2018 International Power Electronics Conference (IPEC-Niigata 2018 –ECCE Asia-)

Niigata, Japan
20-24 May 2018

Pages 2848-3548

IEEE Catalog Number: CFP1854I-POD
ISBN: 978-1-5386-4190-3

Copyright © 2018, IEEJ Industry Applications Society
All Rights Reserved

*** This is a print representation of what appears in the IEEE Digital Library. Some format issues inherent in the e-media version may also appear in this print version.*

IEEE Catalog Number: CFP1854I-POD
ISBN (Print-On-Demand):   978-1-5386-4190-3
ISBN (Online):   978-4-88686-405-5

Additional Copies of This Publication Are Available From:

Curran Associates, Inc
57 Morehouse Lane
Red Hook, NY 12571 USA
Phone: (845) 758-0400
Fax: (845) 758-2633
E-mail: curran@proceedings.com
Web: www.proceedings.com

TABLE OF CONTENTS

THREE-PHASE INDUCTIVE POWER TRANSFER SYSTEM WITH 12 COILS FOR RADIATION NOISE REDUCTION 69
Keisuke Kusaka ; Jun-Ichi Itoh

SECONDARY-SIDE-ONLY CONTROL FOR SMOOTH VOLTAGE STABILIZATION IN WIRELESS POWER TRANSFER SYSTEMS WITH CONSTANT POWER LOAD 77
Giorgio Lovison ; Takehiro Imura ; Hiroshi Fujimoto ; Yoichi Hori

CONSTANT CURRENT CHARGING AND THE MAXIMUM SYSTEM EFFICIENCY TRACKING FOR WIRELESS CHARGING SYSTEMS EMPLOYING DUAL-SIDE CONTROL 84
Zhenjie Li ; Xiaoliang Huang ; Kai Song ; Jinhai Jiang ; Chunbo Zhu ; Zhijiang Du

ELECTRIC FIELD COUPLING TYPE HIGH POWER WIRELESS POWER TRANSFER WITH LEAKAGE ELECTRIC FIELD STRUCURE 88
Mitsuru Masuda

TRANSFER POWER ANALYSIS OF CAPACITIVELY ISOLATED OUTLET AND PLUG (CAPISOP) USING SERIES RESONANCE 94
Hirohito Funato ; Koki Amano ; Takuya Hatsumi ; Junnosuke Haruna

WIDE VOLTAGE GAIN RANGE LLC DC/DC TOPOLOGIES: STATE-OF-THE-ART 100
Qi Cao ; Zhiqing Li ; Haoyu Wang

DUAL HALF-BRIDGE LLC RESONANT CONVERTER WITH HYBRID-SECONDARY-RECTIFIER (HSR) FOR WIDE-OUPUT-VOLTAGE APPLICATIONS 108
Jae-Il Baek ; Chong-Eun Kim ; Keon-Woo Kim ; Min-Su Lee ; Gun-Woo Moon

A STUDY ON THE ANALYSIS AND CONTROL OF NO-LOAD CHARACTERISTICS OF LLC RESONANT CONVERTER FOR PLASMA PROCESS 114
Min-Jun Kwon ; Woo-Cheol Lee

MECHANISM OF CURRENT IMBALANCE IN LLC RESONANT CONVERTER WITH CENTER TAPPED TRANSFORMER 118
Mitsuru Sato ; Shingo Nagaoka ; Takeshi Uematsu ; Toshiyuki Zaitsu

PERFORMANCE STUDY OF HIGH-POWER HALF-BRIDGE INTERLEAVED LLC CONVERTER 123
Hung-I Hsieh ; Hui-Lung Chiu ; Guan-Chyun Hsieh

MULTI-CHIP SIC MOSFET POWER MODULES FOR STANDARD MANUFACTURING, MOUNTING AND COOLING 130
Alberto Castellazzi ; Asad Fayyaz ; Emre Gurpinar ; Abdallah Hussein ; Jianfeng Li ; Bassem Mouawad

AN ALTERNATIVE METHOD TO ACCURATELY DETERMINE THE THERMAL RESISTANCE OF SIC MOSFET STRUCTURES WITH DISCRETE DIODES 137
Andras Vass-Varnai ; Young Joon Cho ; Gabor Farkas ; Marta Rencz

HEAT-RESISTANT PACKAGING TECHNOLOGY FOR WIDE BANDGAP POWER DEVICES AND THERMAL RELIABILITY TESTING 142
K. Suganuma ; H. Zhang ; S. Nagao ; C. Chen ; T. Sugahara ; A. Shimoyama ; A. Suetake

VERIFICATION OF IDENTIFICATION ACCURACY OF LOSS CALCULATED BY INVERSE THERMAL ANALYSIS 148
Yuki Ikari ; Kazushige Nakao

PACKAGING ARCHITECTURES FOR SILICON CARBIDE POWER ELECTRONIC MODULES 153
H. Alan Mantooth ; Simon S. Ang

DEVELOPMENT OF A HOMO-POLAR BEARINGLESS MOTOR WITH CONCENTRATED WINDING FOR HIGH SPEED APPLICATIONS 157
Dai Suzuki ; Takaaki Oiwa

HIGH-SPEED SLOTLESS PERMANENT MAGNET MACHINES: MODELLING AND DESIGN FRAMEWORKS 161
S. Jumayev ; K.O. Boynov ; E.A. Lomonova ; J. Pyrhonen

DEVELOPMENT AND PERFORMANCE OF HIGH-SPEED SPM SYNCHRONOUS MACHINE 169
Kota Kawanishi ; Keisuke Matsuo ; Takayuki Mizuno ; Koji Yamada ; Takashi Okitsu ; Kouki Matsuse

1.2KW 100,000RPM HIGH SPEED MOTOR FOR AIRCRAFT 177
Takehiro Jikumaru ; Gen Kuwata

COMPARATIVE EVALUATION OF Y-INVERTER AGAINST THREE-PHASE TWO-STAGE BUCK-BOOST DC-AC CONVERTER SYSTEMS 181
Michael Antivachis ; Dominik Bortis ; David Menzi ; Johann W. Kolar

DC-POWERED OFFICE BUILDINGS AND DATA CENTRES : THE FIRST 380 VDC MICRO GRID IN A COMMERCIAL BUILDING IN GERMANY 190

Tilo Pueschel

RECENT TREND IN POWER ELECTRONICS FOR ICT SYSTEMS 196

Hiroshi Nakao ; Yu Yonezawa ; Yoshiyasu Nakashima

GREEN BASE STATION USING ROBUST SOLAR SYSTEM AND HIGH PERFORMANCE LITHIUM ION BATTERY FOR NEXT GENERATION WIRELESS NETWORK (5G) AND AGAINST MEGA DISASTER 201

M. Nakamura ; K. Takeno

OPTIMIZATION OF MAINTENANCE BY FAILURE PREDICTION CONSIDERING INSTANTANEOUS AND CUMULATIVE EFFECTS OF EXTERNAL ENVIRONMENTS 207

Kaisei Kanetani ; Masahiro Yamazaki ; Tadatoshi Babasaki ; Hideaki Kim ; Tatsushi Matsubayashi

HYBRID CONVERTERS WITH REDUCED INDUCTOR LOSS FOR INTEGRATABLE POWER CONVERSION 213

Gab-Su Seo ; Hanh-Phuc Le

ENERGY SAVING SYSTEM TREND FOR HARBOR CRANE WITH LITHIUM ION BATTERY 219

Hidemasa Yoshihara

INVERTER DRIVE OF DYNAMOMETERS FORAUTOMOTIVE EVALUATION SYSTEM 227

Shizunori Hamada ; Toshimichi Takahashi ; Nobutaka Kezuka ; Masaju Kouketsu ; Shingo Ishigaki

EXPERIMENTAL INVESTIGATION OF PROTOTYPE ALL-SIC CONVERTER FOR ULTRA-HIGH-SPEED ELEVATOR 233

Kazuhisa Mori ; Kaoru Katoh ; Yohei Matsumoto ; Tatsushi Yabuuchi ; Naoto Ohnuma

HIGH-VOLTAGE, LARGE-CAPACITY CONVERTER TECHNOLOGIES AND THEIR APPLICATIONS 238

Daisuke Yoshizawa ; Paul Bixel ; Masahiko Tsukakoshi

HIGHER RADIAL SUSPENSION FORCE OF MAGNETIC BEARING ON CENTRIFUGAL COMPRESSOR FOR HVAC 244

Yuji Nakazawa ; Yusuke Irino ; Atsushi Sakawaki ; Kazunobu Ohyama

NOVEL SWITCHING CONTROL METHOD FOR FULL-BRIDGE DC-DC CONVERTERS FOR IMPROVING LIGHT-LOAD EFFICIENCY USING REVERSE RECOVERY CURRENT 250

Fumihiro Sato ; Takae Shimada ; Takayuki Ouchi

A 800V/14V SOFT-SWITCHED CONVERTER WITH LOW-VOLTAGE RATING OF SWITCH FOR XEV APPLICATIONS 256

Byeongwoo Kim ; Kangsan Kim ; Sewan Choi

HIGH SPEED CONTROL METHOD FOR SUPERPOSING HIGH-FREQUENCY-HIGH-SINUSOIDAL-CURRENT WITH DC CURRENT TO ANALYZE BATTERY AC IMPEDANCE 261

Jin Xu ; Toshihiko Kishimoto ; Noboru Shimosato

EV BMS WITH TIME-SHARED ISOLATED CONVERTERS FOR ACTIVE BALANCING AND AUXILIARY BUS REGULATION 267

Z. Gong ; B.A.C. Van De Ven ; Y. Lu ; Y. Luo ; K. Gupta ; C. Da Silva ; H.J. Bergveld ; O. Trescases

A DRIVING CIRCUIT WITH PARTIAL POWER REGULATION FOR RGB LED LAMPS 275

You-Chun Huang ; Yu-Jen Chen ; Yong-Jyun Li ; Chin-Sien Moo

FPGA-BASED DYNAMIC DUTY CYCLE AND FREQUENCY CONTROLLER FOR A CLASS-E2DC-DC CONVERTER 282

Sanghyeon Park ; Juan Rivas-Davila

DESIGN METHODOLOGY OF 3 KW INDUCTION HEATING SYSTEM FOR BOTH LOW RESISTANCE AND HIGH RESISTANCE CONTAINERS IN A SINGLE BURNER 289

Si-Hoon Jeong ; Hwa-Pyeong Park ; Jee-Hoon Jung

MULTI-RESONANT INVERTER REALIZING DOWNSIZING AND LOSS REDUCTION FOR ALL-METALLIC IH COOKTOP 296

Takayuki Hirokawa ; Makoto Imai ; Atsushi Fujita

TEMPERATURE ESTIMATION OF ALUMINUM ELECTROLYTIC CAPACITOR UNDER ACTUAL CIRCUIT OPERATION 302

Kazuki Urata ; Toshihisa Shimizu

DESIGN AND EVALUATION OF CURRENT DISTRIBUTION IN POWER MODULE 309

Takaaki Ibuchi ; Eisuke Masuda ; Tsuyoshi Funaki

DEVELOPMENT OF IMPEDANCE-SOURCE INVERTER USING SIC-MOSFET 313

Ryuji Iijima ; Thilak Senanayake ; Takanori Isobe ; Hiroshi Tadano

CONTROL METHODOLOGY FOR REALIZATION OF 100KW HEECS CHOPPER WITH 99.5% EFFICIENCY 318

Yukinori Tsuruta ; Atsuo Kawamura

IRON LOSS REDUCTION IN THE CORES OF INDUCTION HEATING COILS FOR SMALL-FOREIGN-METAL PARTICLE DETECTOR WITH A 400-KHZ SIC-MOSFETS HIGH-FREQUENCY INVERTER 324

Takuya Shijo ; Yuki Uchino ; Yujiro Noda ; Hiroaki Yamada ; Toshihiko Tanaka

FREQUENCY TRACKING BURST-MODE PDM-CONTROLLED CLASS-D ZERO VOLTAGE SOFT-SWITCHING RESONANT CONVERTER FOR INDUCTIVE POWER TRANSFER APPLICATIONS 329

Yoichiro Tabata ; Tomokazu Mishima ; Tatsuya Kido

REDUCED-ORDER DYNAMICAL MODELS OF TUNED WIRELESS POWER TRANSFER SYSTEMS 337

Hongchang Li ; Jingyang Fang ; Yi Tang

DYNAMIC MODELLING AND CLOSED LOOP CONTROL OF TRANSMITTER PARALLEL AND RECEIVER SERIES COMPENSATED IPT TOPOLOGY FOR EV APPLICATIONS 342

Suvendu Samanta ; Akshay Kumar Rathore

DEVELOPMENT OF INDUCTIVE POWER TRANSFER SYSTEM FOR EXCAVATOR UNDER LARGE LOAD FLUCTUATION : CONSIDERATION OF RELATIONSHIP BETWEEN LOAD VOLTAGE AND RESONANCE PARAMETER 348

Jun-Ichi Itoh ; Kent Inoue ; Keisuke Kusaka

WIRELESS POWER TRANSFER SYSTEM USING THREE-PHASE TO SINGLE-PHASE MATRIX CONVERTER 356

Yuji Hayashi ; Hiromasa Motoyama ; Takaharu Takeshita

DESIGN OF A REDUCED-ORDER OBSERVER FOR SENSORLESS CONTROL OF DUAL-ACTIVE-BRIDGE CONVERTER 363

Nguyen Duy Dinh ; Goro Fujita

IMPROVED LOAD TRANSIENT RESPONSE OF A DUAL-ACTIVE-BRIDGE CONVERTER 370

Sheng-Zhi Zhou ; Chuan Sun ; Song Hu ; Guo Chen ; Xiaodong Li

MODULATION AND ACTIVE MIDPOINT CONTROL OF A THREE-LEVEL THREE-PHASE DUAL-ACTIVE BRIDGE DC-DC CONVERTER UNDER NON-SYMMETRICAL LOAD 375

Philipp Joebges ; Anton Gorodnichev ; Rik W. De Doncker

A NOVEL SWITCHING ALGORITHM TO IMPROVE EFFICIENCY AT LIGHT LOAD CONDITIONS FOR THREE-PHASE DAB CONVERTER IN LVDC APPLICATION 383

Hyun-Jun Choi ; Si-Hoon Jung ; Jee-Hoon Jung

DESIGN OF A HIGH-FREQUENCY DUAL-ACTIVE BRIDGE CONVERTER WITH GAN DEVICES FOR AN OUTPUT POWER OF 3.7 KW 388

Philipp Schülting ; Christian Winter ; Rik W. De Doncker

EXPLORATION OF THE DESIGN AND PERFORMANCE SPACE OF A HIGH FREQUENCY 166 KW/10 KV SIC SOLID-STATE AIR-CORE TRANSFORMER 396

Piotr Czyz ; Thomas Guillod ; Florian Krismer ; Johann W. Kolar

NOVEL CALCULATION METHOD OF IRON LOSS OF GAPPED INDUCTORS USING LOSS MAP 404

Yoshihiro Miwa ; Toshihisa Shimizu

VERIFICATION OF THE REDUCTION OF THE COPPER LOSS BY THE THIN COIL STRUCTURE FOR INDUCTION COOKERS 410

Morimasa Hataya ; Koki Kamaeguchi ; Eiji Hiraki ; Kazuhiro Umetani ; Takayuki Hirokawa ; Makoto Imai ; Hideki Sadakata

CONDITION MONITORING OF ELECTROLYTIC CAPACITOR BASED ON ESR ESTIMATION AND THERMAL IMPEDANCE MODEL USING IMPROVED POWER LOSS COMPUTATION 416

Sundararajan Prasanth ; Mohamed Halick ; Mohamed Sathik ; Firman Sasongko ; Tan Chuan Seng ; Peng Yaxin ; Rejeki Simanjorang

TEST SETUP FOR CHARACTERISATION OF BIASED MAGNETIC HYSTERESIS LOOPS IN POWER ELECTRONIC APPLICATIONS 422

Min Luo ; Drazen Dujic ; Jost Allmeling

A FAST OPEN-CIRCUIT FAULT DIAGNOSIS SCHEME FOR MODULAR MULTILEVEL CONVERTERS WITH MODEL PREDICTIVE CONTROL 428

Dehong Zhou ; Shunfeng Yang ; Yi Tang

AN ONLINE OPEN-CIRCUIT FAULT DIAGNOSIS AND FAULT TOLERANT SCHEME FOR THREE-PHASE AC-DC CONVERTERS WITH MODEL PREDICTIVE CONTROL 434

Dehong Zhou ; Yi Tang

THE LIFETIME ASSESSMENT OF A MICRO-INVERTER FOR PV APPLICATIONS 439

Tohihiro Shimao ; Koji Kato ; Youichi Ito ; Akio Iwabuchi ; Yongheng Yang ; Frede Blaabjerg

ONLINE HEALTH MONITORING OF MULTIPLE MOSFETS IN A GRID-TIED PV INVERTER USING SPREAD SPECTRUM TIME DOMAIN REFLECTOMETRY (SSTDR) 446
Sourov Roy ; Faisal Khan

AN IMPROVED EQUIVALENT MODEL FOR A LONG PV STRING UNDER PARTIAL SHADING CONDITIONS 453
Xiaoyang Wang ; Huiqing Wen ; Xingshuo Li

OPTIMIZED FLUX-WEAKENING CONTROL OF INDUCTION MOTOR FOR TORQUE ENHANCEMENT IN VOLTAGE EXTENSION REGION 459
Zhen Dong ; Yong Yu ; Bo Wang ; Qinghua Dong ; Dianguo Xu

IMPROVED PERFORMANCE OF CFTC-BASED DIRECT TORQUE CONTROL OF INDUCTION MACHINES BY INCREASING TORQUE LOOP BANDWIDTH 466
Ibrahim Mohd Alsofyani ; June-Hee Lee ; Byung-Moon Han ; Kyo-Beum Lee

μ-ANALYSIS EVALUATION OF A NOVEL COMBINED CURRENT-AND-SPEED CONTROL FOR INDUCTION MOTORS VIA ILQ DESIGN METHOD 471
Shuto Omori ; Hiroshi Takami ; Masashi Nakamura

LOSS MINIMIZATION CONTROL OF SENSORLESS SCALAR-CONTROLLED INDUCTION MOTOR DRIVES CONSIDERING IRON LOSS 478
Nguyen Anh Tan ; Dong-Choon Lee

TUNING OF INDUCTION MOTOR DRIVE WITH TORQUE SENSOR 483
Hajime Kubo ; Yugo Tadano

QUASI-TWO-LEVEL CONVERTER FOR OVERVOLTAGE MITIGATION IN MEDIUM VOLTAGE DRIVES 488
F. Bertoldi ; M. Pathmanathan ; R. S. Kanchan ; K. Spiliotis ; J. Driesen

A MEDIUM-VOLTAGE THREE-PHASE AC-DC CONVERTER CONSISTING OF CASCADED THREE-LEVEL BOOST-TYPE RECTIFIERS AND AN OPEN-END WINDING TRANSFORMER 495
Ryoji Tsuruta ; Hiromitsu Suzuki ; Ritaka Nakamura

A FAULT TOLERANT CONTROL STRATEGY FOR THE DELTA-CONNECTED CASCADED CONVERTER 503
Ping-Heng Wu ; Po-Tai Cheng

COOLING PERFORMANCE IMPROVEMENT OF HEAT SINK BY OSCILLATING HEAT PIPE ADDITION AND DESIGN FOR ENVIRONMENT OF OSCILLATING HEAT PIPE REFRIGERANT 511
Kuan-Chung Tey ; Kenichiro Suzuki

COMPACT LARGE CAPACITY GAS TURBINE STATIC STARTER 517
Hironori Kawaguchi ; Shigeyuki Nakabayashi ; Akinobu Ando ; Hiroshi Ogino ; Yasuaki Matsumoto ; Ikuto Udagawa ; Takahiro Ohta

VOLTAGE REFERENCE MODIFICATION SCHEME FOR RESONANCE SUPPRESSION IN LCL-FILTERED INVERTERS WITH DISCONTINUOUS PWM METHOD 521
Hyeon-Sik Kim ; Seung-Ki Sul

PARAMETRIC ROBUSTNESS ANALYSIS FOR PARALLEL FEEDFORWARD COMPENSATION BASED ACTIVE DAMPING OF LCL GRID CONNECTED INVERTER 528
Muhammad Talib Faiz ; Muhammad Mansoor Khan ; Xu Jianming ; Muhammad Ali ; Houjun Tang

OPEN-LOOP-BASED ISLAND-MODE VOLTAGE CONTROL METHOD FOR SINGLE-PHASE GRID-TIED INVERTER WITH MINIMIZED LC FILTER 534
Satoshi Nagai ; Jun-Ichi Itoh

EXPERIMENTAL VALIDATION OF ADAPTIVE CURRENT INJECTING METHOD FOR GRID-SYNCHRONIZATION IMPROVEMENT OF GRID-TIED REGS DURING SHORT-CIRCUIT FAULT 542
Shaokang Ma ; Hua Geng ; Geng Yang ; Bo Liu

ADAPTIVE CONTROL OF GRID-VOLTAGE FEEDFORWARD FOR GRID-CONNECTED INVERTERS BASED ON REAL-TIME IDENTIFICATION OF GRID IMPEDANCE 547
Roni Luhtala ; Tuomas Messo ; Tomi Roinila

MODEL BASED TUNING OF PROPORTIONAL RESONANT CONTROLLERS FOR VOLTAGE SOURCE INVERTERS 555
Stefan Almér ; Thomas Besselmann ; Mario Schweizer

AN SOC-BASED PLATFORM FOR INTEGRATED MULTI-AXIS MOTION CONTROL AND MOTOR DRIVE 560
Yongping Sun ; Ming Yang ; Yangyang Chen ; Wangpin He ; Dianguo Xu

VARIABLE SWITCHING FREQUENCY STRATEGY FOR ENHANCED SETTLING PERFORMANCE OF POSITION CONTROL WITHIN INVERTER LOSS LIMIT 565
Choongin Lee ; Jung-Ik Ha

TWO-WHEEL CANE FOR WALKING ASSISTANCE 571
Phi Van Lam ; Yasutaka Fujimoto

FALL PREVENTION AND VIBRATION SUPPRESSION OF WHEELCHAIR USING RIDER MOTION STATE 575
Isseki Takahashi ; Toshiyuki Murakami

STABILIZATION METHOD FOR RESIDENTIAL DC SYSTEM BASED ON PASSIVITY CRITERION 583
Hiroaki Kakigano

A NOVEL CONTROL APPROACH TO MULTI-TERMINAL POWER FLOW CONTROLLER FOR NEXT-GENERATION DC POWER NETWORK 588
Kenji Natori ; Yuta Nakao ; Yukihiko Sato

DC MICROGRID FOR TELECOMMUNICATIONS SERVICE AND RELATED APPLICATION 593
Keiichi Hirose

MVDC DISTRIBUTION GRIDS FOR ELECTRIC VEHICLE FAST-CHARGING INFRASTRUCTURE 598
Marco Stieneker ; Benedict J. Mortimer ; Arne Hinz ; Adolf Müller-Hellmann ; Rik W. De Doncker

REVIEW OF RESONANT GATE DRIVER IN POWER CONVERSION 607
Bainan Sun ; Zhe Zhang ; Michael A.E. Andersen

A LOW PROFILE HIGH FREQUENCY LED DRIVING SYSTEM BASED ON AIRCORE PLANAR INDUCTOR 614
Yueshi Guan ; Xihong Hu ; Shu Zhang ; Yijie Wang ; Dianguo Xu ; Wei Wang

ANALYSIS AND COMPENSATION OF DEAD-TIME EFFECT IN SIC-DEVICE-BASED HIGH-SWITCHING-FREQUENCY INVERTERS 619
Qingzeng Yan ; Xibo Yuan ; Xiaojie Wu ; Yiwen Geng

CONTROL AND PERFORMANCE OF NEW ASYMMETRICAL OPERATION FOR SWITCHED-CAPACITOR-BASED RESONANT CONVERTERS 626
Hadi Setiadi ; Hideaki Fujita

HIGH-FREQUENCY RESONANT CONVERTER WITH SYNCHRONOUS RECTIFICATION FOR HIGH CONVERSION RATIO AND VARIABLE LOAD OPERATION 632
Lei Gu ; Kawin Surakitbovorn ; Juan Rivas-Davila

SMART PV INVERTERS FOR SMART GRID APPLICATIONS 639
Cheng-Jhen Yang ; Terng-Wei Tsai ; Yi-Chan Li ; Cheng-Yu Tang ; Yaow-Ming Chen ; Yung-Ruei Chang

HIGH-VOLTAGE BI-DIRECTIONAL HALF-BRIDGE THREE-LEVEL SERIES RESONANT CONVERTER WITH FREQUENCY MODULATION CONTROL 645
Lee Sih-Yi ; Jhang Jynu-Jhe ; Lin Jing-Yuan ; Hsieh Yao-Ching ; Chiu Haung-Jen

A CONTROL STRATEGY FOR FLYING-START OF SHAFT SENSORLESS PERMANENT MAGNET SYNCHRONOUS MACHINE DRIVE 651
Zih-Cing You ; Sheng-Ming Yang

CONTACTLESS EV POWER TRACK SYSTEM WITH SEGMENT-EXCITED INDUCTIVELY COUPLED STRUCTURE 657
Jia-You Lee ; Yu-Chi Wang ; Chih-Yi Liao

DRIVING TEST EVALUATION OF SENSORLESS VEHICLE DETECTION METHOD FOR IN-MOTION WIRELESS POWER TRANSFER 663
Katsuhiro Hata ; Kensuke Hanajiri ; Takehiro Imura ; Hiroshi Fujimoto ; Yoichi Hori ; Motoki Sato ; Daisuke Gunji

A SYSTEM DESIGN METHOD OF HIGH-FREQUENCY CLASS-D INVERTER FOR WIDEBAND CURRENT CONTROL 669
Hiroki Kurumatani ; Seiichiro Katsura

ANALYSIS OF INTERIOR PERMANENT MAGNET TWO DEGREES OF FREEDOM MOTOR BASED ON CROSS-COUPLED STRUCTURE 675
Yoshiyuki Hatta ; Tomoyuki Shimono

STUDY COMPARISON BETWEEN FIREFLY ALGORITHM AND PARTICLE SWARM OPTIMIZATION FOR SLAM PROBLEMS 681
Mounia Janah ; Yasutaka Fujimoto

BANDWIDTH LIMITATIONS IN FORCE CONTROL OF A SERIES ELASTIC ACTUATOR WITH BACKLASH AND QUANTIZATION 688
Hanul Jung ; Chan Lee ; Sehoon Oh

ROTOR SHAPE OPTIMIZATION OF INTERIOR PERMANENT MAGNET SYNCHRONOUS MOTORS WITH CONCENTRATED WINDINGS BY CONSIDERING END-LEAKAGE FLUX 693
Katsumi Yamazaki ; Hiroki Narushima

LOSS ANALYSIS OF PERMANENT-MAGNET SYNCHRONOUS MACHINES CONSIDERING IN-PLANE EDDY CURRENT IN ELECTRICAL STEEL SHEETS 699

Hideki Ohguchi ; Satoshi Imamori ; Katsumi Yamazaki ; Haiyan Yui ; Masao Shuto

STUDY ON INFLUENCE OF DIFFERENCE IN STRUCTURE OF CONCENTRATED WINDING IPMSMS OBTAINED BY AUTOMATIC DESIGN 704

A. Ura ; M. Sanada ; S. Morimoto ; Y. Inoue

CARRIER HARMONIC LOSS REDUCTION TECHNIQUE ON DUAL THREE-PHASE PERMANENT-MAGNET SYNCHRONOUS MOTORS WITH PHASE-SHIFT PWM 711

Yoshihiro Miyama ; Haruyuki Kometani ; Kan Akatsu

FLUX INTENSIFYING PM-MOTOR WITH VARIABLE LEAKAGE MAGNETIC FLUX TECHNIQUE 718

Masahiro Aoyama ; Toshihiko Noguchi

CONTINUOUS OPERATION CONTROL OF PMSM IN THE CASE OF DC POWER SUPPLY LOSS 726

Jongwon Heo ; Keiichiro Kondo

MODEL PREDICTIVE CONTROL FOR MULTIPHASE MOTOR DRIVES – A TECHNOLOGY STATUS REVIEW 732

A. Tenconi ; S. Rubino ; R. Bojoi

INFLUENCE OF FAST SWITCHING SEMICONDUCTORS ON THE WINDING INSULATION SYSTEM OF ELECTRICAL MACHINES 740

Kay Hameyer ; Andreas Ruf ; Florian Pauli

CENTRALIZED CONTROL OF MODULAR MULTI RECTIFIER FOR MOTOR DRIVE APPLICATIONS UNDER UNBALANCED GRID 746

Yipeng Song ; Pooya Davari ; Frede Blaabjerg

VECTOR CONTROL OF MAGNETICALLY MODULATED MOTOR FOR POWER SPLITTING OF HEV APPLICATION 753

Toshihiko Noguchi ; Sawanth Krishna Machavolu ; Masahiro Aoyama ; Yuto Motohashi

IMPEDANCE-BASED STABILITY EVALUATION OF VIRTUAL SYNCHRONOUS MACHINE IMPLEMENTATIONS IN CONVERTER CONTROLLERS 759

Eneko Unamuno ; Atle Rygg ; Mohammad Amin ; Marta Molinas ; Jon Andoni Barrena

STABLE POWER SUPPLY METHOD FOR HOUSEHOLD APPLIANCES VIA VIRTUAL SYNCHRONOUS GENERATOR IN SINGLE-PHASE THREE-WIRE MICROGRID 767

Yuko Hirase ; Hidehiko Nakagawa ; Eiji Yoshimura ; Shogo Katsura ; Kensho Abe ; Osamu Noro ; Kazushige Sugimoto ; Kenichi Sakimoto

A NOVEL OSCILLATION DAMPING METHOD OF VIRTUAL SYNCHRONOUS GENERATOR CONTROL WITHOUT PLL USING POLE PLACEMENT 775

Jia Liu ; Yushi Miura ; Toshifumi Ise

OPERATION OF A MODULAR MULTILEVEL CONVERTER CONTROLLED AS A VIRTUAL SYNCHRONOUS MACHINE 782

Salvatore D'arco ; Giuseppe Guidi ; Jon Are Suul

ASSESSMENT OF VIRTUAL SYNCHRONOUS MACHINE BASED CONTROL IN GRID-TIED POWER CONVERTERS 790

Chi Li ; Igor Cvetkovic ; Rolando Burgos ; Dushan Boroyevich

RESEARCH ON THE BLOCKCHAIN-BASED INTEGRATED DEMAND RESPONSE RESOURCES TRANSACTION SCHEME 795

Shengnan Zhao ; Yang Li ; Beibei Wang ; Huiling Su

INDIRECT CURRENT CONTROL FOR SEAMLESS TRANSFER OF UTILITY INTERACTIVE INVERTER 803

Kyungbae Lim ; Injong Song ; Jaeho Choi

STUDY OF AC POWER INTERCHANGE AND DC POWER INTERCHANGE FOR MICRO GRID SYSTEMS 809

Kazuto Yukita ; Daiki Owaki ; Shunsuke Horie ; Toshiro Matsumura ; Yasuyuki Goto

STABILITY ENHANCEMENT STRATEGY FOR ISLANDING MICROGRID WITH MULTI-TYPE INVERTERS BASED ON HYBRID IMPEDANCE MODELLING 815

Meiqin Mao ; Yong Ding ; Yatao Shen ; Liuchen Chang

DC POWERED DATA CENTER WITH 200 KW PV PANELS 822

Keiichi Hirose

INFLUENCES OF DETERIORATION IN CAPACITOR AND INDUCTOR ON CURRENT SENSORLESS STATIC MODEL DC-DC CONVERTER 826

Fujio Kurokawa ; Masashi Taguchi ; Jizhe Wang ; Hidenori Maruta ; Nobumasa Matsui

CAPACITIVE DIVIDER BASED PASSIVE START-UP METHODS FOR FLYING CAPACITOR STEP-DOWN DC-DC CONVERTER TOPOLOGIES 831
Michael Halamicek ; Tom Moiannou ; Nenad Vukadinovic ; Aleksandar Prodic

HIGH VOLTAGE GAIN INTERLEAVED ACTIVE-CLAMP FORWARD (IACF) CONVERTER HAVING REDUCED PRIMARY CONDUCTION LOSS 838
Yeonho Jeong ; Mu-Hyun Park ; Gun-Woo Kim ; Byoung-Hee Lee ; Gun-Woo Moon

CONTROL OF SWITCHING-CAPACITOR BASED BUCK-BOOST CONVERTER 845
M. Veerachary ; Vasudha Khubchandani

IMPROVEMENT OF UPLOAD TRANSIENT RESPONSES FOR ULTRA HIGH STEP-DOWN CONVERTER 851
Y.T. Yan ; K.I. Hwu

POWER ELECTRONICS AND CONTROL TECHNOLOGIES FOR HOUSEHOLD WASHER 856
Toru Niki

DEVELOPMENT OF ROOM AIR CONDITIONER WITH TWIN-PROPELLER FANS 860
Takamasa Uemura ; Tomoya Fukui ; Kenichi Sakoda

ELECTROLYTIC CAPACITOR-LESS SINGLE-PHASE TO THREE-PHASE INVERTER WITH HARMONICS SUPPRESSION CONTROL FOR AIR CONDITIONER 866
Nobuo Hayashi ; Takuro Ogawa ; Tomoisa Taniguchi ; Morimitsu Sekimoto

LATEST DEVELOPMENT OF SIC POWER MODULE-BASED SINGLE-STAGE AC-AC RESONANT CONVERTER FOR HIGH-FREQUENCY INDUCTION HEATING APPLICATIONS 872
Tomokazu Mishima

AN OPTIMIZED CONTROL STRATEGY TO IMPROVE THE CURRENT ZERO-CROSSING DISTORTION IN BIDIRECTIONAL AC/DC CONVERTER BASED ON V2G CONCEPT 878
Lei Jing ; Xiaoqing Wang ; Bodong Li ; Maohang Qiu ; Bo Liu ; Min Chen

PER-PHASE CONTROL STRATEGY OF THE THREE-PHASE FOUR-WIRE INVERTER 883
Yi-Chan Li ; Terng-Wei Tsai ; Cheng-Jhen Yang ; Yaow-Ming Chen ; Yung-Ruei Chang

OPPORTUNITIES FOR PERFORMANCE IMPROVEMENT OF SINGLE-PHASE POWER CONVERTERS THROUGH ENHANCED AUTOMATIC-POWER-DECOUPLING CONTROL 889
Huawei Yuan ; Sinan Li ; Wenlong Qi ; Siew-Chong Tan ; S. Y. Ron Hui

ZERO VOLTAGE SWITCHING SCHEME FOR FLYBACK CONVERTER TO ENSURE COMPATIBILITY WITH ACTIVE POWER DECOUPLING CAPABILITY 896
Hiroki Watanabe ; Jun-Ichi Itoh

MODEL PREDICTIVE FAULT TOLERANT CONTROL OF BIDIRECTIONAL AC/DC CONVERTER WITH VOLTAGE BALANCE OF SPLIT CAPACITOR 904
Nan Jin ; Chongyan Zhao ; Leilei Guo

PWM STRATEGY FOR PARALLEL OPERATION OF THREE PHASE CONVERTERS TIED TO GRID 911
Hyun-Sam Jung ; Seung-Ki Sul

PRACTICAL ISSUES AND IMPLEMENTATION CIRCUITS OF THE DIGITAL-ANALOG HYBRID FULL FEED-FORWARD METHOD WITH UNIPOLAR AND BIPOLAR MODULATIONS 917
Xin Zhang ; Henry S. H. Chung ; Zhixun Ma

AN AC-DC POWER CONVERTER FOR ELECTROLYTIC CAPACITOR-LESS LED DRIVER WITH HIGH LUMINOUS EFFICACY 922
Kwon-Sik Park ; Byuong-Jun Seo ; Kyoung-Suk Kang ; Eui-Cheol Nho

AN IMPROVED CASCADED DUAL-BUCK INVERTER 927
Usman Ali Khan ; Honnyong Cha ; Ashraf Ali Khan ; Heung-Geun Kim ; Wilson Eberle ; Liwei Wang

A SINGLE-SWITCH INTEGRATED-STAGE LED DRIVER BASED ON CUK AND CLASS-E CONVERTER 934
Shu Zhang ; Yijie Wang ; Xiaosheng Liu ; Yan Zhou ; Dianguo Xu

A FAULT-TOLERANT PARALLEL INVERTER APPLIED TO MICRO-GRID 939
Xiangyue Shi ; Jinjie Peng ; Zhifeng Qiu ; Wei Xiong

STABILITY ANALYSIS OF GRID-CONNECTED CONVERTERS WITH ADD-ON VOLTAGE SUPPORT FUNCTIONALITY USING REPETITIVE CONTROL 946
Y. Zhang ; M. G. L. Roes ; M. A. M. Hendrix ; J. L. Duarte

ADAPTIVE SERIES STABILIZER MODULE FOR THE GRID CONNECTED INVERTER UNDER VARIABLE GRID CONDITIONS 953
Xin Zhang

AN IMPROVED DROOP CONTROL BASED SMOOTH TRANSFER CONTROL STRATEGY 957
Xin Meng ; Jinjun Liu ; Zeng Liu ; Ronghui An

FREQUENCY RESPONSE ANALYSIS OF LOAD EFFECT ON DYNAMICS OF GRID-FORMING INVERTER 963

Matias Berg ; Tuomas Messo ; Teuvo Suntio

A NEW CONTROL METHOD FOR TRIPLE-ACTIVE BRIDGE CONVERTER WITH FEED FORWARD CONTROL 971

Takanobu Ohno ; Nobukazu Hoshi

ANALYSIS OF PFM OPERATION MODEL FOR CAPACITOR CHARGER RESONANT TOPOLOGY WITH ENERGY DOSAGE 977

Pengyu Jia ; Yiqin Yuan ; Shengwen Fan ; Zhenyu Shan

AN ACTIVE-CLAMPED CURRENT-FED HALF-BRIDGE DC-DC CONVERTER WITH THREE SWITCHES 982

Truong-Duy Duong ; Minh-Khai Nguyen ; Young-Cheol Lim ; Joon-Ho Choi

A HIGH GAIN QUASI SINGLE STAGE LLC RESONANT DC/DC CONVERTER WITH COUPLED INDUCTOR AND PARTIAL ACTIVE CLAMP 987

Chongcan Huo ; Xiaogao Xie ; Shuai Jiang ; Hanjing Dong

SUPPRESSION OF RIPPLE CURRENT IN HIGH STEP-UP DC-DC CONVERTER UTILIZING COCKCROFT-WALTON CIRCUIT WITH INDUCTOR 992

Takumi Yasuda ; Masataka Minami ; Shin-Ichi Motegi ; Masakazu Michihira

AN OPTIMAL DESIGN METHOD CONSIDERING TRANSFORMER PARASITIC CAPACITANCE OF LLC RESONANT CONVERTERS 998

Naizeng Wang ; Xu Yang ; Mofan Tian ; Haiyang Jia ; Guangzhao Xu ; Zhenwei Li

COMPARISON OF HARMONIC LINEARIZATION AND HARMONIC STATE SPACE METHODS FOR IMPEDANCE MODELING OF MODULAR MULTILEVEL CONVERTER 1004

Jing Lyu ; Xin Zhang ; Jingjing Huang ; Jianwen Zhang ; Xu Cai

AN IMPROVED PHASE-SHIFTED PWM FOR A FIVE-LEVEL HYBRID-CLAMPED CONVERTER 1010

Kui Wang ; Nianzhou Liu ; Zedong Zheng ; Yongdong Li

INTEGRATED CONTROL METHODS FOR ASYMMETRICAL CASCADED H-BRIDGE RECTIFIER 1015

Wenjing Dai ; Jie Chen ; Xin Chen ; Chunying Gong

TRANSIENT VOLTAGE STRESS MODELING FOR SUBMODULES OF MODULAR MULTILEVEL CONVERTERS UNDER GRID VOLTAGE SAGS 1021

Zhijian Yin ; Yongheng Yang ; Huai Wang

SVPWM STRATEGY BASED ON MULTILEVEL 3LNPC-CR 1027

Xiaoqiong He ; Pengcheng Han ; Xiaolan Lin ; Yi Wang ; Xu Peng

THE MULTIPLE DEGREE OF FREEDOM BASED NEUTRAL POINT POTENTIAL CONTROL OF THREE LEVEL NEUTRAL POINT CLAMPED CONVERTERS 1032

Bo Guan ; Shinji Doki

A MODIFIED PHASE-SHIFTED PWM TECHNIQUE FOR THE GRID-CONNECTED HYBRID CASCADED CONVERTER 1038

Yu-Chen Su ; Po-Tai Cheng

NOVEL T-TYPE DUAL-BUCK INVERTER WITH MINIMUM NUMBER OF INDUCTORS 1046

Tien-The Nguyen ; Honnyong Cha ; Bang Le-Huy Nguyen ; Heung-Geun Kim

CONTROL OF DIRECT AC/AC MODULAR MULTILEVEL CONVERTER IN RAILWAY POWER SUPPLY SYSTEM 1051

Shuguang Song ; Jinjun Liu ; Shaodi Ouyang ; Xingxing Chen ; Baojin Liu

WIRELESS POWER TRANSFER: CRITICAL REVIEW OF RELATED STANDARDS 1062

Mohamad Abou Houran ; Xu Yang ; Wenjie Chen ; Mehdi Samizadeh

COMPARATIVE STUDY OF SINGLE-PHASE FUNDAMENTAL COMPONENT FREQUENCY ESTIMATION SCHEMES UNDER TIME-VARYING HARMONIC DISTORTION OPERATION 1067

E. B. Kapisch ; J. L. Duarte ; C. A. Duque

A COMPREHENSIVE DEAD-TIME COMPENSATION METHOD FOR A THREE-PHASE DUAL-ACTIVE BRIDGE CONVERTER WITH HYBRID MODULATION SCHEMES 1073

Jingxin Hu ; Zhiqing Yang ; Rik W. De Doncker

EVALUATION OF A HIGH-FREQUENCY REACTOR WITH A NEW WIRE GUIDE FOR A TOROIDAL CORE 1080

Hideki Ayano ; Akira Fujimura ; Yoshihiro Matsui

CORE LOSS EVALUATION IN POWDER CORES: A COMPARATIVE COMPARISON BETWEEN ELECTRICAL AND CALORIMETRIC METHODS 1087

Yuki Ishikura ; Jun Imaoka ; Mostafa Noah ; Masayoshi Yamamoto

MODELING, MAGNETIC DESIGN, AND SIMULATION METHODS CONSIDERING DC SUPERIMPOSITION CHARACTERISTIC OF POWDER CORES USED IN POWER CONVERTERS 1095

Jun Imaoka ; Kenkichiro Okamoto ; Masahito Shoyama ; Yuki Ishikura ; Mostafa Noah ; Masayoshi Yamamoto

MODELLING AND DESIGN OF A MEDIUM FREQUENCY TRANSFORMER FOR HIGH POWER DC-DC CONVERTERS 1103

Miloš Stojadinovic ; Jürgen Biela

EVALUATION OF INDUCTOR LOSSES ON Z-SOURCE INVERTER CONSIDERING AC AND DC COMPONENTS 1111

Ryuji IIjima ; Naoki Kamoshida ; Rene Alexander Barrera Cardenas ; Takanori Isobe ; Hiroshi Tadano

AN INTEGRATING STRUCTURE OF OUTPUT FILTER FOR GRID CONNECTED INVERTER BASED ON FMLF TECHNIQUE 1118

Jie Ma ; Yenan Chen ; Pingping Chen ; Wenxing Zhong ; Dehong Xu

NEW SCREENING METHOD FOR IMPROVING TRANSIENT CURRENT SHARING OF PARALLELED SIC MOSFETS 1125

Junji Ke ; Zhibin Zhao ; Peng Sun ; Huazhen Huang ; James Abuogo ; Xiang Cui

PSPICE MODELING AND APPLICATION FOR SIC POWER MOSFET TO EVALUATE THE POWER LOSS IN FULL-BRIDGE CONVERTER 1131

Juan Wei ; Fei Lin ; Zhongping Yang ; Xianjin Huang ; Chanjuan Xiao ; Hao Zhang ; Wencai Liang

ALL-SIC MODULE PACKAGING TECHNOLOGY 1137

Kento Shirata ; Norihiro Nashida ; Hideyo Nakamura ; Yoshitaka Nishimura

A NEW SMALLEST 1200V INTELLIGENT POWER MODULE FOR THREE PHASE MOTOR DRIVES 1141

Minsub Lee ; Miran Baek ; Junbae Lee ; Daewoong Chung

DESIGN AND ENHANCEMENT OF ESD RELIABILITY IN CIRCULAR UHV 300-V NLDMOS POWER COMPONENTS 1145

Shen-Li Chen ; Yi-Hao Chao ; Chih-Ying Yen ; Jen-Hao Lo ; Chun-Ting Kuo ; Yu-Lin Lin ; Yi-Hao Chiu ; Pei-Lin Wu ; Yu-Lin Jhou

A TECHNOLOGY ANALYSIS OF VOLTAGE SHARING IN SERIES CONNECTED POWER DEVICES 1149

Z Davletzhanova ; O Alatise ; R Bonyadi ; J Ortiz-Gonzalez ; T Dai ; M Jennings ; L Ran ; P Mawby

FAILURE MECHANISM ANALYSIS AND PHYSICS-OF-FAILURE LIFETIME PREDICTION METHOD FOR PRESS-PACK THYRISTOR OF CONVERTER VALVE 1157

Ning Liang ; Zhigang Zhang ; Yating Gou ; Cuicui Liu ; Zebin Yang ; Jiangnan Chen ; Fang Zhuo ; Feng Wang

SURGE VOLTAGE ABSORPTION BY A SILICON CARBIDE AVALANCHE-DIODE WITH P-N STRUCTURE 1162

K. Koseki ; Y. Tanaka

CALCULATION OF THYRISTOR RELIABILITY PARAMETER OF UHVDC CONVERTER VALVE IN HEMP ENVIRONMENT 1167

Zhigang Zhang ; Yating Gou ; Cuicui Liu ; Zebin Yang ; Xiaotong Du ; Jiangnan Chen ; Fang Zhuo ; Feng Wang ; Yuanliang Lan ; Caiwang Sheng

GENERALIZED STACKELBERG GAME-THEORETIC APPROACH FOR JOINTED ENERGY AND RESERVE COORDINATION OF ELECTRIC VEHICLES 1172

Tianyang Zhao ; Xuewei Pan ; Lei Li ; Fei Zhao ; Can Wang

IMPEDANCE INFLUENCE ANALYSIS OF PHASE-LOCKED LOOPS ON THREE-PHASE GRID-CONNECTED INVERTERS 1177

Yuncheng Wang ; Xin Chen ; Yang Zhang ; Jie Chen ; Chunying Gong

PULSE-INJECTION-BASED SENSORLESS CONTROL METHOD WITH IMPROVED DYNAMIC CURRENT RESPONSE FOR PMSM 1183

Hechao Wang ; Kaiyuan Lu ; Dong Wang ; Frede Blaabjerg

INFLUENCE OF PARAMETER VARIATIONS ON OPERATING CHARACTERISTICS OF MTPF CONTROL FOR DTC-BASED PMSM DRIVE SYSTEM 1189

Keisuke Fujii ; Yukinori Inoue ; Shigeo Morimoto ; Masayuki Sanada

A QUIET POSITION SENSORLESS CONTROL FOR AN IPMSM BASED ON EXTENDED EMF AND VOLTAGE INJECTION SYNCHRONIZED WITH PWM CARRIER 1196

Yuki Ishii ; Hiroki Yamashita ; Hisao Kubota

STUDY OF TORQUE RIPPLE REDUCTION AND TORQUE BOOST BY MODIFIED TRAPEZOIDAL MODULATION 1202

Satoshi Joryo ; Kazuto Tatsumi ; Toshimitsu Morizane ; Katsunori Taniguchi ; Noriyuki Kimura ; Hideki Omori

FAULT DIAGNOSIS METHOD OF CURRENT SENSOR FOR PERMANENT MAGNET SYNCHRONOUS MOTOR DRIVES 1206

Guoqiang Zhang ; Guoxin Wang ; Gaolin Wang ; Junya Huo ; Lianghong Zhu ; Dianguo Xu

SENSORLESS SPEED CONTROL OF DIESEL-GENERATOR SYSTEMS BASED ON MULTIPLE SOGI-FLLS 1212
Ngoc Dat Dao ; Dong-Choon Lee ; Dae-Sik Lim

ROBUSTNESS OF SIMPLIFIED SPEED-SENSORLESS VECTOR CONTROL FOR INDUCTION MOTOR 1217
Naoki Akao ; Mineo Tsuji ; Shin-Ichi Hamasaki

MAXIMUM TORQUE CONTROL REFERENCE FRAME BASED ON A TORQUE MAP FOR IPMSMS WITH LARGE INDUCTANCE VARIATION 1223
Kazuki Ohta ; Takumi Ohnuma ; Shinji Doki

PMSM MODEL DISCRETIZATION IN CONSIDERATION OF PARK TRANSFORMATION FOR CURRENT CONTROL SYSTEM 1228
Masamichi Inoue ; Shinji Doki

PSEUDO-RANDOM HIGH-FREQUENCY SINUSOIDAL VOLTAGE INJECTION BASED SENSORLESS CONTROL FOR IPMSM DRIVES 1234
Guoqiang Zhang ; Huiying Wang ; Gaolin Wang ; Junya Huo ; Lianghong Zhu ; Dianguo Xu

AT-NPC 3-LEVEL INVERTER-FED INDUCTION MOTOR VECTOR CONTROL WITH NEUTRAL POINT VOLTAGE CONTROL 1240
K. Sudo ; M. Tsuji ; S. Hamasaki ; T. Fukuoka ; H. Ichinose

INVESTIGATION OF VARIOUS POSITION ESTIMATION ACCURACY ISSUES IN PULSE-INJECTION-BASED SENSORLESS DRIVES 1246
Hechao Wang ; Kaiyuan Lu ; Dong Wang ; Frede Blaabjerg

POSITION SENSORLESS CONTROL OF SWITCHED RELUCTANCE MOTOR USING ESTIMATED PWM PHASE VOLTAGE 1253
Y. Nakazawa ; K. Ohyama ; H. Fujii ; H. Uehara ; Y. Hyakutake

EXPERIMENTAL CONFIRMATION OF THRUST AND ATTRACTIVE FORCE CONTROL OF LINEAR INDUCTION MOTOR BY TWO DIFFERENT FREQUENCY COMPONENTS 1259
Kenta Sannomiya ; Toshimitsu Morizane ; Noriyuki Kimura ; Hideki Omori

GA BASED OPTIMIZED TRAJECTORIES OF ROTATING SPEED AND D-Q AXIS CURRENTS FOR AN IPMSM 1264
Shuta Kumagai ; Kaoru Inoue ; Toshiji Kato

2-DEGREE-OF-FREEDOM DEADBEAT CONTROL WITH DISTURBANCE COMPENSATION FOR PMSM DRIVE SYSTEM USING FPGA 1270
Arata Takahashi ; Shotaro Takakura ; Tomoki Yokoyama

EXTENDED EMF-BASED SIMPLE IPMSM SENSORLESS VECTOR CONTROL USING COMPENSATED CURRENT CONTROLLER 1276
Takatoshi Inoue ; Yasumasa Hamabe ; Mineo Tsuji ; Shin-Ichi Hamasaki

FULL-BAND OUTPUT IMPEDANCE MODEL OF VIRTUAL SYNCHRONOUS GENERATOR IN DQ FRAMEWORK 1282
Li Wenbing ; Wang Jianhua ; Song Jingyu ; Luo Fangfang ; Gao Shang ; Wu Zaijun

AN MTPA CONTROL METHOD OF A PMSM AND A SYNRM BASED ON A DTC IN THE STATOR FLUX LINKAGE SYNCHRONOUS FRAME 1289
Gimpei Itoh ; Yukinori Inoue ; Shigeo Morimoto ; Masayuki Sanada

EEMFS EXCITED BY SIGNAL INJECTION FOR POSITION SENSORLESS CONTROL OF PMSMS AND THEIR PERFORMANCE COMPARISON BY USING IMAGINARY ELECTROMOTIVE FORCE 1295
Takumi Nimura ; Shota Kondo ; Shinji Doki ; Mutuwo Tomita

HARMONIC CURRENT CANCELLATION METHOD FOR PMSM DRIVE SYSTEM USING RESONANT CONTROLLERS 1301
Dongsheng Li ; Yoshitaka Iwaji ; Yasuo Notohara ; Ken Kishita

ESTIMATION ERROR ANALYSIS OF STATOR FLUX OBSERVER FOR DTC-BASED PMSM DRIVES 1308
Atsushi Shinohara ; Kichiro Yamamoto

APPLICATION OF FICTITIOUS REFERENCE ITERATIVE TUNING TO CONTROLLER DESIGN FOR VARIOUS MACHINES 1315
Hidehiro Ikeda ; Kazuya Goto ; Feili Zhang ; Kazuya Kayashima ; Tsuyoshi Hanamoto

HIGH EFFICIENCY CONTROL FOR PERMANENT MAGNET MOTOR DRIVE SYSTEM WITH FUEL CELLS CONNECTED IN SERIES WITH ELECTRIC DOUBLE-LAYER CAPACITORS 1322
Kichiro Yamamoto ; Fumiya Ohdera ; Atsushi Shinohara

COMPARATIVE STUDY OF SPEED RIPPLE REDUCTION BY VARIOUS CONTROL METHODS IN PMSM DRIVE SYSTEMS WITH PULSATING LOAD 1329
Yuma Komaru ; Yukinori Inoue ; Shigeo Morimoto ; Masayuki Sanada

ESTIMATION OF THE PARAMETERS OF THE SERVO DRIVE SYSTEM USING PARTICLE SWARM OPTIMIZATION ALGORITHM 1336

Helin Zhu ; Jae Hyuk Choi ; Sang Uk Park ; Jusuk Lee ; Hyong Gun Lee ; Hyung Soo Mok

A PROGRAMMABLE BATTERY TEST SYSTEM WITH ENERGY RECYCLING FEATURE BASED ON SINUSOIDAL LOADING TECHNIQUE 1341

Chang-Hua Lin ; Guan-Jung Chen ; Hwa-Dong Liu ; Kun-Feng Chen

DEVELOPMENT OF LARGE-CAPACITY CONVERTER FOR BATTERY ENERGY STORAGE SYSTEMS 1346

Hiroyoshi Komatsu ; Tatsuji Katayama ; Noriko Kawakami

ANALYSIS AND COMPARISON OF DC/DC TOPOLOGIES IN PARTIAL POWER PROCESSING CONFIGURATION FOR ENERGY STORAGE SYSTEMS 1351

Maria C. Mira ; Zhe Zhang ; A. E. Michael Andersen

TWO-STAGE PROTECTION FOR MULTI-CHANNEL POWER ELECTRONIC CONVERTERS FED LARGE ASYNCHRONOUS HYDRO-GENERATING UNIT 1358

R. R. Semwal ; Anto Joseph

CURRENT SHARING CONTROL FOR SERIES-PARALLEL CHANGEOVER USING BATTERY AND ELECTRIC DOUBLE-LAYER CAPACITOR BANK 1364

Taisei Nishino ; Keisaku Isozaki ; Naoki Kogai ; Kyungmin Sung

CONTROL METHOD OF ENERGY STORAGE SYSTEM TO IMPROVE OUTPUT POWER OF PCS 1370

Mikiya Ishibashi ; Hitoshi Haga ; Kenji Arimatsu ; Koji Kato

A CONTROL STRATEGY OF MMC BATTERY ENERGY STORAGE SYSTEM BASED ON ARM CURRENT CONTROL 1376

Liu Danqing ; Wang Guangzhu ; Ou Zhujian ; Liu Jiaxing

EQUIVALENT RESISTANCE CONTROL FOR MAXIMUM POWER TRANSFER METHOD OF PIEZOELECTRIC ELEMENT IN VIBRATION POWER GENERATION 1381

Kenya Takamura ; Hiroaki Yamada ; Toshihiko Tanaka ; Tomoharu Yada ; Hajime Fujiwara

DC BUS VOLTAGE STABILIZATION FOR CASCADED POWER CONVERTER BY INTEGRATING AN EXTRA PORT INTO LOAD SIDE PSFB 1386

Jiang You ; Weiyan Fan ; Mengyan Liao

COMMON MODE CURRENT REDUCTION OF THREE-PHASE CASCADED MULTILEVEL TRANSFORMERLESS INVERTER FOR PV SYSTEM 1391

Wenjie Wang ; Ke Chen ; Lijun Hang ; Anping Tong ; Yiliang Gan

CURRENT SHARING/VOLTAGE SHARING CONTROL STRATEGY FOR CASCADED DC/DC CONVERTER IN PHOTOVOLTAIC DC COLLECTION SYSTEM 1397

Bo Chen ; Yi Wang ; Yanjun Tian ; Shilei Wei

PCC VOLTAGE COMPENSATION OF PV INVERTER WITH ACTIVE POWER DECOUPLING CIRCUIT 1403

Duck-Hwan Hwang ; Jung-Yong Lee ; Younghoon Cho

A NOVEL PARTIAL SHADING DETECTION ALGORITHM UTILIZING POWER LEVEL MONITORING OF PHOTOVOLTAIC PANELS 1409

Thusitha Randima Wellawatta ; Sung-Jin Choi

BOOST INTEGRATED THREE-PHASE SOLAR INVERTER USING CURRENT UNFOLDING AND ACTIVE DAMPING METHODS 1414

N. Ha Pham ; Tomoyuki Mannen ; Keiji Wada

LINEAR ACTIVE DISTURBANCE REJECTION CONTROL FOR ISOLATED THREE-PORT CONVERTER 1421

Jiang You ; Mengyan Liao ; Weiyan Fan

STABILITY CONSTRAINED GAIN OPTIMIZATION OF DROOP CONTROLLED CONVERTERS IN DC NANOGRIDS 1426

Soumya Bandyopadhyay ; Laura Ramirez-Elizondo ; Pavol Bauer

SIC BASED SSPC FOR HIGH VOLTAGE SPACE APPLICATIONS 1435

D. Marroquí ; A. Garrigós ; José M. Blanes ; R. Gutiérrez

AN IMPROVED VOLTAGE-TYPE GRID-CONNECTED CONTROL STRATEGY FOR COMPENSATING UNBALANCED VOLTAGE 1442

Liu Hongpeng ; Zhou Jiajie ; Wang Wei

DUAL TWO-STAGE ISOLATED BIDIRECTIONAL DC-DC CONVERTER FOR DC GRID STORAGE 1447

Gabriel Tibola ; Jorge L. Duarte

MODULAR MULTILEVEL CONVERTER WITH CAPACITOR VOLTAGE SELF-BALANCING USING REDUCED NUMBER OF VOLTAGE SENSORS 1455

Taiyuan Yin ; Yue Wang ; Xiaolei Wang ; Shiyuan Yin ; Shumin Sun ; Guanglei Li

PLUG AND OUTLET IN HOUSEHOLD DC LOW VOLTAGE MICRO-GRID POWER DISTRIBUTION 1460
Worapong Pairindra ; Surin Khomfoi

PERFORMANCE PROGRAMMING TECHNIQUE FOR MULTI-STAGE DC POWER DISTRIBUTION SYSTEMS 1465
Syam Kumar Pidaparthy ; Hansang Kim ; Yeonjung Kim ; Byungcho Choi

COORDINATION CONTROL FOR PARALLELED INVERTERS BASED ON VSG FOR PV/BATTERY MICROGRID 1472
Meiqin Mao ; Cheng Qian ; Liuchen Chang ; Yan Du

ADAPTIVE VOLTAGE CONTROL SCHEME FOR DAB BASED MODULAR CASCADED SST IN PV APPLICATION 1478
Tao Liu ; Yang Xuan ; Xu Yang ; Peng Xu ; Yang Li ; Lang Huang ; Xiang Hao

SIX-STEP MMC-BASED HIGH POWER DC-DC CONVERTER 1484
Stefan Milovanovic ; Dražen Dujic

COMBINED DC POWER FLOW CONTROLLER FOR DC GRID 1491
Yongning Chi ; Xizhou Du ; Siqi Liu ; Xu Cai

AN APPROACH FOR THE EMULATION OF DC GRID ADMITTANCES: IMPLEMENTATION ON A BUCK CONVERTER 1498
Enrique Rodriguez-Diaz ; Fracisco D. Freijedo ; Drazen Dujic ; Juan C. Vasquez ; Josep M. Guerrero

A COMPOUND CONTROLLER FOR POWER FLOW AND SHORT-CIRCUIT FAULT IN DC GRID 1504
Han Ye ; Wu Chen ; Pengpeng Pan ; Xiaokun He

DESIGN PROCEDURE AND CONTROL OF A HYBRID CIRCUIT BREAKER WITH ADAPTABLE PULSE CURRENT INJECTION 1509
Andreas Jehle ; Jürgen Biela

A PRAGMATIC SOH AND SOC CO-ESTIMATOR FOR LITHIUM-ION BATTERIES IN SMART GRID APPLICATIONS 1517
Kaiyuan Li ; King Jet Tseng ; Feng Wei ; Boon-Hee Soong

MODELING AND STABILITY ANALYSIS OF PARALLEL DROOP-CONTROLLED AND CURRENT-CONTROLLED INVERTERS 1524
Shike Wang ; Zeng Liu ; Jinjun Liu ; Ronghui An

DIRECT WIRELESS BATTERY CHARGING SYSTEM 1530
Woo-Seok Lee ; Jin-Hak Kim ; Shin-Young Cho ; Il-Oun Lee

AN IMPROVED PWM SCHEME TO ACHIEVE ZERO-VOLTAGE SWITCHING FOR ALL DEVICES IN THREE-PHASE ISOLATED MATRIX RECTIFIER 1537
Xuerui Lin ; Yunwei Ryan Li ; Jahangir Afsharian ; Dewei David Xu

FIXED-FREQUENCY HF GATE DRIVER BY A PUSH-PULL SELF-EXCITATION LC OSCILLATOR HAVING A CAPACITANCE TRANSISTOR 1543
Naoyuki Ishibashi ; Takuya Mizushima ; Masahiko Hirokawa ; Akihiko Katsuki

A FLEXIBLE REDUCED CAPACITOR VOLTAGES STRATEGY FOR VARIABLE-SPEED DRIVES WITH MODULAR MULTILEVEL CONVERTER 1549
Fangzhou Zhao ; Guochun Xiao ; Daoshu Yang ; Zhiqian Wu ; Xin Meng

A LEAKAGE FLUX CANCELLATION TECHNIQUE FOR SERIES-PARALLEL COMBINED RESONANT CIRCUITS WITH ASYMMETRIC ROTARY TRANSFORMERS USED FOR ULTRASONIC SPINDLE DRIVE 1554
Jun Imaoka ; Masahito Shoyama

A NOVEL STRUCTURAL HEALTH MONITORING SYSTEM WITH WIRELESS POWER AND BI-DIRECTIONAL DATA TRANSFER 1562
Yujin Jangs ; Keon-Woo Kim ; Moo-Hyun Park ; Nayoung Lee ; Gun-Woo Moon

CONTROL STRATEGY FOR STARTER GENERATOR IN UAV WITH MICRO JET ENGINE 1567
Jun-Ichi Itoh ; Kazuki Kawamura ; Hiroyuki Koshikizawa ; Kazuyuki Abe

STUDY ON THE INFLUENCE OF VOLTAGE VARIATIONS FOR NON-INTRUSIVE LOAD IDENTIFICATIONS 1575
Yu-Hsiu Lin ; Shun-Kang Hung ; Men-Shen Tsai

BASIC EXPERIMENT OF A MAGLEV SYSTEM FOR A FLEXIBLE STEEL PLATE WITH CURVATURE: FUNDAMENTAL CONSIDERATION ON LEVITATION STABILITY UNDER DISTURBANCE 1580
Makoto Tada ; Kazuki Ogawa ; Takayoshi Narita ; Hideaki Kato ; Hiroyuki Moriyama

PERFORMANCE OF HYBRID MAGNETIC LEVITATION CONTROL SYSTEM FOR THIN STEEL PLATE BY EMS AND PMS: EXPERIMENTAL EVALUATION OF APPLYING OPTIMAL GAP AND ARRANGEMENT OF PMS...... 1586

Yasuaki Ito ; Yoshiho Oda ; Kengo Okuno ; Toshiki Suzuki ; Masahiro Kida ; Takayoshi Narita ; Hideaki Kato ; Hiroyuki Moriyama

A PRACTICAL LITHIUM-ION BATTERY MODEL BASED ON THE BUTLER-VOLMER EQUATION 1592

Kaiyuan Li ; King Jet Tseng ; Feng Wei ; Boon-Hee Soong

BONDING TECHNOLOGY USING COLD-ROLLED AG SHEET IN DIE-ATTACHMENT APPLICATIONS 1598

Seungjun Noh ; Chanyang Choe ; Chuantong Chen ; Hao Zhang ; Katsuaki Suganuma

HIGH-FREQUENCY SELF-DRIVEN SYNCHRONOUS RECTIFIER CONTROLLER FOR WPT SYSTEMS 1602

Akihiro Konishi ; Kazuhiro Umetani ; Eiji Hiraki

AUTOMATIC RESONANCE FREQUENCY TUNING METHOD FOR REPEATER IN RESONANT INDUCTIVE COUPLING WIRELESS POWER TRANSFER SYSTEMS 1610

Masataka Ishihara ; Kazuhiro Umetani ; Eiji Hiraki

INDUCTIVE POWER TRANSFER FOR T5 FLUORESCENT LAMP LIGHTING SYSTEM 1617

Chung-Chuan Hou ; Tang-Jung Chen ; Ching-Chen Chen ; Chen-Wei Chang ; Po-Wei Wang

AN IMPLEMENT 1.5 MHZ OF INDUCTION HEATING FOR ALUMINUM BASED ON VACUUM TUBE OSCILLATOR CIRCUIT 1622

A. Bilsalam ; P. Chanmontree ; S. Supanyapong ; V. Chunkag

SINGLE-INDUCTOR MULTIPLE-OUTPUTS DIMMABLE LED DRIVER WITH BUCK CONVERTER 1626

Ta-Wei Huang ; Wei-Jing Tseng ; Jun-Xian Huang

A SOFT-SWITCHED THREE-LEVEL T-TYPE INVERTER WITH AUXILIARY COMMUTATED POLES 1634

Apollo Charalambous ; Xibo Yuan

CARRIER-BASED REALIZATION OF ARBITRARY SPACE-VECTOR PWM METHODS FOR THREE-LEVEL INVERTERS 1642

Somboon Sangwongwanich ; Supakorn Paiboon

MULTI-LEVEL TOPOLOGY BASED LINEAR AMPLIFIER FAMILY FOR REALIZATION OF NOISE-LESS INVERTERS 1649

Hidemine Obara ; Tatsuki Ohno ; Atsuo Kawamura

A NEW ZERO-VOLTAGE SWITCHING THREE-LEVEL CONVERTER WITH REDUCED RECTIFIER VOLTAGE STRESS 1655

Keon-Woo Kim ; Cheon-Yong Lim ; Dong-Kwan Kim ; Yu-Jin Jang ; Gun-Woo Moon

MODEL PREDICTIVE CONTROL OF A THREE-LEVEL NPC RECTIFIER WITH A SLIDING MANIFOLD TERM 1661

Xiaonan Gao ; Wei Tian ; Xicai Liu ; Zhenbin Zhang ; Ralph Kennel

H∞ CONTROL-BASED VIBRATION SUPPRESSION IN ROBOT ARM WITH STRAIN WAVE GEARING 1666

Tran Vu Trung ; Makoto Iwasaki

FINE FORCE SENSORLESS FORCE CONTROL BASED ON FRICTION-FREE DISTURBANCE OBSERVER 1673

Ohishi Kiyoshi ; Naoki Kamiya ; Toshimasa Miyazaki ; Yuki Yokokura

KINEMATICS AND TRACKING CONTROL OF A FOUR AXIS ANTENNA FOR SATCOM ON THE MOVE 1680

Oguz Kaan Hancioglu ; Mustafa Celik ; Ugur Tumerdem

POSITION SENSORLESS POSITION CONTROL FOR DUAL SOLENOID ACTUATOR 1687

Sakahisa Nagai ; Atsuo Kawamura

CAE TECHNOLOGY APPLICATION TREND FOR LARGE-CAPACITY POWER ELECTRONICS DEVELOPMENT 1692

Teruo Yoshino ; Kuniaki Nagasaka ; Shigeaki Nakabayashi ; Ikuto Udagawa ; Isamu Tominaga ; Junya Konno

XILINX SYSTEM GENERATOR BASED MODELLING OF FINITE STATE MPC 1698

Vijay Kumar Singh ; Ravi Nath Tripathi ; Tsuyoshi Hanamoto

POWER HARDWARE-IN-THE-LOOP SETUP FOR STABILITY STUDIES OF GRID-CONNECTED POWER CONVERTERS 1704

Tommi Reinikka ; Henrik Alenius ; Tomi Roinila ; Tuomas Messo

PASSIVITY-BASED LCL FILTER DESIGN OF GRID-CONNECTED VSCS WITH CONVERTER SIDE CURRENT FEEDBACK 1711

Shih-Feng Chou ; Xiongfei Wang ; Frede Blaabjerg

ADAPTIVE CONTROL OF DC POWER DISTRIBUTION SYSTEMS: APPLYING PSEUDO-RANDOM SEQUENCES AND FOURIER TECHNIQUES 1719

Tomi Roinila ; Hessamaldin Abdollahi ; Silvia Arrua ; Enrico Santi

AN IMPROVED FINITE-SET MODEL PREDICTIVE TORQUE CONTROL FOR INTERIOR PERMANENT MAGNET SYNCHRONOUS MOTOR DRIVES 1724

Xinan Zhang ; Gilbert Foo ; Tung Ngo

PREDICTIVE TORQUE CONTROL FOR FIVE PHASE INDUCTION MOTOR DRIVE WITH COMMON MODE VOLTAGE REDUCTION 1730

Apekshit Bhowate ; Mohan Aware ; Sohit Sharma ; Yogesh Tatte

INDIRECT MATRIX CONVERTER FOR PERMANENT-MAGNET-SYNCHRONOUS-MOTOR DRIVES BY IMPROVED TORQUE PREDICTIVE CONTROL 1736

Yun Jang ; Yeongsu Bak ; Kyo-Beum Lee

PREDICTIVE DC-LINK CURRENT CONTROL BASED ON IPMSM DISCRETE STATE EQUATION FOR INVERTER WITHOUT INDUCTOR OR ELECTROLYTIC CAPACITOR 1741

Yousuke Akama ; Kodai Abe ; Kiyoshi Ohishi ; Yuki Yokokura ; Koji Kobayashi ; Tatsuki Kashihara

NEW SEARCH ALGORITHM OF MODEL PREDICTIVE CONTROL TO REDUCING CALCULATION AMOUNT FOR IMPROVING STEADY CURRENT CONTROL PERFORMANCE 1747

Masahiro Shimaoka ; Shinji Doki

DISTRIBUTED POWER SHARING STRATEGY FOR ISLANDED MICROGRIDS WITHOUT FREQUENCY AND VOLTAGE DEVIATIONS 1752

Tuan V. Hoang ; Hong-Hee Lee

LIFETIME-ORIENTED DROOP CONTROL STRATEGY FOR AC ISLANDED MICROGRIDS 1758

Yanbo Wang ; Dong Liu ; Fujin Deng ; Dao Zhou ; Zhe Chen

EXPERIMENT ON HIERARCHICAL CONTROL BASED POWER QUALITY ENHANCEMENT FOR STANDALONE MICROGRID 1764

Darith Leng ; Sompob Polmai ; Kittichot Soontorntaweesub

A DISTRIBUTED PREDICTIVE CONTROL STRATEGY BASED ON STATE ESTIMATOR FOR ISLANDED MICROGRID 1771

Mi Dong ; Li Li ; Xiaoyu Tian

MAXIMUM POWER POINT TRACKING METHOD FOR PV MODULE UNDER WIDE RANGE VARYING IRRADIANCE LEVELS 1777

Hwa-Dong Liu ; Chang-Hua Lin

DUAL MPPT CONTROL AND FIELD TESTING FOR SWITCHED CAPACITOR-BASED CELL-LEVEL POWER BALANCING UTILIZING DIFFUSION CAPACITANCE OF PHOTOVOLTAIC CELLS 1782

Masatoshi Uno ; Yota Saito ; Masaya Yamamoto ; Shinichi Urabe

SERIES RESONANT DC-DC CONVERTER WITH DUAL-MODE RECTIFIER FOR PV MICROINVERTERS 1788

Yanfeng Shen ; Huai Wang ; Zhan Shen ; Yongheng Yang ; Frede Blaabjerg

VOLTAGE-REFERENCE ACTIVE POWER DECOUPLING BASED ON BOOST CONVERTER FOR SINGLE-PHASE BRIDGE INVERTER 1793

Shuang Xu ; Meiqin Mao ; Riming Shao ; Liuchen Chang

A SINGLE-PHASE COMMON GROUND BOOST INVERTER FOR PHOTOVOLTAIC APPLICATIONS 1799

Tan-Tai Tran ; Minh-Khai Nguyen ; Young-Cheol Lim ; Joon-Ho Choi

STUDY FOR FURTHER INTRODUCTION OF THE ELECTRONIC FREQUENCY CONVERTERS TO THE TOKAIDO SHINKANSEN 1803

Toshimasa Shimizu ; Ken Kunomura ; Masahiko Kai ; Hiroki Miyajima ; Teruhisa Matsui

COUNTERMEASURE FOR PARTIAL TURN-OFF OF THYRISTOR CHANGEOVER SWITCH INTRODUCED TO TOHOKU SHINKANSEN SHIN-YONO SECTIONING POST 1810

Yuki Mizumoto ; Nobuhito Kurosawa

HARDWARE–IN–THE–LOOP REAL–TIME SIMULATION EXPERIMENT PLATFORM FOR TRACTION POWER SUPPLY SYSTEM BASED ON DSPACE-XSIM 1816

Runze Zhang ; Fei Lin ; Zhongping Yang ; Hu Cao ; Yuping Liu

EVALUATING THE NON-SINUSOIDAL AND NON-SYMMETRIC REGIMES FROM A RAILWAY SUPPLYING SUBSTATION 1822

Ileana-Diana Nicolae ; Petre-Marian Nicolae ; Radu-Florin Marinescu

A FUNDAMENTAL TRAIN RUNNING EXPERIMENT FOR A BASIC PERFORMANCE VERIFICATION OF A TRAIN POWER DEMAND CONTROL SYSTEM BY DECENTRALIZED CONTROL ALGORITHM 1828

Yusuke Oki ; Tomoyuki Ogawa ; Yoko Takeuchi ; Tatsuhito Saito ; Jun'ichiro Kawaguchi

VERIFICATION OF SIC BASED MODULAR MULTILEVEL CASCADE CONVERTER (MMCC) FOR HVDC TRANSMISSION SYSTEMS 1834
Y. Ishii ; T. Jimichi

CONTROL OF A 6.6-KV TRANSFORMERLESS STATCOM BASED ON THE MMCC-SDBC USING SIC MOSFETS 1840
Laxman Maharjan ; Toshihisa Tajyuta ; Hiroshi Shinohara ; Akio Suzuki ; Akio Toba

ISOLATED THREE–PHASE AC/DC CONVERTER USING A SOFT–SWITCHING TECHNIQUE FOR BATTERY CHARGER 1847
Yuto Matsui ; Kazuma Suzuki ; Takaharu Takeshita ; Wataru Kitagawa

IMPLEMENTATION OF A MINIATURIZED SIC INVERTER 1854
Hideaki Fujita ; Cristian Andres Garces Guajardo

DESIGN CONSIDERATION OF FLYING CAPACITOR MULTILEVEL INVERTERS USING SIC MOSFETS 1860
Yukihiko Sato ; Kenji Natori

A CONTROL METHOD OF OVERVOLTAGE SUPPRESSION ACROSS THE DC CAPACITOR IN A GRID-CONNECTION CONVERTER USING LEG SHORT-CIRCUIT OF POWER MOSFETS DURING THE INITIAL CHARGE 1866
Tomoyuki Mannen ; Keiji Wada

THE ESSENTIAL RELATIONSHIP BETWEEN DEADBEAT PREDICTIVE CONTROL AND CONTINUOUS-CONTROL-SET MODEL PREDICTIVE CONTROL FOR PWM CONVERTERS 1872
Bi Liu ; Tao Chen ; Wensheng Song

DEADBEAT CONTROL FOR MULTI-LEVEL INVERTER USING 1MHZ MULTISAMPLING METHOD FOR UTILITY INTERACTIVE SYSTEM 1877
Ryosuke Kikuchi ; Ryunosuke Araumi ; Tomoki Yokoyama

1MHZ MULTISAMPLING DEADBEAT CONTROL WITH DISTURBANCE COMPENSATION METHOD FOR THREE PHASE PWM INVERTER 1883
Hiroaki Ueta ; Tomoki Yokoyama

MODULAR MULTILEVEL CONVERTER REPLACED ONE MODULE WITH HIGH VOLTAGE IGBT 1890
Kazunobu Oi ; Kenta Takasho ; Yugo Tadano

INCREASED EFFICIENCY AND REDUCED REALIZATION EFFORT OF DSBC AND DSCC MODULAR MULTILEVEL CONVERTERS (MMCS) 1896
A. Hillers ; J. Biela

COMMON-MODE VOLTAGE INJECTION TECHNIQUES FOR QUASI TWO-LEVEL PWM-OPERATED MODULAR MULTILEVEL CONVERTERS 1904
Jakub Kucka ; Axel Mertens

CURRENT TRACKING AND CELL-VOLTAGE LIMITATIONS OF MODULAR MULTILEVEL CONVERTERS WITH DIRECT DIGITAL CONTROL 1912
T.-F. Wu ; T.-C. Chou ; K.-E. Lin ; T.-Y. Li

SWITCHING LOSS ANALYSIS OF SIC-MOSFET BASED ON STRAY INDUCTANCE SCALING 1919
Keiji Wada ; Masato Ando

MODELING AND OPTIMIZATION OF DISPLACEMENT WINDINGS FOR TRANSFORMERS IN DUAL ACTIVE BRIDGE CONVERTERS 1925
Zhan Shen ; Yanfeng Shen ; Zian Qin ; Huai Wang

OPTIMIZED SELECTION AND UTILIZATION OF DC-LINK CAPACITOR IN A SINGLE-PHASE PV GRID INVERTER SYSTEM 1931
Caspar Collins ; Li Ran

AN EVALUATION CIRCUIT FOR DC-LINK CAPACITORS USED IN A HIGH-POWER THREE-PHASE INVERTER WITH CONDITION MONITORING 1938
Kazunori Hasegawa ; Ichiro Omura ; Shin-Ichi Nishizawa

RECENT MARKET AND TECHNICAL TRENDS IN COPPER ROTORS FOR HIGH-EFFICIENCY INDUCTION MOTORS 1943
Daniel Liang ; Victor Zhou

OVERVIEW OF THE LATEST RESEARCH AND DEVELOPMENT FOR COPPER DIE-CAST SQUIRREL-CAGE ROTORS 1949
Shu Yamamoto

A NOVEL HEAT-RESISTANT INSULATION-PROCESSING AGENT APPLICABLE TO COPPER DIE-CAST SQUIRREL-CAGE ROTORS 1955
Junichi Uchida ; Yuki Sueuchi ; Naosumi Kamiyama

INSULATION-PROCESSING OF COPPER DIE-CAST SQUIRREL-CAGE ROTOR ON MOTOR EFFICIENCY IN HIGH-SPEED OPERATION OVER 10,000 R/MIN 1960
Hideaki Hirahara ; Akira Tanaka ; Shu Yamamoto

HIGH-PRECISION ROTOR POSITION ESTIMATION FOR HIGH-SPEED SPMSM DRIVE BASED ON STATE OBSERVER AND HARMONIC ELIMINATION 1966
Peng Yang ; Xi Xiao ; Meng Zhang ; Shkodyrev Vyacheslav

HARMONIC LOSS REDUCTION IN HIGH SPEED MOTOR DRIVE SYSTEMS BY FLYING CAPACITOR MULTILEVEL INVERTER 1972
Anudari Tumurbaatar ; Sae Mochidate ; Koji Yamaguchi ; Tomohiro Matsuda ; Yukihiko Sato

CURRENT SOURCE TYPE PMSG WIND TURBINE SYSTEM WITH THREE-PHASE THREE-SWITCH BUCK-TYPE RECTIFIER FOR MACHINE-SIDE CONVERTER 1977
Beomseok Chae ; Tahyun Kang ; Yongsug Suh

A STUDY OF 10MW LOAD COMMUTATED INVERTER FOR GAS-TURBINE START-UP 1985
An Hyunsung ; Cha Hanju

PROTOTYPING OF 500 KVA MEDIUM FREQUENCY TRANSFORMER FOR OFFSHORE DIRECT-CURRENT COLLECTION GRID 1991
Tomoyuki Hatakeyama ; Naoyuki Kurita ; Mamoru Kimura

PSCAD/EMTDC AND RTDS SIMULATION ANALYSIS OF MULTIVENDOR MULTI-TERMINAL HVDC SYSTEM CONNECTED TO OFFSHORE WINDFARMS 1997
Hiroshi Suwa ; Takuro Arai ; Takahiro Ishiguro ; Tohru Yoshihara ; Mamoru Kimura ; Tsuneshisa Wachi ; Takahiro Horikoshi ; Tatsuhito Nakajima

INTEROPERABILITY OF MODULAR MULTILEVEL CONVERTERS AND 2-LEVEL VOLTAGE SOURCE CONVERTERS IN A LABORATORY-SCALE MULTI-TERMINAL DC GRID 2003
Salvatore D'arco ; Atsede G. Endegnanew ; Giuseppe Guidi ; Jon Are Suul

PRINCIPLE EXPERIMENT OF CURRENT COMMUTATED HYBRID DCCB FOR HVDC TRANSMISSION SYSTEMS 2011
Ryuta Hasegawa ; Kazuhisa Kanaya ; Yushi Koyama ; Toshiaki Matsumoto ; Takahiro Ishiguro

A THREE-INPUT CENTRAL CAPACITOR DC/DC CONVERTER 2016
Jiaxin Liu ; Feng Gao

SERIES/PARALLEL SWITCHING CIRCUITS USING POWER MOSFETS FOR PHOTOVOLTAIC MODULES 2022
Masamichi Tanemo ; Koki Matsudate ; Shinichi Nomura

MODULARIZED EQUALIZATION ARCHITECTURE BASED ON SWITCHED CAPACITOR CONVERTER TO VIRTUALLY UNIFY MISMATCHED PHOTOVOLTAIC PANEL CHARACTERISTICS 2030
Masatoshi Uno ; Masaya Yamamoto

BUCK-BOOST TYPE MPPT CIRCUIT SUITABLE FOR PHOTOVOLTAIC GENERATION OF VEHICLE INSTALLATION 2036
Fumihisa Kano ; Yuji Kasai ; Hideki Kimura ; Kouhei Sagawa ; Junnosuke Haruna ; Hirohito Funato

VERIFICATION TEST OF ENERGY-EFFICIENT OPERATIONS AND SCHEDULING UTILIZING AUTOMATIC TRAIN OPERATION SYSTEM 2042
Shoichiro Watanabe ; Yasuhiro Sato ; Takafumi Koseki ; Eisuke Isobe ; Jun Kawashita

THE DIRECT BENEFIT OF SIC POWER SEMICONDUCTOR DEVICES FOR RAILWAY VEHICLE TRACTION INVERTERS 2047
Shingo Makishima ; Kazuki Fujimoto ; Keiichiro Kondo

THE LOSS CHARACTERISTICS OF PSFB ZVS DC-DC CONVERTER APPLIED TO THE AUXILIARY POWER SYSTEM 2051
Xianjin Huang ; Juan Zhao ; Fei Lin

SURVEY ON ELECTROMAGNETIC INTERFERENCE ANALYSIS FOR TRACTION CONVERTERS IN RAILWAY VEHICLES 2058
Zhichang Yang ; Hong Li ; Chao Feng ; Yanfeng Jiang ; Fei Lin ; Zhongping Yang

DEVELOPMENT OF TRACTION MOTOR FOR NEW ZERO - EMISSION VEHICLE 2066
Akinobu Iwai ; Satoshi Honjo ; Hirofumi Suzumori ; Toshio Okazawa

EMC DESIGN AND DEVELOPMENT METHODOLOGY FOR TRACTION POWER INVERTERS OF ELECTRIC VEHICLES 2073
Isao Hoda ; Jia Li ; Hiroki Funato

SIMULATION-DRIVEN DESIGN OPTIMIZATION OF A MULTILAYER EMC INPUT FILTER 2078
Fatou Diouf ; Nadim Sakr ; Anna Gheonjian

EV TRACTION INVERTER EMPLOYING DOUBLE-SIDED DIRECT-COOLING TECHNOLOGY WITH SIC POWER DEVICE 2082
Takashi Hirao ; Masami Onishi ; Yusuke Yasuda ; Akihiro Namba ; Kinya Nakatsu

AN OVERVIEW OF STABILITY IMPROVEMENT METHODS FOR WIDE-OPERATION-RANGE FLYBACK CONVERTER WITH VARIABLE FREQUENCY PEAK-CURRENT-MODE CONTROL 2086

Ching-Hsiang Cheng ; Ching-Jan Chen ; Shinn-Shyong Wang

DESIGN AND IMPLEMENTATION OF A HIGH POWER DENSITY ACTIVE-CLAMPED FLYBACK CONVERTER 2092

Yu-Chen Liu ; Bing-Siang Huang ; Cheng-Hung Lin ; Katherine A. Kim ; Huang-Jen Chiu

OPTIMIZED VARIABLE ON-TIME CONTROL FOR LED LIGHTING DRIVER 2097

Jizhe Wang ; Haruhi Eto ; Fujio Kurokawa

DESIGN OF MULTIMODE BATTERY CHARGER WITH DYNAMIC VOLTAGE TRACKING CONTROL 2102

Pang-Jung Liu ; Lin-Hao Chien ; Song-Kai Lee ; Ang-Tung Chen

DUAL-SLOT POWER-PICKUP STRUCTURE FOR CONTACTLESS STRIP INDUCTIVE POWER TRACK SYSTEM 2107

Jia-You Lee ; I-Lin Chen ; Chien-Tzu Ko

DISCONTINUOUS SVM TECHNIQUE FOR THREE-LEG VSI FED BALANCED/UNBALANCED TWO-PHASE LOADS 2113

Supanut Charoensuksirikul ; Yuttana Kumsuwan

REDUCTION OF POWER LOSSES BASED ON GENERALIZED TWO-LEVEL PWM ALGORITHM FOR A NINE-SWITCH VSI 2121

Neerakorn Jarutus ; Yuttana Kumsuwan

SIC-BASED THREE-PHASE QUASI-Z-SOURCE INVERTER VERSUS THE TWO-STAGE TOPOLOGY - A COMPARISON 2129

Kornel Wolski ; Mariusz Zdanowski ; Jacek Rabkowski

DC-SIDE CIRCUIT IMPLEMENTATION OF A THREE-PHASE INVERTER FOR BALANCING PHASE-LEG CAPACITOR CURRENTS 2137

Takashi Hirao ; Keiji Wada ; Toshihisa Shimizu

A THREE-PHASE HYBRID SWITCHED-BOOST INVERTER 2145

Minh-Khai Nguyen ; Tan-Tai Tran ; Hoan-Tien Luong ; Kyoung-Won Lee ; Youn-Ok Choi ; Geum-Bae Cho

THE EFFECT OF BUILT-IN CR SNUBBER CAPACITOR INTO THE POWER MODULE 2149

Ryotaro Hata ; Shigeki Nishiyama

EVALUATION OF NOVEL HYBRID PROTECTION BASED ON PYROSWITCH AND FUSE TECHNOLOGIES 2153

Tomokazu Sakuraba ; Rémy Ouaida ; Song Chen ; Thibaut Chailloux

OPTIMAL DESIGN OF A MAGNETICALLY COUPLED FILTER FOR HIGH EFFICIENCY, LOW COST AND LOW VOLUME DC-DC BATTERY STORAGE CONVERTER 2158

Timothé Delaforge ; Robert Pasterczyk ; Mickaël Robert ; Hervé Chazal ; Jean-Luc Schanen ; Sébastien Mariethoz

HIGH POWER/CURRENT INDUCTOR LOSS MEASUREMENT WITH SHUNT RESISTOR CURRENT-SENSING METHOD 2165

Pin Yu Huang ; Toshihisa Shimizu

SENSITIVITY ANALYSIS OF MEDIUM FREQUENCY TRANSFORMER DESIGN 2170

Marko Mogorovic ; Drazen Dujic

STANDARD MODELS FOR POWER ELECTRONIC SYSTEM SIMULATION 2176

Koichi Shigematsu ; Hiroki Ishikawa ; Taku Noda ; Kentarou Fukushima ; Yoichi Sekiba ; Yusuke Kouno ; Takashi Abe ; Takayuki Sekisue ; Shinji Katoh

MODELING AND MODEL PARAMETER EXTRACTION OF WIDE BANDGAP POWER SEMICONDUCTOR DEVICE, PACKAGE, AND CIRCUIT FOR SIMULATING FAST SWITCHING BEHAVIOR 2181

Tsuyoshi Funaki

STABILITY ANALYSIS METHODS OF A GRID-CONNECTED INVERTER IN TIME AND FREQUENCY DOMAINS 2186

Toshiji Kato ; Kaoru Inoue ; Taiki Sakiyama

FINITE ELEMENT METHODS FOR MULTI-OBJECTIVE OPTIMIZATION OF A HIGH STEP-UP INTERLEAVED BOOST CONVERTER 2193

Wilmar Martinez ; Camilo Cortes ; Ahmad Bilal ; Jorma Kyyra

HIGH FIDELITY REAL-TIME SIMULATION OF MULTI-LEVEL CONVERTERS 2199

Jost Allmeling ; Niklaus Felderer ; Min Luo

AN ENHANCED HIGH FREQUENCY PULSATING VOLTAGE INJECTION METHOD BASED ON IMMUNE ALGORITHM FOR SENSORLESS IPMSM DRIVES 2204

Yanping Zhang ; Zhonggang Yin ; Chao Du ; Youyun Wang ; Xiangdong Sun

POSITION ESTIMATION ACCURACY IMPROVEMENT FOR MAGNETIC SALIENCY BASED SENSORLESS CONTROL INCLUDING CROSS-COUPLING FACTOR 2210

Keita Shimamoto ; Shinya Morimoto

SENSORLESS DRIVE IN THE LOW SPEED REGION AND AUTO-TUNING METHOD FOR PERMANENT MAGNET SYNCHRONOUS MOTORS 2216

Naofumi Nomura ; Shinichi Higuchi

HIGH STABILITY V/F CONTROL OF PMSM USING STATE FEEDBACK CONTROL BASED ON N-T COORDINATE SYSTEM 2224

Yosuke Matsuki ; Shinji Doki

STABILIZATION METHOD USING EQUIVALENT RESISTANCE GAIN BASED ON V/F CONTROL FOR IPMSM WITH LONG ELECTRICAL TIME CONSTANT 2229

Jun-Ichi Itoh ; Takato Toi ; Koroku Nishizawa

SINGLE-PHASE SOLID-STATE TRANSFORMER USING MULTI-CELL WITH AUTOMATIC CAPACITOR VOLTAGE BALANCE CAPABILITY 2237

Jun-Ichi Itoh ; Kazuki Aoyagi ; Keisuke Kusaka ; Masakazu Adachi

A DEVELOPED DUAL MMC ISOLATED DC SOLID STATE TRANSFORMER AND ITS MODULATION STRATEGY 2245

Yan Li ; Chao Liu ; Xu Cai

DC FAULT RIDE-THROUGH OF A THREE-PHASE DUAL-ACTIVE BRIDGE CONVERTER FOR DC GRIDS 2250

Jingxin Hu ; Shenghui Cui ; Rik W. De Doncker

A COMPOUND 10KV DVR SYSTEM BASED ON SOLID STATE TRANSFORMER STRUCTURE 2262

Yaqian Zhang ; Jianzhong Zhang ; Xing Hu ; Zakiud Din

A DUAL-ENERGY-SOURCE UNINTERRUPTIBLE POWER SUPPLY (UPS) 2270

Hao Wang ; Dehong Xu ; Binci Xu ; Haijin Li ; Ye Zhu

INFLUENCE OF WIND POWER FORECASTS ON EQUITABLE DISTRIBUTION METHOD OF WIND POWER CURTAILMENT 2278

Daisuke IIoka ; Hiroumi Saitoh

COMPARISON OF OPTIMIZED DEMAND OF EGS FOR MINIMIZING FUEL CONSUMPTION AND EGS MODEL WITH POWER GRID FREQUENCY USING A HPSPITAL LOAD WITH PV 2283

Yuji Mizuno ; Teppei Baba ; Fujio Kurokawa ; Nobumasa Matsui

COORDINATED DFIG WIND TURBINES AND SOLAR PV GENERATORS FOR INTER-AREA OSCILLATION DAMPING 2287

Tossaporn Surinkaew ; Issarachai Ngamroo

ENERGY MANAGEMENT USING A QUICK CHARGER WITH STORAGE BATTERIES FOR ELECTRIC VEHICLES 2292

Taku Ishibashi ; Toyonari Shimakage ; Norikazu Takeuchi ; Takaaki Kikuchi ; Midori Nonogaki

A METHOD FOR JUNCTION TEMPERATURE ESTIMATION UTILIZING TURN-ON SATURATION CURRENT FOR SIC MOSFET 2296

Hui-Chen Yang ; Rejeki Simanjorang ; Kye Yak See

FIELD BUS FOR DATA EXCHANGE AND CONTROL OF MODULAR POWER ELECTRONIC SYSTEMS WITH HIGH SYNCHRONISATION ACCURACY 2301

Stefan Rietmann ; Simon Fuchs ; André Hillers ; Jürgen Biela

ANALYTICAL INVESTIGATION ON ASYMMETRIC LCC COMPENSATION CIRCUIT FOR TRADE-OFF BETWEEN HIGH EFFICIENCY AND POWER 2309

Kodai Takeda ; Takafumi Koseki

PROBABILISTIC PCA-SUPPORT VECTOR MACHINE BASED FAULT DIAGNOSIS OF SINGLE PHASE 5-LEVEL CASCADED H-BRIDGE MLI 2317

Nagendra Vara Prasad Kuraku ; Yigang He ; Murad Ali

A STUDY ON EDGE SUPPORTED ELECTROMAGNETIC LEVITATION SYSTEM: FUNDAMENTAL CONSIDERATION ON LEVITATION PERFORMANCE OF THIN STEEL PLATE 2324

Yoshiho Oda ; Yasuaki Ito ; Kengo Okuno ; Masahiro Kida ; Toshiki Suzuki ; Takayoshi Narita ; Hideaki Kato ; Hiroyuki Moriyama

APPLICATION OF FACTS DEVICES FOR A DYNAMIC POWER SYSTEM WITHIN THE USA 2329

Jan Paramalingam ; Fuminori Nakamura ; Akihiro Matsuda ; Daisuke Yamanaka ; Taichiro Tsuchiya

CAPACITOR VOLTAGE BALANCING IN SEMI-FULL-BRIDGE SUBMODULE WITH DIFFERENTIAL-MODE CHOKE : (INVITEDPAPER) 2335

Kalle Ilves ; Yuhei Okazaki ; Nan Chen ; Muhammad Nawaz ; Antonios Antonopoulos

RESEARCH ON KEY TECHNOLOGY AND EQUIPMENT FOR ZHANGBEI 500KV DC GRID 2343

Hui Pang ; Xiaoguang Wei

WHAT LED TO SUCCESS IN ACADEMIC RESEARCH ON THE FAMILY OF MODULAR MULTILEVEL CASCADE CONVERTERS? 2352
Hirofumi Akagi

OPERATING PRINCIPLE OF CURRENT RESONANT CONVERTER USING AIR CORE TRANSFORMER FOR ISOLATED POWER SUPPLY ON CHIP 2360
Seiya Abe ; Hikaru Kaishakuji ; Satoshi Matsumoto

ANALYSIS FOR HIGH-FREQUENCY LLC RESONANT CONVERTER WITH PLANAR TRANSFORMER AT LIGHT-LOAD CONDITION 2365
Keon-Woo Kim ; Jae-Il Baek ; Yeonho Jeong ; Ki-Mok Kim ; Gun-Woo Moon

A NOVEL FULL DIGITAL CONTROL H-BRIDGE DC-DC CONVERTER FOR POWER SUPPLY ON CHIP APPLICATIONS 2370
Shigeki Nakano ; Toshiomi Oka ; Seiya Abe ; Satoshi Matsumoto

A HIGH-EFFICIENCY POWER SUPPLY FROM MAGNETIC ENERGY HARVESTERS 2376
Cheon-Yong Lim ; Yeonho Jeong ; Keon-Woo Kim ; Feel-Soon Kang ; Gun-Woo Moon

OPPORTUNITIES FOR LEVERAGING LOW-VOLTAGE GAN DEVICES IN MODULAR MULTI-LEVEL CONVERTERS FOR ELECTRIC-VEHICLE CHARGING APPLICATIONS 2380
Mojtaba Ashourloo ; Mohammad Shawkat Zaman ; Miad Nasr ; Olivier Trescases

A NEW CONTROL STRATEGY FOR MODULAR MULTILEVEL CONVERTER OPERATING IN QUASI TWO-LEVEL PWM MODE 2386
Chao Wang ; Kui Wang ; Zedong Zheng ; Yongdong Li

A CURRENT-SOURCE TYPE MMC WITH DELTA-CONNECTED ARMS FOR SMES 2393
Yushi Miura ; Toshifumi Ise

NEW MODULE WITH ISOLATED HALF BRIDGE OR ISOLATED FULL BRIDGE FOR MODULAR MEDIUM VOLTAGE CONVERTER 2400
Yunpeng Si ; Yifu Liu ; Qin Lei

DEVELOPMENT OF A 700-V-CLASS REVERSE-BLOCKING IGBT FOR ADVANCED T-TYPE NEUTRAL POINT-CLAMPED POWER CONVERSION SYSTEM 2404
Hiroki Wakimoto ; Haruo Nakazawa ; David H. Lu ; Takashi Matsumoto ; Yoichi Nabetani

CERAMIC EMBEDDING AS PACKAGING SOLUTION FOR FUTURE POWER ELECTRONIC APPLICATIONS 2410
Hoang Linh Bach ; Tobias Maximilian Endres ; Daniel Dirksen ; Sigrid Zischler ; Christoph Friedrich Bayer ; Andreas Schletz ; Martin März

MICROELECTROMECHANICAL SYSTEM (MEMS) RESONATOR: A NEW ELEMENT IN POWER CONVERTER CIRCUITS FEATURING REDUCED EMI 2416
A N M Wasekul Azad ; Sourov Roy ; Abu Saleh Imtiaz ; Faisal Khan

A LUMPED THERMAL MODEL INCLUDING THERMAL COUPLING EFFECTS AND BOUNDARY CONDITIONS FOR CAPACITOR BANKS 2421
Qiusheng Wang

HYSTERESIS MODELING OF MAGNETIC DEVICES BASED ON RELUCTANCE NETWORK ANALYSIS 2426
Yoshiki Hane ; Kenji Nakamura

OPTIMAL SIZING AND PLACEMENT OF SOLAR POWERED CHARGING STATION UNDER EV LOADS PENETRATION USING ARTIFICIAL BEE COLONY TECHNIQUE 2430
Yuttana Kongjeen ; Kulsomsup Yenchamchalit ; Krischonme Bhumkittipich

A COMPARISON OF AVERAGE MODEL, SAMPLED-DATA MODEL AND MULTI-FREQUENCY MODEL BASED ON DC/DC CONVERTERS 2435
Xiangpeng Cheng ; Jinjun Liu ; Zeng Liu ; Yiming Tu ; Danhong Xue

SMALL-SIGNAL DISCRETE-TIME MODELING AND DIGITAL CONTROL OF THE BI-DIRECTIONAL DC/DC CONVERTERS 2441
Jia Yaoqin ; Xu Yingchun ; Hou Yijie

ENERGY MANAGEMENT OF HYDROGEN-STORAGE PHOTOVOLTAIC GENERATION SYSTEM WITH A FUNCTION OF SUPPRESSING SHORT-PERIOD COMPONENTS 2449
Yuuki Machida ; Akihisa Goto ; Akiko Takahashi ; Shigeyuki Funabiki

A DYNAMIC BATTERY CHARGING APPROACH FOR ENERGY TRADING IN THE SMART GRID 2456
Avinash Sharma ; Akshay Kumar Rathore ; Rajesh Kumar

A FORCED COMMUTATION METHOD OF THE SOLID-STATE TRANSFER SWITCH IN THE UNINTERRUPTED POWER SUPPLY APPLICATIONS 2462
Meng-Jiang Tsai ; Jiuyang Zhou ; Po-Tai Cheng

ONLINE INTERNAL IMPEDANCE MEASUREMENTS OF LI-ION BATTERY USING PRBS BROADBAND EXCITATION AND FOURIER TECHNIQUES: METHODS AND INJECTION DESIGN 2470
Jussi Sihvo ; Tuomas Messo ; Tomi Roinila ; Roni Luhtala

A DC CURRENT FLOW CONTROLLER FOR MESHED HVDC GRIDS 2476
Viktor Hofmann ; Mark-M. Bakran

AN ISOLATED SOFT-SWITCHING HYBRID-SOURCE DC-DC CONVERTER FOR DC OFFSHORE WIND FARMS 2484
Shenghui Cui ; Jingxin Hu ; Marco Stieneker ; Rik W. De Doncker

A TRANSFORMERLESS MULTI-CELL SOLID-STATE FAULT CURRENT LIMITER FOR MEDIUM VOLTAGE POWER SYSTEM 2490
Pantarote Techama ; Sompob Polmai ; Chanin Bunlaksananusorn

A NOVEL DC POWER FLOW CONTROLLER FOR HVDC GRIDS WITH DIFFERENT VOLTAGE LEVELS 2496
Ya'nan Wu ; Han Ye ; Wu Chen ; Xiaokun He

DESIGN AND CONTROL OF SINGLE-PHASE GRID-CONNECTED PHOTOVOLTAIC MICROINVERTER WITH REACTIVE POWER SUPPORT CAPABILITY 2500
Geon-Hong Min ; Kyung-Hwan Lee ; Jung-Ik Ha ; Myong Hwan Kim

OPTIMAL SIZE AND MULTI-OBJECTIVE CONTROL OF BATTERY ENERGY STORAGES IN DISTRIBUTION SYSTEM WITH HIGH PENETRATION OF DISTRIBUTED PV GENERATORS 2505
Meiqin Mao ; Lei Zhou ; Yangyang Wang ; Liuchen Chang

MISSION PROFILE-ORIENTED CONTROL FOR RELIABILITY AND LIFETIME OF PHOTOVOLTAIC INVERTERS 2512
Ariya Sangwongwanich ; Yongheng Yang ; Dezso Sera ; Frede Blaabjerg

DISCONTINUOUS CURRENT MODE CONTROL FOR MINIMIZATION OF THREE-PHASE GRID-TIED INVERTER IN PHOTOVOLTAIC SYSTEM 2519
Hoai Nam Le ; Jun-Ichi Itoh

A THEORETICAL ANALYSIS ON STATIC CHARACTERISTICS OF VOLTAGE BASED CONTROL METHOD AND CURRENT BASED CONTROL METHOD FOR THE WAYSIDE ENERGY STORAGE SYSTEM IN DC-ELECTRIFIED RAILWAY 2527
Hiroyasu Kobayashi ; Keiichiro Kondo ; Diego Iannuzzi

IMPROVEMENT OF A DC ELECTRICAL RAILWAY SIMULATOR USING ARTIFICIAL INTELLIGENCE 2534
Alvaro J. Lopez-Lopez ; Ramon R. Pecharroman ; Antonio Fernandez-Cardador ; Asuncion P. Cucala

FEEDING-LOSS REDUCTION BY HIGHER-VOLTAGE DC RAILWAY FEEDING SYSTEM WITH DC-TO-DC CONVERTER 2540
Hidenori Shigeeda ; Hiroaki Morimoto ; Kazuhiko Ito ; Toshiyuki Fujii ; Naoki Morishima

MODELING AND SIMULATION OF NOVEL RAILWAY POWER SUPPLY SYSTEM BASED ON POWER CONVERSION TECHNOLOGY 2547
Minwu Chen ; Ruofei Liu ; Shaofeng Xie ; Xiaofang Zhang ; Yimin Zhou

COMPARATIVE STUDY ON FRONT-END PARAMETER IDENTIFICATION METHODS FOR WIRELESS POWER TRANSFER WITHOUT WIRELESS COMMUNICATION SYSTEMS 2552
Sinan Li ; S. Y. Ron Hui

A NEW TYPE OF WIRELESS V2X SYSTEM WITH A DUAL-ACTIVE BIDIRECTIONAL SINGLE-ENDED CONVERTER AND OPTIMIZED SIC-MOSFET 2558
Hideki Omori ; Aoto Yamamoto ; Naoki Mukaiyama ; Masahito Tsuno ; Kenji Fukuda ; Hisato Michikoshi ; Noriyuki Kimura ; Toshimitsu Morizane

METAL OBJECT DETECTION SYSTEM WITH PARALLEL-MISTUNED RESONANT CIRCUITS AND NULLIFYING INDUCED VOLTAGE FOR WIRELESS EV CHARGERS 2564
Seog Y. Jeong ; Van X. Thai ; Jun H. Park ; Chun T. Rim

WIRELESS EV CHARGING SYSTEM WITHOUT AIR-GAP AND MISALIGNMENT 2569
Wenxing Zhong ; Dehong Xu

FIXED SLOPE CARRIER PWM FOR INDIRECT MATRIX CONVERTER 2576
Tzung-Lin Lee ; Chun-Yao Hung ; Yen-Wen Chen ; Wen-Mei Huang

CARRIER-BASED OVERMODULATION STRATEGY FOR MATRIX CONVERTERS 2581
Paiboon Kiatsookkanatorn ; Somboon Sangwongwanich

THREE-PHASE TO HIGH-FREQUENCY SINGLE-PHASE MATRIX CONVERTER : A FREQUENCY CONTROL SUITABLE FOR SOFT SWITCHING 2589
Wataru Kodaka ; Satoshi Ogasawara ; Koji Orikawa ; Masatsugu Takemoto ; Takashi Hyodo ; Hiroyuki Tokusaki

TWO-STEP COMMUTATION FOR ISOLATED DC-AC CONVERTER WITH MATRIX CONVERTER 2596
Shunsuke Takuma ; Jun-Ichi Itoh

A DC-LINK CAPACITOR VOLTAGE OSCILLATION REDUCTION METHOD FOR A MODULAR MULTILEVEL CASCADE CONVERTER WITH SINGLE DELTA BRIDGE CELLS (MMCC-SDBC) 2604

Takaaki Tanaka ; Huai Wang ; Frede Blaabjerg

OPTIMIZED DECOUPLING CONTROL OF FLYING CAPACITOR IN ANPC FIVE-LEVEL INVERTER 2611

Fusheng Wang ; Deyou Zheng ; Jianing Wang ; Fei Li ; Fang Liu ; Shuying Yang ; Zhen Xie

CASCADED DUAL-BUCK AC-AC CONVERTER USING COUPLED INDUCTORS 2619

Sanghun Kim ; Duekjin Jang ; Heung-Geun Kim ; Honnyong Cha

INSTANTANEOUS POWER LOSS CALCULATION FOR MMC BASED ON VIRTUAL ARM MATHEMATICAL MODEL 2625

Yin Shiyuan ; Wang Yue ; Yin Taiyuan ; Nie Cheng ; Duan Guozhao ; Wang Zhang

COMPARISON OF CURRENT CONTROL STRATEGIES IN MODULAR MULTILEVEL CONVERTER 2630

Jianzhao Wei ; Anirudh Budnar Acharya ; Lars Norum ; Pavol Bauer

MODEL PREDICTIVE CONTROL OF A MODULAR MULTILEVEL CONVERTER WITH AN IMPROVED CAPACITOR BALANCING METHOD 2638

Shichong Zhang ; Baodong Bai ; Dezhi Chen

HIGH STEP-UP DC-DC CONVERTER BASED ON MULTI-CELL COUPLED INDUCTOR DIODE-CAPACITOR NETWORK 2646

Xinying Li ; Yan Zhang ; Jinjun Liu ; Pengxiang Zeng

NOVEL ACTIVE CLAMPING STEP-DOWN DC-DC CONVERTER WITH LOWER VOLTAGE STRESS 2653

Chi-Hsuan Hsu ; Jun-Min Jian ; Jiann-Fuh Chen ; Hsuan Liao

DESIGN AND EVALUATION OF A MAGNETICALLY-LOOSELY-COUPLED INDUCTOR FOR A FOUR-PHASE INTERLEAVED BOOST CHOPPER 2660

Hiroki Kowatari ; Toshinori Kitamura ; Nobukazu Hoshi

A SYNCHRONOUS-REFERENCE-FRAME I-V DROOP CONTROL METHOD FOR PARALLEL-CONNECTED INVERTERS 2668

Mingshen Li ; Yonghao Gui ; Zheming Jin ; Yajuan Guan ; Josep M. Guerrero

TRANSIENT STABILITY IMPACT OF THE PHASE-LOCKED LOOP ON GRID-CONNECTED VOLTAGE SOURCE CONVERTERS 2673

Heng Wu ; Xiongfei Wang

COMPREHENSIVE ANALYSIS OF VIRTUAL IMPEDANCE-BASED ACTIVE DAMPING FOR LCL RESONANCE IN GRID-CONNECTED INVERTERS 2681

Teng Liu ; Zeng Liu ; Jinjun Liu ; Yiming Tu ; Zipeng Liu

A COMPARATIVE STUDY OF THE TRADITIONAL FS-MPC AND THE PROPOSED CSF-PCC FOR THE THREE-PHASE GRID-CONNECTED INVERTERS 2688

Zhixun Ma ; Xin Zhang ; Jingjing Huang

CONSTANT SWITCHING-FREQUENCY PREDICTIVE- CURRENT-CONTROL METHOD WITH A DICHOTOMY SOLUTION FOR THE GRID-TIED INVERTERS 2692

Zhixun Ma ; Xin Zhang ; Jingjing Huang ; Zhao Bin ; Lyu Jing

OBSERVER-BASED ACTIVE DAMPING FOR GRID-CONNECTED CONVERTERS WITH LCL FILTER 2697

Y. Zhang ; M. G. L. Roes ; M. A. M. Hendrix ; J. L. Duarte

CONDUCTION LOSS ANALYSIS AND OPTIMIZATION DESIGN OF FULL BRIDGE LLC RESONANT CONVERTER 2703

Yugang Yang ; Lifei Zhang ; Tianshu Ma

FULL-BRIDGE T-TYPE ISOLATED DC/DC CONVERTER WITH WIDE INPUT VOLTAGE RANGE 2708

Dong Liu ; Yanbo Wang ; Fujin Deng ; Zhe Chen

RESEARCH ON HIGH EFFICIENCY LLC DC-DC CONVERTER BASED ON SIC MOSFET 2714

Pengcheng Han ; Xiaoqiong He ; Haijun Ren ; Zhiqing Zhao ; Xu Peng

AN IMPROVED DUAL PHASE SHIFT CONTROL STRATEGY FOR DUAL ACTIVE BRIDGE DC-DC CONVERTER WITH SOFT SWITCHING 2718

Miao Hong ; Gao Xuanjie ; Zeng Chengbi ; Duan Shujiang

DEVELOPMENT OF AN SIC HIGH-FREQUENCY PWM INVERTER USING A THICK MULTILAYER PCB TO MINIMIZE STRAY INDUCTANCE 2725

Kohsuke Ishikawa ; Satoshi Ogasawara ; Masatsugu Takemoto ; Koji Orikawa

FAST SWITCHING PLANAR POWER MODULE WITH SIC MOSFETS AND ULTRA-LOW PARASITIC INDUCTANCE 2732

Arash Edvin Risseh ; Hans-Peter Nee ; Konstantin Kostov

EXPERIMENTAL EVALUATION OF INVERTER SYSTEM CONSISTING OF 4-PARALLEL GAN DEVICES UNIT..........2738
Yoshiya Ohnuma ; Satoshi Miyawaki ; Fumiya Hattori ; Masayoshi Yamamoto

IMPACT OF THE THERMAL-INTERFACE-MATERIAL THICKNESS ON IGBT MODULE RELIABILITY IN THE MODULAR MULTILEVEL CONVERTER..........2743
Yi Zhang ; Huai Wang ; Zhongxu Wang ; Yongheng Yang ; Frede Blaabjerg

NANOSCALE INVESTIGATION OF THE POWER MOSFET BY THE AFM/KFM/SCFM..........2750
Mizuki Nakajima ; Yuuki Uchida ; Nobuo Satoh ; Hidekazu Yamamoto

SIMULATION ANALYSIS OF OPTIMUM GATE DRIVING CONDITIONS OF IGBTS..........2756
Satoshi Sugahara ; Masaki Kawakami ; Kousuke Kamakura

IMPROVEMENT OF THE I2T CAPABILITY FOR XEV ACTIVE SHORT CIRCUIT PROTECTION BY COMBINATION OF RC-IGBT AND LEADFRAME TECHNOLOGIES..........2764
Keiichi Higuchi ; Hayato Nakano ; Akihiro Osawa ; Akio Kitamura ; Shunji Takenoiri ; Daisuke Inoue ; Souichi Yoshida ; Hiromichi Gohara

INVESTIGATION OF SWITCHING BEHAVIOR OF AN IGBT UNDER SOFT TURN-OFF IN APPLICATION FOR DUAL-ACTIVE BRIDGE CONVERTERS..........2768
Eri Ogawa ; Yuichi Onozawa ; Rik W. De Doncker

600 V HIGH VOLTAGE GATE DRIVER IC (HVIC) WITH 1.0 MHZ HIGH FREQUENCY OPERATION FOR LLC CURRENT RESONANT POWER SUPPLY..........2774
Masaharu Yamaji ; Masashi Akahane ; Takahide Tanaka ; Akihiro Jonishi ; Hidetomo Ohashi ; Masahiro Sasaki ; Hitoshi Sumida

AN INTEGRATED VOLTAGE AND CURRENT BALANCING STRATEGY OF SERIES-PARALLEL CONNECTED IGBTS..........2780
Xiaotong Du ; Fang Zhuo ; Haotian Sun ; Hao Yi ; Yanlin Zhu

THERMAL DESIGN AND ANALYSIS OF A CABLE CHARGER USED FOR PORTABLE ELECTRONICS..........2785
Mofan Tian ; Xu Yang ; Naizeng Wang ; Yang Chen ; Laili Wang

PARASITIC INDUCTANCE DESIGN CONSIDERATIONS TO SUPPRESS GATE VOLTAGE OSCILLATION OF FAST SWITCHING POWER SEMICONDUCTOR DEVICES..........2789
Yusuke Sugihara ; Kimihiro Nanamori ; Masayoshi Yamamoto ; Yasuki Kanazawa

THE EXAMINATION OF INCREASING OPERATION SPEED OF CONSEQUENT POLE TYPE AXIAL GAP MOTOR FOR HIGHER OUTPUT POWER DENSITY..........2796
Toru Ogawa ; Tomohira Takahashi ; Masatsugu Takemoto ; Satoshi Ogasawara ; Hideaki Arita ; Akihiro Daikoku

BASIC STUDY OF PMASYNRM WITH BONDED MAGNETS FOR TRACTION APPLICATIONS..........2802
Marika Kobayashi ; Shigeo Morimoto ; Masayuki Sanada ; Yukinori Inoue

STUDY ON ROTOR STRUCTURE SUITABLE FOR IMPROVING POWER DENSITY AND EFFICIENCY IN IPMSMS FOR AUTOMOTIVE APPLICATIONS..........2808
R. Imoto ; M. Sanada ; S. Morimoto ; Y. Inoue

EXAMINATION OF THE DEMAGNETIZATION SUPPRESSION EFFECT OF PLACING FLUX BARRIERS IN AN IPMSM USING RARE-EARTH BONDED MAGNETS..........2814
Takashi Umeda ; Masayuki Sanada ; Shigeo Morimoto ; Yukinori Inoue

A NOVEL POLE-CHANGING METHOD WITH A MULTIPLE THREE-PHASE INVERTER..........2820
Yuki Hidaka ; Taiga Komatsu ; Hideaki Arita

STARTING CHARACTERISTICS OF AN ULTRA-LIGHTWEIGHT MOTOR USING MAGNETIC RESONANCE COUPLING..........2826
Kenta Takishima ; Kazuto Sakai

DESIGN AND BASIC CHARACTERISTICS ANALYSIS OF TOROIDAL WINDING AXIAL GAP INDUCTION MOTOR..........2832
Ryosuke Sakai ; Yukihiro Yoshida ; Katsubumi Tajima

MAGNET ARRANGEMENT SUITABLE FOR LARGE AIR GAP LENGTH IN LINEAR PM VERNIER MOTOR..........2836
Tatsuya Ninomiya ; Abdulaziz Gasim ; Shoji Shimomura

MICRO ELECTROMAGNETIC VIBRATION ENERGY HARVESTER WITH MECHANICAL SPRING AND IRON FRAME FOR LOW FREQUENCY OPERATION..........2842
Yecheng Shen ; Kaiyuan Lu ; Yongming Xia

MEASUREMENT OF TWO-LEVEL INVERTER INDUCED CURRENT SLOPES AT HIGH SWITCHING FREQUENCIES FOR CONTROL AND IDENTIFICATION ALGORITHMS OF ELECTRICAL MACHINES..........2848
Simon Decker ; Andreas Liske ; Daniel Schweiker ; Johannes Kolb ; Michael Braun

A NEW TOPOLOGY OF SWITCHED-CAPACITOR MULTILEVEL INVERTER FOR SINGLE-PHASE GRID-CONNECTED WITH ELIMINATING LEAKAGE CURRENT......2854

Mehdi Samizadeh ; Xu Yang ; Bagher Karami ; Wenjie Chen ; Mohamad Abou Houran ; Adib Abrishamifar ; Abdolreza Rahmati

AN INTERLEAVED BUCK-CASCADED BUCK-BOOST INVERTER FOR PV GRID-CONNECTION APPLICATIONS......2860

Chien-Hsuan Chang ; Chun-An Cheng ; Hung-Liang Cheng

A NOVEL PV ARRAY CONNECTION STRATEGY WITH PV-BUCK MODULE TO IMPROVE SYSTEM EFFICIENCY......2866

Chi Shao ; Wenjie Wang ; Lijun Hang ; Anping Tong ; Shitao Wang

A COMMON-MODE VOLTAGE REDUCTION FOR TWO-STAGE THREE-PHASE TRANSFORMERLESS PV INVERTERS......2871

Adisak Promyoo ; Surapong Suwankawin

A GRID-CONNECTED PV-ENERGY STORAGE SYSTEM WITH SYNCHRONOUS GENERATOR CHARACTERISTICS......2877

Huadian Xu ; Jianhui Su ; Ning Liu ; Yong Shi ; Yan Du

A TRANSFORMERLESS BIDIRECTIONAL DC-DC CONVERTER BASED ON POWER UNITS WITH UNIPOLAR AND BIPOLAR STRUCTURE FOR MVDC INTERCONNECTION......2882

Lejia Sun ; Fang Zhuo ; Feng Wang ; Hao Yi ; Baohui Ma

NEW MODULATION CONTROL OF CONVERTER SYSTEM APPLIED FOR OFFSHORE WIND FARMS......2887

Naoki Kawabata ; Noriyuki Kimura ; Toshimitsu Morizane ; Hideki Omori

SPHERE DECODING BASED LONG-HORIZON PREDICTIVE CONTROL OF THREE-LEVEL NPC BACK-TO-BACK PMSG WIND TURBINE SYSTEMS......2895

Ferdinand Grimm ; Zhenbin Zhang ; Ralph Kennel

BASED ON PCHD AND HPSO SLIDING MODE CONTROL OF D-PMSG WIND POWER SYSTEM......2901

Lijun Hou ; Xuemei Zheng ; Chao Wang ; Yangman Li ; Haoyu Li

ESTABLISHMENT AND DYNAMIC CONTROL OF WIND INDUCTION GENERATOR......2907

M. Z. Lu ; V. K. Ganisetti ; C. M. Liaw

MIDDLE FREQUENCY SOLID STATE TRANSFORMER FOR HVDC TRANSMISSION FROM OFFSHORE WINDFARM......2914

Noriyuki Kimura ; Toshimitsu Morizane ; Isao Iyoda ; Kazushige Nakao ; Tomoki Yokoyama

SIMULATION OF WIND POWER GENERATION SYSTEM USING SWITCHED RELUCTANCE GENERATOR AND CAPACITOR-LESS AC-AC CONVERTER......2921

Guyuan Ji ; Kazuhiro Ohyama

VARIABLE FREQUENCY CONTROL AND FILTER DESIGN FOR OPTIMUM ENERGY EXTRACTION FROM A SIC WIND INVERTER......2932

Abdallah Hussein ; Alberto Castellazzi

EXPERIMENTAL VERIFICATIONS OF UPFC USING DEADBEAT CONTROL WITH 3-PHASE UNBALANCED COMPENSATION......2938

Shin-Ichi Hamasaki ; Hiroto Fukuda ; Syohei Tokumaru ; Mineo Tsuji

A CONTROL METHOD FOR TWO TYPES OF THREE-PHASE TRANSFORMERLESS UNIFIED POWER QUALITY CONDITIONER......2944

Fujian Li ; Guochun Xiao ; Fangzhou Zhao ; Shuai Zhang ; Baojin Liu

DESIGN OF CUSTOMER-END CONVERTER SYSTEMS FOR LOW VOLTAGE DC DISTRIBUTION FROM A LIFE CYCLE COST PERSPECTIVE......2948

A. Mattsson ; P. Nuutinen ; T. Kaipia ; P. Peltoniemi ; J. Karppanen ; V. Tikka ; A. Lana ; P. Pinomaa ; P. Silventoinen ; J. Partanen

A CONTROL METHOD OF DC CAPACITOR VOLTAGE IN MMC FOR HVDC SYSTEM USING NEGATIVE SEQUENCE CURRENT......2956

Hanis Afiqah Binti Jaffar ; Ahmad Arif Bin Abd Rahman ; Hiroaki Kakigano

A COORDINATE AND DISTRIBUTED CONTROL SCHEME FOR MULTILEVEL AND MULTI-STAGE MEDIUM VOLTAGE SOLID STATE TRANSFORMER......2963

Jintong Nie ; Liqiang Yuan ; Qing Gu ; Jianning Sun ; Zhengming Zhao

AN IMPROVED HARMONIC POWER SHARING SCHEME OF PARALLELED INVERTER SYSTEM......2969

Liu Hongpeng ; Liu Xiaoxi ; Zhang Wei ; Wang Wei

THE GRID IMPEDANCE ADAPTATION DUAL MODE CONTROL STRATEGY IN WEAK GRID......2973

Ming Li ; Xing Zhang ; Ying Yang ; Pengpeng Cao

TRANSMISSION POWER ANALYSIS AND CONTROL OF THE DC TRANSFORMER IN HYBRID AC/DC MICROGRID 2980

Jingjin Huang ; Xin Zhang ; Tengfei Zhang

A NOVEL FLEXIBLE INTERCONNECTION SCHEME FOR MICROGRID TO OPTIMIZE THE CAPACITY OF ENERGY STORAGE SYSTEM (ESS) 2986

Zhou Jianqiao ; Zhang Jianwen ; Cai Xu ; Li Zhuyong ; Wang Jiacheng ; Zang Jiajie

VSC CONTROL AND PARAMETERS DESIGN BASED ON VIRTUAL SYNCHRONOUS GENERATOR 2992

Fang Liu ; Meng Wang ; Zhen Xie ; Fusheng Wang ; Jinxin Deng ; Xing Zhang

MULTI-TARGET VIRTUAL RESISTANCE CONTROL STRATEGY IN A 400 HZ LOW VOLTAGE MICROGRID 2997

Yuze Li ; Xuejun Pei ; Zhi Chen ; Hanyu Wang ; Yong Kang

AN ADAPTIVE POWER COMPENSATION STRATEGY FOR THE VOLTAGE STABILIZATION OF LCL-VSC BASED MICROGRIDS 3002

Sheng Xu ; Wu Cao ; Dongchen Fan ; Jianfeng Zhao ; Shunyu Wang

RESONANCE DETECTION STRATEGY FOR MULTIPLE GRID-CONNECTED INVERTERS-BASED SYSTEM USING CASCADED SECOND-ORDER GENERALIZED INTEGRATOR 3010

Wu Cao ; Dongchen Fan ; Kangli Liu ; Jianfeng Zhao ; Liheng Ruan ; Xiaojun Wu

HARMONIC STABILITY ASSESSMENT BASED ON GLOBAL ADMITTANCE FOR MULTI-PARALLELED GRID-CONNECTED VSIS USING MODIFIED NYQUIST CRITERION 3015

Wu Cao ; Dongchen Fan ; Kangli Liu ; Jianfeng Zhao ; Liheng Ruan ; Xiaojun Wu

THE AC TRACTION POWER SUPPLY SYSTEM FOR URBAN RAIL TRANSIT BASED ON NEGATIVE SEQUENCE CURRENT COMPENSATOR 3020

Tianshu Zhao ; Xu Peng

GRID CONNECTED POWER GENERATION CONTROL METHOD FOR Z-SOURCE INTEGRATED BIDIRECTIONAL CHARGING SYSTEM 3025

Xu Jia ; Guoming Chuai ; Haonan Niu ; Qianfan Zhang

AN ISOLATED PFC CONVERTER WITH HARMONIC MODULATION TECHNIQUE FOR EV CHARGERS 3030

Byung-Kwon Lee ; Jun-Young Lee ; Dong-Hun Kang

HIGHLY DYNAMIC SWITCHING FREQUENCY-BASED CALCULATION OF POWER QUANTITIES, FUNDAMENTAL WAVEFORMS, AND RMS VALUES OF INVERTER-FED ELECTRICAL MACHINES 3034

Alexander Stock ; Johannes Teigelkötter ; Johannes Büdel

DESIGN AND ANALYSIS OF HIGH VOLTAGE POWER SUPPLY FOR INDUSTRIAL ELECTROSTATIC PRECIPITATORS 3040

Shengwen Fan ; Yiqin Yuan ; Pengyu Jia ; Zhigang Chen ; Haisi Li

LOAD SHARING OPERATION IN N+1 UPS SYSTEM BY USING HARMONIC SHARING CONTROL METHOD 3046

Prashant Patel ; Sagar Naina ; Utsav Patel ; Premal Patwa

RESEARCH ON CAPACITY OPTIMIZATION OF PV-WIND-DIESEL-BATTERY HYBRID GENERATION SYSTEM 3052

Cailing Zhu ; Furong Liu ; Sheng Hu ; Shu Liu

A NUMERICAL ANALYSIS AND IMPROVEMENT OF OUTPUT CHARACTERISTICS IN DIFFERENT PASSIVE RECTIFIERS BASED ON VIBRATION GENERATORS 3058

Tomoki Sakabe ; Masataka Minami ; Shin-Ichi Motegi ; Masakazu Michihira

CIRCUIT MODELING APPROACH FOR ANALYZING TRIBOELECTRIC NANOGENERATORS FOR ENERGY HARVESTING 3063

Bo-Kyung Yoon ; Jeong Min Baik ; Katherine A. Kim

GENERAL POWER ELECTRIC CONVERTER MODEL 3069

Jingwen Xie

A MODULAR CONVERTER- AND SIGNAL-PROCESSING-PLATFORM FOR ACADEMIC RESEARCH IN THE FIELD OF POWER ELECTRONICS 3074

Rüdiger Schwendemann ; Simon Decker ; Marc Hiller ; Michael Braun

CONTROL IC FOR BOOST-FLYBACK CONVERTER FOR ENERGY HARVESTING APPLICATIONS 3081

Jhih-Sian Li ; Kai-Hui Chen ; Jui-Hung Lai ; Jun-Xian Huang

NEW CONCEPT OF THE DC-DC CONVERTER CIRCUIT APPLIED FOR THE SMALL CAPACITY UNINTERRUPTIBLE POWER SUPPLY 3086

Dang Minh Huynh ; Yoichi Ito ; Shinji Aso ; Koji Kato ; Kenji Teraoka

COMPARATIVE STUDY ON THE PERFORMANCE OF DUAL-PHASE TAPPED-INDUCTOR BOOST CONVERTER AND INTERLEAVED BOOST PARALLEL-INPUT SERIES-OUTPUT CONVERTER IN 40 TO 400V APPLICATIONS 3092
Niño Christopher Ramos ; Tsuyoshi Funaki

A NEW STANDBY STRUCTURE INTEGRATED WITH BOOST PFC CONVERTER FOR SERVER POWER SUPPLY 3100
Jae-Il Baek ; Jae-Kuk Kim ; Jae-Bum Lee ; Moo-Hyun Park ; Gun-Woo Moon

NONISOLATED TWO-CHANNEL LED DRIVER WITH SIMPLE SNUBBER 3107
Jong-Woo Kim ; Jung-Kyu Han ; Jih-Sheng Lai

DESIGN AND IMPLEMENTATION OF SINGLE-PHASE ASYMMETRIC MULTILEVEL STATCOM 3112
Hao Chen ; Yang Han ; Ping Yang ; Congling Wang ; Josep M. Guerrero

SUBMODULE VOLTAGE BALANCING AND LOSS EQUALISATION IN ALTERNATE ARM CONVERTERS BASED ON VIRTUAL VOLTAGES 3117
Georgios Konstantinou ; Harith R. Wickramasinghe ; Salvador Ceballos ; Josep Pou

BALANCED CONDUCTION LOSS DISTRIBUTION AMONG SMS IN MODULAR MULTILEVEL CONVERTERS 3123
Zhongxu Wang ; Huai Wang ; Yi Zhang ; Frede Blaabjerg

SIMPLIFICATION OF MODEL PREDICTIVE CONTROL FOR MODULAR MULTILEVEL CONVERTER THROUGH DIRECT VOLTAGE LEVEL SELECTION 3129
Xingxing Chen ; Jinjun Liu ; Shaodi Ouyang ; Shuguang Song ; Rui Luo

FAMILY OF INTEGRATED MULTI-INPUT MULTI-OUTPUT DC-DC POWER CONVERTERS 3134
Bang Le-Huy Nguyen ; Honnyong Cha ; Tien-The Nguyen ; Heung-Geun Kim

LOW-COMPLEXITY STATE-SPACE BASED SYSTEM IDENTIFICATION AND CONTROLLER AUTO-TUNING METHOD FOR MULTI-PHASE DC-DC CONVERTERS 3140
Marc Kanzian ; Harald Gietler ; Christoph Unterrieder ; Matteo Agostinelli ; Michael Lunglmayr ; Mario Huemer

A PHASE-SHIFT DOUBLE FULL-BRIDGE (PSDB) CONVERTER WITH THREE SHARED LEADING-LEGS 3145
Junjie Zhu ; Qinsong Qian ; Shengli Lu ; Weifeng Sun ; Le Zhang

DUAL ACTIVE BRIDGE SYNCHRONOUS RECTIFIED STEP-DOWN CONVERTER 3151
Chien-Chun Huang ; Chang-Lin Tsai ; Tsung-Lin Tsai ; Yao-Ching Hsieh ; Huang-Jen Chiu ; Jing-Yuan Lin

ACCURATE IMPEDANCE MODEL OF GRID-CONNECTED INVERTER FOR SMALL-SIGNAL STABILITY ASSESSMENT IN HIGH-IMPEDANCE GRIDS 3156
Tuomas Messo ; Roni Luhtala ; Aapo Aapro ; Tomi Roinila

MODELING OF UNBALANCED THREE-PHASE GRID-CONNECTED CONVERTERS WITH DECOUPLED TRANSFER FUNCTIONS 3164
Wei Liu ; Xiongfei Wang ; Frede Blaabjerg

PREDICTING VOLTAGE CHARACTERISTIC OF CHARGING MODEL FOR LI-ION BATTERY WITH ANN FOR REAL TIME DIAGNOSIS 3170
Minella Bezha ; Naoto Nagaoka

IMPEDANCE MODELING AND STABILITY ANALYSIS OF THE CASCADED THREE-PHASE SYMMETRIC SYSTEMS USING COMPLEX TRANSFER FUNCTIONS 3176
Teng Liu ; Zeng Liu ; Jinjun Liu ; Yiming Tu ; Zipeng Liu

ACOUSTIC NOISE REDUCTION OF 12/8 POLES SRM WITHOUT EFFICIENCY DROP USING SIMPLE CURRENT WAVEFORMS 3182
Kyohei Kiyota ; Kenji Amei ; Takahisa Ohji ; Jun Jisaki ; Masanobu Nakai

STUDY OF SWITCHED RELUCTANCE MOTOR DIRECTLY DRIVEN BY COMMERCIAL THREE-PHASE POWER SUPPLY 3186
Masaki Takahashi ; Kohei Aiso ; Kan Akatsu

DOUBLE STATOR AXIAL-FLUX SWITCHED RELUCTANCE MOTOR FOR ELECTRIC CITY COMMUTERS 3192
Hiroki Goto

TORQUE RIPPLE REDUCTION USING ASYMMETRIC FLUX BARRIERS IN SYNCHRONOUS RELUCTANCE MOTOR 3197
Yuuto Yamamoto ; Shigeo Morimoto ; Masayuki Sanada ; Yukinori Inoue

ON-BOARD SINGLE-PHASE ELECTRIC VEHICLE CHARGER WITH ACTIVE FRONT END 3203
Theodore Soong ; Peter W. Lehn

A BIDIRECTIONAL BUFFERED CHARGING UNIT FOR EV'S (BBCU) 3209
Gabriel Fernandez

RECONFIGURABLE CONVERTER WITH MULTIPLE-VOLTAGE MULTIPLE-POWER FOR E-MOBILITY CHARGING 3215
Mohamed S A Dahidah ; He Liu ; Vassilios G. Agelidis

DEVELOPMENT OF A SERIES HYBRID ELECTRIC VEHICLE LABORATORY TEST BENCH WITH HARDWARE-IN-THE-LOOP CAPABILITIES..3223
Poria Fajri ; Nima Lotfi ; Mehdi Ferdowsi

NEW THREE-PHASE STATIC TRANSFER SWITCH USING AC SSCB................................3229
Seung-Min Song ; Jin-Young Kim ; In-Dong Kim

HARMONICS COMPENSATION IN HIGH FREQUENCY RANGE OF ACTIVE POWER FILTER WITH SIC-MOSFET INVERTER IN DIGITAL CONTROL SYSTEM..................................3237
Shin-Ichi Hamasaki ; Kengo Nakahara ; Mineo Tuji

CONTROL OF BUCK-BOOST DIRECT MATRIX CONVERTER WITH LOW VOLTAGE RIDE-THROUGH CAPABILITY..3243
Nico Remus ; Martin Leubner ; Wilfried Hofmann

AN IMPROVED PLL BASED SEAMLESS TRANSFER CONTROL STRATEGY...................3251
Xin Meng ; Jinjun Liu ; Zeng Liu ; Ronghui An

EFFICIENT URBAN RAILWAY DESIGN INTEGRATING TRAIN SCHEDULING, ONBOARD ENERGY STORAGE, AND TRACTION POWER MANAGEMENT....................................3257
Warayut Kampeerawar ; Takafumi Koseki ; Fulin Zhou

OPTIMAL CONTROL METHOD OF AN ENERGY STORAGE SYSTEM FOR ENERGY SAVING.................3265
Yoko Takeuchi ; Tomoyuki Ogawa ; Keisuke Sato ; Hiroaki Morimoto ; Tatsuhito Saito

START-UP AND TRANSIENT OPERATION OF A BIDIRECTIONAL CHOPPER WITH AN AUXILIARY CONVERTER...3273
Hamzeh J. Ahmad ; Haruna Ohnishi ; Makoto Hagiwara

EXPERIMENTAL RESULTS OF QUASI-OPTIMAL CHARGING CURRENT PATTERNS TO REDUCE THE INTERNAL HEAT GENERATION OF THE LITHIUM-ION BATTERY..................3280
Yoshiaki Taguchi ; Gaku Yoshikawa

DEVELOPMENT OF TEST METHODS AND EVALUATION RESULTS FOR 500KV HVDC CONVERTER...3286
Keisuke Hattori ; Asuka Ohtake ; Takayoshi Kamejima ; Haruhisa Wada

DISSIPATION LOOP FOR SHOOT-THROUGH FAULTS IN HVDC CONVERTER CELLS...................3292
Keijo Jacobs ; Staffan Norrga ; Hans-Peter Nee

A SUPPRESSION METHOD OF HARMONIC INSTABILITY IN LINE-COMMUTATED CONVERTERS APPLYING ACTIVE HARMONIC FILTERS.......................................3299
Kenichiro Sano ; Toshiaki Kikuma ; Tatsuhito Nakajima ; Junya Kanno

EXPERIMENT OF SEMICONDUCTOR BREAKER USING SERIES-CONNECTED IEGTS FOR HYBRID DCCB..3304
Kazuyasu Takimoto ; Hiroshi Takenaka ; Toshiaki Matsumoto ; Takahiro Ishiguro

STUDY OF EMI CAUSED BY BUCK CONVERTER ON CONTROLLER AREA NETWORK....................3309
Ryo Shirai ; Toshihisa Shimizu

A STUDY ON REDUCTION TECHNIQUES OF A WIDEBAND COMMON-MODE VOLTAGE PRODUCED BY A PWM INVERTER...3315
Shotaro Takahashi ; Satoshi Ogasawara ; Masatsugu Takemoto ; Koji Orikawa ; Michio Tamate

A MODIFIED DISCONTINUOUS PWM FOR COMMON-MODE VOLTAGE ELIMINATION IN 3-LEVEL 4-LEG PWM CONVERTER SYSTEM...3323
Seon-Ik Hwang ; Jun-Hyung Jung ; In-Ho Cho ; Jang-Mok Kim ; Yung-Deug Son

EMI ANALYSIS OF FULL-SIC INTEGRATED POWER MODULE.....................................3329
Xiliang Chen ; Wenjie Chen ; Yu Ren ; Liang Qiao ; Yilin Sha ; Xu Yang

EXPERIMENTAL VERIFICATION OF COUPLING EFFECT AND POWER TRANSFER CAPABILITY OF DYNAMIC WIRELESS POWER TRANSFER......................................3332
Chan Anyapo ; Nithiphat Teerakawanich ; Chowarit Mitsantisuk ; Kiyoshi Ohishi

NEIGHBORING EFFECTS ON THE DEACTIVATED INVERTER IN A SEGMENTED DYNAMIC WIRELESS EV CHARGING SYSTEM...3338
Qingwei Zhu ; Yanjie Guo ; Lifang Wang ; Shufan Li ; Chenglin Liao

MULTIPLE EXCITING VOLTAGE CONTROL FOR MAXIMIZATION OF MULTI-HOP WIRELESS POWER TRANSFER EFFICIENCY...3344
Masato Sasaki ; Masayoshi Yamamoto

GENERAL ANALYTICAL MODEL FOR INDUCTIVE POWER TRANSFER SYSTEM WITH EMF CANCELING COILS..3349
Keita Furukawa ; Keisuke Kusaka ; Jun-Ichi Itoh

STABILITY INFLUENCE OF FILTER COMPONENTS PARASITIC RESISTANCE ON LCL-FILTERED GRID CONVERTERS..3357
Hiroaki Matsumori ; Toshihisa Shimizu ; Frede Blaabjerg ; Xiongfei Wang ; Dongsheng Yang

REAL-TIME ESTIMATION CONTROL OF INDUCTANCE PARAMETERS USING DUST CORE MATERIALS FOR PWM INVERTER...3363
Kazu Imai ; Takuma Yoshino ; Ohasi Shunsuke ; Tomoki Yokoyama

CONTROL DESIGN OF OUTPUT-STAGE FILTERLESS SINUSOIDAL-WAVE INVERTER........................3369
Shinichi Hiroshige ; Kenji Yamanaka ; Masahide Hojo

SERIES REACTIVE POWER COMPENSATOR WITH REDUCED CAPACITANCE FOR HYBRID TRANSFORMER...3375
Yuki Takahashi ; Takanori Isobe ; Hiroshi Tadano

AN INSIGHT INTO THE VOLTAGE RISING BEHAVIOR DURING TURN-OFF PROCESS OF SERIES CONNECTED SIC MOSFETS ON CIRCUIT LEVEL..3383
Panrui Wang ; Feng Gao ; Yang Jing ; Yufeng Chen ; Lei Zhang

PARALLELING SIX 320A 1200V ALL-SIC HALF-BRIDGE MODULES FOR A LARGE CAPACITY POWER STACK..3390
David Hongfei Lu ; Hiromu Takubo ; Sho Takano ; Yuhei Suzuki

3.3KV ALL-SIC MODULE FOR ELECTRIC DISTRIBUTION EQUIPMENT....................................3396
Ryohei Takayanagi ; Katsumi Taniguchi ; Satoshi Kaneko ; Naoyuki Kanai ; Keishirou Kumada ; Motohito Hori ; Yoshinari Ikeda ; Kouji Maruyama ; Itsuo Kawamura

PRESENT STATUS OF SIC BASED POWER CONVERTERS AND GATE DRIVERS – A REVIEW..............3401
Abhijit Choudhury

METHOD OF APPLYING FORCE DISTRIBUTION FUNCTION FOR LINEAR SWITCHED RELUCTANCE MOTOR DRIVEN BY CURRENT SOURCE INVERTER..3406
Tadashi Hirayama ; Shuma Kawabata

A NOVEL DRIVE CIRCUIT FOR SWITCHED RELUCTANCE MOTORS WITH BIPOLAR CURRENT DRIVE...3412
Hiroki Ishikawa ; Yuma Uesugi ; Seiya Sakurai

TORQUE RIPPLE MINIMIZATION CONTROL OF SRM BASED ON NOVEL MOTOR MODEL CONSIDERING MUTUAL COUPLING EFFECT...3418
Sungyong Shin ; Naruse Hikaru ; Takashi Kosaka ; Nobuyuki Matsui

COMPARISON OF HIGH FREQUENCY VOLTAGE INJECTION METHODS FOR SHAFT SENSORLESS CONTROL OF WOUND-FIELD FLUX SWITCHING MACHINE...................................3426
Hong-Quan Nguyen ; Sheng-Ming Yang

DESIGN AND EXPERIMENTAL VERIFICATION OF A DAB MEDIUM FREQUENCY TRANSFORMER FOR A 6.6KV/200V SOLID STATE TRANSFORMER...3431
Rene Barrera-Cardenas ; Takanori Isobe ; Terazono Katsushi ; Tadano Hiroshi

RESEARCH ON THE UNBALANCED COMPENSATION RANGE OF DELTA-CONNECTED CASCADED H-BRIDGE MULTILEVEL SVG..3439
Rui Luo ; Yingjie He ; Yiming Tu ; Xingxing Chen ; Jinjun Liu

STATIC SYNCHRONOUS COMPENSATOR TO STABILIZE GRID VOLTAGE FOR WIND AND PHOTOVOLTAIC POWER PLANT..3450
Ryota Okuyama ; Naoki Morishima ; Yusuke Ashizaki ; Yohei Itaya

LARGE EQUALIZATION CURRENT CONTROL STRATEGY FOR SERIES CONNECTED BATTERY PACKS BASED ON BUCK-BOOST CONVERTER..3455
Xinbo Liu ; Zhuo Gao ; Xuehao Huang ; Yaohan Zou

A MULTI-PORT BIDIRECTIONAL POWER CONVERSION SYSTEM FOR REVERSIBLE SOLID OXIDE FUEL CELL APPLICATIONS..3460
Xiang Lin ; Kai Sun ; Jin Lin ; Zhe Zhang ; Wei Kong

SELF-PREHEATING METHOD FOR LI-ION BATTERY USING BATTERY IMPEDANCE ESTIMATOR...3466
Dong-Kwan Kim ; Young-Dal Lee ; Sang-Hyun Ha ; Yu-Jin Jang ; Gun-Woo Moon

ACTIVE ANTI-ISLANDING TECHNIQUE WITH REDUCED NON-DETECTION ZONE FOR CENTRALIZED INVERTERS...3471
Prashant Jain ; Vivek Agarwal ; Bishnu Prasad Muni ; Eswar Rao ; Deepak Gehlot ; S. Gautam Kumar

DEVELOPMENT OF SIC APPLIED TRACTION SYSTEM FOR SHINKANSEN HIGH-SPEED TRAIN..3478
Kenji Sato ; Hirokazu Kato ; Takafumi Fukushima

DEVELOPMENT OF A HIGH POWER DENSITY AUXILIARY CONVERTER BASED ON 1700V 225A SIC MOSFET FOR TRAMS...3484
Liu Hao ; Fei Lin ; Zhongping Yang ; Hu Cao ; Meng Xia

EXPERIMENTAL TESTS RESULTS OF DAMPING CONTROL WITH OVER VOLTAGE RESISTOR FOR REGENERATIVE BRAKE CONTROL OF RAILWAY VEHICLE.............................3490
Natsuki Kawagoe ; Febry Pandu Wijaya ; Hiroyasu Kobayashi ; Keiichiro Kondo ; Tetsuya Iwasaki ; Akihiko Tsumura ; Takumi Nagashima ; Yoshinori Yamashita ; Ryota Gondo

COILS LAYOUT OPTIMIZATION OF DYNAMIC WIRELESS POWER TRANSFER SYSTEM TO REALIZE OUTPUT VOLTAGE STABLE 3495
Yi Wang ; Fei Lin ; Zhongping Yang ; Panpan Cai ; Zhiyuan Liu

QUICK CHARGER FOR A BATTERY USING MODULAR MATRIX CONVERTER (MMXC) 3501
Kazuma Suzuki ; Takaharu Takeshita

VARIABLE OUTPUT VOLTAGE CONTROL OF AN ISOLATED BI-DIRECTIONAL AC/DC CONVERTER WITH A SOFT-SWITCHING TECHNIQUE 3507
Takumi Hamaguchi ; Kazuma Suzuki ; Wataru Kitagawa ; Takaharu Takeshita

A NEW MODULATION METHOD APPLYING OPTIMAL DUTY CYCLE AND PHASE SHIFT FOR BIDIRECTIONAL ISOLATED THREE-PHASE AC/DC CONVERTER BASED ON MATRIX CONVERTER 3514
Koji Shigeuchi ; Jin Xu ; Noboru Shimosato ; Yukihiko Sato

DECOUPLING CONTROL METHOD FOR ELIMINATING DC BIAS FLUX OF HIGH FREQUENCY TRANSFORMER IN A BIDIRECTIONAL ISOLATED AC/DC CONVERTER 3522
Kensuke Sakuma ; Koji Shigeuchi ; Jin Xu ; Noboru Shimosato ; Yukihiko Sato

INTERLEAVED VOLTAGE-DOUBLER BOOST CONVERTER FOR POWER FACTOR CORRECTION 3528
Bo-Jia Huang

ZVS INTERLEAVED TOTEM-POLE BRIDGELESS PFC CONVERTER WITH PHASE-SHIFTING CONTROL 3533
Moo-Hyun Park ; Jae-Il Baek ; Jung-Kyu Han ; Cheon-Yong Lim ; Gun-Woo Moon

A ZERO-VOLTAGE-SWITCHING TOTEM-POLE BRIDGELESS BOOST POWER FACTOR CORRECTION RECTIFIER HAVING MINIMIZED CONDUCTION LOSSES 3538
Young-Dal Lee ; Chong-Eun Kim ; Jae-Il Baek ; Dong-Kwan Kim ; Gun-Woo Moon

POWER-FACTOR-CORRECTION WITH POWER DECOUPLING FOR AC-TO-DC CONVERTER 3544
Wan-Jung Chen ; Tsung-Hsi Wu ; Yao-Ching Hsieh ; Chin-Sien Moo ; Po-Hsiang Wen

DESIGN AND ANALYSIS OF THE DISTRIBUTED CONTROLLER FOR THE MODULAR MULTILEVEL CASCADED CONVERTER 3549
Ping-Heng Wu ; Yu-Chen Su ; Po-Tai Cheng

ASYMMETRIC MIXED MODULAR MULTILEVEL CONVERTER TOPOLOGY IN HYBRID BIPOLAR HVDC TRANSMISSION SYSTEMS 3557
Joon-Hee Lee ; Jae-Jung Jung ; Seung-Ki Sul

HIGH POWER MEDIUM VOLTAGE 10 KV SIC MOSFET BASED BIDIRECTIONAL ISOLATED MODULAR DC–DC CONVERTER 3564
Sayan Acharya ; Ritwik Chattopadhyay ; Anup Anurag ; Satish Rengarajan ; Yos Prabowo ; Subhashish Bhattacharya

MULTI-LEVEL POWER CONVERTER USING SERIES-CONNECTED SOLID-STATE TRANSFORMERS 3572
Yuichi Mabuchi ; Yuki Kawaguchi ; Kimihisa Furukawa ; Mitsuhiro Kadota ; Mizuki Nakahara ; Akihiko Kanoda

CAPACITOR VOLTAGE CONTROL OF MMC-STATCOM DURING UNBALANCED AC SYSTEM FAULT 3578
Kaho Nada ; Takeshi Kikuchi ; Tsuguhiro Takuno ; Toshiyuki Fujii ; Ryosuke Uda ; Takashi Sugiyama

SIC BASED POWER SEMICONDUCTOR IN APPLICATIONS - ASPECTS AND PROSPECTS 3584
Peter Friedrichs

ELECTROMAGNETIC MODELING APPROACHES TOWARDS VIRTUAL PROTOTYPING OF WBG POWER ELECTRONICS 3588
Ivana Kovacevic-Badstübner ; Daniele Romano ; Giulio Antonini ; Jonas Ekman ; Ulrike Grossner

SILICON BASED DEVICES FOR DEMANDING HIGH POWER APPLICATIONS 3596
A. Kopta ; J. Vobecky ; M. Rahimo ; T. Wikström ; U. Vemulapati ; C. Papadopoulos ; C. Corvasce ; M. Andenna ; F. Dugal ; F. Fischer ; S. Hartmann

RECENT PROGRESS IN HIGH TO ULTRA-HIGH-VOLTAGE SIC POWER DEVICES: DEVELOPMENT AND APPLICATION 3603
Y. Yonezawa

DYNAMIC DRIFT EFFECTS IN GAN POWER TRANSISTORS: CORRELATION TO DEVICE TECHNOLOGY AND MISSION PROFILE 3607
Joachim Würfl ; Eldad Bahat-Treidel ; Oliver Hilt ; Maria Troppenz ; Mihaela Wolf ; Jan Böcker ; Carsten Kuring ; Sibylle Dieckerhoff

COMPENSATION METHOD OF RADIAL UNBALANCE FORCE AT FAILURE OF A MOTOR SECTION IN A D-Q AXIS CURRENT CONTROL BEARINGLESS MOTOR 3613
Masahide Ooshima

A BEARINGLESS SYNCHRONOUS RELUCTANCE SLICE MOTOR WITH ROTOR FLUX BARRIERS 3619
Thomas Holenstein ; Thomas Nussbaumer ; Johann W. Kolar

PARAMETER IDENTIFICATIONS OF CURRENT-FORCE FACTOR AND TORQUE CONSTANT IN SINGLE-DRIVE BEARINGLESS MOTORS 3627
Hiroya Sugimoto ; Akira Chiba

DAMPENING OF AXIAL VIBRATIONS IN A BEARINGLESS FLUX-SWITCHING SLICE MOTOR BY FIELD CURRENT REGULATION 3632
Bianca Klammer ; Karlo Radman ; Wolfgang Gruber

ANALYSIS AND DESIGN OF A BEARINGLESS AXIAL-FORCE/TORQUE MOTOR WITH FLEX-PCB WINDINGS 3640
Nobuyuki Kurita ; Walter Bauer ; Gerald Jungmayr ; Wolfgang Gruber ; Wolfgang Amrhein

A PLOTTER-BASED AUTOMATIC MEASUREMENT AND STATISTICAL CHARACTERIZATION OF MULTIPLE DISCRETE POWER DEVICES 3644
Michihiro Shintani ; Benjamin Dauphin ; Kazuki Oishi ; Masayuki Hiromoto ; Takashi Sato

A NOVEL HIGH-SPEED SIC MOSFET DRIVER WITH A LOW SWITCH-VOLTAGE STRESS 3650
Xiuqin Wei ; Yuchong Sun ; Hiroo Sekiya

ENHANCEMENT OF DRIVING CAPABILITY OF GATE DRIVER USING GAN HEMTS FOR HIGH-SPEED HARD SWITCHING OF SIC POWER MOSFETS 3654
Takafumi Okuda ; Takashi Hikihara

DESIGN AND EXPERIMENTAL VERIFICATION OF ROBOT ARM OPERATION FOR POWER PACKET DISPATCHING SYSTEM 3658
Tomoki Yokoyama ; Ryunosuke Araumi ; Kazunori Asada ; Takashi Ando

A RESOURCE SHARING MODEL IN A POWER PACKET DISTRIBUTION NETWORK 3665
H. Ando ; R. Takahashi ; S. Azuma ; M. Hasegawa ; T. Yokoyama ; T. Hikihara

DECOUPLED DSOGI-PLL FOR IMPROVED THREE PHASE GRID SYNCHRONISATION 3670
A. A. Nazib ; D. G. Holmes ; B. P. Mcgrath

A DEVIATION ELIMINATION CONTROL BASED ON AUTONOMOUS CURRENT-SHARING CONTROLLER FOR THE PARALLEL-CONNECTED INVERTERS IN AC MICROGRIDS 3678
Yajuan Guan ; Wei Feng ; Baoze Wei ; Wenzhao Liu ; Mingshen Li ; C. Juan Vasquez ; M. Josep Guerrero

SISO TRANSFER FUNCTIONS FOR STABILITY ANALYSIS OF GRID-CONNECTED VOLTAGE-SOURCE CONVERTERS 3684
Hongyang Zhang ; Lennart Harnefors ; Xiongfei Wang ; Jean-Philippe Hasler ; Hans-Peter Nee

A COMMUNICATION-INDEPENDENT REACTIVE POWER SHARING SCHEME WITH ADAPTIVE VIRTUAL IMPEDANCE FOR PARALLEL CONNECTED INVERTERS 3692
Ronghui An ; Zeng Liu ; Jinjun Liu ; Shike Wang

DESIGN AND INTEGRATION OF THE BI-DIRECTIONAL ELECTRIC VEHICLE CHARGER INTO THE MICROGRID AS EMERGENCY POWER SUPPLY 3698
Yang Song ; Pengcheng Li ; Yuanliang Zhao ; Shuai Lu

STABILITY IMPACT OF PV INVERTER GENERATION ON MEDIUM VOLTAGE DISTRIBUTION SYSTEMS 3705
Ye Tang ; Rolando Burgos ; Chi Li ; Dushan Boroyevich

1MW POWER CONDITIONING SYSTEM WITH MULTIPLE DC INPUTS FOR PVS AND BATTERIES 3711
Yasuaki Furusho ; Yasuyuki Noto ; Kansuke Fujii

A ROBUST AND FLEXIBLE DC-LINKED 3-PHASE ENERGY MANAGEMENT SYSTEM WITH ADAPTIVE DROOP CONTROL STRATEGY 3717
Yue Ma ; Yuki Ishikura ; Hitoshi Tsuji ; Kazuaki Mino

MAXIMUM POWER POINT TRACKING CONTROL FOR SMALL HYDROELECTRIC GENERATION 3723
Kazuya Azegami ; Masashi Takiguchi ; Junya Yano ; Hirohiko Tsutsumi ; Toshitake Masuko

DESIGN AND EXPERIMENTAL VERIFICATION OF A THREE-PHASE DUAL-ACTIVE BRIDGE CONVERTER FOR OFFSHORE WIND TURBINES 3729
Takushi Jimichi ; Murat Kaymak ; Rik W. De Doncker

OPTIMIZED BIDIRECTIONAL PFC RECTIFIERS & INVERTERS - SI VS. SIC VS. GAN IN 2L AND 3L TOPOLOGIES - 3734
Jonas Wyss ; Jürgen Biela

A STANDARD BLOCK OF "SERIES CONNECTED SIC MOSFET" FOR MEDIUM/HIGH VOLTAGE CONVERTER 3742
Qin Lei ; Chunhui Liu ; Yunpeng Si ; Yifu Liu

DESIGN AND TESTING OF 1 KV H-BRIDGE POWER ELECTRONICS BUILDING BLOCK BASED ON 1.7 KV SIC MOSFET MODULE 3749
Jun Wang ; Rolando Burgos ; Dushan Boroyevich ; Zeng Liu

A FLYBACK CONVERTER WITH SIC POWER MOSFET OPERATING AT 10 MHZ: REDUCING LEAKAGE INDUCTANCE FOR IMPROVEMENT OF SWITCHING BEHAVIORS 3757
Kazuki Hashimoto ; Takafumi Okuda ; Takashi Hikihara

A STUDY ON LOAD FLUCTUATION OF ISOLATED DC-DC CONVERTER WITH CLASS PHI-2 INVERTER USING GAN-HFET 3762
Yuta Yanagisawa ; Yushi Miura ; Hiroyuki Handa ; Tetsuzo Ueda ; Toshifumi Ise

SINGLE-INDUCTOR MULTIPLE-OUTPUT CURRENT-SOURCE CONVERTER WITH IMPROVED CROSS REGULATION AND SIMPLE CONTROL STRATEGY 3768
Zheng Dong ; Xiaolu Lucia Li ; Chi K. Tse

LIMIT OPERATING FREQUENCY OF PEAK CURRENT-MODE CONTROL DC-DC CONVERTER CONSIDERING TURN-OFF DELAY TIME 3773
Ryo Ute ; Kazuya Fujiwara ; Jun Imaoka ; Masahito Shoyama

A NOVEL SINGLE SWITCH HIGH FREQUENCY DC/DC CONVERTER AND ITS MATHEMATIC MODEL 3780
Yueshi Guan ; Xihong Hu ; Shu Zhang ; Yijie Wang ; Dianguo Xu ; Wei Wang

ANALYSIS OF CLOSED LOOP OPERATION OF AN ISOLATED BIDIRECTIONAL DAB DC-DC CONVERTER WITH LC COUPLING 3785
Bruno Yukio Enomoto ; Kelly C. M. Carvalho ; Lourenço Matakas Junior ; Wilson Komatsu

ISOLATED AC/DC CONVERTER USING SIMPLE PWM STRATEGY 3791
Naoki Hirose ; Yuto Matsui ; Takaharu Takeshita

ANALYSIS OF ONE PHASE LOSS OPERATION OF THREE-PHASE ISOLATED BUCK MATRIX-TYPE RECTIFIER WITH EIGHT-SEGMENT PWM SCHEME 3797
Jahangir Afsharian ; Dewei David Xu ; Bin Wu ; Bing Gong ; Zhihua Yang ; Jun-Ichi Itoh

NOVEL ISOLATED BIDIRECTIONAL INTEGRATED DUAL THREE-PHASE ACTIVE BRIDGE (D3AB) PFC RECTIFIER 3805
F. Krismer ; E. Hatipoglu ; J. W. Kolar

LOAD VOLTAGE REGULATION METHOD FOR AN ISOLATED AC-DC CONVERTER WITH POWER DECOUPLING OPERATION 3813
Shohei Komeda ; Hideaki Fujita

OPTIMAL DESIGN OF A LOW COST 20KW 99.1% EFFICIENCY ACTIVE ZCS ISOLATED DC-DC CONVERTER 3820
Timothé Delaforge ; Sébastien Mariéthoz

SOFT-SWITCHING ANALYSIS AND PFM CONTROL METHOD OF BIDIRECTIONAL DC/DC CONVERTER TOPOLOGY 3825
Yijie Wang ; Haoyu Wang ; Hongyu Song ; Dianguo Xu

A FULLY SOFT-SWITCHED PWM DC-DC CONVERTER USING AN ACTIVE-SNUBBER-CELL 3833
Hai N. Tran ; Adhistira M. Naradhipa ; Sunju Kim ; Ali Tausif

FLYING CAPACITOR RESONANT POLE INVERTER WITH DIRECT INDUCTOR CURRENT FEEDBACK 3840
Sjef J. Settels ; Jorge L. Duarte ; Jeroen Van Duivenbode

DESIGN OF A GAN-BASED WIRELESS POWER TRANSFER SYSTEM AT 13.56 MHZ TO REPLACE CONVENTIONAL WIRED CONNECTION IN A VEHICLE 3848
Kawin Surakitbovorn ; Juan Rivas-Davila

EFFICIENCY MAXIMIZATION OF INDUCTIVE POWER TRANSFER SYSTEM BY IMPEDANCE AND SWITCHING FREQUENCY CONTROL IN SECONDARY-SIDE CONVERTER 3855
Ryosuke Ota ; Dannisworo S. Nugroho ; Nobukazu Hoshi

ANALYSIS OF OPTIMAL OPERATION FREQUENCY RANGE FOR BATTERY CHARGING IN WPT SYSTEM 3863
Yongbin Jiang ; Min Wu ; Junwen Liu ; Yue Wang ; Laili Wang ; Hailong Zhang

INITIAL CURRENT INJECTION METHOD OF A DIRECT THREE-PHASE TO SINGLE-PHASE AC/AC CONVERTER FOR INDUCTIVE CHARGER 3870
Ferdi Perdana Kusumah ; Jorma Kyyrä

MISSION PROFILE EMULATOR FOR PERMANENT MAGNET SYNCHRONOUS MACHINE BASED ON THREE-PHASE POWER ELECTRONIC CONVERTER 3877
Yubo Song ; Ran Cheng ; Ke Ma

A VARIABLE DC BUS VOLTAGE BASED POWER HARDWARE-IN-THE-LOOP EMULATION OF ELECTRIC MOTORS WITH WIDE VARIATION IN INTERFACE FILTER INDUCTANCE 3884
Tsai-Fu Wu ; Mitradatta Misra ; Ying-Yi Jhang ; Chang-Jun Yang ; Yin-Chi Xu

COPPER LOSS MINIMIZATION CONTROL AT ZERO OUTPUT VOLTAGE FOR ELECTROLYTIC CAPACITOR-LESS INVERTER 3890

Kodai Abe ; Haruya Kada ; Kiyoshi Ohishi ; Hitoshi Haga ; Yuki Yokokura

ARMATURE TEMPERATURE ESTIMATION INSENSITIVE TO ROTOR FLUX VARIATION FOR SPMSM 3896

Toshiki Sano ; Kiyoshi Ohishi ; Yuki Yokokura ; Hiroki Iwata ; Yuji Ide ; Daigo Kuraishi ; Akihiko Takahashi

VIRTUAL SYNCHRONOUS GENERATOR CONTROL WITH RELIABLE FAULT RIDE-THROUGH CAPABILITY BY ADOPTING MODEL PREDICTIVE CONTROL 3902

Jonggrist Jongudomkarn ; Jia Liu ; Toshifumi Ise

RESHAPING QUADRATURE-AXIS IMPEDANCE OF THREE-PHASE GRID-CONNECTED CONVERTERS FOR LOW-FREQUENCY STABILITY IMPROVEMENT 3910

Yi Tang ; Jingyang Fang ; Xiaoqiang Li ; Hongchang Li

COMPARISON BETWEEN TRADITIONAL DROOP AND A NEW AUTONOMOUS CONTROL SCHEME FOR PARALLEL INVERTERS 3916

Mohammad Bani Shamseh ; Teruo Yoshino ; Atsuo Kawamura

A NOVEL MICROGRID POWER SHARING SCHEME ENHANCED BY A NON-INTRUSIVE FEEDER IMPEDANCE ESTIMATION METHOD 3924

Baojin Liu ; Zeng Liu ; Jinjun Liu ; Ronghui An ; Shuguang Song

DEVELOPMENT OF A 3.2MW PHOTOVOLTAIC INVERTER FOR LARGE-SCALE PV POWER PLANTS 3929

Naoya Shibata ; Tsuguhiro Tanaka ; Masahiro Kinoshita

IMPEDANCE-BASED STABILITY ANALYSIS OF LARGE-SCALE PV STATION UNDER WEAK GRID CONDITION CONSIDERING SOLAR RADIATION FLUCTUATION 3934

Yiming Tu ; Jinjun Liu ; Teng Liu ; Xiangpeng Cheng

EXPERIMENTAL VERIFICATION OF GRID-CONNECTION OF A PV CONVERTER USING A SYMMETRICALLY CONNECTED BOOST CONVERTER FOR A HIGH-LEG DELTA TRANSFORMER 3940

Daiki Yamaguchi ; Hideaki Fujita

A NOVEL SINGLE- STAGE HIGH-FREQUENCY BOOST INVERTER CASCADED BY RECTIFIER-INVERTER SYSTEM FOR PV GRID-TIE APPLICATIONS 3945

Hamdy Radwan ; Mahmoud A. Sayed ; Takaharu Takeshita ; Adel A. Elbaset ; G. Shabib

NINE SWITCHES MATRIX CONVERTER USING BI-DIRECTIONAL GAN DEVICE 3952

Takashi Hirota ; Kentaro Inomata ; Daisuke Yoshimi ; Masato Higuchi

A MODEL PREDICTIVE DUAL CURRENT CONTROL METHOD FOR INDIRECT MATRIX CONVERTER FED INDUCTION MOTOR DRIVES 3958

Mei Yang ; Chen Lisha ; Liang Wang ; Yunwei Li

FAULT TOLERANT PREDICTIVE CONTROL OF THREE-LEVEL NEUTRAL-POINT-CLAMPED BACK-TO-BACK POWER CONVERTERS 3965

Zhenbin Zhang ; Xicai Liu ; Kejun Cai ; Feng Gao ; Ralph Kennel

TWO-STAGE OPTIMIZATION BASED PREDICTIVE TORQUE CONTROL WITH REDUCED COMPLEXITY FOR A THREE-LEVEL INVERTER DRIVEN INDUCTION MOTOR 3971

Ilham Osman ; Dan Xiao ; Faz Rahman

DESIGN CHALLENGES OF SIC DEVICES FOR LOW- AND MEDIUM-VOLTAGE DC-DC CONVERTERS 3979

Georges Engelmann ; Alexander Sewergin ; Markus Neubert ; Rik W. De Doncker

DESIGN AND TESTING OF 6 KV H-BRIDGE POWER ELECTRONICS BUILDING BLOCK BASED ON 10 KV SIC MOSFET MODULE 3985

Jun Wang ; Slavko Mocevic ; Jiewen Hu ; Yue Xu ; Christina Dimarino ; Igor Cvetkovic ; Rolando Burgos ; Dushan Boroyevich

HIGH POWER MEDIUM VOLTAGE CONVERTERS ENABLED BY HIGH VOLTAGE SIC POWER DEVICES 3993

Sanket Parashar ; Ashish Kumar ; Subhashish Bhattacharya

SOFT-SWITCHING – THE KEY TO HIGH POWER WBG CONVERTERS 4001

Deepak Divan ; Zheng An ; Prasad Kandula

SIC: TECHNOLOGY ENABLER FOR MV DC/DC GALVANICALLY INSULATED MODULAR CONVERTERS 4009

S. Alvarez ; M. Bellini ; U. Vemulapati ; F. Canales ; M. Rahimo

A BEARINGLESS SLICE MOTOR WITH A SOLID IRON ROTOR FOR DISPOSABLE CENTRIFUGAL BLOOD PUMP 4016

Tadahiko Shinshi ; Ryo Yamamoto ; Yoshiki Nagira ; Junichi Asama

REDUCED HARDWARE PARALLEL DRIVE FOR NO VOLTAGE BEARINGLESS MOTORS 4020

Eric L. Severson

DUAL FIELD-ORIENTED CONTROL OF BEARINGLESS MOTORS WITH COMBINED WINDING SYSTEM 4028
Wolfgang Gruber ; Siegfried Silber

OPEN-CIRCUIT FAULT TOLERANT STUDY OF BEARINGLESS MULTI-SECTOR PERMANENT MAGNET MACHINES 4034
G. Valente ; L. Papini ; A. Formentini ; C. Gerada ; P. Zanchetta

BALANCE CONTROL OF SPLIT CAPACITOR POTENTIAL FOR MAGNETICALLY LEVITATED MOTOR SYSTEM USING ZERO-PHASE CURRENT 4042
Takaaki Oiwa

ASYMMETRICAL HALF-BRIDGE CONVERTER WITH ZERO DC-OFFSET CURRENT IN TRANSFORMER USING NEW RECTIFIER STRUCTURE 4049
Jung-Kyu Han ; Jong-Woo Kim ; Seung-Hyun Choi ; Jih-Sheng Lai ; Gun-Woo Moon

CIRCULATING CURRENT-LESS PHASE-SHIFTED FULL-BRIDGE CONVERTER WITH NEW RECTIFIER STRUCTURE 4054
Jung-Kyu Han ; Gun-Woo Moon

A BI-DIRECTIONAL CURRENT DETECTION USING CURRENT TRANSFORMERS FOR BI-DIRECTIONAL DC-DC CONVERTER 4059
Seiji Iyasu ; Yuji Hahashi ; Yuuichi Handa ; Kimikazu Nakamura ; Keiji Wada

A 10 MHZ GANFET BASED ISOLATED HIGH STEP-DOWN DC-DC CONVERTER 4066
Prasanth Thummala ; Dorai Babu Yelaverthi ; Regan Zane ; Ziwei Ouyang ; Michael A. E. Andersen

ANALYSIS AND DESIGN OF A PARALLEL RESONANT CONVERTER FOR CONSTANT CURRENT INPUT TO CONSTANT VOLTAGE OUTPUT DC-DC CONVERTER OVER WIDE LOAD RANGE 4074
Tarak Saha ; Hongjie Wang ; Baljit Riar ; Regan Zane

NOVEL SINUSOIDAL INPUT CURRENT SINGLE-TO-THREE-PHASE Z-SOURCE BUCK+BOOST AC/AC CONVERTER 4080
M. Haider ; D. Bortis ; J. W. Kolar ; Y. Ono

SIMPLE PWM STRATEGY OF A MATRIX CONVERTER FOR MINIMIZING OUTPUT VOLTAGE HARMONICS 4088
Takuya Oshima ; Takaharu Takeshita

NOVEL THREE-LEVEL BACK-TO-BACK CONVERTERS: STRUCTURE, MODULATION METHOD, AND EXPERIMENT 4096
S. Sangwongwanich ; K. Niyomsatian ; S. Samermurn ; S. Nuchnoi ; S. Suwankawin

MODEL PREDICTIVE CONTROL USING SUBDIVIDED VOLTAGE VECTORS FOR CURRENT RIPPLE REDUCTION IN AN INDIRECT MATRIX CONVERTER 4104
Keon Young Kim ; Yeongsu Bak ; Jin-Hyuk Park ; Kyo-Beum Lee

DC-LINK RIPPLE CURRENT REDUCTION IN BACK-TO-BACK CONVERTERS WITH DPWM 4109
Anatolii Tcai ; Kyo-Beum Lee

AN ANALYSIS OF CLASS DE VOLTAGE-SOURCE PARALLEL RESONANT INVERTER 4114
Takeshi Kondo ; Tsuyoshi Inaba ; Yoshikazu Sakai ; Hirotaka Koizumi

AN IMPROVEMENT ON EXTENDED IMPEDANCE METHOD TOWARDS EFFICIENT STEADY-STATE ANALYSIS OF HIGH-FREQUENCY CLASS-E RESONANT INVERTERS 4122
Junrui Liang

OUTPUT POWER CAPABILITY COMPARISONS OF CLASS-E POWER AMPLIFIERS WITH HARMONIC RESONANCE 4127
Hiroo Sekiya ; Xiuqin Wei ; Yuchong Sun

A CLASS Φ2 RESONANT BUCK CONVERTER WITH RIPPLE INJECTION BURST CONTROL METHOD 4133
Min Lin ; Masahiko Hirokawa

PRACTICAL DESIGN TECHNIQUE FOR HIGH POWER DENSITY LLC RESONANT CONVERTER 4139
Shingo Nagaoka ; Hiroyuki Onishi ; Koji Takatori ; Toshiyuki Zaitsu ; Takeshi Uematsu

OPERATIONAL STUDY AND PROTECTION OF A SERIES RESONANT CONVERTER WITH DC CURRENT INPUT APPLIED IN DC CURRENT DISTRIBUTION SYSTEMS 4145
Hongjie Wang ; Tarak Saha ; Baljit Riar ; Regan Zane

A STUDY ON IMPROVEMENT OF POWER UTILIZATION RATE OF ENERGY SYSTEMS WITH PVS AND BATTERIES 4151
Hiroaki Endo ; Masakatsu Kurisaka ; Tsutomu Ueno ; Yusuke Yoshioka ; Kaoru Inoue ; Toshiji Kato

A NOVEL DC DISTRIBUTION NETWORK WITH MULTI-LEVEL BUS VOLTAGES AND ITS ENERGY MANAGEMENT SYSTEM DESIGN 4157
Jingjin Huang ; Xin Zhang ; Zhixun Ma ; Jianfang Xiao

A NOVEL DC-SIDE-PORT IMPEDANCE MODELING OF MODULAR MULTILEVEL CONVERTERS BASED ON HARMONIC STATE SPACE METHOD 4162

Jing Lyu ; Xin Zhang ; Zhixun Ma ; Xu Cai

AN IMPROVED MASTER-SLAVE CONTROL FOR THREE-PORT CONVERTER BASED DISTRIBUTED DC GRID-CONNECTED PV SYSTEM 4168

Siyue Jiang ; Kai Sun ; Hongfei Wu ; Haixu Shi ; Xiaofeng Dong ; Syed Muhammad Raza Kazmi

SENSORLESS POSITION ESTIMATION, PARAMETER IDENTIFICATION AND CONTROL INTEGRATION FOR PERMANENT MAGNET SYNCHRONOUS MACHINES USING CURRENT DERIVATIVE MEASUREMENTS 4174

M.X. Bui

DYNAMIC PERFORMANCE IMPROVEMENT OF BIDIRECTIONAL SWITCHED-CAPACITOR DC/DC CONVERTER BY RIGHT-HALF-PLANE ZERO ELIMINATION 4181

Ding Kaicheng ; Zhang Yan ; Liu Jinjun ; Zeng Pengxiang ; Zhang Jinshui

A MATRIX BASED ISOLATED BIDIRECTIONAL AC-DC CONVERTER WITH LCL TYPE INPUT FILTER FOR ENERGY STORAGE APPLICATION 4186

Prathamesh Pravin Deshpande ; Amit Kumar Singh ; Sanjib Kumar Panda

ON A STUDY OF VOLTAGE DIVIDING CLASS Φ AMPLIFIER 4193

Katsutoshi Hirayama ; Tadashi Suetsugu ; Yudai Furukawa ; Fujio Kurokawa

A DPWM BASED CONTROL STRATEGY TO INTEGRATE PHOTOVOLTAIC SYSTEM AND BATTERY STORAGE USING GRID CONNECTED THREE-LEVEL T-TYPE INVERTER 4198

Mohammad M. Hashempour ; Yue-Ting Tsai ; T. L. Lee

IMPEDANCE MEASUREMENT OF MEGAWATT-LEVEL RENEWABLE ENERGY INVERTERS USING GRID-FORMING AND GRID-PARALLEL CONVERTERS 4205

Matias Berg ; Tuomas Messo ; Tomi Roinila ; Henrik Alenius

IMPROVED VIRTUAL INDUCTANCE BASED CONTROL STRATEGY OF DFIG UNDER WEAK GRID CONDITION 4213

Ran Fang ; Wenjia Chen ; Xueguang Zhang ; Dianguo Xu

CONTROL OF VSC-HVDC FOR WIND FARM INTEGRATION WITH REAL-TIME FREQUENCY MIRRORING AND SELF-SYNCHRONIZING CAPABILITY 4220

Renxin Yang ; Chen Zhang ; Xu Cai ; Gang Shi ; Jing Lyu

A STUDY ON STEADY-STATE CHARACTERISTICS OF SERIES-CONNECTED WIND FARM USING AN EXPERIMENTAL SET OF LABORATORY SIZE 4227

Fujio Tatsuta ; Shoji Nishikata

A NOVEL ISLANDING DETECTION METHOD WITH TWO-PHASE MAGNIFICATION INSPECTION 4233

Jian-Tang Liao ; Shun-Hao Yeh ; Hong-Tzer Yang

Author Index

The 2018 International Power Electronics Conference

Measurement of Two-Level Inverter Induced Current Slopes at High Switching Frequencies for Control and Identification Algorithms of Electrical Machines

Simon Decker[1]*, Andreas Liske[1], Daniel Schweiker[1], Johannes Kolb[2] and Michael Braun[1]

1 Institute of Electrical Engineering, Karlsruhe Institute of Technology, Karlsruhe, Germany
2 SHARE at KIT, Schaeffler Technologies AG & Co. KG, Karlsruhe, Germany
*E-mail: simon.decker@kit.edu

Abstract— **Several modern control and online identification algorithms for electrical machines are based on fast current slope detection. This paper shows and compares several identification methods for the inverter induced current slopes at high switching frequencies and high bandwidth of the measured signal. Test bench measurements with a SiC-MOSFET-inverter with switching frequencies up to 60 kHz and an RL-load are used to compare the different identification methods. Best results among the investigated methods have been achieved with an easy implementable printed circuit board (PCB) design of a planar Rogowski coil.**

Keywords— current slope measurement, online parameter identification, PCB integrated planar Rogowski coil, sensorless control.

I. Introduction

Precise and fast identification of the inverter induced current slopes is mandatory for many sensorless or self-sensing control schemes [1], adaptive control algorithms [2], online parameter identification [3], rotor-temperature identification [4] or inner fault detection [5] of electrical machines.

Higher switching frequencies of actual silicon-carbide semiconductors are desired to reduce weight and space and increase power density compared to the regular IGBT inverters. This requires a higher bandwidth for the measurement- and signal processing equipment. Especially the requirements for the fast current slope identification are increasing.

This paper describes and compares some measurement principles and sensors for fast current slope identification at high switching frequencies for future power electronic devices and control algorithms.

Section II. of this paper gives an overview of the different implemented identification methods. The next chapters discusses the implementation of the investigated methods in hard- and software. Section V. shows the experimental setup of the power electronics and signal-processing test-bench hardware. Then the evaluation and comparison of measurements results of the different identification methods are illustrated in section VI. In the last section a short summary is given.

II. Current Slope Sensing

Current slope detection can be done with several methods, which can be categorized into direct and indirect current slope identification.

Direct current slope identification means direct sensing of the current slopes ahead the analog-digital converter. This can be realized e.g. with a hardware circuit which differentiates the absolute sensor signal of a current transducer, as proposed in [6]. Another possibility is the use of Rogowski coil based current transducers which has the advantage of high bandwidth. But the disadvantage is the necessary integrator circuit for absolute current measurement [7]. By using Faradays law of induction the Rogowski coil yields direct to a $\frac{di}{dt}$ value, which has to be integrated to get the absolute current value. Without the integration, only the raw current slope can be measured and evaluated.

Indirect current slope identification means, that the current slopes are calculated in the signal-processing hardware after the analog digital conversion. This is done by evaluating the measured absolute current values for example with least-squares estimators [2], linear regression or current deviation principles [1].

The absolute current measurement values can be determined with different physical principles. Ohm's law can be evaluated by shunt resistors, Faradays law of induction is used e.g. in current transformers. Magnetic field sensors based on the Hall-effect or the Fluxgate principle as well as magneto resistive effects like the Anisotropic Magneto Resistance (AMR). Several other principles like Faraday effect sensors, etc. are described in [8].

III. Hardware Implementation

Galvanic insulated current sensors are typical for power electronic applications. Hence, these sensors are used to compare the different current slope identification methods. Typical sensors for power electronic applications are magnetic field sensors like Hall-effect sensors and magneto resistive effect sensors like AMR sensors. Therefore the Hall-effect sensor LAH 25-NP from LEM and the AMR sensor CMS3025 from Sensitec are chosen for the indirect current slope identification methods. In Tab. I. both sensor characteristics are listed.

TABLE I
INVESTIGATED CURRENT SENSORS FOR INDIRECT CURRENT SLOPE DETECTION

	Hall-effect sensor	AMR sensor
Model	LAH 25-NP	CMS3025
Rated current	25 A	25 A
Bandwidth	200 kHz	2 MHz
Delay	0.5 µs	0.06 µs
Accuracy	± 0.3 %	± 1 %
Sensing	$i(t)$	$i(t)$
Sensitivity	$1\frac{mA}{A}$	$40\frac{mV}{A}$

The in Tab. I. listed sensors are mounted on a PCB and provided with filters according to the reference design from the manufacturer. The current sensor signal of the LAH 25-NP is converted to a voltage signal with a $100\ \Omega$ measuring resistor, and amplified to ± 10 V at ± 50 A for the signal-processing hardware.

The CMS3025 voltage output signal is directly amplified to ± 10 V at ± 50 A.

For the direct measurements an analog differentiator circuit is implemented, according to [6]. A planar Rogowski coil for a PCB is designed, based on the "pick up coil" as described in [9]. Additionally a prototype di/dt-sensor for direct current slope measurement from Sensitec, based on a Rogowski coil, is investigated. In Tab. II. the investigated sensors are listed.

TABLE II
INVESTIGATED AND DEVELOPED CURRENT SLOPE SENSORS FOR DIRECT CURRENT SLOPE DETECTION

	di/dt-sensor	Planar Rogowski coil	Analog Differentiator
Model	SENSITEC PROTOTYPE	PROTOTYPE	PROTOTYPE
Rated current	25 A	-	-
Bandwidth	n.a.	n.a.	100 kHz
Delay	n.a.	n.a.	n.a.
Accuracy	n.a.	n.a.	n.a.
Sensing	$\frac{di}{dt}$	$\frac{di}{dt}$	$\frac{dv}{dt}$
Sensitivity	$41\frac{mV}{kA/s}$	$0.6\frac{mV}{kA/s}$	n.a.

The Sensitec di/dt-sensor Prototype is evolved from a CMS3025 current sensor with an additional output for the current slope signal. The signal is generated by a Rogowski coil mounted opposite to the busbar followed by an integrated amplifier.

In [9] and [10] a planar Rogowski coil for current sensing and its implementation is introduced. The planar Rogowski coil is integrated in the top and bottom layer of the PCB and connected through vias. The conductor is an isolated external mounted busbar soldered on the top layer. The top and side view of the design is shown in Fig. 1 and Fig. 2. The busbar is made of copper with a thickness of 0.5 mm and a distance $z_0 = 0.25$ mm to the planar Rogowski coil.

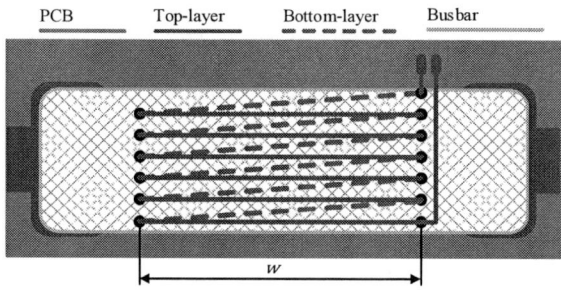

Fig. 1. Top view of the PCB integrated planar Rogowski coil for one of the three-phases.

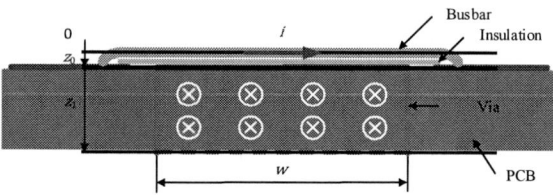

Fig. 2. Side view of the PCB integrated planar Rogowski coil for one of the three-phases.

The flux density B induced in the sensor coil by the busbar can be calculated according to equation (1).

$$\vec{B}(r) = \mu_0 \frac{i}{2\pi r} \cdot \vec{e}_\varphi \tag{1}$$

Assuming a homogenous field of the conductor, the flux inside the planar Rogowski coil can be determined with equation (2).

$$\Phi = \int_{z_0}^{z_1} \mu_0 \frac{i \cdot w}{2\pi r} dr = \mu_0 \frac{i \cdot w}{2\pi} \cdot \ln\left(\frac{z_1}{z_0}\right) \tag{2}$$

With Faraday's law of induction the induced voltage $v(t)$ defined in equation (3) can be derived to the mutual-inductance M multiplied with the current slope $\frac{di}{dt}$. N is thereby the number of turns, $w = 50$ mm the width of the planar Rogowski coil.

$$v(t) = N \cdot \frac{d\Phi}{dt} = M \cdot \frac{di}{dt} \qquad (3)$$

The self-inductance, significant for the dynamic behavior, is analytically calculated with Niwa's method for rectangular coil inductances [11].

The required sensitivity for equal conditions between the compared direct current slope identification methods is thereby adopted from the di/dt-sensor to $41\frac{mV}{kA/s}$. The planar Rogowski design sensitivity has to be amplified with an operational amplifier circuit to the required sensitivity.

The analog differentiator circuit is based on a design presented in [6] and can be assembled to the current sensors. The circuit is shown in Fig. 3. It consists of a passive, first order high-pass filter and a two-stage operational amplifier (op-amp) circuit. The high speed op-amps are ADA49898-1 from Analog Devices. The input voltage v_{in} is generated by the current transducer from the absolute measured value. The output voltage v_{out} is described with equation (4).

The response time $\tau = R_1 C_1$, directly influences the dynamics of the voltage v_{out}. Hence, for the dimensioning of τ, the switching frequency of the inverter has to be taken into account. It is assumed that the final value of v_{out} is reached after 3τ. The time constant τ is set to $\tau = \frac{0.02}{3f}$, for a minimum pulse-width of $\frac{0.02}{f}$ to be able to measure the current slopes at small duty cycles. This has to represent a duty cycle of about 2%.

$$v_{out} = \left(R_1 C_1 \frac{dv_{in}}{dt}\right) \cdot \left(1 + \frac{R_{21}}{R_{22}}\right) \cdot \left(1 + \frac{R_{31}}{R_{32}}\right) \qquad (4)$$

Adding the capacitors C_{2x}, C_{3x} results in a low-pass filter and reduces the noise at higher frequencies.

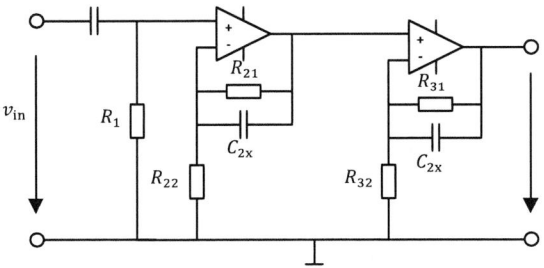

Fig. 3. Schematic of the analog differentiator circuit.

IV. SOFTWARE IMPLEMENTATION

The signal processing system calculates the necessary duty cycles for a given sine wave form which are the reference values for the measurement.

Several methods of current slope evaluations are described in the literature. Common methods are e.g. moving average or least-squares algorithms, artificial neuronal networks and several other principles for software-based current slope identification [12].

In this paper an easy software-based identification method for the current slope will be introduced and discussed. The inverter induced current ripple depends on the switching states of the pulse width modulation. With the knowledge of the applied switching states and the assumption of linearly rising and falling currents between those, the current slopes can be calculated. Two measured current values and the time between these, are sufficient. This in principle, is similar to the described method in [1]. Both necessary current values are generated from the average value of a sequence of oversampled current samples during the switching state. Due to the averaging for the two measured current values, a better signal to noise ratio is achieved especially at higher duty cycles. Samples close to the switching transitions are neglected in the calculation. In Fig. 4 the current slopes and the switching states are displayed. The blue and black circles and the green squares mark the oversampling current. The current slope of III. is calculated with the average value, marked with a star, of the sequence of the blue circles and the average value of the sequence of the green squares. The difference $\Delta \bar{I}$ among the current values and the time Δt between these yields to the current slope during this switching state. This procedure is repeated each switching state.

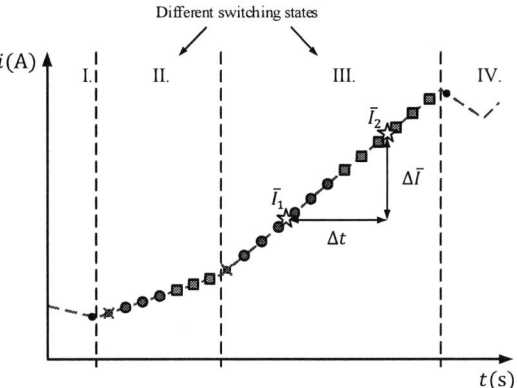

Fig. 4. Principle of the sequential current slope calculation with oversampled absolute current values.

V. EXPERIMENTAL SETUP

The single PCB, Silicon-Carbide MOSFET voltage source inverter consists of a CCS050M12CM2 Six-Pack Modul from Wolfspeed, the DC-link capacitors, the current transducers CMS3050 from Sensitec, an internal fault management and protection system. The inverter switching frequency is up to 100 kHz. The calculation of the reference values and the evaluation of the Analog-Digital Converters (ADCs) is implemented on a real-time system based on a Xilinx Zynq-7030 System on Chip (SoC) with a dual-core ARM Cortex A9 processor and a Kintex-7 Field-programmable gate array (FPGA).

The measured current values are sampled with a 15-bit and five Mega samples per second (Msps) ADC from Linear Technology LTC2323-14. For reference measurements the Agilent DSOX3024A oscilloscope with 200 MHz bandwidth and the Agilent current probe 1147 A are used. The RL-load is a power yoke with shared iron core which has an inductance of 230 µH and a resistance of 300 mΩ.

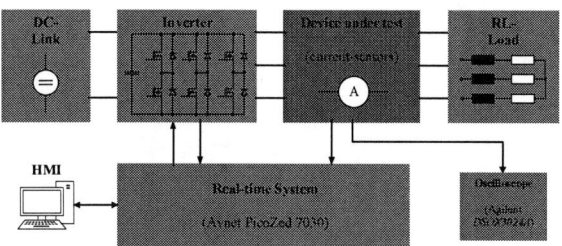

Fig. 5. Experimental setup of the test-bench. The real-time system needs measurement of the currents for the software based evaluation of the current slopes and the control of the RL-load.

VI. MEASUREMENT RESULTS

The measurement results in this section are used for a qualitative analysis of the investigated current slope identification methods. The measured time variant current slope signals can thereby compared to each other with regard to control and identification methods usability.

For the measurements a fundamental sine wave current with an RMS value of 25 A and a frequency $f_0 = 450$ Hz is generated with a DC-link voltage of 80 V. The switching frequency of the two-level inverter for the first measurement is 10 kHz, for the second measurement 60 kHz. For the evaluation of the investigated current sensors, the outputs from the LEM LAH 25-NP and the Sensitec CMS3025 are compared with a current probe from Agilent. In the first measurement with 10 kHz switching frequency, the signals of the current transducers are similar and not displayed this paper. In the second measurement (Fig. 6.) at 60 kHz the delay in the signal from the LAH 25-NP sensor is clearly visible due to the limited bandwidth. The signal from the CMS3025 conform to the signal from the current probe from Agilent. The signal from the Agilent current probe is displayed black, the one from the CMS3025 sensor red and the LAH 25-NP sensor green.

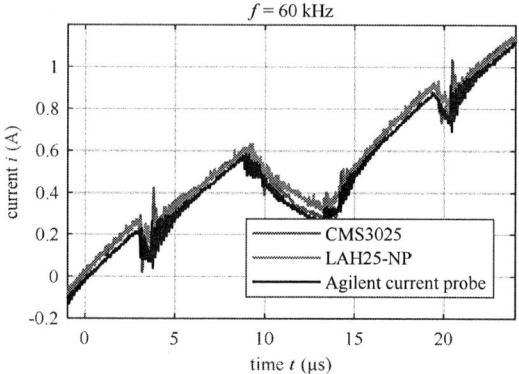

Fig. 6. Reference measurement of the absolute value of the investigated current sensors compared to the current probe.

In the following subsections the current slope detection methods measurements are discussed. In each section first the measurement with 10 kHz switching frequency is displayed, followed by the measurement with 60 kHz switching frequency. Thereby the grey current slope reference signal is generated from the absolute measured values of the current sensors which are logged by the oscilloscope and derived afterwards. In each figure also the measured current value is shown. The signal form of the investigated current slope detection sensor/method is displayed in blue.

A. di/dt-sensor

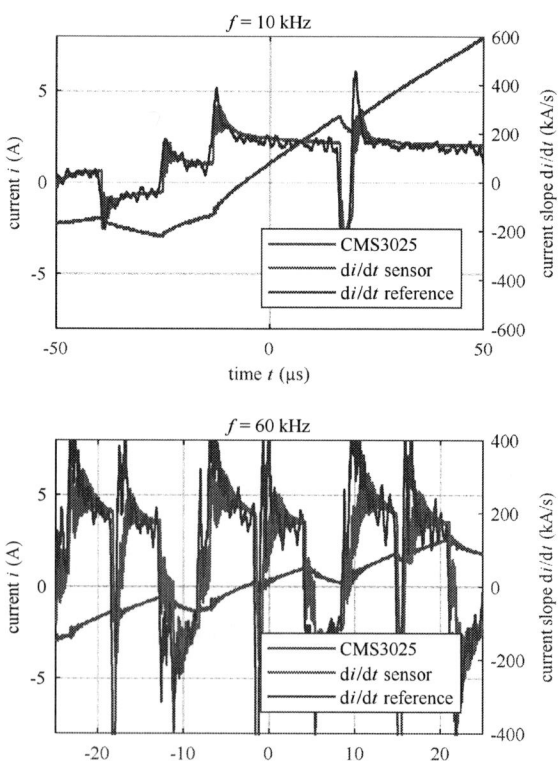

Fig. 7. Measurement of the di/dt-sensor prototype with 10 kHz and 60 kHz switching frequency.

At 10 kHz the d*i*/d*t*-sensor in blue shows the expected waveform. At small pulse width of approximately 15 µs a delay is visible. But due to the switching transitions these detected current slopes of these small duty cycles cannot be evaluated anyway.

At 60 kHz the signal is of high noise and the sensor shows delays. The current slopes at small duty cycles cannot be measured due to the limited bandwidth probably because of the high self-inductance of the used Rogowski coil. But the reference value in gray is noisy as well due to the calculation of the derivative from the measured absolute value of the CMS3025 sensor.

B. PCB planar Rogowski coil

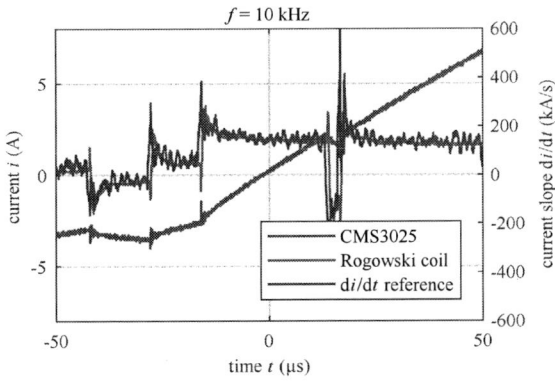

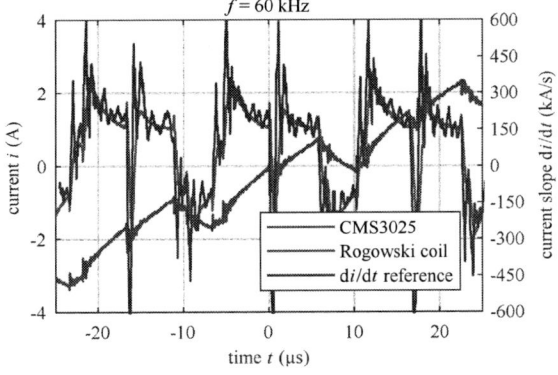

Fig. 8. Measurement of the self-designed planar PCB Rogowski coil prototype with 10 kHz and 60 kHz switching frequency.

The PCB integrated planar Rogowski coil delivers better results compared to the current slope reference measurement, generated from the CMS3025 signal in red. Small delays at small duty cycle indicate a limited bandwidth, but even at these small duty cycle the average current slope is identified precise and correctly.

C. Analog differentiator

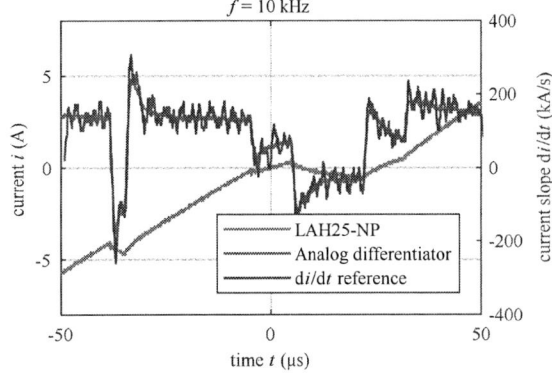

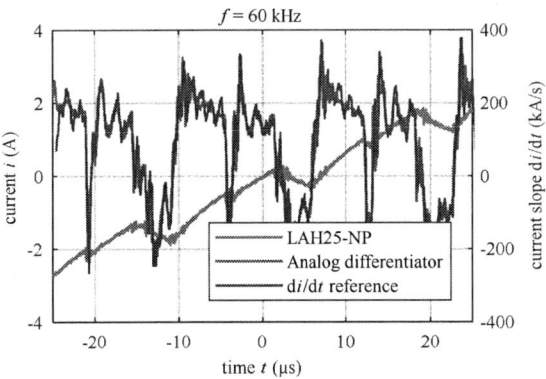

Fig. 9. Measurement of the analog differentiator circuit in combination with the LAH 25-NP with 10 kHz and 60 kHz switching frequency.

The CMS3025 sensor in combination with the designed analog differentiator circuit did not lead to good results. Due to the high bandwidth of the CMS3025 the circuit was too sensitive for valid results. Therefore, the LAH 25-NP signal with a lower bandwidth was evaluated with the designed analog differentiator circuit. The quality of the analog differentiator circuit output depends thereby on the signal quality from the LAH 25-NP sensor.

D. Software current slope identification

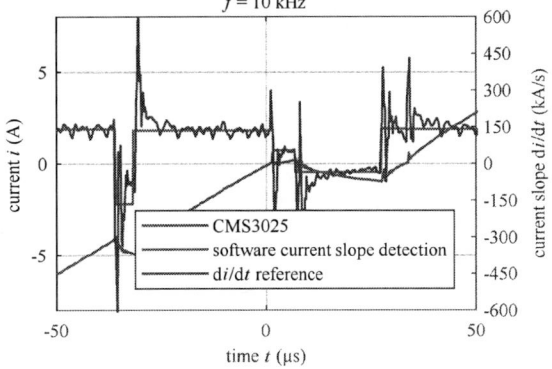

Fig. 10. Measurement of the proposed software based current slope detection with 10 kHz switching frequency.

2852

The online calculation of the actual current slope is processed on the introduced real-time systems programmable logic. Therefore the measured value from the CMS3025 current sensor is digitized and evaluated with the presented algorithm above.

The mean value of the current slope of the 10 kHz switching frequency measurement is thereby identified correctly, seen in Fig. 10. However small duty cycles cannot be evaluated correctly, because of the limited bandwidth of the AD-conversion. The second measurement with 60 kHz is not evaluated due to the limited bandwidth and is not displayed here.

VII. CONCLUSION

This paper investigates different identification methods for the inverter induced current slope. The principles of different methods for current slope detection are explained. The design of an analog differentiator, a planar PCB-integrated Rogowski coil and a software based current slope identification is shown. The presented techniques are implemented on a test-bench and compared to each other. Additionally a di/dt-sensor prototype is used for parallel measurement. Every tested sensor and principle work well so far for standard two level inverters with a switching frequency below 10 kHz. The quality of the output signal of the analog differentiator depends thereby highly on the input signal. Increasing the switching frequency up to 60 kHz leads to high-noise signals and increases the requirements of filter circuits and EMC of all sensors. The developed and designed simple and cheap planar Rogowski coil offers the best signals at these high switching frequencies.

REFERENCES

[1] E. Robeischl and M. Schroedl, "Optimized INFORM Measurement Sequence for Sensorless PM Synchronous Motor Drives With Respect to Minimum Current Distortion," in *IEEE Transactions on Industry Applications*, 2004, pp. 591–598.

[2] A. Liske and M. Braun, "Direct adaptive current control a universal current control scheme for electrical machines," in *Industrial Electronics Society, IECON 2014 - 40th Annual Conference of the IEEE*, 2014, pp. 514–520.

[3] S. Decker, J. Richter, and M. Braun, "Predictive current control and online parameter identification of interior permanent magnet synchronous machines," in *18th European Conference on Power Electronics and Applications (EPE'16 ECCE Europe)*, 2016, pp. 1–10.

[4] M. Ganchev, C. Kral, and T. Wolbank, "Identification of sensorless rotor temperature estimation technique for Permanent Magnet Synchronous Motor," in *International Symposium on Power Electronics Power Electronics, Electrical Drives, Automation and Motion*, 2012, pp. 38–43.

[5] T. M. Wolbank, P. Nussbaumer, H. Chen, and P. E. Macheiner, "Non-invasive detection of rotor cage faults in inverter fed induction machines at no load and low speed," in *IEEE International Symposium on Diagnostics for Electric Machines, Power Electronics and Drives*, 2009, pp. 1–7.

[6] P.-H. Liu, Y. Yan, F. C. Lee, and P. Mattavelli, "External ramp autotuning for current mode control of switching converters," in *Twenty-Eighth Annual IEEE Applied Power Electronics Conference and Exposition (APEC)*, 2013, pp. 276–280.

[7] Texas Instruments Incorporated, *High Accuracy AC Current Measurement Reference Design Using PCB Rogowski Coil Sensor (Rev. A)*. [Online] Available: http://www.ti.com/tool/TIDA-01063.

[8] S. Ziegler, R. C. Woodward, H. H.-C. Iu, and L. J. Borle, "Current Sensing Techniques: A Review," in *IEEE Sensors Journal*, 2009, pp. 354–376.

[9] L. Zhao, J. D. van Wyk, and W. G. Odendaal, "Planar embedded pick-up coil sensor for integrated power electronic modules," in *Applied Power Electronics Conference and Exposition, 2004. APEC '04. Nineteenth Annual IEEE*, 2001, pp. 945–951.

[10] Y. Xue *et al.*, "A compact planar Rogowski coil current sensor for active current balancing of parallel-connected Silicon Carbide MOSFETs," in *2014 IEEE Energy Conversion Congress and Exposition (ECCE)*, 2014, pp. 4685–4690.

[11] F. Grover, *Inductance calculations: Working formulas and tables*. Research Triangle Park: Instrument Society of America, 1982-1946.

[12] D. Hind, C. Li, M. Sumner, and C. Gerada, "Realising robust low speed sensorless PMSM control using current derivatives obtained from standard current sensors," in *2017 IEEE International Electric Machines and Drives Conference (IEMDC)*, 2017, pp. 1–6.

A New Topology of Switched-Capacitor Multilevel Inverter for Single-Phase Grid-Connected with Eliminating Leakage Current

Mehdi Samizadeh[1], Xu Yang[1]*, Bagher Karami[2], Wenjie Chen[1], Mohamad Abou Houran[1],
Adib Abrishamifar[2] and Abdolreza Rahmati[2]

1 Department of Electrical Engineering, Xi'an Jiaotong University, Xi'an, China
2 Department of Electrical Engineering, Iran University of Science & Technology, Tehran, Iran
*E-mail: yangxu@xjtu.edu.cn

Abstract—A new switched-capacitor multilevel inverter (SCMLI) topology for grid-connected photovoltaic (PV) systems with eliminating the leakage current is presented in this study. Due to directly connecting the negative pole of the PV cells to the neutral line of the grid, the leakage current will be decreased to zero. The newest topologies for SCMLI are based on H-bridge stage; that means the common mode voltage in output is variable and leads to leakage current. The main advantages of this new topology are: (1) It does not need the boost stage; thus, it has light weight and small size. (2) The negative pole of the PV is directly connected to the neutral line of the grid, so the leakage current is eliminated. (3) Has good capability to conduct the reverse current from the inductive load to the dc source. (4) By using the Sinusoidal Pulse-Width Modulation (SPWM) technique the voltage of capacitors are self-balanced, thus this structure has the ability to be extended as n-level SCMLI and reduces the output filter size. The 9-level structure of the proposed inverter is simulated. The performance of which will be verified by MATLAB/SIMULINK software.

Keywords—*Multilevel inverter; switched capacitor; leakage current; SPWM modulation technique; step up.*

I. INTRODUCTION

In the recent years, because of the global warming and decreasing the fossil fuels the renewable energy sources has become interesting topic in research and industry. Inverters are indeed necessary for these renewable power systems; and, multilevel inverters due to low cost, high efficiency and quality of output waveform are very popular. These multilevel inverters are useful to inject the energy conversion system such as fuel cells, wind turbines and photovoltaic systems to the grid [1-2]. Due to the low output voltage in renewable energy systems, thus it needs to be boosted to inject to the grid [3]. In MLIs, the output voltage is same as staircase waveforms with the high quality and acceptable total harmonic distortion (THD) which decrease the cost and voltage stresses on switches and output filter size compare with the classical inverters [4-6]. Generally, the conventional topologies of MLIs have been divided into three main types including diode clamped (DCMLIs) [7], flying

capacitors (FCMLIs) [8-9] and cascade H-bridge (CHB) which typically classified into two basic types including symmetric and asymmetric values of input dc power supplies [10]. However, increasing the output levels in these topologies, needs isolated dc power supplies and large capacitor bank. In addition, due to discharging the voltage of capacitors, charge balancing control topology is required. The grid-connected inverters are divided into two types, without and with transformer structure.

Transformerless inverters have a great benefit to further reduce size, cost and improve the efficiency [11-13]. However, in these systems, there is no galvanic isolation which leads to leakage current due to the parasitic capacitor between the PV panels and ground [14]. By maintaining the common mode voltage (CMV) constant, the leakage current is minimized [15-16]. However, to achieve the constant CMV, it requires extra circuitry elements.

Nowadays, many researchers have presented numerous developed topologies of MLIs based on the switched-capacitor structure, which can produce more output voltage levels and decrease the required power supplies [17-19]. All of these Structures require H-bridge inverter to generate the negative output levels. Thus, common mode voltage in output is variable and leads to leakage current. Nowadays, some common ground transformerless grid-connected inverters are presented [20-21]. In these topologies, the negative polarity of the PV is directly connected to the grid, which eliminates leakage current. Nevertheless, a boost converter circuit is used here to raise the input voltage. In addition, these inverters only can make three voltage levels in output.

To overcome aforementioned drawbacks, a novel switched-capacitor multilevel inverter for single-phase grid-connected PV system is presented in this study. The proposed topology is the first structure of SCMLI, which can eliminate leakage current. It includes several capacitors and series/parallel power switches, which have the ability to generate high-quality staircase voltage waveform. Furthermore, due to using the capacitors as a virtual dc source, it can boost input voltage with ease. And by using SPWM technique, the requirement to the charge control circuit is eliminated. Moreover, it has good capability to conduct the reverse current for inductive load. Several simulation results by MATLAB/SIMULINK software are given to acknowledge the validation of proposed topology.

II. PROPOSED SCMLI

A. Proposed general SCMLI topology

The extendable structure for proposed SCMLI is shown in Fig. 1. As can be seen, this structure includes similar basic circuit that to achieve the high output levels, it needs to connect k units together. Each unit contains one capacitor. If the number of capacitors is assumed n, thus, the number of required switches, capacitors, maximum output voltage and the number of the levels which be created are acquired in following equations respectively:

$$N_{Switches} = 6n - 1 \qquad (1)$$

$$N_{Capacitor} = n \qquad (2)$$

$$N_{Level} = 2n + 1 \qquad (3)$$

$$V_{Out(max)} = nV_{dc} \qquad (4)$$

In this structure, without the requirement to the extra dc power supply, by adding the basic unit to the circuit, the output voltage levels will be increased. In addition, boosting of voltage is enhanced. In comparison with the structure of [19] and [20], which only can make three output levels, thus, proposed topology has the acceptable THD and the output filter size is decreased.

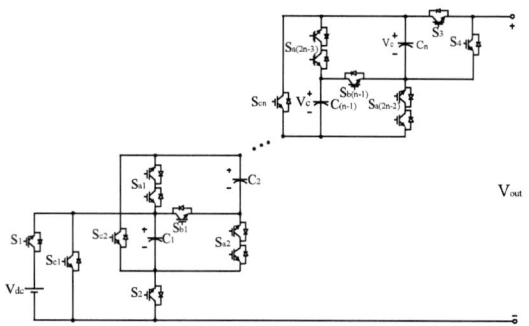

Fig. 1. Proposed general SCMLI topology.

In order to demonstrate these performances, a 9-level inverter based on proposed topology is analyzed in following.

B. Proposed 9-Level Inverter

Fig. 2 shows the 9-level structure of proposed topology that consists of one dc power supply, 4 capacitors and 23 power switches. In this case, if the bidirectional switches of each unit (S_a) become ON, the other switches of this unit, S_b and S_c will be OFF and vice versa. By turning ON the bidirectional switches S_a, the path to charging the capacitors is acquired. In the other state, when S_b turns ON, the capacitors will be series together and discharge in load to generate greater positive

output level. The switch S_c becomes ON to generated negative output voltage levels.

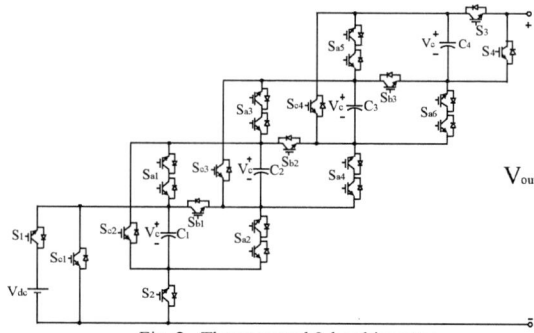

Fig. 2. The proposed 9-level inverter.

The different operating modes of inverter are shown in Fig.4. As can be seen, it is necessary to mention that the all bidirectional switches of each unit -(S_{a1}, S_{a2}), (S_{a3}, S_{a4}) and (S_{a5}, S_{a6})- will be ON or OFF simultaneously. In fact, each pair of bidirectional switches is controlled with the same pulse signal, which makes the control circuit to be simpler. According to Fig. 4, at zero level, all bidirectional switches along with S_1, S_2 and S_4 become ON and charge all capacitors as shown in Fig. 3(a). To generating the first positive level (+V_{dc}), only the switch S_3 becomes ON instead of switch S_4, while the state of other switches is same as zero level (Fig. 3(b)).

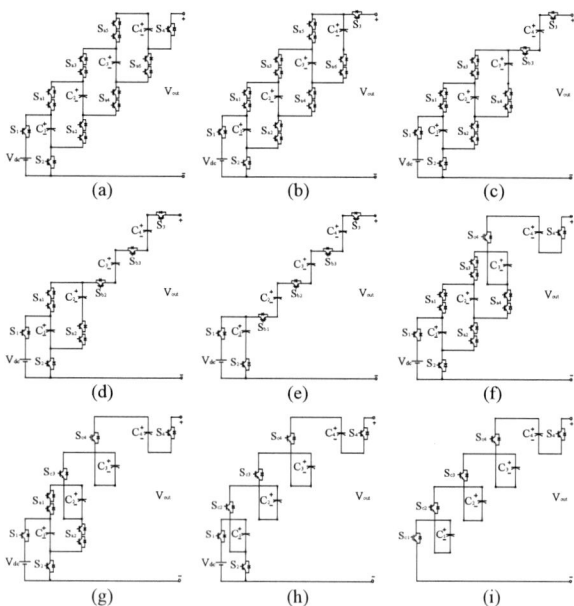

Fig. 3. Different switching states of proposed inverter.

By using the capacitor C_4 as series with dc power supply, +$2V_{dc}$ level can be built. To achieve this state, the switches S_{a5} and S_{a6} turn OFF and switch S_{b3} becomes ON (Fig. 3(c)). Here, dc power supply is charging other capacitors (C_1, C_2, C_3) simultaneously. The same situation occurs for the other positive levels which, by turning ON the switch S_b of each unit and turning OFF the switches S_a, the capacitor of that unit (C_i) as series is

2855

added to the last series capacitor (C_{i+1}) and the next level will be made (Fig. 3(d), (e)).

The first negative level ($-V_{dc}$) is obtained by turning ON the switch S_{c4} and turning OFF the switches S_{a5}, S_{a6} and S_{b3} as shown in (Fig. 4(f)). In this state, the capacitor C_4 from the path, which is created by switches S_2, S_{a2}, S_{a4}, S_{c4} and S_4, is parallel with the load in reverse polarity and the other capacitors (C_1, C_2, C_3) are charging. By turning OFF switches S_{a3}, S_{a4}, S_{b2} and turning ON switch S_{c3}, C_3 and C_4 are series and other negative level ($-2V_{dc}$) can be built (Fig. 3(g)). Continuing this process generates $-3V_{dc}$ level (Fig. 3(h)). To generate $-4V_{dc}$, the all switches S_c and S_4 become ON and other switches were OFF as shown in (Fig. 3(i)).

C. Challenge the Proposed Topology

Fig. 4 demonstrate the reason why have to use two bidirectional switches. According to Fig.4 (a), if bidirectional switches Sa1 is substituted with the single switch, which the anode of diode is connected to positive polarity of capacitor C_1, by turning ON the switch S_{c2} to generate the negative level, thus, the capacitor C_1 can forward bias the diode of switch Sa1 and leads to short circuit. Fig. 4(b) shows that if bidirectional switches Sa1 is replaced with the single switch, which the anode of diode is connected to positive polarity of capacitor C_2, while switch Sb1 becomes ON, to connect two capacitors as series together, the diode of switch Sa1 will be biased by capacitor C_2 and occur to short circuit. According to the same circumstances for bidirectional switch S_{a2}, the diode of switch will be forward bias by capacitor C_1 and C_2 respectively and leads to short circuit as shown in Fig.4 (c), (d). According to aforementioned states, using the single switches instead of the bidirectional switches makes short-circuit and disturbs the Performance of inverter.

According to aforementioned states, using the single switches instead of the bidirectional switches makes short-circuit and disturbs the performance of inverter.

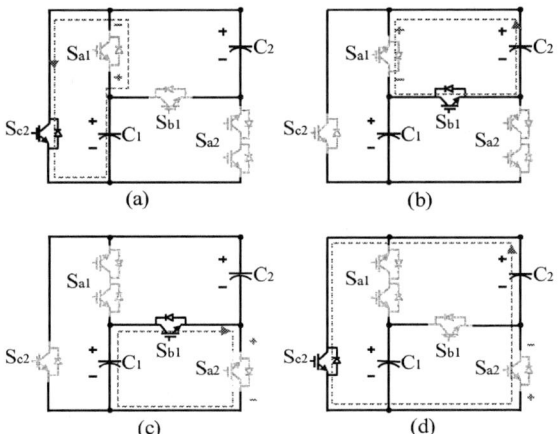

Fig. 4. Short circuit problem by replacing single switch instead of bidirectional.

III. PROPOSED SPWM MODULATION STRATEGY

Table I demonstrates the ON state switches in each level. From this table, it is obvious that, the switch S_3 only in positive levels becomes ON. Furthermore, the switch S_{c1} just to generate $-4V_{dc}$ level will be ON and during the generating other levels is OFF. In addition, the switches S_1 and S_2 become ON in generating all levels except $-4V_{dc}$ level simultaneously.

Fig. 5 illustrates the modulation method of the proposed inverter. As can be seen, a level-shifted PWM (LSPWM) technique is used to make the gate control signals for proposed 9-level inverter. With respect to this figure, a perfect sinusoidal reference signal with amplitude of A_r and frequency of 50Hz is compared with four triangular high frequency carrier waveforms, which have the same amplitude of A_c. Here, the modulation index M is determined by:

$$M = \frac{A_r}{4A_c} \tag{5}$$

To generate the required switching pulse in negative cycle, the absolute value of reference will be compare with carriers.

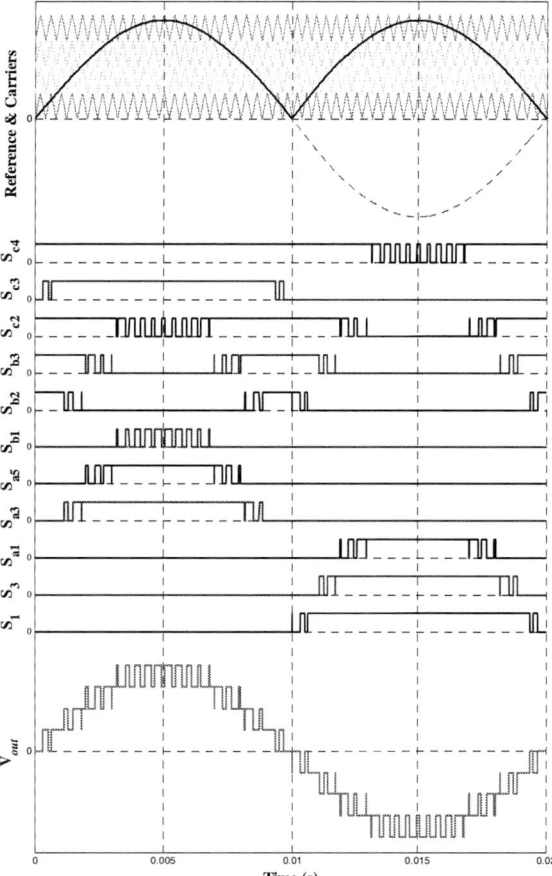

Fig. 5. Proposed SPWM modulation and corresponding gate pulses.

TABLE I
LIST OF ON-STATE SWITCHES IN EACH LEVEL

Reference		ON State Switches	V_{out}
Positive ref	$ref > e_4$	$S_1,S_2,S_3,S_{b1},S_{b2},S_{b3}$	$4V_{dc}$
	$e_4 \geq ref > e_3$	$S_1,S_2,S_3,S_{a1},S_{a2},S_{b2},S_{b3}$	$3V_{dc}$
	$e_3 \geq ref > e_2$	$S_1,S_2,S_3,S_{a1},S_{a2},S_{a3},S_{a4},S_{b3}$	$2V_{dc}$
	$e_2 \geq ref > e_1$	$S_1,S_2,S_3,S_{a1},S_{a2},S_{a3},S_{a4},S_{a5},S_{a6}$	V_{dc}
	$|ref| \leq e_1$	$S_1,S_2,S_4,S_{a1},S_{a2},S_{a3},S_{a4},S_{a5},S_{a6}$	0
Negative ref	$e_2 \geq |ref| > e_1$	$S_1,S_2,S_4,S_{a1},S_{a2},S_{a3},S_{a4},S_{c4}$	$-V_{dc}$
	$e_3 \geq |ref| > e_2$	$S_1,S_2,S_4,S_{a1},S_{a2},S_{c3},S_{c4}$	$-2V_{dc}$
	$e_4 \geq |ref| > e_3$	$S_1,S_2,S_4,S_{c2},S_{c3},S_{c4}$	$-3V_{dc}$
	$|ref| > e_4$	$S_4,S_{c1},S_{c2},S_{c3},S_{c4}$	$-4V_{dc}$

IV. THE CAPACITOR VOLTAGE DROP CALCULATION

In this part, the capacitors voltage drop will be estimated. When the capacitors are used as a virtual dc power supply (are series with the load), their voltage level will be decreased. As expected, the voltage drop of C_4 is much more than other capacitors. In each of voltage level, some capacitors are charging while the other capacitors are discharging. TABLE II shows the charging and discharging states for capacitors in each of output voltage levels. In this table C and D refer to charging and discharging states for capacitors.

TABLE II
CHARGING AND DISCHARGING STATES OF CAPACITORS FOR
PROPOSED 9-LEVEL INVERTER

	-4	-3	-2	-1	0	1	2	3	4
C_1	D	C	C	C	C	C	C	C	C
C_2	D	D	C	C	C	C	C	C	D
C_3	D	D	D	C	C	C	C	D	D
C_4	D	D	D	D	C	C	D	D	D

When the switching operation is between $\pm n$ and $\pm(n+1)$ levels, which, the capacitor C_i will be charged and discharged in $\pm n$ and $\pm(n+1)$ levels respectively, hence the charging and discharging occur in switching frequency, thus, it does not have the effective voltage drop. If the capacitor is discharging in two sequential levels, thus the voltage drop of the capacitor will be effective. The effectual charge and discharge modes of capacitors are shown in TABLE III. For example, the capacitor C_1, while the output is switching between $-3V_{dc}$ and $-4V_{dc}$, is in charge/discharge mode (CD), thus, does not create effective voltage drop.

TABLE III
EFFECTUAL CHARGE AND DISCHARGE MODES OF
CAPACITORS

				N Level				
-4	-3	-2	-1	0	1	2	3	4
CD	C	C	C	C_1	C	C	C	C
D	CD	C	C	C_2	C	C	C	CD
D	D	CD	C	C_3	C	C	CD	D
D	D	D	CD	C_4	C	CD	D	D
t'_3	t'_2	t'_1			t_1	t_2	t_3	

As the worst condition for the voltage drop of capacitors occurs in the resistive load, thus, the calculations are based on resistive load.

Fig. 6 illustrates the determination of time interval in different output voltage levels for one-half cycle. According to this figure, in the time interval between t_2 and t_3, and also between t'_3 and t'_2, the reference wave will be compared with the third carrier while output waveform is switching between $2V_{dc}$ and $3V_{dc}$ levels. According to this pattern, in the other time interval, the reference wave will be compared to the other carriers and will create different output levels. t_i in defined by:

$$t_i = \frac{2\pi}{T} \arcsin(\frac{i}{A_r}) \Big|_{i=1,2,...,(n-1)}$$

$$t'_i = \frac{T}{2} - t_i \qquad (6)$$

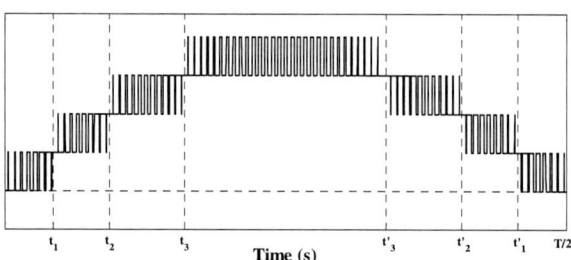

Fig. 6. Determination of time interval in different output voltage levels.

For example, if, $A_r = 3.67$ is assumed, according to (6) t_1, t_2 and t_3 are given 0.88 ^{ms}, 1.83 ^{ms} and 3.05 ^{ms} respectively.

The voltage drop of capacitor is obtained by:

$$\Delta V = \frac{1}{C} \int i\, dt \qquad (7)$$

Since in the pure resistance load, the current waveform is same as a staircase, such as the voltage waveform, then, calculation of voltage drop of the capacitor will be complicated. Thus, by using an accurate assumption, the current can be expressed by the following equations:

$$i = I_m \sin\omega t \qquad (8)$$

$$I_m = \frac{nMV_{DC}}{R} \qquad (9)$$

With respect to (9) and for $n=4$, $M=0.92$, $V_{DC}=85$ [V] and $R=100[\Omega]$, the value of I_m is 3.12 [A]. Due to the modulation technique is used; the voltage drop of the one-half cycle is compensated in the same one-half cycle. According to the TABLE II and III, the states of capacitors for positive and negative cycles are not same. Thus, the voltage drop of capacitors for each cycle is independent of each other and will not sum together.

$$\Delta V^+_{Ci} = \frac{I_m}{C} \int_{t_{n+2-i}}^{t'_{n+2-i}} \sin\omega t\, dt = \frac{2I_m \cos\varphi_{n+2-i}}{C\omega} \qquad (10)$$

$$\Delta V^-{}_{Ci} = \frac{I_m}{C} \int_{t_{n+1-i}}^{t_{n+1-i}} \sin \omega t \, dt = \frac{2I_m \cos \varphi_{n+1-i}}{C\omega} \qquad (11)$$

For example, according to (10) and (11) the voltage drop of the C_4 for positive and negative cycle is obtained by following equations:

$$\Delta V^+{}_{C4} = \frac{I_m}{C} \int_{t_2}^{t_2} \sin \omega t \, dt = \frac{2I_m \cos \varphi_2}{C\omega} = 7.57 \,[V]$$

$$\Delta V^-{}_{C4} = \frac{I_m}{C} \int_{t_1}^{t_1} \sin \omega t \, dt = \frac{2I_m \cos \varphi_1}{C\omega} = 8.64 \,[V] \qquad (12)$$

V. SIMULATION RESULT

To prove the validity of proposed SCMLI structure, several simulations results for the proposed 9-level inverter by using MATLAB/SIMULINK software, is presented as shown in Fig. 7 and 8. Here, the circuit parameters were set to be as follows. The voltage of dc power supply $V_{dc}=85$ *[V]*, the capacitance of all capacitors $C=2.2$ *[mF]* and the all semiconductor devices are supposed ideal with internal resistance $R_{on}=1$ *[mΩ]*. The load is assumed Inductive-Resistive *(R-L)* with $R_{on}=80$ *[Ω]* and $L_{Load}=191$ *[mH]* *(cosφ=0.8)*. According to the proposed modulation strategy, $f_{sw}=10$ *[KHz]* and the modulation index *M=0.92*. The maximum value of output voltage and current waveform for proposed method is 340 volts and 3.12A, respectively. The observed voltage and current in circuit simulation are shown in Fig. 7(a,b) *(cosφ=0.8)* and Fig. 7(c,d) *(cosφ=1)* respectively. The THD of the output waveform without assuming the filter is 16.70% as shown in Fig.8.

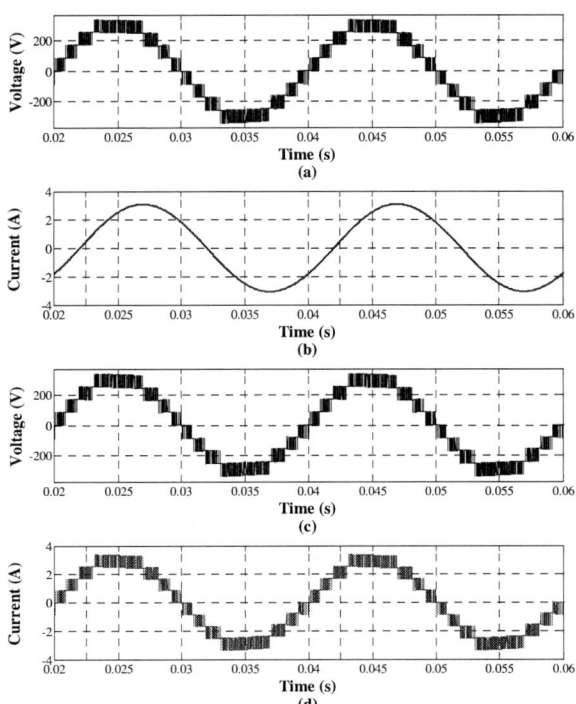

Fig. 7. Output 9-level voltage and current waveform (a, b) cosφ=0.8 and (c, d) cosφ=1.

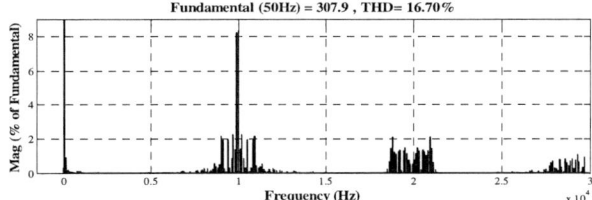

Fig. 8. Frequency spectrum of the output.

In table IV, the proposed topology is compared with two structures [20-21], which have the capability to eliminate the leakage current, and are mentioned to some performance of them.

TABLE IV
COMPARISON OF PROPOSED TOPOLOGY WITH STRUCTURES IN [20] AND [21]

Parameter/ Description	Reference [20] (PROTOTYPE)	Reference [21] (PROTOTYPE)	Proposed Inverter (SIMULATION)
Rated Power (*P*)	1 kW	500 W	500W
Input voltage V$_{dc}$	168 V	400 V	85 V
Output voltage (*RMS*)	230 V	220 V	220 V
Carrier frequency (*f$_s$*)	24.4 kHz	24 KHz	10 KHz
Line frequency	50 Hz	50 Hz	50 Hz
Capacitors	C_{FC}=470 μF, 400 V C_{FC}=100 μF, 400 V	C_{FC}=220 μF, 500 V C_{FC}=320 μF, 500 V	C_{FC}=2200 μF, 100 V
Ability to be extended	No	No	Yes
Boosting capability	No	No	Yes
Load	45 Ω + 67 mH	100 Ω + 150 mH	80 Ω + 191 mH

In the TABLE V, the simulation results of voltage drop of capacitors are compared with the calculation result.

TABLE V
COMPARISON OF SIMULATION AND CALCULATION RESULT OF VOLTAGE DROP

		ΔV_{C1}	ΔV_{C2}	ΔV_{C3}	ΔV_{C4}
Positive	calculation	0	0	5.7	7.57
	simulation	0	0	5.35	7.58
	error	-	-	6.5%	0.1%
Negative	calculation	0	5.7	7.57	8.64
	simulation	0	5.31	7.48	8.46
	error	-	7.3%	1.2%	2%

VI. CONCLUSION

In this study, a novel switched-capacitor multilevel inverter for single-phase grid-connected PV system is presented. The proposed topology is the first structure of SCMLI (does not need boost circuit) which can eliminate leakage current. Due to direct connecting the negative pole of the PV cells to the neutral line of the grid, the leakage current will be decreased to zero. The H-bridge inverter is removed from the end of proposed inverter; thus, to generate the negative output levels, the stored voltage in capacitors, is pumped to the output in reverse polarity. The proposed inverter it has the good capability to conduct the reverse current for inductive load. Due to the output is a 9-level staircase waveform, thus, THD of the output waveform of the proposed inverter is reduced compared to the other topologies, which can eliminate leakage current. Finally, to acknowledge the validity of proposed topology, simulation results were shown.

REFERENCES

[1] A. M. Mahfuz-Ur-Rahman, Md. Mazharul Islam, Md. Rabiul Islam, "Performance analysis of modulation techniques in multilevel inverters for direct grid connected photovoltaic arrays, " 9th International Conference on Electrical and Computer Engineering (ICECE), pp.66-69, 2016.

[2] V Ravi Kumar; A. Sreedevi, "Design and development of Z-source multi-level inverter for solar energy, " International Conference on Applied and Theoretical Computing and Communication Technology (iCATccT), pp.229-232, 2015.

[3] M. Narimani and G.Moschopoulos, "A novel single-stage multilevel type full-bridge converter," IEEE Trans. Ind. Electron., vol. 60, no. 1, pp. 31–42, Jan. 2013.

[4] M. Malinowski, K. Gopakumar, 1. Rodriguez, and M. A. Perez, "A survey on cascaded multilevel inverters," IEEE Trans. Ind. Electron., vol. 57,no. 7, pp. 2197-2206, Jul. 2010.

[5] L. G. Franquelo, 1. Rodriguez, 1. I. Leon, S. Kouro, R. Portillo, and M.A. M. Prats, "The age of multilevel converters arrives," IEEE Ind. Electron., vol. 2, no. 2, pp. 28-39, Jun. 2008.

[6] J. Rodriguez, L. J. Sheng, and P. Fang Zheng, "Multilevel inverters: A survey of topologies, controls, and pplications," IEEE Trans. Ind. Electron., vol. 49, no. 4, pp. 724–738, Aug. 2002.

[7] M. M. Renge and H. M. Suryawanshi, "Five-Level Diode Clamped Inverter to Eliminate Common Mode Voltage and Reduce dv/dt in Medium Voltage Rating Induction Motor Drives," IEEE Trans. Power Electron. vol. 23, no. 4, pp. 1598-1607,2008.

[8] M. Khazraei, H. Sepahvand, and M. Ferdowsi, "Active capacitor voltage balancing in single-phase flying capacitor multilevel power converters," IEEE Trans. Ind. Electron., vol. 59, no. 2, pp. 769- 778, Feb. 2012.

[9] Jing Huang and K. A. Corzine, "Extended operation of flying capacitor multilevel inverters," IEEE Trans. Power Electron., vol. 21, no. I, pp.140-147,2006.

[10] A. Ajami, "Developed cascaded multilevel inverter topology to minimize the number of circuit devices and voltage stresses of switches," IET Power Electron., vol. 7, no. 2, pp. 459–466, Feb. 2014.

[11] T. Kerekes, R. Teodorescu, and U. Borup, "Transformerless photovoltaic inverters connected to the grid," in Proc. IEEE APEC 2007, pp. 1733 – 1737, 2007.

[12] R. González, E. Gubía, J. López, and L. Marroyo, "Transformerless single-phase multilevel-based photovoltaic inverter," IEEE Trans. Ind. Electron, vol. 55, no. 7, pp. 2694-2702, Jul. 2008.

[13] H. Xiao, and S. Xie, "Leakage current analytical model and application in single-phase transformerless photovoltaic grid-connected inverter," IEEE Trans. on Electromagnetic Compatibility, vol.52, no. 4, pp. 902-913, Nov 2010.

[14] Yunjie Gu, Wuhua Li, Bo Yang, Jiande Wu, Yan Deng, Xiangning He, "A transformerless grid connected photovoltaic inverter with switched capacitors" Twenty-Sixth Annual IEEE Applied Power Electronics Conference and Exposition (APEC), pp. 1940 – 1944, 2011.

[15] C. Woo-Jun, "Evaluation and analysis of transformerless photovoltaic inverter topology for efficiency improvement and reduction of leakage current," IET Power Electronics, vol. 8, no. 2, pp. 255–267, 2015.

[16] Venu Sonti; Sachin Jain; Subhashish Bhattacharya, "Analysis of the Modulation Strategy for the Minimization of the Leakage Current in the PV Grid-Connected Cascaded Multilevel Inverter," IEEE Trans. Power Electron., vol. 32, no. 2, pp. 1156–1169, Feb. 2017.

[17] Elyas Zamiri, Naser Vosoughi, Seyed Hossein Hosseini, Reza Barzegarkhoo, and Mehran Sabahi, "A New Cascaded Switched-Capacitor Multilevel Inverter Based on Improved Series–Parallel Conversion With Less Number of Components," IEEE Trans. Ind. Electron., vol. 63, no. 6, pp. 3582-3594, Jun. 2016.

[18] Bagher Karami, Mehdi Samizadeh, " A switched-capacitor multilevel inverter for high AC power systems with reduced ripple loss using SPWM technique," The 6th Power Electronics, Drive Systems & Technologies Conference (PEDSTC2015), pp. 627-632, 2015.

[19] Asuka Tsunoda, Youhei Hinago, and Hirotaka Koizumi," Level- and Phase-Shifted PWM for Seven-Level Switched-Capacitor Inverter Using Series/Parallel Conversion," IEEE Trans. Ind. Electron., vol. 61, no. 8, pp. 4011-4021, Aug. 2014.

[20] Yam P. Siwakoti; Frede Blaabjerg, "A novel flying capacitor transformerless inverter for single-phase grid connected solar photovoltaic system" in IEEE 2016 7th International Symposium on Power Electronics for Distributed Generation Systems (PEDG), 2016, pp. 1-6.

[21] Jaber Fallah Ardashir; "A Single-Phase Transformerless Inverter With Charge Pump Circuit Concept for Grid-Tied PV Applications," IEEE Trans. Ind. Electron., vol. 64, no. 7, pp. 5403-54015, Jul 2017.

An Interleaved Buck-Cascaded Buck-Boost Inverter for PV Grid-Connection Applications

Chien-Hsuan Chang, Chun-An Cheng, and Hung-Liang Cheng
Department of Electrical Engineering, I-Shou University,
Kaohsiung, Taiwan
Email: chchang@isu.edu.tw

Abstract— To simplify circuit configuration and increase system efficiency, a two-switch buck-boost (TSBB) inverter has been proposed by connecting a buck-cascaded buck-boost (BuCBB) converter in series with a line-frequency unfolding circuit. In accordance with the level of input voltage and sinusoidal output voltage, this inverter can work with buck or boost operation principle. Wherefore, efficiency can be increased by delivering partial energy directly to output. This inverter is suitable for developing photovoltaic grid-connection systems with wide voltage-variation of input. In order to further increase its output power, this paper proposes to develop a high-efficiency inverter by connecting two BuCBB converters in interleaving parallel. Due to the cancelation of inductor ripple-currents, size and cost of input and output filters can be significantly reduced. Since the output power is shared equally by two parallel dc-dc converters, the overall conduction losses of power switches and inductors can also be reduced to improve efficiency. Finally, an experimental prototype is implemented with proper analysis and design. The experimental results are measured to verify the feasibility and correctness of the proposed inverter.

Keywords— *buck-boost; inverter; photovoltaic; grid-connection, interleave.*

I. INTRODUCTION

Recently, because of the seeking for the alternative energy resource, the development of the photovoltaic (PV) power applications has been valued greatly [1]–[5]. As inverters are required in PV grid-connected systems, there are many investigator committing to the studies of inverter circuit architecture and their control strategies [6]–[13]. Since typical inverter is buck-type and the PV output voltage varies widely with insolation and temperature changes, a non-isolated boost-type dc-dc converter is usually adopted between the PV panel and the inverter, as shown in Fig. 1. The functions of boost-type converter are to step-up PV voltage as well as to track maximum-power point [14]–[18]. The main advantage of this two-stage system is the independent control of each stage, so that it is easy to achieve optimal designs of all functions. However, except of the complexity of the circuit structure, high cost, and lower reliability, the main disadvantage is high power dissipations caused by two-stage power conversions.

In order to overcome these drawbacks, high-efficiency inverters are developed by using a dc-dc converter to generate a rectified sinusoidal voltage and an H-bridge unfolding circuit with line-commuted operation to change its polarity [19], [20]. The block diagram is shown in Fig. 2, called pseudo dc-link structure. Because the power switches of the unfolding circuit only switch with line-frequency, it could achieve extremely high efficiency.

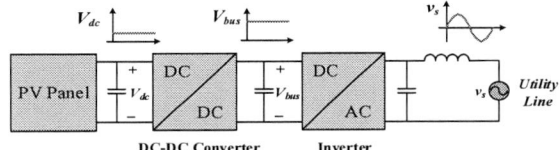

Fig. 1. Block diagram of a two-stage PV power system.

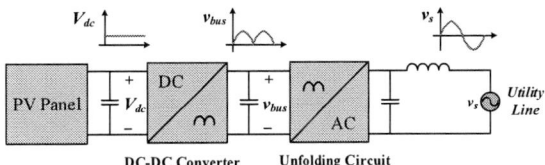

Fig. 2. Block diagram of a PV power system with pseudo dc-link.

Taking into account the wide variations in PV voltage, a dc-dc converter with both step-down and step-up functions is more suitable to generate the rectified sinusoidal voltage. The circuit structure of conventional single-switch buck-boost converter is simple, but the voltage stress of power switch is higher, and entire input power has to store in the inductor and then transfer to output, resulting in low efficiency [21]. If the switches of the two-switch buck-boost (TSBB) converter are operated independently, the TSBB converter can work with buck or boost operation principle. Partial input power can be transferred directly to the output to improve converter efficiency[22]–[24]. The buck-cascaded buck-boost (BuCBB) converter, which is one of the TSBB converters, has been used for developing high efficiency inverter [25]. By controlling the two switches independently, this inverter can work with either buck or boost operation principle. Therefore, efficiency can be improved by delivering partial energy to output directly. This inverter is suitable for developing PV grid-connection systems with wide voltage-variation of input.

In order to further increase the output power of BuCBB inverter, this paper proposes to develop a high-efficiency inverter by connecting two BuCBB dc-dc converters in interleaving parallel [26]–[30]. As the inductor ripple-currents are with phase-difference of 180 degrees and can be canceled by each other, size and cost of input and output filters can be significantly reduced. Since the output power is shared equally by two parallel DC-DC converters, the overall conduction losses of power switches and inductors can also be reduced to improve efficiency. Finally, an experimental prototype is implemented with proper analysis and design. The experimental results are measured to verify the feasibility and correctness of the proposed inverter.

II. CIRCUIT CONFIGURATION

Fig. 3 shows the circuit configuration of the proposed interleaved inverter for PV grid-connection applications. We use two BuCBB dc-dc converters with interleaved operation to generate output rectified sinusoidal voltage, and then connect with an H-bridge unfolding circuit to change the output polarity. Two interleaved BuCBB converters are mainly formed by the MOSFETs S_{Bu1}, S_{Bu2}, S_{Bo1}, S_{Bo2}, the diodes D_{Bu1}, D_{Bu2}, D_{Bo1}, D_{Bo2}, and the inductors L_1, L_2. The two converters can work with either buck or boost operation principle, so that partial energy can be directly transferred between input and output to improve efficiency. In addition, the H-bridge unfolding circuit is formed by four MOSFETs (S_1, S_2, S_3, S_4) and a low-pass filter (L_s & C_s). Because the four MOSFETs operate with low frequency, their switching losses are nearly negligible. This line-frequency unfolding circuit can achieve extremely high efficiency. Furthermore, the currents of inductors L_1, L_2 are interleaving, so that inductance can be reduced to increase resonant frequency and to improve system stability.

Because the proposed interleaved BuCBB inverter has both step-down and step-up functions, it can overcome the difficulty of wide voltage variation in PV applications. While the inverter works in step-up mode, the higher input current can be shared by two inductors, resulting in lower inductor conduction losses and high system efficiency. The two dc converters are paralleled and with interleaving operation, which helps to reduce the ripple of input or output currents and to shrink the volume of filters. Additionally, since only two MOSFETs S_{Bu1}, S_{Bu2} (or S_{Bo1}, S_{Bo2}) are simultaneously switched with high frequency, the switching losses can be significantly reduced, leading to high efficiency.

III. OPERATION PRINCIPLE

Fig. 4 shows the conceptual diagram of PV voltage V_{PV}, line voltage $v_s(t)$, operation modes and duty ratios ($d_{Bu1}(t)$, $d_{Bu2}(t)$, $d_{Bo1}(t)$, $d_{Bo2}(t)$), in which V_{PV} is lower than the peak of $v_s(t)$, and T_{ac} is line period. As shown in this figure, the proposed inverter can work in step-down or step-up modes according to the level of V_{PV} and $v_s(t)$.

A. Step-down Mode

While PV voltage V_{PV} is above the instantaneous voltage of $v_s(t)$，the proposed inverter works in step-down mode. In this mode, the MOSFETs S_{Bo1}, S_{Bo2} are never conducted, and their duty ratios are both 0%. Therefore, the operation principle of inverter is similar to buck converter. The MOSFETs S_{Bu1}, S_{Bu2} high-frequency switching with following duty ratio:

$$d_{Bu1}(t) = d_{Bu2}(t) = \frac{V_M \sin \omega t}{V_{PV}} \tag{1}$$

It should be noted that there is a phase difference of 180 degrees between the gate signals of MOSFETs S_{Bu1}, S_{Bu2}.

B. Step-up Mode

While PV voltage V_{PV} is below the instantaneous voltage of $v_s(t)$，the proposed inverter works in step-up mode. In this mode, the MOSFETs S_{Bu1}, S_{Bu2} are always in conduction state, and their duty ratios are both 100%. Therefore, the operation principle of inverter is similar to

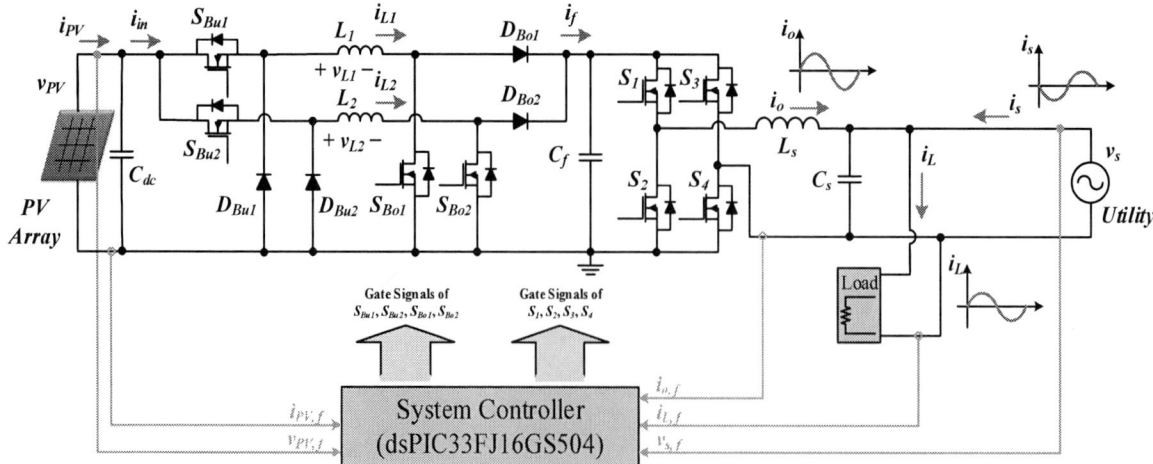

Fig. 3. Schematic diagram of the proposed interleaved inverter for PV grid-connection applications.

boost converter. The MOSFETs S_{Bo1}, S_{Bo2} are high-frequency switching with following duty ratio:

$$d_{Bo1}(t) = d_{Bo2}(t) = 1 - \frac{V_{PV}}{V_M \sin \omega t} \qquad (2)$$

There is also a phase difference of 180 degrees between the gate signals of MOSFETs S_{Bo1}, S_{Bo2}. The statues of switching components are listed in Table I for reference.

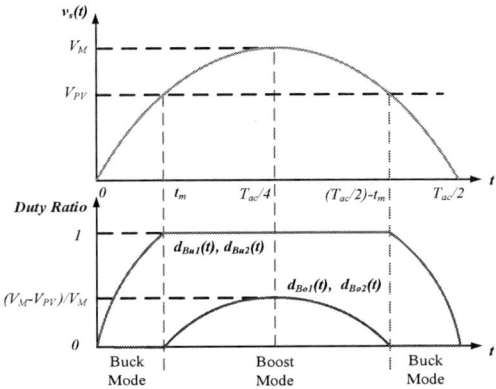

Fig. 4. The conceptual diagram of operation modes and duty ratios.

TABLE I
STATUES OF SWITCHING COMPONENTS OF THE PROPOSED INVERTER

	Step-down Mode $(V_{PV} > v_s(t))$	Step-up Mode $(V_{PV} < v_s(t))$
S_{Bu1}, S_{Bu2}	switching with $d_{Bu1}(t)$ ($d_{Bu2}(t)$)	always on
D_{Bu1}, D_{Bu2}	switching	always off
S_{Bo1}, S_{Bo2}	always off	switching with $d_{Bo1}(t)$ ($d_{Bo2}(t)$)
D_{Bo1}, D_{Bo2}	always on	switching
Gain (M)	$d_{Bu1}(t)$	$1 / (1- d_{Bo1}(t))$

IV. EXPERIMENTAL RESULTS

According to the electrical specifications shown in Table II and the components parameters shown in Table III, we built an experimental prototype as a validation example. The experimental results are used to confirm the feasibility of the proposed circuit structure and the correctness of the theoretical analysis.

The proposed interleaved BuCBB inverter is controlled by sinusoidal pulse-width modulation. At the conditions of 100 V_{DC} input and full load, the measured waveforms of output voltage and output current are shown in Fig. 5. The results are proved that the proposed inverter can effectively convert the DC voltage to AC output. The measured waveforms of inductor voltage v_{L1}, v_{L2} are shown in Fig. 6. As can be seen, when input voltage is above instantaneous output voltage, the proposed inverter works in step-down mode. Conversely, the inverter works in step-up mode.

TABLE II
ELECTRICAL SPECIFICATIONS OF THE INVERTER

Electrical Specifications	
PV Voltage, V_{PV}	80 ~ 200 V_{dc}
Line Voltage, v_s	110 V_{rms}
Line Frequency, f_{ac}	60 Hz
Rated Power, P_o	800 W
Switching Frequency, f_s	40 kHz

TABLE III
COMPONENT PARAMETERS OF THE INVERTER

Component Parameters	
Input Capacitor, C_{dc}	2000 μF
MOSFETs, S_{Bu1}, S_{Bu2}, S_{Bo1} and S_{Bo2}	47N60C3
Diodes, D_{Bu1}, D_{Bu2}, D_{Bo1} and D_{Bo2}	C3D10060A
Inductors, L_1, L_2	1 mH
Capacitor, C_f and C_S	4.7 μF

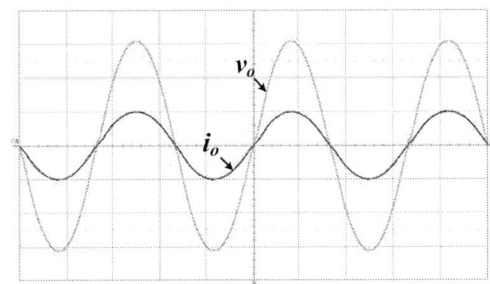

(v_s: 50 V/div; i_o: 10 A/div; time: 5 ms/div)

Fig. 5. Measured waveforms of output voltage v_o and output current i_o at the conditions of 100 V_{DC} input and full load.

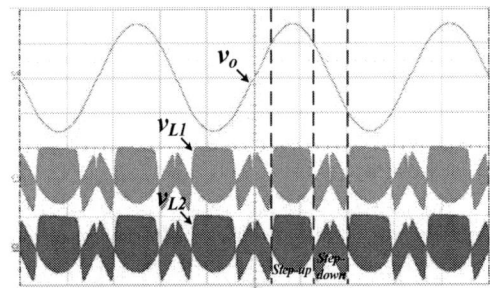

(v_o: 100 V/div; v_{L1}, v_{L2}: 100 V/div; time: 5 ms/div)

Fig. 6. Measured waveforms of output voltage v_o and inductor voltages v_{L1}, v_{L2} at the conditions of 100 V_{DC} input and full load.

Fig. 7 shows the measured waveforms of output voltage, output current, and inductor currents i_{L1}, i_{L2} with 100 V_{DC} input and full load. It can be observed that currents can be evenly distributed on two inductors, resulting in low conduction losses. Fig. 8 shows the zoom-in waveforms of peak inductor currents. Obviously, the average value of i_{L1}, i_{L2} are close, and their ripples are with phase difference of 180 degrees. Since the ripples can be counteracted after merged, the ripple of the current i_{in} is very small, which can significantly narrow the size of the input capacitor. At the same condition, Fig. 9 shows the zoom-in waveforms of step-down mode. As shown in the results, the ripple of the currents i_{L1}, i_{L2} are also out of phase. The ripple of the combined current i_f is

2862

extremely low, helping to reduce the size of the output filter.

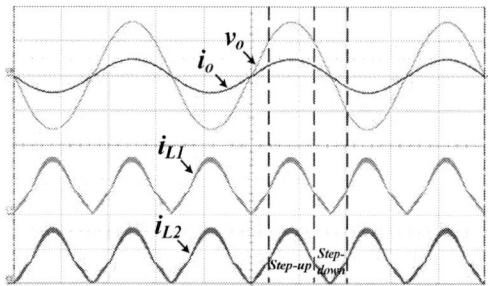

(v_o: 100 V/div; i_o: 20 A/div; i_{L1}, i_{L2}: 5 A/div; time: 5 ms/div)
Fig. 7. Measured waveforms of output voltage v_o, output current i_o and inductor currents i_{L1}, i_{L2} with full load and 100 V_{DC} input.

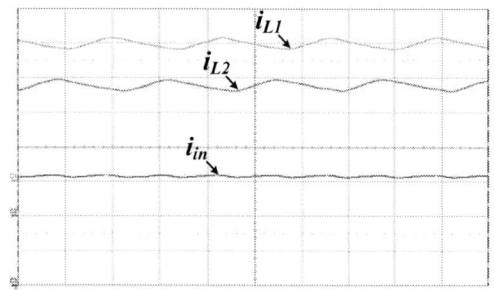

(i_{in}: 5 A/div; i_{L1}, i_{L2}: 2 A/div; time: 10 μs/div)
Fig. 8. Measured waveforms of inductor currents i_{L1}, i_{L2} and their merged current i_{in} with full load and 100 V_{DC} input.

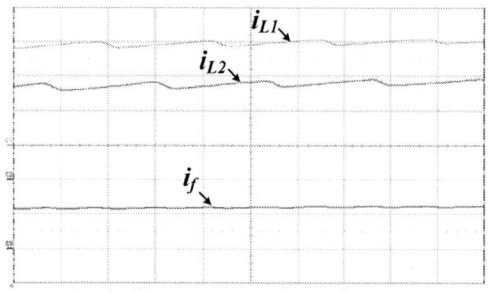

(i_{in}: 5 A/div; i_{L1}, i_{L2}: 1 A/div; time: 10 μs /div)
Fig. 9. Measured waveforms of inductor currents i_{L1}, i_{L2} and their merged current i_f with full load and 100 V_{DC} input.

Furthermore, the prototype is tested at the conditions of 200 V_{DC} input and full load, which results are used to verify the feasibility of the proposed inverter for the applications with wide voltage variation. Fig. 10 shows the measured waveforms of output voltage and output current. It can be seen that the output voltage and output current are sinusoidal waveforms with almost no distortion. The measured waveforms of inductor voltage v_{L1}, v_{L2} are shown in Fig. 11. Because input voltage is always above the sinusoidal output voltage, the proposed inverter only operates in step-down mode. Fig. 12 shows the measured waveforms of output voltage, output current, and inductor currents i_{L1}, i_{L2} at the same condition. The current is also evenly distributed on two inductors to reduce conduction losses of inductors.

The measured total harmonic distortion (T.H.D.) and odd harmonics of output voltage are listed in Table IV. The results meet the requirements of electrical standard, but also verify the correctness of the theoretical analysis. The measured efficiencies are up to 96.2% at input voltage of 100 V_{DC} and 97.2% at input voltage of 200 V_{DC}.

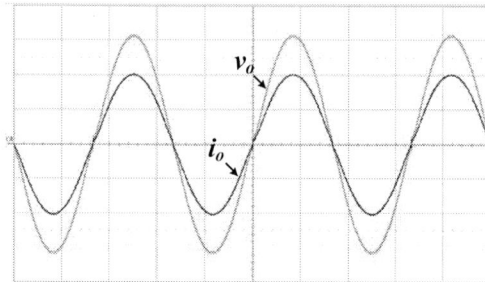

(v_s: 50 V/div; i_o: 5 A/div; time: 5 ms/div)
Fig. 10. Measured waveforms of output voltage v_o and output current i_o at the conditions of 200 V_{DC} input and full load.

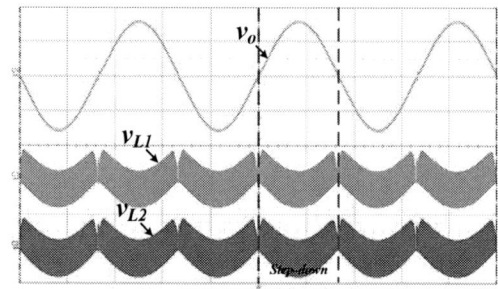

(v_o: 100 V/div; v_{L1}, v_{L2}: 200 V/div; time: 5 ms/div)
Fig. 11. Measured waveforms of output voltage v_o and inductor voltages v_{L1}, v_{L2} at the conditions of 200 V_{DC} input and full load.

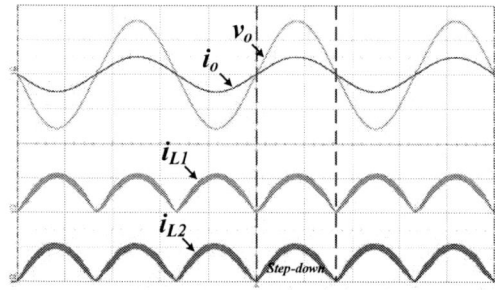

(v_o: 100 V/div; i_o: 20 A/div; i_{L1}, i_{L2}: 5 A/div; time: 5 ms/div)
Fig. 12. Measured waveforms of output voltage v_o, output current i_o and inductor currents i_{L1}, i_{L2} at the conditions of 200 V_{DC} input and full load.

TABLE IV
MEASURED RESULTS OF TOTAL HARMONIC DISTORTION AND ODD HARMONICS OF OUTPUT VOLTAGE

Input Voltage / Odd Harmonics	100 V_{dc}	200 V_{dc}
T.H.D.	1.38 %	0.88%
Harmonic of 3rd	1.10 %	0.12%
Harmonic of 5th	0.50 %	0.45%
Harmonic of 7th	0.22 %	0.22%
Harmonic of 9th	0.11 %	0.28%
Harmonic of 11th	0.10 %	0.21%

At the condition of 100 V_{DC} input voltage, the dynamic tests are performed with step load changes of 50% to 100% and 100% to 50%. Fig. 13 shows the measured transient responses of output voltage v_o. As shown in the waveforms, the output voltage can quickly recover to steady state after step load changes, which verifies that good dynamic performance can be achieved by the proposed inverter. Besides, inductor currents i_{L1}, i_{L2} are also measured during the dynamic tests. It is obvious that the current can be evenly distributed on two inductors whether at half load or full load.

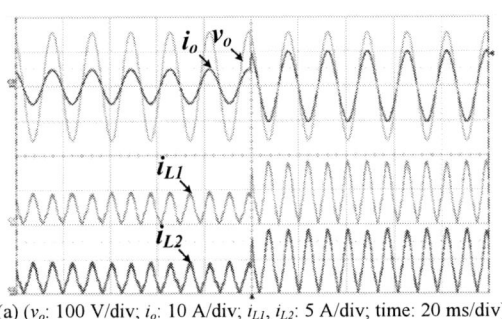

(a) (v_o: 100 V/div; i_o: 10 A/div; i_{L1}, i_{L2}: 5 A/div; time: 20 ms/div)

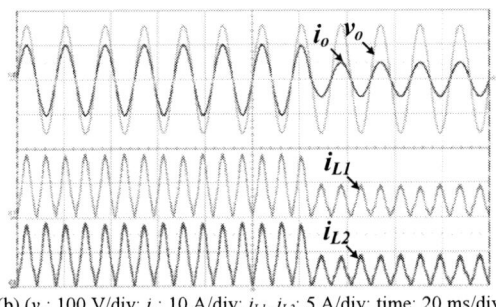

(b) (v_o: 100 V/div; i_o: 10 A/div; i_{L1}, i_{L2}: 5 A/div; time: 20 ms/div)

Fig. 13. Measured transient responses of v_o with 100 V_{DC} input-voltage and (a) 50% to 100%, and (b) 100% to 50% step load changes.

V. CONCLUSION

This paper proposes an interleaved BuCBB inverter for the PV grid-connection applications. Only two MOSFETs switch with high-frequency, which can effectively reduce the switching losses to improve efficiency. With the use of interleaved parallel structure, the input and output current ripple can be significantly canceled, leading to lower conduction loss of inductors and smaller size of the filters. The digital signal processor dsPIC33FJ16GS504 is the core of control circuit. It can provide multiple sets of SPWM control signals and implement complex mathematical calculations, so that the hardware circuit can be simplified and cost can be reduced. The measured experimental results have confirmed the feasibility of the proposed interleaved BuCBB inverter.

ACKNOWLEDGMENT

The authors would like to express their appreciation for grant support from the Ministry of Science and Technology (MOST) of Taiwan, under its grant number MOST 106-2221-E-214-032.

REFERENCES

[1] J. P. Benner and L. Kazmerski, "Photovoltaics gaining greater visibility," *IEEE Spectrum*, vol. 29, no. 9, pp. 34–42, Sep. 1999.

[2] J. T. Bialasiewicz, "Renewable energy systems with photovoltaic power generators: Operation and modeling," *IEEE Trans. Ind. Electron.*, vol. 55, no. 7, pp. 2752–2758, Jul. 2008.

[3] Y.-M. Chen, H.-C. Wu, Y.-C. Chen, K.-Y. Lee, and S.-S. Shyu, "The AC line current regulation strategy for the grid-connected PV system," *IEEE Trans. Power Electron.*, vol. 25, no. 1, pp. 209–218, Jan. 2010.

[4] X. Guan, Z. Xu, and Q. S. Jia, "Energy-efficient buildings facilitated by microgrid," *IEEE Trans. Smart Grid*, vol. 1, no. 3, pp. 243–252, Dec. 2010.

[5] Chien-Hsuan Chang, En-Chih Chang, and Hung-Liang Cheng, "A high-efficiency solar array simulator implemented by an LLC resonant DC-DC converter," *IEEE Trans. Power Electron.*, vol. 28, no. 6, pp. 3039–3046, Jun. 2013.

[6] M. Calais, J. Myrzik, T. Spooner, and V. G. Agelidis, "Inverter for single-phase grid connected photovoltaic systems–An overview," in *Proc. IEEE 33rd Annu. Power Electron. Spec. Conf.*, Jun. 2002, vol. 4, pp. 1995–2000.

[7] D. Casadei, G. Grandi, and C. Rossi, "Single-phase single-stage photovoltaic generation system based on a ripple correlation control maximum power point tracking," *IEEE Trans. Energy Convers.*, vol. 21, no. 2, pp. 562–568, Jun. 2006.

[8] Y. Huang, M. Shen, F. Z. Peng, and J. Wang, "Z-source inverter for residential photovoltaic systems," *IEEE Trans. Power Electron.*, vol. 21, no. 6, pp. 1776–1782, Nov. 2006.

[9] N. Kasa, T. Iida, and L. Chen, "Flyback inverter controlled by sensorless current MPPT for photovoltaic power system," *IEEE Trans. Ind. Electron.*, vol. 52, no. 4, pp. 1145–1152, Aug. 2005.

[10] F. Shinjo, K. Wada, and T. Shimizu, "A single-phase grid-connected inverter with a power decoupling function," in *Proc. IEEE Power Electron. Spec. Conf.*, 2007, pp. 1245–1249.

[11] S. Jain and V. Agarwal, "A single-stage grid connected inverter topology for solar PV systems with maximum power point tracking," *IEEE Trans. Power Electron.*, vol. 22, no. 5, pp. 1928–1940, Sept. 2007.

[12] H. Patel and V. Agarwal, "A single-stage single-phase transformer-less doubly grounded grid-connected PV interface," *IEEE Trans. Energy Convers.*, vol. 24, no. 1, pp. 93–101, Mar. 2009.

[13] D. Cao, S. Jiang, X. Yu, and F. Z. Peng, "Low-cost semi-Z-source inverter for single-phase photovoltaic systems," *IEEE Trans. Power Electron.*, vol. 26, no. 12, pp. 3514–3523, Dec. 2011.

[14] J. M. Kwon, K. H. Nam, and B. H. Kwon, "Photovoltaic power conditioning system with line connection," *IEEE Trans. Ind. Electron.*, vol. 53, no. 4, pp. 1048–1054, Aug. 2006.

[15] P. G. Barbosa, H. A. C. Braga, M. do Carmo Barbosa Rodrigues, and E. C. Teixeira, "Boost current multilevel inverter and its application on single-phase grid-connected photovoltaic systems," *IEEE Trans. Power Electron.*, vol. 21, no. 4, pp. 1116–1124, Jul. 2006.

[16] M. Armstrong, D. J. Atkinson, C. M. Johnson, and T. D. Abeyasekera, "Auto-calibrating dc link current sensing technique for transformerless, grid connected, H-bridge inverter systems," *IEEE Trans. Power Electron.*, vol. 21, no. 5, pp. 1385–1396, Sep. 2006.

[17] L. Asiminoaei, R. Teodorescu, F. Blaabjerg, and U. Borup, "Implementation and test of an online embedded grid impedance estimation technique for PV inverters," *IEEE Trans. Ind. Electron.*, vol. 52, no. 4, pp. 1136–1144, Aug. 2005.

[18] B. Yang, W. Li, Y. Zhao, and X. He, "Design and analysis of a grid-connected photovoltaic power system," *IEEE Trans. Power Electron.*, vol. 25, no. 4, pp. 992–1000, Apr. 2010.

[19] X. Yuan and Y. Zhang, "Status and opportunities of photovoltaic inverters in grid-tied and micro-grid systems," in *Proc. Power Electron. Motion Control Conf.*, 2006, pp. 1–4.

[20] B.S. Prasad, S. Jain, and V. Agarwal, "Universal single-stage grid-connected inverter," *IEEE Trans. Energy Convers.*, vol. 23, no. 1, pp. 128–137, Mar. 2008.

[21] J. Chen, D. Maksimovic, and R. Erickson, "Buck-boost PWM converters having two independently controlled switches," in *Proc. IEEE 32nd Annu. Power Electron. Spec. Conf.*, Jun. 2001, vol. 1, pp. 736–741.

[22] Y.-C. Chang, C.-L. Kuo, K.-H. Sun, and T.-C. Li, "Development and operational control of two-string maximum power point trackers in DC distribution systems," *IEEE Trans. Power Electron.*, vol. 28, no. 4, pp. 1852–1861, Apr. 2013.

[23] C.-L. Wei, C.-H. Chen, K.-C. Wu, and I-T. Ko, "Design of an average-current-mode noninverting buck–boost DC–DC converter with reduced switching and conduction losses," *IEEE Trans. Power Electron.*, vol. 27, no. 12, pp. 4934–4943, Dec. 2012.

[24] J. Chen, D. Maksimovic, and R.W. Erickson, "Analysis and design of a low-stress buck-boost converter in universal-input PFC applications," *IEEE Trans. Power Electron.*, vol. 21, no. 2, pp. 320–329, Mar. 2006.

[25] Chien-Hsuan Chang, Chun-An Cheng, En-Chih Chang, and Hung-Liang Cheng, "Design and implementation of a two-switch buck-boost typed inverter with universal and high-efficiency features," in *Proc. 9th International Conference on Power*

Electronics – ECCE Asia, Jun. 2015, pp. 2737–2743.

[26] Feng Hong, Jun Liu, Baojian Ji, Yufei Zhou, Jianhua Wang, and Chenghua Wang, "Interleaved dual buck full-bridge three-level inverter," *IEEE Trans. Power Electron.*, vol. 31, no. 2, pp. 964–974, Feb. 2016.

[27] B. Tamyurek and B. Kirimer, "An interleaved high-power flyback inverter for photovoltaic applications," *IEEE Trans. Power Electron.*, vol. 30, no. 6, pp. 3228–3241, Jun. 2015.

[28] Young-Ho Kim, Jin-Woo Jang, Soo-Cheol Shin, and Chung-Yuen Won, "Weighted-efficiency enhancement control for a photovoltaic AC module interleaved flyback inverter using a synchronous rectifier," *IEEE Trans. Power Electron.*, vol. 29, no. 12, pp. 6481–6493, Dec. 2014.

[29] Udupi R. Prasanna and Akshay K. Rathore, "Current-fed interleaved phase-modulated single-phase unfolding inverter: analysis, design, and experimental results," *IEEE Trans. Ind. Electron.*, vol. 61, no. 1, pp. 310–319, Jan. 2014.

[30] Mohammad A. Abusara and Suleiman M. Sharkh, "Design and control of a grid-connected interleaved inverter," *IEEE Trans. Power Electron.*, vol. 28, no. 2, pp. 748–764, Feb. 2013.

A Novel PV Array Connection Strategy with PV-buck Module to Improve System Efficiency

Chi Shao[1], Wenjie Wang[1], Lijun Hang[1], Anping Tong[2], Shitao Wang[3]

1 School of automation, Hangzhou Dianzi University, Hangzhou, China, 310016

2 Dept. of Electrical and Electronic Engineering, Shanghai Jiao Tong University, Shanghai, China, 200240

3 State Grid of China Technology College , er-huan south RD, Jinan, China, 250002

lijunhang.hhy@aliyun.com

Abstract-The traditional directly connected PV array has low output power and efficiency under partially shaded condition. This paper proposes a new PV array connection method by installing a buck converter for each PV panel before being connected as novel PV array. More power can be extracted from novel PV array than the traditional PV array by using perturb and observer (P&O) MPPT (Maximum Power Point Tracking) algorithm under any uneven irradiance condition. The simulation results show that the proposed method is feasible and efficient, meanwhile the complex GMPPT (Global MPPT) algorithm is not required.

Keywords-PV; MPPT; partially shaded condition; distributed converter PV array

I. INTRODUCTION

The PV power system becomes popular in this new era because the solar energy is renewable and environmental friendly. Although the research and development on solar cell design and fabrication is carried out continuously to reduce the high capital cost, the improvement of overall PV system performance is equally important. One of the interesting areas is by implementing MPPT technique to control the operating condition of the PV system. This approach is to track the maximum available output power of the PV system and hence to ensure that the maximum power can be extracted regardless the changes of environmental conditions such as solar irradiance level and ambient temperature.

In order to increase the extracted power from PV system, a bunch of individual PV panels (it can be defined as PV-diode Module shown in Fig. 1 (a)) are connected in series and parallel to construct a PV array and enhance the output capacity.   However, this kind of connection has a serious drawback under partially shaded condition. The maximum power cannot be acquired when the irradiance or temperature on each PV panel are not the same in the PV array system. Bruendlinger and et al. have tested various commercially available inverters in partially shaded conditions and have found that the power loss due to shading can be as high as 70% [2]. Even using accurate and efficient GMPPT algorithm, it cannot ensure that each PV-diode module can produce maximum power under partially shaded condition because of the existence of bypass diode.

To improve the output power of PV array under partially shaded condition, this paper proposes a novel method that let every PV module output maximum power, meanwhile to avoid using the complex GMPPT algorithm.

II. NOVEL STRUCTURE OF PV ARRAY

In a traditional PV-diode array, if there are two PV-diode modules of column jth under different irradiance (PV1:1000 W/m^2 ; PV2:800 W/m^2), then their I-V characteristic curve would be shown in Fig. 1(b) and their MPPs are point a and point b. GMPPT will be used in this PV-diode array system. The operating current I may be equal to I_1 or I_3, and I_2 is the shaded PV-diode module's short - circuit current. If $I = I_1$, PV1 would work at MPP, PV2's would not output power, because $I > I_2$ and its bypass diode would turn on. If $I = I_3$, PV2 would work at MPP and PV1 work at point d. It is obvious that the power at point d is smaller than MPP(point a). Thus, wherever the column operating current is, the system would not make sure each PV-diode module produce maximum power under partially shaded condition.

In order to increase the extracted power for the PV array, a buck converter is proposed to parallel with each PV panels before being connected in series, as shown in Fig. 2 (a). The novel structure of PV panel can be called PV-buck module. The proposed PV-buck module can be used to construct the required PV array with certain voltage and power.

Assuming the buck converter has no conversion loss, the individual PV-buck module's output characteristic of P-V and I-V with different irradiance condition are presented in Fig. 2(b) and Fig.2 (c). From Fig. 2(c), it can be concluded that the individual PV-buck module has different output characteristic of I-V under different duty ratio according to its irradiance. From Fig. 2(b), the characteristic of P-V for each PV-buck module is shown under different duty ratio according to different irradiance, meanwhile the maximum power can be extracted from the PV panel under different irradiance. We can find that the maximum output power is not changed under different duty ratios. Meanwhile, it is known that the traditional PV-diode module has only one output characteristic curve.

It is impossible for the PV array to make each PV panel work at its own MPP under partially shaded condition. However, if the PV-buck module is used, it is easy to make sure each PV-buck module work at MPP by controlling the buck converter as showed in Fig.1(c). The two dotted line are PV-buck module I-V curves, point a' and point b' are their MPP and column operating current $I = I_4$.

The 2018 International Power Electronics Conference

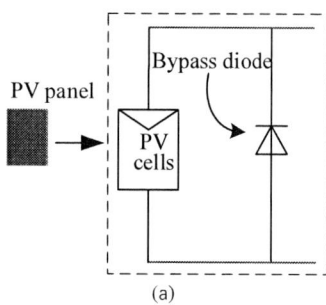

(a)

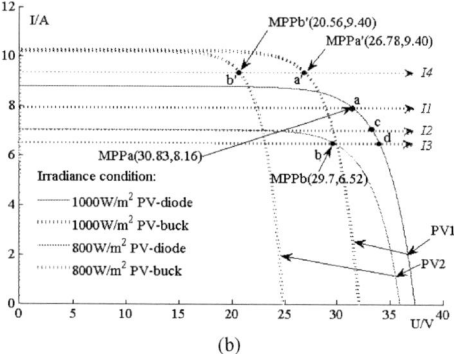

(b)

Fig. 1. (a) Traditional PV-diode module. (b) I-V curves of two PV modules in series. (Data is from the PV panel of CSUN 250-60P.)

It is impossible for the PV array to make each PV panel work at its own MPP under partially shaded condition. However, if the PV-buck module is used, it is easy to make sure each PV-buck module work at MPP by controlling the buck converter as showed in Fig.1(c). The two dotted line are PV-buck module I-V curves, point a' and point b' are their MPP and column operating current $I = I_4$.

A k*n PV array is shown in Fig. 3 by using the proposed PV-buck module. The valuables of current and voltage are marked in the figure. The voltage and current of the buck converter in each column can be derived as (1),

$$\begin{cases} I_{1j} = I_{2j} = \cdots = I_{ij} = \cdots = I_{kj} (j = 1,2 \cdots n) \\ \sum_{i=1}^{k} U_{t1} = \sum_{i=1}^{k} U_{i2} = \cdots = \sum_{i=1}^{k} U_{ij} = \cdots = \sum_{i=1}^{k} U_{in} \end{cases} \quad (1)$$

Where I_{sj} is the current of jth column.

The mean value of voltage and current in each PV-buck module can be deduced as follows:

$$\begin{cases} U_{pvij} \times d_{ij} = U_{ij} \\ I_{pvij} \div d_{ij} = I_{ij} \end{cases} (i = 1,2,\cdots k; j = 1,2,\cdots n) \quad (2)$$

U_{pvmij} and I_{pvmij} are assumed to be the corresponding panel voltage and current of the PV-buck module of the ith row and the jth column in the array, meanwhile the output voltage and current of the corresponding buck converter are assumed to be U_{mij} and I_{mij}. The

maximum power of the column j is $P_{mj} = \sum_{i=1}^{k} P_{mij}$, where, P_{mij} is the maximum output power of the PV-buck module in the ith row of column j. If every PV component work at MPP, the voltage of parallel column must satisfy the rule of as shown in (1), then the power relation of each column can be calculated as:

$$P_{m1} : P_{m2} \cdots P_{mj} \cdots P_{m(n-1)} : P_{mn} = I_{sm1} : I_{sm2} \cdots I_{smj} \cdots I_{sm(n-1)} : I_{smn} \quad (3)$$

I_{smj} is the current of jth column when each PV-buck operates under MPP. Therefore, the duty ratio d_{mij} of each buck converter can be written as:

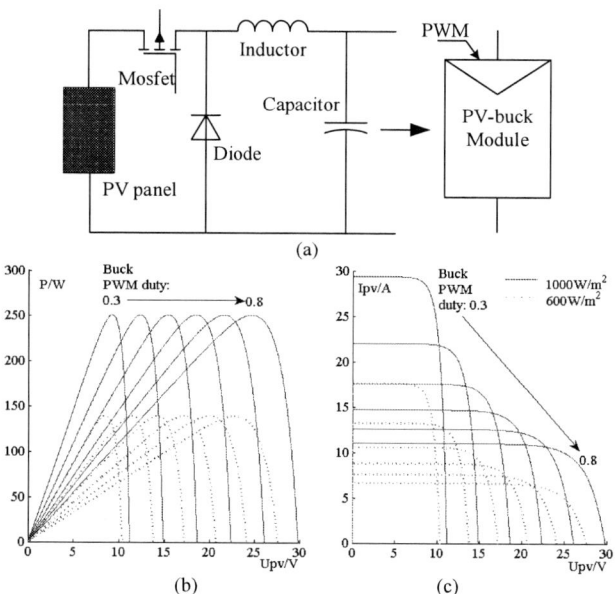

(a)

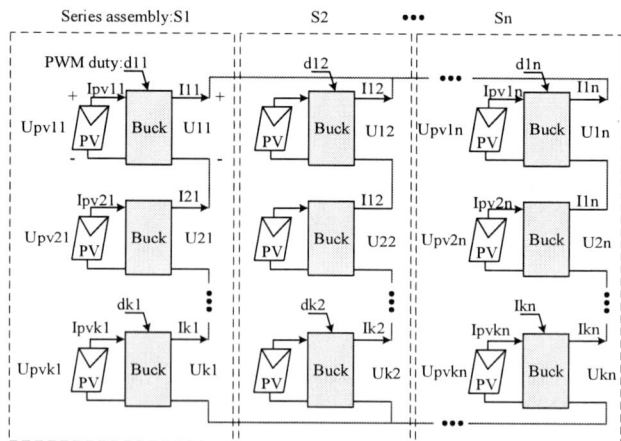

(b)   (c)

Fig. 2. (a) PV-buck module. (b) and (c) are P-V and I-V curves for PV buck module under different duty ratio. (Data is from the PV panel of CSUN 250-60P.)

Fig. 3. k*n PV array with PV-buck module

$$d_{mij} = I_{pvmij} \div I_{ij} = I_{pvmij} \div I_{smj} (j = 1,2,\cdots n) \quad (4)$$

Therefore, in the jth column, d_{mij} can be achieved in

2867

each buck converter to extract the maximum power from the ith row PV panel. d_{mij} can be found by the traditional P&O[3].

III. SIMULATION RESULTS

In order to verify the feasibility and high efficiency of PV array constructed by PV-buck modules, two simulation models are built in Matlab/Simulink, as shown in Fig. 4(a) and Fig. 4(b). CSUN 250-60P polysilicon PV component is used. Under standard test condition, the key parameters of PV are: Maximum Power P_{max} =250W; Open circuit voltage U_o =37.3V; Short-circuit current I_{sso} =8.81A; Voltage at MPP U_{max} =29.9V; Current at MPP I_{max} =8.36A. In addition, the practical parameters of the active switches and diodes are used in the simulation.

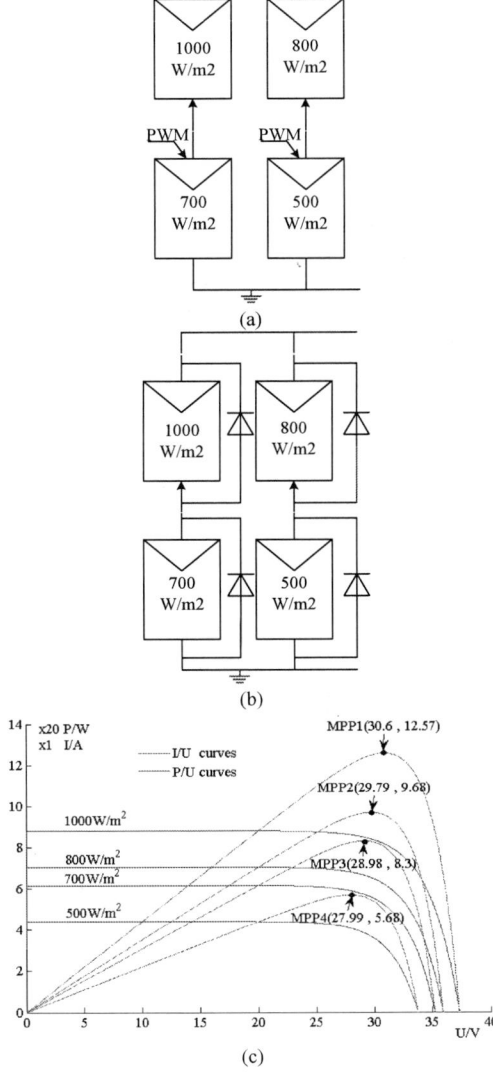

(a)

(b)

(c)

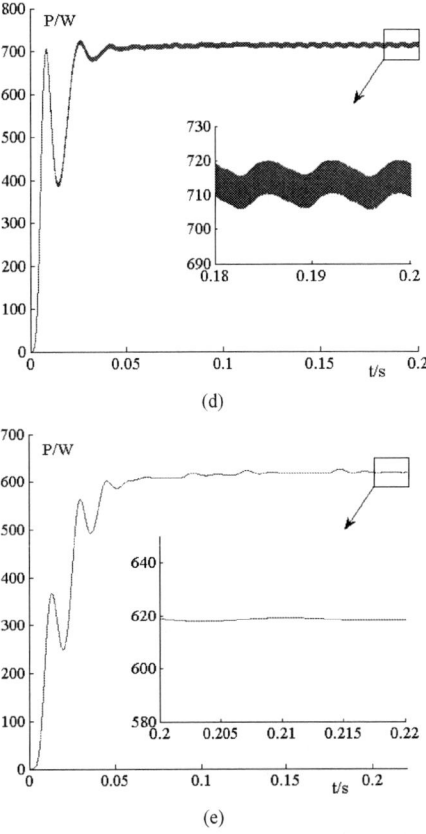

(d)

(e)

Fig. 4. (a) 2 series & 2 parallel PV-buck array. (b) 2 series & 2 parallel PV-diode array. (c) Four theoretical I-V & P-V curves under different isolation. (d) The output power of PV-buck array power. (e) The output power of PV-diode array power.

Fig. 4. (c) shows the theoretical I-V and P-V curves for the PV panel under four different irradiance conditions.

From Fig. 4(c), the theoretical maximum power of four PV panels with different irradiance condition can be calculated, which is P_{max} = (12.57+9.68+8.3+5.68) × 20=724.6W. From Fig. 4(d), the maximum power from the PV-buck module array is $P_{max}' \approx 715$ W and it is almost the same as the calculated value. However, from Fig. 4(e), the maximum power from the PV-diode module array is about 620W. The PV array constructed by the PV-buck module can produce 15% more power than the traditional PV array in the above assumed insolation condition.

IV. EXPERIMENTAL RESULTS

In order to verify the effectiveness of PV array improving output power and efficiency under partially shaded condition, we built the experimental platform based on the AMETEK TerraSAS ETS60 PV simulator. Due to the limit of laboratory condition, we just prepared two simulators. The largest open voltage of single machine is 60V. The resistor is the load.

This control software of simulator which called TerraSAS can record the data of output power in real time. All of the following power waveforms are the data recorded in the background. The simulator setting parameters are shown in Tab.1.

Tab.1   The key parameters of PV Simulator

parameters	value	
	1000W/ m²	800W/ m²
open-circuit voltage U_{oc}/V	35.53	34.11
short-circuit current I_{sc}/A	9.653	7.415
maximum power point U_m/V	28.55	28.21
maximum power current I_m/A	8.933	6.862
maximum power P/W	255.04	193.58
fill factor FF	0.816	
influence of voltage and temperature factor β_V / (% / °K)	-0.36	
the temperature effect of power factor β_P / (% / °K)	-0.5	

According to Tab.1, when the irradiance is 1000W/m2 and 800W/m2, the maximum output power of PV simulator is 255.04W and 193.58W, the parameters are basically consistent with those of CSUN 250-60P model polysilicon subassembly.

The array structure shown in above was initiated experimental verification under two conditions: no shadow and partial shadow. Fig.5 shows the experimental platform.

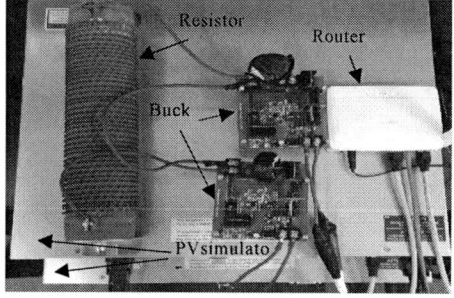

Fig.5.   The photo of experimental platform

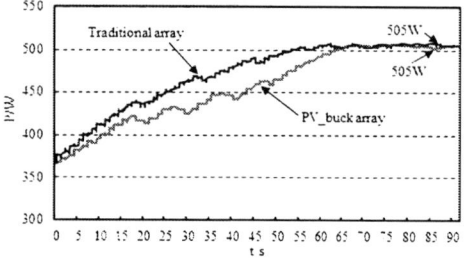

Fig.6.   The output power of 2*1 string PV-buck array and the traditional array without partially shaded condition

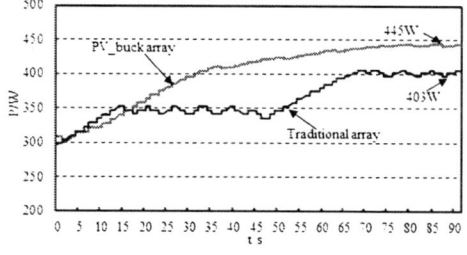

Fig.7.   The output power of 2*1 string PV array under partially

shaded condition

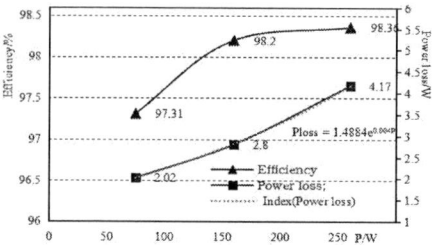

Fig.8.   The efficiency of buck converter

Under no shaded condition, S1=S2=1000W/m2, the theory maximum output power of simulator is 510.08W , we can know from Fig.6, two kinds of array output power are 505W, MPPT precision is 99%. Under partially shaded condition, S1=1000W/m2, S2=800W/m2, the theory maximum output power of simulator is 448.62W, Fig.7 show us that the output power of PV-buck photovoltaic array is 445W, MPPT precision is 99.2%. While the traditional array output power is only 403W, the power promotion is about 10.42%.

The experimental results show that the conclusions are basically consistent with the simulations, moreover in accordance with the results of theoretical analysis. When partial shadow appears, PV-buck array will output more power than traditional string array, which can ensure that the maximum power can be extracted from all PV component. However, the traditional array cannot simultaneously ensure individual PV component of different radiation intensity to achieve the maximum power output. Fig.8 shows that when the buck works at rated power, the efficiency can reach 98.3%. In this experiment, when the power loss of the buck converter is about 4.1W, the group array of PV-buck can effectively improve the 33.6W power, which is about 8.4%.

The experimental results verify the effectiveness of the PV-buck PV array in improving the power generation efficiency of PV panels in local shadows. It should be pointed that it is important to reduce the cost of the buck converter.

V. CONCLUSIONS

According to the theoretical analysis, simulation and experimental results, the advantage of the new PV array connection method using PV-buck is verified. The maximum output power of PV array under partially shaded condition is increased greatly than the traditional PV array structure. Although the use of this method will increase the cost, but after mass production, the cost will be greatly reduced.

VI. ACKNOWLEDGMENT

This paper and its related research are supported in part by the National Natural Science Foundation of China (NSFC) (Grant No. 51777049 and 51707051), in part by the State Grid Electric Power Company of Jiangsu province science and technology project (J2017073).

REFERENCES

[1]   Hiren Patel, Vivek Agarwal, "Maximum power point tracking scheme for PV system operating under shaded condition," IEEE Transactions On Industrial Electronics, vol. 55, no4, pp. 1689-1698, April, 2008.

[2]   R. Bruendlinger, B. Bletterie, M. Milde, and H. Oldenkamp, "Maximum power point tracking performance under partially shade PV conditions," in Proc. 21st EUPVSEC, Dresden, Germany, Sept. 2006.

[3] N. Fernia, G. Petrone, G. Spagnuolo, and M. Vitelli, "Optimization of perturb and observe maximum power point tracking method," IEEE Trans. Power Electron., vol. 20, no. 4, pp. 963–973, Jul. 2005.

[4] M. A. S. Masoum, H. Dehbonei, and E. F. Fuchs, "Theoretical and experimental analyses of photovoltaic systems with voltage- and current-based maximum power-point tracking," IEEE Trans. Energy Convers., vol. 17, no. 4, pp. 514–522, Dec. 2002.

[5] N. Mutoh, M. Ohno, and T. Inoue, "A method for MPPT control while searching for parameters corresponding to weather conditions for PV generation systems," IEEE Trans. Ind. Electron., vol. 53, no. 4, pp. 1055–1065, Aug. 2006.

[6] A. Pandey, N. Dasgupta, and A. K. Mukerjee, "Design issues in implementing MPPT for improved tracking and dynamic performance," in Proc. IEEE IECON, pp. 4387–4391, 2006.

[7] Sera, D., Teodorescu, R., Hantschel, J., Knoll, M., "Optimized maximum power point tracker for fast changing environmental conditions", IEEE Trans. Ind. Electron., 55, (7), pp. 2629–2637, 2008.

[8] Femia, N., Gianpaolo, L., Giovanni, P., Spagnuolo, G., Vitelli, M., "Distributed maximum power point tracking of photovoltaic arrays novel approach and system analysis", IEEE Trans. Ind. Electron., 55, (7), pp.2610–2621,2008.

A Common-Mode Voltage Reduction for Two-Stage Three-Phase Transformerless PV Inverters

Adisak Promyoo and Surapong Suwankawin*
Department of Electrical Engineering, Faculty of Engineering,
Chulalongkorn University, Bangkok, Thailand
*E-mail: surapong.su@chula.ac.th

Abstract- This paper presents a common-mode voltage cancellation method for two-stage transformerless string PV inverters. The exact common-mode equivalent circuit is firstly modeled to exhibit simultaneously the instantaneous common-mode voltages of boost DC-DC converter and three-phase two-level inverter. The selection of zero voltage for PWM inverter is then proposed in order to pave the way for common-mode voltage cancellation. The simulation results show the accuracy of exact common-mode model. In addition, the common-mode voltages can moderately cancel each other out, and the leakage current of the overall system is significantly attenuated.

I. INTRODUCTION

Transformerless string inverters are the promising technology for the grid-connected PV generation systems so far. Two-stage topology; DC-DC boost converter and DC-AC three-phase inverter, governs the major share of commercial PV inverters. The main drawback of this topology is the significant rise of EMI problem in the whole system. This is due to great common-mode voltages generated by both DC-DC boost converter and DC-AC inverter and the leakage current flowing throughout the transformerless PV system is increasing accordingly.

There are several approaches to reduce the common-mode voltage and/or leakage current, e.g. employing EMI filters, modifying modulation techniques or adaptation of converter topology. String PV inverters normally equip with the EMI filters on both sides; DC PV-string-connected and AC grid-connected. Off-the-shelf DC and AC EMI filters for PV systems are also provided [1]. Nevertheless, the required footprint for EMI filters makes the PV inverters bulkier. The modeling of common-mode circuit and the design of EMI filters are also investigated in [2]-[5]. Besides the EMI filters, various modulation techniques of AC side inverters are introduced for both 2-level and 3-level topologies [6]-[8] with the conceding of maximum modulation index. On the contrary, the structure of 3-level inverter is modified in [9] to succeed in another modulation technique without compromising modulation index. Other research works also alter the topology of 2-level three-phase inverter, e.g. the using of four-leg structure in [10] or the addition of neutral point clamp circuit in [11]. On the other hand, the cascaded multilevel H5 inverter is proposed in [12] for high-power application.

In [13], the boost converter is taken into consideration with the so-called coupled inductor boost inverter. This topology is a kind of Z-source inverter, the converter is modulated to not only boost the input DC voltage and simultaneously control the output AC voltage, but also to help reduce the common-mode voltage of the whole PV system.

Nevertheless, some restrictions from the above literature survey are as follows:
1) Apart from the aforementioned research works [5] and [13], almost all solutions consider only the common-mode voltage generated by grid-connected inverter at AC side. This is perhaps not sufficient for the two-stage string inverter scheme.
2) The structure of PV inverters is altered from the common building blocks of DC-DC boost converter, 2-level and/or 3-level inverters. Probably, the industrialists are reluctant to put these solutions into practice.
3) The design of EMI filters is likely to treat DC and AC sides separately. In fact, there is the interaction among both sides through the DC Bus. Considering DC and AC sides together can help to design the filters efficiently.

In this paper, the elementary two-stage transformerless string PV inverter is considered; DC-DC boost converter with 2-level three-phase inverter (Fig. 1). Firstly, the exact model of two-stage PV inverters is derived and the common-mode equivalent circuit of PV system is given. Secondly, the instantaneous common-mode voltages of both DC-DC boost converter and the two-level three-phase inverter are simultaneously investigated. Thirdly, a cancellation of common-mode voltages among of DC-DC boost converter and of two-level inverter is elaborated. Proposed cancellation method is conducted by the selection of zero voltage for modulation of 2-level inverter. Finally, the simulation demonstrates the effectiveness of the proposed concept.

II. COMMON-MODE MODEL OF TWO-STAGE PV SYSTEM

In [14], the rigorous approach is employed to obtain the exact model for AC drive system; both rectifier and 2-level PWM inverter are considered instantaneously with the switching function. In this paper, the same approach is applied to PV system instead; both DC-DC boost converter and grid-connected inverter are taken into account.

The 2018 International Power Electronics Conference

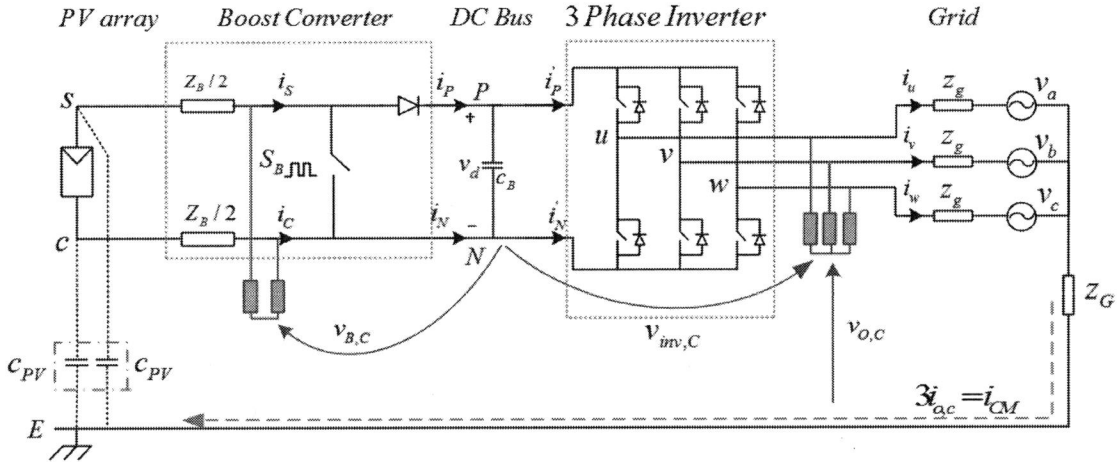

Fig. 1 Standard grid-connected PV system with two-stage three-phase two-level transformerless inverter.

Basically, common-mode and differential-mode quantities X_C, $X_{j,D}$ are defined in (1) and (2) respectively

$$X_C \equiv \sum_{j=1}^{P} X_j \Big/ P \qquad)1($$

$$X_{j,D} = X_j - X_C \quad ; \quad \sum_{j=1}^{P} X_{j,D} = 0 \qquad)2($$

where X_j is the phase variable referenced to ground (E in Fig. 1). Equations (1) and (2) can be written by the matrix formation in (3)-(4)

$$\begin{bmatrix} X_{1,D} \\ X_{2,D} \\ \overline{X_C} \end{bmatrix} = T_{1\phi} \begin{bmatrix} X_1 \\ X_2 \end{bmatrix} \quad ; T_{1\phi} = \frac{1}{2} \begin{bmatrix} 1 & -1 \\ -1 & 1 \\ \overline{1} & \overline{1} \end{bmatrix} \qquad (3)$$

$$\begin{bmatrix} X_{1,D} \\ X_{2,D} \\ X_{3,D} \\ \overline{X_C} \end{bmatrix} = T_{3\phi} \begin{bmatrix} X_1 \\ X_2 \\ X_3 \end{bmatrix} \quad , T_{3\phi} = \frac{1}{3} \begin{bmatrix} 2 & -1 & -1 \\ -1 & 2 & -1 \\ -1 & -1 & 2 \\ \overline{1} & \overline{1} & \overline{1} \end{bmatrix} \qquad (4)$$

$T_{1\phi}$ and $T_{3\phi}$ are the transformation matrix for single-phase and three-phase cases correspondingly.

The procedures for modeling the PV system in Fig. 1 are as follows:
- Firstly, partition the whole system into sub-circuits and label the phase variable of the input and output, for instance, DC-DC boost converter, DC bus and two-level inverter.
- Secondly, express the relationships among the above phase variable in the matrix formation.
- Thirdly, use the transformation matrixes in (3) and (4) to separately express the common-mode component X_C and differential-mode component X_{jD}.
- Finally, after all sub-circuits are completely expressed by the third step, the common-mode and differential-mode equivalent circuits are attained.

For instance, the below contents give some ideas how the above procedures apply to the DC-DC boost converter and inverter sub-circuits. By considering the DC-DC boost converter, the relationship between input voltages (v_{SE}, v_{CE}) and output voltages (v_{PE}, v_{NE}); including the currents (i_S, i_C), can be expressed in (5)

$$\begin{bmatrix} v_{SE} \\ v_{CE} \end{bmatrix} = \frac{Z_B}{2} \begin{bmatrix} 1 & 0 \\ 0 & 1 \end{bmatrix} \begin{bmatrix} i_S \\ i_C \end{bmatrix} + \begin{bmatrix} 1-s_B & s_B \\ 0 & 1 \end{bmatrix} \begin{bmatrix} v_{PE} \\ v_{NE} \end{bmatrix} \qquad (5)$$

where s_B is the switching function of the boost converter as indicated in (6)

$$s_B = \begin{cases} 1 \, ; \; switch = on \\ 0 \, ; \; switch = off \end{cases} \qquad (6)$$

Using the transformation $T_{1\phi}$ in (3) for (5), the common-mode and differential-mode components can be decomposed in (7)

$$\begin{bmatrix} v_{S,D} \\ v_{C,D} \\ \overline{v_{S,C}} \end{bmatrix} = \frac{Z_B}{4} \begin{bmatrix} 1 & -1 & | & 0 \\ -1 & 1 & | & 0 \\ \overline{0} & \overline{0} & | & \overline{3} \end{bmatrix} \begin{bmatrix} i_{S,D} \\ i_{C,D} \\ i_{1,C} \end{bmatrix} + \frac{1}{2} \begin{bmatrix} (1-s_B) & -(1-s_B) & | & 0 \\ -(1-s_B) & (1-s_B) & | & 0 \\ \overline{-s_B} & \overline{s_B} & | & \overline{2} \end{bmatrix} \begin{bmatrix} v_{P,D} \\ v_{N,D} \\ v_{BUS,C} \end{bmatrix} \qquad (7)$$

Two-level inverter can be conducted with the same procedures as well as shown in the following.

Relationship between the phase variables:

$$\begin{bmatrix} v_{uE} \\ v_{vE} \\ v_{wE} \end{bmatrix} = \begin{bmatrix} s_u & 1-s_u \\ s_v & 1-s_v \\ s_w & 1-s_w \end{bmatrix} \begin{bmatrix} v_{PE} \\ v_{NE} \end{bmatrix} \qquad (8)$$

where s_u, s_v, s_w are the switching functions for each phase as indicated in (9)

$$s_u, s_v, s_w = \begin{cases} 1 \, ; \; upper \; switch = on \\ 0 \, ; \; lower \; switch = on \end{cases} \qquad (9)$$

Expression of common-mode and differential-mode components:

2872

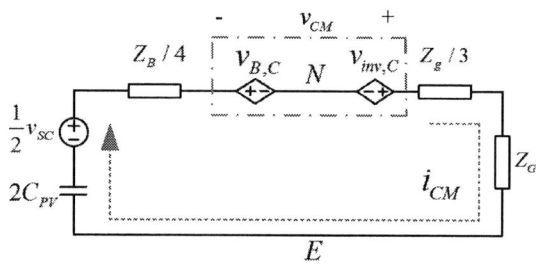

Fig. 2 Common-mode equivalent circuit for PV system in Fig. 1.

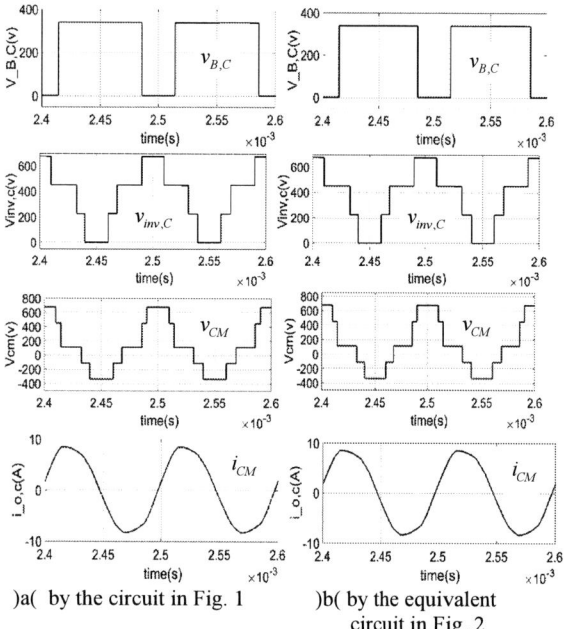

)a(by the circuit in Fig. 1   )b(by the equivalent circuit in Fig. 2

Fig. 3 Simulation results showing the comparison of the common-mode signals between Fig. 1 and Fig. 2.

Using the three-phase transformation matrix in (4), the common-mode and differential-mode components of inverter can be separately expressed in (10)

$$
\begin{bmatrix} v_{u,D} \\ v_{v,D} \\ v_{w,D} \\ \hline v_{O,C} \end{bmatrix} = \begin{bmatrix} s_u - \sum_{n=u,v,w} s_n/3 & -\left(s_u - \sum_{n=u,v,w} s_n/3\right) & 0 \\ s_v - \sum_{n=u,v,w} s_n/3 & -\left(s_v - \sum_{n=u,v,w} s_n/3\right) & 0 \\ s_w - \sum_{n=u,v,w} s_n/3 & -\left(s_w - \sum_{n=u,v,w} s_n/3\right) & 0 \\ \hline \left(\sum_{n=u,v,w} s_n/3 - \frac{1}{2}\right) & -\left(\sum_{n=u,v,w} s_n/3 - \frac{1}{2}\right) & 1 \end{bmatrix} \begin{bmatrix} v_{P,D} \\ v_{N,D} \\ \hline v_{BUS,C} \end{bmatrix} \quad (10)
$$

The rest of the sub-circuits can be easily derived by the same procedure; hence it is omitted in this extended summary. Considering the common-mode components in (7) and (10) in association with the common-mode components from the rest of sub-circuits, the common-mode equivalent circuit for PV system in Fig. 1 can be depicted in Fig. 2. And the common-mode voltages of

boost converter and inverter; $v_{B,C}, v_{inv,C}$, are derived in (11)-(12).

$$
v_{B,C} = \frac{1}{2}(1 - S_B)v_d \quad (11)
$$

$$
v_{inv,C} = \sum_{n=u,v,w} (S_n/3)v_d \quad (12)
$$

Simulation results in Fig. 3 compare the common-mode signals from the circuit in Fig. 1 and the equivalent circuit in Fig. 2. It can be seen that the proposed equivalent circuit is accurate by which all resultant common-mode signals, e.g. common-mode voltages $v_{B,C}$, $v_{inv,C}$ and the leakage current i_{CM}, have the same waveforms. With regard to the employment of the same switching frequency for both boost converter and three-phase inverter, both common-mode voltages $-v_{B,C}$ and $v_{inv,C}$ are in phase and consequently, the total common-mode voltage v_{CM} is larger and includes 8-step change of common-mode voltage in one switching period.

The benefit of the equivalent circuit in Fig. 2 is that it can reflect clearly the instantaneous interaction between the common-mode voltages of DC side from DC-DC boost converter and of AC side from three-phase inverter. Such understanding helps to conduct a reduction of overall common-mode voltage; much more details are available in the next section.

III. COMMON-MODE VOLTAGE REDUCTION

Reduction of the total common-mode voltage v_{CM} is based on the voltage cancellation [15]. From Fig. 2, the common-mode voltages of boost converter $v_{B,C}$ and the inverter $v_{inv,C}$ are connected in series, and with regard to the degree of freedom of zero voltage of three-phase inverter, the proper selection of this zero voltage can alter $v_{inv,C}$ to cancel the common-mode voltage of boost converter $v_{B,C}$ and help to reduce the total common-mode voltage v_{CM} accordingly.

There are 2 steps to achieve the aforementioned scheme. Firstly, the carrier signal of the boost converter is turned out of phase with that of the inverter. Secondly, the proper selection of zero voltage is proceeded by considering the nearest value among of the three-phase voltage commands of inverter and the DC voltage command of boost converter.

Figure 4 provides a comparison of common-mode voltages among various modulation methods. Fig. 4(a) shows the result from the convention modulation method which is used as a benchmark. Fig. 4(b) is the case where the carrier signal of boost converter is turn out of phase only. Though Fig. 4(b) shows less amplitude of total common-mode voltage than the benchmark in Fig. 4(a), the number of step change of common-mode voltage is still eight steps in one switching period.

The 2018 International Power Electronics Conference

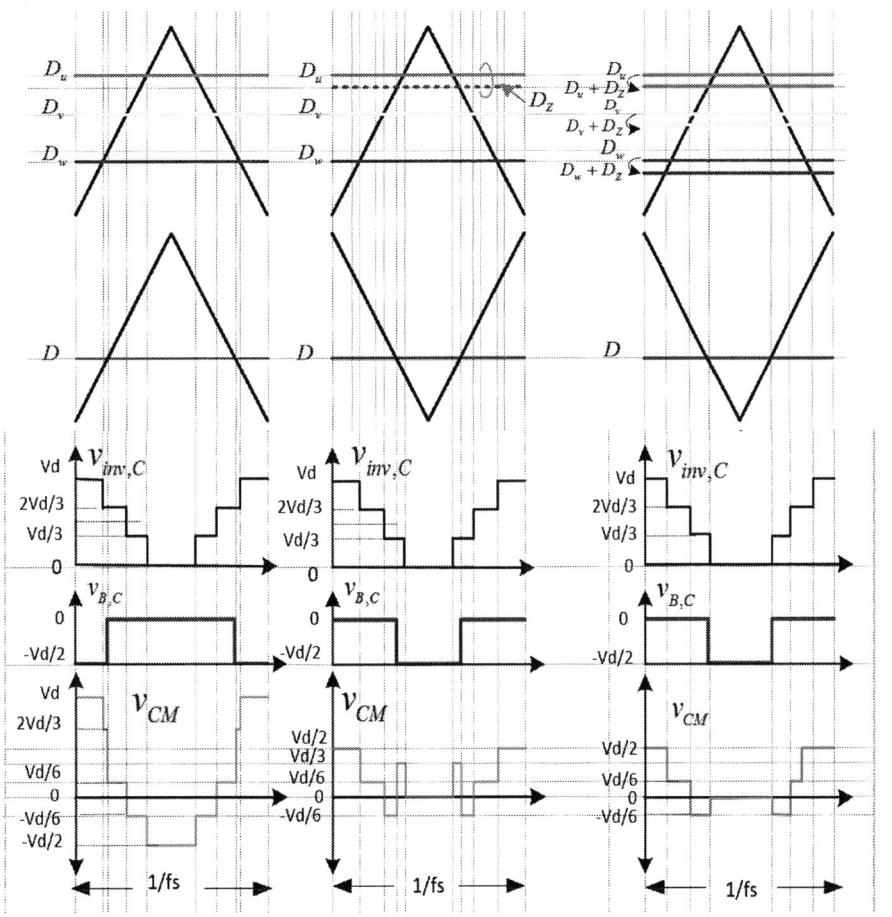

(a) Conventional Modulation method.   (b) Modulation method with   (c) Proposed Modulation method.
out-of-phase carrier signal.

Fig. 4 Concept of common-mode voltage reduction.

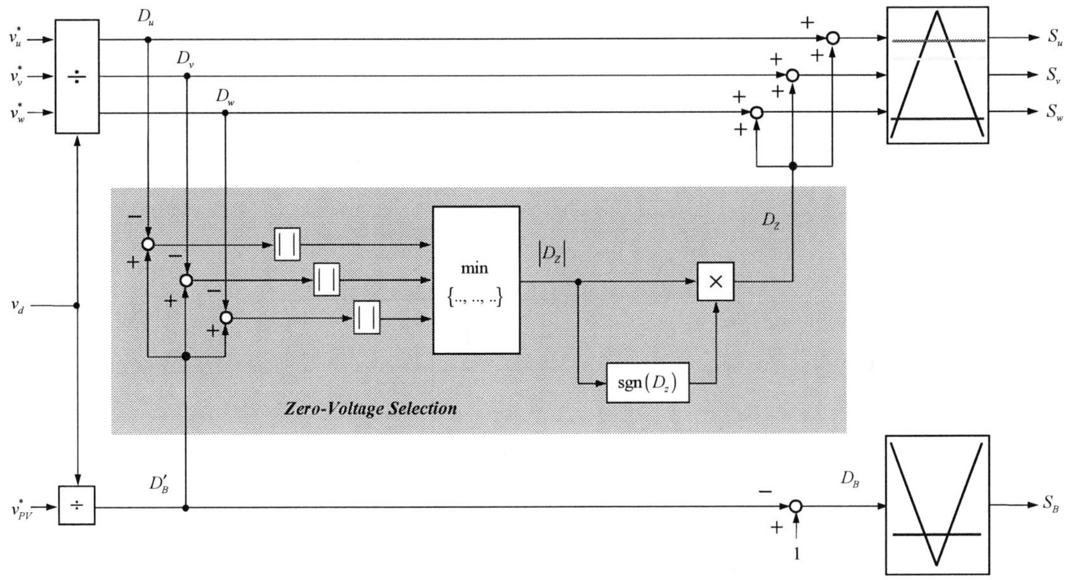

Fig. 5 Selection of zero voltage for common-mode voltage reduction.

2874

Fig. 4(c) demonstrates the mechanism of the proposed modulation method, the selection of zero voltage makes the timing of the step changes of $v_{B,C}$ and $v_{inv,C}$; common-mode voltage from one of the three-phase voltage, to happen at the same moment. In comparison with Figs 4(a) and 4(b), not only the amplitude of the step change of total common-mode voltage can be far more reduced, but the number of the step change also decreases down to six steps in one switching period.

Figure 5 shows the block diagram of the selection of zero voltage. The duty cycle of zero voltage D_z is selected by considering the duty cycle of three-phase voltage (D_u, D_v, D_w) which is nearest to the complementary duty cycle of boost converter $1 - D_B$. Equation (13) indicates the determination of duty cycle for zero voltage D_z .

$$|D_z| = \min\left\{ |D_B' - D_u|, |D_B' - D_v|, |D_B' - D_w| \right\} \quad (13)$$

IV. SIMULATION

Simulation results in Fig. 6 are carried out by the PV system in Fig. 1 with the proposed selection of zero voltage. In comparison with Fig. 3, the common-mode voltage can be dramatically reduced especially the switching frequency component (10 kHz) as shown in the spectrum of common-mode voltage in Fig. 7. In addition, the leakage current i_{cm} is significantly attenuated.

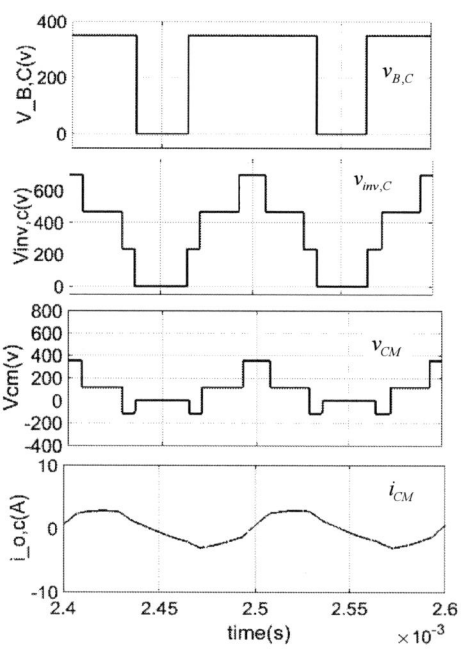

Fig. 6 Simulation result of 2-stage PV system with the proposed zero-voltage selection.

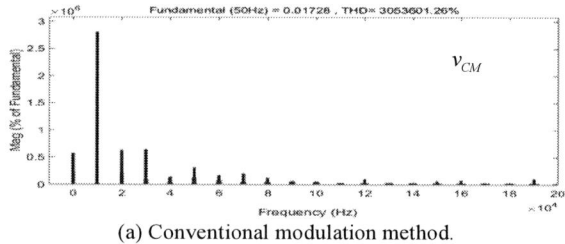

(a) Conventional modulation method.

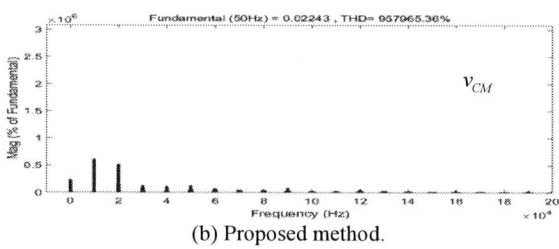

(b) Proposed method.

Fig. 7 Comparison of spectrum of total common-mode voltage.

V. CONCLUSIONS

In this paper, a common-mode voltage cancellation method is proposed for 2-stage 3-phase transformerless grid-connected PV inverters. The exact model of common-mode circuit is derived. Then, the selection of zero voltage of PWM inverter allows one phase of inverter and boost converter to switch simultaneously and consequently conducts the common-mode voltage cancellation. The proposed method is verified by simulation and the results demonstrate that the overall common-mode voltage and the leakage current can be significantly reduced.

REFERENCES

[1] Schaffner, "EMC/EMI Filter for PV Inverters," DC Filters FN 2200 Datasheet.

[2] R. Burkart, J. W. Kolar and G. Griepentrog, "Comprehensive Comparative Evaluation of Single- and Multi-Stage Three-Phase Power Converters for Photovoltaic Applications," *in Proc. IEEE Intelec* 2012.

[3] B. Cougo, T. Friedli, D. O. Boillat and J. W. Kolar, "Comparative Evaluation of Individual and Coupled Inductor Arrangements for Input Filters of PV Inverter Systems," *in Proc. IEEE CIPS* 2012.

[4] H. Zhang, S. Wang and J. Puukko, "Common Mode EMI Noise Modeling and Prediction for a Three-phase, Three-level, Grid Tied Photovoltaic Inverter," *in Proc. IEEE APEMC* 2016, vol. 1, pp. 1188-1194.

[5] D. Dong, X. Zhang, F. Luo, D. Boroyevich and P. Mattavelli, "Common-Mode EMI Noise Reduction for Grid-interface Converter in Low-voltage DC Distribution System," *in Proc. IEEE APEC* 2012, pp. 451-457.

[6] M. C. Cavalcanti, K. C. de Oliveira, A. M. de Farias, F. A. S. Neves, G. M. S. Azevedo and F. C. Camboim, "Modulation Techniques to Eliminate Leakage Currents in Transformerless Three-Phase Photovoltaic Systems," *IEEE Trans. Ind. Electron.*, vol. 57, no.4, pp. 1360-1368, Apr. 2010.

[7] C-Chuan Hou, C-Chung. Shih, P-Tai Cheng and A. M. Hava, "Common-Mode Voltage Reduction Pulsewidth Modulation Techniques for Three-Phase Grid-Connected Converters," *IEEE Trans. Power Electron.*, vol. 28, no. 4, pp. 1971-1979, Apr. 2013.

[8] J-Seok Lee and K-Beum Lee, "New Modulation Techniques for a Leakage Current Reduction and a Neutral-Point Voltage Balance in Transformerless Photovoltaic Systems Using a Three-Level

Inverter," *IEEE Trans. Power Electron.*, vol. 29, no. 4, pp. 1720-1732, Apr. 2014.

[9] Z. Chen, W. Yu, X. Ni and A. Huang, "A New Modulation Technique to Reduce Leakage Current without Compromising Modulation Index in PV Systems," *in Proc. IEEE ECCE* 2015, pp. 460-465.

[10] X. Guo, R. He, J. Jian, Z. Lu, X. Sun and J. M. Guerrero, "Leakage Current Elimination of Four-Leg Inverters for Transformerless Three-Phase PV Systems," *IEEE Trans. Power Electron.*, vol. 31, no. 3, pp. 1841-1846, Mar. 2016.

[11] L. Zhou, F. Gao, G. Shen, T. Xu and W. Wang, "Low Leakage Current Transformerless Three-Phase Photovoltaic Inverter," *in Proc. IEEE ECCE* 2016.

[12] X. Guo, R. He, X. Jia and C. A. Rojas, "Leakage Current Reduction of Transformerless Three-Phase Cascaded Multilevel PV Inverter," *in Proc. IEEE ISIE* 2015, pp. 1110-1114.

[13] Y. Zhou, W. Huang, P. Zhao and J. Zhao, "A Transformerless Grid-Connected Photovoltaic System Based on the Coupled Inductor Single-Stage Boost Three-Phase Inverter," *IEEE Trans. Power Electron.*, vol. 29, no. 3, pp. 1041-1046, Mar. 2014.

[14] P. Pairodamonchai and S. Sangwongwanich, "Exact Common-Mode and Differential-Mode Equivalent Circuits of Inverters in Motor Drive Systems Taking into Account Input Rectifiers," *in Proc. IEEE PEDS* 2011, pp. 278-285.

[15] H-Dong Lee and S-Ki Sul, "A Common-Mode Voltage Reduction in Boost Rectifier/Inverter System by Shifting Active Voltage Vector in a Control Method," *IEEE Trans. Power Electron.*, vol. 15, no. 6, pp. 1094-1101, Nov. 2000.

A Grid-Connected PV-energy Storage System with Synchronous Generator Characteristics

Huadian Xu*, Jianhui Su, Ning Liu, Yong Shi and Yan Du
School of Electrical Engineering and Automation, Hefei University of Technology, Hefei, China
*E-mail:xuhuadian@163.com

Abstract—The technology of the virtual synchronous generator can efficiently improve the distributed generator in term of the capability to support frequency and the compatibility with the main grid. However, inertia of synchronous generators (SGs) cannot be introduced into photovoltaics (PV) generation systems in the absence of energy buffers. In this paper, a PV-energy storage (ES) system capable of mimicking SGs is proposed. The topology, control schemes and small-signal analysis for performance are investigated. Moreover, the designed coordination control enables the inverter to cooperate with the ES device in the system. The proposed PV-ES system is verified with simulations.

Keywords—coordination control, energy storage, photovoltaics generation, virtual synchronous generator.

I. INTRODUCTION

Increasing penetration of inverter-based distributed generator (DG) leads to reducing inertia of the power system, which plays an important role in short term system stability [1]. To support frequency and enhance the compatibility between DGs and the power system, the technology, virtual synchronous generator (VSG), is proposed [2-4]. Nonetheless, it is almost impossible to introduce inertia, the indispensable property for VSG, into the inverter only powered by photovoltaics (PV). To enable the inverter to emulate the inertia of SGs, an energy buffer, whose function is identical to the rotor for kinetic energy, is necessarily installed at the dc link of the inverter.

In [3], inverters that mimic SGs are proposed with an assumption that inverters are supplied by stiff DC voltage sources, not taking PV generation into account. In [5, 6], energy storage (ES) is applied to PV generation, but the interface inverter possesses no characteristics of SGs due to lack of VSG control scheme. In [7], battery is used as the backup for the PV inverter that employs PQ control or droop control, causing the inverter incapable of behaving like SGs. In [4,8], although it is pointed out that ES should be installed for emulating the kinetic energy

stored in rotating rotors of SGs, the detailed system topology and the coordination control scheme for the system are not concerned.

In this paper, a grid-connected PV-ES system capable of mimicking SGs is proposed. The content of this paper is organized as follows. Section II introduces the topology and control schemes. In Section III, a small-signal model of the system is established and performance analysis is conducted. Verifications with simulation results are presented in Section IV. Section V draws conclusions.

II. TOPOLOGY AND CONTROL SCHEMES

Fig. 1 shows the topology and control schemes of the proposed system. The hardware consists of an ES device, a bidirectional DC-DC converter, a set of PV array and an inverter. The PV array is tied directly to the dc link, sharing the same dc bus with the DC-DC converter and the inverter. The typical DC-DC converter, buck/boost converter, is adopted in this paper.

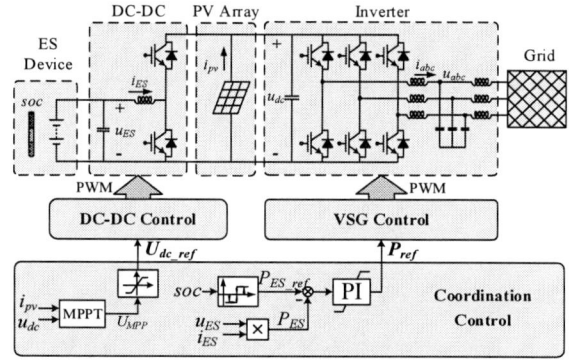

Fig. 1. Topology and control schemes of the proposed grid-connected PV-ES system.

In addition to DC-DC control and VSG control, a coordination control with two tasks is proposed to attune the system as shown in Fig. 1. First is to ascertain the value of dc link voltage reference U_{dc_ref} utilizing a maximum power point tracking (MPPT) algorithm. U_{dc_ref} is provided for DC-DC control, which performs the regulation of dc link voltage u_{dc}. Second is to obtain the inverter active power reference P_{ref} according to the state of charge (SOC) of the ES device. P_{ref} is delivered to

---

This work was supported partly by the National Key Research and Development Program of China (No. 2017YFB0903503), Grants from the Power Electronics Science and Education Development Program of Delta Environmental & Educational Foundation (No. DREM2015002); and the Fundamental Research Funds for the Central Universities (No. JZ2016HGBZ1023).

VSG control, which controls the inverter to emulate the characteristics of SGs.

A. Coordination Control for the PV-ES System

To control dc link voltage u_{dc}, the SOC of the ES device must be kept within a proper range so that the ES device can output energy when u_{dc} falls and store the absorbed energy when u_{dc} rises. To regulate the SOC, the proposed coordination control should be competent to manage the exchanged power P_{ES} between the ES device and the dc link.

Fig. 2 shows the exchanged power reference P_{ES_ref} with respect to the SOC, where SOC_M is the mean of lower limit SOC_L and higher limit SOC_H of the range, P_0 is the absolute value of the charge power and discharge power. The ES device will start to be charged once the SOC is less than SOC_L and to be discharged when the SOC is more than SOC_H. Both charging and discharging are terminated when the SOC reaches SOC_M. Applying the curve shown in Fig. 2 to specify P_{ES_ref}, frequent operations of charge/discharge near SOC_L/SOC_H can be avoided by the coordination control. In this work, the certain SOC(%) range of the ES device is set from 40% to 80%.

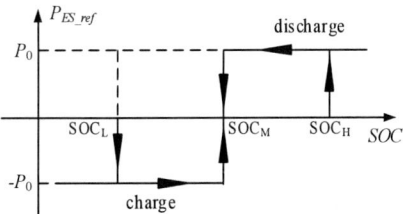

Fig. 2. Relationship between exchanged power reference and SOC.

According to the law of conservation of energy, the following equation is obtained when the energy change of capacitor at the dc link is ignored:

$$P_{PV} + P_{ES} = P_e \qquad (1)$$

where P_{PV} is the power generated by the PV array, P_e is the output power of the inverter in the PV-ES system.

A proportional-integral (PI) regulator, whose input is the error between P_{ES_ref} and P_{ES}, is used for generating the inverter active power reference P_{ref} to regulate P_e with VSG control. Accordingly, P_{ES} can be tuned under the condition that P_{PV} fails to be adjusted due to the PV array operates at the maximum power point.

In coordination control as depicted in Fig. 1, dc link voltage reference U_{dc_ref} is set equal to U_{MPP}, the voltage at the maximum power point of the PV array, in order to draw the maximum power from the PV array. U_{MPP} is calculated by the MPPT algorithm of incremental conductance method.

B. DC-DC Control

As illustrated in Fig. 3, the control scheme of the buck/boost converter incorporates an outer voltage loop with an inner current loop, which ensures dc link voltage u_{dc} to track U_{dc_ref}.

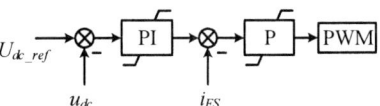

Fig. 3 Block diagram of DC-DC control.

C. VSG Control for the Inverter of the PV-ES System

Fig. 4 shows the VSG control for the inverter of the proposed system. A fifth-order model of a SG, including third-order electrical equations and second-order mechanical equations, is used to calculate the reference currents i_{a_ref}, i_{b_ref}, and i_{c_ref}. The electrical equations set, which reproduces the stator circuit of the SG, is given by

$$L_s \frac{di_{abc\_ref}}{dt} = e_{abc} - u_{abc} - R_s i_{abc\_ref} \qquad (2)$$

where $u_{abc}=[u_a, u_b, u_c]^T$ is the phase terminal voltages of the stator windings; R_s and L_s are, respectively, the stator resistance and the stator inductance, $i_{abc\_ref}=[i_{a\_ref}, i_{b\_ref}, i_{c\_ref}]^T$ embodies the calculated reference currents, which is also the currents of the stator windings, $e_{abc}=[e_a, e_b, e_c]^T =E[\sin\theta, \sin(\theta-2\pi/3), \sin(\theta+2\pi/3)]^T$ is the induced phase electromotive forces in the stator windings。

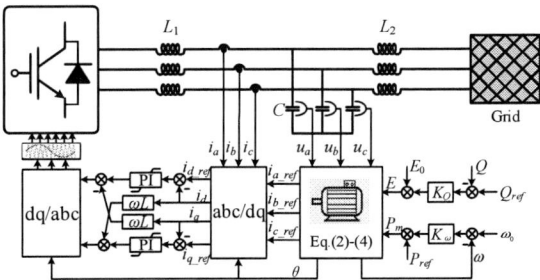

Fig. 4. VSG control for the inverter of the PV-ES System.

The electromechanical characteristics of the SG can be described, neglecting the mechanical losses and considering the effect of damper windings, as

$$J \frac{d\omega}{dt} = \frac{P_m}{\omega} - \frac{P_e}{\omega} - D(\omega - \omega_0) \qquad (3)$$

$$\frac{d\theta}{dt} = \omega \qquad (4)$$

where J is the inertia, P_m is the mechanical power, D is the damping coefficient, ω and ω_0 are, respectively, the actual and nominal angular frequency, θ is the electrical rotation angle.

To emulate the droop characteristics of primary frequency control and primary voltage control, P_m and E can be expressed as

$$P_m = P_{ref} + K_\omega(\omega_0 - \omega) \qquad (5)$$

$$E = E_0 + K_Q(Q_{ref} - Q) \qquad (6)$$

where K_ω is the unit power regulation, E_0 is the no-load electromotive force, Q_{ref} and Q are, respectively, the reference value and the actual value of the inverter output reactive power, K_Q is the voltage droop coefficient.

III. ANALYSIS OF PERFORMANCE

A small-signal model of the proposed PV-ES system is established and corresponding performance analysis is conducted in this section.

Normalizing equations (3)-(5) and linearizing them around an initial operating condition represented by $\delta=\delta_0$, $\omega=\omega_0$ yields

$$\frac{d\Delta\delta}{dt} = \frac{d\Delta\theta}{dt} - \omega_0 = \omega_0\Delta\omega \tag{7}$$

$$\frac{d\Delta\omega}{dt} = -\frac{1}{J}\Big[(D+K_\omega)\Delta\omega + S_E\Delta\delta\Big] \tag{8}$$

where $\Delta\delta$ is the deviation of power angle, $\Delta\omega$ is the angular frequency deviation, S_E is the synchronous torque coefficient given by

$$S_E = \frac{dP_e}{d\delta}\Big|_{\delta=\delta_0} \tag{9}$$

according to (1) and (9), it can be deduced that

$$\Delta P_{pv} + \Delta P_{ES} = \Delta P_e = S_E\Delta\delta \tag{10}$$

To describe the small-signal performance, the block diagram is obtained in Fig. 5, where the PI regulator is the one in coordination control.

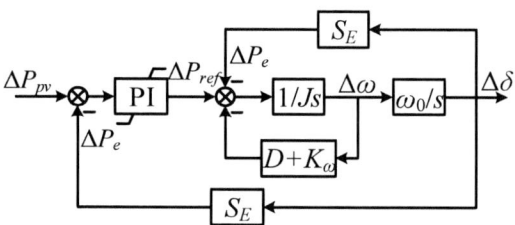

Fig. 5. Small-signal model of the proposed system.

From the block diagram, the transfer function can be expressed as following,

$$\frac{\Delta\omega}{\Delta P_{pv}} = \frac{(K_p s + K_i)s}{Js^3 + (D+K_\omega)s^2 + \omega_0 S_E(K_p+1)s + \omega_0 S_E K_i} \tag{11}$$

where K_p and K_i is, respectively, the proportional coefficient and integral coefficient of the PI regulator.

According to (11), the root loci family of the proposed system can be obtained as Fig. 6(a), where $J=1$, 3 and 15 and $(D+K_\omega)$ changes from 10pu to 200pu. Among three poles depicted in the plane, s_3 is real root and its position depends on K_p and K_i. As $(D+K_\omega)$ increases, the conjugate complex roots s_1 and s_2 evolve in the direction of the arrows. With larger J, s_1 and s_2 approach to imaginary axis, resulting in the reduced frequency response speed and the increased frequency damping. The dynamic performance and stability worsen when J increases, and excessively large J may lead to system instability. With the increase of $(D+K_\omega)$, the conjugate complex roots s_1 and s_2 move away from the imaginary axis at first, which leads to the damping rising and the dynamics and stability improving. And then, root loci of s_1 and s_2 separate after converging on the real axis, which causes the frequency response overdamped. When $(D+K_\omega)$

further increases, s_1 moves toward to the origin, resulting in the decreased system stability margin.

Fig. 6(b) illustrates the root loci family considering variation of K_i from 0.1 to 50 and $K_p=0.3$, 1.1 and 2. As Fig. 6(b) shows, the damping drops, the overshoot and the undamped natural frequency rise when K_p becomes larger. With K_i increasing, s_3 moves away from the origin, meanwhile, s_1 and s_2 close toward to the imaginary axis contributing to the improved dynamics. As K_i increases further, s_1 and s_2 become closer to the imaginary axis, and even turn to be the dominant poles, indicating that the system stability margin and the response speed decrease. Instability may happen with excessively large value of K_i, which causes s_1 and s_2 to cross the imaginary axis into the right half plane.

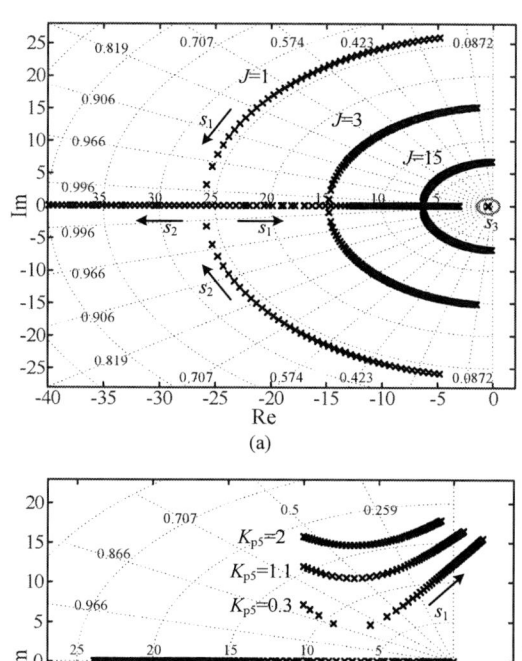

(a)

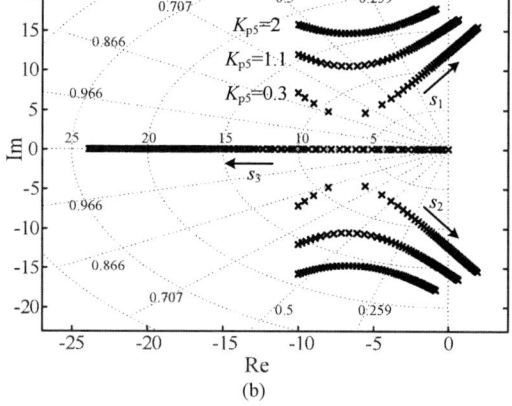

(b)

Fig. 6. Family of root loci with different values of: (a) J and D; (b) K_p and K_i.

IV. VALIDATIONS

Simulations with parameters listed in Table I are performed to verify the proposed PV-ES system. A step of light intensity on the PV array is exerted to validate coordination control and MPPT. Additionally, load at the point of common coupling (PCC) is switched to confirm the frequency support capability of the PV-ES system.

The 2018 International Power Electronics Conference

TABLE I
PARAMETERS FOR SIMULATION

Symbol	Meaning	Value
S_n	Rated power	20kVA
f_n	Nominal frequency	50Hz
V_n	Nominal voltage	380 V
f_s	Carrier frequency	10kHz
C_{dc}	Capacitance at dc link	300μF
$L_1/C/L_2$	Filter parameters	1.2mH/20μF/0.8mH
R_s	Stator resistance	0. 002p.u.
L_s	Stator inductance	0. 5p.u.

A. *Responses in the Case of Light Intensity Stepping*

The responses of the inverter output frequency f, the inverter output power P_e, the PV array output current i_{pv}, dc link voltage u_{dc} and the ES device output current i_{ES} are shown in Fig. 7 when the light intensity on the PV array steps at 1s. Fig. 7 indicates that the system is capable of providing a stable dc link voltage to realize the emulation of SGs characteristics and is able to complete MPPT after the stepping. At the same time, the exchanged power P_{ES} can reach the reference value of 0 and thus the SOC can be regulated. It is worth mentioning that the system power fed to the grid can be smoothed when the PV array output power fluctuates sharply.

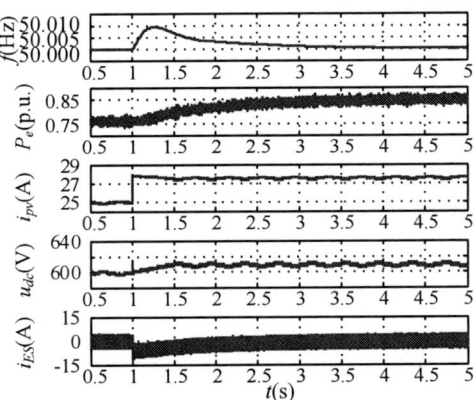

Fig. 7. Responses when the light intensity steps.

B. *Responses in the Case of Load Switching*

As illustrated in Fig. 8, the system output power increases after the load being switched on at 1.05s. On the contrary, the system output power decreases after the load being switched off at 2.5s. As to the inverter output frequency f, it returns to the nominal value after a short time smooth change in aforementioned load switching cases. Therefore, the proposed PV-ES system emulates the inertia of SGs and possesses the capability of supporting frequency.

C. *Frequency Response with Parameters Variation*

As shown in Fig. 9, the increment of J decelerates the response, causes larger overshoot and even results in oscillation. In contrast, the increment in D can enhance the damping of frequency and reduce the oscillation,

leading to the improvement of dynamics and stability. Furthermore, increments of K_p and K_i deteriorate dynamics due to the induced damping ratio reduction and oscillation frequency intensification.

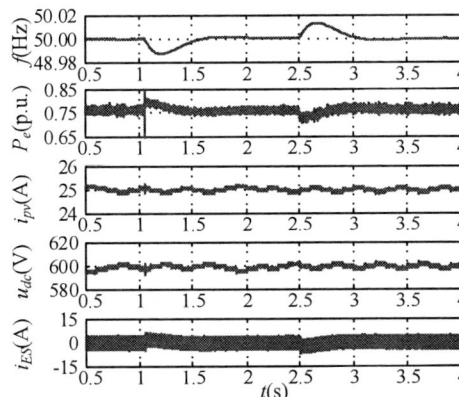

Fig. 8. Responses when 10kW load is switched at PCC.

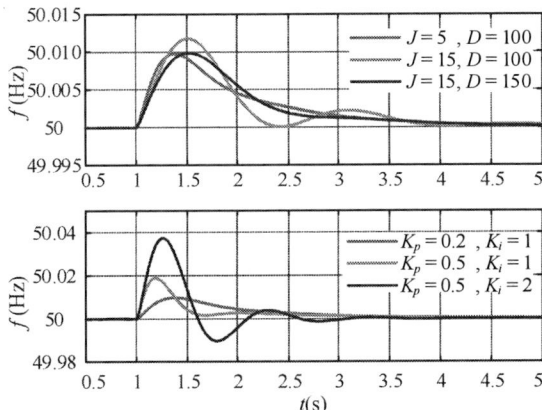

Fig. 9. Frequency response with parameters variation.

V. CONCLUSION

This paper proposes a grid-connected PV-ES system with synchronous generator characteristics. The topology, control schemes and performance analysis are investigated. Applying the developed topology and the designed control schemes, the inertia of SGs can be introduced into the PV-ES system. The PV-ES system is capable of mimicking SGs, managing the exchanged energy to regulate the SOC of the ES device, implementing MPPT and smoothing the power fed to the grid. The proposed system is verified with simulations.

REFERENCES

[1] M. P. N van Wesenbeeck, S. W. H. de Haan, P. Varela and K. Visscher "Grid tied converter with virtual kinetic storage," in *IEEE Bucharest PowerTech, Bucharest, Romania*, pp. 1-7, 2009.

[2] European FP6 project workshop, "VSYNC project". [Online] {Available} http://www.vsync.eu/ (Accessed: June 2011).

[3] Q. C. Zhong and G. Weiss. "Synchronverters: Inverters that mimic synchronous generators," *IEEE Trans on Industrial Electronics*, vol. 58, no. 4, pp. 1259-1267, 2011.

[4] K. Sakimoto, Y. Miura, and T. Ise. "Stabilization of a power system with a distributed generator by a virtual synchronous

generator function," in *IEEE 8th International Conference on Power Electronics and ECCE Asia (ICPE & ECCE), Jeju, South Korea*, pp. 1498-1505, 2011.

[5] M. J. E. Alam, K. M. Muttaqi and D. Sutanto, "A novel approach for ramp-rate control of solar PV using energy storage to mitigate output fluctuations caused by cloud passing," *IEEE Trans. on Energy Conversion*, vol. 29, no.2, pp. 507-518, 2014.

[6] J. Traube, F. Lu, D. Maksimovic, J. Mossoba, M. Kromer, P. Faill, S. Katz, B. Borowy, S. Nichols and L. Casey, "Mitigation of solar irradiance intermittency in photovoltaic power systems with integrated electric-vehicle charging functionality," *IEEE Trans. on Power Electronics*, vol. 28, no. 6, pp. 3058-3067, 2013.

[7] D. V. de la Fuente, C. L. Trujillo Rodríguez, G. Garcerá; E. Figueres and R. O. Gonzalez, "Photovoltaic power system with battery backup with grid-connection and islanded operation capabilities," *IEEE Trans. on Industrial Electronics*, vol. 60, no. 4, pp. 1571-158, 2013.

[8] J. Liu, Y. Miura, H. Bevrani and T. Ise, "Enhanced virtual synchronous generator control for parallel inverters in microgrids," *IEEE Trans. on Smart Grid*, vol. 8, no. 5, pp. 2268-2277, 2017.

A Transformerless Bidirectional DC-DC Converter based on Power Units with Unipolar and Bipolar structure for MVDC Interconnection

Lejia Sun[1], Fang Zhuo[1], Feng Wang[1], Hao Yi[1] and Baohui Ma[2]

1 School of Electrical Engineering, Xi'an Jiaotong University, Xi'an, China

2 State Key Laboratory of Large Electric Drive System and Equipment Technology，Tianshui, gansu, China

sunlejia@163.com

Abstract —In this paper, a novel bidirectional DC-DC converter based on series-connected power units (BPUC) for unipolar and bipolar MVDC interconnection is proposed. Compared with traditional grid-connecting converters, the proposed converter has a transformerless structure to reduce size, cost, and power loss, being more suitable for MVDC systems emphasizing on efficiency, small volume, and low cost. In addition, the proposed converter can be applied as both a unipolar and a bipolar converter in MVDC networks, which greatly extends its scope of application and improves its transmission efficiency. Its control system can realized precise power interaction between DC grids. Simulations of the 4-level unipolar BPUC and bipolar BPUC were implemented, and a 1 kW experimental implementation of the 4-level unipolar BPUC was developed to verify the converter effectiveness.

Index Terms—unipolar，bipolar, bidirectional, multilevel

I. INTRODUCTION

Owing to the increasing installation of DC transmission lines, DC-DC converters for DC grid-connection applications will play a decisive role in future energy grids. In recent years, the increasing amount of renewable energy sources stimulates the development of medium-voltage DC (MVDC) transmission lines[1]. It has been shown that large renewable energy sources, such as windfarms and photovoltaic (PV) power plants, operate more efficiently when connected to an MVDC grid instead of an AC grid [2] and All these renewable energy systems that provide different voltage DC levels, can be efficiently integrated using DC conversion in MVDC lines [3]. In order to establish MVDC networks and efficiently transfer power between unipolar and bipolar MVDC systems, many researches have focused on the medium-voltage (MV) DC-DC converter, used as an interface between MVDC systems [4].

Some researchers propose that the dual-active bridge (DAB) converter and the face-to-face modular multilevel

This work was supported by the National Research Program of China (973 Program) under project 2015CB251001 and the National High Technology Research and Development Program of China (863 Program) 2015AA050606

converter (MMC) can be applied in MVDC connection applications [5]. However, they share some disadvantages:

1) Transformers are required. Their required transformer having adequate isolation and high-frequency operation is difficult to manufacture and faces some technical drawbacks. Furthermore, the weight, parasitic capacitance, losses, and cost associated with transformers add to the system complexity.

2) The topology and control strategy are highly complex. In the DAB and the MMC, a large number of switches need to be controlled and cooperate to accomplish voltage conversion.

With the development of switched-capacitor (SC) circuits, novel DC-DC converters with high conversion ratio have emerged by combining SC and boost circuits [6]. A typical example is the MBC mentioned in [7]. However, the MBC cannot be applied to connect different MVDC systems due to its inherent defects: 1) the power flow cannot reverse; 2) The SC circuit is not suitable for high-power applications, since the required charging current of capacitors would be excessively large. 3) It is difficult to realize bipolar conversion.

In this paper we present a novel kind of bidirectional power-unit-based DC-DC converters with unipolar and bipolar structure for MVDC connection applications, named the BPUC. Topologies of both unipolar and bipolar BPUCs are shown in Fig.1 and Fig.2 and excels with the following characteristics:

1) Simple structure and smaller size. The proposed BPUC is a transformerless topology. Thus, the weight, parasitic capacitance, losses, and cost associated to transformers, which add to the system complexity, are removed. In addition, since issues related to transformers such as cooling and insulation are avoided, the overall size of the device is reduced.

2) Realizing bipolar DC conversion. There is evidence that bipolar MVDC systems are advantageous due to higher reliability and increased power transmission capability if compared with unipolar MVDC systems.The BPUC can be applied as a bipolar DC-DC converter.

3) More intelligent control system. By applying a hierarchical and local control system, the control system not only enhances converter's dynamic performance and control flexibility, but also improves the accuracy of power interaction between connected MVDC systems.

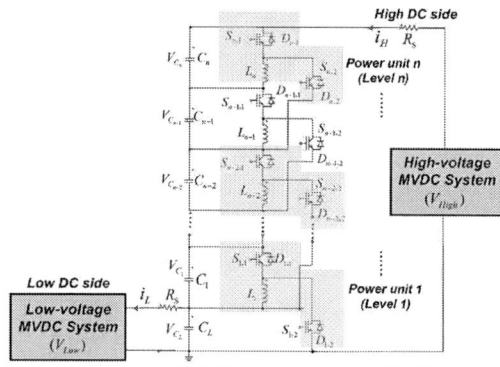

Fig.1 The n-level unipolar BPUC

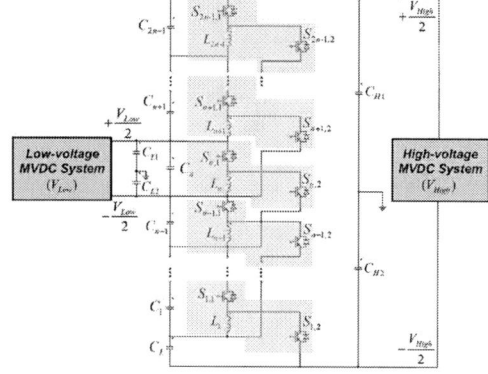

Fig.2 The bipolar BPUC

II. OPERATION PRINCIPLE OF THE BPUC

A. Unipolar BPUC

The diagram of the unipolar BPUC is shown in Fig. 1. Each power unit includes two switching devices $S_{i,1}$ and $S_{i,2}$, and one inductor L_i, $i=1,2,...,n$. There are two operation modes of the unipolar BPUC based on the power flow direction: buck mode and boost mode. For a given power flow direction, in each power unit, one switch is used as power switch and the other as freewheeling diode. The BPUC utilizes L-C resonance between power units and capacitors to transmit power and preserve voltage balance between series capacitors. A high voltage conversion ratio can be obtained between the series capacitors and the bottom level capacitor. In this section, only the power transferring process between the two bottom levels (i.e., C_L and C_1) in boost mode are discussed in detail, since the power transferring process between other adjacent levels is analogous.

For unipolar BPUC, in boost mode, The power unit can pump power from capacitor C_L up to capacitor C_n step-by-step and keep the voltage of every series capacitor at U. The high-voltage output is obtained from the series capacitors output. Fig. 3(c) shows the theoretical waveform when power is pumped from capacitor C_L to

capacitor C_1. The power transferring process can be divided into two stages, illustrated in Fig. 3(a) and Fig. 3(b), respectively. $V_{G(1,2)}$ is the driving signal of $S_{1,2}$, f is the switching frequency, and i_{1-MAX} is the maximum value of inductor current.

Stage □ (t_0-t_1): From time t_0 to t_1, $S_{1,2}$ is turned on and $S_{1,1}$ is turned off; hence, power is transferred from C_L to L_1. At the same time, voltage V_{CL} decreases and inductor current i_1 increases. The rated voltage of $S_{1,1}$ is the sum of V_{CL} and V_{C1}, which is $2U$. Variable k is the increasing gradient of i_1, and i_{1-MAX} is the maximum value of i_1,

Stage □ (t_1-t_2): From time t_1 to t_2, $S_{1,2}$ is turned off. Current i_1 flows through freewheeling diode $D_{1,1}$ and power is transferred from L_1 to C_1. Then, i_1 decreases and V_{C1} increases. Voltage V_{CL} increases charged by input voltage V_{Low}. The rated voltage of $S_{1,2}$ is the sum of V_{CL} and V_{C1}, which is $2U$. The declining gradient of i_1 is equal to k in (1). Finally, at time t_2, $S_{1,2}$ is turned on, and the process repeats from ***Stage □***.

When the converter operates in buck mode, the converter power flow is reversed and the power is transferred from top capacitor C_n to bottom capacitor C_L step by step. The working principle is similar with the boost mode.

B. Bipolar BPUC

Worldwide, most operating commercial MVDC networks are bipolar systems. The bipolar BPUC which is used in bipolar MVDC system connection is shown in Fig. 2. The part of the converter above the middle level (i.e., from C_{n+1} to C_{2n-1}) is named positive pole, whereas the part below the middle level (i.e., from C_1 to C_n, including C_L) is named negative pole. Both pole circuits can be considered as two independent unipolar BPUC converters. According to the power flow direction, the bipolar BPUC also has two operation modes: up-conversion and down-conversion mode. The operation principle of the bipolar BPUC converter is shown in Fig.4.

When the bipolar BPUC works in up-conversion mode, the positive pole circuit works under boost mode, and the negative pole works under buck mode. The power transferring route is represented by blue arrows in Fig.4. When the bipolar converter works in down-conversion mode, the power flow is reversed. The positive pole works under buck mode, and the negative pole works under boost mode. The power transferring route is represented by red arrows in Fig.4. By using the bipolar BPUC, the current level in inductors can be reduced to 50% and the rated voltage of switching devices can also be reduced to 50% of those present in the unipolar BPUC. Therefore, the efficiency and reliability of transmission are both improved.

The 2018 International Power Electronics Conference

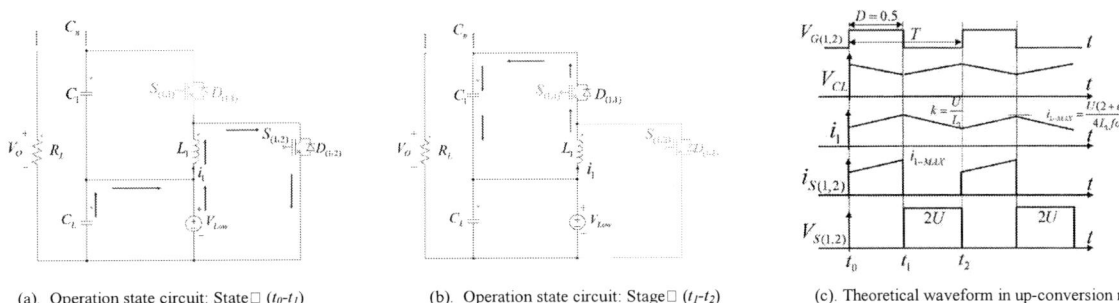

(a). Operation state circuit: State□ (t_0-t_1)   (b). Operation state circuit: Stage□ (t_1-t_2)   (c). Theoretical waveform in up-conversion mode

Fig. 3. Transferring process of the n-level unipolar BPUC in the boost mode when the power transferred from the C_L to C_1

III. CONVERTER DESIGN

A. Inductor Design

In the n-level unipolar BPUC, the conversion ratio is $V_{High}/V_{Low}=N=n+1$. The power transferred by the i-th level can be expressed by $(N-i)P/N$, $i=1,2,...,n$, and α is the current ripple of inductor current i_k. To maintain the energy balance of inductor L_k, the following equation can be used:

$$\frac{N-k}{N} \cdot P = \frac{1}{2} \cdot L_k \cdot f \cdot [(i_k+\frac{\alpha}{2}i_k)^2 - (i_k-\frac{\alpha}{2}i_k)^2]. \quad (1)$$

From (1),the L_k can be calculated as

$$L_k = \frac{U^2}{2f\alpha P(N-k)}. \quad (2)$$

B. Capacitor Design

In order to keep the voltage exchange between series capacitors, the capacitance of C_L and C_n should be the same. The voltage ripple of capacitor C_L is expressed as β and the following equation can be obtained:

$$\frac{1}{2} \cdot C_L \cdot [(U+\frac{\beta}{2}U)^2 - (U-\frac{\beta}{2}U)^2] = \frac{D \cdot P}{f}. \quad (3)$$

From (3), capacitance C_n can be obtained as

$$C_n = C_L = \frac{P}{2f\beta U^2}. \quad (4)$$

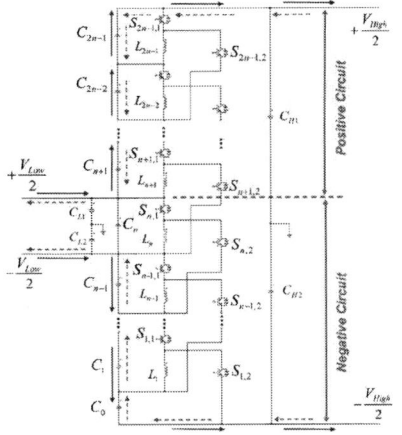

Fig. 4. Operation principle of the bipolar BPUC converter

IV. POWER CONTROL SYSTEM

A proposed hierarchical and local control system is used in the BPUC to achieve more flexible and precise power control between interacting MVDC systems. The general control block diagram of the BPUC is presented in

Fig.5. The control system consists of two levels: the central controller and local power unit controllers. The BPUC provides two power control strategies according to its application: unipolar power control for the unipolar BPUC and bipolar power control for the bipolar BPUC.

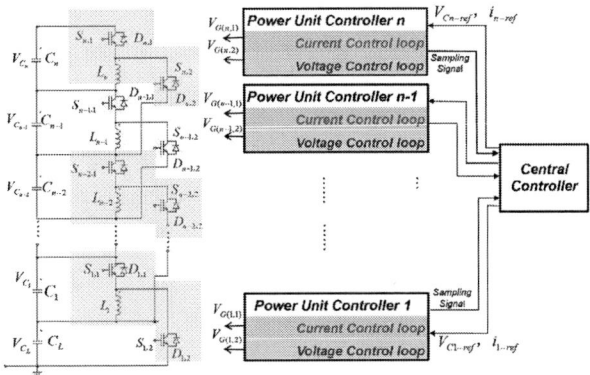

Fig. 5. General control scheme for the BPUC

A. Unipolar Power Control

The unipolar power control is used in unipolar applications, where the converter is required to connect unipolar MV systems such as battery storage systems, PV systems, or distributed generation systems. In the unipolar power control strategy, either the top power unit controller or the bottom power unit controller is set to current control loop to perform constant power interaction between two DC grids. The remaining power unit controllers are set to voltage control loop. Take the unipolar BPUC in buck mode for example, and its control block diagram is shown in Fig.6. Based on the required constant output power P_{ref}. Reference input current i_{H-ref} can be expressed as

$$i_{H-ref} = \frac{P_{ref}}{V_{High}}. \quad (5)$$

The reference voltage for the remaining power unit controllers is given by

$$V_{C-ref} = V_{C(n-1)-ref} \cdots\cdots = V_{C1-ref} = \frac{V_{high}}{n+1} \quad (6)$$

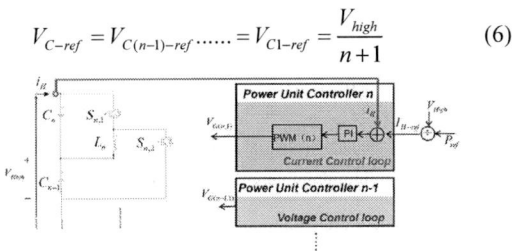

Fig. 6. Control block diagram of the unipolar BPUC for unipolar power control in buck mode

B. Bipolar power control

2884

When the BPUC works as a bipolar converter to connect bipolar MVDC networks and perform precise power interaction, bipolar power control is used. Take the bipolar BPUC in down-conversion as an example,and its block diagram is shown in Fig.7. The positive pole of the converter works under unipolar power control in buck mode and the power unit controller (2n-1) is set to current control loop. The negative pole of the converter works under unipolar power control in boost mode, and Power unit controller 1 is set to current control loop.

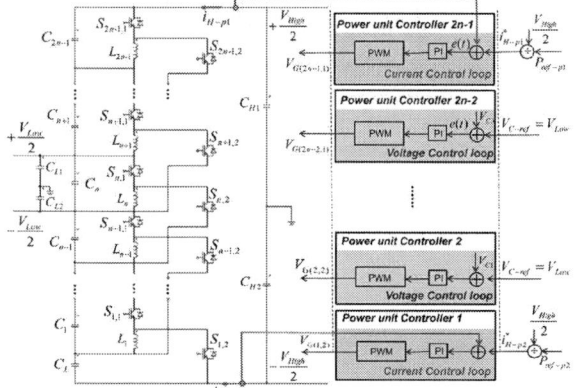

Fig. 7. Control block diagram of the bipolar BPUC for bipolar power control in down-conversion mode

To verify the feasibility of the proposed converter, a simulation model based on a 4-level unipolar BPUC was implemented using PSIM. The simulation parameters are given as follows: V_{High}=5 kV, V_{Low}=1 kV, f=10 kHz, P=1 MW, N=5, α=50%, C_L=C_i=1mF, (i=1,2,3,4), L_i, (i=1,2,3,4)= 25 μH, 33 μH, 50 μH,100 μH

A. Simulation for Unipolar Power Control

Fig. 8 shows the simulation waveforms of the 4-level unipolar BPUC for unipolar power control in the boost mode (V_{in}=V_{Low}=1 kV, V_O=V_{High}=5 kV). It can be seen that an equal voltage is present in both capacitors (i.e., V_{C1}=V_{C2}), with an obtained ripple below 5% and inductor currents approximately equal to the theoretical values calculated by (10). Next, the dynamic step response of the proposed converter was investigated in a simulation of the converter with unipolar power control in buck mode, whose outputs are shown in Fig. 9, and a slightly lower damping is observed during the transient process.

B. Simulation for Bipolar Power Control

Fig. 10 shows the waveforms of the 4-level bipolar BPUC for bipolar power control in up-conversion mode ($\pm V_{in}/2 = \pm V_{Low}/2 = \pm 500$ V, $\pm V_O/2 = \pm V_{High}/2 = 2.5$ kV, P_{ref}= 1 MW). Compared with the outcomes of Fig.12, the inductor current level and its ripple are greatly reduced.

V. SIMULATION RESULTS

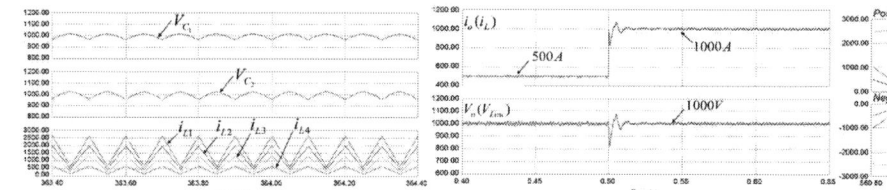

Fig. 8. Simulation of the 4-level unipolar BPUC for unipolar power control in boost mode (V_{in}=V_{Low}=1kV, V_O=V_{High}=4 kV)

Fig. 9. Simulation of dynamic step response of the 4-level BPUC for unipolar power control in buck mode (the load R varies from 2 Ω to 1 Ω)

Fig. 10. Simulation waveform of inductor currents and output voltage of bipolar 4-level BPUC in bipolar power control under up-conversion mode

VI. EXPERIMENTAL RESULTS

Owing to laboratory conditions, a small-scale (i.e., *1 kW*) 4-level prototype of the unipolar BPUC was developed, as shown in Fig.11. The prototype parameters are: *n=4, P=1kW, V_{High}=500V, V_{Low}=100V, f=20kHz*. Fig. 12 shows the the waveforms for inductor currents for unipolar power control in boost mode. Fig. 13 shows the input and the output waveform of the prototype. Fig. 14 shows the startup process of the converter with a slightly lower damping during this process, until the output voltage finally reaches the steady-state rated voltage of 500 *V*.

Fig. 11. Photograph of the 1 kW 4-level unipolar BPUC prototype

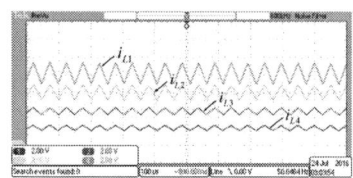

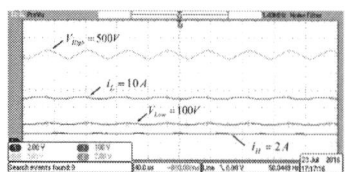

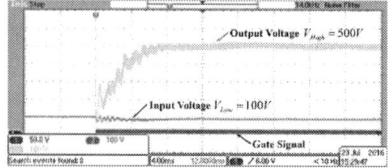

Fig.12 inductor currents in power units(2A/1V)

Fig.13 Waveform of the input and output (2A/1V)

Fig.14 Waveform of automatic starting-up process

VII. CONCLUSION

In this paper, a novel bidirectional DC-DC converter based on series power units for unipolar and bipolar MVDC connection and its corresponding control strategies are proposed. Compared with traditional DC grid-connecting converters, the proposed converter presents the following improvements:

--Transformerless.

--Bipolar MVDC connection with simple structure.

--Precise power control between MVDC systems.

Finally, a 1 kW implementation of the 4-level BPUC is developed. The experiment results validate the operation principle and feasibility of the proposed converter.

ACKNOWLEDGMENT

This work was supported by the National Research Program of China (973 Program) under project 2015CB251001 and the National High Technology Research and Development Program of China (863 Program) 2015AA050606

REFERENCE

[1] St´ephane Vighetti and Jean-Paul Ferrieux "Optimization and Design of a Cascaded DC/DC Converter Devoted to Grid-Connected Photovoltaic Systems," in *IEEE transaction on power electronics,* vol.27, no.4, pp.2018-2027, April. 2012.

[2] S. L undberg, "Wind farm configuration and energy efficiency studies— Series dc versus ac layouts," *P h.D. dissertation,* Chalmers University of Technology, Sweden, 2006.

[3] M. Bragard, N. Soltau, S. Thomas, and R. W . De Doncker, "The balance of renewable s ources and user demands in grids: Power electronics for modular battery energy storage systems," *IEEE Trans. Power Electron,* vol. 25, no. 12, pp. 3049–3056, Dec. 2010.

[4] F. Mura and R. W. De Doncker,"Design aspectsofa medium-voltage directcurrent(MVDC) grid for a university campus,"in *Proc. IEEE EighthInt.Conf. Power Electron. ECCE Asia* ,2011,pp.2359–2366.

[5] Stefan P. Engel and Marco Stieneker, "Comparison of the Modular Multilevel DC Converter and the Dual-Active Bridge Converter for Power Conversion in HVDC and MVDC Grids" *IEEE transaction on power electronics,* vol.30, no.1, pp.124-137, January. 2012.

[6] Dong Cao, and Shuai Jiang, "Optimal Design of a Multilevel Modular Capacitor-Clamped DC–DC Converter," *IEEE transaction on power electronics,* vol.28, no.8, pp.3816-3826, August. 2013.

[7] J.C. Rosas-Caro and J.M. Ramirez, "A DC–DC multilevel boost converter," *IET Power Electronics,* vol.3, no.1, pp129-137, November 2008.

The 2018 International Power Electronics Conference

New Modulation Control of Converter System Applied for Offshore Wind Farms

Naoki Kawabata[1], Noriyuki kimura[1], Toshimitsu Morizane[1] and Hideki omori[1]

1 Electrical and Electronic Engineering, Osaka Institute of Technology, Osaka, Japan

Abstract—This paper proposed the modified trapezoidal modulation (MTM) as a new modulation technique of a voltage source converter (VSC) for an offshore wind farm connected to a HVDC transmission. This system is especially suitable for the series connected offshore windfarm. Experimental results show that MTM is more effective for the reduction of the cost and loss of the VSC than conventional sinusoidal modulation (SM). In this paper, Power Factor Corrected Converter (PFCC) is installed parallel with VSC to reduce the total converter system cost. The simulation and experimental results proved that the selection of the exciting voltage and the slip of the induction generator (IG) can make the VSC dc side current to zero. These facts verified that the capacity of the VSC can be minimized and the system cost can be reduced by the proposed system configuration.

Keywords— modified trapezoidal modulation, offshore windfarm, power factor corrected converter, wind power generator system

I. INTRODUCTION

Recently, the offshore wind farm is actively developed. Especially in Europe, several projects to develop large offshore wind farm are founded, since it is effective compared with the other configurations of power generation [1]. For long distance power transmission from the offshore to the mainland, it is well known that the high-voltage-DC (HVDC) transmission has lower cost than the AC transmission [2]. Fig.1 is an example of the offshore wind farm configuration with the HVDC transmission. It has a converter system with voltage source converter (VSC) and DC/DC converter to connect the induction generator (IG) to the HVDC transmission system.

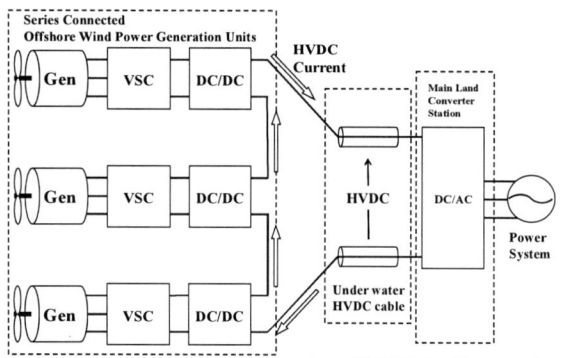

Fig.1. Offshore Wind Farm Configuration with HVDC grid connection

The cost and efficiency is an important issue for the wind power system. In this paper, new modulation technique, Modified Trapezoidal Modulation (MTM), is proposed for cost reduction of VSC. This technique makes the VSC more efficient by reducing switching loss. And as the output power is greater than the sinusoidal modulation (SM) at the same DC voltage, the rated voltage of the switching device can be lowered and gives a better cost performance. Additionally, by using a Power Factor Corrected converter (PFCC) in parallel with the VSC, the total cost of the converter system can be reduced more. The good feasibility of the proposed modulation and configuration is shown by the simulation and experiments.

II. PROPOSED GENERATION SYSTEM

Fig. 2(a) shows the synchronous generator (SG) having the diode rectifier with dc/dc chopper. The SG with permanent magnets has been already installed in Japan. However the cost is too much higher than the induction machine. The induction generator (IG) needs the voltage source converter (VSC) to supply the exciting current as shown in Fig. 2(b). Although the efficiency of the IG is not good as the SG, the cost of the IG is much lower than the SG. To investigate the better cost performance, decrease of the rating of the VSC used for exciting the induction generator is desirable and we proposed the system shown in Fig. 2(c) [3].

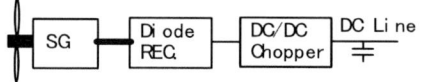

(a) SG system (SG: synchronous generator)

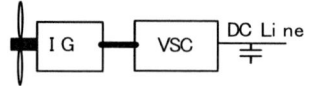

(b) IG system (IG: induction generator)

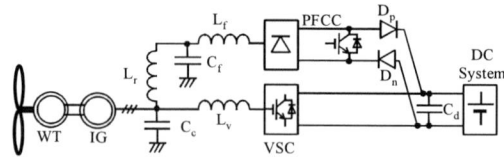

(c) Proposed system with IG (IG: induction generator)

Fig.2. Wind power generation system for dc transmission

III. Modified Trapezoidal Modulation

The modified trapezoidal waveform is shown in Fig.3(a) [4, Fig. 4]. This waveform can be obtained by superimposing a 120 degree rectangular wave on a trapezoidal wave which has a 120 degree flat part. γ is superposition ratio of rectangular amplitude to the maximum value. When $\gamma = 0$, the waveform appears as a trapezoidal wave which has a 120 degree flat part. When $\gamma = 1$, the waveform is a rectangular wave with a 120 degree width. Additionally, Fig.3 (b)-(e) shows the output waveform when modified trapezoidal wave as a modulating signal on a triangle comparison based PWM for a two-level three-phase voltage source inverter. Here E_d, v_u, v_v, v_w, and v_{uv} are inverter DC voltage, U-phase voltage, V-phase voltage, W-phase voltage and line voltage of U-phase to V-phase, respectively.

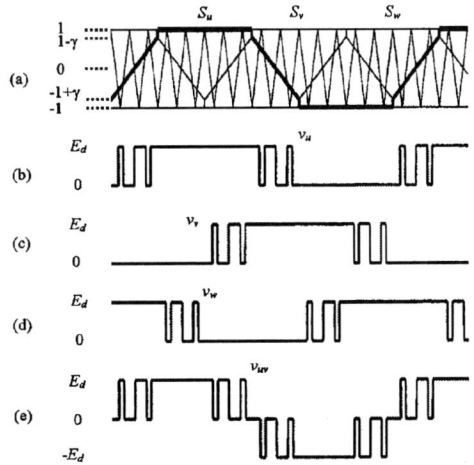

Fig.3. Modified Trapezoidal Waveform and PWM output voltage

IV. Introduction Reduction of Torque Ripple

The amplitude of the fundamental component of modified trapezoidal modulation (MTM) is 17% higher than that of a sinusoidal modulation (SM). However, it is a non sinusoidal waveform and it contains harmonics. The large amplitude components are 5th and 7th harmonics in AC voltage. They appear in the AC current and generate the 6th harmonic torque in an induction machine. At the induction machine, harmonic torque is approximately defined with equation (1). From equation (1), the 6th harmonic torque $T_6 \cong 0$ when $I_5 = I_7$. Here, $\gamma = 0.38$ satisfies $I_5 = I_7$. Fig.4 shows the experimental results and calculation results of the 6th harmonic torque of an induction machine for γ[5]. The amplitude represents the proportion of the 6th harmonic relative to the fundamental component. Whether the slip $s = 0.03$ or $s = 0.05$, the 6th harmonic torque reaches a minimum when $\gamma = 0.38$.

$$T_{6k} \cong \frac{3}{2\omega_s} R_1 I_1 \left(I_{6k+1} - I_{6k-1} \right) \sin 6k\omega_s t \qquad (1)$$

$k = 1, 2, \cdots$

$R_1 = \left(1 - s_1\right) R_r / s_1$

s_1 : Slip

R_r : Rotor resistance

I_1 : Rotor fundamental current

ω_s : Fundamental angular frequency

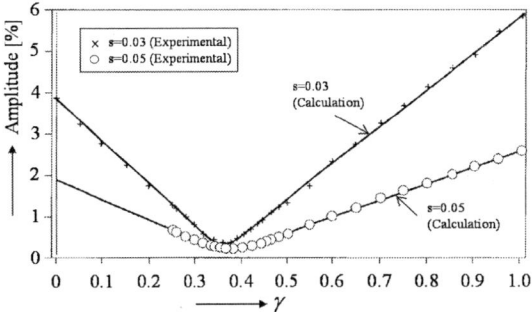

Fig.4. 6th harmonic torque of induction machine for γ

V. Experimental Results of MTM

To compare the modified trapezoidal modulation (MTM) and the sinusoidal modulation (SM) in an induction generator system, the following experiment was conducted. Fig.5 shows the experimental setup of the induction generator system (Here, Cage-rotor IG). The parameters of the IG and the VSC are shown in Table 1 and Table 2.

First, the IG was driven by a three-phase voltage source converter (inverter) for the synchronous speed (50Hz, 1500 rpm). Next, the DC Motor (DCM) was controlled to the regulated revolutions (1618rpm, s=-7.87%). Then, the power analyzer (HIOKI PW6001) recorded the three-phase output voltage, three-phase output current, DC output voltage, and DC output current. The same experiment was conducted while changing the DC voltage and the modulation method.

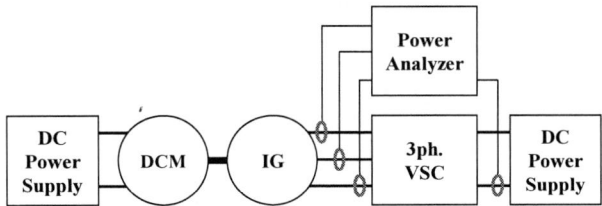

Fig.5. Experimental platform of cage-rotor induction generator connected to DC bus

Table.I   PARAMETERS OF THE CAGE-ROTOR INDUCTION GENERATOR

Parameter	Value
Phase	3
Pole	4
Stator resistance r_1 (Ω)	1.8
Rotor resistance r'_2 (Ω)	2.81
Mutual resistance r_m (Ω)	1500
Mutual inductance L_m (H)	0.309
Stator inductance L_1 (mH)	1.82
Rotor inductance L'_2 (mH)	19.82
Rated Voltage (V)	200
Rated Power (VA)	800

Table.II   Parameters of VSC

Parameter	Value
DC Voltage (V)	140-230
Modulation Rate	1.0
Switching Frequency (kHz)	10

Fig.6 shows the experimental and the simulated results of the DC voltage vs. the DC output power. As the DC voltage increases, the DC output power increases. This is because the AC excitation voltage grows proportionally to the DC voltage. The DC output power of MTM has increased by an average of 42% from the SM. The converter loss is shown in Fig.7. As for the average of converter loss when the DC output power between 300W and 500W, MTM is reduced by 30% when compared to SM.

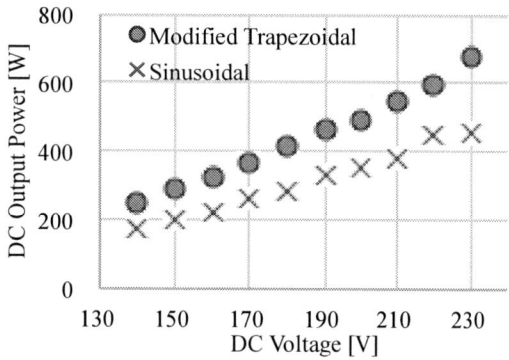

Fig.6. Comparison of DC Voltage and DC output Power by experiment

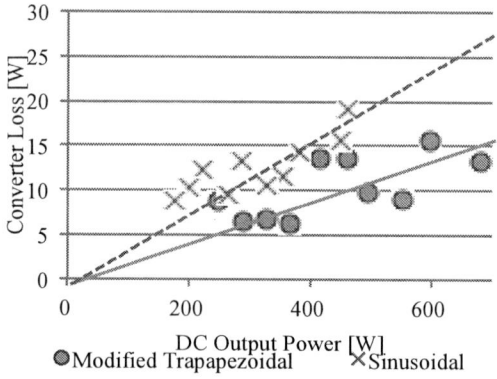

Fig.7. Comparison of DC output power vs. Converter loss by experiment

The waveforms of three-phase line voltage, three-phase output current and DC output current case of DC voltage is 230V acquired by the power analyzer is shown in Fig.8. Fig.8(a) shows the results of MTM and Fig.8(b) shows that of SM.

The period of no switching appears on the three-phase line voltage at MTM, due to the 120 degree flat part in the modulation signal. The three-phase current distortion by the harmonics can be observed at MTM. Also, the DC output current at MTM has a larger current ripple than that of SM.

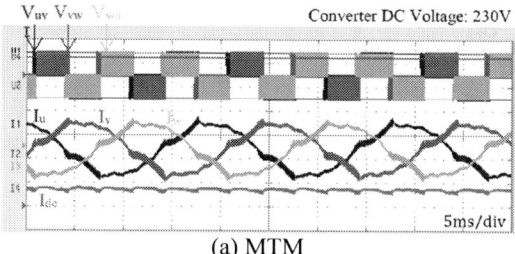

(a) MTM

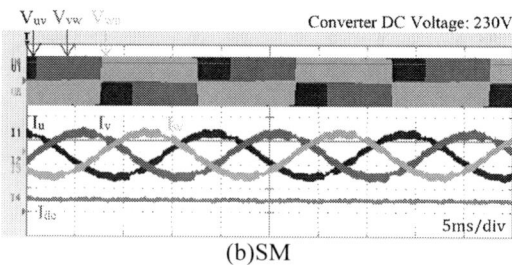

(b)SM

Fig.8. Experimental Waveforms recorded by Power Analyzer

The amplitude of the fundamental component of the AC output voltage with MTM and SM was provided by the Discrete Fourier Transform. When the DC voltage is 230V, the amplitude with MTM is 233V. At the same time, the amplitude with SM is 198V. The voltage with MTM is 17.7% higher than that of SM. The result fits the theoretical value.

VI. SIMULATION OF PROPOSED SYSTEM

Simulated system is shown in Fig.9. It is same as the system shown in Fig. 1(c). The VSC is modulated by the MTM (Modified Trapezoidal Modulation). DC voltage must be changed according to the MPPT (Maximum Power Point Tracking). Therefore it is necessary to confirm the controllability of the PFCC (Power Factor Corrected Converter) to have the zero real power output from the VSC. Simulation software PSIM has been used [6].

The induction generator (IG) generates the electric power to extract the maximum power from the wind turbine. In this simulation, the IG is set to have the constant rotating speed which coincides the maximum power point. Then IG generates the real power Pg and consume the lag reactive power Qg.

$$S_g = V_g I_g = P_g + j Q_g \qquad (2)$$

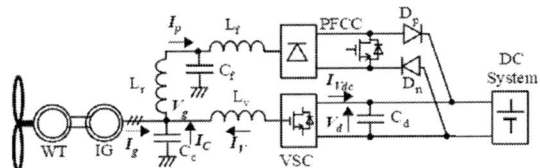

Fig.9. Simulated proposed system with IG (IG: induction generator)

The PFC converter (PFCC) absorb the real power Pp only. If the Pp = Pg, the voltage source converter (VSC) supplies the excitation reactive power Qg only, and has the minimum power rating. The reactive power of the VSC is decreased since Qg is also partly supplied from the capacitors Cc connected at the IG terminal.

$$Q_g = Q_v + Q_c = V_g I_v + \omega C V_g^2 \qquad (3)$$

The circuit diagram of the PFCC is shown in Fig.10. The parameters of PFCC are shown in Table. 3.

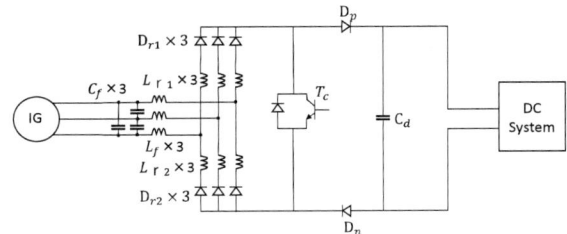

Fig.10. Schematic of PFCC

Table.III   Parameters of PFCC

Parameter	Value
Filter reactor Lf[μH]	100
Reactor Lr[μH]	57
Filter capacitor Cf[μF]	10
Smoothing capacitor Cd[μF]	1000

VII. PAM CONTROL OF VSC WITH MTM

Fig. 11 shows the relation between the voltage Vac at the IG terminal and the slip of the IG to have the same output power, when the mechanical input power and the rotating speed of the IG are constant. This relation is used to have the proper real power absorption to the PFCC. There is a freedom of the choice of the voltage and the slip. Previously the authors proposed to select the loss minimization set for IG depending on the wind speed[8].

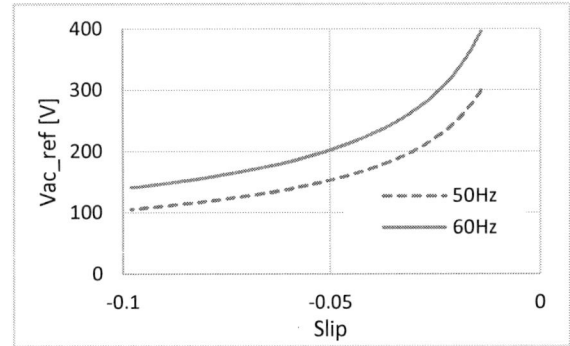

Fig.11. AC exciting voltage at induction generator terminal

To change the IG terminal voltage Vac, PAM (Pulse Amplitude Modulation) control is necessary since the modulation factor of the MTM cannot be changed. For MTM, the IG terminal voltage Vac is determined as following equation by using the dc voltage Vdc.

$$V_{ac} = V_{dc} \left(\frac{4\gamma}{n\pi} \cos\frac{n}{6}\pi + \frac{24(1-\gamma)}{n^2\pi^2} \sin\frac{n\pi}{6} \right) \cos\frac{n}{6}\pi \qquad (4)$$

The relation between the Vdc and the Vac of the VSC is shown in Fig. 12. As shown in Fig. 11, the peak value of the IG terminal voltage exceeds the dc voltage and the current flow into the PFC converter when the duty factor *Df* is zero, that is the diode rectifier.

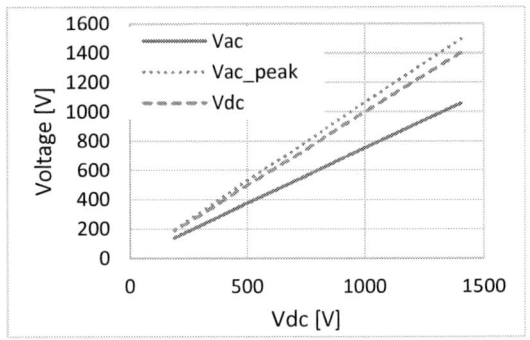

Fig.12. AC exciting voltage at induction generator terminal Vac, its peak value Vac_peak and DC voltage Vdc

If the voltage difference Vdif=Vac_peak – Vdc becomes smaller, the current flow into the PFCC becomes smaller. Then the absorbed real power to the PFCC, Ppfc, also becomes smaller. Therefore, increasing the slip and

2890

decreasing the dc voltage is meaningful to minimize the VSC rating. Feasibility study of the PAM control is investigated by using PSIM ver.10 software. Slip is set to be -4.5% and -6% when the mechanical input power is constant depending on the rotating speed. Simulation results are shown in Fig. 13 and 14 when the duty factor Df is zero. In Fig. 12, Pg=P0, Pv=P2, Pp=P2. Even though the Df is zero, the absorbed Ppfc is larger than the output power from the IG. The mechanical input power is set to be 700 W and the IG output power is around 600W. In Fig. 14, the slip is changed to -6%, and the IG terminal voltage is lower to 180V. The absorbed Ppfc is smaller than the IG output power and the VSC also absorbs the real power, Pvsc, from the IG. Fig. 15 shows that the optimum duty factor can set the Pvsc zero. In this case, the duty factor Df of the PFCC is 0.03.

Fig. 16 and 17 show the change of Pvsc when the slip is set to be -4.5% and -6%, respectively. The synchronous frequency is 60Hz, 50Hz and 40Hz. And the input mechanical power is set to be 700W, 400W 300W depending on the assumed wind speed.

In Fig. 16, the voltage difference between the peak value of the IG terminal voltage Vac_peak and the dc voltage Vdc, that is Vdif, becomes larger and the Ppfc exceeds the IG output power, Pg, even when the duty factor Df is zero. Pvsc is increased as the Df increases when the synchronous frequency is 60Hz.

In Fig. 17, the voltage difference between the peak value of the IG terminal voltage Vac_peak and the dc voltage Vdc, that is Vdif, becomes smaller and the Ppfc decreased to less than the Pg when the duty factor Df is zero. The Pvsc becomes zero at the optimu Df for all synchronous frequency. These results indicates the good feasibility of the stable operating ability of the proposed system.

VIII. Experiment of proposed system

Experiments of the system in Fig.9 were conducted. The parameters of the experiment shown in Table 4. Other values are those in Table 1 and Table 3.

Table.IV parameter of experiment

Parameter	Varue
Rotor speed[rpm]	1500
Rated frequency[Hz]	50
Input power P_{in}[W]	460
Slip[%]	-5, -8

The IG terminal voltage Vac was obtained from equations (5) and the dc voltage Vdc was obtained from equations (6).[7]

$$V_{ac} = \sqrt{\left| \frac{P_{in}}{(1-s)r_2'/s} \times \left\{ \left[r_1 + \frac{r_2'}{s} \right]^2 + \left[2\pi f_{ref}(L_1 + L_2') \right]^2 \right\} \right|} \quad (5)$$

$$V_{dc} = \frac{\sqrt{2}V_{ac}}{\left(\frac{4\gamma}{n\pi} \cos\frac{n}{6}\pi + \frac{24(1-\gamma)}{n^2\pi^2} \sin\frac{n\pi}{6} \right)\cos\frac{n}{6}\pi} \quad (6)$$

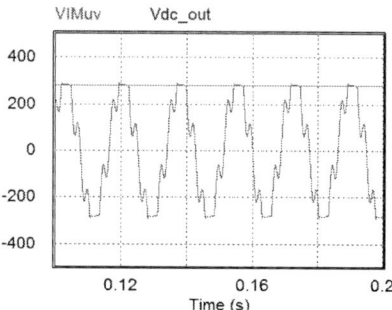

(a)IG voltage V_{IM_UV}, DC voltage Vdc_out

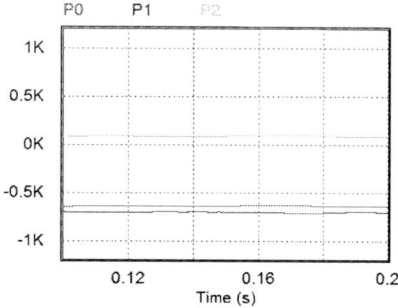

(b)IG power P0, PFCC power P1, VSC power P2

Fig.13. Simulation results of proposed system (Vac_ref=200V, slip=-4.5%, Df=0.00, fs=60Hz)

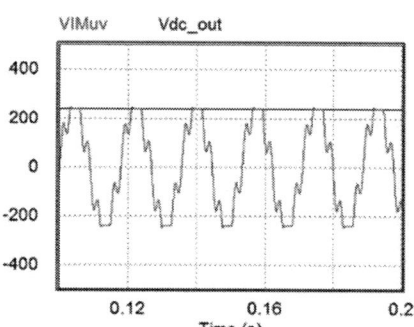

(a)IG voltage V_{IM_UV}, DC voltage Vdc_out

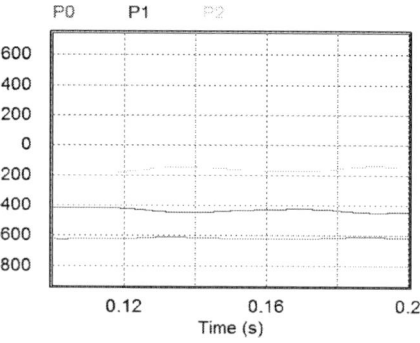

(b)IG power P0, PFCC power P1, VSC power P2

Fig.14. Simulation results of proposed system (Vac_ref=180V, slip=-6%, Df=0.00, fs=60Hz)

The 2018 International Power Electronics Conference

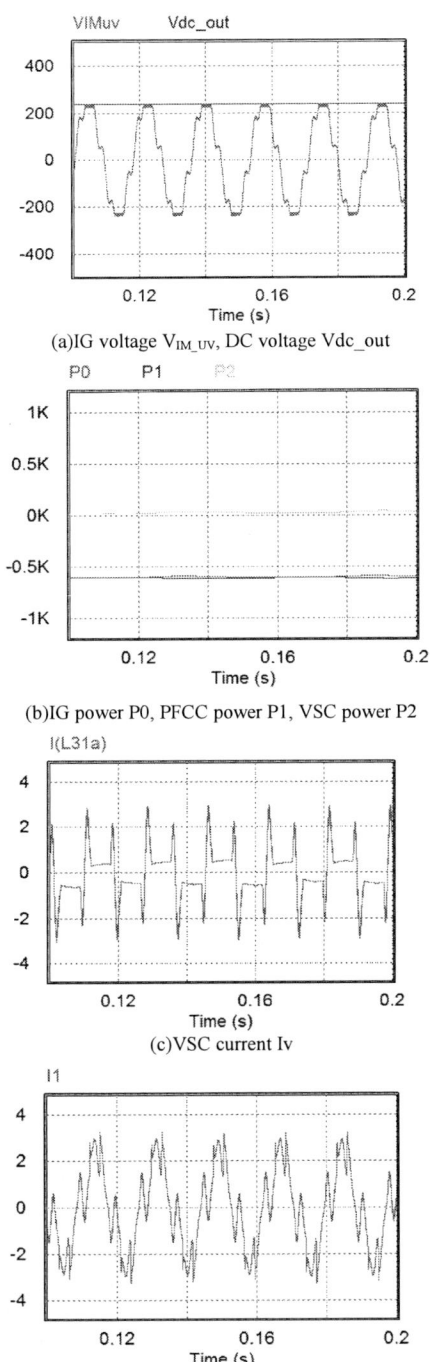

(a)IG voltage V_{IM_UV}, DC voltage Vdc_out

(b)IG power P0, PFCC power P1, VSC power P2

(c)VSC current Iv

(d)PFCC current Ip

Fig.15. Simulation results of proposed system (Vac_ref=200V, slip=-6%, Df=0.03, fs=60Hz)

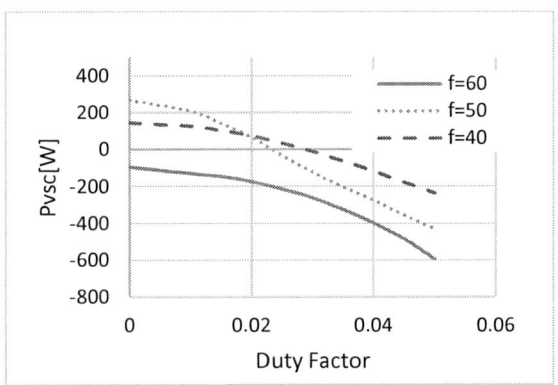

Fig.16. Duty Factor of PFCC vs. Real Power of VSC. (Vac_ref=200V, slip=-4.5%)

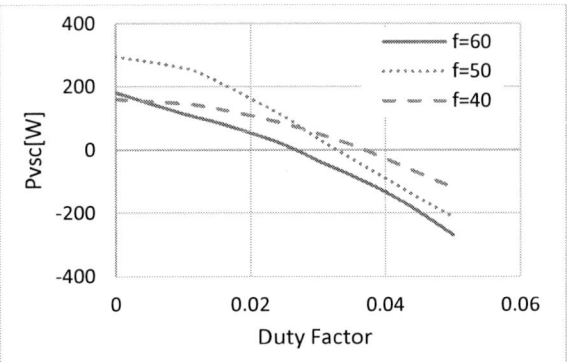

Fig.17. Duty Factor of PFCC vs. Real Power of VSC. (Vac_ref=200V, slip=-6%)

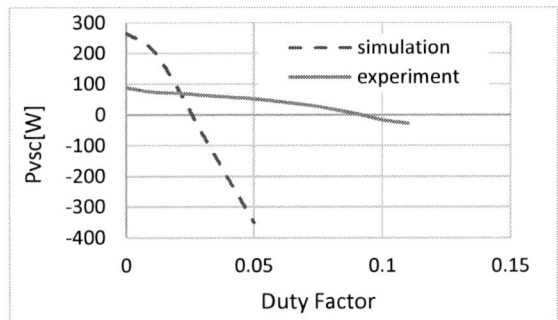

Fig.18. Duty Factor of PFCC vs. Real Power of VSC. (Vac_ref=152.96V, slip=-5%)

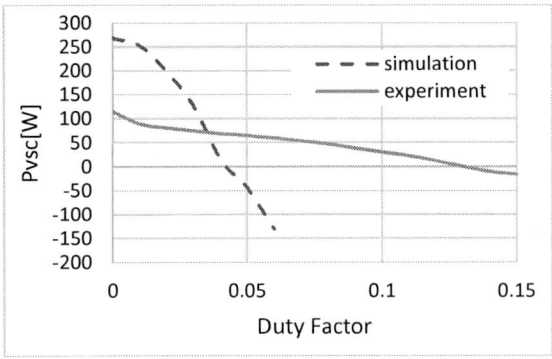

Fig.19. Duty Factor of PFCC vs. Real Power of VSC. (Vac_ref=118.1V, slip=-8%)

Fig. 18 and 19 show the change of the real power of the VSC, Pvsc, in the experiment and simulation when the slip is set to -5% and -8%, respectively.

In Fig. 18 and 19, they were confirmed that there is duty factor Df that can make Pvsc zero. It was also confirmed that Df which makes Pvsc of the experiment zero is larger than simulation.

The reason why Pvsc in the experiment is smaller than the simulation is considered to be the difference in waveforms of the Vac. Fig.20 and 21 show the waveform of the Vac in experiment and simulation, respectively.

From Fig. 20 and 21, it was seen that the waveform of Vac in the experiment was more distorted than the simulation. It means that the maximum value of the Vac of the experiment was larger than the simulation. As a result, the difference between the maximum values of Vac and Vdc becomes smaller in the experiment compared with the simulation. Then, the current flowing in the PFCC, Ipfc, becomes larger, and the PFCC absorbs larger Ppfc.

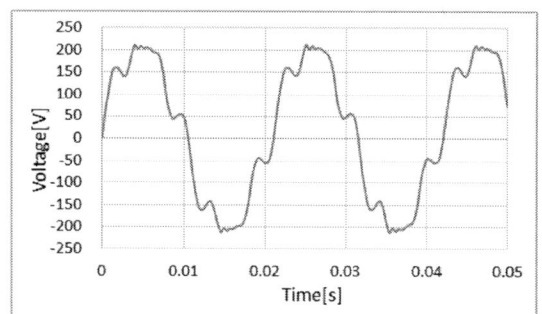

Fig.20. waveform of the Vac in experiment (slip=-5%, Df=0)

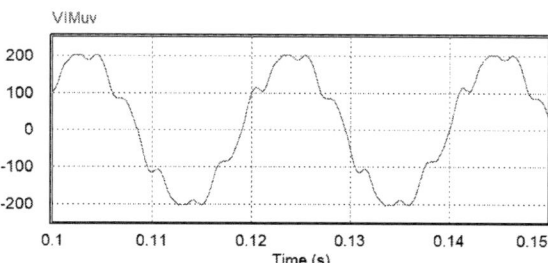

Fig.21. waveform of the Vac in simulation (slip=-5%, Df=0)

Since the exciting voltage of the induction generator (IG) is defined by the VSC ac side voltage which depending on the dc side voltage directly, it is impossible to make the PFCC dc side current to zero.

Fig.22 shows the waveform of the Ipfc in the experiment. Fig.23 shows the waveform of the Ipfc in the simulation. From Fig. 22 and Fig. 23, it can be confirmed that the distortion and the amplitude of the Ipfc in the experiment are larger than the simulation. Therefore, Ppfc absorbed by PFCC in the experiment became larger than the simulation.

Fig. 24 and 25 show the change of the real power of the PFCC, Ppfc, in the experiment and the simulation when the slip is set to -5% and -8%, respectively. From Fig. 24 and Fig. 25, when *Df* is small, Ppfc in the experiment becomes larger than the simulation. This makes Pvsc in the experiment smaller than the simulation. However, it is important to investigate the reason why the reduction rate of Pvsc and the increase rate of Ppfc in the simulation are larger than the experiment.

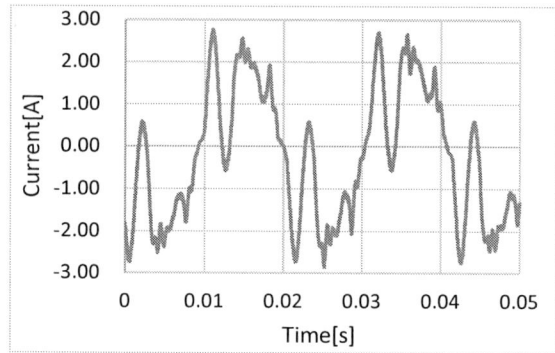

Fig.22. waveform of the Ipfc in experiment (slip=-5%, Df=0)

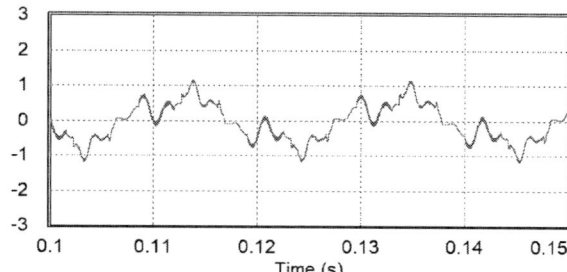

Fig.23. waveform of the Ipfc in simulation (slip=-5%, Df=0)

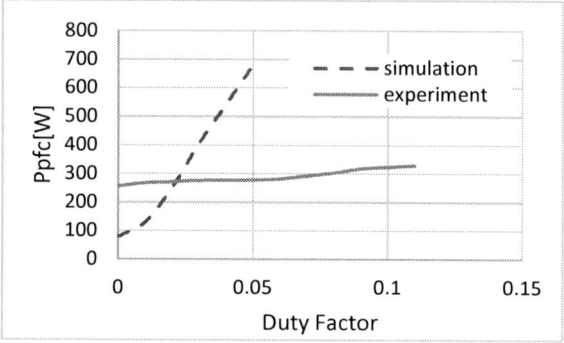

Fig.24. Duty Factor of PFCC vs. Real Power of PFCC (Vac_ref=152.96V, slip=-5%)

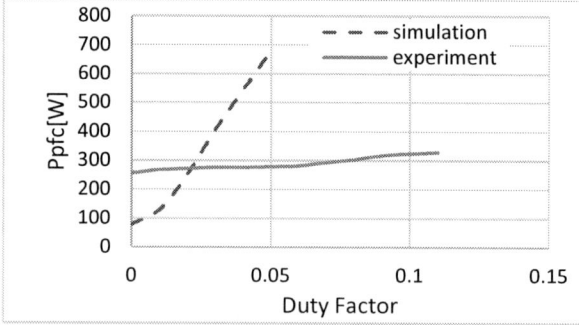

Fig.25. Duty Factor of PFCC vs. Real Power of PFCC (Vac_ref=1V, slip=-8%)

A similar experiments were conducted for MTM and SM. DC voltage Vdc in SM was obtained from equations(7).

$$V_{dc} = \frac{2\sqrt{2}V_{ac}}{\sqrt{3}} \tag{7}$$

Fig. 26 and Fig.27 show the results. Fig. 28 shows waveform of the Vac in SM in experiment.

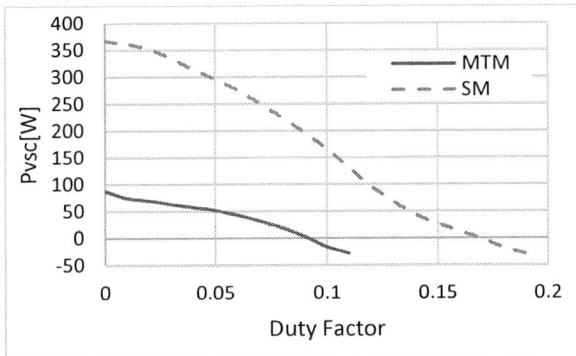

Fig.26. Duty Factor of PFCC vs. Real Power of VSC.
(Vac_ref=152.96V, slip=-5%)

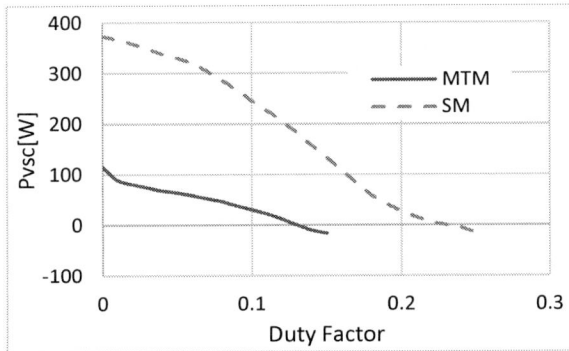

Fig.27. Duty Factor of PFCC vs. Real Power of VSC.
(Vac_ref=118.1V, slip=-8%)

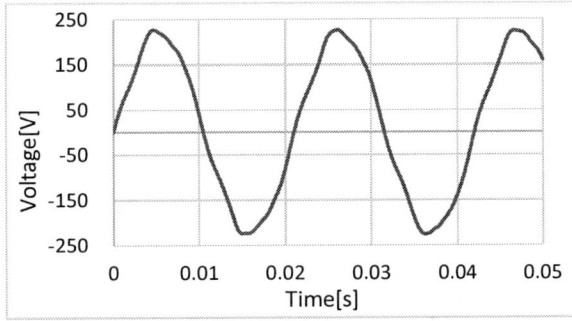

Fig.28. waveform of the Vac in SM in experiment (slip=-5%, Df=0)

In Fig. 26 and Fig. 27, it was confirmed that the duty factor Df that can make Pvsc of the MTM zero is smaller than SM. From equations (6) and (7), to obtain the same amount of Vac, Vdc of about 1.44 times is required for MTM and about 1.63 times is required for SM. In case of MTM, the difference between the maximum value of the amplitude of the Vac and the Vdc is smaller than that of SM. Also, from Fig. 28 and Fig. 20, the Vac waveform of MTM has greater distortion than SM, and the period of the larger value than the Vdc is longer. Therefore, the control range of Df may be smaller with the MTM. From this fact, MTM is effective for reducing VSC loss, but SM is considered to be effective for widening the control

range of Df. Though there is some trade-off between MTM and SM, MTM is more attractive with lower loss and better cost performance.

IX. CONCLUSION

In this paper, new modulation technique for inverter excited induction generator (IG) for DC output wind power generation has been shown. It has a system configuration using PFCC for cost reduction of the voltage source converter (VSC).

The new modulation technique called modified trapezoidal modulation (MTM) has been introduced. It can have a 2/3 non-switching period in the one fundamental cycle. This fact leads to a large reduction of switching loss in the VSC.

The experimental results show that the reduction rate of the converter loss reaches almost 30%. In the AC current from the IG, a certain amount of higher harmonics exists. However, the 6th harmonic torque ripple in the IG can be suppressed by using the modification factor $\gamma = 0.38$. The output power of MTM is larger than that of SM at the same DC voltage range. This fact can reduce the voltage rating of the converter switching device. These facts can reduce the cost of offshore wind farm by cost reduction of the VSC with the MTM.

The proposed system with the power factor corrected converter (PFCC) parallel with the VSC is confirmed that Pvsc can be reduced to zero by controlling the Df of PFCC from simulation and experiment. This leads to the minimization of the VSC rating and the cost of the converter system can be reduced. It was confirmed that the loss of VSC can be reduced and capacity can be reduced by using the proposed system with MTM.

References

[1] A. R. Henderson, C. Morgan, H. C. Sørensen, R. J. Barthelmie and B. Boesmans, "Offshore wind energy in Europe-A review of the state-of-the-art.," Wind energy, vol. 6, no. 1, pp. 35-52, 2003.

[2] K. Søbrink, P. L. Sørensen, E. Joncquel and D. Woodford, "Feasibility study regarding integration of the Læsø Syd 160 MW wind farm using VSC transmission," CIGRE SC14 Colloquium, 2001.

[3] N. Kimura, M. Hirao, T. Morizane, K. Taniguchi, "Wind Power Generation System with Induction Machine and Diode Rectifier", EPE-Power Electronics and Motion Control '06 Conference, Paper No.T12-115, pp.1580-1584, 2006.

[4] H. Yonezawa, K. Taniguchi, T. Morizane and N. Kimura, "Modified Trapezoidal Modulating Signal suitable for PM Synchronous Motor Drives," IEEJ Trans. IA, vol. 125, no. 1, pp. 46-53, 2005.

[5] K. Taniguchi, T. Morizane, "Characteristics of PAM Inverter System for Electric Vehicle," in IEE-Japan Industry Applications Society Conference, 2016.

[6] Powersim Inc: http://powersimtech.com/

[7] T. Masuda, K. Taniguchi, No. Kimura and T. Morizane, "PM Motor Drives by Modified Trapezoidal Modulating Signal," Journal of the Japan Institute of Power Electronics, Vol. 29, No. 2, pp.69-72, 2003.

[8] N. Kimura, S. Kashiwagi, T. Morizane and H. Omori, "Loss Minimization For MPPT Of Converter Excited Induction Generator For Wind Power Generation ", 6th International Power Electronics and Motion Control Conference and Exposition, PEMC, Paper ID-80, 2014.

The 2018 International Power Electronics Conference

Sphere Decoding Based Long-Horizon Predictive Control of Three-level NPC Back-to-Back PMSG Wind Turbine Systems

Ferdinand Grimm * [†], Zhenbin Zhang *[‡] and Ralph Kennel [†]

*Institute for Sustainable Energy and Smart Grid, Shandong University, Jinan, China

[†] Institute for Electrical Drive Systems and Power Electronics, Technische Universitat Munchen, Munich, Germany

[‡]Corresponding author, E-mail: zbz@sdu.edu.cn

Abstract—We propose a computationally efficient control scheme for model predictive control of a three level-neutral point clamped (NPC) inverter permanent magnet synchronous generator (PMSG) back-to-back system. In order to increase the performance a multistep prediction of the system behaviour is applied. Therefore, multistep prediction models for the machine side as well as the grid side are introduced. Since the computational complexity would be growing exponentially with increasing horizon length we apply a sphere decoding algorithm to reduce the computational complexity. The sphere decoding algorithm can only be used for linear systems while the 3L-NPC BTB system is nonlinear. Therefore different simplification methods are studied in this paper.

I. INTRODUCTION

Back-to-back permanent magnet synchronous generator (PMSG) systems with 3 level neutral point clamped (NPC) converter [1] have shown to be a promising technology due to its advantages. Namely the decoupling ability between the network and the generator and the low-voltage ride-through ability showed to be advantageous compared to the Double-Fed-Industrial-Generator.

A promising type of control strategies for renewable energies is discrete control set model predictive control (MPC) [2]. Contrary to continuous control set model predictive control in discrete control set MPC it is not required to use pulse width modulation since each input is applied to the system for the same duration [3]. Discrete control set MPC can again be divided into several subcategories. The most popular discrete control set MPC strategy is predictive current control (PCC). The goal of PCC is to control the stator currents of the machine by solving a linear mixed integer least squares optimization problem during each sampling step [4]. Promising PCC-based schemes that are tailored to the 3 Level-NPC PMSG back-to-back system have been proposed in [5]–[8].

An increased performance of model predictive control schemes can be achieved by using multistep model predictive control [9]. The performance increase comes at a price however, having an exponential increase in computational complexity with increasing horizon. Recently, there have been promising new algorithms for multistep model predictive control in power electronics based on branch-and-bound [10] which aim to exclude certain switching state sequences based on the performance of their prefixes. A special case of

branch-and-bound for linear systems is the sphere decoding algorithm [11], [12].

Modifications of the sphere decoding algorithm that achieves multistep model predictive control for nonlinear converter models were presented e.g. in [13] and [14].

In this paper we apply the sphere decoding algorithm to the 3-level NPC PMSG back-to-back system to obtain a computationally efficient framework that achieves multistep model predictive control. There fore we generalize the results from [13] and [14].

II. SYSTEM MODEL

Figure 1 shows the block diagram of a 3 level-NPC PMSG back-to-back converter. In this section the system equations

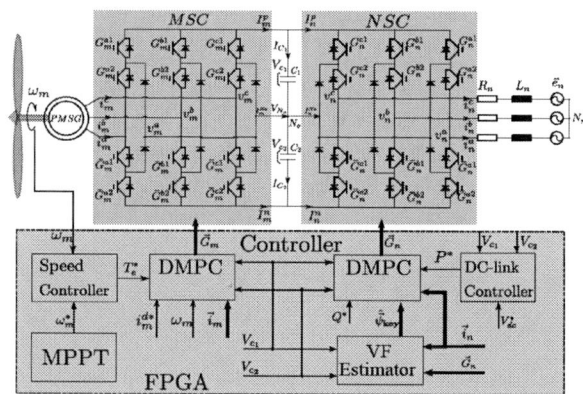

Fig. 1. Block diagram of a 3L-NPC PMSG back-to-back system consisting of the wind turbine, the machine side converter, the DC-link, the grid side converter and the net side RL-filter [15]. The parameters are explained in [15] and table II.

shall be defined starting from the wind turbine.

A. Wind Turbine

The mechanical power $P_w(t)$ which is absorbed by the wind turbine is given by [16], [1]:

$$P_w[k] = \frac{1}{2}\rho A v_w^3[k] C_p[k], \tag{1}$$

2895

where ρ is the wind density, A the turbine cross section, v_w the translational wind speed and C_p the power coefficient of the wind turbine. We assume that $P_w(t)$ is ideally transformed to mechanical power for the generator $P_g(t)$:

$$P_w[k] = P_g[k] = T_m[k]\omega_m[k], \tag{2}$$

where $\omega_m(t)$ is the mechanical generator speed and $T_m(t)$ is the torque. With the help of the theorem of angular momentum it is possible to formulate a relationship between $\omega_m(t)$ and the torque $T_m(t)$

$$J\dot{\omega}_m(t) = T_m(t) - N_p\psi_{pm}i_m^q(t) + (\omega_m)(t). \tag{3}$$

The mechanical generator speed $\omega_m(t)$ is related to the electric speed $\omega_e(t)$ by the number of pole pairs $\omega_e = N_p\omega_m$.

B. Permanent magnet synchronous generator

The current which is produced by the PMSG is given by [15]

$$
\begin{aligned}
\vec{i}_m^{dq}[k+1] &= \begin{bmatrix} 1 - \frac{T_s R_s}{L_s} & T_s\omega_e \\ -T_s\omega_e & 1 - \frac{T_s R_s}{L_s} \end{bmatrix} \vec{i}_m^{dq}[k] + \\
&\quad + \begin{bmatrix} \frac{T_s}{L_s} & 0 \\ 0 & \frac{T_s}{L_s} \end{bmatrix} \vec{u}_m^{dq}[k] - \frac{T_s\psi_{pm}}{L_s}\begin{bmatrix} 0 \\ \omega_e \end{bmatrix} \\
&= \mathbf{A}_m\mathbf{x}_m + \mathbf{B}_m\mathbf{u}_m + \mathbf{h}_m, \tag{4}
\end{aligned}
$$

where $\mathbf{x}_m = \vec{i}_m^{dq}[k]$ is the system state in direct-quadrature coordinates and $\mathbf{u}_m = \vec{S}_m^{abc}[k]$ is the machine side switching state which serves as system input. ω_e is the electric speed which is assumed to be slowly changing and T_s the sampling time. The machine side system output shall be chosen identical to the machine side system state $\mathbf{y}_m = \mathbf{x}_m$.
$\vec{S}_m^{abc}[k]$ is the machine side switching state in abc$-$coordinates. The machine side voltage in $dq-$coordinates $\vec{u}_m^{dq}[k]$ can be transferred to abc$-$coordinates using the Park transform $\mathbf{P}$ and the Clarke transform $\mathbf{T}$:

$$\mathbf{T} = \sqrt{\frac{2}{3}} \cdot \begin{bmatrix} 1 & -\frac{1}{2} & -\frac{1}{2} \\ 0 & \frac{\sqrt{3}}{2} & -\frac{\sqrt{3}}{2} \end{bmatrix}, \mathbf{P} = \begin{bmatrix} \cos(\theta) & \sin(\theta) \\ -\sin(\theta) & \cos(\theta) \end{bmatrix}. \tag{5}$$

θ denotes the electric phase angle. Using (5) yields:

$$\vec{u}_m^{abc}[k] = \mathbf{T}^\dagger\mathbf{P}^T\vec{u}_m^{dq}[k]. \tag{6}$$

The relationship between the switching state $\vec{S}^{abc}$ and the input voltage $\vec{u}_m^{abc}$ is determined by the capacitor voltages V_{c1}, V_{c2}

$$
\begin{aligned}
\mathbf{u}^{abc}[k] &= \frac{V_{dc}[k]}{3}\begin{bmatrix} 2 & -1 & -1 \\ -1 & 2 & -1 \\ -1 & -1 & 2 \end{bmatrix} \vec{S}^{abc}[k] \\
&= \mathbf{T}_l\vec{S}^{abc}[k]. \tag{7}
\end{aligned}
$$

We can then define the machine side input matrix $\mathbf{B}_m = \frac{T_s}{L_s}\mathbf{P}^T\mathbf{T}^\dagger\mathbf{T}_l$. In this paper $V_{dc} \gg V_O$ is assumed such that $\mathbf{u}^{abc}[k]$ can be modeled as independent of V_O.

C. Net-side RL-filter

The mathematical model of the filter and the net (grid) side in stator fixed frame is given by [15]

$$
\begin{aligned}
\vec{i}_n^{\alpha\beta}[k+1] &= (1 - T_s\frac{R_n}{L_n})\vec{i}_n^{\alpha\beta}[k] + \frac{T_s}{L_n}\vec{u}_n^{\alpha\beta}[k] - \frac{T_s}{L_n}\vec{e}_n^{\alpha\beta} \\
&= \mathbf{A}_n\mathbf{x}_n[k] + \mathbf{B}_n\mathbf{u}_n + \mathbf{h}_n. \tag{8}
\end{aligned}
$$

where the grid side system state is defined as $\mathbf{x}_n = \vec{i}_n^{\alpha\beta}[k]$, and the grid side input is the switching state $\mathbf{u}_n = \vec{S}_n^{abc}[k]$. $\vec{e}$ denotes the grid voltage vector. The net side input matrix $\mathbf{B}_n$ is given as $\mathbf{B}_n = \frac{T_s}{L_n}\mathbf{T}^\dagger\mathbf{T}_l$. The output of the net side of the system is the active $P[k]$ and reactive power $Q[k]$ which are defined as follows:

$$
\begin{aligned}
\begin{bmatrix} P[k] \\ Q[k] \end{bmatrix} &= \begin{bmatrix} e_n^\alpha[k] & e_n^\beta[k] \\ e_n^\beta[k] & -e_n^\alpha[k] \end{bmatrix} \vec{i}_n^{\alpha\beta}[k] \tag{9} \\
&= \mathbf{C}_n\vec{i}_n^{\alpha\beta}[k]. \tag{10}
\end{aligned}
$$

D. DC-link

In order to keep the converter working properly it is important to keep the dc-link balance V_O close to zero. Therefore we need to observe the dynamics of V_O [15]:

$$
\begin{aligned}
V_O[k+1] &= V_O[k] + \frac{T_s}{C}|\vec{S}_m^{abc}|^T\vec{i}_m^{abc} - \frac{T_s}{C}|\vec{S}_n^{abc}|^T\vec{i}_n^{abc} \tag{11} \\
&= V_O[k] + \frac{T_s}{C}(|\mathbf{u}_m|^T\mathbf{T}^\dagger\mathbf{P}^T\mathbf{x}_m - |\mathbf{u}_n|^T\mathbf{T}^\dagger\mathbf{x}_n).
\end{aligned}
$$

Since the states $\mathbf{x}_{m,n}$ depend on the neutral point voltage again, equation (12) is nonlinear.
Additionally to the DC-link balance, it is furthermore required to control the DC-link voltage V_{dc}. In this paper the DC-link voltage is controlled with the help of the net-side power. For this reason it is necessary to formulate a reference active power P^{ref} and a reference reactive power Q^{ref}. P^{ref} consists of two parts, see e.g. [15]: First, a reference DC-link current I_g^{ref} is first estimated using a PI-controller [17]: Using I_g^{ref} we can now formulate the reference powers:

$$P^{ref} = V_{dc} \cdot I_g^{ref} + \omega_m^{ref}T_e^{ref}. \tag{12}$$

Moreover, in this paper $Q^{ref} = 0$ is used for the simulations.

III. MULTISTEP PREDICTION MODEL

The multistep prediction model is set up for the machine side and the grid side separately. Since both system equations (8) and (4) have the same structure, we introduce the index $i \in \{m, n\}$ which indicates that the following derivation is similar for the machine side and the net side.

Thus using the input-output relationships (8) and (4) we formulate the output prediction for the grid side as

$$\mathbf{y}_i[k+1] = \mathbf{C}_i\mathbf{A}_i\mathbf{x}_i[k] + \mathbf{C}_i\mathbf{B}_i\mathbf{u}_i[k] + \mathbf{C}_i\mathbf{h}_i, \tag{13}$$

for $i \in \{m, n\}$. For the machine side it further holds $\mathbf{C}_m = \mathbf{I}$ where $\mathbf{I}$ denotes an identity matrix. Moreover a prediction model of the DC-link balance depending on the input is given by

$$V_O[k+1] = V_O \pm \frac{T_s}{C}|\mathbf{u}_i^T[k]|\mathbf{x}_i[k]. \tag{14}$$

Note that the sign of $\pm$ is "+" for the machine side and "$-$" for the net side.

In order to achieve multistep prediction it is necessary to formulate a prediction model over the next N sampling steps. The multistep input, sytem- and output variables $\mathbf{U}_i[k]$, $\mathbf{X}_i[k]$ and $\mathbf{Y}_i[k]$, $\mathcal{V}_O[k]$ are formed by stacking the corresponding single step quantities, $\mathbf{u}_i[k]$, $\mathbf{x}_i[k]$, $\mathbf{y}_i[k]$ and $V_O[k]$ over N prediction steps:

$$\mathbf{X}_i[k] = \begin{bmatrix} \mathbf{x}_i^T[k] & \mathbf{x}_i^T[k+1] & \dots & \mathbf{x}_i^T[k+N] \end{bmatrix}^T . \quad (15)$$

A relationship between the multistep input $\mathbf{U}_i[k]$ and the multistep system state $\mathbf{X}_i[k]$ can now be formulated using the matrices from the appendix:

$$\mathbf{X}_i[k] = \tilde{\mathbf{\Xi}}_i \mathbf{U}_i[k] + \tilde{\mathbf{\Gamma}}_i \mathbf{x}_i[k] + \tilde{\mathbf{H}}_i . \quad (16)$$

The idea of model predictive control is to predict the system output at certain time instants in the future. Therefore it is required to formulate a relationship between the future system output and the input at the current time. Using (32e) the relationship between $\mathbf{U}_i[k]$ and $\mathbf{Y}_i[k]$ for the given system is

$$\mathbf{Y}_i[k+1] = \mathbf{\Theta}_i \left(\mathbf{\Xi}_i \mathbf{U}_i[k] + \mathbf{\Gamma}_i \mathbf{x}_i[k] + \mathbf{H}_i \right) . \quad (17)$$

Another important goal for the control of electrical drives is the limitation of the switching frequency. This can be achieved by minimizing the the control effort $\mathbf{\Delta U} = \mathbf{U}[k] - \mathbf{U}[k-1]$. In [11] it was shown that the multistep control effort can be expressed as

$$\mathbf{\Delta U}_i = \mathbf{S} \mathbf{U}_i[k] - \mathbf{E} \mathbf{u}_i[k-1] . \quad (18)$$

Both matrices $\mathbf{S}$ and $\mathbf{E}$ are given in the appendix and can be found e.g. in [11].

The DC-link voltage prediction $V_O[k+1]$ is depending on the system states $\mathbf{x}_i[k]$ of the previous time step. A relationship between the multistep control input $\mathbf{U}_i[k]$ and the multistep DC-link balance $\mathcal{V}_O$ can be formulated similarly to (17): Using the matrices defined in (34) it is now possible to formulate a relationship between the multistep input and the multistep DC-link balance:

$$\mathcal{V}_O[k+1] = \mathbf{V}_O[k] \pm \frac{T_s}{C} \left(\mathbf{L}_i^{\text{abc}} \mathbf{X}_i (\mathbf{U}_i[k]) \right) \circ |\mathbf{U}_i[k]| . \quad (19)$$

$\mathbf{L}_i^{\text{abc}}$ includes the coordinate transform and is defined as $\mathbf{L}_m^{\text{abc}} = \mathbf{L} \mathbf{T}^{\alpha\beta \mapsto \text{abc}} \mathbf{T}^{dq \mapsto \alpha\beta}$ for the machine side or $\mathbf{L}_n^{\text{abc}} = \mathbf{L} \mathbf{T}^{\alpha\beta \mapsto \text{abc}}$ for the net side. The definitions of $\mathbf{L}$, $\mathbf{T}^{\alpha\beta \mapsto \text{abc}}$ and $\mathbf{T}^{dq \mapsto \alpha\beta}$ are given in the appendix. Again, "+" is associated with the machine side and "$-$" with the grid side.

IV. SOLUTION TO THE OPTIMIZATION PROBLEM

Due to the nonlinearity of (19) the computation of the next input is difficult. In this paper we present two approaches how the nonlinearity can be bypassed.

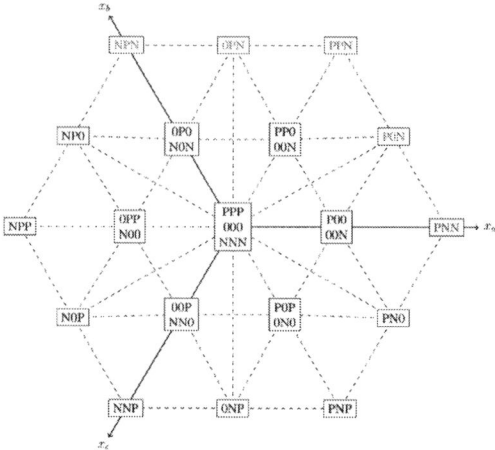

Fig. 2. Applicable converter states, [14]

A. Linearization of the DC-Link Balance

In order to obtain a computationally efficient solution (19) is linearized [14] in both cost functions and a modified sphere decoding algorithm is applied to the linear least squares problem. The linearization point $\mathbf{U}_0$ is selected using a deadbeat control formulation [6] and extrapolated it through the prediction horizon. For example, if the current state is "PNN" a possible linearization point is shown in Figure 2. The advantage of this method is that the quality of the linearization point and thus of the approximation will be independent of the horizon length. With the help of $\mathbf{U}_0$ it is possible to compute the gradient of (19). Since $\mathcal{V}_O$ is not differentiable with respect to $\mathbf{U}_0$ at $\mathbf{U} = 0$, a weak derivative is computed [14]. Thus, the obtained linearization $\mathcal{V}_{\text{lin},i}$ for $\mathcal{V}_O$ can be expressed as

$$\mathcal{V}_{O,c,i} = \mathcal{V}_O(\mathbf{U}_{0,i}) - D_{\mathbf{U}_i} \mathcal{V}_O(\mathbf{U}_{0,i}) \mathbf{U}_{0,i} \quad (20)$$

$$\mathcal{V}_{O,l,i} = D_{\mathbf{U}_i} \mathcal{V}_O(\mathbf{U}_{0,i}) \quad (21)$$

$$\mathcal{V}_{\text{lin},i} = \mathcal{V}_{O,c,i} + \mathcal{V}_{O,l,i} \mathbf{U}_i , \quad (22)$$

where $D_{\mathbf{U}_i}$ denotes the weak derivative with respect to $\mathbf{U}_i$ for $i \in \{m, n\}$. Using the linearized framework, it is possible to define two cost function for the machine side and the net side respectively:

$$J_i = \|\mathbf{Y}_i(\mathbf{U}_i) - \mathbf{Y}_i^{\text{ref}}\|_2^2 + \lambda \|\mathbf{\Delta U}_i\| + \lambda_v \|\mathcal{V}_{\text{lin},i}(\mathbf{U}_i)\|_2^2 . \quad (23)$$

The parameters λ and λ_v are used to set a trade-off between the different control parameters.

B. A priori state selection

Another possible approach suggests that only those states that increase the DC-link balance shall be chosen over the complete horizon length if V_O is below a certain threshold V_C [13]. Similarly, only states that decrease the DC-link balance shall be selected if V_O exceeds V_C. Figure 2 shows the converter states that are applicable to the 3L-NPC converter. An overview over the effect of the selected swithcing state on

V_O can be seen in table I. From the table it is visible that all the

TABLE I
EFFECT OF THE CONVERTER STATES ON V_O

State Effect	States S_i^{abc}		$i \in m, n$		
Increase of V_O	ONN,	OON	ONO	NOO	NON
Decrease of V_O	POO	PPO	POP	OPP	OPO
Unknown effect	PON	PNO	ONP	NOP	NPO	OPN
No effect	PPN	PNN	PNP	NNP	NPN	NNN
	PPP	OOO				

DC-link balance increasing states do not include the position P. On the other hand side, the DC-link balance decreasing states do not include the position N. This information is used to narrow down the number of applicable states. if $|V_O| > V_C$, only those states that change the value of V_O in the desired direction are chosen. In that case either only the switching positions O, N or only the positions O, P will be considered. After that the value for S_i that optimizes the control performance for machine side and grid side respectively:

$$\mathbf{U}_i^{\mathrm{opt}} = \arg\min \|\mathbf{Y}_i(\mathbf{U}_i) - \mathbf{Y}_i^{\mathrm{ref}}\|_2^2 + \lambda\|\varDelta\mathbf{U}_i\|. \quad (24)$$

Equation (24) can for example be solved using a modified sphere decoding algorithm 1 ([11], [12]). The solution $\mathbf{U}_i^{\mathrm{opt}}$ is the optimal switching state that would achieve both, DC-link balancing as well as control performance on grid side and machine side respectively. In case the DC-link current is negative, V_O-increasing states will decrease V_O and vice versa. For this reason it is improtant to furthermore predict the sign of the DC-link current. In the following, $S_i^{\mathrm{abc}+}$ shall be the candidate solution that has only positive and neutral switching states and $S_i^{\mathrm{abc}-}$ the one that only has negative or neutral switching states. As the DC-link balance depends on both, the machine side states as well as the grid side states, we have four possibilities:

$$V_{O1,2,3,4} = V_O + \mathbf{x}_n^{\mathrm{abc}}|S_n^{\mathrm{abc}\pm}| - \mathbf{x}_m^{\mathrm{abc}}|S_m^{\mathrm{abc}\pm}| \quad (25)$$

Note that all four combinations would have the same effect on the control performance. For the next switching state we select the combination of $S_i^{\mathrm{abc}+}$ and $S_i^{\mathrm{abc}-}$ that brings the DC-link balance closest to zero:

$$S_n^{\mathrm{abc}}, S_m^{\mathrm{abc}} = \arg\min \left[\; |V_{O1}| \quad |V_{O2}| \quad |V_{O3}| \quad |V_{O4}| \; \right] \quad (26)$$

The whole framework is summarized in the following:

1) Compute the reference values. $\mathbf{y}_i^{\mathrm{ref}}$.
2) Assemble the multistep prediction model (i.e. $\boldsymbol{\Xi}_i, \boldsymbol{\Gamma}_i,...$).
3) If $|V_O| > V_C$: Set allowed inputs to $\mathcal{S} = \{0, 1\}$.
4) Solve problem (24) with algorithm 1.
5) Post-processing: Solve (26) by exhaustive search.

C. Sphere decoding algorithm

Since each cost function J_i only depends on the corresponding inputs $\mathbf{U}_i$ both cost functions can be minimized in parallel. The goal is to find an input $\mathbf{U}_i^{\mathrm{opt}}$ that minimizes the cost function (23) and furthermore can be produced by the

inverter. Mathematically this is expressed by an optimization problem

$$\mathbf{U}_i^{\mathrm{opt}}[k] = \arg\min J_i(\mathbf{U}_i) \quad \text{s.t.} \quad \mathbf{U}_i \in \mathcal{S}^{3N}, \quad (27)$$

for $i \in \{m, n\}$ and either $\mathcal{S} = \{-1, 0, 1\}$ or $\mathcal{S}$ being determined by a priori state selection. The optimization problem (23) is a mixed integer linear least squares problem and can be solved using a modified sphere decoding algorithm [11]. Therefore, (23) has to be transformed to a triangular structure first:

$$\mathbf{Q}_i = \boldsymbol{\Xi}_i^T\boldsymbol{\Xi}_i + \lambda\mathbf{S}^T\mathbf{S} \pm \lambda_V \mathcal{V}_{O,l,i}^T\mathcal{V}_{O,l,i}, \quad (28)$$

$$\mathbf{Q}_i = \mathbf{H}_i^T\mathbf{H}_i, \quad (29)$$

$$\mathbf{U}_i^{\mathrm{opt}} = \arg\min \|\mathbf{H}\mathbf{U}_{\mathrm{unc},i} - \mathbf{H}_i\mathbf{U}_i\| \quad \text{s.t.} \quad \mathbf{U}_i \in \mathcal{S}^{3N}(30)$$

where $\mathbf{U}_{\mathrm{unc},i}$ is the unconstrained solution of (27). For a priori state selection it holds $\lambda_V = 0$. To solve the triangular system (30) we apply the modified sphere decoding algorithm given in algorithm 1.

Algorithm 1 Sphere decoding algorithm.

1: **function** $\mathbf{U}^{\mathrm{OPT}} = \mathrm{SPDEC}(\tilde{\mathbf{U}}, \mathbf{H}, d^2, k, \rho^2, \check{\mathbf{U}}_{\mathrm{unc}}, \mathcal{U})$
2:     **for** $\mathbf{u} \in \mathcal{U}$ **do**
3:         $\tilde{\mathbf{U}}_k = \mathbf{u}$
4:         $\tilde{d}^2 = \|\check{\mathbf{U}}_{\mathrm{unc}} - \mathbf{H}_{k,1:k}\tilde{\mathbf{U}}_{1:k}\|$
5:         **if** $\tilde{d}^2 \leq \rho^2$ **then**
6:             **if** $k > 1$ **then**
7:                 $\mathbf{U}^{\mathrm{opt}} = \mathrm{SpDec}(\tilde{\mathbf{U}}, \mathbf{H}, \tilde{d}^2, k{+}1, \rho^2, \check{\mathbf{U}}_{\mathrm{unc}}, \mathcal{U})$
8:             **else**
9:                 $\mathbf{U}^{\mathrm{opt}} = \tilde{\mathbf{U}}$
10:                $\rho^2 = \tilde{d}^2$
11:             **end if**
12:         **end if**
13:     **end for**
14:     **return** $\mathbf{U}^*$
15: **end function**

The algorithm is initialized by selecting $k = 1$, $d = 0$, $\mathcal{U} = \mathcal{S}^3$. Furthermore, an initial radius ρ_0 and initial solution which are proposed e.g. in [11] are necessary For the back-to-back PMSG system it is necessary to call the algorithm twice, first for the machine side and parallel to that for the net side:

$$\mathbf{U}_i^{\mathrm{opt}}[k] = \mathrm{SpDec}(\tilde{\mathbf{U}}_i, \mathbf{H}_i, 0, 1, \rho_{0,i}, \mathbf{U}_{\mathrm{unc},i}, \mathcal{S}^3) \quad (31)$$

After that the first three entries of $\mathbf{U}_n^{\mathrm{opt}}[k]$ and $\mathbf{U}_m^{\mathrm{opt}}[k]$ are selected respectively and applied to the machine- and net side of the converter. This procedure is repeated during each sampling interval.

V. SIMULATION RESULTS

For the simulation a 3L NPC back-to-back PMSG system with the parameters given in table II was used: In order to evaluate the performance of the framework, we compare the proposed frameworks with a horizon length $N = 3$ to the performance of a single step prediction with $N = 1$. From

The 2018 International Power Electronics Conference

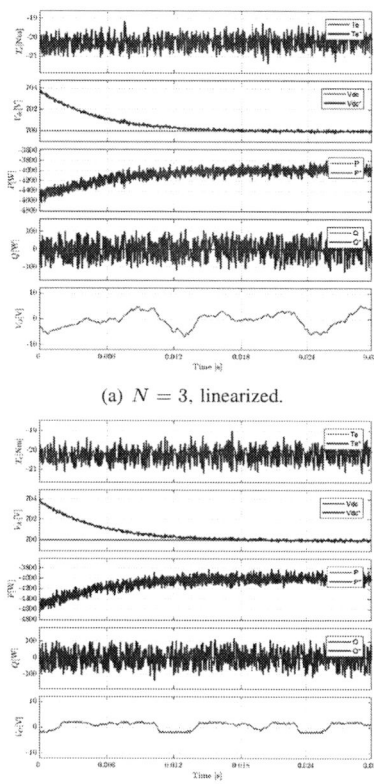

(a) $N = 3$, linearized.

(b) $N = 3$, A Priori selection.

Fig. 3. Torque, power and voltage comparison for long horizon.

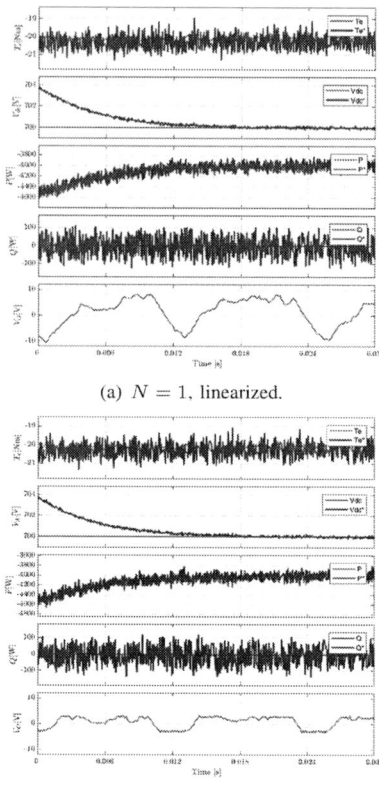

(a) $N = 1$, linearized.

(b) $N = 1$, A Priori selection.

Fig. 4. Torque, power and voltage comparison for short horizon.

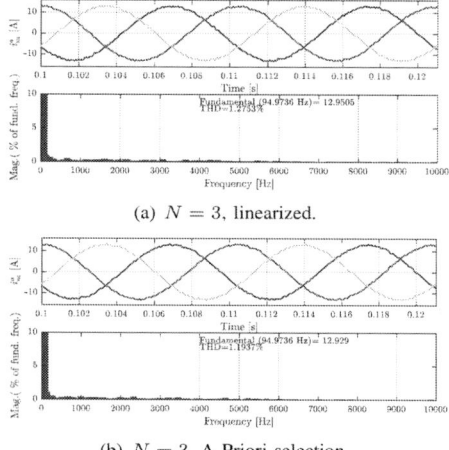

(a) $N = 3$, linearized.

(b) $N = 3$, A Priori selection.

Fig. 5. Current spectrum comparison for long horizon.

TABLE II
MACHINE, GENERATOR AND DC-LINK PARAMETERS

Parameter	Symbol	Simulation value
Sampling Time	T_s	50 µs
Generator/Net side resistance	R_s, R_n	0.1379Ω , 0.156 Ω
Generator/Net side inductance	L_s, L_n	0.019 H, 0.020 H
Rotor permanent magnet flux	ψ_{pm}	0.426 75 Wb
DC-link voltage	V_{dc}	700 V
DC-link capacitance	C	1100 µF
Grid side voltage peak	$\bar{e}_n^{abc}$	250 V
Grid side voltage frequency	ω_n	100π Hz
Generator pole pairs	N_p	3
Weighting factors	λ, λ_V	0.1, 0.02

figure 4 it can be seen that under the same conditions, a priori state selection has a distinctively better DC-link balancing performance. The other parameters such as active and reactive power, torque and DC-link voltage have approximately equal performance. The current spectrum in figure 6 show that the total harmonic distortion (thd) is almost equal for both methods with a slight edge for a priori state selection. Comparing both methods for the long prediction horizon, it is visible that compared to the short prediction horizon the same tendency holds. Figure 3 shows that the DC-link balancing is visibly better for A priori selection compared to linearization and the same holds for the current spectrum as seen in figure 5.

VI. CONCLUSION

Since the systems is nonlinear and the sphere decoding algorithm requires a linear system, it is necessary to bypass the nonlinearity of the system first. Therefore two frameworks were proposed. The first one linearized the nonlinear equation of the DC-link balance while the second one reduces the allowed input states to influence the nonlinearity. The simulation results showed a significant improvement for the long horizon

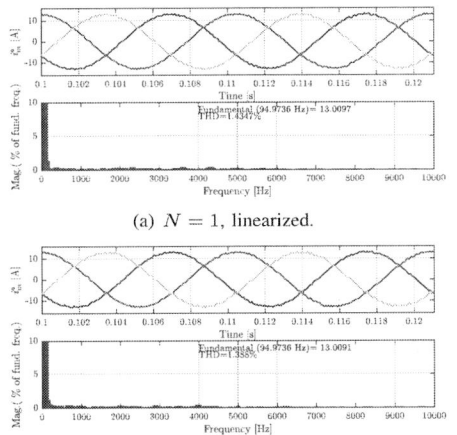

(a) $N = 1$, linearized.

(b) $N = 1$, A Priori selection.

Fig. 6. Current spectrum comparison for short horizon.

performance.

APPENDIX

The matrices given in equation (17) and (18) are defined as:

$$\Xi_i = \begin{bmatrix} \mathbf{B}_i & \mathbf{0} & \dots & \mathbf{0} \\ \mathbf{A}_i\mathbf{B}_i & \mathbf{B}_i & \dots & \mathbf{0} \\ \vdots & \vdots & \ddots & \vdots \\ \mathbf{A}_i^{N-1}\mathbf{B}_i & \mathbf{A}_i^{N-2}\mathbf{B}_i & \dots & \mathbf{B}_i \end{bmatrix}, \quad (32a)$$

$$\Gamma_i = \begin{bmatrix} \mathbf{A}_i^T & \left(\mathbf{A}_i^2\right)^T & \dots & \left(\mathbf{A}_i^N\right)^T \end{bmatrix}^T, \quad (32b)$$

$$\mathbf{H}_i = \begin{bmatrix} \left(\mathbf{h}_i\right)^T & \left(\mathbf{A}_i\mathbf{h}_i\right)^T & \dots & \left(\mathbf{A}_i^{N-1}\mathbf{h}_i\right)^T \end{bmatrix}^T \quad (32c)$$

$$\Theta_i = \mathbf{I} \otimes \mathbf{C}_i, \quad (32d)$$

$$\mathbf{S} = \begin{bmatrix} \mathbf{I} & \mathbf{0} & \dots & \mathbf{0} \\ -\mathbf{I} & \mathbf{I} & \dots & \mathbf{0} \\ \vdots & \vdots & \ddots & \vdots \\ \mathbf{0} & \mathbf{0} & \dots & -\mathbf{I} \end{bmatrix}, \mathbf{E} = \begin{bmatrix} \mathbf{I} \\ \mathbf{0} \\ \vdots \\ \mathbf{0} \end{bmatrix}. \quad (32e)$$

$\otimes$ denotes the Kronecker product in this case.
The time shifted system matrices from (16) are e.g.:

$$\tilde{\Xi}_i^T = \begin{bmatrix} \mathbf{0} & \Xi_{i_{1:2N-2,:}}^T \end{bmatrix}, \dots \quad (33)$$

The matrices for the multistep DC-link balance from (19) are

$$\mathbf{L} = \begin{bmatrix} 1 & 0 & 0 & \dots & 0 \\ 1 & 1 & 0 & \dots & 0 \\ \vdots & \vdots & \vdots & \ddots & \vdots \\ 1 & 1 & 1 & \dots & 1 \end{bmatrix} \otimes \begin{bmatrix} 1 \\ 1 \\ 1 \end{bmatrix}^T, \quad (34a)$$

$$\mathbf{T}^{\alpha\beta\mapsto abc} = \mathbf{I} \otimes \mathbf{T}^\dagger, \quad \mathbf{T}^{dq\mapsto\alpha\beta} = \mathbf{I} \otimes \mathbf{P}^T, \quad (34b)$$

$$\mathbf{V}_O[k] = V_O[k] \cdot \begin{bmatrix} 1 & 1 & \dots & 1 \end{bmatrix}^T. \quad (34c)$$

An analytic expression for the derivative is (compare [14]):

$$D_{\mathbf{U}_i}\mathcal{V}_O = \mathbf{L} \cdot \operatorname{diag}\left(\tilde{\Gamma}^{abc} \circ \operatorname{sign}(\mathbf{U})\right) +$$

$$+ \mathbf{L} \cdot \left(\tilde{\Xi}^{abc} \cdot \operatorname{diag}(|\mathbf{U}|)\right)^T +$$

$$+ \mathbf{L} \cdot \operatorname{diag}\left(\tilde{\Xi}^{abc}\mathbf{U} \circ \operatorname{sign}(\mathbf{U})\right). \quad (35)$$

In this paper $\circ$ denotes the elementwise product.

ACKNOWLEDGEMENT

This research was financially supported by Qilu Young Scholar program of Shandong University

REFERENCES

[1] D. Schröder and D. Schröder, "Modellierung und regelung von modernen windkraftanlagen: Eine einführung," *Elektrische Antriebe-Regelung von Antriebssystemen*, pp. 1540–1614, 2015.

[2] M. Morari and J. H. Lee, "Model predictive control: past, present and future," *Computers and Chemical Engineering*, vol. 23, no. 45, pp. 667 – 682, 1999.

[3] P. Cortes, M. Kazmierkowski, R. Kennel, D. Quevedo, and J. Rodriguez, "Predictive control in power electronics and drives," *IEEE Transactions on Industrial Electronics*, vol. 55, no. 12, pp. 4312–4324, Dec 2008.

[4] J. Rodriguez, J. Pontt, C. A. Silva, P. Correa, P. Lezana, P. Cortes, and U. Ammann, "Predictive current control of a voltage source inverter," *IEEE Transactions on Industrial Electronics*, vol. 54, no. 1, pp. 495–503, Feb 2007.

[5] Z. Zhang and R. Kennel, "Fpga based direct model predictive power and current control of 3l npc active front ends," in *PCIM Europe 2016; International Exhibition and Conference for Power Electronics, Intelligent Motion, Renewable Energy and Energy Management; Proceedings of.* VDE, 2016, pp. 1–8.

[6] Z. Zhang, X. Cai, R. Kennel, and F. Wang, "Fully fpga based predictive control of back-to-back power converter pmsg wind turbine systems with space vector modulator," in *2016 IEEE 8th International Power Electronics and Motion Control Conference (IPEMC-ECCE Asia)*, May 2016.

[7] J.-Z. Zhang, T. Sun, F. Wang, J. Rodríguez, and R. Kennel, "A computationally efficient quasi-centralized dmpc for back-to-back converter pmsg wind turbine systems without dc-link tracking errors," *IEEE Transactions on Industrial Electronics*, vol. 63, no. 10, pp. 6160–6171, 2016.

[8] Z. Zhang, C. M. Hackl, and R. Kennel, "Computationally efficient dmpc for three-level npc back-to-back converters in wind turbine systems with pmsg," *IEEE Transactions on Power Electronics*, vol. 32, no. 10, pp. 8018–8034, 2017.

[9] P. Karamanakos, T. Geyer, N. Oikonomou, F. Kieferndorf, and S. Manias, "Direct model predictive control - a review of strategies that achieve long prediction intervals for power electronics," *IEEE Industrial Electronics Magazine*, pp. 32–43, March 2014.

[10] T. Geyer, G. Papafotiou, and M. Morari, "Model predictive direct torque control; part i: Concept, algorithm, and analysis," *IEEE Transactions on Industrial Electronics*, vol. 56, no. 6, pp. 1894–1905, June 2009.

[11] T. Geyer and D. E. Quevedo, "Multistep direct model predictive control for power electronics; part 1: Algorithm," in *2013 IEEE Energy Conversion Congress and Exposition*, Sept 2013, pp. 1154–1161.

[12] ——, "Multistep direct model predictive control for power electronics; part 2: Analysis," in *2013 IEEE Energy Conversion Congress and Exposition*, Sept 2013, pp. 1162–1169.

[13] F. Grimm, Z. Zhang, and R. Kennel, "Computationally efficient predictive control of three-level NPC converters with DC-link voltage balancing: A priori state selection approach," in *2017 IEEE International Symposium on Predictive Control of Electrical Drives and Power Electronics (PRECEDE)*, Sep 2017.

[14] F. Grimm, Z. Zhang, F. Wang, and R. Kennel, "Multistep predictive control of three-level npc converters using weak derivative linearization," in *2017 Chinese Automation Congress (CAC)*, Oct 2017, pp. 4672–4677.

[15] Z. Zhang, "On control of grid-tied back-to-back power converters and permanent magnet synchronous generator wind turbine systems," Dissertation, Technische Universitt Mnchen, Mnchen, 2016.

[16] E. Yang and G. Chen, "Research on direct-drive wind generation power converter control without pmsg parameters," in *IECON 2007 - 33rd Annual Conference of the IEEE Industrial Electronics Society*, Nov 2007, pp. 2087–2091.

[17] Z. Zhang, X. Cai, R. Kennel, and F. Wang, "Fully fpga based predictive control of back-to-back power converter pmsg wind turbine systems with space vector modulator," in *2016 IEEE 8th International Power Electronics and Motion Control Conference (IPEMC-ECCE Asia)*, May 2016, pp. 1468–1473.

Based on PCHD and HPSO sliding mode control of D-PMSG wind power system

Lijun Hou, Xuemei Zheng*, Chao Wang, Yangman Li, Haoyu Li
Electrical Engineering Department, Harbin Institute of Technology, Harbin, China
*E-mail: xmzheng@hit.edu.cn

Abstract— In order to simplify the model of Direct-drive Permanent Magnet Synchronous Generator (D-PMSG) wind power system and improve the robustness, the variable Speed Constant Frequency (VSCF) D-PMSG wind power system is studied from the perspective of energy and robustness. Firstly, the Port Control Dissipative Hamiltonian (PCHD) model is established for D-PMSG, which simplifies the design of controller. Then, a global High-order Non-singular Terminal Sliding Mode (HNTSM) controller is designed to realize the Maximum Power Point Tracking (MPPT) below the rated wind speed, and it has much faster response speed and better robustness compared with PI controller. However, in the parameters design of PCHD controller, there is no specific selection criterion, so it is difficult for the parameters selected by experience to achieve the ideal effect. Therefore, the hybrid particle swarm optimization algorithm (HPSO) is used to optimize the parameters of PCHD controller, so that the control effect is faster and more accurate. Simulations validate the proposed control.

Keywords— HPSO, global HNTSM, PCHD, D-PMSG

I. INTRODUCTION

Wind power as a clean and reliable renewable energy has been rapid development in recent years. Direct-driven Permanent Magnet Synchronous Generator (D-PMSG) is a typical nonlinear, multivariable, strongly coupled system. As the traditional vector control method is sensitive to system parameters, uncertainties in wind speed changes can even lead to system instability. At present, the Passivity-Based Control (PBC) method [1], showing superior control performance. It does not depend on the precise linearization of the object model and can simplify the controller design effectively and improve the robustness of the system. In paper [2], the Port Control Dissipative Hamiltonian （PCHD）is applied to the stator current optimum control of the D-PMSG wind power system, and the control effect is achieved. However, for PCHD controller, parameters selection is the key factor, and the normal experienced selected parameters are difficult to achieve optimal control effect.

Particle Swarm Optimization [3] (PSO) is a new intelligent algorithm for computing intelligence. It has the characteristics of simple algorithm and strong searching ability to find the best value. In the paper [4] and [5], the PSO algorithm is applied to the PID and the more complex fractional PID parameters design. In paper [6], the PSO algorithm is applied to the parameter design of the Sliding Mode Controller (SMC), and achieved the

desired effect. However, when PSO is to optimize complex objects, it will fall into the local optimal region easily and influence the speed of convergence. In paper [7], combining natural selection mechanism with the PSO algorithm, the HPSO algorithm is proposed. HPSO algorithm can balances the global and local search ability dynamically and overcome the shortcoming of falling into local optimum.

This paper combines the PCHD method with global High-order Non-singular Terminal Sliding Mode (HNTSM) control and applies it to D-PMSG wind power system to realize the maximum power point tracking (MPPT) control under the rated wind speed. In addition, the HPSO algorithm is applied to optimize the PCHD controller parameters to improve controller performance effectively. Finally, the simulation model is built in MATLAB/SIMULINK, and the comparing optimum results are shown in the simulation. The simulation results are verified that the proposed optimum PCHD SMC had the excellent dynamic and static performance.

II. THE MODEL OF D-PMSG

A typical D-PMSG wind turbine system is shown in Fig. 1. Due to the back-to-back dual PWM converter topology, the Machine-side Converter (MSC) and the Grid-side Converter (GSC) can be controlled independently. In this paper, the control strategy of the machine-side is mainly studied, and the MSC is used to realize the control of D-PMSG.

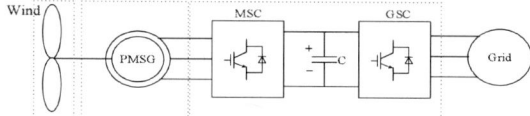

Fig. 1. A direct-drive D-PMSG wind turbine system

A. Wind energy conversion model

According to the aerodynamics principle of the wind turbine, the mechanical torque T_ω of the wind turbine can be expressed as [8]

$$T_\omega = \frac{1}{2}\rho\pi R^2 v_{\text{wind}}^3 C_p(\beta,\lambda)/\lambda \qquad (1)$$

where ρ is the air density, usually $\rho = 1.25 kg/m^3$; R is the blade radius of the wind turbine; v_{wind} is wind speed; $C_p(\beta,\lambda)$ is wind energy utilization factor; β is the pitch angle of the blade of the wind turbine; λ is the tip speed ratio, the expression is shown as followed:

$$\lambda = \frac{R\omega_v}{v_{\text{wind}}} = \frac{\pi R n_v}{30 v_{\text{wind}}} \quad (2)$$

where ω_v is the angular velocity of the blade rotation of the wind turbine; v_{wind} is wind speed; n_v is the speed of the blade of the wind turbine and satisfied $n_v = 30\omega_v / \pi$.

According to experience, the expression of $C_p(\beta, \lambda)$ is:

$$C_p(\beta, \lambda) = 0.5176 \cdot (116 \cdot (\frac{1}{\lambda_i}) - 0.4\beta - 5) \cdot e^{-21(\frac{1}{\lambda_i})} + 0.0068 \cdot \lambda \quad (3)$$

where $\dfrac{1}{\lambda_i} = \dfrac{1}{\lambda + 0.08\beta} - \dfrac{0.035}{\beta^3 + 1}$.

Below the rated wind speed, in order to make full use of wind energy, β should be $0°$. Therefore, there is always an optimum blade tip speed ratio λ_{opt} to make the $C_p(\beta, \lambda)$ maximum, which makes the system achieve MPPT.

B. Port Controlled Hamiltonian Model of D-PMSG system

The form of a PCHD system can generally be expressed as[9]

$$\begin{cases} \dot{x} = [J(x) - R(x)]\dfrac{\partial H(x)}{\partial x} + g(x)u \\ y = g^T(x)\dfrac{\partial H(x)}{\partial x} \end{cases} \quad (4)$$

where the Hamiltonian function $H(x)$ represents the total energy of the system; the interconnected matrix $J(x)$ is an anti-symmetric matrix, which is used to represent the interconnected structure inside the system; the damping matrix $R(x)$ and the input matrix $g(x)$ are both positive semi-definite symmetric matrices which reflect the additional resistive structure of the port and the port input characteristics of the system respectively.

If the friction coefficient of the system is ignored, the mathematical model of D-PMSG in the d-q rotating coordinate system can be expressed as

$$\begin{cases} L_d \dfrac{di_d}{dt} = -R_s i_d + p_n \omega L_q i_q + u_d \\ L_q \dfrac{di_q}{dt} = -R_s i_q - p_n \omega L_d i_d - p_n \omega \psi_f + u_q \\ J \dfrac{d\omega}{dt} = T_e - T_\omega = p_n[\psi_f i_q + (L_d - L_q)i_d i_q] - T_\omega \end{cases} \quad (5)$$

where ψ_d, ψ_q are d-q axis component of permanent magnet motor flux; L_d, L_q are equivalent inductance of stator three-phase winding in d-q coordinate system; i_d, i_q are stator current d-q axis component of motor; p_n is motor pole logarithm; u_d, u_q are stator voltage d-q axis component of motor; J is equivalent value of moment of inertia of the system; ω is mechanical angular velocity of motor rotor; T_ω is load torque; ψ_f is flux chain of permanent magnet of motor.

In the D-PMSG wind power system (5), through the analysis of the energy transfer structure of the machine-side port of the wind power system, the system (5) can be transformed into the PCHD system model shown by Eq.(4). In this paper, the state vector x, the input vector u and the output vector y are selected as follows:

$$x = \begin{bmatrix} x_1 \\ x_2 \\ x_3 \end{bmatrix} = \begin{bmatrix} L_d i_d \\ L_q i_q \\ J\omega \end{bmatrix} = \begin{bmatrix} L_d & 0 & 0 \\ 0 & L_q & 0 \\ 0 & 0 & J \end{bmatrix} \begin{bmatrix} i_d \\ i_q \\ \omega \end{bmatrix} = D \begin{bmatrix} i_d \\ i_q \\ \omega \end{bmatrix} \quad (6)$$

$$u = \begin{bmatrix} u_d & u_q & -T_\omega \end{bmatrix}^T \quad (7)$$

$$y = \begin{bmatrix} i_d & i_q & \omega \end{bmatrix}^T \quad (8)$$

The Hamiltonian energy function of the system is chosen as the sum of electrical energy and mechanical kinetic energy as followed:

$$H(x) = \frac{1}{2}(\frac{1}{L_d}x_1^2 + \frac{1}{L_q}x_2^2 + \frac{1}{J}x_3^2) \quad (9)$$

After transforming through Eq. (6)~(9), D-PMSG system (5) is rewritten in form of PCHD system (4), then we can get the PCHD model of system (5) as:

$$\begin{cases} \begin{bmatrix} \dot{x}_1 \\ \dot{x}_2 \\ \dot{x}_3 \end{bmatrix} = [J(x) - R(x)] \begin{bmatrix} i_d \\ i_q \\ \omega \end{bmatrix} + g(x) \begin{bmatrix} u_d \\ u_q \\ -T_\omega \end{bmatrix} \\ y = g^T \dfrac{\partial H(x)}{\partial x} = [i_d \quad i_q \quad \omega]^T \end{cases} \quad (10)$$

where interconnect matrix is

$$J(x) = \begin{bmatrix} 0 & 0 & p_n x_2 \\ 0 & 0 & -p_n(x_1 + \psi_f) \\ -p_n x_2 & p_n(x_1 + \psi_f) & 0 \end{bmatrix}; \text{ damping matrix}$$

is $R(x) = \begin{bmatrix} R_s & 0 & 0 \\ 0 & R_s & 0 \\ 0 & 0 & 0 \end{bmatrix}$; R_s is stator three-phase winding resistance; input matrix is $g(x) = I_{3\times3}$;

III. CONTROLLER DESIGN OF MACHINE-SIDE CONVERTER

The whole control scheme block is shown in Fig.2, the MPPT control is applied in order to capture more energy through controlling the speed of the wind turbine. The outer loop of optimum speed regulation is carried out by the global HNTSM control, while the inner loop is designed to a PCHD system, and the parameters in the PCHD are optimized by HPSO algorithm.

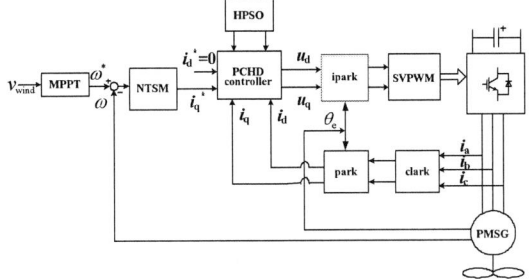

Fig. 2. The control block of the whole system

The control objection is to achieve the MPPT control under the rated-speed, combined with $i_d=0$ vector control method to configure the balance point of the system states and PCHD loop is to guarantee the power is minimum and be stable automatically.

A. Control of Port-controlled Dissipative Hamiltonian

First, let the balance point of the system state are:

$$x_0 = \begin{bmatrix} L_d i_{d0} & L_q i_{q0} & J\omega_0 \end{bmatrix}^T \tag{11}$$

where $i_{d0}=0$ is the expected value of the d-axis current, i_{q0} is the expected value of the q-axis current obtained by the output of the speed outer loop controller, ω_0 is the optimal speed at the corresponding wind speed.

The overall design idea of the PCHD controller is to construct a feedback control $u = \alpha(x)$ to converge the system at the desired balance point x_0.

Now, define:

$$H_d(x) = H(x) + H_a(x) \tag{12}$$

where $H_a(x)$ is the pending energy function in the closed-loop system, reflecting the feedback control into the energy of the closed-loop system.

Through the feedback control $u = \alpha(x)$, the system (4) will form a closed-loop system:

$$\dot{x} = [J_d(x) - R_d(x)]\frac{\partial H_d(x)}{\partial x} \tag{13}$$

where $J_d(x)$ is the desired interconnect matrix, $R_d(x)$ is the desired damping matrix, and satisfied $J_d(x) = J(x) + J_a(x)$, $R_d(x) = R(x) + R_a(x)$.

The expected Hamiltonian energy function of permanent magnet wind power system is designed as:

$$H_d(x) = \frac{1}{2}(x-x_0)^T D^{-1}(x-x_0) \tag{14}$$

The interconnection and damping matrices selected as following:

$$J_a(x) = \begin{bmatrix} 0 & -J_{12} & J_{13} \\ J_{12} & 0 & -J_{23} \\ -J_{13} & J_{23} & 0 \end{bmatrix}, R_a(x) = \begin{bmatrix} r_1 & 0 & 0 \\ 0 & r_2 & 0 \\ 0 & 0 & 0 \end{bmatrix} \tag{15}$$

where J_{12}, J_{13}, J_{23} are the pending matrix of the interconnected matrix; r_1, r_2 are the undetermined parameter of the damping matrix.

After injecting the energy function (14), the controller based on PCHD algorithm is obtained as follows:

$$u_d = -r_1 i_d - J_{12} i_q + J_{13}\omega + (R_s + r_1)i_{d0} + J_{12}i_{q0} \\ - (J_{13} + p_n i_q)\omega_0 \tag{16}$$

$$u_q = J_{12}i_d - r_2 i_q - J_{23}\omega - J_{12}i_{d0} + (R_s + r_2)i_{q0} \\ + (J_{23} + p_n(i_d + \psi_f))\omega_0 \tag{17}$$

$$-T_\omega = -J_{13}i_d + J_{23}i_q + (J_{13} + p_n x_2)i_{d0} \\ - (J_{23} + p_n(i_d + \psi_f))i_{q0} \tag{18}$$

If the system is in equilibrium, means:

$$T_e = T_\omega \tag{19}$$

Substituting (19) into (18), we can get:

$$J_{13} = -\frac{L_d}{L_q}p_n x_{20} \tag{20}$$

$$J_{23} = -p_n x_{10} \tag{21}$$

$$J_{12} = 0 \tag{22}$$

Substituting equations (20)~ (22) into equations (16) and (17), the controller of the PWM rectifier is further obtained as follows:

$$\begin{cases} u_d = -r_1 i_d - L_d p_n i_q \omega - p_n i_q L_q \omega_0 + L_d p_n i_q \omega_0 \\ u_q = -r_2 i_q + L_d p_n i_d \omega + (R_s + r_2)i_{q0} + p_n \psi_f \omega_0 \end{cases} \tag{23}$$

where r_1, r_2 are the key parameters that affects the performance of controller, so they are necessary to be optimized further.

B. Control of Global HNTSM

In order to guarantee the whole system robustness to disturbances, the global HNTSM control is used in the outer loop, and q axis current i_q^* is the input signal of the inner current loop. The mechanical angular velocity ω is to track the given signal ω^* which is obtained by MPPT control.

The paper selected the following speed error:

$$e_\omega = \omega^* - \omega \tag{24}$$

$$\dot{e}_\omega = \dot{\omega}^* - \dot{\omega} = \dot{\omega}^* - \frac{p_n \psi_f i_q - T_\omega}{J} \tag{25}$$

The global HNTSM surface design is as follows:

$$s_\omega = e_\omega + \gamma_1 \dot{e}_\omega^{p_1/q_1} - \lambda_1 \tag{26}$$

where λ_1 is global slip factor, $\lambda_1 = (e_{\omega(0)} + \gamma_1 \dot{e}_{\omega(0)}^{p_1/q_1})e^{-t/\tau_1}$, $e_{\omega(0)}$ and $\dot{e}_{\omega(0)}$ are the initial values of e_ω and $\dot{e}_\omega$, τ_1 is the time constant; γ_1, p_1, q_1 are design parameters, and satisfied $\gamma_1>0$, p_1, q_1 are odd, satisfied $1<p_1/q_1<2$.

Theorem 1: For the system (5), if we selected global HNTSM surface s_ω is selected as (26), speed error e_ω as (24), and designed the control as followed, then the whole D-PMSG system can be guaranteed to be stable :

$$i_q^* = i_{qeq} + i_{qn} \tag{27}$$

$$i_{qeq} = \frac{J}{p_n \psi_f}(\dot{\omega}^* + \frac{T_\omega}{J}) \tag{28}$$

$$i_{qn} = \frac{J}{p_n \psi_f}\int_0^t [\frac{q_1}{\gamma_1 p_1}\dot{e}_\omega^{1-\frac{p_1}{q_1}}(\dot{e}_\omega - \dot{\lambda}_1) + \cdots \\ (k_{10} + \eta_{10})\text{sgn}(s_\omega) + \eta_{11}s_\omega]d\tau \tag{29}$$

where $k_{10} > |\dot{T}_\omega / J| > 0$, $\eta_{10} > 0$, $\eta_{11} > 0$ are the designed constants.

Proof:

Selected the following Lyapunov function:

$$V_\omega(t) = 0.5s_\omega^2 \tag{30}$$

Derivative Eq. (30):

$$\dot{V}_\omega(t) = s_\omega(t)\dot{s}_\omega(t)$$

$$= s_\omega\left[\dot{e}_\omega + \frac{\gamma_1 p_1}{q_1}\dot{e}_\omega^{\frac{p_1}{q_1}-1}\ddot{e}_\omega - \dot{\lambda}_1\right]$$

$$= \frac{\gamma_1 p_1}{q_1}\dot{e}_\omega^{\frac{p_1}{q_1}-1}\left[\left(-k_{10}|s_\omega| + \frac{s_\omega \dot{T}_\omega}{J}\right) - \left(\eta_{10}|s_\omega| + \eta_{11}s_\omega^2\right)\right]$$

$$\leq -\frac{\gamma_1 p_1}{q_1} \dot{e}_\omega^{\frac{p_1}{q_1}-1} \left(\eta_{10} \left| s_\omega \right| + \eta_{11} s_\omega^2 \right) \leq 0 \qquad (31)$$

The proof is ended. That means the global sliding surface s_ω, e_ω will converges to zero within finite-time along $s_\omega = 0$, which can guarantee the speed tracking.

IV. PARAMETERS OPTIMIZATION OF PCHD CONTROLLER BASED ON HPSO ALGORITHM

Since the parameters r_1, r_2 involved in Eq. (23) have a direct impact on the performance of the control system. In order to get a more perfect control strategy, this paper uses HPSO algorithm to design the control parameters.

In the basic PSO algorithm, N particles search for the optimal solution in D-dimensional search space according to certain rules. Each particle has its own position x_i, velocity v_i and fitness value f_i. Among them, the position represents the possible solution of the optimization problem, which is composed of D optimization variables, $x_i = \begin{bmatrix} x_{i1} & x_{i2} & \cdots & x_{iD} \end{bmatrix}$; the velocity represents the distance that the particle needs to move in each iteration, $v_i = \begin{bmatrix} v_{i1} & v_{i2} & \cdots & v_{iD} \end{bmatrix}$; the fitness value is used to measure the current position of the particle ith and is defined according to the specific optimization objective. The own best position of the ith particle (pbest) so far is $P_i = \begin{bmatrix} P_{i1} & P_{i2} & \cdots & P_{iD} \end{bmatrix}$, and the global best position (gbest) of the whole particle swarm is $G = \begin{bmatrix} G_1 & G_2 & \cdots & G_D \end{bmatrix}$.

The d dimensional velocity and position of the ith particle in the global PSO model are updated as follows:

$$v_{id}^{k+1} = w v_{id}^k + \zeta_1 c_1 \left(p_{id}^k - x_{id}^k \right) + \zeta_2 c_2 \left(p_{gd}^k - x_{id}^k \right) \quad (32)$$

$$x_{id}^{k+1} = x_{id}^k + v_{id}^{k+1} \qquad (33)$$

where k is the current iteration number, w is the inertia weight, representing the trade-off between the global and local search ability of the population; c_1 and c_2 are the weighting factors known as the cognitive learning parameter and the social learning parameter, ζ_1 and ζ_2 are two uniformly distributed random numbers from the interval [0,1].

In order to balance the global search ability and the local improvement ability of the PSO algorithm, the nonlinear dynamic inertial weight coefficient formula can also be used. The expressions are as follows[10]:

$$w = \begin{cases} w_{\min} - \dfrac{(w_{\max} - w_{\min}) * (f - f_{\min})}{(f_{avg} - f_{\min})}, f \leq f_{avg} \\ w_{\max} \qquad\qquad\qquad\qquad , f > f_{avg} \end{cases} \quad (34)$$

where $w_{\max}$ and $w_{\min}$ are the maximum and minimum values of w respectively, f is the current objective function value of the particle, f_{avg} and $f_{\min}$ represent the average and minimum target value of all current particles respectively. In the upper formula, the inertial weight w changes automatically with the object function of the particle, so it is called adaptive weight.

By combining the natural selection mechanism with the PSO algorithm, the HPSO algorithm is proposed. The basic idea is to sort the whole particle swarm according to the fitness value in each iteration process, replacing the worst half velocity and position with the velocity and position of the best half of the particles, while preserving the historical optimal value of each individual's original memory.

In this paper, due to r_1 and r_2 these two parameters need to be optimized, particles move randomly over a 2-dimensional search space in order to optimize an objective function as (35), which is used as the criterion of fitness of each particle.

$$f = \int_0^\infty k_3 t \left| e_3(t) \right| + k_2 t \left| e_2(t) \right| + k_1 t \left| e_1(t) \right| dt \quad (35)$$

where $e_1 = i_{d0} - i_d$, $e_2 = i_{q0} - i_q$, $e_3 = \omega^* - \omega$, $k_j (j=1,2,3)$ is the weight, you can choose the value according to the different performance indicators, and satisfied $\sum_{j=1}^n k_j = 1$.

The basic steps of HPSO are as follows:
1) The position and velocity of each particle in the population are initialized randomly;
2) The fitness of each particle is evaluated, the current position and fitness of each particle is stored in the particle's pbest, and the position and fitness of the individual with the best fit in pbest are stored in the gbest;
3) Update the velocity and position of each particle according to the formula (32) and (33);
4) Update weight w according to the formula (34);
5) For each particle, compare its fitness values calculated according to formula (35) with the best values it has experienced and, if it is better, use it as its current pbest;
6) Compare all current pbest and gbest values and update gbest;
7) Ranking the whole particle swarm according to fitness value, replacing the worst half position and velocity with the velocity and position of the best half particle in the population, keeping the pbest and gbest unchanged;
8) If the stop condition is met (usually the preset position accuracy or number of iterations), the search is stopped and the result is output. Otherwise, return to step 3) to continue the search.

V. SIMULATIONS

Based on the proposed method, the paper carried out the simulation of the maximum wind energy tracking system in MATLAB / SINULINK. In the paper, the wind speed is lower than the rated wind speed stage, using the fixed pitch control, so that $\beta = 0°$, if the corresponding maximum wind energy utilization coefficient $C_{pmax} = 0.411$, then the corresponding optimal tip speed ratio $\lambda_{opt} = 8.1$, The desired optimal angular speed is the control target $\omega^* = \lambda_{opt}/R = 8.1v/R$, and the maximum wind energy tracking is achieved by controlling the ω to track ω^*. Assuming the air density is 1.25 kg/m³, take the wind turbine blade radius of 5m, D-PMSG parameters were

$P_N = 1k\text{W}$, $p_n = 6$, $L_d = L_q = 33\text{mH}$, $\psi_f = 0.8\text{Wb}$, $R_s = 2.872\Omega$, $B = 0N \cdot m \cdot s$, $J = 0.011kg \cdot m^2$.

The design parameters of global HNTSM controller are: $\gamma_1 = 0.01$, $p_1 = 5$, $q_1 = 3$, $\tau_1 = 0.016$, $k_{10} + \eta_{10} = 10^5$, $\eta_{11} = 20$.

For the current loop PCHD controller design parameters: r_1, r_2 are optimized by HPSO, r_1, r_2 are in the range of [0, 500], group number of particles $N = 50$, number of iterations is 100, $w_{max} = 1.3$, $w_{min} = 0.4$, $c_1 = c_2 = 2.05$, $k_1 = 0.25$, $k_2 = 0.25$ $k_3 = 0.5$. After iterative calculation of simulation, the curve of fitness and parameters can be obtained as shown in Fig. 3:

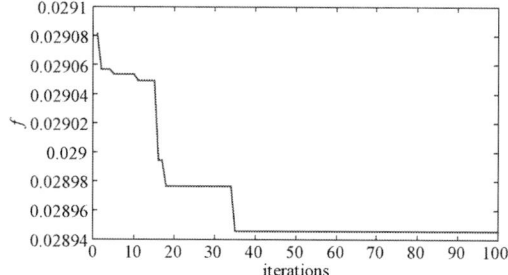

a) Fitness value curve

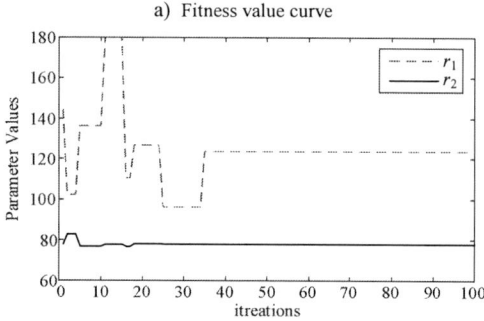

b) Variation curve of Optimal Parameters

Fig. 3. Parameter optimization result

Finally, the optimal controller parameters and performance indicators are: $r_1 = 123.85$, $r_2 = 77.98$, $f = 0.0289$.

Based on the above results, apply it to the PCHD controller. At the simulation process, the value of the wind speed is shown in Fig. 4. The simulation results are shown in Fig. 5. It can be seen from in Fig. 5a), if the wind speed changed, the controller can track the desired angle speed well, and achieved MPPT control. It can be seen from Fig. 5b)~ Fig. 5g), the three-phase current, voltage of wind power system are adjusted in the process of maximum wind energy tracing, and have good dynamic and static performance.

In order to prove that the system has good dynamic and robust dynamic performance, the paper compared the PCHD and global HNTSM control with the traditional PI control. The compared results are shown in Fig.6. From the compared simulation, the PCHD with global HNTSM control strategy is faster than that of PI control. And the prior control also has better dynamic and static performance than later, and is more suitable for wind power system with real-time wind speed change machine-side control. Fig. 6 also compared the results to the parameters of PCHD optimized by HPSO with those

without optimization. After compared, finding the control speed using the optimum parameters obtained by HPSO is faster and more accurate shown in Fig.6a) and Fig. 6b).

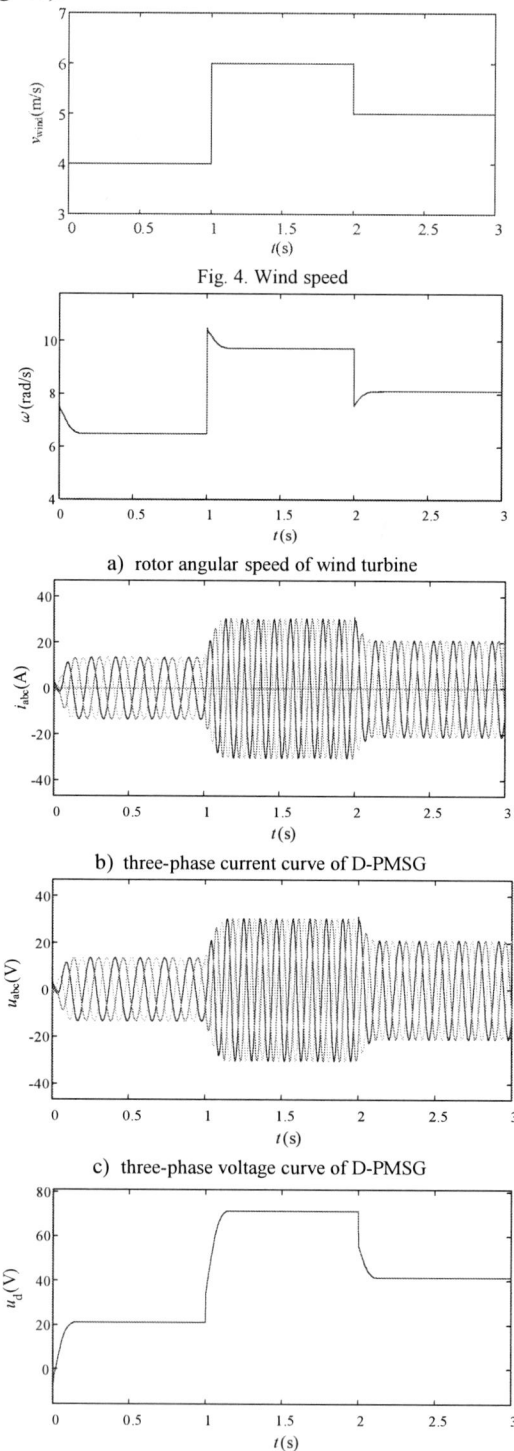

Fig. 4. Wind speed

a) rotor angular speed of wind turbine

b) three-phase current curve of D-PMSG

c) three-phase voltage curve of D-PMSG

d) d-axis voltage curve of D-PMSG

The 2018 International Power Electronics Conference

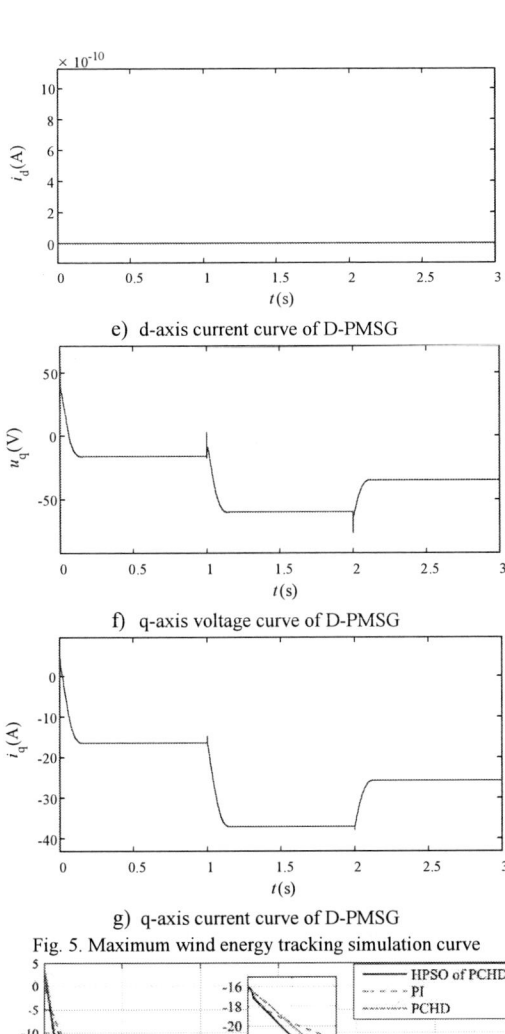

e) d-axis current curve of D-PMSG

f) q-axis voltage curve of D-PMSG

g) q-axis current curve of D-PMSG

Fig. 5. Maximum wind energy tracking simulation curve

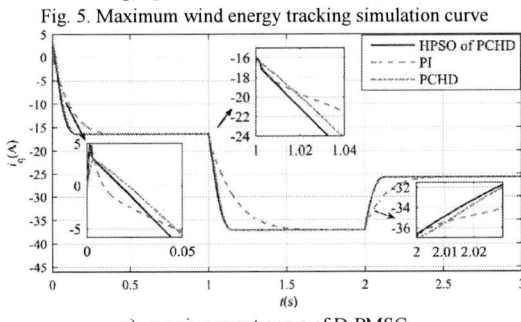

a) q-axis current curve of D-PMSG

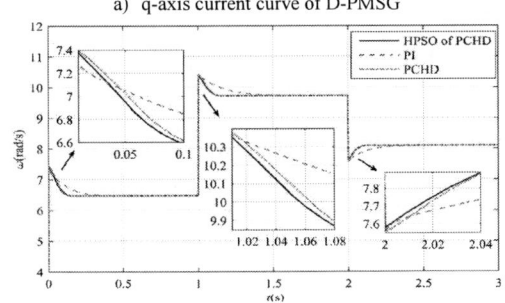

b) mechanical angular speed curve of D-PMSG

Fig. 6. Comparison between PCHD control and PI control

VI. CONCLUSIONS

Based on the PCHD model of D-PMSG wind power system, the PCHD with global HNTSM dual closed-loop controller is designed. The inner loop adopts PCHD

controller. The outer loop of the mechanical angular velocity is controlled by global HNTSM. The application of the proposed HPSO method provides a simple and effective method for parameters design of PCHD controller. The simulation results show that the controller design is more effective than traditional PI and good dynamic and static performance in the case of wind speed change. And the control effect is better than that of the controller without parameter optimization, and the maximum wind energy tracking control under the rated wind speed is realized.

ACKNOWLEDGMENT

This work was supported by the National Natural Science Foundation of China (51577039).

REFERENCES

[1] C CECATI, N ROTONDALE, "Torque and speed regulation of induction motors using the passivity theory approach," *IEEE Trans. on Industrial Electronics.* 1999, 46 (1): 119-127

[2] Ren Yi. "Research on passivity and sensorless control of direct-driven permanent magnet synchronous wind generator system,"*Harbin Institute of Technology*,2013.

[3] Y. Rahmat-Samii, Dennis Gies, Jacob Robinson, "Particle swarm optimization (PSO): A novel paradigm for antenna designs," *URSI Radio Science Bulletin*, 2003: 14–22.

[4] Solihin M I, Tack L F, Kean M L. "Tuning of PID controller using particle swarm optimization (PSO)." International Journal on Advanced Science, *Engineering and Information Technology*, 2011, 1(4): 458-461.

[5] Bingul Z, Karahan O. "Tuning of fractional PID controllers using PSO algorithm for robot trajectory control//Mechatronics (ICM)," *2011 IEEE International Conference on. IEEE*, 2011: 955-960.

[6] Feng Y, Zheng J, Yu X, et al. "Hybrid terminal sliding-mode observer design method for a permanent-magnet synchronous motor control system," *Industrial Electronics, IEEE Transactions on*, 2009, 56(9): 3424-3431.

[7] P. J. Angeline, "Using selection to improve particle swarm optimization", *IEEE International Conference on Evolutionary Computation Proceedings*, 1998:84-89.

[8] R. Jones and G. A. Smith. High Quality Mains Power form Variable-Speed Wind Turbines [J]. Wind Eng, 1994,18 (1):45-49.

[9] Rodrfguez H，Ortega R. "Stabilization of electromechanical systems via interconnection and damping assignment". Int. J. Robust Nonlinear Control，2003，13 (12)：1095-1111.

[10] GONG Chun, WANG Zhenglin, "Well versed in MATLAB optimization calculation (Fourth Edition),"*Publishing House of Electronics Industry*, 2016.11:293,313-314.

Establishment and Dynamic Control of Wind Induction Generator

M. Z. Lu[1], V. K. Ganisetti[1] and C. M. Liaw[1*]

1 Department of Electrical Engineering, National Tsing Hua University, Hsinchu, Taiwan, ROC

*E-mail: cmliaw@ee.nthu.edu.tw

Abstract— The establishment and dynamic control of a wind squirrel-cage induction generator (SIG) generating system are presented in this article. First, a three-phase wind SIG with suited excitation capacitor bank is established to yield stable three-phase generated voltages. Then the IG followed three-phase single-switch (3P1SW) continuous-current switch-mode rectifier (SMR) is developed to establish the boosted inverter DC-link voltage with suited magnitude. Well-regulated voltage against large line and load variations is achieved by the proposed adaptive feedforward controller and feedback controller. Meanwhile, good IG armature line supplying power quality is obtained to reduce the derate problem. Finally, a single-phase load inverter is designed and implemented. The proportional-resonant (PR) feedback controller is designed to yield AC 110V/60Hz voltage with good waveform quality under unknown and nonlinear loads. Successful microgrid-to-home (M2H) and microgrid-to-grid (M2G) operations are demonstrated experimentally.

Keywords— *Induction generator, switch-mode rectifier, boost converter, feedforward control, inverter, proportional-resonant control.*

I. INTRODUCTION

As generally recognized, the extensive use of microgrid [1,2] with renewable energy source is helpful in reducing fossil energy utilization. Moreover, the incorporated operations between the microgrid, the traditional grid and the electric vehicles [3-5] can further facilitate the accomplishment of this goal. Wind generator [6-8] is one of the most popularly employed RES, and squirrel-cage induction generator (SIG) is the simplest one [8-12]. It is adopted here for developing its performance enhancement control approaches.

In DC microgrid, all kinds of sources, storage devices and load converters must be connected to the common DC bus via suited interface DC/DC converters [13-15]. For a wind generator, its generated output voltage is usually highly fluctuated. In the developed wind IG, a followed interface boost DC/DC converter is employed. To quickly and effectively reject the input voltage variation, a voltage feedback controller is augmented with a input feedforward controller, which is adapted to the sensed input voltage. The well-regulated DC output voltage is established for the followed load inverter.

On the other hand, PWM inverters [13,16-21] are needed to generate AC output voltages for powering the loads or performing the grid-connected operations. The inverter output voltage waveform quality is affected by many aspects, such as the adopted switching approaches, the output filter design, and the dynamic control methods, etc. For the existing advanced control approaches considering the effects of nonlinear load effects, some typical ones include composite observer based control [17], the simple robust control [19] and the PR feedback controllers [19-21]. In the proposed IG system, a single-phase inverter is established to generate AC 110V/60Hz voltage. Good voltage waveforms under varied and non-linear loads are achieved using PR control.

An SIG-based wind generating system is developed in this paper. First, the stable DC output voltage is generated by the IG via its properly equipped excitation capacitor bank and followed 3P1SW CCM boost SMR. The adaptive feedforward controller and feedback controller are proposed to yield good DC output voltage regulation response against the converter input voltage and load fluctuations. Finally, a single-phase load inverter is designed and implemented. And the PR feedback controller is developed to yield the AC 110V/60Hz voltage with good waveform quality under unknown and nonlinear loads. Some measured results are provided to demonstrate the normal M2H and M2G operations.

II. SYSTEM CONFIGURATION

Fig. 1 shows the system configuration and control schemes of the developed induction generator system. It consists of a DC motor (DCM) driven IG with self-excited capacitor bank, a 3P1SW SMR consisting of a three-phase diode rectifier and a DC/DC boost converter, and a H-bridge load inverter. The DCM is served as the alternative of wind turbine. The wind IG generated three-phase voltages are converted by its followed SMR to yield the DC output voltage in the range of $v_d = 115V \sim 170V$ ($\omega_r = 1543$rpm ~ 1951rpm, $v_{dm} = 150 \sim 200V$). The DC-link voltage ($v_{dc} = 200V$) is then established by the DC/DC boost converter. Finally, the 110V/60Hz AC voltage is generated by the load inverter for powering home appliances (M2H) or conducting M2G operation.

As well known, the wind generator output is highly fluctuated with speed. To effectively and quickly counteract this effect, an adaptive feedforward controller is proposed. On the other hand, the inverter output voltage waveform will be highly distorted under nonlinear loads. The PR voltage feedback controller is applied in the load inverter to improve its waveform quality.

To form a complete home DC microgrid, an additional photovoltaic (PV) renewable energy system and a battery energy storage system (BESS) as shown in Fig. 1(a) can further be added. Moreover, the battery bank in an electric scooter or electric vehicle can also be incorporated into the developed system to achieve the vehicle-to-microgrid and microgrid-to-vehicle operations.

Due to the limit of scope, the expectedly added parts indicated in Fig. 1(a) are not included in this article; rather they will be conducted and presented in the near future.

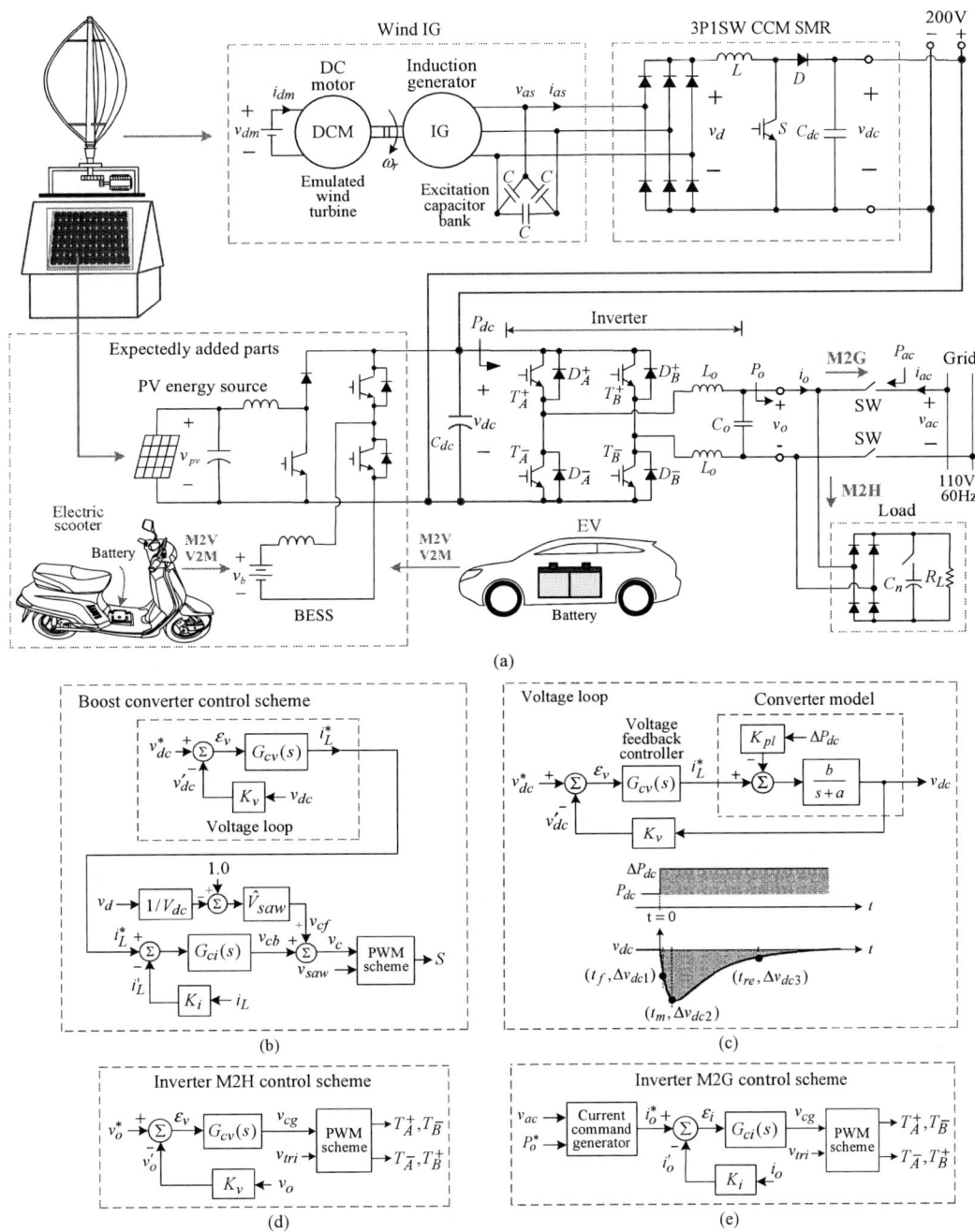

Fig. 1. System configuration of the developed wind induction generator system: (a) power circuit; (b) control scheme of the 3P1SW boost SMR; (c) equivalent v-loop control block and the sketched desired step load regulation response of the 3P1SW boost SMR; (d) H-bridge inverter M2H control scheme; (e) H-bridge inverter M2G control scheme.

2908

III. WIND INDUCTION GENERATOR WITH FOLLOWED INTERFACE CONVERTER

A. Wind Induction Generator

(a) Power circuit

- DCM: 220V, 0.88kW, 2100rpm.
- IG: 120V, 0.8kW, 2000rpm (Nikki Denso).
- Self-excited capacitor bank: three-phase delta connected, using single-phase capacitor of $100\,\mu F/450V$

(b) Measured result

For an induction generator, the proper self-excited capacitance must be determined considering the saturated machine magnetizing characteristic, the stable generation criterion, the generated voltage level, and the loading effect, etc. Fig. 2 shows the output characteristics of the established SIG with the capacitor bank of $C=100\mu F/450V$. It is obvious that the heavier load is, the harder voltage can be generated. By proper consideration, the IG generated output voltage range is specified as $v_d = 115V \sim 170V$.

B. Three-phase Single-switch Boost SMR

(a) Power circuit

- Voltage rating: $v_d = 115V \sim 170V, v_{dc} = 200V$.
- Power rating: $P_{dc} = 300W$.
- Switching frequency: $f_s = 30kHz$, CCM operation.
- Energy storage inductor: $L = 1.7028mH$, $ESR = 1.95\Omega$ at 30kHz.
- DC bus filtering capacitor: $C_{dc} = 2200\mu F/250V$.
- Power device: (i) Three-phase diode rectifier: IXYS VUO 62-08 NO7 (600V, 63A); (ii) IGBT module CM100DY-12H (Mitsubishi Electric, 600V, 100A DC).

(b) Control scheme

(1) Current control scheme

The current sensing factor is set as:

$$K_i = 0.5V/A \tag{1}$$

The proposed current control scheme shown in Fig. 1(b) consists of a feedback controller $G_{ci}(s)$ and a feed-

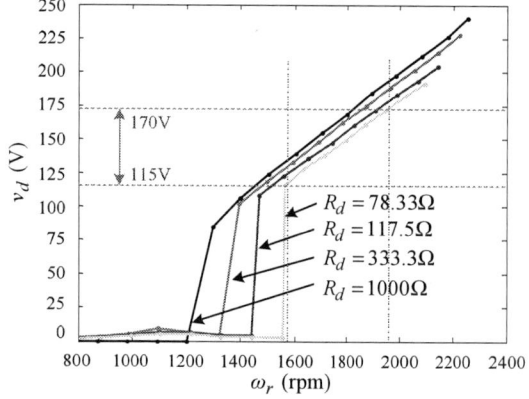

Fig. 2. The output voltage generation characteristics of the developed wind SIG.

forward controller $G_{cf}(s)$ to yield the composite PWM control signal v_c:

$$v_c = v_{cb} + v_{cf} \tag{2}$$

(i) Feedforward controller

The proposed feedforward control methodology can be comprehended from the mechanism depicted in Fig. 3. The control voltage is generated to adaptively reject the highly varied input voltage v_d ($\underline{v}_d \le v_d \le \bar{v}_d$) (115V ~ 170V) for establishing the well-regulated output voltage $v_{dc} = V_{dc} = 200V$. To achieve this goal, one can derive to yield the adaptive control voltage v_{cf} adapted to the sensed input v_d as:

$$v_{cf} = (1 - \frac{v_d}{V_{dc}}) \hat{V}_{saw} \tag{3}$$

where $\hat{V}_{saw}$ denotes the amplitude of saw-tooth carrier.

(ii) Feedback controller

The control signal v_{cb}, which is yielded by $G_{ci}(s)$ is in charge of maintaining the closed-loop stability and also regulating the tracking error due to load variations and non-ideal feedforward control. Thanks to the feedforward control, the less stringent feedback control can be applied. For simplicity, the PI controller is adopted and its parameters are set as:

$$G_{ci}(s) = 8 + \frac{14.1844}{s} \tag{4}$$

(2) Voltage control scheme

The voltage sensing factor is set as:

$$K_v = 0.05V/V \tag{5}$$

Under the chosen operating point with ($v_{dm} = 162V$ ($\omega_r = 1696rpm$), $v_d = 155V$, $v_{dc} = 200V$, $R_{dc} = 509\Omega$), and the PI feedback controller being arbitrarily set as:

$$G_{cv}(s) = 0.5 + \frac{7.5947}{s} \tag{6}$$

Fig. 4(a) shows the measured v_{dc} due to a step load change of $R_{dc} = 509\Omega \rightarrow 202\Omega$ ($\Delta P_{dc} = 119.43W$). By choosing three response points as indicated in Fig. 1(c), the converter dynamic model parameters (a, b, K_{pl}) are estimated. The measured and simulated results shown in Fig. 4(b) demostarte their correctness.

Then, by specifying the desired step load voltage regulation control specifications at the same case as above to be (maximum dip $\Delta v_{dc,max} = 5V$ and restore time $t_{re} = 0.35sec$, $\Delta v_{dc}(t = t_{re}) \equiv 0.1\Delta v_{dc,max}$), the controller G_{cv} can be derived as:

$$G_{cv}(s) = 1.7206 + \frac{23.3314}{s} \tag{7}$$

Figs. 5(a) and 5(b) show the measured and simulated v_{dc} of the developed 3P1SW CCM SMR by the

The 2018 International Power Electronics Conference

quantitatively designed voltage feedback controller under the same condition. The results verify the correctness of the designed controller and the closeness between the measuered and the simulated rtesults.

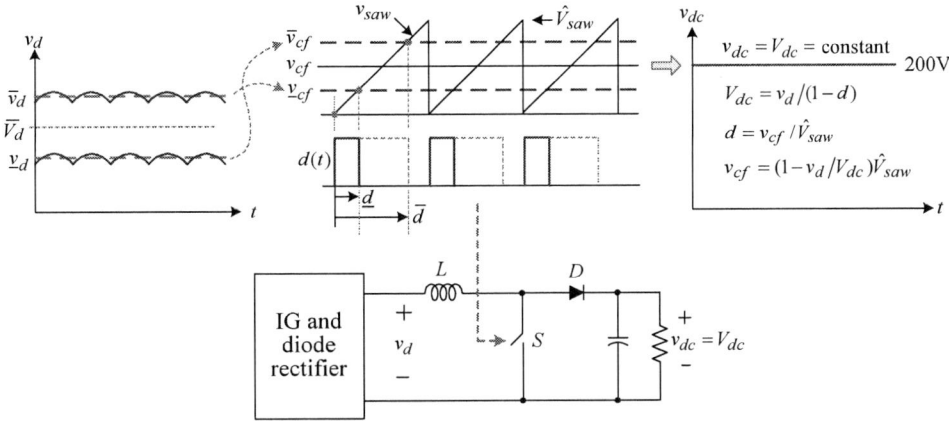

Fig. 3. The proposed adaptive input feedforward control methodology for wind generator 3P1SW CCM boost SMR.

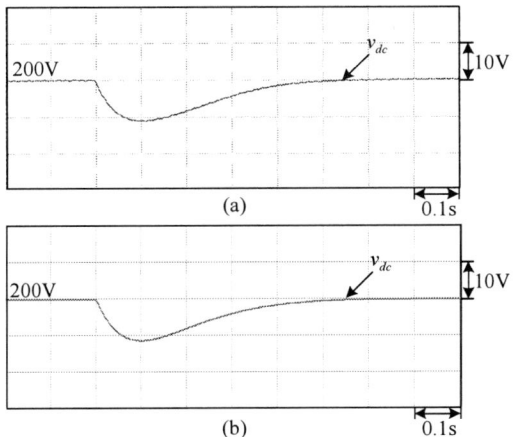

Fig. 4. The output voltages v_{dc} of the developed 3P1SW CCM SMR by the arbitrarily set controller due to a step load change of $R_{dc} = 509\Omega \rightarrow 202\Omega$ ($\Delta P_{dc} = 119.43\text{W}$) : (a) measured result; (b) simulated result.

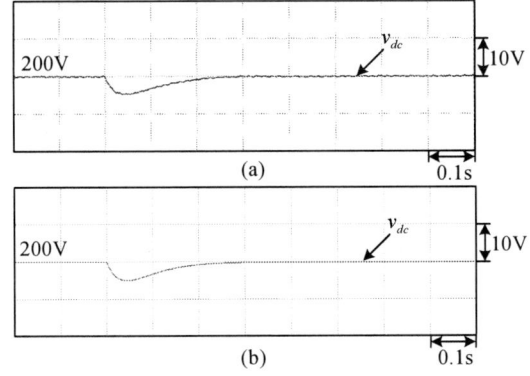

Fig. 5. The output voltages v_{dc} of the developed 3P1SW CCM SMR by the quantitatively designed controller due to a step load change of $R_{dc} = 509\Omega \rightarrow 202\Omega$ ($\Delta P_{dc} = 119.43\text{W}$) : (a) measured result; (b) simulated result.

(c) Measured results

(1) Effectiveness of feedforward control

Now at $R_{dc} = 250\Omega$, the DC motor terminal voltage is changed from 170V to 190V to vary the IG driving speed, the measured v_{dm}, v_{dc} and v_c with and without feedforward controller are compared in Fig. 6. The results indicate that the proposed feedforward controller can effectively reject the input voltage disturbance to let the output voltage v_{dc} be alomost unchanged.

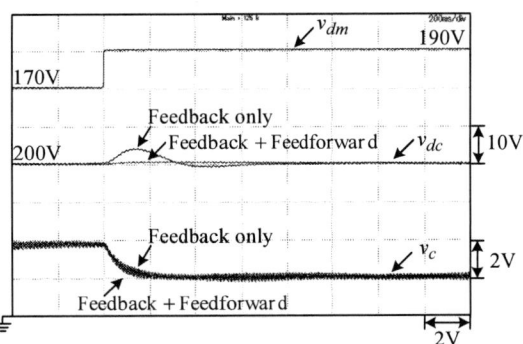

Fig. 6. Measured v_{dm}, v_{dc} and v_c of the 3P1SW CCM SMR due to the step voltage change of DCM terminal voltage v_{dm} with and without feedforward controller at $R_{dc} = 250\Omega$.

(2) Evaluation of wind IG followed 3P1SW SMR

The DC motor driving speed is ajusted to let the DC/DC converter input voltage be $v_d = 155\text{V}$, Figs. 7(a) and 7(b) plot the measured steady-state ((v_{as}, i_{as}), v_{dc}, i_L) and ((v_{dc}^*, v_{dc}'),(i_L^*, i_L')). And Fig. 7(c) depicts the measured v_{dc} due to a step load change of $R_{dc} = 338\Omega$ to $R_{dc} = 202\Omega$ ($\Delta P_{dc} = 79.6\text{W}$). Normal operation and good dynamic as well as static operating characteristics can be observed from the results.

2910

The 2018 International Power Electronics Conference

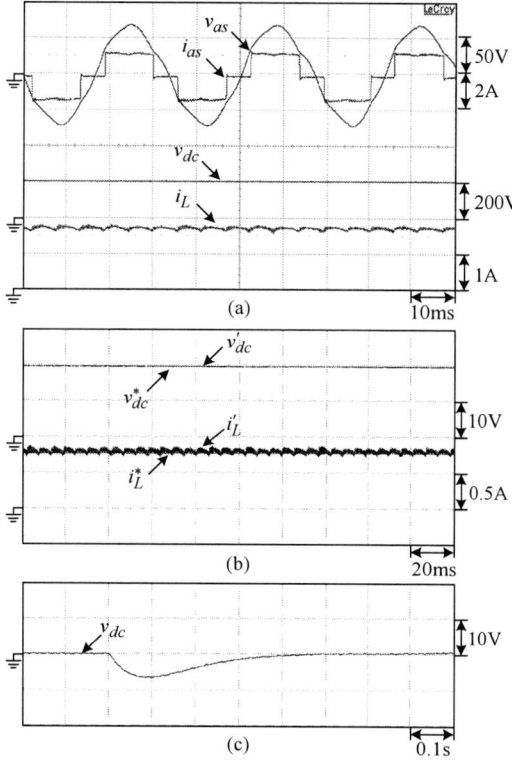

Fig. 7. Measured results of the developed IG followed by 3P1SW CCM SMR under $v_d = 155V$: (a) steady-state ((v_{as}, i_{as}), v_{dc}, i_L); (b) steady-state ((v_{dc}^*, v_{dc}'),(i_L^*, i_L')); (c) v_{dc} due to a step load change of $R_{dc} = 338\Omega$ to 202Ω ($\Delta P_{dc} = 79.6W$).

IV. SINGLE-PHASE H-BRIDGE INVERTER

A. Power Circuit

The specifications and the system components are listed as follows:

- Output voltage: $V_o = 110V/60Hz$.
- Power rating: $P_o = 300W$.
- Switching frequency: $f_s = 20kHz$.
- Output filter: $C_o = 3.3\mu F$, $L_o = 1.274mH$.
- Power devices: IPM PS21454-E (Powerrex).
- Load: a diode rectified load with paralleled R_L and C_n is used.

B. Control Scheme

To preserve better inverter output waveform under varied and nonlinear loads, the PR feedback controller is applied. The sensing factors and the chosen controller parameters are:

- *M2H operation:*

$$K_v = 0.0258 \text{ V/V}, \quad G_{cv}(s) = 0.05 + \frac{16.61}{s^2 + 377^2} \quad (8)$$

- *M2G operation:*

$$K_i = 0.4 \text{V/A}, \quad G_{ci}(s) = 2.41 + \frac{18.32}{s^2 + 377^2} \quad (9)$$

C. Measured Results

(a) M2H operation

(1) Linear load

The established induction generator system is stably operated under ($v_{dm} = 186V, \omega_r = 1870rpm$, $v_{dc} = 200V, R_L = 58.75\Omega$), and the inverter output voltage command of $v_o^* = k_v\sqrt{2}110\sin 2\pi 60t$ is set. The load resistor R_L (C_n is removed) shown in Fig. 1 is used as the test load of the inverter. The measured ((v_{as}, i_{as}), v_{dc}, v_d) , (v_o, i_o) and (v_o^*, v_o') are plotted in Figs. 8(a) to 8(c). Besides, Figs. 9(a) and 9(b) show the measured (v_o, i_o) and (v_o^*, v_o') due to a step load change of $R_L = 117.5\Omega \rightarrow 58.75\Omega$. As the results indicated that the established single-phase H-bridge inverter using PR controller has good tracking responses.

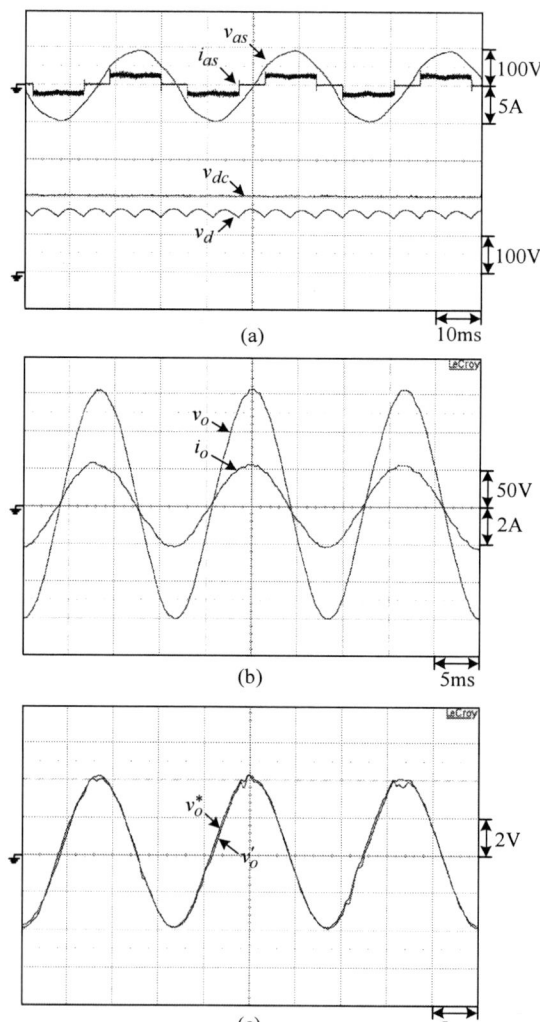

Fig. 8. Measured results of the established induction generator system under ($v_{dm} = 186V, \omega_r = 1870rpm$, $v_{dc} = 200V$, $R_L = 58.75\Omega$): (a) ((v_{as}, i_{as}), v_{dc}, v_d); (b) (v_o, i_o); (c) (v_o^*, v_o').

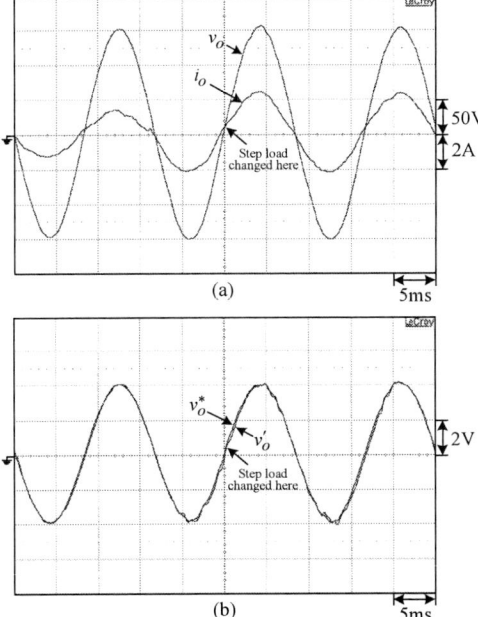

(a)

5ms

(b)

5ms

Fig. 9. Measured (v_o, i_o) and (v_o^*, v_o') of the established induction generator system due to a step load resistance change of $R_L = 117.5\Omega \rightarrow 58.75\Omega$.

(2) Nonlinear load

At ($v_{dm} = 186$V, $\omega_r = 1870$rpm, $v_{dc} = 200$V, $R_L = 58.75\Omega$), the measured (v_o, i_o) and (v_o^*, v_o') under nonlinear rectified load with ($R_L = 117.5\Omega$, $C_n = 2200\,\mu$F/250V) are plotted in Figs. 10(a) and 10(b). Rather good voltage waveform tracking characteristics can be seen.

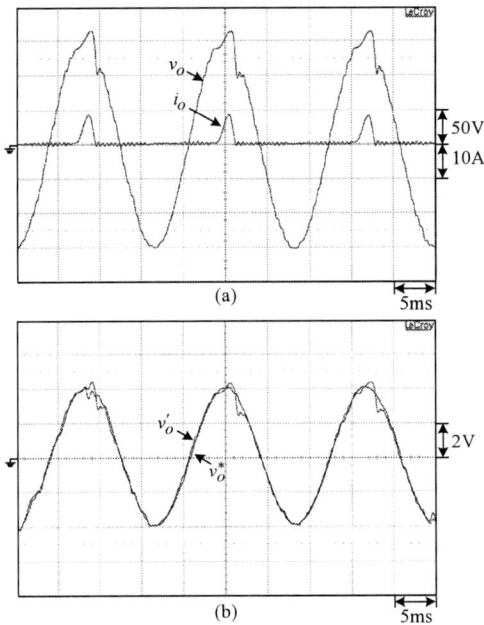

(a)

5ms

(b)

5ms

Fig. 10. Measured (v_o, i_o) and (v_o^*, v_o') of the established established induction generator system under nonlinear rectified load with ($v_{dc} = 200$V, $R_L = 117.5\Omega$, $C_n = 2200\,\mu$F) : (a) (v_o, i_o); (b) (v_o^*, v_o').

(b) M2G operation

The inverter of the established induction generator system is connected to the mains. The power sent back to the grid is set by the current command i_o^*, which is synchronized to the sensed grid voltage v_{ac}. At ($v_{dm} = 186$V, $\omega_r = 1870$rpm, $v_{dc} = 200$V), Fig. 11 and Fig. 12 show the measured ($v_{as}, i_{as}), v_{dc}, v_d$), ($i_o^*, i_o'$) and ($v_o, i_o$) under 100W and 200W, respectively. The results indicate that the established H-bridge inverter can achieve grid-connected operation successfully.

V. CONCLUSION

The establishment and control of a wind IG system have been presented in this paper. By equipping a suited self-excited capacitor bank and a three-phase diode rectifier, the generated DC output voltage in the range of 115V~170V is yielded. The varied IG generated voltage is boosted and regulated by a followed DC-DC boost converter to establish the inverter DC-link voltage (200V). Thanks to the developed input feedforward controller, good voltage regulation response against the input voltage fluctuation is achieved. Finally, a single-phase H-bridge load inverter is developed, and the PR feedback controller is applied to yield AC 100V/60Hz voltage with good waveform quality under nonlinear loads. The inverter M2H and M2G operations with satisfactory performances have been verified by some measured results.

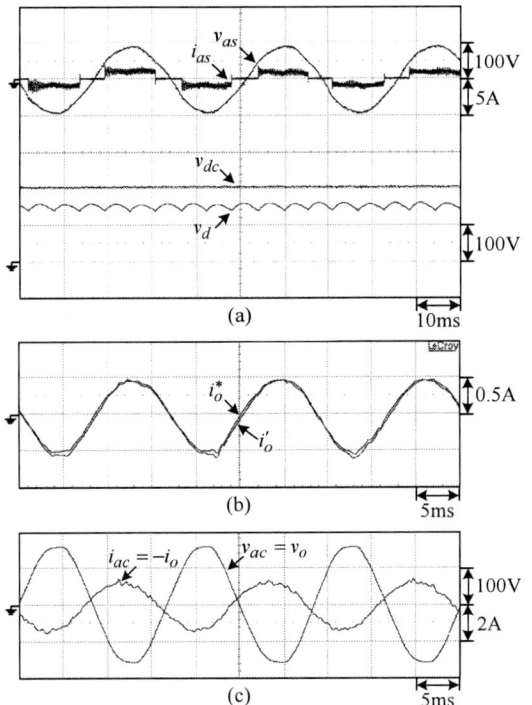

(a)

10ms

(b)

5ms

(c)

5ms

Fig. 11. Measured results of the established induction generator system under ($v_{dc} = V_{dc} = 200$V, $P_o = 100$W): (a) ((v_{as}, i_{as}), v_{dc}, v_d); (b) (v_o, i_o); (c) (v_o^*, v_o').

2912

The 2018 International Power Electronics Conference

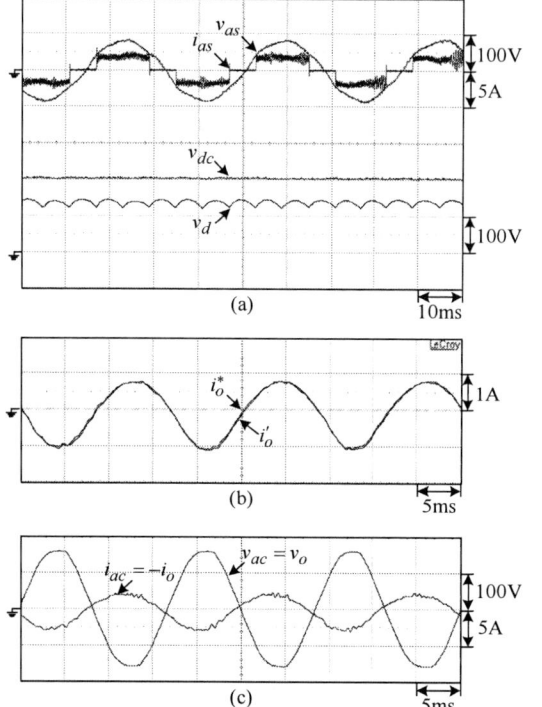

Fig. 12. Measured results of the established induction generator system under ($v_{dc} = V_{dc} = 200V$, $P_o = 200W$): (a) ((v_{as}, i_{as}), v_{dc}, v_d); (b) (i_o^*, i_o'); (c) (v_o, i_o).

REFERENCES

[1] N. Hatziargyriou, H. Asano, R. Iravani, and C. Marnay, "Microgrids," *IEEE Power Energy Mag*, vol. 5, no. 4, pp. 78-94, 2007.

[2] T. Ma, M. H. Cintuglu and O. A. Mohammed, "Control of a hybrid AC/DC microgrid involving energy storage and pulsed loads," *IEEE Trans. Ind. Appl.*, vol. 53, no. 1, pp. 567-575, 2017.

[3] K. W. Hu and C. M. Liaw, "Incorporated operation control of DC microgrid and electric vehicle," *IEEE Trans. Ind. Electron.*, vol. 63, no. 1, pp. 202-215, 2016.

[4] C. Liu, K. T. Chau and S. Gao, "Opportunities and challenges of vehicle-to-home, vehicle-to-vehicle, and vehicle-to-grid technologies," in *Proc. IEEE*, vol. 101, no. 11, November, 2013.

[5] S. S. Williamson, A. K. Rathore and F. Musavi, "Industrial electrics for electric transportation: current state-of-the-art and future challenges," *IEEE Trans. Ind. Electron.*, vol. 62, no. 5, May, 2015.

[6] G. M. Master, *Renewable and Efficiency Electric Power Systems*, New York: John Wiley & Sons Ltd, 2004.

[7] F. Blaabjerg, F. Iov, R. Teodorescu and Z. Chen, "Power electronics in renewable energy systems," in *Proc. IEEE EPE-PEMC*, 2006, pp. 1-17.

[8] P. C. Sen, *Principles of Electric Machines and Power Electronics*, 3rd Edition, John Wiley & Sons, 2014.

[9] G. Raina and O. P. Malik, "Wind energy conversion using a self-excited induction generator," *IEEE Trans. Power App. Syst.*, pp. 3933-3936, 1983.

[10] S. M. Alghuwainem, "Steady-state analysis of an isolated self-excited induction generator driven by regulated and unregulated turbine," *IEEE Trans Energy Convers.*, vol. 14, pp. 718-723, 1999.

[11] T. Ahmed, O. Noro, E. Hiraki, and M. Nakaoka, "Terminal voltage regulation characteristics by static var compensator for a three-phase self excited induction generator," *IEEE Trans. Ind. Appl.*, vol. 40, pp. 978-988, 2004.

[12] Hao Chen and Dionysios C. Aliprantis, "Analysis of squirrel-cage induction generator with Vienna rectifier for wind energy conversion system," *IEEE Trans Energy Convers.*, vol. 26, no.3, pp. pp. 967-975, 2011.

[13] N. Mohan, T. M. Undeland, and W. P. Robbins, *Power Electronics: Converters, Applications, and Design*, 3rd ed., New Jersey: John Wiley & Sons, Inc., 2003.

[14] L. Palma and P. N. Enjeti, "A modular fuel cell, modular DC-DC converter concept for high performance and enhance reliability," *IEEE Trans. Power Electron.*, vol. 24, no. 6, pp. 1437-1443, 2009.

[15] A. A. Fardoun, E. H. Ismail, A. J. Sabzali, and M. A. Al-Saffar, "Bi-directional converter with low input/output current ripple for renewable energy applications," in *Proc. ECCE*, 2011, pp. 3322-3329.

[16] M. Castilla, J. Miret, J. Matas, L. G. de Vicuña, and J. M. Guerrero, "Control design guidelines for single-phase grid-connected photovoltaic inverters with damped resonant harmonic compensators," *IEEE Trans. Ind. Electron.*, vol. 56, no. 11, pp. 4492-4500, 2009.

[17] K. Selvajyothi and P. A. Janakiraman, "Reduction of voltage harmonics in single phase inverters using composite observers," *IEEE Trans. Power Del.*, vol. 25, no. 2, pp. 1045-1057, 2010.

[18] K. W. Hu and C. M. Liaw, "On an auxiliary power unit with emergency AC power output and its robust controls," *IEEE Trans. Ind. Electron.*, vol. 60, no. 10, pp. 4387-4402, 2013.

[19] R. Teodorescu, F. Blaabjerg, M. Liserre and P.C. Loh, "Proportional-resonant controllers and filters for grid-connected voltage-source converters," *IEE Proc.-Electr. Power Appl.*, vol. 153, no. 5, pp. 750-762, 2006.

[20] M. C. Chou and C. M. Liaw, "Dynamic control and diagnostic friction estimation for a PMSM driven satellite reaction wheel," *IEEE Trans. Ind. Electron.*, vol. 58, no. 10, pp. 4693-4707, October 2011.

[21] W. L. Malan, D. M. Vilathgamuwa, and G. R. Walker, "Modeling and control of a resonant dual active bridge with a tuned CLLC network," *IEEE Trans. Power Electron.*, vol. 31, no. 10, pp. 7297-7310, Oct. 2016.

Middle Frequency Solid State Transformer for HVDC Transmission from Offshore Windfarm

Noriyuki kimura[1], Toshimitsu Morizane[1] Isao Iyoda[2],
Kazushige Nakao[3], Tomoki Yokoyama[4]
1 Osaka Institute of Technology, Osaka, Japan;   2 Osaka Electro-Communication University, Osaka, Japan;
3 Fukui Institute of Technology, Fukui, Japan; 4 Tokyo Denki University, Tokyo, Japan

Abstract- **This paper investigates the solid state transformer applied for the HVDC transmission from the offshore windfarm. The 5kVA experimental setup is developed and comparison of up to 1200Hz operation. Dependency of the loss in the transformer with different core materials to frequency is investigated. Frequency loss dependency of nano-crystaline and amorphous core materials are compared in experimental measurement. Nano-crystaline material called "FINEMET" is the best performance, although amorphous seems to have better cost performance. It is also indicated that the converter voltage causes the larger loss in the transformer than the sinusoidal power supply.**

I. INTRODUCTION

Wind power is clean, renewable and plentiful amount and expected to be the major resource of generating electric power. Recently, large scale offshore windfarms are constructed or planned mainly in European region. In 2015, NEDO (New Energy and Industrial Technology Development Organization) in Japan launched the new project for the HVDC transmission from the offshore windfarm[1].

II. HVDC TRANSMISSION USING SST

Some configurations for the HVDC transmission system from the offshore windfarm are under investigation. One of them is using the solid-state transformer (SST). The SST uses the middle frequency (MF) transformer with the MF inverter and rectifier[2, 3]. Fig. 1 shows the schematic diagram of HVDC transmission using SST. It is expected to have 1/100 volume at 100kHz frequency compared with 50/60 Hz conventional transformer. In this NEDO project, the converter system and the middle frequency (MF) transformer will be developed by our institute to realize the large rating solid-state transformer (SST).

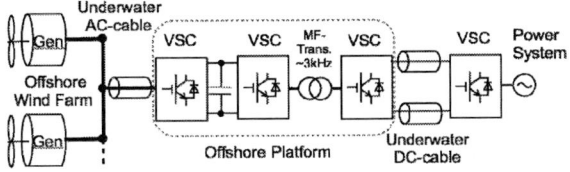

Fig. 1. DC transmission using offshore substation with middle frequency (MF) transformer

Kolar et. al. showed the expected size reduction of middle frequency transformer as Fig. 2. Some experimental results are compared for 1MW ratings. Authors are aiming to reduce the size of MF transformer to 1/10 by using 1-3 kHz frequency. Reduction of size and weight of the transformer at HVDC converter station can reduce the construction cost of the offshore platform. However the efficiency of the SST cannot be as better as the 50/60 Hz transformer since it has the converters. Therefore the efficiency of the converters are important factor of the SST. To have smaller converter loss, using modular multi-level converter structure is used for the middle frequency converter for HVDC side.

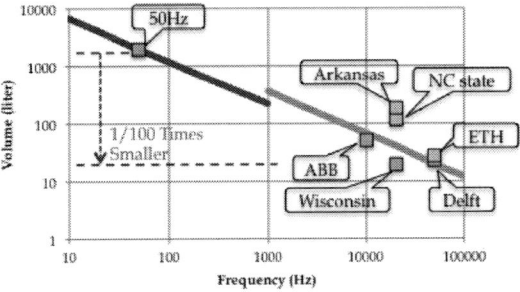

Refer to Kolar et.al [2]

Fig. 2. Expected size reduction of middle frequency (MF) transformer

III. SST EXPERIMENTAL RESULTS

Fig. 3 shows the configuration of the solid state transformer (SST) experimental setup.

Fig. 4 shows the experimental results of the SST experimental setup. HIOKI PW6001 power meter is used for measurement. Fig. 5 shows the results of 400Hz, 10.7 A ac current. Stable operation of the SST with MMC can be seen. Table 1 shows the comparison of the results of 50Hz and 400Hz of about 3.8 kW. The waveforms of the voltage seems to be almost sinusoidal and the ripple in the currents are small enough.

The efficiency of the normal transformer becomes very low at 400Hz as less than 80%. The reactive power is increased largely at 400Hz when the current increases. This may be caused by the leakage inductance of the transformer. The loss in the transformer is also increased largely when current increases. The causes may be the eddy current loss, the hysteresis loss, the skin effect, the proximity effect, etc.

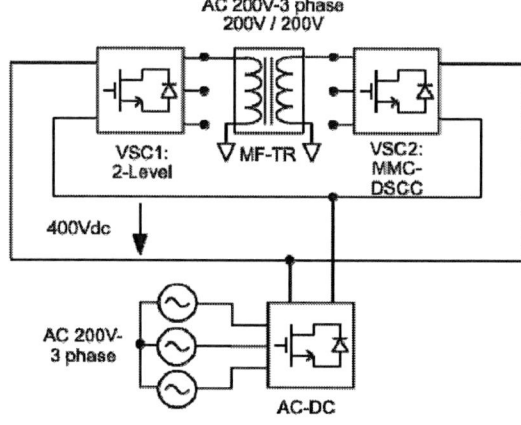

(a) Configuration of the SST

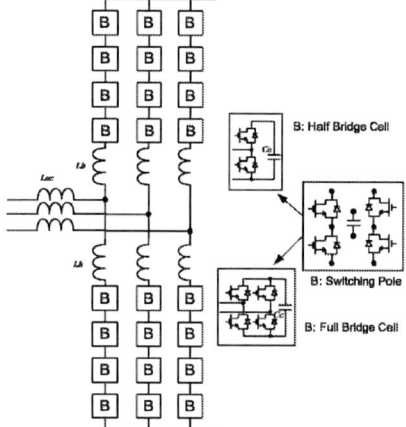

(b) VSC2 (Modular Multilevel Converter)

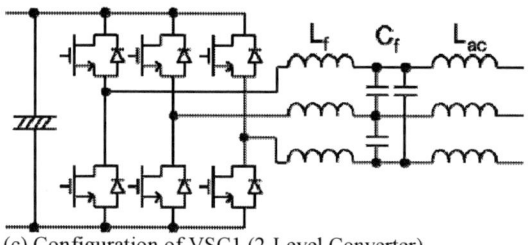

(c) Configuration of VSC1 (2-Level Converter)

Fig. 3. Configuration of the solid state transformer (SST)

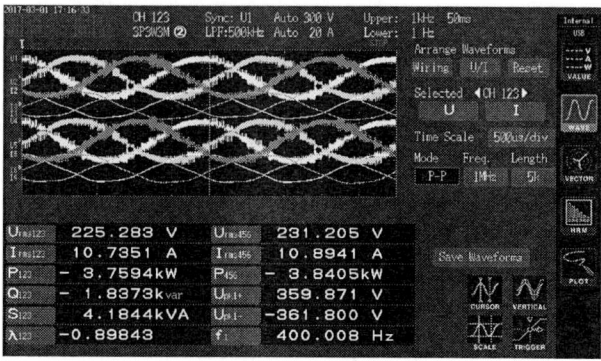

Fig. 4. Waveforms of currents and voltages at both ends of transformer

Table I

COMPARISON OF EXPERIMENTAL RESULTS OF SST AT 50HZ AND 400HZ

Frequency (Hz)	50.0	400.0
AC Voltage (V)	186.2	206.4
AC Current (A)	12.03	11.88
Pinv1 （W）	3844	3183
Reactive Power (Var)	458.4	2812
Pmmc2 （W）	3717	3075
Efficiency (%)	96.0	72.4

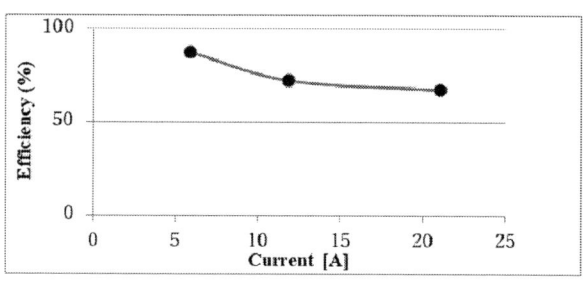

Fig. 5. Efficiency of transformer (Silicate steel) at 400Hz

IV. MIDDLE FREQUENCY TRANSFORMER

To decrease the loss of the SST, core materials of the transformer are investigated. Core materials of the transformers are super-core[4], amorphous, nano-crystaline (called Finemet[5]). Two different manufacturer (called A and B here) made 5kVA MF-transformer. Table II shows the rating of the transformers.

Fig. 6(a) shows the experimental setup of core loss measurement with the sinusoidal power supply. Fig. 6(b) shows the frequency dependency of the core loss. In measurement, the excitation current was set to be the same. Then the loss increases as the frequency increases. The lowest loss material is Finemet as expected. The next lowest is amorphous. These results mean that Finemet is the most attractive core material. However, amorphous is next to Finemet and the cost performance is much higher since the cost is much less than the Finemet. Super-core is not good enough for both manufacturers. So only amorphous and Finemet are compared in this paper.

It is also interesting that the transformers made by the manufacturer-A show the larger loss than the manufacturer-B. One of the causes may be the structure of the windings. It is notified again that this measurement is made by tuning the excitation current same and it may not be the same at the normal operating condition.

Table II

RATING OF MF TRANSFORMER

Rating	Manufacturer A	Manufacturer B
Primary Voltage	200[V]	200[V]
Primary Current	1.45[A]	1.45[A]
Rated Power	5000[VA]	5000[VA]
Frequency	3000[Hz]	3000[Hz]

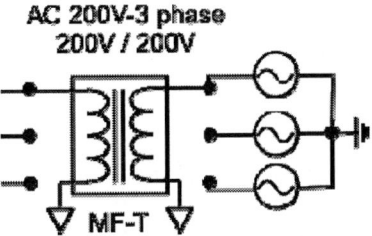

AC 200V-3 phase
200V / 200V

(a) Experimental schematic circuit

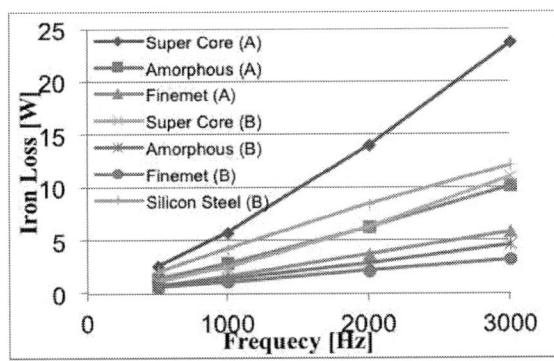

(b) Experimental results of core loss

Fig. 6. Experimental setup and results of several kinds of MF transformer core

The experimental setup of SST is shown in Fig. 3(a). The operating frequency is set at 750 -1200 Hz. This criteria is depending on the L-C values in SST circuit. Enhancement of operating range is under progress.

Fig. 7 shows the experimental results of the SST experimental setup. HIOKI PW6001 power meter is used for measurement. Fig. 7 shows the results of 1000Hz, 3.0 A ac current at the primary side of the MF-transformer. Stable operation of the SST with MMC can be seen. However the ripple of the voltage seems larger than the

400Hz operation, though the ripples in the currents are still small enough.

At this frequency the core loss in the FineMet and the amorphous is not different much. Table III shows the relative cost of the core materials. Around 1 kHz, the amorphous has good cost performance.

Table III

RELATIVE COST OF MF TRANSFORMER

Core Material	Relative cost[%]
FINEMET (Nanocrystaline)	100
Amorphous	30-40
Silicon steel (Super Core, 6.5%)	20-30
Silicon steel (3%)	10-40

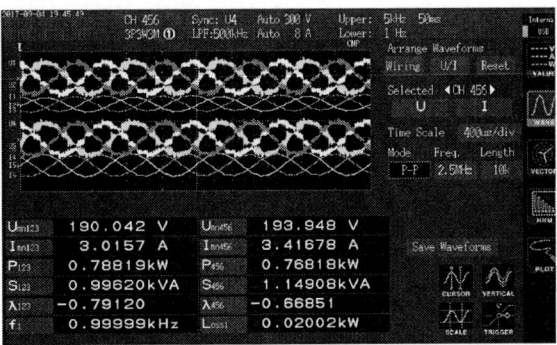

Fig. 7. Waveforms of currents and voltages at both ends of transformer at 1000Hz operation (Upper: Primary side, Lower: Secondary side)

Fig. 8 and 9 show the loss of the transformer used in the SST. Fig. 8 shows the results of FINEMET (nano-crystaline) core and Fig. 9 shows the results of amorphous core.

The higher the frequency is, the lower the loss is in these results. The reason is that the excitation current is smaller for the higher frequency at the same voltage.

The copper loss increases when the input power increases since the current is larger. The loss in SST is larger than the loss with the sinusoidal power supply. The effect of the higher harmonic components and the minor loop of the hysteresis of the transformer may cause this.

In Fig. 9, the increase rate of the loss in SST is larger than the loss with sinusoidal power supply. The reason may be the skin effect caused by the higher harmonic components. Further investigation is required.

The 2018 International Power Electronics Conference

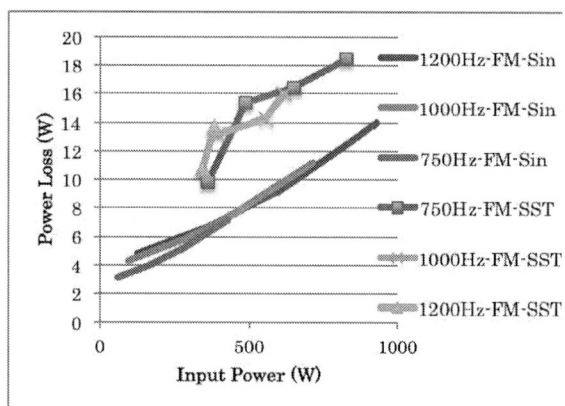

Sin: Sinusoidal Power Supply, SST: Solid State Transformer

Fig. 8. Experimental results of loss of MF-transformer with FINEMET (FM) core (Manufacturer-B) (750-1200Hz)

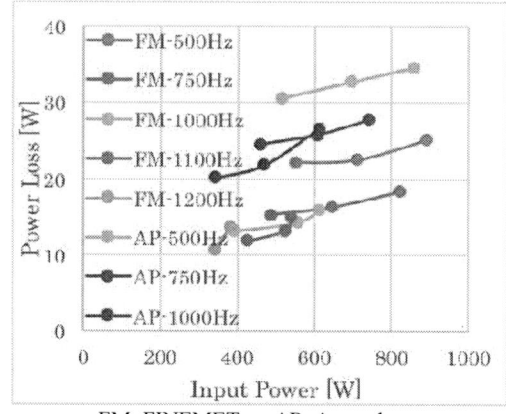

FM: FINEMET : 、AP: Amorphous

Fig. 10. Comparison of experimental results of loss of MF-transformer (Manufacturer-B) in SST (500-1200Hz)

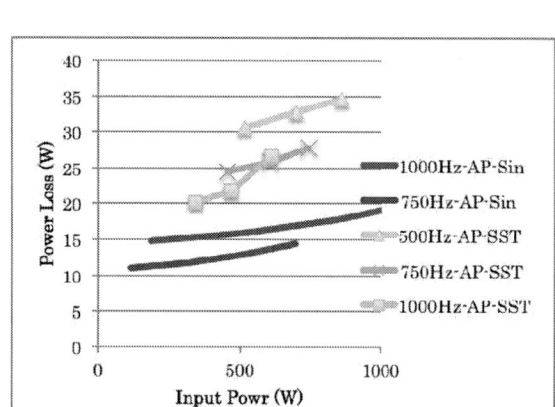

Sin: Sinusoidal Power Supply, SST: Solid State Transformer

Fig. 9. Experimental results of loss of MF-transformer with Amorphous (AP) core (Manufacturer-B) (500-1000Hz)

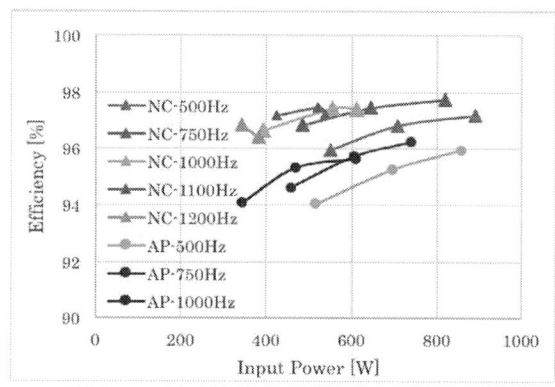

(NC: Nano-crystaline core、AP: Amorphous core; Manufacturer B)

Fig. 11. Frequency dependency of efficiency of the MF-transformer

Fig. 10 shows the loss of MF-transformer dependency on the input power. The core materials are FINEMET (FM) and amorphous (AP). In both materials, the higher frequency shows smaller loss. The FINEMET core can reduce the loss to half of the amorphous core.

Fig. 11 shows the efficiency of MF-transformer in SST depending on the input power. Nano-crystaline (NC) core material shows the better efficiency than amorphous core material. It is also shown that the higher frequency shows better efficiency as expected from the loss characteristics.

Since these results are the primary report of the project, the range of the power is smaller than the rating. The large power experiments are under preparation.

Fig. 12 shows the efficiency of MF-transformer in SST depending on the input power at 1000Hz. FINEMET (Nano-crystaline) core material shows the better efficiency than the amorphous core material. Both results of manufacturer A and B show the similar dependency.

The 2018 International Power Electronics Conference

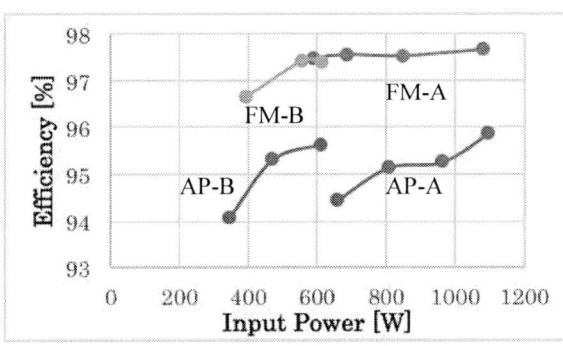

(FM: FINEMET core、AP: Amorphous core;
A: Manufacturer A, B: Manufacturer B) 1000Hz

Fig. 12. Output power dependency of efficiency of the MF-transformer

V. CHARACTERISTICS OF CORE MATERIALS

In this chapter, the hysteresis characteristics of the middle frequency transformer are compared.

The fundamental frequency is 1kHz and the carrier frequency of the converters is 10kHz.

Fig. 13 and 14 show the hysteresis characteristics of the FINEMET and the amorphous core material excited by the sinusoidal amplifier. The transferred power is 2kVA. It can be seen that the area of the FINEMET is much smaller than the amorphous. This obviously leads to the smaller core loss.

Fig. 15 and 16 show the hysteresis characteristics of the FINEMET and the amorphous core material excited by the MMC based SST. The transferred power is 1kVA and 2kVA for Fig. 15 and 16, respectively. It is seen that there is many deviation in the FINEMET as shown in Fig. 15. This phenomenon is well known as the increment cause of the core loss. It is not clear for the amorphous core as shown in Fig. 16. The cause of this difference shall be investigated more in future.

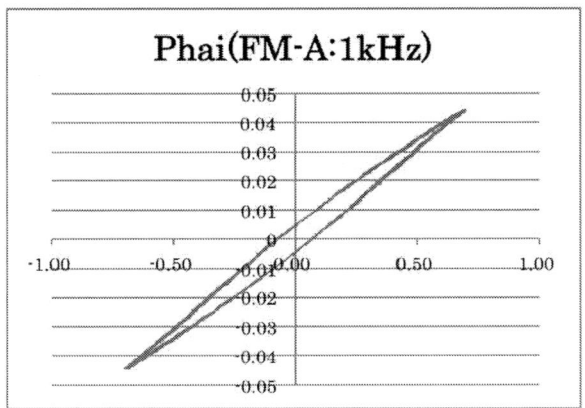

Fig. 13. Hysteresis characteristics of MF-transformer with FINEMET (FM) core (Manufacturer-A)

(Excited by Sinusoidal-Amp; Vac=200V; Sout=2kVA; FINEMET; Manufacturer-A; fsys=1kHz; fcar=10kHz)

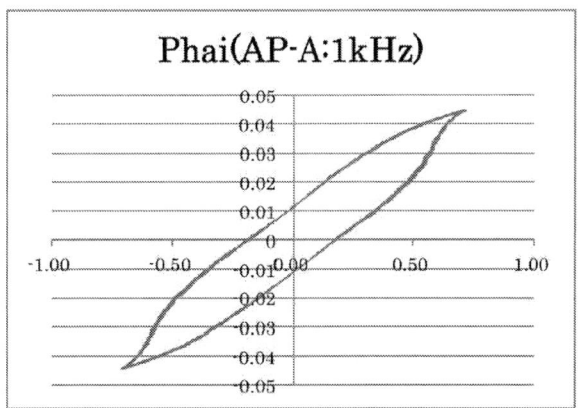

Fig. 14. Hysteresis characteristics of MF-transformer with FINEMET (FM) core (Manufacturer-A)

(Excited by Sinusoidal-Amp; Vac=200V; Sout=2kVA; Amorphous; Manufacturer-A; fsys=1kHz; fcar=10kHz)

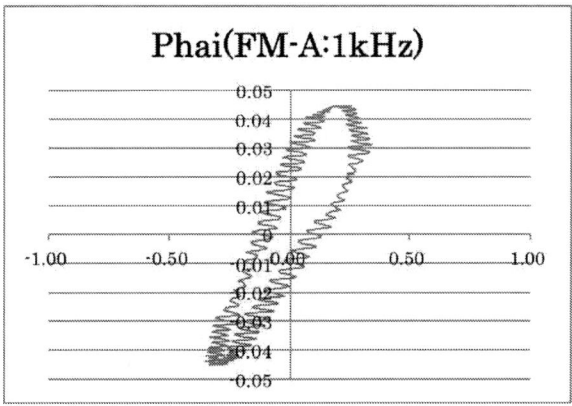

Fig. 15. Hysteresis characteristics of MF-transformer with FINEMET (FM) core (Manufacturer-A)

(Excited by MMC-SST; Vac=200V; Sout=1kVA; FINEMET; Manufacturer-A; fsys=1kHz; fcar=10kHz)

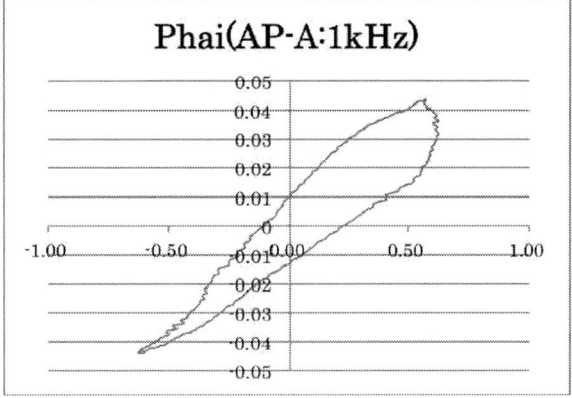

Fig. 16. Hysteresis characteristics of MF-transformer with Amorphous (FM) core (Manufacturer-A)

(Excited by MMC-SST; Vac=200V; Sout=2kVA; Amorphous; Manufacturer-A; fsys=1kHz; fcar=10kHz)

Asymmetrical shapes of both materials are also seen. The excitation current which is calculated by subtracting the secondary current from the primary current of the transformer. The waveform of the calculated excitation current is shown in Fig. 17. The negative side of the excitation current is obviously small. This is caused by the waveforms of the secondary side of the MF-transformer. The secondary side is connected to the MMC. Therefore the control of the MMC may have some trouble. However the voltages of the primary and the secondary side of the voltage do not have obvious difference as shown in Fit. 19(a) and (b). The more investigation will be done in future.

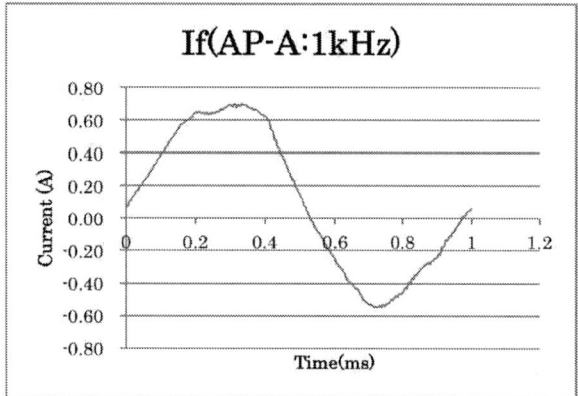

Fig. 17. Excitation current waveform of MF-transformer with Amorphous (FM) core (Manufacturer-A)

(Excited by MMC-SST; Vac=200V; Sout=2kVA; Amorphous; Manufacturer-A; fsys=1kHz; fcar=10kHz)

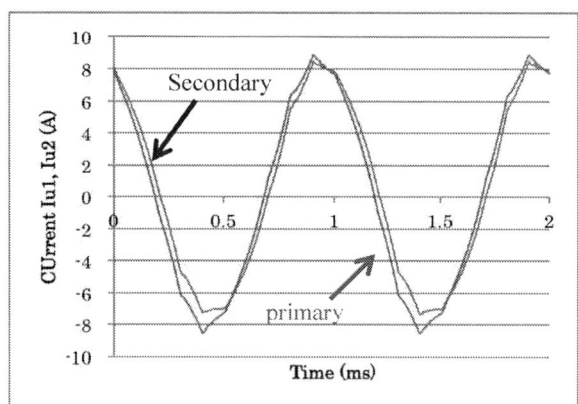

Fig. 18. Current waveform of primary and secondary side of MF-transformer with Amorphous (FM) core (Manufacturer-A)

(Excited by MMC-SST; Vac=200V; Sout=2kVA; Amorphous; Manufacturer-A; fsys=1kHz; fcar=10kHz)

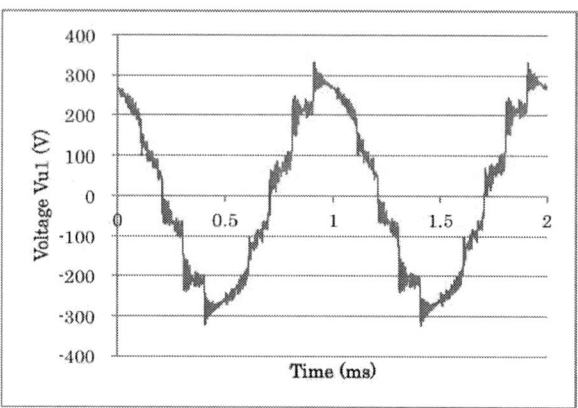

(a) Current waveform of primary side

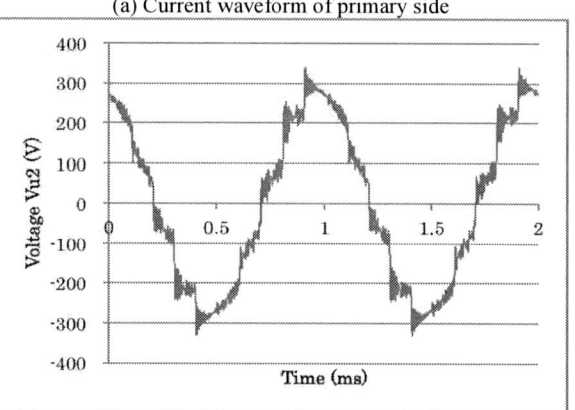

(b) Current waveform of secondary side

Fig. 19. Current waveform of primary and secondary side of MF-transformer with Amorphous (FM) core (Manufacturer-A)

(Excited by MMC-SST; Vac=200V; Sout=2kVA; Amorphous; Manufacturer-B; fsys=1kHz; fcar=10kHz)

VI. ACKNOWLEDGEMENT

This paper is based on the results obtained from the project commissioned by the New Energy and Industrial Technology Development Organization (NEDO) in Japan.

VII. CONCLUSIONS

This paper shows the possibility of smaller and lighter HVdc converter station using the technology of the solid state transformer (SST).

Experimental results show that the MMC based SST can be operated stably at the middle frequency.

Core materials are compared in efficiency and the cost. It is interesting that the higher frequency shows the better efficiency around 1kHz, 1kW operation.

The amorphous has good cost performance, though the nano-crustaline "FINEMET" has the better efficiency than the amorphous. However, the difference is not so large at 1kHz, and the amorphous core may have the better cost performance.

It is also indicated that the higher harmonic components of the converter current causes the larger loss in the transformer than the sinusoidal power supply.

REFERENCES

[1] New Energy and Industrial Technology Development Organization, Japan. (March, 2015); in Japanese, http://www.nedo.go.jp/koubo/FF2_100136.html , accessed on June 2017.

[2] Johann W. Kolar, "Intelligent Solid State Transformers (SSTs), A Key Building Block of Future Smart Grid Systems", Keynote speech at China Power System Society Conference 2011.

[3] J.W. Kolar, G.I. Ortiz, "Solid State Transformer Concepts in Traction and Smart Grid Applications", EPE-PEMC-2012-Tutorial, (2012)

[4] http://www.jfe-steel.co.jp/en/products/electrical/supercore/, accessed on June 2017.

[5] http://www.hilltech.com/pdf/Hitachi/Datasheets/FINEMET_CMC_Core_FT-3KM_F_Series.pdf, accessed on June 2017.

[6] H. Tanaka, K. Nakamura, and O. Ichinokura: "Winding Arrangement of High-frequency Amorphous Transformers for MW-class DC-DC Converters", J. Magn. Soc. Jpn., Vol.40, pp.35-38 (2016)

[7] J. E. Huber and J. W. Kolar, "Solid-State Transformers: On the Origins and Evolution of Key Concepts," in IEEE Industrial Electronics Magazine, vol. 10, no. 3, pp. 19-28, Sept. 2016.

Simulation of Wind Power Generation System Using Switched Reluctance Generator and Capacitor-less AC-AC converter

Guyuan Ji[1] and Kazuhiro Ohyama[1]

[1] Electrical Engineering, Fukuoka Institute of Technology, Japan

*E-mail: ohyama@fit.ac.jp

Abstract— The purpose of this paper is to analyze the possibility of applying capacitor-less AC-AC converters and a switched reluctance generator (SRG) to the wind power generation system (WPGS). The simulation system with Matlab Simulink composed of the SRG, controllers, and capacitor-less AC-AC converters is utilized for this purpose. In this paper, the each element that structures the simulation system is discussed. It also proposes the new concept of converter which can be applied to the WPGS using SRG. Based on the new concept taken a hint from the idea of a matrix converter, this paper presents a new control method for improving the efficiency of SRG by utilizing voltages between lines. Using the capacitor-less AC-AC converter enables to use a bipolar voltage to boost and extract the stator current quickly. This feature of converter will improve the efficiency of SRG. So, we propose the WPGS using the capacitor-less AC-AC converter and SRG.

Keywords— *Switched Reluctance Generator, Capacitor-less converter, Wind Power Generation System, Matlab simulink*

I. INTRODUCTION

Recently, due to the environmental problems, people has been drawing attention to the utilization of the natural resources to solve the problems. A wind energy is a kind of natural source that is difficult to control and low cost. For the sake of recover energy from the wind, the system to generate electricity is controlled by the generator and converter to ensure optimum performance at variable speed. Originally, doubly fed induction generator (DFIG), induction generator (IG), permanent magnet synchronous generators (PMSG) have been used for the WPGS in the past. The potential development among the common types of generators would be the SRG since it has some suitable feature for the small size WPGS. Also, this paper focuses attention on the capacitor-less AC-AC converter which can output arbitrary voltage and frequency from an AC input. Figure 1 is showing the wind power generator system which comprises of the SRG, controllers, and capacitor-less AC-AC converter.

The SRG has attracted a lot of researches to survey the possibility as the generator mainly because of its attractive advantages exceeding traditional generators as follows:

- Simple construction of stator and rotor using silicon steel plates
- Only the stator has windings
- The rotor has no magnet, i.e. the cost is low
- The each phase is independent electrically and magnetically, so it has higher reliability
- The generator can operate at low wind speed because it does not have windings or magnets on the rotor [1]. Also it permits fast response to load changes due to the low inertia of rotor [2]

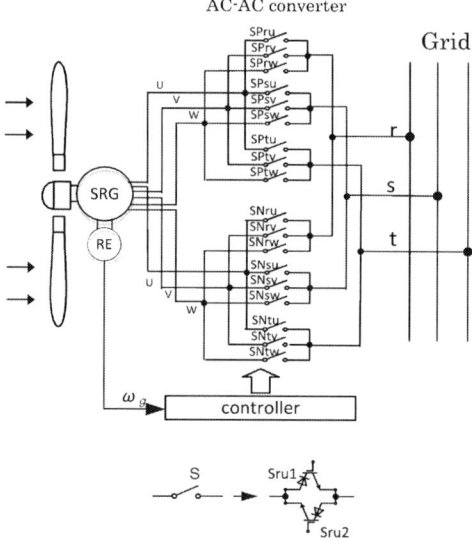

Fig. 1. WPGS using SRG and Capacitor-less AC-AC Converter.

The capacitor-less AC-AC converter has attracted a lot of researches to survey the possibility of application for the WPGS mainly because of its attractive advantages exceeding traditional converters as follows:

- It can output arbitrary voltage and frequency from AC input
- Absence of DC-link capacitors which need maintenances, space, and costs.

• It can use bipolar voltage to control bi-directional stator coils' voltages

In survey of reference literatures for converters of SRG, there are a few papers have discussions concerning power converters [3-5] and their structures [6-9]. Around the same time, the wind energy application using SRG is still in progress, and it needs to make further research on it in the future. Therefore this paper considers the SRG and capacitor-less AC-AC converter which are the elements of WPGS, and the total system of WPGS is verified by using the Matlab Simulink.

II. SWITCHED RELUCTANCE GENERATOR

The SRG is operated by controlling the switching on-off sequence of the power converter which is synchronized with an inductance curve of SRG. The stator and rotor of SRG have salient poles. And they are composed of magnetic material (generally silicon steel plates are utilized). Each tooth of stator has a concentrated coil winding. The coil windings for one phase are connected in series. Its configuration of coil windings can make two or more than two stator poles. The common structure of the SRG have 3 phase, 12 stators and 8 rotors (12/8) poles or 6 stators and 4 rotors (6/4) poles. The example of 3 phase SRG with 12/8 poles structure is shown in figure 2.

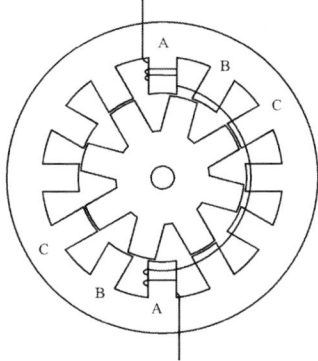

Fig. 2. Configuration of 12/8 SRG.

A. SRG Operation Principles

The operation principle of switched reluctance motor (SRM) as a motor is well-known. However the operation method of SRG is developing slowly.

The continuous rotation of SRG depends on the force of wind turbine connected to the rotor. The inductance of the generator changes periodically from maximum to minimum due to the positional relation between a rotor and the relevant stator phase. The inductance is maximum when the position of rotor and stator are aligned. When the non-aligned position, inductance is minimum.

The operating principle of SRG is similar to that of SRM. The figure 3 is showing the inductance change for the rotor position and stator current examples of generating and motoring modes. The excitation of generating mode make the flux pass through the rotor when the inductance is decreasing. When the rotor and stator poles reach to the non-aligned position, the flux path changes. It creates conversion of energy during the process of variation. In each cycle, the flux has to be returned to zero before starting the excitation of next phase.

The equation (1) is showing the torque T_{EM}. It is generated by stator current and inductance change caused due to the movement of the rotor for the excitation stator winding. The inductance change is independent to the direction of current i as follow:

$$T_{EM} = \frac{1}{2}i^2 \frac{\partial L}{\partial \theta} \qquad (1)$$

where, L is the phase inductance, θ is the rotor position.

In order to produce regenerating torque, the phase current of stator has to synchronize with the position of the rotor. In figure 3, the SRG will not produce the torque when the stator and rotor poles are aligned position since the $\partial L/\partial \theta$ is zero. However, when the rotor pole passes the stator pole, the phase inductance starts to decrease. If the stator current are flowing when the inductance is decreasing, the negative torque will be generated. Then the SRG absorbs the mechanical energy, and it converts mechanical energy to electrical energy.

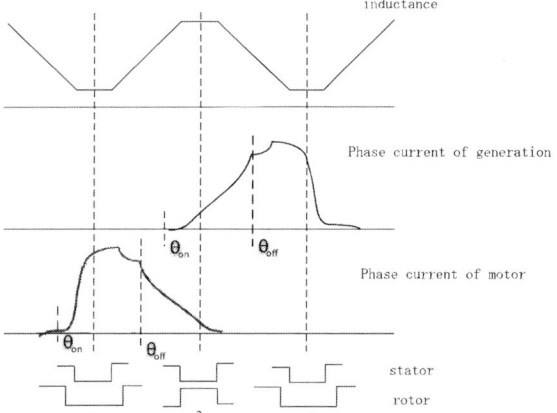

Fig. 3. Variation of inductance.

III. CAPACITOR-LESS AC-AC CONVERTER

In the generation system of SRG, the converter is an important part of generating and outputting the electric energy. Originally, a half-bridge converter is used for the wind power generation system using SRG. But, there is a DC-link capacitor between the half-bridge inverter and grid-connected converter, and it requires regular maintenance. Thus, we focus attention for a capacitor-less AC-AC converter which can be applied to a wind power generator system using SRG. The coils of SRG are independent, so that the capacitor-less AC-AC converter structure as shown in Figure 4. It composed of two matrix converters.

The capacitor-less converter which can output arbitrary voltage and frequency from alternating current (AC) input. It consists of bidirectional switch which consists of reverse blocking of Anti-parallel reverse blocking IGBTs with built-in diode. The SPij (i= r, s, t;

2922

j=u, v, w) is descried as the positive terminal of the excitation coil. Also, the SNij (i= r, s, t; j=u, v, w) is described as the negative terminal of the excitation coil.

The switch on/off, current flow from the upper switch through the SRG to the below switch. According to the position of the rotor correspondence with stator, the excitation and generation are decided.

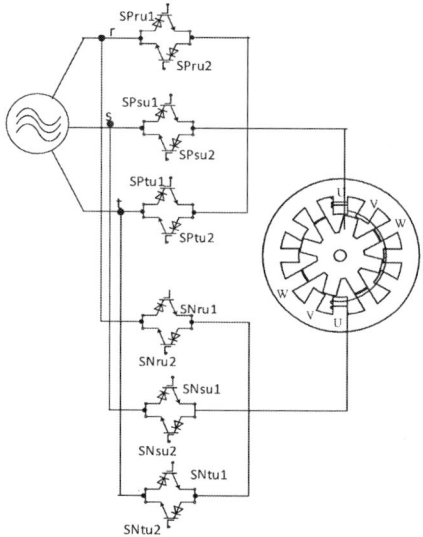

Fig. 4. Capacitor-less AC-AC Converter.

IV. CONTROL METHODS FOR SRG

In the system, the most important thing is the control for SRG depends on switching of power devices as shown in figure 4. As observed in the generation system, the control method to operate the capacitor-less AC-AC converter has to be universal for both high and low speeds. However the control method has to consider the operating conditions of SRG.

SRG control method based on the turn on and turn off angles could be divided into three groups. These groups include:

- adjust the turn on angles and turn off angle
- one of the angles may be fixed, another one is changed
- change the angles of the turn on and turn off

Due to the variation of wind speed, the fixed turn on angle and turn off angles to control the SRG in order to achieve the best performance. The position of rotor must be careful of placement to ensure the operation of the SRG, which can be continuous and improve its maximizing efficiency. Figure 5 shows the control method for SRG. The position sensor block's operation is to estimate the position of the rotor θ from output signal of the rotary encoder which is installed on rotation axis of SRG. The excitation timing block's operation is to estimate the excitation timing for position of rotor θ. According to the excitation timing, the system will choose the maximum line voltage from detectable value of line voltages. Also, the minimum line voltage will be chosen, during the generation time. The 4 steps

commutation is applied to change the line voltage for the excitation timing.

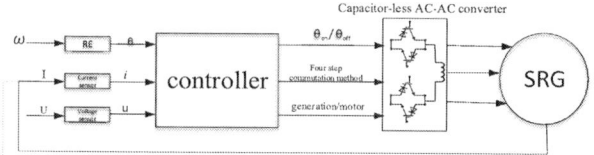

Fig. 5. Control method for SRG

In order to prevent a short circuit of the power and the open circuit of the load when SRG is controlled. During the capacitor-less AC-AC converter is operating, only one of the three switching of the upper arm turns on, and only one of the three switching of lower arm turns on at the same time.

Figure 6 shows a schematic of a two phases U and V commutation method of the matrix converter. In a steady state, both of the devices in the bi-directional switch are gated to allow both directions of current flow. The following explanation assumes that the current is from U phase to V phase and that the upper bi-directional switch SP_{rv1}, SP_{rv2} are closed.

① When Iu>0, the commutation for switch SP_{ru} to SP_{rv}, the switch SP_{ru2} is turned off, only the switch SP_{ru1} is turned on. The current will continue to flow from the switch SP_{ru1}.

② The switch SP_{rv1} is turned off. when $V_u<V_v$, switch SP_{rv1} on. The current will flow from switching SP_{rv2}.

③ The switching SP_{ru1} has no current, switching SP_{ru1} is turn off.

④ Switching SP_{rv2} turning on, finish the commutation.

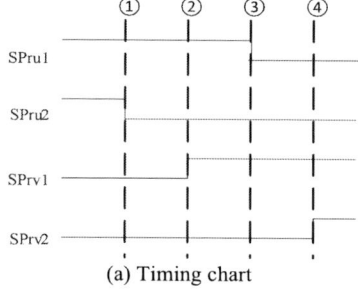

(a) Timing chart

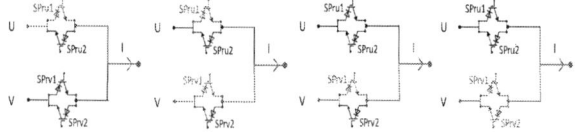

(b) switching chart

Fig. 6. 4 step commutation method.

The control method of the capacitor-less AC-AC converter for SRG is based on the 4 steps commutation method of matrix converter as shown in figure 7. The picture shows the excitation process of the lines voltage from the V_{rr} to the V_{rs}.

① When lines voltage is V_{rr} and Iu>0, the commutation for switch SP_{ru} to SN_{ru}, the switch

2923

The 2018 International Power Electronics Conference

SP$_{ru2}$ and the SN$_{ru1}$ are turned off, the switch SP$_{ru1}$ and the switch SN$_{ru2}$ are turned on. The current will continue to flow from the SP$_{ru}$ to SN$_{ru}$.

② When lines voltage is V$_{rs}$ and Iu>0, the commutation for switch SP$_{ru}$ to SN$_{su}$, the switch SN$_{ru2}$ is turned off, the switch SN$_{su2}$ is turned on. The current will flow from rom the SP$_{ru}$ to SN$_{su}$.

③ The switching SP$_{ru1}$ has no current, switching SP$_{ru1}$ is turn off. The switch SP$_{su1}$ is turned on.

④ Switching SP$_{su2}$ and the SN$_{su1}$ are turning on, finish the commutation.

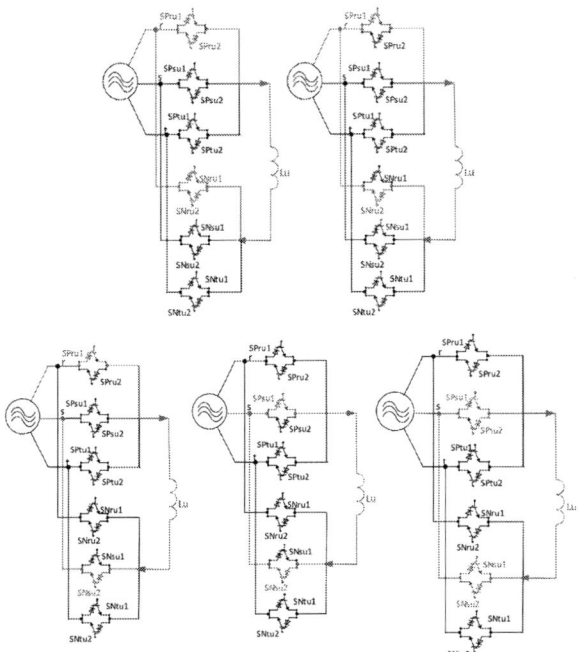

Fig. 7. Control method for SRG

SIMULATION

An output of SRG depends on a lot of various parameters for example: excitation voltage, position of rotor, excitation current, and rotor speed. These parameters may optimize the SRG's operation to get best efficiency. However, the challenge is how to optimize the parameters to achieve the optimum value that will produce the best power. As an initial stage of optimization, the modeling of wind power generator system using SRG and capacitor-less AC-AC converter is verified by Matlab Simulink. The software can simulation the dynamics of machines using all the tools that it has. The simulation is executed by using these blocks and Matlab functions.

The completed Matlab Simulink blocks are shown in figure 8. The simulation model is constituted of some modules such as SRG, capacitor-less AC-AC converter, and controller.

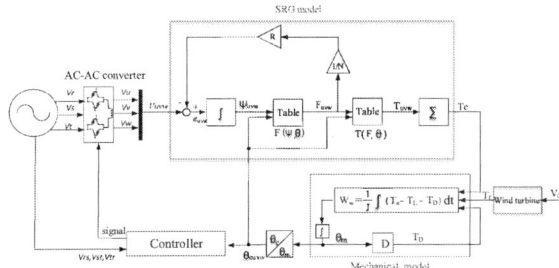

Fig. 8. Simulink system.

A. SRG Block

Figure 9 shows the SRG block that as a resistor in tandem with an inductor.

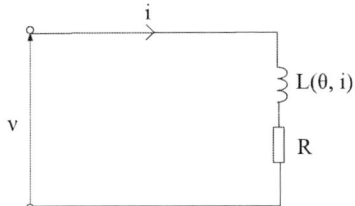

Fig.9. One phase circuit of SRG

The magnetic flux ψ obtained from the integral equation (2) of the induced voltage of stator. The SRG model is comprised of the magnetic force and the toque based on magnetic field analysis using finite element method (FEM). The magnetic force is referred to the table $F(\psi,\theta)$, which can estimate F from the rotor position θ and magnetic flux linkage ψ. Also, the toque is referred to the table $T(F,\theta)$, which can estimate T from the magnetomotive force F and rotor position θ.

$$\psi = \int (v - Ri)dt \qquad (2)$$

where, ψ is the flux linkage, v is the terminal voltage, i is the phase current, R is the phase resistance.

B. Capacitor-less AC-AC converter

The converter used is the capacitor-less AC-AC converter which consists of IGBTs and diodes. Figure 10 is showing an example of the capacitor-less AC-AC converter. The phase winding of SRG is activated by the capacitor-less AC-AC converter. The received switching signal from the controller controls the gates of the IGBTs.

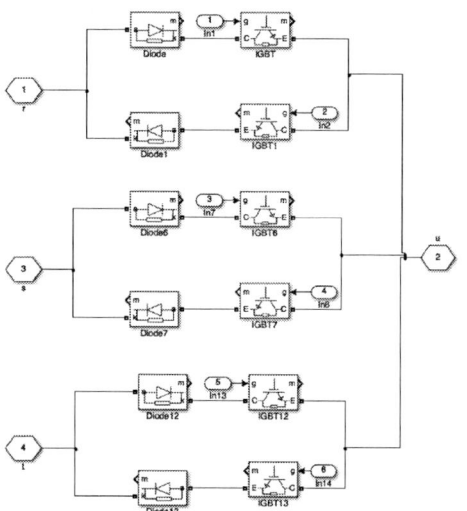

Fig.10. Configuration of capacitor-less AC-AC converter simulation block.

C. Controller

The function of the controller is make sure that the power switches can performed by the switch function between the turn on and turn off angles in figure 5.

RESULTS

Fig.11(a) shows the rotor position. In this paper, the 3 phase 12/8 SRG is used. Therefore, due to the 8 poles of rotor, aligned and unaligned rotor positions iterate every 45 degrees of rotation. And because of three phase stator windings, each phase shifts by 15 degrees.

Figure 11(b) shows the gate signals of U phase. The turn on and turn off angles are fixed. According to the operation state of the corresponding rotor position, the successful on/off sequence of switching are shown. The switching signals are input to the capacitor-less AC-AC converter.

Figure 11(c) shows the excitation voltage of capacitor-less AC-AC converter. Figure 11(d) shows the output phase current of SRG. From the figures, during the voltage is positive, the stator current starts to flow (excitation time), and the phase current is going up. When the stator voltage is switched to the negative voltage (generation time), the stator current flows continuously to the opposite polarity until the phase current becomes zero. This is the generating condition of SRG.

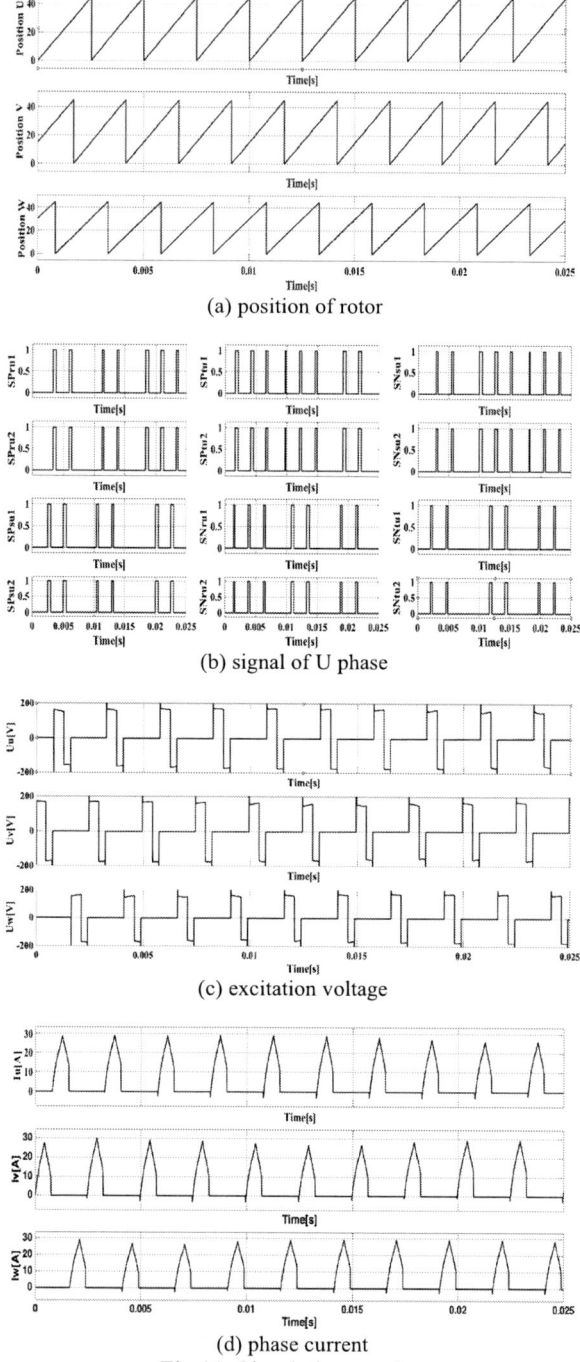

(a) position of rotor

(b) signal of U phase

(c) excitation voltage

(d) phase current

Fig.11. Simulation results.

CONCLUSIONS

The attention is focused on the potential of SRG as the wind generator and capacitor-less AC-AC converter as the power converter. This paper proposed the WPGS using SRG and capacitor-less AC-AC converter. the system are analyzed through the structure, operation principle, and control method of SRG and capacitor-less AC-AC converter. The potential of the system has been discussed. The SRG has advantages of the simple structures, ability to control the wind turbine speed at low speed, and low maintenance cost. Also, the capacitor-less AC-AC converter can be applied to the SRG, and can be used for the wind power generation system. According to the proper control method, it can make the power to be maximized. In the future, the proposing system will be verified by experiments.

REFERENCES

[1] H. Chen, "Implementation of a Three-Phase Switched Reluctance Generator System for Wind Power Applica- tions," 14th Symposium on Electromagnetic Launch Technology, Victoria, 2008, pp. 1-6. doi:10.1109/ELT.2008.104

[2] E. Darie and C. Cepisca, "The Use of Switched Reluc- tance Generator in Wind Energy Applications," 13th Power Electronics and Motion Control Conference, Poz- nan, 2008, pp. 1963-1966. doi:10.1109/EPEPEMC.2008.4635553

[3] A. Fleury, D. Andrade, E. S. L. Oliveira, G. A. Fleury- Neto, T. F. Oliveira, R. J. Dias and A. W. F. V. Silveira, "Study on an Alternative Converter Performance for Switched Reluctance Generator," Industrial Electronics, 2008. IE-CON 2008. 34th Annual Conference of IEEE, Orlando, 2008, pp. 1409-1414.

[4] M. Lipták, V. Hrabovcová, P. Rafajdus and B. Zigmund, "Switched Reluctance Machine with Asymmetric Power Converter in Generating Mode," Acta Electrotechnica et Informatica, Vol. 7, No. 1, 2007, pp. 5-10.

[5] N. K. Singh, J. E. Fletcher, S. J. Finney, D. M. Grant and B. W. Williams, "Evaluation of Sparse PWM Converter for Switched Reluctance Generator," International Con- ference on Power Electronics and Drives Systems (PEDS), Kuala Lumpur, 2005, pp. 721-725.

[6] A. Takahashi, H. Goto, K. Nakamura, T. Watanabe and O. Ichinokura, "Characteristics of 8/6 Switched Reluctance Generator Excited by Suppression Resistor Converter," IEEE Transactions on Magnetics, Vol. 42, No. 10, 2006, pp. 3458-3460. doi:10.1109/TMAG.2006.880388

[7] H. C. Lovatt and J. M. Stephenson, "Influence of Number of Poles per Phase in Switched Reluctance Motors," IEE Proceedings-B Electric Power Applications, Vol. 139, No. 4, July 1992, pp. 307-314. doi:10.1049/ip-b.1992.0037

[8] M. A. Mueller, "Design of Low Speed Switched Reluc- tance Machines for Wind Energy Converters," 9th Inter- national Conference on Electrical Machines and Drives, Canterbury, 1999, pp. 60-64.

[9] M. A. Mueller, "Design and Performance of a 20 kW, 100 rpm, Switched Reluctance Generator for a Direct Drive Wind Energy Converter," IEEE International Conference on Electric Machines and Drives, San Antonio, 2005, pp. 56-63. doi:10.1109/IEMDC.2005.195701

Gap in pagination due to withheld paper.

Pages 2927-2931

The 2018 International Power Electronics Conference

Variable frequency control and filter design for optimum energy extraction from a SiC wind inverter

Abdallah Hussein, and Alberto Castellazzi
The University of Nottingham
Power Electronics, Machines and Control (PEMC) Group

Abstract—This paper proposes the optimised control and filter design of a 12 kW 3-phase 2-level wind inverter specifically taking into account the intermittent nature of the input power. In particular, the applied control scheme aims at optimising the low-load efficiency, which corresponds to the most frequent operational condition in time, by varying the switching frequency. Specifically, a silicon carbide (SiC) converter is addressed, which operates at relatively high frequency, thus enabling a significant reduction of the filter elements. So, the output filter design also needs to be optimised to ensure that the inverter electro-magnetic performance and the size reduction enabled by SiC are kept. That is achieved by designing a variable inductor based on soft saturation core material.

Index Terms—metaloxide semiconductor field-effect transistors (MOSFETs), two-level voltage source converter (2L-VSC), small-scale wind turbine, variable inductor, closed control loop.

I. INTRODUCTION

Wind power converters have to function under intrinsically intermittent input power availability conditions and, in particular, operate most of their lifetime at relatively low load conditions. So, it is important to optimise their design and behaviour taking the intermittent nature of the source into consideration. Power semiconductor devices based on SiC exhibit many advantages in industrial applications power electronics: higher switching frequency, higher temperature capability, higher power density and higher reliability [1]. Nowadays, the commercially available power MOSFETs that satisfy the above benefits are with voltage ratings of 400, 650, 900, 1200 and 1700V for both discrete and module packages, and Schottky diodes with ratings up to 8kV [1]. This recent commercialization is expected to have the potential to deliver revolutionary impact on power electronics industry in the future. According to the availability of SiC power MOSFETs on the markets; the main areas that can be utilized by this technology are home appliances, switching power supply, PV inverters, speed drives, solid state transformers and small-scale wind turbines.

Most published studies have focused particularly on the benefits of SiC over Si in wind turbines [2]–[6], indicating the benefits to improve efficiency and higher switching frequency capability and higher power densities due to the heat-sink and inductor sizes reduction. However, in this study the aim at optimising the low load operation and energy efficiency rather than only the power efficiency taking advantage of the characteristics of SiC and the intermittent nature of wind power availability.

Small-scale wind turbines are mainly used to supply electricity to homes, farms, and small business in rural areas and developing countries. They differ from their large-scale wind turbine in their rating, size, generator type, and the topology of the power electronics conversion system [7]. For small-scale wind conversion system in the power ranges up to 100kW, it is a trend to use a PMSG in industry applications especially for a directly driven wind turbine with benefits of size and weight reduction. As there is no reactive power needed, the conversion system is a simple 3-phase diode rectifier with dc-dc boost converter connected to 2-level 3-phase inverter as shown in Fig.1 [8], [9].

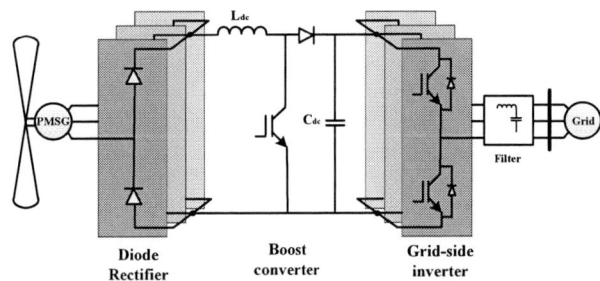

Fig. 1: Small scale wind turbine power conversion system.

The focus is on the inverter stage in this work, so that, the wind turbine model under the maximum power point tracking is simplified by a current source connected in parallel with the DC link capacitor of the grid-side inverter as shown in Fig. 2

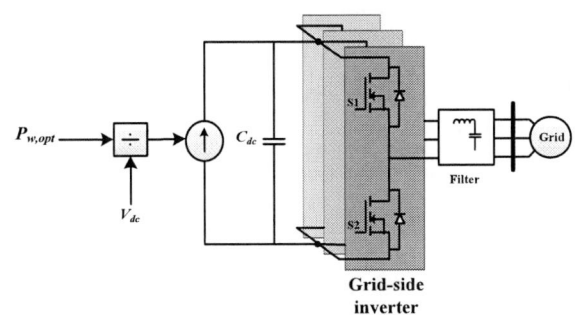

Fig. 2: Simplified model of small-scale wind turbine under MPPT.

The input signal of the current source is:

$$I_w = \frac{P_{w,opt}}{V_{dc}} \tag{1}$$

2932

Where, $P_{w,opt}$ is the optimal wind turbine output power and V_{dc} is the DC link voltage of the converter.

A commercial wind turbine system and probability based on actual measured wind data are considered in this study. Fig.3 shows the output power curve of a 10kW small-scale wind turbine, and the wind speed distribution versus wind speed. It is seen that wind turbines spend most of their operational time working at low output power (the region of interest).

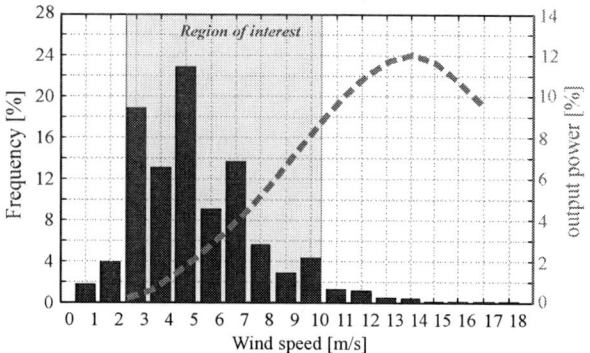

Fig. 3: The wind speed distribution (blue data) and wind output power (red data) versus wind speed.

II. INVERTER EFFICIENCY AT DIFFERENT LOADS

In this section, a SiC-based three-phase inverter has been designed and built in the lab to deliver a through characterization performance of 1200V/20A SiC MOSFETs (CCS020M12CM2 from CREE). The inverter was 12kW 3-phase 2-level used for small-scale wind turbines, and its parameters are shown in Table I. To deliver a demonstration of the inverter power cell characterization, a series of measurements of the power losses and efficiency are performed for different power loading upto 12kW input power at heatsink temperature of $60^{o}C$, which is a reference temperature for some industrial applications , and different switching frequency from 16kHz to 32kHz. The dead-time of 500ns and DC-link voltage of 720V are considered. The test was made with a variable three-phase resistive load to emulate the grid connection at different power loading.

TABLE I: Converter parameters and test conditions

Parameter	Value
Input power rating, P_{in}	12kW
DC-link voltage, V_{DC}	720V
DC-link capacitor C_{DC}	56μF
Grid phase voltage, V_g	230V
Switching frequency, f_{sw}	16-32kHz
Dead-time	500ns
Devices	CREE CCS020M12CM2

The test setup schematic is shown in Fig 4: the converter is powered by a dc power supply, and the power losses are measured by Yokogawa WT3000E precision power analyser. In this stage, to assess the performance of the power cell based on three-phase SiC power module, the voltage measurements are set before the filter inductor L_f to exclude the inductor losses from the assessment.

The DC-link PCB is designed based on planar structure to minimize parasitic inductance in the commutation loop at high switching speed, and two sets of DC link is to have a asymmetric parasitic inductance between the individual half-bridge sections. High frequency ceramic capacitors are fixed very close to inverter legs used to provide minimum voltage overshoots. The carrier and modulating signals are performed using a control platform based on Texas Instrument TMS320C6713 floating point digital signal processor (DSP) and Actel ProAsic3 field programmable gate array (FPGA), and then processed to the gate driver to generate the required switching sequence. The gate driver board is kept very close and directly connected to the power cell in order to minimize the driving loop and reduce the noise pick-up at the gate signals.

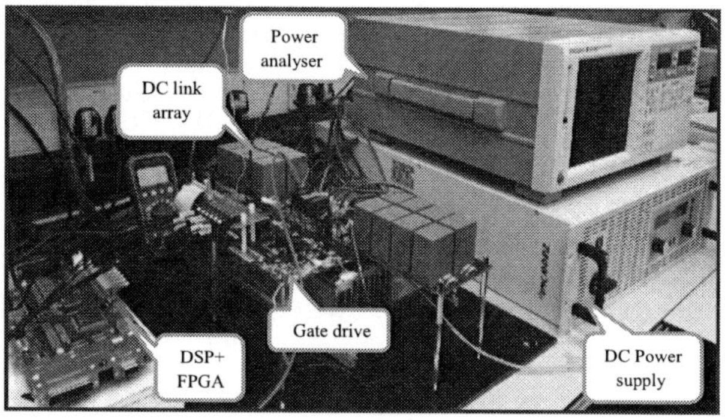

Fig. 4: Three-phase Inverter test setup.

The power cell measured efficiency against the output power at different switching frequency and at $60^{o}C$ heat sink temperature, is shown in Fig. 5. At higher output powers, the performance difference between different switching frequencies is insignificant, it is clear by increasing the switching frequency by a factor 2 implies a reduction in efficiency of less than 0.35% when working around 11.5kW output power and $60^{o}C$ heat sink temperature (less than 40 W difference in total power dissipation). At light loads (the region of interest), in which the wind turbines spend most of their operational time, however, the efficiency is reduced gradually by increasing the switching frequency. The reason for the efficiency reduction is due to the domination of switching losses instead of conduction losses. For example, at 400W output power which corresponds to the starting wind speed, about 5% efficiency drop in the performance curve is observed when moving from 16kHz to 32kHz switching frequencies.

Fig. 6 presents the annual lost energy calculated based on the experimental measured power losses of the converter power cell at different power loading and the hourly wind speed distribution throughout the year in Nottingham. It is clear that lost energy difference at high wind speeds, which has a chance of small fraction of wind turbine operational time, is almost negligible. The significant reduction of efficiency at lighter loads in the region of highly probable wind speed distribution, on the other hand, leads to a significant difference in the lost energy throughout the whole year. To minimize the lost energy at partial loads, one can consider the reduction of

The 2018 International Power Electronics Conference

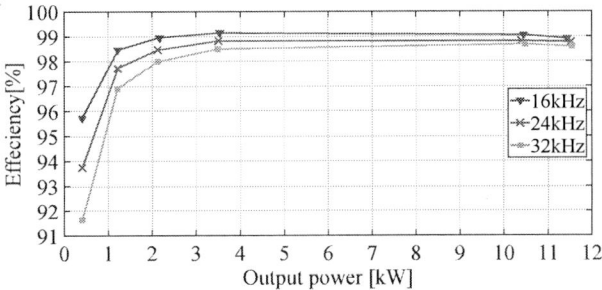

Fig. 5: Power cell measured efficiency vs. output power.

the lost energy by changing the switching frequency in the region of light loading.

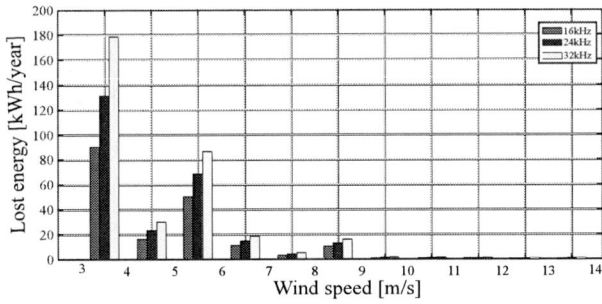

Fig. 6: Annual lost energy vs. wind speeds at different switching frequencies.

III. DESIGN OF VARIABLE FREQUENCY CONTROL

To improve the converter efficiency over a wide range of operation, one important parameter is the switching frequency. The switching frequency is varied by tracking the peak load current based on the relation shown in Fig.7, which is extracted based on the filter-inductor requirement to ensure the harmonic distortion limits and the nature of wind speed distribution. This relation is implemented in a C script in DSP of the control platform to drive the operating switching frequency. Fig.8 shows an implementation flow chart for the variable switching frequency algorithm. The algorithm can be run continuously, the sensed input current are taken from the Analog-to-Digital Converter(ADC). Next, the algorithm replaces the previous sample of the current with the present sample, and averages the number of samples per cycle to generate the rms and peak current. Based on the current value, the switching frequency ($f_{sw}[kHz]$) will updated.

$$f_{sw} = \begin{cases} 16, & 1A \leq I \leq 8A \\ 0.003I^2 + 0.002I + 13.86, & 8A < I \leq 17.5A \\ 32, & 17.5A < I \leq 24.5A \end{cases}$$

IV. DESIGN OF VARIABLE INDUCTOR-FILTER

The minimum inductance value of the inductor-filter for various frequencies, i.e 16kHz to 32kHz, are calculated based on

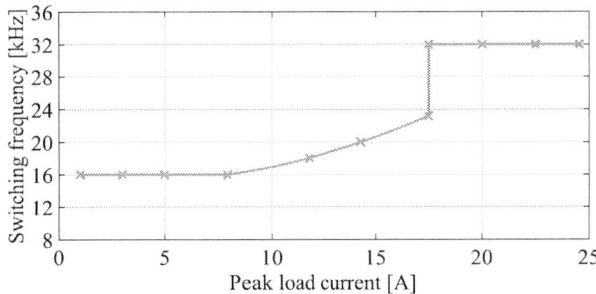

Fig. 7: Switching frequency and output current relationship implemented in the control platform.

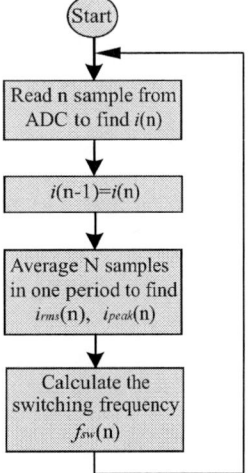

Fig. 8: Variable switching frequency control algorithm flowchart.

(2) [10], taken into account the maximum output ripple current of 20% of the peak current [11]. The switching frequencies and the corresponding minimum required inductance are listed in Table II.

$$L_{f(min)} = \frac{V_{dc}}{4f_{sw}\triangle I_{PP}} \qquad (2)$$

TABLE II

$f_s(kHz)$	$L_{min}(mH)$
16	2.3
24	1.5
32	1.1

Here, we investigate the use of Kool Mμ core material for the design of variable inductor with SiC switches operating at variable switching frequency, i.e 16kHz to 32kHz. This material has a soft saturation behaviour: it permits lower permeability at high currents, while higher permeability values at low currents. However, inductors designed using ferrite cores exhibit constant permeability over the dc bias current range and the permeability sharply dropped when the core operating close to saturation as shown in Fig.9.

2934

The 2018 International Power Electronics Conference

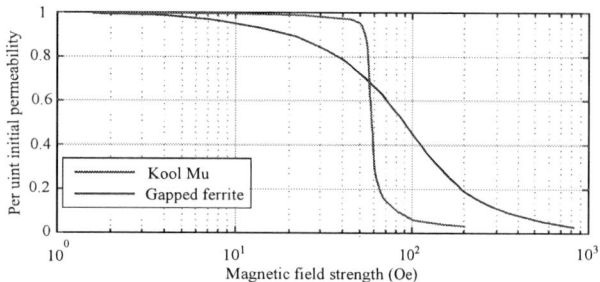

Fig. 9: The permeability of 60μ Kool Mμ powdered iron core and gapped ferrite core versus magnetic field strength.

The material properties of the Kool Mμ LE114 E-core are mentioned in [12]. It has a permeance of 445 for 60μ KoolMμ and a relative permeability of 60 at no load and characterized by a distributed air gap which has no issues with the fringing flux.

Three single-phase Kool Mμ inductors were designed with 85 turns using 11×0.63mm round conductor resulting in 1.1mH inductance at the maximum current of 24.5A and switch frequency of 32kHz. Some design simulations are performed to calculate the variable inductance value at different dc offset currents as shown in Table III, where the inductance value is 2.97mH at 1A current which reduces to 1.07mH at 25A.

TABLE III

$\%\mu_r$	μ_r	$I(A)$	$L(mH)$
100	60	1	2.97
85	51	5.7	2.73
70	42	11	2.25
55	33	15.9	1.77
35	21	25	1.07

The designed single-phase 60μ Kool Mμ inductor was tested by impedance spectroscopy up to 20A dc bias current (the current limit of the impedance analyser). The calculated results are in agreement with measurements over the dc offset current range as shown in Fig. 10.

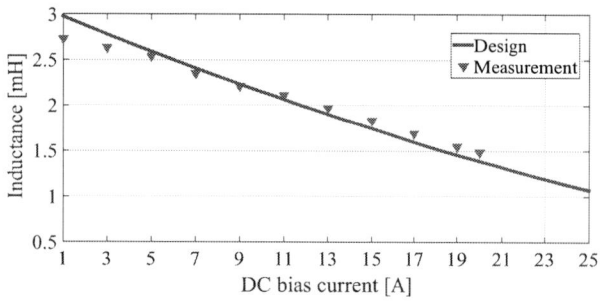

Fig. 10: Calculated variable inductance and measured values vs. dc offset current

V. EXPERIMENTAL VERIFICATION

The three-phase two-level inverter with the technical parameters given in Table I will be used with the variable inductor for the verification of the variable switching frequency technique at low power demands. The final design of variable inductor-filter together with the variable inductor are shown in Fig. 11.

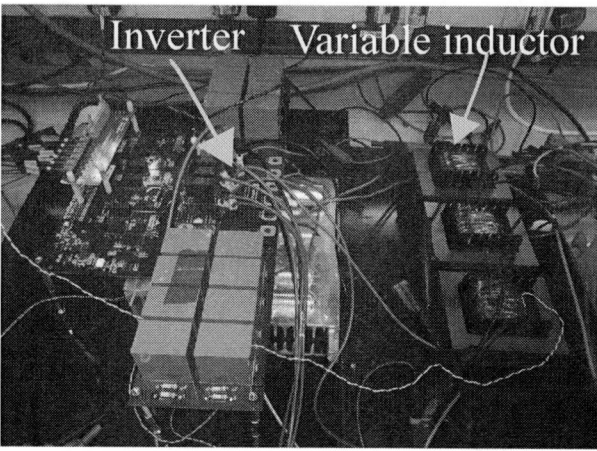

Fig. 11: Hardware of variable inductor-filter with 3-phase 2-level inverter.

Fig 12 shows the experimental phase current waveform at full load: the current is processed in the fast fourier transform (FFT) and its spectrum is shown in Fig. 13 and Fig. 14, which indicates that the total harmonic distortion (THD), which is less than 5%, and the first 40th current harmonics orders meet the standard requirements [13], [14].

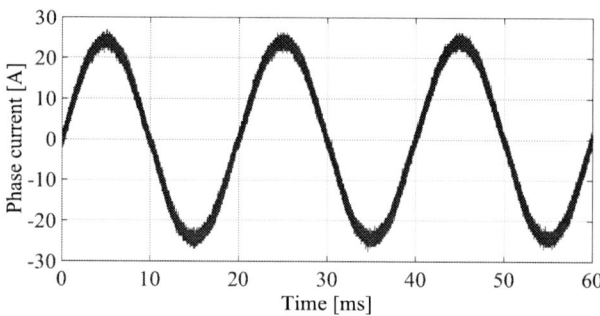

Fig. 12: Experimental current waveform of the inverter at full power.

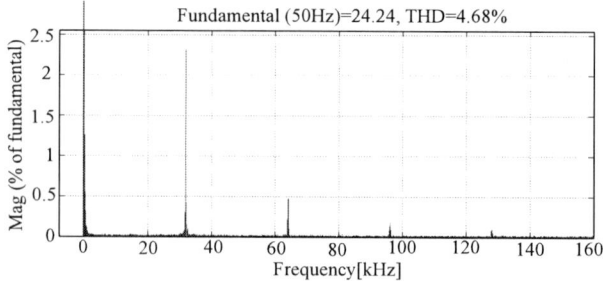

Fig. 13: Current spectrum at full power.

The design also ensures that the power quality requirements is attained over the whole range of wind turbine power loading.

2935

The 2018 International Power Electronics Conference

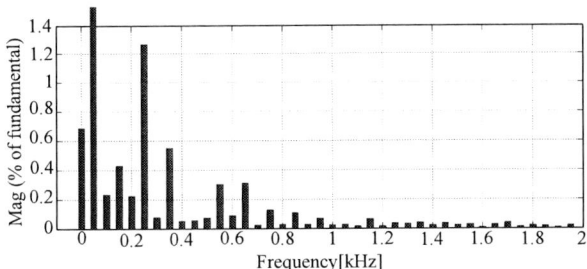

Fig. 14: Current harmonic limits for the first 40th harmonics at full power.

The total demand distortion (TDD) gives better insight about the impact of harmonic distortion at different loads, i.e the THD could be very high at low loads, but its impact on the system is low. Fig 15 shows the TDD is comply with the grid requirements, which is less than 5% [13] over the whole range of loading even at high THD at low power demands.

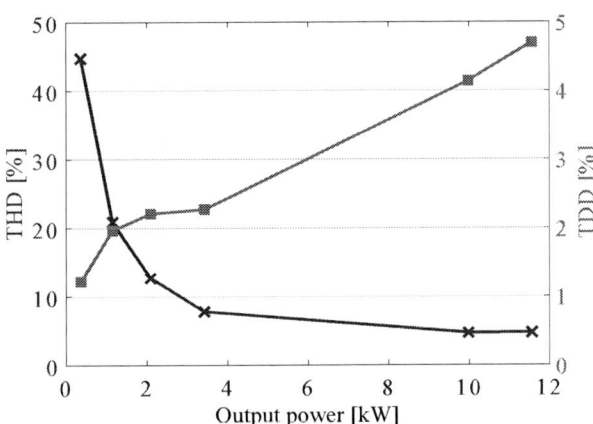

Fig. 15: Measured current THD and TDD for various output power.

Yokogawa WT3000E precision power analyser is used to measure the overall efficiency. The voltage output is measured after the filter-inductor to include the inductor losses together with the power cell losses. The gate driver and the control signal losses are excluded from the measurements. The overall efficiency with fixed switching frequency (32kHz) and variable switching frequency (16kHz-32kHz) is presented in Fig. 16. The performance difference between fixed and variable switching frequency becomes clearer at low power demand, which corresponds to the most frequent operational time of wind turbine. The converter achieved peak efficiency of 99% with variable switching frequency.

For grid-connected applications, the annual cumulative lost energy for SiC inverter with fixing switching frequency is about 227.5kWh, and in contrast, 111.2kWh for inverter with variable switching frequency as shown in Fig 17. This results in a total lost energy saving up-to 51% by adapting variable switching frequency concept.

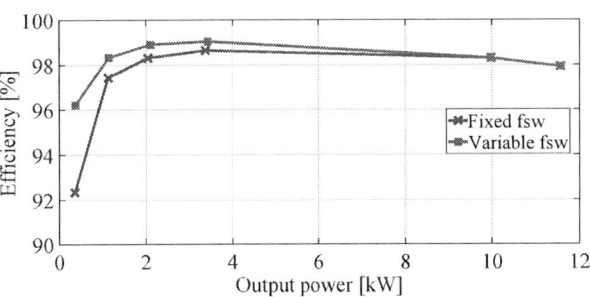

Fig. 16: Measured overall efficiency vs. Load operating at fixed and variable switching frequency.

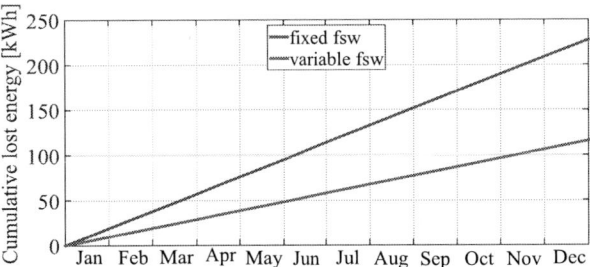

Fig. 17: Cumulative energy loss of the two-level grid-connected inverter based on fixed and variable switching frequency.

VI. CONCLUSIONS

Given the fact that wind turbines work at their low output power for most of their operational time, the authors have presented an approach to maximize the converter efficiency by varying the switching frequency at the region of highly probable wind speed distribution. This corresponding to an increase in the annual energy delivered to the grid (about 51% saving in the lost energy), while keeping the decrease in the size of passive filter elements due to high frequency capability of SiC MOSFETs. Finally, the electro-magnetic performance of the inverter was experimentally demonstrated, meeting grid quality requirements at full and partial loads.

REFERENCES

[1] A. Castellazzi, A. Fayyaz, G. Romano, L. Yang, M. Riccio, and A. Irace, "Sic power mosfets performance, robustness and technology maturity," *Microelectronics Reliability*, vol. 58, pp. 164–176, 2016.

[2] H. Zhang and L. M. Tolbert, "Efficiency impact of silicon carbide power electronics for modern wind turbine full scale frequency converter," *IEEE Transactions on Industrial Electronics*, vol. 58, no. 1, pp. 21–28, Jan 2011.

[3] M. Adamowicz, S. Giziewski, J. Pietryka, M. Rutkowski, and Z. Krzeminski, "Evaluation of sic jfets and sic schottky diodes for wind generation systems," in *2011 IEEE International Symposium on Industrial Electronics*, June 2011, pp. 269–276.

[4] W. L. Erdman, J. Keller, D. Grider, and E. VanBrunt, "A 2.3-mw medium-voltage, three-level wind energy inverter applying a unique bus structure and 4.5-kv si/sic hybrid isolated power modules," in *2015 IEEE Applied Power Electronics Conference and Exposition (APEC)*, March 2015, pp. 1282–1289.

[5] A. Blinov, A. Chub, D. Vinnikov, and T. Rang, "Feasibility study of si and sic mosfets in high-gain dc/dc converter for renewable energy applications," in *IECON 2013 - 39th Annual Conference of the IEEE Industrial Electronics Society*, Nov 2013, pp. 5975–5978.

[6] A. Hussein, A. Castellazzi, P. Wheeler, and C. Klumpner, "Performance benchmark of si igbts vs. sic mosfets in small-scale wind energy conversion systems," in *2016 IEEE International Power Electronics and Motion Control Conference (PEMC)*, Sept 2016, pp. 963–968.

[7] D. M. Whaley, G. Ertasgin, W. L. Soong, N. Ertugrul, J. Darbyshire, H. Dehbonei, and C. V. Nayar, "Investigation of a low-cost grid-connected inverter for small-scale wind turbines based on a constant-current source pm generator," in *IECON 2006 - 32nd Annual Conference on IEEE Industrial Electronics*, Nov 2006, pp. 4297–4302.

[8] F. Blaabjerg, M. Liserre, and K. Ma, "Power electronics converters for wind turbine systems," *IEEE Transactions on Industry Applications*, vol. 48, no. 2, pp. 708–719, March 2012.

[9] M. Malinowski, A. Milczarek, R. Kot, Z. Goryca, and J. T. Szuster, "Optimized energy-conversion systems for small wind turbines: Renewable energy sources in modern distributed power generation systems," *IEEE Power Electronics Magazine*, vol. 2, no. 3, pp. 16–30, Sept 2015.

[10] Y. Liu, H. A. Mantooth, J. C. Balda, and C. Farnell, "Realization of high-current variable ac filter inductors using silicon iron powder magnetic core," in *2017 IEEE Applied Power Electronics Conference and Exposition (APEC)*, March 2017, pp. 855–860.

[11] T. C. Y. Wang, Z. Ye, G. Sinha, and X. Yuan, "Output filter design for a grid-interconnected three-phase inverter," in *Power Electronics Specialist Conference, 2003. PESC '03. 2003 IEEE 34th Annual*, vol. 2, June 2003, pp. 779–784 vol.2.

[12] P. P. U. . Magnetics Powder Core Catalog, MAGNETICS.

[13] "Ieee recommended practice and requirements for harmonic control in electric power systems - redline," *IEEE Std 519-2014 (Revision of IEEE Std 519-1992) - Redline*, pp. 1–213, June 2014.

[14] "Photovoltaic (pv) systems characteristics of the utility interface," *International Advances in Surgical Oncology Electrotechnical Commission, Geneva, Switzerland, IEC 61727*, 2004.

Experimental Verifications of UPFC using Deadbeat Control with 3-phase Unbalanced Compensation

Shin-ichi Hamasaki[1][*], Hiroto Fukuda[1], Syohei Tokumaru[1] and Mineo Tsuji[1]

1 Electrical and Electronic Engineering, Nagasaki University, Nagasaki, Japan
*E-mail: hama-s@nagasaki-u.ac.jp

Abstract-- Unified Power Flow Controller (UPFC) is one of the Flexible AC Transmission System (FACTS). The UPFC has been developed to improve the power quality of the power systems in order to keep conditions of the power line without increasing the number of transmission branches. The UPFC is divided into two parts. One is the parallel system and the other is the series system. The parallel system can compensate the reactive power and 3-phase unbalance due to the load in the power line to improve the power factor. In addition, the active power is controlled to maintain voltage of the common DC capacitor and unbalance among 3-phase active power flow. The series system can control active and reactive power flow. In addition, the voltage drop at load side can be compensated to maintain proper line voltage. In both systems, Output current in parallel system and output voltage control in series side are reqired accurate and quick response in adopt to variable situations. The deadbeat (DB) control is applied to realize good performance. Experiment is demonstrated to investigate performance of the proposed method. The theory of the control are described and experimental results in same tipical situations are presented for verification of its characteristics .

Keywords— power flow controller, reactive power compensation, compensation, deadbeat control

I. INTRODUCTION

In the current power system, a drop in the power factor due to the reactor component of the power transmission line and a reduction in voltage due to an unexpected accident or disturbance causesvarious problems.Also distributed generators using the renewable energy are increasing all over the world. This causes increase of power transmission capacity, power factor drop, harmonics problem and other unexpected accidents in the power transmission system. Such a decline of the power quality should be improved to obtain the stability of the power transmission system. However, when trying to send more electric power to solve this problem, it is necessary to newly add power transmission lines. As a solution to these problems, the introduction of UPFC has attracted attention and development is being carried out.The Unified Power Flow Controller (UPFC)[1]-[11] is a kind of FACTS components. This fills a role of management of the active power and the reactive power flow and maintenance of proper line voltage. UPFC is able to improve the total power quality and

optimize the power flow in the power transmission system.UPFC is reguired to obtain quick and accurat responce for these operations.

In this research, the deadbeat (DB) control[12][13] is introduceed to realize accurate and quick response of output current in parallel system and voltage in series system. The proposed UPFC gives compensation of the reactive power and 3-phase unbalance in parallel system and control of power flow and voltage drop compensation on series system. In this paper, control theory of the proposed system is explained. Experimental results in some tipical situations are presented to confirm characteristics of the proposed UPFC. Also, confirm the transient response of the DB control due to load fluctuation.

II. CONTROL SCHEME

A. Configuration of UPFC

The configuration of UPFC in this study is illustrated in Fig.1. The UPFC has two inverters with a common capacitor on DC bus. The proposed block diagram in the parallel system with the DB control is shown in Fig.2. The parallel system is able to compensate the reactive power and 3-phase unbalance to improve the power factor. In addition, the active power of output current of the parallel system is controlled in order to maintain voltage of the DC capacitor constant. The DB control can realize accurate and quick output following to the current commnad for compensations.

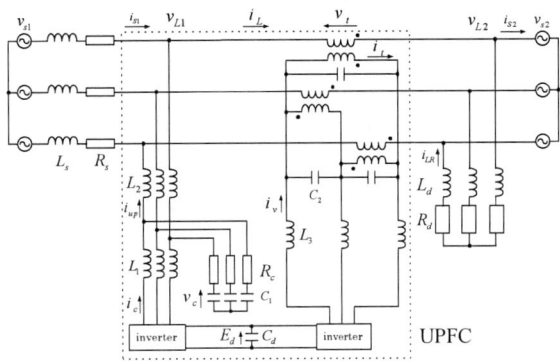

Fig. 1. Circuit configuration of UPFC.

The proposed block diagram in the series system with the DB control is shown in Fig.3. The series system can control active and reactive power flow and compensate voltage drop to maintain line voltage properly. Control performance with quick and accurate response is required in both system as the control performance directly affects compensation performance.

All calculations are executed in α-β coordinates. After calculation, states in α-β are transformed into 3-phase component. The DB control is applied to regulate the output current in parallel system and output voltage in series system. The DB control can follows the controlled object to the command value by a sampling period. Fig.4 shows the sequence image of the DB control. In the proposed scheme, output is controlled by changing the output pulse width $\Delta T(k)$ of the inverter. It is calculated from the detected current and voltage at a sampling period using the theoretical equation obtained from the state equation.

B. 3-phase unbalance compensation

The 3-phase unbalance compensation is performed by the parallel system in Fig.2. The compensation current reference of the parallel side is determined from 3-phase unbalance currents to cancel the line current unbalance.Current reference $i_{upx}{}^{*}$ for 3-phase unbalanced compensation is calculated by (1).

$$I^{*}_{upx} = I^{*}_{sx} - I_{Lx} \quad (x = u, v, w) \tag{1}$$

Average amplitude of the current is calculated from the detected load 3-phase load current in order to keep the active power balance into the parallel side inverter. After that, Line current reference for balanced 3-phase can be obtained from the average amplitude and phase angle of line current. Difference between the line current reference and the load current becomes the output reference of the parallel side for the unbalance compensation.

C. Power flow control

The power flow control is performed by the series system in Fig.3. The active power and the reactive power can be controlled independently by using the p-q theory. The active p component and the reactive q component are calcurated by the p-q transformaton. The power flow control is realized by PI controller for p-component and q-component respectively.

D. DB control in parallel system

Fig.5 shows an equivalent circuit for the DB control. Theory of single phase circuit is enhanced to 3-phase system using α-β transformation. The output pulse width ΔT can be calculated from current and voltage value. When four variables i_c, v_c, i_{up} and v_{L1} are selected as the states, the state equations are derived in (2) and (3).

$$\frac{d}{dt} \boldsymbol{x_I} = \boldsymbol{A_I}\boldsymbol{x_I} + \boldsymbol{b_I} v_{i1} \tag{2}$$

$$i_{up} = \boldsymbol{c_I}\boldsymbol{x_I} \tag{3}$$

$$\boldsymbol{x_I} = \begin{bmatrix} i_c \\ v_c \\ i_{up} \\ v_{L1} \end{bmatrix} \quad \boldsymbol{A_I} = \begin{bmatrix} -\dfrac{R_c}{L_1} & -\dfrac{1}{L_1} & \dfrac{R_c}{L_1} & 0 \\ \dfrac{1}{C_1} & 0 & -\dfrac{1}{C_1} & 0 \\ \dfrac{R_c}{L_2} & \dfrac{1}{L_2} & -\dfrac{R_c}{L_2} & -\dfrac{1}{L_2} \\ 0 & 0 & 0 & 0 \end{bmatrix} \quad \boldsymbol{b_I} = \begin{bmatrix} \dfrac{1}{L_1} \\ 0 \\ 0 \\ 0 \end{bmatrix}$$

$$\boldsymbol{c_I} = \begin{bmatrix} 0 & 0 & 1 & 0 \end{bmatrix}$$

(2) and (3) are expression of the continuous time system. These can be converted into the sample value system using the output pulse width ΔT as shown in (4) and (5) respectively.

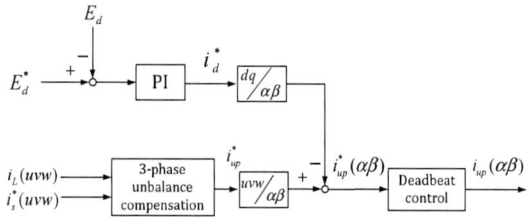

Fig. 2. Block diagram of parallel system of UPFC.

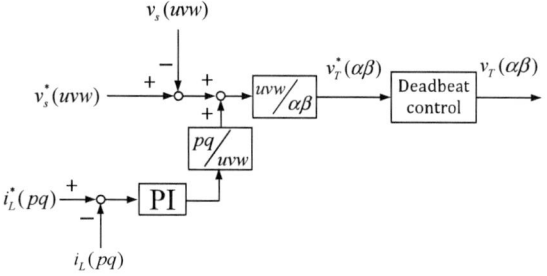

Fig. 3. Block diagram of series system of UPFC.

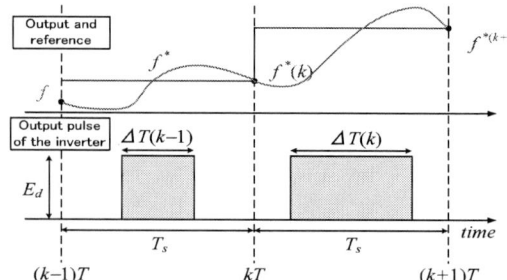

Fig. 4. Sequence of DB control.

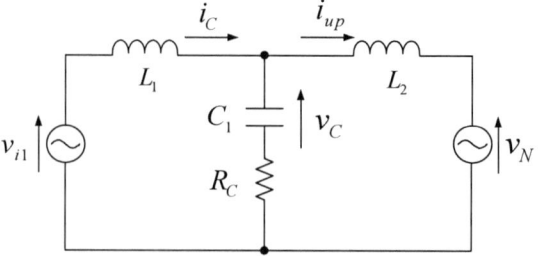

Fig. 5. Equivalent circuit of parallel system for DB control (single phase in α,β components).

2939

$$x_1(k+1) = Fx_1(k) + g\Delta T_1(k) \tag{4}$$

$$i_{up}(k) = c_1 x_1(k) \tag{5}$$

$$F = e^{A_1 T_s} \tag{6}$$

$$g = e^{\frac{A_1 T_s}{2}} b_1 E_d \tag{7}$$

$$F = \begin{bmatrix} F_{11} & F_{12} & F_{13} & F_{14} \\ F_{21} & F_{22} & F_{23} & F_{24} \\ F_{31} & F_{32} & F_{33} & F_{34} \\ 0 & 0 & 0 & 1 \end{bmatrix} \quad g = \begin{bmatrix} g_1 \\ g_2 \\ g_3 \\ g_4 \end{bmatrix}$$

F and g are the matrixes, which have constant parameters calculated from A and b by (6) and (7).

(8) is obtained from (4) and (5) as the discrete time system.

$$i_{up}(k+1) = F_{31}i_c(k) + F_{32}v_c(k) + F_{33}i_{up}(k) + F_{34}v_N + g_3\Delta T_1(k) \tag{8}$$

The pulse width $\Delta T_1(k)$ can be derived in (9) by rewriting $i_{up}(k+1)$ to $i_{up}^*(k+1)$.

$$\Delta T_1(k) = \frac{i_{up}^*(k+1) - F_{31}i_c(k) - F_{32}v_c(k) - F_{33}i_{up}(k) - F_{34}v_N(k)}{g_3} \tag{9}$$

When $\Delta T_1(k)$ is negative, the DC voltage is iven as $-E_d$ with $|\Delta T_1|$.

E. DB control in series system

Fig.6 illustrates a single phase equivalent circuit for DB control. The output pulse width is obtained by calculating from current and voltage value. When three variables i_v, v_t, and I_t are selected as the states, the state equations are derived in (10) and (11).

$$\frac{d}{dt}x_2 = A_2 x_2 + b_2 v_{i2} \tag{10}$$

$$v_t = c_2 x_2 \tag{11}$$

$$x_2 = \begin{bmatrix} v_t \\ i_v \\ I_t \end{bmatrix} \quad A_2 = \begin{bmatrix} 0 & \frac{1}{3C_2} & -\frac{1}{C_2} \\ -\frac{1}{L_3} & 0 & \frac{1}{L_3} \\ 0 & 0 & 0 \end{bmatrix} \quad b_2 = \begin{bmatrix} 0 \\ \frac{1}{L_3} \\ 0 \end{bmatrix}$$

$$c_2 = \begin{bmatrix} 1 & 0 & 0 \end{bmatrix}$$

Deriving as the same as the parallel system, These can be converted into the sample value system in (12) and (13).

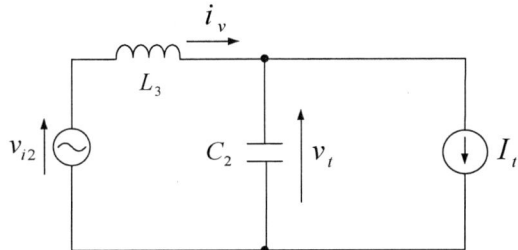

Fig. 6. Equivalent circuit of series system for DB control (α,β single phase).

$$x_2(k+1) = Mx_2(k) + n\Delta T_2(k) \tag{12}$$

$$v_t(k) = c_2 x_2(k) \tag{13}$$

$$M = e^{A_2 T_s}, \quad n = e^{\frac{A_2 T_s}{2}} b_2 E_d \tag{14}$$

$$M = \begin{bmatrix} M_{11} & M_{12} & M_{13} \\ M_{21} & M_{22} & M_{23} \\ 0 & 0 & 1 \end{bmatrix} \quad n = \begin{bmatrix} n_1 \\ n_2 \\ n_3 \end{bmatrix}$$

The matrixes M and n calculated by (14) have constant values. (15) is derived from (12) and (13).

$$v_t(k+1) = M_{11}v_t(k) + M_{12}i_v(k) + M_{13}I_t(k) + n_1\Delta T_2(k) \tag{15}$$

The pulse width $\Delta T(k)$ can be derived in (16) by rewriting $v_t(k+1)$ to $v_t^*(k+1)$.

$$\Delta T_2(k) = \frac{v_t^*(k+1) - M_{11}v_t(k) - M_{12}i_v(k) - M_{13}I_t(k)}{n_1} \tag{17}$$

the calculated $\Delta T_2(k)$ outputting, $v_t(k+1) = v_t^*(k+1)$ can be realized in one sample later.

III. SIMULATION

Simulation is performed in Fig.1. Circuit parameters in the simulation are listed in Table I. 3-phase load resistances of R_d are selected to be 3-phase unbalance situation. Figs.7-9 show transient response without grid connection and power flow control when the load increases at 0.05s. Figs.7 and 8 show that improvement of power factor and compensation of voltage drop can be performed properly by the UPFC operation without the power flow control. Furthermore, Fig.9 shows that the line current by 3-phase unbalance compensation. Even though the load is 3-phase unbalance, the line current is kept in balance. In transient response, the UPFC is able to work quickly for variations in load. In spite of load change, the line volgate in Fig.8 can be kept constant and 3-phase unbalance compensation in Fig.9 flexibly corresponds and the current is stable after about 0.03 s.

The 2018 International Power Electronics Conference

TABLE I
CIRCUIT PARAMETERS OF SIMULATION

L_s : 0.4 mH	L_1 : 1.5 mH	L_3 : 4.0 mH
R_s : 0.1 Ω	L_2 : 2.0 mH	C_{c2} : 50.0 μF
L_d : 10 mH	R_c : 10.0 Ω	E_d : 200 V
R_{du} : 13.3 Ω	C_c : 5.0 μF	C_d : 2200 μF
R_{dv} : 20 Ω	R_{dw} : 30 Ω	N_1 / N_2 : 2 / 1

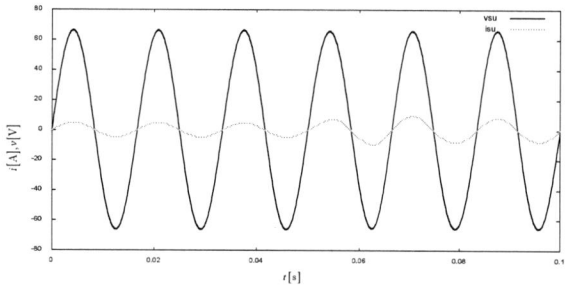

Fig. 7. Line voltage and line current in u-phase with reactive power compensation.

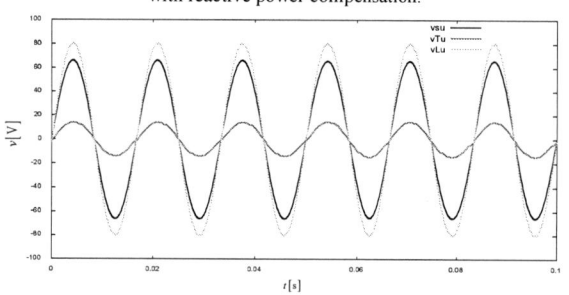

Fig. 8. Voltages with voltage drop compensation.

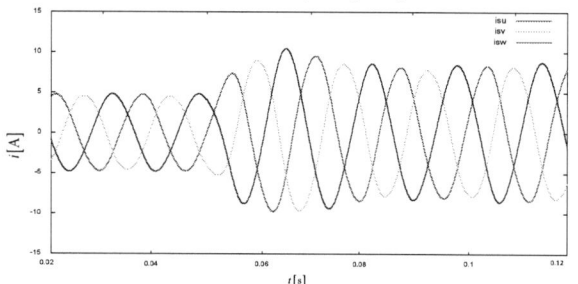

Fig. 9. Line current with the 3-phase unbalance compensation.

Figs.10-14 show the simulation results of the UPFC with power flow control. In this situation, the current command of the active current I_{Lp} is changed from 5A to 10A at 0.05s. The current command of the reactive current I_{Lq} is constant −1A. The active current and the reactive current in series system are regurated by the power flow control accurrately. As a result, the active power flows from feeder 1 to the load befere changing and flows from feeder 1 to the load and feeder 2 after changing.

On the other hand, the reactive power flows from the parallel system of UPFC to the load via the series system.

As a result, the power factor of feeder 1 and 2 become almost 1.0 respectively in this case.

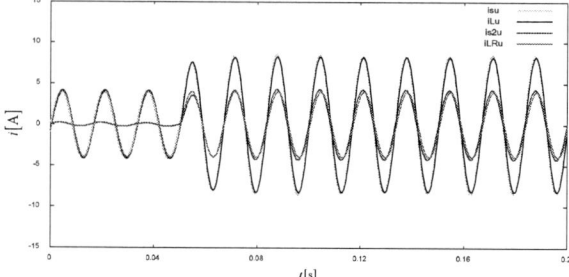

Fig. 10. Each current with power flow control.

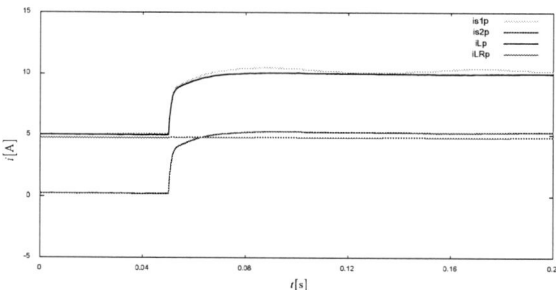

Fig. 11. Each active current with power flow control.

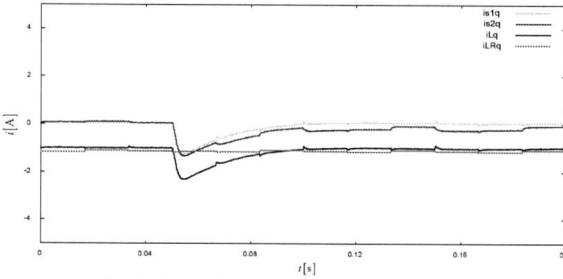

Fig. 12. Each reactive current with power flow control.

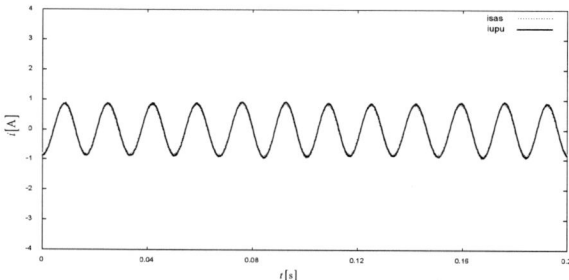

Fig. 13. Output current and command in the parallel syatem

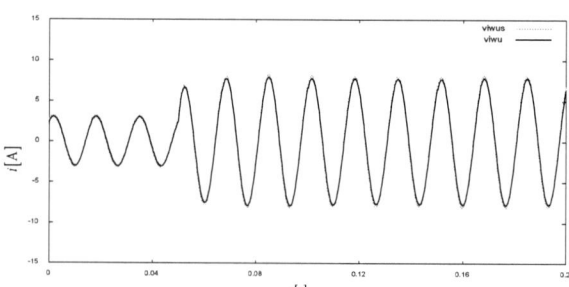

Fig. 14. The output voltage and command in the series system

IV. EXPERIMENT

Experimental equipment is shown in Fig.15 and parameters are listed in Table II. The control is realized by the DSP system. The sampling time of digital control is 50μF.

Experimental results in steady state are shown in Figs.16-21. Fig.16 shows result of the reactive power compensation. The line voltage and current become same phase Thus the power factor is 1.0 and the compensation can be performed properly. Fig.17 shows a result of voltage drop compensation. In this case, the line voltage decreases around 10V but the UPFC can operate by compensation voltage outputting. Thus the load voltage is properly kept in constant AC voltage. Figs.18-19 show results with 3-phase unbalanced compensation. The line current in Fig.19 become 3-phase balanced waveform by operaton of the UPFC in spite of 3-phase unbalanced load current in Fig.18. Figs.20-21 show the followability of the output with respect to the command. It is confirmed that both output of current and voltage accurately follow the command value without delay by the DB control.

Experimental results in transient response are shown in Figs.22-24. Fig.22 shows a result of the transient response of line voltage and line current in u-phase with reactive power compensation. In transient response, the line voltage and current can keep same phase and the power factor is 1.0 as the same as Fig.16. Fig.23 shows a result of transient response of voltage drop compensation. As with in Fig.17, the line voltage always can be kept constant without fluctuation in spite of load changing. It is confirm that the voltage drop compensation of the proposed control is able to keep proper voltage flexibly even in the case of transient response. Fig. 24 shows a result of the transient response of line current with the 3-phase unbalance compensation. It is confirmed that 3-phase unbalance compensation can flexibly correspond for 3-phase balanced current on the line.

It is verified that the proposed control of the UPFC is able to perform all the operation quickly, accurately and stabley in the experiment.

TABLE II
CIRCUIT PARAMETERS OF EXPERIMENT

L_s :	0.4	mH	L_1 :	1.5	mH	L_3 :	4.0	mH
R_s :	0.1	Ω	L_2 :	2.0	mH	C_{c2} :	50.0	μF
L_d :	10	mH	R_c :	10.0	Ω	E_d :	200	V
R_{du} :	13.3	Ω	C_c :	5.0	μF	C_d :	2200	μF
R_{dv} :	20	Ω	R_w :	30	Ω	N_1 / N_2 :	5/1	

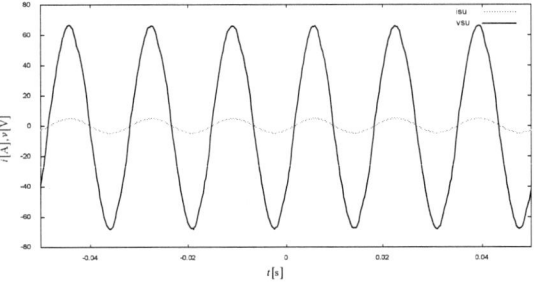

Fig. 16. Line voltage and line current in u-phase
with reactive power compensation.

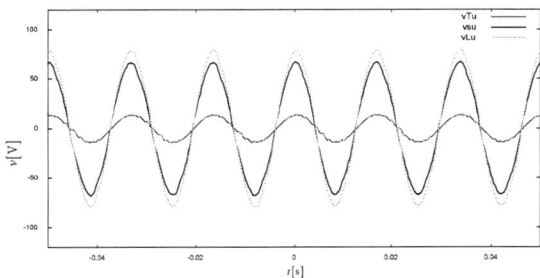

Fig. 17. Voltages with voltage drop compensation.

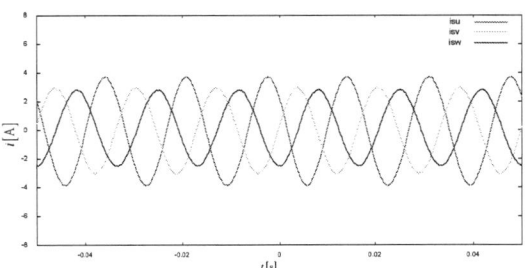

Fig. 18. Load current in 3-pahse unbalance load.

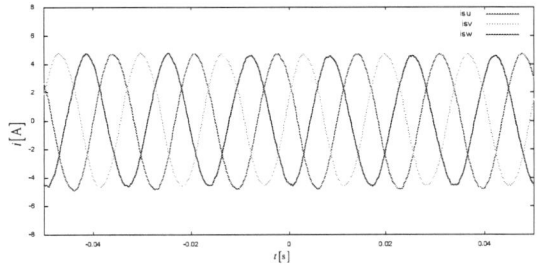

Fig. 19. Line current with 3-phase unbalance compensation .

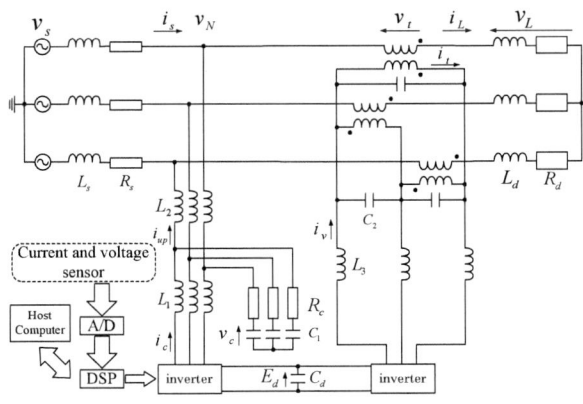

Fig. 15. Experimental equipment

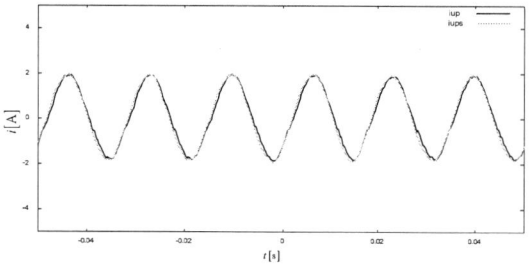

Fig.20. Output current i_{up} and command i_{ups} in the parallel system

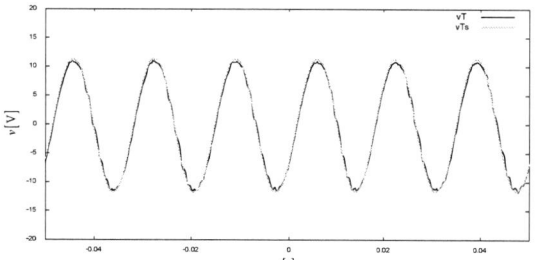

Fig.21. Output voltage v_t and reference v_{ts} in the series system

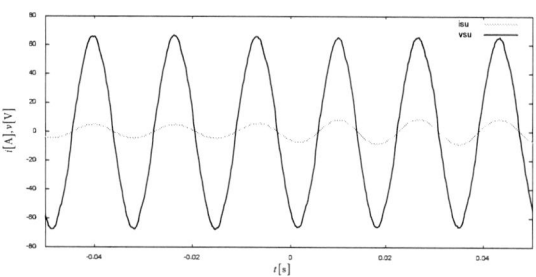

Fig. 22. The transient response of Line voltage and line current in u-phase with reactive power compensation.

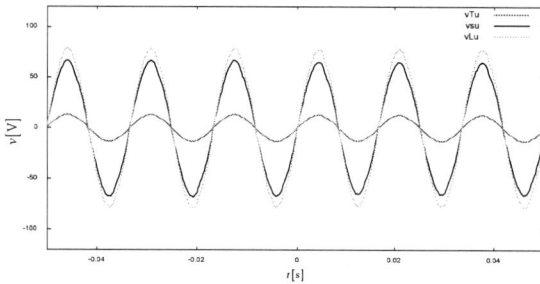

Fig. 23. The transient response of Voltages with voltage drop compensation.

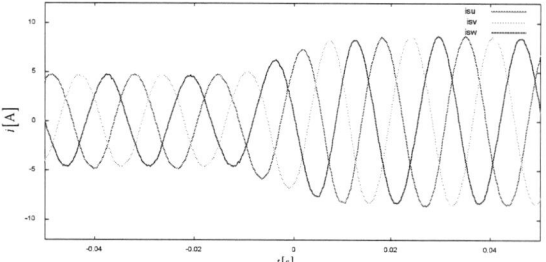

Fig. 24. The transient response of Line current with the 3-phase unbalance compensation .

V. CONCLUSIONS

The DB control for the UPFC with compensation of reactive power, voltage drop, 3-phase unbalance and power flow control is investigated. This control can obtain quick and accurate performance for current and voltage regulation by the DB control. With respect to the control of the UPFC, the reactive power, 3-phase unbalance can be compensated in the parallel syatem of UPFC. The voltage drop can be compensated in series system of UPFC. The power factor and 3-phase balance in the line can be improved by the parallel system. The power flow control can be performed and the voltage drop can be compensated to maintain voltage properly by the series system. It is cleared that the proposed UPFC is able to perform objective compensation appropriately in the experiment.

REFERENCES

[1] H. Fujita, Y. Watanabe and H. Akagi, "Control and Analysis of a Unified Power Flow Controller", *IEEE Trans. on Power Electronics*, Vol.14, No.6, pp.1021-1027 , 1999.

[2] H. Fujita, H. Akagi and Y. Watanabe, "Dynamic Control and Performance of a Unified Power Flow Controller for Stabilizing an AC Transmission System", *IEEE Trans. on Power Electronics*, Vol.21, No.4, pp.1013-1020, 2006.

[3] H. Fujita, Y. Watanabe and H. Akagi, "Transient Analysis of a Unified Power Flow Controller and its Application to Design of the DC-Link Capacitor", *IEEE Trans. on Power Electronics*, Vol.16, No.5, pp.735-740, 2001.

[4] Y. Chen, B. Mwinyiwiwa, and et. al, "Unified Power Flow Controller (UPFC) Based on Chopper Stabilized Diode-Clamped Multilevel Converters", *IEEE Trans. on Power Electronics*, Vol.15, No.2, pp.258-267, 2000.

[5] M.A. Sayed and T. Takeshita, "Line Loss Minimization in Isolated Substations and Multiple Loop Distribution Systems Using the UPFC", *IEEE Trans. on Power Electronics*, Vol.29, No.11, pp.5813-5822, 2014.

[6] J. Wang and F.Z. Peng, "Unified Power Flow Controller Using the Cascade Multilevel Inverter", *IEEE Trans. on Power Electronics*, Vol.19, No.4, pp.1077-1084, 2004.

[7] S. Dasgupta, S.N. Mohan, S.K. Sahoo and et. al, "Lyapunov Function-Based Current Controller to Control Active and Reactive Power Flow From a Renewable Energy Source to a Generalized Three-Phase Microgrid System", *IEEE Trans. on Industrial Electronics*, Vol.60, No.2, pp-799-812, 2013

[8] J. Monteiro, F. Silva, S. F. Pinto and J. Palma, "Linear and Sliding-Mode Control Design for Matrix Converter-Based Unified Power Flow Controllers", *IEEE Trans. on Power Electronics*, Vol.29, No.7, pp.3357-3367, 2014.

[9] U. Malhotra and R. Gokaraju, "An Add-On Self-Tuning Control System for a UPFC Application", *IEEE Trans. on Industrial Electronics*, Vol.61, No.5, pp.2378-2388, 2014.

[10] F.M. Albatsh, S. Mekhilef, S. Ahmad and H. Mokhlis, "Fuzzy-Logic-Based UPFC and Laboratory Prototype Validation for Dynamic Power Flow Control in Transmission Lines", *IEEE Trans. on Industrial Electronics*, Vol.64, No.12, pp.9538-9548, 2017

[11] S. Hamasaki, S. Miyazaki, T. Takaki and M. Tsuji, "Improvement of Responsivity of Unified Power Flow Controller in Digital Control System", *Journal of International Conference on Electrical Machines and Systems*, Vol.3, No.3, pp.354-361, 2014.

[12] S. Hamasaki, T. Kusaba, Y. Mazaki and M. Tsuji, "A Novel Method for Active Filter Applying The Deadbeat Control and The Repetitive Control", *International Conference on Electrical Machines and Systems(ICEMS)*, No. LS5B-2, 2009.

[13] S. Hamasaki and A. Kawamura, "Improvement of Current Regulation of Line-Current-Detection-type Active Filter based on Deadbeat Control", *IEEE Trans. on Industrial Application*, Vol.39, No.2, pp.536-541, 2003.

The 2018 International Power Electronics Conference

A Control Method for Two Types of Three-Phase Transformerless Unified Power Quality Conditioner

Fujian Li, Guochun Xiao, Fangzhou Zhao, Shuai Zhang and Baojin Liu
School of Electrical Engineering, Xi'an Jiaotong University, Xi'an, China
shen2012@stu.xjtu.edu.cn

Abstract-This paper compares two types of three-phase transformerless unified power quality conditioner (UPQC) and analyzes their performance under different grid or load conditions. Unbalanced grid voltage or load current will cause the DC-side voltage imbalance with the traditional control strategy. Thus an improved control method is proposed to suppress the DC-side voltage imbalance, which is analyzed in detail. Finally, the correctness of the proposed method is verified by PLECS simulation.

I. INTRODUCTION

With the development of the power electronics industry, the research on power quality has been continuously deepened. UPQC can solve the problems of voltage quality of power supply and the current quality of load that is a research hotspot [1, 2]. The three-phase transformerless UPQC is built on the basis of single-phase UPQC [3]. Compared to the traditional UPQC, its special topology design can inherently avoid the AC or DC system short circuit, so that no need extra isolating circuits; the UPQC can achieve reactive and negative sequence current compensation.

According to the relative position of series compensator and parallel compensator, UPQC will be divided into two categories, parallel compensator in series compensator left called UPQC-L, otherwise known as UPQC-R [1]. The power flow of two UPQCs in the same grid and load condition is analyzed in detail in [4], which provides reference for cascade mode determination of UPQC system. Theoretically, the connection mode of UPQC-R is ideal, but the practical implementation of UPQC-L is relatively simple. For the UPQC-R, the series compensator is on the grid side, and the compensator energy is taken from the load side. So, there is an energy circulation, which increases the capacity of the device, and the control of UPQC-R is more complex. So, the UPQC-R is not widely used. However, if DC-link capacitor energy is taken from photovoltaic array or wind power generation, UPQC-R is much better [5].

The single-phase half-bridge transformerless UPQC topology needs to maintain balance of the two capacitors voltage [6]. Aiming at the unbalanced DC-link voltage, zero sequence current is injected in [7]. In this paper, a simple and effective method based on individual phase control strategy is proposed for grid imbalance and load imbalance. All the compensators in the traditional UPQC share one DC-link capacitor, while the three-phase UPQC uses three DC-link capacitors. When the grid voltage and load current are unbalanced, the negative parts of

compensation current and voltage will cause DC voltage imbalance, thereby affecting the compensation effect of the device. Based on the steady-state analysis, by adding zero sequence current in each phase of the compensation current, we can use three-phase split-control to realize the redistribution of active power in DC-link capacitors. Finally, three DC-link capacitors voltage reach a steady-state and become balanced. The above control methods are verified by PLECS simulation, and the compensation performance of two UPQCs are compared when the grid voltage is unbalanced or depth drop.

II. THREE-PHASE THREE-WIRE TRANSFORMERLESS UPQC

A. Topologies of UPQC

Fig. 1 and Fig. 2 are UPQC-R and UPQC-L topologies. In Fig. 1, u_{sa}, u_{sb} and u_{sc} are the instantaneous value of three-phase grid voltage, respectively (reference point is N); V_{ca}, V_{cb} and V_{cc} are the instantaneous value of series converter output voltage, respectively; i_{sa}, i_{sb} and i_{sc} are the instantaneous value of three-phase grid current, respectively; i_{la}, i_{lb} and i_{lc} are the instantaneous value of the load current respectively; i_{ca}, i_{cb} and i_{cc} are the instantaneous value of parallel converter output current respectively; i_{c1}, i_{c2} and i_{c3} are the inductor current value of three series converter filter respectively.

V3, V4, V5 and V6 constitute a series compensator for voltage sag and voltage harmonic compensation; V1, V2, V3 and V4 constitute a parallel compensator for current harmonics and reactive power compensation. It can be seen that V3 and V4 are used for current compensation and voltage compensation, namely the coupling problem existing in the three legs. [3] combining with SVPWM technology, the space vector decoupling is realized on the three legs.

Fig. 3 are equivalent circuits of two UPQC topologies. Three-phase three-wire transformerless UPQC is a topology construction based on single-phase three-leg transformerless UPQC. It has inherited advantage of the single-phase transformerless UPQC. Compared with the traditional UPQC, there is no need to introduce the isolation link, and no AC short and DC short problems.

UPQC-R under the condition of normal operation, load current harmonics and reactive power are compensated by the shunt part, not through the series part; voltage harmonic is compensated by the series part, not through the shunt part. So, UPQC-R has the advantages of small harmonic content and reactive power loss. When the reactive power and harmonic content are large, the compensation performance of UPQC-R is better. When the grid voltage drops, the grid current RMS is greater than the load current RMS. For two UPQCs have the same compensation voltage, the power of series part of UPQC-L is smaller. If the DC power supply is connected to the new energy, there will be no energy circulation problem, and the UPQC-R topology is better.

In the single-phase system, current compensation only involves the load current harmonics and reactive power, and DC voltage can be controlled through the PI controller to maintain a stable value. But, the three-phase UPQC cannot. For the three-phase transformerless UPQC, negative sequence of current compensation will make the three-phase DC sides exchange active power. And the three-phase DC-link voltage will deviate from reference value, resulting in DC voltage imbalance, voltage and current compensation performance getting worse. The control strategy of three-phase transformerless UPQC is equivalent to the control of three single-phase transformerless UPQCs and the control of DC voltage.

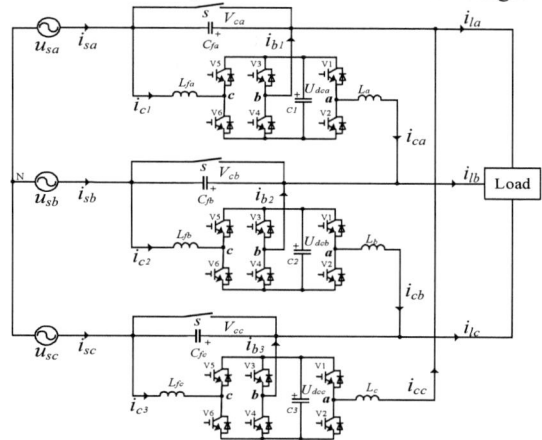

Fig. 1. Three-phase three-wire transformerless UPQC-R.

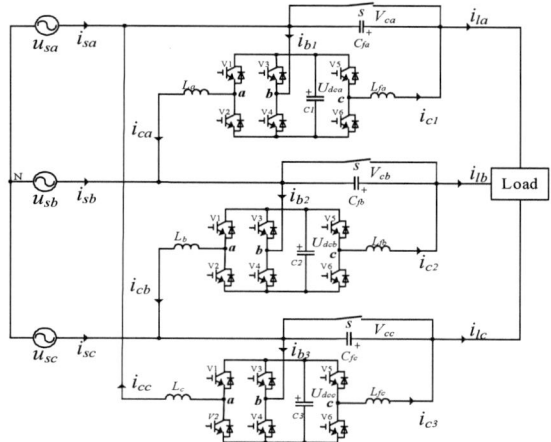

Fig. 2. Three-phase three-wire transformerless UPQC-L.

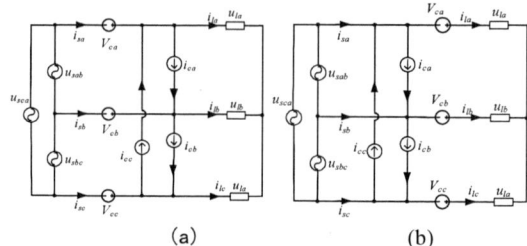

Fig. 3. Equivalent circuit of two UPQCs. (a) UPQC-R. (b) UPQC-L.

B. Three-phase Split-control of UPQC

The compensating current reference (i_{aref}, i_{bref}, i_{cref}) can be obtained by the i_p-i_q method based on the instantaneous reactive power theory. Fig. 4 is a block diagram of a single shunt compensator, which mainly includes the DC-link voltage control part i_{ry}, the harmonic and negative sequence current compensation part i_{xref}, the parallel repetitive controller and the parallel system transfer function $G_{p(z)}$. The i_0 is used for DC-link balance control, and latter section will describe its derivation process in detail. In Fig. 4, x represents a, b and c, y represents ab, bc and ca, i_{labc} represents three-phase load current, θ_y is the initial phase of grid current.

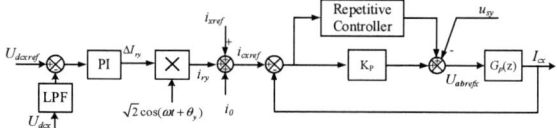

Fig. 4. Single-phase current compensation control block diagram ($x = a, b, c$).

Fig. 5 is single-phase voltage compensation control block diagram. The outer loop is the voltage loop and the inner loop is the inductor current feedback. U_{cbrefx} is the output of inner current loop. V_{cxref} represents the compensation reference value of three-phase compensation voltage.

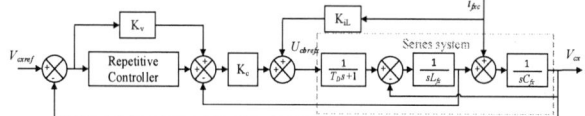

Fig. 5. Single-phase voltage compensation control block diagram ($x = a, b, c$).

According to [3], the six switch control signals of three legs can be gotten from U_{abrefx} and U_{cbrefx}.

III. DC VOLTAGE BALANCE CONTROL METHOD

A. Compensation Performance of Zero Sequence Current

Effect of positive sequence voltage and negative sequence compensation current, and effect of negative sequence voltage and positive sequence compensation current will cause the active power exchange among three DC-link capacitors and the DC voltage imbalance. The imbalance of DC-link voltage will cause the difference of the active power among the three compensators and the grid, resulting in the unbalanced grid current. When the active power flowing into the compensator is equal to the outflow of the compensator, the system will be stable. Because the zero sequence current causes active power transfer among the three without absorbing active power from power grid, we can add the zero sequence current into compensation current to offset all the imbalance.

Fig. 6(a) is the compensation performance of compensating current without adding zero sequence current. The blue, black and green solid line represent the negative sequence of the three-phase compensation current; the blue, black and green dashed line represent ΔI_{rab}, ΔI_{rbc} and ΔI_{rca} respectively. Three red dotted lines are perpendicular to the corresponding line in voltage direction. It indicates that the current compensation system is stable. The DC-link capacitor voltages increase or decrease to different value. That is to say, the DC-link voltages are unbalanced.

Fig. 6(b) is the compensation effect diagram of compensating current with adding zero sequence current. When the grid voltage is unbalanced, the voltage compensation will also lead to different changes of the three-phase DC-link voltage. After the zero sequence current is added, the system enters the steady state again, and the DC-link voltages rise or fall in same degree. Finally, the system is balanced.

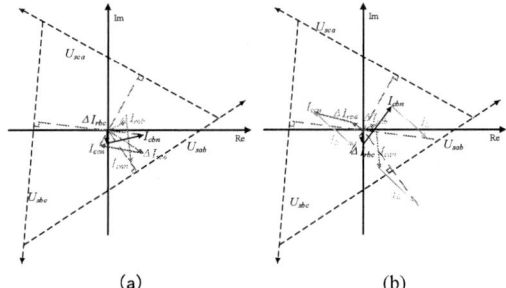

(a)   (b)

Fig. 6. Compensation performance. (a) Without adding zero sequence current. (b) Adding zero sequence current.

B. Reference Value of Zero Sequence Current

When the grid voltage is asymmetrical, three-phase line voltages are:

$$
\begin{cases}
u_{sab} = A_a \sin(\omega t + \varphi_1) + \sum_{i=3,5\cdots} A_{ai} \sin(i\,\omega t + \varphi_{ai}) \\
u_{sbc} = A_b \sin(\omega t + \varphi_2) + \sum_{i=3,5\cdots} A_{bi} \sin(i\,\omega t + \varphi_{bi}) \\
u_{sca} = A_c \sin(\omega t + \varphi_3) + \sum_{i=3,5\cdots} A_{ci} \sin(i\,\omega t + \varphi_{ci})
\end{cases} \quad (1)
$$

Where, A_a, A_b and A_c are three-phase line voltage amplitude, respectively; φ_1, φ_2 and φ_3 are three-phase fundamental phase, respectively; A_{ai}, A_{bi} and A_{ci} are three-phase harmonic voltage amplitude, respectively; φ_{ai}, φ_{bi} and φ_{ci} are initial phase of harmonic voltage, respectively.

Because the zero sequence current harmonic will not cause the power exchange among three phases, suppose the zero sequence is:

$$
\begin{aligned}
i_0 &= \sqrt{2} I_0 \sin(\omega t + \theta_0) \\
&= \sqrt{2}(I_0 \cos\theta_0 \sin\omega t + I_0 \sin\theta_0 \cos\omega t)
\end{aligned} \quad (2)
$$

Combine (1) and (2), the change of the power of the three-phase DC caused by the zero sequence current are obtained.

$$
\begin{cases}
P_{ab}^0 = u_{sab} i_0 = \dfrac{\sqrt{2}}{2} A_a I_0 \cos(\varphi_1 - \theta_0) - \dfrac{\sqrt{2}}{2} A_a I_0 \cos(2\omega t + \varphi_1 + \theta_0) \\
P_{bc}^0 = u_{sbc} i_0 = \dfrac{\sqrt{2}}{2} A_b I_0 \cos(\varphi_2 - \theta_0) - \dfrac{\sqrt{2}}{2} A_b I_0 \cos(2\omega t + \varphi_2 + \theta_0) \\
P_{ca}^0 = u_{sca} i_0 = \dfrac{\sqrt{2}}{2} A_c I_0 \cos(\varphi_3 - \theta_0) - \dfrac{\sqrt{2}}{2} A_c I_0 \cos(2\omega t + \varphi_3 + \theta_0)
\end{cases} \quad (3)
$$

The average value of three-phase DC-link power variation caused by zero sequence current are:

$$
\begin{cases}
\overline{P_{ab}^0} = \dfrac{\sqrt{2}}{2} A_a I_0 \cos(\varphi_1 - \theta_0) \\
\overline{P_{bc}^0} = \dfrac{\sqrt{2}}{2} A_b I_0 \cos(\varphi_2 - \theta_0) \\
\overline{P_{ca}^0} = \dfrac{\sqrt{2}}{2} A_c I_0 \cos(\varphi_3 - \theta_0)
\end{cases} \quad (4)
$$

Negative sequence of compensation current and grid negative voltage will cause three-phase DC-link voltage imbalance and active power exchange. Suppose $\overline{\Delta P_A}$, $\overline{\Delta P_B}$ and $\overline{\Delta P_C}$ as the average value of the three-phase DC-link capacitor power variation as:

$$
\begin{cases}
\overline{\Delta P_A} = \dfrac{\sqrt{2}}{2} A_a \Delta I_{rab} \cos(\varphi_1 - \theta_{ab}) \\
\overline{\Delta P_B} = \dfrac{\sqrt{2}}{2} A_b \Delta I_{rbc} \cos(\varphi_2 - \theta_{bc}) \\
\overline{\Delta P_C} = \dfrac{\sqrt{2}}{2} A_c \Delta I_{rca} \cos(\varphi_3 - \theta_{ca})
\end{cases} \quad (5)
$$

To make the DC-link capacitor voltage increase or decrease in the same extent, we can obtain:

$$
\begin{cases}
\overline{\Delta P_{ABC}} = \overline{\Delta P_A} + \overline{\Delta P_B} + \overline{\Delta P_C} \\
\overline{P_{ab}^0} = \dfrac{1}{3} \overline{\Delta P_{ABC}} - \overline{\Delta P_A} \\
\overline{P_{bc}^0} = \dfrac{1}{3} \overline{\Delta P_{ABC}} - \overline{\Delta P_B} \\
\overline{P_{ca}^0} = \dfrac{1}{3} \overline{\Delta P_{ABC}} - \overline{\Delta P_C}
\end{cases} \quad (6)
$$

The degrees of freedom of the above equation are 2, and the equation can be solved by any two of (6).

$$
\begin{aligned}
I_0 \cos(\varphi_1 - \theta_0) &= (\tfrac{1}{3} \overline{\Delta P_{ABC}} - \overline{\Delta P_A}) / \dfrac{\sqrt{2}}{2} A_a \\
&= \left[\tfrac{1}{3}(A_a \Delta I_{rab} + A_b \Delta I_{rbc} + A_c \Delta I_{rca}) - A_a \Delta I_{rab} \right] / A_a
\end{aligned} \quad (7)
$$

$$
\begin{aligned}
I_0 \cos(\varphi_2 - \theta_0) &= (\tfrac{1}{3} \overline{\Delta P_{ABC}} - \overline{\Delta P_B}) / \dfrac{\sqrt{2}}{2} A_b \\
&= \left[\tfrac{1}{3}(A_a \Delta I_{rab} + A_b \Delta I_{rbc} + A_c \Delta I_{rca}) - A_b \Delta I_{rbc} \right] / A_b
\end{aligned}
$$

So, the expression of zero sequence current can be obtained as shown in (8).

$$
\begin{aligned}
i_0 &= \frac{\Delta I_{0ra} \sin\varphi_1 - \Delta I_{0rb} \sin\varphi_2}{\sqrt{2}(\sin\varphi_1 \cos\varphi_2 - \sin\varphi_2 \cos\varphi_1)} \sin\omega t \\
&\quad + \frac{\Delta I_{0ra} \cos\varphi_2 - \Delta I_{0rb} \cos\varphi_1}{\sqrt{2}(\sin\varphi_1 \cos\varphi_2 - \sin\varphi_2 \cos\varphi_1)} \cos\omega t
\end{aligned} \quad (8)
$$

Where,

$$
\Delta I_{0ra} = \left[\tfrac{1}{3}(A_a \Delta I_{rab} + A_b \Delta I_{rbc} + A_c \Delta I_{rca}) - A_a \Delta I_{rab} \right] / A_a
$$

$$
\Delta I_{0rb} = \left[\tfrac{1}{3}(A_a \Delta I_{rab} + A_b \Delta I_{rbc} + A_c \Delta I_{rca}) - A_b \Delta I_{rbc} \right] / A_b
$$

IV. SIMULATION RESULTS

The simulation parameters are shown in table I.

TABLE I
SIMULATION PARAMETERS

Symbol	Meaning	Value
U_{sa}, U_{sb}, U_{sc}	Rated phase rms voltage	110V
f_s	Switching frequency	15kHz
L_a, L_b, L_c	Filter inductance	4.6mH
L_{fa}, L_{fb}, L_{fc}	Filter inductance	1.5mH
C_{fa}, C_{fb}, C_{fc}	Filter capacitor	20µF
U_{dca}, U_{dcb}, U_{dcc}	DC-link voltage	380V
$C1$, $C2$, $C3$	DC-link capacitor	6600µF

A. Shallow Drop of Grid Voltage

In 0.35s, the three-phase grid voltage fall to 90%, 80% and 70% of the rated value, respectively, and the 3rd and 5th harmonics are injected. Fig. 7 (a) is the voltage and current waveform without adding the zero sequence current in the compensation current. The grid current and the DC-link voltage are not balanced. While the zero sequence current is added in Fig. 7 (b). Grid current and DC-voltage in balance, grid current and grid voltage in same phase, show that the reactive power compensation performance is good.

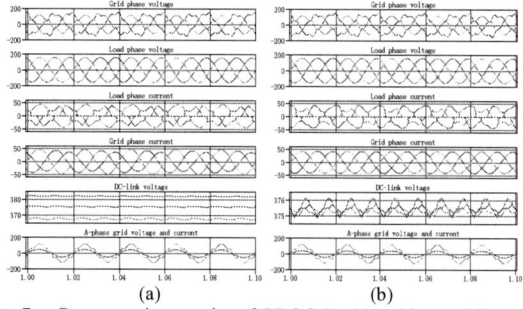

(a)   (b)

Fig. 7. Compensation results of UPQC-R. (a) Without adding zero sequence current. (b) Adding zero sequence current.

B. Deep Drop of Grid Voltage

In 0.35s, the three-phase grid voltage fall to 50%, 40% and 30% of the rated value respectively, and the 3rd and 5th harmonics are injected. Fig. 8 is compensation waveform of UPQC-R. The load voltage and grid current waveform are sinusoidal approximatively; three-phase DC voltage are approximately equal to 349V, basically reaching equilibrium. So, the compensation performance is good.

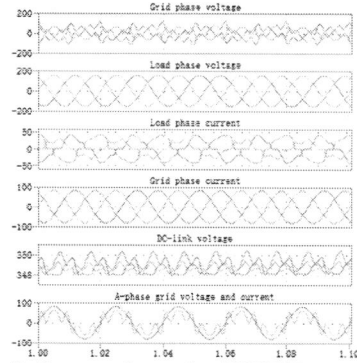

Fig. 8. Compensation results of UPQC-R.

The current inflowing or outflowing from three legs (**a**, **b** and **c** in Fig. 1 or Fig. 2) are obtained in table Ⅱ. Compared with UPQC-L, the inflow or outflow of **c** point in UPQC-R is very large. So, when the grid voltage drops badly, the flow capacity of UPQC-R switch should be taken more seriously than UPQC-L.

TABLE Ⅱ
MAX VALUE AND RMS

Symbol (A)	UPQC-R		UPQC-L	
	Max	RMS	Max	RMS
i_{ca}	37.7	24.1	32.8	21.3
i_{cb}	27.9	20.3	27.2	19.4
i_{cc}	28.5	17.8	34.2	22.4
i_{b1}	52.0	35.1	53.6	37.2
i_{b2}	64.6	42.9	66.2	44.2
i_{b3}	66.6	48.3	60.6	45.2
i_{c1}	82.9	57.6	31.9	18.2
i_{c2}	84.7	57.9	42.8	25.8
i_{c3}	82.2	57.4	43.1	29.5

V. CONCLUSION

The active power of three DC-link capacitors is redistributed with the injection of zero sequence current, which, as a result, keeps the DC-side voltage balanced. The simulation verifies the accuracy of the control method, and the compensation performance between UPQC-R and UPQC-L when the grid voltage is unbalanced and drops badly. It can be used for scheme design and device selection in practical application.

REFERENCES

[1] V Khadkikar. Enhancing Electric Power Quality Using UPQC: A Comprehensive Overview[J]. IEEE Transactions on Power Electronics, 2012, 27(5):2284-2297.

[2] A Teke, L Saribulut and M Tumay. A Novel Reference Signal Generation Method for Power-Quality Improvement of Unified Power-Quality Conditioner[J]. IEEE Transactions on Power Delivery, 2011, 26(4):2205-2214.

[3] Y Lu, GC Xiao, XL Wang, et al. Control Strategy for Single-Phase Transformerless Three-Leg Unified Power Quality Conditioner Based on Space Vector Modulation[J]. IEEE Transactions on Power Electronics, 2016, 31(4):2840-2849.

[4] Q Wang, YL Geng and YH He. Power analysis on universal power quality controller under different system connections[J]. Proceedings of the CSEE, 2005.

[5] M Davari, SM Aleemran, H Nafisi, et al. Modeling the combination of UPQC and photovoltaic arrays with Multi-Input Single-Output DC-DC converter[C]. IEEE International Conference on Industrial Technology, 2009: 1-6.

[6] VSP Cheung, SCR Yeung, HSH Chung, et al. A Transformer-less Unified Power Quality Conditioner with Fast Dynamic Control[J]. IEEE Transactions on Power Electronics, 2017, PP(99):1-1.
JJ Jung, JH Lee, SK Sul, et al. DC Capacitor Voltage Balancing Control for Delta-Connected Cascaded H-Bridge STATCOM Considering Unbalanced Grid and Load Conditions[J]. IEEE Transactions on Power Electronics, 2017, PP(99):1-1.

Design of Customer-End Converter Systems for Low Voltage DC Distribution from a Life Cycle Cost Perspective

A. Mattsson*, P. Nuutinen, T. Kaipia, P. Peltoniemi, J. Karppanen, V. Tikka, A. Lana, P. Pinomaa,
P. Silventoinen, J. Partanen
LUT School of Energy Systems
Lappeenranta University of Technology, Lappeenranta, Finland
*aleksi.mattsson@lut.fi

Abstract- **This paper presents an approach to power electronic converter design in which the objective is to minimize the life cycle cost when the converter is used to supply a residential customer. In the paper, the life cycle cost is defined as the sum of the cost of the main components of the power stage and the cost of the losses during the utilization period. The semiconductor switches, output filtering, heat sink, gate drivers, and DC link capacitance are included in the analysis, whereas their parameters are freed in the optimization process. The behavior of the load of an average residential customer is taken as one of the inputs in the calculation. It is shown that parameters such as the optimal silicon area, filter inductance, and switching frequency can differ significantly from the industry norm in which a high weight factor is given to the performance near the nominal power.**

I. INTRODUCTION

The low voltage DC (LVDC) distribution has been researched for many use cases such as microgrids, data centers, commercial and telecommunication buildings, and public utility grid distribution [1]–[10]. Even though all these areas use DC as the medium for supplying energy, the main focus and optimization targets in the system design may vary, and consequently, the same approaches and solutions may not be optimal when moving from one application to another. For example, a preferable DC voltage level may be a result of some application-specific advantages, which is one of the main reasons for the popularity of the 380–400 VDC range in data centers as it allows the removal of the power factor correction stage from the server power supply while maintaining the same internal DC voltage level and structure of the original power supply. The main focus of this paper is on public utility grid distribution [7]–[10] in which DC is used as a part of the low voltage network replacing parts of the 20 kV medium voltage and 400 V low voltage AC networks. However, owing to standards and regulations, the residential customers are still supplied with 3-phase 230/400 VAC$_{RMS}$, and no modifications are made to the electrical installations on the customer's premises. A simplified diagram of the LVDC system under study is depicted in Fig. 1. As shown in Fig. 1, the LVDC network connects to the 20 kV medium voltage AC network through a double-tier transformer, and the stepped-down 3-phase AC voltage is fed to a rectifier substation. The rectifier substation is

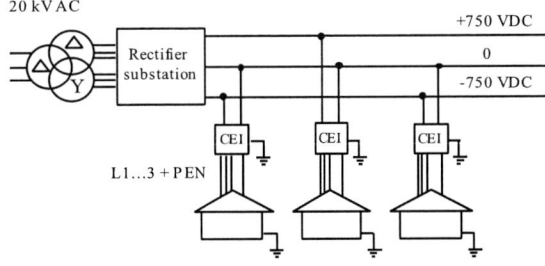

Fig. 1. Simplified diagram of the LVDC system under study. The LVDC network connects to the medium voltage network through a step-down transformer and a bidirectional rectifier. The rectifier can feed either a bipolar (as depicted) or a unipolar DC network and the customers are supplied with 3-phase inverters.

implemented with two series-connected line converters, which generate an isolated three-wire bipolar DC network. The customers are connected either between +750 VDC and 0 or between 0 and −750 VDC, and the 230/400 VAC 3-phase supply is produced using a customer-end inverter (CEI). The low voltage directive [11] allows the usage of DC voltages up to 1500 VDC while the values of 400 VDC, 750 VDC, and 1500 VDC have mainly been considered in previous research regarding this type of a system [9],[10],[12]. However, to the author's knowledge, little effort has been put to investigate the load profile of a power electronic converter in the case of the LVDC system used to supply residential customers and the question of how the load profile could be used to optimize the design of the converter. Therefore, the aim of this paper is to (1) analyze how typical loads, in this case a residential customer, behave as a load for a power electronic converter and (2) show how the profile of the load affects the design of the converter when the objective is to minimize the life cycle cost. The example cases in the paper are calculated for a customer-end inverter (CEI), whereas the major power stage components such as the semiconductor switches, heat sink, output filtering, and the DC capacitance are included in the optimization process. The maintenance cost is evaluated as the cost for a replacement converter unit after its estimated lifetime is exceeded. Major design variables such as the selection of the switching frequency, the key components, and the nominal power of the CEI are addressed. The option of dividing the nominal power among several parallel CEI modules is also considered.

The 2018 International Power Electronics Conference

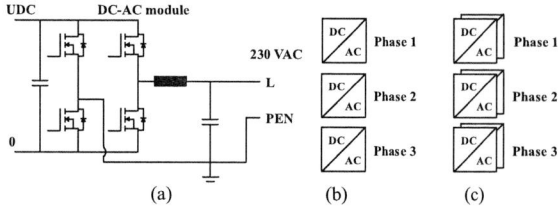

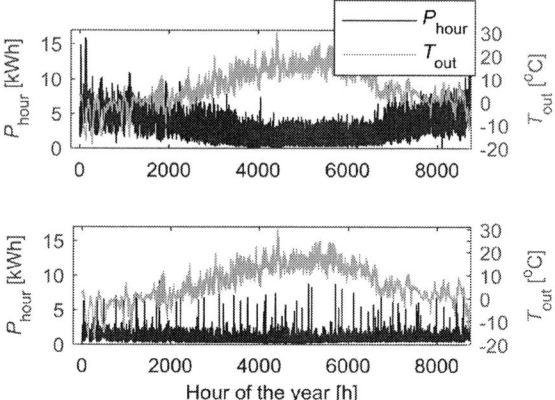

Fig. 2. Simplified diagram of the example case in which a full-bridge inverter (a) is used as a building block for the CEI (b) and the option of paralleling several CEI modules (c) is also considered.

The example cases are calculated for a CEI structure in which each of the three phases of the CEI is implemented using a full-bridge inverter with an isolated DC input. The isolation stage is discussed in more detail in [13]. A simplified diagram of the CEI is depicted in Fig. 2. A preliminary analysis of the CEI structure in Fig. 2 was provided in [14],[15]. The key differences to the previous work are as follows. The design methodology of [14],[15] is extended to incorporate the effect of an overload situation. Requirements of the protection devices on the customer's premises and the dynamic properties of the load are included in the analysis. Therefore, a more detailed study of the load is performed using recent AMR data, averaged load profile models, and measurements are made on typical household appliances. The calculation algorithms are updated to include methods with improved accuracy for the calculated losses of the various components of the power stage. Additional components not yet available during the writing of the previous publications are considered in the analysis. A sensitivity analysis is conducted on the main input variables to determine their significance in the final result. The results are compared with a case in which the CEI is designed to run at its rated power to illustrate how using the load profile of a residential customer as one of the inputs for the design affects the parameters such as the silicon area, output filter inductance, and switching frequency of the optimized converter. Two distinct types of customer load profiles are used in the analysis to determine if a similar CEI design can be used to supply a variety of customers without major drawbacks. Even though the emphasis is laid on the design from the perspective of an LVDC system, the methodology is still valid in other applications if application-specific parameters such as the load profile are adjusted accordingly.

II. RESIDENTIAL CUSTOMER AS A LOAD FOR A POWER ELECTRONIC CONVERTER

In order to fully understand the key factors in the design optimization of the CEI, we must first understand how a residential customer behaves as a load for the CEI. In Finland, LVDC is considered a feasible alternative to AC in rural areas [7],[8]. Typical customers in these potential areas are households residing in detached dwellings (cf. Fig. 1) and having either electricity, oil, a ground source heat pump, or district heating as their main source of heating. The average annual energy consumption of a single customer in the case of electrical

Fig. 3. Hourly AMR measurements of the load P_{hour} and outside temperatures T_{out} for a period of one year (1 Jan. 2015–31 Dec. 2015) from a customer (CUST1) that has electrical heating and a yearly energy consumption of 23.9 MWh (upper graph) and from a customer (CUST2) that has modern facilities and uses a ground source heat pump as the source for heating and has a yearly energy consumption of 7.9 MWh (lower graph)

heating is $E = 20$ MWh (from here on *type 1*), being $E = 7$–11 MWh in the cases in which the customer has some other heating system (from here on *type 2*) than electrical heating [16]. When we consider the customer as a load for a power electronic converter, we are also interested in what the peak powers of the customers are and how the load behaves as a function of time, or to be more exact, the amount of time that is spent at different load values. The peak power determines the minimum requirement for the nominal power of the converter, whereas the load behavior as a function of time defines at what percentages of the nominal power most of the energy is being supplied and thus, which the most important areas are regarding the energy efficiency of the converter. The type 1 and type 2 customers typically have 25 A main fuses, which technically allow a constant three-phase power draw of $P_{max} = 17.25$ kW, and therefore, this would be the dimensioning power for the converter. Even though the actual peak powers of the customers would not be as high as $P_{max} = 17.25$ kW, the customers are paying for the ability to utilize the whole capacity as the electricity tariff is determined by the main fuse rating. Fig. 3 depicts an example of the behavior of a customer's load in two cases in which the customers CUST1 and CUST2 have different types of heating systems. CUST1 uses electricity for heating, whereas CUST2 has a ground source heat pump. The graphs of Fig. 3 are based on automatic reader metering (AMR) measurements made on the customer's premises. As can be seen in Fig. 3, the load P_{hour} varies considerably during the year, and a very small amount of time is spent near P_{max}. The graphs of Fig. 3 are meant to serve as an example of a single customer, whereas averaged load profiles [17] representing a larger sample size of similar customers as in Fig. 3 are used in the further analysis of this paper. The profiles contain averaged load values for each hour of the year relative to the average hourly power

2949

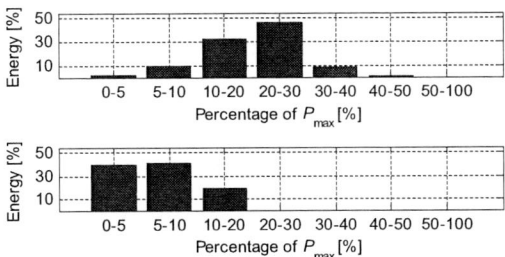

Fig. 4. Distribution of the energy supplied using the CEI as a function of percentage of the peak power $P_{\max}$ for averaged load profiles representing a larger sample size of customer types (upper graph: type 1, lower graph: type 2) similar to Fig. 3.

of the customer, whereas the average hourly power is calculated by

$$P_{\text{hour}} = \frac{E}{8760}. \qquad (1)$$

The load value P_{hour} for each hour of the year can then be calculated by multiplying the hourly values of the load profile by the value of P_{avg}. To better understand how the energy supplied using the CEI is distributed between 0–100% of $P_{\max}$, the energy supplied at different load values was calculated as a percentage of the total supplied energy during the year. The averaged load profiles [17] and annual energy consumptions of $E = 20$ MWh (for type 1) and $E = 9$ MWh (for type 2), both values in good agreement with [16], were used in the calculation. The results are depicted in Fig. 4. When we consider the results in Fig. 4 from the viewpoint of the converter that would supply the customer, we can conclude that in both cases, type 1 and type 2, the nominal power $P_{\text{nom}} = P_{\max}$ of the CEI would seldom be used as most of the energy is being supplied between the load values 0–40% of P_{nom}.

Because of the hour-based resolution of the data in Fig. 3 and [17], they do not adequately represent the dynamic properties of the load. Based on a set of measurements made on typical household appliances, turn-on currents of up to 40 A_{pk} were observed. Another thing to consider is the protection devices that are used on the customer's premises and need a high current in order to operate in the required time in the case of a short-circuit fault in the customer's AC network.

III. Design Optimization of the CEI

The results and load profiles discussed in Section II were taken as one of the inputs for an optimization process, which calculates the design parameters of the CEI. Because the LVDC system under study is considered to be a part of the distribution network, in which the decisive factor for feasibility is the life cycle cost of the system, we have to take into account the different cost factors involved when evaluating whether a particular converter design is optimal or not. As a result, the objective function for the optimized design is similar to the one that is commonly used in the techno-economic analyses of the distribution network and is written as

$$\min \sum_{t=1}^{t_u} (C_{\text{inv}}(t) + C_{\text{loss}}(t) + C_{\text{int}}(t) + C_{\text{main}}(t)), \qquad (2)$$

where C_{inv} is the investment cost, C_{loss} the cost of the losses, C_{int} the interruption cost, C_{main} the maintenance cost, and t_u the utilization period [18]. A typical value for t_u in the distribution network analysis is 40 years [19]. However, if we compare this value with the expected lifetimes of power electronic converters, which can range from 5 to 30 years depending on the application [20], we can see that the converters in the LVDC system probably have to be replaced at least once during a period of $t_u = 40$ years. The cost of the replacement unit can be calculated using the present value as defined by

$$C_{\text{inv}}(t) = C_{\text{rep}}(1 + p)^{-t_{\text{rep}}}, \qquad (3)$$

where C_{rep} is the reference cost of the replacement unit and p the interest rate. A replacement interval of $t_{\text{rep}} = 20$ years and a value of $p = 5\%$ were used in the calculations, when the total cost was calculated for a period of $t_u = 40$ years. Only the replacement of whole converter units was considered and thus, the effect of C_{main}, which indicates for example the cost of replacing a single component of the converter, and C_{int}, which refers to the cost of an unexpected interruption in the electricity supply, were not included in the analysis. Because we are also projecting future costs for the losses over a period of several years, C_{loss} was calculated as the present value of an annuity by

$$C_{\text{loss}} = P_{\text{tot}} C_e \frac{1 - (1 + p)^{-t_u}}{p}, \qquad (4)$$

where P_{tot} is the power loss over a period of one year and C_e the cost of electricity. A value of $C_e = 40$ €/MWh was used in the calculations as it well represents an average market price for electricity in Finland during the past ten years [21]. However, because the electricity price can be considered one of the main input variables affecting the result, a higher electricity cost of $C_e = 80$ €/MWh was also considered. The general design constraints used for the CEI were the voltage quality [22], the minimum nominal power $P_{\text{nom}} = 17.25$ kW, and the maximum ambient temperature inside the CEI cabinet, which was considered to be $T_{\text{amb}} = 50$ °C. The requirements set by the protection devices at the customer-end electrical installations [23],[24] were used as the constraints for the overload capability resulting in values of 110 A_{RMS} for 5 s in the case of a 25 A gG-type main fuse and 160 A_{RMS} for 0.4 s in the case of a 16 A C-type circuit breaker. A value of 160 A_{RMS} for 5 s was taken as the constraint as it satisfies both conditions. The semiconductor-related constraints are the voltage rating and the maximum junction temperature $T_{\text{j,max}}$. A selection of state-of-the-art 1200 V and 900 V SiC MOSFETs [23] and 650 V GaN HEMTs [26] were used in the analysis, whereas fit functions of the switching energies E_{on}, E_{off}, and on-state resistance $R_{\text{ds,on}}$ were generated based on the datasheet curves and parameters. The datasheets of the GaN switches do not give sufficient data of the switching energies as a function of switched current, and therefore, the switching instants were simulated in Orcad PSpice using the device models provided by the manufacturer [26] and a double pulse circuit. As pointed out in [27], the effect of the ripple current in the output filter on the power stage losses is often omitted in the power stage analysis. However, if we consider a typical ripple factor of $RF = 10$–30% and compare it with the load introduced in Section II, we can conclude that in the studied application, omitting the ripple could result in an underestimation of the losses over a typical load cycle. Therefore, the switching losses are calculated by sampling the current waveform in the inverter bridge and by calculating the time-average switching loss using

$$P_{sw} = \frac{1}{T} \sum_{n=1}^{m} \left(E_{on}(I_{sw,n}) + E_{off}(I_{sw,n}) \right), \qquad (5)$$

where I_{sw} is the switched current at the nth switching instant and f_{sw} the switching frequency. The values of I_{sw} as a function of time can be determined by analyzing the inductor current, which is defined by

$$I_{sw}(\omega t) = \sqrt{2} I_{out} \sin(\omega t) \pm \frac{\Delta i_{pp}(\omega t)}{2}, \qquad (6)$$

where I_{out} is the output current of the CEI and Δi_{pp} the peak-to-peak value of the inductor ripple current, which is defined by

$$\Delta i_{pp}(\omega t) = \frac{U_{DC}}{2Lf_{sw}} (1 - M \sin(\omega t)) M \sin(\omega t), \quad (7)$$

where M is the modulation index, L the filter inductance and U_{DC} the input voltage of the CEI [28]. During the positive half cycle of the line current, one pair of switches in the H-bridge turns on at the peak of the inductor current and turns off at the valley of the inductor current, whereas another pair of switches turns on at the valley of the inductor current and turns off at the peak of the inductor current. The order is then reversed during the negative half-cycle. The conduction losses are calculated by

$$P_{cond} = I_{RMS,FET}^2 R_{ds,on}(T_j), \qquad (8)$$

where $I_{RMS,FET}$ is the RMS current of the semiconductor and T_j is the junction temperature of the semiconductor. The expression for the RMS ripple current through a single switch can be derived similarly to the equation for the filter ripple current in [28], and it is defined by

$$I_{ripple,sw} = \frac{U_{DC}}{4Lf_{sw}} \sqrt{\frac{M^2}{3\pi} \left(\frac{\pi}{4} \left(1 + \frac{3}{4} M^2 \right) - \frac{4}{3} M \right)}, (9)$$

where M is the modulation index, L is the filter inductance, and U_{DC} the input voltage of the CEI. The RMS current through a single switch including the ripple current can then be calculated by combining the value of $I_{ripple,sw}$ into the simplified RMS current equations for MOSFETs in [29], and it is defined by

$$I_{RMS,FET} = \sqrt{\frac{I_m^2}{4} + I_{ripple,sw}^2}, \qquad (10)$$

where I_m is the peak of the sinusoidal output current. The expressions of (9) and (10) were verified by simulating the inverter in Fig. 2 while varying the values of L and the output power P_{out} and a good agreement between the calculated and simulated values was observed. In the parallel module case of Fig. 2 (c), a maximum of six modules in parallel was considered. Only a minimum number of parallel modules required to supply the load were switched on at a given time as defined by

$$N_{op} = \left\lceil \frac{P_{hour}}{3P_{nom,n}} \right\rceil, \quad N_{op} \in [3,6,9...18], \qquad (11)$$

where $P_{nom,n}$ is the nominal power of a single module, defined by

$$P_{nom,n} = \frac{P_{max}}{3n}, \quad n \in [1,2,3,...6]. \qquad (12)$$

The RMS load current of a single inverter module can be calculated by

$$I_{RMS,hour,n} = \frac{P_{hour}}{3n \cdot 230}. \qquad (13)$$

The result of (13) was used to calculate the value of I_m in (10), and the value of I_{out} in (6).

Fit functions for the thermal resistance, thermal capacitance, and price of heat sinks were generated based on the heat sink data in [30],[31] to perform the necessary thermal calculations. The thermal pathway from the semiconductor junction to the air cooling the heat sink should satisfy the constraint

$$\begin{aligned} T_{j,max,nom} &- P_{loss,nom} \left(R_{th,j-c} + R_{th,c-hs} \right) \\ &- P_{loss,nom,tot} R_{th,hs} - T_{amb} \geq 0 \end{aligned}, \qquad (14)$$

where $T_{j,max,nom}$ is the maximum allowed junction temperature at the nominal load, $P_{loss,nom}$ is the power loss of a single semiconductor at the nominal load, $R_{th,j-c}$ is the junction-to-case thermal resistance, $R_{th,c-hs}$ is the case to heat sink thermal resistance, $P_{loss,nom,tot}$ is the total heat load on the same heat sink and $R_{th,hs}$ is the thermal resistance of the heat sink. However, in a dynamic overload condition, caused either by a load connected to the customer's electrical installations, or by a short-circuit fault, the temperature of the heat sink will not necessarily reach thermal equilibrium because of its high thermal capacitance C_{th}. In the worst-case scenario, the CEI is operating at its rated power P_{max} when the overload occurs. The total temperature rise of the heat sink during the overload is, therefore, calculated by

$$\Delta T_{hs,o}(t_o) = P_{loss,nom,tot} R_{th,hs} + P_{loss,o,tot} R_{th,hs} \left(1 - e^{-\frac{t_o}{R_{th,hs}C_{th}}} \right), \qquad (15)$$

where $P_{loss,o,tot}$ is the power loss during the overload situation, and t_o is the duration of the overload. Therefore, the constraint for the junction temperature during the overload becomes

$$T_{j,\max} - P_{\text{loss,o}}\left(R_{\text{th,j-c}} + R_{\text{th,c-hs}}\right)$$
$$- \Delta T_{\text{hs,o}}(t_o) - T_{\text{amb}} \geq 0 \quad , \qquad (16)$$

where $T_{j,\max}$ is the maximum allowed junction temperature at the overload situation, and $P_{\text{loss,o}}$ is the power loss of a single semiconductor during the overload. Values of $T_{j,\max,\text{nom}} = 100$ °C, and $T_{j,\max} = 150$ °C were used in the thermal calculations. Although the previously discussed value of $T_{\text{amb}} = 50$ °C was set as the constraint for the dimensioning in (14) and (16), the average value of T_{amb} during the year is much lower, and therefore, a fit for $R_{\text{ds,on}}$ between $T_j = 25$ °C at 0% load and $T_j = 75$ °C at 100% load was applied when calculating the costs for the losses C_{loss} in (2). The output LC filter was designed using amorphous C-cores and N87 ETD ferrite cores [32],[33], whereas solid round winding wires with diameters of 1–6 mm were used for the winding [34]. A constraint for the maximum current density $J_m = 5$ A/mm², which can be considered a reasonable limit for natural convection [35], was used to filter the winding wire data. The values of $B_{\max} = 1.5$ T for amorphous and $B_{\max} = 0.39$ T for ferrite cores, both at a temperature of 100 °C, were taken as the constraints for the maximum flux density [32],[33]. The core losses were calculated using the Improved Generalized Steinmetz Equation (iGSE) [36] and the current waveform in the inductor. In the case of the CEI, the high-frequency component of the inductor current is triangular, and the integral expression of iGSE reduces to

$$P_v = \frac{k_i}{T}\left(DT\left(\frac{\Delta B}{DT}\right)^a (\Delta B)^{b-a} + (1-D)T\left(\frac{\Delta B}{(1-D)T}\right)^a (\Delta B)^{b-a} \right), \quad (17)$$

where T is the switching period, a and b the Steinmetz coefficients, k_i a parameter of the iGSE method, and D the ratio between the rising and falling slopes of the triangle wave. D is defined by

$$D = M\sin(\omega t). \qquad (18)$$

Therefore, the ith value of D for a full-bridge inverter over a half cycle of the 50 Hz output is calculated by

$$D_i = M\sin\left(\frac{\omega i}{2 f_{sw}}\right), \; i = 1,2,3...2T_s f_{sw} \qquad (19)$$

where T_s is the period of a half cycle, in this case 10 ms. The peak-to-peak value of the inductor ripple current $I_{\text{p-p}}$ is calculated with (7) and is then used to calculate the value of ΔB in (17). The result of (17) is calculated at every switching instant over a half cycle, and thus, the average power loss can be calculated by

$$P_{v,\text{avg}} = \sum_i P_{v,i} \frac{T}{T_s}, \qquad (20)$$

where $P_{v,i}$ is the value of the ith instantaneous power loss [36]. The copper losses were calculated using the

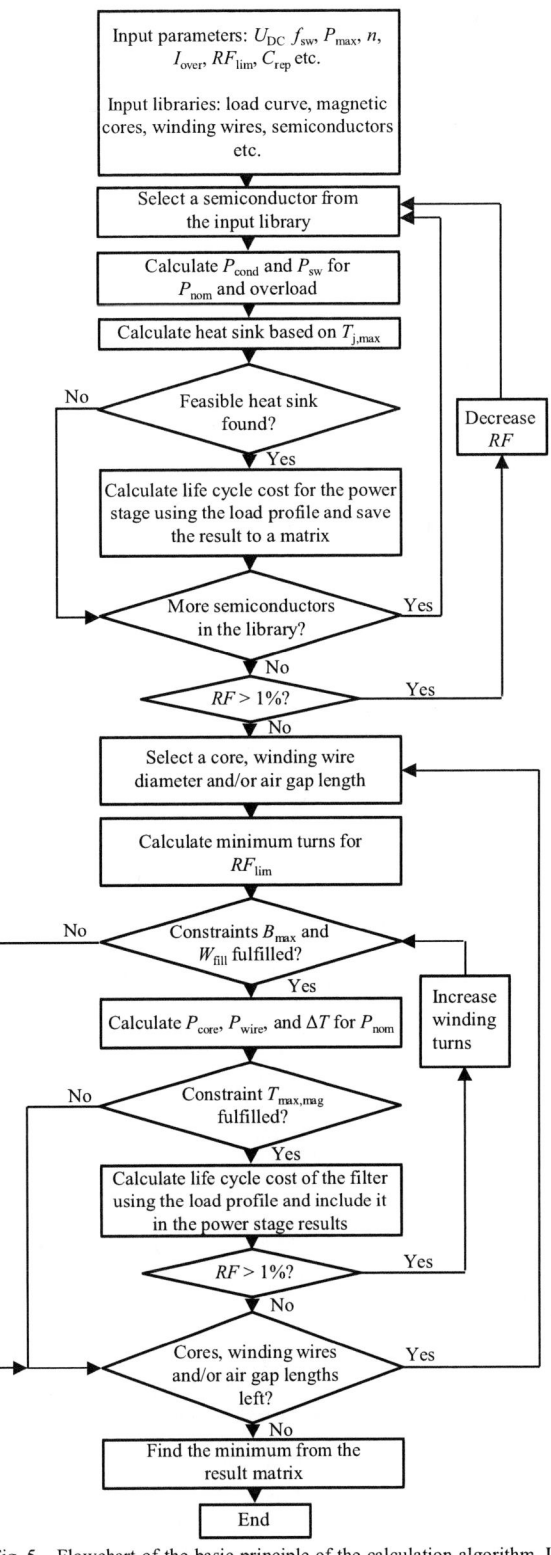

Fig. 5. Flowchart of the basic principle of the calculation algorithm. If the user inputs a number of i alternative values for U_{DC} f_{sw}, n, I_{over}, and RF_{lim}, the flowchart is run i times.

AC resistance of the winding [37] and the RMS current values at the line frequency and the f_{sw}-related harmonic

TABLE I. LIFE CYCLE COST MINIMIZED DESIGNS OF THE CEI FOR THE TWO CUSTOMER TYPES (TYPE 1 AND 2) AND FOR A CASE IN WHICH THE CEI IS DESIGNED TO OPERATE AT ITS RATED POWER. PROTECTION I DENOTES THE USAGE OF AN INTELLIGENT PROTECTION, WHEREAS M DENOTES THE USAGE OF MECHANICAL CIRCUIT BREAKERS, WHICH SET FAR MORE DEMANDING REQUIREMENTS [23],[24] FOR THE CURRENT RATING OF THE CEI.

Load type	Protec-tion	n	C_{DC} [μF]	f_{sw} [kHz]	Main transistors	Filter core	d_{wire} [mm]	L [μH]	η [%]	LCC
Type 1	I	2	976	10	C2M0160120D	AMCC63	2.5	2496	98.1	1
Type 1	M	5	6247	40	C2M0080120D	AMCC80	1.5	1353	97.0	2.98
Type 2	I	2	976	10	C2M0160120D	AMCC63	2.5	2496	97.3	0.92
Type 2	M	5	6247	40	C2M0080120D	AMCC80	1.5	1353	95.4	2.88
P_{max}	I	2	976	10	C2M0040120D	AMCC63	3	1870	98.6	2.52

frequencies in the inductor. The harmonic content was determined by simulating the inverter in Fig. 2 and by calculating the FFT of the inductor current. A fit function representing the RMS values at the different harmonic frequencies $f_{h,m}$ as a function of f_{sw} and the inductor parameters was then generated and used in the calculations. The fit function for the harmonic currents is defined by

$$I_{h,m} = a(f_{sw},m)L^{b(f_{sw},m)}, \quad m = 1,2,3, \quad (21)$$

where L is the inductance of the filter inductor, and a and b are the model coefficients expressed as a function of f_{sw} and m. The three first harmonic groups, that is, $m = 1$–3, are included in the analysis as the RMS value of the harmonics at $m > 3$ is already insignificant. A value of $T_{max,ind} = 80\ ^\circ$C was used as the constraint for the maximum temperature of the inductor, whereas the temperature was calculated by

$$T_{ind} = T_{amb} + \left(\frac{P_{core} + P_{copper}}{SA}\right)^{0.833}, \quad (22)$$

where P_{core} is the core loss, P_{copper} the copper loss, and SA the surface area of the inductor [38],[39]. The copper loss and the inductor temperature were determined using an iterative process, in which the copper resistivity is multiplied by a temperature correction factor, which is defined by

$$T_{corr} = 1 + \alpha(T_{calc} - 20), \quad (23)$$

where α is the temperature coefficient of copper and T_{calc} the temperature to which the resistivity is corrected. The constraint for the DC link capacitance is the maximum allowed peak-to-peak voltage ripple ΔU_{ripple}, whereas the required DC capacitance can be calculated by

$$C_{DC} = \frac{P}{2\pi f U_{DC} \Delta U_{ripple}}, \quad (24)$$

where P is the output power, and f is the output frequency [40]. A value of $\Delta U_{ripple} = 0.1 U_{DC}$ was used in the calculations. The gate driver and control IC cost was estimated at 15 € for a single CEI module when a low-power isolated DC-DC per each switch is assumed [41],[42]. The gate driver power loss is calculated by

$$P_{gate} = Q_g V_g f_{sw}, \quad (25)$$

where Q_g is the gate charge and V_g the driver output voltage.

In order to solve the objective function (2), an optimization algorithm was written in Matlab®. The algorithm takes the necessary data of the main components of the CEI [23]–[34], the price data in 250–1000 quantities [21],[41]–[43], the load profiles of the type 1 and type 2 customers [17], and the main design variables as inputs. The previously described constraints and (1)–(25) are used in the algorithm to evaluate the different combinations of the input data and the design variables $U_{DC} = 400$–750 V, $f_{sw} = 10$–100 kHz, $P_{max} = 17.25$ kW, the number of the parallel CEI modules $n = 1$–6, overload requirement $I_{over} = 50$–160 A$_{RMS}$, and the upper limit for the ripple factor $RF_{lim} = 10$–30%. A simplified flowchart of the algorithm is depicted in Fig. 5. The algorithm evaluates a total of 1.5–3.1 billion possible designs (depending on some of the user-defined input variables) and selects the life-cycle-cost-minimized design as the solution. The algorithm is able to minimize the optimization problem in 3–15 minutes, depending on the user-defined input variables, using a computer equipped with a core i7-3770 processor and 16 GB of RAM. The results with input parameters $U_{DC} = 750$ V, $f_{sw} = 10$–100 kHz, $RF_{lim} = 10\%$, $n = 1$–6, and $C_e = 40$ €/MWh are depicted in Table I, where C_{DC} is the DC link capacitance, d_{wire} is the winding wire diameter, L is the filter inductance, η is the energy efficiency calculated using the load profile in question (type 1, type 2 or the rated power P_{max}), and LCC is the life cycle cost calculated as a relative value. Two distinct types of protection systems, an intelligent one not requiring the converter to exceed its nominal rating and mechanical circuit breakers requiring a high value of overload current $I_{over} = 160$ A$_{RMS}$ [23],[24] to operate, were considered in the analysis. Looking at the results in Table I, we can see that the life-cycle-cost-optimized designs of the CEI are identical for the both customer types (type 1 and type 2) despite the significant differences in their load profiles. Therefore, it can be concluded that a similar design of the CEI could be used to supply both types of customers without major drawbacks. However, the energy efficiency in the case of the type 2 customer will be somewhat lower compared with the type 1 customer as a result of the higher unbalance between the nominal power of the CEI and the actual load (cf. Section II). When the optimization is carried out using the rated power P_{max} instead, the design favors a significantly lower on-state resistance (40 mΩ vs. 160 mΩ), a lower inductance in the filter inductor, and a larger cross-sectional area in the winding wire. If we analyze a larger sample size of the results outside the minima, the behavior is very similar. Therefore, it can be concluded that it is clearly beneficial to use the load behavior of residential customers as one of the inputs in the design process of a converter that is used to supply a residential customer if the target is to find a cost-optimal design. If we compare the results between the two protection systems, we can see that the requirements set by the mechanical protection devices (I_{over} = 160 A_{RMS}) have a significant impact on the CEI parameters. Because of the high overload requirement, the results show a decrease in the energy efficiency and an almost threefold cost compared with the cases in which no overload capability is needed. However, some of the drawbacks of the high overload rating such as a higher cost and lower efficiency with the load profile can be at least to some extent mitigated by implementing the CEI using parallel modules of a lower nominal power (n = 5) instead of a single high-power module. A second set of results were calculated by changing the input voltage from U_{DC} = 750 V to U_{DC} = 400 V. This allows the usage of the switches with a lower voltage rating of 650–900 V [23],[26] and reduces the amount of output filtering, which can potentially reduce the costs. On the other hand, more DC capacitance is needed to suppress the voltage ripple in the DC link as can be seen from (24). Based on the results, the average improvement in the energy efficiency is 1%. Nevertheless, the costs are very similar when no overload capability is required, whereas the results for the overload now show significantly higher costs compared with the cases calculated for U_{DC} = 750 V. Therefore, it can be concluded that there can be some merit in reducing the value of U_{DC} when no overload capability is needed but a higher value of U_{DC} should be chosen if the CEI has to supply an overload situation far exceeding its nominal rating.

A sensitivity analysis was also done on the main design parameters, which can potentially affect the final result. Out of the variables C_e = 40–80 €/MWh, the component price discount rate of 1–0.5, the overload rating I_{lover} = 50–160 A_{RMS}, and the maximum value of the ripple current in the inductor RF_{lim} = 10–30%, the overload rating was found to have the most significant effect on the design and the life cycle cost. When all of the component prices were multiplied by 0.5 to better approximate the volume pricing, only minor changes towards improved energy efficiency were observed in the results. Similar behavior was observed when increasing the value of C_e from C_e = 40 €/MWh to C_e = 80 €/MWh, whereas the life cycle cost increased by 11% on average. When the maximum allowed ripple factor was increased from RF_{lim} = 10% to RF_{lim} = 30% the results stayed relatively unchanged if no overload capability is required or when I_{over} = 50 A_{RMS}. However, a reduction of up to 8% in the life cycle cost was observed when I_{over} = 160 A_{RMS}. It is pointed out that the value of RF_{lim}

should not be increased considerably unless a grid-side inductor is used because a part of the ripple current will begin to flow to the customer network if a highly capacitive load is connected.

IV. CONCLUSIONS

Using the load behavior of actual customers as one of the inputs in the design process is clearly beneficial if the target is to find a cost optimal design of a converter used to supply a residential customer. A preference for relatively high values of inductance in the output LC filter and relatively low values of f_{sw} is observed even when using the current state-of-the-art semiconductor switches such as silicon carbide and gallium nitride ones. Even though increasing the switching frequency further would result in a reduction in the size and cost of the output filter, the cost of the additional switching losses over the utilization period was found to outweigh the advantage in the studied application. Significant differences in the optimal silicon area, filter inductance, and winding wire cross-sectional area were observed compared with a case in which the converter was designed using only the rated power even though the rated power was the same in both cases. When load profiles of two different types of residential customers were used as the input in the design process, the resulting designs for both customers were nearly identical, whereas the energy efficiency was somewhat lower in the case of the customer having a lower annual energy consumption. Even though the energy efficiency for the other customer could be improved by altering the design of the CEI, this can be cost prohibitive. It can be concluded that a similar converter could be used without major drawbacks to supply a variety of residential customers all having their individual differences in their electricity usage profiles. Finally, a sensitivity analysis was carried out on the main parameters and constraints that were initially fixed in the design process. When the electricity cost, component price discount rate, overload rating, and the upper limit for the ripple current in the inductor were varied using realistic ranges for the values, the overload rating was found to have the most significant impact on the results. However, some of the drawbacks of the high overload rating such as a higher cost and lower energy efficiency with the load profile can be, at least to some extent, mitigated by implementing the CEI using parallel modules of a lower nominal power instead of a single high-power module.

REFERENCES

[1] T. Babasaki *et al.*, "Developing of higher voltage direct-current powerfeeding prototype system," in *Proc. INTELEC*, Incheon, Korea, 2009, pp. 1–5

[2] A. Fukui, T. Takeda, K. Hirose, and M. Yamasaki, "HVDC power distribution systems for telecom sites and data centers," in *Int. Power Electron. Conf. (IPEC)*, 21–24 June 2010, pp. 874–880

[3] H. Kakigano, Y. Miura, and T. Ise, "Low-Voltage Bipolar-Type DC Microgrid for Super High Quality Distribution," *IEEE Trans. Power Electron.*, vol. 25, no.12, pp. 3066–3075, Dec. 2010

[4] D. Boroyevich, I. Cvetkovic, Dong Dong, R. Burgos, Fei Wang, and F. Lee, "Future electronic power distribution systems a contemplative view," in *12th Int. Conf. Optimization of Electrical and Electronic Equipment (OPTIM)*, 2010, pp.1369–1380, 20–22 May 2010

[5] Xunwei Yu; Xu She; Xiaohu Zhou; Huang, A.Q., "Power Management for DC Microgrid Enabled by Solid-State Transformer," in *IEEE Trans. Smart Grid*, vol.5, no.2, pp.954–965, March 2014

[6] T. Sakagami, A. Werth, M. Tokoro, Y. Asai, D. Kawamoto and H. Kitano, "Performance of a DC-based microgrid system in Okinawa," *2015 International Conference on Renewable Energy Research and Applications (ICRERA)*, Palermo, 2015, pp. 311-316

[7] T. Hakala, T. Lähdeaho and P. Järventausta, "Low-Voltage DC Distribution—Utilization Potential in a Large Distribution Network Company," in *IEEE Transactions on Power Delivery*, vol. 30, no. 4, pp. 1694-1701, Aug. 2015.

[8] T. Kaipia, P. Salonen, J. Lassila, and J. Partanen, "Possibilities of the low voltage DC distribution systems," in *Proc. NORDAC '06*, Stockholm, Sweden, Aug. 20–21, 2006

[9] Kaipia, T.; Karppanen, J.; Mattsson, A.; Lana, A.; Nuutinen, P.; Peltoniemi, P.; Salonen, P.; Partanen, J.; Lohjala, J.; Wookyu Chae; Juyong Kim, "A system engineering approach to low voltage DC distribution," in *Electricity Distribution (CIRED 2013), 22nd International Conference and Exhibition on* , vol., no., pp.1–4, 10–13 June 2013

[10] Nuutinen, P.; Kaipia, T.; Peltoniemi, P.; Lana, A.; Pinomaa, A.; Mattsson, A.; Silventoinen, P.; Partanen, J.; Lohjala, J.; Matikainen, M., "Research Site for Low-Voltage Direct Current Distribution in a Utility Network—Structure, Functions, and Operation," in *IEEE Trans. Smart Grid*, vol.5, no.5, pp.2574–2582, Sept. 2014

[11] LVD, Low Voltage Directive 2014/35/EU, Std., 2014, [Online]. Available: http://eur-lex.europa.eu/legal-content/EN/TXT/?uri= CELEX:32006L0095

[12] Karppanen, J.; Kaipia, T.; Mattsson, A.; Lana, A.; Nuutinen, P.; Pinomaa, A.; Peltoniemi, P.; Partanen, J., " Selection of Voltage Level in Low Voltage DC Utility Distribution System", in *Proc. CIRED '15*, Lyon, France, June 15–18, 2015

[13] Mattsson, A., Väisänen, V., Nuutinen, P., Peltoniemi, P., Kaipia, T., Silventoinen, P., Partanen, J., "Evaluation of isolated converter topologies for low voltage DC distribution," *Industrial Electronics Society, IECON 2015 - 41st Annual Conference of the IEEE*, Yokohama, 2015, pp. 003301–003307.

[14] Mattsson, A., Lana, A., Nuutinen, P., Väisänen, V., Peltoniemi., Kaipia, T., Silventoinen., and Partanen, J., "Galvanic Isolation and Output LC Filter Design for the Low-Voltage DC Customer-End Inverter," *IEEE Trans. Smart Grid.* vol. 5, no. 5, Sept. 2014, pp. 2593–2601.

[15] Mattsson, A., Nuutinen P., Peltoniemi, P., Kaipia, T., Karppanen, J., Väisänen, V., Partanen, J., and Silventoinen, P., "Life-cycle cost analysis for the customer-end inverter used in low voltage dc distribution," in *2015 IEEE First International Conference on DC Microgrids (ICDCM)*. Atlanta, GA, 7–10 June 2015, pp. 148-153.

[16] Household electricity usage 2011, research report Adato Energia Oy, Finland, 2011. [Online], Available: http://www.motiva.fi/files/ 8300/Kotitalouksien_sahkonkaytto_2011_Tutkimusraportti.pdf

[17] Suomen Sähkölaitosyhdistys ry. Sähkön käytön kuormitustutkimus 1992 (Finnish Electricity Association Sener, research on electricity consumption), in Finnish, SLY 7103, Helsinki 1992, ISSN 0786-7905.

[18] E. Lakervi; J. Partanen, *Sähkönjakelutekniikka* (Electricity distribution technology), Helsinki, Finland, 2008 (in Finnish)

[19] H. L. Willis, "A look at failure and age in a utility system," in *Power Distribution Planning Reference Book*, 2nd ed. New York: Marcel Dekker, 2004

[20] Chung, Henry Shu-hung; Wang, Huai; Blaabjerg, Frede; Pecht, Michael, *Reliability of Power Electronic Converter Systems*, London, United Kingdom, The Institution Of Engineering And Technology, 2016

[21] Nord Pool Spot, "Average yearly Elspot prices in Finland," [Online]. Available: www.nordpoolspot.com

[22] SFS-EN 50160, *Voltage characteristics of electricity supplied by public electricity networks*. SESKO standardization, Finland, 2010

[23] SFS 60269-1, *Low-voltage fuses. Part 1: General requirements*, SESKO standardization, Finland, 2008.

[24] SFS 6000-4-41, *Low-voltage electrical installations. Part 4-41: Protection for safety. Protection against electric shock*. SESKO standardization, Finland, 2012.

[25] C2M, C3M and CAS120 SiC MOSFETs, Wolfspeed, [Online]. Available: http://www.wolfspeed.com/power/products/sic-mosfets/table

[26] GS66506T, GS66508T, GS66516T GaN E-HEMT, GaN Systems, [Online]. Available: http://www.gansystems.com/transistors.php

[27] C. L. Nge, O. M. Midtgård, L. Norum and T. O. Sætre, "Power loss analysis for single phase grid-connected PV inverters," *INTELEC 2009 - 31st International Telecommunications Energy Conference*, Incheon, 2009, pp. 1-5.

[28] Hyosung Kim and Kyoung-Hwan Kim, "Filter design for grid connected PV inverters," *2008 IEEE International Conference on Sustainable Energy Technologies*, Singapore, 2008, pp. 1070-1075.

[29] Bin Gu; Dominic, J.; Jih-Sheng Lai; Chien-Liang Chen; LaBella, T.; Baifeng Chen, "High Reliability and Efficiency Single-Phase Transformerless Inverter for Grid-Connected Photovoltaic Systems," in *IEEE Trans. Power Electron.*, vol.28, no.5, pp.2235–2245, May 2013

[30] Fischer Elektronik SK157/150, SK56/150, SK57/150, SK04/100, SK514, SK185, SK104-38, SK09 heat sinks, [Online]. Available: http://www.fischerelektronik.de/en/service-en/downloads-en/katalog-download-en/

[31] Fischer Elektronik SK157/150, SK56/150, SK57/150, SK04/100, SK514, SK185, SK104-38, SK09 heat sinks, [Online]. Available: www.tme.eu/en/katalog/#id_category=100095&search=fischer+el ektronik

[32] Hitachi Metals America Ltd., "Metglas POWERLITE C-Cores technical bulletin" [Online]. Available: http://www.hitachimetals.com/materials-products/amorphous-nanocrystalline/powerlite-c-cores/documents/POWER LITE_C_opt.pdf

[33] TDK Europe - EPCOS Ferrites and accessories, N87 ETD cores, Epcos, [Online], Available: http://en.tdk.eu/tdken/529424/products/product-catalog/ferrites-and-accessories/epcosferrites-and-accessories/er-etd-eq-cores-and-accessories-

[34] Wires.co.uk, "Enamelled copper magnet winding wire," [Online]. Available: http://wires.co.uk/.acatalog/cu_enam.html

[35] M. K. Kazimierczuk, *High-Frequency Magnetic Components*, 2nd ed. Hoboken, NJ, USA: Wiley, 2013.

[36] K. Venkatachalam, C. R. Sullivan, T. Abdallah and H. Tacca, "Accurate prediction of ferrite core loss with nonsinusoidal waveforms using only Steinmetz parameters," *Computers in Power Electronics, 2002. Proceedings. 2002 IEEE Workshop on*, 2002, pp. 36-41

[37] R. P. Wojda and M. K. Kazimierczuk, "Analytical optimisation of solid-round-wire windings conducting dc and ac non-sinusoidal periodic currents," in *IET Power Electron.*, vol. 6, no. 7, pp. 1462-1474, August 2013.

[38] Metglas Inc. Application Guide: *Power Factor Correction Inductor Design For Switched Mode Power Supplies Using Metglas Powerlite C-Cores*. 2009.

[39] Magnetics® Powder Core Catalogue, *Temperature Rise Calculation*, p. 20. [Online]. Available: https://www.mag-inc.com/Media/Magnetics/File-Library/Product%20Literature/ Powder%20Core%20Literature/MagenticsPowderCoreCatalog201 3Update.pdf?ext=.pdf

[40] S. B. Kjaer, J. K. Pedersen and F. Blaabjerg, "A review of single-phase grid-connected inverters for photovoltaic modules," in *IEEE Trans. Ind. Appl.*, vol. 41, no. 5, pp. 1292-1306, Sept.–Oct. 2005.

[41] Mouser Electronics, [Online], Available: http://www.mouser.com/

[42] Digi-key Electronics, [Online], Available: http://www.digikey.com

[43] TME® Electronic Components, [Online], Available: http://www.tme.eu/en/

The 2018 International Power Electronics Conference

A Control Method of DC Capacitor Voltage in MMC for HVDC System using Negative Sequence Current

Hanis Afiqah binti Jaffar[1], Ahmad Arif bin Abd Rahman[1]*and Hiroaki Kakigano[1]
1 Graduate School of Science and Engineering, Ristumeikan University, Kusatsu, Japan
*E-mail: re0099ee@ed.ritsumei.ac.jp

Abstract— The application of Modular Multilevel Converter (MMC) for High Voltage Direct Current (HVDC) system is expected to enhance power transmission capacity and improve power system stability. To control an MMC-HVDC system properly, the ac current, circulating current among arms and submodule (SM) capacitor voltages are considered. In the conventional balance control method of the dc capacitor voltage, the reference value of the circulating current control includes ac component as same frequency as ac grid. In this case, it is difficult to follow the reference value perfectly by proportional-integral (PI) controller. This article presents a new control method of dc capacitor voltage using negative sequence of ac grid current and circulating current. The validity of the proposed control method is verified by a numerical simulation.

Keywords— *HVDC transmission, modular multilevel converter (MMC), voltage-source converter, negative sequence current.*

I. INTRODUCTION

The renewable energy topic is one of the hottest issues in the entire world today due to the fast and huge consumption of fossil fuels in recent years. Renewable energy can be derived from various sources, such as solar, wind, biological sources, and so on [1]. Due to the growth in power demand, power systems are being expanded and power stations are being located further away from load. The application of HVDC systems is the transmission of bulk power over long distances for both overhead line and sea cable. The overall cost for the transmission system is less and the losses are lower than ac transmission.

New converter topologies and lower priced fast-switching semiconductors have recently made it possible to build voltage source converter (VSC)-based HVDC transmission systems. The VSC-HVDC system has the advantage of independent control of the active and reactive power, fast transient response characteristics owing to pulse-width modulation (PWM) method, and reduced size of the ac-side filter. For these reasons, many investigations related to this subject have been carried out [2]-[4].

Currently, a modular multilevel converter (MMC) is being studied as a self-commutated converter applied to a HVDC power transmission system [5]-[6]. MMC consists of single submodules (SMs) of half-bridge converters or full bridge converters which are connected in series. For the MMC to give power output stably, it is necessary to balance the dc capacitor voltage SMs and control the dc

capacitor voltage in an appropriate value range. The voltage and current of the modular multilevel converter can be converted into three main components as follows:
1) dc current component
2) 3-phase ac current component
3) Arm circulating current component.

Under unbalanced conditions, however, positive and negative sequence components also exist in circulating current and 3-phase ac current. The variation of circulating current components and the dc voltage/current ripple with different current control schemes remain unknown [7]. Therefore, using the main components, dc voltage regulation, average voltage regulation of the capacitors and arm balancing control can be performed.

For arm balancing control method, there are interphase balancing control for suppressing deviation between arms of each phase. In addition, there is also upper and lower arm balancing control for suppressing deviation of dc capacitor voltage between upper and lower arms, as well as, SM balancing control for suppressing deviation between SM of each arms. It is assumed that the circulating current is composed of dc component, positive sequence components, and negative sequence components. The reference values are then generated on the dq frame.

However, the reference values of the positive and negative sequence components include ac components. Since proportional-integral (PI) controller is used for current controller cannot always follows the fluctuating reference values, it is undesirable to contain ac components in reference value.

In this paper, power flow in MMC is analyzed using positive and negative sequence components of not only circulating current but also 3-phase ac current. From this study, dc current reference is generated from negative sequence components of circulating current and 3-phase ac current. Thus, arm balancing control between each phase and arm balancing control between upper and lower sides of converter have been proposed using this method. The effectiveness of the proposed method has verified through simulation.

II. MMC SYSTEM CONFIGURATION

Fig. 1 illustrates a circuit configuration of an MMC, consisting of three series-connected SMs in each arm. The

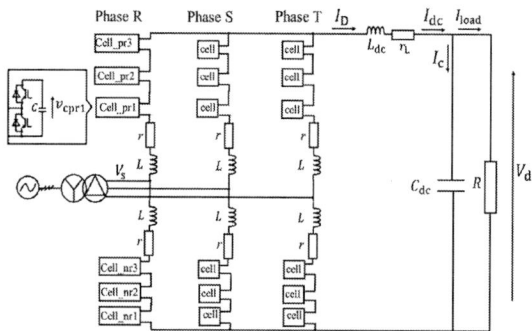

Fig. 1. Circuit configuration of the MMC with 3 half bridge circuit per arm used for the simulations

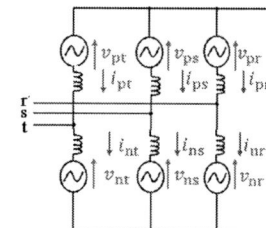

Fig. 2. MMC equivalent circuit.

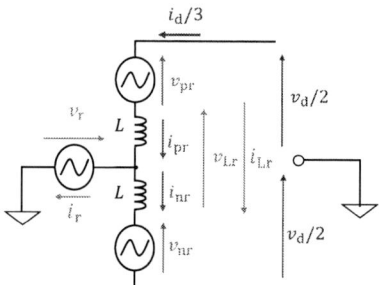

Fig.3. Equivalent circuit of R-phase.

SM is typically a chopper topology made up from two IGBTs/diodes and an energy storage dc capacitor. The point of the inductors connects to the ac grid while two outer phase legs are connected to the dc bus terminals. A $Y-\Delta$ transformer configuration is used in this work.

III. CONTROL VARIABLES

A. Current Control Variables

MMC can be represented as an equivalent circuit depicted in Fig. 2, where the output voltage from the SM is viewed as a voltage source. Meanwhile, the equivalent circuit in Fig. 3 showing voltages, currents, components of ac grid, dc-link components and circulating components of R-phase.

Based on Kirchhoff's current law, the ac grid current of each phase can be analytically defined as follows in terms of positive and negative sequence arm currents.

$$
\begin{aligned}
i_r &= i_{pr} - i_{nr} \\
i_s &= i_{ps} - i_{ns} \\
i_t &= i_{pt} - i_{nt}
\end{aligned} \tag{1}
$$

Next, the ac grid current i_r is considered flows evenly

from the positive and negative sequence for each arm in the r phase and suppose that the alternating current flows equally into each phase. At this time, the currents flowing through the upper arm i_{pr} and the lower arm i_{nr} are expressed by the following equation in term of the alternating current i_r, direct current i_d, and circulating current i_{Lr}.

$$
\begin{aligned}
i_{pr} &= i_{Lr} + 1/3 \cdot i_d + 1/2 \cdot i_r \\
i_{nr} &= i_{Lr} + 1/3 \cdot i_d - 1/2 \cdot i_r
\end{aligned} \tag{2}
$$

Similarly, to phase s and t,

$$
\begin{aligned}
i_{ps} &= i_{Ls} + 1/3 \cdot i_d + 1/2 \cdot i_s \\
i_{ns} &= i_{Ls} + 1/3 \cdot i_d - 1/2 \cdot i_s \\
i_{pt} &= i_{Lt} + 1/3 \cdot i_d + 1/2 \cdot i_t \\
i_{nt} &= i_{Lt} + 1/3 \cdot i_d - 1/2 \cdot i_t.
\end{aligned} \tag{3}
$$

Meanwhile, the dc current i_d is expressed by the following equation

$$
i_d = \frac{1}{2} \cdot \left(i_{pr} + i_{nr} + i_{ps} + i_{ns} + i_{pt} + i_{nt} \right) \tag{4}
$$

In addition, the components of the circulating current of each phase i_{Lr}, i_{Ls}, i_{Lt} are expressed as follows

$$
\begin{aligned}
i_{Lr} &= 1/3 \cdot (i_{pr} + i_{nr}) - 1/6 \cdot (i_{ps} + i_{pt} + i_{ns} + i_{nt}) \\
i_{Ls} &= 1/3 \cdot (i_{ps} + i_{ns}) - 1/6 \cdot (i_{pr} + i_{pt} + i_{nr} + i_{nt}) \\
i_{Lt} &= 1/3 \cdot (i_{pt} + i_{nt}) - 1/6 \cdot (i_{pr} + i_{ps} + i_{nr} + i_{ns})
\end{aligned} \tag{5}
$$

From the above, the transformation matrix for deriving each component of the current is given by the following equation

$$
\begin{bmatrix} i_r \\ i_s \\ i_t \\ i_{Lr} \\ i_{Ls} \\ i_{Lt} \\ i_d \end{bmatrix} = \begin{bmatrix} 1 & -1 & 0 & 0 & 0 & 0 \\ 0 & 0 & 1 & -1 & 0 & 0 \\ 0 & 0 & 0 & 0 & 1 & -1 \\ 1/3 & 1/3 & -1/6 & -1/6 & -1/6 & -1/6 \\ -1/6 & -1/6 & 1/3 & 1/3 & -1/6 & -1/6 \\ -1/6 & -1/6 & -1/6 & -1/6 & 1/3 & 1/3 \\ 1/2 & 1/2 & 1/2 & 1/2 & 1/2 & 1/2 \end{bmatrix} \begin{bmatrix} i_{pr} \\ i_{nr} \\ i_{ps} \\ i_{ns} \\ i_{pt} \\ i_{nt} \end{bmatrix} \tag{6}
$$

B. Voltage Control Variables

Next, it is assumed that the neutral point of the dc voltage and the neutral point of the three-phase ac voltage are at the same potential in the equivalent circuit of Fig. 3. Hence, the output voltages of the upper arm and the lower arm of the r phase v_{pr} and v_{nr} can be expressed by the following equation in the term of dc voltage, circulating voltage, and ac voltage.

$$
v_{pr} = -v_r + \frac{v_d}{2} - \frac{v_{Lr}}{2} \tag{7}
$$

$$
v_{nr} = v_r + \frac{v_d}{2} - \frac{v_{Lr}}{2} \tag{8}
$$

The modulated wave in each cell is obtained by dividing these two equations by the number of cells. When the 3-phase ac voltage is represented by the voltage of the upper and lower arms, it is expressed by the following equation

$$
\begin{aligned}
v_r &= \frac{1}{2}(-v_{pr} + v_{nr}) \\
v_s &= \frac{1}{2}(-v_{ps} + v_{ns}) \\
v_t &= \frac{1}{2}(-v_{pt} + v_{nt})
\end{aligned} \tag{9}
$$

The dc output voltage of the converter becomes as follows

$$
v_d = 1/3 \cdot (v_{pr} + v_{nr} + v_{ps} + v_{ns} + v_{pt} + v_{nt}) \tag{10}
$$

Furthermore, the circulating voltages of each phase v_{Lr}, v_{Ls}, v_{Lt} are expressed by the following equations

$$v_{Lr} = -\frac{2}{3}(v_{pr} + v_{nr}) + \frac{1}{3}(v_{ps} + v_{ns} + v_{pt} + v_{nt})$$

$$v_{Ls} = -\frac{2}{3}(v_{pr} + v_{nr}) + \frac{1}{3}(v_{ps} + v_{ns} + v_{pt} + v_{nt})$$

$$v_{Lt} = -\frac{2}{3}(v_{pr} + v_{nr}) + \frac{1}{3}(v_{ps} + v_{ns} + v_{pt} + v_{nt}) \quad (11)$$

From the above, the voltage conversion matrix is expressed by the following equation

$$\begin{bmatrix} v_r \\ v_s \\ v_t \\ v_{Lr} \\ v_{Ls} \\ v_{Lt} \\ v_d \end{bmatrix} = \begin{bmatrix} -1 & 1 & 0 & 0 & 0 & 0 \\ 0 & 0 & -1 & 1 & 0 & 0 \\ 0 & 0 & 0 & 0 & -1 & 1 \\ -2/3 & -2/3 & 1/3 & 1/3 & 1/3 & 1/3 \\ 1/3 & 1/3 & -2/3 & -2/3 & 1/3 & 1/3 \\ 1/3 & 1/3 & 1/3 & 1/3 & -2/3 & -2/3 \\ 1/3 & 1/3 & 1/3 & 1/3 & 1/3 & 1/3 \end{bmatrix} \begin{bmatrix} v_{pr} \\ v_{nr} \\ v_{ps} \\ v_{ns} \\ v_{pt} \\ v_{nt} \end{bmatrix} \quad (12)$$

Also, its inverse matrix can be drawn as following

$$\begin{bmatrix} v_{pr} \\ v_{nr} \\ v_{ps} \\ v_{ns} \\ v_{pt} \\ v_{nt} \end{bmatrix} = \begin{bmatrix} -1 & 0 & 0 & -1/2 & 0 & 0 & 1/2 \\ 1 & 0 & 0 & -1/2 & 0 & 0 & 1/2 \\ 0 & -1 & 0 & 0 & -1/2 & 0 & 1/2 \\ 0 & 1 & 0 & 0 & -1/2 & 0 & 1/2 \\ 0 & 0 & -1 & 0 & 0 & -1/2 & 1/2 \\ 0 & 0 & 1 & 0 & 0 & -1/2 & 1/2 \end{bmatrix} \begin{bmatrix} v_r \\ v_s \\ v_t \\ v_{Lr} \\ v_{Ls} \\ v_{Lt} \\ v_d \end{bmatrix} \quad (13)$$

In the control block, each arm current is detected and then it is converted to each current component by the equation (6), before it used for current control. The reference values of the 3-phase ac voltages, circulating voltages, and the dc voltages are calculated by using the matrix of expression (12). Lastly, the gate signal is produced by the PWM control.

IV. CONTROL SCHEMES OF THE MMC

Fig. 4 shows the configuration of the control block. The three main components; alternating current, circulating current, and direct current are controlled, respectively. Voltage output by each arm is calculated by equation (12).

A. Analysis of 3-Phase AC Grid Current and Active Power

The positive and negative sequence components of the active power of the 3-phase ac grid current and voltage are extracted by means of park transformation using the phase angle determined by a PLL. Let the ac grid voltage be expressed by the following equation.

$$\begin{bmatrix} v_r \\ v_s \\ v_t \end{bmatrix} = \sqrt{\frac{2}{3}} V_s \begin{bmatrix} \cos \omega t \\ \cos(\omega t - \frac{2}{3}\pi) \\ \cos(\omega t + \frac{2}{3}\pi) \end{bmatrix} \quad (13)$$

Then, the 3-phase ac grid currents i_r, i_s, and i_t are expressed by the following equation as the sum of the current of the positive sequence component and negative

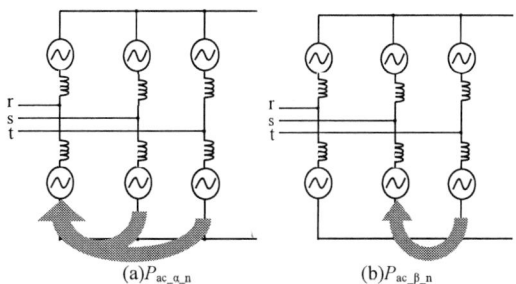

(a)$P_{ac\_\alpha\_n}$ (b)$P_{ac\_\beta\_n}$

Fig. 5. Power-flow directions of $P_{ac\_\alpha\_n}$ and $P_{ac\_\beta\_n}$

sequence component in each phase,

$$\begin{bmatrix} i_r \\ i_s \\ i_t \end{bmatrix} = \sqrt{\frac{2}{3}} I_p \begin{bmatrix} \cos(\omega t + \varphi_p) \\ \cos(\omega t + \varphi_p - \frac{2}{3}\pi) \\ \cos(\omega t + \varphi_p + \frac{2}{3}\pi) \end{bmatrix} + \sqrt{\frac{2}{3}} I_n \begin{bmatrix} \cos(\omega t + \varphi_n) \\ \cos(\omega t + \varphi_n + \frac{2}{3}\pi) \\ \cos(\omega t + \varphi_n - \frac{2}{3}\pi) \end{bmatrix} \quad (14)$$

where φ_p and φ_n denote phase differences of the ac grid voltage. The first term of this equation indicates the positive sequence and the second term indicates the negative sequence of the ac current, respectively. The active power P_{ac_r}, P_{ac_s}, P_{ac_t} in one period of the ac grid voltage and the current can be drawn as follows:

$$\begin{bmatrix} P_{ac\_r} \\ P_{ac\_s} \\ P_{ac\_t} \end{bmatrix} = \frac{1}{T} \int_{-\frac{T}{2}}^{\frac{T}{2}} \begin{bmatrix} v_r \cdot i_r \\ v_s \cdot i_s \\ v_t \cdot i_t \end{bmatrix} dt \quad (15).$$

Here, the first term of the equation (14) is assumed to operate with $\varphi_p = 0$, meaning the power factor is equal to 1. At this time, by controlling the active power input from the ac grid system to the converter, the average reference value of the dc capacitor voltages of all the SMs can be controlled. In this paper, we also consider the active power $P_{ac\_r\_n}$, $P_{ac\_s\_n}$, $P_{ac\_t\_n}$ of the negative sequence ac grid current in the second term. The equation of the active power is expressed by the following equation.

$$\begin{bmatrix} P_{ac\_r\_n} \\ P_{ac\_s\_n} \\ P_{ac\_t\_n} \end{bmatrix} = \frac{1}{T} \int_{-\frac{T}{2}}^{\frac{T}{2}} \sqrt{\frac{2}{3}} V_s \cdot \sqrt{\frac{2}{3}} I_n \begin{bmatrix} \cos \omega t \cdot \cos(\omega t + \varphi_n) \\ \cos(\omega t - \frac{2}{3}\pi) \cdot \cos(\omega t + \varphi_n + \frac{2}{3}\pi) \\ \cos(\omega t + \frac{2}{3}\pi) \cdot \cos(\omega t + \varphi_n - \frac{2}{3}\pi) \end{bmatrix} dt \quad (16)$$

Hence, the active power $P_{ac\_\alpha\_n}$, $P_{ac\_\beta\_n}$ on the $\alpha\beta$ frame can be expressed by the following equation.

$$\begin{bmatrix} P_{ac\_\alpha\_n} \\ P_{ac\_\beta\_n} \end{bmatrix} = \frac{1}{\sqrt{6}} \begin{bmatrix} 2P_{ac\_r\_n} - P_{ac\_s\_n} - P_{ac\_t\_n} \\ \sqrt{3}P_{ac\_s\_n} - \sqrt{3}P_{ac\_t\_n} \end{bmatrix} = \frac{1}{\sqrt{6}} V_s I_n \begin{bmatrix} \cos \varphi_n \\ \sin \varphi_n \end{bmatrix} \quad (17)$$

Fig. 5 shows the power flow in the converter derived from equation (17). Meanwhile, in the dq frame the negative sequence components i_{d_n} and i_{q_n} are expressed as

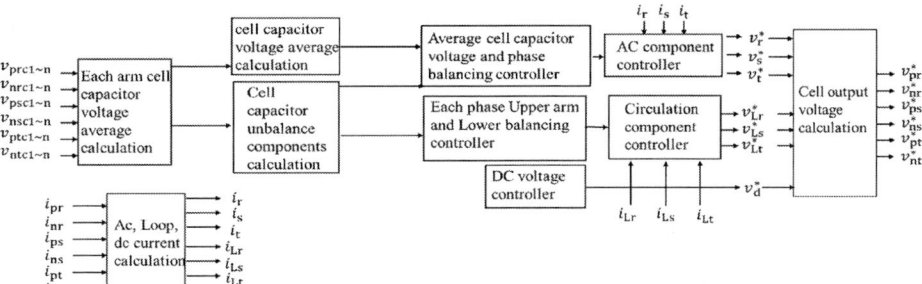

Fig. 4. Overall control block diagram.

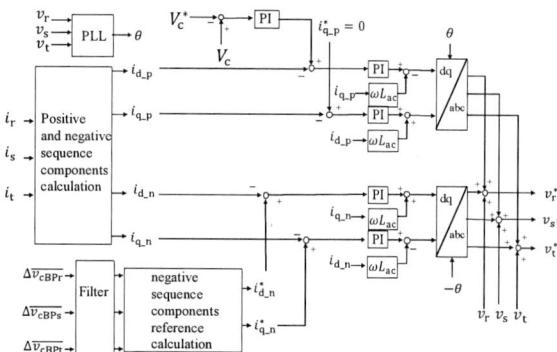

Fig. 6. AC components, dc capacitor average voltage reference and arm balancing controller block diagram

$$\begin{bmatrix} i_{d\_n} \\ i_{q\_n} \end{bmatrix} = I_n \begin{bmatrix} \cos\varphi_n \\ -\sin\varphi_n \end{bmatrix} \tag{18}.$$

Therefore, based on equations (17) and (18), arm balancing control between r-phase, s-phase and t-phase can be perform by manipulating i_{d_n}, and arm balancing control between s-phase and t-phase can be controlled by manipulating i_{q_n}.

B. DC Capacitor Average Voltage Reference and Phase Balancing Control

The average voltage of the dc capacitor V_c can be expressed as

$$V_c = (v_{cpr} + v_{cnr} + v_{cps} + v_{cns} + v_{cpt} + v_{cnt})/6 \tag{19}.$$

The positive sequence ac grid current is applied to the proportional integral control and its reference values $i^*_{d\_p}$, $i^*_{q\_p}$ are given as below using the proportional gain K_{cp} and the integral gain K_{ci}.

$$i^*_{d\_p} = K_{cp}(-V_c^* + V_c) + \int K_{ci}(-V_c^* + V_c)dt \tag{20}$$

$$i^*_{q\_p} = 0 \tag{21}.$$

Then, the deviation between the upper and lower average value of the dc capacitor voltage of the r-phase $\overline{v_{cpr}}$, $\overline{v_{cnr}}$ and the reference value of the dc capacitor voltage V_c^* are defined as follows:

$$\Delta\overline{v_{cpr}} = V_c^* - \overline{v_{cpr}}$$
$$\Delta\overline{v_{cnr}} = V_c^* - \overline{v_{cnr}} \tag{22}.$$

The deviation of the dc capacitor voltage of the r phase can be expressed as

$$\Delta\overline{v_{cBPr}} = \Delta\overline{v_{cpr}} + \Delta\overline{v_{cnr}} \tag{23}.$$

Similarly, the expressions of other phases are also available to obtain

$$\Delta\overline{v_{cBPs}} = \Delta\overline{v_{cps}} + \Delta\overline{v_{cns}}$$
$$\Delta\overline{v_{cBPt}} = \Delta\overline{v_{cpt}} + \Delta\overline{v_{cnt}} \tag{24}.$$

Hence, the reference values $i^*_{d\_n}$, $i^*_{q\_n}$ on the dq frame are given by the following equation using the proportional gain K_n.

$$i^*_{d\_n} = -K_n(2\Delta\overline{v_{cBPr}} - \Delta\overline{v_{cBPs}} - \Delta\overline{v_{cBPt}}) \tag{25}$$

$$i^*_{q\_n} = K_n(\Delta\overline{v_{cBPs}} - \Delta\overline{v_{cBPt}}) \tag{26}$$

From the equations above, a controller block diagram for ac grid current, dc capacitor average voltage reference and arm balancing controller are illustrated in Fig. 6.

A moving average filters are utilized in order to eliminate the dc capacitor voltage ripple of 60 Hz due to the active power from the ac grid system. Therefore, before calculating the equations (25) and (26), the filter is inserted. Hence, the frequency component of the ac current is removed from the deviation of the dc capacitor voltage of each phase. In the decoupling control section, L_{ac} consists of arm reactors and leakage inductance of the transformer that connected the converter and ac grid system, and it is expressed as $L_{ac} = 1/2 \cdot L + L_L$, where L denotes the arm reactor and L_L denotes leakage inductance of the transformer, respectively.

C. Analysis of the Circulating Current and the Active Power

For the analysis of the circulation model, the general equation of circulating current can be drawn as

$$\begin{bmatrix} i_{Lr} \\ i_{Ls} \\ i_{Lt} \end{bmatrix} = \sqrt{\frac{2}{3}}I_{Lp}\begin{bmatrix} \cos(\omega t + \varphi_{Lp}) \\ \cos(\omega t + \varphi_{Lp} - \frac{2}{3}\pi) \\ \cos(\omega t + \varphi_{Lp} + \frac{2}{3}\pi) \end{bmatrix} + \sqrt{\frac{2}{3}}I_{Ln}\begin{bmatrix} \cos(\omega t + \varphi_{Ln}) \\ \cos(\omega t + \varphi_{Ln} + \frac{2}{3}\pi) \\ \cos(\omega t + \varphi_{Ln} - \frac{2}{3}\pi) \end{bmatrix} \tag{27}$$

where φ_{Lp} and φ_{Ln} denote phase differences from the ac grid voltage. The first term of this equation indicates the positive sequence and the second term indicates the negative sequence of the ac current, respectively. The active power P_{L_pr}, P_{L_ps}, P_{L_pt} for one cycle of this current and the upper arm voltage of each phase is expressed by the following equation.

$$\begin{bmatrix} P_{L\_pr} \\ P_{L\_ps} \\ P_{L\_pt} \end{bmatrix} = \frac{1}{T}\int_{-\frac{T}{2}}^{\frac{T}{2}}\begin{bmatrix} v_p \cdot i_{Lr} \\ v_p \cdot i_{Ls} \\ v_p \cdot i_{Lt} \end{bmatrix}dt \tag{28}.$$

Similar to equation (17), expressing the active power $P_{L\_p\alpha}$, $P_{L\_p\beta}$ on the $\alpha\beta$ frame are expressed as follows:

$$\begin{bmatrix} P_{L\_p\alpha} \\ P_{L\_p\beta} \end{bmatrix} = -\sqrt{\frac{3}{2}}V_sI_{Lp}\begin{bmatrix} \cos\varphi_{Lp} \\ \cos\varphi_{Lp} \end{bmatrix} - \frac{1}{\sqrt{6}}V_sI_{Ln}\begin{bmatrix} \cos\varphi_{Ln} \\ \sin\varphi_{Ln} \end{bmatrix} \tag{29}$$

On the other hand, the active power $P_{L\_n\alpha}$, $P_{L\_n\beta}$ on the $\alpha\beta$ axis flowing in the lower arm is

$$\begin{bmatrix} P_{L\_n\alpha} \\ P_{L\_n\beta} \end{bmatrix} = \sqrt{\frac{3}{2}}V_sI_{Lp}\begin{bmatrix} \cos\varphi_{Lp} \\ \cos\varphi_{Lp} \end{bmatrix} + \frac{1}{\sqrt{6}}V_sI_{Ln}\begin{bmatrix} \cos\varphi_{Ln} \\ \sin\varphi_{Ln} \end{bmatrix} \tag{30}$$

Comparing these two equations, the polarity of the active power flowing in the upper and lower arm are opposite. Therefore, it is possible to control the balancing between upper and lower arms by controlling the circulating current. The active power $P_{L\_pr\_n}$, $P_{L\_ps\_n}$, $P_{L\_pt\_n}$ caused by the negative sequence circulating currents $i_{L\_pr\_n}$, $i_{L\_ps\_n}$, $i_{L\_pt\_n}$ of each phase in the upper arm are expressed by the following equations.

$$\begin{bmatrix} P_{L\_pr\_n} \\ P_{L\_ps\_n} \\ P_{L\_pt\_n} \end{bmatrix} = -\frac{1}{T}\int_{-\frac{T}{2}}^{\frac{T}{2}}\begin{bmatrix} v_{pr} \cdot i_{L\_r\_n} \\ v_{ps} \cdot i_{L\_s\_n} \\ v_{pt} \cdot i_{L\_t\_n} \end{bmatrix}dt \tag{31}$$

Hence, the active power on the $\alpha\beta$ frame is described as

$$\begin{bmatrix} P_{L\_p\alpha\_n} \\ P_{L\_p\beta\_n} \end{bmatrix} = \frac{1}{\sqrt{6}}\begin{bmatrix} 2P_{L\_pr\_n} - P_{L\_ps\_n} - P_{L\_pt\_n} \\ \sqrt{3}P_{L\_ps\_n} - \sqrt{3}P_{L\_pt\_n} \end{bmatrix} = -\frac{1}{\sqrt{6}}V_sI_{Ln}\begin{bmatrix} \cos\varphi_{Ln} \\ \sin\varphi_{Ln} \end{bmatrix} \tag{32}$$

On the other hand, in the lower arm, the active power $P_{L\_nr\_n}$, $P_{L\_ns\_n}$, $P_{L\_nt\_n}$ are generated by the negative sequence circulating current and the active power $P_{L\_n\alpha\_n}$ and $P_{L\_n\beta\_n}$ are obtained by converting into $\alpha\beta$ frame is obtained as

$$\begin{bmatrix} P_{L\_n\alpha\_n} \\ P_{L\_n\beta\_n} \end{bmatrix} = \frac{1}{\sqrt{6}}\begin{bmatrix} 2P_{L\_r\_n} - P_{L\_s\_n} - P_{L\_t\_n} \\ \sqrt{3}P_{L\_s\_n} - \sqrt{3}P_{L\_t\_n} \end{bmatrix} = \frac{1}{\sqrt{6}}V_sI_{Ln}\begin{bmatrix} \cos\varphi_{Ln} \\ \sin\varphi_{Ln} \end{bmatrix}$$

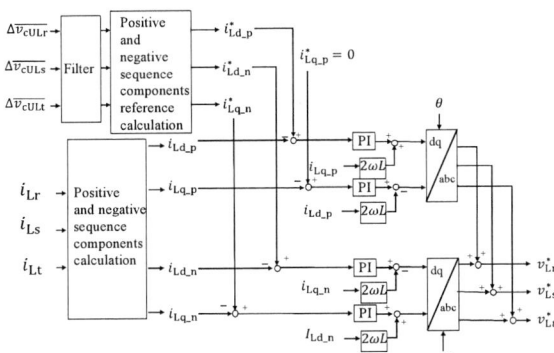

Fig. 7. Circulation components and upper-lower balancing

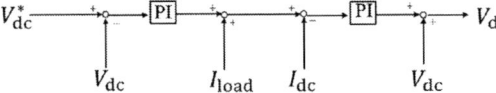

Fig. 8. DC voltage controller

(33).

Through expressions (32) and (33), the balancing among the phase is possible because the form is same as the case of the active power generated by the negative sequence of the ac current in the equation (17). On the right side of these equations, the polarities are different in these two formulas. Here, the negative sequence components i_{Ld_n} and i_{Lq_n} on the dq frame of the negative sequence of the circulation current are expressed by the following equations:

$$\begin{bmatrix} i_{Ld\_n} \\ i_{Lq\_n} \end{bmatrix} = I_{Ln} \begin{bmatrix} \cos \varphi_n \\ -\sin \varphi_n \end{bmatrix} \quad (22).$$

Therefore, from the equations (20), (21), and (22) the balance control between the upper and lower arms becomes possible in each phase by controlling the negative sequence of the circulating current on the dq frame.

D. Upper and Lower Balancing Control

Referring to the first term equations (17) and (18), it is shown that the active power generated by the positive sequence circulating current flows into the dc capacitor voltage in each phase equally. However, since the polarity between the upper and lower arms is different, the power factor is set to 1 for the positive sequence circulating current. Therefore, active power flowing into each arm and maintaining the average value of the unbalance between the upper and lower arms can be controlled easily.

By using the dc capacitor voltage of each arm's SM and the deviations from the reference value $\Delta \overline{v_{cpr}}$, $\Delta \overline{v_{cps}}$, $\Delta \overline{v_{cpt}}$, $\Delta \overline{v_{cnr}}$, $\Delta \overline{v_{cns}}$, $\Delta \overline{v_{cnt}}$, the unbalance components between upper and lower arms $\Delta \overline{v_{cULr}}$, $\Delta \overline{v_{cULs}}$, $\Delta \overline{v_{cULt}}$ are described as follows

$$\Delta \overline{v_{cULr}} = -\Delta \overline{v_{cpr}} + \Delta \overline{v_{cnr}}$$
$$\Delta \overline{v_{cULs}} = -\Delta \overline{v_{cps}} + \Delta \overline{v_{cns}}$$
$$\Delta \overline{v_{cULt}} = -\Delta \overline{v_{cpt}} + \Delta \overline{v_{cnt}} \quad (23).$$

From these equation, the reference values $i^*_{Ld\_p}$ and $i^*_{Lq\_p}$ of the positive sequence circulating current on the dq

frame are set to the following equation using the proportional gain K_{pn}.

$$i^*_{Ld\_p} = K_{pn}(\Delta \overline{v_{cULr}} + \Delta \overline{v_{cULs}} + \Delta \overline{v_{cULt}}) \quad (24)$$
$$i^*_{Lq\_p} = 0 \quad (25)$$

The reference values $i^*_{Ld\_p}$ and $i^*_{Lq\_p}$ of the negative sequence circulating current on the dq frame are obtained by the unbalance component in the upper and lower arms and the proportional gain K_{Ln}.

$$i^*_{Ld\_n} = K_{Ln}(2\Delta \overline{v_{cULr}} - \Delta \overline{v_{cULs}} - \Delta \overline{v_{cULt}}) \quad (34)$$
$$i^*_{Lq\_n} = -K_{Ln}(\Delta \overline{v_{cULs}} - \Delta \overline{v_{cULt}}) \quad (35)$$

From the equations above, the block diagrams of the positive sequence and negative sequence of the circulating current control and the upper and lower arm balancing control are shown in Fig. 7.

E. DC Voltage Control

The dc voltage controller regulates the output of the dc voltage V_d to control the dc output voltage on the dc system. The block diagram is shown in Fig. 8. The input of voltage controller takes the error between the reference voltage and the measured voltage before it is fed to the PI regulator and current loop. V_{dc} is added to prevent an overcurrent generated by a sudden difference between V_{dc} and v_d.

TABLE I
SIMULATION PARAMETERS.

Parameters		Values
Rate power	P	150 MW
AC grid frequency		60 Hz
Transformer rate voltage	V_s	66 kV/66 kV
Transformer inductor	L_{Tr}	3.85 mH (5%)
Arm inductor	L	7.7 mH (10%)
Arm equivalent lost resistor	r	150 mΩ
DC bus voltage	V_{dc}	150 kV
Cell capacitor	C	400 µF
Cell capacitor voltage	V_c^*	50 kV
Cell number	n	3
Carrier frequency	f_c	2 kHz
DC inductor	L_{dc}	3.85 mH
DC coil resistance	r_L	50 mΩ
DC capacitor	C_{dc}	20 µF
PI parameters of the ac current control	K_{p_ac}, K_{i_ac}	14, 143
PI parameters of the circulating current control	K_{p_loop}, K_{i_loop}	14, 143
PI parameters of the cell voltage average control	K_{p_cell}, K_{i_cell}	0.4, 5
PI parameters of the dc current control	K_{p_dcc}, K_{i_dcc}	7, 70
PI parameters of the dc voltage control	K_{pdcv}, K_{idcv}	0.018, 0.2

V. SIMULATION RESULTS

Simulation of the MMC circuit as shown in Fig. 1 was carried out using PLECS. System parameters are listed in Table 1. The converter is operated as the rectifier with the rated value of the active power. The arm balancing controller is applied to the circuit when there was an imbalance between the arms. Then, the response is studied and analyzed.

1) Arm Balancing Controller

Fig. 9, Fig. 10 and Fig. 11 presents the simulation results of the upper and lower arm balancing control. As shown in Fig. 9, the phase balancing control was started at 2 s from the steady-state where the unbalance dc capacitor

2960

The 2018 International Power Electronics Conference

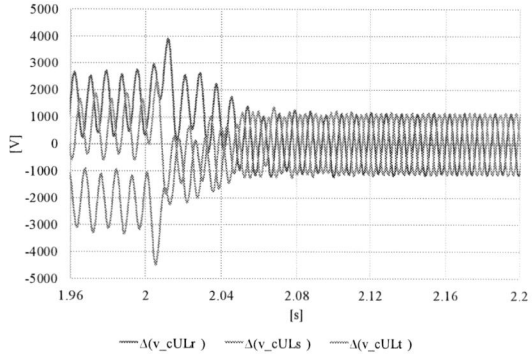

Fig. 9. Upper and lower voltage difference.

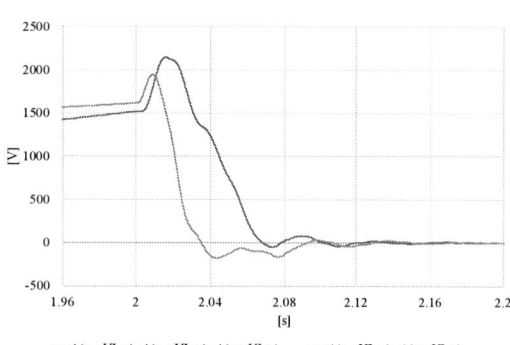

Fig. 10. Upper and lower voltage unbalance.

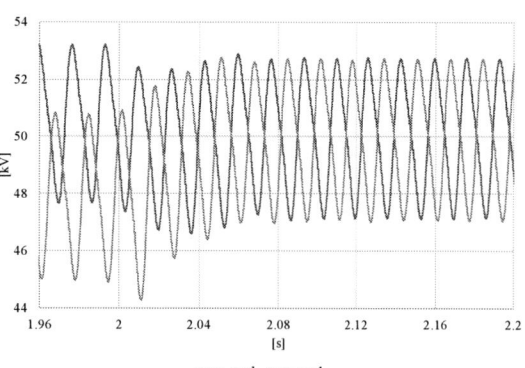

Fig. 11. R-phase upper and lower cell capacitor voltage.

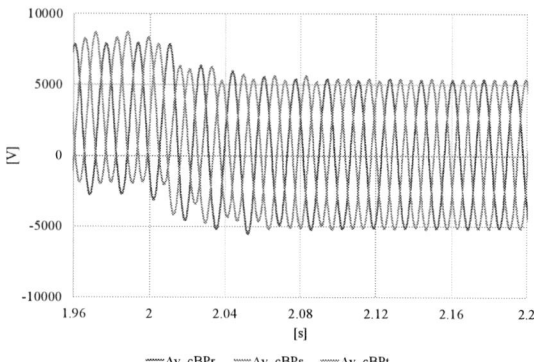

Fig. 12. Interphase voltage difference.

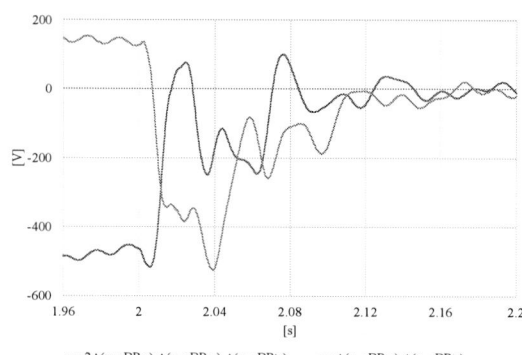

Fig. 13. Voltage unbalance.

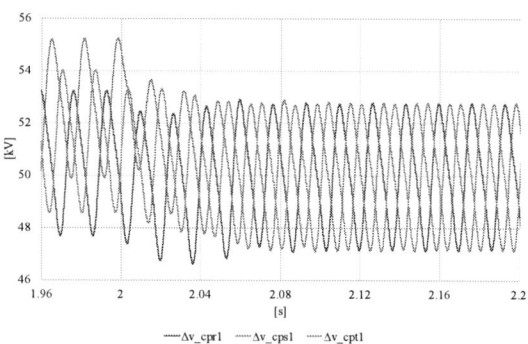

Fig. 14. Upper arm cell capacitor voltage.

voltage occurred between the upper and lower arms. Since the unbalanced voltage is controlled by the negative sequence currents of the equations (34) and (35), the unbalance component has converged to 0 as shown in Fig. 10. Hence, the imbalance between the upper and lower arms has been suppressed using this method.

There is a deviation from the upper and lower arms of the dc capacitor voltages in the r-phase in Fig. 11 before applying upper and lower arm balancing controller. However, after the application of the controller, the amplitudes of the voltages of the upper and lower arms became equal to each other. The same results were obtained for s-phase and t-phase.

2) Interphase Balancing Controller

The simulation result of the interphase balancing control is shown in Fig. 12, Fig. 13 and Fig. 14. As shown in Fig. 12, the phase balance control was started at 2 s from the state where the DC capacitor voltage of each phase was unbalanced.

Since the unbalanced voltage is controlled by the negative sequence currents of the equations (25) and (26), the unbalance component has converged to 0 as shown in Fig. 13. Hence, the imbalance between the phases has been suppressed using this method. On the other hand, the dc capacitor voltages of the upper arm in Fig. 14 differ from phase to phase before applying the phase arm balance control, but after the application, the amplitudes are equalized in each phase.

3) Effect of negative sequence current on active power

2961

The 2018 International Power Electronics Conference

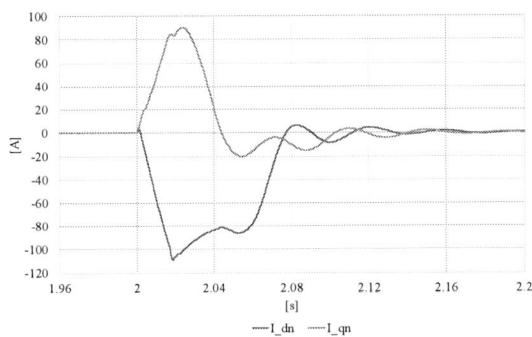

Fig. 15. Negative sequence of ac current on dq axis

Fig. 15 shows the dq axis of negative sequence current components. For the 2400 A of positive sequence of ac current on d axis, the negative sequence on dq axis increase instantly after the application of balancing control. It reaches about ± 110 A at the maximum (4.58% of the total current flow), but it reduced within 0.16 s immediately. From this, it was confirmed that the effect due to the negative sequence component on the system is negligible.

VI. CONCLUSIONS

This paper evaluates power flow in the MMC-HVDC transmission system by separating the ac grid current and the circulating current controller into positive and negative sequence components. An enhance control method has been proposed to maintain the balance control between the upper and lower arms in each phase and also maintain the balance control interphase by controlling the negative sequence circulating current on the dq axis. The validity and effectiveness of this method is confirmed by the simulation results.

REFERENCES

[1] Li Wang and Mi Sa Nguyen Thi, "Comparative Stability Analysis of Offshore Wind and Marine-Current Farms Feeding into a Power Grid Using HVDC Links and HVAC Line," *IEEE Trans. Power Del.*, vol. 28, no. 4, Oct. 2013.

[2] Q. Tu, Z. Xu, Y. Chang, and L. Guan, "Suppressing DC voltage ripples of MMC-HVDC under unbalanced grid conditions, "*IEEE Trans. Power Del.*, vol. 27, no. 3, pp. 1332-1338, Jul. 2012.

[3] Q. Tu, Z. Hu, H. Huang, and J. Zhang, "Parameter design principle of the arm inductor in modular multilevel converter based HVDC," in *Proc. Int. Conf. Power Syst. Technol.*, Hangzhou, China, pp. 1-6, 2010.

[4] N. Flourentzou, V. G. Agelidis, and G. D. Demetriades, "VSC-based HVDC power transmission system: An overview," *IEEE Trans. Power Electron.*, vol. 24, no. 3, pp. 592-602, Mar. 2009.

[5] Q. Tu, Z. Xu, Y. Chang, L. Guan: "Suppressing DC Voltage Ripples of MMC-HVDC Under Unbalanced Grid Conditions" *IEEE Transactions on Power Delivery*, vol. 27, no. 3, pp. 1332-1338, July 2012.

[6] Q. Tu, Z. Xu, L. Xu, "Reduced Switching-Frequency Modulation and Circulating Current Suppression for Modular Multilevel Converters", *IEEE Transactions on Power Delivery*, vol. 26, no. 3, pp. 4858-4867, December 2012.

[7] Makoto Hagiwara, Ryo Maeda, Hirofumi Akagi, "Control and Analysis of the Modular Multilevel Cascade Converter Based on Double-Star Chopper-Cells (MMCC-DSCC)", *IEEE Transactions on Power Electronics*, vol. 26, no. 6, pp. 1649-1658, June 2011.

[8] Nuntawat Thitichaiworakorn, Makoto Hagiwara, Hirofumi Akagi, "Experimental Verification of a Modular Multilevel Cascade

Inverter Based on Double-Star Bridge Cells", *IEEE Transactions on Industry Applications*, vol. 50, no. 1, pp. 509-519, Feb. 2014.

[9] Yingyu Liang, Jianzheng Liu, Tao Zhang, and Qixun Yang, "Arm Current Control Strategy for MMC-HVDC Under Under Unbalanced Conditions", *IEEE Transactions on Power Delivery*, vol. 32, no. 1, February 2017.

[10] Shunfeng Yang, Peng Wang, Yi Tang, Michael Zagrodnik, Xiaolei Hu, King Jet Tseng, "Circulating Current Suppression in Modular Multilevel Converters with Even-Harmonic Repetitive Control", *IEEE Transactions on Industry Applications*, vol. 54, no. 1, Jan/Feb. 2018.

[11] Hariadi Aji, Mario Ndreko, Marjan Popov, and Mart A. M. M. van der Meijden, "Investigation on Different Negative Sequence Current Control Options for MMC-HVDC During Single Line to Ground AC Faults", in *PES Innovative Smart Grid Technologies Conference Europe*, Ljubljana, Slovenia, Oct. 2016.

[12] Vihan Shahu, Teja Bandaru, Tanmoy Bhattacharya, Dheeman Chatterjee, "Control of hybrid modular multilevel converter based HVDC system under DC short circuit faults", *Industrial Electronics Society, IECON*, pp. 171-176, Nov. 2017.

[13] Eduardo Prieto-Araujo, Adria Junyent-Ferre, Gerard Clariana-Colet, Oriol Gomis-Bellmunt, "Control of modular multilevel converter under singular unbalanced voltage conditions with equal positive and negative sequence components", *PowerTech, IEEE Manchester*, June 2017.

[14] Joao Victor Farias, Allan Fagner Cupertino, Heverton Pereira, Seleme Seleme, Remus Teodorescu, "On the Redundancy Strategies of Modular Multilevel Converters", *IEEE Transactions on Power Delivery*, June 2017.

[15] Xiaojie Shi, Zhiqiang (Jack) Wang, Bo Liu, Yalong Li, Leon M. Tolbert, Fred Wang, "Steady-State Modeling of Modular Multilevel Converter Under Unbalanced Grid Conditions", *IEEE Transactions on Power Electronics*, vol. 32, no. 9, pp. 7306-7324, Nov. 2016.

[16] Maryam Saeedifard, Reza Iravani, "Dynamic Performance of a Modular Multilevel Bak-to-Back HVDC System", *IEEE Transactions on Power Delivery*, vol. 25, no. 4, pp. 2903-2912, September 2010.

A Coordinate and Distributed Control Scheme for Multilevel and Multi-Stage Medium Voltage Solid State Transformer

Jintong Nie, Liqiang Yuan, Qing Gu, Jianning Sun and Zhengming Zhao
State Key Laboratory of Power System, Department of Electrical Engineering
Tsinghua University, Beijing, China
E-mail: njt15@mails.tsinghua.edu.cn

Abstract-Medium voltage (MV) compliant solid state transformer (SST) has attracted lots of research interest in the past years. Multilevel and multiple power conversion stages are the main characteristics of currently applied SST which consists of low voltage switching devices. This presents a challenge to the design and development of control system. A distributed control is developed based on coordinate control of different power stages. Control strategy and modulation calculation, device switching and protection are decentralized to each controller with the proposed scheme. Simulation and experimental results are provided to verify the effectiveness of the control scheme.

Keywords— solid state transformer (SST), distributed control, multilevel and multi-stage converter, modularity.

I. INTRODUCTION

In recent years, Solid state transformer has gained more and more attentions due to lightweight volume and controllable power flow characteristic and is thought to be a promising subject in future smart grid and electric traction systems. In order to connect to the medium voltage grid directly, using multilevel frontend converter is still an effective way for the complexity of serial-connection of power semiconductor devices and immature commercial application of medium voltage SiC devices at present. Cascaded H-bridge (CHB) and modular multilevel converter (MMC) have been reported extensively in medium voltage applications such as high voltage direct current (HVDC) [1], battery energy storage system (BESS) [2][3], static synchronous compensator (STATCOM) [4], PV generator [5], power electronic transformer/power electronic traction transformer (PET/PETT) [6], etc. For CHB and MMC converters, both centralized and distributed control system have been employed to fulfill the control goals [7][8]. Hierarchy and decentralized distributed control is more suitable in the occasion where large number of cascaded power modules exist, for the sample and driver signals are relatively huge in traditional centralized implementation. One main merit of distributed control is that much less wires are needed by digitalizing analog and pulse width modulation (PWM) signals and transferring them through network protocol. Another superiority is that parts of the control

and algorithm code are executed in local processor and fully modularity can be realized by distributed controller. Till now, distributed control has mainly been applied in single stage multilevel converters because of relative low switching frequency and sampling frequency. MV SST always adopts a middle stage which is usually a medium or high frequency input series output parallel dual active bridge (ISOP-DAB) topology to support isolation function. Thus, distributed control in SST is more challenging due to the demands of high speed sampling data transfer network and powerful processors to maintain real-time controllability.

Researches in the control of SST have mainly been focused on system level control strategy realization, few papers put emphasis on coordinate and optimal configuration of different stages and distributed controllers. To the author's knowledge, two approaches can meet the high control performance in high frequency converters by distributed control system: One way is to use high performance FPGA and special high speed intellectual property (IP) core integrated communication protocol, which is rather costly. Another way is by exploiting control strategy to make less data interaction between different processors. Droop control or coordinate control are candidate methods. A latter method based solution including system control strategy, hybrid communication and network control, fault treatment is demonstrated in this paper. Multiple stages of SST and local decentralized control strategy are utilized to achieve global control goals. Then the controllability and control performance can be ensured even with a large number of cascaded modules and higher switching frequency in CHB plus ISOP-DAB type SST.

This paper is organized as follows: Section II introduces generally used topologies and investigates distributed control in multilevel and multi-stage MV SST. Section III presents a coordinate distributed control strategy with decreased data exchange and proposes a simple but fast hybrid communication scheme for single phase three-level power cells (PCs) in MV SST. Section IV gives the simulation and experimental results. Conclusions are summarized in section V.

Supported by the National Natural Science Foundation of China (51577100)

II. DISTRIBUTED CONTROL IN MV SST

A. Multilevel and Multi-Stage MV SST

Most of applied MV SSTs are mainly consists of multilevel frontend-stage, DC-DC stage and inverter stage, which have been reported in[9][10]. The main advantage of CHB and MMC in SST is the capability to constitute a MV compatible bidirectional converter by low voltage switching devices with good performance, generating a stable preset DC voltage and making input current controllable. High DC voltage is usually converted to relatively low DC voltage by ISOP DAB converter, containing high frequency transformers to fulfil isolation function. The last stage is always an inverter and is used to obtain user side residential line frequency low AC voltage.

As is indicated in [11], the CHB cells are usually 12 in 10 kV level medium voltage single phase application. In order to reduce the quantity of high frequency transformers and improve total efficiency of SST, 1200V SiC based three-level CHB and two level DAB topology is adopted, as shown in Fig. 1. The third inverter stage forms an independent part and is implemented with 1200V IGBT, which will not be shown and discussed in this paper.

The cascaded power cell number N is 6, then both the number of distributed controllers and data exchanged can be reduced. This, in turn, allows higher switching frequency and control frequency with limited hardware resources and can save cost.

B. Distributed Control

When using CHB plus ISOP-DAB topology, the main control goals are listed and explained as follows.

1) Power quality of the AC input

With sinusoidal input current control goal, SST brings little harmonics and angular distortion for the grid, which can maintain good power quality even with large user side nonlinear load.

2) Stabilization of the low DC voltage

The output voltage of ISOP-DAB is effectively regulated, then unified DC voltage level equipment can be plugged in directly, including inverters, PV converters, charging-piles and so on.

3) Balancing among all power cells

Voltage balancing and power balancing are very important in series or parallel combination circuits. Total voltage and current are averaged by all power cells and safety operation of switching devices can be guaranteed only in balanced condition.

Goal 1 and goal 2 are achieved by the CHB and ISOP converter respectively, the related control methods have been investigated in many literatures. Mathematical model and control loop design of single phase multilevel active rectifier is analyzed in [12]. CHB converter control containing grid voltage synchronization, output voltage regulation, input current control and DC-link voltage balance control are summarized and an energy based balance control is proposed subsequently in [13]. A novel low frequency and high frequency combined method is

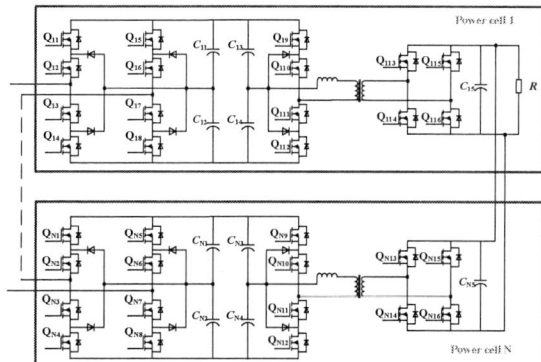

Fig. 1. SST constituted by single phase three level PWM rectifier and half bridge DAB power cell with 1200V SiC MOSFET.

employed in [14] to ensure modular control strategy and voltage balance.For DAB stage, phase shift control method is widely adopted to adjust output power and regulate output DC voltage. An elaborate overview of DAB converter including representative three kinds of control method, that are single phase shift, dual phase shift and triple phase shift methods, are introduced in [15].

In addition to output voltage control, ISOP-DAB need an input voltage sharing control, namely power balance control, as a small mismatch of circuit parameters or phase-shift angle of different power cells will cause the divergency of input voltage, which is modeled and illustrated in [16][17]. In fact, voltage balance control of CHB and power balance control of ISOP DAB have the same goal, that is to ensure the equivalent voltage of each split capacitor in the middle HVDC link. Then the interactive two stages are utilized to simplify balance control in CHB plus ISOP-DAB type SST, which could be implemented in distributed form.

Distributed control in multilevel converter has attracted increasing attention nowadays. A distributed control system consists of multiple digital processors, which are located in every power cell and can communicate with each other through star, multi-tapped or ring network topology[11]. In typical master-slave distributed architecture, one of the local controllers or an additional independent controller is operated as master controller (MC), the others function as slave controllers (SCs). Local analog variables such as sensed voltage and current, state variables such as relay states and fault signals are all digitalized and uploaded to central master controller. PWM signals and synchronization signals are oppositely distributed down to local controllers by digital form. Bidirectional communication is essential in existing distributed control system, however, it is the main reason for transmission delay and limits further increasement of switching frequency and the number of sub-modules. Reference [8] pointed out that capacitor voltage updating would occupy most of the communication bus and execution time in existing MMC distributed control system with a large number of sub-modules. Multilevel SST are faced with the same dilemma, Fig. 2. a) shows that MC needs to allocate other n-1 capacitor voltage to generate common duty cycle m of CHB and common phase shift angle φ of DAB in existing distributed control

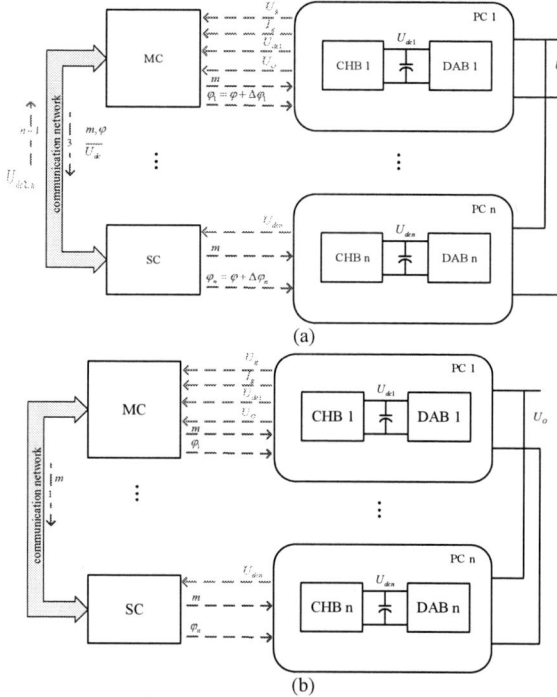

Fig. 2. Distributed control in SST with
a) existing strategy, b) proposed strategy.

strategy, which will be distributed to each SC together with average DC-link voltage $\overline{U_{dc}}$. The proposed distributed strategy aiming at reducing transmission signals in communication network, make the number of power cells independent of network band-width, which is shown in Fig. 2. b).

III. IMPLEMENTATION OF COORDINATE DISTRIBUTED CONTROLLED SST

A. Coordinate Control System

Strongly coupled circuit architecture of CHB plus ISOP DAB SST is first mentioned and a novel control strategy is proposed in [18]. System control under various work condition has been discussed in [19][20]. However, few references have concentrated on coordinate control from distributed perspective, which can reduce hardware resources and increase control performance.

Minimum data exchange oriented master-slave distributed control scheme is proposed and shown in Fig. 3. Master controller is based on TI C6748 DSP and Xilinx XC6SLX16, slave controller is constituted by one XC6SLX16 FPGA and peripheral circuits. Voltage and current control of CHB are performed in master controller, DAB output voltage regulation and power balance control are implemented in each FPGA controller.

B. Hybrid Optical Fiber Communication Network Controller

In conventional master-slave control architecture, most of the algorithm function is executed in master controller [21]. The utilization of slave controllers is not adequate, for they only accomplish the function of AD sample and

DPWM demodulation. A more equilibrium algorithm execution can be launched with the coordinate strategy shown in Fig. 3. One of the power cell controllers is designated as master controller. Main CHB control is carried out in the master controller with only one power cell's information, common duty cycle is distributed to all other slave controllers. Each slave controller has to take charge of DAB control of their own cell.

The corresponding communication topology for control architecture is depicted in Fig. 4. a), with a hybrid combination of ring topology and star topology. Control task can be performed by the alternative two approaches:

1) Ring loop in red dash is used to exchange information for each cell, including sensed voltage, current and DPWM modulation signals. Star topology drawn in solid line is served as fast synchronization and fault operation protection communication link.

2) Ring topology is used to allocate voltage, current and fault state of each cell and synchronize each controller. DPWM is distributed by star topology.

Both of the aforementioned methods can reduce communication pressure but supply one more communication path compared with single topology, as shown in Fig. 4. b) and Fig. 4. c). What is more, the number of optical fibers with N power cells is 2N-1, which is approximate to 2(N-1) in single ring topology or single star topology. Thanks to the parallel capacity of FPGA, ring loop communication through simple but high baud rate universal asynchronous receiver transmitter (UART) protocol is practicable.

C. Operation Principle

In the coordinate condition, capacitor voltage need not to be exchanged among different controllers and the digital control signal is unidirectional, which eases the burden on communication and makes higher frequency real-time control with low speed plastic optical fibers (POF) transmission medium possible. The operation mechanism includes four phases:

1) Preparation phase. SST is in soft start state, each controller is powered on, synchronization signal is distributed to slave controllers from the master controller by star POF links, capacitor voltage and fault state of each power cell are transmitted to the master controller by ring communication network. The total number and state of each cell can be verified, and then the master controller can decide whether or not to continue the operation.

2) Starting phase. With the condition matched, master controller starts the CHB control task. In this phase, master controller needs not to know other cell's capacitor voltage, the corresponding master cell's input current and output capacitor voltage are the main control objects. The total voltage of CHB capacitors will be N times single cell capacitor voltage with common duty cycle control, for the power mismatch is negligible in no load condition. Then the input capacitor voltage of each DAB produces a phase shift reference for output voltage, and the DAB stage starts to work.

The 2018 International Power Electronics Conference

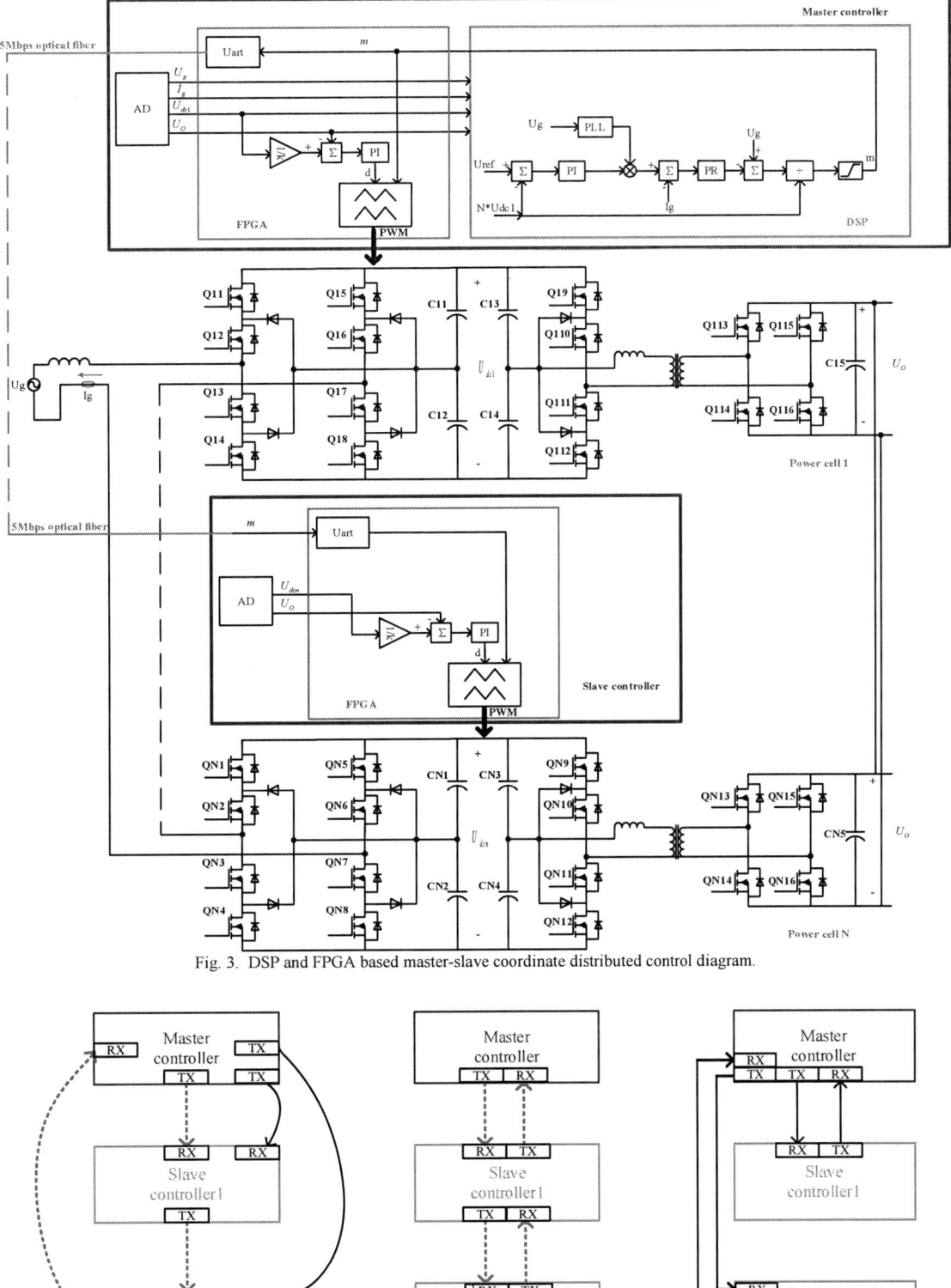

Fig. 3. DSP and FPGA based master-slave coordinate distributed control diagram.

Fig. 4. Different master-slave communication topology with a) proposed hybrid star and ring communication topology architecture, (b) single ring loop communication topology, (c) single star communication topology.

2966

3) Operation phase. Static and dynamic power mismatch exist in running phase, and will cause the divergency of input capacitor voltage. The feedforward voltage control loop of DAB makes input voltage equivalently and constrains the power imbalance.

4) Fault phase. If fault state is generated in any power cell, driver board and local controller will block their own PWM signals immediately, that is usually about several microseconds. The data transmitted in ring loop will be replaced by continual zero state, then the subsequent connected controller in the chain will detect fault state and stop their PWM output orderly and rapidly.

IV. SIMULATION AND EXPERIMENTAL RESULTS

A six power cells single phase SST is constructed in MATLAB/SIMULINK to verify the validity of coordinate control strategy. Line voltage of grid side is 10kV, each power cell's intermediate DC link sustains 1500V voltage and the equivalent HVDC link voltage is 9kV. Transformer ratio of DAB stage is 1:1, and the low DC output voltage is 750V. Single phase total active power is 166.7kW for a three phase 500kW SST system. Switching frequency of the CHB and DAB stage of power cell is 2kHz and 20kHz, respectively. Main control loop parameters are shown in TABLE I.

Main simulation waveforms are shown in Fig. 5 with coordinate control strategy, a FFT analysis is implemented in simulink, the total harmonic distortion (THD) of input current is about 2.83%. The inductor in ISOP-DAB stage of cell 2 and cell 3 are intentionally set to 0.9Pu and 1.1Pu of nominal value to verify balance control in circuit parameter mismatch condition.

TABLE I
CONTROL LOOP PARAMETERS

LOOP	CHB	DAB
Voltage loop PI	Kp=0.2 Ki=0.05	Kp=0.02 Ki=20
Current loop PR	Kp=10 Kr=25	---

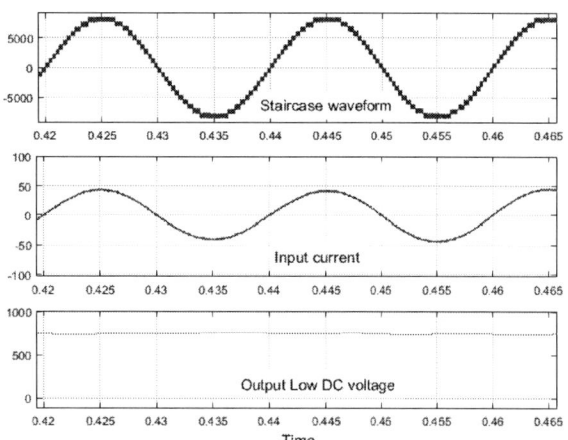

Fig. 5. Input staircase waveform voltage, input current and output low DC voltage with coordinate control strategy.

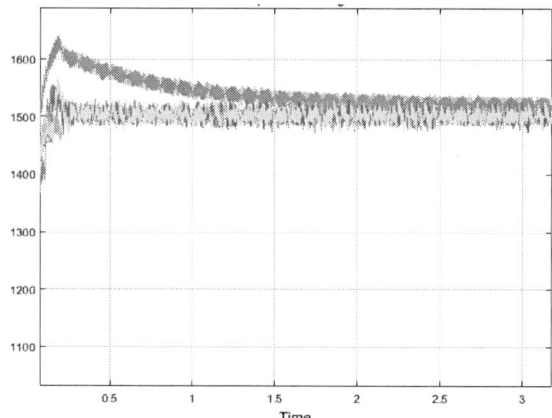

Fig. 6. CHB output DC link voltage of six power cells.

Simulated results in Fig. 6 show that the divergent voltage will be controlled and all the capacitor voltage can be balanced eventually.

Developed power cell and testbench are shown in Fig. 7, the prototype is designed for continuous running with 30kW active power. Main waveforms of CHB and DAB stages are depicted in Fig. 8. Experiments were done at a low voltage and low power condition, due to the experimental condition ristictions.

Fig. 7. Power cell hardware prototype testbench.

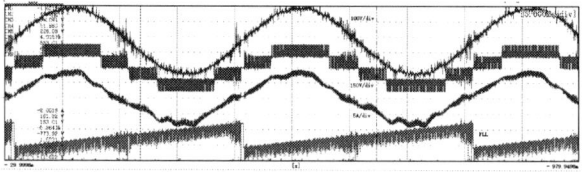

a) Grid voltage, staircase waveform, input current and phase lock loop angle of CHB stage of one power cell.

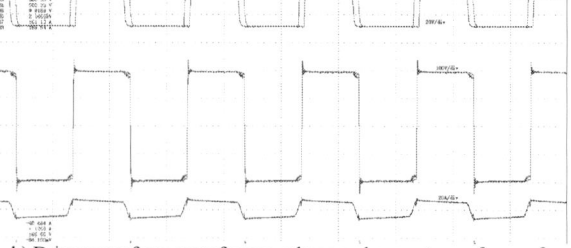

b) Driver waveform, transformer voltage and current waveforms of DAB stage.

Fig. 8. Experimental Results for one SST power cell.

The 2018 International Power Electronics Conference

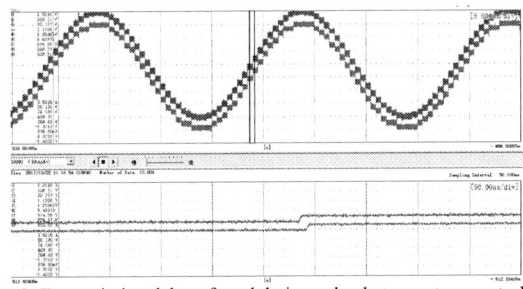

Fig. 9. Transmission delay of modulation value between two controller.

The transmission delay between two adjacent power cell's control system is output by DA and is illustrated in Fig. 9. As can be seen, the delay time is 8us when transmitting 16 bit data. All slave controllers can update modulation value about the same time for the parallel mechnism.

V. CONCLUSIONS

Coordinate control of different power stage in MV SST is utilized to implement a decentralized distributed control. By this approach, hybrid communication topology of distributed control system is proposed, which can relize higher control frequency without the need of complex network protocol and high hardware cost. The scheme can also be applied in other multilevel converter such as MMC system.

REFERENCES

[1]. Li, B., Guan, M., Xu, D., Li, R., Adam, G. P., & Williams, B. "A series HVDC power tap using modular multilevel converters," *Energy Conversion Congress and Exposition.* Milwaukee, 2016, pp.1-7.

[2]. Vasiladiotis, M. and A. Rufer, "Analysis and control of modular multilevel converters with integrated battery energy storage," *IEEE Trans. on Power Electronics*, vol. 30, no. 1, pp. 163-175, 2015.

[3]. Maharjan, L., et al., "State-of-Charge (SOC)-balancing control of a battery energy storage system based on a cascade PWM converter," *IEEE Trans. on Power Electronics*, vol. 24, no. 6, pp. 1628-1636, 2009.

[4]. Akagi, H., S. Inoue and T. Yoshii, "Control and performance of a transformerless cascade PWM STATCOM with star configuration," *IEEE Trans. on Industry Applications*, vol. 43, no. 4, pp. 1041-1049, 2007.

[5]. Coppola, M., et al., "An FPGA-based advanced control strategy of a grid-tied PV CHB Inverter," *IEEE Trans. on Power Electronics*, vol. 31, no. 1, pp. 806-816, 2016.

[6]. Zhao, C., et al., "Power electronic traction transformer—medium voltage prototype," *IEEE Trans. on Industrial Electronics*, vol. 61, no. 7, pp. 3257-3268, 2014.

[7]. Gong, L., et al., "Design and implementation of distributed control system for cascaded H-Bridge multilevel STATCOM," *Annual IEEE Conference on Applied Power Electronics Conference and Exposition (APEC).* Fort Worth, 2011, pp. 1544-1551.

[8]. Yang, S., Y. Tang and P. Wang, "Distributed control for a modular multilevel converter," *IEEE Trans. on Power Electronics*, pp. 1-1, 2018.

[9]. Huang, A.Q., et al., "The future renewable electric energy delivery and management (FREEDM) system," *The Energy Internet. Proceedings of the IEEE*, vol.99, no. 1, pp. 133-148, 2011.

[10].Wang, D., et al., "A 10-kV/400-V 500-kVA electronic power transformer," *IEEE Trans. on Industrial Electronics*, vol. 63, no. 11, pp. 6653-6663, 2016.

[11].Geng, H., et al., "Hybrid communication topology and protocol for distributed-controlled cascaded H-Bridge multilevel STATCOM," *IEEE Trans. on Industry Applications*, vol. 53, no. 1, pp. 576-584, 2017.

[12].Cecati, C., et al., "Design of H-bridge multilevel active rectifier for traction systems," *IEEE Trans. on Industry Applications*, vol. 39, no. 5, pp. 1541-1550, 2003.

[13].Blahnik, V., et al., "Control of single-phase cascaded H-Bridge active rectifier under unbalanced load," IEEE Trans. on Power Electronics: pp. 1-1.

[14].Iman-Eini, H., et al., "A modular strategy for control and voltage balancing of cascaded H-Bridge rectifiers," *IEEE Trans. on Power Electronics*, vol. 23, no. 5, pp. 2428-2442, 2008.

[15].Zhao, B., et al., "Overview of Dual-Active-Bridge isolated bidirectional DC‐DC converter for High-Frequency-Link Power-Conversion system," *IEEE Trans. on Power Electronics*, vol. 29, no. 8, pp. 4091-4106, 2014.

[16].Fan, H.F. and H. Li, "A distributed control of Input-Series-Output-Parallel bidirectional DC-DC converter modules applied for 20 kVA solid state transformer," *Annual IEEE Conference on Applied Power Electronics Conference and Exposition (APEC).* New York, 2011, pp. 939-945.

[17].Ayyanar, R., R. Giri and N. Mohan, "Active Input‐Voltage and Load‐Current sharing in Input-Series and Output-Parallel connected modular DC‐DC converters using dynamic Input-Voltage reference scheme," *IEEE Trans. on Power Electronics*, vol. 19, no. 6, pp. 1462-1473, 2014.

[18].Shi, J., et al., "Research on voltage and power balance control for cascaded modular solid-state transformer," *IEEE Trans. on Power Electronics*, vol. 26, no. 4, pp. 1154-1166, 2011.

[19].Zhang, Z.Y., et al., "Voltage and power balance control strategy for three-phase modular cascaded solid stated transformer," *Annual IEEE Applied Power Electronics Conference and Exposition (APEC).* New York, 2016, pp. 1475-1480.

[20].Liu, J., et al., "Voltage balance control based on dual active bridge DC/DC converters in a power electronic traction transformer," *IEEE Trans. on Power Electronics*, vol. 33, no. 2, pp. 1696-1714.

[21].Wang, L., et al., "Power and voltage balance control of a novel three-phase solid-state transformer using multilevel cascaded H-Bridge inverters for microgrid applications," *IEEE Trans. on Power Electronics*, vol. 31, no. 4, pp. 3289-3301, 2016.

An Improved Harmonic Power Sharing Scheme of Paralleled Inverter System

Liu Hongpeng, Liu Xiaoxi, Zhang Wei and Wang Wei
Department of Electrical Engineering, Harbin Institute of Technology, Harbin, China

Abstract— Due to the growing demand of the electric power and rapid penetration of power electronics equipment, the parallel inverter system which is connected with the nonlinear loads widely exists in islanded micro-grid. The effective harmonic power sharing scheme should be designed to guarantee the system stable operation. In this paper, a negative virtual impedance is introduced to improve the harmonic power sharing precision. The negative virtual impedance is chosen by an improved droop method based on the relationship between virtual inductance and harmonic reactive power. Besides, the voltage drop at PCC can be counteracted. The experimental results validate the correctness and feasibility of the proposed scheme.

Keywords— *Paralleled inverter; Harmonic power sharing; Droop control; Negative virtual impedance; Islanded mode.*

I. INTRODUCTION

Parallel inverter system operating in islanded mode can improve the power capacity of micro-grid. And the parallel inverters should be designed to share active and reactive power in the system, which ensures the system safe and stable operation [1]. Besides, for the parallel inverter system that supplies power for nonlinear loads, harmonic power generated should also be well shared. Meanwhile the harmonic can affect the voltage quality at the point of common coupling (PCC). Therefore, the scheme for the accurate fundamental and harmonic power sharing and better voltage quality at PCC is a key research direction in parallel inverter system [2].

Droop control can provide consistent amplitude and frequency reference for parallel inverter according to the respective power rating [3]. The sharing precision is determined by the inverter output voltage and equivalent impedance. The equivalent impedance of the inverter contains effective line impedance and inverter output impedance. However, the difference of line impedance among inverters is unavoidable and inconvenient to be measured. Thus, in order to change the equivalent impedance, the virtual impedance control strategy is used. In [4], the *G-H* droop control method is applied to change the inverter output impedance. The various harmonic voltage is multiplied by the virtual conductance. Then the product is added to the current reference to modify the harmonic current reference of the inverter. But the virtual conductance may cause the mismatch between the line impedance and output impedance in high-voltage network. The virtual resistance/impedance techniques are proposed

This work was supported by the Lite-On Power Electronics Technology Research Fund (PRC20151382).

to share the fundamental and harmonic power [5]. The virtual inductance changes the effective line impedance at fundamental frequency and the virtual resistance change the effective line impedance at harmonic frequency. But, the introduction of virtual resistance increases the effective line impedance and causes a large voltage drop at PCC. Besides, how to choose the value of virtual impedance is given in this paper. Some variations are reported by [6]-[10]. Though the negative virtual impedance can decrease effective line impedance, the line impedance is unknown. To address the issue above, a *Z-H* droop control method is proposed to improve the harmonic power sharing precision and guarantee the voltage quality at PCC [11]-[13]. The negative virtual impedance can be determined by the linear relationship between its value and various harmonic capacity. However, the calculation of various harmonic power is complicated and it is a burden for the controller. Paper [14] and [15] proposed a secondary voltage control strategy, which the first control strategy is the droop control to achieve the fundamental power sharing and the second control strategy is to achieve the harmonic power sharing and compensate the harmonic voltage. The communication structure is needed to deliver the information of PCC voltage to the central controller. However, the communication structure brings about series of problems such as lower reliability and higher cost.

This paper proposed an improved harmonic power sharing scheme based on negative virtual impedance to optimize the voltage quality at PCC. The choice of the virtual inductance is also provided. The control structure is simple and easy to implement.

II. HARMONIC POWER SHARING IN ISLANDED MODE

Fig. 1 shows a parallel inverter system that consists of two single-phase inverters, nonlinear load and inductive line impedances Z_1 and Z_2. The parallel system is operated in the islanded mode and provide the stable output power for local nonlinear load which can generate large harmonic currents. Based on Thevenin's Theorem, the equivalent

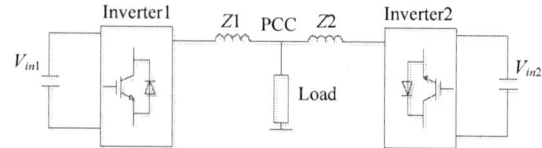

Fig. 1. Configuration of parallel inverter system.

circuits of parallel inverter system connected with the nonlinear loads at fundamental frequency and harmonic frequencies are shown in Fig. 2 and Fig. 3 respectively.

The output impedances of two inverters are mainly inductive and each expressed as jX_{out1} and jX_{out2} because of the existence of the filter inductor. $\vdash\!\!-\!\!\!-$ and $\vdash\!\!-\!\!\!-$ are the equivalent output voltages of two inverters respectively. The voltage at PCC is notated as $U_{PCC}\angle 0$. As shown in Fig. 2, the fundamental active and reactive power sharing precision is determined by the output voltage of the inverter and the equivalent impedance which contains the output impedance of the inverter and line impedance. And the harmonic power sharing precision is mainly determined by the equivalent impedance, as shown in Fig. 3. Droop control can provide stable voltage and frequency by regulating the inverter output reference voltage constantly. The linear droop characteristic when the line impedance is inductive can be represented as

$$\begin{cases} \omega^* = \omega_0 - k_p P \\ U^* = U_0 - k_q Q \end{cases} \quad (1)$$

where ω^* and U^* are the output angular frequency reference and amplitude reference of output voltage; ω_0 and U_0 are its rated angular frequency and output voltage amplitude, k_p and k_q are the active and reactive droop coefficients, P and Q are output active and reactive powers.

In order to modify the equivalent impedance, virtual impedance (jX_{vir1} and jX_{vir2}) are introduced to share the fundamental and harmonic current. By optimizing value of virtual impedance, the equivalent impedances of two inverters are nearly equal. Thus, the nonlinear load power can be distributed evenly between the two inverters.

Essentially, the introduction of virtual impedance can change the output voltage reference. The relationship between output voltage reference and virtual impedance is

$$u_o = G(s)(U^* - Z_{vf}i_f - \sum_{h=3,5,7...} Z_{vh}i_h) - Z_v(s)i_f - \sum_{h=3,5,7...} Z_h i_h \quad (2)$$

where $G(s)$ is open-loop transfer function of the inverter, i_f is the fundamental current and i_h is the harmonic current. Z_{vf} is the fundamental virtual impedance Z_{vh} is the

harmonic virtual impedance. $Z_v(s)$ is the equivalent fundamental output impedance, Z_h is the equivalent harmonic output impedance.

The change of the fundamental and harmonic voltage references can be expressed as (3) and (4), respectively.

$$V_{vf} = ji_f \omega L_v \quad (3)$$

$$\sum_{h=3,5,7...} V_{vh} = \sum_{h=3,5,7...} ji_h \omega L_{vh} \quad (4)$$

where L_v and L_{vh} are fundamental and harmonic virtual inductors, respectively.

In order to calculate V_{vf} and V_{vh}, the various order harmonic current is extracted by applying the fourth order band-pass filter [16], which is consist of two cascading second order band-pass filter. And second-order generalized integrator (SOGI) is utilized to obtain the orthogonal current. As shown in Fig. 4, i_d is same phase with the original current, i_q is orthogonal to i_d and lags behind i_d. The phase difference between $ji_d \omega L$ and $i_q \omega L$ is 180°.

Then, the V_{vf} and V_{vh} can be expressed as

$$V_{vf} = ji_{df} \omega L_v = -i_{qf} \omega L_v \quad (5)$$

$$V_{vh} = ji_{dh} \omega L_{vh} = -i_{qh} \omega L_{vh} \quad (6)$$

The reference of output voltage finally can be obtained as

$$u_o = G(s)(U^* + i_{qf} \omega L_v + \sum_{h=3,5,7...} i_{qh} \omega_h L_{vh}) - Z_o(s)i_o - \sum_{h=3,5,7...} Z_h i_h \quad (7)$$

Based on the analysis above, virtual impedance can decrease the difference of equivalent impedance of parallel inverters.

III. PROPOSED NEGATIVE VIRTUAL IMPEDANCE CONTROL STRATEGY

A. Proposed Control Scheme

Negative virtual impedance can reduce the effective line impedance and counteract the voltage drop at PCC. Fig. 5 shows the overall control diagram of the proposed scheme.

The output voltage u_o and current i_L is measured for computing P and Q. The active and reactive droop expressions can then be used for mapping out the desired voltage and angular frequency. Then, by introducing the negative fundamental and various order harmonic virtual impedances, the voltage reference u^* can be obtained. The

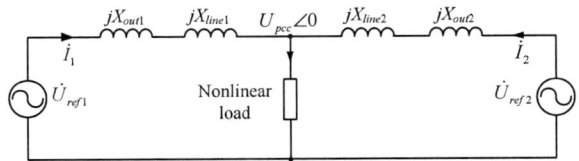

Fig. 2. The equivalent circuit at fundamental frequency.

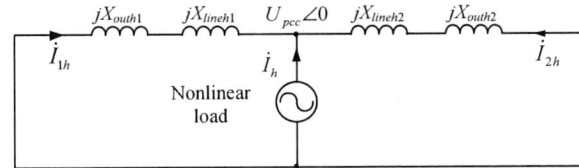

Fig. 3. The equivalent circuit at harmonic frequency.

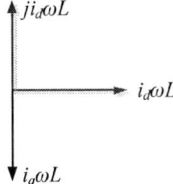

Fig. 4. The diagram of the voltage phase.

output voltage expression of the inverter can thus be expressed as:

$$u_o = G(s)(U^* - i_{qf}\omega L_v - \sum_{h=3,5,7...} i_{qh}\omega_h L_{vh})$$
$$- Z_o(s)i_o - \sum_{h=3,5,7...} Z_h i_h \quad (8)$$

The equivalent impedance can be obtained as

$$Z_{equ} = jX_{out} + jX_{line} - jX_{vir}, X_{vir} = \omega_h L_{vir}, h = 1,2,3...n \quad (9)$$

The negative impedance is smaller than the sum of output impedance and line impedance. But it is large enough to counteract the difference of equivalent impedance between two inverters. According to the linear relationship between the virtual impedance and reactive power, the value of virtual impedance can be determined by

$$L_h = L_{h0} - b(Q_h - Q_{h0}) \quad (10)$$

$$L_v = L_0 - k(Q - Q_0) \quad (11)$$

where L_0 and L_{h0} are the fundamental and harmonic virtual inductance references, Q_0 and Q_{h0} is the rated fundamental reactive power and harmonic reactive power respectively. k and b are the droop coefficients.

Fig. 6 shows the droop curve of negative virtual impedance. Suppose that the equivalent impedance of inverter 1 is larger than that of inverter 2, the reactive power of inverter 1 is less than that of inverter 2. According to the droop curve, the negative virtual inductance introduced by inverter 1 should be larger than that of inverter 2. By regulating the virtual inductance constantly, the fundamental reactive and harmonic power can be shared equally. At the same time, the voltage quality at PCC can be improved because of the negative virtual impedance.

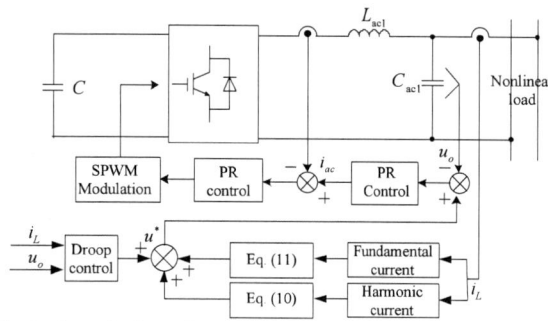

Fig. 5. Control scheme for parallel inverter system.

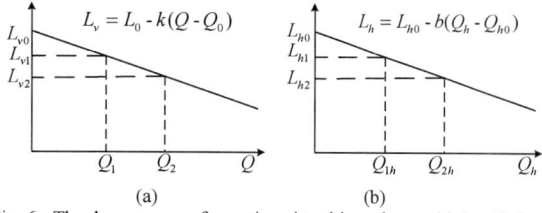

Fig. 6. The droop curve of negative virtual impedance. (a) L_v - Q droop curve, and (b) L_h - Q_h droop curve.

B. Parameters Selection

From (9), it can be seen that the equivalent impedance of the inverter is the sum of output impedance, line impedance and virtual impedance. The minimum value of L_0 and L_{h0} should be greater than the line impedance difference between the inverters to ensure the accuracy of power sharing. One the other hand, in order to avoid generating the negative equivalent impedance, the maximum value of L_0 and L_{h0} should be less than the sum of output impedance and line impedance. Therefore, the range of L_0 and L_{h0} should be selected according to the actual system parameters.

Meanwhile, in order to ensure power sharing of the reactive and the harmonic powers, the parameters k_n and b_n and virtual impedance references also need to meet the following relationship:

$$\frac{b_1}{L_{h01}} = \frac{b_2}{L_{h02}} = \cdots \cdots \frac{b_n}{L_{h0n}}, \frac{k_1}{L_{01}} = \frac{k_2}{L_{02}} \cdots \cdots \frac{k_n}{L_{0n}} \quad (12)$$

In order to speed up the power regulation and to enhance the system response speed, the range of k_n and b_n are given in (13).

$$\begin{cases} 0 < b_n \le \dfrac{L_{h0n} - L_{hn\_min}}{Q_{hn\_max} - Q_{h0n}} \\ 0 < k_n \le \dfrac{L_{0n} - L_{n\text{-}min}}{Q_{n\_max} - Q_{0n}} \end{cases} \quad (13)$$

where L_{hn_min} and L_{n_min} are the minimum value of the harmonic and fundamental virtual impedances, Q_{hn_max} and Q_{n_max} are the maximum value of the harmonic and fundamental reactive powers, respectively. n is the assigned inverter index number.

IV. EXPERIMENTAL RESULTS

Experimental testing with two 1-kW parallel single-phase inverters has been performed for proving the practicality of the proposed scheme, while its parameters are given in Table I. The nonlinear load connecting with inverter 1 consists of a single phase uncontrolled rectifier bridge connected to a 100 Ohms resistor, a 5mH inductor and a 20μF capacitor. And the nonlinear load connecting with inverter 2 consists of a rectifier bridge connected to a 75 Ohms resistor, a 5mH inductor and a 20μF capacitor.

Fig. 7 shows the current and voltage of parallel inverters with the proposed negative virtual impedance. The RMS of u_1 is 219.8V, and THD is 14.6%, The RMS of u_2 is 219.6V, and THD is 15.3%. The RMS of i_1 is 3.29A, and THD is 40.8%, The RMS of i_1 is 3.14A, and THD is 42.7%. It can be seen that the i_1 and i_2 are almost

TABLE I
INVERTER PARAMETERS

PARAMETER	VALUE
Filter inductance (L_{ac})	10mH
Filter capacitance (C_{ac})	10μF
Switching frequency (f_s)	10kHz
Inverter output frequency	50Hz
Active droop coefficient (k_{pi})	0.0001rad/s•W
Reactive droop coefficient (k_{qi})	0.0001V/Var

equal. Two inverters can evenly share the output fundamental and harmonic power. The voltage distortion is compensated by the negative impedance and the voltage quality of PCC is well improved.

Fig. 8 shows the dynamic performance of the system with the proposed harmonic sharing scheme. The capacitor of the load increases from 20μF to 40μF at $t = 0.04$s. During the load switching process, there is no output voltage and current overshoot. And the system has quick response time.

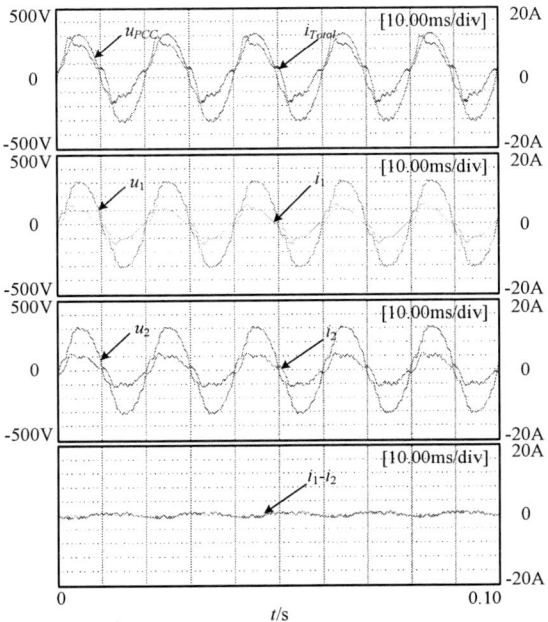

Fig. 7. Output voltage and current of two parallel inverters with negative virtual impedance.

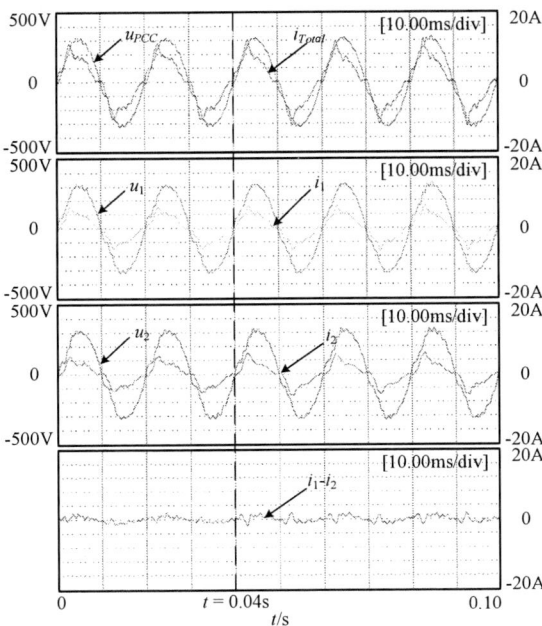

Fig. 8. Dynamic performance of proposed control scheme.

V. CONCLUSIONS

An improved control scheme has proposed to achieve the harmonic power sharing and the voltage quality improvement at PCC in this paper. Based on the analysis of the output impedance of inverter, virtual impedance is introduced to improve the power sharing precision of parallel inverter system. The negative impedance can counteract the difference of equivalent impedance and compensate the voltage drop. And an active method to choose virtual impedances at different order harmonic frequencies is applied to regulate the harmonic power dynamically. Experimental results validate the feasibility and effectiveness of the proposed scheme.

REFERENCES

[1] T. L. Vandoorn, B. Renders, and B. Meersman, et al, "Reactive power sharing in an islanded microgrid," in *Proc. UPEC*, 2010, pp. 1-6.

[2] J. Yang, Y. Chen, M. Dong, et al, "A harmonic and reactive power sharing control strategy for islanded microgrids," in *Proc. CAC*, 2015, pp. 1422-1427.

[3] D. Dipankar, R. Venkataramanan, "Decentralized parallel operation of inverters sharing unbalanced and nonlinear loads," *IEEE Trans. Power Electron.*, vol. 25, no. 12, pp. 3015-3025, 2010.

[4] T. Lee, P. Cheng, "Design of a new cooperative harmonic filtering strategy for distributed generation interface converters in an islanding network," *IEEE Trans. Power Electron.*, vol. 22, no. 5, pp. 1919-1927, 2007.

[5] M. Savaghebi, A. Jalilian, and J. C. Vasquez, et al, "Voltage harmonic compensation of a microgrid operating in islanded and grid-connected modes," in *Proc. ICEE*, 2011, pp. 1-6.

[6] J. M .Guerrero, J. Matas, L. Garcia de Vicuña, et al, "Wireless-control strategy for parallel operation of distributed-generation inverters," *IEEE Trans. Ind. Electron.*, vol. 53, no. 5, pp. 1461-1470, 2006.

[7] M. Savaghebi, J. C. Vasquez, and A. Jalilian, et al, "Selective harmonic virtual impedance for voltage source inverters with LCL filter in microgrids," in *Proc. ECCE*, 2012, pp. 1960-1965.

[8] J. He, Y. Li, and F. Blaabjerg, "An accurate autonomous islanding microgrid reactive power, imbalance power and harmonic power sharing scheme," in *Proc. ECCE*, 2013, pp. 1337-1343.

[9] J. Lu, Y. Zhang, and X. Qian, et al, "A novel virtual impedance method for droop controlled parallel UPS inverters with wireless control," in *Proc. ITEC Asia-Pac.*, 2014, pp. 1-5.

[10] Y. Zhu, B. Liu, F. Wang, et al, "A virtual resistance based reactive power sharing strategy for networked microgrid," in *Proc. ECCE Asia*, 2015, pp. 1564-1572.

[11] J. He, Y. Li, and J. M. Guerrero, et al. "Microgrid reactive and harmonic power sharing using enhanced virtual impedance," in *Proc. APEC*, 2013, pp. 1-4.

[12] J. He, Y. Li, and F. Blaabjerg, "An enhanced islanding microgrid reactive power, imbalance power, and harmonic power sharing scheme," *IEEE Trans. Power Electron.*, vol. 30, no. 6, pp. 3389-3401, 2015.

[13] P. Sreekumar and V. Khadkikar, "A new virtual harmonic impedance scheme for harmonic power sharing in an islanded microgrid," *IEEE Trans. Power Del.*, vol. 31, no. 3, pp. 936-945, 2016.

[14] M. Savaghebi, A. Jalilian, and J. C. Vasquez, et al, "Secondary control scheme for voltage unbalance compensation in an islanded droop-controlled microgrid," *IEEE Trans. Smart Grid*, vol. 3, no. 2, pp. 797-807, 2012.

[15] M. Savaghebi, A. Jalilian, and J. C. Vasquez, et al, "Secondary control for voltage quality enhancement in microgrids," *IEEE Trans. Smart Grid*, vol. 3, no. 4, pp. 1893-1902, 2012.

[16] S. W. Kang and K. H. Kim, "Sliding mode harmonic compensation strategy for power quality improvement of a grid-connected inverter under distorted grid condition," *IET Power Electron.*, vol. 8, no. 8, pp. 1461-1472, 2015.

The Grid Impedance Adaptation Dual Mode Control Strategy In Weak Grid

Ming Li[*], Xing Zhang, Ying Yang and Pengpeng Cao

School of Electrical Engineering and Automation, Hefei University of Technology, Hefei, China

*E-mail: jhuumiuu@163.com

Abstract— **With the increasing penetration of distributed energy resources and the wide distribution of access point, the power grid is becoming more and more weak. For the two main stability control strategies in weak grid, namely the current source grid connected mode (CSM) and the voltage source grid connected mode (VSM), the static characteristics of the power transmission are analyzed in this paper, respectively. In addition, through small signal modeling and analysis of the inverter operating at VSM, the stability and dynamic performance are improved when the grid impedance increases. Furthermore, it is concluded that the inverter is more suitable for operating at VSM in the larger grid impedance occasions. On the basis of the above analysis, this paper proposes the grid impedance adaptation dual mode control strategy in weak grid: when the grid impedance is low, the inverter operates at CSM, and when the grid impedance is large, the inverter operates at VSM. At last, the validity of the above analysis and the feasibility of the proposed scheme are verified by simulation and experiment.**

Keywords— *Weak grid, grid impedance, dual mode control, CCM, VCM.*

I. INTRODUCTION

This With the global environmental pollution and fossil energy crisis intensified, clean and renewable energy has got more and more attention. And the three-phase inverter playing more and more important role [1]. Considering these distributed generation (DG) are mostly in the desert, islands and other remote areas, and because of long transmission lines and transformers, power grid shows weak grid characteristics with high inductance [2-4]. In addition, with the increasing proportion of DG, and the emergence of multi inverter parallel operation conditions, the equivalent grid impedance of a single inverter increases, and the power grid further exhibits the characteristics of weak grid [5].

The stability control strategy of grid connected inverters in weak grid mainly includes two categories:

The first type is a traditional control strategy of grid connected inverter : the active power and reactive power is regulated by the grid current which based on grid voltage orientation, and the phase-locked loop (PLL)

This work is supported in part by the National Key Research and Development Program of China under Project 2016YFB0900300 and in part by the National Natural Science Foundation of China under Project 51677049.

observation grid voltage phase. Because the inverter is equivalent to the current source in this control mode, in this paper is called the current source grid connected mode (CSM). Moreover, most of the existing grid connected inverter operates at CSM, but its current control characteristics deteriorated with grid impedance increases, including trigger voltage distortion, a series of interactive harmonic oscillation, which endangering the safe and stable operation of the inverter. For example, the phase-locked loop (PLL) stability in weak grid is analyzed in [7, 8]. And in [9], the stability of power control loop in weak grid is studied by eigenvalue and transfer function analysis. Based on the linearized state space model, the influence of the PLL on the power transmission limit is studied in [10]. And by using the current cooperative control to improve the stability of power control in weak grid is proposed in [11]. However, these above researches mostly analyze the stability of inverter from the view of small disturbance dynamic characteristics.

The second type is to adjust the active power by adjusting the phase of the output voltage with droop control power loop, and the reactive power is adjusted by the change of amplitude of the output voltage vector. Because the relationship between power and voltage is similar to synchronous generator system, the inverter is equivalent to voltage source under this control mode [12-15], in this paper is called the voltage source grid connected mode (VSM).

Based on the above literature, in this paper, firstly analyzes the characteristics of static relationship of the power transmission between CSM and VSM. In order to analyze the stability of grid connected inverter operates at VSM in weak grid, through small signal modeling and analysis of the inverter operates at VSM, the stability and dynamic performance are concluded when the grid impedance increases, further concluded that it is more suitable for inverter operates at VSM in the larger grid impedance occasions. On the basis of the above analysis, this paper proposes the grid impedance adaptation dual mode control strategy in weak grid: when the grid impedance is low, the inverter operates at CSM, and when the grid impedance is large, the inverter operates at VSM. At last, this paper verifies the correctness and validity of the above analysis by simulation and experiment.

II. CONTROL STRATEGY FOR GRID CONNECTED INVERTERS IN WEAK GRID

Fig. 1 (a) and Fig. 1 (b) represent the control strategy schematic of the grid connected inverter operating at CSM and VSM, respectively.

In Fig. 1 (a): C_{dc} is DC capacitor, L_f represents filter inductance, C_f and R_f respectively represent filter capacitor and damping resistor, R_g and X_g respectively represent impedance component and inductive component of grid impedance. The i_{gdref} and i_{gqref} are the reference values of the d axis and the q axis current, respectively. i_{gd} and i_{gq} are the d axis and q axis component of the output current i_{gabc} in the synchronous rotating coordinate system. θ_{CSM} is the phase obtained by the PLL according to the PCC voltage u_{oabc}. The inverter with CSM is equivalent to the current source.

In Fig. 1 (b), the main circuit of the inverter is the same as that of Fig. 1 (a). VSM control is realized by power control based on droop control. u_{odref} and u_{oqref} are the d axis and q axis components of reference voltage loop , respectively. θ_{VSM} is the output phase of the droop power loop. The u_{od} and u_{oq} are the d axis and q axis components of the PCC voltage u_{oabc}, respectively. The i_{gd} and i_{gq} are the d axis and q axis components of the output current i_{gabc}, respectively. The inverter with VSM is equivalent to the voltage source.

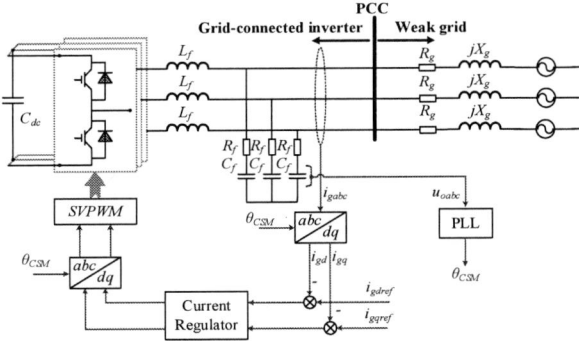

(a) CSM

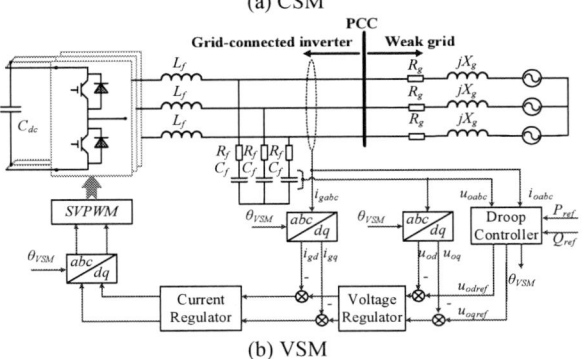

(b) VSM

Fig. 1. The control strategy schematic of the grid connected inverter operating at CSM and VSM.

It is worth mentioning that phase locked loop (PLL) is needed to obtain the phase of grid voltage in CCM control to realize vector control based on grid voltage orientation. However, the bandwidth of PLL will affect the output impedance of grid-connected inverter, which affects the stability of grid-connected inverter in weak grid. In [16], the stability of grid-connected inverters can be improved by adjusting the bandwidth of PLL by identifying the grid impedance is proposed. However, the bandwidth of PLL will be reduced to a very low level when the grid impedance is large, which leads to a significant decline in the dynamic performance of the system under the voltage disturbance.

Different from CCM control, VCM control is similar to synchronous generator control. The phase of grid-connected current is obtained by droop control. It does not need to be orientated by PLL, and there is no problem that PLL affects the stability of inverter mentioned in [16].

III. THE STATIC CHARACTERISTICS OF THE POWER TRANSMISSION

A. Equivalent Model of Grid Connected inverter

Whether CSM or VSM, the equivalent circuit of the inverter connected to the grid can be expressed as Fig. 2. In Fig. 2, X and R are equivalent impedance inductive and resistive components, respectively. The value of the equivalent grid impedance is Z and the impedance angle is φ. P and Q respectively for the inverter output active power and reactive power. The amplitude of grid voltage is U, and the reference phase of grid voltage is 0. The equivalent PCC voltage of grid connected inverter is E and the phase of equivalent PCC voltage is δ. i_g is grid current.

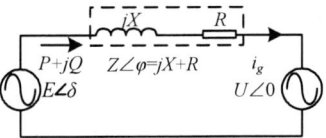

Fig. 2. The equivalent circuit model of grid connected inverter.

B. Power Coupling Characteristics When The Inveter Operates at CSM

The inductive component of the grid impedance is an important reason that affects the stability of the grid connected inverter in weak grid. Considering the worst case, the influence of R is ignored, namely $R=0$.

When the inverter is in steady state, the PLL can accurately track the phase of the voltage E:

$$E\angle\delta = Ee^{j\delta} = U + jX(i_{gd} + ji_{gq})e^{j\delta} \qquad (1)$$

where, i_{gd} and i_{gq} are the d axis and q axis components of the grid current i_g, respectively.

Unfold formula (1), and the two sides of the equation are multiplied by the $e^{-j\delta}$ at the same time. The real and imaginary parts by the left and right side of the equal sign can be equal, respectively:

$$E = U\cos\delta - jU\sin\delta + jXi_{gd} - Xi_{gq} \qquad (2)$$

By (2), the expressions of phase δ and the voltage U of the inverter are shown in (3) and (4), respectively:

$$\delta = \sin^{-1}\left(Xi_{gd}/U\right) \tag{3}$$

$$E = \sqrt{U^2 - X^2 i_{gd}^2} - Xi_{gq} \tag{4}$$

When the phase δ belongs to the range of 0 degrees to 90 degrees, the (3) shows that the larger the grid impedance X, the larger the phase δ. According to (4), with a strong grid condition, the grid impedance value of X is 0, at this time, the output voltage U of the inverter is only related to the reactive current component i_{gq} of the grid connected current. However, when the grid impedance value of X is not 0, the output voltage U are related to both i_{gd} and i_{gq}.

According to Fig. 2, the grid current i_g can be driven as:

$$i_g = \frac{E\angle\delta - U\angle 0}{Z\angle\varphi} = \frac{E\cos\delta - U + jE\sin\delta}{R + jX} \tag{5}$$

So the output of grid connected inverter apparent power S is:

$$\begin{aligned}
S &= E\angle\delta \cdot \left(i_g\right)^* \\
&= \left(E\cos\delta + jE\sin\delta\right) \cdot \\
&\quad \frac{\left[\left(E\cos\delta - U + jE\sin\delta\right)\left(R - jX\right)\right]^*}{R^2 + X^2}
\end{aligned} \tag{6}$$

When $R=0$, the output positive power P and reactive power Q of the inverter are:

$$\begin{cases}
P = \dfrac{EU\sin\delta}{X} \\[2mm]
Q = \dfrac{E^2 - EU\cos\delta}{X}
\end{cases} \tag{7}$$

The (3) and (4) are brought into (7), the following formula can be derived as:

$$\begin{cases}
P = i_{gd}\left(\sqrt{U^2 - X^2 i_{gd}^2} - Xi_{gq}\right) \\[2mm]
Q = i_{gq}^2 X - i_{gq}\sqrt{U^2 - X^2 i_{gd}^2}
\end{cases} \tag{8}$$

Because when the inverter operates at CSM, P and Q are regulated by adjusting i_{gd} and i_{gq}, respectively, the relationship between P and i_{gd} can be plotted according to (8), as shown in Fig. 3. In Fig. 3, the variable uses the unit value, then there is $U=1$, $X=1$.

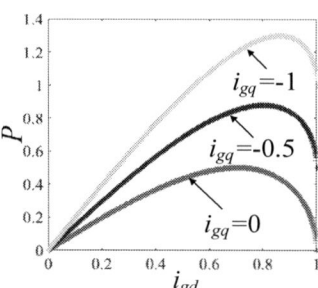

Fig. 3. The relation curve between active power P and grid current i_{gd}.

Fig. 3 shows that in the case of weak grid, when the inverter grid current i_{gd} increases, the output positive power P is not monotonically increasing, but increases first and then decreases. And with the increase of grid current i_{gq}, the output positive power P is also increased.

In order to analyze the coupling degree of active power P and current i_{gq} when the gird impedance increases, according to (8), the partial derivative of output active power P is derived:

$$\begin{cases}
\dfrac{\partial P}{\partial i_{gd}} = \dfrac{U^2 - X^2 i_{gd}^2 + E\sqrt{U^2 - X^2 i_{gd}^2} - U^2}{\sqrt{U^2 - X^2 i_{gd}^2}} \\[4mm]
\dfrac{\partial P}{\partial i_{gq}} = -Xi_{gd}
\end{cases} \tag{9}$$

According to (9), with the increase of the gird impedance X, $\dfrac{\partial P}{\partial i_{gq}}$ increases gradually. It shows that the coupling degree between active power P and grid current i_{gq} is enhanced with the increase of the grid impedance when inveter operates at CSM.

C. Power Coupling Characteristics When the Inveter Operates at VSM

It is different from the inveter operates at CSM, the output power is controlled by controlling the output voltage E and the output angle δ when the inverter operates at VSM.

According to (7), the relationship curve between P and δ as shown in Fig. 4. In Fig. 4, the variable uses the unit value, then there is $U=1$, $X=1$.

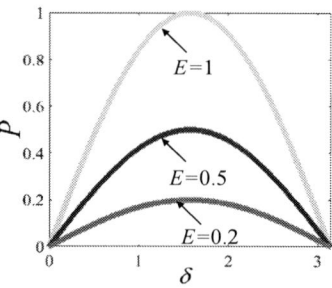

Fig. 4. The relation curve between active power P and angle δ.

Fig. 4 shows that in the case of weak grid, when the the angle δ increases, the output positive power P is not monotonically increasing, but increases first and then decreases. According to Fig. 4, it can be found that the output active power P and angle δ of the inverter operated at VSM have sinusoidal function relationship curve. Besides, with the increase of voltage E, the output positive power P is also increased.

Similar to the inverter operates at CSM, in order to analyze the coupling degree of active power P and voltage E when the grid impedance increases, according to (7), the partial derivative of output active power P is derived:

$$\begin{cases} \dfrac{\partial P}{\partial \delta} = \dfrac{EU}{X}\cos\delta \\[4mm] \dfrac{\partial P}{\partial E} = \dfrac{U}{X}\sin\delta \end{cases} \tag{10}$$

According to (10), with the increase of the grid impedance X, lead to $\dfrac{\sin\delta}{X} \approx 0$, so $\dfrac{\partial P}{\partial E}$ decreased gradually. It shows that the coupling degree between active power P and voltage E is attenuated with the increase of the grid impedance when inverter operates at VSM.

Besides, when the grid impedance X is too small, not only the power coupling degree is very large, but also the power control performance of the inverter is seriously affected. And the angle δ in a small range adjustment case will cause a wide range of output active power P changes, so the inverter control precision of PCC voltage and immunity requirements are significantly improved when inverter operates at VSM, sometimes even lead to system instability. Therefore, in a weak grid case with high grid impedance X, inverter operates at VSM could be helpful to avoid the too high control accuracy requirement, so as to improve the robustness and stability of the system.

In summary, with the increase of the grid impedance, the coupling degree between active power P and current i_{gq} is enhanced when invertes operates at CSM, while the coupling degree between active power P and voltage E is attenuated when invertes operates at VSM. Therefore, it is more suitable for inverter operates at VSM in extremely weak grid with larger grid impedance than operates at CSM. The simulation and experiment in section V verify the conclusion.

D. The Define Margin of X for Identifying the CSM and VSM

The difference between weak grid and strong grid usually defined according to the short circuit ratio (SCR). It is generally considered that the grid with a short circuit ratio less than 2 is a very weak one [17]. The short circuit ratio λ_{SCR} is defined as:

$$\lambda_{SCR} = \frac{S_{sc}}{S_n} \tag{11}$$

where S_n is the rated capacity of the power grid, and the S_{sc} is a short circuit capacity.

If the impedance is larger, the short circuit ratio λ_{SCR} is smaller and grid is weaker. In the section V, a grid-connected inverter with a rated capacity of 100kW is adopted, the rated voltage is 380V, and the short circuit ratio λ_{SCR} takes 2 as the boundary of the VCM and CCM control, and the corresponding grid impedance is $X=0.726\ \Omega$. That is, VSM control is used when the grid impedance X is more than 0.726 Ω, and CCM control is used when the gird impedance X is less than $X=0.726\ \Omega$.

IV. SMALL SIGNAL ANALYSIS OF THE INVERTER OPERATES AT VSM

In this section, through small signal modeling and analysis of the inverter operates at VSM, the stability and dynamic performance are concluded when the grid impedance increases, further concluded that it is more suitable for inverter operates at VSM in the larger grid impedance occasions.

The droop control is a kind of VSM control strategy, and the power control loop is expressed as follows:

$$\begin{cases} \omega = \omega^* + k_p(P^* - P) \\[3mm] E = E^* + \left(k_q + \dfrac{k_i}{s}\right)(Q^* - Q) \end{cases} \tag{12}$$

The voltage integral feedforward control is added to (12), so as to achieve the output of reactive power without steady-state tracking error.

According to (12), the inverter adjust the angle δ by frequency droop curve adjustment. Then according to (7), the output active power P can be adjusted by δ.

The linearization of the (7) and (12) is derived as :

$$\begin{cases} \Delta\omega = -k_p\Delta P \\[3mm] \Delta E = -\left(k_q + \dfrac{k_i}{s}\right)\Delta Q \\[3mm] \Delta P = k_{pe}\Delta E + k_{pd}\Delta\delta \\[3mm] \Delta Q = k_{qe}\Delta E + k_{qd}\Delta\delta \end{cases} \tag{13}$$

where,

$$\begin{cases} k_{pe} = \dfrac{\partial P}{\partial E} = \dfrac{V\sin\delta}{X} \\[3mm] k_{pd} = \dfrac{\partial P}{\partial \delta} = \dfrac{EV\cos\delta}{X} \\[3mm] k_{qe} = \dfrac{\partial Q}{\partial E} = \dfrac{2E - V\cos\delta}{X} \\[3mm] k_{qd} = \dfrac{\partial Q}{\partial \delta} = \dfrac{VE\sin\delta}{X} \end{cases} \tag{14}$$

In addition, the first order low-pass filter is used in power calculation:

$$\begin{cases} \Delta P_{measure} = \dfrac{\omega_f}{s + \omega_f}\Delta P \\[3mm] \Delta Q_{measure} = \dfrac{\omega_f}{s + \omega_f}\Delta Q \end{cases} \tag{15}$$

where, the ω_f is the cutoff frequency of the first order low-pass filter.

According to (12) - (15), the following formula can be derived as:

$$\begin{cases} \Delta\omega = -k_p \Delta P = -\dfrac{k_p \omega_f}{s + \omega_f}\left(k_{pe}\Delta E + k_{pd}\Delta\delta\right) \\[2mm] \Delta E = -\left(k_q + \dfrac{k_i}{s}\right)\Delta Q \\[2mm] = -\dfrac{\left(k_q + \dfrac{k_i}{s}\right)\omega_f}{s + \omega_f}\left(k_{qe}\Delta E + k_{qd}\Delta\delta\right) \end{cases} \quad (16)$$

Consider the $\Delta\omega = s\Delta\delta$, the following formula can be derived as:

$$s^4 + a \cdot s^3 + b \cdot s^2 + c \cdot s + d = 0 \quad (17)$$

where,

$$\begin{cases} a = \left(2 + k_q k_{qe}\right)\omega_f \\[2mm] b = \left(k_i k_{qe} + k_p k_{pd} + k_q \omega_f k_{qe} + \omega_f\right)\omega_f \\[2mm] c = \left(k_i k_{qe} + k_{pd} k_p + k_{qe} k_{pd} k_p k_q - k_p k_{pe} k_q k_{pd}\right)\omega_f^2 \\[2mm] d = \left(k_{qe} k_{pd} - k_{pe} k_{qd}\right)k_p k_i \omega_f^2 \end{cases}$$

$$(18)$$

According to (17) and (18), the root locus diagram is shown in Fig. 5

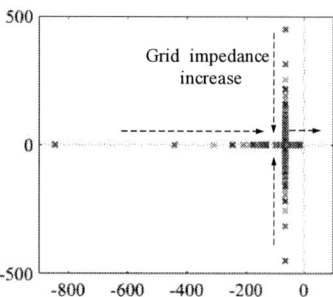

Fig. 5. The root locus diagram of the grid connected inverter with VSM when the grid impedance increase.

Fig. 5 shows that when the grid impedance is very small, the pole damping ratio is also very small. So when the power disturbance, the dynamic response is prone to lead to oscillation, and the inverter output power P fluctuations, even oscillation. With the increase of the grid impedance, the damping ratio increases gradually and the poles tend to the real axis. So the overshoot in the dynamic process can be reduced, and improves the dynamic performance of the system. Also, the simulation and experiment in section V verify this conclusion.

V. SIMULATION AND EXPERIMENTAL VERIFICATION

In order to verify the correctness of the above theoretical analysis, the simulation of three-phase grid connected inverter is constructed by using Matlab/Simulink, and the main parameters are shown in Table I.

TABLE I
PARAMETERS OF THE GRID-CONNECTED INVERTER

Parameters	values
Rated power P_N/kW	100
Rated AC voltage U_N/V	380
DC voltage U_{dc}/V	600
Switch frequency/kHz	5
Inverter side inductance L_1/mH	0.56
Filter capacitor C_f/μF	270
Grid impedance X/Ω	0.726

For illustration, the following four operating cases are defined:

Case I: the inverter operates at VSM in strong grid with very small impedance.

Case II: the inverter operates at VSM in weak grid with large impedance.

Case III: the inverter operates at CSM in strong grid with very small impedance.

Case IV: the inverter operates at CSM in weak grid with very large impedance.

The steady state simulation waveform and the corresponding spectrum analysis of grid current under four different operating cases are presented in Fig.6 and Fig.7, respectively .

As shown in Fig. 6 and Fig.7, it is founded that it is suitable for the inverter operated at CSM in the strong grid condition which the grid impedance is very small, and it is not suitable for the inverter operated at CSM in the weak grid condition which the grid impedance is very large. And it is exactly the opposite when the inverter operates at VSM. This is consistent with the analysis in section II.

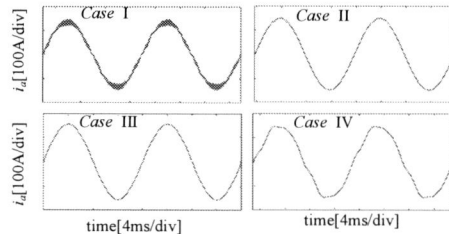

Fig. 6. The steady state simulation waveform of grid current under four different operating cases.

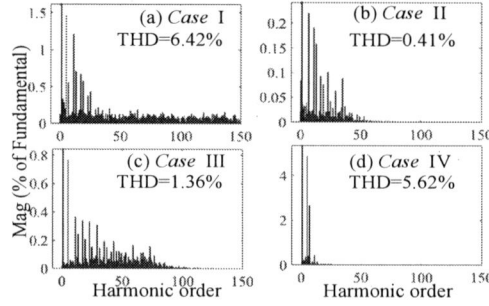

Fig. 7. The corresponding spectrum analysis of simulation waveform of grid current under four different operating cases.

In order to verify the dynamic performance when the inverter operates at VSM under different grid impedance

The 2018 International Power Electronics Conference

conditions, Fig. 8. shows the simulation waveform of the active power amplitude step from 0 to the rated value under the two operating cases: *Case* I, *Case* II. It can be seen from Fig. 8 that with the increase of the grid impedance, the damping ratio of the output active power of the inverter increases, and the power oscillation phenomenon disappears. This is consistent with the small signal model analysis in section III.

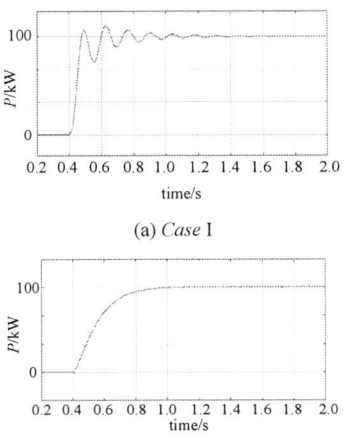

(a) *Case* I

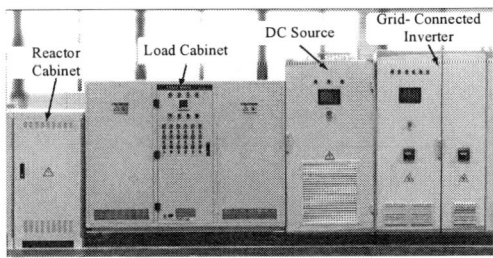

(b) *Case* II

Fig. 8. The simulation waveform of the active power amplitude step from 0 to the rated value under the two operating cases: *Case* I, *Case* II.

In order to validate the proposed theory, experiments have been carried out on a prototype system. The experimental platform is depicted in Fig. 9. The prototype is composed of a 100kW three-phase LCL grid connected inverter and a 100kW DC source. The control circuit is implemented on a DSP chip TMS320F2808. The parameters of the inverter are also shown in Table I. And the grid impedance is achieved by a series of practical reactors.

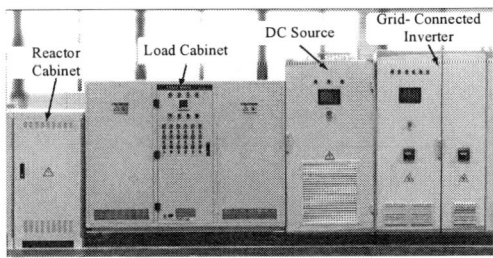

Fig. 9. Experimental platform.

Corresponding to simulation, as shown in Fig. 10 is the grid current steady-state experimental waveform under different four operating cases. The corresponding spectrum analysis is shown in Fig. 11. Comparing with Fig. 7 and Fig. 11, it can be seen that the distribution of grid-connected current spectrum is different in simulation and experiment because of the fact that the filter has some distributed capacitance and the filter parameters are slightly different.

And Fig. 12 shows the experimental waveform of the active power amplitude step from 0 to the rated value under the two operating cases: *Case* I, *Case* II. It can be

concluded from the Fig. 10 to Fig. 12 that the experimental results are in well agreement with the simulation results, the superiority of the proposed control strategy is further verified.

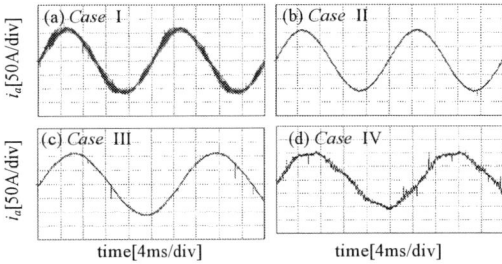

Fig. 10. The steady state experimental waveform of grid current under four different operating cases.

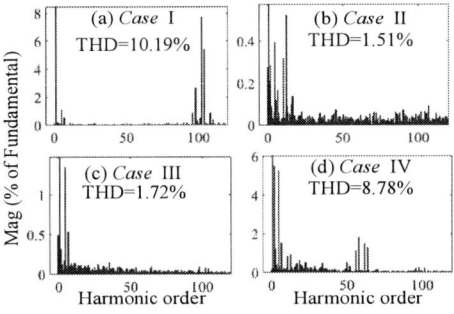

Fig. 11. The corresponding spectrum analysis of experimental waveform of grid current under four different operating cases.

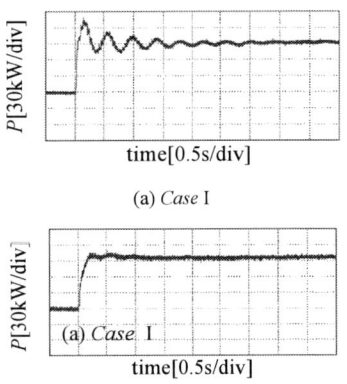

(a) *Case* I

(b) *Case* II

Fig. 12. The experimental waveform of the active power amplitude step from 0 to the rated value under the two operating cases: *Case* I, *Case* II.

VI. CONCLUSION

This paper analyzes the characteristics of static relationship of the power transmission between CSM and VSM. And it is concluded that the inverter is more suitable for operating at VSM in the larger grid impedance occasions. On the basis of the above analysis, this paper proposes the grid impedance adaptation dual mode control strategy in weak grid: when the grid impedance is low, the inverter operates at CSM, and when the grid impedance is large, the inverter operates at VSM.

REFERENCES

[1] F. Blaabjerg, Z. Chen, and S. B. Kjaer, "Power electronics as efficient interface in dispersed power generation systems," *IEEE Transactions on Power Electronics,* vol. 19, pp. 1184-1194, 2004.

[2] A. Etxegarai, P. Eguia, E. Torres, E. Fernandez, A. Etxegarai, P. Eguia, *et al.*, "Impact of wind power in isolated power systems," pp. 63-66, 2012.

[3] M. Liserre, R. Teodorescu, and F. Blaabjerg, "Stability of photovoltaic and wind turbine grid-connected inverters for a large set of grid impedance values," *IEEE Transactions on Power Electronics,* vol. 21, pp. 263-272, 2006.

[4] M. Davari and Y. A. R. I. Mohamed, "Robust Vector Control of a Very Weak-Grid-Connected Voltage-Source Converter Considering the Phase-Locked Loop Dynamics," *IEEE Transactions on Power Electronics,* vol. 32, pp. 977-994, 2017.

[5] C. Yu, X. Zhang, F. Liu, F. Li, H. Xu, R. Cao, *et al.*, "Modeling and Resonance Analysis of Multi-parallel Inverters System under Asynchronous Carriers Conditions," *IEEE Transactions on Power Electronics,* vol. PP, pp. 1-1, 2017.

[6] C. Wan, M. Huang, K. T. Chi, and X. Ruan, "Stability of interacting grid-connected power converters," *Journal of Modern Power Systems & Clean Energy,* vol. 1, pp. 249-257, 2013.

[7] O. Goksu, R. Teodorescu, C. L. Bak, F. Iov, and P. C. Kjaer, "Instability of Wind Turbine Converters During Current Injection to Low Voltage Grid Faults and PLL Frequency Based Stability Solution," *IEEE Transactions on Power Systems,* vol. 29, pp. 1683-1691, 2014.

[8] B. Wen, D. Boroyevich, R. Burgos, P. Mattavelli, and Z. Shen, "Analysis of D-Q Small-Signal Impedance of Grid-Tied Inverters," *IEEE Transactions on Power Electronics,* vol. 31, pp. 675-687, 2015.

[9] Y. Huang, X. Yuan, J. Hu, and P. Zhou, "Modeling of VSC Connected to Weak Grid for Stability Analysis of DC-Link Voltage Control," *IEEE Journal of Emerging & Selected Topics in Power Electronics,* vol. 3, pp. 1193-1204, 2017.

[10] J. Z. Zhou, H. Ding, S. Fan, Y. Zhang, and A. M. Gole, "Impact of Short-Circuit Ratio and Phase-Locked-Loop Parameters on the Small-Signal Behavior of a VSC-HVDC Converter," *Power Delivery IEEE Transactions on,* vol. 29, pp. 2287-2296, 2014.

[11] A. Egea-Alvarez, S. Fekriasl, F. Hassan, and O. Gomis-Bellmunt, "Advanced Vector Control for Voltage Source Converters Connected to Weak Grids," *IEEE Transactions on Power Systems,* vol. 30, pp. 3072-3081, 2015.

[12] X. Yuan, J. Chai, and Y. Li, "Control of variable pitch, variable speed wind turbine in weak grid systems," in *Energy Conversion Congress and Exposition,* 2010, pp. 3778-3785.

[13] X. Yuan, F. Wang, D. Boroyevich, Y. Li, and R. Burgos, "DC-link Voltage Control of a Full Power Converter for Wind Generator Operating in Weak-Grid Systems," *IEEE Transactions on Power Electronics,* vol. 24, pp. 2178-2192, 2009.

[14] R. Piwko, N. Miller, J. Sanchez-Gasca, X. Yuan, R. Dai, and J. Lyons, "Integrating Large Wind Farms into Weak Power Grids with Long Transmission Lines," in *Conference Proceedings of Ces/ieee International Power Electronics and Motion Control Conference,* 2006, pp. 1-7.

[15] E. Muljadi, C. P. Butterfield, B. Parsons, and A. Ellis, "Effect of Variable Speed Wind Turbine Generator on Stability of a Weak Grid," *IEEE Transactions on Energy Conversion,* vol. 22, pp. 29-36, 2007.

[16] M. Cespedes and J. Sun, "Adaptive Control of Grid-Connected Inverters Based on Online Grid Impedance Measurements," in *IEEE Transactions on Sustainable Energy,* vol. 5, no. 2, pp. 516-523, April 2014.

[17] IEEE Guide for Planning DC Links Terminating at AC Locations Having Low Short-Circuit Capacities," *in IEEE Std 1204-1997* , vol., no., pp.1-216, Jan. 21 1997.

The 2018 International Power Electronics Conference

Transmission Power Analysis and Control of the DC Transformer in Hybrid AC/DC Microgrid

Jingjin Huang[1,2], Xin Zhang[2*], Tengfei Zhang[2]

1. Electrical Engineering, Xi'an University of technology, China
2. School of Electrical and Electronic Engineering, Nanyang Technological University, Singapore
E-mail: jackzhang@ntu.edu.sg*

Abstract-In a hybrid AC/DC microgrid, the DC transformer (DCT) is more competitive than the conventional line frequency AC transformer (ACT) to connect the AC and DC bus operating with bidirectional interlinking converter (BIC) due to the high frequency, high power density and adjustable characteristics of the DCT. However, the hybrid AC/DC microgrid application often poses a series of specific challenges to the DCT design, such as the system energy management requirement, efficiency, volume, galvanic isolation and power balance challenges, etc. In this paper, a symmetrical CLLC resonant dual active bridge (DAB) converter has been employed to realize the DCT topology. Meanwhile, a systematic design method of the DCT including transmission power analysis, power control and operation mode analysis have also been presented with the consideration of the hybrid AC/DC microgrid application. Finally, all of the theoretical analysis has been experimentally verified in a 6-kVA hybrid AC/DC microgrid.

I. INTRODUCTION

Nowadays, DC microgrid has become more and more popular due to the rapid development of renewable DC power sources and DC loads [1, 2]. However, in the DC microgrid, the AC power from wind turbine generators should be converted into DC power using AC/DC conversion technique, which needs multiple conversion steps and result in additional power loss to the system and the increasing complexity of the DC microgrid. As a result, the hybrid AC/DC microgrid has come up to facilitate the connection of various renewable AC and DC sources and loads [3, 4].

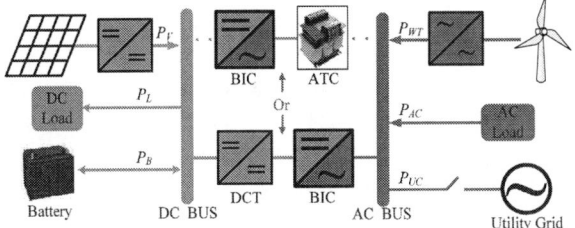

Fig. 1 Configuration of studied hybrid AC/DC microgrid.

In this paper, a typical hybrid AC/DC microgrid as shown in Fig. 1 has been adopted for carrying on the research [3]. As seen, there are two methods to connect the DC and AC bus: a) bidirectional interlinking converter (BIC) + conventional line frequency AC Transformer (ACT in short); b) BIC + DC transformer (DCT). However, comparably, the DCT exhibits competitive advantages by concerning the space occupation, weight, efficiency, *etc.* As a result, the BIC + DCT method is more popular than the BIC + ACT method in the hybrid AC/DC microgrid, where weight and space occupation are of significant concerns [3, 4].

Currently, various DCT topologies have been proposed and studied in the past decades [5, 6]. Among these topologies, the resonant dual active bridge (DAB) converter has attracted lots of research interests due to its simple structure and wide range of soft switching capability [7]. A bidirectional LLC resonant converter is proposed with an extra resonant capacitor [8] to achieve zero voltage switching (ZVS) and zero current switching (ZCS) to reduce the operating power loss. Furthermore, a CLLC-type asymmetric resonant converter is proposed in [9] with additional two capacitors and an inductor. This topology can achieve ZVS for inverter switches and ZCS for rectifier switches with a wide transmission gain. Subsequently, symmetrical CLLC resonant network is also analyzed in several literatures [7, 10]. This topology can operate under high-power conversion efficiency with soft switching capability, as well as keeping the same efficiency bidirectionally.

Though a lot of existing DCT topologies have been studied, they have not been considered at the hybrid AC/DC application. However, the operation environment of the hybrid AC/DC microgrid has its own characteristics, such as the system energy management requirement, efficiency, volume, galvanic isolation and power balance challenges, etc. The above characteristics will bring new challenges to the DCT design.

In this paper, a symmetrical CLLC resonant DCT is implemented in hybrid AC/DC micro-grid as an example to discuss the DCT design issues. The rest of this paper is arranged as follows: in Section II, the general comparison between ACT and DCT has been carried out to visually present the superiority of DCT in the hybrid AC/DC

2980

microgrid; Section III analyses the transmission power features of BIC+DCT at the hybrid AC/DC microgrid; The power control and its corresponding operation modes of the CLLC resonant DCT have been carefully discussed at Section IV. In Section V, experimental prototype has been established to verify the theoretically analysis correctness. Finally, Section VI concludes the whole paper.

II. COMPARISON BETWEEN DCT AND ACT AT HYBRID AC/DC MICROGRID

To visually present the superiority of DCT in the hybrid AC/DC microgrid, the comparison between DCT and ACT has been carried out in this section. In general, the DCT contains two parts: H bridge converters and one high frequency transformer (HFT). Therefore, ACT and HFT with the same rated power are first compared and photographed in Fig. 2. As seen, the volume of HFT is much smaller than ACT, greatly saving the occupied space.

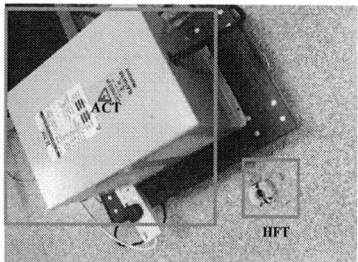

Fig. 2. Transformer with the same rated power.

TABLE I
COMPARISONS BETWEEN DCT AND ACT

Items	ACT	DCT (Two H-bridge converters + HFT)
Rated power	6kVA	6kVA
Weight	64kg	<5kg
Efficiency	96%	97%

Then, according to the datasheet of ACT and the measured characteristics of the tested symmetrical CLLC resonant DCT (See Fig. 3), comparisons are summarized in Table I. It can be observed that the performance of DCT is superior to ACT in terms of weight and efficiency. In addition, it is also expected that the cost of DCT will also be decreased significantly with the popularization of semiconductor device. Furthermore, since the DCT is realized by controllable switches, it also exhibits the advantages as below.

a) It is easier to utilize protection algorithms (over-input/output voltage and current) to ensure the physical isolation instantly, which is one of the impossible functions for the ACT;

b) DCT is feasible for wide-range input / output voltage regulation, however, the voltage regulation ratio of the ACT is unchangeable;

c) DCT has its natural advantages of connecting DC distribution networks than ACT.

Therefore, DCT is more competitive than the ACT in hybrid AC/DC microgrid.

III. TRANSMISSION POWER ANALYSIS OF THE DCT

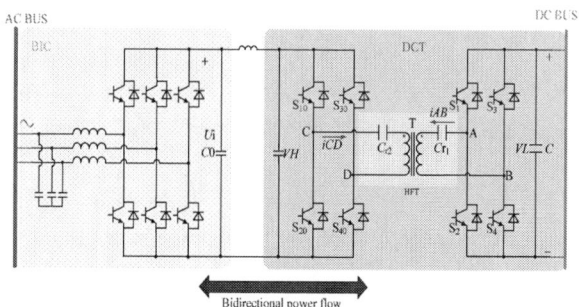

Fig. 3. BIC and DCT circuits in the Hybrid AC/DC microgrid.

The BIC + DCT circuits of this paper have been depicted in Fig. 3. According to Fig. 1, it contains three main power flow scenarios: a) DC bus →AC bus; b) AC bus → DC bus; c) no power flow, whose details are as follows:

Scenario 1: DC bus →AC bus. It indicates the extra DC power is transmitted to the utility grid or to the AC side load.

Scenario 2: AC bus → DC bus. It represents that the DC source hardly satisfies the DC load and battery requirement.

Scenario 3: no power flow exists between AC and DC bus. This means that the DC and AC bus reaches a respective balanced status.

Therefore, BIC and DCT should effectively ensure the bidirectional power flow performance to cooperate with the optimum power dispatch scheme of the Energy Management System (EMS).

A. Reviewer BIC [2 ~ 4]

Fig. 4 shows the control diagram of BIC, which is capable of processing bidirectional power flow efficiently. The BIC is configured to operate in three aforementioned scenarios caused by different distributed sources and loading conditions. By Fig. 4, the specific control mode depends on EMS.

The control mode varies with the operating scenarios, as summarized in Table II. If BIC is connected with the utility grid, the bus voltage and frequency are relatively stable. Therefore, control mode 1 will be selected to transmit power under Scenarios 1 and 2. The desired power P_{EMS} dispatched by EMS is tracked through the BIC, in which the desired active and reactive currents i_d and i_q are sent to the current controller to obtain the modulations. If no power transmission exists as Scenarios 3, mode 1 will be selected. When the system operates under the autonomous condition, scenario 3 will be chosen to stabilize the voltage and frequency.

As seen, control mode 1 is similar to the control of the classical AC/DC rectifier while control model 3 is similar to DC/AC converter. Therefore, for grid-tied mode, the task of BIC is to balance the power; while for the autonomous mode, the BIC main goal is to stabilize its voltage and frequency.

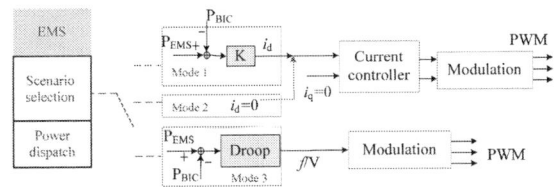

Fig. 4. Control diagram of BIC.

TABLE II
CONTROL MODELS OF THE BIC AT DIFFERENT SCENARIOS

Condition	Scenario	Control mode
Grid-tied condition	1	Mode 1
	2	Mode 1
	3	Mode 2
Autonomous condition	1	Mode 3
	2	Mode 3
	3	Mode 3

B. Transmission power analysis of the DCT

Since the power transmission performance of DCT plays an important role in EMS, the mathematical model with respect to the transmitted power is derived in this sub-section.

If the DCT power is required to be regulated, the phase-shift scheme will be introduced. The input voltage v_{in} consists of three voltage levels $-V_{in}$, 0, and V_{in}, as shown in Fig. 5, which can be expressed using the Fourier series as

$$v_{in} = \frac{4V_{in}}{\pi} \sum_{m=1,3,5...}^{\infty} \frac{1}{m} \cos \frac{m\theta}{2} \sin(m\omega t) \qquad (1)$$

where θ is the phase shift angle of the PS scheme and ω denotes the fundamental angular frequency.

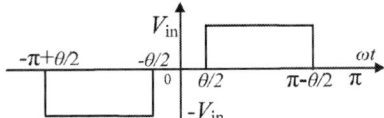

Fig. 5 Input voltage v_{in} with PS scheme.

When DCT operates under the resonant status, the input current i can be expressed as

$$i = I \sin(\omega t - \varphi) \qquad (2)$$

where I is the peak value of input current, and φ indicates the angle difference between the input voltage and current.

The DCT power transferred from the input side $p(t)$ can be calculated as

$$p(t) = v_{in} \cdot i = \overline{p} + \tilde{p} \qquad (3)$$

where

$$\overline{p} = \frac{2V_{in}I}{\pi} \cos \frac{\theta}{2} \cos \varphi$$

$$\tilde{p} = -\frac{2V_{in}I}{\pi} \cos \frac{\theta}{2} \cos(2\omega t - \varphi) + \frac{2V_{in}I}{\pi} \sum_{k=3,5...}^{\infty} \frac{1}{k} \bullet$$

$$\cos \frac{k\theta}{2} \{\cos[(k-1)\omega t + \varphi] - \cos[(k+1)\omega t - \varphi]\}$$

For each operating period T, the DCT average power $p_0(t)$ can be derived as

$$p_0 = \frac{1}{T} \int_0^T \overline{p} \, dt + \frac{1}{T} \int_0^T \tilde{p} \, dt = \overline{p} \qquad (4)$$

$$= \frac{2V_{in}I}{\pi} \cos \frac{\theta}{2} \cos \varphi$$

According to (4), the following DCT transmission power features can be obtained:

a) p_0 can be regulated via changing φ and θ.

b) φ is related with the reactive power, the larger value of which brings higher voltage and current on resonant capacitor. Besides, it is hardly regulated.

c) θ can be adjusted by a simple phase shift scheme to track p_0.

d) The maximum DCT average power $p_{0max} = 2V_{in}I/\pi$ when $\varphi=0$ and $\theta=0$. Thus, 50% duty ratio (DR) scheme can ensure the maximum average power transmission.

e) p_0 is decreased with the increasing $\theta \subseteq [0°, 180°]$, if ignoring the impact of φ.

These features are useful to the power control. Therefore, θ can be regarded as a better candidate to regulate p_0, with the details analyzed as below.

IV. THE POWER CONTROL FOR THE DCT AND ITS OPERATION MODE ANALYSIS

A The power control of the DCT at Hybrid AC/DC microgrid

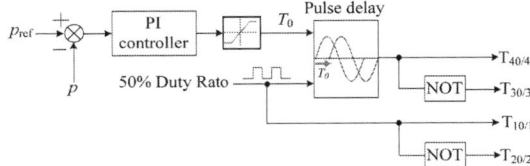

Fig. 6. Power control of the DCT.

DCT may be required to reduce the transmitted power when the renewable energy is insufficient or the market price is high. Thus, a concise and effective closed power loop is introduced by this paper, as shown in Fig. 6, where p_{ref} and $p(\approx p_0)$ indicate the reference and actual output power respectively; $T_{10/20/30/40}$ denotes trigger pulse for the switching state $S_{10/20/30/40}$ in the primary side; 50% DR is set as the reference pulse. T_0 is the delay time of the reference pulse, which is proportional to the phase shift angle θ. The switch $S_{1/2/3/4}$ in the secondary can be

2982

triggered with the same pulse $T_{1/2/3/4}$ correspondingly to facilitate the DCT power control. When regulating the phase shift angle θ, three cases are available in the practical implementation as below.

- Case 1: $\theta > 0$, i.e., the switch $T_{4/40}$ is lag of the switch $T_{1/10}$.
- Case 2: $\theta < 0$, i.e., $T_{4/40}$ is ahead of $T_{1/10}$.
- Case 3: $\theta = 0$, i.e., $T_{4/40}$ is synchronize with $T_{1/10}$.

As discussed in Section III.B, the designed DCT not only can operate under 50% DR scheme ($\theta = 0$) to ensure the maximum transmission power, but also can operate under the power control with the phase shift (PS) scheme to reduce the transmitted power in the hybrid AC/DC microgrid application. Therefore, the specific operating principle of DCT during this power control process will be further discussed in the next sub-section.

B. Operation principle of the DCT with PS scheme

a) Case 1: $\theta > 0^{\circ}$.

The DCT operation principle is analyzed utilizing the power control scheme with the final phase shift angle as $\theta > 0^{\circ}$. The corresponding waveform is shown in Fig. 7. According to Fig. 7(a), the converter operation is repetitive during each switching cycle in the steady state. The steps in the negative part are similar to that in the positive part. Here, take the step in the positive part as an example to aid in understanding the typical operation modes in each switching cycle.

Step 1 ($0-t_0$) (As shown in Fig. 7(a)): S_{10}/S_1 and S_{30}/S_3 are conducting. At $t=0$, $i=0$. The primary/secondary currents of the HFT pass through the antiparallel diodes of S_{30}/S_1. The output capacitor stops charging and the transmission power during this step can be obtained as

$$p_{01} = \int_{0}^{t_0} v_{CD} i_{CD} dt = 0 \tag{5}$$

where v_{CD} and i_{CD} are the input voltage and current at the primary side.

Step 2 (t_0-t_1) (As shown in Fig. 7(b)): At t_0, S_{40} is turned on. S_{10} and the antiparallel diodes of S_1/S_4 are conducting. Combining $i_{CD} = I\sin(\omega t - \varphi)$, the transmitted power can be calculated as

$$p_{02} = \int_{t_0}^{t_1} v_{CD} i_{CD} dt = \int_{t_0}^{t_1} V_{CD} I\sin(\omega t - \varphi) dt$$
$$= \frac{2V_{CD} I}{\omega} \cos\frac{\theta}{2} \cos\varphi \tag{6}$$

where V_{CD} is the magnitude of v_{CD}.

Step 3 (t_1-t_2) (As shown in Fig. 7(c)): At t_1, S_{20} is turned on. However, the current is still positive. Hence, S_{40} and the antiparallel diode of S_{20} are conducting in the primary side. S_2 and the antiparallel diode of S_4 are conducting in the secondary side. Therefore, the transmitted power is

$$p_{03} = \int_{t1}^{t2} v_{CD} i_{CD} dt = 0 \tag{7}$$

The average transmitted power p_t can be obtained by

$$p_t = \frac{p_{01} + p_{02} + p_{03}}{0.5T} = \frac{2V_{CD} I}{\pi} \cos\frac{\theta}{2} \cos\varphi \tag{8}$$

This result is consistent with Eq. (4), i.e., $p_t = p_0$, which further verify the effectiveness of the power control scheme.

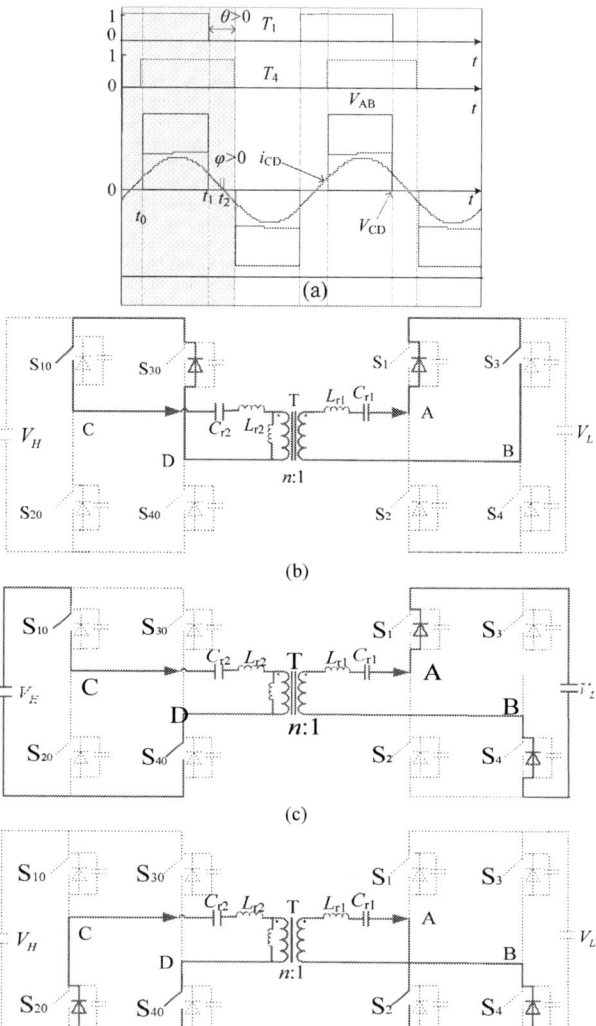

Fig. 7. Operation principle of PS scheme with $\theta > 0$.

Since the operation principle in Case 2 and 3 is similar to Case 1. Only the transmission power is further analyzed in the following.

b) Case 2: $\theta < 0^{\circ}$.

In Fig. 8(a), the input current i_{CD} is a little ahead of v_{CD} when $\theta < 0^{\circ}$, and $i_{CD} = 0$ at $t=0$ and $t=T/2$. Therefore, only two steps are available in the positive part of each cycle in this case.

2983

Step 1 ($0-t_0$): $T_{2/20}$ and $T_{3/30}$ are triggered, which brings $v_{CD}=0$ in this step. Therefore, the transmission power is zero, i.e., $P_{11}=0$.

Step 2 (t_0-t_1): $T_{1/10}$ and $T_{4/40}$ are triggered, and thus the transmission power p_{12} can be calculated as

$$p_{t1} = \frac{2V_{CD}I}{\omega}\cos\frac{\theta}{2}\cos\varphi \qquad (9)$$

Similar to (8), the average transmission power p_{t1} can be calculated as

$$p_{t1} = \frac{2V_{CD}I}{\pi}\cos\frac{\theta}{2}\cos\varphi \qquad (10)$$

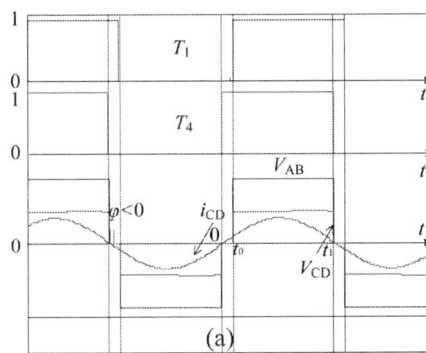

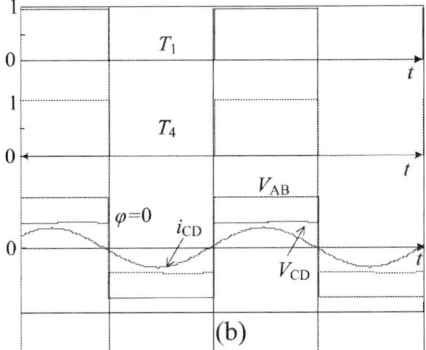

Fig. 8. Operation principle of PS scheme with (a)$\theta<0$ and (b) $\theta=0$.

c) *Case 3: $\theta=0°$.*

As shown in Fig. 8(b), DCT operates with only one step in the positive part of each cycle when $\theta=0°$. Hence, the average transmission power p_{t2} can be calculated as

$$p_{t2} = \frac{1}{0.5T}\int_{t_0}^{\pi} v_{CD}i_{CD}dt = \int_0^{\pi} V_{CD}I\sin(\omega t - \varphi)dt \qquad (11)$$
$$= \frac{2V_{CD}I}{\pi}\cos\varphi$$

Therefore, we can get

$$\begin{cases} p_{t2}(\theta=0°) > p_t(\theta>0°) \\ p_{t2}(\theta=0°) > p_{t1}(\theta<0°) \end{cases} \qquad (12)$$

which further verify that the maximum transmission power is ensured by $\theta=0°$, and both $\theta>0°$ and $\theta<0°$ can be utilized to regulate the transmission power.

According to aforementioned discussion, DCT operation issues at Hybrid AC/DC are summarized as below.

a) DCT input power can maintain its maximum value when no phase shift exists and the HFT input current is kept in phase with the HFT input voltage.

b) The transmission power can be regulated by both θ and φ. Comparably, θ is much easier to be implemented.

c) By Figs. 7 and 8, φ is affected when regulating θ. If the transmission power is greatly reduced, the angle φ must be taken into account to analyze weather the resonant components can withstand the increasing reactive power or not.

V. EXPERIMENTAL VERIFICATION

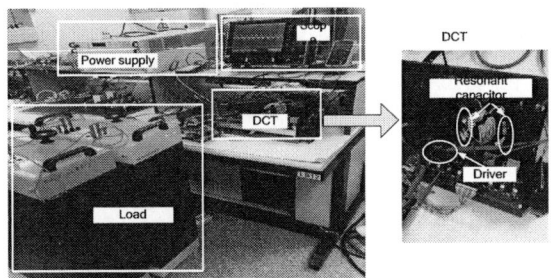

Fig. 9. Experimental prototype of DCT

In this section, a 6 kW DCT prototype (Fig. 9) is established to verify the correctness of the theoretical analysis. Its parameters are listed in Table III.

TABLE III
PARAMETERS OF EXPERIMENTAL SETUPS

Parameters	Values
DC bus nominal voltage	380V
BIC DC bus voltage	760V
BIC and DCT rated power	6kW
Frequency(DCT/BIC)	100kHz(DCT) / 20kHz(BIC)
Designed DCT	L_{r1}=20.4μH, L_{r2}=80.9μH, C_{r1}=0.25μF, C_{r2}=0.07μF

A. Bidirectional power control of DCT

In order to verify the designed power control scheme, the input and output voltages and currents of DCT are tested, the results are shown in Fig. 10. It can be observed that the proposed power control is working well at bidirectional power flow mode.

B. Operating mode verification of DCT

To analyze the operating performance of DCT, the voltage and current of DCT are tested in this sub-section. The results are presented in Fig.11. As seen, the results of the DCT with/without PS scheme agree very well with the theoretical analysis.

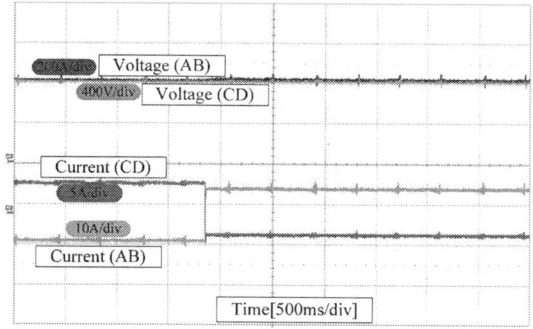

Fig. 10 Currents and voltages with the designed power control.

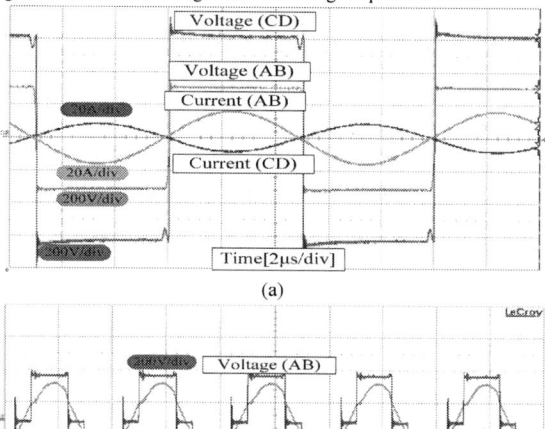

(a)

(b)

Fig. 11 Current and voltage Waveforms: (a) without PS control; (b) with PS control.

VI. CONCLUSION

In this paper, DCT with symmetrical CLLC topology is implemented in hybrid AC/DC microgrid for BIC voltage matching and galvanic isolation. The advantages of DCT over ACT have been thoroughly summarized. The average transmission power is derived and analyzed. Afterwards, a concise and effective power control scheme is employed according to a detailed transmission power analysis. Furthermore, the corresponding operating modes are discussed to facilitate the implementation of the proposed power control. Finally, a 6 kW DCT prototype has been built and various experimental cases have been conducted to confirm the performance of DCT.

REFERENCES

[1] Xiao J, Wang P, Setyawan L. "Hierarchical control of hybrid energy storage system in DC microgrids," *IEEE Transactions on Industrial Electronics*, vol. 62, no. 8, pp. 4915-4924, 2015.

[2] Liu X, Wang P, Loh P C. "A hybrid AC/DC microgrid and its coordination control," *IEEE Transactions on Smart Grid*, vol. 2, no. 2, pp. 278-286, 2011.

[3] Xiao J, Wang P, Huang J, et al. "*Implementation of DC/DC Converter with High Frequency Transformer (DHFT) in Hybrid AC/DC Microgrid*", ACEPT, 2017.

[4] Wang P, Goel L, Liu X, et al. "Harmonizing AC and DC: A hybrid AC/DC future grid solution," *IEEE Power and Energy Magazine*, vol. 11, no. 3, pp. 76-83, 2013.

[5] Jovcic D. "Bidirectional, high-power DC transformer," *IEEE transactions on Power Delivery*, vol. 24, no. 4, pp. 2276-2283, 2009.

[6] Feng W, Mattavelli P, Lee F C. "Pulsewidth locked loop (PWLL) for automatic resonant frequency tracking in LLC DC–DC transformer (LLC-DCX)," *IEEE Transactions on Power Electronics*, vol. 28, no. 4, pp. 1862-1869, 2013.

[7] Zhao B, Song Q, Liu W, et al. "Overview of dual-active-bridge isolated bidirectional DC–DC converter for high-frequency-link power-conversion system," *IEEE Transactions on Power Electronics*, vol. 29, no. 8, pp. 4091-4106, 2014.

[8] Chang C H, Chang E C, Cheng H L. "A high-efficiency solar array simulator implemented by an LLC resonant DC–DC converter," *IEEE Transactions on Power Electronics*, vol. 28, no. 6, pp. 3039-3046, 2013.

[9] W. Chen, P. Rong, and Z. Y. Lu, "Snubberless bidirectional DC-DC converter with new CLLC resonant tank featuring minimized switching loss," *IEEE Trans. Ind. Electron.*, vol. 57, no. 9, pp. 3075–3086, Sep. 2010

[10] J. H. Jung, H. S. Kim, M. H. Ryu, and J. W. Baek, "Design methodology of bidirectional CLLC resonant converter for high-frequency isolation of dc distribution systems," *IEEE Trans. On Power Electron.*, vol. 28, no. 4, pp. 1741–1755, 2013.

[11] Malan W L, Vilathgamuwa D M, Walker G R. "Modeling and control of a resonant dual active bridge with a tuned CLLC network," *IEEE Trans. on Power Electron.*, vol. 31, no. 10, pp. 7297-7310, 2016.

A Novel Flexible Interconnection Scheme for Microgrid to Optimize the Capacity of Energy Storage System (ESS)

Zhou Jianqiao[1], Zhang Jianwen[1*], Cai Xu[1], Li Zhuyong[1], Wang Jiacheng[2], Zang Jiajie[2]

1 Department of Wind Power Research Center, Shanghai Jiao Tong University, Shanghai, China
2 School of Mechatronic Systems Engineering, Simon Fraser University, Surrey, Canada
Email: jianqiaozhou@sjtu.edu.cn

Abstract-A novel flexible interconnection scheme for microgrids has been proposed to optimize the capacity of ESS, which is used to mitigate the power fluctuation of microgrid. Basic structure, function and operation principle of the novel scheme are illustrated, and the effectiveness of flexible interconnection in power fluctuation mitigation has been analyzed mathematically. Based on the existing sizing algorithm of ESS, two cases, including independent microgrid scheme and interconnected microgrid scheme, have been studied. The results demonstrate the benefits of flexible interconnection scheme for microgrid in optimizing the power and energy capacity of ESS.

I. INTRODUCTION

In recent years, numerous photovoltaics (PV) are integrated in distribution grid. The output power of PV is sensitive to climate, temperature and radiation, which leads to the characteristics of randomness and degrades the safety and reliability of power grid. Power fluctuation of PV will affect the operation and planning of distribution grid in different time scale. In terms of grid operation, the short-term fluctuation will deteriorate the distribution grid voltage quality, and cause the wear and tear of voltage control equipment [1]. In terms of grid planning, the long-term fluctuation will interfere the load forecasting and convert the distribution grid structure [2].

The PV power fluctuation can be smoothed by three ways: active power curtailment [3,4], user-side control [5] and microgrid integration with ESS [6]. However, active power curtailment reduces the PV utilization and user-side control requires the high intelligence of local loads. Therefore, in comparison to other ways, microgrid integration with ESS can be used in practice.

Microgrid, which consists of distributed PV, local load, ESS and control & protection platform, has become the desirable solution for PV integration [7]. Microgrid will behave as a controllable power source or load from the superior grid point of view. The power fluctuation arising from PV and loads will be smoothed by controlling flexibly the ESS power. However, considering the high cost of ESS, a new fluctuation mitigation solution is needed to reduce the capacity of ESS.

A novel flexible interconnection scheme for microgrid has been proposed in this paper. Back-to-back voltage source converter (BTB-VSC) is adopted to interconnect multiple microgrid through the common DC bus in this scheme [8-9]. BTB-VSC and corresponding DC bus are used to adjust power flow among microgrids, so that

power fluctuation mitigation can be realized by balancing the complementary microgrid power curves. Therefore, the capacity of ESS will be optimized.

The rest of paper is organized as follows. In section II, the basic structure of flexible interconnection microgrid scheme is introduced. In section III, the function and operation principle of this novel scheme will be illustrated, furthermore, the effectiveness of flexible interconnection in power fluctuation mitigation has been analyzed mathematically. In section IV, two cases will be studied to verify that the capacity of microgrid ESS will be optimized by flexible interconnection. The conclusion will be shown in section V.

II. BASIC STRUCTURE

Flexible interconnection scheme for microgrid is shown as Fig.1. PV, local load and ESS is integrated in each microgrid, which is connected to the superior distribution grid through the point of common coupling (PCC). PV integration is achieved through DC-AC inverter, and ESS integration is achieved through DC-DC converter, common DC bus and BTB-VSC. The power balancing among microgrid could be realized by controlling BTB-VSC. In comparison to existing independent microgrid scheme, only DC common bus and DC breakers should be added in the flexible interconnection microgrid scheme, and distributed ESS should be integrated.

III. FUNCTION, OPERATION PRINCIPLE AND FLUCTUATION MITIGATION ANALYSIS

A. Function and Operation Principle

The function of flexible interconnection scheme for microgrid including:

● Adjusting power flow among microgrids flexibly, mitigating the power fluctuation arising from PV and loads, improving the PV penetration level in microgrids.

● Shifting the peak load to different places, and avoiding the distribution grid transformer overloading.

● Improving the reliability for key loads, and reducing the capacity of back-up supply.

In fact, the ESS in microgrid could be applied in power fluctuation mitigation, peak-load shaving and back-up supply [10]. The flexible interconnection scheme for microgrids could optimize ESS capacity in these situations. Due to limited space, only single situation of

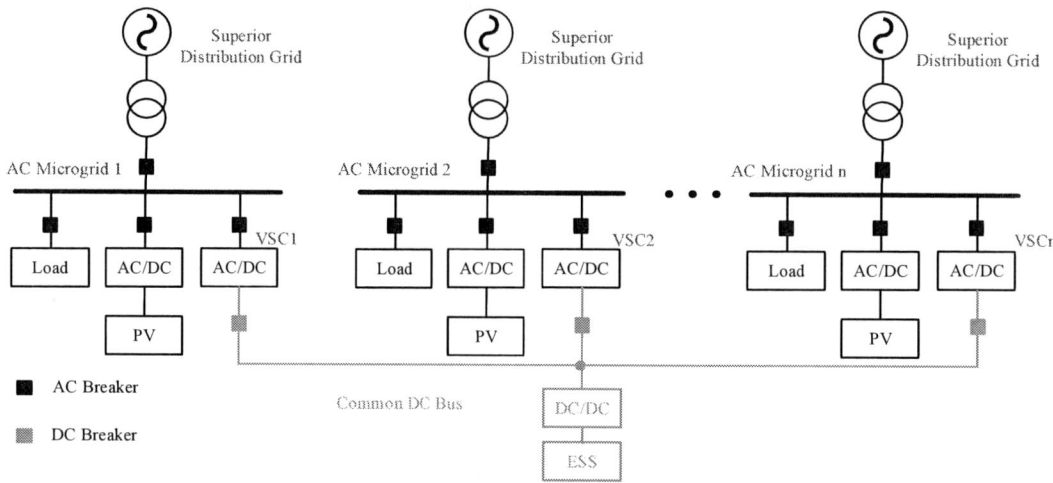

Fig. 1. Basic structure of multi-microgrid flexible interconnection scheme.

power fluctuation mitigation is discussed.

The operation principle of flexible interconnection scheme for microgrid should refer to the control strategy of converter station applied in multi-terminal DC system (MTDC)[11]. The power transfer between microgrids could be achieved by coordinated control of BTB-VSC. In the normal operation mode, one or more VSCs are used to maintain the power balance through controlling the DC bus voltage, and the others will determine the power that need to be transferred. When a fault occurs and results in a microgrid disconnected to superior distribution grid, the operation mode of BTB-VSC should be switched. The voltage and frequency support of disconnected microgrid will be offered by other microgrid through BTB-VSC. In this way, the back-up supply function of flexible interconnection scheme will be achieved.

B. Power Fluctuation Mitigation Analysis

The effectiveness of flexible interconnection scheme for microgrids in power fluctuation mitigation could be demonstrated as follows.

Assuming that the number of microgrid is N, the capacity of each grid is $P_{m1},...,P_{mk},...,P_{mN}$, and the net power (net power = load power consumption – PV power generation) of each microgrid is $P_1(t),...,P_k(t),...,P_N(t)$.

Power fluctuation determined by the net power difference in the constant time interval Δt. Therefore, the power fluctuation of microgrids at the t moment is:

$$P_k(t) - P_k(t-\Delta t) \ , \ k=1,...,N \qquad (1)$$

In the independent microgrid scheme, the sum of power fluctuation that the ESS of microgrids need to deal with at moment t is:

$$\sum_{k=1}^{N} \left| P_k(t) - P_k(t-\Delta t) \right| \qquad (2)$$

In the interconnected microgrid scheme, the whole system power could be allocated to each microgrid via weighted average method, which depends on the capacity of microgrid. That is, the active power that each terminal VSC need to dispatch at t moment is (the power direction that outflow each microgrid is set as positive):

$$-P_k(t) + \frac{P_{mj}}{\sum_{k=1}^{N} P_{mk}} \sum_{k=1}^{N} P_k(t) \ , \ j=1,...,N \qquad (3)$$

The actual net power in each microgrid at t moment is determined by the sum of load consumption, PV generation and VSC dispatching power. The actual net power can be calculated as:

$$\frac{P_{mj}}{\sum_{k=1}^{N} P_{mk}} \sum_{k=1}^{N} P_k(t) \ , \ j=1,...,N \qquad (4)$$

The sum of power fluctuation that the integrated ESS need to deal with at moment t is:

$$\left| \sum_{j=1}^{N} \left[\frac{P_{mj}}{\sum_{k=1}^{N} P_{mk}} \sum_{k=1}^{N} P_k(t) - \frac{P_{mj}}{\sum_{k=1}^{N} P_{mk}} \sum_{k=1}^{N} P_k(t-\Delta t) \right] \right| \qquad (5)$$

The gross power fluctuation that needs to be solved in two schemes could be compared according to absolute value inequality:

$$\left| \sum_{j=1}^{N} \left[\frac{P_{mj}}{\sum_{k=1}^{N} P_{mk}} \sum_{k=1}^{N} P_k(t) - \frac{P_{mj}}{\sum_{k=1}^{N} P_{mk}} \sum_{k=1}^{N} P_k(t-\Delta t) \right] \right|$$
$$= \left| \sum_{k=1}^{N} P_k(t) - \sum_{k=1}^{N} P_k(t-\Delta t) \right| = \left| \sum_{k=1}^{N} \left[P_k(t) - P_k(t-\Delta t) \right] \right| \qquad (6)$$
$$\le \sum_{k=1}^{N} \left| P_k(t) - P_k(t-\Delta t) \right|$$

The inequality (6) indicates that, in the flexible interconnection scheme, the gross power fluctuation could be reduced via equalizing net power of microgrids. Furthermore, the mitigation performance of this scheme is positive correlation to the complementarity of net power among microgrids.

IV. CASE STUDY

To demonstrate the novel scheme performance in optimizing ESS capacity, two cases are studied. As

The 2018 International Power Electronics Conference

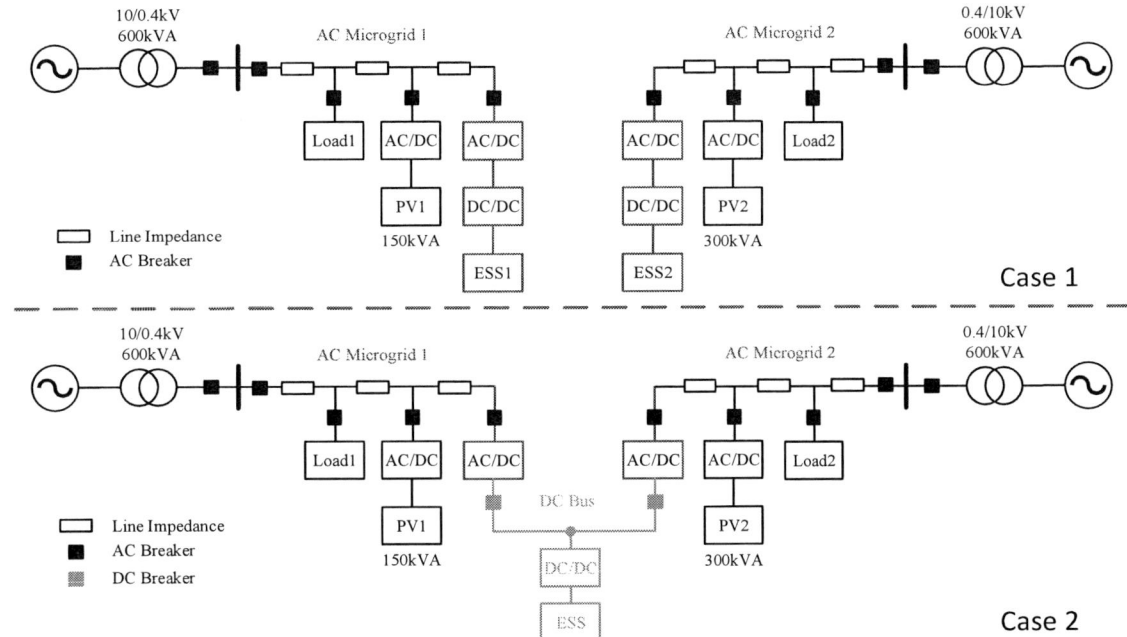

Fig. 2. Basic configuration of case study.

shown in Fig.2, case one has two independent low voltage feeders. Both of them integrate PV and local loads. By configuring the ESS in each feeder separately, two independent microgrid could come into being. On the other hand, in case two, the previous independent microgrids are interconnected by a common DC bus, and the distributed ESSs are integrated. The flexible interconnection microgrids could come into being, so that the integrated ESS capacity is optimized.

The power data of PV and local loads used in case study is from Jinpan Electricity Industrial Park. The data interval is one day, and the sample time is set to 1min. The capacities of both feeders are 600 kVA. PV capacity in feeder 1 is set as 150kVA. PV capacity in feeder 2 is set as 300kVA. The net power curves of two feeders are shown in Fig.3.

To assess the performance of ESS and flexible interconnection in compensating power fluctuation, power fluctuation rate is set as the evaluation index. In the time interval ΔT, the power fluctuation rate could be defined as follow:

$$F_{\Delta T} = \frac{P_{\Delta T}^{\max} - P_{\Delta T}^{\min}}{P_n} \times 100\% \qquad (7)$$

In equation (7), P_n is rated power of the microgrid; $P_{\Delta T}^{\max}$ and $P_{\Delta T}^{\min}$ are the maximum power and minimum power during ΔT separately.

The power fluctuation rate of microgrid should be limited so as to avoid degrading the operation of upper distribution grid. In this paper, two power fluctuation rate limitation objectives in different time scale are applied to comparative analysis.

Objective 1: the system power fluctuation rate is limited as 5% in 1minute interval.

Objective 2: the system power fluctuation rate is limited as 10% in 20 minutes interval.

Without ESS and flexible interconnection, the power fluctuation rate of independent feeder could be calculated by the net power data.

As shown in Fig.4, the maximum power fluctuation rate of feeder 1 in 1 minute interval is 19.33% at 12 o'clock at noon; the maximum power fluctuation rate of feeder 2 in 1 minute interval is 22.67% at 3 o'clock in the afternoon.

As shown in Fig.5, the maximum power fluctuation rate of feeder 1 in 20 minutes interval is 42.67% at 3 o'clock in the afternoon; the maximum power fluctuation rate of feeder 2 in 20 minutes interval is 36.33% at 0 o'clock at midnight.

The net power curves in both feeders cannot satisfy the power fluctuation rate objective requirements. Therefore, ESS installation is imperative to mitigate the power fluctuation in both feeders.

A. ESS Sizing Algorithm Selection

In the application of power fluctuation mitigation, three ESS sizing algorithms could be selected: power margin compensation method, fluctuation analysis method and economic optimization method. Power margin compensation method will calculate the microgrid power difference between actual value and forecasting value, and determine the ESS capacity via probability and statistics theory [12]. Fluctuation analysis method will deal with the microgrid power fluctuation in frequency-domain. The low pass filtering (LPF) [13-14], Discrete Fourier Transform (DFT) [15], or Empirical Mode Decomposition (EMD) could be used to determine the compensation frequency band and capacity of ESS. Economic optimization method will set the capacity of ESS as a variable value, and solve the optimization problem based on economic objective and constraint.

In this paper, the fluctuation analysis method based on DFT would be selected, for this algorithm could achieve

2988

The 2018 International Power Electronics Conference

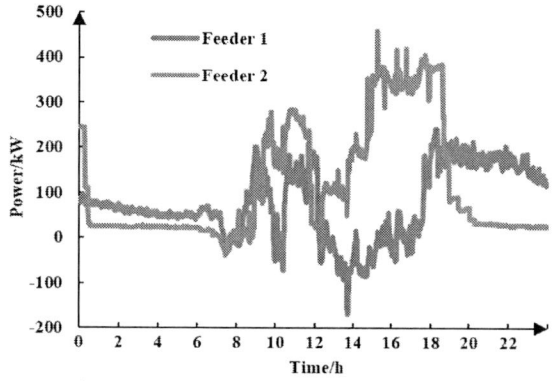

Fig. 3. Actual net load curve of case study.

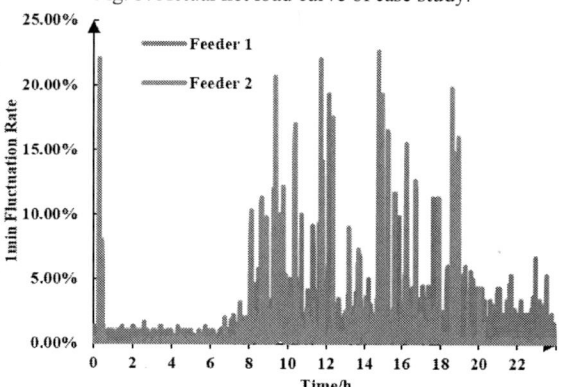

Fig. 4. Power fluctuation rate in 1min.

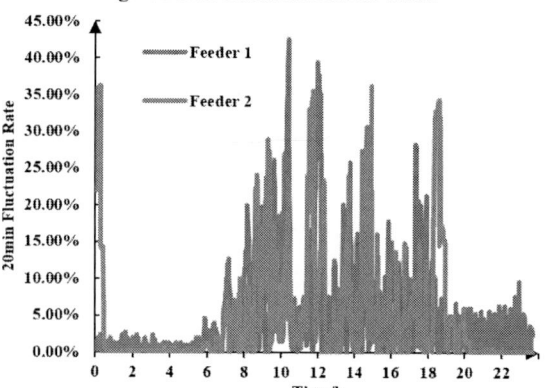

Fig. 5. Power fluctuation rate in 20min.

the precise fluctuation compensation in different frequency band. Under the objective of power fluctuation rate limitation, DFT algorithm could reduce the ESS capacity furthest.

The process of ESS sizing based on DFT algorithm is elaborated in [15]. This paper will directly show the results of ESS sizing in each case.

B. Case one: independent microgrid scheme

In case one, distributed ESSs are installed in two independent microgrids separately, as shown in Fig.2. By the coordinated control of PV, local loads and ESS, the net power fluctuation of microgrid could be smoothed, and the independent microgrid could be integrated to the superior distribution grid friendly.

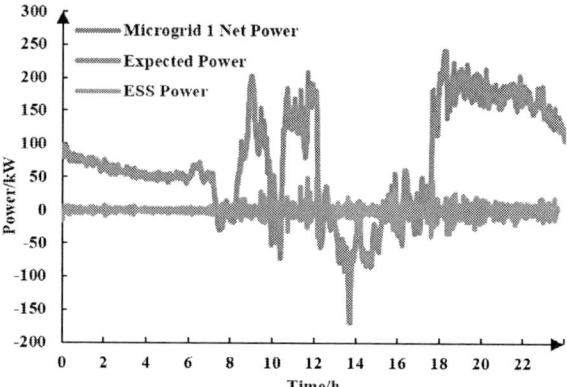

Fig.6. Results of microgrid 1 based on the 1min fluctuation rates control objective.

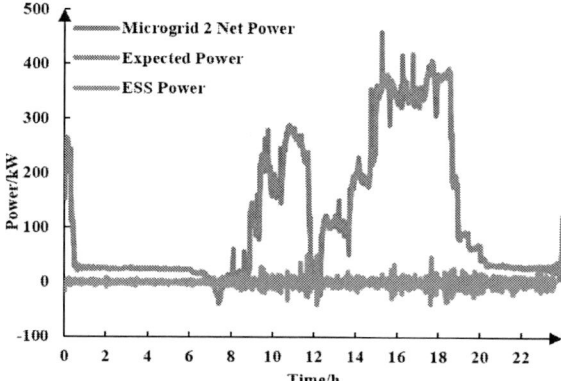

Fig.7. Results of microgrid 2 based on the 1min fluctuation rates control objective.

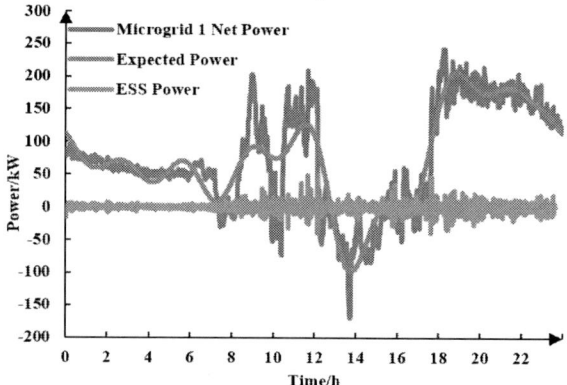

Fig.8. Results of microgrid 1 based on the 20min fluctuation rates control objective.

The capacity of distributed ESS could be sized based on the DFT algorithm. In the control objective 1, the expected net power of microgrid 1 and the corresponding ESS power curve are shown in Fig.6; the expected net power of microgrid 2 and the corresponding ESS power curve are shown in Fig.7. In the control objective 2, the expected net power of microgrid 1 and the corresponding ESS power curve are shown in Fig.8; the expected net power of microgrid 2 and the corresponding ESS power curve are shown in Fig.9.

According to the ESS sizing algorithm and the ESS power curve shown in Fig.6 to Fig.9, the ESS capacity in

2989

The 2018 International Power Electronics Conference

Table 1 Results of energy storage capacity configuration in different schemes

Objective		ESS of Microgrid1	ESS of Microgrid2	ESS of Independent Microgrid Scheme	ESS of Interconnected Microgrid Scheme
$F_{1min} \leq 5\%$	Power Capacity	49kW	112kW	161kW	79kW
	Energy Capacity	1.5kWh	4.4kWh	5.9kWh	1.3kWh
$F_{20min} \leq 10\%$	Power Capacity	150kW	190kW	340kW	206kW
	Energy Capacity	56kWh	79kWh	135kWh	50kWh

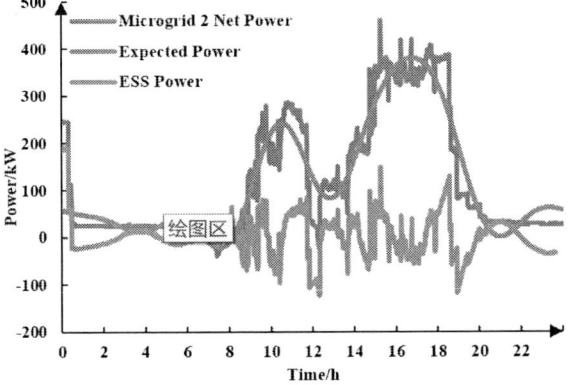

Fig.9. Results of microgrid 2 based on the 20min fluctuation rates control objective.

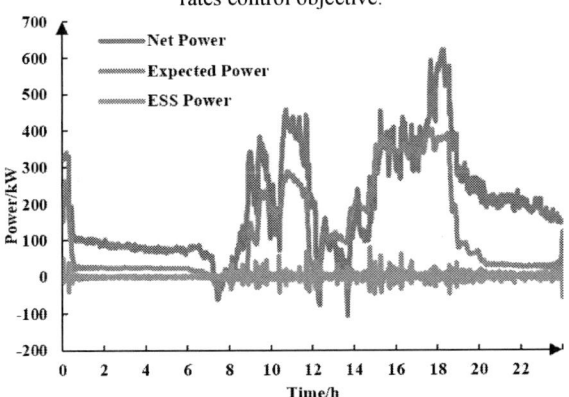

Fig.10. Results of interconnected-microgrid based on the 1min fluctuation rates control objective

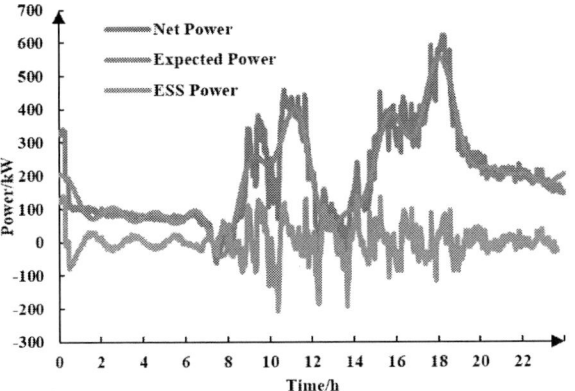

Fig.11. Results of interconnected-microgrid based on the 20min fluctuation rates control objective

different power fluctuation mitigation objectives could be calculated. In the objective 1, the power capacity of ESS in microgrid 1 is 49kW, and the corresponding energy capacity is 1.5kWh; the power capacity of ESS in microgrid 2 is 112kW, and the corresponding energy capacity is 4.4kWh. In the objective 2, the power capacity of ESS in microgrid 1 is 150kW, and the corresponding energy capacity is 56kWh; the power capacity of ESS in microgrid 2 is 190kW, and the corresponding energy capacity is 79kWh.

C. Case two: interconnected microgrid scheme

In case two, integrated ESS is installed in common DC bus of interconnected microgrid, as shown in Fig.2. By the power balancing control between microgrids, the net power fluctuation could be smoothed, and the power and energy capacity of integrated ESS could be optimized.

In the interconnected microgrid scheme, the ESS capacity would be sized based on the sum of net power of two microgrids. In the control objective 1, the expected net power of two microgrids and the corresponding ESS power curve are shown in Fig.10. In the control objective 2, the expected net power of two microgrids and the corresponding ESS power curve are shown in Fig.11.

According to the ESS sizing algorithm and the ESS power curve shown in Fig.10 and Fig.11, the ESS capacity in different power fluctuation mitigation objectives could be calculated. In the objective 1, the power capacity of integrated ESS is 79kW, and the corresponding energy capacity is 1.3kWh. In the objective 2, the power capacity of integrated ESS is 206kW, and the corresponding energy capacity is 50kWh.

D. Results analysis

The overall results of ESS capacity configuration in different objectives are shown in Table I. In the objective 1, the whole power and energy capacity of ESS in independent microgrid scheme are 161kW and 5.9kWh separately; and in interconnected microgrid scheme, these are 79kW and 1.3kWh separately. In the objective 2, the whole power and energy capacity of ESS in independent microgrid scheme are 340kW and 135kWh separately; and in interconnected microgrid scheme, these are 206kW and 50kWh separately.

Some conclusions could be inferred from the results:

● The mitigation of power fluctuation in short time scale should depend on the power capacity of ESS; otherwise, the mitigation of power fluctuation in long time scale should depend on both of the power and the energy capacity of ESS.

● The interconnected microgrid scheme could shift the local loads in spatial scale and balance power among microgrids. Therefore, the power fluctuation will be smoothed and both of the ESS power and energy capacity will be optimized significantly.

2990

V. SUMMARY

To optimize the capacity of ESS which is applied in smoothing the microgrid power fluctuations, a novel flexible interconnection scheme is proposed in this paper. By power balancing among microgrids, the effectiveness of power fluctuation mitigation via flexible interconnection is demonstrated. The complementarity of net power curves among different microgrids will improve the smoothing performance in this scheme. On the basis of DFT algorithm used in ESS capacity sizing, two cases are studied: the independent microgrid scheme and the interconnected microgrid scheme. The analysis results indicate that, in comparison to the existing independent microgrid scheme, the interconnected microgrid scheme will optimize both of the ESS power and energy capacity in microgrids.

In the scenario of optimizing the capacity of ESS in microgrids, the practical application value and benefits of flexible interconnection scheme for microgrids is demonstrated in this paper. In the future work, the key issues to realize the flexible interconnection will be studied intensively.

REFERENCES

[1] Farid Katiraei, Julio Romero Aguero. Solar PV Integration Challenges[J]. IEEE Power and Energy Magazine, 2011, 9(3): 62-71.

[2] Jovan Bebic, Reigh Walling, Kathleen O'Brien, et al. The sun also rises[J]. IEEE Power and Energy Magazine, 2009, 7(3): 45-54.

[3] J. E. S. de Haan, J. Frunt, W. L. Kling. Mitigation of wind power fluctuations in smart grids[C]// Innovative Smart Grid Technologies Conference Europe, 11-13 Oct. 2010, Gothenberg, Sweden: 8p.

[4] Weihao Hu, Zhe Chen, Yue Wang, et al. Wind power fluctuations mitigation by DC-Link voltage control of variable speed wind turbines[C]// Universities Power Engineering Conference, 1-4 Sept. 2008, Padova, Italy: 5p.

[5] Pamela MacDougall, Cor Warmer, Koen Kok. Mitigation of wind power fluctuations by intelligent response of demand and distributed generation[C]// Innovative Smart Grid Technologies, 5-7 Dec. 2011, Manchester, UK: 6p.

[6] Shuaixun Chen, Hoay Beng Gooi, MingQiang Wang. Sizing of energy storage for microgrids[J]. IEEE Transactions on Smart Grid, 2012, 3(1):142-151.

[7] Mike Kleinberg, Jessica Harrison, Niloufar Mirhosseini. Using energy storage to mitigate PV impacts on distribution feeders[C]// Innovative Smart Grid Technologies Conference, 19-22 Feb. 2014, Washington, DC, USA: 5p.

[8] Jeffrey M. Bloemink, Timothy C. Green. Increasing distributed generation penetration using soft normally-open points[C]// Power and Energy Society General Meeting, 25-29 July 2010, Providence, RI, USA: 8p.

[9] Wanyu Cao, Jianzhong Wu, Nick Jenkins, et al. Benefits analysis of Soft Open Points for electrical distribution network operation[J]. Applied Energy, 2016, 165:36-47.

[10] Chee Wei Tan, Tim C. Green, Carlos A. Hernandez-Aramburo. A Stochastic Simulation of Battery Sizing for Demand Shifting and Uninterruptible Power Supply Facility[C]// Power Electronics Specialists Conference, 17-21 June 2007, Orlando, FL, USA: 7p.

[11] Vrana T K, Beerten J, Belmans R, et al. A classification of DC node voltage control methods for HVDC grids[J]. Electric Power Systems Research, 2013, 103(8):137-144.

[12] Xiaoyu Wang, Meng Yue, Eduard Muljadi, et al. Probabilistic Approach for Power Capacity Specification of Wind Energy Storage Systems[J]. IEEE Transactions on Industry Applications, 2013, 50(2): 1215-1224.

[13] K. Yoshimoto, T. Nanahara, G. Koshimizu. Analysis of data obtained in demonstration test about battery energy storage system to mitigate output fluctuation of wind farm[C]// Integration of Wide-Scale Renewable Resources Into the Power Delivery System, 29-31 July 2009, Calgary, AB, Canada: 6p.

[14] Robert B. Bass, Jennifer Carr, José Aguilar, et al. Determining the Power and Energy Capacities of a Battery Energy Storage System to Accommodate High Photovoltaic Penetration on a Distribution Feeder[J]. IEEE Power and Energy Technology Systems Journal, 2016, 3(3):119-127.

[15] Jun Xiao, Linquan Bai, Linquan Bai, et al. Sizing of Energy Storage and Diesel Generators in an Isolated Microgrid Using Discrete Fourier Transform (DFT)[J]. IEEE Transactions on Sustainable Energy, 2014, 5(3):907-916.

VSC Control and Parameters Design Based on Virtual Synchronous Generator

Fang Liu[1*], Meng Wang[1], Zhen Xie[1], Fusheng Wang[1], Jinxin Deng[1] and Xing Zhang[1]

1 School of Electrical Engineering and Automation, Hefei University of Technology, Hefei, China

*E-mail: fragcelau@hfut.edu.cn

Abstract— To cope with the problems that the VSC lacks inertia and synchronous generator controls slowly, the paper combines the advantages of VSC and synchronous generator together to present a control scheme of virtual synchronous generator and a parameters design method. To solve the mutual influence of droop and virtual damping coefficients, a method of combining output frequency and line frequency feedback control is proposed to decouple the two coefficients regulation process while without using the differentiation. Mode switching between constant power and droop control under frequency large disturbance in large scale based on frequency hysteresis control is proposed to maintain system stability. The switching process is realized only through the outer power loop while the inner loop stays the same. At last, a quasi-synchronous-generator parameters design method is established combining the flexible and rapid control characteristics of VSC and the synchronous generator unit's self-droop and self-synchronizing characteristics. The simulation result verified the correctness of theoretical analysis and control scheme.

Keywords— *Decoupling control, quasi-synchronous-generator design, virtual synchronous generator.*

I. INTRODUCTION

In VSC control, the technology of virtual synchronous generator (VSG) [1-2] shows excellent control performance in both grid-connected control mode and islanding control mode. Several key control objectives of VSG-based VSC are: (1) regulate voltage frequency and voltage of grid in grid-connected operation, maintain grid stability, and keep accurate power sharing among VSCs; (2) output constant power when the grid frequency fluctuates violently to exceed limitation, keep stable parallel operation of VSG, and provide frequency and voltage support for the grid.

When VSG-based VSC operates in a grid-connected mode, it can provide virtual inertia, virtual damping, virtual impedance and droop characteristics for the grid to support changes in grid frequency and voltage [3-6]. Literature [4] analyzes distribution and oscillation of instantaneous active power to provide a virtual stator reactance regulator and compensate the voltage drop caused by stator reactance regulator via PCC voltage prediction. Literature [5] introduces design method of virtual damping. Literature [6] presents the evaluation

method for stability of multiple VSGs, and regulates system parameters via particle swarm optimization algorithm and variable virtual inertia. Literature [7] presents an adaptive virtual damping and inertia control scheme to improve stability of frequency. Above literature can improve frequency stability of system. However, the virtual damping does not only influence the frequency stability in dynamic process, but also influences the frequency droop characteristics and the control between the two is not independent.

Literature [8-10] discuss the problem of synchronization stability of power electronic converter under large-disturbance motion. Literature [8] suggests that current limitation in droop control will become asynchronous under large-disturbance and lead to system instability. Literature [2, 11-12] provide VSG parameters design method based on the overall characteristics of synchronous generator unit. Literature [11] shows a quantitative analysis on the influence of perturbation of system model parameters on power tracking of connected grid, and analyzes the setting method for inertia and damping parameters. Literature [12] uses root locus to design VSG closed loop parameters. They both need to take iterative approach to adjust the parameters and get the final optimization result. Literature [2] presents design method of parameters decoupling of outer active and reactive power loop, which can rapidly and accurately get the design parameters without need for iterative approach. Above parameters design methods follow the parameters design idea of synchronous generator which have large moment of inertia and make dynamic response of VSC slowly. Large moment of inertia can make system stable when equivalent output impedance of VSC is large [4, 13]. If equivalent output impedance is small, the value of corresponding moment of inertia should not be too large [13]. Nevertheless, if output impedance is too large, the voltage drop is severe when VSG carries inductive load. Literature [4] combines stator reactance regulator and bus voltage estimator to present a control method of improving output voltage, yet with the need of increasing DC voltage accordingly.

To cope with the problems, this paper provides a control parameters design method and will be discussed in the next sections.

II. CONTROL SCHEME OF VIRTUAL SYNCHRONOUS GENERATOR

The topology of high-power VSC is shown in Fig. 1. VSC is connected to power grid and diesel generator though three-phase half-bridge inverter, LC-filter and Dyn11 transformer.

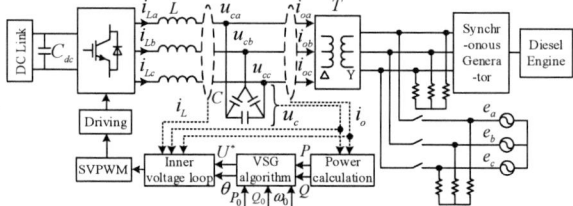

Fig. 1. Topology diagram of VSC

Where, C_{dc} is capacitor of DC side, u_{ca}, u_{cb}, u_{cc} is output capacitor voltage, e_a, e_b, e_c is line voltage at PCC point, i_{La}, i_{Lb}, i_{Lc} is inductive current of VSC-side inductive current , i_{oa}, i_{ob}, i_{oc} is VSC current output , T is transformer, L is VSC-side filter inductance, C is filter capacitor.

The paper adopts VSG-based control scheme, whose basic control equation includes outer power loop [14] and double voltage loop. The outer power loop mainly mimics equation of power angle droop, rotor motion equation and excitation equation. The double voltage loop mainly mimics stator electro-magnetic equation which can be found in [13].

The average active and reactive power are calculated according to output voltage and output current of VSC to establish self-synchronized VSG magnitude and phase. The control block diagram is shown in Fig. 2.

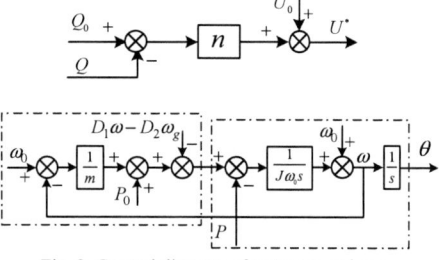

Fig. 2. Control diagram of outer power loop

Where, ω and U^* are respectively angular frequency and voltage magnitude reference, θ is VSC output phase, P, Q are average active and reactive power respectively, ω_0 is angular frequency when MI's active power is P_0, U_0 is output voltage magnitude when MI's reactive power is Q_0, m and n mean $P-\omega$ and $Q-U^*$ droop coefficients respectively, J is virtual inertia. D_1 is VSC frequency feedback coefficient of VSC, D_2 is line frequency feedback coefficient. According to [13, 15], the combination $J\omega_0 m$ in VSG is equal to filtering time constant τ in droop control. While to avoid the large voltage drop of system under reactive power load the

output impedance should be properly decreased. The decrease of equivalent output impedance lowers of the value of virtual inertia accordingly [13]. Now the low-pass filter will be unable to completely filter the low-order harmonic in power. Thus the paper proposes a power calculation method based on notch filter shown as follows:

$$P = (\prod_h \frac{s^2 + \omega_h^2}{s^2 + 2Q_h\omega_h s + \omega_h^2}) \cdot \frac{1.5}{\tau s + 1} \cdot (U_{cq}I_{oq} + U_{cd}I_{od})$$

$$Q = (\prod_h \frac{s^2 + \omega_h^2}{s^2 + 2Q_h\omega_h s + \omega_h^2}) \cdot \frac{1.5}{\tau s + 1} \cdot (U_{cd}I_{oq} - U_{cq}I_{od})$$

(1)

Where, Q_h is Q factor of h-order resonance controller, ω_h is angular frequency of notch filter, τ is time constant of first-order low-pass filter.

Because of the multi-notch filters, the time constant of first-order low-pass filter can be set much smaller and provide fast dynamic response.

For outer power loop, when the large disturbance occurs to line frequency, VSC should adopt constant power control to avoid overcurrent. While the control strategy based on current limitation control will make the system unstable under large disturbances [8].The paper adopts mode switching scheme between constant power and droop control under frequency large disturbance in large scale based on frequency hysteresis control. The control law can be described as follows:

In normal operation, i.e. when $\omega_g > \omega_{g1}$, VSC adopts the control method as shown in Fig. 2. When line frequency falls, i.e. $\omega_g < \omega_{g1}$, assuming $\omega_0 = \omega_g, D_1 = D_2$.

$$\omega = \omega_g + \frac{m}{mJ\omega_0 s + 1 + mD_1}(P_0 - P)$$

(2)

Where, ω_{g1} is low threshold of line frequency, ω_{g2} is high return value of line frequency, $\omega_{g1} < \omega_{g2}$.

When the system is in normal operation, the outer power loop adopts droop control, whose characteristic curve is shown in Fig. 3.

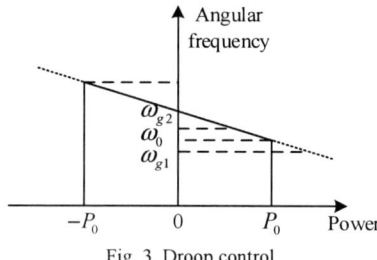

Fig. 3. Droop control

According to Fig. 3, the positive power represents that the inverter delivers power, like the process of power generation by renewable energy such as PV and wind. When the system frequency is too large, the value of power becomes negative, and the system no longer transmits power to grid but absorbs the power from the grid.

The droop characteristic curve as shown in Fig.3 shows that when $\omega_g < \omega_{g1}$, the outer power loop operates

2993

in the mode of constant power. The power reference at this moment is set as rated power P_0; when $\omega_g > \omega_{g1}$, the outer power loop is switched to droop control, and the power reference of droop control is still set as rated power P_0. Although the control modes before and after switching are different, and the processes of achieving steady state are different, the output power in steady state is the same.

III. ANALYSIS ON CHARACTERISTICS OF VIRTUAL SYNCHRONOUS GENERATOR

The equivalent model for outer frequency loop is derived as shown in Fig. 4 according to Fig. 2, where, $K_s = 3EU/X$, E is the effective value for phase voltage of grid, X is equivalent output impedance for each phase of inverter.

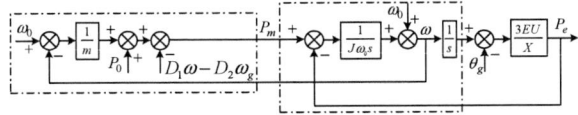

Fig. 4. Equivalent mathematical model of outer loop

According to Fig. 4, it is derived that the mathematical equation for power and frequency is expressed as:

$$P = \frac{mK_s}{Jm\omega_0 s^2 + (1 + mD_1)s + mK_s}P_0 + \frac{K_s(Jm\omega_0 s + 1)}{Jm\omega_0 s^2 + (1 + mD_1)s + mK_s}\omega_0$$
$$- \frac{K_s(J\omega_0 ms + 1 + mD_1 - mD_2)}{Jm\omega_0 s^2 + (1 + mD_1)s + mK_s}\omega_g \quad (3)$$

$$\omega = \frac{ms}{Jm\omega_0 s^2 + (1 + mD_1)s + mK_s}P_0 + \frac{s(Jm\omega_0 s + 1)}{Jm\omega_0 s^2 + (1 + mD_1)s + mK_s}\omega_0$$
$$+ \frac{mK_s + sD_2 m}{Jm\omega_0 s^2 + (1 + mD_1)s + mK_s}\omega_g$$

Assuming $s = 0, D_1 = D_2$ in equation (3) to conclude that when the system is in steady state,

$$P = P_0 + \frac{1}{m}\omega_0 - \frac{1}{m}\omega_g, \omega = \omega_g \quad (4)$$

Equation(4) shows that the frequency droop characteristic of the system is only related to droop coefficient m and unrelated to parameters D_1, D_2.

According to(3), the damping ratio and oscillation frequency of the system can be obtained.

$$\zeta = \frac{(\frac{1}{m} + D_1)}{2\sqrt{J\omega_0 K_s}}, \omega_n = \sqrt{\frac{K_s}{J\omega_0}} \quad (5)$$

According to(5), the damping ratio and oscillation frequency of system can be adjusted via adjusting parameters D_1, X, J while frequency droop coefficient m remains constant. Increasing damping parameter D_1 can increase system damping, and have no influence on oscillation frequency. Increasing equivalent output impedance X can increase system damping and reduce oscillation frequency. Increasing virtual inertia J can decrease system damping but reduce oscillation frequency.

IV. PARAMETERS DESIGN FOR VIRTUAL SYNCHRONOUS GENERATOR

A. Design Principles for Parameters of Droop Coefficients

Frequency droop and voltage droop coefficients are designed according to normalization method based on inverter parameters of different rated power, and conforms to standard of VSCs' grid-connected and islanding parallel operation. The design principle for droop coefficient m of power angle control is that the change of frequency is $\Delta\omega$ when active power changes ΔP. Similarly, the design principle for droop coefficient n of reactive control is that the change of voltage magnitude is ΔU when reactive power changes ΔQ.

$$m = \Delta\omega / \Delta P, n = \Delta U / \Delta Q \quad (6)$$

The angular frequency is set as rated angular frequency of inverter when rated line frequency is 50Hz, i.e. $\omega_0 = 314.1593\text{rad/s}$. VSC's rated power is 500KVA, and rated phase-phase voltage magnitude is 315V. The change in frequency is 1%, and the change in voltage is 2%. The droop coefficient of power angle control is set as $m = \frac{1\% \times 50 \times 2\pi}{500}$rad/KW=0.0063rad/KW , and droop coefficient of reactive control is set as $n = \frac{2\% \times 315}{500}$ V/KVar=0.0126V/KVar .

B. Design Principle for Parameters of Virtual Inertia, Virtual Damping and Impedance

According to(5), the calculation equations for virtual inertia and damping coefficient of system are

$$J = \frac{K_s}{\omega_0 \omega_n^2}$$
$$D_1 = 2\zeta\sqrt{J\omega_0 K_s} - \frac{1}{m}, D_2 = D_1 \quad (7)$$

According to literature [15], there exists a certain relationship between virtual inertia J and time constant τ of first-order low pass filter in droop control, that is, $\tau = J\omega_0 / (1/m + D_1)$. The stability of the VSCs under parallel operation is greatly influenced by this parameter [16-17], especially when switching frequency of high-power VSC is low, and the control delay time is long. In order to avoid the instability caused by this parameter, the paper proposes a quasi-synchronous-generator parameters design method, i.e. designing natural oscillation frequency of outer power loop as about 100 Hz, then equivalent filtering time is in ms grade. Assuming $\omega_n = 2\pi f_n$, where $f_n = 100Hz$, and VSC equivalent output impedance being 5% of rated impedance, then K_s can be set to $K_s \approx 20 \times 500000\text{W}$.

According to(7), $J = 0.08kg \cdot m^2$. If the expected damping ratio is $\zeta = 0.707$, $D_1 = D_2$ =-136313.58 can be calculated though(7). This shows the damping coefficient is positive-negative adjustable, which adds a degree of freedom to the system, and also meets requirements in dynamic response and stability without an influence on the power distribution in steady state.

V. SIMULATION VERIFICATION

A. Simulation Platform Introduction

In order to verify the feasibility and effectiveness of quasi-synchronous generator control scheme proposed in this paper, the simulation platform shown in Fig. 1 is built in MATLAB/Simulink environment. The simulation platform consists of a VSG system that connected to diesel generators and power grids. Setting the per unit values of P_0 and Q_0 to 2.9×10^7 and 0, respectively. Other main parameters are the same as those listed in Table I.

TABLE I
MAIN PARAMETERS OF THE 500KW-VSG PROTOTYPE

Symbol	Meaning	Value
L	Filter induction	0.05mH
C	Filter capacitance	$1200 \, \mu F$
T	Transformer	315/25000V
U_{dc}	Voltage of DC side	650V
k_{vp}	Proportional coefficient in voltage loop	2
k_{vi}	Integral coefficient in voltage loop	200
k_{ip}	Proportional coefficient in current loop	0.2
ω_{g1}	The low threshold of ω_g	311.02rad/s
ω_{g2}	The high return value of ω_g	317.30rad/s
f_{sw}	Switching frequency	3KHz
f_s	Sampling frequency	500KHz

B. Simulation Results

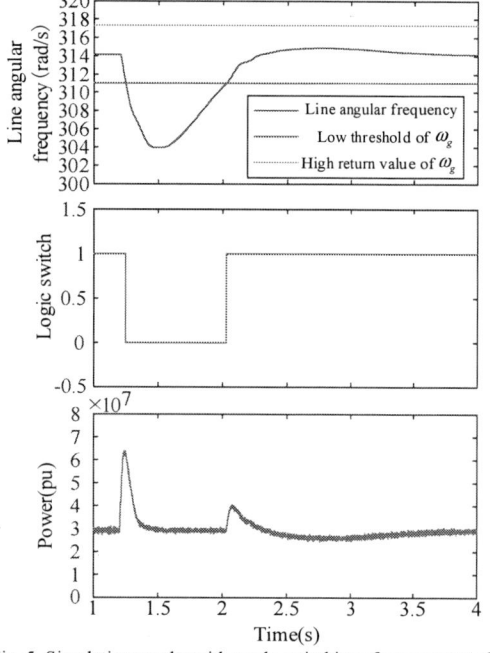

Fig. 5. Simulation results with mode switching of power control

Assuming the virtual inertia $J = 0.03 kg \cdot m^2$, and opening the breaker linking VSC and grid at 1.2s. The system simulation results are shown in Fig. 5.

Fig. 5 illustrates the dynamic performance of active power during line frequency drop at 1.2s. The logic value switches from 1 to 0 due to that the line frequency drops beyond the low threshold at $t = 1.25s$, and the outer power loop begins to adopt constant power control. The output active power is controlled at the set value P_0. At 2.03s, the outer power loop switches to droop control mode while the logic value returns to 1. Although these two control modes are different, the output active powers in steady state are both the set value P_0.

When the breaker is set to be opened at t=2.5s, the VSC system dynamic response with different virtual inertia and damping is shown in Fig. 6.

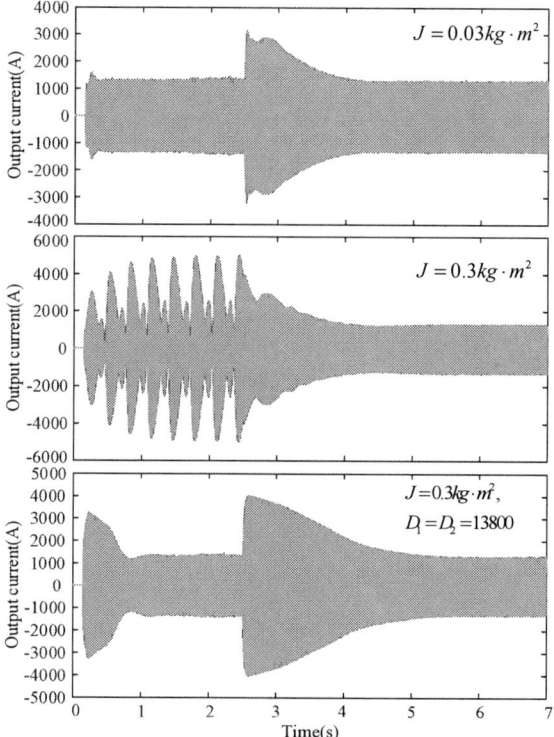

Fig. 6. Dynamic response of VSC output current

As it is illustrated in Fig. 6, the system will be unstable when the virtual inertia is increased. However, the system turns into stable when the value of virtual damping is set properly. Moreover, there exists a corresponding relationship between virtual inertia and damping coefficient according to (7).Thus the virtual damping can be designed based on a definite virtual inertia. Fig. 6 implies that the control method of output frequency and line frequency combined feedback proposed in this paper can greatly improve the system stability and control performance of the system without an influence on droop control.

VI. CONCLUSION

In this paper, a virtual synchronous generator control scheme and a parameters design method are proposed. The proposed control scheme and parameters design method are verified by simulation, and the following conclusions are obtained:(1)When the line frequency exceeds the fluctuation range, the outer power loop operates in constant power control, which effectively reduce overcurrent. But the larger power fluctuation at the moment of mode switching of power control is still worth further discussing and researching;(2)Unlike the SG unit with large inertia, properly parameters design using can maintain system stability, fast dynamic response at the same time;(3)The control method of output frequency and line frequency combined feedback realizes the decoupling control of droop and virtual damping coefficients while without using the differentiation ;(4)When the system is unstable due to excessive virtual inertia, the damping coefficient can be adjusted to stabilize the system.

REFERENCES

[1] J. Liu, Y. Miura, and T. Ise, "Comparison of Dynamic Characteristics between Virtual Synchronous Generator and Droop Control in Inverter-Based Distributed Generators," *IEEE Transactions on Power Electronics*, vol. 31, no. 5, pp. 3600-3611, 2016.

[2] H. Wu, X. Ruan, D. Yang, X. Chen, W. Zhao, Z. Lv, *et al.*, "Small-Signal Modeling and Parameters Design for Virtual Synchronous Generators," *IEEE Transactions on Industrial Electronics*, vol. 63, no. 7, pp. 4292-4303, 2016.

[3] T. Zheng, L. Chen, R. Wang, C. Li, and S. Mei, "Adaptive damping control strategy of virtual synchronous generator for frequency oscillation suppression," in *12th IET International Conference on AC and DC Power Transmission (ACDC 2016)*, 2016, pp. 1-5.

[4] J. Liu, Y. Miura, H. Bevrani, and T. Ise, "Enhanced Virtual Synchronous Generator Control for Parallel Inverters in Microgrids," *IEEE Transactions on Smart Grid*, vol. 8, no. 5, pp. 2268-2277, 2017.

[5] T. Shintai, Y. Miura, and T. Ise, "Oscillation Damping of a Distributed Generator Using a Virtual Synchronous Generator," *IEEE Transactions on Power Delivery*, vol. 29, no. 2, pp. 668-676, 2014.

[6] J. Alipoor, Y. Miura, and T. Ise, "Stability Assessment and Optimization Methods for Microgrid with Multiple VSG Units," *IEEE Transactions on Smart Grid*, vol. 9, no. 2, pp. 1462-1471, 2018.

[7] D. Li, Q. Zhu, S. Lin, and X. Y. Bian, "A Self-Adaptive Inertia and Damping Combination Control of VSG to Support Frequency Stability," *IEEE Transactions on Energy Conversion*, vol. 32, no. 1, pp. 397-398, 2017.

[8] H. Xin, L. Huang, L. Zhang, Z. Wang, and J. Hu, "Synchronous Instability Mechanism of P-f Droop-Controlled Voltage Source Converter Caused by Current Saturation," *IEEE Transactions on Power Systems*, vol. 31, no. 6, pp. 5206-5207, 2016.

[9] H. Lin, C. Jia, J. M. Guerrero, and J. C. Vasquez, "Angle Stability Analysis for Voltage-Controlled Converters," *IEEE Transactions on Industrial Electronics*, vol. 64, no. 8, pp. 6265-6275, 2017.

[10] L. Huang, H. Xin, L. Zhang, Z. Wang, K. Wu, and H. Wang, "Synchronization and Frequency Regulation of DFIG-based Wind Turbine Generators with Synchronized Control," *IEEE Transactions on Energy Conversion*, vol. 32, no. 3, pp. 1251-1262, 2017.

[11] Z. Lu, W. Sheng, Q. Zhong, H. Liu, Z. Zeng, L. Yang, and L. Liu, "Virtual synchronous generator and its applications in micro-grid," *Zhongguo Dianji Gongcheng Xuebao/Proceedings of the Chinese Society of Electrical Engineering*, vol. 34, no. 16, pp. 2591-2603, 2014.

[12] Y. Du, J. M. Guerrero, L. Chang, J. Su, and M. Mao, "Modeling, analysis, and design of a frequency-droop-based virtual synchronous generator for microgrid applications," in *2013 IEEE ECCE Asia Downunder*, 2013, pp. 643-649.

[13] F. Liu, Research on microgrid inverter control strategy based on virtual synchronous generator [D]. Hefei: Hefei University of Technology, 2015.

[14] P. Kundur, N. J. Balu, and M. G. Lauby, *Power system stability and control*. Vol. 7. New York: McGraw-hill, 1994.

[15] S. D'Arco and J. A. Suul, "Equivalence of Virtual Synchronous Machines and Frequency-Droops for Converter-Based MicroGrids," *IEEE Transactions on Smart Grid*, vol. 5, no. 1, pp. 394-395, 2014.

[16] E. A. A. Coelho, P. C. Cortizo, and P. F. D. Garcia, "Small-signal stability for parallel-connected inverters in stand-alone AC supply systems," *IEEE Transactions on Industry Applications*, vol. 38, no. 2, pp. 533-542, 2002.

[17] H. Jinwei and L. Yun Wei, "Analysis, Design, and Implementation of Virtual Impedance for Power Electronics Interfaced Distributed Generation," *IEEE Transactions on Industry Applications*, vol. 47, no. 6, pp. 2525-2538, 2011.

Multi-Target Virtual Resistance Control Strategy in a 400 Hz Low Voltage Microgrid

Yuze Li, Xuejun Pei, Zhi Chen, Hanyu Wang, Yong Kang
State Key Laboratory of Advanced Electromagnetic
Engineering and Technology
School of Electrical and Electronic Engineering
Huazhong University of Science and Technology
Hubei, China
lyzkmust@126.com

Abstract-This paper has proposed a multi-target virtual resistance strategy which can effectively realize targets of unbalanced voltage compensation and unbalanced current (power) sharing among inverters in 400 Hz low voltage microgrid. For unbalanced voltage compensation function, the strategy employs virtual resistance based on unbalanced voltage degree to compensate unbalanced voltage. For unbalanced current sharing function, the strategy uses current sharing degree and virtual resistance to share unbalanced current among inverters in parallel system. The experimental results of two inverters in parallel system are shown to prove the validity of proposed strategy. In addition, this strategy reduces large calculation and is easy to implement.

Keywords— virtual resistance; microgrid; parallel system

I. Introduction

400Hz power systems are often used in aircraft for lighter weight and less cost. Aviation static inverter is important in the power system. Usually the load connected to system is not balanced, so the inverter should have the ability to maintain the balance of load voltage. Traditionally, inverter with △/Y transformer has the ability to maintain the balance of load voltage, but due to its large size and high weight, it is not suitable in aircraft application [1]. Some researchers proposed three-phase inverter with two splitting capacitors at DC link [2]. But the added capacitors should be large enough to guarantee the balance of neutral point. In recent years, three-phase four-leg inverter has more applications [3]-[4]. Compared to three-phase three-leg inverter, it increases a leg to form the neutral line which can flow through zero-sequence current.

Parallel power system can increase capacity of microgrid. When unbalance loads are connected to system, PCC (Point of common coupling) voltage will become unbalanced. To meet grid standard, the unbalanced voltage degree should be limited. Some researchers use negative voltage and negative reactive power to compensate unbalanced voltage at PCC [5]-[6]. But this method needs more voltage sensors to detect PCC voltage, and this method needs a lot of calculation. In parallel system, the power sharing (current sharing) is the another focus. Virtual resistance is employed to improve current sharing performance [5]-[8]. However,

increasing virtual resistance will bring more line voltage drop and higher unbalanced voltage degree of PCC. In this paper a simpler strategy which can compensate unbalanced voltage and share unbalance current without using extra voltage sensors is presented.

II. Inverter Topology And Traditional Distributed Logic Control

A. Topology and Control Method of Single Inverter System

In this paper, three-phase four-leg voltage source inverter (VSI) topology is adopted. Single inverter system with load connection is shown in Fig. 1. The fourth leg can be a neutral wire which can flow through zero sequence current in the three-phase four-wire power system. For four-leg inverter, proportional-resonant (PR) controller as shown in (1) have many advantages, but it cannot fit well when voltage frequency changes [4]. So quasi proportional-resonant controller as shown in (2) is more popular. In this paper, the voltage controller for four-leg inverter is quasi proportional-resonant (PR) controller under αβ0 coordinate system. Fig.2 shows the voltage control strategy for three-phase four-leg inverter under αβ0 coordinate system.

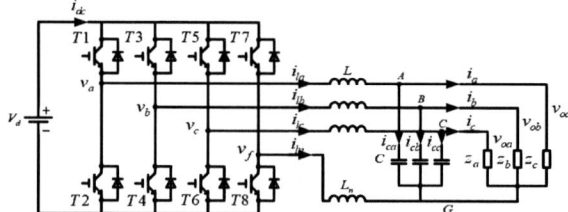

Fig. 1. Three-phase four-leg inverter topology.

$$G_{PR}(s) = k_p + \frac{k_r s}{s^2 + \omega_0^2} \tag{1}$$

$$G_{PR}(s) = k_p + \frac{k_r \omega_c s}{s^2 + 2\omega_c s + \omega_0^2} \tag{2}$$

In (1), k_p is proportional coefficient, and k_r is resonant coefficient. ω_0 is fundamental frequency. In (2), ω_c is cut-off frequency.

Project Supported by the National Natural Science Foundation of China (51577079)

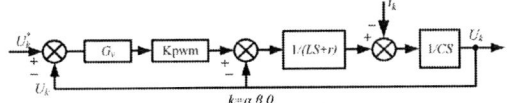

Fig. 2. Voltage control strategy.

B. Distributed Logic Control Strategy of Parallel System

The parallel power system includes two voltage source inverters and uses distributed logic control to share power among inverters. The principle of distributed logic control strategy and distributed logic controller are shown in Fig.3 and Fig.4. Each inverter needs to calculate itself powers and send them to power buses. Then comparing calculated powers with references, and the error of power can be reduced by PI controller. The outputs of PI

controllers are used as voltage reference and phase reference which are added to voltage controller. If one inverter has more active and reactive power output, the distributed logic controller can decrease the amplitude of this inverter's output voltage and change the phase of output voltage. Then the active and reactive power of this inverter can be decreased. In this way, active power and reactive power can be shared among inverters. If unbalance loads connected to PCC, there will be positive, negative and zero sequence current in the system, inverters need more communication lines to exchange unbalance current information for current sharing. So in this paper, distributed logic control is only used to share positive current among inverters.

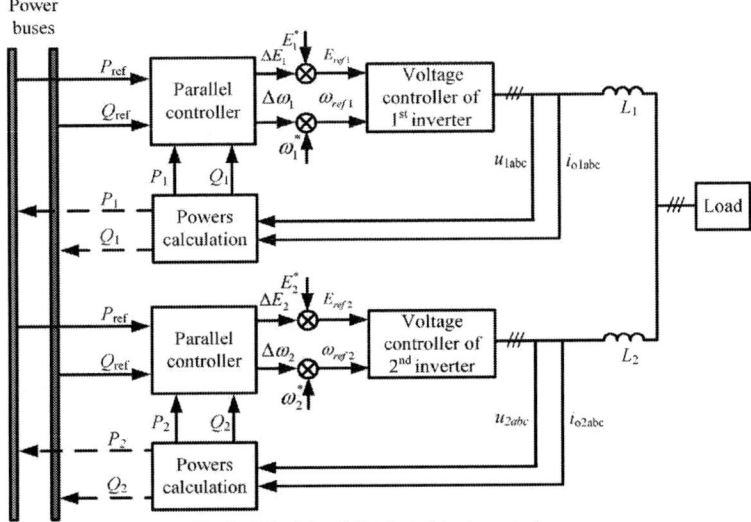

Fig. 3. Principle of distributed logic control

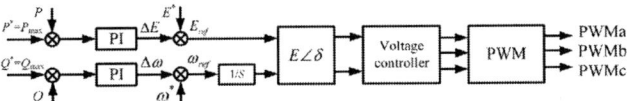

Fig. 4. Distributed logic controller

III. MULTI-TARGET VIRTUAL RESISTANCE CONTROL STRATEGY

A. Unbalanced Voltage Compensation Control Module

Parallel system impedance Z_s includes inverters' output impedance (Z_{out}) and line impedance (Z_{line}). Line impedance contains resistance and inductance.

$$Z_s = Z_{out} + Z_{line} \qquad (3)$$

Where $Z_{line} = R_{line} + jX_{line}$.

It is quite clear that in low voltage microgrid, line impedance mainly consists of resistance [9]. So in this paper, line inductance is ignored.

In IEEE 936-1987 standard, the definition of unbalanced voltage degree is shown as (4), a virtual negative resistance control module for voltage drop decrease and unbalanced voltage compensation is proposed.

$$m_i = \frac{\max(V_{a,i}, V_{b,i}, V_{c,i}) - \min(V_{a,i}, V_{b,i}, V_{c,i})}{(V_{a,i}, V_{b,i}, V_{c,i})/3} \times 100\% \ (i=1,2)$$

$$(4)$$

Parallel system's impedances of positive, negative and zero sequence are in (5). $m_i \times R_{v1,k,i}^{+,-,0}$ is the virtual resistance. The compensation of unbalanced voltage target is $m_i \leq 1.5\%$. The virtual resistance increases, as the unbalanced degree increases, or vice versa.

$$Z_{s,k,i}^{+,-,0} = Z_{out,i}^{+,-,0} + R_{line,k,i}^{+,-,0} + m_i \times R_{v,k,i}^{+,-,0} + jX_{line,k,i}^{+,-,0} \ (k=\alpha,\beta,0 \ i=1,2)$$

$$(5)$$

B. Unbalanced Current Sharing Control Module

Positive, negative and zero sequence equivalent circuits of two four-leg inverters in parallel system are shown in Fig. 5.

From Fig. 5 it is clear that the sharing effect of positive current depends on both inverters' output voltage and system impedance, while negative and zero sequence current sharing only depend on system impedance. According to this mechanism, a virtual resistance control module for unbalanced current sharing is proposed. The definition of the degree of sharing current among two inverters in parallel system is shown in (6)

$$n_{k,i}^{+,-,0} = \frac{i_{k,i}^{+,-,0}}{i_{k,3-i}^{+,-,0}} \times 100\% \quad (k = \alpha, \beta, 0 \ \ i=1,2) \quad (6)$$

And virtual resistance for unbalanced current sharing is $n_{k,i}^{+,-,0} \times R_{v,k2,i}^{+,-,0} (k = \alpha, \beta, 0 \ \ i=1,2)$ .The target of current sharing is $85\% \le n_{k,i}^{+,-,0} \le 115\%$ if inverters in parallel have same rated power. The positive current sharing method consists of distributed logic control and virtual resistance method, while negative and zero sequence current sharing method only use virtual resistance method.

Combining the two control methods proposed in section A and B, the whole system impedance is shown in (7). And the multi-target virtual resistance is $m_i \times R_{v1,k,i}^{+,-,0} + n_{k,i}^{+,-,0} \times R_{v2,k,i}^{+,-,0}$.

$$Z_{s,k}^{+,-,0} = Z_{out,k,i}^{+,-,0} + R_{line,k,i}^{+,-,0} + m_i \times R_{v1,k,i}^{+,-,0} + n_{k,i}^{+,-,0} \times R_{v2,k,i}^{+,-,0}$$
$$+jX_{line,k,i}^{+,-,0} \quad (k = \alpha, \beta, 0 \ \ i=1,2) \quad (7)$$

The whole parallel system control strategy is shown in Fig.6. Inverters collect voltage and current information at PCC and calculate the unbalanced degree m_i. Then add the product of current and multi-target virtual resistance to voltage control loop as a part of voltage reference. The voltage reference consists of three parts, one is rated voltage reference, the other is the output of distributed logic controller, the last one is the product of virtual resistance and current.

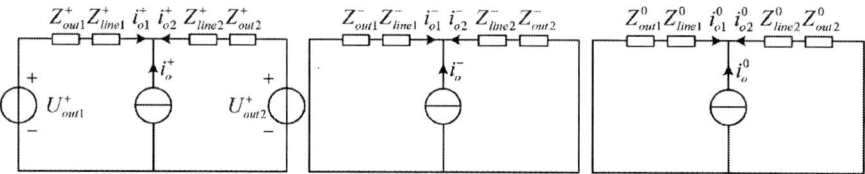

Fig. 5. Positive, negative and zero sequence equivalent circuits of parallel system.

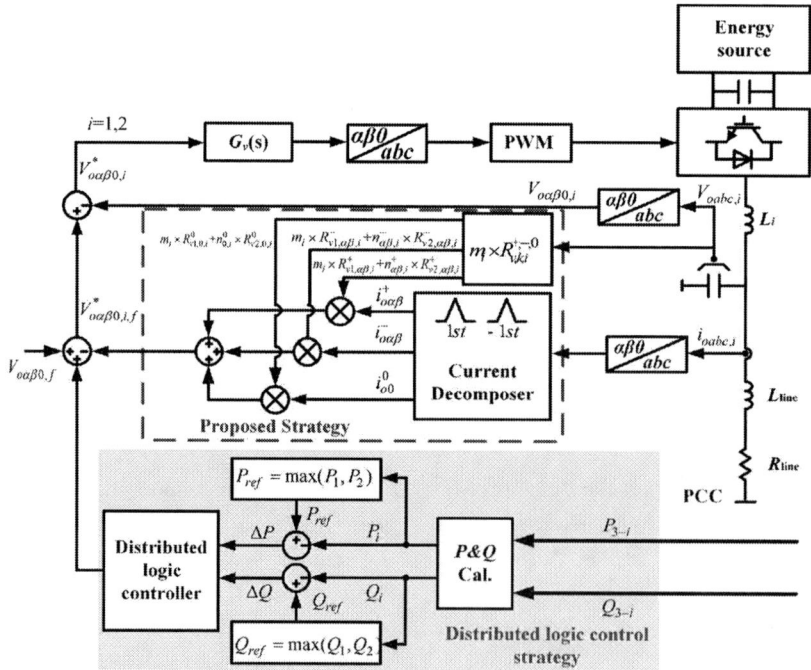

Fig. 6. Block diagram of proposed control strategy for 400Hz low voltage microgird.

IV. EXPERIMENTAL RESULTS

To verify the performance of the proposed approach, two 6 kW 115 V/400 Hz inverters are operated in parallel. In this paper, unbalance load is connected to the system. 3 kW load is connected to phase A, while 6 kW loads are connected to phase B and C. The parameters of system are shown in TABLE I.

Fig. 7 shows the waveforms of two inverters without using multi-target strategy. Because of line impedance,

2999

unbalanced current causes voltage drop on line, so three-phase voltages at PCC are unbalanced. Fig. 8 shows the waveforms of two inverters using the proposed strategy. By further calculation, without using strategy, the unbalanced voltage degree is higher than 2%. While by using the proposed strategy, the unbalanced voltage degree drops to 1.3%. In Fig. 7 and Fig. 8 the current sharing degrees are nearly to 100%. That means load power is equally shared between two inverters.

TABLE I
PARAMETERS OF PARALLEL SYSTEM

Parameters	Values
Rated power	6 kW
Rated voltage	115 V
Rated frequency	400 Hz
Filter inductor	270 μH
Filter capacitor	30μF
Three-phase unbalanced load	50%/100%/100%
Line resistance	0.2Ω
Virtual resistance1	-10Ω
Virtual resistance2	0.2Ω

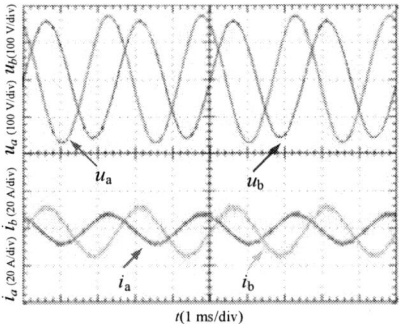 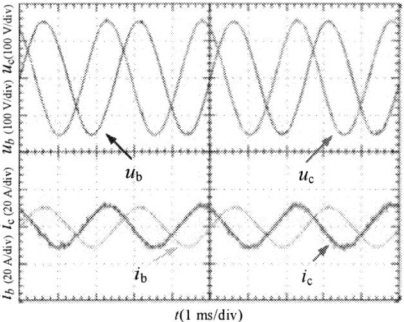

(a)Voltage and current of phase A and B of inverter 1.   (b)Voltage and current of phase B and C of inverter 1.

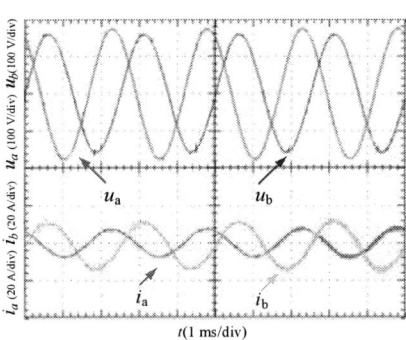 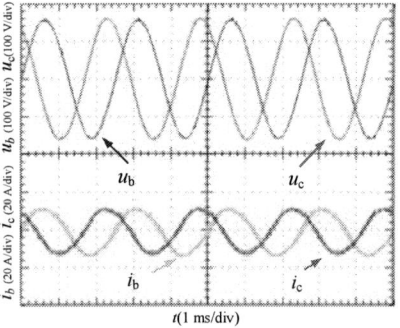

(c)Voltage and current of phase A and B of inverter 2.   (d)Voltage and current of phase B and C of inverter 2.
Fig. 7. Wave forms of two inverters without using the proposed strategy.

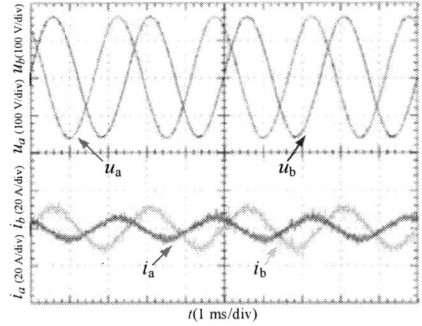 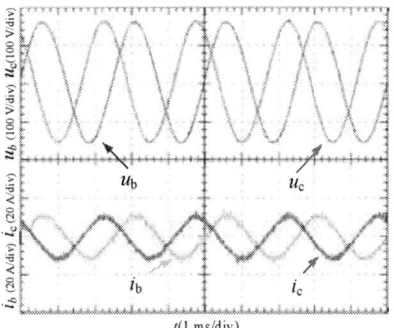

(a)Voltage and current of phase A and B of inverter 1.   (b)Voltage and current of phase B and C of inverter 1.

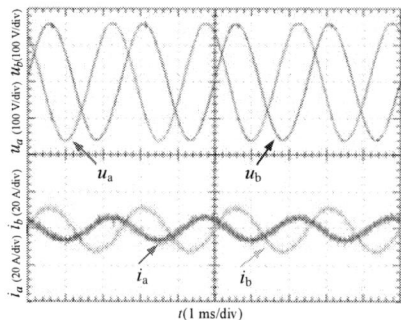

(c)Voltage and current of phase A and B of inverter 2. (d)Voltage and current of phase B and C of inverter 2.
Fig. 8. Wave forms of two inverters using the proposed strategy.

V. CONCLUSION

In 400 Hz low voltage microgrid, loads are sometimes unbalance, and inverter should have the ability to maintain the balance voltage of load. So three-phase four-leg inverter topology is adopted in this paper. The voltage controller of three-phase four-leg inverter is quasi PR controller which can effectively control output voltage under αβ0 coordinate system. To share the power in parallel system, distributed logic control is used. By comparing itself power with power reference, distributed logic controller use the power error to control inverters' output voltage amplitude and phase. So the power can be effectively shared among inverters. The power sharing mechanism of parallel system when unbalance load connected is studied.

Because of line impedance (in low voltage microgrid, resistance is far larger than inductance), unbalanced current generates unbalanced voltage drop on line, the larger impedance, the higher voltage drops. So virtual negative resistance strategy is proposed to compensate unbalanced voltage drop. In order to guarantee the effect of current sharing, the virtual negative resistance should not totally compensate line resistance, so the other virtual resistance for power sharing is added. A two 6 kW 115 V/400 Hz inverters in parallel microgrid system is built in lab. Experimental results verify the proposed strategy.

REFERENCES

[1] Uffe. Burup, Prasad. N. Enjeti, "A New Space Vector based Control Method for UPS System Powering Nonlinear and Unbalanced Loads" *IEEE Trans. Ind. Appl.*, Vol. 37, no. 6, pp. 1864-1870, May. 2001.

[2] Jeong. C. Y, Cho. J. G, "A 1000Kv A Power Conditioner for Three-phase Four-wire Emergency Generators" *IEEE PESC,* Fukuoka, Japan,1998.

[3] A. Mohd, E. Ortjohann, "Control Strategy and Space Vector Modulation for Three-leg Four-wire Voltage Source Inverters under Unbalanced Load Conditions" *IET Power Electron.*, Vol. 3, no. 3, pp. 323-333, May. 2009.

[4] Eyyup Demirkutlu, Ahmet M. Hava, "A Scalar Resonant-Filter-Bank-Based Output-Voltage Control Method and a Scalar Minimum-Switching-Loss Discontinuous PWM Method for the Four-Leg-Inverter-Based Three-Phase Four-Wire Power Supply" *IEEE Trans. Ind. Appl.*, Vol. 45, no. 3, May/June. 2009.

[5] Yang Han, Pan Shen, Xin Zhao, Josep M. Guerrero, "An Enhanced Power Sharing Scheme for Voltage Unbalance and

Harmonics Compensation in an Islanded AC Microgrid" *IEEE Trans. Energy Convers.*, Vol. 31, no. 3, Sep. 2016.

[6] M. Savaghebi, A. Jalilian, J.C. Vasquez, "Autonomous Voltage Unbalance Compensation in an Islanded Droop-Controlled Microgird" *IEEE Trans. Power Electron.*, Vol. 60, no. 4, pp 1390-1402, Apr. 2013.

[7] Xiongfei Wang, Frede Blaabjerg, Zhe Chen, "An Improved Design of Virtual Output Impedance Loop for Droop-Controlled Parallel Three-Phase Voltage Source Inverters" in *IEEE ECCE* 2012, pp 2466-2473.

[8] Xiongfei Wang, Frede Blaabjerg, Zhe Chen, "Autonomous Control of Inverter-Interfaced Distributed Generation Units for Harmonic Current Filtering and Resonance Damping in an Islanded Microgrid" *IEEE Trans. Ind. Appl.*, vol. 50, no. 1, pp. 452-461, Jan /Feb. 2014.

[9] Ping Zhang, Hengyang Zhao, "Power decoupling strategy based on "Virtual Negative Resistor for Inverters in Low-Voltage Microgrids" *IET Power Electron.*, Vol. 9, no. 5, pp.1037-1044, Oct. 2016.

An adaptive power compensation strategy for the voltage stabilization of LCL-VSC based microgrids

Sheng Xu[1], Wu Cao[2], Dongchen Fan[2] ,Jianfeng Zhao[2], Shunyu Wang[2]

Taizhou Electric Power Conversion and Control Engineering Technology Research Center, Taizhou University, Taizhou, China

College of Electrical Engineering, Southeast University, Nanjing, China

xush_00@sina.com

Abstract—In this paper, an adaptive power compensation strategy for the voltage stabilization of LCL-VSC based microgrids is proposed. In this novel method, the load reactive power fluctuation and the reactive power consumption of the microsource output impedances are compensated in real time to adjust the reactive power reference in the droop parameters. First, the voltage and current double closed-loop control system on synchronous rotating frame of the LCL-based microsource inverter is decoupling designed and analyzed systematically. On this basis, the effects of two different feedback voltages on the inverter are studied for the purpose of optimizing the system static and dynamic features of the microsource. Next, via the mathematically modeling of the virtual impedance based droop controller adopted in the microsource, the basic principle of this proposed voltage stabilization method is analyzed in detail. Finally, a simulation experiment is implemented to validate the effectiveness of this new voltage stabilization strategy. The simulation results show that, the two-stage voltage fluctuations caused by the load and the output impedance can be suppressed effectively, and moreover, the smooth transition of the voltage magnitude can be achieved when the microgrids is incorporated into the power networks.

Keywords—*Microgrid; LCL-based inverter; decoupling control; voltage stabilization control; virtual impedance based droop control*

I. INTRODUCTION

It is important to ensure the stable and reliable operation of microgrids, so as to avoid the adverse effects on the power system in grid-connected model and provide excellent power quality for the loads in island mode [1]-[3].

At current control methods for micro-sources, PQ control [4], VF control [5] and droop control [6]-[7] are the prevalent control strategy. By contrast, due to the same external characteristics with synchronous generators, the droop control method can automatically realize the power sharing in terms of their capacity without communications among each microsource, and accordingly, microgrids will have a higher reliability. However, unlike the high-voltage power systems, there is usually a small ratio of X/R in the low-voltage microgrids. Therefore, the active and reactive power are highly coupled, which will result in a poor performance of the conventional P-f/Q-V droop control method. Am at this problem, people introduce several control methods, such as the P-V/Q-f droop control methods [9]-[10] and virtual impedance based control methods [11]-[12]. In order to agree with the traditional power system control theory, and be convenient to plan, build and optimize for microgrids, the virtual impedance based P-f/Q-V droop control method is adopted in this paper.

However, the virtual impedance will contribute to the increasement of the output impedance of microsources, which will result in a ultimately voltage deviation. Furthermore, because of the inherent features of the droop control, the voltage of the islanded microgrids will fluctuate with the load. In view of this, some voltage stabilization methods are put forward in recent years. One is the method of changing the droop coefficients [13]-[14], but this method will affect the power sharing accuracy, dynamic response and even the system stability. Another typical control method is based on the virtual excitation controller [15]-[16], which usually consists of two stage controls, the one-time frequency-modulation control and the integral control of the reactive power deviation. This method can be equivalent to the common droop control with a first-order filter link, so it will cause the dynamic response delay.

In this paper, an adaptive power compensation strategy for the voltage stabilization of microgrids is proposed. In this novel method, the load reactive power fluctuation and the reactive power consumption of the microsource output impedances are compensated in real time to adjust the reactive power reference in the droop parameters.

First, the voltage and current double closed-loop control system on synchronous rotating frame (SRF) of the LCL-based microsource inverter is decoupling designed and analyzed systematically. On the basis of this, for the purpose of optimizing the system static and dynamic features of the microsource, the effects of two different feedback voltages (the capacitor voltage of the LCL filter and the system-side output voltage) on the inverter are studied. Next, via the mathematically modeling of the virtual impedance based droop controller adopted in the microsource, the basic principle of this proposed voltage stabilization method is analyzed in detail. Finally, in order to verify the effectiveness of this proposed control strategy, some simulations are implemented based on MATLAB/Simulink.

II. SYSTEM MODEL OF MICROSOURCES

In this paper, microsources adopt an LCL-type three-phase voltage source converter (VSC) as the grid-connected inverter. In order to simplify the analysis, a DC voltage source is used to substitute for the energy storage system such as batteries in the DC side of the inverter. The circuit and control model of the LCL-VSC based microsource system is depicted in Fig. 1.

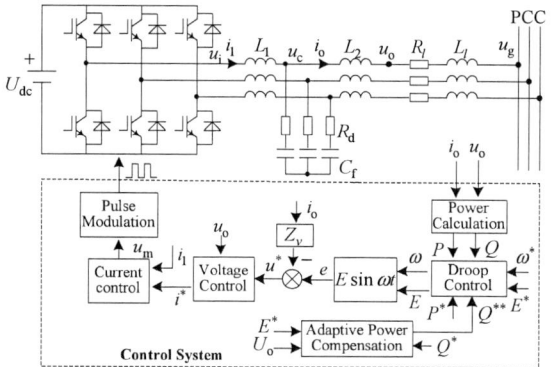

Fig. 1. Circuit and control model of the LCL-VSC based microsources

In Fig. 1, u_g is the system voltage at the point of common coupling (PCC), $Z_l = R_l + j\omega L_l$ is the line impedance from the microsource to PCC, L_1, L_2 and C_f are the components of the LCL filter, R_d is the equivalent resistance of C_f, and u_c is the voltage of C_f. In addition, u_i and i_1 are the device-side output voltage and current, respectively, u_o and i_o are the system-side output voltage and current, respectively, and the DC-side supply is denoted as U_{dc}.

Referring to Fig. 1, the control system consists of several primary parts: power calculation unit, virtual impedance based droop controller, adaptive power compensator, voltage and current controllers and SPWM modulator. The system parameters are defined as follows: ω^* and E^* are the rated frequency and voltage, respectively, P^* and Q^* are the active and reactive power references of the droop controller, respectively, and Z_v is the virtual impedance.

For the aim of zero static deviation of the voltage and current PI controllers, the control system is designed and anlyzed in the synchronous rotating frames (SRF). According to Fig. 1, ignoring the equivalent resistances of all the inductances and capacitances, the frequency-domain mathmatical model of the microsource is described by KVL and KCL equations as follows:

$$\begin{cases} U_{id}(s)-U_{cd}(s)=sL_1I_{1d}(s)-\omega L_1I_{1q}(s) \\ I_{cd}(s)=sCU_{cd}(s)-\omega CU_{cq}(s) \\ I_{1d}(s)=I_{cd}(s)+I_{od}(s) \\ U_{cd}(s)-U_{od}(s)=sL_2I_{od}(s)-\omega L_2I_{1q}(s) \end{cases} \quad (1)$$

$$\begin{cases} U_{iq}(s)-U_{cq}(s)=sL_1I_{1q}(s)+\omega L_1I_{1d}(s) \\ I_{cq}(s)=sCU_{cq}(s)+\omega CU_{cd}(s) \\ I_{1q}(s)=I_{cq}(s)+I_{oq}(s) \\ U_{cq}(s)-U_{cq}(s)=sL_2I_{1d}(s)+\omega L_2I_{1d}(s) \end{cases} \quad (2)$$

III. DESIGN AND ANALYSIS OF THE VOLTAGE AND CURRENT DOUBEL CLOSED-LOOP CONTROL SYSTEM IN SYNCHRONOUS ROTATING FRAME

A. Mathmatical Modeling of the DQ Decoupling Control System

As shown in Fig. 1, the inverter adopts a voltage and current double closed-loop control mode. In general, u_c is

selected as the feedback voltage, and L_2 is considered as the line impedance, which can enhance the inductive characteristics of the line and thereby be beneficial to the conventional P-f/Q-V droop control. However, this method will also further increase the voltage deviation at the PCC. Therefore, in this paper, u_o is chosen as the feedback voltage, and thus L_2 is comprised in the closed voltage control loop. On one hand, the voltage deviation derived from L_2 is eliminated, and the control system has a better high-frequency filtering performance which will be verified in the following analysis. On the other hand, it is possible to avoid the resonance matching between L_2 and the power network parameters.

Analyzing (1) and (2), in the d and q axis, there is coupling of the inverter, which is disadvantageous to the voltage and current independant control. So, it is necessary to design a decoupling control system for the inverter firstly.

According to (1) and (2), the mathmatical relationships between U_i and I_1, and I_1 and U_o can be derived as follows:

$$\begin{cases} U_{id}(s)=sL_1I_{1d}(s)-\omega L_1I_{1q}(s)+U_{cd}(s) \\ U_{iq}(s)=sL_1I_{1q}(s)+\omega L_1I_{1d}(s)+U_{cq}(s) \end{cases} \quad (3)$$

$$\begin{cases} I_{1d}(s)=sCU_{od}(s)-\omega CU_{oq}(s)+ \\ \qquad \left(s^2L_2C-\omega^2L_2C+1\right)I_{od}(s)-2s\omega L_2CI_{oq}(s) \\ I_{1q}(s)=sCU_{oq}(s)+\omega CU_{od}(s)+ \\ \qquad \left(s^2L_2C-\omega^2L_2C+1\right)I_{oq}(s)+2s\omega L_2CI_{od}(s) \end{cases} \quad (4)$$

In Fig. 1, the voltage outer loop utilizes a PI controller to stabilize the output voltage, and the current inner loop adopts a P controller so as to improve the system dynamic response. In the light of (3) and (4), the voltage and current closed-loop controllers are designed as follows:

$$\begin{cases} U_{id}(s)=\left(I_{1d}^*(s)-I_{1d}(s)\right)\times k-\omega L_1I_{1q}(s) \\ U_{iq}(s)=\left(I_{1q}^*(s)-I_{1q}(s)\right)\times k+\omega L_1I_{1d}(s) \end{cases} \quad (5)$$

$$\begin{cases} I_{1d}^*(s)=\left(U_d^*(s)-U_{od}(s)\right)\left(k_P+\dfrac{k_I}{s}\right)- \\ \qquad \omega C\left(U_{oq}(s)-\omega L_2I_{od}(s)-2sL_2I_{oq}(s)\right) \\ I_{1q}^*(s)=\left(U_q^*(s)-U_{oq}(s)\right)\left(k_P+\dfrac{k_I}{s}\right)+ \\ \qquad \omega C\left(U_{od}(s)-\omega L_2I_{oq}(s)+2sL_2I_{od}(s)\right) \end{cases} \quad (6)$$

where k is the proportional coefficient of the current controller, and k_P and k_I represent the proportional and integral coefficients of the voltage PI controller, respectively.

Analyzing (6), the differential terms have little influence on the low-frequency controlled variables in the SRF, and contrarily, the high-frequency interferences may be introduced by them. Therefore, the differential terms are ignoried here, and correspondingly, the voltage controller is simplified as:

3003

$$\begin{cases} I_{1d}^* = \left(U_d^* - U_{od}\right)\left(k_P + \dfrac{k_1}{s}\right) - \omega C\left(U_{oq} - \omega L_2 I_{od}\right) \\ I_{1q}^* = \left(U_q^* - U_{oq}\right)\left(k_P + \dfrac{k_1}{s}\right) + \omega C\left(U_{od} - \omega L_2 I_{oq}\right) \end{cases} \tag{7}$$

According to (5) and (7), the decoupling control scheme is illustrated in Fig. 2.

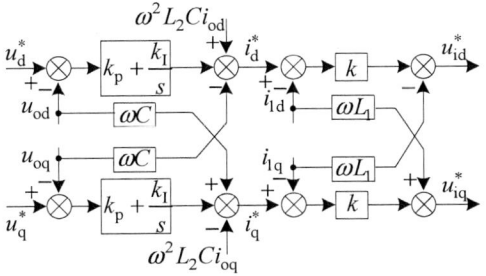

Fig. 2. Block diagram of the voltage and current dq decoupling control

B. Analysis of the Control System

Substituting (5) and (6) into (3) and (4), respectively, it can be seen that the microsource system is decoupled on the d and q axis. In view of the symmetry of the d and q axis, the d axis is considered as an example in this paper. The voltage and current double closed-loop control system on the d axis is shown in Fig. 3.

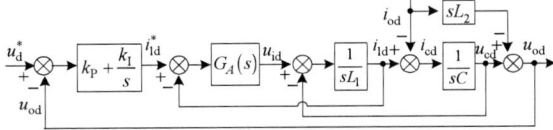

Fig. 3. Block diagram of the voltage and current decoupling control system on the d axis

In Fig. 3, $G_A(s)$ is the current controller considering the total delay of the control system. Defining $G_A(s)$ as:

$$G_A(s) = k \times \frac{k_{PWM}}{T_d s + 1} \tag{8}$$

where, T_d represents the total delay, and k_{PWM} is the equivalent gain of the PWM inverter.

According to Fig. 3, the closed-loop transfer function of the voltage U_{od} can be obtained:

$$U_{od}(s) = G(s)U_d^*(s) - Z(s)I_{od}(s) \tag{9}$$

In (9), $G(s)$ is the voltage gain of $U_{od}(s)$ relative to $U_d^*(s)$, and $Z(s)$ is the impedance of $U_{od}(s)$ relative to $I_{od}(s)$. Ignoring the high-order minimal items, $G(s)$ and $Z(s)$ are given as:

$$G(s) = \frac{U_{od}(s)}{U_d^*(s)} =$$
$$\frac{kk_{PWM}(sk_P + k_i)}{s^3 L_1 C + s^2\left(T_d + kk_{PWM}C\right) + s\left(1 + kk_{PWM}k_P\right) + kk_{PWM}k_i} \tag{10}$$

$$Z(s) = \frac{U_{od}(s)}{I_{od}(s)} =$$
$$\frac{s^3\left((L_1 + L_2)T_d + kk_{PWM}L_2 C\right) + s^2\left(L_1 + L_2\right) + skk_{PWM}}{s^3 L_1 C + s^2\left(T_d + kk_{PWM}C\right) + s\left(1 + kk_{PWM}K_P\right) + kk_{PWM}k_i} \tag{11}$$

According to Thevenin theorem, $G(s)U^*(s)$ is the internal electromotive force, and $Z(s)$ is the internal impedance of the microsource. Analyzing $G(s)$ and $Z(s)$, L_2 exists only in $Z(s)$ while not in $G(s)$. In other words, the different closed-loop feedback voltages do not affect the internal electromotive force, only affect the internal impedance.

Let $L_2 = 0$ in $Z(s)$, that is, u_c is selected as the feedback voltage, and the internal impedance of this closed-loop control system is defined as $Z_1(s)$:

$$Z_1(s) = \frac{U_{od}(s)}{I_{od}(s)} =$$
$$\frac{s^3 L_1 T_d + s^2 L_1 + skk_{PWM}}{s^3 L_1 C + s^2\left(T_d + kk_{PWM}C\right) + s\left(1 + kk_{PWM}k_P\right) + kk_{PWM}k_i} \tag{12}$$

The system parameters are set in Table I.

TABLE I. SYSTEM PARAMETERS

System Parameters	Value
Filter inducanc L_1, L_2	0.3 mH, 0.1 mH
Filter capacitor C	100 μF
Total delay of the control system T_d	100 μs
Equivalent gain of the inverter K_{PWM}	300
Proportional coefficient k	0.5
Proportional and integral coefficient k_P, k_I	5, 100
Switching frequency f_s	10 kHz

Fig. 4 depicts the bode diagrams of $G(s)$.

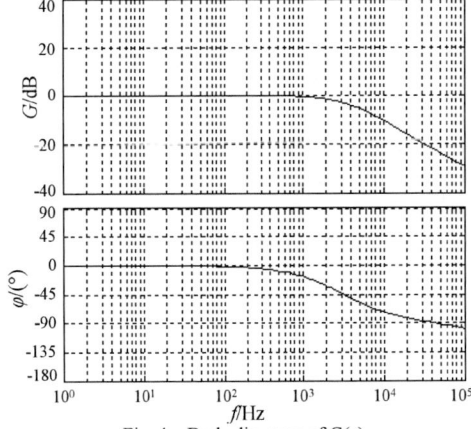

Fig. 4. Bode diagrams of G(s)

Analyzing Fig. 4, it is obvious that, on one hand, U_{od} can keep the accurate track of the command U_d^* in the low-frequency band, and on the other hand, there is a favorable harmonic suppression effect in the high-frequency band.

Fig. 5 depicts the bode diagrams of $Z(s)$ and $Z_1(s)$.

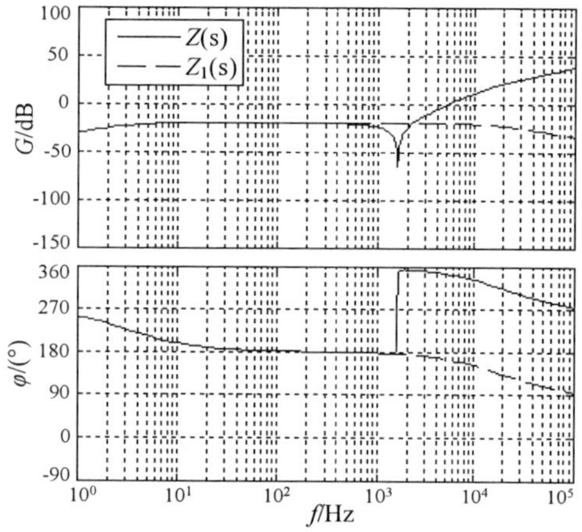

The 2018 International Power Electronics Conference

Fig. 5. Bode diagrams of $Z(s)$ and $Z_1(s)$

Analyzing Fig. 5, in the low-frequency band, $Z(s)$ and $Z_1(s)$ have close amplitude-frequency and phase-frequency characteristics. In the high-frequency band, it can be clearly seen that, there exists a resonance point (depends on the LCL filter parameters) of $Z(s)$, and then has a higher amplitude after the resonance point, which is beneficial for the suppression of high-frequency harmonics. However, the amplitude of $Z_1(s)$ further reduces, and basically no inhibition capability for the high-frequency harmonics.

The above analysis shows that, when u_o is selected as the feedback voltage in the voltage closed loop, compared with u_c, thanks to including all the parameters of the LCL filter, the high-frequency harmonic suppression capability is stronger, and moreover, the low-frequency features is not affected.

IV. MATHMATICAL MODELING OF THE VIRTUAL IMPEDANCE BASED DROOP CONTROLLER FOR THE MICROSOURCE

The virtual impedance can achieve the adjustment of the internal impedance feature of the microsource, so as to meet the needs of varied droop-control strategies. The typical scheme of strategy is described in Fig. 6.

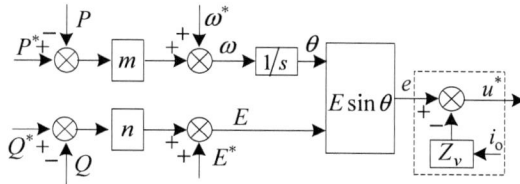

Fig. 6. Schematical diagram of the virtual impedance based droop control

where, Z_v is the virtual impedance.

In Fig. 6, the primary principle of the traditional P-f/Q-V droop controller is depicted, that is:

$$\begin{cases} \omega = \omega^* + m \times \left(P^* - P \right) \\ E = E^* + n \times \left(Q^* - Q \right) \end{cases} \tag{13}$$

where m and n are the frequency and voltage droop coefficients, respectively.

Referring to Fig. 6, a voltage $-i_oZ_v$ is added to the output voltage e of the droop controller. In order to satisfy the P-f/Q-V droop control characteristics, Z_v is set as a pure reactance $j\omega L_v$. Therefore, the mathematical model of the voltage command u^* based on the virtual impedance can be obtained:

$$\begin{cases} U_{\mathrm{d}}^* (s) = E_{\mathrm{d}} (s) - sL_v I_{\mathrm{od}} (s) + \omega L_v I_{\mathrm{oq}} (s) \\ U_{\mathrm{q}}^* (s) = E_{\mathrm{q}} (s) - sL_v I_{\mathrm{oq}} (s) - \omega L_v I_{\mathrm{od}} (s) \end{cases} \tag{14}$$

Combined with (9), the mathematical model of the voltage closed-loop control system can be derived:

$$\begin{cases} U_{\mathrm{od}}(s) = G(s)E_{\mathrm{d}}(s) - (G(s)sL_v + Z(s))I_{\mathrm{od}}(s) \\ \qquad + G(s)\omega L_v I_{\mathrm{oq}}(s) \\ U_{\mathrm{oq}}(s) = G(s)E_{\mathrm{q}}(s) - (G(s)sL_v + Z(s))I_{\mathrm{oq}}(s) \\ \qquad - G(s)\omega L_v I_{\mathrm{od}}(s) \end{cases} \tag{15}$$

According to Fig. 4, in the low-frequency band, $G(s) \approx 1$, and ignoring the differential items including sL_v, formula (15) can be simplified as:

$$\begin{cases} U_{\mathrm{od}}(s) = E_{\mathrm{d}}(s) - Z(s)I_{\mathrm{od}}(s) + \omega L_v I_{\mathrm{oq}}(s) \\ U_{\mathrm{oq}}(s) = E_{\mathrm{q}}(s) - Z(s)I_{\mathrm{oq}}(s) - \omega L_v I_{\mathrm{od}}(s) \end{cases} \tag{16}$$

It is worth noting that, in a low-voltage microgrid, lines are of short length and small impedance [17]. Therefore, it is possible to get a virtual inductance L_v to make $\omega L_v \gg |(Z+Z_l)|$. Accordingly, the influences of Z and Z_l on the droop control can be reduced extremely. Thereby, defining the total output impedance of the microsource $X_o = \omega L_v$, the simplest mathematical model of the micro-source can be given as follows:

$$\begin{cases} U_{\mathrm{od}}(s) = E_{\mathrm{d}}(s) + X_o I_{\mathrm{oq}}(s) \\ U_{\mathrm{oq}}(s) = E_{\mathrm{q}}(s) - X_o I_{\mathrm{od}}(s) \end{cases} \tag{17}$$

In this paper, the power calculation point is chosen at the output terminal (U_o, I_o). As a result, Referring to Fig. 1, the approximate expressions of the output active and reactive power can be obtained:

$$\begin{cases} P \approx \dfrac{U_o E}{X_o} \delta \\ Q = \dfrac{U_o}{X_o} \left(E - U_o \right) \end{cases} \tag{18}$$

V. THE VOLTAGE STABILIZATION CONTROL STRATEGY BASED ON THE ADAPTIVE POWER COMPENSATION FOR MICROGRIDS

A. Primary principle

In general, the voltage and frequency of a microgrid in grid-connected mode are supported by the power system. However, in island mode, the voltage will fluctuate inevitably with the loads under the traditional droop control method, which may result in a adverse condition to the microgrid.

Firstly, according to (13), when Q deviates from Q^*, E will deviate accordingly from E^*, which is determined by the inherent characteristics of the droop control. This voltage deviation can be called the first-stage voltage deviation, which is denoted as $\Delta U_1 = E^* - E$.

Secondly, analyzing (17), X_o will also lead to a voltage deviation between U_o and E. This voltage Deviation can be called the second-stage voltage deviation, which is denoted as $\Delta U_2 = E - U_o$.

Defining the total voltage deviation $\Delta U = \Delta U_1 + \Delta U_2$, in the light of (18), ΔU not only interacts with Q, but also affects P, which will result in a frequency fluctuation.

Based on the simplest equivalent model of the microsource in (17), the proposed voltage stability control strategy based on adaptive power compensation is illustrated in Fig. 7.

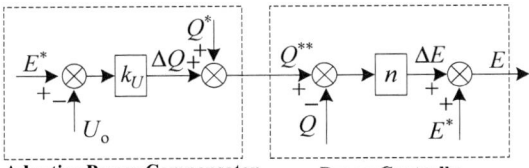

Adaptive Power Compensator   **Droop Controllr**

Fig. 7. Voltage stabilization control strategy based on adaptive power compensation

In Fig. 7, k_U is the power compensation coefficient, and the voltage controller is composed of an adaptive power compensator and a droop controller. According to Fig. 7, The voltage stability control scheme can be described as:

$$\begin{cases} E = E^* + n\left(Q^{**} - Q\right) \\ Q^{**} = Q^* + K_U\left(E^* - U_o\right) \end{cases} \quad (19)$$

Letting $\Delta Q = \Delta Q_1 + \Delta Q_2$, ΔQ_1 and ΔQ_2 is given as follows:

$$\begin{cases} \Delta Q_1 = K_U\left(E^* - E\right) \\ \Delta Q_2 = K_U\left(E - U_o\right) \end{cases} \quad (20)$$

Substituting (13) and (18) into (20), ΔQ_1 and ΔQ_2 are described as:

$$\begin{cases} \Delta Q_1 = -nK_U\left(Q^* - Q\right) \\ \Delta Q_2 = K_U\dfrac{X_o}{U_o}Q \end{cases} \quad (21)$$

Analyzing (19) ~ (21), the compensating power consists of two parts: ΔQ_1 and ΔQ_2, where ΔQ_1 compensates for load fluctuating powers, and ΔQ_2 compensates for the power consumptions of X_o. Therefore, ΔQ_1 and ΔQ_2 are related to the voltage deviations ΔU_1 and ΔU_2, respectively.

Combining (19), (20) and (21), U_o can be obtained:

$$U_o = E^* + \frac{1}{1 + nK_U} \times \left(n\left(Q^* - Q\right) - \frac{X_o}{U_o}Q\right) \quad (22)$$

Analyzing (22), when $nK_U \gg 1$, there is $U_o \approx E^*$. That is, U_o is stabilized near E^*, and the influences of loads, output impedance X_o and the droop coefficient n on U_o are inhibited.

In addition, when the microgrid is connected to the power system, on one hand, since U_o usually equals to the power system voltage U_g, $\Delta Q = 0$ on condition that $E^* = U_g$, and thus, the power compensation link is automatically released, so as not to affect the grid-connected operation of the microgrid. On the other hand, it can achieve the smooth transition of the voltage amplitude while the microgrid is incorporated into the power networks.

Fig. 8 illustrates the schematic diagram of the voltage stability control process based on the adaptive power compensation strategy.

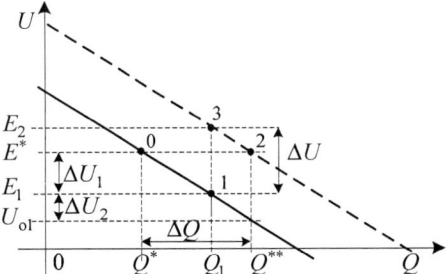

Fig. 8. Schematic diagram of the proposed voltage stability control strategy

Owing to space limitation, the power condition of $Q > Q^*$ is taking as an example in Fig. 8. At the starting time, suppose that the microsource is working at the rated state, i.e. the point of 0 (Q^*, E^*). When Q increases to Q_1, the power deviation ΔQ_1 will result in the voltage error ΔU_1, and accordingly, E drops to E_1. At this time, the working point moves from 0 to 1 (Q_1, E_1). Meanwhile, the power consumption of X_o causes the other voltage deviation ΔU_2. Therefore, the output voltage of the microsource is U_{o1} as a result of $\Delta U = \Delta U_1 + \Delta U_2$. However, under the adaptive power compensation control, the power reference is adjusted to $Q^{**} = Q^* + \Delta Q$. As a result, the droop control line moves to a new position, as the dash line. Accordingly, the new rated working point is in the point of 2 (Q^{**}, E^*), and the current working point of the microsource moves to point 3 (Q_1, E_2). Correspondingly, the final output voltage U_o equals to $E_2 - \Delta U_2$, that is E^*. It is worth noting that, the above process is actually dynamic according to the load power fluctuations.

B. Analysis and Design of the Control System

Combining (15), (18) and (19), and ignoring the internal impedance $Z(s)$, the voltage closed-loop control block diagram of the d axis can be depicted as Fig. 9.

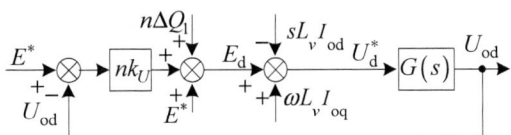

Fig. 9. Block diagram of the voltage closed control loop of the d axis

According to Fig. 9, the mathematical model of the voltage closed-loop control system can be obtained:

$$U_{\text{od}} = \frac{\left(1+nK_{\text{U}}\right)G(s)E^*}{nK_{\text{U}}G(s)+1} + \frac{G(s)\left(n\Delta Q - sL_v I_{\text{od}} + \omega L_v I_{\text{oq}}\right)}{nK_{\text{U}}G(s)+1} \quad (23)$$

Analyzing (23), U_{od} consists of two parts: the first term is the voltage gain of $U_{\text{od}}(s)$ relative to $E^*(s)$, and the second term is the disturbance gain of $U_{\text{od}}(s)$ relative to ΔQ, I_{od} and I_{oq}. Substituting $G(s)$ in (9) into (23), the characteristic equation of the voltage closed-loop system can be obtained:

$$s^3 L_1 C + s^2 \left(T_{\text{d}} + kk_{\text{PWM}}C\right) + s\left(1+\left(1+nK_{\text{U}}\right)\times \right. \\ \left. kk_{\text{PWM}}k_{\text{P}}\right) + \left(1+nK_{\text{U}}\right)kk_{\text{PWM}}k_i = 0 \quad (24)$$

Here, the Rolls criterion is used to analyze the stability of this third-order system, and the ranges of k_{U} can be obtained considering a stable system:

$$K_{\text{U}} > \frac{1}{n}\left(\frac{T_{\text{d}}+kk_{\text{PWM}}C}{\left(L_1 C k_i - \left(T_{\text{d}}+kk_{\text{PWM}}C\right)k_{\text{P}}\right)kk_{\text{PWM}}}-1\right) \quad (25)$$

Substituting the system parameters in Table 1 into (25), it can be seen that, the system is stable as long as $K_{\text{U}}>0$. Therefore, K_{U} has a large stability margin.

VI.   SIMULATION ANALYSIS

In order to verify the effectiveness of the proposed control method, a simple microgrid is modeled in MATLAB, which includes two microsource inverters and three loads to replicate load fluctuations. The inverters are denoted as Inv1 and Inv2 in turn, and their models refer to Fig. 1. The rated line-to-line voltage U_n and frequency f_n are set to 380 V and 50 Hz, respectively. Some of the system parameters refer to Table I, and others are listed in Table II. In addition, the loads are set as: load 1 ($P = 100$ kW and $Q = -100$ kVar), load 2 ($P = 50$kW and $Q = 100$ kVar), and load 3 ($P = 50$kW and $Q = -50$ kVar).

TABLE II.   CONTROL PARAMETERS OF THE INVERTERS

	E^*/V	ω^*/Hz	m ($f_{\min}=1\%f_n$)	n ($\Delta u_{\max}=5\%U_n$)	L_v /mH	k_{U}
Inv1	220	50	6.28e-5	3.8e-4	2	3000
Inv2	220	50	3.14e-5	1.9e-4	1	6000

1) Simulation case 1: island mode

In the initial stage, the microgrid works with the capacitive load 1. At 0.5s, the inductive load 2 is put into working. The simulation waveforms are shown in Fig. 10 and Fig. 11, and P_1 and Q_1 are the active and reactive power of Inv1, P_2 and Q_2 are those of Inv2.

Fig. 10 shows the simulation results of the microgrid without the proposed control strategy. Analyzing the output voltage U_o on the top of Fig. 10, there is a large deviation of U_o, and especially at the time of load changing, the maximum voltage deviation percentage $\Delta u_{\max}/U_n$ reaches to 8.7%. Correspondingly, the active and reactive power also deviate from the expected values because of the voltage deviation.

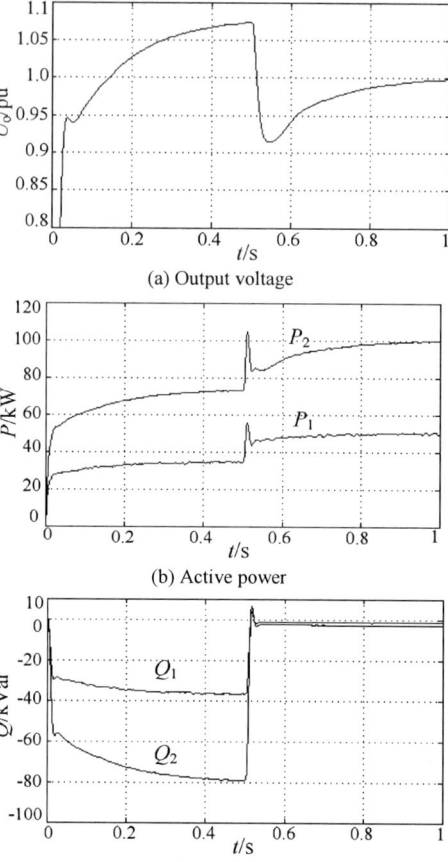

(a) Output voltage

(b) Active power

(c) Rective power

Fig. 10. Simulation waves of the microgrid with the step loads, without the proposed voltage stability control strategy

Fig. 11 shows the simulation results of the microgrid using the proposed control strategy. Analyzing Fig. 11, U_o can be maintained in a narrow range of the rated value even at the time of load changing, and $\Delta u_{\max}/U_n$ only reached to 1.2%. Correspondingly, the output active and reactive power accuracy of each micro-source is relatively higher, and moreover, the load power is shared basically proportional to their ratings.

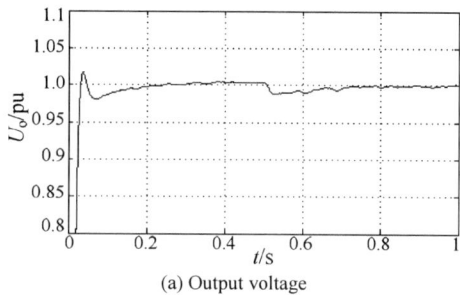

(a) Output voltage

The 2018 International Power Electronics Conference

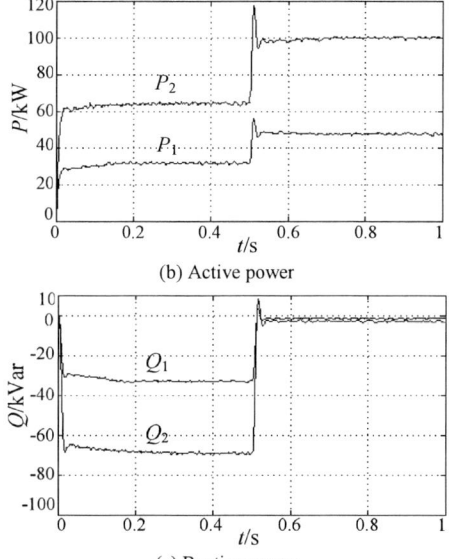

(b) Active power

(c) Rective power

Fig. 11. Simulation waves of the microgrid with the step loads under the proposed voltage stability control strategy

2) Simulation case 2: transition of the operation mode from island mode to grid-connected mode

For the purpose of simplifying the analysis, only a single microsource is used in the microgrid. Setting the reactive power command in grid-connected mode is 60 kVar. In the initial stage, the microgrid works in island mode with the capacitive load 3, and at 0.5 s, is incorporated into the power networks. The simulation waveforms are described in Fig. 12 and Fig. 13.

Fig. 12 shows the simulation results of the microgrid without the proposed control strategy. Analyzing Fig. 12, $\Delta u_{max}/U_n$ reaches to 26% as a result of the large voltage deviation. Therefore, the micro-source output reactive power deviates from the expected values seriously, and furthermore, there is a large power fluctuations when the mode transition.

Fig. 13 shows the simulation results of the microgrid using the proposed control strategy. It can be seen from this figure, that both in islanded and grid-connected mode, there are no obvious voltage and reactive power fluctuations, and $\Delta u_{max}/U_n$ is only 0.7%. In addition, the microsource can achieve the mode smooth transition, and accurately outputs the reactive power according to the command in grid-connected model.

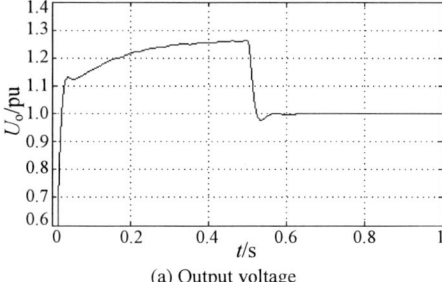

(a) Output voltage

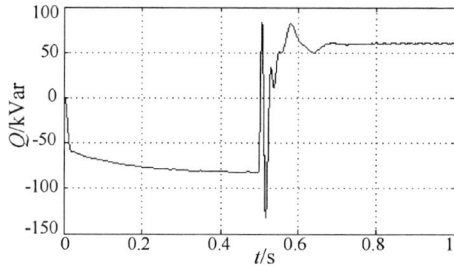

(b) Rective power

Fig. 12. Simulation waves of the microgrid operation model changing without the voltage stability control strategy

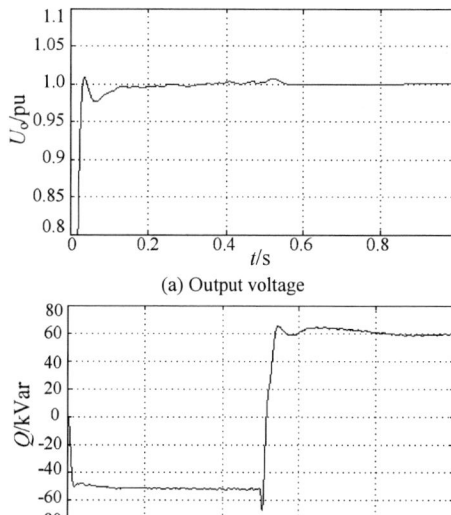

(a) Output voltage

(b) Rective power

Fig. 13. Simulation waves of the microgrid operation model changing under the voltage stability control strategy

In summary, the simulation results above show that the proposed control method can effectively decrease the system voltage deviations caused by load fluctuations and the micro-source output impedance, and do not affect droop control characteristics both in islanded and grid-connected model.

VII. CONCLUSION

An adaptive power compensation strategy for the voltage stabilization of LCL-VSC based microgrids is proposed in this paper. The theoretical analysis and simulation results show that:

Firstly, in the voltage closed-loop system, when the voltage feedback point includes all the parameters of the LCL filter, compared with the point of capacitor voltage of the LCL filter, the low-frequency characteristics of the micro-source internal impedance is similar, but the high-frequency harmonic suppression capability is much stronger. In addition, the different closed-loop feedback voltages do not affect the micro-source internal potential.

Secondly, with the proposed control strategy, the two-stage voltage deviations of the microgrid caused by load fluctuations and the output impedance can be eliminated effectively. Moreover, it can achieve the smooth transition of the voltage amplitude while the microgrid is incorporated into the power networks.

ACKNOWLEDGMENT

The authors wish to thank the National Natural Science Foundation of China (51607037), the "333 Project" Research Program of Jiangsu Province (BRA2016184), the Research Institutes Association Innovative Program of Jiangsu Province (BY2016076-14) and Jiangsu Overseas Research & Training Program for University Prominent Young & Middle-aged Teachers and Presidents

References

[1] Shamshiri M, Chin Kim Gan, Chee Wei Tan, "A review of recent development in smart grid and microgrid laboratories," IEEE International Power Engineering and Optimization Conference, 2012, pp.367-372

[2] Kouluri M K, Pandey R K, "Intelligent agent based microgrid control," International Conference on Intelligent Agent and Multi-Agent Systems, 2011, pp.62-66

[3] Kasem Alaboudy A H, Zeineldin H H, Kirtley J L, "Microgrid stability characterization subsequent to fault-triggered islanding incidents," IEEE Transactions on Power Delivery, 2012, 27(2): 658-669

[4] Qi Li, Weirong Chen, Zhixiang Liu, et al, "Active control strategy based on vectorproportion integration controller for proton exchange membrane fuel cell grid-connected system," IET Renewable Power Generation, 2015, 9(8): 991-1000

[5] Laaksonen H, Saari P, Komulainen R, "Voltage and frequency control of inverter based weak LV network microgrid," International Conference on Future Power System, 2005: 16-18

[6] Gao, Fei, Bozhko, Serhiy, Costabeber, Alessandro, Patel, et al, , "Comparative stability analysis of droop control approaches in voltage-source-converter-based DC microgrids," IEEE Transactions on Power Electronics, 2017, 32(3): 2395-2416

[7] Zhao, Xin, Guerrero, Josep M. , Savaghebi, Mehdi, et al, "Low-Voltage Ride-Through operation of power converters in grid-interactive microgrids by using negative-sequence droop control," IEEE Transactions on Power Electronics, 2017, 32(4): 3128-3143

[8] Majumder R, Ledwich G, Ghosh A, et al, "Droop control of converter-interfaced microsources in rural distributed generation," IEEE Transactions on Power Delivery, 2010, 25(4): 2768-2778.

[9] Tine L, Vandoom, Bart meersman, Guerrero J M, "Automatic power-sharing modification of P/V droop controllers in low-voltage resistive microgrids" IEEE Transactions on Power Electroncs, 2012, 24(4): 2318-2325.

[10] Guerroro J M, matas J, dc Vicuna L G, et al, "Decentralized control for parallel operation of generation inverters using resistive output impedance," IEEE transactions on industrial electronics, 2007, 54(2): 994-54.

[11] Chunxia Dou, Zhanqiang Zhang, Dong Yue, Mengmeng Song, "Improved droop control based on virtual impedance and virtual power source in low-voltage microgrid," IET Generation, Transmission & Distribution, 2017, 11(4): 1046-1055

[12] Zhu, Yixin, Zhuo, Fang, Wang, Feng, et al, "A virtual impedance optimization method for reactive power sharing in networked microgrid," IEEE Transactions on Power Electronics, 2016, 31(4): 2890-2905

[13] de Souza, Wanderson Ferreira, Severo-Mendes, et al, "Power sharing control strategies for a three-phase microgrid in different operating condition with droop control and damping factor investigation," IET Renewable Power Generation, 2015, 9(7): 831-840

[14] Y. A. R. I. Mohamed and E. F. El-Saadany, "Adaptive decentralized droop controller to preserve power sharing stability of paralleled inverters in distributed generation microgrids," IEEE Trans. Power Electron., 2008, 23(6): 2806–2816.

[15] Qingchang Zhong, Philong Nguyen, Zhengyu Ma, Wangxing Sheng, "Self-synchronized synchronverters: inverters without a dedicated synchronization uint," IEEE Transactions on power Electronics, 2014, 29(2): 617-630

[16] Heng Wu, Xinbo Ruan, Dongsheng Yang, et al, "Small- sigal Modeling and parameters design for virtual synchronous generators," IEEE Transactions on Industrial Electronics, 2016, 63(7): 4292-4303

[17] Mohammad S. Golsorkhi, Dylan D. C. Lu, "A control method for inverter-based islanded microgrids based on V-I droop characteristics," IEEE Transactions on power delivery, 2016, 30(3): 1196-1204.

The 2018 International Power Electronics Conference

Resonance Detection Strategy for Multiple Grid-Connected Inverters-based System Using Cascaded Second-Order Generalized Integrator

Wu Cao[1], Dongchen Fan[1], Kangli Liu[1], Jianfeng Zhao[1], Liheng ruan[2], Xiaojun Wu[2]

1 Department of Electrical Engineering

Southeast University, Nanjing, Jiangsu, 210096, China

2 Jiangsu Haihang Electric Technology Co. Ltd

Zhenjiang, Jiangsu, 212200, China

Email: caowu_ee@seu.edu.cn, fdctc @163.com, kangcumt@yeah.net, jianfeng_zhao@seu.edu.cn,

rlh@hhdq.net, wxj@hhdq.net

ABSTRACT— **The mutual interactions in the multi-paralleled grid-connected VSIs coupled through the grid impedance tend to result in the whole system oscillating-formed resonance. Resonance current threatens the stability of grid due to the frequency and magnitude fluctuation. The paper firstly reveals the transformation of resonance point with parallel VSIs. Then, Second Order Generalized Integrator-Frequency Locked Loop (SOGI-FLL) is analyzed to extract resonance current frequency. Secondly, the paper proposes an improved method that two models of SOGI-FLL are cascaded and normalized to eliminate the effect of magnitude and frequency. The fundamental component can be completely filtered by the former part. This method is frequency self-adaptive and faster to extract unknown frequency current. Finally, the simulation results verify the effectiveness of the theoretical analysis.**

Key words— *Grid-connected inverters; Resonance; Frequency estimation; SOGI-FLL*

I. INTRODUCTION

Renewable energy consumption usually uses inverters as the grid interface, which requires long transmission lines and multi-stage

transformers to connect the grid. There exists a complex coupling among inverters and the power grid, which is prone to brings about the resonance. The resonance in multi-inverters system could not only affect power quality, but also damages the relay protection device, seriously affecting the stability of the grid [7].

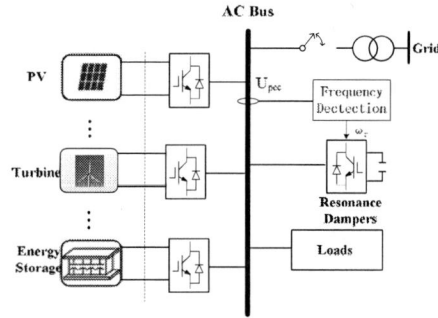

Fig. 1. A micro-grid system with resonance dampers.

Fig.1 illustrates a micro-grid system with photovoltaic, wind turbines, energy storage and resonance dampers. The active damping scheme is an effective resonance suppression method [8]-[9], which suppresses the resonant spike by dampening at the resonant point. However, if the damping broadband is added at the resonant point, the system bandwidth will be reduced. Therefore, in order to suppress the resonance

3010

current, the frequency of the resonant point must be accurately obtained. Discrete Fourier Transformation (DFT) and wavelet transformation are often used to detect resonance [10]. It is appropriate to lose the information of resonance point following the whole frequency band divided evenly in the DFT method. Due to the amplitude distortion and time-varying of the resonance current, the traditional current detection method is limited. SOGI-FLL is widely used in synchronous grid voltage frequency. In [12], the three-phase SOGI-FLL is used to extract the positive and negative sequence components of the current under unbalanced three-phase conditions. The advantage of SOGI-FLL is to simplify the algorithm, which only extract the frequency information, reducing the computation time and space complexity. [13]

In this paper, the impedance model of inverters is establised to analyze the transformation of resoance points, including the number,frequecy and mag. Then, a frequency self-adaptive SOGI-FLL method is presented, which is used to detect the resonance current frequency of multi-inverters grid connected system. The cascaded SOGI is proposed to separate the fundamental frequency component and the resonance component. The proposed method can effectively detect the resonant frequency when the resonant frequency is unknown and time-varying.

II. SYSTEM MODELING AND MECHANISM ANALYSIS

The transformation of resonance point with parallel VSIs could be probed by the impedance model. The open loop impedance model of inverters ignores the influence of the inverter current control strategy.

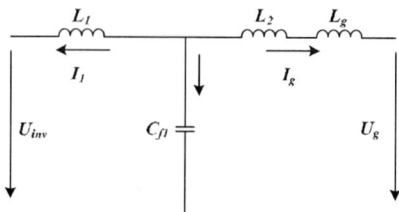

Fig. 2. A double input circuit of a single inverter

A single inverter can be simplified as a double input circuit shown in Fig.2. Assume ω_r is the angular frequency of the LCL inherent resonance peak presented in (4).

$$\omega_r = \sqrt{\frac{L_1 + L_2 + L_g}{L_1(L_2 + L_g)C_{f1}}} \ , \quad f_r = \frac{\omega_r}{2\pi} \quad (4)$$

Assuming that the filter parameters of each inverter are same, the output impedance of each inverter has the same value. N the same paralleled inverter can be equivalent to a single inverter circuit of n times the power grid impedance. Thus the transformation of n to one is completed based AC system in Fig.3 by the above analysis.

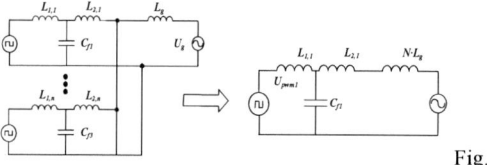

Fig.

3. The equivalent process of N inverters

The new resonance frequency can be derived mentioned above

$$\omega_r = \sqrt{\frac{L_1 + L_2 + n \cdot L_g}{L_1(L_2 + n \cdot L_g)C_{f1}}} \quad (5)$$

Obviously, a new resonance point is added. Meanwhile, the frequency of resonance would change with the increase of inverters from (5).

Different inverter control methods as well as filter parameters could result more complex couplings, even the instability.

III. PROPOSED RESONANCE DETECTION METHOD

A. Frequency self-adaptive SOGI-FLL

Varying resonance frequency and magnitude

are difficult to extract in practical application. To solve this problem, SOGI-FLL is used to extracts resonance components. This has the self-adaptive filter structure, which can extract the current of the unknown frequency.

The structure of this filter is shown in Fig.4, where the resonance frequency is called ω' to difference it from the input frequency ω. The two in-quadrature output signals of the self-adaptive filter in Fig.6, i.e., v' and qv'.

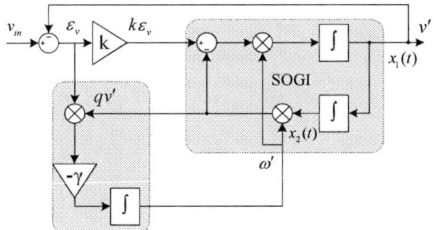

Fig. 4. The structure of SOGI.

The state space method is used to describe the system. Assuming that $x(t)=[x_1(t) \ x_2(t)]^T$. The state equation is shown in (8).

$$\begin{cases} \begin{bmatrix} \dot{x}_1(t) \\ \dot{x}_2(t) \end{bmatrix} = \begin{bmatrix} -k\omega' & -\omega'^2 \\ 1 & 0 \end{bmatrix} \begin{bmatrix} x_1(t) \\ x_2(t) \end{bmatrix} + \begin{bmatrix} k\omega' \\ 0 \end{bmatrix} v(t) \\ \dot{x}_2(t) = x_1(t) \end{cases}$$

$$\dot{\omega}' = -\gamma x_2 \omega'(v - x_1) \quad (8)$$

The lock frequency steady-state error is derived in (10)

$$\varepsilon_v = \omega' x_2 (v - x_1) = \frac{x_2^2}{k}\left(\omega'^2 - \omega^2\right) \quad (10)$$

Substitute (10) into (8),

$$\dot{\omega}' = -\frac{\gamma}{k} x_2^2 \left(\omega'^2 - \omega^2\right) \quad (11)$$

From (11), if the output signal frequency of FLL is lower than the frequency of input resonance signal v_{in} ($\omega'<\omega$), the derivative of output frequency is positive, so the output signal frequency increases until it is equal to the frequency of input resonance signal v_{in}. When $\omega'>\omega$, the derivative of output frequency is negative similarly to realize the self-adaptive matching.

B. The Cascaded SOGI-FLL strategy

Equation (11) reflects the convergence rate of

the signal. Also, in the condition of $\omega' \approx \omega$, $\omega'^2 - \omega^2$ is similar to $2(\omega' - \omega)\omega'$, the (10) can be transformed to (12).

$$\dot{\omega}' = -2\frac{\gamma}{k\omega'}V^2\left(\omega' - \omega\right) \quad (12)$$

It is noticed that the relevance of the frequency ω' as well as magnitude V to the rate of convergence can be eliminated by normalizing the self-adaptive parameter γ. Then, the normalization achieved by multiplying the correlation coefficient could make the derivative function of the output frequency to be related only to the adaptive parameter γ. The normalized derivative of output frequency can be expressed as

$$\dot{\omega}' = -2\gamma\left(\omega' - \omega\right) \quad (13)$$

The cascaded SOGI-FLL is proposed to separate and extract the resonance components in Fig.5. The structure is divided to two parts. The former part SOGI sets its extraction frequency as the fundamental frequency, removing FLL self-adaptive module, by separating the fundamental component of input signal. The latter part is the normalized SOGI-FLL to extract resonance component.

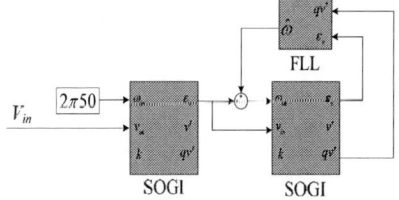

Fig. 5. The structure of the cascaded SOGI-FLL.

$D_b(s)$ and $D_r(s)$ are the transfer function of former part and latter part respectively, the output of the cascaded SOGI is

$$v_r' = D_r(s) \cdot v_{in} \cdot (1 - D_b(s)) \quad (14)$$

Where V_r' is the output of cascaded SOGI-FLL. The Bode of (14) is illustrated in Fig.6. As shown in Fig.6, the gain is -170 dB at 50 Hz, this is because the fundamental wave current can be completely filtered by the former

The 2018 International Power Electronics Conference

part. In addition, of the gain is 0 dB at the resonance point, the resonant frequency of the current information has been retained.

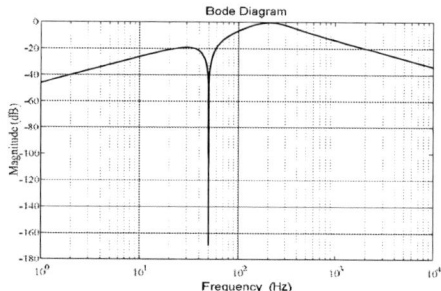

Fig. 6. The Bode of the cascaded SOGI-FLL output.

IV. SIMULATION

A. Resonance extraction effect validation

For verifying the self-adaptive extraction characteristic of the cascaded SOGI-FLL, The simulation of the fundamental wave mixing the resonance frequency resonance signal is set, where the configuration parameters of SOGI-FLL is k=1，γ=-60.

Fig.7 accounts for the FFT window of six cycles with serious distortion, because the current has been mixed by 320 Hz resonance current. Due to the resonance, the input signal of THD and resonant component proportion will be higher.

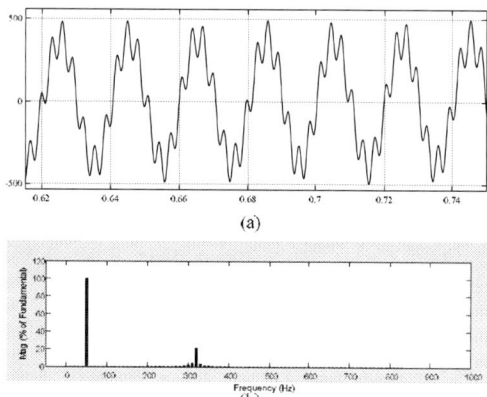

Fig. 7. The FFT window of the resonance current.

Resonance component is extracted by the cascaded SOGI-FLL in Fig.8. From 0.1 s, we can see clearly the resonant component increases from 100A to 200A in Fig.8, where the

resonance envelope is clear.

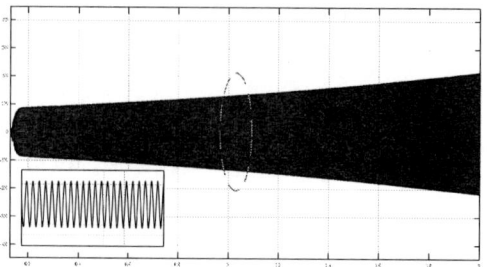

Fig. 8. The extraction of the cascaded SOGI-FLL

B. The comparison of detection speed

The comparison of detection speed results are shown in Fig. 9. The frequency of resonance component is about 320 Hz, and changes to 800 Hz suddenly at 1.5s. Blue curve represents the traditional SOGI-FLL method; the red curve represents the normalized SOGI-FLL. We can see that all the two methods can response to the resonance component quickly, and track the resonant frequency without the static errors. The traditional SOGI-FLL realized on resonance signal tracking in 1.6s, which is lower than the normalized SOGI-FLL tracking at about 1.55s. This is because the normalized SOGI is not affected by the amplitude and frequency. The frequency detection response speed faster when amplification happens.

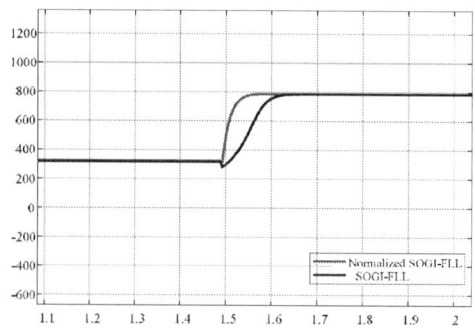

Fig. 9. The extraction effect comparison between the SOGI-FLL and the cascaded SOGI-FLL.

V. CONCLUSION

In this paper, resonance detection principle is analyzed in paralleled grid-connected inverters system. Besides, resonance detection method is proposed based on SOGI-FLL. Moreover, the Cascaded SOGI-FLL algorithm is used to detect and analyze the resonance problem in the grid-connected inverters system. In the process of resonance signal analysis, the influence of the frequency ω' and magnitude V on the rate of convergence is eliminated by normalizing the self-adaptive parameter γ. Simulation results demonstrate that the Cascaded SOGI-FLL can effectively detect the resonance disturbances. Furthermore, resonance analysis based on the proposed method can provide guarantee for the safe and stable operation of the grid.

ACKNOWLEDGMENT

The authors would like to thank the sponsorship of the National Natural Science foundation of China (51607037), Innovation foundation for Combination of Industry and Scientific Research of Jiangsu Province (BY2016076-09) and Jiangsu province ordinary university professional degree graduate practice innovation (SJLX16_0057).

REFERENCES

[1] Yang D, Ruan X, Wu H, "Impedance Shaping of the Grid-Connected Inverter with LCL Filter to Improve Its Adaptability to the Weak Grid Condition," *IEEE Transactions on Power Electronics*, vol. 29, no. 11, pp. 5795-5805, 2014.

[2] Wen B, et al, "Impedance-Based Analysis of Grid-Synchronization Stability for Three-Phase Paralleled Converters," *IEEE Transactions on Power Electronics*, vol. 31, no. 1, pp. 26-38, 2015.

[3] Wang X, et al, "An Active Damper for Stabilizing Power-Electronics-Based AC Systems," *IEEE Transactions on Power Electronics*, vol. 29, no. 7, pp. 3318-3329, 2014.

[4] Kundu U, "Gain-Relationship-Based Automatic Resonant Frequency Tracking in Parallel, LLC Converter," *IEEE Transactions on Industrial Electronics*, vol. 63, no. 2, pp. 874-883, 2016.

[5] Ghadiri-Modarres M, "New Adaptive Algorithm for Delay Estimation of Sinusoidal Signals With Unknown Frequency," *IEEE Transactions on Instrumentation & Measurement*, vol. 64, no. 9, pp. 2360-2366, 2015.

[6] Mansouri M, "Estimation of Electromechanical

Oscillations From Phasor Measurements Using Second-Order Generalized Integrator," *IEEE Transactions on Instrumentation & Measurement*, vol. 64, no. 4, pp. 943-950, 2015.

[7] Reza M S, et al, "Accurate Estimation of Single-Phase Grid Voltage Parameters Under Distorted Conditions," *IEEE Transactions on Power Delivery*, vol. 29, no. 3, pp. 1138-1146, 2014.

[8] J. H. R. Enslin, "Harmonic interaction between a large number of distributed power inverters and the distribution network," *IEEE Trans. Power Electronics*, vol. 19, no. 6, pp. 1586-1593, 2004.

[9] L. Harnefors, L. Zhang and M. Bongiorno, "Frequency-domain passivity-based current controller design," *IET Power Electronics*, vol. 1, no. 4, pp. 455-465, December 2008.

[10] J. Zeng, Z. Zhang and W. Qiao, "An interconnection-damping-assignment passivity-based controller for a DC-DC boost converter with a constant power load," in *2012 IEEE Industry Applications Society Annual Meeting*, Las Vegas, NV, 2012, pp. 1-7.

[11] L. Harnefors, X. Wang, A. G. Yepes and F. Blaabjerg, "Passivity-Based Stability Assessment of Grid-Connected VSCs—An Overview," *IEEE Journal of Emerging and Selected Topics in Power Electronics*, vol. 4, no. 1, pp. 116-125, March 2016.

[12] Y. Liao, Z. Liu, G. Zhang and C. Xiang, "Vehicle-Grid System Modeling and Stability Analysis With Forbidden Region-Based Criterion," *IEEE Transactions on Power Electronics*, vol. 32, no. 5, pp. 3499-3512, May 2017.

[13] Q. Ye, R. Mo, Y. Shi and H. Li, "A unified Impedance-based Stability Criterion (UIBSC) for paralleled grid-tied inverters using global minor loop gain (GMLG)," in *2015 IEEE Energy Conversion Congress and Exposition (ECCE)*, Montreal, QC, 2015, pp. 5816-5821.

The 2018 International Power Electronics Conference

Harmonic Stability Assessment based on Global Admittance for Multi-Paralleled Grid-Connected VSIs using Modified Nyquist Criterion

Wu Cao[1], Dongchen Fan[1], Kangli Liu[1], Jianfeng Zhao[1], Liheng ruan[2], Xiaojun Wu[2]
1 Department of Electrical Engineering
Southeast University, Nanjing, Jiangsu, 210096, China
2 Jiangsu Haihang Electric Technology Co. Ltd
Zhenjiang, Jiangsu, 212200, China
Email: caowu_ee@seu.edu.cn, fdctc @163.com, kangcumt@yeah.net, jianfeng_zhao@seu.edu.cn, rlh@hhdq.net, wxj@hhdq.net

ABSTRACT-The mutual interactions in the multi-paralleled grid-connected VSIs coupled through the grid impedance tend to result in the whole system oscillating-formed resonance, namely the harmonic stability issue. To address them, the traditional impedance-based stability criterion (IBSC) evaluates the stability by the impedance ratio, also known as minor loop gain (MLG). In this paper, the novel global admittance is proposed to assess the harmonic stability using the modified Nyquist criterion. Compared to the IBSC, the computation efforts would be reduced significantly and the resonance damping factor as the stability margin index can be obtained to assess the stability degree directly. The simulation results verify the effectiveness of the theoretical analysis.

Keywords—Harmonic Stability, Global Admittance, Voltage Source Inverters, Modified Nyquist Criterion..

I. INTRODUCTION

Grid-connected VSIs are generally applied in these systems as energy interaction applications with grid to enhance system efficiency and power quality. However, there exists a complex coupling among inverters and the power grid when plenty of inverters connected to a weak grid via a common connection point (PCC) [1].Such couplings and interactions are known as harmonic stabilities [2]. The existence of these problems not only affects the distributed generation (DG) system reliability and power quality, but also seriously can lead to regional power grid paralysis. Many methods are proposed to analysis harmonic stability. Impedance-based method evaluates stability by the minor loop gain [3]. The system is stable only if the impedance ratio manifests the Nyquist stability criterion [4]-[5]. While the impedance ratio is hard to obtain with the large number of inverters of different control and filter parameters. Modal analysis serves as an effective tool to address the system stability. Different from frequency domain analysis method, we can find the useful resonance information by analyzing the characteristics of the Nodal matrix in [6]-[7]. A number of PCC admittance constitutes the nodal matrix.

However, the admittance is the s function, so that nodal matrix is complicated and the matrix order is high [8]. Passivity-based method is very useful for controller design of linear systems [9]. All grid-connected inverters can be designed such that the system output admittance becomes passive [10]. The system will be stale and the passivity-based inverter provides damping of an oscillation of any frequency. This method sacrifices the bandwidth in the purist of stability [11]. Therefore, the modified Nyquist criterion is proposed based on Nyquist criterion to assess the harmonic stability.

In this paper, a Norton admittance model is established to describe multi-paralleled Grid-Connected VSIs in detailed firstly. Secondly, the impedance-based stability criterion (IBSC) is elaborated and Y_{total} defined global admittance is proposed to assess the harmonic stability by modified Nyquist criterion. Also, resonance damping factor is proposed as the stability margin index. The criterion is realized and validated by simulation.

II. SYSTEM DESCRIPTION

The topology of multi-paralleled grid-connected VSIs with LCL filters is illustrated in Fig. 1. Since this research is concerned with the harmonic instability owing to the dynamics of inner control loops, the dc-link voltages of inverters are assumed to be constant.

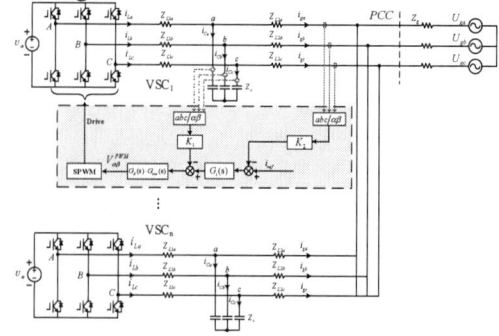

Figure 1: Multi-paralleled grid-connected VSIs system.

3015

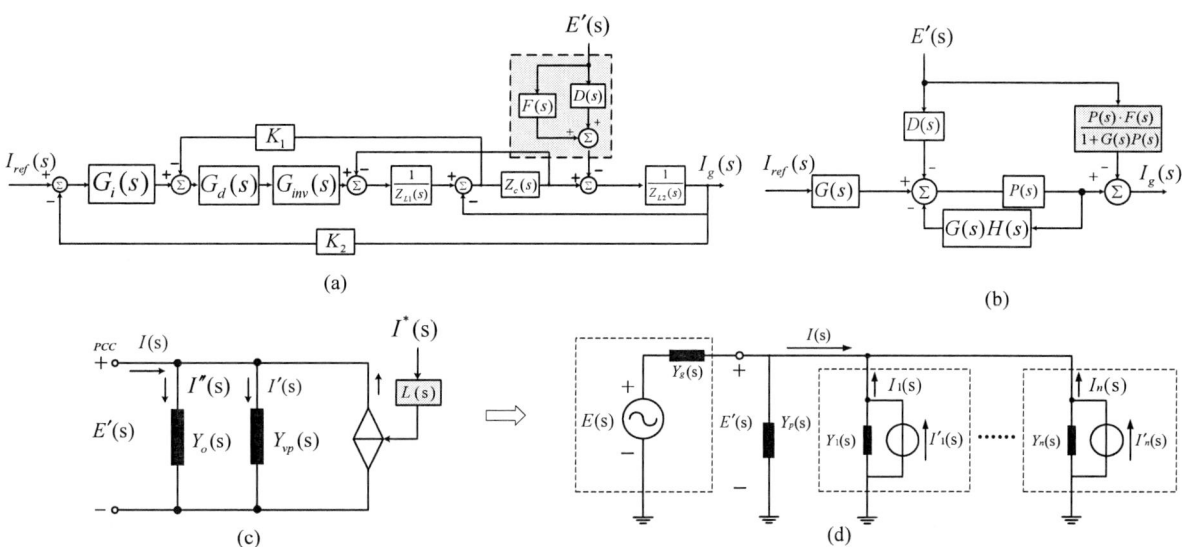

(a)

(b)

(c) ⟹ (d)

Figure 2: (a) Control block diagram of dual current loop LCL inverter (b) The equivalent transformation (c) The Norton equivalent model of single inverter (d) The admittance model of multi-parallel VSIs.

Taking single phase as example, Fig. 2(a) shows the block diagram of dual-loop control strategy based on the feedback of capacitor current and injected current and its equivalent transformation is illustrated in Fig. 2(b) by moving the comparison points or extraction points. According to the circuit theory, the Norton equivalent model of single inverter can be concluded from Fig. 2(c). $L(s)$ is defined as the controlled factor of controlled current source, which could be any current controller such as $Y_o(s)$ is output admittance, $Y_{vp}(s)$ is feedforward channel equivalent admittance.

Fig. 2(d) depicts the admittance model of multi-parallel VSIs, which is combined of single inverter Norton models. $Y_n(s)$ is the sum of the $Y_o(s)$ and $Y_{vp}(s)$.

III. THE PROPOSED STABILITY CRITERION

A. Impedance-Based Stability Criterion(IBSC)

With several grid-connected inverters modeled as a current source in parallel with output impedance, the overall system can be represented by the equivalent circuit shown in Fig. 3.

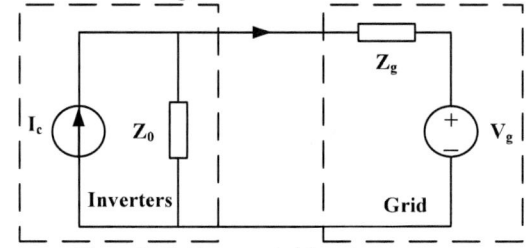

Figure 3: Grid-connected inverters system.

The output current of inverters is:

$$I(s) = \left[I_c(s) - \frac{V_g(s)}{Z_0(s)} \right] \cdot \frac{1}{1 + Z_g(s) / Z_0(s)} \quad (1)$$

It can be assumed that the inverters are stable individually and the grid voltage is stable when unloaded. The system is stable if the ratio of the grid impedance to the inverters output impedance Z_g/Z_0 satisfices the Nyquist criterion. The ratio Z_g/Z_0 is also known as minor loop gain (MLG).

IBSC has been utilized in AC power system, DC distributed power system and inverter-based power system successfully [12]. However, due to different parameters inside the complicated system, the parallel output impedance $Z_1//Z_2//\cdots Z_n$ is hard to obtain. The computation effort could be multiplied if more parallel inverters added.

B. Modified Nyquist Criterion

$Y_{total}(s)$ defined global admittance is assumed as the sum of all admittances in the equivalent circuit in (2), including the grid admittance, passive admittance and inverters output admittance, generally influenced by the long export ac cables and the power transformers, reactive compensation components, and the quantity of inverters.

$$Y_{total}(s) = \left(Y_g(s) + Y_p(s) + \sum_{m=1}^{n} Y_m \right) \quad (2)$$

The total current of PCC is derived in from Fig. 2, where $n=1$, 2, and 3..., represents the number of inverters:

$$I(s) = \sum_{i=1}^{n} I'_i(s) - \frac{Y_g(s)}{Y_{total}(s)} E(s) \sum_{i=1}^{n} Y_i - \frac{\sum_{i=1}^{n} Y_i}{Y_{total}} \sum_{i=1}^{n} I'_i(s) \quad (3)$$

It can be concluded that the total current $I(s)$ is divided into three components. To evaluate the system stability, it is essential to ascertain all three components stability.

Given that the single inverter is designed to be stable, the sum of equivalent controlled current source $\Sigma I'_i(s)$ and the sum of admittance $\Sigma Y(s)$ must be stable. And because the grid admittance $Y_g(s)$ is consists of passive elements, it behaves stable. Therefore, the stability of current depends on $Y_{total}(s)$. Modified Nyquist criterion has been proposed to assess the stability.

Fig.5 depicts bode diagram and the Nyquist plots for global admittance $Y_{total}(s)$. Applied Cauchy's argument principle, global admittance corresponds to the denominator $1+Z_g(s)/Z_0(s)$ in impedance-based stability criterion. The number of encirclements of the origin $(0, j0)$ should be zero when the system is stable.

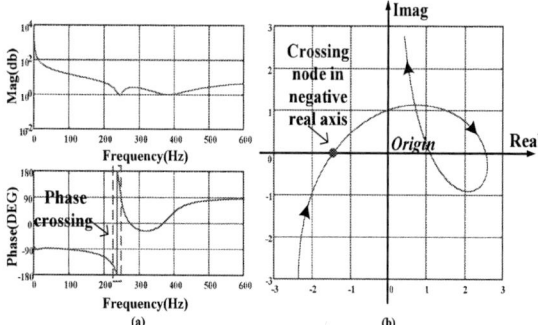

Figure 5: Global admittance trans function bode and Nyquist plot.

The crossing nodes in real axis mean that the magnitude of imaginary part is zero that reveals the condition of system resonant. Thus, the crossing nodes in real axis are resonant nodes.

The real part value of resonant point is defined as resonance damping factor, where

$$R_d = Y_{total}\left(j\omega\right)\Big|_{Im=0} \qquad (4)$$

If $R_d<0$, it represents a positive crossing in negative real axis at resonant nodes and the laps of Nyquist plot surround the origin isn't zero, causing the instability. If $R_d>0$, it represents a crossing in real axis that the number of Nyquist plot surrounded the origin is zero, the system is stable. That is, $R_d>0$ is the sufficient condition of the system stability. In addition, the value of R_d could reflect stability margin index. The absolute value of R_d represents the degree of stability or instability.

This proposed modified stability criterion can be concluded as follow:

(1) Whether global admittance Nyquist encircled the origin is applicable to estimate the stability;

(2) The criterion can be extended that R_d must be larger than zero;

The MLGs seen from each inverter reveal different system stability status. Therefore, it is necessary to calculate all the MLGs to address paralleled grid-tied inverters system stability. Compared to the proposed modified Nyquist criterion, it requires much more computation efforts when n becomes large. Based on the accessible admittance of individual inverter, the complicated computation of parallel impedance can be reduced, under VSIs with internal different parameters and much number.

TABLE 1
COMPARISON WITH DIFFERENT STABILITY CRITERION

Method	Judgement Range	Computational Complexity
Modal analysis	Global	high
IBSC	Local	high
Modified Nyquist Criterion	Global	low

The comparison of judgement range and computational complexity is shown in table 1.

IV. SIMULATION VERIFICATION

Under the condition of weak gird, the grid-side impedance is not a constant value and varies rapidly. Besides, the number of inverters affects the stability of multiple inverters gird-tied system. The more inverters, the possibility of resonance is become serious. In this section, two simulation cases are presented to verity the consequence of the modified stability criterion.

The filter parameters, control parameters and grid parameters of single inverter are shown in table 2. The current controller is designed by Vector proportion integral (VPI) method.

TABLE 2
ELECTRICAL AND CONTROLLER PARAMETERS

Parameter	Value
Rated line voltage (motor)	380VAC
dc-link voltage	800V
Inductance	$L_1=0.1$mH $L_2=0.3$mH
Capacitance	$C_1= 1$uF
Switching Frequency	10kHz
VPI current controller	$Kp_1=Kp_2=18$
	$Ki_1=Ki_2=600$

The global admittance Bode diagram of a 3 parallel gird-connected VSIs system is depicted in Fig. 6(a). A sudden peak in the magnitude of the global admittance causes the system unstable. Furthermore, the resonance damping factor R_d defined in (3) is less than zero shown in Fig. 6(b). R_d represents the degree of instability. The Nyquist plot of Y_{total} shown in Fig. 6 (e) encircles the origin and crosses through negative real axis. The amplitude of PCC voltage is enlarged due to the instability in Fig.7 (a). Theses mean the system is unstable and consistent with the proposed criterion. The modified Nyquist criterion is only used once based on $Y_{total}(s)$. With the traditional IBSC, the MLGs of the three inverters have to be derived individually in expression (5) or calculate the parallel impedance.

$$MLG_1 = \frac{Y_{o,1}}{\dfrac{1}{Z_g} + Y_{o,2} + Y_{o,3}} \dots MLG_i \qquad (5)$$

Another case is about a 2 parallel grid-connected VSIs system. The global admittance bode diagram is depicted in Fig. 6(b), that the resonance peak is weakened. The resonance damping factor R_d is larger than zero shown in Fig. 6 (d). It can be illustrated that system meet the stability requirement. Here, R_d

represents the degree of instability The Nyquist plot of Y_{total} doesn't encircle the origin and crosses through passive real axis in Fig. 6(f). The PCC voltage in time domain keeps stable in Fig. 7(b). It proves that the number of inverters affects the stability by the modified Nyquist criterion.

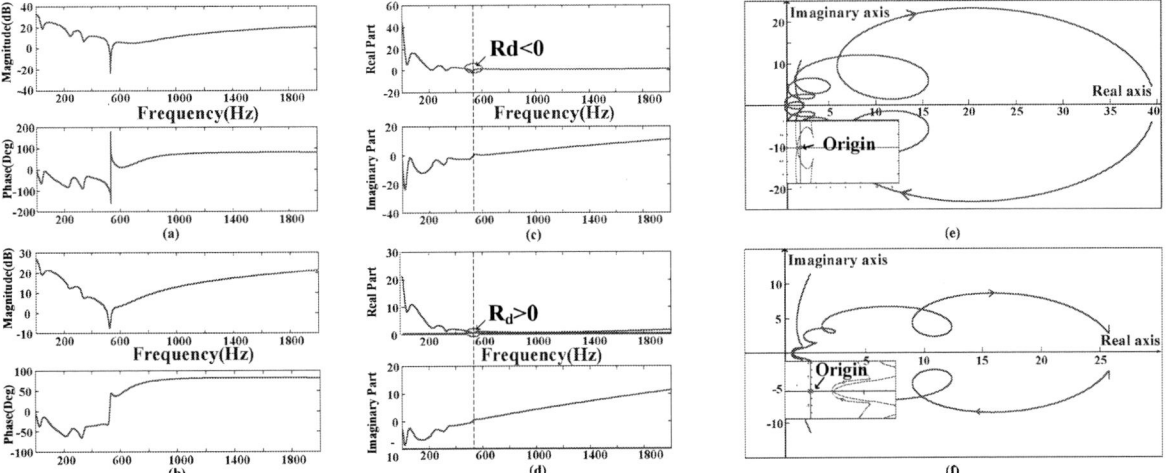

Figure 6: (a) (b) Global admittance bode diagrams (c) (d) Real part curves and imaginary part curves (e) (f) Nyquist plots(Amplitude-phase curve) (n=2,3 respectively).

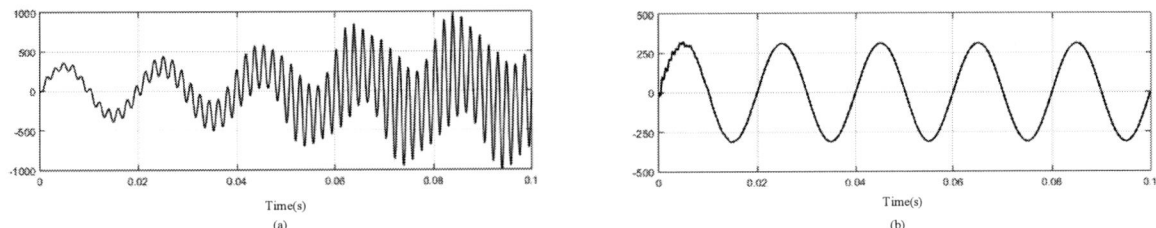

Figure 7: (a) (b) Waveform of PCC voltage (n=3, 2 respectively).

V. CONCLUSIONS

In this paper, a Norton admittance model is established to describe multi-paralleled Grid-Connected VSIs in detailed. Based on the model, the global admittance is introduced to assess the system harmonic stability using modified Nyquist criterion. Compared to the traditional IBSC, the modified stability criterion reduces computation efforts. Also, resonance damping factor as the stability margin index is obtained to assess the stability degree directly. The criterion is realized and validated by simulation. In the future work, the deep meaning of R_d would be explored. Some work about the contribution of each inverter to stability is going to study.

ACKNOWLEDGMENT

The authors would like to thank the sponsorship of the National Natural Science foundation of China (51607037), Innovation foundation for Combination of Industry and Scientific Research of Jiangsu Province (BY2016076-09).

REFERENCES

[1] J. L. Agorreta, M. Borrega, J. López and L. Marroyo, "Modeling and Control of N-Paralleled Grid-Connected Inverters With LCL Filter Coupled Due to Grid Impedance in PV Plants," *IEEE Transactions on Power Electronics*, vol. 26, no. 3, pp. 770-785, March 2011.

[2] C. Yoon, X. Wang, F. M. F. D. Silva, C. L. Bak and F. Blaabjerg, "Harmonic stability assessment for multi-paralleled, grid-connected inverters," in *2014 International Power Electronics and Application Conference and Exposition*, Shanghai, 2014, pp. 1098-1103.

[3] J. Sun, "Impedance-Based Stability Criterion for Grid-Connected Inverters," *IEEE Transactions on Power Electronics*, vol. 26, no. 11, pp. 3075-3078, Nov. 2011.

[4] M. Cheah-Mane; L. Sainz; J. Liang; N. Jenkins; C. E. Ugalde Loo, "Criterion for the Electrical Resonance Stability of Offshore Wind Power Plants Connected through HVDC Links," *IEEE Transactions on Power Systems*. (to be published)

[5] T. Liu, Z. Liu, J. Liu and Q. Dou, "Mechanism analysis and mitigation of instability in grid-connected Voltage Source Inverter with LCL filters based on terminal impedance,"in *2016 IEEE Applied Power Electronics Conference and Exposition (APEC)*, Long Beach, CA, 2016, pp. 2272-2277.

[6] C. Yang, K. Liu and Q. Zhang, "An improved modal analysis method for harmonic resonance analysis," in *2008 IEEE International Conference on Industrial Technology*, Chengdu, 2008, pp. 1-5.

[7] T. Ngo and S. Santoso, "Modal-based voltage stability analysis of low frequency AC transmission systems," in *2016 IEEE Power and Energy Society General Meeting (PESGM)*, Boston, MA, 2016, pp. 1-5.

[8] F. A. El-Sheikhi, Y. M. Saad, S. O. Osman and K. M. El-Arroudi, "Voltage stability assessment using modal analysis of power systems including flexible AC transmission system (FACTS)," in *Large Engineering Systems Conference on Power Engineering, 2003*, 2003, pp. 105-108.

[9] J. Siegers, S. Arrua and E. Santi, "Stabilizing controller design for multi-bus MVDC distribution systems using a passivity based stability criterion and positive feed-forward control," in *2015 IEEE Energy Conversion Congress and Exposition (ECCE)*, Montreal, QC, 2015, pp. 5180-5187.

[10] L. Harnefors, L. Zhang and M. Bongiorno, "Frequency-domain passivity-based current controller design," *IET Power Electronics*, vol. 1, no. 4, pp. 455-465, December 2008.

[11] J. Zeng, Z. Zhang and W. Qiao, "An interconnection-damping-assignment passivity-based controller for a DC-DC boost converter with a constant power load," in *2012 IEEE Industry Applications Society Annual Meeting*, Las Vegas, NV, 2012, pp. 1-7.

[12] L. Harnefors, X. Wang, A. G. Yepes and F. Blaabjerg, "Passivity-Based Stability Assessment of Grid-Connected VSCs—An Overview," *IEEE Journal of Emerging and Selected Topics in Power Electronics*, vol. 4, no. 1, pp. 116-125, March 2016.

[13] Y. Liao, Z. Liu, G. Zhang and C. Xiang, "Vehicle-Grid System Modeling and Stability Analysis With Forbidden Region-Based Criterion," *IEEE Transactions on Power Electronics*, vol. 32, no. 5, pp. 3499-3512, May 2017.

[14] Q. Ye, R. Mo, Y. Shi and H. Li, "A unified Impedance-based Stability Criterion (UIBSC) for paralleled grid-tied inverters using global minor loop gain (GMLG)," in *2015 IEEE Energy Conversion Congress and Exposition (ECCE)*, Montreal, QC, 2015, pp. 5816-5821.

[15] X. Wang, F. Blaabjerg and W. Wu, "Modeling and Analysis of Harmonic Stability in an AC Power-Electronics-Based Power System," *IEEE Transactions on Power Electronics*, vol. 29, no. 12, pp. 6421-6432, Dec. 2014.

The AC Traction Power Supply System for Urban Rail Transit Based on Negative Sequence Current Compensator

Tianshu Zhao[1], Xu Peng[2*]
1 Chengdu NO. 7 High School, Chengdu, China
2 School of Electrical Engineering, Southwest Jiaotong University, Chengdu, China
*Email: pengxuswjtu@foxmail.com

Abstract-The main problems of stray current and the low availability of the regenerative braking energy widely exist in DC traction power supply system which is currently widely used in China. The AC traction power supply system for Urban Rail Transit based on negative sequence current compensator is proposed as a kind of novel power supply scheme to solve these problems in DC system. And the construction of the AC traction power supply system is designed in detail. In the meantime, the system of negative sequence compensator based on three - phase three - level rectifier parallel is designed to compensate for the negative sequence generated by the proposed system in the paper. The negative sequence current compensation algorithm of the AC traction power supply system and the control method of the negative sequence current compensator are deeply analyzed and designed, and the proposed system is feasible to connect with three-phase power grid. Meanwhile, the equalization of the AC traction power supply system was also achieved. Finally, the performance of the power electronic compensator to eliminate the negative sequence current was verified by simulations and experiments.

Keywords: Traction Power Supply System for Urban Rail Transit; Three-level Neutral Point Clamped Convertor; Negative Sequence Current Compensator; Control Strategy

I. INTRODUCTION

With the rapid development of urban railway transit, the researches of the traction power system for urban railway transit are also broadly launched. At present, the DC traction power system is adopted in all the equipped urban railway transit in China, and the 750V or 1500V is used as the main voltage level. The no neutral section devices and the streamlined operation of urban rail transit train are recognized as the main strength of the DC traction power system. However, the problem of stray current and the low availability of regenerative braking energy widely exist in the DC system [1]-[3]. The prevention and control measures of stray current are required to invest a large number of construction funds and complicated researches, nevertheless, its effect is proved to be limited. So it is difficult to control the stray current in general. In addition, the regenerative braking energy can be used only if the extra equipment is designed and connected with the existing systems, while the new utilization programs of regenerative braking energy are also complex and at a great cost.

The single-phase AC traction power supply system is designed as the new traction power system in this paper.

In comparison with the DC system, the improved power supply capacity, the lower construction cost, and the higher carrying capacity are the characters of the new AC traction power supply system. Meanwhile, the AC traction system designed for the urban railway transit not only maintains no neutral section devices and the streamlined operation, but also be capable to solve the stray current and the low availability of regenerative braking energy.

The construction of the AC main traction substation, which has the highest capacity, is composed of the main step-down traction transformer and the negative sequence current compensator. The negative sequence current compensator is mainly composed of several power electric converters [4]-[6]. The negative sequence current compensator can relieve the negative sequence current produced by the traction power system and make use of the regenerative braking energy efficiently. And the other traction substations are based on the three-phase to single –phase cascaded converters to cancel the neutral section device [7-10].

The structure of the novel AC main traction substation, the principle of negative sequence current compensator and the control strategy of the power electric compensator will be described and analyzed amply in this paper.

II. THE STRUCTURE OF THE AC MAIN TRACTION SUBSTATION

The novel AC traction power supply system is exhibited in Figure 1. The construction of the main AC traction substation is composed of the main step-down traction transformer and negative sequence current compensator. The electricity is taken from the 110kV or 220kV three - phase grid and transformed into 27.5kV by the step-down transformer. Phase A and phase B is selected to connect with the catenary of the traction power system. The voltage value of the catenary is same as the line voltage of the phase A and phase B. Meanwhile, phase C is used as the alternate phase. When a single-phase fault occurs in phase A or phase B, phase C can be put into operation to protect the traction power system from power cut.

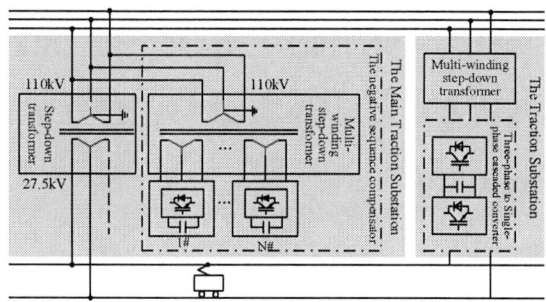

Figure 1. The AC traction power supply system for Urban Rail Transit Based on Negative Sequence Current Compensator

The structure of the new AC traction power system is concise and easy to be applied. Because the catenary of traction substation is only connected with phase A and phase B, the unbalanced three-phase current contained with harmonic and negative sequence current will be produced by the AC traction power system. If the three-phase current can't be compensated, the power quality of the grid will be seriously influenced by it.

The negative sequence current compensator is mainly composed of power electric converters. The negative sequence current compensation system is constitute of multi-winding step-down transformer and several power electronic converters connected in parallel. The primary side of multi-winding step-down transformers is also connected with the primary side of the main traction transformer, while the secondary sides are linked with the compensators.

The compensator operates as a rectifier and the output is connected to a capacitor for energy storage. And the structure of the compensator is shown as Figure 2. Each compensator would accomplish several proportions of the negative sequence current compensation. The capacity of the system can be up to 1MW through the parallel architecture.

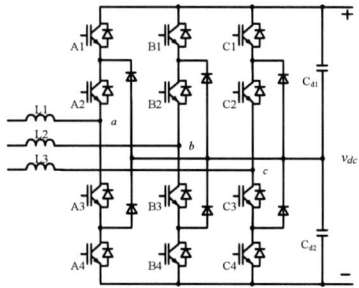

Figure 2. The three-level structure of the compensator

III. THE CONTROL STRATEGY

The negative sequence current compensator of the AC traction power system is completed by the control of the converters. The control strategy of a single compensator is taken as an example.

First, the coordinate of the normalized three-phase voltage and current is transformed respectively from three phase static coordinate (*abc*) to synchronously rotating reference frame (*dq*). The transformation is based on the phase of the grid-side voltage distilled through the three-

phase phase-locked loop. The output of the coordinate transformation which is processed through the low pass filter is divided into the active component and the reactive component of the positive and negative sequence voltage and current. The block diagram of extraction method is shown in Figure 3.

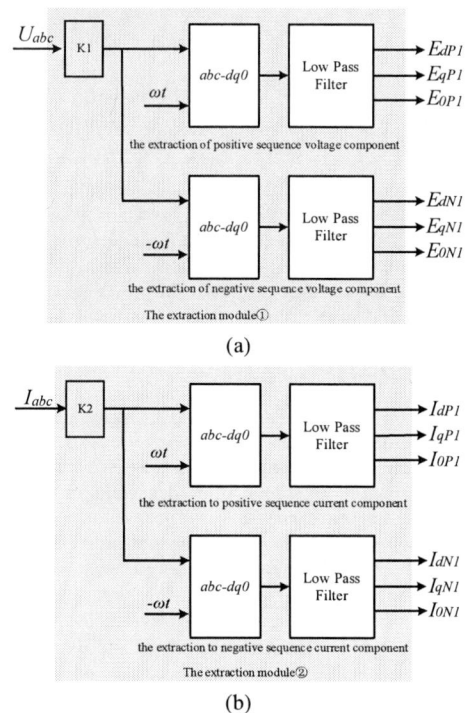

(a)

(b)

Figure 3. The component extraction method.
(a) The extraction of the voltage component;
(b) The extraction of the current component.

Each of the current component can be separated by the above-stated method. These components can be respectively marked as I_{dP} (active component of the positive current), I_{qP} (reactive component of the positive current), I_{dN} (active component of the negative current), I_{qN} (reactive component of the negative current).

In order to eliminate the negative sequence current of the grid side, the reference of each component needs to be given or calculated. The reference of active component $I_{dN}*$ and reactive component $I_{qN}*$ of the negative current is given as zero to clear up the negative sequence current IN of the grid side. While the reference of active component $I_{dP}*$ and reactive component $I_{qP}*$ of the positive current require to be calculated.

Firstly the voltage across the DC capacitor of output side V_{dc} should to be detected and compared with the given value $V_{dc}*$. The difference between V_{dc} and $V_{dc}*$ need to be regulated by the PI controller. The value multiplied by the given voltage $V_{dc}*$ and the output value of the PI controller is recognized as the reference active power P_0 of a single negative sequence current compensator. The calculation method of P_0 is shown in Figure 3. Then connected with the detected active component E_{dP} and reactive component E_{qP} of the

3021

positive voltage, the excepted value of active component $I_{dP}{}^*$ and reactive component $I_{qP}{}^*$ of the positive current can be figured out as the formula (1) and (2).

$$\begin{bmatrix} i_d^{P*} \\ i_q^{P*} \end{bmatrix} = \frac{2P_0}{3D} \begin{bmatrix} e_d^P \\ e_q^N \end{bmatrix} = KP_0 \begin{bmatrix} e_d^P \\ e_q^N \end{bmatrix} \qquad (1)$$

Among the formula (1), parameter D is calculated as formula (2):

$$D = \left[(e_d^P)^2 + (e_q^P)^2 \right] - \left[(e_d^N)^2 + (e_q^N)^2 \right] \neq 0 \qquad (2)$$

The calculation process is illustrated by Figure 4.

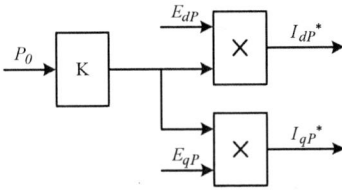

Figure 4. The calculation method of $I_{dP}{}^*$ and $I_{qP}{}^*$

A double closed-loop control algorithm is adopted to eliminate the negative sequence current of the grid side and control the active and reactive power of the grid side. The active current component of positive sequence, for example, the difference between it and its reference current along with the reactive current component and active voltage component of positive sequence can be calculated and superimposed as the new component. The new component is the positive sequence active component of modulation wave in synchronously rotating reference frame. Others component of modulation wave can be calculated by the similar terminology. The Overall control strategy is exhibited by the Figure 5.

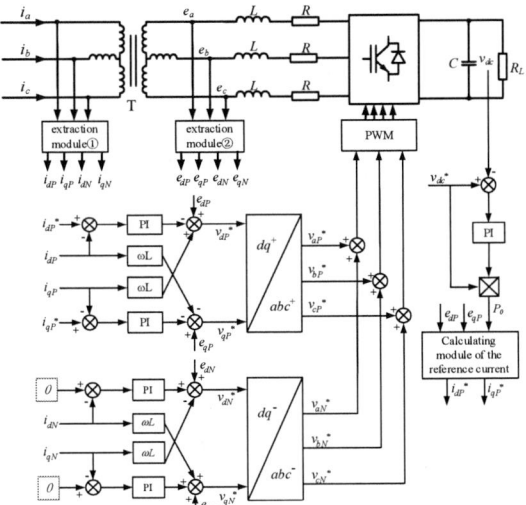

Figure 5. The control strategy of the negative sequence current compensator

IV. SIMULATION AND VERIFICATION

In this section, simulation model of the new AC traction power system based on negative sequence current compensators is built by Matlab/Simulink to verify the feasibility and stability of the AC traction power system.

The simulation model is built as the structure exhibited by Figure 1. The control algorithm is built as the scheme shown in Figure 5. The parameters of simulation are shown in Table 1.

TABLE I
THE PARAMETERS OF THE SIMULATION

Parameter	Value
Reference of DC voltage	1500V
DC capacitor C	1mF
Initial value of capacitor C voltage	1200V
Filter inductance	5mH
AC load	1000Ω
Sampling frequency	2kHz

Figure 6 to Figure 8 are the results of simulation. The results of compensation can be seen through the compare between the (a) and (b) of Figure 6. It is obviously to know that the current are nearly balanced after the compensation in Figure 6 (b). Figure 7 shows that the output DC voltage can be stabilized at the reference voltage 900V. Figure 8 shows the active power waveform and reactive power waveform of grid side.

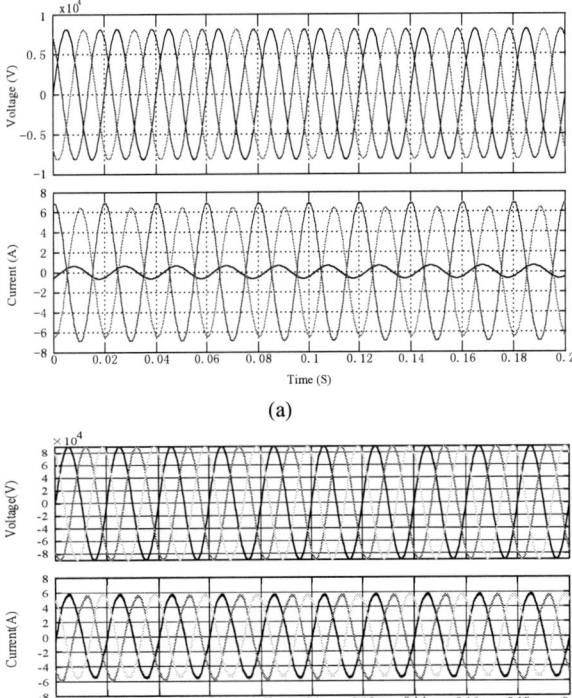

(a)

(b)

Figure 6. The voltage and current waveform of the grid side.
(a)The voltage and current waveform before the compensation;
(b)The voltage and current waveform after the compensation.

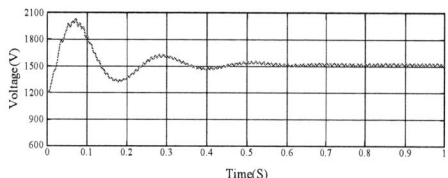

Figure 7. Output DC voltage waveform of compensator

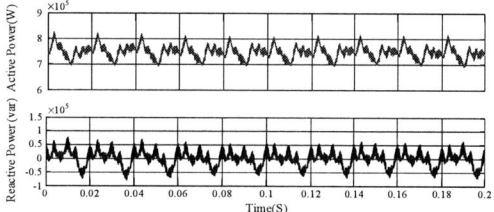

Figure 8. Active power and reactive power of grid side

The feasibility and stability of the AC traction power system have been verified by experiment as well. In the experiment, the parameters are show in Table 2. The FPGA-based controller implements the PWM algorithm, and the version of FPGA is EP3C55F484C7N. The controller is totally coded using Verilog-HDL, synthesized by Quartus II.

TABLE II
THE PARAMETERS OF THE EXPERIMENT

Parameter	Value
Reference of DC voltage	50V
Capacitor of DC side	1mF
Filter inductance	5mH
Sampling frequency	2kHz

The current waveform and output DC voltage waveform is shown in Figure 9. It can be known that the current of grid side is balanced and stable from the Figure 9(a). And it can be known that the Output DC voltage waveform of a compensator can maintain stable when load mutation or reference is changed from the Figure 9(b).

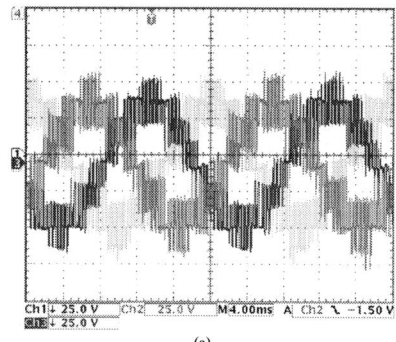

(a)

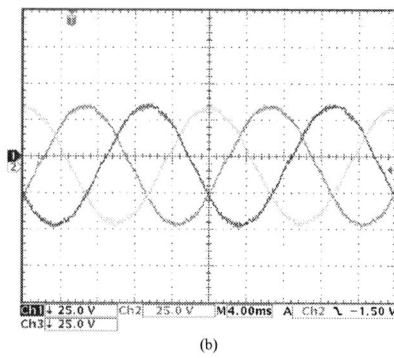

(b)

Figure 9. Comparison of current waveform.
(a) Input current of compensator. (b) Current of grid side.

V. CONCLUSION

A kind of novel AC traction power supply system for urban rail transit, which is based on negative sequence current compensator is proposed in this paper. The structure of the main traction substation is introduced in detail, and the negative sequence current problem, which results from the structure of the main traction substation, is analyzed theoretically. Meanwhile, the corresponding algorithm and control strategy of the negative sequence compensator are also illustrated elaborately in this paper. Finally, the feasibility and stability of the AC traction power system are verified by simulation and experiment.

VI. ACKONWLEDGEMENT

This work was supported by the National Natural Science Foundations of China (Grant Nos. 51477144).

REFERENCES

[1] Fujun Ma, Zhixing He, Qianming Xu, An Luo, Leming Zhou, Mingshen Li, "Multilevel Power Conditioner and its Model Predictive Control for Railway Traction System," *IEEE Trans. on Industrial Electronics*, vol.63, no.11, pp.7275-7285, 2016.

[2] Kejian Song, Georgios Konstantinou, Wu Mingli, etc, "Windowed SHE-PWM of Interleaved Four-Quadrant Converters for Resonance Suppression in Traction Power Supply Systems, " *IEEE Trans. on Power Electronics*, vol.32, no.10, pp. 7870-7881, 2017.

[3] Shibin Gao; Xin Li, Xiaolan Ma, Haitao Hu, Zhengyou He, Jianwei Yang, "Measurement-Based Compartmental Modeling of Harmonic Sources in Traction Power Supply System, " *IEEE Trans. on Power Delivery*, vol.32, no.2, pp. 900-909, 2016.

[4] Xiaoqiong He, Zeliang Shu, Xu Peng, Qi Zhou, Yingying Zhou, Qijun Zhou, Shibin. Gao, "Advanced Co-phase Traction Power Supply System Based on Three-Phase to Single-Phase Converter," *IEEE Trans. on Power Electron*, vol.29, no.10, pp.5323-5333, 2014.

[5] S. Wensheng, F. Xiaoyun, and K.M. Smedley, "A Carrier-Based PWM Strategy With the Offset Voltage Injection for Single-Phase Three-Level Neutral-Point-Clamped Converters, " *IEEE Trans. on Power Electronic*, vol.28, no.3, pp.1083-1095, 2013.

[6] Lu Fang, Xianyong Xu, Houhui Fang, Yuhu Xiao, "Negative-sequence current compensation of power quality compensator for high-speed electric railway," *2014 IEEE Conference and Expo Transportation Electrification Asia-Pacific (ITEC Asia-Pacific)*, pp. 1-5.

[7] Yanjun Tian, Poh Chiang Loh, Fujin Deng, etc, "Impedance Coordinative Control for Cascaded Converter in Bidirectional Application," *IEEE Trans. on Industrial Application*, vol. 52, no.5, pp. 4084-4095, 2016.

[8] Xu Peng, Xiaoqiong He, Pengcheng Han, Aiping Guo, Zeliang Shu, Shibin Gao, "Smooth Switching Technique for Voltage Balance Management Based on Three-Level Neutral Point Clamped Cascaded Rectifier," *Energies*, vol. 9, pp. 1-16, 2016.

[9] Gregory Arthur de Almeida Carlos; Cursino Brandão Jacobina; João Paulo Ramos Agra Méllo; Euzeli Cipriano dos Santos, "Shunt Active Power Filter Based on Cascaded Transformers Coupled With Three-Phase Bridge Converters" *IEEE Trans. on Industrial Application*, vol. 53, no.5, pp. 4673-4681, 2017.

[10] Amritesh Kumar; Vishal Verma, "Analysis and control of improved power quality single-phase split voltage cascaded converter feeding three-phase OEIM drive," *IET Power Electronics*, vol.10, no.8, pp. 903-910, 2017.

The 2018 International Power Electronics Conference

Grid Connected Power Generation Control Method for Z-Source Integrated Bidirectional Charging System

Xu Jia, Guoming Chuai, Haonan Niu and Qianfan Zhang*

Department of Electrical Engineering, Harbin Institute of Technology, Harbin, 150001, China

E-mail: zhang_qianfan@hit.edu.cn

Abstract—When electric vehicles stop to charge the batteries, the motor drive in vehicles do not work. The motor drives can be reconstructed into the charging system of the vehicle in an appropriate way. With a few devices added, it can possess the capacity of fast charging and the function of V2G(Vehicles to grid). In this paper, a Z-source integrated bidirectional charging system with LCL filter is proposed. An active damping method based on the integrated charging system is used to reduce the harmonic current at the resonant frequency of the output current. The mathematical model of proposed system is established, and the control methods of the DC side and the AC side are put forward. Finally, the Z-source experimental platform under the operation of grid connected is set up. The experiment was carried out to verify the feasibility of system.

Keywords—Bidirectional charging, Electric vehicles, Grid connected, Z-Source..

I. INTRODUCTION

The popularity of electric vehicles has contributed to the continuous development of Vehicle to Grid (V2G) technology [1]. In the concept of V2G, electric vehicles act as distributed energy storage resources to feed power to the grid through bidirectional charging devices. When the batteries need to be charged, the power grid supplies energy to the electric vehicles; when the electric vehicles dock, as distributed energy storage sources, they can quickly respond to the power requirements of the grid, and supply the power to the grid. Under the restrictions of volume and weight, the onboard charger on vehicle usually can not achieve high-power energy conversion [2-4]. On the condition of docking, if the motor drives components with higher power levels are reconstructed into a charging system, it can not only save limited space, but also improve the power. This is integrated charging. According to the number of power conversion stages under the mode of V2G, grid connected inverter can be divided into single-stage and double-stage. Single-stage inverter needs high DC link voltage to feed power to the grid, which needs to increase the serial number of batteries. In double-stage inverter, DC/DC converter is added to boost battery voltage to required value. In this

paper, a Z-source converter is used. Compared with the traditional voltage source inverter, Z-source inverter with a DC link voltage boost function tolerates the bridge leg shoot-through to improve system robust. Without dead-time, output voltage THD is decreased [5-8]. In the process of feeding power to grid, a wide power output range, faster response speed, and low current harmonics are needed. LCL filter is applied [8-11]. An active damping method with additional feedback variables can effectively suppress the generation of resonant spikes of the filter.

II. Integrated bidirectional charger scheme

The concept of integrated bidirectional charging is illustrated in Fig.1. Parts of onboard drives include Z source inverter and control circuits are used as bidirectional charging system to complete battery charging and generating back to grid during parking.

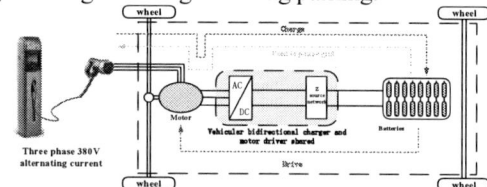

Fig. 1 Integrated bidirectional charging system.

Fig.2(a) shows the diagram of tranction motor drives main circuit. DC link voltage can be boost in this Z source inverter. In battery charging mode, the motor drive part is working as a controlled Z source rectifier. In V2G mode, the motor drives with additional LCL filter is working as a grid connecting Z source inverter. This paper mainly focuses on grid connected power generation.

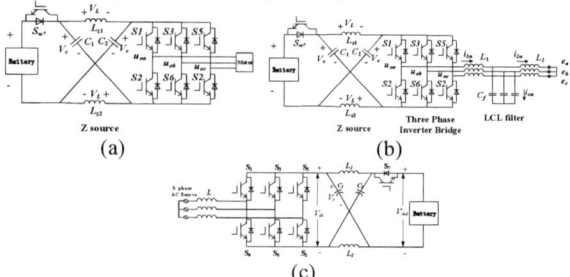

(a)

(b)

(c)

Fig.2 Three modes of integrated charger system. (a) Driving mode. (b) V2G mode. (c) Battery charging mode.

III. Mathematical Model of Three-phase Z-source Inverter with LCL Filter

The three-phase Z-source inverter topology with LCL filtering is shown in Fig. 3, where u_{on} represents the electric potential difference between the output of the inverter and the neutral point of the filter capacitor (n = a, b, c).

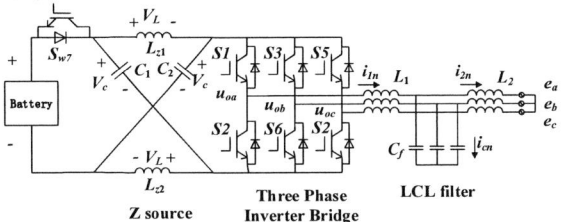

Fig. 3 Three phase Z source inverter topology with LCL filter.

The Z source network is modeled and analyzed, and the capacitor voltage and the inductor current are selected as the state variables to obtain the mathematical model of the Z source network in the two-phase rotating coordinate system, When the system is in the non-direct state shown in Fig. 4(a), the equation of state can be obtained as following formula (1). When the system is in the direct state shown in Fig. 4(b), the equation of state is obtained as following formula (2).

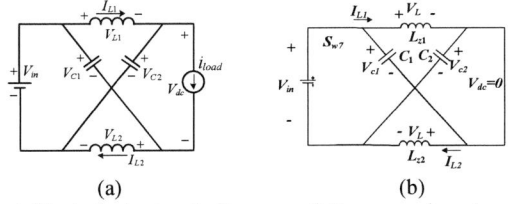

(a)   (b)

Fig. 4 Equivalent circuit diagram of Z source inverter. (a) Non-direct state. (b) Direct state.

$$\begin{cases} C \cdot \dfrac{dV_C}{dt} = i_L - i_{load} \\[2mm] L \cdot \dfrac{di_L}{dt} = V_{in} - V_C \end{cases} \tag{1}$$

$$\begin{cases} C \cdot \dfrac{dV_C}{dt} = -i_L \\[2mm] L \cdot \dfrac{di_L}{dt} = V_C \end{cases} \tag{2}$$

Where i_L, i_{load} are inductance current and DC bus current of inverter; C and L are inductor value and capacitance value of Z source. The variable $S_o = 0$ is defined to express the direct state, and the non-direct state is expressed by the variable $S_o = 1$. So the Z source network mathematical model is obtained by switching function as following formula (3).

$$\begin{cases} C \dfrac{dV_C}{dt} = (1-2S_0)i_L - \dfrac{3}{2}(i_d S_d + i_q S_q) \\[2mm] L \cdot \dfrac{di_L}{dt} = (1-S_0) \cdot V_{in} - (1-2S_0) \cdot V_C \end{cases} \tag{3}$$

Because the DC bus current of three phase inverter is related to the conduction state of switch tube, the switch function S_a, S_b, S_c are used to indicate the conduction or turn off state of three-phase bridge (1 indicates conduction and 0 indicates turn off), and the DC side current i_{load} can be described as following formula (4).Where i_a, i_b, i_c are inverter output three-phase current.

$$i_{load} = i_a S_a + i_b S_b + i_c S_c \tag{4}$$

The above models are obtained in three phase static coordinate system, so the variables are all AC flow, which is not conducive to the design of the control system. It needs coordinate transformation is shown in Fig. 5, and the transformation way is shown in formula (5).

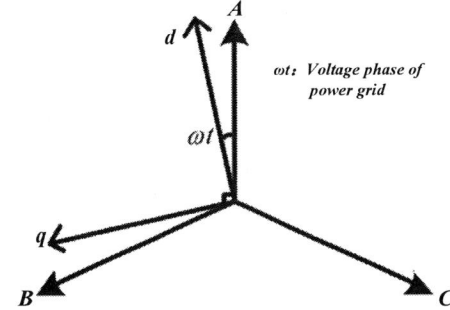

Fig. 5 d-q coordinate transformation system

$$\begin{bmatrix} X_d \\ X_q \\ X_0 \end{bmatrix} = \frac{2}{3} \begin{bmatrix} \cos\omega t & \cos(\omega t - 120°) & \cos(\omega t + 120°) \\ \sin\omega t & \sin(\omega t - 120°) & \sin(\omega t + 120°) \\ 1/2 & 1/2 & 1/2 \end{bmatrix} \begin{bmatrix} X_a \\ X_b \\ X_c \end{bmatrix} \tag{5}$$

According to the derivation process above, the model structure of the Z source converter in the synchronous rotating coordinate system is shown in Fig. 6.

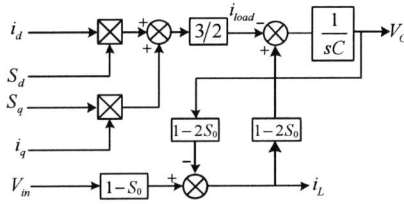

Fig. 6 Model of Z source converter.

According to the state equation of inverter with LCL filter, inverter side current i_{1n}, grid connected current i_{2n} and filter capacitor voltage u_{cn} are selected as the state variables, then the mathematical model of three-phase inverter under rotating coordinate system through d-q transform are as following formulas (6).

$$\begin{cases} L_1 \dfrac{di_{1d}}{dt} = V_C S_d - \omega L_1 i_{1q} - r_1 i_{1d} - u_{cd} \\[2mm] L_1 \dfrac{di_{1q}}{dt} = V_C S_q + \omega L_1 i_{1d} - r_1 i_{1q} - u_{cq} \\[2mm] L_2 \dfrac{di_{2d}}{dt} = -e_d - \omega L_2 i_{2q} - r_2 i_{2d} + u_{cd} \\[2mm] L_2 \dfrac{di_{2q}}{dt} = -e_q + \omega L_2 i_{2d} - r_2 i_{2q} + u_{cq} \\[2mm] C_f \dfrac{du_{cd}}{dt} = i_{1d} - i_{2d} - \omega C_f u_{cq} \\[2mm] C_f \dfrac{du_{cq}}{dt} = i_{1q} - i_{2q} + \omega C_f u_{cd} \end{cases} \tag{6}$$

Where i_{1d}, i_{1q} are d and q axis components of inverter output current vectors; i_{2d}, i_{2q} are d and q axis components of grid current vectors; u_{cd}, u_{cq} are d and q axis components of filter capacitor voltage vectors; e_d, e_q are d and q axis components of power grid voltage vectors. According to the formula (2), the model structure of three-phase grid connected Z source inverter with LCL filter in two-phase rotating coordinate system can be obtained, as shown in Fig. 7.

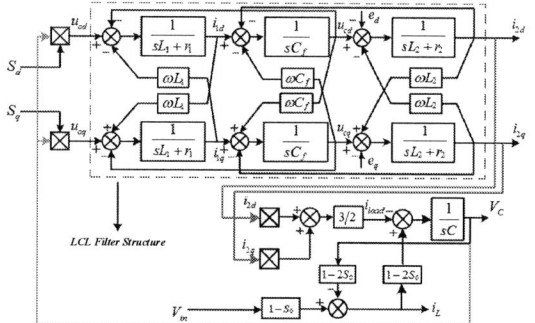

Fig. 7 Model of three phase grid connected Z source inverter with LCL filter in d-q coordinate system.

IV. Grid connected power generating control strategy

Based on the system structure model, control system can be divided into two parts: one is the control of DC side voltage, aiming to ensure the Z-source capacitance voltage stabilized in an appropriate range by algorithm, to meet the requirements of DC link voltage of 3-phases inverter; the other part is the control of AC side voltage, aiming to actualize the independent control of d-q axis current, adjusting the gird-connected reactive power, to actualize sinusoidal output current.

SVPWM with a high voltage utilization ratio can reduce harmonic content and switching losses effectively, which is a huge advantage in improving the quality of gird-connected electricity. The Z-source SVPWM based on the traditional SVPWM inserts the short-through vector integrally or dividedly into traditional zero-vector. 6-stage SVPWM strategy is adopted [12].

According to the topology of three-phase inverter, ignoring the influence of filter capacitor, that is, the grid connected current is equal to the output current of inverter, and get the equation of state under three dimensional static coordinate system as following formula (7).

$$
\begin{cases}
L_f \dfrac{di_{2a}}{dt} + r_f i_{2a} = u_{oa} - e_a \\[2mm]
L_f \dfrac{di_{2b}}{dt} + r_f i_{2b} = u_{ob} - e_b \\[2mm]
L_f \dfrac{di_{2c}}{dt} + r_f i_{2c} = u_{oc} - e_c
\end{cases}
\tag{7}
$$

$$
\begin{cases}
u_{od} = L_f \dfrac{di_{2d}}{dt} + r_f i_{2d} - \omega L_f i_{2q} + e_d \\[2mm]
u_{oq} = L_f \dfrac{di_{2q}}{dt} + r_f i_{2q} + \omega L_f i_{2d} + e_q
\end{cases}
\tag{8}
$$

Where $L_f = L_1 + L_2$, $r_f = r_1 + r_2$. Then the transformation of the formula (7) can obtain the following formula (8).

Because of the coupling between the axis components, if the axial current and the axial current are independent of each other, a feedforward decoupling control strategy can be used. The output of the PI regulator is the equivalent variable, and the control equations are as following formula (9).

$$
\begin{cases}
u_{od} = (K_P + \dfrac{K_I}{s})(i^*_{2d} - i_{2d}) - \omega L_f i_{2q} + e_d \\[2mm]
u_{oq} = (K_P + \dfrac{K_I}{s})(i^*_{2q} - i_{2q}) + \omega L_f i_{2d} + e_q
\end{cases}
\tag{9}
$$

The battery voltage will be changed with time. In order to ensure the stability of inverter DC link-voltage, adjusting the short-through duty cycle d_0 is needed. The battery output voltage will not change suddenly, so the quick response of inductor current loop is not necessary. In order to reduce the amount of sensors and simplify the control structure, single voltage close-loop structure is adopted to control the short-through duty cycle. The control structure is shown as Fig. 8.

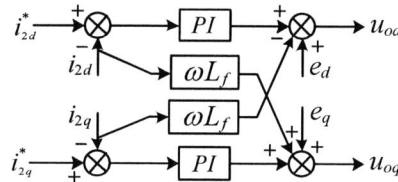

Fig.8 Single close-loop short-through duty control structure.

In order to adjust the gird-connected power, gird-connected current control strategy is used in the current control of Z-source gird-connected system. A LCL filter is applied to suppress high frequency harmonic. Current loop design considering LCL filter is conducted to actualize the decoupling between d, q axis current. The current close-loop model is shown as Fig. 9.

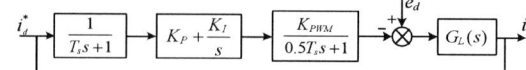

Fig. 9 Inner current loop control structure.

To suppress a resonant peak at the resonant frequency, an active dumping method with additional feedback variables is applied to offset the positive resonant peak at the resonant frequency. From the previous analysis, the current loop structure is shown in Fig. 10, where T_s is the switching period. K_p, K_i are the PI regulator parameters. K_{pwm} is the equivalent gain of the inverter, and $G_{L(s)}$ is the transfer function of the three order filter structure LCL.

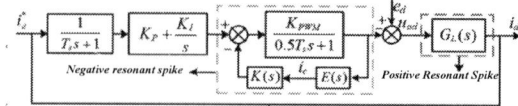

Fig.10 Structure diagram of current loop

Adding the additional feedback variables to current control loop to form a close-loop, close-loop can create negative resonant peak to offset the positive resonant peak. Control block diagram is shown as Fig. 11.

Fig. 11 Current loop structure with additional feedback variables.

3027

A capacitance current estimation algorithm instead of current sensor is proposed to reduce the cost. The block diagram of Z-source gird-connected control system is shown as Fig. 12

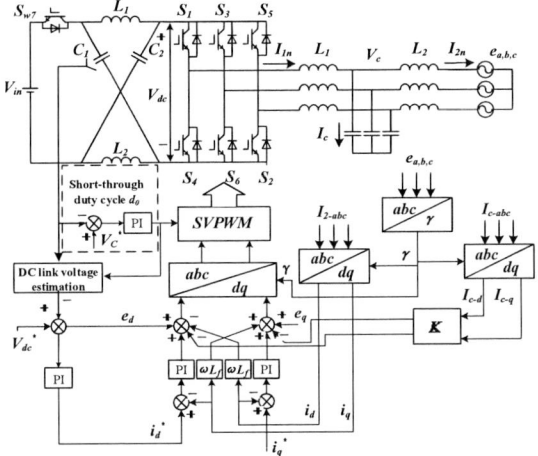

Fig. 12 Block diagram of Z-source gird-connected control system.

V. Simulation Result

According to the above theoretical analysis, the system simulation model of grid connected Z source inverter system is built, and the simulation results are obtained by selecting the appropriate circuit parameters.

Fig. 13 shows the steady state waveform of grid side voltages and currents, the Fig. (a), (b), (c), (d) respectively is the unit power factor operation voltage and current waveform, inductive power feedback operation voltage and current waveform, capacitive power feedback operation voltage and current waveform and current waveform of active and reactive power. Through the simulation waveform, it can be seen that changing the reference i_d and i_q can adjust the reactive power to the grid.

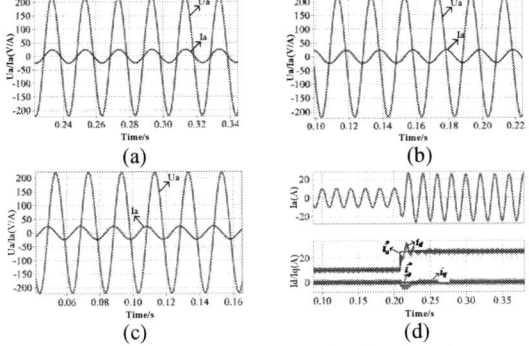

Fig. 13 Waveforms of grid connected voltage and current in steady state. (a)Unit power factor operation. (b)Inductive power feedback operation. (c)Capacitive power feedback operation. (d) Active and reactive currents.

V. Experimental Verification

The experimental platform of grid connected system of is designed as shown in Fig. 14. After grid connected operation, the waveforms of the correlation variables of the Z source converter are obtained as shown in Fig. 15.

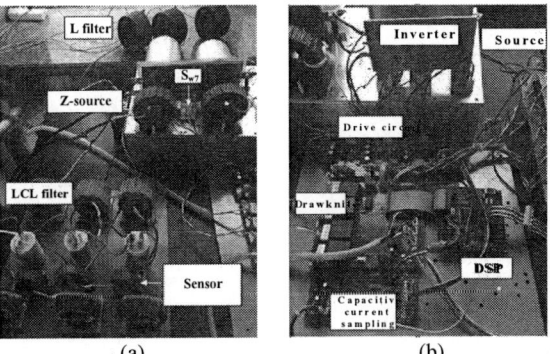

Fig. 14 Experimental platform of Z source grid connected system. (a) Z source converter and filter device. (b) Drive and control circuit.

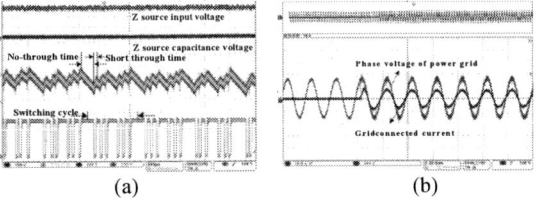

Fig. 15 Grid connected waveform based on Z source. (a) Correlation variable waveform Z source converter. (b) Output waveform of Z source system grid-connected.

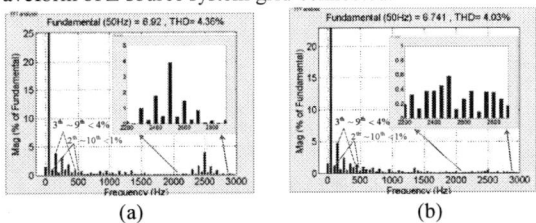

Fig. 16 Harmonic analysis of grid connected current based on LCL filter output. (a) Capacitor current loop not added. (b) Capacitance current loop added.

The experimental results verify the effectiveness of the control method, and the grid connected experiment achieves the unit power feedback operation. It can be seen from Fig. 16 that the method effectively reduces the harmonic content at the resonance point.

VI. Conclusion

This paper proposes a grid connected power generation control scheme of Z source bidirection charger with LCL filter. By deducing the mathematical model of inverter operation, the voltage and current-dual closed loop control strategy is given, and the active damping control method of integrated system is put forward to restrain the resonant spike.

After the completion of the model building and simulation verification, the Z source inverter experimental platform is built. Through the voltage and current closed-dual loop control, the fast response of the grid current command is realized, the operation under unit power factor is realized, and the reactive power demand of the power grid can be satisfied. Adding a filter capacitor current loop in the current loop can effectively suppress the resonant spike at the resonant point. The experimental results verify the effectiveness of the proposed control method.

REFERENCES

[1] H. Sekyung, H. Soohee, Sezaki K. "Development of an optimal vehicle-to-grid aggregator for frequency regulation" [J]. *IEEE Transactions on Smart Grid*, 2010, 1(1): 65-72.

[2] Y. Ota, H. Taniguchi, T. Nakajima, et al. "Effect of autonomous distributed vehicle-to-grid (V2G) on power system frequency control"[C]. *Industrial and Information Systems (ICIIS), 2010 International Conference*, Mangalore, India, July. 2010: 481-485.

[3] Lacroix S, Laboure E, Hilairet M. "An integrated fast battery charger for Electric Vehicle." *Vehicle Power and Propulsion Conference (VPPC), 2010 IEEE*. 2010: 1-6.

[4] Anderson J, Peng F. Four quasi-Z-Source inverters. Power Electronics Specialists Conference, *2008 PESC 2008 IEEE*. 2008: 2743-2749.

[5] Strzelecki R, Adamowicz M, Strzelecka N, et al. New type T-Source inverter. Compatibility and Power Electronics, *2009 CPE '09*. 2009: 191-195.

[6] Wei Q, Fang Zheng P, Honnyong C. Trans-Z-Source Inverters [J]. Power Electronics, *IEEE Transactions on*, 2011, 26(12): 3453-3463.

[7] Hanif M, Khadkikar V, Xiao W, et al. Two degrees of freedom active damping technique for LCL filter-based grid connected PV systems[J]. *IEEE Trans. on Industrial Electronics*, 2014, 61 (6):2795-2803.

[8] He N, Xu D, Zhu Y, et al. Weighted Average Current Control in a Three-Phase Grid Inverter With an LCL Filter [J]. *IEEE Transactions on Power Electronics*, 2013, 28(6):2785-2797.

[9] Tian-Hua L, Pei-Heng Y, Jui-Ling C. Implementation of an integrated battery-charger for an electric-propulsion system. Industrial Electronics Society, IECON 2014 - 40th Annual Conference of the IEEE. 2014: 1526-1531.

[10] Liangrong W, Jianing L, Guoqing X, et al. A novel battery charger for plug-in hybrid electric vehicles. Information and Automation (ICIA), 2012 International Conference on. 2012: 168-173.

[11] Lacroix S, Laboure E, Hilairet M. An integrated fast battery charger for Electric Vehicle. Vehicle Power and Propulsion Conference (VPPC), 2010 IEEE. 2010: 1-6.

[12] Dong Shuai. Research on ripple characteristics of Z source inverter [D]. Harbin Institute of Technology, 2016.

No.0385

An Isolated PFC Converter with Harmonic Modulation Technique for EV Chargers

Byung-Kwon LEE [1], Jun-Young LEE [1*] Dong-Hun KANG[1]

1 Myongji University, Young-in, Republic of Korea

Abstract— **This paper suggests an Single-Phase isolated PFC converter and Non-isolated Buck converter for general-purpose EV chargers with a wide-output voltage range. Harmonic regulation are accomplished by PFC stage and charging control is performed by the simple non-isolated buck converter. Output controls and harmonic controls are performed by secondary switches and primary switches only operated 50% duty-ratio.**

The proposed charger has been verified with a 3.3kW prototype.

Index Terms— **Isolated PFC, Battery charger, converter, Electric vehicles.**

I. INTRODUCTION

RECENTLY, Industrial development has caused serious damage to the environment. We already know the seriousness of the global warming. Global warming problem for the whole world, and Researchers are still studying global warming. But near future Fuel prices will also go up, and future generations will experience food shortages and rising global temperatures. So we find a new transportation methods have been required to solve air pollution and greenhouse gasses so that Vehicle electrification has been spotlighted as a solution.

However, in the old days because petroleum was plentiful and the environmental damage from car usage was largely unknown, the technology was neglected.

But recently electric cars are the inevitable future. They aren't as noisy as regular cars and they emit less pollution. The environmental standards around the world are changing, which means there will be more demand for electric cars in the future. Recently new transportation methods have been required to solve air pollution and greenhouse gasses, so that vehicle electrification has been spotlighted as a solution. There are various kinds of xEVs such as battery electric vehicles, hybrid electric vehicles, and plug-in hybrid electric vehicles, and fuel-cell electric vehicle. A battery charger is a device used to put energy into a secondary cell or rechargeable battery by forcing an electric current through it. The charging protocol depends on the size and type of the battery being charged.

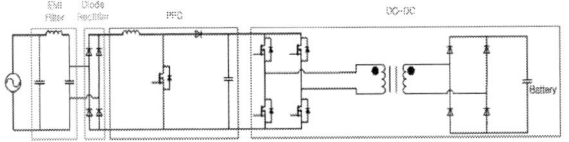

Fig. 1. General battery charger circuit.

Fig. 1 shows the general battery charger circuit. The conventional OBC has a two-stage structure of a power factor correction (PFC) converter for harmonic reduction. It has the two-stage structure of a power factor correction (PFC) converter for harmonic reduction followed by a dc/dc converter for output control and electrical isolation [1], [2]. The total efficiency of OBC is dependent on the efficiency of the dc/dc stage, and there are several requirements for dc/dc converter topology to meet in OBC use. In the PFC stage, there are many papers to improve the efficiency, and over 99% has been reported in prototype phase. The efficiency of OBC is related to mileage so that it is an important issue to improve efficiency of EV. For AC/DC converter, H-bridge-based converters such as single-phase half-bride, single-phase full-bride, and three-phase full-bride converters are general used for grid-connected converters and multilevel topologies are also considered in high power chargers [3~5]. Since all the chargers have high efficiency, many developers are studying the efficiency increase due to the development of semiconductor devices rather than the development of the hardware technology. Recently, many efforts have been focused on improvement of the efficiency in conventional AC/DC converter structures and good efficiencies over 97~99% have been reported based on new devices such as silicon carbide (SIC) devices. Some isolated PFC approaches have been reported using asymmetrical dual active bridge (ADAB) converter and it has various operational modes [6]. Among them, "Boost-DCM (Discontinuous Conduction Mode) operation" is adequate to reduce the switching loss of the primary switch due to ZV-ZCS switching characteristics, which can provides the availability of IGBT as well as MOSFET as switching devices.

However, this operational mode has the disadvantages of low power factor and high switching loss of the secondary switch, which is the obstacle for high power design such as EV battery charger.

In this paper, a design of the isolated PFC converter based on ADAB structure is suggested by operating the converter under the Boost-DCM operation, primary switches can be operated under a fixed frequency with 50% duty-ratio so that it leads good switching conditions of zero-voltage turn-on and zero-current turn-off.

All of the harmonic and output controls are performed by secondary switches and harmonic modulation technique [7] is also adopted to obtain near unity power factor without input current monitoring. It provides a simple control scheme so that a simple digital controller is available.

To verify the feasibility, the proposed circuit has been implemented and tested using a 3.3kW charger.
[8-14].

No.0385

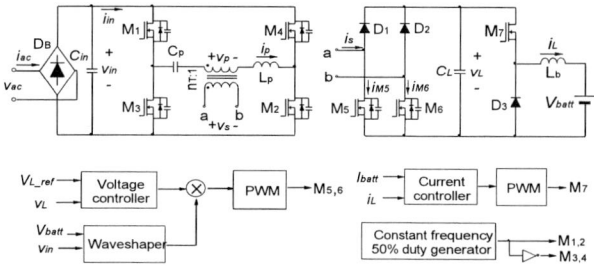

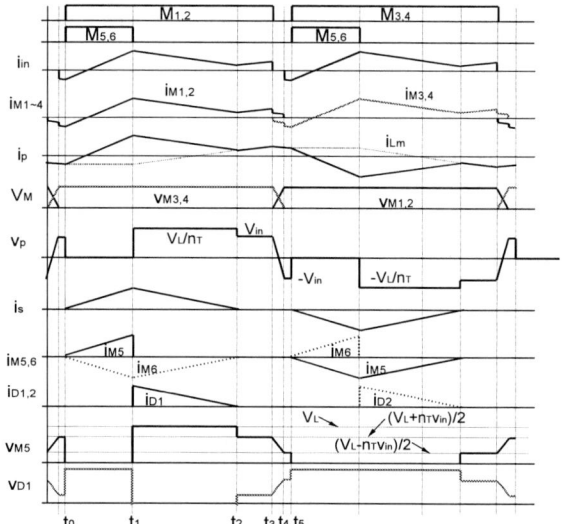

Fig. 2. Schematic diagram of the charger comprised of the proposed isolated PFC converter with harmonic modulation technique and buck converter

Fig. 3. Key waveforms of the proposed PFC converter

II. PROPOSED ISOLATED PFC CONVERTER

Fig. 2 is the schematic diagram of the charger comprised of the proposed isolated PFC converter with harmonic modulation technique and buck converter for battery charging control. The primary switches of $M_1 \sim M_4$ are fixed frequency and 50% duty-ratio. Isolation and harmonic regulation control are accomplished by PFC stage and charging control is performed by the simple non-isolated buck converter. All controls of the output voltage and the line current regulations and harmonic modulation of PFC stage are performed only by the secondary switches of M_5 and M_6.

Detailed operational analysis is as follows: Mode analysis of the proposed PFC converter

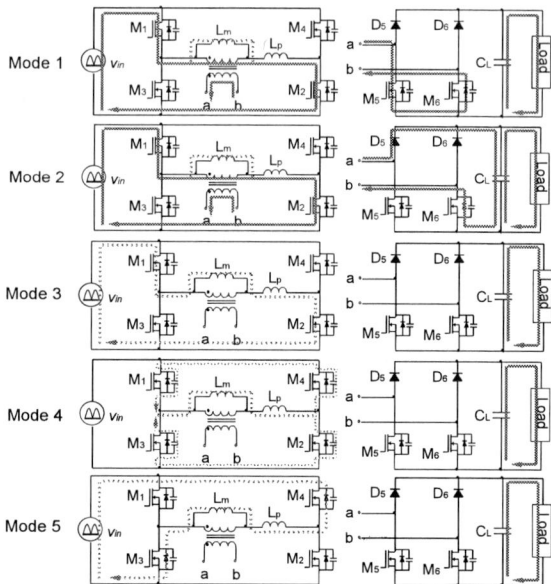

Fig. 4. Operational mode diagrams of the proposed PFC converter

Mode 1 (t0 $\leq$ t< t1): When M_1, M_2, M_5, and M_6 are turned on at t_0 simultaneously, the primary current i_p, flows through M_1, M_2, the inductor L_p, and transformer primary side. The secondary current is flows through M_5, M_6-body diode, and transformer secondary side. i_p is equal to the sum of i_{Lm} and the secondary current referred to the primary $n_T i_s$.

They can be expressed as

$$i_s(t) = (v_{in}/n_T L_p)(t - t_0) \quad (1)$$

$$i_{Lm}(t) = i_{Lm}(t_0) \quad (2)$$

$$i_p(t) = i_{Lm}(t_0) + (v_{in}/L_p)(t - t_0) \quad (3)$$

Mode 2 (t1 $\leq$ t< t2): When M_5, and M_6 are turned off, mode 2 start. Primary current path same mode 1 but the secondary side current path only through D_5 and M_6-body diode.

So, begins mode 2, is, i_{Lm}, and i_p can be expressed as follows:

$$i_s(t) = \frac{v_{in}}{n_T L_p}(t_1 - t_0) + \frac{v_{in} - V_L/n_T}{n_T L_p}(t - t_1) = \frac{v_{in}}{n_T L_p}d_p T_s + \frac{v_{in} - V_L/n_T}{n_T L_p}(t - t_1) \quad (4)$$

$$i_{Lm}(t) = i_{Lm}(t_1) + \frac{V_L/n_T}{L_m}(t - t_1) = i_{Lm}(t_0) + \frac{V_L/n_T}{L_m}(t - t_1) \quad (5)$$

$$i_p(t) = i_{Lm}(t_0) + \frac{V_L/n_T}{L_m}(t - t_1) + \frac{v_{in}}{L_p}d_p T_s + \frac{v_{in} - V_L/n_T}{L_p}(t - t_1) \quad (6)$$

$$t_2 - t_1 = d_{p2}T_s = (v_{in}/(V_L/n_T - v_{in}))d_p T_s \quad (7)$$

Mode 3 (t2 $\leq$ t < t3): This mode only the magnetizing current circulates through M_1 and M_3 and that time v_p is changed to vin and so that exchanged i_{Lm}.

So i_p can be expressed as

$$i_s(t) = 0 \quad (8)$$

$$i_{Lm}(t) = i_{Lm}(t_2) + (v_{in}/L_m)(t - t_2) = i_p(t) \quad (9)$$

$$i_{Lm}(t_2) = i_{Lm}(t_0) + v_{in}V_L/(L_m(V_L - n_T v_{in}))d_p T_s \quad (10)$$

Mode 4, 5 (t3 $\leq$ t < t5): The magnetizing current to flow through M_3, M_4 body-diodes.

3031

And this time zero-voltage-switching (ZVS) condition can be achieved. When the of M_1 and M_2 (Primary Switchs) turn-off. Assuming that total duration of modes 4 and 5,ignore the dead-time
iLm during these two modes can be written as

$$i_{Lm}(t) \approx -i_{Lm}(t_0) \tag{11}$$

Under the above conditions $i_{Lm}(t_0)$ can be derived using eq. (9), (10), and (11). It is as follows:

$$i_{Lm}(t_0) = -(v_{in}/2L_m)T_s \tag{12}$$

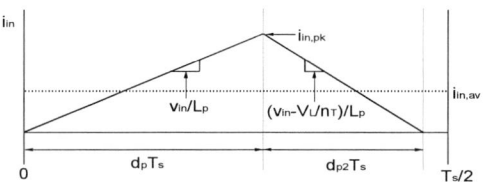

Fig. 5. Input current depicted excluding the magnetizing current during one switching cycle

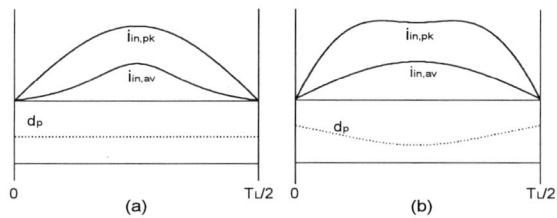

Fig. 6. Inductor current envelope and line current waveform of conventional DCM operation (a) and those adopting harmonic controller (b)

The magnetizing inductance current (i_{Lm}) is determined by the ZVS condition and has no influence on the average current value of the input current (i_{in}) as described in Mode 4 described above. Therefore, the waveform of (i_{in}) can be expressed as shown Fig. 5 by removing (i_{Lm}). Where I_ ($i_{in,pk}$) is the maximum value of the input current and I_ ($i_{in,av}$) is the average value of the input current. the peak value of input current $i_{in,pk}$ and the switching average value of input current $i_{in,av}$ can be derived as

$$i_{in,pk}(t) = (v_{in}/L_p)d_pT_s \tag{13}$$

$$i_{in,av}(t) = (d_p^2 v_{in}T_s/L_p)(V_L/(V_L - n_T v_{in})) \tag{14}$$

Fig. 6(a) is the conventional DCM operation line current waveform d_p is maintained at a constant value during control,
Power factor control is required to obtain a pure sinusoidal current waveform.
This can be expressed as

$$P_o = I_{ac,rms}V_{ac,rms} \tag{15}$$

Since $I_{ac,rms}V_{ac,rms}$ is equal to $I_{in,pk}V_{in,pk}/2$, $I_{in,pk}$ can be found as

$$I_{in,pk} = 2P_o/V_{in,pk} \tag{16}$$

Thus, the line current for unity power factor is derived as follows:

$$i_{ac} = (2P_o/V_{in,pk})\sin \omega_L t \tag{17}$$

If the switching average current value of I_ ($i_{in,av}$) is equal to the absolute value of i_{ac}, power factor control by d_p is required. The formula for d_p can be expressed by the following equation (18) through equations (14) and (17).

Where d_p is a constant value for voltage control and M_f is a modulation value for power factor control.

$$d_p = \sqrt{2L_pP_o/T_sV_{in,pk}^2} \times \sqrt{1 - n_T v_{in}/V_L} \equiv D_p \times M_f \tag{18}$$

TABLE I
Key components list for prototype implementation

Component	Value
$M_1 \sim M_7$	IPW65R041CFD×1
$D_1 \sim D_3$	FFH50US60S×1
C_L	1020μF EL capacitor//50uF film capacitor
C_p	8.8μF
f_s	50kHz
L_p	16.3μH (Core: PQ40/40)
L_b	900μH (Core: CH467125×2)
Transformer	L_m =501μH/L_{lkg}=3.7μH/n_T =0.89 (Core: EER6062)
Controller	DSP TMS320F28335

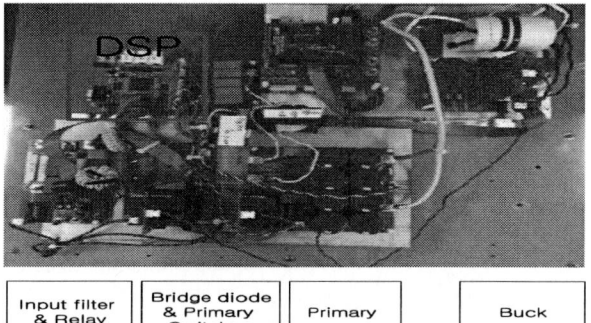

Fig. 7. Experimental prototype for performance validation

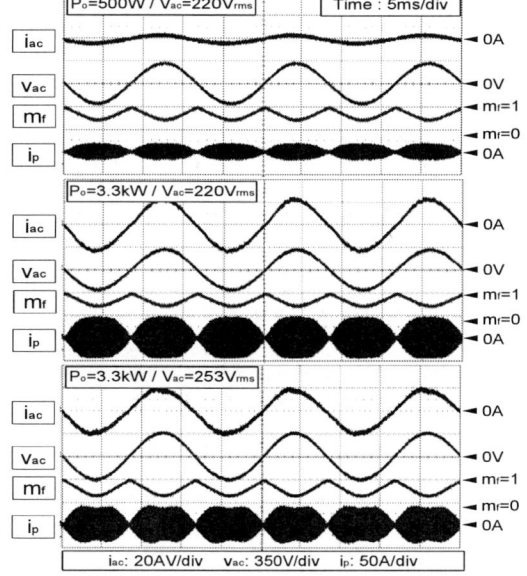

Fig. 8. Measured waveforms according to load and line voltage variations

3032

No.0385

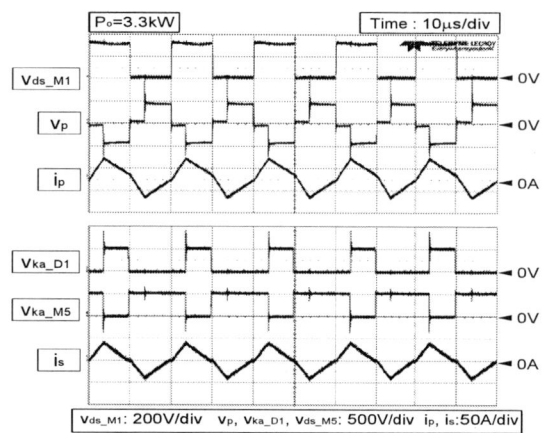

Fig. 9. Switching waveforms at V_{ac}=220V$_{rms}$/V_L=500V

I. EXPERIMENTAL RESULTS

The prototype PFC converter has been designed with the following specifications:

◆ Grid characteristics: 3.3kW/30A/60Hz
◆ Maximum output power: 3.3kW
◆ Input voltage: 220V$_{rms}$±15%
◆ Output voltage: 500V

Fig. 7 is the experimental prototype for validation excluding the output electrolytic capacitor and Table 1 is the key component list of the prototype PFC converter.

Fig. 8 is the measured waveforms of line voltage v_{ac}, line current i_{ac}, primary current i_p, and modulation factor M_f according to load and line voltage variations. The modulation factor M_f is plotted using DAC (Digital-to-Analog Converter) in DSP controller. This figure shows that clean sinusoidal current waveforms can be obtained under wide load ranges by changing operational duty-ratios continuously according to M_f. Also, the prototype design guarantees the normal operation under the given line voltage range. Fig. 9 is the switching waveforms at the rated condition. They are well agreed with theoretical analysis except for the parasitic resonance due to pattern inductances in the printed-circuit-board (PCB).

A high power factor can be obtained under wide load range and the power factor is over 0.983 at the output power more than 1kW. Also, the efficiency is recorded as about 95.6% at V_{ac}=220V$_{rms}$ and P_o=3.3kW and the maximum efficiency of 96.2% occurs at V_{ac}=253V$_{rms}$ and P_o=3.0kW.

II. CONCLUSIONS

Another candidate for an isolated PFC converter that can be used for EV chargers has been suggested based on ADAB structure. It is not affected by the parasitic inductance of the transformer due to the voltage-fed structure. The operation is based on DCM and the harmonic modulation method is applied to overcome the low power factor featured in general DCM operation. All operations including harmonic regulation and output control are performed only by secondary switches with simple control algorithm, which provides another advantage in the aspect of controller and circuit design. As a result of analyzing the operating characteristics of the 3.3 kW prototype, the power factor of over 0.983 was recorded at over 1kW and the efficiency was recorded as about 95.6% at V_{ac}=220V$_{rms}$ and P_o=3.3kW. Therefore, the proposed circuit could be used in the general-purpose battery chargers by configuring it with a simple non-isolated converter.

REFERENCES

[1] M. Pahlevaninezhad, J. Drobnik, P. K. Jain, and A. Bakhshai, "A load adaptive control approach for a zero-voltage-switching dc/dc converter used for electric vehicles," IEEE Trans. Ind. Electron., vol. 59, no. 2, pp. 920–933, Feb. 2012.

[2] D. C. Erb, O. C. Onar, and A. Khaligh, "Bi-directional charging topologies for plug-in hybrid electric vehicles," in Proc. IEEE APEC, 2010, pp. 2066–2072.K. M. Rahman, S. Jurkovic, S. Hwakins, S. Tarnowsky, and P. Savagian, "Propulsion System Design of a Battery Electric Vehicle," IEEE Ind. Electron. Mag., vol. 2, no. 2, pp. 14-24, Jun. 2014.

[3] D. C. Erb, O. C. Onar, and A. Khaligh, "Bi-directional Charging Topologies for Plug-In Hybrid Electric Vehicles," in Proc. IEEE APEC, 2010, pp. 2066-2072.

[4] M. Yilmaz and P. T. Krein, "Review of Battery Charger Topologies, Charging Power Levels, and Infrastructure for Plug-In Electric and Hybrid Vehicles," IEEE Trans. Power Electron., vol. 28, no. 5, pp. 2151-2169, May 2013.

[5] B.Singh, B. N. Singh, A. Chandra, K. Al-Haddad, A. Pandey, and D. P. Kothari, "A Review of Three-Phase Improved Power Quality AC–DC Converters," IEEE Trans. Ind. Electron., vol. 51, no. 3, pp. 641-660, Jun. 2004.

[6] J. Y. Lee, Y. D. Yoon, and J. W. Kang, "A Single-Phase Battery Charger Design for LEV Based on DC-SRC With Resonant Valley-Fill Circuit," IEEE Trans. Ind. Electron., vol. 62, no. 4, pp. 2195-2205, Apr. 2015. 2004.J. Schmenger, S. Zeltner. R. Kramr, S. Endres, and M. Marz, "A 3.7 kW On-board Charger Based on Modular Circuit Design," in Proc. IEEE IECON, 2015, pp. 1382-1387.

[7] J. Y. Lee and H. J. Chae, "6.6-kW Onboard Charger Design Using DCM PFC Converter With Harmonic Modulation Technique and Two-Stage DC/DC Converter," IEEE Trans. Ind. Electron., vol. 61, no. 3, pp. 1243-1251, Mar. 2014.

[8] Y. D. Kim, K. M. Cho, D. Y. Kim, G. W. Moon, :Wide-Range ZVS Phase-Shift Full-Bridge Converter with Reduced Conduction Loss Caused by Circulating Current," IEEE Trans. Power Electron., vol. 28, no. 7, pp. 3308-3316, Jul. 2013.

[9] D. Guatam, F. Musavi, M. Edington, W. Eberle, and W. G. Gunford, "An Automotive On-Board 3.3 kW Battery Charger for PHEV Application," in proc. IEEE VPPC, 2011, pp. 1-6.

[10] I. O. Lee and G. W. Moon, "Half-Bridge Integrated ZVS Full-Bridge Converter with Reduced Conduction Loss for Electric Vehicle Battery Chargers," IEEE Trans. Ind. Electron., vol. 61, no. 8, pp. 3978-3988, Aug. 2014.

[11] R. Beiranvand, M. R. Zolghadri, B. Rashidian, and S. M. H. Alavi, "Optimizing the LLC-LC Resonant Converter Topology for Wide-Output-Voltage and Wide-Output-Load Applications," IEEE Trans. Power Electron., vol. 26, no. 11, pp. 3192-3204, Nov. 2011.

[12] Y. K. Lo, C. Y. Lin, M. T. Hsieh, and C. Y. Lin, "Phase-Shifted Full-Bridge Series-Resonant DC-DC Converters for Wide Load Variations," IEEE Trans. Ind. Electron., vol. 58, no. 6, pp. 2572-2575, Jun. 2011.

[13] H. Wang, S. Dusmez, and A. Khaligh, "Design and Analysis of a Full-Bridge LLC-Based PEV Charger Optimized for Wide Battery Voltage Range," IEEE Trans. Veh. Technol., vol. 63, no. 4, pp. 1603-1613, May 2014.

[14] J. Park and S. Choi, "Zero-Current Switching Series Loaded Resonant Converter Insensitive to Resonant Component Tolerance for Battery Charger," IET Power Electron., vol. 7, no. 10, pp. 2517-2524, Oct. 2013.

3033

The 2018 International Power Electronics Conference

Highly Dynamic Switching Frequency-Based Calculation of Power Quantities, Fundamental Waveforms, and RMS Values of Inverter-fed Electrical Machines

Alexander Stock[1]*, Johannes Teigelkötter[1], Johannes Büdel[1]
1 Laboratory of Electrical Machines, Power Electronics and Drives, University of Aschaffenburg, Germany
*E-mail: alexander.stock@h-ab.de

Abstract—This contribution presents an innovative approximation method for the estimation of power quantities, fundamental waveforms, and RMS values of inverter-fed electrical machines. Typically, all these parameters are calculated as cyclically linked to the fundamental cycle of the voltages and currents of the inverter output or machine input. In order to increase the dynamic of this calculation, a suitable evaluation method based on the inverter switching cycle has been presented in previous publications. The present paper extends this switching frequency-based analysis by introducing additional quantities such as apparent and reactive power. Moreover, the paper describes the utilization of this analysis method for a power control of a dual two-level inverter-fed application.

Keywords—*Active/Reactive Power Measurement, Pulse Width Modulation Converters, Electrical Machines, Power Control*

I. INTRODUCTION

In order to characterize the behaviour and efficiency of inverters and electrical machines, several quantities, such as power and RMS values, have to be measured and evaluated. These measurements provide information about the performance and quality of the inverter control as well as pertaining to the efficiency of the electrical machine, the inverter, and the overall system. In general, these quantities are calculated periodically based on the fundamental period of the electrical voltages and/or currents. For this reason, in a strict sense, all the relevant quantities can only be calculated during steady state conditions for exactly periodical waveforms. During balancing processes, the results of the fundamental cycle-based evaluations may lead to unrealistic and non-useful results. Therefore, in [1], a highly dynamic inverter switching frequency-based calculation method for active power and fundamental waveforms has been introduced. In the current paper, this method will be extended by further switching cycle-based quantities, such as the RMS values of voltages or currents, reactive power, and apparent power. This would allow one to define a related analysis of these conventional fundamental-based quantities during highly dynamic balancing processes. Nevertheless, it approximates the well-known fundamental-based evaluation during steady state operation. The performance of this analysis will be demonstrated by measurements

on a dual two-level inverter-fed permanent magnet synchronous machine (PMSM). In this configuration, the PMSM is an open winding machine, where both winding ends are connected to two galvanically separated two-level inverters (see Fig 6). The inverters themselves are fed by two independent energy storage systems (batteries, fuel cells, supercaps, etc.). Comparable systems have also been presented in [2]. The following section presents a summarizing overview of the switching frequency-based analysis in reference to [1].

II. SWITCHING FREQUENCY-BASED CALCULATIONS

A. Active Power and Fundamental Component Approximation

The conventional RMS values, fundamental waveforms, and power values are calculated by a cyclic averaging window with the size of the fundamental period T of the electrical quantities. In contrast, the switching frequency-based quantities are calculated by the running mean value over the switching cycle. The following example should illustrate the correlations of these definitions in an exemplary manner considering the active power. As is well-known from scientific literature and standards, the active power P is defined as the mean value of the instantaneous power $p(t)$ over the fundamental cycle T (see [3], for example):

$$P = \frac{1}{T} \cdot \int_{t-T}^{t} p(\tau)\,\mathrm{d}\tau \tag{1}$$

However, according to [1], the switching frequency-based active power can be calculated by the mean value of the instantaneous power over the switching cycle T_{sw}:

$$P_{\mathrm{sw}} = \frac{1}{T_{\mathrm{sw}}} \cdot \int_{t-T_{\mathrm{sw}}}^{t} p(\tau)\,\mathrm{d}\tau \tag{2}$$

Reference [1] further describes the significance and legitimization of this definition in relation to the conventional definition (1). It also proves the following equivalence

$$\lim_{T_{\mathrm{sw}} \to 0} P_{\mathrm{sw}} = P \tag{3}$$

keeping in mind some constraints concerning the inverter control, especially the pulse width modulation (PWM) method. Obviously, (3) describes a solely ideal inverter operation with an infinitely high switching frequency $f_{sw} = \frac{1}{T_{sw}} \to \infty$. In practical applications, the accuracy of the approximation

$$P_{sw} \approx P \qquad (4)$$

is optimized by increasing the T/T_{sw}-ratio—that is, the higher the switching frequency compared to the fundamental frequency the better the results of the approximation. The switching frequency-based active power approaches the fundamental active power of the inverter-fed system. An exemplary measurement at one of the two 2-level inverter outputs is shown in Fig 1. The power quantities are normalized to the reference power

$$P_r = \frac{3}{2} \cdot \hat{u}_r \cdot \hat{i}_r \qquad (5)$$

with the reference voltage amplitude $\hat{u}_r = E_d$ and the reference current amplitude $\hat{i}_r$, where E_d represents half the DC-link voltage of the corresponding two-level inverter. Obviously, the switching frequency-based active

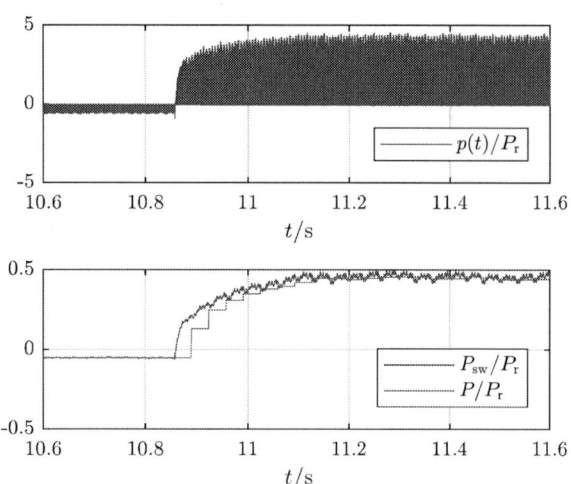

Fig. 1. Instantaneous power $p(t)$, conventional active power P, and switching frequency-based active power P_{sw} during dynamic load change of one inverter output of a dual two-level inverter-fed PMSM system; $f = 30\,\mathrm{Hz}$, $f_{sw} = 8\,\mathrm{kHz}$

power contains the dynamic information of the instantaneous power during balancing processes, whereas it approximates the conventional active power during steady state. Thus, it combines the advantages of both conventional power quantities in one novel power quantity, which is therefore very suitable for control algorithms and real-time monitoring systems. Analogously, in [1], the switching frequency-based fundamental components have been introduced: By calculating the switching cycle-based mean value of the inverter output voltages $u_{\nu r}$ (whereby the ν-th phase is measured against an arbitrary potential r) and currents i_ν and by additionally subtracting a possibly overlayed zero system, the fundamental voltages $u_{\nu r,h1}$ and currents $i_{\nu,h1}$ can be approximated (see Fig. 2,

including the reference potential $r = 0$):

$$
\begin{aligned}
u_{\nu sw,h1}(t) &= u_{\nu sw}(t) - u_{sw0}(t) \\
&= \frac{1}{T_{sw}} \cdot \int_{t-T_{sw}}^{t} u_{\nu r}(\tau)\,d\tau - \ldots \\
&\ldots \frac{1}{n} \cdot \sum_{\mu=1}^{n} \left(\frac{1}{T_{sw}} \cdot \int_{t-T_{sw}}^{t} u_{\mu r}(\tau)\,d\tau \right) \\
&\approx u_{\nu r,h1}(t)
\end{aligned} \qquad (6)
$$

Supposing that there is no neutral connector and no parasitic leakage current, the current zero system is zero. This leads to:

$$i_{\nu sw,h1}(t) = \frac{1}{T_{sw}} \cdot \int_{t-T_{sw}}^{t} i_\nu(\tau)\,d\tau \approx i_{\nu i,h1}(t) \qquad (7)$$

It must be considered that these approximations are also limited to several restrictions, considering the PWM of the inverter (in analogy to the active power approximation; see [1]).

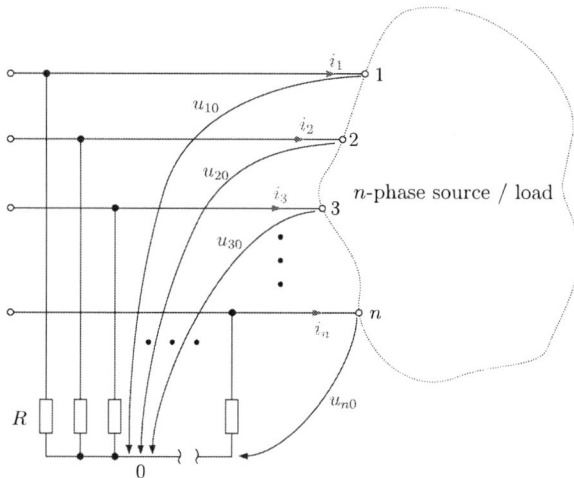

Fig. 2. Electrical n-phase system (see [1], interpreting [5])

Fig. 3 illustrates the switching frequency-based fundamental voltage for the same operating point of the dual two-level inverter-fed three-phase PMSM as Fig. 1. All the depicted voltage values are normalized to the half DC-link voltage $\hat{u}_r = E_d$ of the corresponding two-level inverter. The reference potential r of the measured phase voltages $u_{\nu r}(t)$ with the phases $\nu \in \{1,2,3\}$ is the negative DC-link potential of this inverter. For this, the normalized phase voltages vary in the interval $[0,2]$. The switching frequency-based mean value $u_{1sw}(t)$ of the voltage of Phase 1 shows the waveform of the underlying PWM modulation scheme of the inverter—in this case, the symmetrical PWM (SYPWM) (see [4]). Subtracting the switching frequency-based zero-system voltage

$$u_{0sw}(t) = \frac{1}{3}\big(u_{1sw}(t) + u_{2sw}(t) + u_{3sw}(t)\big) \qquad (8)$$

from the switching frequency-based mean value $u_{1sw}(t)$, the fundamental voltage of Phase 1 is approximated: $u_{1sw,h1}(t) \approx u_{1r,h1}(t)$. This is visualized in the bottom plot of Fig. 3. This plot, moreover, displays the zero-sum voltage of Phase 1 (in accordance with Fig. 2):

$$u_{10} = u_{1\mathrm{r}} - u_0 \quad \text{with} \quad u_0 = \frac{1}{3}(u_{1\mathrm{r}} + u_{2\mathrm{r}} + u_{3\mathrm{r}}) \quad (9)$$

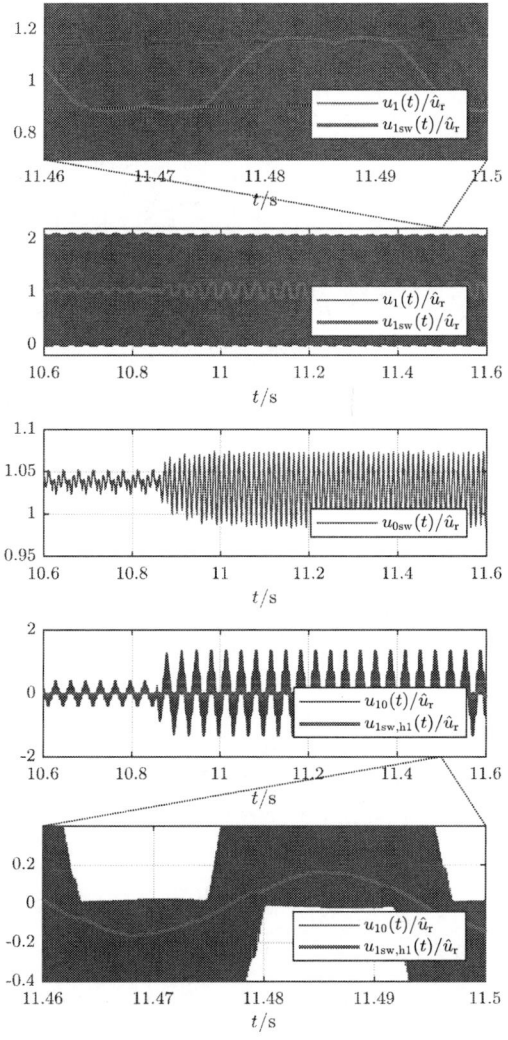

Fig. 3. Switching frequency-based fundamental voltage (the top and bottom plots represent the zoomed areas of the inner ones)

However, the conventional fundamental waveform calculation is not defined for this dynamic balancing process. The switching frequency-based method, moreover, includes transient peaks of the inverter control, which are not visible in the conventional fundamental signal. In real-time applications, the latency period of the fundamental waveform is the fundamental period, whereas the switching frequency-based fundamental waveform value is updated in every switching cycle (in many typical applications, this can be up to several hundred times faster).

Considering a symmetrical three-phase system, the switching frequency-based fundamental amplitude of the

voltages $\hat{u}_{\nu\mathrm{sw},\mathrm{h1}}$ or currents $\hat{i}_{\nu\mathrm{sw},\mathrm{h1}}$ can be calculated applying the well-known space vector theory to the results of (6) and (7) (see [6]):

$$u_{\alpha\mathrm{sw},\mathrm{h1}}(t) = \frac{2}{3}u_{1\mathrm{sw},\mathrm{h1}}(t) - \frac{1}{3}u_{2\mathrm{sw},\mathrm{h1}}(t) - \frac{1}{3}u_{3\mathrm{sw},\mathrm{h1}}(t)$$

$$u_{\beta\mathrm{sw},\mathrm{h1}}(t) = \frac{1}{\sqrt{3}}u_{2\mathrm{sw},\mathrm{h1}}(t) - \frac{1}{\sqrt{3}}u_{3\mathrm{sw},\mathrm{h1}}(t) \quad (10)$$

$$\hat{u}_{\mathrm{sw},\mathrm{h1}} = \sqrt{(u_{\alpha\mathrm{sw},\mathrm{h1}}(t))^2 + (u_{\beta\mathrm{sw},\mathrm{h1}}(t))^2}$$

Since the space vector transformation eliminates the zero system of the phase quantities, it would be alternatively possible to calculate (10) by using $u_{\nu\mathrm{sw}}$ instead of $u_{\nu\mathrm{sw},\mathrm{h1}}$ with $\nu \in \{1, 2, 3\}$. According to (10), the fundamental amplitude of the current is calculated by:

$$i_{\alpha\mathrm{sw},\mathrm{h1}}(t) = \frac{2}{3}i_{1\mathrm{sw},\mathrm{h1}}(t) - \frac{1}{3}i_{2\mathrm{sw},\mathrm{h1}}(t) - \frac{1}{3}i_{3\mathrm{sw},\mathrm{h1}}(t)$$

$$i_{\beta\mathrm{sw},\mathrm{h1}}(t) = \frac{1}{\sqrt{3}}i_{2\mathrm{sw},\mathrm{h1}}(t) - \frac{1}{\sqrt{3}}i_{3\mathrm{sw},\mathrm{h1}}(t) \quad (11)$$

$$\hat{i}_{\mathrm{sw},\mathrm{h1}} = \sqrt{(i_{\alpha\mathrm{sw},\mathrm{h1}}(t))^2 + (i_{\beta\mathrm{sw},\mathrm{h1}}(t))^2}$$

For an ideal symmetrical system in the steady state, the fundamental voltage or current amplitudes of the individual phases are equal and, moreover, equivalent to the constant magnitude of the corresponding space vectors: $\hat{u}_{1\mathrm{sw},\mathrm{h1}} = \hat{u}_{2\mathrm{sw},\mathrm{h1}} = \hat{u}_{3\mathrm{sw},\mathrm{h1}} = \hat{u}_{\mathrm{sw},\mathrm{h1}}$ and $\hat{i}_{1\mathrm{sw},\mathrm{h1}} = \hat{i}_{2\mathrm{sw},\mathrm{h1}} = \hat{i}_{3\mathrm{sw},\mathrm{h1}} = \hat{i}_{\mathrm{sw},\mathrm{h1}}$. For an asymmetrical system, by implication, the corresponding voltage and/or current fundamental amplitudes are time-dependent values $\hat{u}_{\mathrm{sw},\mathrm{h1}}(t)$, $\hat{i}_{\mathrm{sw},\mathrm{h1}}(t)$. This information can be used as an indicator to characterize the symmetry of the monitored system.

B. RMS Value, Reactive and Apparent Power Approximation

Based on the evaluations in the previous sections, several additional quantities, such as RMS voltages or currents, and reactive and apparent power, can be determined. Considering a symmetrical system, as mentioned, the magnitude of the fundamental voltage or current space vector is equivalent to the fundamental phase voltage or current amplitudes. As is well-known from the scientific literature (see [3], for example), the RMS values of the sinusoidal waveforms are equal to the product of their amplitude multiplied by $1/\sqrt{2}$:

$$U_{\mathrm{sw},\mathrm{h1}} = \frac{\hat{u}_{\mathrm{sw},\mathrm{h1}}}{\sqrt{2}} \qquad I_{\mathrm{sw},\mathrm{h1}} = \frac{\hat{i}_{\mathrm{sw},\mathrm{h1}}}{\sqrt{2}} \quad (12)$$

Fig. 4 depicts the switching frequency-based RMS values of the fundamental voltage $U_{1\mathrm{sw},\mathrm{h1}}$ and current $I_{1\mathrm{sw},\mathrm{h1}}$ as well as the conventional RMS values of the fundamental voltage $U_{10,\mathrm{h1}}$ and current $I_{1,\mathrm{h1}}$ (since a symmetrical system has been assumed to calculate the switching frequency-based RMS values, $U_{1\mathrm{sw},\mathrm{h1}} = U_{\mathrm{sw},\mathrm{h1}}$; $U_{10,\mathrm{h1}}$ is the RMS value of the fundamental of Phase 1 measured against an ideal star point 0, which indicates the zero-sum voltage of Phase 1).

Obviously, in analogy to the switching frequency-based active power in Fig. 1, the advantage concerning

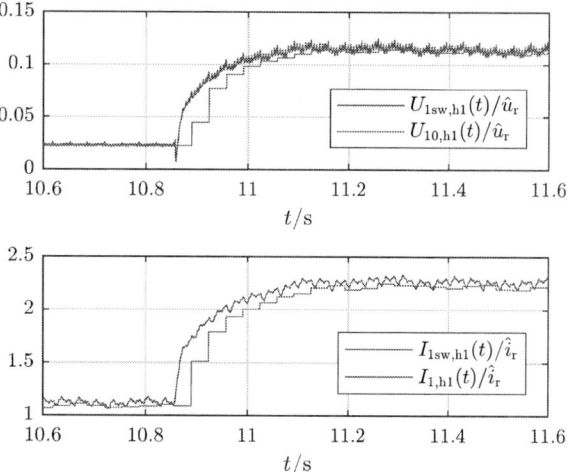

Fig. 4. Switching frequency-based and conventional fundamental RMS values

the dynamic of this analysis is demonstrated. By using the switching frequency-based RMS values, the apparent power can be calculated in accordance with the scientific literature (e.g. [3]). In analogy to the switching frequency-based active power, the apparent power also represents an approximation of the conventional fundamental apparent power of a symmetrical system during steady state:

$$S_{\text{sw}} = 3 \cdot U_{\text{sw,h1}} \cdot I_{\text{sw,h1}} \tag{13}$$

In addition, the reactive power is calculated by (e.g. [3]):

$$Q_{\text{sw}} = \sqrt{S_{\text{sw}}^2 - P_{\text{sw}}^2} \tag{14}$$

The conventional and switching frequency-based power quantities are illustrated in Fig. 5. Furthermore, several other conceivable switching frequency parameters can be calculated by further evaluations based on the ones that have already been presented. It is possible to calculate machine or inverter parameters (e.g. THD values) highly dynamic in dependence of the switching cycle of the inverter. This is a very powerful method for approximating these parameters in real time for control algorithms or monitoring systems. One great advantage of the switching frequency-based quantities is the combination of instantaneous information with the steady state information of the fundamental cycle-based quantities in one unified quantity.

III. Benefits and Suitability of the Switching Frequency-based Analysis for the Dual Two-level Inverter-fed PMSM

In this section, the benefits and the suitability of the switching frequency-based analysis is shown by using the example of a dual two-level inverter-fed PMSM. The major advantage of this compared to the conventional fundamental cycle-based analysis is the enhancement of the dynamic of the voltage, current, and power calculation. This increased dynamic allows the utilization of this analysis in real-time control algorithms—for example, for active and/or reactive power control. The power control

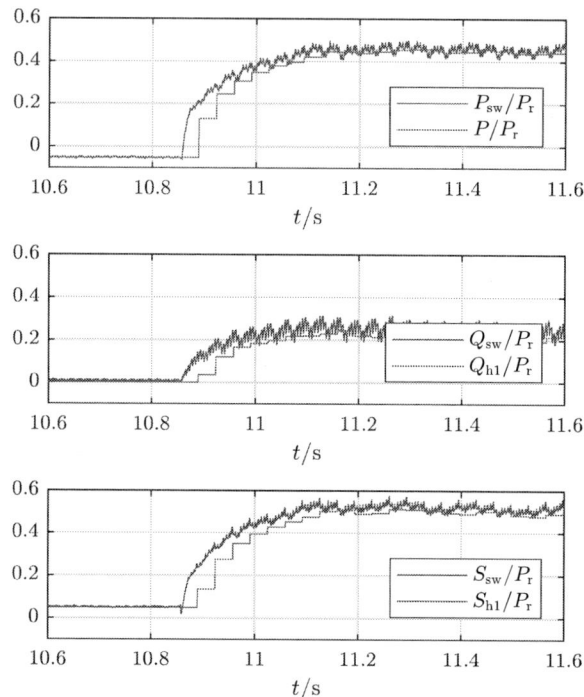

Fig. 5. Switching frequency-based active, reactive, and apparent power

is essential for an efficient operation of the dual two-level inverter application. Fig. 6 shows the basic circuit diagram of this topology:

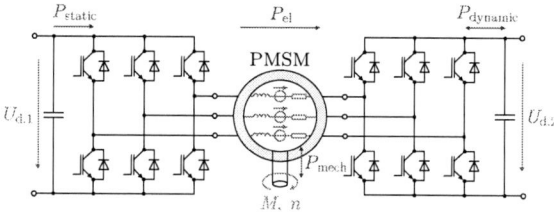

Fig. 6. Block diagram of the dual two-level inverter-fed PMSM

The two DC-links are connected to independent DC energy sources or storage systems. It is not necessary to utilize energy sources of the same type. On the contrary, applying completely different types with diverse voltage levels may lead to several advantages; it may open up the possibility to mutually compensate complementary attributes of the individual sources. Depending on the operating point, the energy flow of the overall system has to be controlled. During the motor operating mode, the machine is fed by one or both of the energy sources, whereby the energy may be transferred to one or both energy storage systems during generator operation. Additionally, it is possible to exchange energy between the two sources [7]. In order to realize reliable and efficient energy transfer between the sources and the machine, a robust and highly dynamic power control is essential. The percentage distribution of the energy on two separate energy storage and power electronics systems reduces

3037

the performance requirements for each of the two partial systems compared to a single two-level inverter topology. The DC-link voltage of both parts of the dual two-level inverter could, for example, be halved compared to the single two-level inverter in order to realize the same resulting machine voltage and power classification. As a result, the breakdown voltage of the IGBTs can be drastically reduced. Furthermore, the control is able to compensate for fluctuating DC-link potentials by adapting the energy flow of the individual energy storage systems and the corresponding partial inverters. A further idea is the separation of the energy flow in a stationary and a highly dynamic component. Each component is related to a corresponding energy storage. Therefore, completely different types of energy storage systems, such as a fuel cell combined with a battery (or a battery and a supercap), could be used. The power during transient balancing processes between two different operating points is delivered or recuperated by the highly dynamic energy storage, whereas the mean power demand during steady state is transferred by the stationary energy storage. Typically, the stationary source has a low reaction speed but a high continuous output power while the highly dynamic source is able to provide very high but short peak power. Usually, its capacity is much smaller.

Fig. 7 shows the basic principle of separating transient and stationary power: Due to mechanical load changes

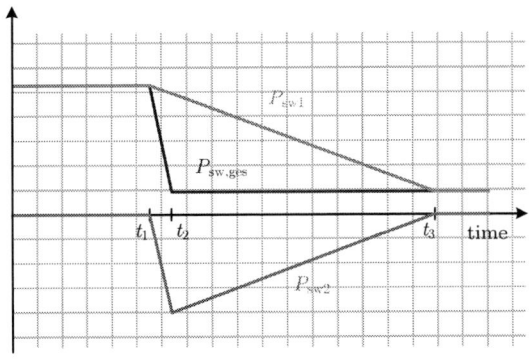

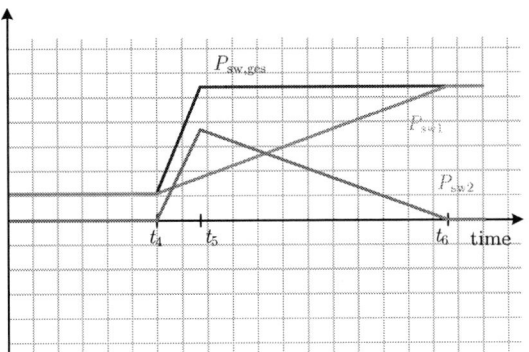

Fig. 7. Qualitative diagram of the possible active power distribution of a dual two-level inverter-fed PMSM

of the electrical machine, power varies rapidly from one

to another operating point. Since the maximum energy transfer ratio of Inverter 1 is strictly limited by its internal structure, the lag time to provide the requested energy of the new operating point is relatively high. In order to guarantee a quick transfer to the new operating point, Inverter 2 delivers the power difference of the demanded machine power and the power delivered by Inverter 1. This realization is based on a highly dynamic electrical power calculation that is used to control the power distribution of the two energy sources in the requested modality.

The authors' test bench includes two batteries with equal dynamic behaviour and capacity. In order to demonstrate the behaviour of the previously mentioned system, including energy storage systems with different dynamic characteristics, the inverter software limits the reaction time of the power supply connected to Inverter 1. The measurement results are displayed in Fig. 8.

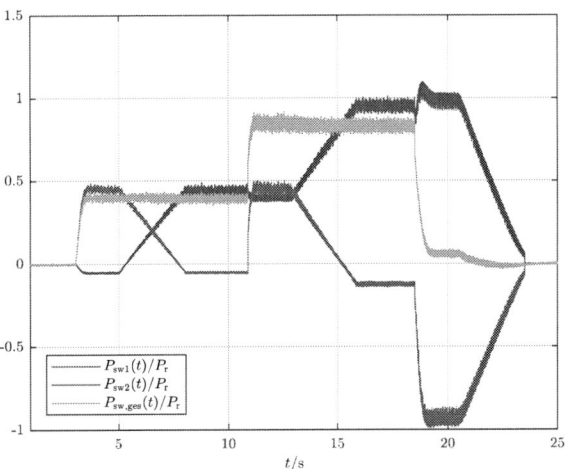

Fig. 8. Measurement of the switching frequency[c2]-based active power of both partial inverters and the overall system

At $t \approx 2{,}5\,\mathrm{s}$, the operating point jumps from no load to the first load operating point. The total switching frequency-based active power $P_{\mathrm{sw,ges}}$ is at first delivered by Inverter 2 $P_{\mathrm{sw,ges}} \approx P_{\mathrm{sw2}}$. At the beginning, the power supply of Inverter 1 remains more or less unaffected. After a given delay time, the power of Inverter 1 is increased; in the same way, the power of Inverter 2 is decreased. After this balancing time, the lagging Inverter 1 provides the total amount of active power $P_{\mathrm{sw,ges}} \approx P_{\mathrm{sw1}}$. At $t \approx 11\,\mathrm{s}$, the same procedure is repeated before the load is reduced to no load at $t \approx 18\,\mathrm{s}$. Since there is no mechanical load, in this case the energy overshoot of the battery of Inverter 1 is transferred to the other battery connected to Inverter 2. The robust functionality of this power control is based on the evaluation of the switching frequency-based power analysis. This analysis represents the essential precondition for a dependable realization of this application.

3038

IV. CONCLUSIONS

In this contribution, the switching frequency-based analysis of power quantities, fundamental waveforms, and RMS voltages and currents is presented. It is a powerful method for approximating the corresponding conventional fundamental cycle-based quantities during steady state. For dynamic balancing processes, it additionally contains transient information of instantaneous quantities. The switching frequency-based quantities therefore combine the steady state information of the conventional fundamental cycle-based quantities with the transient information of instantaneous values in one novel quantity. Owing to the dramatically increased update rate of this analysis, it is very suitable for real-time applications, such as control algorithms and real-time measurement. The relevance and importance of this novel analysis is underlined by measurements at an active power-controlled dual two-level inverter-fed PMSM. The switching frequency-based analysis provides the basis for the realization of the robust and highly dynamic power control of this application.

REFERENCES

[1] A. Stock, J. Teigelkötter, S. Staudt, T. Kowalski, "Highly Dynamic Analysis of Active Power and Fundamental Component Approximation of Inverter-Fed Applications", IEEE 12th International Conference on Power Electronics and Drive Systems (PEDS), Honolulu, USA, 2017, pp. 291-296.

[2] A. Amerise, M. Mengoni, L. Zarri, A. Tani, S. Rubino, R. Bojoi, "Open-Ended Induction Motor Drive with a Floating Capacitor Bridge at Variable DC Link Voltage", IEEE Energy Conversion Congress and Exposition (ECCE), Cincinnati, Ohio, 2017, pp. 3591-3597.

[3] IEEE Standard 1459-2010, Definitions for the Measurement of Electric Power Quantities Under Sinusoidal, Nonsinusoidal, Balanced, or Unbalanced Conditions, IEEE, New York, USA, 2010

[4] K. Zhou, D. Wang, "Relationship between space-vector modulation and three-phase carrier-based PWM: a comprehensive analysis [three-phase inverters]", IEEE Transactions on Industrial Electronics, vol. 49, no. 1, pp. 186-196, 2002.

[5] V. Staudt, "Fryze - Buchholz - Depenbrock: A time-domain power theory", International School on Nonsinusoidal Currents and Compensation, Łagów, Poland, 2008, pp. 1-12

[6] J. Teigelkötter, Energieeffiziente elektrische Antriebe - Grundlagen, Leistungselektronik, Betriebsverhalten und Regelung von Drehstrommotoren, Springer Vieweg: Wiesbaden, Germany, 2013.

[7] G. Grandi, C. Rossi, A. Lega, D. Casadei, "Multilevel Operation of a Dual Two-Level Inverter with Power Balancing Capability", Conference Record of the 2006 IEEE Industry Applications Conference, 2006. 41st IAS Annual Meeting, 2006.

Design and Analysis of High Voltage Power Supply for Industrial Electrostatic Precipitators

Shengwen Fan[1], Yiqin Yuan[2]*, Pengyu Jia[2], Zhigang Chen[1], Haisi Li[2]

1 College of Automation, University of Science and Technology Beijing, Beijing, China
2 Power Electronics and Motor Drives Engineering Center, North China University of Technology, Beijing, China
*E-mail: knightsam@outlook.com

Abstract— Electrostatic precipitator (ESP) is an efficient dust removal equipment. High voltage power supply is the most important part of ESP. In this paper, a 60kV/60kW high voltage power supply for ESP is designed. It uses a multi module half-bridge quasi resonant circuit, and could offer a wide range of output voltage under the soft switch which could reducing the energy consumption. The working states of the circuit are analyzed in detail, and the rough calculation process of the parameters is given in this paper. In addition, the design essentials of high voltage transformer are also given. Finally, the simulation results verify the correctness of the theory. A 60kV/6kW prototype is designed, which proves the correctness of the simulation.

Keywords—Electrostatic precipitator, High voltage converter , LCC half-bridge resonant converter , resonance parameters

I. INTRODUCTION

Electrostatic precipitator (ESP) is a device which uses high voltage electric field to remove solid particles in the air. And it is widely used in industrial field. The power supply is the most important part, which determines the efficiency of ESP [1][2][3]. Due to the characteristics of load, the high voltage power supply of ESP needs to provide a wide range of power output.

In the traditional design, the high voltage power supply usually adopts the hard switch, which will increase the loss and reduce the efficiency. In order to improve the efficiency of high voltage power supply and reduce energy loss, resonant soft switch technology has been applied in these years. LCC series parallel resonant converters combined the advantages of series resonant converters and parallel resonant converters, which have been widely used in high voltage power supply [4][5]. In practical application, LCC resonance needs to use leakage inductance and distribution capacitor of transformer to participate in resonance. Therefore, the parasitic parameters of transformer largely determine the performance of high voltage power supply. For these reasons, the design of transformer is demanding.

In this design, we use multi-module parallel half-bridge resonant converters. As the power is dispersed, the circuit's requirement for leakage inductance of transformer is reduced. Therefore, the design of transformer becomes much simpler. At the same time, the design of multi-module converters is helpful to reduce the requirement of power module, which is beneficial to reduce the loss.

II. ANALYSIS OF PRINCIPLE

The structure of the multi-module parallel half-bridge resonant converters is shown in Fig. 1 [6][7].

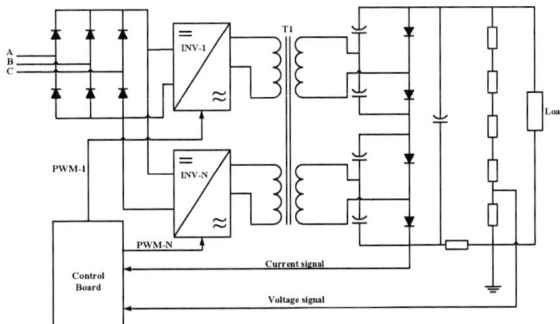

Fig. 1. The high voltage power supply system structure of ESP.

There are three half-bridge inverters in this design. These inverters share one transformer. This transformer has three pairs of windings. The three secondary sides are connected in series by the voltage-multiplier circuit.

The topology of the inverter is shown in Fig. 2.

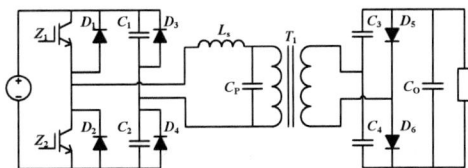

Fig. 2. Half-bridge resonant circuit.

Z_1 and Z_2 are the switch transistors. D_1, D_2, D_3 and D_4 are the freewheel diode. C_1 and C_2 are the resonant capacitors. L_s is leakage inductance of transformer, and C_p is distributed capacitance of transformer. C_3, C_4, D_5 and D_6 constitute voltage-multiplier circuit. C_o is output capacitor.

Unlike the traditional half-bridge circuit, C_1 and C_2 are replaced with small capacitors to participate in resonance which can limit the total energy in the resonant network and control the maximum output power by adjusting C_1 and C_2. D_3 and D_4 parallel with C_1 and which can keep the voltages of C_1 and C_2 not-negative.

This circuit can be divided into two different working states depending on the load.

A. Heavy-load state

The switching modes are shown in Fig.3.

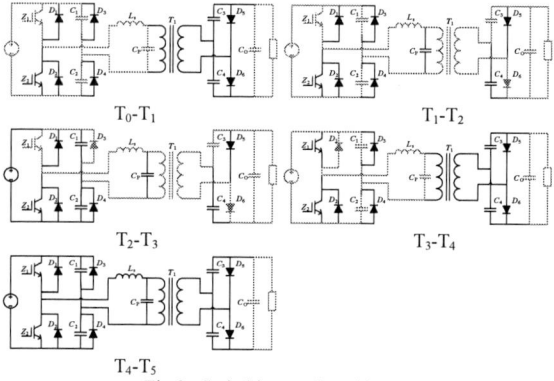

T₀-T₁ T₁-T₂

T₂-T₃ T₃-T₄

T₄-T₅

Fig.3. Switching modes of heavy-load.

[T₀-T₁]

At T_0 time, the switch transistor Z_1 is turn-on. The current of inductance i_{Ls} begin to increase from zero. in this case, Z_1 is zero current switching(ZCS). The resonant capacitance C_1 and the power supply V_g simultaneously discharge into the resonant circuit, and the voltage of C_2 gradually increases with the decrease of the voltage of C_1. The voltage of the distributed capacitance (V_{CP}) rises from $V_{CP}(t_0)$ to V_{out}/n (V_{out} is the output voltage, n is the ratio of transformer, V_{out}/n is denoted as V_L). During this time, the circuit is in LCC resonant state. Its equivalent circuit is shown in Fig. 4.

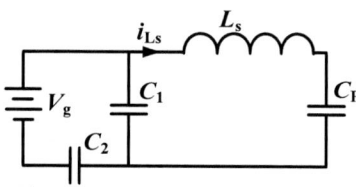

Fig. 4. Equivalent circuit during T₀-T₁.

The voltage of C_1, C_P and the current of inductor are:

$$i_{Ls}(t) = \frac{V_g - V_{C2}(t_0) - V_{CP}(t_0)}{\rho_r} \sin \omega_r(t - t_0) \quad (1)$$

$$V_{C1}(t) = V_{C1}(t_0) - \frac{1}{C_S} \int_{t_0}^{t} i_{Ls}(t)dt \quad (2)$$

$$V_{CP}(t) = V_{CP}(t_0) + \frac{1}{C_P} \int_{t_0}^{t} i_{Ls}(t)dt \quad (3)$$

Where $\omega_r = 1/\sqrt{L_s C_r}$, $\rho_r = \sqrt{L_s/C_r}$, $C_S = C_1 + C_2$, $C_r = C_P C_S/(C_P + C_S)$.

[T₁-T₂]

At T_1 time, the voltage of C_P reaches to V_L. After this point, the voltage-multiplier circuit is connected, and the primary side of the transformer transmits energy to the secondary side. During this time, the circuit is in LC resonant state. Its equivalent circuit is shown in Fig. 5.

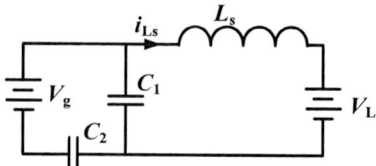

Fig. 5. Equivalent circuit during T₁-T₂.

The voltage of C_1 and the current of inductor are:

$$i_{Ls}(t) = \frac{V_g - V_{C2}(t_1) - V_L}{\rho_s} \sin \omega_s(t - t_1) + i_{Ls}(t_1)\cos \omega_s(t - t_1) \quad (4)$$

$$V_{C1}(t) = V_{C1}(t_1) - \frac{1}{C_S} \int_{t_1}^{t} i_{Ls}(t)dt \quad (5)$$

Where $\omega_s = 1/\sqrt{L_s C_S}$, $\rho_s = \sqrt{L_s/C_S}$.

[T₂-T₃]

At T_2, the resonance capacitor C_1 completes the discharge process. The voltage of C_1 become to zero, and the voltage of C_2 increase to the bus voltage V_g. Because the current cannot be mutated, so the current flows through the diode D_3. The power supply is separated from the resonant circuit and no energy is provided. The energy which stored in the leakage inductance continues to release to the secondary side, and the current decreases linearly. Its equivalent circuit is shown in Fig. 6.

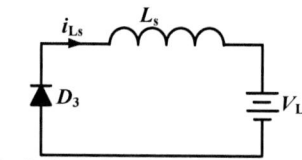

Fig. 6. Equivalent circuit during T₂-T₃.

The current of inductor is:

$$i_{Ls}(t) = i_{Ls}(t_2) - \frac{V_L}{L_s}(t - t_2) \quad (6)$$

[T₃-T₄]

At T_3 time, the current i_{Ls} is reduced to zero. The current continues to decrease through D_1 which parallels with Z_1. So, the energy which stored in C_P starts to release. Most of the energy in C_P is returned to the power supply, and the small part charges for C_1. In this case, the voltage of C_1 increase tiny. During this time, the switch transistor Z_1 is turn-off. There is no voltage at the switch transistor and no current flowing inside, so it is zero voltage and zero current switching(ZCZVS). The voltage of C_P decreases. The load is powered by output capacitor C_O. The circuit is in LCC resonant state. Its equivalent circuit is shown in Fig. 7.

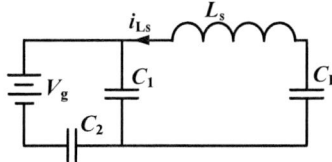

Fig. 7. Equivalent circuit during T_3-T_4

The voltage of C_1, C_P and the current of inductor are:

$$i_{Ls}(t) = \frac{V_L}{\rho_r}\sin\omega_r(t-t_3) \tag{7}$$

$$V_{C1}(t) = -\frac{1}{C_S}\int_{t_3}^{t} i_{Ls}(t)dt \tag{8}$$

$$V_{CP}(t) = V_L + \frac{1}{C_P}\int_{t_3}^{t} i_{Ls}(t)dt \tag{9}$$

[T_4-T_5]

All switch transistors and diodes are closed, the current is zero, and the voltages of C_1, C_2 and C_P remain constant. The load is powered by output capacitor C_O.

When Z_1 is opened, the switching modes are the same as Z_2.

The working state curve is shown in Fig. 8.

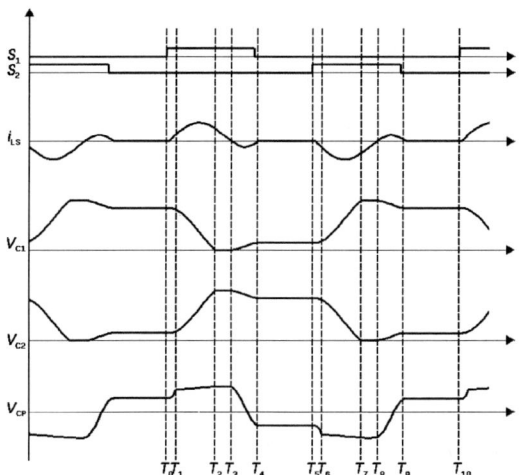

Fig. 8. Working state curve in heavy-load.

B. *Light-load state*

The switching modes are shown in Fig. 9.

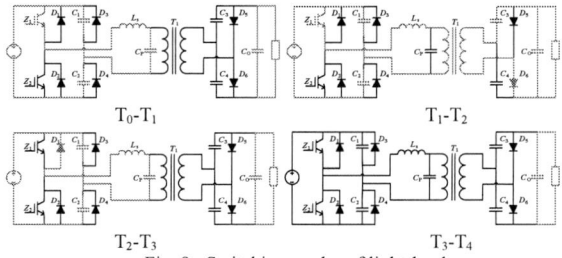

Fig. 9. Switching modes of light-load.

[T_0-T_1]

In this interval, the circuit operation state is consistent with the heavy-load. At T_0 time, the switch transistors Z_1 is turn-on. The current of inductance i_{Ls} begins to increase from zero. The voltage of C_2 increases, and the voltage of C_1 decreases. The voltage of the distributed capacitance rises from $V_{CP}(t_0)$ to V_L. The circuit is in LCC resonant state. Its equivalent circuit is shown in Fig. 10.

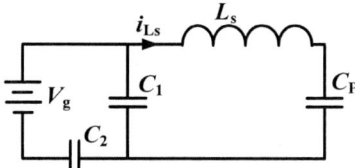

Fig. 10. Equivalent circuit during T_0-T_1.

[T_1-T_2]

At T_1 time, the voltage of C_P reaches to V_L. After this point, the voltage-multiplier circuit is connected. During this time, the circuit is in LC resonant state. Its equivalent circuit is shown in Fig. 11.

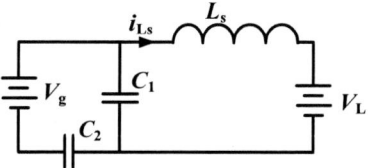

Fig. 11. Equivalent circuit during T_1-T_2.

[T_2-T_3]

At T_2 time, the current reverses, and C_1 stops releasing energy. At this point, D_1 is opened, Z_1 is clamped. The voltage of C_P is decreased. Z_1 needs to be turn-off during this time. The load is powered by output capacitor C_O. Its equivalent circuit is shown in Fig. 12.

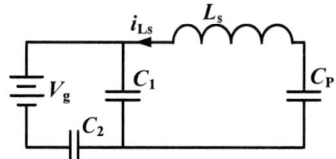

Fig. 12. Equivalent circuit during T_2-T_3.

[T_3-T_4]

All switching transistors and diodes are closed, the current is zero, and the voltages of C_1, C_2 and C_P remain constant. The load is powered by output capacitor C_O.

The working state curve is shown in Fig. 13.

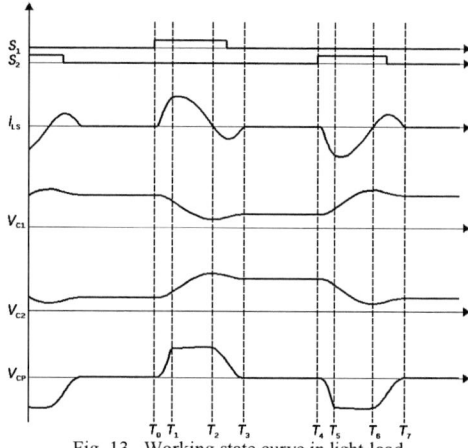

Fig. 13. Working state curve in light-load

In this circuit, the difference between heavy-load and light-load is that when the current decreases to zero, whether the energy stored in the resonant capacitor is fully released. In light-load state, the voltage of resonant capacitors will not be reduced to zero, So the diodes D_3, D_4 don't work. However, it should be pointed out that when the power is starting, the load can be regarded as infinite, so D_3, D_4 cannot be omitted. When the circuit enters steady state and also is in a light-load state, the system has no voltage regulating characteristic. The change in load or frequency within a certain range does not cause a voltage change. This is very harmful. To avoid this, you need to limit the system work state to heavy-load state.

III. CALCULATION OF PARAMETER

In engineering, for simple parameter design process, we have made a significant simplification of the resonance cycle calculation. In this case, the output power is 60kW, the bus voltage is about 530V, and the working frequency is about 40kHz.

According to the formula of capacitance transferable energy: $P = C_s V_r^2 f_s$. Here $C_s = C_1 + C_2$, P is the expected output power, V_r is the change voltage of the resonant capacitor, f_s is the switching frequency. First, we need to assume the number of modules. here, we use three windings to output the full power. A single module needs to output 20kW. Take a 15% allowance, the individual resonant capacitor is about 1 μF.

The resonance period can be easily obtained when all parameters are known. However, at the beginning of the design, the design requirements of the transformer are not determined, so some parameters need to be ignored. We exclude the effect of load on the resonance period and treat the entire resonance process as an ideal state. We use LC resonance to replace the positive working cycle of the actual current, and LCC resonance replaces the reverse working cycle of the actual work. In order to reduce the switching loss, we hope to turn off the switch transistors during the current reverse flow. Therefore, the actual opening time of switch transistors is more than half

of the LC resonance cycle. When the rated switching frequency is 40kHz, the LC resonant frequency must be greater than 40kHz. The reverse flow time is less than the forward working time, so the LCC resonant frequency must be less than 80kHz. Here, choose about 60kHz, according to the formula: $f_o = 1/\sqrt{L_s C_s}$, The leakage inductance value is about 3.5 μH.

According to the gain characteristics of LCC, the gain in DCM is nearly 0.47. The three secondary sides are connected in series by the voltage-multiplier circuit, So, the gain of rectifier is 6. The bus voltage is 530V, and the output voltage is 60kV, the ratio of transformer is about 40.

According to above methods, we can get the design requirements of the transformer.

When the transformer parameters are known. The maximum voltage and power can be calculated according to the set switch frequency, and the appropriate number of modules can be selected.

IV. DESIGN OF TRANSFORMER

Transformer design is the most important part of high voltage power supply. This part briefly introduces the method of transformer design. In this circuit, three inverters are connected in parallel. The output power of the single inverter is reduced, and the resonant capacitor is also reduced. It makes the value of the leakage inductance higher. The design of the transformer is much easier. The transformer structure is shown in Fig. 14.

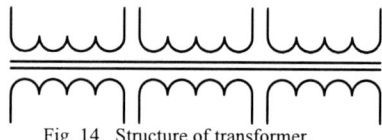

Fig. 14. Structure of transformer.

In order to reduce the leakage inductance of the transformer, the coupling degree should be increased. Therefore, the EE type magnetic core is chosen. The secondary sides are wound outside the primary sides.

The distributed capacitance C_p can partially consume the energy of the resonant capacitor C_s and reduce the output power. The distributed capacitance need be reduced. Two kinds of winding methods of transformer are shown in Fig. 15.

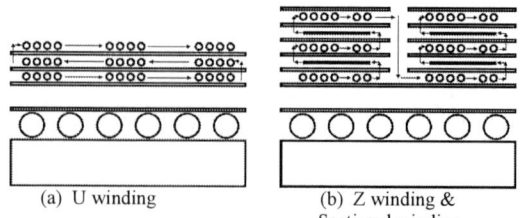

(a) U winding   (b) Z winding &
Sectional winding

Fig. 15. Winding methods of transformer.

The parameters of the transformer in two methods are as TABLE1.

3043

TABLE1
TRANSFORMER PARAMETERS UNDER DIFFERENT WINDING METHOD

	U winding	Z winding & Sectional winding
Magnetic induction	80 μH	100 μH
Leakage inductance	1.3 μH	1.7 μH
Distributed capacitance	600 μF	100 μF

Using 'Z winding' and 'Sectional winding' to replace 'U winding', the distributed capacitance is reduced by six times. However the 'Z winding' makes the distance larger between each layer. The coupling degree of the two sides of transformer is decreased, so the leakage inductance is slightly increased. But it is still within acceptable limits. The real picture of the transformer is shown in Fig. 16.

Fig. 16. Transformer of ESP power supply.

V. SIMULATION AND RESULTS

The Saber simulation software is used to verify the simulation according to the design requirements. The simulation parameters are as TABLE2.

TABLE2
60kW SIMULATION PARAMETERS

Parameter name	Value
Bus voltage	530V
Resonant capacitance	1 μF
Number of inverters	3
Transformer ratio	1:40*3
Leakage inductance	3.5 μH
Distributed capacitance	100 μF
Load resistance	60 kΩ
Switch frequency	<40kHz

The simulation results are shown in Fig.17.

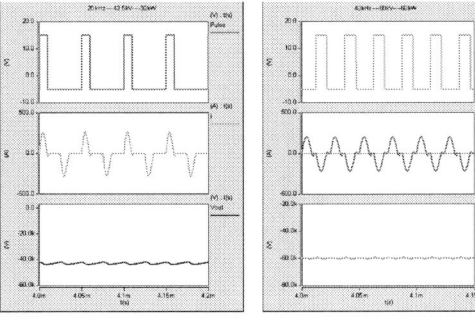

(a) 20kHz   (b) 40kHz
Fig.17. Output waveforms at different frequency.

When the switching frequency is 20kHz, the output voltage is 42.5kV, and the output power is 30kW. When the switching frequency is 40kHz, the output voltage is 60kV, and the output power is 60kW. The output power is proportional to the frequency.

This circuit uses the method of energy dosage modulation. When the system starts to work, the load can be regarded as infinite, so the dynamic response at startup is consistent. By reducing the bus voltage, the maximum output voltage at light load reduce to 55kV.

The simulation results are shown in Fig.18.

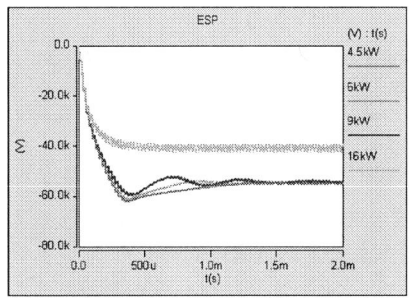

Fig.18. Output voltage at different load.

As shown in Fig.18, in light-load state which power are 4.5kW, 6kW and 9kW, the time from the beginning to the minimum is the same under the same switching frequency and different load resistance. And the final output voltage is consistent. when the system is in heavy-load state which power is 16kW, the energy stored in the resonant capacitors is completely released and the voltage drops. This proves that under light-load state, the change of load will not cause the change of output voltage, and the circuit has no voltage regulating characteristic.

Due to limited conditions, we cannot obtain a stable 60kW load. Therefore, we limit the output power to 6kW by reducing the switching frequency and the size of the resonant capacitors. The switching frequency is 20kHz, and resonant capacitor is 0.2 μF. Using above design method, we redesigned the parameters. The inductor is connected with the leakage inductor in series to increase the inductance of the circuit. The simulation parameters are as TABLE3.

TABLE3
6KW SIMULATION PARAMETERS

Parameter name	Value
Bus voltage	530V
Resonant capacitance	0.2 μF
Number of inverters	3
Transformer ratio	1:40*3
Inductor	60 μH
Distributed capacitance	100 μF
Load resistance	600 kΩ
Switch frequency	<20kHz

The simulation results are shown in Fig.19.

3044

The 2018 International Power Electronics Conference

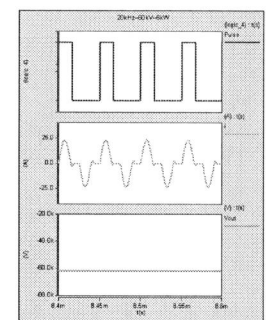

Fig.19. 20kHz/6kW output waveform.

As shown in Fig.19, the new resonance parameters meet the system requirements. In 20kHz, the output voltage is 60kV, the output power is 6kW and the peak point current is 25A.

The experimental prototype is shown in Fig. 20.

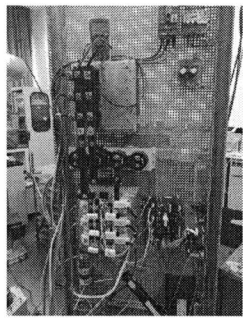 

(a) Main circuit   (b) Tank and load

Fig.20. Experimental prototype

The experimental waveform is shown in Fig. 21.

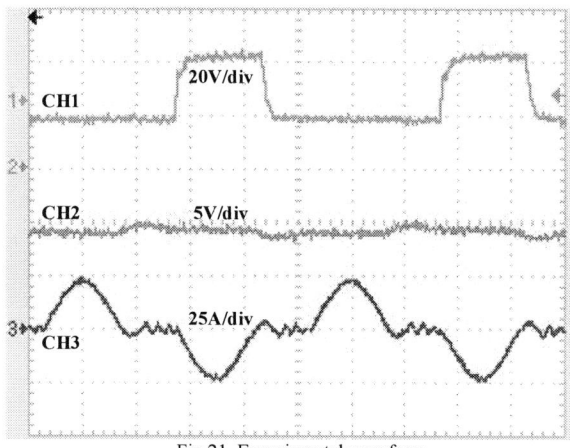

Fig.21. Experimental waveform

CH1 is the waveform of drive pulse. CH2 is the output voltage (10000 times attenuation). CH3 is the waveform of current. The experimental waveform is consistent with the simulation waveform. The correctness of the theory is verified.

VI. CONCLUSION

A 60kV/60kW high voltage power supply for ESP is designed in the paper which uses a multi-module half-bridge quasi resonant circuit. The working states of this circuit under different loads are analyzed in detail, which should be limited to the heavy load mode in order to make the output power adjustable and proportional to switching frequency. On above basis, a design method based on power is proposed to design the resonance parameters of high voltage power supply. In addition, the transformer is also designed to match with the resonance parameters, and its accuracy is verified by simulation. Finally, the simulation and experimental results are given to verify the correctness of the theory with a good dynamic response and adjustability.

REFERENCES

[1] Ye Z, Yong J K, Tai G L, et al. Experimental and theoretical studies of ultra-fine particle behavior in electrostatic precipitators[J]. Journal of Electrostatics, 2000, 48(3):245-260.

[2] Grass N, Hartmann W, Klockner M. Application of different types of high-voltage supplies on industrial electrostatic precipitators[J]. IEEE Transactions on Industry Applications, 2004, 40(6):1513-1520.

[3] Leonard G, Mitchner M, Self S A. Particle transport in electrostatic precipitators ☆[J]. Atmospheric Environment, 1980, 14(11):1289-1299.

[4] Soeiro T B, Mühlethaler J, Linnér J, et al. Automated Design of a High-Power High-Frequency LCC Resonant Converter for Electrostatic Precipitators[J]. IEEE Transactions on Industrial Electronics, 2013, 60(11):4805-4819.

[5] Pawellek A, Bucher A, Duerbaum T. Analysis and design of a resonant LCC converter for low-profile applications[C]// Energy Conversion Congress and Exposition. IEEE Xplore, 2010:3333-3339.

[6] Pokryvailo A, Carp C, Scapellati C. High-Power High-Performance Low-Cost Capacitor Charger Concept and Implementation[J]. IEEE Transactions on Plasma Science, 2010, 38(10):2734-2745.

[7] Pokryvailo A, Carp C, Scapellati C. A High-Power High-Voltage Power Supply for Long-Pulse Applications[J]. IEEE Transactions on Plasma Science, 2010, 38(10):2604-2610.

Load Sharing Operation in N+1 UPS System by Using Harmonic Sharing Control Method

Prashant Patel[1], Sagar Naina[2], Utsav Patel[2] and Premal Patwa[2]

1 Research and Development Centre, Hitachi India Pvt. Ltd., Bangalore-560052, India
2 Research and Development Centre, HHPE Pvt. Ltd., Gujarat-382044, India
E-mail: prashant@hitachi.co.in, utsav_patel@hitachi-hirel.com, premal_patwa@hitachi-hirel.com

Abstract- **Due to the rapid expansion of infrastructure, the demand for Uninterruptible Power Supply (UPS) is gradually increasing. For good redundancy and less downtime, the parallel operation (N+1) is a special feature of high performance UPS system. The main objective of this configuration is to operate all the UPSs independently. The parallel operation of UPS is a challenging task as the synchronization among all parallel connected inverters has to be maintained. In this work, a Master-Slave (M-S) based concept is used for N+1 configuration and the problem associated with parallel operation, such as unequal load sharing and high voltage THD, are addressed. In this method, a modified PI based control is used for both Master and Slave units to achieve the desired voltage THD and load sharing accuracy. In addition, experimental results from a setup with 8 parallel 60kVA UPS verify the effectiveness of the proposed M-S based control.**

Keywords— Master-Slave Configuration, Parallel Operation, PI Controller, THD (Total Harmonic Distortion)

I. INTRODUCTION

UPS is generally used to supply power to critical loads like datacenters, hospitals and industries. In order to achieve high power rating several UPSs are connected in parallel. UPS units are generally connected in N+1 configuration to achieve more reliability, redundancy and modularity. In 'N+1' configuration, 'N' corresponds to the number of UPSs which are supplying the load, '+1' refers to one redundant (standby) UPS. The control of parallel UPS has challenges like equal power sharing among parallel units, lower load voltage THD and minimum circulating currents. There are several control methods are available for parallel operation of UPS system such as Master-Slave[M-S] based control method, Droop control method, Average load sharing, Circular Chain control and so on [1].

The main aim for any parallel connected systems is that every system should play dynamic role to maintain the load voltage THD within permissible range and also have to share the load according to their capacities. In master-slave configuration the conventional fundamental component based PI control is not sufficient to maintain the load voltage THD within permissible range. In addition, it is also incapable to share load equally among the parallel connected system. The reason behind this issue is that the fundamental component based voltage control method is unable to control load voltage THD while the fundamental component based current control method is unable to share harmonic load current equally

as load current depends on the type of load, impedance between source to load and so on [2], [3]. Hence to resolve this issue, voltage based harmonic control loops are added in both master and slave units. The overall control logic of Master and Slave units with modified PI control is explained and validated in this paper for N+1 UPS systems (Here N=7).

This paper explains the complexity involved in control method for parallel connected UPS system and the problem involved in paralleling operation. It addresses the fundamental component based control loop with modified PI controllers for both master and slave units. Further, it covers the development of digital harmonic control method to overcome load sharing problem. The rest of the paper is organized as follows: The concept of N+1 UPS control is discussed in Section II. Implementation of fundamental and harmonics control loops for master and slave unit are focused in section III. The hardware based results are shown in Section IV. Finally in section V, conclusions are carried out.

II. CONCEPT OF N+1 UPS CONTROL

The developed configuration of the N+1 UPS system is shown in Fig. 1. Fig.1 depicts the simplified single line diagram of the N+1 UPS configuration, where one voltage controlled and seven current controlled UPSs are interconnected as a meshed power network through power cables. Each UPS in Fig.1 consists of a PWM rectifier, PWM inverter with output LC-L filter. The system is operated in master-slave control mode. The voltage-controlled inverter (master unit) regulates the system voltage amplitude and frequency. The current-controlled inverters (slave units) operate in current control mode to share the power based on the given reference value [4].

The basic requirements for the parallel operation of UPSs are that all the UPS units should operate at same frequency, phase and voltage level. Further, to prevent the overloading of any UPS, it is required to share the load power equally among the parallel units. Here, the information of inverter current (I_{inv}), load current (I_L) and load voltage (V_L) are used by all UPSs as inputs. It also clears from Fig.1. In Fig.1, switch CBI (Circuit Breaker Input) is used to provide input supply to the rectifier and switch MIS (Mains Isolator Switch) is used to connect inverter to the load. Moreover, the switch CBA (Circuit Breaker Alternate) is used to supply the load directly-

The 2018 International Power Electronics Conference

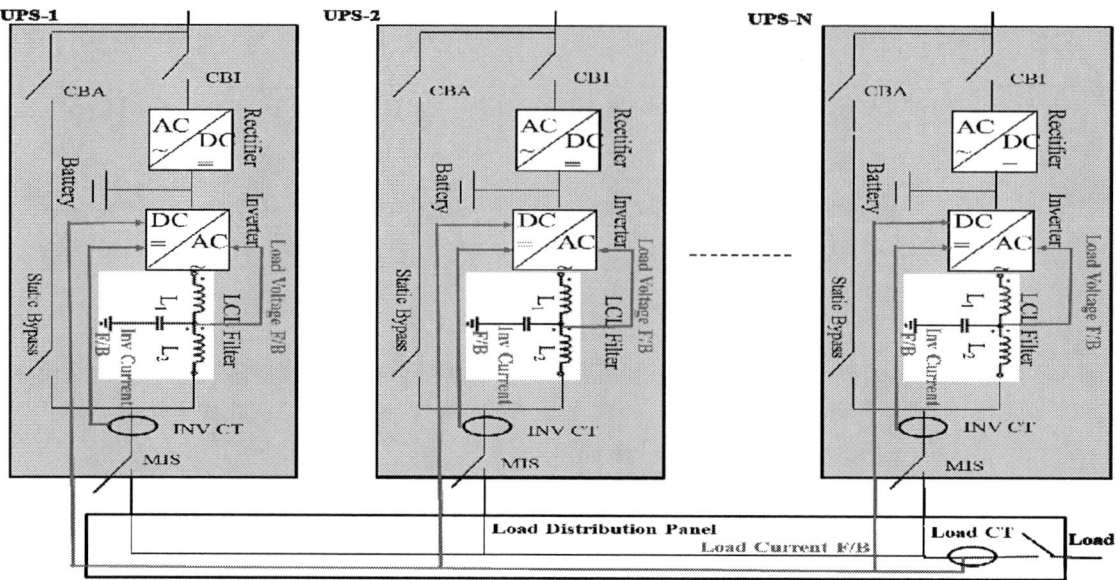

CBI: Circuit Breaker Input CBA: Circuit Breaker Alternate MIS: Mains Isolator Switch

Fig. 1. Single line diagram of 7+1 distributed UPS system

from the bypass supply source.

A. Development of Load Sharing Technique for Parallel Inverter System

The developed Master-Slave based load sharing technique in this work ensures proper load-sharing with two control methods: fundamental current sharing control and harmonic current sharing control. Moreover, the system reliability can be maintained by master-slave toggling functionality in case of failure of master UPS unit to avoid single point of failure. In this work a modified PI based per phase dq0 reference frame based control strategy is implemented. Because the per-phase voltage control method can independently control all the phases, this is preferred over three-phase $dq0$ reference frame based control [5], [6]. Also, it can maintain exact balanced voltage against unbalanced loading. Further, the detail descriptions of master and slave controls are given in next subsections.

B. Fundamental Component Based Load Sharing Technique for Parallel Inverter System

The block diagram of a master control contains a per phase $dq0$ based voltage control loop for load voltage regulation while the slave control contains a per phase $dq0$ based current control loop for proper load sharing operation. The orthogonal component of phase voltage and phase current are generated by applying αβ→dq transformation and its reverse transformation dq→αβ [5], [6].

The fundamental component based voltage controlled method for master R-phase is shown in Fig.2. Here the phase voltage is considered as a α component. Next, the dq components are filtered through the low

pass filter (LPF) that has cut-off frequency at $1Hz$. This digital LPF allows only the fundamental components to pass out as it is converted to DC quantity by $\alpha\beta$-to-dq transformation. Here, $\alpha\beta$-to-dq and dq-to-$\alpha\beta$ blocks require θ which is generated from the PLL block. The comparison of d and q components of the phase voltage V_{dr} and V_{qr} respectively with the respective reference values V_{dref} and V_{qref} give errors and those are being used in PI controllers. The PI controllers in turn generate-

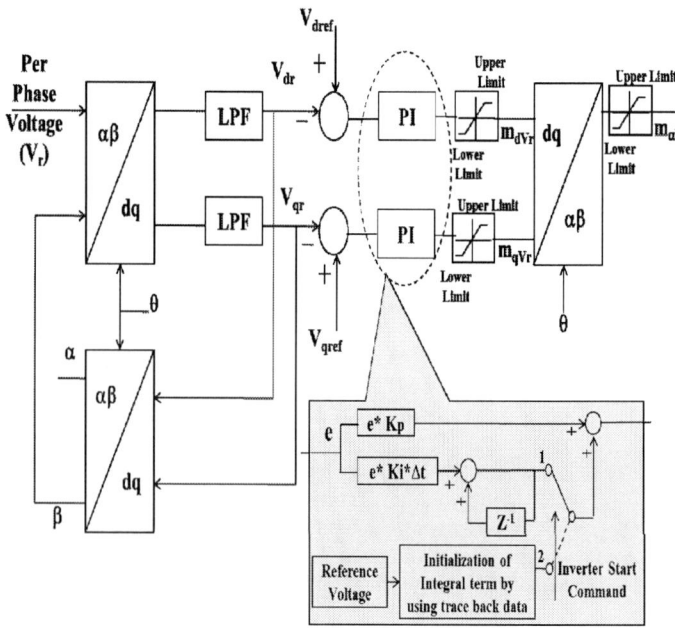

Fig. 2. Per phase fundamental component based voltage control with modified PI controller

The 2018 International Power Electronics Conference

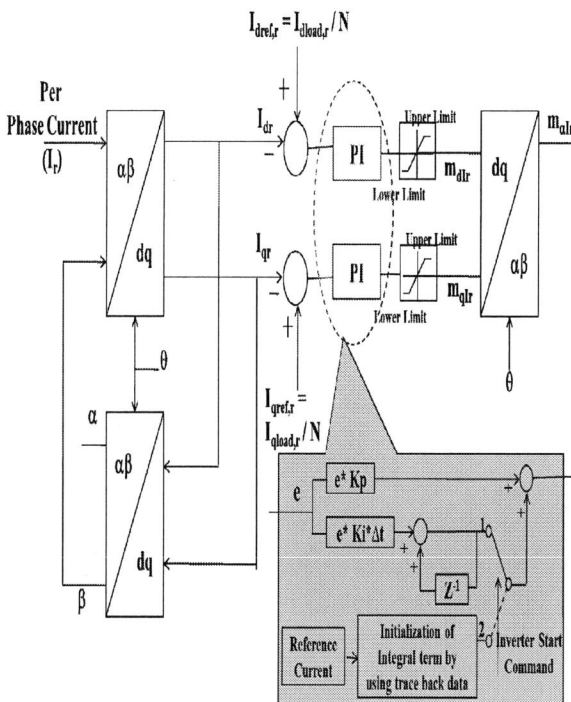

Fig. 3. Per phase fundamental component based current control with modified PI controller

respective modulating waveforms m_{dVr} and m_{qVr} which are fed to dq -to- $\alpha\beta$ to give out modulation signal $m_{\alpha Vr}$ for the given phase. Here, the integral terms of PI controllers are initialized by the particular value that is calculated by using the trace-back data of the system. As shown in Fig.2 the reference voltage signal is used to calculate the initial value of integral term by using trace-back record. As indicated in Fig.2, once inverter start command is generated the integral loop comes to position 1 from position 2 with an initialized value. This concepts helps an individual unit to do plug and play operation from the distributed UPS network.

The fundamental component based current sharing control for slave R-Phase is shown in Fig.3. Current control method is similar to voltage control method except the reference value given to PI controller are I_{dref} and I_{qref} which are derived from the sensed load current. Here the sensed load current is divided by the number of UPS units those are connected in parallel. Moreover, the d and q components of inverter current I_{dr} and I_{qr} respectively are not filtered out through digital low pass filter to avoid the LPF delay in current control. Further, the modulating waveform that is generated by current control loop is marked as $m_{\alpha dr}$ in Fig.3. In this control the integral terms of PI controllers are initialized from the trace-back data by using the reference current value. The same control strategy is used in other two phases Y and B but the

angles of rotation for reference frames are $\theta - 120$ and $\theta - 240$ respectively.

III. IMPLEMENTATION OF HARMONIC SHARING CONTROL

Fig.4 shows the overview of control block diagram of the N+1 distributed UPS system, where the overall system decomposes into N subsystems by inverters and the connection network. There is one voltage controlled system and the others are current controlled system. To achieve correct power sharing, an essential voltage harmonic control method is suggested for both master and slave units. Fig.4 shows the overview of master-slave based 7+1 UPS control method that involve both fundamental component based control method and harmonic components based control methods.

The prime interest for any parallel connected system is that every system should share power in equal proportion, failing which shall lead to overloading of some units. It is therefore required to share not only fundamental component but also necessary to share harmonic components along with fundamental. Moreover, it is also required to maintain load bus voltage THD within permissible limit and hence voltage harmonic cancelation control is required.

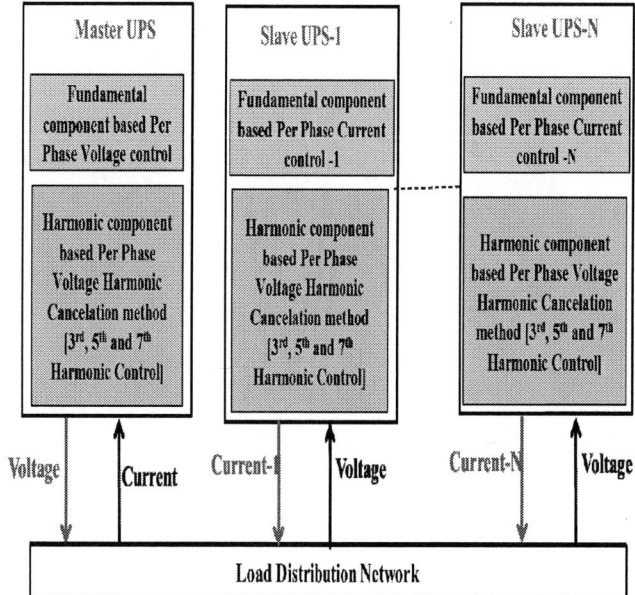

Fig. 4. Overview of master-slave based 7+1 UPS control method

Once master unit starts with fundamental voltage control method, the slave unit with fundamental current conrol method starts to sync its output with master unit. After synchronising is done slave unit connected with master unit and start to work in current sharing mode by taking reference from load feedback current transformer (CT). The N+1 unit maintains the load voltage THD within permissible limit and also shares equal current current with linear load but fails to do the same with non linear load. Here, the problem is not only with higher

3048

load voltage THD but also the current shared by slave units was delayed by master unit and as a result at step loading condition the master unit has to provide more load current compare to slave. As this condition can create overload condition for master unit during transient or step loading, it is recommended to share equal load at transient condition also.

Further, to resolve this issue the FFT analysis was carried out on the traceback data of load voltgae, load current and individual UPS inverter current. From FFT analysis it was observed that the dominant harmonics- 3rd, 5th and 7th - were presents in load bus voltage. It was also observed that the slave was sharing only fundamental current but they don't play any role for harmonics current sharing. Consequently, only master unit has to suffer from high current because it has to give harmonics currents to the load. To mitigate this higher voltage THD problem and to add harmonic sharing features, the fundamental component based control loop is modified by adding voltage harmonic control method in both master and slave units.

The per phase voltage harmonic control loop is shown in Fig.5. Here $dq0$ reference frame is selected to extract the harmonic information in terms of magnitude and phase. Such harmonic correction method cancels the voltage harmonics and support slave units to share harmonic currents.

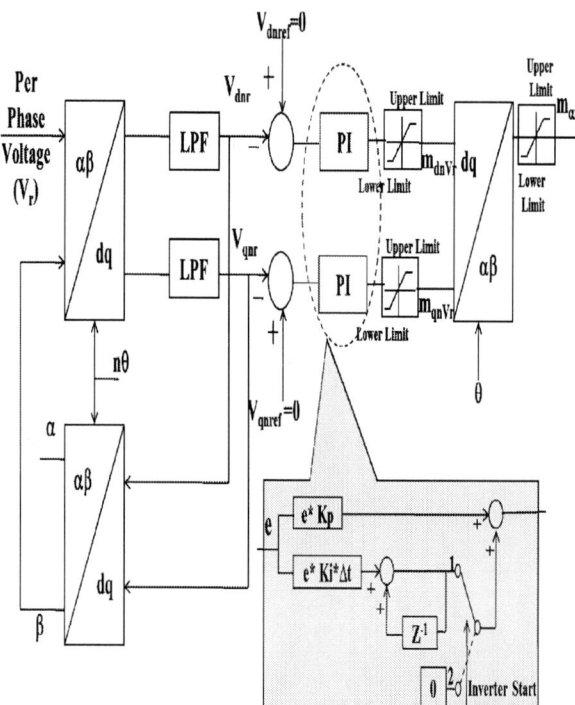

Fig. 5. Per phase voltage harmonic cancelation control with resettable integrator based PI control

As shown in Fig.5 the load R-phase voltage V_r is fed to αβ-to-dq transformation block. To extract the nth harmonic (where n = 3, 5, and 7) information from load voltage waveform the αβ-to-dq transformation is done at $\theta = n\omega$ rotational speed (where ω is the angular frequency of the fundamental component in rad/sec). As a result the nth harmonic component shall convert to a constant DC and other components will be appeared as AC. The constant DC quantity is extracted by using the LPF (with a cut-off at $10Hz$) that suppresses the AC quantity as well. The outputs of d and q axis LPF filters are marked as V_{dnr} and V_{qnr} respectively. These V_{dnr} and V_{qnr} are compared with the d and q axis references, $V_{dnref} = 0$ and $V_{qnref} = 0$, respectively. Further, the errors are being used by PI controllers. Here PI controllers with resettable integrator are used to generate respective modulating waveforms m_{dnVr} and m_{qnVr} for n^{th} harmonic as shown in Fig.5. Here, the resettable integrators reset the integral terms to zero once inverter stops command is generated. These m_{dnVr} and m_{qnVr} are converted to αβ components by dq-to-αβ transformation as displayed in Fig.5. Now this $m_{\alpha nVr}$ component is added with the modulating waveform generated by fundamental component based control loop, $m_{\alpha Vr}$ and $m_{\alpha dr}$, as marked in Fig.2 and Fig.3. By modifying the modulating waveform for both master and slave units the master unit improves the load voltage THD and the slave units start to share n^{th} harmonic current in equal proportion. As the proposed method is based on the per phase control, the correction of nth harmonic in other phases Y and B are given in same manner except the reference for αβ-to-dq and dq-to-αβ transformation are done at $(n\theta - 120)$ and $(n\theta - 240)$ respectively. Moreover, because of the impedance offered by LC-L filter, the interaction of voltage harmonic control loops in parallel units do not play any role for circulating current. Here the load side inductor (L₂ as marked in Fig.1) nullifies the mismatch of cable impedance from source to load as L₂ is far greater in value compare to the cable impedances. Moreover, L2 also helps to reduce harmonic circulating current that is caused due to mismatch in carrier frequency of DSP (Digital Signal Processor).

IV. RESULTS AND DISCUSSION

To validate above described fundamental component based control loop along with harmonics control loop in Master-Slave concept, the hardware testing was carried out on eight parallel connected three phase 60kVA UPS units. The test set up has been done as shown in Fig.1. The UPS input is taken from the utility grid of 400V line to line rms voltage. The DC bus voltage, output voltage and frequency of each UPS are adjusted to 783V, 240Vrms (Ph-N) and 50 Hz respectively. The UPS output is fed to the nonlinear load which is made of three phase diode bridge whose output is connected with parallel connected capacitors (C= 67200 μF) and resistors(R=0.04444 Ω).

The aim of this hardware testing is to show the efficacy of the proposed method of harmonics control in Master-Slave (M-S) based parallel connected UPS system. The results are divided and presented in following two subsections:

A. M-S Concept with Non-linear Load and with only Per Phase Fundamental Component Based Control Loop with Modified PI Controller
B. M-S Concept with Non-linear Load and with Per Phase Fundamental Component Based Control Loop with Modified PI Controller and Per Phase Voltage Harmonic Control Method with Resettable Integrator based PI Controller.

A. M-S Concept with Non-linear Load and with only Per Phase Fundamental Component based Control Loop with Modified PI Controller

To verify the effectiveness of the per phase voltage and per phase current control method as explained in section II, the 7+1 system was tested with non-linear load. Here, 7+1 indicates that there are total eight UPS systems are connected in parallel and any one of them acts as a redundant system. Fig.6 shows the load bus voltage waveform (ornge color), Master unit inverter current (violet color), Slave-1 inverter current (sky blue color) and Slave-7 inverter current (green color). Here for simplicity only two slave current is captured and details of all units current sharing are provided in Table.I. Right hand side of Fig.6, indicates that the load bus voltage THD is 5.01%. Further, as per the Table.I, the current sharing was also unequal. Here the per phase fundamental component based control method was unable to maintain voltage THD as per the desired specification. Hence, the adopted control method for master-slave concepts require revising to improve the parallel performance with nonlinear load.

B. Master-Slave M-S Concept with Non-linear Load and with Per Phase Fundamental Component based Control Loop with Modified PI controller and Per Phase Voltage Harmonic Control Method with Resettable Integrator based PI Controller

The master-slave units control is modified by adding per phase voltage harmonic cancelation method with resettable integrator based PI controller. The resettable integrator makes integral term zero during inverter stop operation hence it can be used for plug and play operation of individual UPS unit from the distributed network. Here the per phase voltage harmonic cancelation loops were added for 3rd, 5th and 7th harmonics as they were the dominant harmonics in load voltage waveform. With modified M-S control loop the N+1 system were again connected with nonlinear load. Fig.7 shows that the UPS output voltage waveform is improved and the measured output voltage THD at common load bus is 2.90% (with nonlinear load crest factor of 3:1). Fig.7 shows the load bus voltage waveform (orange color), Master unit inverter current (violet color), Slave-1 inverter current (sky blue color) and Slave-7 inverter current (green color). The load sharing information with this control method is provided in Table. I. Table.1 gives the load sharing information of master and all slave units. Further, it also shows the effect of harmonic control for load current sharing among parallel connected units. From Table.1 it clears that without harmonic control the current sharing among master and slave units are unequal while with the harmonic control the current distributions are equal.

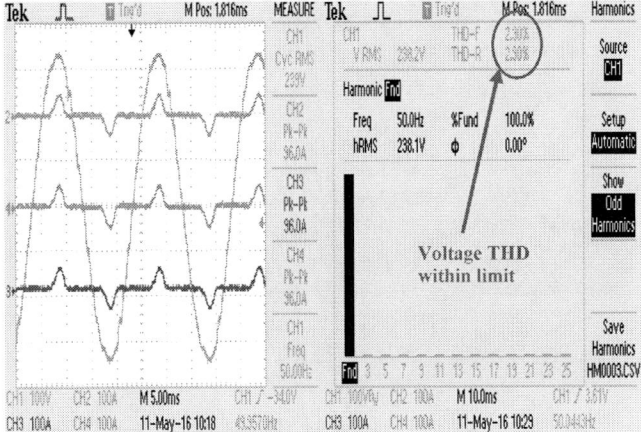

Fig. 7. Voltage/ Current/ THD waveforms of 7+1 UPS connections with nonlinear load for fundamental component based control with harmonic control method

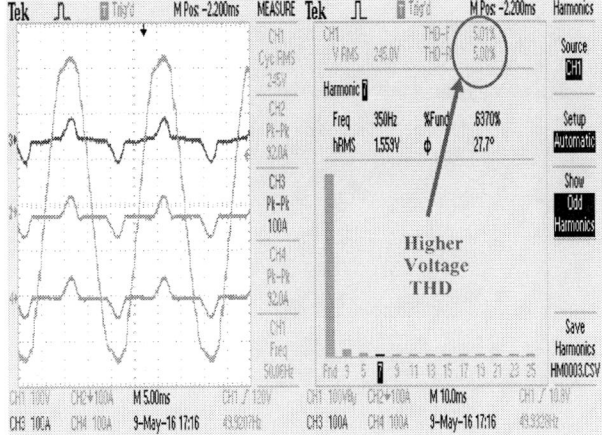

Fig. 6. Voltage/ Current/ THD waveforms of 7+1 UPS connections with nonlinear load for fundamental component based control

Further, to get clarity of current sharing during transient condition the step response of the 7+1 system is shown in Fig.8. In Fig.8 the orange color waveform indicates the load bus voltage waveform, violet color waveform shows the master unit inverter current, sky blue and green color waveforms indicate the slave-1 and slave-7 inverters' currents respectively. To show the sharing accuracy all the current waveforms are made coincide with others and it shows that all the current waveforms are exactly coincide with others (right hand side image) hence they are exactly equal in proportion.

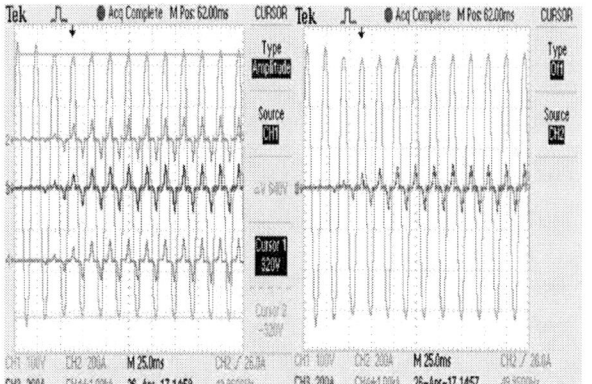

Fig. 8. Step response of 7+1 UPS system with non linear load

V. CONCLUSION

A load voltage THD and current sharing behavior of master-slave based parallel UPS units are studied and the poor voltage THD and unequal power sharing problem with only fundamental component based control is explained. Further, these problems are resolved by addressing per phase voltage harmonic control along with per phase fundamental component control. The working confirmation of parallel UPS units with proposed control is demonstrated on 60kVA UPS systems for 7+1 configuration. The result concludes that master and slave units are sharing current in exact proportion as per their specified ratings and also maintains the load voltage THD as per the standard IEC 62040.

REFERENCES

[1] Josep M. Guerrero, Lijun Hang, and Javier Uceda, "Control of Distributed Uninterruptible Power Supply Systems," IEEE transaction on Industrial Electronics, Vol.55, No.8, August 2008

[2] U. Borup, F. Blaabjerg, and P. N. Enjeti, "Sharing of nonlinear load in parallel-connected three-phase converters," *IEEE Trans. Ind. Appl.*, vol. 37, no. 6, pp. 1817–1823, Nov./Dec. 2001.

[3] Mohammad Nanda R. Marwali, "Digital Control of Pulse Width Modulated Inverters for High Performance Unitnterruptible Power Supplies", thesis, Graduate School of the Ohio State University

[4] A. P. Martins, A. S. Carvalho, and A. S. Araújo, "Design and implementation of a current controller for the parallel operation of standard UPSs," in Proc. IEEE IECON, 1995, pp. 584–589.

[5] A. Yazdani and R. Iravani, Voltage-Sourced Converters in Power System- Modelling, Control, and Applications. New Jersy: John Willey and Sons, 2010

[6] Paul C.Krause, Analysis of Electric Machinery and Drive Systems.U.S.. John Willey and Sons, 2002

TABLE I

POWER SHARING DATA OF 7+1 UPS SYSTEM

Unit-Order of Harmonic Current	Load Sharing Without Harmonic Control (A-rms)			Load Sharing With Harmonic Control (A-rms)		
	Ir	Iy	Ib	Ir	Iy	Ib
Master	71.5	70.2	71.9	64.4	64.5	65.3
3rd	170.7	169.6	170.2	40.7	41.2	40.1
5th	126.7	125.5	125.9	27.1	26.7	26.8
7th	57.4	56.8	56.1	12.8	12.2	11.9
Slave-1	61.8	62.2	61.6	63.2	64.4	62.9
3rd	20.5	19.4	18.9	38.6	38.2	39.1
5th	10.4	9.6	11.1	25.7	25.1	26.2
7th	4.4	4.9	5.1	10.7	10.2	10.9
Slave-2	62.2	62.9	61.6	64.2	64.1	63.8
3rd	21.4	20.1	21.8	38.6	37.8	38.2
5th	11.2	10.4	9.8	26.3	25.8	26.9
7th	4.2	3.5	3.9	10.7	10.2	10.4
Slave-3	61.8	62.5	63.1	63.7	63.4	64.2
3rd	19.2	20.8	21.2	39.5	39.1	39.8
5th	11.5	9.5	10.8	24.7	23.9	24.4
7th	4.1	3.4	4.7	10.9	9.9	10.2
Slave-4	61.2	62.6	61.9	62.2	62.8	63.2
3rd	19.4	20.5	18.7	38.7	38.2	37.9
5th	10.9	11.3	9.5	24.3	24.9	25.1
7th	4.6	5.4	5.1	10.5	9.8	10.1
Slave-5	62.8	62.1	63.2	62.7	63.6	62.5
3rd	19.8	20.4	18.5	38.6	37.8	38.3
5th	12.2	11.2	10.8	26.1	26.6	25.9
7th	3.9	4.4	3.2	10.7	10.2	10.5
Slave-6	62.7	62.9	63.1	64.1	63.9	64.2
3rd	20.3	21.6	22.1	39.6	39.2	40.1
5th	11.7	9.8	10.8	26.2	25.8	26.6
7th	5.1	4.4	4.9	10.7	9.9	10.4
Slave-7	63.2	63.7	63.2	62.9	63.2	63.5
3rd	21.7	19.7	22.1	39.1	39.8	40.1
5th	12.1	11.7	10.8	25.2	24.6	24.9
7th	4.3	3.7	4.1	11.3	10.8	11.6
Load	508.2	510.1	510.6	508.4	510.9	510.6
3rd	304.3	313.1	314.5	314.4	312.3	314.6
5th	197.7	201.1	201.5	206.6	204.4	207.8
7th	79.8	88.5	88.1	89.3	84.2	87.1

Research on capacity optimization of PV-wind-diesel-battery hybrid generation system

Cailing Zhu[1], Furong Liu[1],Sheng Hu[1] and Shu Liu[2]
1College of Automation, Wuhan university of technology, Wuhan, China
2 SINOMACH Intelligence Technology Research Institute, Wuhan,China
*E-mail: 991318311@qq.com

Abstract—For small scale isolated PV-wind-diesel-battery hybrid generation system, the capacity optimization needs to determine capacity of each component. This paper proposes an optimization method that takes economic and technical indexes into account, which optimize the capacity of the battery and diesel generators under the premise of satisfying the technical indexes, and also optimize capacity of wind turbine (WT) and photovoltaic array (PV) by considering the economic index. Among them, the economic index of the system is to minimize the life-cycle cost (LCC). According to the wind-solar resources of the installed site, the mathematical model of the system is built after analyzing the output characteristics of each micro-source and the energy dispatching strategy. Finally, the particle swarm optimization (PSO) algorithm is adopted to obtain the minimum cost and the optimal capacity configuration scheme by MATLAB simulation software. The influence of capacity of WT and PV on LCC is analyzed too.

Keywords—*PV-wind-diesel-battery system; Economic and technical indexes; Capacity optimization; PSO algorithm.*

I. INTRODUCTION

In recent years, with depletion of traditional energy, wind and solar energy as the ideal clean energy have been widely used, which are also random and intermittent and can not be a stable power supply. Adding energy storage system and diesel generator, the weakness of WT and PV can be balanced and a more stable power output that ensure the reliability of the system can be obtained [1]. In order to guarantee that system can make full use of wind-solar source and save cost, a reasonable capacity configuration is required.

At present, the capacity optimal configuration of PV-wind-diesel-battery hybrid generation system has been studied by researchers. In reference [2], annual minimum cost is regarded as the optimal objective and the number of different power sources as variables, and battery power and pollution emissions as constraints. Based on the mathematical model of the system, the genetic algorithm (GA) is used to solve the capacity optimization configuration of each component. In reference [3-4], life cycle cost (LCC) and reliability index, namely the loss of power supply probability (LPSP) are considered as objective functions. But the objective function in these studies does not take the charge current of battery impracticability of capacity allocation results of battery and and other technical indexes into account, leading to diesel

generator. Charge current will make big difference to the life cycle of battery, whether it's too large or too small. If charge current is too large, the temperature of battery will raise and the internal resistance will decline, resulting in oversize of the thermal runaway current, thus influencing the cycle life of battery. While the charge current is too small, the negative plate will be in under-charging state for a long-term,leading to the occurrence of negative plate Sulfuric acid evolution and early damage to the battery [5]. Because the battery has a great impact on the cost of system, it is essential for the battery to choose a suitable charge current.

This paper proposes an optimization method which considers economy indexes under the premise of satisfying the technical indexes. The capacity of the battery can be configured on the basis of satisfying 24 hours power supply for the loads. And then the capacity of the diesel generator can be obtained according to the charge current of the battery and the load demand. According to the wind-solar resources of the installed site, the mathematical model of the system is built by analyzing the output characteristics of each micro-source and the energy dispatching strategy of the system., the economy model is established with the LCC as objective function and the number of WT and PV as variables. The minimum cost and the optimal capacity configuration results are obtained by using the PSO.

II. SYSTEM ARCHITECTURE

The architecture of the system studied in this paper is showed in Fig.1.

As shown in Fig. 1, the system mainly consists of WT, PV module, battery, diesel generator, loads and converter, and the DC bus voltage is 48V. When renewable energy is abundant, WT and PV will charge the battery while supplying power to the load .When the voltage of DC bus becomes high and reach a preset value ,the dump load [6] will be put into operation to prevent overcharging of the battery. If the state of charge (SOC) reaches lower limit, the controller start up the diesel generator, which supply the load while charging the battery.

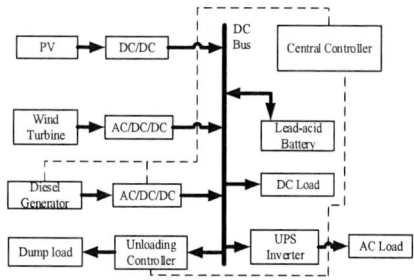

Fig. 1. Architecture of DC micro-grid system

III. MATHEMATICAL AND ECONOMIC MODEL

A. Mathematical Model of system

1) Mathematical Model of WT

The relationship between the output of WT and wind speed can be expressed by the following formula:

$$P_{WT}(v) = \begin{cases} 0 & v < v_{ci} \ or \ v > v_{co} \\ \dfrac{P_r(v^3 - v_{ci}^3)}{v_r^3 - v_{ci}^3} & v_{ci} \le v < v_r \\ P_r & v_r \le v \le v_{co} \end{cases} \quad (1)$$

where V_r、V_{ci}、V_{co} are rated、cut-in and cut-out wind speed respectively, m/s; P_r is the rated power, kW.

2) Mathematical Model of PV

Temperature of PV module is estimated by testing the ambient temperature.

$$T_c = T + 30G_c / 1000 \quad (2)$$

where T_c and T are PV module temperature and ambient temperature respectively, °C; G_c is the solar radiation, kW/m^2.

$$P_{PV} = P_{STC} \frac{G_c}{G_{STC}} [1 + m(T_c - T_{STC})] \quad (3)$$

where G_{STC}、T_{STC}、P_{STC} are the solar radiation、temperature and rated power under standard test conditions respectively; m is the power temperature coefficient. The standard test conditions are $AM1.5, 1 \ kW/m^2, 25 \ ℃$.

3) Mathematical Model of battery

Charging and discharging process are given in (4) and (5)

$$SOC(t) = (1-\delta)SOC(t-1) + P_{bat}(t) \cdot \Delta t \cdot \eta_c / E_c \quad (4)$$

$$SOC(t) = (1-\delta)SOC(t-1) - \frac{|P_{bat}(t)| \cdot \Delta t}{\eta_d \cdot E_c} \quad (5)$$

where $SOC(t)$ and $SOC(t-1)$ are the SOC at the beginning and the end of interval t respectively; δ is the self-discharge rate of battery; P_{bat} is charge and discharge power, kW, if $P_{bat}(t) > 0$, the battery is charging, otherwise it is discharging[7]; η_c and η_d are the battery charge and discharge efficiency respectively; E_c is the rated capacity, kWh.

4) Mathematical Model of diesel generator

$$V_f = \sum_{t=1}^{8760} P_d(t).ge \quad (6)$$

where V_f is the fuel consumption of diesel generator, ge is the fuel consumption rate.

B. Economic Model

5) Objective function

The optimization goal of this paper is to minimize LCC of the system, including the purchase cost C_I, the replacement cost C_R, the maintenance cost C_M and the fuel cost C_F.

$$f(x) = \min LCC \quad (7)$$

$$LCC = C_I + C_R + C_M + C_F \quad (8)$$

$$C_I = N_{pv} \cdot C_{pv} + N_{wt} \cdot C_{wt} + C_{bs} + C_g \quad (9)$$

$$C_R = N_{bs} \cdot C_{bs} \cdot k \quad (10)$$

$$C_M = [N_{pv} \cdot C_{pv\_r} + N_{wt} \cdot C_{wt\_r} + C_{g\_r}] \cdot L_f \quad (11)$$

$$C_F = L_f \cdot pr \cdot V_f \quad (12)$$

Where N_{pv} and N_{wt} are the number of PV and WT respectively; C_{pv}、C_{wt}、C_{bs}、C_g are the price of PV, WT, battery, diesel generator; If the life of WT, PV and diesel generator are 20 years, the life cycle of the system is set to 20 years, the cost of replacement for other devices other than the battery may not be considered, L_f is the life cycle of the system; C_{pv_r}、C_{wt_r}、C_{g_r} are the maintenance of the PV、WT and diesel generator; pr is the price of diesel oil.

6) Cosntraints

$$SOC_{\min} \le SOC(t) \le SOC_{\max} \quad (13)$$

where $SOC_{\min}$ and $SOC_{\max}$ are the minimum and maximum SOC of the battery.

$$P_{b.\min} \le P_{bat} \le P_{b.\max} \quad (14)$$

$$1 \le N_{pv} \le N_{pv.\max} \quad (15)$$

$$1 \le N_{wt} \le N_{wt.\max} \quad (16)$$

where $P_{b.\min}$ and $P_{b.\max}$ are the minimum and maximum power of the battery; $N_{pv.\max}$ and $N_{wt.\max}$ are the largest installed numbers of WT and PV modules respectively.

IV.　THE ENERGY DISTRIBUTION OF THE SYSTEM

The output power of WT and PV is random and non-dispatchable. The battery can either charge or discharge., which plays a role of energy buffer in the whole system. The power difference [8] between the wind-PV and the load is as follows:

$$\Delta P(t) = P_{wt}(t) + P_{PV}(t) - P_L(t) \qquad (17)$$

（1）When $\Delta P(t) > 0$, wind-PV as the main power supplies to the load power as well as charge the battery, if SOC of the battery is within the maximum

$$P_{bat}(t) = \Delta P(t) \qquad (18)$$

When the battery is fully charged, the rest of the energy is consumed by the dump load.

$$P_{bat}(t) = 0 \qquad (19)$$

When $\Delta P(t) < 0$, the battery discharge and priority to provide the power shortage, if it can be satisfied only by the battery, then

$$P_{bat}(t) = \Delta P(t) \qquad (20)$$

$$P_d(t) = 0 \qquad (21)$$

If SOC of the battery is less than or equal to the minimum, the diesel generator starts charging the battery and supplying power to the load

$$P_{bat}(t) = P_b \qquad (22)$$

$$P_d(t) = P_{bat}(t) + |\Delta P(t)|$$

where P_b is the charging power of the battery by diesel generator.

V.　SIIMULATION AND RESULT ANALYSIS

A. Wind and Solar Resources

In this paper, Jakarta, the capital city of Indonesian, where the wind-solar source is very rich, is chosen for the study site. And the products of a domestic company which produces this system sell well in Jakarta. Jakarta is located at east longitude 106°45', south latitude 6°08'. It is perennial high temperature and rainy and monthly average temperature is supposed to be 27°C.

The local solar radiation can be searched by entering the geographical coordinates directly in the HOMER software, and the local average daily solar radiation is 4.76 kWh/m²/d, and the solar is the most plentiful in September. If the height

of blades of WT is selected as 10 meters, the average wind speed per hour is 6.66m / s.

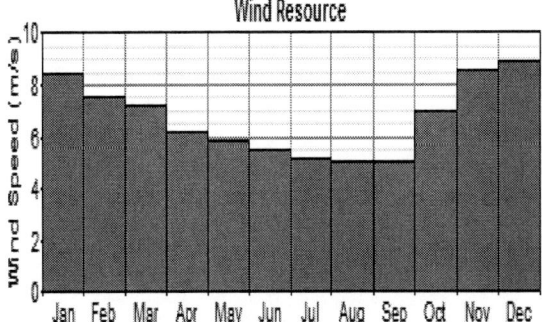

Fig. 2. annual average monthly wind speed in Jakarta

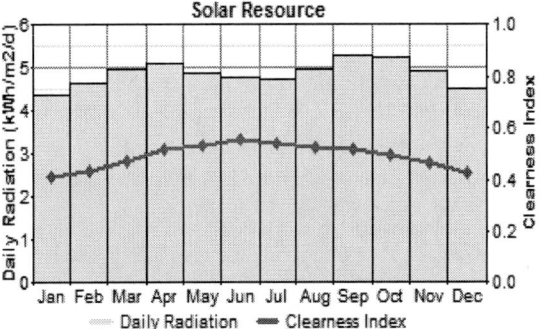

Fig. 3. annual average solar radiation year in Jakarta

B. PSO Algorithm

The basic idea of PSO algorithm comes from the imitation of the social behavior of birds, where each particle represents a possible solution to the problem, and the fitness function is used to evaluate the quality of these particles. Each particle has its own direction and velocity in the contraction space, and adjusts the probabilistic search strategy according to its past experience and group behavior. In PSO algorithm, the particle updates its velocity and position according to the following formula:

$$\begin{cases} v_i \leftarrow \omega \cdot v_i + c_1 \cdot rand() \cdot (pbest_i - x_i) + c_2 \cdot rand() \cdot (g_i - x_i) \\ x_i = x_i + v_i \end{cases} \quad (23)$$

where ω is the inertia weight, c_1 and c_2 are the learning factor, and rand() is the random number, which obeys uniform distributed in [0,1].

The particles are constantly tracking the personal and global best values in the solution space until the termination condition is satisfied. The flow chart of PSO is shown in Fig 4.

C. Capacity optimization configuration

For the $3kW$ rated load power, in order to enable the system to meet the load demands and maintain a stable operation for 24 hours when the battery is powered alone, so the battery

The 2018 International Power Electronics Conference

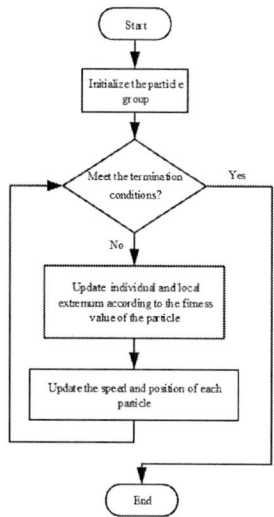

Fig. 4. The flow chart of PSO

is configured as 1500*Ah*, 48*V*. The limited charge current is $I_{min} = 0.15 \cdot C = 225A$, the charging power of the battery is $P_b = U_b \cdot I_{lim} = 48 \cdot 225 = 10.8kW$ .Considered 3*kW* load power, rated power of diesel generator is 15kW, the number of PV modules and WT are set as variables.

The simulation duration is 8760 h, i.e., the whole year, and the time step is 1h. Firstly, import wind speed, solar radiation and load data to the system model, and the output power of each micro-source is obtained. Secondly, calculate the value of objective function with constraints according to the economic data. Finally, apply the PSO algorithm to solving the objective function, and then search configuration results and the minimum cost by iterative optimization. The required parameters are shown in TABLE 1. Costs and cost factors for each power supply is shown in TABLE 2.

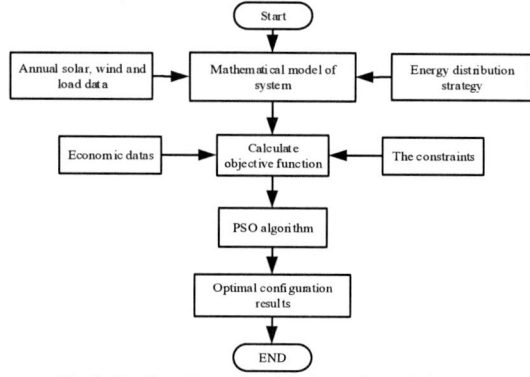

Fig. 3. The flow chart of model of capacity optimization

TABLE 1
SIMULATION PARAMETERS

Component	Parameters	Value
	Cut-in speed	2m/s

WT	Rated speed	8m/s
	Cut-out speed	30m/s
battery	Maximum SOC	0.9
	Minimum SOC	0.1
	Initial SOC	0.5
	Charge/discharge efficiency	90%
Diesel generator	Fuel consumption rate	250 g/kWh
	Diesel price	5.85RMB/L

TABLE 2
COST AND COST FACTORS FOR EACH POWER

Power types	Rated capacity	Purchase cost/RMB	Maintenance cost/RMB
WT	1kW	4250	200/y
PV	1kW	5000	200/y
Battery	1500Ah	36000	0
Diesel generator	15kW	8000	200/y

Cycle life of the battery is affected by many factors, such as temperature, peak current and the depth of discharge (DoD), etc [8]. This paper only considers the effect of DoD on number of cycles of the battery.

The relationship between cycle life and DoD is as shown as follows:

$$N_{cyc}(DoD) = a_1 + a_2 e^{-a_3 \cdot DoD} + a_4 e^{-a_5 \cdot DoD} \qquad (24)$$

where $N_{cyc}(DoD)$ is the number of cycles, $a_1 \sim a_5$ are the fit factors. According to the actual data provided by the battery manufacturer, the fitting curve is as follows:

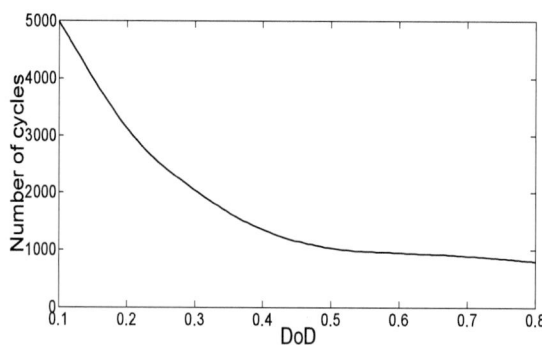

Fig. 5. Fitting curve between number of cycles and DoD

A complete cycle consists of a discharge half cycle and a charging semi-cycle . As is shown in Fig.5, when DoD is 0.8, the cycle life of the battery is around 850 times. When the power of WT and PV are zero, SOC that varies with time is shown in Fig.6. As can be seen in Fig.6, a complete cycle is about 30 hours, thus there are 290 cycles per year and cycle life cycle life will be prolonged greatly. So the cycle life of the battery is supposed to be 4 years.

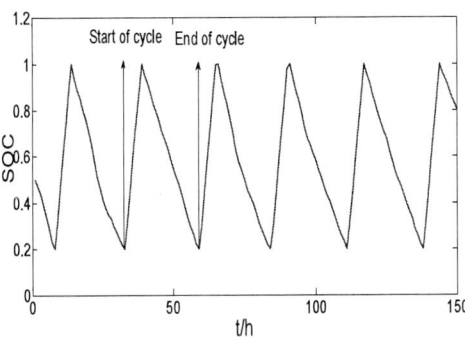

Fig. 6. SOC in battery-diesel system

D. Discuss of optimization results

Based on mathematical and economic model established in this paper, the computational results by PSO algorithm are shown in Table 3. The relationship between LCC and number of iterations is shown in Fig 8.

TABLE 3
CAPACITY OF OPTIMIZATION CONFIGURATION RESULTS

Case	Pwt /kW	Ppv/ kW	Battery/ Ah	Diesel generator/kW	C_F/ $\times 10^4 RMB$	LCC/ $\times 10^4 RMB$
Case1	7	1	1500	15	4.350	30.225
Case 2	8	3	1500	15	2.725	31.230
Case 3	7	6	1500	15	1.385	31.760
Case 4	6	7	1500	15	1.373	31.823
Case 5	0	0	1500	15	110.582	130.58

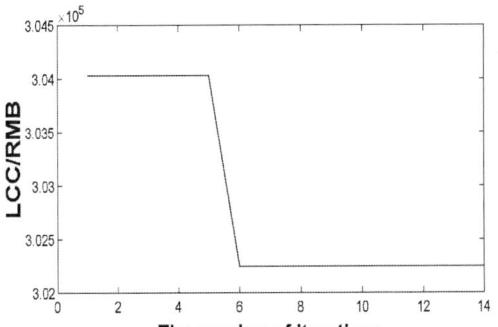

Fig .7. The relationship between LCC and number of iterations

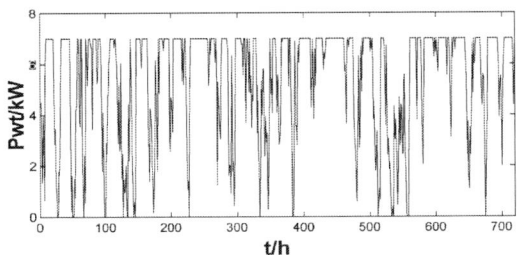

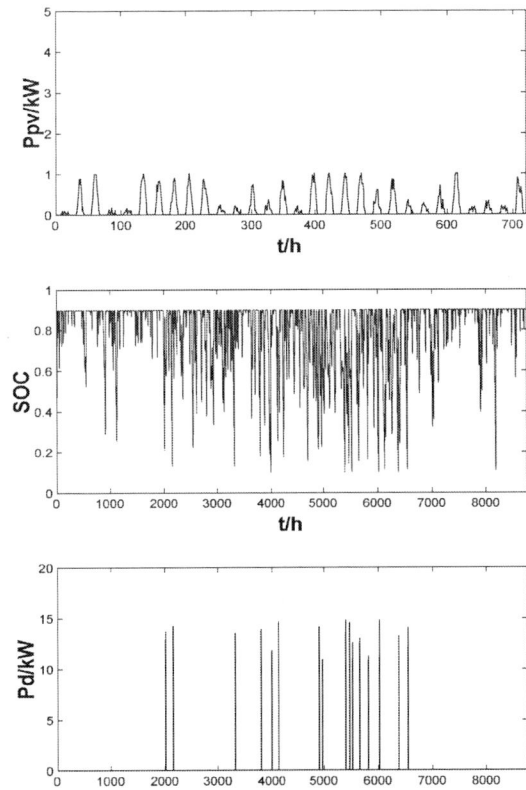

Fig. 8. The outut of micro-sources of case 1

As shown in Table 3, the optimal sizing is Case 1 that wind turbine is 7 kW, PV is 1 kW, battery is 1500 Ah and diesel generator is 15 KW, the minimum cost is 30.225 RMB. The output of micro-sources of case 1 is shown in Fig 8, P_{wt}, P_{pv} and P_d are the output power of WT,PV and diesel genotor. It can be seen form the picture, P_{pv} and P_{wt} are complementary [10] and P_{wt} is lager than P_{pv} because PV can only generate power in the daytime when the solar radiation is zero and can't match with load all the day, let alone the worse continuty of power generation compared with WT. Especially on rainy days, the output power of PV is almost zero, so the PV capacity configuration is relatively small. The SOC of battery is between $0.7 \sim 0.9$ most of the time, which can ensure the normal operation of the system without wind and solar. The battery dose not charge or discharge too much which can extend the life cycle of batteries.

As seen from Fig.9 and Fig.10, in the early, LCC declines sharply because of WT and PV provide some energy to system and reduce the number of start-up times for diesel generator, thus the fuel cost reduce. The capacity of WT has a greater impact on cost than PV. In the medium term, after the capacity of WT and PV arrive at the optimal sizing [7, 1], LCC increases slowly. In the later, when diesel generator does not need to start, fuel cost is zero, increasing the cost of equipment for WT and PV leads to increase of cost of the system.

The 2018 International Power Electronics Conference

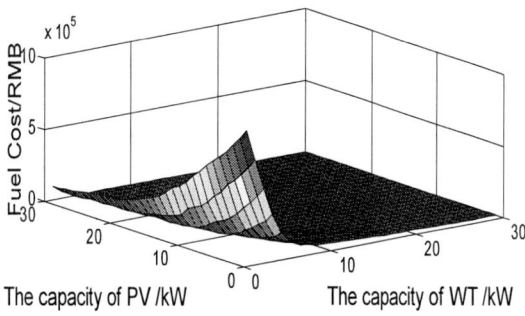

Fig .9. Fuel cost

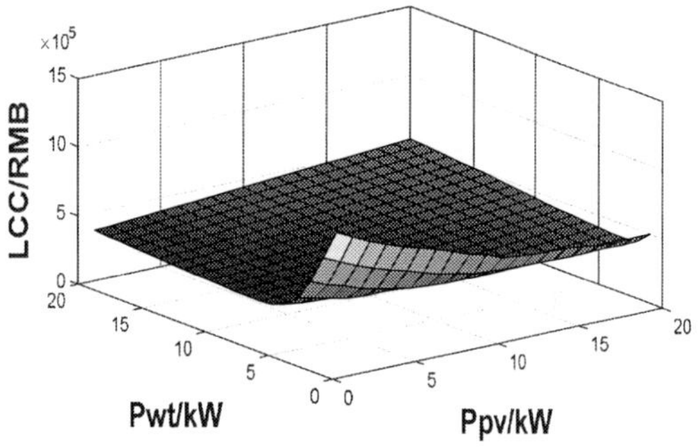

Fig. 10. The life cycle cost

VI. CONCLUSION

This paper proposes a capacity optimization method which takes economic and technical indexes into account. The influence of charge current on cycle life of battery is analyzed and the technical index of system is given priority to optimize the capacity of battery and diesel generator. Then optimize capacity of WT and PV by considering the economic index. And the impact of capacity of WT and PV on LCC is also analyzed. Not only can equipment costs of WT and PV have a direct impact on the LCC, but also the capacity of WT and PV can make indirect impact on LCC by affecting the fuel cost.

ACKNOWLEDGMENT

This work is supported by the project of Wuhan University of Technology(Grant No.23050201). The authors would like to thank the reviewers for their invaluable comments and advice.

REFERENCES

[1] Yang, J., et al. (2012). "Energy storage capacity optimization for independent PV power system based on improved PSO algorithm." East China Electric Power 40(8): 1370-1374.
[2] Ding , M., et al. (2013). "Configuration optimization of capacity of standalone PV-wind-diesel-battery hybrid microgrid." Power System Technology 37(3): 575-581.
[3] Liu, Z. and M. Huang (2014). "Optimal energy allocation of microgrid based on reliability and economy by considering characteristics of micro source." Power System Technology 38(5): 1352-1357.
[4] Rhein, A., et al. (2017). Reliability-based improvement of life-cycle maintenance and replacement strategies in transmission systems using ant colony optimization. International Youth Conference on Energy IYCE.
[5] Chen, J.-j. and N.-m. Yu (2004). "Study on the multi-stage constant current charging method for VRLA battery system." Chinese Journal of Power Sources 28(1; ISSU 160): 32-33.
[6] Wies, R. W., et al. (2005). "Simulink model for economic analysis and environmental impacts of a PV with diesel-battery system for remote villages." IEEE Transactions on Power Systems 20(2): 692-700.
[7] Dagdougui, H., et al. (2016). Power management strategy for sizing battery system for peak load limiting in a university campus. Smart Energy Grid Engineering (SEGE), 2016 IEEE, IEEE.
[8] Wei, L., et al. (2015). Optimal Capacity Allocation of Large-Scale Wind-PV-Battery Hybrid System. Intelligent Human-Machine Systems and Cybernetics (IHMSC), 2015 7th International Conference on, IEEE.
[9] Cheng, L. I., et al. (2016). "Optimization allocation of standalone microgrid system capacity based on complementary characteristic of wind and solar." Chinese Journal of Power Sources.
[10] Hao, W., et al. (2016). The research on adaptability evaluation of wind power integration capacity in power grid considering wind-solar hybrid complementary characteristics and flexible loads interference. Power and Energy Engineering Conference (APPEEC), 2016 IEEE PES Asia-Pacific, IEEE.

A Numerical Analysis and Improvement of Output Characteristics in Different Passive Rectifiers based on Vibration Generators

Tomoki Sakabe[1], Masataka Minami[1], Shin-ichi Motegi[1] and Masakazu Michihira[1]

1 Department of Electrical Engineering, Kobe City College of Technology, Kobe, Japan
*E-mail: r113214@g.kobe-kosen.ac.jp

Abstract—**Vibration generators based on piezoelectric elements have been attracted attention energy harvesting. This paper investigates different passive rectifiers for utilization of vibration generators. In addition, a novel rectifier is proposed to improve the output voltage and power by LC resonance. The validity of the proposed circuit is numerically verified.**

I. INTRODUCTION

Recently, the energy harvesting system is attracted attention [1]. The application of the system determines which energy sources are available in the environment to power it [2]. The ambient energy sources include vibration, solar, and thermal energy [3]. The system is expected to become practical application [2]. This paper focuses on vibration generators based on piezoelectric elements. The vibration generators convert vibration energy into electric energy [4]. As well as the solar generator and electromagnetic generator [5]. The output power of the piezoelectric elements is small, and since the output voltage is low, vibration generators do not maintain large power [6][7]. Therefore, the vibration generators are not generally used.

This paper investigates different passive rectifiers for utilization of the vibration generators. In addition, a novel rectifier is proposed to improve the output voltage and power by LC resonance. The validity of the proposed circuit is numerically verified.

II. PIEZOELECTRIC ELEMENTS AND RECTIFIER CIRCUIT

Piezoelectric elements convert vibration energy into electric energy [8]. The piezoelectric elements are expressed as shown in Fig. 1 [9]. The vibration energy and the piezoelectric ceramic are expressed in AC current source and an internal impedance as a capacitor and a resistor [9].

Since the output power is generally used as DC power, rectifiers are necessary to convert the output power of the piezoelectric elements into DC power. The various rectifiers, such as Fig. 2 (a) and (b), are widely used in many systems [10][11]. Fig. 2 (a) describes a diode bridge rectifier (DB), and Fig. 2 (b) describes a voltage doubler rectifier (VD). The output voltage of DB is lower than VD. On the other hand, the output power of DB is larger than VD. These circuits have such problems. Prof. K. Fujiwara proposed a novel rectifier

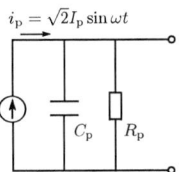

Fig. 1. Equivalent circuit of piezoelectric elements [9].

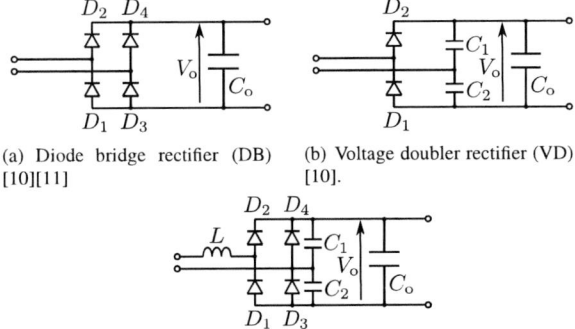

(a) Diode bridge rectifier (DB) [10][11].

(b) Voltage doubler rectifier (VD) [10].

(c) Diode rectifier with improved input current waveform (IW) [12]

Fig. 2. Various rectifier circuits using passive devices.

TABLE I. PARAMETERS OF THE PIEZOELECTRIC ELEMENTS AND THE RECTIFIERS

$I_p = 9.2\,\mathrm{mA}$	$C_o = 22\,\mu\mathrm{F}$
$\omega = 2\pi \times 120\,\mathrm{rad/s}$	$L = 150\,\mathrm{mH}$
$C_p = 1.06\mu\mathrm{F}$	$R_L = 10.4\,\Omega$
$R_p = 2.8\,\mathrm{k}\Omega$	$C_1 = C_2 = 0.33\,\mu\mathrm{F}$

as shown in Fig. 2 (c) [12]. The rectifier improves the output power and voltage by using diode rectifier with improved input current waveform (IW) [12].

In the next section, the output characteristics of these circuits are numerically analyzed.

III. NUMERICAL ANALYSIS OF RECTIFIER CIRCUIT

In this section, DB, VD, and IW are numerically analyzed and compared about the output characteristics. Operations of the rectifier circuits are described every mode.

The 2018 International Power Electronics Conference

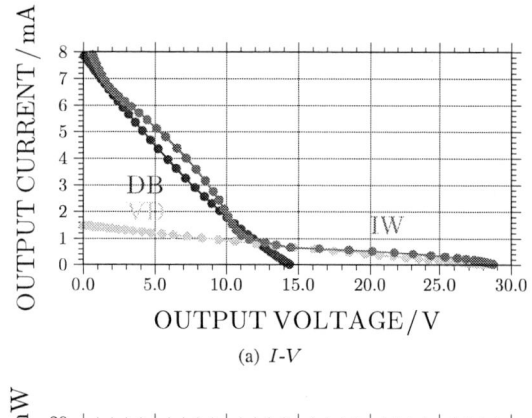

(a) *I-V*

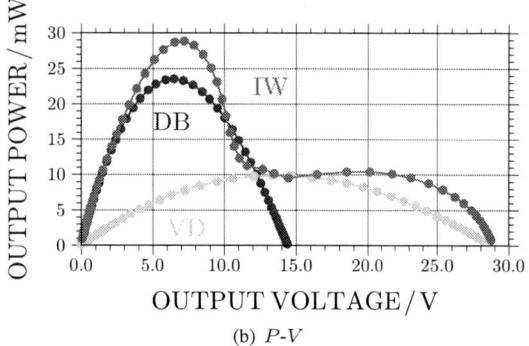

(b) *P-V*

Fig. 3. *I-V* and *P-V* output characteristics of DB, VD, and IW.

A. Numerical conditions and parameters

A circuit simulator called LTspice IV are used for numerical analysis. Table I gives the parameters of the piezoelectric elements and the rectifiers. The parameters are set by the results of preliminary experiments. The diodes of the rectifiers are used schottky barrier diodes 1N5818.

B. I-V and P-V characteristics

Fig. 3 represents the numerical results of the three rectifier circuits. Fig. 3 (a) and (b) describe *I-V* and *P-V* characteristics. At first, in Fig. 3 (b), the output voltage of DB is lower than VD. On the other hand, the output power of DB is larger than VD. The features of DB and VD are implied. Secondly, the IW is compared with DB and VD. The maximum output voltage of the IW is as high as VD. Furthermore, the maximum output power of the IW is larger than DB. Therefore, the IW improves the output voltage and power.

C. Operation modes

In this section, the operation modes of the rectifier circuits are divided and verified. Fig. 4 represents the time waveforms of the rectifier circuits. The operation modes based on reference [10][11][12] are described.

Fig. 5 illustrates the operation modes of Fig. 4 (a). In mode 1, all diodes are turned off. Therefore, the input current i_i does not flow. After that, mode 1 shifts mode 2. In mode 2,

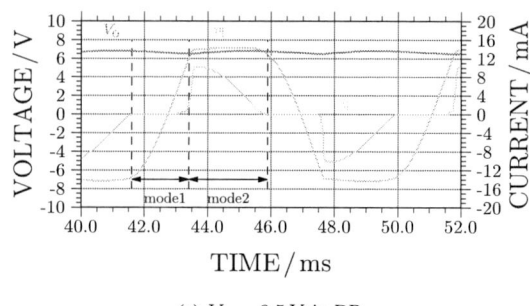

(a) $V_o = 6.5$ V in DB

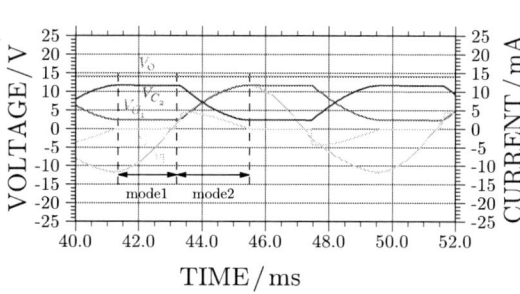

(b) $V_o = 14$ V in VD

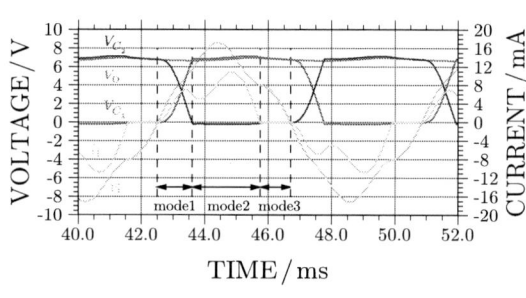

(c) $V_o = 7$ V in IW

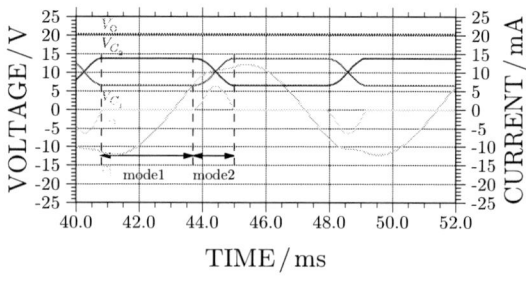

(d) $V_o = 20$ V in IW

Fig. 4. The time waveforms of three rectifier circuits.

the diodes D_2 and D_3 conduct. Therefore, the input current i_i flows to the load R_o.

Fig 6 illustrates the operation modes of Fig. 4 (b). Since the each capacitance of C_1 and C_2 is small in this paper, the each voltage of the capacitors is fluctuation. In mode 1, all the

3059

The 2018 International Power Electronics Conference

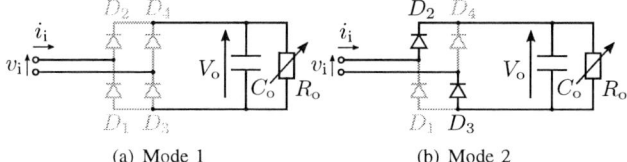

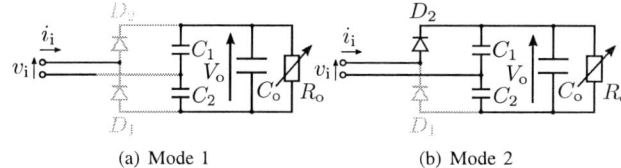

(a) Mode 1   (b) Mode 2

Fig. 5. Operation modes of the circuit with diode bridge rectifier (DB).

(a) Mode 1   (b) Mode 2

Fig. 6. Operation modes of the circuit with voltage doubler rectifier (VD).

diodes are turned off. Therefore, the input current i_i does not flow. When mode 1 shifts mode 2, The diode D_2 conducts. Therefore, the input current i_i flows to C_1 and C_2.

Fig. 7 illustrates the operation modes of Fig. 4 (c) at low voltage region ($V_o = 7\,\text{V}$). At low voltage region, the IW has the three operation modes. In mode 1, only the diode D_2 conducts. Therefore, the charging current flows to C_1, and the discharging current flows through the load R_o to C_2. When the capacitors finish charging and discharging, mode 1 shifts mode 2. In mode 2, The diodes D_2 and D_3 conduct. Therefore, the input current i_i flows to the load R_o. At first, since the inductor stores the energy, the input current i_i continues to flow. However, when the input voltage v_i is lower than the output voltage V_o, the input current i_i decreases. Secondly, when the input voltage v_i is higher than the output voltage V_o, the input current i_i increases. Finally, when the input voltage v_i is lower than the output voltage V_o, the input current i_i decreases again. Since the input current i_i becomes zero, mode 2 shifts mode 3. In mode 3, The diode D_3 conducts, and D_2 does not conduct. Since the input side does not supply power, C_o and C_1 supply power to the load R_o. Here, the interval during flowing the input current i_i is focused on. In Fig. 5, the input current i_i flows in only mode 2. On the other hand, in Fig. 7, the input current current i_i flows in mode 1 and mode 2. In addition, the output voltage of Fig. 7 is higher than Fig. 5. For these reasons, at low voltage region, the output power of the IW is larger than the DB.

Fig. 8 illustrates the operation modes of Fig. 4 (d) at high voltage region ($V_o = 20\,\text{V}$). At high voltage region, the IW has the two operation modes. In mode 1, all the diode are turned off. Therefore, the input current i_i does not flow. In mode 2, the diode D_2 conducts. At that time, the charging current flows to C_1, and the discharging current flows through the load R_o to C_2. At high voltage region, the operation modes of the IW is similar to VD. In addition, the input current i_i increases by the inductor. Therefore, at high voltage region, the maximum output voltage of the IW is higher than the DB, and the output power of the IW is larger than the VD. For these reasons, IW improves the output voltage and power.

IV. PROPOSED CIRCUIT

In the previous section, the IW improves the output voltage and power. However, the output voltage is not high enough to use vibration generators, and the output power is not large enough too. Then, we propose a novel circuit for the vibration generators.

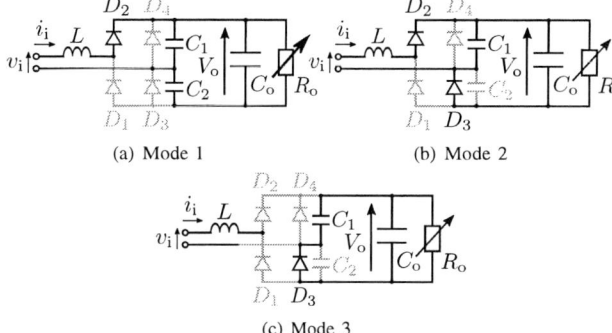

(a) Mode 1   (b) Mode 2

(c) Mode 3

Fig. 7. Operation modes of IW at low voltage region ($V_o = 7\,\text{V}$).

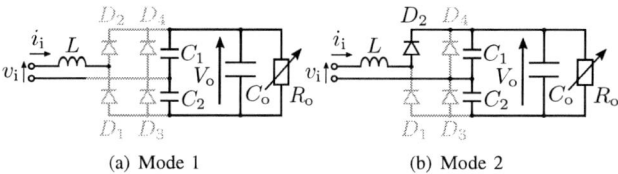

(a) Mode 1   (b) Mode 2

Fig. 8. Operation modes of IW at high voltage region ($V_o = 20\,\text{V}$).

Fig. 9 (a) represents Fig. 1 connected with the load R_o. The output current I_o of Fig. 9 (a) is expressed as,

$$I_o = \sqrt{I_p^2 - (\omega C_p)^2 V_o^2} - \frac{V_o}{R_p}. \qquad (1)$$

The output current I_o is depend on the internal capacitance C_p from this equation. Therefore, LC resonance equivalently reduces the internal capacitance C_p. Fig. 9 (b) represents the circuit using the LC resonance. A parallel inductor L_A and capacitor C_A are connected with the piezoelectric elements. The parallel inductor L_A and capacitor C_A resonate with the internal capacitor C_p. The parallel capacitor adjusts the LC resonance. The LC resonance equivalently enables to decrease the internal capacitance C_p. The output current of Fig. 9 (b) is expressed as [13],

$$I_o = \sqrt{I_p^2 - \left\{\omega(C_p + C_A) - \frac{\omega L_A}{R_A^2 + (\omega L_A)^2}\right\}^2 V_o^2} \\ - \left(\frac{1}{R_p} + \frac{R_A}{R_A^2 + (\omega L_A)^2}\right) V_o. \qquad (2)$$

Fig. 10 (a) and (b) describe the characteristics of Eqs. (1) and (2). Since the inductor L_A neutralizes the internal capacitance, the output current I_o increases as shown in Fig. 10 (a). In addition, the maximum output voltage of Fig. 9

3060

The 2018 International Power Electronics Conference

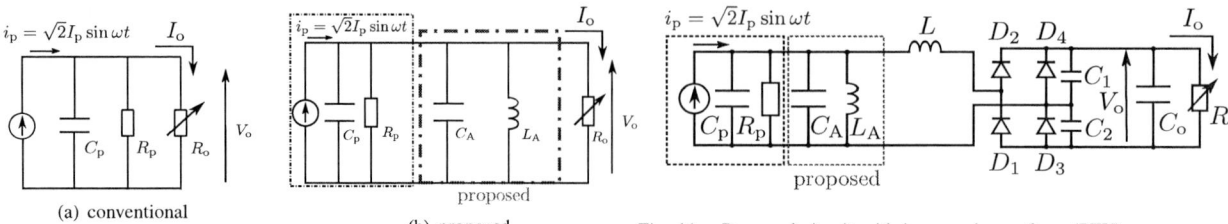

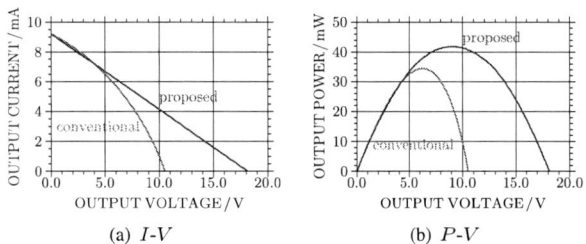

(a) conventional

(b) proposed

Fig. 9. Reference circuit of piezoelectric elements

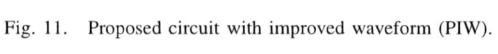

Fig. 11. Proposed circuit with improved waveform (PIW).

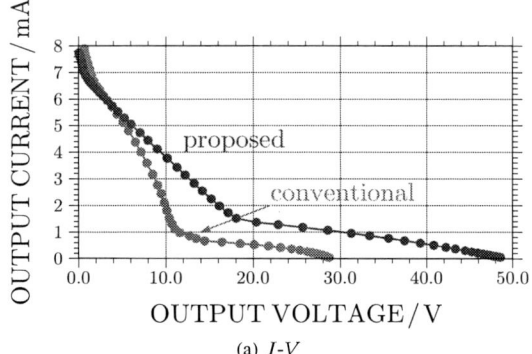

(a) *I-V*

(b) *P-V*

Fig. 10. *I-V* and *P-V* characteristics of Eqs. (1) and Eqs. (2)

(b) is higher than Fig. 9 (a), and the output power of Fig. 9 (b) is larger than Fig. 9 (a) in Fig. 10 (b). For these purposes, a decrease of the internal capacitance improves the output voltage and power. Fig. 11 represents the proposed circuit with improved waveform (PIW). In the next section, the output characteristics of the PIW is numerically analyzed, and compared with the IW in Sec. III.

V. NUMERICAL ANALYSIS OF PROPOSED CIRCUIT

In this section, the PIW is numerically analyzed. In addition, the output characteristics of the PIW are compared with the IW in Sec. III, and the effect of the PIW is verified.

A. Numerical conditions and parameters in PIW

The parameters of the piezoelectric elements and the rectifiers are the same as parameters of section III. A parallel inductor and capacitor are set by the results of preliminary experiments. The inductor is determined $L_A = 816.3$ mH, $R_A = 56.4\,\Omega$. The capacitor is determined $C_A = 1.0\,\mu$F.

B. I-V and P-V characteristics in PIW

Fig. 12 describes the numerical results of the IW and PIW. The output characteristics of the PIW are compared with the IW. The maximum output voltage of PIW is 1.6 times as high as IW, and the maximum output power is 1.35 times as large as IW. At low voltage, the output power is improved. Therefore, the vibration generators may be used for many systems. At high voltage, since the output voltage and power are improved, the vibration generators are able to maintain large power. Therefore, vibration generators may be used for many devices. For these reasons, PIW improves the output voltage and power, and vibration generators are expected to be used for more and more devices and systems.

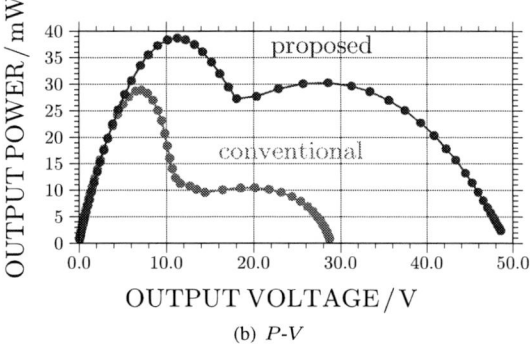

(a) *I-V*

(b) *P-V*

Fig. 12. *I-V* and *P-V* output characteristics of IW and PIW.

VI. CONCLUSIONS

This paper IW improves the output voltage and power of vibration generators with piezoelectric elements. In addition, the proposed circuit using LC resonance improves the output voltage and power. Since two points of the output voltage and power are improved, vibration generators are expected to be used for a wide range of devices and systems.

In future work, the results represented in this paper are experimentally verified.

ACKNOWLEDGMENTS

This research is partially supported by Power Academy in Japan.

REFERENCES

[1] L. Mateu, M. Echeto, and F. de Borja, "Review of energy harvesting techniques and applications for microelectronics." International Society for Optical Engineering, 2005.

[2] P. Spies, M. Pollak, and L. Mateu, *Handbook of Enetgy Harvesting Power Supplies and Applications*, 1st ed. Pan Stanford, 2015, ch. 1.

[3] P. Glynne-Jones, S. P. Beeby, and N. M. White, "Towards a piezo-electric vibration-powered microgenerator," *IEE Proceedings-Science, measurement and technology*, vol. 148, no. 2, pp. 68–72, 2001.

[4] A. Erturk and J. Inman, "An experimentally validated bimorph can-tilever model for piezoelectric energy harvesting from base excitations," *Smart Materials and Structures*, vol. 18, no. 2, p. 025009, 2009.

[5] C. B. Williams and R. B. Yates, "Analysis of a micro-electric generator for microsystems," *Sensors and Actuators A: Physical*, vol. 52, no. 1–3, pp. 8–11, 1906.

[6] F. Lu, H. P. Lee, and S. P. Lim, "Modeling and analysis of micro piezoelectric power generators for micro-electromechanical-systems ap-plications," *Smart Materials and Structures*, vol. 13, no. 1, pp. 57–63, 2003.

[7] P. Becker, B. Folkmer, and Y. Manoli, "The hybrid vibration generator, a new approach for a high efficiency energy scavenger." International workshop on micro and nanotechnology for power generation and energy conversion applications. PowerMEMS, Wahington, DC, 2009, pp. 439–442.

[8] H. A. Sodano, D. J. Inman, and G. Park, "A review of power harvesting from vibration using piezoelectric materials," *The Shock and Vibraion Diggest*, vol. 36, no. 3, pp. 197–205, 2004.

[9] S. Roundy and P. K. Wright, "A piezoelectric vinration based generator for wireless electrinics," *Institute of Physics Publishing Smart Materials Structures*, vol. 13, pp. 1131–1142, 2004.

[10] N. Mohan, T. M. Undeland, and W. P. Robbins, *Power Electronics: Conveters, Applications, and Design*, 3rd ed. John Wiley & Sons, Inc., 2003, ch. 6.

[11] J. G. Kassakian, M. F. Schlecht, and G. C. Verghese, *Principles of Power Electronics*, 1st ed. Addison-Wesley, 1991, ch. 4.

[12] K. Fujiwara and H. Nomura, "A new operating principle of voltage-doubler diode rectifiers to meet the harmonic guide lines (in Japanese)," *IEEJ Transactions on Industry Applications*, vol. 119, no. 1, pp. 103–108, 2008.

[13] M. Minami, T. Sakabe, S. Motegi, and M. Michihira, "Improvement in output power of vibration generators based on piezoelectric elements using passive devices (in Japanese)," *IEEJ Transactions on Industry Applications*, vol. 137, no. 12, 2017.

Circuit Modeling Approach for Analyzing Triboelectric Nanogenerators for Energy Harvesting

Bo-Kyung Yoon*, Jeong Min Baik[†] and Katherine A. Kim*
*School of Electrical and Computer Engineering
[†]School of Materials Science and Engineering
Ulsan National Institute of Science and Technology
Ulsan, South Korea 44919
Email: kkim@unist.ac.kr

Abstract—**Triboelectric Nanogenerators (TENGs) are relatively new energy harvesting devices that show promise for effectively capturing vibrational energy. However, they have unique capacitive characteristics unlike many existing energy harvesting transducers. In order to develop proper power conversion circuits for TENGs, they must be first modeled and analyzed. A TENG is emulated using numerical simulation and analyzed to determine the optimal output resistance and its dependance on frequency. Simulation results indicate that lower frequency and lower amplitude of the plate movement requires very high optimal output resistance, while higher frequency and higher amplitude decreases the optimal output resistance to a more reasonable value.**

I. INTRODUCTION

Energy harvesting is the concept of capturing energy from ambient sources (e.g. vibration, light, heat difference, or wireless signals) to power electronic devices, generally used for low power devices. As applications like internet of things (IoT) and wearable healthcare devices are gaining popularity, energy harvesting solutions are being developed to replace batteries in many of these kinds of applications. To this end, new energy harvesting generators are also being developed. In particular, the triboelectric nanogenerator (TENG) is an emerging type of energy harvesting technology that can be used to capture electricity from vibrational motion [1]–[5]. Compared to more traditional energy harvesting transducers that capture vibration energy, such as piezoelectric and electromagnetic generators [6], TENGs generate power in a fundamentally different way.

The analytical model for ideal TENGs has been provided in [7], which is based on material properties of the device itself. The ideal equivalent circuit model is then given in [7], but focuses on the mathematical derivation of power optimization rather than the circuit implementation. A circuit simulation of a TENG was implemented in SPICE and results were given in [8], but the specific circuit implementation was not given in detailed. Another theoretical study is provided in [9], which provides time domain plots of a TENG based on calculation, but only for one single movement of the TENG, rather than repeated movements of the device. As this technology moves from improving the performance of the TENG device to implementing it in energy harvesting applications, an accurate circuit model is needed for developing the associated power conversion and management circuits.

This paper gives the mathematical equations for a TENG with two dielectrics and discusses the equivalent circuit model. The two-plate TENG is modeled in Matlab using numerical simulation methods. Then, the operational waveforms of the two-plate TENG are simulated to understand the ideal operation waveforms. Finally, the optimal resistance for an ideal sinusoidal plate distance function at various amplitudes of the input are investigated and discussed.

II. TENG ANALYTICAL MODEL

A TENG consists of two charged plates that move back and forth towards each other. The TENG is capacitive in nature, with an inherent voltage that changes with the distance between plates. The voltage is directly proportional to distance between plates while the capacitance is inversely proportional. The structure of the two-plate TENG with two dielectrics is shown in Fig. 1, where the distance $x(t)$ is the distance between the two dielectric layers.

The 2018 International Power Electronics Conference

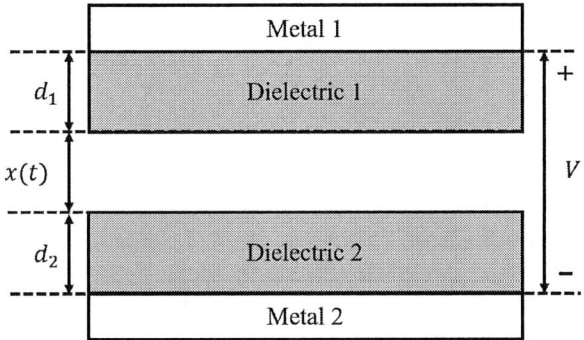

Fig. 1: Structure of two-plate TENG with two dielectrics.

The analytical model for the voltage between the plates V_{OC} is

$$V_{OC}(x) = \frac{\sigma}{\epsilon_0}x \qquad (1)$$

where σ is the charge surface density, ϵ_0 is the permittivity of free space, and x is the distance between plates. The equivalent capacitance of the TENG depends on the plate surface area A and the distance between plates x, according to

$$C_{TENG}(x) = \frac{A\epsilon_0}{d_0 + x} \qquad (2)$$

where d_0 is based on the thickness of the two plates d_1 and d_2, and the relative permittivity of the dielectrics ϵ_{r1} and ϵ_{r2}, respectively, which is defined as

$$d_0 = \frac{d_1}{\epsilon_{r1}} + \frac{d_2}{\epsilon_{r2}} \qquad (3)$$

Equations (1) (2), and (3) define the characteristics of a TENG based on material properties [7], [9].

To generate electrical energy from the TENG, its voltage and current must also be modeled. By adding the open-circuit and capacitor voltage, the voltage V of the TENG is defined as

$$V = -\frac{1}{C_{TENG}(x)}Q + V_{OC} \qquad (4)$$

and the current I of the TENG depends on the change in the capacitor charge according to

$$I = \frac{dQ}{dt} \qquad (5)$$

From equations (4) and (5), the basic equivalent circuit model of the TENG is shown in Fig. 2. It is important to note that the capacitance and open-circuit voltage depend on the distance between the two plates, which

will change during energy harvesting operation. The current and voltage of the TENG also depend on the load characteristics.

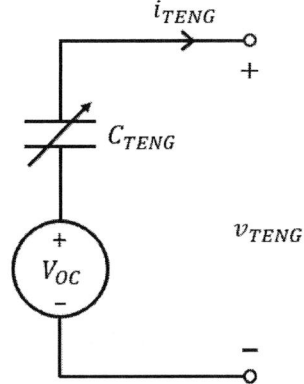

Fig. 2: Equivalent circuit model of a TENG.

The capacitive nature of the TENG is different than many of the other ac transducers for energy harvesting, such as electromagnetic or piezoelectric transducers. The basic circuit model for many ac transducers consist of an ac source and an equivalent resistance, where the optimal output impedance for power generation is the same as the equivalent resistance. However, the optimal output impedance for a TENG is less obvious, such that simulation can be utilized to better understand the operational waveforms and output impedance that results in optimal power generation.

TABLE I: TENG Model Parameters

Variable	Value
A	4×10^{-4} m^2
ϵ_0	8.854×10^{-12} F/m
ϵ_{r1}	2.1
ϵ_{r2}	10
d_1	80×10^{-6} m
d_2	30×30^{-6} m
σ	300×30^{-6} C/m^2

III. NUMERICAL SIMULATION

First, the TENG model is investigated through numerical simulation that is implemented in Matlab. The parameters for the modeled TENG are given in Table I. It is assumed that the TENG output is connected to an impedance R_{eq}, as shown in Fig. 3. Using equations (1) to (5) for the TENG and equivalent impedance R_{eq}, the current is expressed according to

$$I = \frac{dQ}{dt} = -\frac{Q(d_0 + x)}{A\epsilon_0 R_{eq}} + \frac{V_{OC}}{R_{eq}} \qquad (6)$$

The differential equation in (6) can be coded to investigate the time domain waveforms of the TENG based on a function that defines the distance between the plates $x(t)$. In this simulation, the Euler method was used for the numerical integration method but other numerical methods could also be implemented. In this work, sinusoidal waveforms for $x(t)$ are investigated.

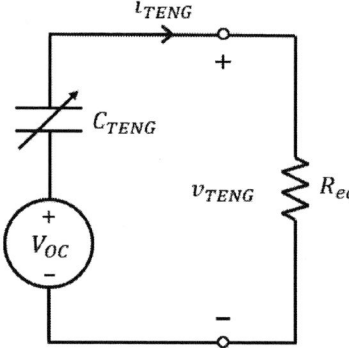

Fig. 3: TENG Equivalent circuit model with equivalent output impedance.

A. Sinusoidal Input Waveforms

First, it is assumed that the input mechanical motion to the TENG is sinusoid which compressed the plates from their maximum distance X_{max} to their minimum distance of 0 at a frequency of 10 Hz. The numerical simulation results for the distance between plates x, open-circuit voltage V_{OC}, and capacitance C_{TENG} over one period are shown in Fig. 4. As shown, V_{OC} changes proportionally with x, but the capacitance is inversely proportional and spikes in capacitance sharply as x approaches 0. In the simulation results, which do not take into account non-idealities, the open-circuit voltage varies from 0 to over 33,880 V—a large voltage range. The capacitance varies from about 3.4 pF to 86 pF.

To investigate the TENG's operation, there must be a load connected across the TENG. The equivalent resistance R_{eq} of the connected circuit will affect the current and voltage characteristics, and the resulting average power and energy capture. From preliminary numerical

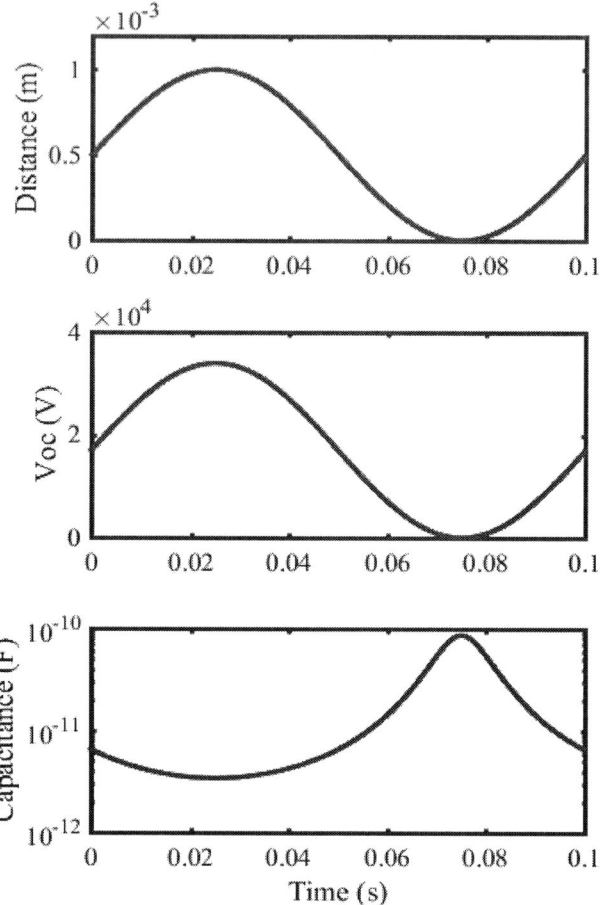

Fig. 4: Plate distance, open-circuit voltage, and capacitance of the TENG over one period.

analysis, it was determined that 200 MΩ yields the highest output power at 10 Hz. To see how the R_{eq} value affects the time domain waveforms, 20 MΩ, 200 MΩ, and 2000 MΩ were simulated. Results for the capacitor charge, TENG current, and TENG voltage are shown in Fig. 5. Although the mechanical input waveform is sinusoidal, the current and voltage waveforms take a non-sinusoidal form.

Using numerical simulation, the power over time and delivered energy is also be calculated. The simulation results from the TENG power and delivered energy are shown in Fig. 6. In the power-over-time waveform, it can be seen that there are two peaks in the power related to when the voltage crosses 0. Among the three tested R_{eq} values, it is clear that 200 MΩ reaches a

The 2018 International Power Electronics Conference

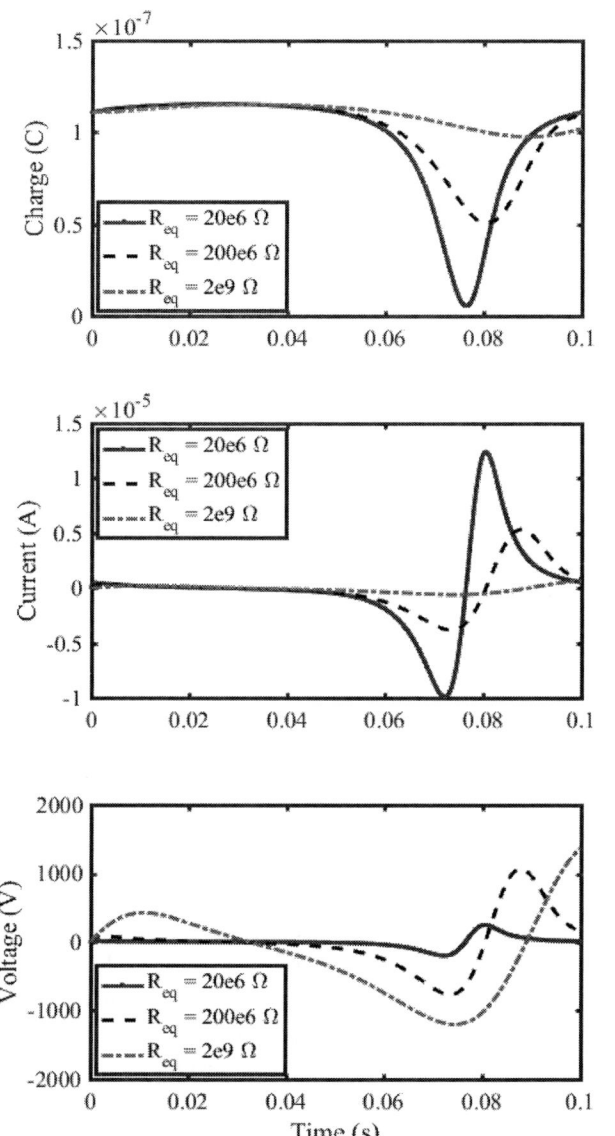

Fig. 5: Charge, current, and voltage waveforms for a TENG.

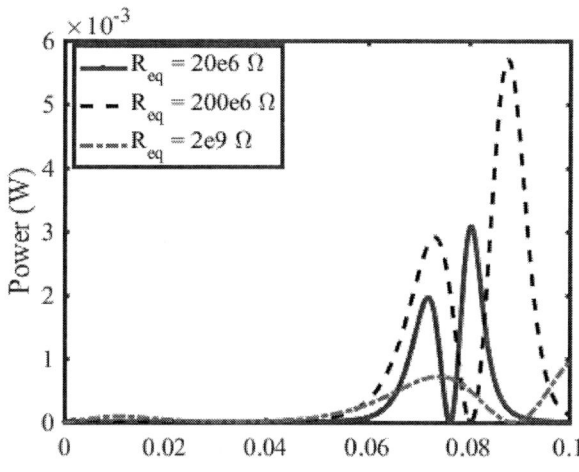

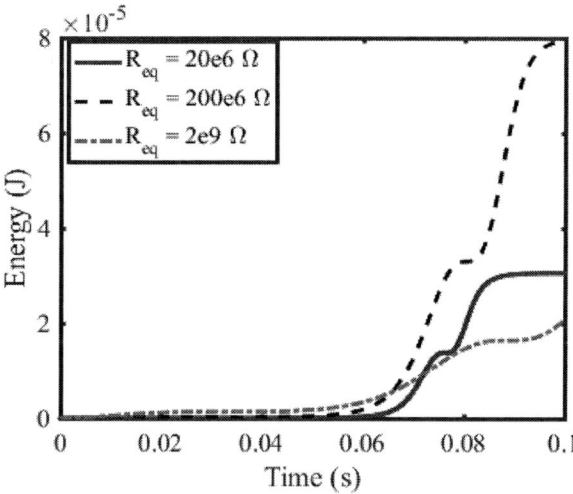

Fig. 6: Power and energy waveforms for a TENG.

significantly higher power level. The most important aspect is the energy capture, where $R_{eq} = 200$ MΩ has the highest energy capture after one period of about 80 μJ, which is an average power of 800 μW. This means that to optimize energy, the equivalent resistance seen by the ideal TENG operating at 10 Hz should have an impedance of approximately 200 MΩ. In practice, this is a very high impedance value that is difficult to implement in a circuit due to various sources of leakage.

B. Optimal Output Impedance

Next, the numerical simulation is used to investigate the optimal output impedance for the TENG at a frequency range of 1 to 100 Hz. The results of optimal impedance over frequency are shown as a logarithmic-scale plot in Fig. 7. The optimal resistance is not a constant value; the optimal value decreases as frequency increases. This indicates that the optimal output impedance should be tuned to a target frequency value and that frequencies far away from the target value will not be captured as effectively. Also, the optimal impedance values at lower frequencies are so high that it is difficult to implement in a real circuit, whereas the impedance

3066

value at higher frequencies become more reasonable. This indicates that operation at higher frequencies may be more effective for a given TENG or that the design of the TENG should be tailored to achieve a reasonable optimal output impedance at the target frequency.

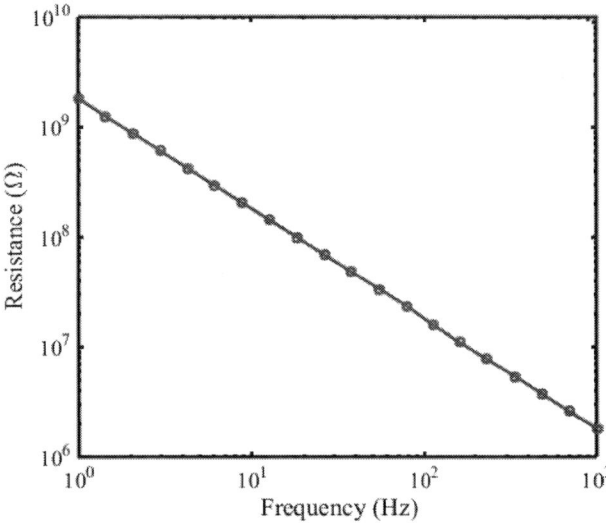

Fig. 7: Optimal resistance of a TENG at various frequencies.

C. Varying Amplitude

The characteristics of the vibrational input may vary, resulting in the movement of the plate distance $x(t)$ to have different amplitudes, which then affects the optimal output resistance. To investigated this, the amplitude of the sinusoidal function used as $x(t)$ was varied from 100% of X_{max} to 80%, 60%, 40%, and 20%. The results of the optimal impedance over frequency are shown in Fig. 8 as a logarithmic-scale plot. The trends over frequency are the same for all the amplitudes, but the optimal impedance value increases as amplitude decreases. This indicates that using the full amplitude of the plate distance will help keep the optimal impedance values in a more reasonable range.

This numerical simulation approach is effective for simulating a TENG with a simple sinusoidal input and an purely resistive output. Using the simulation for the ideal TENG, the range of V_{OC} and C_{TENG} can be calculated, along with the output impedance that gives maximum output power. However, nonidealities and leakage are not

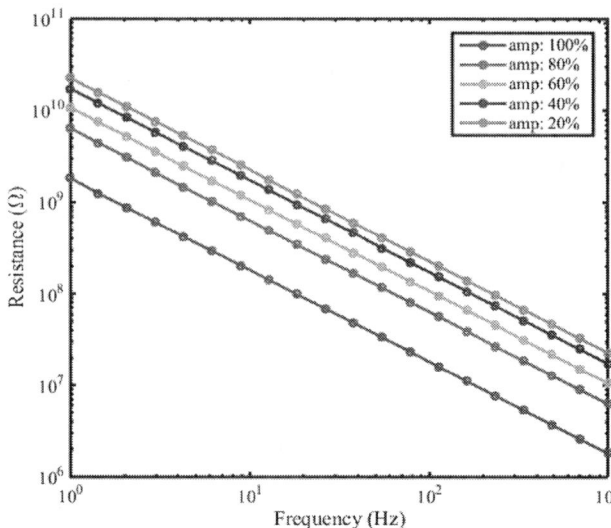

Fig. 8: Optimal resistance of a TENG versus frequency for various plate distance amplitudes.

yet considered in the model, which do substantially affect the operating conditions.

IV. DISCUSSION

Because the TENG generates ac power and most loads in energy harvesting applications are dc, a rectifier must be used. The TENG circuit model with a diode rectifier, capacitor, and load resistance are shown in Fig. 9. After the rectifier, a capacitor is needed to hold up the dc voltage. After this capacitor a dc-dc converter is likely to be used, but the converter can be simplified as an equivalent impedance. Mathematically, an ideal diode rectifier and capacitor do not affect the optimal output resistance [8], such the analysis of the optimal resistance is also true for the circuit with a rectifier and output capacitor.

The purpose of modeling the TENG is to accurately model a real TENG. However, because the optimal output resistance is over 10 MΩ voltage measurement proper measurement can be a challenge. Most oscilloscope probes have an impedance of 10 MΩ which prevents measurements above this impedance value. Another challenge is that the actual plate movement is not likely to be a perfect sinusoid. Future work will focus on modeling a more realistic vibrational input and comparing the simulation results to the experimental results to validate and improve the TENG model.

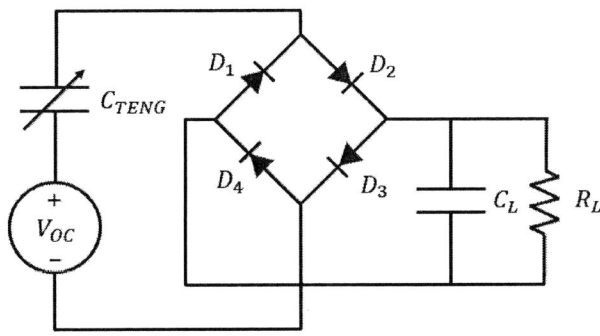

Fig. 9: Basic converter circuit model for a TENG with a diode rectifier.

V. CONCLUSION

The TENG operation principle has been introduced and its analytical model has been presented. The numerical simulation shows the effect of mechanical oscillation of the plates on V_{OC} and C_{TENG}, along with waveforms in the time domain. Then, the numerical simulation was used to determine the optimal output resistance for the TENG at a variety of frequencies and plate-movement amplitudes. The simulation results at lower frequencies and lower amplitudes give unrealistically high optimal output resistances, while higher frequencies and full amplitude give more reasonable optimal output resistance values. This work focused on the ideal model of the TENG with a pure sinusoidal input, but future work will investigate realistic inputs and improvements to the TENG circuit model.

ACKNOWLEDGMENT

This research was supported by the MSIT (Ministry of Science and ICT), Korea, under the ITRC (Information Technology Research Center) support program (IITP-2017-2017-0-01635) supervised by the IITP (Institute for Information & Communications Technology Promotion).

REFERENCES

[1] A. Khaligh, P. Zeng, and C. Zheng, "Kinetic energy harvesting using piezoelectric and electromagnetic technologies-state of the art," *IEEE Transactions on Industrial Electronics*, vol. 57, no. 3, pp. 850–860, Mar. 2010.

[2] S. Niu, Y. Liu, Y. S. Zhou, S. Wang, L. Lin, and Z. L. Wang, "Optimization of triboelectric nanogenerator charging systems for efficient energy harvesting and storage," *IEEE Transactions on Electron Devices*, vol. 62, no. 2, pp. 641–647, Feb. 2015.

[3] D. Bhatia, J. Lee, H. J. Hwang, J. M. Baik, S. Kim, and D. Choi, "Design of mechanical frequency regulator for predictable uniform power from triboelectric nanogenerators," *Advanced Energy Materials*, p. 1702667.

[4] J. P. Lee, B. U. Ye, K. N. Kim, J. W. Lee, W. J. Choi, and J. M. Baik, "3d printed noise-cancelling triboelectric nanogenerator," *Nano Energy*, vol. 38, pp. 377 – 384, 2017.

[5] J. W. Lee, B. U. Ye, and J. M. Baik, "Research update: Recent progress in the development of effective dielectrics for high-output triboelectric nanogenerator," *APL Materials*, vol. 5, no. 7, p. 073802, 2017.

[6] J. Moon and S. B. Leeb, "Power electronic circuits for magnetic energy harvesters," *IEEE Transactions on Power Electronics*, vol. 31, no. 1, pp. 270–279, Jan 2016.

[7] Z. L. Wang, *Triboelectric Nanogenerators*. Springer International, 2016.

[8] S. Niu, Y. S. Zhou, S. Wang, Y. Liu, , L. Lin, Y. Bando, and Z. L. Wang, "Simulation method for optimizing the performance of an integrated triboelectric nanogenerator energy harvesting system," *Nano Energy*, vol. 8, no. Supplement C, pp. 150 – 156, 2014.

[9] A. Abdelwahed, M. Amin, M. Elosairy, and N. Abbasy, "Theoretical modelling for enhancing contact-separation triboelectric nanogenerator performance," in *2016 Annual Connecticut Conference on Industrial Electronics, Technology Automation (CT-IETA)*, Oct. 2016, pp. 1–5.

The 2018 International Power Electronics Conference

General Power Electric Converter Model

Jingwen Xie

R&D, Schneider Electric, Shanghai, China

E-mail: jingwen.xie2@Schneider-electric.com

Abstract— According to the traditional structure and functions, we classified the converters, but they have been confusing and vague. In this paper, an universal model and expression is proposed, which can help to reclassify the converters considering relations among them. Firstly, based on the basic Buck converter, the general model of the converter is built according to the volt-second balance principle. And then a unified mathematic expression of the converters is put forward. Within the appropriate boundary conditions and some possible variations of this unified expression, the expressions of the specific converters can be deduced, such as Boost, Buck-boost, Inverter and Rectifier. Using the general model and expression, multilevel converters are also easier constructed. The proposed analysis method reveals the consistency principle behind different converters, also can help engineers to development converters.

Keywords—Five in one converter model, General converter principle, Multilevel, Theory basis of converter classification.

I. INTRODUCTION

In traditional power electronics theories, the switching power supplies are classified by referring to DC/DC converters. DC/DC converters are divided into six types, including Buck, Boost, Buck-boost, Cuk, Zeta and SEPIC converters. Among these converters, Buck and Boost converters are the basic converters, and the others can be derived from them.

However, is such classification really logical? What is its theoretical basis? This paper is intend to classify and research switching power supply by the relations between the converters, and try to answer these questions. Based on the Buck circuit, a general model of switching power supply is built, and then a unified mathematic expression of all kinds of converters is put forward. Within the appropriate boundary conditions and some possible variations of this unified expression, the expressions of all the specific switching power supplies can be deduced. Each possible deduced expression will be corresponding to one kind of switching power converter; and also, each kind of converters is one special expression in different condition. This can be a theoretical basis for switching power supply classification. According to the proposed classification method, the DC and AC converters can be involved in one mathematic expression, and all the Buck, Boost, Buck-boost, Inverter and Rectifier are derived from a common expression.

II. GENERAL EXPRESSION OF SWITCHING POWER CONVERTER

Figure 1 shows a typical diagram of a buck converter. It is composed of switch transistor, fly-wheel diode, choke and filter capacitor

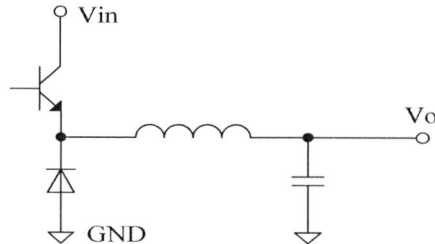

Fig. 1. Buck converter

This converter can convert the DC input voltage to a lower DC voltage required. It's known as the DC chopper, and its work waveform is shown in Figure 2

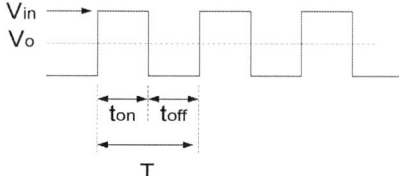

Fig. 2. Waveform of transistor switch

In Figure 2, T is the period of switch, t_{on} is the turn-on time, t_{off} is the turn-off time. According to the volt-second balance principle of the choke, the following relational expression is expressed:

$$(V_{in}-V_o) \times t_{on}-V_o \times t_{off}=0 \quad (1)$$

V_{in} and V_o are the input and output voltages respectively, transpose and get:

$$V_o=V_{in} \times t_{on}/ \left(t_{on}+t_{off} \right) \quad (2)$$

Definition $D= t_{on}/T$ is the ratio of turn-on time in one switch period, which is known as duty cycle, and then (2) can be rewritten as:

$$V_o=V_{in} \times D \quad (3)$$

(3) is the expression of buck converter between input and output voltages. From (3), it is obvious that V_o can be controlled by D. The capacitor on output side in Figure 1 is to filter the high-frequency components so that the smooth output voltage can be achieved. As a result, (3) will not be influenced if this capacitor is taken out.

3069

The 2018 International Power Electronics Conference

Transistor and diode are switches, these two devices can be represented by the ideal switch symbols. Therefore, the ideal buck converter can be changed as figure 3.

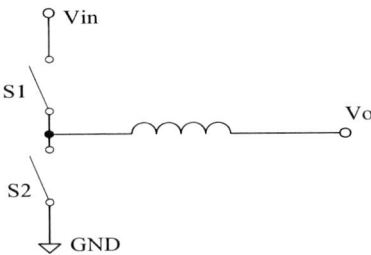

Fig. 3. Ideal switch symbols represent electric switches

Furthermore, V_{in} and V_o can be regarded as voltage supply sources. GND reference ground is essentially also a voltage source, it is endowed with zero as the reference ground for better understanding of the circuit. These three power source will be endowed with general voltage source symbols as V_1, V_2, V_3. Therefore a general model of switching power converter can be achieved as Figure 4.

Figure 4 is the general model of switching power converter, and the basic topology transformation of switching power converter is the research of transformation between the three voltage sources.

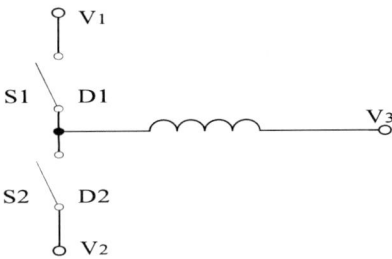

Fig. 4. General model of switching power converter

D_1 in Figure 4 is the duty cycle (normalized conduction time of the S_1 switch) of the S_1 switch, D_2 is the duty cycle of the S_2 switch. According to the volt-second balance principle, the relationship of the three voltage sources in Figure 4 can be expressed as:

$$(V_1 - V_3) \times D_1 + (V_2 - V_3) \times D_2 = 0 \tag{4}$$

(4) can be sorted out as:

$$V_1 \times D_1 + V_2 \times D_2 = V_3 \times (D_1 + D_2) \tag{5}$$

The formula is derived from the principle of volt-second balance which is applicable in both inductance continuous current mode (CCM) and discontinuous current mode (DCM). For easier understanding, the following analysis will be mainly based on the CCM mode.

In CCM, the follow expression can be gotten:

$$D_1 + D_2 = 1 \tag{6}$$

If $D_1 + D_2$ is larger than 1, that means S1 and S2 have common turn on time, which means that there is short current between V_1 and V_2 and an infinite current will be resulted. And if $D_1 + D_2$ is less than 1, that means S_1 and S_2 have common turn off time, which means that the

current of the reactor is completely cut off at same time, resulting in an infinite high voltage. Therefore, the establishment of $D_1 + D_2 = 1$ is inevitably determined by nature.

The combination of (6) and (5) is a description of the switching power converter in CCM. (6) is called as the constraint or boundary condition.

Put the constraint into (5), then (5) becomes:

$$V_1 \times D_1 + V_2 \times D_2 = V_3 \tag{7}$$

Formula (7) is called as the general formula of switching power converter, which is an expression of transformation between the three voltage sources. Certainly, it can be extended to the relationship between numbers of voltage sources, as

$$V_1 \times D_1 + V_2 \times D_2 + \cdots V_n \times D_n = V_o \tag{8}$$

And the constraint is:

$$D_1 + D_2 + \cdots D_n = 1 \tag{9}$$

III. RELATIONSHIP BETWEENT DIFFERENT CONVERTERS

Several major variations of (7) are being discussed below.

A. If V2=0

In (7), if $V_2=0$ (or $V_1=0$, because the roles of V_1 and V_2 are identical, see Figure 4), that is, in the case that V_2 is grounded, then,

$$V_3 = V_1 \times D_1 \tag{10}$$

Compared with (3), it is obvious that, it is the expression of the Buck converter. Since $D_1 \leq 1$, so $V_3 \leq V_1$ is being derived, it is the reason why the converter is known as the buck converter. If V_1 is the input voltage and V_3 is the output voltage, you can only get an output voltage that is smaller than the input voltage. (10) also shows that the polarities of output and input voltages of the buck converter is same. This means that a positive voltage source can only be alternate into a positive voltage output. If you want to get a negative voltage output, you must use a negative voltage source to alternate.

(10) can be alternated by a simply shift:

$$V_1 = V_3 \div D_1 \tag{11}$$

Similarly, since $D_1 \leq 1$, so $V_1 \geq V_3$ is being derived. If V_1 is taken as the output voltage and V_3 as the input voltage, the converter will get an output voltage that is higher than the input voltage. This kind of converter is called Boost converter. Similarly, the polarities of input and output voltages of Boost converter are same.

(10) and (11) are one formula, thus the physical objects described by (10) and (11) must be the same one. Therefore, the expressions show that Boost converter and Buck converter is a same converter. In essence, they are the same converter with different perspectives.

And if put the constraint (6) into (11), then:

$$V_1 = V_3 \div (1 - D_2) \tag{12}$$

This is the well know expression of the Boost converter in general textbooks. (11) is consistent with the traditional theory, and their difference lies in different

3070

perspectives. The traditional theory is based on different switching devices to analyze topology, and the expressions of Buck and Boost are not same. This paper is based on the same switch to analyze, and their expressions are same. Boost is Buck, and they are two aspects of the same thing with reciprocal transformation.

B. If $V_3=0$

A In (7), let $V_3=0$, V_3 is grounded, and (7) becomes:

$$V_1= -V_2 \times (D_2 \div D_1) \qquad (13)$$

The result of $D_2 \div D_1$ can be greater or less than or equal to 1

Therefore, V_1 can be greater than V_2, and can also be less than V_2 with opposite symbols. This is the Buck-boost converter. The transformation of Boost and Buck are all integrated

C. If $V_1= -V_2$

In (7), there is also a typically situation that V_2 is equal to V_1 with opposite amplitude. It is the case of $V_1= -V_2$. There are more practical applications in this case, such as power converter with positive and negative DC buses. Then (7) becomes:

$$V_3=V_1 \times (2D_1-1) \qquad (14)$$

And

$$V_1=V_3 \div (2D_1-1) \qquad (15)$$

(14) is using the voltage to multiply by $(2D_1-1)$. Since $-1 \leq (2D_1-1) \leq 1$ and $|2D_1-1| \leq 1$, this transformation of multiplication is also a Buck converter. The output voltage of Buck converter can be either positive or negative, and the variation range of V_3 is from $-V_1$ to $+V_1$. This means that the conversion of DC to AC can be achieved by this converter, such as an inverter. Make $(2D_1-1) =M\sin(\omega t)$, and M is the modulation coefficient,

then (14) becomes:

$$V_3=A\sin(\omega t) \qquad (16)$$

Where $A=V_1 \times M$ is the sinusoid amplitude. This is the expression of the double-bus inverter.

(15) is voltage to divide by $(2D_1-1)$. Since $-1 \leq 2D_1-1 \leq 1$ and $|2D_1-1| \leq 1$, the relationship of division is a boost converter. The output voltage of boost converter can be either positive or negative. From another perspective, the positive and negative polarities of the power sources can be converted to DC output by the converter. This means that AC input source can be converted to DC output, and the converter can be used as a rectifier. If V_3 is the AC power supply with amplitude of A and frequency of ω, i.e., $V_3=A\sin(\omega t)$, meanwhile, carries out phase locked control to be $(2D_1-1)=M\sin(\omega t)$, then (15) is changed to be:

$$V_1=V_3 \div M\sin(\omega t) \qquad (17)$$
$$V_1=A \div M, \qquad (18)$$

Thus, the output voltage V_1 is the DC voltage with a determined transformation relation to the amplitude of the input AC power supply, reaching the purpose of adjusting V_1 by adjusting M; that is the principle of PFC high-frequency rectifier.

From the analysis above, if you wanted to construct an inverter topology, only multiplication transformation relation can be selected, just as the Buck converter. It is determined by the waveform characteristic of the AC power supply. It is the characteristic demand of output voltage from 0 to 1 and can be realized only through multiplication transformation relation. Also, to construct a rectifier (here referring to the one with PFC function), only Boost converter can be selected as only the division relation can convert the sinusoidal wave changed from $0 \sim 1$ to a determined DC voltage.

D. Family tree of power converter

Various converter relations changed from general formula (7) are listed in Figure 5.

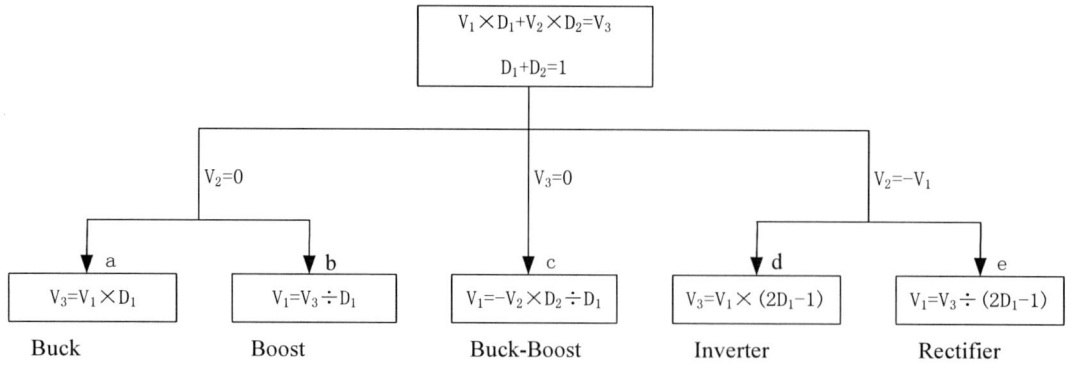

Fig. 5. Family tree of power converter

Figure 5 includes five converters. These converters include three basic DC converters a, b and c and two AC converters d & e. AC and DC converters are from the same formula.

3071

E. Five in one converter model

Mark the five basic transformation relations on common converter model, as shown in Figure 6.

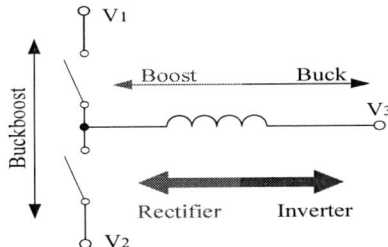

Fig. 6. Five-in-One Converter Model

In Figure 6, from left to right and from right to left are the Buck and Boost converters respectively, Inverter and rectifier regarding V1 and V2 as a group, and the conversion between the upper and lower is the Buck-boost converter.

The physical objects are the same and the five converters are the same converter essentially. Applying this concept in engineering, the design of different converters can be cross-reference. The designed Buck converter is a Boost converter in the opposite way. It can help to have directional reference significance on rapid design of product.

The convenience on derivation of transfer function lies in that, if it is not convenient to derive the Boost transformation relation, one can derive the Buck transformation relation first and then reverse the function; if it is not convenient to derive the control function of switch, one can obtain that of another switch first and then put it in constraint equation $D_1+D_2=1$.

IV. CIRCUIT CONSTRUCTION IN REALITY

A. Commonly used electronic switcher

The inter-relationship between five basic switching power converter is derived at above discussion on switching power converter. In practice, to construct a switching power converter topology, the ideal switching symbols can be replaced in theoretical analysis with specific switching components.

Commonly used representative electronic switches are listed as follows:

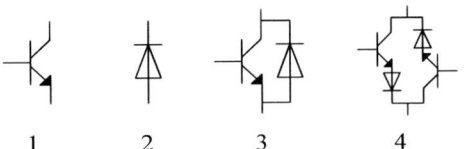

Fig. 7. Commonly used electronic switches

Four types of representative semiconductor switches are listed in Figure 7. No. 1 switch is a symbol of transistor representing three-terminal controllable semiconductor switch, it is a unidirectional switch that has one controllable signal port. Transistor, Mosfet, IGBT, etc. belong to this kind of switch; No. 2 switch is diode; No. 3 switch is semi-controllable bidirectional switch. It is controllable on half of the direction and uncontrollable on the other half direction; No. 4 switch is a full-controllable bidirectional switch, controllable from both directions. The characteristics of No. 4 switch are the closest to the ideal switch. This switch can achieve the function of all the other three switches above and is the most complex. In actual switch selection and application, it is not necessary to always select No. 4 switch. Do selecting devices to fulfill the function by just enough.

For multilevel converter, three and five level T type converters as two examples to show as follow.

B. Construction of three level inverter topology

(8) and (9) are multilevel conversion equation, which are relisted as follows:

$$V_1 \times D_1 + V_2 \times D_2 + \ldots V_n \times D_n = V_o \qquad (8)$$

And constraints:

$$D_1 + D_2 + \ldots \ldots D_n = 1 \qquad (9)$$

For three level inverter, the equation is changed as:

$$V_1 \times D_1 + V_2 \times D_2 + V_3 \times D_3 = V_o \qquad (19)$$

Transfer equation (19) to a diagram as follow:

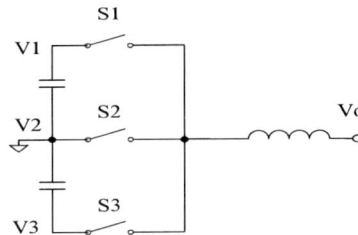

Fig. 8. Three level inverter diagram

Select appropriate semiconductor switches to replace the ideal switch symbol in Figure 8. to obtain the actual three-level topological circuit, as follows:

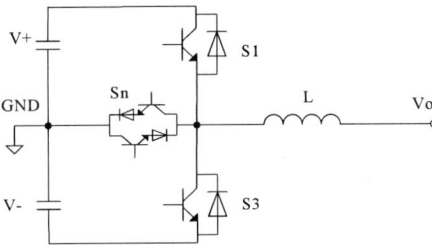

Fig. 9. Three level inverter diagram

According to the theory of reciprocal boost and buck converters introduced above, in order to construct a rectifier, it is only necessary to invert the inverter topology in Figure 9, the rectifier diagram is shown as in Figure 10.

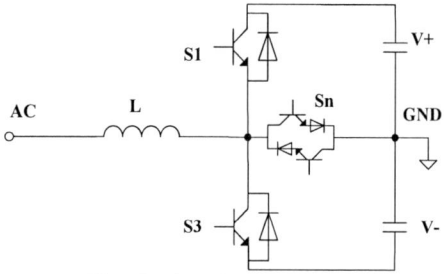

Fig. 10. Three-level rectifier

In Figure 10, a three-level rectifier is shown, the input is AC voltage source and output is positive and negative DC voltage. In actual circuit application, if there is no need for bus energy to feedback the grid, then switches S_1 & S_3 can be simplified as diodes.

C. Construction of five level inverter topology

Another, five level inverter as an example.

(8) is multilevel conversion equation, which to use for five level inverter as follow:

$$V_1 \times D_1 + V_2 \times D_2 + V_3 \times D_3 + V_4 \times D_4 + V_5 \times D_5 = V_o \quad (20)$$

Transfer the expression to diagram as figure 11.

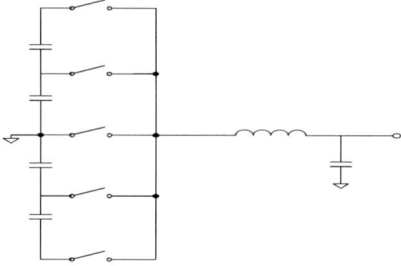

Fig. 11. Five-level inverter

Select appropriate semiconductor switches to replace the ideal switch symbol in Figure 11 to obtain the actual topological circuit of five-level inverter, shown as in Figure 12.

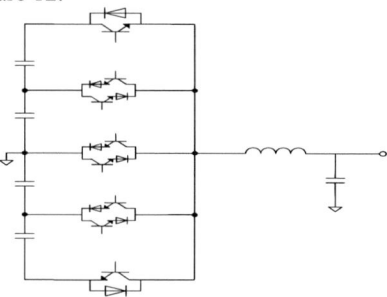

Fig. 12. Actual five-level inverter

Multilevel converter can effectively reduce the voltage stress of switch components, but it doesn't mean that the more levels is the better. When the converter is higher than 6 levels, the relative reduction of voltage is less than 5%(see figure 13) and has bad cost performance at standpoint of engineering. Hence, 3-5 levels are generally used. However, in the application of high-voltage converter, converter with more than 6 levels still has practical significance.

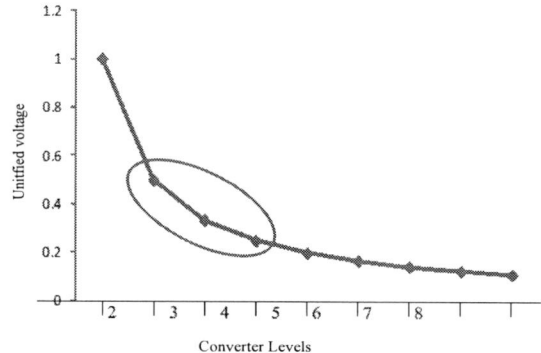

Fig. 13. Voltage stress of multilevel converter switch

V. CONCLUTION

In this paper, basic Buck converter is analyzed firstly, then a universal model of switching power supply and its general expression are constructed, finally the classification method is proposed. Through the investigation of the the general expressions of switching power converter under different conditions, it boils down to five basic power supply topology structures, i.e., Buck, Boost, Buck-boost converters, Inverter and Rectifier. The proposed general power converter model reveals the deep consistency principle behind the different converters and puts forward the theoretical basis of the converter classification. It is clarified that, various converters are connected essentially, buck converter is also a boost converter, they are different applications with the same conversion topology.

From the perspective of a multilevel converter, the more powerful and high voltage converters are, the more multilevel converters are required, and multilevel converter ask for bidirectional semiconductor switching devices.

A Modular Converter- and Signal-Processing-Platform for Academic Research in the Field of Power Electronics

Rüdiger Schwendemann[1*], Simon Decker[1*], Marc Hiller[1], Michael Braun[1]

1 Institute of Electrical Engineering, Karlsruhe Institute of Technology, Karlsruhe, Germany
*E-mail: ruediger.schwendemann@kit.edu, simon.decker@kit.edu

Abstract— For academic research it is mandatory that the theoretical evaluation and modelling of new control methods, modulation schemes, electrical machines, power electronic topologies, etc. is validated with accurate measurements. To guarantee a high quality and high performance research it is necessary to have a modular, scalable, user-frsiendly, adaptable and affordable system. This allows to put the focus on the research topics themselves rather than spending a high effort on the pure implementation of the theoretical research results. The system described in this paper consists of a software environment/toolchain and a hardware platform. The hardware platform can be subdivided into a power electronics platform and a System on Chip based signal processing system. Besides the hardware platform also a user-friendly software environment/toolchain for model-based research is developed and illustrated in this paper. This new system enables rapid-prototyping of new algorithms, hardware and topologies.

Keywords— *Education Tool, Power Electronic Platform, Signal Processing System, rapid control prototyping.*

I. INTRODUCTION

The Institute of Electrical Engineering (ETI) at KIT has three main research topics: "Electrical Drives and Power Electronics", "Hybrid Electric Vehicles" and "Power Electronic Systems". Three professors, 26 scientific assistants and about 50 students deal with issues concerning these research topics. In order to ensure a high quality and fast academic research for such a large number of scientists it is necessary to have a modular and cost-efficient signal processing and power electronics platform. There are several powerful commercial platforms available, but they are expensive and only allow very limited access to the complete system-level, which is a prerequisite in academic research. The described self-developed platform is used for academic research in PhD projects, student laboratories and bachelor- or master theses. So far, the platform has consisted of a modular DSP signal processing system [1] and a single PCB Silicon Insulated Gate Bipolar Transistors (Si-IGBTs) converter with an output rating up to 30 kW. For future academic research, a new modular platform is necessary. This paper describes the next generation of the signal processing system based on a System on Chip (ETI-SoC system), and

the new modular single PCB converter (ETI-combi-EPSR), which can be used with either Silicon Carbide Metal–Oxide–Semiconductor Field-Effect Transistors (SiC-MOSFETs) or Si-IGBTs. Furthermore, the necessary software environment/toolchain for a simple usage of the new hardware platform is described.

II. DEVELOPMENT PROCESS

The development process of new software e.g. control algorithms and modulation schemes or new hardware components like electrical machines or power electronics topologies can be roughly divided in three steps. First, the complete system is analyzed and developed in **simulation**. Model-based development in simulation is essential to enable short research times. Therefore, accurate machine models parametrized by finite element analysis (FEA) or test-bench measurements, exact physical models of the power electronics or precise battery models are available. After the simulation phase, a feasible **hardware** has to be chosen and assembled to confirm the simulation results. Due to the modularity of the ETI platform, a fast test-bench setup is possible. Finally, the hardware has to be commissioned and appropriate tests and measurements can used for the **validation**.

Subsequently, a typical development process of a new electrical machine test-bench is described. This test-bench is used to investigate novel machine control methods.

Simulation:
1. Analysis of an electrical machine with FEA
2. Generation of an electrical machine parameter set
3. Parametrization of a Mathworks Matlab/Simscape electrical machine model
4. Design/Implementation of the converter for the electrical machine in Mathworks Matlab/Simscape
5. Design of the modulator for the converter in Mathworks Matlab/Stateflow
6. Design/Implementation of the machine control in Mathworks Matlab/Simulink
7. Simulation and parametric study of the entire test-bench

8. C-code generation of the developed machine algorithm for the signal processing system
9. HDL-code generation of the modulator for the signal processing system

Hardware:

1. Assembling of the modular power electronics and signal processing components
2. Dimensioning of the necessary passive components, interfaces and peripherals for the test-bench
3. Startup of the test-bench

Validation:

1. Implementation of the developed software on the signal processing system
2. Automated measurements/tests with the Monitor Control Tool based on National Instruments LabView

The time effort for this development process was significant reduced by the usage of the signal processing and the power electronics platform.

III. SIGNAL PROCESSING PLATFORM

System on Chip devices are well-proven in cell phones or software defined radio devices but they are used more and more in automation and control technology [1]. The structural benefit of the System on Chip (SoC) device allows hardware/software co-design and makes them essential for modern automation and control devices.
The proposed system is based on the Xilinx programmable SoC family "Zynq-7000". For the ETI signal-processing platform Avnet's "PicoZed 7030" System on Module (SoM) with the "Zynq Z-7030" and necessary peripheral hardware was chosen (TABLE 1). On the SoM additional hardware components like memory devices or Ethernet driver are already implemented.

TABLE I
AVNET'S PICOZED 7030 SOM [2]

	PicoZed 7030
Processor Core	DUAL CORE ARM CORTEX A9
On-Chip Memory	256 KB
Peripherals	CAN, UART, I2C, ...
Clock Rate	667 MHz
Computer Performance	1334 MFLOPS
Memory	1 GB DDR
Ethernet	10/100/1000 Mbit
Programmable Logic	KINTEX-7
Logic Cells	125 000
PL IOs	250

To achieve downward compatibility with the existing "ETI-DSP system" described in [3] a carrier card for the SoM device is necessary. The existing "ETI-DSP system" is a modular real time measurement and control system in single $(100 \times 160 \, mm^2)$ or double euroboard format

$(233 \times 160 \, mm^2)$. The components are assembled in a subrack with a backplane for power supply and internal communication. The developed carrier card for the "PicoZed 7030" SoM device can be used in the existing system with modular extension boards or as a single system. In Fig. 1 the new ETI-SoC system is illustrated. It was designed for multiple purposes. Therefore a plurality of interfaces was implemented such as Ethernet, Memory card slot, galvanic insulated CAN transceiver, JTAG and USB. A proprietary asynchronous parallel communication bus was connected via backplane connector for internal communication.

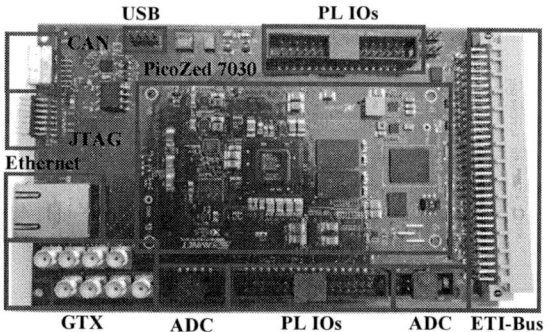

Fig. 1. Topview of the ETI-SoC carrier card design

The SoC system provides another extension ability by the four highspeed Xilinx GTX transceivers, which were attached to SMA connectors. The communication with these transceivers was successfully tested with a data rate of 6.25 Gbit per second. In Fig. 2 the eye pattern is shown. In the middle of the bit cycle the eye is still open with a Bit Error Rate (BER) of 10^{-6}, hence a communication is possible. With this high-speed communication interface an IO-expansion or sharing of computing performance of the SoC device is possible.

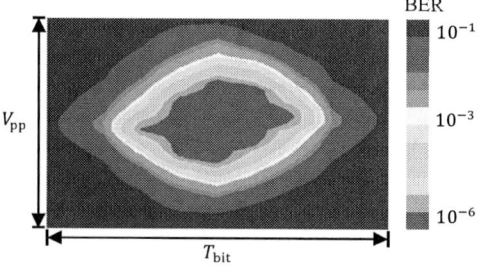

Fig. 2. Eye pattern for a data rate of 6.25 Gbit

On the bottom side of the ETI-SoC system a LVDS connector was attached to facilitate an extension of the carrier card with a second application dependent board.
For control applications it is mandatory to digitize analog measured values. Therefore, six Analog Digital Converters (ADCs) with an analog gain stage were implemented in order to handle ± 10 V input signals. The assembled Linear Technology ADC "LTC2323-14" has a sample rate of five Mega samples per second (Msps) with

3075

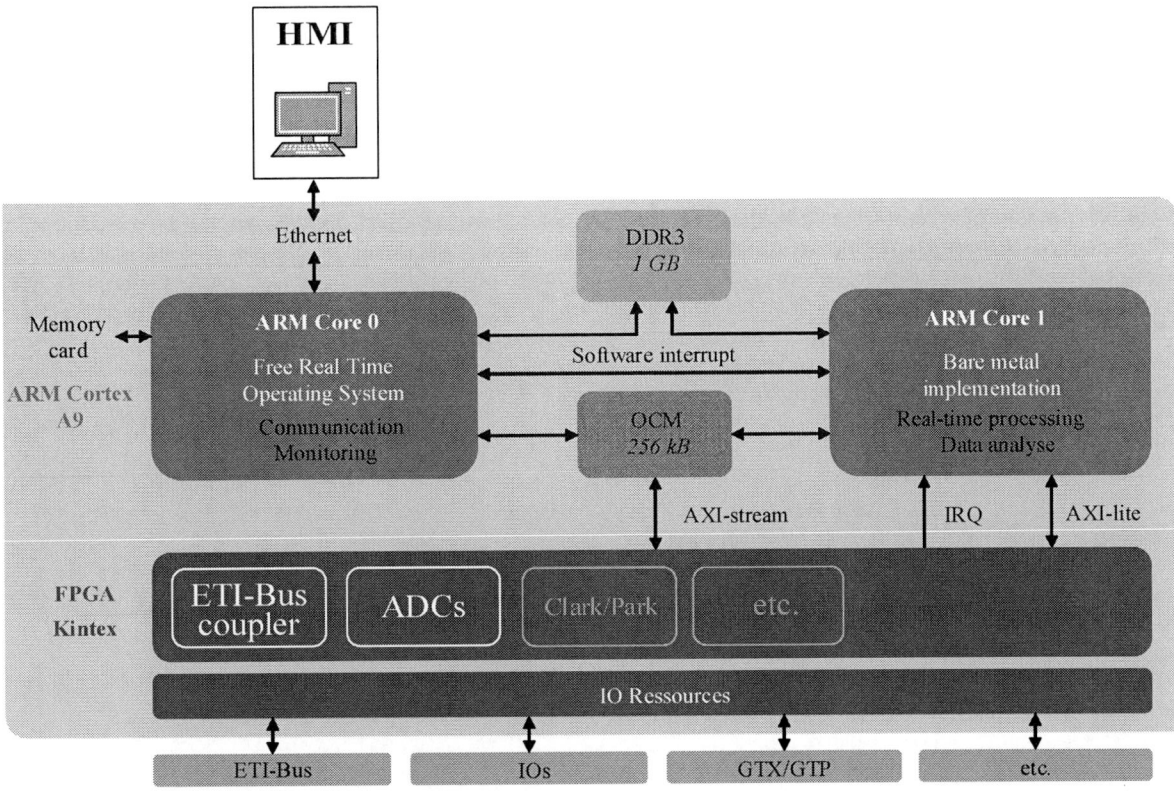

Fig. 3. Block diagram of the ETI-SoC software structure

a resolution of 15 bit. The measured latency of the entire measurement chain was only 400 ns.

To generate the signals for the converter and peripherals also the FPGA programmable logic IOs (PL IOs) of the SoC were attached to two 34 pin connectors.

The eight layer carrier card was designed with ALTIUM Designer during a student's master thesis. Due to the necessary differential routing of the LVDS and the GTX transceivers an accurate design is essential. Therefore, the signal integrity of the whole carrier card design was evaluated with ANSYS SiWave-PI.

IV. SIGNAL PROCESSING SOFTWARE DESIGN

To allow students a fast software development of novel control methods and modulation schemes besides the hardware platform also a software environment and toolchain must be given. The students develop the new algorithms in Matlab/Simulink and use the provided software environment. This allows that only the developed software interfaces in Matlab/Simulink must be addressed to use the particular functions of the SoC without the knowledge of the subordinate code. The developed software environment is described in this section.

To understand the developed software environment the internal system design of the SoC must be known. The used Zynq SoC consists of a dual core ARM Cortex A9

processor, a Kintex-7 FPGA and peripheral units which suit well for real time measurement and control systems. The whole block diagram of the Xilinx Zynq-7000 AP SoC is described in [4]. For the proposed implementation the ARM cores are separately used as an asymmetric multiprocessing system. The whole customized system structure is shown in Fig. 3.

On the first ARM core, ARM core 0, FreeRTOS a slim real-time operating system for embedded devices was implemented. ARM core 0 is responsible for the data exchange of the measured and calculated values with the HMI (host/control computer). For this purpose the FreeRTOS+TCP stack is used. The implemented algorithms, which are developed by the students, are only executed on the second ARM core ARM core 1 due to time-critical latency requirements. The algorithms are realized as raw program code without any operation system, called bare metal implementation. Thereby the program code can be written in C or directly generated with the Matlab/Simulink C-code generation out of the model-based simulation.

Due to this structure, with ARM core 0 as a communication system and ARM core 1 as processing system, the algorithms for the ARM core 1 can be flexibly reconfigured online. The fast execution of these algorithms on ARM core 1 within a defined time up to

10 µs make e.g. the control of power electronics with real time requirements possible.

The software environment also contains the communication between the FPGA and the ARM cores. Hence, measured values from the ADCs, the IO values and control sequences can transmit and receive from the FPGA and the ARM cores. The FPGA also manage the external ETI-Bus and is used as a flexible programmable logic part. For the data transfer between the FPGA and the ARM core 1 the Xilinx AXI4 stream protocol, part of the Advanced Microcontroller Bus Architecture (AMBA), is used. The developed protocol for the AXI4 stream with direct memory access was adapted so that the communication is similar to the existing bus-system described in [3] with a parallel data bus and corresponding address bits. The buffering of the data is utilized in the On-Chip memory (OCM) with a RX- and TX-buffer. The inter processor communication is possible due to synchronized ARM cores via software generated interrupts. The data and control sequences are also transferred between the Ethernet socket/HMI and the ARM core. Due to the easy exchange between the FPGA and the ARM Cores the SoC allow hard/software codesign of the developed functional designs. The executed code on the ARM core 1 is capable to be operated in real time.

On the control computer the developed Monitor Control Tool [3], based on LabView, visualize the received data. For the use of this communication the user only has to import a devised communication block in the designed Matlab/Simulink code model.

V. POWER ELECTRONICS PLATFORM

To evaluate new control methods or modulation schemes besides the signal processing also a power electronics platform is necessary. In Fig. 4 the new ETI-combi-EPSR is depicted. It consists of a local control unit (CU) and the power electronics unit.

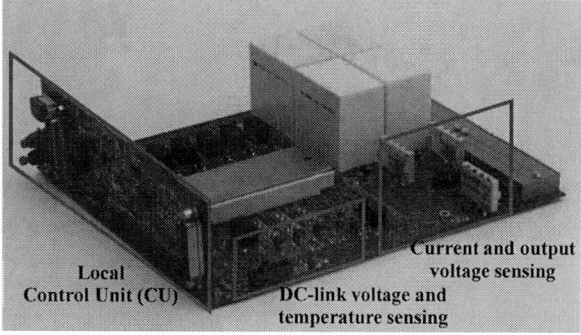

Fig. 4. ETI-combi-EPSR with a Local Control Unit and a power electronics unit

A. Power Electronics Unit

The power electronics unit was designed in such a way that it can either be equipped with Infineon Si-IGBT

module FS75R12KT4 or with Wolfspeed SiC-MOSFET module CCS050M12CM2. The output ratings are 30 kW for the Si-IGBT and 40 kW for the SiC-MOSFET, depending on the switching frequency.

The converter was equipped with measuring systems for the output currents, the heatsink temperature, the DC-link voltage and the output voltages. The combination of switching frequencies in the range up to 100 kHz and new control methods like Direct Adaptive Current Control [5] leads to high demands on the current measurement. Therefore, CMS3000 family current sensors from SENSITEC with a bandwidth of 2 MHz were applied. Since the converter will be used in student laboratories, all measurements are potential free to minimize the risk of accidents.

The converter was analyzed with ANSYS SiWave-DC to check the current density on the PCB [6]. In Fig. 5 the current density on the PCB is illustrated. The PCB was designed in in accordance with the IPC-2221 [7].

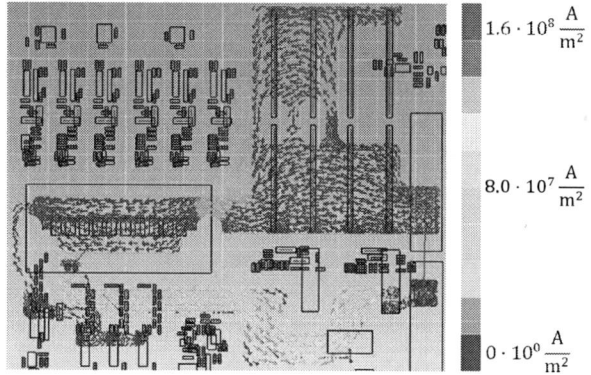

Fig. 5. Simulation with ANSYS SiWave-DC of the current density of one output phase of the three phase converter

B. Local Control Unit

The local control unit consists of a MAX10 FPGA from Altera, an ADS8528SPM Analog-Digital converter from Texas Instruments, digital/analog interfaces and a 50 Mbit/s optical interface. By the use of the CU several tasks of the ETI-SoC system can be carried out. Hence, the resources of the ETI-SoC system are fully available for the superposed control algorithm. Depending on the carried out tasks the CU can be equipped with a MAX10 FPGA from 8000 up to 25000 logic elements. Therefore, it is possible to adapt the CU depending on the application. For the minimal requirements such as monitoring tasks, error management, temperature control and modulation 8000 logic elements are sufficient. If an autarkic system is required, for example an Active Front End (AFE) or a DC/DC converter a MAX10 with 25000 logic elements is necessary. In this case, the ETI-SoC has only to provide the reference values and enable signals and thus no process time of the ETI-SoC is needed. To allow the communication the optical interface is used so the ETI-

SoC system and the CU can be spatially separated. Due to the modular design the CU can also be used for other power electronics system, e.g. the control of one phase of a 3-Level T-Type Neutral Point Clamped Inverter.

VI. EDUCATIONAL USE

At the ETI several lectures cover issues of electric machine design, electrical drive control, power electronics and circuit design. Besides the theoretical knowledge transfer it is also important to teach how to put the theoretical knowledge into practice e.g. in laboratory. Hence, laboratory experiments are necessary which can be built with the described modular platform. Besides that, new control methods or machines developed in PhD programs must be validated. Therefore test benches must be designed with the help of students e.g. as student workers or in different bachelor and master theses by means of the described modular platform.

To allow that universal use of the platform it must be continuously developed. During this development process, several students' projects are supervised in which the students learn to use the knowledge gained from the lectures. Thus, the PCBs for the ETI-combi-EPSR and the ETI-SoC system were developed and validated in one master thesis [8]. The signal processing software design was designed in several master theses [9], [10], [11], [12] and afterward further developed with student workers. Hence, the platform is always used with state of the art devices and can be easily applied for different projects. Three sample projects are subsequently described.

A. Power Hardware in the Loop system

At the ETI a Power Hardware-in-the-Loop (PHIL) system was developed by a PhD student. The system consists of the square-wave powered Modular Multilevel Converter (SPMMC) with a push-pull-converter [13]. The SPMMC can utilize a high-precision 3AC voltage source from DC up to nearly medium frequency [13]. This voltage source is used to emulate permanent magnet synchronous machines with nonlinear magnetics. In Fig. 6 the PHIL and a machine inverter (DUT) are illustrated.

The aim of the PHIL is to show the same behavior as the emulated machine on its terminals. Thus, a DUT can't differentiate whether the PHIL is connected to its terminals or the real machine. For this purpose the SPMMC has to utilize an output voltage v_{ay0} (number of phase: $y \in \{1, 2, 3\}$) that the current i_x is the same as in the real machine. Therefore, the reaction time of the output voltage v_{ay0} must be small in comparison to the switching frequency of the DUT. This is reached with a new hybrid control approach [14], which allows a modulation frequency of the output voltage v_{ay0} of 100 kHz. For the emulation of the machine a model must be calculated. To utilize a correct output voltage v_{ay0} the voltage of the DUT must be measured with a high sample rate and subsequent the machine model must be calculated.

This is done with a rate of 2.5 MHz. The FPGA clock of the ETI-SoC system for the model calculation is 75 MHz. Thus, the system has only 30 cycles for the whole calculation. During that calculation, eight memory accesses to look-up tables are necessary to utilize the correct voltage. Since the calculation itself needs 23 cycles, only 7 cycles are available for the memory accesses. This access time is only possible since the ETI-SoC system has an intern block RAM of 9.3 Mbit with an access time of one cycle [15]. Hence, the new ETI-SoC system made it possible to build up a high-precision PHIL emulation for permanent magnet synchronous machines with nonlinear magnetics.

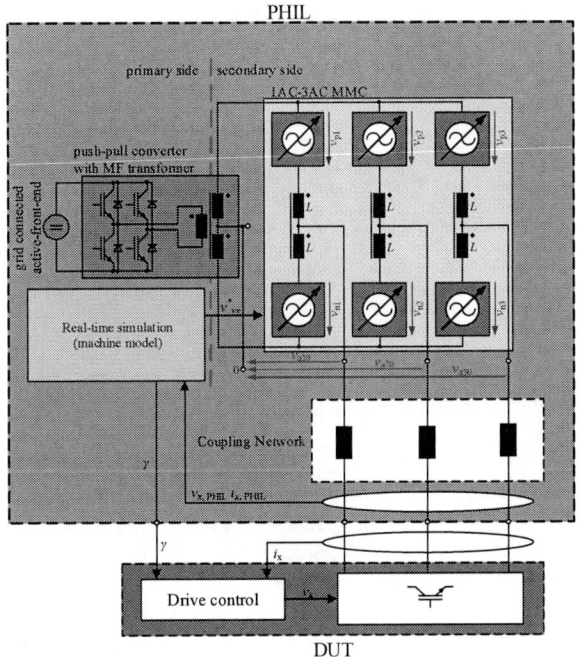

Fig. 6. Power Hardware-in-the-Loop (PHIL) system

B. Motor test-bench

The validation of new electrical machine designs during PhD programs [16] makes extensive measurement of the developed prototypes necessary. Therefore, different motor test-benches with customized signal processing and power electronics hardware are necessary. The buildup of these motor test-benches is often done in student's bachelor and master theses and supervised by the PhD candidates.

For future investigations with high switching frequency inverters for drives and dynamic control of electrical machines a new motor test-bench with Si- and SiC inverter based on the ETI-combi-EPSR was developed in a student bachelor thesis [17].

After the bachelor studies the students have the theoretical knowledge of electrical engineering. Especially within the specialization in power electronics and drives students can enroll lectures about electrical machines and drives, PCB and electrical circuit design and converter control. These lectures are *"Electrical Machines and Power Electronics"*, *"Industrial circuitry"* and *"Converter control technique"*. By building a motor test-bench students can put their learned knowledge into practice.

The described test-bench was developed with the new modular platform. The student started with the design of the cabinet and the power electronic circuit. For this motor test-bench, shown in Fig. 7, the device under test motor and the load motor are coupled directly against each other. For the power supply an active-front (AFE) and a power choke is necessary. For the AFE an ETI-combi-EPSR with SiC-MOSFETs has been used concerning chapter V. The power choke was calculated by the student based on the *"converter control technique"* lecture.

Fig. 7. Motor test-bench setup

The AFE controller was designed based on the "Control of Electrical Drives" lecture. The motor inverters were built up with two ETI-combi-EPSR one with Si-IGBTs and one with SiC-MOSFETs. Hence, a fast buildup of the test bench within one Bachelor thesis was possible. The ETI-SoC system signal processing hardware was used for the motor control and monitoring of the motor test-bench. The modular signal processing system had to be adapted for this purpose. Modular extensions for the rotor angle and temperature measuring boards were assembled. Different customized interface circuit boards were designed with Altium Designer and manufactured. Successive commissioning of the designed boards and the assembled cabinet and programming of the signal processing system with Matlab/Simulink and National Instruments Labview concluded the thesis.

Hence, the student could connect the knowledge of power electronics, drives, information technology, electrical circuit design and control theory from the bachelor studies during the thesis [18].

In Fig. 8 the complete motor test-bench is shown. The control and monitoring computer and the ETI-SoC system can be seen in the upper part. The power electronic consisting of three ETI-combi-EPSR and the power choke is shown in the lower part.

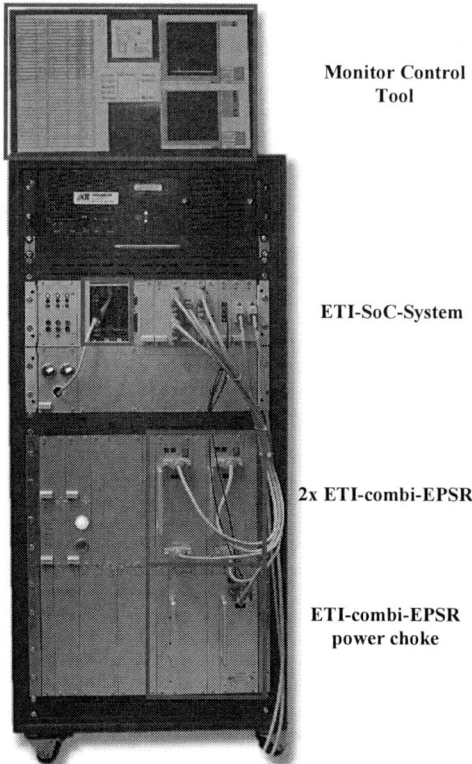

Monitor Control Tool

ETI-SoC-System

2x ETI-combi-EPSR

ETI-combi-EPSR power choke

Fig. 8. Motor test-bench cabinet

C. Formula Student

Another application of the modular signal processing hardware is in the formula student team "ka-racing" (Fig. 9). In the season 2016/17 the signal processing system controlled the drive train of the KIT electric car. For the all-wheel-drive racing car each wheel is powered by a motor over a self-designed axis. One signal processing system thereby executes two predictive motor control algorithms and the required monitoring [19] for one axis with two motors in less than 40 µs.

Fig. 9. Formula Student: *KIT16e* (www.ka-raceing.de)

VII. CONCLUSION

In this paper a modular platform of a signal processing system and a modular converter system for multiple research projects was introduced. With software based development of different algorithms, code generation for C-code and HDL-code fast rapid-prototyping in academic research is possible. Furthermore the modular, flexible converter platform allows fast validation on an experimental set-up for high quality research. Also three sample projects are described which were built with the new modular platform.

VIII. ACKNOWLEDGEMENT

This project has been supported by several generations of research associates, students and technical staff members since about twenty years at the Institute of Electrical Engineering at the KIT.
Special thanks go thereby to the following research associates and as well to their students:

Heinrich Steinhart, Gerhard Clos, Matthias Hauck, Frank Becker, Jörg Weigold, Carsten Ackermann, Klaus-Peter Becker, Alexander Stahl, Alexander Schmitt, Tobias Gemassmer, Mario Gommeringer, Felix Kammerer, Bernd Bohnet, Christian Axtmann, Mathias Schnarrenberger, Patrick Himmelmann, Patrick Winzer, Marc Veigel, Andreas Liske, Lukas Stefanski, Fabian Stamer, Matthias Brodatzki, Stefan Mersche.

REFERENCES

[1]   G. B. Robert Bielby, *Advantages of Xilinx 7 Series All Programmable FPGA and SoC Devices*. [Online] Available: http://www.ni.com/white-paper/14583/en/.

[2]   AVNET, *Picozed Datasheet*. [Online] Available: http://zedboard.org/product/picozed.

[3]   C. Axtmann, M. Boxriker, and M. Braun, "A custom, high-performance real time measurement and control system for arbitrary power electronic systems in academic research and education," in *18th European Conference on Power Electronics and Applications (EPE'16 ECCE Europe)*, 2016, pp. 1–7.

[4]   Xilinx Inc., *Zynq-7000 All Programmable SoC Technical Reference Manual (UG585): Technical Reference Manual*. [Online] Available: https://www.xilinx.com/support/documentation/user_guides/ug5 85-Zynq-7000-TRM.pdf.

[5]   A. Liske, F. Stamer, and M. Braun, "Easy Current Slope Detection for Low Cost Implementation of the Direct Adaptive Current Control for DC-DC-Converters," in *IEEE Energy Conversion Congress and Exposition (ECCE)*, 2015, pp. 180–186.

[6]   ANSYS Inc., *Ensuring the Power and Signal Integrity of Your High-Performance PCBs and Packages Using ANSYS SIwave - Webinar*. [Online] Available: http://www.ansys.com/resource-library/webinar/power-integrity-using-ansys-siwave?tli=en-us. Accessed on: Oct. 24 2017.

[7]   IPC ASSOCIATION CONNECTING ELECTRONICS INDUSTRIES, "Generic Standard on Printed Board Design: IPC-2221A," 1998.

[8]   W. S. Abissi Konga, "Design and Simulation of a novel ETI-System platform Design and Simulation of a novel ETI-System platform composed of power electronic and signal processing board," M.S. thesis, KIT, Karlsruhe, 2016.

[9]   F. Trudel, "Programming and Implementation of a SoC-based Processor System for the Control of Power Electronic Devices," M.S. thesis, KIT, Karlsruhe, 2016.

[10]   B. Schmitz Rhode, "Software Integration of Dynamic Control Algorithms on a System-on-Chip Platform for a Formula Student Electric Driving Chain," M.S. thesis, KIT, Karlsruhe, 2016.

[11]   E. Seidenspinner, "Design and implementation of a fast, efficient communication for the operation of converters," M.S. thesis, KIT, Karlsruhe, 2017.

[12]   B. Baier, "Design of a Motor Control User-Interface for a Formula Student Electric Drive Chain," B.S. thesis, KIT, Karlsruhe, 2017.

[13]   M. Schnarrenberger, F. Kammerer, M. Gommeringer, J. Kolb, and M. Braun, "Current Control and Energy Balancing of a Square-Wave Powered 1AC-3AC Modular Multilevel Converter," in *IEEE Energy Conversion Congress and Exposition (ECCE)*, 2015, pp. 3607–3614.

[14]   M. Schnarrenberger, D. Bräckle, and M. Braun, "A Hybrid Control Approach for Fast Switching Modular Multilevel Converters," in *9th IET International Conference on Power Electronics, Machines and Drives (PEMD 2018)*, 2018.

[15]   Xilinx Inc, *Zynq-7000 AP SoC Family Product Tables and Product Selection Guide*. [Online] Available: https://www.xilinx.com/support/documentation/selection-guides/zynq-7000-product-selection-guide.pdf.

[16]   P. Winzer and M. Doppelbauer, "A hybrid permanent magnet and wound field synchronous machine with displaced reluctance axis capable of symmetric," in *18th European Conference on Power Electronics and Applications (EPE'16 ECCE Europe)*, 2016, pp. 1–11.

[17]   K. Rickert, "Setup and commissioning of a motor test bench with SiC-converter and new SoC-signal processing system," B.S. thesis, KIT, Karlsruhe, 2018.

[18]   KIT, *Department of Electrical Engineering Department of Electrical Engineering and Information Technology Description of Modules*. [Online] Available: https://www.etit.kit.edu/rd_download/ModulbeschrEnglisch.pdf.

[19]   J. Richter, P. Bäuerle, T. Gemassmer, and M. Doppelbauer, "Transient Trajectory Control of Permanent Magnet Synchronous Machines with Nonlinear Magnetics," in *IEEE International Conference on Industrial Technology (ICIT)*, 2015, pp. 2345–2351.

Control IC for Boost-Flyback Converter for Energy Harvesting Applications

Jhih-Sian Li[1], Tsorng-Juu Liang[2], *Fellow, IEEE*, Kai-Hui Chen[3], Jui-Hung Lai[4], and Jun-Xian Huang[5]
Department of Electrical Engineering/Green Energy Electronics Research Center (GREERC)
/ Hierarchical Green-Energy Materials Research Center
National Cheng Kung University, Tainan, Taiwan
Email: tjliang@mail.ncku.edu.tw

Abstract—Conventionally, boost converter is used for high step-up energy harvesting system, but it is needed to be operated in discontinuous conduction mode (DCM) because of the limitation of the topology. Therefore, it causes more conduction loss and current stress on the switch. In this thesis, the control IC for boost-flyback converter in energy harvesting applications is proposed. It can greatly increase the voltage conversion ratio with the cascode of boost converter and flyback converter. In addition, this topology can recycle the energy of the leakage inductance produced from coupled inductor to load. Thus, the efficiency can be increased and the stress on the main switch can be reduced. Finally, The boost-flyback converter with input voltage of 0.2-0.3 V, output voltage of 5 V and output power of 20 mW is implemented to verify the feasibility of the proposed controller.

Index Terms—*Energy harvesting system, high step-up converter, boost-flyback converter.*

I. INTRODUCTION

In recent years, the human health monitoring is developed rapidly, so a lot of sensors are demanded. Fortunately, the human body can generated stable heat, and the TEG can be used to harvest this energy to provide sensors. But the voltage generated by TEG is low [1]-[3]. Therefore, the converter with high step-up and high efficiency is needed to convert the TEG output voltage to designate output voltage for the load.

Traditionally, boost converter can be used for high step-up energy harvesting applications. But considering the parasitic resistances in practical boost converter operated in continuous conduction mode (CCM), the efficiency is very low with extremely high duty ratio. In order to avoid the extremely high duty ratio, boost converter can be operated in discontinuous conduction mode (DCM).

Conventionally, boost converter operated in DCM is designed to increase efficiency by adjusting the switching signal [4]-[8]; however, in the same power condition, the peak current and root-mean-square current in DCM are higher than that in CCM. It causes severe conduction loss. To solve these problems, the topology of high step-up converter can be used for energy harvesting applications to achieve high voltage gain and improve efficiency [9]-[13].

In this thesis, the control IC for boost-flyback converter is proposed. The boost-flyback converter can achieve high step-up gain by cascoding the output of boost converter and flyback converter. This converter can be designed to operate in CCM, and it can reduce the conduction loss; additionally, the voltage stress on power switch is low because the energy stored in the leakage inductance is recycled to the output capacitor of boost stage. Consequently, this topology can achieve high voltage gain but also high efficiency.

II. INTRODUCTION OF BOOST-FLYBACK CONVERTER FOR ENERGY HARVESTING SYSTEM

The equivalent circuit of the boost-flyback converter is shown in Fig. 1, and the characteristics will be analyzed in following.

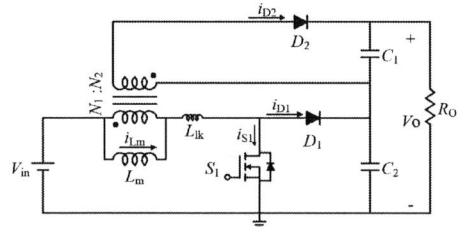

Fig. 1. Boost-flyback converter for energy harvesting applications.

Fig. 2 shows main waveforms of the boost-flyback converter operated in CCM. To simplify the analysis of operating principle, the switch and diodes are assumed to be ideal, and the leakage inductance L_{lk} is considered. There are three operating modes in one switching cycle and are discussed as follows.

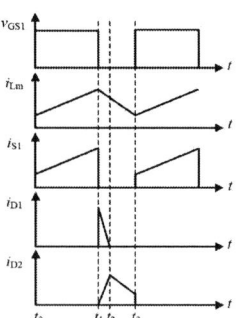

Fig. 2. Key waveforms of the boost-flyback converter operated in CCM.

Mode 1 [$t_0 \sim t_1$], switch S_1 is turned on, and diodes D_1 and D_2 are turned off. Magnetizing inductance L_m and leakage inductance L_{lk} receive energy from input source V_{in}. Thus, the magnetizing inductance current i_{Lm} and leakage inductance current increase linearly. The output capacitor supplies energy to load. This mode ends when switch S_1 is turned off. Mode 2 [t_1~t_2], switch S_1 is turned off, and diodes D_1 and D_2 are turned on. The energy of leakage inductance is recycled to output capacitor C_2, therefore, the voltage stress on the switch S_1 is clamped at V_{C2}, and magnetizing inductance L_m also delivers energy to C_1. This mode ends when the energy of leakage inductance is totally transferred to output capacitor C_2. Mode 3 [t_2~t_3], switch S_1 and diode D_1 are turned off, and diode D_2 is turned on. Magnetizing inductance L_m delivers energy to C_2.

When the boost-flyback converter is operated in steady state condition, the voltage conversion ratio can be derived by using voltage-second balance of the magnetizing inductance L_m.

The voltage gain of the boost-flyback converter can be expressed as Eq. (1).

$$\frac{V_O}{V_{in}} = \frac{1+nD}{1-D} \tag{1}$$

Fig. 3 shows the magnetizing inductance current in boundary conduction mode (BCM). The boundary condition of the magnetizing inductance can be derived by input and output power, and it is showed as Eq. (2).

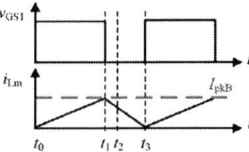

Fig. 3. Magnetizing inductance current of boost-flyback converter in BCM.

$$\frac{\frac{1}{2}L_m I_{pkB}^2}{T_S} = \frac{V_O^2}{R_{O\_BCM}} \tag{2}$$

The boundary condition of magnetizing inductance L_{BCM} can be derived as Eq. (3).

$$L_{BCM} = \frac{1}{2}\frac{(1-D)^2}{(1+nD)^2}R_{O\_BCM}D^2 T_S \tag{3}$$

In order to evaluate the efficiency in practical condition, the ESR and parasitic components are considered. The diodes are replaced with switches. Fig. 4 shows the boost-flyback converter considering ESR and parasitic components.

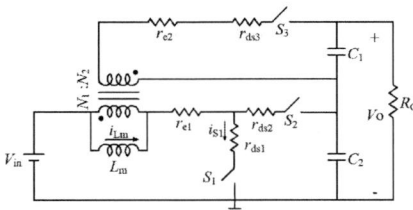

Fig. 4. The boost-flyback converter considering ESR and parasitic components.

By the voltage-second balance rule of magnetizing inductance, the voltage gain can be expressed as Eq. (4).

$$\frac{V_O}{V_{in}} = \frac{1+nD}{1-D} \times \frac{1}{\frac{D(1+n)}{(1-D)^2 R_O}(r_e + r_{ds}) + \frac{2D(r_e + r_D)}{(1-D)R_O} + 1} \tag{4}$$

Then, the efficiency η can be derived by calculating the input and output energy, and expressed as Eq. (5).

$$\eta = \frac{P_{OUT}}{P_{IN}} = \frac{1}{\frac{D(1+n)}{(1-D)^2 R_O}(r_e + r_{ds}) + \frac{2D(r_e + r_D)}{(1-D)R_O} + 1} \tag{5}$$

III. ANALYSIS OF THE PROPOSED CONTROL AND FUNCTION BLOCKS

The proposed control blocks for the boost-flyback converter are shown in Fig. 5. The proposed controller is composed of under voltage lockout (UVLO) [14], protection circuit, oscillator, PWM circuit, compensator, and HV buffer.

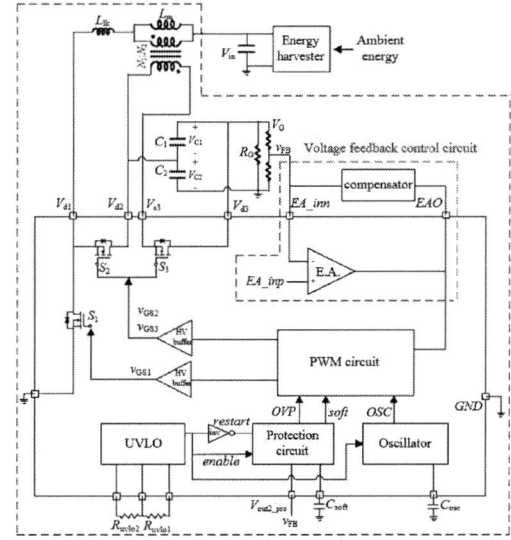

Fig. 5. Proposed control blocks for the boost-flyback converter.

(a) Protection circuit

Fig. 6 (a) shows the schematic of the protection circuit. It is composed of over voltage protection (OVP) and soft start circuit. When the signal *enable* is set to high, the current source starts to charge capacitor C_{soft}. The signal *soft* is sent to PWM circuit and compared with signal *OSC* to achieve the function of soft start. When the output voltage is higher than the voltage which is set to protect system output, the output signal of *OVP* will be low to turn off the main switch. Fig. 6 (b) shows the key waveforms of protection circuit.

3082

The 2018 International Power Electronics Conference

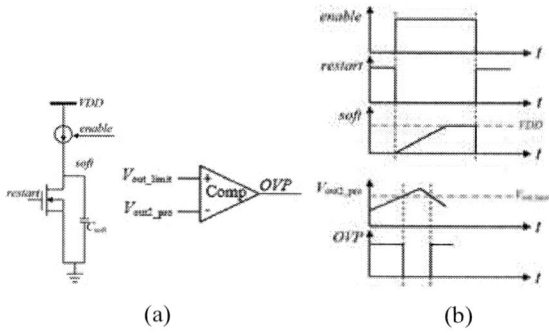

(a) (b)

Fig. 6. (a) Schematic of the protection circuit (b) the key waveforms of protection circuit.

(b) PWM circuit

Fig. 7 shows the schematic and main waveforms of the PWM circuit. This function determines the driving signals of switches. When the output voltage has not been established, the value of the control signal *EAO* is the highest. The duty cycle is determined by the signal *soft* and *OSC*. The slope of the signal *soft* is slower than the signal *OSC*, so the duty cycle increases gradually. The function of soft start can be achieved. When the voltage of the signal *soft* increases to over the voltage $V_{\text{ref\_limit}}$, this duty cycle is limited by the voltage $V_{\text{ref\_limit}}$. The Eq. (6) shows the maximum duty cycle. Finally, when the system is operated in steady state, the duty cycle is determined by the control signal *EAO* and *OSC*.

$$D_{\max} = \frac{V_{\text{ref\_limit}}}{V_{\text{H}}} \quad (6)$$

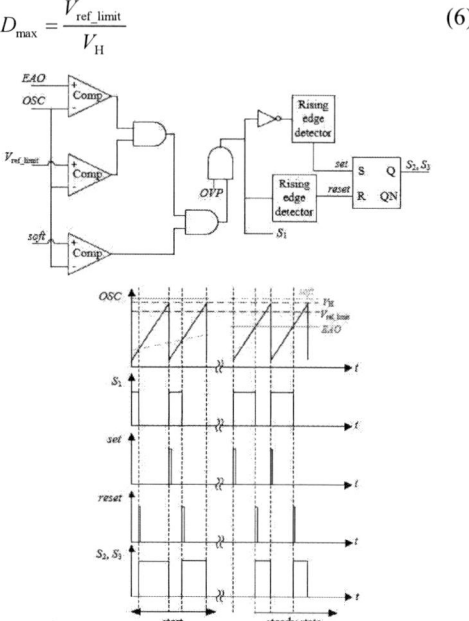

Fig. 7. Schematic and waveforms of the PWM circuit.

IV. EXPERIMENTAL RESULTS

The specifications and parameters of the system are listed in Table I and Table II. This proposed IC is tested by the boost-flyback converter to verify its feasibility.

TABLE I
SYSTEM SPECIFICATIONS

Specifications	Value	Unit
Input DC voltage (V_{in})	0.2~0.3	V
Output voltage (V_O)	5	V
Output power (P_O)	20	mW
Switching frequency (f_s)	40	kHz

TABLE II
CIRCUIT PARAMETERS

Parameters	Value	Unit
Transformer turns ratio ($N_1:N_2$)	1:6	
Magnetizing inductance (L_m)	15	μH
Output capacitor (C_1, C_2)	10	μF
Capacitor of oscillator (C_{OSC})	60	pF
Capacitor of soft start circuit (C_{soft})	2	nF
Resistance of UVLO ($R_{\text{uvlo1}}, R_{\text{uvlo2}}$)	50	kΩ

In 20 mW, input voltage 0.3 V condition, the gate signal is decided by *OSC* and v_{ero}, which is shown in Fig. 8. Although the signal has some noise when the switch turns on and turns off, the hysteresis comparator can avoid the error trigger.

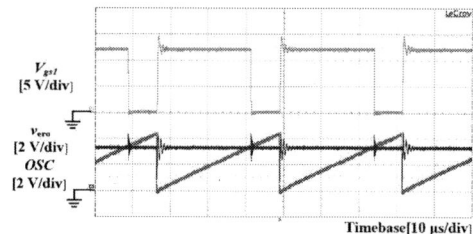

Fig. 8. Measured waveforms of the PWM circuit.

Fig. 9. and Fig. 10. show the measured waveforms of the boost-flyback converter is tested in input power 5 mW and 20 mW, and input voltage 0.2V. When the switch turned on the reversed recovery current i_{D2} cause the ringing on switch V_{ds} and input current i_{in}. When the switch turned off, the energy of leakage inductance is recycled by the current i_{D1}.

The 2018 International Power Electronics Conference

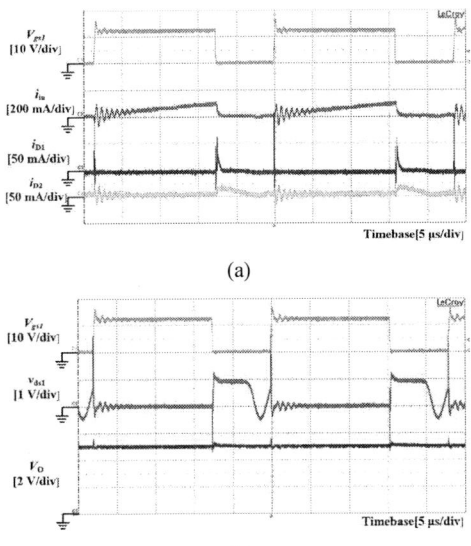

(a)

(b)

Fig. 9. Measured waveforms of (a) V_{gs1}, i_{in}, i_{D1} and i_{D2} (b) V_{gs1}, V_{ds1} and V_O at P_{in}= 5 mW and V_{in}= 0.2 V

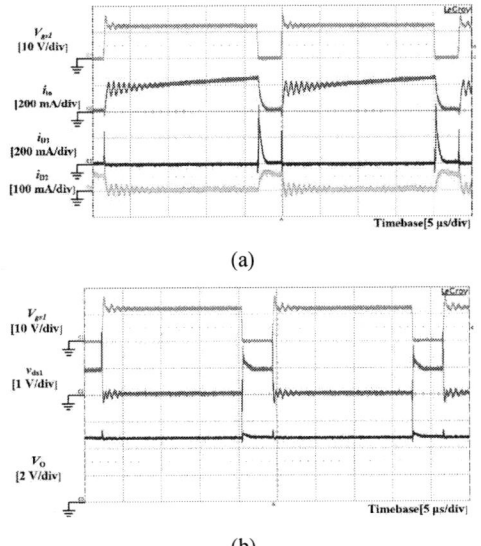

(a)

(b)

Fig. 10. Measured waveforms of (a) V_{gs1}, i_{in}, i_{D1} and i_{D2} (b) V_{gs1}, V_{ds1} and V_O at P_{in}= 20 mW and V_{in}= 0.2 V

In lower input power, the converter operated in DCM. Because the anti-parallel diode of switch also has voltage drop, it allows the V_{DS} resonates to negative.

The measured efficiency curves of the boost-flyback converter in various loads and different input voltages are shown in Fig. 11. The maximum efficiency is 78.7 % and occurs at 10 mW condition.

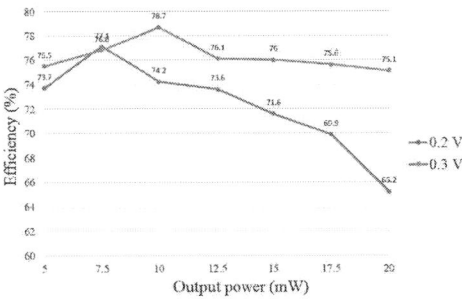

Fig. 11. Efficiency curves of the boost-flyback converter in various output power and input voltage conditions.

V. CONCLUSIONS

A control IC for boost-flyback converter for energy harvesting applications is designed and implemented. Conventionally, boost converter is not suitable for high step-up application according to the voltage gain. With this proposed IC, the boost-flyback converter can be applied in high step-up energy harvesting well. It can not only achieve high voltage gain but also recycle the energy of leakage inductance to output capacitor. So the efficiency can be improved and the voltage stress on the main switch is also reduced.

The hardware prototype is implemented to verify the function of the proposed controller IC. The input voltage is 0.2~0.3 V, the output rated power is 20 mW, and the output voltage is 5 V. The experimental results confirm the feasibility of proposed controller.

REFERENCES

[1] V. Leonov, "Thermoelectric Energy Harvesting of Human Body Heat for Wearable Sensors," *IEEE Sensor Journal*, vol. 13, no. 6, pp. 2284–2291, Jun. 2013.

[2] F. Deng, H. Qiu, J. Chen, L. Wang and B. Wang, and Bo Wang, "Wearable Thermoelectric Power Generators Combined With Flexible Supercapacitor for Low-Power Human Diagnosis Devices," *IEEE Trans. on Industrial Electronics.*, vol. 64, no. 2, pp. 1477–1485, Feb. 2017.

[3] Performance parameters-1MD04-127-15TEG", 2014 RMT

[4] P. H. Chen and P. M. Y. Fan, "An 83.4% Peak Efficiency Single-Inductor Multiple-Output Based Adaptive Gate Biasing DC-DC Converter for Thermoelectric Energy Harvesting," *IEEE Trans. on Circuit and Systems.*, vol. 62, no. 2, pp. 405–412, Feb. 2015.

[5] J. Kim, M. Shim, J. Jung, H. Kim and C. Kim, "A DC-DC Boost Converter with Variation Tolerant MPPT Technique and Efficient ZCS Circuit for Thermoelectric Energy Harvesting Applications," *2014 19th Asia and South Pacific Design Automation Conference (ASP-DAC)*, pp. 35-36, Jan 2014.

[6] Nisha K S and Mini V P, "Battery-less Boost Converter for Thermal Energy Harvesting System," *2015 International Conference on Control Communication & Computing India (ICCC)*, pp. 331-336, Nov 2015.

[7] D. Schillinger, Y. Hu, M. Amayreh, C. Moranz and Y. Manoli, "A 96.7% Efficient Boost Converter with a Stand-by Current of 420 nA for Energy Harvesting Applications," *2016 IEEE International Symposium on Circuits and*

Systems (ISCAS), pp. 654-657, May 2016.

[8] J. Goeppert and Y. Manoli, "Fully Integrated Startup at 70 mV of Boost Converters for Thermoelectric Energy Harvesting," *IEEE Journal of Solid-State Circuits*, vol. 51, no. 7, pp. 1716-1726, Jul 2016.

[9] H. Liu, F. Li and J. Ai, "A Novel High Step-Up Dual Switches Converter With Coupled Inductor and Voltage Multiplier Cell for a Renewable Energy System," *IEEE Trans. on Power Electronics.*, vol. 31, no. 7, pp. 4974-4982, Jul. 2016.

[10] K. I. Hwu and Y. T. Yau, "High Step-Up Converter Based on Coupling Inductor and Bootstrap Capacitors With Active Clamping," *IEEE Trans. on Power Electronics.*, vol. 29, no. 6, pp. 2655-2660, Jun. 2014.

[11] K. I. Hwu and Y. T. Yau, "High Step-Up Converter Based on Charge Pump and Boost Converter," *IEEE Trans on Power Electronics*, vol. 27, no. 5, pp. 2484-2660, May. 2012.

[12] K. C. Tseng and T. J. Liang, "Novel high-efficiency step-up converter," *IEE Proceedings - Electric Power Applications*, vol. 151, no. 2, pp. 182-190, Feb. 2004.

[13] K. C. Tseng and T. J. Liang, "Analysis of integrated boost-flyback step-up converter," *IEE Proceedings - Electric Power Applications*, vol. 152, no. 2, pp. 217-225, May. 2005.

[14] M. H. Cho, W. H. Lee, J. S. Kim, Y. H. Sa, H. S. Kim and H. W. Cha, "Development of Undervoltage Lockout (UVLO) Circuit Configurated Schmitt Trigger," *IEEE International SoC Design Conference (ISOCC)*, pp. 59-60, Nov. 2015.

New Concept of the DC-DC Converter Circuit Applied for the Small Capacity Uninterruptible Power Supply

Dang Minh Huynh[1*], Yoichi Ito[1], Shinji Aso[1], Koji Kato[1], Kenji Teraoka[1]
1 Power System Headquarters, Sanken Electric Co., Kawagoe city, Japan
*E-mail: dangminh@sanken-ele.co.jp

Abstract— This paper discusses about a new concept of the DC-DC converter for the small capacity Uninterruptible Power Supply (UPS) systems. In the UPS system, the DC-DC converter is used to boost battery voltage up to the inverter's input voltage when UPS is operating in the back-up mode. While the conventional small capacity UPS use the push-pull DC-DC converter which has low efficiency and high surge voltage, the new concept which uses the CLC Resonant Type Full-Bridge DC-DC converter is proposed. In this concept, the surge voltage of the main switches and the second-side rectifier diodes is reduced, which helps to improve the efficiency and allows using the low-voltage devices. In the experiment, the efficiency of the DC-AC conversion system which uses the CLC resonant type full-bridge DC-DC converter is measured. As a result, the max efficiency is about 90%, so the usability of the proposed CLC resonant type full-bridge DC-DC converter can be confirmed.

Keywords— *UPS, full-bridge DC-DC converter, snubber circuit, surge voltage.*

I. INTRODUCTION

In recent year, due to the rapid development of the information society and IoT, the demand of Uninterruptible Power Supply (UPS) has been increasing, especially the demand of the UPS applied for the monitoring cameras and telecommunication equipment. UPS is an equipment which ensures the stability of the power supply and avoids data loss due to power failure, has been applied in many data system all over the world [1-2]. At the class of UPS applied for the monitoring cameras and telecommunication equipment, because the load devices is often consumes just tens of watt, the capacity of the UPS of this class is small, about 1kVA. Therefore, the demand of high efficiency at the light load is higher than the other UPS class. Along with this, the miniaturization and low cost are also required.

Up to now, because of the low-cost and simple circuit construction, the push-pull type DC-DC converter circuit is widely used in the class of small capacity UPS [3]. However, recently, because of the requirement of miniaturizing the system, long back-up time and cost saving, an UPS system without cooling fan is considered [4]. Nevertheless, the problem is the push-pull DC-DC

converter has a low efficiency at the light load [5]. Moreover, because the surge voltage of the second-side rectifier diode is not suppressed, so it decreases the efficiency of the DC-DC converter. Therefore, the conventional DC-DC converter is not suitable to use in a natural air-cooling UPS system because it will cause system over heat when backing up in a long time. Hence, in order to satisfy the requirement of the natural air-cooling UPS system, the efficiency of the DC-DC converter need to be improved.

In this paper, in order to solve the above problems, a CLC resonant type full-bridge DC-DC converter is proposed. First, by using the full-bridge DC-DC converter type, the surge voltage of the main switches is reduced when comparing with the conventional push-pull type. By this, it will allow to use the low-voltage FET as the main switches of the DC-DC converter circuit. By reducing the surge voltage of the FET, the switching loss is reduced, and it will improve the efficiency at the light load, where the switching loss is dominant [6].

Second, in the full-bridge DC-DC converter circuit, the conventional problem is the high surge voltage occurring on the second-side rectifier diodes [7-8]. In order to solve this problem, the snubber circuit for the second-side rectifier diodes is optimized. By this, the surge voltage of the rectifier diodes is suppressed more effectively. As a result, the efficiency of the DC-DC converter is improved considerably. Therefore, the volume of the heat sink can be reduced, so it can help to cost-down and minimize the system. Furthermore, in order to optimize the snubber circuit of the second-side rectifier diodes, a capacitor is connected serially with the primary side of the transformer. Because the capacitor can allow only the AC current to flow through, it also help to prevent the biased magnetization of the transformer and limit the current flowing to the transformer. By that, the battery will be kept in safety when the inverter is broken down.

Finally, the efficiency of the CLC resonant Type full-bridge DC-DC converter concept is measured in the experiment. As a result, comparing with the conventional DC-AC conversion system, the efficiency of the system which uses the full-bridge DC-DC converter is higher by maximum 11 points at the light load. Based on this result, it can be confirmed that the UPS system operates at high efficiency when using the CLC resonant type full-bridge DC-DC converter.

II. CIRCUIT CONFIGURATION AND OPERATION MODE OF CLC RESONANT TYPE FULL-BRIDGE DC-DC CONVERTER

A. UPS circuit configuration

Fig.1 shows the circuit configuration of the small capacity UPS. Because the monitoring camera and telecommunication equipments can allow the momentary stop, the standby type UPS is used in this case. In the standby-type UPS, the output is connected directly with the input power supply at the normal operation mode. When the power failure occurs, the delay will switch from a-c contact to b-c contact, and the UPS will operate at the backup mode. In the backup mode, first, the battery voltage 48V will be boosted up to the DC/AC inverter input voltage 180V by the DC/DC converter. After that, the inverter will do the DC/AC conversion to keep the output voltage of the UPS. Because the power conversion at the normal mode is just about to charge the battery, it can be considered that there is no power loss at the normal mode. On the other hand, at the backup mode, there are two steps of the power conversion, so the power loss is considerable. Therefore, in order to improve the efficiency of the UPS, the efficiency improvement at the backup mode is needed. In order to insure the reliability of UPS and the inverter efficiency is about 95%, the efficiency improvement of the inverter without using ZVS is difficult [9]. Therefore, it is needed to improve the efficiency of the DC/DC converter.

Fig.2 shows the conventional push-pull type DC-DC converter circuit. In conventional circuit, the surge voltage of the main switches S_1, S_2 is two times of the input voltage. Therefore, it makes the switching loss become high and decreases the efficiency at the light load. In addition, the snubber circuit including C_{s2}, D_{s1}, D_{s2} is used in order to suppress the surge voltage of the rectifier diodes $D_1 \sim D_4$. However, the snubber circuit design is not optimized, so it does not suppress the surge voltage of the

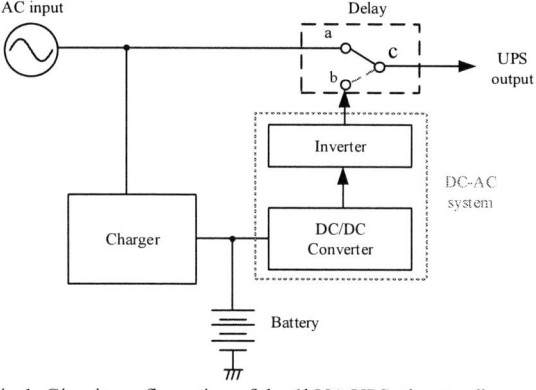

Fig.1. Circuit configuration of the 1kVA UPS, the standby type UPS is used.

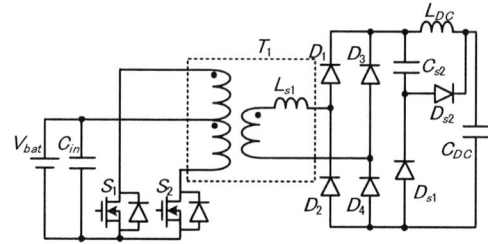

Fig.2. Conventional push-pull DC-DC converter circuit, the capacitor C_{s2} and the leakage inductance L_{s1} are the resonant components.

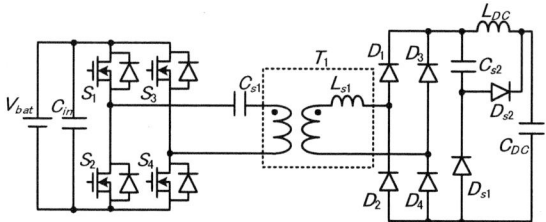

Fig.3. CLC Resonant Type Full-Bridge DC-DC converter circuit, the capacitor C_{s1}, C_{s2} and the leakage inductance L_{s1} are the resonant components.

diodes effectively.

B. Operation switch mode of the CLC resonant type full-bridge DC-DC converter

Fig.3 shows the CLC Resonant Type Full-Bridge DC-DC converter circuit. In this circuit, the snubber circuit including C_{s2}, D_{s1}, D_{s2} is also used to suppress the surge voltage of the rectifier diode $D_1 \sim D_4$. On the other hand, the capacitor C_{s1} is connected serially with the primary side of the transformer in order to adjust the resonant current between the leakage inductance of the transformer and the capacitor C_{s2}.

Fig.4 shows the operation mode of the CLC Resonant Type Full-Bridge DC-DC converter. There are 4 switch modes in 1 carrier period. The current flow in each switch mode is as below.

※Mode 1(S_1, S_4: ON ; S_2, S_3: OFF)

(Primary side) Battery $\rightarrow$ S_1 $\rightarrow$ C_{s1} $\rightarrow$ T_1 $\rightarrow$ S_4 $\rightarrow$ Battery

(Second side) T_1 $\rightarrow$ L_{s1} $\rightarrow$ D_1 $\rightarrow$ L_{DC} // (C_{s2} $\rightarrow$ D_{s2}) $\rightarrow$ C_{DC} $\rightarrow$ D_4 $\rightarrow$ T_1

In mode 1, the drain current of switch S1, S4 is combined from the current which flow through the transformer T_1 and the resonant current between the leakage inductance of the transformer L_{s1} and the total capacity of C_{s1} and C_{s2}.

※Mode 2(S_1, S_2, S_3, S_4: all switches OFF)

(Second side) L_{DC} $\rightarrow$ C_{DC} $\rightarrow$ (D_{s1} $\rightarrow$ C_{s2}) $\rightarrow$ L_{DC}

In mode 2, all switches are OFF, and the reflux current flows through the second-side circuit. In order to suppress the surge voltage of the diodes $D_1 \sim D_4$, it is needed to suppress the reflux current flowing through the diodes in this period. From the current flow, it is

3087

understood that the rectifier diodes $D_1 \sim D_4$ is connected with the snubber circuit C_{s2}, D_{s1} in parallel. Therefore, in order to suppress the reflux current flowing through the rectifier diodes $D_1 \sim D_4$, the voltage of the snubber C_{s2} is hold at non-zero voltage value. It is simple to execute that by increase the capacity of the C_{s2}. However, when increasing the capacity C_{s2} too much, the period of the resonant current will become bigger than the ON period of the switches $S_1 \sim S_4$. In this situation, the peak drain current of the switches $S_1 \sim S_4$ when turn off will become higher, and it will increase the switching loss and noise of the circuit.

In the CLC resonant type full-bridge DC-DC converter, in order to suppress the surge voltage of the rectifier diodes $D_1 \sim D_4$ without increasing the surge voltage of the switches $S_1 \sim S_4$ when turn off, a resonant capacitor C_{s1} is added on the primary side of the transformer T_1 serially. Since the resonant capacitor C_{s1} is connected serially with the leakage inductance and the snubber C_{s2} in the equivalent circuit, the total resonant capacity can be adjusted by changing the capacity of C_{s1}. Therefore, it is possible to increase the capacity of C_{s2} but the resonant frequency is not changed by adjusting the capacity of C_{s1}.

In addition, the resonant frequency f_{rs} between L_{s1} and C_{s1}, C_{s2} can be calculated as (1).

$$ f_{rs} = \frac{1}{2\pi} \sqrt{\frac{C_{s1} + C_{s2}}{C_{s1} \times C_{s2} \times L_{s1}}} \qquad (1) $$

In actual, the resonant frequency f_{rs} is set at 1.5~5 times of the switching frequency, due to the stability of the system.

　※Mode 3(S_2, S_3: ON ; S_1, S_4: OFF)
　(Primary side) Battery $\rightarrow S_3 \rightarrow T_1 \rightarrow C_{s1} \rightarrow S_2 \rightarrow$ Battery
　(Second side) $T_1 \rightarrow D_3 \rightarrow L_{DC} \, // \, (C_{s2} \rightarrow D_{s2}) \rightarrow C_{DC}$

$\rightarrow D_2 \rightarrow L_{s1} \rightarrow T_1$
　In mode 3, the current flow is similar as in mode 1.
　※Mode 4(S_1, S_2,S_3, S_4: all switches OFF)
　(Second side) $L_{DC} \rightarrow C_{DC} \rightarrow (D_{s1} \rightarrow C_{s2}) \rightarrow L_{DC}$
　In mode 4, the current flow is similar as in mode 2.

C. Simulation result

Fig.5 shows the simulation system while table 1 shows the simulation condition which is used to confirm the surge voltage suppression effect when optimizing the capacity of the snubber C_{s2}. In the simulation circuit, a voltage control circuit is used in order to boost up the small input voltage 50 V to higher output voltage 180 V. As the control mechanism, the output voltage is divided by the resistance R_{det1}, R_{det2}, R_{det3}, and the voltage drop v_{outdet} of the resistance R_{det3} is feedback to the control IC in order to execute the voltage control. In addition, the current control is not executed because in actual, it is removed in order to simplify the system and saving cost.

Fig.6 shows the simulation results when changing the capacity of the snubber C_{s2}. In fig.6(a), the capacity of the snubber $C_{s2} = 0.01$uF as the conventional value, while in fig.6(b), the capacity of the snubber C_{s2} is set to 0.47uF. The other parameter is equal as shown in table 1.

Comparing fig.6(a) with fig.6(b), it can be seen that the surge voltage of the diode D_1 is reduced when increasing the capacity of the snubber C_{s2}. When C_{s2}=0.01 uF, the surge voltage of D_1 is 359V, while C_{s2}=0.47uF, the surge voltage of D_1 is reduced by 22%, to 280V. In addition, the resonant period is about 75 kHz, it is equal 1.5 times of the carrier frequency. It is matched with the calculation result.

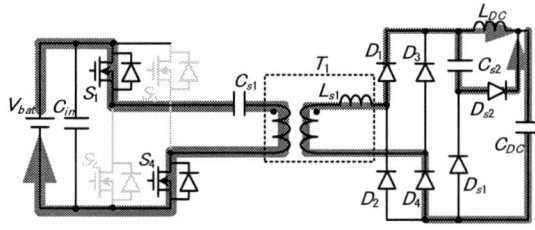

(a) Mode 1: S_1, S_4 ON

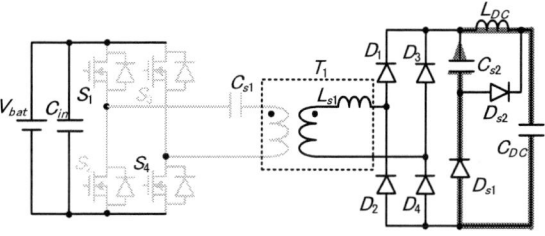

(b) Mode 2: All switches OFF

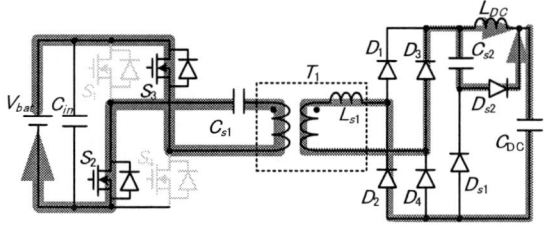

(c) Mode 3: S_2, S_3 ON

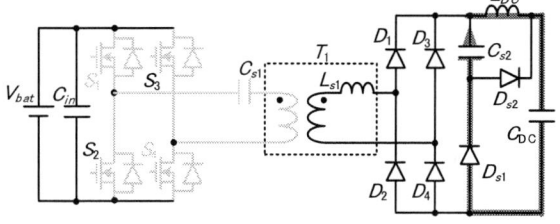

(d) Mode 4: All switches OFF

Fig.4. Operation switch mode of the CLC Resonant Type Full-Bridge DC-DC converter circuit, there are 4 switch modes totally.

The 2018 International Power Electronics Conference

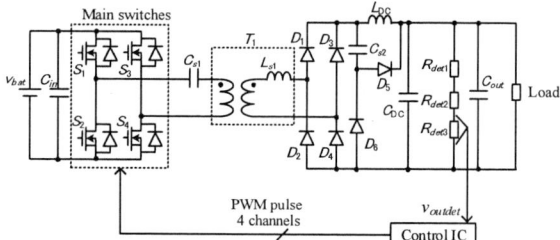

Fig.5. Simulation circuit of the full-bridge DC-DC converter, a simple voltage control circuit is used.

Table 1. Simulation condition

Input voltage	50 V
Output voltage	180 V
Output power	500 W
Output power	500 W
Snubber C_{s1}	7 uF
Switching frequency	50 kHz

III. EXPERIMENTAL RESULT

A. Surge voltage suppression confirmed result

In order to confirm the surge voltage suppression effect when optimizing the capacity of the snubber C_{s2}, an experiment is done in the same condition with the simulation. Fig.7 shows the experimental results in 2 circumstances C_{s2}=0.01 uF and C_{s2}=0.47 uF. Comparing fig.6 to fig.7, it can be seen that the experimental results is matched fairly with the simulation results. Comparing fig.7(a) and fig.7(b), it can be understood that the on duty when C_{s2}=0.47 uF is smaller than that of when C_{s2}=0.01uF. It is because in the resonant period of C_{s2} and L_{s1}, the energy goes directly from transformer's first side to second side without flowing in the transformer core. Therefore, when the resonant energy becomes

bigger, the energy which flows in the transformer core becomes smaller, and the on duty of main switches is also reduced. In addition, in fig.7(a), the surge voltage of the diode D_1 is 420 V, while in fig.7(b), the surge voltage is just about 310 V. By this result, the surge voltage suppression effect when optimizing the capacity of the snubber C_{s2} can be confirmed.

B. Efficiency improvement effect when applying the CLC resonant type full-bridge DC-DC converter into an offline UPS system

In this section, the efficiency improvement effect when applying the CLC resonant type full-bridge DC-DC converter into the offline UPS systems is considered. The efficiency of DC-AC system when using the conventional DC-DC converter circuit and the CLC resonant type full-bridge DC-DC converter will be compared in order to verify the superiority of the CLC resonant type full-bridge DC-DC converter.

Fig.8 shows the conventional DC-AC converter system of the offline UPS, which uses a push-pull DC-DC converter and a single phase full-bridge inverter for converting the battery voltage to the AC output voltage. On the order hand, fig.9 shows the DC-AC converter system of the offline UPS, which use the CLC resonant type full-bridge DC-DC converter.

As the experimental condition, the input voltage is changed from 43V to 54V, corresponding to the actual battery voltage. In addition, the DC-AC converter system output is connected with a resistance load. The total efficiency of both DC-AC converter systems is measured and compared in order to confirm the high efficiency operation of the system which uses the CLC resonant type full-bridge DC-DC converter.

Fig.10 shows the efficiency of the conventional DC-AC conversion system and the DC-AC conversion system which uses the CLC resonant type full-bridge

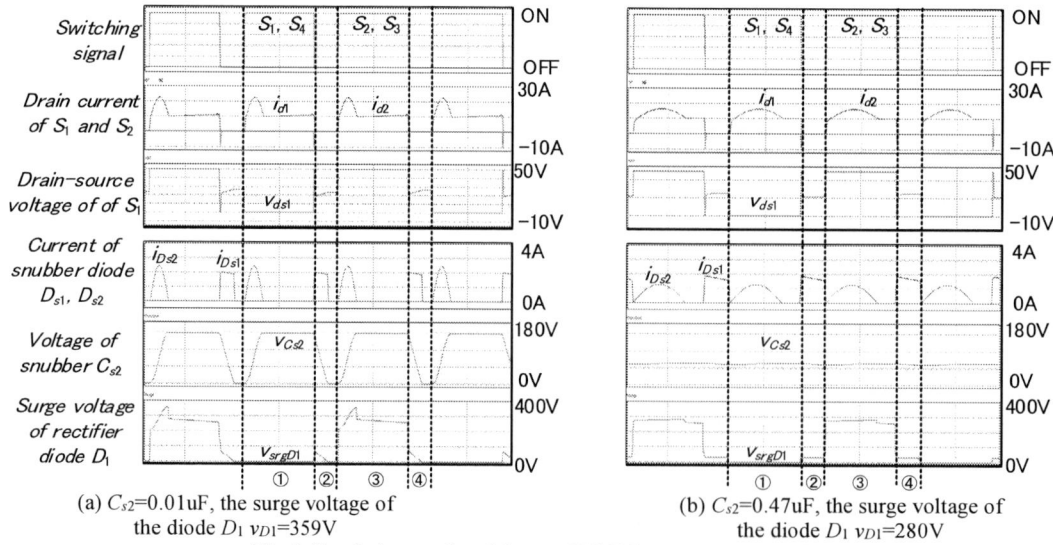

(a) C_{s2}=0.01uF, the surge voltage of the diode D_1 v_{D1}=359V

(b) C_{s2}=0.47uF, the surge voltage of the diode D_1 v_{D1}=280V

Fig.6. Simulation results of the new DC-DC converter concept, the surge voltage of the diodes D_1 is reduced when increasing the capacity of C_{s2}.

3089

The 2018 International Power Electronics Conference

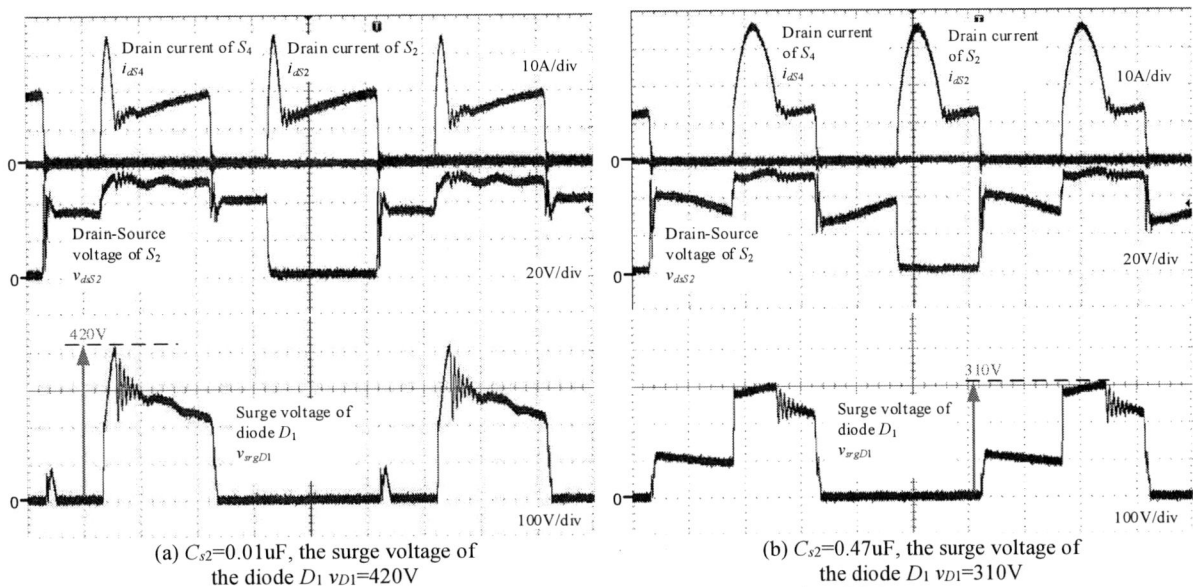

(a) C_{s2}=0.01uF, the surge voltage of the diode D_1 v_{D1}=420V

(b) C_{s2}=0.47uF, the surge voltage of the diode D_1 v_{D1}=310V

Fig.7. Experimental results when changing the capacity of Cs2, it can be confirmed that the surge voltage of the diodes D_1 is reduced when optimizing the capacity of C_{s2}.

DC-DC converter. From fig.10, it can be understood that the max efficiency of the system using the CLC resonant type full-bridge DC-DC converter is about 90%. Due to this result, the high efficiency operation of the system using the CLC resonance type full-bridge DC-DC converter is confirmed. Moreover, comparing the efficiency of the DC-AC conversion system using the CLC resonant type full-bridge DC-DC converter with the conventional system, it can be seen that the efficiency of the system using the CLC resonant type full-bridge DC-DC converter is higher. Especially at the light load below 200 W, the efficiency of the system using the CLC resonant type full-bridge DC-DC converter is higher than that of the conventional system by maximum 11%. It is because the switching loss is reduced significantly when changing from the push-pull DC-DC converter to the CLC resonant type full-bridge DC-DC converter. The loss reduction at the light load is an important improvement for the class of small capacity UPS applied for the low power consumption devices such as monitoring cameras and telecommunication equipment.

Fig.11 shows the actual picture of the UPS unit which uses the CLC resonant type full-bridge DC-DC converter while table 2 shows the product specification. From fig.11, it can be seen that the cooling fan is not used in this UPS system. It is because that from the efficiency improvement, it can be allowed to reduce the large-size circuit parts like the heat-sink or fan.

IV. CONCLUSION

In this paper, a new concept of the DC-DC converter for the small capacity UPS is discussed. By changing the type of the DC-DC converter and optimizing the design of the snubber circuit, the high efficiency DC-DC converter can be achieved. In addition, the efficiency of

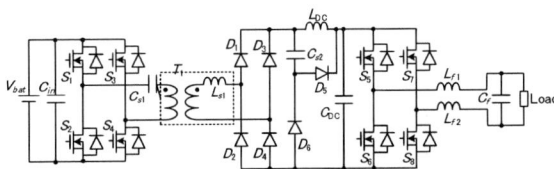

Fig.8. Conventional DC-AC system, which uses the push-pull type DC-DC converter

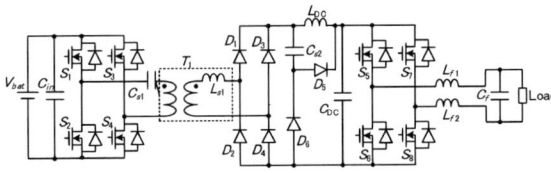

Fig.9. New DC-AC system, which uses the full-bridge type DC-DC converter and optimized snubber circuit

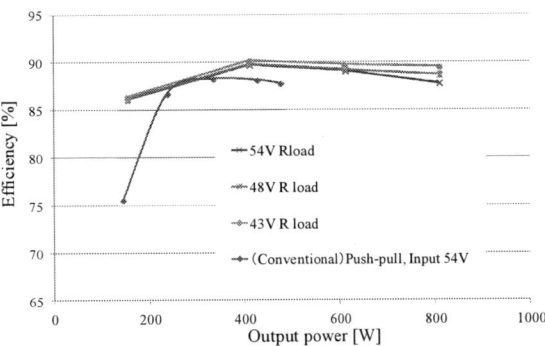

Fig.10. Comparison result between the efficiency of the conventional DC-AC system and new DC-AC system. Comparing with the conventional system, the efficiency of the new system is higher by maximum 11 points at 180W output power.

3090

the DC-AC converter system which uses the CLC resonant type full-bridge DC-DC converter is measured and compared with that of the conventional DC-AC converter system. As a result, it can be confirmed that the system using the CLC resonant type full-bridge DC-DC converter has the higher efficiency, and the efficiency at the light load below 200 W is improved significantly by maximum 11 points when comparing with that of the conventional system. Therefore, the CLC resonant type full-bridge DC-DC converter can satisfies the demand of high efficiency in the natural air-cooling small capacity UPS systems. In the future, in order to achieve a higher efficiency, the other circuit path design will be optimized.

Table 2. Product specification

UPS type		Offline UPS
AC Input	Voltage	90V ~ 110V
	Current	12A
Inverter output	Rated power	800W/1kVA
	Voltage	100V
	Voltage accuracy	Within 10%
Mode-switching time		Less than 10ms
Battery		Lithium-ion battery(*)

(*) Possible to back-up in 3 hours - 418W

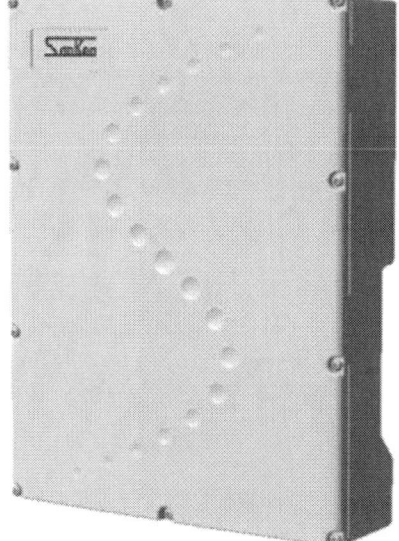

Fig.11. Picture of UPS unit which uses the CLC resonant type full-bridge DC-DC converter, size W300 mm ×D84 mm ×H390 mm

REFERENCES

[1] E. Demirkutlu, S. Cetinkaya, and A. M. Hava, "Output Voltage Control of A Four-Leg Inverter Based Three-Phase UPS by Means of Stationary Frame Resonant Filter Banks", *IEMDC 2007*, pp.880-885, 2007.

[2] H. Li, W. Zhang, and D. Xu, "High-Reliability Long-Backup-time Super UPS with Multiple Energy Sources", *ECCE 2013*, pp. 4926 - 4933, 2013.

[3] Sagar Khare, "Offline UPS Reference Design Using the dsPIC®DSC", http://ww1.microchip.com/downloads/en/AppNotes/01279 B.pdf

[4] T. Izumi, et al., "Development of fanless high efficiency 30 kVA SiC-UPS", 2016 19th International Conference on Electrical Machines and Systems (ICEMS), p.1 – 4, 2016

[5] S. Lee, J. Park, and S. Choi, "A three-phase current-fed push-pull DC-DC converter with active clamp for fuel cell applications", *APEC 2010*, pp. 1934 - 1941, 2010.

[6] K. Shirakawa, etc al., "Method for Reducing Surge Voltage and Turn-on Loss in Push-pull DC-DC Converters", *IEEJ Transactions on Industry Applications*, vol. 137, issue 2, pp. 119-128, 2017 (in Japanese).

[7] K. Domoto, Y. Ishizuka, S. Abe, T. Ninomiya, "A Novel Technique for Control Characteristics Improvement of Full-Bridge Converter with Snubber Capacitor" *IEEJ Journal of Industry Applications*, vol. 5, no. 2 p. 184-190, 2016.

[8] M. Cacciato, A. Consoli, "New Regenerative Active Snubber Circuit for ZVS Phase Shift Full Bridge Converter", *APEC 2011*, pp.1507-1511, 2011.

[9] Y. Chen, G. Hu, C. Du and D. Xu, "A soft-switching full-bridge inverter with high efficiency", *ECCE 2013*, pp. 2737 - 2744, 2013.

The 2018 International Power Electronics Conference

Comparative Study on the Performance of Dual-phase Tapped-inductor Boost Converter and Interleaved Boost Parallel-input Series-output Converter in 40 to 400V Applications

Niño Christopher Ramos[1,2] and Tsuyoshi Funaki[2*]

1 Electrical and Electronics Engineering Institute, University of the Philippines-Diliman, Quezon City, Philippines
2 Division of Electrical, Electronic and Information Engineering, Osaka University, Osaka, Japan
*E-mail: funaki@eei.eng.osaka-u.ac.jp

Abstract— **A comparative study on the efficiency, input current ripple, output voltage ripple and component temperatures for the dual-phase tapped-inductor boost converter (DTBC) and interleaved boost parallel-input series-output converter (IBPSC) is presented. Designs for both converters operating in continuous conduction mode (CCM) with DC-DC voltage conversion from 40V to 400V at 500W rated output power are also presented. The same circuit components are used for both topologies and a multimode board capable of implementing both configurations is used to maintain positions of components and minimize changes in PCB layout.**

Keywords— *interleaved, multiphase, tapped-inductor, voltage lift technique*

I. INTRODUCTION

The boost converter is a basic DC-DC converter topology for stepping up the input voltage to a higher output voltage. However, the boost converter is not suitable for very high voltage gain requirement because of excessive power loss caused by operating at very high duty cycles. Various converter topologies that overcome the limitations of the boost converter without incurring significant power loss were proposed[8-12]. The challenge is to come up with simpler topologies depending on the voltage and power requirement of the target application. This research focuses on the design and analysis of the dual-phase tapped-inductor boost converter(DTBC) and interleaved boost converter parallel-input series output(IBPSC) and their performance in 40V to 400V applications such as grid-tied photovoltaic (PV) systems shown in figure 1.

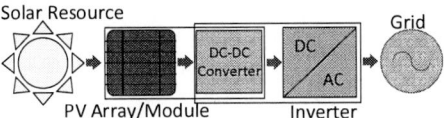

Fig. 1. Solar PV System with DC-DC Converter Module

The converter can function as an intermediate step-up converter between the PV array and the central inverter or become directly integrated within the inverter unit. The converter can also be implemented as a module level power electronics (MLPE)[11] and be paired with a single PV module since PV manufacturers are already producing 96-cell solar panels with up to 58V of maximum power point voltage at standard test conditions [25]. The main objectives for the selection of ITBC and IBCPS as candidate topologies for comparison are:

- To determine the suitability of a topology that relies primarily on the coupled-inductor/transformer (ITBC) and another one that utilizes stacked capacitors (IBPSC) to provide high voltage gain and output. Setting a target application of 40V to 400V, with 500W output power provides a set of specifications where the design and implementation of both topologies are possible.
- The DTBC [1,2,19-20] and IBPSC [3,21] are generally simpler in terms of circuit topology, control and component count over their respective tapped-inductor and charge-pump derived topological counterparts [13-18] and both contain an interleaving/multiphase feature for voltage and current ripple reduction and potential for higher output power production.

II. PRINCIPLE OF OPERATION

A. DTBC

Figure 2 shows the circuit diagram of the DTBC.

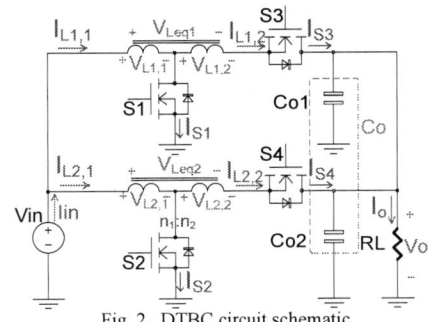

Fig. 2. DTBC circuit schematic

3092

The 2018 International Power Electronics Conference

The circuit is essentially two paralleled tapped-inductor boost converters. However, the corresponding switching signals of the individual modules have a 180 degrees phase-shift (half-period in time domain) relative to each other. The circuit is divided into 4 sub-circuit states as shown in figure 3, and the operation is denoted as sequences of the sub-circuit states in figure 4.

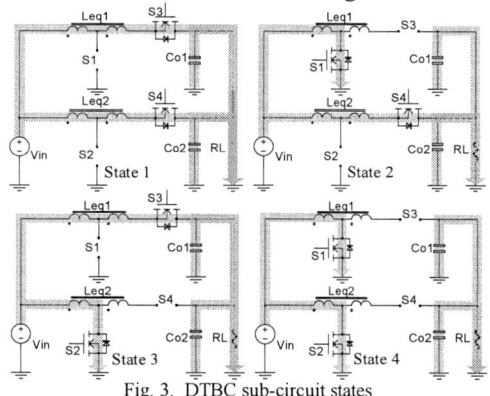

Fig. 3. DTBC sub-circuit states

Figure 4a shows the inductor voltage and current waveforms when the duty cycle (D) is less than 0.5. The switching sequence involves states 1 to 3.

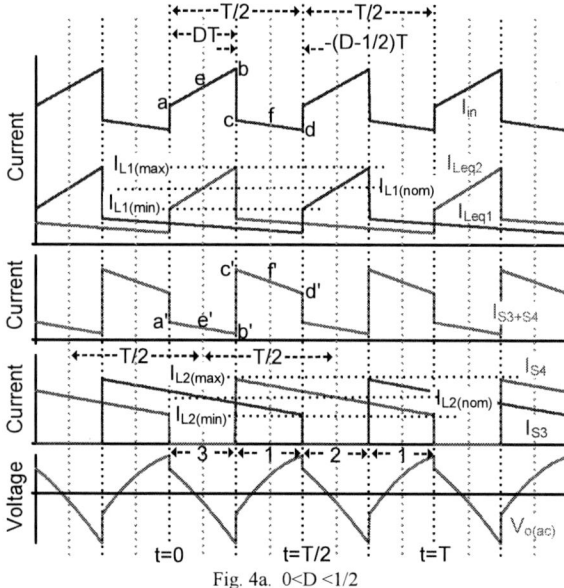

Fig. 4a. 0<D <1/2

For D > 0.5 in figure 4b, the sequence involves states 2 to 4. The shape of the individual inductor current ripple is due to the operation of the tapped-inductor wherein current is transformed from high to low value as power flows from the primary to the secondary side. Conversely, voltage is transformed from low to high value. Higher voltage gain is achieved by adjusting the turns-ratio(**n**) instead of increasing the duty cycle.

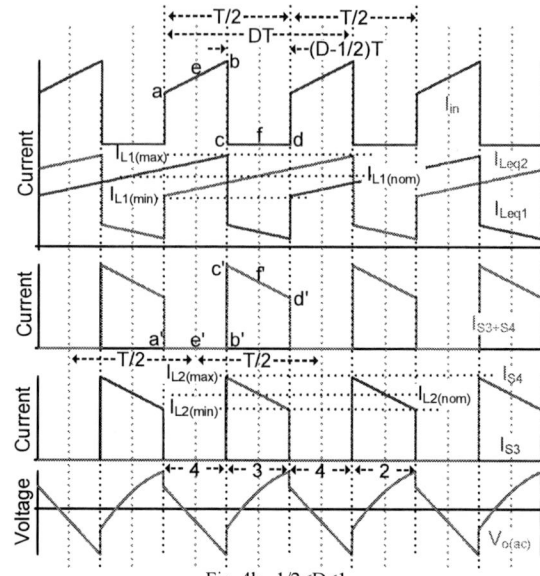

Fig. 4b. 1/2<D<1

Fig. 4. DTBC Current and Voltage Waveforms

For the DTBC, the following relationships are applicable:

$$\frac{V_{Leq1}}{n_1+n_2} = \frac{V_{L1,1}}{n_1} = \frac{V_{L2,1}}{n_2} \tag{1}$$

$$n_1 I_{L1,1} = (n_1 + n_2)I_{L1,2} \rightarrow I_{L1,1} = (1 + n)I_{L1,2} \tag{2}$$

$$\frac{L_{1,1}}{(n_1)^2} = \frac{L_{1,2}}{(n_2)^2} = \frac{L_{eq1}}{(n_1+n_2)^2} \tag{3}$$

where $n = \frac{n_2}{n_1}$.

The individual phases retain their characteristics while working together, therefore the expressions in [1,4] still apply. Since the output of the two phases are in parallel, the ratio of the average output and input voltage can still be expressed as

$$\frac{V_o}{V_{in}} = \frac{1+nD}{1-D} \tag{4}$$

The interleaved operation produces the input current and output voltage waveforms as shown in figure 3. The sum of the high side switch current (I_{S3+S4}) minus the average output current is the effective output capacitor ripple current ($I_{Co}= I_{Co1}+ I_{Co2}$). The corners and midpoints of the input current and the sum of the high-side switch currents are labeled with points **a** to **f** and **a'** to **f'** respectively in figure 4. They are

(For 0<D <1/2)

$$a = I_{L1,min} + I_{L2,min} + \left(\frac{V_o-V_{in}}{L_{eq}}\right)\frac{T}{2}$$

$$a' = I_{L2,min} + \left(\frac{V_o-V_{in}}{L_{eq}}\right)\frac{T}{2}$$

$$b = I_{L1,max} + I_{L2,max} - \left(\frac{V_o-V_{in}}{L_{eq}}\right)\frac{T}{2}$$

$$b' = I_{L2,max} - \left(\frac{V_o-V_{in}}{L_{eq}}\right)\frac{T}{2}$$

$$c = I_{L1,max} + I_{L2,max} - \left(\frac{V_o-V_{in}}{L_{eq}}\right)\frac{T}{2}; c' = I_{L1,max} + b'$$

$$e = \frac{a+b}{2} = I_{L1,nom} + I_{L2,nom}; e' = I_{L2,nom}$$

$$d = I_{L1,min} + I_{L2,min} + \left(\frac{V_o-V_{in}}{L_{eq}}\right)\frac{T}{2}; d' = I_{L1,min} + a'$$

$$f = \frac{c+d}{2} = 2I_{L2,nom}; f' = 2I_{L2,nom}$$

(For 1/2<D <1)

$$a = 2I_{L1,min} + \left(\frac{V_{in}}{L_1}\right)\frac{T}{2}; a' = 0$$

$$b = 2I_{L1,max} - \left(\frac{V_{in}}{L_1}\right)\frac{T}{2}; b' = 0$$

$$e = 2I_{L1,nom}; e' = 0$$

$$c = I_{L1,max} + I_{L2,max} - \left(\frac{V_{in}}{L_1}\right)\frac{T}{2}; c' = I_{L2,max}$$

$$d = I_{L1,min} + I_{L2,min} + \left(\frac{V_{in}}{L_1}\right)\frac{T}{2}; d' = I_{L2,min}$$

$$f = I_{L1,nom} + I_{L2,nom}; f' = I_{L2,nom}$$

where:

$$I_{L1,nom} = \frac{1+n}{1-D}I_o; I_{L2,nom} = \frac{I_{L1,nom}}{1+n}$$

$$\Delta I_{L1,pkpk} = \frac{V_{in}}{L_1}DT; \Delta I_{L2,pkpk} = \left(\frac{V_o-V_{in}}{L_{eq}}\right)DT$$

$$I_{L1,min} = I_{L1,nom} - \frac{\Delta I_{L1,pkpk}}{2}$$

$$I_{L1,max} = I_{L1,nom} + \frac{\Delta I_{L1,pkpk}}{2}$$

$$I_{L2,min} = I_{L2,nom} - \frac{\Delta I_{L2,pkpk}}{2}$$

$$I_{L2,max} = I_{L2,nom} + \frac{\Delta I_{L2,pkpk}}{2}$$

The calculated values are used for capacitor sizing and loss estimation.

B. IBPSC

Figure 5 shows the circuit diagram of the IBPSC. The converter is typically operated at D>0.5 [3].

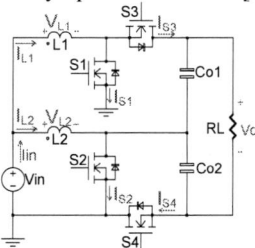

Fig. 5. IBPSC circuit schematic

The converter has 3 sub-circuit states shown in figure 6.

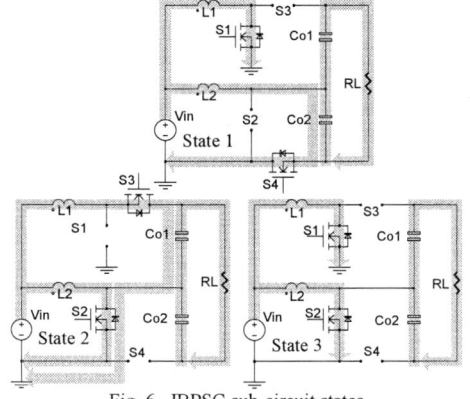

Fig. 6. IBPSC sub-circuit states

The periodic steady-state waveforms in figure 7 shows that the switching sequence creates a drain-source voltage across S3 (V_{S3}) that is twice the voltage of S1, S2 and S4. The switching sequence and stacked configuration of C_{o1} and C_{o2} provides a doubled voltage gain compared to the conventional boost converter.

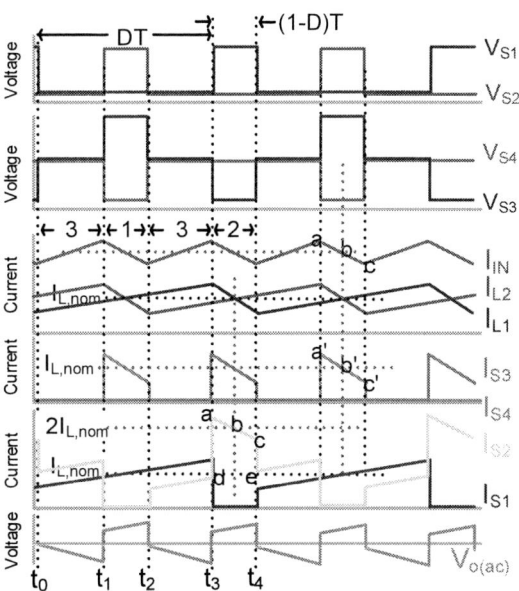

Fig. 7. IBPSC Current and Voltage Waveforms at 1/2<D<1

The voltage gain derived in [3] is given by:

$$\frac{V_o}{V_{in}} = \frac{2}{1-D} \qquad (5)$$

Similar to the analysis of the DTBC, the input current and the sum of the high side switch currents are labeled as points **a** to **c** and **a'** to **c'** respectively in figure 7. They are

$$a = b + \frac{V_{in}(2D-1)T}{2L}; a' = b' + \frac{V_{in}DT}{2L}$$

$$b = 2I_{L,nom}; b' = I_{L,nom}; I_{L,nom} = \frac{I_o}{1-D}$$

$$c = b - \frac{V_{in}(2D-1)T}{2L}; c' = b' - \frac{V_{in}DT}{2L}$$

$$d = I_{L,nom} - \frac{V_{in}(1-D)T}{2L}; e = I_{L,nom} + \frac{V_{in}(1-D)T}{2L}$$

The generated points are used to determine S_2 RMS current in addition to calculation of output capacitor ratings.

III. DESIGN AND IMPLEMENTATION

The respective phases in DTBC is designed to have half of the rated power therefore

$$P_{o,ph} = \frac{P_{o,sys}}{2}; I_{o,ph} = \frac{P_{o,ph}}{V_o}$$

A. INDUCTOR

The inductor design calculations were adapted from [4, 23] and modified accordingly. At the boundary between CCM and DCM, the current at the secondary inductance is given by eq. (6).

$$I_{L2nom,crit} = \frac{\Delta I_{L2,pkpk}}{2} = \frac{I_{o,min}}{1-D} = \frac{I_{ratio} \cdot I_{o,ph}}{1-D} \qquad (6)$$

where I_{ratio} is the ratio of allowable inductor ripple current to rated output current per phase ($I_{o,ph}$). $\Delta I_{L2,pkpk}/2$ which is equal to ($I_{L2,max}$ - $I_{L2,min}$)/2 in figure 4 can be further expressed as

$$\frac{\Delta I_{L2,pkpk}}{2} = \frac{\Delta I_{L1,pkpk}}{2(1+n)} = \frac{V_{in}DT}{2(1+n)L_1} \qquad (7)$$

Eqs. (6) and (7) can be combined to get the minimum required inductance for CCM operation as shown in eq. (8)

$$L_1 = \frac{V_{in}D(1-D)}{2(1+n)I_{ratio} \cdot I_{o,ph} \cdot f}; \; L_2 = n^2 L_1 \quad (8)$$

The design equations for the power components in the IBPS, were adapted from [3] and modified accordingly. At the boundary between CCM and DCM, the nominal inductor current is given by eq. (9)

$$I_{Lnom,crit} = I_{in,ph} = \frac{\Delta I_{L,pkpk}}{2} = \frac{V_{in}DT}{2L} \quad (9)$$

The minimum inductance required for CCM operation can therefore also be obtained from eq. (8) with **n** set to zero and L_1 changed to L.

B. ACTIVE SWITCHES

The maximum voltage across the low-side switch in the DTBC is given by

$$V_{s1max} = V_{s2max} = V_{ds,ls,max} = \frac{nV_{in}+V_o}{n+1} \quad (10)$$

The maximum RMS current through the low-side switch [24] is given by

$$I_{s1rms,max} = I_{s2,rms,max} = \sqrt{\frac{D}{3}\left(3I_{L1,nom}^2 + \frac{\Delta I_{L1,pkpk}^2}{4}\right)} \quad (11)$$

The maximum voltage across the high-side switch is given by

$$V_{s3max} = V_{s4max} = V_{ds,hs,max} = nV_{in} + V_O \quad (12)$$

The maximum RMS current through the high-side switch which corresponds with the current through the secondary inductance is given by

$$I_{s3rms,max} = I_{s4,rms,max} =$$
$$\sqrt{\frac{(1-D)}{3}\left(3I_{L2,nom}^2 + \frac{\Delta I_{L2,pk-pk}^2}{4}\right)} \quad (13)$$

Eqs. (10) and (13) can be proven to be the same as equations in [4].

For the IBPSC, the maximum voltages across the switches occur in the sub-circuit states in figure 6 with each output capacitor charged with half of the output voltage ($V_{co1} = V_{co2} = V_o/2$). Half of the output voltage will therefore be imposed across S1, S2 and S4. S3 will have the highest voltage stress equal to V_o.

The expressions for maximum RMS current through the switches is the same as eq. (11) and (13) with terms associated with primary and secondary inductance treated as equal. The RMS current through S2 is expressed as

$$I_{s2max,rms} = \sqrt{g^2 + h^2} \quad (14)$$

where
$$g = \sqrt{\frac{(1-D)}{3}(a^2 + ac + c^2)}$$

$$h =$$
$$\sqrt{\frac{2D-1}{6}\left(I_{L,min}^2 + I_{L,max}^2 + dI_{L,min} + eI_{L,max} + d^2 + e^2\right)}$$

$$I_{L,min} = I_{L,nom} - \frac{\Delta I_{L,pkpk}}{2}; I_{L,max} = I_{L,nom} + \frac{\Delta I_{L,pkpk}}{2}$$

$$\Delta I_{L,pkpk} = \frac{V_{in}DT}{L}$$

The highest current stress is imposed on S2 as reflected in sub-circuit state 2 of figure 6 and current waveform of figure 7.

C. CAPACITORS

The output capacitor for the two topologies are computed based on the contribution of the capacitance and ESR to the total voltage ripple. The general expression for the ripple voltage is shown in eq. 15.

$$V_{o,rip} = V_{rcpp} + V_{cpp} = V_{ratio} \cdot V_o \quad (15)$$

where $V_{rcpp} = k \cdot V_o$ and $V_{cpp} = (1-k)V_{ratio} \cdot V_o$.
V_{rcpp} represents the voltage ripple caused by ESR while V_{cpp} represents the voltage ripple caused by ideal capacitance. The variables **k** and **(1-k)** assign the percent absorption of the total output voltage ripple. V_{ratio} is the ratio of allowable voltage ripple to the terminal voltage V_t which can be either be the input or output voltage. The voltage ripple contributed by the ESR for the DTBC is

$$V_{rcpp} = \Delta I_{Co,pkpk} \cdot ESR_{Co} \to ESR_{Co} = \frac{V_{rcpp}}{\Delta I_{Co,pkpk}} \quad (16)$$

where $\Delta I_{Co,pkpk} = max(c',d') - min(a',b')$ and $a', b', c' \& d' = points \; in \; I_{S3+S4} \; waveform$.
The voltage ripple contributed by the ideal capacitance is expressed as:

$$V_{cpp} = \frac{Q}{C_o} = \frac{\int I_{Co}dt}{C_o} \quad (17)$$

The integral in eq. (17) is evaluated to obtain C_o as shown in eq. (18)

$$C_o = \frac{-\left(D-\frac{x-1}{2}\right)(e'-I_{out,ave})}{V_{cpp}f} = \frac{-\left(D-\frac{x}{2}\right)(f'-I_{out,ave})}{V_{cpp}f} \quad (18)$$

where x is an integer equal to 1 when D<0.5 and 2 when D>0.5.

The two phases of the IBCPS circuit cannot be separated into independent phases like the ITBC because removing the switches in one phase does not allow the switching sequence to perform the step-up conversion properly. Since the output sides of the phases are in series, I_{S3} and I_{S4} never combine in one branch at the same time. The AC component of I_{S3} flows directly to C_{o1} while the AC component of I_{S4} flows directly to C_{o2}, storing the energies for simultaneous discharge to the load. The voltage of the series combination of C_{o1} and C_{o2} is applied on the load resistor. Assuming that the output capacitors are identical, the total output voltage ripple is expressed as

$$V_{o,rip} = V_{Co1,cpp} + V_{Co1,rcpp} + V_{Co2,cpp} + V_{Co2,rcpp} \quad (19)$$

From figure 7, the ripple components of the currents through S3 and S4 becomes the effective output capacitor current. The corresponding voltage ripple components are expressed as

$$V_{Co1,cpp} = \frac{(I_{L,nom}-I_o)(1-D)T}{C_{o1}} = \frac{I_o DT}{C_{o1}} \quad (20)$$

$$V_{Co1rcpp} = r_{co1}\Delta I_{Co1,pkpk} = r_{co1}\left(I_{L,nom} + \frac{\Delta I_L}{2}\right) \quad (21)$$

$$V_{Co2,rcpp} = \frac{-I_o(1-D)T}{C_{o2}} \quad (22)$$

$$V_{Co2rcpp} = -r_{Co2}I_o \quad (23)$$

The parameters of the output capacitors are then obtained by solving for them using eqs. (20) and (21).

IV. COMPONENT SELECTION

Table 1 shows the additional design specifications for the DC-DC converter and table 2 list of components used in the DTBC and IBPSC. 75KHz was chosen as the switching frequency.

The 2018 International Power Electronics Conference

TABLE I
DTBC/IBPSC Additional Design Specifications

Minimum Output Current for CCM(I_{ratio})	15%
Allowable Input Voltage Ripple ($V_{in,ratio}$)	1%
Allowable Output Voltage Ripple($V_{o,ratio}$)	1%
Input Ripple absorption by ESR$_{Cin}$ (x_{in})	50%
Input Ripple absorption by ESR$_{Co}$ (x_o)	50%

TABLE II
DTBC/IBPSC Power Components

Active Switches	CMF20120D SiC Power MOSFET 1.2kV 80mΩ 42A (C_{oss} = 120pF)
Input Capacitor	Panasonic EZPE Series Film Capacitor 65uF 6.8mΩ 500V
Output Capacitor	Panasonic EZPE Series Film Capacitor 15uF 14.5mΩ, 1.3kV
Inductor	Material: High Flux Ni-Fe Alloy Core: CH467125 (A_L = 281nH/N^2) DTBC (L_1:167µH, L_2:667µH) IBPSC (L:205µH)

Figure 8 shows the overview of constructed inductors that were used for the DTBC and IBPSC respectively.

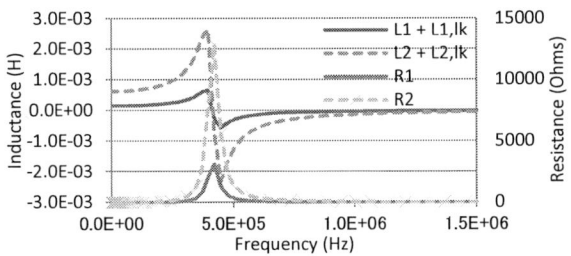

a) DTBC b) IBPSC
Fig. 8. Overview of Constructed Inductors

The tapped-inductor was characterized for snubber sizing and system loss estimation. Figure 9 shows the equivalent circuit model used and the corresponding extracted parameters.

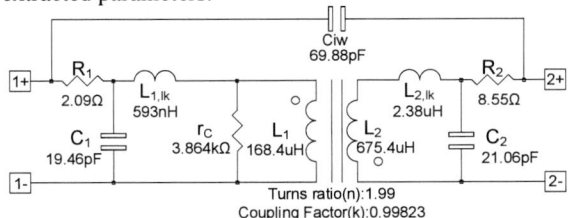

Turns ratio(n):1.99
Coupling Factor(k):0.99823
Fig. 9. Tapped-inductor Equivalent Circuit Model

The ferromagnetic resonance frequency was found to be at 455kHz when measured at either primary or secondary terminals with "1-" and "2-" short-circuited. Figure 10 shows the measured tapped-inductor Ls-Rs characteristics vs. frequency while figure 11 shows the experimental vs. proposed equivalent circuit model primary impedance characteristics vs. frequency.

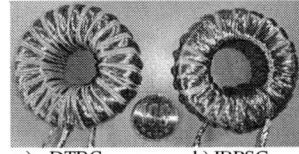

Fig. 10. Tapped-inductor Ls-Rs vs. Freq. Characteristics

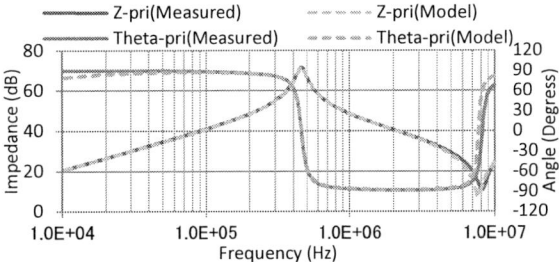

Fig. 11. Tapped-inductor Ls-Rs vs. Freq. Characteristics

Figure 12 shows the DTBC snubbers in one of the phases. RCD voltage clamp and RC snubbers were used for the high and low-side respectively. Snubber designs were adapted from [6,7] for flyback converters and modified accordingly to suit the DTBC topology.

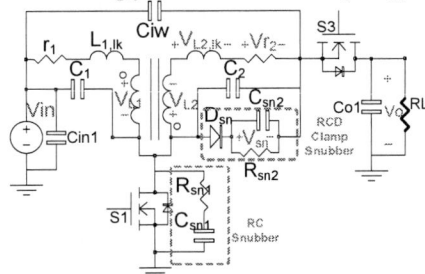

Fig. 12. DTBC Circuit with Snubbers

The RCD snubber loss and the corresponding resistor to dissipate the stored energy are respectively expressed as

$$P_{sn} = \frac{1}{2}\left(L_{2,lk}I_{L2,max}^2 + C_{iw}V_{ciw}^2 + C_2 V_{c2}^2\right)\left(\frac{V_{sn}}{V_{sn}-V_{L2}}\right)f \quad (24)$$

$$R_{sn2} = \frac{V_{sn}^2}{P_{sn}} \quad (25)$$

where $V_{sn} = V_{L2,lk} + V_{L2}$; $V_{L2,lk} = y(V_o + V_{L2})$; $V_{L2} = nV_{in}$ & $V_{ciw} = V_o - V_{in}$; $V_{c2} = (V_o - V_{in})\left(\frac{n}{n+1}\right)$
V_{sn} is the desired clamping voltage and y is the variable that sets the ratio of the allowable voltage that can exceed $V_{ds,hs,max}$ under ideal (no parasitics) conditions.

The characteristic impedance (Z) by the inductor parasitic elements is determined first by measuring the ringing frequency (f_r) and equating it to the snubber resistor as shown in eq. (26). The corresponding capacitor value is then obtained as shown in eq. (27).

$$Z = 2\pi f_r L_{1,lk} = R_{sn1} \quad (26)$$

$$C_{sn1} = \frac{1}{2\pi f_r R_{sn1}} \quad (27)$$

Table 3 lists the specifications of the snubber components used and estimated power loss in suppressing the surge voltage to less than 10% of ideal $V_{ds,hs,max}$.

TABLE III
Selected Snubber Components

Component	Value & Rating	Estimated P$_{loss}$
R$_{sn1}$	62Ω 3W	0.19W/phase
C$_{sn1}$	100pF 1kV	
R$_{sn2}$	5kΩ 35W	2.07W/phase
C$_{sn2}$	1uF 630V Multilayer Ceramic	
D$_{sn}$	2A 1.2kV SiC Schottky	

3096

V. TESTING, RESULTS AND ANALYSIS

Figure 13 shows the overview of developed DTBC/IBPSC in the multimode board. The switches are placed on the bottom side of the PCB to attach to the heatsink base. In the multimode board, S1 to S4 of the DTBC circuit is the same as S1 to S4 of the IBPSC circuit.

Fig. 13. Multimode board with power components

For the testing set-up shown in figure 14, interleaved synchronous PWM signals are generated by FPGA controller. The power source is a DC supply while the load is 320Ω resistor bank. Efficiency is measured with a power analyzer. Component temperatures were measured by thermocouples to observe the thermal stress.

Fig. 14. Experimental Test Set-up

Figure 15 shows the waveforms of one of the phases of the DTBC at rated power. Without the snubber, waveform distortion due to spikes and oscillations is significant.

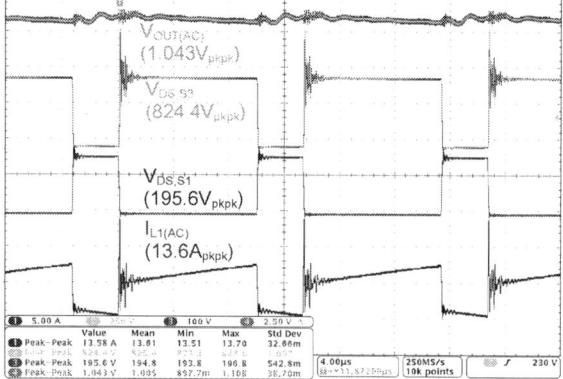

Fig. 15. Single Phase DTBC Waveforms at Rated Power (500W) without Snubber

Figures 16 and 17 show the snubbed DTBC and IBPSC waveforms respectively at 500W loading.

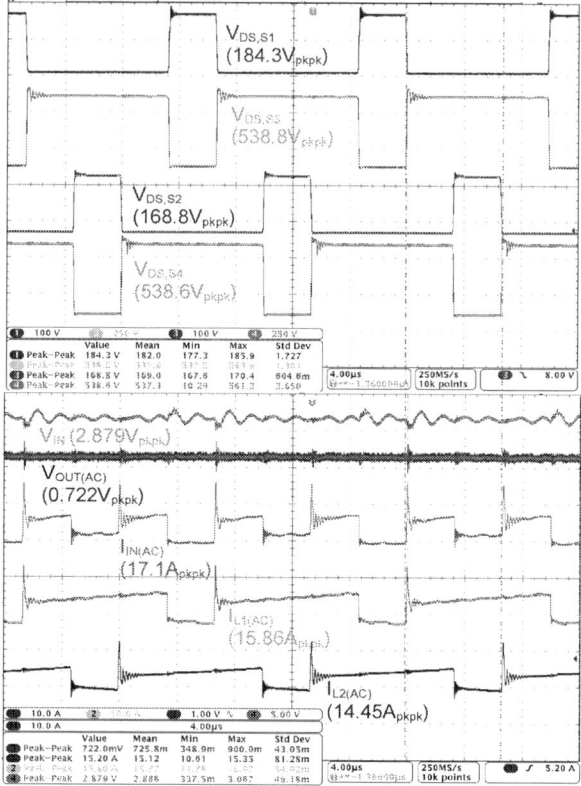

Fig. 16. DTBC waveforms at 500W loading with Snubbers

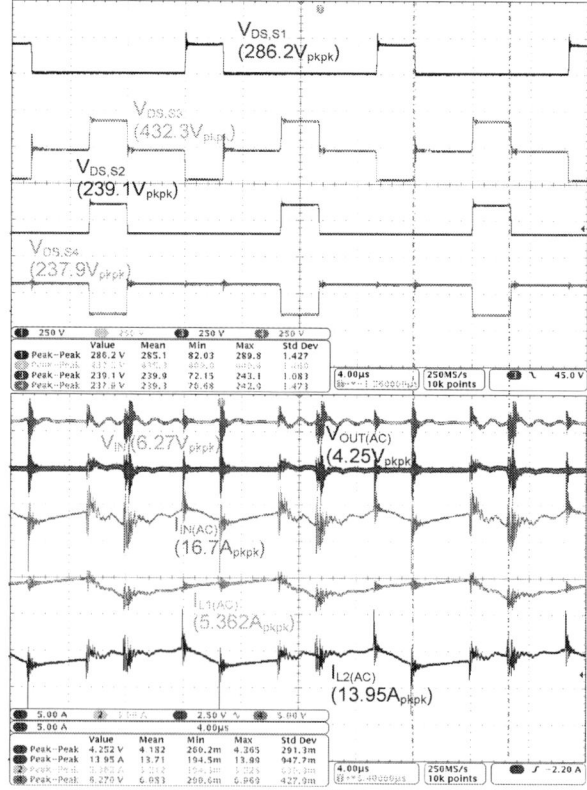

Fig. 17. IBPSC waveforms at 500W loading with Snubbers

Figure 18 shows the input current and output voltage ripple for ITBC and IBCPS at rated power. It can be seen that IBCPS has lower input current ripple and higher voltage ripple than ITBC at rated power.

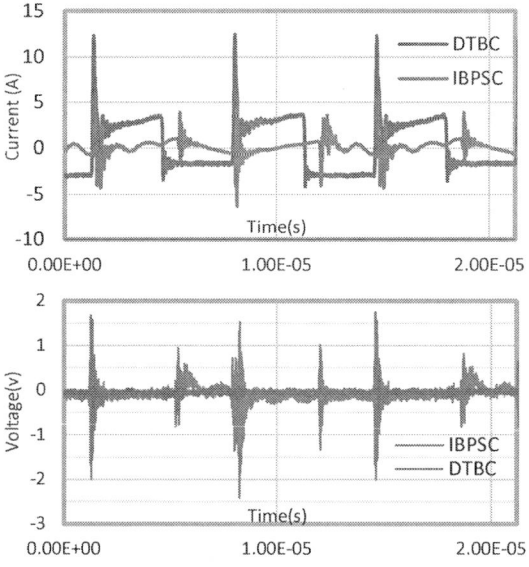

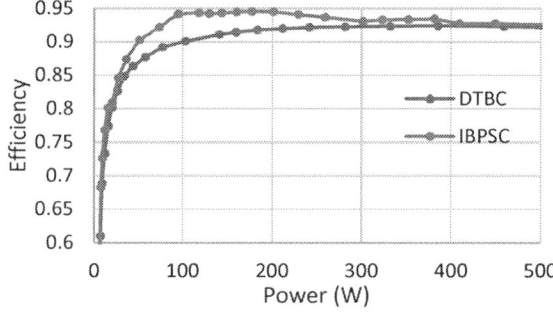

Fig. 18. Input current and voltage ripple at rated power

Figure 19 shows the efficiency vs. output power of DTBC and IBPSC respectively. ITBC and IBCPS obtained an efficiency of 92.15% and 92.53% respectively at rated power.

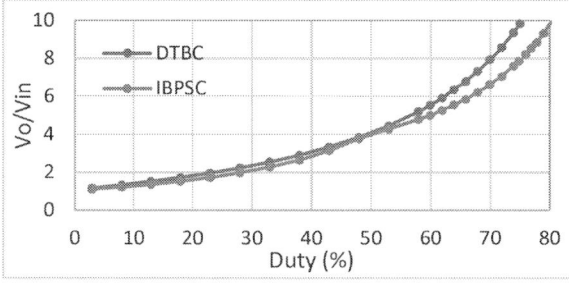

Fig. 19. Efficiency vs. Output power

Figure 20 shows the voltage gain vs. duty cycle of DTBC and IBPSC respectively. It can be seen from the voltage gain curves that the target step-up ratio is achieved by the DTBC at a lower duty cycle compared to the IBPSC.

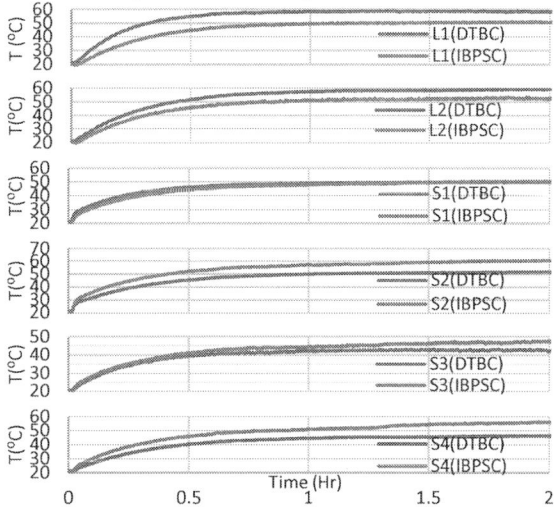

Fig. 20. Voltage Gain vs. Duty Cycle

Figure 21 shows the component temperatures in DTBC and IBPSC when 0 to 100% step loading is applied until temperatures get near the steady-state.

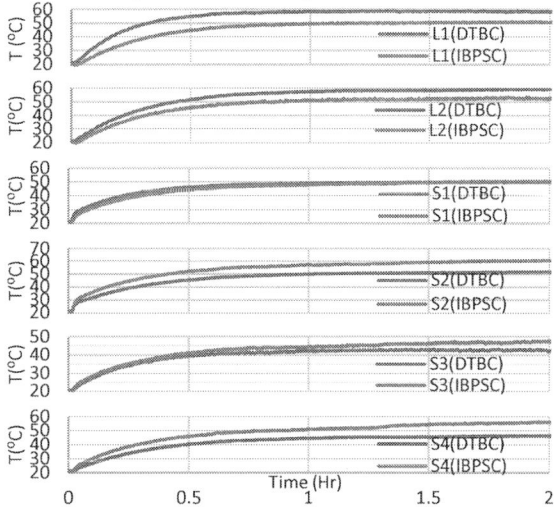

Fig. 21. Component Temperatures vs. Time

The large thermal time constant observed for the switch temperatures is due to the large heatsinks under still air condition. In addition to the heat caused by copper and core loss, the thermal time constant for the inductors is dependent on the volume and spacing of the wires wound on the core. It can be seen that DTBC has obtained higher temperatures for the inductors indicating higher inductor losses compared to the IBPSC. However, for the switches S1 to S4, DTBC obtained lower temperatures compared to the IBPSC ($T_{S1,DTBC}$:49.8□C, $T_{S1,IBPSC}$:50.6□C at approx. steady-state temperature).

Table 4 and 5 contains the summary of the calculated and measured parameters of the DTBC and IBPSC respectively at rated power.

TABLE IV

DTBC CALCULATED & ACTUAL PARAMETER VALUES AT 500W

Parameter	Calculated Value	Actual Value	%error
$I_{L1,min}$	6.5947A	6.6618A	1.0175
$I_{L1,max}$	9.0265A	9.00321A	0.25801
$I_{L2,min}$	2.1982A	2.21259A	0.6546
$I_{L2,max}$ /$\Delta I_{Co,pkpk}$	3.0088A	3.04709A	1.2727
$\Delta I_{in,pkpk}$	6.0601A	5.9649A	1.5699
$V_{ds,hs,clamped}$	528V	535.312	1.3848

TABLE V

IBPSC CALCULATED & ACTUAL PARAMETER VALUES AT 500W

Parameter	Calculated Value	Actual Value	%error
$I_{L,min}$	5.3557A	5.58419A	4.2633
$I_{L,max}$	7.5918A	7.687A	1.25398
$\Delta I_{in,pkpk}$	1.693A	1.6955A	0.15009
$\Delta I_{Co,pkpk}$	2.2361A	2.1201A	5.1843

The designed output voltage ripple for the DTBC and ITBC are 0.195V and 0.896V respectively. The actual output voltage ripple are just shown in figure 18 since it is difficult to determine the percent error due to low signal to noise ratio of the waveforms.

VI. CONCLUSION AND RECOMMENDATIONS

The paper has shown the design and implementation 40V to 400V ITBC and IBCPS. ITBC is the better choice for applications requiring low output voltage ripple, while IBCPS is more suitable for those requiring lower inductor and input current ripple. IBPSC has obtained a higher efficiency than the DTBC in the entire range of output power(94.48% max at 200W loading) . However, the difference in the efficiencies become smaller at rated power. The power loss penalty for operating at higher duty cycle for the IBPSC becomes more apparent as the operation gets near the rated power. DTBC on the other hand was able to operate at lower duty cycles by increasing its turns-ratio. However, the trade-off has manifested in the necessary use of heavy snubbers to suppress the harmful voltage spikes and current surges in the switching nodes. DTBC efficiency can be improved by the use of regenerative snubbers, although it would be better if the leakage inductances and winding capacitances can be further reduced in order to maintain the low complexity of the circuit. The future work will consist of performance comparison of the two topologies using separate PCB designs which cater specifically to their individual characteristics.

REFERENCES

[1] N. Vazquez, L. Estrada, C. Hernandez and E. Rodriguez, "The Tapped Inductor Boost Converter", Industrial Electronics, 2007. ISIE 2007, IEEE International Symposium, 2007

[2] S. Sathyan, H.M. Suryawanshi and A.B. Shitole, "Soft Switched Coupled Inductor based High Step-Up Converter for Distributed Energy Resources," IEEE IECON 2104 Conference Proceedings, Texas, Dallas, USA, Oct. 29-Nov. 1,pp. 1799-1805

[3] Y.Yamamoto, T. Takiguchi, T. Sato and H. Koizumi, "Two-phase Interleaved Bidirectional Converter Input-parallel Output-Series Connection", 2015 9th International Conference on Power Electronics and ECCE Asia, pp. 301-308, 2015

[4] M.N. Gitau, F.M. Mwaniki and I. W. Hofsajer, "Analysis and Design of a Single-Phase Tapped-Coupled-Inductor Boost DC-DC Converter", Journal of Power Electronics, vol. 13, no. 4, July 2013

[5] M. Kazimierczuk, "Pulse-width Modulated DC-DC Converters", Dayton Ohio, Wiley, 2008, chapter 3, no. 3.2.7,pp. 93-95

[6] R. Ridley, "Flyback Converter Snubber Design", Switching Power Magazine, 2005, pp. 1-6

[7] Application Note AN-4147, Fairchild Semiconductor, 2006

[8] T. R. Choudhury and B. Nayak, "Comparison and Analysis of Cascaded and Quadratic Boost Converter", 2015 IEEE Power, Communication and Information Technology Conference, pp. 78-83, 2015

[9] T. J. Liang and K.C. Tseng, "Analysis of Integrated Boost-flyback Step-up Converter", IEE Proceedings – Electric Power Applications, vol. 152, issue 2, pp. 217-225.

[10] C. Shen and P. Chiu, "Buck-boost-flyback Integrated Converter with Single Switch to Achieve High Voltage Gain for PV or Fuel-cell Applications", IET Power Electronics, no. 9, issue 6, pp. 1228-1237, 2016

[11] A. Chub, D. Vinnikov, R. Kosenko and E. Liivik, "Wide Input Voltage Range Photovoltaic Microconverter with Reconfigurable Buck-Boost Switching Stage", IEEE Transactions on Industrial Electronics, vol. 64, no.7, July 2017.

[12] R. Hester, C. Thornton, S. Dhople, Z. Zhao, N. Sridar and D. Freeman. "High Efficiency Wide Load Range Buck/Boost/Bridge Photovoltaic Microconverter", 26th Annual IEEE Applied Power Electronics Conference and Exposition (APEC), pp. 309-313, 2011.

[13] S. –Y. Tseng, J. –Z. Shiang, W. –S. Jwo and C. –M. Yang, "Active Clamp Interleaved Boost Converter with Coupled Inductor for High Step-up Ratio Application", 7th International Conference on Power Electronics and Drive Systems, pp. 1394-1400, 2007

[14] J. Choi, J. Kim, G. Lee, K. Kim and F. Kang, "Cascaded DC-to-DC Converter Employing a Tapped-inductor for High Voltage Boosting Ratio", 41st Annual Conference of the IEEE Industrial Electronics Society, pp. 932-937, 2015

[15] X. Hu and C. Gong, "A High Gain Input-Parallel Output-Series DC/DC Converter with Dual Coupled Inductors, IEEE Transactions on Power Electronics, vol. 3, no. 3, pp. 1306-1317, 2015.

[16] X. Hu, L. Li, Y. Li and G. Wu, "Input-parallel Output-series DC-DC Converter for Non-isolated High Step-up Applications", Electronics Letters, vol. 52, no. 20, pp. 1715-1717, 2016.

[17] P. Wang, L. Zhou, Y. Zhang, J. Li and M. Sumner, "Input-parallel Output-series DC-DC Boost Converter with a Wide Input Voltage Range for Fuel Cell Vehicles", IEEE Transactions on Vehicular Technology, vol. 6, no. 99, 2017.

[18] S. Chen, S. Yang, C. Huang and C. Lin, " Interleaved High Step-up DC-DC Converter with Parallel-input Series-output Configuration and Voltage Multiplier Module", International Conference on Industrial Technology, pp. 119-124, 2017

[19] M.N. Gitau and I.W. Hofsajer, "Analysis of a 4-Phase Tapped-Inductor DC-DC Converter for High Boost Ratio Wide Input Voltage Range Applications," 40th Annual Conference of the IEEE Industrial Electronics Society, Dallas, Texas, USA, Oct. 29-Nov. 1, 2014, pp. 5468-5474

[20] M.N. Gitau, "Analysis of N-phase Tapped-inductor Boost DC-DC Converters", 42nd Annual Conference of the IEEE Industrial Electronics Society, pp. 1294-1300, 2016.

[21] F. Garcia, J. A. Pomilio and G. Spiazzi, "Modeling and Control Design of the Interleaved Double Dual Boost Converter", IEEE Transactions on Industrial Electronics, vol. 60, no. 8, 2013

[22] J.P.Fohringer and F. A. Himmelstoss, "Analysis of a Boost Converter with Tapped- inductor and Reduced Voltage Stress across the Buffer Capacitor", IEEE International Conference on Industrial Technology, pp. 126-131, 2006

[23] A. S. Nastase, "The RMS Value of a Trapezoidal Waveform – Part 2, Internet: https:// masteringelectronicsdesign.com/the-rms-value-of-a-trapezoidal-waveform-part-2/, [Dec. 29, 2017]

[24] HIT Photovoltaic Module N330/N325 Panasonic, Internet: ftp://ftp.panasonic.com/solar/specsheet/n325330-spec-sheet.pdf, [Dec. 29, 2017]

The 2018 International Power Electronics Conference

A New Standby Structure Integrated with Boost PFC Converter for Server Power Supply

Jae-Il Baek[1][*], Jae-Kuk Kim[2], Jae-Bum Lee[3], Moo-Hyun Park[1], and Gun-Woo Moon[1]

1 School of Electronic Engineering, KAIST, Daejeon, Republic of Korea
2 Electrical Engineering, In-ha University, Incheon, Republic of Korea
3 Electrification System Research Team, KRRI, Uiwang-Si, Republic of Korea
*E-mail: dpi1067@kaist.ac.kr

Abstract— In the standby stage of the server power supply, the flyback converter has been widely used due to its simple structure and low cost. However, it suffers from high voltage stress on the primary switch and large transformer size. Thus, the flyback converter in the standby stage degrades the efficiency and power density of the server power supply. To relieve these drawbacks, this paper presents a new standby structure where the standby stage is integrated with the boost PFC stage. In the proposed standby structure, since the primary side of the flyback converter is integrated into the boost PFC stage, the proposed standby structure can eliminate high voltage stress and large transformer. Furthermore, the proposed standby structure helps the boost PFC stage to achieve soft switching operation. Therefore, the proposed standby structure can improve the efficiency and power density of the server power supply. The validity of the proposed standby structure is confirmed by a prototype with 90-264V_{rms} AC line, 750W main output, and 12V/2A standby output.

Keywords— Flyback converter, high efficiency, high power density, standby stage.

I. INTRODUCTION

Recently, as the market of information technology (IT) such as smartphones and cloud services has been dramatically grown all over the world. The number of data centers has been consistently increased. Thus, the server power supplies have been also developed consistently and actively [1]-[2]. In general, the server power supplies need two requirements: 1) high efficiency for energy saving and environmental conservation, 2) high power density for meeting high power capability in the limited space.

Fig. 1(a) shows a typical structure of a server power supply. It is divided into three stages: 1) boost power factor correction (PFC) stage, 2) DC/DC stage, 3) standby stage. In the boost PFC stage, the shape of a input current is controlled to achieve high power quality, and constant link voltage (V_{Link}) is provided for the DC/DC and standby stages [3]-[5]. Next, the DC/DC stage offers galvanic isolation and precisely regulates the main output voltage V_O [6]-[8]. Finally, the main purpose of the standby stage is to provide the standby power for the server computer system. Moreover, as shown in Fig. 1(b), the standby stage should be able to offer the standby output voltage (V_{STB}) before the normal mode and after AC line is lost, i.e., hold-up time.

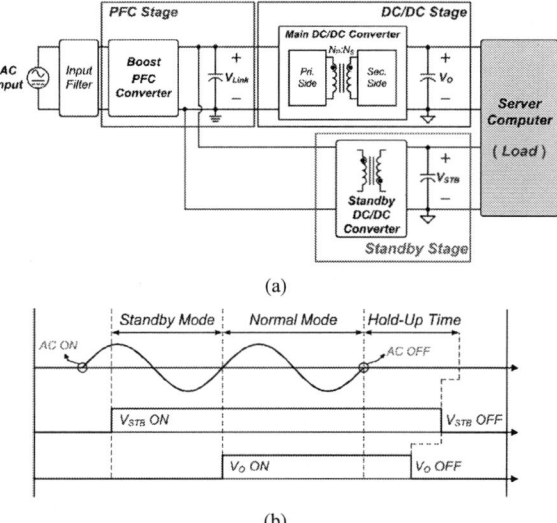

(a)

(b)

Fig. 1. General server power supply. (a) Structure. (b) Operation mode.

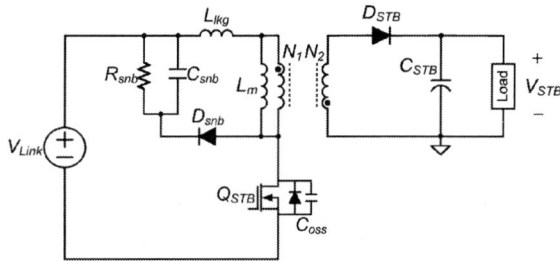

Fig. 2. Circuit diagram of flyback converter in the conventional standby structure.

In the standby stage, the flyback converter has been widely used due to its simple structure and low cost as shown in Fig. 2 [9]-[13]. However, the flyback converter suffers from high voltage stress on the primary switch, which results in large snubber loss. Moreover, due to its hard-switching operation, high voltage stress can cause large switching loss on the primary switch. Thus, the flyback converter can degrade the efficiency of the server power supply. Furthermore, its large transformer size caused by the dc-offset current in the transformer and long creepage distance resulting from high voltage stress decrease the power density of the server power supply. Therefore, in order to improve the efficiency and power

3100

density of the standby stage, many approaches have been researched [9]-[11].

In [9]-[10], the integrated structure between standby and DC/DC stages are proposed to eliminate the switching and snubber losses of the flyback converter. First, in [9], the secondary side of the flyback converter is integrated into the secondary side of the PSFB converter with one additional switch. Thus, in normal mode of Fig. 1(b), it can regulate V_{STB} by controlling one additional switch instead of the flyback switch, which removes the primary loss of the flyback converter. However, in standby and hold-up time modes, since V_{STB} should be regulated without the operation of the DC/DC stage, this structure still has to utilize the conventional flyback structure. Therefore, in spite of high efficiency, it can degrade the power density and increase cost of the sever power supply. Next, in [10], two-switch flyback converter is integrated with the primary side of the PSFB converter by sharing the lagging leg switches of the PSFB converter. Thus, this structure can have the same components count with the conventional flyback converter. Moreover, this structure can achieve high efficiency because the flyback switch can achieve the zero-voltage-switching (ZVS) operation as well as have the clamped voltage stress, i.e., the input voltage of the PSFB converter. However, in the standby mode, the main output voltage cannot be guaranteed as zero because the primary side of the DC/DC stage should be operated to regulate V_{STB}. Moreover, this structure still has large size of the transformer. As a result, it is also difficult to accomplish high power density of the sever power supply.

In this paper, a new standby structure is proposed to achieve high efficiency and high power density of the server power supply. In the proposed structure, the primary side of the flyback converter is integrated with the boost PFC converter. Thus, the proposed standby structure can improve efficiency and power density of the standby stage by eliminating high voltage rated components in the primary side such as the RCD snubber and primary switch. Furthermore, it can help the boost PFC stage to achieve the soft-switching operation, which results in higher efficiency. As a result, the proposed standby structure can improve the efficiency and power density of the server power supply.

II. Concept of Proposed Standby Structure

A. Derivation

Fig. 3(a) shows the conventional boost PFC and flyback converters for the server power supply. From this figure, the boost PFC converter is composed of the input filter capacitor (C_{in}) for the EMI noise, boost inductor (L_B), boost switch (Q_B), and link capacitor (C_{Link}). Moreover, due to the advancement of wide-band gap devices such as SiC and GaN, synchronous rectifier switch (Q_S) can be used instead of the SiC diode to achieve higher efficiency by reducing the conduction loss [18]-[19]. Next, in case of the flyback converter, it is composed of the primary switch (Q_{STB}), RCD snubber for clamping the voltage stress of Q_{STB}, transformer with magnetizing inductor (L_m) and leakage inductor (L_{lkg}),

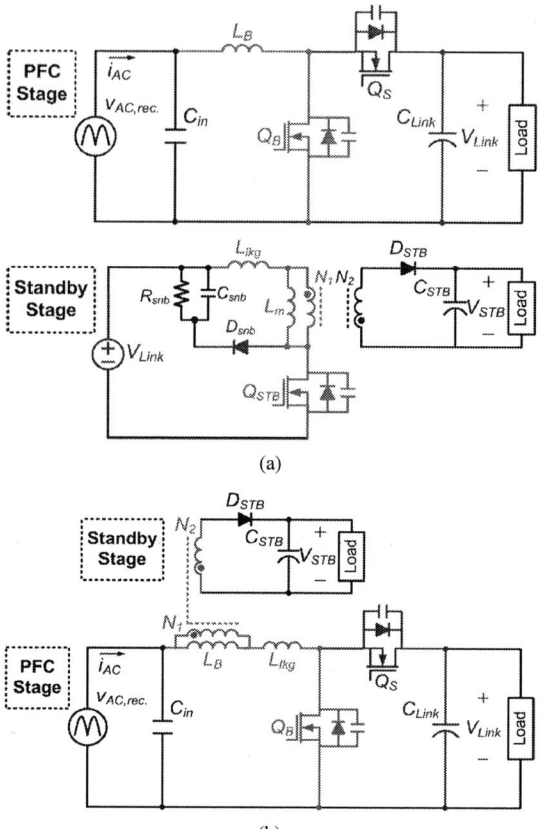

Fig. 3. Derivation of proposed standby structure. (a) Conventional boost PFC and flyback converter. (b) Proposed standby structure

and single-ended rectifier.

As shown in Fig. 3(a), the conventional boost PFC converter has similar structure and components to the conventional flyback converter. For example, L_B and Q_B of the boost PFC converter correspond to transformer and Q_{STB} of the flyback converter. In addition, when R_{snb} and C_{snb} are eliminated, C_{Link} and Q_S of the boost PFC converter are matched to V_{Link} and D_{snb} of the flyback converter. Thus, the proposed standby structure where the flyback converter is integrated into the boost PFC converter is easily derived by replacing L_B with a transformer, as shown in Fig. 3(b). From this figure, the transformer of the flyback converter is eliminated by changing L_B into the transformer. Moreover, since the voltage stress on Q_B, i.e., Q_{STB}, can be clamped to the link voltage (V_{Link}), the proposed standby structure is able to eliminate the RCD snubber and high voltage stress of the conventional flyback converter. Therefore, the proposed standby structure can improve the efficiency and power density of the server power supply.

B. Implementation

For applying the concept of the proposed standby structure to the server power supply, the proposed standby structure should be able to regulate not only the input current (i_{AC}) and V_{Link} as the boost PFC converter but also V_{STB} as the flyback converter. However, unfortunately, it is impossible to meet these requirements

3101

The 2018 International Power Electronics Conference

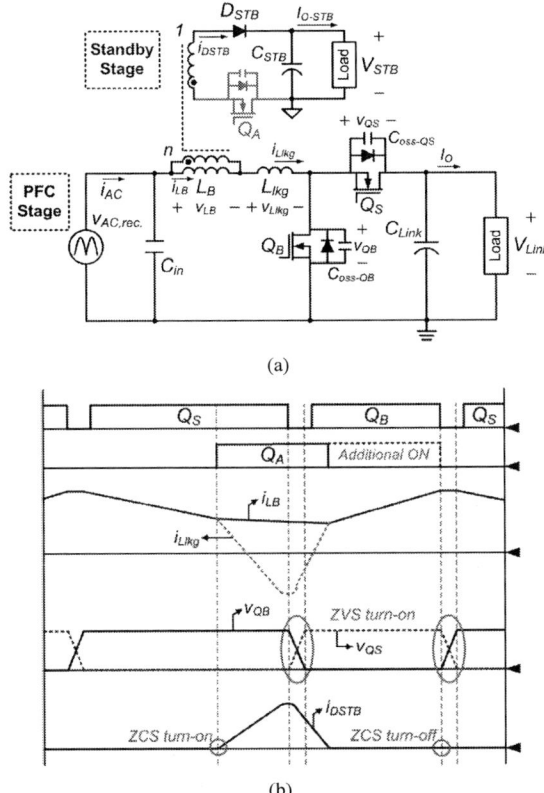

(a)

(b)

Fig. 4. Implementation of proposed standby structure. (a) Circuit diagram. (b) Gate signals and simple key waveforms.

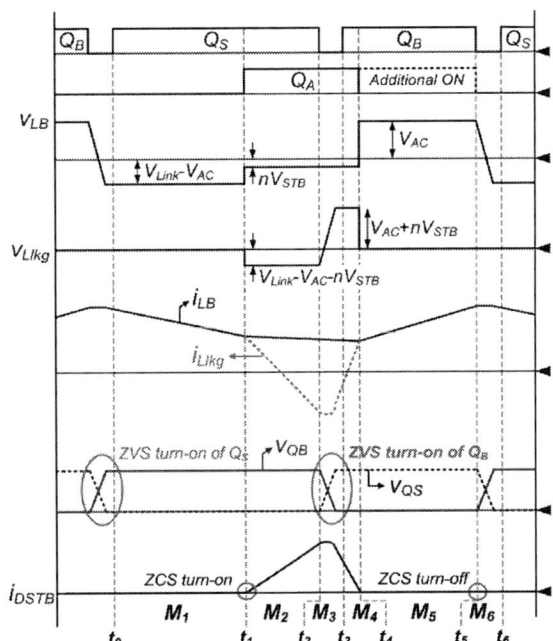

Fig. 5. Operational key waveforms of proposed standby structure.

simultaneously only using Q_B and Q_S because the boost PFC and flyback converters have different voltage gain. Therefore, the proposed standby structure adopts an additional switch Q_A to solve the regulation problem of V_{STB}, as shown in Fig. 4(a). As a result, the proposed structure can regulate i_{AC} and V_{Link} with Q_B and Q_S, and V_{STB} with Q_A having much lower voltage stress compared to the conventional Q_{STB}. From Fig. 4(a), due to the inverse dot structure of the transformer, the power can be transferred from the boost PFC converter to V_{STB} when the voltage across L_B is negative, i.e., the turn-on state of Q_S. However, if Q_A is turned-off at that time, the power cannot be delivered into V_{STB}. Thus, controlling Q_A enables the proposed standby structure to regulate V_{STB} while playing role as the boost PFC converter.

Fig. 4(b) shows the gate signals and simple key waveforms of the proposed standby structure. As abovementioned, when Q_S and Q_A are turned on together, the proposed standby structure can transfer the power into V_{STB}. However, as shown in Fig. 4(b), Q_A is turned on around not the start of the turn-on state of Q_S but the end of that. Therefore, while delivering the power from the boost PFC converter into V_{STB}, the leakage current of the transformer (i_{Llkg}) can have the negative direction at the turn-off state of Q_S, which enables the soft switching operation of the boost PFC converter. Moreover, by synchronizing the turn-off operation of Q_A and Q_B, Q_A can achieve the zero current switching (ZCS) turn-off operation and the control complexity of Q_A can be

simplified. The detailed operation principles of the proposed standby structure are presented in the following section.

III. OPERATIONAL PRINCIPLES OF PROPOSED STANDBY STRUCTURES

Fig. 4(a) and Fig. 5 show the circuit diagram and key waveforms of the proposed standby structure, respectively. To simplify the explanation for the proposed structure, the analysis and the mode operation can be given only for a one switching period of Q_B because the proposed standby structure has the same operational principles over the entire v_{AC} cycles except for the duty-ratio of Q_B and Q_S. Moreover, in order to illustrate the operation of the proposed structure, some assumptions are made as follows: 1) the output capacitor of the boost PFC converter (C_{Link}) is large enough to be considered as a constant voltage (V_{Link}), 2) all parasitic components are ignored except for those specified in Fig. 4(a), 3) the magnetizing inductance of the boost transformer (L_B) is much larger than that of the leakage inductance (L_{lkg}), 4) v_{AC} is constant (V_{AC}) during the one switching period.

Mode 1 [t_0-t_1]: At time t_0, the ZVS operation of Q_S is achieved and Q_S is turned on. Thus, the stored energy in L_B and input power is transferred into V_{Link} through Q_S. In this mode, the currents of L_B and L_{lkg} (i_{LB} and i_{Llkg}) can be expressed as follows:

$$ i_{LB}(t) = i_{Llkg}(t) = i_{LB}(t_0) - \frac{V_{Link} - V_{AC}}{L_B}(t - t_0). \quad (1) $$

Mode 2 [t_1-t_2]: When Q_A is turned on at t_1, mode 2 begins. Since D_{STB} and Q_A are conducted, $-nV_{STB}$ which is the reflected voltage of V_{STB} into the primary side is applied to L_B. Thus, i_{LB} and i_{Llkg} are separated, and they can be

3102

expressed as follows:

$$i_{LB}(t) = i_{LB}(t_2) - \frac{nV_{STB}}{L_B}(t - t_2), \tag{2}$$

$$i_{Llkg}(t) = i_{LB}(t_2) - \frac{(V_{Link} - V_{AC} - nV_{STB})}{L_{lkg}}(t - t_2), \tag{3}$$

$$i_{DSTB}(t) = n(i_{LB}(t) - i_{Llkg}(t)). \tag{4}$$

From (3), i_{Llkg} is decreased linearly so that it can reach negative value, as shown in Fig. 5. In addition, when Q_A is turned on, the initial current value of the switch is zero. It means that Q_A can obtain the ZCS operation, which leads to small switching loss.

Mode 3 [t_2-t_3]: In mode 3, as shown in Fig. 5, since the current flow of Q_S is negative, the body diode of Q_S is not conducted. Thus, the reverse recovery concern of Q_S can be relieved. Moreover, due to the negative direction of i_{Llkg}, the voltage across $C_{oss\text{-}QB}$ (v_{QB}) starts to decrease toward zero, and the voltage across C_{oss_QS} (v_{QS}) is complementally increased to V_{Link}. In this operation, provided that L_{lkg} has larger energy than the stored energy of $C_{OSS\text{-}QB}$, the proposed standby structure can achieve the ZVS operation of Q_B, which results in high efficiency of the boost PFC converter. Moreover, although L_{lkg} has insufficient energy, the proposed standby structure can reduce turn-on overlap switching loss of Q_B and relieve the reverse recovery concerns of Q_S due to the reduced i_{QS} and i_{QB} compared to the conventional boost PFC converter.

Mode 4 [t_3-t_4]: At time t_3, the ZVS operation of Q_B is completed, and the voltage across L_{lkg} (v_{Llkg}) is ($V_{AC}+nV_{STB}$). Thus, i_{Llkg} is linearly increased as in (5) and i_{LB} is continuously decreased as in (2), and the input power is transferred to the standby output until i_{Llkg} meets i_{LB}. This mode ends when D_{STB} is turned off and reverse biased.

$$i_{Llkg}(t) = i_{LB}(t_3) + \frac{(V_{AC} + nV_{STB})}{L_{lkg}}(t - t_3), \tag{5}$$

Mode 5 [t_4-t_5]: At time t_4, D_{STB} is turned off and reverse biased. Thus, the voltage across the magnetizing inductor L_B (v_{LB}) is V_{AC}. As a result, i_{LB} and i_{Llkg} starts to increase linearly, and it can be expressed as follows:

$$i_{LB}(t) = i_{Llkg}(t) = i_{LB}(t_4) + \frac{V_{AC}}{L_{LB}}(t - t_4). \tag{6}$$

Mode 6 [t_5-t_6]: At time t_5, Q_A and Q_B are turned off. Since there is no current flows at Q_A, Q_A can achieve the ZCS operation. In addition, the ZVS operation of Q_S is accomplished with larger energy stored in L_B in this mode. This mode ends when Q_S is turned on.

IV. VOLTAGE GAIN OF PROPOSED STANDBY STRUCTURE

The voltage gain of the proposed standby structure can be derived by using three conditions; 1) the voltage-second balance of L_B, 2) the voltage-second balance of

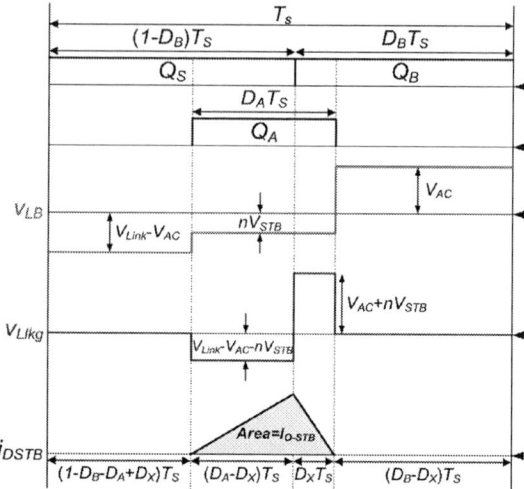

Fig. 6. Simplified key waveforms for voltage gain of proposed standby structure.

L_{lkg}, 3) the average current of i_{DSTB}. Fig. 6 shows the simplified key waveforms for the voltage gain of the proposed standby structure. In this figure, it is assumed that the dead time between Q_B and Q_S, the voltage ripple, and current ripple are negligible. Moreover, the duty-ratios of Q_B and Q_A are D_B and D_A, respectively.

From this figure, the three conditions can be expressed as follows:

$$(V_{Link} - V_{AC})(1 - D_B - D_A + D_X) + nV_{STB}D_A \\ = V_{AC}(D_B - D_X), \tag{7}$$

$$(V_{Link} - V_{AC} - nV_{STB})(D_A - D_X) = (V_{AC} + nV_{STB})D_X, \tag{8}$$

$$\frac{1}{2}D_A \times \frac{n(V_{Link} - V_{AC} - nV_{STB})}{L_{lkg}}(D_A - D_X)T_S = I_{O\text{-}STB}, \tag{9}$$

where $I_{O\text{-}STB}$ is the output current of the standby stage.

Thus, by using (7)-(9), D_X and D_A can be obtained as follows:

$$D_X = \frac{D_A(V_{Link} - V_{AC} - nV_{STB})}{V_{Link}}, \tag{10}$$

$$D_A = \sqrt{\frac{2I_{O\text{-}STB}L_{lkg}V_{STB}}{nT_S(V_{Link} - V_{AC} - nV_{STB})(V_{AC} + nV_{STB})}}, \tag{11}$$

where T_S is the switching period.

First, the boost PFC stage in the proposed standby structure can be obtained by substituting (10) and (11) into (7).

$$V_{Link} = \frac{v_{AC}}{1 - D_B}. \tag{12}$$

From (12), the voltage gain of the boost PFC stage in the proposed standby structure is equal to the conventional one [13]. Therefore, the boost PFC stage in the proposed standby structure can be designed and controlled like the conventional boost PFC converter [14].

Secondly, the voltage gain of the standby stage in the proposed structure can be also achieved by substituting

The 2018 International Power Electronics Conference

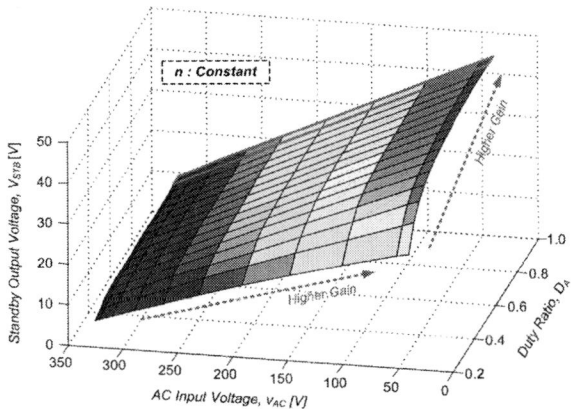

Fig. 7. V_{STB} of proposed standby structure according to v_{AC} and D_A.

TABLE I
DESIGNED PARAMETERS OF PROTOTYPES

Items	Conventional	Proposed
Boost inductor	Core : CH270043*2 L_B : 610μH N_l : 82 (1.0φ)	Core : CH270043*2 L_B : 610μH, L_{lkg} : 35 μH, Main winding (N_l) 82T (1.0φ) Additional winding (N_2) 24T (0.8φ)
Flyback devices	Q_{STB}: SPA08N80C3 (V_{DS}=800V, R_{on}=0.65Ω) D_{STB}: VI20120 (V_{RRM}=120V, V_F=0.51V)	Q_{STB}: IPI110N20N3G (V_{DS}=200V, R_{on}=13mΩ) D_{STB}: VI20202 (V_{RRM}=200V, V_F=0.63V)
Flyback transformer	Core : EED1521 L_m : 1mH $N_l:N_2$=78(0.3φ):10(0.6φ)	-
RCD snubber	R_{snb} : 40kΩ, C_{snb} : 47nF D_{snb} : S1M (V_{RRM}=1200V)	-

(10) and (11) into (8).

$$V_{STB} = \frac{V_{Link}}{n}\left[\frac{1}{2} - \frac{v_{AC}}{V_{Link}} - \frac{A}{D_A^2} + \sqrt{\frac{1}{4} + \frac{A}{D_A^2}\left(\frac{A}{D_A^2} - 1 + \frac{v_{AC}}{V_{Link}}\right)}\right], \quad (13)$$

where A is $L_{lkg}/(n^2T_SR_{STB})$ and R_{STB} is the output resistor of the standby stage.

From (13), since V_{Link} is control output of the boost PFC stage, V_{STB} can be shown according to v_{AC} and D_A, as shown in Fig. 8. From Fig. 7, as D_A increase and v_{AC} decrease, higher V_{STB} can be achieved. This is because the power is transferred from the boost PFC stage to the standby stage when (V_{Link}-v_{AC}) is applied to L_B during D_A. As a result, the standby stage in the proposed standby structure can be designed with (13).

V. EXPERIMENTAL RESULTS

To confirm the validity of the proposed standby structure, a prototype with the specifications of 100-240 V_{rms} AC line (nominal AC line : 115V_{RMS} and 230V_{RMS}), 400V/750W main output, 12V/24W standby output, and 100kHz switching frequency has been built and tested. For the comparison, the prototype of the conventional standby structure is also implemented. The designed parameters are presented in Table I and IPP60R099C6 and GS66508T are used for the boost switch and synchronous rectifier switch for prototypes.

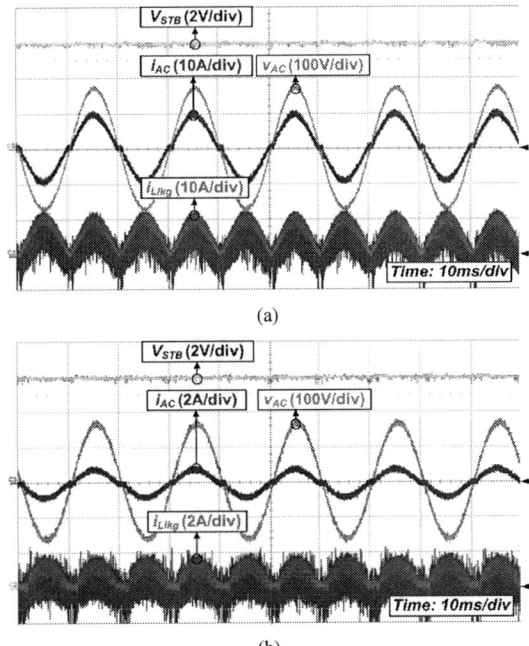

Fig. 8. Experimental key waveforms of proposed standby structure at 115V_{RMS}. (a) 100% load condition. (b) 10% load condition.

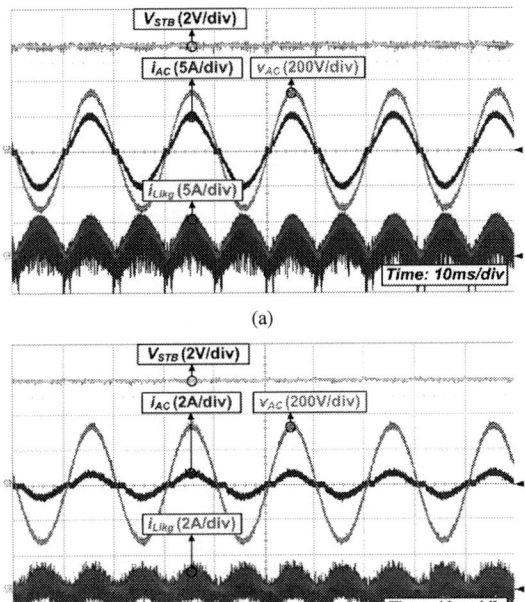

Fig. 9. Experimental key waveforms of proposed standby structure at 230V_{RMS}. (a) 100% load condition. (b) 10% load condition.

Figs. 8 and 9 show the experimental waveforms of the proposed standby structure at 115V_{RMS}/230V_{rms} and 100% /10% load conditions, respectively. From these figures, the shape of the input current (i_{AC}) is well controlled like the input voltage (v_{AC}), and V_{STB} is well regulated under both 115V_{RMS}/230V_{RMS}. Moreover, unlikely to the conventional boost PFC stage, i_{Llkg} of the boost PFC stage in the proposed standby structure can achieve negative value. Thus, it enables the proposed standby structure to

3104

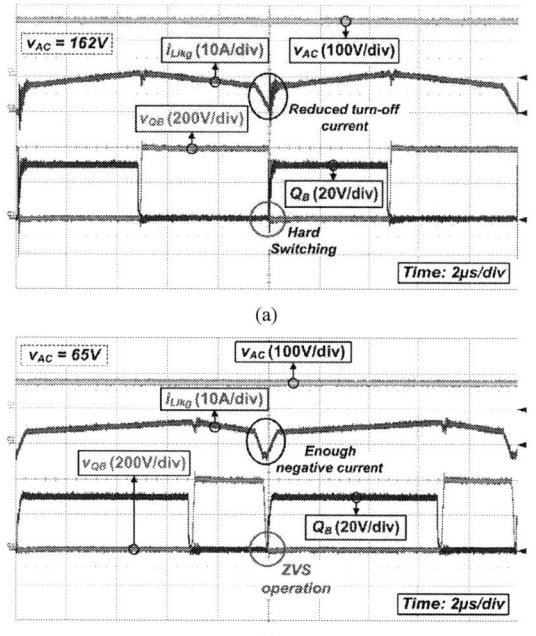

(a)

(b)

Fig. 10. Experimental key waveforms of proposed standby structure at $115V_{RMS}$ and 100% load condition. (a) v_{AC}=162V. (b) v_{AC}=65V.

obtain ZVS operation of Q_B.

Fig. 10 shows the ZVS waveforms of the proposed standby structure under 115VRMS and 100% load condition, which is the worst ZVS case. From Fig. 10(a), when v_{AC} is 162V, since i_{AC} is larger than the required current of the standby stage, i_{Llkg} cannot reach a negative value. Thus, the proposed standby structure is not able to obtain the ZVS operation of Q_B. However, since the turn-off current of switch Q_S and turn-on current of Q_B are significantly reduced, the proposed standby structure can achieve small switching loss and reverse recovery loss. Moreover, as shown in Fig. 10(b), when v_{AC} is smaller than 65V, i_{Llkg} can reach enough negative value to achieve the ZVS operation of Q_B so the proposed converter can eliminate the turn-on switching loss and reverse recovery loss.

Fig. 11 shows the measured system efficiency of the prototypes at $230V_{rms}$ and $115V_{rms}$. To measure the system efficiency, the summation of the output power of the PFC stage and flyback stage is divided by the ac input power. As can be seen in Fig. 11, the proposed standby structure has higher efficiency than the conventional one over the entire AC line and load conditions. This is because the proposed standby structure has the reduced switching loss in the boost PFC stage, and the reduced snubber and switching loss in the standby stage.

VI. CONCLUSIONS

In this paper, a new standby structure where the flyback converter is integrated with the boost PFC converter is proposed to improve the efficiency and power density of the server power supply. Compared to the conventional flyback converter which suffered from large snubber loss, hard switching loss, high voltage stress, and large size of transformer, the proposed

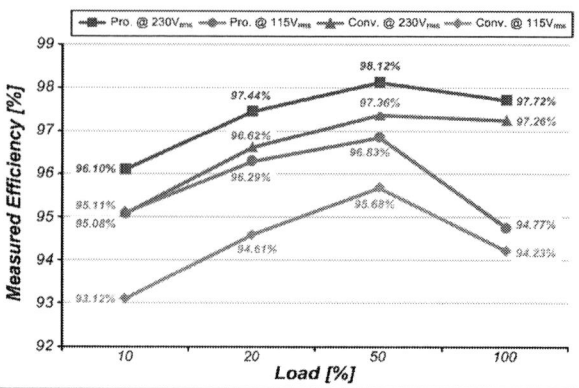

Fig. 11. Measured system efficiency (PFC stage + standby stage).

structure can eliminate the primary side of the flyback converter and large transformer. Thus, the proposed structure can improve the efficiency and power density of the standby stage. Moreover, during the powering period for the standby output, the proposed structure can help the boost PFC converter to achieve soft switching operation, which results in higher efficiency. As a result, the proposed standby structure can improve the efficiency and power density of the server power supply.

ACKNOWLEDGMENT

This work was supported by the National Research Foundation of Kore (NRF) grant funded by the Korea government (MSIP) (No. 2016R1A2B2010328).

REFERENCES

[1] D. Y. Kim, C. E. Kim, and G. W. Moon, "Variable Delay Time Method in the Phase-Shifted Full-Bridge Converter for Reduced Power Consumption Under Light Load Conditions," *IEEE Trans. Power Electron.*, vol. 28, no. 11, pp. 5120-5127, Nov. 2013.

[2] C. Pettey. (2012, 01). [Online]: Gartner Estimates Available : http://www. gartner.com/it/ page.jsp?id=503867.

[3] Y. S. Kim, W. Y. Sung, and B. K. Lee, "Comparative performance analysis of high density and efficiency PFC topologies," *IEEE Trans. Power Electron.*, vol. 29, no. 6, pp. 2666-2679, Jun. 2014.

[4] H. S. Youn, J. S. Park, K. B. Park, J. I. Baek, and G. W. Moon, "A Digital Predictive Peak Current Control for Power Factor Correction With Low-Input Current Distortion," *IEEE Trans. Power Electron.*, vol. 31, no. 1, pp. 900-912, Jan. 2016.

[5] H. S. Youn, J. B. Lee, J. I. Baek, and G. W. Moon, "A Digital Phase Leading Filter Current Compensation (PLFCC) Technique for CCM Boost PFC Converter to Improve PF in High Line Voltage and Light Load Condition," *IEEE Trans. Power Electron.*, vol. 31, no. 9, pp. 6596-6606, Sep. 2016.

[6] J. K. Kim, D. K. Yang, J. B. Lee, and J. I. Baek, "High Light-Load Efficiency Power Conversion Scheme Using Integrated Bidirectional Buck Converter for Paralleled Server Power Supplies, *IEEE Trans. Ind. Electron.*, vol. 64, no. 1, pp. 236-243, Jan. 2017.

[7] J. W. Kim, D. Y. Kim, C. E. Kim, and G. W. Moon, "A Simple Switching Control Technique for Improving Light Load Efficiency in a Phase-Shifted Full-Bridge Converter with a Server Power System," *IEEE Trans. Power Electron.*, vol. 29, no. 4, pp. 1562-1566, Apr. 2014.

[8] J. H. Kim, C. E. Kim, J. K. Kim, J. B. Lee, and G. W. Moon, "Analysis on Load-Adaptive Phase-Shift Control for High Efficiency Full-Bridge LLC Resonant Converter Under Light-Load Conditions," *IEEE Trans. Power Electron.*, vol. 31, no. 7, pp. 4942-4954, Jul. 2016.

[9] J. K. Kim, S. W. Choi, C. E. Kim, and G. W. Moon, "A New Standby Structure Using Multi-Output Full-Bridge Converter

Integrating Flyback Converter," *IEEE Trans. Ind. Electron.*, vol. 58. no. 10, pp. 4763-4767, Oct. 2011.

[10] S. Y. Cho, I. O. Lee, J. K. Kim, and G. W. Moon, "A New Standby Structure Based on a Forward Converter Integrated With a Phase-Shift Full-Bridge Converter for Server Power Supplies," *IEEE Trans. Power Electron.*, vol. 28. no. 1, pp. 336-346, Jan. 2013.

[11] J. P. Hong and G. W. Moon, "A Digitally Controlled Soft Valley Change Technique for a Flyback Converter," *IEEE Trans. Ind. Electron.*, vol. 62. no. 2, pp. 966-971, Feb. 2015.

[12] J. Yang, "Efficiency Improvement with GaN-Based SSFET as Synchronous Rectifier in PFC Boost Converter," *in proc. PCIM Europe 2014*, pp. 1011-1016, May 2014.

[13] R. W. Erickson and D. Maksimovic, *Fundamental of Power Electronics* 2nd ed. Norwell, MA, USA.

[14] S. Abdel-Rahman, "CCM PFC Boost Converter Design (DN 2013-01)," Infineon Technologies North America, Jan. 2013. [Online], Available: http:www.infineon.com.

Nonisolated Two-Channel LED Driver with Simple Snubber

Jong-Woo Kim[1*], Jung-Kyu Han[2] and Jih-Sheng Lai[1]

1 The Bradley Department of Electrical and Computer Engineering, Virginia Tech, Blacksburg, USA

2 Electrical Engineering, KAIST, Daejeon, South Korea

*E-mail: kimjw@vt.edu

Abstract- **In this paper, a new converter topology with a high voltage conversion ratio for non-isolated small power LED drivers is proposed. Conventional converter topologies have limitations such as a large switch turn off loss because of a large turn off current or low power density due to many number of snubber components. The proposed converter utilizes the rectifier diodes and the output capacitors as the snubber components, so it does not require any snubber components. Also, the proposed converter has a small switch turn off loss due to a resonant operation. Furthermore, since the snubber path consists of series-connected output capacitors, the current balancing capability is still preserved for two-channel LED driver. The effectiveness of the proposed driver is validated with 11.55W prototype.**

I. INTRODUCTION

LED is one of the most promising light source nowadays, since it has powerful advantageous characteristics such as high efficiency and long lifetime. Due to its advantages, LED is penetrating the lighting market share from a low to high power lighting applications. Therefore, many researchers have focused on LED drivers achieving a high efficiency and a high power density [1], [2].

The luminance of LED is determined by the current of the LED. A large LED current leads to a brighter light, and whereas a small LED current leads to a darker light. Therefore, in the case many LED are used in a LED driver, it is essential to balance the current in each LED. The easiest and intuitive method to obtain balanced current in each LED is to use series connected LEDs. For these reasons, a high output voltage is required when many LEDs are driven. Furthermore, in outdoor-portable small power LED applications, the input voltage is relatively low since the power source becomes a low voltage battery pack. Therefore, in order to improve the efficiency and high power density, many researchers studied many high step-up converter topologies [3]–[5].

Previous studies have focused on single-switch converter topologies for low power and high step up LED application. Especially, the converter topologies appropriate for the application are based on tapped-boost or flyback converter because they use only one main switch and they can achieve a high step-up voltage conversion ratio with the turns ratio of the transformer.

However, they still have some limitations to achieve a high efficiency and high power density. First, the switch turn off loss is high due to a high input current in a step up application. Secondly, a high voltage stress on the

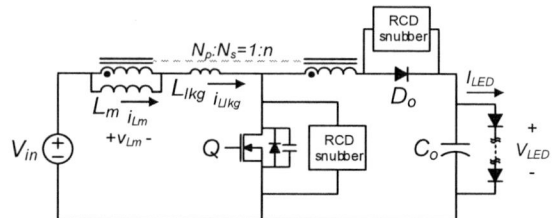

Fig. 1. Tapped-inductor boost converter for a small power and a high voltage conversion ratio application.

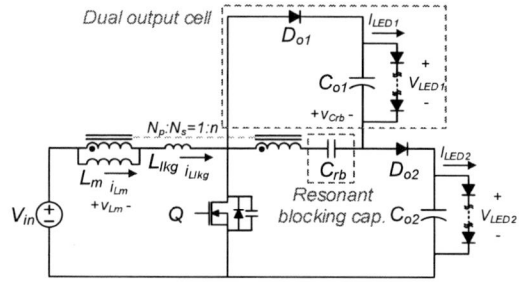

Fig. 2. Proposed converter.

semiconductor devices occur due to the leakage inductor of the transformer. Active/passive clamp circuits can suppress the voltage stress, they result in a low power density and poor cost effectiveness [3]–[5].

A new converter topology for small power and high step-up voltage conversion ratio is proposed in this paper. First, in the proposed converter, the switch turn off loss can be suppressed by using a resonant blocking capacitor with two-channel structure. Secondly, the proposed converter utilizes the rectifier diodes and the output capacitors as the snubber components, it has inherent lossless snubber structure without any additional snubber components.

II. PROPOSED CONVERTER AND ITS OPERATION

The proposed converter is based on the tapped-inductor boost converter shown in Fig. 1. As shown in the figure, the high voltage conversion ratio can be achieved by the turns ratio of tapped-inductor. However, as mentioned before, the tapped-inductor boost converter has a high switching turn off loss due to a large turn off current. Also, the voltage stresses on the semiconductor devices are very high due to the oscillation between the leakage inductor in the tapped-inductor and the parasitic capacitors in the semiconductor devices. In order to suppress that voltage stress, RCD snubber circuits are usually used. However, they result in a large loss since

3107

The 2018 International Power Electronics Conference

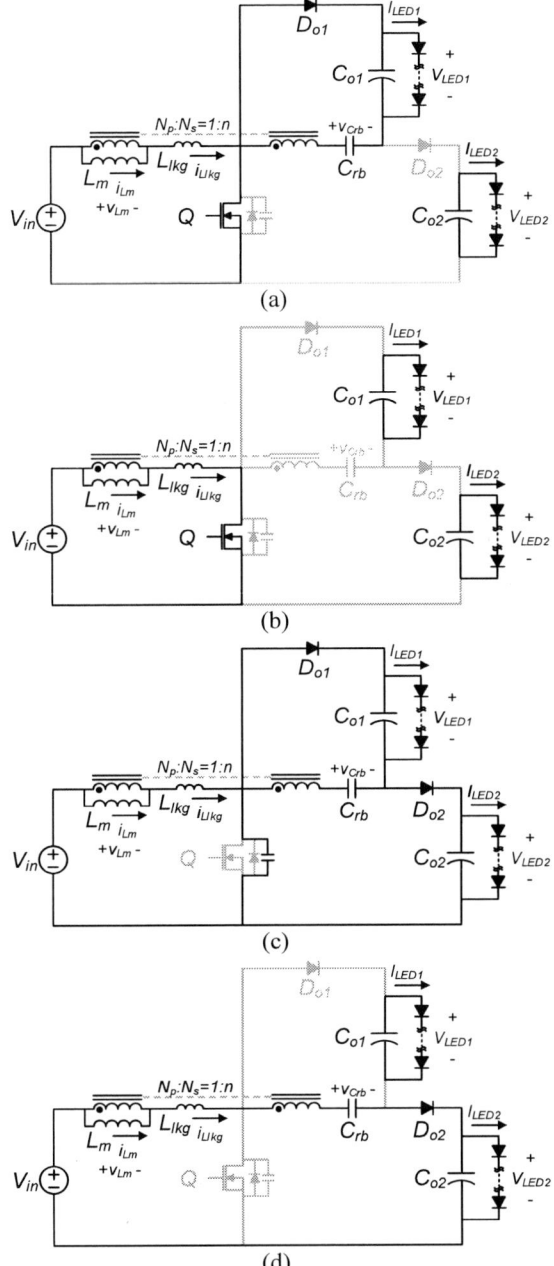

(a)

(b)

(c)

(d)

Fig. 3. Topological stages of the proposed converter during (a) mode 1 $(t_0 \sim t_1)$ (b) mode 2 $(t_1 \sim t_2)$ (c) mode 3 $(t_2 \sim t_3)$ and (d) mode 4 $(t_3 \sim t_0')$

they dissipate the energy stored in the leakage inductor. Although lossless snubber technique can eliminate losses from the snubber circuit, they result in a low power density and high cost problem. Furthermore, although they use snubber circuits, the voltage stress on the boost diode becomes much higher than total output voltage.

Fig. 2 represents the proposed converter. In order to reduce the voltage stress on the semiconductor devices, the proposed converter adapted the two-channel structure. In the tapped-inductor boost converter, the secondary side

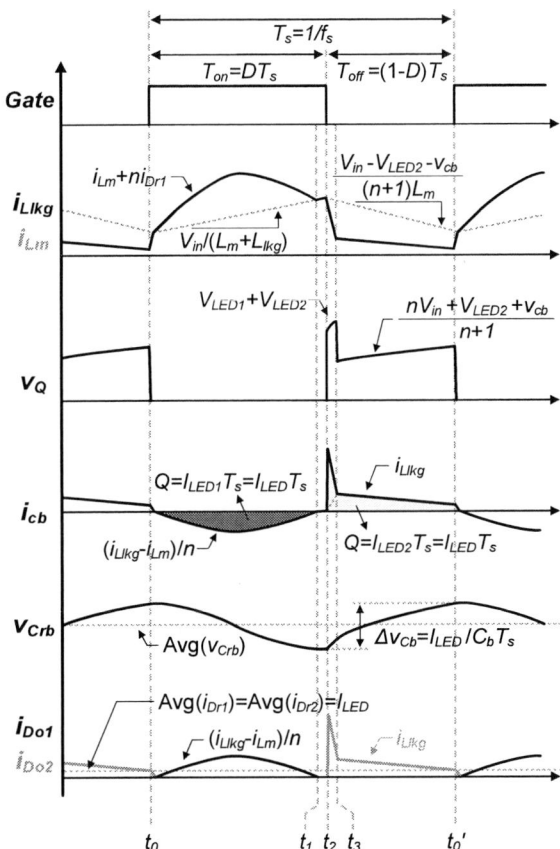

Fig. 4. Key waveforms of the proposed converter.

of the tapped-inductor does not operate when the main switch is turned on. In the proposed converter on the other hand, the tapped-boost converter delivers the output power two the additional output cell. In order to balance the output current for each channel, the blocking capacitor is inserted next to the secondary side of the tapped-inductor. By doing so, the output current in both channel should be equal due to the charge balance condition of the blocking capacitor. Also, the resonant period of the leakage inductor and the blocking capacitor the proposed converter can reduce the turn off switching current of the main switch. Furthermore, it can be seen that the voltage stress on the semiconductor devices are clamped to the sum of the two channel voltages using rectifier diodes so that any snubber component is not required.

Fig. 3 and 4 represent the topological stages and the key waveforms. The proposed converter has a fixed switching frequency and the output is controlled by the duty ratio. During a switching period, 4 topological stages occur.

In mode 1, the main switch is turned on and the power is delivered to the channel 1. Magnetizing inductor current increases linearly, assuming that the magnetizing inductance of the tapped-inductor is much larger than the leakage inductance of it. Mode 1 ends when the resonance ends so that the magnetizing inductor current

3108

TABLE I
COMPONENTS OF THE PROTOTYPE

V_{in}	2.97–3.63V	P_{out}	11.55W

Components	Notation	Design results	
LED	-	XREWHT-L1-0000, 10EA, 5EA in each channel	
Transformer	Core	KoolMu 77059	
	$N_p:N_s$	15(AWG14):61(AWG24), n=4.07, J=400A/cm^2	
	L_m, L_{lkg}	9.8μH, 62nH	
Blocking cap.	C_{rb}	4.4μF MLCC	
Output cap.	C_{o1}, C_{o2}	22μF MLCC 2EA / channel	
Switch	Q	IPB017N08 (70V/1.7mΩ)	
Diode	D_{o1}, D_{o2}	MBRS4201T3G (200V)	

TABLE II
COMPONENTS COMPARISON

Components	[3]	[4]	[5]	Proposed
Magnetic	1	1	1	1
Blocking cap.	1	1	1	1
Switch	1	2	1	1
Rectifier diode	2	2	2	2
Snubber diode	1	-	1	-
Snubber cap.	1	1	-	-
Total	7	7	6	5

and the leakage inductor currents become equal and the rectifier diode D_{o1} is turned off.

In mode 2, the input voltage is applied to magnetizing inductor and the leakage inductor of the tapped-inductor. Mode 2 ends when the main switch is turned off.

In mode 3, the voltage across the main switch increases abruptly by the tapped-inductor current. After its voltage reaches to the sum of two channel voltages, the rectifier diodes D_{o1} and D_{o2} are conducted and they operate as the snubber capacitor. Also, the energy stored in the leakage inductor is delivered to the series connected output capacitors, so that it is lossless structure. Furthermore, the current of the leakage inductor is delivered to each channel equally since two channel are connected in series. Therefore, the proposed converter can suppress the voltage stress just using the existing rectifier diodes and the output capacitors while maintaining the current balancing capability. Mode 3 ends when the current in the leakage inductor is reset enough so that D_{o1} is turned off.

In mode 4, the energy stored in the magnetizing inductor is delivered to the channel 2. Since the magnetizing inductor is much larger than the leakage inductor, the magnetizing inductor current decreases

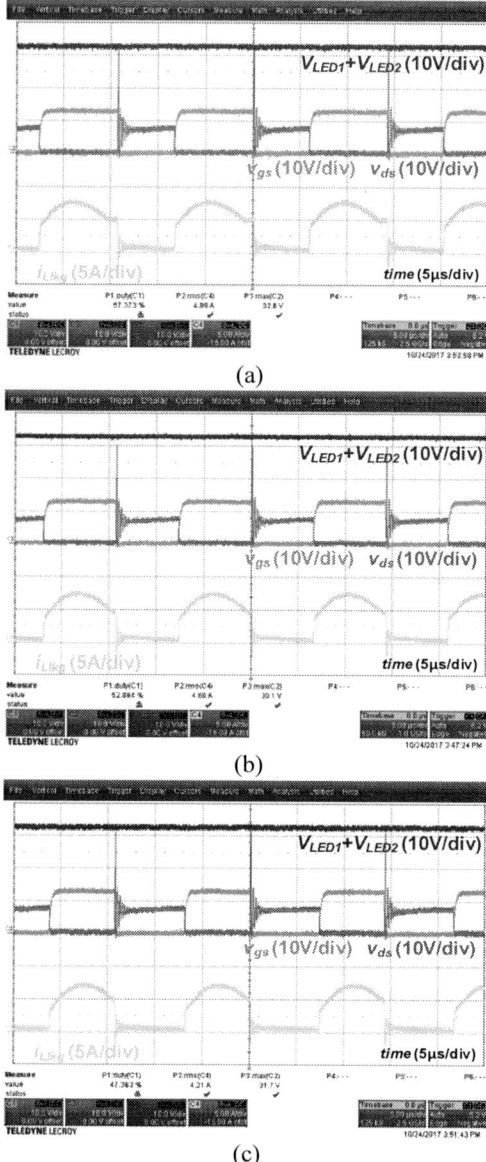

(a)

(b)

(c)

Fig. 5. Key waveforms of the proposed converter at the (a) minimum (b) nominal and (c) maximum input voltage.

linearly. Mode 4 ends when the main switch is turned on and the proposed converter goes back to mode 1.

III. EXPERIMENTAL RESULTS

Based on the operational principles, the prototype of the proposed converter has been constructed with 11.55W rated output. Table I represents the design results of the prototype converter with 70kHz switching frequency. The rated input voltage is 3.3V_{DC} and it has 10% variation. The output voltage/current of each LED is 3.3V/0.35A.

Table II represents the comparison of the number of components with other two-channel LED drivers for similar specification. As shown in the table, the proposed converter has the minimized number of components compared to the other approaches, since the proposed

The 2018 International Power Electronics Conference

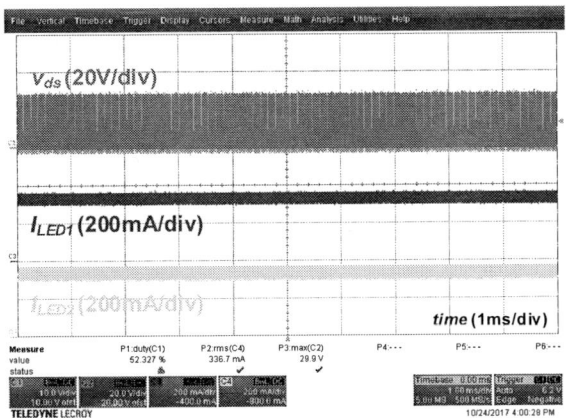

Fig. 6. Current sharing capability of the proposed driver at 3.3V input voltage and the rated output current.

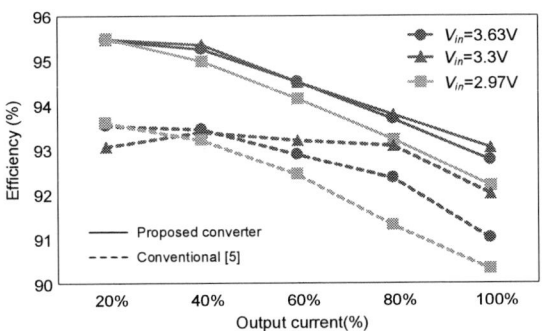

Fig. 7. Efficiency of the proposed converter.

converter does not require any additional snubber diode and snubber capacitor.

Fig. 5 represents the key waveforms of the proposed converter in various input voltages with rated output power. From Fig. 5(a), it can be noted that the voltage stress on the main switch is clamped to the sum of the voltages of each channel, when the input voltage is the minimum. On the other hand, the peak voltage stress on the main switch is lower than the sum of the voltage of each channel in the case of the other input voltages as shown in Fig 5(b) and 5(c). This is because the proposed converter has a small turn off current since it uses a resonant operation.

Also, it can be noted that the duration of Mode 1 in the proposed converter is shorter than the switching on time, so that the proposed converter operates in a below resonance region. By doing so, the rectifier diodes can achieve soft-turnoff. It can be noted from Fig. 5(c) that the on time and the resonant period are the same each other.

Fig. 6 represents the current sharing capability of the proposed converter. As shown in the figure, it can be noted that the proposed converter equally distributes the output current to each channel.

Fig. 7 represents the efficiency of the prototype converter compared to the conventional approach [5]. As shown in the figure, it can be noted that the proposed converter shows a higher efficiency in the entire input

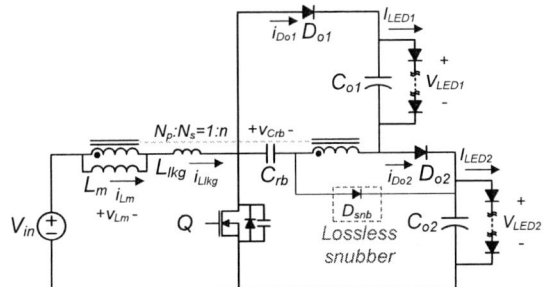

Fig. 8. The proposed converter with a simple lossless snubber.

and output conditions. This is because the proposed converter is based on the tapped-boost converter, whereas the conventional converter in [5] is based on the flyback converter. Since the tapped-boost converter has a higher voltage conversion ratio, the turns-ratio of the transformer can be reduced from 5.08 in [5] to 4.07 in the proposed converter. For these reasons, RMS current in the main switch can be reduced. Also, the core loss of the core can be decreased since the proposed converter can increase the number of turns in the primary side.

IV. SIMPLE LOSSLESS SNUBBER WITH LOWER VOLTAGE STRESS

The proposed driver can be extended with the simple lossless snubber as shown in Fig. 8. In the proposed converter, only one diode with a small current rating can achieve lossless snubber while maintaining the current sharing capability. The equivalent circuits with the proposed converter can be obtained by changing the position of C_r and the secondary side of the transformer. In these cases, the losses snubber can be implemented as shown in Fig. 8 so that the voltage stress on the main switch is clamped to $V_{LED2}+v_{Crb}$ during mode 3. Since v_{Crb} is much lower than V_{LED1}, the voltage can be suppressed effectively. In these configurations, the energy stored in the leakage inductor is delivered to the series connected C_{rb} and C_{o2} during mode 3. According to the charge balance of C_{rb}, the leakage inductance energy absorbed in C_{rb} is delivered to C_{o1} during mode 1 so that the current sharing capability of the proposed converter is preserved. With the lossless snubber, the proposed converter can achieve a higher efficiency since the lossless snubber reduces the switching turn-off loss due to the reduced voltage-current crossover area during the switching turn-off transition. The voltage stress on the snubber capacitor becomes the sum of V_{LED2} and the maximum value of v_{Crb}.

V. CONCLUSION AND FURTHER WORKS

In this paper, a new converter topology for low power two-channel LED driver has been proposed. Since the proposed converter uses the rectifier diodes and the output capacitors as the snubber network also, it can eliminate the snubber components with lossless snubbing. Furthermore, since the proposed converter is based on the tapped-boost converter which has a higher voltage conversion ratio compared to the prior work, it can

3110

achieve a higher efficiency. Therefore, it can be noted that the proposed converter can be a strong candidate for small power LED applications.

In the full paper, the steady state anaylsis, design guidelines, more experimental results, and more references will be added.

References

[1] X. Wu, Z. Wang, and J. Zhang, "Design Considerations for Dual-Output Quasi-Resonant Flyback LED Driver With Current-Sharing Transformer," *IEEE Trans. Power Electron.*, vol. 28, no. 10, Oct. 2013.

[2] X. Wu, J. Zhang, and Z. Qian, "A simple two-channel LED driver with automatic precise current sharing," *IEEE Trans. Ind. Electron.*, vol. 58, no. 10, pp. 4783–4788, Oct. 2011.

[3] K. I. Hwo and W. Z. Jiang, "Single-Switch Coupled-Inductor Based Two-Channel LED Driver with a Passive Regenerative Snubber," *IEEE Trans. Power Electron.*, accepted, vol. 32, no. 6, Dec. 2017.

[4] K. I. Hwo and W. Z. Jiang, "Nonisolated Two-Channel LED Driver With Automatic Current Balance and Zero-Voltage Switching," *IEEE Trans. Power Electron.*, vol. 31, no. 12, Dec. 2016.

[5] J.-W. Kim, J.-M. Choe, and J.-S. Lai, "Non-Isolated Single-Switch Two-Channel LED Driver with Simple Lossless Snubber and Low Voltage Stress," *IEEE Trans. Power Electron.*, accepted.

Design and Implementation of Single-Phase Asymmetric Multilevel STATCOM

Hao Chen[1], Yang Han[*1], Ping Yang[1], Congling Wang[1], Josep M. Guerrero[2]

(1. School of Mechanical and Electrical Engineering, University of Electronic Science and Technology of China,
West High-Tech Zone, Chengdu 611731, China. E-mail: hanyang@uestc.edu.cn
2. Department of Energy Technology, Aalborg University, Aalborg 9220, Denmark, E-mail: joz@et.aau.dk)

Abstract-The design and control of the static synchronous compensator (STATCOM) based on the single-phase asymmetric cascaded H-bridge (ACHB) inverter with different dc voltage values are carried out in this paper. In this paper, the synchronous reference frame (SRF) proportional-integral controller is employed to control the grid current of the ACHB-STATCOM to achieve reactive power compensation. To enhance system stability, an active damping strategy based on the capacitor-current feedback loop is devised to attenuate the resonance of *LCL* filter. Moreover, the hybrid modulation scheme is adopted for the multilevel inverter to generate 19 levels at the output voltage. A reduced-size STATCOM prototype has been built and the experimental results are provided under the inductive and capacitive scenarios, which confirms the effectiveness of the STATCOM based on the designed circuit and the presented control and hybrid modulation scheme.

I. INTRODUCTION

In recent years, multilevel converters are increasingly utilized for STATCOM [1], which shows several benefits such as the multilevel output voltage levels, low switching ripples and reduced-sized output filter. Based on these advantages, the STATCOM based on the asymmetric cascaded H-bridge (ACHB) inverter can be recognized as a promising solution, despite the complex modulation methods and control strategies. In this paper, the ACHB topology with the dc voltage ratio of 1:2:6 is adopted for the STATCOM, where the dc-link voltage ratio is selected by considering a tradeoff between control complexity and the required number of voltage levels [2]. With this ratio, hybrid modulation scheme is established for the STATCOM [3], where the switching signals with different frequency can be applied for each H-bridge module, thus different semiconductor devices can be used and the switching losses can be reduced.

On the other hand, for realizing reactive compensation based on ACHB-STATCOM, the current control scheme is crucially important. In the existing literatures [4]-[6], the various current controllers have been extensively investigated, especially the proportional-resonant (PR) controller, due to its ability to eliminate the steady-state error when tracking the ac signals. However, PR controller suffers the drawbacks of poor dynamic performance and robustness under frequency deviations, which may cause a large tracking error and system instability [6]. For this reason, the synchronous reference frame proportional plus integral (SRF-PI) controller is preferred to overcome these drawbacks for the current

control of single-phase inverters. In [7], the SRF-PI controller was introduced, where the signal of a fictitious phase is generated by the orthogonal signal generator (OSG) technique to emulate a two-phase system, thus the ac signals can be transformed to dc signals in the synchronous frame. Moreover, the conventional PI controller can be used to realize zero steady-state error and fast dynamic response.

This paper introduces the SRF-PI current control scheme with capacitor current feedback based active damping to control the grid current i_g accurately and stably based on the single-phase ACHB-STATCOM. In addition, a reduced-size experimental platform based on the cascaded connection of three H-bridge inverters for ACHB-STATCOM has been designed and established. The system performance with the presented control method and hybrid modulation scheme under different scenarios has been tested experimentally, which validates the effectiveness of the designed circuit and the presented control strategies and hybrid modulation scheme.

II. SYSTEM STRUCTURE AND CONTROL STRATEGY

A. System Structure

The comparison of cascade topologies that consisted of different separated DC voltage cells is illustrated in Table I, where n stands for the number of the cascaded cell. As can be seen, the optimal asymmetry ratio which can give maximal different output levels is 1:3:9, but it's not suitable for the hybrid modulation strategy, because it will lead to an unmodulated part which can't be modulated by the low power cell. And some H-bridge cell may have regenerative power in a lower modulation index, but the inverter outputs real power [11]. Therefore, the mostly used ratio is 1:2:6, which can maximize the number of levels while still being able to apply the hybrid modulation.

TABLE I
THE COMPARISON OF CASCADED TOPOLOGIES ARE CONSISTED OF DIFFERENT SEPARATED DC VOLTAGE CELLS

Ratio of Dc Link	Combination Style	Max Dc Link Voltage	Max Output Voltage	Output Levels
1:1:1	1	E	$n \times E$	$2n+1$
1:2:4	2^{n-1}	$2^{n-1}E$	$(2^n-1)E$	$2^{n+1}-1$
1:2:6	$2 \times 3^{n-2}$	$2 \times 3^{n-2}E$	$0.52(3^n-1)E$	$2 \times 3^{n-1}+1$
1:3:9	3^{n-1}	$2^{n-1}E$	$(3^n-1)E$	3^n

Fig. 1 shows the power stage of the studied single-phase ACHB-STATCOM by using the presented control scheme in the grid-connected mode. As depicted in Fig. 1, the STATCOM is composed of three standard H-bridge cells denoted as Cell1, Cell2, and Cell3 respectively, and connected to the grid through LCL filter, where L_1, C, L_2 constitute the LCL filter; L_g is the equivalent impedance of the grid; v_s is the output voltage of the ACHB inverter. v_g is the grid voltage and each H-bridge cell is provided with a constant and isolated dc source V_{dck} (k=1, 2, 3), where the ratio of dc voltage (V_{dc1} : V_{dc2} : V_{dc3}) is 1:2:6, and each cell generates a three-level output voltage, but at the different switching frequencies.

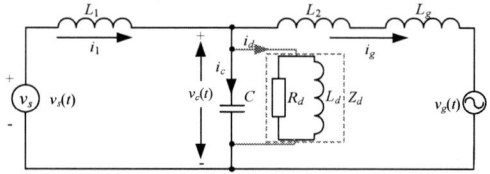

Fig. 1. The diagram of the single-phase ACHB-STATCOM.

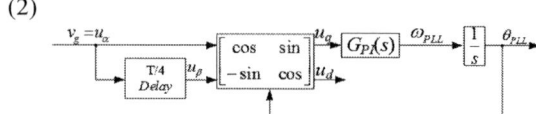

Fig. 2 The simplified circuit for the studied system

Besides, an active damping method based on capacitor current feedback is introduced to suppress the resonance produced by LCL filter, which is equivalent to place an impedance (Z_d) in parallel with filter capacitor C [12]. According to Fig. 1, Z_d can be written as $L_1/M(s)G_d(s)CK$, where the $M(s)$ is the transfer function of the hybrid modulation and can be assumed equal to one [8], and the system delay $G_d(s)$ is set as $e^{-1.5sTs}$ [12]. Thus, the system equivalent and simplified circuit can be depicted, as shown in Fig. 2, and the system-model of the grid-connected ACHB inverter in the s-domain also can be derived as.

$$i_g(s) = G_v(s)v_s(s) + G_u(s)u_g(s)$$

$$= \frac{L_d R_d \sum_{k=1}^{n} \eta_k V_{dck}}{D_0} v_s(s) - \frac{D_1}{D_0} u_g(s) \tag{1}$$

where
$D_0 = s^3 CL_1 L_d(L_2+L_g)R_d + s^2 L_1 L_d(L_2+L_g) + s[L_1(L_2+L_g)+L_d(L_2+L_g)+L_1 L_d]R_d$, $D_1 = s^2 CL_1 L_d R_d + sL_1 L_d + (L_d+L_1)$, and the η_1 .. η_k are the relation between the module terminal voltage. In this paper, n is set as 3, and η_2 =2, η_3=6 [9].

B. PLL model

In addition, as shown in Fig. 1, a single-phase T/4 delayed phase-locked loop (PLL) using single-phase supply voltage. The detailed diagram of T/4 delayed PLL is shown in Fig. 3, where the $G_{PI}(s)=k_{p,pll}+k_{i,pll}/s$.

From the Fig. 3, the model of the adopted PLL can be established and the transfer function $G_{PLL}(s)$ can be derived as [10]

$$G_{PLL}(s) = \frac{k_{p,pll}s + k_{i,pll}}{s^2 + u_d\left(k_{p,pll}s + k_{i,pll}\right)}$$
(2)

Fig. 3 The diagram of T/4 delayed PLL.

C. SRF-PI controller model

Fig. 1 depicts the block diagrams for the current control implemented by the stationary reference frame equivalent of the SRF-PI, where two identical conventional PI controllers are introduced in the rotating (dq-) frame, which can be written as $PI(s)=k_p+k_i/s$.

In the time domain, the model of the SRF-PI controller can be written as [7]:

$$\begin{bmatrix} i^*_{c,\alpha}(t) \\ i^*_{c,\beta}(t) \end{bmatrix} = \begin{bmatrix} \cos\theta_{PLL} & -\sin\theta_{PLL} \\ \sin\theta_{PLL} & \cos\theta_{PLL} \end{bmatrix} \cdot$$

$$\left\{ \begin{bmatrix} PI(t) & 0 \\ 0 & PI(t) \end{bmatrix} * \left\{ \begin{bmatrix} \cos\theta_{PLL} & \sin\theta_{PLL} \\ -\sin\theta_{PLL} & \cos\theta_{PLL} \end{bmatrix} \begin{bmatrix} i_\alpha(t) \\ i_\beta(t) \end{bmatrix} \right\} \right\}$$
(3)

By applying the Park transformation, inverse Park transformation and Laplace transform, the s-domain model of (1) can be derived.

$$\begin{bmatrix} i^*_{c,\alpha}(s) \\ i^*_{c,\beta}(s) \end{bmatrix} = \begin{bmatrix} \underbrace{\left(\frac{k_p s^2 + k_p \omega_{PLL}^2 + k_i s}{s^2 + \omega_{PLL}^2} - \frac{k_i \omega_{PLL}^2 - k_i \omega_{PLL}s}{(s^2 + \omega_{PLL}^2)(\omega_{PLL}+s)} \right) i_\alpha(s)}_{G_{SRF-PI\_\alpha}} \\ \underbrace{\left(\frac{k_i \omega_{PLL}^2 + k_i \omega_{PLL}s}{(s^2 + \omega_{PLL}^2)(\omega_{PLL}-s)} + \frac{k_p s^2 + k_p \omega_{PLL}^2 + k_i s}{s^2 + \omega_{PLL}^2} \right) i_\beta(s)}_{G_{SRF-PI\_\beta}} \end{bmatrix}$$
(4)

Finally, the transfer function of the SRF-PI controller in α-axis and β-axis can be obtained. However, only α-axis quantities belong to the real system in a pseudo-two-phase system, just the real controller G_{SRF-PI_α} is applied to the current inner loop, which can be written as:

$$G_{SRF-PI\_\alpha}(s) = \frac{k_p s^2 + k_p \omega_{PLL}^2 + k_i s}{s^2 + \omega_{PLL}^2} - \frac{k_i \omega_{PLL}^2 - k_i \omega_{PLL}s}{(s^2 + \omega_{PLL}^2)(\omega_{PLL}+s)} \tag{5}$$

Fig. 4 Amplitude curves of transfer function $G_{SRF-PI\_\alpha}(s)$ under different values of k_p and k_i.

3113

Fig. 4 illustrates the amplitude curves of transfer function $G_{SRF\text{-}PI\_\alpha}(s)$ under different values of k_p and k_i, where ω_{PLL} is set as $2\pi50$ rad/s. It can be seen that there is always a very high gain at the fundamental frequency with k_p and k_i varied, which not only ensures a zero steady-state error at fundamental frequency, but also simplify the parameters design.

III. SYSTEM OPERATION

As shown in Fig. 1, assuming the grid voltage is $v_g=V_{gm}\sin(\omega_{PLL}t+\varphi)$, thus the reference current of i_g is obtained as $i_{g,ref}=I_{gm}\sin(\omega_{PLL}t+\varphi)$, where the grid angular frequency ω_{PLL} (100π rad/s) is calculated and provided by PLL, and the power factor angle φ is set by researchers. Obviously, when $i_{g,ref}=I_{gm}\sin(\omega_{PLL}t+90°)$ the ACHB-STATCOM operates in the capacitive mode to inject reactive power into the grid. When $i_{g,ref}=I_{gm}\sin(\omega_{PLL}t-90°)$, the ACHB-STATCOM operates in the inductive mode to absorb reactive power from the grid. And the amount of the reactive power Q exchanged between the STATCOM and the grid in steady state can obtained by (6).

$$Q = \frac{1}{2}V_{gm}I_{gm} \qquad (6)$$

where V_{gm} and I_{gm} denote the grid voltage amplitude and the grid current amplitude, respectively.

In current control, the phase positions of i_g and $i_{g,ref}$ are both delayed for $90°$ to generate two fictitious phases to emulate a two-phase system. By using Park's transformation, i_g and $i_{g,ref}$ are transformed to the synchronous frame with the fictitious phase signals. In order to ensure a zero steady-state error, two identical conventional PI controllers in the SRF are employed to regulate the current signals of the d axis and q axis. Subsequently, the output signals of the PI controllers are transformed back to the stationary frame by using the inverse Park's transformation, and the α-axis output current $i_{cref,\alpha}$ is applied for the control system since the β-axis output current $i_{cref,\beta}$ is neglected, because just α-axis quantity corresponds to the single-phase system.

Meanwhile, an active damping method based on capacitor current feedback is adopted to damp the resonance caused by the LCL filter and enhance stability of the system [13]. As shown in Fig. 1, K is the active damping coefficient and the α-axis output $i_{cref,\alpha}$ is used as the reference signal for capacitor current feedback, hence the active damping method has been achieved.

Furthermore, the output signal (u_m) of the current control, as illustrated in Fig. 1, is applied for the hybrid modulation. The detailed principle of hybrid modulation is shown in Fig. 5 and TABLEII, where the simulation is carried out in SIMULINK based on $V_{dc}=1$V and $u_m=9\sin(100\pi\times t)$ to validate the hybrid modulation. In this modulation method, dc values are used as the carriers for Cell3 and Cell2, and only the Cell1 is regulated by the unipolar PWM technique applying the high-frequency triangular carrier. Finally, the 19 levels output voltage can be synthesized by the ACHB inverter and the different semiconductor devices can be used for each

inverter module, thus total switching losses can be reduced and the converter efficiency of ACHB inverter is improved [3].

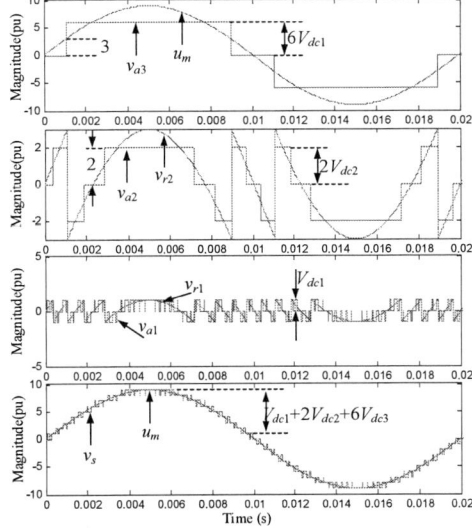

Fig. 5 The generation of output voltage waveforms in the ACHB inverter with dc voltage ratio of 1:2:6.

TABLE II
THE PRINCIPLE OF HYBRID MODULATION

The desired output	The output of Cell3	The output of Cell2	The output of Cell1
9 ◄──► 8	6	2	1 ◄──► 0
8 ◄──► 7	6	2	0 ◄──► -1
7 ◄──► 6	6	0	1 ◄──► 0
6 ◄──► 5	6	0	0 ◄──► -1
5 ◄──► 4	6	-2	1 ◄──► 0
4 ◄──► 3	6	-2	0 ◄──► -1
3 ◄──► 2	0	2	1 ◄──► 0
2 ◄──► 1	0	2	0 ◄──► -1
1 ◄──► 0	0	0	1 ◄──► 0
0 ◄──► -1	0	0	0 ◄──► -1
-1 ◄──► -2	0	-2	1 ◄──► 0
-2 ◄──► -3	0	-2	0 ◄──► -1
-3 ◄──► -4	-6	2	1 ◄──► 0
-4 ◄──► -5	-6	2	0 ◄──► -1
-5 ◄──► -6	-6	0	1 ◄──► 0
-6 ◄──► -7	-6	0	0 ◄──► -1
-7 ◄──► -8	-6	-2	1 ◄──► 0
-8 ◄──► -9	-6	-2	0 ◄──► -1

IV. EXPERIMENTAL RESULTS

To validate the presented system structure and control strategy, a downscaled experimental ACHB-STATCOM has been constructed, as shown in Fig. 6, where L_1, L_2 and C are set as 2mH, 2mH, and 224µF, respectively. The isolated and constant dc-link voltages of the individual H-bridge cells are provided by DC power supplies, and the dc voltage V_{dc1}, V_{dc2}, V_{dc3} are equal to 3V, 6V, 18V, respectively. Besides, V_{gm} is set to be 25V, and L_g is set as 1.78mH. The control algorithm is implemented in TMS320F28335 digital signal processor (DSP), where the sampling frequency and switching frequency of low voltage cell are set as 10 kHz.

The 2018 International Power Electronics Conference

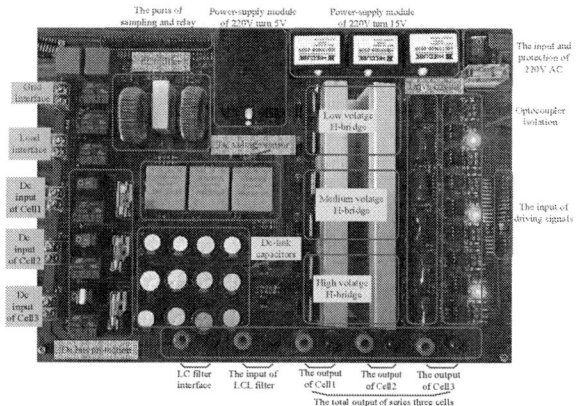

Fig. 6. Photo of the experiment setup for ACHB-STATCOM.

The detailed discussion about parameter design of the SRF-PI controller and capacitor current feedback active damping method for the conventional two-level inverters can be found in [7], [13], [14].

Followed by these parameter design guidelines, the proportional gain k_p and the integral gain k_i of the SRF-PI controller, and the active damping coefficient K are determined experimentally in order to ensure system stability and dynamic performance. Hence, according to the experimental tests, an accurate steady state tracking and fast dynamic response is achieved with k_p=0.3, k_i=10, K=0.8, $k_{p,pll}$=1.3, and $k_{i,pll}$=100.

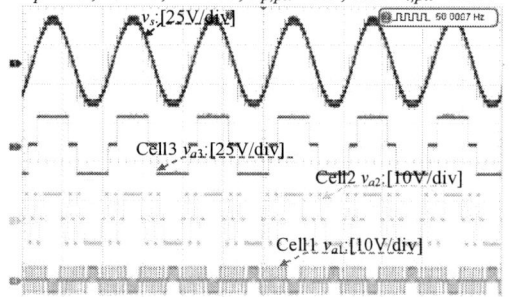

Fig. 7. The experimental waveforms of the hybrid modulation scheme

Fig. 7 shows the experimental results of the presented hybrid modulation scheme, where the output waveforms of each Cell and the synthesized voltage (v_s) are depicted. It can be observed that the synthesized voltage contains 19 levels, which leads to approximately sinusoidal voltage waveform and smaller-sized output filter. Besides, the experimental results under inductive and capacitive operation modes are presented as follows.

A. Experimental verification for the inductive mode of the ACHB-STATCOM

Under this scenarios, the reference $i_{g,ref}$ =$I_{gm}\sin(\omega_{PLL}t$ - 90°), and the experimental results are shown in Fig. 8. It can be observed that i_g is periodic and sinusoidal in the steady state. Moreover, i_g lags v_g by 90° in phase, which indicates that the STATCOM is injecting reactive power into the grid.

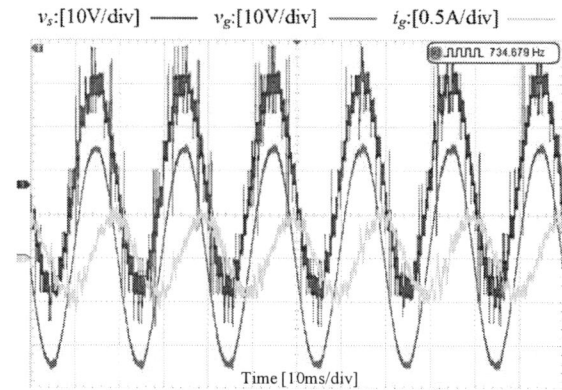

Fig. 8. Experimental waveforms for the ACHB-STATCOM under the inductive operation mode

B. Experimental verification for the capacitive mode of the ACHB-STATCOM

The experimental result in case of $i_{g,ref}$= $I_{gm}\sin(\omega_{PLL}t$ +90°) is shown in Fig. 9. It can be seen that i_g is periodic and sinusoidal in the steady state, additionally, the grid current i_g is leading v_g by 90 degrees in phase, which indicates that the STATCOM absorbs reactive power from the grid.

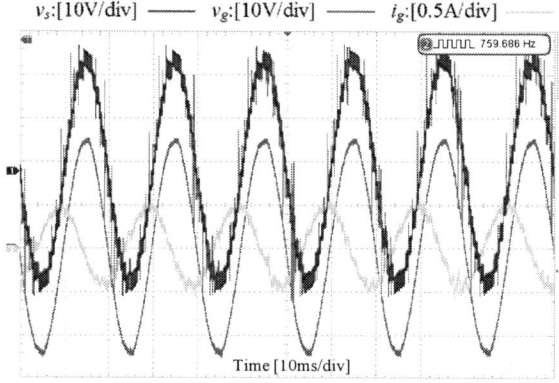

Fig. 9. Experimental waveforms for the ACHB-STATCOM under the capacitive operation mode

To conclude, the experimental results from Fig.8 and Fig.9 suggest that the proposed ACHB-STATCOM is able to provide dynamic reactive power compensation according to the reference current, which validates the effectiveness of the ACHB inverter with SRF-PI current control strategy with inner active damping loop.

V. CONCLUSION

This paper presents the design and control of ACHB-STATCOM based on the single-phase cascaded H-bridge asymmetric cascaded multilevel inverter. To achieve dynamic reactive power compensation, the synchronous reference frame (SRF) PI current control method integrated with inner active damping loop based on capacitor current feedback is adopted to ensure a zero steady-state error and improve dynamic response. Besides, the hybrid modulation method has been employed to maximize the output voltage levels of

3115

ACHB-STATCOM. Moreover, a reduced-size laboratory prototype ACHB-STATCOM is built. The experimental results confirm that the designed system can compensate reactive power stably and effectively. The detailed analysis on controller synthesis, stability, robustness of the ACHB-STATCOM would be presented shortly.

Acknowledgement

This work has been supported by the National Natural Science Foundation of China under grant No.51307015, and in part by the Fundamental Research Fund of Central Universities of China under grant No.ZYGX2015J087.

REFERENCES

[1] Y. Zhang, X. Wu and X. Yuan, "A Simplified Branch and Bound Approach for Model Predictive Control of Multilevel Cascaded H-Bridge STATCOM," *IEEE Trans. Ind. Electron.*, vol. 64, no. 10, pp. 7634-7644, Oct. 2017.

[2] S. Mariethoz and A. Rufer, "Design and control of asymmetrical multilevel inverters," *in Proc. Annu. Conf. IEEE Ind. Electron. Soc., Nov. 2002*, vol. 1, pp. 840–845.

[3] S. Kouro, R. Bernal, H. Miranda, J. Rodriguez, and J. Pontt, "Direct torque control with reduced switching losses for asymmetric multilevel inverter fed induction motor drives," *Conference Record of the 2006 IEEE Industry Applications Conference Forty-First IAS Annual Meeting*, Tampa, FL, 2006, pp. 2441-2446.

[4] H. Komurcugil, N. Altin, S. Ozdemir, and L. Sefa, "Lyapunov-function and proportional-resonant-based control strategy for single-phase grid-connected VSI with LCL filter," *IEEE Trans. Ind. Electron.*, vol. 63, no. 5, pp. 2838-2849, May. 2016.

[5] M. C. Kisacikoglu, M. Kesler, and L. M. Tolbert, "Single-phase on-board bidirectional PEV charger for V2G reactive power operation," *IEEE Trans. Smart Grid*, vol. 6, no. 2, pp. 767–775, Mar. 2015.

[6] X. A. Fu, and S. H. Li, "Control of single-phase grid-connected converters with LCL filters using recurrent neural network and conventional control methods," *IEEE Trans. Power Electron.*, vol. 31, no. 7, pp. 5354-5364, Jul. 2016.

[7] M. Monfared, S. Golestan, and J. M. Guerrero, "Analysis, design, and experimental verification of a synchronous reference frame voltage control for single-phase inverters," *IEEE Trans. Ind. Electron.*, vol. 61, no. 1, pp. 258-269, Jan. 2014.

[8] K. H. Ahmed and G. P. Adam, "New modified staircase modulation and capacitor balancing strategy of 21-level modular multilevel converter for HVDC transmission systems," *in 7th IET International Conference on Power Electronics, Machines and Drives (PEMD 2014)*, 2014, pp.1–6.

[9] T. Busarello, A. Mortezaei, H. Morales-Paredes, A. Al-Durra, J. Antenor Pomilio and M. Simoes, "Simplified small-signal model for output voltage control of asymmetric cascaded H-bridge multilevel inverter," *IEEE Trans. Power Electron.*, vol. PP, no. 99, pp. 1-1.

[10] X. Wang, L. Harnefors and F. Blaabjerg, "A unified impedance model of Grid-Connected voltage-source converters," *IEEE Trans. Power Electron.*, vol. PP, no. 99, pp. 1-1.

[11] S. Wenig, F. Rojas, K. Schönleber, M. Suriyah and T. Leibfried, "Simulation framework for DC grid control and ACDC interaction studies based on modular multilevel converters," *IEEE Trans. Power Delivery*, vol. 31, no. 2, pp. 780-788, April 2016.

[12] D. Pan, X. Ruan, C. Bao, W. Li and X. Wang, "Capacitor-current-feedback active damping with reduced computation delay for improving robustness of LCL-type Grid-Connected inverter," *IEEE Trans. Power Electron.*, vol. 29, no. 7, pp. 3414-3427, July 2014

[13] X. Li, X. Wu, Y. Geng, X. Yuan, C. Xia and X. Zhang, "Wide damping region for LCL-Type Grid-Connected inverter with an improved capacitor-current-feedback method," *IEEE Trans. Power Electron.*, vol. 30, no. 9, pp. 5247-5259, Sept. 2015.

[14] Y. Han, X. Fang, P. Yang, C. Wang, L. Xu, J.M. Guerrero, Stability analysis of digital controlled single-phase inverter with synchronous reference frame voltage control, *IEEE Trans. on Power Electronics*, published online, DOI: 10.1109/TPEL.2017.2746743.

The 2018 International Power Electronics Conference

Submodule Voltage Balancing and Loss Equalisation in Alternate Arm Converters Based on Virtual Voltages

Georgios Konstantinou*, Harith R. Wickramasinghe*, Salvador Ceballos†, and Josep Pou‡

* The University of New South Wales (UNSW Sydney), Sydney, NSW, 2052, Australia
† Tecnalia Research and Innovation, Derio, Spain
‡ School of Electrical and Electronic Engineering, Nanyang Technological University, Singapore
email:g.konstantinou@unsw.edu.au, harith@student.unsw.edu.au, salvador.ceballos@tecnalia.com, j.pou@ntu.edu.sg

Abstract—**Balancing of the submodule (SM) capacitor voltages in the alternate arm converter (AAC) exhibits a great amount of similarities with other modular topologies, especially the modular multilevel converter (MMC). However, the requirements of operation at overmodulation range and the alternate conduction of the phase current further complicate the voltage balancing task. In this paper, a voltage balancing and loss equalisation approach based on virtual SM voltages in combination with sorting algorithms is introduced. The developed approach uses the switching states of the SMs as feedback and provides reduction of the switching frequency of the SMs, further addressing excessive switching due to component parameter variation, tolerances and faults. The effectiveness of the algorithm is demonstrated with simulation results from an equivalent full switching model of an AAC for HVDC applications.**

Index Terms—**Alternate arm converter, modular multilevel converter, multilevel converters, loss equalisation, voltage balancing, virtual voltages**

I. INTRODUCTION

Multilevel voltage source converters (VSCs) have become a mature and proven technology in high-voltage direct current (HVDC) transmission systems due to their prominent features that are highly beneficial for such application [1], [2]. The modular multilevel converter (MMC) [1], [3] is the state-of-the-art VSC, used in the majority of existing and upcoming VSC-HVDC systems. In addition to the common characteristics among VSCs, the unique features of MMCs in high-power applications are: i) modularity, ii) scalability, iii) relatively simple capacitor voltage balancing, iv) ability to handle wide power and voltage ratings, and v) redundant configuration [4]. MMCs provide near sinusoidal output voltage owing to the large number of voltage levels. The typical half-bridge submodule (HB-SM) based MMC is the most commonly used configuration for current HVDC applications, but this configuration suffers from the lack of dc-fault tolerant capabilities [5].

Dc-fault protection, including detection and isolation of transients in a dc system, is among the key requirements of VSC-HVDC converters. This becomes even more important in dc grids, as the faster fault transients may lead to system instability if not treated appropriately. The options to achieve fault-

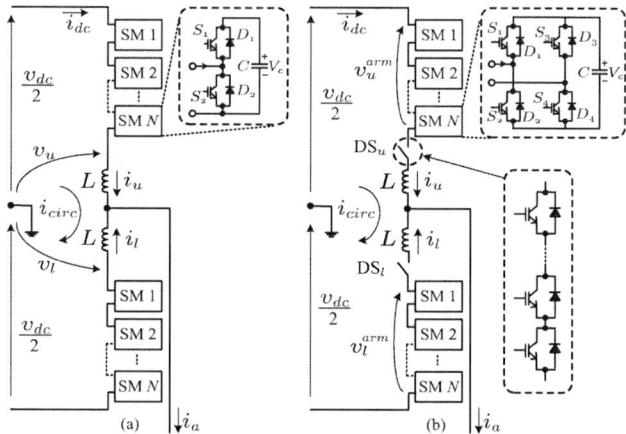

Fig. 1. MMC and AAC circuit configurations; (a) MMC phase-leg and (b) AAC phase-leg.

tolerant operation for the MMC include *i)* bipolar SMs [6] combined with dc isolators, or *ii)* dc-breakers [7] on point-to-point and multiterminal networks. The first option comes at the cost of increased conduction losses and the latter requires a costly and bulky dc-breaker together with a dc-side inductor to limit the rate of rise of current in the dc-side, allowing the protection system to operate effectively [8].

An alternative approach considers modular topologies that have been designed with fault-tolerant characteristics [9]–[12]. The most prominent amongst this family of converters is the alternate arm converter (AAC) [10], [13]–[16]. The AAC shares common aspects with the MMC, especially those related to the modular nature of both topologies. Operational challenges of the MMC can be reflected to the AAC, while system concepts and control approaches that have been successfully applied to one topology can be considered, within certain limitations, for the other.

A key requirement for both the MMC and the AAC is balancing of the SM capacitor voltages within the arms of the converters [17]; an operational requirement that must be fulfilled in steady-state and transient conditions also considering

The 2018 International Power Electronics Conference

converter and SM faults, aging and component deterioration. The objective of this paper is to present a voltage balancing approach for the AAC based on virtual SM capacitor voltages. The proposed approach alters the values of the voltages seen by the sorting stages of the balancing algorithm in order to achieve secondary goals including switching frequency minimisation and power loss equalisation.

The paper is organised in the following manner. Section II provides an introduction to the general operating principles of the AAC and its energy balancing, which is among the key differences between the MMC and the AAC. Section III introduces the voltage balancing algorithm and the virtual voltages including an extended implementation that provides loss equalisation. Extended simulation results are provided in Section IV based on a full switching model of an AAC benchmark topology for HVDC applications while the conclusions of the work will be summarised in Section V.

II. The Alternate Arm Converter

A. Topology and Operating Principles

The AAC [10] is a member of the same family of multilevel converters as the modular multilevel converter (MMC) both in terms of circuit configuration, sharing the two-arm per phase-leg structure, and control requirements; one phase-leg for both the MMC and the AAC are shown in Fig. 1.

A major difference between the two topologies is the presence of the two director switches (DS_u and DS_l), one in the upper and one in the lower arm, leading to the alternate conduction (hence the name of the converter) of the load current from either the upper or the lower arm. The director switches operate at fundamental frequency and are typically switched under zero current (ZCS) in order to avoid increasing the losses of the topology. Additionally, the AAC requires fully controllable bipolar SMs as it needs to generate voltages in the over-modulation range ($M \geq 1$) during its normal operation depending on the stored energy balancing requirements [14]. Also, unlike the MMC, additional complexity in the AAC SMs does not lead to further functionalities and the full-bridge SM offers the optimal topological choice.

B. Energy Balancing

The two director switches, DS_u and DS_l, operate alternatively, during the positive and negative half-cycles of the output reference voltage $v_{am} = m_a \cos(\omega t)$, where m_a is the modulation index. Therefore, the total arm voltage of the AAC ($\sum_1^N V_C$) does not need to be equal to the total dc-link voltage as in the MMC. The number of SMs per arm:

$$N = \left\lceil \frac{\hat{V}_a}{V_C} \right\rceil, \tag{1}$$

is determined based on the requirement for blocking the peak ac-voltage ($\hat{V}_a$), where V_C is the nominal SM capacitor voltage.

Inherent energy balancing exists only when the net exchanged energy of an arm is zero over each half-cycle [10].

Considering the output current $i_a = \hat{I}_a \cos(\omega t + \phi)$, the exchanged energy in the dc-link E_{dc} and the ac-side E_{ac} are:

$$E_{dc} = \int_{-\frac{\pi}{2}}^{\frac{\pi}{2}} \frac{V_{dc}\hat{I}_a}{2} \cos(\omega t + \phi) \, d\omega t, \tag{2}$$

and

$$E_{ac} = \int_{-\frac{\pi}{2}}^{\frac{\pi}{2}} \frac{m_a V_{dc}\hat{I}_a}{2} \cos(\omega t) \cos(\omega t + \phi) \, d\omega t. \tag{3}$$

The net energy of the SM capacitors within an arm can be determined by (2) and (3) as:

$$\Delta E_{arm} = E_{dc} - E_{ac} = \frac{\pi V_{dc}\hat{I}_a \cos\phi}{4\omega} \left(\frac{4}{\pi} - m_a \right). \tag{4}$$

Eq. (4) shows that the operating point for the natural energy balance ("sweet-spot") is $m_a = M_a = 4/\pi$ where ΔE_{arm} becomes zero at this operating point [10].

Unlike the MMC, there is no continuous energy exchange between the upper and lower arm of the AAC due to the alternate operation of the arms. Instead, an overlap period can be utilized around the zero crossing points of v_{am} by allowing both DSs to conduct at the same time and the energy is exchanged using a circulating current (i_{circ}). The overlap operation of the AAC mimics the MMC operation. Hence, the two arm currents (i_u and i_l) during the overlap period can be expressed as,

$$i_u = \frac{i_a}{2} + i_{circ}, \tag{5}$$

$$i_l = \frac{i_a}{2} - i_{circ}. \tag{6}$$

Circulating current dynamics during the overlap periof depend on the arm inductance and the differential voltage across the inductors as:

$$v_{diff} = L\frac{d}{dt}i_{circ}, \tag{7}$$

applied across the arm inductor.

The amount of energy exchanged within an arm by the circulating current depends on the duration of overlap period, and the magnitude and direction of the circulating current. Hence, energy balancing can be achieved by appropriately controlling the circulating current during the overlap period, while maintaining ZCS. Methods of overlap control for the AAC include *i)* symmetric/asymmetric overlap period, *ii)* fixed overlap period, *iii)* regulation of a variable overlap period, *iv)* resonance based circulating current control or *v)* active utilisation of redundant SMs for hysteresis current control to meet the energy balancing goals.

III. Voltage Balancing for the AAC

A. Balancing Algorithm with Virtual Voltages

The concept of virtual voltages for voltage balancing relies on the modification of the value of the capacitor seen by the sorting stage in order to achieve secondary control and regulation goals. The direct implementation for the AAC is shown in Fig. 2. Instead of using the measured capacitor

3118

voltages, i.e. v_{sm} directly in the sorting stages of the balancing algorithm, their sign is modified based on both the arm current direction ($sgn(i_{arm})$) as well as the modulating signal. This is a key difference in the implementation between the AAC and the MMC as the AAC normally operates in the overmodulation region where the connection of SMs leads to an opposite effect on the capacitor voltages. This logic defines the "Arm Current Logic" block in Fig. 2. Reduction in the switching frequency can be achieved by adding an offset K to the voltages of the SMs that are connected to each arm based on the switching states of the SMs, S_w [18].

B. Loss Equalisation based on Switching Transitions

As the derivation of the switching signals directly depends on the capacitor voltages, any deviation between the values of the capacitors due to tolerances, component aging or partial SM faults will have a detrimental effect on the switching frequency of individual SMs [19] leading to unequal transitions and unequal loss distribution between the different SMs.

In order to address this issue, a loss equalisation loop is included in the voltage balancing algorithm, considering the average transitions of the SMs within a given time interval. The virtual voltages v''_{sm} are further modified with the addition of a second offset to generate the final virtual voltages, v'''_{sm}, as shown in Fig. 3(a). The aim of the additional offset is to change the activation priority of SMs depending on the average deviation of their switching frequency from the average across the whole arm. This means that SMs that have a higher switching frequency compared to the average will drop in priority over those that have switched less. The offset is calculated through a cumulative calculation of transitions in the whole arm, shown in Fig. 3. Unlike the offset for switching frequency reduction that has a constant value, the value of the offset is defined by the deviation of each SM from the average switching frequency; the lower part of the algorithm determines the sign at which the offset is added depending on the switching state of the SM.

C. Loss Equalisation based on Loss Modeling

The method introduced in sub-section III-B provides a generic approach to the balancing of the switching losses of the AAC based on the assumption that every transition is weighted equally; this is a result of considering the arithmetic average of transitions. This leads to satisfactory results considering the simplicity of the method. A more accurate reflection can

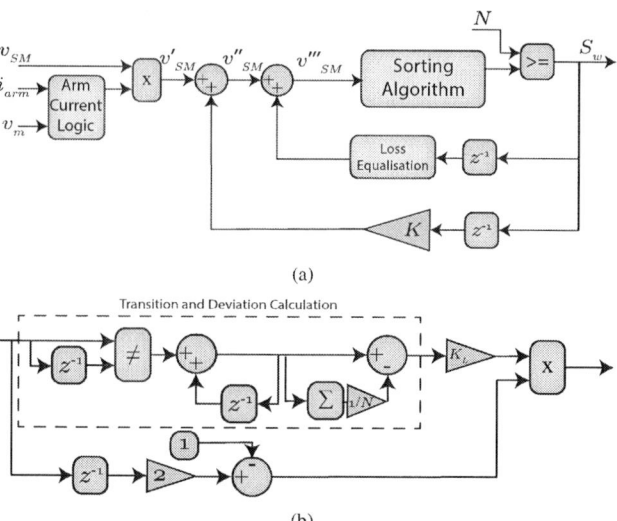

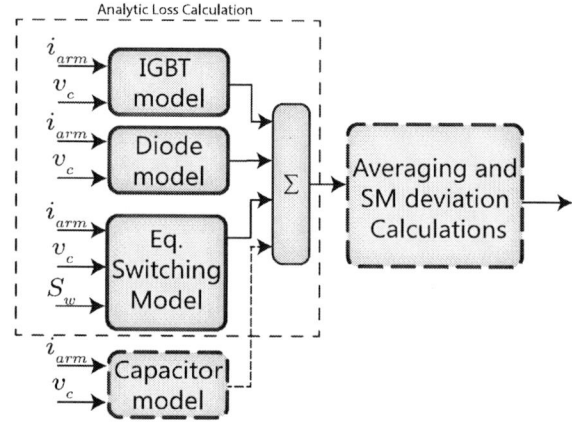

Fig. 3. Extension of the virtual voltage concept for loss equalisation, (a) extended implementation and, (b) loss equalisation algorithm.

Fig. 4. SM voltage balancing for the Alternate Arm Converter based on virtual voltages.

be made assuming an accurate loss model of the AAC that considers both the characteristics of the switching devices as well as the instant of each individual transition. In this case, the weighing and compensation is done of the basis of actual losses, as shown in Fig. 4, rather than switching transitions. Both detailed and simplified loss models [20] can also be considered for this calculation However, the computational requirements of the loss estimation method are relatively high for the FB-SMs of the AAC and this method will not be considered at this stage in the following analysis.

D. Loss Equalisation based on Capacitance Values

Assuming accurate knowledge of the SM capacitor values and operation with phase-shifted PWM (PS-PWM), feed-forward compensation methods that modify the individual duty cycles of each SM can also be considered. Here, following

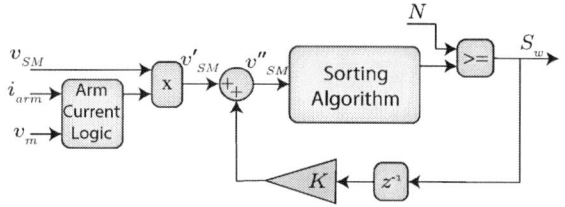

Fig. 2. SM voltage balancing for the Alternate Arm Converter based on virtual voltages.

TABLE I
PARAMETERS OF THE AAC BENCHMARK CONVERTER STATION

Parameter	AAC
Rated Power	400 MVA
DC Voltage	±200 kV
Number of SMs per Arm	8
SM Voltage	35 kV
SM Capacitance	150 μF
Arm Inductance (p.u.)	0.0019
Fundamental Frequency	50 Hz
Transformer Resistance (p.u.)	0.0015
Transformer Leakage Inductance (p.u.)	0.045
Transformer Ratio	1.218
Grid Voltage	380 kV

an estimation [21] of the capacitor in each SM, a gain can be calculated for the reference of each SM to account for the unbalance due to capacitance variation achieving voltage or losses balancing. However, the relatively high switching frequency of each SM, the need for continuous estimation of the capacitance values and the overmodulation operation of the AAC further complicate the practical application of such an approach. The issues become more pronounced as the number of SMs within the arms of the converter increase.

IV. SIMULATION RESULTS

The operation of the algorithm is demonstrated through simulation results from an AAC model for a ±200 kV HVDC application. The parameters of the converter have been derived from the conversion of an AAC with 255 SMs per arm [22] to an equivalent one with 8 SMs per arm, so that the impact of the algorithm for relatively small deviations can be more pronounced. The parameters of the AAC converter station are given in Table I.

Energy balancing of the AAC is achieved through a typical fixed overlap period, described in Section II-B at an operating point of $m_a = 1.15$, which is slightly lower than the sweet-spot, and with the reactive power reference set to 0 MVAr. The steady-state results (steady-state output voltages and currents of the AAC) for this operating point are given in Fig. 5. The currents through the arms with the overlap period that provide energy balancing and the corresponding states of the director switches are also shown in Fig. 6.

Two distinct cases of SM capacitance deviations are considered in this article; deviation of a single SM capacitance and deviation of multiple capacitors within an arm of the converter. These represent a general case of a single faulted SM as well as an extreme case of variability for capacitances within the arm.

A. Single SM Capacitor Deviation

In the first case, a single SM in the upper arm of phase a is assumed to have 50% of the rated capacitance leading to a larger SM capacitor voltage deviation. As the switching frequency is already restricted, the SM with the lower capacitance

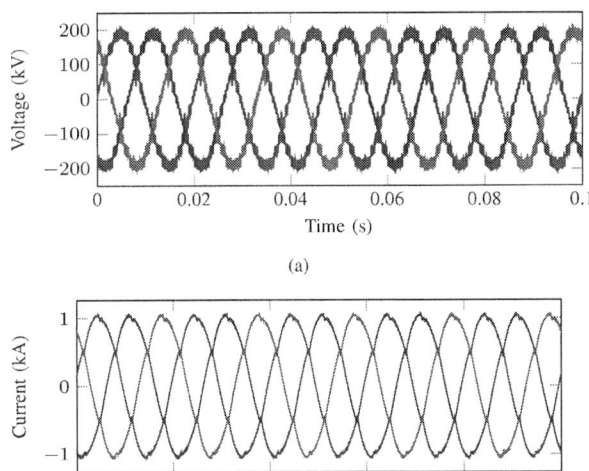

Fig. 5. Output waveforms, (a) converter voltages and (b) output currents of the AAC in steady state operation.

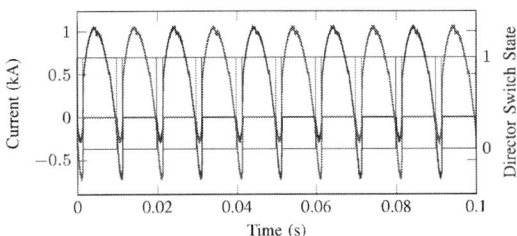

Fig. 6. Arm currents and state of the director switches under steady-state operation.

will not switch continuously (something that would happen if the first feedback loop is not implemented), but it will switch more than the other SMs.

The SM capacitor voltages are shown in Fig. 7, where the greater deviation of SM_{1a} can be clearly observed. The total losses (calculated instantaneously over a 20-ms moving window) of each SM in the upper arm of phase a of the AAC are shown in Fig. 8. Firstly, the alternate operation of the arms in the AAC, leads to the losses remaining relatively constant for approximately half of a fundamental period (minus the overlap period). The maximum deviation in the losses of the SMs (the normalised different between the maximum and minimum average losses) is approximately 21%.

The application of the loss equalisation method reduces the switching of the SM due to the voltage deviations. This leads to a greater deviation of the SM capacitor voltage of the SM with the lower capacitance, as shown in Fig. 9. The implementation of the algorithm leads to a reduction in the average losses (through a reduction in the switching losses of the arm by eliminating additional transitions between SMs), and a maximum deviation of approximately 11%, a 50%

The 2018 International Power Electronics Conference

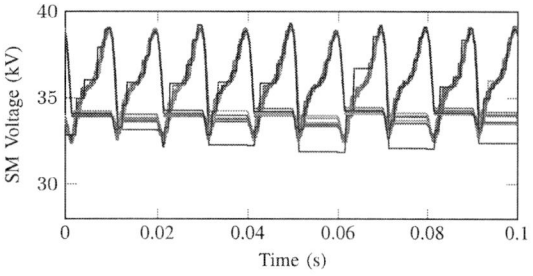

Fig. 7. Single capacitor deviation: SM capacitor voltages with the virtual voltages balancing algorithm of Fig. 2.

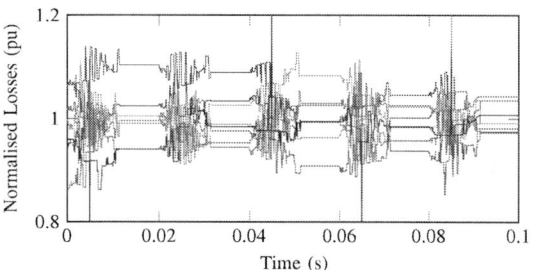

Fig. 8. Single capacitor deviation: Normalised average losses of all SMs in the upper arm of phase a with the balancing algorithm of Fig. 2.

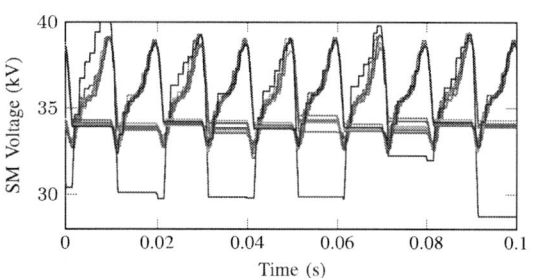

Fig. 9. Single capacitor deviation: SM capacitor voltages with the loss equalisation loop of Fig. 3.

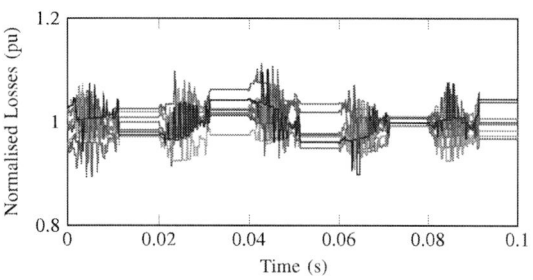

Fig. 10. Single capacitor deviation: Normalised average losses of all SMs in the upper arm of phase a with the balancing algorithm of Fig. 3.

reduction compared to the case where the loss equalisation loop is not implemented.

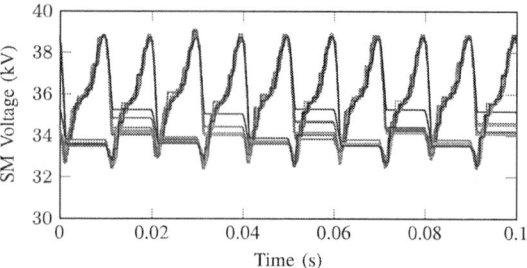

Fig. 11. Multiple SM capacitor deviations: SM capacitor voltages with the virtual voltages balancing algorithm of Fig. 2.

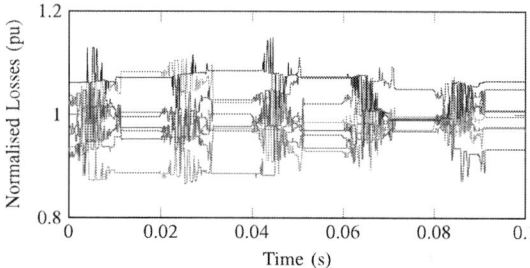

Fig. 12. Multiple SM capacitor deviations: Normalised average losses of all SMs in the upper arm of phase a with the balancing algorithm of Fig. 2.

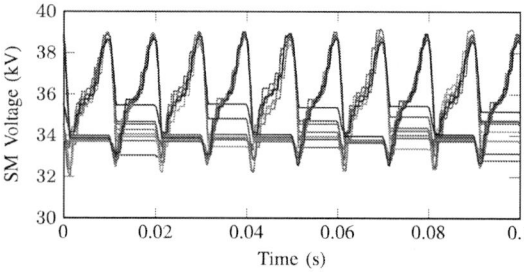

Fig. 13. Multiple SM capacitor deviations: SM capacitor voltages with the loss equalisation algorithm of Fig. 3.

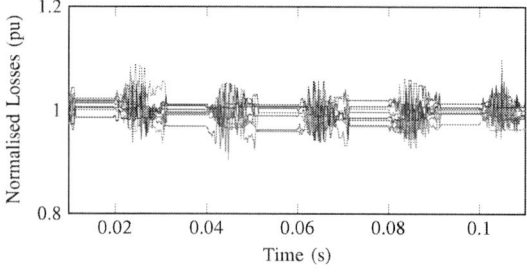

Fig. 14. Multiple SM capacitor deviations: Normalised average losses of all SMs in the upper arm of phase a with the balancing algorithm of Fig. 3.

B. Multiple SM Capacitor Deviations

In this case, the impact of multiple capacitor deviations within one arm is investigated as an extreme case that leads

3121

to major unbalances in the switching frequency, and therefore, switching losses amongst multiple submodules of the AAC. Again focusing on the upper arm of phase A, step deviations of $\pm 5\%$ in the values of the capacitors (from 85% of the original capacitance to 115% of the original capacitance) in seven SMs of the arm and a capacitance of 150% in the eighth SM capacitor are considered.

Without the loss equalisation algorithm, the SM capacitor voltages are well regulated – to the extend this can be achieved considering the low switching frequency of each individual SM – as shown in Fig. 11. The key objective of the balancing algorithm is achieved at the cost of higher number of transitions for the SMs and an unequal distribution of losses increasing the thermal stress of specific SMs within the arms. The results of the normalised losses for the SM of the upper arm of the AAC are provided in Fig. 12. Since capacitances are considered equal in the other five arms, no additional results are provided.

Considering the additional equalisation loop for the balancing algorithm, the SM capacitor voltage deviation increases amongst the SMs of the arm with the capacitance variations (Fig. 13) while the spread of the losses in each of the SMs improves (Fig. 14) leading to improved distribution of losses and thermal management for the SMs of the AAC.

V. CONCLUSION

This paper has presented a comprehensive voltage balancing algorithm for the AAC based on the concept of virtual voltages. The modification of the values of the voltages that are used by the sorting stages of the algorithm provide an effective way to achieve voltage balancing while limiting the effective switching frequency of an individual SM. An additional feedback loop based on switching count or accurate loss estimation of the SMs for the calculation of the virtual voltages can help achieve further loss equalisation and improve thermal management in the arms of the AAC, considering component parameter deviations. The restriction imposed by the algorithm can reduce the loss deviation among the SMs (up to 50% in certain cases such as that of a single SM) while leading to an increase in the SM capacitor voltage ripples. The implementation through the virtual voltages does not increase the complexity of the algorithms while gain selection for the feedback loops is relatively straightforward.

ACKNOWLEDGMENT

This research was supported under Australian Research Council's Discovery Early Career Research Award (DECRA - DE170100370).

REFERENCES

[1] H. Akagi, "Multilevel converters: Fundamental circuits and systems," *Proceedings of the IEEE*, vol. 105, no. 11, pp. 2048–2065, Nov. 2017.

[2] M. Barnes, D. V. Hertem, S. P. Teeuwsen, and M. Callavik, "Hvdc systems in smart grids," *Proceedings of the IEEE*, vol. 105, no. 11, pp. 2082–2098, Nov. 2017.

[3] A. Lesnicar and R. Marquardt, "An innovative modular multilevel converter topology suitable for a wide power range," in *Porc. IEEE Bologna Power Tech Conf.*, 2003, pp. 1–6.

[4] G. Konstantinou, J. Pou, S. Ceballos, and V. G. Agelidis, "Active redundant submodule configuration in modular multilevel converters," *IEEE Trans. Power Del.*, vol. 28, no. 4, pp. 2333–2341, Oct. 2013.

[5] A. Nami, J. Liang, F. Dijkhuizen, and G. D. Demetriades, "Modular multilevel converters for HVDC applications: Review on converter cells and functionalities," *IEEE Trans. Power Electron.*, vol. 30, no. 1, pp. 18–36, Jan. 2015.

[6] G. Konstantinou, J. Zhang, S. Ceballos, J. Pou, and V. G. Agelidis, "Comparison and evaluation of sub-module configurations in modular multilevel converters," in *2015 IEEE 11th International Conference on Power Electronics and Drive Systems*, June 2015, pp. 958–963.

[7] O. Cwikowski, A. Wood, A. Miller, M. Barnes, and R. Shuttleworth, "Operating dc circuit breakers with mmc," *IEEE Trans. Power Del.*, vol. PP, no. 99, pp. 1–1, 2017.

[8] O. Cwikowski, H. R. Wickramasinghe, G. Konstantinou, J. Pou, M. Barnes, and R. Shuttleworth, "Modular multilevel converter dc fault protection," *IEEE Transactions on Power Delivery*, vol. PP, no. 99, pp. 1–1, 2017.

[9] X. Yu, Y. Wei, Q. Jiang, X. Xie, Y. Liu, and K. Wang, "A novel hybrid-arm bipolar mmc topology with dc fault ride-through capability," *IEEE Trans. Power Del.*, vol. 32, no. 3, pp. 1404–1413, June 2017.

[10] M. M. C. Merlin, T. C. Green, P. D. Mitcheson, D. R. Trainer, R. Critchley, W. Crookes, and F. Hassan, "The alternate arm converter: A new hybrid multilevel converter with dc-fault blocking capability," *IEEE Trans. Power Del.*, vol. 29, no. 1, pp. 310–317, Feb. 2014.

[11] Y. Hu, G. Chen, Y. Liu, L. Jiang, P. Li, S. J. Finney, W. Cao, and H. Chen, "Fault-tolerant converter with a modular structure for hvdc power transmitting applications," *IEEE Trans. Ind. Appl.*, vol. 53, no. 3, pp. 2245–2256, May 2017.

[12] M. B. Ghat and A. Shukla, "A new h-bridge hybrid modular converter (hbhmc) for hvdc application: Operating modes, control and voltage balancing," *IEEE Trans. Power Electron.*, vol. PP, no. 99, pp. 1–1, 2017.

[13] H. R. Wickramasinghe, G. Konstantinou, J. Pou, and V. G. Agelidis, "Asymmetric overlap and hysteresis current control of zero-current switched alternate arm converter," in *IECON 2016*, Oct 2016, pp. 2526–2531.

[14] H. R. Wickramasinghe, G. Konstantinou, and J. Pou, "Comparison of bipolar sub-modules for the alternate arm converter," *Electric Power Systems Research*, vol. 146, pp. 115–123, May. 2017.

[15] V. Najmi, R. Burgos, and D. Boroyevich, "Design and control of modular multilevel alternate arm converter (aac) with zero current switching of director switches," in *2015 IEEE Energy Conversion Congress and Exposition (ECCE)*, Sept 2015, pp. 6790–6797.

[16] M. M. C. Merlin and T. C. Green, "Cell capacitor sizing in multilevel converters: cases of the modular multilevel converter and alternate arm converter," *IET Power Electronics*, vol. 8, no. 3, pp. 350–360, 2015.

[17] A. Dekka, B. Wu, R. L. Fuentes, M. Perez, and N. R. Zargari, "Evolution of topologies, modeling, control schemes, and applications of modular multilevel converters," *IEEE Journal of Emerging and Selected Topics in Power Electronics*, vol. 5, no. 4, pp. 1631–1656, Dec 2017.

[18] R. Darus, J. Pou, G. Konstantinou, S. Ceballos, R. Picas, and V. G. Agelidis, "A modified voltage balancing algorithm for the modular multilevel converter: Evaluation for staircase and phase-disposition pwm," *IEEE Trans. Power Electron.*, vol. 30, no. 8, pp. 4119–4127, Aug 2015.

[19] R. Picas, J. Pou, J. Zaragoza, A. Watson, G. Konstantinou, S. Ceballos, and J. Clare, "Submodule power losses balancing algorithms for the modular multilevel converter," in *IECON 2016*, Oct 2016, pp. 5064–5069.

[20] A. Christe and D. Dujic, "Virtual submodule concept for fast semi-numerical modular multilevel converter loss estimation," *IEEE Trans. Ind. Electron.*, vol. 64, no. 7, pp. 5286–5294, July 2017.

[21] F. Deng, D. Liu, Y. Wang, Z. Chen, M. Cheng, and Q. Wang, "Capacitor monitoring for modular multilevel converters," in *IECON 2017 - 43rd Annual Conference of the IEEE Industrial Electronics Society*, Oct 2017, pp. 934–939.

[22] H. R. Wickramasinghe, G. Konstantinou, Z. Li, and J. Pou, "Development of an alternate arm converter benchmark model for hvdc applications," in *IECON 2017*, Oct 2017, pp. 4506–4511.

The 2018 International Power Electronics Conference

Balanced Conduction Loss Distribution among SMs in Modular Multilevel Converters

Zhongxu Wang, Huai Wang, Yi Zhang, and Frede Blaabjerg

Department of Energy Technology, Aalborg University, Aalborg, Denmark

E-mail: zho@et.aau.dk, hwa@et.aau.dk, yiz@et.aau.dk, and fbl@et.aau.dk

Abstract—Due to the parameter mismatch, the unbalanced power loss distribution among SMs in the modular multilevel converter (MMC) can be introduced and further deteriorated by the low-frequency asynchronous switching transients related to no-carrier modulation techniques. The unbalanced thermal stress can reduce the reliability of the MMC and increase the complexity of cooling system design. Nevertheless, an internal balance mechanism exists in the MMC thanks to the capacitor voltage balancing. It contributes to an even conduction loss dissipation among SMs, which is studied and revealed in this paper. Moreover, a computationally light conduction loss estimation method is proposed correspondingly relying on the characteristics of semiconductors and the arm current only. Simulations and experiments are conducted to verify the effectiveness the proposed method.

Index Terms—Modular multilevel converter (MMC), balanced conduction loss distribution, conduction loss estimation, semiconductor.

I. INTRODUCTION

The modular multilevel converter is an emerging and attractive voltage-source converter (VSC) topology for high-voltage direct current (HVDC) transmission systems due to its modularity, scalability to different voltage levels, high output quality and no high voltage dc-link capacitor [1]–[3].

Reliability is one of the major concerns for the MMC because of the large cost investment and the large number of semiconductors, which are the weakest components in power converters [4]. Thus, it is necessary to fulfill the full potential of submodules (SMs) by posing even thermal stress on the devices. However, unbalanced power loss behavior (component-level and submodule-level) in the MMC can lead to various thermal stresses, which brings a challenge to the cooling system design and the converter reliability. The component-level unbalance is caused by a dc bias in the arm current when the active power is transferred through the MMC. The four semiconductors (taking the half bridge SM for example) undertake different thermal stress [5]. Submodule-level uneven power loss dissipation mainly results from the parameter mismatch among SMs and the low switching frequency for the MMC based on nearest level modulation (NLM) [6].

To address above problems, some research efforts have been made. An explanation about the loss unbalance for both conduction loss and the switching loss is detailed, and a two-dimension sorting method is proposed for a balanced junction temperature behavior [7]. Experiment validations and reliability assessment are further conducted on a down-scale

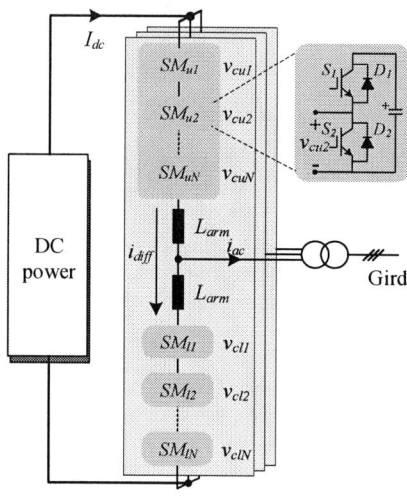

Fig. 1. Circuit configuration of a typical three-phase MMC. (I_{dc} is the dc-bus current, i_{ac} is the ac output current, and i_{diff} is the differential current.)

bench [6]. Various active thermal balancing methods based on the circulating current are explored for component-level power loss balancing, but the effectiveness is limited [8]. [4] focuses on the submodule-level power loss balancing integrated with the capacitor voltage balancing, but the loss model for MMC with a large number of SMs is computationally burdened.

In fact, an internal power loss balancing mechanism already exists in the MMC thanks to the capacitor voltage balancing as mentioned in [6]. However, no detailed explanation has been given to the phenomenon until now, to which the attention will be paid in this paper. An analytical derivation will be given to confirm this. In addition, a very computationally light conduction loss estimation method is proposed and verified through simulations and experiments.

The rest of this paper is organized as follows: Section II gives the introduction of the basic operation principles of the MMC. In Section III, the proposed submodule-level conduction loss estimation method is introduced followed by the full-scale simulation validation in Section IV. Experimental validation based on a down-scale test bench is conducted and describes in Section V. Section VI gives the conclusions.

3123

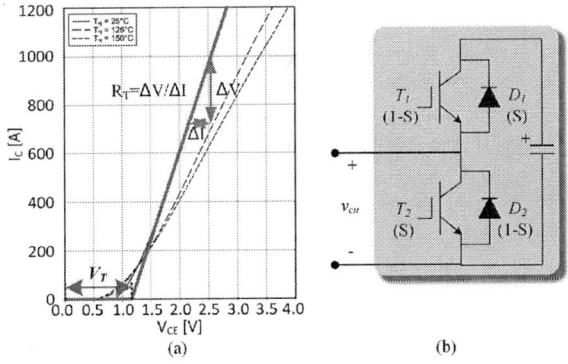

Fig. 2. (a). Curve fitting of IGBT datasheet, and (b). Gate signal relationship between four different semiconductors in a half-bridge SM.

II. OPERATING PRINCIPLE OF MMC

A typical circuit configuration of three-phase MMCS is presented in Fig.1. The MMC is composed of three phases, which can be divided into the upper arm and the lower arm. Each arm includes N series-connected SMs, and an arm inductor to restrain the circulating current within the phase leg. Half-bridge SM with four semiconductors is adopted in this paper. Generally, the arm current can be divided into two parts, namely the common component i_{comm} and the differential component i_{diff} [9]. A dc bias is an essential part of the differential component for the active power transfer, and to maintain the SM's voltage at the rated value. In addition, harmonic currents can be injected into the differential current to achieve certain objectives as well [10]. If the dc bias differential current is only considered here neglecting other harmonics, the upper arm current i_{up} of phase A can be expressed as

$$i_{up} = \frac{I_{dc}}{3} + \frac{I_{ac}}{2}\cos(\omega t + \varphi_1), \quad (1)$$

where I_{dc} and I_{ac} are the amplitude of dc-bus current and ac output current; φ_1 is the power factor angle, and ω is the angular frequency.

Assuming a lossless MMC system, the relationship between I_{dc} and I_{ac} can be derived [11] as

$$I_{ac} = \frac{4I_{dc}}{3m\cos(\varphi_1)}. \quad (2)$$

III. SUBMODULE-LEVEL BALANCED CONDUCTION LOSS DISTRIBUTION

A. Submodule-level Conduction Loss Calculation

Considering the equivalence among the six arms in three-phase MMC system, the analysis in the following will only take the upper arm of phase A for example. The upper arm current i_{up} is first divided into the positive part i_p and the negative part i_n respectively for an easy loss calculation.

$$i_p = \frac{|i_{up}| + i_{up}}{2}, \quad i_n = \frac{|i_{up}| - i_{up}}{2}. \quad (3)$$

The conduction loss averaged in one fundamental period of IGBT and diode can be calculated by

$$P_{con\_T2/D1} = \frac{1}{T}\int_0^T \left(V_{T/D}i_p + R_{T/D}i_p^2\right) S_{T2/D1}dt$$
$$P_{con\_T1/D2} = \frac{1}{T}\int_0^T \left(V_{T/D}i_n + R_{T/D}i_n^2\right) S_{T1/D2}dt \quad (4)$$

where P_{con_x} is the average conduction loss of device x in one fundamental period T; V_T, V_D, R_T and R_D are the on-state voltage and the on-state resistance of IGBT and diode obtained by curve fitting of the data-sheet as shown in Fig. 2; S_i is the time-dependent gate signal of the ith SM, equal to 1 or 0, where $S_{T2} = S_{D1} = S_i$, $S_{T1} = S_{D2} = (1 - S_i)$.

Adding up the equations in (4) and combining with (3), the total conduction loss of one SM can be derived as

$$P_{total} = \frac{1}{T}\int_0^T (V_D i_p + R_D i_p^2 + V_T i_n + R_T i_n^2)dt$$
$$+ \frac{1}{T}\int_0^T \Delta V i_{up}S_i dt + \frac{1}{T}\int_0^T \Delta R i_{up}|i_{up}|S_i dt \quad (5)$$

where $\Delta V = V_T - V_D$ and $\Delta R = R_T - R_D$, are the parameter differences between IGBT and diode.

In the normal operation of the MMC, the capacitor voltages among SMs are balanced in the steady-state, which can be achieved by various voltage balancing control methods [12], [13]. Therefore, it is reasonable to assume that the increase and decrease of the SM's capacitor voltage (charged by D_2 and discharged by T_1) are equal during one fundamental period, and the relationship can be expressed as

$$\Delta U^+ = \int_0^T \frac{i_p(1-S_i)}{TC_i}dt = \Delta U^- = \int_0^T \frac{i_n(1-S_i)}{TC_i}dt \quad (6)$$

where $\Delta U^{+/-}$ is the voltage increase/decrease during one fundamental period; C_i is the capacitance of the ith SM.

The relationship in (6) can be further simplified as

$$\int_0^T i_{up}S_i dt = \int_0^T i_{up}dt. \quad (7)$$

Substituting (7) into (5), the total conduction loss per SM can be re-expressed as

$$P_{total} = \underbrace{\frac{1}{T}\int_0^T (V_D i_p + R_D i_p^2 + V_T i_n + R_T i_n^2)dt}_{P_{com1}}$$
$$+ \underbrace{\frac{1}{T}\int_0^T (\Delta V + \Delta Rk)dt}_{P_{com2}} + \underbrace{\frac{\Delta R}{T}\int_0^T i_{up}(|i_{up}| - k)S_i dt}_{\Delta P_i} \quad (8)$$

where P_{com1} and P_{com2} are the common conduction loss components for all SMs , they are unrelated to the switching actions; ΔP_i is the specific conduction loss component of the ith SM; k is a constant related to the arm current.

The 2018 International Power Electronics Conference

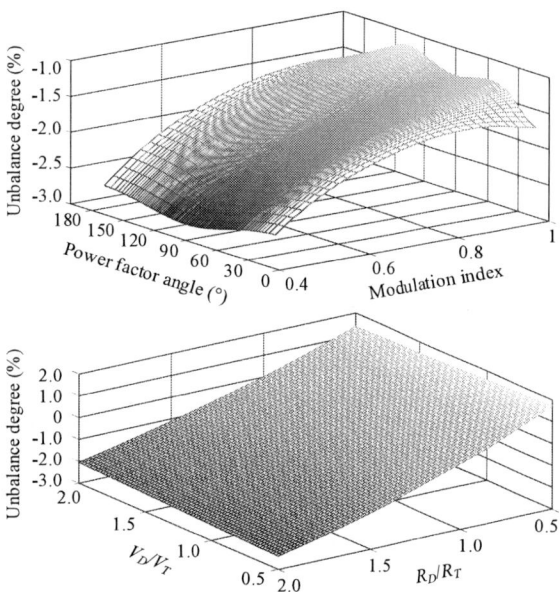

TABLE I
SIMULATION PARAMETERS FOR FULL-SCALE AND DOWN-SCALE MMC

Parameter	Fullscale	Downscale
Power rating	30 MVA	24 kVA
Dc-link voltage V_{dc}	50 kV	2 kV
SM number N	20	20
Arm inductor L_{arm}	13 mH	20 mH
Arm capacitor C_{arm}	3 mF	0.22-0.26 mF
V_T	1.5 V	1.9 V
V_D	2.5 V	1.36 V
R_T	0.52 mΩ	31.6 mΩ
R_D	0.94 mΩ	13.8 mΩ

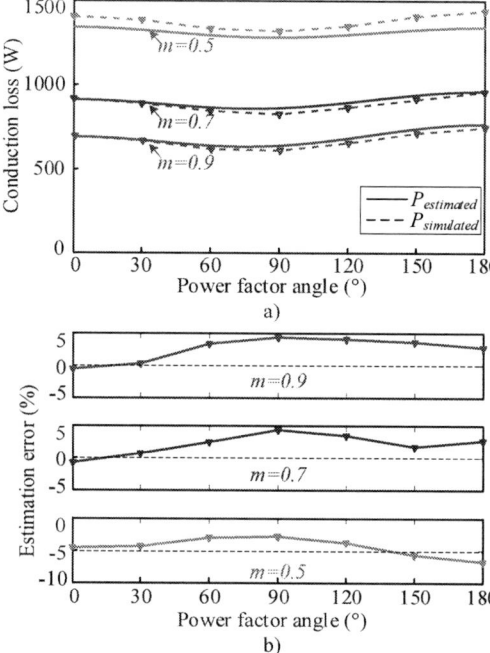

Fig. 3. Unbalance degree regarding different modulation index, power factor and semiconductor parameters.

ΔP_i can be estimated by $P(k)$, whose minimum value can be achieved when (10) holds.

$$|\Delta P_i| \le \left| \frac{\Delta R}{T} \int_0^T |i_{up}(|i_{up}| - k)| dt \right| = P(k) \qquad (9)$$

$$k = \frac{\int_0^T |i_{up}|^2 dt}{\int_0^T |i_{up}| \, dt}. \qquad (10)$$

B. Conduction Loss variation

The conduction loss variation among SMs is cause by ΔP_i, which is dependent on the switching actions. Its impact on the total conduction loss of one SM can be evaluated by a defined parameter, unbalance degree $e_{SM} = P(k)/P_{total}$. It can be affected by several parameters, such as the MMC operation conditions (modulation index and power factor) and the semiconductor on-state characteristics. Their effects are illustrated by full-scale simulation results in the following based on the IGBT module 5SNA-1200G450350 from ABB, whose on-state parameters are listed in Table I.

Fig. 3 shows the unbalance degree under different power factor, modulation index, and power device characteristics. It can be seen that e_{SM} is always within $\pm4\%$ when the modulation index ranges from 1 to 0.4, which covers the normal operating range of MMC system. The impact of parameter differences between IGBT and diode are also evaluated in Fig. 3, and the unbalance degree within 3% can be achieved as well. Note that the unbalance degree here is overestimated, and its actual value should be less than that in Fig. 3. Based on the results above, two preliminary conclusions which will be validated in the following sections can be achieved:

Fig. 4. Conduction loss per SM during one fundamental period regarding different power factor angles. a) Conduction loss, and b) conduction loss estimation error.

1). The total conduction loss of one SM can be estimated by P_{total}. This method is independent on the gate signal, and is computationally light with the need of on-state semiconductor characteristics and the arm current information only.

2). Different SMs share a balanced submodule level conduction loss regardless of the switching transient or the modulation strategies when the capacitor voltage of SMs are well balanced.

IV. FULL-SCALE SIMULATION VALIDATION

To validate the proposed conduction loss estimation approach, and the balanced SM-level conduction loss distribution, simulations based on a three-phase MMC as shown in Fig. 1 are conducted. IGBT module 5SNA-1200G450350 from ABB is used in this paper. Other system parameters are listed in Table I. Fig. 4 and Fig. 5 illustrate the conduction loss of

3125

The 2018 International Power Electronics Conference

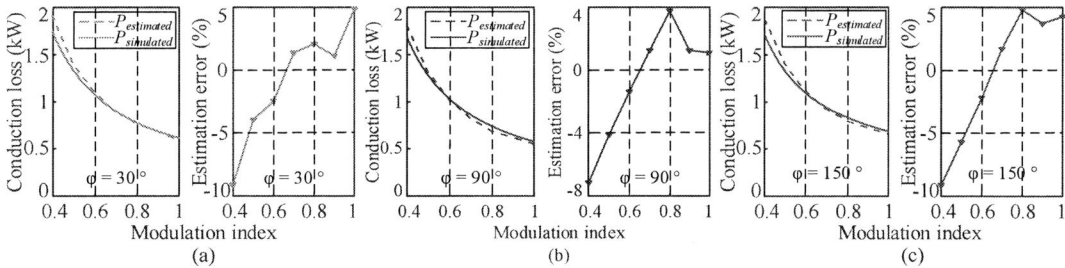

Fig. 5. The total conduction loss and the estimation error of an SM during one fundamental period regarding different modulation indexes. a) $\varphi = 30°$, b) $\varphi = 90°$, and c) $\varphi = 150°$.

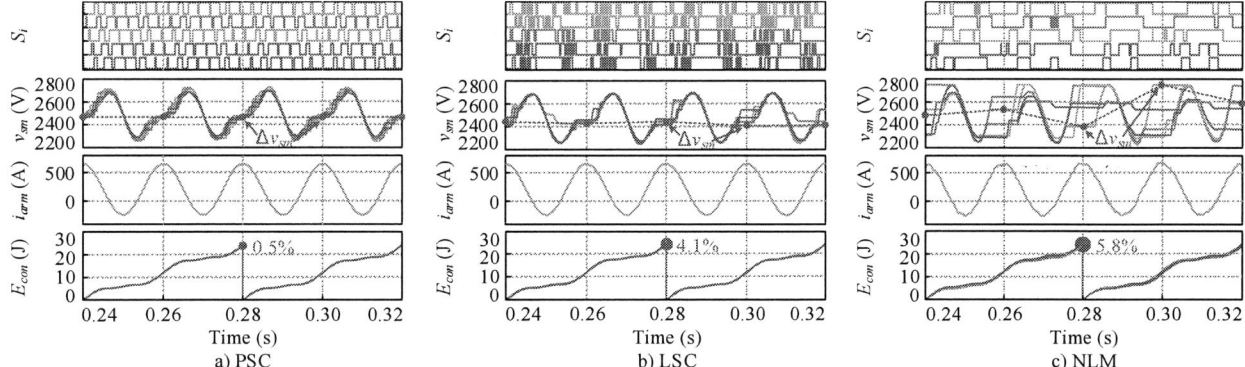

Fig. 6. Conduction loss of five SMs in the upper arm of phase A with modulation index $m = 0.8$ and power factor angle $\varphi = 30°$: a) PSC, b) LSC and c) NLM. (Sub-graphs from top to bottom are: the gate signal, the arm current, the SM capacitor voltage, and the conduction loss per SM averaged in 0.02 s.)

one SM under different power factor angles and different modulation indexes. It can be seen that conduction loss increases greatly with the decrease of the modulation index, and, in contrast, the power factor has a small impact on the conduction loss. The estimation error remains acceptable with the value being around 5% when the modulation index is larger than 0.5. However, it increases sharply for the modulation index less than 0.5. Nevertheless, the proposed method is still valid since MMCs operate in a high modulation index (e.g., around 0.9) condition in most cases.

To illustrate the balanced conduction loss distribution, a series of simulations are done with regards to different modulation strategies and different SM capacitances based on the scaled-down three-phase MMC in Fig. 1. Three commonly-used modulation methods, namely phase-shifted Carrier (PSC) modulation, level-shifted carrier (LSC) modulation and nearest level modulation (NLM) are validated respectively. The capacitance mismatch introduced by manufacturing process, degradation and maintenance of a broken SM [7] is taken into account by evenly setting its value from 2.2 mF to 2.6 mF for SM1 to SM20 with the variation of 18%.

Fig. 6 shows the simulation waveforms of 5 SMs (SM1, SM5, SM10, SM15 and SM20) in the same arm in two fundamental periods with the modulation index and power factor being 0.9 and 1 respectively. Different switching patterns for the three modulation methods can be clearly observed. The

capacitor voltages are well regulated averaging at 2500 V. The current waveform for NLM contains more harmonic components compared with that of PSC due to the lower switching frequency. The average accumulated conduction losses are 23.9 J, 24.2 J, and 24.2 J for PSC, LSC and NLM respectively, and it can be seen that the loss unbalance degree increases from 0.5%, 4.1% to 5.8%. The reason, as mentioned in Section III, is that the capacitor voltage balancing performance gets worse as shown in Fig. 6. Δv_{sm} gets larger and larger, and equations (6) and (7) are not hold perfectly. Nevertheless, the small loss difference can still confirm that the modulation method and capacitance mismatch have a negligible effect on the balanced conduction loss of one SM.

V. DOWN-SCALE SIMULATION AND EXPERIMENT VALIDATION

In addition to the full-scale simulation validation, a three-phase MMC with scaled down system parameters is simulated as well. The arm current contains dc and ac components with the peak value being 4 A and 10 A respectively. Unity power factor is used, and the modulation index is set at 0.8. Moreover, the same IGBT module F4_50R12KS4 from Infineon with the experiment is used in the simulation. Thermal profiles of both IGBT and diode in the simulation are tested through Curve Tracer B1506A under various temperatures ranging from $25°C$ to $125°C$. Meanwhile, a down-scale experiment with the same

3126

The 2018 International Power Electronics Conference

Fig. 7. Scaled down experiment test bench.

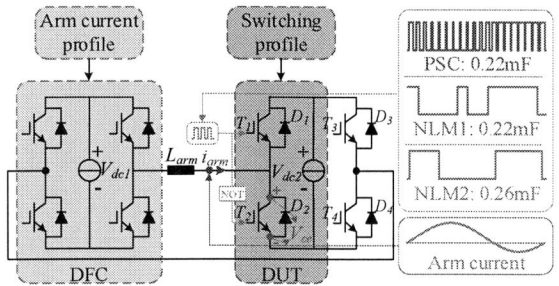

Fig. 8. Circuit topology of the down-scaled experimental bench.

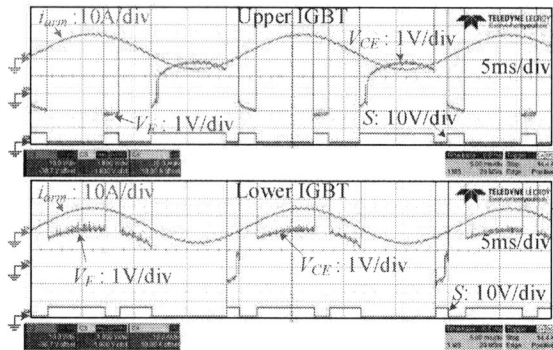

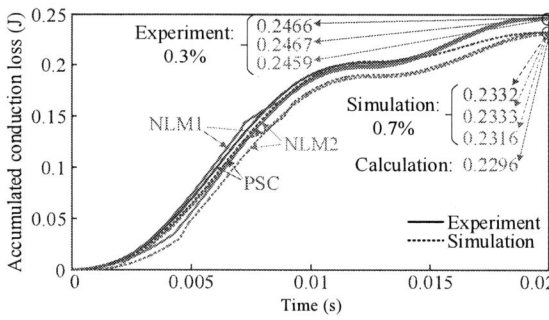

Fig. 9. Experiment waveforms of the arm current, the on-state voltage and the gate signal of upper and lower IGBTs under the unity power factor and the modulation index of 0.8.

Fig. 10. Simulation and experiment results of the accumulated conduction loss of one SM under three different modulation methods.

system parameters in the simulation is conducted based on the prototype in Fig. 7. Fig. 8 shows the circuit scheme where two full bridge converters are used. One is used as the device for control (DFC) regulating the inductor current to track the current profile from simulation. Another one is divided into the DUT and the auxiliary half bridge. The DUT is controlled by the switching profile. Besides, two capacitors (0.22 mF and 0.26 mF) and two modulation methods (PSC and NLM) are used in this paper for validation. Other parameters are listed in Table I.

Fig. 9 shows the experiment waveforms of the arm current, the on-state voltages of both IGBT and diode, and the gate signals under the condition of NLM. It can be seen that the arm current is well regulated, and the on-state voltages of both IGBT and diode are sampled in two fundamental periods. By using the waveform data exported from the oscilloscope, the accumulated conduction loss of the four semiconductors can be calculated in Matlab. The experimental results are compared with the simulation as shown in Fig. 10. The simulated total conduction losses of one SM in one fundamental period are 0.2332 J, 0.2333 J and 0.2316 J for PSC, NLM1 and NLM2 respectively with the variation of 0.7%. The average accumulated conduction losses from the experiment are around 0.2464 J with the variation as low as 0.3%. In addition, the conduction loss calculated by the proposed method is 0.2296 J with the error of 1.4% and 6.8% compared with the simulation and experiment results respectively.

The small errors between the simulations, the experiments

and the calculation confirm the balanced conduction loss distribution and validate the effectiveness of the proposed conduction loss estimation method.

VI. CONCLUSION AND FUTURE WORK

Different SMs share a balanced conduction loss regardless of the operation condition, the modulation techniques and the parameter mismatch related to the capacitor and the semiconductor when the SM capacitor voltages are well balanced. The conclusion is helpful to guide the practical cooling design and the active thermal balanced control of the MMC system, where more attention should be paid to the switching loss. Moreover, a computationally light conduction loss estimation method is proposed correspondingly, which depends on the on-state characteristics of the semiconductors and the arm current only without considering the switching transients. The validity of the conclusion is verified through both full-scale and down-scale simulation and experiment.

REFERENCES

[1] A. Lesnicar and R. Marquardt, "An innovative modular multilevel converter topology suitable for a wide power range," in *Proc. IEEE Conf. Power Tech.*, vol. 3, 2003, p. 6.

[2] M. Hagiwara and H. Akagi, "Control and experiment of pulsewidth-modulated modular multilevel converters," *IEEE Trans. Power Electron.*, vol. 24, no. 7, pp. 1737–1746, Jul. 2009.

3127

[3] M. Saeedifard and R. Iravani, "Dynamic performance of a modular multilevel back-to-back hvdc system," *IEEE Trans. Power Del.*, vol. 25, no. 4, pp. 2903–2912, Sep. 2010.

[4] R. Picas, J. Pou, J. Zaragoza, A. Watson, G. Konstantinou, S. Ceballos, and J. Clare, "Submodule power losses balancing algorithms for the modular multilevel converter," in *Proc. 42nd Conf. Ind. Electron. (IECON)*, 2016, pp. 5064–5069.

[5] S. Rohner, S. Bernet, M. Hiller, and R. Sommer, "Modulation, losses, and semiconductor requirements of modular multilevel converters," *IEEE Trans. Ind. Electron.*, vol. 57, no. 8, pp. 2633–2642, Sep. 2010.

[6] F. Hahn, M. Andresen, G. Buticchi, and M. Liserre, "Thermal analysis and balancing for modular multilevel converters in hvdc applications," *IEEE Trans. Power Electron.*, Apr. 2017.

[7] A. Sangwongwanich, L. Mathe, R. Teodorescu, C. Lascu, and L. Harnefors, "Two-dimension sorting and selection algorithm featuring thermal balancing control for modular multilevel converters," in *Proc. 18th European Conf. Power Electron. and Applications (ECCE Europe)*, Sep. 2016, pp. 1–10.

[8] M. M. Merlin and P. D. Mitcheson, "Active power losses distribution methods for the modular multilevel converter," in *Proc. IEEE 17th Conf. Control and Modeling for Power Electron. (COMPEL)*, 2016, pp. 1–6.

[9] R. Darus, J. Pou, G. Konstantinou, S. Ceballos, R. Picas, and V. G. Agelidis, "A modified voltage balancing algorithm for the modular multilevel converter: Evaluation for staircase and phase-disposition pwm," *IEEE Trans. Power Electron.*, vol. 30, no. 8, pp. 4119–4127, Sep. 2015.

[10] S. Debnath, J. Qin, B. Bahrani, M. Saeedifard, and P. Barbosa, "Operation, control, and applications of the modular multilevel converter: A review," *IEEE Trans. Power Electron.*, vol. 30, no. 1, pp. 37–53, Mar. 2015.

[11] H. Wang, G. Tang, Z. He, and J. Cao, "Power loss and junction temperature analysis in the modular multilevel converters for hvdc transmission systems," *Journal of Power Electron.*, vol. 15, no. 3, pp. 685–694, Nov. 2015.

[12] H. Peng, R. Xie, K. Wang, Y. Deng, X. He, and R. Zhao, "A capacitor voltage balancing method with fundamental sorting frequency for modular multilevel converters under staircase modulation," *IEEE Trans. Power Electron.*, vol. 31, no. 11, pp. 7809–7822, Jan. 2016.

[13] F. Deng and Z. Chen, "Elimination of dc-link current ripple for modular multilevel converters with capacitor voltage-balancing pulse-shifted carrier pwm," *IEEE Trans. Power Electron.*, vol. 30, no. 1, pp. 284–296, May 2015.

The 2018 International Power Electronics Conference

Simplification of Model Predictive Control for Modular Multilevel Converter through Direct Voltage Level Selection

Xingxing Chen, Jinjun Liu, Shaodi Ouyang, Shuguang Song, Rui Luo

Power Electronics and Renewable Energy Research Center
The University of Xi'an Jiaotong, School of Electrical Engineering
Xi'an, Shaanxi, China
Email: xingxingchen@stu.xjtu.edu.cn

Abstract- Model predictive control (MPC) method has attracted the attention of researchers for modular multilevel converter (MMC). However, the computation burden of the controller can be very high by using a direct MPC (D-MPC) method, especially when the submodule (SM) number of MMC becomes large. A stepwise indirect MPC (SI-MPC) strategy is first presented in this paper combined with capacitor voltage sorting algorithm to reduce the considered switching states. Ac side current control and circulating current control are separated of SI-MPC. The capacitor voltage ripple is neglected in the discrete-time mathematical model of MMC with SI-MPC, therefore, a direct voltage level selection (DVLS) method can be further derived with the simplification of SI-MPC. The current control process of MMC is implemented by comparing the current control error with the predesigned error bands. Simulation results verified the theoretical analysis and effectiveness of the proposed methods.

I. INTRODUCTION

Modular multilevel converter (MMC) is a promising topology for medium and high voltage applications [1]-[4], such as motor drives, static synchronous compensator (STATCOM), battery energy storage system (BESS) and high-voltage direct current (HVDC) transmission. It exhibits expandable and redundant configuration, reduced switching frequency and voltage stress of power switches, high quality of output voltage and current waveforms. The circuit configuration of MMC is shown in Fig. 1, which consists of six arms, each being composed of N SMs and one arm inductor.

The conventional system controller of MMC is relatively complicated and can be separated as three parts: ac side controller, inner difference current controller and capacitor voltage controller. The first two controllers can be realized by PI, PR or repetitive control [2]-[4]. The third controller can be further separated as four parts, i.e. sorting algorithm, the total energy control, the horizontal, and vertical energy balancing [5]. The control parameters of these controllers are difficult to design and could affect the system performance if they are not optimal. In addition, an extra modulator is necessary to generate the switching signals. There are mainly two types of modulation methods of MMC, i.e. carrier phase-shifted pulse width modulation (CPS-PWM) [2] and nearest level modulation (NLM) [1]. The former is applicable for

This work was supported by the State Key Laboratory of Electrical Insulation and Power Equipment under Grant EIPE14112.

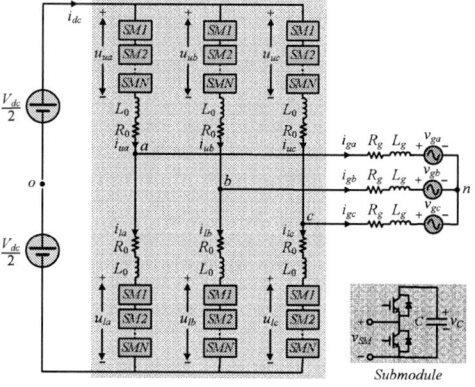

Fig. 1. Circuit configuration of modular multilevel converter.

medium voltage applications, and the latter is more suitable for high voltage applications.

Model predictive control is suitable for multi-input-multi-output system attributing to the following benefits:
· Simple control structure.
· Handling multiple control objectives.
· Nonlinear characteristic of the converters is considered.
· Dead-time compensation.
· High dynamic performance.

Among model predictive control methods, the finite control set MPC (FS-MPC) is more attractive for power electronic converters. The computation burden can be very high with the increasing SM number of MMC through direct FS-MPC [6]. For a MMC with N SMs in each arm, the required switching states is C_{2N}^N. In [7], an indirect FS-MPC method is proposed combined with SM capacitor voltage sorting strategy. The available switching states can be reduced to $N+1$. Considering the dv/dt problem, a fast MPC strategy is presented in [8] to reduce the control options to 3, and only the nearest voltage levels around the previous optimal voltage level are considered.

In this paper, a SI-MPC strategy is first presented combined with capacitor voltage sorting algorithm which can significantly reduce the computation burden of the controller. Different from D-MPC, the ac side current and circulating current control are no long realized by one cost function, but are separated in two steps. Since the capacitor voltage ripple in the discrete-time model of

3129

MMC is neglected with SI-MPC, a proposed DVLS method can be further derived through the simplification of SI-MPC. Cost function minimization process is substituted by comparing the current control error with the predesigned error bands, but won't significantly affect the steady state performance of MMC. Benefit of the high dynamic performance is preserved. The implementation process of DVLS is much simpler than D-MPC and SI-MPC, and the computation burden is also reduced.

The rest of this paper is organized as follows. Section II derives the mathematical model and discrete-time mathematical model of MMC. Section III first presents the implementation process of SI-MPC step by step, and then explains the simplification method from SI-MPC to DVLS. Section IV reports the simulation results and Section V concludes this paper.

II. OPERATING PRINCIPLE AND MODELING OF THE MMC

A. Operating Principle

Fig.1 shows the three-phase MMC structure, which consists of six arms (three phase upper arms and lower arms), each being composed of N SMs and one arm inductor L_0. R_0 represents the equivalent resistor of switching and conduction losses. u_{uj} and u_{lj} represents the output voltage of upper arm and lower arm cascaded SMs respectively ($j=a,b,c$). i_{uj} and i_{lj} represents the upper arm current and lower arm current respectively. The dc side voltage is generated by a dc source V_{dc}. The ac side system is connected to the grid v_{gj} through inductor L_g and equivalent resistor R_g. i_{gj} is the ac side current. There are two states of MMC SM under normal operating principle, i.e. inserted and bypassed states. When the upper switch is turned on and the lower switch is turned off, the SM is inserted, otherwise, the SM is bypassed. S_{xjy} is the SM switching signal, which is given as:

$$S_{xjy} = \begin{cases} 1 & \textit{SM is inserted} \\ 0 & \textit{SM is bypassed} \end{cases} \quad (1)$$

where $x=u,l$ and $y=1,2,3…2N$.

B. Mathematical Model

According to Fig.1, the voltage equation of the upper arm and lower arm can be given as:

$$\frac{V_{dc}}{2} - u_{uj} - L_0\frac{di_{uj}}{dt} - R_0 i_{uj} = L_g\frac{di_{gj}}{dt} + R_g i_{gj} + v_{gj} + v_{no} \quad (2)$$

$$\frac{V_{dc}}{2} - u_{lj} - L_0\frac{di_{lj}}{dt} - R_0 i_{lj} = -L_g\frac{di_{gj}}{dt} - R_g i_{gj} - v_{gj} - v_{no} \quad (3)$$

the three phases of MMC are controlled separately in this paper and v_{no} is neglected. But in some cases, v_{no} can be used to improve the performance of MMC [9]. Through the addition and subtraction of (2) and (3), the mathematical model of MMC can be given as follows [2]:

$$\frac{u_{\Delta j}}{2} = (\frac{L_0}{2} + L_g)\frac{di_{gj}}{dt} + (\frac{R_0}{2} + R_g)i_{gj} + v_{gj} \quad (4)$$

$$\frac{u_{\Sigma j}}{2} = \frac{V_{dc}}{2} - L_0\frac{di_{diffj}}{dt} - R_0 i_{diffj} \quad (5)$$

where $u_{\Delta j}$, $u_{\Sigma j}$ are difference and sum voltage of lower and upper arm, which are given as:

$$u_{\Delta j} = u_{lj} - u_{uj} = \sum_{y=1}^{N}(S_{cljy}v_{cljy} - S_{cujy}v_{cujy}) \quad (6)$$

$$u_{\Sigma j} = u_{lj} + u_{uj} = \sum_{y=1}^{N}(S_{cljy}v_{cljy} + S_{cujy}v_{cujy}) \quad (7)$$

where v_{cxjy} is SM capacitor voltage. i_{diffj} is the inner difference current of phase j and can be expressed as:

$$i_{diffj} = \frac{i_{uj} + i_{nj}}{2} = \frac{i_{dc}}{3} + i_{cirj} \quad (8)$$

which consists of dc component and other even order circulating current components i_{cirj}. Under balanced grid condition, i_{diffj} should be set to $i_{dc}/3$ to eliminate the circulating current.

The SM capacitor voltage dynamic is described as:

$$C\frac{dv_{cxjy}}{dt} = S_{xjy}i_{xj} \quad (9)$$

The sum and difference SMs energy of upper arm and lower arm can be obtained as:

$$\begin{cases} W_{\Sigma j} = \frac{C}{2}\sum_{y=1}^{N}(v_{cujy}^2 + v_{cljy}^2) \\ W_{\Delta j} = \frac{C}{2}\sum_{y=1}^{N}(v_{cujy}^2 - v_{cljy}^2) \end{cases} \quad (10)$$

C. Discrete-Time Mathematical Model

By using forward Euler method, the discrete-time mathematical model of MMC can be obtained according to (1)-(10). The ac side current and inner difference current are given as follows:

$$i_{gj}(k+1) = Ai_{gj}(k) + Bu_{\Delta j}(k) + Cv_{gj}(k) \quad (11)$$

$$i_{diffj}(k+1) = Di_{diffj}(k) - \frac{u_{\Sigma j}(k)T_s}{2L_0} + \frac{T_s V_{dc}}{2L_0} \quad (12)$$

where $A=1- (R_0+2R_g)B\approx 1$, $B=-T_s/ (L_0+2L_g)$, $C=-2B$ and $D=1- T_s R_0/L_0\approx 1$. T_s is the sampling period. The capacitor voltage is expressed as:

$$v_{cxjy}(k+1) = v_{cxjy}(k) + \frac{1}{C}S_{xjy}i_{xj}(k) \quad (13)$$

Therefore, the sum and difference SM energy of upper arm and lower arm can be obtained as:

$$\begin{cases} W_{\Sigma j}(k+1) = \frac{C}{2}\sum_{y=1}^{N}(v_{cujy}^2(k+1) + v_{cljy}^2(k+1)) \\ W_{\Delta j}(k+1) = \frac{C}{2}\sum_{y=1}^{N}(v_{cujy}^2(k+1) - v_{cljy}^2(k+1)) \end{cases} \quad (14)$$

III. PROPOSED MPC OF MMC

A. Implementation process of SI-MPC

The ac side current control, circulating current control and capacitor voltage control are usually realized by one cost function ([7], [10]) given as follows:

$$J = \lambda_1 | i_{gj}^*(k+1) - i_{gj}(k+1)| + \lambda_2 | i_{diffj}^*(k+1) - i_{diffj}(k+1)| + \lambda_3 | W_{\Delta j}^*(k+1) - W_{\Delta j}(k+1)| + \quad (15)$$
$$\lambda_4 | W_{\Sigma j}^*(k+1) - W_{\Sigma j}(k+1)|$$

3130

where i_{gj}^* is the ac side current reference, i_{diffj}^* is the inner difference current reference, $W_{\Delta j}^*$ is the difference SM energy of upper arm and lower arm reference, $W_{\Sigma j}^*$ is the sum SM energy of upper arm and lower arm reference, and λ_1, λ_2, λ_3 and λ_4 are weighting factors. By neglecting the capacitor voltage ripple in the discrete-time model of MMC and using the nominal voltage V_{dc}/N to calculate the next step predicted values, the cost function can be simplified into:

$$J' = \lambda_1 \mid i_{gj}^*(k+1) - i_{gj}(k+1) \mid + \lambda_2 \mid i_{diffj}^*(k+1) - i_{diffj}(k+1) \mid \tag{16}$$

and the capacitor voltage balancing relies on the inherent characteristic of MMC [10]. As the presented SI-MPC method, J' is further separated into two parts:

$$J_1 = \mid i_{gj}^*(k+1) - i_{gj}(k+1) \mid \tag{17}$$

$$J_2 = \mid i_{diffj}^*(k+1) - i_{diffj}(k+1) \mid \tag{18}$$

ac side current and circualting current control are realized in two steps. Capacitor voltage balancing algorithm is introduced to balance the capacitor voltage in each arm. The implementation process of SI-MPC is given as bellow:

1) Measure the ac side current i_{gj} and the inner difference current i_{diffj}.

2) First generate a nearest ac level based on v_{gj} [1]:

$$u_{\Delta j} = \frac{2V_{dc}}{N} round(\frac{Nv_{gj}}{V_{dc}}) \tag{19}$$

Then adjust $u_{\Delta j}$ by adding up an additional voltage level: $-2V_{dc}/N$, 0 or $2V_{dc}/N$, therefore three control options can be derived to control the ac side current. Calculate the next step ac side current of the three control options according to (11) and the ac side cost function J_1 according to (17).

The option with the smallest cost function will be chosen as the optimal voltage level to control the ac side current. If the optimal voltage level is $u_{\Delta j}^*(k)$, then the upper arm and lower arm inserted SM number are obtained as:

$$n_{uj1}^*(k+1) = \frac{N}{2} - \frac{Nu_{\Delta j}^*(k)}{2V_{dc}} \tag{20}$$

$$n_{lj1}^*(k+1) = \frac{N}{2} + \frac{Nu_{\Delta j}^*(k)}{2V_{dc}} \tag{21}$$

3) Under normal operating condition, the inserted SM number of MMC one phase-leg equals to N. By inserting or bypassing an additional SM in the upper arm and lower arm at the same time, $N+2$ and $N-2$ SM states can be introduced to control the circulating current [11] and the ac side control is unaffected. The corresponding phase-leg voltage levels of the three SM states ($N-2$, N, $N+2$) are given as:

$$u_{\Sigma j} \in \{\frac{(N-2)V_{dc}}{N}, \frac{NV_{dc}}{N}, \frac{(N+2)V_{dc}}{N}\} \tag{22}$$

Calculate the next step inner difference current of the three options according to (12) and the cost function J_2 according to (18).

The option with the smallest cost function will be chosen as the optimal voltage level $u_{\Sigma j}^*(k)$ to control the inner difference current. Combined with (20) and

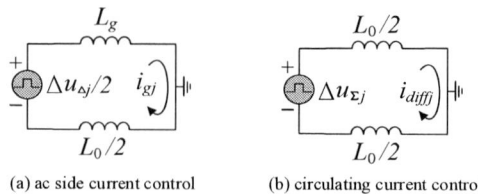

(a) ac side current control   (b) circulating current control

Fig. 2. Equivalent control circuits of SI-MPC.

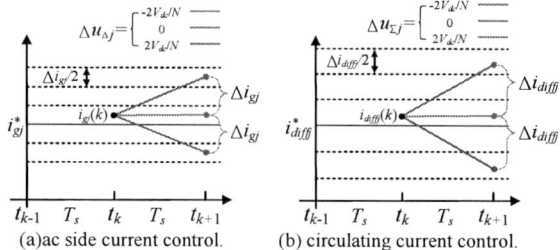

(a) ac side current control.   (b) circulating current control.

Fig. 3. Simplification method from SI-MPC to DVLS.

(21), the upper arm and lower arm inserted SM number are obtained as:

$$n_{uj2}^*(k+1) = n_{uj1}^*(k+1) + \frac{(u_{\Sigma j}^*(k) - V_{dc})N}{2V_{dc}} \tag{23}$$

$$n_{lj2}^*(k+1) = n_{lj1}^*(k+1) + \frac{(u_{\Sigma j}^*(k) - V_{dc})N}{2V_{dc}} \tag{24}$$

4) The capacitor voltage sorting algorithm is executed to determine the final switching signals. The SMs are arranged as descending order. Taking the upper arm as an example, if $i_{uj}>0$, then the last n_{uj2}^* SMs will be inserted and if $i_{uj}<0$, then the first n_{uj2}^* SMs will be inserted.

B. The proposed DVLS method

As mentioned before, the capacitor voltage ripple is neglected in the discrete-time model of MMC with SI-MPC. The ac side current control is realized by adding up an additional voltage level and the inner difference current control is realized by the three SM states. The equivalent control circuits are given in Fig. 2, where $\Delta u_{\Delta j}$, $\Delta u_{\Sigma j} = -2V_{dc}/N$, 0 or $2V_{dc}/N$. Equation (11) and (12) can be simplified into:

$$i_{gj}(k+1) = i_{gj}(k) - \frac{T_s \Delta u_{\Delta j}}{L_0 + 2L_g} \tag{25}$$

$$i_{diffj}(k+1) = i_{diffj}(k) - \frac{\Delta u_{\Sigma j}(k)T_s}{2L_0} \tag{26}$$

The simplification method from SI-MPC to DVLS is shown in Fig. 3. Δi_{gj} and Δi_{diffj} are expressed as:

$$\Delta i_{gj} = \frac{V_{dc}T_s}{N(L_g + L_0/2)} \tag{27}$$

$$\Delta i_{diffj} = \frac{V_{dc}T_s}{NL_0} \tag{28}$$

Take the ac side current control as an example. The purple line is the current reference which can be regarded

3131

as a straight line within two adjacent sampling period. $i_{gj}(k)$ locates within $i_{gj}^*(k)-\Delta i_{gj}/2$ and $i_{gj}^*(k)+\Delta i_{gj}/2$. The next step predicted values are represented by green, red and blue dots of the three control options. According to the diagram, the red dot more close to the reference, therefore, $\Delta u_{\Delta j}=0$ is chosen as optimal control option to control the ac side current. An important law can be discovered, that is when $i_{gj}(k)$ locates within $i_{gj}^*(k)-\Delta i_{gj}/2$ and $i_{gj}^*(k)+\Delta i_{gj}/2$, the optimal $\Delta u_{\Delta j}$ always equals to 0, when $i_{gj}(k)$ locates within $i_{gj}^*(k)+\Delta i_{gj}/2$ and $i_{gj}^*(k)+3\Delta i_{gj}/2$, the optimal $\Delta u_{\Delta j}$ always equals to $2V_{dc}/N$, and when $i_{gj}(k)$ locates within $i_{gj}^*(k)-3\Delta i_{gj}/2$ and $i_{gj}^*(k)-\Delta i_{gj}/2$, the optimal $\Delta u_{\Delta j}$ always equals to $-2V_{dc}/N$. Thus, the cost function minimization process of SI-MPC can be substituted by comparing the current control error with the error bands $\{-3\Delta i_{gj}/2,$ $-\Delta i_{gj}/2, \Delta i_{gj}/2, 3\Delta i_{gj}/2\}$. And control principle of the proposed DVLS method can be given as:

$$
\Delta u_{\Delta j}=\begin{cases}\dfrac{2V_{dc}}{N} & \Delta i_{gj}/2 < i_{gj}(k)-i_{gj}^*(k) < 3\Delta i_{gj}/2 \\[2mm] 0 & -\Delta i_{gj}/2 < i_{gj}(k)-i_{gj}^*(k) < \Delta i_{gj}/2 \\[2mm] -\dfrac{2V_{dc}}{N} & -3\Delta i_{gj}/2 < i_{gj}(k)-i_{gj}^*(k) < -\Delta i_{gj}/2 \end{cases}
$$

(29)

Similarly, the circulating current control principle can be obtained as:

$$
\Delta u_{\Sigma j}=\begin{cases}\dfrac{2V_{dc}}{N} & \Delta i_{diffj}/2 < i_{diffj}(k)-i_{diffj}^*(k) < 3\Delta i_{diffj}/2 \\[2mm] 0 & -\Delta i_{diffj}/2 < i_{diffj}(k)-i_{diffj}^*(k) < \Delta i_{diffj}/2 \\[2mm] -\dfrac{2V_{dc}}{N} & -3\Delta i_{diffj}/2 < i_{diffj}(k)-i_{diffj}^*(k) < -\Delta i_{diffj}/2 \end{cases}
$$

(30)

The optimal voltage levels are directly selected after the comparisons. The traditional capacitor voltage balancing control strategy [5] is introduced to adjust the reference of inner difference current, and to improve the steady state and dynamic performance of capacitor voltage control.

IV. SIMULATION RESULTS

The performance of the proposed two control methods are verified on a three-phase MMC system. The system parameters are given in TABLE I. Fig.4 shows the simulation results. Since the three phases of MMC are controlled separated and the simulation performance are similar, therefore, only the simulation results of phase a are taken as an example. The simulation waveforms of SI-MPC and DVLS are almost the same which verify the control performance of SI-MPC is unchanged after simplification. The ac side currents and inner difference currents are effectively controlled. The second order circulating currents are suppressed. The SM capacitor voltages are well balanced with the capacitor voltage balancing strategy. In Fig. 4, $S_{\Sigma j}$ and $S_{\Delta j}$ are expressed as:

$$
S_{\Sigma j}=\sum_{y=1}^{N}(S_{cujy}+S_{cljy})
$$

(22)

$$
S_{\Delta j}=\sum_{y=1}^{N}(S_{cujy}-S_{cljy})
$$

(23)

TABLE I
SIMULATION PARAMETERS

Parameters	Values
Rated active power P	1.95 MW
Grid voltage frequency f_0	50 Hz
Peak value of v_{gj}	4 kV
Ac side inductor L_g	5 mH
Ac side resistor R_g	0.01 Ω
DC side voltage V_{dc}	10 kV
DC side current reference	195 A
Submodule capacitor C	5000 uF
Arm SM number N	10
Rated capacitor voltage	1 kV
Arm inductance L_0	10 mH
Arm resistor R_0	0.02 Ω
Sampling frequency	10 kHz

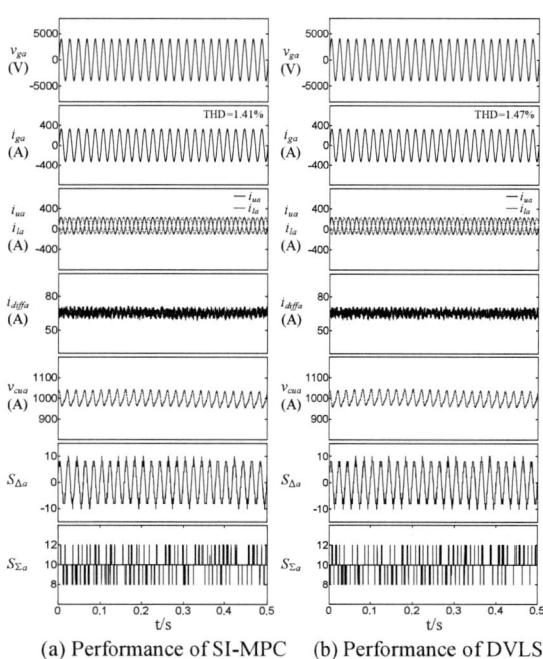

(a) Performance of SI-MPC   (b) Performance of DVLS

Fig. 4. Simulation results of two MPC methods.

V. CONCLUSIONS AND FUTURE WORK

A SI-MPC method is first presented in this paper. Ac side current control and circulating current control are separated. Compared with traditional D-MPC method, the computation burden is reduced. The capacitor voltage ripple is neglected in the discrete model of MMC with SI-MPC, therefore, a proposed DVLS method can be further derived through the simplification of SI-MPC. In this paper, only $-2V_{dc}/N$, 0 or $2V_{dc}/N$ voltage levels are used to control i_{gj} and only three SM states (N-2, N, N+2) are

3132

used to control i_{cirj}. But actually, more voltage levels and more SM states can be easily introduced if necessary. The implementation process of DVLS is much simpler, especially when the control options are increased (more voltage levels and SM states), dichotomy method can be used to reduce the system computation burden.

REFERENCES

[1] Lei Lin, Yizhe Lin, Zhen He, Yu Chen, Jiabing Hu and Wuhua Li, "Improved nearest-level modulation for a modular multilevel converter with a lower submodule number," *IEEE Trans. Power Electron.*, vol. 31, no. 8, pp. 5369-5377, Aug. 2016.

[2] Qingrui Tu, Zheng Xu and Lie Xu, "Reduced switching-frequency modulation and circulating current suppression for modular multilevel converters, " *IEEE Trans. Power Delivery*, vol. 26, no. 3, pp. 2009-2017, Jul. 2016.

[3] Shaohua Li, Xiuli Wang, Zhiqing Yao, Tai Li and Zhong Peng, "Circulating current suppressing strategy for MMC-HVDC based on nonideal proportional resonant controllers under unbalanced grid conditions," *IEEE Trans. Power Electron.*, vol. 30, no. 30, pp. 387-397, Jan. 2015.

[4] Liqun He, Kai Zhang, Jian Xiong and Shengfang Fan, "A repetitive control scheme for harmonic suppression of circulating current in modular multilevel converters," *IEEE Trans. Power Electron.*, vol. 30, no. 1, pp. 471-481, Jan. 2015.

[5] Andres E. Leon and Santiago J. Amodeo, "Energy balancing improvement of modular multilevel converters under unbalanced grid conditions," *IEEE Trans. Power Electron.*, vol. 32, no. 8, pp. 6628-6637, Agu. 2015.

[6] Jiangcha Qin and Maryam Saeedifard, "Predictive control of a modular multilevel converter for a back-to-back HVDC system," *IEEE Trans. Power Delivery*, vol. 27, no. 3, pp. 1538-1547, Jul. 2012.

[7] Pu Liu, Yue Wang, Wulong Cong and Wanjun Lei, "Grouping-sorting-optimized model predictive control for modular multilevel converter with reduced computational load," *IEEE Trans. Power Electron.*, vol. 31, no. 3, pp. 1896-1907, Mar. 2016.

[8] Zheng Gong, Peng Dai, Xibo Yuan, Xiaojie Wu and Guosheng Guo, "Design and experimental evaluation of fast model predictive control for modular multilevel converters," *IEEE Trans. Ind. Electron.*, vol. 63, no. 6, pp. 3845-3856, Jun. 2016.

[9] Apparao Dekka, Bin Wu, Venkata Yaramasu and Navid Reza Zargari, "Model predictive control with common-mode voltage injection for modular multilevel converter," *IEEE Trans. Power Electron.*, vol. 32, no. 3, pp. 1767-1778, Mar. 2017.

[10] Fei Zhang, Wei Li and Géza Joós, "A voltage-level-based model predictive control of modular multilevel converter," *IEEE Trans. Ind. Electron.*, vol. 63, no. 8, pp. 5301-5312, Aug. 2016.

[11] Xingxing Chen, Jinjun Liu, Shaodi Ouyang and Shuguang Song, "An improved circulating current suppressing strategy for modular multilevel converters through redundant voltage levels," in *Proc. IEEE 2nd Annual Southern Power Electron. Conf.* 2016, pp. 1-4.

The 2018 International Power Electronics Conference

Family of Integrated Multi-Input Multi-Output DC-DC Power Converters

Bang Le-Huy Nguyen[1][*], Honnyong Cha[1], Tien-The Nguyen[1] and Heung-Geun Kim[2]

[1] School of Energy Engineering, Kyungpook National University, Daegu, Korea
[2] Department of Electrical Engineering, Kyungpook National University, Daegu, Korea
*E-mail: bangnguyen@ieee.org

Abstract— **This paper explores a family of integrated multiport converters using three-switch which can provide single-input dual-output (SIDO) or dual-input single-output (DISO) with bidirectional power flow between any two ports. The concept can be extended to the *n*-switch converters to achieve more inputs and/or outputs. The proposed converters can be applied to interfacing sources, loads and storage elements having different voltage levels in applications such as dc nanogrids, electric vehicle, multiport power supplies, distributed generation systems. Various topological configurations of the integrated multiport *n*-switch converter are investigated. The operating principles and PWM control strategy of these converters are analyzed in detail. A universalized hardware prototype is built, experimental results are provided for verification.**

Keywords— DC-DC power converter, integrated, multi-input, multi-output.

I. INTRODUCTION

The dc-dc power conversion plays a major role in the power electronics field spreading from low to high power range. Recently, the development of distributed generation systems, dc nanogrids, and electric vehicle, which are included many energy sources, loads, and storage elements, leads to the strong demand for compact integrated converter systems. Conventionally, to supply to different loads, to obtain energy from different sources, and to provide bidirectional power transfer for storage components, discrete power converters are implemented and regulated separately. These converters are coordinated to manage the power flow between them through a common control system via communication channels.

For more effectively monitor and cooperative control these converter systems, the concept of multiport power electronic interface (MPEI) [1]–[3] was introduced and implemented. This architecture considers all related converters as a united converter with a dynamic model [1] to enhance the transient stability and reliability. Fig. 1(a) shows the MPEI using separated dc-dc converters. For further improvements, the integrated multiport dc-dc converters as in Fig. 1(b) appears to reduce the number of redundant components and to make the system more compact and lower cost.

The integrated multiport converter topologies are more and more attractive with many topologies introduced in

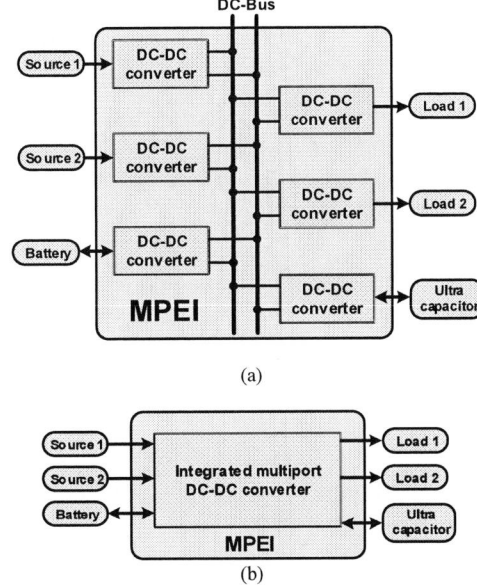

Fig. 1. Multiport power electronics interface architecture with separated dc-dc converters (a) and integrated multiport converter (b).

the literature. The paper [4]–[6] presented two families of multi-input single-output (MISO) converter topologies by combining some pulsating source cells with one output filter. Whereas, in [7], one pulsating source cell was combined with different filter cells to make single-input multi-output (SIMO) converter topologies. In [8], The family of three ports converters is introduced. However, some of these converters cannot provide bidirectional power flow which is crucial for systems involving energy storage elements. Besides, the complexity and inflexibility in structure make these converters ineffective. In [9], the multi-input bidirectional converter is introduced to interface a battery, an ultra-capacitor, and the dc-link of the hybrid electric vehicle/ fuel cell vehicle (HEV/FCV) application. This converter is derived by combining two non-inverting buck/boost converter with a sharing leg to provide both buck and boost function between any two ports. However, it uses more switches and the current in the battery is discontinuous. A bidirectional SIDO buck converter using only three switches were proposed in [10]–[12]. Similarly, the paper

3134

The 2018 International Power Electronics Conference

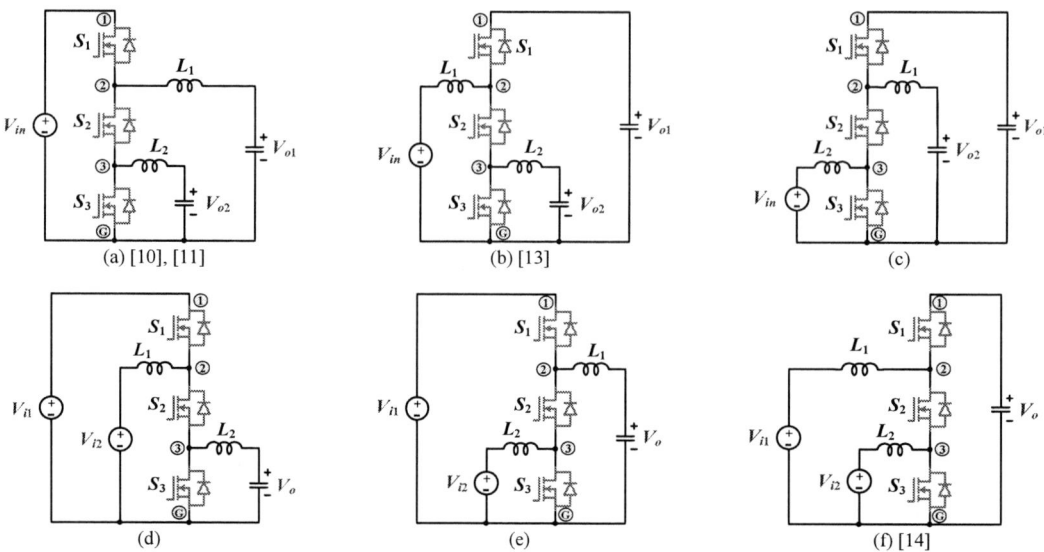

Fig. 2. Topological configurations of three-switch converter for SIDO and DISO converters. (a) two buck outputs [10], [11] (b) one boost and one buck outputs [13], (c) two boost outputs, (d) two buck inputs, (e) one buck and one boost inputs, and (f) two boost inputs [14].

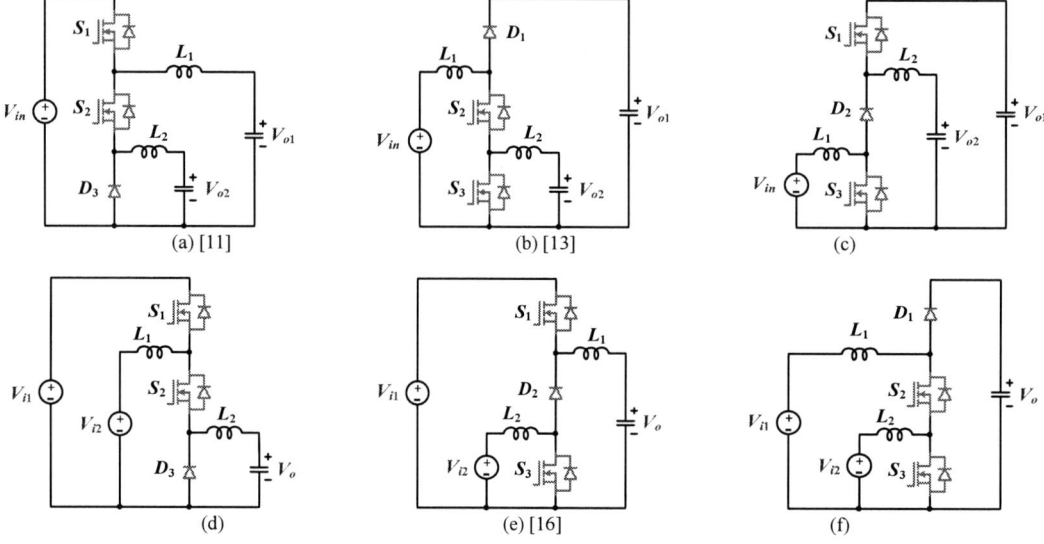

Fig. 3. Nonsynchronous derivations of SIDO and DISO converters. (a) two buck outputs [11], (b) one boost and one buck outputs [13], (c) two boost outputs, (d) two buck inputs, (e) one buck and one boost inputs [16], and (f) two boost inputs.

[13] introduced another configuration of three-switch to achieve one buck and one boost outputs. These converters are also extended to more switches to achieve more buck outputs. In [14]–[15], a DISO using three-switch was introduced. More generally, in [16], a topological synthesis method is proposed to derive nonsynchronous SIDO and DISO three-switch converters from conventional SISO converters. These above converters have benefits of the minimum number of switches and the integration in both hardware and control system.

Interestingly, other SIDO and DISO converters using three switches are still veiled in these papers [10]–[17]. This paper explores more topological configurations of three-switch converters. Gathering with the existed three-

switch converters, now the whole family of the integrated multiport converter which can provide bidirectional power between three ports is fully explored. The nonsynchronous versions of this converter family are also derived. Intensively, the principle and synthesis method for various multiport *n*-switch converter configurations are proposed and investigated. The operating principle and PWM control strategy are also analyzed in detail.

II. TOPOLOGICAL CONFIGURATIONS

In this section, firstly, the bidirectional three-switch SIDO and DISO converters with the common-ground ports are described. Secondly, the nonsynchronous derivations of these converters are considered. The extension to the *n*-switch converter is also discussed.

3135

Then, other derivations of three-switch converters with different-ground ports are investigated. Finally, the universalized configurations which are the multi-purposes converters are proposed and analyzed.

A. Common-Ground Topologies

By adding a switch to the conventional buck converter to have one more output, the first SIDO converter using three-switch were proposed in 2006 [10] as shown in Fig. 2(a). In [11], this converter topology is analyzed in detail, its nonsynchronous version shown in Fig. 3(a) is also introduced. By adding a switch to conventional boost converter to get one more buck output, other SIDO converter is proposed in [13] as shown in Fig. 2(b). In [14], by combining two boost converters, a DISO converter is proposed as shown in Fig. 2(f). In the paper [16], the basic source and load cells are defined. Inserting these cells to the basic dc-dc converter topologies (buck, boost, buck-boost, Cuk, sepic, zeta), six SIDO and six DISO converters are proposed included those shown in Figs.2 (a) and (e).

More simply, there are four nodes in the three-switch converter as denoted in Figs. (a)–(f). Except for the ground node G, any other nodes can be connected to the input source or output load. When one node is connected to the input and the remained nodes are the outputs, the three-switch SIDO converter family can be achieved as in Figs. 2(a)–(c). The converter shown in Fig. 2(c) is a new member with two boost outputs. On the other hand, if two nodes are used for the inputs and the remained node is connected to the output, the three-switch DISO converter family is created as in Figs. 2(d)–(f).

B. Nonsynchronous Derivations

The three-switch converters in Figs. 2(a)–(f) have the bidirectional power transfer between any two ports. However, if only unidirectional power flow is required, the nonsynchronous derivations are considered to reduce the number of active switches and accompanied circuits as gate driver power supply and gate driver circuit. Moreover, there is no shoot-through path in the nonsynchronous converters, then the dead-time protection can be removed.

The nonsynchronous derivations of the above three-switch converter can be achieved from the synchronous ones by replacing the switch with a diode as shown in Figs. 3(a)–(f), respectively. Where the switches, which are possible to be replaced, always conduct current in its diode.

Summarily, the family of SIDO and DISO dc-dc converters with common-ground ports using three-switch included the nonsynchronous derivations is fully explored.

C. n-Switch Extension

Extensively, the three-switch converters can be extended to n-switch as shown in Fig. 4. A similar approach of synthesis principle can be applied. Except for the ground node G, the remained n nodes of this n-switch converter can be connected to inputs or outputs. Thus, any k-input and p-output converters can be derived,

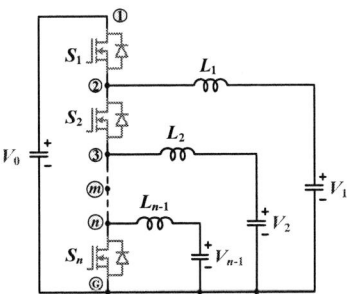

Fig. 4. The common-ground multiport n-switch converter.

TABLE I
THE NUMBER OF N-SWITCH CONVERTERS

Number of input ports	Number of output ports	Number of converters
1	$n-1$	C_n^1
2	$n-2$	C_n^2
...	....	....
$n-1$	1	C_n^{n-1}
$C_n^1 + C_n^2 + ... + C_n^{n-1} = 2^n - 2$		

where $n = (k+p)$. Entirely, the total number of different multiport converters are $(2^n - 2)$ as described in Table I. Notably, all those multiport converters have the same ground of inputs and outputs.

D. Different-Ground Topologies

The paper [17] introduced a SIDO using three-switch with different-ground outputs as shown in Fig. 5(a). Where V_0 is connected to the input, whereas V_1 and V_2 are connected to outputs. Reversely, using V_1 and V_2 as inputs and V_0 as output, the paper [15] formed a DISO converter. Where the port V_1 is connected to nodes 2 and 3. Whereas the port V_2 is connected to node 3 and ground (G).

Seemingly, except for the fixed node V_0, the ports V_1 and V_2 can be connected to other two nodes. More topologies, therefore, can be achieved as described in Figs. 5(b)–(g). In Figs. 5(a)–(c), the ports are connected to the adjacent nodes, whereas the further-distance nodes are combined in Figs. 5(d)–(g).

Similar to the analysis of the common-ground multiport converter topology, each topology shown in Fig. 5 also possess abilities as follows:

1) They can form three SIDO and three DISO converters by arranging the inputs and outputs by the same way shown in Fig. 2.

2) The nonsynchronous versions of these SIDO and DISO converters can also be derived.

3) They can be extended to the n-switch converters.

Considerably, when extending to n-switch, the number of nodes becomes $(n+1)$. Beside the port V_0 is connected to nodes 1 and G, other $(n-1)$ ports can be connected to any two nodes. A myriad of topological configurations can be formed. As a result, various integrated MIMO dc-dc power converters can be achieved.

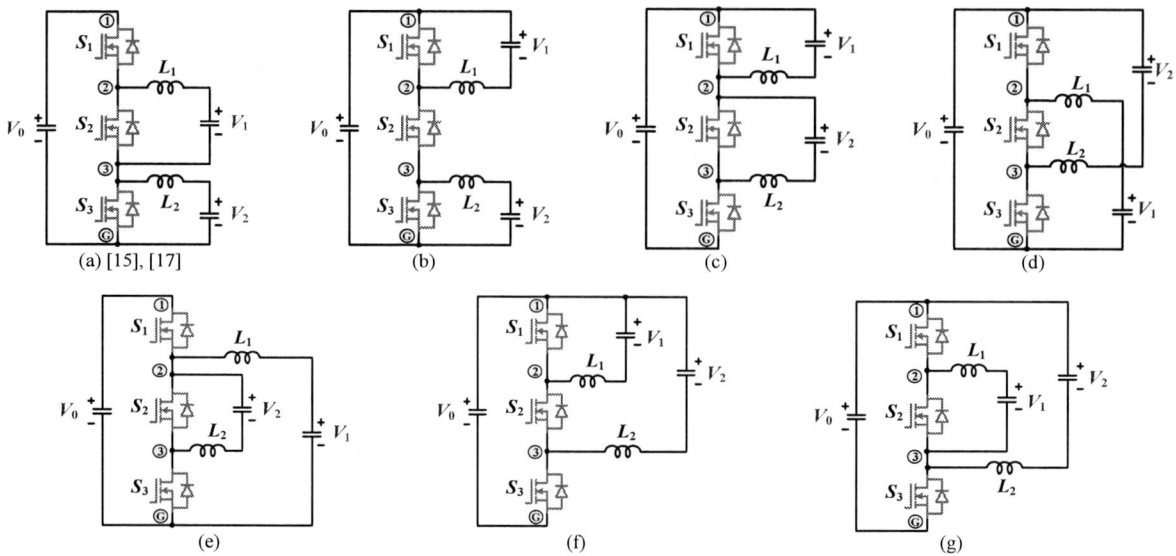

Fig. 5. Topological configurations of the three-switch converters with different-ground ports.

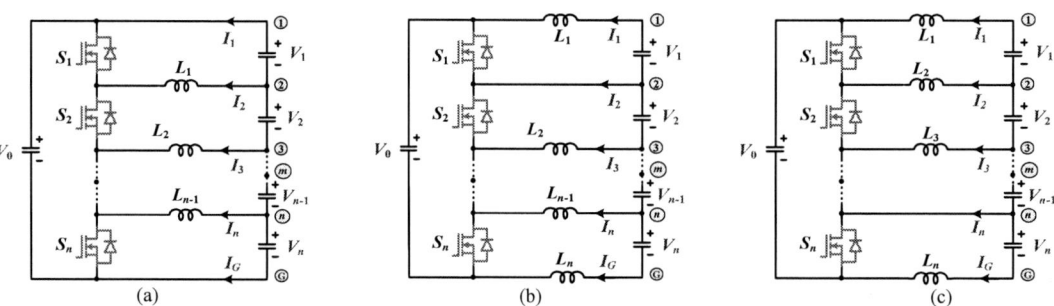

Fig. 6. The multi-purpose integrated *n*-switch multiport converters.

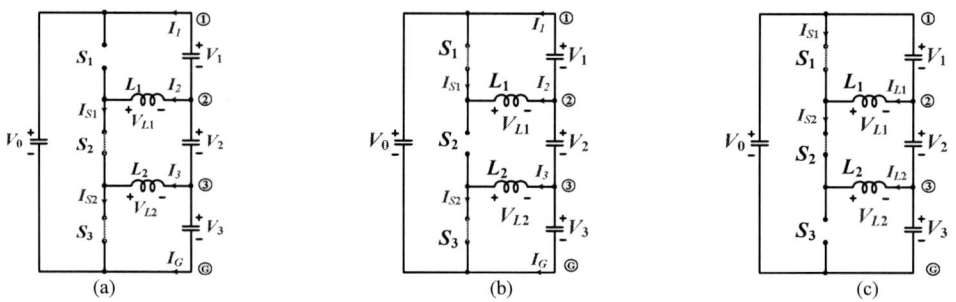

Fig. 7. The operating modes of the multi-purpose three-switch converter (a) Mode 1. (b) Mode 2. (c) Mode 3.

E. Universalized Configurations

The universalized configurations of the integrated multiport converter are derived as shown in Fig. 6. They are the multi-purpose *n*-switch converters which can replace equally all above topologies. Where the pulsating voltage across each switch is filtered to the port by the *LC* filter. All ports are connected in series and the nodes now are placed between the ports. Hence, the input sources and/or output loads can feed and draw the power by connecting to any two nodes.

The *n*-switch multiport converter now becomes an energy router which diverts and allocates the power from and to its $(n+1)$ ports. The total voltages of ports V_1–V_n is the port voltage V_0. The port voltages can be controlled by the duty ratios of the responding switches, respectively, as in equation (9) ($V_i = D_i V_0$, *i*: 1–*n*) which is derived in Section III. Where the duty ratios are defined as the proportion of the period when turning off the switch over the total duty cycle time.

Fig. 6(a) shows the configuration in which the currents $I_2 - I_n$ are continuous owing to the presence of inductors. Whereas the currents I_1 and I_G are discontinuous. In some applications, if the currents I_1 and I_G are required to be continuous. The inductors can be placed in their paths as shown in Figs. 6(b)–(c). Where one inductor in any middle branches can be removed to reduce the number of inductors.

TABLE II
CURRENT FLOWING THROUGH SWITCHES

Mode	S_1	S_2	S_3	S_4	...	S_n
1	0	I_2	$\Sigma I_{2\text{-}3}$	$\Sigma I_{2\text{-}4}$	...	$\Sigma I_{2\text{-}n}$
2	$-I_2$	0	I_3	$\Sigma I_{3\text{-}4}$	...	$\Sigma I_{3\text{-}n}$
3	$-\Sigma I_{2\text{-}3}$	$-I_3$	0	I_4	...	$\Sigma I_{4\text{-}n}$
...	...	...	...	...	....	....
$n\text{-}1$	$-\Sigma I_{2\text{-}(n\text{-}1)}$	$-\Sigma I_{3\text{-}(n\text{-}1)}$	$-\Sigma I_{4\text{-}(n\text{-}1)}$	$-\Sigma I_{5\text{-}(n\text{-}1)}$	....	I_n
n	$-\Sigma I_{2\text{-}n}$	$-\Sigma I_{3\text{-}(n\text{-}1)}$	$-\Sigma I_{4\text{-}(n\text{-}1)}$	$-\Sigma I_{5\text{-}(n\text{-}1)}$	....	0

III. STEADY-STATE ANALYSIS AND PWM CONTROL STRATEGY

For simplicity, the port voltages are assumed to be stable at a fixed value in steady-state and the inductor currents are continuous. In the operation of these multiport converters, only one switch is turned off at the time, whereas all remained switches are turned on. Thus, there are n modes for the n-switch converters. Figs. 7(a)–(c) show the equivalent circuits in the operating modes of the integrated multi-purpose three-switch multiport converter. The analysis of three modes is discussed below.

1) Mode 1 (duty ratio D_1): the switch S_1 is off, while both switch S_2 and S_3 are on. The voltage imposing on inductors L_1, L_2 are as follows:

$$V_{L1}=V_1-V_0 \ , \ V_{L2}=-V_3 \quad (1)$$

2) Mode 2 (duty ratio D_2): the switch S_2 is off, while both switch S_1 and S_3 are on. The voltage imposing on inductors L_1, L_2 are as follows:

$$V_{L1}=V_1 \ , \ V_{L2}=-V_3 \quad (2)$$

3) Mode 3 (duty ratio D_3): the switch S_3 is off, while both switch S_1 and S_2 are on. The voltage imposing on inductors L_1, L_2 are as follows:

$$V_{L1}=V_1 \ , \ V_{L2}=V_0-V_3 \quad (3)$$

Apply the flux (or volt-sec) balance condition for inductor L_1 and L_2, respectively, we have:

$$D_1(V_1-V_0)+(D_2+D_3)V_1=0 \quad (4)$$

$$-(D_1+D_2)V_3+D_3(V_0-V_3)=0 \quad (5)$$

Where,

$$D_1+D_2+D_3=1 \quad (6)$$

$$V_1+V_2+V_3=V_0 \quad (7)$$

According to equations (4)–(7), the gains between the port voltages can be determined as:

$$V_1=D_1V_0 \ ; \ V_2=D_2V_0 \ ; \ V_3=D_3V_0 \quad (8)$$

Extensively, the port voltage V_i can be determined by the duty ratio D_i (i: 1–n) as follows.

$$V_i=D_iV_0 \ , \ i\text{: }1\text{–}n \quad (9)$$

Using the duty ratios D_1–D_n of switches S_1–S_n, all port voltages and inductor current can be regulated independently. The general gate drive signals and duty ratio control diagram are given as illustrated in Figs. 8 and 9, respectively.

The voltage stresses on switches are the voltage across nodes 1 and G (V_0). The currents flowing through the switches in the operating modes are described as in Table II. Where all inductor currents have the direction as denoted in Fig. 6. Apparently, the total current stresses on

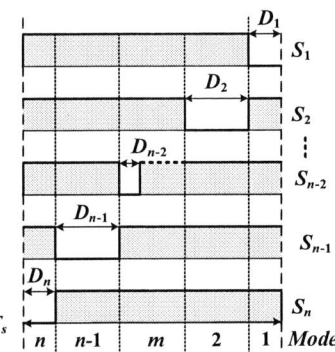
Fig. 8. Gate drive signals

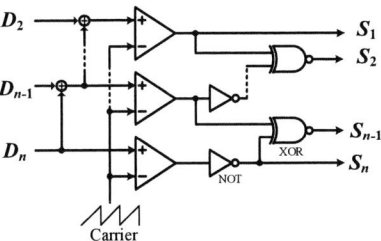
Fig. 9. The duty ratio control diagram

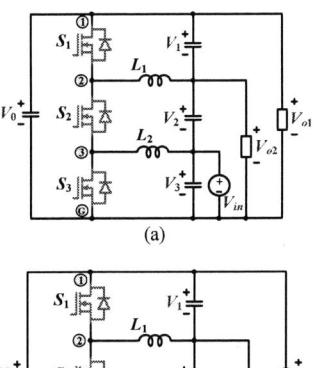
(a)

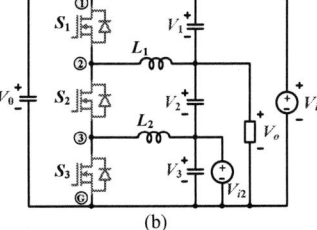
(b)

Fig. 10. The generalized three-switch multiport converter with sources and loads connected as SIDO in test 1 (a) and DISO in test 2 (b).

switches will be reduced provided that the inductor current is different in direction or the middle nodes (except for nodes 1 and G) are connected to inputs and outputs at the same time.

IV. EXPERIMENTAL VERIFICATION

The universalized hardware prototype of the generalized three-switch multiport converter using MOSFETs 47N60CFD of Infineon is built and tested. The inductors are of 0.72 mH inductance value and the capacitors are of 560 μF capacitance value. The switching frequency is 30 kHz.

Firstly, one source and two loads are connected to form a SIDO converter as shown in Fig. 10(a). Where, the input voltage V_{in} = 40 V and the duty ratios D_1 = 0.35,

3138

$D_2 = 0.25$, and $D_3 = 0.4$. Two outputs are the resistive loads of 50 Ω. The experimental results of this test are given in Fig. 11. The output voltages are calculated as $V_{o1} = =V_3/D_3 = 40/0.4 =100$ V; $V_{o2} =V_2+V_3=D_2V_3/D_3 +V_3=65$ V. These calculated values are verified by the voltage waveforms of V_{o1} and V_{o2} shown in Fig. 11. The load currents are the average values of the inductor currents as the I_{L1} and I_{L2} waveforms.

In the second test, two sources and one load are used to achieve a DISO converter as shown in Fig. 10(b). Where the input voltages are $V_{i1} = 50$ V, $V_{i2} = 20$ V and the duty ratios are the same as the first test. The output is the resistive load of 50 Ω. Where, the output voltage V_o can be calculated as $V_{o1}= V_2+V_3 = D_2V_{i1}+V_3 = 32.5$ V. The experimental inductor currents and output voltage waveforms shown in Fig. 12 are agreed with the theoretical analysis. The output load and input source currents of V_{i2} are the average value of the inductor currents I_{L1} and I_{L2} waveforms, respectively.

V. CONCLUSIONS

In this paper, the topological configurations for SIDO and DISO three-switch converters with both common and different ground of ports are explored and analysis in detail. The extension to n-switch is explained. The multi-purpose topologies which can be employed as an energy router are proposed. The steady-state analysis and PWM control strategy for the n-switch converters are provided. The gate signals and duty ratio control diagram are given. The hardware prototype is built and tested, the experimental results are agreed with the theoretical analysis.

ACKNOWLEDGMENT

This work was supported by the Korea Institute of Energy Technology Evaluation and Planning (KETEP) and the Ministry of Trade, Industry & Energy (MOTIE) of the Republic of Korea (No. 20174030201490) and was also supported by Basic Science Research Program through the National Research Foundation of Korea (NRF) funded by the Ministry of Education (NRF-2016R1D1A1B03934577).

REFERENCES

[1] W. Jiang and B. Fahimi, "Multiport power electronic interface—concept modeling and design," *IEEE Trans. Power Electron.*, vol. 26, no. 7, pp. 1890-1900, Jul. 2011.

[2] P. Shamsi and B. Fahimi, "Dynamic behavior of multiport power electronic interface under source/load disturbances," *IEEE Trans. Ind. Electron.*, vol. 60, no. 10, pp. 4500-4511, Oct. 2013.

[3] M. McDonough, "Integration of inductively coupled power transfer and hybrid energy storage system: a multiport power electronics interface for battery-powered electric vehicles," *IEEE Trans. Power Electron.*, vol. 30, no. 11, pp. 6423-6433, Nov. 2015.

[4] Y. Li, X. Ruan, D. Yang, F. Liu, and C. K. Tse, "Synthesis of multiple-input DC/DC converters," *IEEE Trans. Power Electron.*, vol. 25, no. 9, pp. 2372-2385, Sep. 2010.

[5] Y. C. Liu and Y. M. Chen, "A systematic approach to synthesizing multi-input DC–DC converters," *IEEE Trans. Power Electron.*, vol. 24, no. 1, pp. 116-127, Jan. 2009.

[6] P. Yang, C. K. Tse, J. Xu, and G. Zhou, "Synthesis and analysis of double-input single-output DC/DC converters," *IEEE Trans. Ind. Electron.*, vol. 62, no. 10, pp. 6284-6295, Oct. 2015.

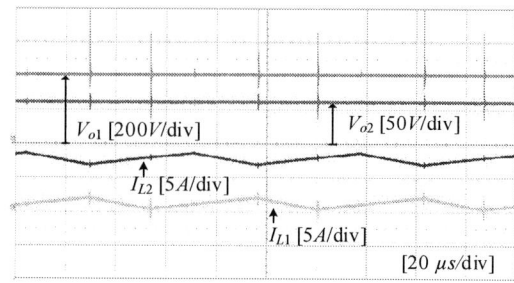

Fig. 11. The experimental waveforms for the SIDO converter in Fig. 10(a) with $D_1 = 0.35$, $D_2 = 0.25$, and $D_3 = 0.4$.

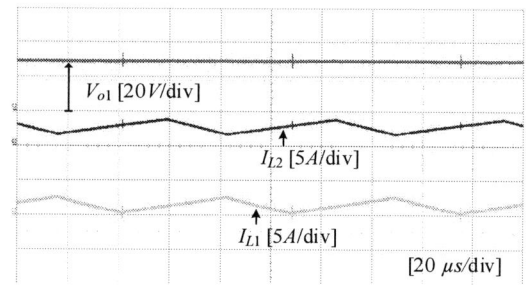

Fig. 12. The experimental waveforms for the DISO converter in Fig. 10(b) with $D_1 = 0.35$, $D_2 = 0.25$, and $D_3 = 0.4$.

[7] M. B. Ferrera Prieto, S. P. Litran, E. D. Aranda, and J. M. E. Gomez, "New single-input, multiple-output converter topologies: combining single-switch nonisolated DC-DC converters for single-input, multiple-output applications," *IEEE Ind. Electron. Mag.*, vol. 10, no. 2, pp. 6-20, Jun. 2016.

[8] H. Wu, K. Sun, S. Ding, and Y. Xing, "Topology derivation of nonisolated three-port DC–DC converters from DIC and DOC," *IEEE Trans. Power Electron.*, vol. 28, no. 7, pp. 3297-3307, Jul. 2013.

[9] A. Hintz, U. R. Prasanna, and K. Rajashekara, "Novel modular multiple-input bidirectional DC–DC power converter (MIPC) for HEV/FCV application," *IEEE Trans. Ind. Electron.*, vol. 62, no. 5, pp. 3163-3172, May 2015.

[10] P. Kumar and M. Rojas-Gonzalez, "Novel 3-switch dual output buck voltage regulator," in *Proc. IEEE Appl. Power Electron. Conf. Expo.*, 2006, pp. 467–473.

[11] E. C. dos Santos, "Dual-output dc-dc buck converters with bidirectional and unidirectional characteristics," *IET Power Electron.*, vol. 6, no. 5, pp. 999-1009, May 2013.

[12] G. Chen, Y. Deng, J. Dong, Y. Hu, L. Jiang, and X. He, "Integrated multiple-output synchronous buck converter for electric vehicle power supply," *IEEE Trans. Veh. Technol.*, vol. 66, no. 7, pp. 5752-5761, Jul. 2017.

[13] O. Ray, A. P. Josyula, S. Mishra, and A. Joshi, "Integrated dual-output converter," *IEEE Trans. Ind. Electron.*, vol. 62, no. 1, pp. 371-382, Jan. 2015.

[14] Jun Cai and Qing-Chang Zhong, "Compact bidirectional DC-DC converters with two input sources," *2014 IEEE 5th International Symposium on Power Electronics for Distributed Generation Systems (PEDG)*, Galway, 2014, pp. 1-5.

[15] R. J. Wai, C. Y. Lin, J. J. Liaw, and Y. R. Chang, "Newly designed ZVS multi-input converter," *IEEE Trans. Ind. Electron.*, vol. 58, no. 2, pp. 555-566, Feb. 2011.

[16] G. Chen, Z. Jin, Y. Deng, X. He, and X. Qing, "Principle and topology synthesis of integrated single-input dual-output and dual-input single-output DC–DC converters," *IEEE Trans. Ind. Electron.*, vol. 65, no. 5, pp. 3815-3825, May 2018.

[17] G. Chen, Y. Deng, J. Dong, Y. Hu, L. Jiang, and X. He, "Integrated multiple-output synchronous buck converter for electric vehicle power supply," *IEEE Trans. Veh. Technol.*, vol. 66, no. 7, pp. 5752-5761, Jul. 2017.

Low-Complexity State-Space Based System Identification and Controller Auto-Tuning Method for Multi-Phase DC-DC Converters

Marc Kanzian[1*], Harald Gietler[2], Christoph Unterrieder[1],
Matteo Agostinelli[1], Michael Lunglmayr[3], Mario Huemer[3]

1 Infineon Technologies Austria AG, Villach, Austria
2 Institute of Smart Systems Technologies, Alpen-Adria Universität Klagenfurt, Klagenfurt, Austria
3 Institute of Signal Processing, Johannes Kepler University Linz, Linz, Austria
*E-mail: marc.kanzian@infineon.com

Abstract—The importance of online system identification (SI) in power electronics is ever increasing. It enables the tracking of system parameters, which in turn can be used for online controller tuning. Hence, SI is a key element for improving a converter's dynamic performance, stability and reliability. In this paper, a state-space based SI approach utilizing the step-adaptive least squares (SALS) estimation algorithm with observation matrix randomization is proposed. The presented concept yields an accurate state-space model of the converter while simultaneously achieving a fast convergence rate and low computational complexity. Consequently, the estimated state-space model is utilized to automatically tune a full state feedback (FSF) controller, resulting in an improved converter performance. The proposed concept is verified by a prototype system comprised of a two-phase buck converter and a field-programmable gate array (FPGA). The provided measurement results highlight the effectiveness and benefits of the presented method over state of the art z-domain estimation. It is shown that the number of required iterations is more than halved, while accuracy is improved.

Keywords—*Adaptive Control, DC-DC power converters, Digital Control, System Identification*

I. INTRODUCTION

Today's electronic devices require highly efficient, accurate and reliable switched-mode power supplies (SMPSs). An insufficiently performing SMPS might have a negative impact on the system's operating range, stability and reliability. In order to provide a well-regulated output voltage under all operating conditions, sophisticated control concepts are essential [1,2]. Manufacturing tolerances, temperature dependency and long-term aging effects can degrade the performance of an SMPS. For this reason, controller coefficients have to be conservatively chosen, in order to cover worst case parameter spreads or different set of passive components, leading to an unsatisfactory performance. By employing online system identification (SI), the actual converter parameters can be estimated [3,4]. Subsequently, these estimates are used in order to auto-tune the controller, yielding improved stability and dynamic performance [4].

SI can be divided into two main approaches: non-parametric and parametric [5]. On the one hand, a fixed system model is not required in non-parametric approaches but they typically require long measurement data sets and are computationally complex [6,7]. On the other hand, parametric SI is typically computationally less complex, but additional uncertainties are introduced, since a fixed system model has to be assumed [3].

In recent parametric SI approaches for SMPS a stimulus, such as a pseudo random binary sequence (PRBS), is superimposed on the steady state duty cycle [8]. Then, the stimulus and the measured output voltage perturbation are used to estimate the coefficients of a discrete-time transfer function (TF) model of the converter. Due to their long stimulus injection times and high computational complexities, these approaches contradict the usual low power and small area requirements. In [3], a concept using the step-adaptive least squares (SALS) estimator [9], an algorithm similar to the Randomized Kaczmarz algorithm [10], has been proposed. The method introduced therein can operate at higher switching frequencies and is computationally less complex than previously presented SI approaches. Nevertheless, there, only an accurate estimate of the converter's natural frequency is obtained and no controller auto-tuning is performed.

In this paper, a state-space based parametric SI approach is proposed. The benefits of this approach are demonstrated on the basis of a two-phase buck converter. In contrast to state of the art concepts, which only estimate the duty cycle to output voltage TF, the presented method yields a model of the converter in the discrete-time state-space. Therefore, auto-tuning of more sophisticated control structures than a standard proportional-integral-derivative (PID) controller is straightforward. Exemplarily, a full state feedback (FSF) controller is employed in this paper. Moreover, the improved accuracy, convergence speed and reduced complexity of the proposed concept compared to state of the art approaches is demonstrated.

The rest of this paper is structured as follows: In Section II, the discrete-time modeling of a two-phase buck converter is reviewed. Then, the applied digital FSF controller is addressed in Section III-A. Subsequently, Section III-B covers the proposed SI concept. Measurements results are reported in Section IV. Finally, the major findings are summarized in Section V.

II. SYSTEM MODEL

In Fig. 1, the schematic of a two-phase buck converter is reported. Each phase m includes a power stage com-

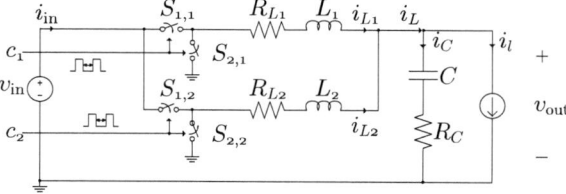

Fig. 1. Schematic of a two-phase synchronous buck converter.

prising the switches $S_{1,m}$ and $S_{2,m}$ and an inductor with inductance L_m. The power switches $S_{1,m}$ and $S_{2,m}$ are toggled by the control signals $c_m \in \{0,1\}$ and $(1-c_m)$, respectively. The output capacitor with capacitance C is shared by the phases. Furthermore, the parasitic resistance of the output capacitor is modeled by R_C, while R_{L_m} models the resistances in the power paths. The input and output voltages are denoted by v_{in} and v_{out}, respectively. Finally, the load is modeled by the current sink i_l. In order to facilitate the following derivations, it is assumed that $L_1 = L_2$ and $R_{L_1} = R_{L_2}$ hold. Typically, the individual power stages of a multi-phase converter are designed to be equal [11]. Hence, these simplifications are reasonable.

From Fig. 1, a continuous-time state-space model of the form

$$\dot{x}(t) = Ax(t) + Bu(t) \tag{1a}$$

$$y(t) = Cx(t), \tag{1b}$$

with the system matrix $A \in \mathbb{R}^{2\times 2}$, the input matrix $B \in \mathbb{R}^{2\times 2}$, the output matrix $C \in \mathbb{R}^{2\times 2}$, the state vector $x(t) = \begin{bmatrix} v_{\text{out}}(t) & i_L(t) = i_{L_1}(t) + i_{L_2}(t) \end{bmatrix}^T$, the input vector $u(t) = \begin{bmatrix} (c_1(t) + c_2(t)) \cdot v_{\text{in}}(t) & i_l(t) \end{bmatrix}^T$, and the output vector $y(t) = \begin{bmatrix} v_{\text{out}}(t) & i_L(t) \end{bmatrix}^T$ can be derived. By state-space averaging (SSA) and subsequent linearization, a linear time-invariant (LTI) small-signal model given as [4]

$$\dot{\tilde{x}}(t) = A\tilde{x}(t) + B\begin{bmatrix} V_{\text{in}} \\ 0 \end{bmatrix} \tilde{d}(t) \tag{2a}$$

$$\tilde{y}(t) = C\tilde{x}(t), \tag{2b}$$

is obtained. In (2a), the small-signal duty cycle $\tilde{d}(t)$ represents the new control input and the steady state input voltage is given by V_{in}. Furthermore, all small-signal quantities are marked by a tilde. It should be noted that, both phases use the same duty cycle as their control input in the small-signal model. Moreover, instead of each individual phase current, only the sum $i_L(t)$ of the inductor currents represents a state variable. As will be later shown, this is fully justified for the presented controller, which calculates a common duty cycle for both phases.

Furthermore, the delays caused by the digital pulse-width modulation (DPWM) are appropriately modeled. In the employed trailing-edge modulation scheme, the duty cycle is updated at the beginning of the first phase's switching period. Hence, the DPWM delays of the first and second phase can be approximated by $t_{d,P1} = e^{-sDT_{\text{sw}}}$ and $t_{d,P2} = e^{-s\left(DT_{\text{sw}} + \frac{T_{\text{sw}}}{2}\right)}$, respectively. Here, the steady state duty cycle D and the switching period T_{sw} have been used for the derivation. Therefore, the overall output delay is given by $t_d = \frac{t_{d,P1} + t_{d,P2}}{2}$ [1].

Then, applying the impulse-invariant transform with sampling period T_{sw} to (2), with the output delay t_d considered, yields the discrete-time converter model [4]

$$x_{k+1} = \Phi x_k + \gamma d_k \tag{3a}$$

$$y_{k+1} = Cx_k. \tag{3b}$$

The subscript k denotes the kth switching period, and the discrete-time system and input matrices are given by Φ and γ, respectively. If the component values of the converter are exactly known, (3) represents an accurate discrete-time model of the converter. This model can be used to design a digital FSF controller, which will be briefly outlined in the next section.

III. PROPOSED CONCEPT

A. Controller Concept

One of the SMPS's main objectives is to regulate the output voltage to V_{ref}. To that end, (3a) is first augmented by an additional integrator state. The integrator removes any remaining steady state error $e_k := V_{\text{ref}} - v_{\text{out},k}$ in the output voltage. The augmented system is therefore governed by

$$\underbrace{\begin{bmatrix} x_{k+1} \\ x_{i,k+1} \end{bmatrix}}_{x_{\text{aug},k+1}} = \underbrace{\begin{bmatrix} \Phi & \begin{matrix} 0 \\ 0 \end{matrix} \\ \begin{bmatrix} -1 & 0 \end{bmatrix} & 1 \end{bmatrix}}_{\Phi_{\text{aug}}} \underbrace{\begin{bmatrix} x_k \\ x_{i,k} \end{bmatrix}}_{x_{\text{aug},k}} + \underbrace{\begin{bmatrix} \gamma \\ 0 \end{bmatrix}}_{\gamma_{\text{aug}}} d_k. \tag{4}$$

In the FSF controller, the duty cycle satisfies the equation

$$d_k = -kx_{\text{aug},k}, \tag{5}$$

with the gain vector k. Substituting the right-hand side of (5) for d_k in (3a), yields the closed-loop system

$$\begin{aligned} x_{\text{aug},k+1} &= \Phi_{\text{aug}}x_{\text{aug},k} - \gamma_{\text{aug}}kx_{\text{aug},k} \\ &= \underbrace{(\Phi_{\text{aug}} - \gamma_{\text{aug}}k)}_{\Phi_{\text{cl}}} x_{\text{aug},k}. \end{aligned} \tag{6}$$

The required gain vector k for the desired closed-loop dynamics can be readily obtained by pole placement [12]. Note that, although (5) includes the summed inductor currents $i_{L,k}$, current sensing is actually not necessary. Instead, a current observer can be used to obtain a current estimate $\hat{i}_{L,k}$ [13,14].

3141

B. System Identification

As outlined in the previous section, the state-space model allows for a straightforward controller design if $\boldsymbol{\Phi}$ and $\boldsymbol{\gamma}$ are precisely known. Model limitations and other uncertainties, e.g. component tolerances, as discussed in Section I, introduce mismatches between the converter model and the actual SMPS. Depending on the choice of $\boldsymbol{k}$, these mismatches may result in instabilities or degraded performance of the real system. Therefore, SI is employed to obtain accurate estimates $\hat{\boldsymbol{\Phi}}$ and $\hat{\boldsymbol{\gamma}}$ of the real converter's system and input matrix, respectively. Consequently, these estimates are used to automatically tune the controller, yielding an improved performance.

Figure 2 illustrates the proposed SI concept. The

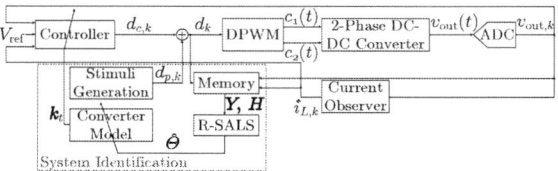

Fig. 2. Proposed closed-loop system identification and controller auto-tuning concept.

duty cycle $d_{c,k}$ is calculated by the controller according to (5), whereby the current estimate $\hat{i}_{L,k}$ is obtained by a current observer. During SI, $d_{c,k}$ is superimposed by the PRBS stimulus $d_{p,k}$, which is generated by a linear-feedback-shift register (LFSR). The output voltage $v_{\text{out},k}$, current observer output $\hat{i}_{L,k}$, and duty cycle d_k samples obtained during SI are stored until the stimulus injection is finished. Then, the measurement matrix $\boldsymbol{Y} = \begin{bmatrix} \boldsymbol{v}_{\text{out}} & \hat{\boldsymbol{i}}_L \end{bmatrix} \in \mathbb{R}^{(M-1)\times 2}$ and the observation matrix $\boldsymbol{H} \in \mathbb{R}^{(M-1)\times \frac{P}{2}}$ are composed, with M denoting the perturbation length in switching periods and P being the number of parameters to be estimated. The $(k+1)^{\text{th}}$ row $\boldsymbol{h}_{(k+1)}^T = \begin{bmatrix} v_{\text{out},k} & \hat{i}_{L,k} & d_k \end{bmatrix}$ of $\boldsymbol{H}$ contains the delayed input and output samples. For the converter model (3), $P = 6$, since $\boldsymbol{\Phi} \in \mathbb{R}^{2\times 2}$ and $\boldsymbol{\gamma} \in \mathbb{R}^{2\times 1}$. Note that, in a TF estimation approach, $P = 8$ parameters have to be estimated, since z-transforming (3) yields two second-order TFs, each with two poles and zeros [4]. Hence, the overall complexity is reduced by using the proposed state-space based approach.

The least squares (LS) solution of

$$\boldsymbol{Y} = \boldsymbol{H} \underbrace{\begin{bmatrix} \boldsymbol{\Phi} & \boldsymbol{\gamma} \end{bmatrix}^T}_{\boldsymbol{\Theta}} + \boldsymbol{N}, \qquad (7)$$

where the measurement noise is combined in the matrix $\boldsymbol{N}$, is obtained by minimizing the cost function

$$J(\boldsymbol{\Theta}) = \text{tr}\left((\boldsymbol{Y} - \boldsymbol{H}\boldsymbol{\Theta})^T (\boldsymbol{Y} - \boldsymbol{H}\boldsymbol{\Theta}) \right). \qquad (8)$$

In other words, the minimization of (8) yields the parameter matrix estimate $\hat{\boldsymbol{\Theta}} = \begin{bmatrix} \hat{\boldsymbol{\Phi}} & \hat{\boldsymbol{\gamma}} \end{bmatrix}^T$ comprising the estimates $\hat{\boldsymbol{\Phi}}$ and $\hat{\boldsymbol{\gamma}}$ of the system matrix $\boldsymbol{\Phi}$ and input matrix $\boldsymbol{\gamma}$, respectively. Although an exact solution of

(8) can be found, the required matrix inversion poses a significant challenge for many applications due to its high computational complexity. Hence, the objective of the R-SALS estimation algorithm is now to approximate the exact solution of (8). In each iteration step n, an update [9]

$$\hat{\boldsymbol{\Theta}}^{(n)} = \hat{\boldsymbol{\Theta}}^{(n-1)} + \mu^{(n)} \boldsymbol{h}_l \left(\boldsymbol{y}_l^T - \boldsymbol{h}_l^T \hat{\boldsymbol{\Theta}}^{(n-1)} \right), \qquad (9)$$

of the parameter matrix estimate is calculated. In (9), $\boldsymbol{h}_l^T$ and $\boldsymbol{y}_l^T$ denote the lth row of $\boldsymbol{H}$ and $\boldsymbol{Y}$, respectively. Furthermore, a dynamic step size calculated by [9]

$$\mu^{(n)} = \frac{1}{\|\boldsymbol{h}_l^T\|_2^2} \qquad (10)$$

is used, which improves both convergence speed and accuracy. In contrast to increasing the row index l in each iteration step, l is randomly selected using the selection strategy of the Randomized Kaczmarz algorithm [10]. For this selection, a cumulative sum $F(i) = \sum_{j=1}^{i} \|\boldsymbol{h}_j^T\|_2^2 \ \forall i \in \{1, 2, ..., M-1\}$ is calculated once, before the first iteration step. Then, the row index l in the nth iteration step is selected such that $F(l-1) \leq r^{(n)}$ and $F(l+1) \geq r^{(n)}$ are fulfilled, using a uniformly distributed random number $r^{(n)}$. This results in row of $\boldsymbol{H}$'s selection probability proportional to its squared norm and further accelerates the convergence of the estimation [10].

After N iterations, the estimator output $\hat{\boldsymbol{\Theta}}^{(N)}$ is used as the converter model parameters, yielding the estimated closed-loop system matrix $\hat{\boldsymbol{\Phi}}_{\text{cl}} = \left(\hat{\boldsymbol{\Phi}}^{(N)} - \hat{\gamma}^{(N)} \boldsymbol{k} \right)$. Finally, the updated and tuned gain vector $\boldsymbol{k}_t$ is calculated by placing the poles of $\hat{\boldsymbol{\Phi}}_{\text{cl}}$ at the desired locations.

IV. MEASUREMENT RESULTS

In order to verify the proposed SI and auto-tuning method, a prototype comprised of a two-phase buck converter with the parameters reported in Table I has been built. Furthermore, the digital controller and SI

TABLE I. TWO-PHASE BUCK CONVERTER PROTOTYPE PARAMETERS

Parameter	Value	Parameter	Value
V_{in}	3.3 V	V_{ref}	1.26 V
L_1, L_2	3.3 μH	R_{L_1}, R_{L_2}	320 mΩ
C	20 μF	R_C	10 mΩ
i_l	1 A	f_{sw}	1 MHz

have been implemented on an Altera Cyclone IV E field-programmable gate array (FPGA). The digital clock rate of the FPGA has been set to $f_{\text{digi}} = 100\,\text{MHz}$, resulting in a DPWM resolution of 100 steps per switching period. Furthermore, a 25 MHz, 8 bit tracking analog-to-digital converter (ADC) with a resolution of 5 mV has been used to obtain $v_{\text{out},k}$. Note that, both the digital controller and the SI approach only require one output voltage sample per switching period. In other words, the proposed concept is also applicable if $v_{\text{out}}(t)$ is sampled with the switching frequency f_{sw}, e.g., by a flash ADC.

During the SI phase, the duty cycle $d_{c,k}$ is superimposed by the perturbation $d_{p,k}$, which is generated by a 9 bit LFSR. Since the amplitude of $d_{p,k}$ determines the magnitude of the output voltage and inductor current perturbations, a trade-off between estimation accuracy and acceptable excursion around the nominal value of $v_{\mathrm{out}}(t)$ has to be made. An amplitude of ± 5 DPWM steps has been selected for $d_{p,k}$. This choice yields a deviation of approximately $\pm 6.5\,\%$ around the nominal output voltage value V_{ref}, while achieving an excellent estimation accuracy. Finally, the injection time has been set to $0.511\,\mathrm{ms}$, which corresponds to one LFSR cycle. Similar to the amplitude of $d_{p,k}$, the stimuli injection time constitutes a design parameter. On the one hand, longer injection times typically improve the estimation accuracy. On the other hand, the number of rows of the observation matrix H is determined by M, hence, more multiplications are required.

To demonstrate the controller auto-tuning, an initial gain vector k has been chosen such that a wide parameter range of $\pm 50\%$ of the nominal values of L and C is supported. For the measured parameters listed in Table I, this results in a closed-loop system response with a bandwidth of $w_n = 2\pi \cdot 15\,\mathrm{kHz}$ and a damping factor of $\zeta = 1$. In Fig. 3, the perturbed output voltage $v_{\mathrm{out}}(t)$ during SI can be seen. Moreover, a sub-harmonic oscillation in

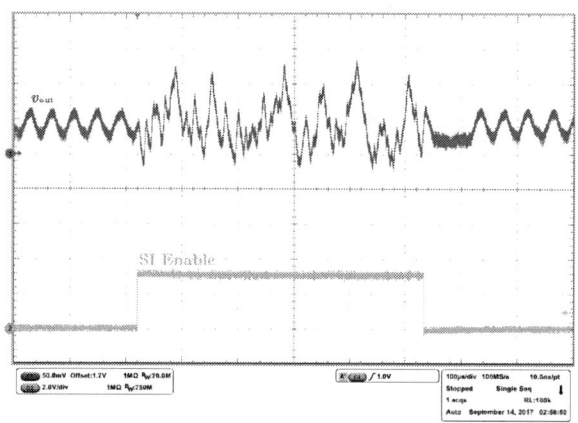

Fig. 3. Output voltage perturbation during the SI phase. The reported waveforms are the output voltage (v_{out}) and the SI enable signal (SI Enable). The untuned controller coefficients are used before, during and after the SI phase.

steady state is observed, caused by the low bandwidth of the initial controller and quantization effects. After the stimulus injection has finished, the observation matrix H and measurement matrix Y are processed by the proposed state-space estimation algorithm, which yields the parameter matrix estimate $\hat{\Theta}$.

In Fig. 4, the duty cycle to output voltage TF $\hat{G}_{vd,\mathrm{ss}}(z)$, obtained by z-transforming the estimated state-space model, is reported. Additionally, the estimate $\hat{G}_{vd,\mathrm{tf}}(z)$, directly obtained in the z-domain by applying the estimator presented in [3], is shown for comparison. Finally, the converter's TF $G_{vd}(z)$ with the measured component values is also included in Fig. 4. As can be

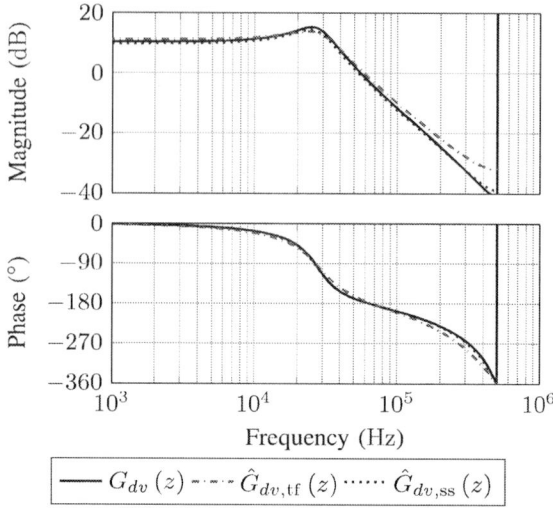

Fig. 4. Bode plots of the duty cycle to output voltage TF $G_{vd}(z)$ with the measured component values and its estimates $\hat{G}_{vd,\mathrm{tf}}(z)$ and $\hat{G}_{vd,\mathrm{ss}}(z)$ obtained in the z-domain and state-space, respectively.

seen, similar results can be achieved with both methods, whereby for higher frequencies the state-space approach achieves a better fitting to the reference $G_{vd}(z)$. It should be noted that the TF $\hat{G}_{vd,\mathrm{ss}}(z)$ is not required for the proposed auto-tuning process, which is carried out in state-space and is only shown as an illustrative example.

The convergence speed of the proposed state-space estimator in comparison with a state of the art z-domain approach is shown in Fig. 5. A significant improvement

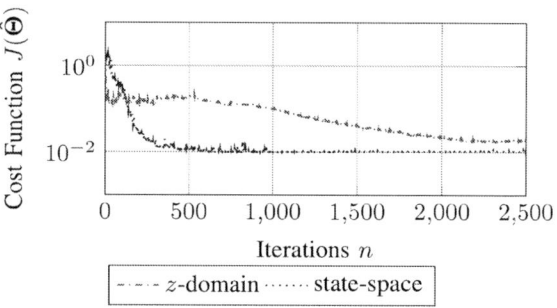

Fig. 5. Comparison of $J(\hat{\Theta})$ as a function of iterations for a state of the art z-domain estimation and the proposed state-space approach.

in convergence speed as well as a higher accuracy are achieved. In other words, the proposed approach requires less iterations to obtain an accurate model of the converter. Table II lists the required number of iterations for a conventional z-domain TF estimation and the proposed approach. The latter one also requires less multiplications per iteration, since fewer parameters have to be estimated. Moreover, the total number of multiplications required until $J(\hat{\Theta})$ converges to its final value is significantly reduced by the state-space estimation.

Ultimately, the estimated state-space model is used for automatic controller tuning. The target closed-loop

TABLE II.   COMPARISON OF THE REQUIRED NUMBER OF MULTIPLICATIONS

Method	Per Iteration	Total
z-domain	28	57000
state-space	22	22000

bandwidth and damping factor have been specified as $w_n = 2\pi \cdot 30\,\text{kHz}$, and $\zeta = 0.7$, respectively. Then, the required gain vector k_t for fulfilling the specification is calculated by Ackermann's formula [12]. In Fig. 6, the output voltage response to a load jump from $0\,\text{A}$ to $1\,\text{A}$ followed by a load drop back to $0\,\text{A}$ is shown. For both the load jump and load drop a reduced under-

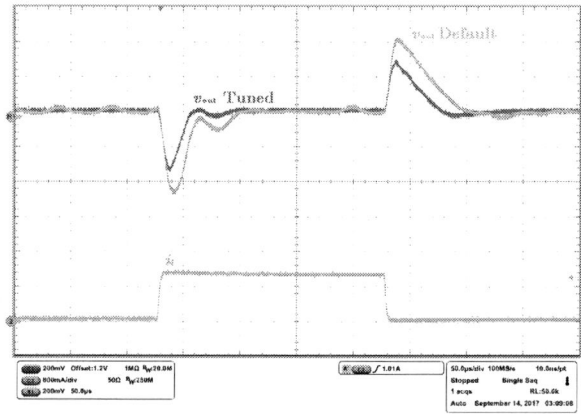

Fig. 6.   Output voltage response to a load jump from $1\,\text{A}$ to $1\,\text{A}$ and a load drop from $1\,\text{A}$ to $0\,\text{A}$. The reported waveforms are the output voltage for the standard controller (v_{out} Default), the output voltage for the auto-tuned controller (v_{out} Tuned), and the load current (i_l).

and overshoot, respectively, is achieved with the SI based auto-tuned controller coefficients. Moreover, the steady state oscillation is suppressed by the updated controller, which further highlights the improved performance.

V. Conclusion

In this paper, a parametric SI concept for DC-DC converters has been presented. The proposed approach combines the efficient SALS algorithm with prior randomization of the observation matrix. Furthermore, in contrast to state of the art solutions, the SI is carried out in state-space instead of the z-domain. This enables a fast and accurate estimation of the plant with low computational complexity. Subsequently, the estimated model is used to automatically tune an FSF controller, improving both its static and dynamic performance. Experimental results for a two-phase buck converter highlight the fast convergence speed and accuracy of the proposed SI concept as well as the performance gain of the auto-tuning.

References

[1] M. Kanzian, M. Agostinelli, and M. Huemer, "Modeling and Simulation of Digital Control Schemes for Two-Phase Interleaved

Buck Converters," In Proceedings of the *2016 Austrochip Workshop on Microelectronics (Austrochip 2016)*, pp. 7–12, Villach, Austria, October 2016.

[2] R. Priewasser, M. Agostinelli, C. Unterrieder, S. Marsili, and M. Huemer, "Modeling, Control, and Implementation of DC-DC Converters for Variable Frequency Operation," In *IEEE Transactions on Power Electronics*, Vol. 29, No. 1, pp. 287–301, January 2014.

[3] H. Gietler, C. Unterrieder, A. Berger, R. Priewasser, and M. Lunglmayr, "Low-Complexity, High Frequency Parametric System Identification Method for Switched-Mode Power Converters," In Proceedings of the *2017 IEEE Applied Power Electronics Conference and Exposition (APEC)*, pp. 2004–2009, Tampa, FL, USA, March 2017.

[4] L. Corradini, D. Maksimović, P. Mattavelli, and R. Zane, *Digital Control of High-Frequency Switched-Mode Power Converters*. John Wiley & Sons, Piscantaway, NJ, USA, 2015.

[5] L. Ljung, *System Identification*. Birkhäuser, Boston, MA, USA, 1998, pp. 163–173.

[6] B. Miao, R. Zane, and D. Maksimovic, "System Identification of Power Converters with Digital Control Through Cross-Correlation Methods," In *IEEE Transactions on Power Electronics*, Vol. 20, No. 5, pp. 1093–1099, September 2005.

[7] M. Shirazi, J. Morroni, A. Dolgov, R. Zane, and D. Maksimovic, "Integration of Frequency Response Measurement Capabilities in Digital Controllers for DC-DC Converters," In *IEEE Transactions on Power Electronics*, Vol. 23, No. 5, pp. 2524–2535, September 2008.

[8] M. Algreer, M. Armstrong, and D. Giaouris, "Active Online System Identification of Switch Mode DC-DC Power Converter Based on Efficient Recursive DCD-IIR Adaptive Filter," In *IEEE Transactions on Power Electronics*, Vol. 27, No. 11, pp. 4425–4435, November 2012.

[9] M. Lunglmayr, C. Unterrieder, and M. Huemer, "Step-Adaptive Approximate Least Squares," In Proceedings of the *2015 23rd European Signal Processing Conference (EUSIPCO 2015)*, pp. 1108–1112, Nice, France, August 2015.

[10] L. Dai, M. Soltanalian, and K. Pelckmans, "On the Randomized Kaczmarz Algorithm," In *IEEE Signal Processing Letters*, Vol. 21, No. 3, pp. 330–333, March 2014.

[11] V. Michal, "Optimal Peak-Efficiency Control of the CMOS Interleaved Multi-Phase Step-Down DC-DC Converter with Segmented Power Stage," In *IET Power Electronics*, Vol. 9, No. 11, pp. 2223–2228, 2016.

[12] C.-T. Chen, *Analog and Digital Control System Design: Transfer-Function, State-Space, and Algebraic Methods*. Oxford University Press, Inc., Oxford, UK, 1995.

[13] M. Kanzian, M. Agostinelli, and M. Huemer, "Sliding Mode Control with Inductor Current Observer for Interleaved DC-DC Converters," In Proceedings of the *2017 IEEE 18th Workshop on Control and Modeling for Power Electronics (COMPEL 2017)*, pp. 1–7, Stanford, CA, USA, July 2017.

[14] K. Y. Cheng, F. Yu, F. C. Lee, and P. Mattavelli, "Digital Enhanced V^2-Type Constant On-Time Control Using Inductor Current Ramp Estimation for a Buck Converter with Low-ESR Capacitors," In *IEEE Transactions on Power Electronics*, Vol. 28, No. 3, pp. 1241–1252, March 2013.

A Phase-shift Double Full-bridge (PSDB) Converter with Three Shared Leading-legs

Junjie Zhu, Qinsong Qian, Shengli Lu, Weifeng Sun and Le Zhang

National ASIC System Engineering Research Center, Southeast University, Nanjing, Jiangsu, PRChina

E-mail: ipec2018@jtbcom.co.jp

Abstract— This paper proposes a novel **PSDB** converter with three shared leading-legs. The ZVS of all the primary-side switches can be realized with wide conversion range (55%~100%) and full load range. Moreover, the effective duty cycle has nothing to do with the dead time. Thus, the **PSDB** would be useful for high switching frequency application. A 250W PSDB converter has been made with GaN transistors, SiC diodes, and planar transformers to verify the characteristics of PSDB. The experimental results show that the PSDB converter can achieve a peak efficiency of 90.6% when the switching frequency is 300kHz.

Keywords— *Phase-shift Double Full-bridge, ZVS, wide conversion range, shared leading leg.*

I. Introduction

Phase-shift full-bridge (PSFB) converters have been paid great attention [1-2]. Since all the primary-side switches are able to be turned on with ZVS. Nevertheless, the lagging-leg switches cannot realize ZVS without the help of a series inductance under conditions of high input voltage and light load. People tried a lot of methods to solve these problems, such as Zero-Voltage Zero-Current-Switching (ZVZCS) PSFB converter and "shared leg" technique [3~8]. However, there are always lagging-legs in these converters. And it is still hard to realize the ZVS of all the switches when the load is light and the switching frequency is high.

In this paper, a phase-shift double full-bridge (PSDB) converter is proposed. And all the bridge-legs in it are both leading-legs and lagging-legs. Thus, the ZVS of the primary-side switches can be realized easily with full load range and wide conversion range. Moreover, the effective duty cycle has nothing to do with the dead-time. As a result, the efficiency and the switching frequency of the proposed converter is quite high among phase-shift converters.

II. Circuit Description and Operation Principle

Fig.1 shows the circuit configuration of the proposed converter which includes triple full-bridge. The turns ratios of TR1 and TR2 are k_1:1:1 and k_2:1:1. L_{lk1} and L_{lk2} are the leakage inductors of TR1 and TR2. L_{m1} and L_{m2} are the magnetizing inductors of TR1 and TR2. R_{on} is the on-resistance of S_1-S_6, v_{on} is the voltage drop of all the anti-

parallel diodes of S_1-S_6, C_{oss} is the output capacitance of S_1-S_6.

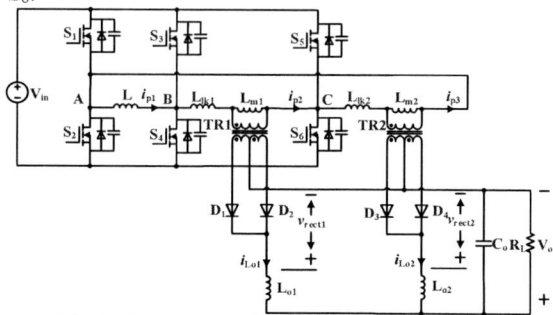

Fig. 1. Circuit diagram for the proposed converter.

Assume that L_{m1} and L_{m2} are large enough, the primary-side magnetizing currents are nearly zero. If $(di_{Lon})/dt \geq 0$ and $i_{Lon} \neq 0$, then $|i_{p2}| = i_{Lo1}/k_1$, $|i_{p3}| = i_{Lo2}/k_2$.

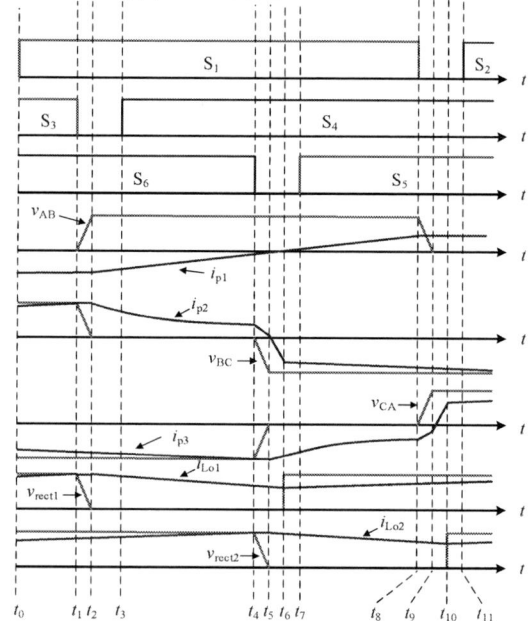

Fig. 2. Key waveforms of the proposed converter.

The proposed converter goes through eleven stages during a half-switching cycle. The key waveforms are shown in Fig. 2. The control order of the next half switching cycle is as follows: the turn-off of S_4, the turn-on of S_3, the turn-off of S_5, the turn-on of S_6, the turn-off of S_2, the turn-on of S_1.

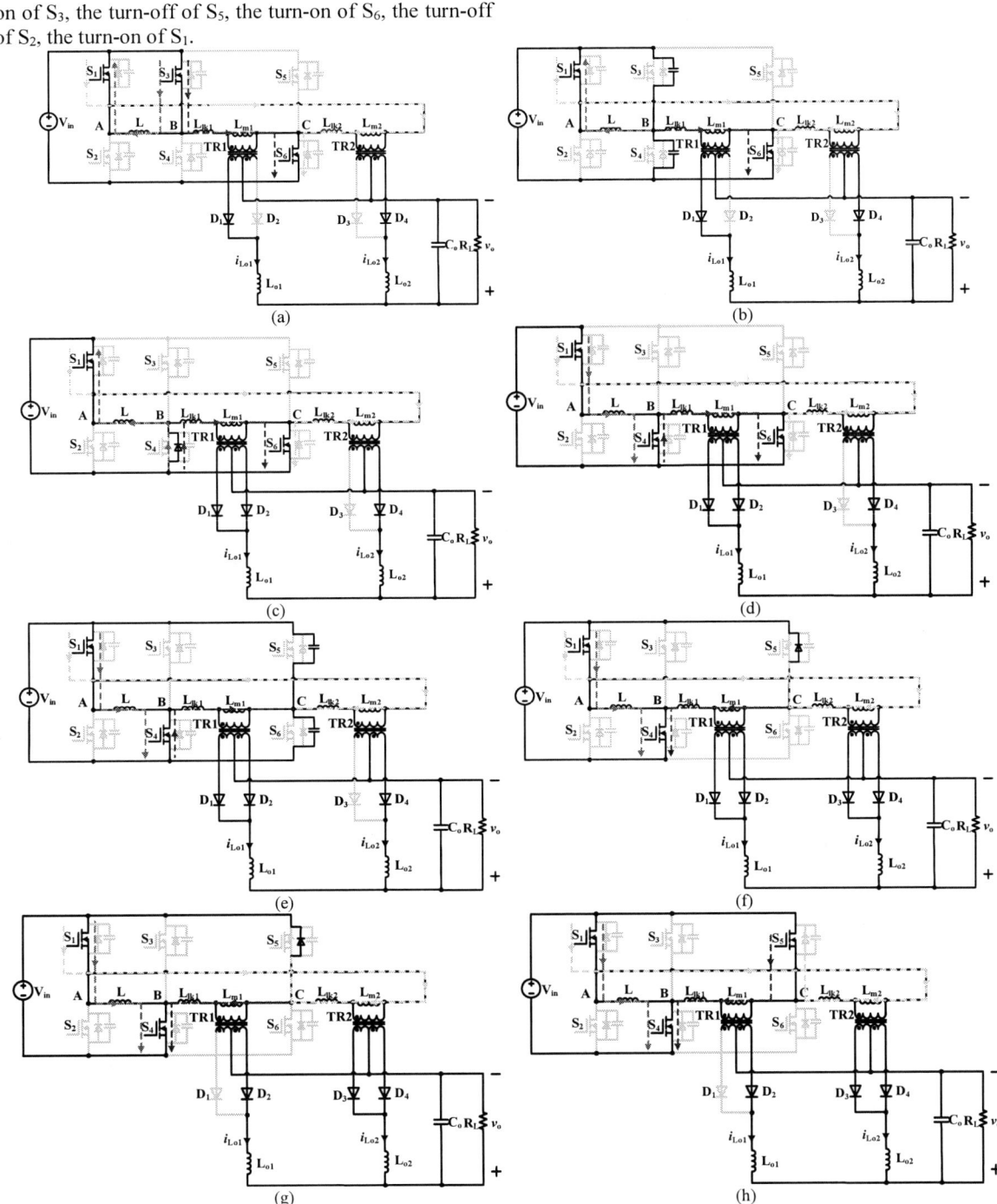

The 2018 International Power Electronics Conference

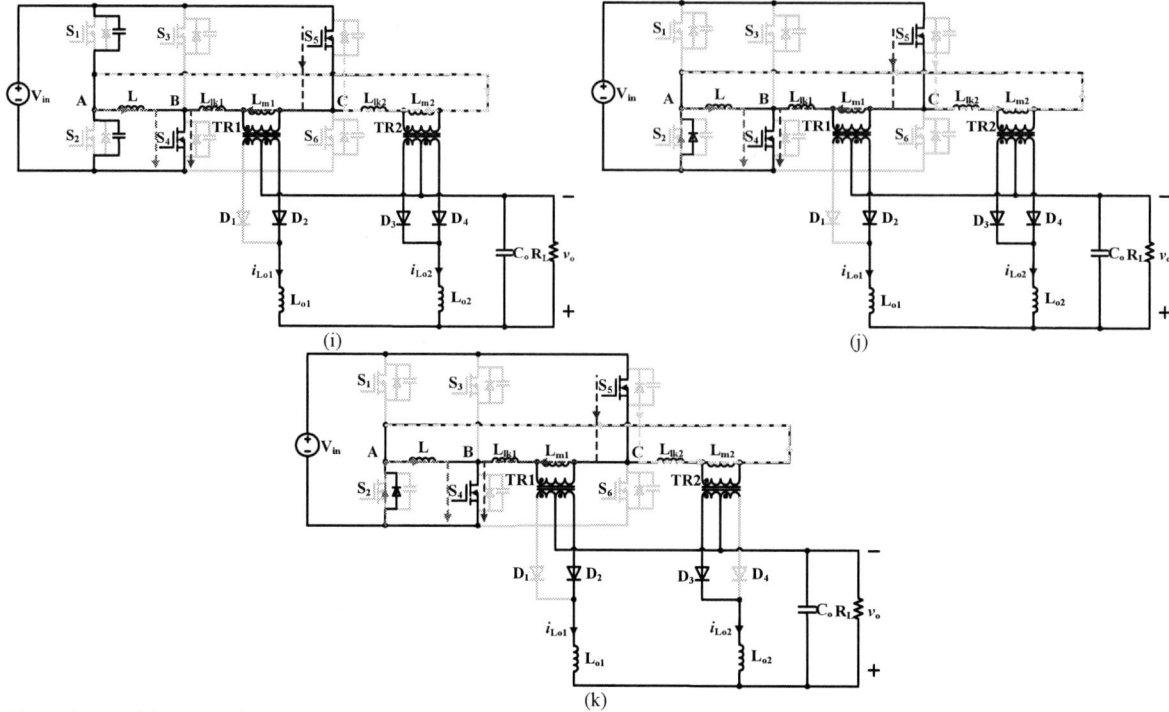

Fig. 3. Stages of the proposed converter. (a) Stage 1, (b) Stage 2, (c) Stage 3, (d) Stage 4, (e) Stage 5, (f) Stage 6, (g) Stage 7, (h)Stage 8, (i) Stage 9, (j) Stage 10, (k) Stage 11.

The equivalent operation circuits for different stages are shown in Fig. 3. To explain the stages well, in Fig. 3, the color lines show the real current following directions. In Fig. 2, the current following directions are the same as Fig. 1.

Stage 1 $[t_0 < t < t_1]$: Before t_0, the anti-parallel diode of S_1 is on. Thus, S_1 has been turned on with ZVS at t_0. i_{Lo1} and i_{Lo2} increase with a constant slope as:

$$\frac{di_{Lo1}(t)}{dt} = \frac{\frac{V_{in}}{k_1} - V_o}{L_{o1}} \quad (1)$$

$$\frac{di_{Lo2}(t)}{dt} = \frac{\frac{V_{in}}{k_2} - V_o}{L_{o2}} \quad (2)$$

The voltage v_{AB} is zero, and L is large enough. Thus, i_{p1} stays the same.

Stage 2 $[t_1 < t < t_2]$: At t_1, S_3 is turned off. During this stage, the output capacitor of S_3 is charged and that of S_4 is discharged. Also, i_{p1}, i_{p2} and i_{Lo1} are considered to be constant during this stage. Since L and L_{o1} are large enough. So, v_{BC} is given by

$$v_{BC}(t) = \frac{V_{in} \times 2C_{oss} - (i_{p2} - i_{p1}) \times (t - t_1)}{2C_{oss}} \quad (3)$$

Stage 3 $[t_2 < t < t_3]$: At t_2, v_{BC} reaches zero. Then the anti-parallel diode of S_4 and D_2 are conducted. During this stage, v_{AB} is V_{in}, hence i_{p1} increases.

$$\frac{di_{p1}(t)}{dt} = \frac{V_{in}}{L} \quad (4)$$

Since v_{rect1} is zero during this stage, i_{p2} and i_{Lo1} decrease as follows:

$$\frac{di_{p2}(t)}{dt} \times L_{lk1} = R_{on} \times \left(i_{p3}(t) - i_{p2}(t) \right) - v_{on} \quad (5)$$

$$\frac{di_{Lo1}}{dt} = -\frac{V_o}{L_{o1}} \quad (6)$$

Stage 4 $[t_3 < t < t_4]$: At t_3, S_4 is turned on with ZVS since v_{BC} maintains zero. During this stage, i_{p2} decreases as depicted as follows:

$$\frac{di_{p2}(t)}{dt} \times L_{lk1} = R_{on} \times (i_{p1}(t) + i_{p3}(t) - 2i_{p2}(t)) \quad (7)$$

The following direction of i_{p1} may change during Stage 4 to Stage 8.

Stage 5 $[t_4 < t < t_5]$: At t_4, S_6 is turned off. During this stage, the output capacitor of S_6 is charged and that of S_5 is discharged. And i_{p3} and i_{Lo2} are considered to be constant. So, v_{CA} is given by

$$v_{CA}(t) = \frac{-V_{in} \times 2C_{oss} - i_{p3}(t) \times (t - t_4) + \int_{t_4}^{t} i_{p2}(t)dt}{2C_{oss}} \quad (8)$$

$$\frac{di_{p2}(t)}{dt} = \frac{v_{BC}(t)}{L_{lk1}} = \frac{-V_{in} - v_{CA}(t)}{L_{lk1}} \quad (9)$$

Stage 6 $[t_5 < t < t_6]$: At t_5, v_{CA} and v_{rect2} reach zero, then the anti-parallel diode of S_5 and D_3 are conducted. During this stage, the voltage v_{BC} is $-V_{in}$. Thus, there are relationships.

$$\frac{di_{p3}(t)}{dt} \times L_{lk2} = R_{on} \times \left(i_{p1}(t) - i_{p3}(t) \right) + v_{on} \quad (10)$$

$$\frac{di_{Lo2}}{dt} = -\frac{V_o}{L_{o2}} \quad (11)$$

Stage 6 ends when $-k_1 i_{p2}$ reaches the output inductance current i_{Lo1}.

Stage 7 $[t_6 < t < t_7]$: At t_6, the diode D_1 is off. Thus, i_{Lo1} increases with a constant slope as (1).

Stage 8 $[t_7 < t < t_8]$: At t_7, S_5 is turned on with ZVS since v_{CA} maintains zero. During this stage, i_{p3} increases as follows:

$$\frac{di_{p3}(t)}{dt} \times L_{lk2} = R_{on} \times (i_{p1}(t) + i_{p2}(t) - 2i_{p3}(t)) \quad (12)$$

Stage 9 $[t_8 < t < t_9]$: At t_8, S_1 is turned off. During this stage, the output capacitor of S_1 is charged and that of S_2

3147

is discharged. Also, ip1 is considered to be constant during this stage, v_{AB} is given by

$$v_{AB}(t) = \frac{V_{in} \times 2C_{oss} - i_{p1} \times (t-t_8) + \int_{t_8}^{t} i_{p3}(t)dt}{2C_{oss}} \quad (13)$$

$$\frac{di_{p3}(t)}{dt} = \frac{v_{CA}(t)}{L_{lk2}} = \frac{V_{in} - v_{AB}(t)}{L_{lk2}} \quad (14)$$

Stage 10 [$t_9 < t < t_{10}$]: At t_9, v_{AB} reaches zero, then the anti-parallel diode of S_2 is conducted. Since L is large enough, ip1 keeps the same. During this stage, the voltage v_{CA} is V_{in}. Thus, there is relationship as follows:

$$\frac{di_{p3}(t)}{dt} = \frac{V_{in}}{L_{lk2}} \quad (15)$$

This stage ends when $k_2 i_{p3}$ reaches i_{Lo2}.

Stage 11 [$t_{10} < t < t_{11}$]: During this stage, D_4 turns off. The change of i_{Lo2} is the same as which shows in (2).

III. ANALYSIS OF THE PSDB CONVERTER

A. Analysis of steady-state operation

To give the DC gain characteristic of PSDB, the calculation is based on the following assumptions:

1) It takes no time to achieve ZVS comparing to a full-switching cycle. Thus, $t_1=t_2$, $t_4=t_5$, $t_8=t_9$.

2) TR1 and TR2 are the same. Thus, $k=k_1=k_2$, $L_{lk}=L_{lk1}=L_{lk2}$, $L_m=L_{m1}=L_{m2}$.

3) The system works in Continuous Conduction Mode (CCM) and Boundary Conduction Mode (BCM).

4) The phase-shift duty cycles of TR1 and TR2 are the same. Thus, $t_5-t_2=t_9-t_5=(1-D)*(t_{11}-t_0)=(1-D)T/2$. D is the duty cycle of PSDB converter. D is from 0.5 to 1. T is the time of a full duty cycle.

5) Assume that the values of L_{o1} and L_{o2} are the same. Thus, $L_o=L_{o1}=L_{o2}$.

Then, the proposed converter can be seen as two same PSFB converters working in parallel. The DC gain characteristic of the proposed converter is the same as that of one PSFB converter. According to the analysis above, the DC gain characteristic is given by

$$G = \frac{V_o}{V_{in}} = \frac{D}{k} \quad (16)$$

Since D has nothing to do with the dead time, the effective duty cycle would not be influenced by the dead time.

B. Soft-switching conditions for the switches of PSDB

The soft conditions for switches depend on the peak values of the primary-side currents. I_{p1} is the peak value of i_{p1}, I_{p2} and I_{p3} are the peak values of i_{p2} and i_{p3}. I_{Lo} are the peak value of i_{Lo1} and i_{Lo2}.

$$I_{p1} = \frac{V_{in} \times T \times (1-D)}{L} \quad (17)$$

$$I_{p2} = I_{p3} = k \times I_{Lo} \quad (18)$$

$$I_{Lo} = \frac{V_o}{2R_L} + \frac{(1-D)V_o T}{2L_o} \quad (19)$$

The energy used to realize the ZVS of S_3 and S_4 is always enough. Since L and L_{o1} are large enough. The

matter is how long it takes to achieve ZVS. Assume the time needed to realize the ZVS of S_3 and S_4 is $\Delta t_{1,2}$.

$$\Delta t_{1,2} = \frac{2V_{in} \times C_{oss}}{I_{p1} + I_{p2}} \quad (20)$$

During Stage 5, if L_{lk1} is too small or C_{oss} is too large, $i_{p2}(t)$ will reach zero even $-i_{Lo1}(t_6)/k$ before v_{CA} reaches zero. Then, it may take longer time to realize the ZVS of S_5. Nevertheless, even this happens, the energy used to realize the ZVS of S_5 is always enough. Since L_{o2} is large enough, and I_{p3} is always larger than $i_{Lo1}(t_6)/k$. Assume the time needed to realize the ZVS of S_5 and S_6 is $\Delta t_{4,5}$, the longest needed to realize the ZVS of S_5 and S_6 is $\Delta t_{4,5max}$, $\Delta t_{4,5}$ should be smaller than $\Delta t_{4,5max}$.

$$\Delta t_{4,5max} = \frac{2V_{in} \times C_{oss}}{I_{p3} + i_{p2}(t_6)} = \frac{6k \times V_{in} \times L_o \times C_{oss}}{(\frac{V_{in}}{k} - v_o) \times T} \quad (21)$$

During Stage 9, if I_{p1} is smaller than $i_{p3}(t_{10})$, $i_{p3}(t)$ may reach I_{p1} before v_{AB} reaches zero, then the ZVS of S_2 can never be realized. As a result, I_{p1} should be larger than $i_{p3}(t_{10})$. Assume the time needed to realize the ZVS of S_1 and S_2 is $\Delta t_{8,9}$, the longest needed to realize the ZVS of S_1 and S_2 is $\Delta t_{8,9max}$, $\Delta t_{8,9}$ should be smaller than $\Delta t_{8,9max}$.

$$\Delta t_{8,9max} = \frac{2V_{in} \times C_{oss}}{I_{p1} - i_{p3}(t_{10})} = \frac{2V_{in} \times C_{oss}}{I_{p1} - \frac{V_o}{2R_L} + + \frac{(1-D)V_o T}{2L_o}} \quad (22)$$

IV. EXPERIMENT VERIFICATION

A 250W hardware prototype of the proposed PSDB converter has been made and tested to verify the circuit operation principles. The specifications of the converter are as follows: input voltage $V_{in}=220V\sim400V$; output voltage $V_o=24V$; output power $P_o=250W$; switching frequency $f_s=300kHz$. This converter is made with GaN transistors, SiC diodes and planar transformers.

Fig. 4 shows the key waveforms of the PSDB converter when $V_{in}=375V$, $P_o=250W$. It verifies the analysis above. Moreover, the effective duty cycle is not infected by the dead-time. Fig. 5 shows the waveforms of S_2, S_4 and S_6 when $V_{in}=375V$, $P_o=25W$.

In Fig. 5, V_{ds2} is the drain-source voltage of S_2, V_{gs2} is the gate-source voltage of S_2, V_{ds4} is the drain-source voltage of S_4, V_{gs4} is the gate-source voltage of S_4, V_{ds6} is the drain-source voltage of S_6, V_{gs6} is the gate-source voltage of S_6.

It shows that the ZVS of switches of PSDB can be realized no matter how light the load is. The peak efficiency of the PSDB converter is 90.6% when $V_{in}=220V$ and $P_o=250W$. The main loss of the proposed converter is the loss of the diodes, which is almost 8% of the input power. Since the conduct voltage of SiC diodes is 1.5V, and the secondary-side diodes cannot turn off with ZCS. The conduction loss of diodes can be calculated as 1.5V/(24+1.5)V=5.8% roughly, which is inevitable.

The 2018 International Power Electronics Conference

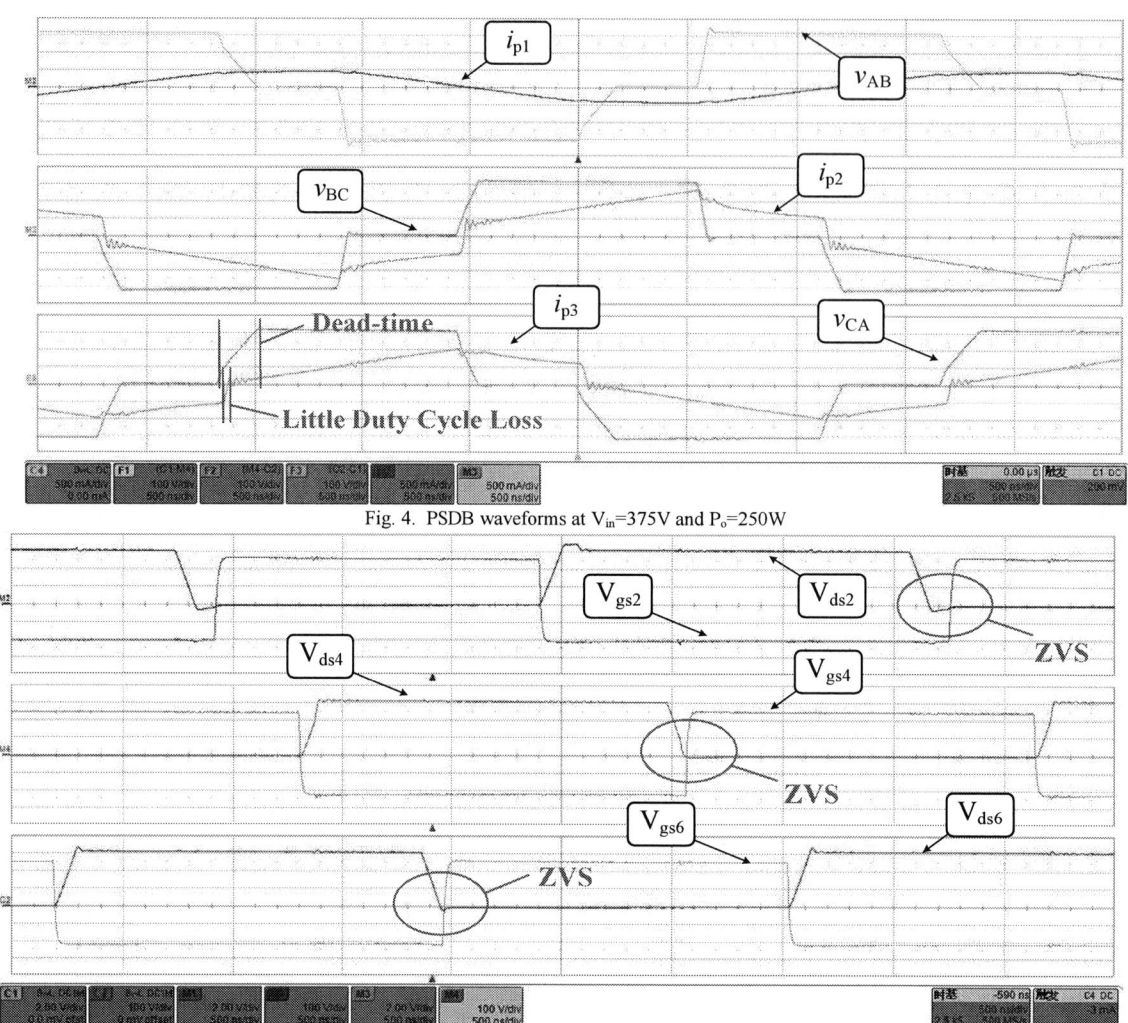

Fig. 4. PSDB waveforms at V_{in}=375V and P_o=250W

Fig. 5. Waveforms of S_2, S_4 and S_6 at V_{in}=375V and P_o=25W

TABLE I
EFFICIENCY COMPARISON OF SEVERAL PHASE-SHIFT CONVERTERS

PS Converters	ZVZCS FB [5]	ZVZCS PWM [3]	PSDB[8]	Proposed PSDB
Efficiency (%)	96.8	91.4	95.3	90.6
Output Voltage (V)	300	25	800	24
Switching Frequency(kHz)	220	40	50	300

Table I shows the efficiency comparison of several phase-shift converters. In Table I, the proposed PSDB has the lowest efficiency among Phase-shift converters. However, it is because the proposed PSDB has the highest switching frequency and the lowest output voltage. When diodes are used in rectifier, the voltage drop of diodes influence the efficiency. Assume the voltage drop of diodes is not change. When the output voltage is lower, the efficiency is lower. According to the analysis above, the conduction loss in diodes in the proposed PSDB converter is 5.8% of the output power. It is inevitable. However, the conduction loss in diodes in [5] and [8] may be less than 1%. Moreover, the efficiency decreases when the switching frequency is higher. Considering the output voltage and the switching frequency, PSDB topology has higher efficiency than most PS topologies. ZVZCS FB converter in [5] may have higher efficiency than PSDB converter when their switching frequencies and output voltages are same. Since the diodes in ZVZCS FB converter can be turned off with ZCS. It is the improvement direction of the proposed PSDB converter.

V. CONCLUSION

This paper proposes a novel PSDB converter with three shared leading-legs. The ZVS of all the primary-side switches can be realized with wide conversion range and full load range, and the effective duty cycle has nothing to do with the dead-time. Thus, this converter would be useful for high switching frequency application. The proposed PSDB converter also has high efficiency among phase-shift converters. In the end, experimental results are presented to verify the performance of this converter.

REFERENCES

[1] B. Chen and Y. Lai, "Switching control technique of phase-shift-controlled full-bridge converter to improve efficiency under light-load and standby conditions without additional auxiliary components," *IEEE Trans. Power Electron.*, vol. 25, no. 4, pp. 1001-1011, Apr. 2010.

[2] A. J. Mason, D. J. Tschirhart, and P. Jain, "New ZVS phase shift modulated full-bridge converter topologies with adaptive energy storage for SOFC application," *IEEE Trans. Power Electron.*, vol. 23, no. 1, pp. 332-342, Jan. 2008.

[3] E. Chu, X. Hou, H. Zhang, M. Wu, and X. Liu, "Novel zero-voltage and zero-current switching (ZVZCS) PWM three-level DC/DC converter using output coupled inductance," *IEEE Trans. Power Electron.*, vol. 29, no. 3, pp. 1082-1093, Mar. 2014.

[4] C. Zhao, X. Wu, P. Meng, and Z. Qian, "Optimum design consideration and implementation of a novel synchronous rectified soft-switched phase-shift full-bridge converter for low-output-voltage high-output-current applications," *IEEE Trans. Power Electron.*, vol. 24, no. 2, pp. 388-397, Feb. 2009.

[5] M. Pahlevaninezhad, P. Das, J. Drobnik, P. K. Jain, and A. Bakhshai, "A novel ZVZCS full-bridge DC/DC converter used for electric vehicles," *IEEE Trans. Power Electron.*, vol. 27, no. 6, pp. 2752-2769, Jun. 2012.

[6] W. Yu, J. Lai, W. Lai, and H. Wan, "Hybrid resonant and PWM converter with high efficiency and full soft-switching range," *IEEE Trans. Power Electron.*, vol. 27, no. 12, pp. 4925-4933, Dec. 2012.

[7] C. Liu, B. Gu, J. Lai, M. Wang, Y. Ji, G. Cai, Z. Zhao, C. Chen, C. Zheng, and P. Sun, "High-efficiency hybrid full-bridge-half-bridge converter with shared ZVS lagging-leg and dual outputs in series," *IEEE Trans. Power Electron.*, vol. 28, no. 2, pp. 849-861, Feb. 2013.

[8] Kaimin Shi, Donglai Zhang, Zhicheng Zhou, Mengqiao Zhang, Di Zhang, and Yu Gu, "A Novel Phase-shift Dual Full-bridge Converter with Full Soft-switching Range and Wide Conversion Range," *IEEE Trans. Power Electron.*, vol. 31, no. 11, pp. 7747 - 7760, Nov. 2016.

Dual Active Bridge Synchronous Rectified Step-Down Converter

Chien-Chun Huang[1] Chang-Lin Tsai[1] Tsung-Lin Tsai[1] Yao-Ching Hsieh[2] Huang-Jen Chiu[1]

Jing-Yuan Lin[1]

[1]Department of Electronic and Computer Engineering, National Taiwan University of Science and Technology, Taipei, Taiwan

[2]Department of Electrical Engineering National Sun Yat-Sen University, Kaohsiung, Taiwan

E-mail: ychsieh@mail.ee.nsysu.edu.tw

Abstract-This paper presents a novel high-efficiency bidirectional zero-voltage-transition (ZVT) pulse-width-modulation (PWM) converter. By controlling the direction and timing of the resonant current injection, the main power switches can achieve zero voltage switching under various loads and various output voltage. Moreover, zero voltage switching eliminates the switching noise on the output voltage. Experimental results based on a 400W prototype are presented to demonstrate the performance of the proposed converter.

I. Introduction

With the increasingly stringent requirements on the instrument power supply for its output quality, the noise of the output voltage should be minimized. Conventional linear regulators have excellent performance in terms of the noise concern, but cannot handle high power due to efficiency problems. Switching power converters, while improving efficiency and can cope with high-power applications, the switching elements cause switching noise on the output voltage and degrade the power quality of the load. Synchronous rectified step-down converter operation in the triangular-current mode (TCM) [1]-[3] is a widely known way to achieve natural zero voltage switching (ZVS). But for operation in TCM mode, the output inductor current is usually designed to operate in boundary conduction mode (BCM) at full load conditions to reduce the output ripple current. In other words, the peak inductor current is twice of the output current. Besides, the frequency modulation operation brings difficulty on passive components design for the wide output range of the instrument power supply. However, there are many auxiliary ZVT switching converters proposed [4]-[10], through the auxiliary switch and the resonant circuit to achieve zero voltage switching on the main switch. However, the ZVS feature cannot be maintained within the full load and voltage range. LCC synchronous rectified step-down converter [11], with additional resonant inductor and two capacitors, achieves ZVS at wider load range; however, the current in the resonant inductor causes the main switch current to increase to twice the full-load current at heavy load. Because the resonant inductor peak current must be greater than the output inductor current to procure the ZVS merit, the resonant inductor design value is better to have just little more resonant inductor current than full load current. However, the resonant inductor peak

current in LCC architecture is dependent on the duty cycle. With a wide range of output voltage applications, the fixed resonant inductance value can only be at a specific output voltage where the resonant inductor current is at its best value; otherwise, the resonant inductor current will be much larger than the full load current. This extra-large peak current not only increases the current stress upon the switch, but also makes it difficult for magnetic component realization. In this paper, a new circuit architecture is proposed to meet the functions of wide output voltage regulation, and ZVS over wide load range.

II. Dual Active Bridge Synchronous Rectified Buck Converter

Since the circuit operations in the forward and reverse directions are similar, the following will be explained only for the forward direction. In order for the high side switch of the synchronous rectified buck converter to achieve ZVS, the C_{oss} of the upper arm switch must be discharged until the body diode is turned on within the dead time after the turn-off of the low side switch. To achieve this purpose, we add a current source to the synchronous rectified switch connection point and control the current source injection timing. The switch Q_{aux1} is turned on after the dead time following the cutoff of the switch Q_{sr2}, the external current source is injected into the connection point of high / low side switches and linearly rises until the current exceeds the output inductor current, and the C_{oss} of the high side switch Q_{sr1} is discharged so that its body diode is forward-biased to achieve zero voltage switching.

After the high side switch turns on with zero voltage switching, it is necessary to disconnect the injection path of the external current source and provide a freewheeling path for demagnetization. Thus, the power diodes D_{1b} and D_{2a} are added, and the energy stored in the resonant inductor vents back to the input source. The resonant inductor current is reset to zero per cycle.

In order to allow the switches to meet the ZVS requirements in both directions, the reverse auxiliary switch Q_{aux2} is added to control the injection timing of the applied current source and the switching Q_{2b} in the reverse operation to provide the exciting path of the resonant inductor.

Ideally, in order to reset the resonant inductor current and clamp at zero, only Q_{1a} and Q_{aux1} need to be turned off and Q_{2b} remains on. Resonant inductor current flows through freewheeling path of Q_{2b} and D_{1b}, and the current is reset and clamped at zero amperes by using the forward voltage of the D_{1b}. However, the energy stored in the parasitic capacitance of the switch is transferred to inductive current in the resonant interval, if the energy is only consumed by D_{1b}, Q_{2b} path (forward voltage of D_{1b}, R_{DSON} of Q_{2b}), it cannot be consumed completely in the limited demagnetization time. In practice, another resistance R_D and Q_D to replace the original consumption path, which shortens the time required for the resonant inductor current to fall to zero. Fig. 1 shows the final circuit configuration proposed in this paper.

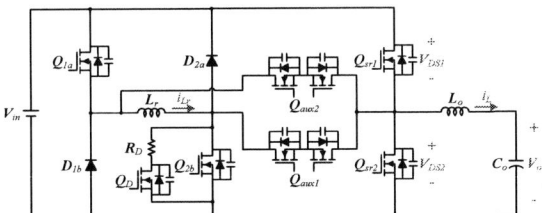

Fig. 1 Dual active bridge synchronous rectified step-down converter

III. Circuit operation interval analysis

In order to simplify the analysis, the circuit operation analysis is based on the following assumptions:

(1) All components on the circuit are considered ideal components.

(2) Trace impedance is omitted.

(3) The parasitic capacitance on the auxiliary switches Q_{aux1}, Q_{aux2}, Q_D are negligible.

Forward mode is analyzed to illustrate the operation of this circuit, and the reverse auxiliary switches Q_{aux2} and Q_{2b} are remained off.

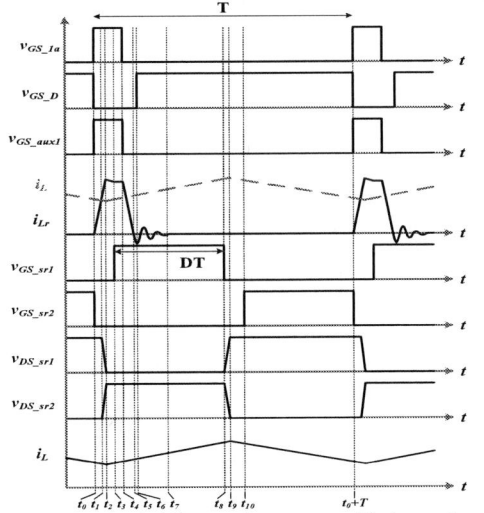

Fig. 2 Dual active bridge synchronous rectified step-down converter operation waveform (forward mode)

(i) Interval 1 ($t_0 \sim t_1$)

At t = t_0, the low side switch Q_{sr2} is turned off and the output inductor current flows through the body diode of Q_{sr2}. Q_{1a} and the forward auxiliary switch Q_{aux1} are turned on, speedup demagnetizing switch Q_D is cutoff. At this time, the voltage across the resonant inductor is V_{in}, the resonant inductor current rises linearly.

(ii) Interval2 ($t_1 \sim t_2$)

At t = t_1, the resonant inductor current is equal to the output inductor current, and the body diode of Q_{sr2} turns off. Resonant inductor and output inductance begins to resonate with C_{oss1} and C_{oss2}. The resonant current charges C_{oss2} and discharges C_{oss1}.

(iii) Interval3 ($t_2 \sim t_3$)

At t = t_2, the voltage on the C_{oss1} of the switch Q_{sr1} drops to zero and the body diode of Q_{sr1} turns on. The resonant inductor current is then clamped by the body diode forward voltage V_F of the Q_{sr1} switch.

(iv) Interval4 ($t_3 \sim t_4$)

At t = t_3, the body diode of the switch Q_{sr1} has been turned on before Q_{sr1} is turned on to achieve ZVS. While the resonant inductor current continues to circulate in this interval. The power is passed from the input side to the output, and the output inductor current rises linearly.

(v) Interval5 ($t_4 \sim t_5$)

At t = t_4, the switch Q_{1a} is turned off simultaneously with the forward auxiliary switch Q_{aux1}, where the resonant inductor current returns the energy to the input power supply via the power diodes D_{2a} and D_{1b}. When the resonant inductor current drops to zero, this interval ends.

(vi) Interval6 ($t_5 \sim t_6$)

At t = t_5, D_{1b}, D_{2a} naturally turns off because the resonant inductor current is zero . The parasitic capacitors of power switches Q_{1a}, Q_{2b}, Q_D and the power diodes D_{1b}, D_{2a} begin to resonant with the resonant inductor L_r.

(vii) Interval7 ($t_6 \sim t_7$)

At t = t_6, a zero-current-detection (ZCD) trigger circuit turns Q_D on. The parasitic capacitance C_{oss2a} of the power diode D_{2a} has been charged by the resonant inductor current, and the initial value of the resonant inductor current at t= t_6 is negative. The power switch Q_D turns on at this time not only reduces the time inductor current falling to zero, but also prevents the diode D_{2a} reverse recovery current which will cause unnecessary switching noise.

(viii) Interval8 ($t_7 \sim t_8$)

At t = t_7, the resonant inductor current is no longer resonant and stable at zero. The interval continues until the switch Q_{sr1} turns off.

(ix) Interval9 ($t_8 \sim t_9$)

At t = t_8, the switch Q_{sr1} is turned off and the output inductor current charges the C_{oss1} of the switch Q_{sr1} and discharges the C_{oss2} of the switch Q_{sr2}.

(x) Interval10 ($t_9 \sim t_{10}$)

At t = t_9, the C_{oss2} of the switch Q_{sr2} is discharged to zero and the body diode turns on.

(xi) Interval11 ($t_{10} \sim t_0+T$)

At t = t_{10}, since the body diode of the switch Q_{sr2} has been turned on, then open Q_{sr2} can reach ZVS.

The 2018 International Power Electronics Conference

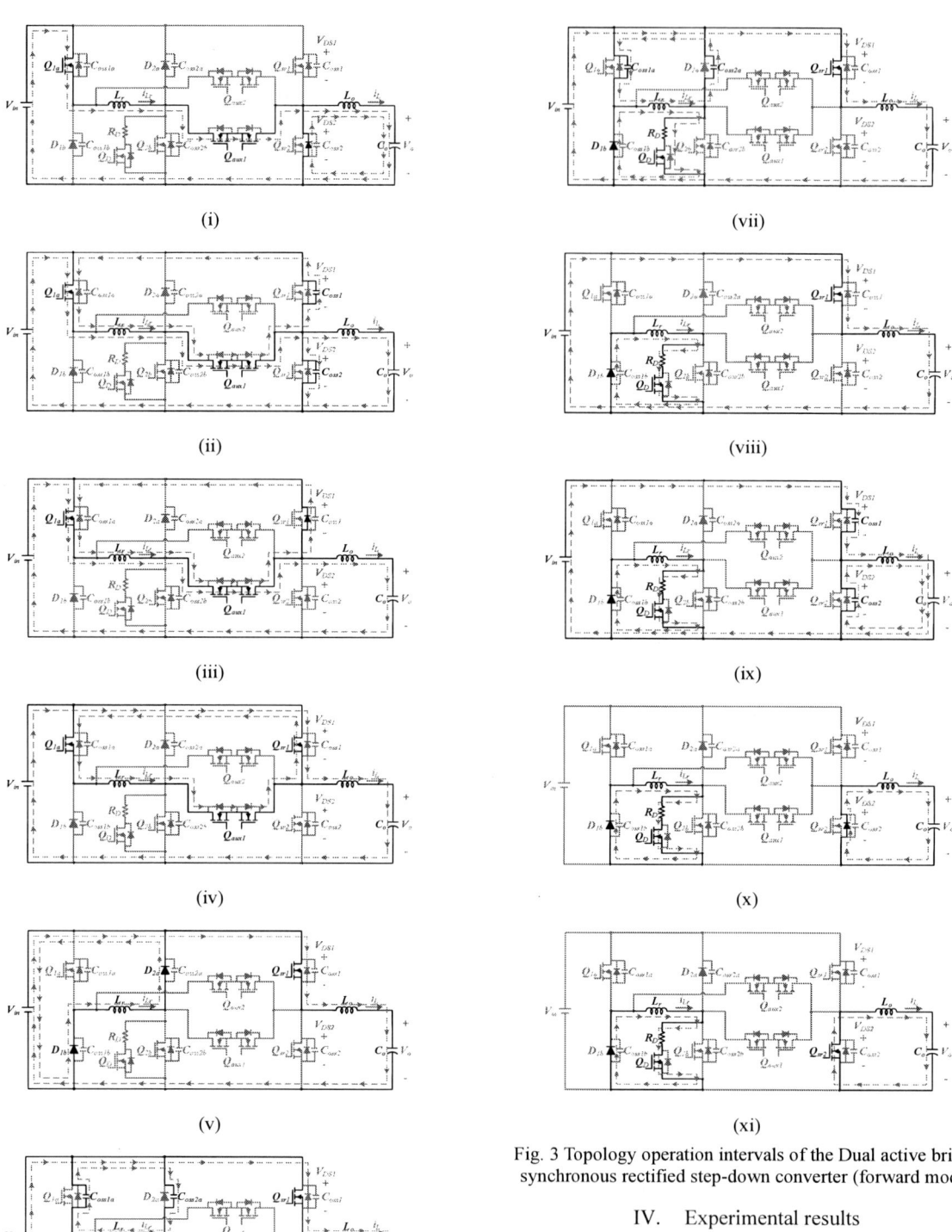

(i)

(ii)

(iii)

(iv)

(v)

(vi)

(vii)

(viii)

(ix)

(x)

(xi)

Fig. 3 Topology operation intervals of the Dual active bridge synchronous rectified step-down converter (forward mode)

IV. Experimental results

In order to verify the theoretical analysis, an implementation of input voltage $V_{IN} = 100V$, output voltage $V_{OUT} = 20 \sim 80V$, output power $P_O = 400W$, maximum load current $I_{O\_MAX} = 10A$ is performed.

Fig. 4 shows the various switching signals and resonant inductor current waveforms under the test conditions for the output voltage is 80V and the output power is 400W.

3153

The 2018 International Power Electronics Conference

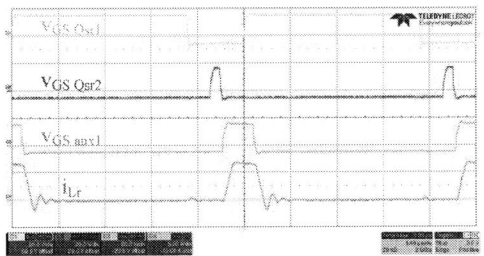

Fig. 4 Switch signal and resonant inductor current waveform
CH1: v$_{GS\_Qsr1}$ (20V/div); CH2: v$_{GS\_Qsr2}$ (20V/div);
CH3: v$_{GS\_aux1}$ (20V/div); CH4: i$_{Lr}$ (5A/div); Time:1μs/div

Fig. 5 shows the Q$_{sr1}$ switch's v$_{GS}$ signal, Q$_{sr1}$ switch's v$_{DS}$ voltage, resonant inductor current and output inductor current waveform under the test conditions for the output voltage is 80V and the output power is 400W.

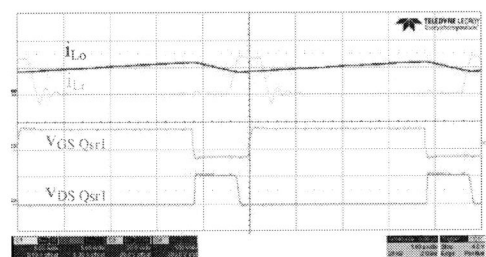

Fig. 5 Resonant inductor current and output inductor current waveform, Q$_{sr1}$ switch's v$_{GS}$ signal, Q$_{sr1}$ switch's v$_{DS}$ voltage
CH1: i$_{Lr}$ (5A/div); CH2: i$_{Lo}$ (5A/div);
CH3: v$_{GS\_Qsr1}$ (20V/div); CH4: v$_{DS\_Qsr1}$ (100V/div); Time:1μs/div

Fig. 6 shows the output voltage ripple when the auxiliary switching circuit does not operate, Q$_{sr1}$ switch is hard switching. Test conditions is the output voltage is 80V, the output watt is 400W and the output inductor current operation in continuous conduction mode. And Fig. 7 shows the output voltage ripple when the auxiliary switching circuit operates. Q$_{sr1}$ is soft switching and the test condition is the same. The switching noise on the output voltage ripple in hard switching mode is 208mV, and the switching noise on the output voltage ripple in soft switching is 86mV. It can be seen from the Fig. that the switching noise on the output voltage ripple is greatly reduced by the soft switching.

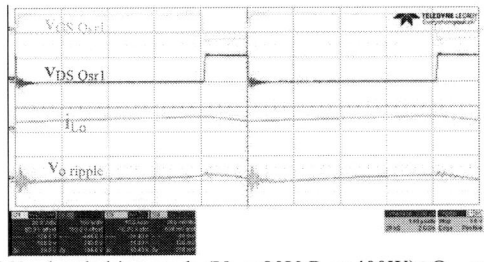

Fig. 6 Hard switching mode (V$_O$ = 80V, P$_O$ = 400W)：Q$_{sr1}$ switch's v$_{GS}$ and v$_{DS}$, output inductance current and output voltage ripple.
CH1: v$_{GS\ Qsr1}$ (20V/div); CH2: v$_{DS\ Qsr1}$ (100V/div);
CH3: i$_{Lo}$ (10A/div); CH4: v$_{o\ ripple}$ (200mV/div); Time:1μs/div

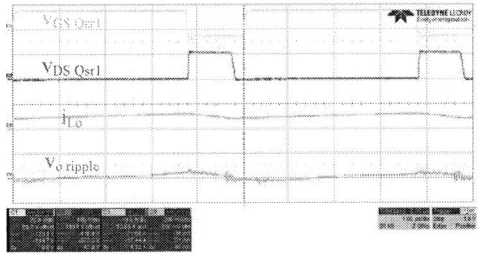

Fig. 7 Soft switching mode (V$_O$ = 80V, P$_O$ = 400W)：Q$_{sr1}$ switch's v$_{GS}$ and v$_{DS}$, output inductance current and output voltage ripple.
CH1: v$_{GS\ Qsr1}$(20V/div); CH2: v$_{DS\ Qsr1}$ (100V/div);
CH3: i$_{Lo}$ (10A/div); CH4: v$_{o\_ripple}$ (200mV/div); Time:1μs/div

Fig. 8 shows the output voltage ripple when the auxiliary switching circuit does not operate, Q$_{sr1}$ switch is hard switching. Test conditions is the output voltage is 20V, the output watt is 200W and the output inductor current operation in continuous conduction mode. And Fig. 9 shows the output voltage ripple when the auxiliary switching circuit operates. Q$_{sr1}$ is soft switching and the test condition is the same. The switching noise on the output voltage ripple in hard switching mode is 310mV, and the switching noise on the output voltage ripple in soft switching is 88mV. It can be seen from the Fig. that the switching noise on the output voltage ripple is greatly reduced by the soft switching. Combined with the results of Fig. 7 and Fig. 9, it is verified that the auxiliary switching circuit can switch to the flexible switching function under different output voltage and current state.

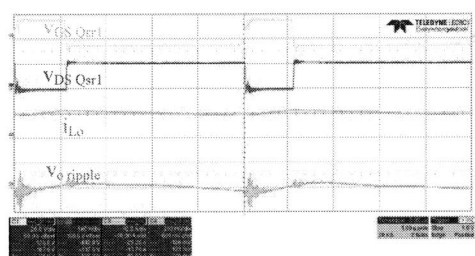

Fig. 8 Hard switching mode (V$_O$ = 20V, P$_O$ = 200W)：Q$_{sr1}$ switch's v$_{GS}$ and v$_{DS}$, output inductance current and output voltage ripple.
CH1: v$_{GS\ Qsr1}$(20V/div); CH2: v$_{DS\ Qsr1}$ (100V/div);
CH3: i$_{Lo}$ (10A/div); CH4: v$_{o\_ripple}$ (200mV/div); Time:1μs/div

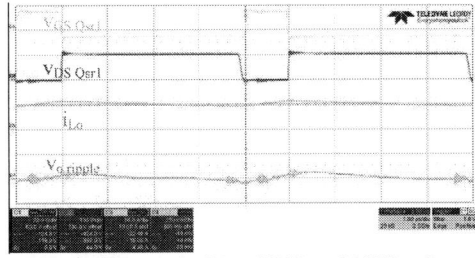

Fig. 9 Soft switching mode (V$_O$ = 20V, P$_O$ = 200W)：Q$_{sr1}$ switch's v$_{GS}$ and v$_{DS}$, output inductance current and output voltage ripple.
CH1: v$_{GS\ Qsr1}$ (20V/div); CH2: v$_{DS\ Qsr1}$ (100V/div);
CH3: i$_{Lo}$ (10A/div); CH4: v$_{o\_ripple}$ (200mV/div); Time:1μs/div

Table I lists the output voltage ripple value for hard switching and soft switching modes under different output voltage and load conditions.

3154

Table I Output voltage ripple values under different output voltages and powers and switching modes

Vo (V)	Io (A)	V_{Ripple_HS}(mV)	V_{Ripple_SS}(mV)
80	1.25	213	68
80	2.5	233	69
80	3.75	200	72
80	5	208	75
20	2.5	207	98
20	5	227	103
20	7.5	240	109
20	10	310	114

It can be seen from Table I that the soft switching can drastically reduce the switching noise on the output voltage

V. Conclusion

The purpose of this paper is to propose a bi-directional converter operating at wide output voltage and load application. By the auxiliary circuit, main switches can achieve zero voltage switching and the switching noise on the output voltage can be eliminated.

Reference

[1] Junhong Zhang, Jih-Sheng Lai, Rae-Young Kim, Wensong Yu, "High-Power Density Design of a Soft-Switching High-Power Bidirectional dc–dc Converter", IEEE Transaction on Power Electronics, Vol. 22, No. 4, pp. 1145 - 1153 (2007).

[2] Christoph Marxgut, Florian Krismer, Dominik Bortis, Johann W.Kolar, "Ultraflat Interleaved Triangular Current Mode (TCM) Single-Phase PFC Rectifier", IEEE Transaction on Power Electronics, Vol.29, No. 2, pp. 873 - 882 (2014).

[3] Xiucheng Huang, Fred C. Lee, Qiang Li, Weijing Du, "High-Frequency High-Efficiency GaN-Based Interleaved CRM Bidirectional Buck/Boost Converter with Inverse Coupled Inductor" ,IEEE Transaction on Power Electronics, Vol. 31, No. 6, pp. 4343 -4352 (2015).

[4] J. Yang and H. Do, "High-efficiency bidirectional dc–dc converter with low circulating current and ZVS characteristic throughout a full range of loads," IEEE Trans. Ind. Electron., vol. 61, no. 7, pp. 3248–3256, Jul. 2014.

[5] K. Chao and C. Huang, "Bidirectional dc–dc soft-switching converter for stand-alone photovoltaic power generation systems," IET Power Electron., vol. 7, no. 6, pp. 1557–1565, Jun. 2014.

[6] I. Lee, J. Kim, T. Lee, Y. Jung, and C.Won, "A new bidirectional DC-DC converter with ZVT switching," in Proc. IEEE Veh. Power Propulsion Conf., 2012, pp. 684–689.

[7] J. Lee et al., "Auxiliary switch control of a bidirectional soft-switching dc/dc converter," IEEE Trans. Power Electron., vol. 28, no. 12, pp. 5446–5457, Dec. 2013.

[8] M. Pavlovsky, G. Guidi, and A. Kawamura, "Buck/boost dc–dc converter topology with soft switching in the whole operating region," IEEE Trans. Power Electron., vol. 29, no. 2, pp. 851–862, Feb. 2014.

[9] I. Kim, S. Paeng, J. Ahn, E. Nho, and J.Ko, "New bidirectional ZVS PWM Sepic/Zeta DC-DC converter," in Proc. IEEE Int. Symp. Ind. Electron., 2007, pp. 555–560.

[10] M. Song, Y. Son, and K. Lee, "Non-isolated bidirectional soft-switching SEPIC/ZETA converter with reduced ripple currents," J. Power Electron., vol. 14, no. 4, pp. 649–660, Jul. 2014.

[11] O. Abdel-Rahman, J. Liu, L. Yao, I. Batarseh, and H. Mao, "LCC zero voltage-switching buck converter with synchronous rectifier," in Proc. IEEE IAS 2006, pp. 2150–2156.

The 2018 International Power Electronics Conference

Accurate Impedance Model of Grid-Connected Inverter for Small-Signal Stability Assessment in High-Impedance Grids

Tuomas Messo[1*], Roni Luhtala[2], Aapo Aapro[1], Tomi Roinila[2]

1 Electrical Energy Engineering, Tampere University of Technology, Tampere, Finland

2 Hydraulics and Automation, Tampere University of Technology, Tampere, Finland

*E-mail: tuomas.messo@tut.fi

Abstract—Power quality problems caused by grid-connected renewable energy inverters have been reported in recent literature. Excessive harmonics and interharmonics may arise when the inverter starts to interact with the grid impedance. Small-signal impedance models have been proven to be useful tools to analyze the stability margins. However, most often the grid voltage feedforward loop employed by the inverter is not included in impedance-based analysis. To fill this gap, this paper presents an impedance model, which includes the effect of feedforward, to analyze impedance-based stability in the presence of large grid impedance. The model is verified by impedance measurements from a laboratory prototype. The model is shown to give accurate prediction of small-signal stability when the Nyquist stability-criterion is applied. Thus, the model can be used to re-shape the inverter impedance to avoid stability problems. The developed impedance model will also provide a useful tool to monitor stability margins online, which necessitates adaptive impedance-shaping of grid-connected inverters.

Keywords—DC-AC power converters, Impedance, Power conversion harmonics, Renewable energy sources, Stability analysis.

I. INTRODUCTION

The amount of grid-connected inverters is increasing as a consequence of the ongoing renewable energy boom [1] which creates unforeseen challenges for grid operators and inverter designers [2], [3]. Inverters have been shown to cause resonance phenomena when connected to a weak or series compensated line [4], [5]. Unexplained resonance in a photovoltaic power plant has been reported recently in [3]. In general, the instability occurs when the inverter control system interacts with grid impedance.

The precondition for instability can be identified from the ratio of inverter and grid impedance. The inverter becomes unstable when the impedance ratio does not satisfy the Nyquist stability criterion [6], [7], [8]. Stable operation can be guaranteed if the inverter impedance has much larger magnitude than grid impedance [9] or if both impedances resemble passive circuits [10], [11]. The inverter impedance can be re-shaped by changing the current control parameters [12] or by applying feedforward measurement from grid voltages [13].

Several studies have proposed small-signal models which consider the dynamic effect of grid voltage feedfor-

ward [21], [22], [18], [19], [20]. Impedance model with proportional grid voltage feedforward has been derived and used in stability analysis in [23] in the sequence-domain. In [24], an impedance model based on single-phase equivalent circuit was applied for stability analysis of parallel converters. In the present paper, however, the dq-domain model is used because it provides numerous benefits, such as the possibility to include the dynamic effects of source and load impedances [25], which is an important feature to consider in microgrids and power-electronics-based power systems.

High-impedance grids have been reported to challenge the stability of grid-connected inverters [2], [3], [4], [5], [6]. One example of a high-impedance grid is a "weak" grid characterized by a small short-circuit-ratio (SCR) which is commonly emulated in the laboratory by using large series-connected inductance or a series RL-circuit [26]. This paper extends the analysis of [19] and [20] to consider the use of proportional grid voltage feedforward in the presence of high grid impedance. It is shown that the feedforward has a significant effect on inverter impedance and even a small control delay can cause instability in the presence of high grid impedance. Therefore, impedance models that do not consider the effect of delay in the feedforward path will arrive at incorrect conclusions regarding stability. The model is verified by impedance measurements and it is shown to give accurate prediction on stability. Applicability of the impedance model is demonstrated by re-shaping the inverter impedance to stabilize the grid interface. The developed impedance model can be further used to necessitate adaptive control of feedforward to enable stable operation during changing grid impedance [38].

II. SMALL-SIGNAL STABILITY OF GRID-CONNECTED INVERTER

A grid-connected inverter can be depicted as in Fig. 1, where the Thevenin and Norton equivalent circuits in dq-domain are used to model the grid and the inverter.

$$\mathbf{Z}_{\mathrm{grid}} = \left[\begin{array}{cc} Z_{\mathrm{gd}} & Z_{\mathrm{gqd}} \\ Z_{\mathrm{gdq}} & Z_{\mathrm{gq}} \end{array} \right] \tag{1}$$

$$\mathbf{Y}_{\mathrm{inv}} = \left[\begin{array}{cc} Y_{\mathrm{od}} & Y_{\mathrm{oqd}} \\ Y_{\mathrm{odq}} & Y_{\mathrm{oq}} \end{array} \right] \tag{2}$$

3156

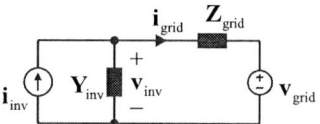

Fig. 1. Inverter connected to a high-impedance grid.

Stability can be evaluated by applying the Nyquist stability criterion to impedance ratios $Z_{gd}Y_{od}$ and $Z_{gq}Y_{oq}$. The stability can be also evaluated by examining the full-order return-ratio matrix $\mathbf{Z}_{grid}\mathbf{Y}_{inv}$, although this is laborious process since it requires solving of eigenvalues for each frequency point. Using the simplified stability criterion provides an accurate stability analysis in most cases [28], [29], [30], [31]. In short, the inverter is deemed unstable if either $Z_{gd}Y_{od}$ or $Z_{gq}Y_{oq}$ encircles -1 on the real axis on complex plane.

The grid impedance is assumed to resemble a series RL-branch [26]. D and q-components of grid impedance, Z_{gd} and Z_{gq} can be defined as in (3) and (4) when cross-couplings between d and q-components are neclegted.

$$Z_{gd} = R_g + sL_g \tag{3}$$

$$Z_{gq} = R_g + sL_g \tag{4}$$

III. IMPEDANCE MODEL OF GRID-CONNECTED INVERTER

Grid-connected three-phase inverter is shown in Fig. 2, where L_{tf} denotes the equivalent inductance of an isolation transformer. Therefore, the inverter effectively employs an LCL-type output filter. The impedance model is developed first for the inverter employing an L-type filter. Subsequently the AC capacitor is added in the impedance model as a load-effect. In this work, the transformer inductance is analyzed as part of the grid impedance. This procedure makes it possible to use the same impedance model for inverters with L, LC or LCL-type filters, given that the feedforward is taken from capacitor voltage.

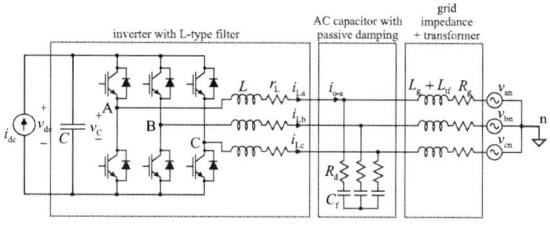

Fig. 2. Grid-connected three-phase inverter.

The average model of the inverter with an L-type filter can be derived in the dq-domain [33], [20] and is given by (5) – (8). Variables in the dq-domain are defined as follows: i_{Ld} and i_{Lq} are the inverter-side inductor currents; v_C is the DC capacitor voltage; i_{dc} is the DC current; v_{od} and v_{oq} are the voltages at inverter output (that is, over

the passively-damped AC capacitors); d_d and d_q are the three-phase duty ratios; v_{dc} is the DC voltage; i_{od} and i_{oq} are the inverter output currents (before the transformer); and ω_s is the angular frequency of the grid.

$$\frac{d\langle i_{Ld}\rangle}{dt} = \frac{1}{L}\left(-r_L\langle i_{Ld}\rangle + \omega_s L\langle i_{Lq}\rangle + d_d\langle v_C\rangle - \langle v_{od}\rangle\right) \tag{5}$$

$$\frac{d\langle i_{Lq}\rangle}{dt} = \frac{1}{L}\left(-\omega_s L\langle i_{Ld}\rangle - r_L\langle i_{Lq}\rangle + d_q\langle v_C\rangle - \langle v_{oq}\rangle\right) \tag{6}$$

$$\frac{d\langle v_C\rangle}{dt} = \frac{1}{C}\left(\langle i_{dc}\rangle - \frac{3}{2}\left(d_d\langle i_{Ld}\rangle + d_q\langle i_{Lq}\rangle\right)\right) \tag{7}$$

$$\langle v_{dc}\rangle = \langle v_C\rangle, \langle i_{od}\rangle = \langle i_{Ld}\rangle, \langle i_{oq}\rangle = \langle i_{Lq}\rangle \tag{8}$$

Steady-state operating point, around which the state-space is linearized can be given as

$$D_d = \frac{V_{od} + \sqrt{\left(V_{od}\right)^2 + \frac{8}{3}r_L V_{dc} I_{dc}}}{2V_{dc}}, \tag{9}$$

$$D_q = \frac{2\omega_s L I_{dc}}{3D_d V_{dc}}, I_{Ld} = \frac{2I_{dc}}{3D_d}. \tag{10}$$

The linear state matrices are given by (11) and (12).

$$\frac{d}{dt}\overbrace{\begin{bmatrix} \hat{i}_{Ld} \\ \hat{i}_{Lq} \\ \hat{v}_C \end{bmatrix}}^{\mathbf{x}} = \overbrace{\begin{bmatrix} -\frac{r_L}{L} & \omega_s & \frac{D_d}{L} \\ -\omega_s & -\frac{r_L}{L} & \frac{D_q}{L} \\ -\frac{3D_d}{2C} & -\frac{3D_q}{2C} & 0 \end{bmatrix}}^{\mathbf{A}}\overbrace{\begin{bmatrix} \hat{i}_{Ld} \\ \hat{i}_{Lq} \\ \hat{v}_C \end{bmatrix}}^{\mathbf{x}}$$

$$+ \overbrace{\begin{bmatrix} 0 & -\frac{1}{L} & 0 & \frac{V_{dc}}{L} & 0 \\ 0 & 0 & -\frac{1}{L} & 0 & \frac{V_{dc}}{L} \\ \frac{1}{C} & 0 & 0 & -\frac{I_{dc}}{D_d C} & 0 \end{bmatrix}}^{\mathbf{B}}\overbrace{\begin{bmatrix} \hat{i}_{dc} \\ \hat{v}_{od} \\ \hat{v}_{oq} \\ \hat{d}_d \\ \hat{d}_q \end{bmatrix}}^{\mathbf{u}} \tag{11}$$

$$\overbrace{\begin{bmatrix} \hat{v}_{dc} \\ \hat{i}_{od} \\ \hat{i}_{oq} \end{bmatrix}}^{\mathbf{y}} = \overbrace{\begin{bmatrix} 0 & 0 & 1 \\ 1 & 0 & 0 \\ 0 & 1 & 0 \end{bmatrix}}^{\mathbf{C}}\overbrace{\begin{bmatrix} \hat{i}_{Ld} \\ \hat{i}_{Lq} \\ \hat{v}_C \end{bmatrix}}^{\mathbf{x}}. \tag{12}$$

Transfer functions can be solved from (13), where $\mathbf{Y}$ and $\mathbf{U}$ are the Laplace-transformed output and input vectors and state matrices $\mathbf{A}$, $\mathbf{B}$, $\mathbf{C}$, and $\mathbf{D}$ are defined according to (11) and (12). The element (i,j) in matrix (14) corresponds to the open-loop transfer function from input variable on row j and output variable on row i [35]; e.g.. $\hat{i}_{od}/\hat{d}_d = \mathbf{G}(2,4) = G_{cod\text{-}o}$.

$$\mathbf{Y}(s) = \left[\mathbf{C}\left(s\mathbf{I} - \mathbf{A}\right)^{-1}\mathbf{B}\right]\mathbf{U}(s) = \mathbf{G}\mathbf{U}(s) \tag{13}$$

$$\overbrace{\begin{bmatrix} \hat{v}_{dc} \\ \hat{i}_{od} \\ \hat{i}_{oq} \end{bmatrix}}^{\mathbf{Y}} = \overbrace{\begin{bmatrix} Z_{in\text{-}o} & T_{oid\text{-}o} & T_{oiq\text{-}o} & G_{cid\text{-}o} & G_{ciq\text{-}o} \\ G_{iod\text{-}o} & -Y_{od\text{-}o} & -Y_{oqd\text{-}o} & G_{cod\text{-}o} & G_{coqd\text{-}o} \\ G_{ioq\text{-}o} & -Y_{odq\text{-}o} & -Y_{oq\text{-}o} & G_{codq\text{-}o} & G_{coq\text{-}o} \end{bmatrix}}^{\mathbf{G}}\overbrace{\begin{bmatrix} \hat{i}_{dc} \\ \hat{v}_{od} \\ \hat{v}_{oq} \\ \hat{d}_d \\ \hat{d}_q \end{bmatrix}}^{\mathbf{U}} \tag{14}$$

The 2018 International Power Electronics Conference

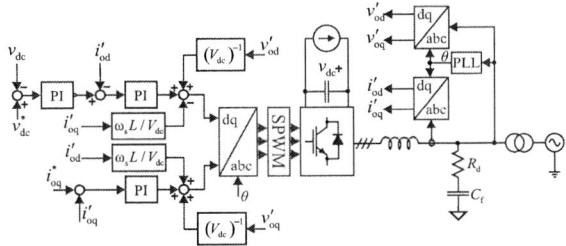

Fig. 3.   Illustration of inverter control system.

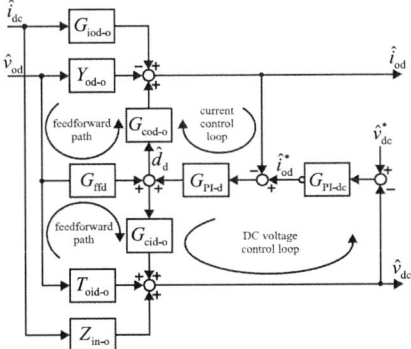

Fig. 4.   Block diagram representing the dynamics of current d-component.

The inverter is controlled with a cascaded control scheme as shown in Fig. 3. Measured grid voltage d and q-components v'_{od} and v'_{oq} are scaled and used as feedforward signals, which is referred as proportional grid voltage feedforward in literature [21]. PI-controller ($G_{PI} = K_P + K_I/s$) parameters are given in Table I. Many excellent papers, such as [33], discuss the current control of photovoltaic inverters. In this paper the loop gains were designed using the well-known loop-shaping method [34].

Output admittance d-component Y_{od} is derived in two stages. First, control functions, such as AC current control, feedforward and DC voltage control, are included in the model by solving the control block diagram in Fig. 4. Secondly, the admittance of the AC capacitor C_f with passive damping resistor R_d is added as a load effect. It is assumed that cross-coupling transfer functions G_{coqd-o}, G_{codq-o}, Y_{oqd-o} and Y_{odq-o} in (14) can be neglected due to decoupling in the current control. Moreover, it is assumed that input dynamics depend mainly on the d-components [36]. Thus, the effect of transfer functions T_{oiq-o} and G_{ciq-o} in (14) can be neglected.

D-component of inverter output admittance Y_{od-c} can be solved from the block diagram and given as in (15). The current control loop gain L_{cc-d}, DC voltage control loop gain and the special parameter $Y_{od-\infty}$ are given in the Appendix. G_{del} is the delay transfer function, according to second-order Pade approximation.

$$Y_{od-c} = \frac{Y_{od-o} - G_{cod-o}G_{ffd}G_{del}}{(1 + L_{cc-d})(1 + L_{dc})} + \frac{L_{dc}}{(1 + L_{dc})}Y_{od-\infty}$$

(15)

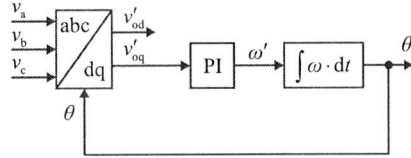

Fig. 5.   Control diagram of conventional PLL.

The effect of the output capacitor can be added in the model as a load-effect, according to [25], as

$$Z_{od} = (Y_{od})^{-1} = \left(Y_{od-c} + \frac{sC_f}{(1 + R_dC_fs)}\right)^{-1}.$$

(16)

TABLE I.   THREE-PHASE INVERTER PARAMETERS.

V_{dc}	414.3 V	L	2.2 mH
I_{dc}	6.58 A	r_L	100 mΩ
$V_{a,b,c}^{rms}$	120 V	C	1.95 mF
ω_s	$2\pi \cdot 60$ rad/s	f_{sw}	20 kHz
C_f	25 μF	R_d	1 Ω
K_p^{dc}	0.096	K_p^d	0.015
K_i^{dc}	1.2	K_i^d	23.4
K_p^{pll}	0.67	K_p^q	0.015
K_i^{pll}	38	K_i^q	23.4
L_g	5.6 mH	R_g	0.4 Ω

The phase-locked-loop in Fig. 5 affects the low-frequency behavior of impedance q-component by introducing negative resistance region below the PLL bandwidth [8], [37], [23]. The control block diagram of output dynamics (q-component) can be depicted as in Fig. 6, from which the admittance q-component Y_{oq-c} can be solved and given as in (17); the current control and PLL loop gains are given in the Appendix.

$$Y_{oq-c} = \left\{ \frac{\left(Y_{oq-o} - \frac{G_{coq-o}G_{ffq}G_{del}}{(1+L_{pll})}\right)}{(1+L_{cc-q})} \right\}$$
$$- \frac{I_{od}}{V_{od}}\frac{L_{cc-q}}{(1+L_{cc-q})}\frac{L_{pll}}{(1+L_{pll})}$$
$$- \frac{D_d}{V_{od}}\frac{G_{coq-o}}{(1+L_{cc-q})}\frac{L_{pll}}{(1+L_{pll})}$$

(17)

The effect of output capacitor can be added in the model as

$$Z_{oq} = (Y_{oq})^{-1} = \left(Y_{oq-c} + \frac{sC_f}{(1 + R_dC_fs)}\right)^{-1}.$$

(18)

IV.   IMPEDANCE MODEL VERIFICATION

A four-quadrant power amplifier from Spitzenberger & Spies (PAS15000) was used to emulate a three-phase system according to Fig. 7. The parameters of the prototype are given in Table I. Inverter output impedance was measured by injecting a small-signal perturbation to grid voltages, which enables measuring the frequency response between three-phase voltage v_o and current i_o (in the dq-domain). The angle of grid voltages was generated inside dSPACE, allowing the impedances to be measured in a

3158

The 2018 International Power Electronics Conference

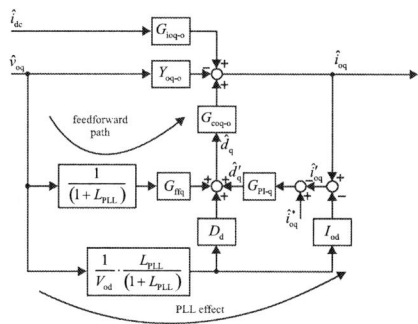

Fig. 6. Block diagram representing the dynamics of current q-component.

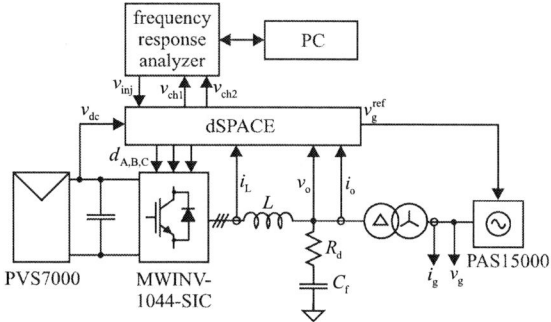

Fig. 7. Measurement setup to extract inverter output impedance.

synchronous reference frame that is fixed to grid voltages. Thus, the effect of the PLL is included in the measured impedances.

The measured impedance d-component and the impedance given by the model are shown in Fig. 8. The impedance d-component was measured at the MPP at 414.3 V. Feedforward gains G_{ffd} and G_{ffq} were selected equal to inverse of DC voltage; that is, 0.0024. The control delay was approximated as 1.5 times the switching period ($T_d = 1.5/f_{sw}$) [33]. The feedforward increases the impedance magnitude significantly between 30 and 600 Hz as can be seen in Fig. 8. At high frequencies the impedance is determined by the AC capacitor. Thus, proportional feedforward does not increase impedance at frequencies higher than the resonant frequency of LC-filter. In fact, the impedance magnitude has a slightly smaller value around the resonant frequency (approx. 1 kHz) when proportional feedforward is used.

The impedance q-component is shown in Fig. 9 at the MPP, both with and without feedforward. A significant increase in impedance magnitude can be noticed between 20 and 600 Hz. The impedance model matches with sufficient accuracy with the measured impedances in all cases, and is therefore suitable for use in stability analysis.

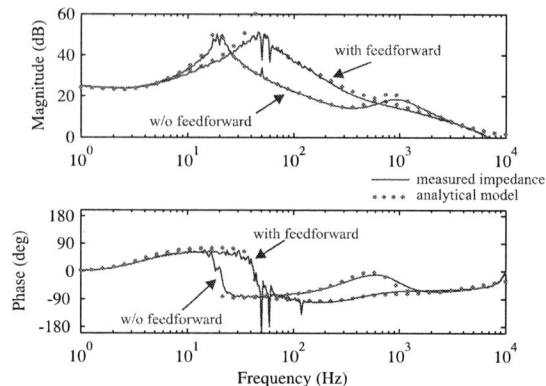

Fig. 8. The effect of proportional grid voltage feedforward on impedance d-component.

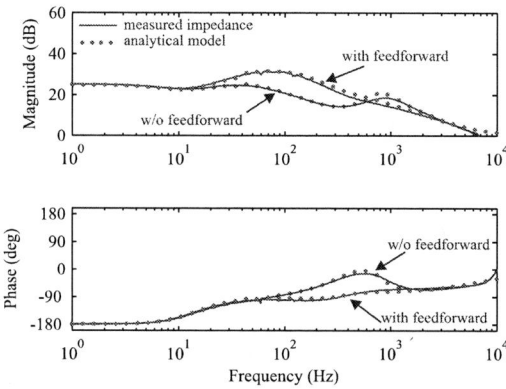

Fig. 9. The effect of proportional grid voltage feedforward on impedance q-component.

V. IMPEDANCE-BASED STABILITY IN THE PRESENCE OF LARGE GRID IMPEDANCE

A. Stability analysis without background harmonics

The grid impedance is high at the end of a long MV line where transformers with low power rating are used [26]. A small SCR is generally emulated in laboratory using a large series inductor. Grid impedance was artificially increased by using a series inductance of 5 mH. The analytical grid impedance was approximated as a series connection of 0.4 Ω resistor and 5.6 mH inductance ($Z_g = R + Ls$), which includes the impedance of the isolation transformer (verified in an offline measurement). The equivalent SCR is 2 for a 10 kVA inverter. The grid impedance was measured to verify that the grid impedance model is accurate and can be used in stability analysis. The measured grid impedance and the analytical model are shown in Fig. 10.

The inverter was observed to become unstable when switching frequency is reduced from 20 kHz to 16 kHz. Impedance d-components in the unstable condition are shown in Fig. 11 where the blue curve depicts the inverter output impedance in unstable case and the black curve is

3159

The 2018 International Power Electronics Conference

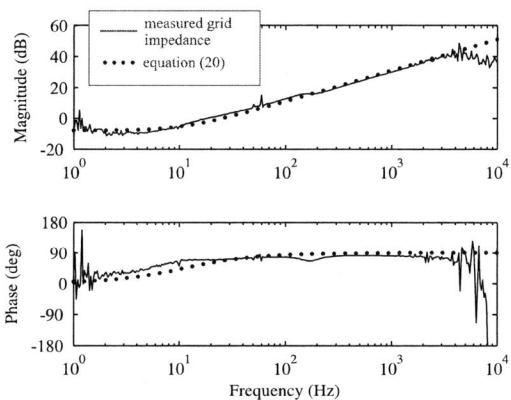

Fig. 10. Measured system impedance and its analytical model.

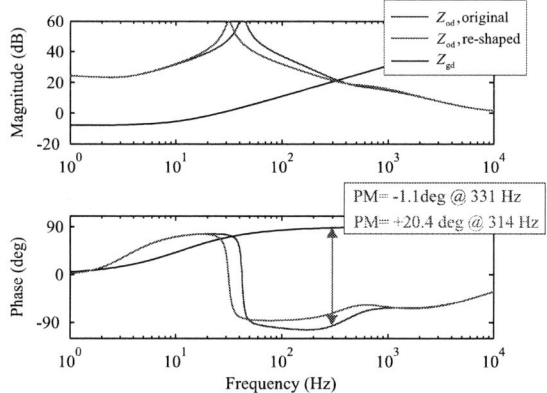

Fig. 11. D-components of original inverter impedance, re-shaped inverter impedance and grid impedance.

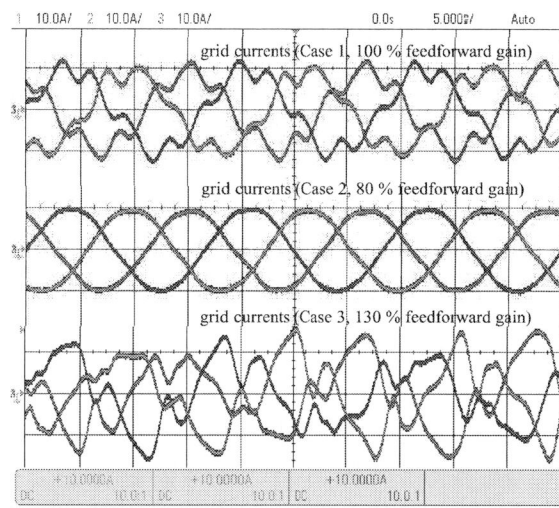

Fig. 12. Instability due to impedance-based interactions.

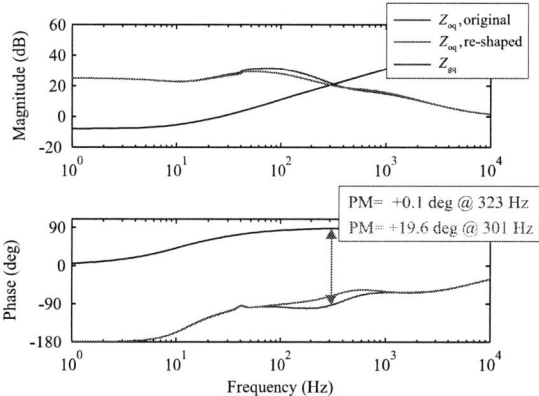

Fig. 13. Q-components of original inverter impedance, re-shaped inverter impedance and grid impedance.

the grid impedance. The impedances overlap at 331 Hz with a phase difference of more than 180 degrees, which suggests that the inverter output impedance and grid impedance form an under-damped resonance. In fact, the impedance model gives slightly negative phase margin, -1.1 degrees. The inverter output currents are shown in Fig. 12 as 'Case 1' and it is clear that the inverter is unstable.

The impedance model was used to re-shape the inverter impedance in order to boost the phase around the critical frequency of 325 Hz. Phase boost was achieved by reducing the feedforward gain to $0.8/V_{dc}$, since feedforward inherently introduces a phase drop in inverter impedance originating from control delay. The green curve in Fig. 11 illustrates the impedance after re-shaping the impedance. The phase margin is 20.4 degrees and the grid currents in the corresponding 'Case 2' are stable, as can be seen in Fig. 12.

Fig. 13 shows the impedance q-components, where the blue curve is the inverter impedance in the unstable case and green curve the re-shaped impedance. The impedance model gives positive phase margin in both cases. In fact, it was observed that the q-component was not the reason for

instability, because the inverter becomes stable (Case II), when only the feedforward gain affecting d-component is reduced. However, to guarantee sufficient phase margin, the feedforward gain affecting the q-component is also reduced.

The model can be used to evaluate stability by examining the ratio of impedance d and q-components Z_{gd}/Z_{od} and Z_{gq}/Z_{oq}, as shown in Figs. 14 and 15. The inverter is stable when the ratio satisfies the Nyquist stability criterion; i.e., the curve does not encircle the critical point (-1,0). Thus, in 'Case 1', the inverter is unstable due to negative phase margin caused by impedance d-component. However, after re-shaping the impedance the inverter is stable, as can be seen by studying the impedance ratio in 'Case 2'. Moreover, increasing the feedforward gain makes the situation even worse, as illustrated by 'Case 3'. Increasing the feedforward gain beyond $1.3/V_{dc}$ would trip the over-current protection of the inverter, which was set to 25 A.

3160

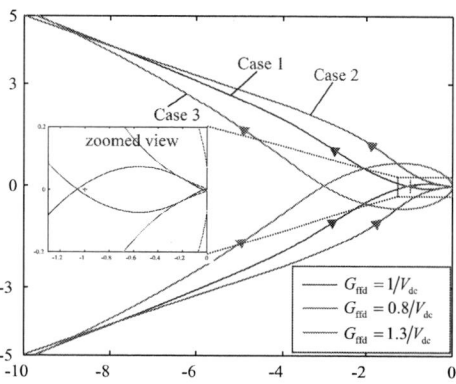

Fig. 14. Nyquist diagrams derived based on d-components.

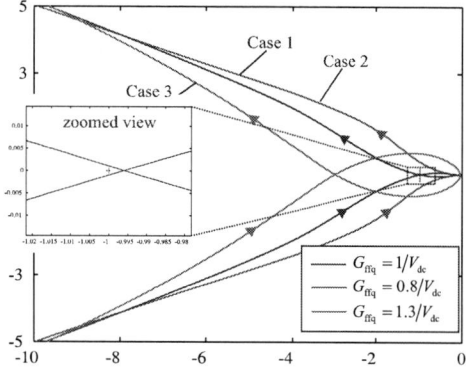

Fig. 15. Nyquist diagrams derived based on q-components.

B. Stability analysis with considerable amount of background harmonics

The effect of feedforward was studied when the inverter is switched at 12 kHz and is connected to a 230 V, 50 Hz public low-voltage grid. A different inverter is used, which is based on IGBT switches requiring dead-time of 4 μs (IPM module 7MBP50RJ120 by Fuji Electric). Moreover, the grid voltages contain initially significant amount of background harmonics at the fifth and seventh harmonics, as shown in Fig. 16. This test was done to demonstrate the accuracy of stability analysis in more realistic conditions.

The spectrum of grid current is shown in Fig. 17, both with and without the proportional grid voltage feedforward. The current has slightly better harmonic spectrum when feedforward is activated, which supports the use of feedforward, especially due to the lower seventh harmonic. In this case the grid impedance is the true impedance seen at the PCC without any additional modifications other than the isolation transformer and can be considered small.

The grid impedance was artificially increased, to emulate a weak grid, by adding a 6.2 mH inductor between

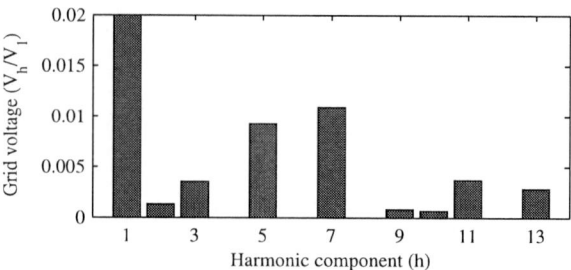

Fig. 16. Initial background harmonics in grid voltage.

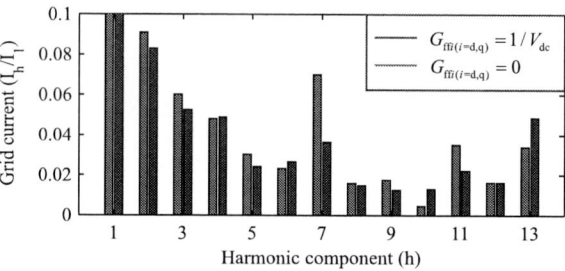

Fig. 17. Harmonic components of inverter current without (red) and with (blue) proportional grid voltage feedforward.

the inverter and the grid (SCR 7.4). Figs. 18a-c show the grid currents when proportional feedforward gain is increased from zero to 80 percent of the ideal value ($1/V_{dc}$). A significant increase in harmonic currents can be noticed by studying the spectrum in Fig. 19. Dead-time causes significant odd harmonics (mainly fifth and seventh), which distort the grid voltage at the inverter-grid interface. Moreover, the isolation transformer was observed to saturate and cause even harmonics. The harmonic components caused by dead-time and the saturating transformer are amplified by the feedforward. Thus, the harmonic currents generated by the inverter are increased rather than decreased when the grid impedance is high. Therefore, the feedforward gains should be selected carefully in grids with large impedance. However, the inverter does not suffer from instability in Figs. 18a-c since all the harmonics can be explained by grid background harmonics or by non-idealities, such as dead-time and saturating transformer.

Fig. 20 shows the ratios of impedance d and q-components when the proportional feedforward gain is selected as $0.8/V_{dc}$ and after increasing the feedforward gain to the nominal value of $1/V_{dc}$. The ratio encircles the (-1,0) -point with the larger feedforward gain, which suggests instability. Fig. 18d shows the grid currents in the unstable case, while Fig. 21 shows the corresponding spectrum of phase current. The spectrum indicates that the current is not distorted only by harmonics in grid voltage, but also by interharmonics that appear due to impedance-based interaction.

VI. FUTURE WORK

The impedance model was found out to give correct prediction on stability in the presence of background harmonics and non-linearities arising from dead-time and a saturating transformer. Thus, the impedance model would be suitable for online stability analysis of grid-connected inverters. Therefore, as a continuation of this study the impedance model has been applied for adaptive control of grid voltage feedforward. Interested readers are urged to see [38] for further reference and demonstration of the adaptive control method.

VII. CONCLUSIONS

This paper presents an impedance model of grid-connected inverter with proportional grid voltage feedforward. The model is verified by measuring the output impedance of a prototype inverter. The model is shown to give accurate assessment of stability in the presence of large grid impedance and considerable amount of background harmonics and non-linearities caused by dead-time and saturating transformer. The model shall be further used in developing active output impedance -shaping algorithms for grid-connected three-phase converters.

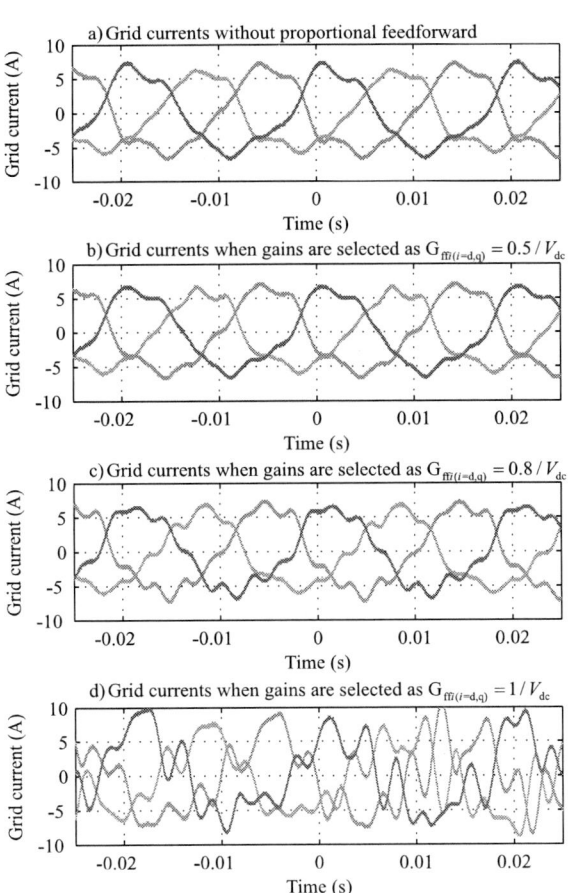

Fig. 18. Grid currents when grid has high inductance (6.2 mH).

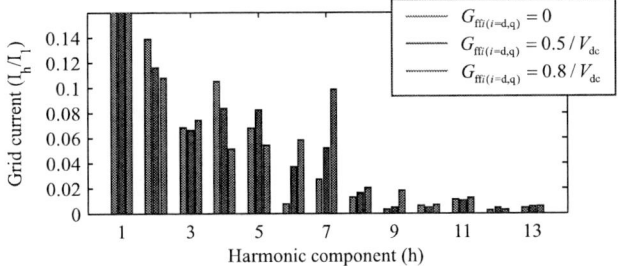

Fig. 19. Harmonic components corresponding to grid currents in cases a), b) and c) of Fig. 18.

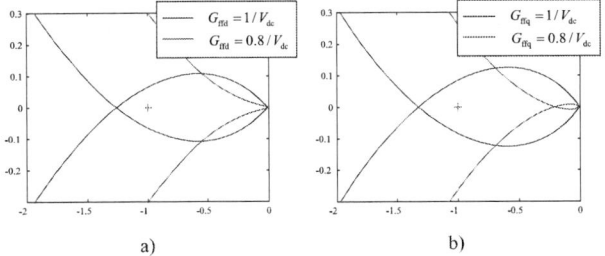

Fig. 20. Ratio of impedance (a) d-components and (b) q-components obtained by using the impedance model.

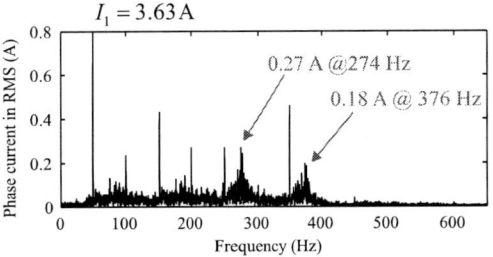

Fig. 21. Spectrum of inverter current in the unstable case (Fig. 18 (d)).

APPENDIX

$$L_{\text{cc-d}} = G_{\text{cod-o}}G_{\text{PI-d}}, \; L_{\text{dc}} = -\frac{G_{\text{cid-o}}}{G_{\text{cod-o}}}\frac{L_{\text{cc-d}}}{(1 + L_{\text{cc-d}})}G_{\text{PI-dc}} \quad (19)$$

$$Y_{\text{od}-\infty} = Y_{\text{od-o}} + \frac{G_{\text{cod-o}}T_{\text{oid-o}}}{G_{\text{cid-o}}} \approx \frac{I_{\text{od}}}{V_{\text{od}}} \quad (20)$$

$$L_{\text{pll}} = \frac{V_{\text{od}}}{s}G_{\text{PI}}^{\text{pll}}, \; L_{\text{cc-q}} = G_{\text{PI-q}}G_{\text{coq-o}} \quad (21)$$

REFERENCES

[1] B. K. Bose, "Global energy scenario and impact of power electronics in 21st century", *IEEE Trans. Ind. Electron.*, vol. 60, no. 7, pp. 2638-2651, Jul 2013.

[2] J. H. R. Enslin and P. J. M. Heskes, "Harmonic interaction between a large number of distributed power inverters and the distribution network", *IEEE Trans. Power Electron.*, vol. 19, no. 6, pp. 1586-1593, Nov 2004.

[3] C. Li, "Unstable Operation of Photovoltaic Inverter from Field Experiences", *IEEE Trans. Power Deliv.*, vol. 8977, no. c, pp. 1-1, 2017.

[4] J. Huang, and X. Yuan, "Impact of the voltage feed-forward and current decoupling on VSC current control stability in weak grid based on complex variables", IEEE Energy Conversion Congress and Exposition, pp. 6845-6852, 2015.

[5] P. Belkin. "Event of 10-22-09", CREZ Technical Conference, Electrical Reliability Council of Texas, 2010.

[6] X. Wang, F. Blaabjerg, and W. Wu, "Modeling and analysis of harmonic stability in an AC power-electronics-based power system", IEEE Trans. Power Electron., vol. 29, no. 12, pp. 6421-6432, Dec 2014.

[7] H. Liu, and J. Sun., "Voltage stability and control of offshore wind farms with AC collection and HVDC transmission", IEEE Trans. Emerg. Sel. Topics Power Electron., vol. 2, no. 4, pp. 1181-1189, Dec 2014.

[8] B. Wen et al, "Analysis of D-Q small-signal impedance of grid-tied inverters", IEEE Trans. Power Electron., vol. 31, no. 1, pp. 675-687, Jan 2016.

[9] D. Yang, X. Ruan and H. Wu. "Impedance shaping of the grid-connected inverter with LCL filter to improve its adaptability to the weak grid condition", IEEE Trans. Power Electron., vol 29, no 11, pp. 5795-5805, Nov 2014.

[10] L. Harnefors et al, "Passivity-based stability assessment of grid-connected VSCs - an overview", IEEE Trans. Emerg. Sel. Topics Power Electron., vol. 4, no. 1, pp. 116-125, Mar. 2016.

[11] K.M. Alawasa, Y. A.-R. I. Mohamed and W. Xu, "Active mitigation of subsynchronous interactions between PWM voltage-source converters and power networks", IEEE Trans. Power Electron., vol 29, no 1, pp. 121-134, Jan 2014.

[12] L. Harnefors, M. Bongiorno and S. Lundberg, "Input-admittance calculation and shaping for controlled voltage-source converters", IEEE Trans. Ind. Electron., vol. 54, no. 6, pp. 3323-3334, Dec 2007.

[13] X. Zhang et al, "Influence of voltage feed-forward control on small- signal stability of grid-tied inverters", IEEE Applied Power Electronics Conference and Exposition, pp. 1216-1221, 2015.

[14] M. Rasheduzzaman, J. A. Mueller and J. W. Kimball, "Reduced-order small-signal model of microgrid systems", IEEE Trans. Sustain. Energy, vol. 6, no. 4, pp. 1292-1305, Oct 2015.

[15] L. Kunjumuhammed et al, "Electrical oscillations in wind farm systems: analysis and insight based on detailed modeling", IEEE Trans. Sustain. Energy, vol. 7, no. 1, pp. 51-62, Jan 2016.

[16] J. Hu et al, "Modeling of grid-connected DFIG-based wind turbines for dc-link voltage stability analysis", IEEE Trans. Sustain. Energy, vol. 6, no. 4, pp. 1325-1336, Oct 2015.

[17] Q. Yan et al, "An improved grid-voltage feedforward strategy for high-power three-phase grid-connected inverters based on the simplified repetitive predictor", IEEE Trans. Power Electron., vol. 31, no. 5, pp. 3880-3897, May 2016.

[18] S. Y. Park et al, "Admittance compensation in current loop control for a grid-tie LCL fuel cell inverter", IEEE Trans. Power Electron., vol. 23, no. 4, pp. 1716-1723, vol. 23, no. 4, Jul. 2008.

[19] T. Messo, A. Aapro, T. Suntio and T. Roinila, "Design of grid-voltage feedforward to increase impedance of grid-connected three-phase inverters with LCL-filter," 2016 IEEE 8th International Power Electronics and Motion Control Conference, pp. 2675-2682, 2016.

[20] T. Messo, J. Jokipii and T. Suntio, "Effect of conventional grid-voltage feedforward on the output impedance of a three-phase photovoltaic inverter," 2014 International Power Electronics Conference, pp. 514-521, 2014.

[21] X. Wang et al, "Full feedforward of grid voltage for grid-connected inverter with LCL filter to suppress current distortion due to grid voltage harmonics", IEEE Trans. Power Electron., vol 25, no 12. pp. 3119-3127, Dec 2010.

[22] X. Wu et al, "Grid harmonics suppression scheme for LCL-type grid-connected inverters based on output admittance revision", IEEE Trans. Sustain. Energy, vol. 6, no. 2, pp. 411-421, Apr 2015.

[23] M. Cespedes and J. Sun, "Impedance modeling and analysis of grid-connected voltage-source converters", IEEE Trans. Power Electron., vol. 29, no. 3, pp. 1254-1261, Mar 2014.

[24] X. Wang, F. Blaabjerg, and P. C. Loh, "An impedance-based stability analysis method for paralleled voltage source converters", in 2014 International Power Electronics Conference (IPEC-Hiroshima 2014 - ECCE ASIA), pp. 1529-1535, 2014.

[25] J. Puukko and T. Suntio, "Modelling the effect of non-ideal load in three-phase converter dynamics", Electronics Letters, vol. 48, no. 7, pp. 402-404, Mar 2012.

[26] M. Liserre, R. Teodorescu and F. Blaabjerg, "Stability of photovoltaic and wind turbine grid-connected inverters for a large set of grid impedance values", IEEE Trans. Power Electron., vol. 21, no. 1, pp. 263-272, Jan. 2006.

[27] L. Jessen and F. W. Fuchs, "Modeling of inverter output impedance for stability analysis in combination with measured grid impedances". IEEE 6th International Symposium on Power Electronics for Distributed Generation Systems, 2015, pp. 1-7.

[28] J. Sun, "Impedance-based stability criterion for grid-connected inverters", IEEE Trans. Power Electron., vol 26, no 11, pp. 3075-3078, Nov. 2011.

[29] T. Messo, A. Aapro and T. Suntio. "Generalized multivariable small-signal model of three-phase grid-connected inverter in DQ-domain", IEEE 16th Workshop on Control and Modeling for Power Electronics, pp. 1-8, 2015.

[30] R. Burgos et al, "On the Ac stability of high power factor three-phase rectifiers", IEEE Energy Conversion Congress and Exposition, pp. 2047-2054, Sep. 2010.

[31] B. Wen, R. Burgos, D. Boroyevich, P. Mattavelli, and Z. Shen, "AC Stability Analysis and DQ Frame Impedance Specifications in Power Electronics Based Distributed Power Systems", IEEE J. Emerg. Sel. Top. Power Electron., vol. 6777, no. c, pp. 1-1, 2017.

[32] T. Messo, R. Luhtala, D. Yang, T. Roinila, X. Wang and F. Blaabjerg, "Real-Time Impedance-Based Stability Assessment of Grid Converter Interactions", IEEE 18th Workshop on Control and Modeling for Power Electronics, pp. 1-8, 2017.

[33] E. Figueres et al, "Sensitivity study of the dynamics of three-phase photovoltaic inverters with an LCL grid filter", IEEE Trans. Ind. Electron., vol. 56, no. 3, pp. 706-717, Mar 2009.

[34] T. Suntio, T. Messo and J. Puukko, Power Electronic Converters: Dynamics and Control in Conventional and Renewable Energy Applications, John Wiley & Sons Ltd., Chichester, 2017.

[35] J. Puukko, T. Messo and T. Suntio, "Effect of photovoltaic generator on a typical VSI-based three-phase grid-connected photovoltaic inverter dynamics", IET Conference on Renewable Power Generation, 2011, pp. 1-6.

[36] H. Mao, D. Boroyevich and F.C. Lee, "Novel reduced-order small-signal model of three-phase PWM rectifiers and its application in control design and system analysis", IEEE Trans. Power Electron., vol. 13, no. 3, May 1998.

[37] T. Messo et al, "Modeling the grid synchronization induced negative-resistor-like behavior in the output impedance of a three-phase photovoltaic inverter", 4th IEEE International Symposium on Power Electronics for Distributed Generation Systems, . pp. 1-7, 2013.

[38] R. Luhtala, T. Roinila and T. Messo, "Adaptive Control of Grid-Voltage Feedforward for Grid-Connected Inverters based on Real-Time Identication of Impedance-Based Stability Margins", International Power Electronics Conference. (IPEC-Niigata 2018 -ECCE Asia-), pp. 1-8, 2018.

The 2018 International Power Electronics Conference

Modeling of Unbalanced Three-Phase Grid-Connected Converters with Decoupled Transfer Functions

Wei Liu, *Student Member, IEEE*, Xiongfei Wang, *Senior Member, IEEE*, and Frede Blaabjerg, *Fellow, IEEE*
Department of Energy Technology, Aalborg, Denmark
E-mail:liuwei.lock@gmail.com, xwa@et.aau.dk, fbl@et.aau.dk

Abstract—The unbalanced three-phase passive filters or grid impedances lead to a multiple-input multiple-output (MIMO) dynamic model for grid-connected converters in the stationary *αβ*-frame. The MIMO system with the couplings between the *α*-axis and *β*-axis components complicates the stability analysis and current controller design. This paper thus presents a modeling method that decomposes the MIMO system into two single-input single-output (SISO) transfer functions, which enables to use the classical SISO system tools for the dynamic analysis and controller design. Time-domain simulations and experiments are performed and the results validate the effectiveness of the approach.

Keywords—Current control, three-phase unbalance, stability,transfer matrices, transfer functions

I. INTRODUCTION

There have been increasing application of power electronic converters in power grids. Three-phase grid-connected power converters are commonly used in renewable energy generation [1], high-voltage direct-current transmission systems [2] and static synchronous compensators [3], etc. In order to meet the grid code, numerous research efforts have been made on the modeling and control of grid-connected converters [4].

The current control of grid-connected converters in the unbalanced three-phase system has been developed [5]-[7]. The proportional + resonant (PR) controller, among other alternatives, provides a computationally efficient approach to control both the positive- and negative-sequence current [5]. However, in most research works, only the unbalanced three-phase grid voltage is considered. In contrast, the impact of three-phase unbalanced control plant, including the converter filter (*L*- or *LCL*-filter), transformer, and the grid impedance, on the stability of current control loop is often overlooked. In practice, the three-phase *L*-filter is often unbalanced due to the parameter variation, and the grid impedance tends to be unbalanced in distribution grids [8]. Moreover, during the transient conditions, e.g. grid faults, the magnetic core saturation in transformers and filter inductors may also cause the phase unbalance in the control plant of the current loop [9].

A balanced three-phase system can be represented by two decoupled single-input single-output (SISO) transfer functions in the stationary *αβ*-frame [10]. However, the unbalanced three-phase control plant leads to a multiple-input multiple-output (MIMO) model with the couplings between the *α*- and *β*-axis. This MIMO model complicates

the stability analysis and controller design of the current loop. To overcome this challenge, this paper proposes a modeling method that can decompose the MIMO dynamic system into two decoupled SISO transfer functions, and thus the classical SISO system tools can be used to design the current controller for unbalanced three-phase converters. Simulations and experimental tests on a three-phase grid converter with the unbalanced L-filter are performed. The results confirm the effectiveness of the proposed method.

II. CONVERTER MODELING UNDER UNBALANCED CONDITION

A general diagram of an L-filtered three-phase grid converter is depicted in Fig. 1. The output current is sensed for the current control. The voltage at the point of common coupling (PCC) is measured for synchronizing the converter with the grid using the synchronous reference frame phase-locked loop (SRF-PLL). The constant dc-link voltage and an ideal grid voltage are assumed. The phase inductors L_a, L_b, L_c are not equal under unbalanced condition. Fig. 2 shows the equivalent circuit of the three-phase converter in the *abc*-frame.

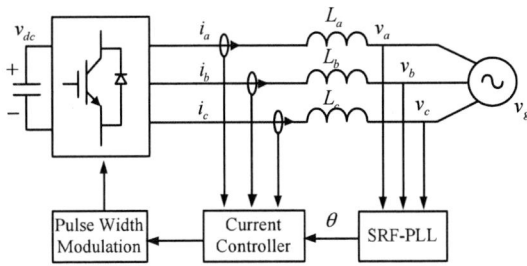

Fig. 1. General diagram of a three-phase grid converter with *L*-filter.

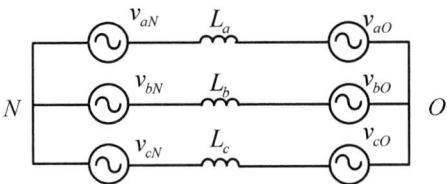

Fig. 2. Equivalent circuit of the three-phase grid converter.

3164

The 2018 International Power Electronics Conference

Based on the equivalent circuit, the model of the three-phase L-filtered converter in the *abc*-frame can be derived as follows:

$$
\begin{bmatrix} v_{aN}(t) \\ v_{bN}(t) \\ v_{cN}(t) \end{bmatrix} - \begin{bmatrix} v_{aO}(t) \\ v_{bO}(t) \\ v_{cO}(t) \end{bmatrix} - \begin{bmatrix} v_{ON}(t) \\ v_{ON}(t) \\ v_{ON}(t) \end{bmatrix} = \begin{bmatrix} L_a & 0 & 0 \\ 0 & L_b & 0 \\ 0 & 0 & L_c \end{bmatrix} \frac{d}{dt} \begin{bmatrix} i_a(t) \\ i_b(t) \\ i_c(t) \end{bmatrix} \quad (1)
$$

where v_{aN}, v_{bN}, v_{cN} are output voltages of the converter, v_{aO}, v_{bO}, v_{cO} are grid voltages, i_a, i_b, i_c are output currents of the converter, v_{ON} is the voltage between neutral point O and N.

Substituting Clarke transformation matrix into (1), the converter model in the stationary $\alpha\beta\gamma$-frame is derived as:

$$
\begin{bmatrix} v_{\alpha N}(t) \\ v_{\beta N}(t) \\ v_{\gamma N}(t) \end{bmatrix} - \begin{bmatrix} v_{\alpha O}(t) \\ v_{\beta O}(t) \\ v_{\gamma O}(t) \end{bmatrix} = T_{abc/\alpha\beta\gamma} \begin{bmatrix} L_a & 0 & 0 \\ 0 & L_b & 0 \\ 0 & 0 & L_c \end{bmatrix} T_{\alpha\beta\gamma/abc} \frac{d}{dt} \begin{bmatrix} i_\alpha(t) \\ i_\beta(t) \\ i_\gamma(t) \end{bmatrix} \quad (2)
$$

where $v_{\alpha N}$, $v_{\beta N}$, $v_{\gamma N}$ are output voltages of the converter in $\alpha\beta\gamma$-frame, $v_{\alpha O}$, $v_{\beta O}$, $v_{\gamma O}$ are grid voltages in the $\alpha\beta\gamma$-frame, i_α, i_β, i_γ are output currents of the converter in the $\alpha\beta\gamma$-frame. There is no zero sequence current in a three-phase three-line system, so only the parameters in the α-axis and β-axis are considered. According to the Laplace transformation, the model of the three-phase converter in the $\alpha\beta$-frame in frequency domain can be formulated as:

$$
\begin{bmatrix} v_{\alpha N}(s) \\ v_{\beta N}(s) \end{bmatrix} - \begin{bmatrix} v_{\alpha O}(s) \\ v_{\beta O}(s) \end{bmatrix} = \begin{bmatrix} sL_{\alpha\alpha} & sL_{\alpha\beta} \\ sL_{\beta\alpha} & sL_{\beta\beta} \end{bmatrix} \begin{bmatrix} i_\alpha(s) \\ i_\beta(s) \end{bmatrix} \quad (3)
$$

where $L_{\alpha\alpha}=2L_a/3+L_b/6+L_c/6$, $L_{\beta\beta}=L_b/2+L_c/2$, $L_{\alpha\beta}=L_{\beta\alpha}=\sqrt{3}(L_b-L_c)/6$.

According to (3), the equivalent inductances $L_{\alpha\beta}$, $L_{\beta\alpha}$ are not zero when the system is unbalanced. $L_{\alpha\beta}$, $L_{\beta\alpha}$ contribute to the coupling items between α-axis and β-axis. The SISO system under the balanced condition results in an MIMO system in the unbalanced condition. Fig. 3 shows the block diagram of the current control plant in the unbalanced condition in $\alpha\beta$-frame. The coupling items make it complex to analyze the stability of the system.

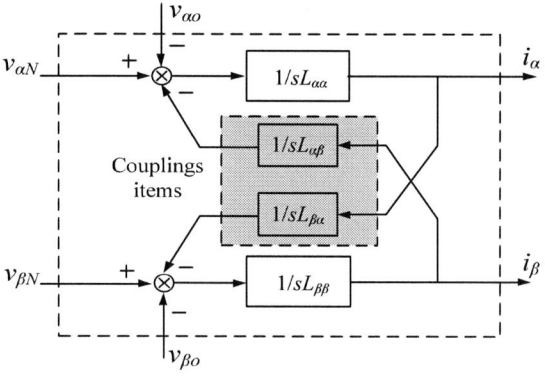

Fig. 3. Block diagram of converter model in $\alpha\beta$-frame.

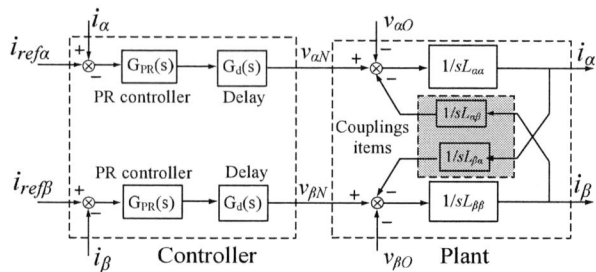

Fig. 4. Block diagram of the current control loop.

III. PROPOSED MODELING METHOD

This section introduces the proposed modeling method that transforms the MIMO system as two decoupled SISO transfer functions.

A. Current control loop

Fig. 4 illustrates the block diagram of the current control loop with an unbalanced control plan in the $\alpha\beta$-frame. The PR controller is used to control both the positive-sequence and negative-sequence components with zero steady-state errors. $G_d(s)$ is the time delay incurred from digital control system, which is composed by one sampling period of the computational delay and half sampling period of the PWM delay [11], and is given by:

$$
G_d(s) = e^{-1.5T_s s} \quad (4)
$$

$G_{PR}(s)$ is the PR controller which can be expressed as:

$$
G_{PR}(s) = k_p + \frac{k_i s}{s^2 + \omega_1^2} \quad (5)
$$

According to the current controller and unbalanced control plant as Fig. 4 shows, the relationship between the references and the output currents are derived as:

$$
\begin{bmatrix} G_{PR}(s)G_d(s) & 0 \\ 0 & G_{PR}(s)G_d(s) \end{bmatrix} \left(\begin{bmatrix} i_{ref\alpha}(s) \\ i_{ref\beta}(s) \end{bmatrix} - \begin{bmatrix} i_\alpha(s) \\ i_\beta(s) \end{bmatrix} \right)
$$
$$
= \begin{bmatrix} sL_{\alpha\alpha} & sL_{\alpha\beta} \\ sL_{\beta\alpha} & sL_{\beta\beta} \end{bmatrix} \cdot \begin{bmatrix} i_\alpha(s) \\ i_\beta(s) \end{bmatrix} \quad (6)
$$

The closed loop transfer function matrix can be given:

$$
\begin{bmatrix} i_\alpha(s) \\ i_\beta(s) \end{bmatrix} = \begin{bmatrix} \dfrac{sL_{\alpha\alpha}+G_{PR}(s)G_d(s)}{G_{PR}(s)G_d(s)} & \dfrac{sL_{\alpha\beta}}{G_{PR}(s)G_d(s)} \\ \dfrac{sL_{\beta\alpha}}{G_{PR}(s)G_d(s)} & \dfrac{sL_{\beta\beta}+G_{PR}(s)G_d(s)}{G_{PR}(s)G_d(s)} \end{bmatrix}^{-1} \begin{bmatrix} i_{ref\alpha}(s) \\ i_{ref\beta}(s) \end{bmatrix} \quad (7)
$$

From the closed-loop transfer function matrix, it is clearly seen that the system is a two-input two-output system (MIMO), and the non-diagonal elements are nonzero, which imply that the MIMO system tools are required for the dynamic analysis and the controller design.

B. Proposed Decoupled SISO Transfer Function

In order to simplify the method of analyzing the system, the proposed modeling method with the decoupled transfer function is derived in this part.

According to (6), the relationship between the references and the outputs can be derived and rewritten as:

$$
\begin{bmatrix} i_\alpha(s) \\ i_\beta(s) \end{bmatrix} = \left([\mathbf{I}] + \begin{bmatrix} sL_{\alpha\alpha} & sL_{\alpha\beta} \\ sL_{\beta\alpha} & sL_{\beta\beta} \end{bmatrix}^{-1} \begin{bmatrix} G_{PR}(s)G_d(s) & 0 \\ 0 & G_{PR}(s)G_d(s) \end{bmatrix} \right)^{-1}
$$
$$
\cdot \begin{bmatrix} sL_{\alpha\alpha} & sL_{\alpha\beta} \\ sL_{\beta\alpha} & sL_{\beta\beta} \end{bmatrix}^{-1} \begin{bmatrix} i_{ref\,\alpha}(s) \\ i_{ref\,\beta}(s) \end{bmatrix}
\tag{8}
$$

Thus, the open-loop transfer function matrix of the MIMO system can be expressed as below:

$$
\mathbf{G(s)} = \mathbf{L^{-1}(s) \cdot G_c(s)} \tag{9}
$$

$$
\mathbf{G_C(s)} = \begin{bmatrix} G_{PR}(s)G_d(s) & 0 \\ 0 & G_{PR}(s)G_d(s) \end{bmatrix} \tag{10}
$$

$$
\mathbf{L^{-1}(s)} = \begin{bmatrix} sL_{\alpha\alpha} & sL_{\alpha\beta} \\ sL_{\beta\alpha} & sL_{\beta\beta} \end{bmatrix}^{-1} \tag{11}
$$

where $\mathbf{G_c(s)}$ is the transfer function matrix for the PR controller, $\mathbf{L^{-1}(s)}$ is the control plant matrix under unbalanced condition.

According to the generalized Nyquist stability criterion, the eigenvalues $\lambda_1(s)$, $\lambda_2(s)$ of $\mathbf{G(s)}$ can be used to analyze the stability of the system [12]. And if the Nyquist curves of $\lambda_1(s)$, $\lambda_2(s)$ do not have the encirclement of the critical point (-1, 0), the system is stable.

For the matrix $\mathbf{G(s)}$, the eigenvalue can be calculated as below:

$$
\det\left[\lambda \mathbf{I} - \mathbf{G(s)}\right] = 0 \tag{12}
$$

As the $\mathbf{G_c(s)}$ is a diagonal matrix, (12) can be simplified as:

$$
\det\left[\lambda \mathbf{I} - G_{PR}(s)G_d(s)\mathbf{L^{-1}(s)}\right] = 0 \tag{13}
$$

Solve the equation, the eigenvalues can be expressed as:

$$
\lambda_1(s) = G_{PR}(s)G_d(s)Y_1(s) \tag{14}
$$

$$
\lambda_2(s) = G_{PR}(s)G_d(s)Y_2(s) \tag{15}
$$

where $Y_1(s)$ and $Y_2(s)$ are the eigenvalues of $\mathbf{L^{-1}(s)}$.

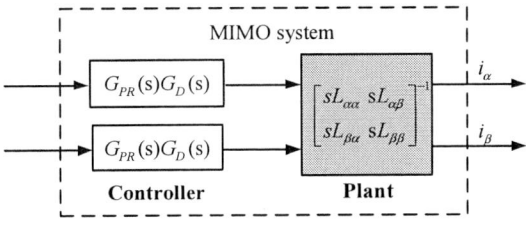

(a)

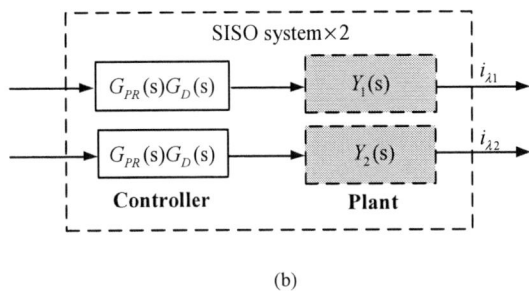

(b)

Fig. 5. Equivalent block diagram of the current control loop. (a) MIMO model. (b) Equivalent SISO models.

According to formula (13), (14), (15) and generalized Nyquist stability criterion, it can be seen that the stability of the system is decided by the eigenvalues of $\mathbf{L^{-1}(s)}$ and the parameters of current controller. As Fig. 5 illustrates, the MIMO control plant can be replaced by two decoupled SISO transfer functions, which are the eigenvalues of the control plant matrix.

Based on (14), (15), once the equivalent model is derived according to the eigenvalues of the matrix $\mathbf{L^{-1}(s)}$, the stability of the unbalanced system can be analyzed using SISO system tools.

C. Stability Analysis for current controller

As shown in Fig. 3 and formula (3), the control plant is the same as the phase parameter without couplings between α-axis and β-axis only when the system is balanced, in which case the root locus analysis can be used to analyze the stability and design the current controller. Otherwise when analyzing the unbalanced control plant, the eigenvalue of the control plant should be used, which also means there are two equivalent plants reflecting the stability of the system as (14) and (15).

In this work, the three-phase unbalance in the L-filter are assumed as $L_a=L_b=3$mH, $L_c=1$mH. Based on the proposed modeling method for the unbalanced three-phase grid converter, the equivalent control plants can be derived as (16) and (17).

$$
Y_1(s) = \frac{1}{0.003s} \tag{16}
$$

$$
Y_2(s) = \frac{1}{0.0017s} \tag{17}
$$

3166

The 2018 International Power Electronics Conference

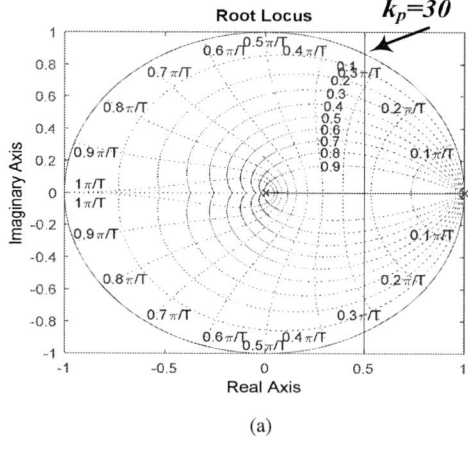

(a)

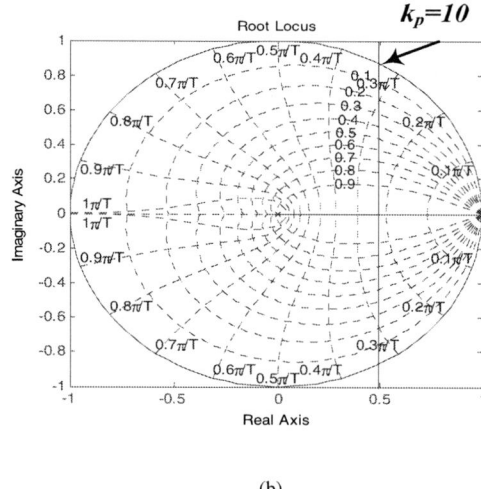

(b)

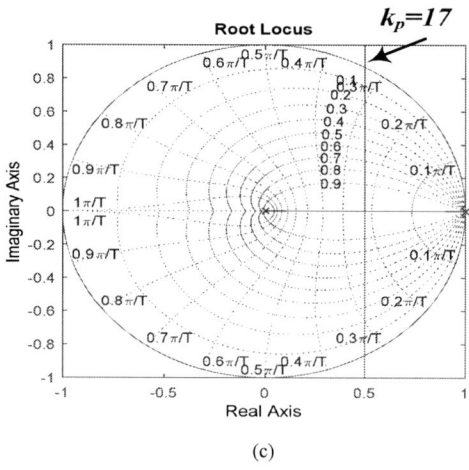

(c)

Fig. 6. Root loci analysis of current control with balanced and unbalanced L-filter. (a) Balanced condition with L=3mH. (b) Balanced Condition with L=1mH. (c) Unbalanced condition with $L_a=L_b=$3mH, $L_c=$1mH.

To analyze the system, the discrete z-domain root loci for the equivalent plants under the unbalanced condition

and the normal plant under the balanced condition are obtained by z-transformation with the Zero Order Hold (ZOH), of which the sampling period T_s is 0.1ms (f_s=10kHz). The time delay for the digital control is $1.5T_s$. Three loci are given in Fig. 6, which illustrate the stable region for both the balanced and unbalanced systems, respectively. Fig. 6(a) shows the boundary of stable region when the system is balanced with the control plant L=3mH. The stable region for the balanced system with the control plant L=1mH is given in Fig. 6(b). Two equivalent control plants for the unbalanced condition is derived as (16), (17), which is the eigenvalue of the matrix $\mathbf{L^{-1}(s)}$. And the critical one (17) is used to analyze the stability of the unbalanced system, of which the root locus is given in Fig. 6(c).

To validate the correctness of the z-domain root locus analysis, Nyquist diagrams of the equivalent control plants for the unbalanced system in continuous s-domain is also plotted in Fig. 7. It is clearly seen that the stable region is $k_p<17$ shown in Fig. 7(b), which matches the result of the root locus given by Fig. 6(c). The effectiveness and correctness of the z-domain model is validated by the Nyquist digrams.

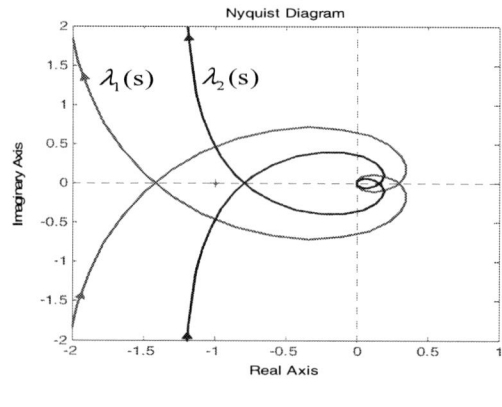

(a)

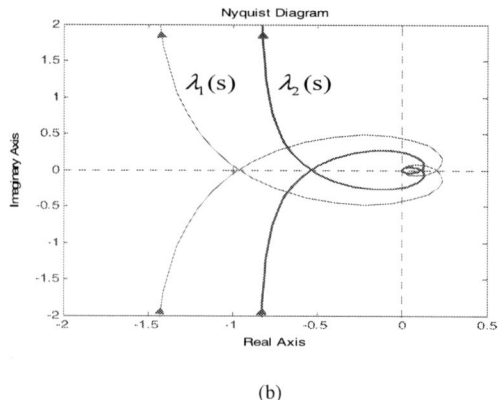

(b)

Fig. 7. Nyquist diagrams of eigenvalues for the unbalanced system with L_a=3mH, L_b=3mH, L_c=1mH. (a) k_p=25 unstable. (b) k_p=17 stable.

3167

Table I
Stable region analysis under unbalanced condition

Method	Control plant	Stable region
I	$L_a=L_b=L_c=1\text{mH}$	$K_p<10$
II	$L_a=L_b=L_c=3\text{mH}$	$K_p<30$
Proposed method	$L_{\gamma 1}=3\text{mH}, L_{\gamma 2}=1.7\text{mH}$	$K_p<17$

Table I concludes the different results when analyzing the stability region of the unbalanced system by using different control plants. Under the unbalanced condition with filter parameters as $L_a=3\text{mH}$, $L_b=3\text{mH}$, $L_c=1\text{mH}$, if the control plant is assumed as $L=1\text{mH}$, the predicted stable region will be $k_p<10$. On the other hand, if the control plant is assumed as $L=3\text{mH}$, the stable region will be $k_p<30$. However, according to the proposed method, the equivalent control plants should be the eigenvalue of the matrix $\mathbf{L^{-1}(s)}$, of which the stable region is kp<17.The proposed method can predict the stable region more accurately and effectivity.

IV. SIMULATION AND EXPERIMENTAL RESULTS

In order to verify the proposed modeling method, time domain simulations using the PSCAD/EMTDC are carried out. Experiments are accomplished in the lab using a three-phase frequency converter with an L-filter and the grid voltage is provided by a Chroma 61845 grid simulator. The controller is designed in the dSPACE DS1007. The simulation and experimental parameters of the system are shown in Table II. The same parameters are used in both cases. The unbalanced condition is built up by changing the inductance of L_c to 1mH.

TABLE II
SYSTEM PARAMETERS FOR SIMULATIONS AND EXPERIMENTS

Symbol	Meaning	Value
f_0	Fundamental frequency	50Hz
f_{sw}	Switching frequency	10kHz
f_s	Sampling frequency	10kHz
V_{dc}	DC link voltage	700Vdc
V_g	Grid voltage (RMS L-L)	230Vac
L_a	Phase a inductance	3mH
L_b	Phase b inductance	3mH
L_c	Phase c inductance	3mH

Fig. 8 shows the simulated waveforms. The results with the balanced L-filter is given in Fig. 8(a). It can be seen that at the time instant of 0.3s the loop gain k_p changes from 30 to 32, the system get into the unstable region under the balanced condition, which matches the root locus analysis in Fig. 6(a). In contrast, with the unbalanced α, when the loop gain k_p changes from 16 to 18 at the time constant of 0.3s, the system becomes unstable as shown in Fig. 8(b), which agrees with the root locus analysis in Fig. 6(c).

Fig. 9 shows the measured waveforms obtained in the experimental test. The PCC voltage and the converter current are measured respectively. The system is stable when the loop gain k_p is 30 under balanced conditions and the system is unstable when the loop gain k_p changes from 30 to 33 as Fig. 9(a) shows, which closely correlate with the simulation results in Fig. 8(a). Fig. 9(b) illustrates the

experimental result of the unbalanced condition with the filter parameters as $L_a=3\text{mH}$, $L_b=3\text{mH}$, $L_c=1\text{mH}$. The system is stable with the loop gain $k_p=17$, yet, the system becomes unstable when the loop gain k_p changes from 17 to 19. The result matches with the time domain simulation shown by Fig. 8(b). Simulation and experimental results both validate the effectiveness of the proposed method.

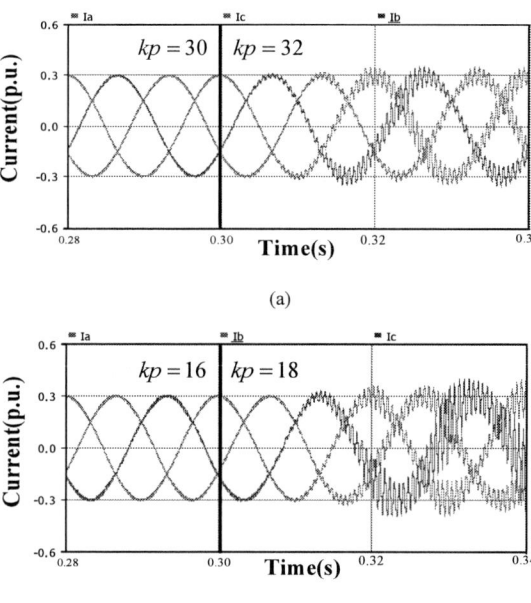

(a)

(b)

Fig. 8. Simulation results. (a) Change loop gain k_p at t=0.3s under balanced condition with L=3mH. (b) Change loop gain k_p at t=0.3s under unbalanced conditions with $L_a=L_b$=3mH, L_c=1mH.

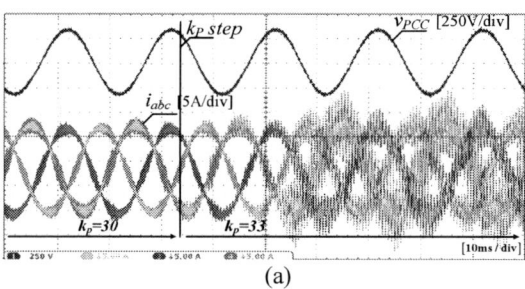

(a)

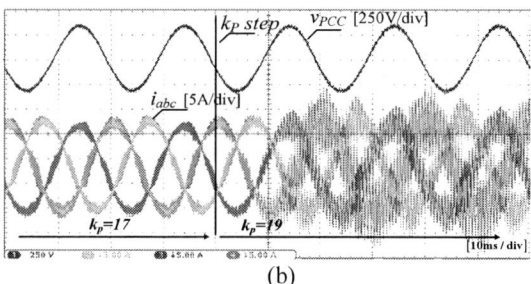

(b)

Fig. 9. Experimental waveforms with balanced inductance and unbalanced inductances. (a) Balanced condition with L=3mH. (b) Unbalanced condition with $L_a=L_b$=3mH, L_c=1mH.

V. CONCLUSIONS

This paper has discussed a modeling method with decoupled transfer functions for the stability analysis and controller design of unbalanced three-phase converters. With the method, the MIMO system with cross couplings between two axes in the $\alpha\beta$-frame can be transformed as two SISO transfer functions. Simulations and experiments confirm the effectiveness of the method.

REFERENCES

[1] F. Blaabjerg, R. Teodorescu, M. Liserre, and A. V. Timbus, "Overview of control and grid synchronization for distributed power generation systems," *IEEE Trans. Ind. Electron.*, vol. 53, no. 5, pp. 1398-1409, Oct. 2006.

[2] A. Lesnikar, R. Marquardt, J. Hildinger, "Modulares Stromrichter Concept für Netzkupplungsanwendung bei hohen Spannungen," ETG, 2002.

[3] F. Peng and J. Wang, "A universal STATCOM with delta-connected cascade multilevel inverter," *Proceedings of the IEEE 35th Annual Power Electronics Specialists Conference* vol. 5, 2004, pp. 3529–3533.

[4] E. Twining and D. G. Holmes, "Grid current regulation of a three-phase voltage source inverter with an LCL input filter," *IEEE Trans. Power Electron.*, vol. 18, no. 3, pp. 888–895, May 2003.

[5] R. Teodorescu, M. Liserre, and P. Rodr´ıguez, *Grid Converters for Photovoltaic and Wind Power Systems*. Hoboken, NJ, USA: John Wiley & Sons, 2011.

[6] Lie Xu, Andersen, B. R. and Cartwright, P., "VSC Transmission Operating under Unbalanced AC Conditions–Analysis and Control Design," *IEEE Transactions on Power Delivery*, vol. 20, no. 1, pp. 427-434, January 2005.

[7] Rodriguez, P., Timbus, A., Teodorescu, R., Liserre,M. and Blaabjerg, F., "Reactive Power Control for Improving Wind Turbine System Behavior Under Grid Faults," *IEEE Transactions on Power Electronics*, vol.24, no. 7, pp. 1898-1810, July 2009.

[8] Mauricio Cespedes, Jian Sun, "Methods for Stability Analysis of Unbalanced Three-Phase Systems," *Proceedings of IEEE Energy Conversion Congress and Exposition (ECCE)*, 2012, pp. 3090-3097.

[9] Bo Liu, Xiaojie Shi , Yalong Li , Fei Fred Wang , Leon M. Tolbert. "A Line Impedance Conditioner for Saturation Mitigation of Zigzag Transformer in Hybrid AC/DC Transmission System Considering Line Unbalances," *IEEE Trans. Power Electron.*, vol. 32, no.7, pp. 5070-5086, 2017.

[10] L. Harnefors, "Modeling of three-phase dynamic systems using complex transfer functions and transfer matrices," *IEEE Trans. Ind. Electron.*, vol. 54, no. 4, pp. 2239–2248, Aug. 2007.

[11] S. Buso and P. Mattavelli, *Digital Control in Power Electronics*. Morgen & Claypool Publishers, 2006.

[12] Mohamed Belkhayat. "Stability Criteria for AC Power System with Regulated loads," Ph.D. Dissertation, Purdue University, USA, Dec. 1997.

Predicting Voltage Characteristic of Charging Model for Li-Ion Battery with ANN for Real Time Diagnosis

Minella BEZHA* and Naoto NAGAOKA

Dept. of Electrical Engineering, Doshisha University, Kyoto, Japan
*E-mail: minella.bezha@yahoo.com

Abstract— An adaptive characteristic of charging for the rechargeable batteries using the artificial neural network (ANN) method is proposed in this study. This model is based on the voltage charging data of a Li-Ion battery. By the voltage characteristic of charging data that have been used as a parameter to describe the actual quantity of energy, which is a key factor in applications. This estimation is an important and challenging task. The upcoming Electric Vehicle (EV) or Hybrid Electric Vehicle (HEV), are becoming the most important technology in transportation, because of the Eco-friendly and its increasing driving autonomy. The battery performance directly influences the total performance and efficiency of the BMS for this kind of vehicles. As already confirmed the importance of the battery state of charge (SOC) prediction and the nonlinear characteristic between the battery SOC and the external variables, the neural network model is proposed in order to investigate further. In this approach, the ANN can predict the characteristic of the charging model from the batteries, with the optimized model it can be simulated within a short time and with a high accuracy. Which is a different type of approach to the difficult task of SOC of the battery.

Keywords— *Artificial neural network (ANN), Lithium-Ion (Li-Ion) battery, state of charge (SOC) estimation, real time estimation.*

I. INTRODUCTION

The usage and the importance rechargeable batteries nowadays are increasing in the portable electronic devices, solar energy industries and the development of the electric vehicles (EV). The improvement of the energy storages system as well as interest to further advance of the battery technology is one of the main purposes of the researchers of this field. The rechargeable batteries have a crucial role in the storage system, especially in the mobile equipment, because the period of its usage and the flexibility of the function are determined by the battery. Despite this, one of the main challenges is the estimation of state of charge (SOC), which expresses the amount of energy available in the battery. Many models have been developed for the estimation of the SOC, and some have good points and bad points. The typical methods to estimate the SOC can be summarized in two groups. The first is based on mathematical relation between different variables and the SOC. These battery variables can be the open circuit voltage (OCV), charging current, internal resistance or AC internal impedance [2,5]. The second approach is based on measuring the energy such as coulomb counting (CC), often known with as current integration. Kalman filter (KF), which is based on the time series approach, mainly based on the step process estimation and the linear quadratic estimation (LQE). Every method, focusing in the estimation of the correct SOC, have positive and negative points, also most of them need an equivalent electrical circuit for the battery model. Although each proposed circuit could explain the behavior of the battery, the accuracy is very low if the model is simple. The accuracy of the complex model may be high, but it will be very difficult to build because the variables in the model are nonlinearly included. Another problem is lacking generality, which lead to wrong SOC estimation in certain battery types. The third method is based on the Artificial Intelligence (AI) by Neural Network (NN) or other machine learning techniques. This paper employs the third method, and proposes an approach, in order to be applicable for commercial applications.

The battery SOC estimation and real time diagnosis is a complex nonlinear system. Although the SOC estimation is essential technology, its accuracy and reliability should be improved still until now.

SOC and deterioration prediction is closely related with the internal characteristics of battery, which could not be directly measured during operation. However, a time series estimation by an appropriate algorithm with a measurement of the indirect parameters of the battery enables the SOC and deterioration estimation. During the operation of EV, the charging process is crucial, because its quality affect the battery state of health (SOH), which means the autonomies in mileage [6]. The charging process can be done during running, but if it's not controlled in appropriate matter, it's easy to damage the battery [3], so the use of motor controllers for achieving a closed loop management in order to charge the batteries at constant current.

The artificial neural network (ANN) does not rely on the mathematical equation or equivalent circuit model to explain the relation between the input variables and output product for resolving the problem. It is because that the ANN characteristics are based on the black box approach, which does not require any knowledge about the system's internal dynamics. In, this paper, a creation of an ANN model based on the voltage at the charging process will be discussed first. In the second, the accuracy of this model will be investigated in order to adapt to this environment of data without any equivalent circuit model. In the final part, this study gives a conclusion about the techniques.

II. OPTIMIZING THE SIMULATED MODEL

A. AI with ANN predicting

Artificial Neural Network is an electronic model based on the imitation of the neural structure of the brain. From this similarity of the learning through experience, NN takes a different approach to solve problems than that of the conventional computers. This approach does not require any knowledge about the system's internal dynamics

Different from the aforementioned methods, the application of the ANN to the estimation of the SOC under the variable of the voltage during charging, it provides a way to deal with the above difficulties. It should be mentioned that ANN does not rely on the explicitly expressed relationship between input variables and the SOC. The relationship between the input variables and the SOC is formed through training. The second feature is the adaptive algorithm, which is an attractive thing of the ANN. Machine Learning (ML) methods are categorized in three methods: Basic ML, Modified ML and Hybrid ML. Each of these methods has their way of application and sub-types of model.

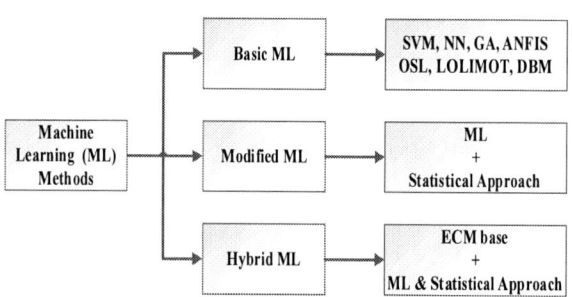

Fig.1. ML methods and usual implementations

B. Training data with the Neural Network

ANN provides algorithms for learning, classification, and optimization. The main ability of this model is defined by the learning process, which is not limited even for nonlinear dynamic system mappings. One of the major advantages of the NN is their ability to generalize. This means that a trained net can make a classification data from the same class as the learning data that it has never seen before. In real applications, developers normally have a small quantity of the possible patterns for the generation of the neural network. In order to achieve the best generalization, the dataset should be split into three parts:

- Training set, used to train the NN. The error is minimized during the training.
- Validation set, used to specify the performance of a NN on patterns that are not used during the learning process.
- Test set, used for the final check during over all the performance of the NN.

It must be mentioned that the learning process should be stopped in the minimum of the validation error. When the learning is not stopped, the overtraining effect occurs and the performance of the NN on all data decreases, despite the fact that the error on the training data is decreasing. In order to check the quality of the estimation, the NN must check the error with third data set, known as the test data after the learning phase.

In the BPNN (back-propagation neural network) model, learning takes place during the propagation of input patterns from the input nodes to the output nodes. To optimize it, the correct values of bias, weights, number of input, hidden and output layers must be found. After obtaining the output values, they are compared with the desired target values. In this moment, the weights are adapted in order to minimize the difference produced between these two variables. The equation of output:

$$Q_i^{(l)} = f_s\left(I_i^{(l)}\right) \tag{1}$$

$$I_i^{(l)} = \sum W_{ij}^{(l)} Q_j^{(l-1)} \tag{2}$$

The function f_s in (1) expresses the response of a neuron.

$$f_s(I) = 1/\left[1 + \exp(-I)\right] \tag{3}$$

The initial values of weights are assumed to be chosen randomly, and the weight between the ith neuron of the $(k-1)^{th}$ layer and the ith neuron of the kth layer is defined as $W_{ij,k}$. The weight adaptation equation is given as below:

$$W_{ij,k}(t_n) = W_{ij,k}(t_{n-1}) - \frac{\alpha E(t_n)}{W_{ij,k}(t_{n-1})} \Delta W_{ij,k}(t_{n-1}) \tag{4}$$

$$E = \frac{1}{2}\sum_{i=1}^{n}(y_i - b_i)^2$$

where $0<\alpha<1$, y_i is i^{th} actual output, and b_i is i^{th} simulated output.

It is not always suitable to increase the number of layers or neurons, because the system can go unstable and will get high error and poor optimization. In addition, the learning rates must be carefully selected as well.

III. SIMULATION AND PREDICTION OF THE NETWORK

A. Creating the model from the experimental data

After the network has been composed for a specific task or application, it is ready to be trained. In the beginning of this process, the initial values of the weights are chosen randomly. The typical patterns or approaches of training are based on the supervised and unsupervised training. The supervised training is based on the learning process from a teaching signal served as a target one. Unsupervised training stands for the network when has to make sense of the inputs without the outside help. It means the outputs are unknown. Unsupervised training often is used to obtain some initial characterization on inputs. In this paper, MATLAB software was used. The training of the network will start by using the supervised model of learning. This training was obtained through the data from the charging process of ICR18650PD Li-Ion battery by using Li-Ion battery test-system As-510-LB4 as shown in Fig.2.

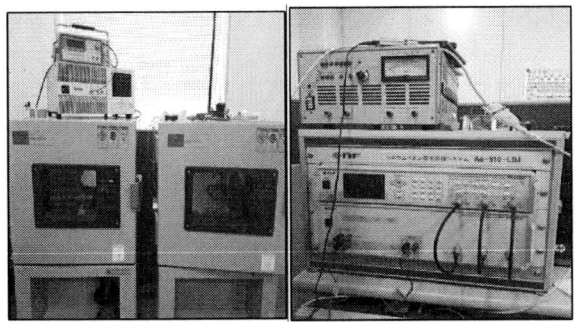

Fig2. Devices used for the retrieving of voltage data.

The charging mode is built by 2635 samples, when every sample is taken every 2 seconds. The temperature of the room is maintained constant at 25°C. All the learning processing time is 1.47 hour. The range of the voltage is from 3.1907 V to 4.199 V. The characteristic is shown in Fig.3. These data will be used for the training and as a target later for the simulation of the network.

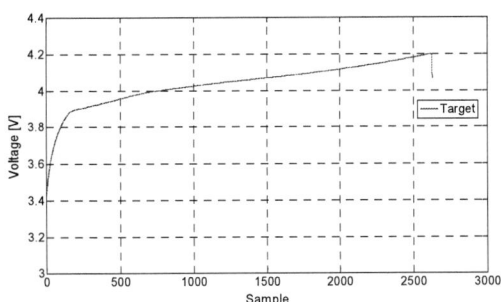

Fig3. Voltage data obtained from a charging process of ICR18650PD

For different type of battery, there are specific equations for charging or discharging equations model. The open voltage source is calculated with a nonlinear equation based on the actual SOC of the battery as [7,8]:

$$V_b = E_o - R \cdot i \qquad (5)$$

During discharge:

$$E_o = E_k - K \frac{Q}{Q - \int i \cdot t} i_t - K \frac{Q}{Q - \int i \cdot t} i^* + A e^{(-B \cdot i \cdot t)} \qquad (6)$$

During charge:

$$E_o = E_k - K \frac{Q}{Q - \int i \cdot t} i_t - K \frac{Q}{\int i \cdot t - 0.1Q} i^* + A e^{(-B \cdot i \cdot t)} \qquad (7)$$

Where E_o is the open circuit voltage, E_k is the battery constant voltage, K is the polarization constant, Q is the battery capacity, i_t is the actual battery current, i^* is the low frequency current dynamics, A is the exponential zone amplitude, B is the inverse of time constant $(Ah)^{-1}$, V_b is the battery voltage, and i is the battery current.

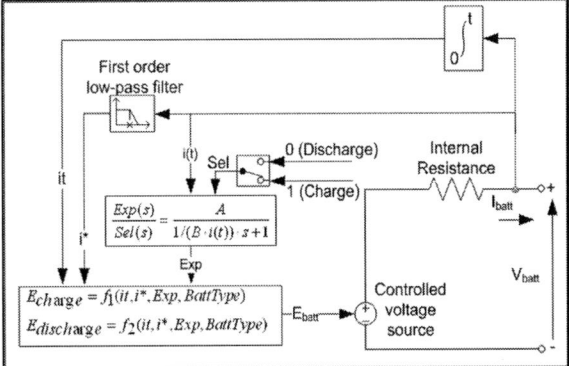

Fig.4. Block diagram of the battery equivalent circuit.

The battery block implements a generic dynamic model parameterized to represent most popular rechargeable batteries. This model will be based in (7), which explain the charging model of the Li-Ion battery.

Fig.4 shows the block diagram of the battery equivalent circuit used in the MATLAB software simulation. In the beginning, the network was started with 15 neurons in the hidden layer and the input layer was composing by a vector with 2635 elements.

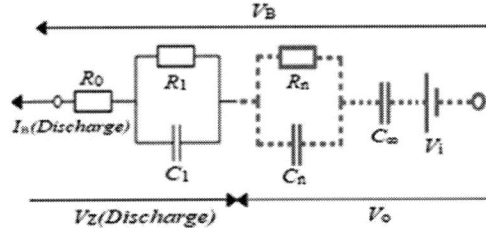

Fig.5. Equivalent circuit of Li-ion battery.

Fig.5 shows the equivalent circuit, which can express the electrical composition of the Li-Ion cell. The parameters related with the capacity fade and the SOC can be defined, and the internal impedance in frequency domain can be calculated using the equivalent circuit [9].

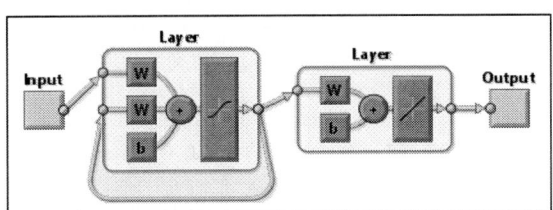

Fig.6. The GDPBNN with adaptive learning rate model

This simulation has been carried out by a personal computer, EPSON ENDEAVOR Model MR4700E with the specs shown in TABLE I.

TABLE I

CPU	i7 6700 3.40 GHz / L3 Cache 8MB
RAM	8GB DDR4 / 2133 MHz
HDD	1 TB

Fig.7 shows the achieved simulated data after few trainings of the network, but the accuracy is not good yet. Even if the waveform of the predicted data is similar, the value is far from the desired target.

Fig.8 shows the changes of the outputs taken the progress of the learning as the parameter. The simulated plots are starting to get near the target value but the still the response is not good. However, it is necessary to tune and calibrate the learning rate and to optimize the error of the output in order to change the bias and weight for the adaptive network. After obtained this characteristic, was realized that the system need to reset the minimum and maximum value, so after doing this thing a new type of model was created and simulated.

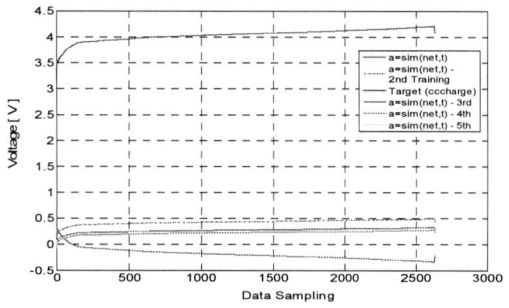

Fig.7. Plot of Target and Simulated data

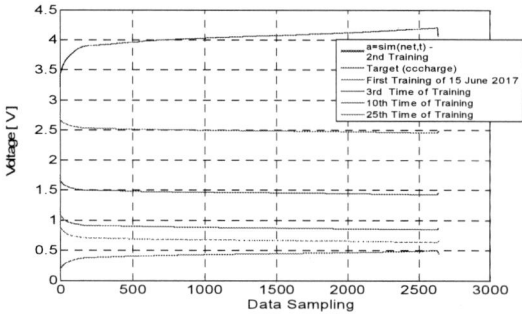

Fig.8. Changed model of NN with simulated data

IV. OPTIMIZING THE SIMULATED MODEL

A. Times Series Estimation

For a complete mathematical model describing a studied phenomenon when is known and not so difficult, forecasting it could be an easy task. However, when an analytical model is unknown, too complex or incomplete, a typical alternative is to try to forecast by building a model that takes into account only previous outcomes of the phenomenon while ignoring any exterior influence. More formally, a time series $\{x(t)\}$ can be defined as a function x of an independent variable t, stemming from a process for which a mathematical description is unknown. A time series is a sequence of vectors, $x(t)$, $t = 0, 1,...n$, where t represents elapsed time. A sequence of scalars will be considered for a simplification in this paper. However, the techniques can be easily extended to a vector series. The series x can be a set of values which continuously vary with t, such as voltage. In practice, for any given physical system, the variable x will be sampled to give a series of discrete data points, equally spaced in time. The rate at which samples are taken dictates the maximum resolution of the model: but, it is not always the situation when the model with the highest resolution has the best predictive power, so that superior results may be obtained by using only every n-th point in the series.

One of the main goals for the neural networks stands to the estimation of future sample $x(t+1)$ of the time-series x up to the actual time $x(t-n) - x(t)$. Formally this can be stated as: find the function $f: R^N \rightarrow R$ such as to obtain an estimate of x at time $t+d$, from the N steps back from time t, so that:

$$x(t+1) = f\left(x(t), x(t-1), \cdots x(t-N+1)\right)$$
$$x(t+1) = f\left(y(t)\right)$$

(8)

where $y(t)$ is the N vector consisting of lagged x values. The function f estimates the next value of x.

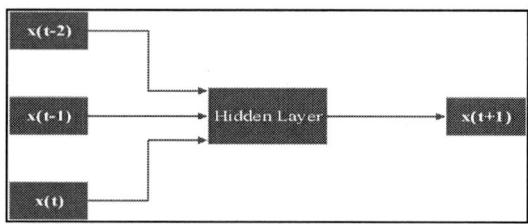

Fig.9. Diagram of the time series processing function

B. Reformulated model of adaptive BFPNN Times Series Estimation

After many training and define the values of bias and weights it was possible to obtain the results as shown below. Fig.10 shows the new simulated data which have been obtained after changed the number of neurons in the hidden layer and changed the value of bias and weights. The updated NN model is with 35 neurons. Now the model is modified with changes in the learning rate, number of epochs and parameter goal.

net.trainParam.lr=0.3; *net.trainParam.epochs*=1500; *net.trainParam.goal*=1×10^{-12}.

Fig.10 shows the state of the trained network after the 1st and 9th learning. The difference in the simulated result is optimized fast and with good accuracy as shown in the Fig.11 and 12. These Figures show the zoomed part in the beginning and in the end of the charging process, by comparing the estimated after every training step.

3173

The 2018 International Power Electronics Conference

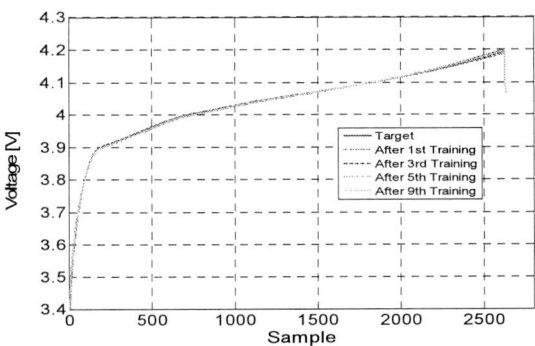

Fig.10. Plots of Target and Simulated data

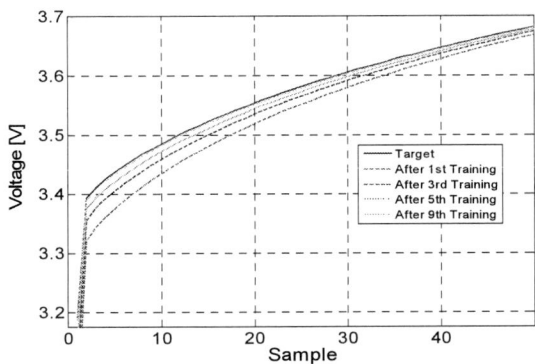

Fig.11. Starting zone of the charging plot

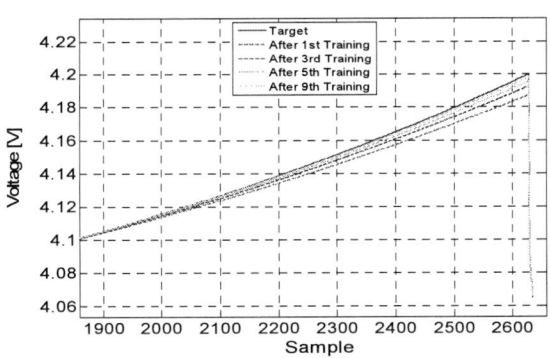

Fig.12. Ending zone of charging plot

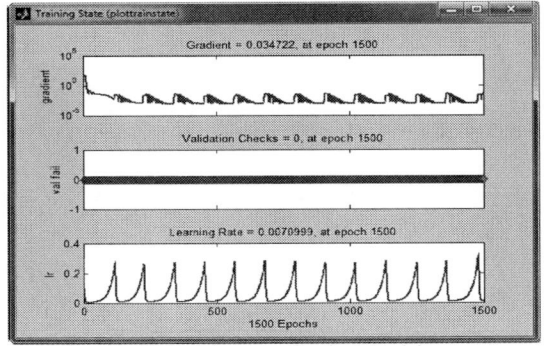

Fig.13. Training state after the 1st training

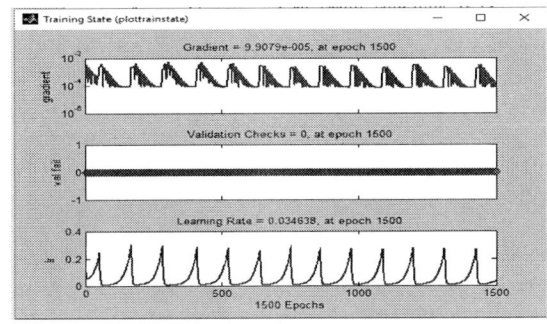

Fig.14. Plots of training state after the 9th

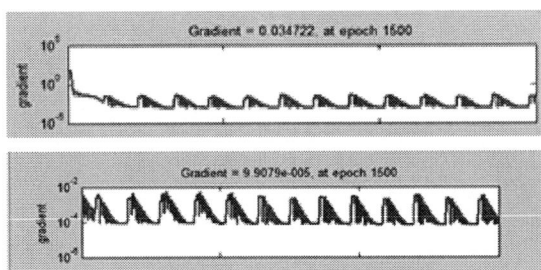

Fig.15. Plots of training state after the 1st and 9th

Fig.15. shows the differences of the gradient between the 1st training and the 9th training. In the second figure, the gradient "9.9079e-005" shows the accuracy is getting very high.

TABLE II

	Min Voltage V]	Max Voltage [V]
Target	**3.1907**	**4.1999**
After 1st Training	3.0603	4.1873
After 3rd Training	3.1519	4.1915
After 5th Training	3.1588	4.1956
After 7thTraining	3.1770	4.1972
After 9thTraining	3.1864	4.198

Table II shows the comparison of the estimated values of the voltage for each estimation after the ith training of the NN. The minimum and maximum values are shown in order to make clear the differences and to observe the stability of the model.

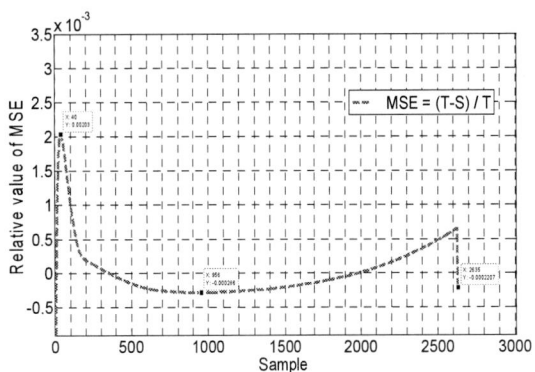

Fig.16. Relative error of the estimated output

3174

Fig.16. shows the relative error calculated for every samples estimated. This plot can be divided in 3 zones, first one regarding to the first peak obtained in the 40th epochs, which the error is increasing because of the learning process and the dynamic changes. The second zone of this plot is the less dynamic characteristic, which continue around the 2000th epochs, when the third zone start to increase the error. By the plot of the relative error with the charging plot, these peaks error accumulation or increases comes because of the fact of the CCCV charging process, by the composition of two processes, when first stand for constant current (CC) and the second for the constant voltage (CV).

V. CONCLUSIONS

In this paper, a voltage charging prediction method was developed using a GD-BFPNN model from the voltage characteristic of charging model of the Li-Ion battery. The calculation is carried out the developed simulation code on MATLAB. In the first part of this study, the experimental data was entered into a typical BP network and the settings of the network were modified to get a high prediction accuracy. The adaptive GD-BFPNN model enables to predict and learn fast. The learning process was finished less than, 1 minute for 2635 cells of the input data. The charging voltage of a rechargeable battery such as Li-Ion can be accurately predicted. It can remove the nonlinear and complex relations between the input variables used in the real-time battery diagnosis for the industrial applications.

The future work is to achieve a good result from the charging process of the battery and to adapt the network to learn also from the discharging process in order to achieve a better understanding of the SOC&SOH estimation and real time estimating diagnosis of battery.

REFERENCES

[1] G. Eason, B. Noble, and I. N. Sneddon, "On certain integrals of Lipschitz-Hankel type involving products of Bessel functions," *IEEE Trans. on Power Electronics*, vol. 247, no. 8, pp. 529-551, 1995.

[2] E. Jensi Miriam, S. Sekar, S. Ambalavanan, "Artificial neural Network Technique for predicting the Lifetime and performance of Lead-Acid Battery", *International Journal (ESTIJ)*, Vol.3, No.2, April 2013.

[3] N. Nagaoka, "A numerical model of Lithium-ion battery for a life estimation," *48th International University Power Engineering Conference*, Dublin, Sep. 2013

[4] S. Andrew Gadsden and Saeid R. Habibi, "Based Fault Detection of a Battery System in a Hybrid Electric Vehicle," *Journal of Energy and Power Engineering 7*, pp. 1344-1351, 2013

[5] M. Negnevitsky, "Artificial Intelligence: A guide to intelligent systems," Third Edition, 2011

[6] T. Hirai, A. Ohnishi, N. Nagaoka, N. Mori, A. Ametani, and S. Umeda, "Automatic Equivalent-Circuit Estimation System for Lithium-Ion Battery," *43rd International University Power Engineering Conference*, Padova, Sep. 2008

[7] O. Tremblay, L-A. Dessaint, A.-I Dekkiche, "A Generic Battery Model for the Dynamic simulation of Hybrid Electric Vehicles," *IEEE Vehicle Power and Propulsion Conference*, pp. 284-289, 2007

[8] E.M.Natsheh, A.Albarbar, "Hybrid Power Systems Energy Controller Based on Neural Network and Fuzzy Logic," *Smart Grid and Renewable Energy*, Vol.4, pp. 187-197, 2013

[9] Dorffner, G: Neural networks for time series processing, Neural Network World 4/96, 1996

[10] T.Ishii and N.Nagaoka, "A logarithmic segmented Laplace transform and its application to a battery diagnosis" *UPEC 52nd International Universities Power Engineering Conference*, pp. 949-953, 2017

Impedance Modeling and Stability Analysis of the Cascaded Three-phase Symmetric Systems Using Complex Transfer Functions

Teng Liu*, Zeng Liu, Jinjun Liu, Yiming Tu, Zipeng Liu

State Key Laboratory of Electrical Insulation and Power Equipment, Xi'an Jiaotong University
Xi'an, China
*E-mail: teng.liu@stu.xjtu.edu.cn

Abstract—Impedance-based stability analysis method has been widely adopted for addressing the small-signal stability issues of power electronics-based systems. For the three-phase ac systems, the impedance models (IMs) are generally obtained in the rotating *dq*-frame and represented by the 2 × 2 transfer matrices, with which the system stability can be predicted through the Generalized Nyquist Criterion (GNC). However, the computational process of GNC as well as the derivation of the impedance matrices aggravates the complexity of stability analysis. This paper presents a simpler impedance modeling process for the three-phase symmetric systems. The IMs can be directly established by the complex transfer functions (CTFs) in the *dq*-frame, while keeping the equivalent relationship with the original impedance matrices. Consequently, the original multiple-input multiple-output (MIMO) system is simplified to a single-input single-output (SISO) system without losing any accuracy. Further, it is also proved that the Nyquist criterion originally used for dc systems can replace the GNC to predict stability of the cascaded three-phase symmetric system, which thus avoids the massive matrix calculations and leads to the remarkable simplification of the stability analysis. Case study of a three-phase grid-connected voltage source inverter system consisting of the frequency-domain analyses and experimental tests is presented. The obtained results confirm the correctness of the theoretical analyses.

Keywords—complex transfer function, impedance model, stability analysis, symmetric system.

I. INTRODUCTION

With the fast development of the smart grids and the distributed power generation systems (DPGSs), more and more power electronics-based power converters are put into use [1]. Compared with the traditional synchronous generators, the power converters usually have the wider control loop bandwidth leading to the complex dynamic characteristics. Thus, the dynamic interactions among the power converters and the grid tend to cause oscillations in a wide frequency range or even threaten safe operation of the system [2]-[4].

To analyze the unstable phenomena appeared in the multiple power converters system, many research works

---

This work was supported by the National Natural Science Foundation of China under Grant 51437007.

have been conducted in recent decades, among which the impedance-based stability analysis method has drawn the most attention [5]-[9]. The impedance-based method was firstly proposed by Middlebrook to solve the interactions between the input filter and the cascaded buck converter. By dividing the whole system into one source and one load subsystems, the system stability can be predicted by applying the Nyquist criterion to the ratio between the source output impedance and the load input impedance [5]. As for dc system, which is generally described by a single-input single-output (SISO) system, the terminal characteristics of the subsystems can thus be represented by only a single impedance making the stability analysis simple and convenient [6]-[7]. Besides, the impedance can be easily acquired either by the theoretical modeling or practical measurement [8]. Thanks to these merits, the relatively complete impedance-based stability theory has been established for dc multiple converters system.

However, situation becomes more complicated when it comes to the three-phase ac systems. Firstly, impedance modeling gets much complicated due to the periodically time-varying operation trajectories [9]. At present, the most widely used approach to develop impedance models (IMs) of the three-phase system is to transform the model into the synchronous reference frame (SRF or *dq*-frame) [10]-[13], where a three-phase ac system is described as a MIMO system. As a result, the IMs can be represented by the 2 × 2 transfer matrices. Each transfer matrix contains four individual terminal impedances. After obtaining the impedance transfer matrices in the *dq*-frame, the system stability should be predicted by the generalized Nyquist criterion (GNC), where the return ratio matrix $\mathbf{L}(s)$ defined by the product of the source impedance matrix and the load admittance matrix is calculated first. Then, the eigenvalues of $\mathbf{L}(s)$ should also be computed to plot their trajectories on the complex plane [11]-[13]. It is thus clear that applying the GNC for the stability analysis is relatively complicated. Besides, the eigenvalues of $\mathbf{L}(s)$ cannot intuitively reflect the participation factors of the power stage and controllers regarding to the system stability, which causes the inconvenience of the stability design. To sum up, the above limitations to some extent impede the further development of stability theory for the three-phase ac system.

To make the impedance-based stability analysis of the cascaded three-phase system simpler, this paper proposes a novel modeling process, where the IMs are established by the complex transfer functions (CTFs) in the dq-frame. It should be mentioned that only the cascaded three-phase symmetric system is investigated in this paper. As for the three-phase asymmetric systems, the relevant researches will be presented in the future publications. By adopting complex space vectors and CTFs to describe the variables and the transfer matrices in the dq-frame respectively, the original MIMO system can be equivalently represented by only one SISO system, where the IMs can be derived conveniently without applying matrix calculation. As the obtained IMs are described by the one-dimensional CTFs, the traditional Nyquist criterion has the possibility to replace the GNC for predicting the system stability, which will be theoretically proved. Consequently, the impedance-based method using the CTFs will remarkably simplify the stability analysis of the cascaded three-phase symmetric system. Further, a case study of a three-phase grid-connected voltage source inverter (VSI) system aiming at demonstrating the impedance modeling and the relevant stability prediction through using the CTFs is presented. Finally, both the frequency-domain and the experimental results verify the correctness of the above theoretical analyses.

II. BRIEF REVIEW OF CTFs

In recent years, the CTFs are gaining wider use as it can represent the three-phase symmetric systems in an elegant and simple form, which thus bring convenience to theoretical analysis. The complex vector representation was first formally extended to the control system tools in [14], where the authors have related the classical control theory with the CTFs. After that, a tutorial review of modeling and analysis based on CTFs for the three-phase systems was presented in [15]. Herein, a brief review of the CTFs is given to make the proposed modeling process and stability analysis more acceptable.

For illustration, a two-input two-output linear system is taken as an example. Based on the classical control theory, such system can be represented by

$$\begin{bmatrix} y_1(s) \\ y_2(s) \end{bmatrix} = \begin{bmatrix} G_{11}(s) & -G_{12}(s) \\ G_{21}(s) & G_{22}(s) \end{bmatrix} \cdot \begin{bmatrix} x_1(s) \\ x_2(s) \end{bmatrix} \tag{1}$$

where $x_1(s)$ and $x_2(s)$ are the system input signals, $y_1(s)$ and $y_2(s)$ are the system output signals, $G(s)$ represents the system transfer function. When $G_{11}(s) = G_{22}(s) = G_1(s)$ and $G_{12}(s) = G_{21}(s) = G_2(s)$ are satisfied, the system will be symmetric [14]. As a result, (1) can be simplified as

$$\begin{bmatrix} y_1(s) \\ y_2(s) \end{bmatrix} = \begin{bmatrix} G_1(s) & -G_2(s) \\ G_2(s) & G_1(s) \end{bmatrix} \cdot \begin{bmatrix} x_1(s) \\ x_2(s) \end{bmatrix} \tag{2}$$

When the input and output signals are represented by the complex space vectors, which are given as

$$\vec{x}(s) = x_1(s) + j \cdot x_2(s) \tag{3}$$

$$\vec{y}(s) = y_1(s) + j \cdot y_2(s) \tag{4}$$

this symmetric system can thus be equivalently expressed as

$$\vec{y}(s) = [G_1(s) + j \cdot G_2(s)] \cdot \vec{x}(s) \triangleq \vec{G}(s) \cdot \vec{x}(s) \tag{5}$$

where

$$\vec{G}(s) = G_1(s) + j \cdot G_2(s) \tag{6}$$

In (6), $\vec{G}(s)$ is exactly the so-called CTF. Different from real-coefficient transfer functions, the CTFs contain the complex coefficient, which will cause the represented system having different magnitude and phase responses for positive and negative frequencies.

Based on (2) and (5), it is found that a symmetric MIMO system can be equivalently simplified to a SISO system represented by the CTFs and complex space vectors. As a result, the classical control theory applied for SISO system can be adopted for the controller design and stability prediction [15], which makes the analysis process more convenient and elegant for avoiding the use of the multivariable control theory. However, the existing researches have not applied CTFs to the impedance-based stability analysis of the cascaded three-phase ac systems. Meanwhile, the relationship between the CTFs based Nyquist criterion and the GNC has not been revealed yet. To fill in this gap, this paper presents the impedance modeling based on the CTFs and the relevant stability analysis through a case study of a three-phase grid-connected VSI system, which will be discussed in the following parts.

III. CASE STUDY OF A THREE-PHASE GRID-CONNECTED VSI SYSTEM

A. System Structure

The system structure of a three-phase grid-connected VSI is shown in Fig. 1. The dc-link voltage V_{dc} is assumed to be constant for simplicity. v_o is the output voltage of the inverter. L represents the output filter, and a paralleled LC-type grid impedance is considered. v_{pcc} is the point of common coupling (PCC) voltage, whose phase angle θ is acquired by the synchronous reference frame phase-locked loop (SRF-PLL) for synchronizing the VSI with the grid to achieve the unity power factor control. It should be noticed that the bandwidth of SRF-PLL is always designed much slower than that of the inner current loop, which means the effect of SRF-PLL on the high frequency characteristics of the system can be neglected [16]. v_s represents the grid voltage. The main parameters of the system studied in this paper are listed in Table I, where two different capacitances are adopted to intentionally obtain both the stable and unstable cases.

In most cases, the current controller is implemented in the dq-frame, whose block diagram is shown in Fig. 2. In the figure, $G_c(s)$ represents the current regulator which is

3177

The 2018 International Power Electronics Conference

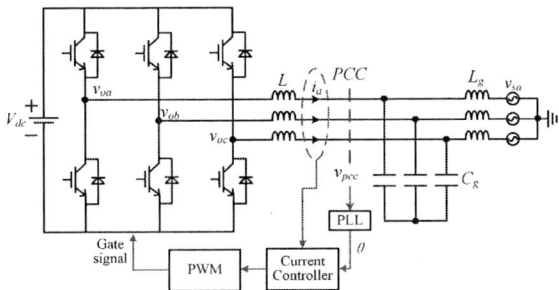

Fig. 1. System structure of a three-phase grid-connected VSI.

TABLE I
MAIN CIRCUIT PARAMETERS

Item	Symbol	Value
DC-link voltage	V_{dc}	400V
Grid voltage (*l-g*, rms)	v_s	120V
Switching frequency	f_{sw}	10kHz
Sampling frequency	f_s	10kHz
VSI filter inductor	L	3.5mH
Grid-side inductor	L_g	1.75mH
Grid-side capacitor	C_g	15μF or 5μF

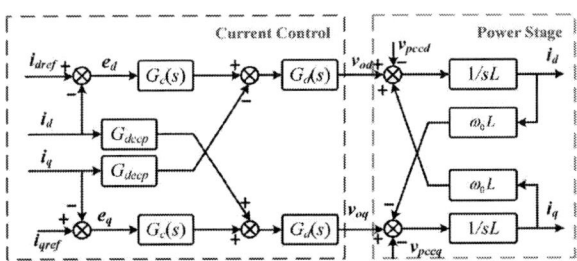

Fig. 2. Block diagram of the current control in the *dq*-frame.

realized by proportional-integral (PI) controller. $G_{decp}(s)$ represents the decoupling control term which is $\omega_0 L$ in this case. ω_0 is the fundamental angular frequency. $G_d(s)$ represents the time delay effects existed in the digital control system, which includes both the computational and pulse width modulation (PWM) delays. When the synchronous sampling scheme is applied, $G_d(s)$ can be expressed as [16]

$$G_d(s) = e^{-1.5T_s s} \tag{7}$$

where T_s is the sampling period.

B. CTFs Based Impedance Modeling

According to Fig. 2, it is clear that the three-phase grid-connected VSI system can be regarded as the three-phase symmetric system if the following two conditions are satisfied: 1) the effect of SRF-PLL is neglected, 2) the parameters of the controllers and the power stage should share the same values for both *d* and *q* axes. Fortunately, these two conditions can be satisfied for a general three-phase grid-connected VSI when only the high frequency instability is investigated. Consequently, the CTFs can be adopted to establish the IMs.

Firstly, the output admittance of the inverter is derived based on the traditional transfer matrix way for showing a clear comparison with the modeling process based on the CTFs. The transfer matrix of the power stage $\mathbf{G_p}(s)$ can be obtained based on Fig. 2, whose expression is given as

$$\begin{bmatrix} i_d \\ i_q \end{bmatrix} = \mathbf{G_p}(s) \cdot \begin{bmatrix} v_{od} \\ v_{oq} \end{bmatrix} = \begin{bmatrix} \dfrac{s/L}{s^2 + \omega_0^2} & \dfrac{\omega_0/L}{s^2 + \omega_0^2} \\ -\dfrac{\omega_0/L}{s^2 + \omega_0^2} & \dfrac{s/L}{s^2 + \omega_0^2} \end{bmatrix} \cdot \begin{bmatrix} v_{od} \\ v_{oq} \end{bmatrix} \tag{8}$$

Then, the transfer matrix of the current control loop gain $\mathbf{T}(s)$ can be expressed as

$$\mathbf{T}(s) = (\mathbf{I} - \mathbf{G_d} \cdot \mathbf{G_p} \cdot \mathbf{G_{decp}})^{-1} \cdot \mathbf{G_d} \cdot \mathbf{G_p} \cdot \mathbf{G_c} \tag{9}$$

where $\mathbf{I}$ is a 2×2 identity matrix. $\mathbf{G_c}(s)$, $\mathbf{G_{decp}}(s)$, and $\mathbf{G_d}(s)$ are expressed in (10), (11) and (12), respectively.

$$\mathbf{G_c}(s) = \begin{bmatrix} k_p + k_i/s & 0 \\ 0 & k_p + k_i/s \end{bmatrix} \tag{10}$$

$$\mathbf{G_{decp}}(s) = \begin{bmatrix} 0 & -\omega_0 L \\ \omega_0 L & 0 \end{bmatrix} \tag{11}$$

$$\mathbf{G_d}(s) = \begin{bmatrix} e^{-1.5T_s s} & 0 \\ 0 & e^{-1.5T_s s} \end{bmatrix} \tag{12}$$

Finally, the transfer matrix of the output admittance of the inverter $\mathbf{Y_{oc}}(s)$ can be derived as

$$\mathbf{Y_{oc}}(s) = [\mathbf{I} + \mathbf{T}(s)]^{-1} \cdot \mathbf{Y_o}(s) \tag{13}$$

where $\mathbf{Y_o}(s)$ is the open-loop output admittance of the VSI. As the output filter is a single inductor, $\mathbf{Y_o}(s)$ is exactly the same with $\mathbf{G_p}(s)$.

Based on the above derivation process, it is clear that obtaining the output admittance of the VSI needs massive matrix calculation, which will cause the stability analysis complicated. For achieving a simpler derivation process, complex space vectors and CTFs can be applied. Firstly, all the state variables in the *dq*-frame can be represented by the complex space vectors. Herein, taking the output current and the PCC voltage as examples, their complex forms are given as

$$\vec{i}_{dq} = i_d + j \cdot i_q \tag{14}$$

$$\vec{v}_{pccdq} = v_{pccd} + j \cdot v_{pccq} \tag{15}$$

Meanwhile, the transfer matrices can be reformulated into the CTF forms based on (2) and (5). Herein, taking $\mathbf{G_c}(s)$ and $\mathbf{G_{decp}}(s)$ as examples, their complex forms are expressed as

$$\vec{G}_c(s) = k_p + k_i/s \tag{16}$$

$$\vec{G}_{decp}(s) = j\omega_0 L \tag{17}$$

Besides, it is worthwhile to mention that the derivation of the power stage model will become much simpler if

3178

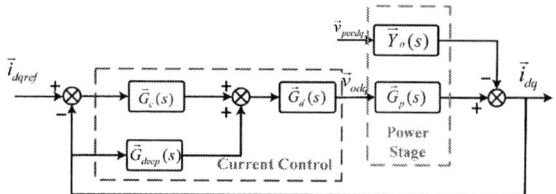

Fig. 3. Block diagram of the CTFs based current control in the *dq*-frame.

the CTF based coordinate transformation is adopted. As the cross coupling terms in $\mathbf{G_p}(s)$ are introduced by the *dq* transformation, these coupling terms can thus be avoided by performing only the stationary *αβ* transformation to the power stage model, whose result is given as

$$G_p^{\alpha\beta}(s) = 1/(sL) \tag{18}$$

Based on the coordinate transformation between the *αβ*-frame and the *dq*-frame [15], the CTF based power stage model in the *dq*-frame can be achieved by directly substituting s with $s + j\omega_0$, whose result is given as

$$\vec{G}_p(s) = \frac{1}{(s+j\omega_0)L} = \frac{s/L}{s^2+\omega_0^2} + j\frac{-\omega_0/L}{s^2+\omega_0^2} \tag{19}$$

Comparing (19) with (8), it can be found that (19) is exactly the CTF form of (8), which verifies the derivation process of the power stage model based on the CTFs.

After obtaining these CTFs, the original 2 × 2 system in the *dq*-frame can be equivalently simplified to a quite simple SISO system represented by the complex space vectors and CTFs, whose control block diagram is shown in Fig. 3. Based on Fig. 3, the current control loop gain $\vec{T}(s)$ and the output admittance of VSI $\vec{Y}_{oc}(s)$ can thus be directly derived as

$$\vec{T}(s) = \frac{\vec{G}_c(s)\cdot\vec{G}_p(s)\cdot\vec{G}_d(s)}{1-\vec{G}_{decp}(s)\cdot\vec{G}_p(s)\cdot\vec{G}_d(s)} \tag{20}$$

$$\vec{Y}_{oc}(s) = \frac{\vec{Y}_o(s)}{1+\vec{T}(s)} \tag{21}$$

Similarly, the grid impedance can also be derived as

$$\vec{Z}_g(s) = \frac{L_g(s+j\omega_0)}{L_gC_g(s+j\omega_0)^2+1} \tag{22}$$

Therefore, the IMs can be obtained without any matrix calculation. Besides, the IMs in the CTF form, different from the impedance matrices, enable the application of the classical control theory, which makes the controller design and stability analysis more convenient.

C. Impedance-based Stability Analyisis Using CTFs

To conduct the impedance-based stability analysis of the grid-connected VSI system, the relationship between the GNC and the Nyquist criterion using the CTFs for the cascaded three-phase symmetric system should be firstly demonstrated. For illustration, the transfer matrices of the

source admittance and load impedance of the three-phase symmetric system in the *dq*-frame are expressed as

$$\mathbf{Y}_s(s) = \begin{bmatrix} Y_d(s) & -Y_q(s) \\ Y_q(s) & Y_d(s) \end{bmatrix} \tag{23}$$

$$\mathbf{Z}_o(s) = \begin{bmatrix} Z_d(s) & -Z_q(s) \\ Z_q(s) & Z_d(s) \end{bmatrix} \tag{24}$$

Based on the GNC, the return ratio matrix $\mathbf{L}(s)$ used for predict the system stability can be defined as

$$\mathbf{L}(s) = \mathbf{Y}_s \cdot \mathbf{Z}_o = \begin{bmatrix} Z_dY_d - Z_qY_q & -(Z_qY_d + Z_dY_q) \\ Z_qY_d + Z_dY_q & Z_dY_d - Z_qY_q \end{bmatrix} \tag{25}$$

As a result, such cascaded system is stable if and only if the characteristic loci of the eigenvalues $\lambda(s)$ of $\mathbf{L}(s)$ do not encircle the critical point (-1 + j0). Based on (25), the eigenvalues of $\mathbf{L}(s)$ can be derived as

$$\lambda_1(s) = (Z_dY_d - Z_qY_q) + j\cdot(Z_qY_d + Z_dY_q) \tag{26}$$

$$\lambda_2(s) = (Z_dY_d - Z_qY_q) - j\cdot(Z_qY_d + Z_dY_q) \tag{27}$$

As all the four elements in the $\mathbf{L}(s)$ are the real-coefficient transfer functions, it can be obtained that the Nyquist curves of $\lambda_1(s)$ and $\lambda_2(s)$ are mirror images in the real axis, which means only $\lambda_1(s)$ or $\lambda_2(s)$ is needed to predict the system stability.

Alternatively, if the impedance matrices are replaced by the CTFs, the minor loop gain applied for evaluating the system stability can be directly obtained as

$$\begin{aligned} \vec{T}_m(s) &= \vec{Y}_s(s)\cdot\vec{Z}_o(s) \\ &= (Z_dY_d - Z_qY_q) + j\cdot(Z_qY_d + Z_dY_q) \end{aligned} \tag{28}$$

Based on (26) and (28), it is obvious that $\vec{T}_m(s) = \lambda_1(s)$, which thus proves that the traditional Nyquist criterion using the CTFs is equivalent to the GNC for predicting stability of the cascaded three-phase symmetric system.

After achieving the equivalent relationship between the GNC and the Nyquist criterion using the CTFs, the stability of the grid-connected VSI system can thus be predicted based on the conventional Nyquist criterion. The parameters listed in Table I are applied. Firstly, the system stability is evaluated by the Nyquist curve of $\vec{T}_m(s)$ shown in Fig. 4, where two different capacitances are used to emulate two different cases. It is clear that the Nyquist curve of $\vec{T}_m(s)$ is totally coincident with that of $\lambda_1(s)$, which further proves the correctness of the above mentioned equivalent relationship. Besides, it can also be seen that the Nyquist curve of $\vec{T}_m(s)$ encircles the critical point (-1, j0) when $C_g = 5\mu\text{F}$. Based on the Nyquist criterion, an unstable system is achieved. On the contrary, a stable grid-connected VSI system can be predicted when $C_g = 15\mu\text{F}$, as the Nyquist curve of $\vec{T}_m(s)$ won't encircle (-1, j0) in this situation.

Further, it is worthwhile to mention that the system stability can also be predicted by the bode diagrams of

3179

The 2018 International Power Electronics Conference

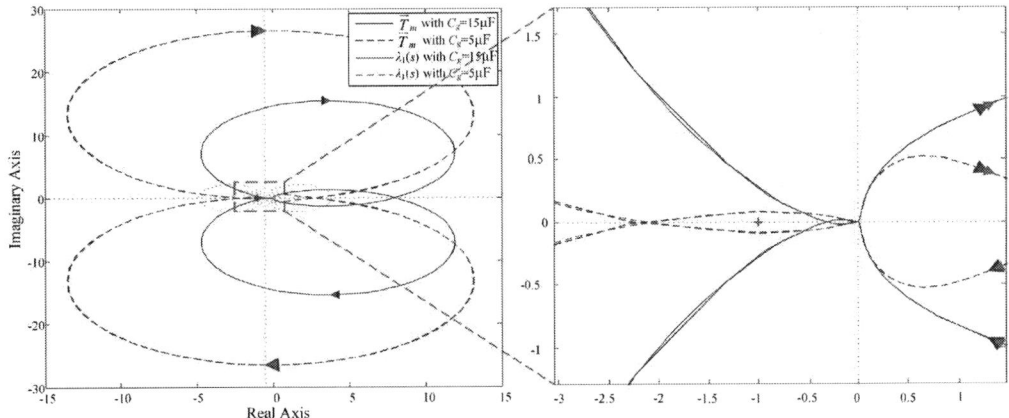

Fig. 4. Nyquist curves of $T_m(s)$ and $\lambda_1(s)$ of $\mathbf{L}(s)$ with different grid-side capacitors: C_g=15μF for the stable case, C_g=5μF for the unstable case.

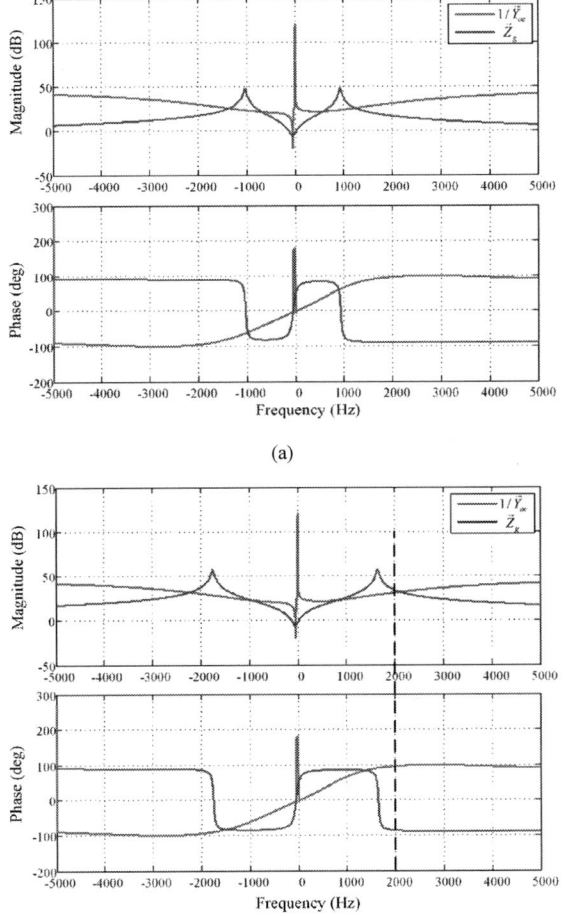

(a)

(b)

Fig. 5. Frequency responses of $1/\vec{Y}_{oc}(s)$ and $\vec{Z}_g(s)$ with different grid-side capacitors: (a) C_g=15μF for the stable case, (b) C_g=5μF for the unstable case.

$\vec{Z}_g(s)$ and $1/\vec{Y}_{oc}(s)$, whose frequency responses can be plotted in Fig. 5. Based on the Nyquist criterion applied for Bode diagrams of the cascaded system, the stability is only related to the frequency ranges where the magnitude of $\vec{Z}_g(s)$ is larger than that of $1/\vec{Y}_{oc}(s)$. Only when the phase angle difference of these two impedances dose not exceed $180°$ in such frequency ranges, a stable system can be guaranteed. Otherwise, the system will be unstable. Due to such conclusion, from Fig. 5(a), it is clear that the phase angle difference is within $180°$ when the magnitude of $\vec{Z}_g(s)$ is larger than that of $1/\vec{Y}_{oc}(s)$, which indicates a stable system. However, when C_g is changed to 5μF, the phase angle difference will exceed $180°$ around 2kHz, and an unstable system can be predicted.

Based on the above analyses, it is found that the same stability results are obtained no matter the Nyquist curves or the Bode diagrams are used. However, it should be noticed that, different from the stability analysis using the eigenvalues of the return ration matrix when the GNC is applied, the Bode diagram of the impedances represented by the CTFs can reflect the stability more intuitively, which can help us have a better insight of the instability mechanism and guide the system design to improve the stability. This advantage distinguish the Nyquist criterion using the CTFs from the GNC to predict the stability of the cascaded three-phase symmetric system, which makes the stability analysis simpler and more convenient.

IV. EXPERIMENTAL RESULTS

A laboratory setup, whose structure is the same with that shown in Fig. 1, is built up to verify the theoretical analyses. The main parameters listed in Table I are used for the experiment. The ac power grid is emulated by a *Chroma Regenerative Grid Simulator* 61860.

Firstly, the grid-side capacitor C_g is set to equal to 15μF. Fig. 6(a) shows the measured phase-*a* PCC voltage v_{pcca}, the inverter output current i_a, and the grid current i_{ga} in the steady state. It is clear that the grid-connected VSI system is stable. However, once C_g is changed to 5μF, the system will become unstable, where the measured results are shown in Fig. 6(b). Hence, the experimental results are consistent with the above stability predictions, which verifies the correctness of the theoretical analyses.

3180

The 2018 International Power Electronics Conference

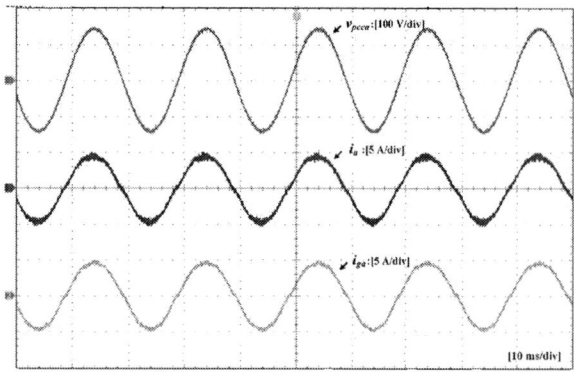

(a)

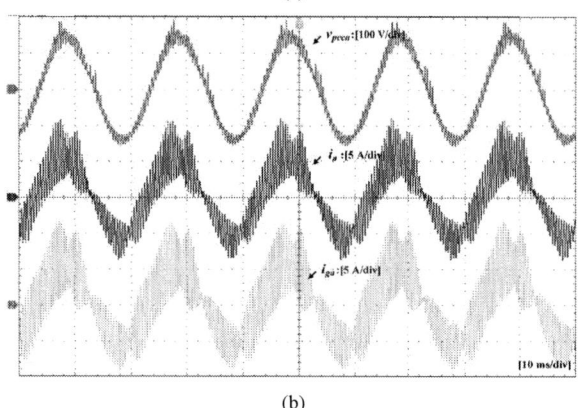

(b)

Fig. 6. Measured waveforms of phase-a PCC voltage v_{pcca}, the inverter output current i_a, and the grid current i_{ga} when (a) $C_g = 15\mu F$, (b) $C_g = 5\mu F$.

V. CONCLUSIONS

This paper proposed a novel impedance modeling process by applying the CTFs, and also conducted the relevant stability analysis of the cascaded three-phase symmetric system. It is shown that applying the CTFs can equivalently change the original MIMO system into the SISO system represented by the CTFs, which will in result simplify the impedance modeling process remarkably. Besides, the obtained IM is represented by only a simple CTF making the possible use of the traditional Nyquist criterion originally applied for dc system to predict the stability of the cascaded system. Then, the equivalent relationship between the GNC and the Nyquist criterion using CTFs was proved. It is obtained that the stability prediction can be performed directly based on the CTFs, which leads to the stability analysis much simpler and more intuitive. Case study of a three-phase grid-connected inverter system was conducted, where the frequency-domain analyses with experimental results were presented for verifying the correctness of the theoretical analyses.

REFERENCES

[1] F. Blaabjerg, R. Teodorescu, M. Liserre, and A. V. Timbus, "Overview of control and grid synchronization for distributed power generation systems," *IEEE Trans. Ind. Electron.*, vol. 53, no. 5, pp. 1398–1409, Oct. 2006.

[2] J. H. R. Enslin and P. J. M. Heskes, "Harmonic interaction between a large number of distributed power inverters and the distribution network," *IEEE Trans. Power Electron.*, vol. 19, no. 6, pp. 1586-1593, Nov. 2004.

[3] L. Harnefors, X. Wang, A. G. Yepes, and F. Blaabjerg, "Passivity-based stability assessment of grid-connected VSCs-An overview," *IEEE J. Emerging Sel. Topics Power Electron.*, vol. 4, no. 1, pp. 116-125, Mar. 2016.

[4] M. Belkhayat, *Stability Criteria for AC Power Systems With Regulated Loads.* West Lafayette, IN, USA: Purdue Univ., 1997.

[5] R. D. Middlebrook, "Input filter consideration in design and application of switching regulators," in *Proc. IEEE Ind. Appl. Soc. Annu. Meeting*, Chicago, IL, USA, 1976, pp. 366-382.

[6] X. Feng, J. Liu, and F. C. Lee, "Impedance specifications for stable dc distributed power systems," *IEEE Trans. Power Electron.*, vol. 17, no. 2, pp. 157-162, Mar. 2002.

[7] S. Vesti, T. Suntio, J. A. Oliver, R. Prieto, and J. A. Cobos, "Impedance-based stability and transient-performance asessment applying maximum peak criteria," *IEEE Trans. Power Electron.*, vol. 28, no. 5, pp. 2099-2104, May. 2013.

[8] J. Liu, X. Feng, F. C. Lee, and D. Borojevich, "Stability margin monitoring for DC distributed power systems via perturbation approaches," *IEEE Trans. Power Electron.*, vol. 18, no. 3, pp. 1254-1261, Nov. 2003.

[9] Z. Liu, J. Liu, W. Bao, and Y. Zhao, "Infinity-norm of impedance-based stability criterion for three-phase ac distributed power systems with constant power loads," *IEEE Trans. Power Electron.*, vol. 30, no. 6, pp. 3030-3043, Jun. 2015.

[10] B. Wen, D. Boroyevich, R. Burgos, and P. Mattawelli, "Analysis of D-Q small-signal impedance of grid-tied inverters," *IEEE Trans. Power Electron.*, vol. 31, no. 1, pp. 675-687, Jan. 2016.

[11] R. Turner, S. Walton, and R. Duke, "A case study on the application of the Nyquist stability criterion as applied to the interconnected loads and sources on grids," *IEEE Trans. Ind. Electron.*, vol. 60, no. 7, pp. 2740-2749, Jul. 2013.

[12] B. Wen, D. Dong, D. Boroyevich, R. Burgos, P. Mattawelli, and Z.Shen, "Impedance-based analysis of grid-synchronization stability for three-phase paralleled converters," *IEEE Trans. Power Electron.*, vol. 31, no. 1, pp. 26-38, Jan. 2015.

[13] V. Valdivia, A. Lazaro, A. Barrado, P. Zumel, C. Fernandez, and M. Sanz, "Black-box modeling of three-phase voltage source inverters for system-level analysis," *IEEE Trans. Ind. Electron.*, vol. 59, no. 9, pp. 3648-3662, Sep. 2012.

[14] S. Gataric and N. R. Garrigan, "Modeling and design of three-phase systems using complex transfer functions," in *Proc. IEEE Power Electron. Spec. Conf.*, Jun./Jul. 1999, vol. 2, pp. 691-697.

[15] L. Harnefors, "Modeling of three-phase dynamic systems using complex transfer functions and transfer matrices," *IEEE Trans. Ind. Electron.*, vol. 54, no. 4, pp. 2239–2248, Aug. 2007.

[16] S. G. Parker, B. P. McGrath, and D. G. Holmes, "Regions of active damping control for LCL filters," *IEEE Trans. Ind. Appl.*, vol. 50, no. 1, pp. 424-432, Jan./Feb. 2014.

The 2018 International Power Electronics Conference

Acoustic Noise Reduction of 12/8 Poles SRM without Efficiency Drop Using Simple Current Waveforms

Kyohei Kiyota[1*], Kenji Amei[1], Takahisa Ohji[1], Jun Jisaki[2] and Masanobu Nakai[2]

1 Graduate School of Science and Engineering for Research, University of Toyama, Toyama, Japan
2 Technology Development Headquarters, Nachi-Fujikoshi Corp., Toyama, Japan
*E-mail: kiyota@eng.u-toyama.ac.jp

Abstract— One of the major challenging issues of switched reluctance motors (SRMs) is improving of the trade-off relation between the acoustic noise and efficiency without the increase of the system cost. In this paper, some simple current waveforms are proposed to reduce both the 3rd and 6th harmonics radial force component and the rms current using simple look up table. The proposed method is adopted to the 12/8 poles SRM for the industrial application.

Keywords— *Acoustic noise reduction, Current control, Efficency, Switched reluctance motor*

I. INTRODUCTION

Switched Reluctance Motors (SRMs) have been attracting attention because of the characteristics such as low cost, robust, and possible high temperature operation as well as an advantage of rare-earth-free. SRMs need only low loss silicon steel to be competitive the Permanent Magnet Synchronous Motors (PMSMs) including rare-earth magnets. Also, SRMs are now achieving to competitive torque density and power density as well as efficiency of PMSMs [1]-[3]. For example, SRMs have been developed by one of the authors with competitive torque, rms current, and efficiency as well as wide operation area with respect to the PMSM installed in the third generation Toyota Prius [4]-[5]. It is noted that the SRM have 1.6 times high power density with respect to that of the PMSM. However, need of a special inverter as well as the acoustic noise and vibration are still the major problem of SRMs. Especially, improve of the trade-off between the acoustic noise reduction and the efficiency improvement is one of the most challenging matter of SRMs.

One of the major causes of the acoustic noise and vibration is caused by an electro-magnet force variation when SRM windings are excited. There are two major classifications of the reduction method of the acoustic noise and the vibration as follows: 1) the mechanical design and 2) the current control. In terms of 1), an enhancement of the stiffness of the stator and frame or a skewing of the stator and rotor has been reported [6], but the effectiveness is limited. On the other hand, this paper deals with the current control method. One of the notable methods is an active vibration cancellation technique [7].

The method can reduce only the vibration at the natural frequency. Although SRMs have the acoustic noise and vibration at not only the natural frequency but also the several harmonics of the current frequency, for example, the third and sixth harmonics of the current frequency. A lot of papers have proposed the flattening torque and radial force sum [8]-[13]. Several papers proposed instantaneously torque and radial force control using predicted motor inductance, torque, or radial force profiles [8]-[10]. Other papers proposed a current waveform which can minimize the variation of the sum of three-phase radial forces [11]-[13]. One of the problems of these methods was the increased rms current and decrease of the efficiency. Also, these methods need precise specifications as well as high performance controller such as FPGAs or high-speed DSPs to realize the flattened torque and radial force, these costs are not ignored in the industrial applications such as compressors, pumps and fans. Some paper proposes current waveform without these precise motor specifications, whereas the trigonometric function is used [14]. However, some low-cost DSPs cannot handle the complex functions including trigonometric functions.

In this paper, some simple current waveforms are proposed to reduce both the 3rd and 6th harmonics radial force component and the rms current of an SRM using simple look up table. The target motor is 12/8 poles 1.5 kW SRM. The effect of the improvement of the efficiency and the radial force ripple is presented. It is found that the triangular waveform can reduce the radial force ripple without the increase of the RMS current.

II. ANALYSIS MODEL

Fig.1 shows the cross-sectional view of the target 3-phase 12/8 poles SRM. Table 1 shows the specification of the target 3-phase 12/8 poles SRM. This SRM is designed for an industrial application. The outer diameter of the stator core is 140 mm. The iron stack length is 60 mm. The DC side of the inverter voltage is 280V. The rated RMS current value is 6.8 A. The rated output torque at the maximum rotational speed is 4 Nm. The maximum output is 1.5 kW at the rotational speed of 3600 r/min. Note that the acoustic noise of the motor is relatively

The 2018 International Power Electronics Conference

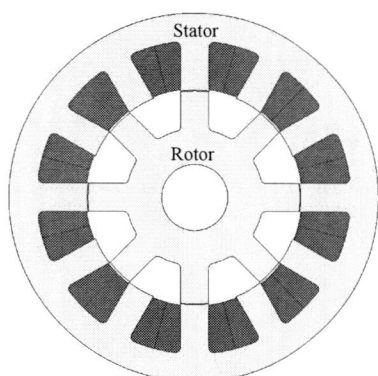

Fig. 1. Structure of the target 12/8 poles SRM.

TABLE I
SPECIFICATIONS OF THE 12/8 POLES SRM

Motor Specifications	
Outer diameter	140 mm
Iron stack length	60 mm
DC side voltage	280 V
Rated rms current	6.8 A
Rated torque	4 Nm
Rated power	1.5 kW
Maximum speed	3600 r/min
Target operation point	
Target speed	650 r/min
Target torque	2 Nm
Target RMS current	3.6 A

significant at the low load operation in some applications, thus, the target operating point is set as low speed and low torque region in this paper. Table 1 also shows the target operating point of this paper. The rotational speed is almost one-sixth of the maximum speed. The target torque is half of the rated torque. The fundamental current frequency is 86.7 Hz; thus, the copper loss is the main loss at this point. In this point, the RMS current of the rectangular waveform is 3.6 A.

Fig. 2 shows the measured acoustic noise of the common rectangular current waveform at the target operating point of the 12/8 SRM. The 3rd, 6th and 9th harmonics of the current frequency is the main acoustic noise in this point.

Fig. 3 shows the analysis model of the 12/8 SRM. As shown in this figure when a SRM generates torque, it also generates radial forces. Both torque and radial force fluctuate significantly in general square current, which results in vibration and acoustic noise. The radial forces of A-phase, B-phase, C-phase are defined as functions of electrical rotor rotational position θ and three-phase currents i_A, i_B, and i_C as $F_{rA}(\theta,i_A)$, $F_{rB}(\theta,i_B)$ and $F_{rC}(\theta,i_C)$. Then sum of radial forces of each phase is given as,

$$F_{rsum} = F_{rA}(\theta,i_A) + F_{rB}(\theta,i_B) + F_{rC}(\theta,i_C). \quad (1)$$

Note that each radial forces $F_{rA}(\theta,i_A)$, $F_{rB}(\theta,i_B)$ and $F_{rC}(\theta,i_C)$ have a phase difference of $2\pi/3$. Then, the radial force vibration can be cancelled except the $3n^{th}$

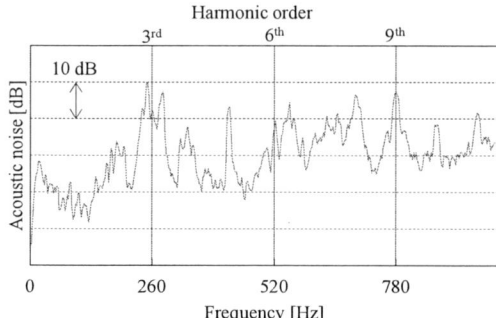

Fig. 2. Measured acoustic noise of the rectangular current waveform at the target point (2 Nm, 650 r/min).

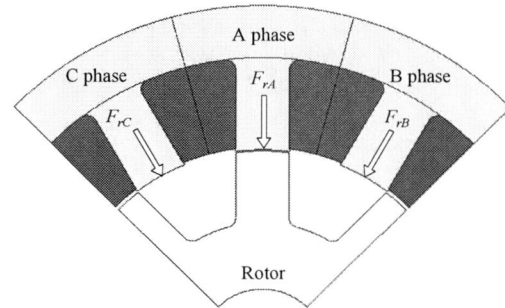

Fig. 3. Analysis model of a 3-phase 18/12 SRM and the radial forces on stator teeth.

harmonics and DC component. In this paper, the 3rd harmonics and 6th harmonics of the radial force are reduced.

III. RADIAL FORCE ANALYSIS

In some industry applications, not only the cost of the motor component, but also the cost of the inverter and the controller should be low. In this paper, the low-cost DSP (DSPIC33EP64MC204) is selected for the SRM controller. This DSP handles only integer values, i.e., the referenced current should be integer value. Thus, the loop up table (LUT) is need when the complicated current waveforms are applied in this system. In this paper, the length of the array is set as 16. Note that the current waveforms in the previous papers [11]-[13] are defined by the function of the sinusoidal waveform with the harmonics components. This method needs precise adjusting of the current amplitude and phase shift of each harmonics components in each motor. Thus, it is difficult to use this method in this system.

This paper compares three simple current arrays: 1) triangular waveform, 2) fundamental sinusoidal waveform with DC offset, and 3) Half-wave rectifier waveform to reduce the radial force sum. Fig. 4 shows the proposed normalized current waveforms. In the triangular waveform, the normalized referenced current $i_{ref}(\theta)$ is given as,

$$i_{ref}(\theta) = \begin{cases} 0 & (0 \le \theta \le \pi) \\ (\pi - \theta)/\pi & (\pi < \theta \le 2\pi) \end{cases}. \quad (2)$$

3183

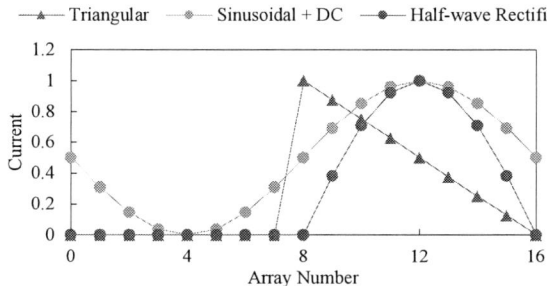

Fig. 4. Proposed current waveforms.

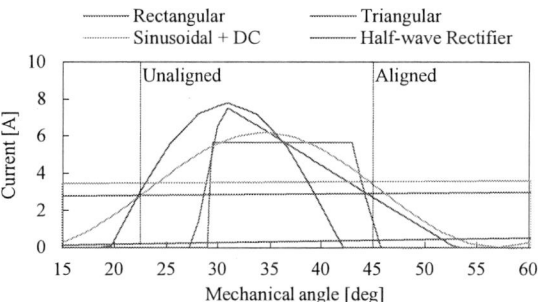

Fig. 5. Current waveforms at 2 Nm, 650 r/min.

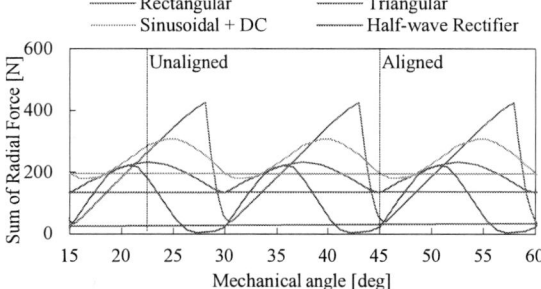

Fig. 6. Sum of Radial force at 2 Nm, 650 r/min.

In the sinusoidal waveform with DC offset, $i_{ref}(\theta)$ is given as,

$$i_{ref}(\theta) = \frac{1 + \sin(\pi + \theta)}{2}. \tag{2}$$

In the triangular waveform, the normalized referenced current $i_{ref}(\theta)$ is given as,

$$i_{ref}(\theta) = \begin{cases} 0 & (0 \le \theta \le \pi) \\ \sin(\pi + \theta) & (\pi < \theta \le 2\pi) \end{cases}. \tag{2}$$

Note that these current waveforms can be shifted. In the later, the shift angles are adjusted to minimize the sum of the 3rd harmonics of radial force. It is also noted that the referenced current is interpolated linearly like lines of fig. 4.

In order to confirm the radial force reduction due to the proposed current waveforms, 2-D FEM analysis is carried out. Fig. 5 shows the current waveforms of the conventional rectangular waveform and three proposed waveforms. The current amplitude of these waveforms is adjusted to set the output torque as 2 Nm. Fig. 6 shows F_{rsum} of the four waveforms. In the conventional rectangular waveform, the radial force ripple is more than 300 N. In the half-wave rectifier waveform, the maximum radial force sum is minimized, whereas the radial force ripple is still high. The radial force ripple is minimized in the triangular waveform.

Table II summarizes the 3rd harmonics F_{r3} and 6th harmonics F_{r6} of radial force and the RMS current I_{rms} as well as the peak current I_{peak}. Although I_{rms} of the triangular waveform is almost identical to that of the rectangular current waveform, F_{r3} and F_{r6} are significantly decreased from the rectangular waveform. Also, the F_{r3} and F_{r6} of the sinusoidal current waveform with DC offset can be reduced from the rectangular waveform, whereas I_{rms} is 14% high.

IV. EXPERIMENTAL VARIDATION

Fig. 7 shows the experimental setup to confirm the proposed current waveforms. There are a fabricated SRM, torque and speed transducer and the generator. A precision sound level meter is located on the side of the fabricated SRM. Fig. 8 shows the experimental control block diagram. The pulse width modulation control is applied in the SRM inverter to fix the switching

TABLE II
RADIAL FORCE RIPPLE AND THE RMS CURRENT OF FOUR CURRENT WAVEFORMS

	T_{ave}	I_{peak}	I_{rms}	F_{r3}	F_{r6}
Rectangular		5.7 A	3.21 A	150 N	65.8 N
Triangular	2.0 Nm	7.7 A	3.33 A	45.0 N	6.94 N
Sinusoidal + DC		6.2 A	3.78 A	64.7 N	3.96 N
Half-wave Rectifier		7.8 A	3.85 A	114 N	7.52 N

frequency as 10 kHz. Note that the SRM inverter has three operation mode: (a) positive voltage mode, (b) negative voltage mode and (c) zero voltage mode. The positive and zero voltage modes are used when the referenced voltage is positive, whereas the negative and zero voltage modes are used when the referenced voltage is negative. Three referenced voltages of each phases $v^*_{a,b,c}$ are generated by the PI control according to the phase currents $i_{a,b,c}$. Note that when the reference current is 0 A, both the high side and low side IGBTs are turned-off, and the integral value of PI controller is reset. The referenced currents of each phase $i^*_{a,b,c}$ are generated by the LUT block like fig. 4 according to the referenced maximum current i^*. Note that in case of rectangular waveform, $i^*_{a,b,c}$ are set as i^* when the rotational angle θ is between the on-angle and off-angle of each phases. The referenced maximum current i^* is generated by the PI control according to the rotational velocity ω. The load torque is fixed to 2.0 Nm by the generator.

Fig. 9 shows the measured current waveform of the triangular waveform. The current waveform is almost identical to the referenced value. The maximum current value is 12 A, that is 58% high with respect to that of the analysis result. Thus, the RMS current should be

3184

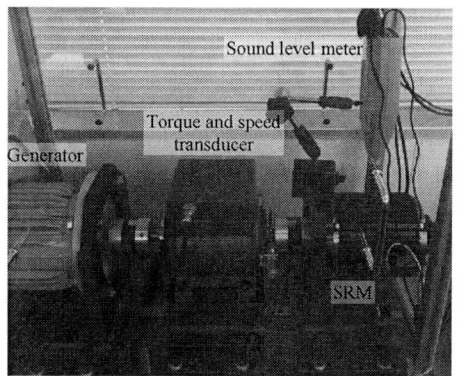

Fig. 7. Experimental setup.

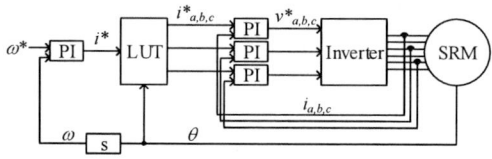

Fig. 8. Control block diagram of the test machine.

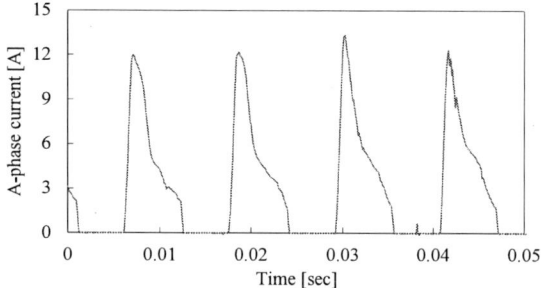

Fig. 9. Current waveform of the triangular waveform of the test machine.

TABLE III
EQUIVALENT NOISE LEVELS OF FOUR CURRENT WAVEFORMS

Rectangular	Triangular	Sinusoidal + DC	Half-wave Rectifier
81.0 dB	77.5 dB	78.6 dB	79.6 dB

increased from the analysis. Table III summarize the equivalent noise levels of each four current waveforms. It is found that the equivalent noise level of the triangular waveform can be 3.4 dB reduced from the rectangular waveform.

V. CONCLUSION

In this paper, three simple current waveforms are proposed to reduce both the 3rd and 6th harmonics radial force component and the rms current of an SRM using simple look up table. In the triangular waveform, the radial force ripple can be significantly decreased from the rectangular current waveform without the increase of the RMS current.

ACKNOWLEDGMENT

The authors acknowledge JMAG software license from J-SOL Corporation.

REFERENCES

[1] X.D. Xue, K.W.E. Cheng, T.W. Ng, N.C. Cheung, "Multi-Objective Optimization Design of In-Wheel Switched Reluctance Motors in Electric Vehicles," *IEEE Transaction on Industrial Electronics*, vol.57, no.9, pp.2980-2987, Sept. 2010.

[2] C.S. Kim, G. Lee, K. Lee, J. Lee, Y. Cho, J. H. Won, H. Shin, C. Choi, H. K. Bae, "Design of Π core and Π2 core PM-aided switched reluctance motors," in *Proc. IEEE IEVC*, Mar. 4-8, 2012, pp.1-6.

[3] I. Boldea, L. N. Tutelea, L. Parsa and D. Dorrell, "Automotive Electric Propulsion Systems With Reduced or No Permanent Magnets: An Overview," *IEEE Transactions on Industrial Electronics*, vol. 61, no. 10, pp. 5696-5711, Oct. 2014.

[4] K. Kiyota, A. Chiba, "Design of Switched Reluctance Motor Competitive to 60-kW IPMSM in Third-Generation Hybrid Electric Vehicle," *IEEE Transactions on Industry Applications*, vol. 48, no.6, pp. 2303-2309, November/December 2012.

[5] K. Kiyota, T. Kakishima, A. Chiba, "Comparison of Test Result and Design Stage Prediction of Switched Reluctance Motor Competitive With 60-kW Rare-Earth PM Motor," *IEEE Transaction on Industrial Electronics*, vol.61, no.10, pp.5712-5721, pp.2564-2575, Oct. 2014.

[6] H. Y. Yang, Y. C. Lim and H. C. Kim, "Acoustic Noise/Vibration Reduction of a Single-Phase SRM Using Skewed Stator and Rotor," *IEEE Transactions on Industrial Electronics*, vol. 60, no. 10, pp. 4292-4300, Oct. 2013.

[7] H. Makino, T. Kosaka and N. Matsui, "Digital PWM-Control-Based Active Vibration Cancellation for Switched Reluctance Motors", *IEEE Transactions on Industry Applications*, vol. 51, no. 6, pp. 4521-4530, Nov.-Dec. 2015.

[8] A. Klein-Hessling, A. Hofmann and R. W. De Doncker, "Direct instantaneous torque and force control: A novel control approach for switched reluctance machines," *2015 IEEE International Electric Machines & Drives Conference (IEMDC)*, Coeur d'Alene, ID, 2015, pp. 922-928.

[9] Cong Ma, Liyan Qu, Rakesh Mitra, Prerit Pramod, Rakib Islam, "Vibration and Torque Ripple Reduction of Switched Reluctance Motors Through Current Profile Optimization", *2016 IEEE Applied Power Electronics Conference and Exposition (APEC)*, pp. 3279 – 3285,2016.

[10] K. Okabayashi, K. Honda, T. Sugiura and H. Kakigano, "Noise reduction using high-switching-frequency operation with SiC MOSFET for switched reluctance motors," *2016 19th International Conference on Electrical Machines and Systems (ICEMS)*, Chiba, 2016, pp. 1-6.

[11] M. Takiguchi, H. Sugimoto, N. Kurihara, A. Chiba, "Acoustic Noise and Vibration Reduction of SRM by Elimination of Third Harmonic Component in Sum of Radial Forces", *IEEE Transactions on Energy Conversion*, pp. 1-9, 2015.

[12] N. Kurihara, J. Bayless, H. Sugimoto, A. Chiba, "Noise Reduction of Switched Reluctance Motor With High Number of Poles by Novel Simplified Current Waveform at Low Speed and Low Torque Region", *IEEE Transactions on Industry Applications*, vol. 52, no.4, pp. 3013 – 3021, July-August 2016.

[13] J. Bayless, N. Kurihara, H. Sugimoto, A. Chiba, "Acoustic Noise Reduction of Switched Reluctance Motor With Reduced RMS Current and Enhanced Efficiency", *IEEE Transactions on Energy Conversion*,vol. 31,no. 2,pp. 627 – 636, June 2016

[14] A. Tanabe and K. Akatsu, "Vibration reduction method in SRM with a smoothing voltage commutation by PWM," *2015 9th International Conference on Power Electronics and ECCE Asia (ICPE-ECCE Asia)*, Seoul, 2015, pp. 600-604.

The 2018 International Power Electronics Conference

Study of Switched Reluctance Motor Directly Driven by Commercial Three-phase Power Supply

Masaki Takahashi[*], Kohei Aiso and Kan Akatsu
Department of Electrical Engineering, Shibaura Institute of Technology, Tokyo, Japan
*E-mail: ma17075@shibaura-it.ac.jp

Abstract— The conventional three-phase induction motors (IMs) have been widely used in various applications because three-phase IMs can be directly driven by commercial three-phase power supply, however, their efficiency are not high due to copper loss in the rotor. To overcome this problem, as an alternative for IMs, a novel three-phase Switched Reluctance Motor (SRM) which can be directly driven by commercial three-phase power supply without converter, inverter and some sensors is proposed. In this paper, it is shown that the proposed SRM can generate the continuous torque and can start by using the simple and low cost drive circuit consisted of only six diodes and no switching devices by the simulation.

Keywords— *Commercial three-phase power supply, Direct drive, Self start-up, Switched reluctance motor*

I. INTRODUCTION

Recently, electric power consumption tends to increase by technological development. Motors account for about 46% of electric power consumption in the world. Besides, three-phase induction motors (IMs) have been widely used in various applications such as fans and pumps because three-phase IMs can be directly driven by commercial three-phase power supply without the complex system such as converter, inverter and sensors, and they realize the compact and low cost of drive system. Therefore, improving their efficiency is very effective for the energy saving.

Some efforts to improve the efficiency of IMs have been researched. These previous researches focus on the techniques to reduce copper loss, core loss, and stray loss in IMs. In [1], the high efficiency is achieved by using low loss materials of the rotor core and die casting cage. The IM consists of a copper squirrel cage and a rotor core made of GO electrical steel sheet to decrease the core losses and the stray load losses. In [2], the winding method mixed the concentric low harmonic windings and Y-Δ connection are applied to reduce the harmonic of magnetomotive force in the air-gap, therefore, the iron loss and the stray load loss can be reduced. However, it is difficult to obtain higher efficiency in three phase IMs because they generate much copper loss in the rotor due to the operation principle. To

achieve higher efficiency, a novel three phase motor which can be directly driven by commercial three-phase power supply without any complex circuits and sensors is an essential requirement as an alternative for IMs in industrial field.

Permanent magnet synchronous machines (PMSMs) are expected as an alternative candidate of IMs because they have high motor efficiency. However, PMSMs generally require converter, inverter and some sensors for driving. To solve the problem, line-start PMSMs (LSPMSMs) are proposed [3][4]. LSPMSMs have a diecasting and permanent magnets in its rotor. It generates the starting torque by utilizing the rotor current excited in die-casting just like IMs and can start directly by applying commercial three-phase power supply. When it reaches the synchronous speed, it is synchronized with the rotating magnetic field and achieve the high performance as a PMSM. However, LSPMSM has not been widely used because the structure is complicated and is not suitable for the mass production.

Furthermore, switched reluctance motors (SRMs) have been researched as a replacement for IM. SRMs consist of only laminated iron core and copper, therefore its structure is simple and robust. In addition, since SRMs do not have permanent magnets, they are inexpensive and suitable for mass production. However, SRMs also need complex drive systems because it is usually driven by using inverter and position sensor. In [5], the single-phase SRM which can be directly driven by a commercial single-phase power supply has been proposed. It has been verified by the experiment that the SRM can start up by using simple circuit consists of only two diodes and achieve the high efficiency compared with a single-phase IM. Although the single-phase SRM is useful as a single-phase machine, three-phase SRM as a replacement for three-phase IM which accounts a large proportion of electric power consumption has not been developed yet.

This research proposes a novel three-phase SRM that can be directly driven by commercial three-phase power supply as a replacement for three-phase IM. The proposed SRM consists of triple-stage structure, the proposed drive

The 2018 International Power Electronics Conference

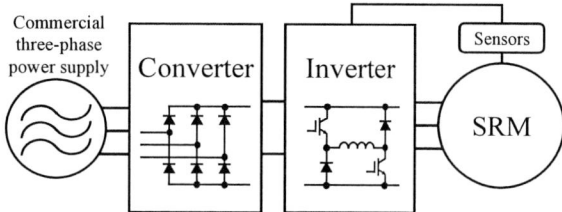

Fig. 1. Conventional drive system of SRM.

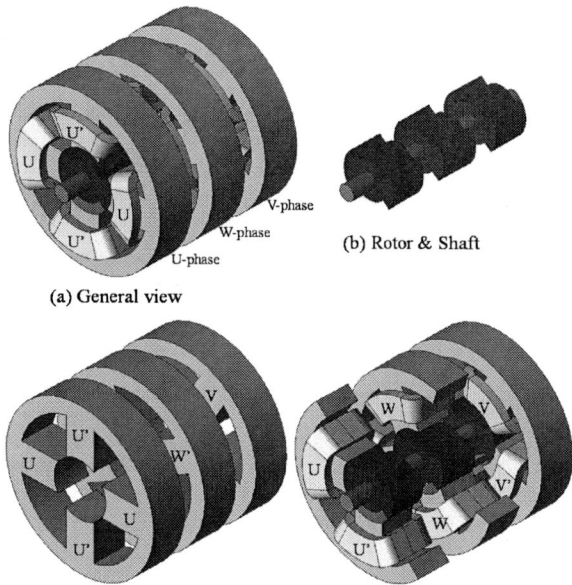

(b) Rotor & Shaft

(a) General view

(c) Stator (d) Internal view

Fig. 2. Structure of proposed three-phase SRM.

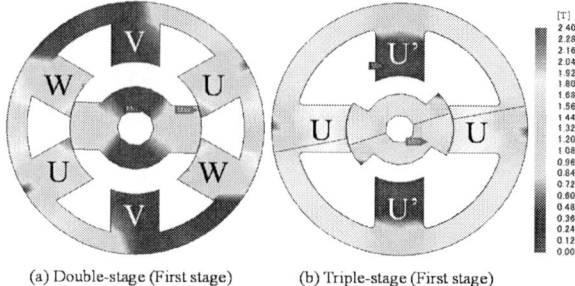

(a) Double-stage (First stage) (b) Triple-stage (First stage)

Fig. 3. Magnetic flux density distribution.

Fig. 4. Double-stage structure.

circuit composed of only six diodes and no switching devices realizes the simple and low cost drive system. The drive circuit can rectifies the commercial three phase power source into the six phase unipolar current. By utilizing the six phase unipolar current, the triple-stage SRM realizes start-up by the commercial three-phase power supply without converter, inverter and some sensors. In this paper, the motor structure, drive circuit and the operation principle are presented and the performances are evaluated by Finite Element Analysis (FEA).

II. PROPOSED THREE-PHASE SRM

As described in chapter I, conventional three-phase SRM requires the complex drive system such as converter, inverter and some sensors for driving, as shown in Fig. 1. Therefore, the size and cost of the system increase compared with IMs. To overcome above problems, the motor structure, drive circuit and principle of torque generation are proposed in this section.

A. Motor structure

The proposed structure of three-phase SRM is shown in

Fig. 2. The specifications of proposed three-phase SRM and conventional high efficient three-phase IM (IE-3) are shown in Table I. As shown in Table I, the design target of this research is to replace 5.5kW three-phase IM with SRM that has same volume (stator diameter, stack length) and output power. As shown in Fig. 2(a)(b), the proposed SRM has triple-stage structure and the combination of 2 pole 4 slot in each stage. As shown in Fig. 2(c), stators of each stage are skewed 30 degrees. The windings of three-phase U-U', V-V', W-W' are separated to each stage, the armature concentrated windings are wound on each stator teeth. By the three-phase windings wound on each stage excited independently, the SRM can generate continuous torque and can start in any rotor position.

Fig. 3 shows the comparison of the magnetic flux density distribution of the double-stage and the triple-stage structure. As shown in Fig. 4, the double-stage structure consists of two stators with skewed 30 degrees, and the windings of the UVW and U'V'W' are separated to each stage. As shown in Fig. 3(a), in the double-stage, the flux interference occurs between different phases U-W. Then, the unintentional current flows between different phases. On the other hand, in the triple stage, as shown in Fig. 3(b), the flux interference does not occur between the different phases because each phase winding is arranged in individual stage. It can avoid the flux interference between three-phase by applying triple-stage structure and the current can flow to an intended phase.

B. Drive circuit

The proposed drive circuit is shown in Fig. 5. As shown in Fig. 5, the drive circuit is composed of the delta

3187

TABLE I
SPECIFICATION COMPARISON OF
PROPOSED SRM AND CONVENTIONAL HIGH EFFICIENT IM

Parameter [unit]	Proposed SRM	Conventional IM
Number of poles	2	4
Number of slots	4	48
Stack length [mm]	44.6×3	134
Stack length including coil-end [mm]	224	224
Stator outer diameter [mm]	213	213
Air gap length [mm]	0.5	0.5
Rated output power [kW]	5.5	5.5
Rated torque [Nm]	35.0	35.9
Rated speed [min⁻¹]	1500	1465
100% load efficiency [%] (50Hz)	72.2	91.2

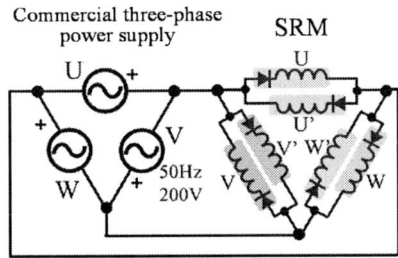

Fig. 5. Proposed drive circuit of three-phase SRM.

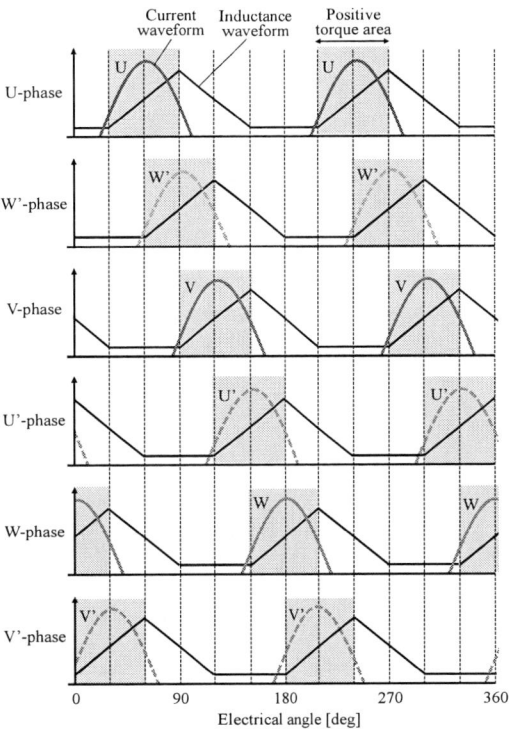

Fig. 6. Current waveform and inductance waveform of the proposed three-phase SRM.

connection which directly connects to commercial three-phase power supply. It includes no switching device and only six diodes. The drive circuit of a phase consists of only a half bridge diode rectifier and it rectifies the commercial AC power source into separated two-phase unipolar voltage and current which have 180 degrees phase difference. That is, six-phase unipolar current are applied to windings on each stator from three-phase commercial AC power source. Therefore, the SRM can be directly driven by the six-phase unipolar current without converter and inverter and realizes the small size and low cost of system.

C. Principles of torque generation

Fig. 6 shows the current waveform and inductance waveform of the proposed SRM. As shown in Fig. 6, the rotor position is expressed as follow:

$$\theta = P\theta_m \tag{1}$$

where θ, P and θ_m are electric angle, pole pairs and mechanical angle, respectively. In this figure, the electric angle is based on U-phase. The unaligned and aligned positions are defined as shown (unaligned: 0 degrees, aligned: 90 degrees in the electric angle). U'-phase leads 90 degrees from U-phase. The output torque of SRMs is expressed as follow:

$$T = \frac{1}{2}\frac{\partial L(i,\theta)}{\partial \theta}i^2 \tag{2}$$

where T, L, and i are the output torque, inductance and phase current, respectively. As shown in (2), the positive

torque is generated when the phase current is applied in positive period of inductance variation. As shown in Fig. 6, the proposed SRM can achieve the continuous torque production by applying unipolar six phase currents shifted 30 degrees each phase during the period of positive inductance variation.

D. Starting method and synchronous speed

The motion equation is expressed as follow:

$$\frac{d\omega}{dt} = \frac{1}{J}(T - T_L) \tag{3}$$

$$T_L = D\omega^2 \tag{4}$$

where ω, J, T_L and D are rotor speed, inertia, load torque and friction, respectively. As shown in (3) and (4), the friction load is assumed as square reduced torque like a fan. The rotor speed quickly rises when inertia is low. The starting method utilizes that the SRM has low rotor-inertia. The proposed SRM can generate continuous torque, therefore, if the torque to reach synchronous speed is generated on every rotor position, the motor can start up without any additional equipment such as magnets and sensors. The synchronous speed of proposed SRM is expressed as follow:

$$N = \frac{f}{P} \times 60 \tag{5}$$

3188

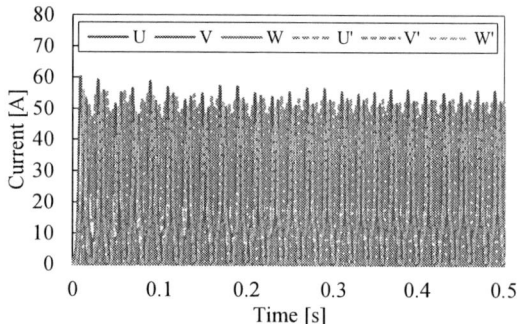

Fig. 7. Current waveform (No load).

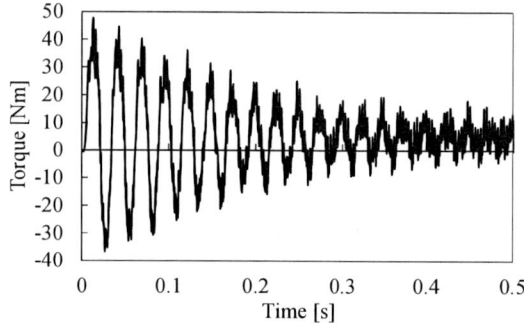

Fig. 8. Torque waveform (No load).

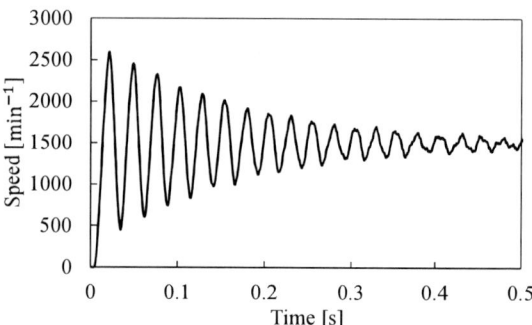

Fig. 9. Motor speed waveform (No load).

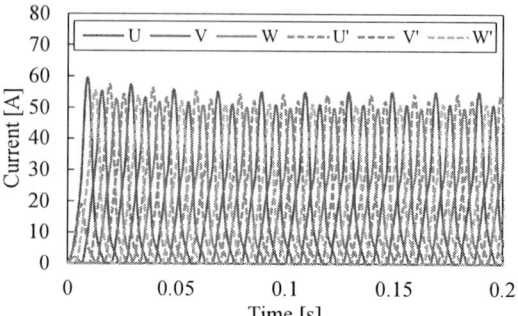

Fig. 10. Current waveform (Load torque 35Nm).

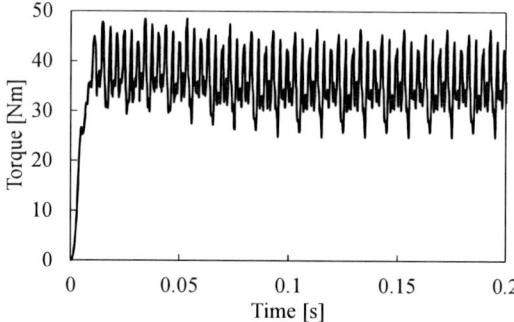

Fig. 11. Torque waveform (Load torque 35Nm).

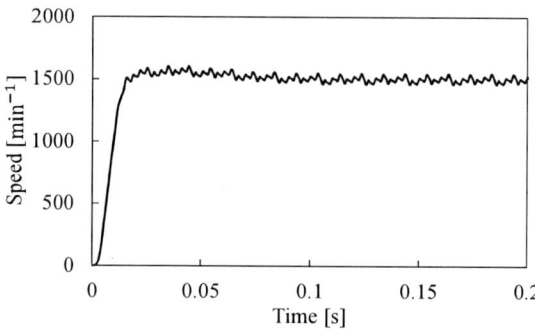

Fig. 12. Motor speed waveform (Load torque 35Nm).

where N is synchronous speed and f is commercial frequency. As shown in (4), the rotation speed depends on commercial frequency and rotor poles. The commercial frequency is 50Hz or 60Hz in the world. Therefore, the synchronous speed is 1500min⁻¹ or 1800min⁻¹.

III. SIMULATION RESULT OF START-UP CHARACTERISTIC

Firstly, in order to realize replacement three-phase IM for proposed motor, it is necessary to reveal whether the proposed motor can start up by commercial three-phase power supply, 200V three-phase AC voltage input. The start-up characteristic is evaluated by the simulation using the motion equation. f, J and D are 50Hz, 0.0015kgm², and 1.556×10^{-5}Nm/(deg/s)², respectively. The J is calculated

based on the rotor volume. The D is determined to give the load torque 35Nm.

Fig. 7, Fig. 8 and Fig. 9 show current, output torque, motor speed under the no load condition, respectively. As shown in Fig. 7 and Fig. 8, the starting torque to reach the synchronous speed is generated by applying the unipolar six phase current. As shown in Fig. 9, the proposed SRM can start up itself and its speed and torque converge to synchronous speed 1500min⁻¹ and 4.5Nm. Fig. 10, Fig. 11 and Fig. 12 show current, output torque, motor speed under the load condition, respectively. As shown in Fig. 10, Fig. 11 and Fig. 12, the starting torque to reach the synchronous speed can be obtained and the proposed motor can start up itself under the load condition. Its speed and torque converge to synchronous speed 1500min⁻¹ and required torque 35.0Nm.

The 2018 International Power Electronics Conference

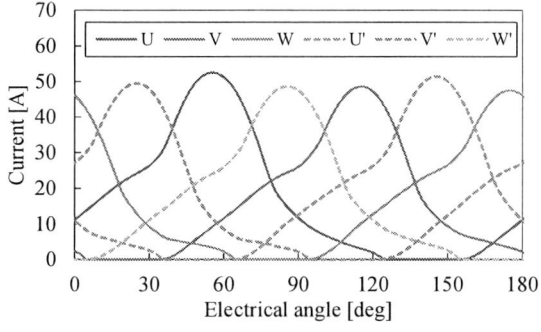

Fig. 13. Current waveform (Constant speed 1500min⁻¹).

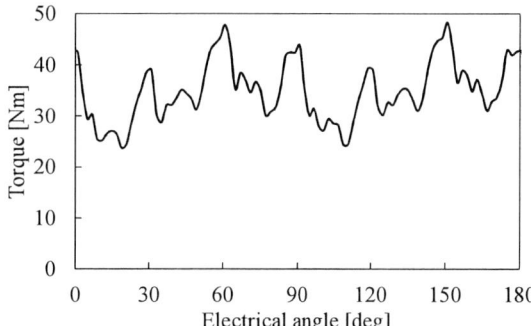

Fig. 14. Torque waveform (Constant speed 1500min⁻¹).

(a) Current

(b) Output power

(c) Copper loss

(d) Efficiency

···△··· d_r : 74.0mm, ϕ_w : 1.7mm
···△··· d_r : 74.0mm, ϕ_w : 1.5mm
──○── d_r : 88.8mm, ϕ_w : 1.4mm
──○── d_r : 88.8mm, ϕ_w : 1.3mm
─▫── d_r : 103.6mm, ϕ_w : 1.2mm
d_r : Rotor diameter
ϕ_w : Winding diameter

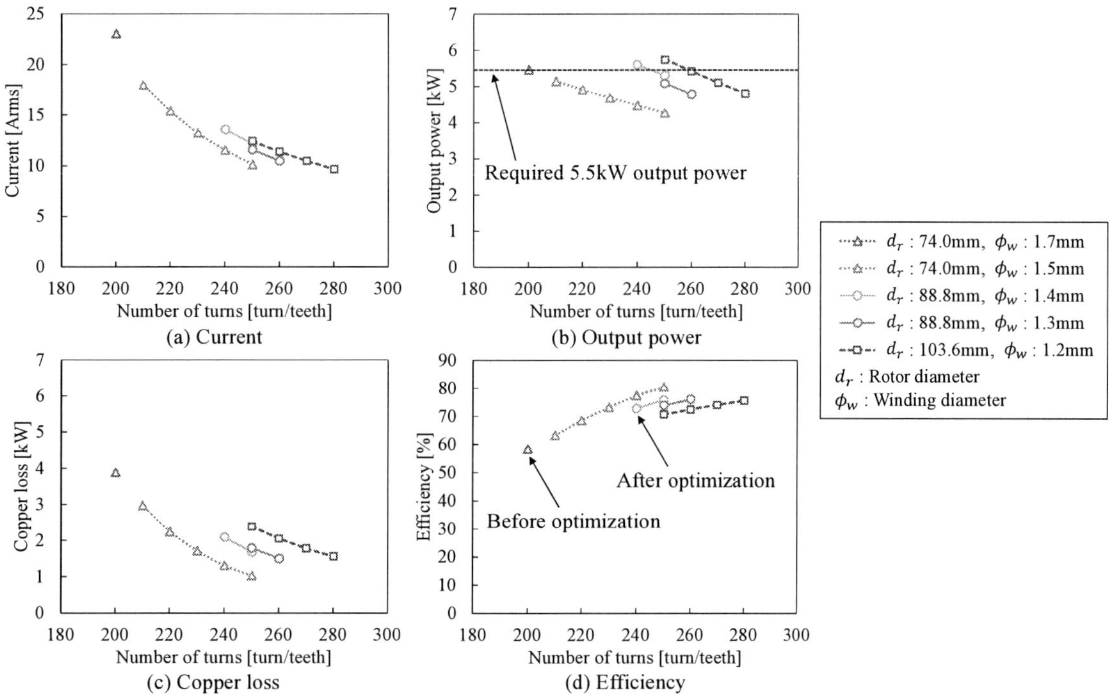

Fig. 15. Characteristic changes depending on the rotor diameter, the winding diameter, and the winding number of turns.

IV. SIMULATION RESULT OF STEADY STATE CHARACTERISTIC

The steady state characteristics of the proposed motor are evaluated by simulation under the condition of the synchronous speed 1500min⁻¹. Current and torque waveforms are shown in Fig. 13 and Fig. 14. As shown in Fig. 13, three-phase input voltage is rectified by diodes and six-phase unipolar current are applied to windings on each stator. As shown in Fig. 14, the proposed motor can achieve the continuous torque and the average torque is satisfied with the required torque 35.0Nm. So, proposed SRM can output 5.5kW. However, the motor efficiency is 56.7% because of a large copper loss. The motor efficiency can be improved if the copper loss can be reduced by optimizing the winding parameters such as number of turns and diameter.

V. DESIGN FOR HIGHER EFFICIENCY MOTOR

In order to improve motor efficiency, it is necessary to optimize parameters by using FEA. The rotor diameter, the winding diameter and the winding number of turns are changed on condition that the winding space factor is 60% or less and the current density is 10A/mm² or less. Phase current, output power, copper loss and the motor efficiency depending on the changes of each parameters are shown in Fig. 15. d_r is the rotor diameter and φ_w is the winding diameter. Those combinations of rotor diameter, winding diameter and winding number of turns are determined by the limitation of space factor and current density. In the relationship between the rotor diameter d_r and the winding diameter φ_w, as the rotor diameter becomes larger, the area of the winding becomes narrower,

3190

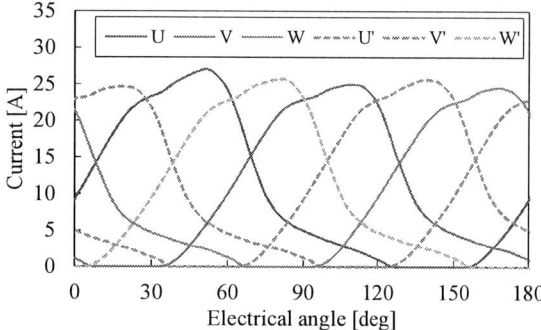

Fig. 16. Current waveform after the optimization (Constant speed 1500min⁻¹).

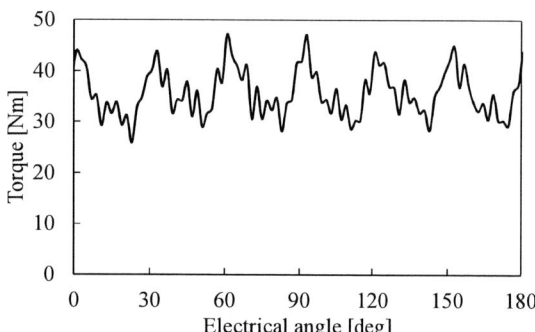

Fig. 17. Torque waveform after the optimization (Constant speed 1500min⁻¹).

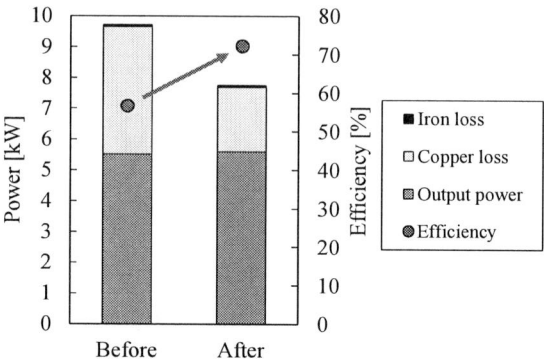

Fig. 18. Output power, copper loss, iron loss and the motor efficiency before and after the optimization.

so the winding diameter becomes thinner. As shown in Fig. 15, as the winding number of turns increases, the motor efficiency increases because the current and the copper loss decrease. However, the output power also decreases as increasing the winding number of turns. Therefore, the winding number of turns which achieves the required output power and the highest motor efficiency need to be selected. On the other hand, when the rotor diameter is larger, the winding diameter is smaller due to space factor. Therefore, as the rotor diameter increases, although the output power is higher, the copper loss increases and the motor efficiency decreases. From these trends of the winding number of turns and the rotor diameter, the winding number of turn and the rotor diameter are determined at the optimal point which achieves the required output power 5.5 kW and the highest motor efficiency. Current and torque waveforms after the optimization are shown in Fig. 16 and Fig. 17. As shown in Fig. 16 and Fig. 17, it is confirmed that the required torque can be achieved by lower current. As shown in Fig. 18, as a result of redesign, the motor efficiency improved from 56.7% to 72.2% due to a significant decrease in copper loss.

VI. CONCLUSION

As an alternative to three-phase IMs, this paper has presented a novel three-phase SRM which can be directly driven by commercial three-phase power supply without converter, inverter and some sensors. The proposed SRM has triple-stage structure, and the proposed drive circuit realizes the simple and low cost drive system because the circuit consists of only six diodes and includes no switching devices. It can rectifies the commercial three-phase power source into the six-phase unipolar current. By exciting the six-phase unipolar current to windings on each stage independently, the triple-stage SRM can generate continuous torque and start-up on any rotor position. It has been shown that the proposed SRM can start up itself and its speed converges to synchronous speed 1500min⁻¹ regardless load condition in the simulation. It has been also shown that the proposed SRM can generate the required torque in the synchronous speed and its maximum motor efficiency is 72.2%.

REFERENCES

[1] V. Mallard, G. Parent, C. Demian, J. F. Brudny, A. Delamotte, Increasing the energy-efficiency of induction machines by the use of grain oriented magnetic materials and die-casting copper squirrel cage in the rotor ", *International Electric Machines and Drives Conference (IEMDC)*, pp. 1-6, 2017.

[2] L. Zhang, Y. Huang, J. Dong, B. Guo and T.Zhou, "Stator Winding Design of Induction Motors for High Efficiency", *International Conference on Electrical Machines and Systems (ICEMS)*, pp. 130-134, 2014.

[3] A. T. de Almeida, F. J. T. E. Ferreira, and G. Baoming, "Beyond Induction Motors - Technology Trends to Move Up Efficiency", *IEEE Trans. on Industry Applications*, vol. 50, no. 3, pp. 2103-2114, 2014.

[4] F. J. T. E. Ferreira, G. Baoming, and A. T. de Almeida, "Stator Winding Connection-Mode Management in Line-Start Permanent Magnet Motors to Improve Their Efficiency and Power Factor", *IEEE Trans. on Energy Conversion*, vol. 28, no. 3, pp. 523-534, 2013.

[5] K. Aiso and K. Akatsu, "A novel Single Phase SRM Simply Driven by Commercial AC Power Supply", *IEEE Energy Conversion Congress and Exposition (ECCE)*, pp. 3878-3885, 2015.

Double Stator Axial-Flux Switched Reluctance Motor for Electric City Commuters

Hiroki Goto[1]

1 Department of Electrical and Electronics Systems Engineering, Utsunomiya University, Utsunomiya, Japan
*E-mail: h_goto@cc.utsunomiya-u.ac.jp

Abstract— **Reluctance Motors which is constructed without rare-earth magnet and robust structure is one of the most suitable in-wheel motors for electric vehicles. In addition, an axial-flux structure has possibility of effective utilization of in-wheel flat motor space and a double stator structure is good at the exhaust heat. This paper presents the designed process of a DSAFSRMs (Double Stator Axial Flux SRMs) and their performance through simulations and experiments. The Manufactured prototype DSAFSRMs have superior torque characteristics depend on rotational speed to a same size geared permanent magnet motor. Finally, a small city commuter vehicle equipped with the prototype DSAFSRMs for in-wheel direct drive and its road test is done successfully.**

Keywords— *switched reluctance motor, electric vehicle, axial flux, in wheel drive*

I. INTRODUCTION

Electric Vehicle (EV) and Hybrid Electric Vehicle (HEV) attract a lot of attention and have been researched and developed for the solution of environmental problems. However, EV needs large capacity rechargeable batteries in their bodies. Now rechargeable battery is still expensive, large and heavy because rechargeable battery has small energy density compared with fossil fuel. Therefore, an electric city commuter which is single or double seated small vehicle and doesn't need large size batteries is effective for popularization of EVs [1].

A small car such as a city commuter doesn't have large space for traction motor and drive train. So, in-wheel drive system that its traction motors are equipped in each wheel is one of the suitable drive systems for city commuters.

Almost in-wheel EVs use PMSMs (Permanent Magnet Synchronous Motors) as its traction motors. However, PMSM has two serious disadvantages. First point, PMSM is expensive than other kinds of motors because rare earth materials used at its rotor magnet. In addition, if PMSMs are used at in-wheel motors for EV, the motors are larger than conventional on-board motors and need plurality. Therefore, costs for the motors become higher increasingly. Second point, PMSM doesn't have robust structure for In-wheel vehicles which motors are directly contracted to ground. There is some possibility that the power converter circuits of PMSMs are broken by high inductive voltage when malfunction of field-weakening

control at high rotational speed. Moreover, the rotation of wheel is stopped suddenly by short-circuit of power wires. Because of these reasons, Switched Reluctance Motor (SRM) is one of the most suitable motors for in-wheel drive systems [2][3]. SRM is the motor which is constructed only iron cores made of electromagnetic steel sheets and windings made of copper. Therefore, SRM has solid structure, is low cost on manufacturing, and no inductive voltage without excitation.

However, SRM also has disadvantages. One of them is that its magnetic saliency causes higher torque ripple, vibration and acoustic noise compared with PMSM. On the other hand, torque density of SRM is smaller than that of PMSM.

To solve these problems, authors have examined flat motor as in-wheel drive systems. So far, torque-volume density can be improved by axial-flux structure at flat space [4]. It can use inner space efficiently because coil end can be placed in plenty radial space. In addition, axial-flux SRM (AFSRM) can be applied to multi-gap structure such as double stator structure or double rotor structure easily. In previous works, it was shown torque of a double rotor type axial-flux SRM (DRAFSRM) is larger than that of a single rotor AFSRM at same volume. A design of DRAFSRM for in-wheel micro bus was examined and its prototype machines were manufactured. The driving test of the micro bus equipped with the DRAFSRMs was done successfully.

However, in double rotor structure, the rotor cores are rotated with outer housing so the part without a rotation is axis only. It means that vibration and acoustic noise caused by stator deformation is directly transferred to outside of the housing [6]. In addition, motor windings which occur heat by copper loss is enclosed in the motor, therefore heat radiation characteristic is not good.

In the paper, focusing on the double stator structure which can improve these problems, a design of DSAFSRM for an electric city commuter with in-wheel direct drive system is presented. The design process of the DSAFSRM is discussed and the measured performance of the prototype motor is shown. Finally, the test driving of a small vehicle equipped with the prototype DSAFSRMs is shown.

II. BASIC CONSTITUTIONS AND DRIVE SYSTEMS OF AFSRM

The basic constitution of one stator-one rotor AFSRM is shown in Fig. 1. AFSRM is constructed by only iron cores made of electromagnetic steel sheet and copper windings.

AFSRM is driven by reluctance torque which caused between stator- and rotor cores. Stator core of AFSRM is formed by reeling toroidal core. Rotor core of SRM is made by cut down laminated steel core. Stator pole of AFSRM is open slot structure therefore winding which is wounded beforehand can be inserted, so winding space factor can be significantly improved. The part of one-phase basic drive circuit of SRM which is an asymmetry half bridge converter is shown in Fig. 2. This type of drive circuit consists of two transistors and two free-wheeling diodes. An arithmetic unit decides a timing to make switching based on rotational angle feedback. Therefore, after the voltage applying to the SRM is end, magnetic energy which is stored in the magnetic circuit regenerates in the power supply.

A phase voltage and a phase current depend on a position of typical AFSRM are shown in Fig. 3. The 'aligned' position is decided that the distance between the rotor pole and the stator pole is minimum, on the other hand, the 'unaligned' position is decided that the distance between the rotor pole and the stator pole is maximum. At an aligned position, the value of the magnetic reluctance is minimum, so its phase inductance is maximum. Then, the phase torque τ_k is produced in the kth-phase can be expressed in

$$\tau_k = \frac{\partial W'(\theta, i_k)}{\partial \theta}. \tag{1}$$

Here $W'(\theta, i_k)$ is the coenergy stored in the winding.

To produce continuous positive torque, windings need applied voltage to the aligned position from the unaligned position. So, AFSRM needs position sensor such as an optical encoder and feedback to arithmetic unit.

On the other hand, AFSRM is often driven at high magnetic flux density, so core material shows high nonlinearity magnetic characteristics. Therefore, above (1) cannot be calculated easily. To confirm the characteristic tendency, 3D-FEM (3-Dimentional Finite Element Method) software to calculate the characteristics of AFSRM model.

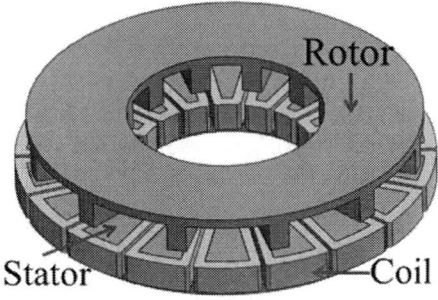

Fig. 1. Basic construction of AFSRM.

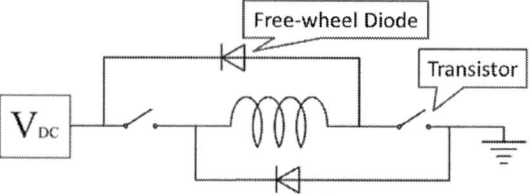

Fig. 2. Asymmetry half bridge converter.

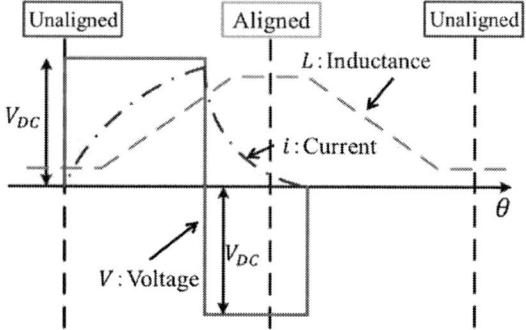

Fig. 3. Change of inductance depends on rotor position.

III. DESIGN OF DOUBLE STATOR AFSRM FOR ELECTRICAL CITY COMMUTER

The overview and specifications of a target electric city commuter is shown in Fig. 4. This vehicle equips two original in-wheel motors beside each rear wheel. These motor' shape and specification are shown in Fig. 5. This motor is an interior permanent magnet synchronous motor (Hereinafter referred to as Base IPMSM) which uses Nd-Fe-B magnet. Base IPMSM is 12-poles / 18-slots and is included a planetary gear (ratio 1/6.267) for speed reduction and torque amplification. The size of housing of Base IPMSM including reduction gear is that the diameter is 260 mm and axial length is 145 mm. Therefore, this housing can be inside of 12 inches wheel. The maximum torque of Base IPMSM is 100 N·m with reduction gear. From the above conditions of Base IPMSM, target of design DSAFSRM is the size which is smaller than that of Base IPMSM and maximum torque is 100 N·m.

The shape of DSAFSRM which designed for electric city commuters is shown in Fig. 6. Windings in facing stator poles are placed in the same rotary direction and are connected in series therefore the magnetic flux works in the direction through facing stator. The number of poles and slots are same as Base IPMSM. The material of stator- and rotor cores is non-oriented electromagnetic steel sheet (35A300). Each rotor pole is independent magnetically and its back yokes are omitted to save the axial length.

The view of prototype DSAFSRM is shown in Fig. 7. A stator core of the DSAFSRM is fixed to the housing made of aluminum (A7075) by inserting 16 support-plates between the stator back yoke and each winding such as Fig. 8. Independent rotor poles are fixed by two support rings (outer and inner) made of stainless steel (SUS303) and non-magnetic bodies made of FRP (Fiber-Reinforced Plastics) are inserted in the vacant space to improve its

Total length	2365 mm
Total width	995 mm
Total height	1600 mm
Vehicle weight	330 kg
Maximum speed	50 km/h
Dimensions of tire	90/90-12

Fig. 4. Specification of Electric City Commuter.

Diameter of case:	260 mm
Axial length of case:	145 mm
Rated output:	290 W
Maximum output:	2 kW
Maximum torque:	100 N·m
Maximum rotational speed:	580 r/min. (= 50 km/h)
Source voltage:	72 V
Weight including case:	11.5 kg

Fig. 5. Specification of Base IPMSM.

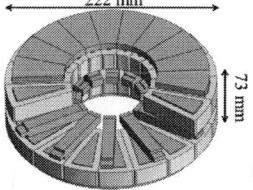

Exciting voltage:	14.4 V
Gap length:	0.3 mm
Winding turns / pole:	99 turns
Winding space factor:	62%
Weight	14.4 kg
Weight with case	32.3 kg
Core material:	35A300

Fig. 6. Structure of prototype DSAFSRM.

Fig. 7. Prototype of the DSAFSRM.

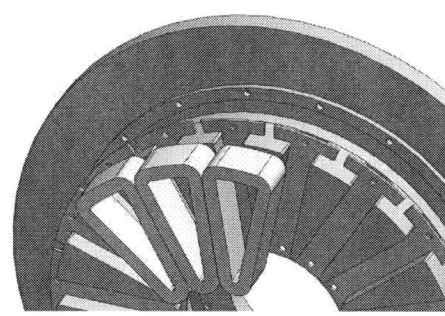

Fig. 8. Support structure of stator pole to case.

strength. In addition to that, two tapered roller bearings made of stainless steel (SUS304) are put inner rotor ring at the axial direction. To detect the rotational angle, a resolver is put in the housing. Finally, a disk brake is put as the brake of the electric city commuter by attaching externally.

IV. PERFORMANCE OF THE PROTOTYPE DRAFSRM

It shows the experimental result of the basic characteristics of two DSAFSRM prototypes which are made for the in-wheel electric city commuter with the characteristics which analysis performed by 3D-FEM.

The current density characteristic depend on torque is shown in Fig. 9. The maximum torque is decided as the torque value at winding current density is 15 A/mm² which is expected value for allowable winding temperature. It shows that the experimental results don't achieve 100 N·m at 15 A/mm² in spite the analysis value. Both of prototype DSAFSRMs achieve 100 N·m at 19 A/mm² at examination. Therefore, the experimental performance of the prototype DSAFSRM is worse than those of simulation.

The rotational speed characteristics depend on torque of two prototype DSAFSRMs with the catalog data of Base IPMSM including gear ratio are shown in Fig. 10. These show that the experimental result and the simulation result are good at agreement. Further, the experimental performance of the prototypes can comparable to Base IPMSM is confirmed.

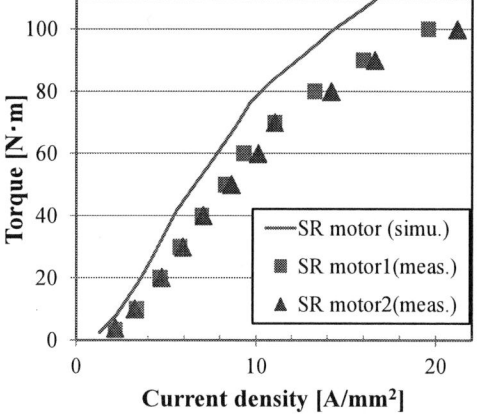

Fig. 9. Current density versus Torque characteristics.

The 2018 International Power Electronics Conference

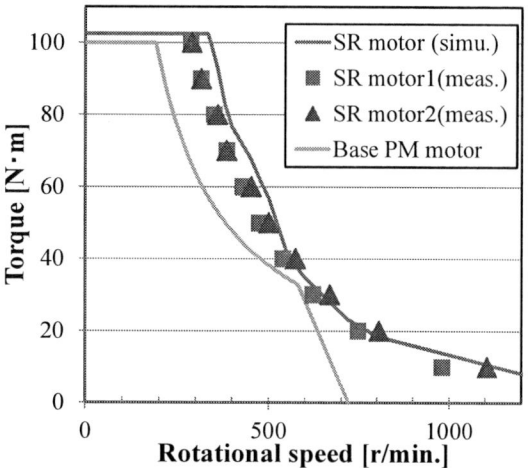

Fig. 10. Rotational speed versus Torque characteristics.

V. PROTOTYPE OF DRIVE SYSTEM FOR ELECTRIC CITY COMMUTER AND DRIVING TEST

The prototype of a drive system of DSAFSRM on the vehicle is shown in Fig. 11. The system is constructed mainly power electronics circuit which is asymmetry half bridge converter and control circuit. Power MOSFETs as switching device and power diodes are used for power electronics circuit and both of them are discrete type module. In addition to this, a FPGA (Field Programmable Gate Array) is used at arithmetic unit and the switching logic is built in internal memory in advance. To detect three phase currents, drive unit has closed—loop hole effect type current sensors. In this way, the drive unit can decide switching command depends on torque command by phase current feedbacks. The power electronics circuit has a smoothing capacitor which is 10,000 μF and snubber circuits to reduce its surge voltage.

To control the torque of the DSAFSRMs depending on accelerator pedal, instantaneous phase torque distribution control is applied [3]. The control logic of FPGA is programmed by model-based development environment using Mathworks Matlab/Simulink and Xilinx System Generator for DSP.

Max voltage (MOSFET)	200 V
Max voltage (Diode)	200 V
Max current (MOSFET)	180 A
Max current (Diode)	240 A
Weight with case	8.0 kg

Fig. 11. The prototype of a drive system.

Using the drive unit, the motor performance is examined. An example of the controlled phase current and the torque waveforms estimated from phase current and rotor angle at torque command given as 15 N·m are shown in Fig. 12 and Fig. 13, respectively. Fig, 12 shows measured phase current waveforms are good agreement with simulated waveforms. And the torque waveforms are almost kept as command value. These results indicate that prototype drive unit functioned normally.

The view of applying the prototype DRAFSRM to city commuter is shown Fig. 14. Motors and drive circuits are totally changed from base vehicle and total weight of vehicle increases 57.6 kg. Therefore, the running resistance depends on angle of inclination and the automobile performance diagram is calculated as shown in Fig. 15. From these characteristics, the maximum speed is expected to 63 km/h at a flat road and the hill-climbing possibility incline is 15 %.

The view of the driving test of the vehicle at hill climbing is shown in Fig. 16 and the waveforms of vehicle velocity and the torque command are shown in Fig. 17. From these results, the driving test is done successfully and the acceleration rate is more than 1.27 m/s² which is enough performance for city vehicle.

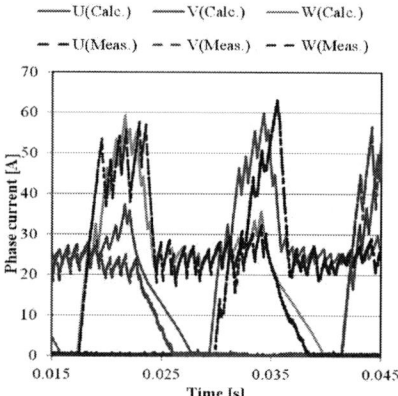

Fig. 12. The example of phase current waveforms at instantaneous phase torque distribution control.

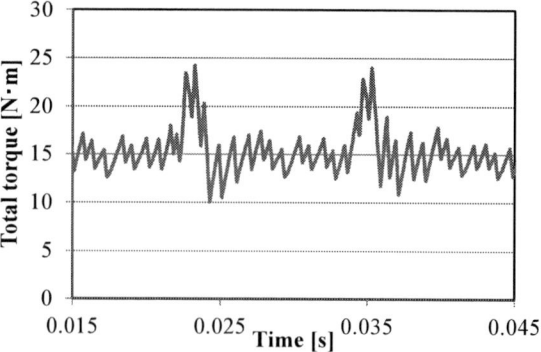

Fig. 13. The example of estimated torque waveforms at instantaneous phase torque distribution control

3195

Fig. 14. Applying DSAFSRMs to electric city commuter.

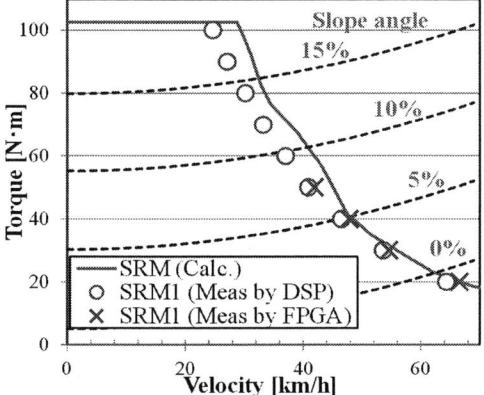

Fig. 15. Expected driving performance of the EV.

Fig. 16. View of the driving test.

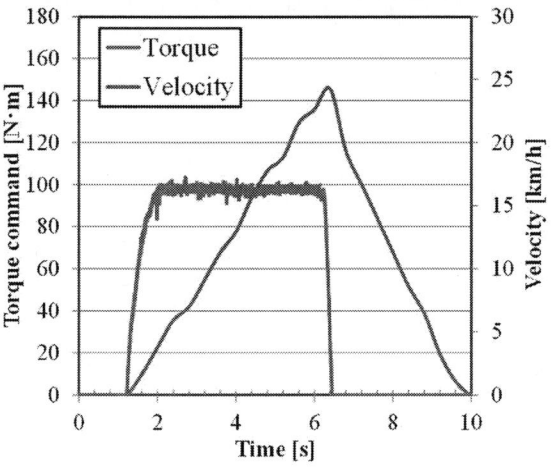

Fig. 17. Driving performance of the EV.

VI. CONCLUSIONS

In the paper, for the purpose of development of the in-wheel motor for an electric city commuter, the double stator structure axial-flux switched reluctance motor (DSAFSRM) was designed, manufactured and tested.

The experimental results indicate the torque of DSAFSRM is comparable to Base IPMSM at same size. At torque characteristics depend on rotational speed, both of simulated value and measured value are superior to Base IPMSM.

In addition, the prototype of drive unit for the prototype DRAFSRMs was designed and manufactured to equip the prototype DRAFSRMs to the target vehicle. As a result, the drive unit provided the result that is comparable to the device which is used conventionally. Therefore, the possibility that we can run as an electric city commuter of the simple substance is made. Finally, the driving test is done successfully.

REFERENCES

[1] Kodai Sone, Masatsugu Takemoto, Satoshi Ogasawara, Kenichi Takezaki, and Hidekatsu Akiyama : "A Ferrite PM In-Wheel Motor Without Rare Earth Materials for Electric City Commuters", IEEE Transactions on Industrial Electronics, Vol.48, No.11, pp.2961-2964 (2012)

[2] X. D. Xue, K. W. E. Cheng, T. W. Ng, and N. C. Cheung : "Multi-Objective Optimization Design of In-Wheel Switched Reluctance Motors in Electric Vehicles", IEEE Transactions on Industrial Electronics, Vol.57, No.9, pp.2980-2987 (2010)

[3] Hiroki Goto, Ayumu Nishimiya, Hai-Jiao Guo, and Osamu Ichinokura : "Instantaneous torque control using flux-based commutation and phase-torque distribution technique for SR motor EV", The international journal for computation and mathematics in electrical and electronic engineering, Vol.29, No.1, pp.173-186 (2010)

[4] Y. Ono, K. Nakamura and O. Ichinokura : "Basic Examination of Congiguration of Axial-gap SR Motor", Journal of the Magnetics Society of Japan, Vol.35, No.2, pp.106-111 (2011)

[5] Tohoru Shibamoto, Kenji Nakamura, Hiroki Goto, and Osamu Ichinokura : "A Design of Axial-gap Switched Reluctance Motor for In-Wheel Direct-Drive EV", The 20th International Conference on Electrical Machines (ICEM2012), FF-001678 (2012)

[6] T. Tokita, H. Goto, and O. Ichinokura : "Reducing Acoustic Noise of Axial-Gap SRMs by Decreasing Axial Electromagnetic Force Ripple (in Japanese)", IEEJ Transaction on Industry Applications , Vol.136, No.4, pp.248-253 (2016)

[7] Keisuke Takase, Hiroki Goto, and Osamu Ichinokura : "Examination of prototype double stator axial-gap SR motor for in-wheel drive (in Japanese)", The Papers of Technical Meeting on Rotating Machinery, RM-15-146 (2015)

[8] Hiroki Goto, Keisuke Takase, and Osamu Ichinokura : "An In-Wheel Axial-Gap SR Motor for Electric City Commuters", EVTeC and APE Japan, 20169099 (2016)

The 2018 International Power Electronics Conference

Torque Ripple Reduction Using Asymmetric Flux Barriers in Synchronous Reluctance Motor

Yuuto Yamamoto[1*], Shigeo Morimoto[1], Masayuki Sanada[1] and Yukinori Inoue[1]
1 Department name, Osaka Prefecture University, Sakai, Japan
*E-mail: sxb01199@edu.osakafu-u.ac.jp

Abstract— **A synchronous reluctance motor is an inexpensive variable-speed motor that is receiving increased attention as a rare-earth-free motor. However, this motor has the problem of having a high torque ripple, which causes unwanted noise and vibration. The focus of this study was to minimize the torque ripple of a synchronous reluctance motor by using an asymmetric flux barrier structure in the rotor. The effects of various flux barrier configurations on torque performance were examined, and two different flux barrier structures were combined by lamination in an asymmetric rotor structure in order to reduce the torque ripple.**

Keywords— *Synchronous reluctance motor, Rare-earth-free motors, Torque ripple reduction, Asymmetric flux barriers*

I. INTRODUCTION

Currently, permanent magnet synchronous motors (PMSMs) with rare-earth permanent magnets are used for various applications because of their high efficiency and wide range of operating speed [1]. However, rare-earth materials are expensive and can be produced in specific areas. Therefore, it is necessary to reduce the use of rare-earth materials [2]-[3]. To replace PMSMs using rare-earth elements, synchronous reluctance motors (SynRMs) are receiving increased attention as rare-earth-free motors [4]-[6].

In general, SynRMs are inexpensive variable-speed motors because they consist of only the core and winding. However, the disuse of permanent magnets causes the torque and efficiency of SynRMs to be inferior to those of PMSMs. In addition, SynRMs, which utilize reluctance torque, have the disadvantage of having a higher torque ripple [7]-[11]. Thus, a suitable structure that achieves a large average torque and the low torque ripple should be examined.

This study investigated how to reduce the torque ripple without also reducing the average torque using an asymmetric flux barrier structure, which is simpler than other torque ripple reduction methods. First, the influence of changes in the flux barrier angle on the torque characteristics was examined. Based on the results, an asymmetric rotor structure combining two kinds of flux barrier shapes was developed and analyzed by the two-dimensional finite element method (FEM). The torque ripple was reduced further when two different models

were combined by stacking and changing their size ratio. The performance of the proposed SynRM using asymmetric flux barriers was compared with that of a reference model.

II. CHARACTERISTICS OF REFERENCE MODEL

Table I shows the common specifications of the analysis models and analytical conditions in this study. Fig. 1 shows the structure of the reference model (Type-R). The stator has 4 poles and a distributed winding with 36 slots. The rotor of the reference model has 6 flux barriers, where the ratio of the slit width to the rib width is 2:1. The wide inner flux barrier of Type R was intended to reduce the *q*-axis inductance.

TABLE I
COMMON SPECIFICATIONS OF ANALYSIS MODELS
AND ANALYTICAL CONDITIONS

Item (Unit)	Value
Number of turns per slot	16
Stator outer diameter (mm)	204
Stator inner diameter (mm)	127
Rotor outer diameter (mm)	126
Rotor inner diameter (mm)	48
Air gap length (mm)	0.5
Stack length (mm)	90
Winding resistance (20°C) (Ω)	0.268
Magnetic steel sheet	50H470
Rated phase current (A)	16.4
Rotational speed (min⁻¹)	3000

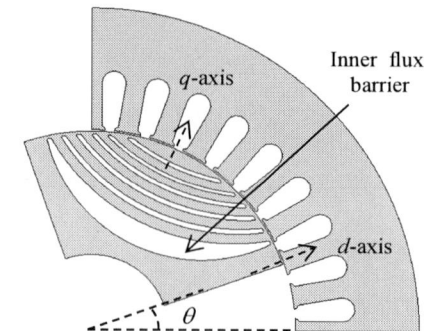

Fig. 1. Structure of reference model.(Type-R)

3197

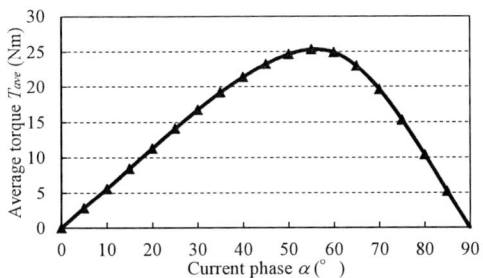

Fig. 2. Average torque versus current phase characteristics of Type-R (I_e =16.4 A).

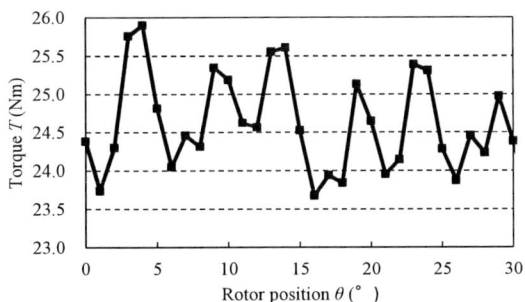

Fig. 3. Instantaneous torque versus rotor position angle in Type-R (I_e =16.4 A, α = 55°).

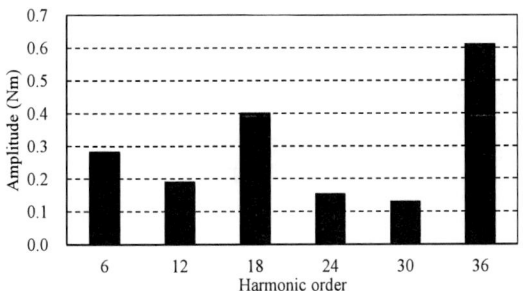

Fig. 4. Amplitude of harmonic components for Type-R (I_e =16.4 A, α = 55°).

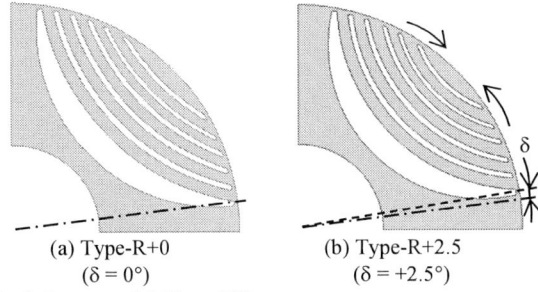

(a) Type-R+0   (b) Type-R+2.5
(δ = 0°)   (δ = +2.5°)

Fig. 5. Rotor model (Type-Rδ).

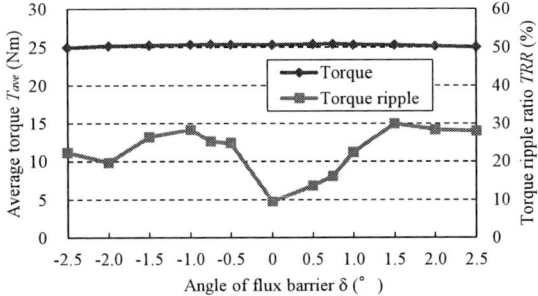

Fig. 6. Average torque and torque ripple versus flux barrier angle (I_e = 16.4 A, α = 55°).

As can be seen, torque ripple occurs approximately at multiples of 6, and in particular, the 6th, 18th, and 36th harmonic components are large. This is attributed to the positional relationship between the slots of the stator and the flux barriers of the rotor.

III. EFFECT OF FLUX BARRIER ANGLE ON TORQUE CHARACTERISTICS

Fig. 5 shows the analytical rotor structure. In this model, all six flux barriers are shifted by δ (°) in a tangential direction with respect to the barrier position of the reference model; we refer to this model as Type-Rδ.

Fig. 6 shows the torque and torque ripple versus shift angle of the flux barrier. The average torque is almost constant over the entire δ region. However, the torque ripple ratio changes depending on δ. It has a minimum value of 9.52 % when δ is 0°. The torque ripple ratio has a maximum value of 29.9% when δ is 1.5°.

Fig. 7 shows the amplitude of the harmonic components versus the flux barrier angle. The amplitude of the 6th harmonic component is almost constant over the entire δ region. However, the changes in amplitudes of the 18th and 36th harmonic components are large.

Fig. 8 shows the phase of the harmonic components versus the angle of the flux barrier. The phase of the 6th harmonic component is almost constant over the entire δ region. However, the phase of the 18th harmonic component increases monotonically, and the phase of the 36th harmonic component changes sharply between δ = -2.5° and δ = -2.0° and between δ = 0° and δ = 0.5°.

Fig. 2 shows the average torque versus the current phase, where the phase current I_e was 16.4 A and the current phase α represents the leading angle of the current vector from the d-axis. As shown in Fig. 2, the maximum torque for Type-R is 25.31 Nm when α = 55°.

Fig. 3 shows the instantaneous torque versus the rotor position angle when I_e = 16.4 A and α = 55°. Here, θ is the geometrical rotor position angle. The torque ripple ratio TRR is defined by

$$TRR = \frac{\Delta T}{T_{ave}} = \frac{T_{MAX} - T_{MIN}}{T_{ave}} \times 100(\%), \quad (1)$$

where ΔT is the difference between the maximum torque T_{MAX} and the minimum torque T_{MIN}, and T_{ave} is the average torque. As shown in Fig. 3, T_{MAX} is 26.7 Nm and T_{MIN} is 24.3 Nm; thus, the torque ripple ratio TRR for Type-R is 9.52 %.

Fig. 4 shows the harmonic components of the instantaneous torque waveform.

3198

The 2018 International Power Electronics Conference

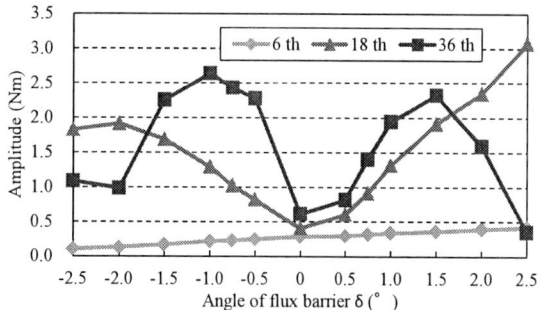

Fig. 7. Amplitude of harmonic components versus flux barrier angle for Type-Rδ (I_e =16.4 A, α = 55°).

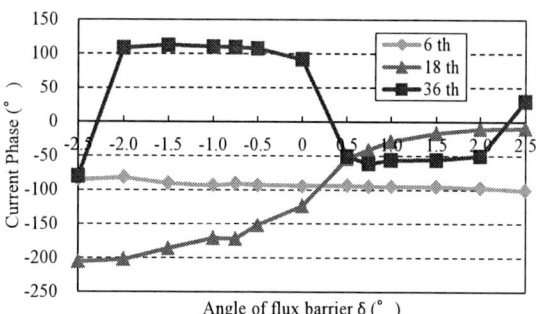

Fig. 8. Phase of harmonic components versus flux barrier angle for Type-Rδ (I_e =16.4 A, α = 55°).

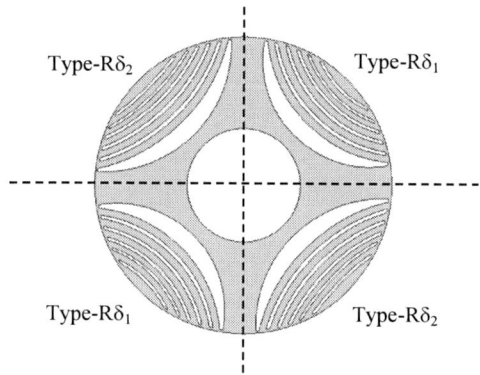

Fig. 9. Rotor structure in asymmetric model (Type-R$\delta_1\delta_2$).

These results confirmed that the flux barrier angle influences the torque ripple and the amplitude and phase of the harmonic components, especially the 18th and 36th harmonic components.

IV. MODEL COMBINING ROTORS WITH DIFFERENT FLUX BARRIER ANGLES

The results in section III suggest that the torque ripple can be reduced by using an asymmetric flux barrier structure and canceling the harmonic components for each model.

Fig. 9 shows the proposed rotor structure formed by combining two flux barrier designs with different angles δ_1 and δ_2 (Type-R$\delta_1\delta_2$).

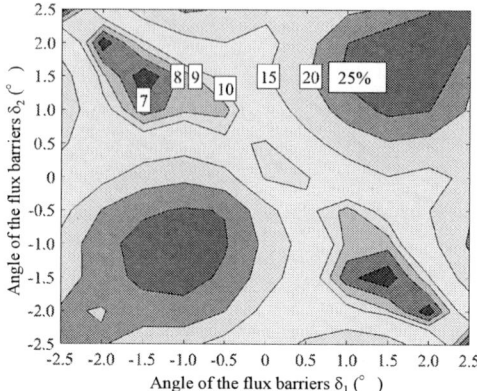

Fig. 10. Torque ripple ratio map for Type-R$\delta_1\delta_2$ with step size of 0.5° (I_e =16.4 A, α = 55°)

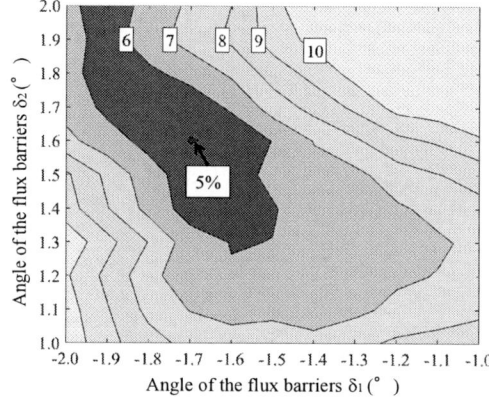

Fig. 11. Torque ripple ratio map for Type-R$\delta_1\delta_2$ with step size of 0.1° (I_e =16.4 A, α = 55°).

Fig. 10 shows the torque ripple ratio map for Type-R$\delta_1\delta_2$ when I_e is 16.4 A and α is 55°. The step size for the flux barrier angle is 0.5°. The torque ripple ratios in Fig. 10 were calculated simply by averaging the instantaneous torques for Type-Rδ_1 and Type-Rδ_2. The model with the lowest torque ripple ratio in Fig. 10 was Type-R-1.5+1.5, with a torque ripple ratio of 6.2 %.

As shown in Fig. 10, the torque ripple for a combination of δ_1 = 1.0 ~ 2.0° and δ_2 = -2.0 ~ -1.0° or δ_1 = -2.0 ~ -1.0° and δ_2 = 1.0 ~ 2.0° is especially low. Hence, the step size for the flux barrier angles was finely adjusted to 0.1° between δ_1 = -2.0 ~ -1.0° and δ_2 = 1.0 ~ 2.0°. Fig. 11 shows the torque ripple ratio map after adjusting the step size of flux barrier angle to 0.1°. The model with the lowest torque ripple ratio is Type-R-1.7+1.6, with a torque ripple ratio of 5.0 %, which is 20.1 % lower than that for Type-R-1.5+1.5.

Fig. 12 shows the instantaneous torque versus rotor position for Type-R-1.7, Type-R+1.6, and Type-R-1.7+1.6. By combining Type-R-1.7 and Type-R+1.6, the instantaneous torque for each model could be canceled, and the torque ripple for Type-R-1.7+1.6 was reduced.

Table II shows that the amplitudes and the phases of the harmonic components for Type-R-1.7, Type-R+1.6,

3199

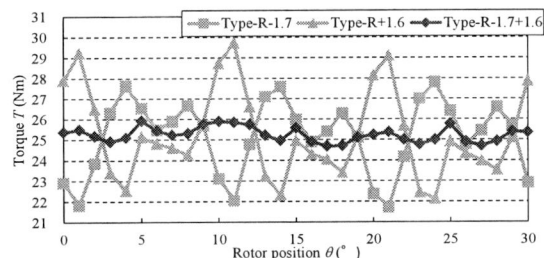

Fig. 12. Instantaneous torque versus rotor position ($I_e = 16.4$ A, $\alpha = 55°$).

TABLE II
AMPLITUDE AND PHASE OF HARMONIC COMPONENTS

		Type-R -1.7	Type-R +1.6	Type-R -1.7+1.6
6th	Amplitude (N·m)	0.16	0.37	0.27
	Phase (°)	-89.09	-96.44	-94.33
18th	Amplitude (N·m)	1.70	1.97	0.15
	Phase (°)	169.33	-14.34	-35.76
36th	Amplitude (N·m)	1.73	1.94	0.30
	Phase (°)	113.20	-53.69	-13.10

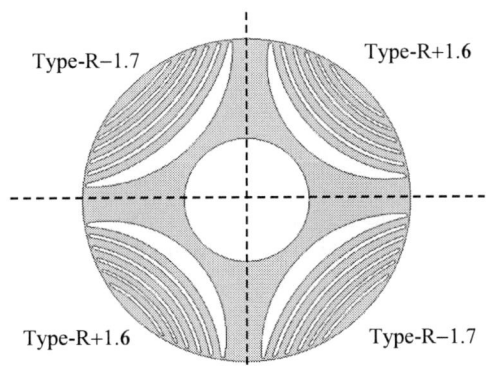

Fig. 13. Proposed rotor structure for Type-R-1.7+1.6.

TABLE III
TORQUE CHARACTERISTICS AT MTPA
($I_e = 16.4$ A)

Item (Unit)	Type-R-1.7+1.6 (calculated)	Type-R-1.7+1.6 (FEM)
α (°)	55	55
T_{ave} (N·m)	25.24	25.29
T_{MAX} (N·m)	25.93	26.08
T_{MIN} (N·m)	24.68	24.77
ΔT (N·m)	1.25	1.31
TRR (%)	4.95	5.20

and Type-R-1.7+1.6, respectively. Comparing the 18th and 36th harmonic components, the amplitudes of Type-R-1.7 and Type-R+1.6 are almost equal, with a phase

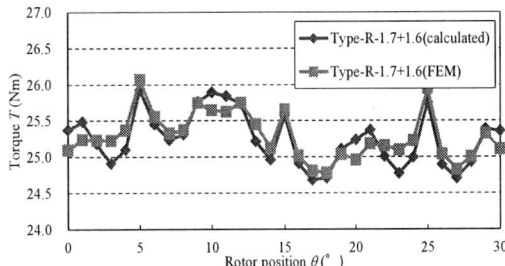

Fig. 14. Instantaneous torque versus rotor position ($I_e = 16.4$ A, $\alpha = 55°$).

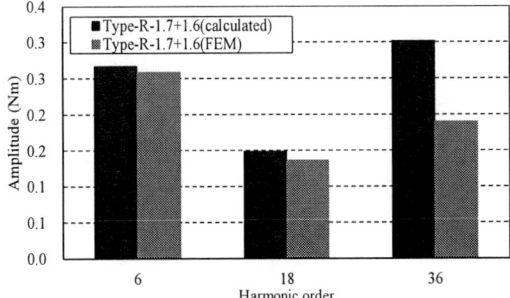

Fig. 15. Amplitude of harmonic components ($I_e = 16.4$ A, $\alpha = 55°$).

difference of approximately 180°. The torque ripple was reduced by canceling out of the 18th and 36th harmonic components of the instantaneous torques for Type-R-1.7 and Type-R+1.6.

V. EVALUATION OF ASYMMETRIC MODEL BY FINITE ELEMENT METHOD

The torque ripple ratios in Fig. 10 and Fig. 11 were calculated by simply averaging the instantaneous torques for each model. Therefore, in this section, we describe an analysis of the proposed asymmetric model (Type-R-1.7+1.6) by the two-dimensional FEM and compare the results with those in the previous section. Fig. 13 shows the proposed rotor structure for Type-R-1.7+1.6.

Table III presents the characteristics of Type-R-1.7+1.6. Type-R-1.7+1.6 (calculated) was calculated simply by averaging the instantaneous torques for each model of Type-R-1.7 and Type-R+1.6. Type-R-1.7+1.6 (FEM) was calculated by the FEM under maximum torque per ampere (MTPA) control. Fig. 14 shows the instantaneous torque versus rotor position angle for Type-R-1.7+1.6 (calculated) and Type-R-1.7+1.6 (FEM). As shown in Table III, the average torque for Type-R-1.7+1.6 (FEM) is almost equal to that for Type-R-1.7+1.6 (calculated), and the torque ripple for Type-R-1.7+1.6 (FEM) is lower by 4.8 % than that for Type-R-1.7+1.6 (calculated).

As shown in Fig. 14, the waveforms of the instantaneous torque for Type-R-1.7+1.6 (calculated) and Type-R-1.7+1.6 (FEM) are different. Fig. 15 shows the amplitude of the 6th, 18th, and 36th harmonic components for Type-R-1.7+1.6 (FEM). As shown in Fig.

3200

The 2018 International Power Electronics Conference

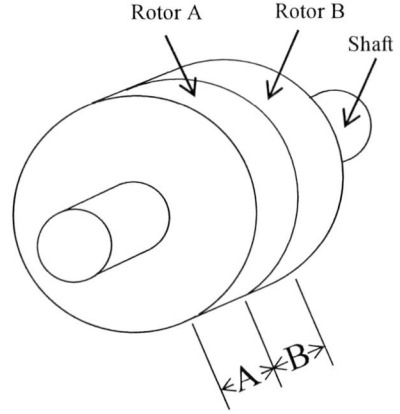

Fig. 16. Structure of rotor formed by stacking two designs.

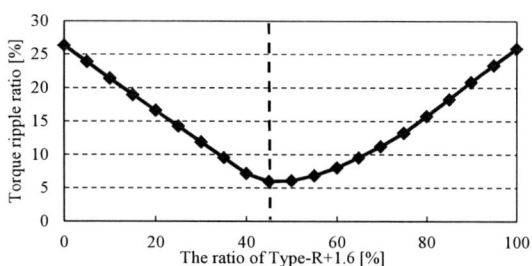

Fig. 17. Torque ripple ratio versus ratio for Type-R+1.6
(I_e = 16.4 A, α = 55°).

TABLE IV
TORQUE CHARACTERISTICS UNDER MTPA CONTROL
(I_e = 16.4 A)

Item (Unit)	Type-R-1.7+1.6	Type-LM
α (°)	55	55
T_{ave} (N·m)	25.24	25.24
T_{MAX} (N·m)	25.93	25.99
T_{MIN} (N·m)	24.68	24.75
ΔT (N·m)	1.25	1.24
TRR (%)	4.95	4.90

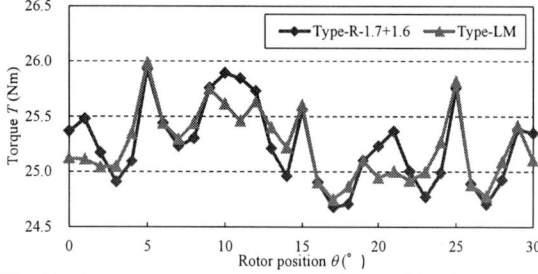

Fig. 18. Instantaneous torque versus rotor position for Type-
R-1.7+1.6 and Type-LM (I_e = 16.4 A, α = 55°).

15, the 6th, 18th, and 36th harmonic components for
Type-R-1.7+1.6 (FEM) are lower than those for Type-R-

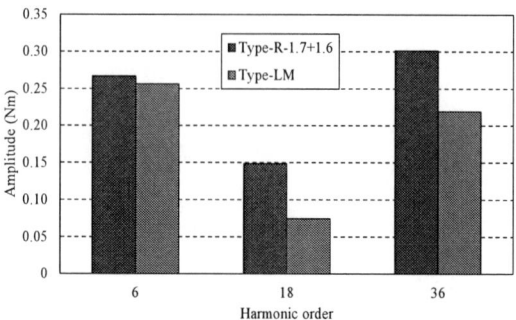

Fig. 19. Amplitude of harmonic components for Type-R-
1.7+1.6 and Type-LM (I_e = 16.4 A, α = 55°).

1.7+1.6 (calculated). Though the torque ripple ratio for
Type-R-1.7+1.6 (FEM) is higher than that for Type-R-
1.7+1.6 (calculated), the harmonic components for Type-
R-1.7+1.6 (FEM) are lower than those for Type-R-
1.7+1.6 (calculated).

VI. COMBINATION OF TWO DESIGNS IN ONE ROTOR

In section IV, an asymmetric flux barrier structure was
used to reduce the torque ripple. In this section, we
discuss how the torque ripple can be reduced by stacking
two different rotor designs on the shaft and changing
their size ratio. Fig. 16 shows the structure of the rotor
stack. In the previous section, the combination of Type-
R-1.7 and Type-R+1.6 showed the lowest torque ripple.
Therefore, our next step was to reduce the torque ripple
further by combining Type-R-1.7 (Rotor A) and Type-
R+1.6 (Rotor B) as shown.

Fig. 17 shows the torque ripple ratio versus the ratio
for Type-R+1.6. The torque ripple ratio is the lowest, at
4.90%, when the size ratio of Type-R-1.7 is 55 % and the
size ratio of Type-R+1.6 is 45 %. This optimal model is
called Type-LM.

Table IV shows the calculated torque characteristics
for Type-R-1.7+1.6 and Type-LM under MTPA control.
Fig. 18 shows the instantaneous torque versus rotor
position. As shown in Table IV, the average torques for
Type-R-1.7+1.6 and Type-LM are equal and the torque
ripple and torque ripple ratio for Type-LM is lower than
that for Type-R-1.7+1.6.

Fig. 19 shows the amplitude of the harmonic
components. The 6th, 18th, and 36th harmonic
components for Type-LM are lower than those for Type-
R-1.7+1.6. In particular, note that the 18th harmonic
component for Type-LM is 49.8 % lower than that for
Type-R-1.7+1.6.

As shown in Table II, the stacking ratio of the
amplitude of 18th and 36th harmonic component of
Type-R-1.7 to Type-R+1.6 is about 45% and 55% and the
18th harmonic component and 36th harmonic component
are canceled.

VII. COMPARISON OF TYPE-R AND TYPE-LM

In this section, the analysis results for the final model
(Type-LM) are compared with those for the reference

3201

TABLE V
TORQUE CHARACTERISTICS FOR TYPE-R AND TYPE-LM
UNDER MTPA CONTROL
(I_e = 16.4 A)

Item (Unit)	Type-R	Type-LM
α (°)	55	55
T_{ave} (N·m)	25.31	25.24
T_{MAX} (N·m)	26.69	25.99
T_{MIN} (N·m)	24.27	24.75
ΔT (N·m)	2.41	1.24
TRR (%)	9.54	4.90
η (%)	94.92	94.96

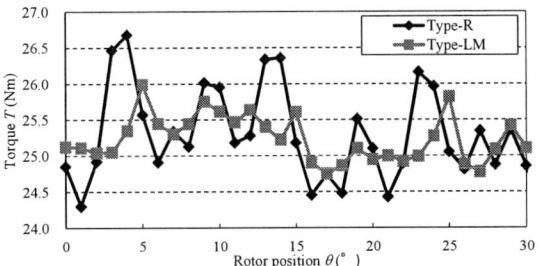

Fig. 20. Torque versus rotor position for Type-R and Type-LM (I_e = 16.4 A, α = 55°).

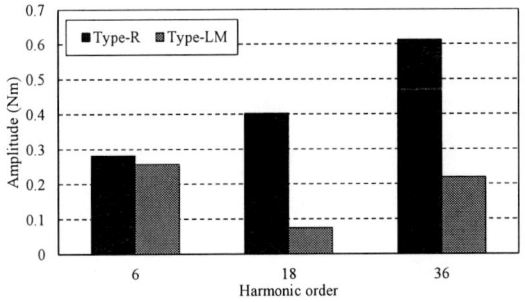

Fig. 21. Amplitude of harmonic components for Type-R and Type-LM (I_e = 16.4 A, α = 55°).

model (Type-R). Table V shows the characteristics for Type-R and Type-LM under MTPA control, and Fig. 20 shows the instantaneous torque versus rotor position angle for Type-R and Type-LM.

As shown in Table III and Fig. 20, the average torque and efficiency for Type-LM are the same as those for Type-R, and the torque ripple for Type-LM is 44.7 % lower than that for Type-R. Fig. 21 shows the amplitudes of the 6th, 18th, and 36th harmonic components for Type-R and Type-LM. The 18th and 36th harmonic components for Type-LM are lower than those for Type-R because of the asymmetric structure.

VIII. CONCLUSIONS

This study examined how to reduce the torque ripple in an SynRM without reducing the average torque by using an asymmetric flux barrier structure.

The analysis results showed that the flux barrier angle influences the torque ripple and the amplitude and phase of the harmonic components, especially the 18th and 36th harmonic components. Consequently, an asymmetric rotor structure having two flux barrier designs (Type-R-1.7+1.6) was proposed. By combining two different flux barrier structures, which cancelled the 18th and 36th harmonic components for each model, the torque ripple was greatly reduced. The torque ripple was further reduced by adjusting the size ratio of the Type-R-1.7 and Type-R+1.6 rotor components. The torque for Type-LM, the proposed asymmetric model with optimal stacking ratio, was equal to that for the reference model Type-R, and the torque ripple for Type-LM was 48.6 % lower than that for Type-R.

REFERENCES

[1] S. Morimoto, Y. Asano, T. Kosaka, and Y. Enomoto, "Recent Technical Trends in PMSM," 2014 International Power Electronics Conference, pp. 1997-2003, 2014.

[2] T. Ota, "Rare Earth Resources and Related Industries in Japan," in Journal of MMIJ, vol. 127, pp. 549-557, 2011.

[3] I. Boldea, L. N. Tutelea, L. Parsa, and D. Dorrell, "Automotive Electric Propulsion Systems With Reduced or No Permanent Magnets: An Overview," IEEE Trans. on Industry Electronics, vol. 61, no. 10, 2014.

[4] R. R. Moghaddam, F. Magnussen, and C. Sadarangani, "Theoretical and Experimental Reevaluation of Synchronous Reluctance Machine," IEEE Trans. on Industry Electronics, vol. 57, no. 1, pp. 6-13, 2010.

[5] K. Kim, J. S. Ahn, S. H. Won, J. Hong, and J. Lee, "A Study on the Optimal Design of SynRM for the High Torque and Power Factor," IEEE Trans. on Magnetics, vol. 43, no. 6, pp. 2543-2545, 2007.

[6] A. Vagati, M. Chiampi, M. Pastorelli, and M. Repetto, "Design Refinement of Synchronous Reluctance Motors Through Finite-Element Analysis," IEEE Trans. on Industry Applications, vol. 36, no. 4, pp. 1094-1102, 2000.

[7] A. Vagati, M. Pastorelli, G. Franceschini, and S. C. Petrache,"Design of Low-Torque-Ripple Synchronous Reluctance Motors, " IEEE Trans. on Industry Applications, vol. 34, no. 4, pp. 758-765, 1998.

[8] E. Howard, M. J. Kamper, and S. Gerber, "Asymmetric Flux Barrier and Skew Design Optimization of Reluctance Synchronous Machines," IEEE Trans. on Industry Applications, vol. 51, no. 5, pp. 3751-3760, 2015.

[9] N. Bianchi, S. Bolognani, D. Bon, and M. D. Pré, "Torque Harmonic Compensation in a Synchronous Reluctance Motor," IEEE Trans. on Energy Convention, vol. 23, no. 2, pp. 466-473, 2008.

[10] N. Bianchi, S. Bolognani, D. Bon, and M. D. Pré, "Rotor Flux-Barrier Design for Torque Ripple Reduction in Synchronous Reluctance and PM-assisted synchronous Reluctance Motors," IEEE Trans. on Industry Applications, vol. 45, no. 3, pp. 921-928, 2009.

[11] M. Sanada, K. Hiramoto, S. Morimoto, and Y. Takeda, "Torque Ripple Improvement for Synchronous Reluctance Motor Using an Asymmetric Flux Barrier Arrangement," IEEE Trans. on Industry Applications, vol. 40, no. 4, pp. 1076-1082, 2004.

The 2018 International Power Electronics Conference

On-board Single-Phase Electric Vehicle Charger with Active Front End

Theodore Soong[1*], Peter W. Lehn[1]

1 Electrical and Computer Engineering, University of Toronto, Toronto, Canada

*E-mail: theodore.soong@utoronto.ca

Abstract—Mass adoption of electric vehicles (EVs) is contingent on the availability of charging infrastructure. One solution to this issue is the introduction of on-board chargers, but such solutions typically require the installation of additional magnetic components that increase EV mass. An alternative approach is the dynamic re-deployment of drivetrain components for stationary charging. This work proposes an on-board single-phase charger that re-uses the traction inverter and motors. The system consists of a dual inverter drivetrain, which affords higher voltage charging compared to other systems. In addition, the system is able to operate bidirectionally and operate at any power factor for grid support services with real and reactive power without subjecting the motor to low frequency harmonic currents. The principle of operation and control of the proposed system is presented and demonstrated by simulation results.

Keywords—*Electric Vehicle, Charging Infrastructure for EVs, Power converters for EV, Battery charger*

I. INTRODUCTION

A major limitation to the mass adoption of electric vehicles (EVs) is the availability and rate of charging infrastructure. A convenient solution for EV owners is the installation of on-board chargers. However, these are typically separate systems installed into the vehicle with a low power rating [1]. An alternative approach to installing separate on-board chargers is to re-deploy drivetrain components to limit weight, volume, and cost while simultaneously increasing charging power [1]–[8]. The traction converter and motor would normally only be used as a drive when the vehicle is in motion, but is also re-deployed to enable single-phase charging when stationary.

Examples of single-phase on-board chargers are shown in Fig. 1 where they all utilize both the traction converters and motor to achieve higher power charging. For single-phase applications, [5], shown in Fig. 1(a), uses a diode rectifier connected to a motor with an accessible neutral point and the negative terminal of the battery. An alternative approach is applicable to systems with two motors [6] as shown in Fig. 1(b) where the single-phase ac grid is directly connected to the neutrals of two motors, while the system shares a single battery.

In all these cases, the operating range of these charging solutions is limited by the battery voltage of the car. These solutions leverage the use of standard drivetrains, which consist of a single battery and traction inverter. However, commercial EVs tend to use higher system

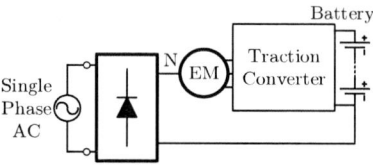

(a) PFC boost charger [5].

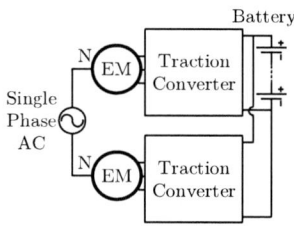

(b) Integrated charger using two-wheel drive [6].

Fig. 1. Existing integrated chargers.

voltages than personal EVs. This allows them to access higher voltage motors, and subsequently results in a higher efficiency system. To leverage passenger vehicle developments, dual inverter drives are of interest due to their ability to drive a higher voltage motor using two lower voltage traction inverters [9], [10], which are effectively operated as a single multilevel drive. This allows EV passenger vehicles access to higher efficiency motors while using standard batteries and traction inverters while also reducing harmonics and lowering $\frac{dv}{dt}$ stresses [9].

A complication of dual inverter drive systems is the charging of two separate batteries. To charge from an ac grid, solutions such as [7], shown in Fig. 2, use a separate on-board charger that charges the first battery while the dual inverters are used to charge the second battery. This implies that the high voltage advantage of the drive system is not leveraged when charging. This can be remedied by connecting the charging terminals across the differential nodes of the dual inverter drive as in [11], which utilizes the full voltage potential of the system both when motoring and charging. However, [11] only showed viability in dc charging. Thus, the purpose of this work is to fully utilize the potential of dual inverter drive systems as part of an on-board ac charger that provides EV owners with more accessible opportunity charging options while reducing components count and weight.

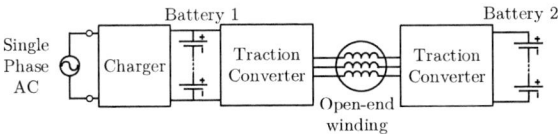

Fig. 2.   Semi-integrated charger using dual inverter drive [7].

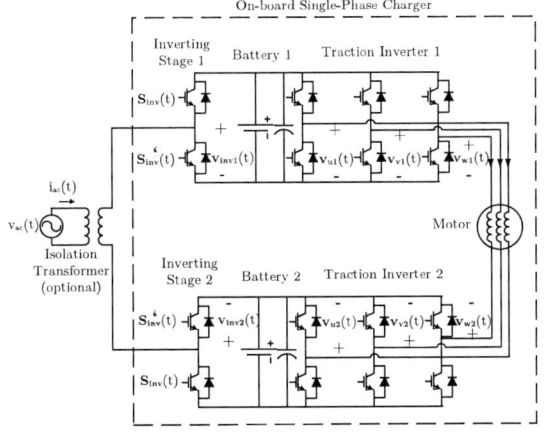

Fig. 3.   Proposed on-board single-phase charger.

II.   PROPOSED TOPOLOGY

The proposed on-board single-phase charger is shown in Fig. 3. The charger is comprised of four major parts, which are the motor, traction inverters, batteries, and inverting stages. In the charging case, the leakage inductance of the motor is used to avoid installation of additional magnetic components for charging. The only additional component are the inverting stages, which would be used in conjunction with the traction inverters to charge the batteries from a single-phase ac grid.

The inverting stages allow the system to produce bipolar voltages, irrespective of the ac grid voltage, provided that the peak ac grid voltage is lower than battery 1 and 2 combined. This implies the charger is capable of bidirectional power flow at any power factor making it suitable for grid support applications in addition to EV charging. The EV charger can therefore provide grid services like load levelling, peak shaving, frequency control, operate as a back-up power supply, etc. In addition, the system is capable of providing reactive power for grid voltage support. As an added benefit, the system is capable of fault blocking capabilities in case of fault events, allowing rapid electronic protection against faults within either the grid or motor. If only unity power factor charging operation were desired, the IGBTs of the inverting stage may be replaced with diodes without changes to the charger's principle of operation.

The AC power outlet is directly connected to the proposed on-board integrated charge and drive system. The isolation transformer may not be necessary depending on requirements. If required, the transformer itself would be

installed at the charging station. This implies that minimal charging infrastructure is needed for the charging station. It would be limited to either a cable or cable with isolation transformer. The proposed on-board single-phase charger serves a dual purpose as both a single-phase charger when the vehicle is stationary, and a traction inverter when the vehicle is in motion.

A.   Principle of Operation

The average model of the on-board charger is provided in Fig. 4, which divides the charger symmetrically into upper and lower charging stages where each charging stage consists of a traction inverter, battery, and inverting stage. Considering first the upper charging stage, each charging stage voltage $v_{chg,u1}(t)$, $v_{chg,v1}(t)$, and $v_{chg,w1}(t)$ represents the average voltage between the inverting stage input and the individual phase outputs of the traction inverter. The charging stage, as a whole, produces an ac voltage. Thus, these dependent voltage sources are defined as

$$v_{chg,u1}(t) = M_{u1}(t)V_{Batt1} \tag{1}$$
$$v_{chg,v1}(t) = M_{v1}(t)V_{Batt1} \tag{2}$$
$$v_{chg,w1}(t) = M_{w1}(t)V_{Batt1} \tag{3}$$

where $M_{u1}(t)$, $M_{v1}(t)$, and $M_{w1}(t)$ are equal to conventional ac modulation signals. They are real valued and range from -1 to 1. The synthesis of $v_{chg,u1}(t)$, $v_{chg,v1}(t)$, and $v_{chg,w1}(t)$ is based on the summation of the inverting stage voltage and traction inverter phase voltages. As the inverting stage is solely used for voltage inversion, the duty cycles sent to the traction inverter phases are found by first defining the inverting stage voltage $v_{inv1}(t)$ as

$$v_{inv1}(t) = S_{inv}(t)V_{Batt1} \tag{4}$$

where $S_{inv1}(t)$ is the switching state of the inverting stage and is an integer of value 0 or 1. Similarly, the output of the traction inverter phases are equal to

$$v_{u1}(t) = d_{u1}(t)V_{Batt1} \tag{5}$$
$$v_{v1}(t) = d_{v1}(t)V_{Batt1} \tag{6}$$
$$v_{w1}(t) = d_{w1}(t)V_{Batt1} \tag{7}$$

where the duty cycles are real valued and range from 0 to 1. The duty cycle values sent to the traction inverter are based on the inverting stage switch state $S_{inv}(t)$ and the modulation signal of the respective phase. Thus, the duty cycles are defined as

$$d_{u1}(t) = S_{inv}(t) - M_{u1}(t) \tag{8}$$
$$d_{v1}(t) = S_{inv}(t) - M_{v1}(t) \tag{9}$$
$$d_{w1}(t) = S_{inv}(t) - M_{w1}(t). \tag{10}$$

An equivalent set of equations may be written for the lower charging stage.

As can be seen from Fig. 4, the upper and lower charging stages are effectively series connected. This implies that the voltage produced by the charger is not limited by an individual battery voltage, but by the sum of both

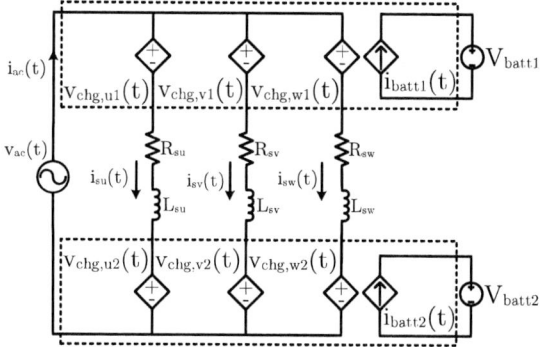

Fig. 4. Average model of proposed system.

batteries in the drivetrain. Therefore, the proposed on-board integrated charger is able to connect to higher voltage systems as compared to on-board integrated chargers based on standard drivetrains, which are limited by their single battery. As both batteries are identical and would be at a nearly equal state of charge, each charging stage would produce approximately half the grid voltage.

Exemplary waveforms of the upper charging stage operating at a power factor of 0.7 leading are provided in Fig. 5. The waveforms show the inverting stage and traction inverter phase voltages, which comprise the overall upper charging stage's ac voltage as described in (8) to (10). As previously discussed, the upper charging stage is expected to output a voltage approximately half that of $v_{ac}(t)$.

The current in each phase of the motor is controlled such that all three-phase currents are equal. Therefore, all phase voltages are equal assuming a symmetrical machine. In consequence, Fig. 5 shows $i_{su}(t)$, $i_{sv}(t)$, and $i_{sw}(t)$ are one third of $i_{ac}(t)$. These equal currents are zero sequence currents and the machine will not produce any average torque when only zero sequence currents are passed through the motor. Thus, this arrangement is suitable for stationary charging of the batteries. In addition to tracking identical currents, phase-shifted modulation may also be applied to the traction inverter to reduce switching harmonics in $i_{ac}(t)$ [11].

Using the system as described offers several benefits in design. As voltage inversion need only occur every half line cycle, the inverting stage may switch at a minimum of 60 Hz. This minimizes conversion loss as the inverting stage can be optimised for conduction losses only, which reduces cooling requirements and system weight. In addition, the traction inverter, which is designed to switch at several kHz as a motor drive, is ideally suited to track the fundamental grid frequency.

B. Energy balancing

In such a drive system, two separate energy storage units are used, which may be batteries, supercapacitors, fuel cells, or any combination thereof. In this work, batteries are used, but energy transfer from one battery to the other is still a necessity to ensure energy balance

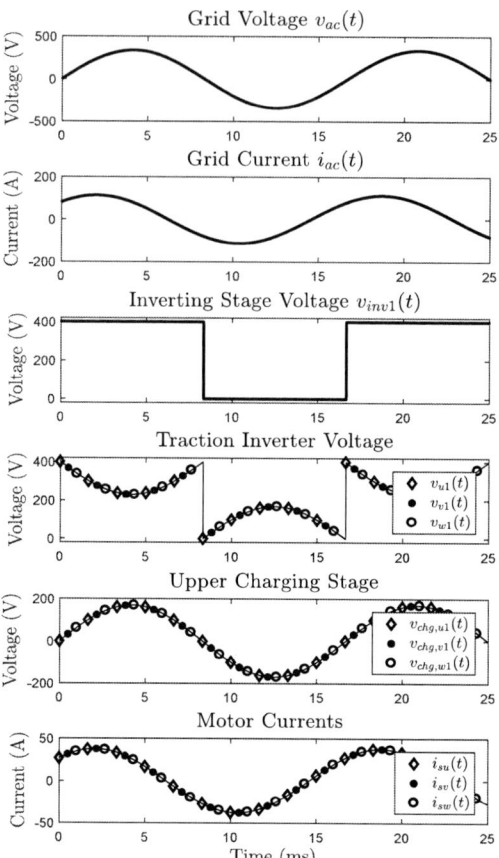

Fig. 5. Exemplary waveforms of the proposed charger operating at a power factor of 0.7 leading.

during charging. This can be achieved by asymmetrically modulating the upper and lower charging stages. However, it is best to first describe the system's nominal operation using the average model shown in Fig. 4. In the model, each phase of the charging stage is represented by a bipolar variable voltage source. In normal operation, as described in Section II-A, the upper and lower charging stage voltages are equal to

$$v_{chg,x1}(t) = M_{x1}(t)V_{batt1} \tag{11}$$
$$v_{chg,x2}(t) = M_{x2}(t)V_{batt2} \tag{12}$$

where subscript $x \in \{u, v, w\}$ denotes the phase. Both the upper and lower charging stage voltages, neglecting voltage drop across the motor impedances, are equal to $\frac{v_{ac}(t)}{2}$ as described in Section II-A. In addition, the grid current $i_{ac}(t)$ is divided equally amongst the three motor phases. Thus,

$$i_{sx}(t) = \frac{i_{ac}}{3} \tag{13}$$

for phase $x \in \{u, v, w\}$. From (11) to (13), the power into the upper and lower charging stages is therefore equal to

$$P_1 = Re\left\{\frac{V_{ac}}{2}I_{ac}^*\right\} \tag{14}$$

$$P_2 = Re\left\{\frac{V_{ac}}{2}I_{ac}^*\right\}. \tag{15}$$

where V_{ac} is the rms phasor of $v_{ac}(t)$ and I_{ac} is the rms phasor of $i_{ac}(t)$. As can be seen, the upper and lower charging stages receive equal power.

The proposed on-board fast charger can balance energy between the batteries of upper and lower charging stages by modifying the relative voltages between the upper and lower charging stages by a given increment $\delta(t)$ as shown in equations (16) to (18).

$$v_{chg,x1}(t) = (1 - \delta(t))M_{x1}(t)V_{batt1} \tag{16}$$

$$v_{chg,x2}(t) = (1 + \delta(t))M_{x2}(t)V_{batt2} \tag{17}$$

$$i_{sx}(t) = \frac{i_{ac}(t)}{3} \tag{18}$$

for phase $x \in \{u, v, w\}$. The resulting power of the upper and lower charging stages are unequal as seen in equation (19) and (20).

$$P_1 = Re\left\{\frac{(1-\delta)V_{ac}}{2}I_{ac}^*\right\} \tag{19}$$

$$P_2 = Re\left\{\frac{(1+\delta)V_{ac}}{2}I_{ac}^*\right\} \tag{20}$$

As the combined voltage of the upper and lower charging stages is still equal to $v_{ac}(t)$, the system is able to transfer energy between the upper and lower charging stages without affecting grid side power transfer.

III. CONTROL STRATEGY

The on-board single-phase charger must enable charging of the EV battery and meet grid requirements. While grid requirements may be met with filter components, the more lightweight solution is to control the on-board single-phase charger to produce voltages and currents with low distortion.

For the on-board single-phase charger, each charging stage may be considered as a bipolar voltage source, which can be used to control the ac grid current with low distortion using the switching method suggested in [11]. A high-level control scheme suitable for this purpose is shown in Fig. 6. To control the currents in each phase of the motor, three individual resonant current controllers are used to ensure the fundamental frequency current is regulated to meet grid code while also dividing the current equally amongst the motor phases so only zero sequence current flows in the motor.

The energy balance controller introduces a $\delta(t)$ reference as described in Section II-B. The $\delta(t)$ reference is used to modify the average power delivered to the upper and lower charging stage batteries, and is used to balance the energy between upper and lower batteries. The $\delta(t)$ reference is created using a PI controller, which regulates

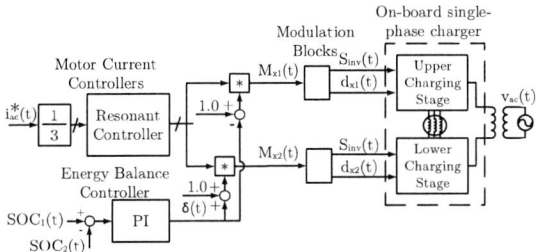

Fig. 6. Control overview where $x \in \{u, v, w\}$ denotes the phase.

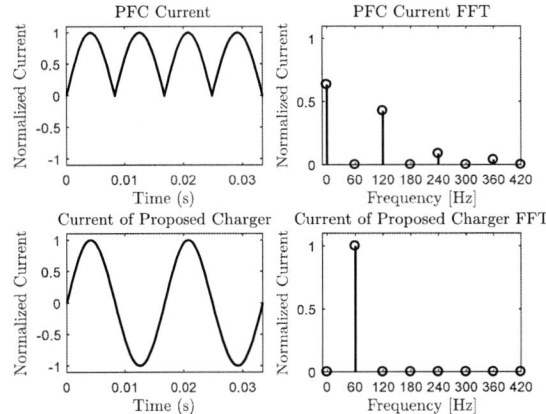

Fig. 7. Comparison of PFC type on-board charger and proposed charger currents.

the difference between the state of charge (SOC) of the upper and lower batteries. The output of the motor current controllers is modified by output of the energy balance controller to regulate the grid current while balancing the energy between the two batteries.

A benefit to the proposed charger is that the current conducted through the motor's leakage inductance is at the grid frequency. In comparison some chargers use a diode rectifier front-end [5], as shown in Fig. 1(a), where the motor must conduct a rectified grid current. Fig. 7 compares the low frequency current harmonics of the two methods. As can be seen, the rectified current contains a fundamental at 120Hz and harmonics while the proposed solution solely contains a fundamental at 60Hz. Therefore, the proposed charger has lower control bandwidth requirements by needing to only track the fundamental frequency current and lower losses as the higher frequency components of the rectified current would incur additional losses in the motor.

IV. RESULTS

A detailed model of the integrated charging topology is simulated in MATLAB using a PLECS toolbox. The high-level diagram of the system is shown in Fig. 3. Two identical battery packs are charged from a 240V single-phase ac source via the on-board single-phase charger and drive system. In this simulation, the battery packs are replaced by ideal voltage sources and the motor is

The 2018 International Power Electronics Conference

TABLE I. SIMULATION PARAMETERS

Grid Parameters	Symbol	Value
Grid voltage	V_{ac}	$240V_{ll,rms}$
Grid current	I_{ac}	$80A_{rms}$
Average Power	P_{avg}	19.2kW
On-board Fast Charger Parameters	Symbol	Value
Inductor resistance	R_s	0.3Ω
Inductance	L_s	0.5mH
Battery Voltage	V_{batt1}, V_{batt2}	400V
Switching frequency	f_{sw}	4.5kHz

assumed symmetric. System parameters are listed in Table I.

Three simulation results are provided to demonstrate system functionality. The first set of results are shown in Figure 8, the voltage and current quantities of the charger when operated at a power factor of 0.7 leading. For the purposes of illustration, the traction inverter voltages and upper charging stage voltages are filtered to show the underlying low frequency components of the voltages and only phase u is shown as the other phases are equal. This figure matches the ideal waveforms as shown in Fig. 5. As expected, the inductor currents per phase track one-third of the input current.

The second set of results are shown in Figure 9 where the charging system is operated at unity power factor while delivering power to the grid. Figures 8 and 9 show the ability of the system to operate at any desired power factor when either charging the EV batteries or providing grid support functionality.

The final result demonstrates the energy balance mechanism as described in Section II-B, which is shown in Fig. 10. Energy balance is achieved by commanding asymmetrical voltage magnitudes from the upper and lower charging stages, which is introduced by a voltage offset $\delta(t)$ between the upper and lower charging stage to charge one battery over the other. From inspection of Fig. 10, $v_{chg,u2}(t)$ is greater in magnitude than $v_{chg,u1}(t)$. Hence the average value of $p_{batt,2}(t)$ is greater than $p_{batt,1}(t)$, and energy balance may be achieved between the upper and lower charging stages.

V. CONCLUSION

Availability and rate of EV charging is a barrier to the mass adoption of EVs. On-board chargers address the availability of EV chargers, but are limited in charging rate. This work presents a single-phase on-board charger, which re-deploys a dual inverter drive and motor system to enable charging with the minimal addition of an inverting stage for each inverter. By utilizing existing high-power electronics, the proposed system is able to surpass the power rating of current state-of-the-art on-board chargers. In addition, a conventional front-end rectifier is avoided with the proposed inverting stages, which avoids unnecessary harmonics. This implies that control bandwidth requirements are reduced and system efficiency is increased. Furthermore, use of the dual inverter drive increases the voltage range of the system, which allows it to connect to higher voltage ac grids in comparison to single inverter systems.

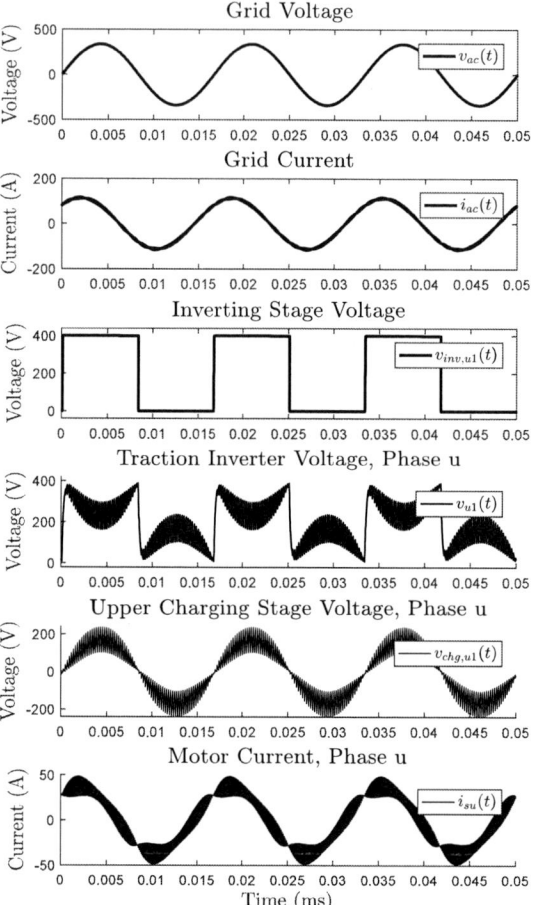

Fig. 8. On-board single-phase charger operating at a power factor of 0.7 leading.

The operating principle of the proposed system was presented with an energy balance mechanism to ensure the energy balance between the two batteries of the dual inverter drive and verified with simulation results. The results show the system is able to operate at any power factor and is even capable of bidirectional power flow while independently transferring unbalanced power to the two batteries. Thus, the vehicle is capable of high-power charging and grid support functionalities, such as voltage support, frequency regulation, and peak shaving.

REFERENCES

[1] M. Yilmaz and P. T. Krein, "Review of Battery Charger Topologies, Charging Power Levels, and Infrastructure for Plug-In Electric and Hybrid Vehicles," *IEEE Transactions on Power Electronics*, vol. 28, no. 5, pp. 2151–2169, may 2013.

[2] I. Subotic, E. Levi, M. Jones, and D. Graovac, "On-board integrated battery chargers for electric vehicles using nine-phase machines," in *2013 International Electric Machines & Drives Conference*. IEEE, may 2013, pp. 226–233.

[3] I. Subotic and E. Levi, "A review of single-phase on-board integrated battery charging topologies for electric vehicles," in

The 2018 International Power Electronics Conference

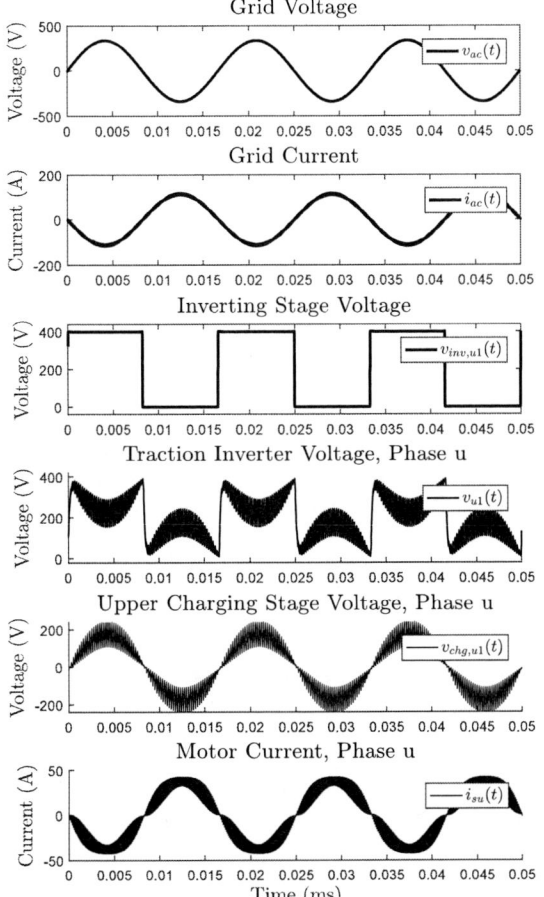

Fig. 9. On-board single-phase charger supplying power to the grid.

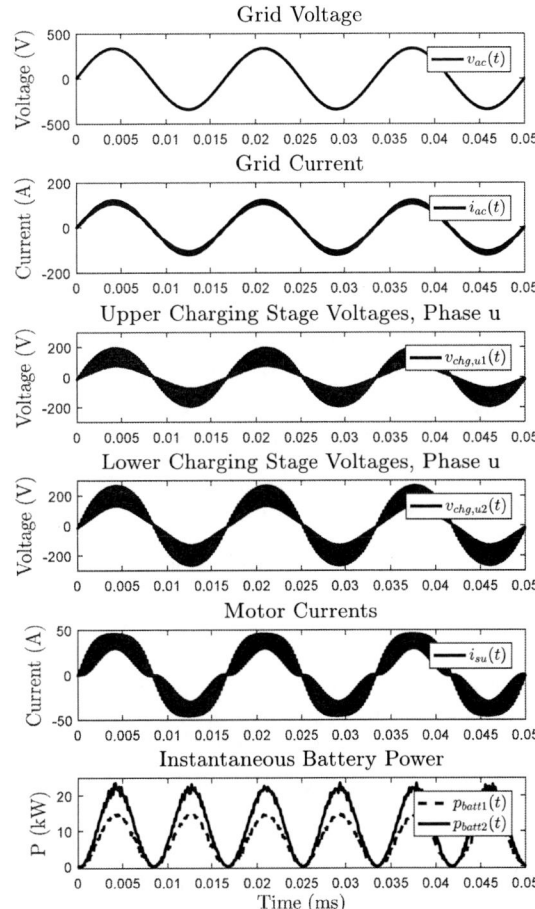

Fig. 10. On-board single-phase charger delivering unequal power to the two battery packs by introducing a voltage difference between traction inverters.

2015 IEEE Workshop on Electrical Machines Design, Control and Diagnosis (WEMDCD). IEEE, mar 2015, pp. 136–145.

[4] L. De-Sousa and B. Bouchez, "Combined electric device for powering and charging," 2010.

[5] G. Pellegrino, E. Armando, and P. Guglielmi, "An integral battery charger with power factor correction for electric scooter," *IEEE Transactions on Power Electronics*, vol. 25, no. 3, pp. 751–759, 2010.

[6] L. Tang and G. J. Su, "A low-cost, digitally-controlled charger for plug-in hybrid electric vehicles," in *2009 IEEE Energy Conversion Congress and Exposition, ECCE 2009*. IEEE, sep 2009, pp. 3923–3929.

[7] J. Hong, H. Lee, and K. Nam, "Charging Method for the Secondary Battery in Dual-Inverter Drive Systems for Electric Vehicles," *IEEE Transactions on Power Electronics*, vol. 30, no. 2, pp. 909–921, feb 2015.

[8] Y. Hu, C. Gan, W. Cao, C. Li, and S. Finney, "Split Converter-Fed SRM Drive for Flexible Charging in EV/HEV Applications," *IEEE Transactions on Industrial Electronics*, vol. 62, no. 10, pp. 6085–6095, 2015.

[9] D. Casadei, G. Grandi, A. Lega, and C. Rossi, "Multilevel operation and input power balancing for a dual two-level inverter with insulated DC sources," *IEEE Transactions on Industry Applications*, vol. 44, no. 6, pp. 1815–1824, 2008.

[10] M. Mengoni, A. Amerise, L. Zarri, A. Tani, G. Serra, and D. Casadei, "Control Scheme for Open-Ended Induction Motor Drives With a Floating Capacitor Bridge over a Wide Speed Range," *IEEE Transactions on Industry Applications*, vol. 53, no. 5, pp. 1–1, sep 2017.

[11] R. Shi, S. Semsar, and P. W. Lehn, "Constant Current Fast Charging of Electric Vehicles via DC Grid Using Dual Inverter Drive," *IEEE Transactions on Industrial Electronics*, vol. 0046, no. c, pp. 1–1, 2017.

3208

A Bidirectional Buffered Charging Unit for EV's (BBCU)

Alfred Rufer[1] Fellow IEEE, Gabriel Fernandez[1]
1 EPFL, Ecole Polytechnique Fédérale de Lausanne, CH 1015 Lausanne, Switzerland
alfred.rufer@epfl.ch

Abstract: **A bidirectional buffered charging unit for EV's is presented, allowing the charge with high power even if the line current capability is limited. The bidirectional buffer is connected to the AC line and is mainly dedicated to vehicles with on-board AC-DC chargers. Extensions are described for the integration of RES as well as for DC charging. The system is also dedicated to operate as a reactive power compensator, or to provide grid system services similar to V2G operation or other power smoothing functions.**
At the side of the buffer battery, the system presented uses a bidirectional DC-DC converter which can be realized in different technologies. A conventional solution with three interleaved channels is compared with a fast switching converter using SiC components.

1. INTRODUCTION

Charging stations for EV's are technically and commercially in strong expansion. Most of them consist today only of AC access points to the electrical distribution grid with simple accounting services. Some of the stations are completed with internet connections with the possibility for reservation of access hours or for remote indication of their occupation. Such systems provide energy to simple vehicles having their single or three phase AC chargers on-board.

Complementarily to such stations more advanced fast chargers with DC connection to the car battery appear more and more, using also "intelligent" connection to the car battery management system [1], [2], [3].

The multiplication of charging ports at the same place addresses the question of the current/power availability of the distribution grid, mainly being a low voltage system. Also load fluctuations for the system operators will be a topic in the future, especially when the fast charging systems are characterized by a high power level.

As a consequence, more and more investigations and proposals are made for the use of local energy storage devices used as power buffers or load equalizing systems, also known as power peak-shaving systems [4], [5].

The local power buffers present simultaneously the property of being able to deliver a power level used for an accelerated charge of an EV even if the local distribution system is not sufficient. The case of local grids powered from renewable sources is a good example of such situations.

Another example will be given by the case of a limited power access by private houses, where the charge of an EV with even a few kW's can bring the consumer in a limited situation when willing simultaneously use powerful devices as cooking or heating apparatus.

Together with the concept of power buffering, the question of the bidirectional power flow will be addressed. More and more investigations in the field of V2G (Vehicle-to-Grid) are made [6]. Controllable reversible EV chargers will be needed in general for such applications.

The local power buffers for EV charging stations can be developed as bidirectional facilities, and in relation with their energy capacity and power ability, such stations present the potential to become interesting players in the context of the exploitation of week grids.

The bidirectional power electronic interfaces to the grid are mostly based on the technique of VSC (Voltage Source Converters), and present in addition the faculty to provide reactive power and to be integrated in the concepts of the voltage support.

As a consequence, large or important swarms of buffered bidirectional charging units for EV's will become interesting partners with the distributors, especially in the context of more and more distributed generation and in the context of smart grids.

The general scheme of a Bidirectional Buffered Charging Unit is represented in Fig. 1.

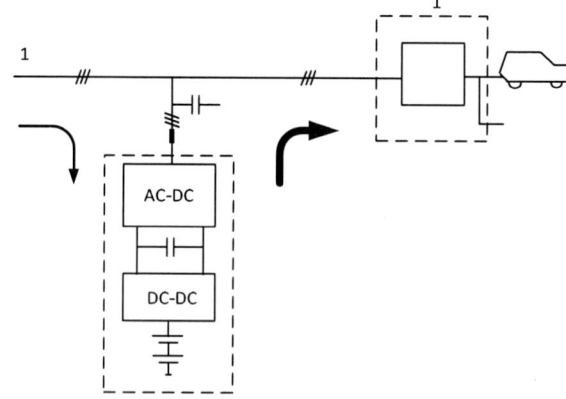

Fig. 1 The concept of the buffered charging unit

1.1 INTEGRATION OF RENEWABLE ENERGY SOURCES

The concept of the Buffered Charging Unit can find its place or have an added function in connection with RES (Renewable Energy Sources) like for example a photovoltaic system in a single family house. The installed power of the panels and its line inverter are not adapted for fast charging. Additionally, the intermediary buffer can play a role for the so-called day-to-night shifting. In such a system, the buffer battery can be charged from the PV panels during the day, and allows charging of the car battery during the end of the day or during the night. A specific design of the buffer battery capacity can, if needed, bridge the power need for charging over one or two days.

Figure 2a) shows a block diagram where the interface to the RES is added to the original system. The design of the buffer energy capacity and power capability in the context of RES should be evaluated accurately, together with the related costs. The number of days to bridge, the number of cars to be charged or the power level of fast charging can be selected as additional parameters for the economic study.

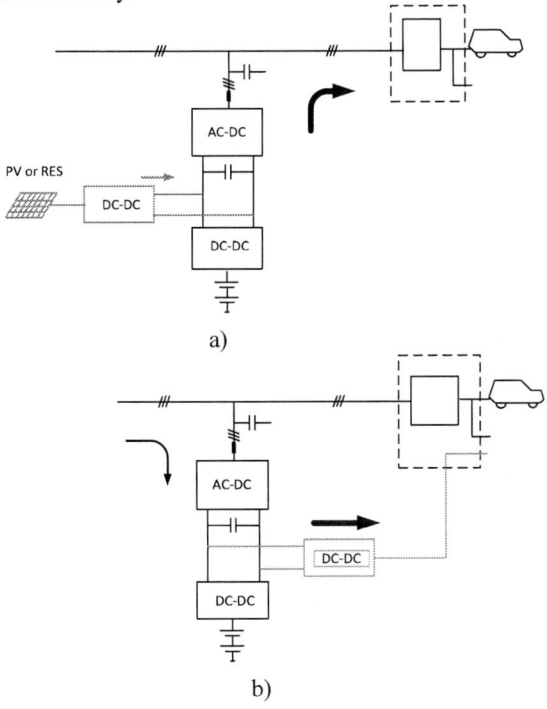

a)

b)

Fig. 2a) Integration with renewable sources, b) Extension for fast DC charging

1.2 A DEDICATED INTERFACE FOR DC CURRENT CHARGING

When the cars to be charged have a charging interface able to interconnect it to a DC current (fast) charging infrastructure, the first system represented in Fig. 1 completed with an AC-DC output converter will lead to an unnecessary cascade of conversions, resulting into higher costs and into reduced energy efficiency (multiple conversions). For such a case, an additional DC-DC converter allowing DC current charging directly from the buffer battery should be added. The block diagram of Fig. 2b) is illustrating this further interface.

1.3 THE POWER CYCLES OF THE BBCU

Figure 3 shows the typical cycles of the BBCU. The upper curve of the figure represents the power delivered by the primary line. In this scenario, the line current is maintained constant and corresponds to a current value of 20 A. The line current is maintained at its nominal value during charging of the buffer (first time-segment of the diagram), as well as during the buffering function when one or two cars are charged from the BBCU at a power level of each 22 kW (second segment of the diagram). The last segment of the diagram illustrates a very fast charge of one vehicle at a typical level of 44 kW.

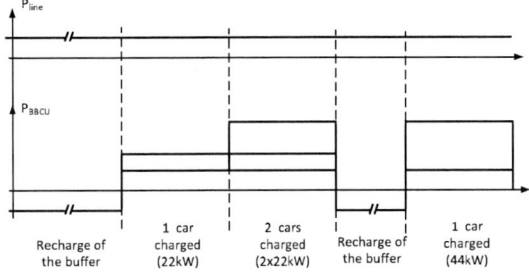

Fig. 3 Typical power cycles of the BBCU

2 THE POWER CIRCUITS OF THE BBCU

Figure 4 shows the power electronic circuits of the proposed BBCU. At the line side of the system, a bidirectional interface is represented, where the topology is based on a 3 Level NPC active rectifier [7]. This circuit is designed for the maximum power to be delivered by the BBCU, and can assume a corresponding high reactive power delivery. The topology of the 3 level NPC-converter is chosen in relation with the expected low distortion of the line current, together with a minimum of costs for the inductors of the output filter.

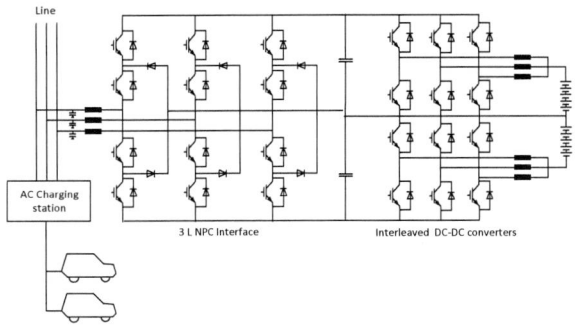

Fig. 4 The power circuits of the BBCU

Additionally, the 3-level converter allows to maintain the switching frequency at a low value, leading to reduced switching losses and to a higher efficiency. Even if the conduction losses of the 3-level inverter are slightly higher than those of a 2 level one, the global efficiency can be maintained at a reduced level [8], [9].

At the battery side of the BBCU, the two half-batteries are interfaced to the positive and to the negative rail of the intermediary circuit.

For a conventional design, circuits can be realized with a pair of three-channel interleaved DC-DC converters (Fig. 4) in order to reach an acceptable low ripple of the battery current with small inductors [10]. The upper and lower half voltages of the DC circuit allow the use of low voltage power devices of lower costs..

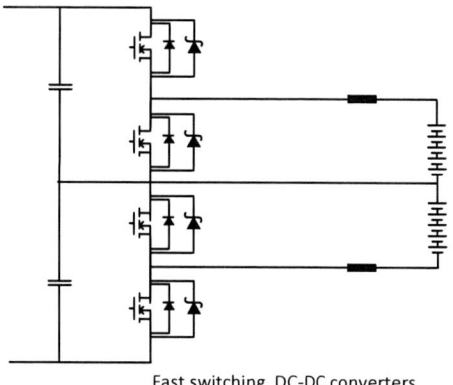

Fast switching DC-DC converters

Fig. 5 Battery interface with SiC components

Type of devices	I= [A]	L= [uH]	Nb. of c.	Unit price	Price
IGBT modules	50		6	34$	204$
Inductors	30	4000	6	173$	1038
Total costs					1242$
SiC moodules	120		2	358$	716$
Inductors	100	250	2	211$	422$
Total costs					1138$

Table 1 Comparison of costs

A more evolved design can be based on the use of modern fast switching SiC components, resulting into a simpler circuit, smaller inductors, but of a higher price of the modules.

The scheme of the SiC based battery interface is represented in Fig. 5.

A simple cost comparison is made in Table 1, based on prices of available components. The table shows that even if the SiC modules are of a considerable price in comparison with conventional IGBT's, the global price of the modern solution is slightly lower.

For this comparison, a switching frequency of 10 kHz is chosen for the IGBT classical solution and 50 kHz for the SiC converters.

3 CONTROL OF THE BBCU

Figure 6 represents a structural diagram of the control functions of the BBCU. A line side vector control of the AC currents is used, with an integrated DC voltage balance control based on carrier modification method [11]. The magnitude of the DC voltage is controlled through the magnitude of the active line current. The imposition of the reactive current component is given by an external reference.

At the side of the buffer battery, classical PI control is used in order to impose the battery current. This control includes also channel balancing strategies for the solution with interleaved channels [12]. The battery charging as the finishing functions are imposed through the battery management system BMS.

The limitation of the grid current and the compensation of the current demand of the car is achieved through the so-called injection control function. This controller is connected to the line active current controller as well to the battery current controllers with feed-forward signals.

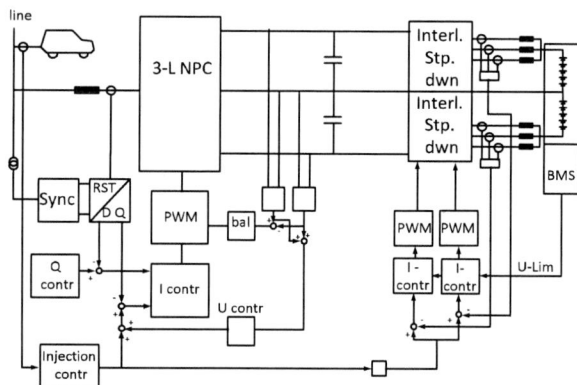

Fig. 6 Control diagram of the BBCU

4 THE BUFFER BATTERY BASED ON LITHIUM TITANATE

The dedicated function of the battery buffer and its capability to provide also power to the grid for compensation or smoothing functions lead to a high number of charges and discharges. As a consequence, a dedicated battery technology is chosen namely the Lithium Titanate technology. The main data of the buffer battery are indicated in Table 2

For the realization of an industrial BBCU facility, the total equipment is expected to be located inside of a standardized underground container as it is used for urban waste collectors (Fig. 8). This type of cabinet is a well-accepted standard and facilitates the specification of the construction.

Battery technology	Lithium Titanate
Type of the modules	Leclanché 936C08Titanate
Rated voltage (module)	46 V
Rated current	90 A
Maximum current	300 A
E (module)	4.2 kWh
Dimensions of one module	463 x 356 x 550 mm
Number of modules	2 x 5
Battery voltage	2 x 230 V
Energy of the buffer	2 x 21 kWh

Table 2 Main data of the buffer battery

The equivalent scheme of the complete buffer battery is given in Fig. 7.

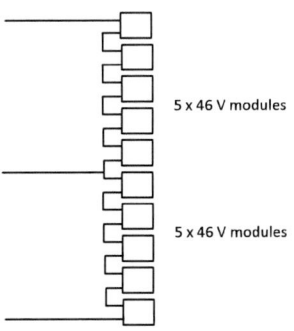

5 x 46 V modules

5 x 46 V modules

Fig. 7 Battery modules

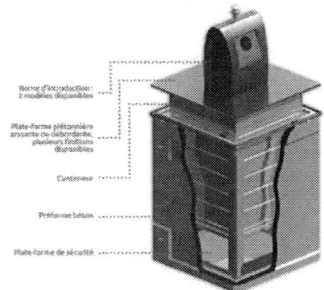

Fig. 8 Underground container construction

4 SIMULATION RESULTS

Figure 9 shows the AC currents at the AC side of the converter when different magnitudes of the consumption must be compensated. The rate of rise of the current is determined by the supervisory control of the car battery charger.

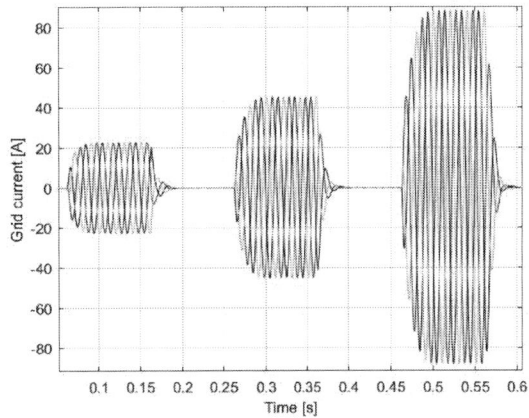

Fig. 9 AC-currents of the BBCU when charging the car at different power magnitudes

For the same magnitude of the AC currents shown in Fig. 9, the current of the battery is represented in Fig. 10. The current in one of the channels of the DC-DC converter is represented in Fig. 11. The comparison of the two solutions for the DC-DC battery interface is given through Fig. 12 and 13. The battery current and the current in the individual channels of the interleaved solution are represented in Fig. 12. The reduced ripple of the battery current is obtained from the superposition of the phase-shifted switching of the individual channel.

The resultant frequency of the ripple of the battery current is equal to three-times the switching frequency of the power devices of the channels (3 x 10 kHz). Then, the battery current of the modern solution using fast switching devices is represented in Fig. 13. For this solution, a switching frequency of 50 kHz is chosen. The values of the inductors are given in Table 1. The design of the inductors has considered an identical value of the ripple of the battery current for both solutions.

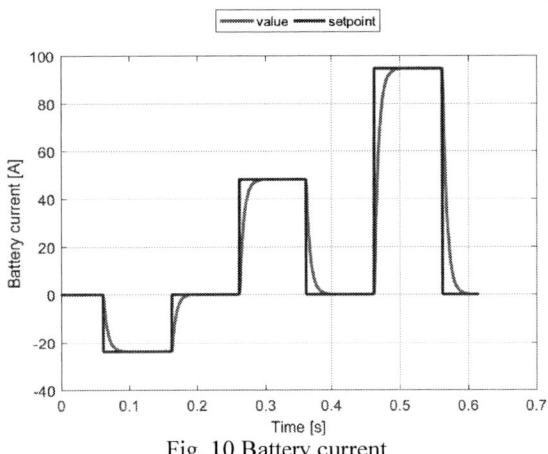

Fig. 10 Battery current

3212

The 2018 International Power Electronics Conference

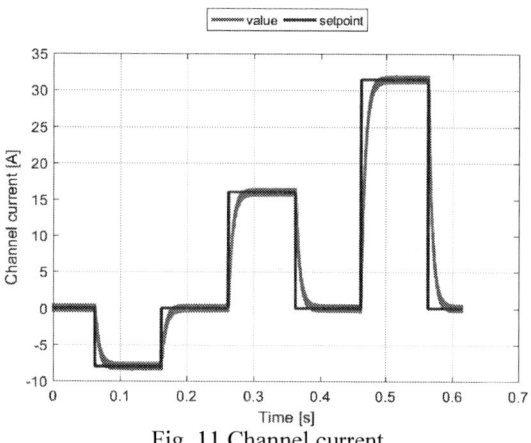

Fig. 11 Channel current

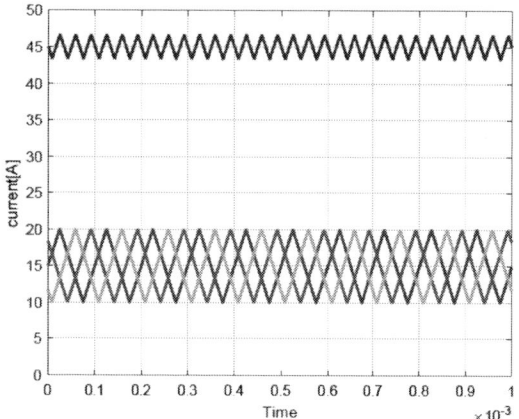

Fig. 12 Current ripples for battery and channels of the interleaved converter

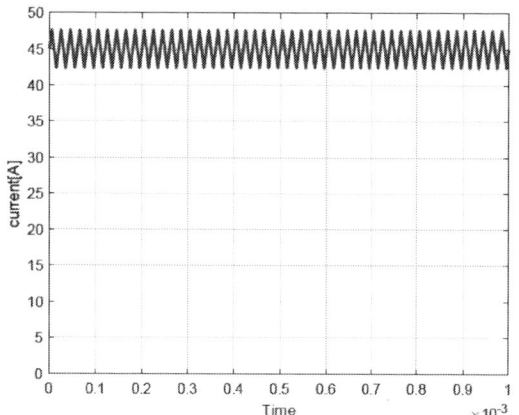

Fig. 11 Ripple of the current of the SiC based solution

5 CONCLUSIONS

A dedicated Bidirectional Buffered Charging Unit BBCU is presented in this paper, allowing a high power / fast charge of EV's even if the line power or current at the point of coupling is limited.

The paper includes simulation results of the 3-level line interface and of two battery-side interfaces realized in 2 different technologies, namely classical interleaved low switching converters with smaller smoothing inductors and a fast switching SiC based solution.

REFERENCES

[1] IEEE Standard Technical Specifications of a DC Quick Charger for Use with Electric VehiclesIEEE Std 2030.1.1-2015 Year: 2016 Pages: 1 - 97

[2] Gautham Ram Chandra Mouli; Johan Kaptein; Pavol Bauer; Miro Zeman, Implementation of dynamic charging and V2G using Chademo and CCS/Combo DC charging standard 2016 IEEE Transportation Electrification Conference and Expo (ITEC) Year: *2016 Pages: 1 - 6*

[3] P. Jampeethong; S. Khomfoi , An EV quick charger based on CHAdeMO standard with grid-support function 2015 18th International Conference on Electrical Machines and Systems (ICEMS) Year: 2015 Pages: 531 - 536

[4] http://ufcev.epfl.ch Ultra Fast Charge of Electric Vehicles, Research project, EPFL

[5] H. Hõimoja and A. Rufer, Infrastructure Issues Regarding the Ultrafast Charging of Electric Vehicles, IAMF 2012 : International Advanced Mobility Forum, Geneva, Switzerland, March 7-8, 2012

[6] Dinh Thai Hoang; Ping Wang; Dusit Niyato; Ekram Hossain , Charging and Discharging of Plug-In Electric Vehicles (PEVs) in Vehicle-to-Grid (V2G) Systems: A Cyber Insurance-Based Model , IEEE Access, Year: 2017, Volume: 5, Pages: 732 - 754 , IEEE Journals & Magazines.

[7] Nabae, A., Takahashi, I., Akagi, H., A new Neutral Point Clamped PWM Inverter, IEEE Trans. On Ind. Applications, Vol. 17, no 5, Sept./Oct. 1981, pp 518-523.

[8] Arifujjaman Md., Shakhawat Md., and Iqbal M. T., Efficiency comparison of 2- and 3-level Inverter Based Power Conditioning System for Grid Connected SOFC Application, IEEE CCECE Canadian Conference on Electrical and Computer Engineering, , May, 4-7, 2014, Toronto, Canada.

[9] Pluschke N., Grasshoff T., More Efficiency for 3-Level Inverters, Power Electronics Europe, Issue 2 2010, www.power-

3213

mag.com/pdf/feature_pdf/1272463225_Semikron_Featur e_Layot_1.pdf, accessed Jan. 16th 2018.

[10] Rufer, A.; Meyer, J.-M., A High Current, Low Ripple, Low Weight PFC Rectifier Using a Standard Power Module, PCIM 98 : International Conference on Power Electronics, Intelligent Motion and Power Quality, Nürnberg, Germany, 25-28 May 1998

[11] Kolomyjski, W., Modulation strategies for Three-Level PWM Converter fed Induction Machine Drive, PhD Thesis, Warsaw University of Technology, Warsaw Poland, 2009.

[12] Fahrni, C., Rufer, A., Bordry, F., Burnet, JP., A novel 60 MW Pulsed Power System based on Capacitive Energy Storage for Particle Accelerators, EPE Journal : European Power Electronics and Drives Association Journal, vol. 18, num. 4, p. 5-13 2008.

Reconfigurable Converter with Multiple-Voltage Multiple-Power for E-Mobility Charging

Mohamed S A Dahidah[1*], He Liu[1] and Vassilios G. Agelidis[2]
[1]School of Engineering, Newcastle University, Newcastle upon Tyne, UK
[2]Department of Electrical Engineering, Technical University of Denmark, Copenhagen, DENMARK
[*]E-mail: mohamed.dahidah@ncl.ac.uk

Abstract— This paper presents a reconfigurable power electronic converter that is able to produce multiple output voltages and multiple output powers to cater for the growing demand of electrified mobility (e-mobility) apparatus. The proposed converter aims to accommodate different charging needs for different e-mobility apparatus, unlike the existing charging infrastructure where it has been designed for a specific type and can only charge one equipment at a time. The proposed charger utilizes the well-known four-leg three-phase converter, three single-phase medium frequency transformers with multiple secondary windings and Active Front-End (AFE) bidirectional PWM rectifier modules. While the high frequency ac enables a compact system, with smaller passive components; the four-leg converter accommodates the potential unbalance caused by simultaneously charging different e-mobility apparatus. The proposed charging system also enables vehicle-to-grid (V2G) operation if required. The feasibility of the proposed reconfigurable power electronic converter is confirmed by intensive simulation studies for different charging scenarios.

Keywords— *E-mobility charger, multiple output power, multiple output voltage, reconfigurable converter.*

I. INTRODUCTION

Electrified mobility (e-mobility) has received unprecedented attention from governments, industry and community in the modern world, where environmental protection and energy conservation are ever growing concerns [1]. However, the wide acceptance of electrified transportation is affected by many barriers, such as driving range and the high cost as compared to the conventional fossil fuel-based competitors. Nevertheless, battery technologies have seen a rapid development and breakthrough improvement resulted in very large batteries, supporting fast charging and enabling longer driving range [2, 3].

Similarly, great efforts have been also put forward towards reducing the cost through improving and integrating various subsystems such as electric motors, power electronics, and electronic controllers as well as the system-level integration and optimization [2].

The charging equipment is the backbone of the e-mobility systems and public charging infrastructure plays an important role in increasing their adaption and for realizing mass commercialization. There have been enormous charging systems developed at both, laboratory and commercial levels, which are broadly classified as either on-boards or off-board chargers [3, 4]. Furthermore, there are several factors influencing the design of these chargers and most importantly are the level of charging power and power flow characteristics (i.e. unidirectional and bidirectional). There are three common charging levels, categorized based on the charging current into [3]:

- Level 1 charging (up to 16A) is the slowest method, which normally integrated with the vehicle or available at the residential properties.
- Level 2 charging (between 16A and 32A) is the primary method for dedicated private and public facilities.
- Level 3 charging, where the current exceeds 32A is introduced to overcome the range limit. This is also known as fast charging [5, 6].

There have been enormous research in the recent years to develop different charging systems for all power levels [3]. A standard on-board EV battery chargers are normally a unidirectional type that uses a simple diode rectifier followed by a boost dc voltage regulator such as [7]. An advanced version that aimed to reduce the current ripple and inductor size via interleaving is presented in [8], where the interleaving reduces the stress on the output capacitor, however, due to the heat management issues, this topology is limited to power levels up to approximately 3.5kW [9]. However these type of chargers do not support injecting power back to the grid and contribute to power stabilization.

Bidirectional chargers on the other hand, support vehicle-to-grid (V2G) operation mode, which contribute to balance the grid power. These chargers are typically have two conversion stages, i.e. active grid-connected ac/dc converter and a bidirectional dc/dc converter to regulate the battery current [10].

Multilevel converters are also explored for such applications due to their features of reducing the switching frequency, enabling lower rating devices and requiring no or small filtering circuits. For example three-level diode-clamp converter is proposed in [11]. However, these converters introduce complexity to the control circuit and the additional components increase the cost.

Fast DC charging have been also widely researched in the recent years to reduce the charging time. The structure of these charging stations can either be with an ac bus, where each charging unit is fed by its independent ac/dc stage or each charging unit connected to a common dc bus formed by a higher power rating ac/dc converter [11-14].

Several power electronic converter topologies have been proposed and documented in the open literature. For instance, [12] proposed three-level dc/dc converters connected to the common dc bus with energy storage integration to help balance the voltages at the dc bus. However, the isolation stage is at the low frequency ac side, which used a bulky transformer. In [14] an ultra-fast charging station for e-buses is proposed where two 2-level voltage source converters are commented to the ac side via three-winding transformer, with a boost interleaved dc/dc converter at the dc side. However, this charging system is also relying on the low-frequency for isolation and it has only one output for a certain type of e-mobility.

The lack of charging infrastructure with more flexibility and choices to consumers still one of the main barriers of wide acceptance of electrified transportation. The approach thus far has followed the development and installation of e-mobility infrastructure suitable only for a single application such as an Electric Vehicle (EV) and for a given type when it comes to charging needs of an EV. For instance, public infrastructure of e-mobility can today only do a particular vehicle charging, leaving out socially important other type of e-mobility equipment such as two- and three-wheel scooters, light delivery vehicles and most importantly socially inclusive equipment such as e-mobility scooters for disabled people.

Therefore, the development of universal, flexible and reconfigurable battery charging systems is extremely important to cater for the future deployment of e-mobility. Hence, this work proposes a reconfigurable, power electronic converter with multiple-voltage, multiple-power outputs to enable simultaneous charging of different e-mobility equipment with different charging needs as envisaged in Fig.1.

The typical charging characteristics of different e-mobility batteries is depicted in Fig.2, where the output voltage and current can essentially move along the Y-axis, via the re-configurability feature of the proposed charger to cater for different charging modes of different e-mobility vehicles.

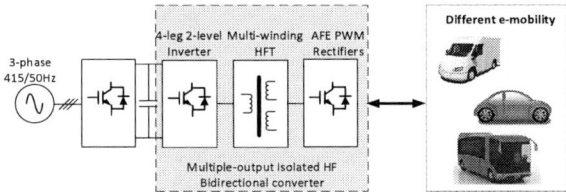

Fig.1. Illustrative block diagram of the proposed e-mobility charger

The rest of the paper is organized as follows: Section II presents the circuit configuration of the proposed reconfigurable converter along with a brief description on the control method of the AFE PWM rectifier. Section III presents some charging scenarios supported by simulation results. The work is concluded in Section IV.

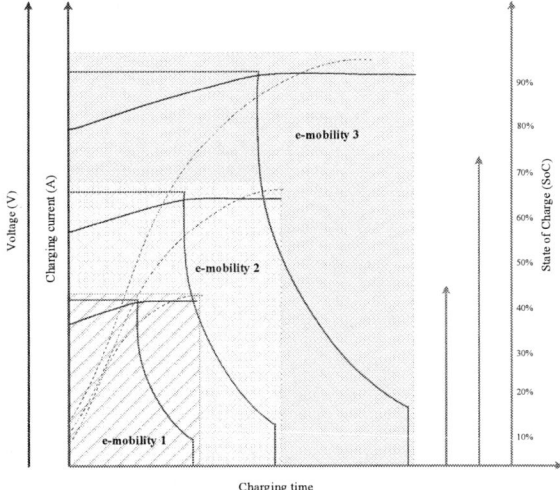

Fig.2. A generalized typical charging characteristics of an e-mobility

II. THE PROPOSED RECONFIGURABLE CONVERTER

Fig. 3 illustrates the generalized schematic diagram of the converter topology of the proposed e-mobility charger.

The charger is fed from three-phase supply through a seven-leg back-to-back two-level converter with 3-leg at the supply side and 4-leg at load side. This configuration is selected to eradicate any imbalance or power quality issues to propagate back to the distribution network. Furthermore, the fourth leg provides a path for the neutral current in case of unequal loading (i.e. if one or more output ports are unloaded) to protect overloading the transformers. Three single-phase, 3-winding, medium-frequency transformers are used to construct the three-phase transformer, where the primary windings are connected in star and the secondary windings are independently decoupled to generate multiple independent outputs. In addition to the galvanic isolation, theses transformers provide also a stepping ratio, if required.

It is worth noting the high-frequency enables smaller passive and filtering components, resulting in a compact and lighter system. As it can be seen from Fig.3, Active-Front-End (AFE) PWM rectifier modules are connected at each secondary winding, forming multiple independent charging ports (output voltages), labelled $V_{o1} - V_{o6}$. Although, only two output ports per ac phase input is considered in this paper, however, in principle, this is expandable to any number.

The re-configurability is achieved by the charging router circuitry (CRC) illustrated in Fig.3. For simplicity, the CRC in this work consists of simple dc connectors arranged in different ways to reconfigure AFE modules in series and/or parallel depending on the charging needs.

The 2018 International Power Electronics Conference

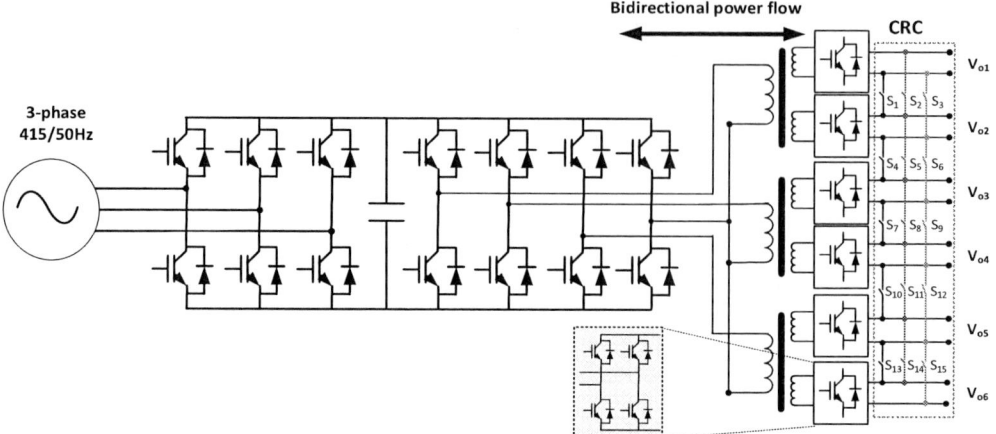

Fig. 3. The schematic diagram of the proposed multiple-voltage multiple-power charger

The Four-leg converter is simply controlled using the method proposed in [19], which regulates the instantaneous voltage and current of the 4-leg converter using dq synchronous reference frame. The control strategy consists of a proportional-integral (PI) voltage controller and a proportional current loop in each phase, generating the required pulse width modulation (PWM) voltage command with zero steady-state tracking error and fast transient response. Details on the control method can be found in [7].

AFE PWM Rectifier and its control:

The proposed reconfigurable charger utilizes the well-known boost active front-end rectifier to provide a controllable output voltage at each port. This enables a simple bidirectional power flow control (if required) with guaranteed unity power factor at the ac side (i.e. the secondary winding of the transformer). Fig.4 illustrates the basic structure of a single-phase bidirectional PWM rectifier, which has been well-documented in the literature, such as [11-14].

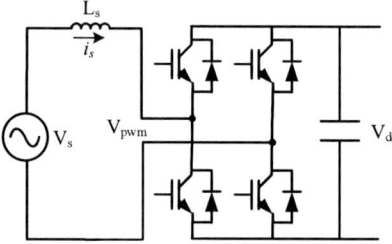

Fig.4 Typical Active-frontend PWM rectifier

The control of the AFE must be able to regulate the output dc voltage to its reference value and also ensures a sinusoidal and unity power factor current at the ac side.
In this paper, the control is developed using the DQ synchronous reference frame method. However, since the AFEs are of a single-phase, which requires two orthogonal components to transform the AC quantities into a DC. Therefore, it is necessary to create a second quantity in quadrature with the real one. This is achieved

by detecting the frequency and phase of the input voltage using a PLL and then generates a fictitious input signal. This method neither requires tuned filters nor store samples to produces a quarter cycle delay.

The purpose of this control is to regulate the inductor current i_s. Considering no losses in the circuit of Fig.4, its average equation through a switching period is given by (1).

$$L\frac{dI_s}{dt} = V_s - V_{pwm} \qquad (1)$$

Fig.5 presents the block diagram of the d-q vector control method applied to the AFEs (i.e. each AFE uses the same control method). The fictitious voltage, V_{rq} is realised by a phase shift of 90 degrees to enable the trasformation of a conventional stationary to rotating frame. I_{ra} and V_{ra} are the actual ac voltage and current of AFE rectifier, respectively; I_{rab} and V_{rab} are the orthogonal current and voltage pairs, respectively, which are realised by the phase shift of 90 degrees . The reference value V^*_{dc} is compared with the measured voltage V_{dc} and the error signal is applied to a PI controller. The d-q transformation matrix of the control system from stationary to rotating reference frame is given as:

$$C_{2s/2r} = \begin{bmatrix} \cos\theta & -\sin\theta \\ \sin\theta & \cos\theta \end{bmatrix} \cdot \begin{bmatrix} V_{rd} \\ V_{rq} \end{bmatrix} \qquad (2)$$

As it can be seen from Fig.5, the q-axis current reference, I^*_{rq} is simply set to zero, to ensure a unity power factor.

The whole control system is realized by two control loops. The outer voltage loop is employed to regulate the output dc voltage of AFE rectifier and the inner current loop regulates the input AC current of AFE rectifier.
It is worth noting that each output port is independently controlled with the same control system.

3217

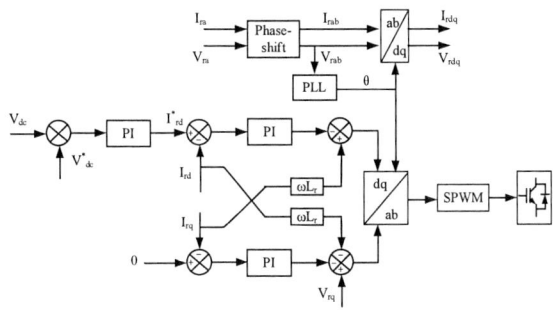

Fig.5. Block diagram of AFE rectifier Control system.

III. SELECTED SIMULATION RESULTS

A simulation model of the proposed converter is developed using MATLAB/SIMULINK software based on the parameters tabulated in Table I. Extensive simulation studies have been are carried out to confirm the feasibility of the proposed solution in accommodating different charging scenarios. Selected cases are presented in the following sub-sections.

Table I: Simulated system parameters

Input AC supply	3-phase 415/50Hz
dc-link voltage	600Vdc
No# of single-phase transformers	3
Transformers turn ratio	2:1
No# output ports (dc)	6
DC output voltage range (each port)	$200 - 400$Vdc
Rated output power	24kW
Rated output power for each port	4kW
Primary AC frequency	2kHz
Switching frequency	20kHz
AFE dc capacitors	300µF

A. Case 1: Multiple outputs of equal voltages and equal powers:

In this case, the charger is configured via the CRC to produce six independent outputs of equal voltages and equal powers. It is worth noting that the use of AFE PWM rectifier modules, each output port draws a close to a sinusoidal current with a unity power factor. This is illustrated in Fig. 6, where the current and voltage waveforms at winding 2 (i.e. as an example) of the secondary side of the transformer. Furthermore, since all ports are equally loaded, drawing equal currents, therefore there will be no imbalance at the primary side of the transformer as it can be observed from Fig. 7, where the neutral current is zero. As shown in Fig.8, all ports are controlled to deliver 4kW power at 250V dc, which confirms the ability of the proposed controller to independently regulate multiple outputs.

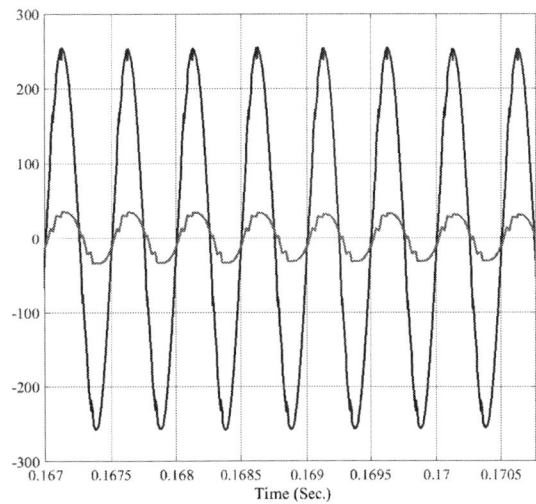

Fig. 6. An example voltage (*blue*) and current (*red*) waveforms at one of the seconday windings

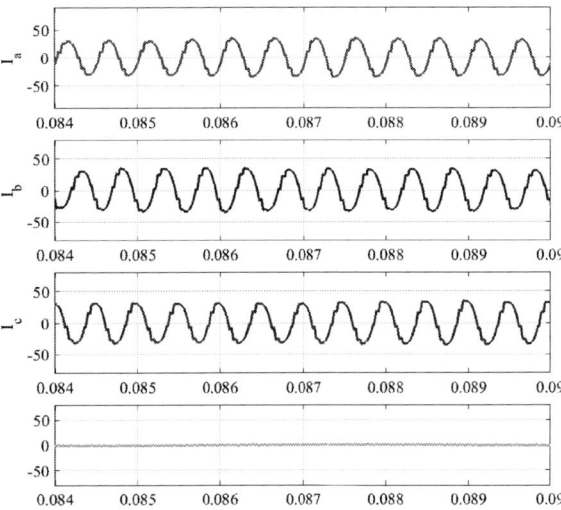

Fig. 7. Primary side input currents for *case 1*. (traces top to bottom, phase a, phase b, phase c, and nutral)

B. Case II: Multiple outputs of different voltages and different powers:

In this case, the charger is reconfigured to provide different outputs. It is assumed that both, output prots 1 and 2 are unloaded and the remaining four ports are equally loaded with 4kW at 250V. As it can be seen from Fig. 9, ports 1 and 2 draw a zero power. The controller still able to achieve equal output voltages across all six output ports.

On the other hand, as the unloaded ports are supplied from phase A of the three-phase input supply, this will casue an imbalance and lead to a high current flowing through the neutral line as shown in Fig.10. However, as the proposed system utilizes a four-leg converter, which will take care of the neutral current and protect the transformer from overloading.

3218

The 2018 International Power Electronics Conference

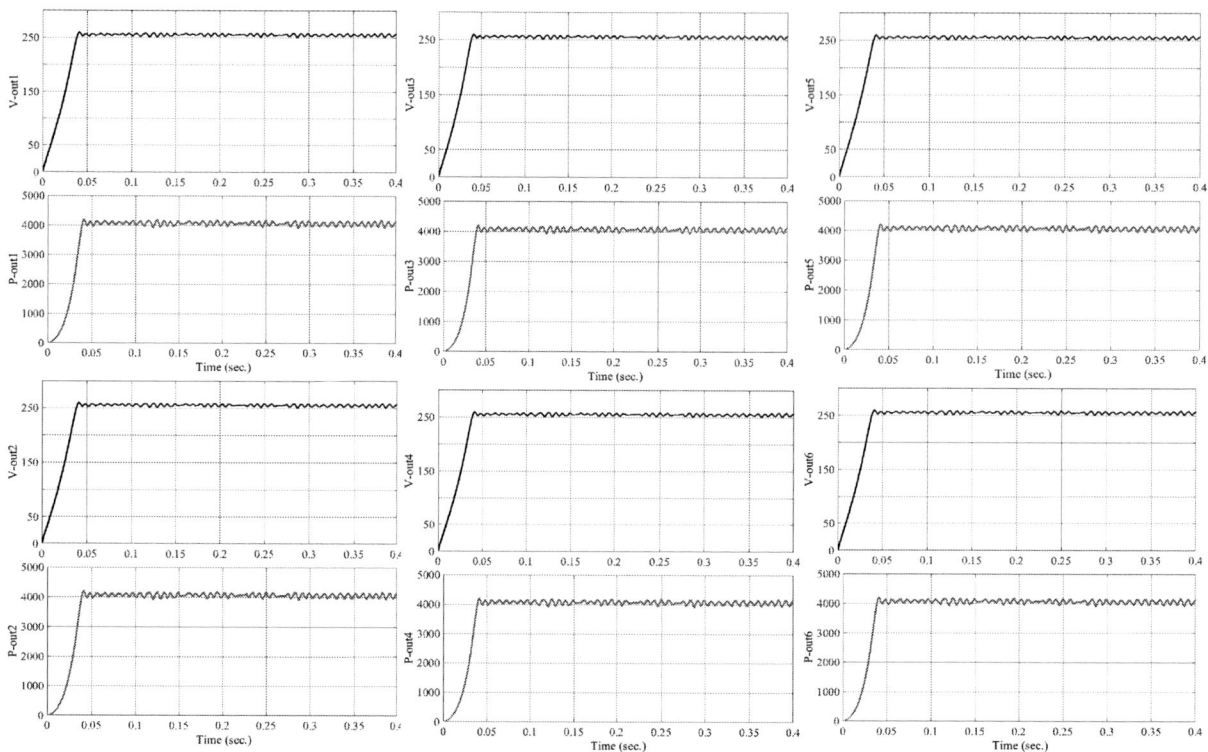

Fig. 8. key-waveforms of *Case I*. upper trace is the output voltage for port-*x* and the lowwer trace is the output power of port-*x*, where x = 1, 2, …, 6.

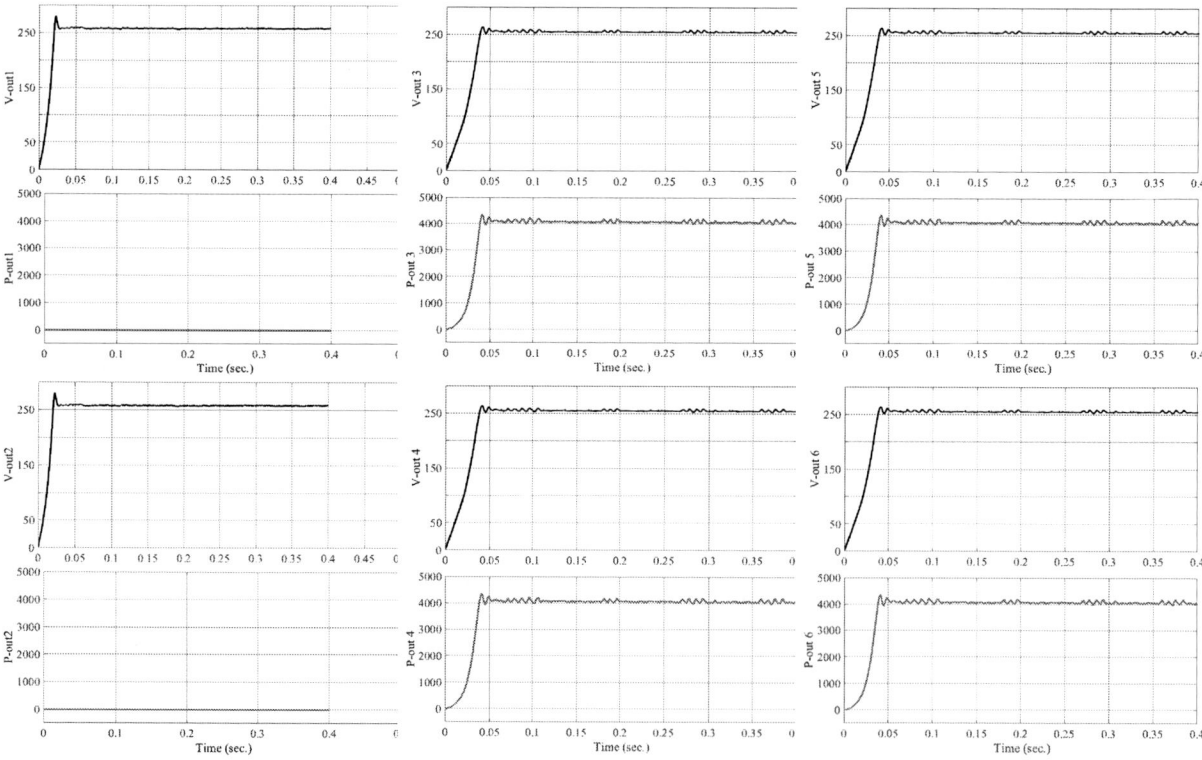

Fig. 9. key-waveforms of *Case II*. upper (*black color*) trace is the output voltage for port-x and the lowwer (*red color*) trace is the output power of port-*x*, where *x* = 1, 2, …, 6.

3219

The 2018 International Power Electronics Conference

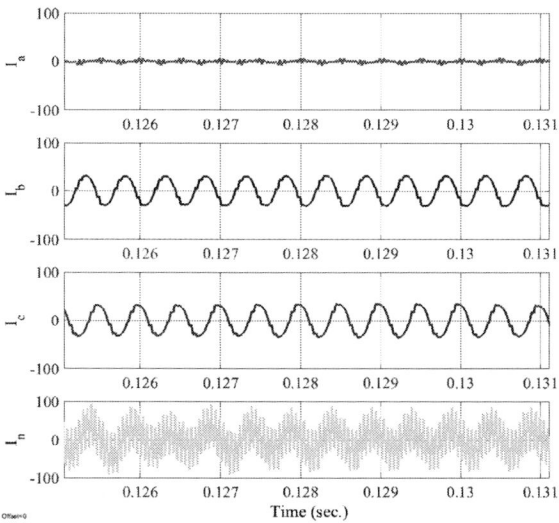

Fig. 10. Primary side input currents for *case II.* (traces top to bottom, phase a, phase b, phase c, and nutral)

C. Case III: Dynamic operation:

The performance of the proposed charger and its control is verified under a step change in both, the output voltage and the output power. As an example, Fig. 11(a) shows the step change introduced to the output voltage of port 3, where the voltage changed from 280V to 320V at time 3 sec.. At the same time, the output voltage of port 6 is increased from 250V to 300V as shown in Fig 11(b).

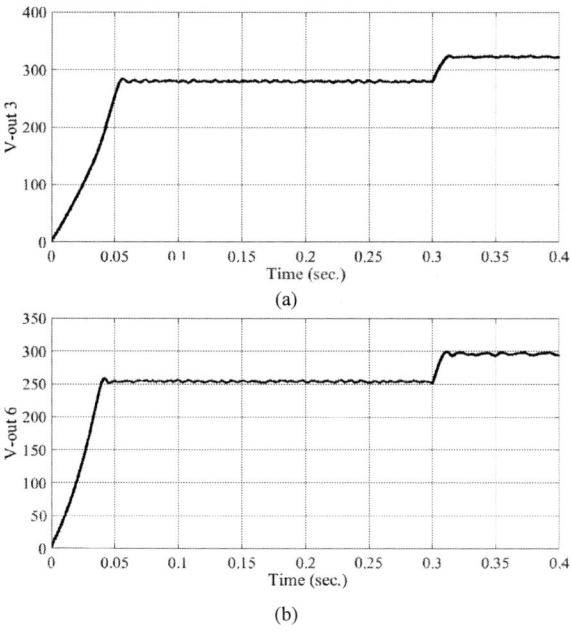

Fig. 11. An example of a step change in the output voltage. (a) port3. (b) port 6.

Another performance validation is confirmed by introducing a step change to the load. Fig. 12(a)

illustrates the operation of the charger when the load at ports 1 and 4 changed at time 2 sec. and 3 sec., respectively. It is worth nothing that the controller was able to maintain the voltage at the reference value despite the change in the load.

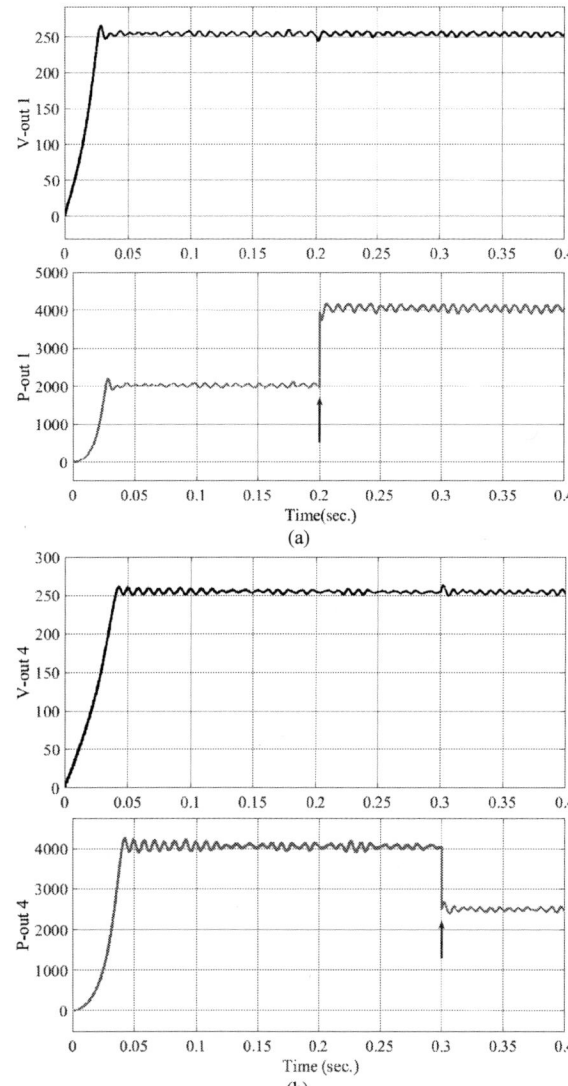

Fig. 12. An example of a step change in the output power. (a) port1. (b) port 4.

D. Case IV: Sereis and parallel operation:

The reconfigurability feature of the proposed system is demonstrated in the case study with different scenarios. Fig.13 (a) illustrate the operation of the proposed charger when ports 1 and 2 are connected in parallel to produce higher power. As it can be noticed, the output voltage is manitaned at 250V but the the output power is doublled. At the same time, ports 5 and 6 and connected in series to deliver higher voltage (i.e. 500V) with boosted output power as illustrated in Fig.13(b).

3220

The 2018 International Power Electronics Conference

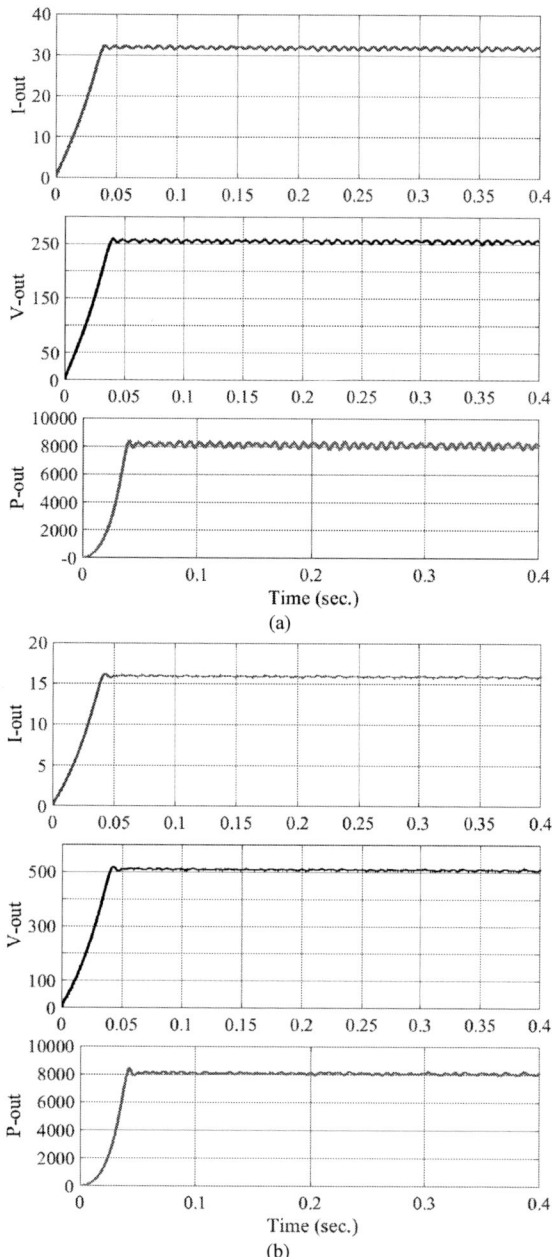

Fig. 13. Key-waveforms of Case IV. (a) output current, voltage and power when ports 1 and 2 are conncted in parallel. (b) output current, voltage and power when ports 5 and 6 are connected in series.

Fig. 14 demonstrates the ability of the proposed charging system to be reconfigured to deliver the maximum power. In this case, all ports are connected in paraller, producing a maximum power of 24kW at an output voltage of 250V.

IV. CONCULSION

The paper presented a power electronic solution that can realize a multiple output voltages and multiple output power EV charger. The charger features modularity and reconfigurability to cater for different charging needs for different mobility devices. The impact of an unequal loading is acommodadted by using a four-leg converter. Active Front-End (AFE) PWM rectifiers are used to regulate the voltage suppiled to the batteries and they also ensure a unity power factor at the ac side. A simple control method based on DQ synchronous reference frame is developed for the single-phase AFE modules. Different cases, emulating different charging scenarios are presented to demonstrate the feasibility of the proposed solution.

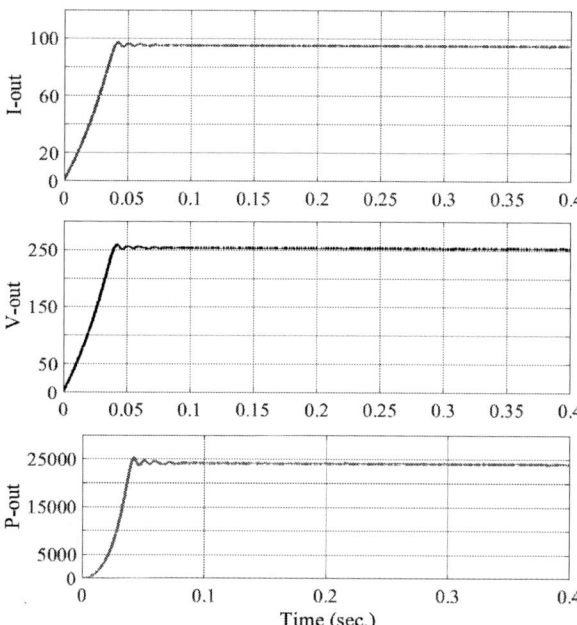

Fig. 12. An example of a step change in the output power. (a) port1. (b) port 4

REFERENCES

[1] P. Bansal, "Charging of electric vehicles: technology and policy implications," *Journal of Science Policy & Governance*, vol. 6, no. 1, February 2015, pp. 1-20.

[2] C.C. Chan, "The state of the art of electric and hybrid vehicles," in *proceedings of the IEEE*, vol.90, no.2, Feb., 2002, pp. 247 – 275.

[3] M. Yilmaz, and P.T. Krein, "Review of battery charger topologies, charging power levels, and infrastructure for plug-in electric and hybrid vehicles," *IEEE Transaction on power Electronics*, vol. 28, no.5, May, 2013, pp. 2151 – 2169.

[4] G.R.C. Mouli, P. Bauer and M. Zeman, "Comparison of system architecture and converter topology for a solar power electric vehicle charging station," *in proceeding of 9th international conference on power electronics –ECCE Asia*, June 1 – 5, 2015, Seoul, Korea, pp. 1908 – 1915.

[5] J.C.G. Justino, T.M. Parreiras and B.J.C. Filho, "Hundreds kW charging stations for e-buses operating under regular ultra-fast charging," *IEEE Transaction on Industry Application*, vol. 52, no.2, March/April, 2016, pp. 1766 – 1774.

[6] S. Rivera, B. Wu, S. Kouro, V. Yaramasu, and J. Wang, "Electric vehicle charging station using a neutral point clamped converter with bipolar dc bus," *IEEE Transactions on Industrial Electronics*, vol. 62, no. 4, April, 2015, pp. 1999 – 2009.

[7] C. Aguilar, F. Canales, J. Arau, J. Sebastian, and J. Uceda, "An integrated battery charger/discharger with power-factor correction," *IEEE Transactions on Industrial Electronics*, vol. 44, no. 5, pp. 597–603, Oct. 1997.

[8] [40] F. Musavi, M. Edington, W. Eberle, and W. G. Dunford, "Evaluation and efficiency comparison of front end AC–DC plug-in hybrid charger topologies," *IEEE Transactions on Smart Grid*, vol. 3, no. 1, pp. 413–421, Mar. 2012

[9] P. Kong, S. Wang, F. C. Lee, and C. Wang, "Common-mode EMI study and reduction technique for the interleaved multichannel PFC converter," *IEEE Transaction on power Electronics*, vol. 23, no. 5, pp. 2576–2584, Sep. 2008

[10] G. Y. Choe, J. S. Kim, B. K. Lee, C. Y. Won, and T. W. Lee, "A bidirectional battery charger for electric vehicles using photovoltaic PCS systems," *in Proc. IEEE Veh. Power Propulsion Conf.*, Sep. 2010, pp. 1– 6

[11] M. Ahmadi, N. Mithulananthan and R. Sharma, "A Review on Topologies for Fast Charging Stations for Electric Vehicles," *proceeding of IEEE International conference on power system technology (POWRCON)*, Wollongong, NSW, Australia, 28 Sep. - 01 Oct., 2016, pp. 1-6

[12] S. Rivera; B. Wu, "Electric Vehicle Charging Station with an Energy Storage Stage for DC Bus Voltage Balancing," *in IEEE Transactions on Power Electronics*, vol.32, no.3, March 2017, pp.2376-2385.

[13] M. Vasiladiotis and R. Alfred "A Modular Multiport Power Electronic Transformer with Integrated Split Battery Energy Storage for Versatile Ultrafast EV Charging Stations," *IEEE Transactions on Industrial Electronics*, vol. 62, no.5, 2015: 3213-3222.

[14] J. C. G. Justino, T. M. Parreiras and B. J. C. Filho, "Hundreds kW Charging Stations for e-Buses Operating Under Regular Ultra-Fast Charging," *IEEE Transactions on Industry Applications*, vol. 52, no.2 March/April 2016, pp. 1766–1774.

[15] B. Singh, B. N. Singh, A. Chandra, K. Al-Haddad, A. Pandey, and D. P. Kothari, "A review of three-phase improved power quality ac–dc converters," *IEEE Transactions on Industrial Electronics*, vol. 51, no. 3, pp. 641–660, Jun. 2004

[16] J. R. Rodríguez, J.W. Dixon, J.R. Espinoza, J. Pontt, and P. Lezana, "PWM Regenerative Rectifiers: State of the Art," *IEEE Transactions on Industrial Electronics*, vol. 52, no. 1, February 2005, pp. 5 – 22.

[17] U.A. Miranda and M. Aredes, and L.G.B. Rolim, "A DQ Synchronous Reference Frame Control for Single-Phase Converters," proceeding of *IEEE 36th Power Electronics Specialists Conference*, Recife, Brazil, 12-16 June, 2005, pp. 1377-1381

[18] B.-T. Ooi and O. Stihi., "A single-phase controlled-current pwm rectifier," *IEEE Transactions on Power Electronics*, 3(4), October 1988, pp. 453–459.

[19] M.R. Miveh, M.F. Rahmat, M. WazirMustafa, A.A. Ghadimi, and A. Rezvani, "An Improved Control Strategy for a Four-Leg Grid-Forming Power Converter under Unbalanced Load Conditions," *Advances in Power Electronics*, 2016, pp. 1 – 14.

Development of a Series Hybrid Electric Vehicle Laboratory Test Bench with Hardware-in-the-Loop Capabilities

Poria Fajri[1*], Nima Lotfi[2] and Mehdi Ferdowsi[3]

1 Electrical and Biomedical Engineering Department, University of Nevada Reno, Reno, Nevada, USA
2 Mechanical and Industrial Engineering Department, Southern Illinois University Edwardsville, Edwardsville, Illinois, USA
3 Department of Electrical and Computer Engineering, Missouri University of Science and Technology, Rolla, Missouri, USA
*E-mail: pfajri@unr.edu

Abstract- **This paper describes the design and development of a laboratory test bench with Hardware-in-the-Loop (HIL) capabilities for the emulation of a Series Hybrid Electric Vehicle (SHEV). The practical challenges of developing a HIL test bench for SHEV emulation are discussed. Furthermore, the overall architecture and system component layout as well as the real-time control algorithm for the test bench are presented. Finally, experimental implementation results on a standard drive cycle are presented to validate the overall operation of the test bench.**

Keywords— Hardware-in-the-Loop, real-time control, Series Hybrid Electric Vehicle, test bench.

I. INTRODUCTION

Transportation electrification is gaining mainstream attention across several sectors, from academia and research laboratories to automotive industries and their suppliers. The automotive industry is leaning increasingly toward developing vehicles that are more fuel efficient and produce less emissions. As a result, electric and hybrid electric vehicles are expected to be the next generation of regular means of transportation due to their relatively high efficiency and low emissions. This has resulted in extensive research efforts in academia and industry on different aspects of these vehicles, such as topology exploration, power electronic components, efficiency maximization strategies [1], and integration into the current power grid [2]. A hybrid vehicle test bench can serve several purposes for research along with providing the perfect educational tool in an academic environment for training engineers in fields related to hybrid vehicle design and control. This kind of laboratory setup gives the engineering students an opportunity to experience the hybrid vehicle technology in practice and allows them to use their knowledge from the classroom and apply it to a physical system that outputs real-time results.

The aim of the work presented in this paper is to demonstrate the development of a Series Hybrid Electric Vehicle (SHEV) test bench with Hardware-in-the-Loop (HIL) capabilities suitable for laboratory environments. The test bench will replace the Internal Combustion

Engine (ICE), one of the main components in SHEVs, with an electric motor to overcome the restrictions of using ICEs in most indoor laboratories, and therefore, is relatively easy to implement.

II. HYBRID VEHICLE TEST BENCH REQUIREMENTS AND DESIGN CHALLENGES

The design and development of a HIL test system which can closely emulate the operating conditions of an actual SHEV is a complex process. In order to best represent the simulated vehicle, it requires accurate selection of components and sub-systems. At the same time, it requires a simulation model capable of controlling all the electric motors simultaneously and in real-time under different driving scenarios. Some of the challenges and considerations that should be taken into account in the design and development of a SHEV test bench are discussed below:

1) Vehicle inertia consideration without using a flywheel

A significant portion of the traction power applied to a vehicle is stored as kinetic energy in its moving mass. Therefore, it is of great importance to carefully take into account this effect while emulating on-road operating conditions on a test bench [3]. Considering vehicle inertia for test bench studies where the actual physical vehicle mass is absent is crucial in obtaining accurate results, especially, when it comes to analyzing energy consumption, motor performance, and regenerative braking capabilities. The use of a flywheel system is an alternative to emulate vehicle's inertia effect. However, due to safety concerns, it is not preferred in many laboratory environments. Another disadvantage of using a flywheel is that each type of vehicle requires a different inertia; therefore, a different flywheel size is needed for different simulation attempts. Additionally, using a flywheel with the same equivalent inertia as a vehicle can occupy a large space, and the weight of such mass can place a great deal of stress on the coupling and motor shafts.

As a more viable alternative, the electric motor used as a load can be controlled to simultaneously emulate both the rotational inertia effect of the flywheel and the resistive vehicle forces. This provides a safer environment for emulating Electric Drive Vehicles (EDVs) and is more suitable in academic and industrial laboratories [3].

2) Limitations on utilizing an ICE in laboratory environments

If the objective is to emulate a hybrid powertrain, due to many limitations on using an ICE in laboratory environments, the best alternative is to use an electric motor in replacement of the ICE. However, the main challenge with using an electric motor is its control to best resemble the operating conditions of an ICE while accurately calculating the fuel consumption to match the consumption of an actual engine based on its given specifications.

3) Communication protocol selection

The electromagnetic noise caused by switching motor drives and switch-mode power supplies are the main sources of disturbance that can interfere in the communication between the controller and components. To ensure fast and accurate data communication between all components of a test bench, special care should be taken in selecting a communication protocol that is standard and can function reliably in a noisy environment. A Controller Area Network (CAN) communication protocol is well-suited for many advanced industrial applications in noisy environments and therefore, is considered as a standard for in-vehicle communication [4].

4) Use of batteries or battery emulator

Although using actual battery packs in a hybrid vehicle test bench is the most straightforward solution, for several reasons such as safety issues, time-consuming charge cycles, and lack of flexibility to facilitate different experiments, it is not typically preferred. A high power converter can be used to emulate the vehicle's battery pack but it should be noted that since regenerative braking is considered, the converter needs to have bidirectional capabilities to satisfy regenerative braking requirements of the simulated vehicle.

5) Vehicle brake emulation and regenerative braking considerations

One inherent feature of electric and hybrid vehicles is regenerative braking [5]. This capability allows the electric motor to operate as a generator during vehicle deceleration and thus absorb the braking energy and store it in the vehicle energy storage unit. Consequently, a need to implement an accurate model of regenerative and friction braking into the test bench control system is of great importance [6]. Proper distribution of the total braking force between the front and rear axle, and proper allocation of brake force between the friction brakes and the regenerative braking should also be considered in the brake controller model [6].

6) Real-time controller and user interface design

In a real SHEV, the vehicle control unit is in charge of energy management and control. It generates the command signals that control all components of the vehicle based on the driver's needs while optimizing the overall energy consumption of the vehicle.

In a hybrid vehicle test bench not only the controller is responsible for accurate control of each component but it also requires real-time computation capability to calculate the resistive forces acting on the vehicle at each time instance. The real-time environment ensures simultaneous interaction between all components of the system and thus provides results that are close to that of the physical vehicle under study.

Another consideration when developing a hybrid vehicle test bench is the design of the user interface. The user interface should be designed such that it would allow the operator to change critical test bench parameters and visually monitor system performance in real-time. It should clearly indicate all the necessary parameters of the system and allow for easy and flexible control of each component.

III. SHEV Test Bench Structure And Component Outline

In a SHEV, the ICE is coupled to the generator and produces the electric power needed to charge the batteries while a Traction Motor (TM) is used to provide traction power to the wheels. In this configuration, the TM is the only source that provides power to the wheels of the vehicle. Since the ICE is only coupled to the generator, the speed of the ICE is independent of that of the vehicle.

Similarly, the test bench architecture shown in Fig. 1 achieves the mentioned goals with the only difference that the ICE is replaced with an electric motor. In this configuration, the TM, generator, engine emulator, Road Load (RL) emulator, and battery emulator are the main physical components, while the real-time simulation of the drive cycle is performed at the software level. The architecture is based on real-time simulation and execution of the HIL system and consists of two sets of 15 kW electric motors on a common shaft. In this setup, the TM represent the vehicle electric motor and the RL emulator is responsible for emulating inertia, braking, and the resistive forces acting on the vehicle. The engine emulator/generator set represent the engine and generator of a SHEV and are responsible for charging the energy storage and maintaining the energy balance of the system.

Tying the TM and generator together is the battery emulator which is a bidirectional programmable battery cycler manufactured by Bitrode Corporation. The incorporation of an external voltage feedback enables the cycler to act as a battery emulator. In order to ensure a high prediction capability for the battery emulator, special care

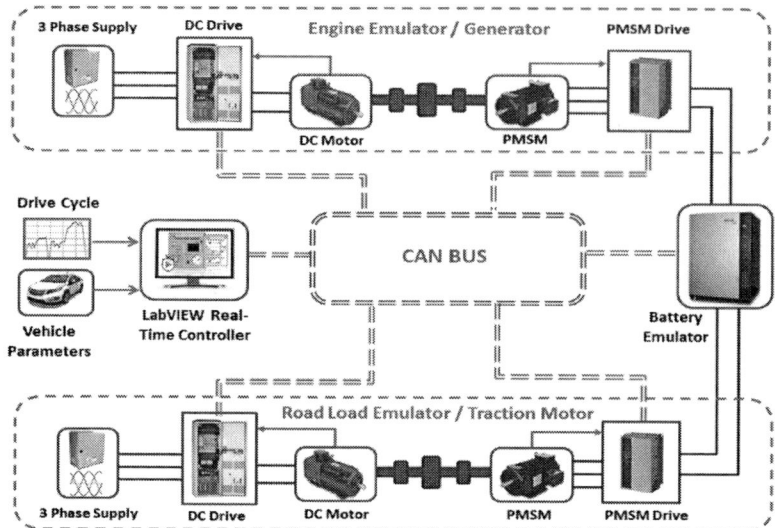

Fig. 1. SHEV test bench architecture and components.

was given to the selection of the model used to emulate the battery performance. Various battery chemistries have been used on-board hybrid vehicles, among which Li-ion batteries have shown to be the most promising solution. There are also a wide variety of literature introducing models with various complexities to describe the performance of the Li-ion batteries [7]. For the simulation purposes in this work, a commonly-used equivalent circuit battery model was employed. Fig. 2 shows a schematic of this battery model in which the battery behavior is emulated using electric circuit elements. The resistor R_s captures the internal resistance of the battery whereas the RC-networks are incorporated to consider the diffusion characteristics of the battery. The battery open-circuit voltage along with all the aforementioned parameters are functions of battery State of Charge (SOC). This model has shown an acceptable performance for hybrid vehicle simulation studies. However, if higher accuracies are needed, the use of such battery emulator can easily facilitate the incorporation of more advanced battery models such as electrochemical models [8].

The commands for output voltage and current of the battery emulator are generated inside the controller using

the discussed model. The need for a real-time communication protocol is satisfied by means of CAN bus which ensures synchronous distribution of reference speed and torque commands as well as real-time monitoring of system critical signals.

The LabVIEW software is used to perform real-time simulation by building the vehicle model and executing the commands in real-time to the battery emulator, TM, RL emulator, engine emulator, and generator. To emulate on-road operation conditions, the LabVIEW real-time controller continuously changes the operating point of the RL emulator and the TM and sends synchronous speed and torque signals to the two motor drives at each instance to allow exact coordination between them. Speed and torque commands for the TM and RL emulator are calculated in real-time from the mathematical models of vehicle dynamics and road load resistance. The required reference speed for the TM is calculated from the vehicle's translational speed by knowing the vehicle wheel radius (r_d) and total gear ratio (G),

$$\omega_{TM} = K_1 \frac{V_{ref}\,G}{r_d} \tag{1}$$

where K_1 is a constant for converting the translational speed of the vehicle from mile/h to m/s.

The required RL emulator torque (T_{RL}) during acceleration is calculated directly from the speed trace using the formula in [3],

$$T_{RL} = \frac{T_R}{\eta G} - (B_{TM} + B_{RL})\omega_{TM}$$
$$+ \left(\frac{J_{ew}}{\eta G^2} - J_{rotation}\right)\left(\frac{d\omega_{TM}}{dt}\right) \tag{2}$$

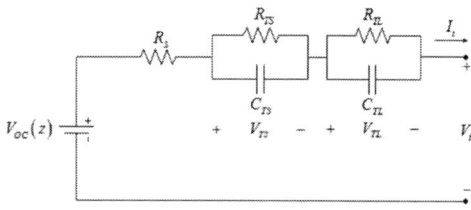

Fig. 2. Equivalent-circuit battery model used in the SHEV test bench.

where J_{ew} represents the equivalent rotational inertia of the vehicle, which is discussed in detail in [9]. T_R is the total resistive torque calculated at the wheels, and is calculated from the resistive forces acting on the vehicle given by,

$$T_R = r_d(mgf_r \cos \alpha + \tfrac{1}{2}\rho_a C_D A_f (V + V_w)^2 + mg \sin \alpha) \quad (3)$$

The definition of parameters used in (2) and (3) are given in Table I.

During deceleration, depending on the vehicle configuration being simulated (rear-wheel drive, front-wheel drive, all-wheel drive), the model takes into account the available brake force on the driving axle and reduces T_{RL} accordingly to mimic the effect of blending friction and regenerative braking in an actual vehicle.

TABLE I
DEFINITION OF PARAMETERS USED IN (2) AND (3)

Parameter	Definition
m	Vehicle mass
ω_{TM}	TM rotational speed
η	Overall efficiency of the vehicle drive train
$J_{rotation}$	Rotating inertia of TM, RL emulator, and coupling
B_{TMr}	Viscous coefficients of the TM
B_{RL}	Viscous coefficients of the RL emulator motor
fr	Vehicle rolling resistance coefficient
C_D	Vehicle aerodynamic drag coefficient
α	Ground slope angle
V	Vehicle speed in m/s
V_w	Wind speed on the vehicle's moving direction
A_f	Vehicle frontal area
ρ_a	Mass density of air

The controller power request signal, dictates the amount of power the engine emulator/generator must transfer to the energy storage. This power request signal is given as a speed command to the engine emulator and a negative torque command to the generator. Since in a SHEV, it is desired to reduce the specific fuel consumption of the engine to achieve the highest economic performance, the speed and torque commands are calculated such that the engine emulator is operating in the most fuel efficient region, having the least fuel consumption for the delivered power. To accomplish this, the torque-speed map of a real engine is employed for calculating the speed and torque commands from the power request signal. In this system, the speed and torque values from the most efficient area of the torque-speed map of a real engine are used to translate the power request signal into the speed and torque commands for the engine emulator/generator component. The controller uses a search algorithm on a lookup table to find the speed and torque values corresponding to a specific power request which minimizes the fuel consumption.

In this setup the LabVIEW software is also utilized as the visual interface between the test bench and the user. The proposed LabVIEW front panel interface designed for this setup is shown in Fig. 3. The interface allows the user to control the start and end of each experiment and graphically indicates the actual speed of each component, distance traveled, battery SOC, and fuel consumption at each instance. The interface also plots the critical system parameters in real-time.

IV. EXPERIMENTAL RESULTS

The developed experimental SHEV test bench is shown in Fig. 4. This arrangement consists of a 15 kW, 6-pole Permanent Magnet Synchronous Motor (PMSM) representing the TM, which is connected to a 15 kW DC

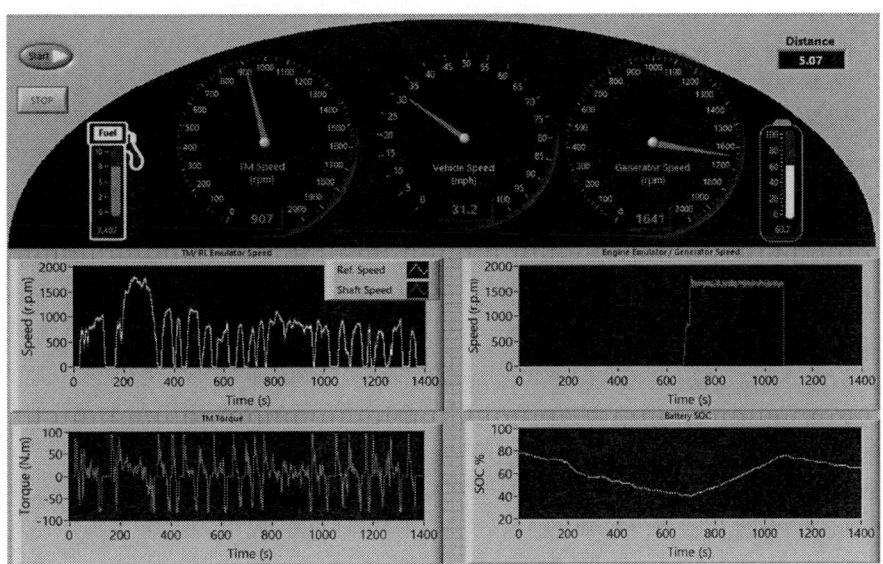

Fig. 3. LabVIEW interface for the SHEV test bench.

The 2018 International Power Electronics Conference

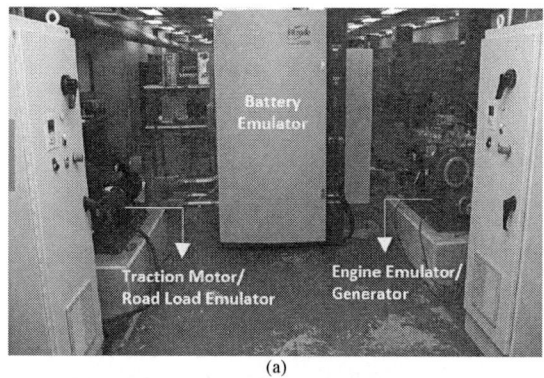

(a)

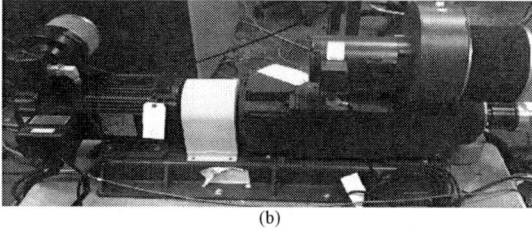

(b)

Fig. 4. Experimental SHEV test bench
(a) Overview of system hardware components
(b) 15 kW Traction Motor/ Road Load Emulator set

machine acting as the RL emulator. To show the application of HIL in emulating a SHEV and to verify the overall performance of the test platform, an experiment was conducted using the test bench of Fig. 4 and considering the vehicle and test bench parameters listed in Table II. The test was performed for one cycle of the Urban Dynamometer Driving Schedule (UDDS).

For this experiment, the control decision utilized for the engine emulator/generator control was based on the engine ON/OFF or thermostat control [10], in which the battery SOC was always maintained between its preset lower and upper limits of 40% and 80%. This was achieved by turning the engine emulator/generator set on when the SOC reached 40% and off when the SOC hit 80%. The results for this experiment are illustrated in Fig. 5.

TABLE II
VEHICLE AND TEST BENCH SPECIFICATIONS

Parameter	Value
m	400 kg
ρ_a	1.22 kg/m^3
C_d	0.19
A_f	1.6 m^2
f_r	0.01
r_d	0.28m
α	0
G	2.3
$J_{rotation}$	0.038 kgm^2
J_{ew}	33.029 kgm^2
η	%90
(B_{TM})	0.0086 N.m/(rad/s)
(B_{RL})	0.0133 N.m/(rad/s)

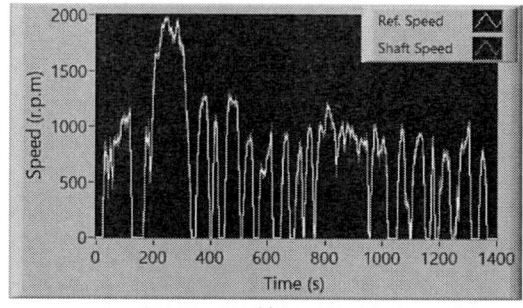

(a)

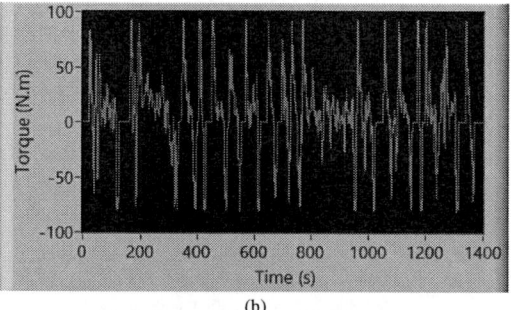

(b)

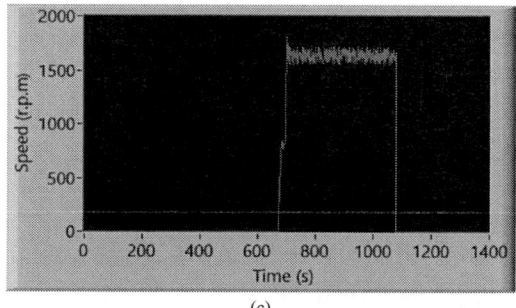

(c)

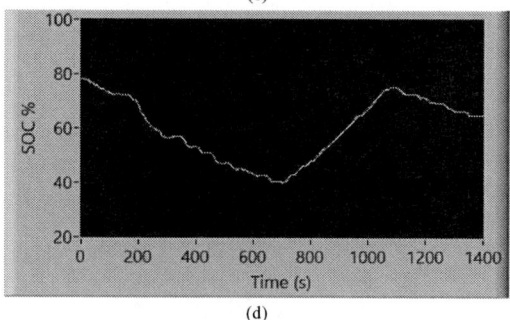

(d)

Fig. 5. SHEV test bench experimental results: (a) Reference and actual shaft speed of TM/RL emulator, (b) TM torque, (c) Engine emulator/ generator speed, (d) Energy storage SOC

Fig. 5 (a) shows the reference and actual shaft speeds of the TM/ RL emulator for a complete drive cycle. Clearly, the speed command is followed throughout the driving cycle with only slight deviations. Fig. 5 (b) shows the actual torque generated by the TM which is used to propel the vehicle and overcome the resistive torque produced by the RL emulator.

3227

Fig. 5 (c) shows the variation in the engine emulator/generator speed and Fig. 5(d) shows the corresponding SOC changes. It can be seen from these figures that the battery SOC experiences both positive and negative slopes during a complete drive cycle while being limited at all times to the preset values of 40% and 80%. As soon as the SOC reaches its lower bound of 40%, the engine/generator is turned on and operated around 1,650 rpm which is the most efficient operating point calculated by the controller. This results in charging the battery and thus an increase in the calculated SOC value is observed during this time.

V. CONCLUSIONS

The challenges of developing a SHEV test platform with HIL simulation capabilities were discussed. A block diagram representation of the test bench was presented and the importance of real-time communication between each component using a reliable communication protocol was emphasized. The developed SHEV platform structure and major components used in the setup as well as the LabVIEW user interface were also discussed. To verify the overall operation of the HIL system, experimental results were presented for a standard UDDS drive cycle.

REFERENCES

[1] W. Deng, Y. Zhao, and J. Wu, "Energy Efficiency Improvement via Bus Voltage Control of Inverter for Electric Vehicles," *IEEE Transactions on Vehicular Technology*, vol. 66, no. 2, pp. 1063-1073, 2017.

[2] K. Clement-Nyns, E. Haesen, and J. Driesen, "The Impact of Charging Plug-In Hybrid Electric Vehicles on a Residential Distribution Grid," *IEEE Transactions on Power Systems*, vol. 25, no. 1, pp. 371-380, 2010.

[3] P. Fajri and M. Ferdowsi, "Emulating On-Road Operating Conditions for Electric-Drive Propulsion Systems," *IEEE Transactions on Energy Conversion*, vol. 31, no. 1, pp. 1-11, March 2016.

[4] Steve Corrigan, Introduction to the Controller Area Network (CAN), *Application Report SLOA101B*, Texas instruments, August 2002.

[5] X. Nian, F. Peng, and H. Zhang, "Regenerative Braking System of Electric Vehicle Driven by Brushless DC Motor," *IEEE Transactions on Industrial Electronics*, vol. 61, no. 10, pp. 5798-5808, Jan. 2014.

[6] P. Fajri, S. Lee, V. A. K. Prabhala, and M. Ferdowsi, "Modeling and Integration of Electric Vehicle Regenerative and Friction Braking for Motor/Dynamometer Test Bench Emulation," *IEEE Transactions on Vehicular Technology*, vol. 65, no. 6, pp. 4264-4273, June 2016.

[7] N. Lotfi, P. Fajri, S. Novosad, J. Savage, R.G. Landers, and M. Ferdowsi, "Development of an experimental testbed for research in lithium-ion battery management systems," *Energies*, vol. 6, pp. 5231 – 5258, 2013.

[8] N. Lotfi, R.G. Landers, J. Li, and J. Park, "Reduced-order electrochemical model-based SOC observer with output model uncertainty estimation," *IEEE Transactions on Control Systems Technology*, vol. 25, no. 4, pp. 1217 – 1230, 2017.

[9] P. Fajri, R. Ahmadi, and M. Ferdowsi, "Equivalent vehicle rotational inertia used for electric vehicle test bench dynamic studies," in *Proc. 38th IEEE Industrial Electronics Society Annual Conference*, Montreal, Canada, Oct. 25-28, 2012, pp. 4115 - 4120.

[10] F. R. Salmasi, "Control strategies for hybrid electric vehicles: Evolution, classification, comparison, and future trends," *IEEE Transactions on Vehicular Technology*, vol. 56, no. 5, pp. 2393-2404, 2007.

New Three-Phase Static Transfer Switch using AC SSCB

Seung-Min Song[1], Jin-Young Kim[1] and In-Dong Kim[1]
1 Power Electronics Energy Lab, Pukyong National University, Busan, Korea
E-mail: thdtmdals92@naver.com
E-mail: jinminerva@naver.com
E-mail: idkim@pknu.ac.kr

Abstract— These days, widespread use of sensitive loads and distributed generators makes static transfer switch (STS) an essential component in power circuits to achieve a good power quality for AC Grids. In case that fault occurs, previous STS cannot break the fault current. However, the proposed STS can break quickly even if a short-circuit fault occurs. Also if there is power quality problems such as Sag/Swell, the proposed STS quickly transfers the load to the good quality source. It is anticipated that the proposed STS may be utilized to realize many reliable AC grid systems.

Keywords— *STS, SSCB, thyristor, circuit breaker*

I. INTRODUCTION

With the development of the IT industry in modern society, a lot of loads that are sensitive to power quality problems are being widely utilized [1].

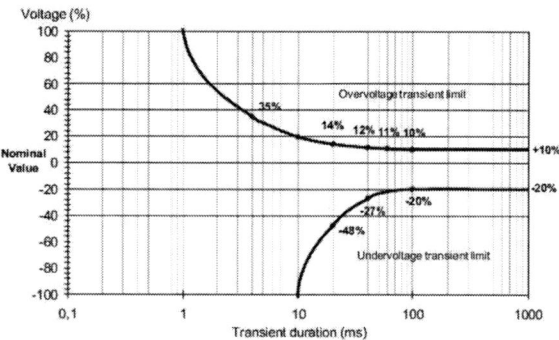

Fig. 1. Output dynamic performance classification.

Fig. 1 shows the regulations for transient duration according to the magnitude of the voltage disturbance in the International Standard Specification (IEC 62040-3). Even if the sag/swell occurs in the power source, if the power source is transferred to another power source quickly within a prescribed time, the load can always be supplied with good quality power. STS (Static Transfer Switch) is essential for sensitive load because it can transfer power source quickly. These days STS is also being applied in high-voltage applications to replace electro-magnetic transfer switch (EMTS) [3]-[4].

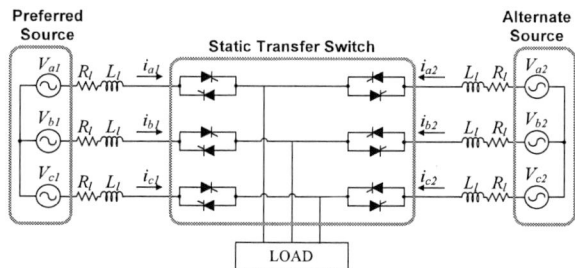

Fig. 2. Previous STS schematic.

Fig. 2 shows the previous STS circuit diagram. SCR is used in the previous STS. When the fault occurs in the preferred source, the SCRs connected to the preferred source are turned off and the SCRs connected to the alternate source are turned on to supply good quality power to the load at all times.

TABLE I
TRANSFER TIMES OF PREVIOUS STS WHEN THE FAULT OCCURS

Case No.	Test Condition	Transfer Time		
		Anal. (ms)	Sim. (ms)	Exp. (ms)
1	Pf=0.8 u=35%	2.89	2.72	3
2	Pf=0.7 single-phase-to ground fault	1.66	1.69	1.83
3	Pf=0.8 phase-to-phase fault	3.40	3.10	3.61
		Total Load-Transfer Time		
1	Pf=0.8 u=35%	3.90	4.10	4.03
2	Pf=0.7 single-phase-to ground fault	4.27	4.67	4.89
3	Pf=0.8 phase-to-phase fault	7.60	7.62	7.91

Table I shows the previous STS transfer time when the fault occurs [5]. Case 1, 2 and 3 show the STS transfer time according to reduced output voltage and load power factor. Case 1 is in case that the voltage of the power source is dropped by 35%. Case 2 is in case that the ground fault occurs on a-phase. And Case 3 is in case that a short fault occurs between a-phase and b-phases. As shown in the Fig 3, the previous STS has very long transfer time. Also, in case of a short circuit fault or

3229

ground fault, previous STS cannot break the fault current rapidly. Since the SCR is not turned off by the gate signal, the commutation circuit is required to turn off the SCR.

As the fault of power grid frequently occurs, if the fault current is not broken quickly, serious damage will occur [6]. Also, in case of short circuit fault for a short time, the power grid has to supply energy stably. But if the broken state is kept for a long time, additional economic loss occurs [7]. For this reason, IEC-62271-100 prescribes the operating duty of circuit breaker that the reclosing and re-breaking operations of circuit breaker should be able to be performed repeatedly [8].

Therefore, this paper proposes a new STS that solves the disadvantages of the previous. The proposed STS performs transfer operation using a solid-state circuit breaker (AC SSCB), so that it can always supply good quality power to the load. Also, AC SSCB used in the proposed STS can quickly break the fault current within 1 [ms], and can perform reclosing and rebreaking operation, so that the operating duty can be performed even in the fault of a short circuit.

II. PROPOSED THREE-PHASE STS USING AC SSCB

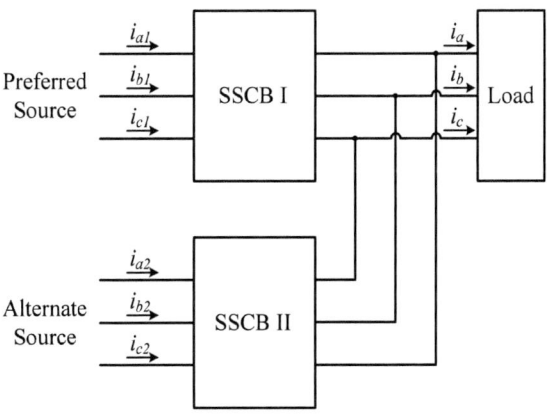

Fig. 4. STS Conceptual diagram.

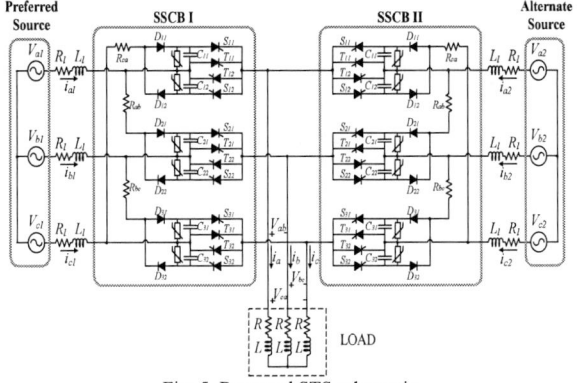

Fig. 5. Proposed STS schematic.

Fig. 4 shows a schematic of STS (Static Transfer Switch). Two independent power sources are connected to each SSCB and the output of each SSCB is connected to the

load. Normally, power is supplied from the preferred source to the load and if a problem occurs in the preferred source the power is transferred to the alternate source.

Fig. 5 is the proposed STS circuit diagram that SSCB is connected instead of antiparallel-connected SCRs of the previous STS. Since the SSCB is connected to each power source, it can break quickly in case of the fault and transfer operation is possible like previous STS.

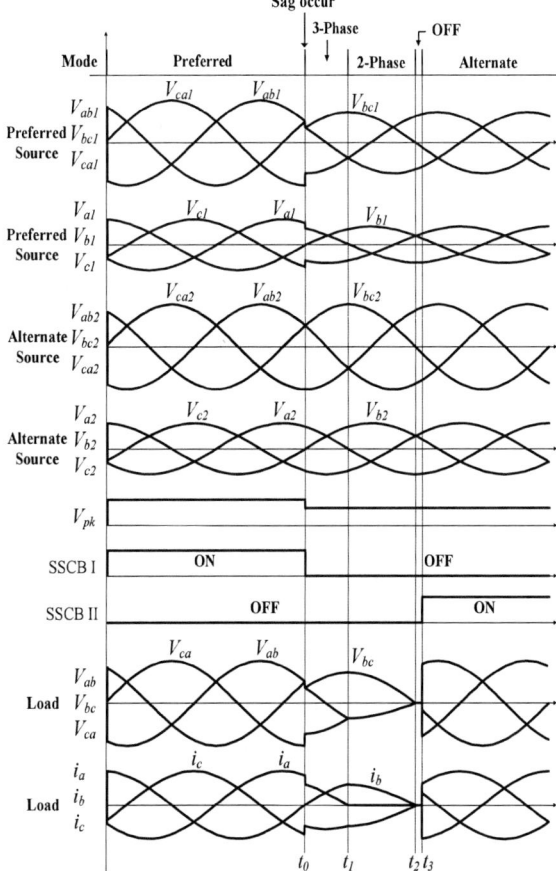

Fig. 6. Proposed STS operation waveforms.

Fig. 6 shows the proposed STS operation waveforms. The STS has five modes.

Four operation modes are the preferred mode to supply power from the preferred source, 3-Phase mode that the gate signal of all switches is off after recognizing the power quality problem, 2-phase mode that current flows only in two phases, off mode that currents in three phases are all 0 [A], and alternate mode that power is supplied through the alternate source.

A. Preferred mode

The preferred mode is to supply power from the preferred source to the load through SSCB I.

In this mode, by detecting the voltage of the preferred source the STS determines whether there is the power quality problem such as the sag/swell or the fault such as the short circuit and the ground fault.

3230

B. Phase mode (t₀~t₁)

The 3-Phase mode is the mode that detects with power quality problem of the preferred source. If the power quality problem is detected, the off signal is applied to the SCR T_{all} (T_{11}, T_{12}, T_{21}, T_{22}, T_{31}, T_{32}) of the SSCB I. However, since the SCR is turned off when the current is 0 [A], the SCR current is still flowing in all three-phases.

C. 2-Phase mode (t₁~t₂)

In the 2-phase mode, while only 2 phase current flows. The phase current of remaining phase is 0 [A] at t_1 in Fig. 6, i_a becomes 0 [A] and only the currents of b-phase and c-phases flow. The currents of b-phase and c-phase are determined by V_{bc}.

D. Off mode (t₂~t₃)

The off mode is the mode that the SCR T_{all} (T_{11}, T_{12}, T_{21}, T_{22}, T_{31}, T_{32}) of SSCB I is turned off. The SCR is completely turned off after tq time from the current becomes zero. Therefore, it is the mode to wait for SCR to turn off completely before turning on the SSCB II.

Since the tq time of SCR is about 100 [us], off mode has a very short time compared with other modes.

E. Alternate mode

The alternate mode is a mode that turns on the SSCB II to supply power from the alternate source to the load. The transfer of the power source is completely done, and the load is supplied with good quality power.

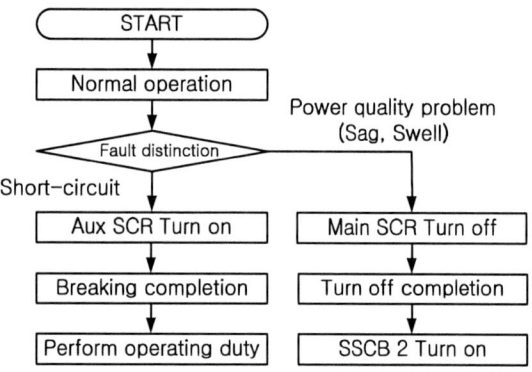

Fig. 7. STS breaking operation algorithm.

Fig. 7 shows the operation algorithm of the STS. Since the proposed STS uses SSCB unlike the previous STS, it should operate differently depending on the problem (short circuit fault and power quality problem such as sag/swell).

In accordance with the IEC-62271-100, SSCB I performs reclosing and rebreaking of the operating duty. The previous STS cannot break short-circuit current in case of short-circuit fault. However if SSCB is used, rapid breaking within 1 [ms] is possible. If power quality problem such as sag/swell occurs in the power supply, SSCB I performs the off operation. Then, when all the

currents of SSCB I become 0 [A], the SSCB II is turned on to transfer the power.

III. THREE-PHASE AC SSCB FOR PROPOSED STS

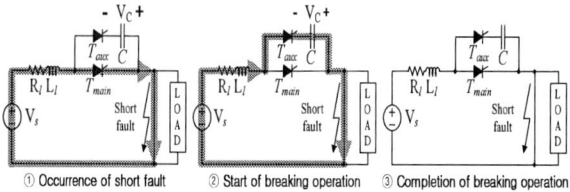

Fig. 8. SCR with commutation circuit.

Fig. 8 shows a circuit with a commutation circuit consisting of a commutation capacitor C to force turn off the SCR in the three-phase AC SSCB for the proposed STS. If the short circuit occurs, SCR T_{aux} turned on the as shown in ② of Fig. 8. When the SCR T_{aux} is turned on, the charging voltage of the commutation capacitor V_C is applied to the SCR T_{main} in the reverse direction. As a result, the current flowing through SCR T_{main} decreases to 0 [A], and after the tq time, the SCR T_{main} is completely turned off and all current flows through the SCR T_{aux}. As the current of grid decreases, the energy stored in the line inductance is moved to the commutation capacitor, and as the current of the line inductance reaches zero, the commutation capacitor is charged to the maximum voltage in the reverse direction.

AC Solid-State Circuit Breaker

Fig. 9. three-phase AC SSCB[9].

Fig. 9 shows the detailed three-phase AC SSCB circuit for the proposed STS.

Fig. 10 shows operation mode of the AC SSCB. The operation of the AC SSCB is divided into six modes. The six operation modes are composed charging mode to charge the commutation capacitor, normal mode to supply energy to the load, breaking mode to break the fault current, recharging mode to recharge commutation capacitor for reclosing, reclosing mode to re-supply energy to the load, and a rebreaking mode to rebreak in case that short circuit faults lasts.

3231

The 2018 International Power Electronics Conference

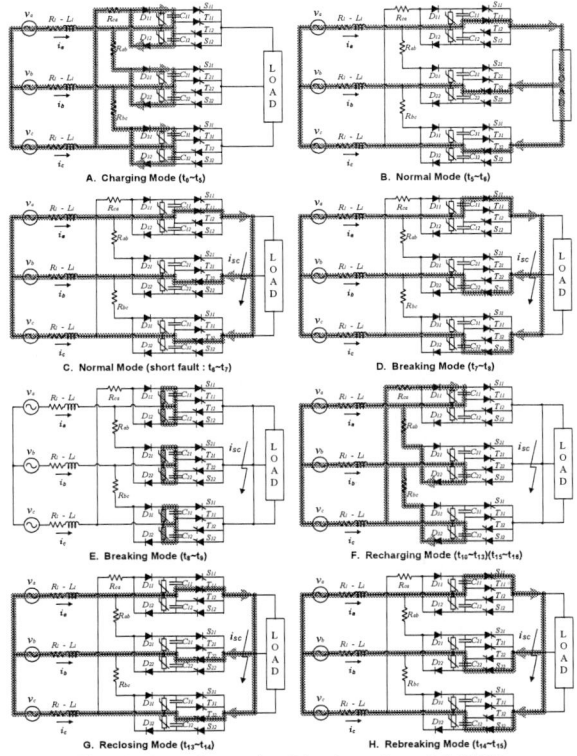

Fig. 10. Operation mode of the three-phase AC SSCB.

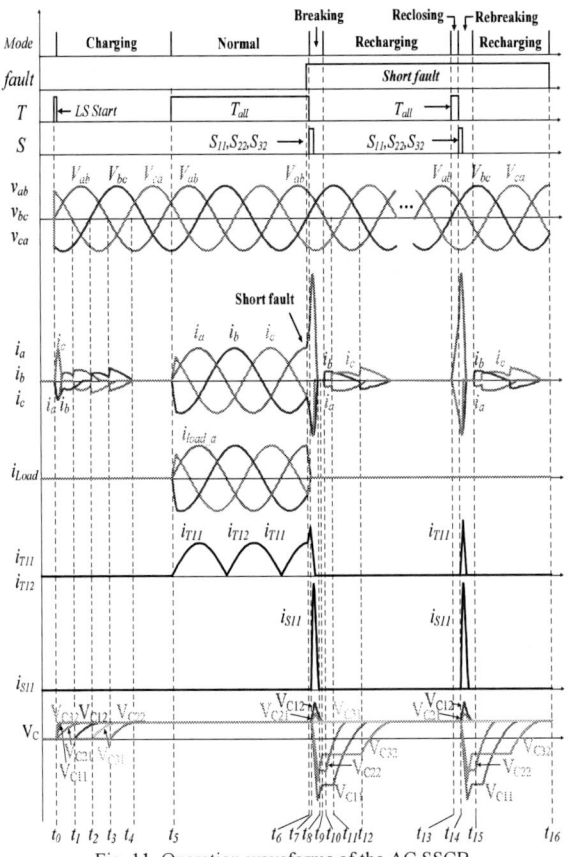

Fig. 11. Operation waveforms of the AC SSCB.

A. Charging mode

All the commutation capacitors in the AC SSCB must be charged at the charging mode because the charging of the commutation capacitors must be preceded in order to perform the breaking operation. Therefore, at the charging mode ($t_0 \sim t_5$) of the AC SSCB, the commutation capacitor is charged to the voltage required for breaking of fault current by using line voltage and charging resistance.

When the LS(line switch) is turned on at t_0 in Fig. 11, it starts to begin the charging of the commutation capacitor. Since all the commutation capacitors have the charge loop that does not include SCR, no additional control is required for charging commutation capacitor. Also, since the commutation capacitor is charged even if the load side is short-circuited, the proposed AC SSCB can perform breaking operation at any time.

B. Normal mode

At the normal mode, by turning on the SCR T_{all} (T_{11}, T_{12}, T_{21}, T_{22}, T_{31}, T_{32}), energy is transferred to the load.

At the normal mode, current and voltage are detected to discriminate accidents such as over current and voltage sag / swell which is the same as preferred mode of STS.

C. Normal mode(short circuit fault)

The mode C in Fig. 10 is the section where the fault current increases due to three-phase short-circuit fault at t_6. Even if a short circuit fault occurs, the magnitude of the fault current is smaller than the reference current judged as an accident. Therefore the AC SSCB operates as the normal mode. As the phase current ia gradually increases, the AC SSCB judges it as a short-circuit fault and starts the breaking mode at t_7.

D. Breaking mode

The breaking mode is the section where break the fault current by using the charged commutation capacitor. When the auxiliary SCR S_{11}, S_{22}, S_{32} corresponding to the direction of each phase current is turned on as shown D circuit in Fig. 10, main SCR T_{11}, T_{22}, T_{32} are turned off by the commutation capacitors C_{11}, C_{22}, C_{32}. The R_1-L_1-C resonant current flows through all phases and the fault current decreases. The commutation capacitors C_{11}, C_{22}, and C_{32} of each phase used for breaking are charged in the reverse direction as the V_C waveform in Fig 8.

E. Breaking mode(all SCR off)

The mode E in Fig. 10 is the section where all fault currents are broken and no current flows in the grid. At t_8 in Fig. 11, the current flowing through the auxiliary SCR becomes zero and all the SCRs of the AC SSCB are turned off, thus the breaking operation of the AC SSCB is complete. Since the reverse-charged voltage of the commutation capacitor is greater than the breakdown voltage of the varistor, the voltage of commutation capacitor discharges through the varistor until t_9.

3232

F. Recharging mode

Recharging mode is the section where the discharged voltage of commutation capacitors at breaking mode is recharged. The recharge loop is the same as the charge loop in charge mode ($t_0 \sim t_5$). The voltage of commutation capacitor can be recharged in case of the short-fault state because recharging operation is performed regardless of the switching state of all the SCR.

G. Reclosing mode

The reclosing mode is the section where the main SCR T_{all} (T_{11}, T_{12}, T_{21}, T_{22}, T_{31}, T_{32}) is turned on. The circuit breaker should be reclosing operation in accordance with the reclosing time of the operating duty. Therefore, all main SCR should be turned on even if short fault is kept on the load side

H. Rebreaking mode

At t_{14}, which is discriminated as short fault, the AC SSCB starts rebreaking operation. The principle of the rebreaking operation is the same as that of breaking mode ($t_7 \sim t_9$).

IV. TRANSFER TIME OF PROPOSED STS

Sag/swell is a representative problem of power quality that lasts more than half cycle (IEC-61000-4-30) [10]. Therefore, if the transfer time of the STS is completed within half cycle, the sag/swell does not occur in the load, so that the load can always be supplied with good quality power.

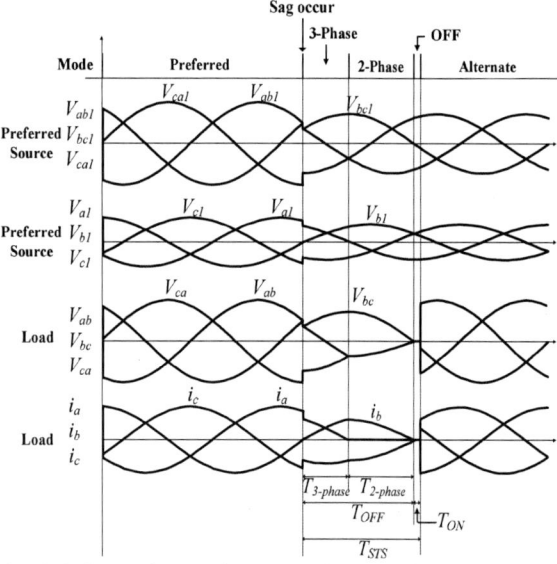

Fig. 12. STS operation waveforms at maximum transfer time (resistance load).

The proposed STS should turn on the main SCR of the SSCB II, confirming that all the currents of SSCB I become 0 [A] because there occurs a short circuit fault

when the main SCR of SSCB II is turned on before the main SCR of SSCB I is turned off. Thus, all main SCRs of SSCB I must be turned off within half cycle, so that the STS can be transferred within half cycle.

Fig. 12 shows the STS operating waveforms at maximum transfer time under resistive load. $T_{3\text{-phase}}$ is time in 3-phase mode and $T_{2\text{-phase}}$ is time in 2-phase mode. T_{off} is the time of until all phase currents become 0 [A] after sag detection T_{on} is the time of off mode, and T_{STS} is the total time after the fault occurs.

As shown in the fig.12, T_{STS} is obtained as follows.

$$T_{STS} = T_{OFF} + T_{ON} \tag{1}$$

T_{OFF} is expressed as follows.

$$T_{OFF} = T_{3\text{-phase}} + T_{2\text{-phase}} \tag{2}$$

Since the point at which the current becomes 0 [A] at three-phases exists every 60°, the range of the $T_{3\text{-phase}}$ is expressed as follows.

$$0 \leq T_{3\text{-phase}} \leq \frac{T}{6} \tag{3}$$

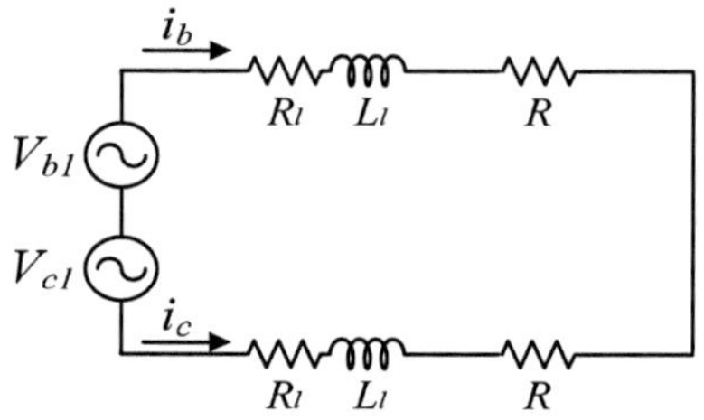

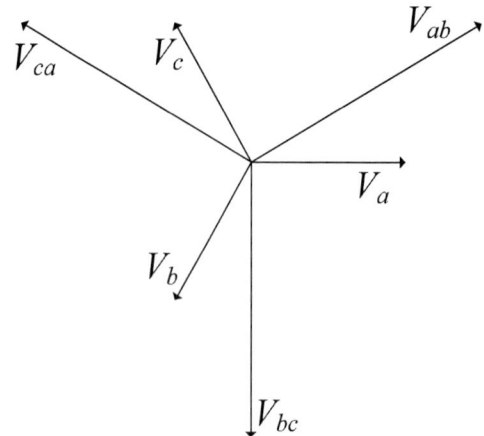

Fig. 13. Equivalent circuit at $T_{2\text{-phase}}$ of Fig 12.

Fig. 13 shows the equivalent circuit consisting of the b and c phase when the a-phase current becomes 0[A] at the resistive load. Fig. 13 shows that the load current i_b and i_c flow by V_{bc}. Therefore, i_b and i_c become 0 [A] when V_{bc} becomes 0[V].

Fig. 14. Line voltage and phase voltage of the power source.

Fig. 14 shows the line voltage and phase voltage phasor of the power supply. As shown in Fig. 14, V_a and V_{bc} have the phase difference of 90°. Since the time from V_a becomes 0 [V] to V_{bc} becomes 0 [V] is $T_{2\text{-phase}}$, $T_{2\text{-phase}}$ is as follows.

$$T_{2\text{-phase}} = \frac{T}{4} \tag{4}$$

Therefore, T_{OFF} is as follows.

$$\frac{T}{4} \leq T_{OFF} \leq \frac{5T}{12} \tag{5}$$

Since the time of T_{ON} is very short compared to T_{OFF}, the time of T_{ON} is ignored.

That is, the maximum transfer time $T_{STS\text{-max}}$ of the STS in the resistive load is as follows.

$$T_{STS\text{-max}} = \frac{5T}{12} \tag{6}$$

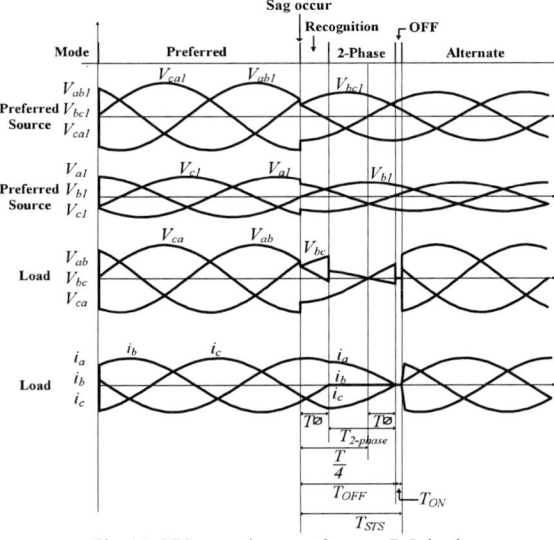

Fig. 15. STS operation waveforms at R-L load.

Fig. 15 shows the STS operating waveforms in the R-L load with a phase delay of Φ °. T_Φ is the phase difference time between load voltage and load current.

As shown in the fig. 15, T_{OFF} is as follows.

$$T_{OFF} = T_\Phi + T_{2\text{-phase}} \tag{7}$$

Fig. 16 shows the equivalent circuit of Fig. 15 at the $T_{2\text{-phase}}$. As shown in Fig. 16, the total impedance is doubled when current flows only in two phases. Therefore, the phase difference between V_{ca} and i_c is the same as the V_b and i_b when current flows through three-phases. Thus, the phase difference between V_{ca} and i_c is Φ°. Therefore, when V_{ca} becomes 0 [V] and after T_Φ, since i_a and i_c become 0 [A], the $T_{2\text{-phase}}$ is as follows.

$$T_{2\text{-phase}} = \frac{T}{4} \tag{8}$$

In R-L load, the time range that the current of one phase becomes zero is the same that as in the case of the resistive load. Thus, regardless of the power factor, the $T_{3\text{-phase}}$ is as follows.

$$0 \leq T_{3\text{-phase}} \leq \frac{T}{6} \tag{9}$$

And $T_{STS\text{-max}}$ is as follows.

$$T_{STS\text{-max}} = \frac{5T}{12} \tag{10}$$

Therefore it is possible to transfer the power supply within half cycle regardless of the power factor.

V. SIMULATION & EXPERIMENT

A. STS Experiment

The experimental parameters of the proposed STS are as shown in Table II.

TABLE II
PROPOSED 3-PHASE STS PARAMETERS

Power	9.63 [kW]
Line Voltage	380 [V]
Full load current	4.8 [A]
Line resistance R_L	100 [mΩ] (0.222%)
Line inductance L_L	35 [uH] (0.0293%)
Load resistance R	45 [Ω]
Range of trip setting	380[V] → 342[V]

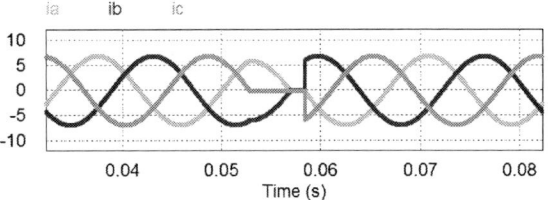

Fig. 17. Simulation i_a, i_b, i_c waveforms when sag occurs.

Fig. 17 shows the simulation waveforms of i_a, i_b, and i_C when the sag occurs, and Fig. 18 shows the experimental waveforms of Fig. 17.

Fig. 16. Equivalent circuit of Fig 15 at $T_{2\text{-phase}}$.

3234

The 2018 International Power Electronics Conference

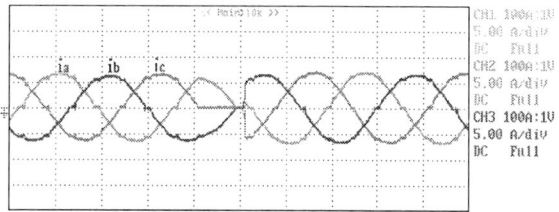

Fig. 18. Measured i_a, i_b, i_c wavefitms when sag occurs.

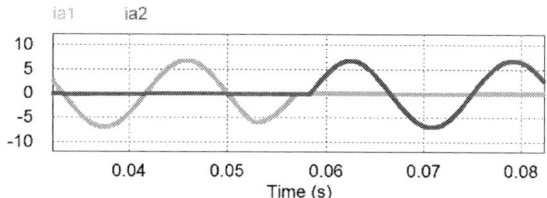

Fig. 19. Simulation i_{a1}, i_{a2} waveforms when sag occurs.

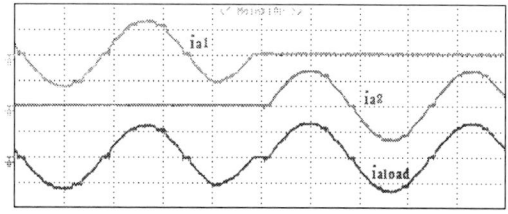

Fig. 20. Measured i_{a1}, i_{a2} waveforms when sag occurs.

Fig. 19 shows the simulation waveforms of the currents i_{a1} and i_{a2} of the AC SSCB when the sag occurs, and Fig. 20 shows the experimental waveforms of Fig. 19.

As shown in Fig. 20, after the current of SSCB I becomes 0 [A], the SSCB II turns on.

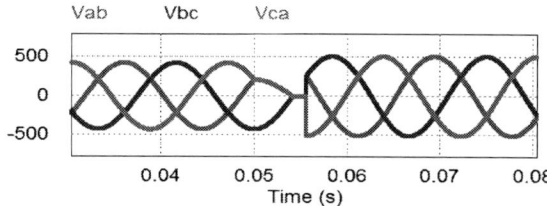

Fig. 21. Simulation V_{ab}, V_{bc}, V_{ca} waveforms when sag occurs.

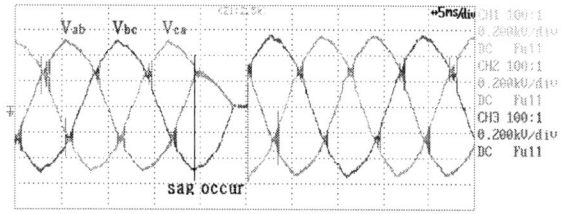

Fig. 22. Measured V_{ab}, V_{bc}, V_{ca} waveforms when sag occurs.

Fig. 21 shows the simulation waveforms of the load voltages V_{ab}, V_{bc} and V_{ca} when sag occurs, and Fig. 22 shows the experimental waveforms of Fig. 21.

As shown in Fig. 21, power supply is transferred within half cycle after sag occurs.

Since the proposed STS is transferred within a half cycle from the three-phase to a distributed power supply, sag/swell does not occur in the load, and the load is always supplied with good quality power.

B. SSCB Experiment

Table III shows the experimental parameters of the AC SSCB.

TABLE III
EXPERIMENT PARAMETER OF THE AC SSCB

Power	46.67 [kW]
Line Voltage	380 [V]
Full load current	70.7 [A]
Line resistance R_L	100 [mΩ] (3.214%)
Line inductance L_L	35 [uH] (0.424%)
Load reisitor R	3 [Ω]
short fault switch resistance	200 [mΩ]
Range of trip setting	100[A_{peak}] $\rightarrow$ 500[A_{peak}]

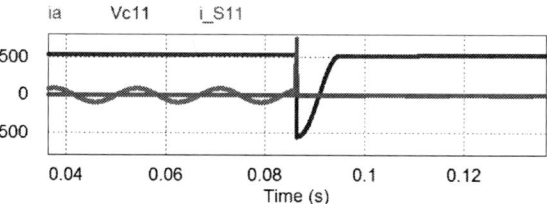

Fig. 23. Simulation V_{C11}, i_a, i_{S11} waveforms at breaking mode.

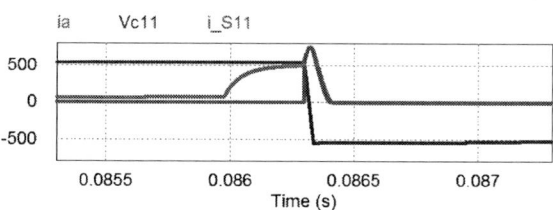

Fig. 24. Magnification waveforms of the Fig 23.

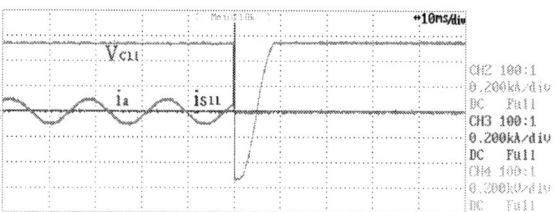

Fig. 25. Measured V_{C11}, i_a, i_{S11} waveforms at breaking mode.

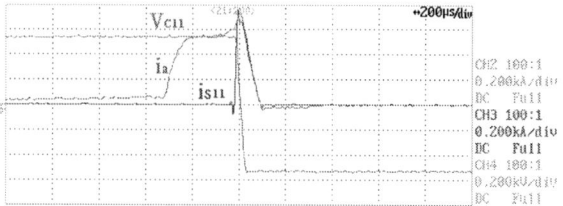

Fig. 26. Magnification waveforms of the Fig 25.

3235

Fig. 23 shows the simulation waveforms of i_a, i_{S11} and V_{C11} at breaking mode, and Fig. 24 shows the experimental waveforms of Fig. 23. As shown in Fig. 23, the AC SSCB breaks the fault current within 100 [us].

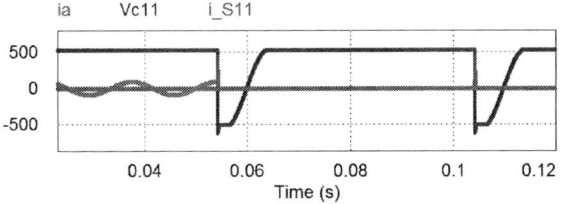

Fig. 27. Simulation V_{C11}, i_a, i_{S11} waveforms in operation duty.

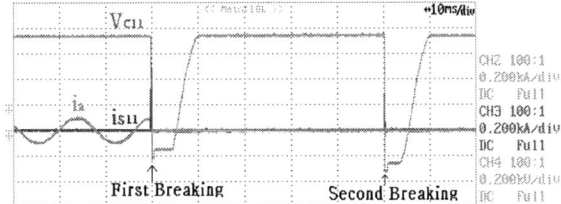

Fig. 28. Measured V_{C11}, i_a, i_{S11} waveforms in operation duty.

Fig. 25 shows the simulation waveforms of i_a, i_{S11}, and V_{C11} when performing operating duty. Fig. 26 shows the experimental waveforms of Fig. 25.

VI. CONCLUSIONS

This paper proposes a new STS using AC SSCB which can perform operating duty of reclosing and rebreaking.

In the event of short fault, the AC SSCB performs rapid breaking operation, and in the event of a problem with power quality such as sag and swell, it performs AC STS operation to replace the power supply. Therefore, regardless of the type of fault in grid, the load is always supplied with good quality power.

The operating characteristics of the proposed STS are verified by simulations and experiments on three - phase short faults and sag.

REFERENCES

[1] Chan-Nan Lu and Cheng-Chieh Shen, "Estimation of Sensitive Equipment Disruptions Due to Voltage Sags," IEEE Trans. Power Delivery, vol. 22, No. 2, pp. 1132–1137, Apr. 2007.

[2]]IEC 62040-3, "Uninterruptible power systems (UPS) – Part 3: Method of specifying the performance and test requirements", International Electotechnical Commission, (1999).

[3] R. W. De Doncker, W. T. Eudy, J. A. Maranto, H. Mehta, and J. W. Schwartzenberg, "Medium voltage subcycle transfer switch," Power Qual. Assur., pp. 46–51, July/Aug. 1995.

[4] J. W. Schwartzenberg and R. W. De Doncker, "15 kV medium voltage static transfer switch," in Proc. IEEE IAS 30th Annu. Meeting, Orlando, FL, pp. 2515–2520, Oct. 1995.

[5] H. Mokhtari, S. B. Dewan, and M. R. Iravani, "Analysis of a static transfer switch with respect to transfer time," in IEEE PES-2000 Summer Meeting, IEEE Trans. on Power Delivery, submitted for publication.

[6] Giuseppe Parise and Luigi Parise, "Unprotected Faults of Electrical and Extension Cords in AC and DC Systems," IEEE Trans. Ind. Appl., vol. 50, No. 1, pp.4–9, Jan. 2014.

[7] C. Abbey, D. Cornforth, N. Hatziargyriou, K. Hirose, A. Kwasinski, E. Kyriakides, G. Platt, L. Reyes and S. Suryanarayanan, "Powering Through the Storm: Microgrids Operation for More Efficient Disaster Recovery," IEEE Power and Energy Magazine, vol. 12, pp. 67–76, May/June 2014.

[8] C37.09, IEEE Standard Test Procedure for AC High-Voltage Circuit Breakers Rated on a Symmetrical Current Basis, 1999

[9] Jin-Young Kim, Seung-Min Song, Seung-Soo Choi, In-Dong Kim and Sun Kyu Choi "New AC solid-state circuit breaker with simple charging and rebreaking capabilities," In Industrial Electronics Society, IECON pp. 3866-3871, Oct.2016

The 2018 International Power Electronics Conference

Harmonics Compensation in High Frequency Range of Active Power Filter with SiC-MOSFET Inverter in Digital Control System

Shin-ichi Hamasaki[1*], Kengo Nakahara[1] and Mineo Tuji[1]
1 Electrical and Electronic Engineering, Nagasaki University, Nagasaki, Japan
*E-mail: hama-s@nagasaki-u.ac.jp

Abstract— Compensation of harmonics in high frequency range by active power filter (APF) using SiC-MOSFET inverter is investigated. A new control method of APF controlled by deadbeat control and repetitive control is proposed. Aiming at frequency, the sampling time is further shortened and increasing the switching frequency is designed in the digital control system. The control of harmonics compensation is realized by the repetitive control, which is adapted for high frequency harmonics by using periodicity. And the current regulator for the output current of the APF is realized by the deadbeat control, which can regulate the output current accurately without delay. SiC-MOSFET inverter is applied to the APF to obtain high frequency switching. Performance of the harmonics compensation by the proposed method is investigated in experiment and excellent compensation is performed in the range to 30th order (1.8kHz) harmonics. A theoretical analysis, simulation results and experimental results using SiC-MOSFET inverter are presented.

Keywords— active power filter, harmonics compensation, SiC-MOSFET inverter, digital control

I.   INTRODUCTION

Problem of the harmonics generated by power converters and other devices may cause serious problem, especially in high frequency. Thus variety kinds of Active Power Filter (APF) have been developed[1]-[10]. The conventional shunt type APF is connected in parallel between the load and the power line. This type is can compensate harmonics current by using the detected current harmonics. The shunt type APF has two types of harmonics detection. One is the load current detection, which is general method for the feedforward control of the APF, and the other is the line current detection. The line current detection type APF[10] detects current of the power line side and is able to compensate by the feedback control. This means that only one current sensor is required even if number of non-linear loads varies.

Generally, performance of harmonics order to be compensated depends on switching frequency of inverter and sampling period of digital control. So that frequency of the harmonics to be compensated is limited to lower frequency. Recently performance of the DSP increases, and SiC devices are developing and practically applied to industrial systems. These can be applied to the APF to

obtain high performance.

SiC-MOSFET inverter is applied to the APF to obtain high frequency switching and the DSP system with high performance is applied to obtain high speed calculation. The repetitive control and the deadbeat (DB) control as a new control method of the APF applied to adapt high frequency. The repetitive control is optimized output current reference of the APF by using periodicity. The harmonics current gradually decreases by repeating the control scheme. In addition, the repetitive control method has the time-leading compensation to cancel the time delay caused by the detection and current regulator. The current regulation of the APF is realized by the DB control, which is able to follow the current command accurately and instantly. Total performance of the APF with the repetitive control and the DB control becomes excellent even if the harmonics is high frequency.

The theory of the proposed method is presented. The sampling periods is shorten and the switching frequency is increased to adapt to high frequency harmonics. When the sampling periods and switching frequency are changed, characteristic of the proposed system is investigated by simulation and experiment.

II.   CONTROL METHOD

A.   Equivalent Circuit

Fig. 1 illustrates the circuit with the APF for analysis.

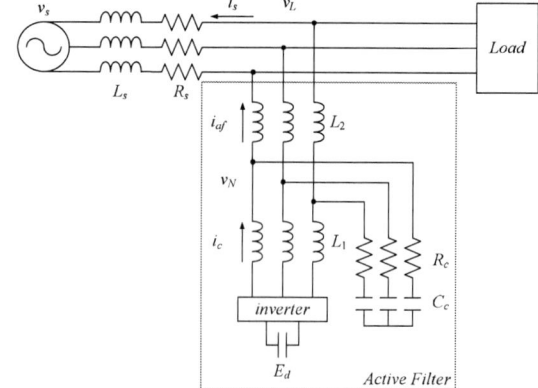

Fig. 1. Circuit for analysis of APF

3237

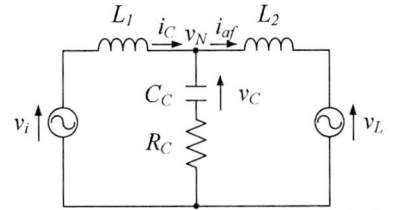

Fig. 2. Equivalent circuit for DB control of APF (single phase)

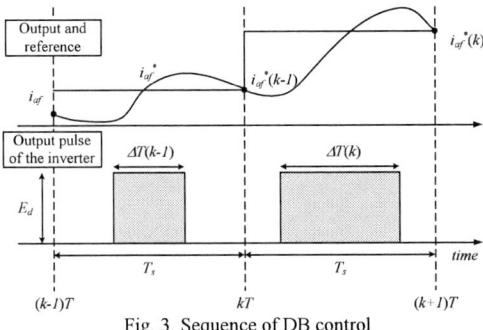

Fig. 3. Sequence of DB control

The APF has the voltage type inverter and output filters (L_1, L_2, C_c and R_c) for switching noise reduction. The load is the harmonics source like the 3-phase diode rectifier. L_s, R_s and v_s are the power line parameters.

Fig.2 illustrates a single phase equivalent circuit for DB control of APF. Based on the circuit, the DB control can be designed.

B. Deadbeat(DB) Control

The DB control can follows the controlled object to the command value by a sampling period. Fig.3 shows the sequence image of the DB control. The APF with the DB control can regulate the output current exactly equal to the current command at next sampling instant. The APF output current i_{af} correspond to the APF current command i_{af}* by calculating optimal pulse width of the inverter by the DB control.

How to calculate the optimal pulse is introduced. The theoretical pulse width can be introduced from Fig.2. When four parameters i_c, v_c, i_{af}, and v_L are selected as the states from Fig.2, the state equations are derived as follows.

$$\frac{d}{dt}\mathbf{x} = \mathbf{A}\mathbf{x} + \mathbf{b}v_i \tag{1}$$

$$i_{af} = \mathbf{c}^T\mathbf{x} \tag{2}$$

$$\mathbf{A} = \begin{bmatrix} -\dfrac{R_c}{L_1} & -\dfrac{1}{L_1} & \dfrac{R_c}{L_1} & 0 \\ \dfrac{1}{C_c} & 0 & -\dfrac{1}{C_c} & 0 \\ \dfrac{R_c}{L_2} & \dfrac{1}{L_2} & -\dfrac{R_c}{L_2} & -\dfrac{1}{L_2} \\ 0 & 0 & 0 & 0 \end{bmatrix}$$

$$\mathbf{x} = \begin{bmatrix} i_c & v_c & i_{af} & v_L \end{bmatrix}$$

$$\mathbf{b} = \begin{bmatrix} \dfrac{1}{L_1} & 0 & 0 & 0 \end{bmatrix}$$

$$\mathbf{c}^T = \begin{bmatrix} 0 & 0 & 1 & 0 \end{bmatrix}$$

(1) and (2) are description of the continuous time system. These can be converted into the sample value sys-

tem using the output pulse width ΔT. (3)-(6) can be obtained.

$$\mathbf{x}(k+1) = \mathbf{F}\mathbf{x}(k) + \mathbf{g}\Delta T(k) \tag{3}$$

$$i_{af}(k) = \mathbf{c}^T\mathbf{x}(k) \tag{4}$$

$$\mathbf{F} = e^{ATs} = \begin{bmatrix} F_{11} & F_{12} & F_{13} & F_{14} \\ F_{21} & F_{22} & F_{23} & F_{24} \\ F_{31} & F_{32} & F_{33} & F_{34} \\ 0 & 0 & 0 & 0 \end{bmatrix}$$

$$\mathbf{g} = e^{\frac{ATs}{2}bE_d} = \begin{bmatrix} g_1 & g_2 & g_3 & 0 \end{bmatrix}^T$$

$\mathbf{F}$ and $\mathbf{g}$ are the matrixes, which have constant parameters calculated from A and b in (1) and (2). (5) is obtained from (3) and (4) as the discrete time system.

$$i_{af}(k+1) = F_{31}i_c(k) + F_{32}v_c(k) + F_{33}i_{af}(k) + F_{34}v_L + g_3\Delta T(k) \tag{5}$$

From (5), the pulse width $\Delta T(k)$ can be derived as shown in (6) by rewriting $i_{af}(k+1)$ to $i_{af}^*(k+1)$.

$$\Delta T(k) = \frac{i_{af}^*(k+1) - F_{31}i_c(k) - F_{32}v_c(k) - F_{33}i_{af}(k) - F_{34}v_L(k)}{g_3} \tag{6}$$

Next, the state observer is designed, which is able to estimate the states on next sampling time previously. So that the APF can reserve the computing time of DSP and calculate $\Delta T(k)$ previously during one previous sampling time. The observer for the estimated values $\hat{i}_c$ and $\hat{v}_c$ are designed in (7).

$$\hat{\mathbf{x}}_c(k) = \mathbf{F}_c\hat{\mathbf{x}}_c(k-1) + \mathbf{F}_a\hat{\mathbf{x}}_a(k-1) + \mathbf{g}_c\Delta T(k-1)$$
$$- \mathbf{L}_c\{i_c(k-1) - \mathbf{c}_c^T\hat{\mathbf{x}}_c(k-1)\} \tag{7}$$

$$\hat{\mathbf{x}}_c(k) = \begin{bmatrix} i_c(k) \\ v_v(k) \end{bmatrix} \quad \mathbf{x}_a(k) = \begin{bmatrix} i_{af}(k) \\ 0 \end{bmatrix}$$

$$\mathbf{F}_c = \begin{bmatrix} F_{11} & F_{12} \\ F_{21} & F_{22} \end{bmatrix} \quad \mathbf{F}_a = \begin{bmatrix} F_{13} & F_{14} \\ F_{23} & F_{24} \end{bmatrix}$$

$$\mathbf{g}_c = \begin{bmatrix} g_1 \\ g_2 \end{bmatrix} \quad \mathbf{c}_c^T = \begin{bmatrix} 1 & 0 \end{bmatrix} \quad \mathbf{L}_c = \begin{bmatrix} l_1 \\ l_2 \end{bmatrix}$$

,where l_1 and l_2 are the observer gain. v_L and i_{af} are estimaed in (8) and (9).

$$\hat{v}_L(k) = v_L(k-1) + K_v\{v_{Lf}(k) - v_{Lf}(k-1)\} \tag{8}$$

$$\hat{i}_{af}(k) = i_{af}(k-1) \tag{9}$$

,where v_{Lf} is the fundamental component of v_L.

The output pulse width $\Delta T(k)$ in (6) is modified to (10) by using the estimated $\hat{i}_c(k)$, $\hat{v}_c(k)$, $\hat{v}_L(k)$ and $\hat{i}_{af}(k)$.

$$\Delta T(k) = \frac{i_{af}^*(k+1) - F_{31}\hat{i}_c(k) - F_{32}\hat{v}_c(k) - F_{33}\hat{i}_{af}(k) - F_{34}\hat{v}_L(k)}{g_3} \tag{10}$$

When $\Delta T(k)$ is negative, the DC voltage of inverter is given as $-E_d$ with $|\Delta T|$.

C. Repetitive Control

The repetitive control aims at periodicity of the harmonics. Data of the detected harmonics current is memorized in a fundamental cycle and used for control repetitively. The harmonics can be compensated by using the memorized signal before one cycle. (11) is the control law in the proposed repetitive control algorithm.

$$i_{af}^{*}(k) = \mu_2 i_{af}^{*}(k-N+L) - \mu_1 i_{sh}(k-N+L) \qquad (11)$$

,where N is number of data in a fundamental cycle and L is number of the time-leading count.

μ_1 is a compensation coefficient to be adjusted to proper ratio to obtain good compensation performance and stability. It is selected within range of 0.0-1.0. μ_2 is an oblivion coefficient, which is introduced to the APF current reference $i_{af}^{*}(k-N)$ to improve stability. μ_2 is selected smaller than 1.0.

Fig.4 shows the whole block diagram of the proposed APF control. Considering that harmonics repeats at constant cycle NT_s, deviation is decreased by using one cycle previous value of the APF current reference $i_{af}^{*}(k-N)$ and adding the inversed phase of harmonics component $\mu_1 i_{sh}(k-N)$. Compensation performance is recursively improved every one fundamental cycle. In addition, the principle of the time-lead is introduced in the repetitive control as the current regulation and the low pass filter cause time delay. The APF current command is improved using the previous data at $k-N+L$ to cancel the time delay.

D. Sampling time

It is possible to improve tracking performance of the APF by shortening the sampling time T_s. Number of date at fundamental cycle N is expressed by fundamental time T_1 and T_s.

Number of date at fundamental cycle N increases by shortening the sampling time T_s. Improving the tracking performance of the APF current regulation, it is possible to compensate higher order harmonics.

III. SIMULATION

Simulation is executed by the circuit in Fig.1. The circuit parameters are listed in Table I and the control parameters are listed in Table II. μ_1, μ_2 and L are selected as shown in Table III. The 3-phase diode rectifier with inductor and capacitor on DC side is installed as the nonlinear load.

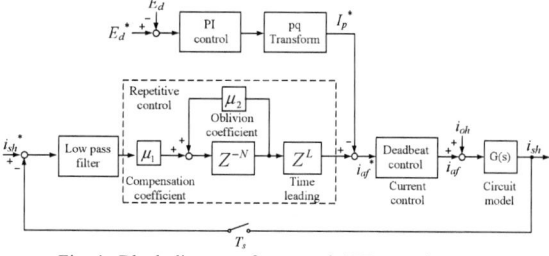

Fig. 4. Block diagram of proposed APF control.

The fundamental frequency is 60Hz. The control cycle of digital control is 50μs (the career frequency is 20kHz) and the control cycle 25μs (the career frequency is 40kHz) are selected to compare the harmonics compensation performance.

Fig.5(a)-(c) show the load current i_L, the APF current i_{af}, the current command i_{af}^{*} and the line current i_s in simulation at T_s=50μs, respectively. About the APF current i_{af} and current command i_{af}^{*}, good performance of following to the command by the DB control can be obtained. Fig.6 shows the FFT analysis of the current. It can be observed that the harmonics of the line current is compensated and the line current becomes sinusoidal waveform by the APF. Even if the load fluctuates, it compensates harmonic components and has high followability. It can be observed that the harmonic is compensated effectively by the repetitive control. Figs. 8 and 9 show the transient response when the harmonics loads increases to double. Even if the load changes, the proposed APF is able to work stably. Simulation result at T_s=25μs is omitted in this paper but the similar result like T_s=50μs can be obtained.

Table IV shows the compensation rate, which is defined by line current RMS per load current RMS and is calculated from FFT analysis at T_s=25μs and T_s=50μs respectively, when the constant AC current source is used as the harmonics load without the diode rectifier. The compensation rate of lower order harmonic is higher at both sampling time 50μs and 25μs. However the compensation rate decreases as the frequency increases. Especially, it can decreases 80% level at T_s=50μs. On the other hand, it keeps around 90% at T_s=25μs.

TABLE I
PARAMETERS IN CIRCUIT

L_s	3.0 (mH)	L_1	1.5 (mH)
L_2	2.0 (mH)	R_c	10.0 (Ω)
C_c	5.0 (μF)	R_d	44.5 (Ω)
L_d	3.0 (mH)	E_d	200 (V)
$v_{s\text{-rms}}$	100 (V)	C_d	5000(μF)

TABLE II
CONTROL PARAMETERS

l_1	−0.75	l_2	5.28	K_v	2.0

TABLE III
COEFFICIENT PARAMETERS

T_s=25μs	μ_1=0.3, μ_2=0.98	L=9
T_s=50μs	μ_1=0.3, μ_2=0.98	L=6

The 2018 International Power Electronics Conference

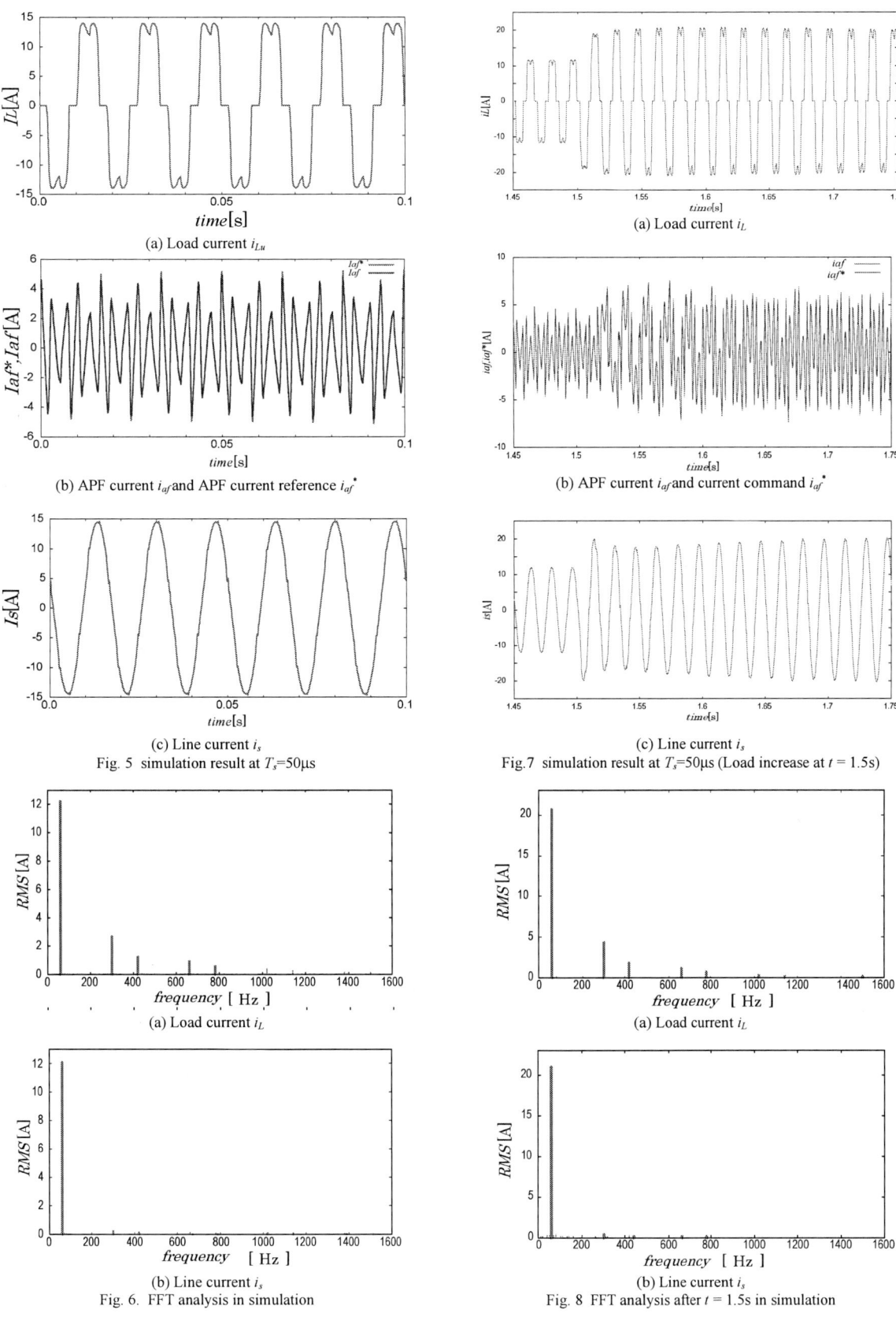

(a) Load current i_{Lu}

(b) APF current i_{af} and APF current reference i_{af}^*

(c) Line current i_s

Fig. 5 simulation result at T_s=50μs

(a) Load current i_L

(b) APF current i_{af} and current command i_{af}^*

(c) Line current i_s

Fig.7 simulation result at T_s=50μs (Load increase at t = 1.5s)

(a) Load current i_L

(b) Line current i_s

Fig. 6. FFT analysis in simulation

(a) Load current i_L

(b) Line current i_s

Fig. 8 FFT analysis after t = 1.5s in simulation

3240

TABLE IV
COMPARISON OF FFT ANALYSIS
(a) T_s=25μs

order	iLh [A]	ish [A]	compensate rate[%]
5	1.000	0.0283	97.17
7	1.000	0.0315	96.85
11	1.000	0.0361	96.39
13	1.000	0.0388	96.12
17	1.000	0.0538	94.62
19	1.000	0.0621	93.79
23	1.000	0.0752	92.48
25	1.000	0.0801	91.99
29	1.000	0.0912	90.88
31	1.000	0.0964	90.36

(b) T_s=50μs

order	iLh [A]	ish [A]	compensate rate[%]
5	1.000	0.0475	95.25
7	1.000	0.0498	95.02
11	1.000	0.0444	95.56
13	1.000	0.0610	93.90
17	1.000	0.081	91.90
19	1.000	0.0949	90.51
23	1.000	0.1064	89.36
25	1.000	0.1066	89.34
29	1.000	0.1016	89.84
31	1.000	0.1147	88.53

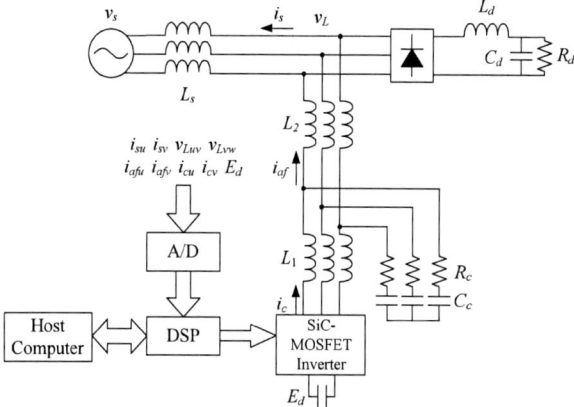

Fig.9 Experimental circuit and control system

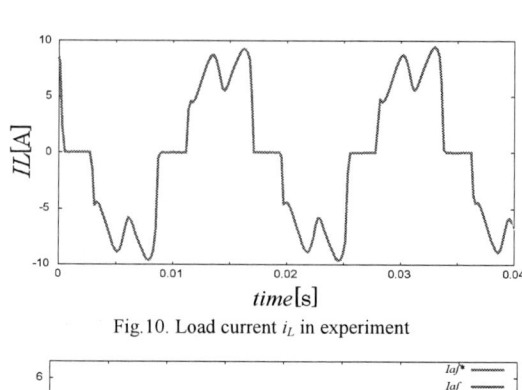

Fig.10. Load current i_L in experiment

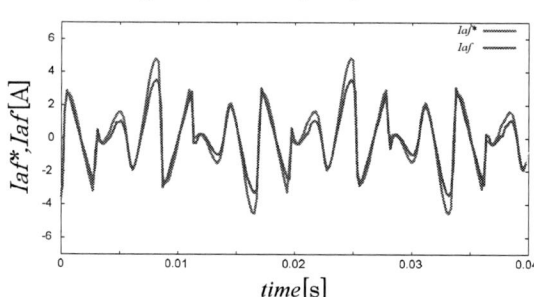

(a) APF current i_{af} and APF current reference i_{af}^*

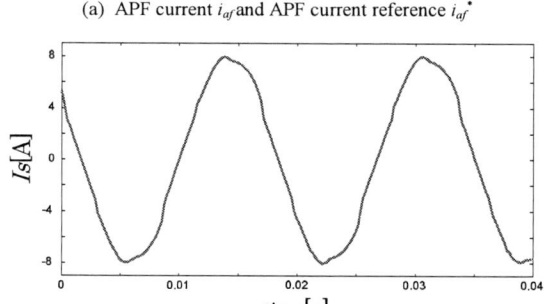

(b) Line current i_s

Fig.11. experimental result at T_s=50μs

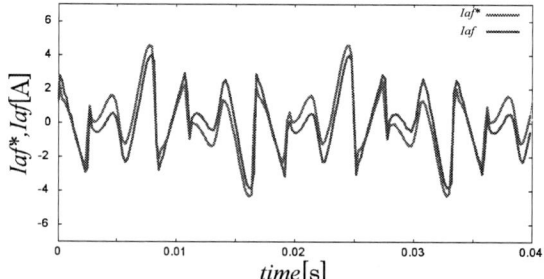

(a) APF current i_{af} and APF current reference i_{af}^*

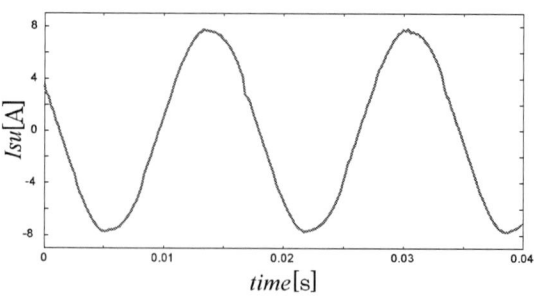

(b) Line current i_s

Fig.12. experimental result at T_s=25μs

V. EXPERIMENT

Fig.9 shows the experimental equipment. Each value (i_{su}, i_{sv}, v_{Luv}, v_{Lvw}, i_{afu}, i_{afv}, i_{cu}, i_{cv} and E_d) is detected by sensors. These signals are sent to DSP through A/D convertor. The control cycle of digital control is 50μs (the career frequency is 20kHz) and the control cycle 25μs (the career frequency is 40kHz) are selected to compare the compensation performance. The parameters are the same as the simulation.

Fig.10 shows the load current from the 3-phase diode rectifier input, Figs.11 and 12 show steady state waveforms at T_s=50μs and 25μs on experiment, respectively.

TABLE V
COMPARISON OF FFT ANALYSIS

(a) T_s=25μs

order	Ilu [A]	Isu [A]	compensate rate[%]
3	0.500	0.040	92.00
5	1.960	0.126	93.57
7	1.085	0.072	93.36
9	0.256	0.018	92.97
11	0.590	0.039	93.39
13	0.521	0.039	92.61
15	0.182	0.013	92.86
17	0.248	0.024	90.32
19	0.338	0.031	90.83
21	0.158	0.014	91.14
23	0.143	0.013	90.91
25	0.212	0.021	90.08
27	0.144	0.012	91.67
29	0.082	0.008	90.24
31	0.152	0.016	89.47

(b) T_s=50μs

order	Ilh [A]	Ish [A]	compensate rate[%]
3	0.460	0.026	94.35
5	2.070	0.154	92.56
7	1.146	0.082	92.84
9	0.193	0.019	90.16
11	0.629	0.044	93.00
13	0.521	0.047	90.98
15	0.139	0.021	84.89
17	0.276	0.022	92.03
19	0.325	0.029	91.08
21	0.125	0.024	80.80
23	0.150	0.030	80.00
25	0.209	0.030	85.65
27	0.110	0.023	79.09
29	0.084	0.016	80.95
31	0.134	0.014	89.55

It can be observed good performance of following to the current command by the DB control in Fig.11(a) and Fig.12(a). Compared to simulation results, small difference between them. It is considered that this is due to parameter error of the circuit and sensors.

From Fig.11(b) and Fig12(b) it can be observed that the waveform at T_s =25μs becomes closer to the sinusoidal waveform than T_s=50μs.

Table V shows the compensation rate calculated by FFT at T_s=50μs and T_s=25μs. The compensation rate of lower order harmonic is higher at both sampling time. However the rates decrease as the order increases. Especially, it decreases 70% level at T_s=50μs. On the other hand, it keeps around 90% at T_s=25μs.

VI. CONCLUSIONS

Aiming at high frequency harmonics, the repetitive control and the DB control of APF is proposed more shortened sampling time. Characteristics of compensation are investigated when the sampling period shorten and switching frequency becomes high by using DSP and SiC-MOSFET inverter. It is able to regulate the APF output current to the command value by the DB control and perform excellent compensation result by the repetitive control. It clarified that the proposed APF is able to compensate the high order harmonics effectively by shortening sampling time. Performance of the proposed system is verified and compared in the experiment. In the future, the sampling will be further shorten time in order to compensate higher order harmonics.

REFERENCES

[1] H. Akagi, "Control Strategy snd Site Selection of a Shunt Active Filter for installation on a Power Distribution System (in Japanese)", *IEEJ Trans. on IA*, No.116, Vol.3, p285-293, 1996.

[2] F. Z. Peng, M. Kohata, H. Akagi, "Compensation Characteristics of Shunt Active and Series Active Power Filters (in Japanese)", *IEEJ Trans.on IA*, No.113, Vol.1, pp.33-39, 1993.

[3] R. R. Pereira, C. Henrique da Silva, et al., "New Strategies for Application of Adaptive Filters in Active Power Filters", *IEEE Trans. on Industry Applications*, No.47, Vol.3, pp.1136-1141, 2011.

[4] A. Bhattacharya, C. Chakraborty, et al., "Parallel-Connected Shunt Hybrid Active Power Filters Operating at Different Switching Frequencies for Improved Performance", *IEEE Trans. on Industrial Electronics*, No.59, Vol.11, pp.4007-4019, 2012.

[5] Q. Trinh and H. Lee, "New Strategies for Application of Adaptive Filters in Active Power Filters", *IEEE Trans. on Industrial Electronics*, No.60, Vol.12, pp.5400-5410, 2013.

[6] P. Acuna, L. Moran, et al., "Improved Active Power Filter Performance for Renewable Power Generation Systems", *IEEE Trans. on Power Electronics*, No.29, Vol.2, pp.687-694, 2014.

[7] G. Carlos, C. B. Jacobina, et al., "Shunt Active Power Filter With Open-End Winding Transformer and Series-Connected Converters", *IEEE Trans. on Industry Applications*, No.51, Vol.4, pp.3273-3283, 2015.

[8] R. Panigrahi and B. Subudhi, "Performance Enhancement of Shunt Active Power Filter Using a Kalman Filter-Based H∞", *IEEE Trans. on Power Electronics*, No.32, Vol.4, pp.2622-2630, 2017.

[9] T. Mannen, H. Fujita, "A DC Capacitor Voltage Control Method for Active Power Filters Using Modified Reference Including the Theoretically Derived Voltage Ripple", *IEEE Trans. on Industry Applications*, No.52, Vol.5, pp.4179-4187, 2016.

[10] S. Hamasaki, M. Cao, and A. Kawamura, "Experimental Verification of Disturbance-Observer-Based Active Filter for Resonance Suppression", *IEEE Trans. on Industrial Electronics*, Vol.50, No.6, 2003

[11] S. Hamasaki and A. Kawamura, "Improvement of Current Regulation of Line-Current-Detection-type Active Filter based on Deadbeat Control", IEEE Trans. on *Industrial Application*, Vol.39, 2002, pp.536-541, 2003.

[12] S. Hamasaki, T. Kusaba, Y. Mazaki and M. Tsuji, "A Novel Method for Active Filter Applying The Deadbeat Control and The Repetitive Control", *The International Conference on Electrical Machines and Systems (ICEMS)*, LS5B-2, pp.5-10, 2009.

The 2018 International Power Electronics Conference

Control of buck-boost direct matrix converter with low voltage ride-through capability

Nico Remus*, Martin Leubner, Wilfried Hofmann

Chair of Electrical Machines and Drives, Elektrotechnisches Institut, TU Dresden, Germany

*E-mail: nico.remus@tu-dresden.de

Abstract—This paper proposes a closed loop grid side control of the direct matrix converter with V-connection AC chopper controlled by the classical field orientated control of an induction machine. The presented approach enables ride-through capability during grid voltage sags. Additionally, the reactive grid current can be controlled independently from the active grid current. Thus, this direct converter becomes an attractive topology for distributed energy systems with low mass and high power density requirements. The control performance at machine operation point changes as well as the transient behaviour at symmetrical and unsymmetrical grid failures is discussed.

Keywords—advanced converter control, AC/AC converter, matrix converter, active damping

I. INTRODUCTION

At fields of applications, where mass and volume are bounded, the direct matrix converter (DMC) is considered as an attractive option to conventional voltage source converters (VSC). Sinusoidal output currents with arbitrary magnitude and frequency together with sinusoidal input currents with controllable power factor in a wide operation range are delivered by the DMC. Due to the absence of large storage elements a high power density can be reached with this topology. The DMC produces lower semiconductor losses in comparison to the VSC at higher switching frequencies. In [1] the converter losses of the DMC were 11 percent lower

than the losses of a VSC with the same chip area at 10 kHz switching frequency and 15 kW rated power. With utilisation of alternative power semiconductors there are further capabilities in loss reduction as denoted in [2]. This thesis states a converter loss reduction up to 53 % with SiC JFET semiconductors in relation to conventional Si-IGBTs with Si-diodes.

The major drawback of DMCs is the limited voltage transfer ratio of 86.6 % which restricts the low voltage ride-through capability. In [3] a ride-through approach for a DMC driving an induction machine (IM) is shown. When a voltage sag occurs, three additional bidirectional switches in the LC filter disconnect the DMC from the grid while the flux linkage is retained with the drive inertia energy. However, no torque can be generated at the IM and neither active nor reactive power can be supplied for grid stabilization. Itoh introduced the DMC with the input side V-connection AC chopper (ACC) in [4]. Four bidirectional switches are added to the LC filter resulting in a boost-up converter as seen in Fig. 1. The new topology can be described as buck-boost direct matrix converter. Consequently, riding through voltage sags without disconnecting the direct converter from the grid is possible. In [5] this topology was classified to the group of matrix-reactance frequency converters (MRFC) and was called cascaded matrix-reactance frequency converter.

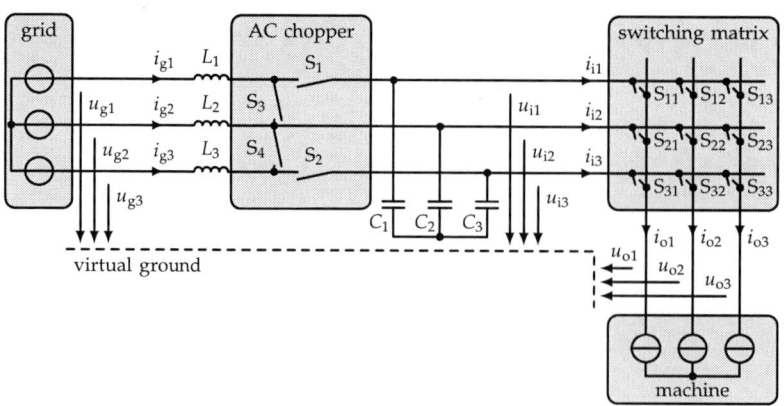

Fig. 1: DMC with input side V-connection AC chopper

In [6] simulative and experimental results, acquired with the buck-boost DMC and the conventional DC link converter, are compared concerning the allocation of the loss. An R-L load was connected to the converter output. In [7] this topology was examined with a permanent magnet synchronous machine and was compared with the conventional DMC. At higher speeds field weakening has to be applied by the conventional DMC while the buck-boost DMC is able to drive the machine with a higher output voltage. Hence, the buck-boost DMC reaches better efficiency at this specific point of operation.

Actually, the low voltage ride-through capability is not considered in former publications. The input side control of the buck-boost DMC is discussed in [4], [6] and [7]. An open loop control strategy for the ACC with a high frequency suppression of the grid current was proposed. This paper gives a first control approach dealing with the low voltage ride-through capability of the DMC with the input side V-connection AC chopper. Additionally, the input side control damps the LC filter actively. Especially for the buck-boost DMC this is essential. Passive damping by resistors within the LC filter would cause high losses [4]. A similar active damping approaches for the DMC is introduced in [8].

After describing the input side control loops of the ACC and DMC, the derivation of the controller parameters is given. Following, the simulation model and test bench for verifying the control approach is depicted. The simulation results for different scenarios are discussed before the experimental results are shown.

II. CONTROL STRUCTURE

A. Field oriented control

For the induction machine a field oriented control (FOC) is utilized as depicted in [9]. The machine current is transformed into the field-forming d'-component $i_{o,d'}$ and the torque-forming q'-component $i_{o,q'}$. With two separated PI current controllers for $i_{o,d'}$ and $i_{o,q'}$ an independent control of the electrical torque is possible while the rotor magnetizing current amplitude is kept constant. The notation of the field synchronous d'q'-components is used to differentiate them from the later introduced grid voltage synchronous dq-components. One objective, which was pursued by developing the input side control, was to remain the structure and functionality of the FOC unaffected.

B. Input side control system

Realizing two switching states $S \in \{0,1\}$ the AC chopper (ACC) fulfils the function of an input side boost converter. Fig. 2 shows the equivalent circuit of the LC filter with ACC. Using the power-variant transformation into space vector components [10], the electrical values are oriented on the

grid voltage vector. Consequently, the grid voltage q-component $u_{g,q}$ is zero. The two ideal switches S_A and S_B are sufficient to represent the ACC functionality.

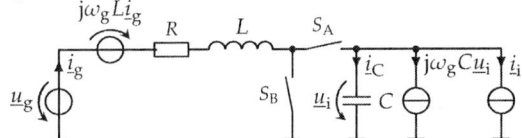

Fig. 2: Equivalent circuit of the LC filter including the AC chopper

If the switch S_A is closed and S_B is open, the filter inductor is connected to the filter capacitor. This combination corresponds to switching state $S = 0$. Otherwise, switch S_A is open and S_B is closed resulting in increasing filter inductor current while the capacitor is disconnected. The switching state is $S = 1$. With help of table I, the position of the simplified equivalent circuit switches can be related to the ACC switches in Fig. 1.

TABLE I: Switch positions of the AC chopper related to the equivalent circuit

S	S_A	S_B	S_1	S_2	S_3	S_4
0	—	╱	—	—	╱	╱
1	╱	—	╱	╱	—	—

The duty cycle d of the applied constant-triangle modulation is defined with switching state S and period time T_s in equation (1).

$$d = \frac{1}{T_s} \int_{t_0}^{t_0+T_s} S(t) \, \mathrm{d}t \tag{1}$$

The averaging of d over the period time T_s removes the high frequency components resulting from the AC chopper switching operation. This corresponds to the averaging approximation discussed comprehensively in [11] and allows the formulation of continuous differential equations for the LC filter system. Hence, the differential equation system only contains the low-frequency components excluding the switching ripple of the inductor current or the capacitor voltage. The equations are determined from the equivalent circuit in Fig. 2 in separated d- and q-components. For a better readability the notation of low-frequency components is omitted.

$$\frac{\mathrm{d}}{\mathrm{d}t}i_{g,d} = \omega_g i_{g,q} - \frac{R}{L}i_{g,d} + \frac{1}{L}u_{g,d} - \frac{1-d}{L}u_{i,d} \tag{2}$$

$$\frac{\mathrm{d}}{\mathrm{d}t}i_{g,q} = -\omega_g i_{g,d} - \frac{R}{L}i_{g,q} + \frac{1}{L}u_{g,q} - \frac{1-d}{L}u_{i,q} \tag{3}$$

$$\frac{\mathrm{d}}{\mathrm{d}t}u_{i,d} = \omega_g u_{i,q} + \frac{1-d}{C}i_{g,d} - \frac{1}{C}i_{i,d} \tag{4}$$

$$\frac{\mathrm{d}}{\mathrm{d}t}u_{i,q} = -\omega_g u_{i,d} + \frac{1-d}{C}i_{g,q} - \frac{1}{C}i_{i,q} \tag{5}$$

Fig. 3 gives the block diagram of the Laplace transformed LC filter system. It is obvious, that the LC filter system gives a cross coupled structure

The 2018 International Power Electronics Conference

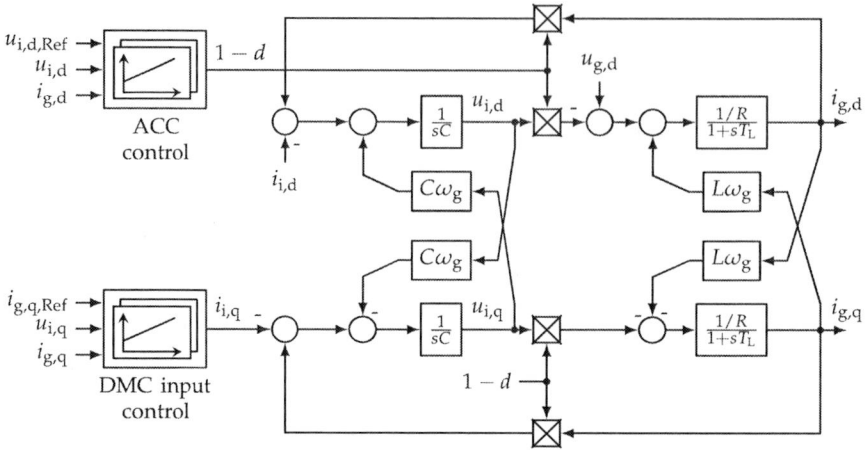

Fig. 3: Block diagram of input filter controlled system

which has to be controlled by both the DMC and the ACC. Considering the fact, that only the input current displacement angle can be manipulated by the DMC without influence on the output voltage [12], the input current $i_{i,q}$ has to be defined as control variable for the q-component of the LC filter system. The ACC is utilized for controlling the d-component. This is implied in Fig 3 with the ACC control and the DMC input control block.

C. Controller structure and tuning

Both ACC control and DMC input control are implemented with cascaded PI controllers. The general controller transfer function is given in (6).

$$G_{\mathrm{R}} = \frac{K_{\mathrm{R}}\left(1 + sT_{\mathrm{N}}\right)}{sT_{\mathrm{N}}} \quad (6)$$

Caused by the multiplication of the inverse modulation factor $(1 - d)$ with the time-variant state variables, the control system described by equations (2) to (5) is non-linear. A common approach to utilize linear controls structures anyway is the

linearisation at an operation point [13]. The process to linearise the system is not shown in this paper. Instead, the controller tuning will be explained with the block diagrams of the already linearised control loops. Fig. 4 (a) shows the inner grid current control loop of the ACC. The AC chopper measurement and modulation time lag is approximated as PT$_1$ element with the time constant T_{ACC}. The grid voltage $u_{\mathrm{g,d}}$ and the cross coupling $\omega_{\mathrm{g}}Li_{\mathrm{g,q}}$ are added to the PI-controller output value to reduce the control system to two PT$_1$ elements from controller view. This simplification enables using the amplitude optimization criterion for the PI-controller tuning [14]. Subsequently, the result is divided by the constant operating point voltage $u_{\mathrm{i,d,OP}}$ resulting from the linearisation to obtain the reverted duty cycle $(1 - d_{\mathrm{Ref}})$ for modulation. The transfer function of the reduced system is:

$$G_{\mathrm{ID}}\left(s\right) = \frac{i_{\mathrm{g,d}}\left(s\right)}{1 - d_{\mathrm{ref}}\left(s\right)} = \frac{1}{1 + sT_{\mathrm{ACC}}} \cdot \frac{1/R}{1 + sT_{\mathrm{L}}} \quad (7)$$

with the inductor time constant $T_{\mathrm{L}} = L/R$. This function is the basis for the amplitude optimization

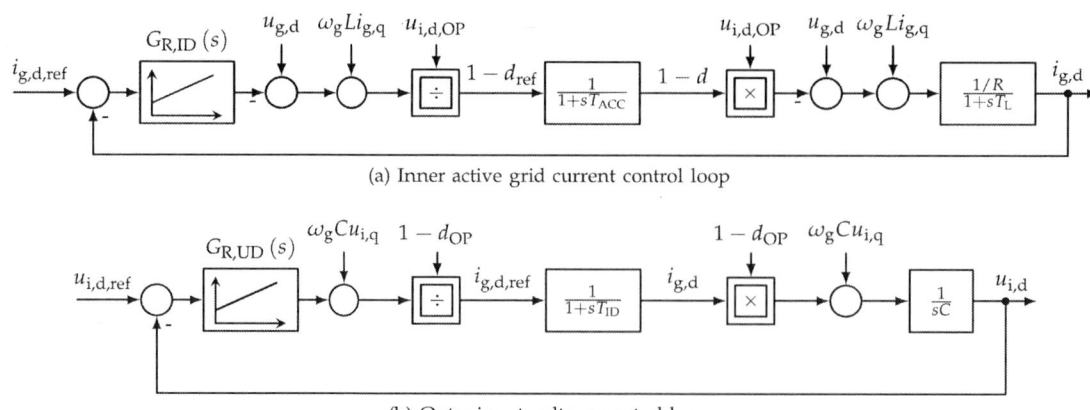

(a) Inner active grid current control loop

(b) Outer input voltage control loop

Fig. 4: Loops of the AC chopper control

3245

The 2018 International Power Electronics Conference

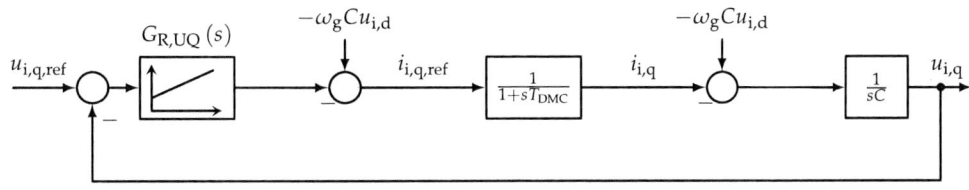

(a) Inner input voltage control loop

(b) Outer active grid current control loop

Fig. 5: Control loops of the DMC

criterion. For compensating the largest time constant T_L of the control system, $T_{N,ID}$ is equal to T_L. The following controller parameters are determined for the AC chopper current control:

$$K_{R,ID} = \frac{L}{2T_{ACC}}, \quad T_{N,ID} = T_L. \tag{8}$$

The amplitude optimization criterion must not be applied for the outer AC chopper voltage control loop in Fig. 4 (b). The control system contains an integral element originated from the filter capacitance. This would result in an unstable behaviour applying this controller tuning. Instead, the symmetrical optimization criterion is used [15]. Adding the cross coupling compensation, the following control system transfer function from controller view is obtained:

$$G_{UD}(s) = \frac{u_{i,d}(s)}{i_{g,d,ref}(s)} = \frac{1}{1 + sT_{ID}} \cdot \frac{1}{sC} \tag{9}$$

$$T_{ID} = 2\sqrt{2}T_{ACC}. \tag{10}$$

For application of the optimization criterion, the inner current control loop transfer function is approximated as PT_1 element with time constant T_{ID}. The resulting voltage controller parameters are:

$$K_{R,UD} = \frac{C}{aT_{ID}}, \quad T_{N,UD} = a^2 T_{ID}. \tag{11}$$

The a factor gives the opportunity to manipulate the phase margin of the voltage control loop. For a stable control this factor must be $a > 1$. For the proposed control scheme $a = 2$ was chosen.

The structure of the DMC input control loops is given in Fig. 5. The procedure from control system up to the optimized controller parameters is broadly similar. Consequently, the controller parameters of the inner voltage control loop are determined using symmetrical optimization criterion:

$$K_{R,UQ} = \frac{C}{aT_{DMC}}, \quad T_{N,UQ} = a^2 T_{DMC} \tag{12}$$

with the time constant T_{DMC} approximating the measurement and modulation time lag of the converter. The outer current control loop parameters are determined using amplitude optimization criterion:

$$K_{R,IQ} = \frac{L}{2T_{UQ}}, \quad T_{N,IQ} = T_L, \quad T_{UQ} = a\sqrt{2}T_{DMC}. \tag{13}$$

EXPERIMENTAL SETUP AND SIMULATION MODEL

Fig. 6 gives the simplified setup. The experimental converter is realized in one enclosed power devise given Fig. 7. The Semikron power modules SKM200GM12T4G are interconnected with copper plates which take the filter capacitors as well. This power devise is installed in an electric cabinet including the 19 inch rack of the control unit and the contactor control. The key parameters of the test bench are given in table II.

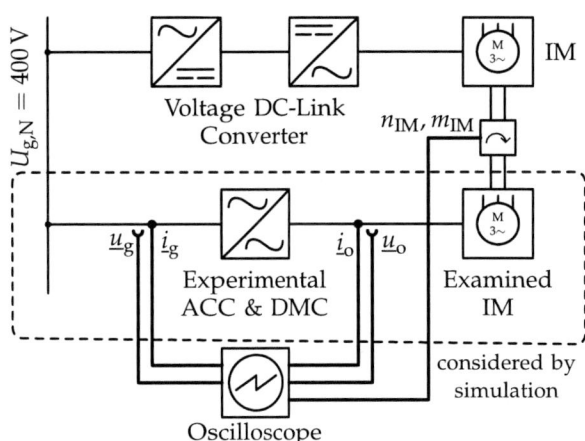

Fig. 6: Overview of the experimental setup

The DMC output voltage demanded from the FOC is modulated with the space vector modulation known from [12]. The AC chopper is controlled with the one-step commutation approach based on U_{CE}-measurement proposed in [16]. For the DMC

3246

TABLE II: Parameters of the test bench

Parameter	Symbol	Value
Filter inductance	L	2.4 mH
Filter capacitance	C	20 µF
Filter inductor resistance	R	30 mΩ
Switching frequency	f_s	10 kHz
ACC lag time	T_{ACC}	150 µs
DMC lag time	T_{DMC}	150 µs
Rated power of IM	$P_{N,IM}$	3.6 kW
Rated torque of IM	$m_{N,IM}$	23 Nm
Rated angular speed of IM	$\omega_{N,IM}$	156 rad/s

TABLE III: Constraints of simulation and experimental study

Parameter	Symbol	Sim./Exp.
Nominal grid voltage	$U_{g,N}$	400/200 V
Initial grid phase voltage	$\hat{U}_{g,0}$	326/163 V
Initial grid current	$\hat{I}_{g,0}$	7.08/3.82 A
Input reference voltage	$U_{i,ref}$	500/250 V
Input phase voltage	$\hat{U}_i$	408/204 V
Initial output current	$\hat{I}_{o,0}$	11.3/7.15 A
IM torque	m_m	-23/11.5 Nm
IM angular speed	ω_m	156/39.3 rad/s

a 2/3/4-step commutation approach was deduced from [17]. The simulation was created modelling

Fig. 7: Power module including IGBTs of DMC and ACC (left) and electric cabinet with complete converter (right)

the existing test bench. The simulation contains the complete buck-boost DMC with discrete switches and the induction machine as well as an ideal grid with voltage sources. A superior speed controller and the proposed control approach are implemented as C-code in the simulation model which is processed with a sampling rate similar to the switching frequency f_s of the converter. Consequently, the controller of the field oriented control and the proposed input side control are working in a time discrete domain as they would do on a microcontroller. The commutation schemes are not implemented in the simulation model. The controller parameters of the ACC as well as the DMC control are determined with the values of table II and the equations (8), (11), (12) and (13). These are given in table IV.

TABLE IV: Controller parameters

ACC control		DMC input control	
$K_{R,ID}$	8.0 V/A	$K_{R,IQ}$	2.83 V/A
$T_{N,ID}$	$80 \cdot 10^{-3}$ s	$T_{N,IQ}$	$80 \cdot 10^{-3}$ s
$K_{R,UD}$	$23.6 \cdot 10^{-3}$ A/V	$K_{R,UQ}$	$66.7 \cdot 10^{-3}$ A/V
$T_{N,UD}$	$1.70 \cdot 10^{-3}$ s	$T_{N,UQ}$	$600 \cdot 10^{-6}$ s

SIMULATION RESULTS

Four different scenarios were examined by simulation. These are a load torque change, a step of the reactive grid current reference $i_{g,q,ref}$, a symmetrical voltage sag and an unsymmetrical voltage sag. At the beginning of every scenario the operating points of IM and DMC are the same resulting in similar electrical and mechanical values given in table III. These values are used for the normalized axis scaling of the diagrams showing the simulation results.

In Fig. 8 the results of a reactive grid current reference step are given. The step time is marked with the grey line at 40 ms. After the reference value has changed from $i_{g,q,ref} = 0$ to $i_{g,q,ref} = 0.5\hat{I}_{g,0}$ the reactive grid current $i_{g,q}$ is followed up within 5 ms.

The results of a load torque change from $m_{load} = -m_{N,IM}$ to $m_{load} = m_{N,IM}$ are shown in Fig. 9. Marked with a grey vertical line, the time of load torque change is 40 ms. The torque-forming component $i_{o,q}$ of the output current rises to its end value within six milliseconds as can be seen in Fig. 9 (e). This current slope is preset from the

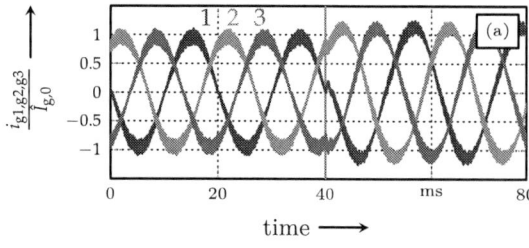 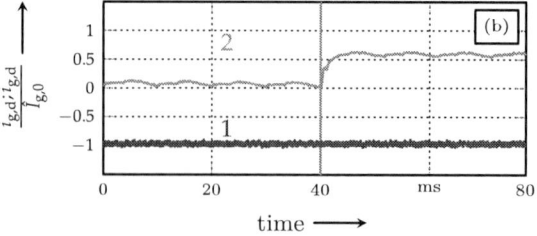

Fig. 8: Simulation results of the new LC filter control combined with FOC of the IM at reactive grid current reference change from $i_{g,q,ref} = 0$ to $i_{g,q,ref} = 0.5\hat{I}_{g,0}$:
(a) grid phase current, 1 - i_{g1}, 2 - i_{g2}, 3 - i_{g3}; (b) grid current in dq-components, 1 - $i_{g,d}$, 2 - $i_{g,q}$;

superior speed control of the induction machine. The proportional relation of the electrical torque m_{el} towards $i_{o,q}$ is observable from Fig. 9 (f). The input voltage d-component $u_{i,d}$ shows a slight sag (Fig. 9 (c)) while the grid current d-component $i_{g,d}$

rises. A stationary operation point is reached in about ten milliseconds after the time of torque change. This shows, that the input side ACC and DMC control loops keep stable and do not restrict the dynamic of the speed controlled IM.

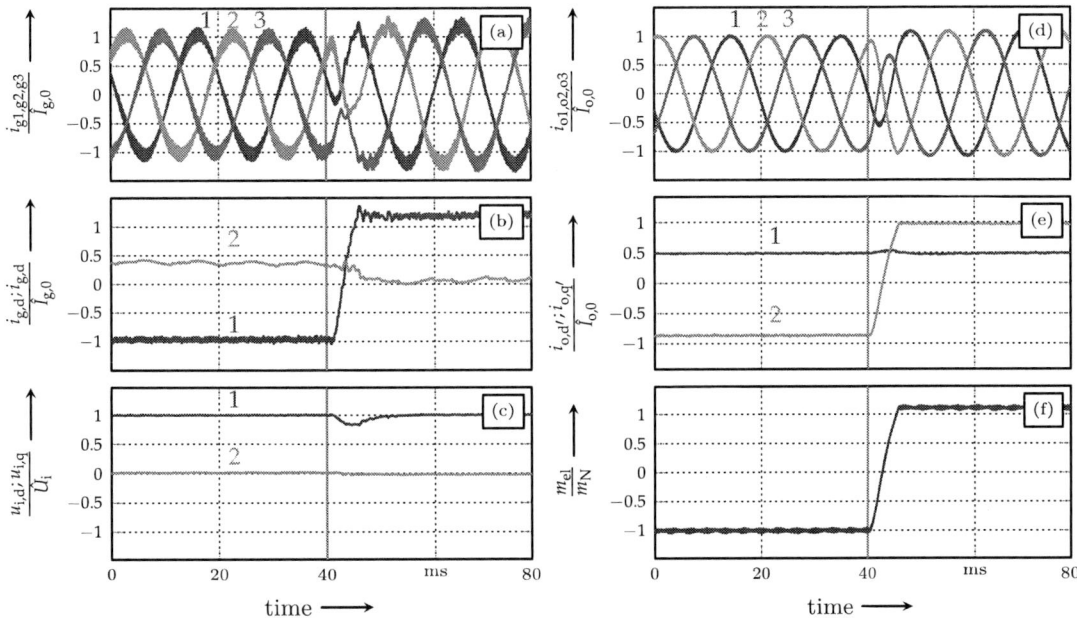

Fig. 9: Simulation results of the new LC filter control combined with FOC at load torque change to $m_l = m_N$: (a) grid phase current, 1 - i_{g1}, 2 - i_{g2}, 3 - i_{g3}; (b) grid current in dq-components, 1 - $i_{g,d}$, 2 - $i_{g,q}$; (c) input voltage in dq-components, 1 - $u_{i,d}$, 2 - $u_{i,q}$; (d) IM stator current, 1 - i_{o1}, 2 - i_{o2}, 3 - i_{o3}; (e) FOR output current, 1 - $i_{o,d'}$, 2 - $i_{o,q'}$; (f) IM electrical torque m_{el}.

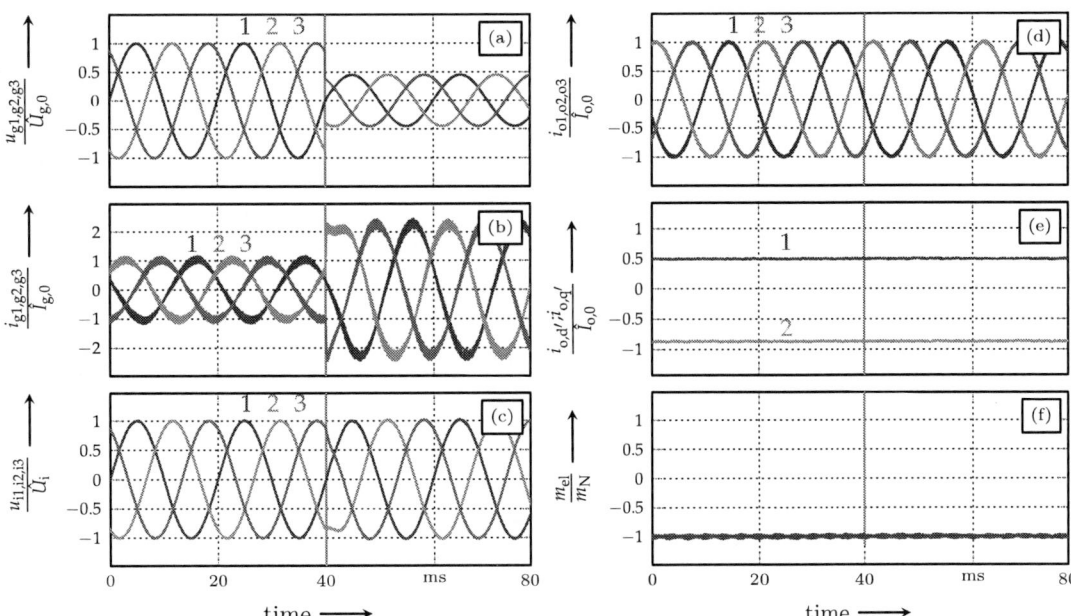

Fig. 10: Simulation results of the LC filter control combined with FOC at grid voltage drop to $U_g = 0.45\,U_{g,N}$: (a) grid voltage, 1 - u_{g1}, 2 - u_{g2}, 3 - u_{g3}; (b) grid phase current, 1 - i_{g1}, 2 - i_{g2}, 3 - i_{g3}; (c) input voltage , 1 - u_{i1}, 2 - u_{i2}, 3 - u_{i3}; (d) IM stator current, 1 - i_{o1}, 2 - i_{o2}, 3 - i_{o3}; (e) FOR output current, 1 - $i_{o,d'}$, 2 - $i_{o,q'}$; (f) IM electrical torque m_{el}.

The 2018 International Power Electronics Conference

A first impression of the low voltage ride-through capability can be obtained in Fig. 10. At 40 ms a symmetrical grid voltage sag to $\hat{U}_g = 0.45\,\hat{U}_{g,0}$ occurs. Due to the immediate raise of the grid current by the ACC control

(Fig. 9 (b)), the input voltage in Fig. 10 (c) shows only a weak distortion. No influence of the voltage sag on the output currents or electrical torque are noticeable at the Fig. 10 (d)-(f).

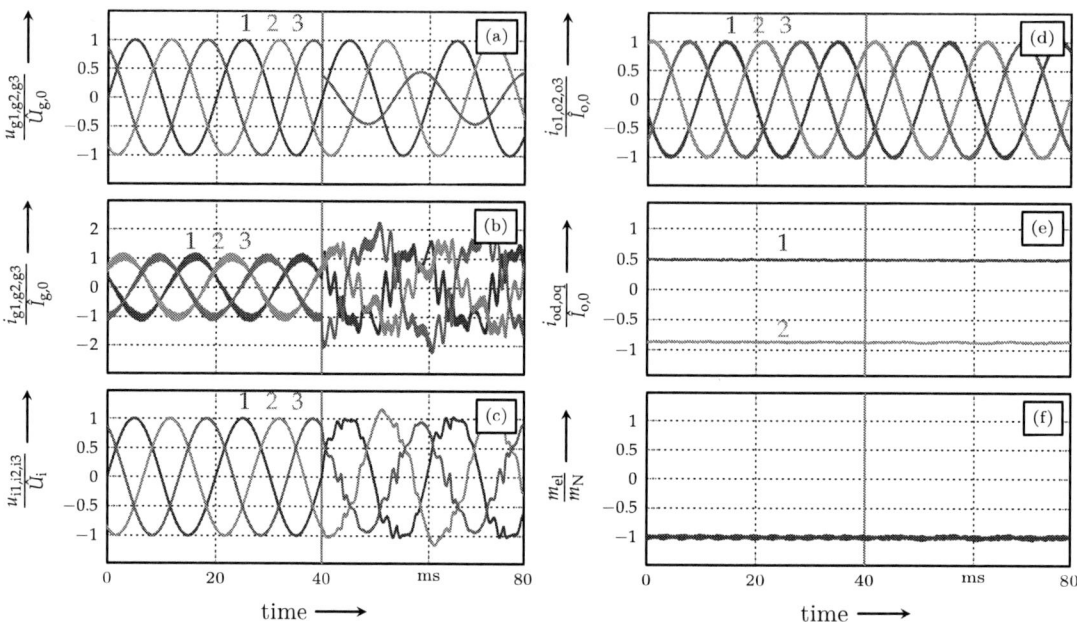

Fig. 11: Simulation results of LC filter control with FOC at one phase grid voltage drop to $U_{g3} = 0.45\,U_{g,N}$:
(a) grid voltage, 1 - u_{g1}, 2 - u_{g2}, 3 - u_{g3}; (b) grid phase current, 1 - i_{g1}, 2 - i_{g2}, 3 - i_{g3};
(c) input voltage , 1 - u_{i1}, 2 - u_{i2}, 3 - u_{i3}; (d) IM stator current, 1 - i_{o1}, 2 - i_{o2}, 3 - i_{o3};
(e) FOR output current, 1 - i_{od}, 2 - i_{oq}; (f) IM electrical torque m_{el}.

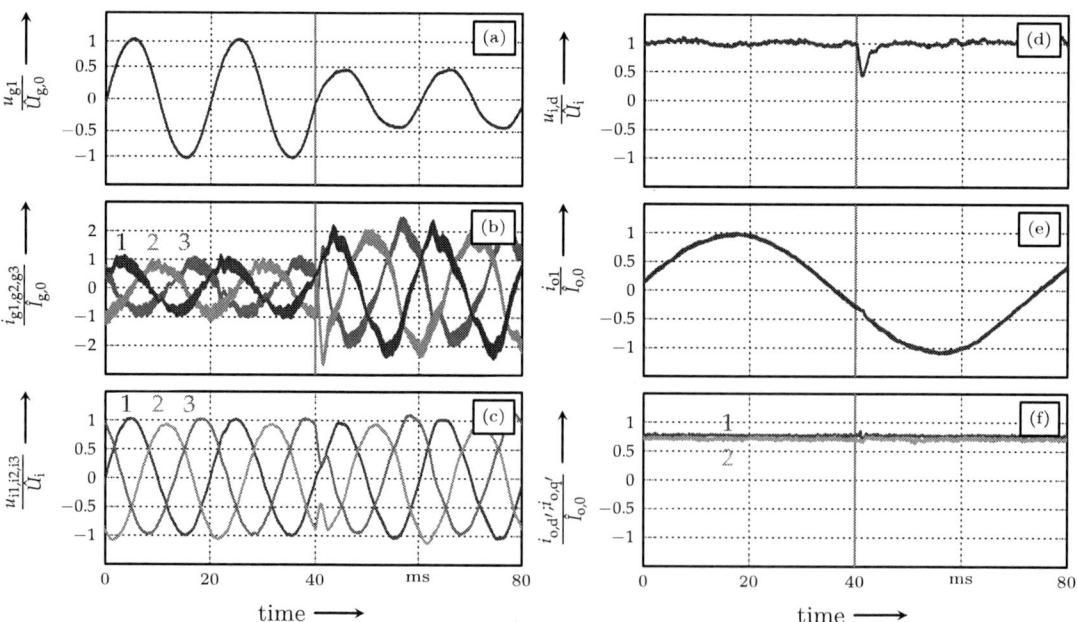

Fig. 12: Experimental results of LC filter control with FOC at grid voltage drop to $U_g = 0.5\,U_{g,N}$:
(a) grid voltage u_{g1}; (b) grid phase current, 1 - i_{g1}, 2 - i_{g2}, 3 - i_{g3};
(c) input voltage , 1 - u_{i1}, 2 - u_{i2}, 3 - u_{i3}; (d) input voltage d-component $u_{e,d}$;
(e) output phase current i_{o1}; (f) FOR output current, 1 - $i_{o,d'}$, 2 - $i_{o,q'}$.

Evaluating the results of an one phase voltage sag to $\hat{U}_{g3} = 0.45\,\hat{U}_{g,0}$ strong distortions of the grid current are apparent in Fig. 11 (b). For reason this, the symmetrical components introduced by [18] has to be considered. At unbalanced grid failures a positive and negative sequence component of the grid voltage occurs. Caused by the space vector transformation of the electrical values the proposed control does not regard the negative sequence. The consequence are undesired low frequency oscillations of the controlled values.

EXPERIMENTAL RESULTS

Fig. 12 shows the experimental results of the proposed control approach at a symmetrical grid voltage sag to $\hat{U}_g = 0.5\,\hat{U}_{g,0}$. As can be seen in table III, the IM operation point and the grid voltage value are different from them used by the simulation model. This difference is reasoned by remaining challenges according to the commutation schemes used by the AC chopper and the matrix converter. For recording the grid phase voltage, the grid phase currents and the input voltages in Fig. 12 (a)–(c) as well as the output phase current in Fig. 12 (e) the Yokogawa DL850E ScopeCoder was used. The d-component $u_{i,d}$ of the input voltage and the FOC dq-current is readout from DACs of the Texas Instruments TMS320F28379D micro-controller executing the same C-code as implemented within the simulation model. In Fig. 12 (b) an immediate raise of the grid current occurs at the time after the grid voltage sag. The input voltage drops shortly as can be seen in Fig. 12 (c) and Fig. 12 (d), but the impact on the IM current barely visible in Fig. 12 (e) and Fig. 12 (f). The grid current and input voltage quality is poor in contrast to the simulation results. This is caused by the commutation schemes which are not considered by the simulation as already mentioned. It has to be emphasised, that no damping resistors are present within the LC filter.

CONCLUSION

A linear control approach for the buck-boost matrix converter is proposed in this paper. No additional damping methods or resistors are necessary to stabilize the LC filter. Consequently, high losses of passive filter damping as mentioned in [4] are avoided. A classical field oriented control is applied to the induction machine without any adaptation. Additionally, the grid side power factor can be controlled directly by the reactive grid current control and symmetrical grid voltage sags are compensated from the AC chopper control without any impact on the machines electrical values. Nevertheless, the low voltage ride-through capability is poor in case of unbalanced grid failures. For a better performance the proposed control approach has to be revised considering the negative sequence components. The effectiveness is actually shown by simulation and experimental result. But for an acceptable performance

in practice, the applied commutation schemes have to be revised.

REFERENCES

[1] S. Bernet, S. Ponnaluri, and R. Teichmann, "Design and loss comparison of matrix converters, and voltage-source converters for modern AC drives," *IEEE Transactions on Industrial Electronics*, vol. 49, no. 2, pp. 304–314, Apr. 2002.

[2] D. Domes, "Untersuchungen zum einsatz von unipolaren sic leistungshalbleiterbauelementen in antriebsstromrichtern," Ph.D. dissertation, TU Chemnitz, 2009.

[3] D. Orser and N. Mohan, "A matrix converter ride-through configuration using input filter capacitors as an energy exchange mechanism," *IEEE Transactions on Power Electronics*, vol. 30, no. 8, pp. 4377–4385, 2015.

[4] J. Itoh, K. Koiwa, and K. Kato, "Input current stabilization control of a matrix converter with boost-up functionality," in *Proc. Int. Power Electronics Conf. (IPEC)*, 2010, pp. 2708–2714.

[5] Z. Fedyczak, P. Szczesniak, G. Tadra, and M. Klytta, "A comparison of basic properties of the integrated and cascade matrix-reactance frequency converters," in *Power Electronics and Motion Control Conference (EPE/PEMC), 2012 15th International*, Sep. 2012, pp. DS1b.8–1–DS1b.8–6.

[6] K. Koiwa and J. Itoh, "Experimental verification for a matrix converter with a v-connection AC chopper," *European Conference on Power Electronics and Applications (EPE 2011)*, pp. 1–10, 2011.

[7] ——, "Verification of effectiveness of a matrix converter with boost-up AC chopper by using an ipm motor," *Applied Power Electronics Conference and Exposition (APEC)*, pp. 2265–2271, 2012.

[8] M. Leubner, N. Remus, M. Stübig, and W. Hofmann, "Active stabilization of direct matrix converter input side filter through grid current control," in *Proc. IEEE Applied Power Electronics Conf. and Exposition (APEC)*, Mar. 2016, pp. 2175–2181.

[9] N. P. Quang and J.-A. Dittrich, *Vector Control of Three-Phase AC Machines*, 2nd ed. Springer-Verlag, 2015.

[10] *IEC 62428:2008 Electric power enginering Modal components in three-phase a.c. systems Quantities and transformations*, DKE Deutsche Kommission Elektrotechnik Elektronik Informationstechnik Std.

[11] R. W. Erickson and D. Maksimović, *Fundamentals of Power Electronics*, 2nd ed. Springer Science+Business Media, LCC, 2001.

[12] L. Huber and D. Borojevic, "Space vector modulated three-phase to three-phase matrix converter with input power factor correction," *IEEE Transactions on Industry Applications*, vol. 31, no. 6, pp. 1234–1246, 1995.

[13] H. Lutz and W. Wendt, *Taschenbuch der Regelungstechnik*. Harri Deutsch, 2007.

[14] C. Kessler, *Über die Vorausberechnung optimal abgestimmter Regelkreise Teil III. Die optimale Einstellung des Reglers nach dem Betragsoptimum*. Regelungstechnik 3, 1955, vol. 3.

[15] ——, "Das symmetrische optimum," *Regelungstechnik 6*, vol. 6, no. 11, pp. 395–400, 1958.

[16] N. Remus, M. Leubner, K. Koethe, and W. Hofmann, "One-step commutation approach for direct converters based on uce-measurement," in *European Conference on Power Electronics and Applications – EPE*, no. 19, 2017.

[17] M. Leubner, N. Remus, S. Schwarz, and W. Hofmann, "Voltage based 2/3/4-step commutation for direct three-level matrix converter," in *Proc. IEEE Applied Power Electronics Conf. and Exposition (APEC)*, 2018.

[18] C. L. Fortescue, "Method of symmetrical co-ordinates applied to the solution of polyphase networks," *Proceedings of the American Institute of Electrical Engineers*, vol. 37, no. 6, pp. 629–716, Jun. 1918.

The 2018 International Power Electronics Conference

An Improved PLL Based Seamless Transfer Control Strategy

Xin Meng, Jinjun Liu, Zeng Liu, Ronghui An
State Key Lab of Electrical Insulation and Power Equipment
School of Electrical Engineering, Xi'an Jiaotong University
Xi'an China
Email: mengxinstar@stu.xjtu.edu.cn

Abstract—This paper proposes a seamless transfer control method for distributed generation units to solve the issue of hybrid voltage and current mode control method and overcome the limitation of direct current control strategy. Upon the occurrence of utility outage, the voltage controller is automatically activated to regulate the load voltage. The inverter can transfer from controlled current source to controlled voltage source automatically. Therefore, the quality of critical load voltage can be maintained during the transferring process and doesn't rely on the islanding detection method. The direct current control strategy can only operate without grid line impedance, but this proposed method can overcome this limitation. When the grid returns normal, the pre-synchronization can be realized automatically without extra infrastructure. The simulation results are provided to verify the proposed control method.

Keywords—seamless transfer; hybrid voltage and current mode; islanding detection;

I. INTRODUCTION

To solve the energy crisis and environmental issues, the distributed generation (DG) unit, such as the solar energy, fuel cell, and the wind power has attracted more and more attentions. When the grid is normal, the DG unit operates in grid-connected (GC) mode, and exchanges power with the utility. When the utility is broken, the DG unit should disconnect with the utility, and operate in islanding mode to provide energy to the critical load. To make sure the critical load voltage quality, the DG unit should provide a seamless transfer between these two modes. Some publication papers proposed several control methods to realize the smooth transfer between GC and SA mode. These control methods can be classified into three categories.

In the first category, all DG units are based on droop control in both GC and SA mode [2-4], thus the seamless transfer from GC mode to SA mode can be achieved. However, on the one hand, if the grid voltage has oscillation in GC mode, the grid current quality is not well and the inverter output power is deviated from its reference value. On the other hand, the dynamic performance is not well because of the low-pass filters (LPFs) in power calculation channel.

The second category named indirect current control is proposed in [1, 5-7]. In the GC mode, the DG unit is controlled as a current source. When islanding happens, the DG unit can transfer from current source to voltage source automatically,

therefore, the quality of load voltage can be improved. However, because of the existence of outer grid current control loop, the system dynamic performance is poor.

In the third category, which named as hybrid voltage and current mode control method, all DG unit are controlled as current sources in the GC mode to exchange power with grid [8-11]. When islanding happens, at least one DG unit transfer from current source to voltage source to provide voltage support for microgrid. From the moment of occurrence of islanding to the moment of switching the controller to voltage mode, the load voltage is neither fixed by the utility, nor regulated by the DG unit [12], therefore, the load voltage quality may be worsen during this transfer period.

On the other hand, when the grid returns normal, before the DG unit transferring from SA mode to GC mode, the pre-synchronization mechanism is necessary. The synchronization is traditionally realized by regulating the phase angle and amplitude of inverter output voltage based on the phase and voltage amplitude difference of two sides of static transfer switch (STS), which needs another phase locking loop (PLL) to detect the phase angle and amplitude of STS grid side voltage and another integral block.

To overcome the drawbacks of the third category control method, this paper proposes an improved PLL based seamless transfer control strategy, which can transfer from current source converter (CSC) to voltage source converter (VSC) automatically and doesn't rely on the speed and accuracy of the islanding detection method. Therefore, the quality of critical load voltage can be maintained during the transferring process. Secondly, the direct current control method in [12] can only operate without grid line impedance, but this proposed method can overcome this limitation. Thirdly, when the grid returns normal, the pre-synchronization can be realized automatically without extra infrastructure.

This paper is organized as follows. Section II introduces the power stage of DG units and the proposed control method. Section III verifies the proposed seamless transfer control method by simulation. Finally, the conclusions are given in Section IV.

This work was supported by the National Natural Science Foundation of China under Grant 51437007, and the Power Electronics Science and Education Development Program of Delta Environmental & Educational Foundation under Grant DREM2014002.

II. POWER STAGE OF DG UNIT AND PROPOSED CONTROL METHOD

The power stage of DG unit is shown in Fig.1. There are two switches between the DG unit and the utility. When the grid is normal, both S_i and S_u are closed. When the grid is broken, the utility protection switch S_u turns off instantly, and then islanding is formed. After the islanding state is confirmed

by islanding detection algorithm of the DG unit, the transfer switch S_i turns off.

The overall control block diagram for the proposed method is described in Fig.2. It includes the capacitor voltage loop and inductor current loop, which function will be analyzed later. Another important control block is the improved PLL block, as shown is Fig.3, which generates the phase angle, the voltage loop reference and the control signal "*flag*".

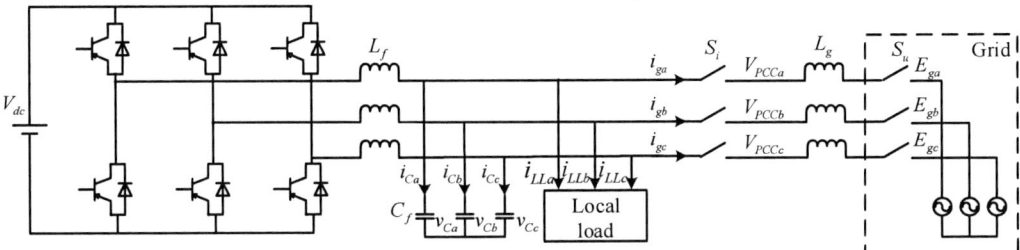

Fig. 1.   Power stage of the DG unit

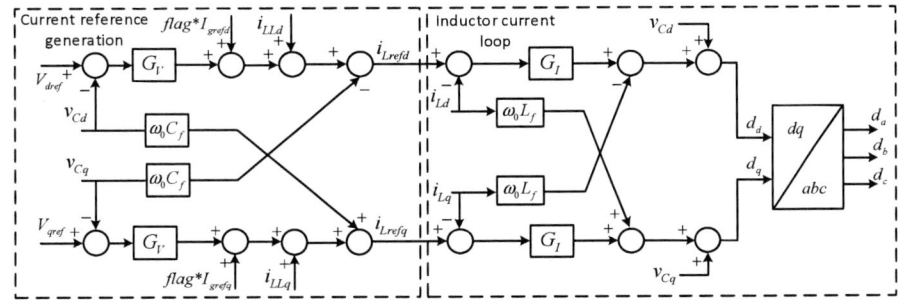

Fig. 2.   The overall control block diagram of the proposed control method

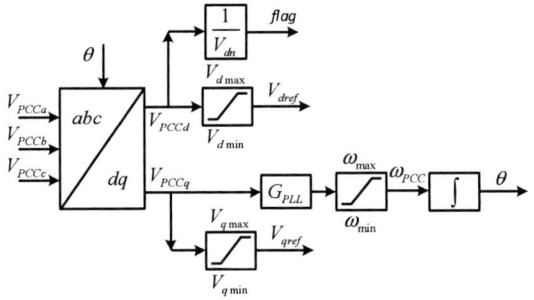

Fig. 3.   The improved PLL block

There are four operating states for the DG unit, including the GC mode, the transition from GC mode to SA mode, the SA mode and the transition from SA mode to GC mode.

1) GC mode:
According to IEEE standard 1547–2003[13], the limiting value V_{dmax}, V_{dmin}, V_{qmax} and V_{qmin} in Fig.3 can be selected as (1)-(4) to make sure the capacitor voltage reference within 0.88 pu -1.1 pu,

$$V_{d\max} = 1.09 \cdot V_{dn} \qquad (1)$$

$$V_{d\min} = 0.9 \cdot V_{dn} \qquad (2)$$

$$V_{q\max} = 0.1 \cdot V_{dn} \qquad (3)$$

$$V_{q\min} = -0.1 \cdot V_{dn} \qquad (4)$$

where V_{dn} represents the nominal amplitude of phase-to-neutral capacitor voltage.

When the DG unit operates in GC mode, the PLL should be utilized to lock the phase angle of the utility, therefore, the angular frequency should not reach the upper value or the lower value of the limiter. Besides, when the DG unit operates in islanding mode, the angular frequency ω is restricted between ω_{max} and ω_{min}, and it should not drift from the normal value too much. So ω_{max} and ω_{min} are selected as follows,

$$\omega_{\max} = 2\pi * 50.5 \, rad \, / \, s, \; \omega_{\min} = 2\pi * 49.5 \, rad \, / \, s \qquad (5)$$

In the GC mode, the voltage of point of common coupling (PCC) equals to the LC filter capacitor voltage, and the relationship $V_{PCCd} = v_{cd}$ and $V_{PCCq} = v_{cq}$ are established. The limiter of D-axis and Q-axis in the PLL block are not function, i.e., $V_{d,qref} = V_{PCCd,q}$. Therefore, the input of voltage controller G_v equals to zero. The voltage controller G_v adopts proportional (P) regulator. The reason for utilizing P regulator in voltage loop is to avoid the integral effect, making sure the output of G_v is zero when the input of G_v is zero. So the voltage control loop is out of function in GC mode. The DG unit is controlled as a current source by the inductor current loop to supply given active and reactive power.

By the Park transformation, the PCC voltage is transformed into the synchronous rotating reference frame (SRF), which is shown as,

$$\begin{cases} V_{PCCd} = V_{PCC} \cos(\theta^* - \theta) \\ V_{PCCq} = V_{PCC} \sin(\theta^* - \theta) \end{cases} \tag{6}$$

where V_{PCC} is the amplitude of the PCC voltage, and θ^* is the actual phase angle. V_{PCCq} is regulated to zero by the PLL, so V_{PCCd} equals to the amplitude of PCC voltage V_{PCC}. Therefore, $flag = V_{PCCd} / V_{dn} \approx 1$.

In the D-axis, the inductor current reference i_{Lrefd} can be expressed as (7) according to Fig.2.

$$i_{Lrefd} = I_{grefd} + i_{LLd} - \omega_0 C_f v_{Cq} \tag{7}$$

The first part I_{grefd} is the grid current reference. The second part i_{LLd} is the load current of D-axis, which is used to compensate the harmonic component in the grid current under nonlinear local load. The third part $-\omega_0 C_f v_{Cq}$ equals to zero due to that $v_{Cq} = 0$, where ω_0 is the rated angular frequency, and C_f is the capacitance of the filter capacitor. Consequently, the current reference i_{Lrefd} is imposed by the given current reference I_{grefd} and the load current i_{LLd}.

In the Q-axis, the inductor current reference i_{Lrefq} is

$$i_{Lrefq} = I_{grefq} + i_{LLq} + \omega_0 C_f v_{Cd} \tag{8}$$

Similarly, I_{grefq} is a given current reference. The second part i_{LLq} is also determined by the characteristic of the local load. The third part $\omega_0 C_f v_{Cd}$ is fixed since v_{Cd} equals to V_{PCC}.

Based on above analysis, the control diagram in GC mode can be simplified as Fig.4 and Fig.5, and the inverter is controlled as a current source by the inductor current loop.

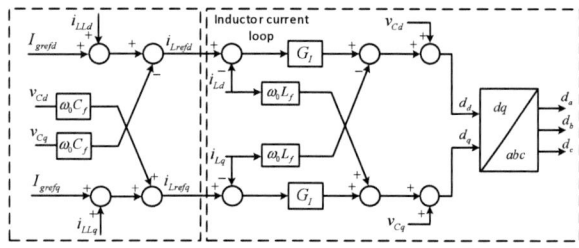

Fig.4 Simplified control block in GC mode

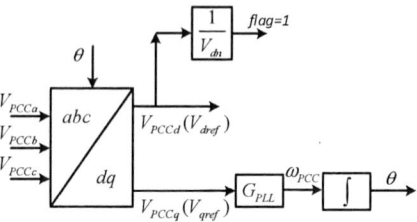

Fig.5 Simplified PLL block in GC mode

2) Transition from GC mode to SA mode:

When the utility protection switch S_u opens, the islanding is formed. After the islanding is detected by the DG unit, the transfer switch S_i turns off. The transition can be divided into two parts as Fig.6 shows. The first time interval starts from the instant of opening S_u to the instant of opening S_i when the islanding is detected. The second time interval starts from the instant of opening S_i.

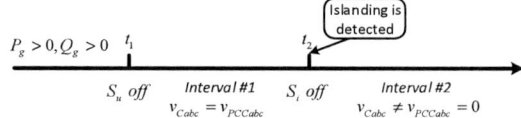

Fig.6 Operation sequence from GC mode to SA mode

To simplify analysis, it is assumed that the local load is represented as a parallel RLC circuit, and the active and reactive power consumed by the load can be expressed by (9) and (10).

$$P_{load} = \frac{3}{2} \frac{V_C^2}{R} \tag{9}$$

$$Q_{load} = \frac{3}{2} V_C^2 \left(\frac{1}{\omega L} - \omega C \right) \tag{10}$$

where V_c and ω represent the load voltage amplitude and angular frequency respectively.

It is also assumed that the DG unit injects active power and reactive power into the utility in the GC mode. Therefore, when islanding happens, the local load must absorb the extra power injected to the grid originally, as the output power of inverter cannot change instantaneously. According to (9), the magnitude of the load voltage will rise with the increase of P_{load}. At the same time, the angular frequency will decrease to consume more reactive power with (10). The amplitude and frequency of load voltage will drift because of the active power and reactive power mismatch between the DG and the load demand.

During t_1-t_2, the PCC voltage V_{PCCabc} is still the same with the capacitor voltage v_{Cabc} because the switch S_i is still closed. Therefore, the amplitude of PCC voltage V_{PCCabc} will also increase, and the D-axis output of PLL block will increase. Because of the existence of D-axis limiter, if the increasing of V_{PCCabc} is large, the limiter will function (V_{dref} will equal to V_{dmax}). The voltage controller will function to regulate the load voltage, and the DG unit can transfer from CSC to VSC automatically. Reversely, if the DG unit absorbs active power

3253

and reactive power from the utility in the GC mode, the analysis is similar.

The second time interval begins from the instant when the switch S_i opens. If the switch S_i opens, the PCC voltage V_{PCCabc} will decrease to zero, and the input voltage of PLL block is zero. Because of the existence of D-axis and Q-axis limiter in the PLL block, the voltage reference V_{dref} equals to the lower value V_{dmin}, and V_{qref} equals to zero. The angular frequency ω_{PCC} equals to the limiter lower value ω_{min}. The control signal "*flag*" equals to zero, which will make the grid current reference out of function.

3) SA mode:

In the SA mode, S_i and S_u are both open. The voltage references in D-axis and Q-axis are V_{dmin} and zero, respectively. The angular frequency ω_{PCC} equals to ω_{min}. The control signal "*flag*" is zero. The voltage controller G_v regulates the load voltage, and the DG unit is worked as a voltage source.

4) Transition from SA mode to GC mode:

Firstly, if the grid is restored and S_u is closed, the PCC voltage equals to the grid voltage. There is no current through the grid inductor because of the opening of switch S_i, and the voltage drop on grid inductor equals to zero. The PLL block will track the phase of grid voltage automatically.

Secondly, the output of PLL block V_{PCCd} and V_{PCCq} will be adjusted within the upper and lower value of the limiter, and the amplitude of load voltage will be regulated to follow the grid voltage. As a result, the synchronization of load voltage and grid voltage is finished. The control signal "*flag*" turns to one automatically, and the grid current reference functions.

Thirdly, the switch S_i turns on, and the DG unit operates in the GC mode.

III. SIMULATION RESULTS AND DISCUSSION

PSCAD simulations are conducted to verify the proposed method. The simulation circuit and parameters are shown in Fig.7 and Table I, respectively. The master DG unit adopts the proposed control method and the slave DG units adopt PQ control.

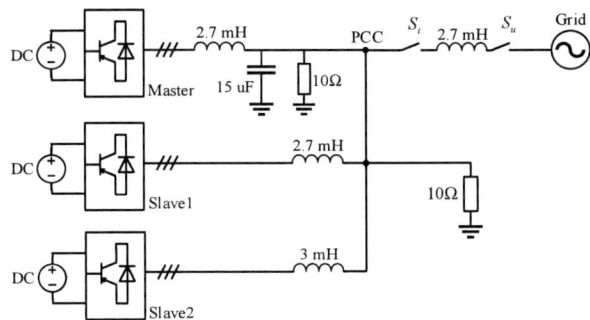

Fig.7 Simulation circuit of microgrid system

Table I Simulation parameters

Parameters	Value
Grid voltage amplitude	162.6V
Grid current reference value I_{grefd}	15A
D-axis limiting value V_{dmax}	177V
D-axis limiting value V_{dmin}	146.5V
Q-axis limiting value V_{qmax}	16V
Q-axis limiting value V_{qmin}	-16V

1) At first, the utility is normal, and the DG unit is connected with the grid.

2) At 2s, the grid is broken and protection switch Su turns off.

3) At 2.02s, the islanding is confirmed and transfer switch Si turns off.

4) During 2.02s-5s, the microgrid operates in islanding mode.

5) At 3.2s, the grid is restored and Su is turned on.

6) At 5s, the transfer switch Si is turned on. The synchronization of load voltage and grid voltage is completed during 3.2s-5s.

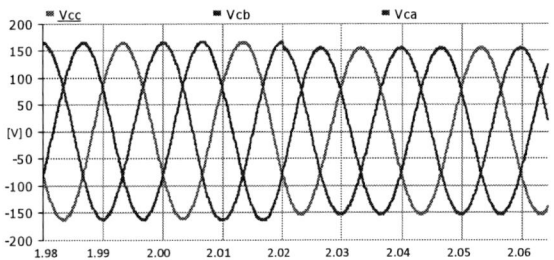

(a)

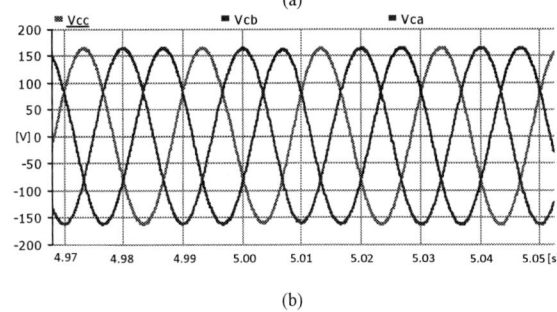

(b)

The 2018 International Power Electronics Conference

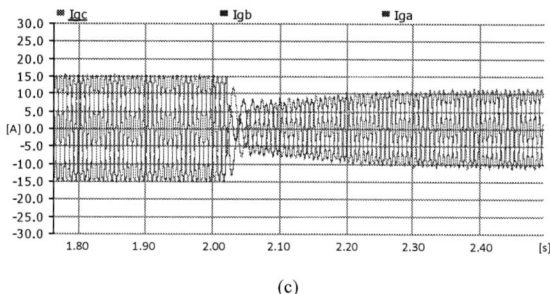

(c)

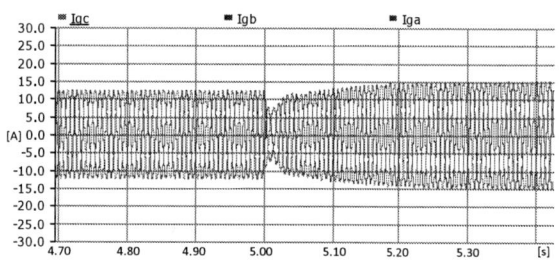

(d)

Fig.8 Simulation results of capacitor voltage and grid current

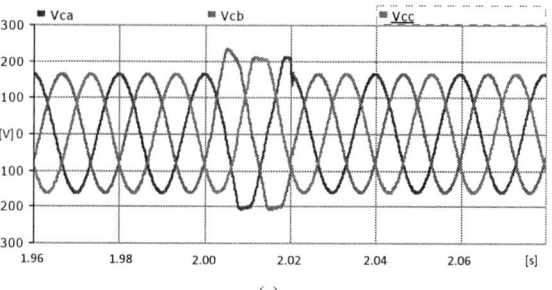

(a)

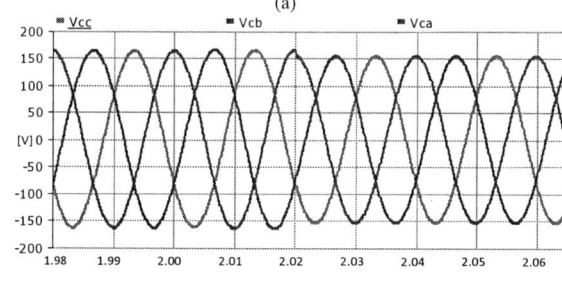

(b)

Fig.9 Simulation results of load voltage with different control strategy (a) Traditional hybrid voltage and current mode control method (b) Improved control method

Fig.8 (a) and (c) shows the capacitor voltage (critical load voltage) and grid current of master DG unit when the microgrid transfers from GC mode to SA mode. Fig.8 (b) and (d) shows the capacitor voltage and grid current when the microgrid transfers from SA mode to GC mode. In Fig.8 (a), the amplitude of capacitor voltage will increase during 2s-2.02s and decrease after 2.02s, which is in accordance with the theoretical analysis above. It can be seen from Fig.8 (a) and (b), there is no voltage distortion during the transferring process, and the change of load voltage is smooth. In Fig.8 (c) and (d), the grid current during the transferring process is also smooth, and there is no apparent inrush current.

In Fig.9 (a), the utility protection switch S_u turns off at 2s, and the transfer switch S_i turns off at 2.02s. During this time interval, the DG unit is disconnected with utility and is still controlled as a current source. The amplitude and frequency of load voltage will drift because of the active power and reactive power mismatch between the DG output and the load demand. Therefore, the quality of load voltage is worsened during this time interval. However, for the improved PLL based control method, when islanding happens, the voltage controller is automatically activated to regulate the load voltage. Therefore, the quality of load voltage can be maintained during the transferring process as shown in Fig.9 (b).

IV. CONCLUSIONS

This paper proposes an improved seamless transfer strategy to solve the issues in the traditional hybrid voltage and current mode control method, which can transfer from CSC to VSC automatically, and doesn't rely on the speed and accuracy of islanding detection method. In GC mode, the inductor current loop is utilized to control the DG unit to act as a current source. The output of voltage controller equals to zero, therefore, the reference of the inner inductor current loop will not be regulated by the voltage loop. Upon the occurrence of utility outage, the voltage controller will be automatically activated to regulate the load voltage. Therefore, the quality of critical load voltage can be maintained during the transferring process. Secondly, the direct current control method in [12] can only operate without grid line impedance, but this proposed method can overcome this limitation. Thirdly, when the grid returns normal, the pre-synchronization can be realized automatically without extra infrastructure.

ACKNOWLEDGMENT

This work was supported by the National Natural Science Foundation of China under Grant 51437007, and the Power Electronics Science and Education Development Program of Delta Environmental & Educational Foundation under Grant DREM2014002.

REFERENCES

[1] Z. Liu and J. Liu, "Indirect Current Control Based Seamless Transfer of Three-phase Inverter in Distributed Generation," IEEE Transactions on Power Electronics, vol. 29, no. 7, pp. 3368-3383, Jul. 2014.

[2] Y. Shi and J. Su, "A Seamless Mode Transfer Method for Microgrid Based on Mode Adaptive Droop Control," in TENCON 2013 - 2013 IEEE Region 10 Conference (31194),2013, pp.1-5.

[3] Y. Jia, D. Liu and J. Liu, "A Novel Seamless Transfer Method for a Microgrid Based on Droop Characteristic Adjustment," in Power Electronics and Motion Control Conference (IPEMC), 2012 7th International , 2012, pp.362-367.

[4] S. Hu, C. Kuo, T. Lee and J. Guerrero, "Droop-Controlled Inverters with Seamless Transition between Islanding and Grid-Connected Operations," in 2011 IEEE Energy Conversion Congress and Exposition, 2011, pp.2196-2201.

[5] T. Yu, S. Choi, and H. Kim, "Indirect current control algorithm for utility interactive inverters for seamless transfer," in Power Electronics Specialists Conference, 2006. PESC '06. 37th IEEE, 2006, pp. 1–6.

[6] J. Kwon, S. Yoon and S. Choi, "Indirect Current Control for Seamless Transfer of Three-Phase Utility Interactive Inverters," IEEE Transactions on Power Electronics, vol. 27, no. 2, pp. 773-781, Feb. 2012.

[7] S. Yoon, H. Oh, and S. Choi, "Controller design and implementation of indirect current control based utility interactive inverter system," in 2011 IEEE Energy Conversion Congress and Exposition, 2011, pp. 955–960.

[8] T.Hwang and S.Park, "A Seamless Control Strategy of Distributed Generation Inverter For Critical Load Safety under Strict Grid Disturbance," in 2012 Twenty-Seventh Annual IEEE Applied Power Electronics Conference and Exposition (APEC) Conference, 2012, pp.254-261.

[9] I. Balaguer, Q. Lei, S. Yang, U. Supatti and F. Peng, "Control for Grid-Connected and Intentional Islanding Operations of Distributed Power Generation," IEEE Transactions on Industrial Electronics,vol.58, no.1, pp.147-157, Jan. 2011.

[10] D. Ochs, B. Mirafzal and P. Sotoodeh, "A Method of Seamless Transitions Between Grid-Tied and Stand-Alone Modes of Operation for Utility-Interactive Three-Phase Inverters," IEEE Transactions on Industrial Applicayions, vol. 50, no. 3, pp. 1934-1941, May. 2014.

[11] Y. Wang and G. Zhao, "Direct Current Control Strategy for Seamless Transfer of Voltage Source Inverter in Distributed Generation Systems," in International Conference on Renewable Power Generation (RPG 2015), 2015, pp.1-5.

[12] Z. Liu and J. Liu, "Unified Control based Seamless Transfer of Microgrids," in 9th International Conference on Power Electronics-ECCE Asia, 2015, pp.1477-1482.

[13] IEEE Standard for Interconnecting Distributed Resources with Electric Power Systems, IEEE Standard 1547-2003, 2003.

Efficient Urban Railway Design integrating Train Scheduling, Onboard Energy Storage, and Traction Power Management

Warayut Kampeerawar[1]*, Takafumi Koseki[1] and Fulin Zhou[2]

1 Department of Electrical Engineering and Information Systems, The University of Tokyo, Tokyo, Japan
2 School of Electrical Engineering, Southwest Jiaotong University, Chengdu, China
*E-mail: kampeeyut@gmail.com

Abstract— This paper presents a design of urban railway operation based on an integrated design approach. The integrated approach aims to integrate the design train schedule, optimizing scenario for installing onboard energy storage, and considering effective power management. The proposed integrated design is formulated as an optimization problem with two alternative form of objective function solved by Genetic Algorithm. First objective function is minimizing energy supplied from substation and energy capacity of energy storage. Alternative form of objective function is minimizing surplus regenerative energy and energy capacity of energy storage. Both objective functions deal with energy-saving and cost-saving concern. To verify the performance of proposed method, the numerical case studies of urban railway system in Thailand and China will be numerically performed. The results shown that the proposed integrated design with the first objective function in off-peak hour can provide up to 16% of energy-saving performance in the case of Chinese Jinan Metro system and 27% of energy-saving performance in the case of Bangkok Metro System.

Keywords— *Onboard Energy Storage, Railway power management, Regenerative Power, Train Scheduling*

I. INTRODUCTION

Reduction of energy consumptions and CO_2 emissions is a prominent trend in transportation system. Because of uncertainty in power consumption in railway systems, power management is a big challenge task. Basically, power management involves how to satisfy the power demand with limited power supply by means of optimizing energy usage at various stages in the system operation in the most efficient way. When a train is operated in powering mode, it will consume power from wayside power substations or onboard-power supply. Due to the advancement of rolling stock and power electronics technology, when a train is operated in braking mode, it can efficiently recycle braking power as regenerative power which can be fed back into catenary, used by train itself or stored in energy storage. By integrating optimization of train operation and applications of power electronics technology, power management cannot only provide balance of demand and supply but also improvement of energy-saving operation.

In modern railway operations, energy-saving operation can be achieved by reducing energy consumption and increasing regenerative energy usage. To reduce energy consumption of a train, various approaches for design of energy-saving driving strategy have been proposed by many researchers. By the advent of rolling stock technology, optimal design of vehicle, advanced technology of traction motor, losses in rolling stocks, transmission line, and relevant system can be considerably reduced. Besides decreasing energy consumption, utilizing regenerative power is one of the most effective way to save energy. Nowadays, the regenerative braking system is commonly employed for railway vehicles. When the regenerative braking system is operated, a traction motor will temporary turn into a generator which generates considerably regenerative power. The regenerative power can be effectively managed by the following ways.

- Powering vehicle itself or accelerating vehicles nearby by adjusting train scheduling or speed pattern for interchanging energy among running trains.
- Being stored and recycled by Energy Storage System (ESS)
- Fed back to utility grid through inverting substation.

In urban railway, regenerative power can be used efficiently due to high frequency of train operating in peak-hour period. However, considerable regenerative power may be unusable in long off-peak period. Therefore, integrating multiple methods to manage regenerative power can increase possibility of utilizing regenerative power.

Many researchers proposed strategies for effective use of regenerative energy by applying the design of timetable, speed profile, energy storage system, or integrating multiple methods as an integrated optimization problem. Integrating multiple methods can provide better improvement of energy-saving operation [1]. An integrated design of speed profile and train scheduling based on energy-efficient algorithm was proposed by [2]. The proposed design provides fast calculation but neglect exchanging power between train.

In addition, a cooperative train control model to design energy-saving train scheduling based on simple search algorithm was developed by [3]. Due to the complexity of integrating various factors and parameters into the same problem, some metaheuristic methods, e.g. Genetic Algorithm, are selected for solving the problem [4-5]. A two-layer optimization including timetable and driving strategy was presented in [4]. The running times of each train were adjusted to minimize energy consumption based on the idea of synchronizing power-time profile by using simple estimation of energy. Moreover, an integrated optimization of driving pattern and timetable was proposed by [5]. They considered optimizing timetable and speed profile in the same problem.

Some integrated design of train operation including installation of energy storage or inverting substation were proposed by [6-7]. An optimal design of speed profiles with consideration of regenerative energy recovery was proposed in [6]. To manage regenerative braking energy, the application of onboard ESS was considered in the design of energy-saving speed profile. There were some evaluations of recovery energy by various methods, but wayside ESS and timetable optimization were not mentioned by this work.

Preceding researches focusing on integrated design of train scheduling and energy storage integrated were found in [8-9]. Design of train scheduling and wayside ESS with minimizing energy supplied from substations was proposed in [8]. The modified objective function, improved solving algorithm and variable weighting factors were evaluated in [9].

In this paper, the design of train scheduling, onboard ESS by means of effective power management was presented. For the proposed design, onboard ESS are included in the optimization of timetable parameters to improve the utilization of regenerative energy. The optimization problem is formulated based on two different objective functions. First objective function deals with minimizing total energy supply and capacity of ESS, while the second one aims to minimize surplus of regenerative energy and capacity of ESS. Evaluation of energy-saving performance and regenerative energy usage are estimated for measuring achievement of the integrated approach. Some numerical case studies based on urban railway system in Thailand and China were performed to explain how the proposed method can improvement of utilizing regenerative energy and energy-saving operation. Furthermore, the effect of traffic condition, variable weighting factor, and different objective function were discussed.

II. PROPOSED METHOD

A. Integrated Design concept

The proposed integrated design aims to design train schedule, onboard ESS, and management strategy of traction power management by simultaneously optimizing timetable parameters and capacity of ESS.

Employing onboard ESS, the proposed management of tractive power and regenerative power can be explained by Fig. 1. In braking mode, regenerative power will be used by onboard auxiliary system, stored in onboard ESS, then the surplus regenerative power will be sent to nearby trains via catenary. If such regenerative power cannot be absorbed by catenary, it will be wasted in resister. The main purpose for increasing the use of regenerative power by train itself and nearby trains is to reduce the requirement of high capacity of onboard ESS. In powering mode, trains mainly consumed power from catenary and additionally from onboard ESS.

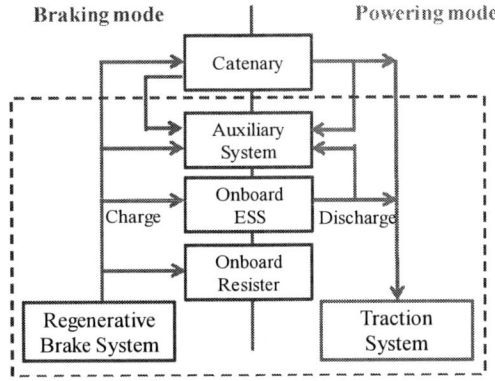

Fig. 1. Power Management Scheme with Onboard ESS

For the proposed integrated design, total energy consumption of railway system will be expected to reduce the total energy supplied from power substations by means of maximizing use of regenerative power. For the design of train schedule, adjusting timetable parameters, e.g. running time and dwell time, are performed to increase the possibility of exchanging power among trains. While timetable is being designed, the appropriate capacity of ESS will be determined based on energy-saving and cost-saving objective.

In addition, the non-integrated design is mentioned as a simplified design compare with the Integrated design. Basically, the non-integrated approach aims to optimize timetable parameters and capacity of ESS sequentially. The detail of Non-integrated design will be explained in the following part. The basic concept of Integrated design and Non-integrated design can be expressed as in Fig.2.

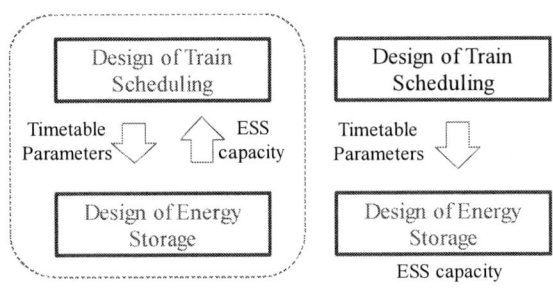

(a) Integrated Design   (b) Non-integrated Design
Fig. 2. Integrated Design vs Non-integrated Design.

B. Problem formulation

The proposed integrated design is formulated as an optimization problem with two alternative form of objective function. First objective function as shown in equation (1) is minimizing energy supplied from traction substations and energy capacity of ESS.

$$\min f(\mathbf{T_r}, \mathbf{T_d}, N_{ess}) = w\left(\frac{\hat{E}_{sup}}{\hat{E}_{sup,base}}\right) + (1-w)\left(\frac{E_{ess}}{E_{ess,max}}\right) \quad (1)$$

Alternative form of objective function defined by equation (2) is minimizing surplus regenerative energy and energy capacity of ESS

$$\min f(\mathbf{T_r}, \mathbf{T_d}, N_{ess}) = w\left(\frac{\hat{E}_{brake} - \hat{E}_{reg}}{\hat{E}_{brake}}\right) + (1-w)\left(\frac{E_{ess}}{E_{ess,max}}\right) \quad (2)$$

The constraints for optimization problem are determine as follows.

Headway limit: $T_{h,min} \leq T_h \leq T_{h,max}$

Dwell time limit: $T_{da,min} \leq T_{da} \leq T_{da,max}$

Running time limit: $T_{r,a \to b,min} \leq T_{r,a \to b} \leq T_{r,a \to b,max}$

Trip time: $T_{trip,min} \leq T_{trip} \leq T_{trip,max}$

Regenerative limit: $V_{tr,reg} \leq V_{reg,max}$

ESS charge and discharge: $SOC_{min} \leq SOC \leq SOC_{max}$

where

$\mathbf{T_d} = [T_{d1}, T_{d2}, ..., T_{dn}]$: Dwell time

$\mathbf{T_r} = [T_{r,1 \to 2}, T_{r,2 \to 3}, ..., T_{r,(n-1) \to n}]$: Running time

T_h : Headway (s), T_{da} : Dwell time at passenger station a (s), $T_{r,a \to b}$: Running time from passenger station a to station b (s), T_{trip} : Trip time for single journey of a train (s)

$\hat{E}_{sup}$: Estimated total energy supplied from substations (kWh), $\hat{E}_{sup,base}$: Estimated total energy supplied from substations (kWh) in case of nominal operating condition

$\hat{E}_{brake}$: Estimated total energy generated from electrical brake system (kWh), $\hat{E}_{reg}$: Estimated total regenerative energy utilized by trains (kWh), E_{ess} : Total energy capacity of energy storage system (kWh), $E_{ess,max}$: Maximum energy capacity of energy storage system (kWh) which can be installed in the system, N_{ess} : Number of energy storage module , SOC: State of charge of ESS, $V_{tr,reg}$: Voltage of train at pantograph in regenerative mode, $V_{reg,max}$: Maximum regenerative voltage, w: weighting factor.

C. Non-integrated design

Non-integrated design is proposed as a simplified version of Integrated design and used for comparing the optimized results. First, the design of timetable parameters is performed, then the capacity of ESS can be determined by a simple search method.

Step 1 Design of timetable parameters

Assume that the effect of ESS on timetable design can be neglected. Therefore, timetable parameters can be optimized without ESS by applying $w=1$ to the proposed

objective function (1) and (2). The optimization problem will be solved by Genetic algorithm (GA).

Step 2 Design of ESS

Use the optimized timetable parameter from previous step as the fixed operating condition, then the ESS capacity will be considered as the only one decision variable. As a result, the optimization problem with objective function (1) or (2) are now easily solved by one-dimensional search method.

D. Solving algorithm

Genetic Algorithm is selected for solving the problem due to the complicated optimization problem including the constraints. The algorithm is shown in Fig.3. The chromosomes are defined as running times, dwell times and energy capacity of ESS. Calculating fitness function is based on the result of power flow calculation of each time step over a specific period. The steps for calculating fitness function is described in Fig 4.

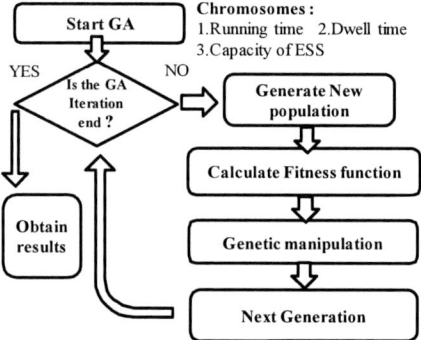

Fig. 3. Flowchart of Genetic Algorithm

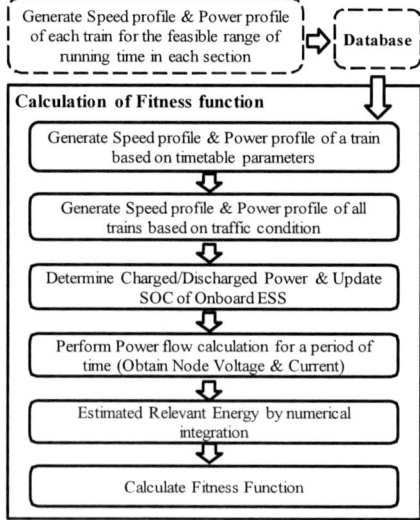

Fig. 4 Calculation of Fitness Function

E. Estimation of Energy

To estimate energy, the power profile versus time of each train and power supply from substations will be first calculated, then the relevant energy will be estimated by numerical integration, i.e. trapezoidal integration. For generating the power profile of each train, train movement calculation including vehicle data and infrastructure data is performed neglecting effect of voltage to train performance. By assuming that all trains have the same power profile, the power profiles of multiple train operation will be generated by corresponding timetable parameters.

To evaluate energy supplied from substation, the power flow calculation implemented based on algorithm proposed by [10] is performed. The power flow calculation considers exchanging regenerative power among trains and charged/discharged power of ESS to determine nodal voltage and current at any point in the system. Accordingly, the power flow results are used for estimating regenerative energy, charged and discharged energy of ESS and Energy supplied by substations. The charged and discharged energy of ESS is estimated based on the strategy mentioned in [11].

For the power flow calculation, the electrical models of system component are explained as follows. The train's power and current are calculated by equation (3) and (4), respectively.

$$P_{tr} = \begin{cases} \dfrac{F_T \times v}{\eta} + P_{aux} & ; \text{Powering mode} \\[2mm] P_{aux} & ; \text{Coasting mode} \\[2mm] \eta \times F_B \times v + P_{aux} & ; \text{Braking mode} \end{cases} \quad (3)$$

$$I_{tr} = \frac{P_{tr}}{V_{pant} - V_{rail}} \quad (4)$$

Where V_{pant}: Nodal voltage at pantograph, V_{rail}: Nodal voltage at rail conductor, F_T: Tractive effort, F_B: Brake effort, P_{aux}: Auxiliary power I_{tr}: Train's current, P_{tr}: Train's power

Power substation is non-inverting substation type which cannot absorb regenerative power from the braking train. When, the negative current is detected, the substation will be modeled as a large resistance or be excluded from the circuit. The amount of power in charging mode or discharging mode will be determined based on State of Charge (SOC) of ESS. The catenary and rail are considered as constant resistance at constant operating temperature.

III. NUMERICAL CASE STUDIES

In this paper, the urban railway system in Thailand named Bangkok Rapid Transit System (BTS-Silom line) and Jinan Metro Railway (Line R1) operated in Jinan City, China, are selected for performing case studies. To verify the performance of proposed integrated design, numerical case studies were implemented on MATLAB r2007b. Non-integrated design and Integrated design were both applied with 2 different objective functions, 2 traffic conditions (Peak hour and Off-peak hour), and 3 scenarios of weighting factors (w=0.5, w=0.8, w=0.99).

For all cases, all relevant energy quantities were evaluated over 1 hour of operating period with a time step of 0.5 seconds as an example shown in Fig.5.

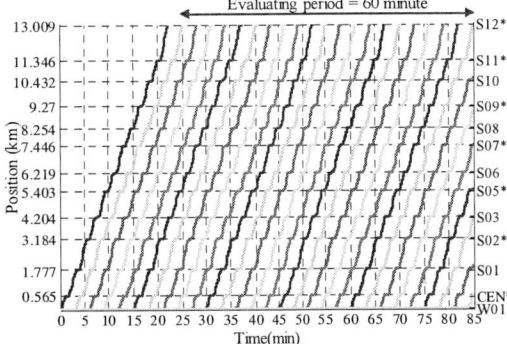

Fig. 5. The position of train versus time of BTS-silom line in peak-hour period

To compare performance of the designed result, the energy-saving percentage (%E_{save}) and the recuperated energy (%E_{recu}) are defined by equation (5) and (6), respectively.

$$\%E_{save} = 100 \times \frac{(\hat{E}_{sup,base} - \hat{E}_{sup,case\,i})}{\hat{E}_{sup,base}} \quad (5)$$

$$\%E_{recu} = 100 \times \frac{(\hat{E}_{brake} - \hat{E}_{reg})}{\hat{E}_{reg}} \quad (6)$$

A. Case studies of BTS

1) System Information and operating condition

BTS-Silom line is operated on 750-Volt D.C. third-rail electrification system. There are 13 passenger stations and 7 traction substations along the total length of 13 km as shown in Fig.6. The system information and the nominal operating condition with evaluation of energy are shown in table I and table II, respectively.

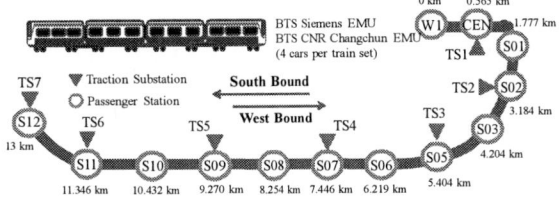

Fig. 6. The Route Map of BTS-Silom line

TABLE I
BASIC INFORMATION OF BTS-SILOM LINE

Data name	Information	
Train's Length	87.25 m (4 cars)	
Voltage Conditions	nominal voltage	750 V-DC
Weight Conditions	tare weight	153 ton
	max. speed	80 km/h
Movement Features	max. acceleration	0.87 m/s²
	max. deceleration	1.00 m/s²
	gear, motor, inverter	98%,88%,98%
Efficiencies	regenerative brake	80%
Max. Auxiliaries	constant load	270 kW
Electrical Resistance	Third-rail	8.23 mΩ/km
	Running rail	40.46 mΩ/km

TABLE II
THE NOMINAL OPERATING CONDITION AND EVALUATION OF ENERGY OF
BTS-SILOM LINE

Traffic volume	T_h (sec)	Load (ton)	T_{trip} (sec)	$E_{sup,base}$ (kWh)	E_{reg} (kWh)	E_{brake} (kWh)	$\%E_{recu}$
Peak	180	75	1338	5753	1662	2294	72.48
Off-peak	300	38	1184	3521	446	1183	37.69

For the nominal operating condition, timetable parameters are assumed as follows. Nominal dwell time ($T_{da,nom}$) in peak hour and off-peak hour are set equally for every station as 30 s and 20 s, respectively. Nominal running time is assumed as 1.05 time of the minimum running time ($T_{r,a \rightarrow b,min}$) which train can be operated. Nominal trip time ($T_{trip,nom}$) is defined as in table II.

To simplify the case studies, all case studies were performed based on the design of trains operated in only one direction, start from station W1, not round trip.

2) Non-integrated design case

In this case, the timetable parameters and ESS capacity were designed by applying the non-integrated design approach. The specification of ESS is Electric double-layer capacitors (EDLC) with the capacity per module of 300 kW, 1 kWh, 428 kg, 750 Vdc.

There are 12 cases for the non-integrated design with different criteria. The timetable was first designed by GA (Variables = 23 (11 dwell times, 12 running times), Population size = 23x50, Crossover probability = 0.8, Mutation probability = 0.2, number of generation = 300, number of stall generations = 30) subjected to the following constraints.

Dwell time limit: $\quad T_{da,nom} \leq T_{da} \leq T_{da,nom} + 5\ sec$

Running time limit: $T_{r,a \rightarrow b,min} \leq T_{r,a \rightarrow b} \leq 1.1 \times T_{r,a \rightarrow b,min}$

Trip time: $\qquad\qquad T_{trip} = T_{trip,nom}$

Regenerative limit: $\quad V_{tr,reg} \leq 900\ V$

After obtaining timetable parameters, the capacity of ESS was optimized for each scenario with the same constraint of $0.25 \leq SOC \leq 1.0$. The results of each case were shown in table III and the corresponding speed profiles were shown in Fig.7 and Fig.8.

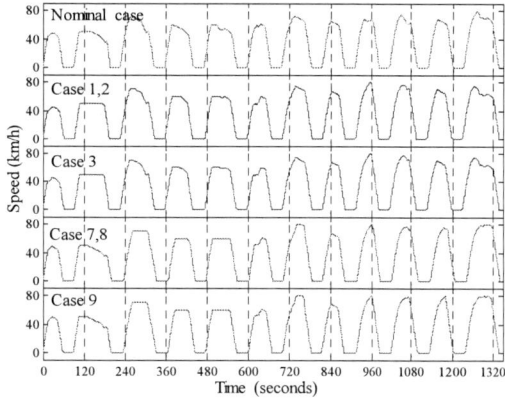

Fig. 7. Speed profile based on optimized timetable parameters designed by Non-integrated design (Peak hour)

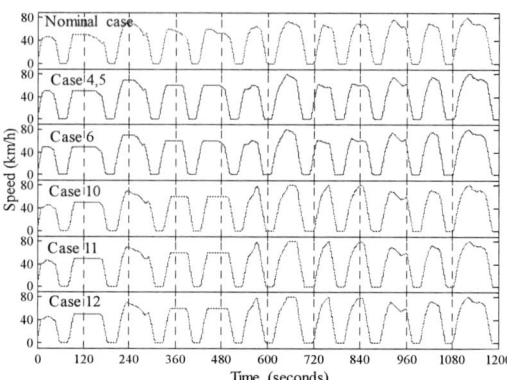

Fig. 8. Speed profile based on optimized timetable parameters designed by Non-integrated design (Off-peak hour)

TABLE III
THE RESULTS OF THE CASE STUDIES OF BTS USING NON-INTEGRATED DESIGN

Case No.	Obj.No. Traffic	w	E_{ess} (kWh)	$\%E_{save}$	$\%E_{recu}$	$\dfrac{\%E_{save}}{E_{ess}}$
1	(1) Peak	0.5	1	17.85	93.56	17.85
2		0.8	1	17.85	93.56	17.85
3		0.99	2	18.08	94.74	9.04
4	(1) Off-peak	0.5	1	18.08	44.92	18.08
5		0.8	1	18.08	44.92	18.08
6		0.99	7	26.10	98.30	3.73
7	(2) Peak	0.5	1	10.97	87.03	10.97
8		0.8	1	10.97	87.03	10.97
9		0.99	10	11.13	99.18	1.11
10	(2) Off-peak	0.5	1	11.50	42.53	11.50
11		0.8	7	22.14	95.45	3.16
12		0.99	10	22.85	99.97	2.29

3) Integrated design case

In this case, 12 scenarios of the same design criteria as previous cases were obtained. Timetable parameters and ESS capacity were designed by GA (Variables = 24 (11 dwell times, 12 running times, 1 ESS capacity), Population size = 23x50, Crossover probability = 0.8, Mutation probability = 0.2, number of generation = 300, number of stall generations = 30) subjected to the same constraints as previous case used. The evaluation of energy and the evaluation of performance were shown in table IV and table V, respectively. The corresponding speed profile generated by designed timetable parameters were shown in Fig.9 and Fig.10.

TABLE IV
THE EVALUATION OF ENERGY FOR THE CASE STUDIES OF BTS USING
INTEGRATED DESIGN

Obj.	Traffic	w	E_{ess} (kWh)	E_{train} (kWh)	E_{sup} (kWh)	E_{reg} (kWh)	E_{brake} (kWh)
(1)	Peak	0.5	1	5062.5	4592.1	1296.1	1419.9
		0.8	2	4922.9	4578.4	1322.0	1426.1
		0.99	4	4664.0	4543.5	1412.5	1434.1
	Off-peak	0.5	1	2782.0	2828.2	355.6	746.2
		0.8	4	2523.1	2587.4	588.9	741.7
		0.99	7	2409.6	2543.9	744.0	752.4
(2)	Peak	0.5	1	5996.0	5140.6	1832.4	2009.4
		0.8	2	5446.4	4862.8	1658.5	1770.1
		0.99	6	4895.5	4817.8	1704.7	1740.2
	Off-peak	0.5	1	3078.9	3122.8	397.8	936.7
		0.8	5	2871.0	2932.4	666.8	1005.3
		0.99	8	2735.3	2833.2	974.1	1045.6

3261

TABLE V
THE EVALUATION OF PERFORMANCE FOR THE CASE STUDIES OF BTS
WITH INTEGRATED DESIGN

Case No.	Obj.No. Traffic	w	E_{ess} (kWh)	$\%E_{save}$	$\%E_{recu}$	$\dfrac{\%E_{save}}{E_{ess}}$
13	(1) Peak	0.5	1	20.19	91.28	20.19
14		0.8	2	20.43	92.70	10.22
15		0.99	4	21.03	98.49	5.26
16	(1) Off-peak	0.5	1	19.68	47.66	19.68
17		0.8	4	26.52	79.40	6.63
18		0.99	7	27.75	98.89	3.96
19	(2) Peak	0.5	1	10.66	91.19	10.66
20		0.8	2	15.49	93.70	7.75
21		0.99	6	16.27	97.96	2.71
22	(2) Off-peak	0.5	1	11.31	42.47	11.31
23		0.8	5	16.72	66.33	3.34
24		0.99	8	19.54	93.16	2.44

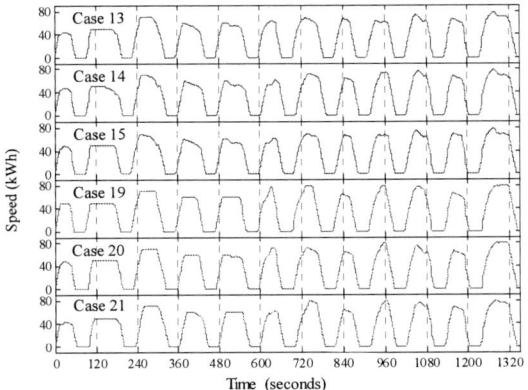

Fig. 9. Speed profile based on optimized timetable parameters designed by Integrated design (Peak hour)

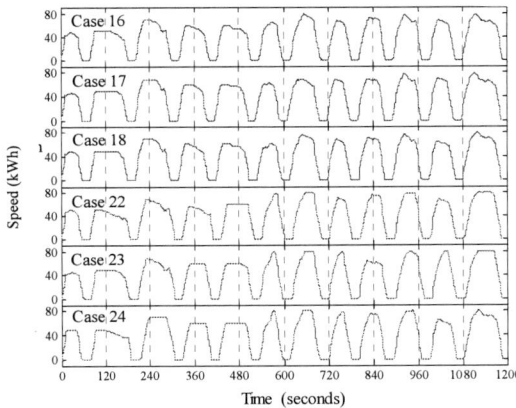

Fig. 10. Speed profile based on optimized timetable parameters designed by Integrated design (Off-peak hour)

B. Case studies of Jinan Metro System

1) System information and operating condition

Presently, Jinan Metro line R1 system is being constructed. It is designed to be operated on 1500-Volt D.C. third-rail electrification system. The system consists of 12 passenger stations and 9 traction substations along the total length of 25.77 km. The nominal operating condition and evaluation of energy are shown in table VI and table VII, respectively.

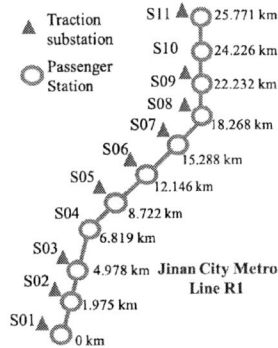

Fig. 11. The Route Map of Jinan metro line R1

TABLE VI
BASIC INFORMATION OF JINAN METRO LINE R1

Data name	Information	
Train's Length	78.29 m (4 cars)	
Voltage Conditions	nominal voltage	1500 V-DC
Weight Conditions	tare weight	140 ton
Movement Features	max. speed	100 km/h
	max. acceleration	1.00 m/s^2
	max. deceleration	1.00 m/s^2
Efficiencies	gear, motor, inverter	98%,88%,98%
	regenerative brake	80%
Max. Auxiliaries	constant load	340 kW
Electrical Resistance	Third-rail	13.8 mΩ/km
	Running rail	36.0 mΩ/km

TABLE VII
THE NOMINAL OPERATING CONDITION AND EVALUATION OF ENERGY OF JINAN METRO

Traffic	T_h (sec)	Load (ton)	T_{trip} (sec)	$E_{sup,base}$ (kWh)	E_{reg} (kWh)	E_{brake} (kWh)	$\%E_{recu}$
Peak	180	83.5	1747	5576	1055	2845	37.08
Off-peak	360	59.6	1719	2776	386	1270	30.42

Compared with BTS, Jinan Metro System is operated on higher voltage level, higher maximum operating speed, and larger average distance between adjacent passenger stations.

2) Integrated design case

In this case, timetable parameters and ESS were designed by using Integrated design with objective function (1) in off-peak hour period. Therefore, only 3 cases with different weighting factors, i.e. Case J1-J3, were performed. The optimization problem was solved by GA (Variables = 20 (9 dwell times, 11 running times, 1 ESS capacity), Population size = 20x50, Crossover probability = 0.8, Mutation probability = 0.2, number of generation = 300, number of stall generations = 30) subjected to the same constraints as previous cases used except for regenerative voltage limit and the specification of ESS.

The regenerative voltage limit is defined that $V_{tr,reg} \leq 1800$ V and the specification of ESS is EDLC with the capacity per module of 300 kW, 1 kWh, 428 kg, 1500 Vdc.

The results and corresponding evaluation were shown in table VIII. The corresponding speed profile based on designed timetable parameters were shown in Fig.12.

3262

The 2018 International Power Electronics Conference

TABLE VIII
THE EVALUATION OF ENERGY AND PERFORMANCE OF DESIGNED CASE
STUDIES OF JINAN METRO SYSTEM USING INTEGRATED DESIGN

w	E_{ess}	E_{sup} (kWh)	E_{reg} (kWh)	E_{brake} (kWh)	$\%E_{save}$	$\%E_{recu}$	$\dfrac{\%E_{save}}{E_{ess}}$
0.5	1	2718.2	505.6	1303.6	2.09	38.79	1.04
0.8	4	2618.8	533.0	1307.0	5.67	40.78	1.42
0.99	10	2325.1	809.4	1319.5	16.25	61.34	1.62

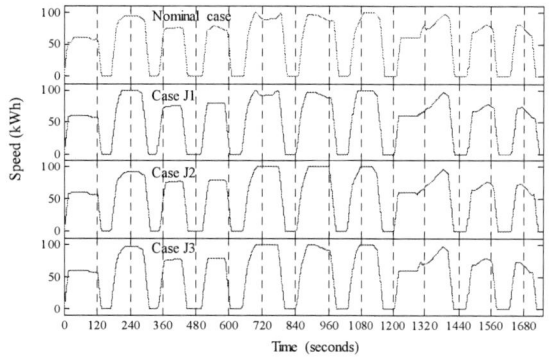

Fig. 12. Speed profile based on timetable parameters designed by Integrated design (Off-peak hour)

IV. RESULT DISCUSSIONS

From the results of case studies, the effect of different objective, weighting factor, and traffic condition can be evaluated. As well as the performance of Integrated design and Non-integrated design was discussed in this section.

A. The effect of different objective function

From comparison of the results of applying different function shown in Fig.13, application of objective function (1) provides considerably better results in energy-saving operation than application of objective function (2) does, while the results in regenerative utilization and capacity of ESS are quite similar for both objective functions. The objective function (1) directly aims to minimize total energy supply, but it can provide good results in both energy-saving operation and regenerative utilization.

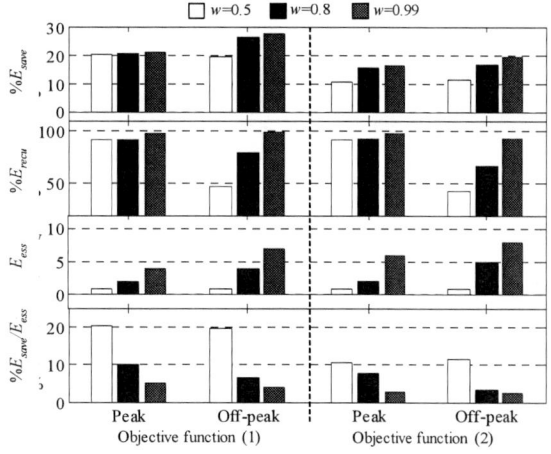

Fig. 13. Comparison of the results of Integrated design applying different objective function

B. The effect of weighting factor

The weighting factor obviously affects the compromise between two sub-objectives in the main objective function. In case of the application of objective function (1), when the weighting factor is increased, the energy-saving performance will increase proportionally but the cost-saving performance significantly decreases as shown in Fig.13. Therefore, the variation of weighting factor may provide the range of appropriate design scenarios considered as the choice for deciding an appropriate solution. However, the logical criteria for selecting the optimal weighting factor has not considered in this paper.

C. The effect of traffic volume

Generally, the traffic volume considerably affects the possibility of exchanging regenerative power among trains. From the results of applying objective function (1) shown in Fig.13, the proposed design can provide effective energy-saving operation and considerable improvement in regenerative utilization in off-peak hour because the regenerative utilization by the system with nominal operating condition in low traffic volume is basically not effective. Moreover, the required capacity of ESS in peak hour is small due to less surplus regenerative energy.

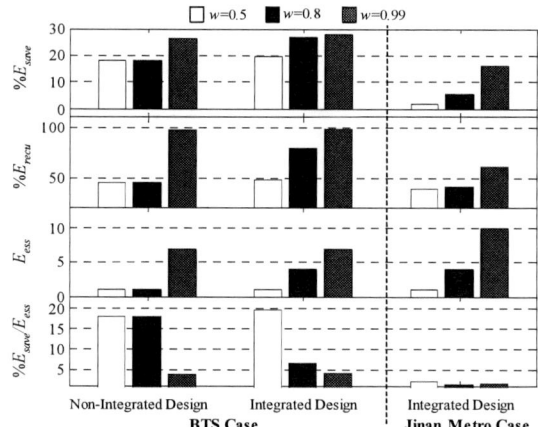

Fig. 14. Comparison of Non-integrated design and Integrated design (Objective function 1 at Off-Peak hour)

D. Comparison of Non-integrated design and Integrated design

The effect of objective function and traffic volume showed that the effective designed results can be obtained by application of objective function (1) in off-peak hour. Therefore, the comparison of the results in such condition was focused. As shown in Fig.14, the Integrated design provided better results in both energy-saving operation and regenerative utilization than the Non-integrated design did, but the Integrated design takes more considerable computation to obtain the optimal result. Although, the Non-integrated design provides the less

3263

effective solution. The non-integrated design may be used as a preliminary design or an alternative design.

E. Comparison between case of BTS and Jinan Metro

Due to some different operating conditions and characteristics of BTS and Jinan Metro System, the design results of BTS were more effective than those of Jinan Metro as shown in Fig.14. Compared with the results of BTS, the case of Jinan Metro System requires larger capacity of ESS, while the energy-saving performance is considerably less, because the average distance between two adjacent passenger stations of Jinan metro is considerably longer than that of BTS. Basically, long distance between two adjacent passenger stations reduces the frequency of braking and accelerating operation leading the less possibility of utilizing regenerative energy.

F. Criteria for selecting an appropriate design

To obtain an effective design with the proposed design, the objective function (1) in off-peak hour is the most appropriated condition. For the case studies of BTS, if the energy-saving operation is considered as the main design objective, the most appropriate designed case should be the case of the maximum $\%E_{save}$, i.e. case 18. If the energy-saving and cost-saving objective are both concerned, the most effective design should be the case of maximum $\%E_{save}$ per kWh of capacity of ESS, i.e. case 16.

For the case studies of Jinan Metro, case J3 is the case of the maximum $\%E_{save}$ and the case of maximum $\%E_{save}$ per kWh of capacity of ESS. Therefore, it is easily selected as the most appropriated design. The capacity of ESS and $\%E_{save}$ were shown in table IX.

TABLE IX
THE MOST APPROPRIATE DESIGN SOLUTION AMONG ALL CASE STUDIED

System Name	Max $\%E_{save}$		Max $\%E_{save}/E_{ess}$	
	E_{ess} (kWh)	$\%E_{save}$	E_{ess} (kWh)	$\%E_{save}$
BTS	7	27.75	1	19.68
Jinan Metro	10	16.25	10	16.25

V. CONCLUSIONS AND FUTURE WORKS

This paper presents an integrated design for designing effective railway operation integrating power management, train scheduling and application of onboard energy storage leading to improve energy-saving operation. The proposed design is formulated as an optimization problem based on compromising of energy-saving purpose and cost-saving purpose in which the timetable parameters and capacity of ESS will be simultaneously optimized by using GA.

The numerical case studies were performed on two urban railway systems, BTS and Jinan Metro System. The results shown that the proposed integrated design can provide effective solution when the objective function consisting of minimizing total energy supplied and

capacity of ESS is applied in off-peak hour condition. To provide alternative solutions with different criteria on energy-saving performance and cost-saving performance, the weighting factor can be adjusted in a specified range. For the case studies of BTS, the maximum energy-saving performance of 27% was obtained by using the optimized timetable with 7 kWh of ESS. In the case studies of Jinan Metro, the maximum energy-saving performance of 16% was obtained by using the optimized timetable with 10 kWh of ESS. However, the cost-saving issue is always concerned in practical implementation. The reduction of capacity of ESS may be required. Compared with the non-integrated design based on sequential optimization, the Integrated design provide better solution.

For the future work, the robustness of the proposed method to traffic variation will be investigated. Furthermore, the criteria for varying weighting factor to obtain optimal solution will be logically investigated. However, the main difficulty of using GA with the proposed design is the considerably large computation time. The parallel computation may be applied for reducing computation time of GA.

REFERENCES

[1] Yang, Xin, Xiang Li, Bin Ning, and Tao Tang. "A survey on energy-efficient train operation for urban rail transit," *IEEE Trans. on Intelligent Transportation Systems*, vol. 17, no. 1, pp. 2-13, 2016.

[2] S. Su, T. Tang, X. Li, & Z. Gao, "Optimization of multitrain operations in a subway system," *IEEE Trans. on intelligent transportation systems*, vol.15, no.2, pp. 673-684, 2014.

[3] S. Su, T. Tang, & C. Roberts, "A cooperative train control model for energy saving," *IEEE Trans. on intelligent transportation systems*, vol.16, no.2, pp.622-631, 2015.

[4] F. Cao, Z. Liao, S. Liu, J. Xun, & X. Luo, "A two-layer construction of energy optimization approach for timetable," In proceedings of IEEE International Conference on Intelligent Rail Transportation (ICIRT), pp. 476-481, 2016.

[5] N. Zhao, C. Roberts, S. Hillmansen, Z. Tian, P. Weston & L. Chen, "An integrated metro operation optimization to minimize energy consumption," *Transportation Research Part C: Emerging Technologies*, vol.75, pp. 168-182, 2017.

[6] M. Domínguez, A. Fernández-Cardador, A. P. Cucala & R. R. Pecharromán, "Energy savings in metropolitan railway substations through regenerative energy recovery and optimal design of ATO speed profiles," *IEEE Trans. on automation science and engineering*, vol.9, no.3, pp. 496-504, 2012.

[7] X. Li, & H. K. Lo, "Energy minimization in dynamic train scheduling and control for metro rail operations," *Transportation Research Part B: Methodological*, vol.70, pp.269-284, 2014.

[8] W. Kampeerawat and T. Koseki, "A strategy for utilization of regenerative energy in urban railway system by application of smart train scheduling and wayside energy storage system", In proceedings of 2017 International Conference on Alternative Energy in Developing Countries and Emerging Economies (AEDCEE-2017), May 25, 2017.

[9] W. Kampeerawat and T. Koseki, "Integrated Design of Train Scheduling and Wayside Energy Storage for Energy-Saving Urban Railway Operation – A Case Study of Bangkok Rapid Transit System", IEEJ Technical meeting, November 30, 2017.

[10] Y. Cai, M. R. Irving, and S. H. Case. "Iterative techniques for the solution of complex DC-rail-traction systems including regenerative braking." *IEE Proceedings-Generation, Transmission and Distribution*, vol. 142, no.5 pp. 445-452, 1995.

[11] Zongyu GA, Jianjun FA, Zhang Y, and Di SU, "Control strategy for wayside supercapacitor energy storage system in railway transit network," *Journal of Modern Power Systems and Clean Energy*, vol.2, no.2, pp. 181-190, 2014.

The 2018 International Power Electronics Conference

Optimal Control Method of an Energy Storage System for Energy Saving

Yoko Takeuchi[1*], Tomoyuki Ogawa[2], Keisuke Sato[1], Hiroaki Morimoto[3] and Tatsuhito Saito[2]

1 Signalling and Transport Information Technology Division, Railway Technical Research Institute, Tokyo, Japan
2 Vehicle Control Technology Division, Railway Technical Research Institute, Tokyo, Japan
3 Power Supply Technology Division, Railway Technical Research Institute, Tokyo, Japan
*E-mail: takeuchi.yoko.49@rtri.or.jp

Abstract — From the viewpoint of energy saving, the optimal control method of stationary energy storage systems is a method of minimizing of the total energy supplied from all the related traction substations. The purpose of this paper is to find the optimal charging/discharging power control method of the energy storage system under the condition in which speed profiles and driving operations of all trains are fixed. We make some assumptions to model this problem as a mathematical optimization problem. In this paper, we introduce some calculation results obtained by varying parameters. We also indicated that this mathematical optimization model can calculate the maximum effect on energy saving which is useful to design the energy storage system. In addition, we confirm the validity of this mathematical optimization model and estimate the effects of the energy storage system on energy saving.

Keywords — *DC feeding system: Optimization: Regenerative energy: Train operation*

I. INTRODUCTION

Recently, several railway companies install stationary energy storage systems on the wayside. For this, there are such reasons as the energy saving, feeding voltage stabilization of each train and the reserving minimal tractive energy for an emergency evacuation from underground tunnels.

In this research, we focus on the control method of the energy storage system and aim to find an optimal method for charging/discharging the system in such a way as to reach the most effective energy saving.

In order to save energy, understanding features of the train operation power is important.

In general, the driving operation which consists of many coasting and less powering can achieve energy saving, although it needs longer running time. In railway system, each train has to run avoiding over the planed running time because of punctuality. It means that the reduction of traction energy is limited.

Many railway companies install rolling stocks which are equipped with regenerative brakes. They can produce regenerative energy in the DC feeding system. If a train powering exist near a train braking, the train braking can provide the train powering with its regenerative energy.

On the other hand, there are no train powering near a train braking, the regenerative energy is wasted, which is called "regenerative energy squeezing" under light load condition. The regenerative energy is calculated by feeding circuit equations of Kirchhoff's low.

Regenerative performance of rolling stocks have been improved by introducing the several recent technologies such as silicon-carbide semiconductor modules and/or high-power motors, in that the amount of regenerative energy increases.

Therefore, the technology with a focus to reduce the regenerative energy squeezing becomes more important. One of the effective technology to reduce regenerative energy squeezing is installing energy storage system. Its control method and performance have a greatly impact on the effect on energy saving.

When the speed profiles and driving operations of all trains are fixed, each train operation power can be determined. In addition, if the charging/discharging power of the energy storage system is given, the supplied power of each rerated traction substation can be calculated by solving feeding circuit equations. The total energy supplied from all substations in a period of time from the first station to the last station is the index of the effect on energy saving. Therefore, there exists the transition of the charging/discharging power which can achieve minimizing of the total energy supplied from all the substations. This transition is called "the optimal control method" for the energy storage system in this paper.

Therefore, a purpose of this paper is to find the optimal control method under the condition which the speed profiles and driving operations of all trains are fixed. In order to model this problem as a mathematical optimization problem, some assumptions are made. This mathematical optimization problem is solved using Gurobi optimizer.

In this paper, we introduce some calculation results obtained by varying parameters. We also indicate that this mathematical optimization model can calculate the maximum effect on energy saving which is useful to design the energy storage system. In addition, using the Train Operation Power Simulator [1], we confirm the

3265

validity of this mathematical optimization model. We also estimate the effect on energy saving when the energy storage system is installed and this optimal control method is applied.

II. CONSTRAINTS OF THE ENERGY STORAGE SYSTEM

In the modelling, the constraints on the performance of the energy storage system should be considered. Concretely, the state-of-energy of the energy storage system should be kept within a certain range (it is called "SOE constraint") because deviating from this range reduces the life of the energy storage system.

For energy saving, the energy storage system should be charged in such a way as to decrease the regenerative energy squeezing as much as possible and be discharged with a proper timing to keep the SOE constraints. In general, since the energy storage system cannot obtain the information of driving operations, it is controlled based on the voltage which is real-time measurable data in the system, while the SOE constraints being kept [2]-[4]. There is another control method based on the real-time train position data[5].

III. MODELLING AS A MATHEMATICAL OPTIMIZATION PROBLEM

Under the condition which the speed profiles and driving operations of all trains are fixed, the other assumptions are needed to model this problem as a mathematical optimization problem.

Assumption (1) : A discrete model is applied.

Each train's powering/regenerative power depends on the interaction between the trains in operation. They also depends on the characteristics of substation rectifiers and rolling stocks, and the resistance of electric wires and rails. Although the train operation power varies continuously, it is enough to find the charge/discharge power of the energy storage system at the short time of each Δt (hereinafter referred to as "each time"), a discrete model is applied to approximate simply. Concretely, the time and the energy are divided into length of time unit Δt.

Assumption (2) : Charging energy of the energy storage system is lost at a fixed rate.

To consider the effectiveness of the transformation of the energy storage system, a fixed efficiency rate η is introduced. The discharging energy used for train operation is equal to η times as much as the charging energy. This assumption represents that not all of the energy stored can be used for the train traction energy.

Assumption (3) : The number and position of energy storage systems are restricted.

To simplify the model, the only one energy storage system is installed inside the only one substation.

From the condition of the fixed speed profiles and driving operations of all trains (Fig.1(a)), all of the train operation power are also fixed (Fig.1(b)) at any time. Therefore, once the value of charging/discharging power

of the energy storage system is given (Fig.2(a)), the total energy supplied from all substations is determined by the calculation of feeding circuit equations (Fig.2(b)). It is

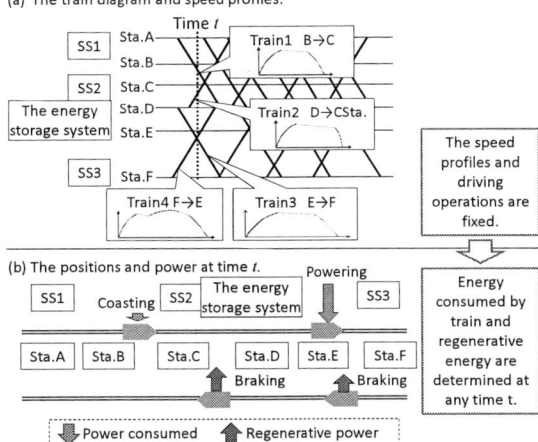

Fig. 1. An example of the precondition of speed profiles and train operation energy[6].

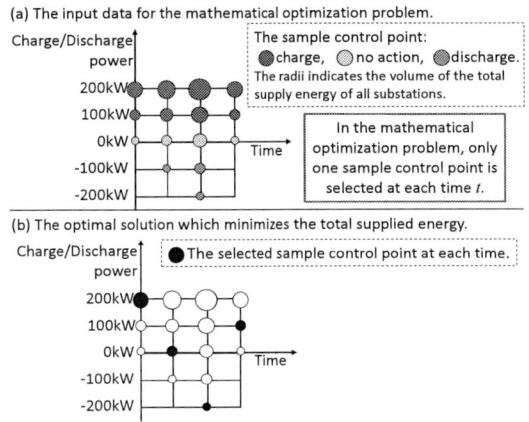

Fig. 2. An example of the sample control point of the energy storage system[6].

(a) The input data for the mathematical optimization problem.

(b) The optimal solution which minimizes the total supplied energy.

Fig. 3. An example of the input data for the mathematical optimization problem and the optimal solution [6].

3266

noticed that the available values of charging/discharging power at time t are changing according to the train operation energy. For example, when all of the trains are braking at time t, the energy storage system cannot be discharged because there are no trains which consume acceleration energy.

The modelling procedures are as follows. At each time, a lot of "sample control points" where some charging/discharging power value is set are prepared. Then, the total energy supplied from all substations for each sample control point is calculated. These data are set as the input data for the mathematical optimization problem (Fig.3(a)). The objective function of this mathematical optimization problem is to minimize the total energy supplied from all substations in a period of time from the first train to the last train (Fig.3(b)). Main constraint is that only one sample control point is selected at each time and the SOE constraint is kept at any time.

An example of the optimal control points is shown in Fig.4 on the left. The transitions of the SOE of the energy storage system and the total energy supplied from all substations are shown in Fig.4 on the right.

IV. PROCEDURES OF THIS RESEARCH

The high estimated accuracy of the total energy supplied from all substations is needed whatever any sample control points is selected, because it has a great impact on the optimization results. In addition, in order to verify this mathematical optimization problem, we should indicate that the optimal objective value approximates to the reality well.

For these, we use the Train Operation Power Simulator which has been developed for the high accuracy estimation of train operation power [1]. Figure 5 shows the input-output parameters, the functions and the components of the Train Operation Power Simulator. This simulator integrated the methods for calculating energy consumed by trains according to the rolling stock characteristics and driving operations, the train control and the power supply. These methods have been developed at RTRI as a result of its researchers drawing on their expertise in their respective fields.

The procedures of this research is shown in Fig.6. Firstly, the total power supplied from all substations by each sample control point are calculated in order to make the input data of the mathematical optimization problem (Fig.6 (1)). Then, we confirm the validity of this mathematical optimization model (Fig.6 (2)) and estimate the effect on energy saving (Fig.6 (3)).

V. FORMULATION AS A MATHEMATICAL OPTIMIZATION PROBLEM

In this section, we formulate the optimal energy storage system control problem as a mathematical optimization problem depending on the assumptions, the constraints and the ideas that are described in the previous sections.

Notation of sets and elements

- $\Delta t \in \mathbb{Z}_+$: The time unit[s].
- $T := \{0, \Delta t, 2\Delta t, \cdots, t_{max}\}$: The time set.

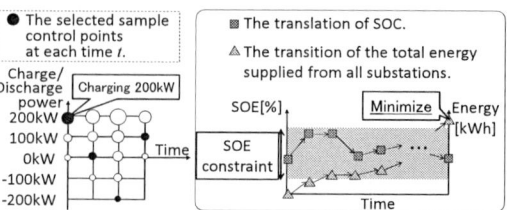

Fig. 4. An example of the optimal solution and the transitions of the SOE and the total energy supplied from all substations.

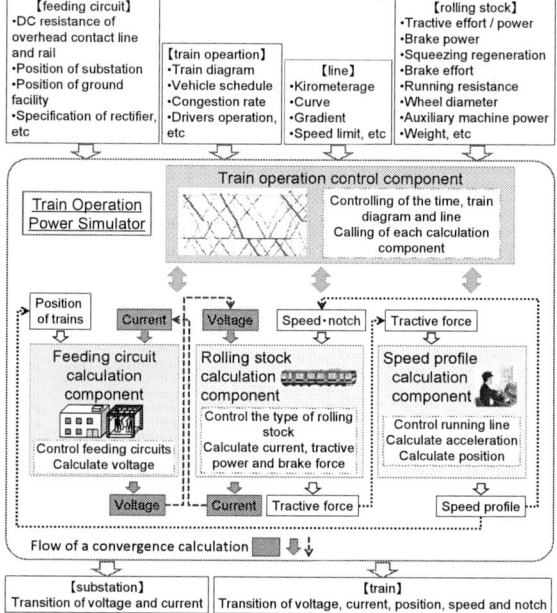

Fig. 5. Input-output parameters, functions and structures of the Train Operation Power Simulator [1].

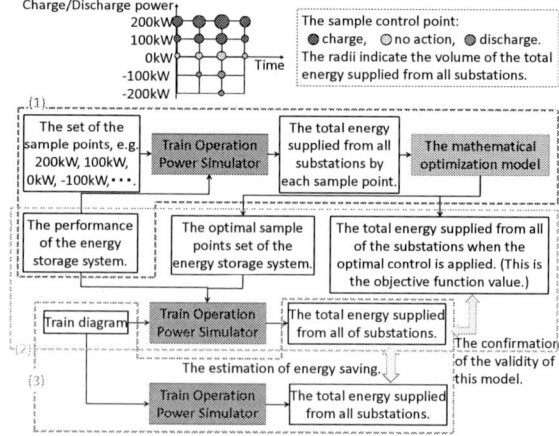

Fig. 6. The procedures of this research.

➢ The element is represented by t.
➢ Time 0 represents the departure time of the first train.
➢ Time t_{max} represents the arrival time of the last train.
- J: The set of substations.
- g: The energy storage system.
- $N := \{1, \cdots, n_{max}\}$: The set of the sample control

3267

points of the energy storage system.

> Its element is represented by n.

Notation of constants

- $P_J^{SS}(t,n) \in R_+$: The total power supplied from all substations J when the sample control point $n \in N$ is selected at time $t \in T \setminus \{t_{max}\}$ [W].
  > They are ordered as $P_J^{SS}(t,1) \geq \cdots \geq P_J^{SS}(t,n_{max})$.

- $P_g^{stg}(t,n) \in R$: The charging/discharging power of the energy storage system when the sample control point $n \in N$ is selected at time $t \in T \setminus \{t_{max}\}$ [W]
  > If $P_g^{stg}(t,n) \geq 0$, the sample point n of the energy storage system represents charging at time t.
  > If $P_g^{stg}(t,n) < 0$, the sample point n of the energy storage system represents discharging at time t.
  > $P_g^{stg}(t,n)$ is called "charging power" whether $P_g^{stg}(t,n)$ is positive or negative.
  > They are ordered as $P_g^{stg}(t,1) \geq \cdots \geq P_g^{stg}(t,n_{max})$.
    ❖ $P_g^{stg}(t,1)$: The maximum instantaneous charging power.
    ❖ $P_g^{stg}(t,n_{max})$: The maximum instantaneous discharging power.

- $E_g^{stg} \in R_+$: The lower limit of the charging capacity of the energy storage system g [J].

- $\overline{E_g^{stg}} \in R_+$: The upper limit of the charging capacity of the energy storage system g [J].

Notation of parameters

- $\alpha^+ \in R_+$: The discharging cost per 1[J] of the energy storage system between the time t_{max} and the next day's time 0.

- $\alpha^- \in R_+$: The charging cost per 1[J] of the energy storage system between the time t_{max} and the next day's time 0.

- $\beta \in R_+$: A coefficient for ordering optimal solutions when there exists multiple such solutions. ($\beta \approx 0$)

- $\eta \in [0,1]$: The energy translation efficiency rate from charging energy to discharging energy of the energy storage system g.

- $f(t,n,\eta) = \begin{cases} \eta & P_g^{stg}(t,n) \geq 0 \\ 1 & otherwise \end{cases}$
  : The efficiency multiplier of the energy storage system g.

Notation of the decision variables

- $x_g(t,n) \in \{0,1\}$: If the sample control point $n \in N$ is selected at time $t \in T \setminus \{t_{max}\}$, $x_g(t,n) = 1$. Otherwise $x_g(t,n) = 0$.

Notation of the other variables

- $e_g^{stg}(t) \in \left[\underline{E_g^{stg}}, \overline{E_g^{stg}}\right]$: The amount of the charging/discharging energy of the energy storage system g at the time $t \in T$ [J].

- $y_g^+ \in R_+, y_g^- \in R_+$: The difference in the amount of the charging energy of the energy storage system g between at the time 0 and the time t_{max} [J]. If $e_g^{stg}(0) \geq e_g^{stg}(t_{max})$, $y_g^+ = 0$. Otherwise $y_g^- = 0$.

The formulation was represented as follows:

$$\text{Min.} \quad \sum_{t \in T \setminus \{t_{max}\}} \sum_{n \in N} P_J^{SS}(t,n)x_g(t,n)\Delta t \quad (1)$$
$$+\alpha^+ y_g^+ + \alpha^- y_g^- + \beta e_g^{stg}(0)$$

$$\text{s.t.} \quad e_g^{stg}(t + \Delta t) = e_g^{stg}(t) \quad (2)$$
$$+ \sum_{n \in N} f(t,n,\eta)P_g^{stg}(t,n)x_g(t,n)\Delta t$$
$$\forall t \in T \setminus \{t_{max}\}$$

$$y_g^+ - y_g^- = e_g^{stg}(t_{max}) - e_g^{stg}(0) \quad (3)$$

$$\sum_{n \in N} x_g(t,n) = 1 \qquad \forall t \in T \setminus \{t_{max}\} \quad (4)$$

$$\underline{E_g^{stg}} \leq e_g^{stg}(t) \leq \overline{E_g^{stg}} \quad (5)$$

$$y_g^+ \geq 0 \quad (6)$$

$$y_g^- \geq 0 \quad (7)$$

$$x_g(t,n) \in \{0,1\} \quad \forall t \in T \setminus \{t_{max}\}, \forall n \in N. \quad (8)$$

The meaning of the terms of the objective function (1) and the constraints (2)-(5) are as follows.

(1) $\sum_{t \in T \setminus \{t_{max}\}} \sum_{n \in N} P_J^{SS}(t,n)x_g(t,n)\Delta t$ is the sum of the total energy supplied from all substations. The main aim of this research is minimizing this value.

(1) $\alpha^+ y_g^+ + \alpha^- y_g^-$ is the cost of charging/discharging energy from $e_g^{stg}(t_{max})$ to $e_g^{stg}(0)$. A smaller value is preferred.

(1) $\beta e_g^{stg}(0)$ is introduced for ordering multiple optimal solutions which possibly exists. Among such solutions one which takes the minimum value of $e_g^{stg}(t_{max})$ are selected.

(2) The storage energy at the time $t + \Delta t$ is equal to the sum of the efficiency multiplier $f(t,n,\eta)$ times the charging power of the selected sample control point and the storage energy at the time t.

(3) The storage energy difference between the operation of the last train and that of the next day's first train is calculated.

(4) Only one sample control point is selected at each time.

(5) The storage energy is kept in the range from $\underline{E_g^{stg}}$ to $\overline{E_g^{stg}}$. This represents the SOE constraint.

VI. CASE STUDY

In this section, we introduce some calculation results, confirm the validity of this mathematical optimization model and estimate the effect on energy saving when the energy storage system is installed and the optimal control method is applied.

A. Input Data and Software

We prepare a sample railway line for the case study. There are neither gradients nor curves, the number of stations is 12, the distance between stations is randomly in the range of 1.5 to 2.5 km. There are 6 substations in every 4.5km and the energy storage system is installed in the substation 9km away from the line end (Fig.7). The train diagram is periodic at a headway of 5 or 10 minutes and all of the trains stop at every station. Figure 8 shows a diagram of a headway of 10 minutes. TABLE I shows the rolling stock elements. Regarding the auxiliary machine power 2 patterns are set according to the outside temperature. One is the smallest power which is correspondent to outside temperature 15[°C] and the other is higher power which is correspondent to 35[°C] [6].

The time unit Δt is 1[s]. As the sample control points of the energy storage system, we present several control points to each of which the charging power is allocated at intervals of 250[kW] from the maximum instantaneous charging power $P_g^{stg}(t,1)$ to the maximum discharging power $P_g^{stg}(t,n_{max})$. We set these sample control points to the Train Operation Power Simulator and obtain the input data of the mathematical optimization problem as previously shown in Fig.3(a). We use Python 3.6 and Gurobi Optimizer 7.5.1 as the interface of the input-output data and the formulation of the mathematical optimization problem.

B. Considering the Optimal Control Method

As an example, we make the assumptions as follows. The train diagram is a headway of 10 minutes (Fig.8), the auxiliary machine power is correspondent to 15[°C], the lower limit of the charging capacity of the energy storage system E_g^{stg} is set at 0[J] and the upper limit $\overline{E_g^{stg}}$ is set at 30[kWh], and the absolute value of the maximum instantaneous charging power $\left|P_g^{stg}(t,1)\right|$ and discharging power $\left|P_g^{stg}(t,n_{max})\right|$ are set at 2000[kW], then we solve

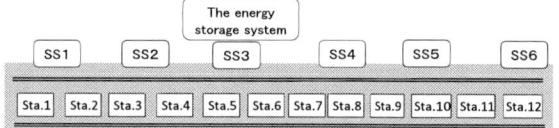

Fig. 7. The position of substations and the energy storage system.

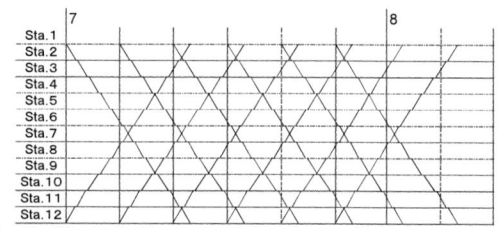

Fig. 8. The train diagram of a headway of 10 minutes.

TABLE I

ROLLING STOCK ELEMENTS

Total mass of the train[ton]	286
Starting acceleration [km/h/s]	3.3
Deceleration of braking [km/h/s]	2.5
The ratio of motored cars and trailer cars	1:1
Traction circuit efficiency	0.855
Auxiliary machine power [kW] corresponding to temperature 15[°C]	39
Auxiliary machine power [kW] corresponding to temperature 35[°C]	284

this mathematical optimization problem. Figure 9 shows the transition of the power and the charging energy of this optimal solution. The storage energy 0[%] indicates 0[kWh] and 100[%] indicates 30[kWh], respectively. The train diagram is a headway of 10 minutes, so that the optimal control method is iterated by 10 minutes from 7:20:00 to 7:50:00, and therefore the time axis is expressed from 7:30:00 to 7:40:00 in Fig.9. There are some considerations as follows.

(a) At the time when the total power consumed by trains is negative (around 7:34:00, 7:36:00 and 7:38:00).

- Figure 10 shows the positions of trains, the power consumed by trains, the output power of substations and the charging power of the energy storage system at the time 7:34:00.
- Trains 5 and 6 are braking near the energy storage system and other trains are coasting.
- The energy storage system is charged with their regenerative energy in order to decrease the regenerative energy squeezing.

(b) At the time except (a) when the energy storage system is charged (around 7:30:00, 7:32:00, 7:37:30 and 7:39:30).

- Figure 11 shows the positions of trains, the power consumed by trains, the output power of substations and the charging power of the energy storage system at the time 7:32:00.
- Trains 5 and 6 are braking near the energy storage system, trains 3 and 4 are powering and other trains are coasting. Trains 5 and 6 are positioned away from trains 3 and 4.
- This optimal solution indicates that there is a case where it is more efficient to charge the energy storage system with the regenerative energy than to provide train 3 or 4 positioned far away with the regenerative energy.

(c) At the time when the energy storage system is discharged (around 7:31:00, 7:33:00, 7:35:00, 7:37:00 and 7:39:00).

- Figure 12 shows the positions of trains, the power consumed by trains, the output power of substations and the charging power of energy storage system at the time 7:37:00.
- In order to be charged as much as possible at the time of (a) and (b), the energy storage system is discharged when the amount of the total energy consumed by trains is large.

C. Comparison of the Claculation Results Obtained by Varying Energy Storage System Performance

We present the following 3 cases; Case 1 where a diagram of a headway of 10 minutes is applied and the auxiliary machine power is corresponding to outside temperature 15[°C], which we call "the basic case", case 2 where a diagram of a headway of 10 minutes is applied and the auxiliary machine power is corresponding to outside temperature 35[°C], which we call "the case No.1" and case 3 where a diagram of a headway of 5 minutes is applied and the auxiliary machine power is corresponding

The 2018 International Power Electronics Conference

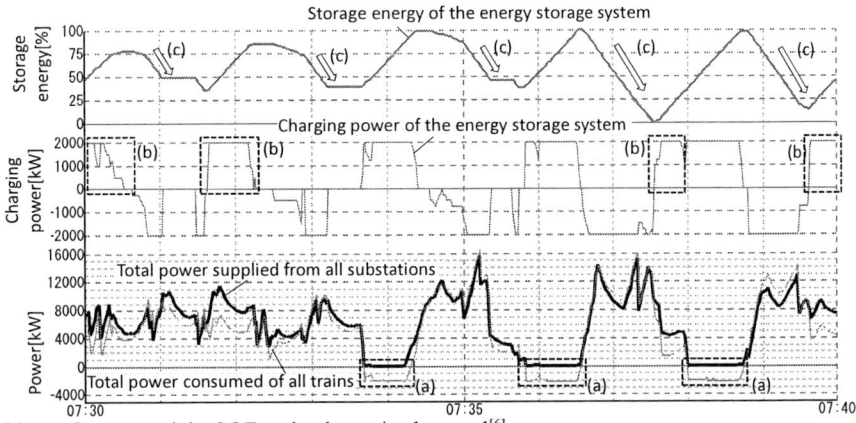

Fig. 9. The transitions of power and the SOE under the optimal control[6].

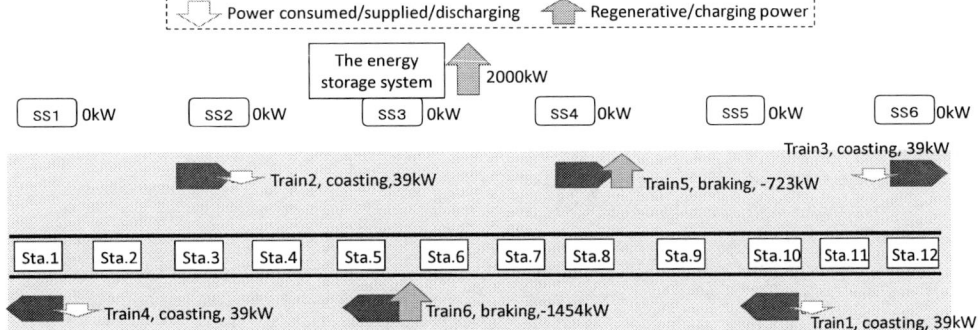

Fig. 10. The positions and power of trains and substations at 7:34:00[6].

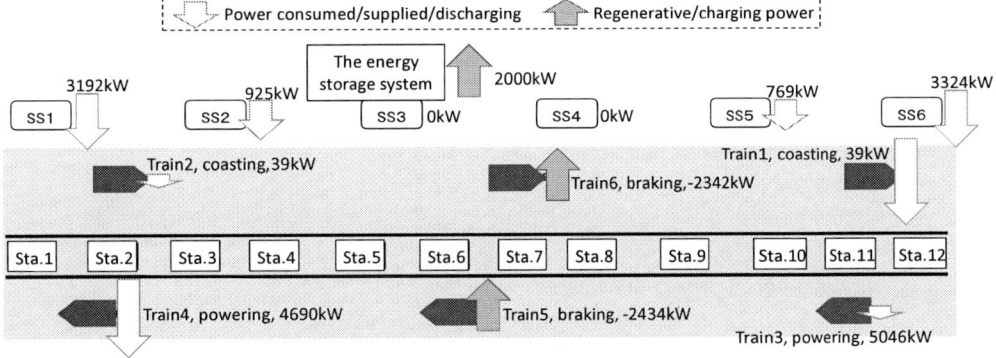

Fig. 11. The positions and power of trains and substations at 7:32:00[6].

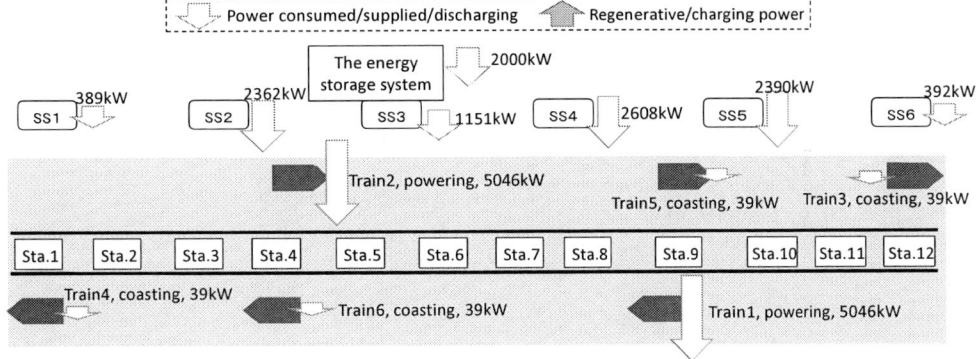

Fig. 12. The positions and power of trains and substations at 7:37:00[6].

3270

to outside temperature 15[°C], which we call "the case No.2".

The lower limit of the charging capacity of the energy storage system E_g^{stg} is set at 0[J] and the upper limit $\overline{E_g^{stg}}$ is varied according to 3 patterns: 10[kWh], 20[kWh] and 30[kWh]. For each upper limit, the absolute value of the maximum instantaneous charging power $\left|P_g^{stg}(t,1)\right|$ and discharging power $\left|P_g^{stg}(t,n_{max})\right|$ are set at the same value (it is called "the maximum instantaneous power") and is varied according to 12 patterns: from 500[kW] to 6000[kW] at 500[kW] intervals. Then the number of calculation patterns is $3 \times 12 = 36$ for each case.

We use a PC with Core i7-5960X (8core 16thread, 3.0~3.5GHz) and 64GB memory. The calculation time to obtain the optimal solution are around from 3[s] to 15[s].

The comparison between the total power supplied from all substations under each case is indicated in Fig.13 - 15. We confirm the inflection points beyond which the effect on energy saving is less increased even if the maximum instantaneous power is increased. Table II shows these inflection points for each case according to the charging capacities. In Table II, the inflection point is defined as the point where the improvement rate is less than 0.2% when the maximum instantaneous power is increased.

Railway companies should take care of the balance between the charging capacity and the maximum instantaneous power when installing the energy storage system. From this case study, we suggest that $\left|P_g^{stg}(t,1)\right| = \left|P_g^{stg}(t,n_{max})\right| = 1500[\text{kW}]$ when $\overline{E_g^{stg}}$ is 10[kWh], $\left|P_g^{stg}(t,1)\right| = \left|P_g^{stg}(t,n_{max})\right| = 2500[\text{kW}]$ or 3000[kW] when $\overline{E_g^{stg}}$ is 20[kWh], and $\left|P_g^{stg}(t,1)\right| = \left|P_g^{stg}(t,n_{max})\right| = 4000[\text{kW}]$ when $\overline{E_g^{stg}}$ is 30[kWh].

Therefore, we indicate that this mathematical optimization model can calculate the maximum effect on energy saving which is useful to design the energy storage system.

D. Confirmation of the Validily of the Mathmatical Optimaization Model

To confirm the validity of the mathematical optimization model, we designate the optimal sample control method of the energy storage system which is obtained by the mathematical optimization problem in the cases indicated in TABLE II, and calculate the total energy supplied from all substations using the Train Operation Power Simulator. As a result, the difference between the objective function value and the simulation result is from 0.1% to 1.8% in terms of relative error. For example, Figure 16 shows the comparison of the total energy supplied from all substations between the simulation result and the objective function value under the basic case. As a result, this mathematical optimization model is high accuracy.

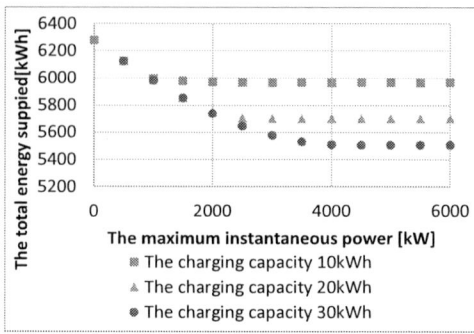

Fig. 13. The comparison of the calculation results obtained by varying the energy storage system performance under the basic case.

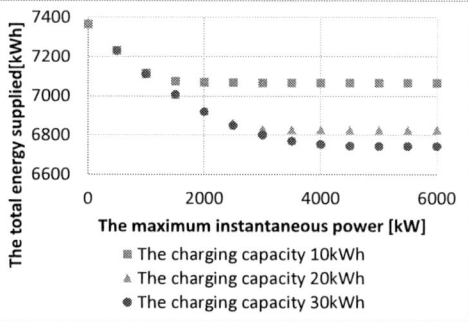

Fig. 14. The comparison of the calculation results obtained by varying the energy storage system performance under the case No.1.

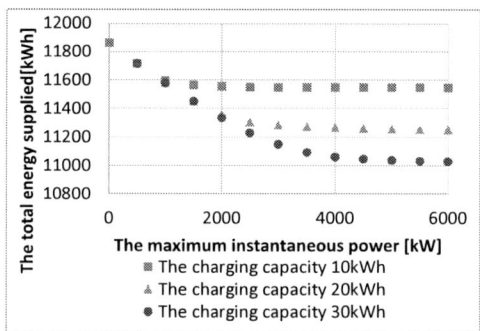

Fig. 15. The comparison of the calculation results obtained by varying the energy storage system performance under the case No.2.

TABLE II
INFLECTION POINTS OF FIG.13 TO FIG.15.

Charging capacity	Basic case	Case No.1	Case No.2
10kWh	1500kW	1500kW	1500kW
20kWh	2500kW	3000kW	2500kW
30kWh	4000kW	4000kW	4000kW

E. Estimation of the Effect of Energy Saveing

We calculate the total energy supplied from all substations in the absence of the energy storage system for each case using the Train Operation Power Simulator. Then, we compare these results with the simulation results of section VI.D and estimate the effect on energy saving when the energy storage system is installed and the optimal control method is applied. Figure 17 shows the effect on energy saving.

When the railway company installs the energy storage system with 30[kWh] and 4000[kW] in this sample line, the effect on energy saving is 12.2% under the basic case, as against 8.3% under the case No.1. There is a large difference between them, so that we should consider the auxiliary machine power to evaluate the effect on energy saving.

Under the case No.2, the high density train diagram causes less regenerative energy canceled, so that the effects on energy saving are about half as compared with the basic case.

VII. CONCLUSION

From a viewpoint of energy saving, the optimal control method of the energy storage system is a method of minimizing of the total energy supplied from all substations. The purpose of this paper is to find the optimal control method provided that all of the train operations are given. We produce the following results.

- The optimal control method of the energy storage system is modeled as the mathematical optimization problem in order to minimize the total energy supplied from all substations.

- We introduce some calculation results obtained by varying parameters. We confirm the inflection points beyond which the effect on energy saving is less increased even if the performance of the energy storage system is increased. This mathematical optimization model can calculate the maximum effect on energy saving which is useful to design the energy storage system.

- We confirm the validity of this mathematical optimization model using the Train Operation Power Simulator. The differences between the simulation result and the objective function value of this mathematical optimization model is from 0.1% to 1.8% in terms of relative error, so that this model is high accuracy.

- We calculate the effect on energy saving under some cases using the Train Operation Power Simulator. The effects on energy saving are within the range of 2.5% to 12.2%. They varies in accordance with not only the train density but also the auxiliary machine power.

This mathematical optimization model enables the estimation of the maximum effect on energy saving. In future, we will focus on the real-time control method of the energy storage system. We will try to develop another control method which is comparable with this optimal control method using the real-time train operation and rolling stock data.

ACKNOWLEDGMENT

Part of the development of the Train Operation Power Simulator was funded by a Railway Technology

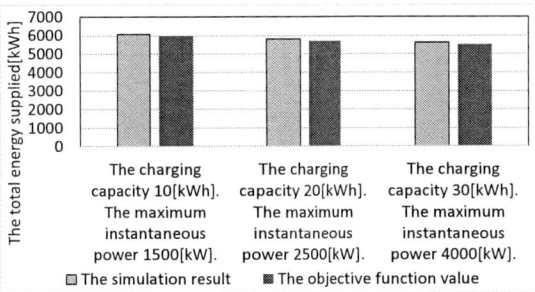

Fig. 16. The comparison of the total energy supplied from all stations between the simulation result and objective function value under the basic case.

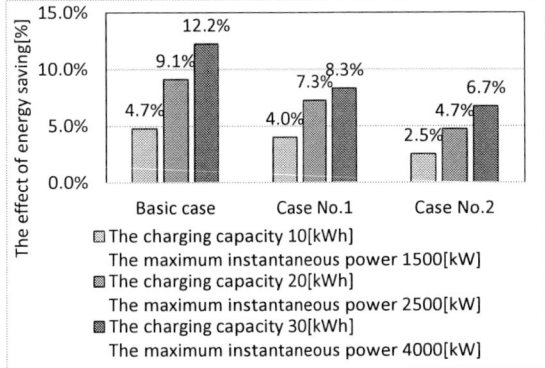

Fig. 17. The comparison of the experimental results obtained by varying the energy storage system performance.

Development Grant from the Ministry of Land, Infrastructure, Transport and Tourism.

REFERENCES

[1] Y. Takeuchi, T. Ogawa, H. Morimoto, Y. Imamura, S. Minobe and S. Sugimoto, "Development of a Train Operation Power Simulator Using the Interaction between the Power Supply Network, Rolling Stock Characteristics and Driving Patterns, as Conditions", *Quarterly Report of RTRI*, Vol.58, No.2, 2017.

[2] H. Konishi, T. Yoshii and H. Shigeeda, "Improvement of Control Method for Fixed Energy Storage System", *Quarterly Report of RTRI*, Vol.55, No.2, 2014.

[3] H. Kobayashi, S. Akita, T. Saito, K. Kondo, "A Voltage Basis Power Flow Control for Charging and Discharging Wayside Energy Storage Devices in the DC-electrified Railway System", *International Conference on Electrical Machines and Systems (ICEMS) 2016, Chiba, JAPAN*, 0458, 2016.

[4] Traction Energy Storage System with SCiB™ For DC Railway Power Supply Systems, https://www.toshiba.co.jp/sis/railwaysystem/en/event/inno-trans2016/pdf/c10TESS.pdf (2017/10/30 access)

[5] K. Minaminosono, M, Hashimoto and D, Yasukochi, "Simplification of Electric Substation System by Utilizing Energy Storage System", *IEE-Japan Industry Applications Society Conference, I.E.E. JAPAN*, 5-62, 2017 (in Japanese).

[6] Y. Takeuchi, T. Ogawa, K. Sato, H. Morimoto and T. Saito, "Energy storage system Control Algorithm for Minimization of Total Train Operation Energy", *Joint Technical Meeting on "Vehicle Technology" and "Transportation and Electric Railway"*, *I.E.E. JAPAN*, VT-17-024 / TER-17-059, 2017 (in Japanese).

[7] H. Kanno, T. Ogawa, S. Manabe, T. Takashige, Y. Imamura, S. Minobe, J. Kawamura, and M. Kageyama, "Effect of seasonal factor and train congestion on the auxiliary power", *J-RAIL2014, S3-3-3, Nigata, JAPAN*, 2014 (in Japanese).

The 2018 International Power Electronics Conference

Start-Up and Transient Operation of a Bidirectional Chopper With an Auxiliary Converter

Hamzeh J. Ahmad*, Haruna Ohnishi, and Makoto Hagiwara
Department of Electrical and Electronic Engineering, Tokyo Institute of Technology, Tokyo, Japan
*E-mail: ahmad.h.ab@m.titech.ac.jp

Abstract—The emerging applications of large-capacity battery-powered motor drive systems, such as those used in dc electric railways in the recent years, have required a dramatic reduction in terms of size and weight of a bidirectional chopper that exists in these drive systems. However, a conventional medium-voltage large-capacity bidirectional chopper is equipped with a heavy and bulky inductor for current smoothing, which would pose many practical challenges when it is installed in a moving train car. The authors of this paper have proposed a bidirectional chopper with an auxiliary converter connected in series with the inductor intended for reducing its size and weight. However, the proposed converter requires the capacitors of the auxiliary converter to be initially charged in order to start its operation. This paper proposes a method for initial charging of the capacitors used in the auxiliary converter without utilization of external charging circuitry. In addition, the auxiliary converter can be operated as a solid-state dc circuit breaker to protect the chopper when a short-circuit fault in the switches of the main converter occurs. This paper also discusses the transient performance of the proposed chopper and its associated control system under current and voltage disturbances. A 150-V, 2-kW downscaled model is used for experimental verification.

Keywords—battery energy storage system, dc circuit breaker, dc-dc converter.

I. INTRODUCTION

Battery energy storage systems (BESS's) have been used recently in the Japanese dc electric railways to power the rolling stock that runs on the non-electrified sections of the track, owing to their benefits in saving energy, minimizing noise and reducing maintenance when compared with the conventional diesel rail-cars [1]. Energy saving can be achieved by utilizing the regenerative braking operation to charge the batteries [2]. It is reported in [3] that the regenerative braking operation can reduce energy consumption by 26%. In order to permit a catenary-free operation, the BESS has to be installed on board, which in turn adds weight to the moving train car [4]. Furthermore, the installation of an on-board BESS permits the realization of pure electric braking [5].

The regenerative energy that is stored in the BESS can be used to support the dc grid, and this operation requires a converter that allows power flow in both directions (i.e., from battery to catenary and vice versa). Moreover, voltage levels at the dc catenary and the battery are different; therefore, a bidirectional chopper (i.e., non-isolated dc-dc converter) is used for voltage conversion (e.g., 1500 V to 600 V) [6]. Fig. 1 shows a typical circuit configuration of the conventional

bidirectional chopper used in dc electric railways [7], where v_{dc1} represents the catenary voltage (high-voltage side), v_{dc2} represents the battery voltage (low-voltage side), DCCB1 and DCCB2 are the high-voltage-side and low-voltage-side dc circuit breakers, respectively. The capacity of a bidirectional chopper can reach as high as 500 kW [6] [8], which makes the associated smoothing inductor heavy and bulky.

The authors of this paper have proposed a bidirectional chopper with an auxiliary converter (BCAC) for the purpose of reducing size and weight of the inductor [9]. It consists of a main converter, an auxiliary converter, a DCCB at the high-voltage side, and an inductor. The auxiliary converter consists of full-bridge cells that are connected in series, which works as an active power filter that absorbs the ac voltage produced by the main converter. It is reported in [9] that the inductance can be reduced to less than 1/60 compared with the required inductance for the conventional bidirectional chopper to achieve the same peak-to-peak current ripple. The validity of the chopper under steady-state operation has been confirmed by an experiment using a 150-V, 2-kW downscaled model.

The initial charging of the capacitors used in the auxiliary converter is indispensable for the operation. In other words, the converter cannot start up if the capacitors are not charged. Additional circuits for initial charging were used in [9]. However, using additional charging circuits can increase cost and complexity of the system. This paper proposes a method for achieving the initial charging of the capacitors used in the auxiliary converter. In this method, the initial charging from zero to the rated voltage is realized by adjusting the duty ratio of the main converter, thus requiring no external circuits.

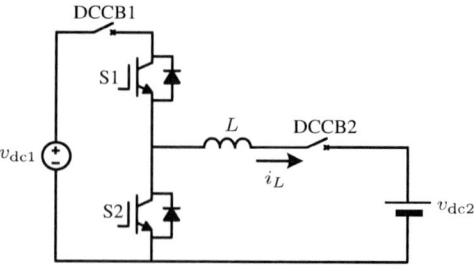

Fig. 1. Circuit configuration of a conventional bidirectional chopper for dc electric railways.

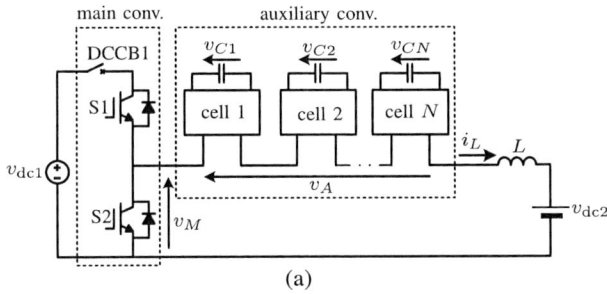

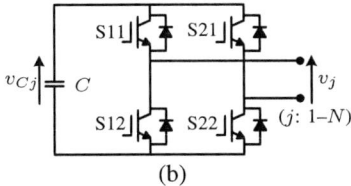

Fig. 2. Circuit configuration of a bidirectional chopper with an auxiliary converter (BCAC).

Beside that, the transient performance of the BCAC under voltage and current disturbances has not been discussed in [9]. This paper investigates the performance of the BCAC and its associated control system during transient operations including a short-circuit fault in the switches of the main converter.

II. CIRCUIT CONFIGURATION

As mentioned above, the BCAC comprises a conventional bidirectional chopper with an auxiliary converter that is connected in series with the smoothing inductor L. Fig. 2 shows the circuit configuration of the proposed bidirectional chopper, where S1 and S2 are the upper and lower switches of the main converter, respectively, v_{dc1} is the dc voltage at the high-voltage side, v_{dc2} is the dc voltage at the low-voltage side, v_M is the voltage produced by the main converter at its low-voltage side, v_{C1}, v_{C2}, and v_{CN} are the dc-capacitor voltages of the bridge cells where N indicates the number of cells in the auxiliary converter, v_A is the sum of the individual cell voltages.

III. PRINCIPLES OF OPERATION

A. Principles of Operation of the Main Converter

A voltage produced by the main converter at its low-voltage side, v_M, can be given by

$$v_M = \begin{cases} v_{dc1} & (S1: \text{on}, S2: \text{off}) \\ 0 & (S1: \text{off}, S2: \text{on}). \end{cases} \quad (1)$$

Here, v_M can be divided into two components, namely, an average dc component $(v_M)_{dc}$ and an ac component $(v_M)_{ac}$. The dc component can be approximated by:

$$(v_M)_{dc} = d v_{dc1}. \quad (2)$$

Here, d is the instantaneous duty ratio of S1, and (2) holds true when the time variation in d during a single carrier period is small enough. The ac component $(v_M)_{ac}$ is a square wave with a frequency of f_{SM}, and it can be calculated from (1) and (2) as

$$(v_M)_{ac} = \begin{cases} (1-d)v_{dc1} & (S1: \text{on}, S2: \text{off}) \\ -d v_{dc1} & (S1: \text{off}, S2: \text{on}). \end{cases} \quad (3)$$

B. Principles of Operation of the Auxiliary Converter

The ac component produced by the main converter, $(v_M)_{ac}$, generates a large amount of switching-ripple current when superimposed on L. Therefore, the auxiliary converter should produce an ac voltage $(v_A)_{ac}$ that is equal to $(v_M)_{ac}$ in order to cancel out the ac voltage component that produces the ripple current. Hence, $(v_A)_{ac}$ is expressed as

$$(v_A)_{ac} = \begin{cases} (1-d)v_{dc1} & (S1: \text{on}, S2: \text{off}) \\ -d v_{dc1} & (S1: \text{off}, S2: \text{on}). \end{cases} \quad (4)$$

In addition, the auxiliary converter can operate as a solid-state DCCB in case of a short-circuit fault occurrence in the switches of the main converter, as discussed in later sections. This can be realized by turning off all IGBTs of the auxiliary converter which in turn electrically disconnects v_{dc2} from the main converter.

IV. CONVERTER CONTROL

The operation of the BCAC can be classified into two modes; the active mode and the idle mode. In the active mode, the reference value of the inductor current is either positive or negative (i.e., $i_L^* \neq 0$), whereas in the idle mode this value equals zero (i.e., $i_L^* = 0$). The idle mode is required to assure that the voltage of each capacitor in the auxiliary converter is kept at its reference value when there is no power flow. Depending on the required mode of operation, either active mode or idle mode is selected.

A. Converter Control during Active Mode

In active mode, the control system of the main converter is responsible for controlling the power flow between v_{dc1} and v_{dc2}, whereas the control system of the auxiliary converter is responsible for producing $(v_A)_{ac}$ and controlling the voltages of the dc capacitors (i.e., v_{C1}, v_{C2}, and v_{CN}).

Fig. 3 shows the control block diagram of the main converter during an active mode, where the current control loop forces i_L to follow a reference current value, i_L^*, producing the duty ratio of S1, d. Here, v_{dc2} and v_B^* are used as feedforward control signals, in which v_B^* is the sum of voltage commands produced by the auxiliary converter for the dc-capacitor voltage control, as described later. v_B^* should be added for achieving decoupling between the main converter and the auxiliary converter in terms of control.

Fig. 4 shows the control block diagram of the auxiliary converter during the active mode, where the switching-ripple current mitigation and the dc-capacitor voltage control are included, whereas a block diagram for protection of the system

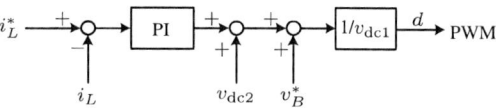

Fig. 3. Control block diagram of the main converter during an active mode.

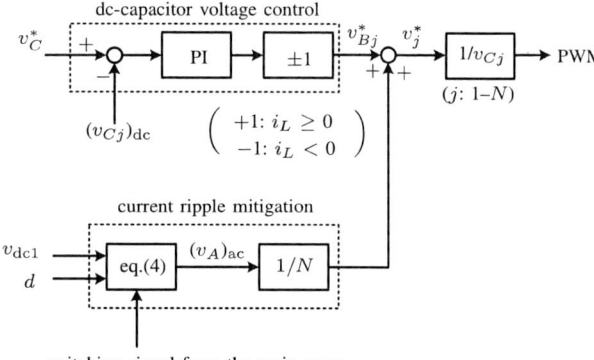

Fig. 4. Control block diagram of the auxiliary converter during an active mode.

is excluded for the convenience of explanation. The ripple current mitigation can be achieved by producing an ac voltage $(v_A)_{\mathrm{ac}}$ as expressed in (4) so that v_{dc1}, d, and the switching signal from the main converter should be used as input signals. Finally, each bridge cell produces an ac voltage $(v_A)_{\mathrm{ac}}/N$.

The dc-capacitor voltage control forces each dc-capacitor voltage $(v_{Cj})_{\mathrm{dc}}$ (j: 1–N) to follow its command value, v_C^*, producing the voltage command v_{Bj}^*. Here, $(v_{Cj})_{\mathrm{dc}}$ can be obtained by applying a moving-average filter with a frequency of f_{SM} to v_{Cj}. When the product of v_{Bj} ($= v_{Bj}^*$) and i_L (i.e., $v_{Bj}i_L$) is positive, a positive active power flows into the capacitor, thus $(v_{Cj})_{\mathrm{dc}}$ increases. On the other hand, when $v_{Bj}i_L$ is negative, a negative active power flows into the capacitor, which causes $(v_{Cj})_{\mathrm{dc}}$ to decrease. It should be noted that the polarity of v_{Bj}^* should be changed properly according to that of i_L as shown in Fig. 4 for achieving the above-mentioned aim.

Let the sum of voltage commands for dc-capacitor voltage control produced by each bridge cell be v_B^*. It is given by

$$v_B^* = \sum_{j=1}^{N} v_{Bj}^*, \qquad (5)$$

where v_B^* is a dc quantity as predicted from Fig. 4. This implies that v_B^* may cause a negative effect on the current control because it is also achieved by adjusting a dc quantity included in d. For achieving decoupling between the main and the auxiliary converters in terms of control, v_B^* should be added to the control block diagram of the main converter.

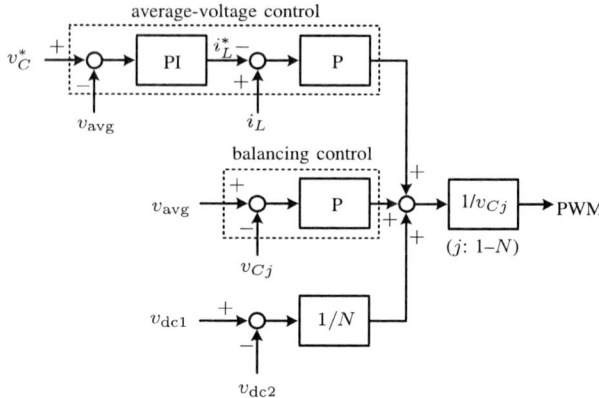

Fig. 5. Control block diagram of the auxiliary converter during an idle mode.

B. Converter Control during an Idle Mode

The idle mode of operation is the case where all capacitors of the auxiliary converter are charged, and the current flowing through the inductor is ideally zero. In order to achieve this mode, either S1 or S2 is kept turned on during the period of operation of this mode. If S1 is on, S2 is off, and i_L is appropriately controlled to zero, the voltage across the terminals of the auxiliary converter will ideally equal $v_{\mathrm{dc1}} - v_{\mathrm{dc2}}$, and $-v_{\mathrm{dc2}}$ if S1 is off and S2 is on, in steady state. Fig. 5 shows the control block diagram of the auxiliary converter during the idle mode when S1 is on and S2 is off. The average-voltage control forces the average value of the capacitor voltages, v_{avg}, to follow v_C^*, and i_L^* which ideally converges to zero when capacitor voltage error converges to zero (i.e., $v_C^* - v_{\mathrm{avg}} = 0$). The value of v_{avg} is given by

$$v_{\mathrm{avg}} = \frac{1}{N} \sum_{j=1}^{N} v_{Cj}, \qquad (6)$$

The balancing control is required to assure that each individual capacitor voltage equals v_{avg}. For a fast-response control, the quantity $(v_{\mathrm{dc1}} - v_{\mathrm{dc2}})/N$ is added to each converter.

V. CONVERTER STARTUP

In order to start up the BCAC, the dc-capacitors of the auxiliary converter must be initially charged. In this paper, an initial-charging method for the capacitors of the auxiliary converter is proposed, in which the capacitors are charged from either v_{dc1} (i.e., $i_L > 0$), or v_{dc2} (i.e., $i_L < 0$). In both cases, the charging process is accomplished by adjusting the switching states of the main and auxiliary converter switches, thus no additional circuits are required. If v_{dc1} is used for charging the capacitors, the main converter is operated as a buck chopper, where S1 is the active switch that controls the charging process and S2 is kept turned off. On the other hand, if v_{dc2} is used, S2 is the active switch and S1 is kept turned off, and the main converter operates as a boost chopper. At this point, all capacitors are charged sequentially in a process that

3275

The 2018 International Power Electronics Conference

TABLE I.   SWITCHING MODES FOR INITIAL CHARGING.

Mode	i_L	S11	S12	S21	S22
Charging mode	$i_L > 0$	on	off	off	on
	$i_L < 0$	off	on	on	off
Short-circuit mode 1		on	off	on	off
Short-circuit mode 2		off	on	off	on

can be called "sequential charging" such that one capacitor is charged at a time. This is achieved by operating one cell of the auxiliary converter cells in "charging mode" while the remaining cells are operated in "short-circuit mode". When the charging process of the capacitor corresponding to the cell is completed, the cell is operated in short-circuit mode, and next cell is operated in charging mode. The process is repeated until all capacitors are charged.

If a certain bridge cell is operated in charging mode and $i_L > 0$, then S11 and S22 are turned on while S12 and S21 are turned off. In this case, $v_j = v_{Cj}$ and the current flows into the positive terminal of the capacitor; as a result the capacitor voltage increases. In the same sense, if $i_L < 0$, then S12 and S21 are turned on while S11 and S22 are turned off. The short-circuit mode can be realized in two ways; short-circuit mode 1 and short-circuit mode 2 (as defined in Table I). In short-circuit mode 1, the switches of the positive-side arm are turned on and the switches of the negative-side arm are turned off. Short-circuit mode 2 is the opposite of short-circuit mode 1. In both modes, $v_j = 0$ and almost no current flows into the capacitor, which makes the voltage of the capacitor unchanged. Table I summarizes the switching modes of a bridge cell for the initial charging.

VI.   CIRCUIT BREAKER OPERATION

In the conventional bidirectional chopper shown in Fig. 1, a short-circuit occurrence in any or both of the switches (i.e., S1 and S2) causes the inductor current i_L to increase or decrease, and shorts the output terminals of v_{dc1} when both switches are closed. Therefore, it is necessary to install a DCCB at both the high-voltage side and the low-voltage side. However, the auxiliary converter in the proposed BCAC can interrupt the inductor current that is caused by a short-circuit fault in the switches of the main converter. This can be realized by turning off all the switches of the auxiliary converter; therefore, DCCB2 can be removed. However, DCCB1 remains indispensable.

The operation of the auxiliary converter as a DCCB holds the advantages of solid-state circuit breakers, such as faster fault clearance when compared to mechanical DCCBs. Moreover, replacement of a mechanical circuit breaker by the auxiliary converter solves the problem of arcing that is associated with mechanical circuit breakers; therefore, an arc extinguishing mechanism is no longer required.

In general, a short-circuit fault in S1 or S2 can be detected by monitoring the voltage and/or current of each switch, which in turn requires additional voltage and current sensors for fault

TABLE II.   CIRCUIT PARAMETERS OF A 150-V, 2-KW DOWN-SCALED MODEL USED FOR EXPERIMENTS.

Rated power	P	2 kW
DC voltage source 1	v_{dc1}	150 V/130 V/60 V
DC voltage source 2	v_{dc2}	95 V/75 V/30 V
Inductance at 0 Hz	L (0 Hz)	0.50 mH (12.6%*)
Inductance at 10.8 kHz	L (10.8 kHz)	0.34 mH (8.5%*)
Bridge-cell count	N	3
DC-capacitor voltage	V_C	45 V/15 V
Capacitance	C	3 mF
Unit capacitance constant [11]	H	4.6 ms
Carrier freq. (main conv.)	f_{SM}	450 Hz
Carrier freq. (auxiliary conv.)	f_{SA}	1800 Hz

*this value is based on a 2 kW, 150 V and 450 Hz base.

detection. In this study, a short-circuit fault is detected by monitoring i_L, and when it reaches a specified threshold value, the circuit breaker operation is triggered. The measurement of i_L is also required for control purposes; therefore, no additional sensors are required.

VII.   EXPERIMENTAL VERIFICATION

A. Experimental Circuit Configuration

Table II summarizes the circuit parameters that were used in the experiments. Each different value of v_{dc1}, v_{dc2} and V_C corresponds to a different experiment as explained in the following subsections. The high-voltage side voltage v_{dc1} and the low-voltage side voltage v_{dc2} of the BCAC were connected to the dc power supplies (DP030 RS) from NF Corporation. The unit capacitance constant H [11] is given by

$$H = \frac{\frac{1}{2}CV_C^2 N}{P}, \qquad (7)$$

where it is defined as the ratio of all electrostatic energy stored in the dc capacitors with respect to the rated active power. Hence, it has a unit of seconds. Substituting the circuit parameters in Table II ($V_C = 45$ V for normal operation) into (7) yields $H = 4.6$ ms, which is a practical value.

The carrier frequency of IGBTs used in the main converter was set as $f_{SM} = 450$ Hz, and the carrier frequency of phase-shifted PWM for bridge cells used in the auxiliary converter was set as $f_{SA} = 1800$ Hz. The inductance at 10.8 kHz (corresponds to $2Nf_{SA}$) is $L = 0.34$ mH, which corresponds to 8.5% based on P, V_{dc1}, and f_{SM}.

The experimental capacitor voltage waveforms were taken by using the Hioki Memory Hicorder 8861-50 with a frequency band of 200 kHz, where the sampling frequency is 1 MHz. The experimental inductor current waveforms were taken by Tektronix oscilloscope DP04104B-L with a frequency band of 1 GHz along with the Tektronix current probe TCP0030A with a frequency band of 120 MHz, where the sampling frequency is 1 MHz.

TABLE III. SEQUENCE OF INITIAL CHARGING.

Period	$t_0 \leq t < t_1$	$t_1 \leq t < t_2$	$t_2 \leq t < t_3$
Process	v_{C3} charging	v_{C2} charging	v_{C1} charging
Cell 1 Mode	short-circuit	short-circuit	charging
Cell 2 Mode	short-circuit	charging	short-circuit
Cell 3 Mode	charging	short-circuit	short-circuit

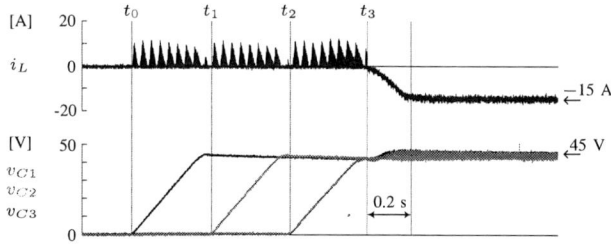

Fig. 6. Experimental waveforms during startup when $i_L^* = -15$ A.

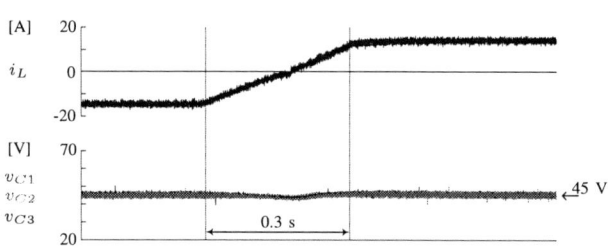

Fig. 7. Experimental waveforms during a ramp change in i_L^*.

B. Converter Operation During Startup

Fig. 6 shows the experimental waveforms of v_{C1}, v_{C2}, v_{C3} and i_L during the startup of the BCAC. In this experiment, V_{dc1} was used for the charging process as can be inferred from the positive direction of i_L during the charging period (i.e., $t_0 \leq t \leq t_3$). Before t_0, all capacitor voltages are set to zero, and at the instant t_0, the charging process starts in a sequence depicted in Table III , where the time instants t_1, t_2 and t_3 represent the end-of-charging instants of v_{C3}, v_{C2} and v_{C1}, respectively. When the initial charging finishes, either idle mode or active mode can be applied. In this experiment, the active mode was applied by switching from the initial charging control to the active mode control. A ramp signal that decreases from 0 to -15 in 0.2 s was selected as a current command signal. It can be shown from Fig. 6 that the initial charging and the switching between the initial charging control and active mode control are successfully achieved without overvoltages or overcurrents.

C. Transient Performance During a Ramp Change in i_L^*

Fig. 7 shows the experimental waveforms when i_L^* increases from -15 to 15 in 0.3 s. The power flow in both directions is achieved smoothly with a very small delay occurs

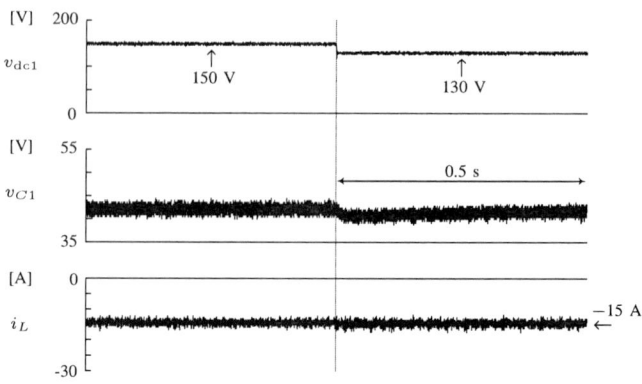

Fig. 8. Experimental waveforms during a step change in v_{dc1}.

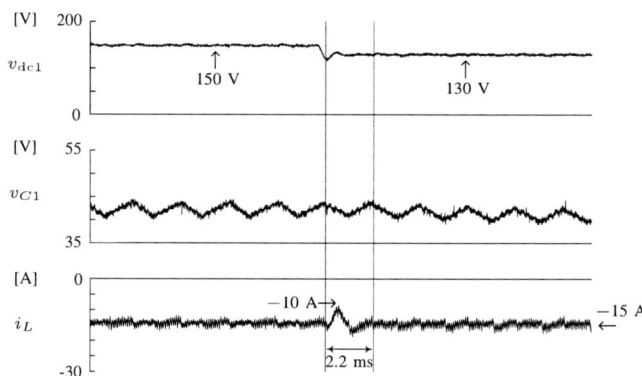

Fig. 9. Time-expanded waveforms of Fig. 8

when i_L approaches 0 A. This delay appears from the effect of the on-state voltage of the switching devices. This effect was partly compensated; therefore, its presence is not clear. It can be also observed that the voltages of all capacitors are very-well regulated around 45 V in steady state.

D. Transient Performance During a Step Change in v_{dc1} and v_{dc2}

In actual systems, catenary and battery voltages fluctuate greatly depending on the operating conditions. For example, if the nominal catenary voltage is 1500 V, the catenary voltage may fluctuate from 900 V to 1850 V. In particular, there is a possibility of a step change occurrence in the catenary voltage at sections that are far away from the substation. Moreover, if the nominal battery voltage is 690 V, the battery voltage may fluctuate from 560 V to 790 V. For this reason, the performance of the BCAC was examined under step changes in v_{dc1} and v_{dc2}. Although a step change in the battery voltage is unlikely to happen in a real operation, the effect of a step change in v_{dc2} was investigated to examine the performance of the BCAC under the worst-case conditions.

3277

The 2018 International Power Electronics Conference

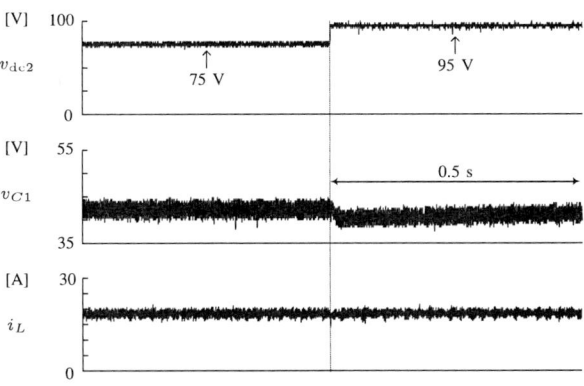

Fig. 10. Experimental waveforms during a step change in v_{dc2}.

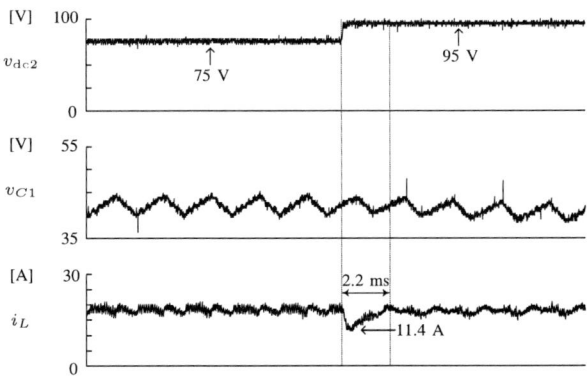

Fig. 11. Time-expanded waveforms of Fig. 10.

Fig. 8 shows the experimental waveforms during a step change in v_{dc1} from 150 V to 130 V. In this experiment, v_{dc2} = 75 V and $i_L = -15$ A. It can be shown that the dc component of v_{C1} during a transient period decreases by approximately 2 V, and then goes back to its value prior to the step change in about 0.5 s. The effect of the step change in v_{dc1} on i_L is barely noticeable from Fig. 8. A time-expanded version of Fig. 8 is shown in Fig. 9 in which a transient in i_L can be easily observed, where it increases from -15 A to -10 A and goes back to -15 A in about 2.2 ms (= 1/450 Hz). This transient is not severe and its period is sufficiently short. In order to study the robustness of the BCAC under fluctuations in the battery voltage, a step change in v_{dc2} was applied. Fig. 10 shows the experimental waveforms during a step change in v_{dc2} from 75 V to 95 V. In this experiment, $v_{dc1} = 150$ V and $i_L = 20$ A. The dc component of v_{C1} experiences a transient period in which a small variation in its value occurs, it recovers from the transient in about 0.5 s. In order to show the effect of this step change on i_L, a time-expanded version of Fig. 10 is shown in Fig. 11. During a transient, i_L decreases from 20 A to 11.4 A and and recovers from the transient after 2.2 ms.

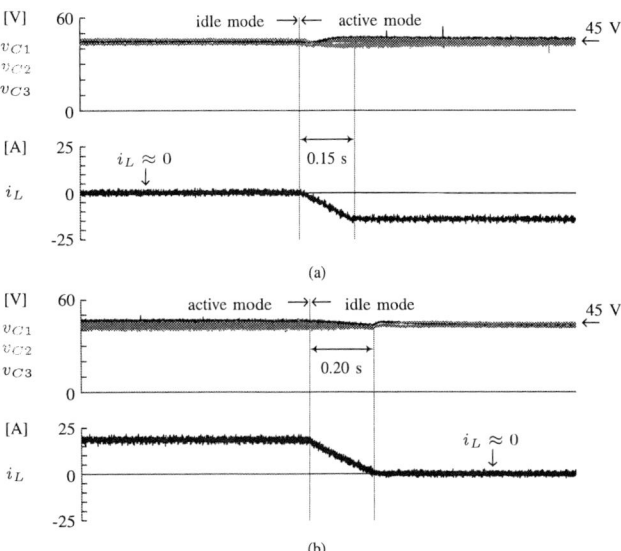

Fig. 12. Experimental waveforms of switch over: (a) From idle mode to active mode. (b) From active mode to idle mode.

This transient can be tolerated provided that it lasts for a short period of time. It should be mentioned that voltage spikes in waveform of v_{C1} that appear in Fig. 11 are not caused by the actual operation of the system, but by the measurement noise.

E. Experimental Verification of an Idle Mode and an Active Mode Controls

Depending on the desired operation, the BCAC can be operated in an active mode or an idle mode. For example, if the BCAC is installed in a dc electric train, it is operated in the active mode when the train is moving. However, when the train is stopped, it is operated in the idle mode. It is necessary to examine the BCAC operation during this transition from the active mode to the idle mode and vice versa. Fig. 12(a) shows the experimental waveforms of a switch over from the idle mode to the active mode. In this experiment, $v_{dc1} = 150$ V and $v_{dc2} = 75$ V. At the instant of the switch over, a ramp current reference i_L^* was applied, such that i_L^* decreases from 0 to -15 in 0.15 s. A transition from the active mode to the idle mode is shown in Fig. 12(b). In this case, i_L^* decreases from 20 A to 0 A in 0.2 s starting from the instant of the switch over. In both cases, it can be observed that the capacitor voltages are well-regulated before, after and at the instant of the switch over. The inductor current i_L follows its ramp reference without an overcurrent.

F. Experimental Verification of Circuit Breaker Operation

As previously mentioned, the auxiliary converter can be used as a solid-state dc circuit breaker when there is a fault in the main converter. In order to test the operation of the auxiliary converter as a circuit breaker, a fault condition can

The 2018 International Power Electronics Conference

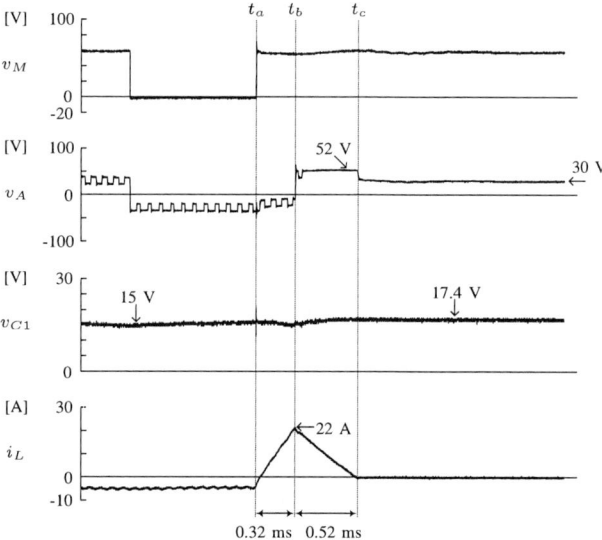

Fig. 13. Experimental waveforms during circuit breaker operation.

be simulated by keeping S1 turned on after a specified time instant. In such a condition, when S2 is turned on during its switching, terminals of v_{dc1} are shorted and DCCB1 disconnects v_{dc1} from the circuit. However, in this experiment, S2 was kept turned-off immediately after the fault occurrence. In this test, $v_{dc1} = 60$ V, $v_{dc2} = 30$ V, $v_{Cj} = 15$ V, and $i_L = -5$ A prior to fault occurrence.

Fig. 13 shows the experimental waveforms during the operation of the auxiliary converter as a circuit breaker. Time instants t_a, t_b and t_c represent fault-inception, fault-detection and fault-clearance time instants, respectively. At t_a, S1 is turned on and i_L starts to increase in a rate given by (8).

$$\frac{di_L}{dt} = \frac{1}{L}(v_{dc1} - v_{dc2} - v_A) \qquad (8)$$

When i_L reaches a threshold value (22 A in this experiment), all switches in the auxiliary converter are turned off at t_b. At this time, i_L starts to flow through the capacitors of the auxiliary converter via the free-wheeling diodes S11 and S22 of each cell. Therefore, capacitor voltages increase and (9) holds.

$$\frac{di_L}{dt} = \frac{1}{L}\left(v_{dc1} - v_{dc2} - \sum_{j=1}^{3} v_{Cj}\right) \qquad (9)$$

Substituting the system parameters in (9) yields a negative value, which means that i_L decreases. At t_c, i_L reaches zero, thus the voltage across the inductor becomes zero. At this condition, $v_A = v_{dc1} - v_{dc2}$, and i_L is interrupted. The fault is cleared by converting the total magnetic energy in the circuit into electrostatic and thermal energies. The high-speed operation of the auxiliary converter as dc circuit breaker is evident, as the time interval between the fault inception and the fault clearance is 0.84 ms as shown in Fig. 13.

VIII. CONCLUSION

This paper has presented a bidirectional chopper with an auxiliary converter (BCAC) that is capable of achieving a significant reduction in size and weight of the current-smoothing inductor. Since the BCAC is unable to start its operation when the capacitors of its auxiliary converter are not initially charged, an initial charging method for the capacitors of the auxiliary converter has been proposed in this paper. Besides that, the transient performance of the BCAC under severe disturbances was experimentally investigated. It has been shown that BCAC can handle severe current and voltage disturbances without interruption. Moreover, the operation of the auxiliary converter as a solid-state dc circuit breaker was demonstrated experimentally. It has been proven that the auxiliary converter of the BCAC can replace the mechanical dc circuit breaker that is equipped with the conventional bidirectional chopper at its low-voltage. The validity of the proposed initial-charging method, the robustness of the BCAC under severe disturbances, and the circuit breaker operation have been confirmed from the experimental results that were obtained by using a 150-V, 2-kW downscaled model.

REFERENCES

[1] Y. Nagaura, R. Oishi, M. Shimada, T. Kaneko, "Battery-powered Drive Systems: Latest Technologies and Outlook," *Hitachi Review.*, vol. 66, no. 2, pp. 138–144, 2017.

[2] A. Rufer, "Energy storage for railway systems, energy recovery and vehicle autonomy in Europe," *International Power Electronics Conference - ECCE ASIA*, Sapporo, 2010, pp. 3124-3127.

[3] M. Ayata et al., "Traction inverter system with Lithium-ion batteries for EMUs," *Proc. 17th Eur. Conf. Power Electron. Appl.*, Geneva, Switzerland, Sep., 2015, pp. 1–9.

[4] D. Ronanki, S. Singh, S. Williamson, "Comprehensive Topological Overview of Rolling Stock Architectures and Recent Trends in Electric Railway Traction Systems," *IEEE Trans. Transportation Electrification.*, vol. 3, no. 3, pp. 724–738, May. 2017.

[5] R. Takagi, and T. Amano, "Evaluating On-board Energy Storage System Using Multi-train Simulator RTSS," *The 4th International Conference on Railway Traction Systems (IET RTS)*, Birmingham, UK, April. 13–15, 2010, pp. 1–3.

[6] Z. Li, S. Hoshina, and M. Nogi, "Development of DC/DC Converter for Battery Energy Storage Supporting Railway DC Feeder Systems," *IEEE Trans. Ind. Appl.*, vol. 52, no. 5, pp. 4218–4224, Sep./Oct. 2016.

[7] "Energy Storage System for DC Electric Railways for TOBU RAILWAY CO., LTD.," *Toyodenki technical report.*, vol. 127, no. 3, pp. 52–55, 2013 (in Japanese).

[8] Z. Li, S. Hoshina, M. Nogi, and N. Satake, "Development of a Chopper for Regenerative Energy Storage System," *IEEJ, annual conf.*, vol. 1, no. 21, pp. I-123–I-126, 2014 (in Japanese).

[9] H. Ohnishi, and M. Hagiwara, "Experimental Verification of a Bidirectional Chopper for Battery Energy Storage Systems Capable of Reduction in Size and Weight of an Inductor," *Energy Conversion Congress and Exposition (ECCE)*, Cincinnati, Ohio, USA, Oct. 1–5, 2017, pp. 197–204.

[10] Z. Huang, S. Wong, and C. Tse, Fellow, "Design of a Single-Stage Inductive-Power-Transfer Converter for Efcient EV Battery Charging," *IEEE Trans. Vehicular Technology*, vol. 66, no. 7, pp. 5808–5821, July 2017.

[11] H. Fujita, S. Tominaga, and H. Akagi, "Analysis and Design of a DC Voltage-controlled Static Var Compensator Using Quad-series Voltage-source Inverters," *IEEE Trans. Ind. Appl.*, vol. 32, no. 4, pp.970–977, Jul./Aug. 1996.

3279

Experimental results of quasi-optimal charging current patterns to reduce the internal heat generation of the lithium-ion battery

Yoshiaki Taguchi[1], Gaku Yoshikawa[1]
1 Traction Control, Railway Technical Research Institute, Japan

Abstract— **Battery powered trains should finish the rapid charge in the scheduled stopping time. It is important to reduce the internal heat generation of the lithium-ion battery during the period of rapid charge, to prevent overheat of the battery. A quasi-optimal current pattern was derived by the developed simple search algorithm to reduce the internal heat generation. This algorithm requires less calculation cost, and is based on a third order equivalent circuit model of a test battery. Experimental results, under some ambient temperatures: 15°C, 25°C, 35°C, demonstrated that the quasi-optimal current pattern reduced 0.3-0.5% of the internal heat generation, and 0.4-0.9% of the temperature rise compared to those of the constant current pattern.**

Keywords— *Rapid charge, Quiasi-optimal current pattern, Internal loss reduction, Lithium-ion battery*

I. INTRODUCTION

In Japan, battery powered trains (contact-wire and lithium-ion battery hybrid trains) have been operating for commercial use since 2014 [1], [2]. As for the traction lithium-ion battery installed to such trains, suppressing the temperature rise is essential to ensure its longer lifetime. Especially, the temperature rise during rapid charge is a significant issue. We have developed the method for estimating the battery temperature easily by using a thermal network model [3], [4]. Improvement of the cooling performance is effective in the temperature reduction. However, the cooling device needs additional cost and energy.

So, we have studied on reducing the internal heat generation of the lithium-ion battery. It is clear that slower charge generates smaller internal heat generation. But, the slow charge isn't allowed when the battery powered train should finish recharging in the limited and scheduled stopping time. Therefore, we searched the effective charging current pattern under the conditions of same average current, charge time, as those of the conventional constant current pattern, aiming to reduce the internal heat generation.

Previously, various optimal charging method has been studied. Some method uses the low order (zero or first) battery equivalent circuit model [5]-[7]. In such studies, in most cases, charging is executed in wide range of SOC (state of charge), and the charging time is 1 h or more with small charging current. Under this charging condition, it is important for optimization to consider the slow change of the internal resistance or capacitance, due to the SOC and/or temperature variation. Therefore, low-order model will be sufficient to estimate the voltage response with practical accuracy.

In the case of battery powered train in Japan [1], [2], the rapid charge is executed with small ΔSOC (change in SOC). Generally, ΔSOC is designed under 40%, to achieve long lifetime of the battery, and to remain enough energy used during unexpected long-time stopping without charging from outside. In many cases, the charging time for battery powered train will not exceed 20 min. As for Series BEC819, typical rapid charging time is approx. 5 min. For the optimization of such a rapid charge, it is important to consider both the fast and slow transient responses of the battery voltage with practical accuracy. Consequently, second or more-order model will be necessary, which contains 2 or more time constants of voltage response.

The study of Ref. [8] deals with the second order equivalent circuit model. The optimized charging current pattern is found by MATLAB-based GPOPS (general pseudo-spectral optimal control software). This requires high calculation cost, hence, the application may be limited.

So the authors developed a simple and low-cost method searching for an optimal charging current pattern, which is effective for the rapid charge of battery powered trains.

In this paper, a developed algorithm for searching an optimal charging current pattern is described. By using the obtained current pattern, some effects are experimentally demonstrated, which is reducing the internal heat generation and suppressing the temperature rise of a test battery module.

II. BATTERY MODEL IDENTIFICATION

A. Tested battery and its application

Table 1 shows the specification of the tested lithium-ion battery. Ratings of cylindrical cell is 3.7V-75Ah. One

TABLE 1	
SPECIFICATIONS OF THE TESTED LITHIUM-ION BATTERY	
Positive electrode active material	LMO
Rated cell voltage	3.7 V
Rated cell capacity	75 Ah
Continuous rating of maximum current	225A
Module ratings	22.2 V-75 Ah (6 cells series connected)
Module weight	Approx. 23 kg

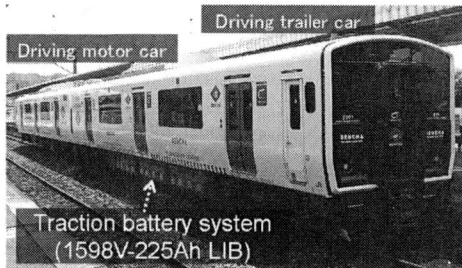

Fig. 1. Overview of the battery powered train, series BEC819.

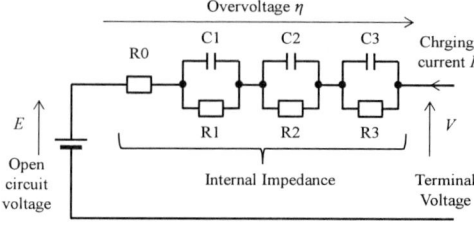

Fig. 2. Employed equivalent circuit model for the test module.

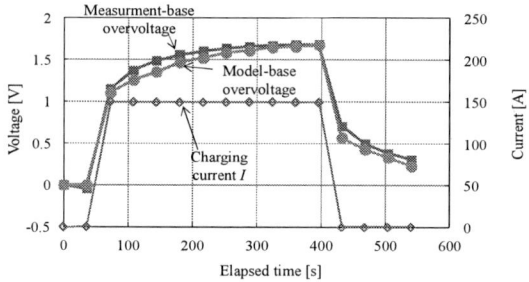

Fig. 3. Waveforms for identification and validation of the equivalent circuit model.

module consists of series-connected 6 cells. In this study, only one module is used for charging tests.

This module is installed to railway vehicles, series BEC819, etc. Series BEC819 is secondly commercialized battery-powered train in Japan, as shown in Fig. 1. This train uses both AC-fed catenary and onboard battery. The ratings of the onboard battery system is 1598V-225Ah. Series BEC819 was developed by Kyushu Railway Company (JR Kyushu), Hitachi Ltd., and Railway Technical Research Institute (RTRI) in 2016. Revenue operation of this train was started on a non-electrified Wakamatsu line in JR Kyushu, in Oct. 2016.

B. Equivalent circuit model

Fig. 2 shows a third order battery equivalent circuit model, contains 3 parallel RC elements, employed for the test module. In former researches, zero to second order models are commonly used [5]-[8]. In this study, we set the model order at 3 for the purpose of small estimation error and stable estimation of the transient voltages.

C. Calculation of measurement-base overvoltage

In this equivalent circuit, overvoltage η is defined as

$$\eta = V - E \tag{1}$$

where, E is open circuit voltage and V is terminal voltage. E is measured only when a value of charging current I is almost zero for long term, at least several hours. Therefore, E is estimated in the charging period, as following sequence.

1) Before charging, current I has been kept to almost zero for 7 h. Hence, the open circuit voltage E is regarded as same as V, which is easily measured.

2) Initial SOC (State of charge) is derived by E/SOC conversion table.
3) While charging, latest SOC is calculated from integrated charging current I and initial SOC.
4) Latest SOC is converted to latest E by SOC/E conversion table.

After E is estimated, overvoltage η is calculated by (1). These calculations is based on measured values of V and I. So, hereafter, we define this overvoltage η as measurement-base overvoltage.

D. Identification of internal impedance

Internal impedance model of the battery should be identified to estimate the overvoltage considering any kind of charging current pattern.

Waveforms for the model identification are shown in Fig. 3. This data is a result of the 150A-360s CC-charging, where the ambient temperature of the battery is kept at 25°C. Both charging current I and measurement-base overvoltage η are measured by 10-Hz sampling, and averaged by each period of 36s. This 36-s period is determined to separate the charging time in 10 steps.

We employ the pulse transfer function to represent the dynamics of the internal impedance, as in (2), where the sampling period is 36s.

$$\frac{\eta(z)}{I(z)} = \frac{b_0 + b_1 z^{-1} + b_2 z^{-2} + b_3 z^{-3}}{a_0 + a_1 z^{-1} + a_2 z^{-2} + a_3 z^{-3}} \tag{2}$$

The order number of (2) is equal to that of the assumed equivalent circuit, shown in Fig. 1. Parameters a_x, b_x are identified by adopting the least squared method to the measured data of I and η. And we assumed that: each

3281

parameter is constant during rapid charge, and independent from current patterns. This assumption is able to be applied because the charging time is short, and the SOC variation and temperature rise are small, and their difference among current patterns are also small, under charging conditions in this paper.

Fig. 3 shows the waveforms of estimated overvoltage η, by using (2) and input data of measured I. Hereafter, this estimated overvoltage is represented as model-base overvoltage.

From the comparison of the measurement-base and model-base overvoltage, it is found that maximum error between them is 10%. It is future task to reduce the estimation error.

III. SIMPLE SEARCH ALGORITHM

A. Calculation of internal heat generation

The objective of this study is reducing the internal heat generation during rapid charge. When considering the rapid charge, the main cause of heat generation is Joule loss and other causes are ignorable. In Fig. 4, the Joule loss W_J is represented as a colored area, calculated as (3).

$$W_J = \int \eta dQ = \int \eta I dt \tag{3}$$

Equation (3) is discretized as

$$W_J = \sum_i \{\eta(i)I(i)T\} \tag{4}$$

where T represents the sampling period of 36s, and i is index number from 1 to 10.

B. Searching algorithm

We developed a simple algorithm that search for a current pattern to reduce the internal heat generation. The developed algorithm is shown as a flowchart in Fig. 5.

On the step 1, initial current pattern is set to 150-A conventional constant current pattern (CCCP). This is similar to the typical charging pattern of series BEC819, whose CC charge is at 150 A/module, and charging time is approx. 5 min. Though charging pattern of BEC819 contains constant-voltage period, it is not considered in this paper.

On the step 2 and 3 perturbation pulse trains are generated, whose amplitude is set to initial value of 5A. Waveform of the perturbation pulse trains are shown in Fig. 6 (a). As shown here, the pulse train have 3 values of $-\Delta I$, 0, ΔI. The number of the pulse train is originally $3^{10}=59049$. After choosing ones whose average value is zero, the number becomes 8953.

On the step 4 and 5, model-base overvoltage is calculated, as is described in section 2 D, where input current pattern is I_{opt} plus perturbation pulse train δI. Then the Joule loss is calculated by (4). This is repeated until the number j reaches 8953.

On the step 7, temporally optimal current pattern $I_{opt}(k)$ is determined as $I_{opt}(k-1)+\delta I(j)$ which minimized the Joule

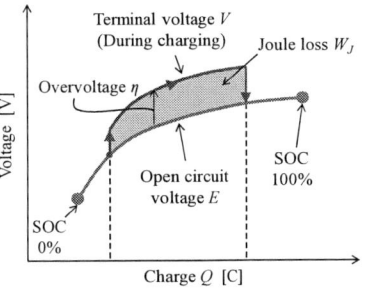

Fig. 4. Conceptual diagram for calculating Joule loss.

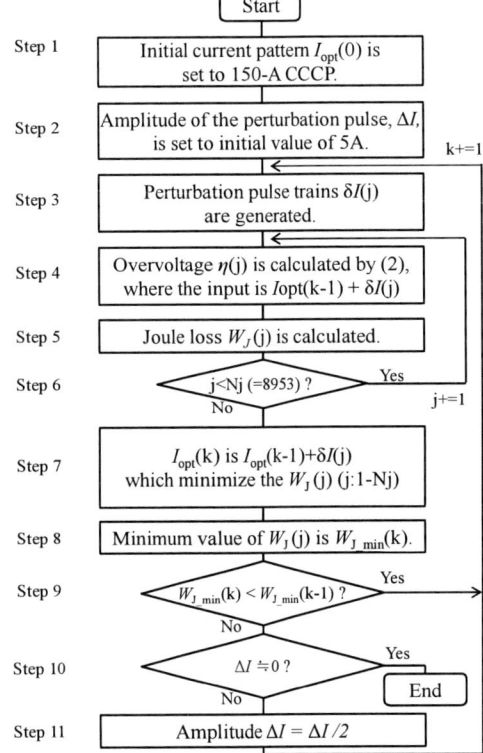

Fig. 5. Flowchart of the developed simple search algorithm.

loss W_J among all of $W_J(j)$. Examples of the step 4 to 7 are shown in Fig. 6 (b), (c).

On the step 8, minimum value of W_J is stored as $W_{J\_min}(k)$.

On the step 9, if W_{J_min} becomes less than former value, index k is incremented by 1, and the step number returns to 3.

On the step 10 and 11, if the value of ΔI is enough small, the pattern $I_{opt}(k)$ is regarded as the final optimal current pattern. Otherwise, the amplitude ΔI is divided by 2, then the index k counts up, and the step number returns to 3.

Thus, the temporally optimal current pattern becomes convergent on a final current pattern.

3282

C. Searched quasi-optimal current pattern

Fig. 7 shows the final current pattern, defined as quasi-optimal current pattern (QOCP). Since the developed algorithm does not assure the precise optimality, we add "quasi" to the name of obtained pattern.

The values of average charging current and charging amount are same between CCCP and QOCP.

The profile of the QOCP is high value at the beginning and the end of the charging, and is symmetrical. In the previous study [8], under some conditions, similar current

profile, excepting symmetry, is reported as an optimal pattern aiming for loss minimization.

IV. EXPERIMENTAL RESULTS

A. Condition of the experiment

We executed the experiment to evaluate the effectiveness of the searched quasi-optimal current pattern (QOCP). Experiment procedure is as follows;

1) Charge at an average current of 150 A for 6 minutes,
2) Rest for 30 minutes,
3) Discharge at a constant current of 37.5A for 24 minutes,
4) Rest for 7 hours.

Procedures 1)-4) are repeated for comparative patterns, which consist one cycle. In each cycle, conventional

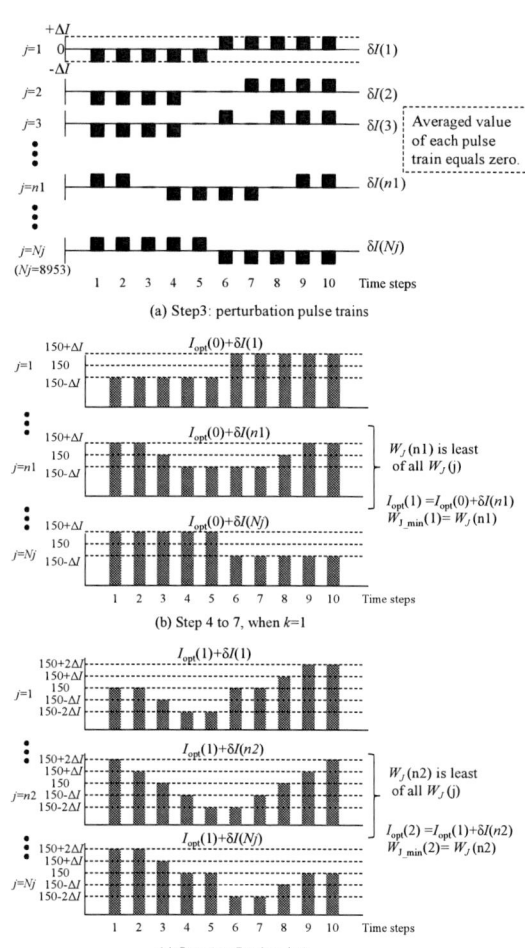

Fig. 6. Typical pulse trains in the searching algorithm.

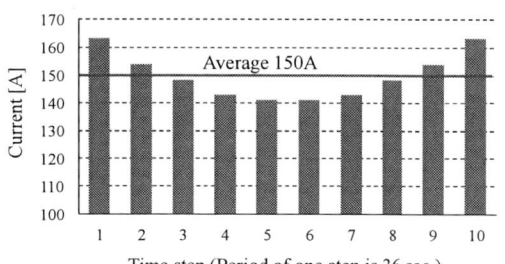

Fig. 7. Searched quasi-optimal current pattern.

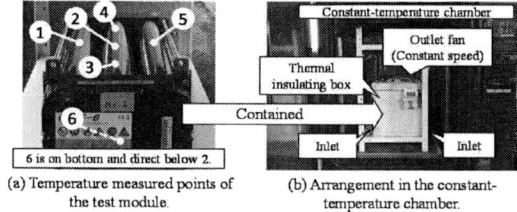

(a) Temperature measured points of the test module.   (b) Arrangement in the constant-temperature chamber.

Fig. 8. Arrangement of experimental devices.

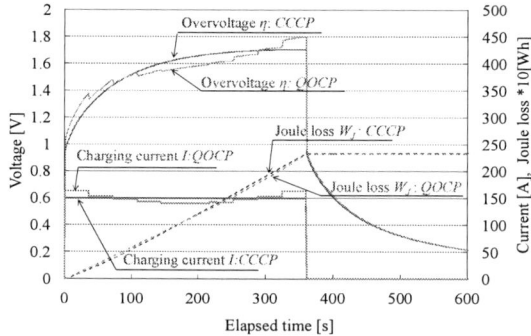

Fig. 9. An example of electrical data from the evaluation experiment (2nd cycle in Table 3).

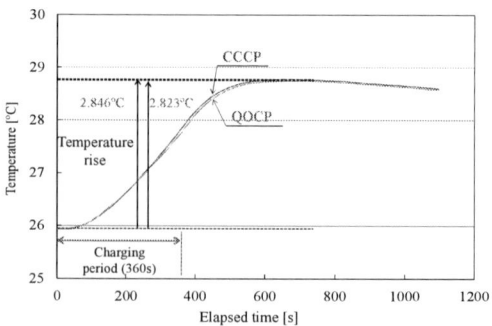

Fig. 10. An example of thermal data from the evaluation experiment (2nd cycle in Table 3).

TABLE 2

EXPERIMENTAL RESULTS UNDER THE AMBIENT TEMPERATURE OF 15°C

Test cycle	Measurement-base Joule loss			Temperature rise		
	CCCP [Wh]	QOCP [Wh]	Ratio [%] (QOCP/CCCP)	CCCP [°C]	QOCP [°C]	Ratio [%] (QOCP/CCCP)
2	30.05	29.98	99.7	3.808	3.788	99.5
3	30.11	30.09	99.9	3.811	3.788	99.4
4	30.36	30.34	99.9	3.833	3.840	100.2
5	30.50	30.30	99.3	3.872	3.841	99.2
6	30.42	30.35	99.7	3.867	3.850	99.6
Average	**30.35**	**30.27**	**99.7**	**3.838**	**3.821**	**99.6**
Average reduction ratio			**0.3**	**Average reduction ratio**		**0.4**

TABLE 3

EXPERIMENTAL RESULTS UNDER THE AMBIENT TEMPERATURE OF 25°C

Test cycle	Measurement-base Joule loss			Temperature rise		
	CCCP [Wh]	QOCP [Wh]	Ratio [%] (QOCP/CCCP)	CCCP [°C]	QOCP [°C]	Ratio [%] (QOCP/CCCP)
2	23.37	23.27	99.6	2.846	2.823	99.2
3	23.34	23.23	99.5	2.846	2.842	99.8
4	23.30	23.20	99.6	2.851	2.848	99.9
5	23.26	23.16	99.6	2.881	2.854	99.0
6	23.23	23.13	99.5	2.891	2.885	99.8
Average	**23.30**	**23.20**	**99.5**	**2.863**	**2.850**	**99.6**
Average reduction ratio			**0.5**	**Average reduction ratio**		**0.4**

TABLE 4

EXPERIMENTAL RESULTS UNDER THE AMBIENT TEMPERATURE OF 35°C

Test cycle	Measurement-base Joule loss			Temperature rise		
	CCCP [Wh]	QOCP [Wh]	Ratio [%] (QOCP/CCCP)	CCCP [°C]	QOCP [°C]	Ratio [%] (QOCP/CCCP)
2	19.02	18.96	99.7	2.211	2.198	99.4
3	19.03	18.95	99.6	2.230	2.225	99.8
4	19.06	19.00	99.7	2.234	2.193	98.2
5	19.17	19.09	99.6	2.195	2.167	98.8
6	19.26	19.21	99.7	2.173	2.159	99.3
Average	**19.11**	**19.04**	**99.6**	**2.208**	**2.188**	**99.1**
Average reduction ratio			**0.4**	**Average reduction ratio**		**0.9**

constant current pattern (CCCP) and the searched quasi-optimal current pattern (QOCP) are executed alternatively. And several cycles are repeated to be evaluated in average.

Fig. 8 shows the arrangement of the experimental devices. The measured points of the battery temperature are shown in Fig. 8(a). Hereafter, we use the average temperature of 6 points for evaluation. The test battery module is contained in the thermal insulating box, and the box is mounted inside the constant-temperature chamber, indicated in Fig. 8(b). The thermal insulating box has ventilating outlet fans, which operate in the constant power and stabilize the cooling condition.

B. Experimental results and consideration

An example of experimental waveforms of current, voltage, and Joule loss are shown in Fig. 9. And thermal waveforms are shown in Fig. 10. Both figures are obtained in a same test cycle.

Left side of each Table 2-4 shows comparison of Joule loss. This measurement-base Joule loss is derived from the measurement-base overvoltage and measured current equation (3). QOCP reduced Joule losses by 0.3%, 0.5%, 0.4% compared to the CCCP on average, when the ambient temperatures are 15°C, 25°C, 35°C respectively.

Right side of each Table 2-4 shows comparison of temperature rise of the test battery. QOCP reduced temperature rises by 0.4%, 0.4%, 0.9% compared to the CCCP on average, when the ambient temperatures are 15°C, 25°C, 35°C respectively.

Reduction ratios of Joule loss were similar value of those of temperature rise, in the conditions of 15°C, 25°C. This implies that the main cause of the temperature reduction is the Joule loss reduction.

In this paper, we employed the same QOCP, which is searched based on the 25-°C battery model, for all ambient temperatures. Nevertheless, close value of reduction ratios are obtained among 3 ambient temperatures. This is desirable advantage for practical use. However, the battery equivalent circuit parameters are affected by the battery temperature and SOC. Therefore, there remains a possibility that some better QOCP is searched by considering the effect of battery temperature and SOC. Searching for better QOCP is one of the future tasks.

V. CONCLUSION

A quasi-optimal current pattern was derived by the developed simple search algorithm, to reduce the internal heat generation during rapid charge. Experimental results, under some ambient temperatures: 15°C, 25°C, 35°C, demonstrated that the quasi-optimal current pattern reduced the measurement-base internal heat generation by 0.3-0.5%, and the temperature rise by 0.4-0.9% compared to those of the constant current pattern.

Since the reduction ratio of the temperature rise is small, this quasi-optimal current pattern (QOCP) will be an ancillary measure for preventing overheat. On the other hand, the summation of saved energy through many times of rapid charge will not be ignorable.

The developed searching algorithm requires little calculation cost. Moreover, charging controller may handle the QOCP with small additional cost. Therefore, the developed searching algorithm and QOCP will potentially provide a fine cost effectiveness for reducing the internal heat generation of the traction lithium-ion battery.

REFERENCES

[1] Y. Kono, N. Shiraki, H. Yokoyama, and R. Furuta, "Catenary and storage battery hybrid system for electric railcar series EV-E301," International Power Electronics Conference (IPEC-Hiroshima 2014 - ECCE-ASIA), 2014, pp.2120-2125 (2014)

[2] Y. Taguchi, et.al, "Development of AC Electric Train Boarding High-Capacity Traction Battery," No. 507, 11th WCRR 2016, Milan, Italy, (2016)

[3] Y. Taguchi, A. Terada, M. Miki, K. Hatakeda, T. Kimura, "Evaluation of a thermal network model for the traction battery of the battery-powered EMU," Vehicle Power and Propulsion Conference (IEEE-VPPC2015), DI-2-1 (2015)

[4] G. Yoshikawa, Y. Taguchi, "Development of a parameter identification method for the thermal circuit model of a lithium-ion battery installed on a battery-powered EMU," Vehicle Power and Propulsion Conference (IEEE-VPPC2017), RT5.1 (2017)

[5] Xiasong Hu, Shengbo Li, Huei peng, Fengchun Sun, "Charging time and loss optimization for LiNMC and LiFePO4 batteries based on equivalent circuit models," J. Power Sources, Vol. 239, pp. 449-457 (2013)

[6] Zheng Chen, Bing Sia, Chunting Chris Mi, Rui Xiong, "Loss-Minimization-Based Charging Strategy for Lithium-ion Battery," IEEE Trans. on Industry Applications, Vol. 51, No.5, pp.4121-4129 (2015)

[7] A. Abdollahi, X. Han, G.V. Avvari, N. Raghunathan, B. Balasingam, K.R. Pattipati, Y. Bar-Shalom, "Optimal battery charging, Part I: Minimizing time-to-charge, energy loss, and temperature rise for OCV-resistance battery model," J. Power Sources, Vol. 303, pp.388-398 (2016)

[8] E. Inoa, J. Wang, "PHEV Charging Strategies for Maximized Energy Saving," IEEE Trans. on Vehicular Technology, Vol. 60, No.7, pp.2978-2986 (2011)

Development of Test Methods and Evaluation Results for 500kV HVDC Converter

Keisuke Hattori[1*], Asuka Ohtake[1], Takayoshi Kamejima[2] and Haruhisa Wada[3]

1 Toshiba Mitsubishi-Electric Industrial Systems Corporation, Tokyo, Japan
2 Toshiba Energy Systems & Solutions Corporation, Kanagawa, Japan
3 Toshiba Corporation, Kanagawa, Japan

Abstract- **An ultra-high-voltage converter rated at 500kV-1200A has been developed for HVDC transmission, a key technology for enhancing operability of electric power systems, a basic infrastructure of the sustainable society. The paper introduces the development of test methods for HVDC converter, the ultra-high voltage power electronics. One is for insulation test performed in the high voltage laboratory and the other is for open line test at site. The converter insulation test was performed successfully with the developed test method. The test method developed for open line test was proven to be feasible by the simulator tests.**

Keywords- HVDC converter, Quadruple valve, Insulation test, Open line test

I. INTRODUCTION

For realizing sustainable electric power systems, renewable generations are increasing rapidly around the world. One of the issues to increase the renewables is the power transmission line from remote renewable power stations to load centers. Sometimes, the AC transmission lines are already heavily loaded. In such case, to add the power flow from large renewables may result in overload the existing AC lines and may result in the AC network instability. When long submarine cables are necessary for transmission, the AC technology cannot send the power effectively. For these cases, the HVDC technology provides solutions. Namely, the HVDC can adjust the power flow by the converter control. It also can send power through very long cables.

This paper introduces a converter rated at 500kV-1200A developed for a submarine cable system, where power is transmitted from remote hydraulic power stations. Section II introduces the system and the converter briefly. Section III describes the development of test method for ultra-high voltage and Section IV describes the insulation test results. Section V describes the back ground of the open line test. Section VI describes the control principle of the open line test, Section VII describes the development of the control and Section VIII describes the evaluation of the developed control by the simulator tests.

II. HVDC TRANSMISSION SYSTEM AND CONVERTER STRUCTURE

A typical HVDC transmission system rated at +/- 500kV-1200A is shown in Fig.1. The converter converts the power from AC to DC or vice versa at the ends of the DC transmission line. In Fig. 1, the transmission line is made of DC cable. Usually, the system consists of two sets of converters and transmission lines. A set of converters and a transmission line is called a pole. Fig. 1 has 2 poles and then the system is called a bipolar system.

The 500kV-1200A converter is the 12-pulse converter and its circuit topology is shown in Fig. 2. In HVDC application, a special mechanical structure is taken considering high voltage application. The four valves in the same column in Fig. 2 are included in a mechanical structure called a quadruple valve. Fig. 3 shows the mechanical structure.

The quadruple valve is suspended from the ceiling of the building. The terminal at the top is connected to the neutral line which is connected to the ground since the top terminal is the nearest to the ground potential in the mechanical structure. On the contrary, the bottom terminal is connected to the DC transmission line through the DC reactor. Then, the highest voltage for insulation tests should be applied to the bottom of the quadruple valve.

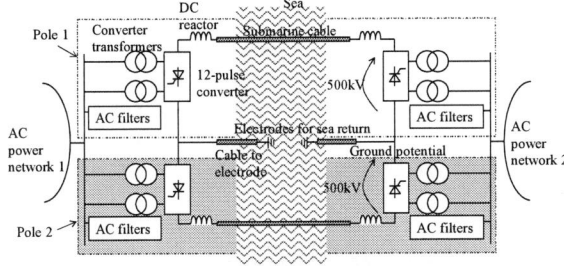

Fig. 1. Typical bipolar HVDC system rated at +/-500kV-1200A.

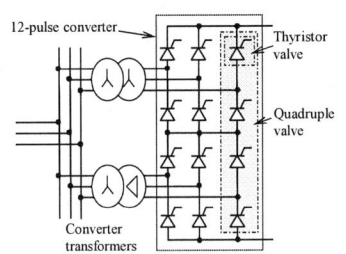

Fig. 2. Converter Circuit Topology for HVDC.

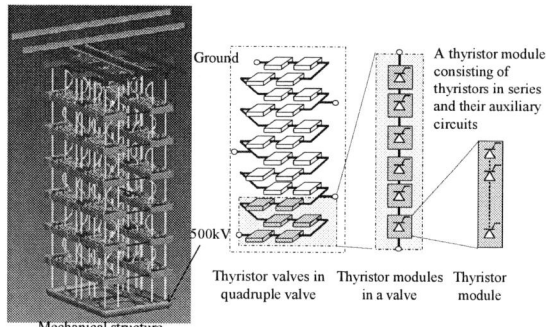

Fig. 3. Quadruple valve mechanical structure and electrical connection.

III. TEST METHOD DEVELOPMENT FOR HIGH VOLTAGE INSULATION TESTS

The electrical tests of the quadruple valve were performed according to the standard IEC 60700-1[1] in the High Voltage Laboratory at Hamakawasaki Factory. The insulation tests for the valves are performed after the thyristor module tests. The quadruple valve is then assembled with the tested modules[2].

A conceptual layout for the insulation tests is shown in Fig. 4. The photograph of the actual test set-up is shown in Fig. 5. In order to make the test condition to be equivalent to the insulation distances in the actual valve building, the distances from the quadruple valve to the supporting structure and to the floor is designed to be equivalent to or more conservative than the actual distances in the valve building. The distance between the 500kV shield of the quadruple valve and the floor was simulated precisely since the highest test voltages are applied between them. The distance between the quadruple valve and the wall is also simulated by the supporting structure.

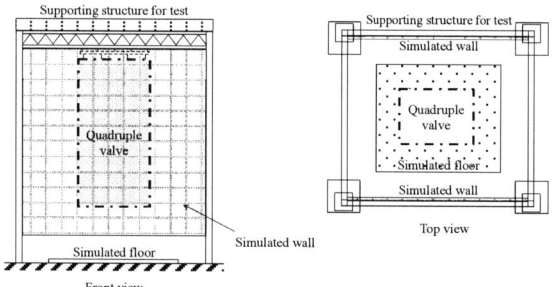

Fig. 4. Layout of developed supporting structure for test.

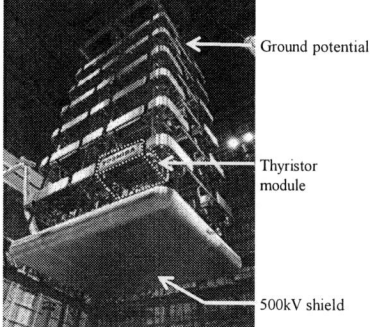

Fig. 5. Quadruple valve under test.

Some of the HVDC valve tests are different from those for the conventional AC equipment as listed below.
a) DC voltage test
b) Insulation tests between valve terminals
c) Non-periodical turn-on test, a dynamic test to turn on the valve while a high voltage is applied.

For item a), the laboratory developed a DC voltage generator which can output DC voltage more than 1000 kV. Considering item b), the power capacity of the test voltage generators should be carefully designed.

In the test of item c), a gate pulse is required to turn on the valve while the switching impulse voltage is applied. Then, a test method is developed to synchronize the switching impulse and the gate signal. Furthermore, in the test, a current of very high di/dt flows in the test circuit. In order to avoid unnecessary voltage rise at the ground potential of the valve, the test ground was carefully designed to minimize the parasitic inductance.

IV. INSULATION TEST RESULTS

The tests were performed successfully. Major test results are summarized in Table 1 for the developed quadruple valve. In the standard, it is called as the MVU (Multiple Valve Unit) tests. The test voltages in Table 1 are the values for the standard environmental conditions. When each test is performed, the test voltages are adjusted considering actual temperature, humidity and atmosphere pressure according to the IEC 60060-1[3] for high voltage test techniques.

Fig. 6 shows the test circuit of the impulse test for the MVU. The positive and negative impulse voltages were applied between the high voltage DC terminal and the earth. The thyristor valve for U1 phase was short-circuited to simulate the actual operation where one of thyristor valves is conducting and the voltage is applied to the other three valves. Fig.7, 8 and 9 show the switching impulse voltage waveforms, lightning impulse voltage waveforms and steep front impulse voltage waveforms for the MVU, respectively. The waveforms are defined by the IEC 60060-1 and IEC 60700-1. It was confirmed the MVU withstood each impulse voltage.

The tests for a valve are also performed. In this case, the tests voltages are applied across the anode terminal and the cathode terminal of the valve. In the tests for the valve, the non-periodical turn-on test is the special test different from the tests for conventional equipment. Fig. 10 shows the test circuit of the non-periodical turn-on test. In this test, thyristor valve is checked to withstand the current and voltage stresses at turn-on under the high voltage conditions. A capacitor was connected in parallel to the thyristor valve in order to simulate the surge current commutated from the arrester.

It is a dynamic test and its test waveform is shown in Fig. 11 as an example. The switching impulse voltage was applied to the X2 phase thyristor valve. The X2 phase thyristor valve was fired at the crest of the impulse voltage by the gate pulse from the Valve Base Electronics (VBE) which generates the gate pulse and detects the thyristor valve failure. When the thyristor valve turns on, the discharging current from the capacitor flows into the

The 2018 International Power Electronics Conference

thyristor valve. The thyristor valve successfully turned on and withstood the high di/dt current and voltage stresses in the test.

TABLE I
TEST RESULTS OF INSULATION TESTS ON MVU

Test item	Test voltages	Test results
DC voltage withstand test	+/- 888 kV DC	Successful
AC voltage withstand test	668 kV AC	Successful
Switching impulse withstand test	+/- 1172 kVp	Successful
Lightning impulse withstand test	+/- 1168 kVp	Successful
Steep front impulse withstand test	+/- 1219 kVp	Successful

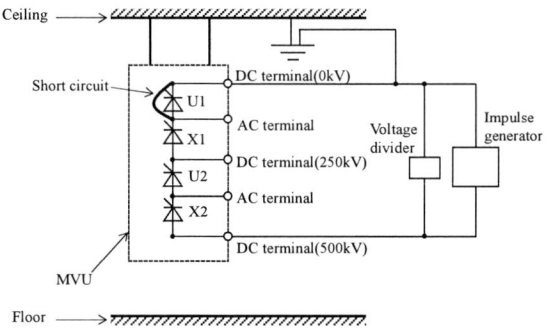

Fig. 6. Test circuit of the impulse test.

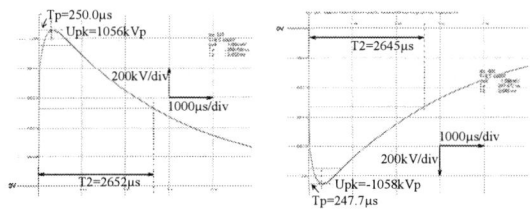

Fig. 7. Waveforms of the switching impulse withstand test.

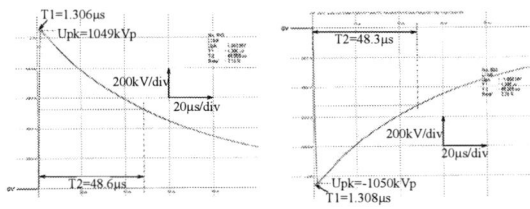

Fig. 8. Waveforms of the lightning impulse withstand test.

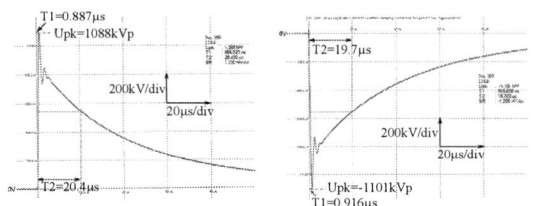

Fig. 9. Waveforms of the steep front impulse withstand test.

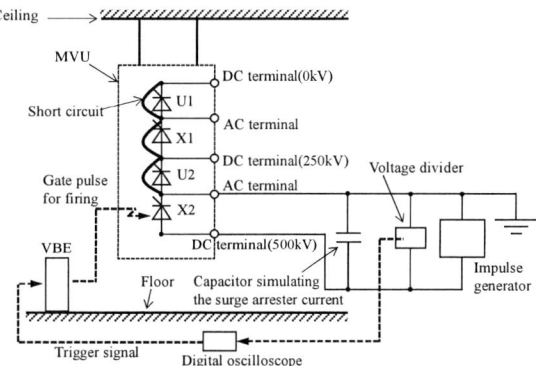

Fig. 10. Test circuit of the non-periodical turn-on test.

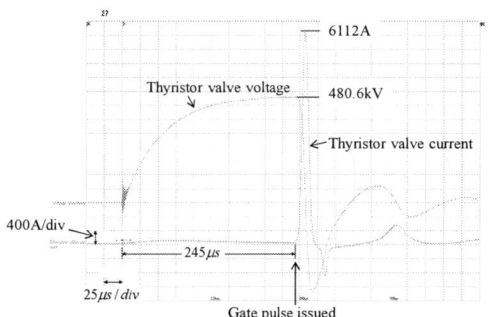

Fig. 11. Waveform of the non-periodical turn-on test.

V. CONTROL METHOD DEVELOPMENT FOR OPEN LINE TEST

At site after the thyristor valves and other components are installed, the energization tests are performed to check the insulation performance of the AC and DC circuits before the operation tests as shown in Fig. 12. The test procedure at site is described in IEC 61975 for HVDC system[4]. For the AC circuits, the energization test can be performed by closing the circuit breakers which connect the converter station to the AC power network.

In contrast, for the DC circuits, no voltage can be applied by just closing the circuit breakers. The converter is required to supply the DC voltage to the DC circuits. In the test, only the DC reactor and the DC lines are connected. Namely, the DC line is open. Then, the test is called as "Open line test."

However, the test includes high risks. If some defects in the DC circuit are hidden and not found before the test, the defects will result in DC short circuit. Then, the converter may suffer from the overcurrent if the converter operates with the conventional DC voltage control.

Then, a control method has been developed in order to perform the open line test with safety procedure. The developed control method is introduced in the next section.

3288

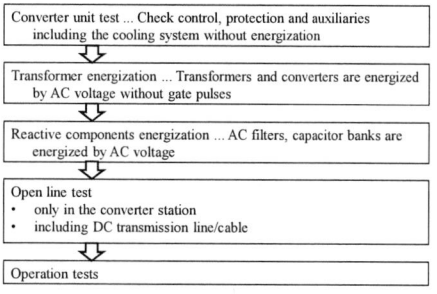

Fig. 12. Typical HVDC test procedure at site for energization tests.

VI. PRINCIPLE OF CONTROL METHOD FOR OPEN LINE TEST

The control method is different from that for ordinary converter operation since it is not based on the conventional theory of line commutated converters. The method is based on the idea, controlling the charging current from the AC inputs of the converter to the capacitances in the DC circuit. The charging is done with the current pulses made each time when gate signals are issued to the converter valves. With this special intermittent operation, the converter can operate in the control angle range of the inverter operation region, greater than 90 degrees, and can also generate the DC voltage required to the open line test.

Because of the control angle range, the control method offers advantage that the converter does not generate large short circuit current even if the DC circuit is short circuited. Then, this control method is suitable for the open line test at site from viewpoint of safety test procedure.

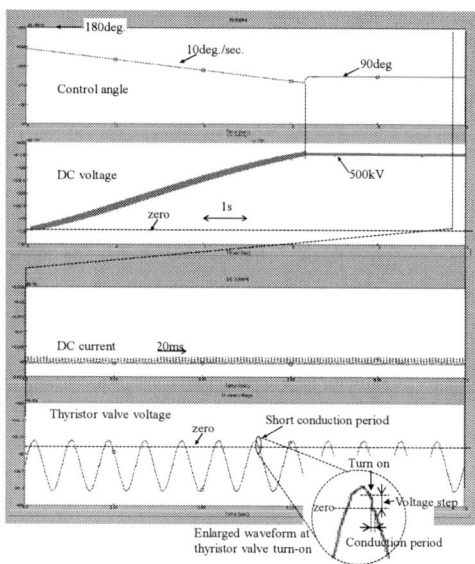

Fig. 13. Simulation of control principle for open line test.

Firstly, the control method was developed through the digital simulation with detailed model of the converter including capacitances in the DC circuit. A simple simulation at the early stage of the development is shown in Fig. 13. It is noted that, in this simulation, only the capacitances and a leakage resistance are connected to the DC circuit. From the simple simulation, it is observed that the DC voltage is established by the intermittent DC current, a train of current pulses.

The DC voltage is kept almost constant, as the capacitance in the DC circuit is charged by train of current pulses. In the steady state, the mean value of the current pulses is considered to be equal to the current discharging through the leakage resistance.

VII. CONTROL METHOD DEVELOPMENT CONSIDERING ACTUAL SITE CONDITIONS

After the basic idea of the control is established, the detailed control method was developed considering the following factors;
a) DC voltage balance among four valve groups in the 12-pulse converter,
b) suitable voltage range of the thyristor valve at the instant of turn-on by the gate pulses and
c) capacitance value change in the DC circuit along with the test procedure[5].
The details of the factor c) are described below.

At the first step of the open line test, only the converter and the DC reactor are connected with the DC bus line in the converter building. Then, the capacitance of the DC circuit is very small. In the next step, the DC cable is connected. When the cable is very long, several hundreds of km, the capacitance value could reach very large, more than tens of micro-Farads.

The capacitance of the DC circuit depends on the DC circuit condition whether the DC cable is connected to the DC circuit or not. Therefore the characteristics of charge and discharge of the DC circuit are different between the first step and the second step. Firstly, the control method for the DC circuit including considerably large capacitance is described below.

As described in the previous section, at the open line test operation, the capacitance of the DC circuit is charged by current pulses made each time when gate signals are issued to the converter valves. The amounts of current depend on the voltage steps when the valves turn on. In other words, the charging speed of capacitance depends on the voltage step. The voltage step is shown in enlarged view of Fig.13. In addition, the voltage steps influence the dv/dt stresses of the thyristor valves. For the above reasons, the voltage step should be kept at an appropriate value in the open line test operation.

If the DC voltage balances among four valve groups in the 12-pulse converter as mentioned in the beginning of this section, the voltage step depends on the DC voltage, the AC voltage and the gate timing. In order to keep the voltage step at the appropriate value, the control method which estimates the gate timing from the DC voltage and the AC voltage was developed. Of course, if the voltage steps were detected by measurement system, such as voltage sensor, the voltage step would be controlled directly. However, such voltage step detection system is not practical to be implemented in the HVDC converter. Therefore, the control method to estimate from the DC voltage and the AC voltage is evaluated to be reasonable.

At the open line test, the DC circuit is charged until the

3289

DC voltage reaches to the reference DC voltage value by the above control method. If converter continued this operation although the DC voltage reached the reference value, the DC voltage would exceed the reference value. In order to hold the DC voltage around the reference value, the converter GB (Gate-Block) logic was added to the control method. The converter GB logic compares the DC voltage and the reference value. When the DC voltage exceeds the reference value, the logic will block the gate signals for the converter. During GB period, the capacitance of the DC circuit discharges via leakage resistance, and the DC voltage falls down. And, when the converter GB logic detects that the DC voltage falls less than the reference value, the logic will DEB (De-Block) the gate signals for the converter. Thus, the DC voltage is held around the reference value by repetition of DEB and GB.

If the converter GB logic just repeats DEB and GB, the counts of valve turn-on are different among 12 valves. The imbalance of the turn-on counts changes the DC voltage balances among four valve groups. In order to prevent the imbalance of the turn-on counts among valves, the periods of DEB and GB are managed to be increments of a cycle. Namely if the AC system frequency is 50Hz, the periods are managed to be increments of 20ms. The control method for the DC circuit with DC cable, including considerably large capacitance, is summarized in Fig. 14.

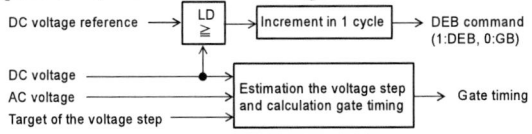

Fig. 14. Open line test control method with DC cable.

Secondly, the other control method was developed for the DC circuit whose capacitance is very small. However, the same basic theory is applied as that for the DC circuit with DC cable. The voltage step is also kept at the appropriate value by the control method. However, because the capacitance is very small, the DC voltage rises a lot by just a single current pulse made by a thyristor valve turn-on. If all of the 12 valves turn on each cycle, the DC voltage rises quickly, and it is difficult to control the DC voltage level. In order to avoid this quick charge, the other gate control method was developed. In this method, only 4 of 12 valves turn on each cycle. When the converter is controlled by this method, the charging speed is 1 of 3 of the first control method.

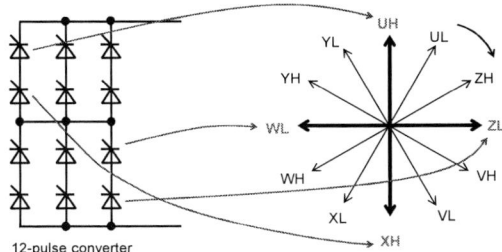

12-pulse converter

Fig. 15. 4 of 12 phases in active at small capacitance case.

Therefore, it is easy to control the DC voltage. The valves turned on in a cycle should be chosen in order to balance the DC voltage among four valve groups. Thus, the combinations of 4 valves are chosen as shown in Fig. 15. In Fig. 15, UH, XH, ZL and WL are chosen for the valves turned on in a cycle. However, the control method can choose not only this combination, but also the other combinations where the phases of the valves are shifted 90 degrees each other.

In the case of small capacitance, it is difficult to keep the DC voltage around the reference value with the repetition of DEB and GB in increments of a cycle, because the ripple of the DC voltage tends to be large. Thus, an auxiliary control function is added to the second control method. The auxiliary control function adjusts the target value of the voltage step of the DC voltage depending on the difference between the DC voltage and the reference value. When the DC voltage approaches the reference value, the method adjusts the voltage step, in other words, the amplitude of current pulse is adjusted. Then, the charging current at turn-on and the discharging current through the leakage resistance can be precisely balanced. And, the DC voltage is held around the reference value. The control method for the DC circuit with small capacitance is summarized in Fig. 16.

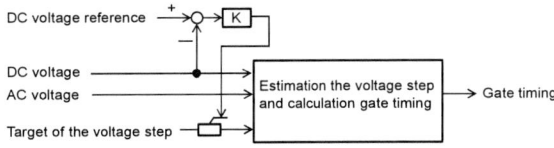

Fig. 16. Open line test control method without DC cable.

As described above, two different control methods for two different capacitance values were developed. These control methods can be implemented in a converter control system and be chosen according to the DC circuit conditions.

VIII. CONTROL PERFORMANCE EVALUATION FOR OPEN LINE TEST

The developed control method was installed in actual digital controllers. The control performance was evaluated with a digital real time simulator. In the simulator, the detailed converter models were prepared and they were controlled by the pulses issued from the actual controllers. The simulator test configuration is shown in Fig. 17.

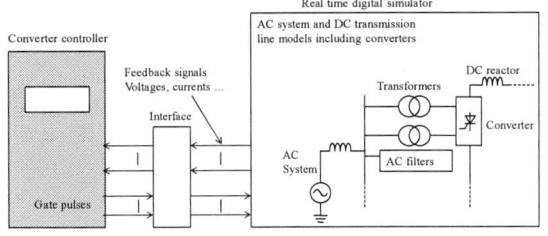

Fig. 17. Simulator test configuration for control performance evaluation.

3290

One of test results is shown in Fig. 18 (a). In this case, the DC cable is connected to the DC circuit. The key point is the voltage step when the valves turn on. Every voltage step is controlled to be almost equal in the range for the thyristor valves to turn on appropriately. Then, each current pulse to the DC circuit is also kept to be almost equal. This behavior keeps the DC voltage balance almost equal among the thyristor valves.

In the simulator test, it takes around 400 seconds to establish the DC voltage to the reference value which is 500kV in Fig. 18 (a). In the simulator model, the leakage resistance in the DC line is also modeled. Then, the DC voltage waveform shows a curve similar to exponential waveform. When the DC voltage reaches the reference value, the controller changes the control mode to keep the DC voltage to be constant by issuing the gate pulses discontinuously.

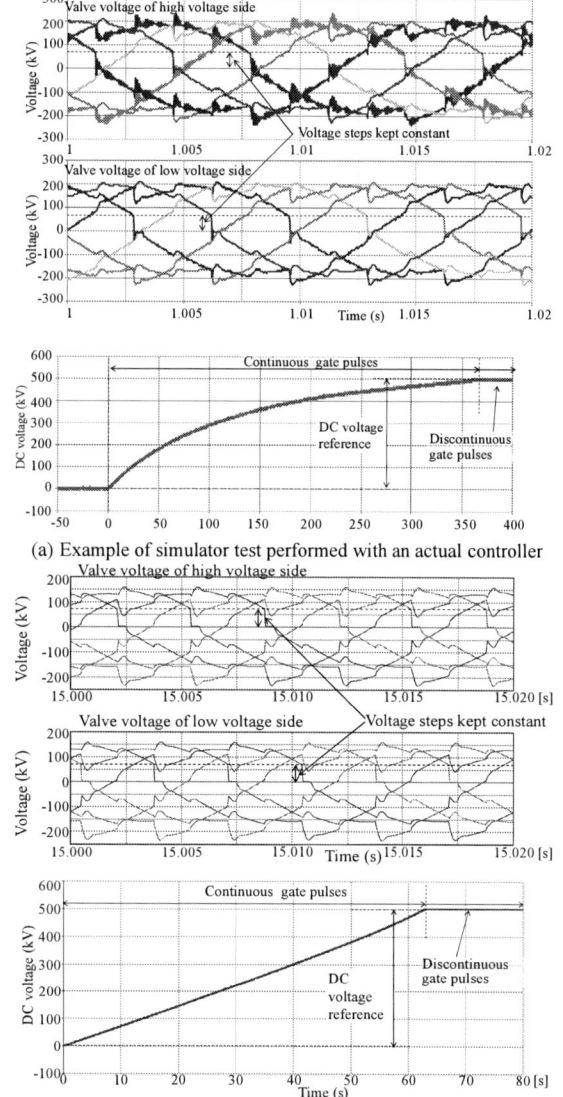

(a) Example of simulator test performed with an actual controller

(b) Example of off-line simulation with analysis tool

Fig. 18. Example of simulator test result and simulation result for case with DC cable.

The digital simulation results are also shown in Fig. 18(b) for comparison. The test results with the actual controller are in good agreement with the simulation results concerning the thyristor valve waveforms which have equal voltage step at turn-on. However, the DC voltage waveform is different because the capacitance of the DC circuit is set smaller in order to reduce the calculation time for simulation. However, the control mode change is confirmed also in the simulation.

The cases for small capacitance without the DC cable were also evaluated in the simulator test and the simulation. In these cases, the voltage steps were adjusted and the DC voltage was kept nearly equal to the reference value. Thus, the second control method was also confirmed to show good control performance as designed.

IX. CONCLUSION

The HVDC transmission has been required for better power network operability. For such purpose, reliable HVDC converters and the controllers are necessary. Then, the test methods to evaluate the converter and the controller performances are also important and essential. From such viewpoints, the paper introduced some of test methods and test results of the thyristor valve and the controller developed for 500kV HVDC transmission. The introduced tests are the special ones for high voltage power application different from the general power electronics applications.

The ultra-high voltage tests were performed to evaluate the quadruple valve insulation design and the performance at the turn-on from very high voltage. For such ultra-high voltage tests in the test laboratory, the test methods were developed and successful test results were obtained.

The insulation test for DC circuit at site requires another test method development. Based on the special intermittent operation of the converter, the control method for the open line test was developed and evaluated successfully through the simulator tests.

HVDC technology is now required furthermore to build future power networks for transmitting power from sustainable generations all over the world. The authors believe that this paper contributes for realizing the future HVDC networks.

REFERENCES

[1] IEC 60700-1, "Thyristor valves for high voltage direct current (HVDC) power transmission - Part 1: Electrical testing"

[2] T. Kamejima et.al, "Development of Suspended-type HVDC Thyristor Valve," *General Meeting of IEE of Japan*, 4-026, 2016.

[3] IEC 60060-1, "High-voltage test techniques - Part 1: General definitions and test requirements"

[4] IEC 61975, "High-voltage direct current (HVDC) installations - System tests"

[5] K. Hattori et.al, "Control Method of Open Line Test for HVDC System," *General Meeting of IEE of Japan*, 4-063, 2016.

3291

Dissipation Loop for Shoot-Through Faults in HVDC Converter Cells

Keijo Jacobs, Staffan Norrga, Hans-Peter Nee
KTH Royal Institute of Technology, EECS, Stockholm, Sweden
E-mail: keijoj@kth.se

Abstract—**Converter cells for HVDC applications store large amounts of energy. This energy might be dissipated in a very short time in case of a shoot-through fault. Measures to avoid shoot-through or handle the extreme currents during a fault and prevent damage from neighboring components are essential to ensure a continued operation of the converter. With future high-voltage silicon carbide semiconductors, cell voltages can be increased leading to higher stored energy per cell. In cells with thyristor-based semiconductors, e.g. IGCTs, a di/dt reactor may have to be employed. This paper presents a method to handle the dissipated energy during shoot-through which makes use of the inherently needed di/dt reactor. The majority of the stored energy in the cell can be dissipated in a dedicated discharge loop formed by the reactor and an antiparallel bypass thyristor. After diverting the fault current into the dissipation loop, there is no current through any other component of the cell.**

Keywords—*Fault currents, short-circuit currents, HVDC transmission, modular multilevel converters, electronic packaging thermal management*

I. INTRODUCTION

High-voltage direct-current (HVDC) converters are built to operate for several decades. A redundancy is needed in order to reliably operate the converter after parts have failed. During maintenance, the failed parts can be repaired or exchanged. In modular multilevel converters (MMCs), which employ cascaded cells, the required redundancy is realized by implementing more converter cells than initially needed. There are concepts using hot-swapping of converter cells [1]; however, they have not been implemented for HVDC converters.

One of the most serious faults in an MMC is the internal cell shoot-through fault. If one power semiconductor of a converter cell fails, the cell capacitor may be short-circuited, resulting in a very fast discharge of the stored energy. The consequences of such a fault are highly dependent on the failure mode of the semiconductor packaging. Insulated-gate bipolar transistors (IGBT) packaged in wire-bonded modules (WBMs) have an open-circuit failure mode, i.e. the bond wires melt in an overcurrent event. The resulting arcing causes the silicone gel in the package to evaporate, leading to an explosion. Therefore, converter cells employing WBM IGBTs require an explosion-proof housing [2] and have to be bypassed as fast as possible at the terminals. Integrated gate-commutated thyristors (IGCT) in press-pack housing have a stable short-circuit failure mode (SCFM). A failed device will melt locally between

two pressured plates and form a low-resistive short-circuit [3]. Additionally, compared to WBM IGBTs, IGCTs are predicted to increase reliability, decrease losses, and offer higher power capability in MMCs [4], [5]. An implementation is presented in [6].

Still, in case of a shoot-through fault, measures have to be taken to protect the components in the cell and in the rest of the converter. The dissipated energy may heat up components to a critical temperature. Furthermore, the forces created by the high surge currents may damage the mechanical construction. In the worst case, the current path is interrupted, resulting in arcing. The effects of the fault should in all cases be contained within the cell, so that damage to neighboring converter cells is avoided. Hence, a strategy to handle these faults is essential for the reliable operation of HVDC converters.

There are measures to reduce the risk of shoot-through by series connecting several semiconductors to a stack, forming a single switch. A redundancy is introduced by rating the series connection for a higher voltage than the operation voltage of the cell. If one semiconductor in the stack fails, the others can still support the necessary voltage. A stable SCFM of the semiconductors is a requirement for this method. The series connection can also be incorporated into a new topology, for example the double module presented in [7]. However, the redundancy in terms of series connected semiconductors comes at the cost of increased conduction losses and complexity. A cell topology reducing conduction losses while having an increased shoot-through protection is presented in [8]. The full-bridge arrangement has two paths for the current during the bypass state of the cell, reducing the effective on-state resistance. This is made possible by an additional switch in series to the capacitor, which also introduces an additional redundancy.

For the future, there are several possible developments leading to higher stored energy (therefore more difficult shoot-through fault handling) in converter cells:

- The maximum cell voltage is limited by today's semiconductor voltage rating (with the exception of cells employing series-connected devices). Future silicon carbide (SiC) power semiconductors may be designed to withstand substantially higher breakdown voltages than current silicon (Si) devices. Increased cell voltage reduces the amount of required cells per arm, thus also reducing converter complexity and possibly volume [9].

- High-voltage cells can be utilized to increase the dc-link voltage. This allows for higher transmission power, or alternatively, for reduced current and losses.

- In order to achieve more transmission power it can be decided to increase the rated current, in which case a higher cell capacitance is needed to maintain a low cell voltage ripple.

In this paper, shoot-through faults for several cell designs are analyzed, and a discharge loop (DL) for handling those faults is presented. Section II presents the characteristics and implications of a shoot-through fault for cells with only parasitic inductance and for cells with a dedicated inductor. Section III describes the DL as a method to handle shoot-through faults. In Section IV, it is analyzed how much energy is dissipated in the half-bridge. Estimations are carried out for the temperature increase of the semiconductors. Conclusions are drawn in Section V.

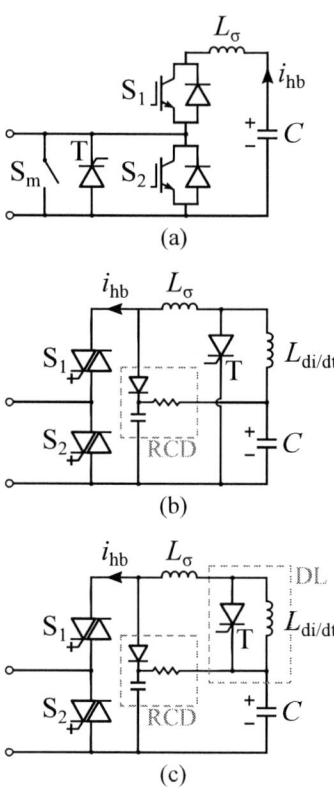

Fig. 1. Half-bridge cell employing IGBTs (a), IGCTs with bypass thyristor (b), and IGCTs with proposed discharge loop (c).

II. UNCONTROLLED SHOOT-THROUGH FAULTS

A shoot-through fault occurs if a low resistive path is provided between the positive and the negative terminal of the cell capacitor. This might happen if one of the switches in the half-bridge fails. The capacitor then discharges through the half-bridge with the characteristic of a damped oscillation. The parameters determining the shape of the oscillation are:

1) cell capacitance C, which is determined by the energy storage requirements of the converter;
2) inductance, which can be parasitic stray inductance of the busbars L_σ or a dedicated inductor $L_{\mathrm{di/dt}}$;
3) resistance, determined by the on-state resistance of the semiconductor devices R_{on}, the busbars R_{bus}, and the ESR of the cell capacitor.

As depicted in Fig. 1 (a), a standard half-bridge cell with IGBTs has only parasitic inductance L_σ. An IGCT implementation (Fig. 1 (b)) requires a di/dt limiting reactor $L_{\mathrm{di/dt}}$ in order to limit turn-on currents through the semiconductors. Without limiting di/dt, the current does not have enough time to spread across the wafer plane, leading to current filamentation. The associated heat dissipation at a certain spot might cause the device to melt locally. Furthermore, an RCD circuit has to be added to discharge $L_{\mathrm{di/dt}}$. This becomes necessary since there is no current path when a positive current through the IGCT S_1 is turned off. A bypass thyristor is added across the half-bridge.

With the circuit shown in Fig. 1 (b), the maximum short-circuit current through the IGCT cell is limited by $L_{\mathrm{di/dt}}$. Hence, the reactor is an inherent protection against high surge currents. This is favourable from a mechanical perspective, since mechanical forces scale quadratically with current. The uncontrolled short-circuit current during a shoot-through fault flows through the half-bridge and can be analytically expressed by solving the differential equation for the 2nd order system formed by R, L, and C.

$$i_{\mathrm{sc}}(t) = \frac{V_{\mathrm{C}}}{L\omega_{\mathrm{o}}} e^{-t/\tau} \sin(\omega_{\mathrm{o}}t), \qquad (1)$$

where the parameters ω_{o}, L, R, and τ are given by

$$\omega_{\mathrm{o}} = \sqrt{\frac{1}{LC} - \frac{R^2}{4L^2}} \qquad (2)$$

$$L = L_{\mathrm{di/dt}} + L_\sigma \qquad (3)$$

$$R = R_{\mathrm{bus}} + \mathrm{ESR} + R_{\mathrm{semicond}} \qquad (4)$$

$$\tau = \frac{2L}{R}. \qquad (5)$$

The peak current can be calculated as

$$i_{\mathrm{sc,max}} = \frac{V_{\mathrm{C}}}{L\omega_{\mathrm{o}}} e^{-\frac{\arctan(\omega_{\mathrm{o}}\tau)}{\omega_{\mathrm{o}}\tau}} \sin[\arctan(\omega_{\mathrm{o}}\tau)] \qquad (6)$$

at the time instant

$$t_{\mathrm{sc,max}} = \frac{\arctan(\omega_{\mathrm{o}}\tau)}{\omega_{\mathrm{o}}}. \qquad (7)$$

Regarding the semiconductors, this analytic approach is subject to the following simplifications:

- The knee voltage of the semiconductors is omitted. In reality, the switching device and the free-wheeling diode have different values; thus, the inserted knee voltage would change with current direction. The knee voltage is relatively small compared to the total voltage drop in the RLC circuit. Therefore, it does not have a major influence for the current characteristic.

- The resistance of the semiconductor is assumed to be constant, also disregarding that switching device and free-wheeling diode have different on-state resistance values. Depending on current direction, this would change the characteristic of the damped oscillation.

The circuit simulations carried out for this paper take the mentioned points into account. Still, there are simplifications which result in differences between the simulations and a real case:

- The conduction characteristics (V-I curves) for the switching device and the free-wheeling diode are linearized, yielding a resistive part (the on-state resistance R_{on}) and a constant potential (the knee voltage V_{k}). Hence, the semiconductors are modeled as a series connection of ideal device, on-state resistance, and knee voltage. In reality, the resistive part changes for high currents outside of the nominal operation. IGBTs, for example, experience a strong increase in on-state resistance when saturation is reached.

- Temperature dependencies are not taken into account. It is difficult to predict the resistivity under the high currents (well beyond the nominal operation) that flow during a shoot-through fault. This is because extreme overheating might melt the metallization of the semiconductors or even the semiconductor material itself.

- Resistances of the package (contact resistance and resistance of the aluminum and other materials) are omitted.

Therefore, care should be taken when interpreting the analytical and simulation results. In the following, simulations carried out with PLECS are presented. For simplicity it is assumed that the energy of the cell dissipates entirely in the semiconductors S_1, S_2, and T (i.e. R_{bus} and ESR are omitted).

Table I lists the four scenarios chosen for comparison. Case 1 is a wire-bonded IGBT module and it can be expected that the wire-bonds melt early during the fault [2]. Hence, it is questionable that the real shoot-through current corresponds to the simulations presented here. Case 2-4 employ semiconductors with stable SCFM. Case 2 and case 3 use data from a currently available Stakpak BIGT and a press-pack IGCT. Both are reverse conducting, i.e. the diode is integrated. The maximum permitted half-sine current surge for both devices is a 10 ms pulse with a peak current of 32 kA. Data for the theoretical SiC GTO and diode (case 4) is generated from the formulae given in [10] and [11] respectively.

The capacitance is calculated for a cell in a modular multilevel converter with the formula derived in [12]. It is proportional to the power rating of the converter and inversely proportional to the cell voltage. Said differently, for increased cell voltage the capacitance decreases by the same factor. This means that in a converter with given power rating and dc voltage, the stored energy in one cell is proportional to the cell voltage. For increased cell voltage the amount of cells required per arm decreases; thus, the energy stored in the whole converter stays the same. For an MMC with 1 GW transmission power and ± 320 kV dc-side voltage, this yields the capacitance values given in Table I. The parasitic inductance of the IGBT cell was assumed to be 200 nH. For the IGCT implementations $L_{\mathrm{di/dt}}$ was calculated to limit the current slope to 1000 A/µs.

TABLE I. SHOOT-THROUGH SIMULATION CASES

Case	1	2	3	4
Semiconductor	IGBT [13]	BIGT [14]	RC-IGCT [15]	SiC GTO** [10][11]
Package	WBM	StakPak	Press-pack	-
Surge current	-	32 kA 10 ms	32 kA 10 ms	-
$R_{on,sw}$ [mΩ]	1.1	0.45	0.43	0.54
$R_{on,d}$ [mΩ]	0.87	0.27	0.96	0.36
$V_{k,sw}$ [V]	1.45	1.8	1.7	4.08
$V_{k,d}$ [V]	1	1.75	2.4	3.68
L [µH]	0.2*	0.2*	2.25	16.5
V_C [kV]	1.65	2.25	2.25	16.5
C [mF]	12.26	8.99	8.99	1.23
E_{tot} [kJ]	16.7	22.8	22.8	167

* parasitic inductance
** theoretical

Fig. 2 illustrates the shoot-through currents for the cases listed in Table I. The converter cells with only parasitic inductance experience a current peak of 280 kA for case 1 and 416 kA for case 2. Both cells have dissipated their entire energy after approximately 1 ms. The maximum current for the cells employing di/dt limiting reactors is lower, around 140 kA. $L_{\mathrm{di/dt}}$ is an inherent protection against high peak currents, but can also be designed to limit the maximum shoot-through current to a desired value at the cost of introducing additional switching losses (when S_1 switches off a positive current, the remaining energy in $L_{\mathrm{di/dt}}$ is dissipated in the RCD circuit). Case 3 and 4 experience a similar shoot-through current, although the total energy in the cells is very different. This is because $L_{\mathrm{di/dt}}$ is designed according to nominal cell voltage V_C. Since the semiconductor characteristics are both similar, case 3 is more damped than case 4. This also becomes clear when determining τ (5) for both cases. All shown cases exceed the permitted surge currents for these devices.

III. DISSIPATION LOOP

To handle the issues described in the last section, converter cells have to be discharged safely, while still providing a path for the converter arm current. In case of a cell employing wire-bonded IGBT modules (in explosion-proof housing), the cell has to be bypassed externally as fast as possible. To bypass the cell for continued operation, a mechanical bypass switch S_m is attached to the cell terminals, as indicated in Fig. 1 (a). For press-pack IGCT cells there are no internal open circuits.

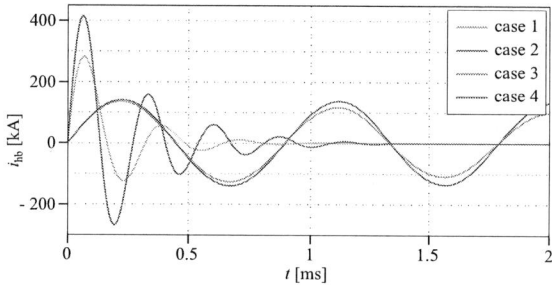

Fig. 2. Shoot-through current for the four cases listed in Table I.

Therefore, a path for the arm current is always available. To distribute the dissipated energy, a bypass thyristor can be arranged in parallel to the half-bridge, as shown in Fig. 1 (b). In a shoot-through fault this thyristor would take over the current (or at least a large share of the current). The capacitor is then discharging through the thyristor. Current will flow through the cell capacitors until all energy is dissipated.

In this paper, we propose to dissipate the cell-capacitor energy in a dedicated dissipation loop (DL). The loop is formed by the di/dt reactor $L_{di/dt}$ and a bypass thyristor T across it, as depicted in Fig. 1 (c). The thyristor is arranged in opposite direction to the initial shoot-through current flow.

The currents and voltages for the uncontrolled case and the proposed DL are shown in Fig. 3. When a shoot-through fault occurs, the stored energy will oscillate from C to $L_{di/dt}$. The thyristor T is reverse-biased until this point. The current path is shown in Fig. 4 (a). When the current reaches its maximum (after approximately 1/4 oscillation period) the remaining total cell energy is stored entirely in $L_{di/dt}$. The voltage V_C has its zero crossing at this point and is about to become negative. Therefore, the thyristor T is now forward-biased. T is fired and the current commutates from the half-bridge to the DL. Parasitic inductances within the circuit limit the speed of this commutation slightly. The current is now circulating in the DL, as depicted in Fig. 3 (b). As soon as T is triggered, there is no zero crossing of the current through the thyristor. The current in the DL will persist until the stored energy is entirely dissipated. S_2 is now permanently turned on (or in a stable SCFM, if destroyed), bypassing the cell.

The instant of triggering the thyristor T has to be chosen carefully. Triggering within the first quarter of the oscillation period will not have any effect, since the thyristor is reverse-biased. This also means that early triggering is not an issue for the functionality of the DL. On the other hand, triggering the thyristor too late, while C is charged negatively (significantly after the instant described above), should be avoided at all cost. The shoot-through current in negative direction is not limited anymore by $L_{di/dt}$ (only by the parasitic inductance); thus, it results in high reverse current peaks. The DL can take over the current at any of the maxima, meaning that it can be decided how much energy is dissipated in the

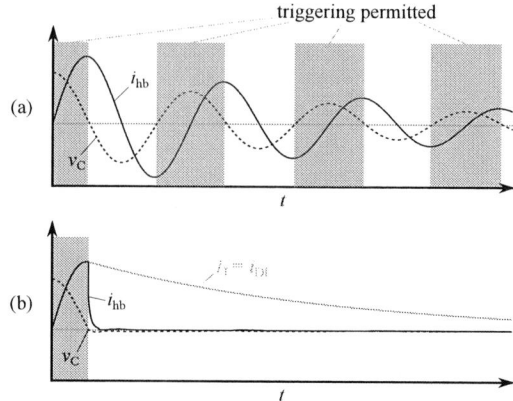

Fig. 3. Comparison of voltages and currents in the cell for uncontrolled shoot-through (a) and the proposed discharge loop mechanism (b).

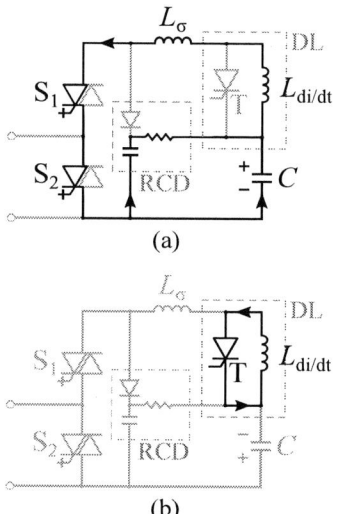

Fig. 4. Current path during shoot-through fault before (a) and after the thyristor T takes over the current (b).

half-bridge before the DL is fired. This will be discussed in more detail in the Section IV. The permitted intervals for triggering the thyristor of the DL are marked in Fig. 3.

With the previously described method, a fraction of the energy dissipation is shifted to the DL. The distribution of the energy among S_1, S_2, and T depends on the characteristic of the oscillation and the trigger instant. Highly damped shoot-through faults (large semiconductor on-state resistance) dissipate a large share of the total energy during the first half-wave. Thus, the DL can only dissipate a smaller fraction of the total energy. For cells with a low damped RLC characteristic (low on-state resistance of the semiconductors), the majority of the energy can be dissipated in the DL. This might be interesting for future cells employing SiC bipolar semiconductors with blocking voltages beyond $15\,kV$. These devices are foreseen to have comparably low on-state resistance [10].

For the shown example of a $2.25\,kV$ cell (case 3) $92\,\%$ of the total energy is dissipated in the DL, when triggering at the first current maximum. By using the same triggering instant for the 16.5 kV cell (case 4), 99 % of the total energy is dissipated in the DL. The DL has to be designed mechanically and thermally robust to withstand the high forces and temperatures. The advantage of the proposed solution is that these high requirements for thermal robustness apply only to a very small part of the cell, namely $L_{\mathrm{di/dt}}$ and T.

The DL can be employed in any cell topology and for any semiconductor device technology with SCFM. However, the DL makes most sense for cell designs that already have a dedicated inductor, e.g. cells with thyristor-based semiconductors.

IV. ENERGY DISSIPATION

The energy dissipation in the semiconductors is determined by their conduction characteristic (V-I curve). Switching instants are either at zero current or zero voltage. Therefore, switching losses can be disregarded. For the sake of simplicity, the V-I curve is linearized, yielding a resistive part R_{on} and the knee voltage of the semiconductor V_{k}. The instantaneous losses are determined by

$$p_{\mathrm{l}} = R_{\mathrm{on}} i_{\mathrm{hb}}^2 + V_{\mathrm{t}} i_{\mathrm{hb}}. \qquad (8)$$

Integration yields the dissipated energy of the switching device and the diode, shown in Fig. 5 (middle).

$$e_{\mathrm{l}} = \int_0^t p_{\mathrm{l}} dt \qquad (9)$$

Since S_1 and S_2 are exposed to the same current (neglecting the arm current, which superposes in S_2) their energy dissipation can be assumed to be equal. Thus, the total dissipated energy in the cell is determined by

$$e_{\mathrm{dis}} = 2 \int_0^t \left(p_{\mathrm{l,S1}} + p_{\mathrm{l,D1}} \right) dt. \qquad (10)$$

In Fig. 5, the shoot-through current (top), the absolute value of the dissipated energy in one press-pack (middle), and the relative value of the dissipated energy for the whole cell (bottom) is shown. The first three possible trigger instants for the DL are denoted t_1, t_2, and t_3. Adding up the energy of the switching device and the free-wheeling diode amounts to the dissipated energy for the semiconductor in one press-pack (marked in Fig. 5 (middle) for t_1, t_2, and t_3. In both cases a similar amount of energy is dissipated during the first few oscillations. As discussed in Section II, the oscillation for case 3 is much more damped than for case 4. Therefore, the total cell energy is dissipated much faster for case 3, as seen in Fig. 5 (bottom).

As mentioned before, it can be decided to trigger the thyristor at any current maximum. This might be interesting for distributing the dissipated energy among the semiconductors in the cell. The energy dissipated in the half-bridge can be tailored to what the press-packs can withstand. At the time instants t_1, t_2, and t_3, the energy in the cell is dissipated by 8.3 %, 48 %, and 70.7 % for case 3. For case 4 it is 1.5 %, 6.5 %, and 11.3 % for the

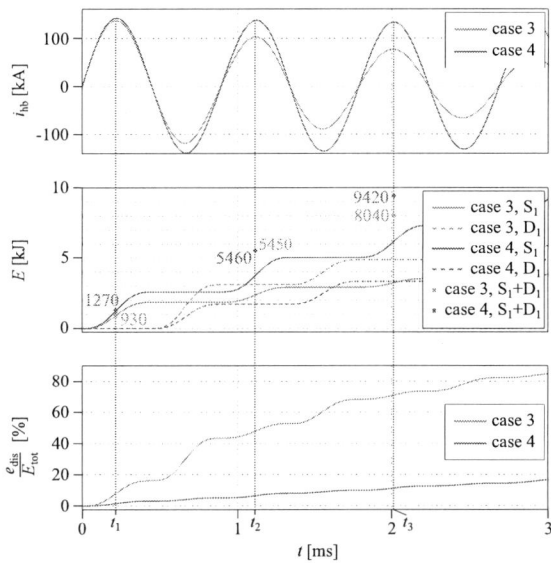

Fig. 5. Comparison of case 3 and 4: Shoot-through current (top), dissipated energy in the press-pack (middle), and relative value of the dissipated energy for the whole cell (bottom).

same time instants. Consequently, the DL has to dissipate the remaining share, after the thyristor is fired. Fig. 6 shows the shoot-through current in the half-bridge for case 3 for the first five possible trigger instants.

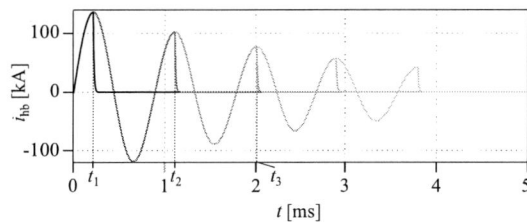

Fig. 6. Simulation of current through half-bridge (case 3) with the DL taking over the shoot-through at different maxima.

It is difficult to predict the consequences of the high current during shoot-through for the circuit. The DL itself has to be built in an extremely robust way. Depending on the characteristic of the bypass thyristor for these high currents beyond 100 kA (including its behavior when materials heat up and melt), the energy dissipation is shared between the parasitic resistance of the inductor and the thyristor press-pack itself. For the semiconductor devices in the half-bridge, simplified predictions about the temperature can be made by assuming adiabatic heating of the materials in the press-pack. The temperature increase for a certain amount of dissipated energy can then be determined by using

$$\Delta T = \frac{E}{c_{\mathrm{p}} m}, \qquad (11)$$

where c_{p} is the specific heat and m the mass of the absorbing medium. As indicated in Fig. 7, the three main materials in the heat conduction path are the semiconductor itself (Si or SiC), the molybdenum (Mo) discs (utilized as intermediate contact material), and the aluminum (Al) package. The material constants are given in Table II and the calculated temperature increase ΔT is given in Table III. The absolute values given in Fig. 5 (middle) serve as input for the temperature calculation. For calculating ΔT, we assume that the dissipated energy is entirely contained in one of these elements (e.g. if 5450 J would be dissipated entirely in the molybdenum discs, they would heat up by 150 K). Thus, the calculations are kept simple, at the cost of slightly overestimating the temperatures. In reality, the semiconductor itself heats up most, but some of the heat is transferred to the adjacent molybdenum discs before the maximum temperature has been reached. The molybdenum discs transfer the heat to the aluminum plates, which form the top and bottom layer of the press-pack. The heat transfer is slow, so it can be expected that the real temperature for the Si or SiC wafers is close to the calculated value. The temperature at the surface of the press-pack is of major interest, since the surfaces are attached to a heatsink. The coolant of the heatsink should be able to handle the temperature increase.

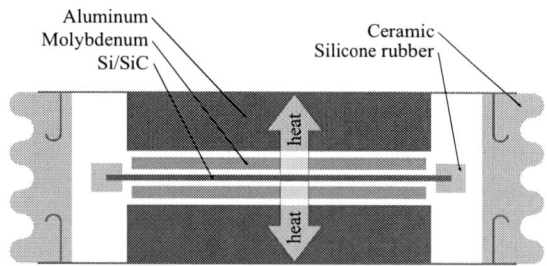

Fig. 7. Simplified cross section of a press-pack device.

TABLE II. MATERIAL CONSTANTS FOR THE MAIN COMPONENTS OF A PRESS-PACK

Material	Si	SiC	Mo	Al
melting temp. [K]	1687	3100	2896	660
spec. heat [J/kg K]	700	690	277	921
density [kg/m3]	2330	3210	10280	2699
volume [m3]	3.18e-6*	3.18e-6*	1.27e-5**	1.27e-4***

* ∅ 90 mm wafer, 500 μm thickness
** two ∅ 90 mm plates, 1 mm thickness
*** two ∅ 90 mm plates, 1 cm thickness

TABLE III. TEMPERATURE INCREASE FOR THE MATERIALS IN THE PRESS-PACK IN KELVIN

case	t	e_{dis} [J]	ΔT_{Si}	ΔT_{SiC}	ΔT_{Mo}	ΔT_{Al}
3	t_1	930	179	132	26	3
3	t_2	5450	1051	774	150	17
3	t_3	8040	1550	1141	222	25
4	t_1	1270	245	180	35	4
4	t_2	5460	1052	775	151	17
4	t_3	9420	1816	1337	260	30

It becomes evident that even for triggering the DL at the first current maximum, the semiconductors in the half-bridge reach temperatures outside of their rating. It can be expected that the semiconductor metallization melts down first, since Si and SiC have higher melting temperatures. The temperatures at the case of the press-pack seem uncritical, even when triggering at the third current maximum. The authors do not doubt that this

kind of stresses can be handled by the press-pack. Shoot-through tests on IGCTs have already been performed with over 60 kJ of energy stored in the capacitor [3]. After a shoot-through, the SCFM was confirmed by subjecting the IGCTs to rated currents for 15 hours. The resistance of the press-packs in SCFM was around 2 Vrms for the duration of the test.

V. CONCLUSION

This paper proposes a discharge loop to handle shoot-through faults in converter cells employing thyristor-based semiconductors, e.g. IGCTs. The dissipation loop, formed by the inherently needed di/dt limiting reactor and a bypass thyristor, can be utilized to relieve the main circuit from electrical, mechanical, and thermal stress. Especially for cells with high stored energy and low resistance, a large portion (> 90 %) of the total cell energy can be dissipated in the dedicated dissipation loop. The remaining stress on the half-bridge is within the range of what current IGCT press-packs can handle. Thus, high requirements for mechanical and thermal robustness have to be fulfilled only for the components forming the discharge loop, i.e. the di/dt limiting reactor and the bypass thyristor. A disadvantage is that the discharge loop can only be triggered at a current maximum, when the cell energy is entirely stored in the reactor. Therefore, it has to be guaranteed that semiconductor packages in the half-bridge can withstand the heat dissipation until that point. This is achieved by using devices with short-circuit failure mode.

Using the discharge loop might be interesting for future cells with a lot of stored energy. These could be cells employing SiC semiconductors with blocking voltages of 15 kV and higher.

ACKNOWLEDGEMENT

The authors would like to acknowledge SweGRIDS for funding this project and ABB Corporate Research Center (SECRC) for their valuable support.

REFERENCES

[1] D. Cottet *et al.*, "Integration technologies for a medium voltage modular multi-level converter with hot swap capability," in *2015 IEEE Energy Conversion Congress and Exposition (ECCE)*, Sep 2015, pp. 4502–4509.

[2] M. Billmann, D. Malipaard, and H. Gambach, "Explosion proof housings for IGBT module based high power inverters in HVDC transmission application," in *Proc. PCIM Eur. 2009 Conf.*, Nuremberg, May 2009, pp. 352–357.

[3] F. Dijkhuizen and S. Norrga, "Fault tolerant operation of power converter with cascaded cells," in *Proceedings of the 2011 14th European Conference on Power Electronics and Applications*, Aug 2011.

[4] P. Ladoux, N. Serbia, and E. I. Carroll, "On the potential of IGCTs in HVDC," *IEEE Journal of Emerging and Selected Topics in Power Electronics*, vol. 3, no. 3, pp. 780–793, Sep 2015.

[5] T. Modeer, H. P. Nee, and S. Norrga, "Loss comparison of different sub-module implementations for modular multilevel converters in HVDC applications," in *Proceedings of the 2011 14th European Conference on Power Electronics and Applications*, Aug 2011, pp. 1–7.

[6] T. Modeer, S. Norrga, and H.-P. Nee, "Implementation and testing of high-power IGCT-based cascaded-converter cells," in *2014 IEEE Energy Conversion Congress and Exposition (ECCE)*, Sep 2014, pp. 5355–5359.

[7] H.-G. Eckel and H. Gambach, "Double module for a modular multi-stage converter (US 9263969 B2)," 2016.

[8] C. Dahmen and R. Marquardt, "Progress of High Power Multilevel Converters: Combining Silicon and Silicon Carbide," in *PCIM Europe 2017; International Exhibition and Conference for Power Electronics, Intelligent Motion, Renewable Energy and Energy Management*, May 2017.

[9] K. Jacobs, D. Johannesson, S. Norrga, and H. P. Nee, "MMC converter cells employing ultrahigh-voltage SiC bipolar power semiconductors," in *2017 19th European Conference on Power Electronics and Applications (EPE'17 ECCE Europe)*, Sep 2017.

[10] D. Johannesson, M. Nawaz, K. Jacobs, S. Norrga, and H. P. Nee, "Potential of ultra-high voltage silicon carbide semiconductor devices," in *2016 IEEE 4th Workshop on Wide Bandgap Power Devices and Applications (WiPDA)*, Nov 2016, pp. 253–258.

[11] N. Kaji, H. Niwa, J. Suda, and T. Kimoto, "Ultrahigh-voltage SiC p-i-n diodes with improved forward characteristics," *IEEE Transactions on Electron Devices*, vol. 62, no. 2, pp. 374–381, Feb 2015.

[12] M. M. C. Merlin and T. C. Green, "Cell capacitor sizing in multilevel converters: cases of the modular multilevel converter and alternate arm converter," *IET Power Electronics*, vol. 8, no. 3, pp. 350–360, Mar 2015.

[13] Mitsubishi, "CM1500HC-66R data sheet, High Voltage Insulated Gate Bipolar Transistor Modules," Dec 2012.

[14] ABB, "5SJA 3000L450300 data sheet, StakPak BIGT Module PRELIMINARY," Mar 2017.

[15] ABB., "5SHX 26L4520 data sheet, Reverse Conducting Integrated Gate-Commutated Thyristor," Apr 2016.

The 2018 International Power Electronics Conference

A Suppression Method of Harmonic Instability in Line-Commutated Converters Applying Active Harmonic Filters

Kenichiro Sano[1*], Toshiaki Kikuma[1], Tatsuhito Nakajima[2] and Junya Kanno[3]

1 System Engineering Research Laboratory, CRIEPI, Yokosuka, Japan
2 Department of Electrical and Electronic Engineering, Tokyo City University, Tokyo, Japan
3 TEPCO Research Institute, TEPCO Holdings, Tokyo, Japan
*E-mail: k.sano@ieee.org

Abstract— Harmonic instability of line-commutated converters (LCC) may cause restriction in grid operations or increase of running costs. Recently, voltage-source converters (VSC) are applied to HVDC transmission system. Some of them are installed neighboring to the existing LCCs. In such case, VSC's fast control characteristics can be utilized to contribute to LCC's stable operation. This paper proposes a novel solution of harmonic instability in LCCs, where adjacent VSC is operated as an active harmonic filter. The VSC improves the high system impedance resulting from parallel resonance in the ac grid. The authors analyze and demonstrate its effectiveness to the harmonic instability based on a stability criterion and electromagnetic transient (EMT) simulations. The required compensation current for the active filter is much smaller than the rated current of the LCC. Moreover, it is confirmed that the active filter control can be applied to normal VSC based BTB, HVDC, or STATCOM systems as an additional function without interrupting their original functions.

Keywords— harmonic instability, active harmonic filter, line-commutated converters, STATCOM.

I. INTRODUCTION

When an ac grid has parallel resonant condition in a specific frequency, harmonic currents from a line-commutated converter (LCC) may gradually increase, cause persistent oscillation, and interrupt normal operation of the LCC. The phenomenon is called "harmonic instability" [1]. There are conventional solutions such as improving control characteristic of the LCC, modifying ac filter design, limiting total capacity of shunt capacitor banks, connecting a synchronous condenser [2][3]. However, it is difficult to modify the controller design or ac filter of the LCCs which are already installed in substations. Limiting shunt capacitor banks causes restrictions on grid operations. Synchronous condenser generates power loss and also requires its maintenance, which will cause the increase of running costs.

Recently, voltage-source converters (VSC) are gradually applied to HVDC transmission system. There are some projects to install new VSC type frequency converters (FC) in parallel with the existing LCC type FCs in Japan. Skagerrak 4 project has installed a VSC HVDC system in parallel with the existing LCC system [4]. Tres Amigas project plans to install VSC and LCC back-to-back (BTB) systems in the same substation [5]. In such cases, VSC's fast control characteristics can be utilized to contribute to LCC's stable operation.

This paper proposes a novel solution of harmonic instability in LCCs, where adjacent VSC is operated as an active harmonic filter to compensate harmonics in ac system. That method provides another option to suppress the harmonic instability, which enables flexible grid operation and reduction of running costs. Among various configurations in active filters [6], this study applies a shunt active filter based on voltage detection [7]. The active filter was originally developed for mitigating harmonic resonance in distribution systems. Although the control method is same as the conventional shunt active filters, this paper investigates the effectiveness of the active filter to suppress the harmonic instability of LCCs. The authors verify the effectiveness applying a stability criterion based on the impedances of the ac grid and the LCC [8]-[11]. The analysis demonstrates its effect to the harmonic instability. Moreover, it is confirmed that the active filter control can be applied to normal VSC based BTB, HVDC, or STATCOM systems as an additional function without interrupting their original functions.

II. SYSTEM CONFIGURATION

Fig. 1 shows the system configuration focusing in this study. The circuit parameters are shown in TABLE I. A 600-MVA BTB system consisting of LCC interconnects between two non-synchronous ac systems. Passive ac filters and shunt capacitors are equipped in the both ac buses. A VSC rated at 150 MVA is connected to one of the ac buses. The VSC applies a neutral-point-clamped (NPC) converter in its circuit topology. Although the VSC consists of single inverter for STATCOM operation in this study, it may be BTB or HVDC transmission system being capable of transferring active power between the two ac systems. Short circuit capacity of the ac system is 1600 MVA, which is modeled by an ac voltage source v_s and an inductance L_{ac}. Power flow of the LCC P_{lcc} is adjusted to 1 pu. The rectifier is controlled to regulate the dc current, and the inverter is controlled to regulate the power flow.

Fig. 2 shows a control block diagram for the LCC. The control block is constructed based on the existing control [12]. Same controller is installed for both rectifier and inverter. Power controller calculates the dc current reference $I_{dc}*$ according to the dc power reference $P_{dc}*$ and detected dc power P_{dc}. Current controller calculates the commutation angle α_c according to the $I_{dc}*$, detected dc

3299

The 2018 International Power Electronics Conference

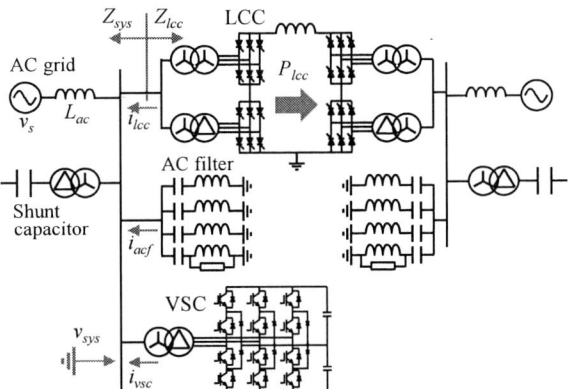

Fig. 1. System configuration.

TABLE I
CIRCUIT PARAMETERS

Line frequency	f	50 Hz
Line-to-line voltage of ac grid	V_s	275 kV
Inductance of ac grid	L_{ac}	150 mH
Short-circuit capacity of ac grid		1600 MVA
Power rating of LCC		600 MVA
Power flow of LCC	P_{lcc}	600 MW
Power rating of VSC		150 MVA
Reactive power rating of shunt capacitor		160 MVA
Reactive power rating of 5th ac filter		59.2 MVA
Reactive power rating of 11th ac filter		43.6 MVA
Reactive power rating of 13th ac filter		31.2 MVA
Reactive power rating of high pass ac filter		67.2 MVA
Gain of active filter	K_{af}	0.006 S (Ω^{-1})

current I_{dc}, and current margin I_m. Voltage controller calculates the commutation angle α_v according to the dc voltage reference V_{dc}^*, and detected dc voltage V_{dc}. Gamma controller calculates the commutation angle α_g to maintain minimum extinction angle γ_{min} according to the detected ac voltage V_{ac}, I_{dc}, and commutation reactance X. The minimum value among α_c, α_v, α_g is selected for the commutation angle α.

III. ACTIVE FILTER CONTROL FOR VSCS

Fig. 3 shows a control block diagram for the VSC. Based on the conventional control [13], the reactive power controller (AQR) and the dc voltage controller (DCAVR) calculate the current references i_d^*, i_q^*. Then the ac current controller calculates the voltage references v_d^*, v_q^*. In this study, active harmonic filter [7] is added to the aforementioned control. The ac bus voltages v_u, v_v, v_w are transformed to rotating d-q frame voltages v_d, v_q by the line frequency f. Then harmonic components are extracted by two-order high-pass filters (HPF) designed to have cutoff frequency of 5 Hz. After multiplying a control gain K_{af}, the values are superimposed to the current references i_d^*, i_q^*. Thus, this active filter control does not respond to 45-55 Hz, which corresponds to line frequency. $1/K_{af}$ has a dimension of resistance. The VSC operates as a resistance of $1/K_{af} = 1/0.006 = 170\ \Omega$ to harmonics. By connecting the active filter to ac system in parallel, impedance of the ac system can be maintained to be low in the harmonic region.

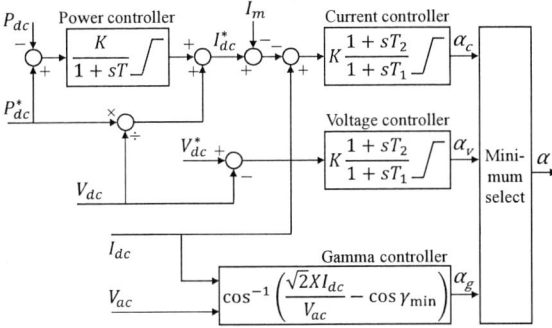

Fig. 2. Control blocks of the LCC.

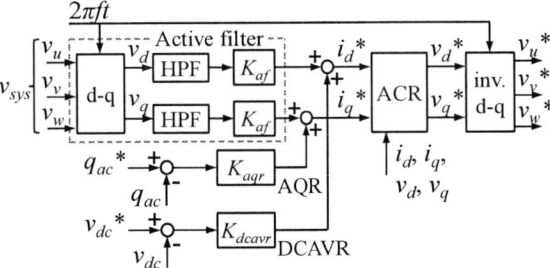

Fig. 3. Control blocks of the VSC.

NPC converter is applied to the VSC in this study. When the active filter control is applied to the NPC converters, their neutral-point voltage will be deviated by compensating even-order harmonic currents. Therefore, neutral-point voltage control by zero-sequence voltages [14] should be also implemented in addition to the control method in Fig. 3.

IV. ANALYTICAL METHOD FOR HARMONIC INSTABILITY ASSESMENT

The impedance based stability criterion was established in 1976 and originally applied to the design of input filters in dc–dc converters [15]. The stability criterion can be also applied to analyses of the harmonic instability of the LCC [8]-[11]. The same method is employed in this study. The outline is as follows.

The target system is divided by the ac bus into ac grid side and LCC side as shown in Fig. 4 (a). The ac grid side includes ac filters and shunt capacitor. Z_{sys} is defined as the combined impedance of the ac grid side obtained from the ac bus. Z_{lcc} is defined as the equivalent impedance of the LCC obtained from the ac bus. Fig. 4 (b) shows the relation between the n-th harmonics in the ac bus voltage V_n, and the n-th harmonics in the current passing through the ac bus I_n. Here, I_{dist} corresponds to the disturbance caused by the LCC or saturation of the transformer. The open-loop transfer function of this system $G_o(f)$ is provided as follows:

$$G_o(f) = Z_{sys} / Z_{lcc}.$$

The following discussion is focusing on the small-signal stability based on the linearized Z_{lcc}. Because actual LCCs are nonlinear system, Z_{lcc} may change according to the operating conditions.

3300

The 2018 International Power Electronics Conference

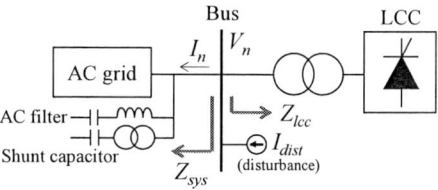

(a) Expression in a circuit diagram.

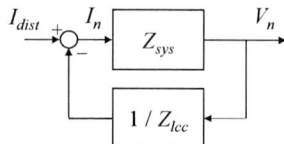

(b) Expression in a block diagram.

Fig. 4. Impedance based stability criterion.

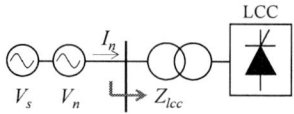

(a) Measurement of the LCC's impedance Z_{lcc}

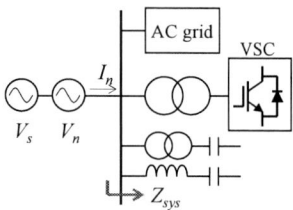

(b) Measurement of the system impedance Z_{sys}

Fig. 5. Configurations for the impedance measurement.

The stability of the feedback system is identified by its open-loop transfer function $G_o(f)$.

- $-|G_o(f_{180°})| < 0$dB (gain margin is negative)
- $\arg[G_o(f_{0dB})] + 180° < 0$ (phase margin is negative).

When the aforementioned conditions are satisfied, the system is instable. Here, $\arg[G_o]$ means the phase angle of G_o, $f_{180°}$ is defined as the frequency where the phase angle of G_o reaches $-180°$, and f_{0dB} is defined as the frequency where the gain reaches 0dB. These conditions can be equivalently transformed as follows:

- $|Z_{sys}(f_{argZ})| > |Z_{lcc}(f_{argZ})|$ (1)
- $\arg[Z_{lcc}(f_{|Z|})] - \arg[Z_{sys}(f_{|Z|})] > 180°$,   (2)

where f_{argZ} is the frequency to be $\arg[Z_{lcc}] - \arg[Z_{sys}] = 180°$, $f_{|Z|}$ is the frequency to be $|Z_{sys}| = |Z_{lcc}|$. Therefore, harmonic instability can be analyzed by the obtained impedances Z_{sys} and Z_{lcc}.

V. VALIDATION BY COMPUTER SIMULATIONS

Effectiveness of the active filter to the harmonic instability is validated by computer simulations. Simulations are carried out by XTAP [16], which is an electromagnetic transient (EMT) analysis program developed by CRIEPI.

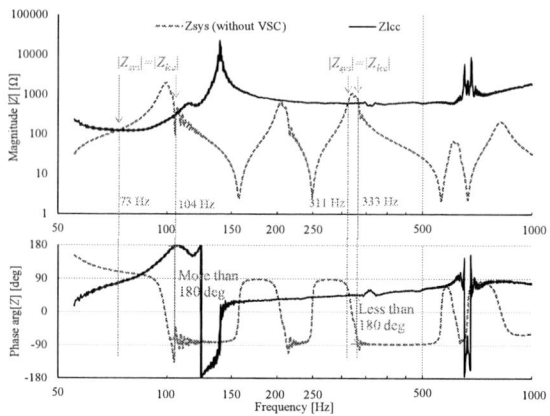

Fig. 6. Impedance characteristics when the VSC is halted.

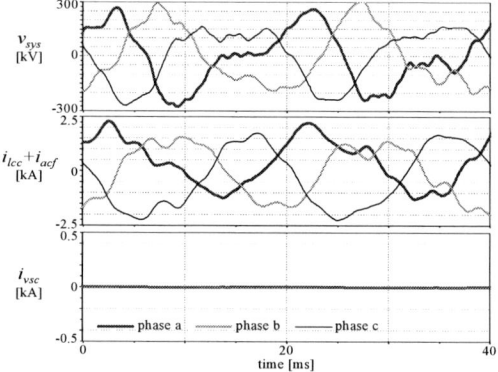

Fig. 7. Simulation waveforms when the VSC is halted.

Fig. 5 depicts a method to measure Z_{lcc} and Z_{sys} by simulations. The target system is divided by the ac bus into ac grid side and LCC side. Each system is connected to a series-connected fundamental voltage source V_s for operating the converter and harmonic voltage source V_n for measuring the impedance. Amplitude and phase of the harmonic current I_n are detected by sweeping the frequency of V_n. Z_{lcc} and Z_{sys} can be calculated by dividing V_n by I_n in each frequency. Z_{sys} is the combined characteristic of the ac grid, VSC, ac filters, and shunt capacitor.

A. Without VSC

As a reference case, an analysis is carried out when the VSC is halted.

1) Impedance based analysis

Fig. 6 shows the frequency characteristics of Z_{sys} and Z_{lcc} in this condition. $|Z_{sys}|$ has some local maximum points resulting from parallel resonance in the system. $|Z_{sys}|$ reaches 1.81 kΩ at 100 Hz and 1.07 kΩ at 320 Hz, whereas $|Z_{lcc}|$ is 231 Ω at 100 Hz and 614 Ω at 320 Hz. "$|Z_{sys}| = |Z_{lcc}|$" is satisfied at 73 Hz, 104 Hz, 311 Hz, and 333 Hz. Phase difference "$\arg[Z_{lcc}(f_{|Z|})] - \arg[Z_{sys}(f_{|Z|})]$" at these frequencies is obtained from the phase diagram, that is $-33°$, $258°$, $-20°$, and $139°$. Since one of the values exceeds $180°$ at 104 Hz, the system is evaluated as instable according to (2).

2) EMT simulation

3301

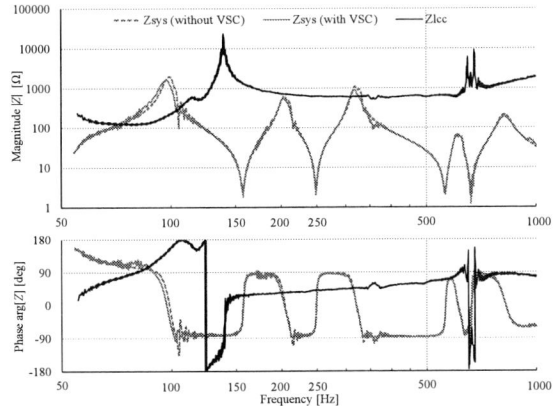

Fig. 8. Impedance characteristics when the VSC operates with reactive power control (without active filter control).

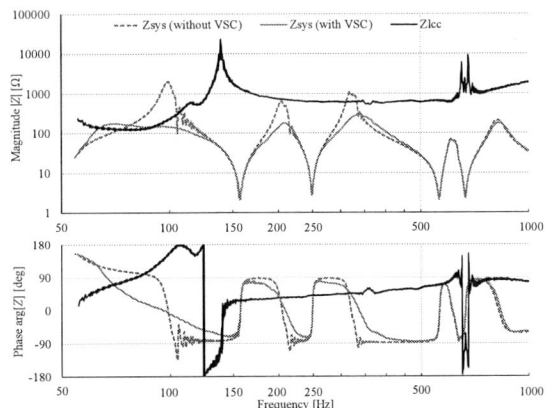

Fig. 10. Impedance characteristics when the VSC operates with active filter control.

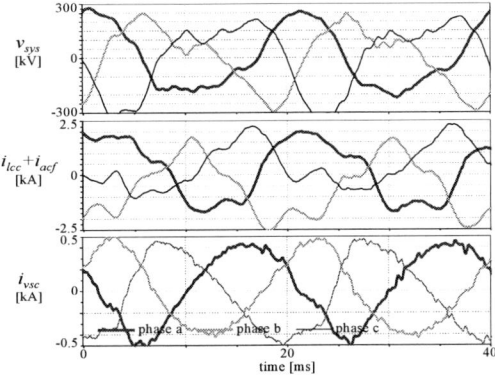

Fig. 9. Simulation waveforms when the VSC operates with reactive power control (without active filter control).

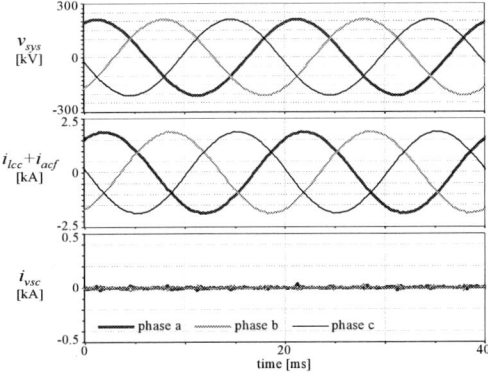

Fig. 11. Simulation waveforms when the VSC operates with active filter control.

Fig. 7 shows the waveforms of ac bus voltage v_{sys}, LCC current after passing passive ac filters $i_{lcc} + i_{acf}$, VSC current i_{vsc}, which are obtained by an EMT simulation. Simulated circuit is same as Fig. 1. The LCC provides highly distorted current, which contains 100 Hz and 320 Hz. Then ac bus voltage is also distorted. The harmonics do not diverge but persist in a specific level. Although this result shows the distorted waveforms in order to observe the harmonic instability, protection relays will detect the malfunction and halt the converter in actual facilities.

B. VSC with reactive power control (conventional control without active filter control)

VSC is operated to compensate the reactive power of the fundamental frequency as a general STATCOM. The reactive power reference $q_{ac}*$ is 1.0 pu of capacitive operation. The active filter control is disabled by setting $K_{af} = 0$.

1) Impedance based analysis

Fig. 8 shows the frequency characteristics of Z_{sys} and Z_{lcc} in this condition. Solid lines show the case with VSC. Dotted lines are the case without VSC as a reference (same result as Fig. 6). Because the VSC operates as a current source, its equivalent impedance is large. Thus, parallel connection of the VSC does not affect the characteristic of Z_{sys}. In this case, the system is evaluated as instable.

2) EMT simulation

Fig. 9 shows the EMT simulation result. Harmonic instability is observed also in this case. The VSC with reactive power control does not suppress the harmonic instability. Because the ac bus voltage is highly distorted, the VSC cannot control its current i_{vsc} to be sinusoidal waveform.

C. VSC with active filter control

VSC is operated with active filter control. The reactive power control is disabled by setting $K_{aqr} = 0$.

1) Impedance based analysis

Fig. 10 shows the frequency characteristics of Z_{sys} and Z_{lcc} in this condition. Since VSC with active filter control operates as a resistance of 170 Ω, local maximums of $|Z_{sys}|$ are decreased to 142 Ω at 100 Hz and 224 Ω at 320 Hz. $|Z_{sys}|$ is smaller than $|Z_{lcc}|$ in these frequency regions. Although there are some points to satisfy "$|Z_{sys}| = |Z_{lcc}|$" below 100 Hz, they does not satisfy the phase condition shown in (2). Thus, the system is evaluated as stable.

There is no change in local minimums of Z_{sys}, which exist in series resonant frequencies of 5th, 11th, and 13th ac filters (250, 550, 650 Hz). It means that the active filter control does not affect the harmonic filtering performance of the passive ac filters.

2) EMT simulation

3302

The 2018 International Power Electronics Conference

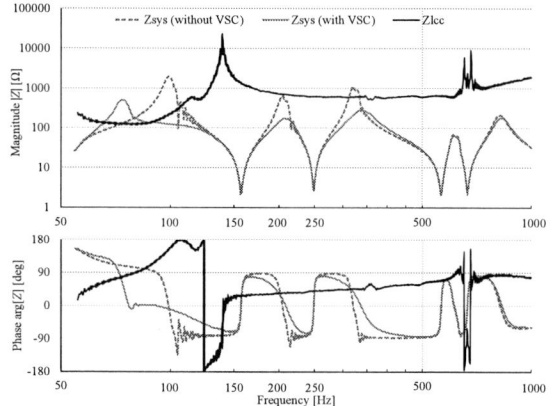

Fig. 12. Impedance characteristics when the VSC operates with active filter control and reactive power control.

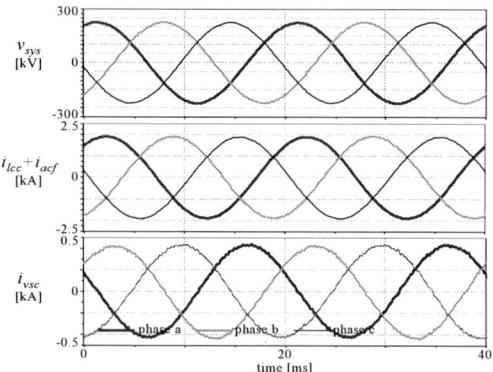

Fig. 13. Simulation waveforms when the VSC operates with active filter control and reactive power control.

Fig. 11 shows the EMT simulation result. Harmonic instability is not observed in this case. Maximum value of the VSC current i_{vsc} is 60 A, which is 13% of rated value of the VSC and 3.4% of rated value of the LCC. Thus, required compensation current for the active filter is much smaller than the rated current of the LCC.

D. VSC with active filter and reactive power control

VSC is operated with active filter control while compensating the reactive power of the fundamental frequency. The reactive power reference $q_{ac}{}^*$ is 1.0 pu of capacitive operation.

1) Impedance based analysis

Fig. 12 shows the frequency characteristics of Z_{sys} and Z_{lcc} in this condition. The characteristic is almost same as the case applying only active filter control. In this case, the system is evaluated as stable.

2) EMT simulation

Fig. 13 shows the EMT simulation result. Harmonic instability is not observed in this case. It is confirmed that the active filter control and the reactive power control can be implemented together without interfering each other. It means that this active filter control can be applied to normal VSC based BTB, HVDC, or STATCOM systems as an additional function without interrupting their original functions.

VI. CONCLUSIONS

This paper proposed a novel solution of harmonic instability in LCCs, where adjacent VSC is operated as an active harmonic filter. Since the VSC operates as a resistance to harmonic region, it can reduce the high impedance resulting from parallel resonance in the ac grid. Simulation results and theoretical analyses demonstrated the effectiveness of a shunt active harmonic filter based on voltage detection to suppress harmonic instability in LCCs. The required compensation current for the active filter is much smaller than the rated current of the LCC. Moreover, the active filter control can be applied to VSC based BTB, HVDC, or STATCOM systems as an additional function.

REFERENCES

[1] J. D. Ainsworth, "Harmonic instability between controlled static convertors and a.c. networks," *Proceedings of the Institution of Electrical Engineers*, vol.114, pp.949-957, Jul. 1967.

[2] P. S. Bodger, G. D. Irwin, and D. A. Woodford, "Controlling harmonic instability of HVDC links connected to weak AC systems," *IEEE Transactions on Power Delivery*, vol. 5, no. 4, pp. 2039-2046, Oct 1990.

[3] A. E. Hammad, "Analysis of second harmonic instability for the Chateauguay HVDC/SVC scheme," *IEEE Transactions on Power Delivery*, vol. 7, no. 1, pp. 410-415, Jan 1992.

[4] J. P. Kjærgaard, K. Søgarrd, S. D. Mikkelsen, T. Pande-Rolfsen, A. Strandem, B. Bergdahl, and H.-O. Bjarme: "Bipolar operation of an HVDC VSC converter with an LCC converter," *CIGRE SC B4 San Francisco Colloquium*, no. B4-10, Mar. 2012.

[5] N. M. Kirby, N. M. Macleod, D. Stidham, and M. Reynolds: "Application of high capacity voltage source converter HVDC technology at the Tres Amigas Superstation," *CIGRE SC B4 San Francisco Colloquium*, no. B4-12, Mar. 2013.

[6] H. Akagi, "Active Harmonic Filters," *Proceedings of the IEEE*, vol. 93, no. 12, pp. 2128-2141, Dec. 2005.

[7] H. Akagi, H. Fujita, and K. Wada, "A shunt active filter based on voltage detection for harmonic termination of a radial power distribution line," *IEEE Transactions on Industry Applications*, vol. 35, no. 3, pp. 638-645, May/Jun. 1999.

[8] A. R. Wood, and J. Arrillaga, "Composite resonance; a circuit approach to the waveform distortion dynamics of an HVdc converter," *IEEE Transactions on Power Delivery*, vol. 10, no. 4, pp. 1882-1888, Oct. 1995.

[9] K. Yamaji, M. Azuma, N. Honjo, Y. Noro, and T. Yoshino, "Study of non-integer harmonic instabilities in HVDC system," *IEEJ Transactions on Power and Energy*, vol. 118, no. 7-8, pp.899-905, Jul./Aug. 1998. (in Japanese)

[10] H. Liu, and J. Sun, "Modeling and analysis of DC-link harmonic instability in LCC HVDC systems," *IEEE Control and Modeling for Power Electronics (COMPEL)*, pp. 1-9. June 2013.

[11] X. Wang, F. Blaabjerg, and W. Wu, "Modeling and analysis of harmonic stability in an AC power-electronics-based power system," *IEEE Transactions on Power Electronics*, vol. 29, no. 12, pp. 6421-6432, Dec. 2014.

[12] "Power electronics handbook," Ohmsha, pp.150-151, July 2010. (in Japanese)

[13] K. Sakamoto, M., Yajima, T. Ishikawa, S. Sugimoto, T. Sato, and H. Abe, "Development of a control system for a high-performance self-commutated AC/DC converter," *IEEE Transactions on Power Delivery*, vol. 13, no. 1, pp. 225-232, Jan. 1998.

[14] S. Ogasawara, and H. Akagi, "Analysis of variation of neutral point potential in neutral-point-clamped voltage source PWM inverters," *IAS Annual Meeting*, pp. 965-970 vol.2. Oct. 1993.

[15] R. D. Middlebrook, "Input filter considerations in design and application of switching regulators," *Proceedings of the IEEE IAS Annual Meeting*, pp. 366-382, 1976.

[16] A. Ametani, "Numerical analysis of power system transients and dynamics," IET, pp. 169-212, 2014.

Experiment of Semiconductor Breaker using Series-Connected IEGTs for Hybrid DCCB

Kazuyasu Takimoto[1*], Hiroshi Takenaka[1], Toshiaki Matsumoto[1]and Takahiro Ishiguro[2]
1 Power and Industrial Systems Research and Development Center,
TOSHIBA CORPORATION, Fuchu-shi, Tokyo, Japan
2 Transmission & Distribution Systems Div.
TOSHIBA ENERGY SYSTEMS & SOLUTIONS CORPORATION, Kawasaki-shi, Kanagawa, Japan
*E-mail: kazuyasu.takimoto@toshiba.co.jp

Abstract-This paper considered application of series-connected press pack Injection Enhanced Gate Transistors (IEGTs) as a semiconductor breaker for hybrid DC circuit breakers (DCCBs) of multi-terminal HVDC transmission systems. In order to interrupt a large current equivalent to fault current without parallel connection of press pack IEGT, the snubber circuit was applied. With the snubber circuit, surge voltages and switching losses can be reduced. Therefore, IEGTs can interrupt lager currents with the snubber circuit. In order to balance voltages across each series-connected IEGT, each IEGT should be selected to have similar properties. The circuit condition and operation timing of each IEGT should be adjusted. In order to confirm the capability of the proposed configuration, a prototype of hybrid DCCB was demonstrated. The semiconductor breaker of hybrid DCCB prototype was composed of four series connected IEGTs (4.5 kV, 2.1 kA) with the snubber circuit. Experimental result shows it successfully interrupted the current of 9 kA. Furthermore, the peak voltage across the semiconductor breaker was 14 kV. Variation of the voltage across each IEGT was less than 5 %. This means each IEGT peak voltage was suppressed under the withstand voltage of IEGT.

I. INTRODUCTION

Recently, as one of measures to promote massive introduction of renewable energy, optimization of a wide area system that transmits electric power from large-scale wind power generations located at offshore to each mainland demand area has been studied. In the case of long distance transmission, DC transmission is better than AC transmission in terms of loss and construction cost [1]. Therefore, it is studied to apply multi-terminal high-voltage DC (HVDC) transmission systems to connect multiple windfarms and multiple onshore stations. In multi-terminal HVDC transmission systems, in order to prevent fault voltage spreading, large capacity DC circuit breakers (DCCBs) are required to disconnect fault points in a few milliseconds [2][3].

In the past several years, hybrid DCCBs which include mechanical and semiconductor switches have been developed for the request of high speed interruption[4][5][6][7]. If those hybrid DCCBs have semiconductors in the normal current path, the transmission losses on the semiconductors will increase.

On the other hand, a low-loss hybrid DCCB has been proposed [8]. The circuit configuration is shown in Fig. 1.

Proposed hybrid DCCB includes mechanical switches, a semiconductor breaker, and a commutation circuit. In the normal mode, the current flows through the mechanical switch. Since there is no semiconductor in the current path, transmission losses can be minimized. When a fault occurs, the commutation circuit operates and the fault current commutates from the mechanical switch to the semiconductor breaker. Then the semiconductor breaker can interrupt the fault current.

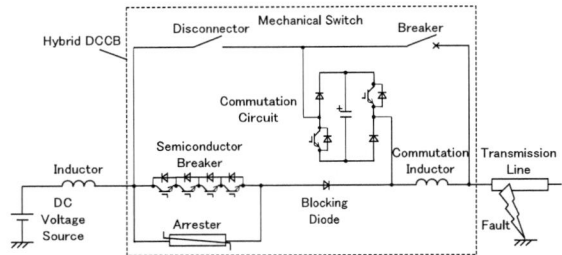

Fig. 1. Proposed hybrid DCCB.

Press pack Injection Enhanced Gate Transistor (IEGT) is a semiconductor suitable for high voltage / large current operation. In general use, press pack IEGT has capability to interrupt twice as large current as rated current. On the other hand, a fault current may be more than twice as large as the rated current. In order to interrupt a fault current without parallel connection of press pack IEGT, press pack IEGT requires a current interrupting capability exceeding normal interruptible current. Furthermore, in order to withstand the high voltage of HVDC transmission systems, it is necessary to connect press pack IEGTs in series. In series-connected IEGTs, each voltage across series-connected IEGTs needs to be balanced so as not to exceed the withstand voltage of IEGT. However, there are no reports that series-connected IEGTs interrupted a current more than twice as large as the rated current of IEGTs.

For the proposed hybrid DCCB, this paper proposed the configuration which can interrupt current more than twice as large as the rated current with series-connected IEGTs. In order to confirm the capability of the proposed configuration, this paper demonstrated using a prototype of hybrid DCCB which include the semiconductor breaker with series-connected IEGTs.

II. BASIC OPERATION PRINCIPLE

An operation procedure of proposed hybrid DCCB is described with Fig. 2.

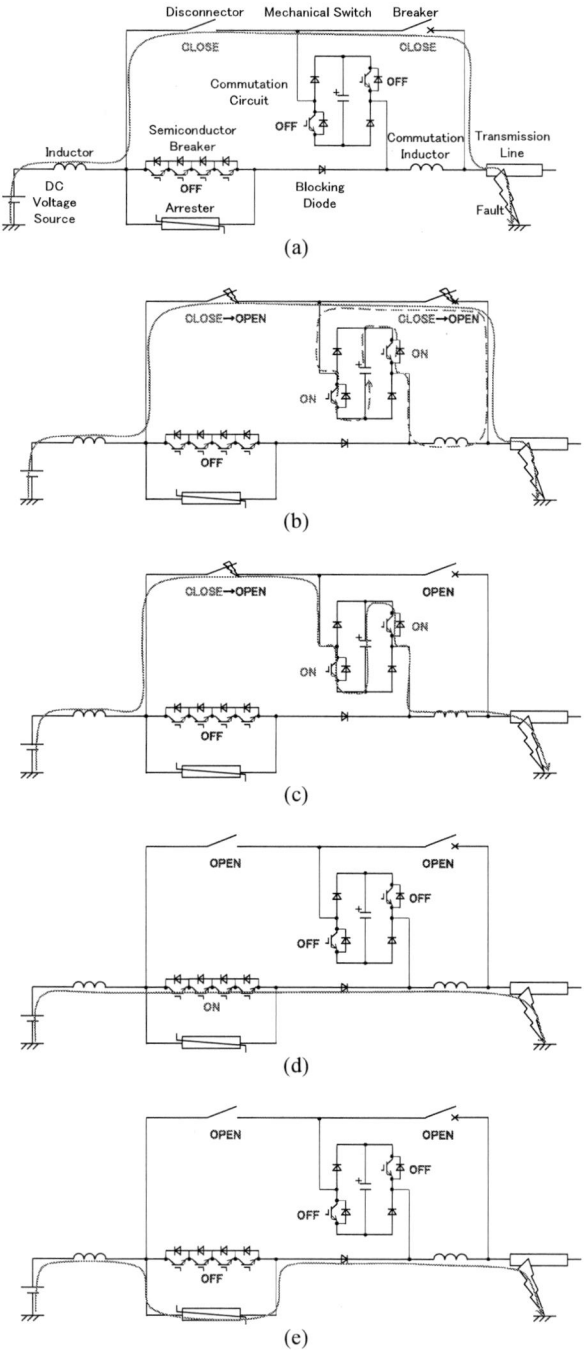

Fig. 2. Operation principle of hybrid DCCB.

(a): A system fault occurs and the fault current increases through the mechanical switch. (b): When the fault is detected, the disconnector and the breaker starts to

open. The semiconductors of the commutation circuit are turned on. Then capacitor of the commutation circuit is discharged and the current of the breaker decreases. (c): When the current of the breaker falls to zero, the arc of the breaker is extinguished. (d): When the semiconductor breaker is turned on and the semiconductor of the commutation circuit is turned off, the fault current commutates to the semiconductor breaker. (e): After insulation recovery of the disconnector, the semiconductor breaker turns off. Then the fault current commutates to the arrester and the fault current decays because the arrester consumes the energy stored in the inductor. Finally the interrupting operation is completed.

Based on the operation above, we consider the duties of the semiconductor breaker in detail. During normal operation, current does not flow through the semiconductor breaker. Only when a system fault occurs, the fault current flows through the semiconductor breaker and interrupted by the semiconductor breaker. When the fault current is interrupted, the overvoltage is suppressed by the arrestor. In addition to the arrester voltage, the voltage across the stray inductance between the arrester and the semiconductor breaker is applied to the semiconductor breaker. As shown in Fig.3, the semiconductor of the semiconductor breaker has the duty for interrupting a large current equivalent to the fault current. Moreover, it is necessary to suppress the overvoltage not to exceed the withstand voltage of the semiconductor breaker.

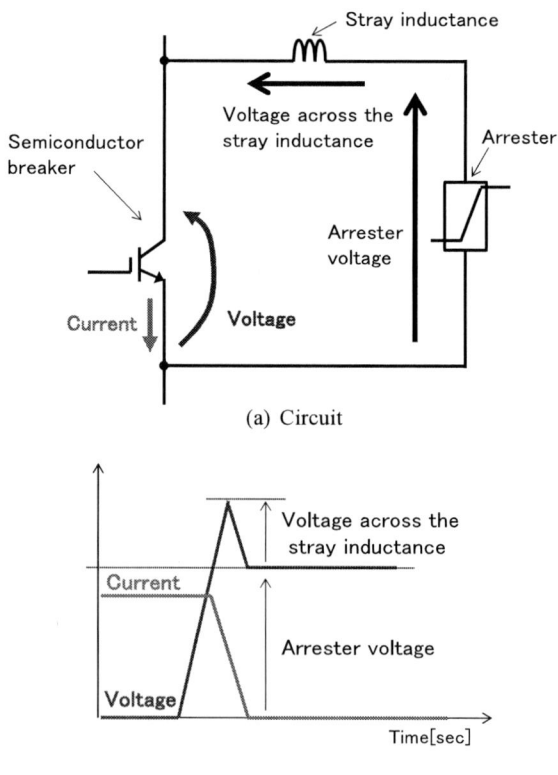

(a) Circuit

(b) Waveform

Fig. 3. Behavior of the semiconductor breaker.

III. CONFIGURATION OF THE SEMICONDUCTOR BREAKER

Based on the duties of the semiconductor breaker, this paper considers IEGTs applied to the semiconductor breaker.

Press pack IEGTs (4.5 kV, 2.1 kA) are used to block high voltage and large current. In general use, press pack IEGT has capability to interrupt twice as large current as rated current (4.2 kA). Since it is extremely short time for the fault current to flow through the semiconductor breaker, it is excessive to use IEGTs with the rated current capability equivalent to the fault current. Therefore, in order to minimize the rated value of IEGTs, this paper apply the snubber circuit to interrupt the fault current more than twice as large as the rated current of the IEGTs. Fig.4 shows diagram of the IEGT and snubber circuit. Applying the snubber circuit makes it possible to suppress the overvoltage as shown in Fig.5. In addition, since the current decreases before the voltage rises as shown in Fig.5, it is also possible to suppress the switching losses of the IEGTs. Therefore, the semiconductors can interrupt the fault current safely.

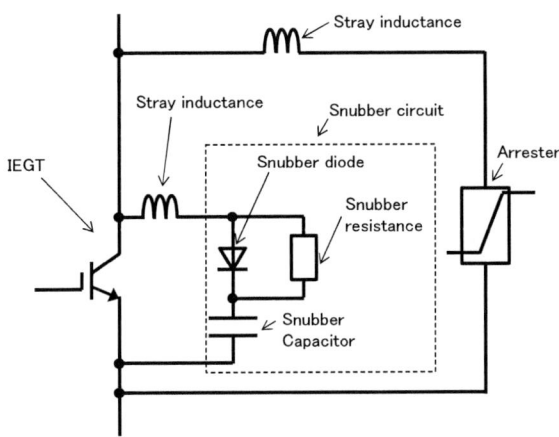

Fig. 4. Diagram of the IEGT and snubber circuit.

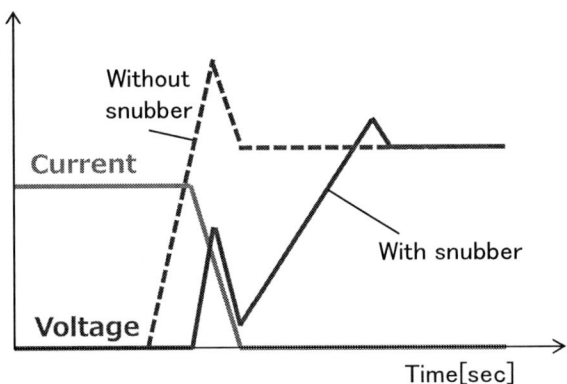

Fig. 5. Effect of the snubber circuit.

The operation of the semiconductor breaker with the snubber circuit will be described in detail. Fig.6 and Fig.7 show the operation of semiconductor breaker with the snubber circuit.

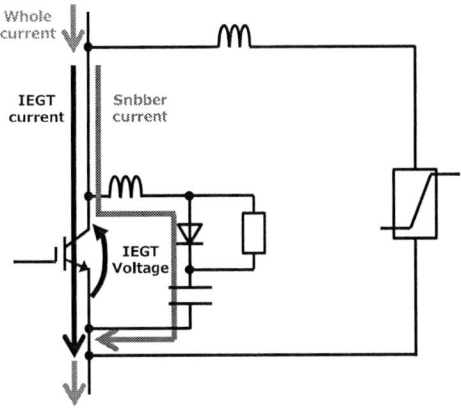

(a) Period 1

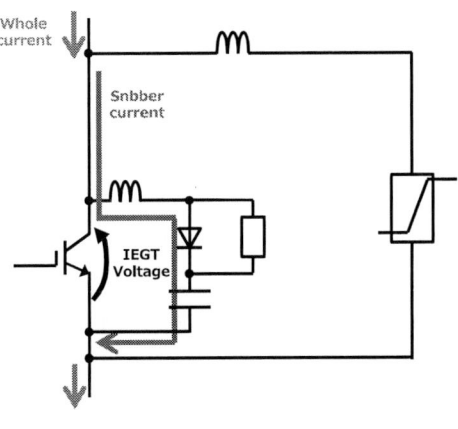

(b) Period 2

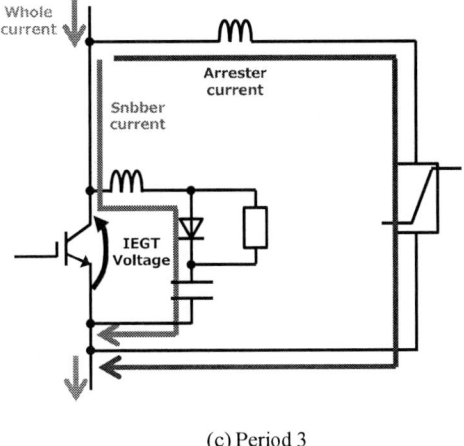

(c) Period 3

Fig. 6. Current flow of semiconductor breaker with the snubber circuit.

The 2018 International Power Electronics Conference

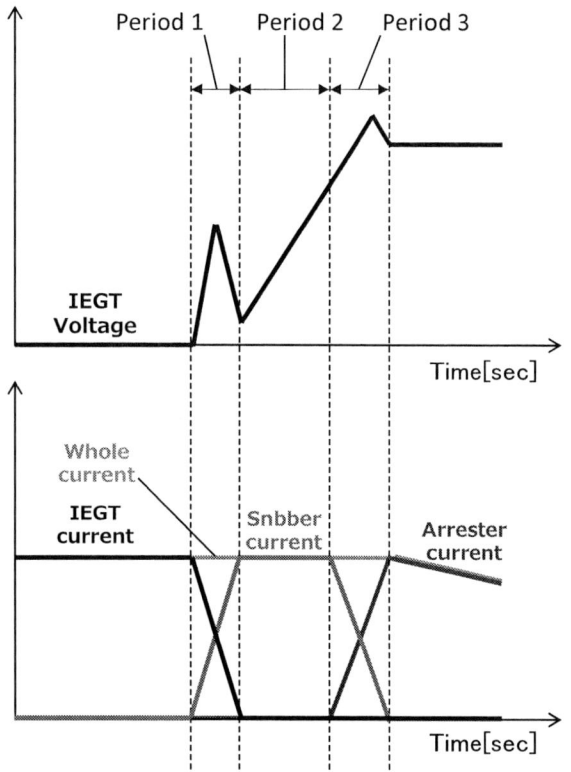

Fig. 7. Schematic diagram of operating waveforms of semiconductor breaker with the snubber circuit.

(Period 1)

When the IEGTs start to turn off, the IEGTs voltage begin to rise, and at the same time the charging current to the capacitor of the snubber circuit starts to flow through the snubber diode and the voltage across the snubber capacitor starts to rise. As the charging current to the snubber capacitor increases, the current of IEGTs decrease. Since there is a stray inductance between the snubber circuit and IEGTs, the voltage across IEGTs becomes higher than the snubber capacitor voltage by the voltage across the stray inductance. When the commutation from the IEGTs to the snubber circuit is completed, the voltage across the IEGTs and the voltage across the snubber capacitor become equal.

In this period, although IEGTs generate switching losses, the voltage is suppressed by the snubber capacitor. Thus, the losses are lower than that without snubber.

(Period 2)

Even after all currents are commutated to the snubber circuit, the snubber capacitor continues to be charged by the current of the snubber circuit, and the capacitor voltage rises. During this period, the voltage across IEGTs is equal to the voltage across the snubber capacitor.

In this period, since current does not flow in IEGTs, losses do not occur in IEGTs.

(Period 3)

When the voltage across the snubber capacitor reaches the operating voltage across the arrester, the current through the snubber circuit starts commutating to the arrester. Since the stray inductance exists between the arrester and the IEGT (snubber circuit), the voltage across IEGTs is the sum of arrester voltage and the voltage across stray inductance. When whole current commutate to the arrester, IEGTs and the snubber circuit complete to turn off.

Also in this period, since current does not flow in IEGTs, losses do not occur in IEGTs.

From the above operation, it is possible to suppress IEGTs overvoltage and losses with the snubber circuit.

Furthermore, in order to withstand the high voltage equivalent to HVDC systems, the semiconductor breaker is constituted with series-connected IEGTs. In order to suppress the voltage across each IEGT under the withstand voltage voltage across IEGT, the voltage balance of each series-connected IEGT is required. It is mainly influenced by characteristics of each IEGT, circuit conditions and driving speed. In order to balance the voltage across each series-connected IEGT, this paper made the characteristics of each IEGT and each circuit condition of IEGT as similar as possible, and adjusted the operation timing of each series-connected IEGT.

IV. EXPERIMENTAL RESULT

This paper built and experimented a prototype of the proposed hybrid DCCB and evaluated the behavior of the semiconductor breaker. The structure of prototype of semiconductor breaker is shown as Fig.8. The semiconductor breaker was composed of four series connected IEGTs (4.5 kV, 2.1 kA) with the snubber circuit.

Fig. 9 shows the waveforms of the experimental result of the semiconductor breaker. The semiconductor breaker with series-connected IEGTs and snubber circuit successfully interrupted large current over 9 kA and peak voltage across the semiconductor breaker was 14 kV.

The waveforms show that the current decreased before the voltage rose, and switching losses of the IEGTs were suppressed.

Fig.10 shows the peak voltage across IEGTs. The result demonstrated that variation of each IEGT peak voltage was less than 5 %, and each IEGT peak voltage was suppressed under the withstand voltage of IEGT.

The 2018 International Power Electronics Conference

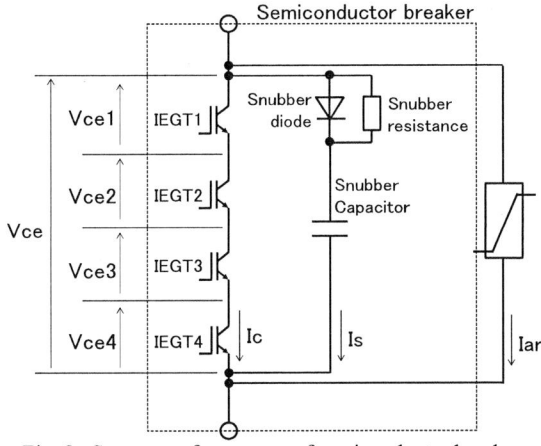

Fig. 8. Structure of prototype of semiconductor breaker.

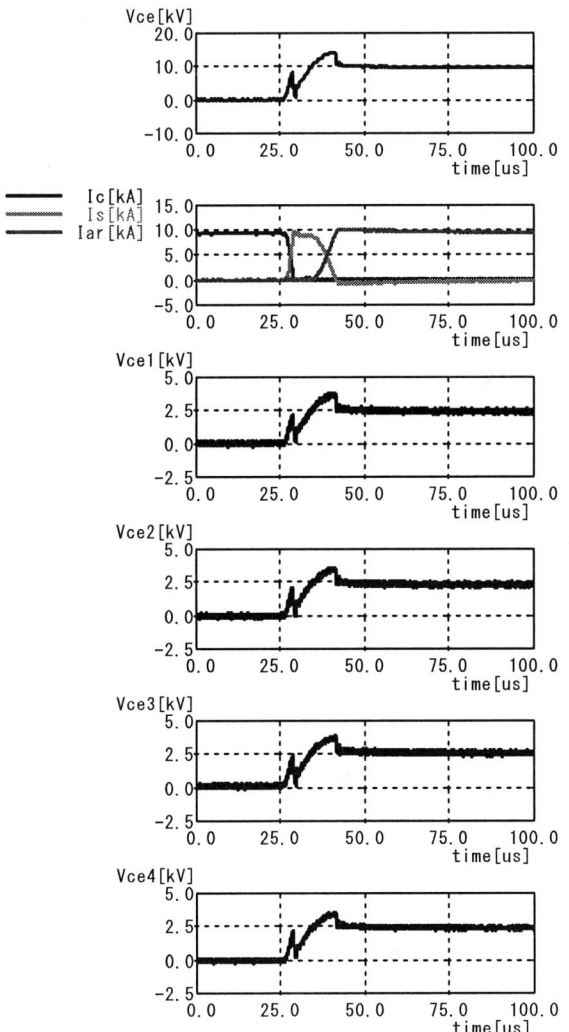

Fig. 9. Experimental result of the semiconductor breaker.

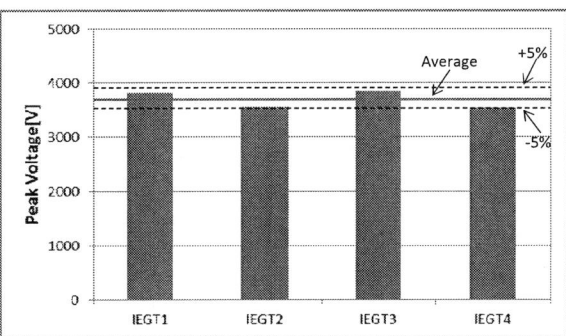

Fig. 10. Peak voltage.

V. CONCLUSION

This paper presented the semiconductor breaker of the proposed hybrid DCCB, and applied series-connected IEGTs with snubber circuits. Experimental results with the prototype of the hybrid DCCB have shown that it successfully interrupt a high voltage and large current. Moreover, it was confirmed that the voltage balance of each series-connected IEGT performed well.

ACKNOWLEDGMENT

This article is based on results obtained from a project commissioned by the New Energy and Industrial Technology Development Organization (NEDO).

REFERENCES

[1] CIGRE TB492 B4-46, "Voltage Source Converter (VSC) HVDC for power Transmission - economic aspects and comparison with other AC and DC Technologies", CIGRE, 2010.

[2] C. Franck, "HVDC circuit breakers: a review identifying future research needs," IEEE Trans. on Power Delivery, vol. 26, no. 2, pp. 998–1007, Apr. 2011.

[3] J. Hafner and B. Jacobson, "A key innovation for reliable HVDC grids", CIGRE symposium, Bologna, September 2011.

[4] A. Hassanpoor, J. Hafner, B. Jacobson, "Technical assessment of load commutation switch in hybrid HVDC breaker", pp. 3667–3673, IPEC, 2014.

[5] W. Grieshaber, J.-P. Dupraz, D.-L. Penache and L. Violleau, "Development and test of a 120kV direct current circuit breaker", B4-301, CIGRE, 2014.

[6] W. Zhou, X. Wei, S. Zhang, G. Tang, Z. He, J. Zheng, Y. Dan, C. Gao, "Development and test of a 200 kV full-bridge based hybrid HVDC breaker", EPE, 2015.

[7] G. F. Tang, X. G. Wei, W. D. Zhou, S. Zhang, C. Gao, Z. Y. He, J. C. Zheng, "Research and Development of a Full-bridge Cascaded Hybrid HVDC Breaker for VSC-HVDC Applications", A3-117, CIGRE, 2016.

[8] R.Hasegawa, K.Takimoto, T.Matsumoto, N.Iio, "Principle verification of Hybrid DCCB for HVDC Transmission System", 6-26, IEEJ Annual Meeting 2017

3308

Study of EMI Caused by Buck Converter on Controller Area Network

Ryo Shirai and Toshihisa Shimizu*
Tokyo Metropolitan University
1-1 Minami-Osawa, Hachioji-shi, Tokyo, Japan
*E-mail: shimizut@tmu.ac.jp

Abstract—The controller area network (CAN) system is one of the serial bus systems used in various applications, such as automobiles, airplanes, and industry. This study describes the analysis of electromagnetic interference (EMI) caused by a buck converter that is installed close to a CAN communication bus, and highlights the problems in a conventional EMI evaluation, followed by the development of an experimental EMI mitigation method. The experiments performed for verification show that the periodic switching noise of a buck converter causes data-transmission failure of CAN. Finally, this study proposes a novel control method for a buck converter, which can reduce the data-transmission failure of CAN.

Keywords—Buck converter, CAN, EMI, Noise, Spread-spectrum technique.

I. INTRODUCTION

The electromagnetic interference (EMI) caused by switching converters has always been a problem [1]. Moreover, communication networks, such as local area networks (LANs), are one of the victims of EMI. The unintentional EMI noise from a power converter, which propagates as radiation noise or a conduction noise, deteriorates the signal integrity in communication networks [2] and causes a data-transmission failure of the controller area network (CAN). To reduce EMI noise, shielding [3]–[5] and filtering [6]–[8] are commonly applied to switching converters or communication networks [9]. However, these conventional EMI reduction methods may not be effective in every situation, and hence, an ad-hoc design that is suitable for each specific application/situation is required. In addition, various international standards are set to ensure electromagnetic compatibility (EMC) between the switching converter and communication network. These standards specify the EMI level of switching converters in the frequency domain, and the EMI reduction methods are generally applied based on these standards in the frequency domain [10].

However, only a few studies are published that mention not only the EMI reduction method of switching converters but also the failure mechanism of EMI victims, such as communication networks. Furthermore, the mechanism of EMI noise propagation from switching converters to its victim is often unknown. To realize the effective noise reduction of the power converter and low noise susceptibility of CAN communication, it is necessary to analyze both simultaneously. Therefore, this study presents the comprehensive analysis of an EMI caused by a buck converter, which is installed close to a CAN communication line. The CAN system is one of the common automotive LAN standards and is widely used in powertrain and power control systems in which high reliability is required [11],[12]. Regarding the electromagnetic susceptibility of CAN, some studies are available that consider the EMI of integrated circuits [13], fast electrical transient, and other such factors [14]–[16]. However, to the best of the authors' knowledge, only a few studies are published that focuses on the periodic switching noise emitted by a switching converter installed close to a CAN communication line. Reference [17] described an improved CAN communication protocol against periodic switching noise of a switching converter, but the physical mechanism of the EMI caused by the switching converter, leading to CAN communication failure was not clarified. Moreover, changing the well-established CAN communication protocol is not a realistic solution when it comes to being applied to an actual system. Therefore, the main purpose of this study is to develop a more reasonable EMI mitigation method without modifying the CAN communication protocol.

This study first presents the configuration of an EMI test bench, which we call a noise injection system that intentionally generates EMI noise in a CAN communication line. This system is effective in realizing a stable noise injection and in ensuring sufficient repeatability of the experimental verification process. In section III, the mechanism of EMI noise propagation is clarified step-by-step through several experiments using the noise injection system. Section IV presents the result of serial bus analysis to evaluate the performance of CAN communication under the EMI noise environment of a buck converter. The signal integrity of a CAN system greatly changes with the input voltage of a buck converter. In sections V and VI, two implemented EMI mitigation methods of the noise injection system are discussed. In the first method, the EMI noise of the buck

The 2018 International Power Electronics Conference

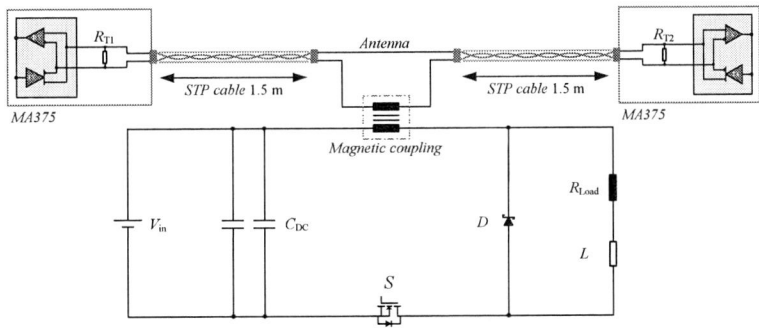

Fig. 1. Circuit diagram of noise injection system.

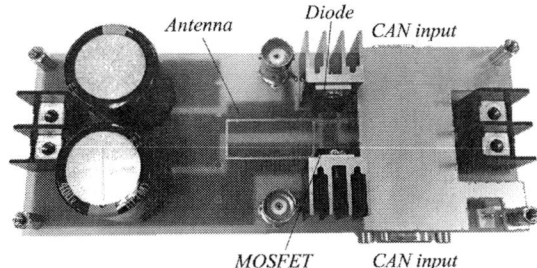

Fig. 2. Picture of noise injection system.

TABLE I
SPECIFICATION OF THE NOISE INJECTION SYSTEM

Buck converter	
DC link capacitor C_{DC}	270 μF, DC 400 $V_{max} \times 2$
Diode D	CREE, C3D 10060 A, 600 V, 14.5 A
MOSFET S	Infineon, SPP20N60C3, 650 V, 20.7 A
Inductor L	810 μH
Load resistor R_{Load}	10 Ω
Gate resistor R_G	6.2 Ω
Switching frequency f_S	20 kHz (Gate driver: 0 V to 17 V)
Input DC voltage V_{in}	0–200 V

CAN system	
CAN node	Micro Application Lab. MA 375 $\times$ 2
Termination resistor R_{T1} / R_{T2}	120 Ω / 120 Ω
CAN communication speed	125 kbps
Frame interval	50 ms

converter is reduced in the frequency domain by applying a spread-spectrum modulation to the buck converter. However, in this method, the data-transmission failure of CAN is drastically increased, which suggests that EMI reduction in the frequency domain is not always effective to suppress the data transmission failure of CAN. In the second method, a novel switching method for the buck converter is proposed and verified, in which the data-transmission failure of CAN is reduced effectively. In section VII, the conclusions of this study are mentioned.

II. NOISE INJECTION SYSTEM

The noise injection system is designed as an EMI test bench to intentionally inject the EMI noise to the CAN bus line. The system consists of a buck converter and a CAN communication system, as shown in Fig. 1 and Table I. As shown in Fig. 2, the parallel-shaped DC bus line and antenna-like CAN bus line are very important features in this system. Moreover, the common-mode noise of the buck converter is suppressed as much as possible by deleting its ground wire in the system. Owing to these features, the differential-mode noise is injected to the CAN bus line by the effect of magnetic coupling between the DC and CAN bus lines [18], and the parasitic-oscillation current flowing through the DC bus line dominantly produces a noise voltage in the CAN bus line [19].

The topology of a CAN system follows the high-speed CAN protocol specified by ISO 11898-2, and the communication speed is set to the lowest speed of 125

kbps. For a CAN communication system, two CAN nodes controlled by programmable interface controllers (PICs) are used in its noise injection system and accordingly connected to the antenna-like CAN bus via shielded twisted pair (STP) cables.

III. MECHANISM OF NOISE PROPAGATION FROM THE BUCK CONVERTER TO THE CAN BUS

In this section, several measurement results are provided to grasp the noise-propagation mechanism in a noise injection system.

A. Parasitic-oscillation waveforms in the DC bus line

The drain-source voltage v_{DS} of the MOSFET when it turns off is measured because the MOSFET is mounted in the parasitic-oscillation path of the mains circuit. In this study, the parasitic-oscillation current in the DC bus when the MOSFET is turned on is almost negligible because a Schottky barrier diode is used to suppress the oscillation current as much as possible.

As shown in Fig. 3, a high-frequency oscillation waveform appears on the measured voltage waveforms of v_{DS}. These undesirable oscillation waveforms are due to the parasitic impedance of the oscillation path. Thus, the parasitic-oscillation waveform in the v_{DS} significantly depends on the parasitic impedance of the oscillation

3310

path. The parasitic impedance of the off-state MOSFET [20] affects the oscillation frequency and damping factor of the oscillation waveform.

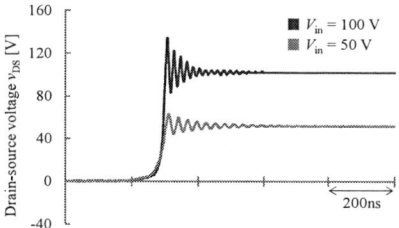

Fig. 3. Measured waveforms of drain-source voltage of the MOSFET when V_{in} is 50 V and 100 V.

B. Noise propagation in the CAN system

The parasitic-oscillation current in the DC bus line is caused when the MOSFET is turned off. Subsequently, as shown in Figs. 4(a) – 4(d), the noise voltage is induced in the CAN bus line by the effect of magnetic coupling between the DC and CAN bus lines. When the noise voltage is higher than the threshold voltage of CAN receiver, the signal output voltage decreases from 5 V to 0 V, which is defined as an error signal in this study. Even though the noise voltage has multiple pulses, the error signal becomes a single pulse because the CAN receiver does not have the required slew rate. In addition, the time length of the error signal gradually becomes longer as V_{in} is increased, and the risk of the bit error rate in CAN communication is increased.

IV. DATA-TRANSMISSION FAILURE OF CAN

The CAN communication process is realized by transmitting data frames at certain intervals. When the EMI noise of a buck converter produces critical error signals in a frame, a frame can fail to transmit its data, which is defined as frame loss, in this study.

Frame losses are caused when the noise voltage is induced in the CAN bus. Hence, some frame errors are observed by serial bus analysis under the operation of a buck converter. The serial bus analysis is carried out by varying V_{in} from 0–200 V by using a serial bus analyzer, Tektronix MDO3054. The transmitting CAN node continuously transmits data frames to the receiving node at 125 kbps. The frame interval is set at 50 ms and the measurement is carried out for 2 s. The measured errors are classified as cyclic-redundancy-check (CRC) error, data error, end-of-frame (EOF) error, and frame error, which are detected by their respective functions of the serial bus analyzer.

Fig. 5 shows the result of serial bus analysis in the range of V_{in} from 0–200 V. The result shows that the number of errors, except for EOF errors, gradually increases as V_{in} is increased, as well as the time width of the error signal becomes longer. The measurement results indicate that the frame error rate is proportional to the total time width of the error signals during one switching period of the buck converter.

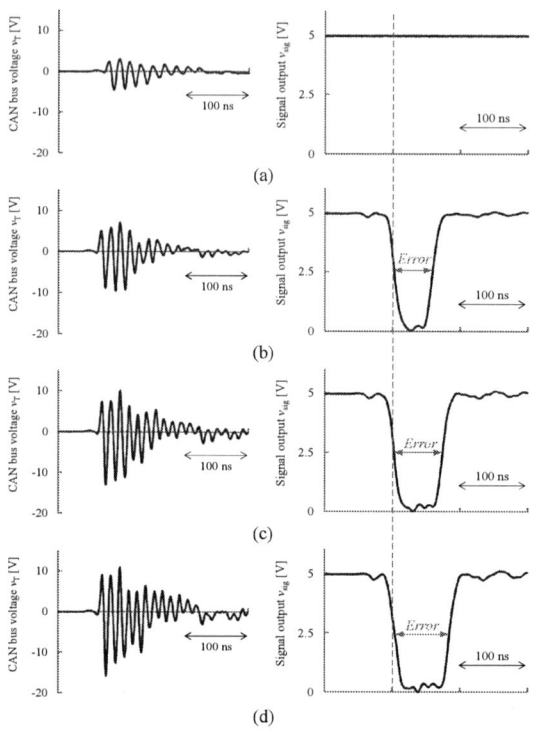

Fig. 4. Measured voltage waveform of CAN bus voltage v_T and signal output v_{sig} when MOSFET is turned off. (a) V_{in} = 50 V, (b) V_{in} = 100 V, (c) V_{in} = 150 V, and (d) V_{in} = 200 V.

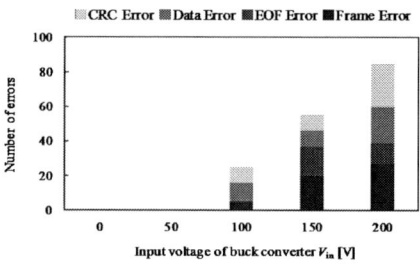

Fig. 5. Measured result of serial bus analysis.

V. EFFECT OF SPREAD-SPECTRUM TECHNIQUE ON A BUCK CONVERTER

The spread-spectrum technique is known as one of the major EMI reduction methods of a switching converter in the frequency domain, which is described in [21]–[24]. By applying this technique to a switching converter, the spectral characteristics can be improved. However, when the spread-spectrum technique is applied to the switching converter, the timing of switching noise is assumed to be possibly more disturbed, rather than operating at a constant frequency. Hence, the effectiveness of the spread-spectrum technique for improving the CAN communication performance is questionable.

To verify the effectiveness of the spread-spectrum technique in improving the performance of CAN communication, frequency-modulation control is applied to the buck converter of the noise injection system. In this

3311

The 2018 International Power Electronics Conference

study, the triangular modulation method is employed to incorporate its simplicity to the system [25]. The switching frequency of the buck converter is linearly modulated from 20–50 kHz in a 0.44 µs cycle, as shown in Fig. 6.

The measurement setup is shown in Fig. 7. The vertical near-magnetic field at the center of the DC bus line is measured by a spectrum analyzer with a 10 mm loop antenna probe. The measurement frequency range is from 10–100 MHz to observe the spectral characteristic of the parasitic-oscillation current in the DC bus. Fig. 8 shows the measurement result of the near-magnetic field of the DC bus line. The result indicates that the frequency-modulation control can reduce the EMI noise of the buck converter in the frequency domain as compared to that with a constant frequency operation at 35 kHz.

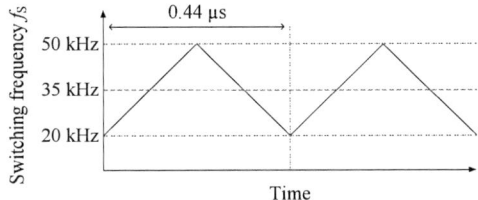

Fig. 6. Modulation pattern of the switching frequency f_s.

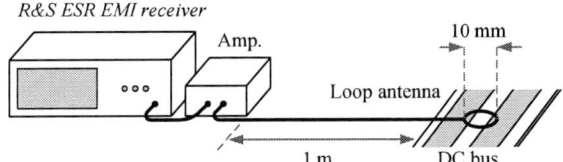

Fig. 7. Measurement setup for spectrum analysis on the vertical near-magnetic field of the DC bus.

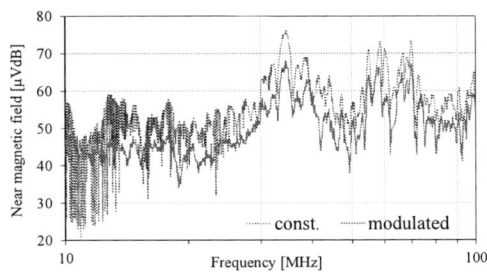

Fig. 8. Measurement result of the vertical near-magnetic field at the center of the DC bus.

Moreover, the result of serial bus analysis indicates that the switching frequency modulation is not always effective for improving the CAN communication performance. The serial bus analysis is carried out by applying frequency modulation to the buck converter

with the measurement same as that of the near-magnetic field.

The measured frame error rates are **13%** when the spread-spectrum technique is applied to the noise injection system and **3%** when the buck converter constantly operates at 20 kHz. The measurement result shows that even though switching frequency modulation is employed, the frame error rate increases drastically. In other words, even though the EMI noise of the buck converter in the frequency domain is reduced by applying the frequency-modulation control, the performance of the CAN communication is deteriorated instead of improving. Therefore, evaluation of the EMI noise propagation in the time domain is strongly needed to suppress the CAN communication failure.

VI. DEVELOPMENT OF AN EMI MITIGATION METHOD BASED ON TIME-DOMAIN ANALYSIS

Because the EMI reduction method based on frequency-domain analysis results in insufficient effect of reducing the EMI in CAN communications, an EMI mitigation method based on time-domain analysis is required. In this section, a novel EMI reduction method is proposed considering the mechanism of CAN communication failure.

A. Basic principle of CAN communication failure

Figs. 9(a) and 9(b) show the simplified schematic waveform of the CAN communication signal and noise voltage. When the noise voltage is induced on a period other than the CAN bit-sampling point as shown in Fig. 9 (a), no error will be happened. However, when the timing of the noise voltage and the CAN bit-sampling point are matched, as shown in Fig 9(b), possibility of the CAN signal error is greatly increased. Therefore, coincidence of the CAN bit-sampling point and noise voltage must be avoided to prevent data-transmission failure of the CAN communication.

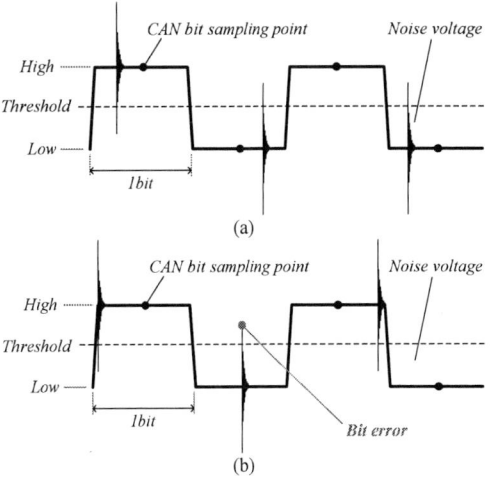

Fig. 9. Simplified schematic waveform of a CAN communication signal and induced noise voltage. (a) No bit errors and (b) with a bit error.

3312

The 2018 International Power Electronics Conference

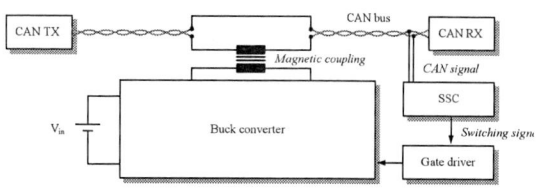

Fig. 10. Noise injection system applying a start-stop synchronization control.

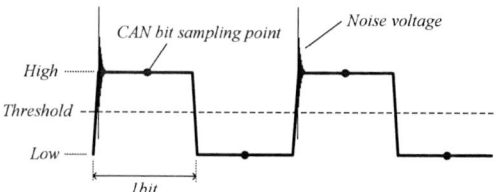

Fig. 12. Simplified schematic waveform of a CAN communication signal and induced noise voltage when the start-stop synchronization control is applied.

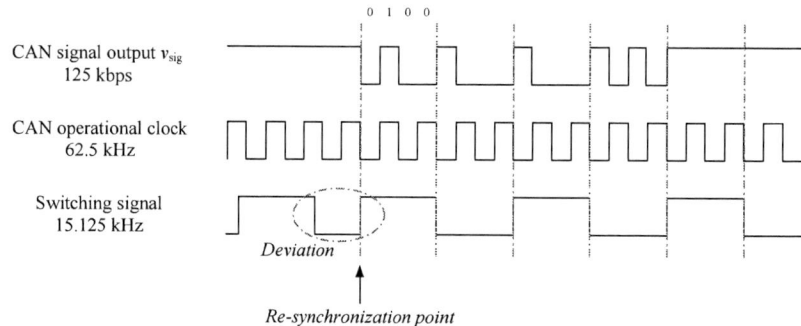

Fig. 11. Timing chart of the start-stop synchronization sequence.

B. Experimental EMI mitigation method

In general, the operation clock of switching converters and CAN systems are not synchronized, and the switching noise is injected randomly to the CAN systems. Hence, the synchronization between the buck converter and the operating clock of CAN is required to control the timing of the noise voltage in the CAN communication signal. To realize synchronization between them, start-stop synchronization method is employed to the noise injection system.

The start-stop synchronization technique is commonly used in various asynchronous communication systems. To employ the control method, a start-stop synchronization circuit (SSC) is needed. Fig. 10 shows the noise injection system embedding to the SSC. The function of the SSC is quite simple. Fig. 11 shows the timing chart to describe how the SSC operates in the time domain. When the start timing of each frame bit is transmitted to the CAN bus, the switching timing of the buck converter is re-synchronized with the operating clock of CAN. In this study, the switching frequency f_s of the buck converter is set to 15.125 kHz, which is one-quarter of the operating-frequency of CAN (62.5 kHz, 125 kbps). Hence, the output signal of the CAN receiver can be observed, as shown in Fig. 12. A noise voltage is simultaneously induced at the timing of voltage transition in the CAN communication signal, and it never coincides with the CAN bit-sampling point.

To verify that the start-stop synchronization control is effective to reduce error frames in the CAN

communication, the serial bus analysis is carried out under applying the start-stop synchronization control to the noise injection system. The result shows that CAN communications are successfully realized, and the frame-error rate is totally **0%** irrespective of DC bus voltage V_{in}.

However, some drawbacks are experienced when the start-stop synchronization control is applied to the noise injection system.

First, when the switching signal is re-synchronized with the operating clock of CAN, the switching signal has a time deviation, which affects the control stability of the output voltage of the buck converter. To solve this problem, an improved switching control method needs to be applied. A detailed explanation of the method will be given in a future study.

Second, the EMI noise emitted by the buck converter is not suppressed even when the start-stop synchronization control is applied. Hence, a combination with other EMI reduction methods is needed to meet the EMI standards.

VII. CONCLUSIONS

In this study, the analysis of EMI caused by buck converters on CAN communication has been presented. The measured voltage waveforms in the noise injection system indicate that the high-frequency oscillation in the DC bus induces noise voltage in the CAN bus, and the noise voltage can cause error signals in the output of the CAN receiver. The results of the serial bus analysis verify that the frame error rate of CAN communication is significantly affected by the operating condition of the buck converter. According to the detailed analysis of the noise injection system, two EMI reduction methods are implemented. The spread-spectrum technique is significantly effective in reducing the EMI in the frequency domain. However, the measurement result indicates that it cannot reduce the transmission failure in CAN communication. Start-stop synchronization control is one of the methods that avoid the noise voltage simultaneously induced with the CAN bit sampling point. The authors have demonstrated that the communication failure of CAN is completely prevented by applying start-stop synchronization.

ACKNOWLEDGMENT

This work was supported by JSPS KAKENHI Grant Number JP16H02328.

REFERENCES

[1] R. Redl, "Power electronics and electromagnetic compatibility," *PESC '96 Record. 27th Annual IEEE Power Electronics Specialists Conference*, vol. 1, pp. 15–21, 1996.

[2] M. Jin, M. Weiming, P. Qijun, K. Jun, Z. Lei, and Z. Zhihua, "Identification of essential coupling path models for conducted EMI prediction in switching power converters," *IEEE Trans. Power Electron.*, vol. 21, no. 6, pp. 1795–1803, 2006.

[3] S. C. Tang, S. Y. R. Hui, and H. S. -H. Chung, "Evaluation of the shielding effects on printed-circuit-board transformers using ferrite plates and copper sheets," *IEEE Trans. Power Electron.*, vol. 17, no. 6, pp. 1080–1088, 2002.

[4] S. Yang, Q. Chen, and W. Chen, "Common mode EMI noise reduction technique by shielding optimization in isolated converters," *2016 7th Asia Pacific International Symposium on Electromagnetic Compatibility (APEMC)*, vol. 1, pp. 622–625, 2016.

[5] S. Stegen and J. Lu, "Shielding effect of high frequency power transformers for DC/DC converters used in solar PV systems," *2010 Asia-Pacific International Symposium on Electromagnetic Compatibility (APEMC)*, pp. 414–417, 2010.

[6] M. Ali, E. Labouré, F. Costa, and B. Revol, "Design of a hybrid integrated EMC filter for a DC-DC power converter," *IEEE Trans. Power Electron.*, vol. 27, no. 11, pp. 4380–4390, 2012.

[7] M. Hartmann, H. Ertl, and J. W. Kolar, "EMI filter design for a 1 MHz, 10 kW three-phase/level PWM rectifier," *IEEE Trans. Power Electron.*, vol. 26, no. 4, pp. 1192–1204, 2011.

[8] F. -Y. Shih, D. Y. Chen, Y. -P. Wu, and Y. -T. Chen, "A procedure for designing EMI filters for AC line applications," *IEEE Trans. Power Electron.*, vol. 11, no. 1, pp. 170–181, 1996.

[9] H. Tanaka, Y. Saito, H. Kumatani, and T. Azuma, "EMC solution for automotive LAN," *IEICE Technical Report*, vol. 104, pp. 11–16, 2009.

[10] K. Mainali and R. Oruganti, "Conducted EMI mitigation techniques for switch-mode power converters: A survey," *IEEE Trans. Power Electron.*, vol. 25, no. 9, pp. 2344–2356, 2010.

[11] S. Corrigan, "Introduction to the controller area network (CAN)," *Application Report, Texas Instruments*, pp. 1–15, 2002.

[12] W. Xing, H. Chen, and H. Ding, "The application of controller area network on vehicle," *Proceedings of the IEEE International Vehicle Electronics Conference, 1999. (IVEC '99)*, vol. 1, pp. 455–458, 1999.

[13] M. Fontana, F. G. Canavero, and R. Perraud, "Integrated circuit modeling for noise susceptibility prediction in communication networks," *IEEE Trans. Electromagn. Compat.*, vol. 57, no. 3, pp. 339–348, 2015.

[14] M. Fontana and T. H. Hubing, "Characterization of CAN network susceptibility to EFT transient noise," *IEEE Trans. Electromagn. Compat.*, vol. 57, no. 2, pp. 188–194, 2015.

[15] F. Ren, Y. R. Zheng, M. Zawodniok, and J. Sarangapani, "Effects of electromagnetic interference on control area network performance," *2007 IEEE Region 5 Technical Conference*, pp. 199–204, 2007.

[16] M. Fontana, F. G. Canavero, and R. Perraud, "Electromagnetic susceptibility assessment of controller area networks," *Proceedings of the 2014 International Symposium on Electromagnetic Compatibility (EMC Europe 2014)*, pp. 795–800, 2014.

[17] M. Nakamura, M. Ohara, M. Arai, K. Sakai, S. Fukumoto, and K. Wada, "Testbeds of a hybrid-ARQ-based reliable communication for CANs in highly electromagnetic environments," *2015 IEEE 2nd International Future Energy Electronics Conference (IFEEC)*, pp. 1–6, 2015.

[18] O. Aouine, C. Labarre, and F. Costa, "Measurement and modeling of the magnetic near field radiated by a buck chopper," *IEEE Trans. Electromagn. Compat.*, vol. 50, no. 2, pp. 445–449, 2008.

[19] R. Shirai and T. Shimizu, "Basic study of a novel EMC solution for the electromagnetic interference of power converter to CAN," *IEEJ Workshop SPC/MD*, 2017.

[20] K. Kam, D. Pommerenke, A. Bhargava, B. Steinfeld, C. -W. Lam, and F. Centelo, "Quantification of self-damping of power MOSFET in a synchronous buck converter," *IEEE Trans. Electromagn. Compat.*, vol. 53, no. 4, pp. 1091–1093, 2011.

[21] R. Mukherjee, A. Patra, and S. Banerjee, "Impact of a frequency modulated pulsewidth modulation (PWM) switching converter on the input power system quality," *IEEE Trans. Power Electron.*, vol. 25, no. 6, pp. 1450–1459, 2010.

[22] A. Bendicks, S. Frei, N. Hees, and M. Wiegand, "Systematic reduction of peak and average emissions of power electronic converters by the application of spread spectrum," *IEEE Trans. Electromagn. Compat.*, no. 99, pp. 1–10, 2017.

[23] K. K. Tse, H. S. -H Chung, S. Y. Huo, and H. C. So, "Analysis and spectral characteristics of a spread-spectrum technique for conducted EMI suppression," *IEEE Trans. Power Electron.*, vol. 15, no. 2, pp. 399–410, 2000.

[24] K. Inoue, K. Kusaka, and J. -I Itoh, "Reduction in radiation noise level for inductive power transfer systems using spread spectrum techniques," *IEEE Trans. Power Electron.*, vol. 33, no. 4, pp. 3076–3085, 2018.

[25] F. Pareschi, R. Rovatti, and G. Setti, "EMI reduction via spread spectrum in DC/DC converters: State of the art, optimization, and tradeoffs," *IEEE Access*, vol. 3, pp. 2857–2874, 2015.

The 2018 International Power Electronics Conference

A Study on Reduction Techniques of a Wideband Common-Mode Voltage Produced by a PWM Inverter

Shotaro Takahashi[1*], Satoshi Ogasawara[1], Masatsugu Takemoto[1], Koji Orikawa[1] and Michio Tamate[2]

1 Graduate School of Information Science and Technology, Hokkaido University, Sapporo, Japan

2 Research & Development Group, Fuji Electric Co., Ltd., Hino, Japan

*E-mail: takahashi@ist.hokudai.ac.jp

Abstract— Switching speed of next-generation power devices such as silicon carbide is about ten times faster than those of silicon devices. This aspect may increase the frequency ranges of common-mode (CM) voltage accompanied by switching operations. Meanwhile, operating frequency ranges of noise filters typically limited by parasitic effects and frequency dependencies of magnetic components. Thus, it is difficult to suppress a wideband noise by installing only conventional passive noise filter. This paper studies about the combination of a conventional low-frequency passive filter and a high-frequency passive filter (HF-PF) which consists of a high-frequency magnetic material such as nickel-zinc ferrite. Furthermore, the proposed active CM voltage feedback circuit is applied for improving the performance of the HF-PF. An insertion effect against CM voltage produced by a pulse-width-modulated inverter is evaluated experimentally. Experimental results show that the constructed system effectively reduce CM voltage in the frequency range from 100 kHz to 100 MHz.

Keywords— Active common-mode filter, Common-mode voltage, EMI, PWM inverter

I. INTRODUCTION

Power electronics equipments based on switching operations of power semiconductors have increased their market share due to its advantages of energy-saving and controllability. To achieve more higher efficiency levels, researches and developments of next-generation power semiconductor devices such as silicon carbide and gallium nitride are actively of late years. Switching speed of these devices are about ten times faster compared to those of conventional silicon devices (e.g. insulated-gate bipolar transistors). This means that the next-generation devices decrease conduction losses dramatically, but they also possess the side effect that electromagnetic noise accompanied by their switching operations has a wider frequency range beyond several tens megahertz. In three-phase adjustable speed motor drive systems, asymmetrical phase voltage pulses of pulse-width-modulation (PWM) generate common-mode (CM) voltage. This voltage causes high-frequency (HF) earth leakage currents [1]−[3] which propagate to a ground plane through stray capacitance, and motor shaft voltages [4] which may damage motor bearings.

To date, various filtering concepts for improving electromagnetic environments have presented [5]−[12].

Most of those are based on passive components. For example, reference [5] improved a differential-mode (DM) output filter (low-pass sinusoidal filter) by connecting a neutral point of the output filter to a dc-link midpoint. The proposed filter configuration can reduce both DM and CM noise at the motor terminals. The detailed design procedure of this filter presented in [6]. Another effective noise filtering techniques are utilizing active devices [9]−[12]. A literature [9] proposed an active common-noise canceller (ACC). The ACC consists of complimentary transistors and four-phase windings CM transformer, and achieve feedforward active CM voltage cancellation at a PWM inverter output.

The main target of these conventional techniques is the reduction of conducted noise with a frequency range from switching frequency up to several megahertz. Meanwhile, the operating frequency ranges of passive components are limited due to the frequency dependencies of magnetic materials and stray effects. As a consequence, performances of conventional filters begin to deteriorate at HF range beyond 10 MHz. This means that conventional filters are not suitable to reduce the wideband electromagnetic noise produced by next-generation devices.

HF CM noise beyond several tens megahertz mainly radiates from power cables [13]−[15]. When grounding condition of a shield sheath has high quality, a shielded cable may become an appropriate solution to suppress radiated noise [14], but, a proper shielding is not always available. Furthermore, in many cases, shield cables are too heavy and too expensive [15], [16]. Based on the aforementioned discussion, this paper studied a combination of a conventional low-frequency passive filter (LF-PF) and a high-frequency passive filter (HF-PF). The HF-PF focuses on the reduction of HF CM noise with frequency beyond several tens megahertz and the suppression of radiated noise from a power cable, and mainly consists of a three-phase HF CM inductor realized by employing a HF magnetic core material. Furthermore, an active CM voltage feedback circuit proposed by the authors in previous work [17], is applied for improving performance of the HF-PF. The detailed filter design procedures are described, and the basic principle of the ACF is analyzed with a simple equivalent circuit model. Insertion effects of the filters against CM voltage produced by a PWM inverter is evaluated in a three-phase motor

3315

The 2018 International Power Electronics Conference

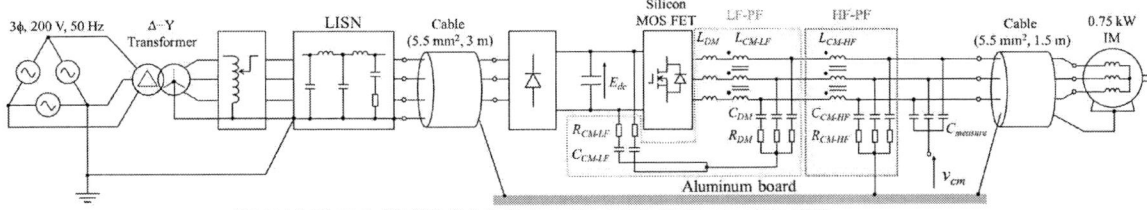

Fig. 1. Experimental system for the measurement of common-mode voltage.

drive system. Experimental results show that the system constructed in this paper effectively suppresses CM voltage at the inverter output over wide frequency range from 100 kHz to 100 MHz.

II. EXPERIMENTAL SYSTEM AND PASSIVE FILTERS

A motor drive system constructed for measurements of output CM voltages is shown in Fig. 1. The system consists of a three-phase 5-kVA PWM inverter and a 0.75 kW induction motor. A 1.5 meters long power cable is used for connecting the inverter to the motor. The cable includes four conductors, three for phases and one for a ground line. An aluminum board which is grounded through a LISN (line-impedance-stabilization-network), imitates a metal box of the inverter. A motor case is connected and grounded to the aluminum board. The LF-PF and the HF-PF are installed at the inverter output. Structures and design procedures of these filters are described as the following subsections respectively.

A. Low-frequency passive filter

As shown in Fig. 1, an LF-PF is a combination of a DM filter and a CM filter. The DM filter which consists of three DM inductors L_{DM}, DM capacitors (commonly called as X-capacitors) C_{DM} and three resistors R_{DM}, eliminates HF switching ripples. As a result of the installation of the DM filter, an output line-to-line voltage of the inverter becomes a sinusoidal waveform, moreover, overvoltage and resonance caused by leakage inductance of CM inductor and cable capacitance are effectively damped. The detailed design procedure of the DM filter is described in [6], thus, the design of the DM filter is not discussed in this paper. A three-phase CM inductor L_{CM-LF}, two capacitors C_{CM-LF}, two damping resistors R_{CM-LF} form the CM filter. Connecting a neutral point of DM capacitors to DC-link constructs a low impedance path of CM current. The specifications of the components selected for the LF-PF are shown in TABLE I.

Firstly, we start to design the three-phase LF CM inductor. We assume that almost all switching frequency component of CM voltage was applied across the CM inductor, under this assumption, the allowable CM current $I_{CM,max}$ flowing to DC-link can be possible to estimate as

$$I_{CM,max} = \frac{E_{dc}}{2\pi f_{sw} L_{CM-LF}} \tag{1}$$

where, E_{dc} is DC-link voltage and f_{sw} is switching frequency. All experiments of the section □ are carried out

TABLE I
SPECIFICATIONS OF SELECTED COMPONENTS FOR THE LF-PF

Symbol	Specification
L_{DM}	Tokyo-Seiden, TSL2T-15A, 1 mH
C_{DM}	KEMET, R47-X2-440, 0.22 µF, 440 VAC
R_{DM}	Ohmite, WH/WN Series, 15 Ω, 5 W
L_{CM-LF}	EPCOS, B64290A0082, OD: 50 mm, ID: 30 mm, HT: 20 mm, $k = 1$, 3×14 turns, ϕ. 1 mm, 1.7 mH
C_{CM-LF}	EPCOS, B32642B, 0.1 µF, 400 VAC
R_{CM-LF}	Ohmite, WH/WN Series, 100 Ω, 5 W

under E_{dc} = 200 V, f_{sw} = 100 kHz, and inverter output frequency is 50 Hz. $I_{CM,max}$ is chosen to be 0.2 A in this paper, so that the designed value of L_{CM-LF} is obtained to be 1.6 mH. A manganese-zinc ferrite N30 is selected as the core material of the LF CM inductor. By using dimensions of the selected core, a relation which the turn number N has to satisfy for specified inductance value L of a toroidal inductor can be described as

$$N > \sqrt{\frac{L}{k \cdot \frac{1}{2\pi} \mu_0 \mu_r HT \cdot \ln\left(\frac{OD}{ID}\right)}} \tag{2}$$

where, OD is the outer diameter, ID is the inner diameter, HT is the height of the selected core, k is the number of the core, μ_0 ($4\pi\times10^{-7}$ H/m) is the permeability of free space, and μ_r is the relative permeability of the magnetic material. In general, each parameter described in the equation (2) is available in datasheets presented from the manufacturers [18], [19].

Magnetic flux density inside the core B_{core} must not exceed the saturation magnetic flux density of the core material B_{sat}. According to [9], the flux density of the core under square-wave magnetization is given by

$$B_{core} = \frac{E_{dc}T}{8A_e k N} \tag{3}$$

where, T is the switching period, A_e is the cross-section area of the selected core. From (2), calculated turn number of the LF CM inductor is 14. Selecting $k = 1$, and using $T = (100\times10^3)^{-1}$ s and $A_e = 195.7$ mm², $B_{core} = 0.09$ T is obtained. This value is much lower than the saturation flux density of the N30 ($B_{sat} = 0.4$ T at 25 °C) [19].

The value of CM capacitor C_{CM-LF} is obtained as

3316

$$C_{CM\text{-}LF} = \frac{1}{2}\left(\frac{1}{2\pi f_{cut,CM\text{-}LF}}\right)^2 \cdot \frac{1}{L_{CM\text{-}LF}} \qquad (4)$$

where, $f_{cut,CM\text{-}LF}$ is the CM cutoff frequency of the LF-PF which should have a frequency range from the inverter output frequency (50 Hz) to the switching frequency (100 kHz), and be chosen to be 10 kHz, so that $C_{CM\text{-}LF} = 0.1$ μF is selected as CM capacitors. In this case, $f_{cut,CM\text{-}LF} = 8.6$ kHz can be calculated.

Taking into account of the allowable power loss of CM $P_{CM,max}$, the value of the CM damping resistor $R_{CM\text{-}LF}$ is given as

$$R_{CM\text{-}LF} = \frac{2P_{CM,max}}{I_{CM,max}^2} \qquad (5)$$

$P_{CM,max} = 2$ W is chosen in this paper, and $R_{CM\text{-}LF} = 100$ Ω is selected. Finally, the CM damping ratio of the LF-PF is calculated to be 0.28.

B. High-frequency passive filter

To suppress electromagnetic noise beyond several tens megahertz, the three-phase HF CM inductor $L_{CM\text{-}HF}$, Y-connected capacitors $C_{CM\text{-}HF}$ and a damping resistor $R_{CM\text{-}HF}$ form the HF-PF. Connecting the neutral point of the capacitors to the aluminum board, CM voltage potential difference between the power cable and aluminum board is reduced, thereby, radiated noise from output power cable is effectively suppressed (the reduction effect of the HF-PF against radiated noise was validated experimentally in an anechoic chamber [17]). A large value of $C_{CM\text{-}HF}$ decreases its self-resonance frequency caused by stray inductance, and limits its operating frequency range. A proper value of $C_{CM\text{-}HF}$ is derived as the relation between the allowable stray inductance L_{stray} and the desiable maximum operating frequency of the capacitor f_{max}. Therefore, $C_{CM\text{-}HF}$ is given by

$$C_{CM\text{-}HF} \leq \frac{1}{3}\left(\frac{1}{2\pi f_{max}}\right)^2 \cdot \frac{1}{L_{stray}} \qquad (6)$$

Frequencies up to 100 MHz are within the CM voltage measurement range of this study. From (6), the value of HF CM capacitor for $f_{max} = 100$ MHz and $L_{stray} = 10$ nH is $C_{CM\text{-}HF} \approx 84.5$ pF. Thus, $C_{CM\text{-}HF} = 33$ pF is selected.

The regulation standard for radiated noise is defined in the frequency range beyond 30 MHz. For the purpose of reducing radiated noise, the cutoff frequency of the HF-PF $f_{cut,CM\text{-}HF}$ should have a frequency range of 1-10 MHz, so that $f_{cut,CM\text{-}HF}$ is designed to be 10 MHz. Thus, the required value of the HF CM inductor $L_{CM\text{-}HF}$ is

$$L_{CM\text{-}HF} = \left(\frac{1}{2\pi f_{cut,CM\text{-}HF}}\right)^2 \cdot \frac{1}{3C_{CM\text{-}HF}} \qquad (7)$$

Thus, $L_{CM\text{-}HF} = 2.6$ μH is obtained.

It can be considered that low-frequency and large

TABLE II
SPECIFICATIONS OF SELECTED COMPONENTS FOR THE HF-PF

Symbol	Specification
$L_{CM\text{-}HF}$	Fair-Rite, 5967001701, OD: 31.8 mm, ID: 19.1 mm, HT: 9.5 mm, $k = 2$, 3×6 turns, ϕ: 1 mm, 2.8 μH
$C_{CM\text{-}HF}$	KEMET, C0G, 0805, 33 pF, 1 kVDC
$R_{CM\text{-}HF}$	VISHAY, RCS e3 Series, 0805, 49.9 Ω, 1%, 0.25 W

amplitude components of output CM voltage is almost reduced by the installation of the LF-PF. Under this assumption, core saturation is negligible in the design of the HF CM inductor, meanwhile an operating frequency range still remain as an important factor at the design-stage. As well known, the complex relative permeability of each magnetic material represents a frequency-dependent behavior, so that the inductance of the fabricated inductor decreases with frequency and the inductor presents a resistive characteristic at HF range. For realizing HF inductors, it is preferable that the real part of the complex permeability is constant over wide frequency range. Thus, we selected Fair-Rite 67 [20] as the core material of the three-phase HF CM inductor.

Self-resonance due to the winding stray capacitance also limits the operational frequency range of an inductor [21]–[24]. The winding stray capacitance can be separated the turn-to-turn capacitance and the turn-to-core capacitance [21]. The turn-to-turn capacitance decreases with the distance between the adjacent turns (the winding pitch) [24], thus, the sufficient winding pitch is preferable for the HF CM inductor. On the other hand, low-permeability magnetic materials such as nickel-zinc have very high resistivity, and their turn-to-core capacitance is almost negligible. Generally, the size of the available magnetic core for HF applications is relatively small. This means that the window area of the core is limited, and an increase in the number of turns directly leads to a decrease in the winding pitch. Based on this, the number of cores is set to 2, in order to obtain the specified inductance without increasing the number of turns. In this case, the required number of turns is derived to be 6 from (2), therefore, the value of $L_{CM\text{-}HF} = 2.8$ μH and the frequency of $f_{cut,CM\text{-}HF} = 9.5$ MHz can be calculated.

The measured CM impedance curve of the inductor fabricated for the three-phase HF CM inductor is presented in Fig. 2. The measurement was conducted using an impedance analyzer (E4990A, Keysight) in the frequency range from 100 kHz to 100 MHz. It can be confirmed that the measured impedance curve has an inductive characteristic over wide frequency up to 70 MHz. The resonance is observed at around 80 MHz caused by the frequency dependency of the material [20], and the inductor shows a resistive behavior at frequencies higher than 90 MHz.

The value of $R_{CM\text{-}HF} = 50$ Ω which is same value as the compensating resistor of the operational amplifier described in the next section, is selected, and the damping ratio of the HF-PF is calculated to be 0.14 in this design. Finally, the specifications of the filter components selected for the HF-PF are also shown in TABLE II.

3317

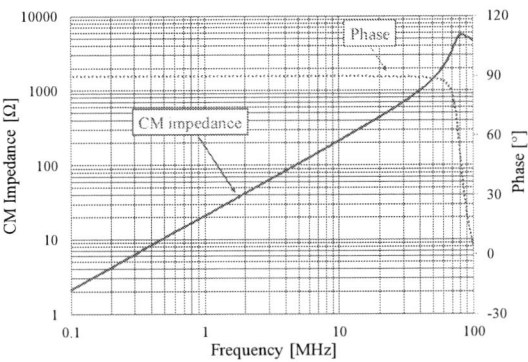

Fig. 2. The CM impedance measurement result of the three-phase high-frequency CM inductor.

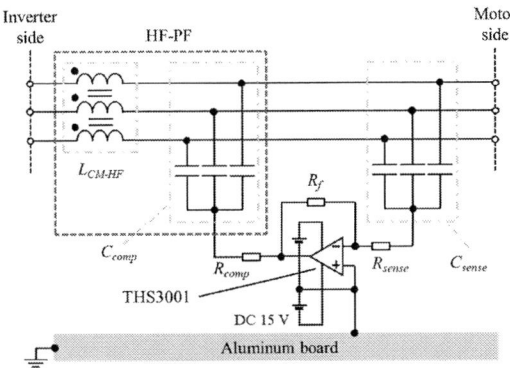

Fig. 3. Circuit configuration of the active common-mode filter applied to a three-phase power cable.

III. ACTIVE COMMON-MODE FILTER

A. Structure of the active common-mode filter

In order to improve the performance of the HF-PF, it is necessary to shift the cutoff frequency $f_{cut,CM-HF}$ to the lower frequency by increasing the values of the filter components. Although the value of L_{CM-HF} increases with the number of cores, the per turn length of the winding increases. This means an increase in the winding stray capacitance, and the maximum operating frequency of the HF CM inductor decreases. Furthermore, the value of C_{CM-HF} is limited by the relation shown in (6).

One conceivable solution for improving the performance of a passive noise filter is utilizing an active-feedback (AF) which enhances the frequency responses of filter components at a specified frequency range [12], [25]. They are commonly called as feed-back type active noise filters which can be classified into four main structures based on types of a sensing and a compensating. Except for the voltage-sensing current-compensating (VSCC) type, the other three types of filter structures need HF transformers [12]. In general, obtaining capacitors with a wide operating frequency range is relatively easy, on the contrary, the realization of the HF transformer is very difficult. Therefore, it can be considered that a transformer-less VSCC type AF structure is an appropriate option for HF applications.

Based on the aforementioned discussion, the authors proposed an active common-mode filter (ACF) which is the VSCC type active noise filter [17]. Fig. 3 shows the circuit configuration of the ACF applied to a three-phase power cable. The ACF consists the HF-PF (three-phase HF CM inductor L_{CM-HF} and a compensating capacitor $C_{comp} = 3 \times C_{CM-HF}$), a sensing capacitor C_{sense}, and a CM voltage-feedback circuit. A high-speed current feedback operational amplifier THS3001 [26] realizes the AF of CM voltage over wide frequency range. The sensing capacitor and a sensing resistor R_{sense} form a CM voltage sensing network and rejects unwanted low-frequency high amplitude voltage for avoiding a saturation of the operational amplifier. The high-pass cutoff frequency of the CM voltage sensing network $f_{cut,HP}$ is derived as

TABLE III
SPECIFICATIONS OF SELECTED COMPONENTS FOR THE ACF

Symbol	Specification
C_{sense}	KEMET, C0G, 0805, 3×33 pF, 1 kVDC
R_{sense}	Stackpole Electronics, RNCP Series, 0805, 499 Ω, 1 %, 0.25 W
R_f	Stackpole Electronics, RPC Series, 0805, 1 kΩ, 5 %, 0.25 W

$$f_{cut,HP} = \frac{1}{2\pi R_{sense} C_{sense}} \quad (8)$$

The sensing resistor and a feedback resistor R_f decide an inverting amplifier gain, and a compensating resistor R_{comp} is installed on the amplifier output for managing stability of the AF system (the compensating resistor also plays as the damping resistor of the HF-PF). The specifications of the selected components for the AF network is shown in TABLE III.

This AF circuit senses the CM voltage between the power cable and a grounded metal box (the aluminum board imitates this in this paper), and feed it back to the compensating capacitor. As a result of this AF, the frequency response of the compensating capacitor is effectively improved, thereby the HF performance of the HF-PF can be enhanced without using larger value of the passive components. This merit of the applying the AF will be analyzed in the following subsection.

B. Analysis of the passive filter performance with the active feedback

The influence of the AF on the compensating capacitor impedance was analyzed in our previous work [27]. However, the impact on the performance of the HF-PF is not presented in [27]. This issue can be analyzed by assuming the ACF as a simplified single-phase equivalent circuit shown in Fig. 4(a). The AF circuit is modeled by considering the inverting amplifier as an ideal voltage-controlled voltage source (v_{noise}) with gain G. CM attenuation by the installation of the ACF (Att) is calculated to be the ratio between v_{noise} and compensated CM voltage v_{cm} as

3318

$$Att = 20 \log_{10} \left(\frac{v_{cm}}{v_{noise}} \right) \qquad (9)$$

From Fig. 4(a), following equations are derived,

$$v_c = G \cdot R_{sense} \cdot I_3 \qquad (10)$$

$$v_{cm} = \left(R_{sense} + \frac{1}{j\omega C_{sense}} \right) \cdot I_3 \qquad (11)$$

$$v_{inv} - j\omega L_{CM\text{-}HF} \cdot I_1 = \left(R_{comp} + \frac{1}{j\omega C_{comp}} \right) \cdot I_2 - v_c \qquad (12)$$

$$\left(R_{comp} + \frac{1}{j\omega C_{comp}} \right) \cdot I_2 - v_c = \left(R_{sense} + \frac{1}{j\omega C_{sense}} \right) \cdot I_3 \qquad (13)$$

$$I_1 = I_2 + I_3 \qquad (14)$$

I_1, I_2, and I_3 are obtained from (10), (12), (13), and (14). Substituting derived I_3 into (11), v_{cm} is obtained, thereby, the CM voltage attenuation caused by the installation of the ACF can be calculated from (9).

Analysis results are shown in Fig. 4(b). For comparison, the CM attenuation curve in the case of only the HF-PF installed (without the AF), was also calculated. Analytical parameters of each component are equal to the values indicated in TABLE II and III, so that the inverting amplifier gain G is 2. As the capacitance of the compensating capacitor is effectively amplified by the AF [27], the cutoff frequency of the HF-PF is shifted to the low-frequency side. Thus, the CM attenuation by the ACF is larger than when the AF is not applied, at HF range beyond 10 MHz. On the other hand, the value of the compensating resistor (the damping resistor of the HF-PF) is equivalently reduced by the AF, thereby, the steep resonance is observed when the AF is applied. It is noted that this resonance will be damped by stray resistances (e.g. the skin effect of a power cable) in practical system.

C. Design of the active common-mode filter

When the inverting amplifier gain is set to be 2, the recommended feedback resistor value R_f for deriving optimum frequency response is 680 Ω (under source voltage is ±15 V) [26]. Taking into account of the stability, the value of 1 kΩ is chosen, so that the value of the sensing resistor R_{sense} is fixed to be 500 Ω, thereby, $f_{cut,HP}$ = 3.2 MHz can be calculated from (8). In order to obtain the proper AF, the high-pass cutoff frequency of the CM voltage sensing network should be designed to be lower than the shifted cutoff frequency of the HF-PF $f_{cut,CM\text{-}HF,shift}$. In [27], the following equation is used for calculating the frequency $f_{cut,CM\text{-}HF,shift}$,

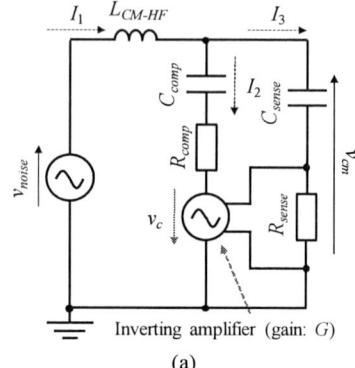

(a)

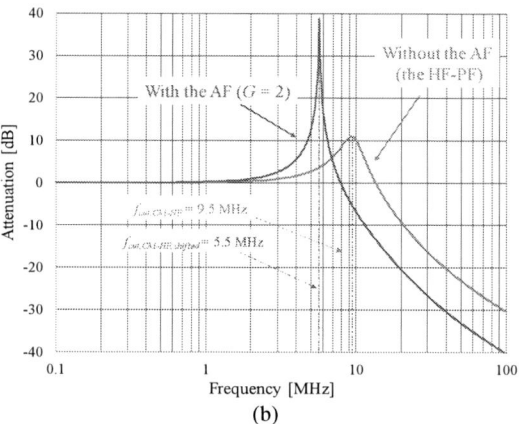

(b)

Fig. 4. Analysis to evaluate the influence of the active-feedback on the high-frequency passive filter. (a) Analytical model which depicts the active common-mode filter as the simplified single-phase equivalent circuit. (b) Analytical attenuation curves.

$$f_{cut,CM\text{-}HF,shift} = \frac{1}{2\pi \sqrt{(1 + G)C_{comp} \cdot L_{CM\text{-}HF}}} \qquad (15)$$

Thus, $f_{cut,CM\text{-}HF,shift}$ = 5.5 MHz is obtained in this design, this specified frequency is lower than the designed high-pass cutoff frequency.

For driving large capacitive loads of greater than 10 pF, it is necessary that the compensating resistor to be placed in series with the output of the operational amplifier. According to the datasheet published from the developer [26], the value of the compensating resistor should have a range from 20 Ω to 75 Ω, so that the value of 50 Ω is chosen.

A stability analysis of the feedback system is important for the design of feedback-type active noise filters [12], however, it is not included in the scope of this paper.

IV. EXPERIMENTAL RESULTS

Figs. 5(a)–(d) show the experimental waveforms of the output CM voltage under the DC link voltage E_{dc} = 200 V,

The 2018 International Power Electronics Conference

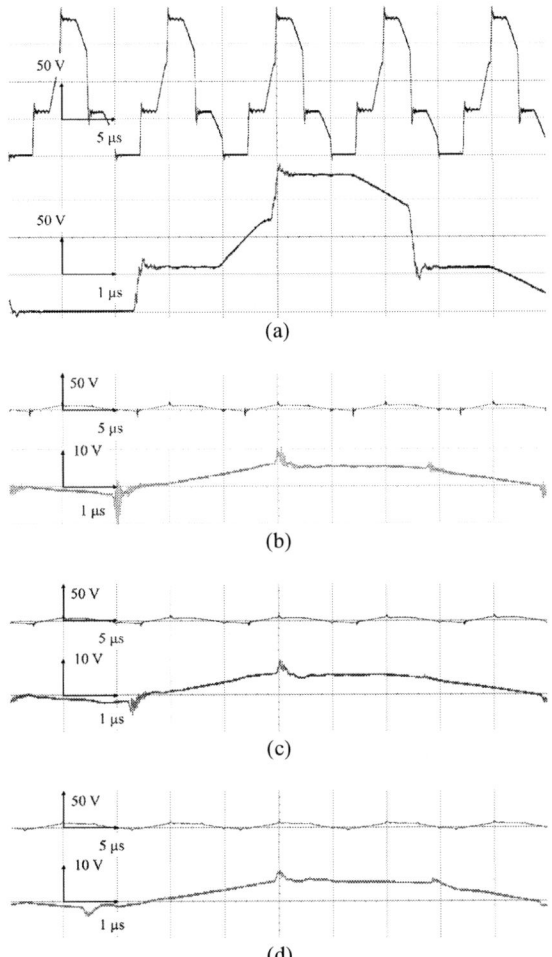

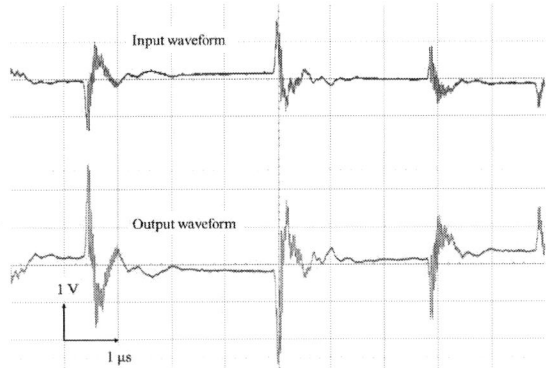

Fig. 6. The input and output waveforms of the operational amplifier.

Fig. 5. Measured output common-mode voltages. (a) No filter is installed. (b) The low-frequency passive filter is installed. (c) The combination of the low-frequency and high-frequency passive filter is constructed at the PWM inverter output. (d) The low-frequency passive filter and the active common-mode filter are installed.

the switching frequency f_{sw} = 100 kHz, and the inverter output frequency is 50 Hz. Note that, in this paper, the CM voltages measured as a voltage potential between the neutral point of the measuring capacitors (3×33 pF) and the aluminum board. It is also noted that the time scales of each figure are different from upper-side and lower-side waveforms. The upper-side waveforms are based on the switching frequency of the PWM inverter, the time scales of the lower-side waveforms are 1μs.

Fig. 5(a) shows the measured waveforms in the case of the inverter is connected to the motor directly. It is observed that this measured CM voltage varies as a stepwise waveform. Additionally, very steep dv/dt which may cause shaft voltages and the damage of motor stator windings, can be observed from the lower-side waveform in Fig. 5(a).

The upper-side waveform in Fig. 5(b) indicates that the switching frequency component of the output CM voltage

is effectively suppressed by installing of the LF-PF at the inverter output. This small CM voltage would not damage to motor bearings and windings. On the other hand, the HF oscillation with frequency around 30 MHz generated by switching operations of power semiconductors, are confirmed from the lower-side waveform. This HF CM voltage may cause radiated noise from power cables, so that it is necessary to reduce this HF oscillation.

Fig. 5(c) shows the experimental waveform of the output CM voltage when the combination of the LF-PF and the HF-PF is constructed at the PWM inverter output. Compared with the lower-side waveform in Fig. 5(b), the HF oscillation is moderated to a certain extent. To accomplish of more CM attenuation, employing the larger values of the filter components is inevitable.

The lower-side waveform in Fig. 5(d) indicates that the ACF effectively suppress the HF oscillation without varying the values of the passive components which form the HF-PF. Furthermore, the measured input and output waveforms of the operational amplifier mounted on the printed-circuit-board, are shown in Fig. 6. It can be observed that the amplifier achieves the AF operation of the inverting amplifier gain of $G = 2$ without the saturation.

The frequency analysis results for each measured output CM voltage waveform are shown in Fig. 7. Due to the dynamic range of the oscilloscope, the measured data when no filter is installed, could not present the analysis result at higher frequencies than 60 MHz. Taking into account of this limitation due to the performance of the measurement equipment, another set of output CM voltage measurements were carried out by using high-pass filter with the cutoff frequency is set to 10 MHz. The CM voltage spectra with frequency up to 100 MHz can be derived from these measured waveforms. Thus, the measured CM voltage spectrum from 100 kHz to 10 MHz and the spectrum with frequency from 10 MHz to 100 MHz are shown in Fig. 7, simultaneously.

It is observed that the switching frequency component and its harmonics are dramatically reduced by the LF-PF. However, the CM attenuation of the LF-PF begins to deteriorate from 10 MHz. And it is also confirmed that HF component with frequency of 30 MHz is not suppressed by installing only the LF-PF. The HF-PF starts to reduce the CM voltage from around 10 MHz, this result agrees

3320

The 2018 International Power Electronics Conference

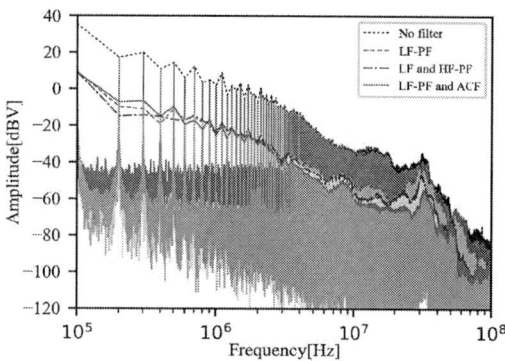

Fig. 7. Frequency analysis results of the measured common-mode voltage waveforms.

with the cutoff frequency of the HF-PF is set to 9.5 MHz. Additionally, the HF component with frequency of 30 MHz is reduced and more HF components with frequency up to 100 MHz is effectively reduced. More CM attenuation is accomplished by the installation of the ACF. Because of the shifted cutoff frequency is 5.5 MHz, the ACF begins to reduce the output CM voltage from around 3 MHz. Due to the frequency response of the AF network, the ACF is not able to enhance the performance of the HF-PF at higher frequencies than 60 MHz. The frequency response of the ACF may be improved by the proper design of the printed-circuit-board.

V. CONCLUSIONS

In order to suppress the wideband output common-mode (CM) voltages produced by switching operations of the PWM inverter, the combination of the conventional low-frequency passive filter (LF-PF) and the high-frequency passive filter (HF-PF) is studied in this paper. Furthermore, an active common-mode filter (ACF) proposed by the authors in the previous work is applied for improving the HF performance of the HF-PF. The structures and the design procedures of each filter are described in detail. And the basic principle of the ACF is analyzed with the simplified equivalent circuit.

The CM voltage attenuation of the filters are evaluated with a 5-kVA motor drive system. Experimental results indicate that high-frequency components of the CM voltages can not be suppressed by the installation of only the LF-PF. Thus, the combination of the LF-PF and the HF-PF is necessary to reduce the wideband CM voltage. Additionally, it is confirmed that the ACF can enhance the performance of the HF-PF at the high-frequency range. Frequency analysis result shows that the combination of the LF-PF and the ACF effectively reduce the output CM voltage from 100 kHz to 100 MHz.

REFERENCES

[1] Y. Murai, T. Kubota, and Y. Kawase, "Leakage current reduction for a high-frequency carrier inverter feeding an induction motor," *IEEE Trans. on Industry Applications*, vol. 28, no. 4, pp. 858-863, 1992.

[2] S. Ogasawara, and H. Akagi, "Modeling and damping of high-frequency leakage currents in PWM inverter-fed AC motor drive systems," *IEEE Trans. on Industry Applications*, vol. 32, no. 5, pp. 1105-1114, 1996.

[3] A. Muetze, "Scaling issues for common-mode chokes to mitigate ground currents in inverter-based drive systems," *IEEE Trans. on Industry Applications*, vol. 45, no. 1, pp. 286-294, 2009.

[4] S. Chen, T. A. Lipo, and D. Fitzgerald, "Modeling of motor bearing currents in PWM inverter drives," *IEEE Trans. on Industry Applications*, vol. 32, no. 6, pp. 1365-1370, 1996.

[5] D. A. Rendusara, and P. N. Enjeti, "An improved inverter output filter configuration reduces common and differential modes *dv/dt* the motor terminals in PWM drive systems," *IEEE Trans. on Power Electronics*, vol. 13, no. 6, pp. 1135-1143, 1998.

[6] H. Akagi, H. Hasegawa, and T. Doumoto, "Design and performance of a passive EMI filter for use with a voltage-source PWM inverter having sinusoidal output voltage and zero common-mode voltage," *IEEE Trans. on Power Electronics*, vol. 19, no. 4, pp. 1069-1076, 2004.

[7] T. Nussbaumer, M. L. Heldwein, and J. W. Kolar, "Common mode EMC input filter design for a three-phase buck-type PWM rectifier system," in *Proc. 21st IEEE Appl. Power Electron. Conf. Expo.*, 2006, pp. 19-23.

[8] H. Akagi, and T. Shimizu, "Attenuation of conducted EMI emissions from an inverter-driven motor," *IEEE Trans. on Power Electronics*, vol. 23, no. 1, pp. 282-290, 2008.

[9] S. Ogasawara, H. Ayano, and H. Akagi, "An active circuit for cancellation of common-mode voltage generated by a PWM inverter," *IEEE Trans. on Power Electronics*, vol. 13, no. 5, pp. 835-841, 1998.

[10] I. Takahashi, A. Ogata, H. Kanazawa, and A. Hiruma, "Active EMI filter for switching noise of high frequency inverters," in *Proc. PCC '97 Conf.*, Nagaoka, Japan, 1997, pp. 331-334.

[11] M. C. Di Piazza, G. Tine, and G. Vitale, "An improved active common-mode voltage compensation device for induction motor drives," *IEEE Trans. on Industrial Electronics*, vol. 55, no. 4, pp. 1823-1834, 2008.

[12] M. L. Heldwein, H. Ertl, J. Biela, and J. W. Kolar, "Implementation of a transformerless common-mode active filter for online converter systems," *IEEE Trans. on Industrial Electronics*, vol. 57, no. 5, pp. 1772-1786, 2010.

[13] S. Ogasawara, H. Ayano, and H. Akagi, "Measurement and reduction of EMI radiated by a PWM inverter-fed AC motor drive system," *IEEE Trans. on Industry Applications*, vol. 33, no. 4, pp. 1019-1026, 1997.

[14] G. L. Skibinski, R. J. Kerkman, and D. Schlegel, "EMI emissions of modern PWM AC drives," *IEEE Industry Applications Magazine*, vol. 5, no. 6, pp. 47-81, 1999.

[15] A. Roc'h, and F. Leferink, "Nanocrystalline core material for high-performance common mode inductors," *IEEE Trans. on Electromagnetic Compatibility*, vol. 54, no. 4, pp. 785-791, 2012.

[16] N. Hanigovszki, J. Landkildehus, G. Spiazzi, and F. Blaabjerg, "An EMC evaluation of the use of unshielded motor cables in AC adjustable speed drive applications," *IEEE Trans. on Power Electronics*, vol. 21, no. 1, pp. 273-281, 2006.

[17] S. Takahashi, S. Ogasawara, K. Orikawa, M. Takemoto, and M. Tamate, "An active common-mode filter for reducing radiated noise from power cables," in *2017 IEEE International Future Energy Electronics Conference (IFEEC-ECCE Asia)*, 2017, pp. 1753-1758.

[18] TDK, "Ferrites and accessories SIFERRIT material N30," 2017. [Online]. Available: https://en.tdk.eu/download/528848/05a0bcaec 69256cbbdbec9fecb243311/pdf-n30.pdf

[19] TDK, "Ferrites and accessories Toroids (ringcores) R50.0×30.0× 20.0, B64290L0082," 2014. [Online]. Available: https://product. tdk.com/info/en/documents/data_sheet/R5000x3000x2000.pdf

[20] Fair-Rite Products Corp., "Toroids (5967001701)," 2013. [Online]. Available: http://www.fair-rite.com/product/toroids-5967001701/

[21] A. Massarini, and M. K. Kazimierzuk, "Self-capacitance of inductors," *IEEE Trans. on Power Electronics*, vol. 12, no. 4, pp. 671-676, 1997.

[22] M. L. Heldwein, L. Dalessandro, and J. W. Kolar, "The three-phase common-mode inductor: modeling and design issues," *IEEE Trans. on Industrial Electronics*, vol. 58, no. 8, pp. 3264-3274, 2011.

[23] S. W. Pasko, M. K. Kazimierzuk, and B. Grzesik, "Self-capacitance of coupled toroidal inductors for EMI filters," *IEEE Trans. on Electromagnetic Compatibility*, vol. 57, no. 2, pp. 216-223, 2015.

[24] A. Ayachit, and M. K. Kazimierczuk, "Self-capacitance of single-layer inductors with separation between conductor turns," *IEEE Trans on Electromagnetic Compatibility*, vol. 59, no. 5, pp. 1642-1645, 2017.

[25] P. C. Murphy, T. C. Neugebauer, C. Brasca, and D. J. Perreault, "An active ripple filtering technique for improving common-mode inductor performance," *IEEE Power Electronics Letters*, vol. 2, no. 2, pp. 45-50, 2004.

[26] Texas Instruments, "420-MHz high-speed current-feedback amplifier," 2009. [Online]. Available: http://www.tij.co.jp/jp/lit/ds/sym link/ths3001.pdf

[27] S. Takahashi, S. Ogasawara, M. Takemoto, K. Orikawa, and M. Tamate, "Common-mode voltage attenuation of an active common-mode filter in a motor drive system fed by a PWM inverter," *in Proc. The 20th International Conference on Electrical Machines and Systems (ICEMS 2017)*, #659, Sydney, NSW, Australia, August. 2017.

A Modified Discontinuous PWM for Common-mode Voltage Elimination in 3-level 4-leg PWM Converter System

Seon-Ik Hwang[1], Jun-Hyung Jung[1], In-ho Cho[2], Jang-Mok Kim[1], and Yung-Deug Son[3]

1 Department of Electrical Engineering, Pusan National University, Busan, Korea
2 Production system research department, Hyundai Heavy Industries, Ulsan, Korea
3 Department of Mechanical Facility Control Engineering, Korea University of Technology and Education, Cheonan, Korea
*E-mail: hsiatop@pusan.ac.kr, jjhyung@pusan.ac.kr, choinho0217@hhi.co.kr, jmok@pusan.ac.kr, ydson@koreatech.ac.kr

Abstract— This paper proposes a pulse width modulation (PWM) method to eliminate common-mode voltage (CMV) in three-level four-leg neutral-point-clamped (NPC) PWM converter system. For the purpose of CMV reduction, several papers have paid attention to the three-level four-leg NPC PWM converter system. However, the conventional methods are unable to totally eliminate the CMV or to use limited voltage vectors according to switching pattern. Therefore, this paper proposes the discontinuous PWM (DPWM) control strategy in each sector and region to eliminate the CMV in three-level four-leg NPC PWM converter system. The DPWM method are used to reduce CMV and the remaining CMV is removed using fourth-phase switching. The simulation results are presented to verify the validity of the proposed method.

Keywords— three-leg four-leg NPC converter, four-leg converter, CMV eleimination, Discontinuous PWM

I. INTRODUCTION

Recently, the use of high-speed switching devices such as IGBTs applied to a power conversion system have been increasing in the industry. These switching devices induce the electromagnetic interference (EMI) noise by their switching operation. This EMI noise leads to system malfunctions and shutdowns and degrades the reliability of equipment [1] – [3]. The EMI noise is classified into differential mode noise and common mode noise. The differential mode noise is a noise flowing between two power transmission conductors, and the common mode noise is a noise flowing between a power conversion system and the ground. The common mode noise is a main element of the EMI noise generated by the CMV [4].

As methods for reducing the CMV, there are methods of using additional hardware such as a common mode filter and changing PWM switching pattern of a converter. A method of reducing CMV by adding hardware such as common mode filter causes an increase in the cost and volume of the system [4]. So, several papers have dealt with the reduction of CMV by changing the switching pattern [5] – [9]. The [5]

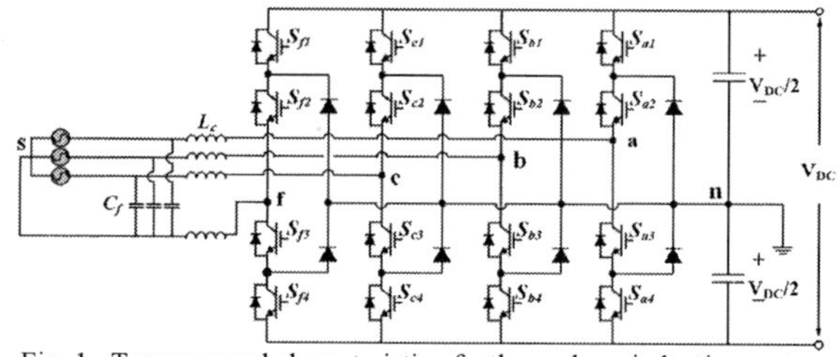

Fig. 1. Torque-speed characteristic of a three-phase induction motor.

proposed the method using the Push-Pull PWM3 (PPPWM3) to reduce CMV in three-level four-leg NPC PWM converter method. However, this method can't perfectly remove the CMV. The [6] propose a PWM method using only a zero vector (O,O,O) and medium vectors which does not generate the CMV. So, the three-level NPC converter does not produce the CMV. But, this method needs a fourth pole to compensate the upper and lower DC link imbalance. Furthermore, this method has a reduced voltage utilization because the use of the voltage vector is limited to eliminate CMV.

Therefore, this paper suggests a method capable of removing the CMV completely. The proposed method can utilize all switching states in the three-level four-leg NPC PWM converter system. The switching patterns of three phase are implemented in a form of modified DPWM to reduce the CMV. And a remaining CMV can be eliminated by using a fourth-phase of the converter system. The validity of proposed method is verified through simulation results.

II. CONVENTIONAL CMV REDUCTION METHOD IN THREE-LEVEL FOUR-LEG TOPOLOGY[5]

Fig. 1 shows the structure of a three-level four-leg NPC PWM converter system. This system consists of a three-phase three-leg converter system and the fourth-phase connected to the neutral point via L filter and LC filter. The CMV of the three-level four-leg NPC PWM converter system can be obtained using (1) ~ (4).

The 2018 International Power Electronics Conference

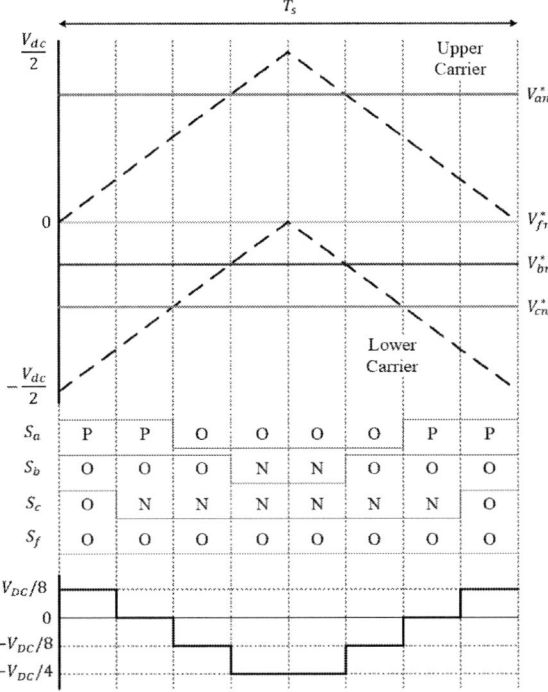

Fig. 2. Switching pattern and CMV in 3-level 4-leg converter system with SPWM

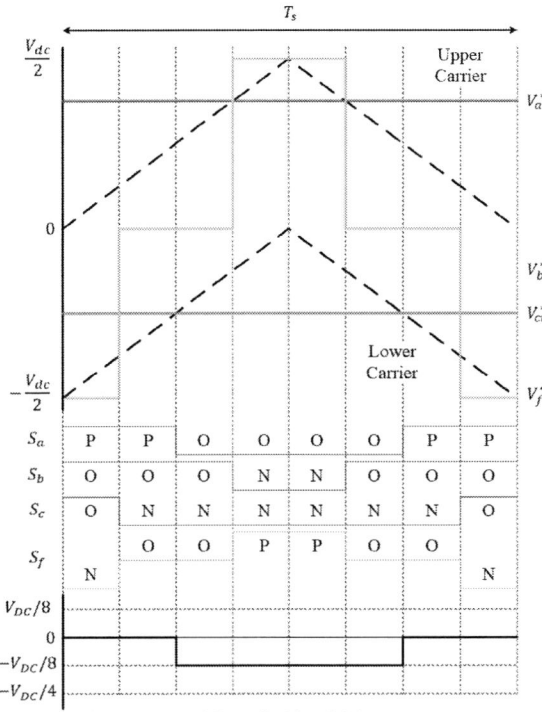

Fig. 3. Switching pattern and CMV in 3-level 4-leg converter system with PPPWM3

$$V_{CMV} = V_{SN} = \frac{V_{an} + V_{bn} + V_{cn} + V_{fn}}{4} \tag{1}$$

$$= \frac{S_a + S_b + S_c + S_f}{4} \times \frac{V_{dc}}{2}$$

$$S_X = \begin{cases} 1 \ or \ P \left(S_{x1}, S_{x2} : ON \right) \\ 0 \ or \ O \left(S_{x2}, S_{x3} : ON \right) \\ -1 \ or \ N \left(S_{x3}, S_{x4} : ON \right) \end{cases} \tag{2}$$

$$S_{sum} = S_a + S_b + S_c \tag{3}$$

$$S_{CMV} = S_{sum} + S_f \tag{4}$$

The S_{x1}, S_{x2}, S_{x3}, and S_{x4} means the switching state of switches in each phase as shown in Fig. 1. The S_{sum} is the sum of the switching states of three phase. The S_{CMV} is the sum of S_{sum} and the switching state of fourth-phase. V_{CMV} is the CMV of converter system according to the switching state.

Fig. 2 shows the CMV of three-level four-leg NPC PWM converter system using sinusoidal PWM (SPWM). The each switching state in single phase has one of -1, 0, and 1 in (2). Since the fourth-phase voltage is not applied, the switching state of fourth-phase is 0. So, the range of S_{CMV} is from -2 to 2 as shown in Fig. 2. Therefore, a range of CMV is from -2Vdc/8 to 2Vdc/8 when applying the SPWM method to the three-level four-

leg NPC PWM converter system. Fig. 3 shows the CMV reduction method using the PPPWM3 in the three-level four-leg NPC PWM converter system [5]. The three phase reference voltages are generated by using the SPWM method, and fourth-phase reference voltage is compensated by applying the PPPWM3 method. In PPPWM3 method, the switching of the fourth-phase is changed at the first and third switching points of three phase as shown in Fig. 3. However, since the fourth-phase switching has three switching states of -1, 0, and 1, it is impossible to set the S_{CMV} to 0 when the S_{sum} is equal to -2 or 2 switching state. When the PPPWM3 is applied, the CMV has a range from -V_{dc}/8 to V_{dc}/8. In this method, the magnitude of the CMV is reduced, but it can't be eliminated. Therefore, additional researches are required to eliminate the CMV.

III. PROPOSED CMV ELIMINATION METHOD

In [5], even if the compensation of the S_{CMV} is performed by using the fourth-phase of the converter, the S_{CMV} isn't 0 in the whole switching state. If the range of S_{sum} can be changed to a range from -1 to 1, the total S_{CMV} can be eliminated by using the compensation of fourth-phase switching. However, the value of S_{sum} has -2 or 2 in some small vectors such as (O,N,N) and (P,P,O) as shown in Fig. 4. The CMV generated by these small vectors remains in spite of using compensation of fourth-phase switching. Fortunately, these small vectors have a small vector whose S_{sum} has a value of 1 or -1 in the redundant vector.

3324

The 2018 International Power Electronics Conference

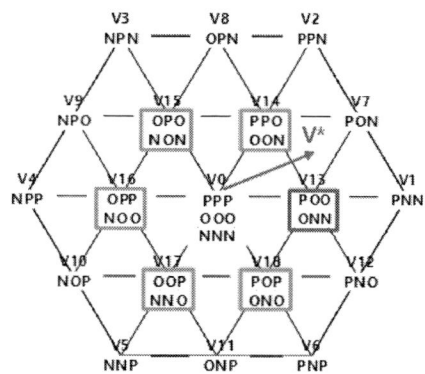

Fig. 4. Vector diagram of three level switching

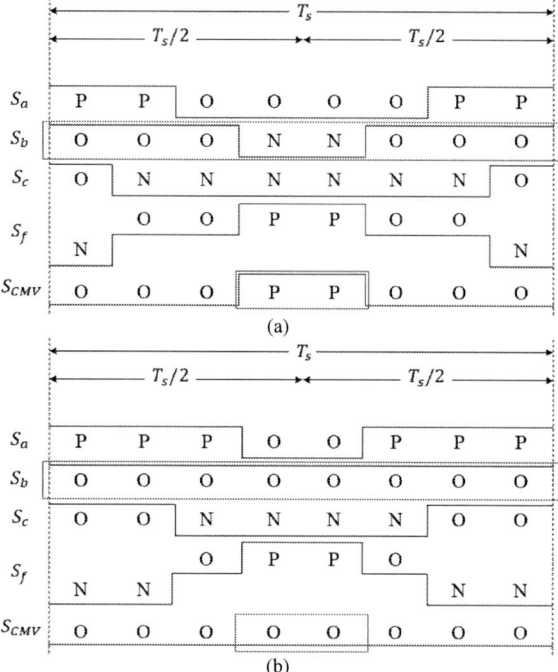

(a)

(b)

Fig. 5. Switching sequence (a)SPWM (b)S_{CMV}=0

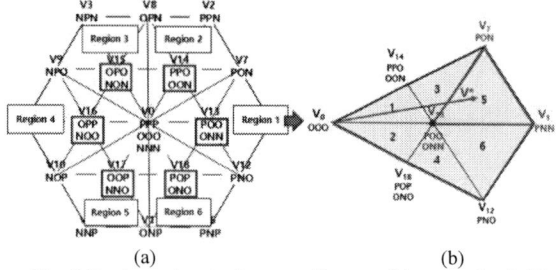

(a)   (b)

Fig. 6. Region and sector in vector diagram of three-level switching

As shown in Fig. 5(a), the value of S_{sum} at switching state (O,N,N) in middle of switching sequence is -2. This switching state have a redundant switching state (P,O,O). If the switching state (O,N,N) can be changed into the switching state (P,O,O), the range of S_{sum} is -1 to 1. So, the S_{CMV} can be 0 by using fourth-phase switching as shown in Fig. 5 (b).

O State DPWM in Type A

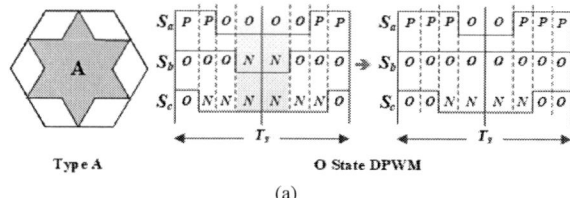

(a)

P State DPWM in Type B

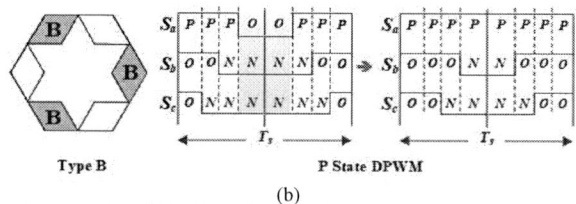

(b)

N State DPWM in Type C

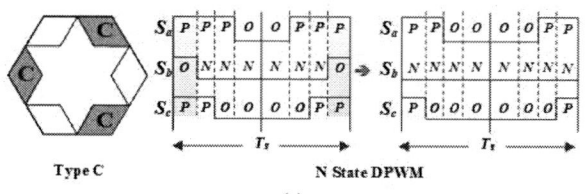

(c)

Fig. 7. Switching type each region and sector

TABLE I
OFFSET VOLTAGE FOR 3-PHASE IN EACH TYPE

	Offset voltage
O state DPWM	$-V_{mid}$
P state DPWM	$V_{DC}/2-V_{max}$
N state DPWM	$-V_{DC}/2-V_{min}$

For change the switching state as shown in Fig. 5, the vector diagram of three-level NPC converter is separated to six regions based on small vector as shown in Fig. 6 (a). Each region is further divided into six sector according to used switching state as shown in Fig. 6 (b). To analyze the whole sectors, the vector diagram of three-level NPC can be divided into three type as shown in Fig. 7.

In the example of Fig. 7 (a), it is the same case as the example in Fig. 5. In this case, if the switching state (O,N,N) is changed to switching state (P,O,O), the range of S_{sum} is -1 to 1. Therefore, CMV can be eliminated to used compensation of fourth-phase switching. To change switching state (O,N,N) from (P,O,O), the offset voltage in table I for O state DPWM is added to each phase. So, the reference voltage of phase b is DPWM to O state. The range of this type is the sector 1 to 4 in whole region.

In case of Fig. 7 (b), the value of S_{sum} at switching state (O,N,N) in middle of switching sequence is -2 same as Fig. 7 (a). So, switching state (O,N,N) is changed into the switching state (P,O,O), the range of S_{sum} is -1 to 1. Therefore, CMV can be eliminated to use compensation

3325

The 2018 International Power Electronics Conference

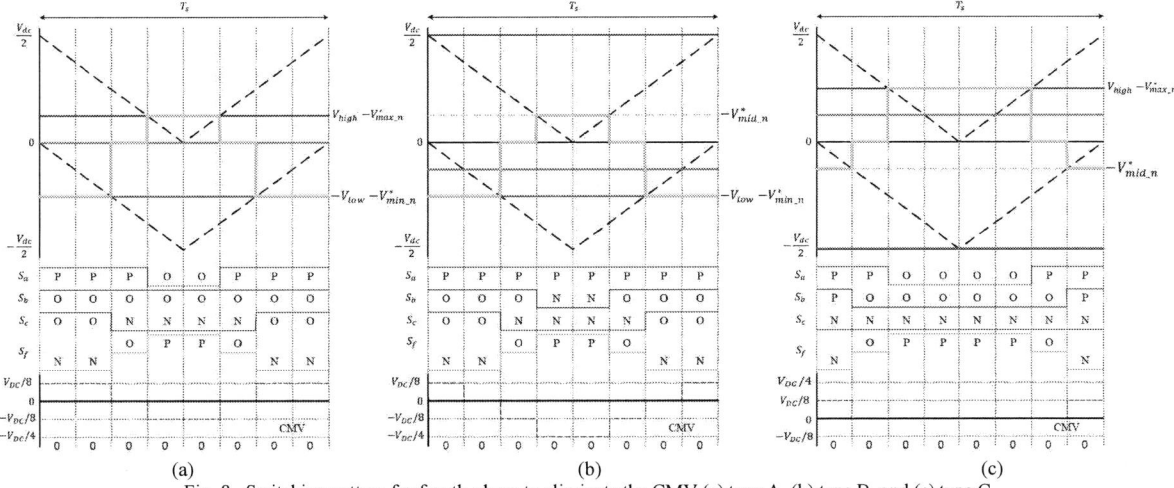

Fig. 8. Switching pattern for fourth-phase to eliminate the CMV (a) type A, (b) type B, and (c) type C

TABLE II
REFERENCE VOLTAGE OF FOURTH-PHASE SWITCHING

		Reference voltage
Type A	Upper	$V_{high}-V^{*}_{max\_n}$
	Lower	$-V_{low}-V^{*}_{min\_n}$
Type B	Upper	$-V^{*}_{mid\_n}$
	Lower	$-V_{low}-V^{*}_{min\_n}$
Type C	Upper	$V_{high}-V^{*}_{max\_n}$
	Lower	$-V^{*}_{mid\_n}$

of fourth-phase switching. To change switching state (O,N,N) from (P,O,O), the offset voltage in table I for P state DPWM is added to each phase. So, the reference voltage of phase a is DPWM to P state. The range of this type is the sector 5, and 6 in region 1, 3, and 5.

In case of Fig. 7 (c), the value of S_{sum} at switching state (P,O,P) in side of switching sequence is 2. So, switching state (P,O,P) is changed into the (O,N,O) in the redundant vector, the range of S_{sum} is -1 to 1. Therefore, CMV can be eliminated to used compensation of fourth-phase switching. To change switching state (P,O,P) from (O,N,O), the offset voltage in table I for N state DPWM is added to each phase. So, the reference voltage of phase b is DPWM to N state. The range of this type is the sector 5, and 6 in region 2, 4, and 6.

After applying the offset voltages in three-phase in each type, the range of S_{sum} becomes -1 to 1 in the range of voltage linearity of SPWM region. So, the S_{CMV} is eliminated by using the fourth-phase switching. The switching pattern for fourth-phase in each type is shown in Fig. 8. In case of type A, compensated switching in fourth-phase to eliminate S_{CMV} is generated to compare carrier with reference voltages. There are two type of reference voltage as shown in table II, the "Upper" is the reference voltage for switches of S_{x1}, S_{x3}, and the "Lower" is the reference voltage for switches of S_{x2}, S_{x4}. So, the S_{CMV} is eliminated as shown in Fig. 8 (a).

In type B and C, compensated switching on fourth-phase to eliminate S_{CMV} is compared with carrier and reference voltage in table II as type A. After compensation of fourth-phase, the S_{CMV} become 0, hence CMV can be eliminated.

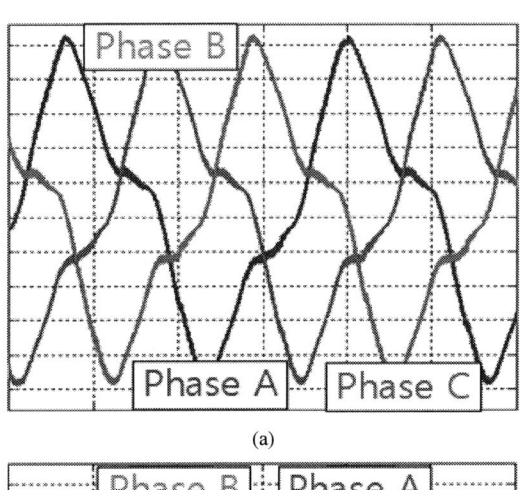

(a)

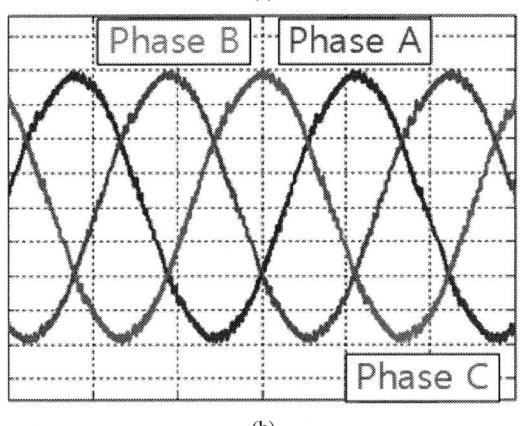

(b)

Fig. 9. Phase current (a) without compensation (b) with compensation

However, due to the voltage of fourth-phase from the compensation switching, the input voltage imbalance occurs. Because of this reason, additional harmonics of input current are generated as shown in Fig. 9 (a). In order to compensate the input voltage imbalance, the compensation voltages are injected onto 2 phases which doesn't use DPWM method as per (5) ~ (6).

The 2018 International Power Electronics Conference

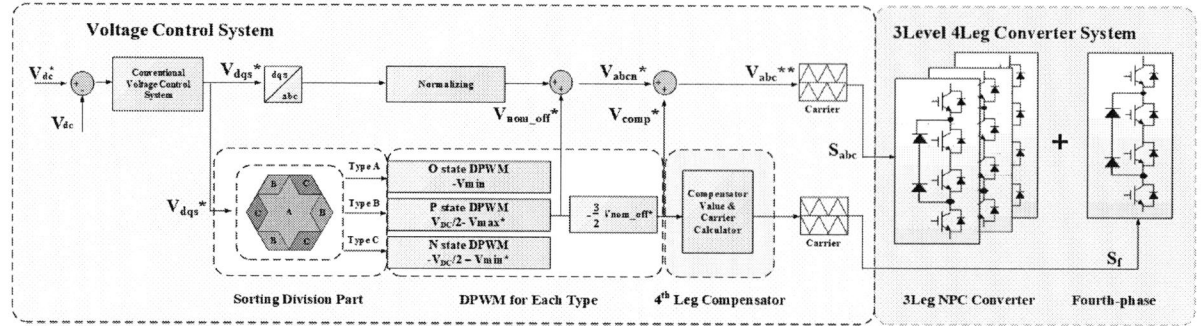

Fig. 10. Proposed control block diagram

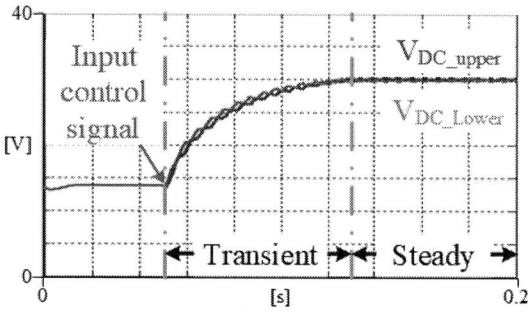

Fig. 11. Upper/Lower DC voltage

TABLE III
THE PARAMETERS OF THE 3-LEVEL 4-LEG CONVERTER SYSTEM

Parameter	Value	Parameter	Value
Switching frequency [kHz]	10	Reference DC-link voltage [V]	60
Grid voltage [V]	22	L_c [mH]	5
Load [Ω]	33	C_f[μF]	5

$$V_{af} + V_{bf} + V_{cf} + 3V^*_{nom\_off} + 2V_{comp} = 0 \quad (5)$$

$$V_{comp} = -\frac{3}{2}V^*_{nom\_off} \quad (6)$$

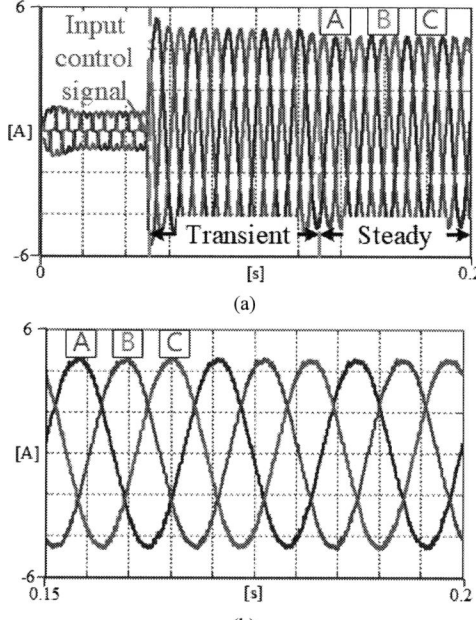

Fig. 12. Phase current (a) waveform(0~0.2s) (b) waveform(0.15~0.2s)

The injection of the compensation voltage is eliminated to the input current imbalance as shown in Fig. 9 (b).

The block diagram for the total proposed method is shown in Fig. 10. The proposed CMV elimination algorithms consist of three parts. The first part is the sorting division part. The reference voltage is separated into type A, B, and C to use d- and q-axis stationary frame. The second part is the calculation of the offset voltage for DPWM and compensation of input voltage imbalance. The third part is the calculation of fourth-phase switching pattern for eliminated to CMV.

IV. SIMULATION RESULTS

To verify the effectiveness of proposed method, MATLAB simulations are carried out in the three-level four-leg PWM converter system. The specifications of the system are listed in table III.

In Fig. 10, before the control signal is input, the three level PWM converter operated as diode rectifier mode,

and the upper and lower dc link voltage is charged to about 13.8V. After input control signal at 0.05s, the upper and lower dc link voltage reaches 30V during transient state. The difference of upper and lower peak to peak capacitor voltages is within ±1V.

The Fig. 12 shows the phase current waveform after applying the DPWM offset voltage. After input control signal, distortion doesn't occur in the phase current.

The Fig. 13 shows the phase voltage and CMV. Because the DC link reference voltage is 60V, phase A, B, and C show three level phase voltages of -30V, 0, +30V. The magnitude of fourth-phase voltage is identical to the A, B and C phase voltages by the fourth-phase switching. The CMV occurs in a transient state for a while, but it is completely eliminated later.

The Fig. 14 shows CMV waveform according to the modulation index (MI). The MI is defined as (7).

$$MI = \frac{V_{phase\_peak}}{V_{dc}/2} \quad (7)$$

The 2018 International Power Electronics Conference

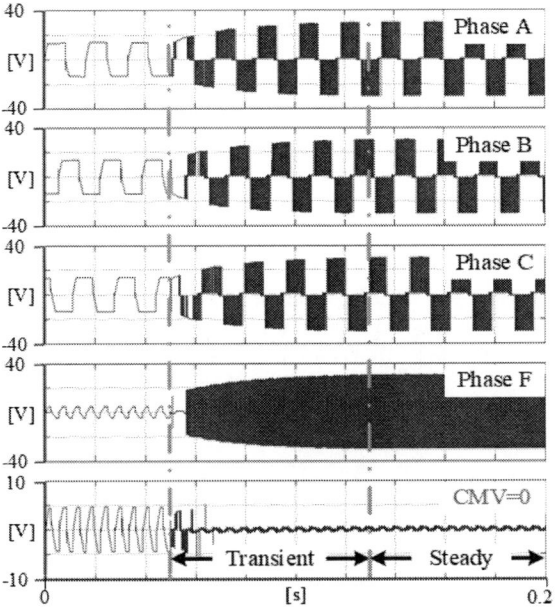

Fig. 13. Phase voltage and CMV

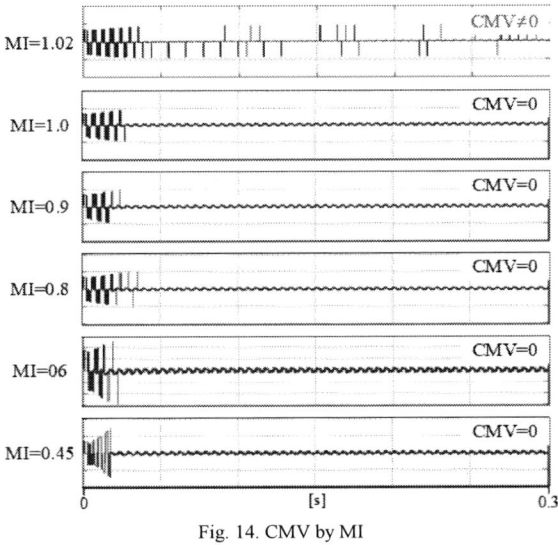

Fig. 14. CMV by MI

The SPWM method has the same voltage linearity characteristics ($0 \le MI \le 1$). If the MI of SPWM is larger than 1, the value of S_{sum} is -2 or 2 due to the influence of OM, it is impossible of compensation of CMV using fourth-phase. So, in this method, CMV doesn't occur until MI reaches 1, but when MI becomes larger than 1, CMV occurs.

V. CONCLUSIONS

This paper proposes the PWM method to eliminate the CMV in three-level four-leg NPC PWM converter system. The switching pattern is combined for output CMV to be zero using the DPWM in each type. The reduced CMV using switching pattern in three phase is

eliminated by using fourth-phase switch. The current imbalance by fourth-phase switching are compensated by additional injected voltage onto 2 phase which doesn't apply DPWM. The CMV are eliminated to use this method. This method can be used to MI of 0 to 1 for elimination of CMV. The validity of the proposed method is verified through the simulation results.

REFERENCES

[1] G. Skibinski, and R. Kerkman, "Emi emissions of modern PWM AC drives", *IEEE Industry Applications Magazine*, vol. 5, no. 6, pp. 47-80, 1999

[2] F. Shih, D. Chen, Y. Wu, and Y. Chen, "A procedure for designing EMI filter for AC line application", *IEEE Trans. On Power Electronics*, vol. 11, no. 1, pp.170-181, 1996

[3] W. S. Im, J. M. Park, D. C. Lee, K. B. Lee, "Diagnosis and Fault-Tolerant Control of Three-Phase AC-DC PWM Converter System", *IEEE Trans. On Industry Applications*, vol. 49, no. 4, pp. 1539-1547, 2013

[4] T. Guo, D. Chen, and F. C. Lee, "Separation of the common-mode- and differential-mode-conducted EMI noise", *IEEE Trans. On Power Electronics*, vol. 11, no. 3, pp.480-488, 1996

[5] Seung-jun Chee, Sanggi Ko, Hyeon-Sik Kim and Seung-Ki Sul, "Common-Mode Reduction of Three-Level Four-Leg PWM Converter," *IEEE Trans. on Industry Application*, vol. 51, no. 5, pp. 4006-4016, Sep/Oct 2015.

[6] Robert M. Cuzner, Ashish R. Bendre, P. J. Faill, and B. Semenov, "Implementation of a Four-pole Dead-Time-Compensated Neutral-Point-Clamped Three-Phase Inverter With Low Common-Mode Voltage Output," *IEEE Trans. on Industry Application*, vol. 45, no. 2, pp. 816-826, Mar/Apr 2009.

[7] N. Nguyen, T. Nguyen, and H. Lee, "A reduced Switching Loss PWM Strategy to Eliminate Common-mode Voltage in Multilevel Inverters," *IEEE Trans. on Power Electronics*, vol. 30, no. 10, pp. 5425-5438, Oct 2015.

[8] Qian. Liu, Fred. Wang, and Dushan. Boroyevich, "Conducted EMI Noise Prediction and Charaterization for Malti-phase-leg Converters Based on Modular-Terminal-Behavioral(MTB) Equivalent EMI Noise Source Model," *in 2006 37th IEEE Power Electronics Specialists Conference*, 2006.

[9] P. C. Loh, D. G. Chen, Y. Fukuta, and T. A. Lipo, "A reduced common mode hysteresis current regulation strategy for multilevel inverter," *IEEE Trans. on Power Electronics*, vol. 19, no. 1, pp. 192-200, Jan 2004

The 2018 International Power Electronics Conference

EMI Analysis of Full-SiC Integrated Power Module

Xiliang Chen, Wenjie Chen*, Yu Ren, Liang Qiao, Yilin Sha, Xu Yang
School of Electrical Engineering, Xi'an Jiaotong University, Xi'an, China
*E-mail: cwj@mail.xjtu.edu.cn

Abstract—Due to the high dv/dt and di/dt slew rates, the electromagnetic interference (EMI) issue in the high power level power module is serious. In this paper, a novel EMI model of full-SiC power module is proposed based on the ANSYS Q3D software and the EMI mechanical is analyzed by discussing the EMI coupling paths both in common-mode (CM) interference and differential-mode (DM) interference. Then, a synchronous buck converter based on the full-SiC power module is simulated in the ANSYS Simplorer software. Finally, a synchronous buck experimental platform based on a 1200V, 288A full-SiC power module is tested. It is found that the intensity of EMI increases as the voltage and current of power module increase. The intensity of the CM EMI presents uniform distribution along with frequency and a resonant peak appears at specific frequency points of the DM EMI.

Keywords— power module; EMI; full-SiC

I. INTRODUCTION

Nowadays, the SiC MOSFET is more frequently used in power electronic applications such as of electric vehicles (EVs) and hybrid electric vehicles (HEVs) [1]. However, just single SiC MOSFET has inadequate power rating to be directly used in high power applications [2]. Therefore, full-SiC power module is really useful in high power electrical systems [3-4].

Due to the high dv/dt slew rate and closely packaged design in these wide-bandgap power semiconductor devices, SiC integrated power module would generate significant EMI concerns [5-6].

In this paper, the EMI mechanism of full-SiC MOSFET integrated power module is analyzed by discussing the EMI coupling paths both in common-mode (CM) interference and differential-mode (DM) interference. Then, a 3D model of full-SiC MOSFET integrated power module is established based on the ANSYS Q3D software. Besides, a synchronous buck converter is simulated by ANSYS Simplorer software which imports the electrical parameters of the full-SiC integrated power module from the Q3D software. Finally, a self-created one which optimize the layout by using split damping capacitors are tested and compared.

II. 3D MODELLING OF POWER MODULE AND EMI MECHANISM ANALYSIS

Structure of the proposed full-SiC power module is shown in Fig.1 by using ANSYS Q3D extractor. In this packaging structure, the adjacent decoupling concept which has several split decoupling capacitors is used. There are three advantages by adding the decoupling

capacitors: (a) Reduce the loop inductance, thus the overshoot voltage of MOSFET would be suppressed and EMI performance could be better. (b) Reduce the coupling effect between the output capacitors of the devices which is good to stability of the power module. (c) The decoupling capacitor could be seen as X capacitor in a filter, so the EMI performance could be improved.

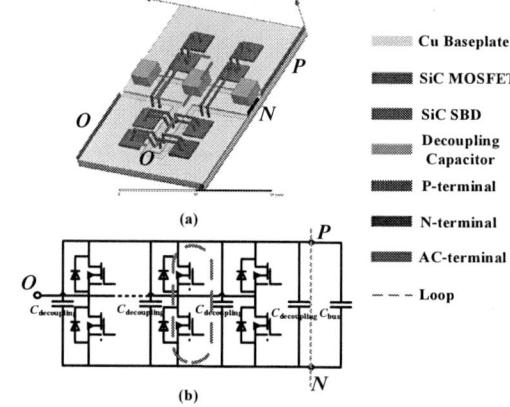

Fig. 1. 3D model of the proposed phase-leg SiC module with adjacent decoupling capacitors and its maximum current looping path.

The high frequency equivalent circuit of synchronous buck converter based on the proposed full-SiC power module is shown in Fig.2. Each MOSFET has a parasitic capacitor C_s between drain and source. And the line impedance stabilization network (LISN) is added to the input side of the circuit for testing the EMI of the synchronous buck converter. The CM conducted EMI coupling paths are shown in Fig.3. The CM noise loops by the LISN and stray capacitance C_{s11}, C_{s12}~C_{sn1}, C_{s11} between MOSFETs' drain and ground. And the DM conducted EMI coupling paths are shown in Fig.4. The DM noise loops by the LISN, MOSFETs and load.

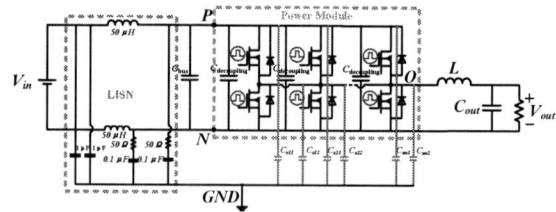

Fig.2. EMI equivalent circuit of synchronous buck converter by using full-SiC integrated power module.

The 2018 International Power Electronics Conference

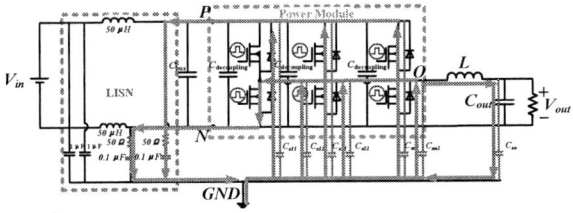

Fig.3. Common-mode conducted EMI coupling path of synchronous buck converter by using full-SiC integrated power module.

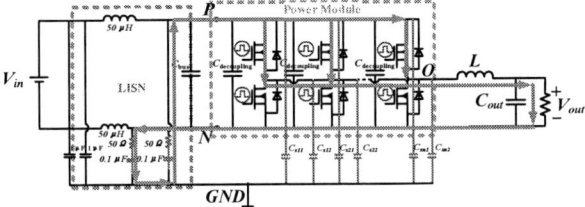

Fig.4. Differential-mode conducted EMI coupling path of synchronous buck converter by using full-SiC integrated power module.

III. SIMULATION ANALYSIS

The synchronous buck converter based on the full-SiC power module is modeled in ANSYS Q3D software as Fig.1 shown. The direct bonding copper (DBC) of the power module has 19.26*30.74mm^2 area and 1.235mm thickness, which is made up by copper. The size of SiC MOSFET is 3.1×3.36mm^2 and the size of SiC schottky-barrier diode (SBD) is 3.08×3.08mm^2.

First, the parameters of the power module are extracted by ANSYS Q3D. Then, these parameters are imported into ANSYS Simplorer software for circuit simulation. Next, the V_{ds} of the MOSFET between drain and source, the output voltage V_{out} and output current I_d as are shown in Fig.5 (a). And the CM and DM EMI amplitude frequency diagram is drawn through MATLAB in Fig.5 (b) by importing data from ANSYS Q3D. It can be seen that the RMS value of I_d is 30A when input voltage is 600V and the output voltage is 150V. The intensity of the CM EMI presents uniform distribution along with frequency. But as to the intensity of DM EMI, a resonant peak appears at specific frequency points.

IV. EXPERIMENT

In this section, by applying to a synchronous buck converter as shown in Fig.2, the characteristic of proposed 1200V, 288A SiC MOSFET module is tested and its EMI performance is also measured.

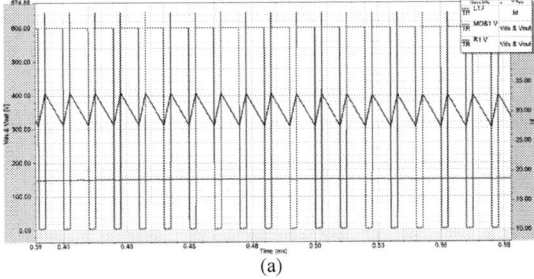

(a)

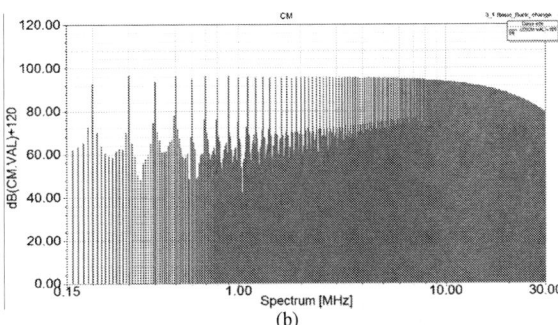

(b)

Fig.5. ANSYS Simplorer simulation waveforms when Vin=500V and Id=30A. (a) Vds (rad), Vout (black) and Id (blue) of the circuit. (b) CM EMI amplitude frequency diagram.

The input voltage and output voltage of the converter is 0-600V DC and 0-150V DC. The maximum output power is 4.5kW. The Switching frequency is 100kHz.

The CM, DM conducted EMI (frequency ranges from 150kHz to 30MHz) and radiated EMI (frequency ranges from 30MHz to 300MHz) are tested as Fig.6, Fig.7 and Fig.8 shown. Also the EMI values are tested separately according to the different working conditions which are V_{ds}=100V, I_d=5A and V_{ds}=600V, I_d=30A. The maximum value of CM EMI when under 100V 5A is approximately 100dBμV, while under 600V 30A is approximately 110dBμV. The maximum value of DM EMI when under 100V 5A is approximately 100dBμV, while under 600V 30A is approximately 120dBμV. The maximum value of radiated EMI when under 100V 5A is approximately 52dBμV, while under 600V 30A is approximately 60dBμV. Some conclusions could be drawn as follow:

a) The intensity of EMI increases as the voltage and current of power module increase. The DM EMI increases more than CM EMI. It is because that DM EMI is affected more obviously than CM EMI when the loop current increasing according to the EMI coupling path in Fig.4. The conducted EMI increases more than radiated EMI. The reason is that the frequency of circuit is only 100kHz, thus the EMI has been reduced largely in the radiated frequency domain.

b) The intensity of the CM EMI presents uniform distribution along with frequency according to Fig.6. However, as to the intensity of DM EMI, a resonant peak appears at specific frequency points.

c) As to the suppression of EMI in power module, the main method to reduce DM EMI may be adjusting the high frequency parameter in EMI coupling path to decrease resonant. And the CM EMI reduction needs to design a filter with high insertion loss.

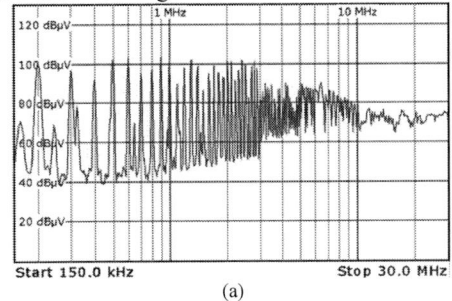

(a)

3330

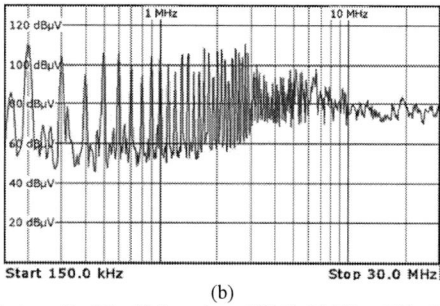

(b)

Fig.6. Test result of the CM conducted EMI. (a) When Vds=100V and Id=5A. (b) When Vds=600V and Id=30A.

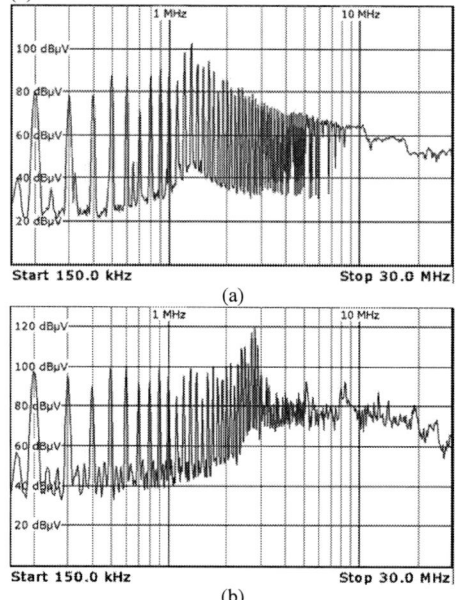

(a)

(b)

Fig.7. Test result of the DM conducted EMI. (a) When Vds=100V and Id=5A. (b) When Vds=600V and Id=30A.

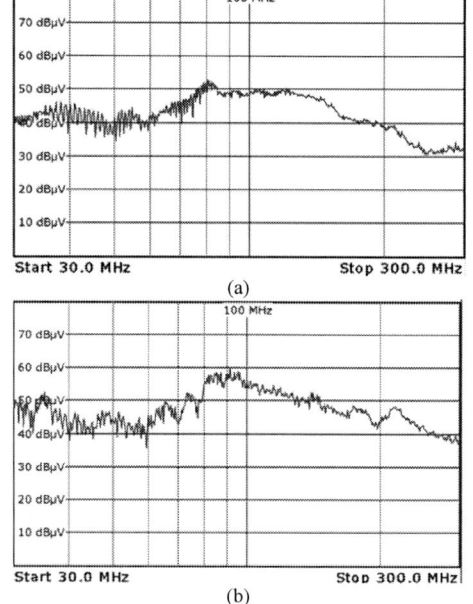

(a)

(b)

Fig.8. Test result of the radiated EMI. (a) When Vds=100V and Id=5A. (b) When Vds=600V and Id=30A.

V. CONCLUSIONS

In this paper, a novel EMI model of full-SiC power module is proposed based on the ANSYS Q3D software and the EMI mechanical is analyzed. Next, a synchronous buck converter based on the full-SiC power module is simulated in the ANSYS Simplorer software. Finally, a synchronous buck experimental platform based on a 1200V, 288A full-SiC power module is tested.

REFERENCES

[1] S. Hazra, A. De, L. Cheng, J. Palmour, M. Schupbach, B. A. Hall, S. Allen, and S. Bhattacharya, "High switching performance of 1700-V, 50-A SiC Power MOSFET over Si IGBT/BiMOSFET for advanced power conversion applications," IEEE Trans. Power Electron., vol. 31, no.7, pp. 4742–4754, Jul. 2016.

[2] Cree Inc, SiC MOSFETs, www.wolfspeed.com/

[3] M. Chinthavali, L.M. Tolbert, H. Zhang, J.H. Han, F. Barlow, and B. Ozpineci, "High power SiC modules for HEVs and PHEVs," in Power Electronics Conference (IPEC), 2010, pp. 1842-1848.

[4] L. Qiao et al., "Performance of a 1.2kV, 288A full-SiC MOSFET module based on low inductance packaging layout," 2017 IEEE Applied Power Electronics Conference and Exposition (APEC), Tampa, FL, 2017, pp. 3038-3042.

[5] C. Yao et al., "Common-mode noise comparison study for lateral wire-bonded and vertically integrated power modules," 2015 IEEE Energy Conversion Congress and Exposition (ECCE), Montreal, QC, 2015, pp. 3092-3098.

[6] A. Dutta and S. S. Ang, "Electromagnetic Interference Simulations for Wide-Bandgap Power Electronic Modules," in IEEE Journal of Emerging and Selected Topics in Power Electronics, vol. 4, no. 3, pp. 757-766, Sept. 2016.

Experimental Verification of Coupling Effect and Power Transfer Capability of Dynamic Wireless Power Transfer

Chan Anyapo[1*], Nithiphat Teerakawanich[1], Chowarit Mitsantisuk[1] and Kiyoshi Ohishi[2]

[1] Department of Electrical Engineering,Faculty of Engineering, Kasetsart University,50 Ngamwongwan Rd.,Ladyao, Jatujak, Bangkok,10900,Thailand, *E-mail: chan.a@ku.th

[2] Motion Control Laboratory, Nagaoka University of Technology, 1603-1,Kamitomioka, Nagaoka, Niigata, Japan 940-2188

Abstract- **This paper presents an evaluation of multi-coils full-bridge resonant inverter for dynamic wireless charging. The validity of coupling effect and power transfer capability of various conditions were also discussed. The newly designed of dynamic wireless charging system uses a sharing of switching branch, which switching devices counts can be reduced. The effect of coupling factor at the different horizontal-axis lateral offsets were studied and investigated. In order to confirm the performance of proposed system, the theoretical analysis and experimental verifications were compared. The comparison results shown that the theoretical calculations and experimental tests were quite similar, which would be acceptable.**

Keywords— Coupling factor, dynamic wireless charging, resonant inverter, wireless power transfer(WPT).

I. INTRODUCTION

Wireless power transfer (WPT) technologies have been gaining worldwide attention since their performances, which electrical power can be transferred without wire [1]. Various applications have been applies these technologies, e.g., the static wireless charging [2]-[4], the dynamic WPT for EVs [5]-[9] and dynamic WPT for railway [10]. In the case of electric vehicles, due to the limitation of energy storage in battery and long-periods in charging time, then to apply this technology is an alternative solution for the future EVs [11],[12]. Thus, many researchers and companies are working to develop the concerning technologies of dynamic WPT for EVs. The example of research topics on dynamic WPT are 1) the new bipolar track pad that can produce a wide power transfer profile to the receiver [13], 2) the dynamic wireless power transfer for electric vehicle developed by ORNL [14], and 3) a proposed dynamic WPT system for wireless charging, which combines the newly designed of pads-array and segmental long coils coupler [15]. However, there have been many interested points should be studied and verified. For instance, the number of switching devices is one of key issues, if a long-distance dynamic-powered roadway is constructed, then the large-number of switching devices counts are required. Most of papers have been proposed the full-bridge topology of resonant inverters, which are typically required 4-switching devices per one transmitter [16]. In previous research, we proposed the multi-coils full-

bridge resonant inverter, which switching devices counts can be reduced [17]. In this study, to evaluate the performance of proposed system, various mathematical calculations of the equivalent circuit were explained. The transferred- power capabilities in various conditions were verified by using the comparison between mathematical calculations and experimental results.

II. SYSTEM OVERVIEW AND EQUIVALENT CIRCUIT DIAGRAM

Fig. 1(a) shows the proposed multi-coils full-bridge resonant inverter for the dynamic WPT. In this study, it is assumed that two-receiving modules (i.e. inverter no.1 and no.3) are detected. Then, the keys operational waveforms are depicted, as shown in Fig.1 (b) and Fig. 1(c) illustrates an equivalent circuit diagram. The advantage of proposed resonant inverter is the reducing in the number of main switching devices, but it still works the same as traditional systems. The comparison of the used devices between conventional full-bridge resonant inverter and proposed multi-coils full-bridge inverter is depicted in table I.

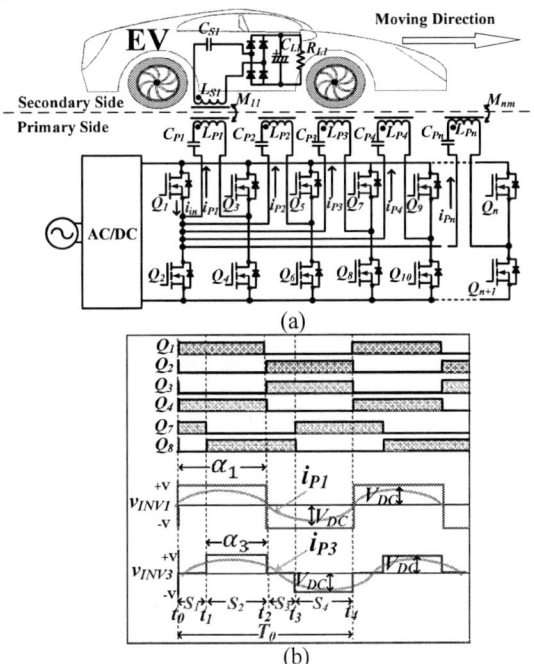

(a)

(b)

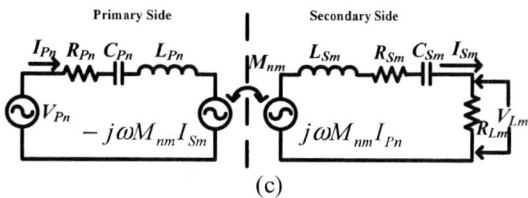

(c)

Fig.1. (a) Multi-coils full-bridge resonant inverter for dynamic WPT (b) keys operational waveforms of the multi-coils full-bridge resonant inverter (c) an equivalent circuit for one-to-one coil.

TABLE I

MAIN DEVICES COMPARISON BETWEEN CONVECTIONAL FULL-BRIDGE RESONANT INVERTER AND PROPOSED MULTI-COILS FULL-BRIDGE RESONANT INVERTER

Devices	Conventional	Proposed
Primary side coils (N)	4	4
switching devices counts	16	10
Switching devices counts calculation	$4 \times N$	$2 + (2 \times N)$

TABLE II

NOMENCLATURE

Symbol	Meaning	Unit
V_{DC}	DC input voltage	volt(V)
V_{Pn}	Output of inverter/primary voltage of Transmitter no.n	volt(V)
L_{Pn}	Primary self-inductance of coil no.n	henry(H)
C_{Pn}	Primary capacitor of inverter no. n	farad(F)
R_{Pn}	Primary resistance of coil no.n	ohm(Ω)
L_{Sm}	Secondary self-inductance of coil no.m	henry(H)
C_{Sm}	Secondary capacitor of receiver no.m	farad(F)
R_{Sm}	Secondary resistance of coil no.m	ohm(Ω)
M_{nm}	Mutual inductance between transmitter No.n and receiver no.m.	henry(H)
R_{Lm}	Load resistance of receiver no.m	ohm(Ω)
V_{Lm}	Load voltage of receiver no.m	volt(V)
I_{Pn}	Primary side current of inverter no. n	ampere(A)
I_{Sm}	Secondary side current of receiver no.m	ampere(A)
f_0	Resonant frequency in hertz	hertz(Hz)
ω_0	Resonant frequency in rad.	radian/sec (rad)
P_{Lm}	Power transferred to the load	watt(W)
η	Efficiency	(%)
K_{nm}	Coupling Coefficient between transmitter no.n and receiver no.m.	-

Note: (n , m =1, 2, 3...).

It is assumed that each resonant inverters will be operated in accordance with the detection of the receiver using the tracking sensors. From the equivalent circuit, the mutual inductance M_{nm} between primary side coil number n (n =1,2,3...) and secondary side coil number m (m =1,2,3...) is defined in (1).Then, the coupling coefficient k_{nm} between transmitter number n and receiver number m are required for calculating the mutual inductance, as described in section III .

$$M_{nm} = k_{nm} \sqrt{L_{Pn} L_{Sm}} \tag{1}$$

From Fig.1 (c), each inverters voltage V_{Pn} can be derived by (2) and each secondary side currents I_{Sm} can be calculated by (3). Then, the transferred power to load can calculated by (4). α_n is the phase-offset of each primary side inverters number n .

$$V_{Pn} = \frac{2\sqrt{2}}{\pi} V_{DC} \sin\left(\frac{\alpha_n}{2}\right) \tag{2}$$

$$I_{Sm} = \frac{j\omega M_{nm} I_{Pn}}{\left(R_{Lm} + R_{Sm} + j\omega L_{Sm} + \frac{1}{j\omega C_{Sm}}\right)} \tag{3}$$

$$P_{Lm} = |I_{Sm}|^2 R_{Lm} \tag{4}$$

To replace I_{Sm} from (3) to (4), then, the power transfer to the resistive load at resonant condition P_{Lm} can be rewritten as (5). The power transfer efficiency η can be calculated by (6), which has been presented in [18].

$$P_{Lm} = \frac{\omega_0^2 M_{nm}^2 R_{Lm} V_{Pn}^2}{\left[\left(R_{Pn}\left(R_{Lm} + R_{Sm}\right) + \omega_0^2 M_{nm}^2\right)\right]^2} \tag{5}$$

$$\eta = \frac{\omega_0^2 M_{nm}^2 R_{Lm}}{(R_{Lm} + R_{Sm})(R_{Pn} R_{Sm} + R_{Pn} R_{Lm} + \omega_0^2 M_{nm}^2)} \tag{6}$$

III. COUPLING COEFFICIENT MEASUREMENTS

To identify the equivalent circuit parameters of dynamic WPT system, the coupling coefficient at different horizontal-axis lateral offsets and varying angles of X-axis θ_X were confirmed by measuring and calculating, which can be drawn its structure, as shown Fig 2. The angles of X-axis θ_X were measured by using the real-time monitoring software, which was described in chapter IV. $L_{Pn(Open)}$ is the measured inductance across the primary side inductance L_{Pn}, while the secondary side inductance L_{Sm} is opened and $L_{Pn(Short)}$ is the same measurement, while the secondary side inductance is shorted. The coupling coefficient K_{nm} is determined by inserting these inductance values into equation (7), which has been presented in [19]. Then, the calculated coupling coefficients and mutual inductance at the different horizontal-axis lateral offsets with 25-mm of height and 0-degree of X-axis angle θ_X are plotted in Fig. 3. With the same measured procedure, but the horizontal-axis lateral offset was fixed to 0-mm, the calculated coupling coefficients and mutual inductance at the different angles of X-axis θ_X are depicted in Fig. 4.

$$k_{nm} = \sqrt{\left(1 - \frac{L_{Pn(Short)}}{L_{Pn(Open)}}\right)} \tag{7}$$

The 2018 International Power Electronics Conference

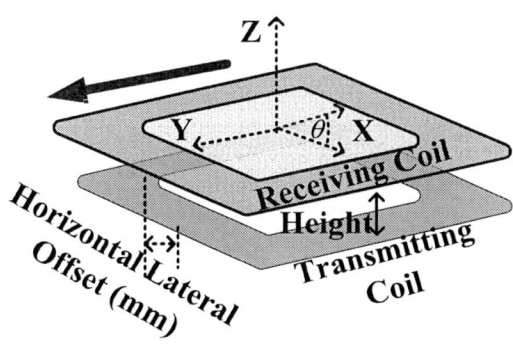

Fig. 2. The horizontal-axis and vertical-axis lateral offsets drawing diagram.

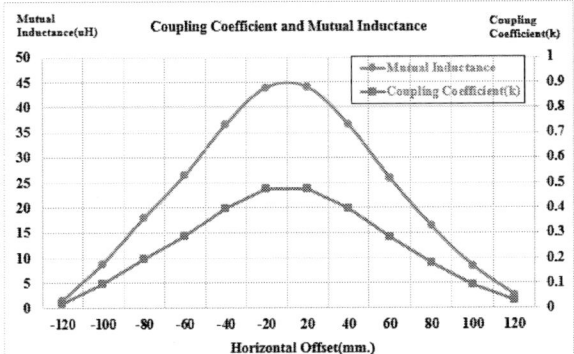

Fig. 3. Coupling coefficient and mutual inductance at different horizontal-axis lateral offsets.

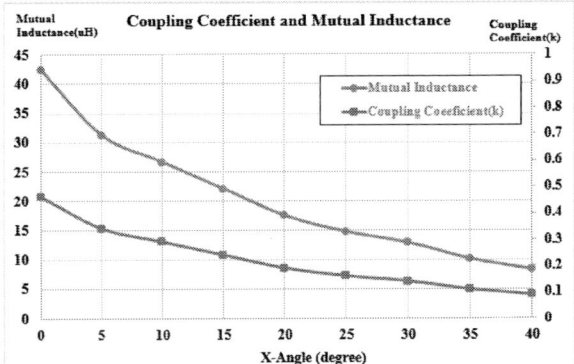

Fig. 4. Coupling coefficient and mutual inductance at different X-axis angles.

IV. EXPERIMENTAL VERIFICATIONS

To verify the theoretical analysis, a scaled-down laboratory prototype of dynamic WPT was built and tested, as depicted in Fig. 5. The detail of experimental parameters are depicted in table III. The transmitting and receiving coils were used a planar-rectangular shape at the width of 95-mm, the length of 160- mm, which were placed 70-mm. of the lateral distance between adjacent coils. The ferrite bars thickness of 5-mm were mounted on both transmitting and receiving coils.

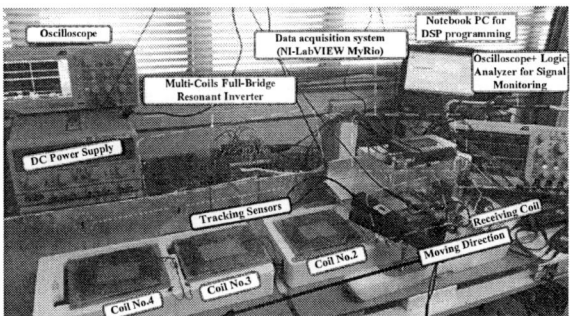

Fig. 5. A scaled-down laboratory prototype of dynamic WPT.

TABLE III
PARAMETERS OF PROPOSED DYNAMIC WPT

Symbol	Meaning	Values
V_{DC}	DC input voltage	12 V
L_{P1}	Primary self-inductance of coil no.1	92.60 μH
L_{P2}	Primary self-inductance of coil no.2	92.17 μH
L_{P3}	Primary self-inductance of coil no.3	93.43 μH
L_{P4}	Primary self-inductance of coil no.4	90.61 μH
$C_{P1}, C_{P2}, C_{P3}, C_{P4}$	Primary side capacitors	1 μF
C_{S1}	Secondary side capacitor of receiver no.1	1.5 μF
C_{S2}	Secondary side capacitor of receiver no.2	1.5 μF
R_{P1}, R_{P3}	Primary side resistances	0.955 Ω
L_{S1}	Secondary self-inductance	103.8 μH
L_{S2}	Secondary self-inductance	63 μH
R_{S1}	Secondary resistance	0.771 Ω
R_{L1}, R_{L2}	Load resistance	10 Ω
f_0	Resonant frequency	15.9 kHz

To confirm the performance of proposed multi-coils full-bridge resonant inverter, the primary side/secondary side voltage and current waveforms were measured, as shown in Fig.6. It was assumed that two-receiving modules were detected at the same time. Then, there were two inverters (i.e. inverter no.1 and inverter no.3) were operated with the difference of phase-offset. The inverter no. 1 was operated at phase-offset of $0°$ ($\alpha_1 = 180°$) while the inverter no. 3 was operated at phase-offset of $60°$ ($\alpha_3 = 120°$). Based on the results, each primary side inverter could operate to generate a high frequency pulse independently. Moreover, each pulse width could be independently adjusted. As a result, the average power could be controlled. To confirm the theoretical calculation of the receiving power, the comparison between theoretical value and measured value have been confirmed. The equation of receiving power from (5) was adopted by using the mutual inductance information from measuring and calculating data.

3334

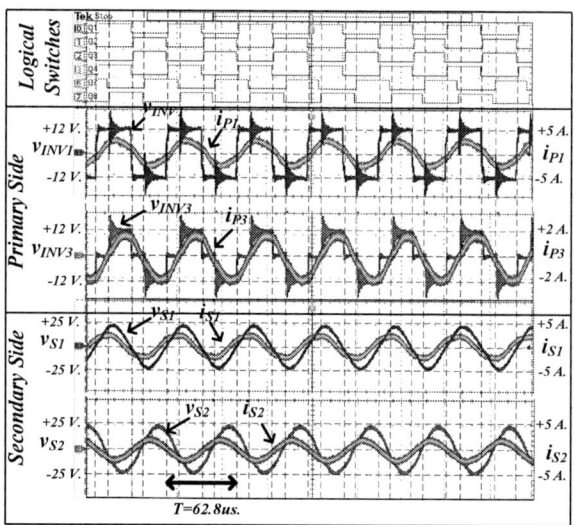

Fig. 6. Experimental results of keys operational waveforms of multi-coils full-bridge resonant inverter.

The receiving power characteristics of both theoretical and experimental values against the position of receiving coils are shown in fig.7. The test was conducted with four inverters, which were sequence operated to transfer power to one receiving module. Based on the results, the experimental result (Square marks) was similarly to the calculated result (Circular marks) with small gap. The maximum power was around 30-W and the minimum power was 0-W. It was observed that the fluctuation of transferred power when the receiver overlaps with the transmitters was occurred. The chipped power is majority problem on the dynamic WPT, which should be solved in the future study.

The commonly used solution was to apply the energy storage device (ESD), which could be compensated the chipped power, as proposed in [10]. In practice, the receiving modules is a moving device, which is quite difficult to monitor electrical parameters and detected conditions. Then, the real-time monitoring system was developed to solve the mentioned problem. The features of proposed system includes the transmitter-coil detectors, the real-time receiving power measuring, and the real-time angle monitoring. The graphic user interface (GUI) of the real-time monitoring system for dynamic WPT is depicted in Fig. 8, which is communicated through WiFi. The most important feature is the real-time acceleration monitoring, which can monitor the XYZ-axis angles.

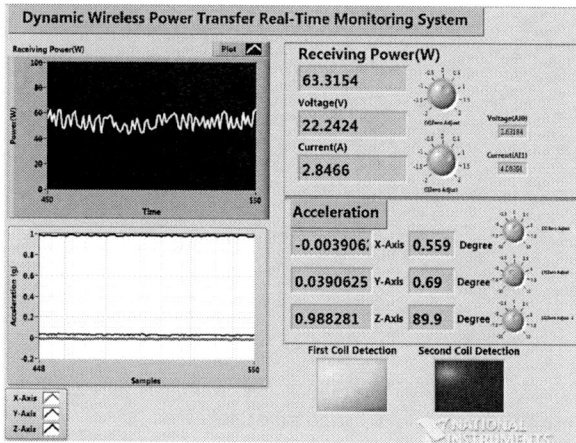

Fig. 8. Graphic user interface of the real-time monitoring system for dynamic WPT.

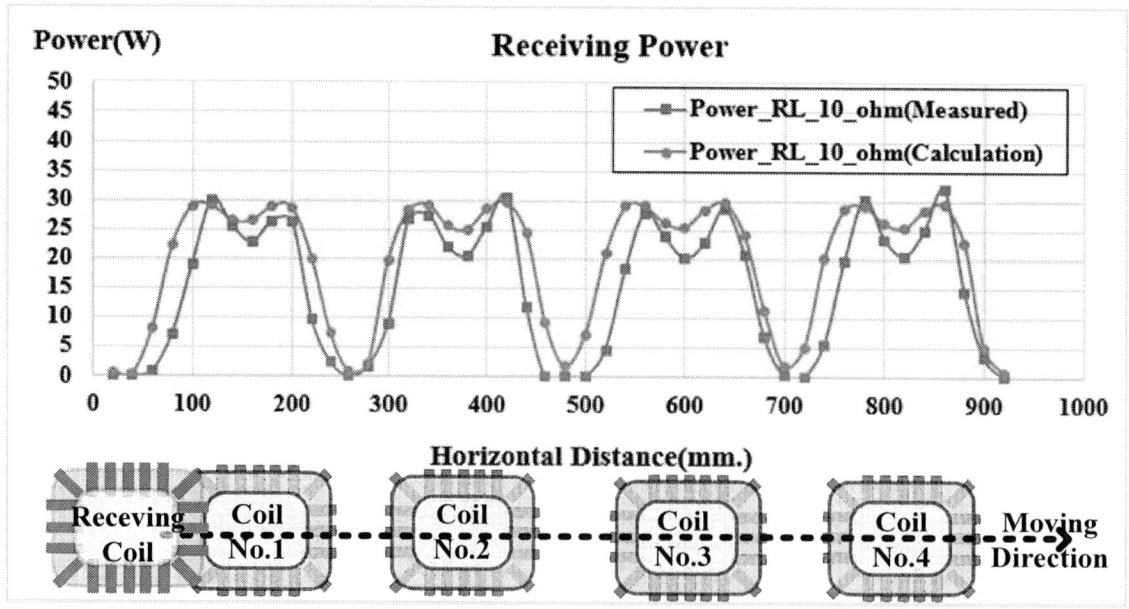

Fig.7. Receiving power characteristic against the position of receiving coil.

3335

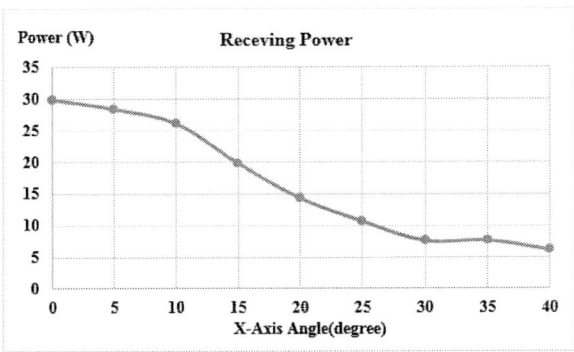
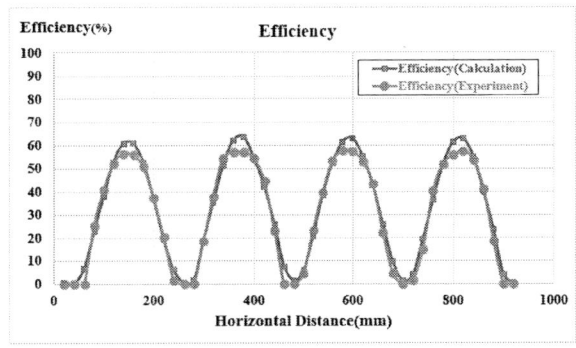

Fig. 9. Power transfer against different X-axis angle that was observed by the real-time monitoring system.

Fig. 10. System efficiency comparison between calculation values and experimental results.

With the feature of real-time monitoring, it could be used the information for to control the secondary side of dynamic WPT in the future. The real-time monitoring results is depicted in fig.9. It was the transferred power against different X-axis angle, which was observed by the real-time monitoring system. The trend of receiving power was dramatically decreased, when the X-axis angle θ_X was increased. This information may be useful in the case of while the vehicle is driven, it may change the horizontal-lateral offsets and XYZ-axis angles at any time. Then, real-time angle's information will be useful for the advance dynamic WPT's controller. However, the real-time angles $\theta_X, \theta_Y, \theta_Z$ will be deeply studied in the future research. The system efficiency comparison between calculation values and experimental results is depicted in Fig.10. As a result, the experimental graph (Circular marks) is trended similarly to the calculation graph (Square marks). The trend of efficiency on both calculation and experimentation are reached at the maximum value when the primary side coil and secondary side coil were completely aligned. According to the calculated results, it was found that the maximum and minimum efficiencies were around 63% and 0%, respectively. From the experimental results, the maximum and minimum efficiencies were around 57% and 0%, respectively.

V. CONCLUSIONS

The aim of this study is to compare the theoretical calculations and experimental tests to confirm the system parameters and to evaluate the performances of proposed dynamic WPT. The newly designed of resonant inverter circuit used the less number of main switching device counts. However, to apply this concept, it was properly performed in the same as a traditional system, which was revealed through the experimental results. We also investigated on the coupling effect and power transfer capability of the dynamic WPT system. The development of real-time monitoring system was also discussed and preliminary tested, which can be applied for dynamic WPT. The theoretical calculations on the key parameters were also examined.

The experimental results affirm the evaluation of theoretical calculations, which could be confirmed the performance of proposed dynamic WPT system. However, several points of the proposed dynamic WPT (e.g., losses analysis and maximum efficiency tracking) should be verified before implementing to a real transportation system.

ACKNOWLEDGMENT

The authors would like to thank the department of electronics and telecommunication engineering, Pathumwan Institute of Technology for supporting PSIM® software, LabVIEW® software, measurements and testing instruments.

REFERENCES

[1] N. Tesla, "Apparatus for transmitting electrical energy," U.S. Patent 1119732, 1914.

[2] H. Abe, H. Sakamoto, and K. Harada, "A noncontact charger using a resonant converter with parallel capacitor of the secondary coil," IEEE Trans. Ind. Appl., vol. 36, no. 2, pp. 444–451, Mar./Apr. 2000.

[3] C. S. Wang, O. H. Stielau, and G. A. Covic, "Design Considerations for a Contactless Electric Vehicle Battery Charger," IEEE Transactions on Industrial Electronics, Vol. 52, No. 5, pp. 1308-1314, Oct. 2005.

[4] K. Kusaka and J. Itoh, "Experimental verification of rectifiers withSiC/GaN for wireless power transfer using a magnetic resonance coupling," in Proc. IEEE Int. Conf. Power Electron. Drive Syst., pp. 1094–1099, 2011.

[5] G. A. Covic, J. T. Boys, M. L. G. Kissin, and H. G. Lu. "A three-phase inductive power transfer system for roadway powered vehicles". IEEE Trans. on Ind. Elect.,Vol. 54 , Iss.6. 2007. pp.3370 – 3378.

[6] C. C. Mi, G. Buja, S. Y. Choi, and C. T. Rim, "Modern advances in wireless power transfer systems for roadway powered electric vehicles," IEEE Trans. Ind. Electron., vol. 63, no. 10, pp. 6533–6545, Oct. 2016.

[7] A. Zaheer, G. A. Covic, and D. Kacprzak, "A bipolar pad in a 10-kHz 300-W distributed IPT system for AGV applications," IEEE Trans. Ind. Electron., vol. 61, no. 7, pp. 3288–3301, Jul. 2014.

[8] J. Huh, S. W. Lee, W. Y. Lee, G. H. Cho, and C. T. Rim, "Narrow-width inductive power transfer system for on-line electrical vehicles (OLEV)," IEEE Trans. Power Electron., vol. 26, no. 12,pp. 3666–3679, Dec. 2011.

[9] Song, K., Zhu, C., Koh, K. E., Kobayashi, D., Imura, T., and Hori, Y. "Modeling and design of dynamic wireless power transfer system for EV applications."2015-41st Annual Conference of the IEEE Industrial Electronics Society, IECON, pp. 005229-005234, 2015.

[10] S. Shibata, F. P. Wijaya and K. Kondo, "A study on the transmission power in dynamic contactless power transmission," 2016 IEEE International Conference on Industrial Technology (ICIT), Taipei, 2016, pp. 245-250.

[11] Y. Guo, L. Wang, Q. Zhu, C. Liao, and F. Li, "Switch-on modeling and analysis of dynamic wireless charging system used for electric vehicles,"IEEE Trans. Ind. Electron., vol. 63, no. 10, pp. 6568–6579, Oct. 2016.

[12] G. Buja, C. T. Rim and C. C. Mi, "Dynamic Charging of Electric Vehicles by Wireless Power Transfer," IEEE Trans. Ind. Electron.,vol. 63, no. 10, pp. 6530-6532, Oct. 2016.

[13] G. A. Covic, L. G. Kissin, D. Kacprzak, N. Clausen, and H. Hao,"A bipolar primary pad topology for EV stationary charging and highway power by inductive coupling," in Proc. IEEE Energy Convers. Congr. Expo. (ECCE), Sep. 2011, pp. 1832–1838.

[14] J. M. Miller, O. C. Onar, C. White, and S. Campbell, "Demonstrating dynamic wireless charging of an electric vehicle: The benefit of electro-chemical capacitor smoothing," IEEE Power Electron. Mag., vol. 1, no. 1,pp. 12–24, Mar. 2014.

[15] S. Zhou and C. C. Mi, "Multi-paralleled LCC reactive power compensation networks and their tuning method for electric vehicle dynamic wireless charging,"IEEE Trans. Ind. Electron., vol. 63, no. 10, pp. 6546–6556, Oct.2016.

[16] Diekhans, T.; Doncker, R.W. A Dual-Side Controlled Inductive Power Transfer System Optimized for Large Coupling Factor Variations and Partial Load. IEEE Trans. Power Electron., 30,pp. 6320–6328,2015.

[17] C. Anyapo, N. Teerakawanich and C. Mitsantisuk ,"Development of multi-coils full-bridge resonant inverter for dynamic wireless power transfer" in Proc. ECTI-CON 2017. June 27-30, Thailand,2017, pp.588-591.

[18] K. Aditya and S. S. Williamson, "A Review of Optimal Conditions for Achieving Maximum Power Output and Maximum Efficiency for a Series–Series Resonant Inductive Link," IEEE Trans. Transp. Electrification, vol. 3, no. 2, pp. 303-311, June 2017.

[19] B. Hesterman. Analysis and modelling of magnetic coupling. Technical report, Advanced Energy Industries,www.denverpels.org, 2007.

Neighboring Effects on the Deactivated Inverter in a Segmented Dynamic Wireless EV Charging System

Qingwei Zhu[1], Yanjie Guo[2], Lifang Wang[2*], Shufan Li[2], and Chenglin Liao[2]

1 School of Electrical and Electronic Engineering, University of Manchester, Manchester, UK

2 Institute of Electrical Engineering, Chinese Academy of Sciences, Beijing, China

*E-mail: wlf@mail.iee.ac.cn

Abstract— In practical dynamic wireless electric vehicle (EV) charging (DWEVC) applications, the wireless charging roadway is often split into short individually controllable segments; only those segments which there are EVs running above are activated. The activated segments often impact on their adjacent deactivated segments due to the inter-coupling between neighboring segments. In particular, the neighboring effects on the inverter of a deactivated segment are studied in this paper. Characteristics of the deactivated segmental inverter under neighboring effects are studied and estimation of its DC-link voltage is derived. Moreover, the often overlooked self-inductance variation of the transmit coils with respect to the position of the receive coil or EV is investigated. Further, a 2.4-kW dynamic wireless EV charging prototype was established and tested; testing results validated the proposed DC-link voltage evaluation method and demonstrated the non-negligible influence of the self-inductance variation. Finally, segmental transition with zero DC-link voltage pulsation has been achieved on our prototype by using the proposed voltage estimation method.

Keywords— *DC-link voltage estimation, dynamic wireless EV charging, neighboring effects, self-inductance variation.*

I. INTRODUCTION

Applications of the emerging wireless power transfer (WPT) technology on electric vehicle (EV) charging, especially the eye-catching dynamic wireless EV charging [1-5] technology or roadway powered EV [6-8], provide a new promising form of green transportation and could bring evolutions on our future transportation. To enable wirelessly charging a dynamic EV running on the lane, firstly, EVs need to be installed with a WPT receiver to harvest the wireless power delivered from the wireless charging roadway. Also, wireless power transfer related infrastructure needs to be established along the roadway, serving as the groundside transmitting system.

Fig. 1 depicts the configuration of a typical dynamic wireless EV charging (DWEVC) system. For efficiency and safety, the wireless charging lane is often split into a number of WPT segments [2, 4-6, 8-10], each having their individual compensation network, inverter, and individual DC bus switch. State and operation current of each segment can be freely controlled by tuning its own segmental inverter and bus switch. In practice, real time detection of the electric vehicles running on the DWEVC roadway is conducted; only those segments on which there are EVs which have sent out charging requests running are energized while all the other segments on which there are no EVs at all or there are EVs with no charging demands running are all deactivated. This vehicle state dependent operation mode eliminates the unnecessary electromagnetic radiation (EMR) and power loss of the uncoupled segments, could significantly improve the safety and systematic efficiency of the dynamic wireless EV charging system.

But the price is that it unavoidably brings to the repeatedly switching of the state of the WPT segments, the uncertainty of real traffic even adds to the complexity of the WPT segments' function state. Interactions between the neighboring segments which are of volatile functioning states, together with the dynamically moving WPT receiver could cause reliability issues to these multi-segmented DWEVC systems, in particular to those that utilize the preferable LCC compensation topology [3-5, 11-13], which is gaining popularity due to its advantages in primary current regulation and systematic control. One particular issue is the so called neighboring effects [4], which result from the interaction between adjacent segments due to the weak but non-negligible coupling of the closely mounted transmit coils. Neighboring effects induces unwanted current in the seemingly deactivated segment, and produces unexpected DC-link voltage over the deactivated segmental inverter [4]. Excessive DC-link voltage could occur and break down the segmental inverter, and unmatched voltages between the DC bus and the DC-link of the deactivated inverter could cause reliability issues like EMI problems when turning on [5] the about-to-activate inverter. But this issue found in DWEVC systems unfortunately remains less investigated to date.

This paper focuses on the study of the aforementioned neighboring interactions between DWEVC segments, in particular on the operation characteristic of the switched-off inverter of the deactivated segments. Estimation on the DC-link voltage of the deactivated inverter under

This work was supported by International Science & Technology Cooperation Program of China under contract No. 2016YFE0102200.

neighboring effects is theoretical derived. Influence of the often overlooked self-inductance variation of the transmit coils and its impact on neighboring effects are also studied. A 2.4-kW dynamic wireless EV charging prototype is build up; simulations and experiments are conducted to verify the theoretical analyses and the DC-link voltage estimation method.

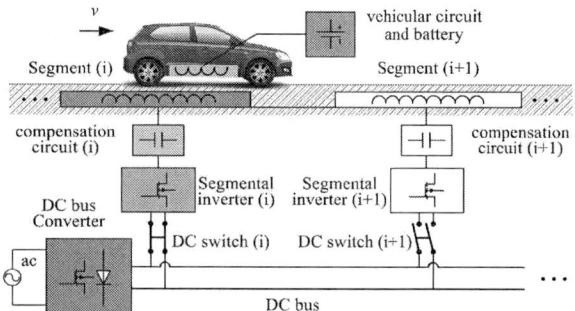

Fig. 1. Diagram of segmented dynamic EV wireless charging system with an EV traversing from one arbitrary functioning segment to its adjacent about-to-activate segment.

II. DEACTIVATED INVERTER UNDER NEIGHBORING EFFECTS AND EVALUATION OF ITS DC-LINK VOLTAGE

A. Deactivated Inverter under Neighboring Effects

For the deactivated DWEVC segments, the segmental switch is open and the segmental inverter is shut down, as depicted with the unshaded segment (i+1) on the right of Fig. 1. In this section, characteristics of the shut-down inverter of a deactivated segment in a LCC-compensated dynamic charging system are investigated. Simulations on a typical voltage source inverter with RC snubber circuit, which is the most commonly used in EV wireless charging application, are conducted under the contest of dynamic wireless EV charging, with neighboring effects from the adjacent functioning segment considered.

Output/input characteristics of the normally working inverter and the switched-off inverter are compared in Fig. 2. The current flowing in the snubber circuit I_{snb}, and the current in the reverse diode of a power switch I_D, are looked into and plotted in Fig. 3. Fig. 2(b) shows that the supposedly idle inverter behaves as a reactive power source with great inductive impedance. The deactivated inverter actually operates as a backwards rectifier as described in [4], which is demonstrated by the reverse diode voltage U_D, diode current I_D and the DC-link capacitor voltage U_C waveforms in Fig 3(a). And the two legs of snubber resistors and capacitors become the load of this rectifier. Resultantly, the output current I_{inv} is about twice of the snubber circuit current I_{snb}, as seen in Fig. 3(b). The input of this rectifier is more like a current source after the transformation [4] of the LCC compensation network connecting the output ends of this segmental inverter and the deactivated transmit coil, on which a certain voltage is induced by the neighboring segment. Since the capacitance of the snubber capacitors is often as small as several nF or hundreds of pF for wireless charging systems operating at tens of kHz, the

'output' DC-link voltage of this reversely rectifying inverter could be even higher than the normal DC bus input voltage and cause unexpected problems.

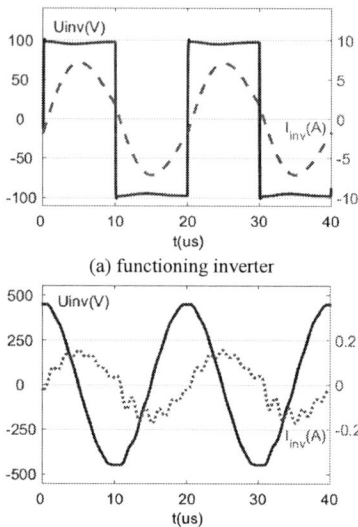

Fig. 2. Output/input voltage and current waveforms of the inverter of the functioning segment and the deactivated segment.

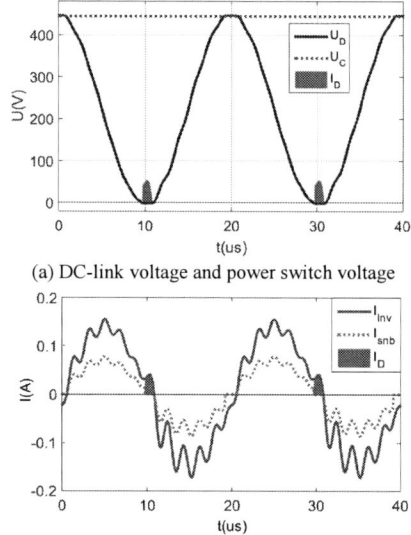

Fig. 3. Operation characteristics of the deactivated inverter.

B. DC-Link Voltage Estimation for Deactivated Inverter

As aforementioned, only the segments with EVs running above are activated. In practice, one of the most typical scenarios is that an EV is leaving one arbitrary functioning segment S1 for its adjacent segment S2, which is still idle at the moment but is about to be energized once the vehicle arrives S2. Fig. 4 and Fig. 5 show the equivalent circuits for a typical primary side LCC-compensated DWEVC system which is undergoing the described segmental transition scenario. This segment transition process is divided into two phases according to

3339

the position of the vehicle. Fig. 4 is for phase I when the EV is still moving on the domain of the working segment S1; Fig. 5 is for phase II that the EV has entered S2. The coupling strength of the receive coil L_R with the functioning segment S1 and that with the idle segment S2 varies with the two operation phases. Operation mode of the DWEVC system varies as a result, in particular the power flow among the three coils, as indicated by the arrow and dashed circle in the two figures.

In the equivalent circuits, impedances of the left hand, bottom hand, and the right hand of the LCC compensation networks are defined as $j\alpha X_P$, $-jX_P$, and $j\beta X_P$, respectively. The DC bus input voltage of the system is U_{dc}, the voltage over the DC-link capacitor C_{dc} for the deactivated inverter of segment S2 is noted with U_C. Z_{snb} is the overall impedance of the snubber circuit for each power switch. Z' is the resultant impedance of Z_{snb} after the impedance transformation [4] of the LCC network for the deactivated segment. Expression of Z' is present in (1). R_{eq} is the equivalent resistance of the load R seen from the inputs of the vehicle side rectifier, $R_{eq} = 8R/\pi^2$. The activated segment S1, the idle segment S2, and the receive coil are nominated with subscripts 1, 2, and R, respectively in parameters of self and mutual inductance.

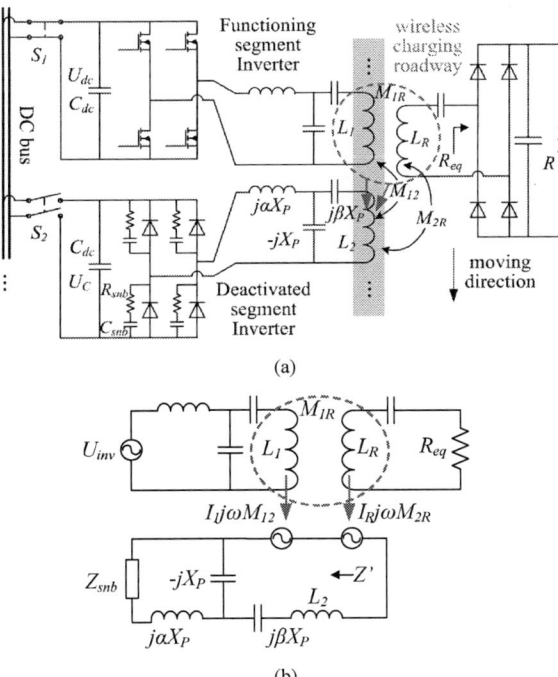

Fig. 4. Equivalent circuit of a typical dynamic charging system when an EV is traversing from the functioning segment S1 to the deactivated segment S2 next to S1. Phase I, the vehicle is still within the domain of segment S1, $M_{1R} > M_{2R}$. (a) equivalent circuit and power flow of the DWEVC system (b) simplified circuit.

In phase I as shown in Fig. 4, the EV is still above the domain of segment S1, the on-board receive coil has a much stronger coupling with the functioning transmit coil L_1 of S1 compared to that with transmit coil L_2 of the about-to-activate segment S2. In this situation, receive coil current I_R is proportional to the transmit coil current I_1 in the functioning segment S1 while almost indifferent to L_2, due to the negligible coupling of the deactivated segment S2 to the receive coil. Therefore, the functioning transmit coil L_1 and the receive coil L_R can be regarded as two individual current source with current I_1 and I_R, simultaneously working on the deactivated transmit coil L_2. Correspondingly, the induced voltage U_{ind1} in L_2 is made up of two portions as seen in (2) and Fig. 4(b), representing the influence of the functioning transmit coil L_1 and receive coil L_R, respectively. The voltage U_{snb} of the virtual overall impedance Z_{snb} of the deactivated inverter could be derived based on the simplified circuit shown in Fig. 4(b), then the DC-link voltage U_C of the deactivated inverter, which approximates the amplitude of U_{snb}, could be estimated as in (3).

$$Z' = \frac{X_P^2}{Z_{snb} + j(\alpha-1)X_P} + j(\beta-1)X_P \tag{1}$$

$$U_{ind1} = I_1 j\omega M_{12} + I_R j\omega M_{2R}$$
$$\approx I_1 j\omega M_{12} + \frac{I_1 j\omega M_{1R}}{R_{eq}} \cdot j\omega M_{2R} \tag{2}$$

$$U_C = \sqrt{2}\left| \frac{U_{ind1}}{Z'} \cdot \frac{-jX_P(Z_{snb} + j\alpha X_P)}{-jX_P + (Z_{snb} + j\alpha X_P)} \cdot \frac{Z_{snb}}{Z_{snb} + j\alpha X_P} \right|$$
$$= \sqrt{2}\left| \frac{Z_{snb}U_{ind1}}{-Z_{snb}(\beta-1) + jX_P(\alpha+\beta-\alpha\beta)} \right| \tag{3}$$
$$\approx \sqrt{2}\left| \frac{I_1 Z_{snb}(j\omega M_{12} - \omega^2 M_{1R}M_{2R}/R_{eq})}{-Z_{snb}(\beta-1) + jX_P(\alpha+\beta-\alpha\beta)} \right|$$

As the vehicle proceeds to move forward and enters segment S2, the receive coil begins to have a much stronger coupling with the transmit coil L_2 of the about-to-activate segment S2 than with the previous segment S1. Accordingly the DWEVC system switches to phase II as shown in Fig. 5. In this situation, since the receive coil current drops to a small level, the receiver acts as more of a strongly coupled load than a current source as in phase I to the deactivated segment S2. The power flow of the DWEVC system changes and is indicated in Fig. 5(a). The functioning transmit coil L_1 acts as the single current source and the induced voltage U_{ind2} in L_2 becomes (4).

The whole receiver can be equaled as a reflected impedance R_{ref}, as expressed in (5) and indicated in Fig. 5(b). The overall impedance seen from the deactivated transmit coil L_2 is composed of two parts, the first is the overall impedance Z' of the deactivated inverter, and the second is the reflected impedance R_{ref} of the receiver. The induced voltage U_{ind2} in L_2 are divided between Z' and R_{ref}. Resultantly, in phase II the DC-link voltage U_C of the deactivated inverter is estimated as (6), which has a similar form to (3) of phase I except the additional factor corresponding to voltage division. Z' is relatively small compared to R_{ref} due to the high impedance of Z_{snb} and the fact that α, β are usually close to 1; on the other hand,

3340

R_{ref} is proportional to the square of M_{2R}, which increases as the vehicle moves from the edge to the center of S2. Therefore, U_C decrease rapidly once the EV enters into the about-to-activate segment S2.

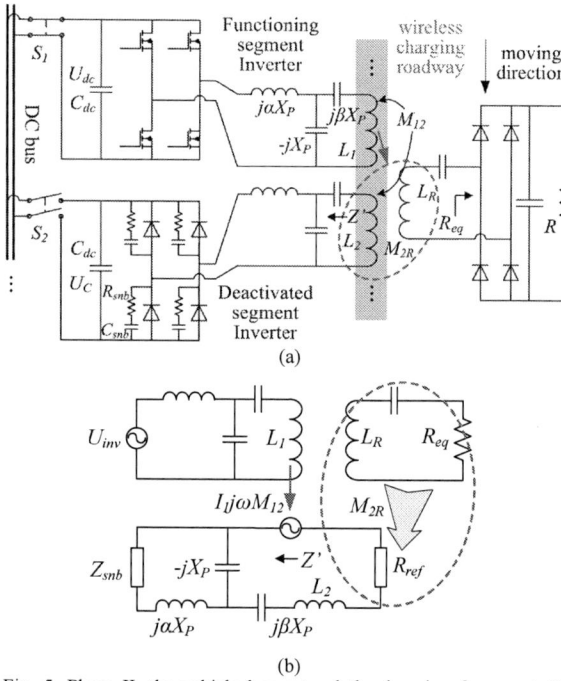

(a)

(b)

Fig. 5. Phase II, the vehicle has entered the domain of segment S2, $M_{1R} < M_{2R}$. (a) equivalent circuit and power flow of the DWEVC system (b) simplified circuit.

$$U_{ind2} \approx I_1 j\omega M_{12} \tag{4}$$

$$R_{ref} = \frac{\omega^2 M_{2R}^2}{R_{eq}} \tag{5}$$

$$
\begin{aligned}
U_C &= \sqrt{2} \left| \frac{U_{ind2}}{R_{ref} + Z'} \cdot \frac{-jX_P(Z_{snb} + j\alpha X_P)}{-jX_P + (Z_{snb} + j\alpha X_P)} \cdot \frac{Z_{snb}}{Z_{snb} + j\alpha X_P} \right| \\
&= \sqrt{2} \left| \frac{Z_{snb} \cdot \dfrac{Z' U_{ind2}}{R_{ref} + Z'}}{-Z_{snb}(\beta - 1) + jX_P(\alpha + \beta - \alpha\beta)} \right| \\
&\approx \frac{\sqrt{2}|Z'|}{|R_{ref} + Z'|} \cdot \left| \frac{I_1 j\omega M_{12} Z_{snb}}{-Z_{snb}(\beta - 1) + jX_P(\alpha + \beta - \alpha\beta)} \right|
\end{aligned}
$$

$$\tag{6}$$

III. Effects of Self-Inductance Variation

Another factor which is often overlooked in the studies of dynamic wireless EV charging systems is the self-inductance variation phenomenon. Due to the relatively small WPT gap between the vehicle side receive coil and the ground side transmit coils, the ferrite of the receive coil actually has a positive contribution to the inductance

of the transmit coils as opposite to what is often thought independent, especially on those multi-coil [3-6, 8] powered DWEVC systems where the dimension of the transmit coils is comparable to that of the receive coil. Fig. 6 below shows the transmit coil self-inductance variation with respect to the position of the receive coil measured from a predesigned dynamic EV charging prototype. In this system, the transmit coil is 72 cm*27 cm, the receive coil is 36 cm*36 cm, and the WPT gap is 10.5 cm. In Fig. 6, $x=80$ cm corresponds to the scenario that the receive coil is perfectly aligned with the transmit coil, while $x=0$ cm corresponds to that the centers of these two coils are horizontally separated by 80 cm.

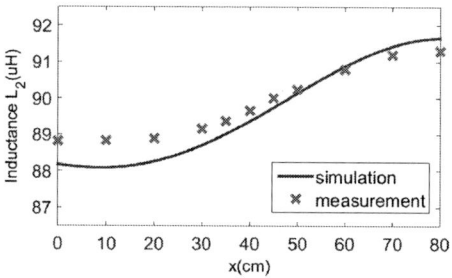

Fig. 6. Transmit coil self-inductance variation corresponding to the position of the receive coil. The transmit coil is fixed at $x=80$ cm.

Data shown in Fig. 6 indicates a variation around $\pm 2\%$ on the self-inductance of the transmit coil. The more the two coils overlaps, the larger the self-inductance gets. In practice, the mean value of the transmit coils' inductance with the receive coil moving over the whole transmit coil is used for parameter design. The value of β, which is one of the most important parameters in LCC network design, is dependent on the self-inductance of the transmit coil, as expressed in (7). C_{PS} is the capacitor in the right hand of the LCC network, directly connecting to the transmit coil in series. Due to the described self-inductance variation phenomenon, the actual value of β varies with the position x of the moving receive coil as opposite to remains its designed value β, as indicated in (8). For a functioning segment like S1, β decreases as the vehicle leaves that segment; for an about-to-activate segment like S2, β rises gradually as the vehicle enters that segment.

$$\beta = \frac{j\omega L_{i-mean} - 1/j\omega C_{PS}}{X_P} \tag{7}$$

$$\beta(x) = \frac{j\omega L_i - 1/j\omega C_{PS}}{X_P} = \frac{j\omega L_i(x) - \omega L_{i-mean}}{X_P} + \beta \tag{8}$$

Primary study in [4] shows that the DC-link voltage U_C due to the neighboring effects has an approximately reciprocal relationship to the factor $|\beta - 1|$. If the designed value of β was smaller than 1, which is often the case, β would get closer to 1 as the EV moves towards the center of the about-to-activate segment. As a result, U_C of that deactivated segment could increase dramatically in this process. Therefore, it is important to take account of the inductance variation and ensure that β does not exceed 0.98 wherever the vehicle position is when designing an LCC-compensated DWEVC system. Otherwise an excessively high DC-link voltage on the deactivated inverter could be produced and breakdown the power switches. Also, variation of β should be considered when using (3) and (6) for DC-link voltage estimation.

IV. EXPERIMENTAL VALIDATION

A predesigned miniature dynamic wireless EV charging prototype was used to validate the analyses on neighboring effects and the proposed DC-link voltage estimation for the deactivated segmental inverter. The prototype is operated at 50 kHz, the rated DC bus voltage is 300V. The system consists of a cart-carried receive coil and two identical DWEVC segments, each segment is 80 cm long, including one bundled transmit coil (72 cm*27 cm) and eight assembled ferrite stripes (36.5 cm*5.2 cm), as shown in Fig. 7. Each segment has its own segmental inverter and DC bus switch, which are fitted in the controller box at the bottom left of the photograph. The two WPT segments or transmit coils are compensated with LCC networks as marked in the photo, whereas the receive coil is simply compensated with a series-connected capacitor. Electrical connections of the bus switch, segmental inverter, LCC compensation networks and the transmit coils are exactly the same to the circuits shown in Fig. 4 and Fig. 5. Parameter design is based on [4], the value of α and β is designed to be 1.238 and 0.955, X_P is 12.3, snubber capacitor C_{snb} for the inverters is 0.66 nF, transmit coil current is 21.65A, and the average output power over one segment unit is designed to be 2.4 kW. The cart moves along x axis and the center of segment S1 is defined as the origin, coordinates of the center of the two segments as well as the middle point between them are marked in Fig. 7 for reference.

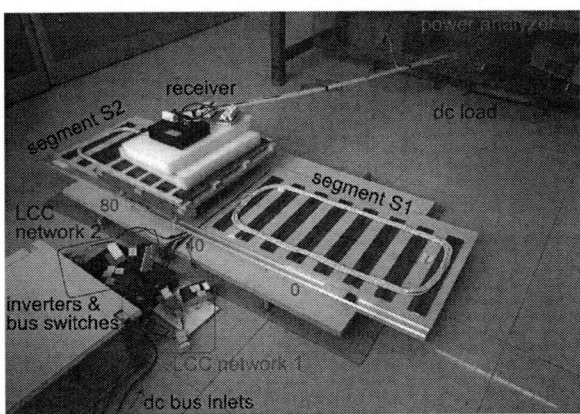

Fig. 7. Photograph of our dynamic EV charging prototype.

During the experiments, segment S2 was set as the functioning segment while segment S1 was deactivated with its segmental inverter switched off and bus switch open; the cart was initially put at x=80 cm, and was slowly moved towards the center of S1. The DC bus

input voltage was reduced to half of the rated voltage for safety reason. DC-link voltage of the segmental inverter for the deactivated S1 was recorded and compared with the simulation data as well as the analytical estimation results based on (3) and (6) using the measured self-inductance and mutual inductance data of the coils. Analytical estimations were conducted for both the case that self-inductance variation is accounted (shown as the thick dashed curves in Fig. 8), and the case that the inductance variation was neglected (shown as the thin solid curves in Fig. 8). For 80 cm $\geq x \geq$ 40 cm, $M_{2R} > M_{1R}$, the system is operated in phase I and (3) was referred to; for 40 cm $\geq x \geq$ 0 cm, $M_{1R} > M_{2R}$, the system is operated in phase II and (6) was used.

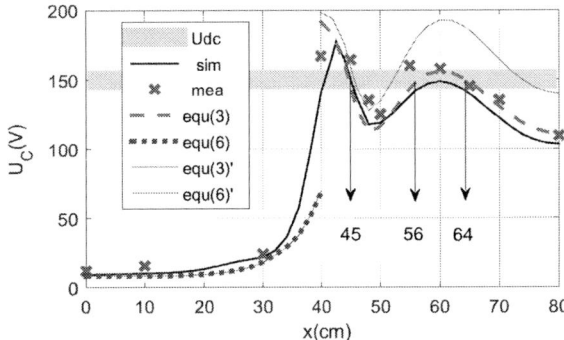

Fig. 8. DC-link voltage of the switched-off inverter of the deactivated segment S1.

Fig. 8 clearly shows that the estimated results when considering the self-inductance variation of the transmit coil matches well with both the simulations and the measured data, which validates the effectiveness of the proposed operation phase dependent DC-link voltage evaluation method for LCC-compensated dynamic wireless charging system. It also shows that it is much more essential to account for the self-inductance variation in phase I than in phase II for precise DC-link voltage prediction. As mentioned, the DC-link voltage U_C due to backward rectifying operation of the deactivated inverter is almost reciprocal to the factor $|\beta-1|$. In phase I, as the value of β is close to 1, a little variation of β even less than 2% caused by the self-inductance variation could result in a big difference in the result of U_C, as indicated with the deviation of curve equ(3) and equ(3)' in the plot. Whereas in phase II, with the mechanism of voltage division between R_{ref} and the relatively smaller Z' in the deactivated segment, sensitivity to β of U_C is greatly reduced, resulting the overlapping of equ(6) and equ(6)'. In real practice of DWEVC applications, to maximize the energy wirelessly delivered to the receiver over the whole charging roadway, the deactivated segments are often to be activated in advance before the EV fully entering in, i.e. at somewhat time when the system is operating in late phase I. Thus, the phase-I voltage estimation is of more practical significance. The results indicate the necessity of taking the self-inductance variation, which is often neglected in literature, into account in LCC-compensated dynamic wireless EV charging systems.

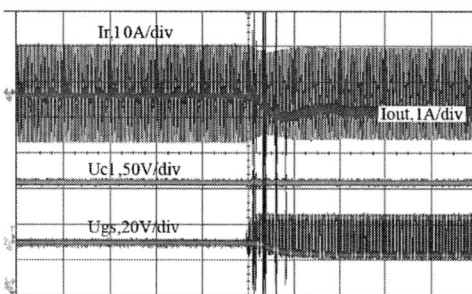

Fig. 9. Captured voltage and current transient when activating the segmental inverter of S1 as the cart just arrives at the position of $x = 64.5$cm. $Uc1$ is the DC-link voltage of the inverter for segment S1. Ir is the output dc current.

Based on the equ(3) estimation of the DC-link voltage of the deactivated inverter, we can see that when the vehicle moves to the positions near x=45 cm, 56 cm, 64 cm, the DC-link voltage of the inverter for the about-to-activate segment S1 would be perfectly matched with the DC bus input voltage. Voltage pulsation and the related EMI issues could be minimized if the activation of segment S1 were executed at these particular points. To testify this, a set of S1-activation experiments has been further conducted with S1 being energized at different positions. It turned out that zero-pulsation of the DC-link voltage was achieved at the position of x = 49 cm, 53 cm, and 64.5 cm, corresponding to the three predicted positions, respectively. The captured waveform for the zero voltage pulsation activation at x=64.5cm is shown in Fig. 9. As mentioned, in practical DWEVC applications, segments are to be repetitively switched on and off according to the traffic, the realization of zero-voltage-pulsation segment activation/deactivation here also could provide guideline for selection of the segment switching position to improve systematic reliability.

V. CONCLUSIONS

In this paper, focused on the dynamic wireless EV charging systems which adopt the more and more popular LCC compensation network, the neighboring effects a functioning segment exerts on its adjacent deactivated segment, particularly on the switched-off segmental inverter is investigated. DC-link voltage estimation for the switched-off inverter under two different operation phases of the DWEVC system is proposed. The often overlooked self-inductance variation of the transmit coils with respect to the position of the receive coil is studied as well. A 2.4-kW LCC compensated miniature dynamic wireless EV charging prototype has been setup and tested. Experimental results verified the proposed DC-link voltage estimation method and demonstrated the necessity of considering the often overlooked inductance variation. Finally, the DC-link voltage matching points of our DWEVC prototype were found and segment switching with zero-voltage-pulsation was achieved on our dynamic wireless EV charging prototype.

REFERENCES

[1] G. Buja, C. T. Rim, and C. C. Mi, "Dynamic Charging of Electric Vehicles by Wireless Power Transfer," *IEEE Trans. Ind. Electron.*, vol. 63, pp. 6530-6532, 2016.

[2] A. Zaheer, M. Neath, H. Z. Z. Beh, and G. A. Covic, "A Dynamic EV Charging System for Slow Moving Traffic Applications," *IEEE Trans. Transport. Electrific.*, vol. 3, pp. 354-369, 2017.

[3] S. Zhou and C. C. Mi, "Multi-Paralleled LCC Reactive Power Compensation Networks and Their Tuning Method for Electric Vehicle Dynamic Wireless Charging," *IEEE Trans. Ind. Electron.*, vol. 63, pp. 6546-6556, 2016.

[4] Q. Zhu, L. Wang, Y. Guo, C. Liao, and F. Li, "Applying LCC Compensation Network to Dynamic Wireless EV Charging System," *IEEE Trans. Ind. Electron.*, vol. 63, pp. 6557-6567, 2016.

[5] Guo, Y., Wang, L., Zhu, Q., Liao, C., and Li, F., 'Switch-on Modeling and Analysis of Dynamic Wireless Charging System Used for Electric Vehicles', *IEEE Trans. Ind. Electron.*, 2016, vol. 63, pp. 6568-6579.

[6] H. Liu, X. Huang, L. Tan, J. Guo, and W. Wang, "Switching control optimisation strategy of segmented transmitting coils for on-road charging of electrical vehicles," *IET Power Electron.*, vol. 9, pp.2282-2288, 2016.

[7] C. C. Mi, G. Buja, S. Y. Choi, and C. T. Rim, "Modern Advances in Wireless Power Transfer Systems for Roadway Powered Electric Vehicles," *IEEE Trans. Ind. Electron.*, vol. 63, pp. 6533-6545, 2016.

[8] R. Tavakoli and Z. Pantic, "Analysis, Design and Demonstration of a 25-kW Dynamic Wireless Charging System for Roadway Electric Vehicles," *IEEE J. Emerg. Select. Topics Power Electron.*, in press, 2017.

[9] Deng, Q., Liu, J., Czarkowski, D., Bojarski, M., Chen, J., Hu, W., and Zhou, H., 'Edge Position Detection of on-Line Charged Vehicles with Segmental Wireless Power Supply', *IEEE Trans. Veh. Technol.*, 2017, vol. 66, pp. 3610-3621.

[10] Mou, X. and Sun, H., 'Analysis of Multiple Segmented Transmitters Design in Dynamic Wireless Power Transfer for Electric Vehicles Charging', *Electronics Letters*, 2017, vol. 53, pp. 941-943.

[11] G. Zhu, D. Gao, S. Wang, and S. Chen, "Misalignment tolerance improvement in wireless power transfer using LCC compensation topology," in *IEEE PELS Workshop Emerg. Technol., Wireless Power Transfer* (WoW 2017), Chongqing, China, 2017.

[12] C. Xiao, D. Cheng, and K. Wei, "An LCC-C Compensated Wireless Charging System for Implantable Cardiac Pacemakers: Theory, Experiment and Safety Evaluation," *IEEE Trans. Power Electron.*, in press, 2017.

[13] H. Feng, T. Cai, S. Duan, J. Zhao, X. Zhang and C. Chen, "An LCC-Compensated Resonant Converter Optimized for Robust Reaction to Large Coupling Variation in Dynamic Wireless Power Transfer," *IEEE Trans. Ind. Electron.*, vol. 63, pp. 6591-6601, 2016.

Multiple Exciting Voltage Control for Maximization of Multi-hop Wireless Power Transfer Efficiency

Masato Sasaki[*] and Masayoshi Yamamoto

Department Electrical Engineering, Nagoya University, Nagoya, Japan

*E-mail: sasaki.masato@h.mbox.nagoya-u.ac.jp

Abstract— **Recently multi-hop wireless power transfer (WPT) systems for the area extension of a wireless battery charging have been addressed by several research groups. This paper describes a discrete exciting voltage control technique for the multi-hop WPT system via magnetic resonant coupling which can maximize the power transfer efficiency in response to the change of coupling status. The theory allows the equations of the wireless power transfer efficiency of the system to be determined at all the ratio of the mutual inductance. The calculated results are included to confirm the advantage to single transmitting multi-hop WPT systems.**

Keywords— wirless power transfer, efficiency, multi-hop

I. INTRODUCTION

Wireless power transfer Systems via magnetic resonant coupling (MRC-WPT) have attracted a great deal of attention in recent years and have been reported in [1]-[3]. MRC-WPT systems allow power to traverse large air gaps with high efficiency. MRC-WPT research and developments have been applied to wireless charging systems for electric vehicles and portable equipment such as mobile phones and electric vehicles.

The studies about multi-hop WPT systems to improve the convenience of WPT systems are presented in [4]-[8]. However, most of the previous studies focus on single transmitting.

In this paper, the method for controlling the exciting voltage of transmitter coils is proposed. First, the characteristic of relation between power transfer efficiency and ratio of mutual inductance and ratio of exciting voltage is derived. Next, the conditions of the ratio of exciting voltage for the wireless power transfer efficiency's maximization are derived. Finally, the advantage of the multiple transmitting multi-hop WPT systems which can maximize the wireless power transfer efficiency is confirmed.

II. TECHNICAL WORK PREPARATION

Fig. 1 shows single transmitting multi-hop WPT system. Coil1 is transmitter coil with a resonant capacitor, Coil2 and Coil3 are the repeater coils with a resonant capacitor and Coil4 is the receiver coil loaded with a resonant capacitor and a resistive load R_L. This system assumes wireless charge systems for moving EV. So, the number of repeater coils repeater coils is not limited to three.

We study multiple transmitting techniques for the minimum configuration of multi-hop WPT systems. Fig. 2

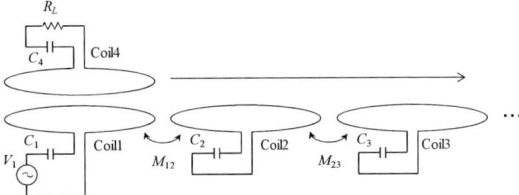

Fig. 1. Multi-Hop WPT System (Conventional)

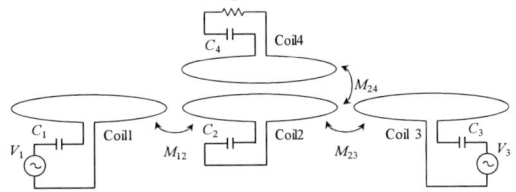

(a) Receiver Coil:Center

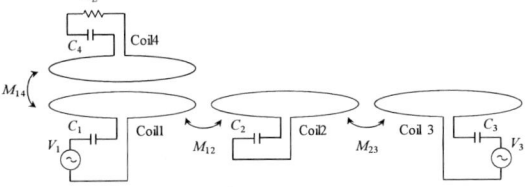

(b) Receiver Coil:Side 1

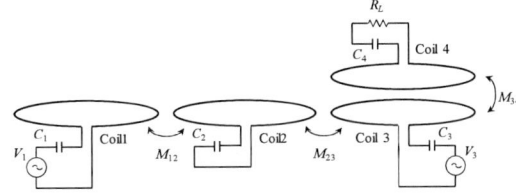

(c) Receiver Coil:Side 2

Fig. 2. Multiple Transmitting Multi-Hop WPT Systems. (Proposal)

shows multiple transmitting multi-hop WPT system for analysis. The transmitter coils are Coil1 and Coil3.The two transmitter coils are excited with two ac voltage source V_1 and V_3 with the same frequency and phase. M_{12} is mutual inductance of Coil1 and Coil2, M_{23} is mutual inductance of Coil2 and Coil3, M_{14} is mutual inductance of Coil1 and Coil4 and M_{24} is mutual inductance of Coil2

3344

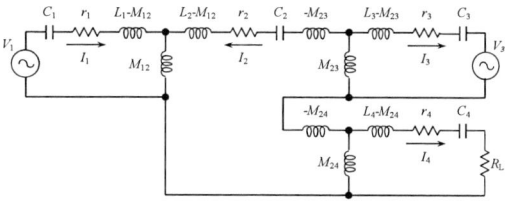

(a) Receiver Coil:Center

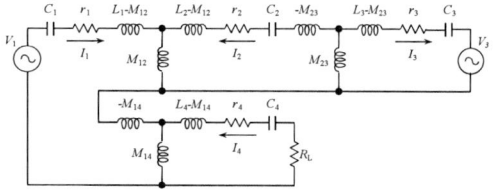

(b) Receiver Coil:Side 1

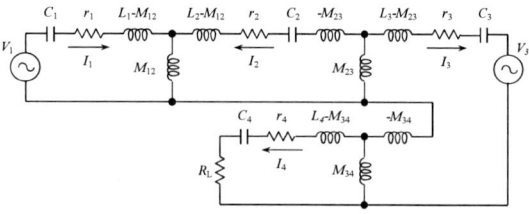

(c) Receiver Coil:Side 2

Fig. 3. Equivalent Circuit of Proposal Systems.

and Coil4. In Fig. 2(a) the receiver coil Coil4 is positioned in the center, in Fig. 2(b) and Fig. 2(c) the receiver coil Coil4 is positioned in the side. The position state is divided to mainly these three states.

Fig. 3 shows equivalent circuit of multi-hop WPT systems shown in Fig. 2. r_1, r_2, r_3 and r_4 are respectively internal resistance of Coil1, Coil2, Coil3 and Coil4, L_1, L_2 L_3 and L_4 are respectively self-inductance of Coil1, Coil2, Coil3 and Coil4. I_1, I_2, V_1 and V_2 are respectively RMS value.

III. ANALYSIS OF MULTIPLE TRANSMITTING MULT-HOP WPT SYSTEM (RECEIVER COIL:CENTER)

A. Power and Wireless Power Transfer Efficiency

According to the equivalent circuit shown in Fig. 3(a), equation (1) - (4) are given.

$$V_1 = \left\{ r_1 + j\left(\omega L_1 - \frac{1}{\omega C_1} \right) \right\} I_1 + j\omega M_{12} I_2 \quad (1)$$

$$0 = \omega M_{12} I_1 + \left\{ r_2 + j\left(\omega L_2 - \frac{1}{\omega C_2} \right) \right\} I_2 + j\omega M_{23} I_3 + j\omega M_{24} I_4 \quad (2)$$

$$V_3 = j\omega M_{23} I_2 + \left\{ r_3 + j\left(\omega L_3 - \frac{1}{\omega C_3} \right) \right\} I_3 \quad (3)$$

$$0 = j\omega M_{24} I_2 + \left\{ r_4 + R_L + j\left(\omega L_4 - \frac{1}{\omega C_4} \right) \right\} I_4 \quad (4)$$

Inductance and capacitance of receiver coil satisfies (5)

because WPT via magnetic resonant coupling send the power by electrical resonance.

$$\omega L_1 = \frac{1}{\omega C_1} \ , \ \omega L_2 = \frac{1}{\omega C_2} \ , \ \omega L_3 = \frac{1}{\omega C_3} \ , \ \omega L_4 = \frac{1}{\omega C_4} \quad (5)$$

From (1)-(5), we can obtain

$$\begin{pmatrix} V_1 \\ 0 \\ V_3 \end{pmatrix} = \begin{pmatrix} r_1 & j\omega M_{12} & 0 \\ j\omega M_{12} & r_2 + \dfrac{(\omega M_{14})^2}{r_4 + R_L} & j\omega M_{23} \\ 0 & j\omega M_{23} & r_3 \end{pmatrix} \begin{pmatrix} I_1 \\ I_2 \\ I_3 \end{pmatrix} \quad (6)$$

The current of coils I_1, I_2, I_3 and I_4 are expressed as:

$$I_1 = \frac{1}{\Delta_1} \left[\left\{ r_2 r_3 + \frac{r_3 (\omega M_{24})^2}{r_4 + R_L} + (\omega M_{23})^2 \right\} V_1 - \omega^2 M_{12} M_{23} V_3 \right] \quad (7)$$

$$I_2 = -\frac{1}{\Delta_1} \left\{ j\omega M_{12} r_3 V_1 + j\omega M_{23} r_1 \cdot V_3 \right\} \quad (8)$$

$$I_3 = \frac{1}{\Delta_1} \left[-\omega^2 M_{12} M_{23} V_1 + \left\{ r_1 r_2 + \frac{r_1 (\omega M_{24})^2}{r_4 + R_L} + (\omega M_{12})^2 \right\} V_3 \right] \quad (9)$$

$$I_4 = -\frac{j\omega M_{24}}{r_4 + R_L} I_2 \quad (10)$$

$$\Delta_1 = r_1 r_2 r_3 + \frac{r_1 r_3 (\omega M_{25})^2}{r_5 + R_L} + r_1 (\omega M_{23})^2 + r_3 (\omega M_{12})^2 \quad (11)$$

We define the ratio of mutual inductance K_M and the ratio of driving voltage K_V as:

$$K_M = \frac{M_{23}}{M_{12}} \quad (12)$$

$$K_V = \frac{V_3}{V_1} \quad (13)$$

The current of coils I_1, I_2, I_3 and I_4 are expressed as:

$$I_1 = \frac{V_1}{\Delta_1} \left\{ r_2 r_3 + \frac{r_3 (\omega M_{25})^2}{r_5 + R_L} + (\omega M_{12})^2 \left(K_M^2 - K_M K_V \right) \right\} \quad (14)$$

$$I_2 = -\frac{V_1}{\Delta_1} \left(j\omega M_{12} \right) \left(r_3 + r_1 K_M K_V \right) \quad (15)$$

$$I_3 = \frac{V_1}{\Delta_1} \left\{ r_1 r_2 K_V + \frac{r_1 (\omega M_{24})^2}{r_4 + R_L} K_V + (\omega M_{12})^2 \left(K_V - K_M \right) \right\} \quad (16)$$

$$I_4 = -\frac{V_1}{\Delta_1} \cdot \frac{\omega^2 M_{12} M_{24} \left(r_3 + r_1 K_M K_V \right)}{r_4 + R_L} \quad (17)$$

From (13)-(18), output power P_o, input power P_{in} and the power transfer efficiency are expressed as:

$$P_o = |I_4|^2 R_L = \frac{V_1^2 R_L}{\Delta_1^2} \frac{\left\{ \omega^2 M_{12} M_{24} \left(r_3 + r_1 K_M K_V \right) \right\}^2}{(r_4 + R_L)^2} \quad (18)$$

$$P_{in} = \frac{V_1^2}{\Delta_1} \left[r_2 r_3 + r_1 r_2 K_V^2 + \frac{(\omega M_{24})^2 \left(r_3 + r_1 K_V^2 \right)}{r_4 + R_L} + \left\{ \omega M_{12} \left(K_M - K_V \right) \right\}^2 \right] \quad (19)$$

$$\eta = \frac{V \alpha \left(r_1 K_M K_V + r_3 \right)^2}{r_1 r_2 K_V^2 + \dfrac{\omega^2 M_{24}^2 \left(r_1 K_V^2 + r_3 \right)}{r_4 + R_L} + (\omega M_{12})^2 \left(K_M - K_V \right)^2 + r_2 r_3} \quad (20)$$

$$\alpha = \frac{\omega^4 M_{12}^2 M_{24}^2 R_L}{\Delta_1 (r_4 + R_L)^2} \quad (21)$$

B. Maximize Wireless Power Transfer Efficiency

The partial derivative of the power transfer efficiency η with respect to K_V is expressed as:

$$\frac{\partial \eta}{\partial K_V} = 2\alpha\beta\left(r_1 K_M K_V + r_3\right)\left(K_M - K_V\right) \tag{22}$$

$$\beta = \frac{r_1 r_2 r_3 + \dfrac{\left(\omega M_{24}\right)^2 r_1 r_3}{r_4 + R_L} + \left(\omega M_{12}\right)^2\left(K_M^2 r_1 + r_3\right)}{\left\{r_1 r_2 K_V^2 + \dfrac{\omega^2 M_{24}^2\left(r_1 K_V^2 + r_3\right)}{r_4 + R_L} + \left(\omega M_{12}\right)^2\left(K_M - K_V\right)^2 + r_2 r_3\right\}} \tag{23}$$

From (23) and (24), the optimal ratio of driving voltage K_V which maximizes the power transfer efficiency is derived as:

$$K_{V\_OPT} = K_M \tag{24}$$

By substituting (24) into (20), the optimal power transfer efficiency can be calculated and by substituting (24) into (14) and (16), the current of transmitting coils I_1 and I_3 are derived as:

$$I_1 = \frac{V_1}{\Delta_1}\left(r_2 r_3 + \frac{r_3\omega^2 M_{24}^2}{r_4 + R_L}\right) \tag{25}$$

$$I_3 = \frac{V_3}{\Delta_1}\left(r_1 r_2 + \frac{r_1\omega^2 M_{24}^2}{r_4 + R_L}\right) \tag{26}$$

The transmitting current is independent from the other exciting voltage. Therefore, the interpretation of the optimal ratio of exciting voltage K_V is that the exciting voltage does not interfere with the other transmitter current.

IV. ANALYSIS OF MULTIPLE TRANSMITTING MULT-HOP WPT SYSTEM (RECEIVER COIL:SIDE 1)

A. Power and Wireless Power Transfer Efficiency

According to the equivalent circuit shown in Fig. 3(b), equation (27) - (30) are given.

$$V_1 = \left\{r_1 + j\left(\omega L_1 - \frac{1}{\omega C_1}\right)\right\}I_1 + j\omega M_{12}I_2 + j\omega M_{14}I_4 \tag{27}$$

$$0 = \omega M_{12}I_1 + \left\{r_2 + j\left(\omega L_2 - \frac{1}{\omega C_2}\right)\right\}I_2 + j\omega M_{23}I_3 \tag{28}$$

$$V_3 = j\omega M_{23}I_2 + \left\{r_3 + j\left(\omega L_3 - \frac{1}{\omega C_3}\right)\right\}I_3 \tag{29}$$

$$0 = j\omega M_{14}I_1 + \left\{r_4 + R_L + j\left(\omega L_4 - \frac{1}{\omega C_4}\right)\right\}I_4 \tag{30}$$

From (5) and (27)-(30), we can obtain

$$\begin{pmatrix} V_1 \\ 0 \\ V_3 \end{pmatrix} = \begin{pmatrix} r_1 + \dfrac{\left(\omega M_{14}\right)^2}{r_4 + R_L} & j\omega M_{12} & 0 \\ j\omega M_{12} & r_2 & j\omega M_{23} \\ 0 & j\omega M_{23} & r_3 \end{pmatrix}\begin{pmatrix} I_1 \\ I_2 \\ I_3 \end{pmatrix} \tag{31}$$

The current of coils I_1, I_3 and I_4 are expressed as:

$$I_1 = \frac{1}{\Delta_2}\left[\left\{r_2 r_3 + \left(\omega M_{23}\right)^2\right\}V_1 - \omega^2 M_{12}M_{23}V_3\right] \tag{32}$$

$$I_3 = \frac{1}{\Delta_2}\left[-\omega^2 M_{12}M_{23}V_1 + \left\{r_1 r_2 + \frac{r_2\left(\omega M_{14}\right)^2}{r_4 + R_L} + \left(\omega M_{12}\right)^2\right\}V_3\right] \tag{33}$$

$$I_4 = -\frac{j\omega M_{14}}{r_4 + R_L}I_1 \tag{34}$$

$$\Delta_2 = r_1 r_2 r_3 + \frac{\omega^2 M_{14}^2\left(r_2 r_3 + \omega^2 M_{23}^2\right)}{r_4 + R_L} + \omega^2\left(M_{23}^2 r_1 + M_{12}^2 r_3\right) \tag{35}$$

By using the ratio of mutual inductance K_M and the ratio of driving voltage K_V, the current of coils I_1, I_3 and I_4 are expressed as:

$$I_1 = \frac{V_1}{\Delta_2}\left[r_2 r_3 + \left(\omega M_{12}\right)^2\left(K_M^2 - K_M K_V\right)\right] \tag{32}$$

$$I_3 = \frac{V_1}{\Delta_2}\left[-\left(\omega M_{12}\right)^2 K_M + \left\{r_1 r_2 + \frac{r_2\left(\omega M_{14}\right)^2}{r_4 + R_L} + \left(\omega M_{12}\right)^2\right\}K_V\right] \tag{33}$$

$$I_4 = \frac{-j\omega M_{14}V_1}{\Delta_2\left(r_4 + R_L\right)}\left[r_2 r_3 + \left(\omega M_{12}\right)^2\left(K_M^2 - K_M K_V\right)\right] \tag{34}$$

From (32)-(34), output power P_o, input power P_{in} and the power transfer efficiency are expressed as:

$$P_o = \frac{\omega^2 M_{14}^2 R_L\left\{r_2 r_3 + \left(\omega M_{12}\right)^2\left(K_M^2 - K_M K_V\right)\right\}^2}{\Delta_2^2\left(r_4 + R_L\right)^2}V_1^2 \tag{35}$$

$$P_{in} = \frac{V_1^2}{\Delta_2}\left[r_2 r_3 + \omega^2 M_{12}^2\left(K_M - K_V\right)^2 + \left\{r_1 r_2 + \frac{r_2\omega^2 M_{14}^2}{r_4 + R_L}\right\}K_V^2\right] \tag{36}$$

$$\eta = \frac{\omega^2 M_{14}^2 R_L}{\Delta_2\left(r_4 + R_L\right)^2}\cdot\gamma \tag{37}$$

$$\gamma_1 = \frac{\left\{r_2 r_3 + \left(\omega M_{12}\right)^2\left(K_M^2 - K_M K_V\right)\right\}^2}{r_2 r_3 + \omega^2 M_{12}^2\left(K_M - K_V\right)^2 + \left\{r_1 r_2 + \dfrac{r_2\omega^2 M_{14}^2}{r_4 + R_L}\right\}K_V^2} \tag{38}$$

B. Power and Power Transfer Efficiency

By solving

$$\frac{\partial \eta}{\partial K_V} = 0 \tag{39}$$

the optimal ratio of driving voltage K_V which maximizes the power transfer efficiency is derived as:

$$K_{V\_OPT} = 0 \tag{40}$$

Equation (40) represents that the optimal exciting voltage is the single transmitting. By substituting (40) into (37), the optimal power transfer efficiency is expressed as:

$$\eta_{KV\_OPT} = \frac{\omega^2 M_{14}^2 R_L\left(r_2 r_3 + \omega^2 M_{12}^2 K_M^2\right)}{\Delta_2\left(r_4 + R_L\right)^2} \tag{41}$$

V. ANALYSIS OF MULTIPLE TRANSMITTING MULT-HOP WPT SYSTEM (RECEIVER COIL:SIDE 2)

A. Power and Wireless Power Transfer Efficiency

According to the equivalent circuit shown in Fig. 3(c), equation (42) -(45) are given.

$$V_1 = \left\{ r_1 + j \left(\omega L_1 - \frac{1}{\omega C_1} \right) \right\} I_1 + j\omega M_{12} I_2 \qquad (42)$$

$$0 = \omega M_{12} I_1 + \left\{ r_2 + j \left(\omega L_2 - \frac{1}{\omega C_2} \right) \right\} I_2 + j\omega M_{23} I_3 \qquad (43)$$

$$V_3 = j\omega M_{23} I_2 + \left\{ r_3 + j \left(\omega L_3 - \frac{1}{\omega C_3} \right) \right\} I_3 + j\omega M_{34} I_4 \qquad (44)$$

$$0 = j\omega M_{34} I_1 + \left\{ r_4 + R_L + j \left(\omega L_4 - \frac{1}{\omega C_4} \right) \right\} I_4 \qquad (45)$$

From (5) and (42)-(45), we can obtain

$$\begin{pmatrix} V_1 \\ 0 \\ V_3 \end{pmatrix} = \begin{pmatrix} r_1 & j\omega M_{12} & 0 \\ j\omega M_{12} & r_2 & j\omega M_{23} \\ 0 & j\omega M_{23} & r_3 + \dfrac{(\omega M_{34})^2}{r_4 + R_L} \end{pmatrix} \begin{pmatrix} I_1 \\ I_2 \\ I_3 \end{pmatrix} \qquad (46)$$

The current of coils I_1, I_3 and I_4 are expressed as:

$$I_1 = \frac{1}{\Delta_2} \left[\left\{ r_2 r_3 + \frac{r_2 (\omega M_{34})^2}{r_4 + R_L} + (\omega M_{23})^2 \right\} V_1 - \omega^2 M_{12} M_{23} V_3 \right] \quad (47)$$

$$I_3 = \frac{1}{\Delta_3} \left[-\omega^2 M_{12} M_{23} V_1 + \left\{ r_1 r_2 + (\omega M_{12})^2 \right\} V_3 \right] \qquad (48)$$

$$I_4 = -\frac{j\omega M_{34}}{r_4 + R_L} I_3 \qquad (49)$$

$$\Delta_3 = r_1 r_2 r_3 + \frac{\omega^2 M_{34}^2 (r_1 r_2 + \omega^2 M_{12}^2)}{r_4 + R_L} + \omega^2 \left(M_{23}^2 r_1 + M_{12}^2 r_3 \right) \quad (50)$$

We define the ratio of mutual inductance K_M and the ratio of driving voltage K_V as:

$$K_{V2} = \frac{V_1}{V_3} \qquad (51)$$

The current of coils I_1, I_3 and I_4 are expressed as:

$$I_1 = \frac{V_3}{\Delta_3} \left[-(\omega M_{12})^2 K_M + \left\{ r_2 r_3 + \frac{r_2 (\omega M_{34})^2}{r_4 + R_L} + (\omega M_{12} K_M)^2 \right\} K_{V2} \right] \quad (52)$$

$$I_3 = \frac{V_3}{\Delta_3} \left[r_1 r_2 + (\omega M_{12})^2 (1 - K_M K_{V2}) \right] \qquad (53)$$

$$I_4 = \frac{-j\omega M_{34} V_3}{\Delta_3 (r_4 + R_L)} \left[r_1 r_2 + (\omega M_{12})^2 (1 - K_M K_{V2}) \right] \qquad (54)$$

From (52)-(54), output power P_o, input power P_{in} and the power transfer efficiency are expressed as:

$$P_o = \frac{\omega^2 M_{34}^2 R_L \left\{ r_1 r_2 + (\omega M_{12})^2 (1 - K_M K_{V2}) \right\}^2}{\Delta_3^2 (r_4 + R_L)^2} V_3^2 \qquad (55)$$

$$P_{in} = \frac{V_3^2}{\Delta_3} \left[r_1 r_2 + \omega^2 M_{12}^2 (K_M K_{V2} - 1)^2 + \left\{ r_2 r_3 + \frac{r_2 \omega^2 M_{34}^2}{r_4 + R_L} \right\} K_{V2}^2 \right] \quad (56)$$

$$\eta = \frac{\omega^2 M_{34}^2 R_L}{\Delta_3 (r_4 + R_L)^2} \cdot \gamma_2 \qquad (57)$$

$$\gamma_2 = \frac{\left\{ r_1 r_2 + (\omega M_{12})^2 (1 - K_M K_{V2}) \right\}^2}{r_1 r_2 + \omega^2 M_{12}^2 (K_M K_{V2} - 1)^2 + \left\{ r_2 r_3 + \frac{r_2 \omega^2 M_{34}^2}{r_4 + R_L} \right\} K_V^2} \quad (58)$$

B. Power and Power Transfer Efficiency

By solving

$$\frac{\partial \eta}{\partial K_{V2}} = 0 \qquad (59)$$

the optimal ratio of driving voltage K_V which maximizes the power transfer efficiency is derived as:

$$K_{V2\_OPT} = 0 \qquad (60)$$

Equation (60) represents that the optimal exciting voltage is the single transmitting. By substituting (60) into (57), the optimal power transfer efficiency is expressed as:

$$\eta_{K_{V2\_OPT}} = \frac{\omega^2 M_{34}^2 R_L (r_1 r_2 + \omega^2 M_{12}^2)}{\Delta_3 (r_4 + R_L)^2} \qquad (61)$$

VI. Evaluation of Power Transmitter Efficiency

We evaluate the power transfer efficiency of multi transmitting multi-hop WPT system described in the previous section. Table I shows the specifications of the proposal multi-hop WPT system.

Fig. 4 and Fig. 5 show the calculated results of the wireless power transfer efficiency and output power. Fig. 4 shows evaluate results of the proposal system shown in Fig. 2(a). When K_v is 1, the power transfer efficiency is maximized and the value of the power transfer efficiency is 79.9%. When K_v is -1, the output power and the power transfer efficiency are minimized and the value is zero. Fig. 5 shows evaluate results of the proposal system shown in Fig. 2(b). When K_v is zero, the power transfer efficiency is maximized and the value of the power transfer efficiency is 97.99%. When K_v is 1, the output power and the power transfer efficiency are minimized and the value is zero. From these results, the validity of the derived formulas is confirmed.

Table II shows the comparison result of the power transfer efficiency with the conventional method. When the receiver coil Coil4 is positioned in the side 1 as shown in Fig. 2(b), the values of the power transfer efficiency of each method are equal. When the receiver coil Coil4 is positioned in the center and the side 2, the power transfer efficiency is improved. From these results, the advantage to single transmitting multi-hop WPT systems is confirmed.

Table I. Specifications of the System.

Parameter		Value
Operating Frequency	f_s	6.78MHz
Self-Inductance	$L_1 = L_2 = L_3 = L_4$	99.93μH
Resonant capacitance	$C_1 = C_2 = C_3 = C_4$	5.51pf
Resistance of Coil	$r_1 = r_2 = r_3 = r_4$	8.51Ω
Mutual Inductance	$M_{12} = M_{23}$	3.00μH
	M_{14}, M_{34}	49.97μH
Road Resistance	R_L	5kΩ

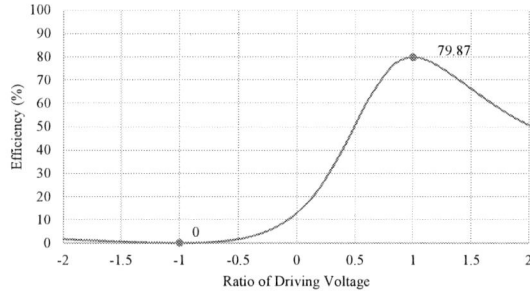

(a) Efficiency versus Ratio of Driving Voltage

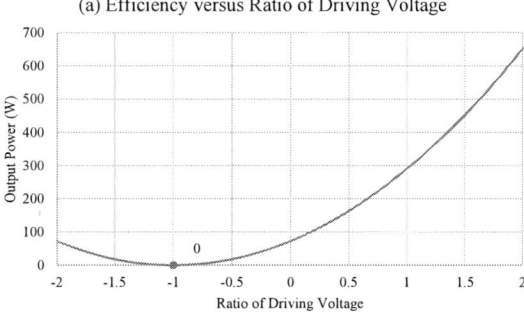

(b) Output Power versus Ratio of Driving Voltage

Fig. 4. Efficiency and Output Power (Receiver Coil:Center)

(a) Efficiency versus Ratio of Driving Voltage

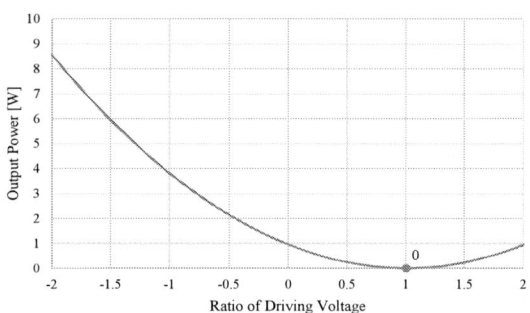

(b) Output Power versus Ratio of Driving Voltage

Fig. 5. Efficiency and Output Power (Receiver Coil:Side 1)

Table II. Comparicon of Efficiency.

	Receivre Coil's Position		
	Side 1	Center	Side 2
Proposal	97.99%	79.87%	97.99%
Conventional	97.99%	12.89%	66.08%

VII. CONCLUSIONS

In this paper, the theoretical power transfer efficiency formula of multiple transmitting multi-hop WPT system is derived. And the exciting voltage control method for maximizing the power transfer efficiency is proposed by analyzing the equivalent circuit. The exciting voltage control can realize a higher power transfer efficiency WPT system than single transmitting WPT system. We studied multiple transmitting techniques for the minimum configuration of multi-hop WPT systems. Therefore future work includes analysis of the larger multi-hop WPT systems and experiments to verify the mathematical analysis in this paper.

REFERENCES

[1] A. Kurs, A. Karalis, R. Moffat, J. D. Jonnopoulos and M. Soljačić : "Wireless Power Transfer via Strongly Coupled Magnetic Resonances", in Science Express, Vol.317, No.5834, pp.83-86 (2007)

[2] A. Karalis, J. D. Jonnopoulos and M. Soljačić : "Efficient wiewless non-radiative mid-range energy transfer", Annals of Physics, Vol.323, No.1, pp.34-48 (2008)

[3]

Q. Chen, L. Li, and K. Sawaya : "Numerical Analysis on Transmission Efficiency of Evanescent Resonant Coupling Wireless Power Transfer System", IEEE Trans. Antennas Propag., Vol.58, No.5, pp.1751-1758 (2010)

[4] Y. Naruse, et al. : "Impedance Matching Method for Any-Hop Straight Wireles Power Transmission Using Magnetic Resonance", in *Proc. IEEE RWS*, Jan. 2013, pp.20-23.

[5] S. Ohtake, et al. : "Location Estimation Considering Receiver Height in 2D Multi-hop Wireless Power Tranfer", *Technical Report of IEICE*, WPT 2013-21, Nov. 2013 (in Japanese)

[6] W. X. Zhong, et al. : "Wireless power domino-resonator systems with non-coaxial axes and circular structuer", *IEEE Trans. Power Electron.*, vol. 27, no. 11, pp.4750-4762, Nov. 2012

[7] K. L. Hhi, et al. : "Effects of magnetic coupling of nonadjacent resonators on wireless power domino-resonator systems", IEEE Trans. Power Electron., vol. 27, no. 4, pp.1905-1916, Apr. 2012.

[8] T. Imura : "Equivalent Circuit for Repeater Antenna for Wireless Power Transfer via Magnetic Resonant Coupling Considering Signed Coupling ", in *Proc.* ICIEA, 2011 pp.1501-1506

The 2018 International Power Electronics Conference

General Analytical Model for Inductive Power Transfer System with EMF Canceling Coils

Keita Furukawa, Keisuke Kusaka, and Jun-ichi Itoh
Department of Electrical Engineering, Nagaoka University of Technology, NUT
Niigata, Japan

*E-mail: itoh@vos.nagaokaut.ac.jp, kusaka@vos.nagaokaut.ac.jp, archer_FK@stn.nagaokaut.ac.jp

Abstract- **This paper provides an analytical model and design criteria of additional windings for reducing electromagnetic field (EMF) generated from inductive power transfer (IPT) systems. In particular, the canceling coils are connected to the main transmission coils with common-mode connection or differential-mode connection. Otherwise, the canceling coils are short-circuited. Parameter variation, which may degrade the system efficiency, occurs in the IPT system due to the unwanted coupling between the main transmission coils and the canceling coils. Therefore, theoretical analysis of the IPT system with the canceling coils is conducted in order to evaluate the effects of the EMF shielding methods on the system parameter variation. The calculation results of the system parameters agree to the measurement results in the prototype of the four-winding transformer with the error of lower than 5%.**

Keywords— active shielding, electromagnetic field, inductive power transfer, multiple magnetic coupling

I. INTRODUCTION

In recent years, inductive power transfer (IPT) systems have attracted much attention in terms of safety and convenient battery chargers for electrical vehicles (EVs) [1–5]. The IPT systems achieve power transmission using a magnetic coupling without electrical contacts. Electromagnetic field (EMF) should be considered because EMF may cause health impairment or malfunction of electrical equipment [5–7]. In particular, high-power IPT systems such as quick battery chargers for EVs generate higher level of EMF. Therefore, it is necessary to reduce EMI noise to widespread use of the IPT system.

In order to reduce EMF, many studies have been conducted on circuit topologies, modulation methods and configurations of transmission coils [5–13]. In particular, non-resonant reactive shield and active shield, which are focused on the configurations of the transmission coils, have been proposed [11–13].

Non-resonant reactive shielding methods require additional short-circuited coils in order to reduce EMF [11]. The magnetic flux induces a current which cancels the leakage magnetic field in the short-circuited coils when a leakage magnetic flux crosses the short-circuited coils. However, the EMF-reduction effectiveness is low at the place far from the canceling coils because the magnetomotive force of the canceling coil is limited by the interlinkage magnetic flux [11].

On the other hand, the active shielding method employs canceling coils which are connected to the main transmission coils or additional power sources [11–13]. In this method, the current flowing in the canceling coils is controllable with the main coils or the power sources. Therefore, the cancellation of the magnetic field becomes more effective compared to the non-resonant reactive shielding method because the magnetic field generated by the canceling coils are controllable [11].

The reduction effect of EMF by adding coils has been reported in several papers [11–13]. Nevertheless, the influences on parameter variations of the transmission coil due to the additional coils have not been analytically discussed. In particular, the magnetic flux generated by the canceling coils not only reduces the EMF but also crosses the main coils as an interlinkage flux. Thus, the self-inductance and the mutual inductance of the entire transmission coil are changed due to the additional coils. This variation of the self-inductance changes the resonant frequency. Thus, the transmission power or the efficiency might be decreased compared that of the non-canceling coils. This problem in the past work is that the influences of attaching the canceling coils to the entire IPT systems are not discussed enough.

In this paper, the effect of the canceling coils is revealed as focusing on the variations of the equivalent self-inductance and the equivalent mutual inductance of the entire transmission coil, which is theoretically analyzed using a model of a four-winding transformer. The new contribution of this paper is providing a general analytical model and design criteria in order to reduce the parameter verification for the EMF canceling coils. In particular, the canceling coils are connected in parallel to the main transmission coils when the non-resonant reactive shielding method or the active shielding method are introduced in order to reduce EMF of transmission coils system. The equivalent self-inductances and the equivalent coupling coefficient taking into account the influence of the canceling coils are calculated from a view of the entire transmission system. Then, the calculation results are confirmed with measurement of the equivalent self-inductances and the equivalent coupling coefficient in

The 2018 International Power Electronics Conference

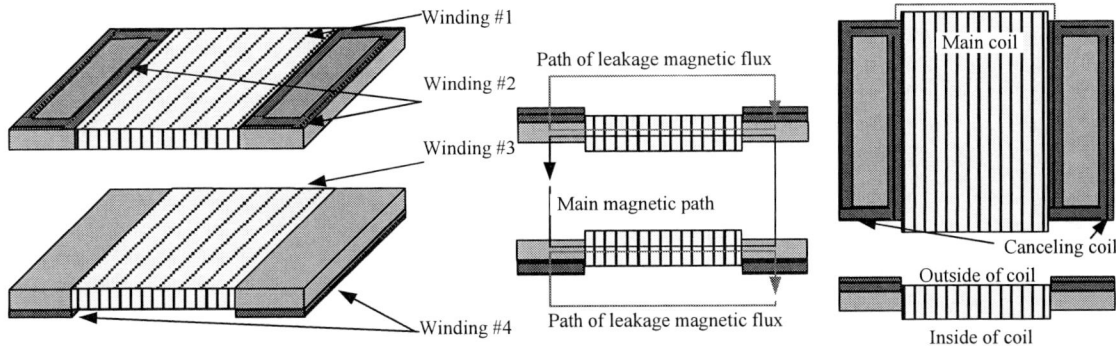

(a) Outline of transmission coil.　　　(b) Front view with magnetic paths.　　　(c) Configuration of coil at one side.

Fig. 1. Investigated model of transmission coil as example. The solenoid-type transmission coil with the canceling coils (winding #2 and winding #4). EMF upper and under cores is reduced by the canceling coils.

prototypes of the four-winding transformer. Moreover, the effect of EMF reduction is simulated when the prototypes are installed to the 1-kW IPT systems.

II. CONFIGURATION OF TRANSMISSION COILS

In this section, the equivalent self-inductance and the equivalent coupling coefficient are formulated by a model of the multi-winding transformer when the canceling coils are connected to the main coils in parallel or short-circuited. The equivalent self-inductance is defined as self-inductances from a view of a primary side or secondary side of the entire transmission coil. The equivalent coupling coefficient is defined as a coupling coefficient between the primary side and the secondary side. Both of the equivalent values are essential to design a resonant frequency and a transmission power for IPT systems.

A. Analytical Model with Four-winding Transformer

Figure 1 shows the schematic of the analyzed transmission coils. Although following analysis is possible to be applied to general transmission coils, which fulfill some conditions as mentioned later, the solenoid-type transmission coil is analyzed as an example. The transmission coil behaves as the four-winding transformer. In order to avoid the additional core, the canceling coil is wired on the same core of the main coil. The main coils, e.g., winding #1 and winding #3, are wired on the cores to the form of the solenoid coils, whereas the canceling coils, e.g., winding #2 and winding #4, form a pair of the serial rectangular coil. As shown in Fig. 1(a), the canceling coils are placed on the outside of the primary and secondary cores in order to reduce EMF which emits in the direction of the upper and below the transmission coils.

Due to the placement of the canceling coils, the mutual inductances between following windings, e.g., winding #2 and winding #3, winding #2 and winding #4, and winding #4 and winding #1, are negligibly weak. Thus, the relationship between current and voltage of each coil in Fig. 1 is expressed by the four-order inductance matrix as

$$\begin{pmatrix} v_1 \\ v_2 \\ v_3 \\ v_4 \end{pmatrix} = \begin{pmatrix} L_{11} & L_{12} & L_{13} & 0 \\ L_{12} & L_{22} & 0 & 0 \\ L_{13} & 0 & L_{11} & L_{12} \\ 0 & 0 & L_{12} & L_{22} \end{pmatrix} \frac{d}{dt} \begin{pmatrix} i_1 \\ i_2 \\ i_3 \\ i_4 \end{pmatrix}$$

$$= L \begin{pmatrix} 1 & k_c\sqrt{\alpha} & k_M & 0 \\ k_c\sqrt{\alpha} & \alpha & 0 & 0 \\ k_M & 0 & 1 & k_c\sqrt{\alpha} \\ 0 & 0 & k_c\sqrt{\alpha} & \alpha \end{pmatrix} \frac{d}{dt} \begin{pmatrix} i_1 \\ i_2 \\ i_3 \\ i_4 \end{pmatrix} \quad (1),$$

where v_m and i_m ($m = 1, 2, 3,$ and 4) are the input voltage and the current of the winding m, L_{mm} is the self-inductance of the winding m, and L_{mn} ($m \neq n$, $n = 1, 2, 3,$ or 4) is the mutual inductance between the winding m and the winding n, respectively. Here, L is equal to L_{11}. Note that k_M and k_c are the coupling coefficients between the main-coil to the main-coil or the canceling-coil, respectively. Moreover, α is the inductance ratio L_{13} / L.

Figure 2 shows the connection diagrams of the non-resonant reactive shielding method and the active shielding method. In the active shielding method, both of the cases, where the canceling coils are connected as the common-mode coils and the differential-mode coils, are considered. In particular, the common mode coils result in the positive mutual inductance between the parallel-connected coils, whereas the negative mutual inductance between the parallel-connected coils occurs in the differential mode coils.

III. CALCULATION OF EQUIVALENT PARAMETERS

A. Inverse Matrix of Inductance Matrix

In order to clarify the equivalent self-inductance and the equivalent coupling coefficients, the inductance matrix is calculated. It is convenient to calculate the current of the coils from the input voltage with the inverse matrix of the inductance matrix when conditions of the input voltage is decided as shown in Fig. 2. The relationship between current and voltage of each coil in Fig. 1 is also expressed by a four-order inverse matrix in (2) (bottom of next page) and (3)

$$\det L = \alpha L \left\{ \left(1 + k_c\right)^2 \left(1 - k_c\right)^2 - k_M^2 \right\} \quad\quad\quad (3).$$

where $\det L$ is the determinant of the four-order inductance matrix in (1).

3350

B. Short-circuited Coil

When the canceling coils are shorted as shown in Fig. 2(a), the input voltage is $v_1 = v_p$, $v_3 = v_s$ and $v_2 = v_4 = 0$. Hence, the input current is expressed by

$$
\begin{pmatrix} i_1 \\ i_2 \\ i_3 \\ i_4 \end{pmatrix} = \frac{1}{\det L} \begin{pmatrix} \alpha\left(1-k_c^2\right)\int v_p\,dt - \alpha k_M \int v_s\,dt \\ -\sqrt{\alpha}\,k_c\left(1-k_c^2\right)\int v_p\,dt + \sqrt{\alpha}\,k_c k_M \int v_s\,dt \\ -\alpha k_M \int v_p\,dt + \alpha\left(1-k_c^2\right)\int v_s\,dt \\ \sqrt{\alpha}\,k_c k_M \int v_p\,dt - \sqrt{\alpha}\,k_c\left(1-k_c^2\right)\int v_s\,dt \end{pmatrix}
$$

....(4).

In addition, the conditions of the input current $i_1 = i_p$, $i_3 = i_s$ are considered. The relationship between the voltage and the current of the entire transmission coils is shown as

$$
\begin{pmatrix} i_p \\ i_s \end{pmatrix} = \begin{pmatrix} i_1 \\ i_3 \end{pmatrix} = \frac{\alpha}{\det L} \begin{pmatrix} 1-k_c^2 & -k_M \\ -k_M & 1-k_c^2 \end{pmatrix} \begin{pmatrix} \int v_p\,dt \\ \int v_s\,dt \end{pmatrix} \quad ......(5),
$$

$$
\begin{pmatrix} v_p \\ v_s \end{pmatrix} = \begin{pmatrix} L_{peq\_short} & M_{eq\_short} \\ M_{eq\_short} & L_{seq\_short} \end{pmatrix} \frac{d}{dt}\begin{pmatrix} i_p \\ i_s \end{pmatrix}
$$

$$
= L \begin{pmatrix} 1-k_c^2 & k_M \\ k_M & 1-k_c^2 \end{pmatrix} \frac{d}{dt}\begin{pmatrix} i_p \\ i_s \end{pmatrix} \quad ....(6),
$$

where L_{peq_short} is the equivalent self-inductance of the primary side, L_{seq_short} is the equivalent self-inductance of the secondary side, and M_{eq_short} is the equivalent mutual inductances between the primary side and the secondary side.

Through the calculation, the self-inductance and the coupling coefficient of the entire transmission coil are changed from the original values L and k_M, respectively. The equivalent self-inductance $L_{eq\text{-}short}$ ($= L_{peq\_short} = L_{seq\_short}$) and the equivalent coupling coefficient k_{eq_short} are expressed in (7) and (8).

$$
L_{eq\_short} = L\left(1-k_c^2\right) ....(7)
$$

$$
k_{eq\_short} = \frac{M_{eq\_short}}{L_{eq\_short}} = \frac{k_M}{1-k_c^2} ....(8)
$$

C. Commom-mode Coil

The conditions of the input voltage is $v_1 = v_2 = v_p$ and

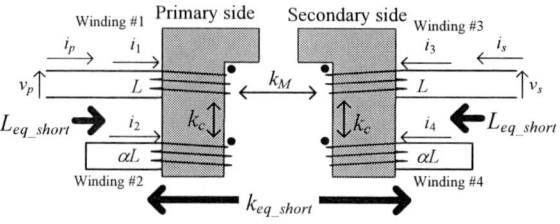

(a) Short-circuited canceling coils.

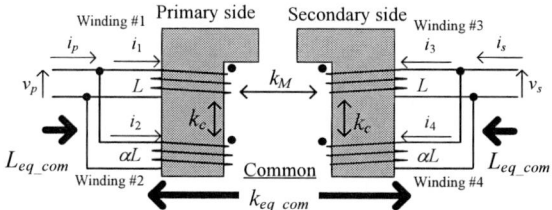

(b) Canceling coils connected as common-mode coils.

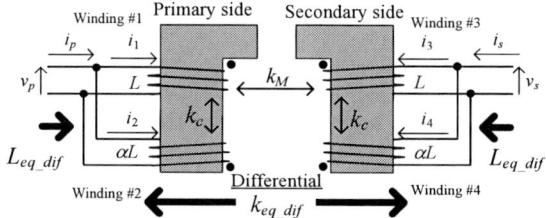

(c) Canceling coils connected as differential-mode coils.
Fig. 2. Connection diagrams of main coils and canceling coils.

$v_3 = v_4 = v_s$. Hence, when the canceling coils are connected to the main coils as the common-mode coils as shown in Fig. 2 (b), the input current is expressed by (9).

In addition, the conditions of the input current $i_p = i_1 + i_2$, $i_s = i_3 + i_4$ are considered. The relationship between the voltage and the current of the entire transmission coils is shown in (10) and (11) (next page), where L_{peq_com} is the equivalent self-inductance of the primary side, L_{seq_com} is the equivalent self-inductance of the secondary side, and M_{eq_com} is the equivalent mutual inductances between the primary side and the secondary side. The equivalent self-inductance L_{eq_com} ($= L_{peq\_com} = L_{seq\_com}$) and the equivalent

$$
\begin{pmatrix} i_1 \\ i_2 \\ i_3 \\ i_4 \end{pmatrix} = \begin{pmatrix} L_{11} & L_{12} & L_{13} & 0 \\ L_{12} & L_{22} & 0 & 0 \\ L_{13} & 0 & L_{11} & L_{12} \\ 0 & 0 & L_{12} & L_{22} \end{pmatrix}^{-1} \begin{pmatrix} \int v_1\,dt \\ \int v_2\,dt \\ \int v_3\,dt \\ \int v_4\,dt \end{pmatrix} = \frac{1}{\det L} \begin{pmatrix} \alpha\left(1-k_c^2\right) & -\sqrt{\alpha}\,k_c\left(1-k_c^2\right) & -\alpha k_M & \sqrt{\alpha}\,k_c k_M \\ -\sqrt{\alpha}\,k_c\left(1-k_c^2\right) & \left(1-k_c^2\right)-k_M^2 & \sqrt{\alpha}\,k_c k_M & -k_c^2 k_M \\ -\alpha k_M & \sqrt{\alpha}\,k_c k_M & \alpha\left(1-k_c^2\right) & -\sqrt{\alpha}\,k_c\left(1-k_c^2\right) \\ \sqrt{\alpha}\,k_c k_M & -k_c^2 k_M & -\sqrt{\alpha}\,k_c\left(1-k_c^2\right) & \left(1-k_c^2\right)-k_M^2 \end{pmatrix} \begin{pmatrix} \int v_1\,dt \\ \int v_2\,dt \\ \int v_3\,dt \\ \int v_4\,dt \end{pmatrix} \quad (2)
$$

$$
\begin{pmatrix} i_1 \\ i_2 \\ i_3 \\ i_4 \end{pmatrix} = \frac{1}{\det L} \begin{pmatrix} \sqrt{\alpha}\left(\sqrt{\alpha}-k_c\right)\left(1-k_c^2\right)\int v_p\,dt - \sqrt{\alpha}\left(\sqrt{\alpha}-k_c\right)k_M \int v_s\,dt \\ \left\{\left(1-\sqrt{\alpha}\,k_c\right)\left(1-k_c^2\right)-k_M^2\right\}\int v_p\,dt + \left(\sqrt{\alpha}-k_c\right)k_c k_M \int v_s\,dt \\ -\sqrt{\alpha}\left(\sqrt{\alpha}-k_c\right)k_M \int v_p\,dt + \sqrt{\alpha}\left(\sqrt{\alpha}-k_c\right)\left(1-k_c^2\right)\int v_s\,dt \\ \left(\sqrt{\alpha}-k_c\right)k_c k_M \int v_p\,dt + \left\{\left(1-\sqrt{\alpha}\,k_c\right)\left(1-k_c^2\right)-k_M^2\right\}\int v_s\,dt \end{pmatrix} \quad (9)
$$

3351

coupling coefficient k_{eq_com} are expressed in (12) and (13).

$$L_{eq\_com} = \frac{\alpha L\left\{\left(1-k_c^2\right)\left(1+\alpha-2k_c\sqrt{\alpha}\right)-k_M^2\right\}}{\left(\alpha-2\sqrt{\alpha}k_c+1\right)^2-k_M^2} \quad ....(12)$$

$$k_{eq\_com} = \frac{M_{eq\_com}}{L_{eq\_com}} = \frac{k_M\left(\sqrt{\alpha}-k_c\right)^2}{\left(1-k_c^2\right)\left(1+\alpha-2k_c\sqrt{\alpha}\right)-k_M^2}$$

$$..(13)$$

D. Differential-mode Coil

The conditions of the input voltage is $v_1 = -v_2 = v_p$ and $v_3 = -v_4 = v_s$ when the canceling coils are connected to the main coils as the differential-mode coils as shown in Fig. 2 (c). Hence, the input current is expressed by (14).

In addition, the conditions on the input current $i_p = i_1 - i_2$, $i_s = i_3 - i_4$ are considered. The relationship between the voltage and the current of the entire transmission coils is shown in (15) and (16), where L_{peq_dif} is the equivalent self-inductance of the primary side, L_{seq_dif} is the equivalent self-inductance of the secondary side, and M_{eq_dif} is the equivalent mutual inductances between the primary side and the secondary side. The equivalent self-inductance L_{eq_dif} ($= L_{peq\_dif} = L_{seq\_dif}$) and the equivalent coupling coefficient k_{eq_dif} are expressed in (17) and (18).

$$L_{eq\_dif} = \frac{\alpha L\left\{\left(1-k_c^2\right)\left(1+\alpha+2k_c\sqrt{\alpha}\right)-k_M^2\right\}}{\left(\alpha+2\sqrt{\alpha}k_c+1\right)^2-k_M^2} \quad ....(17)$$

$$k_{eq\_dif} = \frac{k_M\left(\sqrt{\alpha}+k_c\right)^2}{\left(1-k_c^2\right)\left(1+\alpha+2k_c\sqrt{\alpha}\right)-k_M^2} \quad ....(18)$$

E. Design Criteria of Canceling Coils

Figures 3 and 4 show the contour diagrams of the equivalent coupling coefficients and the equivalent self-inductances derived from Eqs. (7–8, 12–13, 17–18), respectively. It is noted that the coupling coefficient k_M is fixed at 0.2, whereas the coupling coefficient k_c and the self-inductance ratio α are variables. In particular, k_c is adjusted by changing installation location and configuration of the canceling coils, whereas α is also adjusted by changing the number of the turns of the canceling coils.

Figure 3(a) shows that k_{eq_short} is improved by increasing k_c, whereas k_{eq_short} is not varied by α. Meanwhile, Fig. 4(a) shows that the L_{eq_short} is decreased by increase k_c, and L_{eq_short} is not also varied by α. Hence, the parameter variation with the short-circuited canceling coils is not influenced by the number of the turns of the canceling coils.

$$\begin{pmatrix} i_p \\ i_s \end{pmatrix} = \begin{pmatrix} i_1+i_2 \\ i_3+i_4 \end{pmatrix} = \frac{1}{\det L}\begin{pmatrix} \left(\alpha-2\sqrt{\alpha}k_c+1\right)\left(1-k_c^2\right)-k_M^2 & -\left(\sqrt{\alpha}-k_c\right)^2 k_M \\ -\left(\sqrt{\alpha}-k_c\right)^2 k_M & \left(\alpha-2\sqrt{\alpha}k_c+1\right)\left(1-k_c^2\right)-k_M^2 \end{pmatrix}\begin{pmatrix} \int v_p dt \\ \int v_s dt \end{pmatrix} \quad (10)$$

$$\begin{pmatrix} v_p \\ v_s \end{pmatrix} = \begin{pmatrix} L_{peq\_com} & M_{eq\_com} \\ M_{eq\_com} & L_{seq\_com} \end{pmatrix}\frac{d}{dt}\begin{pmatrix} i_p \\ i_s \end{pmatrix}$$

$$= \frac{\alpha L}{\left(\alpha-2\sqrt{\alpha}k_c+1\right)^2-k_M^2}\begin{pmatrix} \left(\alpha-2\sqrt{\alpha}k_c+1\right)\left(1-k_c^2\right)-k_M^2 & \left(\sqrt{\alpha}-k_c\right)^2 k_M \\ \left(\sqrt{\alpha}-k_c\right)^2 k_M & \left(\alpha-2\sqrt{\alpha}k_c+1\right)\left(1-k_c^2\right)-k_M^2 \end{pmatrix}\frac{d}{dt}\begin{pmatrix} i_p \\ i_s \end{pmatrix} \quad (11)$$

$$\begin{pmatrix} i_1 \\ i_2 \\ i_3 \\ i_4 \end{pmatrix} = \frac{1}{\det L}\begin{pmatrix} \sqrt{\alpha}\left(\sqrt{\alpha}+k_c\right)\left(1-k_c^2\right)\int v_p dt - \sqrt{\alpha}\left(\sqrt{\alpha}+k_c\right)k_M\int v_s dt \\ \left\{-\left(1+\sqrt{\alpha}k_c\right)\left(1-k_c^2\right)+k_M^2\right\}\int v_p dt + \left(\sqrt{\alpha}+k_c\right)k_c k_M\int v_s dt \\ -\sqrt{\alpha}\left(\sqrt{\alpha}+k_c\right)k_M\int v_p dt + \sqrt{\alpha}\left(\sqrt{\alpha}+k_c\right)\left(1-k_c^2\right)\int v_s dt \\ \left(\sqrt{\alpha}+k_c\right)k_c k_M\int v_p dt + \left\{-\left(1+\sqrt{\alpha}k_c\right)\left(1-k_c^2\right)+k_M^2\right\}\int v_s dt \end{pmatrix} \quad (14)$$

$$\begin{pmatrix} i_p \\ i_s \end{pmatrix} = \begin{pmatrix} i_1-i_2 \\ i_3-i_4 \end{pmatrix} = \frac{1}{\det L}\begin{pmatrix} \left(\alpha+2\sqrt{\alpha}k_c+1\right)\left(1-k_c^2\right)-k_M^2 & -\left(\sqrt{\alpha}+k_c\right)^2 k_M \\ -\left(\sqrt{\alpha}+k_c\right)^2 k_M & \left(\alpha+2\sqrt{\alpha}k_c+1\right)\left(1-k_c^2\right)-k_M^2 \end{pmatrix}\begin{pmatrix} \int v_p dt \\ \int v_s dt \end{pmatrix} \quad (15)$$

$$\begin{pmatrix} v_p \\ v_s \end{pmatrix} = \begin{pmatrix} L_{peq\_dif} & M_{eq\_dif} \\ M_{eq\_dif} & L_{seq\_dif} \end{pmatrix}\frac{d}{dt}\begin{pmatrix} i_p \\ i_s \end{pmatrix}$$

$$= \frac{\alpha L}{\left(\alpha+2\sqrt{\alpha}k_c+1\right)^2-k_M^2}\begin{pmatrix} \left(\alpha+2\sqrt{\alpha}k_c+1\right)\left(1-k_c^2\right)-k_M^2 & \left(\sqrt{\alpha}+k_c\right)^2 k_M \\ \left(\sqrt{\alpha}+k_c\right)^2 k_M & \left(\alpha+2\sqrt{\alpha}k_c+1\right)\left(1-k_c^2\right)-k_M^2 \end{pmatrix}\frac{d}{dt}\begin{pmatrix} i_p \\ i_s \end{pmatrix} \quad (16)$$

The 2018 International Power Electronics Conference

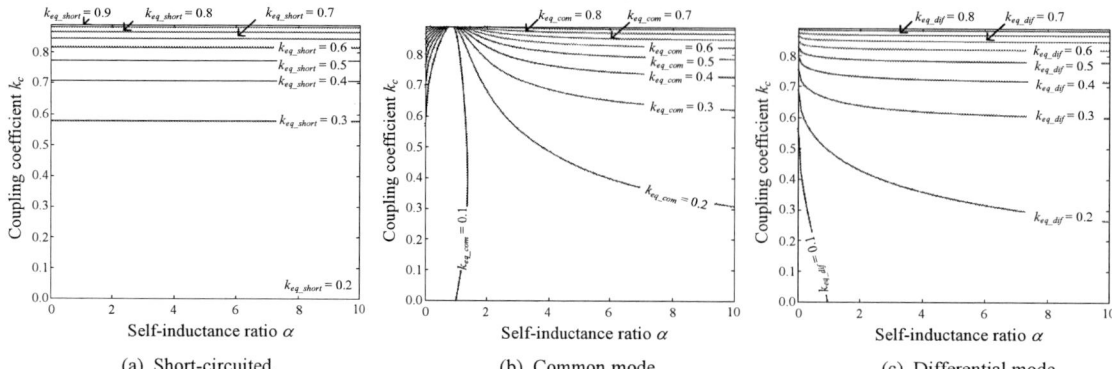

(a) Short-circuited.   (b) Common mode.   (c) Differential mode.

Fig. 3. Relationship among equivalent coupling coefficients, coupling coefficient between main coil and canceling coil, and self-inductance ratio. The coupling coefficient k_M is 0.2. Variation range of the coupling coefficient between the main coil and the canceling coil k_c is from 0 to 0.89, whereas range of the self-inductance ratio α is from 0 to 10.

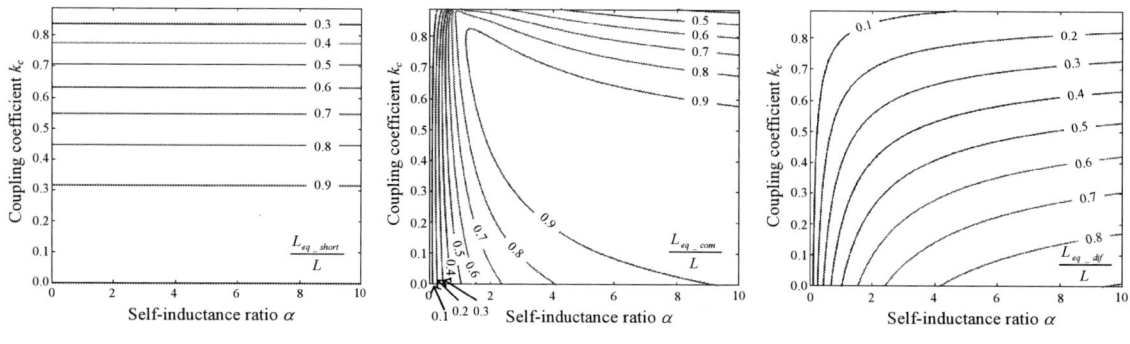

(a) Short-circuited.   (b) Common mode.   (c) Differential mode.

Fig. 4. Relationship among ratio of equivalent self-inductances to self-inductance of main windings, coupling coefficient between main coils and canceling coils, and self-inductance ratio. Conditions of k_M, k_c, and α are same as in Fig. 3.

Figure 3(b) shows that k_{eq_com} is improved by increasing k_c or α. Figure 4(b) shows that L_{eq_com} is decreasing due to the increase in k_c or the decrease in α. Thus, in order to avoid the parameter variation which is caused by the common-mode-connected canceling coils, not only the number of the turns of the canceling coils should be designed larger than the number of the turns of the main coils, but also the canceling coils have to be placed close to the main coils.

Figure 3(c) shows that k_{eq_dif} is improved in the range of high k_c and high α. Meanwhile, Fig. 4(c) shows that L_{eq_dif} is decreasing due to the high k_c and the decrease in α. Therefore, not only the winding turn should be much larger than the main coils, but also the canceling coils have to be installed apart from the main coil in order to avoid the parameter variation which is caused by the differential-mode-connected canceling coils.

However, setting the canceling coils apart from the main coils degrades a canceling performance of EMF. There is the trade-off between the parameter variation (decreasing of the equivalent self-inductance) and the canceling performance. Thus, an operation frequency or the construction of a transmission coil should be redesigned, when the canceling coils are installed in order to reduce EMF.

As a conclusion, it is shown that the design criteria for the three connection methods of the canceling coils at the view point of avoiding the parameter variations as follows:

1) the short-circuited connection (non-resonant reactive shield)
- the long distance between the main coils and the canceling coils.
- the low self-inductance of the canceling coils.

2) the common-mode connection
- installation of the canceling coils by the main coils.
- twice times or more the number of the turn of the canceling coils compared with the main coils.

3) the differential-mode connection
- the long distance between the main coils and the canceling coils.
- twice or more the number of the turn of the canceling coils compared with the main coils.

The advantage of avoiding the parameter variations is reducing the mismatch between the operation frequency and the resonant frequency, which is necessary to operate IPT systems under the conditions of the high efficiency and the high power transmission.

IV. Experimental Verification with Prototype Transmission Coil

The self-inductances and the mutual inductances of the wired coils are measured in order to confirm Eqs. (7–8, 12–13, 17–18) with the prototype four-winding transmission coil.

3353

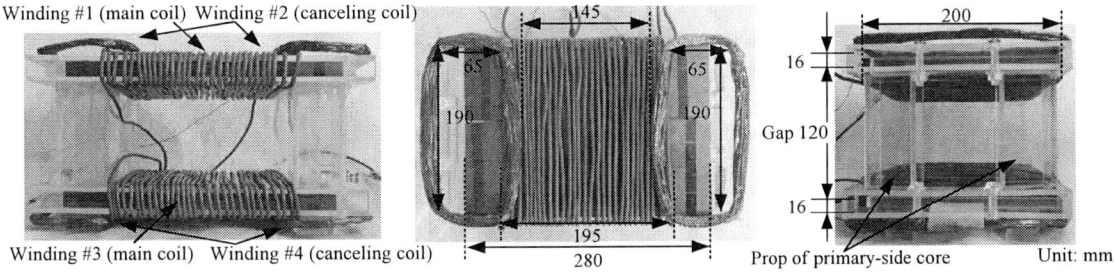

Winding #1 (main coil) Winding #2 (canceling coil)

Winding #3 (main coil)   Winding #4 (canceling coil)

| (a) Front view. | (b) Top view. | (c) Side view. |

Gap 120
Prop of primary-side core   Unit: mm

Fig. 5. Outline of prototype four-winding transmission coils. The main coils are wired on center of the cores (winding #1 and winding #3). The canceling coils are put on top of the primary core and bottom of the secondary core (winding #2 and winding #4). The canceling coils construct the serial two coils. The four props support the primary-side core.

Table I. Experimental results: equivalent coupling coefficient and inductances of prototype four-winding transformer.

Connection of canceling coils		Short-circuited	Common mode	Differential mode
Inductance matrix as 4-winding transformer			$\begin{pmatrix} 399 & 432 & 70.4 & 12.2 \\ 432 & 3120 & 12.8 & 6.7 \\ 70.5 & 12.8 & 393 & 422 \\ 12.2 & 6.7 & 422 & 3010 \end{pmatrix}$ [μH]	
Self-inductance of main winding L	Measured value	396 μH		
Self-inductance ratio α	Measured value	7.94		
Coupling coefficient k_c	Measured value	0.382		
Coupling coefficient k_M	Measured value	0.178		
Equivalent primary self-inductance	Measured value	334 μH	396 μH	236 μH
	Calculated value	338 μH	394 μH	241 μH
Error of calculated equivalent primary self-inductance		1.3%	0.6%	2.3%
Equivalent secondary self-inductance	Measured value	332 μH	391 μH	235 μH
	Calculated value	338 μH	394 μH	241 μH
Error of calculated equivalent secondary self-inductance		1.9%	0.7%	2.7%
Equivalent mutual inductance	Measured value	67.5 μH	73.0 μH	43.3 μH
	Calculated value	70.4 μH	72.1 μH	46.6 μH
Equivalent coupling coefficient	Measured value	0.203	0.185	0.184
	Calculated value	0.208	0.183	0.193
Error of calculated coupling coefficient		2.7%	1.2%	4.9%

Figure 5 shows the prototype of the transmission coil. In order to shield EMF on the top and below, the canceling coils shaped double-D are put on the outside cores. The core material is ferrite (TDK Corp., N87). The number of turns of the main coils is 30 with 3.5-mm² insulated wires, whereas the number of turn of the canceling coils is 130 with enameled wires. Note that the number of turns of the primary side and the secondary side are the same.

Table I shows the measurement results of the four-order inductance matrix, the equivalent self-inductances and the equivalent coupling coefficients in the each connection. In order to compare the equivalent self-inductance, the equivalent mutual inductance, and the equivalent coupling coefficient, both the measured values and the calculated values are shown.

In particular, the calculated values of the equivalent self-inductances correspond to the measured values with a maximum error of 2.7%. The self-inductance is the important factor because the IPT system should be designed to resonate at the transmission frequency. Thus, a precise calculation is crucial for the design of the IPT system. Besides, the maximum error of the equivalent coupling coefficients is 4.9%, which is larger than the error of the equivalent self-inductance because of the influence of the ignored magnetic coupling between following

windings, i.e., winding #2 to winding #3, winding #2 to winding #4, and winding #4 to winding #1.

V. EMF REDUCTION WITH CANCELING COILS

A. Circuit and Model Configuration

In order to confirm the effect of the EMF reduction, the prototype transmission coil is simulated with JMAG (JSOL Corporation). JMAG is a software for the electromagnetic field analysis with a finite element method.

Figure 6 shows the circuit configuration in the simulation model, whereas Table II shows the specification of the circuit. The IPT system is constructed with S/S topology, which has the resonant capacitors connected to both of the primary side and secondary side of the transmission coil in series. The input voltage is the sinusoidal wave for focusing on fundamental frequency. The capacitances of the resonant capacitors C_{s1}, C_{s2} are decided in order to resonate with the considered L at a resonance frequency of 84.75 kHz. Noted that operating frequency is decided by the resonant conditions of the resonance capacitances and the equivalent self-inductance such as L_{eq_short}, L_{eq_com}, or L_{eq_dif}. The output power is 1 kW by adjusting the value of the equivalent load resistance

The 2018 International Power Electronics Conference

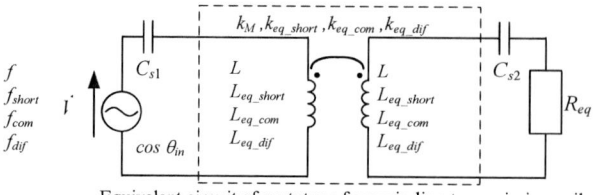

Equivalent circuit of prototype four-winding transmission coil

Fig. 6. Circuit configuration of simulation model. The prototype four-winding transformer is equivalent to the two-winding transformer when the canceling coils are shorted or connected to the main coils in parallel. For individual connection such as when the canceling coils are shorted, the equivalent self-inductance, the equivalent coupling coefficient and the operation frequency are chosen L_{eq_short}, k_{eq_short} and f_{short} differently.

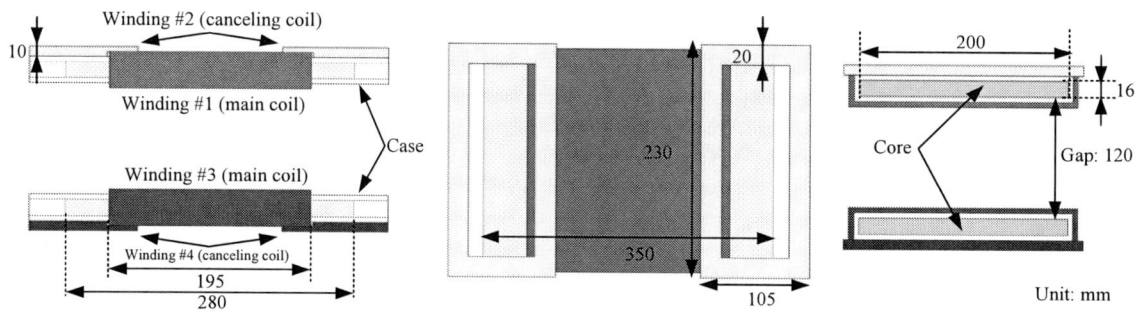

(a) Front view.   (b) Top view.   (c) Side view.

Fig. 7. Outline of CAD model of prototype four-winding transmission coil. The size of the cores and the cases matches the prototype four-winding transformer; the coils are constructed as the solid model, and the material of the cores is PC40. The cases are treated as air, whereas the thickness of the cases and the main coils is 5 mm.

Table II. Simulation conditions.

Parameter	Symbol	Value	
Input AC voltage	V_{in}	252	Vrms
Rated power	P	1.0	kW
Operation frequency without canceling coils	f	84.75	kHz
Operation frequency of short-circuited operation	f_{short}	95.00	kHz
Operation frequency of common-mode operation	f_{com}	85.40	kHz
Operation frequency of differential-mode operation	f_{dif}	115.5	kHz
Resonant capacitors	C_{s1}, C_{s2}	8.56	nF

R_{eq} because the output current of the S/S resonant circuit is inversely proportional to the equivalent mutual inductance of the transmission coil at the LC resonance.

Figure 7 shows the computer-aided-design (CAD) model of the transmission coil on JMAG. The structure and the size of the CAD model are based on the prototype four-winding transmission coil as shown in Fig. 5. The wires are expressed as the colored solid models. Noted that the eddy current and hysteresis loss are not considered.

B. EMF Reduction Effect with Canceling Coils

Figure 8 shows the simulation results of the magnetic flux distribution under the each condition. The result is obtained on the cross-section of the center of the transmission coil as a representative case. Note that the input power factor cos θ_{in} is unity when the operation frequencies of Fig. 8(a)–(d) are 84.75 kHz, 95.00 kHz, 85.40 kHz, 115.5 kHz, respectively.

The flux distributions on the top and bottom of the transmission coils decreases with the short-circuited connection and differential-mode connection in

comparison with the flux distribution without the canceling coils. The magnetic flux distribution of Fig. 8(a)–(d) are 13.8 µT, 9.41 µT, 12.6 µT, and 8.06 µT at the 50-cm bottom of the secondary core as the representative values, respectively. In addition, the magnetic density at the outside of the canceling coil decreases by the canceling coils. Moreover, the effect of EMF reduction is more effective with the differential operation than that with the short-circuited connection.

VI. CONCLUSION

In this paper, the effect of the canceling coils on the parameter variation of the IPT system was considered regarding to the equivalent self-inductance and the equivalent coupling coefficient. Three connection methods of the canceling coils, i.e. the short-circuited connection, the common-mode connection and the differential-mode connection, were evaluated.

The calculated equivalent values agreed with the measured values through the inductance measurement using the prototype transmission coil attached the canceling coils. The relative error between the calculated values and measured values was 4.9%.

In addition, EMF reduction near the canceling coils were confirmed with the short-circuit connection (by 32%) and the differential-mode connection (by 41%) in the simulation using JMAG.

Above of the results, the design criteria focusing on avoiding the parameter verifications are as follows:
- the canceling coils far from the main coils at the short-circuited connection.
- the canceling coils closed to the main coils with

3355

The 2018 International Power Electronics Conference

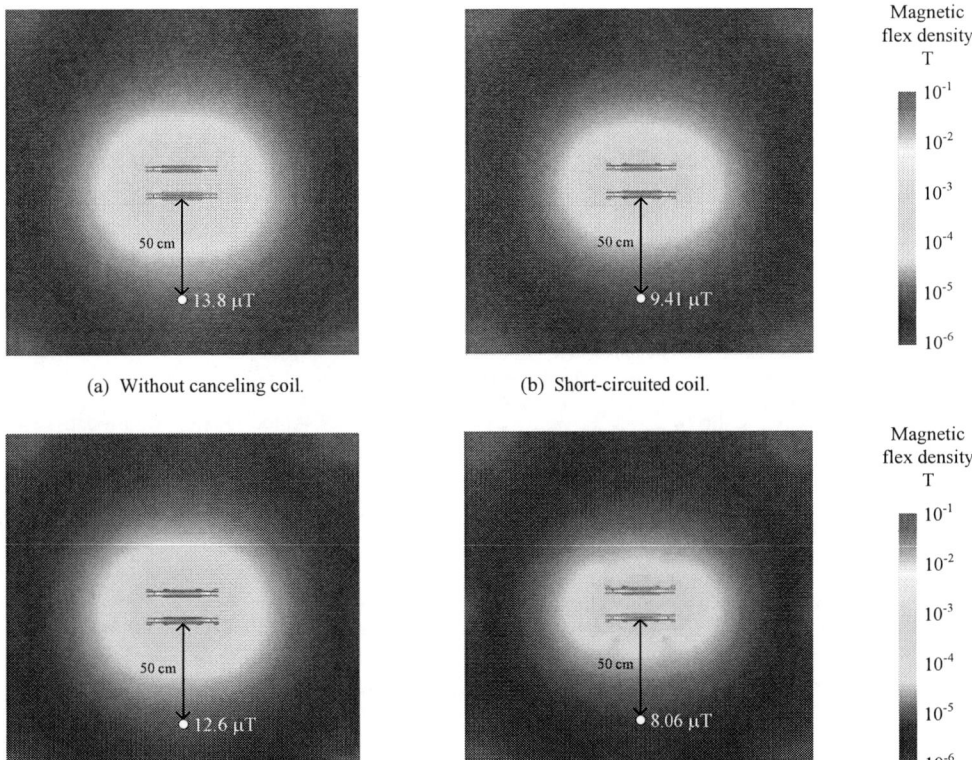

(a) Without canceling coil. (b) Short-circuited coil.

(c) Common-mode coil. (d) Differential-mode coil.

Fig. 8. Magnetic flux distribution of prototype transmission coil. The transmission coils are placed at the center of the contour plots (orange or yellow area). The upper-side core is the primary side. The bottom-side core is the secondary side. Note that the operation frequency is different for each the simulation conditions, which are shown in Table II. Transmission powers of without-canceling coil, short-circuited connection, common-mode connection and differential-mode connection are 991 W, 1.16 kW, 1.03 kW and 1.27 kW, respectively.

the large number of the turn at the common-mode connection.

- the canceling coils far from the main coils with the large number of the turn at the differential-mode connection.

Future plans are considerations of the parameter verifications introduced cross couplings between canceling coils.

REFERENCES

[1] T. Mizuno, T. Ueda, S. Yachi, R. Ohtomo and Y. Goto: "Dependence of Efficiency on Wire Type and Number of Strands of Litz Wire for Wireless Power Transfer of Magnetic Resonant Coupling", IEEJ Journal of Industry Applications, Vol. 3, No. 1, pp. 35-40 (2014)

[2] H. Ishida, H. Furukawa and T. Kyoden: "Development of Design Methodology for 60 Hz Wireless Power Transmission System", IEEJ Journal of Industry Applications, Vol. 5, No. 6, pp. 429-438 (2016)

[3] K. Kusaka, and J. Itoh: "Development Trends of Inductive Power Transfer Systems Utilizing Electromagnetic Induction with Focus on Transmission Frequency and Transmission Power", IEEJ Journal of Industry Applications, Vol. 137, No. 5, pp. 328-339 (2017)

[4] R. Ota, N. Hoshi and J. Haruna: "Design of Compensation Capacitor in S/P Topology of Inductive Power Transfer System with Buck or Boost Converter on Secondary Side", IEEJ Journal of Industry Applications, Vol. 4, No. 4, pp. 476-485 (2015)

[5] Su Y. Choi, Beom W. Gu, Seog Y. Jeong and Chun T. Rim: "Advances in Wireless Power Transfer Systems for Roadway-Powered Electric Vehicles", IEEE Trans. PE, Vol.3, No.1 pp.18 - 36 (2015)

[6] D. Shimode, T. Murai and S. Fujiwara: "A Study of Structure of Inductive Power Transfer Coil for Railway Vehicles", IEEJ Journal of Industry Applications, Vol. 4, No. 5, pp. 550-558 (2015)

[7] T. Watanabe and M. Ishida: "Study on the influence of the magnetic field and the induced electrical field in human bodies by wireless charging systems", EVTec and APE 2016, (2016)

[8] K. Kusaka, K. Inoue and J. Itoh: "Radiation Noise Reduction using Spread Spectrum for Inductive Power Transfer Systems considering Misalignment of Coils", Energy Conversion Congress and Exposition, pp. 5507-5514 (2017)

[9] T. Campi and Silvano Cruciani Mauro Feliziani: "Magnetic Shielding of Wireless Power Transfer Systems", Institute of Electronics, Information and Communication Engineers, 15A-H1, pp. 422-425 (2014)

[10] S. Kim, H. Park, J. Kim, J. Kim, and S. Ahn: "Design and Analysis of a Resonant Reactive Shield for a Wireless Power Electric Vehicle", IEEE TRANSACTIONS ON MICROWAVE THEORY AND TECHNIQUES, Vol. 62, No. 4, pp. 1057-1066 (2014)

[11] J. Park, D. Kim, K. Hwang, H. Ho Park and S. Il Kwak: "A Resonant Reactive Shielding for Planar Wireless Power Transfer System in Smartphone Application", IEEE TRANSACTIONS ON ELECTROMAGNETIC COMPATIBILITY, Vol. 59, No. 2, pp. 695-703 (2017)

[12] T. Shijo, K. Ogawa, M. Suzuki, Y. Kanekiyo and M. Ishida: "EMI Reduction Technology in 85 kHz Band 44 kW Wireless Power Transfer System for Rapid Contactless Charging of Electric Bus", IEEE Energy Conversion Congress and Exposition (2016)

[13] S. Lee et al., "Active EMF cancellation method for I-type pickup of online electric vehicles," in Proc. IEEE Appl. Power Electron. Conf. Expo., pp. 1980-1983 (2011)

The 2018 International Power Electronics Conference

Stability Influence of Filter Components Parasitic Resistance on LCL-Filtered Grid Converters

Hiroaki Matsumori[1], Toshihsia Shimizu[1*], Frede Blaabjerg[2], Xiongfei Wang[2] and Dongsheng Yang [2]

1 Electrical and electricity department, Tokyo metropolitan university, Tokyo, Japan
2 Department of energy technology, Aalborg university, Aalborg, Denmark
*E-mail: shimizut@tmu.ac.jp

Abstract— This paper presents the equivalent series resistance (ESR) modeling method of filter components and its stability impact of *LCL* filtered grid-connected voltage-source converters (VSCs). First, the time-domain models of filter inductor and capacitor are developed for modeling the ESR effect, which are used in simulations. Then, the current control loop gains of the grid-connected three-phase VSC, which include the ideal *LCL* filter and the ESR-modeled *LCL* filter are compared. The result shows that the ESR affects both the resonance peak and the phase characteristic in the frequency domain. Finally, in order to verify accuracy of proposed model, the loop gain is measured by network analyzer in the experimental set up, which shows that the analytical loop gain with proposed ESR modeled inductor agrees very well with experimental value. This verification means that the ESR modeling method is useful for an accurate stability analysis.

Keywords— *LCL-Filters, Stability assessment, equivalent series resistance (ESR), Grid Converters.*

I. INTRODUCTION

Advances in power semiconductor devices have significantly enhanced the power-density of power electronic converters, and the voltage-source converters (VSCs) are widely used in renewable power generation sources, energy-efficient power loads, and flexible ac/dc transmission power systems, etc [1] [2]. For VSCs, an *L* filter or an *LCL* filter has to be used at the output to filter the pulsed voltages and limit the harmonic contents. The *LCL* filter can realize very effective switching harmonic attenuation with much reduced inductance requirement [3]. And so, various stability analysis of *LCL* filtered VSCs which including time delay of digital control system is already performed [3]-[16]. The time delay of the digital control system, which, in the worst cases, includes one sampling period (T_s) of computational delay [4] and half sampling period ($0.5\,T_s$) of pulse-width modulation (PWM) delay [5], has been found to have a stabilizing effect within the single-loop grid-side current control [6], [7]. Given the time delay of $1.5T_s$, whereas computational delay plus PWM delay, the one-sixth of the sampling frequency ($f_s/6$) is known as the critical frequency below the Nyquist frequency ($f_s/2$) [16]. A stable grid-side current control can be achieved for the *LCL* resonance

frequency between $f_s/6$ and $f_s/2$. However, this delay-dependent stability assessment assumes that the impedance of passive component is ideal, i.e. *LCL* filter resonance gain is infinity, while the effect of equivalent series resistance (ESR) of impedance is overlooked. In [17], the ESR is considered and set to constant value. But, the ESR of frequency-dependent characteristic such as skin effect, proximity effect, etc. have yet to considered. Considering the further implementation of renewable energy [18], more practical stability assessment is necessary.

This paper thus considers practical ESR model of filter components such as inductor and capacitor, and stability impact of the ESR of filter components is assessed. First, the time domain model of inductor is considered in order to address the frequency-dependent characteristic of the ESR. Then, the loop gains for grid-side current controlled three-phase VSC which using the ideal inductor and the ESR modeled inductor are calculated and compared under the stable grid-side current control condition, i.e., the *LCL* resonance frequency between $f_s/6$ and $f_s/2$. The calculation result of loop gain shows that the ESR can affect resonance peak damping and phase characteristic over a wide frequency range. Finally, in order to verify proposed model, the loop gain of VSC is measured by network analyzer in the experimental set up, and it is compared to calculated value. The experiment result agrees very well with simulation result, and it means the ESR modeling method of inductor is useful for accurate stability analysis. The proposed ESR modeling method can be also useful for other power electronics system analysis, such as EMI filter design, PI controller design, etc.

II. PARASITIC RESISTANCE MODELING

Normally, the inductor and capacitor are modeled by three circuit elements, *R*, *C* and *L*. The frequency domain model is easy to derive by curve fitting based on the measured data, yet it is difficult used for circuit simulator. Hence, the time domain models of filter inductor and capacitor are developed. For the stability analysis of the current-controlled VSCs, the precise parasitic resistance modeling below the Nyquist frequency ($f_s/2$) is important.

3357

A. Inductor modeling method

The conventional time domain model is shown in Fig.1 (a) which consists of L, $R_{winding}$, $R_{iron\,loss}$, C. However, the winding resistance $R_{winding}$ is set as a constant value in the whole frequency range and does not take the skin effect and the proximity effect into consideration.

For the stability analysis of VSCs, the modeling of the iron loss and the self-capacitance is not so important because the iron loss $R_{iron\,loss}$ is only affected by the switching ripple and fundamental-frequency component [19] and the self-capacitance C which is between windings is very small and affected over the Nyquist frequency ($f_s/2$).

From those reasons, the modeling of the inductance L and the winding resistance is important for the stability analysis of the current-controlled VSCs. The inductance L can be modeled by conventional circuit. For the winding resistance, the resistance value become increased by skin effect and proximity effect. In this case, LR ladder circuit, as shown in Fig. 1 (b), can be used [20].

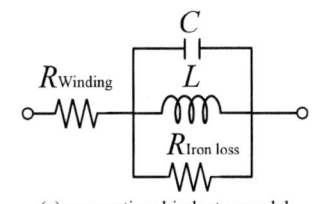

(a) conventional inductor model

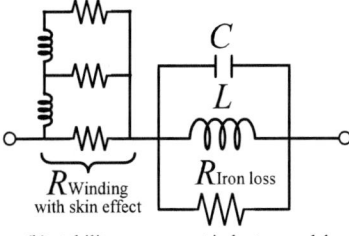

(b) stability assessment inductor model

Fig. 1 The equivalent circuit of inductor.

B. Impedance measurement and inductor modeling

In order to model inductor, the frequency response of the inductor is measured by using the impedance analyzer (Keysight: E4990A). Then the inductor is modeled by least squares fitting from measured data. Fig. 2 shows gapped amorphous inductor (Shinenergy: MPR10322-15-01) and equivalent circuit of the inductor. Figs. 3 to 6 show measured and calculated frequency responses of the inductor in terms of ESR, impedance, inductance, and phase. These values are modelled up to 50MHz. The value of LR ladder circuit is calculated by curve fitting from measured ESR data in target frequency. As seen in Fig. 6, the maximum phase difference is only 1.1 degree at 50Hz. Even if winding resistor part of proposed model contains inductors, the values of inductors do not have much influence on the phase and impedance, etc. That's because

inductance of resistor part is much smaller than main inductance L. It should be mentioned that the ESR can be modeled more precisely by increasing LR ladder. In case of bias characteristic for inductance, the LCR meter (WK: 3260B) is used. Fig. 7 shows inductance value for DC bias current. The test inductor keeps inductance value for DC bias current. The inductor is modelled very well.

(a) test inductor

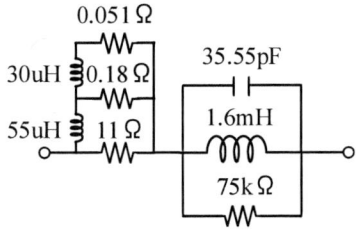

(b) modeled equivalent circuit of inductor

Fig. 2 The test inductor and the equivalent circuit.

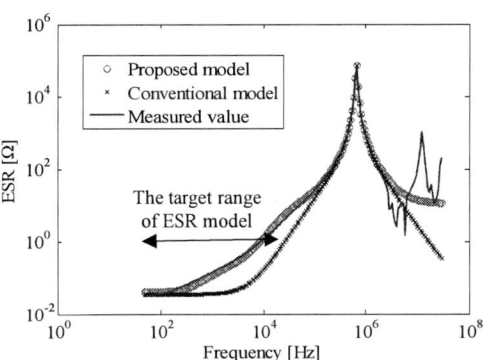

Fig. 3 The ESR characteristic.

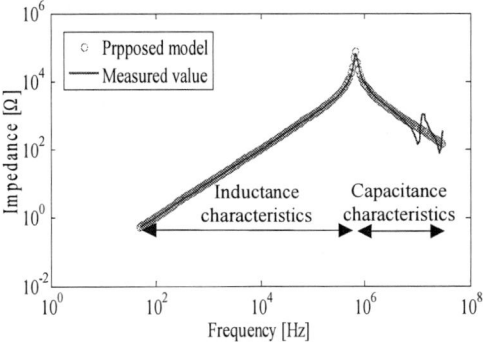

Fig. 4 The impedance characteristic.

The 2018 International Power Electronics Conference

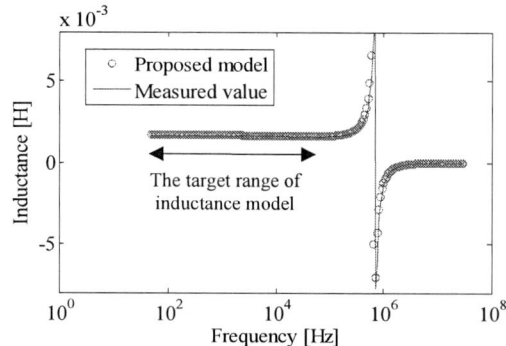

Fig. 5 The inductance characteristic.

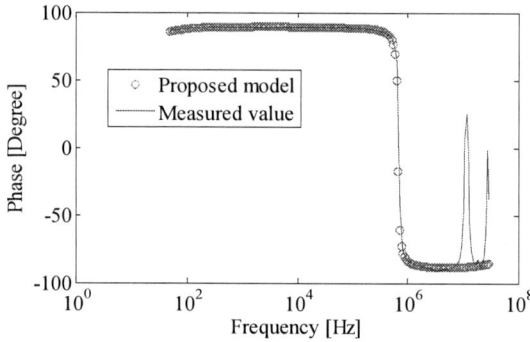

Fig. 6 The phase angle characteristic.

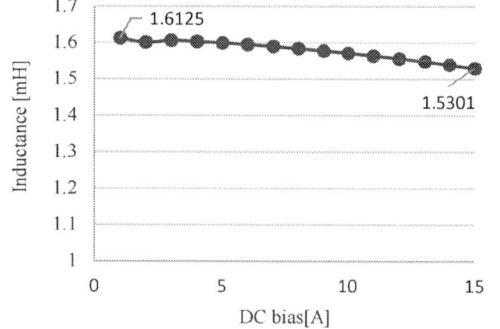

Fig. 7 The DC bias current characteristic.

(a) conventional equivalent circuit model of capacitor

(b) stability assessment model

Fig. 8 The equivalent circuit of capacitor

case, CR ladder circuit, as shown in Fig. 8 (b), can be used in order to implement the ESR in low frequency region [22].

Note: it is possible to model the increment of ESR due to skin effect and proximity effect in high frequency range by adding LR ladder circuit such as ESR modeling of inductor.

D. mpedance measurement and modeling result

The frequency response of the capacitor is also measured by using the impedance analyzer (Keysight: E4990A). Then the capacitor is modeled by least squares fitting from measured data.

Fig. 9 is shows three-phase capacitor and modeled value (line to line) of the capacitor. Figs. 10 to 13 show measured and calculated frequency responses of the capacitor in terms of ESR, impedance, capacitance, and phase. These values are modeled up to 50MHz. The value of CR ladder circuit is calculated by curve fitting from measured ESR data in target frequency. The maximum phase difference in low frequency region is only 0.09° at 13kHz. The capacitor is modelled very well.

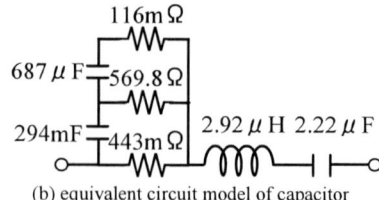

Fig. 9 The test capacitor and the equivalent circuit.

C. Capacitor modeling method

The conventional time domain model of capacitor is shown in Fig.8 (a) which consists of L, R, C. The L and C can be modeled by constant value. But, the ESR value is not considered for frequency-dependence characteristics.

For the stability analysis of VSCs, the modeling of the parasitic inductance L is not so important because its affected over the Nyquist frequency ($f_s/2$). The ESR of the capacitor generally shows a high value in the low frequency region and decreases as it approaches the resonance frequency of capacitor. After the resonance frequency of capacitor, the ESR is increased by the skin effect [21],[22]. However, the modeling of the skin effect of ESR in capacitor is not considered because its normally not affected below the Nyquist frequency ($f_s/2$). In this

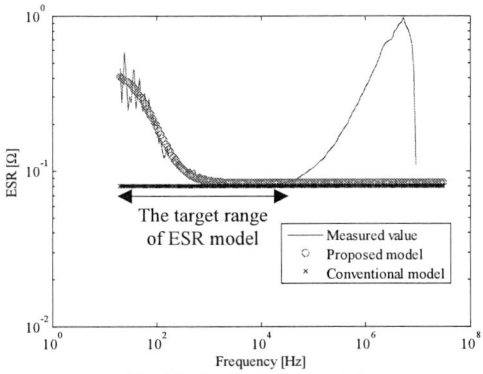

Fig. 10 The ESR characteristic.

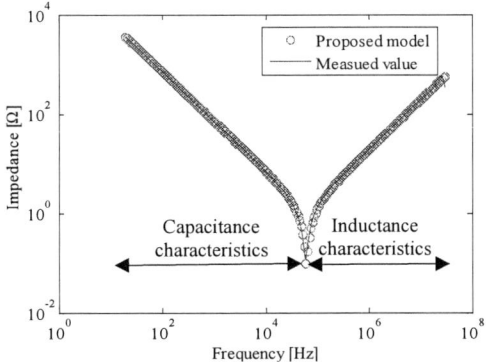

Fig. 11 The impedance characteristic of capacitor

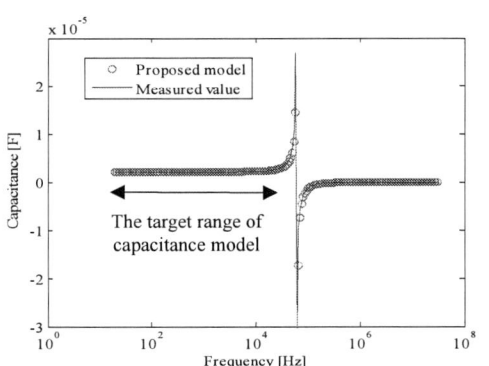

Fig. 12 The capacitance characteristic.

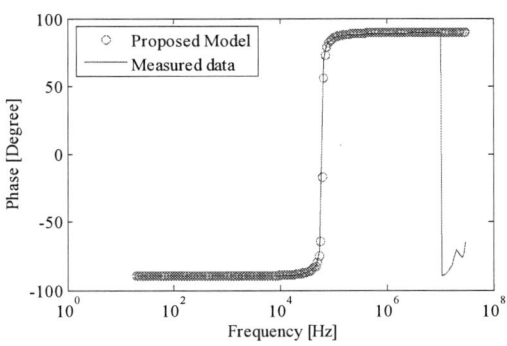

Fig. 13 The phase angle characteristic.

III. System Modeling and Stability Assessment

The stability of grid side current controlled voltage sauce inverter is analyzed. The simulation circuit of a grid-connected three-phase VSC is shown in Fig. 14. The dc-link voltage control is not discussed in this paper, because it is usually designed to be a slow control loop, and will not interact with the inner current control as long as a proper control bandwidth is chosen. By considering the equivalent series resistances (ESRs) of passive components, the plant-equivalent block diagram is shown in Fig. 15. From Fig. 15, the loop gain T can be obtained as follows [23]:

$$T = \frac{G_c G_d Z_C}{Z_{L1} Z_{L2} + Z_{L1} Z_C + Z_{L2} Z_C}$$ (1)

where Z_{L1} and Z_{L2} are the converter-side impedance of inductor L_1 and the grid-side impedance of inductor L_2 plus grid inductance, respectively. L_g is the grid inductance, and Z_C is the impedance of filter capacitor C_f, G_c is $K_p + \frac{K_i T_s}{1-e^{-sT_s}}$ for PI controller, G_d is $e^{-1.5sT_s}$ for time delay of system. The grid inductance uses measured value by pseudo-random binary sequence (PRBS), expressed in [24].

Then, the grid side control gain T is calculated and compared between ideal LCL filter and ESR-modeled LCL filter. The circuit parameter is shown in table I. In the calculation of ESR modeled LCL filter case, the equivalent circuit of inductor, as shown in Fig. 2, are used for impedance of converter and grid side inductor. Also, the equivalent circuit of three-phase filter capacitor, as shown in Fig. 9, is used for impedance of filter capacitor. And the PI controller G_c is designed as $G_c = 11.17 + \frac{0.3899}{1-e^{-1\times10^{-4}s}}$.

The calculation result of the inner loop gain T and phase of three-phase VSC is shown in Fig. 16. The ESR affect the resonance damping, resonance frequency and the phase characteristic. When the ESR is considered, the resonance peak is damped from infinity to 6.98dB, the resonance frequency is changed from 2363kHz to 2297kHz, and the crossover frequency is changed from 1645kHz to 1635kHz. The phase margin at cross over frequency is increased from 65 degree to 85 degree. The PI controller G_c gain can be increased. The gain margin at phase crossover frequency is little bit reduced from 13.10 dB to 12.95 dB.

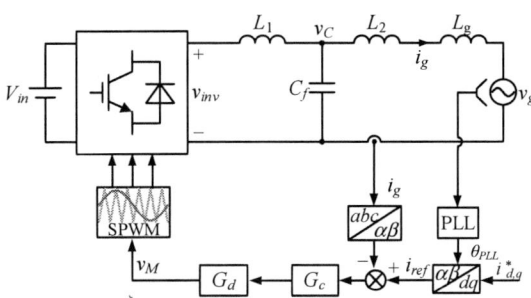

Fig. 14 The simulation circuit of grid-connected three-phase VSC

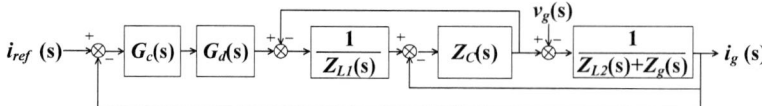

Fig. 15 The block diagram of Fig. 14

TABLE I
MAIN CIRCUIT PARAMETER

Grid voltage : V_g	400V
Grid fundamental frequency : f_1	50 Hz
Grid inductance : L_g	1.8 mH
Converter switching frequency : f_{sw}	10 kHz
Converter sampling frequency: f_s	10 kHz
Converter dc-link voltage : V_{dc}	730 V
LCL-filter inductor : L_1	1.6 mH
LCL-filter inductor : L_2	1.6 mH
LCL-filter capacitor : C_f	4.4 µF

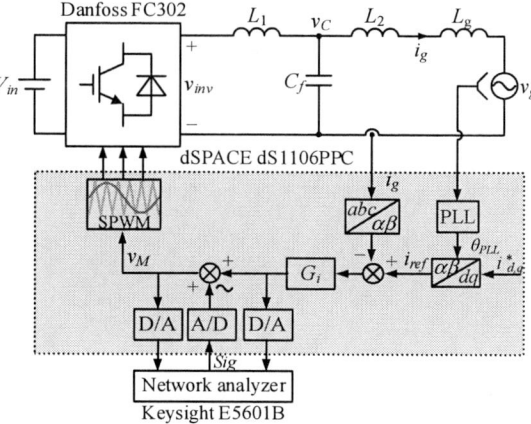

Fig. 17 The measurement system.

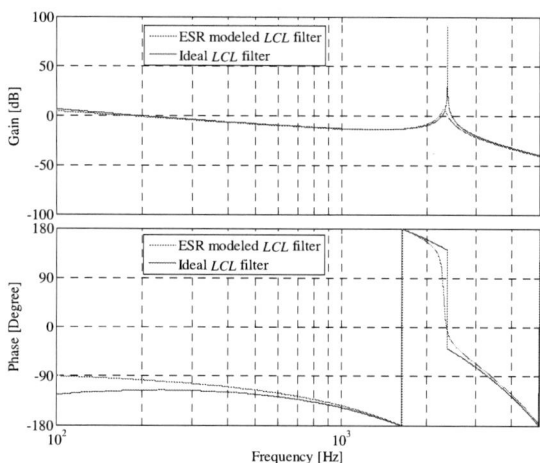

Fig. 16 The calculation result of loop gain and phase.

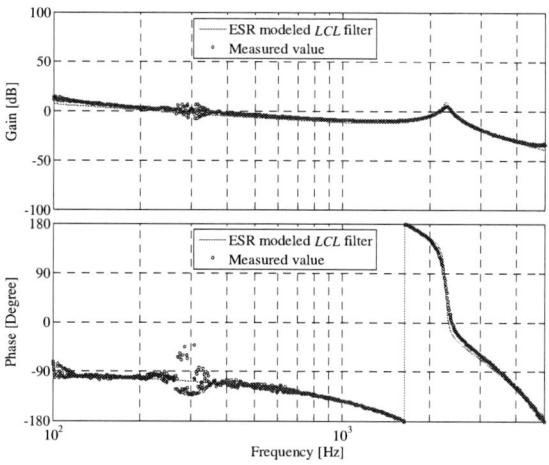

Fig. 18 The measurement result of loop gain and phase.

IV. EXPERIMENTAL VERIFICATION

In order to verify the proposed inductor model, the inner loop gain T and phase of three-phase VSC is evaluated. In the present study, three single-phase ac filter inductors (Shinenergy: MPR10322-15-01) are used for converter and grid side inductor. The circuit parameter is same as simulation. The inner loop gain T is measured by the network analyzer (Keysight: E5601B) and the network analyzer is connected to loop gain controller through D/A and A/D converter, as shown in Fig. 17. Fig. 18 shows measurement value and calculation value of gain loop gain T. The loop gain T is agree very well and the phase crossover frequency and the resonance frequency are coinciding very well. If the ESR is considered, the peak value of resonance frequency deference is 5.11dB, the resonance frequency deference is only 15Hz, the crossover frequency deference is only 5Hz. By using this model, the

simulation is more accurately. The deviation around 300 Hz is caused by 5th and 7th order harmonics of fundamental frequency from grid voltage.

From these results, the accurate prediction of control loop gain and phase can be possible by considering ESR of filter components, grid impedance and time delay.

V. CONCLUSIONS

This paper presents the equivalent series resistance (ESR) modeling method of filter components and its impact on stability for grid connected voltage-source converters (VSCs) with LCL filter. Firstly, the time domain models of filter components are presented in order to address the frequency-dependent characteristic of the ESR by LR and CR ladder circuit and the ladder circuits does not much affect for impedance and phase. Then, the loop gains of the grid connected three-phase VSC which using the ideal *LCL* filter and the ESR modeled *LCL* filter are calculated and compared. The calculation result shows that the ESR affect the resonance peak and the phase characteristic of the loop gain. Finally, in order to verify accuracy of the proposed model, the loop gain of VSC is measured by network analyzer in the experimental set up. The simulation and experiment result show good agreement with each other, and it means the ESR modeling method of *LCL* filter is useful for accurate stability analysis, PI controller design, *LC* trap filter design, etc.

ACKNOWLEDGMENT

This work was supported by JSPS KAKENHI Grant Number 17J06592.

REFERENCES

[1] F. Blaabjerg, Z. Chen, and S. B. Kjaer, "Power electronics as efficient interface in dispersed power generation systems," *IEEE Trans. Power Electron.*, vol. 19, no. 5, pp. 1184-1194, Sept. 2004.

[2] N. Flourentzou, V. G. Agelidis, and G. D. Demetriades, "VSC-based HVDC power transmission systems: an overvew," *IEEE Trans. Power Electron.*, vol. 24, no. 3, pp. 592-602, Mar. 2009.

[3] M. Huang, S. Member, X. Wang, and P. C. Loh, "LLCL - Filtered Grid Converter With Improved Stabilit and Robustness," vol. 31, no. 5, pp. 3958–3967, 2016.

[4] S. Buso and P. Mattavelli, *Digital Control in Power Electronics*, San Francisco, CA: Morgan & Claypool Publ., 2006.

[5] D. M. Van de Sype, K. D. Gusseme, F. D. Belie, A. P. Van den Bossche, and J. A. Melkebeek, "Small-signal *z*-domain analysis of digitally controlled converters," *IEEE Trans. Power Electron.* vol. 21, no. 2, pp.470-478, Mar. 2006.

[6] J. Yin, S. Duan, and B. Liu, "Stability analysis of grid-connected inverter with *LCL* Filter adopting a digital single-loop controller with inherent damping characteristic," *IEEE Trans. Ind. Inform.*, vol. 9, no. 2, pp. 1104-1112, May. 2013.

[7] S. G. Parker, B. P. McGrath, and D. G. Holmes, "Regions of active damping control for *LCL* Filters," *IEEE Trans. Ind. Appl.*, vol. 50, pp. 424-432, Jan./Feb. 2014.

[8] J. H. Enslin and P. J. Heskes, "Harmonic interaction between a large number of distributed power inverters and the distribution network," *IEEE Trans. Power Electron.*, vol. 19, no. 6, pp. 1586-1593, Nov. 2004.

[9] X. Wang, F. Blaabjerg, and W. Wu, "Modeling and analysis of harmonic stability in an ac power-electronics-based power system," *IEEE Trans. Power Electron.*, vol. 29, no. 12, pp. 6421-6432, Dec. 2014.

[10] M. Liserre, R. Teodorescu, and F. Blaabjerg, "Stability of photovoltaic and wind turbine grid-connected inverters for a large set of grid impedance values," *IEEE Trans. Power Electron.*, vol. 21, no. 1, pp. 263-272, Jan. 2006.

[11] F. Wang, J. Duarte, M. Hendrix, and P. Ribeiro, "Modelling and analysis of grid harmonic distortion impact of aggregated DG inverters," *IEEE Trans. Power Electron.*, vol. 26, no. 3, pp. 786-797, Mar. 2011.

[12] P. Brogan, "The stability of multiple, high power, active front end voltage sourced converters when connected to wind farm collector system," *in Proc. EPE WECS* 2010, pp. 1-6.

[13] C. Zou, B. Liu, S. Duan, and R. Li, "Influence of delay on system stability and delay optimization of grid-connected inverters with LCL filter," *IEEE Trans. Ind. Inform.*, vol. 10, no. 3, pp. 1775-1784, Aug. 2014.

[14] J. Wang, J. Yan, L. Jiang, and J. Zou, "Delay-dependent stability of single-loop controlled grid-connected inverters with *LCL* filters," *IEEE Trans. Power Electron.*, vol. 31, no. 1, pp. 743-757, Jan. 2016.

[15] Y. Tang, W. Yao, P. C. Loh, and F. Blaabjerg, "Design of *LCL* Filters With *LCL* Resonance Frequencies Beyond the Nyquist Frequency for Grid-Connected Converters," IEEE J. Emerg. Sel. Top. Power Electron., vol. 4, no. 1, pp. 3–14, 2016.

[16] M. Huang, S. Member, X. Wang, and P. C. Loh, "LLCL - Filtered Grid Converter With Improved Stability and Robustness," vol. 31, no. 5, pp. 3958–3967, 2016.

[17] A. Reznik, M. G. Simoes, A. Al-Durra, and S. M. Muyeen, "LCL Filter design and performance analysis for grid-interconnected systems," *IEEE Trans. Ind. Appl.*, vol. 50, no. 2, pp. 1225–1232, 2014.

[18] Danish Energy Agency, "Energy and Climate Policies beyond 2020 in Europe -Overall and selected countries," pp. 1-50, 2015.

[19] H. Matsumori, T. Shimizu, K. Takano and H. Ishii, "Evaluation of Iron Loss of AC Filter Inductor Used in Three-Phase PWM Inverters Based on an Iron Loss Analyzer," *IEEE Trans. Power Electron.* vol. 31, no. 4, pp. 3080–3095, 2016.

[20] S. Jazebi and F. de León, "Duality-Based Transformer Model Including Eddy Current Effects in the Windings," *IEEE Trans. Power Delivery.* vol. 30, no. 5, 2015.

[21] H. Nagasaki, P. Y. Huang, and T. Shimizu, "Characterization of power capacitors under practical current condition using capacitor loss analyzer," *IEEE Energy Convers. Congr. Expo (ECCE)*, 2016.

[22] S. Koyama and T. Goutsu, "Proposal of precise SPICE model of conductive polymer aluminum solid capacitors," *IEEE Appl. Power Electron. Conf. Expo (APEC)*, 2016.

[23] X. Wang, X. Ruan, S. Liu, and C. K. Tse, "Full Feedforward of Grid Voltage for Grid-Connected Inverter With," *IEEE Trans. Pe.* vol. 25, no. 12, pp. 3119–3127, 2010.

[24] T. Messo, R. Luhtala, T. Roinila, D. Yang, X. Wang, and F. Blaabjerg, "Real-Time impedance-based stability assessment of grid converter interactions," *2017 IEEE 18th Work. Control Model. Power Electron. COMPEL 2017*, 2017.

Real-time estimation control of inductance parameters using dust core materials for PWM inverter

Kazu Imai[1], Takuma Yoshino[1], Ohasi Shunsuke[2]**, Tomoki Yokoyama[1]*

1 Tokyo Denki University, Tokyo, Japan
2 Fuji Electric Co.,Ltd, Hyogo, Japan
*E-mail: yoko@fr.dendai.ac.jp
**E-mail: ohasi-shunsuke@fujielectric.com

Abstract—In recent years, the switching frequency is increased in power converters due to improvement of the switching devices, which result in the reduction of the size of the passive components. The dust core material is used for the inductance core of the PWM inverter to downsize the filter component. In that case, the flux density is saturated for the large inductance current and the inductance parameter is varied widely due to the inductance current variations. The output characteristics is deteriorated with the conventional control. The deadbeat control method combined with the disturbance compensation and the inductance estimation methods were proposed. Also 1 MHz multisampling method was applied to adjust the control gains for the inductance parameter variations. The advantage of the proposed method was verified through simulations and experiments.

Keywords—Single-phase inverter, Inductor, FPGA, Dust core, Inductance estimation

I. INTRODUCTION

In recent years, the progress of power semiconductor devices enables increasing of the switching frequency in power converters, which result in the reduction of the passive components[1], [2], [3]. The dust core material is one approach to downsize the inductor component for the core materials[4], [5]. The flux density can be improved in the dust materials compared with the ferrite materials[6]. However, the density of magnetic flux becomes small in the case of the numbers of coil turns is not enough because of the core size[5], [7]. In that case, the magnetic flux is saturated for the large inductance current and the inductance parameter is varied widely during the converter operations[8]. As the result, the harmonic distortion is increased in the output waveforms.

In this study, the digital control method for the single phase PWM inverter using the estimation control for the inductance parameter was proposed based on the inductance characteristics for the current variations. The deadbeat control method with disturbance compensation is applied for the inverter control method. The output current error is compensated by the disturbance compensation method.

For the other approach, the estimation method for the inductance parameter is also proposed to improve the output waveforms when the inductance parameter is varied due to inductor current. The inductance is estimated based on the MHz sampling data of the inductor current, and the control gains of the deadbeat control are recalculated using the estimated inductance parameters for every sampling period. In addition, the multisampling method is applied which result in the real time compensation for the inductance variations. In this method, the PWM pulse width is recalculated in MHz order sampling frequency[9], [10], [11]. The inductance estimation is carried out in every MHz sampling to adjust the control gains, and real time compensation of the PWM pulse width is realized by adopting the estimation method and multisampling mecthod. The robustness of the proposed methods were verified when the inductance parameters were varied during the inverter operation. The advantage of the proposed method was verified through simulations and experiments.

II. CONTROL METHOD

A. Deadbeat control with disturbance compensation method

The proposed inverter system was indicated in Fig.1.

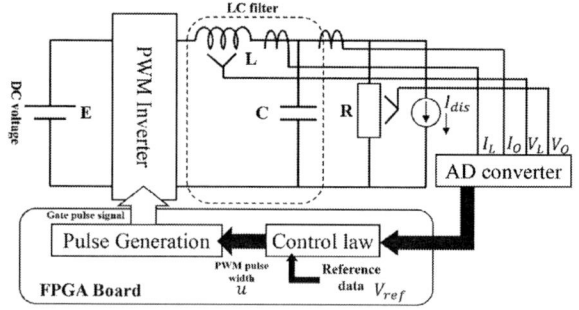

Fig. 1. Single-phase PWM inverter system

$$\dot{x}(t) = A_c x(t) + B_c u(t) \qquad (1)$$

$$, where \ A_C = \begin{bmatrix} -1/RC & 1/C & -1/C \\ -1/L & 0 & 0 \\ 0 & 0 & 0 \end{bmatrix}$$

$$, B_C = \begin{bmatrix} 0 \\ -1/L \\ 0 \end{bmatrix}, x(t) = \begin{bmatrix} V_O \\ I_L \\ I_{dis} \end{bmatrix}.$$

3363

$$I_{dis} = I_O - \frac{V_O}{R}. \tag{2}$$

$$x[k+1] = A_d x[k] + B_d u_c[k] \tag{3}$$

$$, where\ A_d = e^{A_c T_u} = \begin{bmatrix} F_{11} & F_{12} & F_{13} \\ F_{21} & F_{22} & F_{23} \\ 0 & 0 & 1 \end{bmatrix}$$

$$, B_d = e^{\frac{A_c T_u}{2}} B_c E = \begin{bmatrix} g_1 \\ g_2 \\ 0 \end{bmatrix}$$

$$, x[k] = \begin{bmatrix} V_O[k] \\ I_L[k] \\ I_{dis}[k] \end{bmatrix}, x[k+1] = \begin{bmatrix} V_{ref}[k] \\ I_{Lref}[k] \\ 0 \end{bmatrix}.$$

$$u[k] = \frac{V_{ref}[k] - F_{11}V_O[k] - F_{12}I_L[k] - F_{13}I_{dis}[k]}{g_1}. \tag{4}$$

The continuous time model of the proposed system becomes as (1)[11]. The disturbance current is defined as $I_{dis}(t)$ for the error current compared with the nominal resistance current[11], [12]. $I_{dis}(t)$ can be calculated as (2). The discrete time model of (1) becomes as (3). (4) is obtained by solving the first line of (3), and the pulse width $u[k]$ can be calculated.

B. Inductance estimation method with dust core characteristics

The inductance characteristics of the dust core for the current variation is shown in Fig.2, which is supplied by the dust core supplier.

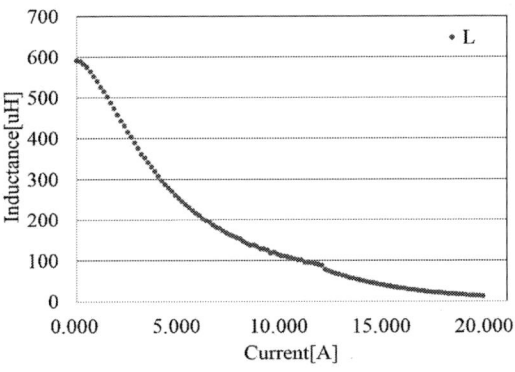

Fig. 2.   Inductance characteristics for the current variations

When the small core is used for the filter inductor, the flux density is saturated for the large inductor current. In this case, the inductance value is varied widely due to the inductor current variation. When the inductor current is 10[A], the inductance value becomes one sixth of the nominal value. In the proposed method, the inductance is estimated based on the current-inductance characteristics in Fig.2 according to the inductor current variations.
Also the control gains are recalculated using the estimated inductance in every sampling period. Fig.3 shows the control gains calculated using the inductance characteristics. The blue lines of Fig.3 indicate the control gains calculated using the inductance characteristics as shown in Fig.2.

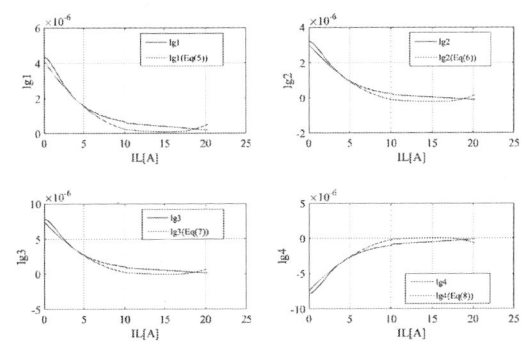

Fig. 3.   Variations of the deadbeat control gains for the inductor current variation

The control gains can be obtained from the lookup tables for the inductor current variation. To reduce the logic element region of FPGA consumed by the lookup tables, these gains can be expressed using approximate expression from (5) to (8). The red lines of Fig.3 indicates that the control gains are calculated using (5)~(8), and the calculation of the control gains using estimated inductance can be realized without lookup tables.

$$lg1 = 2.19e-08 * i_L^2[k] - 6.57e-07 * i_L[k] + 5.03e-06. \tag{5}$$
$$lg2 = 1.87e-08 * i_L^2[k] - 5.57e-07 * i_L[k] + 3.94e-06. \tag{6}$$
$$lg3 = 3.32e-08 * i_L^2[k] - 9.96e-07 * i_L[k] + 7.45e-06. \tag{7}$$
$$lg4 = -3.32e-08 * i_L^2[k] + 9.96e-07 * i_L[k] - 7.45e-06. \tag{8}$$

C. Inductance estimation method using MHz sampled inductance current

The inductance can be calculated using the inductor voltage and current by (9). And (9) can be obtained as discrete model of (10). The inductance estimation can be realized by calculating (10) in FPGA controller without the inductance characteristic in the case that the sampling time is short enough to suppress the calculation error.

$$L = \frac{V_L}{\frac{dI_L}{dt}}. \tag{9}$$

$$L[k] = \frac{V_L[k]}{\frac{|I_L[k] - I_L[k-1]|}{sampling\ time}}. \tag{10}$$

The deadbeat control gains for the inductance variations are shown in Fig.4. The control gains is liner for the inductance variation as shown in Fig.4. The blue lines of Fig.4 indicate the control gains calculated using (11)~(14).

$$lg1 = 8.9718e-03 * L[k] + 3.090e-08. \tag{11}$$
$$lg2 = 7.6102e-03 * L[k] - 3.022e-07. \tag{12}$$
$$lg3 = 1.3611e-02 * L[k] - 1.393e-07. \tag{13}$$
$$lg4 = -1.3611e-02 * L[k] + 1.393e-07. \tag{14}$$

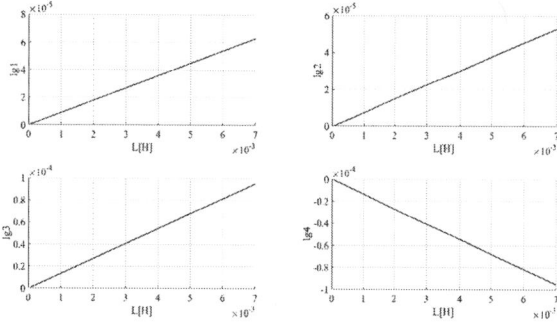

Fig. 4.   Variations of the deadbeat control gains for the inductance

D. Multisampling method

The timing chart of the multisampling method is shown for one carrier period in Fig.5.

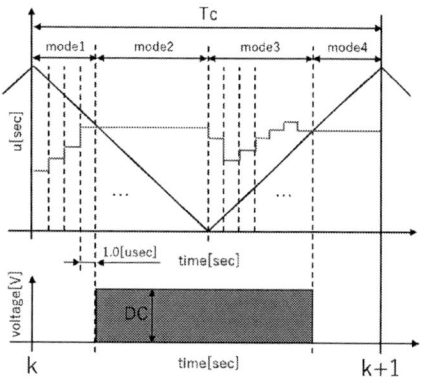

Fig. 5.   Timing chart of multisampling method

Multisampling is a method which the PWM pulse width is recalculated in MHz order sampling frequency[10][11]. The one carrier interval is divided into four modes to derive the PWM pulse width.

In the mode 1, the PWM pulse width calculation is carried out using every 1MHz sampling data until the instant that PWM pulse is turned on, and shifts to mode 2. In the mode 2, the calculation is not carried out, and maintain PWM pulse state on until the center of the carried interval. In the mode 3, the PWM pulse width calculation is carried out using every 1MHz sampled data. When the instant that PWM pulse is turned off, shifts to mode 4. In the mode 4, the PWM pulse is maintained state off until the next carrier interval. As the result, the PWM pulse width can be adjusted for the inductance variations in real time operations.

III.   SIMULATION RESULT

The proposed method is verified through the simulations. The parameter conditions are shown in Table.I. The simulations were carried out in the resistive load conditions and the nonliner load conditions. In this verification, the capacitor input rectifier load is used as the nonliner

load.

The output waveforms of the voltages, resistance current and inductor current are indicated from Fig.6 to Fig.11. The waveforms without the inductance estimation method and without the multisampling method are shown in Fig.8 and Fig.9. The waveforms using the inductance estimation method with dust core characteristics are shown in Fig.8 and Fig.9. The waveforms for the inductance estimation method using Mhz sampled inductance current are shown in Fig.10 and Fig.11. Table.II shows the THD characteristics in each condition.

TABLE I.   SIMULATION CONDITIONS

Parameter	Value
Control cycle	20[kHz]
Sampling frequency(with multisampling)	1[MHz]
Sampling frequency(without multisampling)	20[kHz]
Input DC voltage	141[V]
Output voltage	70.2[Vrms]
Inductance	591(1.86%) 84.4(0.782%)[μH]
Capacitance	30.4[μF](9.54%)
Rated Resistance	10.0[Ω]
nonliner load R	50[Ω]
nonliner load C	660[μF]
nonliner load L	0.6[mH]

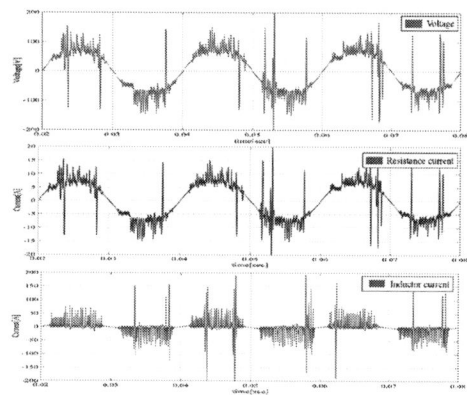

Fig. 6.   Simulation result(without inductance estimation, fixed gain, resistive load)

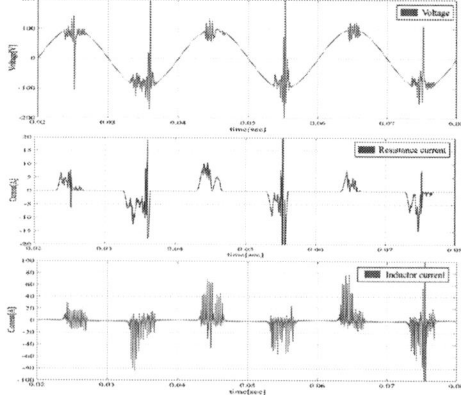

Fig. 7.   Simulation result(without inductance estimation, fixed gain, nonliner load)

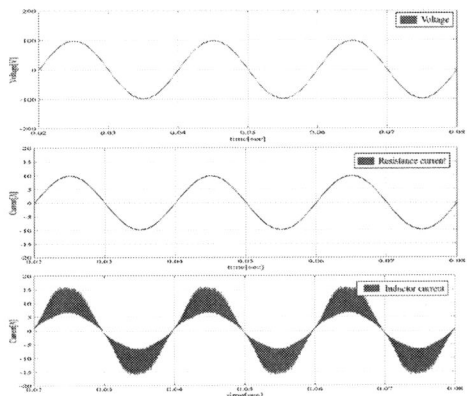

Fig. 8. Simulation result(inductance estimation with dust core characteristics, resistive load)

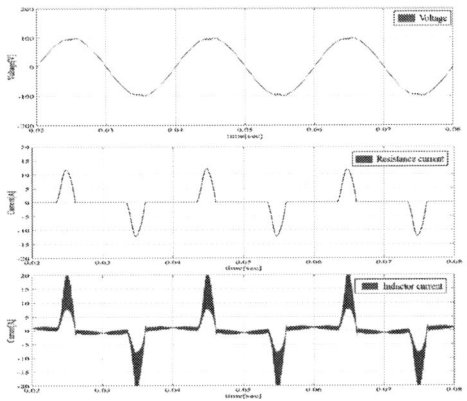

Fig. 9. Simulation result(inductance estimation with dust core characteristics, nonliner load)

TABLE II. THD(SIMULATIONS)

Parameter	Resistive load[%]	Nonliner load[%]
without L estimation	80.68	71.79
with L estimation (with dust core characteristics)	0.96	1.62
with L estimation (using MHz sampled inductance current)	0.78	0.73

Fig.6 shows the waveform of the resistive load condition. The inductor current is varied around 0 to ±15[A] in the resistive load condition. As shown in Fig.2, the inductance parameter is varied between 591 to 43[μH]. The output voltage was distorted because of the rapid inductance variation for the inductor current. Fig.7 shows the waveform for the nonliner load condition. In this case, the current flows when the diode turns on around the peak voltage duration, so the load current is rapidly changed. The time duration of the inductance variation is narrow compared with Fig.6. In Fig.8 and Fig.9, the output characteristics of the voltages and the currents were improved compared with Fig.6 and Fig.7, and it was found that the output characteristics were improved using the real time inductance estimation and the multisampling method. In Fig.10 and Fig.11, the voltages and the currents are also improved. The inductance estimation using inductance current is verified through the simulations.

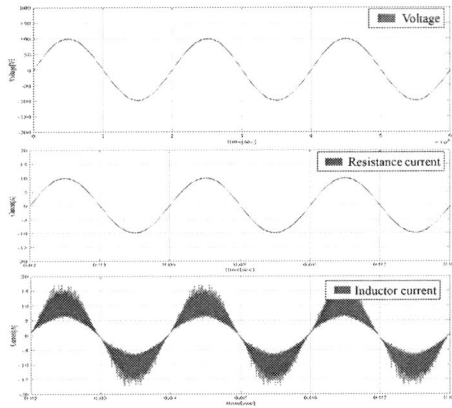

Fig. 10. Simulation result(inductance estimation using MHz sampled inductance current, resistive load)

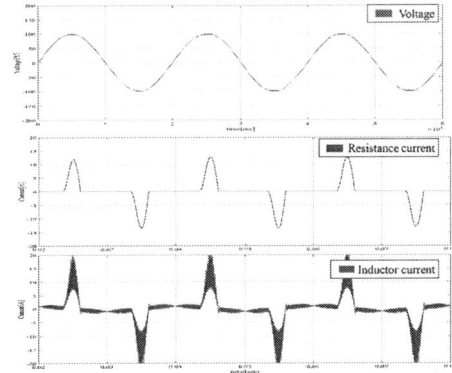

Fig. 11. Simulation result(inductance estimation using MHz sampled inductance current, nonliner load)

IV. EXPERIMENTAL RESULT

The proposed method is verified through the experiments. Parameters and load conditions are same with the simulations. The output waveforms without the estimation control and the multisampling method are shown in Fig.12 and Fig.13. Fig.14 and Fig.15 indicates the experimental results in the case of the inductance estimation method using dust core characteristics and the multisampling method are adopted to the deadbeat control method with disturbance compensation. Fig.16 and Fig.17 show the experimental results in the case of the inductance estimation method using MHz sampled inductance currrent and the multisampling method are adopted to the deadbeat control method with disturbance compensation. Table.III shows the THD characteristics in each condition.

TABLE III. THD(EXPERIMENTS)

Parameter	Resistive load[%]	Nonliner load[%]
without L estimation	20.36	7.843
with L estimation (with dust core characteristics)	7.392	5.117
with L estimation (using MHz sampled inductance current)	5.24	4.46

The 2018 International Power Electronics Conference

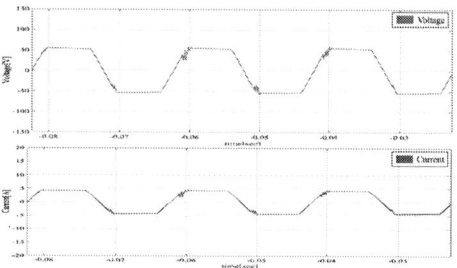

Fig. 12. Experimental result(without inductance estimation, fixed gain, resistive load)

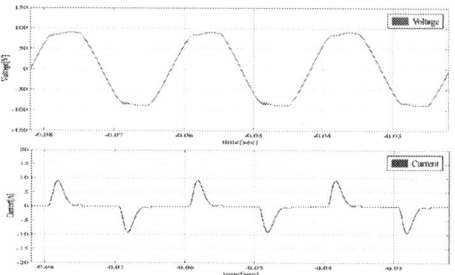

Fig. 13. Experimental result(without inductance estimation, fixed gain, nonliner load)

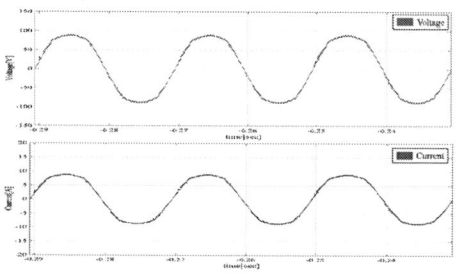

Fig. 14. Experimental result(inductance estimation with dust core characteristics, resistive load)

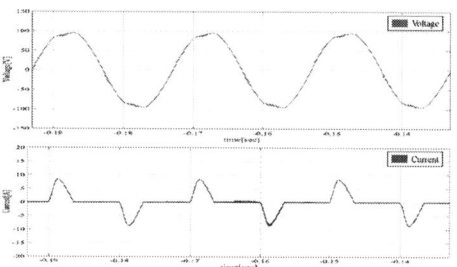

Fig. 15. Experimental result(inductance estimation with dust core characteristics, nonliner load)

Fig.12 shows the waveforms in the case of the resistive load condition when the input DC voltage is 90[V], and Fig.13 shows the waveforms in the case of noliner load condition when the input DC voltage is 120[V]. The harmonic distortion was increased when the input DC voltage was increased in each condition. From Fig.14 to Fig.17 show the waveforms in the case of the rated dc input voltage condition. It was found that the stable operation was realized using the inductance estimation method and the multisampling method. In Fig.16 and

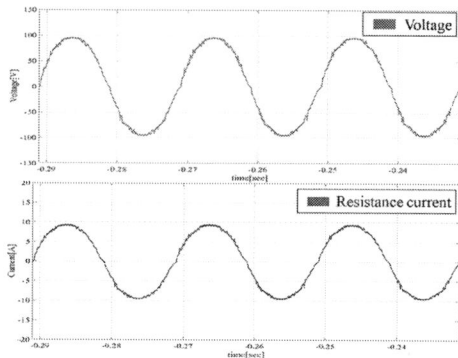

Fig. 16. Experimental result(inductance estimation using MHz sampled inductance current, resistive load)

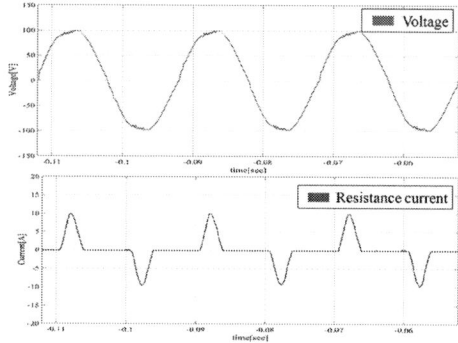

Fig. 17. Experimental result(inductance estimation using MHz sampled inductance current, nonliner load)

Fig.17, the stable operation was realized. The inductance estimation using MHz sampled inductance current is effective for the dust core materials.

V. CONCLUSION

The digital control method for single phase PWM inverter using the estimation method for the inductance parameter of the dust core materials was proposed. To improve the output characteristics, the real time adjustment of the control gains for the inductance variation was realized by using the inductance estimation method and the multisampling method. Also the inductance estimation method using MHz sampled inductance current was proposed. The effectiveness of the both proposed method were verified through the simulations and the experiments, and the output characteristics were improved.

REFERENCES

[1] H.Matumori, T.Shimizu, K.Takano, H.Ishii, "Iron loss calculation of AC filter inductor for three phase PWM inverter",*Energy Conversion Congress and Exposition*, Raleigh NC USA, 2012

[2] A.B.Ponniran, K.Orikawa, J.Ito, "Minimization of Passive Components in Multi-level Flying Capacitor DC-DC Converter", *IEEJ Journal of Industry Applications*, vol. 5, no. 1, pp. 10-11, 2015

[3] A.Hino, K.Wada, "Resonance Analysis Focusing on Stray Inductance and Capacitance of Laminated Bus Bars", *IEEJ Journal of Industry Applications*, vol. 5, no. 6, pp. 407-412, 2016

[4] K.Kabeya, S.Yanase, Y.Okazaki, K.Yun, "Magnetic Property of Iron-Dust Cores With Mixture of Ferromagnetic Ferrite Powder and Alumina Powder", *IEEE TRANSACTIONS ON MAGNETICS*, vol. 50, no. 4,2014

[5] A.Hirai, B.Cougo, "Optimal Inductor Design and Material Selection for High Power Density Inverters Used in Aircraft Applications", *Electrical Systems for Aircraft, Railway, Ship Propulsion and Road Vehicles & International Transportation Electrification Conference*, Toulouse France, 2016

[6] E.Otsuki, K.Ishii, S.Nakano, "SMD Inductors Based on Soft-Magnetic Powder Compacts", *IEEE Applied Power Electronics Conference and Exposition*, pp. 74-78, Palm Springs CA USA, 2010

[7] Y.Li, Y.Xie, R.Chen, L.Han, D.Chen, H.Su, "A layer Power Inductor Fabricated by Cofirable Ceramic/Ferrite Materials With LTCC Technology", *IEEE TRANSACTIONS, PACKAGING AND MANUFACTURING TECHNOLOGY*, vol. 7, no. 9, 2017

[8] X.Ji, T.Noguchi, "Online q-axis Inductance Identification of IPM Synchronous Motor Based on Relationship between Its Parameter Mismatch and Current", *IEEJ Journal of Industry Applications*, vol. 4, no. 6, pp. 730-731, 2015

[9] A.Kawamura, H.Fujimoto, T.Yokoyama, "Survey on the real time digital feedback control of PWM inverter and the extension to multi-rate sampling and FPGA based inverter control", *Annual Conference of the IEEE Industrial Electronics Society*, Taipei Taiwan, 2007

[10] R.Saito, T.Yokoyama, "Digital Control of PWM Inverter Using Ultrahigh-Speed Network for Feedback Signals with Communication Disturbance Observer Based on Rocket I/O Protocol", *IEEJ Journal of Industry Applications*, vol. 4, no. 6, pp.752-757, 2015

[11] M.Ito, R.Fujiwara, "3MHz multi sampling deadbeat control of single phase PWM inverter using FPGA based hardware controller", *IEEE International Future Energy Electronics Conference*, Taipei Taiwan, 2015

[12] T.Fujii, T.Yokoyama, "FPGA based Deadbeat Control with Disturbance Compensator for Single Phase PWM Inverter", *Power Electronics Specialists Conference*, South Korea, 2006

Control Design of Output-Stage Filterless Sinusoidal-Wave Inverter

Shinichi Hiroshige[1*], Kenji Yamanaka[1] and Masahide Hojo[1]

1 Dept. of Electrical and Electronic Engineering, Tokushima University, Tokushima, Japan

*E-mail: hiro41@ee.tokushima-u.ac.jp

Abstract— **A power supply unit with its output voltage sinusoidal is extremely useful not only for general use but also in an emergency case. In order to operate various apparatuses stably, the feedback control system of sinusoidal-wave inverter should be carefully designed. In this paper, feedback control system is designed for a sinusoidal-wave inverter without filter circuit at its output. In addition, its effectiveness is verified in both theoretical and detail simulation study.**

Keywords— *sinusoidal-wave inverter, voltage feedback control, feedback control design.*

I. INTRODUCTION

Recently, stand-alone sinusoidal-wave inverter is noticed as a power supply for portable and emergency use such as automobile interior electrical outlet of electric vehicle or fuel cell vehicle. In addition, sinusoidal-wave inverter will be useful in case that the utility cannot supply electric power because of a severe disaster.

A concept of a sinusoidal-wave inverter which consists of a buck converter and an inverter without LC output filter is proposed in Ref. [1]. A prototype of the proposed converter was constructed and its operating characteristics and efficiency was verified by experimental study. It can be also extended to a three-phase inverter by multiplying the units[2].

A conventional sinusoidal-wave inverter can be used for many apparatuses as a general-purpose power supply because it can output a high-quality waveform as well as the commercial grid. However, a feedback control design of the sinusoidal inverter is used to be difficult because of the secondary delay element of the LC low-pass filter existing in its output-stage.

In this paper, a converter controller for power supply without output LC filter is theoretically designed and verified by detail simulation study.

II. CIRCUIT CONFIGURATION AND PRINCIPLE

Sinusoidal-wave inverter is generally controlled by a pulse width modulation (PWM). In this circuit topology, LC filter is necessary at the output-stage to remove voltage harmonics included in the PWM waveform. In the proposed inverter circuit topology shown in Fig. 1(a), there is not LC filter at the output terminal and a buck converter is connected at the input to regulate the voltage amplitude.

Fig. 2 shows operating waveform of proposed inverter.

Switching command of buck converter is produced by the traditional triangular-wave PWM. Absolute value of the sinusoidal-wave is compared with a triangular wave and the switching command is generated. Output voltage of the buck converter is regulated as absolute value of the sinusoidal wave like $v_{out(abs)}$ in Fig. 2. Inverter is operated as shown in Figs. 1 (b) or (c), where the electrical angle is $0 \leq \omega t < \pi$ or $\pi \leq \omega t < 2\pi$, respectively. The absolute value of the sinusoidal wave is alternately inverted by these operation to achieve a sinusoidal wave v_{out} as shown in Fig. 2.

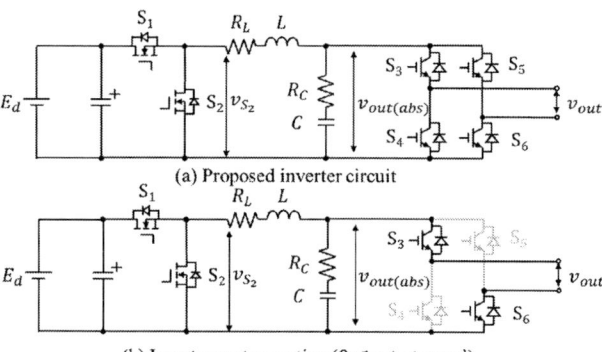

(a) Proposed inverter circuit

(b) Inverter part operation ($0 \leq \omega t < \pi$ rad)

(c) Inverter part operation ($\pi \leq \omega t < 2\pi$ rad)

Fig. 1. Proposed inverter circuit.

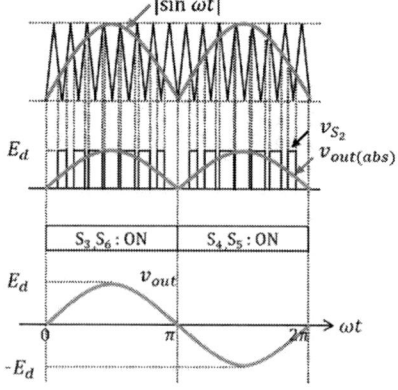

Fig. 2. Operating waveform of proposed inverter.

The absolute value of the sinusoidal wave is outputted by buck converter at the input-stage of inverter, and it is inverted alternately by the inverter. Therefore, a sinusoidal wave of 60 Hz is achieved at the output terminal. The inverter works for only inversion alternately, and absolute value of the sinusoidal wave is outputted by buck converter. Therefore, the proposed inverter can be design of the control system same as buck converter.

III. CONTROL DESIGN OF THE PROPOSED INVERTER

A. Derivation of Transfer Functions

In order to design a feedback control system, we need to derive a transfer function. In this paper, a control system to keep a constant voltage by feedback control of the output voltage as shown in Fig. 3 is considered under a specification of power supply and circuit parameters summarized in Table 1.

In the control circuit, the proportional gain multiplied by the error between absolute value of the sinusoidal wave as a reference and the one of output voltage. The error is modulated by PWM as shown in Fig. 4. Buck converter in main circuit is operated by the PWM signal. The absolute value of sinusoidal-wave is outputted by an LC filter, and feedback to control circuit. It is inverted alternately by the inverter, and a sinusoidal wave is achieved at the output voltage terminal.

A mathematical model of a DC-DC converter is generally obtained by a state-space averaging method. In this paper, a mathematical model of the proposed inverter is obtained by the classical control theory. As a result, the control system can be easily designed.

Output voltage $v_{S_2}(t)$ of the buck converter can be represented as

$$v_{S_2}(t) = d(t)E_d,$$

where $d(t)$ is output signal of PWM modulation circuit. On-state during a switching period is represented by duty ratio D. Switching command is assumed as an input of switching part, transfer function of the switching part is

$$G_{SW}(s) = \frac{V_{S_2}(s)}{D(s)} = E_d \qquad (1)$$

LC filter characteristic is important in feedback control design. Therefore, a resistance component of inductor and ESR of capacitor should be considered in the transfer function of LC filter as shown in Fig. 5. Voltage and current in the LC filter can be represented as follows;

$$\begin{cases} v_{S_2} = R_L i_L + L\dfrac{di_L}{dt} + v_{out(abs)} \\ v_{out(abs)} = R i_R \\ v_{out(abs)} = R_C i_C + \dfrac{1}{C}\displaystyle\int i_C\, dt \\ i_L = i_C + i_R \end{cases}$$

Laplace transform of these equations results in the following equations.

$$\begin{cases} V_{S_2} = R_L I_L + sLI_L + V_{out(abs)} \\ V_{out(abs)} = R I_R \\ V_{out(abs)} = R_C I_C + \dfrac{1}{sC} I_C \\ I_L = I_C + I_R \end{cases}$$

I_L, I_C, I_R is removed from these equation, take the ratio of input voltage and output voltage. Transfer function of LC filter is

$$G_{filter}(s) = \frac{V_{out(abs)}(s)}{V_{S_2}(s)}$$
$$= \frac{R}{R_C + R}\cdot\frac{\frac{1}{LC}(1 + sCR_C)}{s^2 + s\left(\frac{1}{L}\frac{R_C R_L + R_L R + RR_C}{R_C + R} + \frac{1}{C}\frac{1}{R_C + R}\right) + \frac{1}{LC}\frac{R_L + R}{R_C + R}}$$

Fig. 6 is a bode diagram when the load resistance R is varied under constant LC filter parameter. In case of low-load state, quality factor Q is large at cut-off frequency (resonance frequency). Feedback control system is likely unstable in case of no load, because the phase angle curve sharply delays around the resonant frequency. Hence, proposed inverter control system design is assumed in the no-load state. Transfer function of LC filter in case of no-load can be represented as follows.

$$G_{filter}(s) = \frac{\frac{1}{LC}(1 + sCR_C)}{s^2 + s\frac{1}{L}(R_C + R_L) + \frac{1}{LC}} \qquad (2)$$

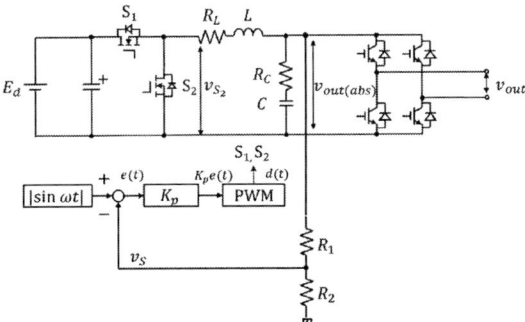

Fig. 3. Voltage feedback system of proposed inverter.

Table 1. Design specification and circuit parameters.

Input d.c. voltage E_d		150V	
Output ac voltage v_{out}		100Vrms±5%, 60Hz	
Rated output power		1500W	
L	180µH	R_1	100 k Ω
R_L	0.01Ω	R_2	3k Ω
C	7µF		
R_C	0.02Ω		
Cut-off frequency		4.48kHz	
Switching frequency		100kHz	

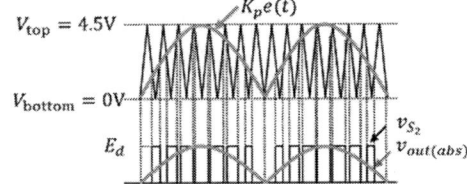

Fig. 4. PWM modulation waveforms in the proposed inverter.

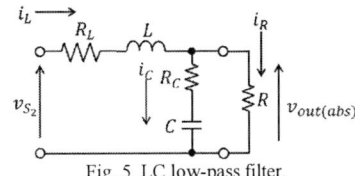

Fig. 5. LC low-pass filter.

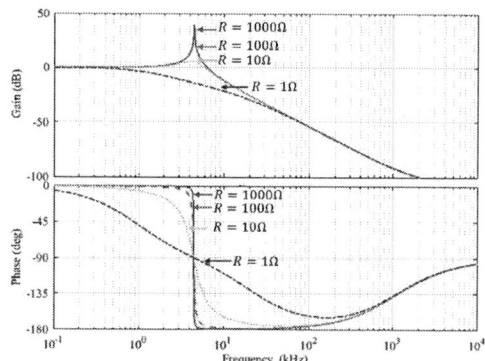

Fig. 6. Bode diagram of LC filter in each case of load resistance.

Output voltage $v_{out(abs)}$ of the buck converter is divided by resister R_1 and R_2 as follows, and it is applied to the control circuit.

$$v_s(t) = \frac{R_2}{R_1 + R_2} v_{out(abs)}(t)$$

Transfer function of voltage sensing part is

$$G_{sensor}(s) = \frac{V_s(s)}{V_{out(abs)}(s)} = \frac{R_2}{R_1 + R_2}. \quad (3)$$

PWM modulation signal $d(t)$ is obtained as follows using Fig. 7.

$$d(t) = \frac{K_p e(t)}{V_{top} - V_{bottom}}$$

Transfer function of PWM modulation is as follows.

$$G_{PWM}(s) = \frac{D}{K_p E(s)} = \frac{1}{V_{top} - V_{bottom}} \quad (4)$$

Obtained transfer function of each part is same as the one of the buck converter. Proposed inverter feedback control system can be designed by these equations.

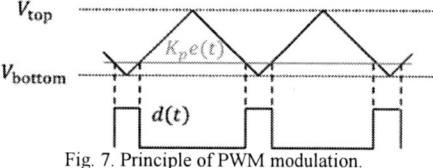

Fig. 7. Principle of PWM modulation.

B. Design of LC Fillter in the Buck Converter

In case of general buck converter, frequency spectrum of switching waveform includes switching frequency, its harmonics and d.c. component. Cut-off frequency of the LC filter is decided so that the switching frequency component is attenuated by the LC filter. In this case, lower limit of the cut-off frequency is not defined. However, the buck converter part of the proposed inverter outputs absolute value of sinusoidal-wave. It includes d.c. component, fundamental wave (120 Hz) and harmonics. Therefore, LC filter of proposed inverter should be considered based on this characteristic.

Fig. 8 shows the amplitude spectrum of switching waveform that the absolute value of sinusoidal-wave is modulated by PWM at 100 kHz switching frequency. Fig. 9 is the amplitude spectrum of absolute value of sinusoidal-wave. Harmonics of absolute value of

sinusoidal wave is very small at range over 2 kHz. Hence, cut-off frequency is necessary to design such that larger than 2 kHz and attenuates the switching frequency.

Fig. 10 is the bode diagram of LC filter in case of no load where its parameters are shown in Table 1. The absolute value of sinusoidal-wave component is passed through LC filter without resonance and is attenuated around the switching frequency of 100 kHz. Fig. 11 shows output voltage waveform when an actual circuit is operated by open-loop control. Sinusoidal-wave is outputted by the proposed inverter.

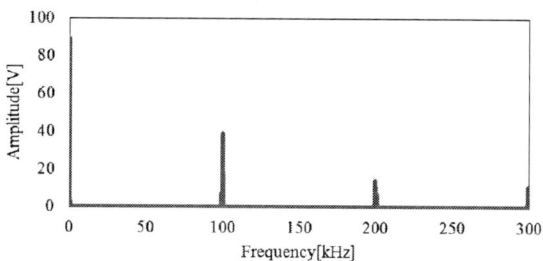

Fig. 8. Output voltage amplitude spectrum of proposed inverter.

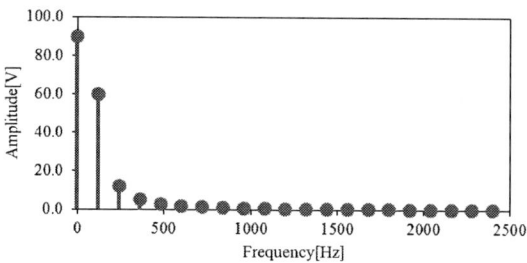

Fig. 9. Amplitude spectrum of absolute value of the sinusoidal wave.

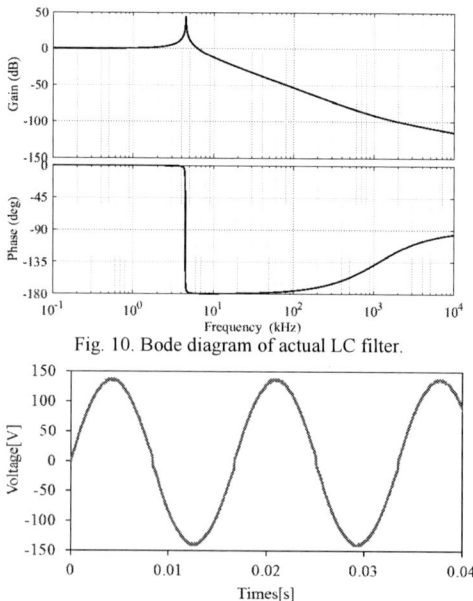

Fig. 10. Bode diagram of actual LC filter.

Fig. 11. Actual output voltage waveform of proposed inverter (open-loop system).

3371

C. Example of Feedback Control Design

Block diagram of the part of the buck converter in the proposed inverter is shown in Fig. 12(a). Fig. 12(b) shows the result of calculating the gain of each blocks in the block diagram for d.c. component and the sinusoidal absolute value. Table 2 shows the propotional gain and the output voltage amplitude at each open-loop gain. In oder to output the voltage amplitude satisfying the inverter design specification, the proportional gain is required to 20dB (10 times).

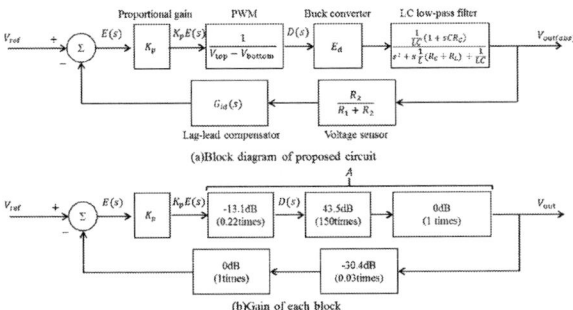

(a)Block diagram of proposed circuit

(b)Gain of each block

Fig. 12. Block diagrams of proposed inverter.

Table 2. Each gain and output voltage amplitude.

Open-loop gain[dB] (times)	K_p[dB] (times)	A[dB] (times)	Sensor gain [dB](times)	Amplitude[V]
0(1)	0(1)			74.9
10(3.16)	10(3.16)			113.9
20(10)	20(10)	30.4(33.3)	-30.4(0.03)	136.4
30(31.6)	30(31.6)			145.4

Since an LC filter exists at the output of buck converter, the phase lags 180 deg. around its cutoff frequency. Therefore, output may become unstable due to a load fluctuation and the lag of control circuit. Fig. 13 shows a bode diagram of open-loop transfer function. Phase margin is 0.86 deg. and there is almost no stability margin.

Phase margin θ_M [rad] have relation to harmful effect of output voltage by lagging in feedback loop. The margin for the delay in the feedback loop is called delay margin. The delay margin T_{DM} [s] is defined by the following equation.

$$T_{DM} = \frac{\theta_M}{\omega_{(0dB)}} \qquad (5)$$

Where $\omega_{(0dB)}$ is angular frequency that is 0dB at open-loop gain. Delay margin of this feedback control system is

$$T_{DM} = \frac{\frac{0.86}{180}\pi}{9.34 \times 10^4} = 0.16\mu s.$$

and there is almost no stability margin for the delay in the feedback loop.

Therefore, a phase lag-lead compensator is used to avoid the instability problem. The transfer function of the phase lag-lead compensator is represented as

$$G_{id}(s) = G_i(s)G_d(s) = \frac{\frac{s}{\omega_{zi}} + 1}{\frac{s}{\omega_{pi}} + 1} \frac{\frac{s}{\omega_{zd}} + 1}{\frac{s}{\omega_{pd}} + 1}. \qquad (6)$$

This compensator is combined a phase lead compensator and a phase lag compensator. A phase lead compensator cancels the phase delay due to a LC filter, and improves the stability. Although the lead compensator leads the phase, it maintain the gain in the high frequency range more than necessary. In this case, output volatage is influenced by swhiching frequency and noise. Therefore, the gain in the high frequency is attenuated by a phase lag compensator, and the influence of noise is suppressed. Stability of the feedback control system is ensured by arranging the pole and zero. Fig. 14 shows bode diagram of lag-leed compensator when the arrangement of poles and zeros.

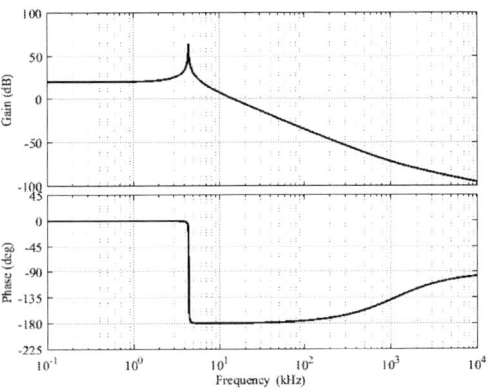

Fig. 13. Bode diagram of open-loop gain.

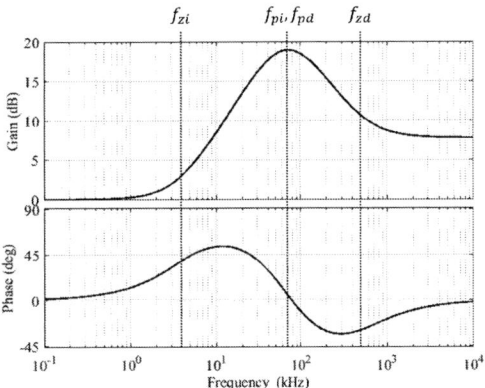

Fig. 14. Bode diagram of the phase lag-lead compensation.

Table 3. Pole and zero of lag-lead compensator.

Pole of Lead Compensator	f_{pi}	70kHz
Zero of Lead Compensator	f_{zi}	4kHz
Pole of Lag Compensator	f_{pd}	70kHz
Zero of Lag Compensator	f_{zd}	500kHz

Pole and zero of the compensator is placed suitably, stable margin can be improved. Fig. 15 shows a bode diagram when the pole and zero is placed as summarized in Table 3. After the phase compensator is added to the control system, the gain margin is 12.3dB and phase margin is 32.2 deg.

In case of phase compensation is performed, the delay margin is

$$T_{dM} = \frac{\frac{32.2}{180}\pi}{2.46 \times 10^5} = 2.29\mu s$$

The margin of stability is ensured for the delay in the feedback loop, compared case of without the phase compensation.

The above is the control system design of a general buck converter; the proposed inverter can design the control system using this method.

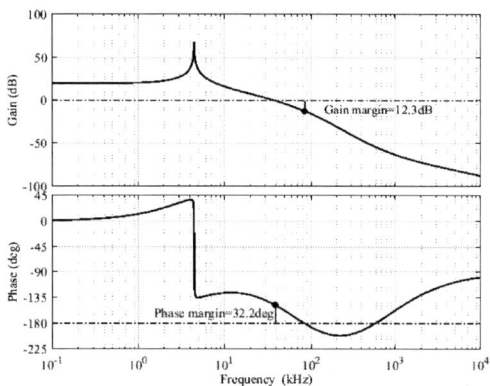

Fig. 15. Bode diagram with phase compensation.

IV. RESULTS OF COMPUTER SIMULATION

Computer simulation by PSIM is executed to confirm the above control design method. Fig. 16 shows the output voltage waveform when the proportional gain is varied. Increment of the proportional gain increases amplitude of output voltage. When the proportional gain is 20 dB (10 times), amplitude of the output voltage is 136V which satisfies the design specification of 100Vrms±5% at 60Hz. Output voltage amplitude is matched with the calculated value.

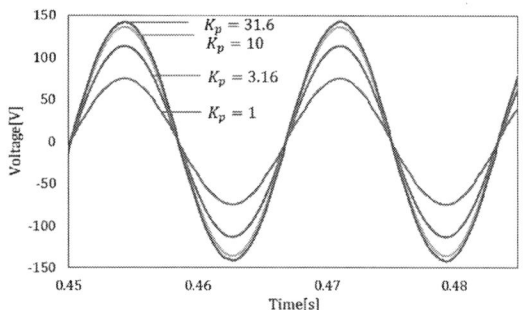

Fig. 16. Output voltage changes due to the proportional gain.

Fig. 17 shows the output voltage waveform when a load is disconnected from the output of the inverter. The output voltage waveform in case that the phase compensation is not applied, is oscillated just after the load disconnection, as shown in the middle figure. However, as shown in the top figure, the proposed phase compensation can prevent the output voltage from such oscillation and realize a sinusoidal output voltage, continuously.

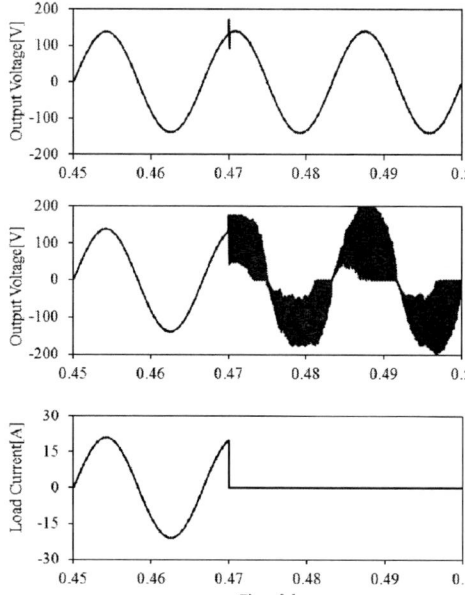

Fig. 17. Comparison of the output voltage and current in case of a sudden load variation; (top) voltage waveform with the proposed controller design, (middle) voltage waveform without the controller, and (bottom) its current waveform.

Fig. 18 is output voltage waveforms in case of condition of delay of 1μs in the control circuit. The output voltage waveform is oscillated when the phase compensation is not applied. However, the feedback system with phase compensation can prevent voltage waveform from oscillation.

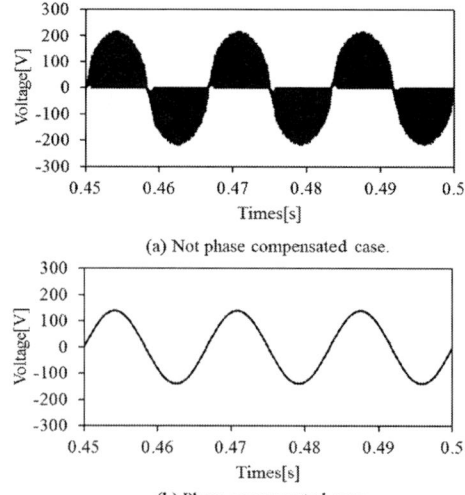

Fig. 18. Output voltage in case of control circuit including delay time.

V. CONCLUSION

The proposed feedback control for the sinusoidal voltage inverter without output filter can be designed similarly as the one of the conventional buck converter. Therefore, the control design of sinusoidal-wave inverter becomes easy as compared with conventional circuits. From results of computer simulation, proposed inverter and a design method of its controller are useful as a power supply of general use. In this paper, an analog feedback system is utilized but the configuration of the system with operational amplifier is complicated. Future issue is extension to digital control system and verification by the actual circuit.

REFERENCES

[1] Daishi Shimakawa, Kenji Yamanaka, and Masahide Hojo, "Experimental Study of Filter-Less Single-Phase Sine Wave Inverter," NCSP'15 ,28PM2-2-3, 2015.

[2] Shinichi Hiroshige, Kenji Yamanaka, and Masahide Hojo," Three-Phase Sinusoidal-wave Inverter without Output Filter Circuit," IEE-Japan Industry Applications Society Conference, 2017.

[3] Keng C. Wu, "Switch-Mode Power Converters Design and Analysis," Academic Press, 2005.

[4] Ned Mohan, Tore M. Undeland and William P. Robbins, "Power Electronics Third Edition," Wiley, 2003.

[5] R. D. Middlebrook and Slobodan Cuk, "A General Unified Approach to Modeling Switching-Converter Power Stages," IEEE Power Electronics Specialists Conference (PESC), 1976.

[6] Shi-Ping Hsu, Art Brown, Loman Rensink and R.D.Middlebrook, "Modelling and Analysis of Switching DC-to-DC Converters in Constant-Frequency Current-Programmed Model," IEEE PESC, 1979.

[7] Kaiwei Yao, Yuancheng Ren and Fred C. Lee, "Critical Bandwidth for the Load Transient Response of Voltage Regulator Modules," IEEE Transactions on Power Electronics, 2004.

The 2018 International Power Electronics Conference

Series Reactive Power Compensator with Reduced Capacitance for Hybrid Transformer

Yuki Takahashi[1], Takanori Isobe[2], Hiroshi Tadano[2]

1 Graduate School of Pure and Applied Sciences, University of Tsukuba, Ibaraki, Japan
2 Faculty of Pure and Applied Sciences, University of Tsukuba, Ibaraki, Japan
*E-mail: 1720373@s.tsukuba.ac.jp

Abstract—This paper discusses a hybrid transformer which consists of a conventional line frequency transformer with comparatively high leakage impedance and a series reactive power compensator. Several topologies and control of series compensator are reviewed in terms of losses and harmonics. The required capacitance, which can be important design factor in the phase separated configuration, is also discussed. Experimental demonstrations with a laboratory prototype, and loss evaluations with practical scale voltage compensators designed for a 30 kVA hybrid transformer are reported.

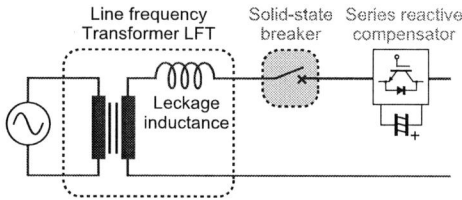

Fig. 1. Conceptual diagram of the hybrid-transformer using a series reactive power compensator.

I. INTRODUCTION

For more flexible, controllable and functional power systems, several power control devices achieved by power-electronics technology are studied and have already applied to actual power grids. One of challenging applications is a solid-state circuit breaker, which can be applied in both of low and medium voltage distribution network. The solid-state circuit breaker enables comparative high speed fault current shut down after detection than conventional mechanical circuit breakers; therefore, design of power system protection can be changed. However semiconductor switches have a drawback of much poor short-time over-current capability compared with conventional grid control devices like mechanical switches and transformers, and this makes the power electronics applications applied to power system controllers difficult.

A concept of solid-state-transformer (SST) has been proposed and studied to replace or eliminate conventional line frequency transformers connected to the medium voltage distribution line [1], [2], [3]. The SST provides some controllability and functionality in addition to possible size and weight reduction; however, it has not yet used widely in practical applications due to cost and loss increases. On the other hand, a concept of hybrid transformers has been proposed [4], [5]. It consists of a line frequency transformer and a partial rated power converters; for instance, a series voltage injector in the secondary side of the transformer. This concept does not aim at size and weight reductions but prioritizes cost and efficiency therefore is more practical for some applications which does not require DC terminals and frequency conversion. This concept is attractive for high voltage (HV) to medium voltage (MV) transformers in power substation, and MV to low voltage (LV) transformers installed on poles and in buildings.

This paper proposes to apply the hybrid transformer concept to reduce possible fault current by designing leakage inductance of the transformer to be comparatively high, but without increase of voltage fluctuation. The reduced maximum fault current makes the concept of the solid-state circuit breaker feasible, and results in more safe and flexible power distribution system. This paper mainly discusses series reactive power compensators to be used in the hybrid transformer, comparison of topologies, and its design and control principles.

II. HYBRID TRANSFORMER FOR POWER DISTRIBUTION NETWORK

The basic idea of the hybrid transformer is using a series voltage injection in either winding. The third winding can be equipped to provide power to the series voltage injection; however, this can be eliminated by achieving it with only reactive power handling. This paper proposes to use series reactive power compensators as shown in Fig. 1.

Fault current caused by possible short circuit in the load side is determined by an equivalent transformer series impedance and line impedance. High transformer impedance results in a reduced fault current amplitude; however, voltage drop caused by the load current in regular operation will be high. The hybrid transformer can be designed to have a comparatively high impedance to reduce the fault current since the series voltage injection can compensate for the comparatively high voltage drop. The voltage rating of the series compensators is required to be higher than the maximum voltage drop; however, it is expected to be around 10 to 15% of the rated voltage of the winding to achieve a sufficient fault current reduction. Therefore, widely available 1200 V IGBTs or SiC-MOSFETs can be used for MV winding with a high level of insulation to ground, and a low voltage (less than

3375

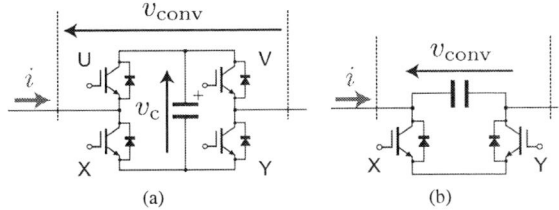

Fig. 2. Series reactive power compensators. (a)Full-bridge voltage source converter type. (b)Gate controlled series compensator (GCSC).

100 V) and high current Si-MOSFETs can be used for LV winding. For both cases, the partial voltage ratings lead to comparatively low losses and therefore sufficiently high overall efficiency, which should be closer to that of conventional transformers.

On the other hand, the series compensator is controlled to be inactive (shorted) when the over-current is detected so that the fault current is limited by its high equivalent series impedance. The reduced fault current flows through the series compensator; therefore, its over current capability is still important.

III. SERIES REACTIVE POWER COMPENSATORS FOR HYBRID TRANSFORMER

A. Topologies

Two possible topologies exist for the series reactive compensators as shown in Fig. 2. Most natural way is full-bridge voltage source converter connected in series as shown in Fig. 2(a) [6]. Another way is gate controlled series capacitor (GCSC) [7], [8] shown in Fig. 2(b), and its schematic operating waveforms are shown in Fig. 3(a). GCSC has a reduced number of switches and conduction loss can be reduced; however, it can operate only with line frequency switching. The compensation degree is controlled by the phase angle of 180° width gate signals to the current phase, and its equivalent reactance can varies from zero to $1/\omega C$, where C is the capacitance of the actually equipped capacitor.

GCSC also has an advantage of reduced capacitance compared to typical full-bridge voltage source converters, although an AC capacitor is needed. Since series compensators are usually achieved by phase-separated configurations, considerably high capacitance is needed to avoid high capacitor voltage fluctuation caused by double line frequency instantaneous power oscillation. Similar operation to GCSC is possible in the full-bridge compensator with line frequency switching operation [9]. Same operating waveforms with GCSC, as shown in Fig. 3(a), are available with the full-bridge configuration with a reduced capacitance. The distinguished capacitor voltage waveform, which has a zero voltage period in every half line cycle due to the reduced capacitance, can be seen. Basic functionality of the full-bridge compensator with line frequency switching is not different from that of GCSC; however, the full-bridge compensator has increased number of devices and increased losses; there-

fore, GCSC is advantageous for line frequency switching operation, in terms of loss and simplicity.

PWM modulations are also applicable for the full-bridge reactive power compensators with reduced capacitor with considering the highly oscillating capacitor voltage and voltage level control techniques [10], [11]. Its schematic waveforms are shown in Fig. 3. The highly oscillating capacitor voltage should be taken into account for modulation. There are two control strategies: one is fixing the voltage level as usual converters do as shown in Fig. 3(b), and the other is to achieve the minimum capacitor voltage as shown in Fig. 3(c), which is a distinctive operation mode and applicable only for series compensation.

Line frequency switching results in a distorted injecting voltage but switching losses can be negligibly low. On the other hand, PWM modulations do not generate low order harmonic voltages but a filter is needed to remove high order components. Selection depends on harmonic requirement in applications. In general, injecting voltage in the hybrid transformer application is expected to be 10 to 15% of voltage applied to the load; therefore, the voltage distortion generated by series compensators has low impact to the total voltage. On the other hand, PWM modulations can introduce waveform compensation for distorted source voltage as a series connected active filter.

B. Control Principle

For PWM operated full-bridge series compensators, capacitor voltage level with high oscillation should be controlled. A peak control technique for highly oscillating capacitor voltage has been proposed in [12] for STATCOMs. The peak and bottom voltage estimation block is shown in Fig. 4(a). The square of the capacitor voltage is used as a processed signal since it contains pure sinusoidal component with an offset. A second-order generalized-integrator (SOGI) configured as a quadrature signal generator (QSG) detects its average and amplitude of the sinusoidal component dynamically, then the peak and bottom of the signal can be estimated.

Fig. 4(b) shows the overall control block diagram with the peak voltage estimation. A voltage reference, V_q^*, is given to control injecting voltage as reactive power component, and a simple PI control of active power is applied for the capacitor voltage control.

For bottom voltage control, same structure with some modification can be applied. The estimation process of the bottom voltage, $\check{v}_c$, is also described in Fig. 4(a), and the bottom voltage can also be controlled by the same PI control.

C. Design Principle

Ratings of all the components for series compensators are basically considered to be based on partial series injecting voltage and rated system current. The PWM controls mentioned above require a LC filter; however, the voltage rating of the filter inductor can be several percents

3376

The 2018 International Power Electronics Conference

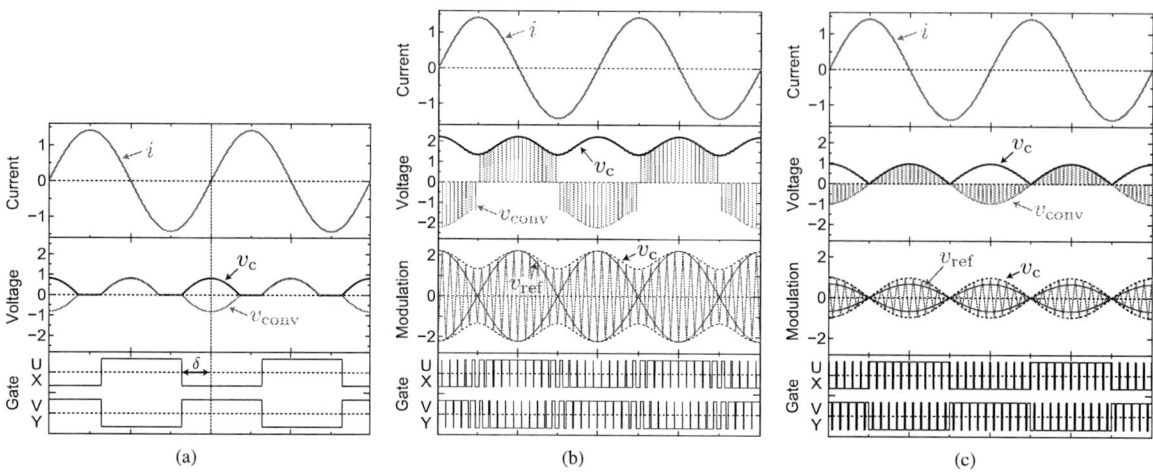

Fig. 3. Schematic waveforms of compensators. (a)GCSC and the full-bridge compensator operated in line frequency switching. (b)Full-bridge compensator with PWM switching operation and a constant peak voltage control. (b)With Minimum capacitor voltage control.

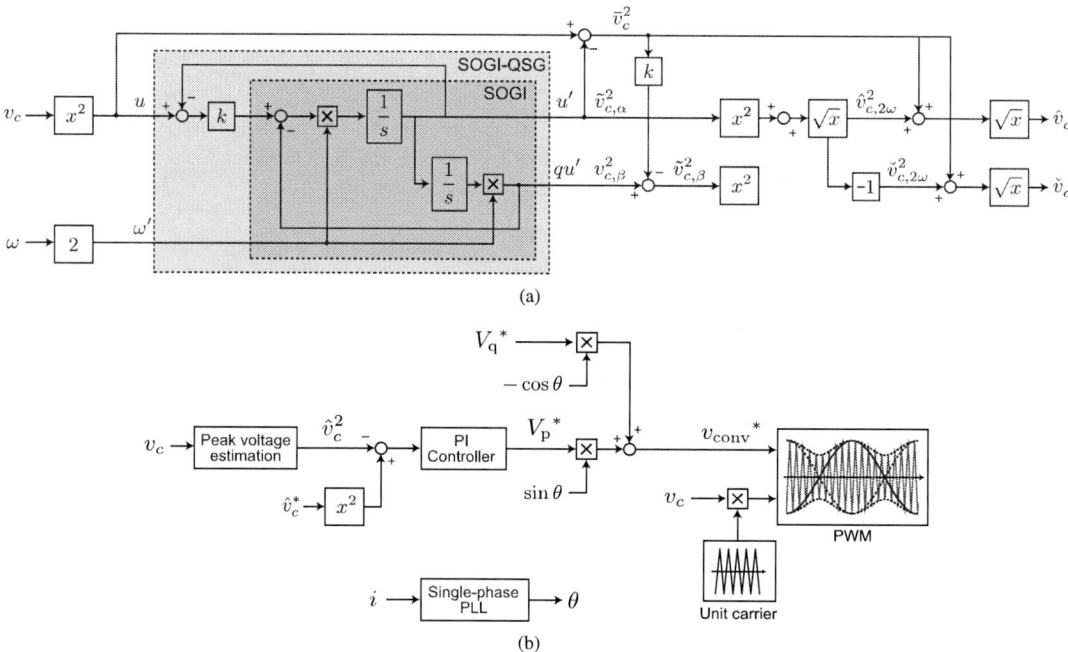

Fig. 4. Control block diagrams for PWM controlled full-bridge series compensator. (a)Dynamic peak voltage and bottom voltage estimation block. (b)Overall control block diagram with the peak voltage estimation block.

of the rated series injecting voltage although full current rating is required. The voltage rating of the filter capacitor is corresponding to the rated series injecting voltage; and its current rating is several percents of the rated system current. These features result in considerably low volume and loss of the filter.

However, a short-time over current must be taken into account for current rating considerations of the filter inductor and semiconductors. In fault conditions, the series compensator is switched to a shorting mode and does not operate in PWM; therefore, the filter indcutor is not required to have enough inductance for the over

current but should have conducting capability.

In the particular application of the hybrid transformer, the maximum reactance of the series compensator must be higher than the reactance of transformer leakage, ωL_{leak} to compensate fully for the voltage drop; therefore, the main capacitor should be designed to satisfy

$$C \leq \frac{1}{\omega^2 L_{\text{leak}}}, \qquad (1)$$

in case of GCSC. The full-bridge configuration can select much higher capacitance freely; however, same low capacitance is enough for the operation by using the

3377

TABLE I. PARAMETERS OF LABORATORY-SCALE EXPERIMENTS

(a)Voltage Supply

V_{in}	200 V	
I_{rated}	2.5 A	
f	50 Hz	
L_{leak}	24 mH	0.094 p.u.

(b)Load

R_{load}	60~210 Ω
L_{load}	165 mH

(c)Compensator

V_{rated}	26 V	
I_{rated}	2.5 A	
C_{conv}	280 μF	1.37 p.u.
C_{filter}	23 μF	0.0545 p.u.
L_{filter}	25 μH	0.00071 p.u.
MOSFET	IRFP4568	(150 V/171 A)
f_{sw}	5 kHz	

above proposed control techniques. The voltage rating of the capacitor is corresponding to the rated series injecting voltage with some margin, as same as for the semiconductor devices.

IV. LABORATORY SCALE DEMONSTRATIONS

Demonstrations with a laboratory-scale prototype were conducted. Schematic diagram of the circuit for the demonstrations is shown in Fig. 6, and its parameters are listed in Table I. As shown in Table I(a) and (b), comparatively high transformer impedance was assumed and inductive loads were applied. As shown in Table I(c), the compensator is designed to have capability to compensate for possible maximum voltage drop but with some margin since non-negligible resistance of the inductance introduces additional voltage drop in this small scale setup. MOSFETs with sufficiently low voltage rating than the system voltage, but higher than the maximum series injecting voltage, was selected. A LC filter was used for PWM switching operation but not used for line frequency switching control.

Resulting waveforms with three different control modes and two current levels are shown in Fig. 5. When the current was close to the rated current, waveforms with three control modes were similar, which have strong fluctuating capacitor voltage due to the proposed capacitance design. Difference could be observed in low current operations. PWM operation with the bottom voltage control and the line frequency switching control had low capacitor voltage; on the other hand, it was kept to be high with the peak voltage control.

Fig. 7 shows load voltages in r.m.s values as function of the load current with and without the compensator. The voltage highly dropped in the case without compensation; however, it was kept at constant in cases with the compensator regardless of control strategies. In all the cases, voltage generation in the compensator was supported from the load inductance. If the load power factor is unity, complete voltage compensation is not possible due to source side resistance; however, it does not matter since loads are usually inductive, and the transformer resistance can be negligibly low in real scale systems.

Efficiencies of compensators including the filter based on the system handling power are shown in Fig. 8. Very high efficiencies over 99% were observed even in this small scale setup since the series compensation is a kind of partial power processing. The efficiency of the line frequency switching control was highest and its reason can be thought as almost zero switching losses and no filter losses. In comparison between PWMs, the minimum voltage control achieved higher efficiencies in light load condition. The reason for that can be thought as low switching losses due to the low capacitor voltage, and low filter inductor losses due to low amplitude of the voltage applied to the inductor. The efficiencies of two controls became almost equal at rated current since the waveforms became almost same.

Total harmonic distortions (THDs) are also shown in Fig. 8. It is natural that the PWM controls achieved lower THDs than that with the line frequency control; however, THDs with the line frequency control were still low. This is because of the partial voltage injection.

V. EVALUATION OF A PRACTICAL SCALE COMPENSATORS

TABLE II. PARAMETERS OF FABRICATED PRACTICAL-SCALE COMPENSATORS AND EXPERIMENTAL SETUP

(a)Assumed Transformer

Number of phases	3 phase	
Capacity	30 kVA	
Line-to-Line Voltage	400 V	
f	50 Hz	
I_{rated}	43.3 A	
L_{leak}	1.7 mH	0.10 p.u.

(b)Voltage Supply

V_{variable}	0~10 V

(c)Load

L_{load}	1 mH

(d)Compensator

V_{rated}	23.1 V	
I_{rated}	43.3 A	
C_{conv}	4.2 mF	1.42 p.u.
C_{filter}	560 μF	0.0939 p.u.
L_{filter}	100 μH	0.0535 p.u.
R_{filter}	1.25 Ω	
MOSFET	IXFN200N07	(70 V/200 A)
f_{sw}	20 kHz	

To verify its feasibility in terms of size and loss of the series compensators, practical-scale prototypes were fabricated and evaluations were conducted. An overviews of the fabricated compensators are shown in Fig. 10. Schematic circuit diagrams for the evaluations are shown in Fig. 9, and its parameters are listed in Table II. In addition to the full-bridge series compensator, GCSC is also fabricated and tested in this evaluation.

The ratings of the fabricated series compensators are based on the current rating of the assumed transformer, that is 43.3 A, and the maximum voltage drop caused by the assumed leakage inductance, that is 23.1 V. Based on the nominal ratings, relatively low capacitance was selected for the capacitor in the compensators based on the proposed design strategy, that is 560 μF. Relatively low inductance and capacitance for filter were

The 2018 International Power Electronics Conference

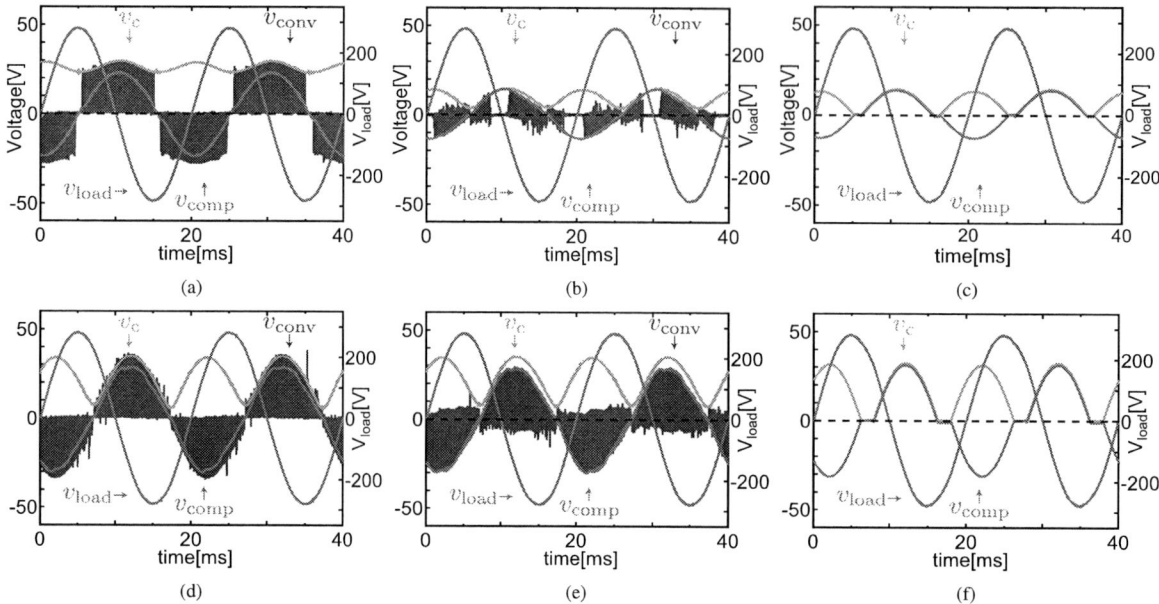

Fig. 5. Experimental waveforms. (a)PWM with peak voltage control and 0.4 p.u current. (b)PWM with minimum voltage control and 0.4 p.u current. (c)Line frequency switching with 0.4 p.u current. (d)PWM with peak voltage control and 1.0 p.u current. (e)PWM with minimum voltage control and 1.0 p.u current. (f)Line frequency switching with 1.0 p.u current.

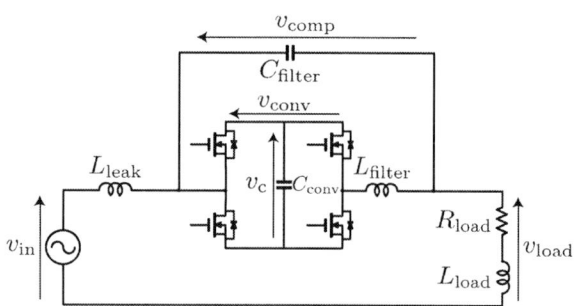

Fig. 6. Circuit diagram of laboratory-scale experiments. A voltage source and a series connected inductor, L_{leak}, were used in stead of using an actual transformer.

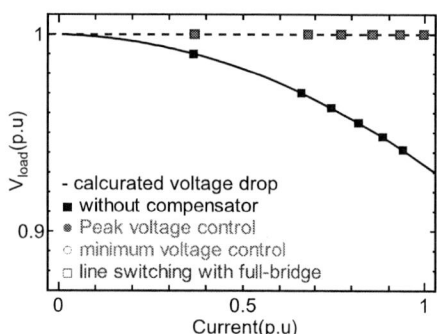

Fig. 7. Load voltage characteristics normalized by the no-load voltage as function of the load current, with and without compensators.

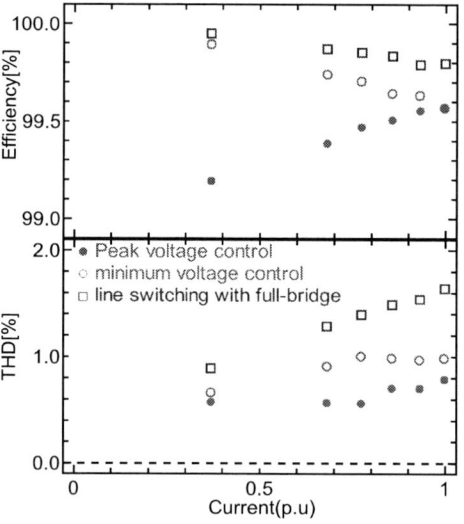

Fig. 8. Efficiencies of compensators based on the system handling power (top) and THDs of the load voltage (bottom).

also selected and a damping resistor was connected in series to the filter capacitor, to avoid possible resonance between the filter inductor and capacitor. Voltage ratings for the semiconductor switches and the capacitor are sufficiently low, which is corresponding to the voltage rating of the series compensators. The current rating of the semiconductor devices was relatively high to ensure the capability of conducting possible fault current; however, that is not discussed in this paper.

Only the series compensators were evaluated and

3379

The 2018 International Power Electronics Conference

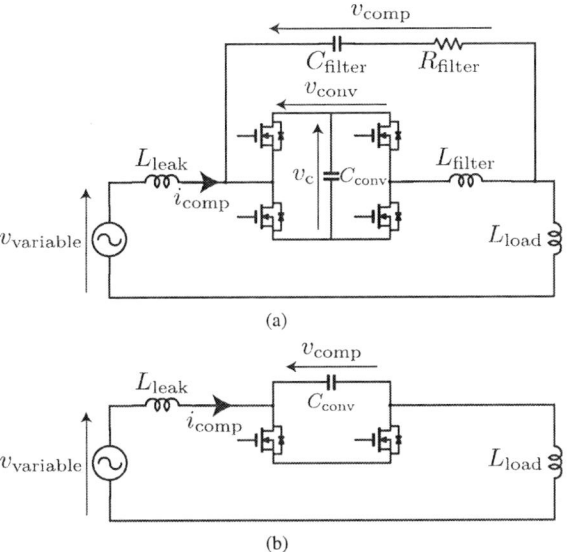

(a)

(b)

Fig. 9. Circuit diagram for evaluating practical scale compensators. (a)Full-bridge series compensator. (b)Gate controlled series compensator (GCSC).

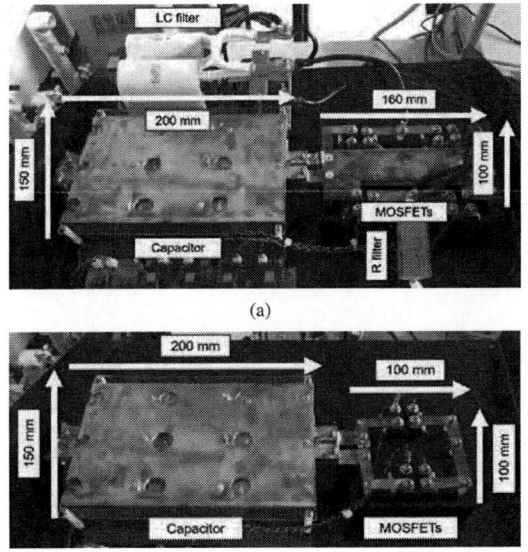

(a)

(b)

Fig. 10. Circuit overview of evaluated practical scale compensators. (a)Full-bridge series compensator. (b)Gate controlled series compensator (GCSC).

actual transformer was not used; therefore, the circuit configuration for the evaluation is not same with that for the actual application. To apply the nominal current and voltage to the fabricated series compensators, only an inductor, L_{load}, is connected as load and a variable voltage source was used. The source voltage was controlled to achieve a given current set-point, and the compensation degree of the compensators were controlled to achieve a given voltage set-point. The current and voltage were controlled to simulate actual series compensator for a

particular leakage inductance, that means the injecting voltage was controlled to be linear to the current.

Resulting waveforms with three different control modes of the full-bridge series compensator and the GCSC are shown in Fig. 11. As shown in Fig. 11(a) to (f), waveforms of three control modes of the full-bridge compensator were similar to those in the small-scale demonstrations. The waveforms of the full-bridge compensator with line frequency switching were similar to those of the GCSC, which are shown in Fig. 11(g)(h); however, a slight difference in zero voltage period was observed. The reason for that can be thought as a slight difference in compensation degree.

Measured losses of the compensators including the filter are shown in Fig. 12. All the losses were lower than 20 W. Those losses are considerably low since the total rating of the assumed system is 30 kVA; therefore, the efficiency based on the system ratings will be higher than 99.8%, by considering that the total loss of the compensators for three phases can be up to 60 W for 30 kVA system.

The losses with line frequency switching were lower than those with PWM again. The losses of the GCSC were lower than those of the full-bridge compensator with line frequency switching. The reason for that is the reduced number of switches to conduct the line current. In GCSC, the current flows through capacitor without semiconductor switches when the capacitor is charging and discharging, and the current flows through the semiconductor switches only during the zero voltage period.

Total harmonic distortions (THDs) of the injecting voltage, v_{comp}, are also shown in Fig. 12. It is natural that PWM controls achieved lower THDs than those with line frequency switching and GCSC; however, amplitude of the injecting voltage is about 10% even in the maximum current in normal operation; therefore, the worst THD of the load voltage can be lower than 3%.

VI. CONCLUSION

This paper proposed a hybrid transformer concept with series reactive compensation, and reviewed series compensation topologies and controls. The capacitance design was discussed since the phase-separated configuration can increase the size of the capacitor.

In the discussion, GCSC configuration with line frequency switching was concluded to be the best choice in terms of loss and simplicity, and possible harmonic distortion generated by the line frequency switching is expected to be low-enough for the hybrid transformer application. However, further discussion is still needed about capability of the fault current handling and switch operation to fault current limiting.

REFERENCES

[1] S. Xu, A. Q. Huang, and R. Burgos, "Review of solid-state transformer technologies and their application in power distri-

3380

The 2018 International Power Electronics Conference

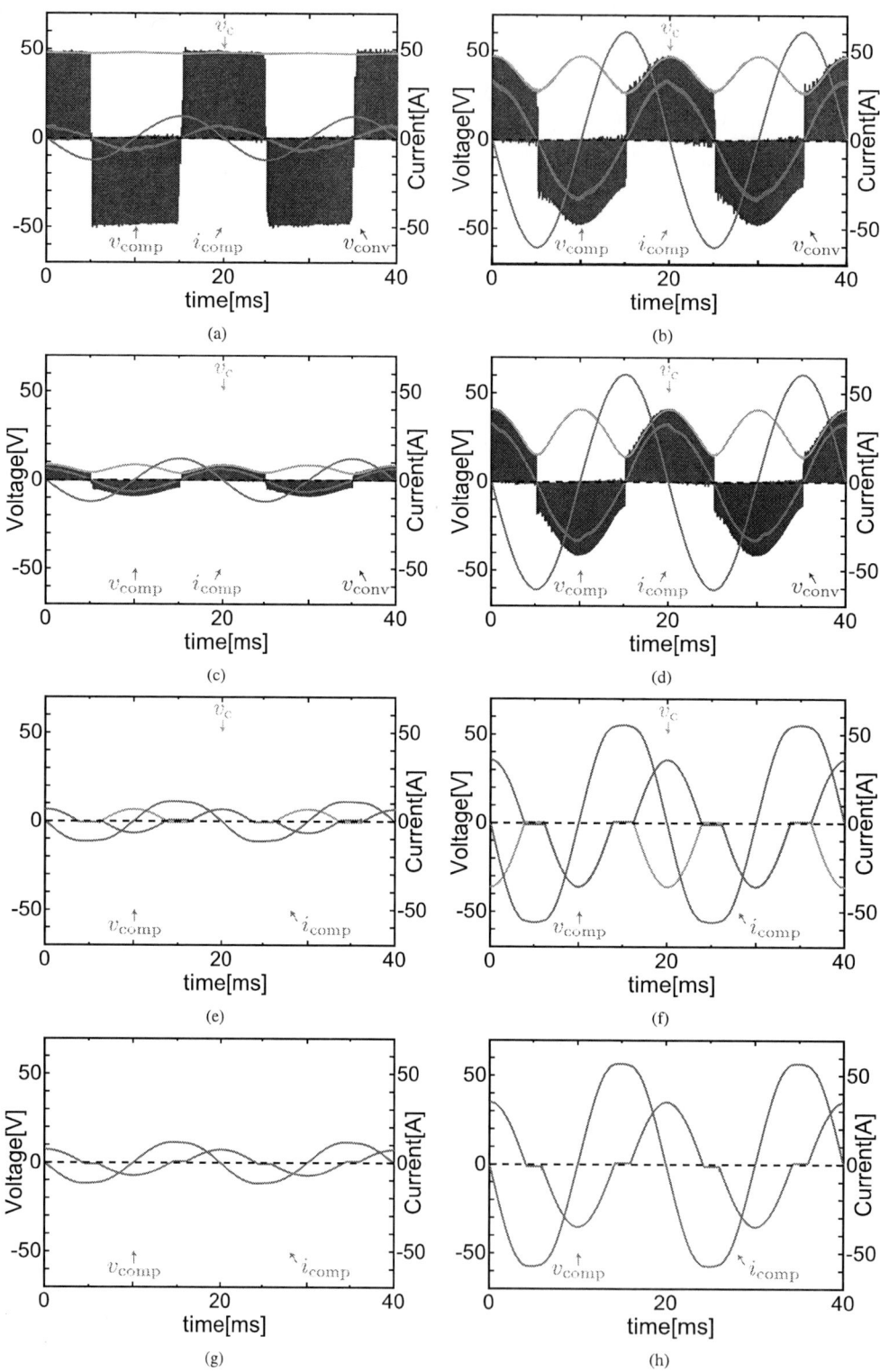

Fig. 11. Experimental waveforms. (a)PWM with peak voltage control and 0.2 p.u current. (b)PWM with peak voltage control and 1.0 p.u current. (c)PWM with minimum voltage control and 0.2 p.u current. (d)PWM with minimum voltage control and 1.0 p.u current. (e)Line frequency switching of full-bridge compensator with 0.2 p.u current. (f)Line frequency switching of full-bridge compensator with 1.0 p.u current. (g)Line frequency switching of GCSC with 0.2 p.u current. (f)Line frequency switching of GCSC with 1.0 p.u current.

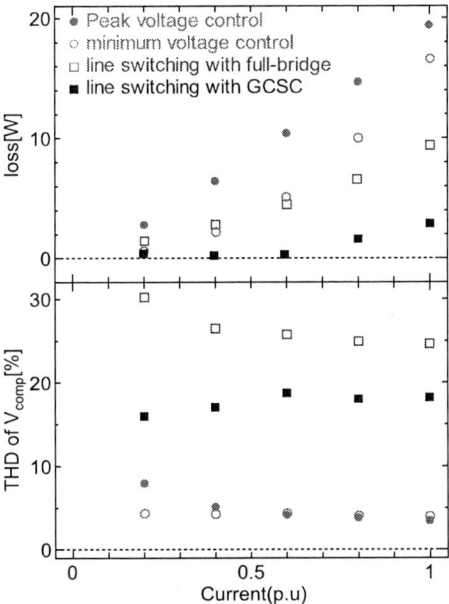

Fig. 12. Losses of compensators (top) and THDs of the injecting voltage (bottom).

bution systems", *IEEE Journal of Emerging and Selected Topics in Power Electronics*, Vol. 1, No. 3, pp. 186–198, 2013.

[2] J. W. Kolar, and G. Ortiz, "Solid-State-Transformers: Key components of future traction and smart grid systems", *Proceedings of the International Power Electronics Conference – ECCE Asia (IPEC 2014)*, Hiroshima, Japan, May 18–21 (2014)

[3] J. E. Huber and J. W. Kolar, "Optimum number of cascaded cells for high-power medium-voltage multilevel converters," *2013 IEEE Energy Conversion Congress and Exposition (ECCE 2013)*, Denver, CO, 2013, pp. 359–366.

[4] S. Bala, D. Das, E. Aeloiza, A Maitra and S. Rajagopalan, "Hybrid distribution transformer: Concept development and field demonstration," *IEEE Energy Conversion Congress and Exposition (ECCE2012)*, Raleigh, NC, 2012, pp. 4061–4068.

[5] J. Sastry and S. Bala, "Considerations for the design of power electronic modules for hybrid distribution transformers," *IEEE Energy Conversion Congress and Exposition (ECCE2013)*, Denver, CO, 2013, pp. 1422–1428.

[6] L. Gyugyi, C. D. Schauder, and K. K. Sen, "Static synchronous series compensator: A solid-state approach to the series compensation of transmission lines," *IEEE Transactions on Power Deliverly*, vol. 12, no. 1, pp. 406–417, (1997)

[7] G. G. Karady, T. H. Ortmeyer, B. R. Pilvelait and D. Maratukulam, "Continuously regurated series capacitor," *Transactions on Power Deliverly*, vol. 8, No. 3, pp. 1348–1355, (1993)

[8] E. H. Watanabe, L. F. W. de Souza, F. D. de Jesus, J. E. R. Alves and A. Bianco, "GCSC - gate controlled series capacitor: a new facts device for series compensation of transmission lines," *2004 IEEE/PES Transmission and distribution conference and exposition: Latin america*, pp. 981–986, (2004)

[9] J. A. Wiik, F. D. Wijaya, and R. Shimada:, "Characteristics of the magnetic energy recovery switch (MERS) as a series FACTS controller", *IEEE Trans. on Power Delivery*, Vol.24, No.2, pp.828–836 (2009)

[10] T. Isobe, D. Shiojima, K. Kato, Y. R. R. Hernandez, and R. Shimada, "Full-bridge Reactive Power Compensator with Minimized-Equipped Capacitor and its Application to Static Var Compensator," *IEEE Transactions on Power Electronics*, vol. 31, No. 1, pp. 224 – 234, 2016.

[11] T. Isobe, "A full-bridge AC power flow controller with reduced capacitance operated with both FFS (fundamental frequency switching) and PWM," *IEEE energy conversion congress and exposition (ECCE 2014)*, Pittsburgh, PA, United States, 14 –18 September 2014.

[12] T. Isobe, L. Zhang, H. Tadano, J. A. Suul, M. Molinas "Control of DC-capacitor Peak Voltage in Reduced Capacitance Single-Phase STATCOM," *in Proceedings of the Seventeenth IEEE Workshop on Modeling and Control for Power Electronics*, COMPEL 2016, Trondheim, Norway, 27–30, June 2016.

The 2018 International Power Electronics Conference

An Insight into the Voltage Rising Behavior during Turn-off Process of Series Connected SiC MOSFETs on Circuit Level

Panrui Wang[1]*, Feng Gao[1], Yang Jing[1], Yufeng Chen[2] and Lei Zhang[2]
1 School of Electrical Engineering, Shandong University, Jinan, China
2 State Grid Shandong Electric Power Research Institute, Jinan, China
*E-mail: qaz113259@163.com

Abstract— Current commercial available SiC MOSFET can hardly meet the requirement of high-voltage electrical application. Connecting SiC MOSFETs in series is a feasible solution. In order to utilize series connected SiC MOSFETs better, a comprehensive understanding of its switching process is of great necessity, especially turn-off process when huge different voltage sharing may occur. This paper focus on two factors, the nonlinearity of SiC MOSFET junction capacitance C_j and the existence of freewheeling diode junction capacitance C_f. A circuit analysis of series connected SiC MOSFETs is proposed. This paper analyzes the turn-off process of series connected SiC MOSFETs qualitatively and mathematically, and the expressions of key parameters, such as dv_{ds}/dt during turn-off process, are derived. The characteristics of series connected SiC MOSFETs during turn-off process are revealed. Finally, the accuracy of the analysis is verified by experimental results.

Keywords— equivalent model, series connected, SiC MOSFET

I. INTRODUCTION

Silicon carbide (SiC) device, as a new generation of power semiconductor device, has been showing excellent performance in power converters and applications [1][2]. Among the family of SiC devices, SiC MOSFETs are widely used because of its normally-off property. Series connected SiC MOSFET is a feasible way to apply low rated voltage devices to high voltage applications. Similar with series connected IGBTs, SiC MOSFETs in series connection still face the problem of unequal voltage sharing in turn-off process. The approaches for analyzing semiconductor switches can be divided into three main categories: physical theory, switching behavior, and equivalent circuit. Reference [3] created a physical model for single SiC MOSFET for PSpice simulation and reference [4] built a MATLAB/Simulink model of series connected SiC MOSFET based on McNutt/Hefner model. Physical models are relatively accurate but to some extend complicated, not clear and time consuming when used for calculation. Reference [5] established a behavior model of series connected IGBTs in PSIM simulation environment, which is still difficult to find out the internal correlation between semiconductor switches in series connection. Also, the two methods mentioned above rely on simulation and cannot find out the

switching patterns of series connected SiC MOSFETs directly. However, the analysis from the perspective of equivalent circuit are more straight-forward and can yield results with acceptable error within much shorter time. Reference[6]-[9] analyzed the switching transient and switching losses of single power MOSFET thoroughly on circuit level, while the circuit analysis of series connected MOSFETs have not been explored deeply. When SiC MOSFETs are series connected, delicate distinction of device parameter or very tiny signal delay (several nanoseconds) can trigger huge voltage difference, sometimes even causes device failure because of its fast switching transient. Thus, a comprehensive discussion of series connected SiC MOSFETs on circuit level is of importance and necessity. This paper proposes a clear insight into the turn-off behavior of series connected SiC MOSFETs, which can become guidance when series connected SiC MOSFETs are utilized in high voltage and high power applications. The analysis can also help to design SiC MOSFET driver and evaluate its performance such as switching speed and electromagnetic interference(EMI) problem. The final voltage distribution can be predicted from the perspective of circuit. Experimental results are presented for verifying the theoretical findings.

II. QUALITATIVE DEMONSTRATION ABOUT THE TURN-OFF TRANSIENT OF SERIES CONNECTED SiC MOSFETs

In this chapter, the nonlinearity of SiC MOSFET junction capacitance C_j and the existence of freewheeling diode junction capacitance C_f are emphasized. Fig. 1 is

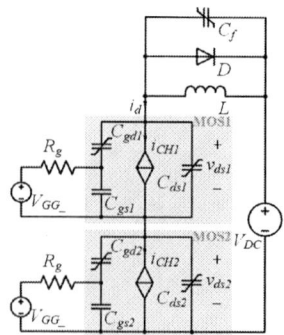

Fig. 1. Equivalent circuit of series connected SiC MOSFETs

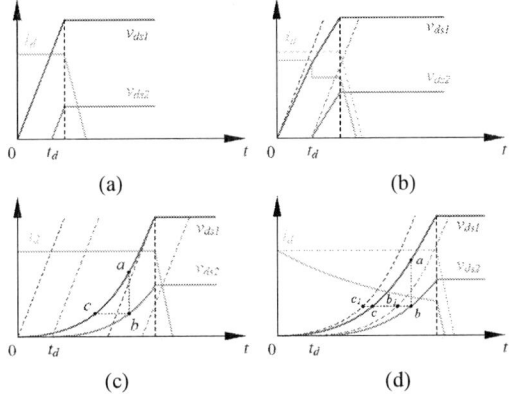

Fig. 2. Analytical waveforms of turn-off transient under four situations (a) ignore both the nonlinearity of C_j and the existence of C_f; (b) ignore the nonlinearity of C_j but consider the existence of C_f;(c) consider the nonlinearity of C_j but ignore the existence of C_f;(d) consider both the nonlinearity of C_j and the existence of C_f.

the equivalent circuit of series connected SiC MOSFETs. Fig. 2 is the analytical waveforms to illustrate the effects of C_j and C_f. For simplification, only two switches are series connected, but the conclusion drawn from this analysis can be extended directly when more switches are series connected together. The detailed discussion is shown as follows.

A. Ignore both the nonlinearity of C_j and the existence of C_f

Fig. 2(a) is an ideal turning off waveform of series connected MOSFETs when both the nonlinearity of C_j and the existence of C_f are ignored. The turning off process, in a sense, can be treated as a charging process of the junction capacitance C_j, or C_{oss}, specifically. The junction capacitance used here is the value when the MOSFETs operates in cutoff region ($v_{ds} = 1000V$ in this paper). The slew rate of v_{ds} is a function of load current i_d and apparently, the larger the load current i_d is, the faster the v_{ds} rises. The slew rate of v_{ds1} and v_{ds2} here are constant and same during the voltage rising transient because in this process, the load current i_d and C_j keeps constant. The final voltage difference value of v_{ds} is the delay time t_d multiplies the slew rate of v_{ds}.

B. Ignore the nonlinearity of C_j but consider the existence of C_f

Fig. 2(b) considers the existence of C_f but ignores the nonlinearity of C_j. When v_{ds} rises, the DC power supply voltage is constant. In this way, the voltage on C_f decreases and thus draws a displacement current from the load current. In Fig. 2(b), the dashed line is the waveform in Fig. 2(a). Compared with the situation (a), when v_{ds1} starts rising, i_d drops a little and the charging speed of C_j decreases. After the time point of t_d, both v_{ds1} and v_{ds2} enter the process of voltage rising, the total slew rate of v_{ds} becomes larger and much more displacement current is drawn from i_d, thus the slew rate of v_{ds} decreases more. However, as the junction capacitance C_j keeps constant, the slew rate of v_{ds1} and v_{ds2} are still same after t_d, and the

final voltage difference value of v_{ds} is equal to that in situation (a).

C. Consider the nonlinearity of C_j but ignore the existence of C_f

Fig. 2(c) is the situation where the nonlinearity of C_j is considered but the existence of C_f is ignored. The dashed line is the waveform of situation (a). It is noted that the voltage rising transient is no longer a straight line but a curve, which is caused by the nonlinearity of C_j. Because C_f does not exist in this situation, the load current keeps constant. The junction capacitance decreases with the increase of v_{ds} and smaller capacitance means faster charging speed at same load current. That is why the slew rate of v_{ds} is small when v_{ds} is small and vice versa. At a same time point, consider points of a and b in Fig. 2(c), v_{ds1} is larger than v_{ds2}, thus C_{j1} is smaller than C_{j2}, which causes the slew rate of v_{ds1} to be larger than that of v_{ds2}. When it comes to points of b and c, v_{ds1}, v_{ds2} and i_d are all same, yielding same voltage slew rate of v_{ds1} and v_{ds2}, which means the switch process of S_2 is only a repetition of that of S_1 with a time delay of t_d.

D. Consider both the nonlinearity of C_j and the existence of C_f

Fig. 2(d) is the turning off waveform when both the nonlinearity of C_j and the existence of C_f are considered. The dashed line is the waveform of situation (c). According the analysis in situation (c), at a same time point, the slew rate of v_{ds1} is bigger than that of v_{ds2} because of the larger v_{ds} value. When the existence of C_f is taken into account, the value of i_d decreases gradually, which is because the slew rate of v_{ds} increases as v_{ds} gets larger, resulting in an increasing displacement current of C_f. This effect causes the slew rate at point c in Fig. 2(d) to be smaller than that at point c_1 and the slew rate at point b to be smaller than that at point b_1, too. The waveform curves skew to right entirely and the slew rate at point b is smaller than that at point c because the load current i_d drops as time goes on.

If the slew rate of v_{ds} is referred to k, then in Fig. 2(d), the relationship of $k_b < k_a$, $k_c < k_a$ and $k_b < k_c$ can be derived. Base on the analysis above, the following conclusions can be concluded:

1. The nonlinearity of the junction capacitance C_j makes the voltage rising transient not a straight line but a curve. The slew rate of v_{ds} increases with the time going on., and thus $k_c < k_a$. At a same time point, the switch that turns off later has lower slew rate of v_{ds}, and thus $k_b < k_a$;

2. The existence of C_f slows down the slew rate of v_{ds} and correlates the turning off process of each switch with each other. The switching behavior of one switch can influence that of other switches by the bridge of load current i_d, and in this way, the switches no longer operate independently with each other;

3. The total effect of the nonlinearity of the junction capacitance C_j and the existence of C_f makes $k_b < k_c$, which means that the later the switch turns off, the slower the slew rate of its v_{ds} is.

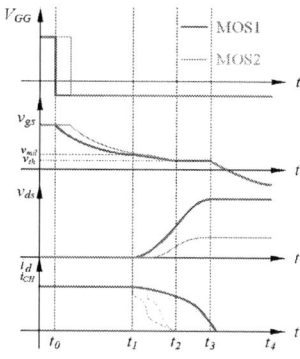

Fig. 3. Turn-off transient waveforms of series connected SiC MOSFETs.

III. MATHEMATICAL ANALYSIS OF TURN-OFF PROCESS OF SERIES CONNECTED MOSFETS

In this chapter, a mathematical analysis is implemented. For simplification, two switches connected in series are mainly discussed, but the results when n switches connected in series are also provided. The whole turn-off waveforms of two series connected MOSFETs is shown in Fig. 3, where the red thick line refers to S_1 and blue thin line refers to S_2. S_2 is assumed to turn off later than S_1 for time t_d. It is noted that the switching transient can be divided into 4 time intervals. The first three time intervals, from t_0 to t_3, will be discussed in detail, while the last one will be neglected because there is no tail current at the end of turn-off process like IGBT and the final voltage distribution will keep constant after t_3 [10].

Time interval 1 [$t_0 - t_1$] Delay time: The equivalent circuit in this period is shown in Fig. 4(a). The input capacitance of the MOSFETs start to discharge through the drive circuit when the driver source V_{GG} flips from positive value to negative value. Here, v_{ds} can be considered as 0 and thus C_{gs} and C_{ds} are parallel connected with each other. In this process, the v_{ds} and i_d of two MOSFETs keep constant and two MOSFETs operate independently.

Time interval 2 [$t_1 - t_2$] Voltage rising time I: After v_{gs1} reaches v_{mil}, S_1 enters its Miller plateau region and v_{ds1} starts to rise, while v_{gs2} have not reached v_{mil} and S_2 still operate in the same pattern with the last time interval. The equivalent circuit in this period is shown in Fig. 4(b).

For S_1, in this time interval, both C_{gd1} and C_{ds1} start to be charged. C_{gs1} can be considered as open circuit and all current that charges C_{gd1} flows into gate driver. The next equations can be derived for S_1 from Fig. 1(a).

$$I_L = i_{CH1} + i_{oss1} + i_{Cf} \tag{1}$$

$$i_{CH1} = g_{fs}\left(v_{gs1} - v_{th}\right) \tag{2}$$

$$v_{gs1} = R_g i_{g1} + V_{GG\_} \tag{3}$$

$$i_{g1} = C_{gd1}\frac{dv_{ds1}}{dt} \tag{4}$$

$$i_{oss1} = \left(C_{ds1} + C_{gd1}\right)\frac{dv_{ds1}}{dt} \tag{5}$$

$$i_{Cf} = C_f \frac{d\left(v_{ds1} - V_{DC}\right)}{dt} = C_f \frac{dv_{ds1}}{dt} \tag{6}$$

where I_L is the current flowing through the load inductor, which can be treated as a constant value during the turn-off transient, i_{CH1} is channel current, v_{th} is gate threshold voltage and g_{fs} is the transconductance of the SiC MOSFET.

Equations (1) – (6) can be arranged as:

$$A_1\left[\frac{dv_{ds1}}{dt}\right] = \left[a_1 + C_f\right]\left[\frac{dv_{ds1}}{dt}\right] = [c] \tag{7}$$

where $a_1 = g_{fs}R_g C_{gd1} + C_{oss1}$, $c = I_L - V_{GG}\_g_{fs} + g_{fs}v_{th}$. The slew rate of v_{ds1} can be solved as

$$\frac{dv_{ds1}}{dt} = \frac{I_L + g_{fs}v_{th} - g_{fs}V_{GG\_}}{C_f + C_{oss1} + g_{fs}R_g C_{gd1}} \tag{8}$$

When v_{gs2} falls to v_{mil}, both v_{ds1} and v_{ds2} start to rise. The equivalent circuit in this period is shown in Fig. 4(c). Equations (1) – (5) can still be used here. There is also another set of equations for S_2 in the same form of equations (1) – (5), with an only change of subscript. Here, C_f induces a coupling between S_1 and S_2 and the equation showing the effect of C_f is written as follows

$$I_{Cf} = C_f \frac{d\left(v_{ds1} + v_{ds2} - V_{DC}\right)}{dt} = C_f \frac{dv_{ds1}}{dt} + C_f \frac{dv_{ds2}}{dt} \tag{9}$$

Then, combining all equations mentioned above can write them into the form of matrix as follows

$$A_2\begin{bmatrix}\dfrac{dv_{ds1}}{dt}\\[2mm]\dfrac{dv_{ds2}}{dt}\end{bmatrix} = \begin{bmatrix}a_1 + C_f & C_f\\ C_f & a_2 + C_f\end{bmatrix}\begin{bmatrix}\dfrac{dv_{ds1}}{dt}\\[2mm]\dfrac{dv_{ds2}}{dt}\end{bmatrix} = \begin{bmatrix}c\\ c\end{bmatrix} \tag{10}$$

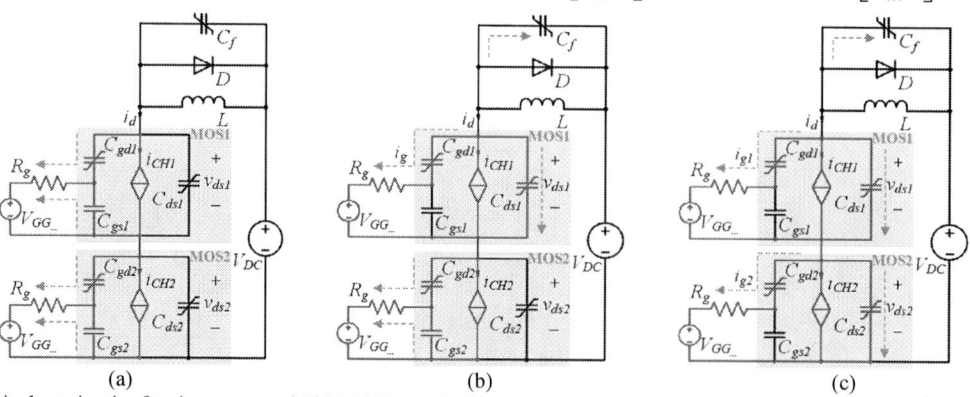

(a) (b) (c)

Fig. 4. Equivalent circuit of series connected SiC MOSFETs in different time intervals (a) time interval 1; (b) time interval 2, when only S_1 starts to turn off; (c) time interval 2, when both S_1 and S_2 start to turn off

where $a_1 = g_{fs}R_gC_{gd1} + C_{oss1}$, $a_2 = g_{fs}R_gC_{gd2} + C_{oss2}$, $c = I_L - V_{GG\_}g_{fs} + g_{fs}v_{th}$. By solving (10), the results are expressed as

$$\begin{bmatrix} \dfrac{dv_{ds1}}{dt} \\ \dfrac{dv_{ds2}}{dt} \end{bmatrix} = \frac{\mathbf{A}_2^*}{\det(\mathbf{A}_2)}\begin{bmatrix} c \\ c \end{bmatrix} \quad (11)$$

$$= \frac{I_L - V_{GG\_}g_{fs} + g_{fs}v_{th}}{\det(\mathbf{A}_2)}\begin{bmatrix} C_{oss2} + C_{gd2}R_gg_{fs} \\ C_{oss1} + C_{gd1}R_gg_{fs} \end{bmatrix}$$

It is interesting to find out from the equation (11) that the slew rate of v_{ds1} and v_{ds2} in this time interval is positively related to the junction capacitance value of its complementary switch. Specifically, at the same time point of this time interval, v_{ds1} is larger than v_{ds2}, thus C_{j1} is smaller than C_{j2}, yielding a larger slew rate of v_{ds1}.

When the number of MOSFETs connected in series increases to n ($n \geq 2$), there will be n sets of equations in the same form of equations (1)-(5) with an extra equation as follows

$$I_{Cf} = C_f\frac{d\left(\sum\limits_{i=1}^{n}v_{dsi} - V_{DC}\right)}{dt} = C_f\sum\limits_{i=1}^{n}\frac{dv_{dsi}}{dt} \quad (12)$$

Combining the $5n+1$ equations mentioned above and write them into the form of matrix as follows

$$\mathbf{A}_n\begin{bmatrix} \dfrac{dv_{ds1}}{dt} \\ \dfrac{dv_{ds2}}{dt} \\ \vdots \\ \dfrac{dv_{dsn}}{dt} \end{bmatrix} = \begin{bmatrix} a_1 + C_f & C_f & \cdots & C_f \\ C_f & a_2 + C_f & \cdots & C_f \\ \vdots & \vdots & \ddots & \vdots \\ C_f & C_f & \cdots & a_n + C_f \end{bmatrix}\begin{bmatrix} \dfrac{dv_{ds1}}{dt} \\ \dfrac{dv_{ds2}}{dt} \\ \vdots \\ \dfrac{dv_{dsn}}{dt} \end{bmatrix} = \begin{bmatrix} c \\ c \\ \vdots \\ c \end{bmatrix} \quad (13)$$

By solving (13), the results are expressed as

$$\begin{bmatrix} \dfrac{dv_{dsk}}{dt} \end{bmatrix} = \frac{\mathbf{A}_n^*}{\det(\mathbf{A}_n)}\begin{bmatrix} c \\ c \\ \vdots \\ c \end{bmatrix} \quad (14)$$

$$= \frac{I_L - V_{GG\_}g_{fs} + g_{fs}v_{th}}{\det(\mathbf{A}_n)}\left[\prod\limits_{\substack{i=1 \\ i \neq k}}^{n}\left(C_{ossi} + C_{gdi}R_gg_{fs}\right)\right]$$

The equation (14) features in the same way of (11). From (14), it can be inferred that the slew rate of v_{ds} is decided by the order of turn-off. If a MOSFET turns off later, the v_{ds} of the others will be larger and the junction capacitance of the others will be smaller, resulting in a slower slew rate of this MOSFET.

It can be learnt from equations (1), (5) and (12) that the channel currents, i_{CH1} and i_{CH2}, are decreasing gradually in the turn-off transient, which are drawn in dashed lines in Fig. 3. The results derived above are valid only at the time when i_{CH1} and i_{CH2} are positive because the current flowing through the channel cannot be negative. If the load current is small, sometimes i_{CH} will fall to 0 during the voltage rising transient because of the higher v_{ds} slew rate of SiC MOSFET, especially when series connected. Fig. 5 elaborates the calculated and experimental waveforms, where the transient of i_{CH} is

depicted. This phenomenon has also been pointed out in [8] but have not been analyzed in detail. Usually, i_{CH1} and i_{CH2} drop to 0 almost at the same time. Here for simplification, the time point when either i_{CH1} or i_{CH2} reaches 0 is considered as the end of this time interval.

Time interval 3 [$t_2 - t_3$] Voltage rising time II: This time interval only occurs when the channel current of the MOSFET falls to 0, which is widely observed in SiC MOSFET. In this time interval, the channel current i_{CH} is no longer controlled by v_{gs} and stays at 0. In this way, the equivalent circuit can be simplified into a capacitor charging circuit in Fig. 6. In Fig. 6(b), the drive side is replaced with a voltage source v_{gs} and the value of the v_{gs} should be v_{th} because the channel current i_{CH} has dropped to 0. Fig. 6(c) further simplifies the circuit by omitting the drive side because v_{gs}, usually several volts, is very small when compared with v_{ds}. The final equivalent circuit of Fig. 6(c) is two nonlinear capacitors connected in series.

In this circumstance, the next equations can be employed for finding out the slew rate of v_{ds1} and v_{ds2}

$$C_{oss1}\frac{dv_{ds1}}{dt} = C_{oss2}\frac{dv_{ds2}}{dt} \quad (15)$$

$$C_{oss1}\frac{dv_{ds1}}{dt} + C_f\frac{d\left(v_{ds1} + v_{ds2} - V_{DC}\right)}{dt} = I_L \quad (16)$$

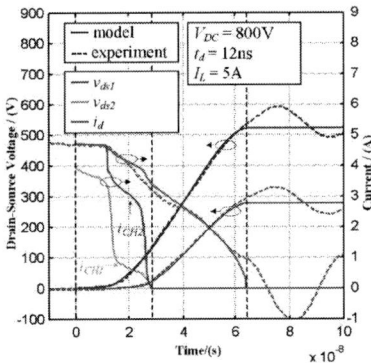

Fig. 5. The decrease of i_{CH} during v_{ds} rising transient.

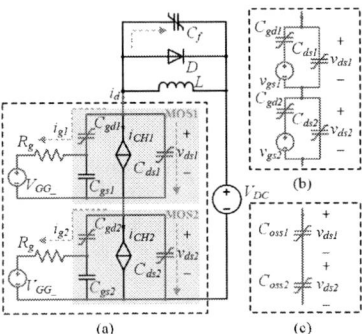

Fig.6. Equivalent circuit of series connected SiC MOSFETs (a) the equivalent circuit; (b) the simplified equivalent circuit; (c) the final equivalent circuit

3386

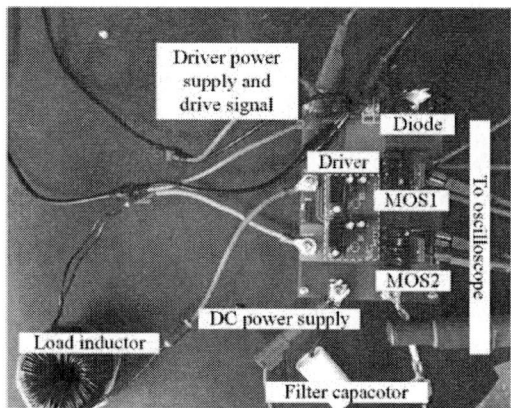

Fig. 7. Photo of experimental platform.

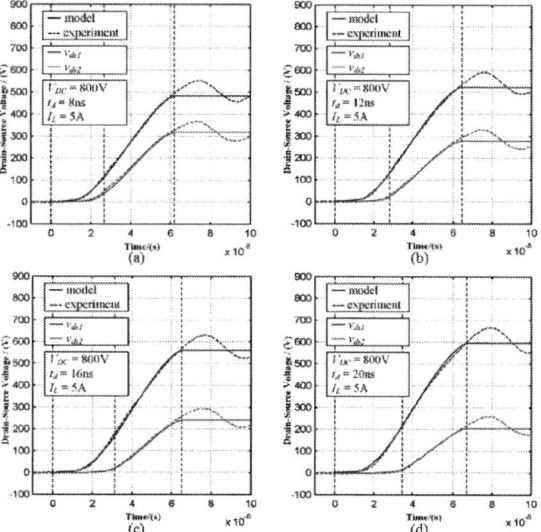

Fig. 8. Experimental and simulated waveforms of v_{ds1} and v_{ds2} with different drive signal delay.

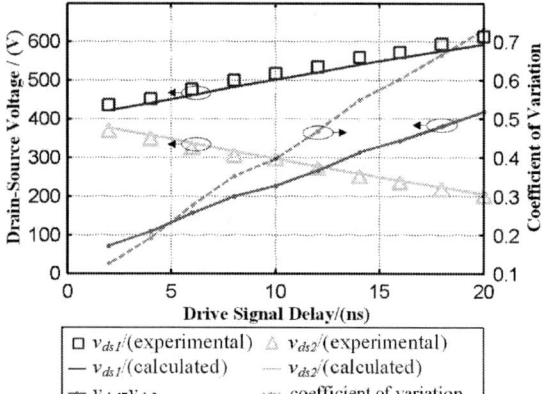

Fig. 9. Relationship between final voltage distribution and drive signal delay.

$$C_{oss2}\frac{dv_{ds2}}{dt}+C_f\frac{d\left(v_{ds1}+v_{ds2}-V_{DC}\right)}{dt}=I_L \tag{17}$$

The calculated slew rate of v_{ds} are expressed as follows

$$\frac{dv_{ds1}}{dt}=\frac{I_L}{C_{oss1}+C_f\left(1+\dfrac{C_{oss1}}{C_{oss2}}\right)} \tag{18}$$

$$\frac{dv_{ds2}}{dt}=\frac{I_L}{C_{oss2}+C_f\left(1+\dfrac{C_{oss2}}{C_{oss1}}\right)} \tag{19}$$

From equations (18) and (19), it is noted that the slew rate of v_{ds} is negatively related with its own junction capacitance and positively related with the junction capacitance of the other MOSFET. If S_1 turns off earlier and v_{ds1} is higher than v_{ds2}, thus C_{oss1} is smaller than C_{oss2}, yielding a faster slew rate of v_{ds1}, which is still in consistent with the conclusion drawn above. Similarly, if n MOSFETs connected in series, equation (15) – (17) will become

$$C_{oss1}\frac{dv_{ds1}}{dt}=C_{oss2}\frac{dv_{ds2}}{dt}=\cdots=C_{ossn}\frac{dv_{dsn}}{dt} \tag{20}$$

$$C_{ossk}\frac{dv_{dsk}}{dt}+C_f\frac{d\left(\sum_{i=1}^{n}v_{dsi}-V_{DC}\right)}{dt}=I_L \tag{21}$$

The result of equation (20) – (21) is

$$\frac{dv_{dsk}}{dt}=\frac{I_L}{C_{ossk}+C_f\left(1+\displaystyle\sum_{\substack{i=1\\i\neq k}}^{n}\frac{C_{ossk}}{C_{ossi}}\right)} \tag{22}$$

Another point needed attention is that when the sum of v_{ds} approaches V_{DC}, the slew rate of v_{ds} slows down and the curve of voltage rising transient seems S-shaped. That is because the junction capacitance of the freewheeling diode C_f increases significantly and slows down dv_{ds}/dt when the voltage of C_f decrease to near 0.

IV. EXPERIMENTAL VERIFICATION

An experimental platform of a chopper circuit in Fig. 1 is established to verify the equivalent circuit model of series connected SiC MOSFETs proposed in this paper. The main switch is formed by two series connected SiC MOSFETs C2M0040120D manufactured by Cree. The key parameters of the MOSFET is shown in Table I. Each of the MOSFET has its own isolated driver with an integrated driver chip IXDD614 from IXYS. An

TABLE I MAIN PARAMETERS OF C2M0040120D

Parameters	Values
Minimum/Maximum Allowed Gate Source Voltage	-10V/+25V
Gate Threshold Voltage	2.6V
Internal Gate Resistance	1.8Ω
Drain-Source Breakdown Voltage	1200V
Continuous Drain Current	60A
Transconductance	15.1S
Input Capacitance (V_{ds}=1000V)	1893pF
Output Capacitance (V_{ds}=1000V)	150pF
Reverse Transfer Capacitance (V_{ds}=1000V)	10pF

The 2018 International Power Electronics Conference

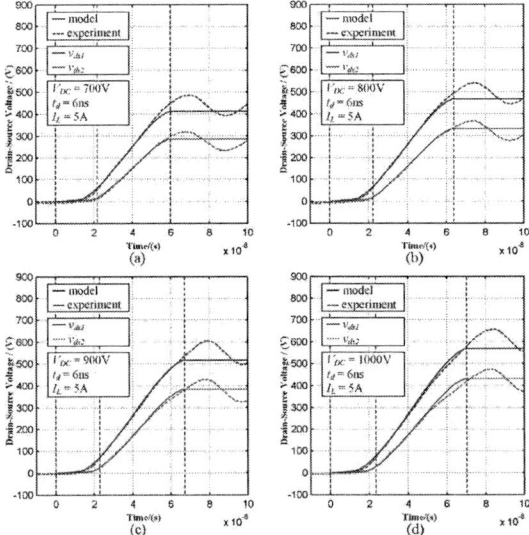

Fig. 10. Experimental and simulated waveforms of v_{ds1} and v_{ds2} with different DC bus voltage.

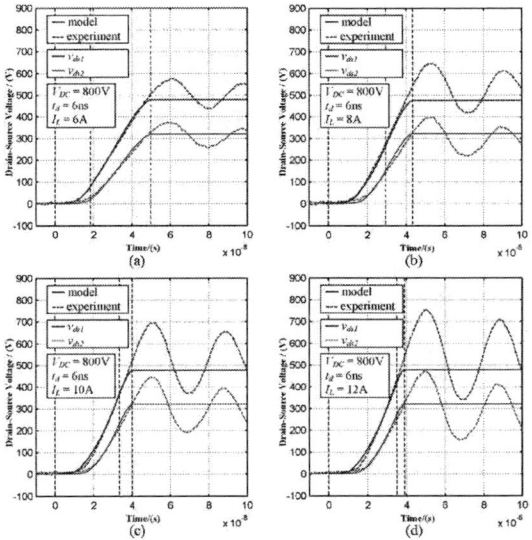

Fig. 11. Experimental and simulated waveforms of v_{ds1} and v_{ds2} with different load current.

asymmetric drive voltage of +20V/-4V is adopted and the gate series resistor is 39 Ω. The drive signal is generated by a dual channel signal generator AFG 3022B made by Tektronix and the oscilloscope is 104MXi-A from LeCroy with a bandwidth of 1GHz. The photo of the test platform is shown in Fig. 7.

Three sets of experiments with different drive signal delay, different DC bus voltage and different load current were implemented. The experimental and simulated waveforms are presented in Fig. 8, Fig. 10 and Fig. 11, respectively. The vertical dashed lines are the boundaries of different time intervals. It is noticed that the calculated results agree well with the experimental results. Fig. 7 is the graph of statistic data from calculation and experiment. The dashed line refers to the coefficient of

variation V_σ to evaluate the degree of voltage unbalance, which is defined as follows

$$V_\sigma = \sqrt{\frac{\sum_{i=1}^{n}\left(v_{dsi} - \overline{v}_{ds}\right)^2}{(n-1)\overline{v}_{ds}^2}} \tag{23}$$

In Fig. 9, it is noticed that the difference between v_{ds1} and v_{ds2} increase almost linearly along with the drive signal delay. The coefficient of variation also rises from 0.13 to 0.73 linearly when the drive signal delay increases from 2ns to 20ns. Although some discrepancy appears, which is mainly caused by the neglected elements such as parasitic inductance, the analysis proposed in this paper can reveal the characteristics of turn-off behavior of series connected SiC MOSFETs. The final voltage distribution is also predicted accurately.

V. Conclusion

This paper provides an insight into the voltage rising transient of series connected SiC MOSFETs during turn-off transient based on circuit theory. The nonlinearity of SiC MOSFET junction capacitance C_j and the existence of freewheeling diode junction capacitance C_f are emphasized and their effects on turn-off behavior of series connected SiC MOSFETs are fully elaborated. The switching process of series connected SiC MOSFETs is analyzed qualitatively and mathematically. The expressions of key parameters, such as dv_{ds}/dt during voltage rising time, are derived and the characteristics of series connected SiC MOSFETs turn-off process are revealed. Finally, the accuracy of the analysis is verified by the experimental results.

Acknowledgement

This work was supported by State Grid Corporation of China under Grant SGSDDK00KJJS1600066.

References

[1] J. Wang et al., "Characterization, Modeling, and Application of 10-kV SiC MOSFET," in *IEEE Transactions on Electron Devices*, vol. 55, no. 8, pp. 1798-1806, Aug. 2008.

[2] K. Chen, Z. Zhao, L. Yuan, T. Lu and F. He, "The Impact of Nonlinear Junction Capacitance on Switching Transient and Its Modeling for SiC MOSFET," in *IEEE Transactions on Electron Devices*, vol. 62, no. 2, pp. 333-338, Feb. 2015.

[3] L. Ceccarelli, F. Iannuzzo and M. Nawaz, "PSpice modeling platform for SiC power MOSFET modules with extensive experimental validation," *2016 IEEE Energy Conversion Congress and Exposition (ECCE)*, Milwaukee, WI, 2016, pp. 1-8.

[4] G. Tsolaridis, K. Ilves, P. D. Reigosa, M. Nawaz and F. Iannuzzo, "Development of Simulink-based SiC MOSFET modeling platform for series connected devices," *2016 IEEE Energy Conversion Congress and Exposition (ECCE)*, Milwaukee, WI, 2016, pp. 1-8.

[5] Yu Sang, T. Lu, Zhengming Zhao and Shiqi Ji, "Behavior model for series connected high voltage IGBTs," *2014*

IEEE Conference and Expo Transportation Electrification Asia-Pacific (ITEC Asia-Pacific), Beijing, 2014, pp. 1-5.

[6] Yuancheng Ren, Ming Xu, Jinghai Zhou and F. C. Lee, "Analytical loss model of power MOSFET," in *IEEE Transactions on Power Electronics*, vol. 21, no. 2, pp. 310-319, March 2006.

[7] J. Wang, H. S. h. Chung and R. T. h. Li, "Characterization and Experimental Assessment of the Effects of Parasitic Elements on the MOSFET Switching Performance," in *IEEE Transactions on Power Electronics*, vol. 28, no. 1, pp. 573-590, Jan. 2013.

[8] M. Rodríguez, A. Rodríguez, P. F. Miaja, D. G. Lamar and J. S. Zúniga, "An Insight into the Switching Process of Power MOSFETs: An Improved Analytical Losses Model," in *IEEE Transactions on Power Electronics*, vol. 25, no. 6, pp. 1626-1640, June 2010.

[9] I. Castro *et al.*, "Analytical Switching Loss Model for Superjunction MOSFET With Capacitive Nonlinearities and Displacement Currents for DC–DC Power Converters," in *IEEE Transactions on Power Electronics*, vol. 31, no. 3, pp. 2485-2495, March 2016.

[10] P. R. Palmer, Xueqiang Zhang and Jin Zhang, "The direct series connection of SiC MOSFETs," *IECON 2016 - 42nd Annual Conference of the IEEE Industrial Electronics Society*, Florence, 2016, pp. 1171-1176.

The 2018 International Power Electronics Conference

Paralleling six 320A 1200V All-SiC Half-bridge Modules for a Large Capacity Power Stack

David Hongfei Lu[1*], Hiromu Takubo[1], Sho Takano[1] and Yuhei Suzuki[2]

1 Advanced Technology Laboratory, Corporate R&D Headquarters, Fuji Electric Co., Ltd., Hino City, Japan
2 Development Division, Power Electronics Systems Business Group, Fuji Electric Co., Ltd., Suzuka City, Japan
*E-mail: Lu-David@fujielectric.com

Abstract— **We developed a 400V/400kW power stack by paralleling six 320A/1200V half-bridge all-SiC modules. Structural imbalance of parasitic inductance is contained within 3% in the laminated bus-bar with the aid of electromagnetics simulation. Switching dynamics in normal operation conditions and arm short circuit condition were investigated under a hard-switching driving scheme. Current imbalance less than 12% was ensured during both steady-state conduction and dynamic switching. Compared with conventional IGBTs, the switching loss of the all-SiC modules is reduced by more than 60%, and the switching frequency can be increased more than twice.**

Keywords— ***Power Stack, Parallel Connection, Short-Circuit, All-SiC Module***

I. INTRODUCTION

As one of its prominent features attributed to wide band gap semiconductor material, SiC power MOSFET is capable of achieving high-speed switching and low power loss at the same time. By adopting SiC MOSFETs, improvements of power electronics equipment in terms of size, weight and power efficiency are expected relative to their IGBT-based counterparts. Extensive investigations have been conducted for applications with small power capacity. Meanwhile, it is required to increase the capacity of the SiC devices to meet the requirement of larger power capacity applications. Paralleling modules, with multiple chips paralleled within in turn, is considered as one of the effective methods [1-3].

We have successfully parallelized six 320A/1200V half-bridge non-wire bonded modules with largest total current capability to our knowledge, and reported preliminary results in [4]. In this paper, we further report specifics on bus-bar design, inductive switching dynamics up to 2700A, and short-circuit behaviors.

II. GENEAL CONSIDERATIONS UPON PARALLELING MOSFETS

An illustrative circuit diagram of a single phase inverter in which six modules paralleled in each arm is shown in Fig. 1. To achieve large total controllable current, it is important to equalize the current conduction of each module during steady state and switching

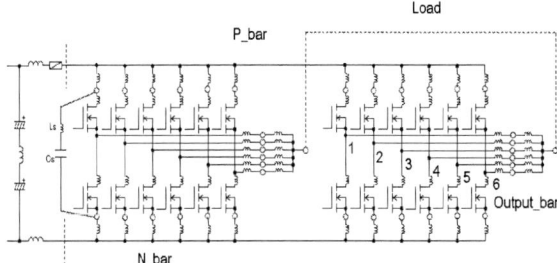

Fig. 1. An illustrative diagram of single-phase inverter with six paralleled modules in each arm.

Fig.2. The outlook of 1200V/320A all-SiC half-bridge power module with a new packaging concept.

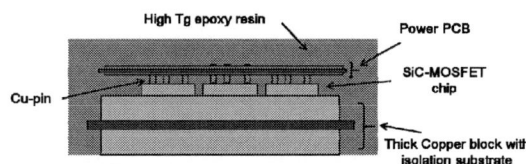

Fig. 3. A schematic internal structure of module in Fig. 2.

operation. Main factors with negative impact on current balance are (1) the scattering of modules' electric characteristics, notably, the gate threshold voltage V_{th} and on-state voltage V_{on} (2) bus bar structural difference resulting in insufficiently low and unmatched inductance and resistance in each current circulating loop, and (3) unmatched gate driving signal distribution.

III. SIC MODULE SPECIFICS

Fig. 2 shows the appearance of the all-SiC module, with dimensions of W126×D45× H13 in mm. Fig. 3 shows its internal structure. SiC chip was soldered with copper pin rather than Aluminum wire on the source and

3390

gate side, internal power PCB board was used to achieve high current density. The pattern on the PCB is optimized to reduce parasitic inductance in the path of the main current and to avoid EMI issues. Furthermore, screw terminals are used in main terminals to accommodate large current.

Relative to a wire-bonded IGBT (300A / 1200V) module, a volume reduction of 55%, a footprint reduction of 15%, and a wiring inductance reduction of 25% between the P-N terminals are achieved. At the same time, its power cycle capability is significantly improved [5-6].

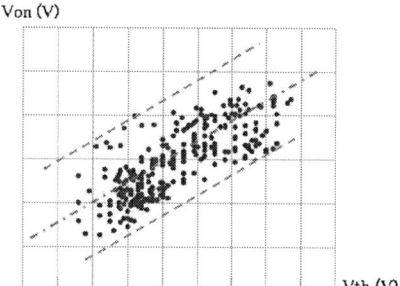

Fig.4. Von-Vth distribution of all-SiC modules.

The V_{on} - V_{th} distribution of the all-SiC module is shown in Fig. 4, with scattering generally in line with literature report [7]. In this work, common-practice is followed to sort V_{th} and V_{on} in different ranks, six modules from the same rank is paralleled in each arm to ensure no deviation from guaranteed reverse-biased safe operating area (RBSOA) of the module.

IV. BUS-BAR DESIGN

As mentioned in section II, it is important to equalize laminated P, N bars from DC-link capacitors to each module in the same arm, and output bars from each module to the load for power delivery, as well as equalize and minimize the loop inductance from each module through the path of P bar, short-circuited DC-link capacitors, and N bars, or vice versa, to suppress transient voltage stress during switching [8].

Electromagnetic field analysis with ANSYS Q3D was carried out to optimize a bus-bar structure shown in Fig. 5. The maximum inductance imbalance ratio among module mounting points, $(L_{max}-L_{min})/L_{av}$, was designed to be within 3%. L_{max}, L_{min}, L_{av} are maximum, minimum, and average value among six positions, respectively.

Table I shows the parasitic inductance imbalance ratio $(L-L_{av})/L_{av}$ at each position in turn-off loop for modules in the same phase-leg farthest away from DC-link capacitors.

With bypass capacitors connected in parallel with modules, the imbalance ratio could be further lowered.

TABLE I.

SIMULATED LOOP INDUCTANCE IMBALANCE RATIO

Module position	1	2	3	4	5	6
Imbalance ratio	0.93%	-0.10%	-0.86%	-0.88%	-0.06%	0.96%

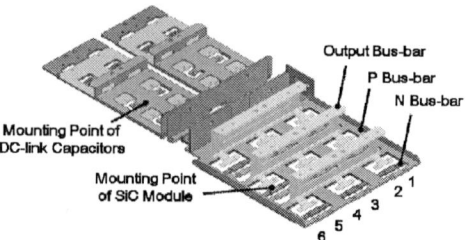

Fig. 5. A three-phase bus-bar structural model.

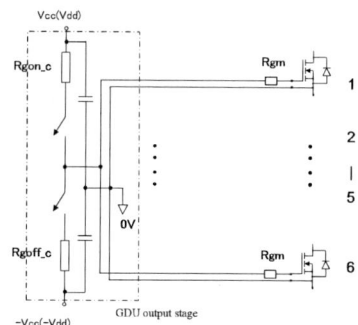

Fig. 6. . Gate resistance allocation.

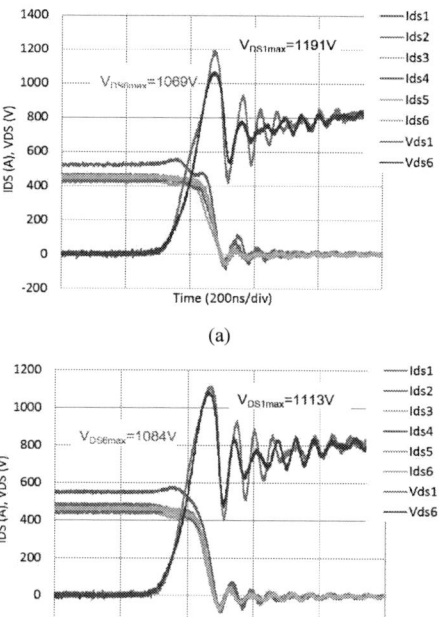

(a)

(b)

Fig. 7. Room temperature turn-off waveforms at a total current of 2700A without (a) and with (b) a gate ballasting resistance on each module, V_{DC}=700V.

V. EXPERIMENTAL RESULTS

A. Gate Driving Conditions

A ballasting resistance R_{gm} of 5-10% total gate resistance is placed on each module's gate terminal [9], Fig. 6. It leads to a relatively stable dynamic transition as compared in Fig.7 with room temperature turn-off waveforms at a current of 2700A. Furthermore, the spread of turn-off surge voltage is suppressed by the presence of R_{gm}, Fig. 7(b). Total Turn-on/off gate resistances are decided to keep transient voltage within RBSOA under application-specific overvoltage and overcurrent conditions upon hard-switching. In Fig. 7, DC-link voltage V_{DC} is limited to 700V due to parasitic inductance associated with spacers on module terminals surrounding which current is measured with current transformer.

B. Switching Waveforms at Normal Operation Conditions

Typical turn-on waveforms and turn-off waver forms are shown in Fig. 8 and Fig. 9, respectively. The load current is 1900A. V_{DC} is 700V. In Fig. 8, 9, the modules have a 0.36V spread in V_{th} and a 0.28V spread in V_{on} at room temperature.

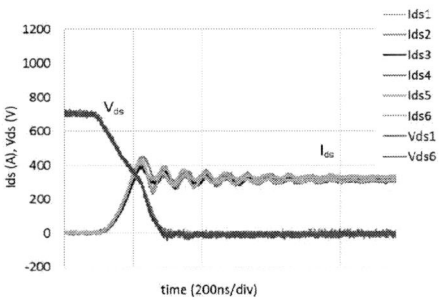

Fig. 8. Typical turn-on waveforms at a current of 1900A and V_{DC}=700V, room temperature.

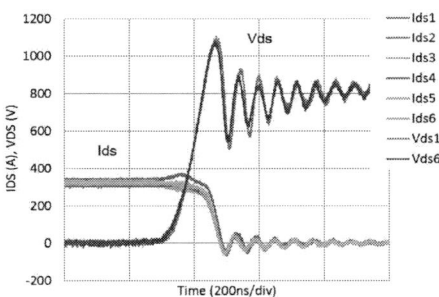

Fig. 9. Typical turn-off waveforms at a current of 1900A and V_{DC}=700V, room temperature.

The current imbalance ratio of each module is defined as

$$\beta \equiv (I_M - I_{av})/I_{av} \qquad (1)$$

β can be further classified as β_{SS} in the steady state and as β_D during switching transition. Variance in V_{th} and V_{on}, current imbalance ratios are tabulated in Table II for turn-off waveform shown in Fig. 9. Maximum β_{SS} is 5.55%, and maximum β_D is below 12%, which could be considered as a good result.

TABLE II
VARIANCE IN MODULE PARAMETERS AND CURRENT IMBALANCE

Module position	1	2	3	4	5	6
ΔVth (V)	-0.26	0.06	0.06	0.05	-0.01	0.10
ΔVon (V)	-0.21	-0.01	0.05	0.07	0.07	0.03
β_{SS}(%)	5.55	0.27	-3.48	-4.20	2.83	-0.99
β_D (%)	11.65	-1.64	-5.39	-5.48	1.47	-0.61

The di/dt during turn-on is below 10kA/μs, and dV/dt during turn-off is below 18kV/μs for current up to rated current (total 320×6=1920A) and for temperature up to 175°C.

C. Dynamic Behavior at Short-Circuit Condition

Short circuit failures can be categorized either as a hard switching failure (HSF) or a fault under load failure(FUL) according to their occurrence timing relative to gate driving signal. HSF refers to devices'

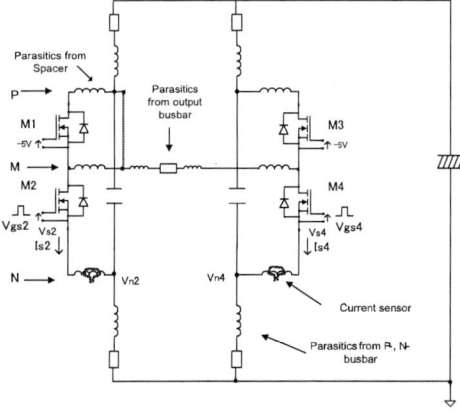

Fig. 10. Simplified circuit diagram with short-circuit in the upper arm of module at position 1 in Fig. 1.

turning on into a short circuit, while FUL refers to an occurrence of circuit shorting during fully-on period of the devices. Short-circuit could occur either at switch arms or a load. The severer one is arm short-circuit, in which dc-link voltage is mainly sustained by the other devices in the same phase-leg, with less parasitic inductance transiently sustaining a portion of the voltage.

As arm short-circuit is more prone to occur during assembling, and HSF type experiment is more controllable. Dynamic balance of current and voltage is investigated by shorting the upper arm of module at position 1 in Fig. 1. A more detailed circuit diagram with various parasitic inductances and resistances is shown in Fig. 10 (two modules in parallel for simplicity).

A set of representative waveforms is shown in Fig. 11 for a group of modules having a spread of 0.3V in V_{th} and a spread of 0.15V in V_{on} at room temperature. Among which the module with lowest V_{th} and V_{on} is placed at position 1 in Fig. 1. Short-circuit time is set as 2μs. The current at position 1 I_{ds1} was measured with a 3000A Rogowski probe. Other currents were measured with current transformers.

The 2018 International Power Electronics Conference

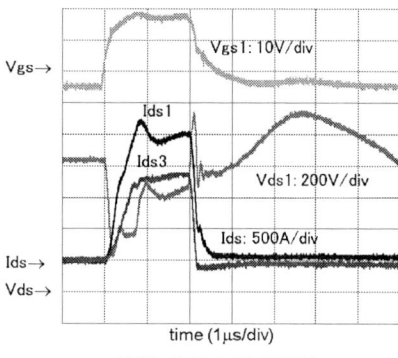

(a) Vgs1, Ids1, Ids3, Vds1

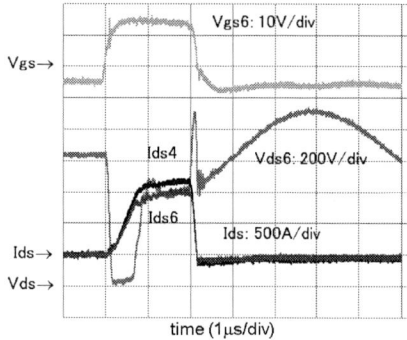

(b) Vgs6, Ids4, Ids6, Vds6

Fig. 11. Waveforms upon a 2μs short-circuit of the upper arm of module at position 1 in Fig. 1 (M1 in Fig. 10), V_{DC}=830V, room temperature.

With the existence of parasitic inductance of the output bus-bar as shown in Fig. 10, the waveforms of modules away from position 1 bears the characteristics of inductive switching. Peak Ids1 reaches 2300A (7 times of rated current I_0) at room temperature, and 2500A at 175°C (Fig. 12). These current levels are similar to reports on different modules relative to their rated current [10]. The peak current at position 6 I_{ds6}, which is farthest away from position 1, reaches only 1070A (3.4×I_0) at room temperature and 1100A at 175°C. Large peak current at 175°C is considered resulting from a quicker initial turn-on due to a V_{th} lowering of 0.8V from room temperature to 175°C.

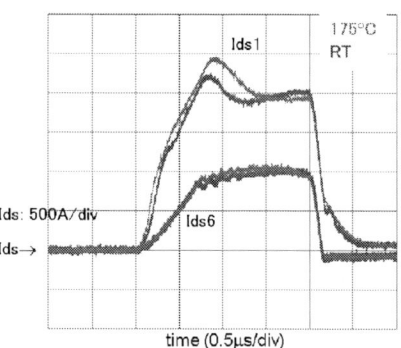

Fig. 12. Comparison of short-circuit currents at RT and 175°C.

There are other two prominent features in Fig. 11, one is gate voltage difference between position 1 and 6, and another is a slugging turn-off behavior for both V_{gs} and I_{ds} at position 1.

The gate voltage difference can be explained by gate charge behavior due to difference in miller capacitance between short-circuit and inductive load conditions [11-12]. In the period from gate voltage rising to the timing that I_{ds1} and I_{ds6} reach their separate peaks, the lower arm switch at position 1 is in a direct short-circuit situation that sustains a higher V_{ds} and owns a shorter or zero miller period, while its counterpart at position 6 is in a situation of inductive turn-on with a normal miller period, due to parasitic inductance on the output bar (Fig. 10). In the same period, V_{gs} at positon 1 can rise to a level higher than that at positon 6, and kept higher in accordance with its current level afterwards.

The slugging turn-off behavior can be explained with current circulation in gate driver's ground wires. During the initial phase of turn-off, current transition is more abrupt at position 1 than at other positions, 10.8kA/μs at position 1 versus 7.75kA/μs at position 6. Parasitic inductance associated with module's N terminal could pull down the potential of auxiliary source at position 1 more lower due to its larger negative di/dt (V_{s2}<V_{s4} in Fig.10). Current circulation could be initialized from auxiliary source terminals at other five positions to position 1 through gate driver's ground wires (Fig. 6), and appeared as source tail current even after channel current turned off.

For V_{DC} of 600-800V, the minimum V_{ds} during short-circuit is at least 10V higher than the Von after normal inductive turn-on even to an over-current(OC) level which is 1.5 times of total module rating, Fig. 13. This allows a voltage window for adopting conventional Desat short circuit protection methods [13-14].

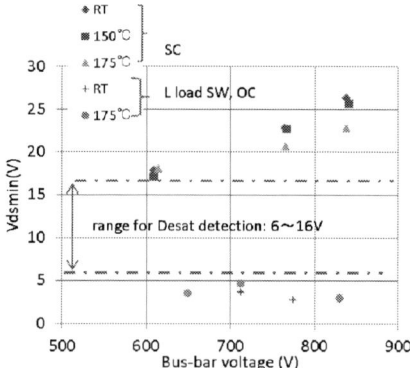

Fig. 13. Drain-source voltage difference between normal turn-on and short circuit situations.

A set of demonstrative waveforms is shown in Fig. 14, obtained with a gate driver equipped with Desat circuitry. Desat diode is connected to module's output terminal M at position 6, short-circuit period is set as 3.5μs, and the mask time of Desat circuit is set as 2μs. The current are safely turned off after the mask period.

3393

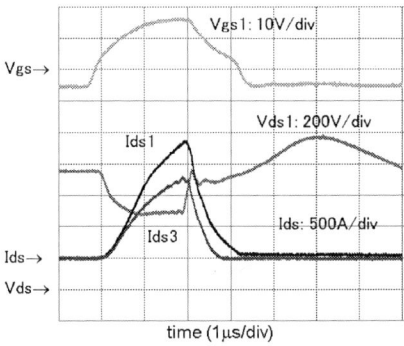

Fig. 14. A demonstration for the effectiveness of a conventional Desat circuit at room temperature, V_{DC}=750V.

D. Effect on Inverter Loss Reduction

Compared with Si-IGBT module, the switching loss of all-SiC module is reduced by 62% for turn-on switching and 74% for turn-off switching, Fig. 15. For a 400V/400kW 3-phase inverter, its semiconductor loss is reduced by 20% even if its PWM carrier frequency is increased from 4 kHz to 15 kHz, Fig. 16.

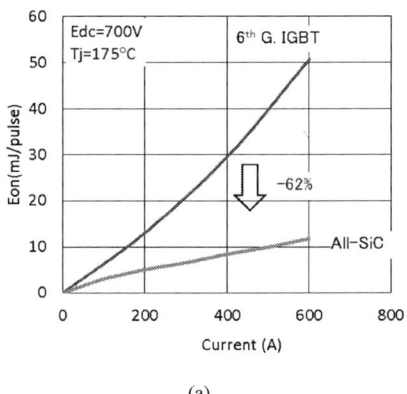

(a)

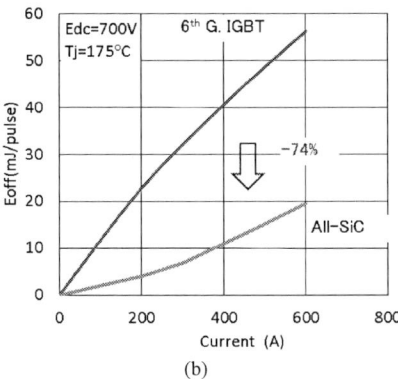

(b)

Fig. 15. Switching loss comparison between all-SiC and 6th generation IGBT modules (a) turn-on (b) turn-off.

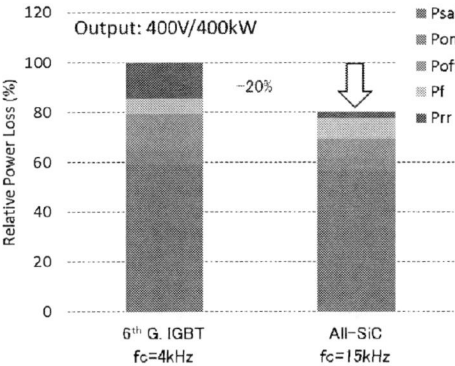

Fig. 16. Semiconductor loss comparison for a 400V/400kW 3-phase inverter.

A prototype 400V/400kW three-phase inverter is shown in Fig. 18. Along with double side cooling, its volume reaches half of the conventional one with IGBT modules, and its efficiency is improved by 1.3%.

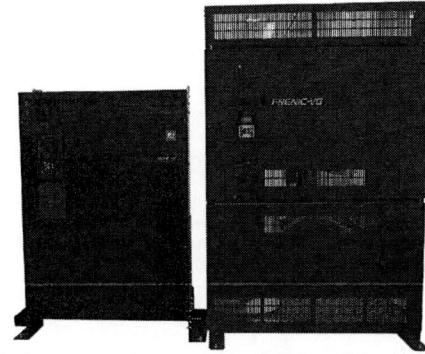

Fig. 18: A prototype inverter with all-SiC modules (left) along a conventional one with IGBT modules (right).

VI. CONCLUSION

We developed a 400V/400kW power stack by paralleling six 320A/1200V half-bridge all-SiC modules. Investigations on structural imbalance of bus bar, switching dynamics in normal operation condition and arm short circuit condition, were conducted to ensure current imbalance less than 12% during both steady-state conduction and dynamic switching. Short-circuit protection is demonstrated possible by conventional Desat method.

Compared with inverter with IGBTs modules, total semiconductor loss is reduced by 20% while the switching frequency increased more than twice. Along with double side cooling, a prototype inverter achieves 50% volume reduction and 1.3% efficiency improvement.

REFERENCES

[1] J. Fabre, P. Ladoux, "Parallel Connection of 1200-V/100-A SiC-MOSFET Half-Bridge Modules", IEEE Trans. on Industrial Applications, Vol.52, No. 2, pp. 1669-1676, 2016.

[2] J. Colmenares, D. Peftitsis, J. Rabkowski et al., "High-efficiency 312-kVA three-Phase inverter using parallel connection of Silicon Carbide MOSFET power modules", IEEE Trans. on Industrial Applications, Vol.51, No. 6, pp. 4664-4676, 2015.

[3] Ruxi Wang, Juan Sabate, Fengfeng Tao et al., "H-Bridge Building Block with SiC Power MOSFETs for Pulsed Power Applications", In IEEE Energy Conversion Congress and Exposition (ECCE-US), 2016.

[4] H. Takubo, D. H. Lu, and Y. Suzuki, "Development of a Large Capacity Power Stack by Paralleling All-SiC Modules", IEEJ Industry Applications Society Conference, pp. I-583-584, 2017.

[5] M. Chounabayashi, Y. Otomo, T. Karasawa, "All-SiC 2-in-1 Module", Fuji Electric Review. vol. 89, no.4. pp.238-241, 2016.

[6] Y. Iwasaki, M. Chounabayashi, M. Nakazawa et al., "All-SiC Module with 1st Generation Trench Gate SiC MOSFETs and New Concept Package", Proceeding of PCIM Europe 2017. pp.651-657.

[7] G. Wang, J. Mookken, J. Rice et al., "Dynamic and Static Behavior of Packaged Silicon Carbide MOSFETs in Paralleled Applications", in IEEE 29th Applied Power Electronics Conference and Exposition (APEC), pp1478-1483, 2014.

[8] A. Callegaro, J. Guo, M. Eull, et al., "Bus bar design for high power inverters", IEEE Trans. on Power Electronics, accepted for publication, 2017.

[9] J. B. Forsythe, "Paralleling of Power MOSFETs for Higher Power Output", IEEE-IAS Conference Record (1981).

[10] P. Reigosa, H. Luo, F. Iannuzzo et al., "Investigation on the short circuit safe operation area of SiC MOSFET power modules", IEEE Energy Conversion Congress and Exposition (ECCE-US), 2016.

[11] K. Yuasa, S. Nakamichi, and I. Omura, "Ultra High Speed Short Circuit Protection for IGBT with Gate Charge Sensing", Proc. of the 22th International Symposium on Power Semiconductor Devices IC's (ISPSD), pp. 37-40, 2010.

[12] R. Bayerer, and S. Suleri, "Low Impedance Gate Drive for full Control of Voltage Controlled Power Devices", Proc. of the 26th International Symposium on Power Semiconductor Devices IC's(ISPSD), pp.438-441, 2014.

[13] S. Musumeci, R. Pagano, A. Raciti et al., "A new gate circuit performing fault protections of IGBTs during short circuit transients," Conference Record of the 2002 IEEE Industry Applications Conference. 37th IAS Annual Meeting, 2002, pp. 2614-2621, vol.4.

[14] T. Bertelshofer, A. März and M. M. Bakran, "A temperature compensated overcurrent and short-circuit detection method for SiC MOSFET modules," 2017 19th European Conference on Power Electronics and Applications (EPE'17 ECCE Europe), 2017.

3.3kV All-SiC Module for Electric Distribution Equipment

Ryohei Takayanagi, Katsumi Taniguchi, Satoshi Kaneko,
Naoyuki Kanai, Keishirou Kumada, Motohito Hori, Yoshinari Ikeda,
Kouji Maruyama and Itsuo Kawamura

Fuji Electric Co., Ltd.
4-18-1 Tsukama, Matsumoto, Nagano, 390-0821 Japan
takayanagi-ryouhei@fujielectric.com

Abstract -- We developed 3.3kV All-SiC power module on which SiC-MOSFET and SiC-SBD were applied for electric distribution equipment. This power module has characteristic structure of pin-connection and resin-molded package. These features lead to long-term reliability. In addition, low static loss and low switching loss is achieved by using SiC devices. Total switching loss of All-SiC module in the inverter driving is 64% lower than a module which uses Si-IGBT and Si-FWD.

Index Terms-- *All-SiC module, inverter loss, module structure, electric distribution equipment, power module, reliability*

I. INTRODUCTION

To prevent global warming, reduction in greenhouse gas emission, especially CO_2, is required. Active utilization of the renewable energy is desired, and then it is necessary to use power electronics equipment in order to convert the generated electric energy efficiently. Power semiconductor devices play an important role in power conversion for power electronics equipment. Mainstream Si (silicon) devices have been improved; however power devices generally have the trade-off relationship between the breakdown voltage and the resistivity and the performance of Si devices is already approaching the limit of its physical properties. To realize the devices with lower resistance, in other words lower energy loss, much attention has been paid to SiC (silicon carbide), which is expected as the next-generation semiconductor material. SiC has the great physical properties for power devices with high withstand voltage; especially has around 10 times higher electric field strength than Si [1]. Thus SiC devices are expected to achieve both high withstand voltage and low resistance, and allow power electronics equipment to be more efficient, compact and lightweight.

Nowadays we are faced with many technical problems about electric distribution because of the increase of decentralized power generation systems (mainly solar power generation systems). Specifically, the production of excess power and lack of ability to adjust frequency may cause voltage rise or frequency variation in power lines. To deal with them, next-generation decentralized

electric distribution equipment with high efficient SiC devices is currently being developed.

II. SiC POWER SEMICONDUCTOR MODULE FOR ELECTRIC DISTRIBUTION EQUIPMENT

In Japan, electric distribution equipment should be introduced to 6.6 kV electric distribution systems, because many residential houses have solar power generators and the voltage rise by them cause reverse power flow in 6.6 kV power line. In order to install them in a residential area, it is demanded that the equipment is compact, self-air-cooled and lightweight in order to be mounted on utility poles. In addition, it is required to utilize single existing utility poles for reduction of the construction cost. However, downsizing or weight reduction is difficult for conventional equipment with Si devices because it has larger loss and then needs a larger heat sink, and it need to be mounted on double poles [2], [3]. Therefore high cost and lack of installation space have prevented the equipment from wide introduction. Application of the newly developed All-SiC semiconductor power module enables to realize small and light electric distribution equipment, and this equipment can be mounted on a single utility pole. It can also be operated at high frequency of 13 kHz or more, which is inaudible region. The operation at the inaudible frequency is important in order to avoid noise pollution

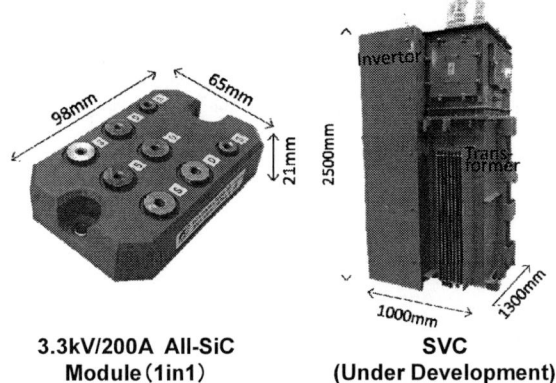

3.3kV/200A All-SiC Module (1in1)   **SVC (Under Development)**

Fig. 1. The developed 3.3 kV All-SiC module mounted with electric distribution equipment (SVC)

because it can be set up in residential areas. Fig. 1 shows the appearance of the developed 3.3 kV All-SiC module applied with electric distribution equipment (SVC: Static Var Compensator). This paper describes the structure and characteristics of the 3.3 kV All-SiC 1in1 module developed for the electric distribution equipment.

The All-SiC module contains SiC-MOSFET (Metal Oxide Semiconductor Field Effect Transistor) and SiC-SBD (Schottky Barrier Diode). In conventional 3.3kV class module, Si-IGBT (Insulated Gate Bipolar Transistor) and Si-PND (P-N Diode, often referred as FWD: Free-Wheeling Diode) has been often applied, but IGBT and PND have the characteristics of high switching loss because they are bipolar devices. On the other hand, SiC MOSFET and SBD are unipolar devices; therefore they can lead to low loss in high-voltage application.

III. MODULE STRUCTURE

Fig. 2 shows a comparison of the schematic cross-sectional structures of the conventional module and the the developed module. The structure of the 3.3 kV All-SiC module differs significantly from that of Si-IGBT modules [4]-[6]. It has mainly two structual features, pin-wiring structure and epoxy resin molding.

In the developed structure, semiconductor devices and power board were connected with copper pins in place of conventional aluminum wire bonding. More current can be applied at pins than Al wire, and the bonding area on the substrate is not necessary for 3-dimensional layout by using the power board. For this reason, pin-wiring structure allows a large current flow and enables high-density assembly of SiC devices. Moreover, this structure contributes to enhancing long-term reliability since lift-off hardly occurs in pins connected by solder, unlike in wire bonding.

Epoxy resin is applied in place of conventional silicone gel as the module internal sealing material. The

structure without the metal base and the resin case contributes to weight reduction. Epoxy resin has higher breakdown voltage than the gel, and then it is expected that higher long-term reliability is ensured at the point of insulation performance [7]. Epoxy resin can prevent the degradation of solder joint or insulating characteristics at high temperature operation. Epoxy resin, however, may cause cracks at the insulated substrate by bending stress when temperature is changed for thermosetting of resin. AlN (aluminum nitride) is the representative material for the insulating substrate of high capacity power module, because it has great thermal conductivity. Meanwhile, AlN dose not have high bending strength enough to withstand the stress by resin molding. It is found that high-strength Si_3N_4 (silicon nitride) with thick copper plates is suitable for the insulating substrate of All-SiC module. This material can withstand the bending stress from the resin molding, and thick copper plates enhance thermal conductivity.

These features lead to higher reliability. Fig. 3 shows the thermal cycling capacity. In the thermal cycling test, thermal stress from -40 °C to 150 °C is applied to the modules. We can decide whether the module is degraded or not by partial discharge test, in which high voltage is applied between all terminals (gate, source, and drain terminals are short-circuited) and back copper (outside of the module). It is decided that the upper limit level of

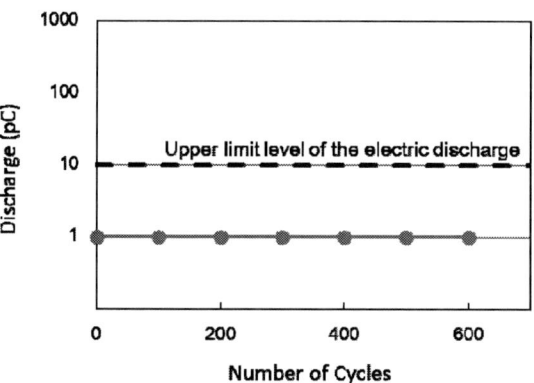

Fig. 3. Results of the electric charge amount in the partial discharge test after the thermal cycling test

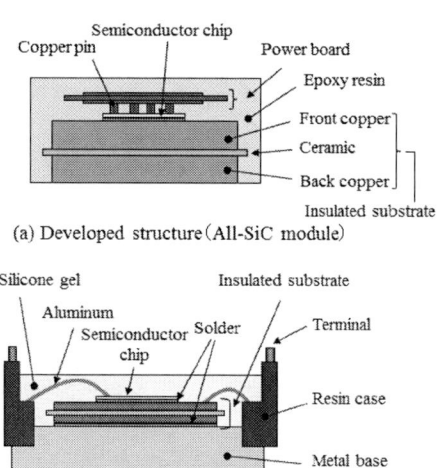

(a) Developed structure (All-SiC module)

(b) Conventional structure (Si-IGBT module)

Fig. 2. Comparison of the schematic structure of the cross section of the existing module

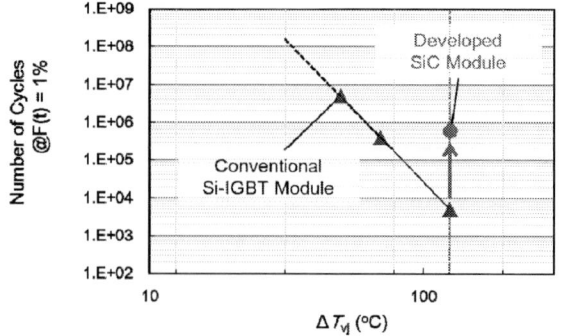

Fig. 4. Reliability of $\triangle T_{vj}$ power cycling test of the All-SiC module and conventional Si-IGBT module

3397

the electric discharge is 10 pC by IEC standard (IEC 60664-1) [8]. After 600 cycles progress, the result of partial discharge test did not change from initial characteristics, which is lower than a detection limit. Thus we concluded that no degradation of insulating performance occurs and the module satisfies the standard after the thermal cycles.

Fig. 4 shows the results of the $\triangle T_{vj}$ (T_{vj} : virtual junction temperature) power cycling test for the All-SiC module and conventional Si-IGBT module. The current stress is applied continually in this power cycling test. All-SiC module has greater torelance of power cycling test, because the rigid resin molding holds internal structure stable and threfore lift-off does not happened. The All-SiC module has 120 times higher power cycle capability than a conventional Si-IGBT module at $\triangle T_{vj}$ = 150 °C.

The required lifetime of electric distribution equipment like SVC is at least 20 years, twice as ordinary industrial equipment. All-SiC module has higher reliability as mentioned above, and therefore is suitable for electric distribution equipment.

IV. ELECTRIC CHARACTERISTICS

Current capacity of the newly developed 3.3 kV All-SiC 1in1 module is 200 A. We fabricated a Si-IGBT module of the same current capacity class, in which SiC MOSFET and SiC SBD chips were replaced with Si-IGBT and Si-FWD. Then the electric characteristics of these modules were compared in terms of energy loss.

A) I-V characteristics

I-V characteristics determine static loss of the power module. Fig. 5 shows I-V characteristics of the All-SiC module and the conventional Si-IGBT module in ON-state (V_{GS} = +15V). As shown in Fig. 5, V_{DS} at the rated current (200A), meaning ON-state resistance, is 48 % smaller in the All-SiC module than the Si-IGBT module at 25 °C and 20% smaller at 150 °C. In addition, the reduction ratio is greater at lower current. IGBT is a bipolar device, so it has the "knee" voltage, the threshold voltage from which the drain current increases. On the

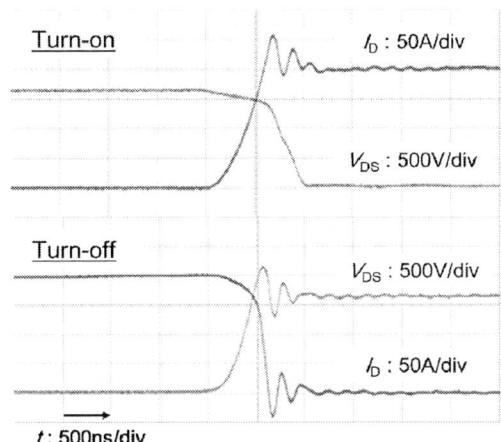

Fig. 6. Switching waveform of All-SiC module at turn-on and turn-off

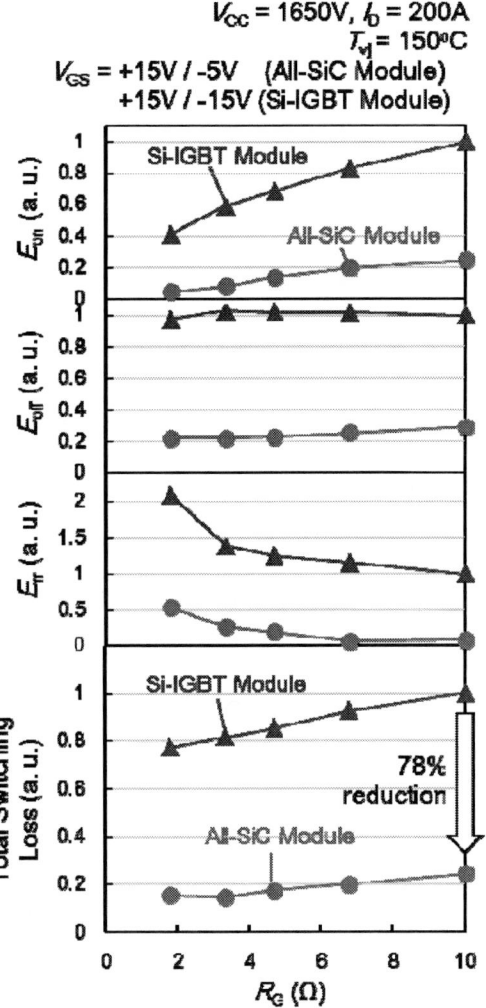

Fig. 7 The comparison of the R_G dependence of E_{on}, E_{off}, E_{rr} and total switching loss between All-SiC module and Si-IGBT module. All data are normalized by each datum of Si-IGBT module at R_G = 10 Ω.

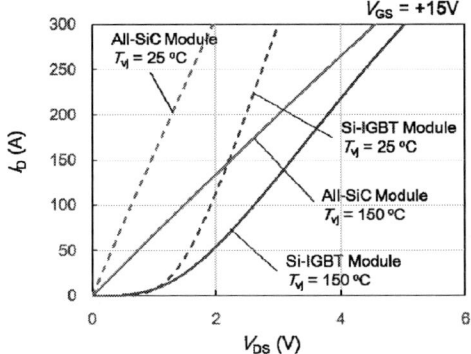

Fig. 5. I-V characteristics at T_{vj} = 25°C and T_{vj} = 150°C for the All-SiC module and Si-IGBT module at ON-state

3398

other hand, MOSFET has the linear *I-V* curve. This is the reason that the SiC-MOSFET resistance is smaller than that of Si-IGBT at relatively lower current. Thus, it is found that reduction of static loss is achieved by using All-SiC module.

B) Switching characteristics

Fig. 6 shows the switching waveform of the All-SiC module at turn-on and turn-off. We achieved faster switching and smaller tail current in All-SiC module by using unipolar devices (SiC MOSFET and SBD). These features contribute to low switching loss because switching loss is culculated by the multipulication of I_D and V_{DS}. Switching loss generates at the timing of turn-on, turn-off, and reverce recovery in All-SiC module and Si-IGBT module.

Fig.7 shows R_G dependence of E_{on}, E_{off}, E_{rr} and total switching loss of All-SiC module and Si-IGBT module. Here, E_{on} is the loss at turn-on, E_{off} is that of turn-off and E_{rr} is that of reverse recovery. The total switching loss is the sumention of them. Note that the loss of Si-IGBT module at $R_G = 10\ \Omega$ is normalized to 1 in each graph of Fig. 7. E_{on} increased as R_G increases, and E_{rr} indicated the opposite tendency. E_{off} had small R_G dependence. Then the total switching loss increases slightly as R_G increases. These tendencies are observed in both All-SiC module and Si-IGBT module. In all R_G, E_{on}, E_{off} and E_{rr} were reduced significantly by using SiC devices. The All-SiC module achieved 78% of reduction of total switching loss at $R_G = 10\ \Omega$ compared to the conventional Si-IGBT module.

C) Loss simulation during inverter operation

Fig. 8 shows the circuit diagram of the 3-level inverter for the SVC we are developing for electric distribution equipment. Here, All-SiC clamp diode module, which has only SiC-SBD chips, is used as a clamp diode. The circuit of Si-IGBT module is constituted by Si-IGBT module and Si-PND clamp diode module, instead of All-SiC module and All-SiC clamp diode module shown in Fig. 8. Loss simulation of 3-level inverter operation was performed for All-SiC and Si-IGBT modules. The

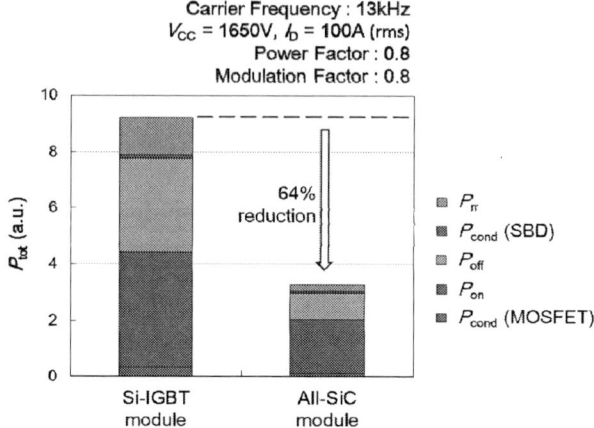

Fig. 9. Result of loss simulation during inverter operation at the usage conditions of the SVC 3-level inverter

simulation condition of inverter operation is set upon the ordinary driving condition of SVC equipment. The carrier fleqency is 13 kHz, which is higher than that of general driving condition for Si-IGBT module. $V_{cc} = 1650V$, and RMS (Root Mean Square) of I_D is 100A. Power factor and modulation factor are set for 0.8.

The results of simulation are shown in Fig. 9. The switching losses (P_{on}, P_{off} and P_{rr}) occupies almost all of the total power dissipation P_{tot}. In Si-IGBT module, 94% of the total loss is due to the switching losses. Because of its low switching losses, the All-SiC module performed 64% reduction of P_{tot}. This low-loss characteristics allow the cooling system of the equipment to be self-air-cooling, which is simple, small and lightweight. Then it contribute to downsizing of SVC, and we achieved to develop the compact and lightweight SVC equipment enough to be mounted on a single utility pole.

V. CONCLUSION

This paper describes the structural features and the electrical characteristics of the 3.3 kV All-SiC 1in1 module developed for SVC, a kind of electric distribution equipment.

The All-SiC module we developed has the structural features. Copper pins are used for electric connection. This structure enables high-density assembly of SiC devices, so the compact module is achieved. Molding with epoxy resin prevents degradation of solder joint and insulation performance at high temperature. High-strength Si_3N_4 is employed for insulated substrate to enhance resistance against bending stress due to epoxy resin molding. These features provide the module with high reliability.

This All-SiC module takes advantage of low loss and can drive at high-frequency because of the great electric characteristics of SiC MOSFET and SiC SBD. Both the static loss and the switching loss are reduced significantly, and 64% reduction of the total power dissipation in inverter operation is achieved compared to Si-IGBT

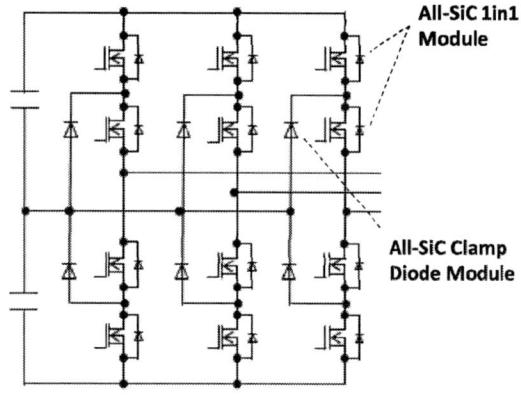

Fig. 8. The circuit diagram of the 3-level inverter using All-SiC module

module. Thus the All-SiC module contributes to development of compact and lightweight electric distribution equipment which has a simple self-air-cooling system, and this SVC equipment can be mounted on a single utility pole.

It is expected that the decentralized electric distribution systems will spread widely by using SVC equipment with All- SiC module and the low-carbon society will be realized.

VI. ACKNOWLEDGMENT

These results were obtained as a result of the assistance program "Project to verify decentralized energy next-generation power grid construction" funded by the New Energy and Industrial Technology Development Organization (NEDO).

REFERENCES

[1] A. O. Konstantinov, Q. Wahab, N. Nordell and U. Lindefelt," Study of avalanche breakdown and impact ionization in 4H silicon carbide", J. ELECTRON. MATER, vol. 27, pp. 335-341, Apr. 1998

[2] Ohinata, T. et. al., "A study on cooperative operation of the Magnetic Flux Contorol Type Variable Inductor Voltage Control Integration System", *In Proc. the Twenty-fifth Annual Conference of Power and Energy Society, IEE of Japan*, 2014, pp. 3-3-19, 3-3-20. (in Japan)

[3] Ohinata, T et. al., "A study on cooperative operation of the Magnetic Flux Contorol Type Variable Inductor Voltage Control Integration System (2)", *In Proc. the Twenty-sixes Annual Conference of Power and Energy Society. IEE of Japan*, 2015, pp. 4-3-5, 4-3-6. (in Japan)

[4] Horio, M. et. al., "New Power Module Structure with Low Thermal Impedance and High Reliability for SiC Device", *In Proc. PCIM*, 2011, pp.229-234.

[5] Ikeda, Y. et. al., "Investigation on Wirebond-less Power Module Structure with High-density Packaging and High Reliability", *In Proc. International Symposium on Power Semiconductor Devices and ICs*, 2011, pp.272-275

[6] Hinata, Y. et. al., "Full SiC Power Module with Advanced Structure and its Solar Inverter Application", Presented at The Applied Power Electronics Conference and Exposition, 2012.

[7] Hori, M. et. al, "Enhanced Breakdown Voltage for All-SiC Modules", *In Proc. IEEE CPMT Symposium Japan*, 2017, pp.127-130

[8] *Insulation coordination for equipment within low-voltage systems - Part 1: Principles, requirements and tests*, IEC 60664-1:2007

The 2018 International Power Electronics Conference

Present Status of SiC based Power Converters and Gate Drivers – A Review

Abhijit Choudhury, *Member, IEEE*

Experimental Power Grid Centre (EPGC), Institute of Chemical and Engineering Sciences (ICES)
Agency for Science Technology and Research (A*STAR), Singapore
Email : choudhurya@epgc.a-star.edu.sg

Abstract–**Wide band gap (WBG) based semiconductor devices are considered to be the next generation of power electronic devices due to their higher switching frequency of operation with reduced device losses compared to their strong competitor Si based switches. Among the WBG devices (SiC, GaN), SiCs (Silicon Carbide) are considered to be the most preferable choice for the medium and higher power levels due to their reliability in physical construction. The major part of the loss reduction in SiC devices come from the lower on state resistance and the reduction in the reverse recovery losses associated with the antiparallel diodes. The total estimated loss reduction can be around 70% in SiC based converters compared to the Si based ones. However, there are issues related to driving these switches at safer operating area (SOA) with reduced switching losses at higher switching frequencies (20 kHz – 100 kHz), which also leads to a significant EMI/EMC issues in comparison to Si device (Operates at 20 kHz maximum). Higher switching frequency of operation further increases the effect of stray capacitances and associated oscillation which might not be acceptable for stable system operation. This paper will present different applications of SiC based inverters, challenges associated with high switching frequency SiC driver design and the problem with commercially available drivers.**

Index Terms—SiC Device; Gate Drive; Cross Talk.

I. INTRODUCTION

POWER converters/inverters are the common link to convert the AC/DC power to the DC/AC power with desired voltage level. The total investment on this types of systems are keep on increasing nowadays due to the penetration of different renewable energy based power sources like solar, wind, electric vehicles to the power grid to reduce the carbon emission. Even in micro grids, inverters are also used to interconnect different power sources, which need a lot of investments on the converter side. Fig. 1 shows the interconnection of different power sources (From different renewable energy resources) to the grid by power inverters. The grid network will then be connected to the industries and domestic household loads through transmission and distribution lines.

To connect solar system to grid a DC-DC converter will initially boost the voltage to the desired level and will also make sure that the maximum power can be extracted from the available solar energy. It will be then converted to a AC power with a DC-AC converter. In case of wind energy, the generated AC power will be converted to DC and then it will be connected to the grid system through a DC-AC converter. To take care of the intermittent power fluctuations related to the renewable energy systems, an energy storage is essential to supply power to the grid when required power levels can't be meet from the available wind or solar power and to store the excess power whenever it is not required by the load.

On the load side, major parts of the industrial loads are electrical machines. In this type of system grid power is supplied to the machine through inverters to improve the efficiency and power factor. For household applications starting from mobile or laptop charger to drive LED/ LCD TV or air-condition system, all will need some sort of power converter systems.

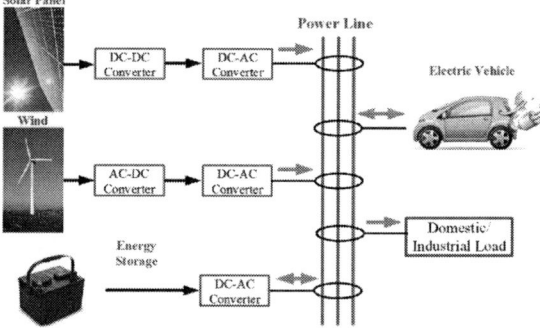

Fig. 1. Renewable energy integration to power system

Another major application area for power converters nowadays are electric vehicles (EV). In EVs the high power battery (Li/NiCd) is charged from grid power through a battery charger and then it is connected to the drive system with a three phase inverter. From all of the above mentioned applications it is clear that power electronic converters are one of the most critical elements for interconnecting different available resources to meet the consumer's requirements. However, the main challenges associated with the converters are their efficiency, size, reliability and the overall system cost. The conventional inverters use silicon (Si) based power switches,

3401

which have their limitation in terms of efficiency and switching frequency of operation. All the grid connected systems needs some sort of isolation, which is generally provided by transformer. If the system operates at higher switching frequency, the magnetic component size can be reduced, which will reduce the system size and cost associated with it. With the Si based IGBTs it is not possible and also there are limitations in the maximum voltage that these high power switches can block at medium and high power applications. The switching losses associated with the Si based inverters are also considerable which further increases the size of cooling system. Hence, where system needs higher switching frequencies to reduce the magnetic component size for reduced foot print area, Si based converter is not the best solution. There are quite a few soft switching strategies available in the literature to reduce the switching losses [1], however they all have some limitations in term of system performance range, reliability and cost. Different multilevel inverter topologies are also available in the literature for medium and high power applications to reduce the device voltage stress; however it will increase the system cost and introduce other control challenges [2].

Due to the advent of more efficient power semiconductor devices like WBD (Wide band gap device), it might be possible to solve the challenges associated with the converter switching frequency and at the same time to have a higher power density. Among the commercially available WBD devices, SiCs (Silicon Carbide) are the most preferable choice because of their higher power rating, faster switching frequency, significantly lower switching losses and ability to withstand higher junction temperature compared to the Si based devices.

This paper is organised as follows. Section II shows the physical property of wide band gap devices (WBD) and their associated driving challenges. Section III explains about different proposed gate driver topologies, their advantages and disadvantages with different commercially available gate drivers.

II. WBG DEVICES PHYSICAL PROPERTY AND ASSOCIATED CHALLENGES

Table I shows a performance comparison between the Si and SiC based device characteristics. Because of the significantly higher break down voltage compared to the Si based switches, SiC can withstand higher DC-link voltage. It also helps to reduce the footprint size for similar rated power converters. In addition higher band gap helps to operate at higher junction temperature. At present the commercially available SiC devices can operate up to 150°C - 175°C. For this reason the cooling system requirement or size will be greatly reduced compared to the Si based devices and more power can be push through the power switches compared to similar rated Si IGBTs. It helps to save the additional space

and cost on cooling equipment. As, these devices can operate at higher switching frequencies, the temperature ripple is low, which in turn help further improving the device reliability. Junction capacitance of SiC MOSFET is also lower than that of the Si IGBT, resulting in a faster switching speed.

Table I: Physical property comparison between SiC and Si based semiconductor devices

	Bandgap Energy (eV)	Dielectric Constant (dimension)	Electron Mobility (cm^2/Vs)	Break Down Electric Field (10^6 V/cm)
SiC	3.25	9.7	1140	3
Si	1.1	11.8	1500	0.3

Table II: Device parameter comparison for SiC MOSFET and Si based IGBT and MOSFET

Parameter	SiC MOSFET	Si IGBT	Si MOSFET
Name	C2M0040120D	IHW30N120R3	SPP17N80C3
Breakdown Voltage (V)	1200	1200	800
Conduction current (at 25°C) (A)	60	60	51
$R_{ds,on}$ (mΩ)	40	V_{cesat} = 2.2 V	290
Fall Time (ns)	60	365	80
Turn off Loss (mJ)	0.4	1.47	Not Available
Input Capacitance (pF)	1893	2038	2320
V_{th}	2 – 4 V	5.1 - 6.4 V	2.1 – 3.9 V

Table II shows the comparison of SiC, Si MOSFET and Si IGBT. From the details it is clear that SiC will show lower on state resistance, hence less conduction loss, has shorter turn off time, and input capacitance. All these properties will help to switch at faster switching frequency and with reduced loss (Conduction and switching). Moreover, SiC is a majority carrier device, which does not have the 0.7 V on state voltage drop and also reverse recovery charges from the antiparallel diodes. These two properties help to reduce significantly the switching losses. However, SiC MOSFETs on state resistance ($R_{ds,on}$) is higher than the normal MOSFET. Hence, to keep it low its gate driver voltage needs to keep high compared to the conventional MOSFET. Furthermore, gate resistance for SiC MOSFET is higher than Si MOSFET. Hence, a small external resistance can be used, with a surge current taken into consideration.

Due to higher cost of the SiC switches compared to Si, the main focus areas of applications are for high switching frequencies (> 20.0 kHz) and high blocking voltage (> 12.0 kV) in hard switch applications. Hybrid packages (Si switch + SiC diode) are good alternative to pure SiC based modules for reduced system cost and high efficiency with lower switching loss. However, the switching speed and the maximum allowable device junction temperature will be limited by the Si device. Fig. 2 shows the three phase inverter loss comparison between SiC, Si and hybrid models [3]. As, expected a pure SiC module has the highest loss reduction compared to the Si based IGBT module (70%). Hence, a three phase two level hard switched inverter can show a similar loss reduction like three-level NPC inverter [4] with comparatively simpler hardware configuration. Significant part of the loss reduction comes from the reduction in antiparallel diode revere recovery loss (E_{rr}).

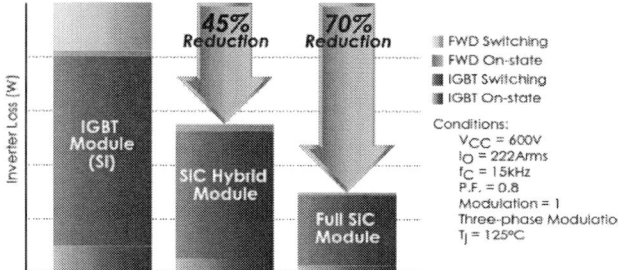

Fig. 2. Total inverter loss comparison for Si, SiC and Hybrid module [3]

Fig. 3 shows more detailed performance difference in between SiC and Si device during turning on operation. Two devices of similar rating of 1.7 kV is considered to compare the switching characteristics. From this figure it is clear that, during short circuit where IGBT enters into ohmic region, which somehow limits the short circuit current, SiC still stays in saturation; which makes the fault current to increase rapidly. Hence, during the short circuit condition, special attention needs to be provided for SiC devices because the current will quickly shoot-up compared to Si based IGBTs and will damage the device. Unlike the Si devices where desaturation is considered as a feedback to know about the fault condition, for SiC it can't be used in similar way, to get a fast feedback on the device state. Moreover, compared to Si switches SiC based power switches have much lower short circuit withstand capability.

There are other critical factors like DC-link voltage and operating temperature which are also considerably involved in estimating the fault clearing time for these types of power switches as shown in Fig. 4. From the results is can be observed that SiC switches from different vendors and generation show different short circuit withstand time (SCWT) with change in case temperature and DC-Link voltage. Hence system specification is very important in terms of driver design, to accurately estimate the SCWT and design accordingly the gate driver. As shown, with increase in junction temperature and applied DC voltage the fault tolerant time goes down.

Hence, while designing a gate driver, device operating temperature and applied voltage also need to be taken care of.

Another issue with SiC devices is that, they are more prone to cross talk at higher switching frequencies, which may create shoot trough fault and damage the devices.

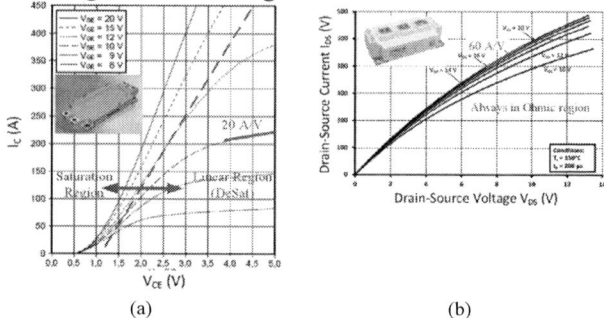

(a)   (b)

Fig. 3. Device output characteristics comparison between (a) Si IGBT (Infineon FF225R17ME4 IGBT, 1.7 kV) and (b) SiC MOSFET (SiC MOSFET CAS300M17BM2, 1.7 kV)

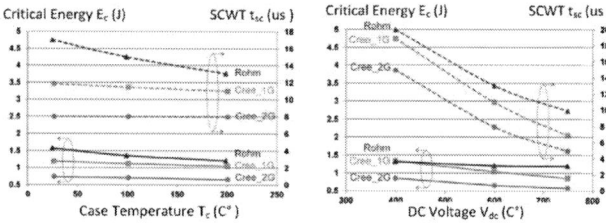

Fig. 4. Comparison of short circuit withstand time (SCWT) based on case temperature and DC bus voltage.

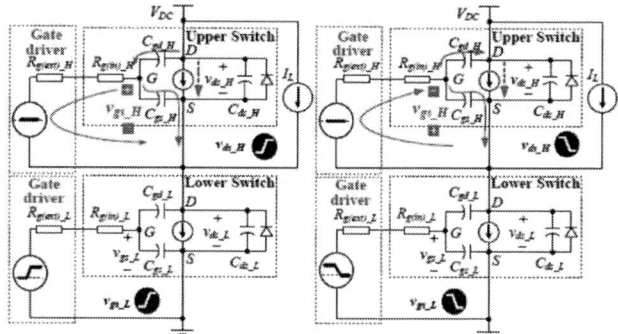

Fig. 5. Mechanism of cross talk in power switches [5]

As shown in Fig. 5, when the lower switch tries to turn on, the voltage across the upper switch (V_{ds_H}) increases. It creates a current through the miller capacitor (C_{gd_H}). If the voltage drop across C_{gs_H} goes higher than the threshold voltage, it will turn on the upper switch and a shoot through current will flow through both the switches. This will in turn damage the power switches. Now when the lower devices will try to turn off, the voltage across the upper switch (V_{ds_H}) will go down, which will force to discharge the two capacitors C_{gd_H} and C_{gs_H}, and

3403

will again overstress the upper switch. For the SiC MOSFETs due to lower threshold voltage and high gate resistance the phenomenon of cross talks are more severe especially at higher switching frequency of operation and requires special gate driver design to mitigate this issue [5].

III. AVAILABLE GATE DRIVING TOPOLOGIES AND THEIR PERFORMANCE COMPARISON

Due to the lower transconductance of the SiC devices compared to the IGBTs, the gate driving voltage is higher (20 - 25V) than IGBTs, which makes the driving circuit requirements for SiCs different from Si based switches. The lower gate-source threshold voltage ($V_{gs,th}$) (~2.5 V) makes the device more prone to the unwanted turn on due to the increased di/dt [6]. This voltage even further goes down with increase in device temperature. As the negative gate diving voltage is limited, a miller clamping circuit is proposed in [7] to neutralize the miller effect. Connecting a capacitor across gate-source terminal might solve the problem; however it will make the system slow.

Table III shows the commercially available SiC modules from three major suppliers and their key specifications. The mentioned part numbers are collected based on their maximum available drain source voltage (V_{ds}) and drain current ratings (I_d). From the comparison study it can be observed that, performance of SiC modules will differ based on the manufacturer and the power ratings. Some of the key performance differences are the driving voltage, switching time and the on state resistance. These are also the key performance parameters for the SiC modules to operate safely with reduced switching losses. Hence, it is difficult to find a universal gate driver which can be a best fit for all SiC modules. Moreover detailed analysis on the safe operating area (SCWT capability), switching time and allowable switching voltage ringing across SiC modules based on gate driving condition needs also to be evaluated based on operating condition beforehand. There are various literature available on the SiC gate driving topologies and few of them are included in this paper. In [8] three different configurations are proposed to keep the SiC device safe from over current. The first one is based on the solid state circuit breaker (SSCB). In this strategy an IGBT/MOSFET based switch is connected in series with SiC device. In case of fault SSCB is going to protect the SiC. This solution needs more power electronics based components and will increase the cost and complexity of the system. Hence it can't be consider as a smart solution. The second proposed one is based on the desaturation technique as generally being used for IGBT protection. As shown in Fig 3, unlike the Si based IGBTs, SiC MOSFETs do not have any clear transition from ohmic to active region especially at lower voltage level. Where in case of IGBT the increased current will be limited by the saturation region to some extent, which increases the sort circuit

withstand time a bit higher, for SiC it is not possible. Hence, desaturation technique can't be used reliably for SiC module if the voltage is not large enough and the data sheet of the device does not show similar kind of V_{CE} vs I_C characteristics as of IGBT. In the third strategy a kelvin source based over current protection scheme is proposed. In this strategy an additional terminal is connected to the SiC source and a parallel RC circuit is connected across it. This passive circuit directly measures the over current without any delay generally proposed by other di/dt based over current protection schemes [9], [10].

Table III: Performance comparison between three different SiC power module manufacturers

Manufacturer	CREE	Infineon	Semikron
Part Number	CAS300m17BM2	FF11mR12w1m1	SKM500Mb120
V_{ds} (V)	1700.0	1200.0	1200.0
I_d (A)	325.0	100.0	541.0
T_j (°C)	150.0	150.0	175.0
V_{gs} (V)	-10.0/+25.0	-10.0/+20.0	-6.0/+22.0
T_{on} (ns)	177.0 @ 900V, 300 A	41.5 @ 600 V, 100 A	320.0 @ 600 V, 250 A
T_{off} (ns)	267.0 @ 900V, 300 A	72.5 @ 600 V, 100 A	465.0 @ 600 V, 250 A
R_{dson} (mΩ)	8.0	11.0	3.75

To reduce the effect of cross talk shown in Fig. 5, a gate assisted circuit is proposed in [5]. In the proposed circuit the gate voltage of the non-operative switch is maintained into a desired level to avoid the cross talk due to the switching of the active switching devices. There are few transformer based gate driving topologies also proposed by researcher in [11]. However, these topologies increase the complexity and cost of the system. Moreover, detailed performance of the system is also not demonstrated.

An active gate driver (AGD) is proposed in [12]. The proposed strategy reduces the gate energy by increasing the gate resistance in the miller plateau region of the gate source voltage. The proposed strategy reduces the overshoot, oscillation and losses.

Compared to the Si based IGBTs, SiC MOSFETs can't withstand lower negative gate voltage (below -10.0 V). When SiC switches operate at higher switching frequencies, during the turning off of one of the switches in half bridge, the gate drain voltage (V_{GD}) of the lower switch will have a negative voltage induced. If these voltage goes lower than the allowable limit (-10.0 V), it will damage the switch. To reduce the effect of this negative voltage, a transistor based gate driver is proposed in [13]. In the proposed configuration a capacitor is connected in series with the transistor. Detailed experimental analysis is shown to verify the design concept.

An improved desaturation based SiC gate driver topology for short circuit protection is proposed in [14]. The proposed response time is 600 ns and it is a two stage based soft turned off technique. The total response time includes the IC delay time and the blanking time. An additional extra charging capacitor is introduced in the proposed circuit to reduce the blanking time.

For Si MOSFETs there is a blanking time introduced in the driver with DeSet protection to allow the switch to completely turn on. In case of SiC, due to the ringing of current during turning on, the delay time may need to increase, which may allow the switch current to increase in case of short circuit. Temperature variations also affect the MOSFET characteristics and in turn the protection. To overcome this issue a rogowski coil based short circuit protection technique is implemented in [15]. This coil directly senses the device current and if it exceeds the set value, the gate driver will turned off the triggering pulses to the devices. Detailed experimental work is carried out to show the efficacy of the proposed system. However, the gate driver circuit becomes complicated and needs a lot of conditioning to filter out the sensed current signal.

Table IV: Commercially available SiC gate driver comparison

Set parameters	CGD15HB62P1	SKYPER 32 R/ 32 PRO R	62EM1
Maximum Switching Frequency	64 kHz	50 kHz	200 kHz
Short circuit response time	1.5 μsec	Not Specified	Not Specified
Gate driver resistance selection based on EMI/EMC suppression	Not Specified	Not Specified	Not Specified
Eliminate the effect of cross talking at higher switching frequency	Not Specified	Not Specified	Not Specified

Table IV shows few commercially available SiC gate drivers and their performance based on few specific criterias required to maintain the device in the SOA. All the commercially available drivers are specifically designed for a particular SiC module and may not be able to show similar efficient performance for other modules as well. This is because of different driving profile requirements from different SiC module supplier. Hence, an application and device specific driver design is essential for an efficient and safer operation of the SiC devices.

IV. CONCLUSIONS

A detailed review on different commercially available SiC MOSFET modules and the challenges associated with driving them at higher switching frequencies are presented. It can be observed that the driving characteristics for SiC modules are quite different from Si based IGBTs. Hence, a direct replacement of IGBT gate drivers is not possible for SiC. Moreover, for more efficient design of the gate drivers a detail study of the device operating range is essential, which will also help to set the fault protection logic in place to operate the devices in the required SOA.

V. REFERENCES

[1] M. Reza and H. Farzanehfard, "Family of soft-switching bidirectional converters with extended ZVS range," *IEEE Trans. on Industrial Electronics*, vol. 64, no. 9, pp. 7000-7008, Sept. 2017.

[2] A. Choudhury, P. Pillay, and S. S. Williamson, "Modified DC-bus voltage balancing algorithm for three-level neutral point clamped PMSM traction inverter drive with reduced common mode voltage," *IEEE Trans. on Industry Application*, vol. 52, no. 1, pp. 278-292, Jan. 2016.

[3] Powerex, "Full SiC & Hybrid SiC IGBTs & IPMs," Internet: http://www.pwrx.com/Promotion/FullSicHybridDesign, Feb. 02, 2018.

[4] A. Choudhury, P. Pillay, and S. S. Williamson, "Comparative analysis between two-level and three-level DC/AC electric vehicle traction inverters using a novel DC-link voltage balancing algorithm," *IEEE Journal of Emerging and Selected Topic in Power Electronics*, vol. 2, no. 3, pp. 529-540, Sept. 2014.

[5] Z. Zhang, Z. Wang, F. Wang, L. M. Tolbert, and B. J. Blalock, "Reliability-oriented design of gate driver for SiC devices in voltage source converter," Published in *IEEE Int. Workshop on Integrated Power Packaging*, Chicago. IL, USA, May 2015, 20-23.

[6] D. Peftitsis and J. Rabkowski, "Gate and base drivers for silicon carbide power transistors: An overview," *IEEE Trans. on Power Electronics*, vol. 31, no. 10, pp. 7194-7214, Oct. 2016.

[7] Z. Chen, M.Danilovic, D. Boroyevich, and Z. Shen, "Modularized design consideration of a general-purpose, high-speed phase-leg PEBB based on SiC MOSFETs," in *Proc. 14th Eur. Conf. Power Electron. Appl.*, Aug. 2011, pp. 1–10.

[8] Z. Wang, X. Shi, Y. Xue, L. M. Tolbert, F. Wang, and B. J. Blalock, "Design and performance evaluation of overcurrent protection schemes for silicon carbide (SiC) power MOSFETs," *IEEE Trans. on Industrial Electronics*, vol. 61, no. 10, pp. 5570-5581, Oct. 2014.

[9] L. Pierre, B. Dominique,M. Herve, A. Bruno, and R. Jean-Francois, "Fast over-current protection of high power IGBT modules," in *Proc. Eur. Conf. Power Electron. Appl.*, 2005, pp. 1–10.

[10] L. Chen, "Intelligent gate drive for high power MOSFETs and IGBTs," Ph.D. dissertation, Dept. Elect. Comput. Eng., Michigan State Univ., East Lansing, MI, USA, 2008.

[11] Q. Qian, J. Yu, J. Zhu, W. Sun, and Y. Yi, "Isolated gate driver for SiC MOSFETs with constant negative off voltage," *in Proc. IEEE Applied Power Electronics Conf. and Exposition*, Tampa,USA, Mar. 2017.

[12] A. P. Camacho, V. Sala, H. Ghorbani, and L. Romeral, "A noval active gate driver for improving SiC MOSFET switching trajectory" *IEEE Trans. on Industrial Electronics*, vol. 99, no. 99, pp. 1-10, Sept. 2017.

[13] F. Gao, Q. Zhou, P. Wang, and C. Zhang, "A gate driver of SiC MOSFET for suppressing the negative voltage spike in a bridge circuit" *IEEE Trans. on Power Electronics*, vol. 33, no. 3, pp. 2339-2353, Mar. 2018.

[14] Y. Shi, R. Xie, L. Wang, Y. Shi, and H. Li, "Switching characterization and short-circuit protection of 1200 V SiC MOSFET T-type module in PV inverter application," *IEEE Trans. on Industrial Electronics*, vol. 64, no. 11, pp. 9135-9143, Nov. 2017.

[15] J. Wang, Z. Shen, R. Burgos, and D. Boroyevich, "Integrated switch current sensor for short circuit protection and current control of 1.7-kV SiC MOSFET modules," in proc. of *IEEE Energy Conv. Congress and Expo.*, 2016, pp. 1-7.

Method of Applying Force Distribution Function for Linear Switched Reluctance Motor Driven by Current Source Inverter

Tadashi Hirayama[1][*]and Shuma Kawabata[1]
1 Graduate School of Science and Engineering, Kagoshima University, Kagoshima, Japan
*E-mail: hirayama@eee.kagoshima-u.ac.jp

Abstract— This paper presents a thrust ripple reduction method for a linear switched reluctance motor (LSRM) driven by a current source inverter (CSI). For thrust ripple reduction, a control method using a force distribution function (FDF) is applied. A new circuit configuration is proposed in order to apply the FDF to the LSRM driving system with the CSI. Performances of the LSRM with the proposed driving system are investigated by a dynamic simulation and experiments. As a result, it was confirmed that the thrust ripple reduction method using the FDF could be applied to the LSRM driven by the CSI.

Keywords— *Linear switched reluctance motor, Thrust ripple reduction, Force distribution function, Current source inverter, Circuit-magnet coupled analysis*

I.   INTRODUCTION

A linear switched reluctance motor (LSRM) has distinctive advantages such as simple and robust structure, inexpensive price, maintenance free, high fault tolerance, and so on. The LSRM has a potential to realize carrier devices with superior performance compared to other linear machines such as a linear induction motor (LIM) and linear synchronous motor (LSM). However, the LSRM has a problem of large thrust ripple because both the stator and mover iron core have a salient pole structure. In general, the thrust ripple is reduced by a design or a control. In LSM, a cogging force and the thrust ripple can be reduced by optimizing shapes of permanent magnets and iron cores and skewing [1]. However, it is not easy to completely eliminate the cogging force by the permanent magnet. A linear flux switching permanent magnet motor [2] which combines the merits both the LSRM and the LSM also has the cogging force by the permanent magnet. On the other hand, as a thrust ripple reduction method for the LSRM, a control method using a force distribution function (FDF) has been proposed [3]-[6]. The FDF is a control method for reducing thrust ripple by distributing a force to a plurality of phases of the LSRM in order to generate commanded thrust. Since the LSRM doesn't have the permanent magnet, there is no problem of the cogging force, and the thrust ripple can be easily eliminated by applying the FDF.

For a high performance drive of the LSRM or a SRM, the driving system with a current source inverter (CSI) has been proposed [7] [8]. The LSRM driven by the CSI is high efficiency compared with a conventional voltage

source inverter (VSI). In the LSRM drive using the FDF, there is a section in which current flows in a single phase and a section in which current flows simultaneously in two phases. In the conventional driving system with the VSI, it is possible to easily supply command currents generated by the FDF to the LSRM because each phase current can be controlled independently. In the driving system with the CSI, the current waveform is controlled by a chopper circuit and the excited phase is switched by the asymmetric H-bridge inverter. Therefore, it is impossible to independently control current waveforms of respective phases during the interval in which two phases are energized. Although the efficiency of the LSRM is improved by using the CSI, the LSRM drive using the FDF can't be realized.

This paper proposes a new driving system for applying the FDF to the LSRM driven by CSI. First, we explain the command current generation method by the FDF. Next, we show the conventional driving system with the VSI and the proposed driving system with the CSI for applying the FDF. In the proposed driving system, IGBTs are added between each output terminal of the asymmetric H-bridge inverter. Then, we confirm the usefulness of the thrust ripple reduction effect and the proposed driving system by performing a dynamic simulation. Finally, the characteristic performances when the FDF is applied to the proposed driving system with the CSI are verified by performing experiments.

II.   THRUST RIPPLE REDUCTION METHOD BY FDF

The LSRM drive using the FDF reduces the thrust ripple by using all forces positively generated at each phase with respect to the moving direction. In this paper, command currents using FDF are derived by a method shown in reference [3]. Here, our LSRM has 3-phase excitation windings. First, the rate of change, $g_k(x)$, to the mover position x of self-inductances is given by the following equation.

$$g_k(x) = \frac{\partial L_k(x)}{\partial x} \tag{1}$$

where, $k = a, b, c$. $L_a(x)$, $L_b(x)$, and $L_c(x)$ are A, B, and C-phase self-inductance, respectively.
Although self-inductances depend on the current value, it is assumed to be a function only of the mover position in

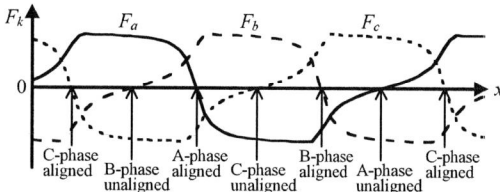

Fig. 1. Fundamental thrust waveforms of 3-phase LSRM.

TABLE I
RELATIONSHIP BETWEEN MOVER POSITION AND FDFS

x	$f_k(x)$
C-phase aligned $\leq x <$ B-phase unaligned	$f_a(x) = 1$ $f_b(x) = 0$ $f_c(x) = 0$
B-phase unaligned $\leq x <$ A-phase aligned	$f_a(x) = \|g_a(x)\| / (\|g_a(x)\| + \|g_b(x)\|)$ $f_b(x) = \|g_b(x)\| / (\|g_a(x)\| + \|g_b(x)\|)$ $f_c(x) = 0$
A-phase aligned $\leq x <$ C-phase unaligned	$f_a(x) = 0$ $f_b(x) = 1$ $f_c(x) = 0$
C-phase unaligned $\leq x <$ B-phase aligned	$f_a(x) = 0$ $f_b(x) = \|g_b(x)\| / (\|g_b(x)\| + \|g_c(x)\|)$ $f_c(x) = \|g_c(x)\| / (\|g_b(x)\| + \|g_c(x)\|)$
B-phase aligned $\leq x <$ A-phase unaligned	$f_a(x) = 0$ $f_b(x) = 0$ $f_c(x) = 1$
A-phase unaligned $\leq x <$ C-phase aligned	$f_a(x) = \|g_a(x)\| / (\|g_c(x)\| + \|g_a(x)\|)$ $f_b(x) = 0$ $f_c(x) = \|g_c(x)\| / (\|g_c(x)\| + \|g_a(x)\|)$

this paper.

Next, FDFs of each phase are determined by the mover position and equation (1) as the following: 1) all sections where thrust occurs in the moving direction are used; 2) in the section where two phases are simultaneously energized, the FDF $f_k(x)$ is given by following equation.

$$f_k(x) = \frac{|g_k(x)|}{|g_k(x)| + |g_{k\pm1}(x)|} \quad (2)$$

Figure 1 shows fundamental thrust waveforms generated by exciting each phase and Table I shows FDFs to the mover position x. Here, F_a, F_b, and F_c are the generated thrust by A, B, and C-phase excitation, respectively. All positive thrusts in Fig. 1 are used. In the case of utilizing two phases, the ratio of the thrust generated at each phase is determined by (2).

When a command thrust of the LSRM is F_x^*, command thrusts of each phase, $F_k(x)$, are given by following equation.

$$F_k^*(x) = F_x^* \cdot f_k(x) \quad (3)$$

Assuming that the magnetic circuit of the LSRM is linear, the thrust generated in each phase is expressed as follow;

$$F_k(x) = \frac{1}{2} i_k^2 \frac{\partial L_k(x)}{\partial x} \quad (4)$$

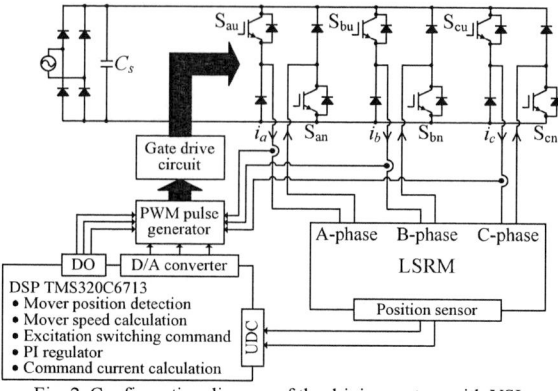

Fig. 2. Configuration diagram of the driving system with VSI.

The command currents $i_k^*(x)$ of each phase are given as follow from (1) and (4);

$$i_k^*(x) = \sqrt{\frac{2F_k^*(x)}{g_k(x)}} \quad (5)$$

III. INVERTER CIRCUIT FOR LSRM DRIVING SYSTEM

A. Voltage Source Inverter

Figure 2 shows the conventional driving system with the VSI. This driving system consists of an asymmetric H-bridge inverter system and a control system. In the control system, command currents are calculated by a DSP. Signals from a position sensor attached to the LSRM are counted by an up / down counter (UDC) and that value is taken into the DSP to obtain the mover position. The excitation section commands of each phase according to the mover position are calculated and outputted from the digital output (DO). The actual velocity of the mover is calculated from the mover position and the command thrust is calculated by PI control. Furthermore, $f_k(x)$ according to the mover position is calculated from $g_k(x)$ and command currents are calculated from (5). The command currents are outputted from the D/A converter. The PWM signals for IGBT control of the asymmetric H-bridge inverter are generated by comparing actual currents from the current sensor with the command currents.

B. Current Source Inverter

Figure 3 shows the proposed driving system with the CSI. This inverter circuit consists of the chopper circuit and the general asymmetric H-bridge inverter. Compared with the conventional driving system with the CSI [7], IGBTs S_{ab}, S_{bc}, S_{ca} and capacitors C_g are newly added in order to actualize the current control using the FDF. S_{ab}, S_{bc}, S_{ca} are used to energize two phases simultaneously. Capacitors C_g are used to improve the fall time of the current when the excitation is switched off.

In the control system, the command currents for each phase and the excitation section command for the asymmetric H-bridge inverter are calculated. The

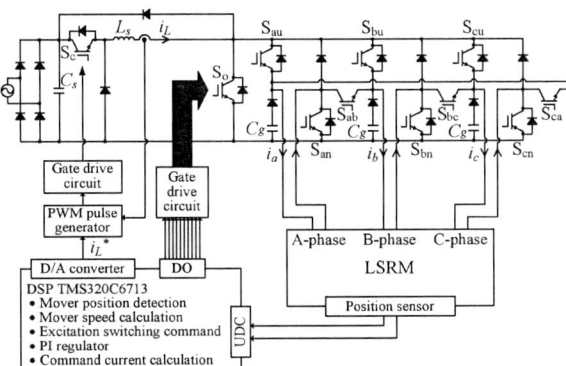

Fig. 3. Configuration diagram of the proposed driving system with CSI.

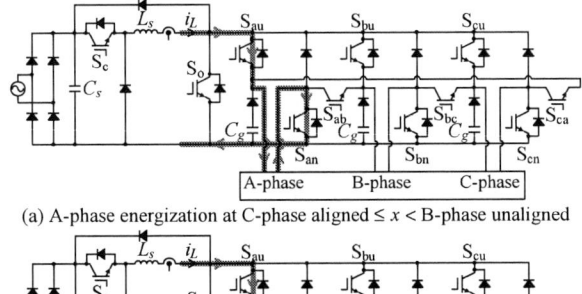

(a) A-phase energization at C-phase aligned $\leq x <$ B-phase unaligned

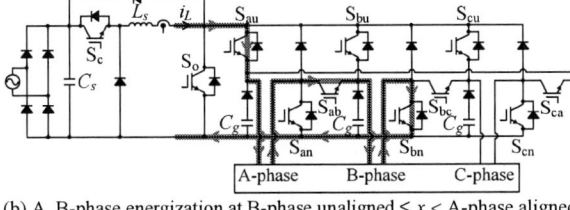

(b) A, B-phase energization at B-phase unaligned $\leq x <$ A-phase aligned

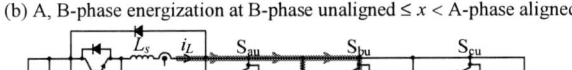

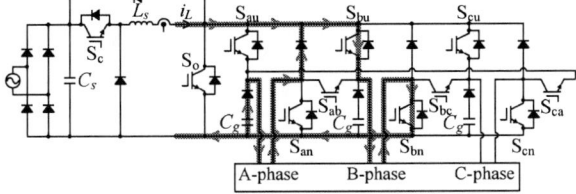

(c) Discharge of accumulated energy of A-phase coil at A-phase aligned $\leq x <$ C-phase unaligned

Fig. 4. Flow of the current for the proposed driving system with CSI.

command current i_L^* of the reactor L_s outputted from D/A converter is compared with the actual current i_L from the current sensor, and the PWM signal for IGBT control of the chopper circuit is generated. In addition, the excitation section commands outputted from DO are input to the gate drive circuit for IGBTs of the asymmetric H-bridge inverter.

In the current control using FDF, the command current at the simultaneous energization of two phases is given as follow from (2), (3) and (5);

$$i_k^*(x) = \sqrt{\frac{2F_x^*}{|g_k(x)| + |g_{k+1}(x)|}} \tag{6}$$

Note that, since $g_k(x)$ is only used in positive section, $g_k(x) = |g_k(x)|$.

Figure 4 shows the flow of current for the proposed driving system with CSI. At the single phase energization,

TABLE II
SWITCHING SIGNALS AND COMMAND CURRENT FOR FDF

x	Switching signals / Command current
C-phase aligned $\leq x <$ B-phase unaligned	ON : S_{au}, S_{an} OFF : Others $i_L^* = i_a^*$
B-phase unaligned $\leq x <$ A-phase aligned	ON : S_{au}, S_{ab}, S_{bn} OFF : Others $i_L^* = i_a^*$
A-phase aligned $\leq x <$ C-phase unaligned	ON : S_{bu}, S_{bn} OFF : Others $i_L^* = i_b^*$
C-phase unaligned $\leq x <$ B-phase aligned	ON : S_{bu}, S_{bc}, S_{cn} OFF : Others $i_L^* = i_b^*$
B-phase aligned $\leq x <$ A-phase unaligned	ON : S_{cu}, S_{cn} OFF : Others $i_L^* = i_c^*$
A-phase unaligned $\leq x <$ C-phase aligned	ON : S_{cu}, S_{ca}, S_{an} OFF : Others $i_L^* = i_c^*$

the current i_L generated by the chopper circuit is supplied to one phase of LSRM as shown in Fig. 4 (a). In the case of two phases simultaneous energization, i_L controlled by (6) is supplied to two phases by connecting in series two phases as shown in Fig. 4 (b). Before switching to the next single phase energization, the A-phase current flows as shown in Fig. 4 (c) until the accumulated energy of the A-phase coil is discharged and becomes zero.

The switching signals for the IGBT control of the asymmetric H-bridge inverter and the command current i_L^* according to the mover position are summarized as shown in Table II.

IV. SIMULATION AND EXPERIMENTAL RESULTS

A. Experimental Machine

The physical dimensions and appearance of our experimental machine are shown in Fig. 5. The experimental machine is 6/4 type LSRM and a long stator type linear motor with primary side on vehicle. The mover has six coils of concentrated winding and the stator is only iron core. The number of turns of excitation winding is 180 turns per pole, the stack length is 60 mm, the air gap length is 1 mm, and the rated current is 3 A in effective value (5 A in amplitude value). We designed the experimental machine with the rated velocity 1 m/s and the rated thrust 50 N. The stator length is approximately 1.2 m. The position information of the mover is derived from an optical linear sensor which has a resolution of 2 μm. The mass including the iron core, windings, and the base of the mover is 5.9 kg.

B. Simulation Model

The simulation is performed by a circuit and magnetic field coupled analysis. For the simulation model of our LSRM, flux linkages of each winding and thrusts at various mover positions and currents are calculated by FEM analysis and these data tables are created. In this paper, the motion equation of mover propulsion direction for the dynamic simulation is defined as follow;

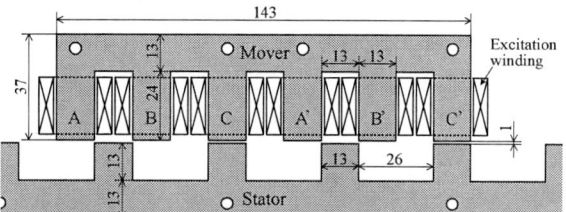

(a) Dimensions of the experimental machine (unit : mm)

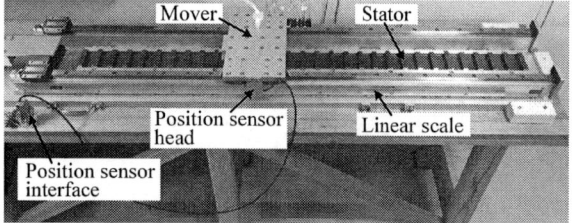

(b) Appearance of the experimental machine

Fig. 5. Experimental machine of the LSRM.

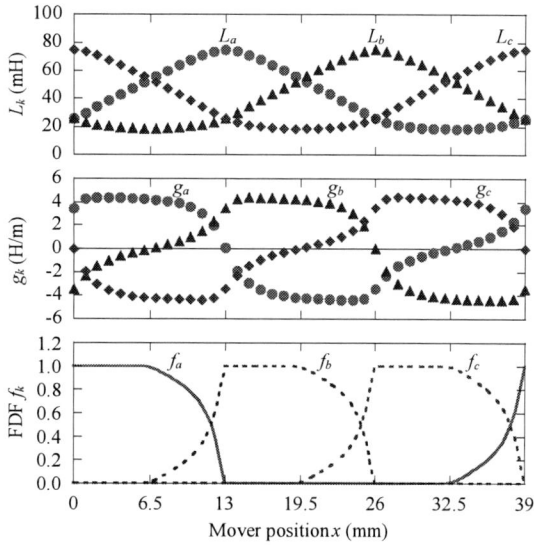

Fig. 6. Calculated results of self-inductances $L_k(x)$, $g_k(x)$, and $f_k(x)$ using simulation.

$$M\frac{dv_s}{dt} = F_x - F_L \qquad (7)$$

where, M is the mass of mover, v_s is the velocity of mover, F_x is the thrust force, F_L is the load force.

For the simulation by circuit and magnetic field coupled analysis, a circuit simulator "SIMPLORER Ver. 7.0 S. V." and a FEM software "JMAG-Designer Ver. 16.1" are used.

C. Simulation Results

Figure 6 shows calculated results of self-inductances $L_k(x)$ and $g_k(x)$. The self-inductances are calculated at 5 A and intervals of 1 mm by using FEM analysis. From (1), $g_k(x)$ at intervals of 1 mm is calculated by a three-point differential approximation method. $g_k(x)$ at intervals of 1 mm is given in the simulation model. $f_k(x)$ and command

currents are calculated by performing a linear interpolation of $g_k(x)$.

Figure 7 shows the simulation results of command currents $i_a{}^*$, $i_b{}^*$, $i_c{}^*$, actual currents i_a, i_b, i_c and thrust waveforms at $v_s = 5$ mm/s and the command thrust $F_x{}^* = 50$ N. Figs. 7 (a) and (b) are the driving system with VSI and the proposed driving system with CSI, respectively. The DC link voltage is 90 V. In the proposed driving system with CSI, L_s and C_g are 100 mH and 0.1 μF, respectively. As shown in Fig. 7 (a), the actual currents are in good agreement with the command currents and almost constant thrust is obtained. Therefore, the usefulness of the thrust ripple reduction by FDF is confirmed. As shown in Fig. 7 (b), in the proposed driving system with CSI, it is confirmed that the actual currents can be flowed as well as the driving system with VSI of Fig. 7 (a), and the thrust ripple reduction by FDF can be realized.

Figure 8 shows the simulation results of velocity control. The command velocity $v_s{}^*$ is 50 mm/s and the load force F_L is 40 N. Figs. 8 (a) and (b) are the driving system with VSI and the proposed driving system with CSI, respectively. As shown in Fig. 8 (a), it is confirmed that there is almost no thrust ripple in both the acceleration and constant velocity section, also there is almost no velocity ripple. As shown in Fig. 8 (b), it is confirmed that the same drive as the driving system with VSI can be realized under the velocity control.

The LSRM drive using the FDF can be performed in the driving system with the CSI by the proposed method. In the proposed driving system with the CSI compared with the driving system with the VSI, the pulsation of actual current is reduced because the current waveform is controlled by the chopper circuit with a large reactor. Therefore, it is confirmed that the pulsation of thrust due to the pulsation of current is reduced.

D. Experimental Results

Figure 9 shows the measured results of self-inductances $L_k(x)$ of the experimental machine at various current values. The self-inductances were measured at intervals of 1 mm by a current saturation method [9]. Although the self-inductances are changed by the current value, the self-inductances at 5 A are used for calculation of $g_k(x)$ in this paper. In the experimental system as with the simulation, $g_k(x)$ at intervals of 1 mm is given into the DSP control system. $f_k(x)$ and command currents are calculated from $g_k(x)$.

Figures 10 and 11 show measured results of command currents $i_a{}^*$, $i_b{}^*$, $i_c{}^*$, actual current i_a, i_b, i_c, and thrust waveforms at $v_s = 5$ mm/s and $F_x{}^* = 50$ N, respectively. Figs. 10 (a) and (b) are the driving system with the VSI and the proposed driving system with the CSI. The thrust was measured with a load cell. The DC link voltage is 50 V, L_s and C_g are 25 mH and 0.1 μF, respectively. As shown in Fig. 10, in the proposed driving system using the CSI, it is confirmed that actual currents can follow command currents well as with the driving system using the VSI. From Fig. 11, the thrust waveform of the

3409

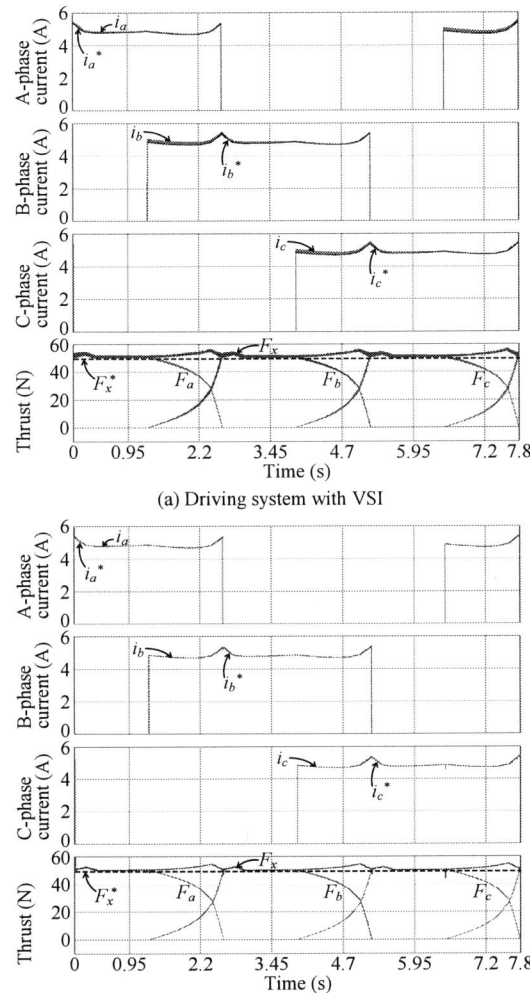

(a) Driving system with VSI

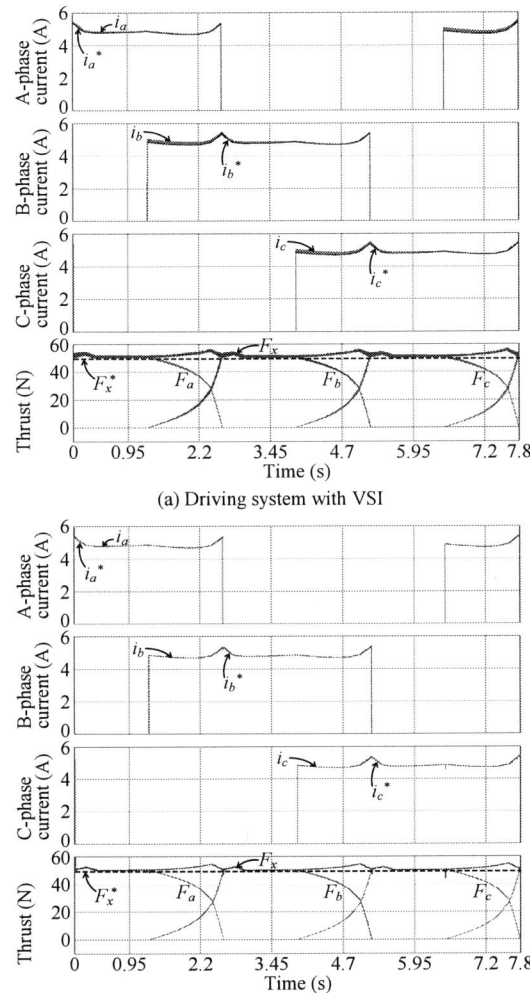

(b) Proposed driving system with CSI
Fig. 7. Simulation results at F_x^*=50 N, v_s=5 mm/s.

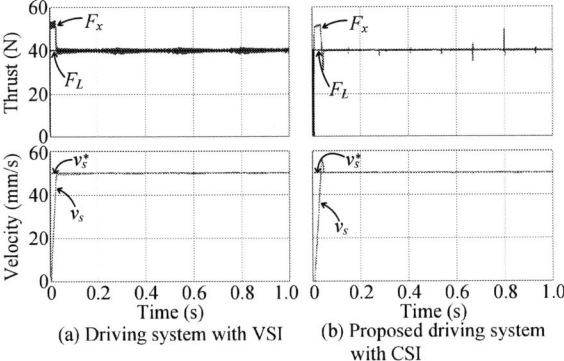

(a) Driving system with VSI   (b) Proposed driving system with CSI

Fig. 8. Simulation results of velocity control at v_s^*=50 mm/s, F_L=40 N.

proposed driving system with the CSI can be obtained almost the same characteristics as the driving system with the VSI. A large thrust ripple is confirmed because the thrust ripple reduction by the FDF is unsatisfactory. However, the application of the FDF to LSRM driven by the CSI which is the aim of this paper could be realized.

Figure 12 shows the velocity response waveforms

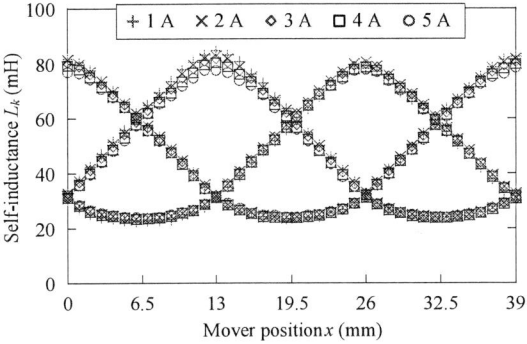

Fig. 9. Measured results of self-inductances at various current values.

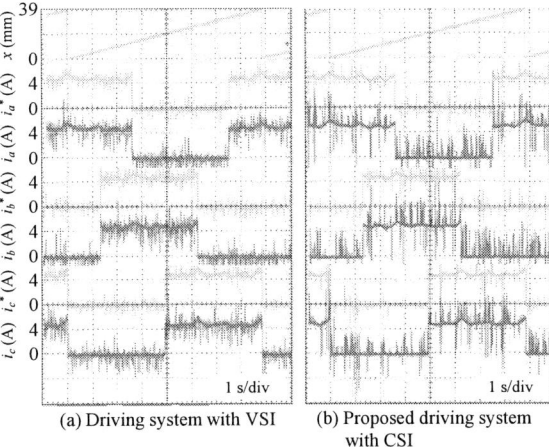

(a) Driving system with VSI   (b) Proposed driving system with CSI

Fig. 10. Measured results at F_x^*=50 N, v_s=5 mm/s.

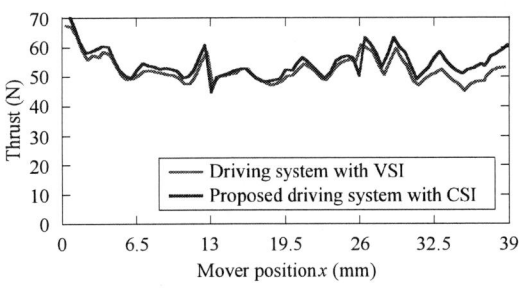

Fig. 11. Measured results of thrust vs. mover position characteristics.

under the velocity control at v_s^*= 50 mm/s and no load. As shown in Fig. 12, there is velocity ripple from -15 mm/s to +15 mm/s. The large velocity ripple occurs because the thrust ripple reduction by the FDF is not sufficient. However, since the actual currents closely follow command currents and both the rise and fall of the current can be switched at high speed, the LSRM driving using the FDF can be applied to the driving system with the CSI by the proposed method.

V. CONCLUSIONS

In this paper, we proposed the new driving system for applying the driving method using the FDF to the driving system with the CSI. First, the usefulness of the driving

3410

The 2018 International Power Electronics Conference

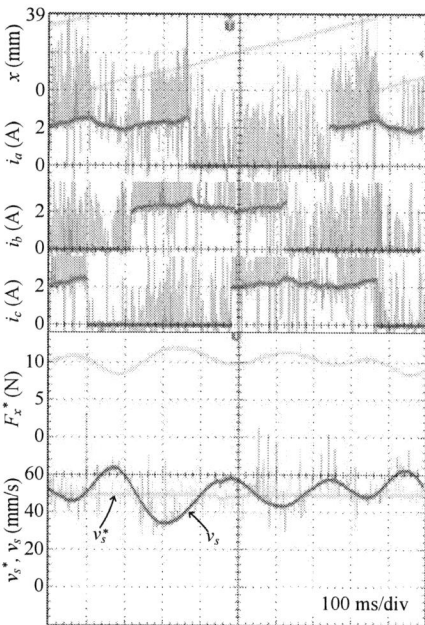

Fig. 12. Measured results of velocity control at v_s^*=50 mm/s, no load.

method using the FDF in our experimental machine of LSRM was confirmed by the dynamic simulation. Next, we measured performance characteristics of the proposed circuit for applying the FDF to the driving system with the CSI, and confirmed that the same current and thrust waveforms and velocity control characteristics as the conventional driving system with the VSI can be obtained. From these results, it was possible to achieve both the high efficiency driving of the LSRM by the CSI and the thrust ripple reduction of the LSRM by the FDF. In the future, we will improve the performance of the FDF and clarify the characteristic performances such as the efficiency and operating characteristics of the LSRM driven by the proposed driving system with the CSI.

ACKNOWLEDGMENT

This work was supported in part by JSPS KAKENHI Grant Number 16K18063.

REFERENCES

[1] M. Wang, L. Li, and R. Yang, "Overview of Thrust Ripple Suppression Technique for Liner Motors," *Chinese Journal of Electrical Engineering*, vol. 2, no. 1, pp. 77-84, 2016.

[2] W. Hao, Z. Deng, and Y. Wang, "Cogging Force Reduction for the Linear Flux-Switching Permanent Magnet Machines Based on a Twisted Stator," *Proc. of 18th International Conference on Electrical Machines and Systems*, pp. 1147-1150, 2015.

[3] H. S. Lim, R. Krishnan, and N. S. Lobo, "Design and Control of a Linear Propulsion System for an Elevator Using Linear Switched Reluctance Motor Drives," *IEEE Trans. on Industrial Electronics*, vol. 55, no. 2, pp. 534-542, 2008.

[4] J. F. Pan, N. C. Cheung, and Y. Zou, "An Improved Force Distribution Function for Linear Switched Reluctance Motor on Force Ripple Minimization With Nonlinear Inductance Modeling," *IEEE Trans. on Magnetics*, vol. 48, no. 11, pp. 3064-3067, 2012.

[5] H. K. Bae, B. S. Lee, P. Vijayraghavan, and R. Krishnan, "A Linear Switched Reluctance Motor: Converter and Control," *IEEE Trans. on Industry Applications*, vol. 36, no. 5, pp. 1351-1359, 2000.

[6] X. D. Xue, K. W. E. Cheng, and S. L. Ho, "Optimization and Evaluation of Torque-Sharing Functions for Torque Ripple Minimization in Switched Reluctance Motor Drives," *IEEE Trans. on Power Electronics*, vol. 24, no. 9, pp. 2076-2090, 2009.

[7] T. Hirayama, R. Matsumoto, T. Hiraishi, and S. Kawabata, "Drive Performance Evaluation of Linear Switched Reluctance Motor using a Current Source Inverter," *Proc. of 18th International Conference on Electrical Machines and Systems*, pp. 285-289, 2015.

[8] Y. Suzuki and K. Akatsu, "High efficiency SRM drive using a Quasi-Current Source Inverter," *IEEE 2nd International Future Energy Electronics Conference 2015*, pp. 1-5, 2015.

[9] P. Zhang, P. A. Cassani, and S. S. Williamson, "An Accurate Inductance Profile Measurement Technique for Switched Reluctance Machines," *IEEE Trans. On Industrial Electronics*, vol. 57, no. 9, pp. 2972-2979, 2010.

The 2018 International Power Electronics Conference

A Novel Drive Circuit for Switched Reluctance Motors with Bipolar Current drive

Hiroki Ishikawa[1*], Yuma Uesugi[2], and Seiya Sakurai[1]

1 Dept. of Electrical, Electronic and Computer Engineering, Gifu University, Gifu, JAPAN
2 Dept. of Human and Information Systems, Gifu University, Gifu, JAPAN
*E-mail: ishikawa@gifu-u.ac.jp

Abstract—This paper proposes a novel drive circuit for three-phase switched reluctance motors (SRMs) with bipolar current drive. Asymmetric H-bridge drive circuit is often used as a SRM drive circuit, and only positive current flows through windings in the SRMs. The SRMs can also generate positive torque even if the motor current is negative during positive torque generation angles.

In this paper, a bipolar current drive and a suitable drive circuit for SRMs are discussed, and the effectiveness of the proposed drive system is confirmed by some simulations and experimental results.

Keywords—*Bipoler current drive, single-phase voltage source inverters, and switched reluctans motors.*

I. INTRODUCTION

Switched reluctance motors (SRMs) are often driven in a different way from AC motors such as induction motors and synchronous motors. Motor current through a stator winding in the SRMs is controlled to be positive pulses as a function of the position of salient rotor poles in positive torque generation angles. The current control is defined as unipolar current drive in this paper. Asymmetric H-bridge drive circuit in Fig. 1 is used as SRM drive circuit frequently to realize such current waveforms, and the unique configuration can realize to control phase current independently.

The three-phase full bridge voltage source inverter (3p-VSI) is also applied to the drive circuit for the SRMs[1]-[11] under some limitations of the winding voltages and the phase current conduction. The 3p-VSI fed SRMs with the star connection[1]-[6] requires two-phase conduction at any time because two windings are connected in series through the neutral point of the star connection. Independent phase current control cannot be achieved, and all phase current cannot contribute to the same direction of the torque generation. For the SRMs with the delta connection including a series diode for each phase winding in Fig. 2[7]-[9], the winding voltages are limited during two-phase conduction between the incoming and the outgoing phases.

In order to generate torque in the same direction, the polarity of the phase current is not important because the rotor poles have neither permanent magnet nor winding. Not only positive current but also negative current is allowed[10][11]. In reference [10], two capacitors are added

to the three-phase full bridge inverter to divide dc source voltage in half as shown in Fig. 3, and the middle point of the capacitors and the neutral point in the star connection are connected. Though the magnitude of each phase voltage becomes half of the dc source voltage, both polarity of the current can be realized. In reference [11], IGBTs are applied in spite of the capacitors, and full-dc source voltage can be supplied to the windings. However, the selection of the phase current polarity is decided by

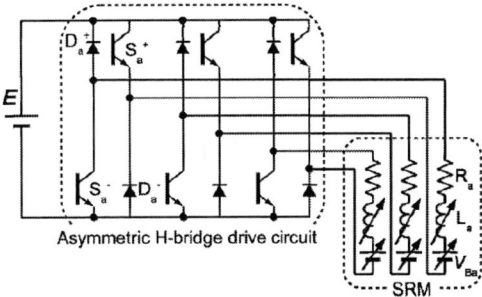

Fig. 1. Conventional asymmetric H-bridge drive circuit.

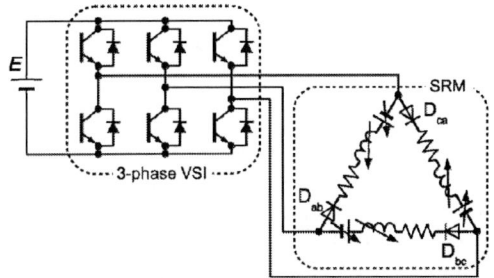

Fig. 2. Three-phase voltage source inverter fed SRMs with delta connection.

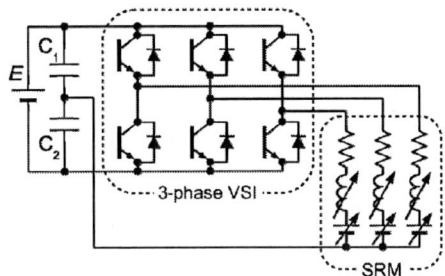

Fig. 3. Three-phase voltage source inverter fed SRMs with a neutral line.

the suitable condition for the drive circuits such as the suppression of the capacitor voltage fluctuation and the routine for the switching sequence.

This paper discusses the positive and the negative pulse current alternately flowing for three-phase SRMs as a selection of the phase current polarity. The phase current control to allow both polarities of the current pulses is defined as bipolar current drive in this paper. The phase voltage under bipolar current drive can be obtained from the division of single-phase ac voltage into each phase. Based on the consideration of the phase voltage waveforms, a suitable drive circuit for three-phase SRMs to realize the bipolar current drive is also proposed. The proposed circuit consists of a single-phase full bridge inverter (1p-VSI) and switches to distribute the output voltage of the 1p-VSI to each phase. The effectiveness of the proposed circuit and bipolar current drive is clarified by some simulation and experimental results in this paper.

II. OVERVIEW OF THREE-PHASE SRMs AND CONVENTIONAL DRIVE CIRCUITS

Figure 4 shows the cross sectional view of a 12/8 SRM as an example of three-phase SRMs. The SRM has 12 stator poles and 8 rotor poles. 6/4 SRMs are also one of three-phase SRMs. In the figure, stator windings on stator poles A_1, A_2, A_3, and A_4 are connected in series, and it is called as phase-a winding. The phase-b and –c windings have similar connections, respectively.

Supposing that the initial rotor position as shown in Fig. 4 and the rotor rotates in clockwise, the phase current flows through phase-a winding until rotor poles X_1, X_2, X_3, and X_4 are aligned with stator poles A_1, A_2, A_3, and A_4, respectively. In order to apply the phase current to a winding of phase-a, the switches S_a^+ and S_a^- of the conventional asymmetric H-bridge drive circuit in Fig. 1 are turned on at turn-on position θ_{on}, and the phase current starts flowing. After S_a^+ and S_a^- are turned off at turn-off position θ_{off}, the phase current decreases to reach zero by the aligned position. The θ_{on} and θ_{off} are designed to generate the positive torque from the torque characteristics of the SRMs with respect to the magnitude of the phase current and the rotor position. The phase current can be also controlled by PWM from θ_{on} to the aligned position, and the waveform shaping of the phase current is available[12]-[14].

The general SRM model of phase-a is expressed by;

$$v_a = R_a i_a + \frac{\partial \lambda_a}{\partial i_a}\frac{di_a}{dt} + \frac{\partial \lambda_a}{\partial \theta}\frac{\partial \theta}{\partial i_a}$$
$$= R_a i_a + L_a \frac{di_a}{dt} + V_{Ba}, \tag{1}$$

where, v_a is voltage across the phase-a winding, λ_a is flux linkage, i_a is the current of phase-a, θ is the rotor position, L_a is the inductance of the phase windings, R_a is winding resistance, and V_{Ba} is the speed EMF. Supposing that magnetization characteristics of the stator and the rotor cores have linearity, the torque equation in the range of the rotor position for the positive torque is given by;

$$Tq = \frac{dL_a}{dt}i_a^2 + \frac{dL_b}{dt}i_b^2 + \frac{dL_c}{dt}i_c^2, \tag{2}$$

where, L_b and L_c are the inductances of phase- b and –c windings, respectively, and i_b and i_c are the phase currents of phase-b and –c, respectively. The magnitude of the current pulses is regulated to achieve the required torque although the torque characteristic with respect to the phase current has non-linearity. On the other hand, the polarity of the generated torque has no relation with the direction of the motor current. Not only the positive current but also the negative current can flow through the phase windings to generate the positive torque [10][11].

III. STRATEGY FOR BIPOLAR CURRENT DRIVE OF SRMs

Figure 5 illustrates difference of the phase current and phase winding voltage waveforms between unipolar and bipolar current drive without PWM control, which is called one-pulse control. Figure 5(a) shows phase current and winding voltage waveforms with the conventional unipolar current drive. Only positive phase current flows, and each voltage is positive from θ_{on} to θ_{off}, and negative from θ_{off} to zero current position θ_0. The corresponding phase is distinguished by means of the suffixes -a, -b, and -c. During two-phase conduction from θ_{off} to θ_0, the incoming phase requires positive phase voltage to increase current, and the outgoing phase requires the negative phase voltage to decrease phase current. In this case, difference polarities of the phase voltage are needed.

Figure 5(b) shows the waveforms of phase currents and winding voltages with the bipolar current drive. The positive and the negative current pulses flow alternately. Figure 6 shows the experimental waveforms of the phase current and voltage driven by the 3p-VSI in Fig. 3. The experimental conditions are listed in Table 1. The waveforms in Fig. 6 are similar to those in Fig. 5(b). From the waveforms, both of incoming and outgoing phase during the two-phase conduction require the same polarity of the phase voltage as the hatched and gray area in Fig. 5(b).

Such winding voltage during two-phase conduction can be realized, if the single-phase ac voltage v_d as shown in the bottom waveform of Fig. 5(b) is applied across two windings connected in parallel.

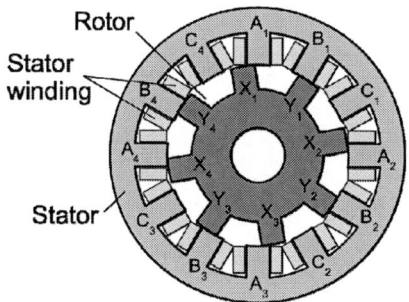

Fig. 4. Cross sectional view of a 12/8 SRM.

IV. CONFIGULATION OF THE PROPOSED DRIVE CIRCUIT

Figure 7 shows the proposed drive circuit for the bipolar current drive for the three-phase SRMs. The circuit consists of 1p-VSI and phase selector switches. The 1p-VSI generates the single-phase ac voltage v_d. Output frequency f_O of the 1p-VSI is calculated by

$$f_O = \frac{N}{60} \frac{p_S p_R}{2 n_S} \qquad (3)$$

where N[rpm] is the rotational speed, n_S is number of winding on the stator pole in series, p_S is number of the stator poles, and p_R is number of the rotor poles.

The v_d is applied across appropriate phase windings by the phase selector switches. Examples of the

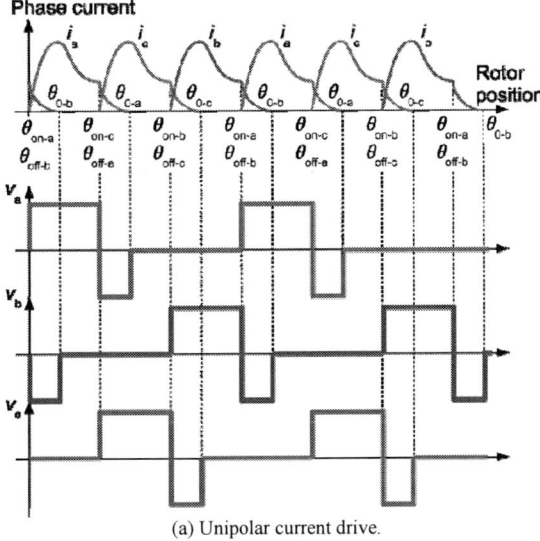

(a) Unipolar current drive.

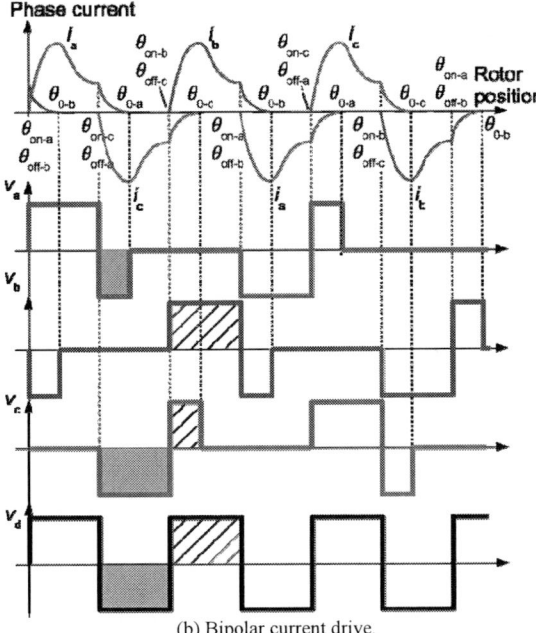

(b) Bipolar current drive.

Fig. 5. Difference between unipolar and bipolar current drive.

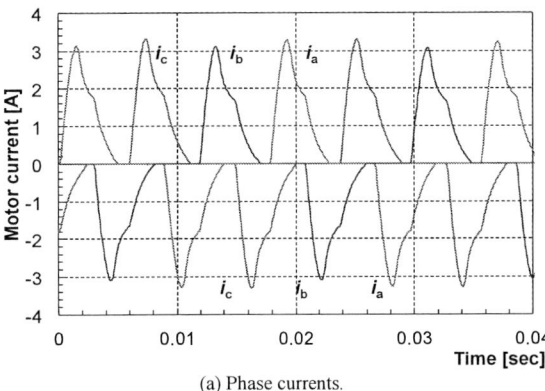

(a) Phase currents.

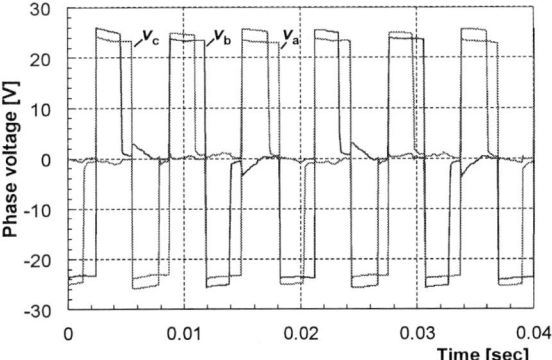

(b) Phase voltages across each winding.

Fig. 6. Experimental waveforms of phase current and phase voltages of 12/8 SRM with the bipolar current drive.

TABLE I
EXPERIMENTAL CONDITIONS FOR THE 3P-VSI-12/8 SRM DRIVE SYSTEM WITH BIPOLAR CURRENT DRIVE

Symbol	Value
Source voltage E	50 V
Capacitors C_1 and C_2	1000 μF
Turn-on angle	-20°
Turn-off angle	-5°
Rotor speed	850 rpm

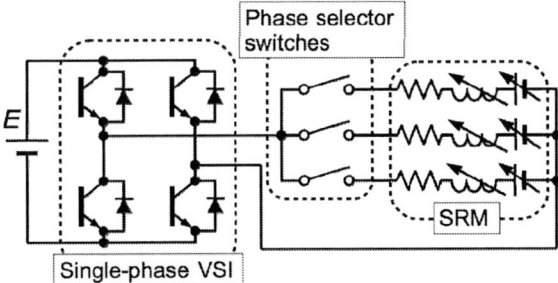

Fig. 7. The proposed drive circuit for bipolar current drive.

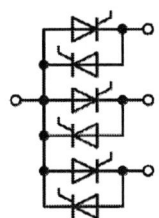

Fig. 8. Examples of the phase selector switches configuration.

3414

The 2018 International Power Electronics Conference

configuration for the phase select switches are shown in Fig. 8. The switches need to allow the bidirectional current flowing. for the phase selector switches are shown in Fig. 8. The Each switch requires reverse blocking characteristics, and IGBTs with reverse blocking capability and thyristors are available.

V. SIMULATION AND EXPERIMENTAL RESULTS

A. Simulation Results

The simulation results are shown in Figs. 9 and 10. Figure 9 shows simulated waveforms of the phase current, the phase voltage, and the output voltage and current of 1p-VSI in the proposed drive circuit fed three-phase SRM with the one-pulse control. The source voltage E is

143V, the rotational speed of the SRM is 1,500rpm, θ_{on} is -40°, θ_{off} is -10°, and the load torque is 9.5Nm, which is the rated torque of the SRM. The f_0 in Fig. 9(c) is 150Hz because the tested SRM is 6/4 SRM.

Figure 10 shows examples of the phase current waveform shaping. The phase current is controlled to be flat-topped in Fig. 10 (a) and to reduce the torque ripples[12][13] in Fig. 10 (b). The proposed drive circuit can control the phase current waveforms from Fig. 10 although the accuracy of the waveform shaping for the out going phase in Fig. 10(b) should be improved.

B. Experimental Results

The experimental setup in Fig. 11 has been built based on Figs 7 and 8, and the verification by experiment for above theoretical discussion has been carried out. The tested SRM is a 12/8 SRM with Y-connection of the stator windings, and the rated power is 750 W. For the phase selector switches in the proposed drive circuit, the IGBTs with reverse blocking capability IXRH40N120 are used. The rating voltage and current are 1200V and 55A, respectively. All switching signals are generated by FPGA (Spartan-3AN, XILINX) from the rotor position θ. Compared with the proposed drive circuit, the SRM has been driven by the conventional asymmetric H-bridge drive circuit in Fig. 1 with unipolar current drive. The experimental conditions are listed in Table II.

Figure 12 shows voltage and current waveforms of the conventional system with unipolar current drive. The rotational speed is 1,040 rpm under no load condition.

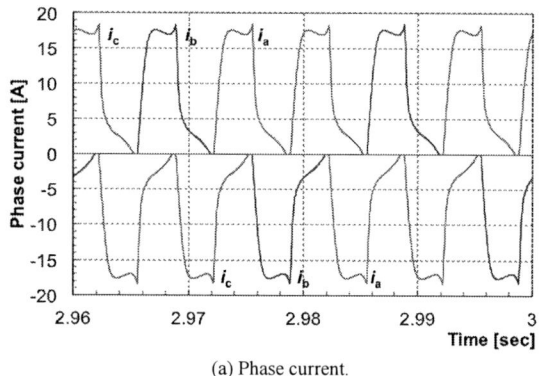

(a) Phase current.

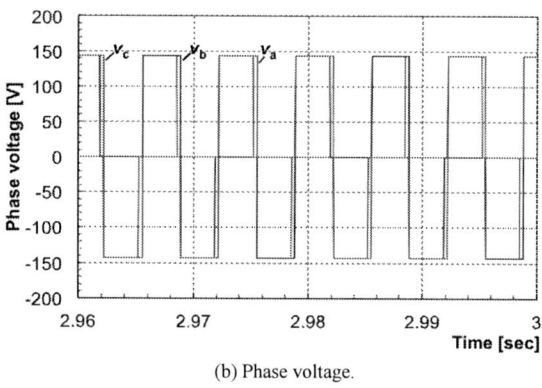

(b) Phase voltage.

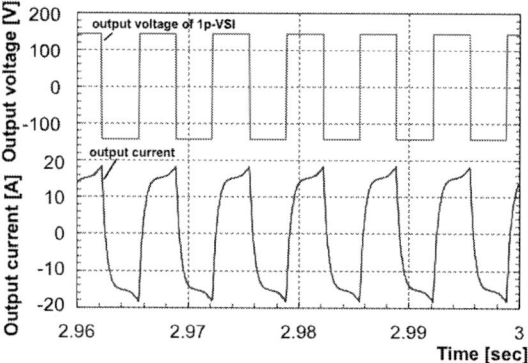

(c) Output voltage and current of 1p-VSI.
Fig. 9. Waveforms of the proposed drive circuit with one-pulse control.

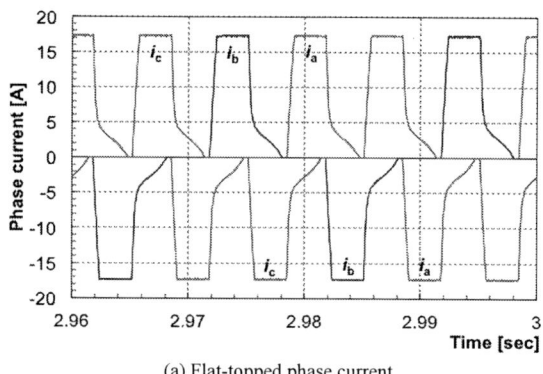

(a) Flat-topped phase current.

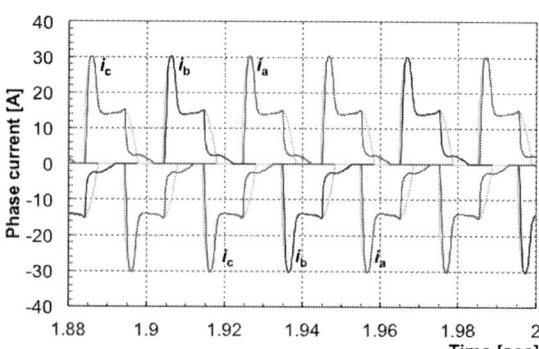

(b) Phase current for torque ripple reduction.
Fig. 10. Examples of the phase current waveform shaping.

3415

The 2018 International Power Electronics Conference

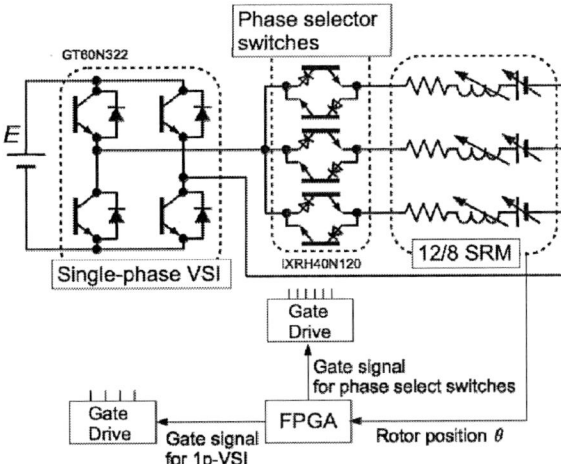

Fig. 11. Experimental setup for the proposed drive circuit.

TABLE II
EXPERIMENTAL CONDITIONS FOR THE PROPOSED DRIVE CIRCUIT-12/8
SRM DRIVE SYSTEM WITH BIPOLAR CURRENT DRIVE

Symbol	Value
Source voltage E	50 V
Turn-on angle	-20°
Turn-off angle	-5°

The waveforms of the source voltage E, the winding voltages v_a, v_b, and v_c in Fig. 12(a) and current i_{DC} through E, phase currents i_a, i_b, and i_c in Fig. 12(b) are typical waveforms with unipolar current drive.

The voltage and current waveforms in the proposed drive system with bipolar current drive are shown in Fig. 13. The rotational speed is 1,035 rpm under no load condition, and the speed is as almost same as the conventional system. The waveforms of the output voltage v_{inv} of the 1p-VSI, the winding voltages v_a, v_b, and v_c in Fig. 13(a) and the output current i_{INV} of the 1p-VSI, phase currents i_a, i_b, and i_c in Fig. 13(b) are obtained. In the proposed system, the bipolar current drive is achieved.

I. CONCLUSIONS

This paper discussed bipolar current drive for three phase SRMs, and proposed a suitable drive circuit to realize the bipolar current drive.

The proposed drive circuit is consists of a single-phase voltage source inverter and phase selector switches. The windings of the tested three phase SRM is Y-connected, and the neutral point is connected to the inverter. IGBTs with reverse blocking capability are applied to the phase selector switches, and all switching signals are generated by FPGA based on the rotor position.

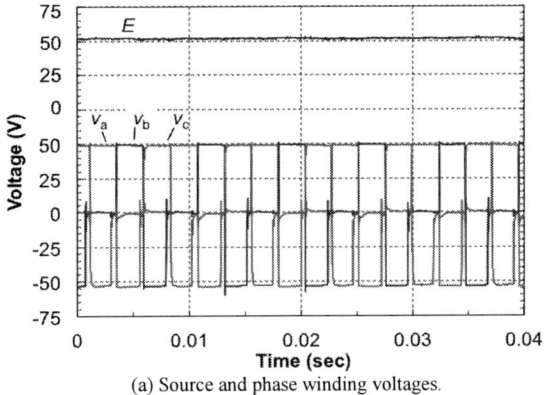

(a) Source and phase winding voltages.

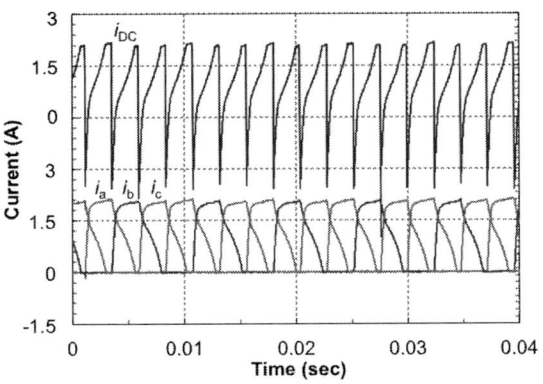

(b) Source and phase currents.

Fig. 12. Waveforms of the conventional asymmetric H-bridge drive circuit with unipolar current drive.

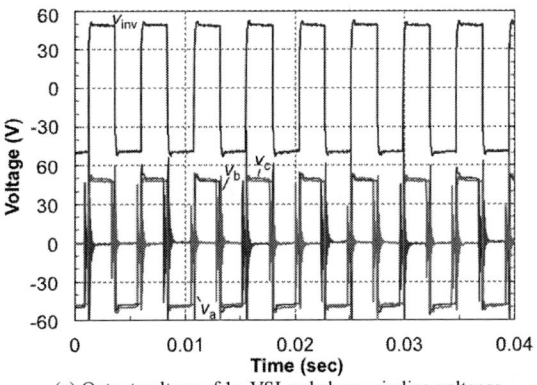

(a) Output voltage of 1p-VSI and phase winding voltages.

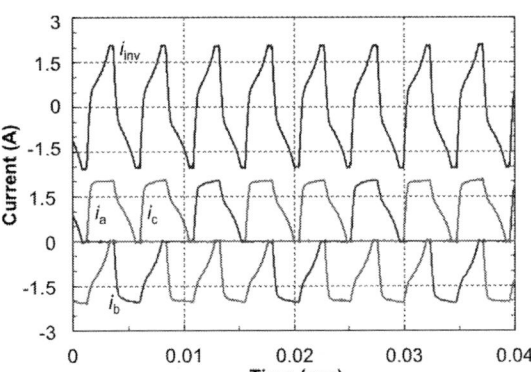

(b) Output current of 1p-VSI and phase currents.

Fig. 13. Waveforms of the proposed drive circuit with bipolar current drive.

Some simulations and experiments have been carried out to confirm the effectiveness of the proposed drive circuit. From the results, it is clarified that the proposed drive circuit realized bipolar current drive for three phase SRMs.

REFERENCES

[1] Dan Ilea, Mircea M. Radulescu, Frédéric Gillon, Pascal Brochet, "Multi-objective optimization of a switched reluctance motor for light electric traction applications," *2010 IEEE Vehicle Power and Propulsion Conference*, pp.1-6. 2010.

[2] Dan Ilea, Mircea M. Radulescu, Frédéric Gillon, Pascal Brochet, "Dynamic model of a three-phase full-bridge inverter-fed Switched Reluctance Motor for optimization purposes," *International Aegean Conference on Electrical Machines and Power Electronics and Electromotion, Joint Conference*, pp.260-265, 2011

[3] Suebsuang Kachapornkul, Pakasit Somsiri, Nattapon Chayopitak, Kanokvate Tungpimolrut, Ruchao Pupadubsin, Prapon Jitkreeyan, "Sensorless control of switched reluctance motor for three-phase full-bridge inverter drive," *2008 International Conference on Electrical Machines and Systems*, pp.3321-3326, 2008

[4] Yuen-Chung Kim, Yong-Ho Yoon, Byoung-Kuk Lee, Jin Hur, Chung-Yuen Won, "A new cost effective SRM drive using commercial 6-switch IGBT modules," *2006 37th IEEE PESC*, pp.1-7, 2006.

[5] Kui-Jun Lee, Nam-Ju Park, Kyung-Hwan Kim, Dong-Seok Hyun, "Simple fault detection and tolerant scheme in VSI-fed switched reluctance motor," *2006 37th IEEE PESC*, pp.1-6, 2006

[6] Xu Liu, Zaiping Pan, "Study on switched reluctance motor using three-phase bridge inverter: Analysis and comparison with asymmetric bridge," *2008 International Conference on Electrical Machines and Systems*, pp.1354-1358, 2008.

[7] B. C. Mecrow, A. C. Clothier, P. G. Barrass, C. Weiner, "Drive configurations for fully-pitched winding switched reluctance machines," *Conference Record of 1998 IEEE Industry Applications Conference*, Vol.1, pp.563-570, 1998.

[8] A. C. Clothier, B. C. Mecrow, "Inverter topologies and current sensing methods for short pitched and fully pitched winding SR motors," *APEC '99*, Vol. 1, pp.416-423, 1999.

[9] Mohammed Elamin, Yusuf Yasa, Ali Elrayyah, Yilmaz Sozer, "Performance improvement of the delta-connected SRM driven by a standard three phase inverter," *2017 IEEE International Electric Machines and Drives Conference (IEMDC2017)*, pp.1-7, 2017.

[10] A. C. Oliveira, C. B. Jacobina, A. M. N. Lima, F. Salvadori, "Startup and Fault Tolerance of the SRM Drive with Three-Phase Bridge Inverter," *2005 36th IEEE PESC*, pp. 714-719, 2005.

[11] Hiroaki Makino, Takashi Kosaka, Shinya Morimotoi, "Instantaneous Current Profile Control of Four-Phase SR Motor for Industrial Servo Drives," *IEEJ Transactions on Industry Applications*, Vol.135, No.6, pp.711–717, 2015 (in Japanese).

[12] Hiroki Ishikawa, Paudel Rishab, Akinori Tsutsumi, and Haruo Naitoh, "Novel Speed Control System with Flat Torque Control for Switched Reluctance Motor Drives," *The 12th International Conference on Electrical Machines and Systems(ICEMS2009)*, 2009.

[13] Hiroki Ishikawa, Yuki Kobayashi, and Haruo Naitoh, "A Novel Position Control System with Torque Ripple Reduction for SRMs," *Proceedings of the 37th IECON*, pp. 1651-1656, 2011.

[14] Hiroki Ishikawa, Masahide Uraji, and Haruo Naitoh, "A New Current Control for Eddy Current Loss Reduction of Switched Reluctance Motors," *IEEE-COMPEL 2013*, 2013.

Torque Ripple Minimization Control of SRM Based on Novel Motor Model Considering Mutual Coupling Effect

Sungyong Shin[1]*, Naruse Hikaru[1], Takashi Kosaka[1,2], and Nobuyuki Matsui[1]

1 Dept. of Electrical and Mechanical Eng., Nagoya Institute of Technology, Nagoya, Japan
2 Frontier Research Institute for Information Science, Nagoya Institute of Technology, Nagoya, Japan
*Email: 28513003@stn.nitech.ac.jp

Abstract- **In order to minimize torque ripple in switched reluctance motor (SRM), many kinds of methods have been proposed. One of them is current profile tracking control based on a motor model, torque-position-current model, which can be derived from magnetizing curves. However, the current profile tracking control has a problem which is an increase in torque ripple at the current overlapping region under high load condition. In this paper, through magnetic analyses using 2D-FEA and mathematical analyses regarding 4-phase 8/6 SRM, the reason of the increase in torque ripple is revealed and a novel torque- position-current model taking mutual coupling effect into account is proposed for better torque ripple minimization control.**

Keywords— SRM, torque ripple, mutual magnetic flux, current tracking

I. INTRODUCTION

Generally, for industrial servo system, permanent magnet synchronous motor (PMSM) is most popular because that it has strong points such as high efficiency, high torque density and sophisticated control performance. However, it has essential problems in terms of demagnetization at high temperature use, mechanical strength at high speed operation and a cost increase due to supply anxiety of permanent magnet concerned with rare-earth materials as well known. On the other hand, switched reluctance motor (SRM) is not only non-permanent magnet machine, but also simple and robust structure [1], so that it is expected as a good candidate for servo drive applications. However, vibration, noise and high torque ripple make SRM difficult to widely apply for servo drive applications. Particularly in servo drive applications, a precise torque control is indispensable. To minimize torque ripple, many kinds of control methods have been proposed [2]-[8].

In order to achieve flat torque control in SRM, the instantaneous current profile tracking control is needed accordance with target torque and rotor position. The tracking control consists of two parts: one is current profiling part and the other is current profile tracking control part. For a precise current profiling, many kinds of methods were proposed [2-8]. One of them was the method to use the torque-rotor position-current (here after T-θ-i) function obtained from measured self-inductance profile [2]. Others were the things to use the T-θ-i function obtained from FEA [5] and direct torque measurement [7].

However, above methods had some errors because that in [2] the effect of magnetic saturation was not considered, that in [5] had torque measurement error, that in [7] had some error between real motor and simulation. In order to overcome this problem, the current profiling method using T-θ-i function calculated from the measured magnetizing curves was proposed in [8]. For a precise current profile tracking control, PWM voltage control method was also proposed in [8]. This current profiling method worked well at light load condition. However, it did not work well at high load condition over rated load especially at current overlapping region.

In this paper, a reason why this method in [8] (here after, conventional method) does not work is investigated through the magnetic analyses using 2D-FEA. Then a novel current profiling method considering mutual coupling effect is proposed. Finally, through the simulation and experiment of a 400W 4-phase 8/6 SRM that has shape of Fig.1, it is proven that the torque ripple minimization by using the proposed method is working well at wide load condition.

II. PRINCIPLE OF TORQUE RIPPLE MINIMIZATION IN CONVENTIONAL METHOD

The process of torque ripple minimization is explained. Firstly, by using torque contour function, the target phase torque is generated. Then, through the current-torque-position (here after, i-T-θ) function, the target phase current profile required is computed. Finally, through PWM voltage control, the target current profile is realized.

A. Torque Contour Function [2][8]

As can be seen Fig. 2, the torque contour function selected among many kinds of possible functions [2] as an

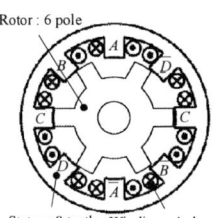

Fig. 1. Photograph and cross-sectional view of test motor

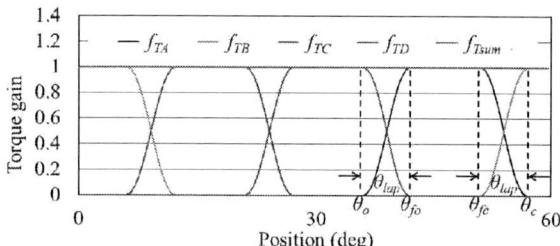

Fig. 2. Torque contour function in conventional method [2][8].

example is explained for 4-phase SRM. The torque contour function per phase can be expressed as $f_{Tx}(\theta)$ with generalized form. The total summation of contour function with respect to 4-phase can be rewritten below,

$$f_{Tsum} = \sum_{x=A}^{D} f_{Tx}(\theta) = 1 \tag{1}$$

where, x denotes the phase number. The target phase torque T_x^* with respect to the rotor position is given in,

$$T_x^*(\theta) = T_t^* \times f_{Tx}(\theta) \tag{2}$$

where T_t^* is the total target torque command. Then, the total torque can be controlled to be flat shape like below,

$$\sum_{x=A}^{D} T_x^*(\theta) = T_t^* \tag{3}$$

In this torque contour function, two kinds of angle θ_{fo}, θ_{lap} can be adjustable parameters. The former represents single phase conduction start angle and the latter is two phase overlapping conduction angle, respectively. As the treated motor has a symmetric structure and the phase shift between adjacent phases is 15 degrees, another angle parameters can be given in the following fashion,

$$\theta_o = \theta_{fo} - \theta_{lap}, \theta_{fc} = \theta_o + 15, \theta_c = \theta_{fo} + 15 \tag{4}$$

where, θ_o is a turn-on angle of the excited phase conduction, θ_{fc} is a termination angle of the single phase excitation, θ_c is termination angle of the excited phase conduction. To get unity gain at overlapping region, the contour function of A-phase can be expressed like below.

$$f_{TA}(\theta) = \frac{1}{2}\left[\cos\left(\frac{\theta - \theta_{fo}}{\theta_{lap}}\pi\right) + 1\right] \tag{5}$$

B. Torque-Position-Current Characteristics Modeling and Target Current Calculation

In conventional method, for calculation of the target current profile to produce target torque at a given rotor position, the T-θ-i model obtained from magnetizing curve is used. Fig. 3 shows the magnetizing curve of A-phase. The flux linkage λ_x can be expressed by using a polynomial equation at every fixed rotor position θ_x and given in,

$$\lambda_x(i_x)\big|_{\theta=\theta_x} = \sum_{n=1}^{n_{max}} L_n\big|_{\theta=\theta_x} \cdot i_x^n \tag{6}$$

where, n_{max} denotes the highest order of polynomial equation, i_x is the phase current. The coefficient L_n can be obtained by using DFT (Discrete Fourier Transform) and given in,

$$L_n(\theta) = \sum_{k=0}^{k_{max}} L_{nk} \cos(k\alpha\theta) \tag{7}$$

where, L_{nk} and k_{max} denote the coefficients and the highest order of cosine series function. α is the number of rotor poles. As a result, the magnetizing curves can be derived as the following fashion.

$$\lambda_x(i_x, \theta) = \sum_{n=1}^{n_{max}} L_n(\theta) \cdot i_x^n \tag{8}$$

By using the magnetic co-energy W_{mx}' expression, torque expression can be deduced as,

$$W_m' = \int_0^{i_x} \lambda_x(i_x, \theta) \implies T_x(i_x, \theta) = \frac{\partial w_m'}{\partial \theta}\bigg|_{i_x=const} \tag{9}$$

Fig. 4 shows the T-θ-i characteristic calculated from (9). Through this T-θ-i characteristic, i-T-θ characteristics model can be also obtained by using a polynomial equation and given in,

$$i_x(T_x, \theta) = \sum_{m=1}^{m_{max}} K_m(\theta) \cdot T_x^m \tag{10}$$

where, K_m and m_{max} denote the coefficients and the highest order of the polynomial equation.

C. Current Profile Tracking Control

The target phase torque can be calculated from (2) with respect to a given rotor position. Then, by using i-T-θ model in (10), the target phase current profile can be computed. Through PWM voltage control based on the magnetizing curves model and voltage equation of SRM [8], the real phase current can track the target profile. Fig.

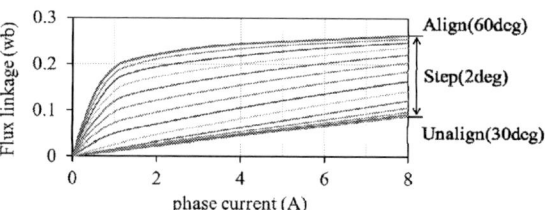

Fig. 3. Magnetizing curve of A-phase in conventional method.

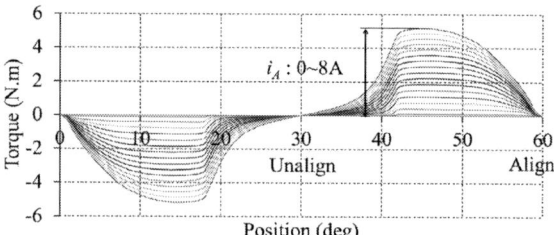

Fig. 4. T-θ-i model of A-phase in conventional method.

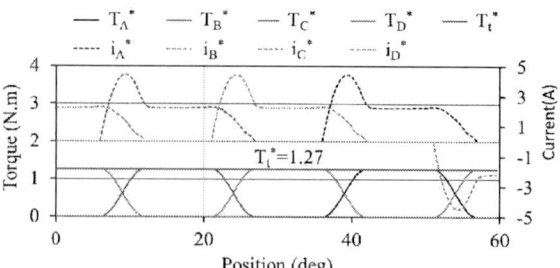

Fig. 5. Taget torque contour and current in conventional method.

5 illustrates the target phase torque and current profiles.

III. REVIEW OF CONVENTIONAL TORQUE RIPPLE MINIMIZATION CONTROL METHOD

A. Problem of Conventional Method

As a weak point, the conventional torque minimization method in [8] does not work well in some operating condition. Especially in high load condition, torque ripple increases at the overlapping region in which adjacent two phases are simultaneously excited. As an example, it can be seen from Fig. 6 that the torque ripple at 200% load condition (the rated torque 1.27Nm) is much bigger than that at 100% load condition. The value in the bracket ( ) of ΔT means the fluctuation ratio with respect to T_t^*.

B. Detailed Reason of Problem

In order to reveal the reason of the increase in torque ripple mentioned above, magnetic analyses using 2D-FEA and mathematical analyses are conducted. In fact, at the overlapping region, the magnetic flux linkage of one phase is affected by mutual flux generated from another exciting phases. In the case that the A- and D-phase are simultaneously energized in the overlapping region, the total flux linkage can be expressed as,

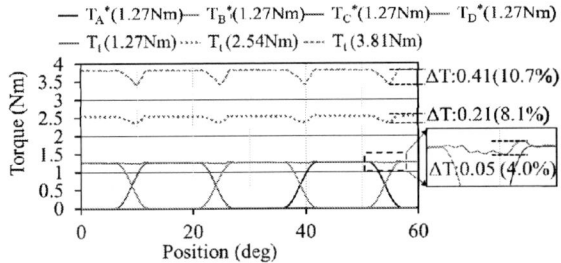

Fig. 6. Torque ripple ΔT in different load conditions, 1.27N·m (100%), 2.54N·m (200%) and 3.81N·m (300%) at 100r/min.

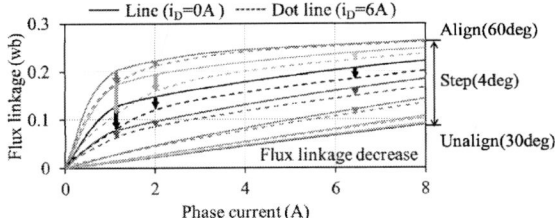

Fig. 7. Magnetizing curves of A-phase considering mutual coupling effect by D-phase.

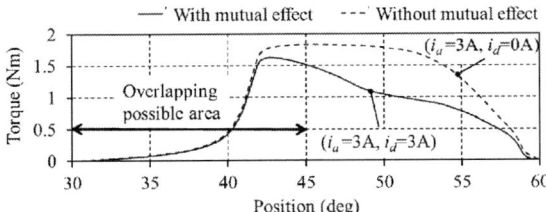

Fig. 8. T-θ curves of A-phase with and without mutual coupling effect.

$$\lambda_A(i_A, i_D, \theta) = \lambda_{AA}(i_A, i_D, \theta) + \lambda_{AD}(i_A, i_D, \theta) \quad (11)$$

where, λ_{AA} is self-induced flux linkage in A-phase, λ_{AD} is flux linkage in A-phase mutually-induced by D-phase. As a result, the magnetic co-energy of A-phase W_{mA}' is affected by λ_{AD}. The magnetic co-energy considering the mutual coupling effect is given in,

$$W_{mA}' = \int_0^{i_A} \lambda_{AA}(i_A, i_D, \theta) di_A + \frac{1}{2}\int_0^{i_A} \lambda_{AD}(i_A, i_D, \theta) di_A \quad (12)$$

Since the conventional method has no consideration on mutual coupling effect in the co-energy calculation, T-θ-i model also has some errors for the real machine. As can be seen from Fig. 7, the magnetizing curve is obviously affected by the mutual coupling effect. As a consequence, the infinitesimal variance of the magnetic co-energy for the position variance $\Delta\theta$ at a given rotor position decreases due to the mutual coupling effect, resulting in torque reduction. As can be seen from Fig. 8, in the possible overlapping region from 30 to 45degrees, the torque of A- phase decreases due to the influence of the mutual coupling effect by D-phase.

IV. TORQUE RIPPLE MINIMIZATION BY USING NOVEL T-θ-I MODEL CONSIDERING MUTUAL COUPLING EFFECT

A. A Novel T-θ-i Model

In order to minimize the torque ripple, a novel T-θ-i model taking the mutual coupling effect into account is examined. For that, the magnetizing curves including the mutual coupling effect is needed. Fig. 9 demonstrates the magnetizing curves of A-phase considering the mutual coupling effect of D-phase. These magnetizing curves also can be measured based on equations below.

$$\lambda_A(i_A, i_D, \theta)\big|_{i_A=pulse, i_D=const} = \int_0^t (v_A - Ri_A) dt + M_{AD} i_D \quad (13)$$

$$M_{AD} i_D = \lambda_{AD}(i_A, i_D, \theta)\big|_{i_A=0, i_D=pulse} = \int_0^t (v_A - Ri_A) dt \quad (14)$$

where, M_{AD} is the mutual inductance between A- and D-phase, v_A, i_A and R are the applied A-phase voltage, the corresponding phase current and the phase resistance, respectively. The total flux linkage of A-phase is calculated via two steps. Firstly, under the fixed rotor position θ and D-phase current, through the voltage pulse injection into A-phase and an integration of v_A -Ri_A, the first term in right hand side of (13) can be calculated. Secondly, the initial mutual flux $M_{AD} i_D$ at $t=0$ is computed by (14) and the measured v_A and i_A when the voltage pulse is applied to D-phase. Through the summation of the first and the second calculation results, the total flux linkage can be obtained experimentally. This magnetizing curve also can be expressed by using a polynomial equation and given in,

$$\lambda_x(i_x, i_y)\big|_{\theta=\theta_x} = \sum_{n=0}^{n_{max1}} \lambda_{x\_n}(i_y)\big|_{\theta=\theta_x} \cdot i_x^n \quad (15)$$

$$\lambda_{x\_n}(i_y)\big|_{\theta=\theta_x} = \sum_{m=0}^{m_{max1}} L_{x\_nm}\big|_{\theta=\theta_x} \cdot i_y^m \quad (16)$$

where, λ_{x_n}, n_{max1}, and m_{max1} denote the coefficients and the highest order of the polynomial equations, respectively. The coefficient L_{x_nm} can be obtained by DFT,

3420

$$L_{x\_nm}(\theta) = \sum_{k=0}^{k_{max1}} L_{x\_nmk} \cos(k\alpha\theta) \qquad (17)$$

where, L_{x_nmk} and k_{max1} denote the coefficients and the highest order of cosine series function. Finally, the total flux linkage can be given in,

$$\lambda_x(i_x,i_y,\theta) = \sum_{n=0}^{n_{max1}} \left[\sum_{m=0}^{m_{max1}} L_{x\_nm}(\theta) \cdot i_y^m \right] \cdot i_x^n \qquad (18)$$

At this point, in order to use the magnetic co-energy W_{mx}' expression (12) for the torque calculation, the total flux linkage must be divided into the self-induced and the mutually-induced flux components. In this study, the equation below is introduced [9].

$$\lambda_A(i_A,i_D,\theta) = \lambda_A(i_A,0,\theta) + \lambda_{AD}(i_A,i_D,\theta) \qquad (19)$$

Assuming that $\lambda_A(i_A,0,\theta)$ is self-induced flux linkage and $\lambda_{AD}(i_A, i_D,\theta)$ is the mutually induced flux linkage, the magnetic co-energy can be derived as,

$$W_{mA}'(i_A,i_D,\theta) = \int_0^{i_A} \lambda_A(i_A,0,\theta) di_A$$
$$+ \frac{1}{2}\int_0^{i_A}\left(\lambda_A(i_A,i_D,\theta)-\lambda_A(i_A,0,\theta)\right)di_A \qquad (20)$$

By using (21) below,

$$T_x(i_x,i_y,\theta) = \left.\frac{\partial w_{mx}'(i_x,i_y,\theta)}{\partial \theta}\right|_{i_x,i_y=const} \qquad (21)$$

the novel T-θ-i model considering the mutual coupling effect between two phases can be obtained and given in,

$$T_{xy}(i_x,i_y,\theta) = T_x(i_x,i_y,\theta) + T_y(i_y,i_x,\theta) \qquad (22)$$

Fig. 10 illustrates the novel T-θ-i curves under A-and D-phase excitation with respect to i_A, i_D and θ.

B. Target Current Profiling and Control

Based on the proposed T-θ-i model, a novel i-T-θ model can be obtained by using the polynomial equation

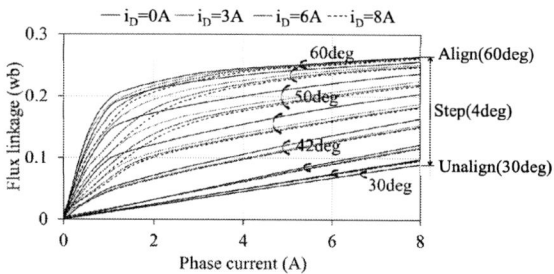

Fig. 9. Magnetizing curves of A-phase considering mutual coupling effect of D-phase.

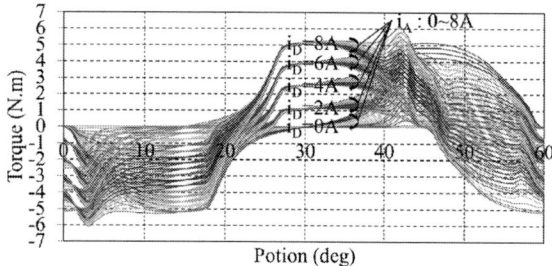

Fig. 10. T-θ-i curves considering the mutual coupling effect of A- and D-phase.

and given in,

$$i_x(T_{xy},i_y,\theta) = \sum_{n=0}^{n_{max2}} \left[\sum_{m=0}^{m_{max2}} K_{x\_nm}(\theta)\cdot i_x^m \right] \cdot T_{xy}^n \qquad (23)$$

where, K_{x_nm}, n_{max2}, and m_{max2} are the coefficient, the highest orders of the polynomial equation, respectively. According to this equation, we can determine the target current profile i_x^* for a given total torque $T_{xy}^* = T_t^*$, the detected adjacent phase current i_y^* at a detected rotor position θ. As can be seen Fig. 11, in the overlapping region I, the target current profile i_A^* is calculated considering the pre-excited phase current i_D^*. After the overlapping area I, the target current profile i_A^* must be determined so as to achieve the phase target torque T_A^* by only single phase similar to conventional method.

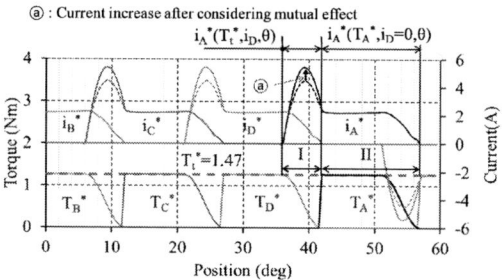

Fig. 11. Torque contour functions and current profiles in the proposed method.

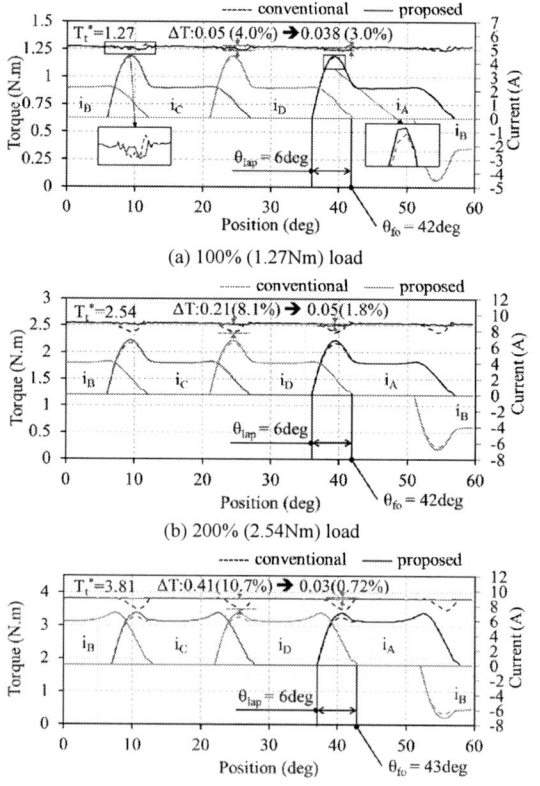

(a) 100% (1.27Nm) load

(b) 200% (2.54Nm) load

(c) 300% (3.81Nm) load

Fig. 12. Comaprisons of torque and current waveforms under the conventional and the proposed control methods for different load conditions at 100r/min.

(a) Equipment for magnetizing curve

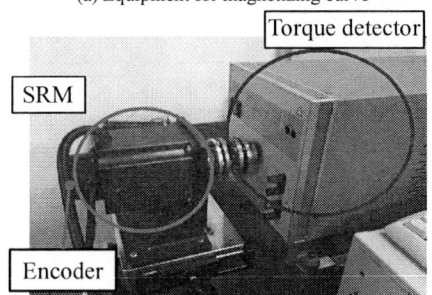

(b) Equipment for torque ripple test

Fig. 13. Experimental setup

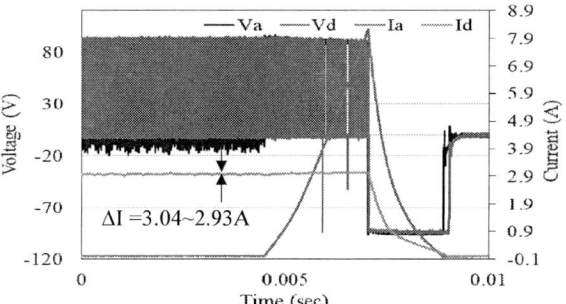

Fig. 14. A-phase impulse current and D-phase DC current to be controlled to 3A using 30kHz PWM and DC-link 100V

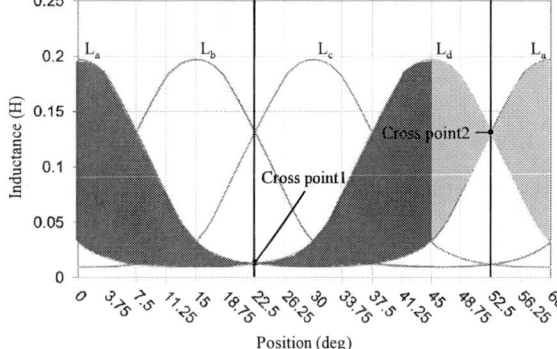

Fig. 15. A symmetricity of between A-phase and D-phase

V. SIMULATION RESULTS

To prove the proposed current profiling being effective for torque ripple minimization, some simulation based on 2D-FEA is executed. Fig. 12 demonstrates torque and current waveforms under the conventional control and the proposed control methods for 100%, 200% and 300% load conditions at 100r/min. As can be seen in the figure, the torque ripple is minimized.

VI. EXPERIMENT RESULTS

A. Experimental setup

To get a magnetizing characteristic of tested motor and evaluate a performance of torque ripple minimization, the experimental equipment is set up as can be seen Fig. 13. Fig.13 (a) is a equipment to obtain magnetizing curve. To obtain magnetic flux linkage data at fixed position, the voltage and current data through the pulse voltage injection is needed. To fix the position exactly for measurement, rotary encoder (Tamagawa Seiki, TS5667N420, 131072 ppr, 0.0027/pulse) and stepping motor (oriental motors, UPH599HG2-B2, 0.0072deg/pulse) with harmonic gear (oriental motors, parking torque 360kg.fcm, gear ratio 1:100) to have high resolution are used. Fig. 13 (b) is a equipment to take torque ripple experiment. The PE-Expert III (Myway Plus Corporation) was used as a controller to generate gate signals for the asymmetric H-bridge converter with a PWM frequency of 10 kHz. For position sensing, a encoder (same with above) was attached. As a torque meter and load, torque detector (ONO SOKKI, MT-6254A, Max. torque : 5Nm, resolution : 0.01Nm) consisting of a geared DC motor and torque sensor was coupled to the tested SRM and used as a constant speed dynamo.

B. Magnetizing curve and torque curve

By using equipment, magnetizing curve is obtained through integration of measured voltage and current as mentioned at chapter IV-A every angle with step 0.5deg from 22.5deg to 52.5deg. At first, $\lambda_{AA}(i_A,i_D,\theta)$ and $\lambda_{DA}(i_A,i_D,\theta)$ is obtained from v_A, i_A, v_D, i_D that measured when impulse voltage is inputted A-phase while D-phase is controlled to DC current- i_D (0.5A step from 0 to 8A)by using PI controller as can be seen in Fig. 14. And then $\lambda_{AD}(i_A,i_D,\theta)$ and $\lambda_{DD}(i_A,i_D,\theta)$ is obtained from v_A, i_A, v_D, i_D that measured when impulse voltage is inputted D-phase while A-phase is zero. As a result, the total flux linkage of phase is calculated by using below equation.

$$\lambda_A(i_A,i_D,\theta) = \lambda_{AA}(i_A,i_D,\theta) + \lambda_{AD}(i_A,i_D,\theta) \tag{24}$$

$$\lambda_D(i_A,i_D,\theta) = \lambda_{DD}(i_A,i_D,\theta) + \lambda_{DA}(i_A,i_D,\theta) \tag{25}$$

And from the inductance profile of 8/6 SRM as can be seen Fig. 15, inductance of A and D-phase are crossed at 22.5 and 52.5deg because of symmetric structure. Through symmetricity, the data from 0 to 22.5deg can be mirrored by below relation,

$$\lambda_A(i_A,i_D,\theta) = \lambda_D(i_D,i_A,45-\theta), \lambda_D(i_A,i_D,\theta) = \lambda_A(i_D,i_A,45-\theta)$$

and also the data from 52.5 to 60deg can be mirrored by using below relationship.

$$\lambda_A(i_A,i_D,\theta) = \lambda_D(i_D,i_A,52.5-\theta), \lambda_D(i_A,i_D,\theta) = \lambda_A(i_D,i_A,52.5-\theta)$$

By using the calculated magnetic flux linkage and it's

The 2018 International Power Electronics Conference

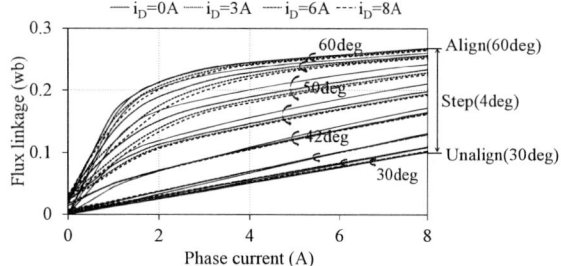

Fig. 16. Measure magnetizing curves of *A*-phase considering mutual coupling effect of *D*-phase.

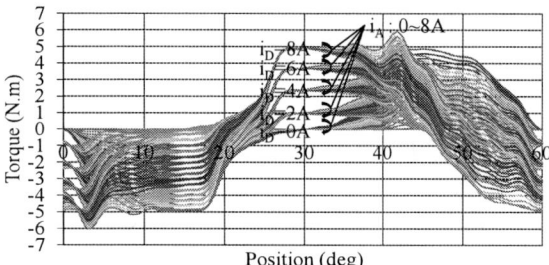

Fig. 17. *T-θ-i* curves considering the mutual coupling effect of *A*- and *D*-phase from test result.

symmetric data, magnetizing curve is obtained. Fig. 16 shows magnetizing curves of A-phase that consider mutual flux effect by D-phase current. And by using these flux linkage curve and equation from 20) to 22), torque curve (*T-θ-i*) is obtained as can be seen Fig. 17. In this calculation, as the order of polynomial, n_{max1}, $m_{max1} = 10$ is used and as the order of DFT, $k_{max1} = 19$ is used.

C. Torque ripple calculation results

From torque curve, the *i-T-θ* model is obtained by using equation 23). Through this *i-T-θ* model, the target current to meet the target torque at current overlap area is calculated while considering the mutual magnetic flux by adjacent phase current. Fig. 18 shows the calculated torque about target current profiles obtained from the *i-T-θ* model that mutual effect is considered and not considered. The current and torque profile in current overlap region are not same with simulation results because the magnetizing characteristics obtained from tested motor are different with simulated one. In Fig. 18 the reduction of torque ripple is small due to the small affection of mutual flux because the current overlap is finished before 42deg that the stator teeth overlap with rotor teeth. In this paper, for evaluation of the reduction performance of torque ripple when the affection of mutual flux is high, the experiment in the condition of $\theta_{fo} = 44$deg is conducted. Fig. 19 shows the calculated torque about target current profile calculated from the *i-T-θ* model for $\theta_{fo} = 44$deg. As can be seen figures, the torque ripple can be reduced effectively through proposed method especially in rated torque condition (1.27Nm).

D. Experimental results

To evaluate the effectiveness of torque ripple reduction, a experiment was conducted about the current profile ob-

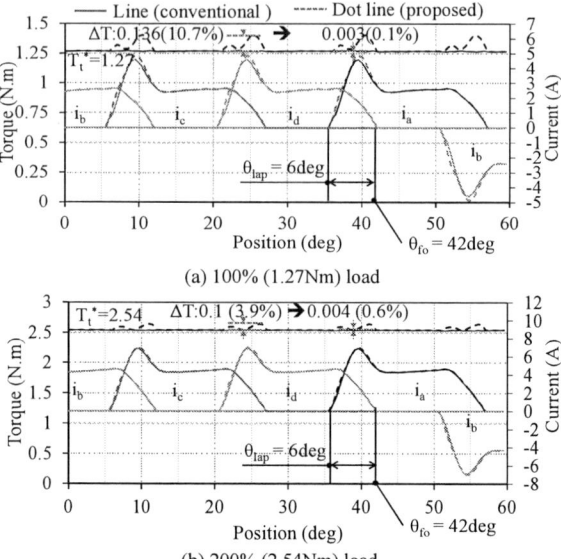

(a) 100% (1.27Nm) load

(b) 200% (2.54Nm) load

Fig. 18. Torque and current calculated from the *i-T-θ* model under the conventional and the proposed methods for different load conditions at $\theta_{fo} = 42$deg

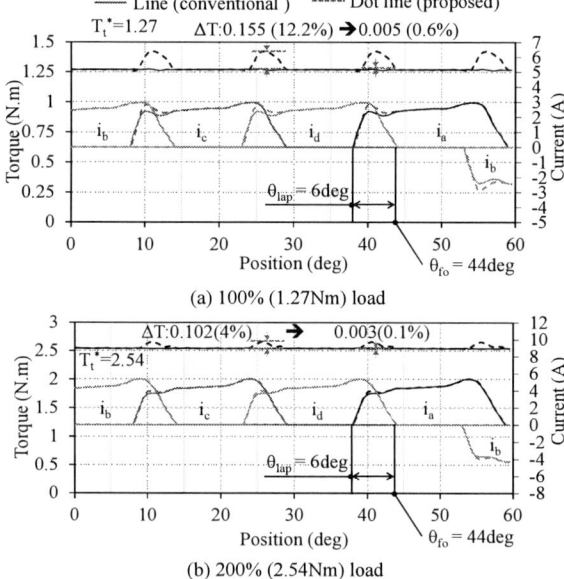

(a) 100% (1.27Nm) load

(b) 200% (2.54Nm) load

Fig. 19. Torque and current calculated from the *i-T-θ* model under the conventional and the proposed methods for different load conditions at $\theta_{fo} = 44$deg

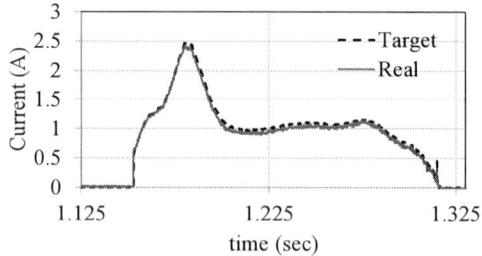

Fig. 20. Experimental result of current tracking with target value

The 2018 International Power Electronics Conference

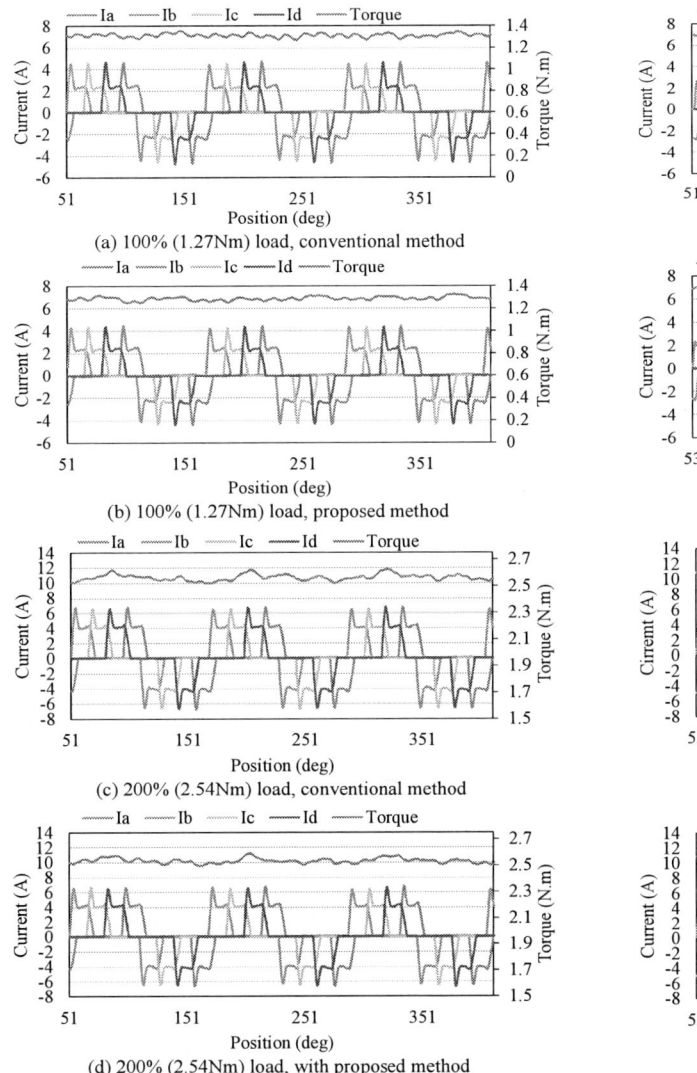

Fig. 21. Experimental torque and current result under the conventional and the proposed control methods for different load conditions at θ_{fo} = 42deg, speed 20r/min

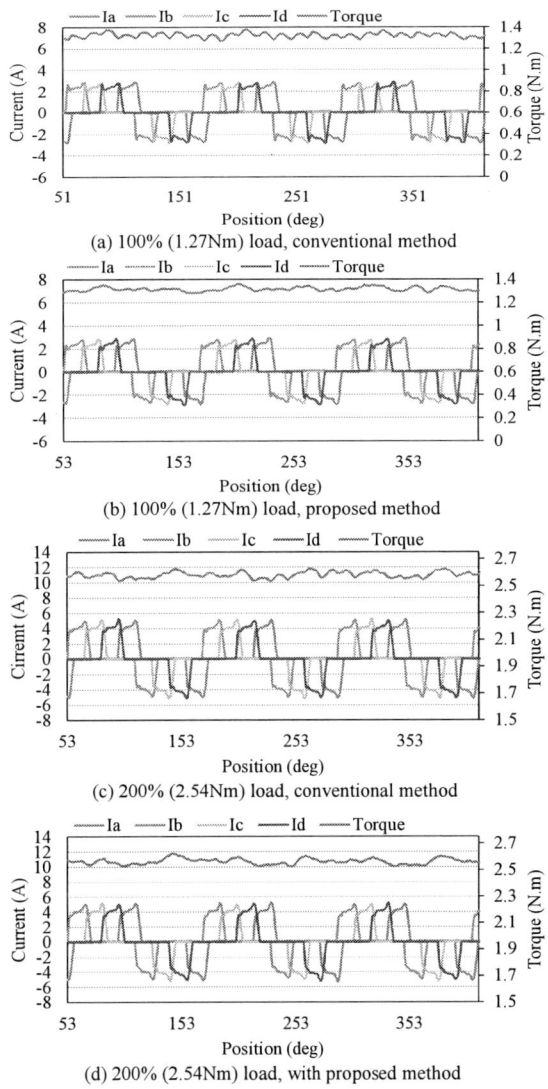

Fig. 22. Experimental torque and current result under the conventional and the proposed control methods for different load conditions at θ_{fo} = 44deg, speed 20r/min

tained above. For safety, the test is conducted about maximum 100% and 200% of rated torque. For current control tracking the target current, active current regulator (ACR) based on the voltage equation of motor is used [8]. Fig. 20 shows the real current is tracking with target current well. As can be seen Fig. 21 and 22, torque ripple became smooth at both load condition of 1.27 and 2.54Nm when the proposed method with the consideration of mutual effect is used. Especially, the torque ripple reduction performance is high at θ_{fo} = 44deg that mutual affection is high as can be seen Fig 22-(a) and (b). it is difficult to recognize how much a torque ripple is reduced. In this paper, through the DFT about the torque data of one rotation-360deg mechanical angle, the performance of torque ripple reduction was evaluated. Because a phase is turned on 24times every rotation, it is not difficult to figure out that 24th order of DFT is strong relation with torque ripple produced by mutual magnetic flux. Fig. 23

shows the DFT result about torque results at each operation condition. 24th and 48th component of torque when using the proposed method is reduced compared with that when using the conventional method in every operating condition. As a result, variance of torque-ΔT also reduced except for the condition of 2.45Nm, θ_{fo} = 44deg. In this condition, because 6th order components of torque is increased, variance of torque is increased. However, it can be evaluated that the reduction of torque ripple component by mutual effect is well conducted through experiment. Table I shows the summary of experiment result. T^*, T_{ave} is target torque and average torque of one rotation respectively.

VII. CONCLUSION

To overcome the torque ripple increase under the control method based on the conventional modelling approach, in this paper the reason has been revealed through

The 2018 International Power Electronics Conference

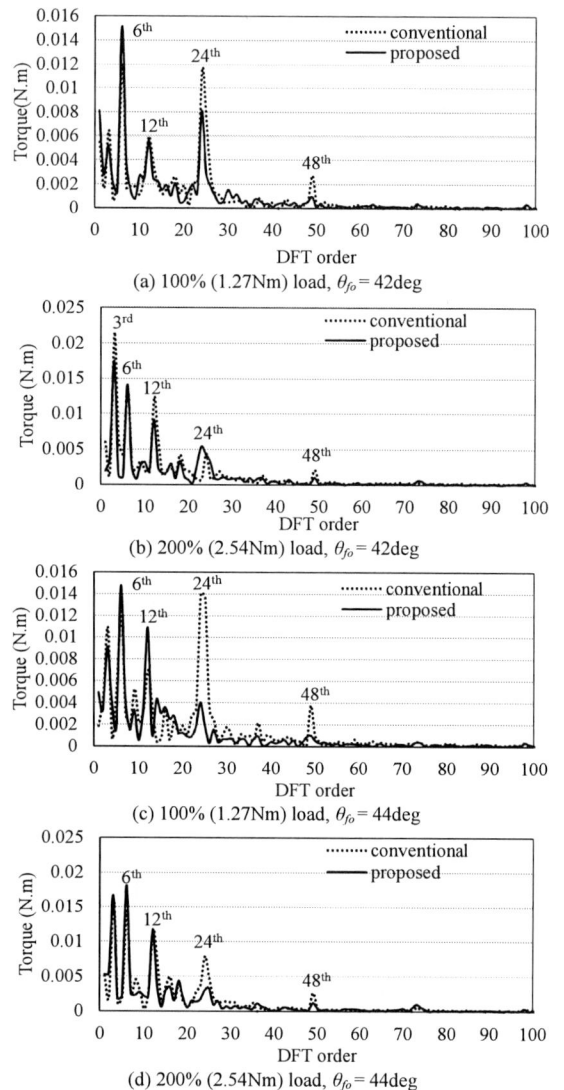

Fig. 23. DFT result of torque under the operation condition at speed 20r/min

TABLE I
SUMMARY OF EXPERIMENTAL RESULTS

θ_{fo} (deg)	T^{*} (N.m)	method	T_{ave} (N.m)	ΔT (N.m)	DFT 24th (N.m)	DFT 48th (N.m)
42	1.27	conventional	1.313	0.084	0.0116	0.0028
		proposed	1.287	0.077	0.0081	0.0010
	2.54	conventional	2.564	0.105	0.0044	0.0021
		proposed	2.535	0.096	0.0043	0.0010
44	1.27	conventional	1.324	0.107	0.0140	0.0038
		proposed	1.316	0.08	0.0040	0.0010
	2.54	conventional	2.582	0.093	0.0080	0.0027
		proposed	2.563	0.101	0.0029	0.0012

[3] K. M. Rahman, A. V. Rajarathnam, and M. Ehsani, "Optimized Instantaneous Torque Control of Switched Reluctance Motor by Neural Network", Conf. Rec. of IEEE-IAS Annual Meeting, Vol. I, pp.556-563, 1997.

[4] S. Mir, M. Elbuluk, and I. Husain, "Torque Ripple Minimization in Switched Reluctance Motors using Adaptive Fuzzy Control", *IEEE Trans. on IA.*, Vol. 35, pp. 461–468, Mar./Apr., 1999.

[5] H. Ishikawa, Y. Kamada and H. Naito, "Instantaneous Current Profile Control for Flat Torque of Switched Reluctance Motors", *T. IEE Japan*, Vol. 125-D, No. 12, pp.1113-1121, 2005. (in Japanese)

[6] I. Husain, "Minimization of torque ripple in SRM drives", IEEE Trans. on PE, Vol. 49, No.1, pp.28-39, 2010.

[7] Y. Niwa, T. Abe and T. Higuchi: "A Study of Rotor Position Control for Switched Reluctance Motor", Proc. of 10th IEEE International Conference on Power Electronics and Drive Systems (PEDS), pp.1039-1044, 2013.

[8] H. Makino, T. Kosaka, N. Matsui, M. Hirayama and M. Ohoto: "PWM-based Instantaneous Current Profile tracking control for Torque Ripple Suppression in Switched Reluctance Servomotors", Proc. of 10th IEEE International Conference on Power Electronics and Drive Systems (PEDS), pp. 1055-1060, 2013

[9] Bingni Qu, Jiancheng Song, Tao Liang, Hongda Zhang, "Mutual Coupling and Its Effect on Torque Waveform of Even Number Phase Switched Reluctance Motor", 2008 International Conference on Electrical Machines and Systems, pp. 3405-3410, 2008.

the magnetic analyses using 2D-FEA and mathematical analyses theoretically. For the torque ripple minimization, the novel current profiling technique has been proposed by employing the novel T-θ-i model considering the mutual coupling effect. To verify the effectiveness of the proposed control method, simulation and experiment have been conducted. As a result, compared with the conventional method, the torque ripple has been minimized effectively.

REFERENCES

[1] T. J. E. Miller, "Switched Reluctance Motors and their Control", Magna Physics Publishing and Clarendon Press Oxford, 1993.

[2] I. Husain and M. Ehsani, "Torque Ripple Minimization in Switched Reluctance Motor Drives by PWM Current Control", *IEEE Trans. on PE*, Vol.11, No.1, Jan/Feb, 1996.

Comparison of High Frequency Voltage Injection Methods for Shaft Sensorless Control of Wound-Field Flux Switching Machine

Hong-Quan Nguyen and Sheng-Ming Yang*
Department of Electrical Engineering
National Taipei University of Technology, Taipei, Taiwan, R.O.C.
Email: smyang@ntut.edu.tw

Abstract- This paper compares three shaft sensorless control schemes for wound-filed flux switching machines based on high frequency square-wave voltage injection. Because the machine has armature and field windings, high frequency voltage can be injected and processed in the armature winding, as the conventional PMSM, or injecting in one winding and processing the signals from the other winding. The analytic result shows that the scheme with *d*-axis voltage injection and *q*-axis current processing is generally similar to that used for PMSMs. An additional polarity identification is required to prevent phase error in the estimated position. However, schemes with injection and processing at separate windings have superior performance, and do not require polarity identification. In addition to the theoretical analysis, experimental verifications of the proposed sensorless control schemes are also presented.

I. INTRODUCTION

Shaft sensorless control techniques have been developed and applied for permanent magnet synchronous machines (PMSMs), especially the saliency-based algorithms, extensively recently [1-7]. Eliminating the position sensor not only increases system reliability but also reduces system volume, cost, and noise effects. Among the existing methods, high frequency (HF) voltage injection based schemes are particularly suited for standstill and low speed applications. These algorithms are divided into three groups: (a) Rotating voltage injection, extracts rotor position through injection of a rotating voltage vector in both *d*- and *q*-axes on the stationary reference frame [2]. (b) Pulsating voltage injection, injects a carrier voltage to either *d* or *q*-axis in the rotor reference frame [3]. (c) Square-wave voltage injection, injects a square voltage to either *d* or *q*-axis in the rotor reference frame [4-5]. The last method demonstrated a larger position and speed loop bandwidth than the other methods [4]. For all of the above methods, however, polarity identification is required because the induced HF current has two cycles per electrical period [6-7].

The flux switching machine (FSM) is a relatively new class of motor that has gain considerable research attentions in recent years. Among various FSM topologies, the wound-field FSM (WF-FSM) has the advantages of low cost, appropriate for harsh environment, and it is easy to regulate field for higher operating speeds or larger starting torque [8-9]. Due to the similarity of the operational principle, the conventional vector control and HF voltage injection methods developed for PMSMs can also be employed for flux switching machines [10]. Because a WF-FSM has armature and filed windings, this leads to the possibility of injecting and processing high frequency signals in the armature winding, as the conventional PMSM, or injecting and processing high frequency signals in separate windings.

This paper presents three motor position estimation and shaft sensorless control schemes with high frequency square-wave voltage injection for WF-FSM. These schemes are: voltage injection in the *d*-axis and the induced *q*-axis current is processed, voltage injection in the *q*-axis and the induced field current is processed, and voltage injection in the field and the induced *q*-axis current is processed. Implementations of the rotor position estimation with these schemes are analyzed, and their performances are evaluated experimentally.

II. HIGH FREQUENCY MODEL

The WF-FSM investigated in this paper is a three-phase, external rotor machine with 24-stator slot and 14-rotor pole. Figure 1 shows the cross-section view of the machine, and Table I lists its parameters. Detail design and performances of this machine can be found in [11].

Fig. 1. Cross-section view of the WF-FSM.

TABLE I   Parameters of the WF-FSM

Symbol	Meaning	Value
N_s	Stator slot	24
N_p	Rotor pole	14
ω_m	Rated speed	600 rpm
T	Rated torque	5.70 Nm
L_q/L_d	*q-d* axis inductance	13.32/14.56 mH
L_{fs}	Field self-inductance	36.0 mH
L_{mf}	Mutual inductance	9.6 mH

The mathematical model of the WF-FSM expressed in the rotor reference frame is

$$\begin{bmatrix} v_{qs}^r \\ v_{ds}^r \\ v_f \end{bmatrix} = \begin{bmatrix} r_s + L_{qs}s & \omega_r L_{ds} & \omega_r L_{mf} \\ -\omega_r L_{qs} & r_s + L_{ds}s & L_{mf}s \\ 0 & (3/2)L_{mf}s & r_f + L_{fs}s \end{bmatrix} \begin{bmatrix} i_{qs}^r \\ i_{ds}^r \\ i_f \end{bmatrix} \quad (1)$$

where r_s and r_f are the armature and field resistance, respectively, ω_r is the electrical speed, v_{qs}^r, v_{ds}^r, v_f and i_{qs}^r, i_{ds}^r, i_f are the q-d axes field voltages and currents, respectively, L_{qs} and L_{ds} are the q and d axes inductances, L_{mf} is the mutual inductance between the field and armature windings, and "s" is the differential operator. From Eq. (1), the high frequency voltage model of the WF-FSM can be found as

$$\begin{bmatrix} v_{qsi}^r \\ v_{dsi}^r \\ v_{fi} \end{bmatrix} = \begin{bmatrix} L_{qs}s & 0 & 0 \\ 0 & L_{ds}s & L_{mf}s \\ 0 & (3/2)L_{mf}s & L_{fs}s \end{bmatrix} \begin{bmatrix} i_{qsi}^r \\ i_{dsi}^r \\ i_{fi} \end{bmatrix} \quad (2)$$

where the subscript 'i' represents the corresponding high frequency component. Rewrite Eq. (2) as

$$s\begin{bmatrix} i_{qsi}^r \\ i_{dsi}^r \\ i_{fi} \end{bmatrix} = \begin{bmatrix} \dfrac{1}{L_{qs}} & 0 & 0 \\ 0 & \dfrac{2L_{fs}}{2L_{ds}L_{fs}-3L_{mf}^2} & \dfrac{-2L_{mf}}{2L_{ds}L_{fs}-3L_{mf}^2} \\ 0 & \dfrac{-3L_{mf}}{2L_{ds}L_{fs}-3L_{mf}^2} & \dfrac{2L_{ds}}{2L_{ds}L_{fs}-3L_{mf}^2} \end{bmatrix} \begin{bmatrix} v_{qsi}^r \\ v_{dsi}^r \\ v_{fi} \end{bmatrix} \quad (3)$$

III. Armature Winding HF Voltage Injection

Because WF-FSM has armature and filed windings, the HF voltage can be injected in the d- or q-axis of the armature winding, or the field winding. This section analyzes the armature winding injections, and the next section presents the filed winding injection.

A. d-axis HF square-wave voltage injection

The position estimation method presented in this subsection is denoted as <u>Method 1</u> for convenience. Rewrite Eq. (3) without the field component as

$$s\begin{bmatrix} i_{qsi}^r \\ i_{dsi}^r \end{bmatrix} = \begin{bmatrix} \dfrac{1}{L_{qs}} & 0 \\ 0 & \dfrac{2L_{fs}}{2L_{ds}L_{fs}-3L_{mf}^2} \end{bmatrix} \begin{bmatrix} v_{qsi}^r \\ v_{dsi}^r \end{bmatrix} \quad (4)$$

Transform the above equation into the estimated rotor reference frame, and let $\Delta\theta_r$ be the error between the actual and the estimated rotor positions, the induced currents in the estimated rotor reference frame can be found as

$$s\begin{bmatrix} i_{qsi}^{re} \\ i_{dsi}^{re} \end{bmatrix} = \frac{1}{L_1^2-L_2^2}\begin{bmatrix} L_1-L_2\cos(2\Delta\theta_r) & L_2\sin(2\Delta\theta_r) \\ L_2\sin(2\Delta\theta_r) & L_1+L_2\cos(2\Delta\theta_r) \end{bmatrix}\begin{bmatrix} v_{qsi}^{re} \\ v_{dsi}^{re} \end{bmatrix} \quad (5)$$

where the superscript 're' indicates the variable is in the estimated rotor reference frame, and L_1 and L_2 are

$$L_1 = \frac{L_{qs}+L_{ds}}{2} - \frac{3L_{mf}^2}{4L_{fs}}, \quad L_2 = \frac{L_{qs}-L_{ds}}{2} + \frac{3L_{mf}^2}{4L_{fs}} \quad (6)$$

Note that as shown in Eq. (5), the induced HF currents for the WF-FSM is similar to the induced HF currents for PMSM except L_1 and L_2 [4]. By letting $v_{qsi}^{re}=0$ and v_{dsi}^{re} to a square voltage with magnitude V_{di} and the differential currents replaced with difference currents, a current error can be deduced from the induced q-axis current as follows for small $\Delta\theta_r$,

$$i_{err\_dq} = \Delta i_{qsi}^{re} = \frac{L_2}{L_1^2-L_2^2}V_{di}\Delta T\sin(2\Delta\theta_r)$$
$$\cong \frac{2L_2}{L_1^2-L_2^2}V_{di}\Delta T\Delta\theta_r = K_{err\_dq}\Delta\theta_r \quad (7)$$

where both i_{err_dq} and Δi_{qsi}^{re} represent the q-axis current error, ΔT is the period of the injection voltage, and K_{err_dq} is a factor representing the sensitivity of i_{err_dq} to the position error. Therefore, i_{err_dq} is equivalent to a position error signal. Express K_{err_dq} with the actual inductances,

$$K_{err\_dq} = \frac{2L_{fs}(L_{qs}-L_{ds})+3L_{mf}^2}{L_{qs}(2L_{ds}L_{fs}-3L_{mf}^2)}V_{di}\Delta T\Delta\theta_r \quad (8)$$

It shows that both the machine saliency level and the inductances of the windings affect K_{err_dq}.

Figure 2 shows the rotor position estimator based on the d-axis HF square-wave voltage injection. The measured q-axis current is processed with a high-pass filter (HPF) first to separate the high frequency and the fundamental components. Then, the difference current is calculated and its sign corrected. This yields the current error i_{err_dq} shown in Eq. (7). Finally, the PI controller forces i_{err_dq} to zero and tracks the rotor position. Note that as shown in Eq. (7), current error is proportional to twice the position error. Therefore, correction of the polarity of the estimated rotor position is required to prevent phase error. Polarity identification procedures similar to those used for PMSM machines can be used for WF-FSM [6].

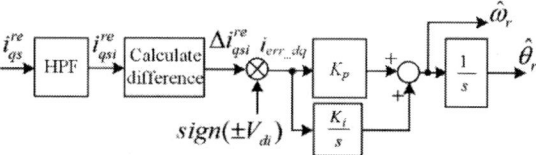

Fig. 2. Rotor position estimator for the d-axis square wave voltage injection.

B. q-axis HF square-wave voltage injection

The position estimation method presented in this subsection is denoted as <u>Method 2</u>. If injecting HF voltage into the q-axis of the armature winding, the induced field current can also be processed for a position error signal. Rewrite Eq. (3) to the following form and set $v_{fi}=0$, the induced HF field current is

$$s\,i_{fi} = \begin{bmatrix} 0 & \dfrac{-3L_{mf}}{2L_{ds}L_{fs}-3L_{mf}^2} \end{bmatrix}\begin{bmatrix} v_{qsi}^r \\ v_{dsi}^r \end{bmatrix} \quad (9)$$

Transform the voltages into the estimated rotor reference frame, field current becomes

3427

$$s\,i_{fi} = \frac{-3L_{mf}}{2L_{ds}L_{fs} - 3L_{mf}^2}\begin{bmatrix} \sin(\Delta\theta_r) & \cos(\Delta\theta_r) \end{bmatrix}\begin{bmatrix} v_{qsi}^{re} \\ v_{dsi}^{re} \end{bmatrix} \quad (10)$$

By letting $v_{dsi}^{re} = 0$ and v_{qsi}^{re} to a square voltage with magnitude V_{qi}, a current error can be deduced from the induced field current as follows for small $\Delta\theta_r$,

$$i_{err\_qf} \cong \frac{-3L_{mf}}{2L_{ds}L_{fs} - 3L_{mf}^2}V_{qi}\Delta T\Delta\theta_r = K_{err\_qf}\Delta\theta_r \quad (11)$$

where i_{err_qf} and K_{err_qf} are the field current error and the position error sensitivity factor, respectively. The rotor position estimator for q-axis square-wave voltage injection is identical to the estimator shown in Fig. 2 except for the input becomes i_{err_qf}. As shown in Eq. (11), there is only one current error cycle per electrical period in i_{err_qf}. Therefore, this scheme does not require polarity identification. Besides, i_{err_qf} is negatively proportional to the mutual inductance, and is independent to the saliency of the machine.

IV. FIELD WINDING HF VOLTAGE INJECTION

The position estimation method presented in this section is denoted as <u>Method 3</u>. If injecting HF voltage into the field winding, the induced q-axis current can be processed for a position error signal. Rewrite Eq. (3) to the following form and set both v_{qsi}^{re} and v_{dsi}^{re} to zero, then the induced armature current is

$$s\begin{bmatrix} i_{qsi}^r \\ i_{dsi}^r \end{bmatrix} = \begin{bmatrix} 0 \\ \frac{-2L_{mf}}{2L_{ds}L_{fs} - 3L_{mf}^2} \end{bmatrix}v_{fi} \quad (12)$$

Transform the currents into the estimated rotor reference frame, and the current become

$$s\begin{bmatrix} i_{qsi}^{re} \\ i_{dsi}^{re} \end{bmatrix} = \frac{-2L_{mf}}{2L_{ds}L_{fs} - 3L_{mf}^2}\begin{bmatrix} \sin(\Delta\theta_r) \\ \cos(\Delta\theta_r) \end{bmatrix}v_{fi} \quad (13)$$

By letting v_{fi} to a square voltage with magnitude V_{fi}, a current error can be deduced from the induced q-axis current as follows for small $\Delta\theta_r$,

$$i_{err\_fq} \cong \frac{-2L_{mf}}{2L_{ds}L_{fs} - 3L_{mf}^2}V_{fi}\Delta T\Delta\theta_r = K_{err\_fq}\Delta\theta_r \quad (14)$$

where i_{err_fq} is the q-axis current error with the field voltage injecting, and K_{err_fq} is the position error sensitivity factor. It can be seen that i_{err_fq} is very similar to i_{err_qf} shown in Eq. (11) except for its magnitude is 1.5 times smaller. Therefore, the estimator shown in Fig. 2 can also be used to estimate rotor position with i_{err_fq} as the input. Note that i_{err_fq} also has one current error cycle per electrical period, therefore does not require polarity identification. It is also interesting to note that in the case of field voltage injection, field current does not contain any position error information.

Talking all the relevant factors into consideration, the q-axis and the field HF voltage injection methods are much simpler than the d-axis HF voltage injection because they do not need polarity identification. Besides, among these

three methods, the q-axis HF voltage injection has the highest position error sensitivity.

V. CONTROL SYSTEM

Figure 3 shows the block diagram for the shaft sensorless control of WF-FSM based on the square-wave HF voltage injection. To compare the performances of the HF voltage injection methods presented in the previous sections, the input to the rotor position estimation is set to either i_{qs}^{re} or i_f, depending on the type of voltage injection used. The output of the position estimator is motor electrical position and speed. The estimated electrical position is used for the coordinate transformations of the current controller, and the speed is used as the feedback of the speed controller.

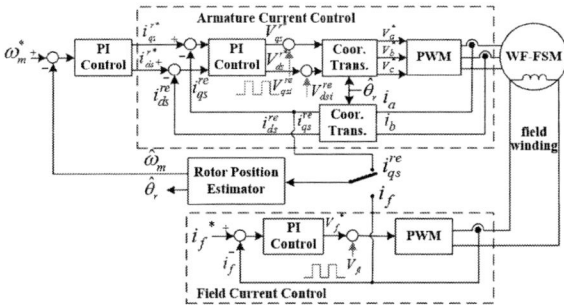

Fig. 3. Block diagram of the shaft sensorless control based on the square-wave HF voltage injection.

VI. EXPERIMENTAL VALIDATIONS

The proposed control schemes are implemented with a TI-TMS320F28335 digital signal processor and verified experimentally. The execution rate for the current and speed control loops are 18.31 and 2.28 kHz, respectively. Figure 4 shows the experimental system. The WF-FSM is mounted on a dynamometer. A hysteresis brake attached to the rotor shaft delivers the required load torque. An encoder mounted on the same shaft provides the actual rotor position for position error assessments.

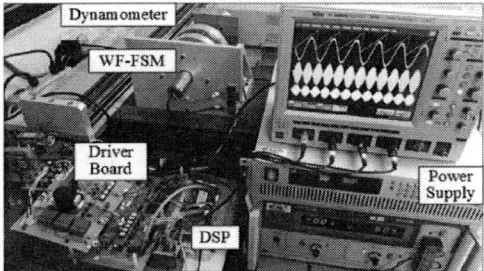

Fig. 4. Experimental system.

Figure 5 compares the measured high frequency currents for the proposed methods when the motor is at standstill and the error between the actual rotor position and the estimated position ($\Delta\theta_r$) varied from zero to 360 degrees. The frequency and magnitude of the injection voltage is 2.28 kHz and 20 V, respectively. It is clear that, the position error signal has two cycles per electrical period for Method 1, while only one cycle per electrical period for Method 2 and 3. Moreover, the peak of the position error signals i_{err_dq}, i_{err_qf}, and i_{err_fq} are 45mA, 125mA, and 90mA, respectively. The sensitivity factor for Method 2 and 3 are significantly higher than that of the Method 1. These results are consistent with the analysis in Section III and IV. It is also apparent that the difference current Δv_{qsi}^{re} shown in Fig. 5(a) is asymmetric for the adjacent half cycles. This property causes dc offset to i_{err_dq}, and consequently adds error to the estimated rotor position. Figure 6 compares the calculated and the measured induced position error signals: i_{err_dq}, i_{err_qf}, and i_{err_fq} when the injection voltage is 20V.

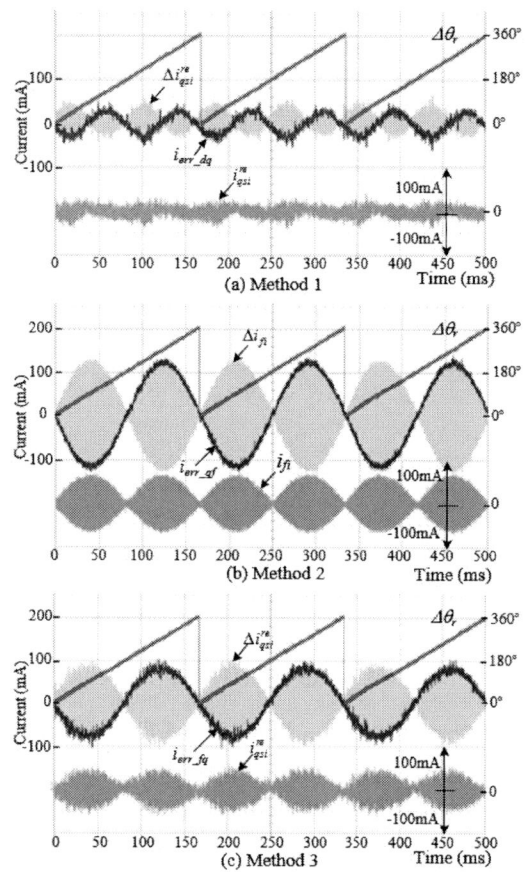

Fig. 5. Measured high frequency currents when the motor is at standstill and the position error varied from zero to 360 degrees.

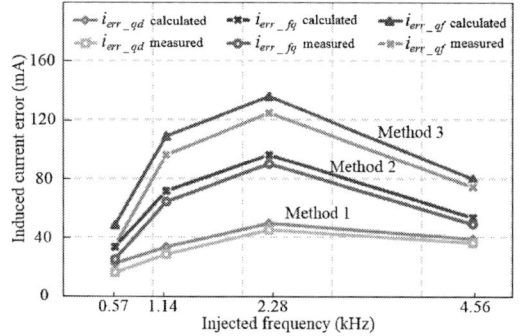

Fig. 6. Comparison of the calculated and the measured induced position error signals with injected voltages = 20V.

Figure 7 compares the dynamic responses of shaft sensorless control under no load and 40% load with the proposed methods. The external load is applied when the motor is running at constant speed. As shown in the figure that all the methods display good dynamic performances. However, the position errors are relatively different. The position error for Method 1 in Fig. 7(a) is approximately 18° without load, and dropped down to 7° after the load is applied. The high estimated position error is due to the low sensitivity of the position error signal i_{err_dq}. Besides, the asymmetric differential HF current also contributed some error to the estimated position. Moreover, the asymmetric level is related to the load torque. Higher q-axis current reduces the asymmetric level of the differential HF current, consequently, reduces the estimation errors.

As shown in Fig. 7(b), the position error for Method 2 is 4°, 5°, and 8° during acceleration, constant speed, and deceleration period, respectively. The position error ripple is low, approximately 1° and is insensitive to load variation. Fig. 7(c) shows the response for Method 3. It can be seen that the position error is within 5° to -5° when the motor is running at constant speed, with or without load. During the acceleration and deceleration period the error is slight higher, approximately 8°. The position error ripple is higher than the other two methods. The position error of Method 2 and 3 are more consistent than Method 1 with respect to load.

To summarize the above results, the q-axis and field voltage injection (Method 2 and 3) have better performance at the experimental conditions. However, the q-axis injection voltage will ultimately limit the operating speed range of the machine. In this regard, Method 3 is more advantageous for high speed operations because the HF voltage is injected in the field winding.

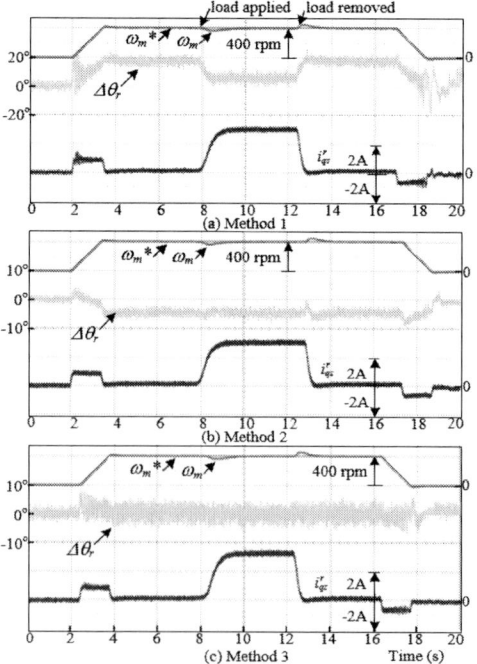

Fig. 7. Responses of shaft sensorless speed control, 40% load is applied when the motor is running at constant speed.

3429

VII. CONCLUSION

This paper investigated three high frequency square-wave voltage injection schemes for shaft sensorless control of WF-FSM drives. The experimental results show that all of the three methods can track rotor position reasonably well. However, the d-axis voltage injection method exhibits the highest position errors due to its low sensitivity of the position error signal and asymmetric differential HF current. Both the q-axis and the field voltage injections are simpler for implementation since they do not require polarity identification. Moreover, these two methods are more robust due to the high sensitivity of the induced current errors and independent to the saliency of the machine. Among the three methods, the field HF voltage injection is more advantageous for high speed operations because the HF voltage is injected in the field winding.

ACKNOWLEDGEMENTS

This work was supported by the Ministry of Science and Technology, Taiwan, R.O.C., under grant MOST 105-2221-E-027 -066 -MY2.

REFERENCES

[1] D. Raca, M. C. Harke, and R. D. Lorenz, ''Robust magnet polarity estimation for initialization of PM synchronous machines with near-zero saliency,'' *IEEE Trans. on Industry Applications*, vol. 44, no. 4, pp. 1199–1209, 2008.

[2] F. Gabriel, F. De Belie, X. Neyt, and P. Lataire, "High-frequency issues using rotating voltage injections intended for position self-sensing," *IEEE Trans. on Industrial Electronic*, vol. 60, no. 12, pp. 5447–5457, 2013.

[3] X. Luo, Q. Tang, A. Shen, and Q. Zhang, "PMSM sensorless control by injecting HF pulsating carrier signal into estimated fixed-frequency rotating reference frame," *IEEE Trans. on Industrial Electronic*, vol. 63, no. 4, pp. 2294–2303, 2016.

[4] Y. D. Yoon, S. K. Sul, S. Morimoto, and K. Ide, "High-bandwidth sensorless algorithm for ac machines based on square-wave-type voltage injection," *IEEE Trans. on Industry Applications*, vol. 47, no. 3, pp. 1361–1370, 2011.

[5] D. Kim, Y. C. Kwon, S. K. Sul, J. H. Kim, and R. S. Yu, "Suppression of injection voltage disturbance for high-frequency square-wave injection sensorless drive with regulation of induced high-frequency current ripple," *IEEE Trans. on Industry Applications*, vol. 52, no. 1, pp. 302–312, 2016.

[6] J. Holtz, "Acquisition of position error and magnet polarity for sensorless control of PM synchronous machines," *IEEE Trans. on Industry Applications*, vol. 44, no. 4, pp. 1172–1180, 2008.

[7] Y. S. Jeong, R. D. Lorenz, T. M. Jahns, and S. K. Sul, "Initial rotor position estimation of an interior permanent-magnet synchronous machine using carrier-frequency injection methods," *IEEE Trans. on Industry Applications*, vol. 41, no. 1, pp. 38–45, 2005.

[8] Y. Tang, J. J. H. Paulides, T. E. Motoasca, and E. A. Lomonova, "Flux-switching machine with DC excitation," *IEEE Trans. on Magnetics*, vol. 48, no. 11, pp. 3583–3586, 2012.

[9] H. Q. Nguyen, J. Y. Jiang, and S. M. Yang, "Design of a Wound-field Flux Switching Machine with Dual-stator to Reduce Unbalanced Shaft Magnetic Force", *Journal of The Chinese Institute of Engineers*, Vol. 40, No. 5, May 2017, pp. 441-448.

[10] T. C. Lin, Z. Q. Zhu, K. Liu, and J. M. Liu, "Improved sensorless control of switched-flux permanent-magnet synchronous machines based on different winding configurations," *IEEE Trans. on Industrial Electronic*, vol. 63, no. 1, pp. 123–132, 2016.

[11] S. M. Yang, J. H. Zhang, and J. Y. Jiang, "Modeling Torque Characteristics and Maximum Torque Control of a Three-Phase, DC-Excited Flux Switching Machine", *IEEE Transactions on Magnetics*. Vol. 52, No. 7, Jul. 2016, pp.1-4.

Design and Experimental Verification of a DAB Medium Frequency Transformer for a 6.6kV/200V Solid State Transformer

Rene Barrera-Cardenas[*], Takanori Isobe[*], Terazono Katsushi[†], Tadano Hiroshi[*]
[*]Faculty of Pure and Applied Sciences, University of Tsukuba, Japan, barrera.rene.fm@u.tsukuba.ac.jp
[†]Tsukuba Research Laboratory, Yaskawa Electric, Japan

Abstract—The optimal design and experimental verification of a medium frequency transformer as key element in a 6.6kV/200V modular solid state transformer is reported. The optimization algorithm is based on brute-force approach, which explore the complete design space by parametric sweeping. A 15kVA 20kHz Transformer has been built based on the outputs of the proposed optimal design algorithm and the experimental verification been performed.

I. INTRODUCTION

Research in direction of Solid State Transformer (SST) technology has been intensified recently in the area of power electronics [1], [2]. The SST architecture normally provides high flexibility, easy integration of renewable energy sources and energy storages facilities, which is essential for implementation of smart grids and island grids [1]. However, compared to a conventional low-frequency transformer, the SST increased functionality is contrasted by its higher cost and lower efficiency in medium voltage (MV) to low voltage (LV) AC-AC applications [1].

In a full AC-AC MV/LV SST system, isolated DC-DC converters are used to link two AC-DC stages, which provide the interfaces to single- or three-phase grids. The DC-DC converters are mostly based on Dual Active Bridge (DAB) or Resonant converter topologies, which both feature a Medium Frequency Transformer (MFT) as key component for isolation and step-down in voltage level. Demands on high performance from semiconductor devices and specially from MFT because the combination of medium frequency and MV, makes the isolated DC-DC converter the major challenge in the realization of the SST concept[3], [4].

Despite the increased worldwide research effort, the SST technology is still on a level of academic and advanced industry prototypes [2]. It should be mentioned that this work is linked to an academy-industry collaborative research project between University of Tsukuba and Yaskawa Electric Corporation, where a 6.6kV/200V SST system has been investigated. In the earliest stages of this research project, a 200V/200V 6kVA prototype (Generation 1) has been built to research in the SST operation using an oscillating power control concept. Then, a 6.6kV/200V 20kVA SST (Generation 2) has been designed based on DAB converter and research has been focus on 1.7kV SiC device implementation and power loss analysis for main system components. Different system parameters

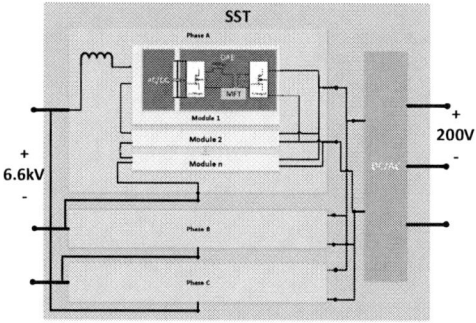

Fig. 1. Configuration of the 6.6kV/200V SST

has been investigated and the DAB operative frequency has been detected as one of the main parameters affecting system performance. Different 900V/900V 1kW DC-DC converter prototypes has been researched based on 1kHz, 10kHz and 100kHz DAB frequencies with three different MFTs prototypes, however no special attention was paid on MFT optimal design and therefore one of the critical components regarding power losses were the DAB MFT. Currently, the third stage of this project is under going with a 6.6kV/200V 300kVA SST (Generation 3) as main target with the MFT design and optimization as one of the main requirements to improve system performance.

The optimal design and experimental verification of a MFT as key element in a Dual Active Bridge (DAB) in a 300kVA 6.6kV/200V modular SST is reported. The optimization algorithm is based on brute-force approach, which explore the complete design space by parametric sweeping. A 15kVA 20kHz Transformer has been built based on the outputs of the proposed optimal design algorithm and the experimental verification has been performed.

II. SOLID STATE TRANSFORMER AND MFT: SPECIFICATIONS AND CONSTRAINTS

Fig. 1 shows the schematic diagram of a three-phase 6.6kV/200V SST considered for discussion. The primary side (6.6kV) consists of several series connected single-phase AC/DC converters per phase. The DC terminals of the AC/DC converters are connected to Dual Active Bridge

The 2018 International Power Electronics Conference

TABLE I
SPECIFICATIONS OF THE TARGET SST

System	Output Capacity	300 kVA
	Input Voltage	6.6kV
	Output Voltage	200 V
	Line Frequency	50 Hz
AC/DC	Number of Cascade	10
	Configuration	Δ
	Grid Connecting inductance	69.3 mH
	Nominal Cell DC voltage	855 V
	Carrier Frequency	4 kHz
DAB	Nominal Primary Voltage	855 V
	Nominal Secondary Voltage	855 V
	Equivalent series inductance	50-80 μH
	Switching Frequency	20 kHz
DC/AC	Smoothing Capacitor	300 μF
	Grid connecting inductance	2.22 mH
	Carrier Frequency	4 kHz

TABLE II
MFT SPECIFICATIONS AND CONSTRAINTS

General	Nominal power	15 kVA
	Input/Output Voltage	855V/855V
	Operation Frequency	20 kHz
	Leakage inductance	55μH
	Overload factor	30 %
Insulation	Insulation voltage	6.6 kV
	Insulation level	40 kV
	Insulation safety margin	100%
	Insulation type	Dry / Tape
Thermal	Cooling system	Natural convection
	Cooling type	dry-type
	Ambient Temperature	40 °C
	Maximum temperature rise	100 °C
Others	Minimum efficiency	99%
	Maximum current density	1.2 A/mm2
	Maximum flux density	0.8* Bsat

(DAB) converters for isolation and voltage adaptation by a Medium Frequency Transformer (MFT), with a smoothing capacitor C_1. DAB converter is controlled based on phase-shift modulation. The secondary side of the DAB converters are all connected in parallel and fed to a three-phase DC/AC inverter with 200V output voltage. The main specifications of the target SST system are shown in Table I. The SST general design is beyond of the scope of this paper, but it should be noted that the system parameters will determine the MFT specifications. MFT specifications and main constraints are presented in Table II.

Despite the power density reduction on final design, a thermal management based on dry-type natural convention has been selected because high reliability, simplicity and lower cost compared with other solutions. A comparison of power-frequency-isolation specification for different dry-type isolation, air cooling MFT reported in literature [5] is presented in Fig. 2. The reported MFT design is based on shell-type core structure. A cross-section view of the considered structure with the main core dimensions is presented in Fig. 3. Also, a cross-section view of the considered winding structure for MFT is shown in Fig. 4, which is according with the specified insulation type discussed as following. First, it should be noted that although the intended applied voltage to the MFT windings is 855 V and the MFT voltage ratio is 1:1, the insulation voltage between primary and secondary windings is determine by the SST voltage (6.6kV) and therefore an insulation level of 40 kV between windings has been selected following IEC 60076-3 standard recommendations [6]. The insulation level together with the insulation material define the minimum insulation distances between low voltage (LV) and high voltage (HV) windings, and between the core and the HV winding. Following the main project frame, a simple, reliable and low cost MFT design is preferred, therefore a dry-type insulation consists of insulation tape has been chosen. The insulation material used between windings is an insulation tape based on Poly-Tetra-Fluoro-Ethylene (PTFE) film, NITOFLON No.900UL from Nitto manufacturer, with 0.5mm thickness and using many layers to achieved the

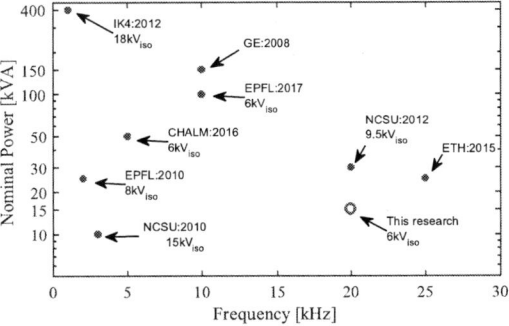

Fig. 2. Power-Frequency specifications for Dry-type isolation and Air cooling MFT reported in literature [5]. (Label= Institution:Year, Isolation voltage). GE[7], IK4[8], EPFL [9], [2], CHALM [10], NCSU [11], [12], ETH[13].

desired insulation distance. For winding layers, an insulation tape based on polyester film, CT289 from Yahua manufacturer, with 0.08mm thickness and breakdown voltage of 7kV, as for winding voltage lower than 1.1 kV, an insulation level of 6 kV is suggested [6].

In order to estimated the minimum distance between primary and secondary winding, the practical expression for the breakdown voltage in solid materials proposed in [14], can be used:

$$V_{BD} = V_{BD.ref} \cdot \left(\frac{d_{BD}}{d_{ref}} \right)^{\alpha_{isoM}} \quad (1)$$

where V_{BD} is the breakdown voltage, $V_{BD.ref}$ is a reference breakdown voltage, d_{BD} is the thickness of the sample, d_{ref} is the reference thickness and α_{isoM} is a fitting exponent, $\alpha_{isoM} = 0.5$ according to [14]. Then, for the case in hand, the minimum insulation thickness between primary and secondary winding ($d_{ins1.min}$) can be estimated by:

$$d_{ins.min} = d_{ref} \cdot \left(\frac{K_{SFI} \cdot V_{iso}}{V_{BD.ref}} \right)^{\frac{1}{\alpha_{isoM}}} \quad (2)$$

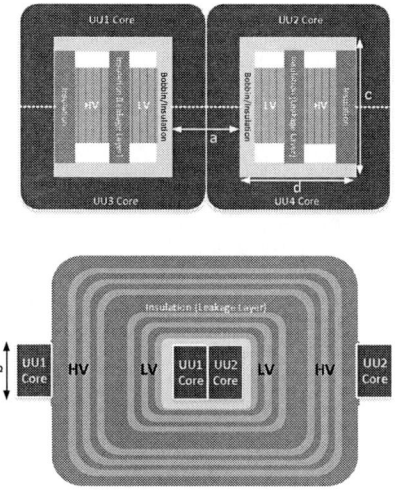

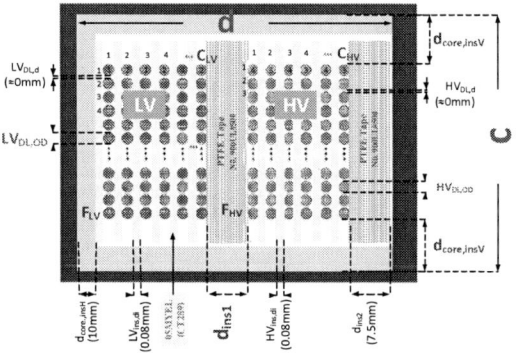

Fig. 3. Cross-section view of the considered Shell-type MFT. Top: frontal view, bottom: top view

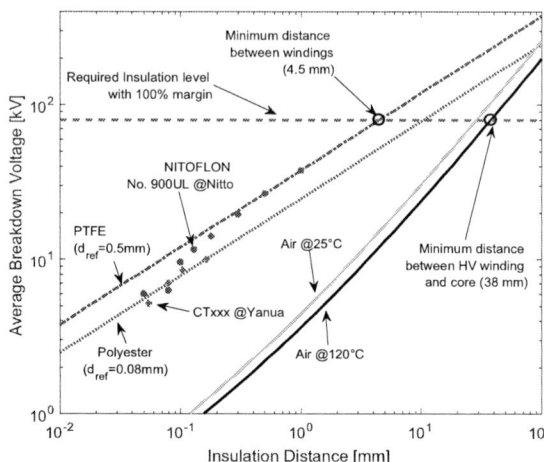

Fig. 5. Average Breakdown Voltage vs. Insulation thickness/distance for the considered insulation materials

Fig. 4. Considered winding structure for the MFT

where K_{SFI} is the insulation safety factor and V_{iso} is the required insulation level. In the case of vertical space between HV winding and core ($d_{core,insV}$), not insulation material has been used, only mainly air (plus the bobbin), so the expression used in [14] for the breakdown voltage of air as function of distance can be used to find the minimum $d_{core,insV}$:

$$V_{BD.air} = 112.4 \cdot \left(\frac{p_r \cdot d_{core,insV}}{0.0159 \cdot T_{air}} + \sqrt{\frac{p_r \cdot d_{core,insV}}{T_{air}}} \right) \tag{3}$$

where p_r is the air pressure in [kPa], T_{air} is the air temperature in K, and the voltage and distance are expressed in kV and m, respectively. Fig. 5 shows the average breakdown voltage versus the insulation material thickness/distance for the considered insulation materials, also includes air breakdown voltage for 25°C and 120°C with 101.3 kPa pressure.

III. MFT OPTIMAL DESIGN

The proposed MFT optimal design algorithm is presented in Fig. 6. The system specifications and the minimum insulation distances are the main inputs of the optimization algorithm. Then, a geometry parametrization is perform for a selected core material and winding conductor, so all the possible combination of design parameters are evaluated. Calculations are focus on power losses, box volume, outer area and temperature rise.

Considering the phase shift angle (γ) between voltage V_1 and V_2, the transformer RMS current can be estimated by

$$I_p = \frac{(V_{DC1} + V'_{DC2}) \cdot \gamma}{4 \cdot \pi \cdot L_\delta \cdot f_s} \cdot \sqrt{\frac{\pi - \gamma}{\pi}} \tag{4}$$

where L_δ is the additional MFT leakage inductance and f_s is the operating frequency. The maximum flux density applied to a coil with effective cross section A_e and number of turns N_{LV} is given by

$$B_{pk} = \frac{V_{DC1} \cdot (\pi - \gamma)}{4 \cdot \pi \cdot N_{LV} \cdot A_e \cdot f_s} \tag{5}$$

The MFT power losses are mainly core and winding losses. Core Losses are estimated by improved generalized Steinmetz equation (iGSE)[15] with the
effective core volume ($Vol_{core} = A_e \cdot l_e$):

$$P_{core} = k_i \cdot \left(\frac{V_{DC1}}{N_{LV} \cdot A_e} \right)^\beta \cdot \left(\frac{\pi - \gamma}{2\pi} \right)^{\beta - \alpha + 1} \cdot \left(\frac{1}{f_s} \right)^{\beta - \alpha} \cdot A_e \cdot l_e \tag{6}$$

The winding losses are calculated based on the equivalent frequency dependent AC resistance from referred to primary side ($R_{ac1}(f)$) :

$$P_w = \sum_{h=1}^{h_n} R_{ac1}(h \cdot f_s) \cdot I_{1h}^2 \tag{7}$$

where I_{1h} is the RMS value of the primary current trough the transformer winding at h^{th} harmonic and h_n is the highest considered harmonic component. In order to estimated the current harmonic components the approach reported in [16]. Then, the RMS value of the h_{th} current harmonic can be calculated by

$$I_{1h} = \frac{1}{\sqrt{2}} \cdot \frac{\Delta V_h}{2\pi \cdot f_s \cdot h \cdot L_\delta} \qquad (8)$$

d+

$$\Delta V_h = \sqrt{V_{AC1h}^2 + V_{AC2h}^2 - 2 \cdot V_{AC1h} \cdot V_{AC2h} \cdot \cos(h \cdot \gamma)}$$

$$V_{ACXh} = V_{DCX} \cdot \frac{2}{h \cdot \pi} \cdot (1 - \cos(h \cdot \pi))$$

The frequency dependent AC resistance, which includes skin and proximity effects, is estimated following the Dowell's models [17], [18], [19]. The frequency depended leakage inductance is estimated based on hybrid model presented in [20]. Then, d_{ins1} can be estimated to fulfill the required leakage inductance value at a given frequency.

Dielectric losses are calculated but it should be mention that are very low compared with other losses and therefore could be neglected for this case study. Only the design which fulfill the considered constraints are storage and the performance space for a given core and winding conductor type could be obtained.

Transformer temperature estimation is needed during the optimization process to verify that temperature specifications are not exceeded. In a transformer with natural air cooling, as considered is in this paper, the dominant heat-transfer mechanism is by convection [21]. The Newton's equation of convection is therefore used to determine the temperature rise (ΔT_{Tr})of the magnetic component [19], [18]:

$$P_{core} + P_w = h_{Tr} \cdot A_t \cdot \Delta T_{Tr} = \frac{1.42 \cdot A_t}{H_{Tr}^{0.25}} \cdot \Delta T_{Tr}^{1.25} \qquad (9)$$

where A_t is the external surface area of the core and windings, and h_{Tr} is the convection heat transfer coefficient ($h_{Tr} = 1.42 \cdot \left(\Delta T_{Tr}/H_{Tr}\right)^{0.25}$ [19]), and H_{Tr} is the height of the transformer ($H_{Tr} = c + a$, see Fig. 3).

The MFT loss-volume performance space for the considered case is presented in Fig. 7. Two core materials has been considered, ferrite LP9 and Finemet F3CC. Because waiting time for customize core dimensions want to be avoid, only standard and available cores has been considered in this case study. Under that conditions, Ferrite LP9 cores are the best option, and only two standard sizes can be used for the given constraints: UU120/118/40 and UU160/120/40. The selected design is also shown in Fig. 7 with a start mark. A summary of the main design parameters are shown in Table III. In order to simplify the winding design, same winding parameters have been used for both windings (primary and secondary side).

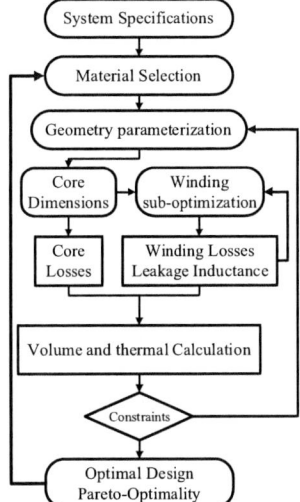

Fig. 6. MFT Optimal Design Algorithm

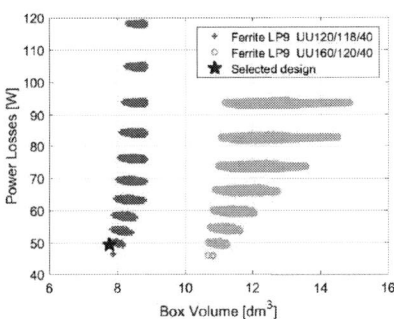

Fig. 7. MFT Losses-Volume Performance space for the considered case

IV. MFT PROTOTYPE AND EXPERIMENTAL VERIFICATION

The built MFT prototype based on the selected design parameters from MFT optimal design algorithm is shown Fig. 8. First, the estimation of MFT electrical parameters is verified by compared them with measured values from short circuit test. The MFT AC to DC resistance ratio and leakage inductance referred to low voltage side versus frequency are shown in Fig. 9, where measurements and calculations based on Dowell's model [17], [19] are compared. High deviation between calculated and measured values has been found, about

TABLE III
MFT SELECTED DESIGN

Core: Ferrite LP9			Winding: TEX-E conductor		
Dimensions	a	60 mm	Cond. diameter	d_s	0.8 mm
	b	40 mm	Number of turns	N_{LV}	30
	c	175 mm	Parallel cond.	n_p	33
	d	59 mm	Number of layers	C_{LV}	11
	B_{pk}	0.2 T			

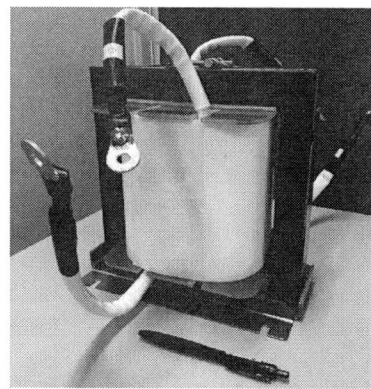

Fig. 8. MFT Prototype

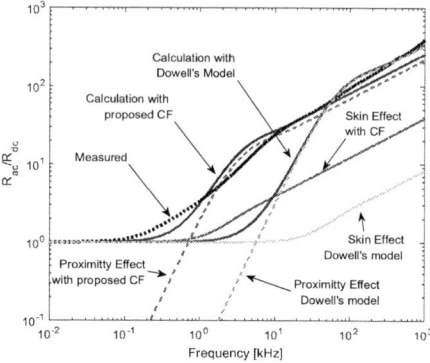

(a) AC Resistance to DC resistance ratio

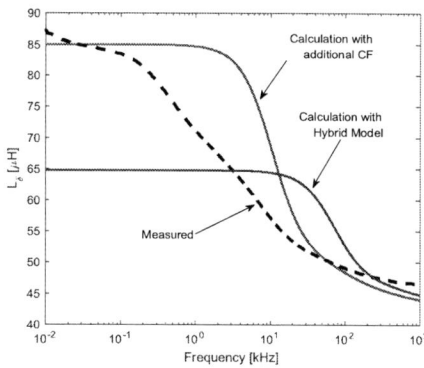

(b) Leakage Inductance

Fig. 9. MFT Electrical parameters

60% lower R_{ac} estimation and 20% higher L_δ estimation at 20 kHz, compared with the measured values from the MFT prototype. It can be observed that measured data shows frequency effects (increments on R_{ac} and decrement on L_δ) from frequency above 100 Hz, however frequency effects on calculated data becomes evident only beyond 2 kHz, with proximity effects as dominant source of losses in the selected design. It is clear that the Dowell's models can not estimated accurately the main MFT electrical parameters for the considered winding structure. The insulation requirement for this application combined with the selected insulation method forces to consider designs with very low winding fill factor, which differs from ideal conditions considered in Dowell's model. In this case, the winding height (h_w) is very small compared with the core window height (c), so the magnetic field distribution along core window is not longer unidimensional, as it is the main assumption in Dowell's model. The magnetic field around the top and bottom of the winding has an additional horizontal component in this case, increasing the magnetic field magnitude and therefore the penetration on the conductor as well as a higher increment on R_{ac}.

Here, a Correction Factor (CF) is proposed to calculate an equivalent conductor diameter with similar penetration effects based on Dowell's model. Then, considering the winding structure presented in Fig. 4, the AC resistance of low voltage side winding ($R_{ac,LV}$) can be calculated by

$$\frac{R_{ac,LV}}{R_{DC,LV}} = \frac{2\left(C_{LV}^2 - 1\right)}{3 \cdot CF} \cdot G_R\left(f_s, d_{eq,LV}\right) + 2 \cdot F_R\left(f_s, d_{eq,LV}\right) \tag{10}$$

$$d_{eq,LV} = LV_{DL,OD} \cdot CF \tag{11}$$

$$CF = \left(\frac{c}{h_w}\right)^2 \tag{12}$$

$$h_w = F_{LV} \cdot LV_{DL,OD} + (F_{LV} - 1) \cdot LV_{DL,d} \tag{13}$$

where $F_R\left(f_s, d_s\right)$ and $G_R\left(f_s, d_s\right)$ are the Dowell's skin and proximity factors as function of frequency f_s and conductor diameter d_s, $d_{eq,LV}$ is the equivalent conductor diameter of low voltage side winding. Similar expressions can be used for estimation of high voltage side ac resistance ($R_{ac,HV}$). In the case of leakage inductance, the CF is applied to the hybrid model presented in [20], by simply using the equivalent conductor diameter ($d_{eq,LV}$) from equation 11 in the calculations. Complete hybrid model description can be found in [20]. Fig. 9 also shows estimations based on the proposed CF. It can be observed that by applying the proposed CF with the Dowell's model, the increments on R_{ac} and decrement on L_δ start from frequency above 200 Hz, and the estimations are close to the measured data. However, it can be noted that the tendency of calculated values are not the same as the measured data, because limitation on the model. Table IV shows the measured and calculated DC resistance, AC resistance and leakage inductance referred to low voltage side at 20 kHz. Estimation errors with proposed CF have been reduced to 13% and 8.3%, in R_{ac} and L_δ, respectively, compared with Dowell's model estimation in this case (60% and 20%).

For the MFT experimental verification, a 15-kW class DAB converter, which was designed according with the specifica-

3435

TABLE IV
MFT ELECTRICAL PARAMETERS

	Measured	Calculated	Error
R_{dc}	23.79 $m\Omega$	26.66 $m\Omega$	12 %
R_{ac} @ 20kHz	0.854 Ω	0.963 Ω	13 %
L_{δ} @ 20kHz	53.39 μH	57.83 μH	8.3 %

tions of a single cell for the target 300-kVA SST (see Table I) , was fabricated. The 1.7kVx72A SiC MOSFET devices (C2M0045170D) from Cree manufacturer have been used for the fabricated DAB. A dead time of $0.5\mu s$ is considered for the DAB operation. Fig. 10 shows statically measured losses and efficiencies of the fabricated DAB converter. The loss characteristics are well agreed to the expected ones. There is a comparatively high offset loss which comes from transformer core loss and switching losses at light load due to shorting of the device output capacitance.

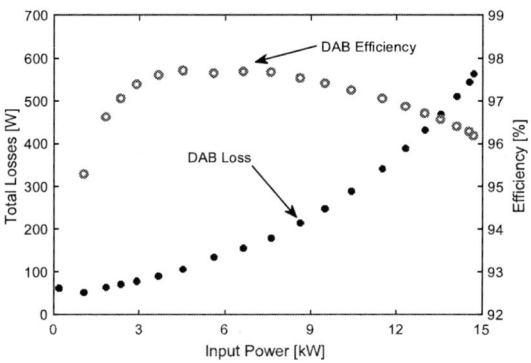

Fig. 10. Power losses and efficiency of the fabricated DAB converter

Fig. 11a shows the experimental waveforms for the primary/secondary voltages (v_p, v_s) and currents (i_p, i_s) at nominal power (15kW). The DAB had been operated with voltage ratio different than one ($V_{in} > V_{out}$) in order to compensate the unexpected high winding resistance of the MFT prototype and reduce the transitory peak current on MFT terminals.

In order to estimated MFT core losses from measurements, the magnetizing current (i_m) and the induce voltage need to be measured. The magnetizing current is estimated from the difference between measured i_p and i_s, ($i_m = i_p - i_s$). On the other hand, an auxiliary winding with 3 turns has been included in one of the outer core legs of the MFT and the open circuit voltage in the auxiliary winding (v_m) has been measured, so the induce voltage referred to the primary side (v_{p*}) can be estimated, $v_{p*} = N_{pm} \cdot v_m$, using the measured effective voltage turn ratio N_{pm} between primary winding and auxiliary winding. Fig. 11b shows the measured i_m and v_m at nominal power. Then, the average core losses can be calculated by

$$P_{core} = f_s \cdot A_e \cdot l_e \cdot \oint H \cdot dB = f_s \cdot N_{pm} \cdot \int v_m \cdot i_m \cdot dt \quad (14)$$

Additionally, the magnetic field (H) and the magnetic flux density (B) can be estimated from magnetizing inductance and induce voltage, respectively:

$$H = N_{LV} \cdot i_m / l_e \quad (15)$$

$$B = \frac{N_{pm}}{A_e \cdot N_{LV}} \cdot \int v_m dt \quad (16)$$

where N_{LV} is the number of turn of primary side winding. Fig. 11c shows the measured B-H hysteresis curve at nominal power. It can be observed that i_m shows a DC component, which can be explain because the operation of DAB with voltage ratio major than one.

Fig. 12a shows a comparison between measured and calculated phase shift angle ($\gamma_{mea}, \gamma_{cal}$), MFT input RMS current ($I_{p,mea}, I_{p,cal}$) and peak flux density ($B_{pk.mea}, B_{pk.cal}$) as function of the input power. Measured and calculated phase shift angle are close for input power beyond 5 kW. However, for low power, γ_{mea} differs from γ_{cal}, mainly because influences of the device output capacitance are more significant at at light load. Estimation of MFT input RMS current agree with measurements for all the power range. The measured peak flux density decreases faster than estimated values between 0 and 3 kW, and keeps almost constants beyond 3 kW. Even considering the measured phase shift angle and input voltage into the calculations from equation 5, the estimations are not agree with the measurements. This can be related to the H_{DC} component introduced during the DAB operation with voltage ratio major than one.Fig. 12b shows the measured H_{DC}. It is expected that analytical core losses estimation by equation 6 also differ from measured core losses if the DC magnetic field (H_{DC}) influence on core loss parameters (β, k_i) is neglected. There is not reported core parameter dependences with H_{DC} for the considered core material, ferrite LP9. In order to account for those variations, the parameter dependences with H_{DC} reported in [22] for a similar core material (Ferrite N27) has been used. Fig. 12c shows the considered core loss parameters variation based on Ferrite N27 material dependences, and these parameter variation can be used to estimated core losses by equation 6. Fig. 13a shows the measured core losses and the estimations with and without accounting β and k_i dependences with H_{DC}. Measured core losses increases as power increases. It is mainly due to the H_{DC} component as it can be noted from calculated core losses taking into account β and k_i dependences in Fig. 13a.

On the other hand, the winding losses can be estimated from measured primary current and the measured AC resistance referred to primary side by equation 7 and getting primary current frequency components using fast-Fourier transformer. Fig. 13b shows the measured and calculated winding losses. Calculations and measurements agree very well and winding losses are dominated by proximity effects. Additionally, Fig. 13c shows the total MFT losses obtained from added the measured core and winding losses. MFT estimation agrees

The 2018 International Power Electronics Conference

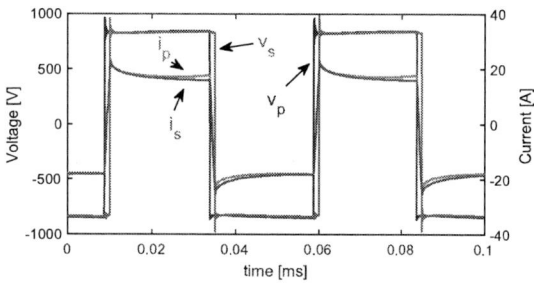

(a) Primary/Secondary voltages and currents

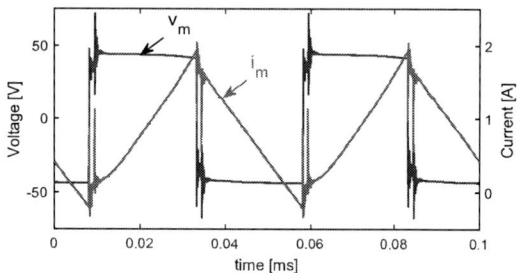

(b) Magnetizing current (i_m) and auxiliary winding voltage (v_m)

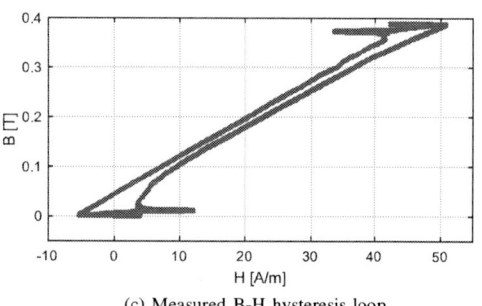

(c) Measured B-H hysteresis loop

Fig. 11. Experimental Waveforms at nominal power (15 kW)

with the measurements for all the power range except between 3 kW and 5 kW, where underestimation was obtained.

Finally, Fig. 14 shows a possible DAB loss breakdown based on transformer measured data and the calculation of MOSFET conduction loss by:

$$P_{MOS.Cond} = 8 \cdot R_{ds} \cdot \frac{I_p^2}{2} \qquad (17)$$

where R_{ds} is the MOSFET ON-resistance (assumed to be $60 m\Omega$ for C2M0045170D devices).

V. CONCLUSIONS

In this paper, a MFT optimal design algorithm has been proposed and its application and validation has been reported to fulfill design requirements linked to a 6.6kV/200V SST application. A 15kVA- 20kHz MFT has been built based on the

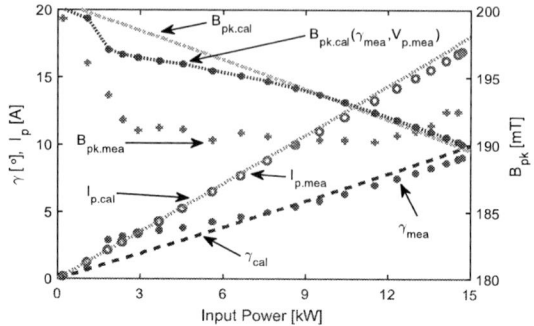

(a) Comparison between measured (mea) and calculated (cal): Phase shift angle(γ), MFT input rms current (I_p) and peak flux density (B_{pk})

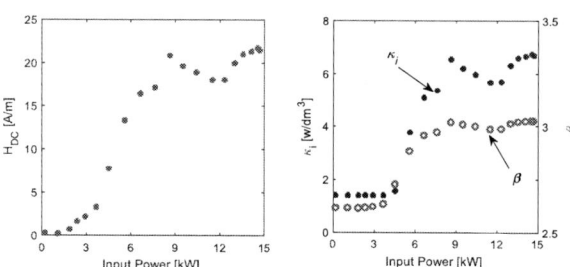

(b) Measured DC Magnetic Field (c) Core loss parameters variation based (H_{DC}) on Ferrite N27 material dependences

Fig. 12. Measured and estimated MFT variables as function of input power

selected design parameters from the proposed optimization algorithm. Validation of electrical parameter estimation based on a proposed modification of the Dowell's model has been performed. Also, power loss measurements has been performed to determine the core and winding losses as function of the input power. By considering influences of DC magnetic field introduced in the DAB operation, the considered models has been validated. Future work will be focus on proximity loss reduction, as it is the main loss contributor in the fabricated MFT.

REFERENCES

[1] M. Leibl, G. Ortiz, and J. W. Kolar, "Design and Experimental Analysis of a Medium-Frequency Transformer for Solid-State Transformer Applications," *IEEE Journal of Emerging and Selected Topics in Power Electronics*, vol. 5, pp. 110–123, Mar. 2017.

[2] M. Mogorovic and D. Dujic, "Medium Frequency Transformer Design and Optimization," in *PCIM Europe 2017; International Exhibition and Conference for Power Electronics, Intelligent Motion, Renewable Energy and Energy Management*, pp. 1–8, May 2017.

[3] S. Inoue and H. Akagi, "A Bidirectional Isolated DC-DC Converter as a Core Circuit of the Next-Generation Medium-Voltage Power Conversion System," *IEEE Transactions on Power Electronics*, vol. 22, pp. 535–542, Mar. 2007.

[4] H. Akagi and S. Inoue, "Medium-Voltage Power Conversion Systems in the Next Generation," in *Power Electronics and Motion Control Conference, 2006. IPEMC 2006. CES/IEEE 5th International*, vol. 1, pp. 1 –8, Aug. 2006.

[5] Peng Shuai, *Optimal Design of Highly Efficeinct, compact and Silent Medium Frequency Transformer for Future Solid State Transformers*. PhD Thesis, ETHZ, 2017.

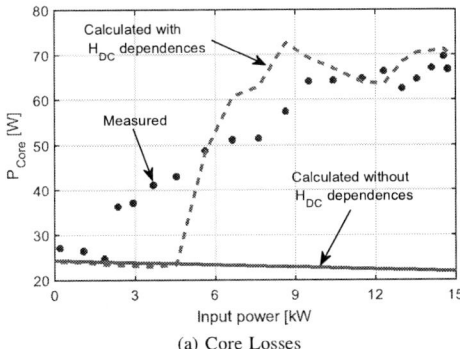

(a) Core Losses

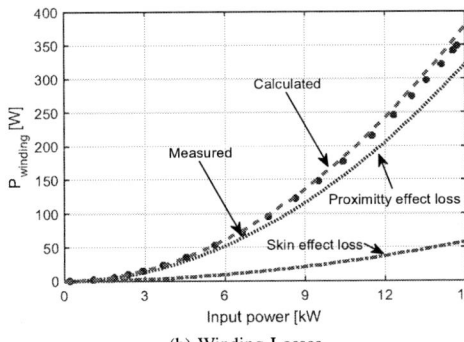

(b) Winding Losses

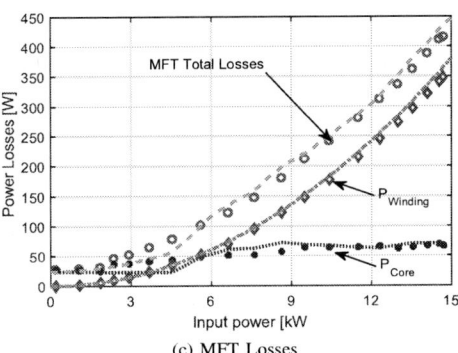

(c) MFT Losses

Fig. 13. Comparison between measured and estimated MFT Power Losses

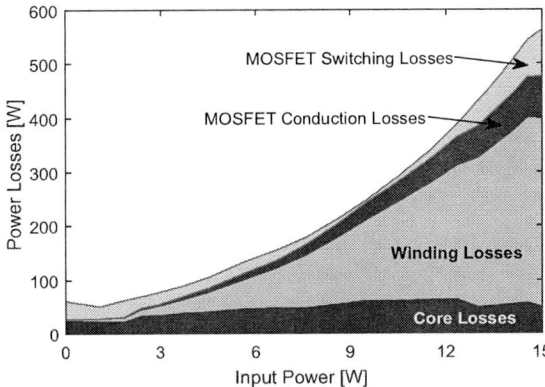

Fig. 14. Possible DAB loss breakdown based on measured data

frequency transformer design for a 7.2kv to 120v/240v 20kva solid state transformer," in *IECON 2010 - 36th Annual Conference on IEEE Industrial Electronics Society*, pp. 493 –498, Nov. 2010.

[12] S. S. Baek, S. Bhattacharya, B. Cougo, and G. Ortiz, "Accurate equivalent circuit modeling of a medium-voltage and high-frequency coaxial winding DC-link transformer for solid state transformer applications," in *2012 IEEE Energy Conversion Congress and Exposition (ECCE)*, pp. 1439–1446, Sept. 2012.

[13] D. Rothmund, G. Ortiz, T. Guillod, and J. W. Kolar, "10kv SiC-based isolated DC-DC converter for medium voltage-connected Solid-State Transformers," in *2015 IEEE Applied Power Electronics Conference and Exposition (APEC)*, pp. 1096–1103, Mar. 2015.

[14] Salmon E.R., "Dielectric strength of an insulation materialâĂŤIs it a constant?," *IEEE Electrical Insulation Magazine*, vol. 5, Feb. 1989.

[15] K. Venkatachalam, C. R. Sullivan, T. Abdallah, and H. Tacca, "Accurate prediction of ferrite core loss with nonsinusoidal waveforms using only Steinmetz parameters," in *2002 IEEE Workshop on Computers in Power Electronics, 2002. Proceedings.*, pp. 36–41, June 2002.

[16] I. Villar, U. Viscarret, I. Etxeberria-Otadui, and A. Rufer, "Global Loss Evaluation Methods for Nonsinusoidally Fed Medium-Frequency Power Transformers," *IEEE Transactions on Industrial Electronics*, vol. 56, pp. 4132–4140, Oct. 2009.

[17] P. L. Dowell, "Effects of eddy currents in transformer windings," *Proceedings of the Institution of Electrical Engineers*, vol. 113, pp. 1387–1394, Aug. 1966.

[18] W. G. Hurley, "Optimizing core and winding design in high frequency transformers," in *Power Electronics Congress, 1996. Technical Proceedings. CIEP '96., V IEEE International*, pp. 2–13, IEEE, Oct. 1996.

[19] W. G. Hurley, W. H. Wolfle, and J. G. Breslin, "Optimized transformer design: inclusive of high-frequency effects," *IEEE Transactions on Power Electronics*, vol. 13, pp. 651–659, July 1998.

[20] M. Mogorovic and D. Dujic, "Medium Frequency Transformer Leakage Inductance Modeling and Experimental Verification," 2017.

[21] R. A. Jabr, "Application of geometric programming to transformer design," *IEEE Transactions on Magnetics*, vol. 41, pp. 4261– 4269, Nov. 2005.

[22] J. Muhlethaler, J. Biela, J. Kolar, and A. Ecklebe, "Improved Core-Loss Calculation for Magnetic Components Employed in Power Electronic Systems," *IEEE Transactions on Power Electronics*, vol. 27, pp. 964 –973, Feb. 2012.

[6] International Standard IEC 60076-3, "Power Transformers: Insulation levels, dielectric tests and external clear- ances in air," 2000.

[7] S. Prabhakaran, C. Stephens, F. Johnson, and L. Iorio, "High frequency power transformers for future military propulsion and energy systems," 2008.

[8] I. Villar, L. Mir, I. Etxeberria-Otadui, J. Colmenero, X. Agirre, and T. Nieva, "Optimal design and experimental validation of a Medium-Frequency 400kva power transformer for railway traction applications," in *2012 IEEE Energy Conversion Congress and Exposition (ECCE)*, pp. 684–690, Sept. 2012.

[9] I. Villar, *Multiphysical Characterization of medium frequency power electronic transformers*. PhD Thesis, Ecole Polytechnique Federale de Lausanne (EPFL), 2010.

[10] M. A. Bahmani, *Design and Optimization Considerations of medium frequency power transformers in high-power DC-DC Applications*. PhD Thesis, Chalmers University of Technology, 2016.

[11] Y. Du, S. Baek, S. Bhattacharya, and A. Huang, "High-voltage high-

Research on the Unbalanced Compensation Range of Delta-connected Cascaded H-bridge Multilevel SVG

Rui Luo[*], Yingjie He, Yiming Tu, Xingxing Chen, Jinjun Liu
School of Electrical Engineering, Xi'an Jiaotong University, Xi'an, China
*E-mail: luorui0324@foxmail.com

Abstract— **Delta-connected cascaded H-bridge multilevel SVG can compensate the reactive and negative-sequence current synthetically and it is one of the most effective solutions for improving power quality in high-voltage and large-power applications. Focusing on the dc-link voltage control problems for SVG under unbalanced conditions, the asymmetric active powers between three-phase clusters produced by line voltages and phase currents are analyzed and the expression of zero-sequence current for cluster voltage control is also derived on this paper. On the basis of this, the quantitative relationship between the unbalance degree of the supply voltage, the unbalance degree of the compensation current and the compensation ability is analyzed. Finally, the correctness of the relevant theoretical analysis is verified by simulation. These quantitative analyses have directive significance for the application and selection of the delta-connected cascaded H-bridge multilevel SVG.**

Keywords— *Delta-connected cascaded H-bridge multilevel SVG, zero-sequence current, unbalance degree.*

I. INTRODUCTION

In recent years, static var generator (SVG), due to good compensation, fast response, small energy storage component size and low harmonic content and so on, has been widely applied to improve power quality problems [1-3]. Among the SVGs family, the cascaded H-bridge multilevel SVG has been receiving considerable attention due to the advantages of easy modular expansion, independent inverter units, no need for multiple transformer access and less switching devices required at the same output level [4-7]. The cascaded H-bridge multilevel SVG has two different structures, namely star and delta. When the asymmetrical loads are compensated, the star-connected SVG is required to output both reactive and negative-sequence currents, and offset occurs at neutral-point of the star-connected SVG. Therefore, it is not suitable for the star-connected SVG to compensate the severely asymmetrical load. For the delta-connected SVG, the H-bridge modules in each cluster can be independently controlled, which is equivalent to the change of the asymmetrical load structure. The delta-connected SVG can quickly and accurately compensates symmetrical and asymmetrical loads well, which is impossible for the star-connected SVG to realize. For above reasons, the delta-connected SVG is chosen as the emphases for research in this paper.

Delta-connected cascaded H-bridge multilevel SVG has many advantages that many other topologies don't have, but there are also some defects in the delta-connected SVG. Among them, cluster voltage control is difficult for the design of SVG control system. In practical applications, the parallel loss and the switching loss of the H-bridge module, and the conduction inconsistency between the trigger pulses of the switching device will lead to the imbalance of the dc-link voltage. If corresponding measures are not taken to control the dc-link voltage, the imbalance of the dc-link voltage will not only affect the compensation effect of SVG, but also threaten the safe and stable operation of the SVG. Taking aim at the above mentioned problem, many control strategies have been put forward by experts and scholars. The split phase control method [8-9], the equilibrium component method [10] and the zero-sequence current injection method [11-15] are three main methods currently used for cluster voltage control in the delta-connected SVG. Among the methods, zero-sequence current injection method has received extensive attention because the zero-sequence current only flows in the loop and does not affect the output compensating line current of the device.

The control strategies based on the zero-sequence current superposition introduce zero-sequence current into the output phase current of the delta-connected cascaded H-bridge multilevel SVG to guarantee dc-link voltage balancing, which increase the output phase current. So the unbalanced range in which SVG can operate reliably is limited when the rated output phase current is decided. At present, the most of the literature focuses on researching the control strategies, and only a few literatures have given a quantitative analysis on the compensating capability of the star-connected cascaded H-bridge multilevel SVG, and there is rarely introduction about quantitative analysis on the compensating capability of the delta-connected cascaded H-bridge multilevel SVG. For the star-connected SVG, in the work presented in [16], the compensation ability of the negative-sequence current was analyzed under the symmetrical grid voltage, but the compensation ability under the asymmetrical grid voltage was not studied. In [17], the compensation capacity under the condition of simple negative-sequence current was quantitatively analyzed, but the research on the whole compensation domain was lacking. For the delta-connected SVG, in

[18], it was discussed that the reactive power rating is related to the voltage unbalance factor and phase angle of negative-sequence voltage, but ignored the effect of the unbalanced compensation currents. In addition, in the work presented in [19], the star-connected SVG and delta-connected SVG compensation ability were investigated, but it only considered the conditions of symmetrical supply voltage for star structure and symmetrical compensation current for delta structure.

At present, there is no comprehensive quantitative analysis on compensation ability of the delta-connected cascaded H-bridge multilevel SVG. Therefore, the aim of this paper is to analyze the key factors that affect the compensation ability of the SVG under unbalanced conditions, and give the compensation range where the device can be reliably operated and its analytical method. Finally, the results of the analysis are verified by simulation.

II. CLUSTER VOLTAGE CONTROL UNDER UNBALANCED CONDITIONS

The system configuration of the delta-connected cascaded H-bridge multilevel SVG is depicted in Fig.1.

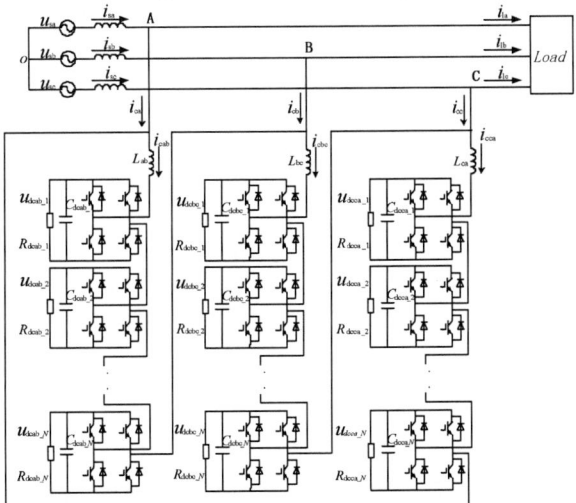

Fig.1 Delta-connected cascaded H-bridge multilevel SVG system configuration

Three phases of A, B and C are in delta configuration, and each phase consists of N identical H-bridge modules in series, and then is connected to the grid by interface inductance $L_i(i=ab,bc,ca)$ The voltages of the three-phase system are u_{sa}, u_{sb}, u_{sc} and the currents of the three-phase system are i_{sa}, i_{sb}, i_{sc}. The output line currents of SVG are i_{ca}, i_{cb}, i_{cc}, and the output phase currents of SVG are i_{cab}, i_{cbc}, i_{cca}, and the load currents are i_{la}, i_{lb}, i_{lc}, u_{dci_k} ($i=ab,bc,ca$, $k=1,2,3,...,N$) represents dc-link voltage of H-bridge module, and R_{dci_k}, C_{dci_k} ($i=ab,bc,ca$, $k=1,2,3,...,N$) represent the resistance equivalent to the power loss and DC capacitor of each H-bridge module respectively.

Referring to Fig.1, the voltage of each phase leg of delta-connected cascaded H-bridge multilevel SVG under unbalanced supply voltage can be written as

$$u_{scb}=\sqrt{2}U_p\sin(\omega t)+\sqrt{2}U_n\sin(\omega t+\varphi)$$
$$u_{sbc}=\sqrt{2}U_p\sin(\omega t-120°)+\sqrt{2}U_n\sin(\omega t+\varphi+120°) \quad (1)$$
$$u_{sca}=\sqrt{2}U_p\sin(\omega t+120°)+\sqrt{2}U_n\sin(\omega t+\varphi-120°)$$

Where, U_p, U_n are the effective values of positive and negative-sequence components of voltage, and φ is the initial phase of negative-sequence voltage. Similarly, the current flowing through each phase leg is shown as follows:

$$i_{cab}=\sqrt{2}I_p\sin(\omega t+\gamma)+\sqrt{2}I_n\sin(\omega t+\phi)$$
$$i_{cbc}=\sqrt{2}I_p\sin(\omega t+\gamma-120°)+\sqrt{2}I_n\sin(\omega t+\phi+120°) \quad (2)$$
$$i_{cca}=\sqrt{2}I_p\sin(\omega t+\gamma+120°)+\sqrt{2}I_n\sin(\omega t+\phi-120°)$$

Where, I_p, I_n are the effective values of positive and negative-sequence components of current, and γ, ϕ are the initial phase of positive and negative-sequence current respectively.

According to the equation (1) and (2), the average power flow at each phase in one cycle can be calculated by equation (3).

$$\overline{p_{ab}}=U_pI_p\cos\gamma+U_nI_n\cos(\phi-\varphi)$$
$$+U_pI_n\cos\phi+U_nI_p\cos(\gamma-\varphi)$$
$$\overline{p_{bc}}=U_pI_p\cos\gamma+U_nI_n\cos(\phi-\varphi) \quad (3)$$
$$+U_pI_n\cos(\phi-120°)+U_nI_p\cos(\gamma-\varphi+120°)$$
$$\overline{p_{ca}}=U_pI_p\cos\gamma+U_nI_n\cos(\phi-\varphi)$$
$$+U_pI_n\cos(\phi+120°)+U_nI_p\cos(\gamma-\varphi-120°)$$

It can be seen from the equation(3) that the average power absorbed by each phase can be divided into two parts: $\overline{p}$ is the same part of the phase cluster power flows and $\overline{\Delta p_i}$ is the deviation of the phase cluster power flows, and the relationship between two parts is shown as follow,

$$\overline{p_i}=\overline{p}+\overline{\Delta p_i}(i=ab、bc、ca) \quad (4)$$

Where,

$$\overline{p}=U_pI_p\cos\gamma+U_nI_n\cos(\phi-\varphi) \quad (5)$$

$$\begin{cases}\overline{\Delta p_{ab}}=U_pI_n\cos\phi+U_nI_p\cos(\gamma-\varphi)\\ \overline{\Delta p_{bc}}=U_pI_n\cos(\phi-120°)+U_nI_p\cos(\gamma-\varphi+120°)\\ \overline{\Delta p_{ca}}=U_pI_n\cos(\phi+120°)+U_nI_p\cos(\gamma-\varphi-120°)\end{cases} \quad (6)$$

Assuming that there is no converter power loss, which is reasonable because power loss is so small that it can be ignored in the real installation. In order to maintain cluster voltage balancing, it is necessary to superimpose the zero-sequence current into the delta-connected cascaded H-bridge multilevel SVG to ensure the zero active power flow. Phasor diagram showing how to find out zero-sequence current is shown in Fig.2, $u_{si}(i=a,b,c)$ and $u_{si}(i=ab,bc,ca)$ are the phase voltage and line voltage on point of common coupling (PCC) respectively, and $i_{ci}(i=ab,bc,ca)$ and $i_{ci}(i=a,b,c)$ refer to the output compensation phase and line currents of SVG respectively, and $i_{ci}(i=ab,bc,ca)$ can be resolved into positive-sequence component (i_{ci}^{p}), negative-sequence component (i_{ci}^{n}) and zero-sequence component (i_{ci}^{0}). Obviously, the phase current $i_{ci}(i=ab,bc,ca)$ and the line voltage $u_{si}(i=ab,bc,ca)$ of each phase leg are not perpendicular without injecting the zero-sequence current component, which will cause some clusters to absorb energy from the grid and dc-link voltage is rising, and the other clusters release energy to the grid and the dc-link is decreasing. After injecting zero-sequence current, the each phase leg current $i_{ci}(i=ab,bc,ca)$ is perpendicular to the corresponding terminal voltage $u_{si}(i=ab,bc,ca)$. In the view of phasor, zero-sequence current is introduced to change the phase angle of the fundamental component of the phase current inside the triangle without affecting the line current. By selecting the proper zero-sequence current component, the phase current can be perpendicular to the line voltage respectively, and the active power flowing each phase leg is zero, and cluster voltage is kept balancing.

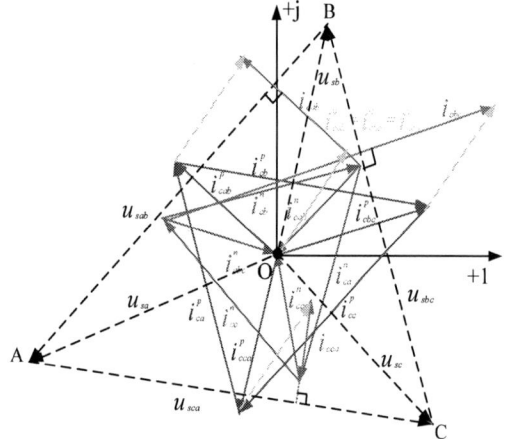

Fig.2 Phasor diagram showing how to find out zero-sequence current

Setting the zero-sequence voltage is

$$i_0 = \sqrt{2}I_0 \sin(\omega t + \delta) \qquad (7)$$

Where, I_0 is the effective value of zero-sequence current; δ is initial phase of zero-sequence current. With injecting zero-sequence current, the average active power absorbed by each phase leg can be expressed by

$$\overline{p_i} = \overline{p} + \overline{\Delta p_i} + \overline{\Delta p_i^0}(i = ab、bc、ca) \qquad (8)$$

Where,

$$\overline{p} = U_p I_p \cos\gamma + U_n I_n \cos(\phi - \varphi) \qquad (9)$$

$$\begin{cases} \overline{\Delta p_{ab}} = U_p I_n \cos\phi + U_n I_p \cos(\gamma - \varphi) \\ \overline{\Delta p_{bc}} = U_p I_n \cos(\phi - 120°) + U_n I_p \cos(\gamma - \varphi + 120°) \\ \overline{\Delta p_{ca}} = U_p I_n \cos(\phi + 120°) + U_n I_p \cos(\gamma - \varphi - 120°) \end{cases} \qquad (10)$$

$$\begin{cases} \overline{\Delta p_{ab}^0} = U_p I_0 \cos\delta + U_n I_0 \cos(\delta - \varphi) \\ \overline{\Delta p_{bc}^0} = U_p I_0 \cos(\delta + 120°) + U_n I_0 \cos(\delta - \varphi - 120°) \\ \overline{\Delta p_{ca}^0} = U_p I_0 \cos(\delta - 120°) + U_n I_0 \cos(\delta - \varphi + 120°) \end{cases} \qquad (11)$$

The power loss of the device is ignored and the overall active power flows is zero, we can get equation (12) easily as follow:

$$\sum_{i=ab,bc,ca} \overline{p_i} = 3\overline{p} = 3(U_p I_p \cos\gamma + U_n I_n \cos(\phi - \varphi)) = 0 \qquad (12)$$

At the same time, in order to maintain the cluster voltages of three phases basically constant, it is necessary to ensure that the active power flowing through each phase leg is zero. Due to the restriction of the existing equation (12), the zero-sequence current only needs to satisfy the equation (13).

$$\begin{cases} \overline{p_{ab}} = 0 \\ \overline{p_{bc}} = 0 \end{cases} \qquad (13)$$

The unbalance degree of supply voltage and compensation current is defined in equation (14) respectively.

$$K_u = U_n / U_p、\ K_i = I_n / I_p \qquad (14)$$

Based on equation (8) ~ (14), the expression of zero-sequence current can be obtained as follows.

$$\begin{aligned} i_0 &= \sqrt{2}I_0 \sin(\omega t + \delta) \\ &= \sqrt{2}I_p \frac{-K_u K_u \cos(\gamma - 2\varphi) + K_i \cos\phi -}{(K_u - 1)(K_u + 1)} \sin\omega t \\ &\quad + \sqrt{2}I_p \frac{K_u K_u \sin(\gamma - 2\varphi) - K_i \sin\phi -}{(K_u - 1)(K_u + 1)} \cos\omega t \end{aligned} \qquad (15)$$

Substituting equation (14) into (12), the relationship between K_u, K_i, γ, ϕ and φ can be obtained as follow:

$$\gamma = \begin{cases} \pi - a\cos(-K_u K_i \cos(\phi - \varphi)) & 0 < \gamma < \pi \\ \pi + a\cos(-K_u K_i \cos(\phi - \varphi)) & \pi < \gamma < 2\pi \end{cases} \qquad (16)$$

Substituting the equation (16) into (15), it is found that the capacitive and inductive characteristics of the positive-sequence current (the value range of γ) have an important impact on the zero-sequence current. In addition, the zero-sequence current of the delta-connected cascaded H-bridge multilevel SVG under unbalanced conditions is determined by the five factors. In addition to the effective value of positive-sequence current I_p, the other four factors are: K_u is the unbalance degree of the supply voltage; K_i is the unbalance degree of compensation current, φ is the initial phase of the negative-sequence voltage component of supply voltage, and ϕ is the initial phase of negative-sequence current component of compensation current. Zero-sequence current is satisfied by

$$i_o = I_p \bullet g(K_u, K_i, \phi, \varphi) \qquad (17)$$

III. UNBALANCED COMPENSATION RANGE

Regardless of line loss, the maximum values of current flowing through three-phase legs of the delta-connected cascaded H-bridge multilevel SVG are shown in equation (18) with injecting zero-sequence current.

$$I_{\max} = \max\{| i_{cab} + i_0 |, | i_{cbc} + i_0 |, | i_{cca} + i_0 |\} \quad (18)$$

Where, '| |' represents phasor modulo operation.

According to equation (1), (2) and (15) ~ (18), the functional relationship between the maximum output current amplitude of the SVG and each variable is obtained by

$$I_{\max} / I_p = G(K_u, K_i, \phi, \varphi) \quad (19)$$

When compensating the negative-sequence current, it is necessary for SVG to inject the zero-sequence current to maintain cluster voltages balance, which is likely to cause the variation of output current of some phase legs. Because the device is prone to overcurrent after injecting the zero-sequence current and I_{max}/I_p can accurately reflect the status of output current of each phase leg, I_{max}/I_p is chosen as the standard to measure the unbalance compensation ability of the SVG in this paper. Only when the value of I_{max}/I_p is smaller than the rated value, the SVG can operate normally.

According to the functional relationship shown in equation (19), it can be seen that the output current level I_{max}/I_p is mainly affected by four factors: K_u, φ, K_i and ϕ. When the other parameters are decided, the quantitative relationship between the output current level I_{max}/I_p, K_u and K_i can be obtained by MATLAB numerical analysis method, as shown in Fig. 3.

When the output voltage level I_{max}/I_p, is determined, there is a plane paralleling to the K_u and K_i axes along the rated value of I_{max}/I_p. Only the unbalanced operating range below the plane is the area where the device can operate reliably. From the point of view of practical application, when the output rated current of the device is determined, in order to ensure the safe and stable operation of the device, it is necessary to control K_u and K_i in the proper range. In addition, the rated output current of the SVG is affected simultaneously by K_u and K_i, and the influence of K_u is more significant.

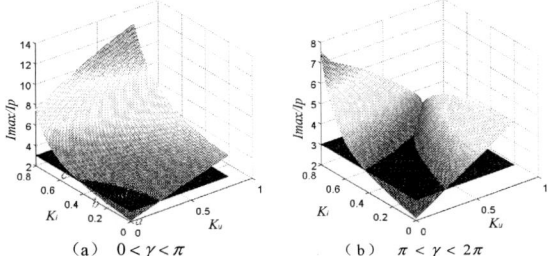

Fig.3 Variation of output current level of delta-connected cascaded H-bridge multilevel SVG with the change of K_u and K_i

IV. SIMULATION RESULTS

In order to verify the accuracy of the theoretical derivation and quantitative analysis in this paper, a simulation model of delta-connected cascaded H-bridge multilevel SVG is built by using Matlab/Simulink. The simulation parameters are shown in Table. I.

TABLE.I
PARAMETERS OF SIMULATION

Variables	Symbol	Value
Line-to-line rms voltage	u_s	10kV
Grid frequency	f	50Hz
Cascaded cell number	N	12
DC bus capacitor	C_{dc}	4700uF
Switching frequency	f_{sw}	10kHz
Couple inductor	L	6mH
Nominal DC voltage	u_{dc}^*	1000V

The simulation results are shown in Fig. 4 and Fig. 5, which are divided into five stages. *Stage I* (t<0.2s): K_u and K_i were zero (point **a** at Fig. 3(a)), only the positive-sequence reactive current was compensated, as illustrated in Fig. 5(a), the dc-link voltage and compensation current also can be kept balance without injecting the zero-sequence current, which was consistent with the theory. *Stage II* (0.2s<t<0.4s): In this stage, K_u was still zero, but K_i changed from 0 to 20% (point **b** at Fig. 3(a)), then the SVG need to compensate reactive and negative-sequence current at the same time. It was necessary for the SVG to inject the corresponding zero-sequence current to achieve the cluster voltage balance control, as shown in Fig. 5(a). *Stage III* (0.4s<t<0.6s): K_i increased from 20% to 50% (point **c** at Fig. 3(a)), which was more than the designed maximum unbalance degree 44%, as shown in Fig. 5(b). After the mutation, the maximum value of the output phase current exceeds its rated value 1000A. *Stage IV* (0.6s<t<0.8s): K_i decreased from 50% to 20%, K_u increased from 0 to 10% (point **d** at Fig. 3(a)). At this stage, the zero-sequence current was adjusted, and the cluster voltage reached stable state again, as shown in Fig. 5(c). *Stage V* (0.8s<t<1s): K_i was still 20%, K_u changed from 10% to 40% (point **e** at Fig. 3(a)), which was more than the designed maximum unbalance degree 21%. SVG cannot maintain cluster voltage balance, and the amplitude of the output phase current is much larger than its rated value, which is verified the correctness of the relevant theoretical analysis. In addition, the output line current does not change after the mutation, indicating that the injected zero-sequence current flows only inside the triangle and has no effect on the output line current, as shown in Fig. 5(d).

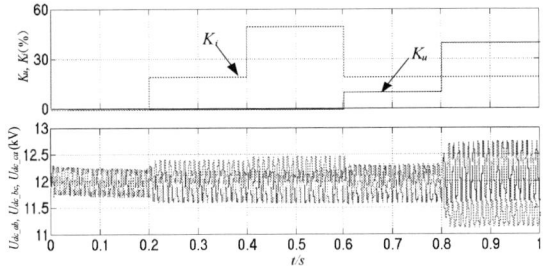

Fig.4 Waveforms during sudden change of unbalance degree

The 2018 International Power Electronics Conference

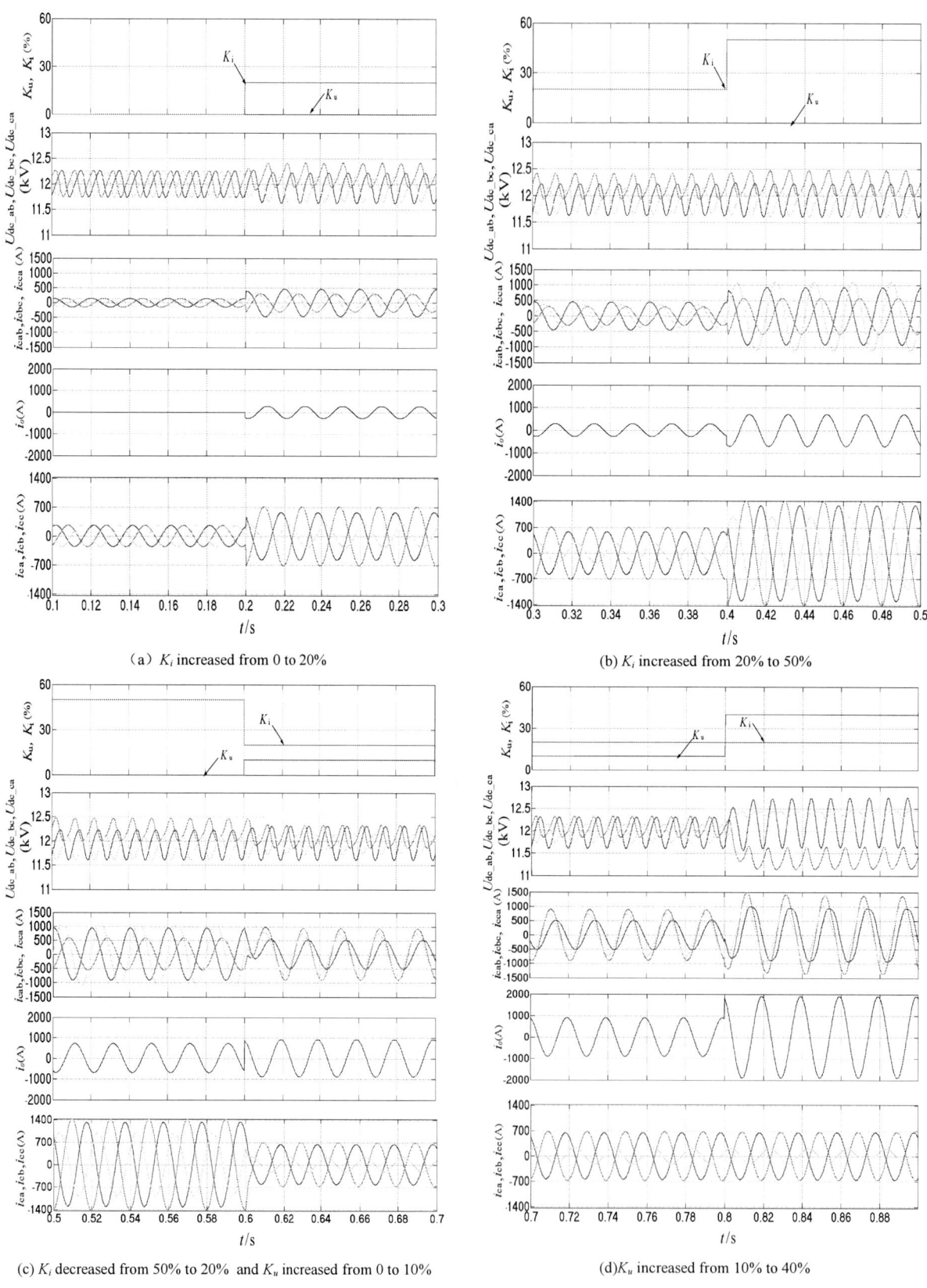

(a) K_i increased from 0 to 20%

(b) K_i increased from 20% to 50%

(c) K_i decreased from 50% to 20% and K_u increased from 0 to 10%

(d) K_u increased from 10% to 40%

Fig.5 Partial enlargement waveforms during sudden change of unbalance degree

V. CONCLUSIONS

This paper considered the unbalance degree of the supply voltage and the compensation current together, gave the detailed expression of zero-sequence current. What's more, the paper presented that the compensation ability of the delta-connected cascaded H-bridge multilevel SVG can be measured by using its maximum output current. Based on the analysis of the factors affecting output current level of the SVG, the unbalanced compensation capability and range of the SVG are given quantitatively. The quantitative equation derived in this paper can provide a good theoretical basis for the parameter design and device selection of the delta-connected cascaded H-bridge multilevel SVG. Finally, the rationality and accuracy of the analysis results is verified by simulation.

REFERENCES

[1] H. Akagi, "Classification, Terminology, and Application of the Modular Multilevel Cascade Converter (MMCC)," *IEEE Transactions on Power Electronics*, vol. 26, no. 11, pp. 3119-3130, Nov. 2011.

[2] F. Z. Peng and Jih-Sheng Lai, "Dynamic performance and control of a static VAr generator using cascade multilevel inverters," *IEEE Transactions on Industry Applications*, vol. 33, no. 3, pp. 748-755, May/Jun 1997.

[3] C. K. Lee, J. S. K. Leung, S. Y. R. Hui and H. S. H. Chung, "Circuit-level comparison of STATCOM technologies," *Power Electronics Specialist Conference, 2003. PESC '03. 2003 IEEE 34th Annual*, 2003, pp. 1777-1784 vol.4.

[4] F. Z. Peng and Jin Wang, "A universal STATCOM with delta-connected cascade multilevel inverter," *2004 IEEE 35th Annual Power Electronics Specialists Conference (IEEE Cat. No.04CH37551)*, 2004, pp. 3529-3533 Vol.5.

[5] L. Maharjan, S. Inoue, H. Akagi and J. Asakura, "A transformerless battery energy storage system based on a multilevel cascade PWM converter," *2008 IEEE Power Electronics Specialists Conference*, Rhodes, 2008, pp. 4798-4804.

[6] N. Hatano and T. Ise, "Control Scheme of Cascaded H-Bridge STATCOM Using Zero-Sequence Voltage and Negative-Sequence Current," *IEEE Transactions on Power Delivery*, vol. 25, no. 2, pp. 543-550, April 2010.

[7] K. Sano and M. Takasaki, "A Transformerless D-STATCOM Based on a Multivoltage Cascade Converter Requiring No DC Sources," *IEEE Transactions on Power Electronics*, vol. 27, no. 6, pp. 2783-2795, June 2012.

[8] Z. Chunyan and L. Zhao, "Advanced compensation mode for cascade multilevel static synchronous compensator under unbalanced voltage," *IET Power Electronics*, vol. 8, no. 4, pp. 610-617, 4 2015.

[9] Y. Feng, X. Wang, J. Wu and H. Zhang, "Research on the cluster current control of delta-connected cascade multilevel STATCOM," *2014 International Conference on Power System Technology*, Chengdu, 2014, pp. 2316-2321.

[10] Wang Baoan, Shang Jiao, Chen Hao and Dai Ningyi, "Steinmetz theory applied to fundamental phase reference current calculation for a STATCOM with delta configuration," *TENCON 2015 - 2015 IEEE Region 10 Conference*, Macao, 2015, pp. 1-5.

[11] M. Hagiwara, R. Maeda and H. Akagi, "Negative-Sequence Reactive-Power Control by a PWM STATCOM Based on a Modular Multilevel Cascade Converter (MMCC-SDBC)," *IEEE Transactions on Industry Applications*, vol. 48, no. 2, pp. 720-729, March-April 2012.

[12] P. H. Wu, H. C. Chen, Y. T. Chang and P. T. Cheng, "Delta-Connected Cascaded H-Bridge Converter Application in Unbalanced Load Compensation," *IEEE Transactions on Industry Applications*, vol. 53, no. 2, pp. 1254-1262, March-April 2017.

[13] M. Nieves, J. M. Maza, J. M. Mauricio, R. Teodorescu, M. Bongiorno and P. Rodríguez, "Enhanced control strategy for MMC-based STATCOM for unbalanced load compensation," *2014 16th European Conference on Power Electronics and Applications, Lappeenranta*, 2014, pp. 1-10.

[14] R. E. Betz, T. Summers and T. Furney, "Symmetry Compensation using a H-Bridge Multilevel STATCOM with Zero Sequence Injection," *Conference Record of the 2006 IEEE Industry Applications Conference Forty-First IAS Annual Meeting*, Tampa, FL, 2006, pp. 1724-1731.

[15] Yang Bo, Zeng Guang and Zhong Yanru, "Research on the control system of cascaded multilevel STATCOM for unbalanced load compensation," *2011 IEEE Power Engineering and Automation Conference*, Wuhan, 2011, pp. 88-91.

[16] Q. Song and W. Liu, "Control of a Cascade STATCOM With Star Configuration Under Unbalanced Conditions," *IEEE Transactions on Power Electronics*, vol. 24, no. 1, pp. 45-58, Jan. 2009.

[17] H. Zhixing et al., "Circulating current derivation and comprehensive compensation of cascaded STATCOM under asymmetrical voltage conditions," *IET Generation, Transmission & Distribution*, vol. 10, no. 12, pp. 2924-2932, 9 2 2016.

[18] Z. He et al., "Reactive Power Strategy of Cascaded Delta-Connected STATCOM Under Asymmetrical Voltage Conditions," *IEEE Journal of Emerging and Selected Topics in Power Electronics*, vol. 5, no. 2, pp. 784-795, June 2017.

[19] E. Behrouzian and M. Bongiorno, "Investigation of Negative-Sequence Injection Capability of Cascaded H-Bridge Converters in Star and Delta Configuration," *IEEE Transactions on Power Electronics*, vol. 32, no. 2, pp. 1675-1683, Feb. 2017.

Gap in pagination due to withheld paper.

Pages 3445-3449

Static synchronous compensator to stabilize grid voltage for wind and photovoltaic power plant

Ryota Okuyama, Naoki Morishima, Yusuke Ashizaki and Yohei Itaya
Toshiba Mitsubishi-Electric Industrial Systems Corporation, Tokyo, Japan

Abstract- The installation of wind and/or photovoltaic energy systems has rapidly increased in Japan. Electric power from renewable energy is always fluctuating owing to its time-varying nature. This weak interconnection of renewable energy in an electrical network affects the power quality and reliability.

A static synchronous compensator (STATCOM) is required to stabilize the voltage fluctuation in a renewable energy power plant. This paper describes the design and test results of a new STATCOM that is suitable for renewable energy. The developed STATCOM has the following features, : a small harmonic distortion, synchronous connection for an interconnection transformer, and continuous operation capability during a grid fault.

Keywords—Converter, IGBT, STATCOM

I. INTRODUCTION

Wind and photovoltaic (PV) power plants are often located in unmanned remote areas. These power plants are connected to a low short circuit capacity grid. Thus, voltage fluctuations occur as a result of variations in the electric power generation. A static synchronous compensator (STATCOM) is required to stabilize such voltage fluctuations [1].

II. REQUIREMENTS

We investigated the specifications of a suitable STATCOM for renewable energy power plants. First, air cooling is recommended because providing cooling water may be difficult in a remote area. Second, self-commutated devices are recommended because a high impedance grid has weak resistance to harmonic currents. Third, a high continuous operation capability is an important factor in an unmanned remote area. If the STATCOM stopped operating, equipment restoration will take much longer than in a manned area. Finally, the inrush current of the interconnection transformer should be suppressed to prevent voltage flicker. A STATCOM should excite the transformer by synchronizing the voltage with the grid before closing the circuit breaker. The inrush current is suppressed when the circuit breaker is closed during excitation.

III. SYSTEM CONFIGURATION

The specifications of the developed STATCOM are listed in TABLE I. The exterior appearance is shown in Fig. 1.

TABLE I
SPECIFICATIONS OF STATCOM

Specifications	
Capacity (kVA)	625
Output voltage (Vrms)	400
Output current (Arms)	902
Dimension [H × W × D] (mm)	2025×2000×700
Power device	IGBT
Circuit	three-phase, three-level
Cooling material	Air

Fig. 1. Exterior appearance.

The STATCOM was designed to have a low voltage classification under Japanese regulations. The power conversion circuit is a three-phase, three-level converter that uses insulated-gate bipolar transistors (IGBTs). The cooling method is forced air cooling using fans. The air intake is at the bottom front, with the exhaust on the top side of the cubicle.

The STATCOM can be connected directly to a 400-V power line. The system capacity can be adjusted using parallel connections. As an example, a system configuration for a total of 6500 kVA is shown in Fig. 2 [2].

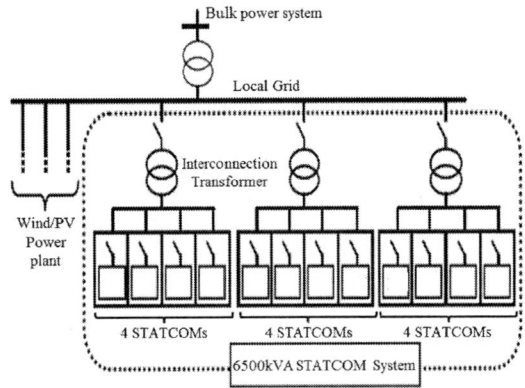

Fig. 2. 6500kVA system configuration.

Four STATCOMs are connected to a single interconnection transformer. Each STATCOM has a circuit breaker at the inlet of the main circuit. If a STATCOM fails, the circuit breaker is opened to disconnect that STATCOM from the circuit, and the other STATCOMs continue to operate without the failed one. This redundant system improves the continuous operation capability.

As an example, a control block diagram for a wind power plant is shown in Fig. 3 [2][3].

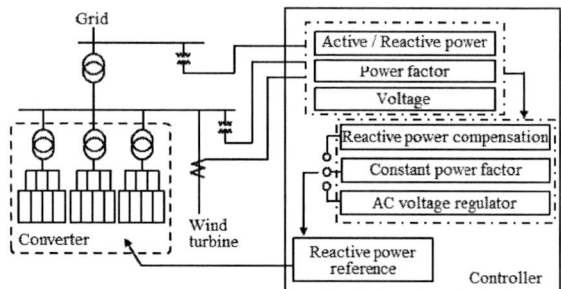

Fig. 3. Control block diagram for a wind power plant.

The controller has a AC automatic voltage regulator (AC-AVR). The AC-AVR calculates reactive power reference depending on a detection voltage of a grid. The grid voltage is controlled by a reactive power that converters generate depending on the reactive power reference.

Fast response is not required because a wind turbine has large moment of inertia [3].

IV. HARMONICS

One of the requirements for the STATCOM is a small harmonic current. We measured the output current of the STATCOM and confirmed the harmonic current.

The configuration of the test equipment is shown in Fig. 4.

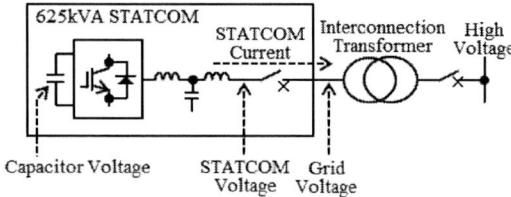

Fig. 4. Test equipment.

One STATCOM was connected to an interconnection transformer, and the transformer was connected to a high-voltage grid through a circuit breaker. Measurement points are indicated by the dotted lines in Fig. 4.

As shown in Fig. 5, the 100% rated current waveforms were measured.

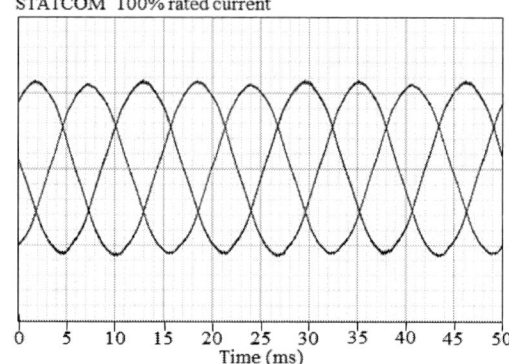

Fig. 5. STATCOM current waveforms.

The total harmonic distortion of the current was less than 3% at the rated current output in each phase. A result of FFT analysis is shown in Fig. 6.

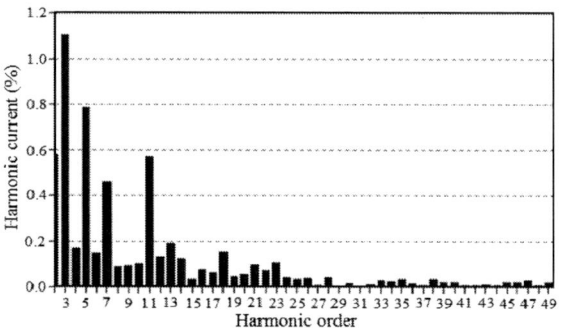

Fig. 6. FFT analysis of a STATCOM current.

It is believed that harmonic filters will not be necessary for the STATCOM.

V. EXCITATION

Synchronous connection by excitation from the STATCOM is a required function. As a feature of the STATCOM, the electric power for excitation is supplied from the electric power stored in DC capacitors.

The power loss during excitation is almost the core loss of the transformer. The excitation voltage is synchronous with the grid voltage, and its amplitude is also adjusted. Under this condition, the relationship between the voltage drop of the DC capacitors and elapsed time is shown as formula (1).

$$W \cdot t = \frac{1}{2} \cdot C \left(V_0^2 - V_t^2 \right) \qquad (1)$$

W is power loss (W)
t is elapsed time (s)
C is capacitance of DC capacitors (F)
V_0 is initial voltage of DC capacitors (V)
V_t is voltage of DC capacitors at the time (V)

From formula (1), we confirmed that the electric power stored in the DC capacitors was sufficient to maintain the excitation over the required duration.

We confirmed that the STATCOM could excite the transformer and connect synchronously.

VI. Continuous Operation Capability

If the circuit breaker of an interconnection transformer is opened suddenly, a STATCOM should withstand this and be able to re-start. We tested it under this condition.

The measured waveforms are shown in Fig. 7. The grid voltage disappeared when the circuit breaker connected to the high voltage was opened, as shown in Fig. 4. The STATCOM also stopped at this time, and the circuit breaker in the STATCOM was opened by an instant over-voltage condition. When the grid voltage came back, the STATCOM recognized the capability of re-starting. Then, the STATCOM re-started.

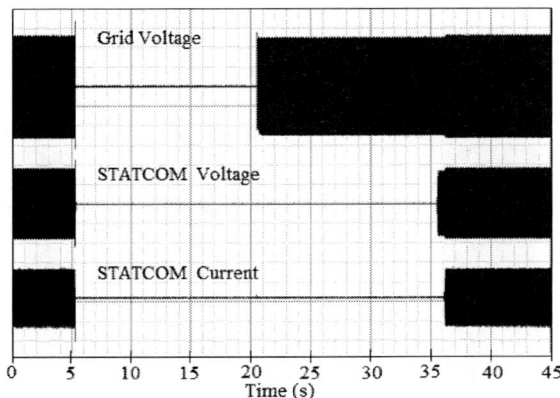

Fig. 7. Sudden circuit breaker opening.

Fig. 8 shows an expanded waveform at the circuit breaker was opened.

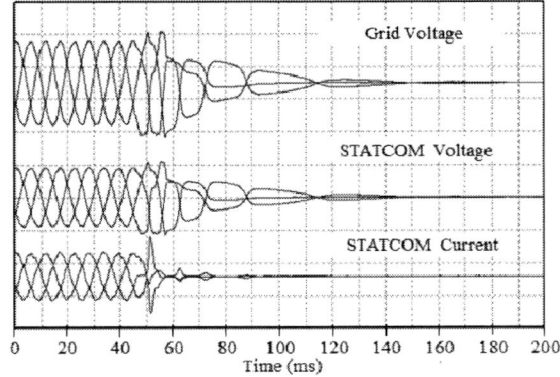

Fig. 8. Expanded waveform at circuit breaker was opened.

It was confirmed that even if an upper circuit breaker opens suddenly, the STATCOM can re-start.

If a grid fault occurs, the grid voltage changes suddenly. Continuous operation capability during a grid fault was confirmed using a small-scale converter (2 kVA). The configuration of the test equipment is shown in Fig. 9. The STATCOM controller was the same as the one that controlled the 625 kVA STATCOM. The grid simulator was able to change the voltage waveform.

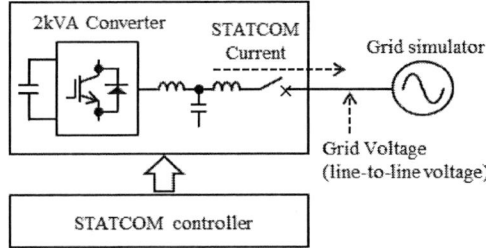

Fig. 9. Test equipment using small-scale converter.

The PV converter has the fault ride through (FRT) function required by the national grid regulation [4][5][6]. The FRT function is not necessary for a STATCOM. However, wind and/or PV power plants should operate continuously. Thus, we confirmed the continuous operation capability of the STATCOM in the case of a grid fault.

Fig. 10 shows the waveforms of a three-phase line-to-ground fault (3LG) when the grid voltage dropped from 100% to 20%. Fig. 11 shows the waveforms at the recovery from the 3LG fault.

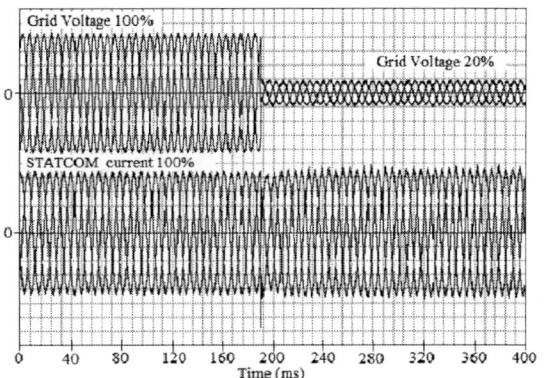

Fig. 10. 3LG fault.

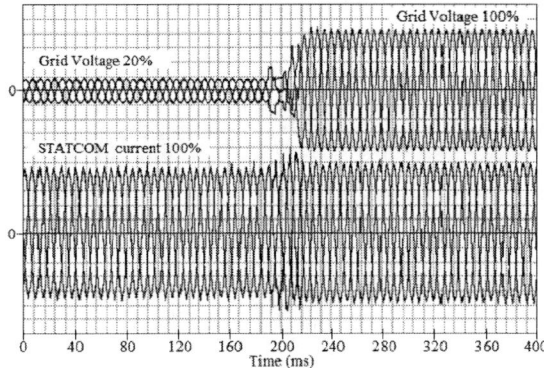

Fig. 11. Recovery from 3LG fault.

At the beginning of the grid fault, the current spiked as a result of the sudden voltage drop. The STATCOM withstood this current spike and operated continuously. At the recovery from the grid fault, the grid voltage was distorted. However, the STATCOM continued to operate.

We also tested one line-to-ground (1LG) and two line-to-ground (2LG) faults. Fig. 12 shows a 1LG fault when a phase voltage dropped from 100% to 20%. Fig. 13 shows the recovery from the 1LG fault.

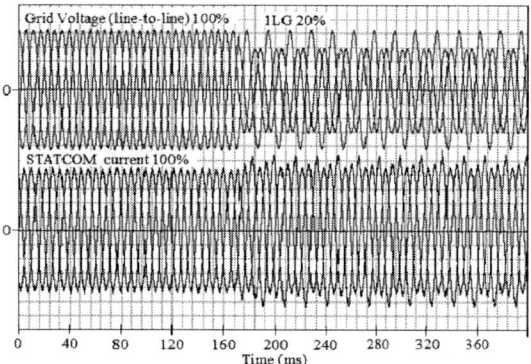

Fig. 12. 1LG fault.

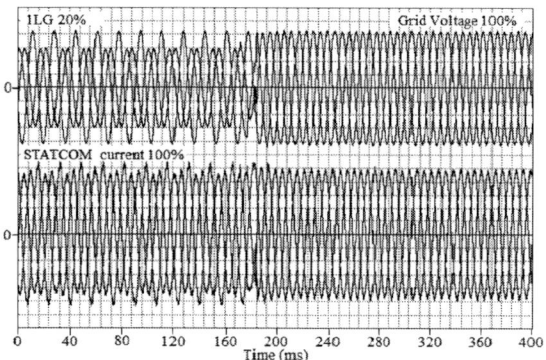

Fig. 13. Recovery from 1LG fault.

The measured grid voltages shown in these figures are line-to-line voltages. Thus, two line-to-line voltages dropped during the 1LG fault.

Fig. 14 shows a 2LG fault when a phase voltage dropped from 100% to 20%. Fig. 15 shows the recovery from the 2LG fault.

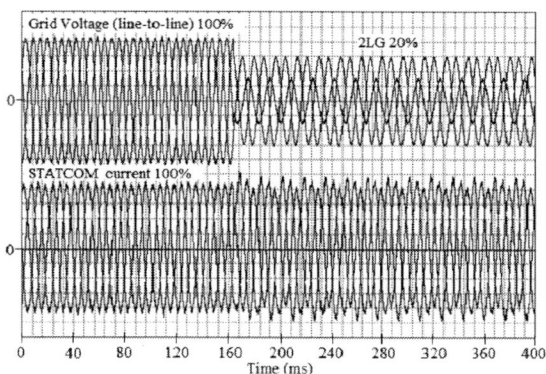

Fig. 14. 2LG fault.

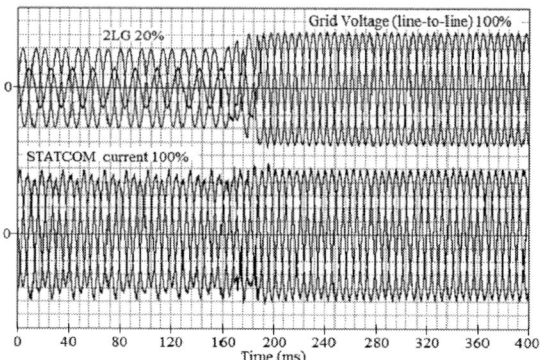

Fig. 15. Recovery from 2LG fault.

The measured grid voltages shown in these figures are line-to-line voltages. Thus, one line-to-line voltage dropped lower than other line-to-line voltages during the 2LG fault.

The STATCOM current in each phase was unbalanced during 1LG and also 2LG fault. However, the STATCOM continued to operate.

It was confirmed that the STATCOM continued to operate despite the distorted voltage under the sudden voltage drop, voltage unbalance, and recovery condition.

The case of a phase shift condition was also tested. The STATCOM continued to operate in the case of a 15 degrees phase shift, as shown in Fig. 16.

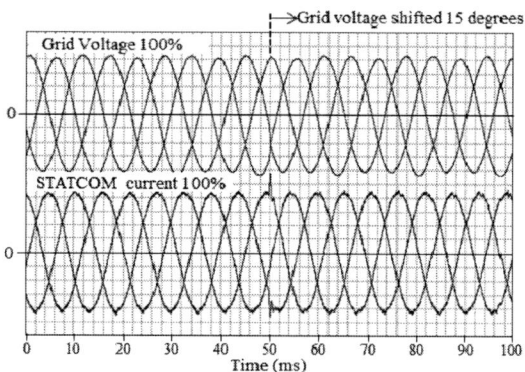

Fig. 16. STATCOM response to phase shift of grid voltage.

The current of the STATCOM fluctuated because of the voltage variation. However, the current fluctuation was immediately reduced by the current controller of the STATCOM.

VII. CONCLUSION

We developed a STATCOM for grid stabilization such as in wind and/or PV power plants.

In order to suit renewable energy power plants, The STATCOM has the following features, : a small harmonic distortion, synchronous connection for an interconnection transformer, and continuous operation capability during a grid fault. It can easily be introduced in a remote area.

The system capacity can be adjusted using parallel STATCOM connections, which also provides redundancy.

The continuous operation capability was confirmed in several grid fault cases.

Renewable energy represents alternative energy sources and its use is increasing. The STATCOM can be expected to assist in grid stabilization.

REFERENCES

[1] M. Sugimoto, R, Okuyama, N. Morishima, Y. Ashizaki "Self-commutated SVC to stabilize Grid voltage of wind and PV power plant," Proceeding of Annual Convention of IEEJ Industrial Applications Society, 1-63, 2016.

[2] Y. Ashizaki, Y. Itaya, N. Morishima, R. Okuyama, "Harmonic-less self-commutated SVC for wind and photovoltaic power plant" The 70th Joint Conference of Electrical, Electronics and Information Engineers in Kyushu, 03-1P-09, 2017.

[3] Y. Ashizaki, N. Tagashira, S. Yasutomi, K. Suzuki, M. Kondo, "SVC Control Panel with High Flexibility to realize various control in wind power generation system" Proceeding of Annual Convention of IEEJ Industrial Applications Society, 1-31, 2010.

[4] JESC E0019 (2012) "Grid-interconnection Code JEAC 9701-2012," appendix-1, 2013.

[5] R. Okuyama, J. Li, E. Ikawa "Certification of Grid regulation for PV inverter, ZVRT test," Proceeding of Annual Convention of IEEJ, 4-142, 2015.

[6] R. Inzunza, E. Ikawa, T. Sumiya, T. Ambo "Development of Grid Interaction Features for a Utility-Interaction Photovoltaic Inverter," Proceeding of Annual Convention of IEEJ Industrial Applications Society, 1-5, 2012.

The 2018 International Power Electronics Conference

Large Equalization Current Control Strategy for Series Connected Battery Packs Based on Buck-Boost Converter

XinBo Liu [123*], Zhuo Gao [123], XueHao Huang[123] and YaoHan Zou[123]

1 Inverter Technologies Engineering Research Center of Beijing
2 Collaborative Innovation Center of Key Power Energy-Saving Technologies in Beijing
3 Collaborative Innovation Center of Electric Vehicles in Beijing, North China University of Technology, China
*E-mail: liuxinbo@ncut.edu.cn

Abstract—Battery imbalance is one of the most serious problems to significantly decrease the battery life time, and Buck-Boost equalization circuits for series connected battery packs are very necessary. This paper proposes a new large equalization current control strategy to achieve fast balancing. The SOC (state of charge), voltages and currents of batteries are all considered in the control loops. The SOC and voltage control loop determines on or off state of the switches, while the current control loop determines the duty cycle of the switches in "on" state based on PI regulator. Depending on the control loops, the equalization currents are controllable, and the maximum values reach the large reference current. The strategy could achieve fast equalization, and is very applicable for large capacity series battery packs. Experimental results verify the validity of the control strategy.

Keywords—Large equalization currents control, series battery packs, Buck-Boost converter, the voltage and SOC control loop

I. INTRODUCTION

Large capacity batteries in series connection are increasingly used in electric vehicles and large scale energy storage power stations. The cost of the batteries is very expensive while the battery life time is limited. Consequently, one important focus is to improve battery utilization and extend the life time. Unfortunately, in practical applications, battery imbalances usually occurs in series packs due to different inherent characteristics, resulting in reducing the capacity of the series battery packs and also achieving lower battery utilization.

To balance the battery voltages and increase life time, many equalization circuits and control methods are introduced. Depending on producing loss or not, these circuits are mainly divided into two categories: dissipative and non-dissipative circuits [2]. Non-dissipative balancing circuits are more attractive for large capacity batteries for the advantages of the energy substantially lossless and high efficiency. Reference [3] presents a novel power inductor-based bidirectional lossless equalization circuit, and obtains a large equalization current. Reference [4] proposed Buck-Boost

equalization circuit to achieve voltage balancing between the series battery packs, but the equalization current is not controllable. Reference [5]-[7] also presents lots of topologies of equation circuits to improve the equalization performances. The Buck-Boost non-dissipative balancing circuit is one of the most commonly used topologies for large capacity batteries in series. Unfortunately, the equalization currents are all coupled together, and are very difficult to be controlled, resulting in uncertain equalization speed and time.

To reduce the equalization time, this paper introduces large equalization current control strategy for series connected battery packs based on buck-boost converter. In the strategy, the voltage and SOC (state of charge) loop defines the switches on or off state, while the current loop determines the switch-on time. Depending on the control loops, the equalization currents are controllable, and the maximum values could reach the large reference current. Experimental results show the proposed control strategy provides more balanced paths and could achieve fast equalization.

II. THE PRINCIPLE BUCK-BOOST EQUALIZATION TOPOLOGY

The Buck-Boost equalization circuit is shown in Fig. 1. $V_{B1} \sim V_{Bn}$ are the voltages of the series batteries, $Q_1 \sim Q_n$, $D_1 \sim D_n$ are fully controlled switches, $L_1 \sim L_{n-1}$ are equalization inductances. Except the first and the last battery, every battery has a path constituted by two equalization inductors, a fully controlled switch and a battery. The adjacent batteries share the same equalization inductors.

The ordinary equalization principle is summarized as follows: if the energy of the battery V_{Bx} is higher, then the energy of V_{Bx} could be transferred to others through equalization inductors. In the first period of a switching cycle, the switch Q_x is on, while $Q_1 \sim Q_{x-1}$ and $Q_{x+1} \sim Q_n$ are all off, then the battery V_{Bx} discharges, and the inductors L_{x-1}, L_{x-2} stores energy. In the second period of the switching cycle, $Q_1 \sim Q_n$ are on, and the stored energy of L_{x-1} and L_{x-2} is transmitted to $V_{B1} \sim V_{Bx-1}$ through the

3455

diodes $D_1 \sim D_{x-1}$, and also transmitted to $V_{Bx+1} \sim V_{Bn}$ through $D_{x+1} \sim D_n$. During these procedures, the equalization currents are not controllable.

The voltage difference between batteries is one of the most important parameters to start equalization. In fact, the combination of voltage differences and SOC differences between batteries are more applicable for large capacity batteries equalization, for the voltage differences are usually very small in most cases. Consequently, the accurate SOC estimation could improve the equalization effect.

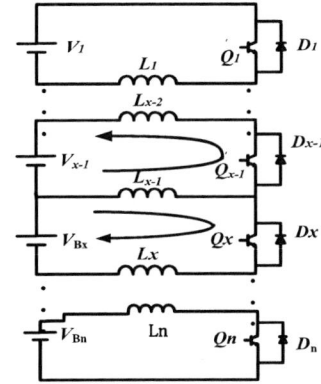

Fig. 1. The Buck-Boost equalization circuit

III. SOC ESTIMATION METHOD

In order to achieve better balance effect, this paper adopts the combination method of current integration method and open-circuit voltage method to estimate SOC of battery :

[1] Current integration method

This method is based on the definition of SOC, and integrates the charge and discharge currents of the battery to obtain the SOC. The calculation formula is shown as :

$$SOC = SOC_0 + \frac{\int i \, dt}{Q_N} \tag{1}$$

Where SOC_0 is the SOC value of the battery at the initial moment, and i is the current through the battery, and Q_N is the rated capacity of the battery (Ah). When the battery is charged, the current is positive, while the discharge current is negative. The new SOC value is the sum of the initial value and the change of battery capacity, which is expressed by the ratio of the current integration to the rated capacity of the battery. The calculation procedure is very simple and fast, and is easy to be implemented in real-time estimation.

[2] Open-circuit voltage method

A lot of experiment results indicate there is a certain relationship between the battery open-circuit voltage (OCV) and SOC. Consequently, the open-circuit voltage value could be used to estimate the battery SOC, especially used in the estimation of the initial SOC_0.

Table I gives the HPPC (hybrid pulse power characteristic) experimental results of iron phosphate battery (3.2V, 200Ah). As the increase of the SOC, the open-circuit voltage (OCV) rises too. When the SOC value is between 0% and 50%, and is also between 80%

and 100%, the OCV significantly changes. On the other sides, when SOC value is between 50% and 80%, the OCV changes slightly. To improve the estimated accuracy of the battery initial SOC_0, three sections estimation method is suitable. Depending on the data in Table I, 1stopt method is used for the polynomial fitting. Three fitting functions are shown as:

$$\begin{aligned} y = & 168665.810338 \times x^{1.5} - 880205.235627 \times x \\ & + 1530495.703109 \times x^{0.5} - 886667.329735 \end{aligned} \tag{2}$$

$$\begin{aligned} y = & 824132.037587 \times x^4 - 7566121.857812 \times x^3 + \\ & 20879647.70109 \times x^2 - 9085395.95715 \times x - 23229313.75581 \end{aligned} \tag{3}$$

$$\begin{aligned} y = & 515893.3857 \times x^{1.5} - 4285106.055188 \times x \\ & + 10506778.50741 \times x^{0.5} - 8038527.27754 \end{aligned} \tag{4}$$

TABLE I

OCV-SOC EXPERIMENTAL RESULTS

SOC	OCV(V)	SOC	OCV(V)	SOC	OCV(V)
0%	2.7706	35%	3.2842	70%	3.3005
5%	3.1479	40%	3.2912	75%	3.3206
10%	3.2088	45%	3.2938	80%	3.3338
15%	3.214	50%	3.2947	85%	3.3346
20%	3.2394	55%	3.2953	90%	3.3355
25%	3.2592	60%	3.296	95%	3.337
30%	3.2721	65%	3.297	100%	3.3758

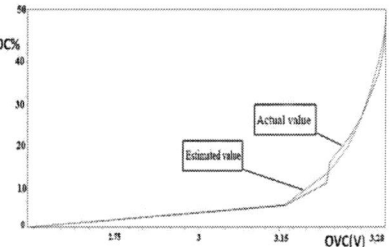

(a)   OCV-SOC0%-50%

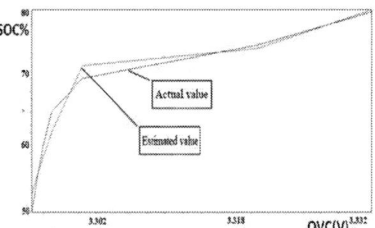

(b)   OCV-SOC50%-80%

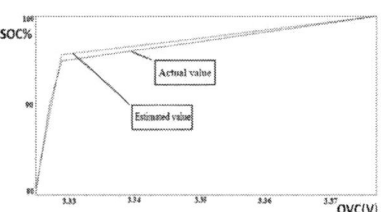

(c)   OCV-SOC80%-100%

Fig. 2.   The relationship curve between SOC and OCV

The 2018 International Power Electronics Conference

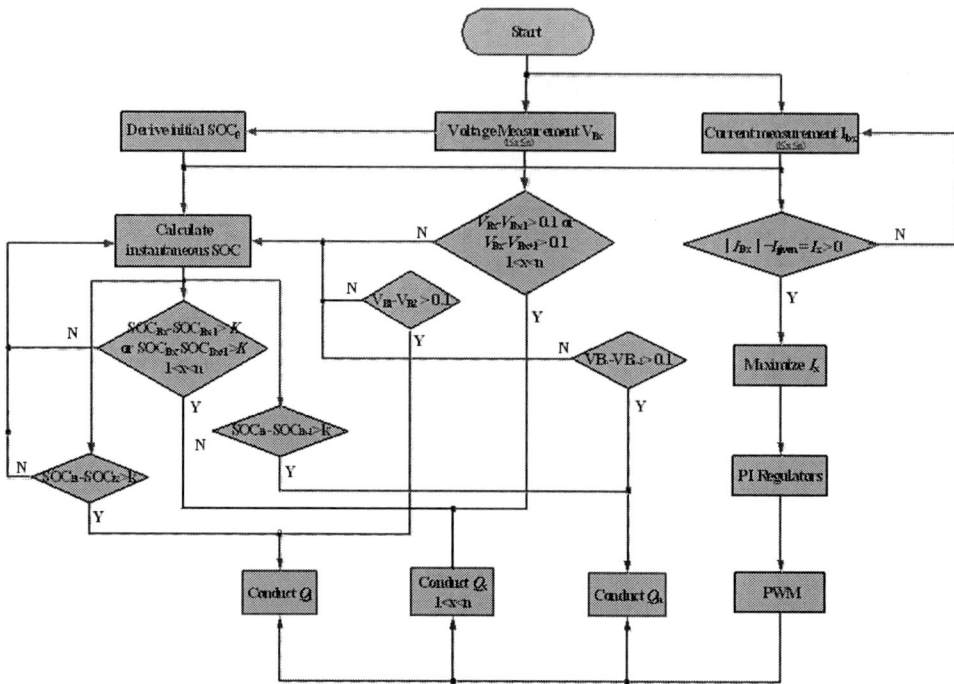

Fig. 3. The equalization principle of the proposed control strategy

On the basis of the three fitting functions, the approximate curves between true values and estimation values of SOC are shown in Fig. 2. Obviously, three fitting functions method is accurate enough to estimate the initial SOC_0 depending on OCV.

Based on the estimated initial SOC_0 and (1), the instantaneous SOC is derived. The combination method of current integration method and open-circuit voltage method could guarantee an accurate estimate of SOC, and simultaneously achieve a rapid speed.

IV. THE PRINCIPLE OF THE NEW CONTROL STRATEGY

A new large equalization current control strategy is presented, and shown in Fig. 3. The SOC、 voltages and currents of batteries are all considered in the control loops. The SOC and voltages control loop determines on or off state of the switches, while the current control loop determines the duty cycle of the switches in "on" state based on PI regulator.

Firstly, the voltage measurement unit detects voltages of every battery, and the current measurement unit detects the currents. If the calculated voltage differences are more than 0.1 V, then the equalization is allowed, otherwise, SOC differences are taken into account.

When $SOC_{B1}-SOC_{B2} > K$ (K is the given SOC threshold), the PWM signal of Q_1 is usable and Q_1 is in "on" state. Similarly, when SOC_{Bx}、 SOC_{Bx-1} and SOC_{Bx+1} ($1<x<n$) satisfy $SOC_{Bx}-SOC_{Bx-1} > K$ or $SOC_{Bx}-SOC_{Bx+1} > K$, Q_x is on. When $SOC_{Bn}-SOC_{Bn-1} > K$, the PWM signal of Q_n is available and Q_n is on.

The PWM signals are generated by the current control loop. In this loop, the first step is to detect currents I_{Bx} of each battery ($1 \le x \le n$), regardless of batteries working conditions. Then all $|I_{Bx}|$ (the absolute value of I_{Bx}) are compared with a given equalization threshold I_{given}, and the difference values are calculated as $I_x = |I_{Bx}| - I_{given}$. The maximum value of I_x are achieved, and put into the PI regulator to generate PWM signals. The procedure is summarized as follows: the maximum difference between measured currents of all batteries I_x is achieved, and taken as the input of the PI regulator to generate variable PWM signals. Based on the control loops, the equalization currents are controllable.

V. EXPERIMENTAL RESULTS

In order to verify the validity of the proposed control strategy, an equalization circuits for three series lithium batteries is constructed, and shown in Fig. 4. The rated capacity of the battery is 200Ah, and the rated voltage is 3.2 V. Based on open-circuit voltage method, the initial SOC of the batteries are shown in Fig. 5. The reference equalization current is 10A, and the equalization currents of the three batteries captured by an oscilloscope are shown in Fig. 6. Depending on the equalization current measurements, the equalization process with calculated SOC is shown in Fig. 7. The results indicate, based on the presented control strategy, three series batteries with different SOC values could achieve a new equilibrium, and simultaneously the equalizing currents are controllable, and could reach as large as the reference value 10 A.

3457

It is concluded that the control strategy could increase the equalization currents, and completes the equalization process as soon as possible. The method is very applicable for large capacity series battery packs.

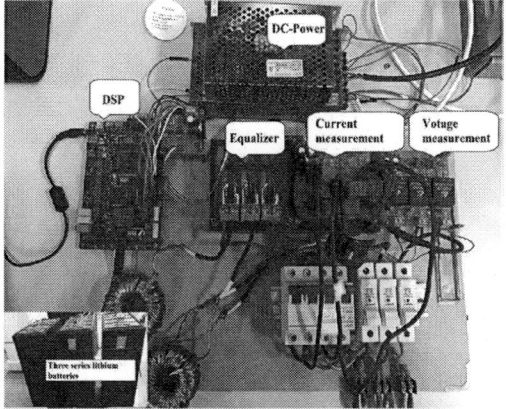

Fig. 4. Experimental platform

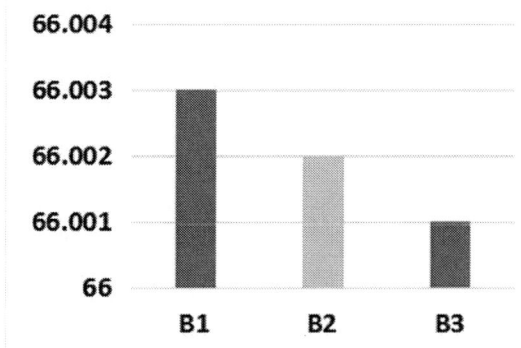

Fig. 5. The initial SOC of three batteries

VI. CONCLUSIONS

This paper presents a new control strategy to increase the equalization current for series battery packs. The characteristics of SOC 、 voltages and currents of batteries are all considered in the control loops. The SOC and voltages control loop determines the state of the switches, while the current control loop determines the duty cycle of the switches in "on" state based on PI regulators. Depending on the strategy, the equalization currents are controllable, and could reach as large as the reference values. Experimental results indicate the proposed control strategy provides more balanced paths and completes the equalization process as soon as possible. The strategy is very applicable for large capacity series battery packs.

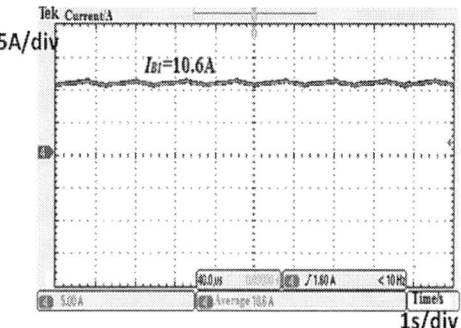

a)   The equalization current of battery B1

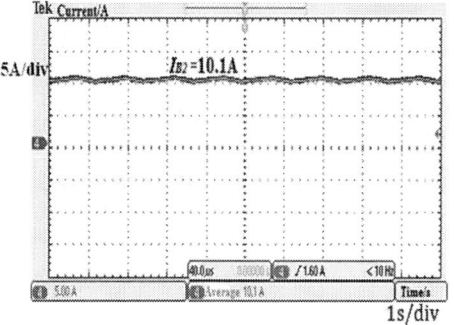

b)   The equalization current of battery $B2$

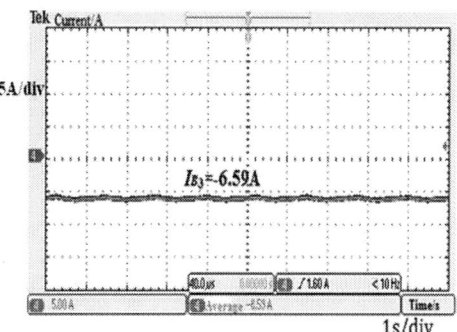

c)   The equalization current of battery $B3$

Fig. 6. When the reference current is 10A, the equalization currents of the three series batteries

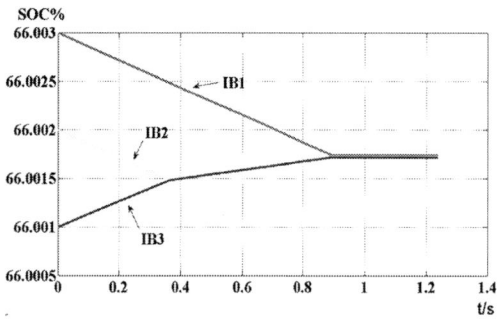

Fig. 7. The calculated SOC of three batteries during the equalization process when reference current is 10A

REFERENCES

[1] Ali M H, Wu B, Dougal R A. An Overview of SMES Applications in Power and Energy System[J]. *IEEE Transactions on Sustainable Energy*, 2010, 1(1): 38-40.

[2] Zhang C, Tseng K J. A Novel Flywheel Energy Storage System with Partially-self-bearing Flywheel Motor[J]. *IEEE Transactions on Energy Conversion*, 2007, 22(2): 477-487.

[3] Xiangwei Guo,Longyun Kang,Zhizhen Huang.Research on a Novel Power Inductor-Based Bidirectional Lossless Equalization Circuit for Series-Connected Battery Packs [J].Energies,2015,8:5555-5576.

[4] F. Mestrallet, L Kerachev, J.C. Crebier, "Multiphase Interleaved Converter for Lithium Battery Active Balancing," *IEEE Transactions on Power Electronics*, vol.29, pp.2874-2881, Aug 2013

[5] Schweitzer M, Lizama I, Friedli T. Comparison of the Chip Area Usage of 2-Level and 3-Level Voltage Source Converter Topologies[J]. *IEEE Transactions on Power Electronics*, 2010, 26(48): 391-396.

[6] Jinlei Sun ,Rengui Lui, Guo Wei, Bingliang Xu,Chunbo Zhu, A High Efficiency Equalizer based on Forward Converter for Series Connected Battery String. *IEEE Vehicle Power and Propulsion Conference*, Oct. 9-12, 2012.

[7] Mestrallet.F, Kerachev L, Crebier.J.C , Multiphase Interleaved Converter for Lithium Battery Active Balancing, *IEEE Transactions on Power Electronics*, 2013,29:2874-2881.

[8] Momayyezan.M, Hredzak.B, Agelidis.V, A Load-Sharing Strategy for the State of Charge Balancing Between the Battery Modules of Integrated Reconfigurable Converter,*IEEE Transactions on Power Electronics*,2016.

[9] Li Linlin, Bian Liangchu, Xing Jing,Ji Xiu. "Research and Design of Battery Initiative Equalization System in Charging and Discharging Period", Control and Decision

[10] Conference (CCDC), 2015 27th Chinese,pp.6237 - 6242. Juan Zhao, Jiuchun Jiang, Liyong Niu. "A Novel Charge Equalization Technique for Electric Vehicle Battery System", ower Electronics and Drive Systems (PEDS), 2003. The Fifth International Conference,Vol.2,pp. 853 - 857

A Multi-Port Bidirectional Power Conversion System for Reversible Solid Oxide Fuel Cell Applications

Xiang Lin[1], Kai Sun[1*], Jin Lin[1], Zhe Zhang[2] and Wei Kong[3]

1 State Key Lab of Power Systems, Department of Electrical Engineering, Tsinghua University, Beijing, China
2 Department of Electrical Engineering, Technical University of Denmark, Kgs. Lyngby, Denmark
3 Department of Electric Power Engineering, Shanghai University of Electric Power, Shanghai, China
*E-mail: sun-kai@mail.tsinghua.edu.cn

Abstract- **Reversible Solid Oxide Fuel Cell/Electrolyser Cell (SOFC/EC) technology is an attractive solution for high energy storage system in the utility grid. However, the wide range of voltage and low power of single SOFC/EC stack make it difficult to design the power conversion system for SOFC/EC storage system. In this paper, a new power multi-port bidirectional conversion system is proposed to connect multiple SOFC/EC stacks with the utility grid. The converter structure contains a multi-port structure with two conversion stages. The two-stage conversion structure is first analyzed to address the wide-range of SOFC/EC stack's voltage. The high-step-down CLLC resonant converter is implemented to achieve efficient voltage transformation, and the interleaved buck converter is employed as the second stage to control the voltage of SOFC/EC stack within a wide range. The derivation of the multi-port structure is introduced, and the control strategy of proposed conversion system is also discussed in this paper. The proposed conversion system enables a flexible control for the application of multiple SOFC/EC stacks. The feature of the proposed system is verified by the experiments from a down-scale prototype.**

Keywords—multi-port bidirectional power conversion system, reversible solid oxide fuel cell, CLLC resonant converter, interleaved buck converter

I. INTRODUCTION

With the advent of renewable energy, there is a strong demand for high power storage system in the power grid. The hydrogen storage system is an attractive solution of the high-power storage system for several reasons: 1) Hydrogen is a clean energy and has the highest energy density (KWh/Kg); 2) Hydrogen storage system is one of few choices for long-term storage; 3) Hydrogen storage system can be extended to high power system [1]. Recently, the Solid Oxide Fuel Cells can operate as Electrolyser Cells to generate hydrogen, which makes this Solid Oxide Fuel Cell/Electrolyser Cell (SOFC/EC) technology more attractive compared to separated water electrolyser and fuel cells [2-3]. To connect the SOFC/EC with the utility grid, high power bidirectional conversion system is needed. Traditionally, the conversion system has two conversion stages consisting of a dc-dc converter to connect distributed storage stacks with a dc bus, and a centralized dc-ac converter to connect the dc bus to the utility grid. This paper focuses on the dc-dc conversion stage which is usually designed with the consideration of SOFC/EC stack's characteristics.

However, the SOFC/EC has certain electrical characteristics which complicate the power conversion system design. Firstly, because of the challenge to equalize fuel/gas-pressure within the cells, a single SOFC/EC stack has a limited number of cells resulting in low stack voltage [4]. So, to improve the system's power capacity, a multi-port conversion system is needed to connect multiple SOFC/EC stacks. Moreover, the conversion system should achieve high step-down voltage transformation. Secondly, the SOFC/EC has a wide operating voltage, and keeping interfacing converter high efficiency over such a wide range is a challenge.

Therefore, to design an efficient power conversion system for SOFC/EC, both the system structure and converter topology need to be addressed. Alternatively, storage stacks can be connected in series directly [5-6]. However, in such a structure, it is difficult to adjust the operation of each stack independently for a higher system efficiency and one stack failure may affect the operation of the whole string. In some literatures, storage stacks are connected in a cascaded structure with isolated dc-dc converters [7]. Such a structure has an improved control flexibility, but the control complexity is high over a wide voltage range. In recent years, multi-port structures also attract an increasing attention because of its ability to connect multiple sources. However, those systems usually have non-modular structure and complex control strategy and thereby are not suitable for utility-scale energy storage system [8]. Regarding to converter topologies, several topologies for unidirectional fuel cell have been proposed [9-13], among which dual-active-bridge (DAB) converter and isolated boost converter are the most popular ones for high-step-down voltage conversion. But auxiliary circuits are usually needed to extend voltage range of the DAB converter or to achieve ZVS for the isolated boost converter.

In this paper, a new multi-port bidirectional power conversion system is proposed as shown in Fig.1. The interleaved buck converter converts the wide-range voltage of SOFC/EC to nearly constant voltage at the middle stage, and the multi-port CLLC resonant bidirectional converter (CLLC-RBC) connects multiple

The 2018 International Power Electronics Conference

sources with the dc bus. With such arrangement, a flexible and simplified control of each stack is achieved, as well as the power capacity can be easily extended.

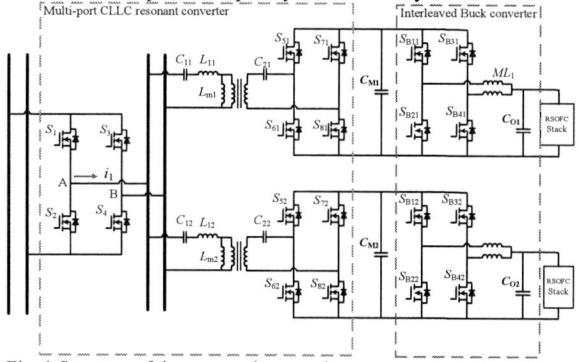

Fig. 1 Structure of the proposed conversion system.

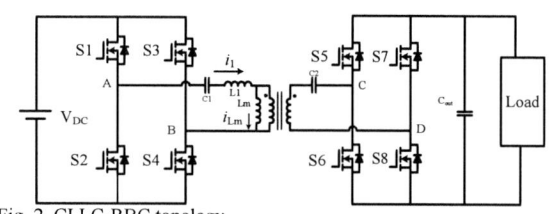

Fig. 2. CLLC-RBC topology.

In the following sections of the paper, the single-input-multi-output CLLC-RBC and interleaved buck converter are discussed. The operation of system is also analyzed. Finally, the experimental results are presented to verify the operation of system.

II PROPOSED CONVERSION SYSTEM

A. Two-stage structure for CLLC-RBC application

In recent years, CLLC-RBC has become a desirable topology for bidirectional dc-dc applications due to its simple control, high-frequency operation and high efficiency [14-15]. A typical CLLC-RBC topology is shown in Fig. 2, and the frequency modulation scheme is usually employed. In the forward mode, in which power is delivered from the high-voltage DC bus to the load, the switches $S1\&S4$ and $S2\&S3$ are driven by complementary signals with 50% duty cycle while all the switches at low voltage side are turned off or are controlled in synchronous rectification mode. However, in the reverse mode, the switches $S5\&S8$ and $S6\&S7$ are driven by complementary signals with 50% duty cycle, and the switches at high voltage side are turned off or are controlled in synchronous rectification mode. The ZVS is achieved with the assistance of the magnetic excitation current i_{Lm}. The voltage gain of the CLLC resonant converter is usually analyzed with the approach of fundamental mode, and a typical voltage gain curve in forward mode (SOEC) is presented in Fig. 3.

While employing CLLC resonant converter for SOFC/EC applications, the wide stack voltage needs to be addressed. Because of the polarization of SOFC/EC, the stack voltage changes with current.

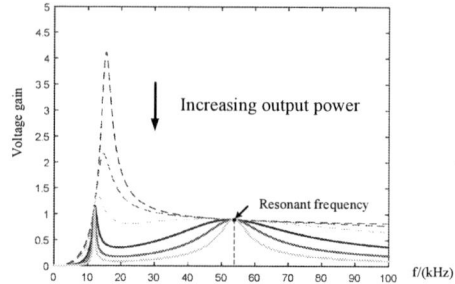

Fig. 3 Typical voltage gain of the CLLC resonant converter in forward mode.

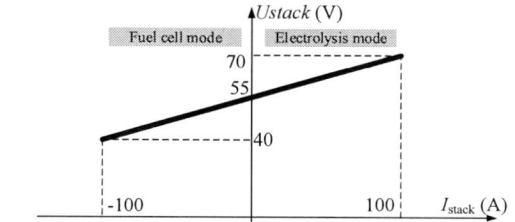

Fig. 4 Simplified U-I curve of SOFC/EC stacks.

A simplified steady-state U-I curve of a SOFC/EC stack is shown as an example in Fig. 4, where the voltage varies from 40 V to 70 V, and the maximum power is 7 kW and 4 kW in SOEC and SOFC modes, respectively. The following evaluation is based on this example. Therefore, such a wide voltage range set a difficulty to design an efficient dc-dc converter.

In the previous literature, the application of resonant dc-dc converter contains two different approaches: single-stage and two-stage structure [14-16]. In single-stage structure, the frequency of CLLC-RBC is adjusted to control the voltage gain, and in two-stage structure, the resonant dc-dc converter is controlled with fixed frequency and a non-isolated dc-dc converter is employed to adjust the voltage gain. But for SOFC/EC application, the single-stage structure is not suitable. The voltage gain curve in Fig.3 indicates that the voltage gain is related to the output power, and in light load, higher voltage gain is easier to be achieved. However, Fig.4 indicates that higher voltage gain is needed in heavy load condition during forward mode. So, if the single-stage structure is employed for SOFC/EC stacks, it is difficult to design the resonant tank and the operation frequency range.

On the other hand, the two-stage structure, as illustrated in Fig.5 is suitable for SOFC/EC application. Fig.4 indicates that the CLLC-RBC has stable voltage gain at resonant frequency in different load condition. Therefore, by controlling frequency of CLLC resonant converter at the fixed resonant frequency, it is much simplified to design the control strategy and converter parameters. The bidirectional interleaved buck converter is chosen to assist the adjustment of operation voltage. Since the CLLC resonant converter has achieved high-step-down voltage transformation, the required voltage-gain (duty cycle) range of interleaved buck converter is

3461

reduced and low voltage-gain is avoided (because the switching loss is related to dc voltage at the high-voltage side and dc current at the low voltage side, and the extreme low voltage gain, for example 0.1, will cause large switching loss). In the meantime, comparing with buck converter, interleaved buck converter has smaller switching and conduction loss with equal output current ripple.

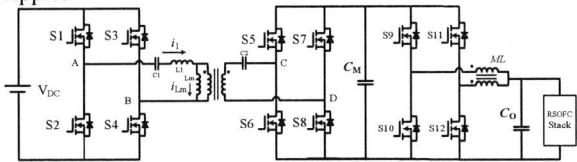

Fig.5 two-stage structure for CLLCBDC application

B. Multi-port structure derivation

By above analysis, the two-stage dc-dc converter is suitable for single SOFC/EC stack. For power storage system, a few of such stacks are needed. Because of the application of CLLC-RBC, a new system structure to connect multiple RSOFC stacks is designed as Fig. 1 shows. Traditionally, to connect several SOFC/EC stacks to the dc bus, several dc-dc converters are needed. But in the proposed conversion system, the multi-port structure is applied instead. In the two-stage dc-dc converter, the CLLC resonant converter is controlled at resonant frequency, so different CLLC resonant converters have same drive signals for corresponding switches while they have same resonant tank parameters. For each converter, the current at the high voltage side is relative low, so it is practical to integrate several full bridge circuits at the high voltage side into one full bridge circuit at the high voltage side as shown in Fig.6. Through this structure, multi-port capacity is achieved with reduced number of active switches. Also comparing with multi-port structure with multi-winding transformer, the design of resonant tank and the isolated transformer are much simplified.

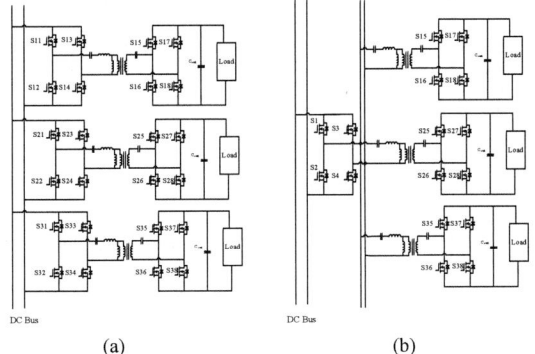

Fig. 6 Comparison of different structures to connect multiple sources. (a) parallel-operated structure, (b) integrated structure

Since the frequency of multi-port CLLC-RBC is fixed, the output power is adjusted by the interleaved buck converter in the two-stage converter. Because the combination of the full bridge circuits does not cause the couple of different stacks power, the proposed multi-port structure enables a flexible control of each stack. The voltage of each stack can be adjusted without affect other

stacks. This is the salient benefit of proposed multi-port conversion system. As shown in Fig.7, in the past research of high power storage system, the storage stacks are usually connected in series or with a cascaded structure. In the series-connected structure, the failure of one storage stack will affect other stacks. In the cascaded structure, each dc-dc converter shares the same current in the high-voltage side, so the power of each stack is coupled. With the constant voltage of dc bus, the power adjustment of one storage stack will affect the common current in the high-voltage side and then the operation of other storage stacks.

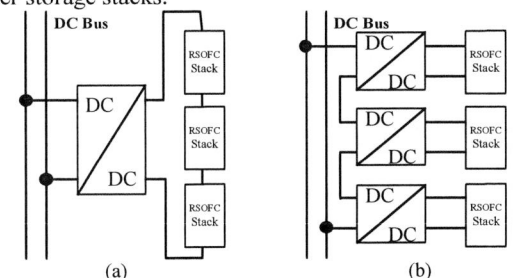

Fig. 7 tradition storage system structure: (a) series-connected structure; (b) cascaded structure.

The proposed conversion system has an extensible structure and is suitable for storage system at different power rate. For high power storage system, the proposed conversion system could be designed as a modular storage block, and the power of storage system can be scaled by connecting several blocks in parallel structure as Fig. 8 shows.

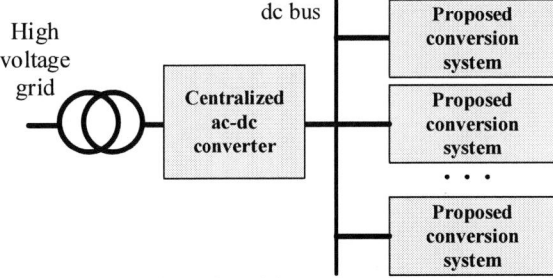

Fig. 8 System topology of modular storage system with proposed conversion system

III CONTROL OF THE PROPOSED CONVERSION SYSTEM

In Section II, the structure of proposed conversion system is fully evaluated and, in this section, the control strategy is further analyzed. The following analysis is based on the single-input-dual-output system as shown in Fig.1.

The switching frequency of multi-port CLLC resonant converter is fixed at the resonant frequency, so the voltage of the middle stage is constant under different load condition. To limit the voltage gain of the interleaved buck converter, the middle stage voltage V_{middle} is designed to be a little higher than the maximum voltage of a single SOFC/EC stack V_{stack_max}: $V_{middle}=1.1V_{stack\_max}$. For example, for the stack shown in Fig.3, the middle stage voltage is designed at 80 V. Then the voltage gain range of interleaved buck converter is

3462

0.5-0.875. Because the middle stage voltage is constant, the control of interleaved buck converter is much simplified. As shown in Fig. 9, by controlling the duty cycle D of interleaved buck converter, the current (or voltage) of a single stack is regulated in forward and reverse mode.

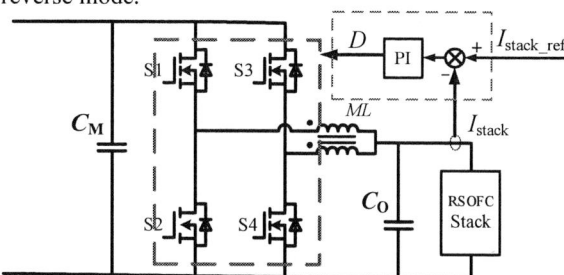

Fig. 9 Control diagram for the interleaved buck converter.

The control diagram of the whole conversion system is presented in Fig.10. The control target is the system power. By measuring the current from dc bus, the power of the entire system can be calculated. Based on an allocation mechanism, the current reference I_{ref} of each stack is calculated. Therefore, the current of each stack is regulated. The allocation of current references can be set to different rules under different load conditions. Because the control of different ports is decoupled, the current references can be adjusted flexibly under different condition to improve the performance of the entire system. For example, in the light load condition, the current reference of the port #2 can be set to 0 to improve the system efficiency. In the heavy load condition, the current reference of each port is set to be equal to realize power requirement.

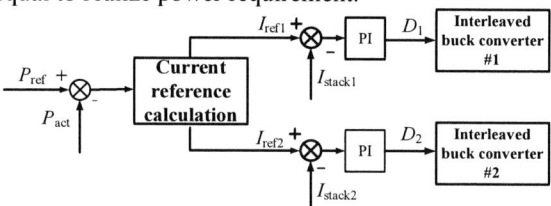

Fig. 10 the control diagram for the conversion system

A single-input-dual-output system (Fig.1) is built in the MATLAB/Simulink to illustrate the flexible operation of proposed power conversion system. The parameters of resonant tanks and mutual inductance in interleaved buck converter are shown in Table 1. The voltage of dc bus is set to 600 V, and the input voltage of middle stage V_{Ck} is designed to around 80 V. The SOFC/EC stacks are simulated with the controllable current sources, and the U-I curve of the controllable current sources is programmed as Fig. 4 to simulate the stacks' characteristics.

TABLE 1 Parameters of resonant tank and mutual inductance

$L_{11\&12}$	$C_{11\&12}$	$C_{21\&22}$	$L_{m1\&m2}$	$L_{ML1\&2}$	$M_{ML1\&2}$
150 uH	100 nF	8000 nF	550 uH	100 uH	-34 uH

The example situation is that both stacks work at the fuel cell mode, and the power demand for the conversion system is increased from light load to heavy load. As illustrated in Fig.11, the current of each stack can be adjusted flexibly without affecting the other stack. The demand of power is addressed with one stack in light load and two stacks in heavy load. So in different load condition, the conversion system can adjust its load distribution for better system efficiency.

The simulation illustrates that the proposed system conversion system enables a flexible control of connected storage stacks. This is an advantage comparing to traditional storage structures as shown in Fig. 7. Such flexible control is achieved because of the application of multi-port CLLC-RBC. In the meantime, the system parameters, especially the parameters of resonant tanks, are easy to design, because no integration of resonant tanks is needed in the proposed structure.

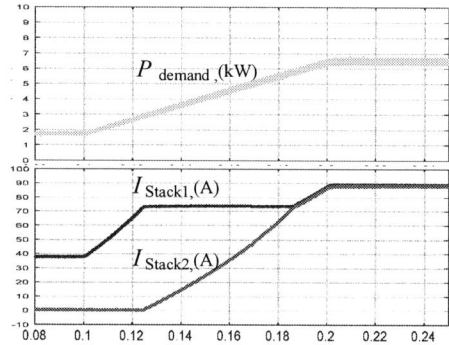

Fig. 11 the control diagram for the conversion system

IV EXPERIMENT RESULTS

To verify the flexible control of the system, a reduced-scale prototypes of proposed conversion system has been designed and built as shown in Fig.12. The conversion system contains two output ports. The SiC MOSFET C2M0080120D and Si MOSFET IRFP4127PbF are used as the switches at high voltage and low voltage side respectively. The design parameters of resonant tanks and mutual inductors are listed in Table 1. The switching frequency of CLLC resonant converter is set to 50 kHz, and the frequency of interleaved buck converter is set to 20 kHz.

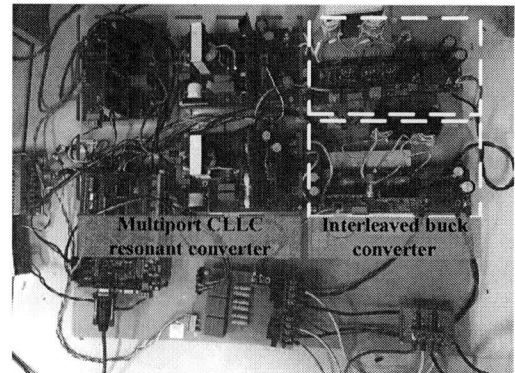

Fig. 12 the prototype of proposed conversion system.

3463

First, the forward-mode operation is verified, and the resistors (5 Ω) are used as the load to simulate the SOFC stacks. The input voltage is set to 200 V, and the voltages of output ports are set as follow: u_{o1} =20 V, u_{o2}=18 V. The steady-state waveforms are shown in Fig. 13. The input current of each resonant tank is shown in Fig.13 (a), and two resonant tanks share the same input ac square voltage waveform. In Fig.13 (b), two interleaved buck converters are controlled with different duty cycles. The experiment waveforms indicate that the output voltages v_{o1}& v_{o2} can be set to different status, and the parameter differences of resonant tanks or mutual inductors don't affect operation of the proposed system. The combination of full bridge circuit at the high voltage side doesn't affect the independent control of u_{o1}& u_{o2}.

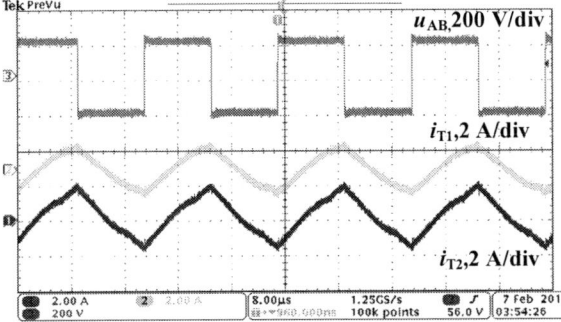

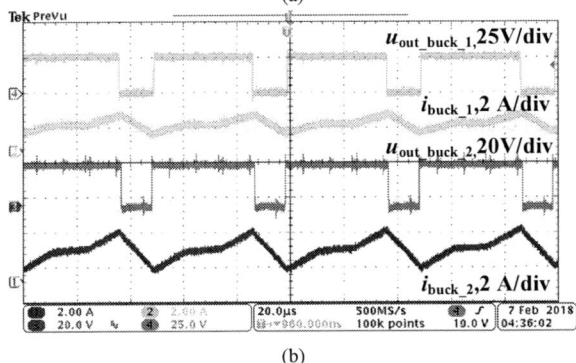

(b)

Fig. 13 Experiment results of steady state: (a) waveforms of input currents of two transformers; (b) waveforms of two interleaved buck converters.

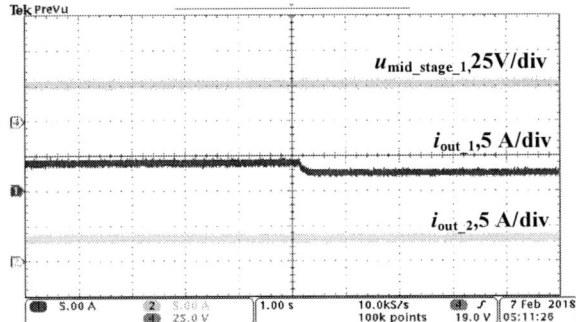

Fig. 14 the experiment results of dynamic process

The experiment of dynamic performance is also conducted. The input voltage is set to 200 V, and u_{o2} is

set to 18 V. The voltage reference of u_{o1} is changed from 20V to 12 V. The experiment waveform is shown in Fig. 13. The output currents are shown in the result. By controlling the duty cycle of the interleaved buck converters, the control of i_{out_1} is achieved and the change of i_{out_1} doesn't affect i_{out_2}. Therefore, in the proposed conversion system, each port at the low-voltage side can be controlled independently.

The back-mode operation is conducted as well. The resistor (500 Ω) is connected to the high-voltage side, and two dc sources are connected to two low-voltage ports. The experiment results of steady state and transient state are presented in Fig. 14 and Fig. 15. The voltage of port #1 is set to 20 V and the voltage of port #2 is set to 15 V. The waveforms indicate that the allocation of each port's power could be adjusted flexibly without affecting the total power of the conversion system.

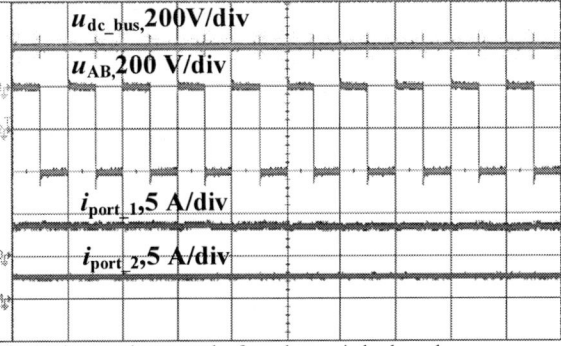

Fig. 14 the experiment result of steady state in back mode

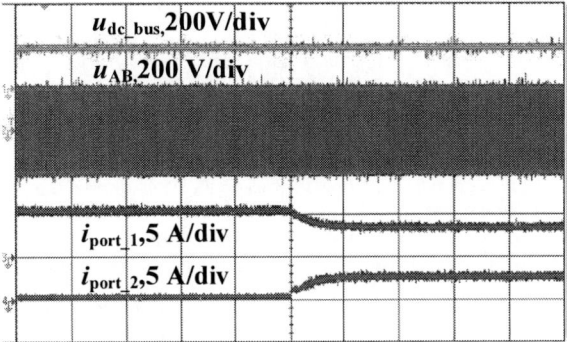

Fig. 15 the experiment result of transient state in back mode

The above experimental results illustrate that the proposed conversion system can achieve flexible control in both forward-mode and back-mode operation. From the angle of control, the proposed conversion system can be seen as two separate converters. And such feature is based on the two-stage conversion structure with CLLC-RBC. In the storage system with SOFC/EC stacks, such ability can enable better performance for the whole conversion system and each stack. This is a salient advantage of proposed conversion system comparing to traditional structure for high power storage system.

V. CONCLUSION

In this paper, a new multi-port bidirectional conversion

system is proposed to connect multiple SOFC/EC stacks with utility grid. With the analysis of CLLC-RBC's feature, the two-stage conversion structure is applied for the application of SOFC/EC. By the combination of full bridge circuit at high-voltage side, the multi-port CLLC-RBC is designed to achieve high-step-down voltage transformation with fixed frequency control, and the interleaved buck converter is employed to control the output voltage. The control of the proposed conversion system is illustrated in this paper, and the propose conversion system enables a more flexible control comparing to traditional storage system structures. A reduced-scale single-input-dual-output prototype is built and the flexible control is verified with the prototype.

ACKNOWLEDGMENT

The authors would like to acknowledge the supports by National Key Research and Development Program (2016YFE0102600) and State Key Lab of Power Systems, Tsinghua University (SKLD16Z04).

REFERENCES

[1] Luo, Xing, et al. "Overview of current development in electrical energy storage technologies and the application potential in power system operation." Applied Energy 137 (2015): 511-536.

[2] Schiebahn S, Grube T, Robinius M, et al. Power to gas: Technological overview, systems analysis and economic assessment for a case study in Germany[J]. International journal of hydrogen energy, 2015, 40(12): 4285-4294.

[3] Ni M, Leung M K H, Leung D Y C. Technological development of hydrogen production by solid oxide electrolyzer cell (SOEC)[J]. International Journal of Hydrogen Energy, 2008, 33(9): 2337-2354.

[4] Pittini, Riccardo, Zhe Zhang, and Michael AE Andersen. "Isolated full bridge boost DC-DC converter designed for bidirectional operation of fuel cells/electrolyzer cells in grid-tie applications." Power Electronics and Applications (EPE), 2013 15th European Conference on. IEEE, 2013.

[5] Lang P, Michel M, Wade N, et al. Dynamic energy storage-A UK first[J]. 2012.

[6] Krishnamoorthy H S, Rana D, Garg P, et al. Wind turbine generator–battery energy storage utility interface converter topology with medium-frequency transformer link[J]. IEEE Transactions on Power Electronics, 2014, 29(8): 4146-4155

[7] Bragard M, Soltau N, Thomas S, et al. The balance of renewable sources and user demands in grids: Power electronics for modular battery energy storage systems[J] IEEE Transactions on Power Electronics, 2010, 25(12): 3049-3056.

[8] Hawke, Joshua T., Harish S. Krishnamoorthy, and Prasad N. Enjeti. "A family of new multiport power-sharing converter topologies for large grid-connected fuel cells." IEEE Journal of Emerging and Selected Topics in Power Electronics 2.4 (2014): 962-971.

[9] Wang, Jin, et al. "Low cost fuel cell converter system for residential power generation." IEEE transactions on Power Electronics 19.5 (2004): 1315-1322.

[10] Zhang, Zhe, et al. "Analysis and design of a bidirectional isolated DC–DC converter for fuel cells and supercapacitors hybrid system." IEEE transactions on Power Electronics 27.2 (2012): 848-859.

[11] Nymand, Morten, and Michael AE Andersen. "High-efficiency isolated boost DC–DC converter for high-power low-voltage fuel-cell applications." IEEE Transactions on Industrial Electronics 57.2 (2010): 505-514.

[12] Prasanna, U. R., and Akshay K. Rathore. "Extended range ZVS active-clamped current-fed full-bridge isolated DC/DC converter for fuel cell applications: analysis, design, and experimental results." IEEE Transactions on Industrial Electronics 60.7 (2013): 2661-2672.

[13] Rathore, Akshay K., Ashoka KS Bhat, and Ramesh Oruganti. "Analysis, design and experimental results of wide range ZVS active-clamped LL type current-fed DC/DC converter for fuel cells to utility interface." IEEE Transactions on Industrial Electronics 59.1 (2012): 473-485.

[14] Chen W, Rong P, Lu Z. Snubberless bidirectional dc–dc converter with new CLLC resonant tank featuring minimized switching loss[J]. IEEE Transactions on industrial electronics, 2010, 57(9): 3075-3086.

[15] Zahid Z U, Dalala Z M, Chen R, et al. Design of bidirectional dc–dc resonant converter for vehicle-to-grid (V2G) applications[J]. IEEE Transactions on Transportation Electrification, 2015, 1(3): 232-244.

[16] Krismer F, Biela J, Kolar J W. A comparative evaluation of isolated bi-directional DC/DC converters with wide input and output voltage range[C]//Industry Applications Conference, 2005. Fourtieth IAS Annual Meeting. Conference Record of the 2005. IEEE, 2005, 1: 599-606.

Self-preheating Method for Li-ion Battery Using Battery Impedance Estimator

Dong-Kwan Kim[1], Young-Dal Lee[2], Sang-Hyun Ha[2,3], Yu-Jin Jang[1], and Gun-Woo Moon[2]

1 Division of Future Vehicle, KAIST, Daejeon, South Koreaw
2 School of Electrical Engineering, KAIST, Daejeon, South Korea
3 Agency for Defense Development, Daejeon, South Korea
*E-mail: pkodower@kaist.ac.kr

Abstract-The performance of Li-ion battery's cell is significantly reduced under subzero temperature. In this paper, a self-preheating method for Li-ion battery using battery impedance estimator is presented. For battery impedance estimation, no information of the actual battery's internal impedance is required, so that the proposed method can be easily and quickly adapted to the commercial Li-ion batteries. In addition, the proposed method includes the optimization of charging/discharging current and preheating frequency using the battery impedance estimator. Moreover, based on DC- discharging estimation in the proposed method, the preheating procedure is automatically shut. Using the presented method the self-preheating method can be widely adapted to commercial applications. The feasibility of the proposed control strategy is verified by the simulation and experimental results.

I. INTRODUCTION

Lithium-ion battery is widely used for many applications requiring mobility such as cellphone, handheld LED lighting devices, and electrical vehicle, which includes smart grid [1]-[4]. Among these applications, smart grid has been an important issue because that are good to reduce the large peak electric energy consumption in a day time using bidirectional energy transferring between EV and grid, so that it prevents the energy shortage problem. Therefore, Lithium-ion battery has several requirements: the large power capability and the wide operating temperature range.

Lithium-ion battery, however, has a critical problem that the capacity and the power capability decreased in sub-zero temperature because the internal impedance of the lithium-ion battery is dramatically increased [5] and [6]. As a result, the lithium-ion battery suffers the reduction of the power ability and capability under the sub-zero temperature.

To overcome the problem, the temperature of Lithium-ion battery should be higher than sub-zero temperature, so that various preheating methods are presented. The very conventional way is resistive heat-up [7]. In this case, large energy can be released from the heater, so that the temperature rises quickly. However, most of the energy of the heating plate is left out to the ambient. Thus, the resistive heat-up method loses large battery energy.

Recently, to increase heat transfer efficiency, self-

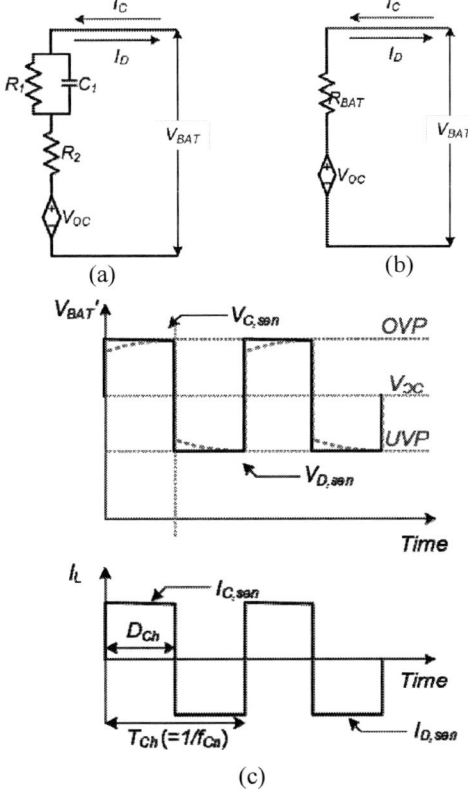

Fig. 1. Battery impedance estimator.
(a). Equivalent internal resistance model.
(b). Modified internal resistance mode.
(c). Equivalent waveforms.

preheating method, also known as internal preheating method, is introduced. [7]-[9]. In internal heating method, the battery is warmed up by the heat energy from the battery's internal resistance. In this case, the current flows from or to the battery, so that the battery's reliability should be high. Therefore, the voltage of battery should be monitored and over voltage protection and under voltage protection circuit prevent the abnormal status. As a result, in the internal heating method, there is tight restriction of the heat-up current, so that the optimization of heat-up current is important to minimize the heat-up time. To fasten preheating time, the heat-up current should be sufficient large to maximize the power dissipation. Fig. 1 shows the RC equivalent circuit of battery. Fig. 1 shows the internal impedance of battery is related with the charging-discharging current period. As shown in fig. 2, the high frequency heat-up has small battery voltage ripple with same RMS current. Thus, high frequency heat-up is required to maximize the current. However, in high frequency heat-up operation, the most amount of current flows not resister but capacitor. Therefore, the heat-up frequency should be minimized to have higher heat-up speed. Therefore, there is a trade-off of the battery reliability problem and the heat-up speed. As a result, the selection of operating current and frequency is important.

In the previous works, there are limitations to adapt the self-preheating method to commercial battery pack directly. [7] and [8] verify the feasibility of the self-preheating method, however, the proper control method was not presented. Recently, Shankar, Youngki, and Anna G. [9] have presented internal heating strategy with the cell's pulse power capability. In [9], the heating strategy is well organized. However, there is a limitation that the algorithm requires information of the battery's internal impedance as a format of lookup table, so that the system is hard to avoid being complex. In addition, the shut-down algorithm is restrictive because it focuses only the peak power capability the availability of the constant current discharging. Moreover, because the heat-up frequency (f_{ch}) is set to be a constant, there are limits to increase the heat-up speed with the limited current conditions.

In this paper, a self-preheating method of Li-ion battery using battery impedance estimator is proposed. The proposed control strategy has simple internal heating method. The proposed method has several advantages. First, by using battery impedance estimator, it does not require the information of battery internal impedance. Second, using the battery impedance estimator, it maximizes the heating speed with given constraints; avoiding overlapping with BMS. Third, in the proposed control method automatic shut-down of the self-preheating is included when the DC discharging is available regardless of the temperature. Therefore, the proposed control method is adaptable to use for the commercial battery packs.

The detailed explanation of the proposed control method will be followed. In additions, the performance of the proposed method is verified by experimental results.

II. PROPOSED CONTROL METHOD

There are requirements to adapt self-heating method to the commercial pack: the adaptable hardware design guide, the battery voltage sensing method, and proper control method.

In section A, the hardware design guide of the self-heating system is introduced. In section B and C, the proposed control method based on the operation mode is explained.

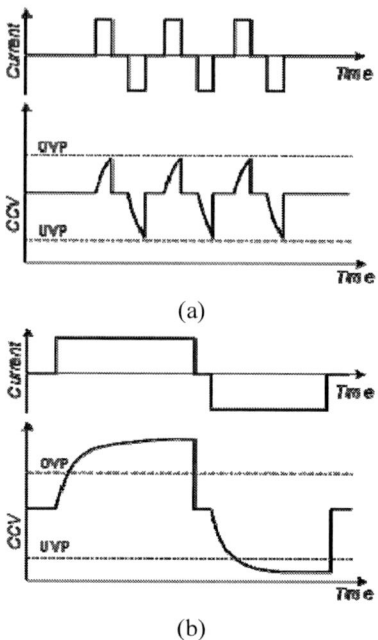

(a)

(b)

Fig.2. Waveform of the battery current and voltage under sub-zero temperature.
(a) With fast heat-up frequency.
(b) With slow heat-up frequency.

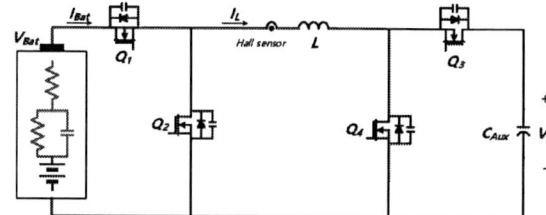

Fig. 3. Non-isolated cascade buck-boost converter.

A. Hardware Configuration

In the self-heating method, the energy in the battery is periodically restored and stored. Thus, sufficient energy storage is required. In this case, the electrolyte capacitor and super capacitor can be used because the capacitors have sufficient large energy and the capability aren't effected strongly by the temperature.

In the self-heating method, the battery charging and discharging current is controlled tightly. Therefore, the input current can be controlled to get constant value without the heat-up frequency. The current-fed type of converter can be good option because the current mode control can be adapted to this converter, and it controls battery charging and discharging current directly. In addition, the input current of the current-fed converter is a constant, so that the input filter size can be minimized. Normally, the current-fed type converter, however, has step-up characteristics, which means the output voltage of converter is always higher than the input voltage. In this case, the relay and PTC is used to prevent inrush current during the start up. However, in the self-heating method, the output capacitance is too large, so that the inrush current and charging time of the output capacitance is too long. Therefore, the relay and PTC cannot be used. As a

3467

result, cascade buck-boost converter is recommended to the self-heating method to overcome the inrush problem.

Fig. 3 shows non-isolated type cascade buck-boost converter. In the start-up period, the cascade buck-boost structure acts as buck, buck-boost, and boost converter sequentially. First, when the output voltage is lower than the input voltage, Q_1 is switching, and the converter operates as buck converter. In this case the output voltage of converter is increased. However, due to the forward voltage drop and loss components the output voltage level is hard to reach the input voltage level. In this case, Q_1 and Q_4 are synchronously controlled, so that the converter operates as buck-boost converter. The buck-boost converter can be both step-down converter and step-up converter, and the current path between the input node and the output node is always disconnected. Thus, there is no inrush problem. When the output voltage is higher than the input voltage level, the Q_1 is fully turned on and Q_4 is switching. In this case the converter is shown as one of the current-fed type converter, boost converter. As a result, using the cascade buck-boost structure the inrush problem is neglected and the requirement of the input current is satisfied.

A. Battery Impedance Estimator

The self-preheating method uses the heat caused by the loss of the internal resistance of battery as shown in fig. 1(a). In this case, to maximize the heating efficiency of the self-preheating method, the charging (or discharging) current is needed to be touch the voltage limits (OVP and UVP) as shown in fig. 1(c). In additions, the open circuit voltage (V_{OC}) is required to check remained battery energy during the operation.

During the heat-up time, the charging and discharging operation of the battery is periodically occurred. Therefore, the battery voltage waves largely in periodical. Thus, both battery voltage during charging time and that during discharging time should be checked separately, and that is not easy work by using analog circuit, so that, in this paper, a digital processor is used.

Actually, the battery's internal impedance is modeled with the various RC component. In addition, the internal resistance and capacitance of the battery are changed by the temperature. Thus, it is hard to get every information of the battery's impedance. Even if the knowing of information of batteries is possible, there exist a problem. During the heat-up time the same energy is take away from the battery and stores to the battery periodically. In this case, some amount of energy is stored to the internal capacitor of battery so that the sensed battery doesn't have correct impedance information. Therefore, the real-time control method with simple battery model is required.

In the proposed battery impedance estimator, the battery internal structure is differently modeled as shown in fig. 1(b). However, in these cases, the internal impedance is required to get current references.

As shown in fig. 1(c), equivalent internal resistance, R_{BAT}, is used to obtain V_{OC} and charging/discharging current references maintaining to utilize nominal battery

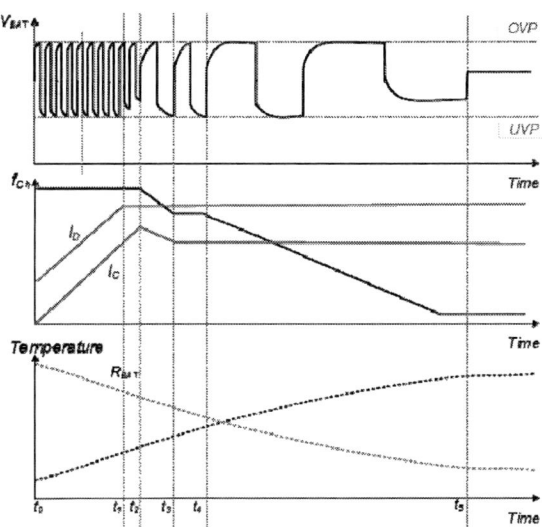

Fig. 4. Control waveforms of
the proposed self-preheating method.

voltage range maximally. In this case, R_{BAT} does not mean the actual internal resistance. However, by formulating R_{BAT} in each charging/discharging period, R_{BAT} is sufficient to replace the information of actual internal resistance. As shown in fig. 1(c), R_{BAT} is calculated by the frequency, sensed battery voltages, and sensed current during charging/discharging period as follows :

$$R_{BAT} = \frac{V_{C,Sensed} - V_{D,Sensed}}{I_{C,Sensed} - I_{D,Sensed}} \qquad (1).$$

B. Control Method

The battery is connected with input of boost converter. In bidirectional boost converter, the power flow is determined by Q_1 and Q_2 operations. When Q_1 is mainly controlled, the converter operates as a boost converter, so that the power is saving to the output capacitor. On the other hand, when Q_2 is main switch, the power is back to the battery. The size of the output capacitor is determined by the stored energy. Normally, the capacity of output capacitance is much smaller than that of the battery, so that the sufficient design conditions of output capacitor is required.

On the other hands, to minimize the output capacitor, the stored energy should be minimized. To achieve it, the energy transferred to output capacitor and from the output capacitor in one charging/discharging period has to be same as follow :

$$V_{C,Sensed} I_{C,Sensed} D_{Ch} = V_{D,Sensed} I_{D,Sensed} (1 - D_{Ch}) \qquad (2),$$

where, D_{ch} is (charging time) $\cdot$ (preheating frequency).

The battery management system is connected to the battery also. Because the Li-ion battery has to be operated in normal operating voltage range due to the safety problem and the lifetime extension, the control method

3468

should keep voltage limits. Based on these constraints, the proposed control method will be explained.

Fig. 4 shows the internal heating strategy of the proposed method. Because the power dissipation on the resistance is proportional to the square of current, maximizing charging/discharging current is mostly important. Thus, from $t_0 \sim t_1$, the charging/ discharging current increase while utilize the nominal battery voltage range.

Using R_{BAT}, V_{OC}, the charging current reference (I_C), and the discharging current reference (I_D) are obtained as follows (1), (2), and (3) :

$$V_{OC} = \frac{V_{C,Sensed} I_{D,Sensed} + V_{D,Sensed} I_{C,Sensed}}{I_{C,Sensed} + I_{D,Sensed}} \quad (3).$$

$$I_C = \frac{V_{OVP} - V_{OC}}{R_{BAT}} \quad (4),$$

where, V_{OVP} is the over voltage limit of battery.

$$I_D = \frac{V_{OC} - V_{UVP}}{R_{BAT}} \quad (5),$$

where, V_{UVP} is the under voltage limit of battery.

From, (4) and (5), the battery voltage always touches voltage limits of battery. However, if one of current reference reaches to the maximum current reference, the battery voltage cannot touch the voltage limits during heating time, so that the heating speed is decreased as shown in $t_1 \sim t_2$ of Fig. 4. To overcome it, proposed control strategy includes frequency modulation period ($t_2 \sim t_3$). Because the frequency is related to RC time constant of the battery internal impedance, the decrease of frequency makes R_{BAT} increased. Thus, the battery voltage touches the voltage limits again ($t_3 \sim t_4$).

In the proposed method, the heating operation is end by checking the RC constant of the battery voltage ($t_4 \sim t_5$). In the frequency modulation period, when the frequency is sufficiently smaller than one of RC constant of the battery, the battery voltage doesn't increase by following the frequency decreases, so that the battery voltage does not touch the limits. In this case, the battery is sufficiently heated to use. Thus, the controller can automatically end up the heating operation by using frequency modulation and checking the control status.

III. EXPERIMENTAL RESULTS

To verify the proposed control strategy, PSIM, a popular power simulation tool, is used. TABLE I shows

TABLE I. USED PARAMETERS IN THE SIMULATION.

	$t_0 \sim t_1$	$t_4 \sim t_5$
V_{OC} [v]	3.7	3.4
R_1 [Ω]	0.19	0.019
R_2 [Ω]	0.125	0.0125
C_2 [F]	1m	1m

TABLE II. EXPERIMENTAL SPECIFICATIONS.

	Parameters
Battery voltage	29V(20~33.6V)
Maximum heat-up current	5A
Minimum heat-up frequency	50Hz
Temperature range	-20~0°C
Switch ($Q_1 \sim Q_4$)	TK100E10N
Inductor(L_B)	CH234125, 22uH
Capacitor(C_O)	20000uF

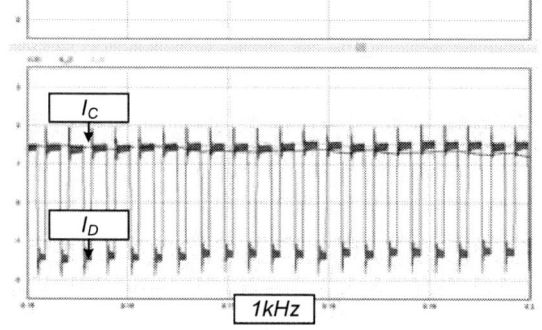

(a)

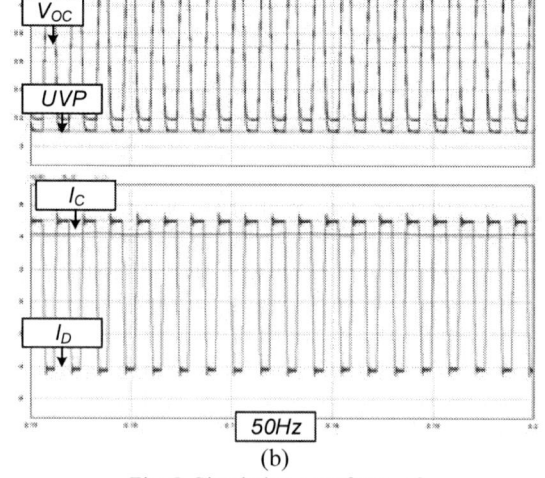

(b)

Fig. 5. Simulation waveforms of the proposed self-preheating method.
(a). $t_0 \sim t_1$. (b). $t_4 \sim t_5$.

the parameters used in the simulation. In addition, TABLE II shows test conditions and used parameter for the converter. Fig. 5(a)-(b) show the simulated waveform of the operation of the proposed method. As shown in fig.

3469

5(a), the current is set to utilize the entire nominal voltage range of the battery. Fig 5(b) represents the automatic frequency decreases to increases used battery voltage range. Finally, the heating method operates well until the end of time. From the simulation results, the proposed method maintains the maximum internal heating speed by touching the voltage limits of the battery. In additions, fig.5 shows that the output voltage regulated exactly from equation (1) during the self-preheating time.

Fig. 6(a)-(b) show the experimental waveform of the proposed method. As shown in fig. 6(a), during the start-up the inrush current is well limited by the cascade buck-boost structure. In addition, fig. 6(a) verifies the $t_0 \sim t_4$ operation. Fig. 6(b) shows the shadowed region of fig. 6(a). By, fig. 6(b), the heat-up current is set to maximum and the heat-up frequency is at the minimum considering OVP and UVP conditions.

IV. CONCLUSION

In this paper, a self-preheating method of bidirectional Li-ion battery charger for low temperature operating is presented. The proposed control strategy covers optimal preheating control by maximizing the nominal battery voltage range utilization. In addition, a proper hardware guide is introduced. Moreover, it doesn't require information of the battery internal impedance. As a result, the proposed method can be widely adapted to various commercialized battery cells in low temperature.

REFERENCES

[1] K. Chung, S.-K. Hong, and O.-K. Kwon, "A Fast and Compact Charger for Li-Ion Battery Using Successive Built-In Resistance Detection," IEEE Trans. Circuits Syst. II Express Briefs, vol. 7747, no. c, pp. 1–1, 2016.

[2] K.-M. Lee, S.-W. Lee, Y.-G. Choi, and B. Kang, "Active Balancing of Li-ion Battery Cells Using Transformer as Energy Carrier," IEEE Trans. Ind. Electron., vol. 64, no. 2, pp. 1–1, 2016.

[3] N. a. Chaturvedi, R. Klein, J. Christensen, J. Ahmed, and A. Kojic, "Algorithms for advanced battery-management systems," IEEE Control Syst. Mag., vol. 30, no. 3, pp. 49–68, 2010.

[4] S. Biswas, L. Huang, V. Vaidya, K. Ravichandran, N. Mohan, and S. V. Dhople, "Universal Current-Mode Control Schemes to Charge Li-Ion Batteries under DC/PV Source," IEEE Trans. Circuits Syst. I Regul. Pap., vol. 63, no. 9, pp. 1531–1542, 2016.

[5] H. Rahimi-Eichi, U. Ojha, F. Baronti, and M. Chow, "Battery Management System: An Overview of Its Application in the Smart Grid and Electric Vehicles," Ind. Electron. Mag. IEEE, vol. 7, no. 2, pp. 4–16, 2013.

[6] Y. Ji, Y. Zhang, and C.-Y. Wang, "Li-Ion Cell Operation at Low Temperatures," J. Electrochem. Soc., vol. 160, no. 4, pp. A636–A649, 2013.

[7] Y. Xing, W. He, M. Pecht, and K. L. Tsui, "State of charge estimation of lithium-ion batteries using the open-circuit voltage at various ambient temperatures," Appl. Energy, vol. 113, pp. 106–115, 2014.

[8] Y. Ji and C. Yang, "Heating strategies for Li-ion batteries operated from subzero temperatures," Electrochim. Acta, vol. 107, pp. 664–674, 2013.

[9] Z. Lei, C. Zhang, J. Li, G. Fan, and Z. Lin, "Preheating method of lithium-ion batteries in an electric vehicle," J.

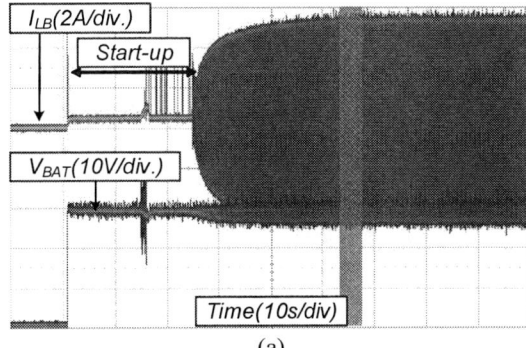

(a)

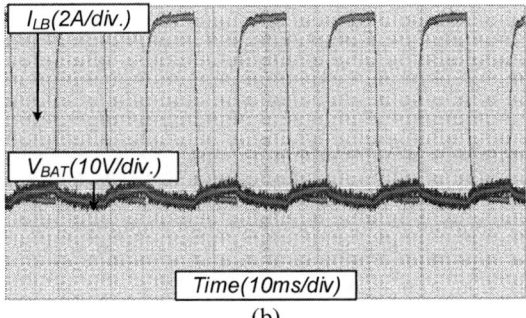

(b)

Fig. 6. Experimental waveforms of the proposed self-preheating method.
(a). *Overall.* (b). *nominal.*

Mod. Power Syst. Clean Energy, vol. 3, no. 2, pp. 289–296, 2015.

[10] S. Mohany, Y. Kim, A. G. Stefanopoulou, and Y. Ding, "Energy-Conscious Warm-Up of Li-ion Cells From Subzero Temperatures," IEEE Trans. Industrial Electron., vol. 63, no. 5, pp. 1547–1552, 2014.

The 2018 International Power Electronics Conference

Active Anti-Islanding technique with reduced Non-Detection Zone for Centralized Inverters

Prashant Jain[1], Vivek Agarwal[2], Bishnu Prasad Muni[3], Eswar Rao[4], Deepak Gehlot[5], S.Gautam Kumar[6]

1,3-6 BHEL R&D, Hyderabad India

2 Fellow, IEEE

Email: prashantjain@bhel.in

Abstract—**This paper presents an active anti-islanding scheme for grid-tied centralized inverters for large Photovoltaic (PV) power plants. The proposed technique involves appropriate reactive power injection into the grid which results in the positive/negative rate of change of frequency. The proposed algorithm is simple and can be easily integrated into DQ based current control technique. It has a smaller Non-Detection Zone (NDZ) and faster response as compared to other active anti-islanding schemes. As the proposed method is based on alarm generation and confirmation, it does not inject any periodical disturbance into the grid unlike other anti-islanding techniques and thus does not cause any power quality issues. The proposed active anti-islanding algorithm is simulated in MATLAB/SIMULINK under different loading conditions as per IEEE/IEC standards. The algorithm is also experimentally verified on a 25 kW centralized inverter prototype using TMS320F2812 DSP controller. The experimental and simulation results are presented to validate the effectiveness of the algorithm.**

Keywords— *Active anti-islanding, Standard IEEE 1547 and IEC 62116, Large PV system.*

I. INTRODUCTION

IEEE 1547.1[1] and IEC 62116[2] standards govern the requirements for anti-islanding in distributed generation systems, particularly in PV fed grid-tied centralized inverters for large PV power plants. As per the standard requirement, it is mandatory to de-energize an islanded electrical network during any intentional or unintentional failure of grid supply. The grid-connected centralized inverter should not energize an islanded network and must deactivate itself within a stipulated time of 2 seconds for the safety of the other equipment connected at the Point of Common Coupling (PCC) and personnel working in the vicinity. The centralized inverter in grid-connected mode as shown in Fig. 1 exports active power P_{pv} into the grid with the Power Factor (PF) close to unity. The reactive power Q_{pv} exported by the inverter is negligible as the power factor is unity. In grid-connected mode, some part of active power required by the load at PCC is supplied by the centralized inverter and remaining part is supplied by the main grid. The reactive power requirement of the load is supplied by the grid. In islanded condition i.e. disconnection from the main grid as shown in Fig. 1, part of an islanded network is energized by PV fed centralized inverter which is undesirable and is hazardous to both personnel working and the local load connected to the islanded network. During islanding condition, there is a deterioration in the quality of power supplied to the load at PCC. Anti-islanding is an important feature for the grid-connected converters, hence it is very important for the inverter to correctly determine the status of the main grid and avoid any faulty trips that may result in loss of active power to the grid.

Many anti-islanding schemes have been proposed in the literature [4]-[14] which are broadly divided into three categories namely passive technique, active technique, and communication-based technique. Passive anti-islanding techniques monitor the grid parameters i.e. voltage, frequency and phase. Any deviation beyond the threshold level in grid parameters as per IEC/IEEE standards [1][2] initiates an

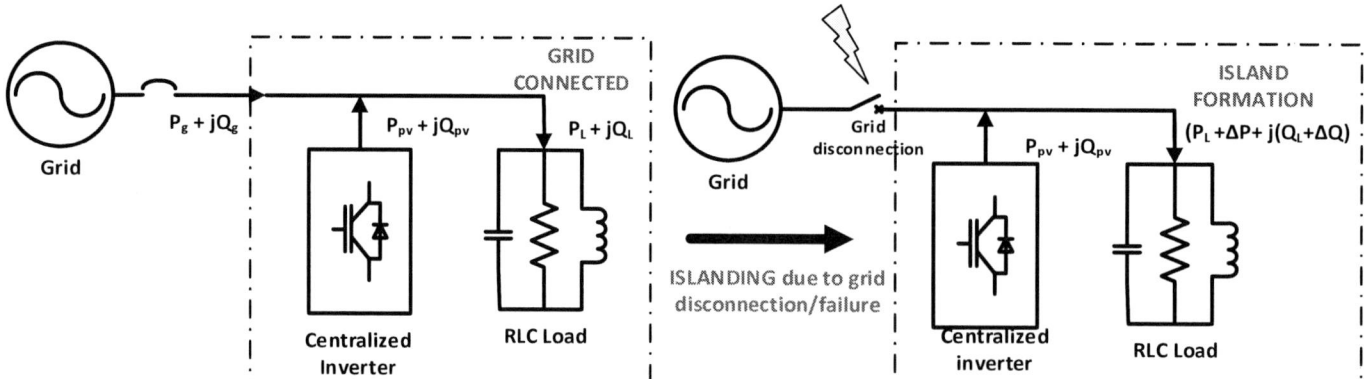

Fig. 1. Operation of centralized inverter in grid connected and islanded modes with RLC Load

3471

islanding condition. A passive anti-islanding technique based on phase shift method [6][7] is presented, where phase shift between inverter voltage and the current is used to detect an islanding condition. This method is preferred over other passive anti-islanding technique and has gained popularity due to its simplicity and smaller NDZ as compared to other passive techniques. The passive technique based on harmonics detection at PCC is presented in [8]. This technique is based on the harmonics detection in the inverter voltage, and hence it is very sensitive to the system parameters and often results in the generation of false alarms and trips. The passive techniques are simpler in implementation but they have bigger NDZ, hence sometimes fail to detect the anti-islanding conditions.

The active anti-islanding techniques are based on the injection of the disturbance in grid voltage and frequency by the PV fed centralized inverter [10]-[14]. The islanding condition is detected by first injecting some disturbance at PCC i.e. (in the active and reactive current reference or in the voltage and current) and then sensing these parameters based on the feedback mechanism. The islanding in the network is detected when the sensed parameters are beyond the threshold values as per IEC/IEEE standards. The active anti-islanding method proposed in [11] is based on harmonics injection in the current. Active Frequency Drift (AFD) based active anti-islanding is proposed in [13], the method based on harmonic injection which results in a small drift in system frequency during islanding condition. A voltage and frequency shift based active anti-islanding technique is proposed in [12]. These techniques are based on positive feedback in voltage and frequency which takes system parameters (i.e. voltage and frequency) to their threshold. The active anti-islanding techniques have lower NDZ as compared to passive anti-islanding techniques but the periodical disturbance in grid parameter causes power quality and stability concerns. Communication-based anti-islanding techniques use Supervisory Control and Data Acquisition (SCADA) and Power Line Carrier Communication (PLCC) [16] based communication infrastructures. These methods have very low NDZ as compared to passive and active technique but need an expensive infrastructure for the implementation.

A hybrid active anti-islanding method for grid-tied centralized inverter is proposed in this paper with almost zero NDZ and no power quality concerns. The proposed active anti-islanding method is a combination of rate of change of frequency and reactive power injection with positive feedback and can be easily integrated with DQ based current control technique. It has a faster response as compared to other active anti-islanding techniques. The proposed technique does not take system parameters to threshold, hence reduces power quality concerns to the loads connected at PCC.

II. PROPOSED ANTI ISLANDING ALGORITHM WITH REDUCED NON-DETECTION ZONE

Non-Detection Zone (NDZ) is a zone/region where anti-islanding schemes fail to detect the grid disconnection or an island formation of the network. Selection criteria for anti-islanding scheme depend on the effectiveness of the scheme to detect the islanding with smaller NDZ [3][4], stability and easier implementation on the microcontroller. NDZ is

determined based on the change in active and reactive power absorbed by the RLC load connected at PCC in grid-connected and islanded modes as shown in Fig. 1. The active power and reactive power of the load in grid-connected mode is given by $P_L = P_g + P_{pv}$ and $Q_L = Q_g + Q_{pv}$ where, P_g, P_{pv} are the grid and inverter active power; Q_g, Q_{pv} are the grid and inverter reactive power and V and f are the voltage and frequency in the grid-connected mode.

$$P_L = \frac{V^2}{R} \tag{1}$$

$$Q_L = V^2 \left(\frac{j}{\omega \times L} - j\omega \times C \right) \tag{2}$$

$$|Z_L| = \frac{V^2}{|P_g + P_{pv} + jQ_g|} \tag{3}$$

$$V' = \frac{V}{\sqrt{\left(1 + \frac{P_g}{P_{pv}}\right)^2 + \left(\frac{Q_g}{P_{pv}}\right)^2}} \tag{4}$$

$$R\sqrt{\frac{C}{L}} \left(\frac{\omega'}{\omega} - \frac{\omega}{\omega'} \right) = \frac{Q_g}{P_{g+pv}} \tag{5}$$

Load impedance Z_L when PV fed inverter is exporting active power at Unity Power Factor (UPF) (i.e. $Q_{pv} = 0$) is given by equation (3). When the grid supply disconnects, the voltage and frequency of the islanded network are given by V' and f' as shown in equation (4) and (5). When the centralized inverter disconnects itself from the main grid, there will be a change in voltage and frequency (i.e. ΔV and Δf) due to change in active and reactive power balance (i.e ΔP and ΔQ). This small change in ΔV and Δf has to be properly captured in the anti-islanding algorithm, otherwise the false detection and tripping may result in reduced power output which affects the overall efficiency of the system.

The proposed active anti-islanding method is a combination of rate of change of frequency and the reactive power injection with the positive feedback i.e. the algorithm detects any small change in the frequency and accordingly injects the reactive power to change the system frequency either in positive or negative direction. The rate of change of frequency i.e. df/dt is monitored both in positive and negative direction and it helps in faster detection of the islanding condition within the safe operating region. The block diagram of proposed anti-islanding algorithm for the grid-tied centralized inverter is shown in Fig.2. In a grid-connected mode, PV fed centralized inverter will export power with unity P.F. and the reactive current reference is controlled to zero as shown in Fig. 2. The current control is implemented in the d-q reference frame where the active current reference is proportional to the d-axis component and the reactive current reference is proportional to the q-axis component [16]. If the anti-islanding algorithm tries to perturb the system frequency by injecting reactive power, due to the presence of the grid the system frequency will not perturb. In islanded mode, the system frequency will perturb in positive or negative direction when the anti-islanding algorithm gives the reactive power reference. The proposed anti-islanding method perturbs the system frequency in positive and negative directions by appropriately injecting reactive power as shown in Fig. 3.At B the system frequency is increased by the injection of reactive power and df/dt is positive. After a time-delay the reactive power injection is reversed and hence at C, the system frequency will decrease and df/dt will become negative. The proposed algorithm monitor the rate of change in frequency in synchronism with the injection of the reactive power and finally at D, the islanding is detected. The flow chart of proposed active anti-islanding are shown in Fig. 4.

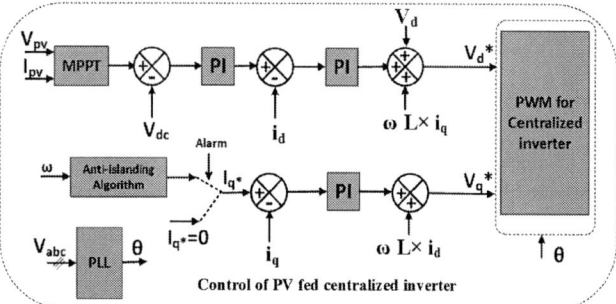

Fig. 2. Control block diagram of 3- φ grid-tied centralized inverter system with proposed active anti-islanding

In an islanded mode, there will be a change in the PCC load characteristic as per the equation 4 and 5 and it will result in a change in voltage and frequency. The small change in frequency is sensed by the master controller which will trigger an alarm to start reactive power injection in positive/negative direction. The reactive power injection in positive/negative direction increases/decreases the system frequency of an islanded network and based on df/dt calculation the islanding is detected.

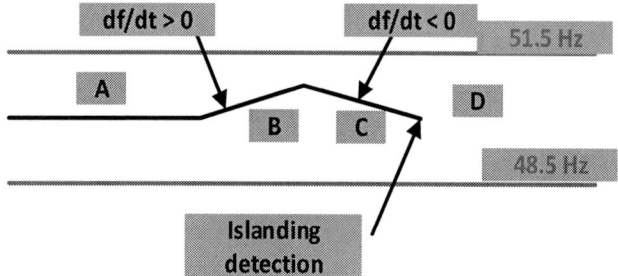

Fig.3. Algorithm of proposed active anti-islanding scheme for PV fed centralized inverter

The flowchart of the proposed anti-islanding is shown in Fig. 4, where the PCC voltage is sensed and the frequency is computed using the Hybrid PLL technique [17]. A small deviation in the grid frequency triggers an alarm, which activates the algorithm and injection of positive/negative reactive power by the PV fed centralized inverter. The rate of change of system frequency is monitored. In grid-connected

mode there won't be any change in frequency but in islanded mode system frequency will change as per the anti-islanding algorithm. The proposed active anti-islanding scheme will be able to detect the islanded condition much faster than other active anti-islanding techniques.

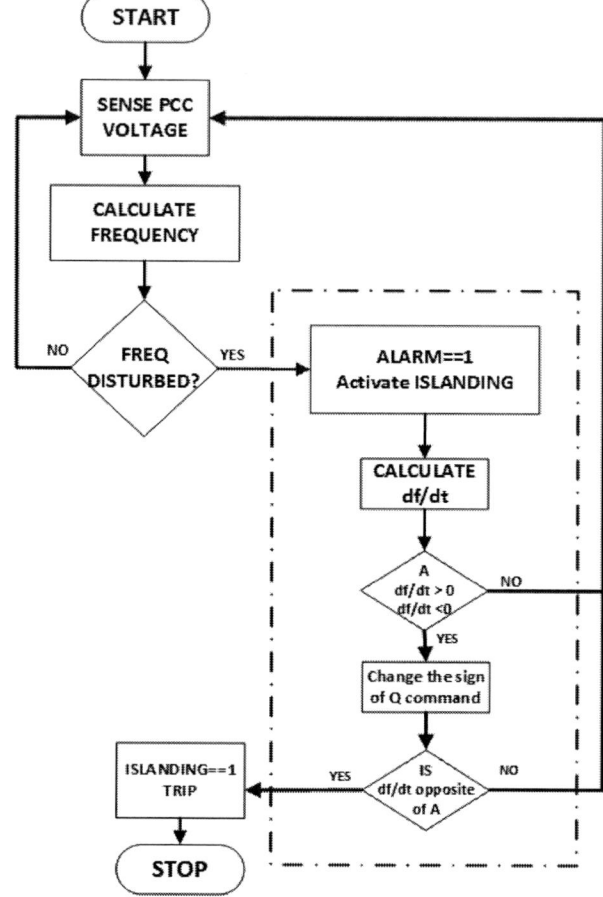

Fig.4. Flow chart of the proposed anti-islanding technique for grid-tied centralized inverter

The features and advantages of proposed technique are as follows.

A. *The proposed anti-islanding algorithm is simple and can be easily applied to DQ based current control of a grid connected centralized inverter.*

B. *This anti-islanding algorithm has almost zero NDZ and can detect anti-islanding even when there is a small change in the system parameter(s).*

C. *Unlike other active anti-islanding scheme, it does not inject periodical disturbance into the grid. Hence, THD is maintained within a tolerable limit.*

D. *The detection is based on the rate of change of frequency within the normal frequency range; hence it does not take system frequency to the positive and negative threshold. Hence, there is no abrupt change in system frequency*

3473

E. The reactive power requirement is smaller as compared to other active anti-islanding schemes. Hence the algorithm is more effective and has a faster response.

III. TEST SETUP FOR ANTI-ISLANDING DETECTION AS PER STANDARDS

The overall arrangement and test setup for testing the proposed active anti-islanding algorithm for grid-tied centralized inverter is shown in Fig. 5. The DC power source can be either a PV source or a programmable PV simulator where irradiation profile can be pre-programmed as per the user requirement. The DC programmable PV simulator is connected at the input of the grid-tied inverter which is the Equipment Under Test (EUT).

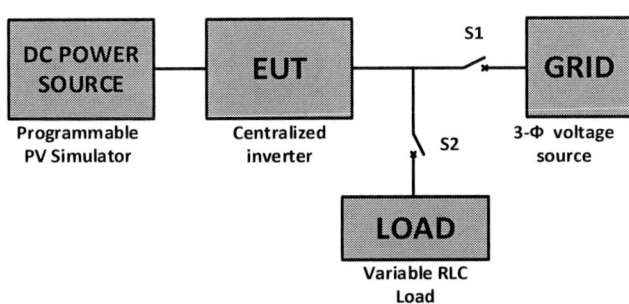

Fig.5. Overall arrangement of experimental test setup for testing the proposed anti-islanding algorithm

The generalized test procedure is adopted for testing anti-islanding for centralized grid-tied inverter which is in line with IEC and IEEE standards. It includes the following steps:

1) Connect all the sub-systems i.e. DC source, EUT, AC load and grid as per the arrangement shown in Fig. 5.
2) Switch S1 is closed and EUT is switched ON. EUT is controlled to export the power at UPF into the grid as per Table I.
3) EUT DC voltage is maintained as per MPPT range of PV fed centralized inverter (i.e. approximately at 75% to 100 % of nominal MPPT range).
4) EUT frequency trip setting is kept at ± 1.5 Hz of the nominal value and voltage trip setting is kept at 0.85 to 1.15 of the nominal value.
5) S2 is closed and variable RLC load is adjusted to keep the active power and reactive power as per Table II.
6) Once all the conditions are achieved as per Table II, the islanded mode is simulated by the opening of the switch S1 to disconnect the grid supply. In this condition, the centralized inverter energizes only the AC loads at the PCC.
7) The time taken for tripping is recorded if the EUT trips with an active anti-islanding error.

8) Steps 2 to 7 are repeated for other conditions listed in Table I and Table II.

TABLE I

S.No	EUT Output Power in percentage
1	100 % of rated power
2	Between 66 % to 33 % of rated power
3	Less than 33 % of rated power

TABLE II

S.No	Load Active Power	Load Reactive Power
1	100 % of rated power	less than 5 % (Inductive)
2	100 % of rated power	less than 5 % (Capacitive)
3	Between 66 % to 33 %	less than 5 % (Inductive)
4	Between 66 % to 33 %	less than 5 % (Capacitive)
5	Less than 33 %	less than 5 % (Inductive)
6	Less than 33 %	less than 5 % (Capacitive)

IV. SIMULATION RESULTS

PV source, centralized inverter, RLC load, and grid are modeled in MATLAB as per the arrangement shown in Fig. 5. The centralized inverter is connected to the grid via controllable switch S1 and a variable RLC load is connected at PCC.The proposed active anti-islanding detection based on the rate of change of frequency and appropriate reactive power injection is simulated with different conditions as per Tables I and II. It can be seen from the simulation results in Fig. 6 that the proposed algorithm can detect islanding condition even for a small NDZ, i.e. islanding is detected even when the change in frequency and voltage is within the tolerable limits. The detection of islanding condition with 66% / 33% active and 5 % of reactive load (capacitive) is shown in Figs. 6(a) and 6(b). When the load is capacitive, the proposed islanding detection algorithm detects islanded condition by first injecting negative reactive power which results in the decrease of the system frequency. When *df/dt* is negative, the injection of reactive power is reversed which increases the system frequency and *df/dt* becomes positive. The islanding of the system is detected and the trip time is controlled by reactive power injection. The inverter pulses are blocked within stipulated time in compliance with the IEEE/IEC standards. The detection of islanding condition with 66% / 33% active and 5 % of reactive load (inductive) is shown in Figs. 6(c) and 6(d). When the load is inductive, the proposed algorithm detects islanding condition by first injecting positive reactive power which increases of the system frequency and when the *df/dt* is positive, the injection of reactive power is reversed and it decreases the system frequency. The islanding condition is detected when df/dt is negative.

The detection of islanding condition with 66% and 100 % active power is shown in Figs. 6(e) and 6(f). The proposed islanding detection algorithm is able to detect islanding condition for small change in system frequency. The performance of proposed anti-islanding technique is further analyzed and validated on a 25 kW experimental setup with variable RLC Load.

The 2018 International Power Electronics Conference

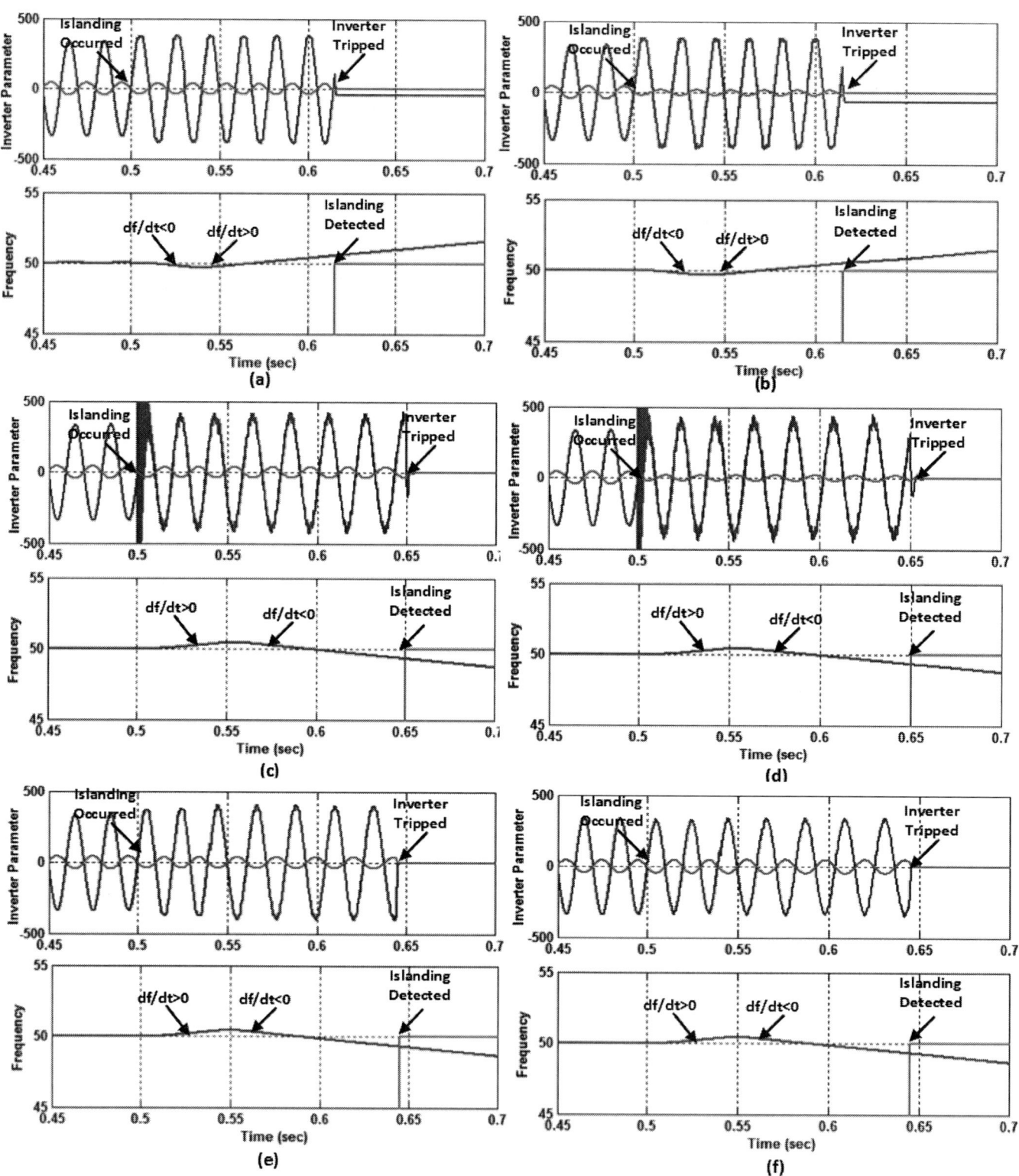

Fig. 6: (a) Simulation results of proposed active anti-islanding scheme with 66 % active and 5 % reactive load (Capacitive); (b) Simulation results of proposed active anti-islanding scheme with 33 % active and 5% reactive load (Capacitive); (c) Simulation results of proposed active anti-islanding scheme with 66 % active and 5 % reactive load (Inductive); (d) Simulation results of proposed active anti-islanding scheme with 33% active and 5% reactive load (Inductive); (e) Simulation results of proposed active anti-islanding scheme with 66 % active and no reactive load; (f) Simulation results of proposed active anti-islanding scheme with 100 % active and no reactive load

3475

V. EXPERIMENTAL RESULTS

The proposed active anti-islanding technique is validated and verified using a 25 kW low power prototype of the grid-connected centralized inverter. The single line diagram (SLD) of the hardware setup is shown in Fig. 7.The PV power plant characteristic is simulated using a PV simulator which is connected to the 3- grid-tied inverter. The output of the 3- inverter is connected to the grid with the help of switch S1. Configurable RLC loads are connected to PCC to simulate various conditions as per Table-I and Table II. The photograph of the experimental test setup is shown in Fig. 8 in which 20 kW PV simulator is connected to an IGBT based inverter stack to 240 V grid via LCL filter. The Proposed active anti-islanding algorithm along with PWM generation, PI current regulator and voltage and current protections are implemented on TMS320F2812 DSP controller [18].

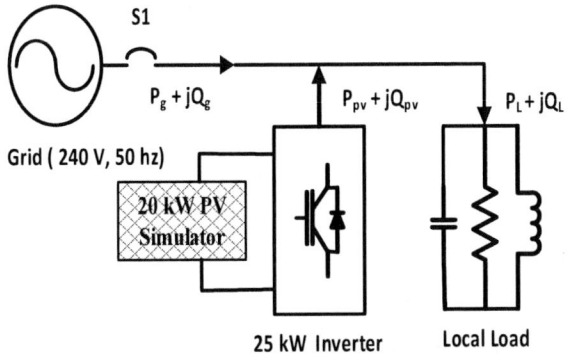

Fig. 7: SLD of 25 kW Centralized inverter with RLC load experimental set-up in the laboratory

Fig. 8: Experimental hardware setup for verification of the proposed active anti-islanding using TMS320F2812 DSP

The R,L,C value of the load at PCC is selected such that the change in frequency and voltage of the islanded system is within NDZ. The switch S1 is used for disconnecting the PV fed centralized inverter system from the main grid to create islanding condition. Experimental results of test setup showing transition from grid-connected mode to islanded mode with a small change in frequency and voltage are shown in Fig. 9. It can be seen that there is a change of ±0.25 Hz in system frequency and ± 5V in the voltage, which is within the tolerable limit as per IEC/IEEE standard. The PV fed centralized inverter system is still feeding the islanded network as system frequency and voltage is within the tolerable limit.

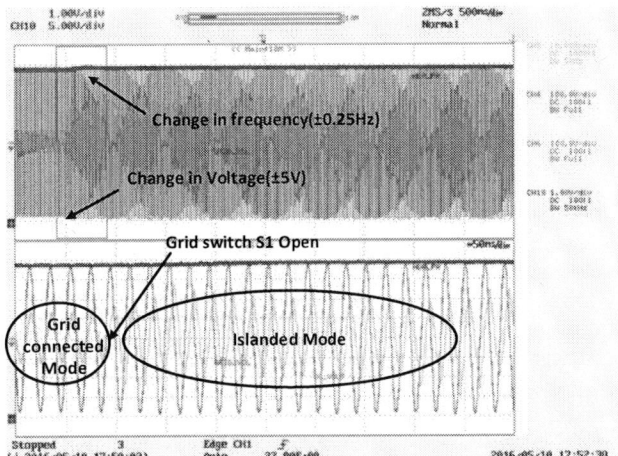

Fig. 9: Experimental results of change in frequency and voltage during islanding condition (voltage grid=blue, inverter current=green, frequency=cyan)

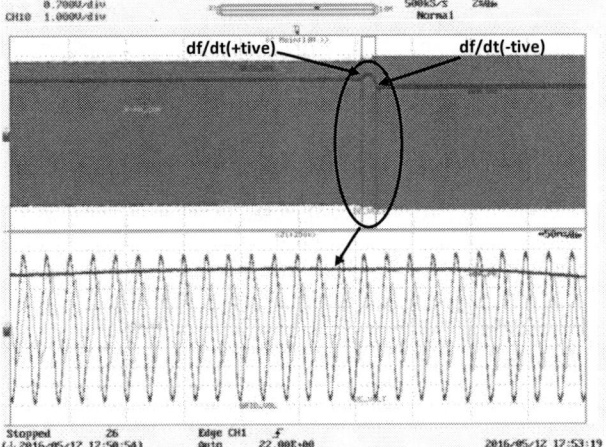

Fig. 10: Experimental results of change in frequency for detection of islanding (voltage grid=blue, inverter current=green, frequency=cyan)

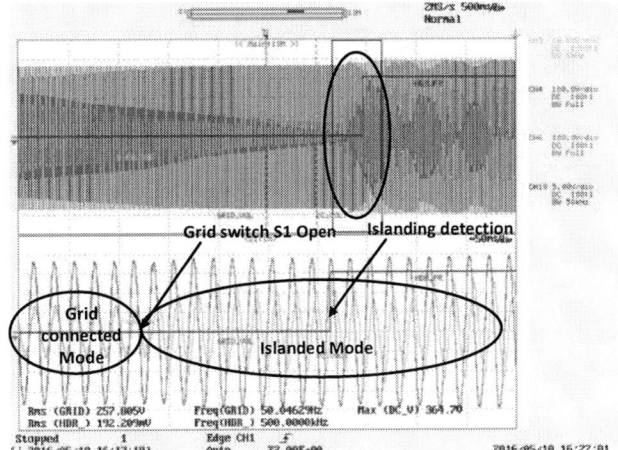

Fig. 11: Experimental results of detection of islanding with df/dt (voltage grid=blue, inverter current=green, frequency=cyan)

The experiment is repeated with the inclusion of the proposed anti-islanding algorithm. It can be observed from the Fig. 10 and Fig. 11 proposed algorithm detects islanding condition by first injecting positive Q, this increases the system frequency and same is detected by the DSP controller. The controller then injects negative Q and this decreases the system frequency. The change in *df/dt* in positive and negative direction enables the controller to detect islanding conditions even with the small change in system frequency and voltage.

VI. CONCLUSION

Grid-tied centralized inverters are preferred in large PV power plant installations. As per the IEC and IEEE standards, a PV fed grid-connected inverter system should not feed an islanded network for the safety of the equipment and personnel. Active and passive anti-islanding schemes are used for detection of islanding condition for the PV fed centralized inverter. The proposed algorithm is simple and can be easily applied to DQ based current control. It has almost zero NDZ and can detect islanding even in the small change of the system frequency and doesn't affect the normal MPPT operation of the centralized inverter. The proposed scheme doesn't cause any power quality issue unlike other active anti-islanding schemes, which inject periodical disturbance into the grid. Hence, THD is maintained within a tolerable limit. The proposed algorithm can detect islanding without taking system frequency to the positive and negative threshold hence it does not require any large reactive power injection. The functioning of the critical loads connected at PCC is not affected as there is no abrupt change in system parameters and the overall stability of the system is also maintained at a safe operating range. The simulations and experimental results on a 25 kW centralized inverter validate the effectiveness of the proposed active anti-islanding scheme.

REFERENCES

[1] IEEE 1547, "Standard for Interconnecting Distributed Resources with Electric Power Systems," 2003.

[2] I.E.C. 62116, "Test procedure of islanding prevention measures for utility-interconnected photovoltaic inverters," 2014.

[3] Zhihong Ye; Kolwalkar, A.; Yu Zhang; Pengwei Du; Reigh Walling. "Evaluation of anti-islanding schemes based on nondetection zone concept", Power Electronics, IEEE Transactions on Volume 19, Issue 5, Sept. 2004, pp.1171-1176.

[4] Z. Ye, R. Walling, L. Garces, R. Zhou, L. Li, T. Wang, Study and development of anti-islanding control for grid-connected inverters, May 2004, [online] Available: http://www.nrel.gov/docs/fy04osti/36243.pdf.

[5] F. De Mango, M. Liserre, A. Dell'Aquila and A. Pigazo, "Overview of Anti-Islanding Algorithms for PV Systems. Part I: Passive Methods," *Power Electronics and Motion Control Conference, EPE-PEMC 2006. 12th International,* Portoroz, 2006, pp. 1878-1883.

[6] Guo-Kiang Hung, Chih-Chang Chang and Chern-Lin Chen, "Automatic phase-shift method for islanding detection of grid-connected photovoltaic inverters," in *IEEE Transactions on Energy Conversion*, vol. 18, no. 1pp. 169-173.

[7] M. E. Ropp, M. Begovic, A. Rohatgi, G. A. Kern, R. H. Bonn, Sr.,and S. Gonzalez, "Determining the relative effectiveness of islanding detection methods using phase criteria and nondetection zones," *IEEE transactions on energy conversion,*September 2000 vol. 15, no. 3, pp.290-296.

[8] Sung-Il Jang and Kwang-Ho Kim, "An islanding detection method for distributed generations using voltage unbalance and total harmonic distortion of current," in *IEEE Transactions on Power Delivery*, April 2004, vol. 19, no. 2, pp. 745-752.

[9] H. Zeineldin, M. J. Marei, E. F. El-Saadany, M. M. A. Salama, "Safe controlled islanding of inverter based distributed generation", IEEE Power Electron. Specialists Conf. Aachen Germany,June 2004 vol. 4, pp. 2515-2520.

[10] F. De Mango, M. Liserre and A. Dell'Aquila, "Overview of Anti-Islanding Algorithms for PV Systems. Part II: Active Methods," *Power Electronics and Motion Control Conference, 2006. EPE-PEMC 2006. 12th International,* Portoroz, 2006, pp. 1884-1889.

[11] M. Ciobotaru, R. Teodorescu, F. Blaabjerg, "On-line Grid Impedance Estimation Based on Harmonic Injection for Grid-Connected PV Inverter", IEEE ISIE, June 2007, pp. 2473-2442.

[12] Z. Ye, L. Li, L. Garces, et al., "A new family of active antiislanding schemes based on DQ implementation for grid-connected inverters," in Power Electronics Specialists Conference, 2004. PESC 04. 2004 IEEE 35th Annual 2004, Vol.1 pp. 235-241.

[13] L. A. C. Lopes and Huili Sun, "Performance assessment of active frequency drifting islanding detection methods," in *IEEE Transactions on Energy Conversion*, March 2006, vol. 21, no. 1, pp. 171-180.

[14] Jin Beom Jeong, Hee Jun Kim, Soo Hyun Back and Kang Soon Ahn, "An improved method for anti-islanding by reactive power control," *2005 International Conference on Electrical Machines and Systems*, Nanjing, 2005, Vol. 2, pp. 965-970.

[15] Remus Teodorescu; Marco Liserre; Pedro Rodrguez, " Grid Converters for Photovoltaic and Wind Power Systems", Wiley-IEEE Press, 2011

[16] W. Xu, G. Zhang, C. Li, W. Wang, G. Wang and J. Kliber, "A Power Line Signaling Based Technique for Anti-islanding Protection of Distributed Generators: Part I: Scheme and Analysis," *Power Engineering Society General Meeting, 2007. IEEE,* Tampa FL, 2007, pp. 1-1.

[17] P. Jain, V. Agarwal, and B. P. Muni, "Hybrid Phase Locked Loop for controlling Centralized inverters in large solar Photovoltaic power plants," 2016 IEEE International Conference on Power Electronics, Drives and Energy Systems (PEDES), Trivandrum, 2016, pp. 1-7.

[18] Instruments, Texas. "TMS320F2812 Digital signal processor data manual."SPRS174M, URL: http://www. ti. Com, 2005

Development of SiC Applied Traction System for Shinkansen High-speed Train

Kenji Sato[1]*, Hirokazu Kato[1] and Takafumi Fukushima[2]

1 Technology Research and Development Department, General Technology Division,
Central Japan Railway Company, Aichi, Japan
2 Rolling Stock Department, Shinkansen Operations Division, Central Japan Railway Company, Tokyo, Japan
*E-mail: ksato@jr-central.co.jp

Abstract— This paper presents the development of a traction system for high-speed trains by adopting SiC power devices to pursue weight reduction and compactness of the system. We found the combination of the SiC applied conversion system with train-draft cooling system and 6-pole induction motors is a suitable approach to highlight the merits of SiC devices. The running tests of a prototype were conducted to confirm its sound performances. The developed traction system is to be installed in the latest-type Shinkansen train, or the Series N700S, which will debut in March 2018, and this SiC application to high-speed train's traction system is the first case over the world.

Keywords— *Conversion Sysem, High-Speed Train, SiC, Traction System*

I. INTRODUCTION

The traction system of Tokido Shinkansen train consists of main transformers, conversion systems and traction motors. Figure 1 shows a typical example of the latest system. The AC 25,000V power is supplied on catenary. The pantograph receives the electricity from the catenary and sends it to the main transformer. The transformer steps down its voltage to AC 1,500V and sends it to the conversion system. The conversion system, which comprise a converter and an inverter, converts the electricity to DC 3,000V once, and then inverts it to three phase AC electricity with changing frequencies and voltages in order to drive induction motors for traction.

By applying the power electronics technology, the traction system for Shinkansen high-speed trains has been improved since its inauguration in 1964 as shown in Table I. In 1992, the Series 300 Shinkansen train adopted the GTO thyristors to realize the PWM Converter Inverter system with the induction motor drives, which achieved the significant weight reduction and compactness of traction systems [1]. The IGBT that was applied to the Series 700 increasingly reduced weight and volume. In addition, low-loss IGBTs of the Series N700 allowed us to develop "the train-draft cooling system," in which the air flow under floors is used for cooling, while the conventional cooling system of Shinkansen train's conversion systems is the forced-ventilation system with cooling blowers as shown in Fig. 2 [2]. Since the conversion system with train-draft cooling abolished the cooling blowers, it realized the further weight reduction, compactness and higher reliability.

Based on these backgrounds, we developed the SiC applied traction system for the latest-type Shinkansen train, whose prototype train for experiments will debut in March, 2018. The key concept of the system is the combination of the SiC applied conversion system with train-draft cooling system and 6-pole induction motors to pursue additional weight-reduction, compactness, and higher reliabilities. We conducted running tests of the prototype of the developed traction system and confirmed its sound performances. This SiC application to high speed trains is the first time in the world.

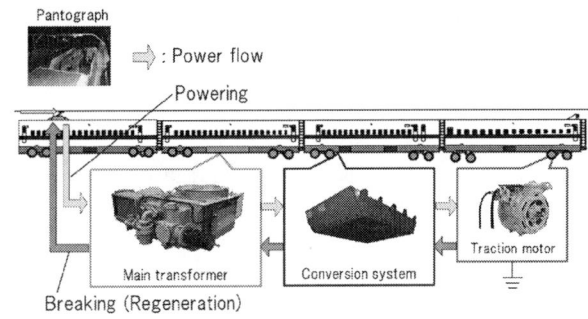

Fig. 1. Traction system of Shinkansen train.

TABLE I
IMPROVEMENT IN THE TRACTION SYSTEM OF TOKAIDO SHINKANSEN

Type	Series 0	Series 100	Series 300	Series700	Series N700 N700A	Series N700S
Year	1964	1985	1992	1999	2007	2018
Semiconductor device	Diode	Thyristor	GTO thyristor	IGBT	Low-loss IGBT	SiC device
Control system	Tap changer control	Thyristor-drived phase control	PWM converter and inverter			
Cooling system	Forced ventilation cooling system				Train draft cooling system	
Traction motor	DC motor		3-phase induction motor			
			4-pole			6-pole
Electric breaking	Rheostatic braking		Regenerative breaking			

The 2018 International Power Electronics Conference

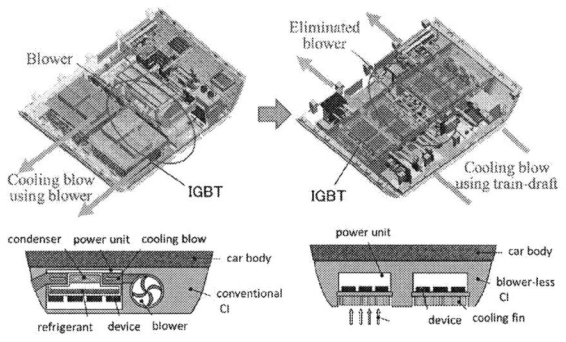

(a)   Forced ventilation   (b)Train-draft cooling

Fig. 2. Conversion system with train draft cooling system.

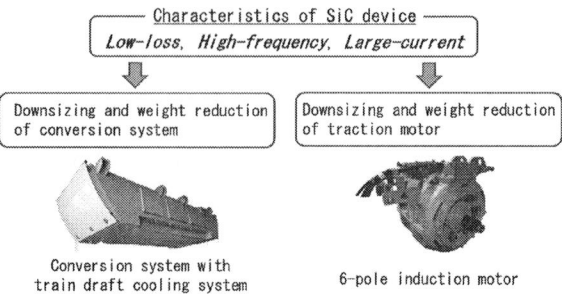

Fig.3. Key concept of SiC application to traction system.

II. KEY CONCEPT OF SiC APPLIED TRACTION SYSTEM

The power devices used in Shinkansen traction system has been the silicon device (Si device) from diodes of the Series 0 in 1964 to low-loss IGBT of the Series N700 in 2007. Though the performance of IGBT has been improved in terms of switching frequency and current capacity, the Si device seems to reach its theoretical limitation because of its indigenous characteristics. On the other hand, the wide-bandgap devices such as a silicon carbide device (SiC device), which is expected to have lower loss and to be resistant to higher temperatures, recently appeared on the market [3]. Considering that the SiC device of 3.3kV-1500A was ready for commercial applications, we decided to adopt the SiC device to the Shinkansen traction system and started developments in 2012 [4, 5].

Through the developments, we thought that our key concept is to take advantage of the merits of the SiC device, which enables lower loss, higher frequency and larger current, in not only the conversion system but also in whole traction system as shown in Fig 3.

After the preliminary study, we found that the merits of SiC device can be utilized in the most effective way by combining the conversion system with train-draft cooling system and 6-pole induction motors. The weight portion of cooling fins in the train-draft cooling system is larger than that of the conventional forced ventilation system. Thus, we found the application of SiC devices to the conversion system with train-draft cooling system could be quite effective because cooling fins can be downsized due to lower switching losses. The larger current capacity enables us to introduce 6-pole induction motors instead of conventional 4-pole induction motors, resulting in drastic weight reduction because of volume reduction of primary cores by increasing poles.

III. DEVELOPMENT OF PROTOTYPE

A. Conversion System

The conversion system of Shinkansen trains consists of a PWM converter and a PWM inverter as shown in Fig. 4. In our study, we tried to apply SiC devices to both of the converter and the inverter, and also to consider the application of two types of SiC devices, which are the IGBT with SiC-SBD (Hybrid SiC) and the SiC-MOSFET with SiC-SBD (Full SiC). Figure 5 and 6 show the schematics of our developing conversion system. We study two types of conversion systems using two different types of SiC devises respectively: Hybrid-SiC and Full-SiC. These are designed to be compatible in terms of basic structural configurations and electrical specifications. Through the bench tests and the running experiments, we compared two types and learned their each characteristics.

Figure 7 shows the developed SiC-applied conversion system with train-draft cooling system for running experiments. The SiC devices are placed on upper side of cooling fins, whereas the lower side of fins faces the under floor and is cooled down by the train-draft airflows. Since the airflow speeds affect cooling capabilities, we had collected the field data in commercial operations which show the relations among train speeds, underfloor airflow and temperature raises of devices.

Figure 8 shows the simulation results of the device energy loss per phase based on data of running experiments, comparing the converter in the conversion system with the IGBT and with the Hybrid-SiC. The results show that not only the loss of the free wheeling diodes (FWD) and the clamp diodes (CDd), or the SiC-SBD, in Hybrid-SiC is reduced mainly because of the decrease of recovery loss but also the loss of IGBTs in Hybrid-SiC is reduced mainly because of the decrease of the currents derived by recovery currents of SiC-SBD. This yields an approximate 30% decrease of total device energy loss per phase of converter. Given those data, we decided the size of fins and positions of devices on the fins to avoid heat spots. This enables us to streamline the cooling fins themselves and then also power unit including cooling fins, devices and switching drive circuits as shown in Fig. 9. Both simulations and running experimental results show that the conversion system with Hybrid-SiC can achieve our original goal for compactness and lightweight. We also found that the Full-SiC has the potential to explore further weight reduction in the future

3479

by optimizing structures of cooling fins and surrounding parts of them, while its cost effectiveness is supposed to be carefully considered in its application. The research bought us to decide that both the Hybrid-SiC and the Full-SiC are to be adopted to our systems to meet the different purposes, depending on the extent of the expected and prospective weight reduction and compactness.

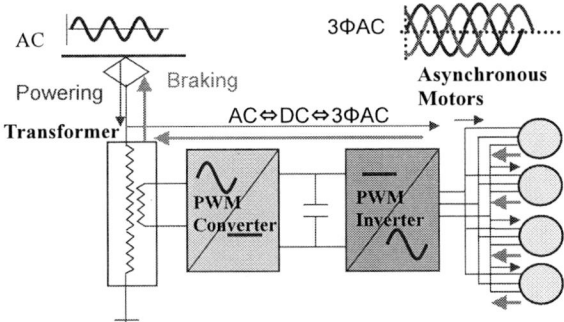

Fig. 4. Simplified circuit of Shinkansen traction system.

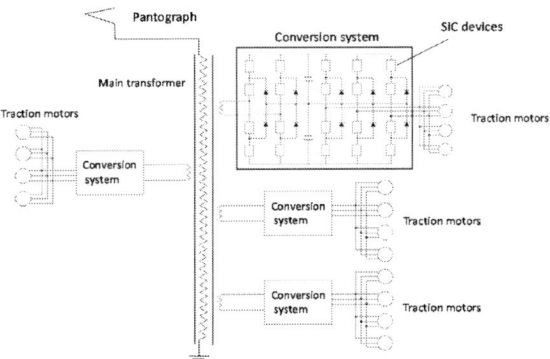

Fig.5. Schematic of the developed conversion system

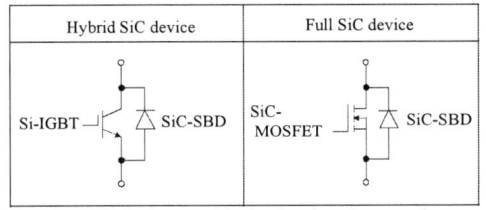

Fig.6. Types of SiC devices used in the developed conversion systems.

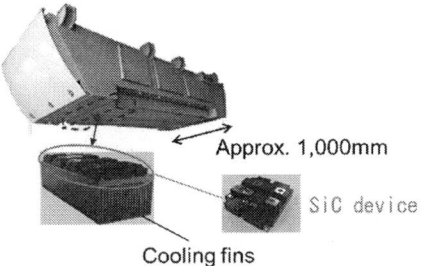

Fig. 7. Development of SiC-applied conversion system with train-draft cooling system (prototype).

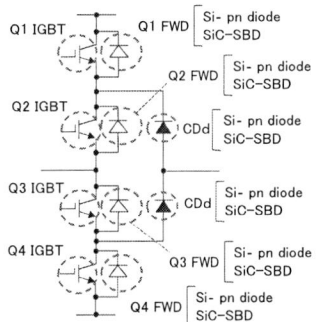

(a)Circuit diagram of one phase in the converter

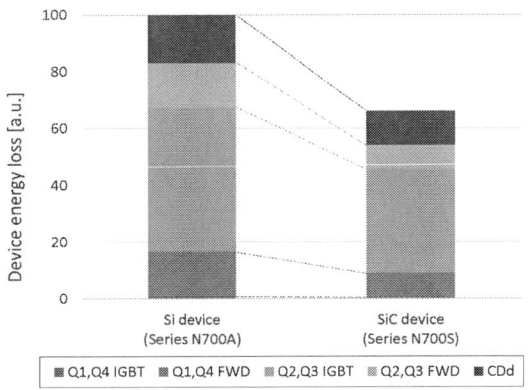

(b) Simulation results of running between Tokyo and Shin-Osaka based on running experimental results

Fig.8. Comparison switching loss per phase of the converter in powering between Si device and SiC device (Hybrid-SiC) .

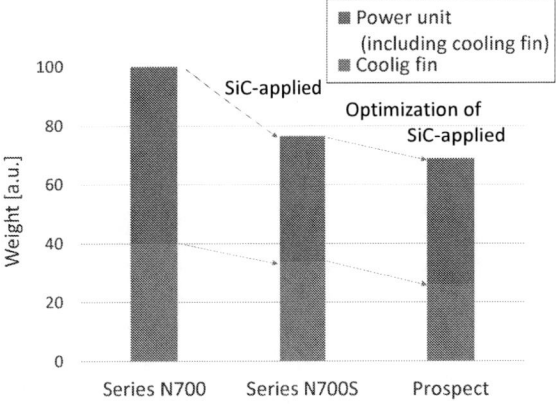

Fig.9. Weight reduction of the cooling fins and power unit including cooling fins, devices and switching drive circuits.

B. Traction Motor

Figure 10 shows our concept in developing the 6-pole traction motor driven by the SiC applied conversion system. On the condition that the total magnetomotive force is kept constant, the increase of poles can make each magnetic circuit smaller while the excitation inductance gets decreased, which leads to decrease of the power factor and need of larger motor currents. The SiC applied conversion system can supply larger current and therefore realize the increase of the number of poles from 4 poles to

3480

6 poles. This enables the volume reduction of the core, subject to keeping the same magnetic density in the core. The 6-pole traction motor also reduces overhangs of the primary coils, resulting in downsizing in the axial direction as shown in Fig 11.

In addition, we developed a novel structure of secondary core, or rotor based on the magnetic field analysis. The secondary core of conventional traction motors of Shinkansen trains have round holes for cooling because the traction motors are more highly power densified for high-speed running. While the core around the rotor bars contributes to the magnetic circuit, the core surrounding the axis, where the magnetic fields don't go through, can be reduced [6]. We conducted magnetic field simulations as shown in Fig. 12, and thus we reached a spoke shape as an optimized secondary core instead of the conventional shape with round holes as shown in Fig. 13.

Thanks to lager current capacity of the SiC applied conversion system, we also changed the motor characteristics from the magnetic-loaded to the electric-loaded. The gap magnetic field is proportional to ratio of the motor input voltage to the motor input frequency (V/f) as in (1), which decides the motor characteristics.

$$V/f \propto \phi \qquad (1)$$

where ϕ is gap magnetic field. If V/f is raised, the input voltage will increase and motor current will decrease. On contrary, if V/f is lowered, the input voltage will decrease and motor current will increased, which lead to changing the motor characteristic to more electric-loaded type. The electric-loaded type motor has an advantage of weight reduction of the motor because it can decrease the volume of iron core which dominates total weight of the motor. However, the margin of stalling torque varies with the square of the V/f as described in (2).

$$T_m \propto \frac{(V/f)^2}{l} \qquad (2)$$

where T_m is a stalling torque and l is a leakage inductance. Therefore, we study the optimum V/f to reduce total weight and motor currents with ensuring the same stalling torque margin as that of Series N700 as shown in Fig. 14. The study leads us to decide that the V/f constant terminal speed is set at a certain speed between 200km/h and 210km so that its weight reduction effect can meet our expectations. Figure 15 shows the motor input voltages and the motor currents comparing the 4-pole motor for Series N700 and the 6-pole motor for N700S. The V/f gets decreased by 15%, while the motor current of N700S gets increased. These designing changes such as 6-poles and electric-loaded characteristics combined with the SiC-applied conversion system can realize drastic weight reduction. Figure 16 summaries the effects of weight reduction by comparing weight/power of motors of Series 300, Series, 700, Series N700, which are conventional 4-pole motors, and Series N700S, which is the developed 6-pole motor. The weight/power of Series N700S is reduced by 20%, which significantly contributes to weight reduction of the tractions system of Series N700S.

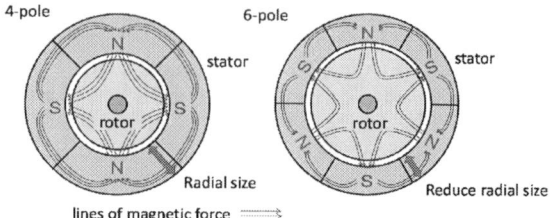

Fig. 10. Concept of developing 6-pole motor instead of 4-pole motor.

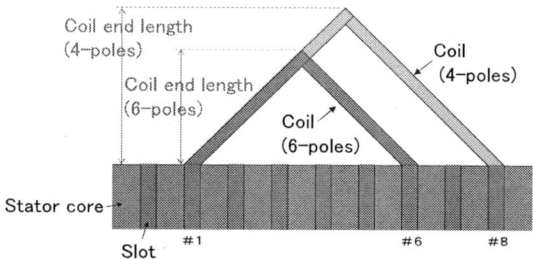

Fig. 11. Overhangs of the primary coils of 4-pole motor and 6-pole motor.

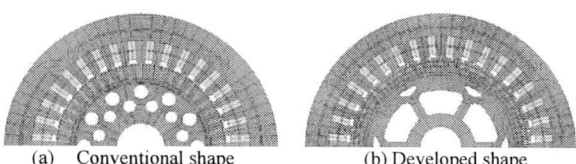

(a) Conventional shape   (b) Developed shape

Fig. 12. Magnetic field simulations of secondary cores.

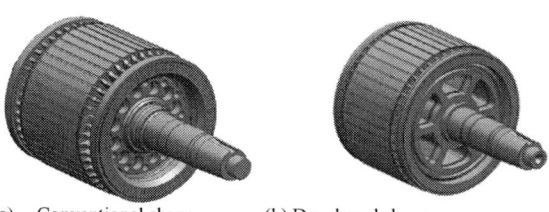

(a) Conventional shape   (b) Developed shape

Fig.13. Secondary core shapes.

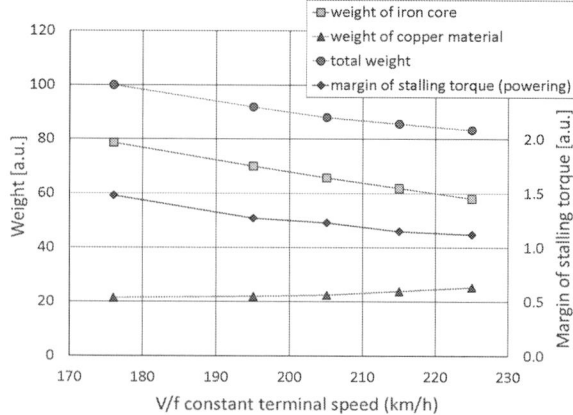

Fig. 14. Weight reduction effect related to V/f, and margin of stalling torques of the developed 6-pole motor in powering.

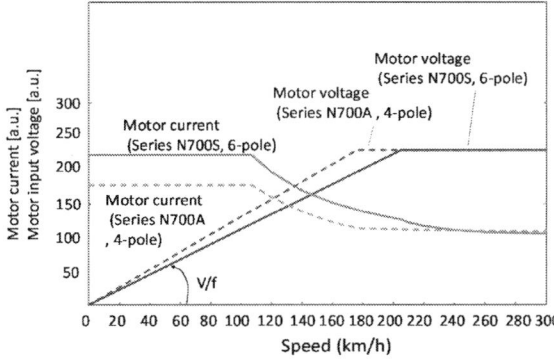

Fig. 15. Comparison of the input voltage and the motor current between motors with 4-poles for Series N700 and those with 6-poles for Series N700S.

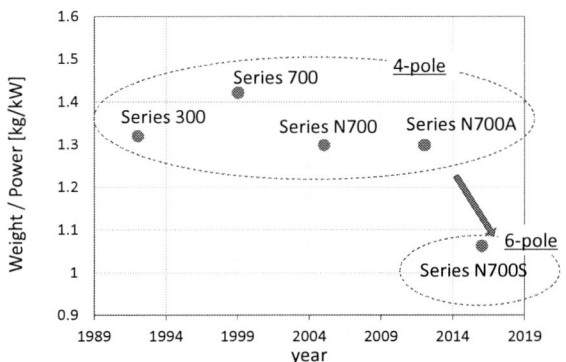

Fig. 16. Summary of weight reduction effect by comparing weight/power of motors of Series 300, Series, 700, Series N700 and Series N700S.

IV. RUNNING TEST RESULTS

We conducted running tests of the developed SiC applied traction system (conversion systems and 6-pole traction motors) by using the N700 Shinkansen train.

Figure 17 shows estimated junction temperatures of SiC device (Hybrid-SiC) which were measured in running tests of 515km between Tokyo and Shin-Osaka. The devise temperature of the inverter raised steeply in train starting because of smaller air flow at lower speed and higher switching pulse modes at low speed range. At high speed ranges, the device temperature rises of the converter are larger than those of the inverter due to larger current with the constant switching pulse mode. Figure 18 shows temperature rises of the stator and rotor cores of 6-pole traction motors related to the running time. The experimental results agreed with what we calculated by considering RMS currents and time constant of heat capacities. They also show that temperature rises are sufficiently blow the limit values and the performances of traction system are satisfactory as we expected.

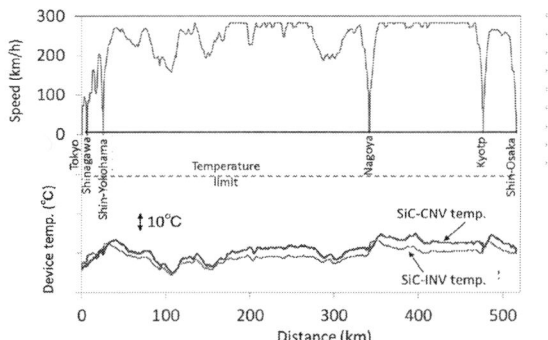

Fig. 17. Running test results of the conversion system.

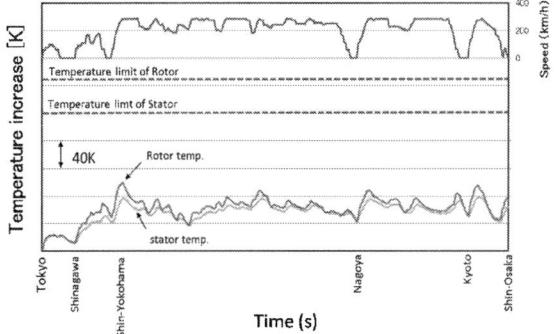

Fig. 18. Running test results of the 6-pole motor.

V. WEIGHT REDUCTION AND DOWNSIZING EFFECTS

The weight reduction and downsizing effects are shown in Fig. 19. The width of developed conversion system for the Series N700S is reduced to a half of conventional one for the Series N700. The axial length of the 6-pole traction motor for N700S is reduced by 10%. In terms of weight, the developed traction system for N700S is lightened by 20%, compared with N700. These results allow us to confirm that our approach to take advantage of SiC merits for weight reduction and compactness is successfully effective.

This weight reduction and compactness can expand the flexibilities of designing the layout of underfloor equipment as shown in Fig. 20. Since a conversion system of Series N700 is not smaller enough to be installed with a main transformer at the same car, the conversion system should be installed at a different car aside, which requires the additional connections from the main transformer via conversion systems to motors bridging between cars. Such installation of a main transformer and a conversion system at different cars restricts the flexibilities of designing the underfloor layout.

The SiC-applied tractions system solves this issue because a main transformer and a conversion system can be installed at the same car due to their compactness and lightweight. These flexibilities enable us to easily redesign different configurations of trains (ex. 8-car train or 12-car train) from the original 16-car train. This concept can be called as "Standard Shinkansen train."

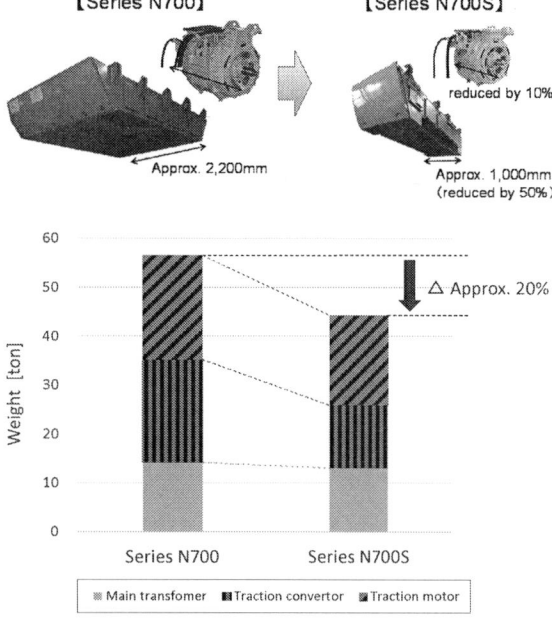

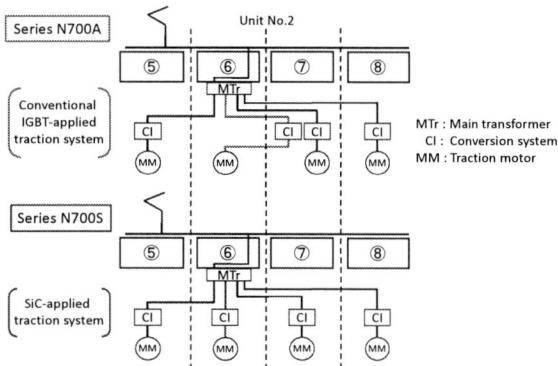

Fig. 19. Weight reduction and downsizing results.

Fig. 20. Flexibilities of designing the layout of underfloor equipment due to compactness and weight reduction of SiC-applied traction system.

VI. CONCLUSIONS

The traction system of Tokaido Shinkansen train has been improved by applying the power electronics. Based on accumulated technologies through the improvements, we developed the SiC applied traction system to pursue further weight reduction and compactness. In the development, we found that the merits of SiC device can be the most effective by combining the conversion system with train-draft cooling system and 6-pole induction motors.

We developed a prototype of the traction system, and conducted running tests. The test results showed the sound performance as we expected. In terms of weight reduction and compactness, the width of the conversion system is reduced to a half, and the axial length of the 6-pole traction motor is reduced by 10%, compared with conventional ones of the Series N700. The total weight of the traction

system is lightened by 20%. Additionally, we can have a prospect on the possibilities of further weight reduction in future research.

This weight reduction and compactness expand the flexibilities of designing the layout of underfloor equipment of the Shinkansen train in that main transformer and a conversion system can be installed at a same car. These flexibilities enable us to easily redesign different configurations of trains (ex. 8-car train or 12-car train) from the original 16-car trains, which can be called as "Standard Shinkansen train."

The developed traction system is to be installed in the latest-type Shinkansen train, or the Series N700S, which will debut in March 2018, and this traction system is the first application of SiC devices to high-speed trains' traction systems in the world. We hope that our development will pave the way for SiC applications in railway at home and abroad.

ACKNOWLEDGEMENT

The authors acknowledge the joint development of Toshiba Infrastructure Systems & Solutions Corporation, Mitsubishi Electric Corporation, Fuji Electric Corporation, Ltd. and Hitachi, Ltd. for this work.

REFERENCES

[1] K. Sato, M. Yoshizawa, and T. Fukushima, "Traction systems using power electronics for Shinkansen high-speed electric multiple units," *IPEC-Sapporo 2010*.

[2] H. Shimoyama, H. Kato, G. Kobayashi, and K. Sakanoue, "Advancement of Shinkansen rolling stock by downsizing and weight reduction of propulsion system," *STECH 2015*

[3] K. Ishikawa, K. Ogawa and M. Nagasu, "Traction Inverter that Applies Hybrid Module Using 3kV SiC-SBDs and High-Speed Drive Circuit," *IEEJ Transactions on Industry Applications,* vol. 135, no. 5, pp. 531-537, 2015.

[4] G. Kobayashi, H. Shimoyama, H. Tanaka, and N. Suyama, "Downsizing and weight reduction of the SiC applied traction system for Shinkansen high speed trains," *R&M*, vol. 23, no. 10, pp. 38-41, 2015.

[5] K. Sato, T. Fukushima, N. Suyama, K. Oda, and M. Kasahara, "Development of the SiC applied traction system for Shinkansen high speed trains," *IEE-Japan Industry Applications Society Conference 2016*, pp. 307-308, 2016.

[6] K. Sato, T. Onuki, S. Wakao, M. Tokuhisa, "A Novel Structure of Induction Motors for Weight Reduction," *Proc. of International Symposium on Theoretical Electrical Engineering (ISTET'99)*, pp. 589-593, 1999.

The 2018 International Power Electronics Conference

Development of a High Power Density Auxiliary Converter Based on 1700V 225A SiC MOSFET for Trams

Liu Hao[1], Fei Lin[1*], Zhongping Yang[1], Hu Cao[2] and Meng Xia[2]
1 School of Electrical Engineering, Beijing Jiaotong University, Beijing, China
2 CCRC Qingdao Sifang Rolling Stock Research Institute Co. Ltd., Qingdao 266000, China
*E-mail: flin@bjtu.edu.cn

Abstract— The application of the SiC MOSFET can improve the efficiency and power density of the auxiliary converter for trams. In this paper, the design process is given. According to the system requirements, the topology, the SiC device and its driver are selected. Then the switching frequency of the high frequency DC/DC module and the three-phase inverter module of the system are determined according to the loss simulation model. At last, the SiC auxiliary converter prototype has been developed whose rated power reaches 40kW and the power density reaches 133VA/L, the output of the system meets the requirement.

Keywords— SiC MOSFET, tram, auxiliary converter, loss model, efficiency

I. INTRODUCTION

The auxiliary converter is one of the most important components in trams, which can supply power to all the auxiliary devices other than the main circuit of the traction system. Therefore, the reliability of the auxiliary converter plays an important role in the smooth operation of the vehicle, and the weight and volume of the auxiliary converter are also important as a vehicle equipment [1-5].At present, the auxiliary converter based on Si devices has reached the bottleneck of development, and it is difficult to break through in efficiency and power density.

With the lower switching losses, the switching frequency of the converter based on SiC devices is higher than those based on Si devices which achieves the benefit of working with smaller filter [6]–[8]. For medium and low voltage applications, scholars have performed a large number of studies. Reference [9] compared SiC devices and Si devices with DC/DC converters. When using Si devices, the efficiency of converters with power of 7 kW, frequency of 50 kHz and temperature of 150℃ was 85%. When using SiC devices, the converter can work at 500 kHz and 300℃, and the efficiency can reach as high as 89%, thus the above facts fully confirmed the advantages of SiC devices. Reference [10] designed a 20 kW three-phase two-level SiC MOSFET V2G inverter. The application of SiC devices allows the efficiency of the system, which operates at 10 kHz, to reach as high as 99.05%, and system efficiency can still be maintained at

97.71% when the switching frequency is increased to 100 kHz. A 50 kW PV string inverter was designed with SiC devices in Reference [11]. However, in the field of rail traffic, few scholars have expressed the overall design method of the auxiliary converter.

In this paper, a SiC converter prototype whose topology is a high-frequency DC/DC converter plus three-phase inverter is designed for the auxiliary converter in trams. In Section II, the system requirements and topology of the auxiliary converter are introduced. In Section III, the SiC device and its drive board used in this system are introduced. In Section IV, the author sets up the loss simulation model for fast calculation and determines the switching frequency. In Section V, how to improve the power density of the auxiliary converter prototype developed in this paper is analyzed. In Section VI, the experimental results of the SiC auxiliary converter prototype are analyzed, and its performance meets the original design goals.

II. AUXILIARY CONVERTER SYSTEM

When designing an auxiliary converter, the circuit topology must be determined firstly. Also, the choice of the topology needs to consider the requirements of the auxiliary converter system. The auxiliary converter system parameters are given in table I.

TABLE I
THE AUXILIARY CONVERTER PARAMETERS

parameter	value
Rated input voltage / V	DC750
Input voltage range/ V	500-900
Rated output power / kW	40
Rated output voltage / V	AC380/3p
Rated output frequency / Hz	50

Due to the fact that the DC bus voltage of the trams is less than 900V and we could get the 1700V/225A SiC MOSFET from Cree , we could adopt two-level topology in the auxiliary converter, and there are two types of commonly used topology, one is the direct inverter, in which DC 750 V is directly inverted into three-phase AC output, and this scheme has the advantages of simple control, and low number of switching devices; at the

3484

same time, its most significant drawbacks are the larger size and weight of its three-phase isolation transformer, and the other one has a high-frequency DC/DC converter and a three-phase inverter, as Figure 1 shows. In this scheme, the number of devices is increased, but the size and weight of the transformer in the previous solution are significantly reduced. The application of SiC devices can improve the efficiency and power density of the converter, which will help greatly improve the performance of the system in this paper, especially for the last topology scheme.

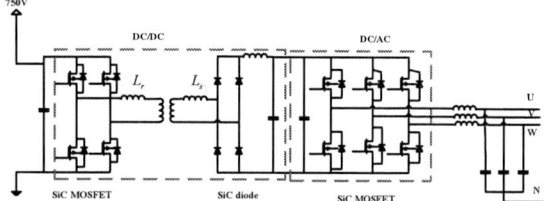

Fig. 1. The topology of SiC auxiliary converter prototype.

III. SiC DEVICES AND DRIVERS

According to the voltage and power demands of the switching device in the system, and considering the safety margin, it is decided that the 1700V/225A SiC MOSFET module (CAS300M17BM2) from Cree can be adopted in the prototype.

The SiC device needs the special drive circuit to fully display its high frequency performance. There are a lot of SiC device drivers, most of whose are for laboratory research. In the market production, we get two drivers, they are PT62SCMD17 from Cree and 62EM1 from AgileSwitch.

These two drivers have different characteristics and functions. Table II shows some features of these two driver boards.

TABLE II
THE FEATURES IN TWO DRIVER BOARDS

Features	Cree PT62SCMD17	AgileSwitch 62EM1
Working voltage	1700V	1700/1200V
Supply Voltage	+15~+24V	+15V
Propagation delay	100ns	250ns
Turn on rise time	85ns	80ns
Turn off fall time	30ns	90ns
Desat Monitor Voltage	6.5V	9V
Dead time	500ns	1000ns
Adjustable dead and blanking time	YES	YES
Gate driving voltage	+20V/-6V	+20V/-6V
Maximum switching frequency	125kHz	125kHz
Under voltage protection	YES	YES
Over voltage protection	YES	YES
Over current protection	YES	YES
Temperature monitoring	NO	YES
DC link voltage monitor	NO	YES

Fig. 2. The SiC device driver from Cree.

Figure 2 shows the driver board (model: PT62SCMD17) from Cree, it has a large range of the input voltage, also has the differential input mode, which can eliminate interference effectively. The primary side and secondary side have under voltage protection function, which can ensure that the SiC device could work in the best condition. It adopts the desaturation over-current detection, by detecting the voltage across the SiC MOSFET with fast response characteristic. In testing, the short-circuit protection time is 1.9us, which can ensure the safety of the SiC module. The dead time can be set according to the actual needs, the default dead time in the circuit is 500ns, the switching frequency is up to 125kHz, the input control signal voltage range is 15~24V, and the output voltage of the gate is -6V/+20V, so that the SiC module can be reliably opened and turned off.

Fig. 3. The SiC device driver from AgileSwitch.

Figure 3 shows the driver board (model: 62EM1) from AgileSwitch, it not only has all the functions of Cree driver board, but also adds temperature monitoring function and DC link voltage monitoring function. However its input control signal voltage needs to be only 15V, which is a drawback. And its price is higher than the PT62SCMD17. Taking into account the functional requirements and price factors of the drivers, we decide to adopt the Cree driver board.

IV. THE SELECTION OF SWITCHING FREQUENCY

After determining the switching device and the driver board, it is necessary to design the parameters of the passive components in the system. The premise of the design is to determine the switching frequency of the DC/DC module and the DC/AC module. Increasing the switching frequency can reduce the passive device parameters, but we can't increase the switching frequency blindly, which will lead to low efficiency. Therefore, the selection of switching frequency needs to consider the efficiency of the system.

A. The SiC device loss in system

In this paper, the topology structure of the SiC auxiliary converter is a high-frequency DC/DC converter plus a three-phase inverter, as shown in Figure 1.

Due to the leakage current of the SiC device in the blocking state being negligible, the loss mainly includes the on-state loss P_{con}, turn-on loss P_{sw-on}, and turn-off loss P_{sw-off}. And then we have some fast loss calculation formulas. In this part, we use the same gate resistance as the manufacturer used in SiC device's datasheet:

$$P_{con} = I_{rms}^2 \cdot R_{ds} \cdot f_s \qquad (1)$$

$$P_{sw-on} = E_{on} \cdot \frac{V_c}{V_n} \cdot \frac{I_c}{I_n} \cdot f_s \qquad (2)$$

$$P_{sw-off} = E_{off} \cdot \frac{V_c}{V_n} \cdot \frac{I_c}{I_n} \cdot f_s \qquad (3)$$

In the formula, I_{rms}^2 is the RMS value of the current flowing through the switch devices; R_{ds} is on-state resistance of switches; E_{on} and E_{off} are the loss of turn on and turn off at the data handbook of standard test conditions; V_n and I_n are the blocking voltage and collector current at the data handbook of standard test conditions; V_c and I_c are the blocking voltage and collector current at the actual conditions; and f_s is the switching frequency.

B. The loss simulation model for fast calculation

To facilitate the selection of the switching frequency and efficiency improvement of the converter, it is necessary to establish an effective loss model which can be easily generalized. The mockup of the converter system is replicated using PLECS simulation software. The software has the advantage of model simulation, which makes it possible and easy to measure the on-state loss and switching loss of the switching device by using the probe module, according to the data of the switching device. Additionally, combining the cycle mean value unit and periodic pulse mean module in the PLECS module library with the probe component, the software can transform the instantaneous value of loss into the average loss and can express the average loss intuitively, which facilitates the generalization of the model.

The high-frequency DC/DC converter simulation model and the three-phase inverter simulation model are shown in Figure 5. The loss of the device is determined by the actual working conditions of the circuit. The average conduction loss and average switching loss of the device can be calculated by the relevant modules in the PLECS module library.

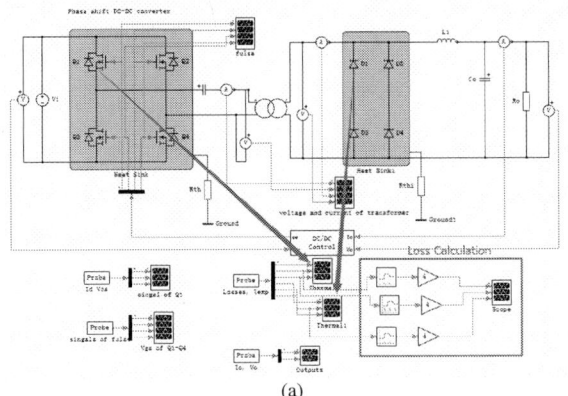

(a)

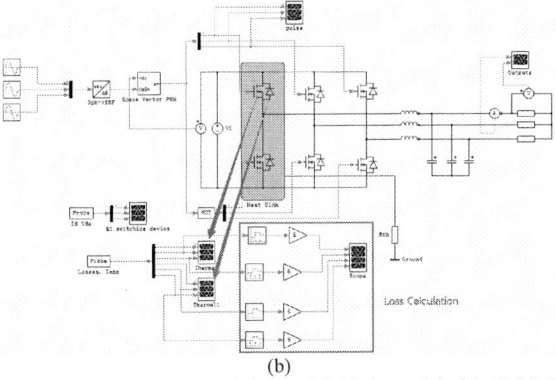

(b)

Fig. 5. The PLECS loss simulation model, (a) the model of the DC/DC converter, (b) the model of the three-phase inverter.

In this simulation model, we need to set up a new thermal description file for the SiC device as Figure 6 shows. And in our paper, we set up this thermal description by formulas (1) ~ (3).

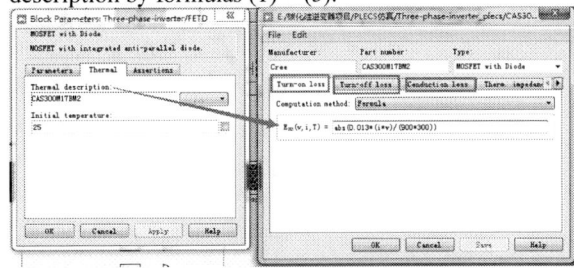

Fig. 6. The device thermal description file.

And the other parameters of the simulation model is calculated according to the system requirements.

C. The switching frequency selection

Then we can obtain the following results by this simulation model. Firstly, we can get the efficiency curve of the DC/DC module as shown in Figure 7.

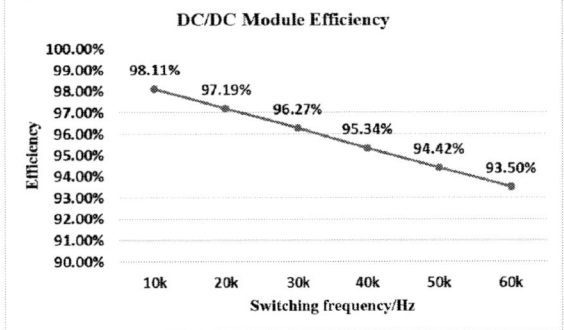

Fig. 7. The efficiency curve of the DC/DC module.

And the efficiency curve of the three-phase inverter module is as shown in Figure 8.

The 2018 International Power Electronics Conference

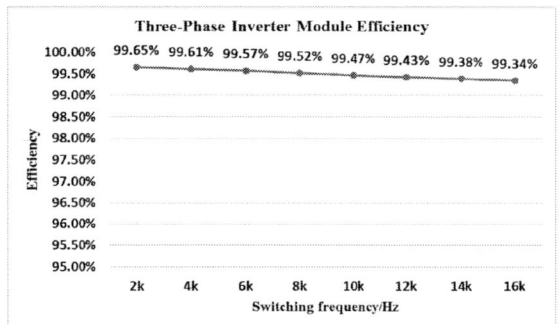

Fig. 8. The efficiency curve of the three-phase inverter module.

From Figure 7 and Figure 8, we can know that the efficiency curves of two modules are decreased, but the overall efficiency of the three-phase inverter module is maintained at more than 99%, and the efficiency of the high frequency DC/DC module decreased more obviously, the reason is that the switching loss will be higher with the higher frequency, which is a significant proportion in all loss. So that the greater switching loss will lead to a greater total loss, when DC/DC module frequency increased significantly. However, increasing the switching frequency is beneficial to reduce the passive device parameters and improve the system power density. In consideration of that the overall efficiency of the system need to meet the requirements of more than 92% (whose efficiency is not lower than that of the original auxiliary converter based on Si devices), the switching frequency of the two modules are defined. At the same time, considering the efficiency of the high frequency transformer is 97.5%, the efficiency of three-phase filter system output is 99.5%, we can calculate the switching frequency limit choices: high frequency DC/DC module works at 40kHz, three-phase inverter module works at 10kHz.

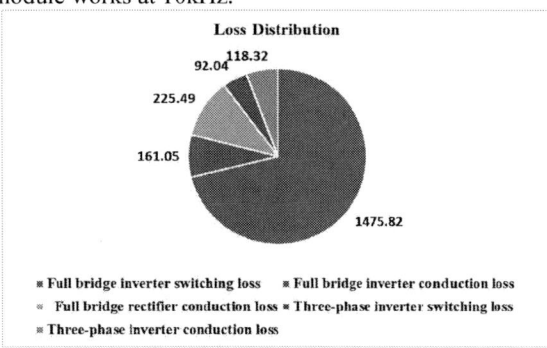

Fig. 9. The loss distribution with selected switching frequency.

At the selected switching frequency, the loss distribution of each module is plotted as shown in Figure 9. We can see that in all the device loss, full bridge inverter switching loss has a more than 1/2 proportion, the main reason is the full bridge inverter switching frequency is higher, at the same time, in order to facilitate the calculation and analysis, results are calculated under the assumption that the full bridge circuit works in hard switching conditions. And in the actual application, the DC/DC converter will work in soft switching condition with high frequency, in order to reduce the switching loss.

V. POWER DENSITY ADVANTAGE OF PROTOTYPE

For vehicle equipment, it is also important to minimize the volume and weight. And we also know that power density and efficiency are two points for a contradictory converter with lots of passive devices and radiators. So we should know that the design objective of the auxiliary converter prototype in this paper is to minimize the volume and weight under the premise of meeting the system efficiency requirement. Next, two aspects are introduced to explain why the auxiliary converter prototype can improve the power density.

First of all, the use of the SiC devices in this prototype reduces the power loss of devices obviously. From table III, we could see that the turn-on loss and turn-off loss of SiC MOSFET module are far less than that in Si IGBT module when we choose two devices with the same power level.

TABLE III
THE SWITCHING LOSS OF TWO DEVICES

Parameter	CAS300M17BM2 1700V/225A SiC MOSFET module	FF300R17KE3 1700V/300A Si IGBT module
Turn-on loss / mJ	13	105
Turn-off loss / mJ	10	94

As mentioned in section II, there are two types of commonly used topology when the supply power is DC750V, named the direct inverter scheme and high-frequency DC/DC converter plus three-phase inverter scheme. So what's the difference between these two schemes? There is a three-phase power frequency transformer in the direct inverter scheme which has a huge volume and weight, which doesn't exist in the latter scheme. And, there is a high-frequency DC/DC converter in the latter scheme.

TABLE IV
THE INFORMATION OF POWER FREQUENCY TRANSFORMER

Three-phase power frequency transformer(50KVA/50Hz)	
Size / mm	800*300*660
Volume / L	158.4
Weight / kg	350

TABLE V
THE INFORMATION OF HIGH FREQUENCY DC/DC CONVERTER

High frequency transformer(50KVA/40kHz)	
Size / mm	160*115*175
Volume / L	3.22
Weight / kg	5.7

TABLE VI
THE INFORMATION OF DC/DC CONVERTER

DC/DC module(radiator included)	
Size / mm	(300+400)*160*300
Volume / L	33.6
Weight / kg	35

Table IV shows the volume and weight information of power frequency transformer while table V shows that of high frequency transformer which is the actual measured

3487

data. We can see that the volume of the high frequency transformer is 2% of that of the power frequency transformer. And the volume of high frequency plus DC/DC module is also less than that of power frequency transformer. So we could know that the power density of the prototype topology scheme is still higher than that of the other one, even though the prototype has a more DC/DC module.

VI. PROTOTYPE EXPERIMENT

According to the above designs, this paper developed the following tram SiC auxiliary converter prototype, the prototype includes main circuit, control circuit, the two controller part, and it can realize the control goal of the DC750V input to 3AC380V output. The power density of the prototype can reach 133VA/L, more than the original power density of auxiliary converter based Si IGBT. The topology is as shown in Figure 1, the working state of each module was tested under rated power of 40kW.

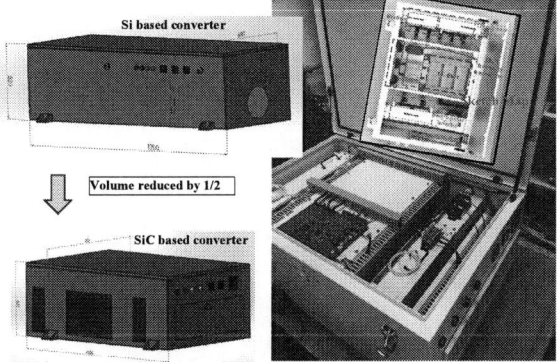

Fig. 10. The SiC auxiliary converter.

Since the switching device is a SiC module, it is impossible to observe the current through the MOSFET, so that the output current of the module is measured when observing the module. When we tested working state of high frequency DC/DC module, three-phase inverter module is blocked, the output of high frequency DC/DC module is connected to the resistive load of 11 ohm. The primary voltage and primary current waveform in high frequency transformer is shown in Figure 11, the output voltage and the output voltage of the DC/DC module is also reflected, we can see that the input voltage of high frequency DC/DC module is 780V, working at the frequency of 40kHz, the prototype removed the filter inductor, caused primary current of transformer is in the on-off state, the DC/DC module in good working condition, the output power is up to 43.25kW.

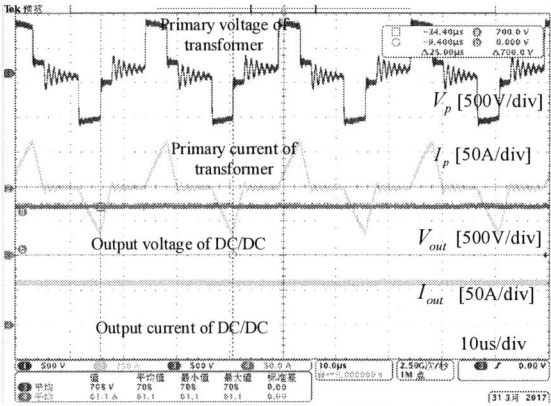

Fig. 11. The waveform of the DC/DC module.

After the test of the high frequency DC/DC module, the full power test of the whole machine is carried out. Figure 12 is the experimental waveform of output phase voltage and output phase current in the three-phase inverter module. The switching frequency is 10kHz, and it can be seen that the output waveform is ideal, and the voltage harmonic content is low.

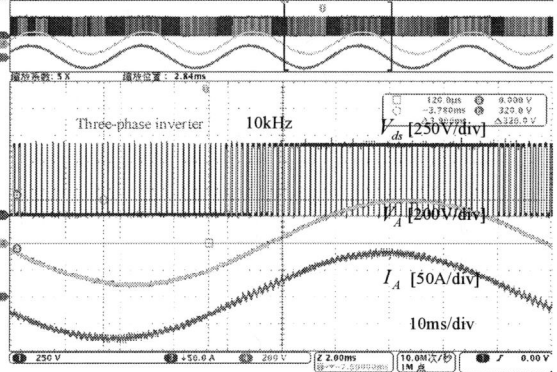

Fig. 12. The waveform of the DC/AC module.

VII. CONCLUSION

This paper presents the design cycle and prototype experimental results of the SiC auxiliary converter system for trams. Firstly, according to the system requirements, the prototype topology is determined, and then the SiC device model is determined. The performance of the two SiC driver boards is simply compared, and the appropriate driver is selected. Then the author set up the loss calculation model by PLECS software and estimated the efficiency of each module in this system and determined that the switching frequency of the high frequency DC/DC is 40kHz and the three-phase inverter module work at 10kHz, which can ensure that the system efficiency is not lower than 92%. At last, the working state of SiC auxiliary converter prototype is tested, the prototype power reaches 40kW, the power density reaches 133VA/L, each module of the system works normally, the output voltage and current waveforms are as expected.

REFERENCES

[1] L. Diao, J. Chen, W. Lin and Z. Liu, "Analysis and design for reducing voltage stress in output rectifier of LFLRV SIV DC-DC

converter," *2009 IEEE International Symposium on Industrial Electronics*, Seoul, 2009, pp. 1077-1080.

[2] Ki-Bum Park, S. Pettersson and F. Canales, "Auxiliary power supply for LV inverter with 1700 V SiC switch," *IECON 2013 - 39th Annual Conference of the IEEE Industrial Electronics Society*, Vienna, 2013, pp. 483-488.

[3] M. Brenna, F. Foiadelli, D. Zaninelli and D. Barlini, "Application prospective of Silicon Carbide (SiC) in railway vehicles," *2014 AEIT Annual Conference - From Research to Industry: The Need for a More Effective Technology Transfer (AEIT)*, Trieste, 2014, pp. 1-6.

[4] D. Murthy-Bellur, E. Ayana, S. Kunin and B. Palmer, "High-frequency split-phase air-cooled SiC inverter for vehicular power generators," *2015 IEEE Transportation Electrification Conference and Expo (ITEC)*, Dearborn, MI, 2015, pp. 1-5.

[5] F. Alkayal and J. B. Saada, "Compact three phase inverter in Silicon Carbide technology for auxiliary converter used in railway applications," *2013 15th European Conference on Power Electronics and Applications (EPE)*, Lille, 2013, pp. 1-10.

[6] A. Kadavelugu, S. Baek, S. Dutta, S. Bhattacharya, M. Das, and A. Agarwal, "High-frequency design considerations of dual active bridge 1200 V SiC MOSFET DC–DC converter," in *Proc. IEEE Appl. Power Electron. Conf. Expo.*, Fort Worth, TX, USA, 2011, pp. 314–320.

[7] Z. Chen, Y. Yao, D. Boroyevich, K. Ngo, P. Mattavelli, and K. Rajashekara, "A 1200 V, 60 A SiC MOSFET multi-chip phase-leg module for high-temperature, high-frequency applications," *IEEE Trans. Power Electron.*, vol. 29, no. 5, pp. 2307–2320, May 2014.

[8] B. Zhao, Q. Song, W. Liu, and Y. Sun, "Dead-time effect of the high frequency isolated bidirectional full-bridge DC–DC converter: Comprehensive theoretical analysis and experimental verification," *IEEE Trans. Power Electron.*, vol. 29, no. 4, pp. 1667–1680, Apr. 2014.

[9] W. Choi, D. Han, C. T. Morris and B. Sarlioglu, "Achieving high efficiency using SiC MOSFETs and reduced output filter for grid-connected V2G inverter," *IECON 2015 - 41st Annual Conference of the IEEE Industrial Electronics Society*, Yokohama, 2015, pp. 003052-003057.

[10] K. Shenai, "Silicon carbide power converters for next generation aerospace electronics applications," in *National Aerospace and Electronics Conference*, 2000. NAECON 2000. Proceedings of the IEEE 2000, 2000, pp. 516–523.

[11] J. Mookken, B. Agrawal, and J. Liu, "Efficient and compact 50kW gen2 SiC device based PV string inverter," in *Proc. PCIM'14 Conf.*, May 2014, pp.1-7.

The 2018 International Power Electronics Conference

Experimental Tests Results of Damping Control with Over Voltage Resistor for Regenerative Brake Control of Railway Vehicle

Natsuki Kawagoe[1]*, Febry Pandu Wijaya[1], Hiroyasu Kobayashi[1], Keiichiro Kondo[1],
Tetsuya Iwasaki[2], Akihiko Tsumura[2], Takumi Nagashima[2], Yoshinori Yamashita[3], Ryota Gondo[3]
1 Chiba University, Chiba, Japan
2 Odakyu Electric Railway Co. Ltd., Tokyo, Japan
3 Mitsubishi Electric Co. Ltd., Tokyo, Japan
*E-mail: nkawagoe@chiba-u.jp

Abstract-**Higher DC-link voltage of the regenerating train under the light-load regenerative brake control increases the regenerative brake power. However, if the regenerating load suddenly changes, the filter capacitor (FC) voltage of the traction inverter rises due to the delay of the motor current control, and the overvoltage protection may be activated. In this paper, damping control method to reduce the peak of FC voltage using the over voltage resistor (OVRe) is examined by experimental test in Odakyu Tama Line. Then, the peak of FC voltage is decreased when the dumping control with OVRe is working.**

Keywords— DC-electrified railway, light-load regenerative brake control, over voltage resistor, load shutdown test

I. INTRODUCTION

One of the beneficial feature of energy saving as well as the reduction of the maintenance work for mechanical brake is highly utilizing the regenerative brake in the electrified railway system [1]. However, in the DC-electrified railway system, when the train uses regenerative brake by the inverter-fed induction motor, as it is shown in Fig. 1, the regenerative energy must be absorbed by the other powering train because the diode rectifier is generally applied to the substation [2]. In this situation, in order to transmit higher regenerative power to the powering train in the distance and reduce the energy supplied from substation to powering train, the filter capacitor (FC) voltage of the vehicle in regenerative brake condition must be controlled at higher value to compensate the voltage drop in the feeding line.

One of the effective ways to increase the FC voltage of the regenerative train by only changing the control is to improve light load regenerative brake control. Where the regenerative brake power and load power are balanced by controlling the q-axis current of the induction motor, i_q, according to the FC voltage, v_c, as shown in Fig. 2(a). In Fig. 2(a), i_{qmax} is the minimum q-axis current corresponding to the maximum regenerative brake force. V_{cmax} is the DC-link voltage to stop the light-load regenerative brake control, where it is usually fixed to avoid over voltage of the traction inverter in the steady state condition. V_{clim} is the DC-link voltage to start

reducing the absolute value of q-axis current, where higher V_{clim} is preferable for saving the substation energy [3], as shown in Fig. 2(b).

By applying a high V_{clim} value in Fig.2(a), there is a possibility that the overvoltage protection (OVD) may operate with the FC voltage reaching the upper limit value. Then, the operation of OVD stops the regenerative brake operation. To cope with these problems, a damping control method that reduces the peak of FC voltage utilizing the over voltage resistor (OVRe) has been proposed [4] and its basic performance to reduce the peak value of FC voltage is verified [5][6].

The load for the regenerative brake sometimes suddenly changes in actual revenue service lines, according to the change of load demand of the powering train. Especially, the load is shutdown as is the worst case of the sudden load change. Thus, a load shutdown test with the resistor to emulate the load as shown in Fig.3 is carried out in the commissioning test at the factory. In this test, V_{clim} is tuned so as to avoid overvoltage of FC [7]. The circuit behavior of the load shut down in the actual line is different from the one at the commissioning test. For instance, the current is shut down by the high speed circuit breaker (HB) on the regenerating vehicle, though by the HB on the powering load vehicle in an actual line test. In this case, the inductance is higher on the route of the shutdown current than the case of the load shut down test at the factory. This results in longer time transient of the current shutdown and it may cause high

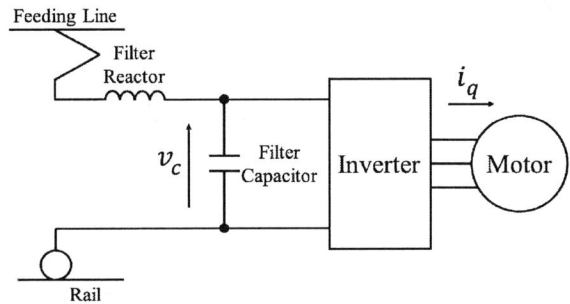

Fig. 1. Traction circuit of the DC-electrified railway vehicles.

3490

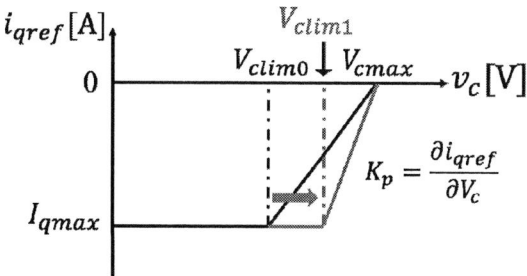

(a) Motor current control under light load condition.

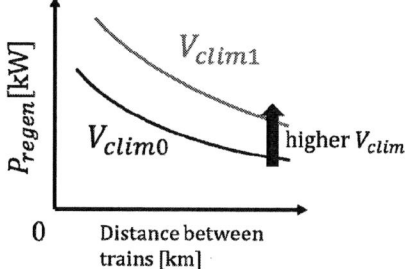

(b) Regenerative brake power characteristics by setting higher V_{clim}.

Fig. 2. Light-load regenerative brake control method.

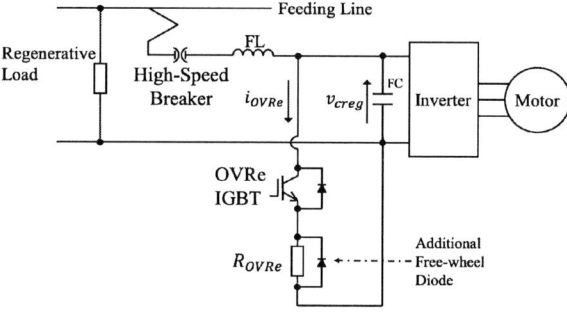

Fig. 3. Circuit configuration of the regenerating train.

spike voltage of the filter capacitor on the regenerative brake vehicle. Therefore, the newly being installed damping control by OVRe [4] must be evaluated under the load shut down test on actually running vehicle condition. In this paper, a damping control method utilizing the over voltage resistor (OVRe) is verified through experimental running tests with an actual load train. The tests evaluate if the damping energy on the OVRe is effective enough and how higher the V_{clim} can be set. As the results, the damping energy on the OVRe is enough much to damp the energy on the capacitor and to increase the V_{clim} 1700V to 1720V in this tests, by which around 20% average higher power of the regenerative brake is expected.

II. THE REGENERATIVE BRAKE CONTROL METHOD UNDER LIGHT LOAD CONDITION [8]

As shown in Fig. 4, the powering and regenerating trains between two substations are assumed, and the suffixes "pow" and "reg" represent powering and regenerating trains, respectively.

$$i_{qref} = \begin{cases} I_{q\max} & (v_c < V_{clim}) \\ k_p(v_c - V_{c\max}) & (V_{clim} \le v_c < V_{c\max}) \\ 0 & (V_{c\max} \le v_c) \end{cases} \quad (1)$$

Afterwards, the DC input side current of the inverter, i_{inv}, can be obtained from (2) by neglecting the inverter loss. In addition, the active power of d-axis current is ignored because it is much lower than the active power of q-axis current.

$$i_{inv} = \begin{cases} v_q I_{q\max} v_c^{-1} & (v_c < V_{clim}) \\ k_p v_q(1 - V_{c\max} v_c^{-1}) & (V_{clim} \le v_c < V_{c\max}) \\ 0 & (V_{c\max} \le v_c) \end{cases} \quad (2)$$

In Fig. 4, the constraint of the equation (4) is given from the voltage drop at the feeding line resistance R_e at the steady state.

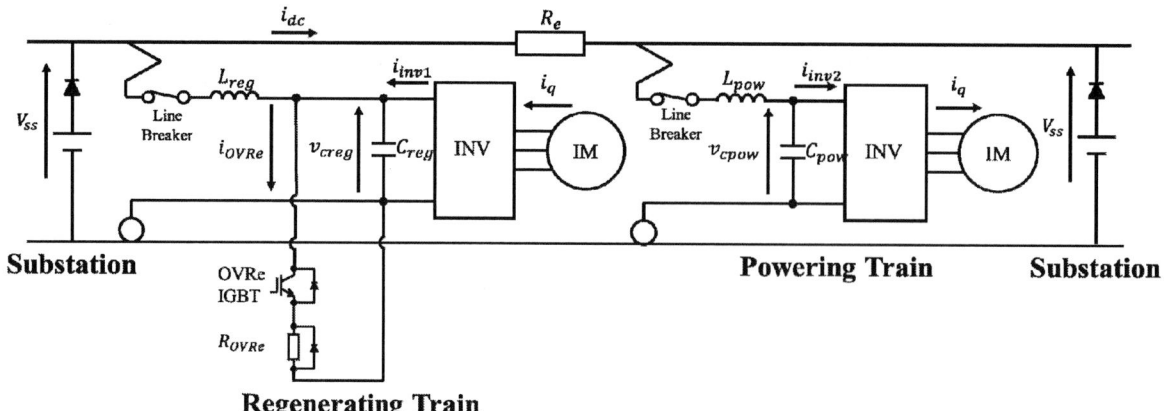

Fig. 4. The typical circuit configuration of the vehicle in the DC-electrified railway network.

3491

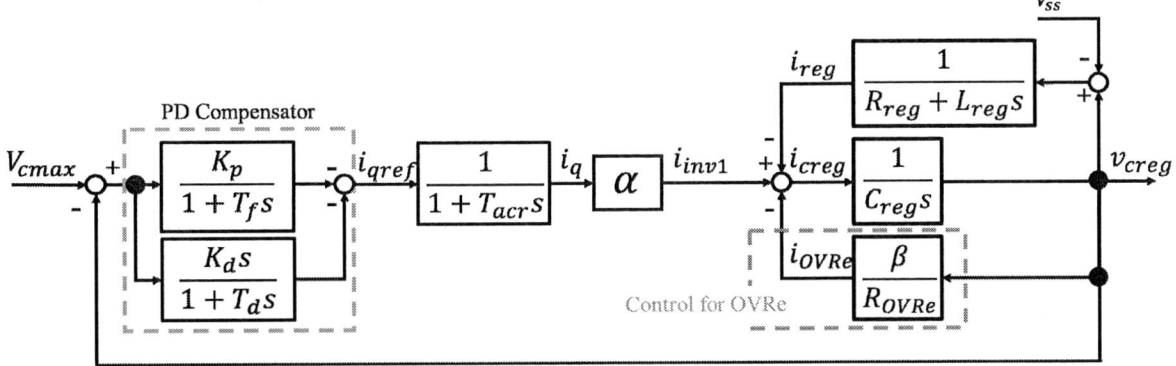

Fig. 5. The block diagram of ligh-load regenerative brake control using OVRe.

$$i_{inv} = R_e^{-1}(v_{creg} - v_{cpow}) \qquad (3)$$

Therefore, by setting the high V_{clim} in equation (2), the operating point of the FC voltage becomes high, and as a result, more regenerative current can be sent to the power running car load by the equation (3), leading to energy saving.

However, as the trade-off of higher energy saving, if the load for the regeneration suddenly changes, the delay of the motor current control causes the FC voltage increases. Then, the FC voltage may reach the upper limit value and activate the OVD by applying higher V_{clim} value.

III. THE CONTROL METHOD OF OVRE SYSTEM [4]

Fig. 3 shows the circuit configuration of the regenerating train. As shown in Fig.3, the OVRe and the IGBT circuit connected in series are generally equipped in the traction inverter. The original purpose of OVRe is to absorb the energy of the filter capacitor in case of OVD or when the inverter is switched-off. Therefore, the heat capacity of the OVRe is not so high. The powering train is represented by the load resistance R_{load} in the load shutdown test. A low-price free-wheel diode is required as an additional circuit to keep the current pass when the OVRe IGBT is turned-off.

Fig. 5 shows the control block diagram of the light load regenerative brake control equipped with the OVRe system. As the controller, a PD compensator is used to compensate the phase delay of the control system due to automatic current regulator (ACR) delay and to avoid sudden change of the FC voltage, v_{creg}.

In Fig. 5, R_{OVRe} represents the value of OVRe resistance. When IGBT is operated by duty ratio β, the current i_{OVRe}, which flows to the OVRe and IGBT, is given as equation (4).

$$i_{OV\,Re} = \frac{\beta}{R_{OV\,Re}} v_{creg} \qquad \left(0 \leq i_{OV\,Re} \leq \frac{v_{creg}}{R_{OV\,Re}}\right) \qquad (4)$$

A high pass filter (HPF) is used to pick-up the FC voltage spike. The output of HPF is reflected as the duty ratio β by multiplying a damping gain K_{OVRe} as follows:

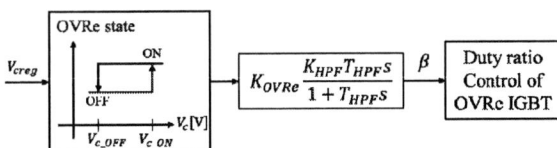

Fig. 6. Duty ratio control of OVRe IGBT.

Table 1 Meanings of symbol of fig. 5,6.

Symbol	Meaning
K_p	Proportional gain of regenerating controller
T_f	Time constant of LPF for proportional controller
K_d	Differential gain of regenerating controller
T_d	Time constant of LPF for differential controller
T_{acr}	Time constant of control system of motor current
α	Modulation ratio of inverter
R_{reg}	Resistance of filter reactor
L_{reg}	Inductance of filter reactor
C_{reg}	Capacitance of filter capacitor
R_{ovre}	OVRe resistance
V_{ss}	Voltage of feeding line
V_{c_ON}	Voltage to start OVRe control
V_{c_OFF}	Voltage to stop OVRe control
K_{ovre}	Damping gain of duty ratio control of OVRe IGBT
T_{HPF}	Time constant of HPF for duty ratio control of OVRe IGBT
K_{HPF}	Damping gain of HPF for duty ratio control of OVRe IGBT
β	Duty ratio of OVRe IGBT

$$\beta = K_{OV \text{Re}} \left(\frac{K_{HPF} T_{HPF} s}{1 + T_{HPF} s} v_{creg} \right) \quad (0 \leq \beta \leq 1) \quad (5)$$

K_{HPF} and T_{HPF} are determined so that the dynamic range of HPF includes 17.75Hz as the resonant frequency of LC filter of the regeneratng vehicle. The time constant of HPF, T_{HPF}, is set to 9ms; whereas the HPF gain, K_{HPF}, is 1.0.

In order to prevent the OVRe from consuming regenerative power more than necessary, a hysteresis control for controlling ON and OFF of the OVRe IGBT according to the FC voltage is incorporated as shown in figure. 6 V_{c_ON} is designed to be as small as possible between V_{cmax} and OVD value. As a result, OVRe can be activated more quickly, and more current can be absorbed. As a result, it can be considered that the peak value of the FC voltage also becomes small.

IV. EXPERIMENTAL RESULT

As an tests condition, Train A to be evaluated with the proposed method and Train B for a powering load are preoared respectively. Train A is the Odakyu Electric Railway type 1000, 4 train-car (2 motor cars), train B is Odakyu Electric Railway type 4000, 10 train-car (6 motor cars). The load condition by the powering train B of type 4000 is high enough to absorb all the regenerative power generated in regenerating train A was set as a condition that can be consumed by train B. The section for the test was Odakyu Tama line. The Tama line running along the Tama hills is a route where a gradual uphill gradient continues from Shin-Yurigaoka station towards Karakida station. Figure. 7 shows the principal stations and substations of the Tama line. The configuration of the feeding circuit is a directional power feeding method.

At the first step of the tests, two trains run on the same track with 1 - 2 km distant each other. At this step, regeneration brake is applied to train A at approximately the maximum power around 90 km/h and train B accelerates at 40 km/h around. It is supposed that the electric power of train A and train B is balanced. The powering operation of the Train B is shut down by opening all of the circuit breaker at the same time to emulate sudden no load status. In this step, the proposed method is verified if it could successfully avoid overvoltage protection of the inverters onboard train A. The proposed method is evaluated by the traction system of which specifications are shown in Table 2. OVRe limits the continuous operation possible time in one operation to 70ms because of heat capacity restriction.

The value of V_{clim} originally set in the Train is 1700 V. In this test, V_{clim} of train A which is a regenerative car was increased to 1720V.

The experimental result is shown in Fig. 8. Fig. 8 (a) shows the waveform of DC-current, i_{inv1} in Fig.4. Fig. 8 (b) shows the waveform of FC voltage, v_{creg} in Fig.4 and duty of OVRe IGBT.

Table 2 Specifications of the traction system for the Train A.

Specification	Values	
Rated power per traction motor	190	kW
C_{reg}	10.05	mF
L_{reg}	8	mH
R_{reg}	0.1	Ohm
I_{qmax}	268	A
V_{clim}	1720	V
V_{cmax}	1750	V
V_{c_ON}	1760	V
V_{c_OFF}	1730	V
Instantaneous OVD value	1900	V
R_{OVRe}	9	Ohm

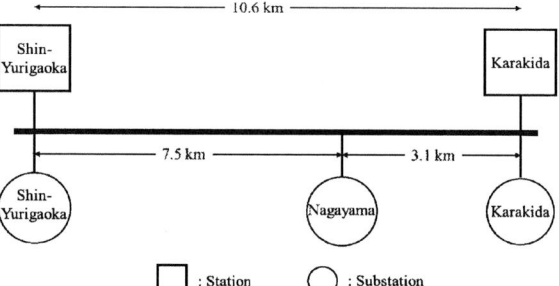

Fig.7. Principal Stations and Substations in Odakyu Tama Line.

According to Fig. 8 (a), the DC current flowing to the load at moment of the load shutdown was reduced to 0 in approximately 0.16seconds. This result is about 6 times longer than the result of the load shutdown test at the commissioning test in Ref. [7]. In addition, the response of the stator q-axis current was about 3 times slower than the result of the q-axis current in the load shutdown test in Ref. [7]. A field orientation control is employed to Odakyu type 1000 for train A. As shown in Fig. 1, the FC voltage at load shutdown is determined in proportion to the difference between the current from the DC side of the inverter and the powering load current in this test condition. The DC current flowing to the load gently decreases than the q-axis current which is almost in proportional to the inverter DC side current. In the test of the factory, the load current decreases from around -500A to 0A with around 10ms. However, it takes more than 100ms in Fig. 8 (a) from -600A to 0A. This may be caused by more storage magnetic energy adjacent to the FLs both on the train A and train B. The slower change of DC current at the load shut down is severe case for avoiding the over voltage in the FC on board regenerative vehicle, because more power flows to the FC and the OVRe must absorb the enough energy to avoid the FC over voltage. From this view point, 70ms of operating duration for OVRe damping control is appropriate for this test condition.

According to Fig. 8(b), the FC voltage temporarily stops rising around 1750 V by operating OVRe IGBT in Fig. 4, and the peak value of FC voltage is successfully reduced.

3493

The 2018 International Power Electronics Conference

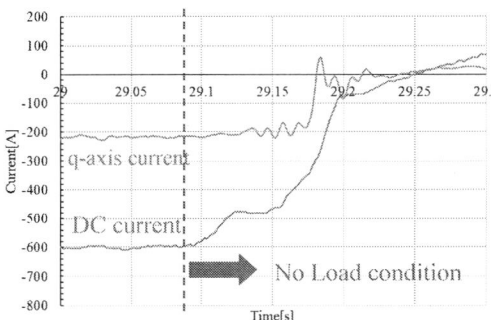

(a)Waveform of DC-current and q-axis current

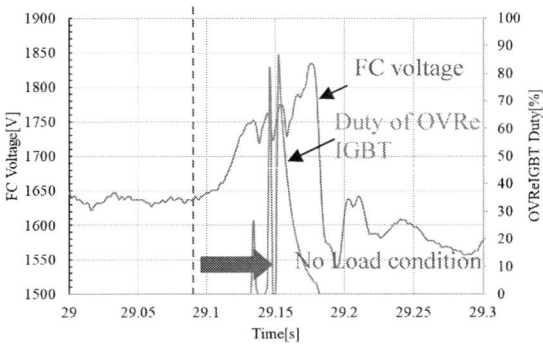

(b)Waveform of FC voltage and duty of OVRe IGBT

Fig. 8. Running test result.

Although the FC voltage sharply rises due to load interruption, it can be seen that OVRe operates around 29.13 s and 29.15 s, and the FC voltage decreases. As a result, since the increase of the FC voltage could be suppressed for about 30ms, it is presumed that the peak value of the FC voltage could be reduced. Because the increase and decrease of the FC voltage repeatedly occurred, the q-axis current command value vibrated by the light load regeneration control.

The peak value of the FC voltage at this time is 1850 V, which is 50 V lower than the reference of the overvoltage in the test vehicle this time. Therefore, it is possible to further increase V_{clim}. According to this result, it is possible to detect an increase in the FC voltage by the HPF even in an actual vehicle. Also, by properly using OVRe, the damping control with OVRe system is effective to reduce the peak of FC voltage and is able to avoid the inverter operation stop by OVD. Therefore, the value of V_{clim} can be increased from 1700V to 1720V so that the regenerative brake power is transmitted to farther powering train.

V. CONCLUSIONS

In this paper, the proposed damping control with OVRe is verified by the load shut down test in an actual rail line with two actual trains. The newly installing OVRe damping control is verified in the actual vehicle running tests. In this test, the value of V_{clim} can be increased from 1700V to 1720V so that the regenerative brake power is transmitted to father powering train. This

leads to the more energy saving in train operation. In addition, it is revealed that the load shutdown current changes slower in the actual vehicle running test than in the case of the test at the factory. From the viewpoint of the overvoltage of FC, the load shout down test in the commissioning tests in a factory is a more severe case. Thus the commissioning test in the factory is enough to evaluate and tune the parameters.

VI. REFERENCES

[1] W. Gunselmann, "Technologies for increased energy efficiency in railway systems," *The 2005 European Conference on Power Electronics and Applications (EPE)*, pp. 1-10, 2005.

[2] M. Ogasa, "Application of energy storage technologies for electric railway vehicle – Example with hybrid electric railway vehicles," *IEEJ Transactions on Electrical and Electronic Engineering*, vol. 5, issue 3, pp. 304-311, 2010.

[3] T. Saito, K. Kondo and T. Koseki, "An analytical design method for a regenerative braking control system for DC-electrified railway systems under light load conditions," *IEEJ Transactions on Industry Application*, vol. 132, no. 2, pp. 268-277, 2012. (in Japanese)

[4] Jun-ichi Asano and Keiichiro Kondo, "A Damping Control Method to Enhance Regenerative Brake Power under Light Load Conditions," *Annual Conference of the IEEE Industrial Electronics Society (IECON2013)*, TT12-4, 2013.

[5] Febry Pandu Wijaya, Hiroyasu Kobayashi, Keiichiro Kondo, Tetsuya Iwasaki and Akihiro Tsumura, "Damping Control Method of Regenerative Brake Control under Light Load Condition Utilizing Over Voltage Resistor," *2017 IEEE 12th International Conference on Power Electronics and Drive Systems (IEEE PEDS 2017)*, 334, 2017

[6] Febry Pandu Wijaya, Hiroyasu Kobayashi, Keiichiro Kondo, Tetsuya Iwasaki and Akihiro Tsumura, "Damping Control Method Utilizing Over Voltage Resistor under Light Load Condition," *IEEJ Transactions on Electrical and Electronic Engineering*, Vol.13, No.2, 2018

[7] Natsuki Kawagoe, Hiroyasu Kobayashi, Keiichiro Kondo, Tetsuya Iwasaki, and Akihiro Tsumura "A Method to Design Light-load Regenerative Braking Control System at Load Shutdown for DC-electrified Railway Systems by Numerical Simulation," *2017 IEE-Japan Industry Applications Society Conference*, Vol. 5, pp. 293-298, 2017 (in Japanese)

[8] H. Kobayashi, T. Saito, K. Matsuo, S. Akita, K. Kondo, T. Suzuki, T. Iwasaki, S. Watanabe, A. Tsumura, S. Yamaguchi, and T. Ando "Experimental Running Test Results of Improving the Regenerative Brake Control Performance to Transmit the Regenerative Energy to the Distant Powering Load", *The Papers of Joint Technical Meeting on Transportation and Electric Railway and Physical Sensor, IEE Japan*, TER-16-16, PHS-16-16, pp.45-50,2016. (in Japanese)

3494

The 2018 International Power Electronics Conference

Coils layout optimization of dynamic wireless power transfer system to realize output voltage stable

Yi Wang*, Fei Lin, Zhongping Yang, Panpan Cai, Zhiyuan Liu
School of Electrical Engineering, Beijing Jiaotong University, Beijing, China
*E-mail: 16121534@bjtu.edu.cn

Abstract- **In order to solve the problem of the received voltage drop caused by the decrease of the coiling coefficient between the primary and secondary coils at the junction of the primary coils of dynamic wireless power transfer (DWPT), a partial overlap arrangement method of primary coil for DWPT system is proposed. The two adjacent primary coils are overlapped partially, and this two coils are used simultaneously to supply power when the secondary coil is located at the junction to ensure less voltage drop without any additional controls. The LCC compensation topology is used to ensure that the output voltage of the system is not affected by mutual inductance between the primary coils. The coil transfer efficiency can be maintained at a high level at the junction. The validity of the proposed arrangement method is verified by experiments.**

Keywords—dynamic wireless power transfer，coupling coil，partial overlap arrangement

I. INTRODUCTION

Compared to the traditional plug-in power supply system, wireless power transfer (WPT) system can achieve mechanical and electrical isolation, minimize cable and socket applications, and ensure safe operation in harsh environments [1-2]. It has the advantages of no exposed wires, no friction, good urban landscape and so on to use dynamic wireless power transfer (DWPT) system instead of catenary for modern tram power supply, and as a result, DWPT has caused widespread concern in recent years [3]. The primary transmitting coil of DWPT can be divided into two types, single-coil type, which is substantially larger than the secondary coil, and segmented coil type, which is commeasurable in size with the secondary coil [4-6]. As the long primary coil has low transmission efficiency, electromagnetic leakage caused by the exposed magnetic field and other issues [7], the DWPT with segmented coils is now widely studied.

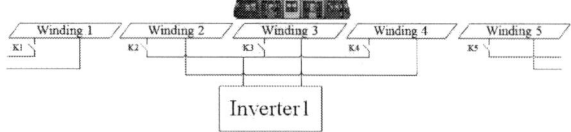

Fig.1. Typical schematic diagram of DWPT with segmented coils supplying modern trams

Fig.1 shows the typical schematic diagram of using DWPT with segmented coils to supply modern trams. The principle of power supply is that when the tram goes

to a power supply coil, the switch corresponding to this coil closes, and this coil can supply the tram with the rest switches off. This method can effectively reduce the electromagnetic leakage, improve the transfer efficiency. When DWPT is applied instead of the catenary for supplying the tram , a stable DC voltage output comparable to the catenary's is desired. However, when the tram is located at the junction of the primary coils, the coupling coefficient of the coupling coils will be greatly weakened, which will seriously affect the pickup of the electric energy. Although the tram can still pass when the supply voltage is low, it increases the difficulty of motor control. It will seriously affect the normal starting of the tram if the tram just right stops at the junction of the primary coils. In view of this problem, many scholars have proposed different solutions. In [8], a cascade energy pickup mechanism with multiple receiving coils is proposed. This structure increases the number of secondary coils to five horizontally arranged double pickup coils, and they are connected in parallel after rectification and filtering to improve the energy transfer capability. As a result, the receiver structure becomes very complex with large volume. A DQ-type power supply rails is proposed in [9] to realize the steady out power, but the phase difference between two power supply inverters must be kept 90 degrees, which means the control is complex, and the circuit device is doubled. The University of Auckland has proposed a variety of magnetic structures to achieve a smooth power transfer profile, such as quadrature pad design [10-11], double-D pad (DDP) primary [12-13], and double-D quadrature pad (DDQP) design [12], bipolar pad (BPP) [14-15]. All of above magnetic structures have multiple coils which help them achieve greater tolerance to lateral displacement by placing the coils in such a way that when one coil is at its null power point, the other coil provides sufficient power to derive the load, thereby eliminating null points from the cumulative output. But these improvement in performance comes out at the cost of an increase in the complexity of both the magnetic and electronic designs on the primary and secondary of the system, and at the same time, the output power still fluctuates.

Aiming at solving the problem of reduced received voltage of the secondary caused by the decrease of coupling coefficient at the junction of the primary coils, this paper proposes a partial overlap arrangement method of the primary coils to guarantee the output voltage stable. At the same time, the LCC compensation topology is

3495

used, this compensation allows the system to use two adjacent coils to supply simultaneously, and the received voltage is the sum of the voltage received by every single coil, independent of the mutual inductance between the primary coils. This method does not need to change the coil shape and control method, with little change to the system circuit. And the coil transfer efficiency can be maintained at a high level. The validity of the arrangement is verified by experiments.

II. ANALYSIS OF LCC TOPOLOGY

The LCC is a constant voltage gain compensation topology. And this compensation topology realizes the current of the primary coil constant, which is beneficial to the circuit protection. At the same time, the compensation parameters are independent of mutual, and the voltage gain is determined by the inductance of the primary side series compensation inductor and the mutual inductance between the primary and secondary coil. That is to say, the voltage gain can be realized by designing the value of the series compensated inductor on the primary side when the coil parameters have been determined [16]. The power supply schematic of DWPT system for the modern tram with LCC compensation topology is shown in Fig.2.

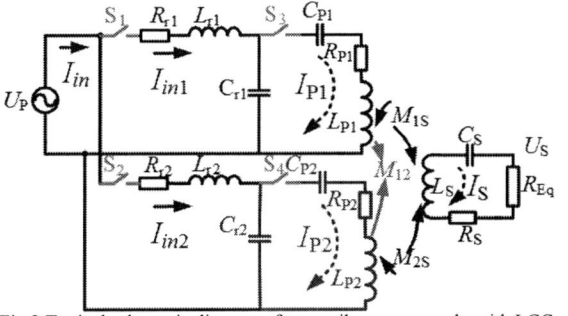

Fig.2 Typical schematic diagram of two coils power supply with LCC compensation topology

U_P and U_S are the RMS values of the output voltage of the high frequency inverter and the input voltage of the diode rectifier filter circuit, and I_{in} is the output current of the inverter. L_{r1} and L_{r2} are primary side series compensation inductors, and I_{in1} and I_{in2} are RMS values of the current of primary side series compensation inductors. L_{P1} and L_{P2} are the self inductance of the primary coils, and L_S is the self inductance of the secondary coil. M_{1S} and M_{2S} are the mutual inductance between the primary coils and secondary coil, and M_{12} are the mutual inductance between the primary coils. C_{r1}, C_{r2}, C_{P1}, C_{P2} and C_S are compensation capacitors. R_{r1} and R_{r2} are parasitic resistances of compensation inductors. R_{P1}, R_{P2} and R_S are parasitic resistances of coils, and R_{Eq} is equivalent load resistance.

The working process is as follows: when the tram is located above the primary coil L_{P1}, the relay S_1, S_3 turn on and S_2, S_4 turn off, so the tram receives the electric energy through the L_{P1} coil. When the tram is located at the junction of the primary coil L_{P1} and L_{P2}, the relay S_1,

S_2, S_3, S_4 are closed, and the L_{P1} and L_{P2} coils feed the tram simultaneously. Similarly, when the tram is located above the primary coil L_{P2}, the relay S_2, S_4 turn on and S_1, S_3 turn off, the coil L_{P2} feeds the tram individually.

According to Kirchhoff's Voltage Law (KVL), the circuit equations when one primary coil feeds the modern tram can be obtained.

$$\begin{bmatrix} U_P \\ 0 \\ 0 \end{bmatrix} = \begin{bmatrix} R_{r1}+j\omega L_{r1} & Z_{P1} & -j\omega M_{1S} \\ -\dfrac{1}{j\omega C_{r1}} & Z_{P1}+\dfrac{1}{j\omega C_{r1}} & -j\omega M_{1S} \\ 0 & j\omega M_{1S} & -\left(Z_s+R_{Eq}\right) \end{bmatrix} \begin{bmatrix} I_{in1} \\ I_{P1} \\ I_S \end{bmatrix} \quad (1)$$

Among them,

$$Z_{P1}=j\omega L_{P1}+\frac{1}{j\omega C_{P1}}+R_{P1} \quad (2)$$

$$Z_S=j\omega L_S+\frac{1}{j\omega C_S}+R_S \quad (3)$$

The ω is the operating angular frequency:

$$\begin{cases} \omega L_{r1}=\dfrac{1}{\omega C_{r1}}=\omega L_{P1}-\dfrac{1}{\omega C_{P1}} \\[3mm] \omega L_S=\dfrac{1}{\omega C_S} \end{cases} \quad (4)$$

When only one primary coil supplies power to the secondary side, taking the coil L_{P1} as an example, the effective value of the voltage received by the secondary side is as follow,

$$U_S=\frac{\omega^2 L_{r1}M_{1S}R_{Eq}U_P}{\left(\omega^2 L_{r1}^2+R_{r1}R_{P1}\right)\left(R_{Eq}+R_S\right)+\omega^2 M_{1S}^2 R_{r1}} \quad (5)$$

In the actual system, the parasitic resistances of coils and compensation inductor are small, which satisfy the following formula.

$$R_{Eq}\gg R_{P1},R_{P2},R_{r1},R_{r2},R_S \quad (6)$$

Neglecting the parasitic resistances of coils and compensation inductor, equation (5) can be simplified as follow.

$$U_S=\frac{M_{1S}}{L_{r1}}U_P \quad (7)$$

When two primary coils supply power to the secondary simultaneously, the voltage received by the secondary side can be calculated by the following formula.

$$U_S=\frac{\left(\dfrac{M_{1S}}{L_{r1}}+\dfrac{M_{2S}}{L_{r2}}+\xi\right)R_{Eq}}{\left(R_{Eq}+R_S\right)\left(1+\dfrac{R_{P1}R_{r1}}{\omega^2 L_{r1}^2}+\dfrac{R_{P2}R_{r2}}{\omega^2 L_{r2}^2}\right)+\dfrac{M_{1S}^2}{L_{r1}^2}R_{r1}+\dfrac{M_{2S}^2}{L_{r2}^2}R_{r2}+\zeta\cdot\dfrac{R_{r1}R_{r2}}{\omega^4 L_{r1}^2 L_{r2}^2}}U_P \quad (8)$$

Among them,

$$\xi=\frac{\left(-j\omega M_{1S}M_{12}+M_{2S}R_{P1}\right)R_{r1}}{\omega^2 L_{r1}^2 L_{r2}}+\frac{\left(-j\omega M_{2S}M_{12}+M_{1S}R_{P2}\right)R_{r2}}{\omega^2 L_{r1}L_{r2}^2} \quad (9)$$

$$\begin{aligned}\zeta=&\omega^2 M_{1S}^2 R_{P2}+\omega^2 M_{2S}^2 R_{P1}-2j\omega^3 M_{1S}M_{2S}M_{12}\\ &+\left(R_{P1}R_{P2}+\omega^2 M_{12}^2\right)\left(R_{Eq}+R_S\right)\end{aligned} \quad (10)$$

When the parasitic resistances are neglected, equation (8) can be simplified as following:

3496

$$U_S = \left(\frac{M_{1S}}{L_{r1}} + \frac{M_{2S}}{L_{r2}} \right) U_P \qquad (11)$$

It can be seen from equation (11) that when two primary coils supply power to the secondary side at the same time with LCC compensation topology, the voltage received by the secondary side is the sum of the voltage received by every single coil and independent of the mutual inductance between the two primary coils.

When the tram is located at the junction of the primary coils, the mutual inductance between the primary and secondary coils is reduced due to the decrease of the coupling coefficient. Assuming that the moving direction of the tram is from the coil L_{P1} to the coil L_{P2}, the mutual inductance M_{1S} decreases continuously with M_{2S} increasing continuously. Once the system parameters are determined, L_{r1} can be considered equal to L_{r2}. If the sum of mutual inductance M_{1S} and M_{2S} can approximately equal to the mutual inductance when a single primary coil is aligned with the secondary coil by adjusting the spacing between two adjacent primary coils, it can ensure that the voltage received by the secondary coil is stable.

III. ANALYSIS OF OVERLAP ARRANGEMENT

As shown in Fig.3, the traditional layout of the primary coils and the scheme of proposed partial overlap arrangement method are compared. According to the above description, the stable voltage output can be achieved by reasonably adjusting the coil spacing, but the overlap distance between coils is a problem that needs to be determined.

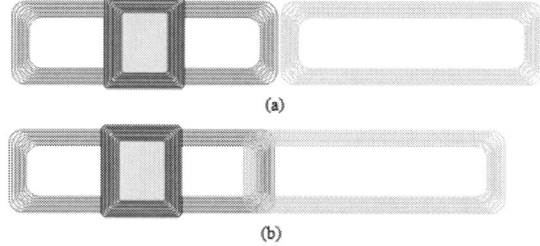

(a)

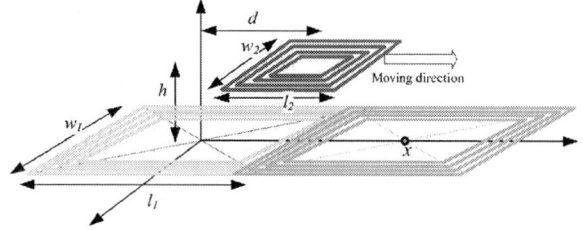

(b)

Fig.3 Coil layout method. (a) Traditional layout method. (b) Proposed partial overlap arrangement method.

The coordinate system is established according to Fig.4, and the Nie Neumann formula is applied to calculate the mutual inductance between square coils.

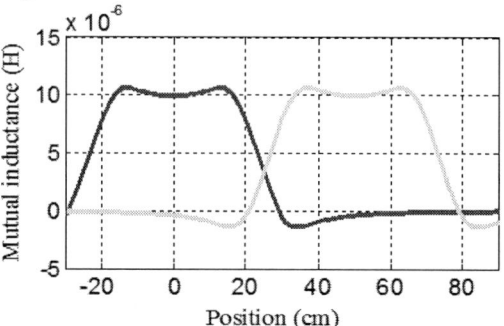

Fig.4 Coordinate system of overlap arrangement

Among them, l_1=50cm, w_1=18cm, l_2=25cm, w_2=18cm, h=5cm, and the number of turns of the primary coil is 10

turns, while the number of turns of the secondary coil is 15 turns. When two primary coils are arranged next to each other, that is, x=50cm, the calculated mutual inductance between the primary and secondary coils varies with the position of the secondary coil, as shown in the Fig.5.

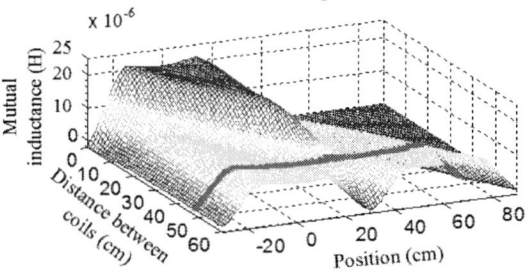

Fig.5 Calculated mutual inductance between the primary and secondary coils

It can be seen from Fig.5 that when the secondary coil is located at the junction, the mutual inductance is lower than when the secondary coil is aligned with a single primary coil. According to the equation (11), it can be ensured that the voltage received by the secondary coil is stable if the sum of mutual inductance M_{1S} and M_{2S} can approximately equal to the mutual inductance when the secondary coil is aligned with a single primary coil by adjusting the spacing between two adjacent primary coils. If the tram is fed only by a single coil at the junction, the distance between the primary coils needs to be reduced further, which will reduce the coefficient of utilization of the primary coils. When the secondary coil is located at the junction of the two coils, two adjacent coils are used to supply power simultaneously, which can ensure that the voltage at the secondary side is stable and the coil utilization is kept high. Combined with Fig.5 and equation (11), it can be seen that the spacing between two adjacent coils directly affects the fluctuation range of output voltage under double coil power supply mode. According to the calculation formula of mutual inductance, in the coordinate system shown in Fig. 4, the schematic diagram of $(M_{1S}+M_{2S})$ varying with the spacing between two adjacent primary coils and the position of the secondary coil is shown in Fig.5.

Fig.6 The schematic diagram of $(M_{1S}+M_{2S})$ varying with the spacing between two adjacent primary coils and the position of the secondary coil

3497

It can be seen from Fig.6 that there is an arrangement interval between the primary coils in dual coils power supply mode, so that $(M_{1S}+M_{2S})$ fluctuates less. The arrangement interval is $x=46\sim47$ cm if the parameters of Fig.4 is applied. As shown in Fig.4, the two primary coils do not share exactly the same plane. The orientation causes a slight difference in the distance between the primary coils and secondary coil, and the mutual inductance is not the same, as a result. A very simple solution is to bend one of the coils only where it overlaps the other coil.

Therefore, after the system parameters are determined, according to the requirements of the fluctuation range of the supply voltage of load, the arrangement spacing of the primary coils can be selected according to Fig.6. At the same time, in order to improve the coefficient of utilization of coils as much as possible, the lower limit of the coil spacing can be selected.

IV. EXPERIMENTS

In order to verify the effectiveness of the proposed partial overlap arrangement method, a low-power experiment was carried out. Experimental verification is performed using the experimental platform as shown in Fig.7. This platform adopts TMS320F28335 DSP as the control core and the primary side inverter switch is FMP65N15T2, MOSFET made by Fuji Electric's. And the system uses the parameters listed in Table 1 for experimental verification.

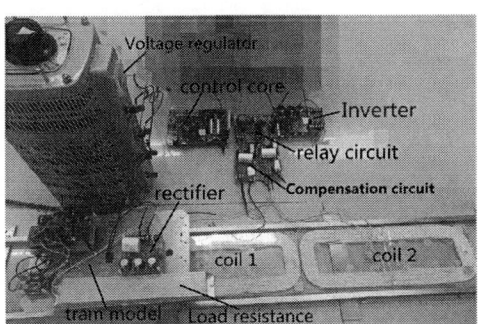

Fig.7 Experiment platform

TABLE I
EXPERIMENTAL PARAMETERS

Symbol	Meaning	Value
U_{in}	Input DC voltage	20 V
f	Frequency	50 kHz
L_{r1}/L_{r2}	Primary compensation inductance	10.0/10.0 µH
L_{P1}/L_{P2}	Primary coil inductcance	59.3/57.6 µH
R_{P1}/R_{P2}	Resistance of primary coils	0.14/0.17 Ω
L_S	Secondary coil inductcance	52.9 µH
R_S	Resistance of secondary coil	0.056 Ω
R_{Eq}	Load resistance	3.0 Ω

The expected secondary output DC voltage is 20V. According to equations (7) and (11) and Fig.5, the compensation inductance is designed to be 10 µH. And the overlap length of the adjacent primary coils is 4 cm. The coils parameters are measured by LCR Bridge KM8118. Single-coil power supply experiments, dual

coil power supply experiments under the traditional adjacent arrangement and dual coil power supply experiments with partial overlap arrangement are carried out respectively. The output voltage is shown in Fig.8.

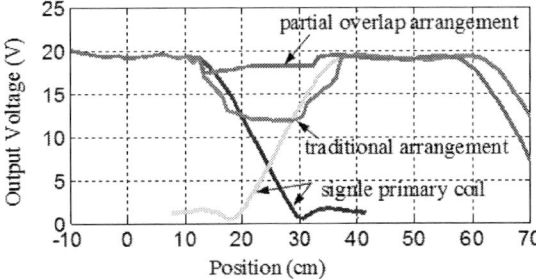

Fig.8 Output voltage

It can be seen from the experimental results that in DWPT system, the coupling coefficient between the primary and secondary coils changes with the position of the secondary coil. At the edge of the primary coils, the voltage received by the secondary coil decreases as the coupling coefficient decreases. The voltage received by the secondary can be increased when the adjacent two coils are used to supply power simultaneously at the junction of the primary coils. But, the received voltage is still lower than when the secondary coil is aligned with a single primary coil if the primary coils are placed in the traditional adjacent arrangement mode. According to the partial overlap arrangement method proposed in this paper, when the center distance of the primary coils is changed to 46 cm, that is, the overlap length of the adjacent primary coils is 4 cm, the voltage received by the secondary is relatively stable at the coils junction, as shown by the red line in Fig. 8. Then, it can be seen that the proposed partial overlap arrangement method of primary coils can well maintain the stability of the output voltage without adding a lot of extra hardware and using complex control methods.

For WPT system, transfer efficiency is a very important indicator, especially for high-power WPT system. In this experiment, PA4000 power analyzer is used to measure the coil transfer efficiency and the measured coil transfer efficiency is shown in Fig. 9.

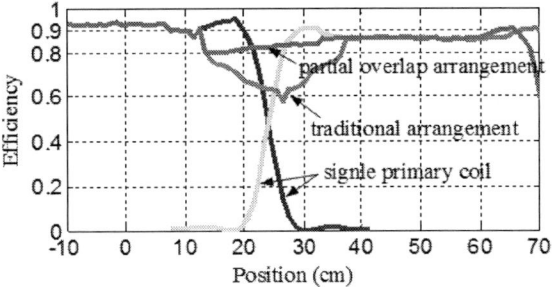

Fig.9 coil transfer efficiency

It can be seen from Fig.9 that, at the junction of the primary coils, when the traditional adjacent arrangement method is adopted, not only the received voltage is decreased, but also the transmission efficiency is reduced.

3498

The 2018 International Power Electronics Conference

With partial overlap arrangement method presented here, the transfer efficiency is kept still high, with only a slight reduction compared to when the secondary and a single primary coil are aligned, which is an additional benefit.

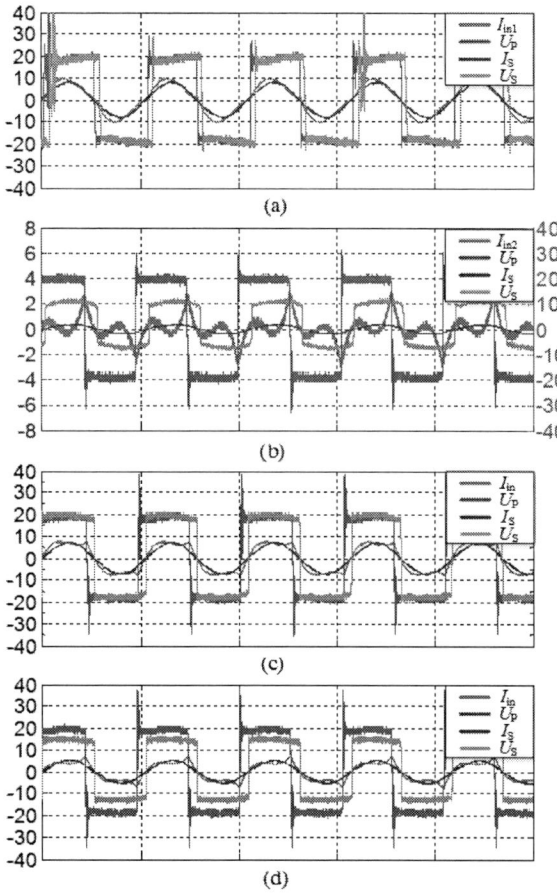

Fig.10 The waveforms of the inverter output voltage and current and received voltage and current. (a) Only the primary coil L_{P1} is powered with secondary coil at 7cm. (b) Only the primary coil L_{P2} is powered with secondary coil at 12cm. (c) Both L_{P1} and L_{P2} are powered and partially overlapped with secondary coil at 21cm. (d) Both L_{P1} and L_{P2} are powered and arranged traditionally with secondary coil at 21cm.

Fig.10 shows the waveforms of the inverter output voltage and current and the received voltage and current under different experimental conditions. Comparing Fig.10 (a) and (b), it can be seen that the inverter output voltage and current are different when the position of the secondary coil is different. When the secondary coil is aligned with a single primary coil, the system works well and the load is small for the inverter, so the loss on the inverter is larger, and as a result, the output voltage of the inverter is lower and the current harmonic content is small. When the secondary coil is longitudinally offset, the load is larger for the inverter and the loss on the inverter is reduced, resulting in an increase in the inverter output voltage but an increase in the harmonic content of the inverter output current. While the secondary coil is located at the junction, the current harmonic content is larger for each WPT system, and the inverter output

current is the superposition of the current of two WPT systems, which means that the harmonic content of the inverter output current is larger, as shown in Fig.10 (c) and (d), which is one of the causes of the experimental error. In addition, due to the internal resistance of the coils, the received voltage is not completely in phase with the inverter output voltage, as shown in the Fig.10, which is also the reason for the experimental error. However, on the whole, the experimental error does not affect the final conclusion and the experimental results validate the validity of the partial overlap arrangement method proposed in this paper.

V. CONCLUSIONS

In this paper, a partial overlap arrangement method of primary coils based on the LCC compensation topology is proposed, which can effectively solve the problem that the rectified output voltage fluctuates greatly at the junction of primary coils. The theoretical and experimental results show that when the spacing between the two adjacent primary coils are selected properly, the rectified output voltage varies very little, and the required power of the load can be ensured. The mutual inductance between the primary coils does not affect the pickup voltage of the secondary when the primary coils are overlapped partially. Besides, the coil transfer efficiency can be maintained at a high level. And the system circuit change is small, so this method has certain practical engineering value.

REFERENCES

[1] J. Shin et al., "Design and Implementation of Shaped Magnetic-Resonance-Based Wireless Power Transfer System for Roadway-Powered Moving Electric Vehicles," *IEEE Trans. on Industry Applications*, vol. 61, no. 3, pp. 1179-1192, March 2014.

[2] W. Zhang, S. C. Wong, C. K. Tse and Q. Chen, "An Optimized Track Length in Roadway Inductive Power Transfer Systems," *IEEE Journal of Emerging and Selected Topics in Power Electronics*, vol. 2, no. 3, pp. 598-608, Sept. 2014.

[3] S. Y. Choi, B. W. Gu, S. Y. Jeong and C. T. Rim, "Advances in Wireless Power Transfer Systems for Roadway-Powered Electric Vehicles," *IEEE Journal of Emerging and Selected Topics in Power Electronics*, vol. 3, no. 1, pp. 18-36, March 2015.

[4] J. Huh, S. W. Lee, W. Y. Lee, G. H. Cho and C. T. Rim, "Narrow-Width Inductive Power Transfer System for Online Electrical Vehicles," *IEEE Trans. on Power Electronics*, vol. 26, no. 12, pp. 3666-3679, Dec. 2011.

[5] C. Park, S. Lee, G. H. Cho, S. Y. Choi and C. T. Rim, "Omni-directional inductive power transfer system for mobile robots using evenly displaced multiple pick-ups," *2012 IEEE Energy Conversion Congress and Exposition (ECCE)*, Raleigh, NC, 2012, pp. 2492-2497.

[6] M. Yilmaz, V. T. Buyukdegirmenci and P. T. Krein, "General design requirements and analysis of roadbed inductive power transfer system for dynamic electric vehicle charging," *2012 IEEE Transportation Electrification Conference and Expo (ITEC)*, Dearborn, MI, 2012, pp. 1-6.

[7] K. Lee, Z. Pantic and S. M. Lukic, "Reflexive Field Containment in Dynamic Inductive Power Transfer Systems," *IEEE Trans. on Power Electronics*, vol. 29, no. 9, pp. 4592-4602, Sept. 2014.

[8] J. Shin *et al.*, "Design and Implementation of Shaped Magnetic-Resonance-Based Wireless Power Transfer System for Roadway-Powered Moving Electric Vehicles," *IEEE Trans. on Industrial Electronics*, vol. 61, no. 3, pp. 1179-1192, March 2014.

[9] C. Park, S. Lee, S. Y. Jeong, G. H. Cho and C. T. Rim, "Uniform Power I-Type Inductive Power Transfer System With DQ-Power Supply Rails for On-Line Electric Vehicles," *IEEE Trans. on Power Electronics*, vol. 30, no. 11, pp. 6446-6455, Nov. 2015.

[10] S. Raabe and G. A. Covic, "Practical design considerations for contactless power transfer quadrature pick-ups," IEEE Trans. Ind. Electron., vol. 60, no. 1, pp. 400–409, Jan. 2013.

[11] G. Elliott, S. Raabe, G. A. Covic, and J. T. Boys, "Multiphase pickups for large lateral tolerance contactless power-transfer systems," *IEEE Trans. Ind. Electron.*, vol. 57, no. 5, pp. 1590–1598, May 2010.

[12] M. Budhia, J. T. Boys, G. A. Covic, and C.-Y. Huang, "Development of a single-sided flux magnetic coupler for electric vehicle IPT charging systems," *IEEE Trans. Ind. Electron.*, vol. 60, no. 1, pp. 318–328, Jan. 2013.

[13] A. Zaheer, D. Kacprzak, and G. A. Covic, "A bipolar receiver pad in a lumped IPT system for electric vehicle charging applications," in *Proc.IEEE ECCE*, 2012, pp. 283–290.

[14] A. Zaheer, M. Budhia, D. Kacprzak, and G. A. Covic, "Magnetic design of a 300 W under-floor contactless Power Transfer system," in *Proc. 37th IEEE IECON/IECON*, 2011, pp. 1408–1413.

[15] A. Zaheer, G. A. Covic, and D. Kacprzak, "A bipolar pad in a 10-kHz 300-W distributed IPT system for AGV applications," *IEEE Trans. Ind. Electron.*, vol. 61, no. 7, pp. 3288–3301, Jul. 2014.

[16] Y Geng, B Li, Z Yang, F Lin, H Sun. "A High Efficiency Charging Strategy for a Supercapacitor Using a Wireless Power Transfer System Based on Inductor/Capacitor/Capacitor (LCC) Compensation Topology." *Energies*, vol. 10, no. 1, pp. 135, 2017.

Quick Charger for a Battery Using Modular Matrix Converter (MMxC)

Kazuma Suzuki, and Takaharu Takeshita
Electrical and Mechanical Engineering
Nagoya Institute of Technology
Gokiso, Showa, Nagoya, Japan
k.suzuki.499@nitech.jp, take@nitech.ac.jp

Abstract—The authors propose the isolated AC/DC converter using modular matrix converter as a quick charger. This converter has a high-frequency transformer instead of a commercial-frequency transformer. The volume and wight of this system become much smaller than a system using the commercial-frequency transformer. Because this converter can directly charge the battery from the input medium voltage such as 6.6 kV, a pole transformer is not required and the total system including the distribution system can be downsized. The soft-switching technique is applied to the secondary converter switching and the switching loss can be reduced. The operation principle and control method of the isolated AC/DC converter are explained. The effectiveness of the proposed circuit was verified by simulations.

Keywords—*distribution system, high-frequency transformer, isolated AC/DC converter, modular matrix converter*

I. INTRODUCTION

In resent years, the demand of the electric vehicle (EV) and plug-in hybrid vehicle (PHV) increases for CO_2 emission reduction and emergency power supply devices. Because the Ministry of Economy, Trade and Industry in Japan announced the rate of EVs and PHVs in the total of new vehicle selling will be increased to 30 % by 2030 [1], the number of EVs and PHVs will further increase. The one of the problems which must be solved for the spread of EVs and PHVs is the charging time for the battery. Therefore, the quick charger is required.

The isolated AC/DC converter for the battery charger consists of commercial transformer, three-phase AC to DC boost converter, and DC to DC buck converter. In this circuit, the volume and weight of the system become large and heavy because of the commercial transformer.

The isolated AC/DC converter with the high-frequency transformer has been proposed for the battery charger. This circuit consists of AC to DC boost converter, DC to high-frequency AC converter, high-frequency transformer, and DC to DC converter [2]-[6]. The volume and weight of the system can be reduced by using the high-frequency transformer. However, because the number of the power conversion is three times, the converter loss increases and the system efficiency has a limit. In the circuit configuration with the high-frequency transformer, many researchers have proposed the circuit systems using a matrix converter, which directly converts an AC voltage to an AC voltage, at the primary side of

the transformer. The circuit consists of matrix converter, high-frequency transformer, AC to DC converter [7]-[13]. Because the number of the power conversion is two times, the volume of the conversion part decreases and the system can be downsized. However, because these circuits [2]-[13] are intended for the low input voltage such as 200 V, the conduction loss becomes large in the case of a quick charge.

The authors propose the isolated AC/DC converter using the Modular Matrix Converter (MMxC) for the input medium voltage such as 6.6 kV. The circuit consists of MMxC, high-frequency transformer without the commercial-frequency transformer, and AC to DC converter. Because this converter can directly charge the battery from the medium input voltage, a pole transformer is not required and the total system including the distribution system can be downsized. The input and output phases of the MMxC are composed of the multiple arm connected in matrix. The one module consists of the capacitor and four switches of a H-bridge. Because the one arm of the MMxC is composed of the several modules connected in series, the MMxC has a high breakdown voltage. Because the MMxC is driven at a high voltage and a low current, the conduction loss can be suppressed and the high system efficiency can be obtained. In addition, the switching loss can be reduced using the soft-switching technique at the secondary converter switching. In this paper, the authors propose the circuit configuration using the MMxC and the control method. The effectiveness is verified by simulations.

II. CIRCUIT CONFIGURATION OF AC/DC CONVERTER

Fig.1 shows the circuit configuration of the isolated AC/DC converter using the MMxC. The source voltages $e_{su}, e_{sv},$ and e_{sw} are connected to each leg of the MMxC through the reactors L_f for suppressing the harmonics current flowing into the power source. The module of the MMxC consists of the capacitor C_m and four switches of the H-bridge. When the module capacitor voltage is v_c, the module can output three level voltages of $-v_c, 0,$ and v_c. Each arm of the MMxC consists of n modules connected in series. The small capacity reactors L_b are connected to each arm to keep the arm current continuity and prevent the short circuit between MMxC arms. The primary side terminals of the transformer are

The 2018 International Power Electronics Conference

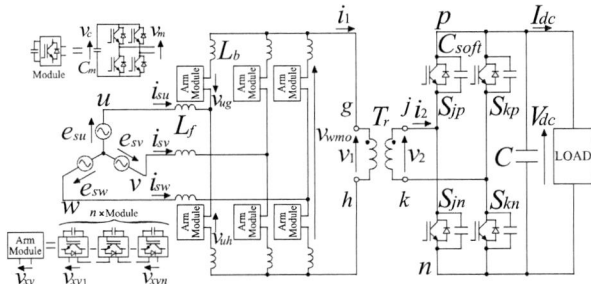

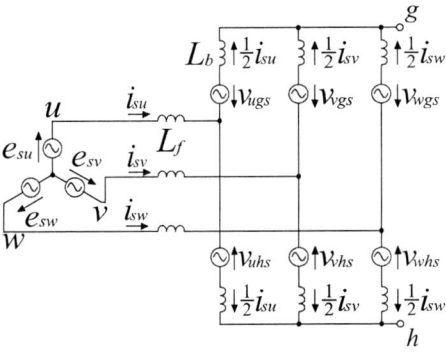

Fig. 1. Proposed circuit configuration

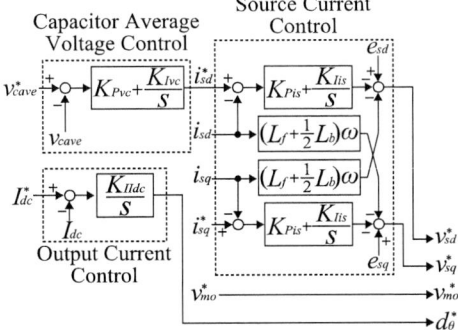

Fig. 2. Block diagram of control system

Fig. 3. Analytical model of source current control

defined as g-phase and h-phase. The g-phase and h-phase are connected to the upper and lower arm sides of the MMxC, respectively. The secondary side terminals of the transformer are defined as j-phase and k-phase. The j-phase and k-phase are connected to the H-bridge composed of four switches $S_{jp} - S_{kn}$. The secondary converter with switches $S_{jp} - S_{kn}$ generates the secondary voltage v_2. For a soft-switching, the parallel capacitors C_{soft} are connected to the secondary converter switches.

III. CONTROL METHOD

This chapter explains the source current control, output current control, and capacitor average voltage control. Fig.2 shows the block diagram of the control system. The control system of the proposed circuit consists of the source current control, output current control, and capacitor average voltage control. Source currents i_{su}, i_{sv}, and i_{sw} and the arbitrary output current I_{dc} can be controlled by source and output current controls. The average value of the MMxC module capacitor voltages v_{cave} is regulated to the reference by the capacitor average voltage control.

A. Source Current Control

Fig.3 shows the analytical model of the source current control. In this analytical model, the modules of the MMxC arms are expressed as the voltage sources $v_{ugs} - v_{whs}$. The voltage sources v_{ugs} and v_{uhs} generate the same voltage, so that the same currents of i_{su} flow into the upper and lower arms. For a simple explanation,

the authors develop the theory focusing on the g-phase. The source voltages $e_{su}, e_{sv},$ and e_{sw} are expressed by (1) using the effective value of the source line voltage E and the phase angle θ $(= \omega t)$.

$$
\begin{bmatrix} e_{su} \\ e_{sv} \\ e_{sw} \end{bmatrix} = \sqrt{\frac{2}{3}} E \begin{bmatrix} \cos\theta \\ \cos(\theta - 2\pi/3) \\ \cos(\theta - 4\pi/3) \end{bmatrix} \tag{1}
$$

The source current references $i_{su}^*, i_{sv}^*,$ and i_{sw}^* are expressed by (2) using the effective value of the source current I, the phase angle θ, and the power factor angle reference φ^*.

$$
\begin{bmatrix} i_{su}^* \\ i_{sv}^* \\ i_{sw}^* \end{bmatrix} = \sqrt{2}I \begin{bmatrix} \cos(\theta + \varphi^*) \\ \cos(\theta - 2\pi/3 + \varphi^*) \\ \cos(\theta - 4\pi/3 + \varphi^*) \end{bmatrix} \tag{2}
$$

In the closed circuit of the three phase power source and the g-phase arms, the voltage equation is obtained as follows:

$$
\begin{bmatrix} e_{su} \\ e_{sv} \\ e_{sw} \end{bmatrix} = \left(L_f + \frac{1}{2}L_b\right)\frac{d}{dt}\begin{bmatrix} i_{su} \\ i_{sv} \\ i_{sw} \end{bmatrix} + \begin{bmatrix} v_{ugs} \\ v_{vgs} \\ v_{wgs} \end{bmatrix} \tag{3}
$$

The transformation matrix C for transforming to the $d-q$ coordinates is defined as follows:

$$
C = \sqrt{\frac{2}{3}} \begin{bmatrix} \cos\theta & \cos(\theta-\frac{2}{3}\pi) & \cos(\theta-\frac{4}{3}\pi) \\ -\sin\theta & -\sin(\theta-\frac{2}{3}\pi) & -\sin(\theta-\frac{4}{3}\pi) \end{bmatrix} \tag{4}
$$

The source voltages, the source currents, and the module output voltages on the $d-q$ coordinates are expressed by (5) using the transformation matrix C and the transposed matrix T.

$$
\begin{aligned}
e_{sdq} &= \begin{bmatrix} e_{sd} & e_{sq} \end{bmatrix}^T = C\begin{bmatrix} e_{su} & e_{sv} & e_{sw} \end{bmatrix}^T \\
i_{sdq} &= \begin{bmatrix} i_{sd} & i_{sq} \end{bmatrix}^T = C\begin{bmatrix} i_{su} & i_{sv} & i_{sw} \end{bmatrix}^T \\
v_{gsdq} &= \begin{bmatrix} v_{gsd} & v_{gsq} \end{bmatrix}^T = C\begin{bmatrix} v_{ugs} & v_{vgs} & v_{wgs} \end{bmatrix}^T
\end{aligned} \tag{5}
$$

In the closed circuit of the three phase power source and the g-phase arms on the $d-q$ coordinates, the voltage equation is expressed by (6) using (3)–(5).

$$
\begin{aligned}
\begin{bmatrix} e_{sd} \\ e_{sq} \end{bmatrix} &= \begin{bmatrix} v_{gsd} \\ v_{gsq} \end{bmatrix} + \left(L_f + \frac{1}{2}L_b\right)\frac{d}{dt}\begin{bmatrix} i_{sd} \\ i_{sq} \end{bmatrix} \\
&\quad + \omega\left(L_f + \frac{1}{2}L_b\right)\begin{bmatrix} 0 & -1 \\ 1 & 0 \end{bmatrix}\begin{bmatrix} i_{sd} \\ i_{sq} \end{bmatrix}
\end{aligned} \tag{6}
$$

3502

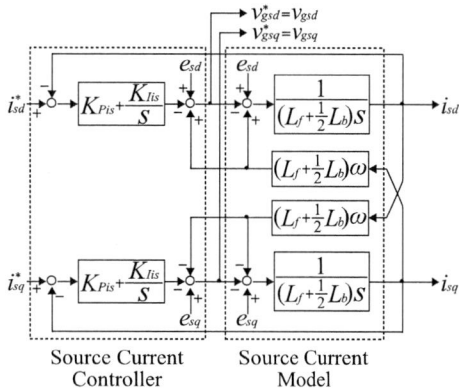

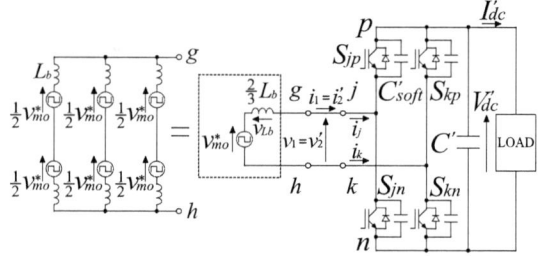

Fig. 5. Analytical model of output current control

Fig. 4. Block diagram of source current control system

The transfer function of the source current is expressed by (7) using (6).

$$
\begin{bmatrix} i_{sd} \\ i_{sq} \end{bmatrix} = \frac{1}{\left(L_f + \frac{1}{2}L_b\right)s} \left\{ \begin{bmatrix} e_{sd} \\ e_{sq} \end{bmatrix} - \begin{bmatrix} v_{gsd} \\ v_{gsq} \end{bmatrix} \right.
$$
$$
\left. - \omega\left(L_f + \frac{1}{2}L_b\right)\begin{bmatrix} 0 & -1 \\ 1 & 0 \end{bmatrix}\begin{bmatrix} i_{sd} \\ i_{sq} \end{bmatrix} \right\} \quad (7)
$$

Because of the coupling terms between d and q axes, the decoupling control system is constructed. Fig.4 shows the block diagram of the source current coltrol system. The module output voltage references v_{gsd}^* and v_{gsq}^* are given by using PI controller as follows:

$$
\begin{bmatrix} v_{gsd}^* \\ v_{gsq}^* \end{bmatrix} = - \left(K_{Pis} + \frac{K_{Iis}}{s}\right)\begin{bmatrix} i_{sd}^* - i_{sd} \\ i_{sq}^* - i_{sq} \end{bmatrix} + \begin{bmatrix} e_{sd} \\ e_{sq} \end{bmatrix}
$$
$$
- \omega\left(L_f + \frac{1}{2}L_b\right)\begin{bmatrix} 0 & -1 \\ 1 & 0 \end{bmatrix}\begin{bmatrix} i_{sd} \\ i_{sq} \end{bmatrix} \quad (8)
$$

The d axis current reference i_{sd}^* is given by the capacitor average voltage control so that the capacitor average voltage of the all modules is equal to the reference. The q axis current reference i_{sq}^* is zero for the unity source power factor.

B. Output Current Control

Fig.5 shows the analytical model of the output current control. In this analytical model, the modules of the MMxC arms are expressed as the voltage sources $\frac{1}{2}v_{mo}^*$. When the output voltage of the MMxC and the reactors L_b of each arm are viwed from the high-frequency transformer side, these are considered as the voltage source v_{mo}^* and the reactor $\frac{2}{3}L_b$. When the high-frequency transformer is assumed to be an ideal transformer at a turn ratio of $a : 1$, the secondary circuit parameters converted on the primary side are defined as follows:

$$
\begin{cases} i_2' = i_2/a, & v_2' = av_2 \\ C_{soft}' = C_{soft}/a^2, & C' = C/a^2 \\ I_{dc}' = I_{dc}/a, & V_{dc}' = aV_{dc} \end{cases} \quad (9)
$$

Fig.6 shows the switching patterns of the secondary

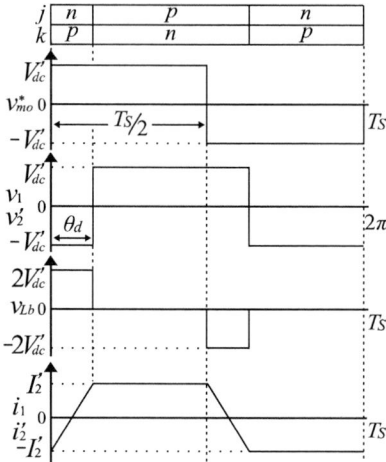

Fig. 6. Switching patterns and waveforms of output current analytical model

converter and waveforms of the output current analytical model during the one period T_s. The output voltage reference v_{mo}^* of the MMxC and the transformer voltages v_1, and v_2' are rectangular waveforms whose peak values are the output voltage V_{dc}'. When the switches S_{jp} and S_{kn} are on-state, the secondary converter outputs the positive voltage V_{dc}'. When the switches S_{jn} and S_{kp} are on-state, the secondary converter outputs the negative voltage $-V_{dc}'$. The phase-difference θ_d between the MMxC output voltage reference and the transformer voltages control the power through the transformer. The ratio of the phase-difference θ_d to the angle of π is defined as the phase-difference ratio $d_\theta = \theta_d/\pi$. When the high-frequency transformer is considered as the ideal transformer, the transformer currents i_1 and i_2' are equal to each other as shown in Fig.6.

In the half period $T_s/2$ under the output voltage $v_{mo}^* = V_{dc}'$, the reactor voltage is $v_{Lb} = 2V_{dc}'$ and the secondary current i_2' changes from $-I_2'$ at the phase zero to I_2' at the phase θ_d. When the instantaneous output power p_2 of the transformer is assumed to be same with the output power p_{out} of the load, the instantaneous output power p_2 can be formulated as follows:

$$
p_2 = V_{dc}'I_2'(1 - d_\theta) = V_{dc}'I_{dc}' = p_{out} \quad (10)
$$

The amplitude of the secondary current I_2' is expressed

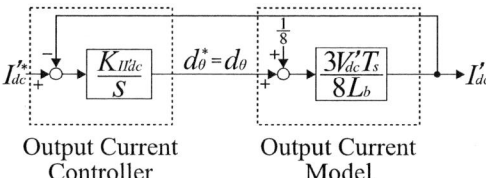

Output Current Controller   **Output Current Model**

Fig. 7. Block diagram of output current control system

by (11) using the reactor $\frac{2}{3}L_b$.

$$I_2' = \frac{3}{4L_b}\int_0^{T_s/2} v_{Lb}\, dt = \frac{3V_{dc}'d_\theta T_s}{4L_b} \quad (11)$$

The output current I_{dc}' is expressed by (12) using (10) and (11).

$$I_{dc}' = \frac{3V_{dc}'T_s}{4L_b}\left\{\frac{1}{4} - \left(d_\theta - \frac{1}{2}\right)^2\right\} \quad (12)$$

Because the range of the phase-difference ratio is $0 \le d_\theta \le 1/2$, the following equation is obtained by performing Taylor expansion around $d_\theta = 1/4$ in the formula (12).

$$I_{dc}' = \frac{3V_{dc}'T_s}{8L_b}\left(d_\theta + \frac{1}{8}\right) \quad (13)$$

Fig.7 shows the block diagram of the output current control system. The reference value of the phase-difference ratio d_θ^* is expressed by (14) using the integration gain $K_{II'dc}$, the output current reference $I_{dc}'^*$, and the detection value I_{dc}'.

$$d_\theta^* = \frac{K_{II'dc}}{s}(I_{dc}'^* - I_{dc}') \quad (14)$$

C. Capacitor Average Voltage Control

The capacitor average voltage v_{cave} of all modules is expressed by (15) using the capacitor voltages of the each module.

$$\begin{cases} v_{cave} = \dfrac{1}{6n}\displaystyle\sum_{x,y,z} v_{cxyz} \\ x = \{u,v,w\}, y = \{g,h\}, z = \{1,2,\dots,n\} \end{cases} \quad (15)$$

The relation between the energy of the all module capacitors and the input-output power energy is obtained as follows:

$$\begin{aligned} 3nC_m v_{cave}^2 &= \int(\sqrt{3}EI\cos\varphi^* - V_{dc}'I_{dc}')dt \\ &= \int Ei_{sd}\,dt - \int V_{dc}'I_{dc}'\,dt \end{aligned} \quad (16)$$

In the case of controling the capacitor average voltage v_{cave} around the capacitor voltage value V_c, the term of v_{cave}^2 is processed by Taylor expansion around $v_{cave} = V_c$ and the following equation is obtained.

$$v_{cave}^2 \simeq V_c^2 + 2V_c(v_{cave} - V_c) = 2V_c v_{cave} - V_c^2 \quad (17)$$

The transfer function of the capacitor average voltage v_{cave} is expressed by (18) using (16) and (17).

$$v_{cave} = \frac{1}{2}V_c + \frac{1}{6nC_m V_c s}\left(Ei_{sd} - V_{dc}'I_{dc}'\right) \quad (18)$$

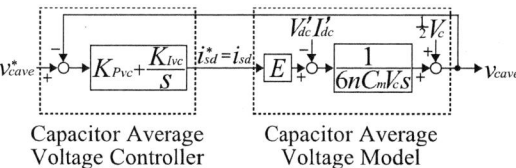

Capacitor Average Voltage Controller   **Capacitor Average Voltage Model**

Fig. 8. Block diagram of capacitor average voltage control system

The capacitor average voltage v_{cave} can be controled by the effective current component i_{sd} of the formula (18). Fig.8 shows the block diagram of the capacitor average control system. The reference value of the effective current i_{sd}^* is expressed by (19) using the proportional gain K_{Pvc}, the integration gain K_{Ivc}, the capacitor average voltage reference v_{cave}^*, and the detection value v_{cave}.

$$i_{sd}^* = \left(K_{Pvc} + \frac{K_{Ivc}}{s}\right)(v_{cave}^* - v_{cave}) \quad (19)$$

D. Module Output Voltage and Capacitor Avelage Voltage

Because the MMxC module generates the total voltage of the source and output current controls, the each module output voltage is obtained as follows:

$$\begin{cases} v_{xgz} = \dfrac{1}{n}\left(v_{xgs} - \dfrac{1}{2}v_{mo}\right) \\ v_{xhz} = \dfrac{1}{n}\left(v_{xhs} + \dfrac{1}{2}v_{mo}\right) \\ x = \{u,v,w\}, z = \{1,2,\dots,n\} \end{cases} \quad (20)$$

The module capacitor voltage must be set higher than the total voltage of the source and output current controls. When the voltage drops through the reactors L_f and L_b by the source current control is assumed to be sufficiently small compared with the source voltages, the condition to be satisfied by the module capacitor voltage V_c is obtained as follows:

$$V_c \ge \frac{1}{n}\left(\sqrt{\frac{2}{3}}E + \frac{1}{2}V_{dc}'\right) \quad (21)$$

IV. SOFT-SWITCHING

A. Soft-Switching Conditions of Secondary Converter

The soft-switching conditions of the secondary converter are determined by the sign of the output currents i_j and i_k. The soft-switching conditions at the one-period T_s are given as follows:

$$\begin{cases} S_{jn} \to S_{jp} & ; & i_j = i_2' > 0 \\ S_{jp} \to S_{jn} & ; & i_j = i_2' < 0 \\ S_{kn} \to S_{kp} & ; & i_k = -i_2' > 0 \\ S_{kp} \to S_{kn} & ; & i_k = -i_2' < 0 \end{cases} \quad (22)$$

The soft-switching from the negative-side switch to the positive-side switch such as S_{jn} to S_{jp} can be realized under the positive output currents i_j, $i_k > 0$. The soft-switching from the positive-side switch to the negative-side switch such as S_{jp} to S_{jn} can be realized under the negative output currents i_j, $i_k < 0$.

3504

The 2018 International Power Electronics Conference

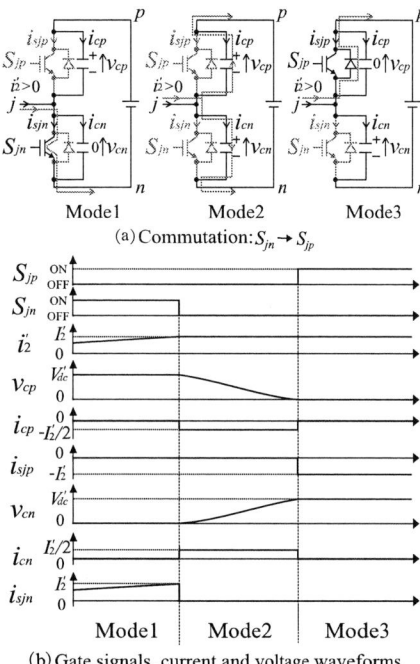

(a) Commutation: $S_{jn} \rightarrow S_{jp}$

(b) Gate signals, current and voltage waveforms

Fig. 9.   Commutation of switch S_{jn} to S_{jp}

B. Soft-Switching Commutations of Secondary Converter

Fig.9 shows the commutation of the negative-side switch S_{jn} to the positive-side switch S_{jp} during the positive half-period $T_s/2$ in Fig.6. Fig.9 (a) shows the commutation of the switch S_{jn} to the switch S_{jp} under the positive secondary current $i_2' > 0$. Fig.9 (b) shows the gate signal, current, and voltage waveforms. In Mode1, the secondary current i_2' flows into the switch S_{jn} and the parallel capacitor voltage of the switch S_{jn} is zero. The parallel capacitor of the off-state switch S_{jp} has the negative charge at the transformer side. When the switch S_{jn} is turned off, Mode1 changes to Mode2 and the secondary current i_2' flows into two capacitors. Because the voltage of the switch S_{jn} gradually changes by the parallel capacitor, the turn-off loss of the switch S_{jn} can be suppressed. When the parallel capacitor of the switch S_{jp} finishes discharging, Mode2 changes to Mode3 and the secondary current i_2' flows into the diode of the switch S_{jp}. The on-signal is given to the switch S_{jp} while the secondary current i_2' flows into the diode of the switch S_{jp}. When the sign of the secondary current i_2' changes from positive to negative, the switch S_{jp} becomes on-state and the secondary current i_2' flows into the switch S_{jp}. Because the parallel capacitor voltage of the switch S_{jp} is zero before and after changing the sign of the secondary current i_2', the soft-switching can be achieved at the switch S_{jp}.

V.   SIMULATION RESULTS

A. System Configuration

Tab.I shows the specification of the simulation system and Fig.10 shows the system configuration. Because the

TABLE I.   SPECIFICATION OF SIMULATION SYSTEM

Source voltage E, ω	200V , $2\,\pi \times 60$ rad/s
Power factor angle φ^*	0 rad
Input filter L_f, L_b	3.0 mH , 0.6 mH
Number of series modules n	3
Module capacitors C_m	300 μF
Module capacitor voltages V_c	100 V
Turn ratio of transformer a	1
Frequency of transformer $1/T_s$	10 kHz
Output capacitor C	1500 μF
Output DC voltage V_{dc}	200 V
Output power P_{out}	1200 W
Capacitor C_{soft}	3 nF

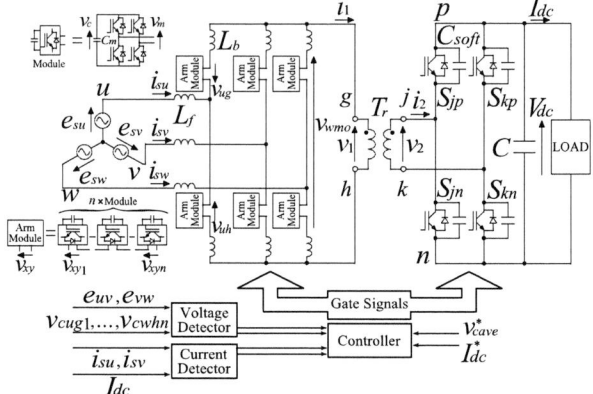

Fig. 10.   System configuration

low input voltage such as 200 V will be used in our laboratory, the effective value of the source line voltage E of 200 V and the frequency of 60 Hz are used in this simulations. The number of series module n is 3 and the arm of the MMxC consists of the three modules connected in series. The output voltage V_{dc} is 200 V and the output power P_{out} is 1200 W. The turn ratio of the high-frequency transformer a is 1 and the frequency of the transformer waveforms are 10 kHz. The duty cycles of the each switch are calculated by detecting the source line voltages e_{uv}, e_{vw}, the source currents i_{su}, i_{sv}, the module capacitor voltages $v_{cug1} - v_{cwh3}$, and the output DC current I_{dc}. The switching signals are generated based on the duty cycles. Because the module capacitor voltage needs to be more than 87.8 V according to the formula (3), the capacitor average voltage reference is set as $v_{cave}^* = V_c = 100$ V in this simulations.

B. Simulation waveforms

Fig.11 shows the simulation waveforms under changing from 600 W to 1200 W of the output power P_{out}. The waveforms are the source voltage e_{su}, the source current i_{su}, the g-phase module output voltage v_{ug}, the module capacitor voltages v_{cug1}, v_{cvg1}, v_{cwg1}, the module output voltage v_{mo} viewed from the transformer side, the transformer voltage v_2, the transformer current i_2, the output DC voltage V_{dc}, and the output DC current I_{dc}. The source current i_{su} can be formed into the sinusoidal waveform. When the output power P_{out} changes from 600 W to 1200 W, the module capacitor voltages v_{cug1},

3505

The 2018 International Power Electronics Conference

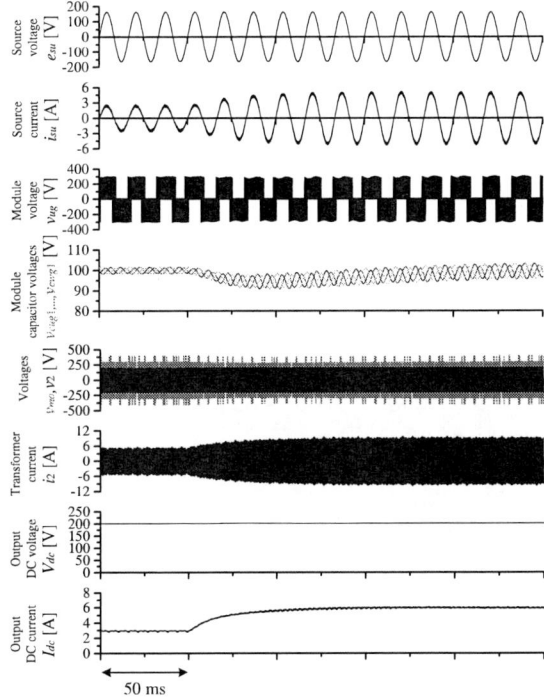

Fig. 11. Simulation waveforms

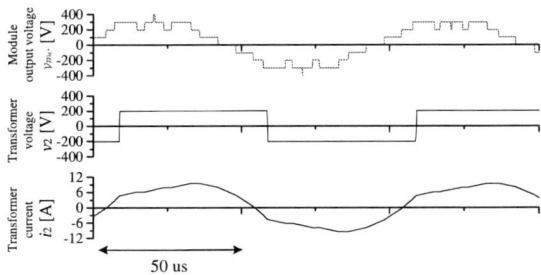

Fig. 12. Magnification waveforms at 1200 W

v_{cvg1}, and v_{cwg1} can be regulated to the reference. Fig.12 shows the magnification waveforms at 1200 W. The waveforms are the module output voltage v_{mo} viewed from the transformer, the transformer voltage v_2, and the transformer current i_2. The each waveform frequency of 10 kHz can be obtained. In the secondary converter, when the transformer voltage v_2 changes higher voltage, the sign of the transformer current i_2 is positive. When the transformer voltage v_2 changes lower voltage, the sign of the transformer current i_2 is negative. The current and voltage waveforms of the transformer i_2 and v_2 satisfy the soft-switching conditions of equation (22).

VI. Conclusion

In this paper, the authors have proposed the circuit configuration and the control method using the MMxC for charging the battery from the input medium voltage such as 6.6 kV. The effectiveness of the proposed circuit and control method was verified by the simulations.

ACKNOWLEDGEMENT

This work was supported by Council for Science, Technology and Innovation (CSTI), Cross-ministerial Strategic Innovation Promotion Program (SIP), "Next-generation power electronics" (funding agency: NEDO).

REFERENCES

[1] Electric Vehicle, Advanced Technology and ITS Promotion Office, Automobile Division, Manufacturing Industries Bureau, "Compilation of the Road Map for EVs and PHVs toward the Dissemination of Electric Vehicles and Plug-in Hybrid Vehicles", Electric Vehicle, Advanced Technology and ITS Promotion Office, Automobile Division, Manufacturing Industries Bureau. Ministry of Economy. Japan (2016)

[2] S. Inoue and H. Akagi, "A Bi-Directional DC/DC Converter for an Energy Storage System", *IEEE Trans. Power Electron.*, Vol.22, No.6, pp.2299-2306 (2007)

[3] Z. Hhang, H. Xu, L. Shi, D. Li and Y. Han, "A Unit Power Factor DC Fast Charger for Electric Vehicle Charging Station", *7th International Power Electronics and Motion Control Conference*, pp.411-415 (2012)

[4] S. Ebrahimi, M. Taghavi, F. Tahami, and H. Oraee, "Integrated Bidirectional Isolated Soft-Switched Battery Charger for Vehicle-to-Grid Technology Using 4-Switch 3Φ-Rectifier", *39th Annual Conference of the IEEE Industrial Electronics Society (IECON)*, pp.906-911 (2013)

[5] R. Huang and K. Mazumder, "A Soft-Switching Scheme for an Isolated DC/DC Converter with Pulsating DC Output for a Three-Phase High-Frequency-Link PWM Converter", *IEEE Trans. Power Electron.*, Vol.24, No.10, pp.2276-2288 (2009)

[6] K. Sekimori, S. Kuroda, Y. Kawabata, and T. Kawabata, "New control method of bidirectional isolated dc/dc converter – Control system which can obtain output voltage much higher than winding ratio of transformer –", *17th European Conference on Power Electronics and Applications (EPE'15 ECCE-Europe)*, pp.1-10 (2015)

[7] S. Manias and P. D. Ziogas, "A Novel Sinewave in AC to DC Converter with High-Frequency Transformer Isolation", *IEEE Trans. Ind. Electron.*, Vol.32, No.4, pp.430-438 (1985)

[8] K. Inagaki, T. Furuhashi, A. Ishiguro, M. Ishida, S. Okuma and Y. Uchikawa, "A Waveform Control Method of AC to DC Converters with High-Frequency Links", *IEEJ Trans. IA*, Vol.110, No.5, pp.525-533 (1990)

[9] Y. Ohnuma and J. Itoh: "Experimental Verification of a 50kVA, 125A High-frequency AC to DC Power Converter Using a Three-phase to Single-phase Matrix Converter", *IEEJ JIASC-2011*, No.1-80, pp.403-406 (2011)

[10] C. Li, Y. Yulin, and D. Xu, "Soft-Switching Three-Phase Matrix based Isolated AC-DC Converter for DC Distribution System", *IEEE Energy Conversion Congress and Exposition (ECCE)*, pp.6755-6761 (2015)

[11] X. Yu, F. Jin, and M. Wang, "A Novel Soft-switching Modulation Scheme for Isolated DC-to-three-phase-AC Matrix-based Converter Using SiC Device", *IEEE Energy Conversion Congress and Exposition (ECCE)*, pp.1-8 (2016)

[12] D. Varajao, L. M. Miranda, R. E. Araujo, and J. P. Lopes, "Power Transformer for a Single-stage Bidirectional and Isolated AC-DC Matrix Converter for Energy Storage Systems", *42nd Annual Conference of the IEEE Industrial Electronics Society (IECON2016)*, pp.1149-1155 (2016)

[13] K. Suzuki, J. Isozaki, W. Kitagawa and T. Takeshita, "Isolated AC/DC Converter Using Soft-Switching Technique", *IEEJ Transactions on Industry Applications*, Vol.136, No.8, pp.540-548 (2016)

Variable Output Voltage Control of an Isolated Bi-directional AC/DC Converter with a Soft-Switching Technique

Takumi Hamaguchi, Kazuma Suzuki, Wataru Kitagawa, and Takaharu Takeshita

Nagoya Institute of Technology

Gokiso, Showa, Nagoya, Japan

ckb14095@stn.nitech.ac.jp, 28513004@stn.nitech.ac.jp, kitagawa.wataru@nitech.ac.jp, take@nitech.ac.jp

Abstract—This paper presents the variable output voltage control method of an isolated bi-directional AC/DC converter. The authors propose the PWM strategy for the both sides converters responding to the output voltage variation. In addition, the proposed method can apply the soft-switching technique. Finally, the experimantal results verify the effectiveness of the control method.

Keywords—AC/DC power converters, soft-switching technique, high-frequency transformer, converter control

I. INTRODUCTION

In recent years, the renewable energy of the photovoltaic and wind turbine generation systems has increased [1]. The output power of these generation systems fluctuates by the weather conditions. The energy storage devices such as batteries and EDLCs are needed for the stable power supply. Also, EVs and PEVs which need the battery systems have increased. Because the battery systems are DC voltage and the distribution systems are AC voltage, the bi-directional AC/DC converter is required for the grid connection. The isolation between AC and DC sides is important to protect the devices. In addition, because these converters are needed to be high-frequency for downsizing the systems, the isolated AC/DC converters using high-frequency transformer have been proposed [2]-[9].

The isolated AC/DC converters, which consists of the boost AC to DC converter, the inverter, the high-frequency transformer, and the AC to DC converter, have been proposed [2]-[5]. Because of three-time power conversions, they have a limit of a circuit improvement for the size and effeciency. For improving these problems, the two-stage AC/DC converters using matrix converter have been proposed [6],[7]. Because of two-time conversions, they have better performance compared with the three-time conversion systems. However, because these systems do not have the ability of the soft-switching, the high switching losses can be generated.

The authors have proposed the circuit configuration and the control method of the isolated bi-directional AC/DC converter [8]-[9]. This AC/DC converter can achieve the soft-switching commutations of all switches without the auxiliary circuit. Because the soft-switching tequnique can reduce the switching losses, the high system efficiency can be obtained under the high switching frequency. However, the authors have mentioned the control method under only constant output voltage. The output voltage usually changes according to the state of charge of the battery. Therefore, this paper presents the variable output voltage control which applys the soft-switching technique to reduce the switching losses. The effectiveness of the proposed method was verified by experiments.

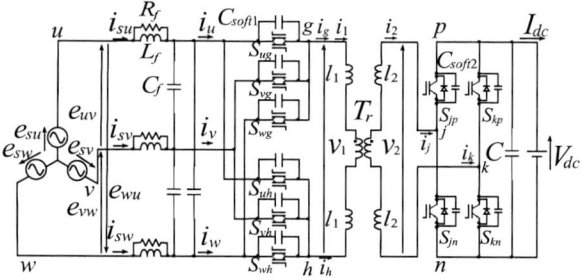

Fig. 1. Proposed circuit configuration

II. CIRCUIT CONFIGURATION AND TRANSFORMER WAVEFORMS

A. Main Circuit Configuration

Fig.1 shows the main circuit configuration. The source voltages e_{su}, e_{sv}, and e_{sw} are connected to the matrix converter through the LC filter. The LC filter, which consists of the reactors L_f, the capacitors C_f, and the resistors R_f, prevents the harmonics current from flowing into the power source. The primary voltage v_1 is generated by the switching of the matrix converter. The matrix converter is composed of the six bi-directional switches $S_{ug} - S_{wh}$. The reactors l_1 and l_2 are connected in series to the both sides of the high-frequency transformer to suppress the changing rate of the currents i_1 and i_2. The secondary voltage v_2 is generated by the switching of the secondary converter. The secondary converter is composed of the four switches $S_{jp} - S_{kn}$. When the turn ratio of the transformer is $a : 1$, the secondary side circuit parameters converted on the primary side are defined as follows:

$$
\begin{cases}
l_2' = a^2 l_2 & , \quad C_{soft2}' = C_{soft2}/a^2 \\
i_2' = i_2/a & , \quad v_2' = av_2 \\
V_{dc}' = aV_{dc}
\end{cases}
\tag{1}
$$

The 2018 International Power Electronics Conference

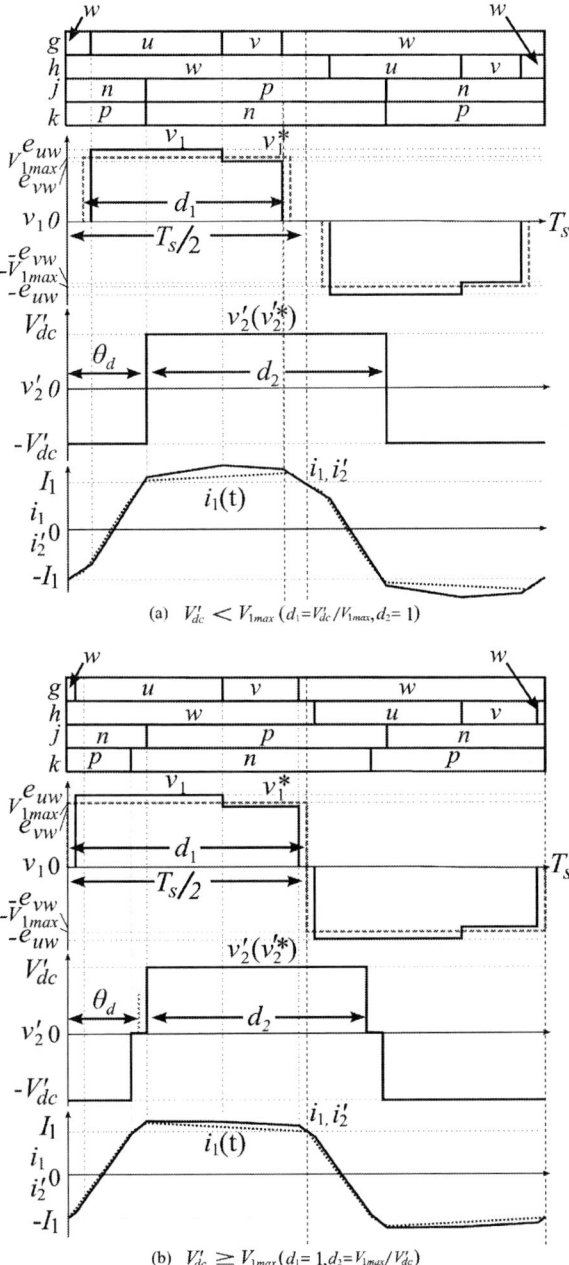

(a) $V'_{dc} < V_{1max}$ ($d_1 = V'_{dc}/V_{1max}, d_2 = 1$)

(b) $V'_{dc} \geq V_{1max}$ ($d_1 = 1, d_2 = V_{1max}/V'_{dc}$)

Fig. 2. Current and voltage waveforms of transformer

B. Transformer Waveforms

Fig.2 shows the switching patterns for both sides converters and the transformer waveforms during one period T_s of the high-frequency transformer under the current and voltage states $i_u^* > i_v^* > 0 > i_w^*$, $e_{su} > e_{sv} > e_{sw}$. Because the matrix converter generates the primary voltage v_1 using the source line voltages, the amplitude of the primary voltage v_1 has a limit. When the maximum output voltage of the matrix converter in the whole phase angle of the source voltage is defined as V_{1max}, the maximum primary voltage V_{1max} is expressed as $V_{1max} = \sqrt{6}E/2$ using the effective value of the source line voltage E. The average value of the primary

voltage v_1 and the secondary voltage v_2 during the half period $T_s/2$ are controlled to be equal because this condition is the most easily to apply the soft-switching technique. Therefore, the PWM strategy is applied to both sides converters of the transformer according to the amplitude relationship between maximum primary voltage V_{1max} and the output voltage V'_{dc}. Fig.2 (a) is under the amplitude relationship of $V'_{dc} < V_{1max}$ and Fig.2 (b) is under the amplitude relationship of $V_{dc}' \geq V_{1max}$. The primary voltage reference v_1^* consists of the three level voltages $-V_{1max}$, 0, and V_{1max}. The secondary voltage reference $v_2^{'*}$ consists of the three level voltages $-V'_{dc}$, 0, and V'_{dc}. The ratios d_1 and d_2 are defined as the time ratios which the voltage references v_1^* and $v_2^{'*}$ outputs non-zero voltage during the half period $T_s/2$. The phase-difference θ_d between the voltage references v_1^* and $v_2^{'*}$ is controlled for the power through the transformer. The ratio of the phase-difference θ_d to the angle of π is defined as the phase-difference ratio $d_\theta = \theta_d/\pi$. The average values of the primary voltage v_1 and the reference v_1^* during a half period $T_s/2$ are controlled to be the same value by the switching of the matrix converter. The swicthing patterns of the secondary converter are $180°$ conduction type. The secondary voltage v_2' is controlled as equal to the secondary voltage reference $v_2^{'*}$ by the switching of the secondary converter. In Fig.2 (a), the ratio d_2 is controlled as maximum of 1 and the ratio d_1 is controlled as V'_{dc}/V_{1max}. Because of the ratio d_2 of 1, the secondary voltage reference $v_2^{'*}$ consists of the two level voltages V'_{dc} and $-V'_{dc}$. In Fig.2 (b), the ratio d_1 is controlled as maximum of 1 and the ratio d_2 is controlled as V_{1max}/V'_{dc}. Because of the ratio d_1 of 1, the primary voltage reference v_1^* consists of the two level voltages V_{1max} and $-V_{1max}$.

C. Power through High-Frequengcy Transformer

In Fig.2, the initial value of the primary current I_1 is expressed by (2) using the series reactors l_1, l_2'.

$$
I_1 = \frac{1}{4(l_1 + l_2')} \int_0^{T_s/2} (v_1^* - v_2^{'*}) \, dt
$$
$$
= \frac{V'_{dc} T_s}{4(l_1 + l_2')}\left(d_\theta - \frac{1 - d_2}{2}\right) \tag{2}
$$

The primary current $i_1(t)$ is defined by (3) using the series reactors l_1, l_2', the voltage references v_1^*, and $v_2^{'*}$.

$$
i_1(t) = \frac{1}{2(l_1 + l_2')} \int (v_1^* - v_2^{'*}) \, dt - I_1 \tag{3}
$$

The instantaneous power of the transformer p_1 during the half period $T_s/2$ is expressed by (4) using the primary voltage reference v_1^* and the primary current $i_1(t)$.

$$
p_1 = \frac{2}{T_s} \int_0^{T_s/2} v_1^* i_1(t) \, dt
$$
$$
= \frac{V_{1max} V'_{dc} T_s}{16(l_1 + l_2')}\{1 - (d_1 - 1)^2
$$
$$
- (d_2 - 1)^2 - 4(d_\theta - \frac{1}{2})^2\} \tag{4}
$$

Because the instantaneous power p_1 is inversely proportional to the sum of the reactors l_1 and l_2', the arbitrary

3508

output power can be obtained by appropriately selecting the series reactors.

III. SOFT-SWITCHING CONDITIONS

A. Soft-Switching Conditions of Primary Converter

In a positive half period $T_s/2$ of the primary voltage v_1 under the current and voltage states of $i_u^* > i_v^* > 0 > i_w^*$ and $e_{su} > e_{sv} > e_{sw}$, the minimum-phase switch S_{wh} is always on-state in the h-phase. The on-state switch of g-phase changes in the order of minimum-phase switch S_{wg}, maximum-phase switch S_{ug}, middle-phase switch S_{vg}, and minimum-phase switch S_{wg}. In a negative half period $T_s/2$ of the primary voltage v_1, the minimum-phase switch S_{wg} is always on-state in g-phase. The on-state switch of h-phase changes in the order of minimum-phase switch S_{wh}, maximum-phase switch S_{uh}, middle-phase switch S_{vh}, and minimum-phase switch S_{wh}. The soft-switching conditions for the six commutations of the primary converter in one period T_s are given as follows:

$$
\begin{cases}
S_{wg} \rightarrow S_{ug}, & i_g = i_1 < 0 \\
S_{ug} \rightarrow S_{vg}, & i_g = i_1 > 0 \\
S_{vg} \rightarrow S_{wg}, & i_g = i_1 > 0 \\
S_{wh} \rightarrow S_{uh}, & i_h = -i_1 < 0 \\
S_{uh} \rightarrow S_{vh}, & i_h = -i_1 > 0 \\
S_{vh} \rightarrow S_{wh}, & i_h = -i_1 > 0
\end{cases}
\tag{5}
$$

The output currents i_g and i_h must be negative during the commutations from the low-voltage phase to the high-voltage phase, such as the commutation from S_{wg} to S_{ug}. The output currents i_g and i_h must be positive during the commutations from the high-voltage phase to the low-voltage phase, such as the commutation from S_{ug} to S_{vg}.

B. Soft-Switching Conditions of Secondary Converter

The soft-switching conditions for the four commutations of the secondary converter in one period T_s are given as follows:

$$
\begin{cases}
S_{jp} \rightarrow S_{jn}, & i_j = i_2 < 0 \\
S_{jn} \rightarrow S_{jp}, & i_j = i_2 > 0 \\
S_{kp} \rightarrow S_{kn}, & i_k = -i_2 < 0 \\
S_{kn} \rightarrow S_{kp}, & i_k = -i_2 > 0
\end{cases}
\tag{6}
$$

The output currents i_j and i_k must be positive during the commutations from the low-voltage phase to the high-voltage phase, such as the commutation from S_{jn} to S_{jp}. The output currents i_j and i_k must be negative during the commutations from the high-voltage phase to the low-voltage phase, such as the commutation from S_{jp} to S_{jn}.

IV. GENERATION OF THE TRANSFORMER WAVEFORMS

A. Duty Cycles of Primary Converter

Because the primary current i_1 must be continuous, the duty cycles $d_{ug} - d_{wh}$ of the matrix converter switches can be formulated based on the following constraints:

$$
d_{ug} + d_{vg} + d_{wg} = 1 \tag{7}
$$
$$
d_{uh} + d_{vh} + d_{wh} = 1 \tag{8}
$$

In the positive half period $T_s/2$, the average value of the primary voltage v_1 can be expressed as follows:

$$
d_1 V_{1max} = (d_{ug} - d_{uh})e_{su} + (d_{vg} - d_{vh})e_{sv} + (d_{wg} - d_{wh})e_{sw} \tag{9}
$$

The matrix converter can generate the seven levels as the primary voltage v_1 and these voltages are expressed as follows:

$$
e_{uw} > e_{vw} > e_{uv} > 0 > -e_{uv} > -e_{vw} > -e_{uw} \tag{10}
$$

In the positive half period $T_s/2$ under the current and voltage states $i_u^* > i_v^* > 0 > i_w^*$, $e_{su} > e_{sv} > e_{sw}$, the three level voltages e_{uw}, e_{vw}, and 0 are used. Because the w-phase switch S_{wh} is always on-state, the duty cycles d_{uh}, d_{vh}, and d_{wh} of the h-phase switches can be given as follows:

$$
d_{uh} = 0, \quad d_{vh} = 0, \quad d_{wh} = 1 \tag{11}
$$

In Fig.2, because the primary current i_1 flows into the u-phase while the switch S_{ug} is on-state, the average value $\bar{i}_u$ of the input current i_u is expressed by (12).

$$
\begin{aligned}
\bar{i}_u &= \frac{2}{T_s} \int_{\frac{T_s}{4}d_{wg}}^{\frac{T_s}{2}(\frac{1}{2}d_{wg}+d_{ug})} i_1(t)\, dt \\
&= \frac{T_s}{8(l_1 + l_2')}(A_u d_{ug}^2 + B_u d_{ug} + C_u) \tag{12}
\end{aligned}
$$

The parameters A_u, B_u, and C_u are given as follows:

$$
\begin{cases}
A_u = (1 + \dfrac{e_{uv}}{e_{vw}})V_{1max} - \dfrac{1}{2}\{(1 + \dfrac{e_{uv}}{e_{vw}})^2 + 1\}V_{dc}' \\
B_u = \{\dfrac{d_1 V_{1max}V_{dc}'}{e_{vw}} + (2d_\theta - 1)V_{dc}'\}\dfrac{e_{uv}}{e_{vw}} \\
\qquad + \dfrac{d_1 V_{1max}}{e_{vw}}(V_{dc}' - V_{1max}) + 2d_\theta V_{dc}' \\
C_u = \{-2d_\theta^2 + 2(1 - \dfrac{d_1 V_{1max}}{e_{vw}})d_\theta \\
\qquad - \dfrac{1}{2}(1 - d_2)^2 - \dfrac{1}{2}(1 - \dfrac{d_1 V_{1max}}{e_{vw}})^2\}V_{dc}'
\end{cases}
\tag{13}
$$

The average value $\bar{i}_v$ of the input current i_v is expressed by (14).

$$
\begin{aligned}
\bar{i}_v &= \frac{2}{T_s} \int_{\frac{T_s}{2}(\frac{1}{2}d_{wg}+d_{ug})}^{\frac{T_s}{2}(1-\frac{1}{2}d_{wg})} i_{1t}\, dt \\
&= \frac{T_s}{8(l_1 + l_2')}(A_v d_{ug}^2 + B_v d_{ug} + C_v) \tag{14}
\end{aligned}
$$

The parameters A_v, B_v, and C_v are given as follows:

$$
\begin{cases}
A_v = (V_{dc}' - V_{1max})(1 + \dfrac{e_{uv}}{e_{vw}}) \\
B_v = \dfrac{d_1 V_{1max}^2}{e_{vw}} - (\dfrac{d_1 V_{1max}}{e_{vw}} + \dfrac{2e_{uw}}{e_{vw}}d_\theta)V_{dc}' \\
C_v = \dfrac{2d_1 V_{1max}}{e_{vw}}d_\theta V_{dc}'
\end{cases}
\tag{15}
$$

For the unity input power factor, the relation between the input current references and the average values is formulated based on the following constraint:

$$
\frac{\bar{i}_v}{\bar{i}_u} = \frac{i_v^*}{i_u^*} \tag{16}
$$

The duty cycles d_{ug}, d_{vg}, and d_{wg} are expressed by (17) − (19) using (7) − (9) and (12) − (16).

$$d_{ug} = \frac{B_{uv} + \sqrt{B_{uv}^2 - 4A_{uv}C_{uv}}}{2A_{uv}} \quad (17)$$

$$d_{vg} = \frac{d_1 V_{1max}}{e_{vw}} - \frac{e_{uw}}{e_{vw}} d_{ug} \quad (18)$$

$$d_{wg} = 1 - \frac{d_1 V_{1max}}{e_{vw}} + \frac{e_{uv}}{e_{vw}} d_{ug} \quad (19)$$

The parameters A_{uv}, B_{uv}, and C_{uv} are given as follows:

$$\begin{cases} A_{uv} = A_u i_v^* - A_v i_u^* \\ B_{uv} = B_v i_u^* - B_u i_v^* \\ C_{uv} = C_u i_v^* - C_v i_u^* \end{cases} \quad (20)$$

In the negative half period $T_s/2$, the three level voltages $-e_{uw}$, $-e_{vw}$, and 0 are used. The duty cycles in the negative half period can be obtained by exchanging the g-phase duty cycles with the h-phase duty cycles in the positive half period. The duty cycles $d_{ug} - d_{wh}$ in the negative half period are expressed by (21) − (24).

$$d_{ug} = 0, \ d_{vg} = 0, \ d_{wg} = 1 \quad (21)$$

$$d_{uh} = \frac{B_{uv} + \sqrt{B_{uv}^2 - 4A_{uv}C_{uv}}}{2A_{uv}} \quad (22)$$

$$d_{vh} = \frac{d_1 V_{1max}}{e_{vw}} - \frac{e_{uw}}{e_{vw}} d_{ug} \quad (23)$$

$$d_{wh} = 1 - \frac{d_1 V_{1max}}{e_{vw}} + \frac{e_{uv}}{e_{vw}} d_{ug} \quad (24)$$

B. Duty Cycles of Secondary Converter

The secondary converter can generate the three level voltages $-V_{dc}'$, 0, and V_{dc}' as the secondary voltage v_2'. The duty cycles of the secondary converter switches $S_{jp} - S_{kn}$ are 0.5. For the positive voltage V_{dc}', the switches S_{jp} and S_{kn} must be on-state. For the negative voltage $-V_{dc}'$, the switches S_{jn} and S_{kp} must be on-state. For the zero voltage, the switches S_{jp} and S_{kp} or S_{jn} and S_{kn} must be on-state. The ratio d_2 is controlled by shifting the switching timings of the j-phase with respect to the k-phase. In addition, the phase of the secondary voltage v_2' is delayed by the phase-difference θ_d with the primary voltage reference v_1^*.

C. Gate Signals Generation

Fig.3 shows the comparison lines with the sawtooth waveform, gate signals of all switches, the input current i_u, and the transformer voltages v_1, v_2 during one cycle of the source voltage. Fig.3 (a) shows those under the amplitude relationship of $V_{dc}' < V_{1max}$. Fig.3 (b) shows those under the amplitude relationship of $V_{dc}' \geq V_{1max}$. The gate signals for the switches S_{ug}-S_{wh} of the matrix converter are generated by the sawtooth carriers which changes from 0 to 1 and the comparison line which uses the duty cycles d_{ug}-d_{wh}. In the primary side, even if the magnitude relationship between the input current references i_u^*, i_v^*, and i_w^* changes, the source voltage

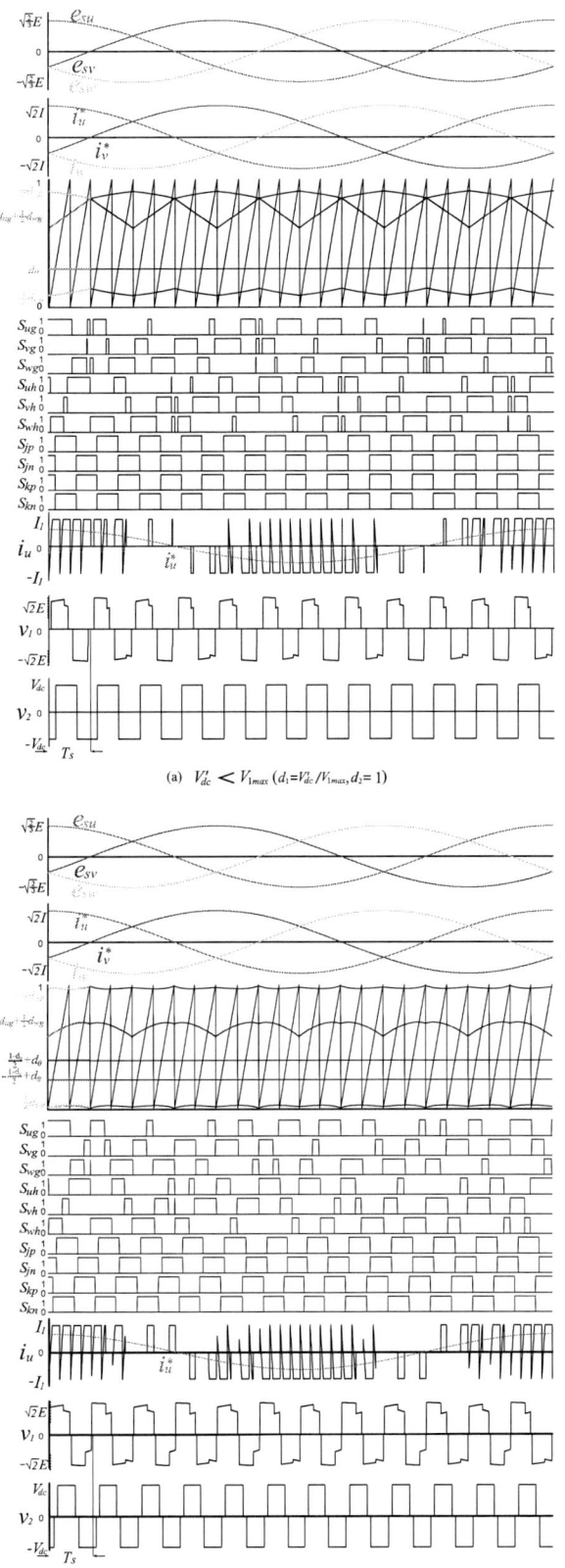

(a) $V_{dc}' < V_{1max} \ (d_1 = V_{dc}'/V_{1max}, d_2 = 1)$

(b) $V_{dc}' \geq V_{1max} \ (d_1 = 1, d_2 = V_{1max}/V_{dc}')$

Fig. 3. Waveforms and gate signals

phase of the maximum absolute value is always connected to g-phase or h-phase during a half period $T_s/2$. The other phase, which is not conneced to the source voltage phase of the maximum absolute value, is connected to each source voltage phase in order. Therefore, the proposed method can apply to the whole phase of the source voltage. The input current i_u is obtained as sinusoidal waveforms including the switching frequency, because it is generated by cutting out the current from the primary current i_1 in the on-state period of the u-phase switches S_{ug} or S_{uh}. The transformer voltages v_1, v_2 are generated by the proposed PWM strategy of each side converter. In Fig.3 (a), the primary voltage v_1 consist of three voltage levels. The secondary voltage v_2 has the maximum pulse width and consists of two voltage levels. In Fig.3 (b), the primary voltage v_1 has the maximum pulse width. The secondary voltage v_2 consists of three voltage levels. The magnitude of the primary voltage v_1 changes in response to the alteration of the source voltage phase θ. Even if the magnitude of the primary voltage v_1 changes, the soft-switching conditions (5),(6) are always satisfied. Because the duty cycles $d_{ug} - d_{wh}$ of the matrix converter are generated to match the average value of the primary voltage v_1 with that of the reference value v_1^* in the half period $T_s/2$.

V. EXPERIMENTAL RESULTS

A. Experimental System

Fig.4 shows the experimental system and Table I shows the experimental conditions. The effective value of the source line voltage E is 200 V and the frequency is 60 Hz. Turn ratio a of the transformer is 1 and the frequency is 10 kHz. The reactors l_1 of 0.1 mH and

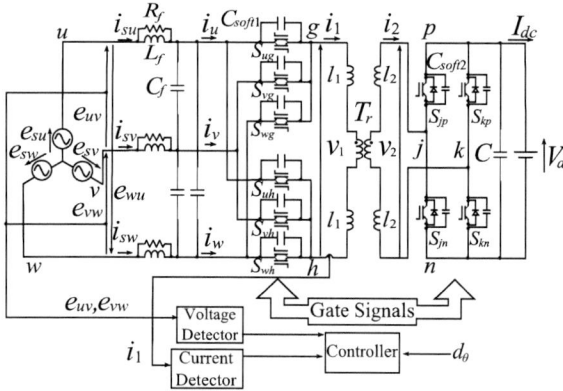

Fig. 4. Experimental Circuit Configuration

TABLE I. SPECIFICATIONS OF EXPERIMENTAL SYSTEM

Source voltage E, ω	200 V, $2\pi \times 60$ rad/s
Output voltage V_{dc}	200,235,280 V
Maximum of primary voltage V_{1max}	235 V
Input filter L_f, C_f	0.6 mH, 5.6 μF
Damping resistor R_f	27 Ω
Inductance l_1, l_2	0.1 mH, 0.1 mH
Capacitors C_{soft1}, C_{soft2}	0.5 nF, 3 nF
Output filter C	300μF
Turn ratio of transformer a	1
Frequency of transformer f_T	10 kHz

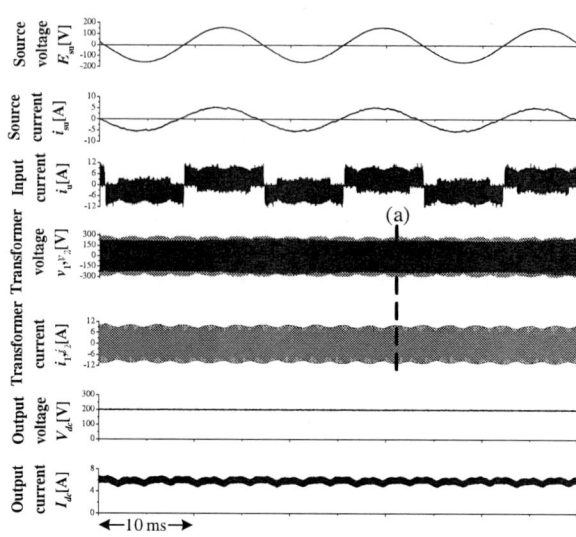

Fig. 5. Experimental waveforms ($V_{dc} = 200$ V, $d_\theta = 0.3$)

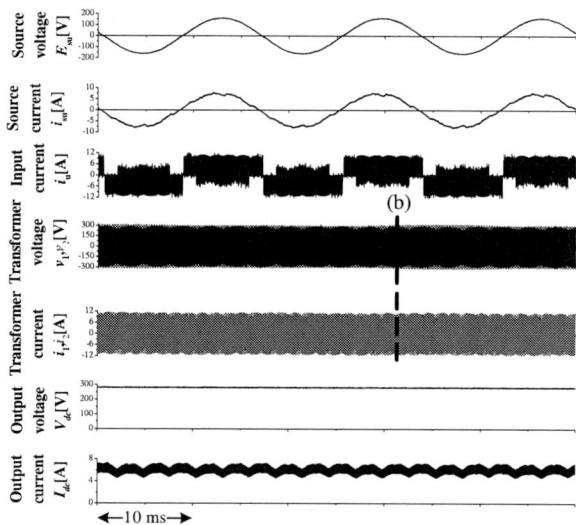

Fig. 6. Experimental waveforms ($V_{dc} = 280$ V, $d_\theta = 0.3$)

l_2 of 0.1 mH are connected to the both sides of the transformer. The capacitors C_{soft1} of 0.5 nF and C_{soft2} of 3 nF for the soft-switching are connected in parallel to the switches of the matrix converter and the secondary converter, respectively. The SiC-MOSFET is used as the switch, and the bi-directional switch is composed of the two switches connected in anti-series. The maximum output voltage V_{1max} is 245 V according to the fomula $V_{1max} = \sqrt{6}E/2$. However, in this experiments, the maximum output voltage V_{1max} is set as 235 V due to the deadtime in the commutation sequences. The experimental characteristics under the three level voltages V_{dc} of 200 V, 235 V, and 280 V are confirmed. The duty cycles are determined by detecting the source line voltages e_{uv}, e_{vw} and the primary current i_1. The phase-difference ratio d_θ is determined by open-roop from 0.1 to 0.5 by every 0.1. The switching signals based on the duty cycles are generated by using the FPGA.

3511

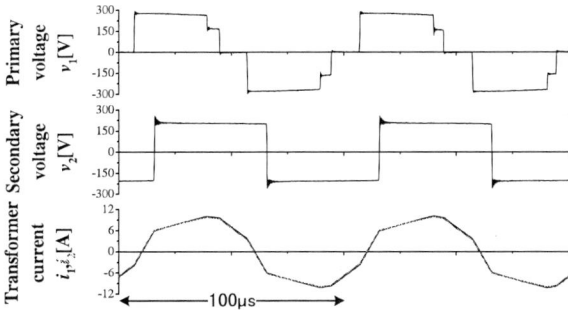

Fig. 7. Magnification waveforms ($V_{dc} = 200\ V, d_\theta = 0.3$)

B. Experimental waveforms

Fig.5 and Fig.6 show the experimental waveforms under the output voltage V_{dc} of 200 V and 280 V, respectively. The phase-difference ratio d_θ is adjusted to 0.3 in the both experiments. The waveforms are the source voltage e_{su}, the source current i_{su}, the input current i_u, the primary and secondary voltages v_1, v_2, the primary and secondary currents i_1, i_2, the output voltage V_{dc}, and the output current I_{dc}. The input current i_u includes the 20 kHz components of twice the switching frequency.. The source current i_{su} is formed into the sinusoidal waveform with the switching frequency components absorbed by the input LC filter. Fig.7 and Fig.8 show the magnification waveforms of the parts in Fig.5 (a) and in Fig.6 (b), respectively. Because the magnetizing current is sufficiently small compared with the transformer currents i_1 and i_2, the primary and secondary currents i_1 and i_2 can be asummed to be equal to each other. In Fig.7, because the output voltage V_{dc} of 200 V is lower than the maximum output voltage V_{1max} of 235 V, the ratios d_1 of 0.851 and d_2 of 1 are obtained. In Fig.8, because the output voltage V_{dc} of 280 V is higher than the maximum output voltage V_{1max} of 235 V, the ratios d_1 of 1 and d_2 of 0.839 are obtained. In Fig.7 and Fig.8, all of the soft-switching conditions (5) − (6) of the transformer currents are satisfied. For example, the primary current i_1 is controlled negative when the primary voltage v_1 changes from 0 to e_{uw} and the secondary current i_2 is controlled positive when the secondary voltage v_2 changes from $-V_{dc}$ to V_{dc}. The phase θ_d of the secondary voltage v_2 is delayed with respect to the primary voltage v_1 to send the power from AC to DC sides.

C. Relation between Output Power and Phase-Difference Ratio

Fig.9 shows the relation between the output power p_{out} and the phase-difference ratio d_θ. The theoretical values of the output power p_{out} are calculated using the formula (4). The experimental values of the output power p_{out} are measured using the precision power analyzer WT1800, Yokogawa. In each case of the output voltage V_{dc} of 200 V, 235 V, and 280 V, the experimental results close to the theoretical values can be obtained.

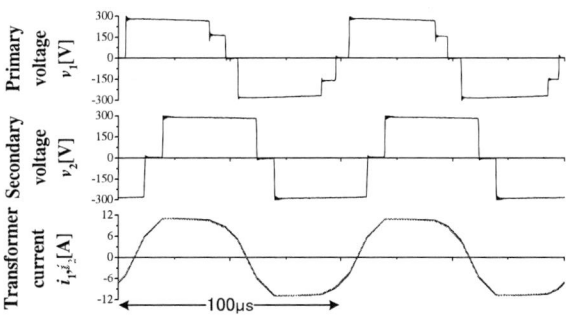

Fig. 8. Magnification waveforms ($V_{dc} = 280\ V, d_\theta = 0.3$)

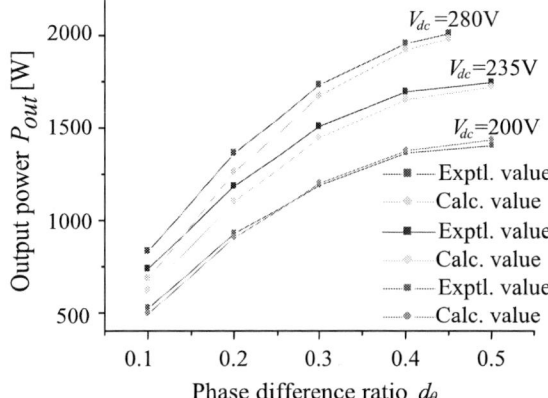

Fig. 9. Relation between output power and phase-difference ratio

VI. CONCLUSIONS

This paper presents novel control method for variable output voltage of an isolated bi-directional AC/DC converter using the high-frequency transformer. The proposed method applys the PWM strategy for both sides converters of transformer in response to the amplitude relationship between the maximum primary voltage and the output voltage to coincide the average value of the primary voltage with the average value of the secondary voltage. In addition, the proposed method applies the commutation patterns for the soft-switching technique which minimizes the switching losses. The output power for the phase-difference ratio of the proposed converter system is theoretically derived. The soft-switching conditions for all commutations of converters are described. The duty cycles for the switches of the both sides converters are theoretically derived. The effectiveness of the proposed control and the output behavior were experimentally verified. In addition, the satisfaction of the soft-switching conditions for all commutation of the both sides converters was verified by experiments.

ACKNOWLEDGEMENT

This work was supported by Council for Science, Technology and Innovation (CSTI), Cross-ministerial Strategic Innovation Promotion Program (SIP), "Next-generation power electronics" (funding agency: NEDO).

REFERENCES

[1] Bahadur Singh Pali, Shelly Vadhera, "Renewable Energy Systems for Generating Electric Power: A Review", *in 2016 IEEE 1st International Conference on Power Electronics, Intelligent Control and Energy Systems(ICPEICES)*, pp.1-6, July 2016.

[2] S. Inoue and H. Akagi, "A Bi-Directional DC/DC Converter for an Energy Storage System", *IEEE Trans. Power Electron*, vol. 22, No. 6, pp. 2299-2306, March 2007.

[3] Jordi Everts, Jeroen Van den Keybus, Florian Krismer, Johan Driesen, andJohann W. Kolar, "Switching Control Strategy for Full ZVS Soft-Switching Operation of a Dual Active Bridge AC/DC Converter", *in 2012 Twenty-Seventh Annual IEEE Applied Power Electronics Conference and Exposition(APEC)*, pp.1048-1055, Feb. 2012.

[4] Qiong Wang, Xuning Zhang, Rolando Burgos, Dushan Coroyevich, Adam White and Mustansir Kheraluwala, ". esign and Optimization of a High Performance Isolated Three Phase AC/DC Converter",*in 2016 IEEE Energy Conversion Congress and Exposition(ECCE)*, pp.1-10, Sept. 2016.

[5] Yu Fang, Songyin Cao, Yong Xie and Pat Wheeler, "Study on Bidirectional-Charger for Electric Vehicle Applied to Power Dispatching in Smart Grid", *in 2016 IEEE 8th International Power Electronics and Motion Control Conference(IPEMC-ECCE Asia)*, pp. 2709-2713, May 2016.

[6] M. Yamada and T. Takeshita: "PWM Strategy of AC to DC Converter with High Frequency Link for Reducing Output Voltage Ripple", *Proc. of 4th IEEE International Conference on Power Engineering, Energy and Electrical Drives (POWERENG) 2013*, pp.846-851, May 2013.

[7] J. J. Sandoval, S. Essakiappan, and P. Enjeti, "A bidirectional series resonant matrix converter topology for electric vehicle dc fast charging", *in 2015 IEEE Applied Power Electronics Conference and Exposition (APEC)*, pp. 3109-3116, March 2015.

[8] K Suzuki, W Kitagawa, and T. Takeshita: " ˇ oft-Switching Three-Phase AC to DC Converter Isolated By High-Frequency Transformer", *18th European Conference on Power Electronics and Applications (EPE) 2016*, pp.1-10, Sept. 2016.

[9] K. Suzuki, W. Kitagawa and T. Takeshita, "Suppression Control of Source Current Harmonics of Bi-Directional Isolated AC/DC Converter using Soft Switching Technique",*in 2017 IEEE 3rd International Future Energy Electronics Conference and ECCE Asia (IFEEC 2017 - ECCE Asia)*, pp57-61, June 2017.

A New Modulation Method Applying Optimal Duty Cycle and Phase Shift for Bidirectional Isolated Three-Phase AC/DC Converter Based on Matrix Converter

Koji Shigeuchi[1], Jin Xu[2], Noboru Shimosato[2], Yukihiko Sato[1*]

1 Graduate School of Science and Engineering, Chiba University, Chiba, Japan
2 Power Supply Systems Department, Myway Plus Corporation, Yokohama, Japan
*E-mail: ysato@faculty.chiba-u.jp

Abstract—This paper presents a new modulation method for high-frequency link bidirectional three-phase AC/DC converter based on matrix converter. Since the matrix converter does not have electrolytic capacitor, it achieves high power density and long lifetime system. The modulation method performs bidirectional active and reactive power control with sinusoidal line current by using optimal duty cycle and phase shift. The optimal duty cycle and phase shift are determined by numerical calculation based on the system mathematical model. Experimental results employing a 1kW laboratory prototype verify the capability to control active and reactive power with sinusoidal line current.

Keywords—AC/DC converter, Bidirectional power flow, High-frequency transformer, Matrix converter

I. INTRODUCTION

Recently, the number of renewable energy system is increasing for solving environmental issues. The renewable energy systems, such as wind power generation and photovoltaic system, have large fluctuation of the power generation. Therefore, the grid is possibly destabilized by those systems. In order to avoid the destabilization, the battery energy storage systems can be used to adjust the active power, and compensate the fluctuation. For connection of the battery energy storage system to the power grid, a bidirectional isolated AC/DC converter, which realize high efficiency, small size, low cost and good power quality, is needed. High-frequency link AC/DC converters are widely investigated because they can reduce the size of transformer. A typical-type high-frequency link AC/DC converter is configured by a non-isolated AC/DC converter and a high-frequency isolated DC/DC converter [1]. However, such converters require a DC link electrolytic capacitor that has problems of large size and short lifetime. To overcome these problems, several types of AC/DC converters applying matrix converter(MC) have been proposed [2]–[9]. Since those converters do not have a DC link, the electrolytic capacitors are not necessary. Therefore, a small-size and long-lifetime system can be realized.

Buck type ([2], [3]), boost type ([4]–[6]) and Dual-Active-Bridge (DAB) type ([7]–[9]) of AC/DC converters based on matrix converter have been proposed. Fig. 1 shows basic topologies of each type. The buck

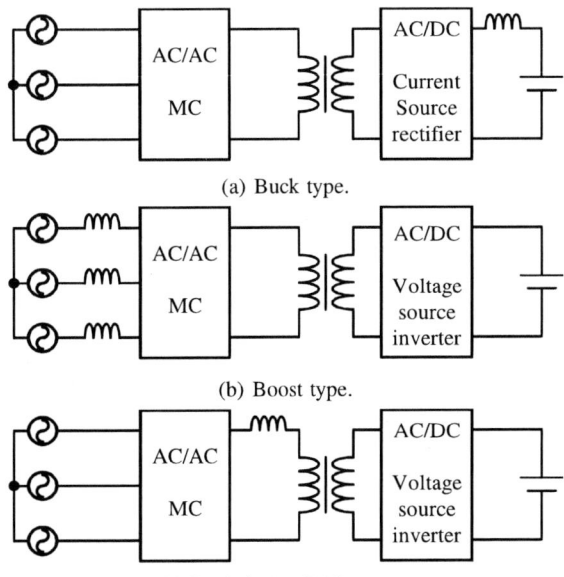

(a) Buck type.

(b) Boost type.

(c) Dual-Active-Bridge type.

Fig. 1: Basic topologies of high-frequency link three-phase AC/DC converter based on matrix converter.

type and boost type topologies are based on current source PWM rectifier and voltage source PWM inverter, respectively. The buck type can realize zero voltage switching (ZVS) for AC to DC conversion, and the boost type can realize ZVS for DC to AC conversion by using a leakage inductance of the high-frequency transformer [5], [6]. However, in case of reverse direction conversion, ZVS in all the power switches cannot be realized. In addition, voltage spike occurs by the leakage inductance. Therefore, those two types are not suitable for bidirectional operation.

In the DAB type topology, both of the MC and secondary voltage source inverter generate high frequency link voltage at the same time. This operation is based on a principle of DC/DC DAB converter [10]. The DC/DC DAB converter is actively investigated because the converter can realize ZVS in all the power switches for bidirectional operation. For the same reason, it is expected that the DAB type AC/DC converter is suitable

The 2018 International Power Electronics Conference

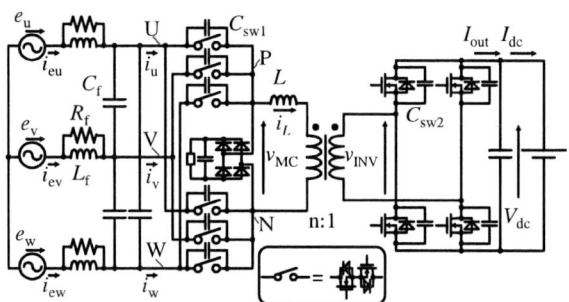

Fig. 2: Bidirectional isolated three-phase AC/DC Dual-Active-Bridge converter based on matrix converter.

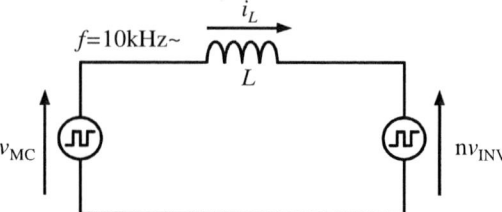

Fig. 3: Equivalent circuit of high-frequency link.

for bidirectional operation.

Several modulation methods for three-phase AC/DC DAB converter have been proposed. However, those methods still have some problems. In [7], a modulation method to realize bidirectional active power flow control with sinusoidal line current under zero current switching (ZCS) has been proposed. However, the power factor at high-frequency link is very low, and conduction losses are increased. For example, when the line current rms is 5.3A, transformer current peak is around 40A in [7]. In [8], the active power control with ZVS is achieved. But, this modulation method uses square-wave approximation to obtain duty cycle, and the approximation increases the line current distortion. The line current Total-Harmonic-Distortion (THD) is higher than 10% under all conditions. In [9], the active power and reactive power flow control based on space vector modulation is realized. However, ZVS in all the power switches are not be realized because the method switches to positive voltage and negative voltage at instants that the transformer current is fixed value.

This paper proposes a new modulation method based on PWM and Phase-Shift-Modulation for bidirectional isolated AC/DC DAB converter. This method realizes bidirectional active and reactive power flow control with sinusoidal line current waveform. The optimal duty cycle and phase shift in order to achieve sinusoidal line current waveform are obtained by numerical calculation based on the system mathematical model. In addition, ZVS in all the power switches can be realized by connecting soft switching capacitor to the power switches.

This paper is organized as follows: section II presents the outline of modulation strategy to realize active and reactive power control with sinusoidal line current waveform. Section III introduces the details of the system mathematical model and explains the numerical calculation method to obtain optimal duty cycle and phase shift. Finally, to demonstrate the validity of the proposed modulation method, the experimental results employing a laboratory 1kW prototype are presented and discussed in section IV.

II. Modulation Method

A. Topology

Fig. 2 shows the circuit configuration of a bidirectional AC/DC DAB converter. The primary side is composed of

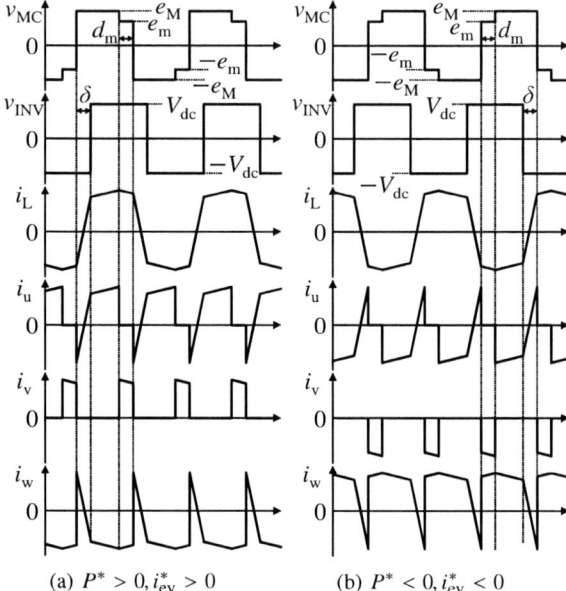

(a) $P^* > 0, i_{ev}^* > 0$ (b) $P^* < 0, i_{ev}^* < 0$

Fig. 4: Proposed modulation waveforms. ($e_u > e_v > e_w$)

a three-phase to single-phase matrix converter (MC), an input LC filter, and a clamp circuit. The secondary side consists of a full-bridge inverter (INV). A high-frequency transformer is connected with an external inductor L in series. The MC input currents i_u, i_v, i_w contain the harmonics of the switching frequency. The input LC filter eliminates these harmonics. Therefore, line current i_{eu}, i_{ev}, i_{ew} do not contain these harmonics. The basic operation of the MC is summarized as below. The MC connects any two of the input phases of U, V, W to the output terminals P and N. For example, when U and W are connected with P and N, respectively, the MC output voltage v_{MC} equals to $e_u - e_w$. Here, e_u, e_v, e_w are the input phase voltages given by comercial line angular frequency ω and line voltage E as follows:

$$
\begin{bmatrix} e_u \\ e_v \\ e_w \end{bmatrix} = \sqrt{\frac{2}{3}} E \begin{bmatrix} \cos(\omega t) \\ \cos(\omega t - 2\pi/3) \\ \cos(\omega t - 4\pi/3) \end{bmatrix}. \tag{1}
$$

At the same time, the transformer current i_L flows through U and W phases, and U and W input currents i_u and i_w are $i_u = i_L$ and $i_w = -i_L$, respectively. In this case, since V phase is not connected to any terminal, V input current i_v equals to 0.

3515

The 2018 International Power Electronics Conference

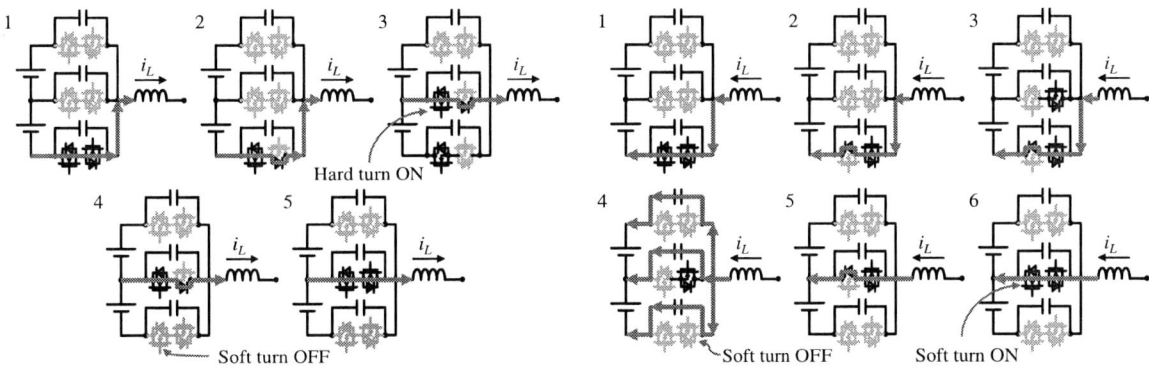

(a) Hard switching condition.

(b) ZVS condition.

Fig. 5: Commutations to high voltage from low voltage phase of the line under ZVS and hard switching conditions.

B. Modulation Strategy

This subsection presents the proposed modulation strategy to realize active and reactive power control with sinusoidal line current waveforms. Fig. 3 shows the equivalent circuit of the high-frequency link. The difference between the MC output voltage v_{MC} and the INV output voltage nv_{INV} is applied to the inductor L. Here, nv_{INV} is referred to the primary side, and n is turns ratio of the high-frequency transformer, and the magnetization current is assumed to be negligible.

In the following explanation, the input phase voltages e_u, e_v, e_w are assumed to have a relationship $e_u > e_v > e_w$. If both the MC and INV generate square wave voltages with a phase shift, the average input power P is obtained as follows:

$$P = \frac{e_M n V_{dc}}{2fL} \frac{\delta}{\pi} \left(1 - \frac{\delta}{\pi}\right). \quad (2)$$

Where δ is phase shift between v_{MC} and v_{INV}, and the voltage e_M is the amplitude of the MC output voltage v_{MC} that is $e_M = e_u - e_w$, and V_{dc} is the amplitude of the INV output voltage v_{INV}. (2) shows that the active power P is controllable by adjusting phase shift δ. However, if the MC generates the square waveform, V phase is not connected with the output of the MC. In this case, the sinusoidal line current waveform cannot be achieved because the current does not flow through V phase. To overcome this problem, PWM is applied to the MC. Fig. 4 shows the outline of the modulation waveforms based on PWM and Phase-Shift-Modulation (PSM). In Fig. 4, the MC and INV output voltages v_{MC} and v_{INV}, the high-frequency transformer current i_L, and the MC input currents i_u, i_v, i_w are shown. The MC generates the voltage e_M and voltage $e_m = e_v - e_w$ or $e_u - e_v$. d_m is defined as the duty cycle when the MC is generating $\pm e_m$ in the switching period. When the MC generates e_m, the transformer current i_L flows through V phase, and the V phase input current i_v equals to i_L. Therefore, the average MC input current of V phase $\bar{i}_v$ is controllable by adjusting the duty cycle d_m. In addition, when the input power P and V phase MC input current i_v are controlled, U and W phase MC input currents are indirectly determined because the relationship of the

following equations exists.

$$\begin{cases} P = \bar{i}_u e_u + \bar{i}_v e_v + \bar{i}_w e_w \\ 0 = \bar{i}_u + \bar{i}_v + \bar{i}_w \end{cases} \quad (3)$$

In this way, the line current i_{eu}, i_{ev}, i_{ew} can become sinusoidal waveforms by controlling P and $\bar{i}_v$ to appropriate values because i_{eu}, i_{ev}, i_{ew} are almost equal to the MC average input currents $\bar{i}_u, \bar{i}_v, \bar{i}_w$. The reference value of the line current $i_{eu}^*, i_{ev}^*, i_{ew}^*$ are determined by given values of a input power reference P^* and a power factor angle reference α^* as follows:

$$\begin{bmatrix} i_{eu}^* \\ i_{ev}^* \\ i_{ew}^* \end{bmatrix} = \sqrt{\frac{2}{3}} \frac{P^*}{E \cos \alpha^*} \begin{bmatrix} \cos(\omega t - \alpha^*) \\ \cos(\omega t - 2\pi/3 - \alpha^*) \\ \cos(\omega t - 4\pi/3 - \alpha^*) \end{bmatrix}. \quad (4)$$

In conclusion, the outline of the proposed modulation method is described as follows:

1) Obtain V phase input current reference i_{ev}^* from input power reference P^* and power factor angle reference α^*.

2) Obtain the optimal duty cycle d_m and phase shift δ from the system mathematical model. In other words, solve $P(\delta, d_m) = P^*$ and $\bar{i}_v(\delta, d_m) = i_{ev}^*$.

3) The MC generates PWM waveform modulated with the duty cycle d_m, and the INV generates square waveform modulated with the phase shift δ.

C. Zero Voltage Switching

The AC/DC DAB converter can realize zero voltage switching (ZVS) in all the power switches as same as DC/DC DAB converters [8], [10]. ZVS is achieved by turn on when the voltage across the power switch is clamped to zero by conducting its antiparallel diode. During turn off, since the soft switching capacitor connected to the power switch suppresses the rising of the voltage, the switching loss is reduced. The ZVS conditions are determined by the direction of i_L. Fig. 5 shows commutations to high voltage from low voltage phase of the line under ZVS and hard switching conditions. The proposed modulation waveforms shown in Fig. 4 meet the ZVS condition at all switching cases. Therefore, the proposed method can realize ZVS in all the power switches.

3516

III. CALCULATION OF DUTY CYCLE AND PHASE SHIFT

A. Mathematical Model

To determine the optimal duty cycle d_m and phase shift δ from $P(\delta, d_\mathrm{m}) = P^*$ and $\bar{i}_\mathrm{v}(\delta, d_\mathrm{m}) = i_\mathrm{ev}^*$, it is necessary to obtain mathematical models of the input power $P(\delta, d_\mathrm{m})$ and V phase average input current $\bar{i}_\mathrm{v}(\delta, d_\mathrm{m})$. Firstly, the model of the input power $P(\delta, d_\mathrm{m})$ is obtained from the system equivalent circuit as shown in Fig. 3 as follows:

$$P = \frac{e_\mathrm{M} n V_\mathrm{dc}}{2fL} \frac{\delta}{\pi}\left(1 - \frac{\delta}{\pi}\right) + \frac{(e_\mathrm{M} - e_\mathrm{m})n V_\mathrm{dc}}{4fL} d_\mathrm{m}\left(1 - 2\frac{\delta}{\pi} - d_\mathrm{m}\right). \tag{5}$$

In the case that the MC generates square wave, the input power $P(\delta, d_\mathrm{m})$ is given by (2). (5) has the extra second term compared to (2). This term corresponds to the effect of PWM. Secondly, the model of the V phase average input current $\bar{i}_\mathrm{v}(\delta, d_\mathrm{m})$ is obtained from the V phase input current i_v with the duty cycle d_m (Fig. 4), as follows:

$$|\bar{i}_\mathrm{v}| = f\left\{\int_{\frac{1-d_\mathrm{m}}{2f}}^{\frac{1}{2f}} i_L(t)dt - \int_{\frac{2-d_\mathrm{m}}{2f}}^{\frac{1}{f}} i_L(t)dt\right\}. \tag{6}$$

The following equation is obtained from substituting the transformer current i_L as shown in the equivalent circuit of Fig. 3.

$$|\bar{i}_\mathrm{v}| = \frac{n V_\mathrm{dc}}{2fL}\frac{\delta}{\pi} d_\mathrm{m} + \frac{e_\mathrm{M} - n V_\mathrm{dc}}{4fL} d_\mathrm{m}(1 - d_\mathrm{m}) \tag{7}$$

(5) and (7) are the system mathematical models. Finally, the optimal duty cycle d_m and phase shift δ are determined by solving equations $P^* = P$ and $i_\mathrm{ev}^* = \bar{i}_\mathrm{v}$ obtained by (5) and (7). The equations are shown as follows:

$$P^* = \frac{e_\mathrm{M} n V_\mathrm{dc}}{2fL}\frac{\delta}{\pi}\left(1 - \frac{\delta}{\pi}\right)$$
$$+ \frac{(e_\mathrm{M} - e_\mathrm{m})n V_\mathrm{dc}}{4fL} d_\mathrm{m}\left(1 - 2\frac{\delta}{\pi} - d_\mathrm{m}\right) \tag{8}$$

$$|i_\mathrm{ev}^*| = \frac{n V_\mathrm{dc}}{2fL}\frac{\delta}{\pi} d_\mathrm{m} + \frac{e_\mathrm{M} - n V_\mathrm{dc}}{4fL} d_\mathrm{m}(1 - d_\mathrm{m}). \tag{9}$$

B. Numerical Calculation

This subsection presents the solving method for equations (8) and (9). It is difficult to solve these equations analytically, because these are nonlinear simultaneous equations. Therefore, to solve the equations, the real-time numerical calculation is introduced. As the numerical calculation method, bisection method is applied due to its high reliability. The solving method is performed as follows:

1) Drop out duty cycle d_m from the equations and obtain one-variable equation $P^* = P(\delta)$.
2) Solve equation $P^* = P(\delta)$ by the bisection method.
3) Determine duty cycle d_m from the phase shift δ.

To apply the bisection method, the nonlinear simultaneous equations are converted to a one-variable equation because bisection method cannot apply

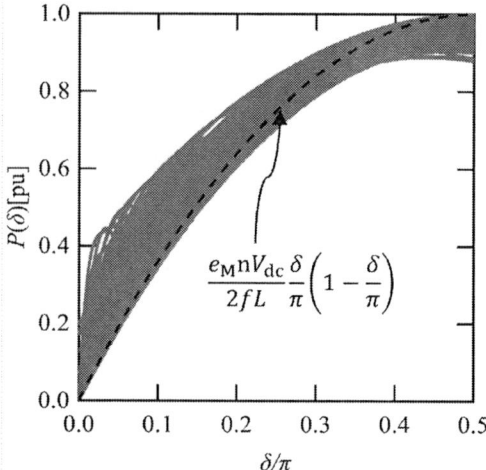

Fig. 6: Function form of $P^*(\delta)$.

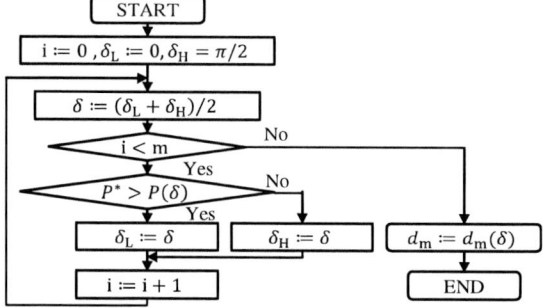

Fig. 7: Flowchart of numerical calculation to determine phase shift δ and duty cycle d_m. (m is a number of iteration)

multivariable equations. The form of the input power mathematical model $P(\delta)$ is similar to a parabola. Therefore, the equation $P^* = P(\delta)$ is suitable to apply the bisection method because the outline of equation can be easy to estimate. Firstly, the duty cycle d_m is dropped out from the (8) and (9) by solving the following equation obtained with dividing (9) by (8).

$$\frac{|i_\mathrm{ev}^*|}{P^*} = \frac{2\frac{\delta}{\pi} d_\mathrm{m} + (e_\mathrm{M}/n V_\mathrm{dc} - 1)d_\mathrm{m}(1 - d_\mathrm{m})}{2e_\mathrm{M}\frac{\delta}{\pi}\left(1 - \frac{\delta}{\pi}\right) + (e_\mathrm{M} - e_\mathrm{m})d_\mathrm{m}\left(1 - 2\frac{\delta}{\pi} - d_\mathrm{m}\right)} \tag{10}$$

The following equation is obtained by solving (10) for the duty cycle d_m.

$$A = 1 - \frac{e_\mathrm{M}}{n V_\mathrm{dc}} + e_\mathrm{M}\frac{|i_\mathrm{ev}^*|}{P^*}\left(1 - \frac{e_\mathrm{m}}{e_\mathrm{M}}\right)$$

$$B = -A + 2\frac{\delta}{\pi}\left\{1 + e_\mathrm{M}\frac{|i_\mathrm{ev}^*|}{P^*}\left(1 - \frac{e_\mathrm{m}}{e_\mathrm{M}}\right)\right\}$$

$$C = -2e_\mathrm{M}\frac{|i_\mathrm{ev}^*|}{P^*}\frac{\delta}{\pi}\left(1 - \frac{\delta}{\pi}\right) \tag{11}$$

$$d_\mathrm{m}(\delta) = \frac{-B + \sqrt{B^2 - 4AC}}{2A}$$

The following equation is obtained by substituting (11) into (8).

$$P^* = P(\delta, d_\mathrm{m}(\delta)) = P(\delta) \tag{12}$$

3517

TABLE I: Specifications of Experimental System

Source line voltage	200V, 50Hz
Secondary DC voltage V_{dc}	60V
External inductor L	440μH
Input filter L_f, C_f, R_f	0.75mH, 10μF, 16.5Ω
Switching frequency f	15.15kHz
Turns ratio n	4
Rated input power	1kW
Dead time of INV	3μs
Commutation time of MC	1.5μs

In (12), $P(\delta)$ contains several parameters. Fig. 6 shows the form of $P(\delta)$ for possible variation in all the parameters. In Fig. 6, 1pu equals to $e_M n V_{dc}/8fL$. Function form of $P(\delta)$ is similar to a parabola in any cases, and it takes the minimum value 0pu at $\delta = 0$, and it takes the maximum value around 1pu near $\delta = \pi/2$. Therefore, the equation $P^* = P(\delta)$ can be solved using the bisection method by setting two initial values δ_L and δ_H to 0 and $\pi/2$, respectively. The flowchart of the solving method is shown in Fig. 7.

C. Operation Range

This section presents possible operation range of the proposed method. The operation range is limited by following conditions.

$$\begin{cases} |P^*| \le P_{max} \approx \dfrac{e_M n V_{dc}}{8fL} \\ 0 \le d_m \le 1 - \dfrac{\delta}{\pi} \end{cases} \tag{13}$$

The first and second equations correspond to the operation range of the active power and reactive power, respectively. In other words, the power reference value P^* and the input power factor angle reference value α^* are limited. The conditions of α^* is obtained by (4), (11) and (13) as follows:

$$|\alpha^*| \le \alpha_{max}$$

$$\tan \alpha_{max} \ge$$

$$\begin{cases} \dfrac{1}{\sqrt{2}} \dfrac{E}{n V_{dc}} & \text{if } \dfrac{E}{n V_{dc}} > \sqrt{\dfrac{2}{3}} \\ \dfrac{\sqrt{3}}{4}\left(1 + \sqrt{\dfrac{3}{2}}\dfrac{E}{n V_{dc}}\right) - \dfrac{1}{\sqrt{3}}\dfrac{1}{1+\sqrt{\frac{3}{2}}\frac{E}{n V_{dc}}} & \text{if } \dfrac{E}{n V_{dc}} \le \sqrt{\dfrac{2}{3}} \end{cases} \tag{14}$$

(14) means that the limitation of α^* depends on E and $n V_{dc}$. When E takes constant value, the limitation of α^* increases if $n V_{dc}$ is low. In the case of $E/n V_{dc} = \sqrt{2/3}$, the limit of α^* exceeds $\pm 30°$. In addition, the minimum value of e_M equals to $n V_{dc}$ in this condition.

IV. Experimental Results

A. System Configuration

A 1kW laboratory prototype was developed to confirm the validity of the proposed method. Fig. 8 shows the system configuration, and Table. I shows specification of the system. The MC and INV are implemented with IGBTs and MOSFETs, respectively. The soft switching capacitors are not implemented because this prototype is focusing on the verification for the principle of the proposed modulation. The controller is implemented by a DSP (TI: TMS320F28335) and a FPGA. The DSP

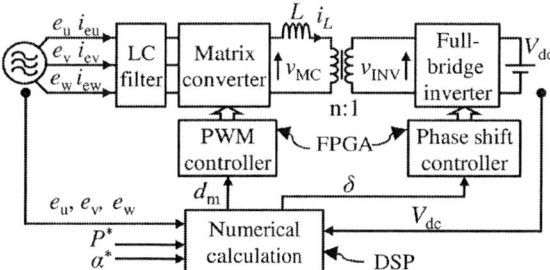

Fig. 8: Experimental system configuration.

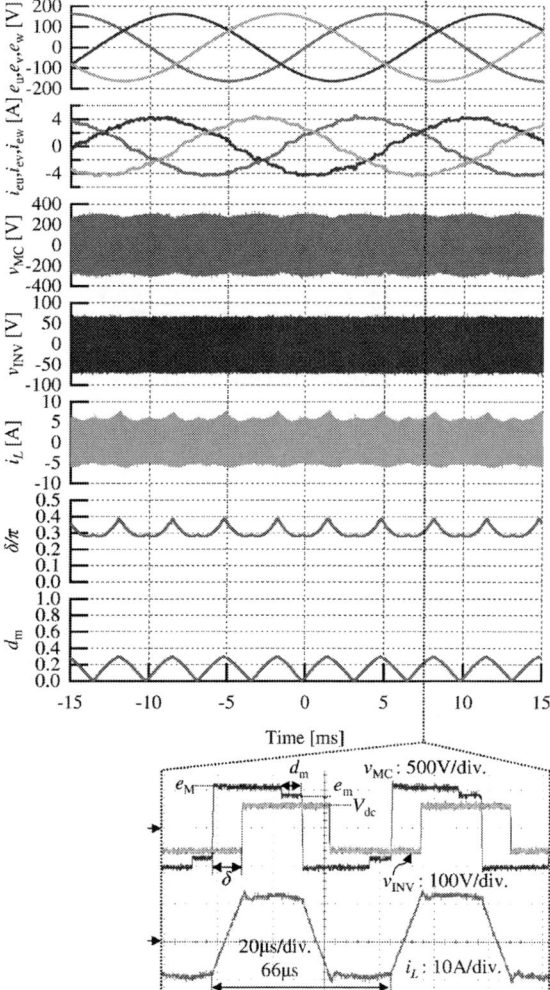

Fig. 9: Experimental waveforms in AC to DC conversion when $P^* = 1$kW and $\alpha^* = 0°$.

calculates the optimal duty ratio d_m and phase shift δ to realize sinusoidal line current by using bisection method. The number of iteration in the numerical calculation is 10, and the calcuration time is less than 8.5μs. The prototype system controls the input power P and the power factor angle α by open-loop control for given reference values P^* and α^*. Here, in AC to DC conversion, P and P^* mean AC input. In DC to AC conversion, they mean DC input.

B. Bidirectional Active Power Flow Control

This section presents the experimental results of bidirectional active power flow control. Fig. 9 shows the experimental waveforms in AC to DC conversion when the input power reference value $P^* = 1$kW and the power factor angle reference value $\alpha^* = 0°$. The line current i_{eu}, i_{ev}, i_{ew} are sinusoidal waveforms. The phase difference between the phase voltage e_u, e_v, e_w and the line current is caused by leading currents of the input LC filter. This phase difference can be decrease by adjusting parameters of the LC filter. The MC and INV output voltages v_{MC}, v_{INV} and the current of the high-frequency transformer i_L are high-frequency waveforms that mainly contain the switching frequency components. The magnified waveforms in the bottom of the figure show the details of the waveforms of v_{MC}, v_{INV} and i_L. From those results, we can confirm that the proposed modulation waveforms as shown Fig. 4 is realized. In this case, THD of the line current is 3.9%.

Fig. 10 shows the experimental waveforms in DC to AC conversion when the input power reference value $P^* = 1$kW and the power factor angle reference value $\alpha^* = 0°$. The sinusoidal line current waveform is also observed in this case and the THD of the line current is 3.7%.

From these results, it is confirmed that the proposed method can realize the sinusoidal line current with THD of less than 4% under the bidirectional operating condition at the rated power.

Fig. 11 shows measured THD characteristics of the line current by using HIOKI PW6001 power analyzer. In the case of AC to DC conversion, the THD is less than 5% in operation range above around 500W. The characteristics of AC to DC conversion and DC to AC conversion are not same, and the line current THD of DC to AC conversion is significantly larger than that in AC to DC conversion in operation range less than 700W. Fig. 12 shows relationship the input power P and reference P^*. Similar to THD characteristics, the characteristics of P in AC to DC conversion and DC to AC conversion are not same. The difference of P and P^* of DC to AC conversion is large in operation range less than 700W.

These results are caused by a difference of a dead time of the INV and a commutation time of the MC. The dead time of the INV is set to 3μs, and the commutation time of one step sequence is set to 1.5μs. If the dead time is short enough, the line current THD and the difference of P and P^* can be reduced in wide operation range.

In [8], it is reported that the line current THD is higher than 10% under all conditions by the modulation using the square wave approximation. Compared with the control method of [8], we can confirm that the proposed method using the optimal duty cycle d_m and phase shift δ is very effective to reduce the line current THD rather than the modulation method using the approximation.

C. Reactive Power Flow Control

This section presents experimental results of reactive power flow control. Fig. 13 shows the experimental waveforms in AC to DC conversion when $P^* = 1$kW, $\alpha^* =$

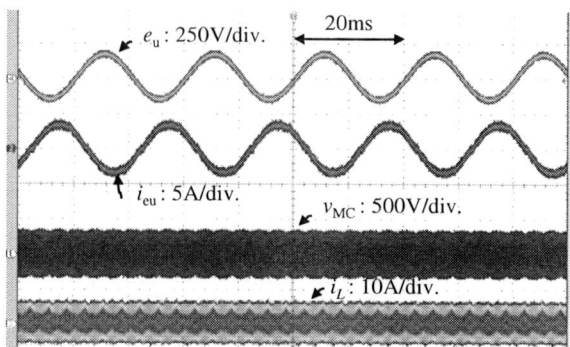

Fig. 10: Experimental waveforms in DC to AC conversion when $P^* = 1$kW and $\alpha^* = 0°$.

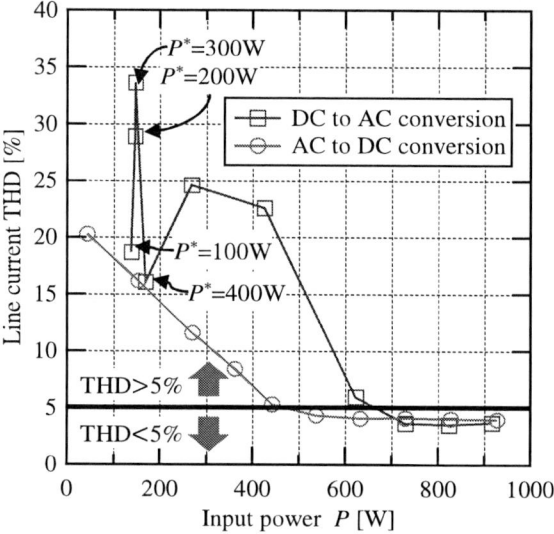

Fig. 11: Measured characteristics of line current THD.

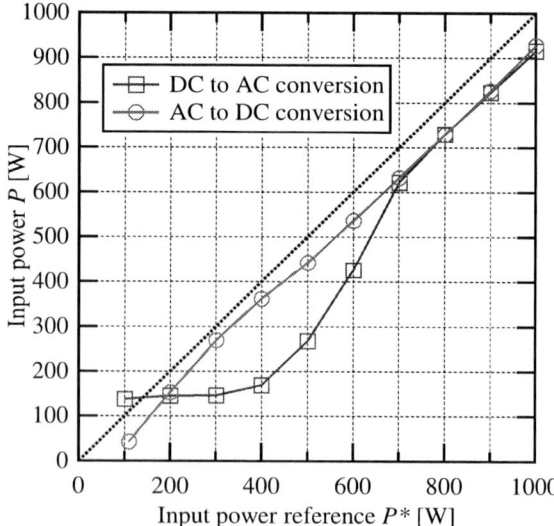

Fig. 12: Characteristics of input power P vs input power reference P^*.

The 2018 International Power Electronics Conference

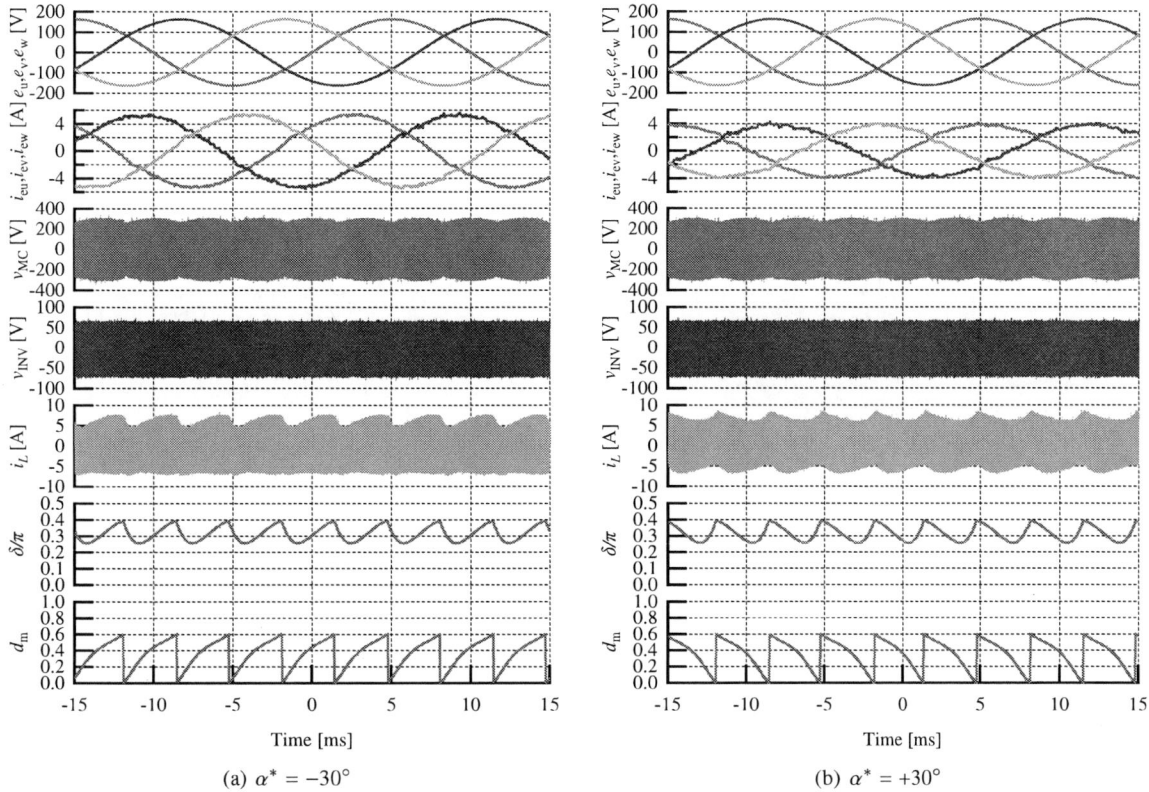

(a) $\alpha^* = -30°$ (b) $\alpha^* = +30°$

Fig. 13: Experimental waveforms when reactive power flow control. (AC to DC conversion, $P^* = 1kW$)

$\pm 30°$. In the case when $\alpha^* = -30°$, as shown in Fig. 13a, the phase of the line current i_{eu}, i_{ev}, i_{ew} leads compared with the case of $\alpha^* = 0°$ (Fig. 9). In the same way, the phase of line current in case of $\alpha^* = 30°$ (Fig. 13b) delays compared with the case of $\alpha^* = 0°$ (Fig. 9). In addition, the waveforms of line currents are sinusoidal in both cases. The THD of line current when $\alpha^* = -30°$ and $\alpha^* = 30°$ are 1.9% and 3.8%, respectively.

The characteristics of the average reactive power Q and the power factor angle α is shown in Table. IIa. α does not coincide with reference value α^* because the leading current of the input LC filter superimposes on the line current. Table. IIb shows the characteristics if the leading currents of the LC filter are neglected. In this case, the power factor angle almost coincide with its reference value.

From these results, it is confirmed that the proposed method can control bidirectional reactive power flow. Actually, in Fig. 13b, the leading current of the LC filter is almost canceled by the lagging current of the converter. Therefore, the proposed method can compensate the leading current of the LC filter by adjusting the power factor reference value α^*.

D. Efficiency and Loss Distribution

Fig. 14 shows the system efficiency in bidirectional operations. The maximum efficiency is 90.3% at $P = 404W$ of AC to DC conversion. Fig. 15 shows measured loss distribution by using HIOKI PW6001 power analyzer

TABLE II: Experimental characteristics of reactive power.

(α^*: power factor angle reference value, P: average input active power, Q: average input reactive power, α: power factor angle) ($P^* = 1kW$)

(a) Contained leading current of LC filter.

$\alpha^*[°]$	$P[W]$	$Q[var]$	$\alpha[°]$
-30	923	-956	-46.0
0	927	-474	-27.1
30	922	-9.98	-0.621

(b) Neglected leading current of LC filter.

$\alpha^*[°]$	$P[W]$	$Q[var]$	$\alpha[°]$
-30	923	-482	-27.6
0	927	0	0
30	922	464	26.7

in AC to DC conversion. The losses of the MC and the external inductor are dominant, and they take 80% of the total loss when $P = 928W$. The reduction of the losses can be realized by appropriate design of system. In addition, applying soft switching capacitor can reduce the losses too.

V. CONCLUSION

This paper proposed a new modulation method for bidirectional isolated AC/DC Dual-Active-Bridge converters based on matrix converter. The proposed method determines the optimal duty cycle d_m and

3520

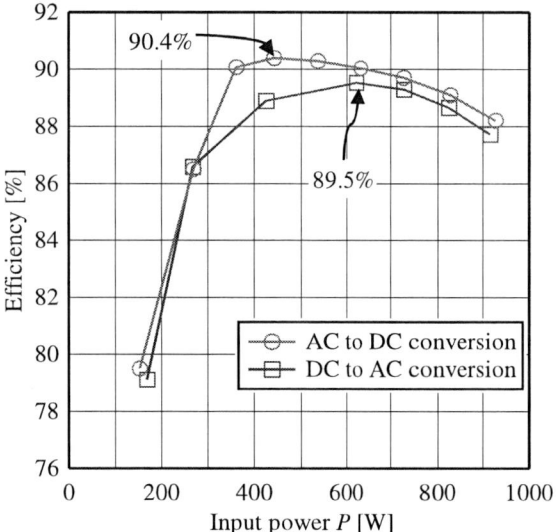

Fig. 14: System efficiency.

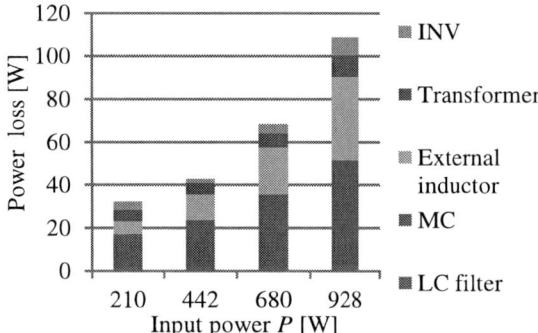

Fig. 15: Measured loss distribution in AC to DC conversion.

the phase shift δ to realize sinusoidal line current by using real time numerical calculation based on the system mathematical model. The experimental results employing the 1kW prototype demonstrated the bidirectional controllability of active power and reactive power with sinusoidal line current by using open-loop control.

The THD of the line current less than 5% was achieved in the operation range above 500W in AC to DC conversion. In DC to AC conversion, the operation range where THD less than 5% can be realized was limited by the dead time of the INV, and the operation range was above 700W. This problem can be improved by shortened dead time of the INV or a new compensation method. The controllability of reactive power was demonstrated at operation points of the power factor angle reference $\alpha^* = \pm 30°$. The results of the power factor angle when the leading current of LC filter is neglected almost coincides with its reference value $(-27.6°, 26.7°)$. In addition, the results of the line current THD were less than 5%.

The maximum system efficiency was 90.4%, and the losses of the MC and the external inductor take 80% of the total loss. These large losses can be reduced by appropriate design of the system and implementation soft

switching discussed in IIc.

REFERENCES

[1] M. Yilmaz, P. T. Krein, "Review of Battery Charger Topologies, Charging Power Levels, and Infrastructure for Plug-In Electric and Hybrid Vehicles," IEEE Trans. on Power Electronics, vol. 28, no. 5, pp. 2151-2169, 2013.

[2] S. Manias and P. D. Ziogas,"A novel sine wave in AC to DC converter with high-frequency transformer isolation,"IEEE Trans. on Industrial Electronics, Vol. 32, No. 4, pp. 430-438, 1985.

[3] M. Yamada, and T. Takeshita,"PWM Strategy of AC to DC Converter with High Frequency Link for Reducing Output Voltage Ripple,"Power Engineering, Energy and Electrical Drives (POWERENG), 2013 Fourth International Conference, pp. 846-851, 2013.

[4] K. Inagaki, S. Okum, "A High Frequency Link DC/AC Converter using a Three-Phase Output PWM Controlled Cycloconverter," IEEJ Trans. on Industry Applications, Vol. 112, No. 6, pp. 545-552, 1992.

[5] S. Norrga, S. Meier, and S. Östlund, "A Three-Phase Soft-Switched Isolated AC/DC Converter Without Auxiliary Circuit," IEEE Trans. on Industrial Electronics, Vol. 44, No. 3, pp. 836-844, 2008.

[6] S. Takuma, K. Orikawa, J. Itoh, R. Oshima, and H Takahashi "Isolated DC to three-phase AC converter using indirect matrix converter with ZVS applied to all switches," IEEE Energy Conversion Congress and Exposition (ECCE), pp. 4678-4689, 2015.

[7] N. Weise, K. Basu, N. Mohan, "Advanced modulation strategy for a three-phase ac-dc dual active bridge for v2g," IEEE Vehicle Power and Propulsion Conference (VPPC), pp. 1-6, 2011.

[8] M. A. Sayed, K. Suzuki, T. Takeshita, W. Kitagawa, "Soft-switching PWM technique for grid-tie isolated bidirectional DC-AC converter with SiC device," IEEE Trans. on Industry Applications, vol. PP, no. 99, pp. 1-1, 2017.

[9] D. Varajão, R. E. Araújo, L. M. Miranda, J. A. P. Lopes, "Modulation Strategy for a Single-Stage Bidirectional and Isolated AC-DC Matrix Converter for Energy Storage Systems," IEEE Trans. on Industrial Electronics, Vol. 65, No. 4, pp. 3458-3468, 2018.

[10] R.W. De Doncker, D.M. Divan, M.H. Kheraluwala, "A three-phase soft-switched high-power-density DC/DC converter for high-power applications," IEEE Trans. on Industry Applications, Vol. 27, No. 1, pp. 63-73, 1991.

Decoupling Control Method for Eliminating DC Bias Flux of High Frequency Transformer in a Bidirectional Isolated AC/DC Converter

Kensuke Sakuma[1], Koji Shigeuchi[1], Jin Xu[2], Noboru Shimosato[2] and Yukihiko Sato[1*]

1 Graduate School of Science and Engineering, Chiba University, Chiba, Japan
2 Power Supply Systems Department, Myway Plus Corporation, Yokohama, Japan
*E-mail: ysato@faculty.chiba-u.jp

Abstract— This paper proposes a new method of eliminating DC bias flux of high frequency transformer in a bidirectional isolated AC/DC converter. The converter treated in this paper has a matrix converter and a full-bridge converter in primary and secondary sides of a high-frequency isolation transformer, respectively. Exciting current of the transformer pulsates when the matrix converter and full-bridge inverter are controlled for eliminating dc bias flux at the same time. Therefore, this paper proposes a decoupling control method considering physical model of the transformer and shows its effectiveness by simulation.

Keywords— *AC/DC converter, DC bias flux, decoupling control, high-frequency transformer*

I. INTRODUCTION

The needs of storage batteries connected to AC systems are increasing with growth of renewable energy and electric vehicles. An AC/DC converter is used when the storage batteries are connected to the AC system. A bidirectional isolated AC/DC converter using high-frequency transformer is investigated for charge or discharge of storage batteries [1]. Fig.1 shows a circuit configuration of the battery systems. This topology has some problems of large size and low efficiency because the topology requires large electrolytic capacitors and three stage conversion. In order to realize a small size and high efficiency bidirectional isolated AC/DC converter, matrix converter (MC) without DC link capacitors as shown in Fig.2 is a promising solution. The authors have proposed the modulation method for the bidirectional isolated AC/DC converter applying MC [2]. But, problem caused by DC bias flux in the high-frequency transformer has not yet been investigated.

In the transformer side, the output voltages of the MC and full-bridge inverter (INV) has DC offset due to the delay in switching signal and individual variation in parameters of the circuit components [1]. As a result, it leads to DC bias flux in the transformer. At the worst case, magnetic saturation occurs and exciting current increases. The larger exciting current increases copper loss and causes fatal damage on the components.

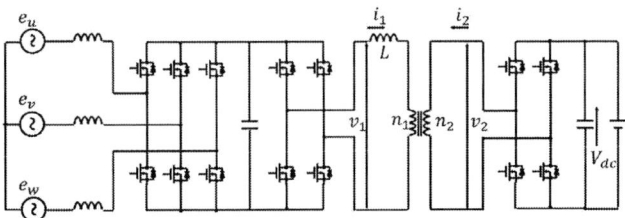

Fig.1. Bidirectional isolated AC/DC converter composed of three-phase PWM xonverter and Dual Active Bridge DC/DC converter.

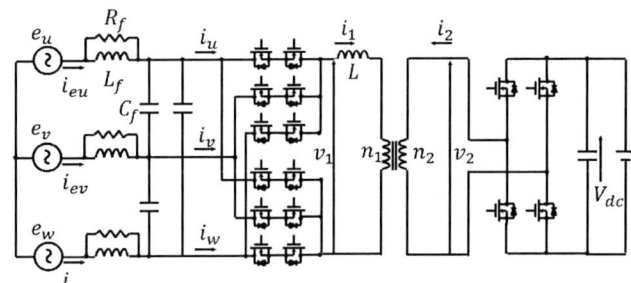

Fig.2. Bidirectional isolated AC/DC converter applying matrix converter.

Increasing the capacity of the transformer and inserting capacitors can prevent the saturation of transformer. Although, these methods have a problem of increasing the volume of the system and number of components. Hence, several methods to eliminate the dc bias flux by feedback control is implemented [3-6]. In [4-5], a PI control based on detection of DC component in the current flowing into the transformer is used for elimination of the DC bias flux. In [6], the DC bias flux of isolated AC/DC converter applying MC is treated. However, the secondary side of the transformer is composed of a diode rectifier. Thus, a different method is required for Dual Active Bridge (DAB) type operation in the secondary side. In particular, the output voltages of the MC and INV must be controlled at the same time so that the DC current flowing into the transformer from the MC side and the INV side becomes around 0 respectively. At this time, if the voltages are not controlled

3522

appropriately, low frequency pulsation appears in the exciting current (detail is described in section III). This is considered to be caused by the cross-coupling between the control of the MC side and INV side. Thus, physical model of the cross-coupling between windings of the transformer should be considered. Therefore, this paper proposes a decoupling control method of primary and secondary windings of the transformer for eliminating DC bias flux in a DAB type bidirectional isolated AC/DC converter.

Firstly, the modulation method of the DAB type bidirectional isolated AC/DC converter is described and the physical model of the transformer is derived. Secondly, a decoupling control method on the physical model is proposed. Finally, simulation and experimental results shows the effectiveness of the proposed control method.

II. MODULATION METHOD OF THREE-PHASE AC/DC DAB CONVERTER

In the following, the modulation method will be described based on Fig.2 and Fig.3. e_u, e_v, e_w are the three-phase voltages of power supply and i_{eu}, i_{ev}, i_{ew} are three phase input currents. f is the switching frequency (= frequency of the high-frequency link) and phase difference δ is normalized with respect to π. It is assumed that $e_u > e_v > 0 > e_w$ and e_H, e_h are defined as follows.

$$e_H = e_u - e_w \quad (1)$$
$$e_h = e_v - e_w. \quad (2)$$

Assuming that the turn ratio of the transformer is $n(= n_1/n_2)$, the equation of the current flowing through the inductor L is obtained as follows:

$$i_1 = \frac{1}{L}\int v_L\,dt = \frac{1}{L}\int \left(v_1 - \frac{n_1}{n_2}v_2\right)dt. \quad (3)$$

The output voltages v_1, v_2 of MC and INV are realized by the duration d_h of e_h and δ as shown in Fig.3. Transmitted power P and average current $(\overline{i_v})$ in a cycle $1/f$ are obtained by calculation of (3) and expressed as

$$P = \frac{e_H n V_{dc}}{2fL}\delta(1-\delta) + \frac{(e_H - e_h)nV_{dc}}{4fL}d_h(1 \quad (4)$$
$$- 2\delta - d_h)$$
$$\overline{i_v} = \frac{nV_{dc}}{2fL}\delta d_h + \frac{e_H - nV_{dc}}{4fL}d_h(1 - d_h) \quad (5)$$

Finally, it is assumed that power reference P^* and v phase current reference i_{ev}^* correspond to (4) and (5) as follows:

$$\begin{cases} P^* = P \\ i_{ev}^* = \overline{i_v}. \end{cases} \quad (6)$$

So, δ and d_h to realize given transmitted power reference and sinusoidal input current can be obtained by solving (4) and (5) [2].

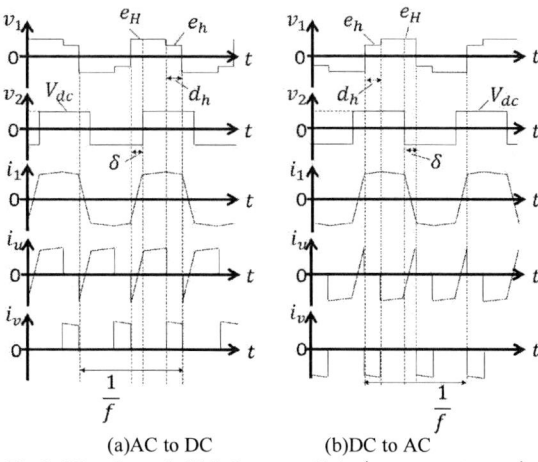

(a)AC to DC   (b)DC to AC

Fig.3. Waveforms in high-frequency link. $(e_u > e_v > 0 > e_w)$

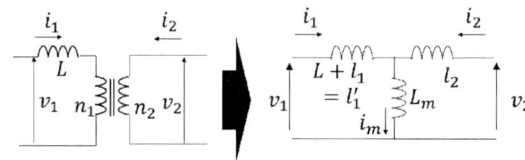

Fig.4. Equivalent circuit of the transformer including external inductor.

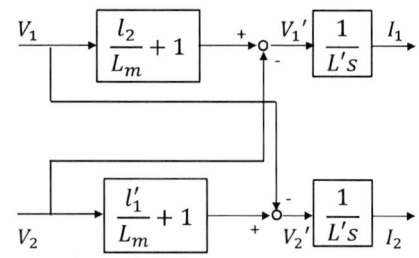

Fig.5. Block diagram of transformer.

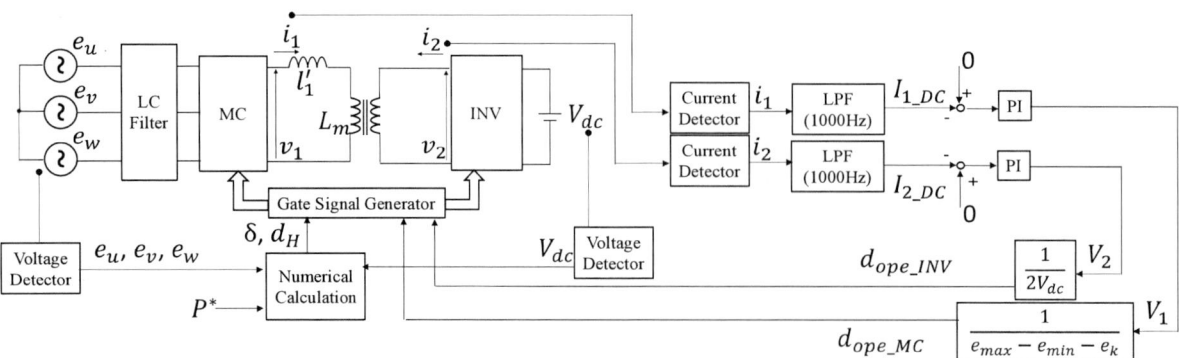

Fig.6. System configuration with control blocks. (without considering cross-coupling)

3523

III. Control Method to Eliminate DC Bias Flux in High-Frequency Transformer

A. Phisical Model of Transformer

An equivalent circuit of the transformer including external inductor L is shown in Fig.4. i_1, i_2, l_1, l_2, and L_m are the currents of the primary and secondary sides, the leakage inductances, and the exciting inductance converted to the primary side. Further, small winding resistances of the transformer are ignored. In order to eliminate DC bias flux in the transformer, it is necessary to control DC offset I_{m_DC} of the exciting current i_m to 0. From Fig.4, i_m can be described as follows;

$$i_m = i_1 + i_2. \tag{7}$$

Thus, DC offset I_{1_DC}, I_{2_DC} of i_1, i_2 should be controlled to 0.

From the equivalent circuit in Fig.4, the voltage equations of the transformer are given by

$$
\begin{cases}
v_1 = l_1' \dfrac{di_1}{dt} + L_m \dfrac{d(i_1 + i_2)}{dt} \\
v_2 = l_2 \dfrac{di_2}{dt} + L_m \dfrac{d(i_1 + i_2)}{dt}.
\end{cases}
\tag{8}
$$

Furthermore, Laplace transform is performed in (8) and to solve for current i_1, i_2. Then the following equation is obtained.

$$
\begin{pmatrix} I_1 \\ I_2 \end{pmatrix}
= \frac{1}{s} \frac{L_m}{(l_1' + L_m)(l_2 + L_m) - L_m^2}
\begin{pmatrix} \frac{l_2}{L_m} + 1 & -1 \\ -1 & \frac{l_1'}{L_m} + 1 \end{pmatrix}
\begin{pmatrix} V_1 \\ V_2 \end{pmatrix}
\tag{9}
$$

In (9), Laplace transforms of v_1, v_2, i_L, and i_2 are V_1, V_2, I_1, and I_2, respectively. In addition, block diagram for V_1, V_2, I_1, and I_2 is described as shown in Fig.5 if $1/L'$ is determined as follows:

$$\frac{1}{L'} = \frac{L_m}{(l_1' + L_m)(l_2 + L_m) - L_m^2}. \tag{10}$$

In Fig.5, V_1' and V_2' are voltages that directly control I_1 and I_2, respectively. From Fig.5, it can be seen that I_1, I_2 are determined by V_1 and V_2 interfering each other. Therefore, in order to control I_1, I_2, separately, appropriate voltage V_1, V_2 must be generated by the MC and INV to obtain required V_1', V_2'.

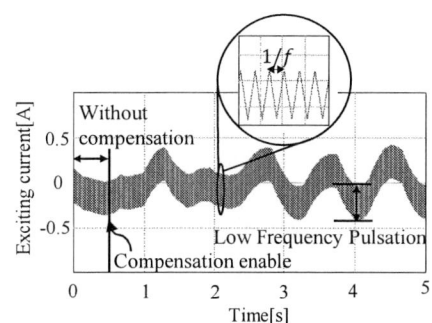

Fig.7. Exciting current in the transformer without decoupling control. (the case of Fig.6)

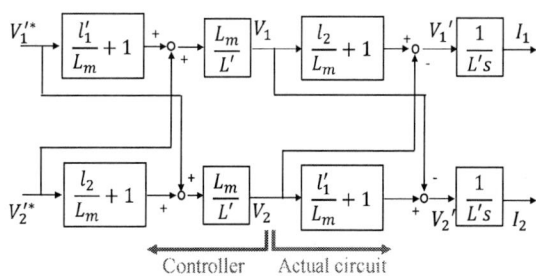

Fig.8. Decoupling control.

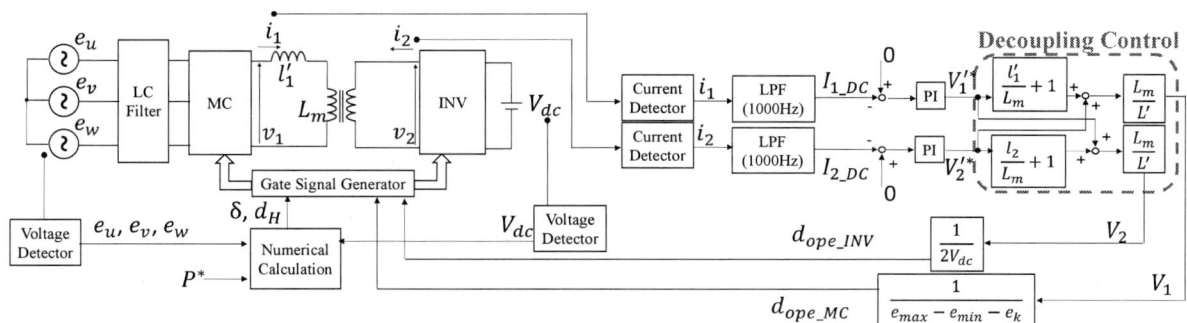

Fig.9. System configuration with control blocks. (with proposed decoupling control)

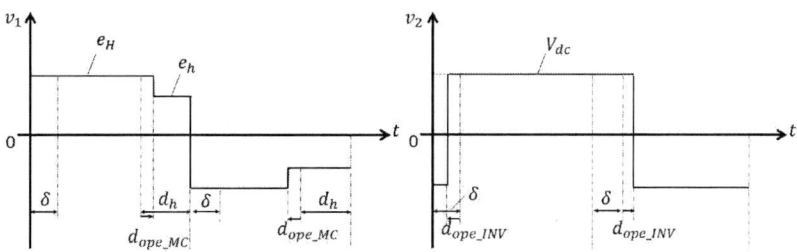

(a) : Output voltage of MC   (b) : Output voltage of inverter

Fig.10. Introduction of d_{ope_MC}, d_{ope_INV}.

3524

B. Decoupling Control method of primaly side and secondaly side

In the previous section, it is shown that the cross-coupling blocks exist in the physical model of the transformer. If the DC offset of I_1, I_2 are controlled to 0 without considering cross-coupling as shown in Fig.6, the waveform as shown in Fig.7 appears as the exciting current. Fig.7 shows waveform of the exciting current i_m in a transient response when the compensation method without consideration of the cross-coupling is enabled. Pulsation with lower frequency than the switching frequency (f=15[kHz]) can be observed. The pulsation is caused by disturbance (the delay in switching signal and individual variation in parameters of the circuit components). In addition, the decay of the pulsation is slow. It is seen that due to the cross-coupling between the primary and secondary sides, appropriate control cannot be realized and the pulsation cannot be suppressed less than the switching frequency component immediately. Hence, in this section, decoupling control is discussed. Here, new variables $V_1'^*$ and $V_2'^*$ are defined as (11) from (9).

$$\begin{pmatrix} \dfrac{l_2}{L_m}+1 & -1 \\ -1 & \dfrac{l_1'}{L_m}+1 \end{pmatrix}\begin{pmatrix} V_1 \\ V_2 \end{pmatrix} = \begin{pmatrix} V_1'^* \\ V_2'^* \end{pmatrix} \tag{11}$$

Then, (11) can be transformed as follows:

$$\begin{pmatrix} V_1 \\ V_2 \end{pmatrix} = \frac{L_m}{L'}\begin{pmatrix} \dfrac{l_1'}{L_m}+1 & 1 \\ 1 & \dfrac{l_2}{L_m}+1 \end{pmatrix}\begin{pmatrix} V_1'^* \\ V_2'^* \end{pmatrix}. \tag{12}$$

Thus, using (10) and (12), equation (9) can be transformed as follows:

$$\begin{pmatrix} I_1 \\ I_2 \end{pmatrix} = \frac{1}{L's}\begin{pmatrix} V_1'^* \\ V_2'^* \end{pmatrix}. \tag{13}$$

From (13), references $V_1'^*$ and $V_2'^*$ of the voltage V_1', V_2' to control I_1, I_2 are obtained. Fig.8 shows relationship among these variables as a block diagram. Further, Fig.9 shows the configuration of whole the system applying the proposed decoupling control.

C. Control Method of V_1 and V_2

In order to realize above control, it is necessary to generate the voltages V_1, V_2 obtained by (12) with MC and INV in every switching cycle $1/f$. Hence, in addition to δ, d_m obtained in section II, new operation variable d_{ope_MC}, d_{ope_INV} as shown in Fig.10 are introduced. In Fig.10 (a), the variables are obtained by the following equations.

$$f\int_0^{\frac{1}{f}} v_1\,dt = V_1 \tag{14}$$

After calculating (14), d_{ope_MC} is obtained as follows:

$$d_{ope\_MC} = \frac{V_1}{e_H - e_h}. \tag{15}$$

In the same way, d_{ope_INV} is obtained as follows:

Table I
Simulation conditions.

AC source(line)	200V$_{rms}$, 50Hz
Input filter L_f, C_f, R_f	0.75mH, 10uF, 16.5Ω
Frequency of transformer f	15kHz
Inductance l_1'	440uH
Secondary leakage inductance l_2	1uH
Excitation inductance L_m	20mH
Turns ratio $n_1:n_2$	4:1
DC source V_{dc}	50V
Rated power P	1kW
PI gain K, T	0.1V/A, 0.1s

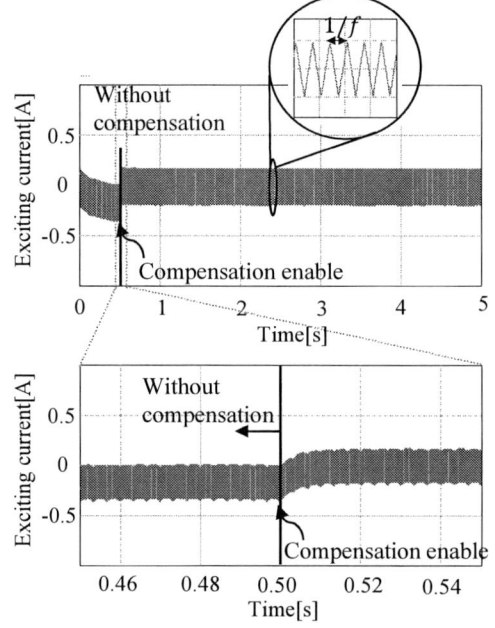

Fig.11. Exciting current in transformer with proposed decoupling control. (the case of Fig.10)

Table II
Experimental conditions.

AC source(line)	200V$_{rms}$, 50Hz
Input filter L_f, C_f, R_f	0.75mH, 10uF, 16.5Ω
Frequency of transformer f	15.15kHz
External inductor L	440uH
Primary leakage inductance l_1	1uH
Secondary leakage inductance l_2 (referred to the primary side)	~1uH
Excitation inductance L_m	20mH
Turns ratio $N_1:N_2$	4:1
Secondary DC voltage V_{dc}	50V
Rated power P	1kW
PI gain K, T	0.1V/A, 0.1s

$$d_{ope\_INV} = \frac{V_2}{2V_{dc}}. \tag{16}$$

On the other hand, there is a concern that d_{ope_MC}, d_{ope_INV} will affect the main converter operation. But, in most case, the DC offset of the output voltage due to the delay in switching signal and variation in circuit parameters is smaller than the AC component. Hence, the effect on the main operation of the converter by adopting the above operation can be negligible.

IV. VERIFICATION OF EFFECTIVENESS

A. Simulation Results

In order to confirm the effectiveness of the proposed method, simulation of the total system shown in Fig.9 was performed by using simulation software PSIM. Table I shows the simulation conditions. Fig.11 shows waveforms of the exciting current i_m in transient response when the proposed decoupling control is enabled. In this case, the pulsation can be suppressed immediately because appropriate control is performed by the proposed method. In the result without the proposed method as shown in Fig.7, low frequency pulsation appeared. Comparing these results, it can be seen that the DC bias flux and low frequency pulsation are effectively eliminated by the proposed method.

B. Experimental Results

This section presents experimental results employing a 1kW laboratory prototype to verify the effectiveness of the proposed decoupling control method. Table II shows the conditions of the experiments. Fig.12 shows experimental waveforms of line current i_{eu} and exciting current i_m in steady state when power reference P^* is 800W. i_m is filtered by a low pass filter with the cut-off frequency f_c is 20Hz because i_m contains switching and commercial frequency components. Fig.12(a) is the waveforms without DC bias compensation, Fig.12(b) is the waveforms with the conventional compensation method without the proposed decoupling control, and Fig.12(c) is the waveforms with the proposed decoupling compensation method. As shown in the lower side of Fig.12(a), the DC bias component appears in i_m if it is not controlled to 0. Also in the case of the conventional method, the DC bias component is not completely eliminated as shown in Fig.12(b). In contrast, in the case of the proposed method, the DC bias component does not appear in i_m as shown in Fig.12(c). In addition, line current i_{eu} shown in the upper side of Fig.12 is sinusoidal waveforms in all cases. From these results, it can be confirmed that the DC bias compensation based on the feedback control can eliminate the DC bias flux achieving sinusoidal line current.

Fig.13 shows waveforms of the exciting current i_m in transient response when the conventional or proposed

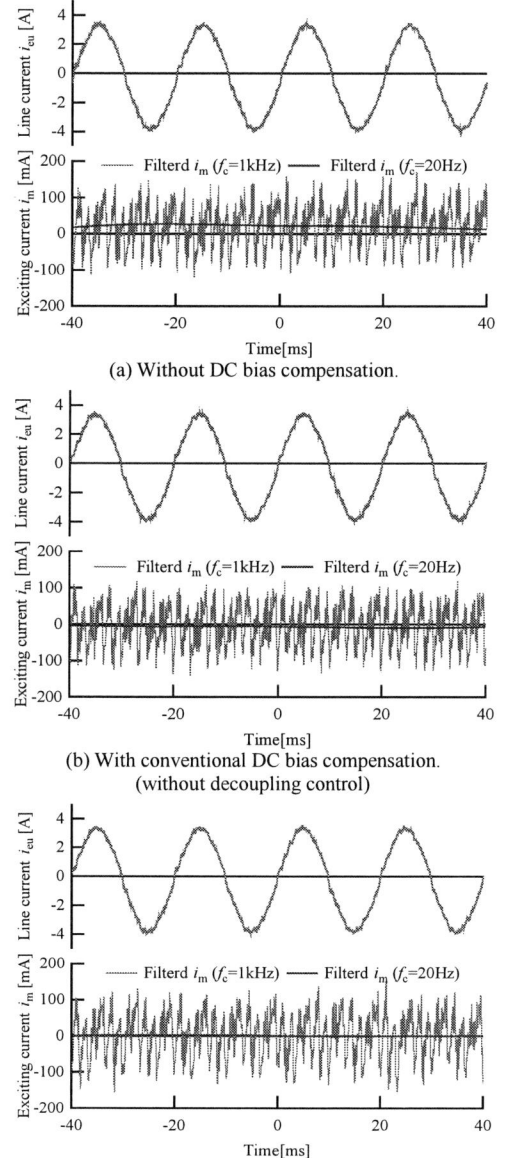

(a) Without DC bias compensation.

(b) With conventional DC bias compensation.
(without decoupling control)

(c) With proposed DC bias compensation.(with decoupling control)
Fig.12. Experimental waveforms of line current and exiting current in steady state. ($P^* = 800$W)

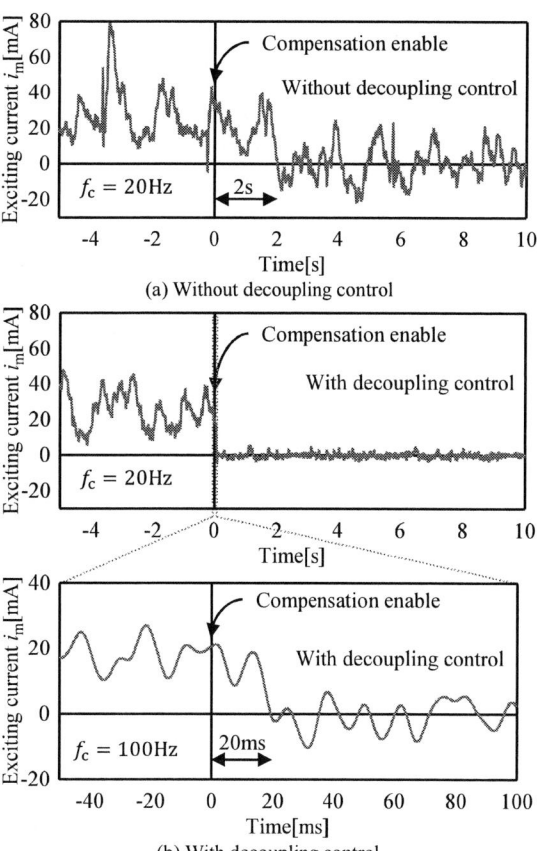

(a) Without decoupling control

(b) With decoupling control
Fig.13. Exciting current waveform in transient state. DC bias compensation is enabled at instant of Time=0. ($P^* = 800$W)

3526

compensation method is enabled. i_m is filtered by a low pass filter with the cut-off frequency $f_c = 20$Hz or 100Hz($f_c = 100$Hz is in order to show a quick response). The transient response of i_m without the decoupling control shown in Fig.13(a) has a long setting time that is around 2s and low frequency pulsation since appropriate control cannot be realized by the cross-coupling between the primary and secondary sides. Compared with the simulation results, the amplitude of the pulsation is small because the dumping components of the system are larger than that in case of the simulation. In contrast, the transient response of i_m with proposed decoupling control shown in Fig.13(b) has a shorter setting time around 30ms. In addition, the proposed method can suppress the pulsation more effective than the conventional method. Therefore, the proposed decoupling control can achieve significant improvement of the exciting current controllability.

Fig.14 shows waveforms of the line current i_{eu} and exciting current i_m in transient response when power reference P^* is changed from 400W to 800W. i_m is filtered by a low pass filter with the cut-off frequency $f_c = 100$Hz. The fluctuation of i_m occurs at instant of steep change in P^*. However, the proposed method eliminates this fluctuation rapidly. The setting time is shorter than 20ms. From this result, it can be confirmed that the proposed method performs the effective compensation method even when P^* is changed steeply.

V. CONCLUSION

This paper proposed the decoupling control method to eliminate the DC bias flux of high-frequency transformer in the bidirectional isolated AC/DC converter. The method decouples the primary and secondary sides of the high-frequency transformer by using the physical model of the transformer, and realizes appropriate control between the primary and secondary side. The simulation and experimental results demonstrated the effectiveness of the proposed method. The proposed method can eliminate the DC bias flux of high-frequency transformer in the bidirectional isolated AC/DC converter and achieve the significant improvement of the exiting current controllability. The experimental results with the proposed method showed the shorter setting time of around 30ms (it is 2s in case of the conventional method). In addition, the conventional method cannot suppress the low frequency pulsation. This is caused by the cross-coupling between the primary and secondary sides of the transformer. In contrast, the proposed method can suppress the low frequency pulsation more effectively than the conventional method because the proposed method decouples the primary and secondary side of the high frequency transformer.

From results demonstrated in this paper, it is confirmed that the proposed decoupling control can eliminate the DC bias flux of the high-frequency transformer in the bidirectional isolated AC/DC converter. In additon, the proposed decoupling control can achieve the significant improvement of the exiting current controllability.

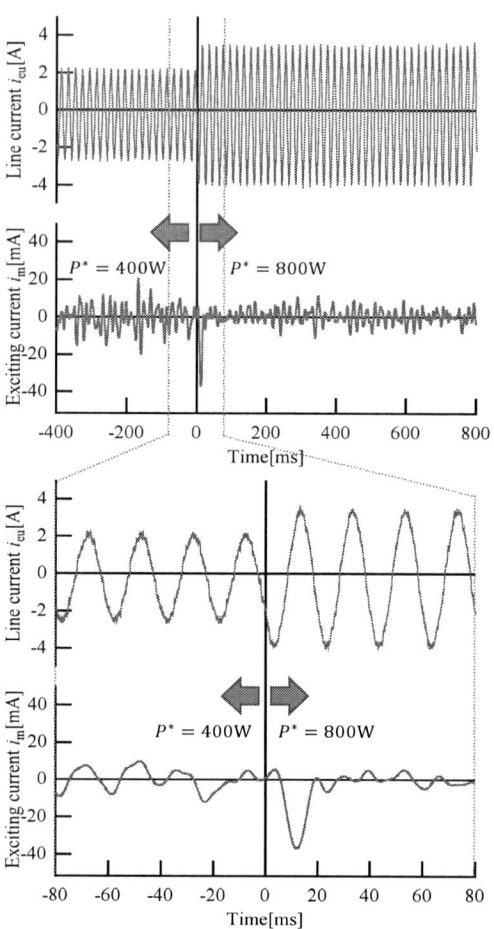

Fig.14. Waveforms of line current and exciting current in transient state. Power reference value P^* is changed to 800W from 400W at instant of Time=0.

REFERENCES

[1] N. M. L. Tan, T. Abe, and H. Akagi, "Design and Performance of a Bidirectional Isolated DC-DC converter for Battery Energy Storage System," *IEEE Transactions on Power Electronics*, vol. 27, no. 3, pp. 1237-1248, 2012.

[2] K. Shigeuchi, K. Sakuma, J. Xu, N. Shimosato, and Y. Sato, "Improvement of Input Current Waveform of an Isolated Bidirectional AC/DC Converter Employing Matrix Converter," *The Papers of Joint Technical Meeting on 'Electron Devices' and 'Semiconductor Power Converter' IEE-Japan*, SPC-16-154, pp. 25-30, 2016.

[3] G. Ortiz, L. Fassler, J. W. Kolar, and O. Apeldoorn, "Flux Balancing of Isolation Transformers and Application of 'The Magnetic Ear' for Closed-Loop Volt-Second Compensation," *IEEE Transactions on Power Electronics*, vol. 29, no. 8, pp. 4078-4090, 2014.

[4] S. Dutta, and S. Bhattacharya, "A Method to Measure the DC Bias in High Frequency Isolation Transformer of the Dual Active Bridge Dc to DC Converter and its removal using Current Injection and PWM switching," *Energy Conversation Congress and Exposition (ECCE)*, pp. 1134-1139, 2014.

[5] B. P. Baddipadiga, and M. Ferdowsi, "Dual Loop Control for Eliminating DC-bias in a DC-DC Dual Active Bridge Converter," *3rd International Conference on Renewable Energy Research and Applications*, pp. 490-495, 2014.

[6] R. Okuoka, M. Ishida, N. Yamamura, and M. Koyama, "Soft-Switching Technique and DC bias suppression control in High Frequency Link Form Matrix Converter System," *IEE-Japan Industry Applications Society Conference*, pp. 181-184, 2017.

Interleaved Voltage-Doubler Boost Converter for Power Factor Correction

Ta-Hsun Lo, *Student Member, IEEE*, Jen-Hao Teng, *Senior Member, IEEE*, Bo-Jia Huang
Department of Electrical Engineering,
National Sun Yat-Sen University,
Kaohsiung, Taiwan

Abstract - An Interleaved Voltage-Doubler Boost Converter (IVDBC) using digitalized average current control to achieve Power Factor Correction (PFC) is proposed in this paper. The basic circuit configuration of proposed IVDBC is a two-phase interleaved boost converter and a full-wave voltage-doubler rectifier in series. The capacitors of the voltage-doubler circuit are charged alternately by the two-phase interleaved operation; therefore, the voltage conversion gain of the IVDBC is twice that of a conventional interleaved boost converter. Consequently, the proposed IVDBC is probable to be used for high-voltage applications. A TI TMS320F28335 is used in this paper to design and implement the average current control scheme for the IVDBC. A prototype circuit with an AC input voltage of 110V, DC output voltages between 400V to 700V and a rated output current of 1.4A is realized in this paper. Experimental results indicate that the Total Harmonic Distortion (THD) and power factor of input current of the proposed IVDBC for PFC can conform the requirements of international regulations. Besides, a maximum conversion efficiency of 94.07% can be achieved.

Keywords—Power Factor Correction, Voltage Doubler Rectifier, Two-Phase Interleaved Operation, Total Harmonic Distortion.

I. INTRODUCTION

With the rise of environmental awareness, higher power quality and energy efficiency are required. International organizations also developed a number of standards [1-2] to regulate harmonics and power factor. Meanwhile, in order to improve the conversion efficiency of active power for power conversion devices, some energy-related programs such as Energy Star, Climate Rescue, and so on postulate power conversion devices to have conversion efficiencies higher than 80% at 20% to 100% of loads and a power factor higher than 0.9 at full load [3]. Power-electronic-based devices can be treated as nonlinear loads and commonly result in harmonic pollution and power factor problems in the power supply side. It also make power supply side must provide more active power to satisfy the load. In order to improve the above problems, Power Factor Correction (PFC) is designed and integrated into most of the power-electronic–based devices [4].

Many different PFC circuits have been proposed and the conventional interleaved boost converter [4-11] is one

---

This work was supported by Ministry of Science and Technology of Taiwan under Contracts MOST 107-3113-E-110-001 – and MOST 104-2221-E-110-042-MY3.

of the commonly-used PFC circuits due to the higher conversion efficiency and lower input current ripple. The output voltage of commonly-used interleaved boost converter should be below 400V due to the voltage stability of boost converter. However, there are many applications that need the output voltage of a PFC circuit to be higher than 400V. For example, the charging voltages of a fast charger for electric vehicles according to CHAdeMO and SAE J1773 [10-11] can be up to 500V and 600V, respectively. The charging voltages for electric buses and electric trucks may be 700V or up. Higher output voltage can usually be realized by transformer; however, it often results in larger size, high voltage stress on output capacitors and diodes and so on [12-14]. Therefore, a Interleaved Voltage-Doubler Boost Converter (IVDBC) is proposed in this paper for high-voltage applications. Experimental results show that the Total Harmonic Distortion (THD) and power factor of the proposed IVDBC for PFC under the rated output current are in accordance with the requirements of international standards.

II. BASIC CONCEPTS OF PROPOSED IVDBC

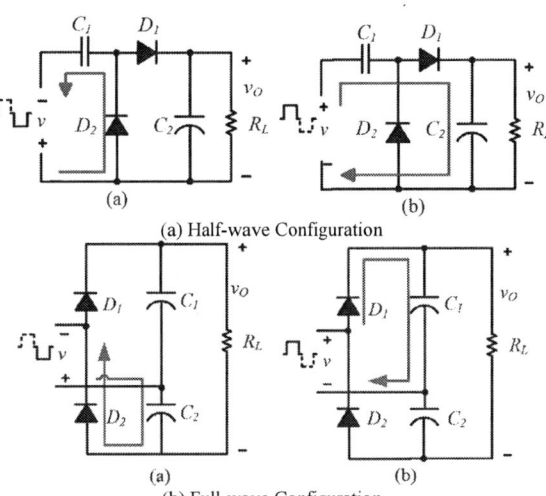

(a) Half-wave Configuration

(b) Full-wave Configuration

Fig. 1: Conventional Voltage Doubler Rectifiers

An interleaved voltage-doubler boost converter is proposed in this paper. The half-wave and full-wave voltage doubler rectifier can be used for step-up voltage applications because the output voltage is twice of the input voltage [13-14]. Fig. 1 illustrates the circuit configurations of two types of voltage doubler rectifiers with the input voltage, v, in one switching cycle being either $+V$, $-V$, or zero. Fig. 1(a) and Fig. 1(b) show the operational modes of the half-wave and full-wave configurations. The output voltage of $2V$ can be achieved. Fig. 2 illustrates the basic configuration of a two-phase interleaved boost converter. The gate driver signals for Q_1 and Q_2 are controlled by two 180 degree shifted signals.

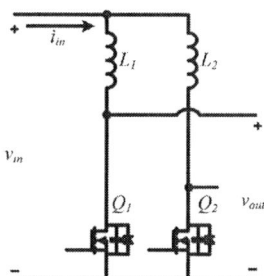

Fig. 2: Basic Circuit Configurations of Interleaved Boost Converter

In this paper, a modified full-wave voltage-doubler rectifier is employed and integrated into Fig. 2. Fig. 3 shows the circuit configuration of proposed IVDBC. From Fig. 3, it can be seen that the proposed IVDBC for PFC is composed of a rectifier, a two-phase interleaved boost converter and a modified full-wave voltage doubler. The modified voltage-doubler circuit with two capacitors in series at the output terminal is cascaded to a two-phase interleaved boost circuit. The capacitors of the voltage-doubler circuit are charged alternately by the two-phase interleaved operation; therefore, the voltage conversion gain of the proposed converter is two-times of a conventional boost converter. One of the advantages of full-wave voltage doubler rectifier is the voltage stresses on capacitors C_1 and C_2 as shown in Fig. 1(b) is equal to the input voltage and is half of the voltage stress on capacitor C_2 of Fig. 1(a). Due to the lower voltage stresses on capacitors, the proposed IVDBC employing the full-wave voltage doubler rectifier is suitable for high-voltage applications.

III. OPERATIONAL PRINCIPLES OF PROPOSED IVDBC

The gate driver signals of proposed IVDBC is similar to the conventional interleaved boost converter. Both of the inductor currents, i_{L1} and i_{L2}, share the input current evenly by controlling two 180 degree shifted signals for the gate driver signals of Q_1 and Q_2. The operational principles of proposed IVDBC can be classified as duty cycle greater than 0.5 and smaller than 0.5, respectively.

A. Mode A : 0.5 ≤ Duty Cycle ≤ 1

Fig. 4 shows the key waveforms for Mode A, i.e. duty cycle is greater than 0.5, where V_{GS1} and V_{GS2} are the gate driver signals for power switches, Q_1 and Q_2, respectively. V_{DS1} and V_{DS2} are the drain-source voltages for power switches, Q_1 and Q_2, respectively. V_{D1} and V_{D2}

are the voltages for diodes, D_1 and D_2, respectively. I_{L1} are I_{L2} are the currents of inductors, L_1 and L_2, respectively. I_{in} is the input current as shown in Fig. 3. I_{D1} and I_{D2} are the currents for diodes, D_1 and D_2, respectively. The proposed IVDBC can be divided into four operational states in one switching cycle and the equivalent circuits are shown in Fig. 5.

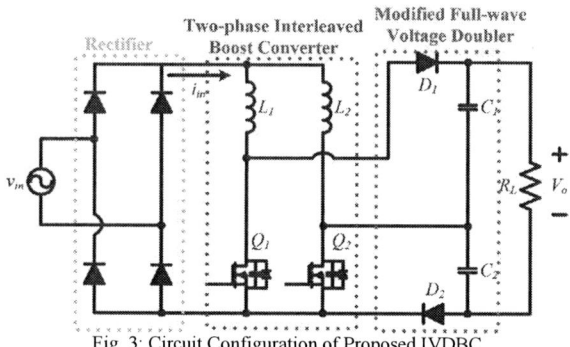

Fig. 3: Circuit Configuration of Proposed IVDBC

Mode A.1 ($t_0 \leq t < t_1$): As shown in Fig. 5(a), power switches Q_1 and Q_1 are in the ON state. The energy from V_{in} is stored in L_1 and L_2 by currents i_{L1} and i_{L2}, respectively. Meanwhile, the energy stored in C_1 and C_2 is transferred to output load R_L. The inductor voltages for L_1 and L_2 can be expressed as

$$V_{L1} = L\frac{di_{L1}}{dt} = V_{in} \tag{1a}$$

$$V_{L2} = L\frac{di_{L2}}{dt} = V_{in}^{'} \tag{1b}$$

Mode A.2 ($t_1 \leq t < t_2$): Power switch Q_1 at t_1 is in the ON state, power switch Q_2 turns OFF. As shown in Fig. 5(b), the energy from V_{in} is stored in inductor L_1 by current i_{L1}. The current i_{L2} transfers energy of V_{in} and inductor L_2 to capacitor C_2. The inductor voltages for L_1 and L_2 can be expressed as

$$V_{L1} = L\frac{di_{L1}}{dt} = V_{in} \tag{2a}$$

$$V_{L2} = L\frac{di_{L2}}{dt} = V_{in} - V_{C2} \tag{2b}$$

Mode A.3 ($t_2 \leq t < t_3$): Power switch Q_1 is in the ON state, power switch Q_2 turns ON at t_2. The current i_{L1} and i_{L2} are same as the Mode A.1 shown in Fig. 5(a).

Mode A.4 ($t_3 \leq t < t_4$): Power switch Q_1 turns OFF and power switch Q_2 is in the ON state at t_3. As shown in Fig. 5(c), energy of V_{in} and L_1 is transferred to capacitor C_1 through diodes D_1 and power switch Q_2. Meanwhile, energy from V_{in} is stored in inductor L_2 by current i_{L2}. The inductor voltages of L_1 and L_2 can be expressed as

3529

$$V_{L1} = L\frac{di_{L1}}{dt} = V_{in} - V_{C1} \tag{3a}$$

$$V_{L2} = L\frac{di_{L2}}{dt} = V_{in} \tag{3b}$$

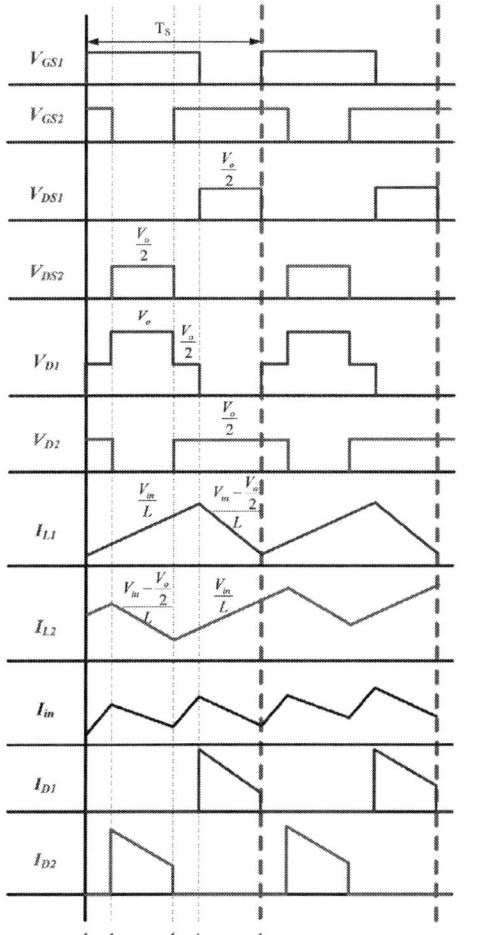

Fig. 4: Key Waveforms of Proposed IVDBC for Mode A

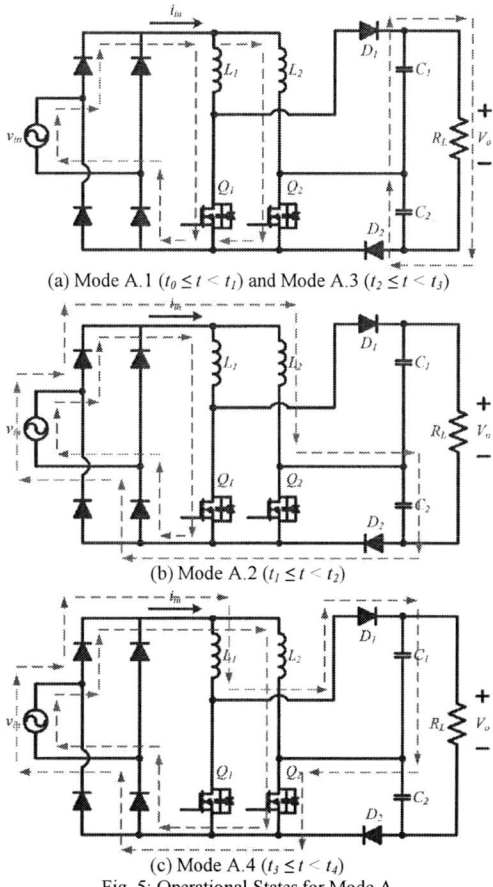

(a) Mode A.1 ($t_0 \leq t < t_1$) and Mode A.3 ($t_2 \leq t < t_3$)

(b) Mode A.2 ($t_1 \leq t < t_2$)

(c) Mode A.4 ($t_3 \leq t < t_4$)

Fig. 5: Operational States for Mode A

The output voltage can be expressed as

$$V_o = V_{C1} + V_{C2} \tag{5}$$

The capacitor voltages can be written as

$$V_{C1} = \frac{1-D_{Q2}}{2-D_{Q1}-D_{Q2}}V_o \tag{6a}$$

$$V_{C2} = \frac{1-D_{Q1}}{2-D_{Q1}-D_{Q2}}V_o \tag{6b}$$

Thus, the voltage conversion ratio, M, calculated from (4) and (6), can be rewritten as

$$M = \frac{V_o}{V_{in}} = \frac{2-D_{Q1}-D_{Q2}}{1-D_{Q1}-D_{Q2}+D_{Q1}D_{Q2}} \tag{7}$$

B. Mode B: 0 < Duty Cycle < 0.5

The key waveforms and operating stages for Mode B, i.e. duty cycle is smaller than 0.5, are similar as Figs. 4 and 5; therefore, they are not shown here due to limited space and will be represented in the final manuscript. The

The voltage of inductors combining with Modes A.1 to A4, when duty cycle is larger than 0.5, can be expressed as

$$V_{L1} = \frac{V_{in}\Delta t_1 + V_{in}\Delta t_2 + V_{in}\Delta t_3 + (V_{in}-V_{C1})\Delta t_4}{T_s} \tag{4a}$$

$$= V_{in} - (1-D_{Q1})V_{C1} = 0$$

$$V_{L2} = \frac{V_{in}\Delta t_1 + (V_{in}-V_{C2})\Delta t_2 + V_{in}\Delta t_3 + V_{in}\Delta t_4}{T_s} \tag{4b}$$

$$= V_{in} - \left(1-D_{Q2}\right)V_{C2} = 0$$

$$\Delta t_n = t_n - t_{n-1} \quad n = 1 \text{L } 4 \tag{4c}$$

where D_{Q1} and D_{Q2} are the duty cycles for the gate driver signals of Q_1 and Q_2.

voltage conversion ratio, M, for Mode B can be derived and expressed as

$$M = \frac{V_o}{V_{in}} = \frac{1 + D_{Q1} - D_{Q2}}{1 - D_{Q1} - D_{Q2} + D_{Q1}D_{Q2}} \qquad (8)$$

The proposed IVDBC can be used for PFC. The commonly-used average current control is employed in this paper. However, due to voltage unbalance for capacitors C_1 and C_2, a separate control scheme for power switches Q_1 and Q_2 as shown in Fig. 6 is designed in this paper to make sure the equal voltages of capacitors C_1 and C_2. Due to limited space, the detailed discussions of Fig. 6 are not described here.

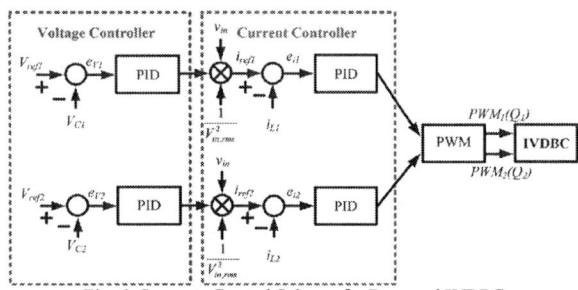

Fig. 6: Separate Control Scheme for Proposed IVDBC

IV. PRELIMINARY EXPERIMENTAL RESULTS

A prototype circuit with an AC input voltage of 110V, DC output voltages between 400V to 700V and a rated output current of 1.4A is implemented to verify the performance of proposed IVDBC for PFC. The TI TMS320F28335 [15] is used to realize the separate control scheme as shown in Fig. 6 and to control the proposed IVDBC. Table I lists the parameters used in the prototype circuit. The inductors, L_1 and L_2, values of 2mH were selected to let the input current more smooth. The capacitors, C_1 and C_2, values of 470μF were selected to reduce the output ripple.

TABLE I: PARAMETERS OF PROPOSED IVDBC

Rated Input Voltge(V_{in})	110 V$_{rms}$
Rated Output Voltage (V_o)	400-700V DC
Rated Output Current	1.4A
Switching Frequency	50 kHz
L_1、L_2	2 mH
C_1、C_2	470 μF
Q_1、Q_2	C2M0025120D
D_1、D_2	DSEP29-12A

Fig. 7 shows the input current I_{in}, input voltage V_{in} and voltages of capacitors C_1 and C_2 under the rated output current of 1.4A and the output voltage of 700V for the proposed IVDBC for PFC. From Fig. 7, it can be observed that the voltages of capacitors C_1 and C_2 are almost equal. Besides, the waveform of input current follows the input voltage and is close to sinusoid. Fig. 8 shows the currents of inductors L_1 and L_2. From Fig. 8, it

can be observed that the currents of inductors L_1 and L_2 are almost equal. The input current can be evenly shared by inductors L_1 and L_2; therefore, the advantages of interleaved boost converter such as lower input current ripple can still be guaranteed. Figs. 9 and 10 are the conversion efficiency and THD of input current under the rated output current of 1.4A and the output voltage of 700V. Fig. 9 indicates that the conversion efficiency of proposed IVDBC is 94.07% with power factor 0.9995. Fig. 10 shows that percentage of harmonic current from the 2nd to 50th orders and THD of the input current is 3.31%.

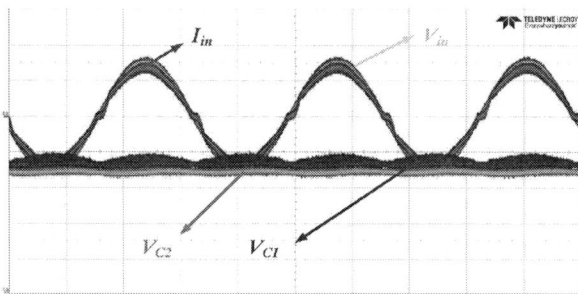

(Vin: 100V/div; Iin: 5A/div; V$_{C1}$, V$_{C2}$: 100V/div; Time: 5ms/div)
Fig. 7: Input voltage, Current and Capacitor Voltages of IVDBC

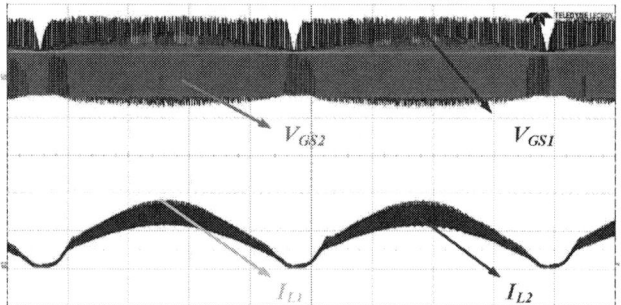

(i$_{L1}$, i$_{L2}$: 2A/div; V$_{GS1}$, V$_{GS2}$: 20V/div; Time: 2ms/div)
Fig. 8: Inductor Currents of IVDBC

U$_{a:1}$	:	109.33	V
I$_{a:1}$	:	9.656	A
U$_{a:2}$	:	0.7033k	V
I$_{a:2}$	:	1.4112	A
P$_1$	:	1.0551k	W
P$_2$	:	0.9926k	W
?$_1$	:	94.07	%
λ$_1$	:	0.9995	

Fig. 9: Conversion Efficiency of IVDBC

From the preliminary experimental results, it can be observed that the THD of input current and power factor of the proposed IVDBC for PFC is less than 5% and larger than 0.99, respectively. These results are in accordance with the requirements of international standards. Therefore, the validity of proposed IVDBC can be demonstrated and should have great potential to be used for high-voltage applications. More experimental results for the proposed IVDBC under different output voltages and currents will be shown in the final manuscript. Besides, the detailed loss analysis for the proposed IVDBC and other PFC circuits will also be investigated and compared.

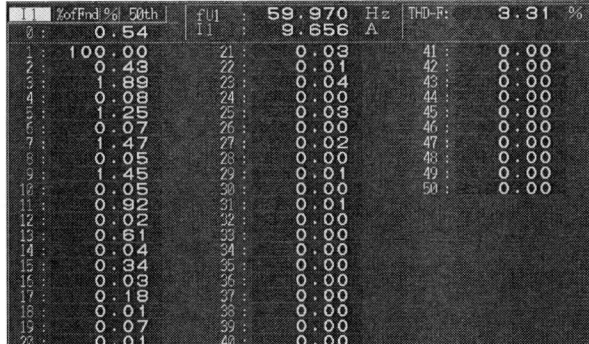

Fig. 10: Harmonic Spectra and THD of IVDBC for PFC

REFERENCES

[1] Schaffner, "IEC 61000-3-2 Harmonics standards overview," May 2006.

[2] M. Thomas, J. Daniel, "Application of IEEE std. 519-1992 harmonic limits" IEEE IAS PULP, 2005.

[3] Eenergy Star Website [On Line]. https://www.energystar.gov/.

[4] "Power Factor Correction (PFC) handbook: Choosing the right power factor controller solution," ON Semiconductor, HBD853/D, Rev. 5, Apr–2014.

[5] Sam Abdel-Rahman, Franz Stückler and Ken Siu, "PFC boost converter design guide," Infineon Application Note, 2016-02-22.

[6] Milan M. Jovanovic and Yungtaek Jang, State-of-the-art, single-phase, active power-factor-correction techniques for high-power applications - an overview, IEEE Trans. on Industrial Electronics, vol. 52, no. 3, June 2005, pp. 701 - 708.

[7] B. A. Miwa, D. M. Otten, M. E. Schlecht, "High efficiency power factor correction using interleaving techniques," in Proc. IEEE APEC, Boston, MA, pp. 557-568, Feb. 1992.

[8] L. Balogh, R. Redl, "Power-factor correction with interleaved boost converters in continuous-inductor-current mode," in Proc. IEEE APEC, San Diego, CA, pp. 168-174, Mar. 1993.

[9] Y. Jang, M. M. Jovanovic, "Interleaved boost converter with intrinsic voltage-doubler characteristic for universal-line PFC front end," IEEE Tran. Power Electronics, vol. 22, pp. 1394-1401, July 2007.

[10] CHAdeMO Association, "Electric vehicle quick charger installation and operation manual," Jul. 2011.

[11] SAE International, http://www.sae.org/

[12] H. D. Thai, J. Barbaroux, H. Chazal, Y. Lembeye, J. C. Crebier, and G. Gruffat, "Implementation and analysis of large winding ratio transformers," in Proc. IEEE Appl. Power Electron. Conf., Feb. 2009, pp. 1039–1045.

[13] J. C. Salmon, " Circuit topologies for single-phase voltage-doubler boost rectifiers," in Proc. IEEE APEC, Boston, MA, pp. 549-556, Feb. 1992.

[14] Yi Zhao, Xin Xiang, Wuhua Li, Xiangning He, and Changliang Xia, "Advanced symmetrical voltage quadrupler rectifiers for high step-up and high output-voltage converters," IEEE Trans. Power Electron., vol. 28, no. 4, pp. 1622–1631, April. 2013.

[15] "TI TMS320F28335 Data Manual," Texas Instruments [Online]. Available: http://www.ti.com/lit/ds/symlink/

The 2018 International Power Electronics Conference

ZVS Interleaved Totem-pole Bridgeless PFC Converter with Phase-shifting Control

Moo-Hyun Park[1*], Jae-Il Baek[1], Jung-Kyu Han[1], Cheon-Yong Lim[1] and Gun-Woo Moon[1]

1 School of Electrical Engineering, Korea Advanced Institute of Science and Technology, Daejeon, Korea

*E-mail: moohyun3@kaist.ac.kr

Abstract— This paper proposes an interleaved totem-pole bridgeless boost power factor correction (PFC) converter employing an additional inductor which has small switching loss. The main problem of conventional interleaved totem-pole bridgeless boost PFC converter is large switching loss due to the hard switching operation. And that problem becomes more serious in high switching frequency operation. In proposed converter, by utilizing the energy of additional inductor, zero-voltage-switching (ZVS) operation of the switches is achieved. Also, by using phase-shifting control between two interleaved PFC units, the flowing current on additional inductor can be controlled and only proper magnitude of current is generated. As a result, the proposed converter has higher efficiency than conventional converter. The feasibility of the proposed converter is confirmed with 50Hz, 230*Vrms* input and 1.6kW (400V / 4A) output prototype.

Keywords— *Totem-pole bridgeless PFC, Phase-shifting control, Zero voltage switching.*

I. INTRODUCTION

Boost power factor correction (PFC) converter is essential for power supply with high power capacity such as on-board charger, server power supply, etc, because high power factor (PF) is required in high power applications. PFC converts AC input voltage to DC voltage of link capacitor with nearly unity PF to satisfy the PF regulation.

Generally, power supply with high power capacity consists of several stages as shown in Fig. 1. After bridge diodes rectification, PFC regulate the voltage of link capacitor. And DC/DC converter supply the tightly regulated output voltage to the load. During this process, power losses occur at each power conversion stages. To increase the efficiency of power supply, there have been many researches to remove the bridge diodes which cause large conduction loss [1]-[3]. By removing the bridge diodes in current flowing path, conduction loss is reduced.

For the past few decades, various bridgeless boost PFC converters were proposed. Many topologies based on basic bridgeless PFC has common mode (CM) noise problem due to the floating input voltage. Dual boost bridgeless PFC with the returned diode can avoid the CM noise problem, however, there are too many circuit elements [1].

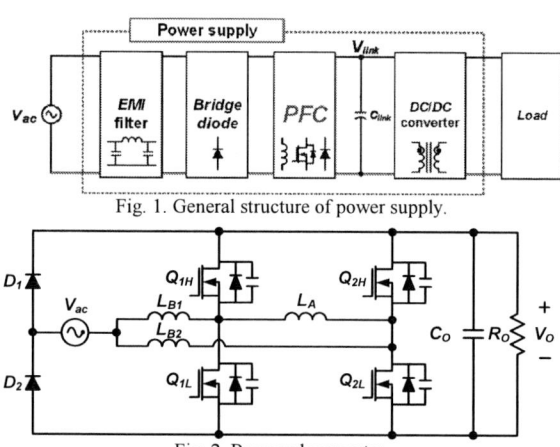

Fig. 1. General structure of power supply.

Fig. 2. Proposed converter.

Totem-pole bridgeless boost PFC has smaller CM noise than other bridgeless topologies, because the input is clamped to the output by diodes during each half-line cycle. Also, it has small counts of circuit elements. However, due to the poor reverse recovery characteristic of intrinsic body diode in general MOSFET switch, totem-pole bridgeless boost PFC had been limited to be used in continuous conduction mode (CCM) operation. Nowadays, due to the development of wide-band-gap switches such as GaN(Gallium nitride), SiC(Silicon carbide) switches, which has good reverse recovery characteristic, more focuses are put on using totem-pole bridgeless boost PFC.

On the other hand, bridgeless boost PFC has large switching loss, because it is based on boost converter which is a hard switching topology. To reduce the switching loss in totem-pole bridgeless PFC, many soft switching converters have been researched [3]-[5]. The basic zero-voltage transition (ZVT) circuit is proposed in [4]. However, too many circuit components are needed to be applied to the totem-pole topologies. The ZVT technique for an interleaved structure is proposed in [3], [5]. A coupled inductor substitutes two boost inductors and zero-current-switching (ZCS) turn off is achieved in the diodes and zero-voltage-switching (ZVS) turn on is achieved in the MOSFETs. However, it needs an additional circuit for valley switching. Also, an inductor is added between two boost inductors in [5]. ZVS is achieved by this additional inductor, however, the duty-ratio should be larger than 0.5. Therefore it is not suitable

3533

for PFC application which has wide duty-ratio range.

In this paper, an interleaved totem-pole bridgeless PFC with reduced switching loss is proposed. Only one inductor is added to the conventional interleaved totem-pole bridgeless PFC converter as shown in Fig. 2. By utilizing the energy of additional inductor, the stored charge at output capacitance of switch can be discharged before being turned on. As a result, ZVS turn on is achieved. And to prevent the excessive energy of the additional inductor, phase-shifting control to make the proper energy without additional active components is proposed .

To verify the feasibility of proposed converter, 1.6kW prototype is experimented with 50Hz, 230$Vrms$ input voltage.

II. PROPOSED CONVERTER

As shown in Fig. 2, boost inductor L_{B1}, MOSFET switches Q_{1L}, Q_{1H}, diode D_1 comprise one PFC converter unit and the components with subscript "2" form the other unit. L_{B1} and L_{B2} have same magnitude. The additional inductor (L_A) is placed between two PFC units. The voltage on L_A (v_{LA}) can be V_O, $-V_O$ or zero depending which switches are turned on. And current on L_A (i_{LA}) changes according to v_{LA}.

Before illustrating the steady-state operation, several assumptions are made as follows:
1) the switches Q_{1L}-Q_{2H} are ideal MOSFET switches except for their output capacitances and body diodes;
2) the diodes D_1-D_2 are ideal components;
3) the output voltage, V_O, is constant.

Because of the symmetric structure and operation of the proposed converter, only the process of positive half-line cycle is explained. There are eight modes in positive half-line cycle and only four modes are explained because they are symmetric. The key waveforms and the equivalent circuits are illustrated in Fig. 3 and Fig. 4. In positive half-line cycle, D_2 is conducted and Q_{1L}, Q_{2L} operate as build-up switches. When i_{LB1} or i_{LB2} changes, the slope is follows:

$$v_{ac} = \sqrt{2}v_{rms}\sin(2\pi f_L t) \qquad (1)$$

$$\frac{di_{LB}}{dt} = \frac{v_{ac}}{L_B} = \frac{\sqrt{2}v_{rms}\sin(2\pi f_L t)}{L_B} \qquad (2)$$

$$\frac{di_{LB}}{dt} = \frac{v_{ac} - V_O}{L_B} = \frac{\sqrt{2}v_{rms}\sin(2\pi f_L t) - V_O}{L_B}, \qquad (3)$$

where v_{ac} is input AC voltage, v_{rms} is RMS input voltage and f_L is line frequency. L_B intends L_{B1} or L_{B2} and i_{LB} intends i_{LB1} or i_{LB2}. i_{LB} increases with slope as (2) and decreases with slope as (3). Q_{1L} and Q_{2L} operate with phase difference as 180 degrees in interleaved structure. The mode analysis is illustrated in case of the duty ratio is less than 0.5. Also, the mode analysis is process in case of i_{LA} is larger than i_{LB1} at t_1. Details of mode analysis are follows.

Mode 1 [t_0-t_1] : Q_{1H} and Q_{2H} are ON state. Therefore, i_{LB1} and i_{LB2} decreases with slope as (3). i_{LA} is maintained

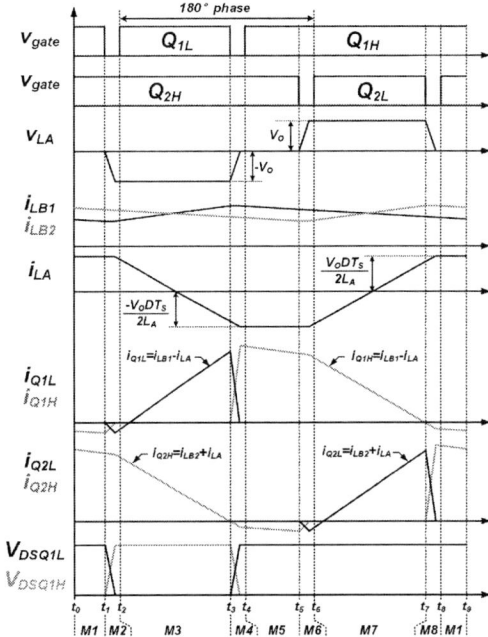

Fig. 3. Key waveforms in positive v_{ac} line cycle.

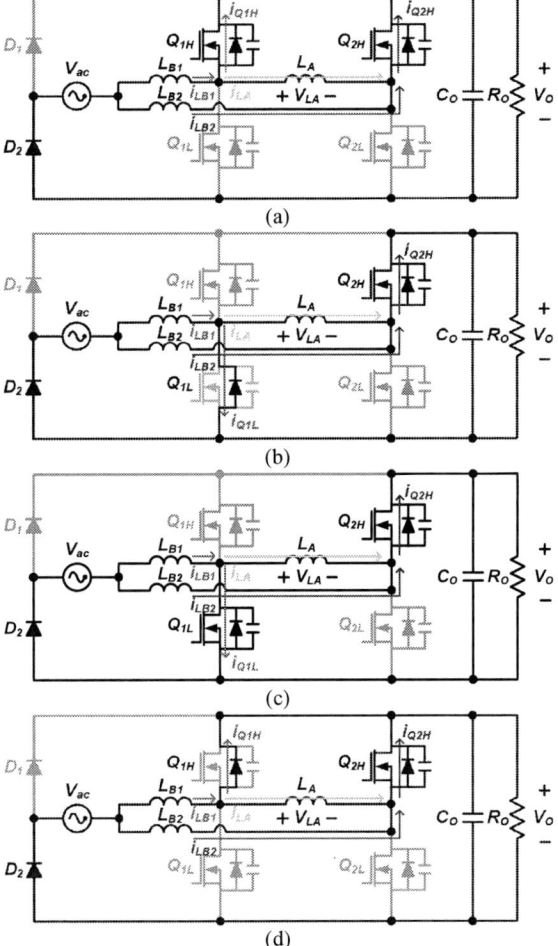

Fig. 4. Equivalent circuit (a) mode 1. (b) mode 2. (c) mode 3. (d) mode 4.

because the both nodes of L_A is connected to V_O and zero voltage is applied on L_A. If there is not L_A, i_{Q1H} equals to i_{LB1} and i_{Q2H} equals to i_{LB2}. However, due to L_A, i_{Q1H} equals to i_{LB1}-i_{LA} and i_{Q2H} equals to i_{LB2}+i_{LA} as illustrated in Fig. 3. Therefore, i_{Q2H} at t_1 has smaller value compared to the case without L_A. This phenomenon reduces the reverse recovery problem of the body diode. At t_1, Q_{1H} is turned off.

Mode 2 [t_1-t_2] : After Q_{1H} is turned off, C_{oss1H} is charged to V_O and C_{oss1L} is discharged to 0 because the magnitude of i_{LA} is larger than i_{LB1}. After C_{oss1L} is totally discharged, D_{1L} is conducted. And the difference between i_{LB1} and i_{LA} flows to D_{1L}.

Mode 3 [t_2-t_3] : Q_{1L} is turned on with ZVS, because the voltage on C_{oss1L} is zero at t_2. Because Q_{1L} and Q_{2H} are ON state, i_{LB1} increases and i_{LB2} decreases. -V_O is applied to L_A, therefore, i_{LA} decreases from $V_O D T_S/2L_A$ to -$V_O D T_S/2L_A$ until Q_{1L} is turned off. Due to L_A, i_{Q1L} equals to i_{LB1}-i_{LA} and i_{Q2H} equals to i_{LB2}+i_{LA} as illustrated in Fig. 3. At t_3, Q_{1L} is turned off.

Mode 4 [t_3-t_4] : After Q_{1L} is turned off, C_{oss1L} is charged to V_O and C_{oss1H} is discharged to 0, because the magnitude of i_{LA} is larger than i_{LB1}. After C_{oss1H} is totally discharged, D_{1H} is conducted. And the difference between i_{LB1} and i_{LA} flows to D_{1H}.

Mode 1-4 describe ZVS turn on process of Q_{1L}. Mode 5-8 are the counterparts for Q_{2L}. Due to the similarity, they are omitted here. And the operations in case of less duty ratio than 0.5 or not are different slightly. When the duty ratio is less than 0.5, i_{LA} increases or decreases during DT_S. In contrary, when the duty ratio is larger than 0.5, i_{LA} increases or decreases during $(1-D)T_S$. However, the magnitude of i_{LA} before the low side switch is turned on is same in both cases. Therefore, ZVS operation is achieved in both cases.

III. DESING GUIDE OF PORPOSED CONVERTER

A. Boost inductor design

Boost inductor (L_B) is designed to have 20% ripple current of maximum current. One PFC unit convers half of output power (P_O) and it has input RMS current (I_{rms}), $P_O/2V_{rms}$. And the minimum duty cycle (D_{min}) is (V_O-$\sqrt{2}$ V_{rms})/V_O. Therefore, the value of LB is as follows:

$$L_B = \frac{V_{rms} D_{min} T_S}{0.4 I_{rms}}. \tag{4}$$

If the ripple current of L_B is large, turn-on switching loss can be reduced. However, the core loss and conduction loss can be larger. In proposed converter, because the switching loss is reduced by additional inductor, L_B can designed to have small ripple current. Therefore, the core loss and conduction loss of L_B can be small.

B. ZVS condition

As illustrated in Fig. 3, the current on low side switch flows in negative direction before the switch is turned on.

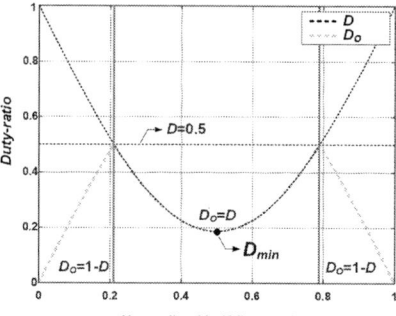

Fig. 5. D_O according to the half-line cycle.

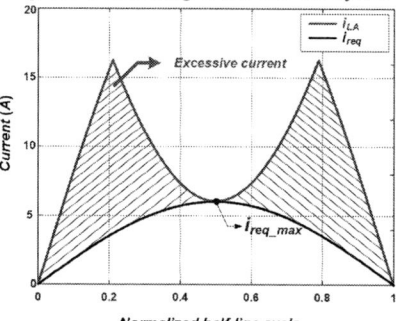

Fig. 6. i_{LA} and i_{req} without phase-shifting control.

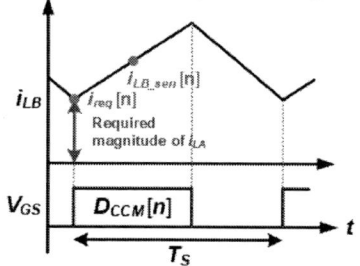

Fig. 7. Sensed current and required current in nth cycle.

And ZVS turn on is achieved. To satisfy this condition, i_{LA} should be larger than i_{LB} at t_1. And i_{LA} needs more margin considering discharging the output capacitor of switch totally. The ZVS condition is as follows:

$$i_{LA}(t_1) - i_{LB1}(t_1) > 0. \tag{5}$$

ZVS is achieved well if i_{LA} is large. However, the conduction loss is increased. Therefore, proper magnitude of i_{LA} is recommended for high efficiency considering both switching loss and conduction loss. i_{LA} can be controlled by phase-shifting and magnitude of L_A.

C. Additional inductor design

The maximum value of i_{LA} is inversely proportional to L_A. And D_O at the middle point of half-line cycle is D_{min} as illustrated in Fig. 5. Therefore, L_A should be designed to achieve ZVS at the middle point of half-line cycle. To achieve aforementioned ZVS condition, i_{LA} at D_{min} should be larger than i_{req_max} as illustrated in Fig. 6. And the available range of L_A is as follows:

$$L_A < \frac{V_O D_{min} T_S}{2 i_{req\_max}}. \tag{6}$$

If the magnitude of L_A is too small, it can be easy to achieve ZVS over all input voltage range and all load condition. However, $\emptyset$ would be small to make proper magnitude of i_{LA} and it causes the larger conduction loss in EMI filter. Therefore, to make $\emptyset$ as large as possible, L_A souled be designed to have maximum magnitude that can achieve ZVS over all input and output condition.

D. Phase-shifting control

Aforementioned before in mode analysis, i_{LA} changes during the overlapping time between the low side switch in a PFC unit and the high side switch in the other PFC unit. Therefore, the ripple of i_{LA} is proportional to the overlapping time, $D_O T$. D_O is D if D is smaller than 0.5 and D_O is $1-D$ if D is larger than 0.5. Therefore, the maximum overlapping time is $0.5T$ in case of D_O is 0.5.

To regulate the constant output voltage, PFC converter has the duty ratio has follows:

$$D = 1 - \frac{v_{ac}}{V_O}. \tag{7}$$

Fig. 5. illustrates D and D_O according to the half-line cycle. And the ripple of i_{LA} and i_{req} without phase-shifting control are illustrated in Fig. 6. i_{LA} has similar shape with D_O. If L_A is designed to achieve ZVS at the middle point of the half-line cycle which has the largest switching loss, i_{LA} has excessive value except middle point. Therefore, the conduction loss on L_A and switches are increased, though ZVS is achieved.

To make i_{LA} to be same with i_{req}, phase-shifting control is proposed.

According to the phase difference between two PFC units ($\emptyset$), the ripple of i_{LA} can be controlled. Fig. 7. illustrates the waveform of i_{LB} at nth switching cycle. The value of i_{LB} is sensed at the center of the switch on-time by digital signal processor (DSP) and sensed value of i_{LB} at nth switching cycle ($i_{LB\_sen}[n]$) is illustrated in Fig. 7.

Also i_{req} at nth switching cycle ($i_{req}[n]$) is illustrated. $i_{req}[n]$ is follows:

$$i_{req}[n] = i_{LB\_sen}[n] - \frac{v_{ac} D T_S}{2 L_B} \tag{8}$$

For PFC operation, i_{LB}, v_{ac} and V_O are sensed in every switching cycle. Therefore $i_{req}[n]$ can be calculated. And i_{LA} changes from $-V_O \emptyset T_S / 2L_A$ to $V_O \emptyset T_S / 2L_A$ with phase-shifting control. To satisfy the ZVS condition, (5) should be larger than $V_O \emptyset T_S / 2L_A$ as (4). And the minimum value of $\emptyset$ is follows:

$$\emptyset > \frac{2 L_A}{V_O T_S} (i_{LB\_sen}[n] - \frac{v_{ac} D T_S}{2 L_B}) \tag{9}$$

By applying calculated $\emptyset$, ZVS is achieved in all CCM operation regions with minimum conduction loss on switches.

And there is no DCM operation, because the synchronous rectifier is turned on in powering period. Therefore, i_{LB} can flow in negative direction and has continuous shape.

TABLE I
SPECIFICATIONS AND PARAMETERS

		Proposed	Conventional
v_{in}	Line input voltage	230V$_{AC}$	
V_O	Output voltage	400V$_{DC}$	
P_O	Maximum output power	1600W(400V/4A)	
f_s	Switching frequency	200kHz	
Q_1-Q_4	MOSFET switch	GS66508T	
L_{B1}, L_{B2}	Boost inductor	123µH (PQ3230, 0.1Φ*100 wire 48Turns)	
L_A	Additional inductor	51µH (PQ35I)	-
C_O	Output capacitor	81µF*10EA	

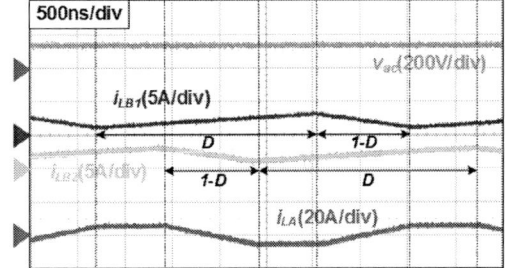

Fig. 8. Phase-shifted waveforms of proposed converter.

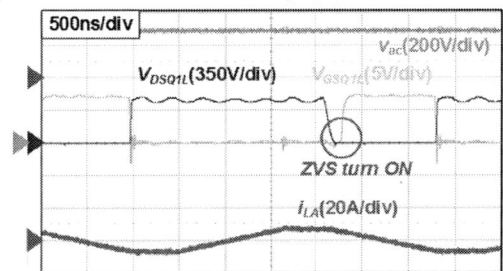

Fig. 9. ZVS operation of proposed converter.

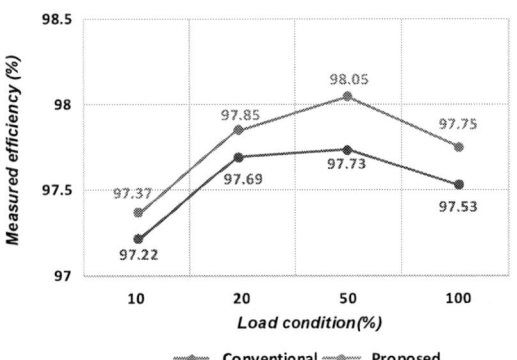

Fig. 10. Measured efficiency of conventional converter and proposed converter.

IV. EXPERIMENTAL RESULTS

The experiment is confirmed with interleaved totem-pole bridgeless boost PFC with 1600W power capacity. Input voltage and output voltage are 230Vrms and 400V each. Switching frequency is 200kHz. Detail specification and parameters are shown in Table I. One prototype has additional inductor and the other prototype does not have. With additional inductor, the proposed converter operates in ZVS condition and has small switching loss. Phase-shifting control is illustrated in Fig. 8 and ZVS turn on is achieved in the proposed converter as illustrated in Fig. 9. The measured efficiency of conventional converter and proposed converter is illustrated in Fig. 10. The proposed converter showed higher efficiency than conventional converter, though it has core loss and conduction loss of additional inductor.

V.   CONCLUSION

In this paper, an interleaved totem-pole bridgeless boost PFC converter with reduced switching loss is proposed. By utilizing the energy of additional inductor, ZVS turn on is achieved. With phase-shifting control, proper energy is generated in additional inductor without excessive current. Therefore, the conduction loss is minimized while ZVS turn on is achieved over all input voltage range and all load condition. Therefore, the proposed PFC converter is expected to be widely used for high power application such as OBC and sever power supply in high switching frequency.

ACKNOWLEDGMENT

This work was supported by the National Research Foundation of Korea (NRF) grant funded by the Korea government (MSIP) (No. 2016R1A2B2010328).

ACKNOWLEDGMENT

This work was supported by Korea Electronic Power Research Institute (KEPRI).

REFERENCES

[1] B. Zhang, Q. Lin, Y. Shimada, J. Imaoka, M. Shoyama, and S. Tomioka, "Analysis and Reduction Method of Conducted Noise in GaN HEMTs based Totem-pole Bridgeless PFC Converter," In Proc. IPEMC-ECCE Asia 2016.

[2] L. Huber, J. Yungtaek, and M. M, Jovanovic, "Performance evaluation of bridgeless PFC boost rectifiers," *IEEE Trans. on Power Electronics*, vol. 23, no. 3, pp. 1381-1390, May 2008.

[3] B. Su, and Z. Lu, "An Interleaved Totem-Pole Boost Bridgeless Rectifier With Reduced Reverse-Recovery Problems For Power Factor Correction," *IEEE Trans. On Power Electronics*, vol. 25, no. 6, pp. 1406-1415, Jun 2010.

[4] H. Guichao, L. Ching-Shan, J. Yimin, and F. C. Y. Lee, "Novel zero-voltage-transition PWM converters," *IEEE Trans. on Power Electron*, vol. 9, no. 2, pp. 213-219, Mar. 1994.

[5] H. Yao-Ching, H. Te-chin, and Y. Hau-Chen, "An interleaved boost converter with zero-voltage transition," *IEEE Trans. on Power Electronics*, vol. 24, no. 4, pp. 973-978, Apr. 2009.

A Zero-Voltage-Switching Totem-pole Bridgeless Boost Power Factor Correction Rectifier having Minimized Conduction Losses

Young-Dal Lee[1*], Chong-Eun Kim[2], Jae-Il Baek[1], Dong-Kwan Kim[1], and Gun-Woo Moon[1]
1 Electrical Engineering, KAIST, Daejeon, Korea
2 Power R&D Group, Solu-m Corp., Yong-in, Korea
*E-mail: youngdal.lee@kaist.ac.kr

Abstract-A new bridgeless totem-pole boost PFC rectifier is proposed to achieve a soft-switching for power factor correction (PFC) circuit. Based on the conventional totem-pole bridgeless rectifier, an auxiliary circuit is applied to retain zero-voltage-switching (ZVS) of main switches under the middle-to-heavy load conditions and minimized conduction loss under the light load conditions utilizing auxiliary switches. Moreover, the proposed converter can reduce additional conduction loss under the light load conditions without complicated control method. Therefore, the proposed converter can have high efficiency at the entire load conditions. The feasibility of the proposed converter is verified by experimental results of prototype configurations with universal AC input and 750W output.

I. INTRODUCTION

Recently, the capability of processing abundant data have been required to satisfy the flow of time such as the Fourth Industrial Revolution and big data analytics. In accordance with the above trends, total power requirements to operate server systems in data centers have been risen dramatically over the past few years for high density. In addition, the efficiency issues under the light load conditions in server power system are becoming more and more important since the server power system operates under the light load conditions due to redundancy operations for high reliability.

Meanwhile, in order to meet the high power factor (PF) and low total harmonics distortion (THD) regulation standard, the conventional power factor correction (PFC) circuit is normally composed of full bridge rectifier and boost converter because of their simple implementation and low cost [1]-[2]. Despite of its simple structure, the conventional PFC rectifier go through three and more semiconductors which result in large conduction loss. So as to simplify the structure and minimize conduction loss, many number of companies have been investigated the bridgeless PFC rectifier. The typical bridgeless PFC rectifier combines a full bridge rectifier and DC/DC PFC circuit as a single stage. Accordingly, the power flow can go through just two semiconductors having reduced conduction loss. Therefore, the overall system efficiency can be improved with the bridgeless PFC rectifier.

There are typical bridgeless PFC rectifiers reported on [3]-[8]. Among them, the totem-pole bridgeless PFC rectifier as shown in Fig. 1 has many advantages such as small conduction components, simple control and better

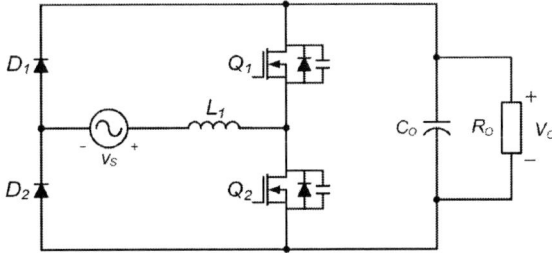

Fig. 1. Circuit diagram of conventional totem-pole bridgeless PFC Rectifier.

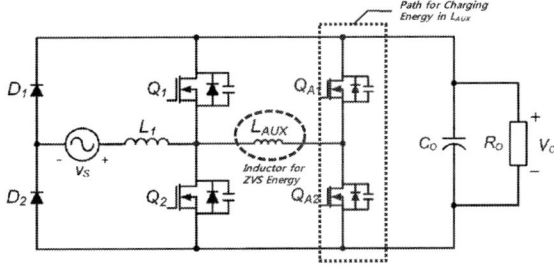

Fig. 2. The Concept and derivation of the proposed soft-switching totem-pole bridgeless PFC Rectifier.

EMI characteristics. Although the totem-pole bridgeless PFC rectifier has been proposed for many years, practical products are rarely reported at the industrial area.

In fact, the totem-pole bridgeless PFC rectifier has some drawbacks such as listed below. The some of the severe problems are hard-switching operations and large current spike due to reverse recovery current. Therefore, utilizing the totem-pole bridgeless PFC rectifier is limited at discontinuous conduction mode (DCM) and critical conduction mode (CRM) due to reverse recovery current and hard-switching operations.

As a result, totem-pole bridgeless PFC rectifier usually uses GaN (Gallium Nitride Power Transistor) devices which have relatively small reverse recovery characteristics than Si-MOSFET. Although GaN devices have relatively small reverse recovery charge characteristics and on-channel resistance, it has some disadvantages like as its much higher cost than Si-MOSFET and less deficient reliability such environmental test, i.e. surge and dynamic conditions.

3538

To overcome the above limitation in the totem-pole bridgeless PFC rectifier, many soft-switching converters have been researched [4]-[8]. In [4]-[8], those converters can reduce switching loss but they have low power density due to many number of components count. Also, the additional circuit for soft-switching makes system efficiency lower due to additional conduction loss under the light load condition. Also, they have complicated control methods for reducing the switching and conduction losses. Therefore, those converters would not meet high power density on account of complex configurations and relative high conduction loss under light load conditions.

So as to solve above-mentioned problems of the earlier researches, this paper presents a new soft-switching totem-pole bridgeless PFC rectifier having small components count without complicated control method as shown in Fig. 2. By applying auxiliary circuits, ZVS operation of two main switches Q_1 and Q_2 can be accomplished under the middle-to-heavy load conditions. Moreover, a minimized conduction loss can be achieved under the light load conditions by utilizing auxiliary switches having relative small output capacitance C_{OSS} characteristics without complex control method. So, the proposed converter can have high efficiency under the entire load conditions. The feasibility of proposed converter is verified by 750W prototype converter.

II. OPERATION PRINCIPLE OF PROPOSED CONVERTER

A. Concept of Proposed Converter

The key of the proposed converter is to achieve high efficiency by adopting soft-switching operation of main switches under the middle and heavy load conditions and delivering the power to output side utilizing auxiliary switches under light load conditions.

Based on the conventional totem-pole bridgeless PFC rectifier as shown in Fig. 1, the proposed converter is comprised with auxiliary circuit as described below. In order to acquire ZVS operation of main switches, the additional current path is required by adopting auxiliary inductor. L_{AUX} is connected between two main switches and two auxiliary switches. Q_1 and Q_2 are main switches for the proposed converter. Q_{A1} and Q_{A2} are auxiliary switches which role as ZVS operations under the middle and heavy load conditions, well as delivering the power to output side under the light load conditions.

B. Operation Principle of middle & heavy load condition

In this section, the operational principle of the proposed converter is going to be analyzed in detail under the middle and heavy load conditions. To describe the operational principle of the proposed converter, Fig. 3 and Fig. 4 shows the key waveforms of the proposed converter.

For the sake of convenience in analysis, only the positive half period of the input voltage is considered. To simplify the analysis, it is assumed that the proposed converter operates in the steady state and the below some suppositions are premised during a few switching cycles.

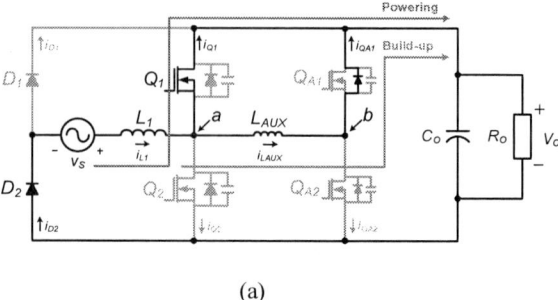

(a)

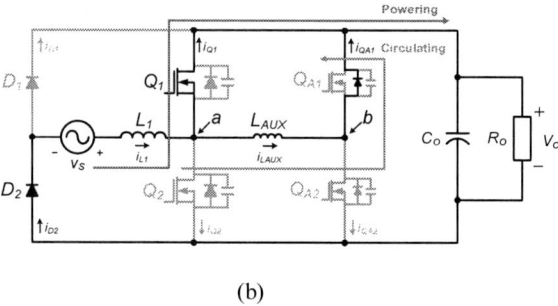

(b)

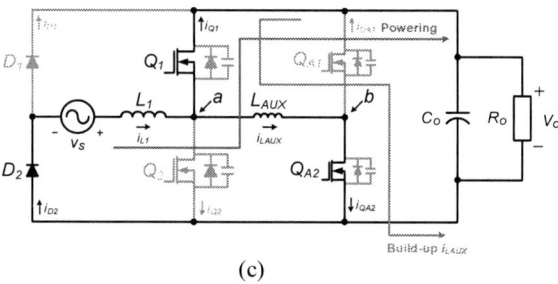

(c)

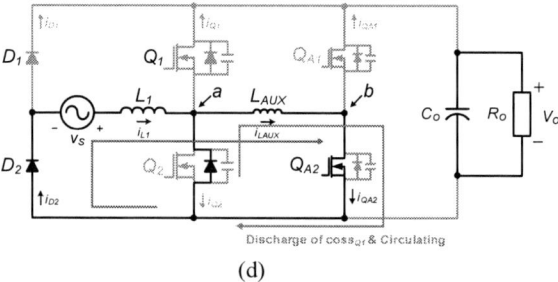

(d)

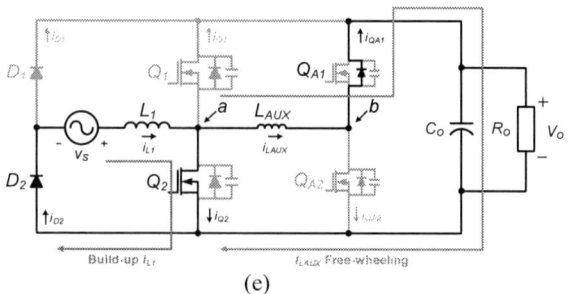

(e)

The 2018 International Power Electronics Conference

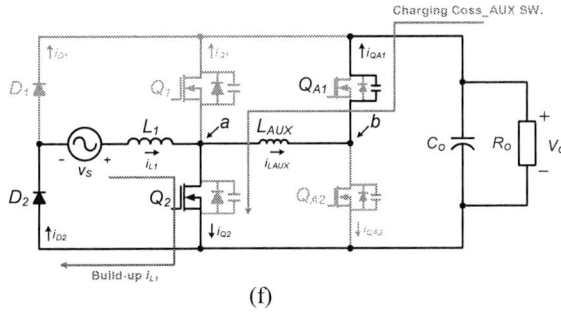

(f)

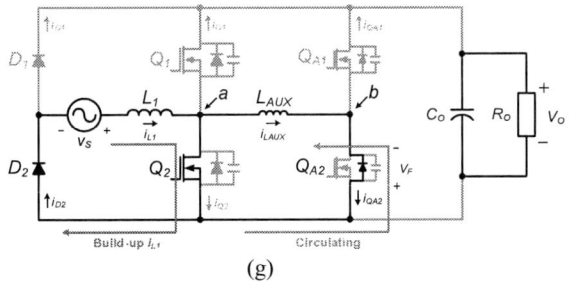

(g)

Fig. 3. The operation principles of the proposed converter under the positive half-line cycle. (@middle and heavy load conditions) (a)*Mode 1*(t_1~t_2), (b)*Mode 2*(t_2~t_3), (c)*Mode 3*(t_3~t_4), (d)*Mode 4*(t_4~t_5), (e) *Mode 5*(t_5~t_6), (f)(g) *Mode 6*(t_6~t_1)

1) All the components and devices are ideal.
2) The output capacitor C_O is large enough to be assumed that output voltage V_O is constant.
3) All the semiconductor, i.e. two main switches and two auxiliary switches are ideal states having parallel body diodes and parasitic capacitance C_{OSS}.

Mode 1 [t_1~t_2]: This mode begins when the main switch Q_2 is turned-off. This mode is similar with the conventional totem-pole bridgeless PFC rectifier. So, the energy storing in main inductor L_1 transfer to output side, going through the body diode of main switch Q_1. The current of the boost main inductor L_1 is governed by the equation (1).

$$i_{L1}(t) = i_{L1}(t_1) + \frac{(V_s - V_0)}{L_1}(t - t_1) \quad (1)$$

During this mode, since the voltage across the auxiliary inductor L_{AUX} is the output voltage V_O, the current of the auxiliary inductor i_{LAUX} starts build-up to i_{peak}, and it can be expressed as below equation (2),(3).

$$\Delta i_{peak}(t) = \sqrt{2} \cdot \sqrt{\frac{COSS_{QA2}}{L_{AUX}}} \cdot V_{OUT} \quad (2)$$

$$i_{peak}(t) = i_{peak}(t_1) + \frac{V_0}{L_{AUX}}(t - t_1) \quad (3)$$

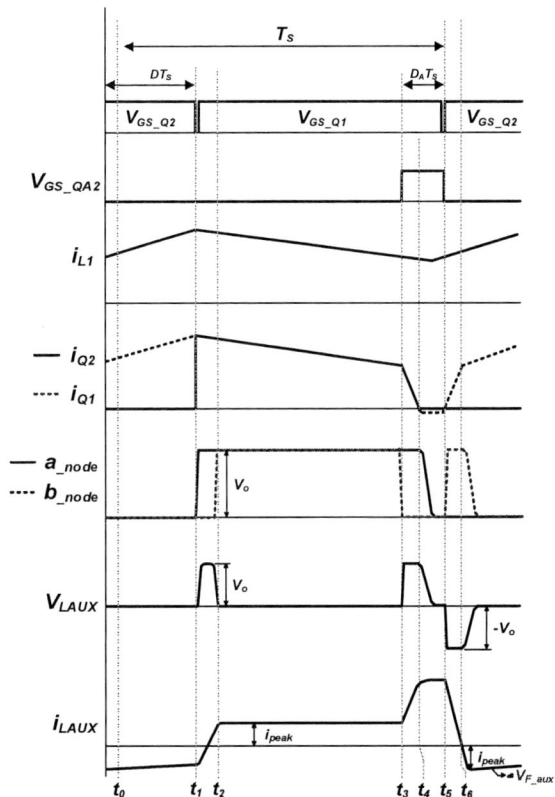

Fig. 4. The key waveforms of the proposed converter during a few switching cycles.

Mode 2 [t_2~t_3]: When Q_{A2_COSS} is fully charged, this mode starts. In this mode, since the voltage across the a_node is the output voltage V_O and the voltage across the b_node is the output voltage V_O, the resulting voltage across the auxiliary inductor L_{AUX} is zero. Therefore, the current of the auxiliary inductor i_{LAUX} is circulating through the main switch $Q1$, auxiliary inductor L_{AUX} and the body diode of the auxiliary switch Q_{A1}.

Mode 3 [t_3~t_4]: This mode begins with turning on auxiliary switch Q_{A2} at the end of mode 2. In accordance with turning on Q_{A2}, the voltage across the auxiliary inductor L_{AUX} induces output voltage V_O. Hence, the current of the auxiliary inductor L_{AUX} increases linearly as expressed in equation (4). As well, the reverse recovery concern can be relieved by controllable di/dt of i_{D1}.

$$\Delta i_{LAUX} = \frac{V_O}{L_{AUX}}(t - t_2) \quad (4)$$

Mode 4 [t_4~t_5]: This mode begins when the current of the i_{Q1} decreases up to zero. As i_{Q1} reaches zero, the junction capacitors of two main switches Q_1 and Q_2 are simultaneously operating as charging and discharging mode. At the end of charging and discharging the junction capacitor of two main switches Q_1 and Q_2, the

3540

voltage across the auxiliary inductor L_{AUX} is zero. Therefore, the circulating current in this mode flows through the body diode of Q_2, auxiliary inductor L_{AUX}, and auxiliary switch Q_{A2}. Since this circulating current can degrade overall system efficiency, this mode should be well designed.

Mode 5 [t_5~t_6]: This mode begins when the main switch Q_2 is turned-on with ZVS operation. And the energy storing in auxiliary inductor L_{AUX} transfers to output side by going through the body diode of auxiliary switch Q_{A1} as expressed in (5).

$$\Delta i_{LAUX} = -\frac{V_O}{L_{AUX}}(t - t_5) \quad (5)$$

Mode 6 [t_6~t_1]: This mode begins when i_{LAUX} reaches zero. Also, the junction capacitor of Q_{A1} is charged up to V_O. And then i_{LAUX} decreases up to i_{peak} as expressed in equation (6). i_{LAUX} flows through the junction capacitor of Q_{A1} as shown in Fig.3 (f).

$$\Delta i_{peak}(t) = \sqrt{2} \cdot \sqrt{\frac{COSS_{QA1}}{L_{AUX}}} \cdot V_{OUT} \quad (6)$$

When the junction capacitor of Q_{A1} is fully charged, i_{LAUX} is circulating through the main switch Q_2, the body diode of auxiliary switch Q_{A2} and auxiliary inductor L_{AUX} as shown in Fig.3 (g).

C. Operation Principle of light load condition

In this section, the operational principle of the proposed converter is described in detail under the light load conditions. For convenience in explanation of this mode, the Fig. 5 shows the operation of the proposed converter under the light load conditions. Fig 5 (a) describes the conventional bridgeless PFC rectifier operation for positive half-line cycle. Whereas, Fig 5 (b) presents the proposed bridgeless PFC rectifier operations under the light load conditions. When the proposed converter is in middle and heavy load conditions, the two main switches Q_1 and Q_2 role as transferring the energy storing in main inductor L_1 to output side.

On the other hand, when the proposed converter is in the light load conditions, the proposed converter transfers the energy storing in inductor to output side not by using two main switches Q_1 and Q_2 but by utilizing two auxiliary switches Q_{A1} and Q_{A2}. The two auxiliary switches Q_{A1} and Q_{A2} used in this paper have relatively small current rating specification.

Hence, the two auxiliary switches have smaller output capacitance C_{OSS} than two main switches of that. Therefore, the efficiency of the system can be improved by utilizing auxiliary switches Q_{A1} and Q_{A2} under the light load conditions.

As a result, the proposed totem-pole bridgeless PFC converter can be improved efficiency at the entire load conditions with utilizing Si-MOSFET having low cost

and higher reliability which is proved sufficiently.

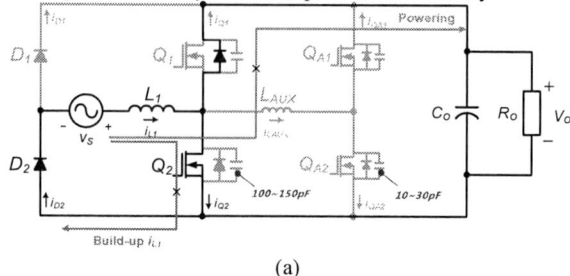

(a)

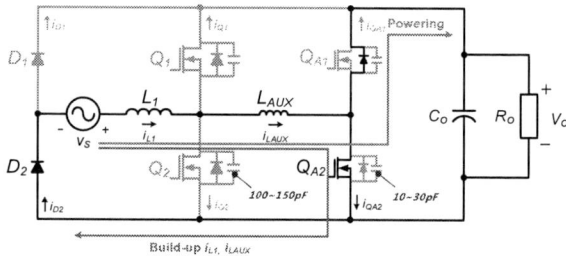

(b)

Fig. 5. The operation principles of the proposed converter under the positive half-line cycle. (@light load conditions) (a) The conventional operation under the light load conditions (b) The proposed operation under the light load conditions

III. ANALYSIS OF THE PROPOSED CONVERTER

A. Concept of Proposed Converter

In the proposed converter, additional losses of the auxiliary circuit should be carefully designed. The additional losses such as switching loss and conduction loss in auxiliary circuit should be considered.

In order to describe the conduction loss of auxiliary circuit, the root mean square value I_{RMS} for i_{LAUX} should be expressed in equation (7).

$$I_{RMS} \approx \sqrt{I_{peak}^2 + \left\{ \left(I_{BMAX} - I_{peak}\right)^2 \cdot \frac{L_{AUX}}{V_{OUT}} \cdot \frac{1}{2} \right\}} \quad (7)$$

Where I_{peak} is offset value due to circulating characteristics. I_{BMAX} is the maximum value for ZVS operations. And L_{AUX} and V_{OUT} are the inductance value for auxiliary circuit and output voltage respectively.

Fig. 6(a) shows the key waveforms for I_{RMS}, which is related to conduction loss of auxiliary circuit. So as to achieve minimize additional loss, the proper inductance should be selected. If the inductance value L_{AUX} is relatively small, the switching loss can be increased since the switching loss is inversely proportional to the inductance value for L_{AUX}. Whereas the inductance value L_{AUX} is relatively large, the conduction loss is increased due to large current flowing through the auxiliary circuit. By considering additional loss in the auxiliary circuit, the proper inductance values which achieves minimized additional loss as shown in Fig 6(b).

to soft-switching characteristics.

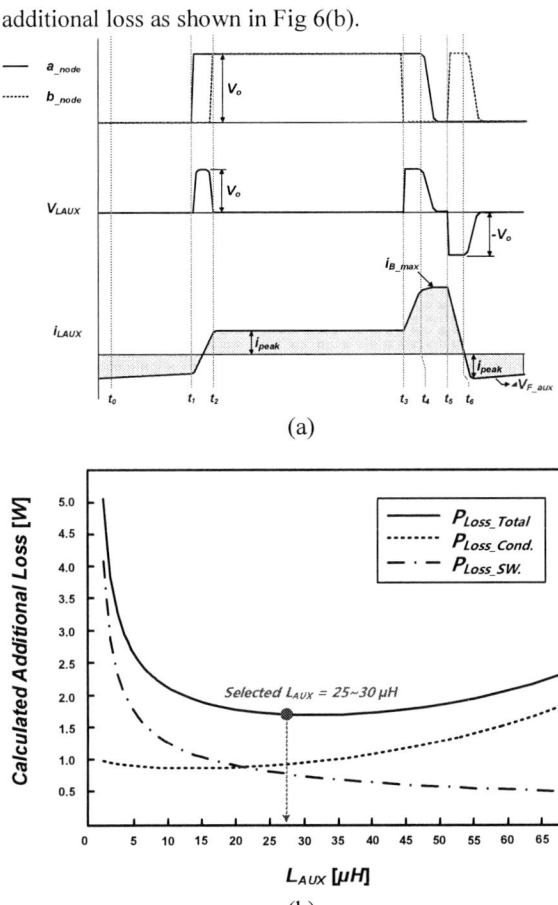

(a)

(b)

Fig. 6. The design for auxiliary inductance value. (a) Key waveforms for RMS value for i_{LAUX}, (b) Design example regarding the change of L_{AUX} at low line $115V_{AC}$ and 100% load conditions.

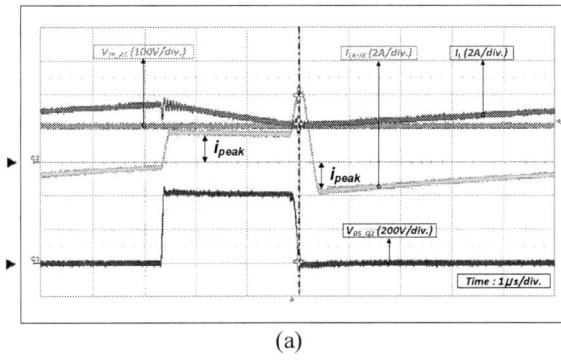

(a)

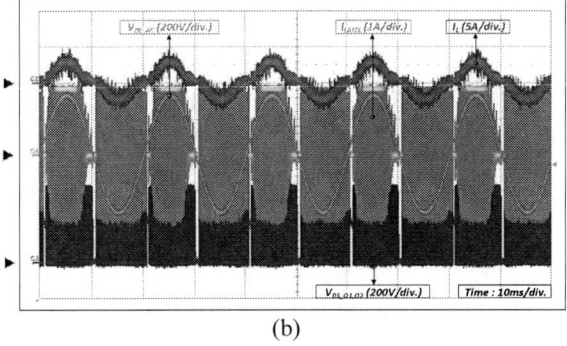

(b)

Fig. 7. The experimental waveforms of the proposed converter for low line input $115V_{AC}$ and 50% load conditions. (a) ZVS waveforms for main switches, (b) Input voltage, Input current and the current flowing L_{AUX}.

IV. EXPERIMENTAL RESULT

In order to verify the feasibility of the proposed totem-pole bridgeless PFC rectifier, the experimental results for the proposed converter is conducted with 750W prototype system. The detail components list is described in Table I.

Fig. 7(a) shows key waveforms for soft-switching operation at the low line input voltage $115V_{AC}$ and 50% load conditions. Fig. 7(a) presents that ZVS operations of the main switch Q_2 can be accomplished in the proposed converter. Also, the proposed converter can be operated in continuous current mode as well.

Fig. 7(b) shows another key waveforms including the input AC voltage, input current and the current flowing through auxiliary inductor for the proposed converter at the rated voltage with 50% load conditions.

By turning-on the auxiliary switch Q_{A2}, the current flowing auxiliary inductor L_{AUX} increase linearly to achieve ZVS operations. Therefore, it is noted that the proposed converter can have small switching losses due

TABLE I
COMPONENTS LIST OF PROTOTYPE

Symbol	Components List
Main switches (Q_1 AND Q_2)	GS66508T (600V/ 30A/ 50mΩ)
Auxiliary switches (Q_{A1} AND Q_{A2})	IPA65R420CFD (650V/ 8.7A/ 420mΩ)
Main Inductor (L_1)	CH270043, 1.0Φ, 82Turns, 610uH
Auxiliary Inductor (L_{AUX})	CH172026, 0.2Φ, 31Turns, 20uH

V. CONCLUSION

A ZVS soft-switching totem-pole bridgeless PFC rectifier for high efficiency has been presented and analyzed. The proposed converter is applying the auxiliary circuit to reduce switching losses and conduction losses in the conventional totem-pole bridgeless PFC rectifier topology. Hence, the proposed converter can have small switching losses due to soft-switching characteristics and small conduction losses for the operations by utilizing auxiliary switches under the light load conditions.

Moreover, the proposed converter overcomes the limitation for totem-pole bridgeless PFC rectifier operation, i.e. CRM and DCM. The proposed converter can operate the system in CCM not only by using GaN devices but also by utilizing Si-MOSFET having lower cost and higher reliability which is proved sufficiently. Therefore, the proposed converter can also attain low cost and high reliability, i.e. environmental test.

REFERENCES

[1] 80Plus Program. [Online]. Available: http://www.plugloadsolutions.com.

[2] Climate Savers Computing Initiative (CSCI). [Online]. Available: http://www.climatesaverscomputing.org

[3] Y. H. Jeong, J. K. Kim, and G. W. Moon, "A Bridgeless Dual Boost Rectifier with Soft-Switching Capability and Minimized Additional Conduction Loss," IEEE Trans. Industrial Electron, Jun. 2017.

[4] Bin Su, Junming Zhang, and Zhengyu Lu, "Totem-Pole Boost Bridgeless PFC Rectifier With Simple Zero-Current Detection and Full-Range ZVS Operating at the Boundary of DCM/CCM," IEEE Trans. Power Electron, vol. 26, no. 2, pp. 427–435, Feb. 2011.

[5] Muntasir Alam, Wilson Eberle, Deepak S. Gautam,and Chris Botting,"A Soft-Switching Bridgeless AC–DC Power Factor Correction Converter," IEEE Trans. Power Electron., vol. 32, no. 10, pp. 7716–7726, Oct. 2017.

[6] Chih-Chiang Hua1, Yi-Hsiung Fang, and Chin-Hsiung Huang,"Zero-voltage-transition bridgeless power factor correction rectifier with soft-switched auxiliary circuit," IET Power Electron., vol. 9, Iss. 3, pp. 546–552, Oct. 2016.

[7] Wei Hu, Yong Kang, and Xuehua Wang,"Novel Zero-Voltage Transition Semi Bridgeless Boost PFC Converter with Soft Switching Auxiliary Switch,"Energy Conversion Congress and Exposition (ECCE), 2014 IEEE , pp. 2707–2712, Sep. 2014.

[8] Khairul Safuan Bin Muhammad, and Dylan Dah-Chuan Lu,"ZCS Bridgeless Boost PFC Rectifier Using Only Two Active Switches,"IEEE Trans. Industrial Electron, vol. 62, no. 5, pp. 2795–2806, May. 2015.

The 2018 International Power Electronics Conference

Power-Factor-Correction with Power Decoupling for AC-to-DC Converter

Wan-Jung Chen[1], Tsung-Hsi Wu[1], Yao-Ching Hsieh[1], Chin-Sien Moo[1], Po-Hsiang Wen[2]

1 Department of Electrical Engineering, National Sun Yat-sen University, Kaohsiung, Taiwan

2 Smart Device SBU, Lite-On Technology Corporation, Taipei, Taiwan

E-mail: cleochen94@gmail.com

Abstract— To prolong the lifetime of electronic circuits by excluding the electrolytic capacitor, this paper attempts to develop a power-factor-correction (PFC) circuit for ac-to-dc converters based on boost conversion with power decoupling. An inductor with two active power switches is introduced to the conventionally used boost converter as a pumped-storage energy tank for storing the energy from the ac input at the higher voltages and pumping it back at the lower voltages to accomplish the functions of PFC and voltage regulation. Experimental results have demonstrated that the PFC circuit with a metalized film capacitor as the output filter can achieve an input power factor higher than 0.99 and a total harmonic distortion (THD) less than 12%.

Keywords— *Boost conversion; Power decoupling; Power factor correction; Pumped-storage energy tank.*

I. INTRODUCTION

Over past few decades, more and more power electronic converters have been used due to the advancement on electronic components and semiconductor technology. These power converters draw distorted currents from the ac source, leading to deterioration on power quality of the ac grid [1, 2]. To mitigate this problem, a power factor corrector (PFC) is inquired in many electronics products to comply with the regulations [3-9]. Conventionally, a diode bridge followed by a boost converter is used as the PFC to shape the waveform of the ac input current. An electrolytic capacitor with large capacitance has to be attached to the PFC to reduce the low-frequency ripple at the dc output [10]. The electrolytic capacitor is not only of large size but also may shorten the overall lifespan of the power electronic converter [11-14].

To exclude the use of the large electrolytic capacitor, this paper attempts to reconfigure the PFC circuit by introducing a pumped-storage energy tank. An additional inductor with associated power switches is implanted in a boost converter as an energy tank for regulating the instant power from the ac source based on power decoupling principle. A high power factor can be realized by properly adjusting the duty-ratios of the active power switches of the energy regulation tank. In this case, only a fraction of the delivered energy has to be processed by the output capacitor for reducing the ripple on the output voltage within an acceptable range. With such a configuration, the converter can achieve a high power

factor and a low total harmonic distortion (THD) at the ac input [15-17]. Moreover, the lifespan of the ac-to-dc converter can be prolonged by replacing the electrolytic capacitor with a metalized film capacitor.

II. POWER DECOUPLING

Fig. 1 illustrates the concept of power decoupling. The ac-to-dc converter is supplied from the ac mains. The rectified input voltage is expressed as

$$v_{rec} = V_p \left| \sin 2\pi f_i t \right| \tag{1}$$

where f_i and V_p are the line frequency and the peak of the ac voltage, respectively.

For a single-phase ac circuit with unit power factor, the input current is sinusoidal and in phase with the voltage of the ac source. The instantaneous input power, p_i, can be represented as

$$p_i = P_o - P_o \cos 4\pi f_i t \tag{2}$$

where P_o is the average output power.

This equation indicates that the input power can be decoupled into a dc component and an alternating power with a frequency twice the line frequency. As shown in Fig.1, the instantaneous power is higher than the average power within the duration from 1/8 to 3/8 and from 5/8 to 7/8 of T_i in an ac line cycle. To sustain a constant output power, an energy tank may be introduced to store the excessive power during these two intervals and pump the stored energy back to the load for the rest time of a cycle.

Ideally, with such a tactic, the output power of the power conversion circuit can be kept at a constant. In practice, only a small ripple presents on the output voltage. As a result, the output filter capacitor can be effectively reduced.

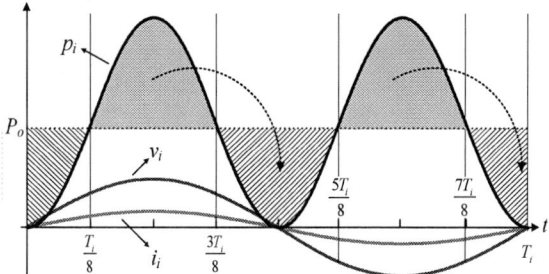

Fig. 1. Concept of pumped-storage energy with power decoupling.

III. Circuit Configuration and Operation

Fig. 2 shows the ac-to-dc converter with the proposed power factor correction circuit consisting of a boost converter with a pumped-storage energy tank. A low-pass filter is used to reduce the high-frequency harmonics at the ac input. The boost converter is formed by an inductor for boost conversion, L_1, with an associated active power switch, S_1, with a freewheeling diode, $\dot{D}_1$. The pumped-storage energy tank is composed of two active power switches, S_2 and S_3, with two diodes, D_2 and D_3, to regulate the amount of the pumped-storage energy in the inductor, L_2. The power decoupling for power factor correction is executed by properly controlling the duty-ratios of these two active power switches. The circuit operation of the pumped-storage energy tank can be illustrated by the energy storing mode and the energy pumping mode according to the power flows into or out from the inductor.

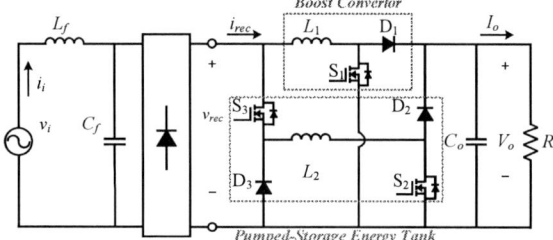

Fig. 2. The ac-to-dc converter with the proposed PFC circuit.

A. Energy Storing Mode

The energy storing mode happens when the input power p_i is higher than the output power P_o. The theoretical waveforms on key components of the energy storing mode are illustrated in Fig. 3. The circuit operation in this mode can be divided into four stages in accordance with the statuses of the power switches, as shown in Fig. 4. In this mode, the active power switch S_2 is always turned on to conduct the freewheeling current of the inductor L_2. There can be two states depending on the extents of the duty-ratios of S_1 and S_3, which are represented by α_1 and α_3, respectively.

Stage I-1

The circuit operation in Stage I-1 is shown in Fig. 4(a). By turning on the active power switches S_1 and S_3, the inductors L_1 and L_2 are both charged by the rectified input voltage, v_{rec}. Both inductor currents, i_{L1} and i_{L2}, increase linearly. Stage I-2 is entered after Stage I-1 in the case that S_3 is switched off first.

Stage I-2

Stage I-2 starts when S_3 is switched off and S_1 remains turned-on. As S_3 is turned off, D_3 is forward-biased to conduct the freewheeling current through S_2. In this stage, only L_1 receives energy from the ac source while the current in L_2 remains unchanged.

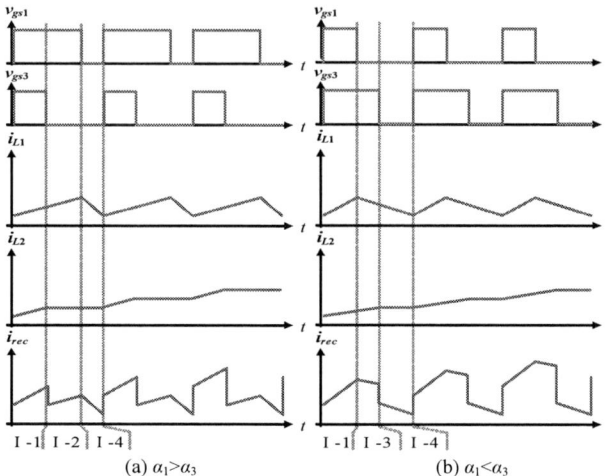

Fig. 3. Theoretical waveforms of energy storing mode.

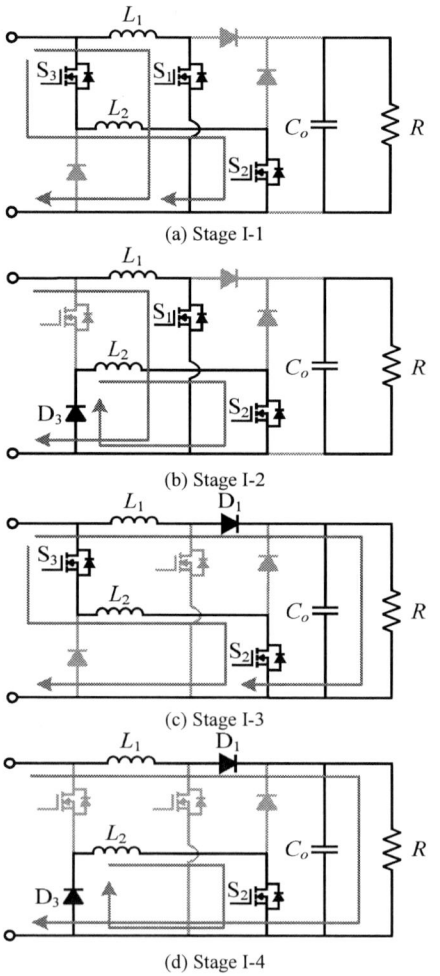

Fig. 4. Operating stages in energy storing mode.

Stage I-3

Stage I-3 comes after Stage I-1 when S_1 is turned off before S_3. In this stage, L_2 receives energy from the ac input, leading to a linearly increasing in i_{L2}. Meanwhile, the load is powered by the ac source and the boost conversion inductor L_1.

Stage I-4

As S_1 and S_3 are both turned off, the operation enters Stage I-4. The ac input source delivers power to the load through the boost conversion inductor L_1 while the current i_{L2} freewheels through S_2 and D_3. When i_{L1} and i_{L2} have reached the designated levels, a period of the energy storing mode is completed.

B. Energy Pumping Mode

The theoretical waveforms of the energy pumping mode are shown in Fig. 5. The energy pumping mode can be divided into four stages as shown in Fig. 6. In this mode, S_3 is always turned off. Similarly, there are two states in this mode depending on the extents of α_1 and α_2, the duty-ratios of the active power switches S_1 and S_2.

Stage II-1

In this stage, the active power switch S_2 is switched on to conduct the freewheeling current as S_1 is switched off. The rectified voltage v_{rec} is applied on the boost inductor L_1, resulting in an increase in i_{L1}, whereas i_{L2} remains nearly constant through the freewheeling path.

Stage II-2

As S_2 is switched off and S_1 remains turned-on, the inductor L_2 releases its stored energy through D_2 and D_3 for sustaining the output voltage while the inductor L_1 is still charged by v_{rec}.

Stage II-3

Stage II-3 comes after Stage II-1, when S_1 has been switched off but S_2 remains at the on state. The inductor L_1 begins to deliver power to the load through D_1. The inductor i_{L1} decreases linearly. In the meanwhile, i_{L2} freewheels through S_2 and D_3 remaining at a constant level. This stage ends when S_2 is switched off.

Stage II-4

In this stage, all the active power switches, S_1, S_2 and S_3 are turned off. Meanwhile, D_1, D_2, and D_3 are forced to be turned on to conduct the freewheeling currents. The ac source supplies the required load current through L_1 and the stored energy in L_2 is pumped back to the loads continuously. The two inductor currents i_{L1} and i_{L2} decreases linearly. The energy pumping mode is completed as i_{L1} and i_{L2} has decreased to the designated levels.

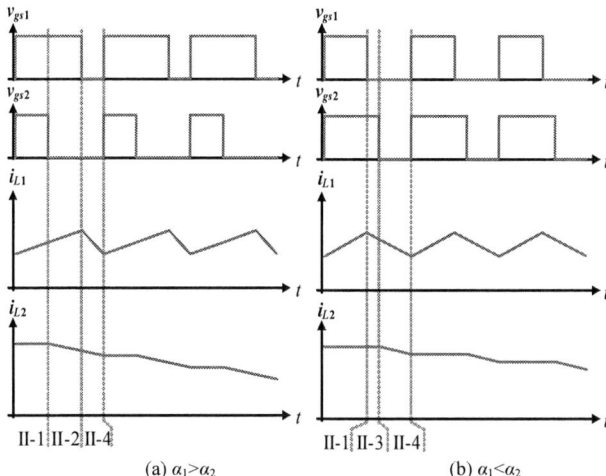

(a) $\alpha_1 > \alpha_2$ (b) $\alpha_1 < \alpha_2$

Fig. 5. Theoretical waveforms of energy pumping mode.

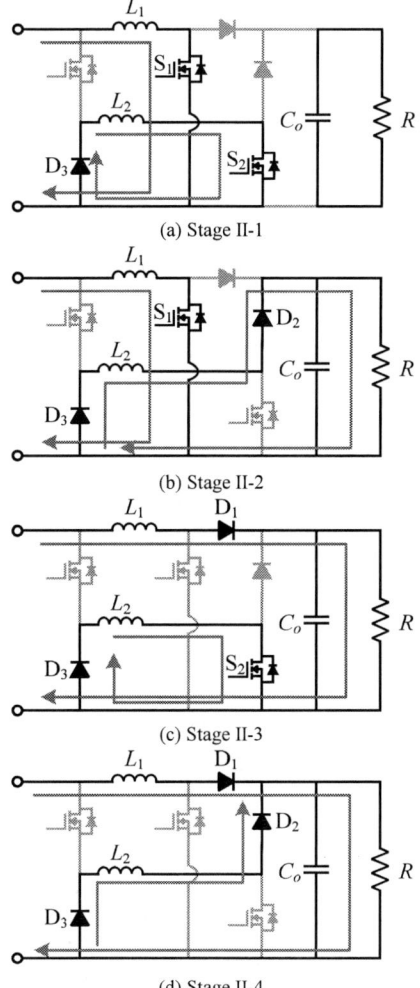

(a) Stage II-1

(b) Stage II-2

(c) Stage II-3

(d) Stage II-4

Fig. 6. Operating stages in energy pumping mode.

IV. Experimental Results

A laboratory circuit is built to verify the theoretical analyses of the proposed PFC circuit. The ac-to-dc converter is supplied from an ac input voltage of 110 V at a frequency of 60 Hz to power a load of 157 W rated at an output voltage of 225 V. The ac voltage may vary in a range between 90 V and 130 V. The power decoupling feature is realized by adjusting the duty-ratios of the active power switches from 0 to 0.85 at an operating frequency of 50 kHz. Since the output voltage ripples can be effectively reduced, a metalized film capacitor of 10 μF can be used as the output capacitor.

Both inductor currents of L_1 and L_2 are designed to be operated at the continuous-conduction-mode (CCM). After passing the peak of the ac voltage, V_p, v_i decreases. At the 1/4 and 3/4 of T_i, the conversion losses are more significant due to the lower input voltage and power. In this case, the inductors may not have enough energy to sustain the load voltage within an acceptable range. To solve the problem, the duty-ratio of active power switch S_1 is increased up to charge more energy to L_1, leading to an input current larger than the sinusoidal waveform at the period around the 1/4 and 3/4 of T_i. As the ac input voltage v_i decreases from the peak, the duty-ratio α_1 is increased to maintain the output voltage. In the case that α_1 has reached the upper limit of 0.85, the input current is not capable of tracking the sinusoidal v_i by the boost conversion circuit, causing deviation on the input current and the output voltage at every zero crossings. A significant distortion happens at the lower input voltage.

Fig. 7 shows the simulated and measured waveforms at the rated ac input voltage of 110 V. The experimental results have shown good agreements with the simulated waveforms. The measured ripple factor is 14.95% with an ac input voltage of 110 V. A high power factor of 0.99 is achieved with a THD of 8.65%.

Fig. 8 shows the measured waveforms at ac input voltages of 90 V and 130 V, respectively. With a lower input voltage of 90 V, a higher input current is needed for fulfilling the load power requirement, resulting in more drastically changes at the zero crossings of the ac input current. The input power factor remains as high as 0.99 but the THD and the ripple factor deteriorates to 12% and 18.83%, respectively.

On the other hand, when the input voltage is increased up to 130 V, conduction losses are reduced with a smaller input current. Besides, the drastic change on the input current can be mitigated around the zero crossings, leading to a less distorted input current and a smaller output voltage ripple. The ac-to-dc converter can achieve a power factor of 0.99 with a THD less than 7%, and a ripple factor of 12.19%.

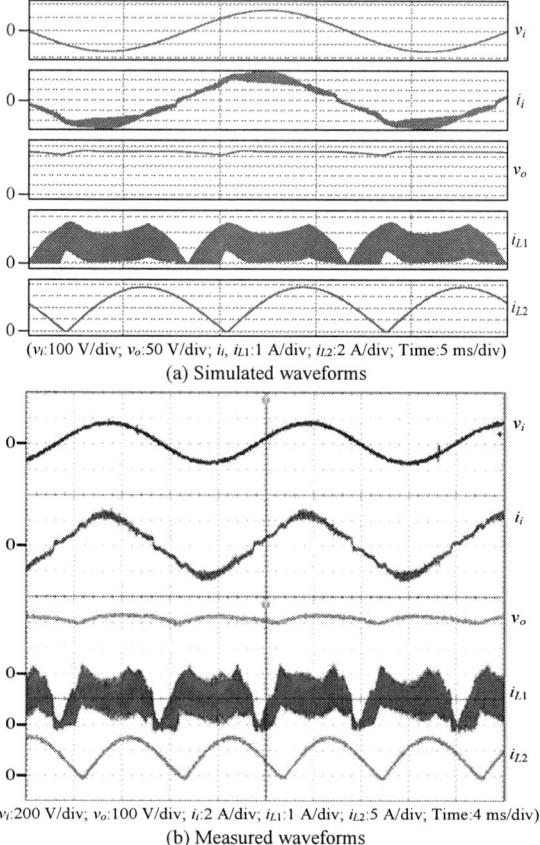

(v_i:100 V/div; v_o:50 V/div; i_i, i_{L1}:1 A/div; i_{L2}:2 A/div; Time:5 ms/div)

(a) Simulated waveforms

(v_i:200 V/div; v_o:100 V/div; i_i:2 A/div; i_{L1}:1 A/div; i_{L2}:5 A/div; Time:4 ms/div)

(b) Measured waveforms

Fig. 7. Simulation and measured waveforms at 110 V.

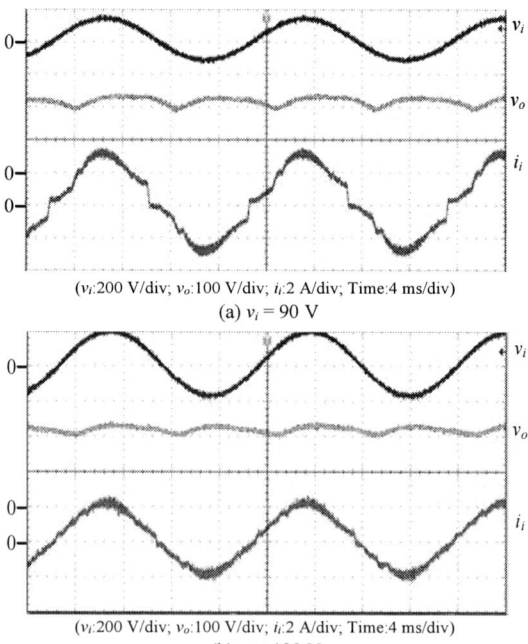

(v_i:200 V/div; v_o:100 V/div; i_i:2 A/div; Time:4 ms/div)

(a) v_i = 90 V

(v_i:200 V/div; v_o:100 V/div; i_i:2 A/div; Time:4 ms/div)

(b) v_i = 130 V

Fig. 8. Ac input current and dc output voltage at 90 V and 130 V.

V. CONCLUSIONS

To reduce the low-frequency ripple on the dc output voltage and hence to exclude the use of an electrolytic capacitor, an auxiliary pumped-storage energy tank is added to the PFC circuit for the ac-to-dc converter. The pumped-storage energy tank is employed to reallocate the excessive power to the insufficient in the ac power by decoupling the input power into an ac component and a dc component. Instead of using a large electrolytic capacitor in the conventional boost or buck-boost typed PFC, the energy tank for power regulation is configured by an inductor with two active power switches. A laboratory circuit is designed for an output power of 157 W with an ac voltage source of 110 V at 60 Hz. In the design case, a relatively large inductor is needed. In addition, more losses are introduced by the additional power switches. These drawbacks, however, can be alleviated by using higher ranked power switches and magnetic cores with higher permeability. Experimental results on a laboratory circuit have demonstrated the effectiveness of the proposed approach. For a designated output power, a better performance is achieved at a higher input voltage.

ACKNOWLEDGMENT

This work was supported by the Ministry of Science and Technology of Taiwan under the grant number of 106-2218-E-006-024.

REFERENCES

[1] J. S. Subjak and J. S. McQuilkin, "Harmonics-Causes, Effects, Measurements, and Analysis: an Update," *IEEE Trans. on Industry Applications*, vol. 26, no. 6, pp. 1034-1042, Nov./Dec. 1990.

[2] C. S. Moo, Y. C. Chuang, and C. R. Lee, "A New Power-Factor-Correction Circuit for Electronic Ballasts with Series-Load Resonant Inverter," *IEEE Trans. on Power Electronics*, vol. 13, no. 2, pp. 273-278, Mar. 1998.

[3] W. L. Qi, H. Wang, X. G. Tan, G. Z. Wang, and K. D. T. Ngo, "A Novel Active Power Decoupling Single-Phase PWM Rectifier Topology," in *Procs. IEEE Applied Power Electronics Conference and Exposition*, pp. 89-95, Mar. 2014.

[4] Y. Ohnuma and J. I. Itoh, "A Novel Single-Phase Buck PFC AC-DC Converter with Power Decoupling Capability Using an Active Buffer," *IEEE Trans. on Industry Applications*, vol. 50, no. 3, pp. 1905-1914, May/June 2014.

[5] B. Singh, B. N. Singh, A. Chandra, K. Al-Haddad, A. Pandey, and D. P. Kothari, "A Review of Single-Phase Improved Power Quality AC-DC Converters," *IEEE Trans. on Industrial Electronics*, vol. 50, no. 5, pp. 962-981, Oct. 2003.

[6] M. Su, P. Pan, X. Long, Y. Sun, and J. Yang, "An Active Power-Decoupling Method for Single-Phase AC-DC Converters," *IEEE Trans. on Industrial Informatics*, vol. 10, no. 1, pp. 461-468, Feb. 2014.

[7] J. P. M. Figueiredo, F. L. Tofoli, and B. L. A. Silva, "A Review of Single-Phase PFC Topologies Based on Boost Converter," in *Procs. IEEE/IAS International Conference on Industry Applications*, pp. 1-6, Nov. 2010.

[8] H. B. Ma, W. S. Yu, C. Zheng, J. S. Lai, Q. Y. Feng, and B. Y. Chen, "A Universal-Input High-Power-Factor PFC Pre-regulator without Electrolytic Capacitor for PWM Dimming LED Lighting Application," in *Procs. IEEE Energy Conversion Congress and Exposition*, pp. 2288-2295, Sep. 2011.

[9] J. M. Alonso, D. Gacio, J. García, M. Rico-Secades, and M. A. D. Costa, "Analysis and Design of The Integrated Double Buck-Boost Converter Operating in Full DCM for LED Lighting Applications," in *Procs. IEEE International Conference on Industrial Electronics*, Nov. 2011.

[10] J. M. Alonso, J. Vina, D. Gacio, G. Martinez, and R. Osorio, "Analysis and Design of the Integrated Double Buck-Boost Converter as A High-Power-Factor Driver for Power-LED Lamps," *IEEE Trans. on Power Electronics*, vol. 59, pp. 1689-1697, Apr. 2012.

[11] M. L. Gasperi, "Life Prediction Model for Aluminum Electrolytic Capacitors," in *Procs. of IEEE Industry Applications Conference*, vol. 3, pp. 1347-1351, Oct. 1996.

[12] P. Spanik, M. Frivaldsky, and A. Kanovsky, "Life Time of the Electrolytic Capacitors in Power Applications," in *Procs. International Conference ELEKTRO*, pp. 233-238, May 2014.

[13] L. L. Gu, X. B. Ruan, M. Xu, and K. Yao, "Means of Eliminating Electrolytic Capacitor in AC/DC Power Supplies for LED Lightings," *IEEE Trans. on Power Electronics*, vol. 24, no. 5, pp.1399-1408, May 2009.

[14] W. Chen and S. Y. R. Hui, "Elimination of An Electrolytic Capacitor in AC/DC Light-Emitting Diode (LED) Driver with High Input Power Factor and Constant Output Current," *IEEE Trans. on Power Electronics*, vol. 27, no. 3, pp. 1598-1607, Mar. 2012.

[15] X. Ruan, B. Wang, K. Yao, and S. Wang, "Optimum Injected Current Harmonics to Minimize Peak-to-Average Ratio of LED Current for Electrolytic Capacitor-Less AC–DC Drivers," *IEEE Trans. on Power Electronics*, vol.26, pp. 1820-1825, July 2011.

[16] S. K. Ki, D. K. W. Cheng, and D. D. C. Lu, "High Efficiency and PF Capability of Single Phase Power Factor Corrector," in *Procs. IEEE Region Ten Conference*, pp. 1-4, Nov. 2006.

[17] K. I. Hwu and W. C. Tu, "Controllable and Dimmable AC LED Driver Based on FPGA to Achieve High PF and Low THD," *IEEE Trans. on Industrial Informatics*, vol. 9, pp. 1330-1342, Aug. 2013.

AUTHOR INDEX

Aapro, Aapo 3156
Abdollahi, Hessamaldin....................1719
Abe, Kazuyuki..................................1567
Abe, Kensho.......................................767
Abe, Kodai 1741, 3890
Abe, Seiya 2360, 2370
Abe, Takashi 2176
Abrishamifar, Adib...........................2854
Abuogo, James 1125
Acharya, Anirudh Budnar 2630
Acharya, Sayan 3564
Adachi, Masakazu.............................2237
Afsharian, Jahangir 1537, 3797
Agarwal, Vivek.................................3471
Agelidis, Vassilios G.........................3215
Agostinelli, Matteo...........................3140
Ahmad, Hamzeh J.............................3273
Aiso, Kohei......................................3186
Akagi, Hirofumi 2352
Akahane, Masashi.............................2774
Akama, Yousuke 1741
Akao, Naoki 1217
Akatsu, Kan 711, 3186
Alatise, O 1149
Alenius, Henrik 1704, 4205
Ali, Muhammad 528
Ali, Murad 2317
Allmeling, Jost..........................422, 2199
Almér, Stefan.....................................555
Alsofyani, Ibrahim Mohd.....................466
Alvarez, S. 4009
Amano, Koki 94
Amei, Kenji......................................3182
Amin, Mohammad 759
Amrhein, Wolfgang...........................3640
An, Ronghui............957, 1524, 3251, 3692, 3924
An, Zheng 4001
Andenna, M......................................3596
Andersen, A. E. Michael 1351
Andersen, Michael A. E. 607, 4066
Ando, Akinobu 517
Ando, H. 3665
Ando, Masato 1919
Ando, Takashi 3658
Ang, Simon S. 153
Antivachis, Michael............................181
Antonini, Giulio................................3588
Antonopoulos, Antonios 2335
Anurag, Anup 3564
Anyapo, Chan 3332
Aoyagi, Kazuki 2237
Aoyama, Masahiro......................718, 753
Arai, Takuro......................................1997
Araumi, Ryunosuke 1877, 3658
Arimatsu, Kenji 1370

Arita, Hideaki 2796, 2820
Arrua, Silvia.................................... 1719
Asada, Kazunori 3658
Asama, Junichi 4016
Ashizaki, Yusuke 3450
Ashourloo, Mojtaba 2380
Aso, Shinji 3086
Aware, Mohan 1730
Ayano, Hideki 1080
Azad, A N M Wasekul......................... 2416
Azegami, Kazuya 3723
Azuma, S. 3665
Baba, Teppei 2283
Babasaki, Tadatoshi 207
Bach, Hoang Linh 2410
Baek, Jae-Il 108, 2365, 3100, 3533, 3538
Baek, Miran 1141
Bahat-Treidel, Eldad.......................... 3607
Bai, Baodong 2638
Baik, Jeong Min 3063
Bak, Yeongsu 1736, 4104
Bakran, Mark-M. 2476
Bandyopadhyay, Soumya 1426
Barrena, Jon Andoni 759
Barrera-Cardenas, Rene..................... 3431
Bauer, Pavol............................ 1426, 2630
Bauer, Walter 3640
Bayer, Christoph Friedrich 2410
Bellini, M. 4009
Berg, Matias........................... 963, 4205
Bergveld, H.J. 267
Bertoldi, F. 488
Besselmann, Thomas......................... 555
Bezha, Minella 3170
Bhattacharya, Subhashish 3564, 3993
Bhowate, Apekshit............................ 1730
Bhumkittipich, Krischonme 2430
Biela, J. 1896
Biela, Jürgen 1103, 1509, 2301, 3734
Bilal, Ahmad 2193
Bilsalam, A. 1622
Bin, Zhao 2692
Bixel, Paul 238
Blaabjerg, Frede................. 439, 746, 1183,
    1246, 1711, 1788, 2512, 2604, 2743, 3123,
    3164, 3357
Blanes, José M. 1435
Böcker, Jan 3607
Bojoi, R.. 732
Bonyadi, R. 1149
Boroyevich, Dushan 790, 3705, 3749, 3985
Bortis, D. 4080
Bortis, Dominik 181
Boynov, K.O..................................... 161
Braun, Michael 2848, 3074

AUTHOR INDEX

Büdel, Johannes 3034
Bui, M.X. 4174
Bunlaksananusorn, Chanin 2490
Burgos, Rolando 790, 3705, 3749, 3985
Cai, Kejun 3965
Cai, Panpan 3495
Cai, Xu 1004, 1491, 2245, 4162, 4220
Canales, F. 4009
Cao, Hu 1816, 3484
Cao, Pengpeng 2973
Cao, Qi 100
Cao, Wu 3002, 3010, 3015
Cardenas, Rene Alexander Barrera 1111
Carvalho, Kelly C. M. 3785
Castellazzi, Alberto 130, 2932
Ceballos, Salvador 3117
Celik, Mustafa 1680
Cha, Honnyong 927, 1046, 2619, 3134
Chae, Beomseok 1977
Chailloux, Thibaut 2153
Chang, Chen-Wei 1617
Chang, Chien-Hsuan 2860
Chang, Liuchen 815, 1472, 1793, 2505
Chang, Yung-Ruei 639, 883
Chanmontree, P. 1622
Chao, Yi-Hao 1145
Charalambous, Apollo 1634
Charoensuksirikul, Supanut 2113
Chattopadhyay, Ritwik 3564
Chazal, Hervé 2158
Chen, Ang-Tung 2102
Chen, Bo 1397
Chen, C. 142
Chen, Ching-Chen 1617
Chen, Ching-Jan 2086
Chen, Chuantong 1598
Chen, Dezhi 2638
Chen, Guan-Jung 1341
Chen, Guo 370
Chen, Hao 3112
Chen, I-Lin 2107
Chen, Jiangnan 1157, 1167
Chen, Jiann-Fuh 2653
Chen, Jie 1015, 1177
Chen, Kai-Hui 3081
Chen, Ke 1391
Chen, Kun-Feng 1341
Chen, Min 878
Chen, Minwu 2547
Chen, Nan 2335
Chen, Pingping 1118
Chen, Shen-Li 1145
Chen, Song 2153
Chen, Tang-Jung 1617
Chen, Tao 1872

Chen, Wan-Jung 3544
Chen, Wenjia 4213
Chen, Wenjie 1062, 2854, 3329
Chen, Wu 1504, 2496
Chen, Xiliang 3329
Chen, Xin 1015, 1177
Chen, Xingxing 1051, 3129, 3439
Chen, Yang 2785
Chen, Yangyang 560
Chen, Yaow-Ming 639, 883
Chen, Yenan 1118
Chen, Yen-Wen 2576
Chen, Yufeng 3383
Chen, Yu-Jen 275
Chen, Zhe 1758, 2708
Chen, Zhi 2997
Chen, Zhigang 3040
Cheng, Ching-Hsiang 2086
Cheng, Chun-An 2860
Cheng, Hung-Liang 2860
Cheng, Nie 2625
Cheng, Po-Tai 503, 1038, 2462, 3549
Cheng, Ran 3877
Cheng, Xiangpeng 2435, 3934
Chengbi, Zeng 2718
Chi, Yongning 1491
Chiba, Akira 3627
Chien, Lin-Hao 2102
Chiu, Huang-Jen 2092, 3151
Chiu, Hui-Lung 123
Chiu, Yi-Hao 1145
Cho, Geum-Bae 2145
Cho, In-Ho 3323
Cho, Shin-Young 1530
Cho, Young Joon 137
Cho, Younghoon 1403
Choe, Chanyang 1598
Choi, Byungcho 1465
Choi, Hyun-Jun 383
Choi, Jae Hyuk 1336
Choi, Jaeho 803
Choi, Joon-Ho 982, 1799
Choi, Seung-Hyun 4049
Choi, Sewan 256
Choi, Sung-Jin 1409
Choi, Youn-Ok 2145
Chou, Shih-Feng 1711
Chou, T.-C. 1912
Choudhury, Abhijit 3401
Chuai, Guoming 3025
Chung, Daewoong 1141
Chung, Henry S. H. 917
Chunkag, V. 1622
Collins, Caspar 1931
Cortes, Camilo 2193

AUTHOR INDEX

Corvasce, C. 3596
Cucala, Asuncion P. 2534
Cui, Shenghui 2250, 2484
Cui, Xiang 1125
Cvetkovic, Igor 790, 3985
Czyz, Piotr 396
D'arco, Salvatore 782, 2003
Da Silva, C. 267
Dahidah, Mohamed S A 3215
Dai, T. 1149
Dai, Wenjing 1015
Daikoku, Akihiro 2796
Danqing, Liu 1376
Dao, Ngoc Dat...1212
Dauphin, Benjamin 3644
Davari, Pooya 746
Davletzhanova, Z 1149
De Doncker, Rik W. 375, 388, 598,
     1073, 2250, 2484, 2768, 3729, 3979
Decker, Simon 2848, 3074
Delaforge, Timothé 2158, 3820
Deng, Fujin 1758, 2708
Deng, Jinxin 2992
Deshpande, Prathamesh Pravin..............4186
Dieckerhoff, Sibylle................................3607
Dimarino, Christina 3985
Din, Zakiud...2262
Ding, Yong...815
Dinh, Nguyen Duy....................................363
Diouf, Fatou 2078
Dirksen, Daniel.......................................2410
Divan, Deepak 4001
Doki, Shinji 1032, 1223, 1228, 1295, 1747, 2224
Dong, Hanjing 987
Dong, Mi 1771
Dong, Qinghua 459
Dong, Xiaofeng.......................................4168
Dong, Zhen 459
Dong, Zheng 3768
Driesen, J. 488
Du, Chao 2204
Du, Xiaotong 1167, 2780
Du, Xizhou 1491
Du, Yan 1472, 2877
Du, Zhijiang...84
Duarte, J. L. 946, 1067, 2697
Duarte, Jorge L. 1447, 3840
Dugal, F..3596
Dujic, Drazen.............. 422, 1484, 1498, 2170
Duong, Truong-Duy..................................982
Duque, C. A. 1067
Eberle, Wilson 927
Ekman, Jonas 3588
Elbaset, Adel A. 3945
Endegnanew, Atsede G. 2003

Endo, Hiroaki 4151
Endres, Tobias Maximilian 2410
Engelmann, Georges 3979
Enomoto, Bruno Yukio 3785
Eto, Haruhi 2097
Faiz, Muhammad Talib 528
Fajri, Poria 3223
Fan, Dongchen 3002, 3010, 3015
Fan, Shengwen 977, 3040
Fan, Weiyan 1386, 1421
Fang, Jingyang........................... 337, 3910
Fang, Ran 4213
Fangfang, Luo 1282
Farkas, Gabor 137
Fayyaz, Asad 130
Felderer, Niklaus 2199
Feng, Chao 2058
Feng, Wei 3678
Ferdowsi, Mehdi 3223
Fernandez, Gabriel...................................3209
Fernandez-Cardador, Antonio 2534
Fischer, F. 3596
Foo, Gilbert 1724
Formentini, A. 4034
Freijedo, Fracisco D. 1498
Friedrichs, Peter 3584
Fuchs, Simon 2301
Fujii, H. 1253
Fujii, Kansuke 3711
Fujii, Keisuke 1189
Fujii, Toshiyuki 2540, 3578
Fujimoto, Hiroshi 77, 663
Fujimoto, Kazuki 2047
Fujimoto, Yasutaka 571, 681
Fujimura, Akira 1080
Fujita, Atsushi 296
Fujita, Goro 363
Fujita, Hideaki.............626, 1854, 3813, 3940
Fujiwara, Hajime 1381
Fujiwara, Kazuya 3773
Fukuda, Hiroto 2938
Fukuda, Kenji 2558
Fukui, Tomoya 860
Fukuoka, T. 1240
Fukushima, Kentarou 2176
Fukushima, Takafumi 3478
Funabiki, Shigeyuki 2449
Funaki, Tsuyoshi 309, 2181, 3092
Funato, Hirohito 94, 2036
Funato, Hiroki 2073
Furukawa, Keita 3349
Furukawa, Kimihisa 3572
Furukawa, Yudai 4193
Furusho, Yasuaki 3711
Gan, Yiliang 1391

AUTHOR INDEX

Ganisetti, V. K. 2907
Gao, Feng 2016, 3383, 3965
Gao, Xiaonan 1661
Gao, Zhuo 3455
Garrigós, A. 1435
Gasim, Abdulaziz 2836
Gehlot, Deepak 3471
Geng, Hua 542
Geng, Yiwen 619
Gerada, C. 4034
Gheonjian, Anna 2078
Gietler, Harald 3140
Gohara, Hiromichi 2764
Gondo, Ryota 3490
Gong, Bing 3797
Gong, Chunying 1015, 1177
Gong, Z. 267
Gorodnichev, Anton 375
Goto, Akihisa 2449
Goto, Hiroki 3192
Goto, Kazuya 1315
Goto, Yasuyuki 809
Gou, Yating 1157, 1167
Grimm, Ferdinand 2895
Grossner, Ulrike 3588
Gruber, Wolfgang 3632, 3640, 4028
Gu, Lei 632
Gu, Qing 2963
Guajardo, Cristian Andres Garces 1854
Guan, Bo 1032
Guan, Yajuan 2668, 3678
Guan, Yueshi 614, 3780
Guangzhu, Wang 1376
Guerrero, Josep M. 1498, 2668, 3112
Guerrero, M. Josep 3678
Gui, Yonghao 2668
Guidi, Giuseppe 782, 2003
Guillod, Thomas 396
Gunji, Daisuke 663
Guo, Leilei 904
Guo, Yanjie 3338
Guozhao, Duan 2625
Gupta, K. 267
Gurpinar, Emre 130
Gutiérrez, R. 1435
Ha, Jung-Ik 565, 2500
Ha, Sang-Hyun 3466
Haga, Hitoshi 1370, 3890
Hagiwara, Makoto 3273
Hahashi, Yuji 4059
Haider, M. 4080
Halamicek, Michael 831
Halick, Mohamed 416
Hamabe, Yasumasa 1276
Hamada, Shizunori 227

Hamaguchi, Takumi 3507
Hamasaki, S. 1240
Hamasaki, Shin-Ichi 1217, 1276, 2938, 3237
Hameyer, Kay 740
Han, Byung-Moon 466
Han, Jung-Kyu 3107, 3533, 4049, 4054
Han, Pengcheng 1027, 2714
Han, Yang 3112
Hanajiri, Kensuke 663
Hanamoto, Tsuyoshi 1315, 1698
Hancioglu, Oguz Kaan 1680
Handa, Hiroyuki 3762
Handa, Yuuichi 4059
Hane, Yoshiki 2426
Hang, Lijun 1391, 2866
Hanju, Cha 1985
Hao, Liu 3484
Hao, Xiang 1478
Harnefors, Lennart 3684
Hartmann, S. 3596
Haruna, Junnosuke 94, 2036
Hasegawa, Kazunori 1938
Hasegawa, M. 3665
Hasegawa, Ryuta 2011
Hashempour, Mohammad M. 4198
Hashimoto, Kazuki 3757
Hasler, Jean-Philippe 3684
Hata, Katsuhiro 663
Hata, Ryotaro 2149
Hatakeyama, Tomoyuki 1991
Hataya, Morimasa 410
Hatipoglu, E. 3805
Hatsumi, Takuya 94
Hatta, Yoshiyuki 675
Hattori, Fumiya 2738
Hattori, Keisuke 3286
Haung-Jen, Chiu 645
Hayashi, Nobuo 866
Hayashi, Yuji 356
He, Wangpin 560
He, Xiaokun 1504, 2496
He, Xiaoqiong 1027, 2714
He, Yigang 2317
He, Yingjie 3439
Hendrix, M. A. M. 946, 2697
Heo, Jongwon 726
Hidaka, Yuki 2820
Higuchi, Keiichi 2764
Higuchi, Masato 3952
Higuchi, Shinichi 2216
Hikaru, Naruse 3418
Hikihara, T. 3665
Hikihara, Takashi 3654, 3757
Hiller, Marc 3074
Hillers, A. 1896

AUTHOR INDEX

Hillers, André 2301
Hilt, Oliver 3607
Hinz, Arne 598
Hirahara, Hideaki 1960
Hiraki, Eiji 410, 1602, 1610
Hirao, Takashi 2082, 2137
Hirase, Yuko 767
Hirayama, Katsutoshi 4193
Hirayama, Tadashi 3406
Hirokawa, Masahiko 1543, 4133
Hirokawa, Takayuki 296, 410
Hiromoto, Masayuki 3644
Hirose, Keiichi 593, 822
Hirose, Naoki 3791
Hiroshi, Tadano 3431
Hiroshige, Shinichi 3369
Hirota, Takashi 3952
Hoang, Tuan V. 1752
Hoda, Isao 2073
Hofmann, Viktor 2476
Hofmann, Wilfried 3243
Hojo, Masahide 3369
Holenstein, Thomas 3619
Holmes, D. G. 3670
Hong, Miao 2718
Hongpeng, Liu 1442, 2969
Honjo, Satoshi 2066
Hori, Motohito 3396
Hori, Yoichi 77, 663
Horie, Shunsuke 809
Horikoshi, Takahiro 1997
Hoshi, Nobukazu 971, 2660, 3855
Hou, Chung-Chuan 1617
Hou, Lijun 2901
Houran, Mohamad Abou 1062, 2854
Hsieh, Guan-Chyun 123
Hsieh, Hung-I 123
Hsieh, Yao-Ching 3151, 3544
Hsu, Chi-Hsuan 2653
Hu, Jiewen 3985
Hu, Jingxin 1073, 2250, 2484
Hu, Sheng 3052
Hu, Song 370
Hu, Xihong 614, 3780
Hu, Xing 2262
Huang, Bing-Siang 2092
Huang, Bo-Jia 3528
Huang, Chien-Chun 3151
Huang, Huazhen 1125
Huang, Jingjin 2980, 4157
Huang, Jingjing 1004, 2688, 2692
Huang, Jun-Xian 1626, 3081
Huang, Lang 1478
Huang, Pin Yu 2165
Huang, Ta-Wei 1626

Huang, Wen-Mei 2576
Huang, Xianjin 1131, 2051
Huang, Xiaoliang 84
Huang, Xuehao 3455
Huang, You-Chun 275
Huemer, Mario 3140
Hui, S. Y. Ron 889, 2552
Hung, Chun-Yao 2576
Hung, Shun-Kang 1575
Huo, Chongcan 987
Huo, Junya 1206, 1234
Hussein, Abdallah 130, 2932
Huynh, Dang Minh 3086
Hwang, Duck-Hwan 1403
Hwang, Seon-Ik 3323
Hwu, K.I. 851
Hyakutake, Y. 1253
Hyodo, Takashi 2589
Hyunsung, An 1985
Iannuzzi, Diego 2527
Ibuchi, Takaaki 309
Ichinose, H. 1240
Ide, Yuji 3896
Iijima, Ryuji 313, 1111
Iioka, Daisuke 2278
Ikari, Yuki 148
Ikeda, Hidehiro 1315
Ikeda, Yoshinari 3396
Ilves, Kalle 2335
Imai, Kazu 3363
Imai, Makoto 296, 410
Imamori, Satoshi 699
Imaoka, Jun 1087, 1095, 1554, 3773
Imoto, R. 2808
Imtiaz, Abu Saleh 2416
Imura, Takehiro 77, 663
Inaba, Tsuyoshi 4114
Inomata, Kentaro 3952
Inoue, Daisuke 2764
Inoue, Kaoru 1264, 2186, 4151
Inoue, Kent 348
Inoue, Masamichi 1228
Inoue, Takatoshi 1276
Inoue, Y. 704, 2808
Inoue, Yukinori 1189, 1289, 1329, 2802, 2814, 3197
Irino, Yusuke 244
Ise, Toshifumi 775, 2393, 3762, 3902
Ishibashi, Mikiya 1370
Ishibashi, Naoyuki 1543
Ishibashi, Taku 2292
Ishigaki, Shingo 227
Ishiguro, Takahiro 1997, 2011, 3304
Ishihara, Masataka 1610
Ishii, Y. 1834
Ishii, Yuki 1196

AUTHOR INDEX

Ishikawa, Hiroki 2176, 3412
Ishikawa, Kohsuke 2725
Ishikura, Yuki 1087, 1095, 3717
Isobe, Eisuke 2042
Isobe, Takanori 313, 1111, 3375, 3431
Isozaki, Keisaku 1364
Itaya, Yohei 3450
Ito, Kazuhiko 2540
Ito, Yasuaki 1586, 2324
Ito, Yoichi 3086
Ito, Youichi 439
Itoh, Gimpei 1289
Itoh, Jun-Ichi 69, 348, 534, 896, 1567,
        2229, 2237, 2519, 2596, 3349, 3797
Iwabuchi, Akio 439
Iwai, Akinobu 2066
Iwaji, Yoshitaka 1301
Iwasaki, Makoto 1666
Iwasaki, Tetsuya 3490
Iwata, Hiroki 3896
Iyasu, Seiji 4059
Iyoda, Isao 2914
Jacobs, Keijo 3292
Jaffar, Hanis Afiqah Binti 2956
Jain, Prashant..................................3471
Janah, Mounia...................................681
Jang, Duekjin 2619
Jang, Yu-Jin 1655, 3466
Jang, Yun 1736
Jangs, Yujin 1562
Jarutus, Neerakorn 2121
Jehle, Andreas 1509
Jennings, M 1149
Jeong, Seog Y 2564
Jeong, Si-Hoon 289
Jeong, Yeonho 838, 2365, 2376
Jhang, Ying-Yi 3884
Jhou, Yu-Lin 1145
Ji, Guyuan 2921
Jia, Haiyang.....................................998
Jia, Pengyu.............................977, 3040
Jia, Xu 3025
Jiacheng, Wang 2986
Jiajie, Zang 2986
Jiajie, Zhou 1442
Jian, Jun-Min 2653
Jiang, Jinhai 84
Jiang, Shuai 987
Jiang, Siyue 4168
Jiang, Yanfeng 2058
Jiang, Yongbin 3863
Jianhua, Wang 1282
Jianming, Xu 528
Jianqiao, Zhou 2986
Jianwen, Zhang 2986

Jiaxing, Liu 1376
Jikumaru, Takehiro 177
Jimichi, T. 1834
Jimichi, Takushi 3729
Jin, Nan 904
Jin, Zheming 2668
Jing, Lei 878
Jing, Lyu 2692
Jing, Yang 3383
Jingyu, Song 1282
Jing-Yuan, Lin 645
Jinjun, Liu 4181
Jinshui, Zhang 4181
Jisaki, Jun 3182
Joebges, Philipp 375
Jongudomkarn, Jonggrist 3902
Jonishi, Akihiro 2774
Joryo, Satoshi 1202
Joseph, Anto 1358
Jumayev, S. 161
Jung, Hanul 688
Jung, Hyun-Sam 911
Jung, Jae-Jung 3557
Jung, Jee-Hoon 289, 383
Jung, Jun-Hyung 3323
Jung, Si-Hoon 383
Jungmayr, Gerald 3640
Junior, Lourenço Matakas 3785
Jynu-Jhe, Jhang 645
Kada, Haruya 3890
Kadota, Mitsuhiro 3572
Kai, Masahiko 1803
Kaicheng, Ding 4181
Kaipia, T. 2948
Kaishakuji, Hikaru 2360
Kakigano, Hiroaki 583, 2956
Kamaeguchi, Koki 410
Kamakura, Kousuke 2756
Kamejima, Takayoshi 3286
Kamiya, Naoki 1673
Kamiyama, Naosumi 1955
Kamoshida, Naoki 1111
Kampeerawar, Warayut 3257
Kanai, Naoyuki 3396
Kanaya, Kazuhisa 2011
Kanazawa, Yasuki 2789
Kanchan, R. S. 488
Kandula, Prasad 4001
Kaneko, Satoshi 3396
Kanetani, Kaisei 207
Kang, Dong-Hun 3030
Kang, Feel-Soon 2376
Kang, Kyoung-Suk 922
Kang, Tahyun 1977
Kang, Yong 2997

AUTHOR INDEX

Kanno, Junya 3299
Kano, Fumihisa 2036
Kanoda, Akihiko 3572
Kanzian, Marc....................................3140
Kapisch, E. B. 1067
Karami, Bagher...................................2854
Karppanen, J.....................................2948
Kasai, Yuji 2036
Kashihara, Tatsuki 1741
Katayama, Tatsuji 1346
Kato, Hideaki 1580, 1586, 2324
Kato, Hirokazu 3478
Kato, Koji 439, 1370, 3086
Kato, Toshiji.................1264, 2186, 4151
Katoh, Kaoru.......................................233
Katoh, Shinji 2176
Katsuki, Akihiko 1543
Katsura, Seiichiro.................................669
Katsura, Shogo 767
Katsushi, Terazono 3431
Kawabata, Naoki 2887
Kawabata, Shuma 3406
Kawagoe, Natsuki 3490
Kawaguchi, Hironori.............................517
Kawaguchi, Jun'ichiro 1828
Kawaguchi, Yuki 3572
Kawakami, Masaki 2756
Kawakami, Noriko 1346
Kawamura, Atsuo 318, 1649, 1687, 3916
Kawamura, Itsuo 3396
Kawamura, Kazuki 1567
Kawanishi, Kota 169
Kawashita, Jun 2042
Kayashima, Kazuya 1315
Kaymak, Murat....................................3729
Kazmi, Syed Muhammad Raza 4168
Ke, Junji 1125
Kennel, Ralph.............1661, 2895, 3965
Kezuka, Nobutaka 227
Khan, Ashraf Ali 927
Khan, Faisal................................446, 2416
Khan, Muhammad Mansoor...................528
Khan, Usman Ali 927
Khomfoi, Surin 1460
Khubchandani, Vasudha........................845
Kiatsookkanatorn, Paiboon..................2581
Kida, Masahiro.........................1586, 2324
Kido, Tatsuya 329
Kikuchi, Ryosuke 1877
Kikuchi, Takaaki 2292
Kikuchi, Takeshi 3578
Kikuma, Toshiaki 3299
Kim, Byeongwoo 256
Kim, Chong-Eun 108, 3538
Kim, Dong-Kwan 1655, 3466, 3538

Kim, Gun-Woo 838
Kim, Hansang 1465
Kim, Heung-Geun 927, 1046, 2619, 3134
Kim, Hideaki....................................... 207
Kim, Hyeon-Sik 521
Kim, In-Dong 3229
Kim, Jae-Kuk 3100
Kim, Jang-Mok 3323
Kim, Jin-Hak 1530
Kim, Jin-Young 3229
Kim, Jong-Woo........................... 3107, 4049
Kim, Kangsan 256
Kim, Katherine A.......................... 2092, 3063
Kim, Keon Young 4104
Kim, Keon-Woo............108, 1562, 1655, 2365, 2376
Kim, Ki-Mok 2365
Kim, Myong Hwan 2500
Kim, Sanghun 2619
Kim, Sunju 3833
Kim, Yeonjung 1465
Kimura, Hideki 2036
Kimura, Mamoru 1991, 1997
Kimura, Noriyuki............1202, 1259, 2558, 2887, 2914
Kinoshita, Masahiro 3929
Kishimoto, Toshihiko 261
Kishita, Ken 1301
Kitagawa, Wataru 1847, 3507
Kitamura, Akio.................................... 2764
Kitamura, Toshinori 2660
Kiyoshi, Ohishi 1673
Kiyota, Kyohei 3182
Klammer, Bianca 3632
Ko, Chien-Tzu 2107
Kobayashi, Hiroyasu 2527, 3490
Kobayashi, Koji 1741
Kobayashi, Marika............................... 2802
Kodaka, Wataru 2589
Kogai, Naoki 1364
Koizumi, Hirotaka 4114
Kolar, J. W. 3805, 4080
Kolar, Johann W. 181, 396, 3619
Kolb, Johannes 2848
Komaru, Yuma 1329
Komatsu, Hiroyoshi 1346
Komatsu, Taiga 2820
Komatsu, Wilson 3785
Komeda, Shohei 3813
Kometani, Haruyuki 711
Kondo, Keiichiro................726, 2047, 2527, 3490
Kondo, Shota 1295
Kondo, Takeshi.................................... 4114
Kong, Wei 3460
Kongjeen, Yuttana 2430
Konishi, Akihiro 1602
Konno, Junya 1692

AUTHOR INDEX

Konstantinou, Georgios 3117
Kopta, A. 3596
Kosaka, Takashi..3418
Koseki, K. 1162
Koseki, Takafumi..................2042, 2309, 3257
Koshikizawa, Hiroyuki 1567
Kostov, Konstantin 2732
Kouketsu, Masaju...227
Kouno, Yusuke 2176
Kovacevic-Badstübner, Ivana 3588
Kowatari, Hiroki...2660
Koyama, Yushi 2011
Krismer, F. 3805
Krismer, Florian 396
Kubo, Hajime 483
Kubota, Hisao...1196
Kucka, Jakub 1904
Kumada, Keishirou 3396
Kumagai, Shuta..1264
Kumar, Ashish 3993
Kumar, Rajesh 2456
Kumar, S. Gautam 3471
Kumsuwan, Yuttana 2113, 2121
Kunomura, Ken 1803
Kuo, Chun-Ting 1145
Kuraishi, Daigo 3896
Kuraku, Nagendra Vara Prasad....................2317
Kuring, Carsten 3607
Kurisaka, Masakatsu 4151
Kurita, Naoyuki 1991
Kurita, Nobuyuki 3640
Kurokawa, Fujio 826, 2097, 2283, 4193
Kurosawa, Nobuhito 1810
Kurumatani, Hiroki......................................669
Kusaka, Keisuke.............69, 348, 2237, 3349
Kusumah, Ferdi Perdana 3870
Kuwata, Gen...177
Kwon, Min-Jun 114
Kyyrä, Jorma 2193, 3870
Lai, Jih-Sheng 3107, 4049
Lai, Jui-Hung 3081
Lan, Yuanliang 1167
Lana, A. 2948
Le, Hanh-Phuc 213
Le, Hoai Nam 2519
Lee, Byoung-Hee...838
Lee, Byung-Kwon..3030
Lee, Chan 688
Lee, Choongin 565
Lee, Dong-Choon 478, 1212
Lee, Hong-Hee 1752
Lee, Hyong Gun 1336
Lee, Il-Oun...1530
Lee, Jae-Bum...3100
Lee, Jia-You...657, 2107

Lee, Joon-Hee 3557
Lee, Junbae 1141
Lee, June-Hee 466
Lee, Jung-Yong 1403
Lee, Jun-Young 3030
Lee, Jusuk 1336
Lee, Kyo-Beum 466, 1736, 4104, 4109
Lee, Kyoung-Won 2145
Lee, Kyung-Hwan 2500
Lee, Min-Su 108
Lee, Minsub 1141
Lee, Nayoung 1562
Lee, Song-Kai...2102
Lee, T. L. 4198
Lee, Tzung-Lin 2576
Lee, Woo-Cheol 114
Lee, Woo-Seok 1530
Lee, Young-Dal 3466, 3538
Lehn, Peter W. 3203
Lei, Qin 2400, 3742
Leng, Darith 1764
Leubner, Martin 3243
Li, Bodong 878
Li, Chi 790, 3705
Li, Dongsheng 1301
Li, Fei 2611
Li, Fujian 2944
Li, Guanglei 1455
Li, Haijin 2270
Li, Haisi 3040
Li, Haoyu 2901
Li, Hong 2058
Li, Hongchang 337, 3910
Li, Jhih-Sian 3081
Li, Jia 2073
Li, Jianfeng 130
Li, Kaiyuan 1517, 1592
Li, Lei 1172
Li, Li 1771
Li, Ming 2973
Li, Mingshen 2668, 3678
Li, Pengcheng 3698
Li, Shufan 3338
Li, Sinan 889, 2552
Li, T.-Y. 1912
Li, Xiaodong 370
Li, Xiaolu Lucia 3768
Li, Xiaoqiang 3910
Li, Xingshuo 453
Li, Xinying 2646
Li, Yan 2245
Li, Yang 795, 1478
Li, Yangman 2901
Li, Yi-Chan 639, 883
Li, Yongdong 1010, 2386

AUTHOR INDEX

Li, Yong-Jyun 275
Li, Yunwei 3958
Li, Yunwei Ryan 1537
Li, Yuze 2997
Li, Zhenjie 84
Li, Zhenwei 998
Li, Zhiqing 100
Liang, Daniel 1943
Liang, Junrui 4122
Liang, Ning 1157
Liang, Wencai 1131
Liao, Chenglin 3338
Liao, Chih-Yi 657
Liao, Hsuan 2653
Liao, Jian-Tang 4233
Liao, Mengyan 1386, 1421
Liaw, C. M. 2907
Lim, Cheon-Yong 1655, 2376, 3533
Lim, Dae-Sik 1212
Lim, Kyungbae 803
Lim, Young-Cheol 982, 1799
Lin, Chang-Hua 1341, 1777
Lin, Cheng-Hung 2092
Lin, Fei 1131, 1816, 2051, 2058, 3484, 3495
Lin, Jin 3460
Lin, Jing-Yuan 3151
Lin, K.-E. 1912
Lin, Min 4133
Lin, Xiang 3460
Lin, Xiaolan 1027
Lin, Xuerui 1537
Lin, Yu-Hsiu 1575
Lin, Yu-Lin 1145
Lisha, Chen 3958
Liske, Andreas 2848
Liu, Baojin 1051, 2944, 3924
Liu, Bi 1872
Liu, Bo 542, 878
Liu, Chao 2245
Liu, Chunhui 3742
Liu, Cuicui 1157, 1167
Liu, Dong 1758, 2708
Liu, Fang 2611, 2992
Liu, Furong 3052
Liu, He 3215
Liu, Hwa-Dong 1341, 1777
Liu, Jia 775, 3902
Liu, Jiaxin 2016
Liu, Jinjun 957, 1051, 1524, 2435, 2646, 2681, 3129, 3176, 3251, 3439, 3692, 3924, 3934
Liu, Junwen 3863
Liu, Kangli 3010, 3015
Liu, Nianzhou 1010
Liu, Ning 2877

Liu, Pang-Jung 2102
Liu, Ruofei 2547
Liu, Shu 3052
Liu, Siqi 1491
Liu, Tao 1478
Liu, Teng 2681, 3176, 3934
Liu, Wei 3164
Liu, Wenzhao 3678
Liu, Xiaosheng 934
Liu, Xicai 1661, 3965
Liu, Xinbo 3455
Liu, Yifu 2400, 3742
Liu, Yu-Chen 2092
Liu, Yuping 1816
Liu, Zeng 957, 1524, 2435, 2681, 3176, 3251, 3692, 3749, 3924
Liu, Zhiyuan 3495
Liu, Zipeng 2681, 3176
Lo, Jen-Hao 1145
Lomonova, E.A. 161
Lopez-Lopez, Alvaro J. 2534
Lotfi, Nima 3223
Lovison, Giorgio 77
Lu, David H. 2404
Lu, David Hongfei 3390
Lu, Kaiyuan 1183, 1246, 2842
Lu, M. Z. 2907
Lu, Shengli 3145
Lu, Shuai 3698
Lu, Y. 267
Luhtala, Roni 547, 2470, 3156
Lunglmayr, Michael 3140
Luo, Min 422, 2199
Luo, Rui 3129, 3439
Luo, Y. 267
Luong, Hoan-Tien 2145
Lyu, Jing 1004, 4162, 4220
Ma, Baohui 2882
Ma, Jie 1118
Ma, Ke 3877
Ma, Shaokang 542
Ma, Tianshu 2703
Ma, Yue 3717
Ma, Zhixun 917, 2688, 2692, 4157, 4162
Mabuchi, Yuichi 3572
Machavolu, Sawanth Krishna 753
Machida, Yuuki 2449
Maharjan, Laxman 1840
Makishima, Shingo 2047
Mannen, Tomoyuki 1414, 1866
Mantooth, H. Alan 153
Mao, Meiqin 815, 1472, 1793, 2505
Mariéthoz, Sébastien 2158, 3820
Marinescu, Radu-Florin 1822
Marroquí, D. 1435

AUTHOR INDEX

Martinez, Wilmar 2193
Maruta, Hidenori 826
Maruyama, Kouji 3396
März, Martin 2410
Masuda, Eisuke 309
Masuda, Mitsuru 88
Masuko, Toshitake 3723
Matsubayashi, Tatsushi 207
Matsuda, Akihiro 2329
Matsuda, Tomohiro 1972
Matsudate, Koki 2022
Matsui, Nobumasa 826, 2283
Matsui, Nobuyuki 3418
Matsui, Teruhisa 1803
Matsui, Yoshihiro 1080
Matsui, Yuto 1847, 3791
Matsuki, Yosuke 2224
Matsumori, Hiroaki 3357
Matsumoto, Satoshi 2360, 2370
Matsumoto, Takashi 2404
Matsumoto, Toshiaki 2011, 3304
Matsumoto, Yasuaki 517
Matsumoto, Yohei 233
Matsumura, Toshiro 809
Matsuo, Keisuke 169
Matsuse, Kouki 169
Mattsson, A. 2948
Mawby, P 1149
Mcgrath, B. P. 3670
Meng, Xin 957, 1549, 3251
Menzi, David 181
Mertens, Axel 1904
Messo, Tuomas 547, 963, 1704, 2470, 3156, 4205
Michihira, Masakazu 992, 3058
Michikoshi, Hisato 2558
Milovanovic, Stefan 1484
Min, Geon-Hong 2500
Minami, Masataka 992, 3058
Mino, Kazuaki 3717
Mira, Maria C. 1351
Mishima, Tomokazu 329, 872
Misra, Mitradatta 3884
Mitsantisuk, Chowarit 3332
Miura, Yushi 775, 2393, 3762
Miwa, Yoshihiro 404
Miyajima, Hiroki 1803
Miyama, Yoshihiro 711
Miyawaki, Satoshi 2738
Miyazaki, Toshimasa 1673
Mizumoto, Yuki 1810
Mizuno, Takayuki 169
Mizuno, Yuji 2283
Mizushima, Takuya 1543
Mocevic, Slavko 3985
Mochidate, Sae 1972

Mogorovic, Marko 2170
Moiannou, Tom 831
Mok, Hyung Soo 1336
Molinas, Marta 759
Moo, Chin-Sien 275, 3544
Moon, Gun-Woo 108, 838, 1562,
    1655, 2365, 2376, 3100, 3466, 3533, 3538,
    4049, 4054
Mori, Kazuhisa 233
Morimoto, Hiroaki 2540, 3265
Morimoto, S. 704, 2808
Morimoto, Shigeo ...1189, 1289, 1329, 2802, 2814, 3197
Morimoto, Shinya 2210
Morishima, Naoki 2540, 3450
Moriyama, Hiroyuki 1580, 1586, 2324
Morizane, Toshimitsu 1202, 1259, 2558, 2887, 2914
Mortimer, Benedict J. 598
Motegi, Shin-Ichi 992, 3058
Motohashi, Yuto 753
Motoyama, Hiromasa 356
Mouawad, Bassem 130
Mukaiyama, Naoki 2558
Müller-Hellmann, Adolf 598
Muni, Bishnu Prasad 3471
Murakami, Toshiyuki 575
Nabetani, Yoichi 2404
Nada, Kaho 3578
Nagai, Sakahisa 1687
Nagai, Satoshi 534
Nagao, S. 142
Nagaoka, Naoto 3170
Nagaoka, Shingo 118, 4139
Nagasaka, Kuniaki 1692
Nagashima, Takumi 3490
Nagira, Yoshiki 4016
Naina, Sagar 3046
Nakabayashi, Shigeaki 1692
Nakabayashi, Shigeyuki 517
Nakagawa, Hidehiko 767
Nakahara, Kengo 3237
Nakahara, Mizuki 3572
Nakai, Masanobu 3182
Nakajima, Mizuki 2750
Nakajima, Tatsuhito 1997, 3299
Nakamura, Fuminori 2329
Nakamura, Hideyo 1137
Nakamura, Kenji 2426
Nakamura, Kimikazu 4059
Nakamura, M. 201
Nakamura, Masashi 471
Nakamura, Ritaka 495
Nakano, Hayato 2764
Nakano, Shigeki 2370
Nakao, Hiroshi 196
Nakao, Kazushige 148, 2914

AUTHOR INDEX

Nakao, Yuta 588
Nakashima, Yoshiyasu 196
Nakatsu, Kinya 2082
Nakazawa, Haruo 2404
Nakazawa, Y. 1253
Nakazawa, Yuji 244
Namba, Akihiro 2082
Nanamori, Kimihiro 2789
Naradhipa, Adhistira M. 3833
Narita, Takayoshi 1580, 1586, 2324
Narushima, Hiroki 693
Nashida, Norihiro 1137
Nasr, Miad 2380
Natori, Kenji 588, 1860
Nawaz, Muhammad 2335
Nazib, A. A. 3670
Nee, Hans-Peter 2732, 3292, 3684
Neubert, Markus 3979
Ngamroo, Issarachai 2287
Ngo, Tung 1724
Nguyen, Bang Le-Huy 1046, 3134
Nguyen, Hong-Quan 3426
Nguyen, Minh-Khai 982, 1799, 2145
Nguyen, Tien-The 1046, 3134
Nho, Eui-Cheol 922
Nicolae, Ileana-Diana 1822
Nicolae, Petre-Marian 1822
Nie, Jintong 2963
Niki, Toru 856
Nimura, Takumi 1295
Ninomiya, Tatsuya 2836
Nishikata, Shoji 4227
Nishimura, Yoshitaka 1137
Nishino, Taisei 1364
Nishiyama, Shigeki 2149
Nishizawa, Koroku 2229
Nishizawa, Shin-Ichi 1938
Niu, Haonan 3025
Niyomsatian, K. 4096
Noah, Mostafa 1087, 1095
Noda, Taku 2176
Noda, Yujiro 324
Noguchi, Toshihiko 718, 753
Noh, Seungjun 1598
Nomura, Naofumi 2216
Nomura, Shinichi 2022
Nonogaki, Midori 2292
Noro, Osamu 767
Norrga, Staffan 3292
Norum, Lars 2630
Noto, Yasuyuki 3711
Notohara, Yasuo 1301
Nuchnoi, S. 4096
Nugroho, Dannisworo S. 3855
Nussbaumer, Thomas 3619

Nuutinen, P. 2948
Obara, Hidemine 1649
Oda, Yoshiho 1586, 2324
Ogasawara, Satoshi 2589, 2725, 2796, 3315
Ogawa, Eri 2768
Ogawa, Kazuki 1580
Ogawa, Takuro 866
Ogawa, Tomoyuki 1828, 3265
Ogawa, Toru 2796
Ogino, Hiroshi 517
Oh, Sehoon 688
Ohashi, Hidetomo 2774
Ohdera, Fumiya 1322
Ohguchi, Hideki 699
Ohishi, Kiyoshi 1741, 3332, 3890, 3896
Ohji, Takahisa 3182
Ohnishi, Haruna 3273
Ohno, Takanobu 971
Ohno, Tatsuki 1649
Ohnuma, Naoto 233
Ohnuma, Takumi 1223
Ohnuma, Yoshiya 2738
Ohta, Kazuki 1223
Ohta, Takahiro 517
Ohtake, Asuka 3286
Ohyama, K. 1253
Ohyama, Kazuhiro 2921
Ohyama, Kazunobu 244
Oi, Kazunobu 1890
Oishi, Kazuki 3644
Oiwa, Takaaki 157, 4042
Oka, Toshiomi 2370
Okamoto, Kenkichiro 1095
Okazaki, Yuhei 2335
Okazawa, Toshio 2066
Oki, Yusuke 1828
Okitsu, Takashi 169
Okuda, Takafumi 3654, 3757
Okuno, Kengo 1586, 2324
Okuyama, Ryota 3450
Omori, Hideki 1202, 1259, 2558, 2887
Omori, Shuto 471
Omura, Ichiro 1938
Onishi, Hiroyuki 4139
Onishi, Masami 2082
Ono, Y. 4080
Onozawa, Yuichi 2768
Ooshima, Masahide 3613
Orikawa, Koji 2589, 2725, 3315
Ortiz-Gonzalez, J. 1149
Osawa, Akihiro 2764
Oshima, Takuya 4088
Osman, Ilham 3971
Ota, Ryosuke 3855
Ouaida, Rémy 2153

AUTHOR INDEX

Ouchi, Takayuki 250
Ouyang, Shaodi 1051, 3129
Ouyang, Ziwei 4066
Owaki, Daiki 809
Paiboon, Supakorn 1642
Pairindra, Worapong 1460
Pan, Pengpeng 1504
Pan, Xuewei 1172
Panda, Sanjib Kumar 4186
Pang, Hui 2343
Papadopoulos, C. 3596
Papini, L. 4034
Paramalingam, Jan 2329
Parashar, Sanket 3993
Park, Hwa-Pyeong 289
Park, Jin-Hyuk 4104
Park, Jun H. 2564
Park, Kwon-Sik 922
Park, Moo-Hyun 1562, 3100, 3533
Park, Mu-Hyun 838
Park, Sang Uk 1336
Park, Sanghyeon 282
Partanen, J. 2948
Pasterczyk, Robert 2158
Patel, Prashant 3046
Patel, Utsav 3046
Pathmanathan, M. 488
Patwa, Premal 3046
Pauli, Florian 740
Pecharroman, Ramon R. 2534
Pei, Xuejun 2997
Peltoniemi, P. 2948
Peng, Jinjie 939
Peng, Xu 1027, 2714, 3020
Pengxiang, Zeng 4181
Pham, N. Ha 1414
Pidaparthy, Syam Kumar 1465
Pinomaa, P. 2948
Polmai, Sompob 1764, 2490
Pou, Josep 3117
Prabowo, Yos 3564
Prasanth, Sundararajan 416
Prodic, Aleksandar 831
Promyoo, Adisak 2871
Pueschel, Tilo 190
Pyrhonen, J. 161
Qi, Wenlong 889
Qian, Cheng 1472
Qian, Qinsong 3145
Qiao, Liang 3329
Qin, Zian 1925
Qiu, Maohang 878
Qiu, Zhifeng 939
Rabkowski, Jacek 2129
Radman, Karlo 3632
Radwan, Hamdy 3945
Rahimo, M. 3596, 4009
Rahman, Ahmad Arif Bin Abd 2956
Rahman, Faz 3971
Rahmati, Abdolreza 2854
Ramirez-Elizondo, Laura 1426
Ramos, Niño Christopher 3092
Ran, L 1149
Ran, Li 1931
Rao, Eswar 3471
Rathore, Akshay Kumar 342, 2456
Reinikka, Tommi 1704
Remus, Nico 3243
Ren, Haijun 2714
Ren, Yu 3329
Rencz, Marta 137
Rengarajan, Satish 3564
Riar, Baljit 4074, 4145
Rietmann, Stefan 2301
Rim, Chun T. 2564
Risseh, Arash Edvin 2732
Rivas-Davila, Juan 282, 632, 3848
Robert, Mickaël 2158
Rodriguez-Diaz, Enrique 1498
Roes, M. G. L. 946, 2697
Roinila, Tomi 547, 1704, 1719, 2470, 3156, 4205
Romano, Daniele 3588
Roy, Sourov 446, 2416
Ruan, Liheng 3010, 3015
Rubino, S. 732
Ruf, Andreas 740
Rygg, Atle 759
Sadakata, Hideki 410
Sagawa, Kouhei 2036
Saha, Tarak 4074, 4145
Saito, Tatsuhito 1828, 3265
Saito, Yota 1782
Saitoh, Hiroumi 2278
Sakabe, Tomoki 3058
Sakai, Kazuto 2826
Sakai, Ryosuke 2832
Sakai, Yoshikazu 4114
Sakawaki, Atsushi 244
Sakimoto, Kenichi 767
Sakiyama, Taiki 2186
Sakoda, Kenichi 860
Sakr, Nadim 2078
Sakuma, Kensuke 3522
Sakuraba, Tomokazu 2153
Sakurai, Seiya 3412
Samanta, Suvendu 342
Samermurn, S. 4096
Samizadeh, Mehdi 1062, 2854
Sanada, M. 704, 2808
Sanada, Masayuki ..1189, 1289, 1329, 2802, 2814, 3197

AUTHOR INDEX

Sangwongwanich, Ariya 2512
Sangwongwanich, S...4096
Sangwongwanich, Somboon 1642, 2581
Sannomiya, Kenta 1259
Sano, Kenichiro..3299
Sano, Toshiki..3896
Santi, Enrico...1719
Sasaki, Masahiro...2774
Sasaki, Masato..3344
Sasongko, Firman 416
Sathik, Mohamed 416
Sato, Fumihiro..250
Sato, Keisuke 3265
Sato, Kenji 3478
Sato, Mitsuru...118
Sato, Motoki..663
Sato, Takashi...3644
Sato, Yasuhiro 2042
Sato, Yukihiko................588, 1860, 1972, 3514, 3522
Satoh, Nobuo 2750
Sayed, Mahmoud A..3945
Schanen, Jean-Luc 2158
Schletz, Andreas...2410
Schülting, Philipp..388
Schweiker, Daniel...2848
Schweizer, Mario 555
Schwendemann, Rüdiger..3074
See, Kye Yak..2296
Sekiba, Yoichi...2176
Sekimoto, Morimitsu 866
Sekisue, Takayuki 2176
Sekiya, Hiroo..3650, 4127
Semwal, R. R. 1358
Senanayake, Thilak..313
Seng, Tan Chuan..416
Seo, Byuong-Jun..922
Seo, Gab-Su...213
Sera, Dezso...2512
Setiadi, Hadi...626
Settels, Sjef J. 3840
Severson, Eric L. 4020
Sewergin, Alexander 3979
Sha, Yilin..3329
Shabib, G. 3945
Shamseh, Mohammad Bani 3916
Shan, Zhenyu..977
Shang, Gao..1282
Shao, Chi...2866
Shao, Riming 1793
Sharma, Avinash 2456
Sharma, Sohit...1730
Shen, Yanfeng...1788, 1925
Shen, Yatao...815
Shen, Yecheng...2842
Shen, Zhan...1788, 1925

Sheng, Caiwang 1167
Shi, Gang...4220
Shi, Haixu..4168
Shi, Xiangyue 939
Shi, Yong...2877
Shibata, Naoya..3929
Shigeeda, Hidenori 2540
Shigematsu, Koichi...2176
Shigeuchi, Koji...3514, 3522
Shijo, Takuya 324
Shimada, Takae 250
Shimakage, Toyonari 2292
Shimamoto, Keita 2210
Shimao, Tohihiro 439
Shimaoka, Masahiro 1747
Shimizu, Toshihisa 302, 404, 2137, 2165, 3309, 3357
Shimizu, Toshimasa 1803
Shimomura, Shoji...2836
Shimono, Tomoyuki 675
Shimosato, Noboru 261, 3514, 3522
Shimoyama, A. 142
Shin, Sungyong 3418
Shinohara, Atsushi 1308, 1322
Shinohara, Hiroshi 1840
Shinshi, Tadahiko 4016
Shintani, Michihiro 3644
Shirai, Ryo...3309
Shirata, Kento 1137
Shiyuan, Yin..2625
Shoyama, Masahito 1095, 1554, 3773
Shujiang, Duan 2718
Shunsuke, Ohasi 3363
Shuto, Masao 699
Si, Yunpeng..2400, 3742
Sihvo, Jussi..2470
Sih-Yi, Lee...645
Silber, Siegfried 4028
Silventoinen, P. 2948
Simanjorang, Rejeki...416, 2296
Singh, Amit Kumar..4186
Singh, Vijay Kumar 1698
Son, Yung-Deug 3323
Song, Hongyu 3825
Song, Injong..803
Song, Kai...84
Song, Seung-Min 3229
Song, Shuguang 1051, 3129, 3924
Song, Wensheng..1872
Song, Yang..3698
Song, Yipeng..746
Song, Yubo..3877
Soong, Boon-Hee 1517, 1592
Soong, Theodore 3203
Soontorntaweesub, Kittichot 1764
Spiliotis, K...488

AUTHOR INDEX

Stieneker, Marco 598, 2484
Stock, Alexander 3034
Stojadinovic, Miloš..................................1103
Su, Huiling 795
Su, Jianhui 2877
Su, Yu-Chen 1038, 3549
Sudo, K. 1240
Suetake, A. 142
Suetsugu, Tadashi 4193
Sueuchi, Yuki 1955
Sugahara, Satoshi....................................2756
Sugahara, T. 142
Suganuma, K. 142
Suganuma, Katsuaki.................................1598
Sugihara, Yusuke 2789
Sugimoto, Hiroya 3627
Sugimoto, Kazushige................................767
Sugiyama, Takashi 3578
Suh, Yongsug 1977
Sul, Seung-Ki 521, 911, 3557
Sumida, Hitoshi 2774
Sun, Bainan 607
Sun, Chuan 370
Sun, Haotian 2780
Sun, Jianning...2963
Sun, Kai 3460, 4168
Sun, Lejia 2882
Sun, Peng 1125
Sun, Shumin 1455
Sun, Weifeng 3145
Sun, Xiangdong 2204
Sun, Yongping 560
Sun, Yuchong 3650, 4127
Sung, Kyungmin 1364
Suntio, Teuvo..963
Supanyapong, S. 1622
Surakitbovorn, Kawin 632, 3848
Surinkaew, Tossaporn 2287
Suul, Jon Are 782, 2003
Suwa, Hiroshi 1997
Suwankawin, S. 4096
Suwankawin, Surapong.............................2871
Suzuki, Akio 1840
Suzuki, Dai..157
Suzuki, Hiromitsu 495
Suzuki, Kazuma 1847, 3501, 3507
Suzuki, Kenichiro......................................511
Suzuki, Toshiki...........................1586, 2324
Suzuki, Yuhei 3390
Suzumori, Hirofumi 2066
Tabata, Yoichiro 329
Tada, Makoto 1580
Tadano, Hiroshi 313, 1111, 3375
Tadano, Yugo 483, 1890
Taguchi, Masashi 826

Taguchi, Yoshiaki 3280
Taiyuan, Yin 2625
Tajima, Katsubumi 2832
Tajyuta, Toshihisa 1840
Takahashi, Akihiko 3896
Takahashi, Akiko 2449
Takahashi, Arata 1270
Takahashi, Isseki 575
Takahashi, Masaki 3186
Takahashi, R. 3665
Takahashi, Shotaro 3315
Takahashi, Tomohira 2796
Takahashi, Toshimichi 227
Takahashi, Yuki 3375
Takakura, Shotaro 1270
Takami, Hiroshi 471
Takamura, Kenya 1381
Takano, Sho 3390
Takasho, Kenta 1890
Takatori, Koji 4139
Takayanagi, Ryohei 3396
Takeda, Kodai 2309
Takemoto, Masatsugu 2589, 2725, 2796, 3315
Takenaka, Hiroshi 3304
Takeno, K. 201
Takenoiri, Shunji......................................2764
Takeshita, Takaharu 356, 1847, 3501, 3507, 3791, 3945, 4088
Takeuchi, Norikazu 2292
Takeuchi, Yoko 1828, 3265
Takiguchi, Masashi 3723
Takimoto, Kazuyasu 3304
Takishima, Kenta 2826
Takubo, Hiromu 3390
Takuma, Shunsuke 2596
Takuno, Tsuguhiro 3578
Tamate, Michio.......................................3315
Tan, Nguyen Anh 478
Tan, Siew-Chong 889
Tanaka, Akira 1960
Tanaka, Takaaki 2604
Tanaka, Takahide 2774
Tanaka, Toshihiko 324, 1381
Tanaka, Tsuguhiro 3929
Tanaka, Y. 1162
Tanemo, Masamichi 2022
Tang, Cheng-Yu 639
Tang, Houjun 528
Tang, Ye 3705
Tang, Yi 337, 428, 434, 3910
Taniguchi, Katsumi 3396
Taniguchi, Katsunori 1202
Taniguchi, Tomoisa 866
Tatsumi, Kazuto 1202
Tatsuta, Fujio...4227

AUTHOR INDEX

Tatte, Yogesh 1730
Tausif, Ali 3833
Tcai, Anatolii 4109
Techama, Pantarote 2490
Teerakawanich, Nithiphat....................3332
Teigelkötter, Johannes........................3034
Tenconi, A. 732
Teraoka, Kenji 3086
Tey, Kuan-Chung.................................511
Thai, Van X. 2564
Thummala, Prasanth 4066
Tian, Mofan.................................998, 2785
Tian, Wei 1661
Tian, Xiaoyu 1771
Tian, Yanjun 1397
Tibola, Gabriel 1447
Tikka, V. 2948
Toba, Akio 1840
Toi, Takato 2229
Tokumaru, Syohei..............................2938
Tokusaki, Hiroyuki.............................2589
Tominaga, Isamu 1692
Tomita, Mutuwo.................................1295
Tong, Anping.............................1391, 2866
Tran, Hai N......................................3833
Tran, Tan-Tai...............................1799, 2145
Trescases, O. 267
Trescases, Olivier...............................2380
Tripathi, Ravi Nath.............................1698
Troppenz, Maria 3607
Trung, Tran Vu 1666
Tsai, Chang-Lin 3151
Tsai, Meng-Jiang...............................2462
Tsai, Men-Shen 1575
Tsai, Terng-Wei 639, 883
Tsai, Tsung-Lin 3151
Tsai, Yue-Ting..................................4198
Tse, Chi K.......................................3768
Tseng, King Jet.............................1517, 1592
Tseng, Wei-Jing 1626
Tsuchiya, Taichiro 2329
Tsuji, Hitoshi 3717
Tsuji, M. 1240
Tsuji, Mineo...................1217, 1276, 2938
Tsukakoshi, Masahiko 238
Tsumura, Akihiko 3490
Tsuno, Masahito 2558
Tsuruta, Ryoji 495
Tsuruta, Yukinori................................318
Tsutsumi, Hirohiko 3723
Tu, Yiming.............2435, 2681, 3176, 3439, 3934
Tuji, Mineo 3237
Tumerdem, Ugur 1680
Tumurbaatar, Anudari.........................1972
Uchida, Junichi.................................1955

Uchida, Yuuki 2750
Uchino, Yuki 324
Uda, Ryosuke 3578
Udagawa, Ikuto 517, 1692
Ueda, Tetsuzo 3762
Uehara, H. 1253
Uematsu, Takeshi 118, 4139
Uemura, Takamasa 860
Ueno, Tsutomu 4151
Uesugi, Yuma 3412
Ueta, Hiroaki 1883
Umeda, Takashi 2814
Umetani, Kazuhiro..............410, 1602, 1610
Unamuno, Eneko 759
Uno, Masatoshi 1782, 2030
Unterrieder, Christoph 3140
Ura, A. 704
Urabe, Shinichi 1782
Urata, Kazuki 302
Ute, Ryo...3773
Valente, G. 4034
Van De Ven, B.A.C. 267
Van Duivenbode, Jeroen 3840
Van Lam, Phi 571
Vasquez, C. Juan 3678
Vasquez, Juan C. 1498
Vass-Varnai, Andras 137
Veerachary, M. 845
Vemulapati, U....................... 3596, 4009
Vobecky, J. 3596
Vukadinovic, Nenad 831
Vyacheslav, Shkodyrev 1966
Wachi, Tsuneshisa 1997
Wada, Haruhisa..................................3286
Wada, Keiji...................1414, 1866, 1919, 2137, 4059
Wakimoto, Hiroki................................2404
Wang, Beibei 795
Wang, Bo 459
Wang, Can 1172
Wang, Chao 2386, 2901
Wang, Congling 3112
Wang, Dong 1183, 1246
Wang, Feng 1157, 1167, 2882
Wang, Fusheng 2611, 2992
Wang, Gaolin 1206, 1234
Wang, Guoxin 1206
Wang, Hanyu 2997
Wang, Hao 2270
Wang, Haoyu 100, 3825
Wang, Hechao 1183, 1246
Wang, Hongjie 4074, 4145
Wang, Huai...........1021, 1788, 1925, 2604, 2743, 3123
Wang, Huiying 1234
Wang, Jianing 2611
Wang, Jizhe.................................826, 2097

AUTHOR INDEX

Wang, Jun 3749, 3985
Wang, Kui 1010, 2386
Wang, Laili 2785, 3863
Wang, Liang 3958
Wang, Lifang 3338
Wang, Liwei 927
Wang, Meng 2992
Wang, Naizeng 998, 2785
Wang, Panrui 3383
Wang, Po-Wei 1617
Wang, Qiusheng 2421
Wang, Shike 1524, 3692
Wang, Shinn-Shyong 2086
Wang, Shitao 2866
Wang, Shunyu 3002
Wang, Wei 614, 3780
Wang, Wenjie 1391, 2866
Wang, Xiaolei 1455
Wang, Xiaoqing 878
Wang, Xiaoyang 453
Wang, Xiongfei 1711, 2673, 3164, 3357, 3684
Wang, Yanbo 1758, 2708
Wang, Yangyang 2505
Wang, Yi 1027, 1397, 3495
Wang, Yijie 614, 934, 3780, 3825
Wang, Youyun 2204
Wang, Yu-Chi 657
Wang, Yue 1455, 3863
Wang, Yuncheng 1177
Wang, Zhongxu 2743, 3123
Watanabe, Hiroki 896
Watanabe, Shoichiro 2042
Wei, Baoze 3678
Wei, Feng 1517, 1592
Wei, Jianzhao 2630
Wei, Juan 1131
Wei, Shilei 1397
Wei, Wang 1442, 2969
Wei, Xiaoguang 2343
Wei, Xiuqin 3650, 4127
Wei, Zhang 2969
Wellawatta, Thusitha Randima 1409
Wen, Huiqing 453
Wen, Po-Hsiang 3544
Wenbing, Li 1282
Wickramasinghe, Harith R. 3117
Wijaya, Febry Pandu 3490
Wikström, T. 3596
Winter, Christian 388
Wolf, Mihaela 3607
Wolski, Kornel 2129
Wu, Bin 3797
Wu, Heng 2673
Wu, Hongfei 4168
Wu, Min 3863

Wu, Pei-Lin 1145
Wu, Ping-Heng 503, 3549
Wu, T.-F. 1912
Wu, Tsai-Fu 3884
Wu, Tsung-Hsi 3544
Wu, Xiaojie 619
Wu, Xiaojun 3010, 3015
Wu, Ya'nan 2496
Wu, Zhiqian 1549
Würfl, Joachim 3607
Wyss, Jonas 3734
Xia, Meng 3484
Xia, Yongming 2842
Xiao, Chanjuan 1131
Xiao, Dan 3971
Xiao, Guochun 1549, 2944
Xiao, Jianfang 4157
Xiao, Xi 1966
Xiaoxi, Liu 2969
Xie, Jingwen 3069
Xie, Shaofeng 2547
Xie, Xiaogao 987
Xie, Zhen 2611, 2992
Xiong, Wei 939
Xu, Binci 2270
Xu, Cai 2986
Xu, Dehong 1118, 2270, 2569
Xu, Dewei David 1537, 3797
Xu, Dianguo 459, 560, 614, 934,
    1206, 1234, 3780, 3825, 4213
Xu, Guangzhao 998
Xu, Huadian 2877
Xu, Jin 261, 3514, 3522
Xu, Peng 1478
Xu, Sheng 3002
Xu, Shuang 1793
Xu, Yin-Chi 3884
Xu, Yue 3985
Xuan, Yang 1478
Xuanjie, Gao 2718
Xue, Danhong 2435
Yabuuchi, Tatsushi 233
Yada, Tomoharu 1381
Yamada, Hiroaki 324, 1381
Yamada, Koji 169
Yamaguchi, Daiki 3940
Yamaguchi, Koji 1972
Yamaji, Masaharu 2774
Yamamoto, Aoto 2558
Yamamoto, Hidekazu 2750
Yamamoto, Kichiro 1308, 1322
Yamamoto, Masaya 1782, 2030
Yamamoto, Masayoshi 1087, 1095, 2738, 2789, 3344
Yamamoto, Ryo 4016
Yamamoto, Shu 1949, 1960

AUTHOR INDEX

Yamamoto, Yuuto 3197
Yamanaka, Daisuke 2329
Yamanaka, Kenji 3369
Yamashita, Hiroki 1196
Yamashita, Yoshinori 3490
Yamazaki, Katsumi.................................693, 699
Yamazaki, Masahiro 207
Yan, Qingzeng 619
Yan, Y.T 851
Yan, Zhang 4181
Yanagisawa, Yuta 3762
Yang, Chang-Jun 3884
Yang, Cheng-Jhen 639, 883
Yang, Daoshu 1549
Yang, Dongsheng 3357
Yang, Geng 542
Yang, Hong-Tzer 4233
Yang, Hui-Chen 2296
Yang, Mei...3958
Yang, Ming 560
Yang, Peng 1966
Yang, Ping 3112
Yang, Renxin 4220
Yang, Sheng-Ming 651, 3426
Yang, Shunfeng 428
Yang, Shuying 2611
Yang, Xu 998, 1062, 1478, 2785, 2854, 3329
Yang, Ying 2973
Yang, Yongheng 439, 1021, 1788, 2512, 2743
Yang, Yugang 2703
Yang, Zebin 1157, 1167
Yang, Zhichang 2058
Yang, Zhihua 3797
Yang, Zhiqing 1073
Yang, Zhongping 1131, 1816, 2058, 3484, 3495
Yano, Junya 3723
Yao-Ching, Hsieh 645
Yaoqin, Jia 2441
Yasuda, Takumi 992
Yasuda, Yusuke 2082
Yaxin, Peng 416
Ye, Han 1504, 2496
Yeh, Shun-Hao 4233
Yelaverthi, Dorai Babu....................................4066
Yen, Chih-Ying..1145
Yenchamchalit, Kulsomsup...............................2430
Yi, Hao 2780, 2882
Yijie, Hou 2441
Yin, Shiyuan...1455
Yin, Taiyuan 1455
Yin, Zhijian 1021
Yin, Zhonggang 2204
Yingchun, Xu...2441
Yokokura, Yuki....................... 1673, 1741, 3890, 3896
Yokoyama, T. 3665

Yokoyama, Tomoki 1270, 1877, 1883,
　　2914, 3363, 3658
Yonezawa, Y. 3603
Yonezawa, Yu 196
Yoon, Bo-Kyung 3063
Yoshida, Souichi 2764
Yoshida, Yukihiro 2832
Yoshihara, Hidemasa 219
Yoshihara, Tohru 1997
Yoshikawa, Gaku 3280
Yoshimi, Daisuke 3952
Yoshimura, Eiji 767
Yoshino, Takuma 3363
Yoshino, Teruo 1692, 3916
Yoshioka, Yusuke 4151
Yoshizawa, Daisuke 238
You, Jiang 1386, 1421
You, Zih-Cing 651
Yu, Yong 459
Yuan, Huawei 889
Yuan, Liqiang 2963
Yuan, Xibo 619, 1634
Yuan, Yiqin 977, 3040
Yue, Wang...2625
Yui, Haiyan..699
Yukita, Kazuto..809
Zaijun, Wu 1282
Zaitsu, Toshiyuki 118, 4139
Zaman, Mohammad Shawkat 2380
Zanchetta, P. 4034
Zane, Regan 4066, 4074, 4145
Zdanowski, Mariusz 2129
Zeng, Pengxiang 2646
Zhang, Chen 4220
Zhang, Feili 1315
Zhang, Guoqiang 1206, 1234
Zhang, H. 142
Zhang, Hailong 3863
Zhang, Hao 1131, 1598
Zhang, Hongyang 3684
Zhang, Jianwen 1004
Zhang, Jianzhong 2262
Zhang, Le 3145
Zhang, Lei 3383
Zhang, Lifei 2703
Zhang, Meng 1966
Zhang, Qianfan 3025
Zhang, Runze 1816
Zhang, Shichong..2638
Zhang, Shu 614, 934, 3780
Zhang, Shuai 2944
Zhang, Tengfei 2980
Zhang, Wang 2625
Zhang, Xiaofang 2547

AUTHOR INDEX

Zhang, Xin 917, 953, 1004, 2688, 2692, 2980, 4157, 4162
Zhang, Xinan 1724
Zhang, Xing 2973, 2992
Zhang, Xueguang 4213
Zhang, Y. 946, 2697
Zhang, Yan 2646
Zhang, Yang 1177
Zhang, Yanping 2204
Zhang, Yaqian 2262
Zhang, Yi 2743, 3123
Zhang, Zhe 607, 1351, 3460
Zhang, Zhenbin 1661, 2895, 3965
Zhang, Zhigang 1157, 1167
Zhao, Chongyan 904
Zhao, Fangzhou 1549, 2944
Zhao, Fei 1172
Zhao, Jianfeng 3002, 3010, 3015
Zhao, Juan 2051
Zhao, Shengnan 795
Zhao, Tianshu 3020
Zhao, Tianyang 1172
Zhao, Yuanliang 3698
Zhao, Zhengming 2963
Zhao, Zhibin 1125
Zhao, Zhiqing 2714
Zheng, Deyou 2611
Zheng, Xuemei 2901
Zheng, Zedong 1010, 2386
Zhong, Wenxing 1118, 2569
Zhou, Dao 1758
Zhou, Dehong 428, 434
Zhou, Fulin 3257
Zhou, Jiuyang 2462
Zhou, Lei 2505
Zhou, Sheng-Zhi 370
Zhou, Victor 1943
Zhou, Yan 934
Zhou, Yimin 2547
Zhu, Cailing 3052
Zhu, Chunbo 84
Zhu, Helin 1336
Zhu, Junjie 3145
Zhu, Lianghong 1206, 1234
Zhu, Qingwei 3338
Zhu, Yanlin 2780
Zhu, Ye 2270
Zhujian, Ou 1376
Zhuo, Fang 1157, 1167, 2780, 2882
Zhuyong, Li 2986
Zischler, Sigrid 2410
Zou, Yaohan 3455

IEEE
445 Hoes Lane
Piscataway, NJ 08854-4141

ISBN 978-1-5386-4190-3